21. $\int \sqrt{a^2+x^2}\,dx = \frac{x}{2}\sqrt{a^2+x^2} + \frac{a^2}{2}\sinh^{-1}\frac{x}{a} + C$

22. $\int x^2\sqrt{a^2+x^2}\,dx = \frac{x(a^2+2x^2)\sqrt{a^2+x^2}}{8} - \frac{a^4}{8}\sinh^{-1}\frac{x}{a} + C$

23. $\int \frac{\sqrt{a^2+x^2}}{x}\,dx = \sqrt{a^2+x^2} - a\sinh^{-1}\left|\frac{a}{x}\right| + C$

24. $\int \frac{\sqrt{a^2+x^2}}{x^2}\,dx = \sinh^{-1}\frac{x}{a} - \frac{\sqrt{a^2+x^2}}{x} + C$

25. $\int \frac{x^2}{\sqrt{a^2+x^2}}\,dx = -\frac{a^2}{2}\sinh^{-1}\frac{x}{a} + \frac{x\sqrt{a^2+x^2}}{2} + C$

26. $\int \frac{dx}{x\sqrt{a^2+x^2}} = -\frac{1}{a}\ln\left|\frac{a+\sqrt{a^2+x^2}}{x}\right| + C$

27. $\int \frac{dx}{x^2\sqrt{a^2+x^2}} = -\frac{\sqrt{a^2+x^2}}{a^2x} + C$

28. $\int \frac{dx}{\sqrt{a^2-x^2}} = \sin^{-1}\frac{x}{a} + C$

29. $\int \sqrt{a^2-x^2}\,dx = \frac{x}{2}\sqrt{a^2-x^2} + \frac{a^2}{2}\sin^{-1}\frac{x}{a} + C$

30. $\int x^2\sqrt{a^2-x^2}\,dx = \frac{a^4}{8}\sin^{-1}\frac{x}{a} - \frac{1}{8}x\sqrt{a^2-x^2}\,(a^2-2x^2) + C$

31. $\int \frac{\sqrt{a^2-x^2}}{x}\,dx = \sqrt{a^2-x^2} - a\ln\left|\frac{a+\sqrt{a^2-x^2}}{x}\right| + C$

32. $\int \frac{\sqrt{a^2-x^2}}{x^2}\,dx = -\sin^{-1}\frac{x}{a} - \frac{\sqrt{a^2-x^2}}{x} + C$

33. $\int \frac{x^2}{\sqrt{a^2-x^2}}\,dx = \frac{a^2}{2}\sin^{-1}\frac{x}{a} - \frac{1}{2}x\sqrt{a^2-x^2} + C$

34. $\int \frac{dx}{x\sqrt{a^2-x^2}} = -\frac{1}{a}\ln\left|\frac{a+\sqrt{a^2-x^2}}{x}\right| + C$

35. $\int \frac{dx}{x^2\sqrt{a^2-x^2}} = -\frac{\sqrt{a^2-x^2}}{a^2x} + C$

36. $\int \frac{dx}{\sqrt{x^2-a^2}} = \cosh^{-1}\frac{x}{a} + C = \ln\left|x+\sqrt{x^2-a^2}\right| + C$

37. $\int \sqrt{x^2-a^2}\,dx = \frac{x}{2}\sqrt{x^2-a^2} - \frac{a^2}{2}\cosh^{-1}\frac{x}{a} + C$

38. $\int (\sqrt{x^2-a^2})^n\,dx = \frac{x(\sqrt{x^2-a^2})^n}{n+1} - \frac{na^2}{n+1}\int (\sqrt{x^2-a^2})^{n-2}\,dx, \qquad n \neq -1$

39. $\int \frac{dx}{(\sqrt{x^2-a^2})^n} = \frac{x(\sqrt{x^2-a^2})^{2-n}}{(2-n)a^2} - \frac{n-3}{(n-2)a^2}\int \frac{dx}{(\sqrt{x^2-a^2})^{n-2}}, \qquad n \neq 2$

40. $\int x(\sqrt{x^2-a^2})^n\,dx = \frac{(\sqrt{x^2-a^2})^{n+2}}{n+2} + C, \qquad n \neq -2$

41. $\int x^2\sqrt{x^2-a^2}\,dx = \frac{x}{8}(2x^2-a^2)\sqrt{x^2-a^2} - \frac{a^4}{8}\cosh^{-1}\frac{x}{a} + C$

42. $\int \frac{\sqrt{x^2-a^2}}{x}\,dx = \sqrt{x^2-a^2} - a\sec^{-1}\left|\frac{x}{a}\right| + C$

43. $\int \frac{\sqrt{x^2-a^2}}{x^2}\,dx = \cosh^{-1}\frac{x}{a} - \frac{\sqrt{x^2-a^2}}{x} + C$

Continued overleaf.

44. $$\int \frac{x^2}{\sqrt{x^2-a^2}}\,dx = \frac{a^2}{2}\cosh^{-1}\frac{x}{a} + \frac{x}{2}\sqrt{x^2-a^2} + C$$

45. $$\int \frac{dx}{x\sqrt{x^2-a^2}} = \frac{1}{a}\sec^{-1}\left|\frac{x}{a}\right| + C = \frac{1}{a}\cos^{-1}\left|\frac{a}{x}\right| + C$$

46. $$\int \frac{dx}{x^2\sqrt{x^2-a^2}} = \frac{\sqrt{x^2-a^2}}{a^2x} + C$$

47. $$\int \frac{dx}{\sqrt{2ax-x^2}} = \sin^{-1}\left(\frac{x-a}{a}\right) + C$$

48. $$\int \sqrt{2ax-x^2}\,dx = \frac{x-a}{2}\sqrt{2ax-x^2} + \frac{a^2}{2}\sin^{-1}\left(\frac{x-a}{a}\right) + C$$

49. $$\int (\sqrt{2ax-x^2})^n\,dx = \frac{(x-a)(\sqrt{2ax-x^2})^n}{n+1} + \frac{na^2}{n+1}\int (\sqrt{2ax-x^2})^{n-2}\,dx,$$

50. $$\int \frac{dx}{(\sqrt{2ax-x^2})^n} = \frac{(x-a)(\sqrt{2ax-x^2})^{2-n}}{(n-2)a^2} + \frac{(n-3)}{(n-2)a^2}\int \frac{dx}{(\sqrt{2ax-x^2})^{n-2}}$$

51. $$\int x\sqrt{2ax-x^2}\,dx = \frac{(x+a)(2x-3a)\sqrt{2ax-x^2}}{6} + \frac{a^3}{2}\sin^{-1}\frac{x-a}{a} + C$$

52. $$\int \frac{\sqrt{2ax-x^2}}{x}\,dx = \sqrt{2ax-x^2} + a\sin^{-1}\frac{x-a}{a} + C$$

53. $$\int \frac{\sqrt{2ax-x^2}}{x^2}\,dx = -2\sqrt{\frac{2a-x}{x}} - \sin^{-1}\left(\frac{x-a}{a}\right) + C$$

54. $$\int \frac{x\,dx}{\sqrt{2ax-x^2}} = a\sin^{-1}\frac{x-a}{a} - \sqrt{2ax-x^2} + C$$

55. $$\int \frac{dx}{x\sqrt{2ax-x^2}} = -\frac{1}{a}\sqrt{\frac{2a-x}{x}} + C$$

56. $$\int \sin ax\,dx = -\frac{1}{a}\cos ax + C$$

57. $$\int \cos ax\,dx = \frac{1}{a}\sin ax + C$$

58. $$\int \sin^2 ax\,dx = \frac{x}{2} - \frac{\sin 2ax}{4a} + C$$

59. $$\int \cos^2 ax\,dx = \frac{x}{2} + \frac{\sin 2ax}{4a} + C$$

60. $$\int \sin^n ax\,dx = \frac{-\sin^{n-1} ax\cos ax}{na} + \frac{n-1}{n}\int \sin^{n-2} ax\,dx$$

61. $$\int \cos^n ax\,dx = \frac{\cos^{n-1} ax\sin ax}{na} + \frac{n-1}{n}\int \cos^{n-2} ax\,dx$$

62. (a) $$\int \sin ax\cos bx\,dx = -\frac{\cos(a+b)x}{2(a+b)} - \frac{\cos(a-b)x}{2(a-b)} + C, \qquad a^2 \neq b^2$$

(b) $$\int \sin ax\sin bx\,dx = \frac{\sin(a-b)x}{2(a-b)} - \frac{\sin(a+b)x}{2(a+b)}, \qquad a^2 \neq b^2$$

(c) $$\int \cos ax\cos bx\,dx = \frac{\sin(a-b)x}{2(a-b)} + \frac{\sin(a+b)x}{2(a+b)}, \qquad a^2 \neq b^2$$

63. $$\int \sin ax\cos ax\,dx = -\frac{\cos 2ax}{4a} + C$$

64. $$\int \sin^n ax\cos ax\,dx = \frac{\sin^{n+1} ax}{(n+1)a} + C, \qquad n \neq -1$$

This table is continued on the endpapers at the back.

CALCULUS AND ANALYTIC GEOMETRY

SIXTH EDITION

GEORGE B. THOMAS, JR.
MASSACHUSETTS INSTITUTE OF TECHNOLOGY

ROSS L. FINNEY
MASSACHUSETTS INSTITUTE OF TECHNOLOGY

CALCULUS AND ANALYTIC GEOMETRY

ADDISON-WESLEY PUBLISHING COMPANY
Reading, Massachusetts • Menlo Park, California • London
Amsterdam • Don Mills, Ontario • Sydney

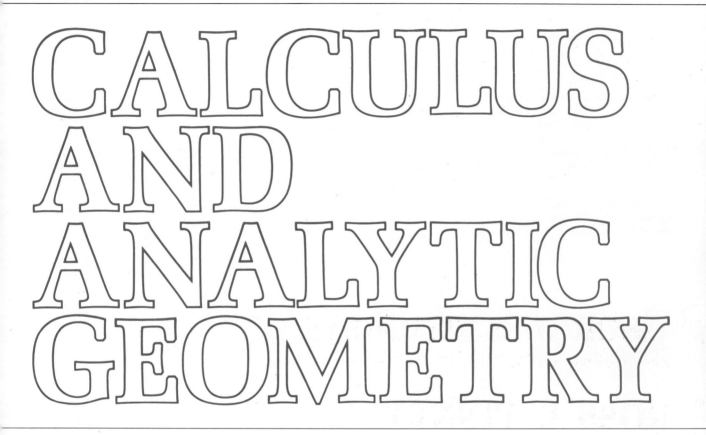

Sponsoring Editor: *Wayne Yuhasz*
Developmental Editor: *Jeffrey Pepper*
Production Manager: *Martha K. Morong*
Production Editor: *Marion E. Howe*
Text Designer: *Margaret Ong Tsao*
Illustrators: *Carmella M. Ciampa*
Dick Morton
Cover Designer: *Richard Hannus, Hannus Design Associates*
Cover Photograph: *Charles Rotmil, Four By Five*
Art Editor: *Dick Morton*
Manufacturing Supervisor: *Ann DeLacey*

Library of Congress Cataloging in Publication Data

Thomas, George Brinton, 1914–
Calculus and analytic geometry.

Includes index.
1. Calculus. 2. Geometry, Analytic.
I. Finney, Ross L. II. Title.
QA303.T42 1984 515′.15 83-2569
ISBN 0-201-16290-3

Reprinted with corrections, May 1984

BCDEFGHIJ-DO-8987654

To the students and teachers,
from all countries, who have used
earlier editions

PREFACE

Calculus, believed by many to be the greatest achievement in all mathematics, was created to meet the pressing mathematical needs of seventeenth century science. Heading the list were the needs to interrelate the accelerations, velocities, and distances traveled by moving bodies, to relate the slopes of curves to rates of change, to find the maximum and minimum values a function might take on (the greatest and least distances of a planet from its sun, for example), and to find the lengths of curves, the areas bounded by curves, the volumes enclosed by surfaces, and the centers of gravity of attracting bodies. Calculus is still the best mathematics for solving problems like these (and many others) and it is now so widely used that there is hardly a professional field that does not benefit from it in some way.

In preparing the *Sixth Edition* from the *Fifth Edition,* we have tried to make calculus easier for people to learn. The number of exercises has nearly doubled. Many of the new exercises are of a basic computational nature, designed to reinforce fundamental skills. There are more examples, three hundred new figures (all captioned), frequent formula summaries, and more (and shorter) applications for people going into engineering and science. Linear and quadratic approximations of functions are introduced early in the book, along with standard error estimates. There continue to be hand-held calculator exercises.

The widespread availability of microcomputers has made it possible for people learning calculus or using it in their professional work to graph functions in the plane or in space, to solve many equations quickly, and otherwise to implement the formulas derived by calculus. This edition therefore contains references at the end of some of the problem sets to a series of twenty-seven Apple II* microcomputer programs appropriate for students enrolled in calculus courses as well as for engineers, physicists, and others who use calculus as a tool in their work. The programs and accompanying manual are available from Addison-Wesley Publishing Company under the title *The Calculus Toolkit.* There is more information about them following this preface. It is to be emphasized, however, that one's progress through the text is in no way dependent upon having access to these or any other computer programs.

The level of rigor of the text is about the same as in earlier editions. For example, we do not prove that a function that is continuous on a

*Apple is a registered trademark of Apple Computer, Inc.

closed and bounded interval has a maximum and a minimum on the interval, but we state that theorem and use it in proving the Mean Value Theorem.

Chapter 1, on the rate of change of a function, has new sections on continuity and on infinity as a limit.

Chapter 2, on derivatives, now has sections on linear (tangent line) approximations, and on inverse functions and the Picard method for finding roots.

Chapter 3, on applications of derivatives, begins with new sections on curve sketching, concavity, and asymptotes. Maxima and minima now precede related rates. The chapter concludes with a section that extends the Mean Value Theorem and develops error estimates for the standard linear and quadratic approximations of functions.

Chapter 4, on indefinite and definite integrals, begins as it has in the past with differential equations of the form $y' = f(x)$, solved by separation of variables. The chapter contains a new development of the two fundamental theorems of integral calculus, however, and devotes a separate section to the technique of integration by substitution (for both definite and indefinite integrals). It is in this section that differentials are first introduced.

Chapter 5, on applications of definite integrals, has more art, problems, and worked examples than before, and frequent formula summaries.

Chapter 6, which introduces the logarithmic, exponential, and inverse trigonometric functions, also discusses relative rates of growth of functions. It concludes with a section on applications of exponential and logarithmic functions to cooling, exponential growth, radioactive decay, and electric circuits, and a section on compound interest and Benjamin Franklin's will.

In Chapter 7, on techniques of integration, the section on improper integrals has been expanded to include comparison tests for convergence. There is also a new section on using integral tables, and the treatment of integration by parts has been moved to the beginning of the chapter.

Chapters 8 (plane analytic geometry) and 9 (hyperbolic functions) have been shortened somewhat and contain additional art and problems.

Chapter 10, on polar coordinates, is shorter than before and contains a new technique for graphing polar equations of the form $r = f(\theta)$.

The presentation of infinite sequences and series has been moved forward in the book and divided into Chapters 11 and 12. Chapter 11 is devoted to sequences and infinite series of constants, Chapter 12 to Taylor's theorem (as an extended mean value theorem) and power series. Series of complex numbers are mentioned briefly. (An introduction to complex number arithmetic and Argand diagrams appears in Appendix 8.)

Chapter 13, on vectors, begins with motion in the plane and moves from there to the study of vector algebra and geometry in space.

Chapter 14, on vector functions and their derivatives, has a new treatment of tangent vectors, velocity, and acceleration, and concludes with a section on Kepler's laws of planetary motion.

Chapter 15, on partial derivatives, has new treatments of limits of functions of two variables, continuity, surfaces, partial derivatives, chain rules, directional derivatives, linear approximation and increment estimation, maxima and minima (both constrained and free), Lagrange multipliers, exact differentials, and least squares. Computer graphics have made it possible to visualize and discuss a number of surfaces that could not have been shown in earlier editions of this book. The chapter also looks briefly at solutions of some of the important partial differential equations of physics (in connection with higher order derivatives) and has a short section on how to apply chain rules when a function's variables are not independent.

Chapter 16, on multiple integrals, contains a new introduction to the subject, along with more examples, problems, and frequent formula summaries. It concludes with a presentation of surface area based on the notion of gradient.

Chapter 17, on vector analysis, begins with vector fields, surface integrals, line integrals, and work, and concludes with Green's theorem, the divergence theorem, and Stokes's theorem. In addition to many new examples and problems, the chapter now contains a brief derivation of the continuity equation of hydrodynamics.

In Chapter 18, on ordinary differential equations, the treatment of linear second order equations with constant coefficients has been expanded to include the method of undetermined coefficients in addition to the method of variation of parameters. The chapter concludes with short sections on power series solutions, direction fields and Picard's theorem, and Euler and Runge–Kutta methods.

The appendixes include expanded sections on determinants and Cramer's rule, and on matrices and linear equations, as well as new sections on mathematical induction and number systems.

The text is available in one complete volume, which can be covered in two or three semesters, depending on the pace, or as two separate parts. Part I treats functions of one variable, analytic geometry in two dimensions, and infinite series (Chapters 1 through 12). It also contains the appendixes on determinants, matrices, and mathematical induction. Part II begins with Chapter 11, on sequences and series, and contains all subsequent chapters, including the appendixes. Both parts contain answers to odd-numbered problems.

We would particularly like to thank Jack M. Pollin and Frank R. Giordano of the Mathematics Department of the U.S. Military Academy at West Point, who generously gave us access to and guided us through their department's carefully kept examination and problem files. Some of the finest problems in this new edition have come from West Point. Likewise, and no less, we would like to thank Carroll O. Wilde of the Mathematics Department of the U.S. Naval Postgraduate School for granting us access to his department's outstanding examination files, and for his many supportive comments and suggestions during the preparation of the present edition.

We would also like to acknowledge the many helpful contributions and suggestions of our colleagues at MIT, particularly Arthur Mattuck and Frank Morgan, and the many students there and elsewhere who have taken the time to share their ideas with us.

Many helpful comments and suggestions were made by people who reviewed the *Fifth Edition* as the *Sixth Edition* was being planned:

Mark Bridger, Northeastern University
Jeff Davis, University of New Mexico
Simon Hellerstein, University of Wisconsin, Madison
Stanley Lukawecki, Clemson University
Ronald Morash, University of Michigan, Dearborn
Harold Oxsen, Diablo Valley College
Walter Read, California State University, Fresno
Michael Shaughnessy, Oregon State University

We would also like to thank and acknowledge the contributions of the people who participated in the preliminary planning sessions that took place in Menlo Park, California, and Cincinnati, Ohio:

Charles Austin, California State University, Long Beach
Douglas Crawford, College of San Mateo
Daniel S. Drucker, Wayne State University
Bruce H. Edwards, University of Florida
Alice J. Kelly, University of Santa Clara
Stanley Lukawecki, Clemson University
Ronald Morash, University of Michigan, Dearborn
Michael Shaughnessy, Oregon State University
Ronald J. Stern, University of Utah
Carroll O. Wilde, Naval Postgraduate School

A great many valuable contributions to the *Sixth Edition* were made by the people who reviewed the manuscript as it developed through its various stages:

Charles Austin, California State University, Long Beach
Mark Bridger, Northeastern University
Stuart Goldenberg, California Polytechnic State University
Simon Hellerstein, University of Wisconsin, Madison
Ronald Morash, University of Michigan, Dearborn
Hiram Paley, University of Illinois at Urbana-Champaign
David F. Ullrich, North Carolina State University

Charles Austin and Mark Bridger made many additional contributions to the chapters on partial derivatives and multiple integration.

Special contributions to the book were also made through correspondence and in conversation by

Harold Diamond, University of Illinois at Urbana-Champaign
Frank D. Faulkner, Naval Postgraduate School
Nathaniel Grossman, University of California, Los Angeles
Richard W. Hamming, Naval Postgraduate School
John P. Hoyt, Franklin and Marshall College
Alice J. Kelley, University of Santa Clara

Ernest J. Manfred, U.S. Coast Guard Academy

Charles G. Moore, Northern Arizona University

Joseph J. Rotman, University of Illinois at Urbana-Champaign

Arthur C. Segal, The University of Alabama in Birmingham

G. Wayne Sullivan, Volunteer State Community College

Joseph D. Zund, New Mexico State University

The answer manuscript for the odd-numbered problems in this edition was developed through the collaborative efforts of

Kenneth R. Ballou, University of California, Berkeley

Lynda Jo Carlson, California State University, Fullerton

J. Howard Jones, Lansing Community College

Chris J. Petti, Massachusetts Institute of Technology

Craig W. Reynolds, Massachusetts Insitute of Technology

Richard J. Palmaccio, Pine Crest School

Richard Semmler, Northern Virginia Community College

Charles Swanson, University of Minnesota

The computer-generated art that appears in Chapters 13, 15, and 18 is the patient and enthusiastic work of John Aspinall of the Plasma Fusion Center at MIT.

To each and every person who has at any time contributed helpful suggestions, comments, or criticisms, whether or not we have been able to incorporate these into the book, we say "Thank you very much."

It is a pleasure to acknowledge the superb assistance in illustration, editing, design, and composition that the staff of Addison-Wesley Publishing Company has given to the preparation of this edition.

Any errors that may appear are the responsibility of the authors. We will appreciate having these brought to our attention.

October, 1983

G. B. T., Jr.
R. L. F.

AVAILABLE SUPPLEMENTS The following supplementary materials are available for use by students:

Self-Study Manual	*Maurice D. Weir* (Naval Postgraduate School)
Student Supplement	*Kenneth R. Ballou* (University of California, Berkeley)
	Charles Swanson (University of Minnesota)

LIST OF PROGRAMS

The Calculus Toolkit

Ross L. Finney *Dale T. Hoffman*
Judah L. Schwartz *Carroll O. Wilde*

Super * Grapher
Name That Function
Secant Lines
Limit Problems
Limit Definition
Continuity at a Point
Derivative Grapher
Function Evaluator/Comparer
Parametric Equations
Root Finder
Picard's Fixed Point Method
Integration
Integral Evaluator
Antiderivatives and Direction Fields
Partial Fraction Integration Problems
Conic Sections
Sequences and Series
Taylor Series
Complex Number Calculator
3D Grapher
Double Integral Evaluator
Scalar Fields
Vector Fields
First Order Initial Value Problem
Second Order Initial Value Problem
Damped Oscillator
Forced, Damped Oscillator

CONTENTS

4 INTEGRATION 250

5 APPLICATIONS OF DEFINITE INTEGRALS 315

6 TRANSCENDENTAL FUNCTIONS 381

7 INTEGRATION METHODS 444

8 PLANE ANALYTIC GEOMETRY 510

APPENDIXES A–1

PROLOGUE

What Is Calculus?

Calculus is the mathematics of motion and change. Where there is motion or growth, where variable forces are at work producing acceleration, calculus is the right mathematical tool. This was true in the beginnings of the subject, and it is true today.

Calculus is used to predict the orbits of earth satellites; to design inertial navigation systems, cyclotrons, and radar systems; to explore problems of space travel; to test theories about ocean currents and the dynamics of the atmosphere. It is applied increasingly to solve problems in biology, business, economics, linguistics, medicine, political science, and psychology. Calculus is also a gateway to nearly all fields of higher mathematics.

Differential calculus deals with the problem of calculating rates of change. When we have a formula for the distance that a moving body covers as a function of time, differential calculus gives us the formulas for calculating the body's velocity and acceleration at any instant. Differential calculus also lets us answer questions about the greatest and smallest values a function can take on. What angle of elevation gives a cannon its greatest range? What is the strongest rectangular beam we can cut from a cylindrical log? When is a chemical reaction proceeding at its fastest rate?

Integral calculus considers the problem of determining a function from information about its rate of change. Given a formula for the velocity of a body as a function of time, we can use integral calculus to produce a formula that tells how far the body has traveled from its starting point at any instant. Given the present size of a population and the rate at which it grows, we can produce a formula that predicts the population's size at any future time or that estimates how big the population was at some time in the past. Given the rate at which carbon-14 disintegrates, we can produce a formula that calculates the age of a sample of charcoal from its present ratio of carbon-14 to carbon-12. The age of Crater Lake in Oregon, 6660 years, was estimated by applying such a formula to the charcoal from a tree that was killed in the volcanic eruption that formed the lake.

Modern science and engineering use both differential and integral calculus to express physical laws in precise mathematical terms and to study the consequences of these laws.

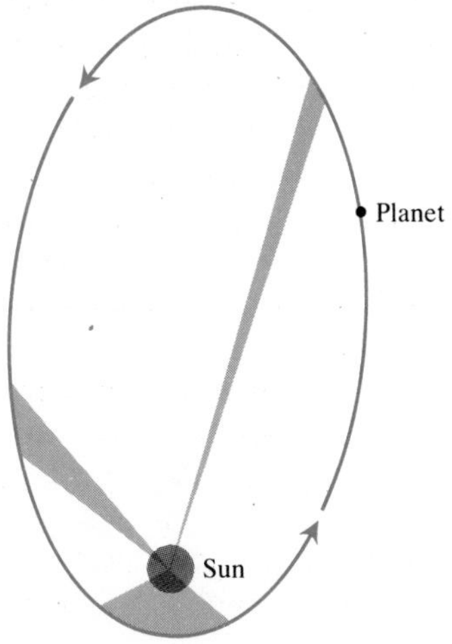

P.1 A planet moving about its sun. The shaded regions shown here have equal areas. According to Kepler's second law, the planet takes the same amount of time to traverse the curved outer boundary of each region. The planet therefore moves faster near the sun than it does farther away.

Before Sir Isaac Newton (1642–1727) was born, Johannes Kepler (1571–1630) spent twenty years studying observational data and using empirical methods to discover the three laws that now bear his name:

1. Each planet moves about the sun in an orbit that is an ellipse with one focus at the sun (Fig. P.1).
2. The line joining the planet and the sun sweeps over equal areas in equal intervals of time.
3. The squares of the periods of revolution of the planets about the sun are proportional to the cubes of their average distances from the sun.

Figure P.1 gives us some idea of what Kepler's first two laws say, but what does his third law say? The earth's period of revolution about the sun is 365 days (close enough for now). Its average distance from the sun is 93 million miles. The ratio of the square of 365 to the cube of 93,000,000 is $(365)^2/(93{,}000{,}000)^3 = 1.66 \times 10^{-19}$. Kepler's third law says that if the period of revolution T of any other planet is measured in earth days, and if the planet's average distance D from the sun is measured in miles, then the numerical value of $T^2/D^3 = (\text{Period})^2/(\text{Distance})^3$ will be 1.66×10^{-19}, just as it was for the earth. The ratio is the same for every planet in the solar system. In particular, the farther a planet is from the sun, the longer it takes to go once about its orbit.

When they were first written down, there can have been no immediately apparent relationship between the measurements Kepler made of T and D for the planets he observed. He must have studied long and hard before discovering the constancy of the ratio T^2/D^3. Only a few years later, Newton was able to use calculus to derive all three of Kepler's laws from the physical assumptions we know as Newton's laws of motion and gravitation. Kepler formulated his laws for our solar system. Newton's work shows that they hold for any orbital system that obeys the laws of motion and gravitation.

The calculus we use today is the contribution of many people. Its roots can be traced to classical Greek geometry, but its invention is chiefly the work of the astronomers, mathematicians, and physicists of the seventeenth century. Among these were René Descartes (1596–1650), Bonaventura Cavalieri (1598–1647), Pierre de Fermat (1601–1665), John Wallis (1616–1703), Blaise Pascal (1623–1662), Christian Huygens (1629–1695), Isaac Barrow (1630–1677), and James Gregory (1638–1675). The work culminated in the great creations of Newton and Gottfried Wilhelm Leibniz (1646–1716).

The efforts of Newton and Leibniz changed the calculus from an appendage of Greek geometry to an independent science that had the power to handle a vast array of new scientific problems. Newton and Leibniz formulated the fundamental theorems of calculus and recognized in the particular works of their colleagues the general methods that applied to all sorts of functions. Many of their operational formulas, methods of calculation, and mathematical symbols are still in use.

The development of the calculus did not stop with the accomplishments of the seventeenth century. Great additions were made by the eighteenth-century mathematicians James Bernoulli (1654–1705) and his brother John Bernoulli (1667–1748); by Leonhard Euler (1707–1783), the key mathematical figure of the century; by Joseph-Louis Lagrange (1736–1813); and many others. The ultimate justification of the procedures of calculus was made by the mathematicians of the nineteenth century, among them Bernhard Bolzano (1781–1848), Augustin-Louis Cauchy (1789–1857), and Karl Weierstrass (1815–1897). The nineteenth century also brought spectacular developments in mathematics beyond the calculus. You can read about them in Morris Kline's magnificent book *Mathematical Thought from Ancient to Modern Times* (New York: Oxford University Press, 1972).

"The calculus was the first achievement of modern mathematics," wrote John von Neumann (1903–1957), one of the great mathematicians of the present century, "and it is difficult to overestimate its importance. I think it defines more unequivocally than anything else the inception of modern mathematics; and the system of mathematical analysis, which is its logical development, still constitutes the greatest technical advance in exact thinking."†

---

† *World of Mathematics*, Vol. 4 (New York: Simon and Schuster, 1960), "The Mathematician," by John von Neumann, pp. 2053–2063.

1 The Rate of Change of a Function

1.1 Coordinates for the Plane

When we were studying geometry in high school, we may not have considered it to be a tool for studying the physical world as the Alexandrian Greeks did, but in their minds the subject was anchored in application. In contrast, the Arab and Hindu mathematicians of the period, who were developing the beginnings of what we now call algebra, were free to carry out their proofs and deductions without the restriction that their results have a physical interpretation. For them it was all right to use zero as a number in arithmetic, but not so for the Greeks, to whom zero was only a notational place holder used to denote the absence of a number. To a great extent, geometry and algebra developed separately until only a few hundred years ago.

In the seventeenth century, Fermat and Descartes made the marriage of algebra and geometry that totally changed the face of mathematics. This marriage, which we call analytic geometry today, provided the tool that the seventeenth-century scientists needed for quantifying their work and laid the foundation for astonishing advances in mathematics, physics, astronomy, and biology.

Analytic geometry begins with the assignment of numerical coordinates to all the points in a plane. These coordinates make it possible to graph algebraic equations in two variables as lines and curves. They also allow us to calculate angles and distances and to write coordinate equations to describe the paths along which objects move. Since most of the theory of calculus can be presented in geometric terms, and since the applications of calculus are chiefly about motion and change, the coordinate plane of analytic geometry is the natural setting for learning about calculus and its applications.

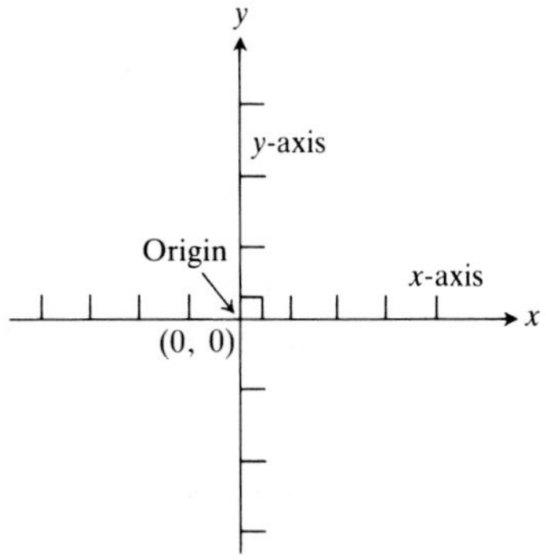

1.1 The points in the plane get their coordinates from scaled axes like the ones shown above.

The first step in assigning coordinates to points in a plane is to choose a pair of lines that cross at right angles, as shown in Fig. 1.1. The point O where the lines cross is called the *origin*. One of these lines is called the *x-axis*, the other the *y-axis*. The directions of the x- and y-axes are referred to as "horizontal" and "vertical," respectively, and they are usually drawn this way.

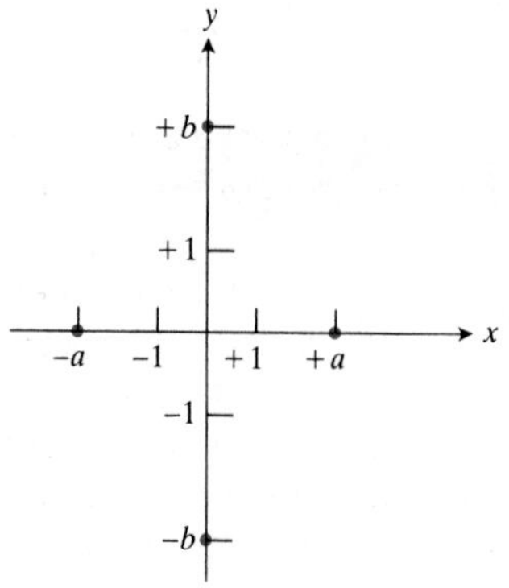

1.2 The scaling on each axis is symmetric with respect to the origin.

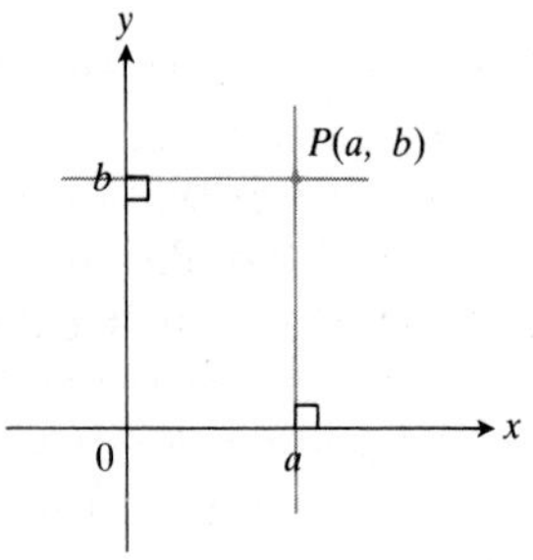

1.3 The point P where the line perpendicular to the x-axis at a crosses the line perpendicular to the y-axis at b is assigned the coordinate pair (a, b).

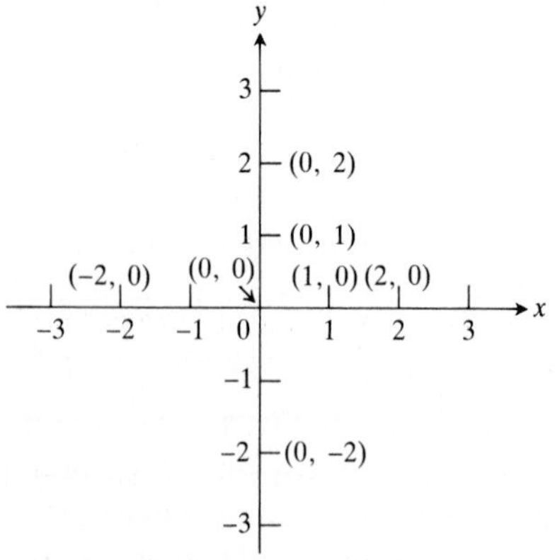

1.4 The points on the axes now have two sets of labels.

The next step is to assign the coordinate pair (0, 0) to the origin and to scale each axis in terms of some unit of length. The units of length on the two axes may be different, as they are in Fig. 1.2.

The scaling on the x-axis assigns the positive number a to the point that is a units to the right of O and the number $-a$ to the point that is a units to the left of O. The positive numbers lie to the right of O, the negative numbers to the left. The scaling completes a one-to-one correspondence between the points on the x-axis and the set of *real numbers* (numbers that may be represented by terminating and nonterminating decimals).

The scaling on the y-axis assigns the positive number b to the point that is b units above the x-axis and the number $-b$ to the point that is b units below the x-axis. With 0 already assigned to the origin, this symmetrical arrangement of the numbers completes a one-to-one correspondence between the points of the y-axis and the real numbers. (See Fig. 1.2.)

We are now ready to assign number pairs to the points in the plane. The first number in the pair (a, b) assigned to a point P is found by dropping a perpendicular from P to the x-axis and taking a to be the number assigned to the point where the perpendicular hits the axis. The second number is found by dropping a perpendicular from P to the y-axis and taking b to be the number assigned to the foot of this perpendicular on the y-axis. The construction is shown in Fig. 1.3.

The number a from the x-axis is the *x-coordinate,* or *abscissa,* of P. The number b from the y-axis is the *y-coordinate,* or *ordinate,* of P. The pair (a, b) is the *coordinate pair* of the point P. It is also an *ordered pair,* listing the x-coordinate first and the y-coordinate second. To call attention to the fact that a point P has been assigned a particular coordinate pair (a, b), we write the symbols P and (a, b) side by side, like this: $P(a, b)$. We did this in Fig. 1.3.

The geometric procedure we just described assigns an ordered pair of numbers to each point in the plane. We can also regard this procedure as one that assigns a point in the plane to each ordered pair of numbers. It assigns to the ordered pair (a, b) the point where the perpendicular to a on the x-axis crosses the perpendicular to b on the y-axis. Thus the assignment of coordinates is a one-to-one correspondence between the points of the plane and the set of all ordered pairs of real numbers.

The coordinates we just defined are called *Cartesian coordinates* in honor of Descartes.

The points on the axes now have two kinds of labels: the numbers they received when the axes were originally scaled, and the number pairs they receive as points in the plane. How do the labels match up? See Fig. 1.4. Note that the y-coordinate of every point on the x-axis is 0, the x-coordinate of every point on the y-axis is 0, and the coordinates of the origin are (0, 0).

If the x-axis is drawn horizontally, as in Fig. 1.5, motion from left to right along the x-axis is said to be motion in the *positive x-direction.* Similarly, motion from right to left is motion in the *negative x-direction.* Along the y-axis, drawn vertically, the positive direction is up, the negative direction down.

The origin divides the x-axis into the *positive x-axis* (the part to the right of O) and the *negative x-axis* (the part to the left of O). Similarly,

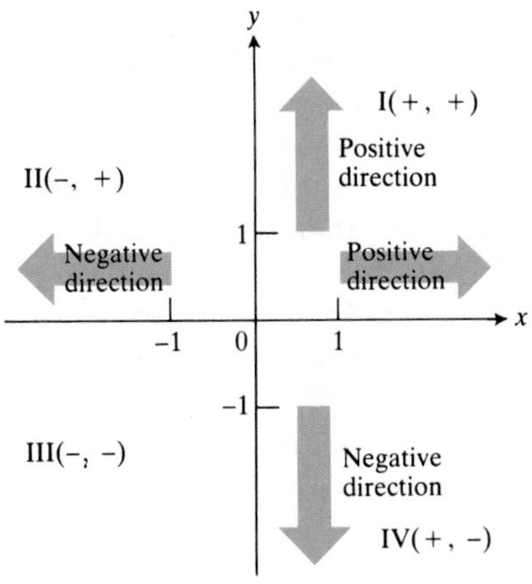

1.5 Directions along the axes: x and y increase in the positive direction and decrease in the negative direction. The roman numerals show the numbering of the four quadrants.

the origin divides the y-axis into the *positive y-axis* and the *negative y-axis*.

The two axes divide the plane into four regions called *quadrants*, numbered first, second, third, and fourth. They are labeled I, II, III, and IV in Fig. 1.5.

A word about scales. When we plot real data in the coordinate plane or graph formulas whose variables have different units of measure, the units shown on the two coordinate axes may have very different interpretations. If we plot data for the size of the national debt at different times, the x-axis might show time in months and the y-axis might show billions of dollars. If we graph steam pressure as a function of boiler temperature, the x-axis might show degrees Celsius and the y-axis pounds per square inch. In cases like these there is obviously no reason to use the same scale when we draw the two axes. There is no need to place the two 1's on the axes the same number of millimeters or whatever away from the origin.

When we graph functions whose variables do not represent physical measurements, however, and when we draw figures in the coordinate plane to learn about their geometry or trigonometry, we will assume that the scales on the axes we draw are the same. One unit of distance up and down in the plane will then look the same as one unit of distance right and left. As on a surveyor's map or a scale drawing, line segments that are supposed to have the same length will look as if they do. Figures that are similar will look similar, and angles that are congruent will look congruent, no matter where they may be in the plane.

PROBLEMS

In Problems 1–12, first draw a pair of coordinate axes and plot the given point $P(a, b)$. Then plot each of the following points:

a) The point Q such that PQ is perpendicular to the x-axis and is bisected by it. Give the coordinates of Q. (P and Q are symmetric with respect to the x-axis.)
b) The point R such that PR is perpendicular to the y-axis and is bisected by it. Give the coordinates of R. (P and R are symmetric with respect to the y-axis.)
c) The point S such that PS is bisected by the origin. Give the coordinates of S. (P and S are symmetric with respect to the origin.)
d) The point T such that PT is perpendicular to and is bisected by the 45°-line L through the origin that bisects the first and third quadrants. (See Fig. 1.6.) Give the coordinates of T, assuming equal units on the axes. (P and T are symmetric with respect to L.)

1. $(1, -2)$ **2.** $(2, -1)$ **3.** $(-2, 2)$
4. $(-2, 1)$ **5.** $(0, 1)$ **6.** $(1, 0)$
7. $(-2, 0)$ **8.** $(0, -3)$ **9.** $(-1, -3)$
10. $(\sqrt{2}, -\sqrt{2})$ **11.** $(-\pi, -\pi)$ **12.** $(-1.5, 2.3)$

1.6 Symmetry with respect to the x-axis, the y-axis, the origin, and the 45°-line L.

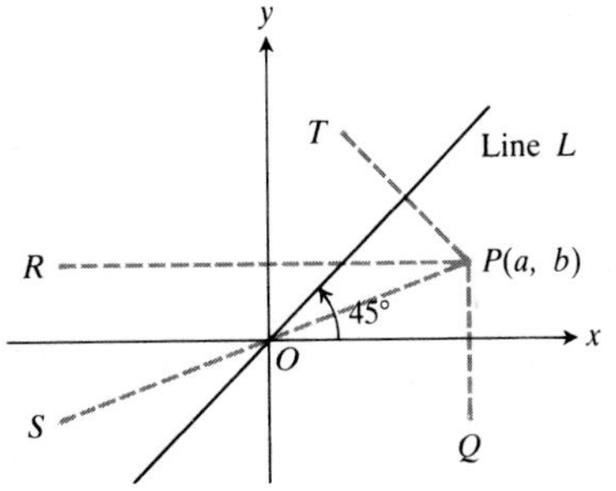

13. If $P = P(x, y)$, then the coordinates of the point Q in (a) above are $(x, -y)$. Give the coordinates of R, S, and T in terms of x and y.

In Problems 14–20, assume that the units of length on the two axes are equal.

14. A line is drawn through the points $(0, 0)$ and $(1, 1)$. What acute angle does it make with the positive x-axis? Sketch.

15. There are three parallelograms with vertices at $(-1, 1)$, $(2, 0)$, and $(2, 3)$. Sketch them and give the coordinate pairs of the missing vertices.

16. A rectangle with sides parallel to the axes has vertices at $(3, -2)$ and $(-4, -7)$.
a) Find the coordinates of the other two vertices.
b) Find the area of the rectangle.

17. The rectangle in Fig. 1.7 has sides parallel to the axes. It is three times as long as it is wide. Its perimeter is 56 units. Find the coordinates of the vertices A, B, and C.

18. A circle in quadrant II is tangent to both axes. It touches the y-axis at $(0, 3)$.
a) At what point does it meet the x-axis? Sketch.
b) Find the coordinates of the center of the circle.

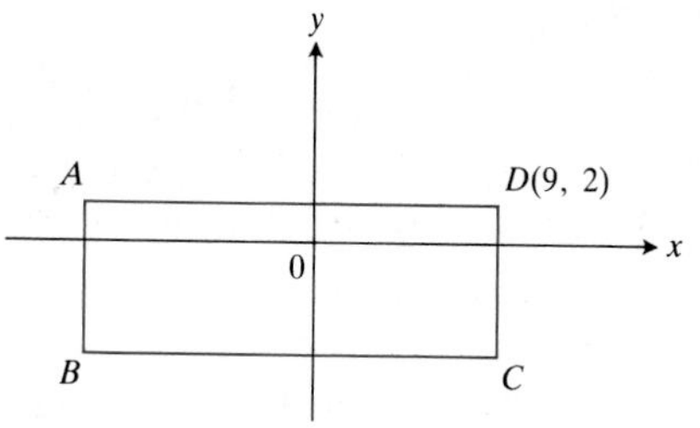

1.7 The rectangle for Problem 17.

19. The line through the points $(1, 1)$ and $(2, 0)$ cuts the y-axis at the point $(0, b)$. Find b by using similar triangles.

20. A 90° rotation counterclockwise about the origin takes $(2, 0)$ to $(0, 2)$ and $(0, 3)$ to $(-3, 0)$, as shown in Fig. 1.8. Where does the rotation take each of the following points?
a) $(4, 1)$ b) $(-2, -3)$ c) $(2, -5)$
d) $(x, 0)$ e) $(0, y)$ f) (x, y)
g) What point is taken to $(10, 3)$?

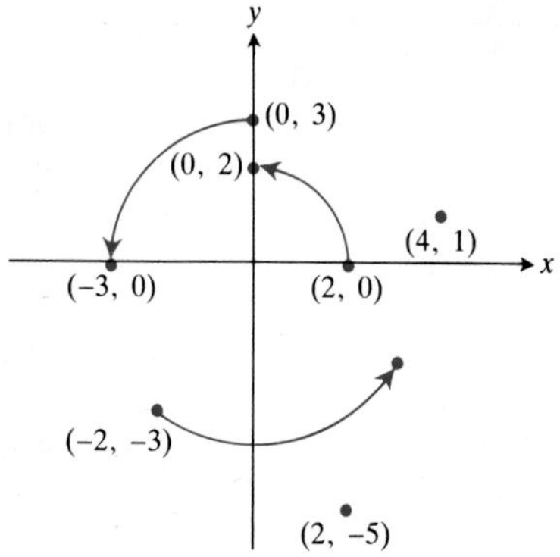

1.8 Points moved to new positions by a counterclockwise 90° rotation about the origin.

1.2 Increments and Distance

Increments

Figure 1.9 reproduces a multiflash photograph of a ball looping the loop on a circular track. The coordinate grid in the picture gives us a way to say where the center of the ball is each time the ball is shown.

When the ball moved down from $B(19, 12)$ and up to $C(39, 18)$, the x-coordinate of the center started at $x = 19$ cm and finished at $x = 39$ cm.

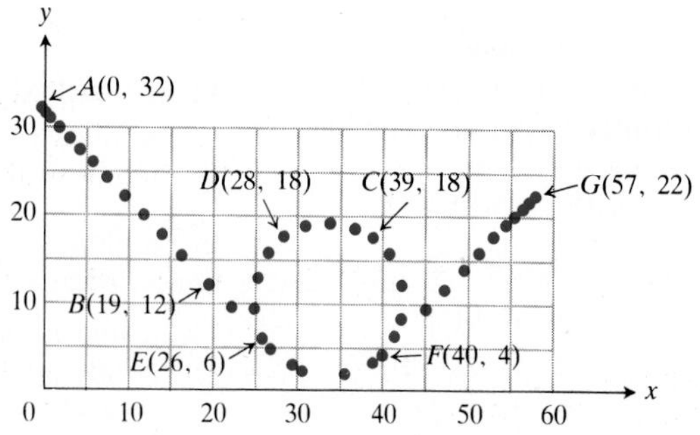

1.9 A reproduction of a multiflash photograph of a ball looping the loop in a vertical circle. The ball was released from rest at $A(0, 32)$ in the upper left corner. The letters label successive positions. The coordinates given are those of the ball's center, rounded to the nearest integer. The axes are scaled in centimeters.

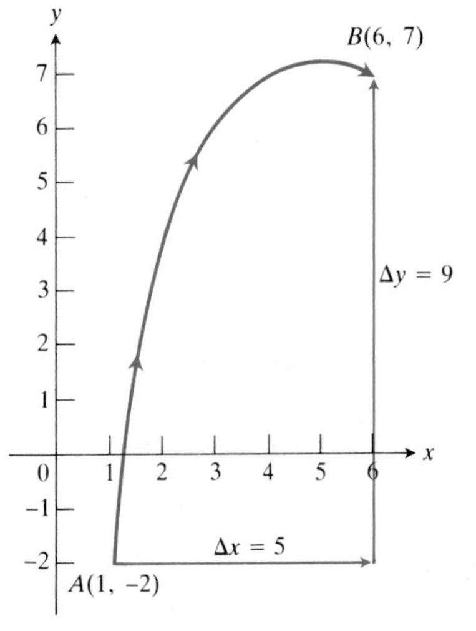

1.10 When a particle moves from one point to another, Δx and Δy are calculated from the coordinates of the initial and terminal positions. When the point moves from A to B, the net coordinate changes are $\Delta x = 6 - 1 = 5$ and $\Delta y = 7 - (-2) = 9$.

The net change Δx (read "delta x") in the coordinate was $\Delta x = 39 - 19 = 20$ cm. The y-coordinate changed from $y = 12$ cm to $y = 18$ cm, for a net vertical change of $\Delta y = 18 - 12 = 6$ cm. We can think of Δx and Δy as measures of net horizontal and vertical change.

EXAMPLE 1 Find the net horizontal and vertical changes that took place in the position of the center of the ball shown in Fig. 1.9 when the ball moved from $C(39, 18)$ to $D(28, 18)$.

Solution The x-coordinate changed from $x = 39$ cm to $x = 28$ cm, for a net change of

$$\Delta x = 28 - 39 = -11 \text{ cm}.$$

The y-coordinate started at $y = 18$ cm and finished at $y = 18$ cm, for a net change of

$$\Delta y = 18 - 18 = 0 \text{ cm}. \quad \square$$

In mathematics, net changes in coordinates are called *increments*. When a particle moves in the plane from a starting position (x_1, y_1) to a final position (x_2, y_2), the increments Δx and Δy in the coordinates are given by the following formulas:

$$\Delta x = x_2 - x_1 \quad \text{and} \quad \Delta y = y_2 - y_1. \qquad (1)$$

To calculate the increment in a coordinate, we subtract the coordinate's initial value from its final value. An increment can be any real number: positive, negative, or zero.

EXAMPLE 2 Figure 1.10 shows the path of a particle that moved from $(x_1, y_1) = (1, -2)$ to $(x_2, y_2) = (6, 7)$. The corresponding increments in the particle's coordinates are calculated from Eqs. (1) to be

$$\Delta x = x_2 - x_1 = 6 - 1 = 5 \quad \text{and} \quad \Delta y = y_2 - y_1 = 7 - (-2) = 9. \quad \square$$

Distance

When the coordinate axes in the plane use the same unit of measure, we can use coordinate increments in combination with the Pythagorean theorem to calculate distances between points in the plane. Figure 1.11 shows how. In applying the Pythagorean theorem to the right triangle P_1CP_2 we see that the distance d between the two points $P_1(x_1, y_1)$ and $P_2(x_2, y_2)$ is given by the following formula:

Distance Formula

$$d = \sqrt{(\Delta x)^2 + (\Delta y)^2} = \sqrt{(x_2 - x_1)^2 + (y_2 - y_1)^2} \qquad (2)$$

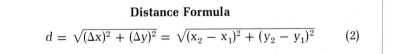

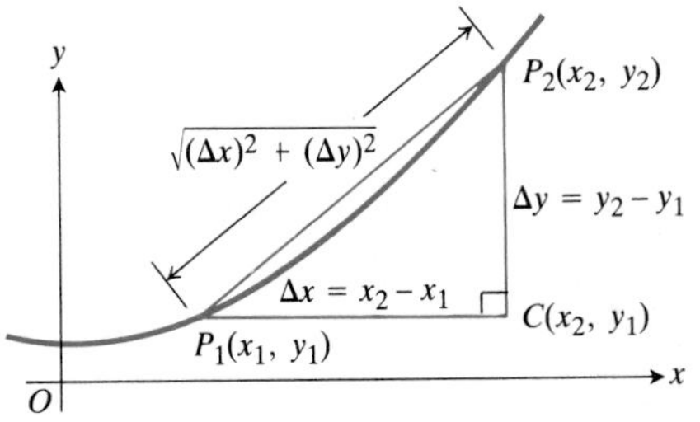

1.11 Distance in the coordinate plane is calculated with the Pythagorean theorem.

Equation (2) is called the *distance formula* for the plane.

EXAMPLE 3 Calculate the distance between the points $(-1, 2)$ and $(2, -2)$.

Solution If we let $(x_1, y_1) = (-1, 2)$ and $(x_2, y_2) = (2, -2)$, then

$$\Delta x = x_2 - x_1 = 2 - (-1) = 3, \qquad \Delta y = y_2 - y_1 = (-2) - 2 = -4,$$

and

$$d = \sqrt{(\Delta x)^2 + (\Delta y)^2} = \sqrt{(3)^2 + (-4)^2} = \sqrt{9 + 16} = \sqrt{25} = 5.$$

If we name the points the other way around, the formula in (2) still gives the distance between them:

$$d = \sqrt{(-1 - 2)^2 + (2 - (-2))^2} = \sqrt{(-3)^2 + (4)^2} = \sqrt{9 + 16} = 5.$$

The increments change sign when we interchange the points, but the squares of the increments remain the same. □

Example 3 illustrates the fact that when we use Eq. (2) to calculate the distance between two points, it does not matter which point we call P_1 and which we call P_2.

Circles

An *equation for a circle* in the coordinate plane is an equation in x and y that is satisfied by the coordinates of the points that lie on the circle and is not satisfied by the coordinates of any of the points that lie elsewhere.

The standard equation for a circle in the plane comes from Eq. (2) in the following way. The circle whose radius is r units and whose center lies at the point (h, k) is the set of all points in the plane whose distance from (h, k) is r units. See Fig. 1.12(a). Accordingly, the coordinates of any point P(x, y) on this circle satisfy the equation

$$\sqrt{(x - h)^2 + (y - k)^2} = r$$

or, if we square both sides,

$$(x - h)^2 + (y - k)^2 = r^2. \tag{3}$$

Conversely, any point in the plane whose coordinates satisfy Eq. (3) is a point whose distance from the point (h, k) is r units. In short, Eq. (3) is an equation for the circle.

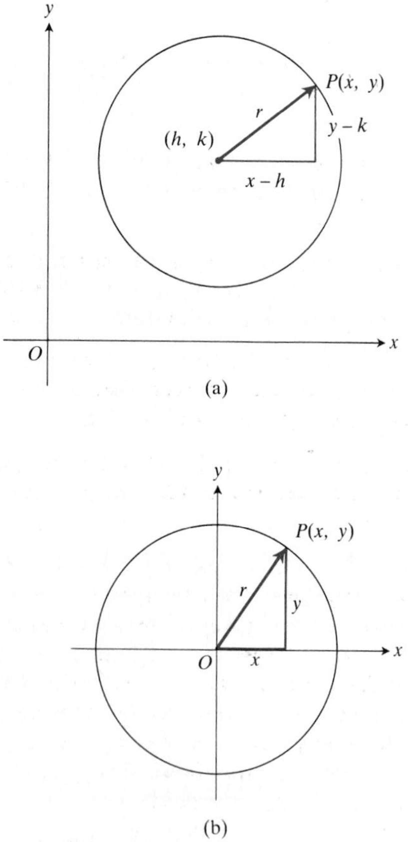

1.12 (a) For the points $P(x, y)$ on the circle of radius r centered at (h, k), $(x - h)^2 + (y - k)^2 = r^2$. (b) For the points $P(x, y)$ on the circle of radius r centered at the origin, $x^2 + y^2 = r^2$.

> The standard equation of a circle in the plane with center at (h, k) and radius r is
>
> $$(x - h)^2 + (y - k)^2 = r^2. \tag{3}$$

If the center of a circle is at the origin, then h and k in Eq. (3) are both equal to 0 and Eq. (3) becomes $x^2 + y^2 = r^2$. (See Fig. 1.12b.)

> The standard equation of a circle in the plane with center at the origin and radius r is
>
> $$x^2 + y^2 = r^2. \tag{4}$$

EXAMPLE 4 Write an equation for the circle whose radius is 7 units and whose center is the point (2, −3).

Solution We take $h = 2$, $k = -3$, and $r = 7$ in Eq. (3) to get

$$(x - 2)^2 + (y + 3)^2 = 49.$$

Since there is no particular reason to square out the quantities on the left, this is the equation we seek. □

REMARK. The distance formula $d = \sqrt{(\Delta x)^2 + (\Delta y)^2}$ in Eq. (2) calculates d in whatever unit the two axes are scaled in. For instance, suppose that x and y are given in meters. Then the increments Δx, Δy are in meters, their squares are in meters squared, the sum of their squares is in meters squared, and the square root of this sum is once again in meters.

PROBLEMS

1. Find the net horizontal and vertical changes that took place in the coordinates of the center of the ball shown in Fig. 1.9 when the ball moved from
a) A to G, b) D to E, c) C to F.

2. By what increments did the coordinates of the center of the ball in Fig. 1.9 change when the ball later rolled backward from G to F?

In Problems 3–8, a particle moves in the plane from A to B. Point C is to be found at the intersection of the horizontal line through A and the vertical line through B. Sketch the lines and find (a) the coordinates of C; (b) Δx; (c) Δy; and (d) the length of AB under the assumption that the x-axis and the y-axis are scaled in the same unit.

3. $A(-1, 1)$, $B(1, 2)$ **4.** $A(1, 2)$, $B(-1, -1)$

5. $A(-3, 2)$, $B(-1, -2)$ **6.** $A(-1, -2)$. $B(-3, 2)$

7. $A(-3, 1)$, $B(-8, 1)$ **8.** $A(0, 4)$, $B(0, -2)$

Suppose that a particle in the plane moves from $P_1(x, y)$ to $P_2(x + \Delta x, y + \Delta y)$ with the increments Δx and Δy given in Problems 9–16. Determine in each problem whether P_2 lies above, below, to the right, or to the left of P_1.

9. $\Delta x = 6$, $\Delta y = 3$ **10.** $\Delta x = 5$, $\Delta y = 0$

11. $\Delta x = -2$, $\Delta y = 0$ **12.** $\Delta x = 0$, $\Delta y = 2$

13. $\Delta x = 3$, $\Delta y = -1$ **14.** $\Delta x = -1$, $\Delta y = -2$

15. $\Delta x = 0$, $\Delta y = -5$ **16.** $\Delta x = -4$, $\Delta y = 0$

In Problems 17–20, write an equation for the circle of radius 5 units whose center is at the given point. Sketch the circle to show its relation to the coordinate axes.

17. $(0, 0)$ **18.** $(5, 0)$ **19.** $(0, 5)$ **20.** $(-3, 4)$

21. Write an equation for the circle that is centered at the origin and passes through the point $(1, 1)$.

22. Let L be the line through the origin that makes a 45° angle with the positive x-axis. Find the coordinates of the points on this line that are 1 unit away from the origin.

23. A particle starts at $A(-2, 3)$ and its coordinates change by increments $\Delta x = 5$, $\Delta y = -6$. Find its new position.

24. A particle starts at $A(6, 0)$ and its coordinates change by increments $\Delta x = -6$, $\Delta y = 0$. Find its new position.

25. The coordinates of a particle change by $\Delta x = 5$ and $\Delta y = 6$ in moving from $A(x, y)$ to $B(3, -3)$. Find x and y.

26. A particle started at $A(1, 0)$, circled the origin once counterclockwise, and returned to $A(1, 0)$. What were the net changes in its coordinates?

27. Find the starting position of a particle whose terminal position is $B(u, v)$ after its coordinates have changed by increments $\Delta x = h$ and $\Delta y = k$.

28. A particle moves from the point $A(-2, 5)$ to the y-axis in such a way that $\Delta y = 3\,\Delta x$. Find its new coordinates.

29. A particle moves at a constant velocity v in a straight line past the origin. The coordinate system and the particle's track are shown in Fig. 1.13. The positions of the particle shown on the line are $\Delta t = 1$ second apart. Why are the areas $A_1, A_2, \ldots, A_5$ in Fig. 1.13 equal? As in Kepler's second law (see the prologue), the line that joins the particle to the origin sweeps out equal areas in equal times. (*Hint:* Drop an altitude from the origin to the line of motion.)

30. A particle moves along the parabola $y = x^2$ from the point $A(1, 1)$ to the point $B(x, y)$, $x \neq 1$. Sketch the parabola and show that

$$\frac{\Delta y}{\Delta x} = x + 1 \quad \text{if } \Delta x \neq 0.$$

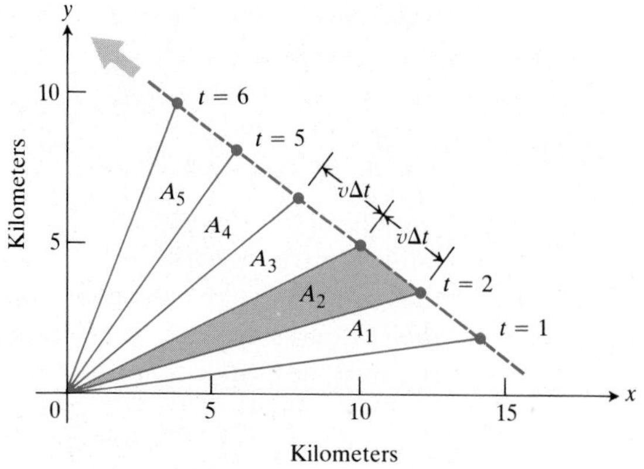

1.13 A particle moving in a straight line past the origin at a constant velocity v sweeps out equal areas $A_1, A_2, \ldots,$ in equal times (see Problem 29). The axes are scaled in kilometers. Time, denoted by t, is measured in seconds.

1.3 The Slope of a Straight Line

If we keep track of the prices of food or steel or computers, we can watch their progress on graph paper by plotting points and fitting them with a curve. We can extend the curve day by day as new prices appear. To what uses can we then put such a curve? We can see what the price was on any date. We can see by the slope of the curve (whose precise meaning will be given later) the rate at which the prices are rising or falling. If we plot other data on the same sheet of paper, we can perhaps see what relation they have to the rise and fall of prices. The curve may also reveal patterns that would help us to forecast the future with more accuracy than someone who has not graphed the data.

One of the many reasons calculus has proved so useful over the years is that it is the right mathematics for making the connection between rates of change and the slopes of smooth curves. Explaining this connection is one of the goals of this chapter. The basic plan is first to define what we mean by the slope of a line, and then to define the slope of a curve at each point on the curve to be the slope of the tangent line there. Just how this is done will become clear as the chapter goes on. Our first step is to find a practical way to calculate slopes of lines.

Civil engineers calculate the slope of a roadbed by calculating the ratio of the distance it rises or falls to the distance it runs horizontally. They call this ratio the *grade* of the roadbed. They usually write grades as percents, as shown in Fig. 1.14. Along the coast, railroad grades are usually less than 2%. In the mountains, they may go as high as 4%. Highway grades are usually less than 5%.

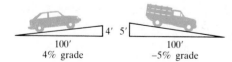

4′
5′
100′
4% grade
100′
−5% grade

Figure 1.14

In analytic geometry we define the slope of a line as a ratio of rise to run, too, but we do not usually express the ratios as percents.

Slope

To begin, let L be a line in the plane that is not parallel to the y-axis. Let $P_1(x_1, y_1)$ and $P_2(x_2, y_2)$ be any two points on L (see Fig. 1.15). We call $\Delta y = y_2 - y_1$ the *rise* from P_1 to P_2, and $\Delta x = x_2 - x_1$ the *run* from P_1 to

rise - y
run - x

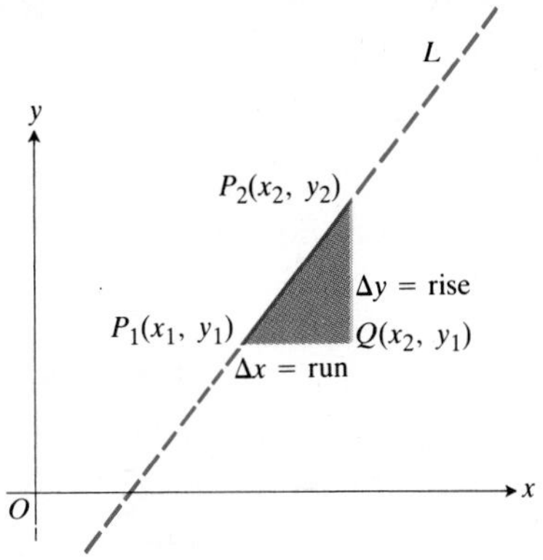

1.15 The slope of the line P_1P_2 is $m = \Delta y/\Delta x = \text{rise/run}$.

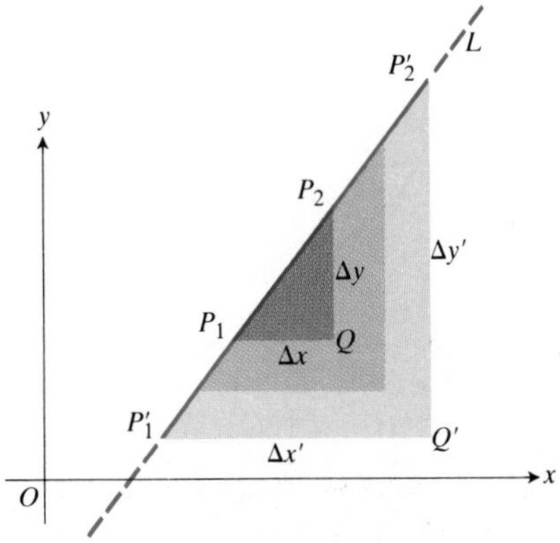

1.16 These triangles are similar. This tells us that $\Delta y/\Delta x = \Delta y'/\Delta x'$, because we know that the ratios of corresponding sides of similar triangles are equal.

P_2. The fact that L is not vertical tells us that Δx is not zero. We define the *slope* of L as the rise per unit of run. Thus if we denote the slope by m, we have the following.

$$\textit{Slope:} \quad m = \frac{\text{rise}}{\text{run}} = \frac{\Delta y}{\Delta x} = \frac{y_2 - y_1}{x_2 - x_1}. \tag{1}$$

Suppose that instead of choosing the points P_1 and P_2 to calculate the slope in Eq. (1), we were to choose a different pair of points P_1' and P_2' on L and calculate

$$m' = \frac{y_2' - y_1'}{x_2' - x_1'} = \frac{\Delta y'}{\Delta x'}. \tag{2}$$

Would we get the same answer for the slope? In other words, would $m' = m$? The answer is yes, as we can see from Fig. 1.16. The numbers m and m' are equal because they are ratios of corresponding sides of similar triangles:

$$m' = \frac{\Delta y'}{\Delta x'} = \frac{\Delta y}{\Delta x} = m. \tag{3}$$

The slope of a line, as we have defined it, depends only on how steeply the line rises or falls. We can calculate the slope from the coordinates of *any two points* on the line.

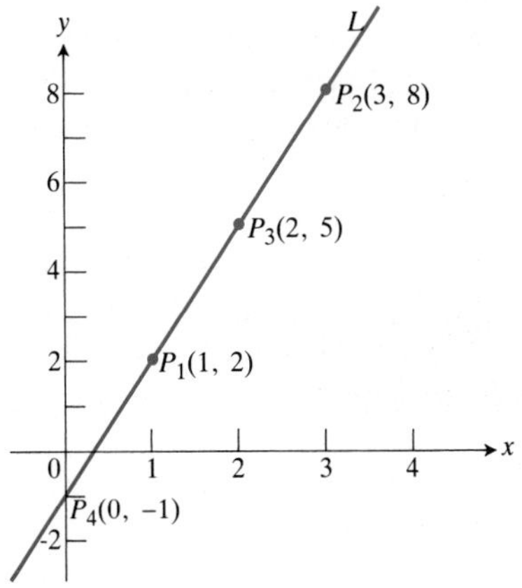

1.17 The slope of this line is $m = (8 - 2)/(3 - 1) = 3$. This means that $\Delta y = 3\,\Delta x$ for every change of position on the line.

EXAMPLE 1 Calculate the slope of the line through the points $P_1(1, 2)$ and $P_2(3, 8)$ in Fig. 1.17.

Solution We use Eq. (1) to get

$$m = \frac{y_2 - y_1}{x_2 - x_1} = \frac{8 - 2}{3 - 1} = \frac{6}{2} = 3.$$

A slope of $m = 3$ means that y increases 3 units every time x increases 1 unit. In other words, the change in y is 3 times the change in x. Along the line, $\Delta y = 3\,\Delta x$. The slope $m = 3$ is a proportionality factor. □

Suppose we take P_2 as (1, 2) and P_1 as (3, 8) instead of the other way around. Does Eq. (1) still give the answer $m = 3$? Yes, it does, as we can see here:

$$m = \frac{y_2 - y_1}{x_2 - x_1} = \frac{2 - 8}{1 - 3} = \frac{-6}{-2} = 3.$$

Renumbering the points P_1 and P_2 changed the signs of the rise and the run but did not change their ratio. When we use Eq. (1) to calculate the slope of a line from two points, it does not matter which point we take first or which we take second. This is true in general because

$$\frac{y_2 - y_1}{x_2 - x_1} = \frac{(-1)(y_2 - y_1)}{(-1)(x_2 - x_1)} = \frac{y_1 - y_2}{x_1 - x_2}.$$

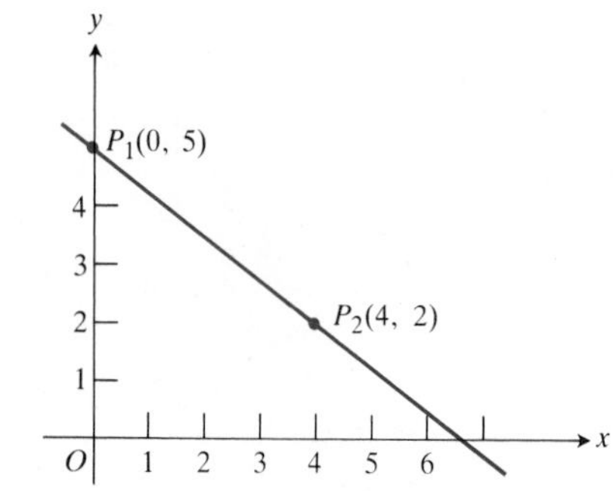

1.18 The slope of this line is $m = \Delta y/\Delta x = (2 - 5)/(4 - 0) = -3/4$. This means that y decreases 3 units every time x increases 4 units.

EXAMPLE 2 Calculate the slope of the line through $P_1(0, 5)$ and $P_2(4, 2)$ in Fig. 1.18.

Solution Equation (1) gives

$$m = \frac{y_2 - y_1}{x_2 - x_1} = \frac{2 - 5}{4 - 0} = \frac{-3}{4} = -\frac{3}{4}.$$

A slope of $m = -\frac{3}{4}$ tells us that y decreases 3 units whenever x increases 4 units. □

Lines that go uphill as x increases, like the line in Fig. 1.17, have positive slopes. Lines that go downhill as x increases, like the line in Fig. 1.18, have negative slopes. The slope of every horizontal line is 0. A horizontal line neither rises nor falls. The points on it all have the same y-coordinate, and the rise $\Delta y = y_2 - y_1$ is always 0.

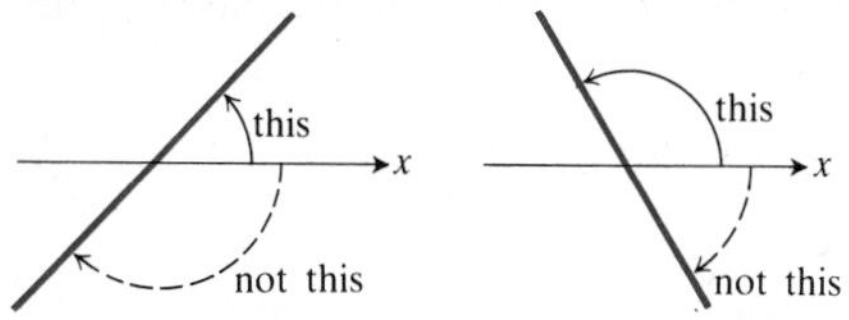

1.19 Angles of inclination are measured counterclockwise from the x-axis.

Angles of Inclination

The *angle of inclination* of a line that crosses the x-axis is the smallest positive angle that the line makes with the positively directed x-axis (Fig. 1.19). The angle of inclination of a line that does not cross the x-axis is taken to be 0°. Thus angles of inclination may have any measure from 0° up to but not including 180°.

> The slope of a line (Fig. 1.20) is the tangent of the line's angle of inclination ϕ (provided $\phi \neq 90°$):
>
> $$m = \tan\phi. \qquad (4)$$

EXAMPLE 3 Let us examine the slopes of lines whose angles of inclination are near 90°, like

$$\phi_1 = 89°59', \qquad m_1 = \tan\phi_1 \approx 3437.7$$

and

$$\phi_2 = 90°01', \qquad m_2 = \tan\phi_2 \approx -3437.7.$$

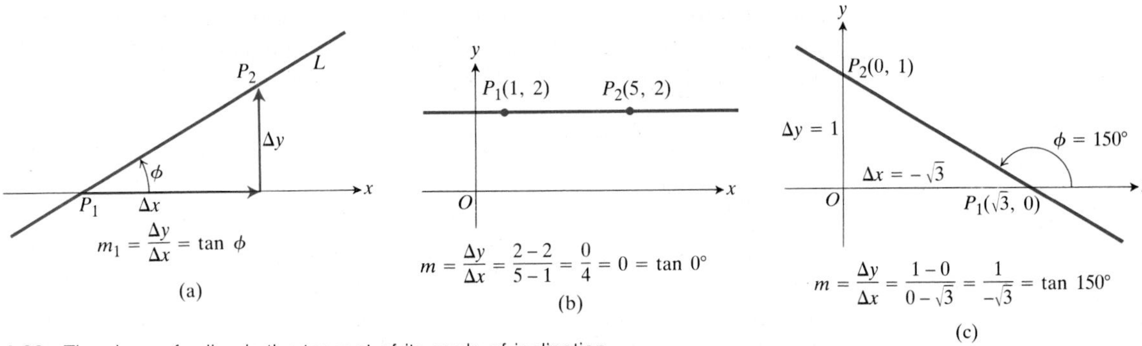

1.20 The slope of a line is the tangent of its angle of inclination.

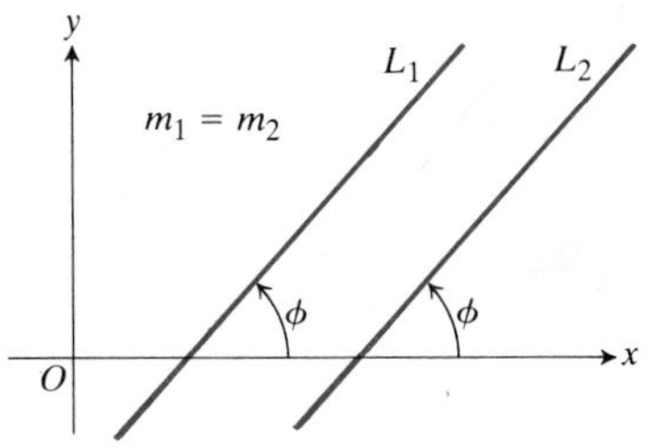

1.21 When parallel lines are not vertical, they have the same slope, $\tan \phi$.

The slopes of such lines are numerically very large. By taking ϕ still closer to $90°$ we can make the slope numerically larger than any number N, no matter how large N may be. This fact can be summarized by saying that the slope of a line "becomes infinite" as its angle of inclination approaches $90°$ or that a vertical line has "infinite slope." But, strictly speaking, we do not assign any real number to be the slope of a vertical line. Vertical lines have no slope. □

The symbol for infinity is ∞. But we should not use this symbol in addition, subtraction, multiplication, and division the way we use ordinary numbers. Infinity is not a number.

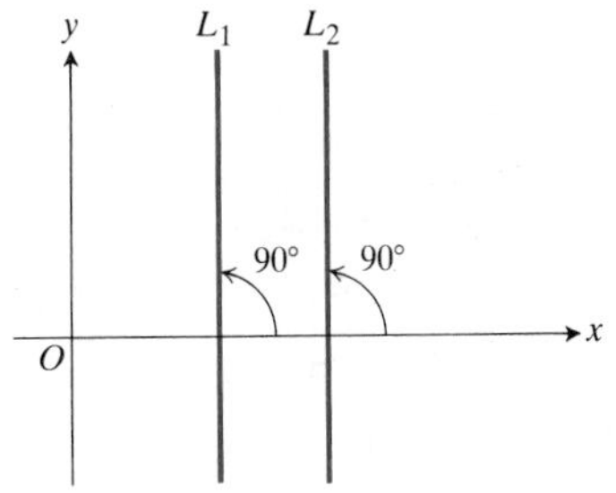

1.22 Vertical lines have equal angles of inclination, but no slope. Note that $\tan \phi$ is not defined when $\phi = 90°$.

Parallel Lines

Parallel lines have equal angles of inclination and hence, if they are not vertical, the same slope. See Figs. 1.21 and 1.22.

Conversely, two lines that have equal slopes $m_1 = \tan \phi_1 = m_2 = \tan \phi_2$ also have equal angles of inclination and are therefore parallel. The reason is this: If

$$0 \le \phi_1 < 180°, \qquad 0 \le \phi_2 < 180°, \tag{5}$$

and

$$\tan \phi_1 = \tan \phi_2, \tag{6}$$

then $\phi_1 = \phi_2$.

Perpendicular Lines

The angles of inclination of perpendicular lines differ by $90°$, so that when ϕ_2 is the larger angle, $\phi_2 = 90° + \phi_1$.

CASE 1. If one line is vertical, then the other must be horizontal.

CASE 2. If neither line is vertical (see Fig. 1.23), then neither ϕ_1 nor ϕ_2 is $90°$ or $0°$, and

$$\begin{aligned} m_2 = \tan \phi_2 = \tan (90° + \phi_1) &= \frac{\sin (90° + \phi_1)}{\cos (90° + \phi_1)} \\ &= \frac{\cos \phi_1}{-\sin \phi_1} = -\frac{1}{\tan \phi_1} = -\frac{1}{m_1}. \end{aligned} \tag{7}$$

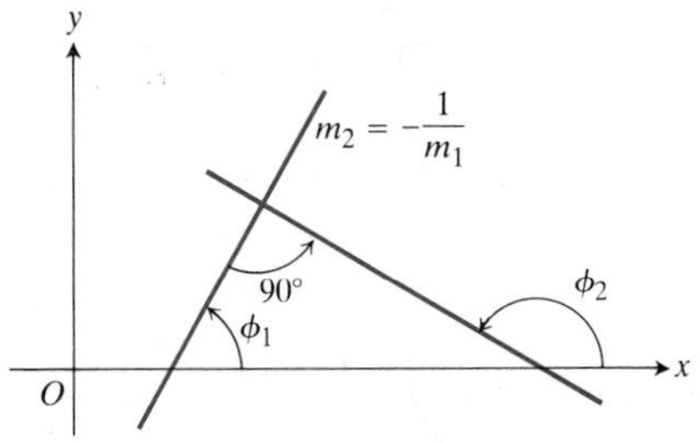

1.23 The exterior angle theorem of geometry says that $\phi_2 = 90° + \phi_1$ in this figure. This in turn means that $m_2 = -1/m_1$, as explained in the text.

Thus the slopes of the lines satisfy the equation

$$m_2 = -\frac{1}{m_1}. \tag{8}$$

The argument can be reversed to show that two lines whose slopes satisfy Eq. (8) are perpendicular. The number $-1/m_1$ in Eq. (8) is called the *negative reciprocal* of m_1.

Summary

1. The slope of the line through $P_1(x_1, y_1)$ and $P_2(x_2, y_2)$, $x_1 \neq x_2$, is
$$m = \frac{\text{rise}}{\text{run}} = \frac{y_2 - y_1}{x_2 - x_1}.$$
2. $m = \tan\phi$ (ϕ is the angle of inclination).
3. Vertical lines have no slope.
4. Horizontal lines have slope 0.
5. For lines that are neither horizontal nor vertical it is handy to remember:
   a) they are parallel $\Leftrightarrow m_2 = m_1$;
   b) they are perpendicular $\Leftrightarrow m_2 = -1/m_1$.

PROBLEMS

In Problems 1–16, plot the points A and B, and find the slope (if any) of the line determined by them. Also find the slope (if any) of a line perpendicular to AB.

1. $A(1, -2)$, $B(2, 1)$ **2.** $A(-1, 2)$, $B(-2, -1)$
3. $A(-2, -1)$, $B(1, -2)$ **4.** $A(2, -1)$, $B(-2, 1)$
5. $A(1, 0)$, $B(0, 1)$ **6.** $A(-1, 0)$, $B(1, 0)$
7. $A(2, 3)$, $B(-1, 3)$ **8.** $A(0, 3)$, $B(2, -3)$
9. $A(0, -2)$, $B(-2, 0)$ **10.** $A(1, 2)$, $B(1, -3)$
11. $A(0, 0)$, $B(-2, -4)$ **12.** $A(\frac{1}{2}, 0)$, $B(0, -\frac{1}{3})$
13. $A(0, 0)$, $B(x, y)$, $(x \neq 0, y \neq 0)$
14. $A(0, 0)$, $B(x, 0)$, $(x \neq 0)$
15. $A(0, 0)$, $B(0, y)$, $(y \neq 0)$
16. $A(a, 0)$, $B(0, b)$, $(a \neq 0, b \neq 0)$

In Problems 17–21, plot the points A, B, C, and D. Then determine whether or not $ABCD$ is a parallelogram. Determine which parallelograms are rectangles.

17. $A(0, 1)$, $B(1, 2)$, $C(2, 1)$, $D(1, 0)$
18. $A(3, 1)$, $B(2, 2)$, $C(0, 1)$, $D(1, 0)$
19. $A(-2, 2)$, $B(1, 3)$, $C(2, 0)$, $D(-1, -1)$
20. $A(-1, -2)$, $B(2, -1)$, $C(2, 1)$, $D(1, 0)$
21. $A(-1, 0)$, $B(0, -1)$, $C(2, 0)$, $D(0, 2)$

Problems 22–24 are about the stairway shown in Fig. 1.24. The slope of the stairway can be calculated from the riser height R and the tread width T as R/T. The

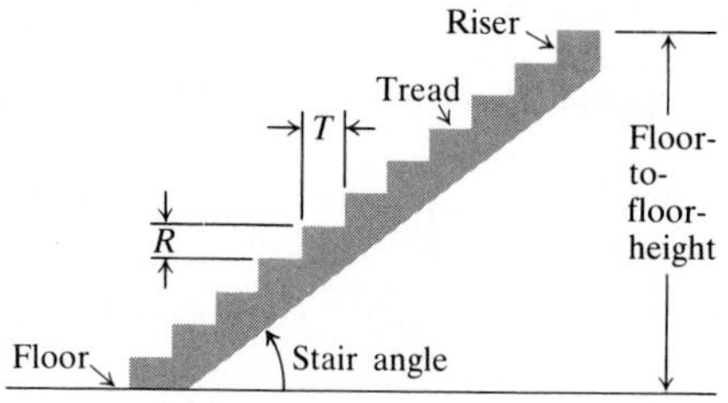

1.24 A stair profile, where R = the riser height and T = the tread width.

manual from which the drawing was adapted defines a *stairway* as a stepped footway having a slope not less than 5:16, or $31\frac{1}{4}$ percent, and not greater than 9:8 or $112\frac{1}{2}$ percent. (The manual goes on to say that below these limits footways become ramps, above them step-ladders!)

22. What are the minimum and maximum stair angles allowed in Fig. 1.24 by the definition of a stairway?

23. A common angle for household stairs is 40°. If the treads on the stairs are 9 in. wide, about how high are the risers?

24. Find the slope of a 40° stairway.

Solve Problems 25 and 26 by measuring slopes in Fig. 1.25.

25. Find the rate of change of temperature in degrees per inch for

a) gypsum wall board,
b) fiberglass insulation, and
c) wood sheathing.

26. Which of the materials just listed is the best insulator? the poorest? Explain.

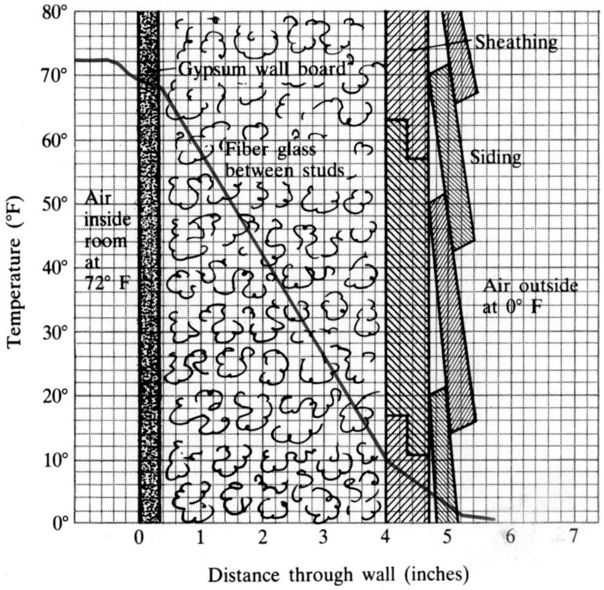

1.25 Temperature gradients in a wall. (*Source: Differentiation*, by W. U. Walton *et al.*, Project CALC, Education Development Center, Inc., Newton, Mass. (1975); p. 25.)

27. The line through the origin and $P(x, y)$ has slope $+2$, and the line through $(-1, 0)$ and $P(x, y)$ has slope $+1$. Find x and y.

28. A line L goes through the origin and has slope $+3$. If $P(x, y)$ is a point on the line, show that $y = 3x$. Sketch the line.

29. Sketch the line L that passes through the origin with slope $-\frac{1}{2}$. Show that if $P(x, y)$ lies on L, then $y = -(x/2)$.

In Problems 30–34, use slopes to determine whether the given points are collinear (that is, lie on a common straight line).

30. $A(1, 0)$, $B(0, 1)$, $C(2, 1)$

31. $A(-2, 1)$, $B(0, 5)$, $C(-1, 2)$

32. $A(-2, 1)$, $B(-1, 1)$, $C(1, 5)$, $D(2, 7)$

33. $A(-2, 3)$, $B(0, 2)$, $C(2, 0)$

34. $A(-3, -2)$, $B(-2, 0)$, $C(-1, 2)$, $D(1, 6)$

35. Given $A(0, -1)$, $B(4, 0)$, and $C(3, 4)$, show that ABC is a right triangle. Find the center and radius of the circumscribed circle. Sketch.

36. Find the coordinates of the point where the line in Fig. 1.17 crosses the x-axis. (*Hint:* On this line, $\Delta y = 3\,\Delta x$.)

37. Find the coordinates of the point where the line in Fig. 1.18 crosses the x-axis.

1.4
Equations of Straight Lines

In this article we show how to write and interpret equations for lines in the coordinate plane.

An *equation for a line* is an equation that is satisfied by the coordinates of the points that lie on the line, and is not satisfied by the coordinates of any points that lie elsewhere.

What do such equations do for us? They tell us when lines are vertical, and what their slopes are when they are not vertical. They show us how to calculate the value of y for any value of x on a nonvertical line or the value of x for any value of y on a nonhorizontal line. They also give us a useful way to summarize numerical data and to predict values of unrecorded data. Equations of lines play an important role in estimating roots of equations that are too complicated to solve directly (we shall have more to say about this in Chapter 2).

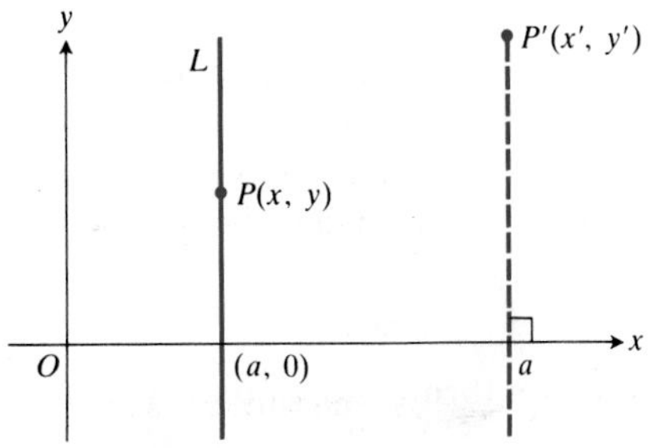

1.26 $x = a$ is an equation for the vertical line through the point $(a, 0)$.

Vertical Lines

Every vertical line L has to cross the x-axis at some point $(a, 0)$ (Fig. 1.26). The other points on L lie directly above or below $(a, 0)$. This means that the first coordinate of a point $P(x, y)$ on L must be $x = a$, whereas y can be

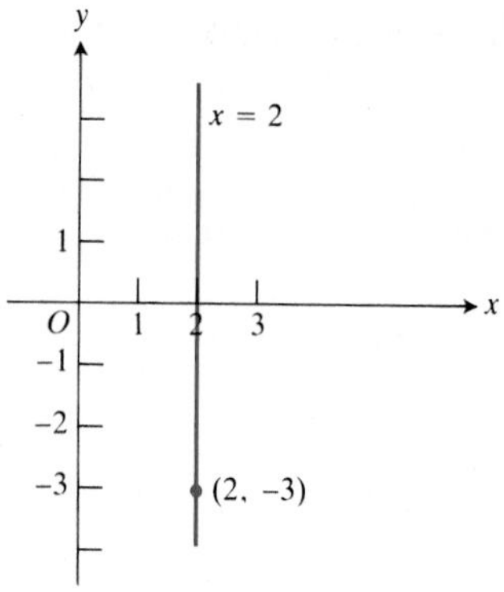

1.27 $x = 2$ is an equation for the vertical line through $(2, -3)$.

any number. In other words, the coordinates of all the points (x, y) on L satisfy the equation $x = a$.

To be sure that $x = a$ is an equation for L it remains to check only that the points not on L all have first coordinates different from a. They do, because the perpendiculars from these points to the x-axis do not hit the x-axis at the point $x = a$. That is, if $P'(x', y')$ does not lie on L, then $x' = a' \neq a$ (see Fig. 1.26).

> An equation for the *vertical line* through the point (a, b) is
>
> $$x = a. \tag{1}$$

EXAMPLE 1 An equation of the vertical line through the point $(2, -3)$ is $x = 2$. See Fig. 1.27. □

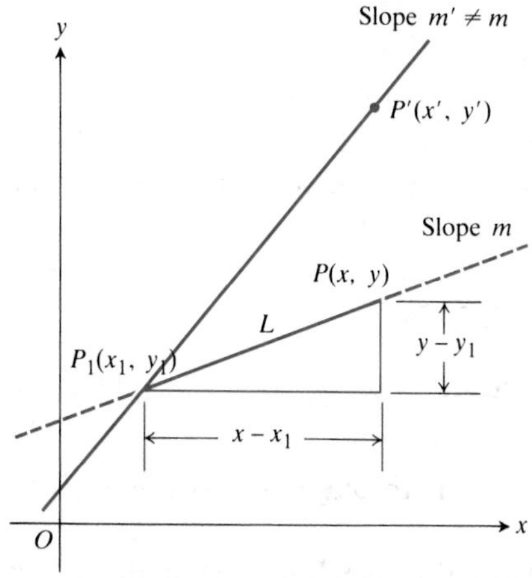

1.28 Suppose L is the line through $P_1(x_1, y_1)$ whose slope is m. Then other points $P(x, y)$ lie on this line if and only if slope $PP_1 = m$. This fact gives us the point–slope equation for L.

Lines That Are Not Vertical

To write an equation for a line L that is not vertical, it is enough to know its slope m and the coordinates of a point $P_1(x_1, y_1)$ on it. If $P(x, y)$ is any other point on L (as in Fig. 1.28), then $x \neq x_1$ and we can write the slope of L as

$$\frac{y - y_1}{x - x_1} \tag{2}$$

and set this expression equal to m to get

$$\frac{y - y_1}{x - x_1} = m. \tag{3}$$

Multiplying both sides of Eq. (3) by $x - x_1$ gives us the more useful equation

$$y - y_1 = m(x - x_1). \tag{4}$$

Equation (4) is an equation for L, as we can check right away. Every point (x, y) on L satisfies the equation—even the point (x_1, y_1). How about the points not on L? If $P'(x', y')$ is a point not on L, then, as Fig. 1.28 shows, the slope m' of $P'P_1$ is different from m and the coordinates x' and y' of P' do not satisfy Eqs. (3) and (4).

Equation (4) is called a *point–slope equation* for L.

DEFINITION

> The *point–slope equation* of the line through the point (x_1, y_1) with slope m is
>
> $$y - y_1 = m(x - x_1). \tag{5}$$

EXAMPLE 2 Write an equation for the line through the point $(1, 2)$ with slope $m = -\frac{3}{4}$. Where does this line cross the x-axis? the y-axis? the line $x = 3$?

Solution To find an equation for the line, we use Eq. (5) with $(x_1, y_1) = (1, 2)$ and $m = -\frac{3}{4}$:

$$y - 2 = -\frac{3}{4}(x - 1)$$

$$y = -\frac{3}{4}x + \frac{3}{4} + 2 \qquad (6)$$

$$y = -\frac{3}{4}x + \frac{11}{4}.$$

To find where the line crosses the x-axis, we set $y = 0$ in Eq. (6) and solve for x:

$$0 = -\frac{3}{4}x + \frac{11}{4} \quad \text{or} \quad x = \frac{11}{3}.$$

This shows that the line crosses the x-axis at $x = \frac{11}{3}$, or at the point $(\frac{11}{3}, 0)$.

To find where the line crosses the y-axis, we set $x = 0$ in Eq. (6) and solve for y:

$$y = -\frac{3}{4}(0) + \frac{11}{4} = \frac{11}{4}.$$

This shows that the line crosses the y-axis at $y = \frac{11}{4}$, or at the point $(0, \frac{11}{4})$.

To find where the given line crosses the line $x = 3$, we set $x = 3$ in Eq. (6) and solve for y:

$$y = -\frac{3}{4}(3) + \frac{11}{4} = -\frac{9}{4} + \frac{11}{4} = \frac{2}{4} = \frac{1}{2}.$$

The coordinates of the point of intersection are therefore $(x, y) = (3, \frac{1}{2})$. □

EXAMPLE 3 Write an equation for the line through $(-2, -1)$ and $(3, 4)$.

Solution We first calculate the slope and then use Eq. (5) with one of the given points as (x_1, y_1). The slope of the line is

$$m = \frac{-1 - 4}{-2 - 3} = \frac{-5}{-5} = 1.$$

Either $(-2, -1)$ or $(3, 4)$ can be the (x_1, y_1) in Eq. (5), as the following calculations show.

Using $(-2, -1)$ we find	Using $(3, 4)$ we find
$y - (-1) = 1 \cdot (x - (-2))$	$y - 4 = 1 \cdot (x - 3)$
$y + 1 = x + 2$	$y - 4 = x - 3$
$y = x + 1.$	$y = x + 1.$

In practice, either one of these calculations will do. It is not necessary to do them both. □

Every line that is not vertical must cross the y-axis at some point $(0, b)$. The number b is called the *y-intercept* of the line. If we take $(x_1, y_1) = (0, b)$ in Eq. (5), we find that the line has the point-slope equation

$$y - b = m(x - 0), \qquad (7)$$

which is equivalent to

$$y = mx + b. \tag{8}$$

The equation $y = mx + b$ is called the *slope-intercept* equation of the line (Fig. 1.29).

DEFINITION

> The *slope-intercept equation* of the line with slope m and y-intercept b is
>
> $$y = mx + b. \tag{9}$$

Note that in Examples 2 and 3 the final form of the equation for each line was its slope-intercept equation.

EXAMPLE 4 The slope-intercept equation of the line whose slope is $m = -\frac{3}{4}$ and whose y-intercept is $b = 5$ is

$$y = -\frac{3}{4}x + 5.$$

This is the line in Fig. 1.18. □

EXAMPLE 5 Find the distance between the point $P(2, 1)$ and the line L: $y = x + 2$.

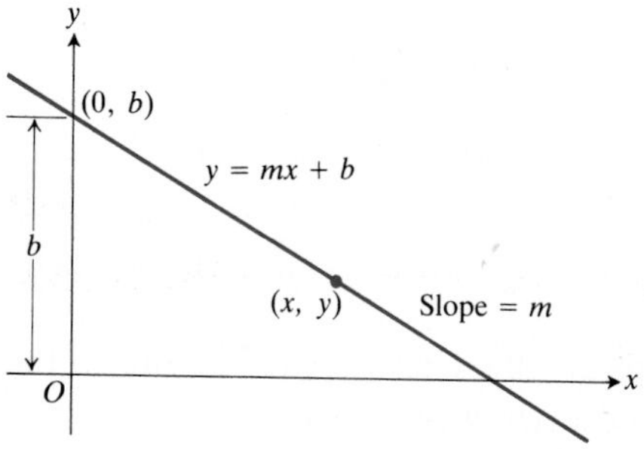

1.29 The line $y = mx + b$ has slope m and y-intercept b.

Solution We solve the problem in three steps: (1) Find an equation for the line L' through P perpendicular to L; (2) find the point Q where L' crosses L; and (3) calculate the distance between P and Q. See Fig. 1.30.

STEP 1. Find an equation for the line L' through $P(2, 1)$ perpendicular to L. The slope of L: $y = x + 2$ is $m = 1$. The slope of L' is therefore $m = -\frac{1}{1} = -1$. We set $(x_1, y_1) = (2, 1)$ and $m = -1$ in Eq. (5) to find L':

$$\begin{aligned} y - 1 &= -1(x - 2) \\ y &= -x + 2 + 1 \\ y &= -x + 3. \end{aligned}$$

STEP 2. Find the point Q where L': $y = -x + 3$ crosses L: $y = x + 2$ by solving their equations simultaneously.

To find the x-coordinate of Q, we equate the two expressions for y:

$$\begin{aligned} x + 2 &= -x + 3 \\ 2x &= 1 \\ x &= \frac{1}{2}. \end{aligned}$$

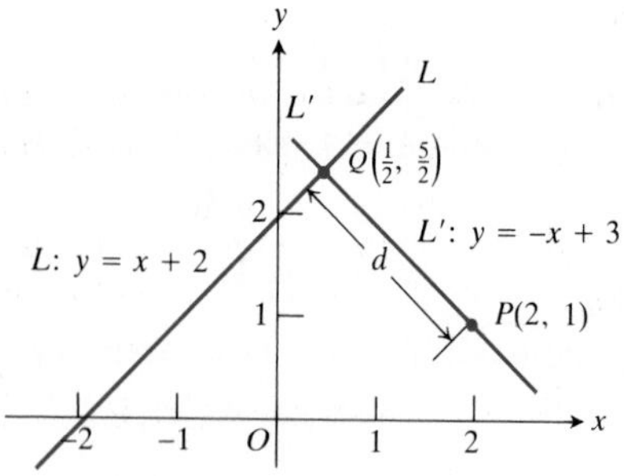

1.30 The distance d between point $P(2, 1)$ and line L is measured along the line L' perpendicular to L. It can be calculated from the coordinates of P and Q.

The x-coordinate of Q is $x = \frac{1}{2}$. The y-coordinate can be obtained by substituting $x = \frac{1}{2}$ in the equation for either line. We chose $y = x + 2$ arbitrarily, and find

$$y = \frac{1}{2} + 2 = \frac{5}{2}.$$

The coordinates of Q are $(\frac{1}{2}, \frac{5}{2})$.

STEP 3. The distance d between $P(2, 1)$ and L is the distance between $P(2, 1)$ and $Q(\frac{1}{2}, \frac{5}{2})$. We calculate d with the distance formula from Article 1.2:

$$d = \sqrt{\left(2 - \frac{1}{2}\right)^2 + \left(1 - \frac{5}{2}\right)^2} = \sqrt{\left(\frac{3}{2}\right)^2 + \left(-\frac{3}{2}\right)^2} = \sqrt{\frac{18}{4}} = \frac{3}{2}\sqrt{2}. \quad \square$$

For a general formula for the distance between a point and a line, see Miscellaneous Problem 8 at the end of this chapter.

When the line $y = mx + b$ is horizontal, its slope is $m = 0$. The equation $y = mx + b$ thus reduces to $y = b$.

> An equation for the *horizontal line* through the point (a, b) is
>
> $$y = b. \tag{10}$$

EXAMPLE 6 The equation of the horizontal line through the point (1, 2) is $y = 2$. See Fig. 1.20(b). □

The General Linear Equation

The equations

$$8x + 5y = 20, \qquad 7x = -14, \quad \text{and} \quad 3y = 6$$

all have the form

$$Ax + By = C \tag{11}$$

for particular values of A, B, and C. Moreover, any point-slope equation

$$y = mx + b$$

can be put in the form $Ax + By = C$ rearranging it as

$$mx - y = b. \tag{12}$$

Equation (12) has the form of Eq. (11), with $A = m$, $B = -1$, and $C = b$. Since the equations of vertical lines like

$$x = 5 \quad \text{and} \quad x = a$$

also have the form of Eq. (11) (with $A = 1$ and $B = 0$), we can see that every line in the coordinate plane has an equation like Eq. (11) with not both A and B equal to zero.

What about the converse? Is the graph of Eq. (11) always a line when at least one of A and B is different from zero? The answer is yes, as we shall now show.

If $B = 0$ in $Ax + By = C$ and $A \neq 0$, then we can solve for x by dividing both sides of the equation by A.

$$\text{If } B = 0: \qquad Ax = C \quad \text{and} \quad x = \frac{C}{A}.$$

The equation is the equation of a vertical line.

If $B \neq 0$ in $Ax + By = C$, we can solve for y by dividing both sides of the equation by B.

$$\text{If } B \neq 0: \qquad \begin{aligned} Ax + By &= C \\ \frac{A}{B}x + y &= \frac{C}{B} \\ y &= -\frac{A}{B}x + \frac{C}{B}. \end{aligned}$$

The equation is the slope-intercept equation of the line with slope $m = -A/B$ and y-intercept $b = C/B$.

An equation like (11) that contains only first powers of x and y is said to be "linear in x and y." Thus we may summarize our discussion of Eq. (11) by saying that every straight line in the plane has a linear equation, and every equation that is linear in x and y represents a straight line.

DEFINITION

> The *general linear equation* is
>
> $$Ax + By = C, \qquad (11)$$
>
> where A, B, and C are constants and not both A and B are zero.

EXAMPLE 7 Find the slope and y-intercept of the line $8x + 5y = 20$.

Solution First we solve the equation for y to obtain the slope-intercept equation of the line. Then we read the slope and y-intercept from the equation:

$$\begin{aligned} 8x + 5y &= 20 \\ 5y &= -8x + 20 \\ y &= -\frac{8}{5}x + 4. \end{aligned}$$

The slope is $m = -\frac{8}{5}$. The y-intercept is $b = 4$. □

Graphing

A quick way to graph a line that crosses both axes is to find the intercepts, mark them on the axes, then draw the line through the marked points. (The method fails only if the line passes through the origin or if the intercepts are hard to plot.)

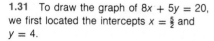

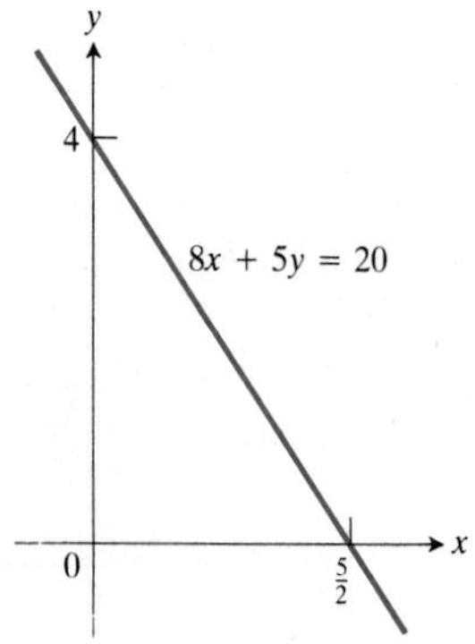

1.31 To draw the graph of $8x + 5y = 20$, we first located the intercepts $x = \frac{5}{2}$ and $y = 4$.

EXAMPLE 8 Graph the line $8x + 5y = 20$.

Solution

1. Find the x-intercept by setting $y = 0$ to obtain $8x = 20$ or $x = \frac{5}{2}$.
2. Find the y-intercept by setting $x = 0$ to obtain $5y = 20$ or $y = 4$.
3. Plot the intercepts and draw the line (Fig. 1.31). □

Important Formulas

$x = a$	Vertical line through (a, b)
$y = b$	Horizontal line through (a, b)
$y = mx + b$	Slope-intercept equation
$y - y_1 = m(x - x_1)$	Point-slope equation
$Ax + By = C$	General linear equation

PROBLEMS

In Problems 1–8, write an equation for (a) the vertical line and (b) the horizontal line through the given point.

1. $(2, 3)$ **2.** $(-7, -7)$ **3.** $(0, 0)$ **4.** $(0, -4)$

5. $(-4, 0)$ **6.** $(a, 0)$ **7.** $(0, b)$ **8.** (x_1, y_1)

In Problems 9–14, write an equation for the line determined by the given point and slope. Then graph the line.

9. $(1, 1),\ m = 1$ **10.** $(1, -1),\ m = -1$

11. $(-1, 1),\ m = 1$ **12.** $(-1, 1),\ m = -1$

13. $(0, b),\ m = 2$ **14.** $(a, 0),\ m = -2$

In Problems 15–28, find an equation for the line determined by the given points.

15. $(0, 0),\ (2, 3)$ **16.** $(1, 1),\ (2, 1)$

17. $(1, 1),\ (1, 2)$ **18.** $(-2, 1),\ (2, -2)$

19. $(-2, 0),\ (-2, -2)$ **20.** $(1, 3),\ (3, 1)$

21. $(T, 0),\ (0, F_0)\quad (T \neq 0, F_0 \neq 0)$

22. $(0, 0),\ (1, 0)$ **23.** $(0, 0),\ (0, 1)$

24. $(2, -1),\ (-2, 3)$ **25.** $(-0.7, 1.5),\ (1.4, 0.8)$

26. $(\sqrt{2}, \sqrt{2}),\ (\sqrt{5}, \sqrt{5})$ **27.** $(x_0, y_0),\ (x_1, y_1)$

28. $(0.5, 10{,}000),\ (2, 35{,}000)$

In Problems 29–34, write an equation for the line with the given slope and y-intercept. Graph the line.

29. $m = 3,\ b = -2$ **30.** $m = -1,\ b = 2$

31. $m = 1,\ b = \sqrt{2}$ **32.** $m = -\frac{1}{2},\ b = -3$

33. $m = -5,\ b = 2.5$ **34.** $m = \frac{1}{3},\ b = -1$

In Problems 35–48, find the slope and intercepts of the line. Then graph the line.

35. $y = 3x + 5$ **36.** $2y = 3x + 5$

37. $x + y = 2$ **38.** $2x - y = 4$

39. $x - 2y = 4$ **40.** $3x + 4y = 12$

41. $4x - 3y = 12$ **42.** $x = 2y - 5$

43. $\dfrac{x}{3} + \dfrac{y}{4} = 1$ **44.** $\dfrac{2x}{5} - \dfrac{y}{3} = 1$

45. $\dfrac{x}{2} - \dfrac{y}{3} = -1$ **46.** $\dfrac{x}{3} + \dfrac{y}{1} = -1$

47. $1.05x - 0.35y = 7$ **48.** $0.98x + 1.96y = 9.8$

49. Find the slope of the line $x/a + y/b = 1$ (a, b constants $\neq 0$). What are the x- and y-intercepts of this line? (This is called the *intercept equation* of the line.)

50. Find the slope of the line $x_1x + y_1y = 1$ (x_1, y_1 constants $\neq 0$).

51. Write an equation for the line (a) in Fig. 1.17; (b) in Fig. 1.20(c).

52. *Lines through the origin.*

a) Write equations for the four lines that pass through the origin and the points $(1, \frac{1}{2})$, $(1, 1)$, $(1, 2)$, $(1, 3)$. Graph the lines. Label each line with its slope.

b) Repeat the instructions in (a) for the points $(-1, \frac{1}{2})$, $(-1, 1)$, $(-1, 2)$, and $(-1, 3)$.

c) Write an equation for the line through the origin with slope m. Find the coordinates of the point of intersection of this line with the line $x = 1$. Illustrate.

In Problems 53–66; find (a) an equation for the line through P parallel to L, (b) an equation for the line L' through P perpendicular to L, and (c) the distance from P to L.

53. $P(2, 1),\quad L: y = x + 2$

54. $P(0, 0),\quad L: y = -x + 2$

55. $P(0, 0),\quad L: x + \sqrt{3}y = 3$

56. $P(1, 2),\quad L: x + 2y = 3$

57. $P(-2, 2),\quad L: 2x + y = 4$

58. $P(3, 6),\quad L: x + y = 3$

59. $P(1, 0)$, L: $2x - y = -2$

60. $P(-2, 4)$, L: $x = 5$

61. $p(3, 2)$, L: $x = -5$

62. $P(3, 2)$, L: $y = -4$

63. $P(a, b)$, L: $x = -1$

64. $P(3, -h)$, L: $y = 4$ $(h > 0)$

65. $P(4, 6)$, L: $4x + 3y = 12$

66. $P(2/\sqrt{3}, -1)$, L: $\sqrt{3}x + y = -3$

In Problems 67–74, find the angle of inclination of the given line.

67. $y = x + 2$ **68.** $y = -x + 2$ **69.** $x + \sqrt{3}y = 3$

70. $x + 2y = 3$ **71.** $2x + y = 4$ **72.** $2x - y = -2$

73. $4x + 3y = 12$

74. $\sqrt{3}x + y = -3$

In Problems 75–78, find the line through the given point with the given angle of inclination ϕ.

75. $(1, 4)$, $\phi = 60°$ **76.** $(-1, -1)$, $\phi = 135°$

77. $(-2, 3)$, $\phi = 90°$ **78.** $(3, -2)$, $\phi = 0°$

79. The pressure p experienced by a diver under water is related to the diver's depth d by an equation of the form $p = kd + 1$ (k is a constant). When $d = 0$ meters, the pressure is 1 atmosphere. The pressure at 100 meters is about 10.94 atmospheres. Find the pressure at 50 meters.

80. A ray of light comes in along the line $x + y = 1$ above the x-axis and reflects off the x-axis. See Fig. 1.32. Write an equation for its new path.

81. The steel in railroad track expands when heated. For the track temperatures encountered in normal outdoor use, the length s of a piece of track is related to its temperature t by a linear equation. An experiment with a piece of track gave the following measurements:

$$t_1 = 65°\text{F}, \qquad s_1 = 35 \text{ ft},$$
$$t_2 = 135°\text{F}, \qquad s_2 = 35.16 \text{ ft}.$$

Write a linear equation for the relation between s and t.

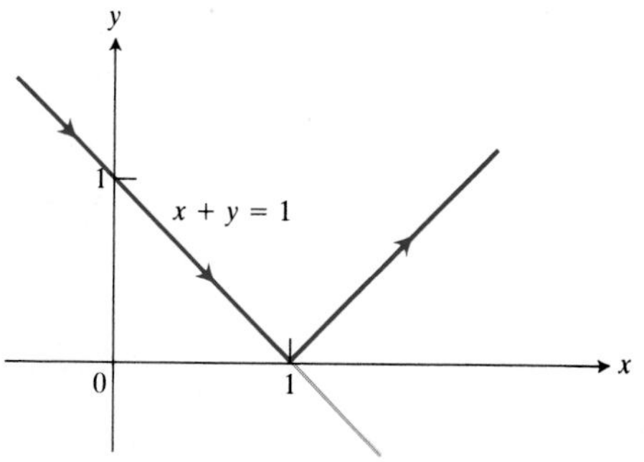

1.32 The path of the light ray in Problem 80.

82. Fahrenheit (F) and Celsius (C) temperature readings are related by a linear equation; that is, the graph of F vs. C is a straight line.
a) Find an equation that relates F and C, given that $C = 0$ when $F = 32$ and $C = 100$ when $F = 212$.
b) Is there a temperature at which $C = F$? If so, what is it?

83. Suppose that the length of the perpendicular ON from the origin to a line L is p, and that ON makes an angle α with the positive x-axis. Show that $x \cos \alpha + y \sin \alpha = p$ is an equation for L. (*Remark:* When $A^2 + B^2 \neq 0$, the general linear equation $Ax + By = C$ can be written $x \cos \alpha + y \sin \alpha = p$, with $\cos \alpha = A/\sqrt{A^2 + B^2}$, $\sin \alpha = B/\sqrt{A^2 + B^2}$, and $p = C/\sqrt{A^2 + B^2}$.)

84. If A, B, C, C' are constants, and not both A and B are zero, show that (a) the lines

$$Ax + By = C, \qquad Ax + By = C'$$

either coincide or are parallel, and that (b) the lines

$$Ax + By = C, \qquad Bx - Ay = C'$$

are perpendicular.

***Toolkit* programs**	
Name That Function	Super * Grapher

1.5
Functions and Graphs

The values of one variable quantity often depend on the values of another; for example:

> The distance it takes to brake a car to a stop is a function of how fast the car is going when you start to brake.

The number of rabbits in a predator–prey food chain is a function of the number of foxes present.

The volume of water remaining in a bathtub is a function of the number of seconds since the plug was pulled.

The amount of time it takes you to decompress from an eighty-foot scuba dive is a function of how long you stayed down.

In all these examples, the values of one variable quantity, which we might call y, depend on the values of another variable quantity, which we might call x. In each case, if it is also true that the value of y is completely determined by the value of x, we can say that y is a function of x.

Euler invented a symbolic way to say "y is a function of x" by writing

$$y = f(x), \tag{1}$$

which we read "y equals f of x." This notation is shorter than the verbal statements that say the same thing. It also lets us give different functions different names by changing the letters we use. To say that braking distance is a function of speed we can write $d = f(s)$. To say that the concentration of a medicine in the bloodstream is a function of time between doses we can write $c = g(t)$ (we use a g here because we just used f for something else). We have to know what the variables d, s, c, and t mean, of course, before these equations make sense.

In mathematics, any rule that assigns to each element in one set some element from another set is called a *function*. The sets may be sets of numbers, sets of number pairs, sets of points, sets of objects of any kind. The sets do not have to be the same. All the function has to do is assign some element from the second set to each element in the first set. Thus a function is like a machine that assigns an output to every allowable input. The inputs make up the *domain* of the function. The outputs make up the function's *range* (Fig. 1.33).

1.33 A flow diagram for a function f.

DEFINITION

A *function* from a set D to a set R is a rule that assigns a single element of R to each element in D.

The word "single" in the definition of function does not necessarily mean that there is only one element in the function's range (although this could, and will, happen from time to time). It means that each input from the domain is assigned exactly one output from the range, no more and no less. In other words, each input appears just once in the list of input-output pairs defined by the function. A function is sometimes defined as a collection of ordered pairs (x, y) in which each element x of the function's domain appears at the first element of a pair exactly once, and each y is some element of the function's range. In such a list, the y's may repeat, but the x's cannot.

The functions in much of our work will have domains and ranges that are sets of real numbers. Such functions, called *real valued functions of a real variable,* are often given by formulas or equations, as in the following examples.

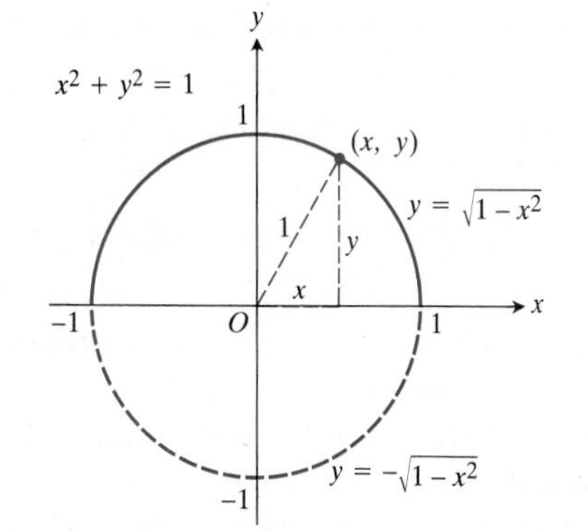

1.34 The circle $x^2 + y^2 = 1$. On the upper half, $y = \sqrt{1 - x^2}$; on the lower half, $y = -\sqrt{1 - x^2}$.

EXAMPLE 1 The formula $A = \pi r^2$ expresses the area A of a circle as a function of its radius r. In the context of geometry, the domain D of the function is the set of all possible radii—in this case, the set of all positive real numbers. The range is also the set of all positive real numbers. □

EXAMPLE 2 The formula $y = x^2$ defines the number y to be the square of the number x. We might call this function of x the "squaring" function because the output numbers are the squares of the input numbers. The general name for the function is "the function $y = x^2$."

The domain of the function $y = x^2$ is the set of allowable values of x—in this case, the set of all real numbers. The range, made up of the resulting values of y, is the set of all nonnegative numbers. □

EXAMPLE 3 If a point (x, y) lies on the circle in the plane that is centered at the origin and has radius 1 unit (see Fig. 1.34), then x and y satisfy the equation

$$x^2 + y^2 = 1.$$

This is equivalent to saying that

$$y^2 = 1 - x^2 \qquad \text{or} \qquad y = \pm\sqrt{1 - x^2},$$

which gives two possible formulas for y.

If the point (x, y) lies on the upper half of the circle, the semicircle with $y \geq 0$, then $y = +\sqrt{1 - x^2}$. If the point lies on the lower semicircle, then $y = -\sqrt{1 - x^2}$. In each case, y is a function of x whose domain is the interval from $x = -1$ to $x = 1$. The range of $y = \sqrt{1 - x^2}$ is $0 \leq y \leq 1$. The range of $y = -\sqrt{1 - x^2}$ is $-1 \leq y \leq 0$. □

REMARK. The equation $x^2 + y^2 = 1$ does not define y as a single function of x because for each x between -1 and $+1$ there are *two* values of y.

The domains and ranges of many functions in mathematics are intervals of real numbers like the ones shown in Fig. 1.35. The set of all real numbers that lie *strictly between* two fixed numbers a and b is called an *open* interval. The interval is "open" at each end because it contains neither of its endpoints. Intervals that contain both endpoints are called *closed*. Intervals that contain one endpoint but not both are called *half-open*. (Half-open intervals could just as well be called half-closed, but no one seems to call them that!)

The domains and ranges of functions can also be *infinite* intervals like the ones in Fig. 1.36.

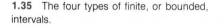

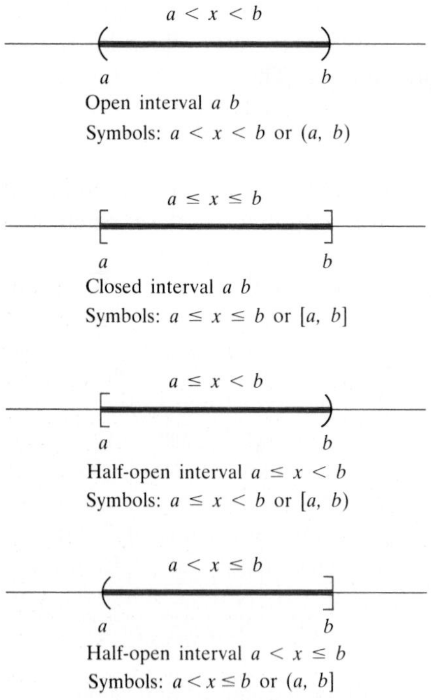

1.35 The four types of finite, or bounded, intervals.

EXAMPLE 4

	Function	Domain	Range
a)	$y = x^2$	$-\infty < x < \infty$	$0 \leq y$
b)	$y = \sqrt{1 - x^2}$	$-1 \leq x \leq 1$	$0 \leq y \leq 1$
c)	$y = \dfrac{1}{x}$	$x \neq 0$	$y \neq 0$
d)	$y = \sqrt{x}$	$0 \leq x$	$0 \leq y$
e)	$y = \sqrt{4 - x}$	$x \leq 4$	$0 \leq y$

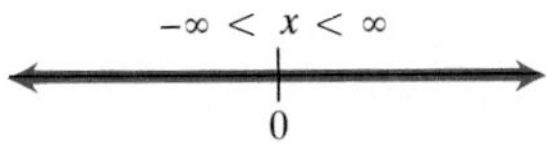

The set of real numbers, $(-\infty, \infty)$

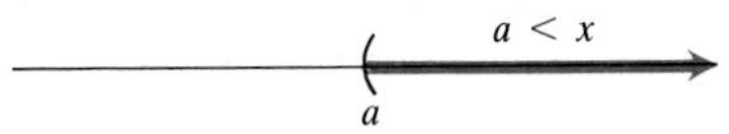

The set of numbers greater than a, (a, ∞)

The set of numbers greater than or equal to a, $[a, \infty)$

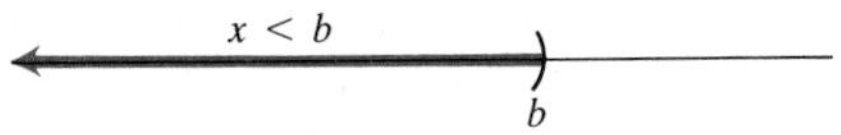

The set of numbers less than b, $(-\infty, b)$

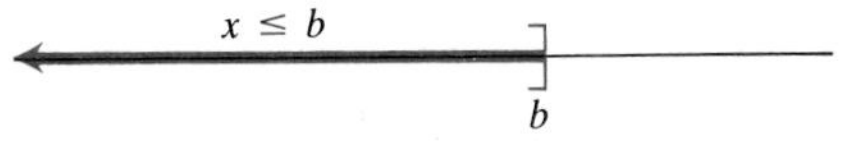

The set of numbers less than or equal to b, $(-\infty, b]$

1.36 Rays on the number line, with or without their endpoints, and the real line itself are called *infinite intervals*. There are five types in all, including the real line.

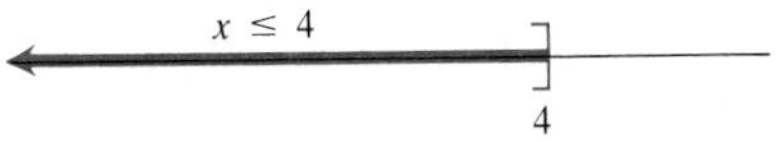

1.37 The domain of $y = \sqrt{4 - x}$.

y - independent
x - dependent

In each case, the domain of the function is taken to be the largest possible set of real x-values for which the formula gives real y-values. □

The formula $y = x^2$ gives real y-values for any real number x.

The formula $y = \sqrt{1 - x^2}$ gives real y-values for every value of x in the closed interval from -1 to 1. Beyond this domain, the quantity $1 - x^2$ is negative and its square root not a real number. (Complex numbers of the form $a + bi$, where $i = \sqrt{-1}$, are excluded from our consideration until Chapter 12.)

The formula $y = 1/x$ gives a real y-value for every x except $x = 0$. We cannot divide 1 (or any other number, for that matter) by 0.

The formula $y = \sqrt{x}$ gives real y-values only when x is positive or zero. The number $y = \sqrt{x}$ is not a real number when x is negative. The domain of $y = \sqrt{x}$ is therefore restricted to the interval $x \geq 0$.

In $y = \sqrt{4 - x}$, the quantity $4 - x$ cannot be negative. That is, $4 - x$ must be greater than or equal to 0. In symbols,

$$0 \leq 4 - x \tag{2}$$

or

$$0 + x \leq 4 - x + x \tag{3}$$

or

$$x \leq 4. \tag{4}$$

The formula $y = \sqrt{x - 4}$ gives a real y-value for any x less than or equal to 4 (Fig. 1.37).

Independent and Dependent Variables. Also, a Warning about Division by 0, and a Convention about Domains.

The variable x in a function $y = f(x)$ is called the *independent variable*, or *argument*, of the function. The variable y, whose value depends on x, is called the *dependent variable* of the function.

We must keep two restrictions in mind when we define functions. First, we *never divide by 0*. When we see $y = 1/x$, we must think $x \neq 0$. Zero is not in the domain of the function. When we see $y = 1/(x - 2)$, we must think "$x \neq 2$." Second, we will deal with real-valued functions only (except for a very short while later in the book). We may therefore have to restrict our domains when we have square roots (or fourth roots, or other even roots). If $y = \sqrt{1 - x^2}$, we should think "x^2 must not be greater than 1. The domain must not extend beyond the interval $-1 \leq x \leq 1$."

There is a convention about domains of functions defined by formulas like those in Example 1. If the domain is not stated explicitly, then the domain is always assumed to be the largest possible set of real x-values for which the formula gives real y-values. Therefore, if we wish to exclude some possible values from the domain of the function, then we must specify the intended domain. For example, writing

$$y = x^2$$

Table 1.1

x	$y = x^2$
−2.0	4.0
−1.75	3.0625
−1.5	2.25
−1.0	1.0
−0.5	0.25
0	0
0.5	0.25
1.0	1.0
1.5	2.25
1.75	3.0625
2.0	4.0

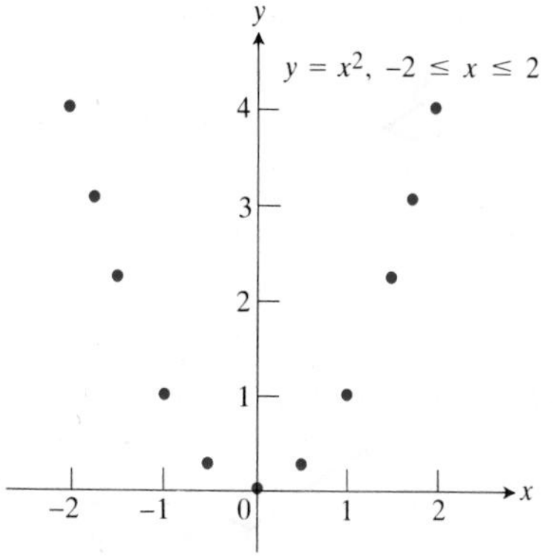

1.38 The points plotted from Table 1.2.

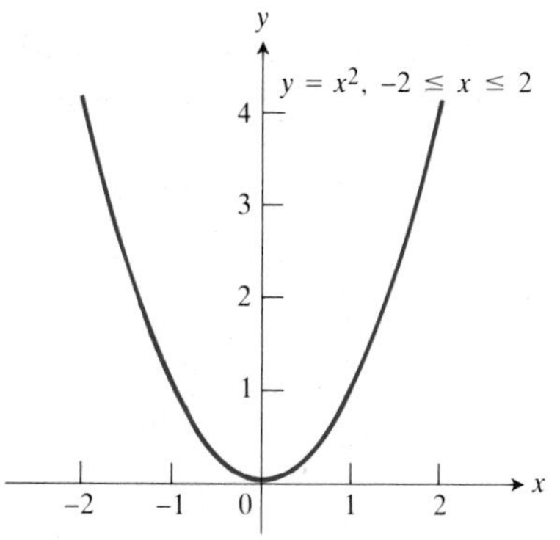

1.39 Figure 1.38 filled in.

without restriction assumes that $-\infty < x < \infty$. If we wish to exclude negative values from consideration, we must write

$$y = x^2, \qquad x \geq 0.$$

REMARK. A given function may be expressed by using different letters for its variables. For example,

$$r = 3s^2 - 7, \qquad y = 3x^2 - 7, \qquad z = 3m^2 - 7,$$

all express the function whose rule is "square, triple, and then subtract seven."

Graphs and Graphing

The straight lines we drew in Article 1.4 were the graphs of functions like $y = x + 1$ and $y = -(\frac{8}{5})x + 4$. To graph one of these functions, we plotted two points whose coordinates satisfied the given equation and drew a line through the plotted points. The line was the set of points in the plane whose coordinate pairs (x, y) were the input-output pairs of the function.

In general, the set of points in the plane whose coordinate pairs are the input-output pairs of a function is called the *graph* of the function. In the following examples we graph functions whose graphs are not straight lines.

EXAMPLE 5 Graph the function $y = x^2$ over the interval $-2 \leq x \leq 2$.

Solution To graph the function, we carry out the following steps:

1. We make a table of input-output pairs for the function (Table 1.1).
2. We plot the corresponding points to learn the shape of the graph (Fig. 1.38).
3. We sketch the graph by connecting the points (Fig. 1.39). □

In Example 5 we graphed the function $y = x^2$ over the interval $-2 \leq x \leq 2$. What about the rest of the graph? The domain and range of $y = x^2$ are both infinite, so we cannot hope to draw the entire graph. But we can imagine what the graph looks like by examining the formula $y = x^2$ and by looking at the picture we already have. As x moves away from the interval $-2 \leq x \leq 2$ in either direction, $y = x^2$ increases rapidly. When x is 5, y is 25. When x is 10, y is already 100. The graph goes up as shown in Fig. 1.40, continuing the pattern we see in Fig. 1.39.

The basic idea for graphing curves that are not straight lines is to plot points until we see the curve's shape, and then to use the formula for y to find out how y changes as x moves between or away from the plotted values. But what points do we plot?

Here are some rules about choosing good points to plot. When we get to Chapter 3, we shall have more to say about using the formula $y = f(x)$ to predict how y changes between plotted points.

Choosing Points for Graphing $y = f(x)$

1. Plot any points where the graph crosses or touches the axes. These points are often easy to find by setting $y = 0$ and $x = 0$ in the equation $y = f(x)$.

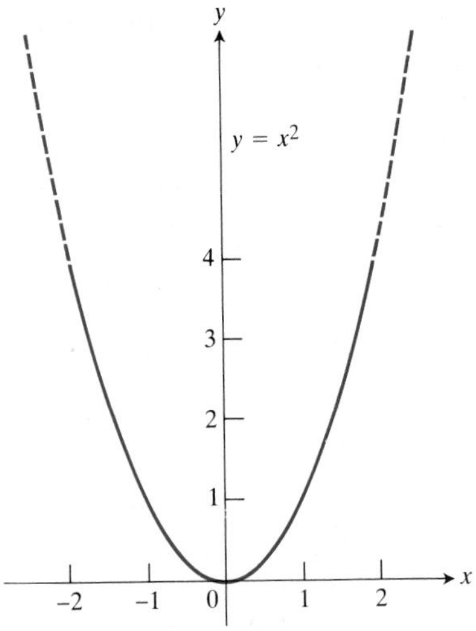

1.40 Figure 1.39 extended.

x	$y = 1/x$
-2	$-\frac{1}{2}$
-1	-1
$-\frac{1}{2}$	-2
$-\frac{1}{3}$	-3
$\frac{1}{3}$	3
$\frac{1}{2}$	2
1	1
2	$\frac{1}{2}$

(a)

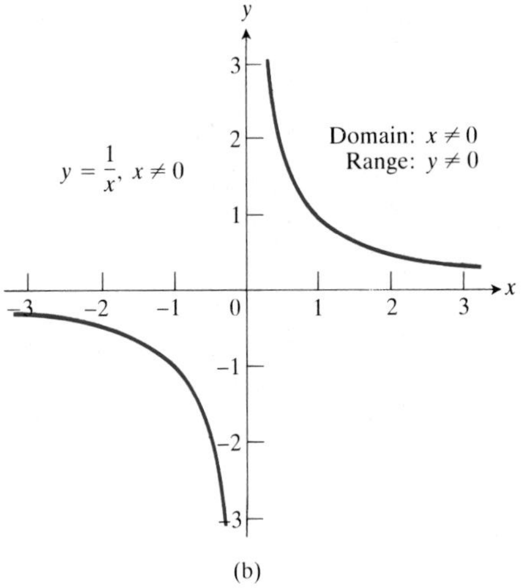

(b)

1.41 (a) Values of $y = 1/x$ for selected values of x. (b) The graph of $y = 1/x$.

2. Plot a few points near the origin. When the values of x are small, the values of y are often easy to compute or estimate.
3. Graph the function at or near any endpoints of its domain.

Figures 1.41, 1.42, and 1.43 show tables of values and graphs for the functions in parts (c), (d), and (e) of Example 4. The graph of the function $y = \sqrt{1 - x^2}$ in Example 4(b) is the upper semicircle in Fig. 1.34.

The functions we have graphed so far have been defined over their domains by single formulas. There is no reason, however, why a function cannot be defined by applying different formulas to different parts of its domain, so long as it is clear which rule applies where. In the following example we define a function $y = f(x)$ by piecing together three different

1.42 (a) Values of $y = \sqrt{x}$ for selected values of x. (b) The graph of $y = \sqrt{x}$.

x	$y = \sqrt{x}$
0	0
$\frac{1}{4}$	$\frac{1}{2}$
1	1
2	$\sqrt{2}$
4	2

(a)

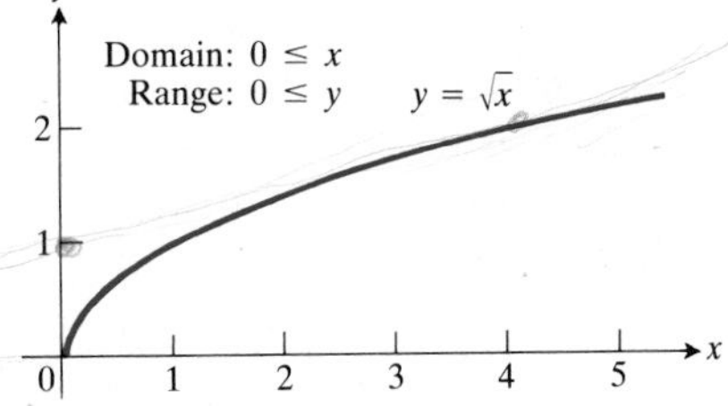

(b)

1.43 (a) Values of $y = \sqrt{4 - x}$ for selected values of x. (b) The graph of $y = \sqrt{4 - x}$.

x	$y = \sqrt{4 - x}$
4.0	0
3.75	0.5
2.0	$\sqrt{2}$
0	2.0
-2	$\sqrt{6}$

(a)

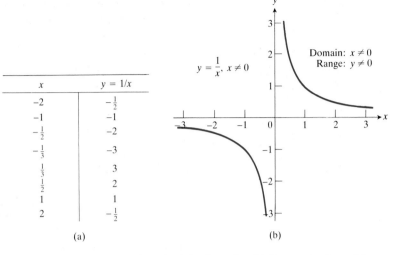

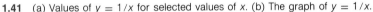

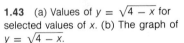

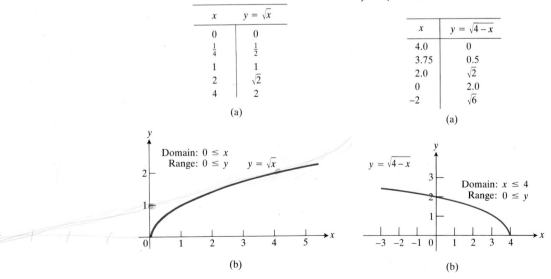

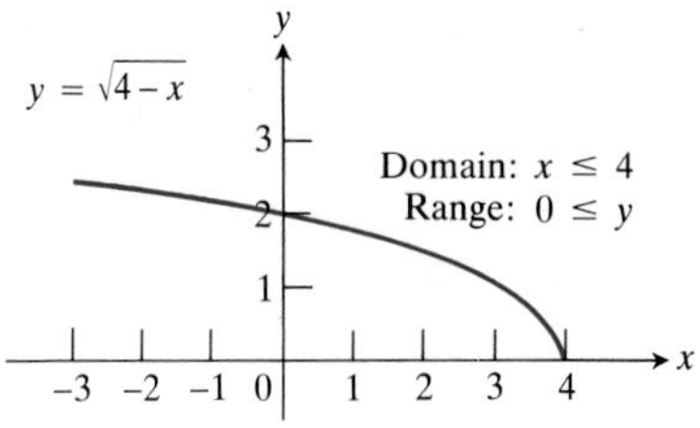

(b)

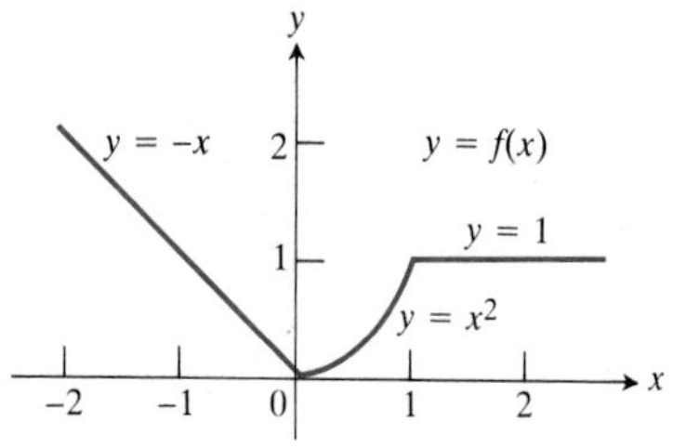

1.44 The function $y = f(x)$ shown here is graphed by applying different formulas to different parts of the function's domain.

rules. One rule applies to the interval $x < 0$, another to the interval $0 \leq x \leq 1$, another to the interval $x > 1$. The function is still *just one function,* whose domain is $-\infty < x < \infty$.

EXAMPLE 6 The values of the function

$$y = f(x) = \begin{cases} -x & \text{if } x < 0, \\ x^2 & \text{if } 0 \leq x \leq 1, \\ 1 & \text{if } x > 1, \end{cases}$$

are given by the formulas $y = -x$ when $x < 0$, $y = x^2$ when $1 \leq x \leq 1$, and $y = 1$ when $x > 1$. See Fig. 1.44 for the graph. □

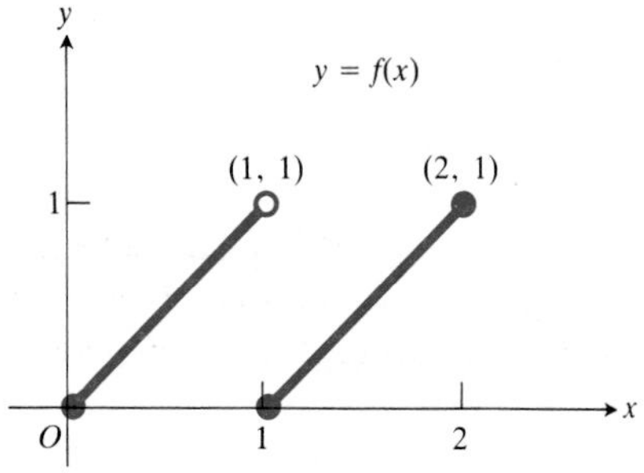

1.45 The graph of the function $y = f(x)$ (Example 7) shown here consists of two line segments. The segment on the left contains the endpoint at the origin (shown by a heavy dot) but does not contain the endpoint (1, 1). The segment on the right contains both endpoints.

EXAMPLE 7 Suppose that the graph of a function $y = f(x)$ consists of the line segments shown in Fig. 1.45. Write a formula for f.

Solution We find formulas for the segments from (0, 0) to (1, 1) and (1, 0) to (2, 1), and piece them together in the manner of Example 6.

Segment from (0, 0) to (1, 1). The line through (0, 0) and (1, 1) has slope $m = (1 - 0)/(1 - 0) = 1$ and y-intercept $b = 0$. Its slope-intercept equation is therefore $y = x$. The segment from (0, 0) to (1, 1) that includes the point (0, 0) but not the point (1, 1) is the graph of the function $y = x$ restricted to the half-open interval $0 \leq x < 1$, namely,

$$y = x, \quad 0 \leq x < 1.$$

Segment from (1, 0) to (2, 1). The line through (1, 0) and (2, 1) has slope $m = (1 - 0)/(2 - 1) = 1$ and passes through the point (1, 0). The corresponding point-slope equation for the line is therefore

$$y - 0 = 1(x - 1), \qquad \text{or} \qquad y = x - 1.$$

The segment from (1, 0) to (2, 1) that includes both endpoints is the graph of $y = x - 1$ restricted to the closed interval $1 \leq x \leq 2$, namely,

$$y = x - 1, \quad 1 \leq x \leq 2.$$

Formula for the function $y = f(x)$ shown in Fig. 1.45. The values of f on the interval $0 \leq x \leq 2$ are given by combining the formulas we obtained for the two segments of the graph:

$$f(x) = \begin{cases} x & \text{for } 0 \leq x < 1, \\ x - 1 & \text{for } 1 \leq x \leq 2. \end{cases} \square$$

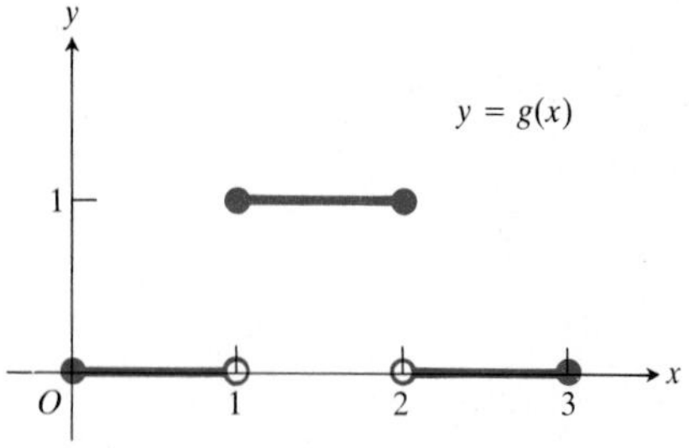

1.46 Functions like the one graphed here are called step functions. Example 8 shows how to write a formula for g.

EXAMPLE 8 The domain of the "step" function $y = g(x)$ graphed in Fig. 1.46 is the closed interval $0 \leq x \leq 3$. Find a formula for $g(x)$.

Solution The graph consists of three horizontal line segments. The left segment is the half-open interval $0 \leq x < 1$ on the x-axis, which we may think of as a portion of the line $y = 0$:

$$y = 0, \quad 0 \leq x < 1.$$

The second segment is the portion of the line $y = 1$ that lies over the closed interval $1 \leq x \leq 2$:

$$y = 1, \quad 1 \leq x \leq 2.$$

Table 1.2 Values of $\sin\theta$, $\cos\theta$, and $\tan\theta$ for selected values of θ.

Degrees	-180	-135	-90	-45	0	45	90	135	180
θ (radians)	$-\pi$	$-3\pi/4$	$-\pi/2$	$-\pi/4$	0	$\pi/4$	$\pi/2$	$3\pi/4$	π
$\sin\theta$	0	$-\sqrt{2}/2$	-1	$-\sqrt{2}/2$	0	$\sqrt{2}/2$	1	$\sqrt{2}/2$	0
$\cos\theta$	-1	$-\sqrt{2}/2$	0	$\sqrt{2}/2$	1	$\sqrt{2}/2$	0	$-\sqrt{2}/2$	-1
$\tan\theta$	0	1		-1	0	1		-1	0

The third segment is the half-open interval $2 < x \leq 3$ on the line $y = 0$:

$$y = 0, \quad 2 < x \leq 3.$$

The values of g are therefore given by the formula

$$g(x) = \begin{cases} 0 & \text{for } 0 \leq x < 1, \\ 1 & \text{for } 1 \leq x \leq 2, \\ 0 & \text{for } 2 < x \leq 3. \ \square \end{cases}$$

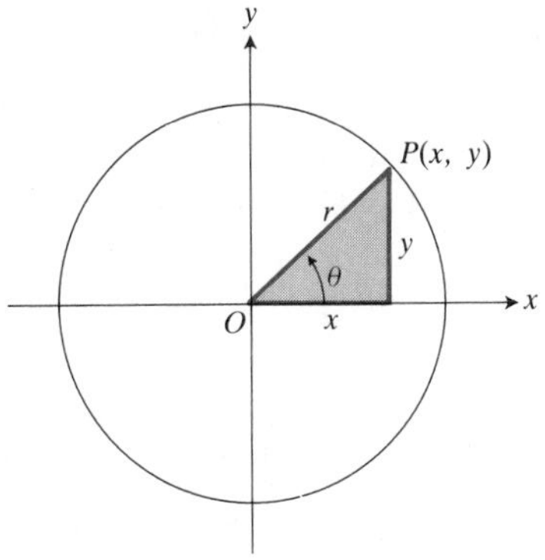

1.47 Angle θ in standard position.

EXAMPLE 9 The sine, cosine, and tangent functions. When an angle of θ degrees is placed in standard position at the center of a circle of radius r as in Fig. 1.47, the values of the sine, cosine, and tangent of the angle are given by the following formulas:

$$\sin\theta = \frac{y}{r}, \qquad \cos\theta = \frac{x}{r}, \qquad \tan\theta = \frac{y}{x}. \tag{5}$$

The function $\tan\theta$ is not defined for angles for which the denominator x is 0. This means that angles of $\pm 90°$, $\pm 270°$, and so on are not in the domain of the tangent function. In radian measure, the excluded angles are $\pm\pi/2$, $\pm 3\pi/2$, ... See Table 1.2 and Fig. 1.48. We shall review radian measure and trigonometric functions in Article 2.7. $\square$

1.48 Graphs of the sine, cosine, and tangent functions.

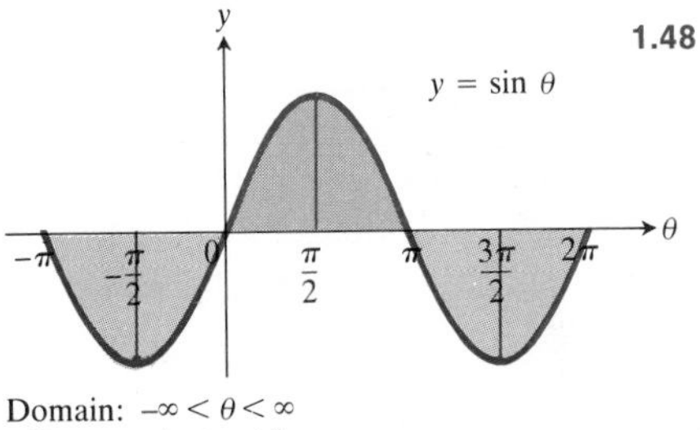

Domain: $-\infty < \theta < \infty$
Range: $-1 \leq y \leq 1$

(a)

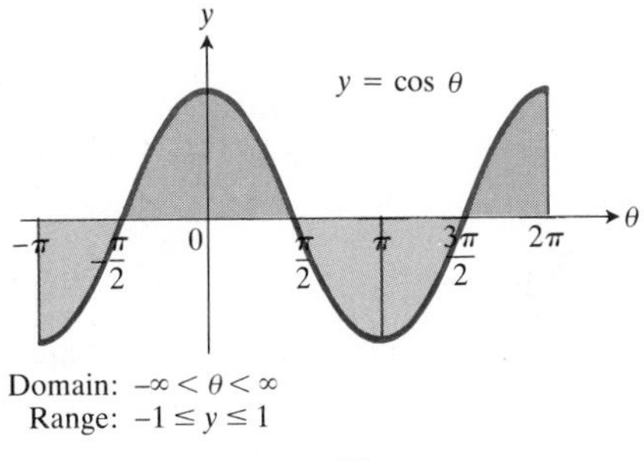

Domain: $-\infty < \theta < \infty$
Range: $-1 \leq y \leq 1$

(b)

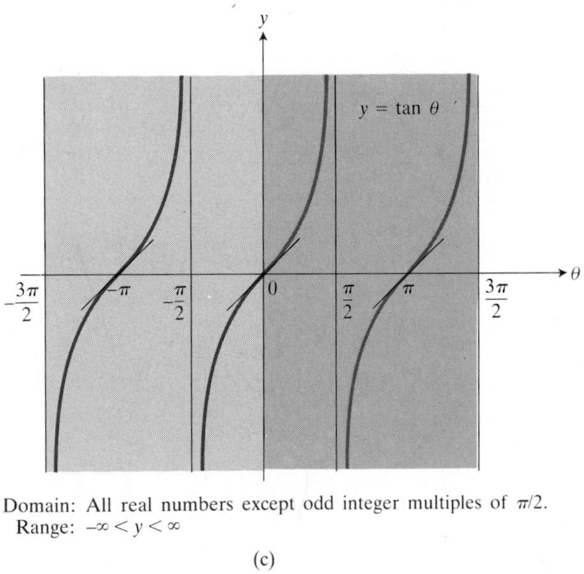

Domain: All real numbers except odd integer multiples of $\pi/2$.
Range: $-\infty < y < \infty$

(c)

Sums, Differences, Products, and Quotients of Functions

If $f(x)$ and $g(x)$ are two functions, with domains D_f and D_g, then the

sum	$f(x) + g(x)$,
differences	$f(x) - g(x)$,
	$g(x) - f(x)$,
product	$f(x) \cdot g(x)$,
quotients	$\dfrac{f(x)}{g(x)}, \quad g(x) \neq 0$,
	$\dfrac{g(x)}{f(x)}, \quad f(x) \neq 0$,

are also functions of x, defined for any value of x that lies in both D_f and D_g. The points at which $g(x) = 0$ must be excluded, however, to obtain the domain of the quotient $f(x)/g(x)$. Likewise, any points at which $f(x) = 0$ must be excluded from the domain of the quotient $g(x)/f(x)$.

EXAMPLE 10 Give the domains of

$$f(x) = \sqrt{x}, \qquad g(x) = \sqrt{1 - x}$$

and the corresponding domains of $f + g$, $f - g$, $g - f$, $f \cdot g$, f/g, and g/f.

Solution See Fig. 1.49. The domains of f and g are

$$D_f = [0, \infty), \qquad D_g = (-\infty, 1].$$

The points common to these domains are the points of the closed interval [0, 1]. On [0, 1], we have

sum $f + g$:	$f(x) + g(x) = \sqrt{x} + \sqrt{1 - x}$,
differences $f - g$:	$f(x) - g(x) = \sqrt{x} - \sqrt{1 - x}$,
$g - f$:	$g(x) - f(x) = \sqrt{1 - x} - \sqrt{x}$,
product $f \cdot g$:	$f(x) \cdot g(x) = \sqrt{x(1 - x)}$,

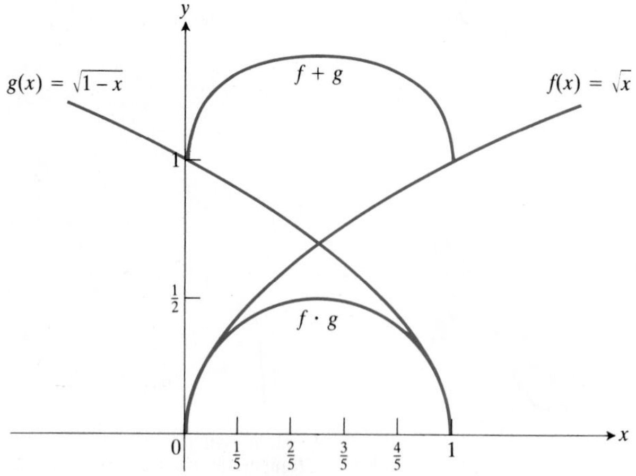

1.49 The domain of the function $f + g$ is the intersection of the domains of f and g, the interval [0, 1] on the x-axis where these domains overlap. This interval is also the domain of the function $f \cdot g$. See Example 10.

quotients:
$$\frac{f(x)}{g(x)} = \sqrt{\frac{x}{1-x}}, \quad x \neq 1,$$
$$\frac{g(x)}{f(x)} = \sqrt{\frac{1-x}{x}}, \quad x \neq 0.$$

The domains of $f + g$, $f - g$, $g - f$, and $f \cdot g$ are all the same, namely, the closed interval $[0, 1]$. The number $x = 1$ must be excluded from the domain of f/g, however, because $g(1) = \sqrt{1 - 1} = 0$. The domain of f/g is therefore the half-open interval $[0, 1)$. Similarly, the number $x = 0$ must be excluded from the domain of g/f because $f(0) = \sqrt{0} = 0$. The domain of g/f is therefore the half-open interval $(0, 1]$. □

Composition of Functions

Suppose that the outputs of a function f can be used as inputs of a function g. We can then hook f and g together to form a new function whose inputs are the inputs of f and whose outputs are the numbers $g(f(x))$, as in Fig. 1.50. We say that the function $g(f(x))$ (pronounced "g of f of x") is a *composite* of f and g. It is made by *composing* f and g in the order first f, then g. The usual "stand alone" notation for this composite is $g \circ f$, which is read as "g of f." Thus, the value of $g \circ f$ at x is $(g \circ f)(x) = g(f(x))$.

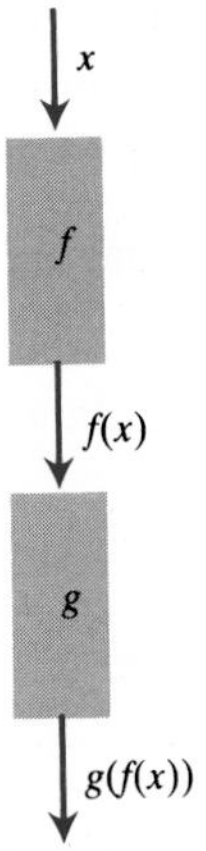

1.50 Two functions can be composed when the range of the first lies in the domain of the second.

EXAMPLE 11 If $f(x) = \sin x$ and $g(x) = -x/2$, write a formula for the composite $g(f(x))$.

Solution As suggested by Fig. 1.50, we can obtain a formula for $g(f(x))$ by substituting $f(x) = \sin x$ for the input variable x in $g(x) = -x/2$:

$$g(x) = -\frac{x}{2},$$
$$g(f(x)) = -\frac{f(x)}{2} = -\frac{\sin x}{2}.$$

The formula for $g(f(x))$ is therefore

$$g(f(x)) = -\frac{\sin x}{2}. \ \square$$

EXAMPLE 12 Find a formula for $g(f(x))$ if $f(x) = x^2$ and $g(x) = x - 7$. Then find the value of $g(f(2))$.

Solution To find $g(f(x))$ we replace the x in the formula for $g(x)$ by the expression given for $f(x)$:

$$g(x) = x - 7,$$
$$g(f(x)) = f(x) - 7 = x^2 - 7.$$

The value of $g(f(2))$ is found by substituting 2 for x in the formula for $g(f(x))$:

$$g(f(x)) = x^2 - 7,$$
$$g(f(2)) = (2)^2 - 7 = 4 - 7 = -3. \ \square$$

The order in which two functions $f(x)$ and $g(x)$ are composed can affect the result of composing them. In Example 12 we composed two functions f and g in the order first f, then g, to obtain the function $g(f(x))$. In the following example we compose the same functions in the reverse order, first g, then f, to obtain the function $f(g(x))$.

EXAMPLE 13 Find a formula for $f(g(x))$ if $f(x) = x^2$ and $g(x) = x - 7$. Then find $f(g(2))$.

Solution To find $f(g(x))$, we replace the x in the formula for $f(x)$ by the expression for $g(x)$:

$$f(x) = x^2,$$
$$f(g(x)) = (g(x))^2 = (x - 7)^2.$$

To find $f(g(2))$, we substitute 2 for x in the expression we just obtained for $f(g(x))$:

$$f(g(x)) = (x - 7)^2,$$
$$f(g(2)) = (2 - 7)^2 = (-5)^2 = 25. \square$$

REMARK. In the notation for composite functions, the parentheses tell which function comes first:

The notation $g(f(x))$ says "first f, then g." To calculate $g(f(2))$, we calculate $f(2)$, then apply g.

The notation $f(g(x))$ says "first g, then f." To calculate $f(g(2))$, we calculate $g(2)$, then apply f.

Absolute Values

Much of calculus concerns the behavior of functions on intervals, and it turns out that the handiest way to describe an interval centered at the origin or any other point is by using absolute values. We first define the absolute value of a real number, then look at some of the properties of the absolute value function that account for its usefulness in calculus.

DEFINITION

The *absolute value* of a number x is the number

$$|x| = \sqrt{x^2} = \begin{cases} x & \text{if } x \geq 0, \\ -x & \text{if } x < 0. \end{cases} \quad (6)$$

The symbol $|x|$ is read "the absolute value of x." The definition says that $|x|$ is equal to x when x is positive or zero and to x with its sign changed when x is negative. The graph of the function $y = |x|$ is shown in Fig. 1.51.

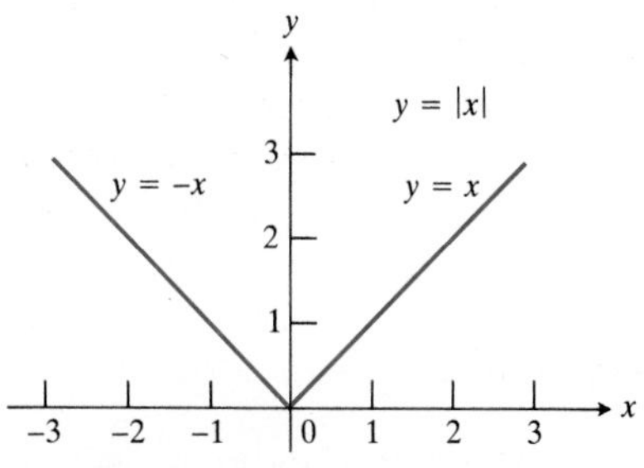

1.51 The absolute value function.

EXAMPLE 14 The absolute value of 3 is $|3| = 3$. The absolute value of 0 is $|0| = 0$. The absolute value of -5 is $|-5| = -(-5) = 5$. $\square$

The absolute value of a product of two numbers is the product of their absolute values. In symbols, we have

$$|ab| = |a||b| \quad \text{for all numbers } a \text{ and } b. \quad (7)$$

EXAMPLE 15 Examples of $|ab| = |a||b|$:

$$|(-1)(4)| = |-1||4| = (1)(4) = 4,$$
$$|3x| = |3||x| = 3|x|,$$
$$|-2(x + 5)| = |-2||x + 5| = 2|x + 5|. \square$$

The explanation of Eq. (7) is that for any numbers a and b the numbers $|ab|$ and $|a||b|$ are connected by the following equalities:

$$|ab| \quad = \quad \sqrt{(ab)^2} \quad = \quad \sqrt{a^2b^2} \quad = \quad \sqrt{a^2}\sqrt{b^2} \quad = \quad |a||b|. \tag{8}$$

⇧	⇧	⇧	⇧
Definition of $\lvert ab\rvert$	Because $(ab)^2 = a^2b^2$	The square root of a product of nonnegative numbers is the product of their square roots.	Definitions of $\lvert a\rvert$ and $\lvert b\rvert$

The absolute value of a sum of two numbers is never larger than the sum of their absolute values. When we put this in symbols, we get an inequality known as the *triangle inequality*.

DEFINITION

The Triangle Inequality

$$|a + b| \leq |a| + |b| \quad \text{for all numbers } a \text{ and } b \tag{9}$$

EXAMPLE 16 Examples of $|a + b| \leq |a| + |b|$:

$$|0 + 5| = 5 \leq |0| + |5| = 0 + 5 = 5,$$
$$|-3 + 0| = 3 \leq |-3| + |0| = 3 + 0 = 3,$$
$$|3 + 5| = 8 \leq |3| + |5| = 3 + 5 = 8,$$
$$|-3 - 5| = 8 \leq |-3| + |-5| = 3 + 5 = 8.$$

In all four cases, $|a + b|$ equals $|a| + |b|$.

On the other hand,

$$|-3 + 5| = |2| < |-3| + |5| = 8,$$
$$|3 - 5| = |-2| < |3| + |-5| = 8.$$

In both cases, $|a + b|$ is less than $|a| + |b|$. $\square$

The general rule is that $|a + b|$ is less than $|a| + |b|$ when a and b differ in sign. In all other cases, $|a + b|$ equals $|a| + |b|$.

Note that the absolute value bars in expressions like $|-3 + 5|$ also work like parentheses: We do the addition *before* taking the absolute value.

The numbers $|a - b|$ and $|b - a|$ are always equal and give the distance between a and b on the number line (Fig. 1.52). This is consistent with the square root formula for distance in the plane, because

$$\sqrt{(a - b)^2 + (0 - 0)^2} = \sqrt{(a - b)^2} = |a - b|, \tag{10}$$
$$\sqrt{(0 - 0)^2 + (a - b)^2} = \sqrt{(a - b)^2} = |a - b|. \tag{11}$$

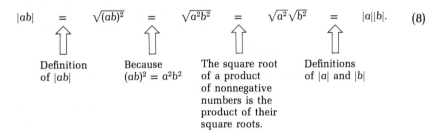

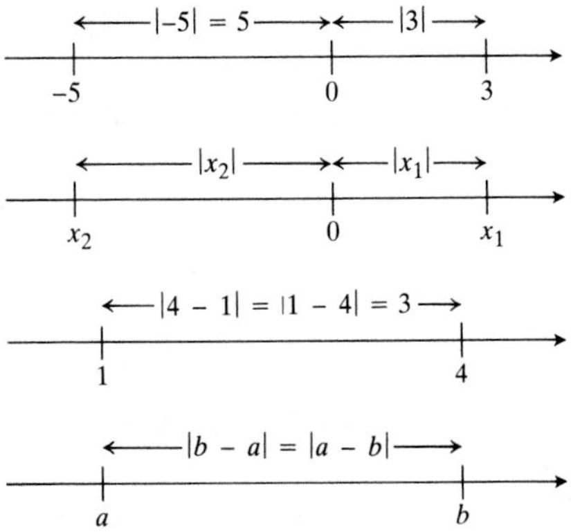

1.52 Absolute values give distances between points on the axes.

$$|a - b| = |b - a| \text{ for all numbers } a \text{ and } b. \tag{12}$$

This number is the distance between a and b on the number line.

The connection between absolute values and distance lets us use absolute value inequalities to describe intervals.

An inequality like $|a| < 5$ says that the distance from a to the origin is less than 5 units. This is equivalent to saying that a lies between -5 and 5. In symbols,

$$|a| < 5 \quad \Leftrightarrow \quad -5 < a < 5. \tag{13}$$

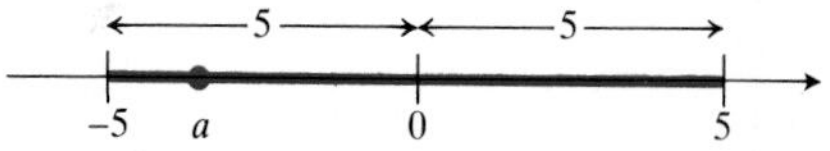

1.53 $|a| < 5$ means $-5 < a < 5$.

The set of numbers a with $|a| < 5$ is the same as the open interval from -5 to 5 (Fig. 1.53).

In general, if c is any positive number, then the absolute value of a is less than c if and only if a lies in the interval between $-c$ and c.

$$|a| < c \quad \Leftrightarrow \quad -c < a < c \tag{14}$$

EXAMPLE 17 Find the values of x that satisfy the inequality $|x - 5| < 9$.

Solution We first use Eq. (14) with $a = x - 5$ and $c = 9$ to change

$$|x - 5| < 9$$

to

$$-9 < x - 5 < 9. \tag{15}$$

Next, we add 5 to all three quantities in Eq. (15). This isolates the x:

$$-9 + 5 < x - 5 + 5 < 9 + 5$$
$$-4 < x < 14.$$

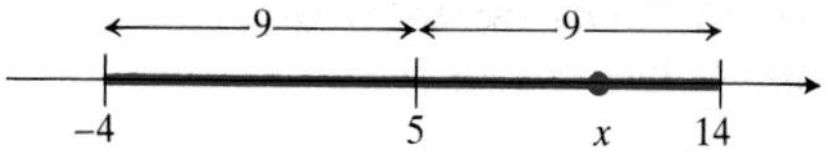

1.54 $|x - 5| < 9$ means $-4 < x < 14$.

These steps show that the values of x that satisfy the inequality $|x - 5| < 9$ are the numbers in the interval $-4 < x < 14$. (See Fig. 1.54.) □

EXAMPLE 18 Find the values of x that satisfy the inequality

$$\left|\frac{3x + 1}{2}\right| < 1.$$

Solution Change

$$\left|\frac{3x + 1}{2}\right| < 1$$

to

$$-1 < \frac{3x + 1}{2} < 1$$

to

$$-2 < 3x + 1 < 2$$

to

$$-3 < 3x < 1$$

to

$$-1 < x < \frac{1}{3}.$$

(See Fig. 1.55.) □

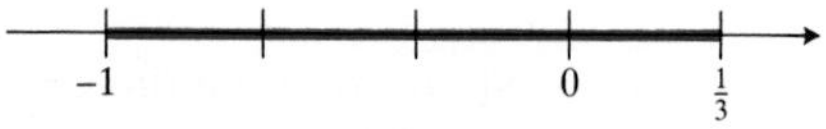

1.55 The inequality

$$\left|\frac{3x + 1}{2}\right| < 1$$

holds on the interval $-1 < x < \frac{1}{3}$.

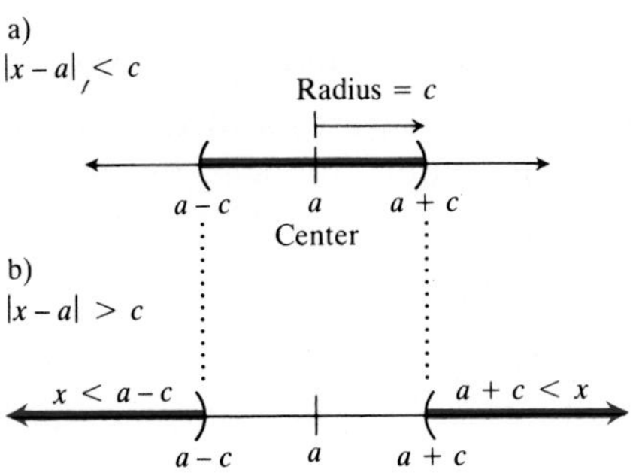

1.56 (a) The open interval $|x-a|<c$ runs from $a-c$ to $a+c$. (b) The graph of the inequality $|x-a|>c$ consists of two infinite intervals: the interval $x<a-c$ and the interval $x>a+c$.

EXAMPLE 19 Find the endpoints of the interval determined by the inequality $|x-a|<c$. What is the geometric meaning of the inequality $|x-a|>c$?

Solution To find the endpoints of the interval $|x-a|<c$ (Fig. 1.56a), we change

$$|x-a|<c$$

to

$$-c<x-a<c$$

to

$$a-c<x<a+c.$$

The endpoints are $a-c$ and $a+c$.

The points that satisfy the inequality $|x-a|>c$ (Fig. 1.56b) are the points on the x-axis whose distances from a are *greater* than c. These are the points *outside* the closed interval $|x-a|\leq c$, the points that lie to the right of $a+c$ and to the left of $a-c$. Geometrically, these points make up the two infinite open intervals $x<a-c$ and $x>a+c$. □

EXAMPLE 20 Graph the function $y=|x^2-1|$.

Solution We first apply the definition of absolute value to rewrite the formula $y=|x^2-1|$ without absolute value bars:

$$y=\begin{cases}(x^2-1) & \text{if } (x^2-1)\geq 0,\\ -(x^2-1) & \text{if } (x^2-1)<0.\end{cases}$$

In other words,

$$y=\begin{cases}x^2-1 & \text{if } x^2\geq 1,\\ 1-x^2 & \text{if } x^2<1,\end{cases}$$

or

$$y=\begin{cases}x^2-1 & \text{if } x\leq -1 \text{ or } x\geq 1,\\ 1-x^2 & \text{if } -1<x<1.\end{cases}$$

The graph of $y=|x^2-1|$ (Fig. 1.57) is the parabola $y=x^2-1$ with the negative part between $x=-1$ and $x=1$ reflected across the x-axis. □

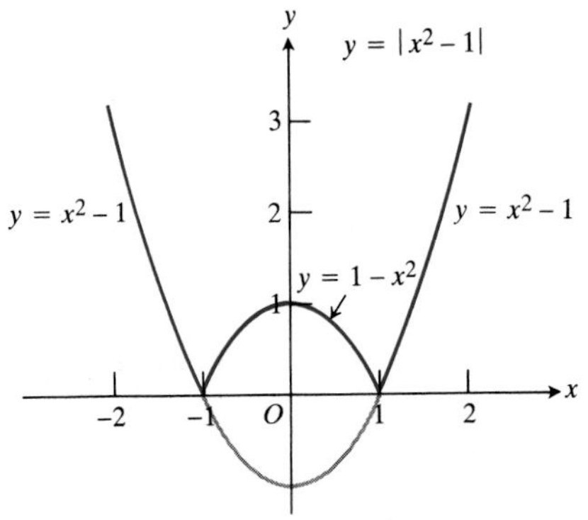

1.57 The graph of $y=|x^2-1|$ is the graph of $y=x^2-1$ if $x\leq -1$ or $x\geq 1$, and it is the graph of $y=1-x^2$ if $-1<x<1$.

Important Properties of Absolute Values

1. $|ab|=|a||b|$
2. $|a-b|=|b-a|$
3. $|a|=|-a|$
4. $|a+b|\leq|a|+|b|$
5. $|a|<c\Leftrightarrow -c<a<c$
6. $|x-a|<c\Leftrightarrow a-c<x<a+c$

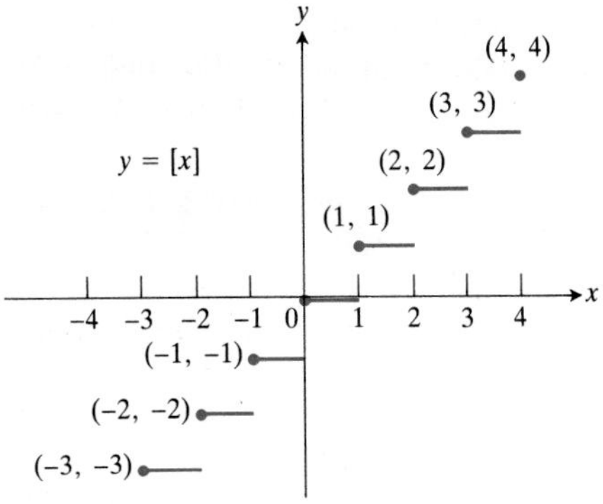

1.58 The graph of $y = [x]$, the greatest integer less than or equal to x. Domain: $-\infty < x < \infty$; range: the integers.

Step Functions

Most of our information about functions comes from graphs and equations. Any rule that gives a single value of y for each value of x, however, expresses y as a function of x. The function in the following example is defined by a verbal statement.

EXAMPLE 21 *The greatest integer function* $y = [x]$. For each real number x the value of $y = [x]$ is the largest integer that is less than or equal to x.

The symbol $[x]$ is read "the greatest integer in x." The function $y = [x]$ is graphed in Fig. 1.58. □

Selected values of $y = [x]$.

Positive values: $[1.9] = 1$, $[2] = 2$, $[3.4] = 3$.

The value 0: $[0.5] = 0$, $[0] = 0$.

Negative values: $[-2.7] = -3$, $[-0.5] = -1$.

Note that if x is negative, $[x]$ may have a larger absolute value than x does.

The greatest integer function is a *step function*. (So is the function in Example 8.) Many things around us can be modeled with step functions; for example,

The price of priority mail, as a function of weight;

The output of a blinking light, as a function of time;

The number displayed by a machine that gives digital outputs, as a function of time.

Step functions exhibit points of *discontinuity*, where they jump from one value to another without taking on any of the intermediate values. As shown in Fig. 1.58, $y = [x]$ jumps from $y = 1$ when $x < 2$ to $y = 2$ at $x = 2$ without taking on any of the values between 1 and 2.

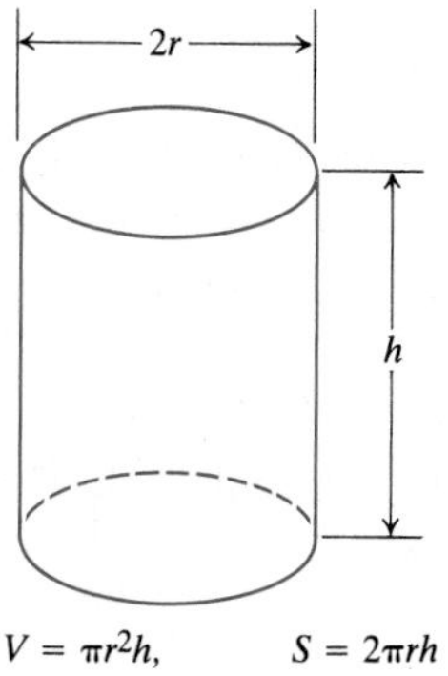

1.59 Right circular cylinder.

Functions of More than One Independent Variable

The values of some functions are calculated from the values of more than one independent variable.

The volume of a right circular cylinder, $V = \pi r^2 h$, is calculated from the values of the two independent variables r and h. So is the cylinder's lateral surface area $S = 2\pi rh$. See Fig. 1.59.

The formula $A = \frac{1}{2}(a + b)h$ calculates the area A of the trapezoid in Fig. 1.60 from the values of the three independent variables a, b, and h.

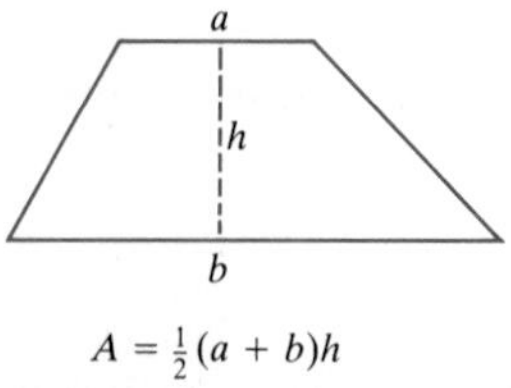

1.60 Trapezoid.

A more modern example is the function that gives the width of the Concorde's sonic boom carpet (Fig. 1.61).

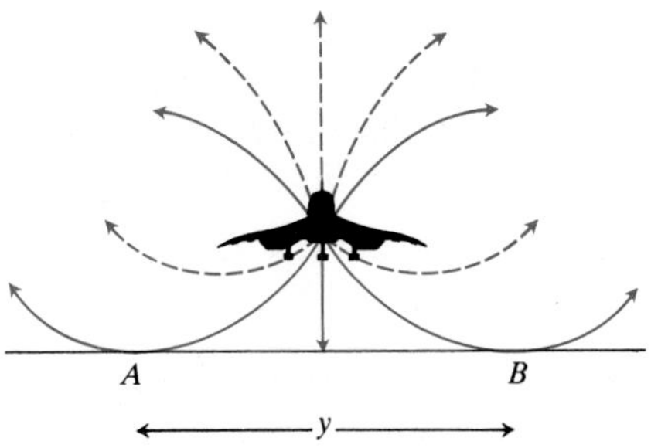

Sonic boom carpet

1.61 Sound waves from the Concorde bend as the temperature changes above and below the altitude at which the plane flies. The sonic boom carpet is the region on the ground that receives shock waves directly from the plane, not reflected from the atmosphere or diffracted along the ground. The carpet is determined by the grazing rays striking the ground from the point directly under the plane.

EXAMPLE 22†

The width y of the region in which people on the ground hear the Concorde's sonic boom directly, and not reflected from a layer in the atmosphere, is a function of

T = the surface temperature (in degrees Kelvin),

h = the elevation of the Concorde (in km),

d = the vertical temperature gradient (temperature drop in degrees Kelvin per km).

The formula for y is

$$y = 4(Th/d)^{1/2}. \tag{16}$$

Evaluated for $T = 296$ K (about 23°C or 73°F), $h = 13$ km, and $d = 6$ K/km, the formula yields $y = 101.3$ km. □

To calculate the width of the Concorde's sonic boom carpet from Eq. (16), we need to know the values of the earth's surface temperature T, the plane's elevation h, and the vertical temperature gradient d. But these values are all we need to know. As soon as we know them, we can determine the value of y. The function $y = 4(Th/d)^{1/2}$ is a function of the three variables T, h, and d.

The domain of the function $y = 4(Th/d)^{1/2}$ is the set of all triples (T, h, d) with T, h, and d nonnegative. The physical meanings of the variables keep their values from being negative. The range of y is the set of numbers $y \geq 0$.

Generally, let's suppose that D is a collection of n-tuples of numbers, written in the form

$$(x_1, x_2, x_3, \ldots, x_n).$$

A function f with domain D is a rule that assigns a number

$$y = f(x_1, x_2, x_3, \ldots, x_n) \tag{17}$$

to each n-tuple. The range of the function is the set of y-values. The number y is called the *output*, or *dependent*, variable of the function. The numbers $x_1, x_2, x_3, \ldots, x_n$ are called the *independent*, or *input*, variables of the function.

In the function $y = 4(TH/d)^{1/2}$, which gives the width of the Concorde's sonic boom carpet, T, h, and d are the independent variables and y is the dependent variable.

PROBLEMS

In Problems 1–12, find the domain and range of each function.

1. $y = \sqrt{x + 4}$ **2.** $y = \sqrt{x^2}$ **3.** $y = 1/(x - 2)$ **10.** $y = \sin(1/x)$ **11.** $y = x \sin x$

4. $y = \sqrt{-x}$ **5.** $y = (\sqrt{x})^2$ **6.** $y = \sqrt{4 - x^2}$

7. $y = 2 \cos x$ **8.** $y = \cos(1/x)$ **9.** $y = -3 \sin x$ **12.** $y = \sqrt{\dfrac{1 + \cos 2x}{2}}$

†From N. K. Balachandran, W. L. Donn, and D. H. Rind, "Concorde Sonic Booms as an Atmospheric Probe," *Science*, 1 July 1977, Vol. 197, p. 47.

In Problems 13–30, find (a) the domain and (b) the range of the function. Then (c) graph the function.

13. $y = x^2 + 1$ **14.** $y = x^2 - 2$ **15.** $y = -x^2$

16. $y = 4 - x^2$ **17.** $y = \sqrt{x + 1}$ **18.** $y = \sqrt{4 - x}$

19. $y = 1 + \sqrt{x}$ **20.** $y = \sqrt{9 - x^2}$ **21.** $y = (\sqrt{2x})^2$

22. $y = \dfrac{2}{x}$ **23.** $y = -\dfrac{1}{x}$ **24.** $y = \dfrac{1}{x^2}$

25. $y = \sin 2x$ **26.** $y = \cos 2x$ **27.** $y = \sin^2 x$

28. $y = \cos^2 x$ **29.** $y = 1 + \sin x$ **30.** $y = 1 - \cos x$

31. Consider the function $y = 1/\sqrt{x}$.
a) Can x be negative?
b) Can $x = 0$?
c) What is the domain of the function?

32. Consider the function $y = \sqrt{2 - \sqrt{x}}$.
a) Can x be negative?
b) Can $\sqrt{x}$ be greater than 2?
c) What is the domain of the function?

33. Consider the function $y = \sqrt{(1/x) - 1}$.
a) Can x be negative?
b) Can $x = 0$?
c) Can x be greater than 1?
d) What is the domain of the function?

34. Which of the graphs in Fig. 1.62 could be the graph of:
a) $y = x^2 - 1$? Why?
b) $y = (x - 1)^2$? Why?

35. Which of the graphs in Fig. 1.62 could *not* be the graph of $y = 4x^2$? Why?

36. By solving $y^2 = x$ for y, replace the equation by an equivalent system of equations each of which determines y as a function of x. Graph these two equations. (*Hint:* See Fig. 1.42.)

Figure 1.62

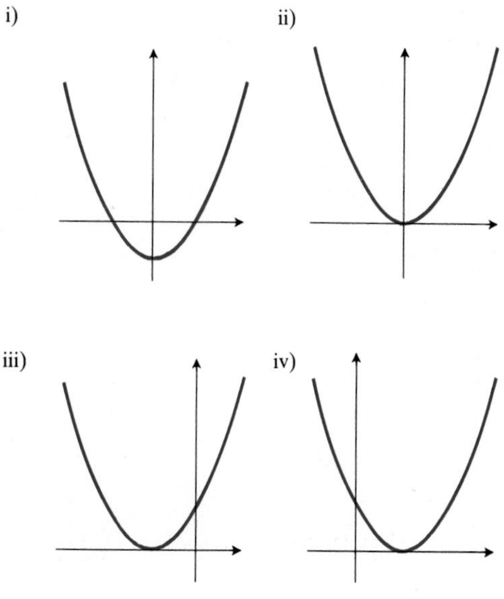

37. Make a table of values with $x = 0$, 1, and 2, and graph the function

$$y = \begin{cases} x, & \text{when } 0 \le x \le 1, \\ 2 - x & \text{when } 1 \le x \le 2. \end{cases}$$

Graph the functions in Problems 38–41.

38. $y = \begin{cases} 3 - x, & x \le 1, \\ 2x, & 1 < x \end{cases}$ **39.** $y = \begin{cases} 1/x, & x < 0, \\ x, & 0 < x \end{cases}$

40. $y = \begin{cases} 1, & x < 5, \\ 0, & 5 \le x \end{cases}$ **41.** $y = \begin{cases} 1, & x < 0, \\ \sqrt{x}, & x > 0 \end{cases}$

42. Find formulas for the functions graphed in the following figures.

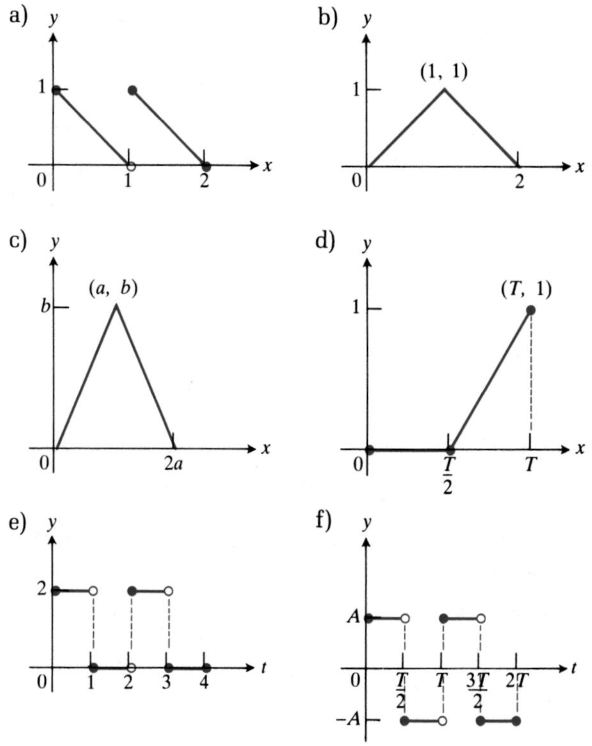

In Problems 43–45, give the domains of f and g and the corresponding domains of $f + g$, $f - g$, $f \cdot g$, f/g, and g/f.

43. $f(x) = x, \quad g(x) = \sqrt{x - 1}$

44. $f(x) = \dfrac{1}{x - 2}, \quad g(x) = \dfrac{1}{\sqrt{x - 1}}$

45. $f(x) = \sqrt{x}, \quad g(x) = \sqrt[4]{x + 1}$

46. If $g(x) = 37 - 3x^3 + x^4 - x$, then $g(-2)$ is:
a) 31;
b) 75;
c) 79;
d) none of the above.

47. If $h(x) = 1 + 5/x$, find:
a) $h(-1)$; b) $h(1/2)$;
c) $h(5)$; d) $h(5x)$;
e) $h(10x)$; f) $h(1/x)$.

48. If $f(x) = x + 5$ and $g(x) = x^2 - 3$, find:
a) $g(f(0))$; b) $f(g(0))$;
c) $g(f(x))$; d) $f(g(x))$;
e) $f(f(-5))$; f) $g(g(2))$;
g) $f(f(x))$; h) $g(g(x))$.

49. Let $f(x) = (x - 1)/x$. Show that $f(x) \cdot f(1 - x) = 1$.

50. If $f(x) = 1/x$, find:
a) $f(2)$; b) $f(x + 2)$;
c) $(f(x + 2) - f(2))/2$.

51. If $F(t) = 4t - 3$, find $(F(t + h) - F(t))/h$.

52. Copy and complete the following table.

	$f(x)$	$g(x)$	$(g \circ f)(x)$
a)	$x - 7$	$\sqrt{x}$	
b)	$x + 2$	$3x$	
c)	$x^2 + 2x + 1$		$\lvert x + 1\rvert$
d)		$\sqrt{x - 5}$	$\sqrt{x^2 - 5}$
e)	$\dfrac{x}{x - 1}$	$\dfrac{x}{x - 1}$	
f)		$\sqrt{x}$	$\lvert x\rvert$
g)		$\sqrt{x}$	x
h)	$\dfrac{1}{x}$		x

53. Find functions $f(x)$ and $g(x)$ whose composites are $(g \circ f)(x) = |\sin x|$ and $(f \circ g)(x) = (\sin \sqrt{x})^2$.

In Problems 54–60, describe the domain of x without absolute values.

54. $|x| < 2$ **55.** $|x| \geq 2$.

56. $|x + 2| < 1$ **57.** $|x - 3| > 3$

58. $|x - 1| \geq 5$ **59.** $\left|\dfrac{1}{x}\right| \leq 1$

60. $\left|\dfrac{x}{2} - 1\right| \leq 1$

In Problems 61–69, use absolute values to describe the given intervals of x- and y-values. It may help to draw a picture of the interval.

61. $-8 < x < 8$ **62.** $-3 < y < 5$

63. $-5 < x < 1$ **64.** $1 < y < 7$

65. $-a < y < a$ **66.** $-1 < x < 2$

67. $L - \varepsilon < y < L + \varepsilon$ (L and ε constant)

68. $1 - \delta < x < 1 + \delta$ (δ constant)

69. $x_0 - 5 < x < x_0 + 5$ (x_0 constant)

In Problems 70–79, match each absolute value inequality with the interval it determines.

70. $|x| < 4$ a. $-2 < x < 1$

71. $|x + 3| < 1$ b. $-1 < x < 3$

72. $|x - 5| < 2$ c. $3 < x < 7$

73. $\left|\dfrac{x}{2}\right| < 1$ d. $-\dfrac{5}{2} < x < -\dfrac{3}{2}$

74. $|1 - x| < 2$ e. $-2 < x < 2$

75. $|2x - 5| \leq 1$ f. $-4 < x < 4$

76. $|2x + 4| < 1$ g. $-4 < x < -2$

77. $\left|\dfrac{x - 1}{2}\right| < 1$ h. $2 \leq x \leq 3$

78. $\left|\dfrac{2x + 1}{3}\right| < 1$ i. $-2 \leq x \leq 2$

79. $|x^2 - 2| \leq 2$

80. Do not fall into the trap $|-a| = a$. The equation does not hold for all values of a.
a) Find a value of a for which $|-a| \neq a$.
b) For what values of a does the equation $|-a| = a$ hold?

81. When does $|1 - x|$ equal $1 - x$ and when does it equal $x - 1$?

82. Which of the graphs in Fig. 1.63 is the graph of $y = |x - 3|$?

Graph the functions in Problems 83–88.

83. $y = -|x|$ **84.** $y = |x - 1|$

85. $y = \dfrac{|x| - x}{2}$ **86.** $y = \dfrac{|x| + x}{2}$

87. $y = \dfrac{1}{|x|}$ **88.** $y = |x - 3| + 2$

Figure 1.63

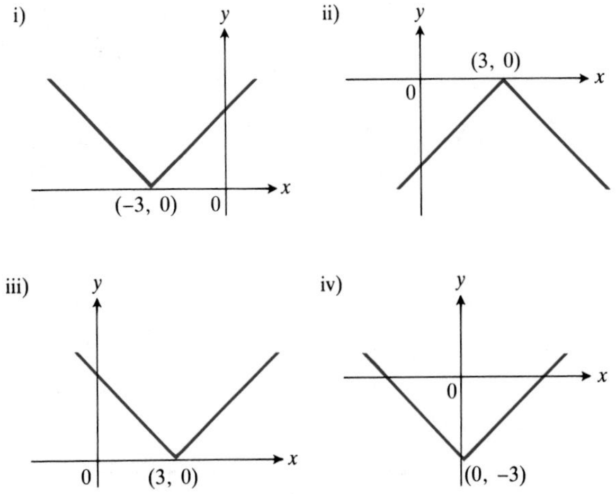

89. Graph each function. Locate any maximum and minimum values it may have.
a) $y = |x + 2| + x$ b) $y = |x| + |x - 1|$

90. a) Graph the function $y = |4 - x^2|$ over the interval $-3 \le x \le 3$.
b) Find the maximum and minimum values of y on the interval.
c) At what values of x does y take on its maximum and minimum values?

91. Graph $|x| - |y| = 1$:
a) in Quadrant I; b) in Quadrant III.

Problems 92–94 are about the greatest integer function.

92. Graph each function.
a) $y = x - [x], \quad -3 \le x \le 3$
b) $y = [x/2], \quad -3 \le x \le 3$
c) $y = [2x] - 2[x]$
d) $y = \frac{1}{2}([x] + x)$ e) $y = \frac{1}{[x]}$

93. For what values of x does $[x] = 0$?

94. When x is positive or zero, $[x]$ is the integer part of the decimal representation of x. What is the corresponding description of $[x]$ when x is negative?

95. The Washington-bound Concorde approaches the United States on a course that takes it south of Nantucket Island at an altitude of 16.8 km. If the surface temperature is 290 K and the vertical temperature gradient is 5 K/km, how far south of Nantucket must the plane be flown to keep its sonic boom carpet away from the island?

96. *Even and odd functions.* A function $y = f(x)$ is an *even* function of x if $f(-x) = f(x)$ for every value of x, and *odd* if $f(-x) = -f(x)$ instead. The function $f(x) = x^2$ is even, because $(-x)^2 = x^2$. The function $f(x) = x^3$ is odd because $(-x)^3 = -x^3$. Which of the functions below are even, and which are odd?
a) $y = x$ b) $y = |x|$
c) $y = x^4$ d) $y = \sin x$
e) $y = \cos x$ f) $y = \tan x$
g) $y = x^2 - 9$ h) $y = x^3 + 5x$
i) $y = x^3 - 5x$ j) $y = \begin{cases} x^2 - 9, & \text{if } x \ne 5, \\ 16, & \text{if } x = 5 \end{cases}$

97. Is $f(x) = x - (1/x)$ even, odd, or neither? (See Problem 96.)

***Toolkit* programs**

Function Evaluator Super * Grapher
Name That Function

1.6

Slopes of Quadratic and Cubic Curves

We shall now get our first view of the role calculus plays in describing change. We begin with the average rate at which a quantity changes over a period of time, and progress toward determining the rate at which a quantity is changing at a specific time.

We encounter average rates of change in such forms as average speeds (distance traveled divided by trip time, say, in miles per hour), growth rates of populations (in percent per year), and average monthly rainfall (in inches per month). The *average rate of change* in a quantity over a period of time is the amount of change in the quantity divided by the length of time in which the change takes place.

Experimental biologists are often interested in the rates at which populations grow under controlled laboratory conditions. Figure 1.64 shows data from a fruit fly growing experiment, the setting for our first example.

EXAMPLE 1 *The average growth rate of a laboratory population.* The graph in Fig. 1.64 shows how the number of fruit flies (*Drosophila*) grew in a controlled 50-day experiment. The graph was made by counting flies at regular intervals, plotting a point for each count, and drawing a smooth curve through the plotted points.

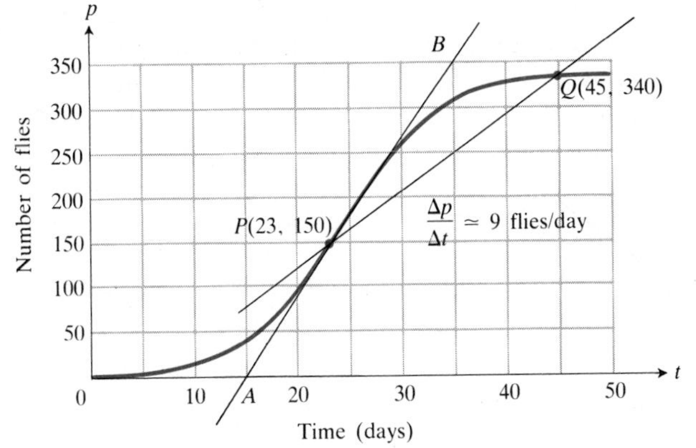

1.64 Growth of a fruit fly population in a controlled experiment. (*Source:* A. J. Lotka, *Elements of Mathematical Biology*. Dover, New York (1956); p. 69.)

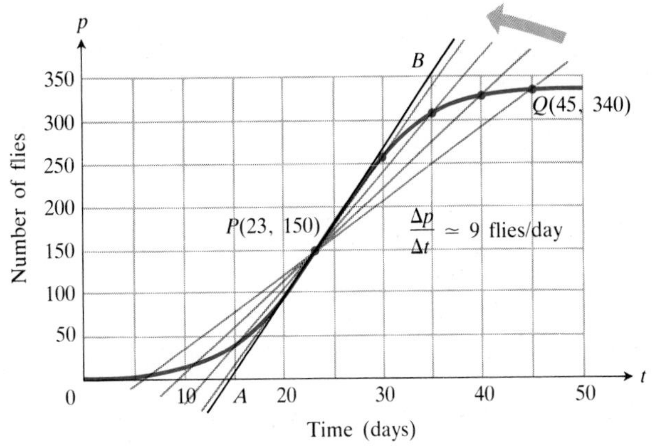

(a)

Q	Slope of $PQ = \Delta p/\Delta t$ (flies/day)
(45, 340)	$(340 - 150)/(45 - 23) \approx 9$
(40, 330)	$(330 - 150)/(40 - 23) \approx 13$
(35, 310)	$(310 - 150)/(35 - 23) \approx 15$
(30, 265)	$(265 - 150)/(30 - 23) \approx 16.4$

(b)

1.65 (a) Four secants to the fruit fly graph of Figure 1.64, through the point $P(23, 150)$. (b) The slopes of the four secants.

There were 150 flies on day 23 and 340 flies on day 45. This gave an increase of $340 - 150 = 190$ flies in $45 - 23 = 22$ days. The average rate of change in the population from day 23 to day 45 was therefore

$$\textit{Average rate of change:} \quad \frac{\Delta p}{\Delta t} = \frac{340 - 150}{45 - 23} = \frac{190}{22} \approx 9 \text{ flies/day.} \tag{1}$$

The average change in Eq. (1) is also the slope of the secant line through the two points

$$P(23, 150) \quad \text{and} \quad Q(45, 340)$$

on the population curve. (A line through two points on a curve is called a *secant* to the curve.) The slope of the secant PQ can be calculated from the coordinates of P and Q:

$$\textit{Secant slope:} \quad \frac{\Delta p}{\Delta t} = \frac{340 - 150}{45 - 23} = \frac{190}{22} = 9 \text{ flies/day.} \tag{2}$$

By comparing Eqs. (1) and (2) we can see that the average rate of change we calculated in (1) is the same number as the slope we calculated in (2), units and all. We can always think of an average rate of change as the slope of a secant line. □

In addition to knowing the average rate at which the population grew from day 23 to day 45, we may also want to know how fast the population was growing on day 23 itself. We can find out by watching the slope of the secant PQ change as we back Q along the curve toward P. The results for four positions of Q are shown in Fig. 1.65.

In terms of geometry, what we see as Q approaches P along the curve is this: The secant PQ approaches the tangent line AB we drew by eye at P. This means that within the limitations of our drawing the slopes of the secants approach the slope of the tangent, which we calculate from the coordinates of A and B to be

$$\frac{350 - 0}{35 - 15} = 17 \text{ flies/day}.$$

In terms of population change, what we see as Q approaches P is this: The average growth rates for the time intervals approach the slope of the tangent to the curve at P (17 flies per day). The slope of the tangent line is therefore the number we take as the rate at which the fly population was changing on day $t = 23$.

The "moral" here is that we should define the rate at which the value of a function $y = f(x)$ is changing with respect to x at any particular value of $x = x_1$ to be the slope of the tangent to the curve $y = f(x)$ at $x = x_1$. But how are we to define the tangent line and to deduce its slope from the formula $y = f(x)$?

The mathematicians of the early seventeenth century devoted a great deal of time to developing methods for finding tangents to curves, a problem that Descartes once said was the most useful and general problem he knew of in geometry. By 1629 Fermat had devised the method we normally use today. We shall use Fermat's method in Examples 2, 3, and 5. We shall then discuss the method's general features in Article 1.7.

EXAMPLE 2 Define the line tangent to the parabola $y = x^2$ at the point P(2, 4), and find its slope.

Solution We begin with a secant line that passes through P(2, 4) and a neighboring point $Q(2 + \Delta x, (2 + \Delta x)^2)$ on the curve (Fig. 1.66). We then write an expression for the slope of the secant, and watch what happens to the slope as Q approaches P along the curve.

The slope of the secant PQ is

$$m_{\text{sec}} = \frac{\Delta y}{\Delta x} = \frac{(2 + \Delta x)^2 - 2^2}{\Delta x} = \frac{4 + 4\,\Delta x + (\Delta x)^2 - 4}{\Delta x}$$

$$= \frac{4\,\Delta x + (\Delta x)^2}{\Delta x} = 4 + \Delta x. \qquad (3)$$

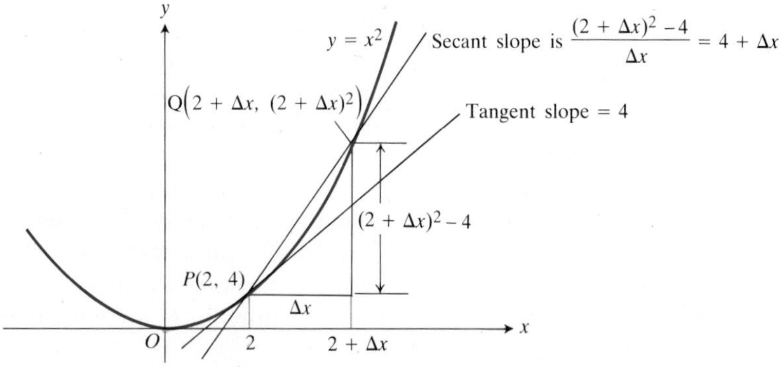

1.66 The slope of the secant line *PQ* approaches 4 as *Q* approaches *P* along the curve.

As Q approaches P along the curve, Δx approaches zero and $m_{sec} = 4 + \Delta x$ approaches $4 + 0 = 4$. We describe this behavior by saying that the limit of $m_{sec} = 4 + \Delta x$ as Δx approaches 0 is 4. What this means geometrically is that as Q moves toward P along the curve, the secant line PQ approaches the line through P whose slope is $m = 4$. This is the line we define to be the tangent to the parabola $y = x^2$ at the point $P(2, 4)$.

The point-slope equation for the tangent line is therefore:

Point:	(2, 4);
Slope:	$m = 4$;
Equation:	$y - 4 = 4(x - 2)$
	$y - 4 = 4x - 8$
	$y = 4x - 4$. □

We can use the method of Example 2 to define the tangent line to any point $P(x, y)$ on the parabola $y = x^2$, no matter what its coordinates are, as shown in the following example.

EXAMPLE 3 Given any point $P(x, y)$ on the curve $y = x^2$, define the line tangent to the curve at P and find its slope.

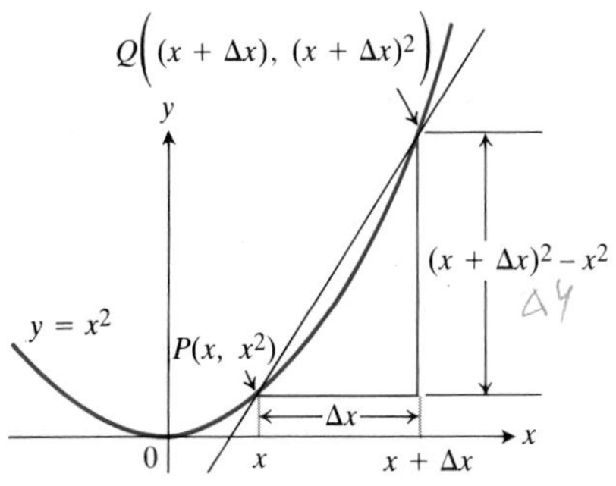

1.67 The slope of the secant *PQ* shown here is $2x + \Delta x$.

Solution As in Example 2, we begin with a secant line PQ that passes through P and a neighboring point Q on the parabola (see Fig. 1.67). We then write an expression for the slope of PQ, and watch what happens to the expression as Q approaches P along the curve.

The coordinates of a typical point $P(x, y)$ on the parabola $y = x^2$ are (x, x^2). We can write the coordinates of any other point Q on the curve as $(x + \Delta x, (x + \Delta x)^2)$, where Δx is the run from P to Q. In terms of these coordinates, the slope of the secant PQ is

$$m_{sec} = \frac{\Delta y}{\Delta x} = \frac{(x + \Delta x)^2 - x^2}{\Delta x} = \frac{x^2 + 2x\,\Delta x + (\Delta x)^2 - x^2}{\Delta x}$$

$$= \frac{2x\,\Delta x + (\Delta x)^2}{\Delta x} = 2x + \Delta x. \tag{4}$$

Now comes an important step. As Q approaches P along the curve, the value of 2x in the expression $2x + \Delta x$ does not change, but the value of Δx approaches 0. Therefore, the value of $2x + \Delta x$ approaches $2x + 0 = 2x$. We describe this behavior of $m_{sec} = 2x + \Delta x$ as Δx approaches 0 by saying that the limit of $m_{sec} = 2x + \Delta x$ as Δx approaches 0 is 2x. Geometrically, the fact that the slope of PQ approaches 2x means that PQ approaches the line through P whose slope is 2x. This is the line we define to be the tangent to the curve $y = x^2$ at the point P. Its slope is

$$m = m_{tan} = 2x.$$

Note that if $x = 2$, then $m_{tan} = 4$ in agreement with the result of Example 2. □

The following example shows how to use the slope formula $m = 2x$ from Example 3 to find equations for the tangents to the curve $y = x^2$.

EXAMPLE 4 Find the equations for the tangents to the curve $y = x^2$ at the points $(-\frac{1}{2}, \frac{1}{4})$ and (1, 1).

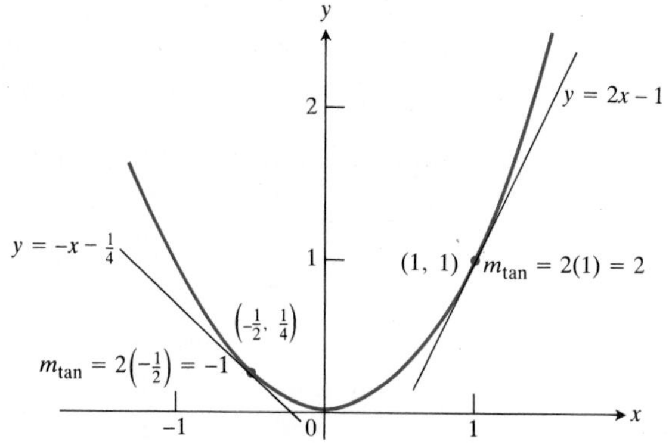

1.68 The slope of the tangent to a point (x, y) on the parabola $y = x^2$ is $m_{\tan} = 2x$.

Solution We use the slope formula $m = 2x$ from Example 3 to find the point-slope equation for each line.

Tangent at $(-\frac{1}{2}, \frac{1}{4})$ Point: $(-\frac{1}{2}, \frac{1}{4})$

Slope: $m = 2x = 2(-\frac{1}{2}) = -1$

Equation:
$$y - \tfrac{1}{4} = -1(x - (-\tfrac{1}{2}))$$
$$y - \tfrac{1}{4} = -x - \tfrac{1}{2}$$
$$y = -x - \tfrac{1}{4}$$

Tangent at $(1, 1)$ Point: $(1, 1)$

Slope: $m = 2x = 2(1) = 2$

Equation:
$$y - 1 = 2(x - 1)$$
$$y - 1 = 2x - 2$$
$$y = 2x - 1$$

See Fig. 1.68. □

DEFINITION

Slope of a Curve

When a curve $y = f(x)$ has a tangent at a point, the *slope of the curve* at the point is the slope of the tangent there.

EXAMPLE 5 Find the slope of the curve $y = x^3 - 3x + 3$ at an arbitrary point $P(x, y)$ on the curve.

Solution We begin by finding an expression for the slopes of the secants through P. From these we find the slope of the tangent at P, which, by definition, is the slope of the curve at P.

Since $P(x, y)$ lies on the curve, its y-coordinate satisfies the equation $y = x^3 - 3x + 3$. In terms of x, then, the coordinates of P are

$$P(x, x^3 - 3x + 3). \tag{5}$$

The coordinates of any point Q on the curve whose first coordinate differs from x by an increment Δx are

$$Q(x + \Delta x, (x + \Delta x)^3 - 3(x + \Delta x) + 3). \tag{6}$$

The run from P to Q is Δx. The rise from P to Q is

$$\begin{aligned}\Delta y &= [(x + \Delta x)^3 - 3(x + \Delta x) + 3] - [x^3 - 3x + 3]\\ &= [x^3 + 3x^2\,\Delta x + 3x\,(\Delta x)^2 + (\Delta x)^3 - 3x - 3\,\Delta x + 3] - [x^3 - 3x + 3]\\ &= 3x^2\,\Delta x + 3x\,(\Delta x)^2 + (\Delta x)^3 - 3\,\Delta x. \qquad (7)\end{aligned}$$

The slope of PQ is therefore

$$\begin{aligned}m_{PQ} &= \frac{\Delta y}{\Delta x} = \frac{3x^2\,\Delta x + 3x\,(\Delta x)^2 + (\Delta x)^3 - 3\,\Delta x}{\Delta x}\\ &= 3x^2 + 3x\,(\Delta x) + (\Delta x)^2 - 3. \qquad (8)\end{aligned}$$

As Q approaches P along the curve, Δx approaches zero and m_{PQ} approaches the number

$$m_{\tan} = 3x^2 + 3x(0) + (0)^2 - 3 = 3x^2 - 3. \qquad (9)$$

The slope of the tangent line at P, and therefore of the curve at P, is

$$m = m_{\tan} = 3x^2 - 3. \qquad (10)$$

We shall graph the curve with the aid of this formula in Example 7. □

x	y	m
−2	1	9
−1	5	0
0	3	−3
1	1	0
2	5	9

(a) Slope table

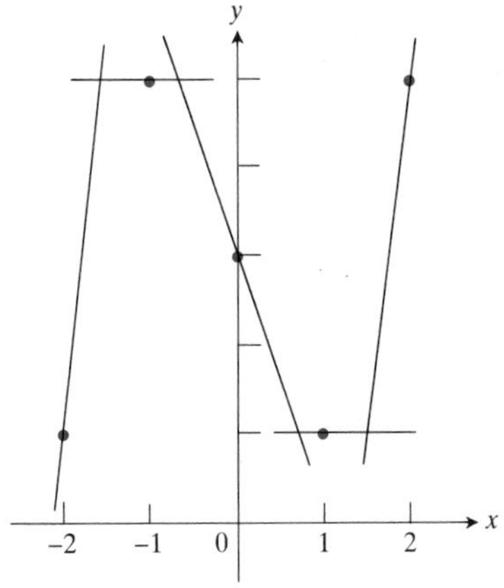

(b) Tangent frame

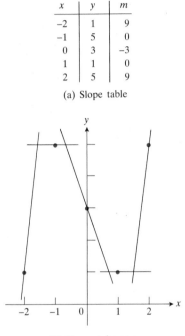

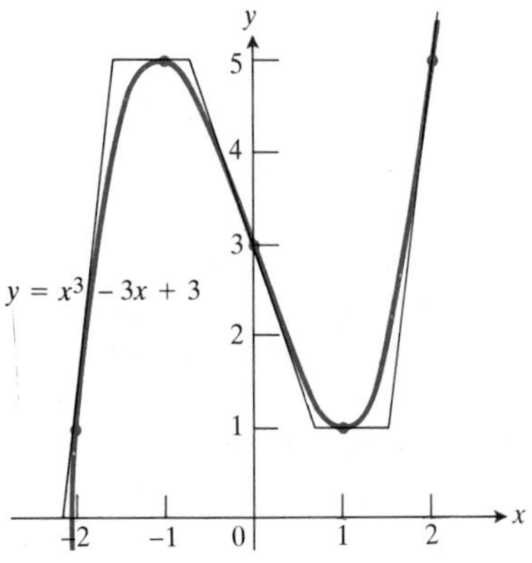

(c) Sketch

1.69 The three steps for sketching the curve $y = x^3 - 3x + 3$ with a tangent frame (see Example 7).

EXAMPLE 6 Find an equation for the line tangent to the curve $y = x^3 - 3x + 3$ at the point (2, 5).

Solution We use the tangent slope formula $m = 3x^2 - 3$ from Example 5 to write a point-slope equation for the tangent line:

Point: (2, 3);

Slope: $m = 3x^2 - 3 = 3(2)^2 - 3 = 3 \cdot 4 - 3 = 9$;

Equation: $y - 3 = 9(x - 2)$

$y - 3 = 9x - 18$

$y = 9x - 15$. □

In graphing, we sometimes use the slope formula for a curve to draw some of the tangents to a curve before we draw the curve itself. This gives us a "frame" within which to draw the curve, and may tell us the shape of the curve after we have plotted only a few points.

EXAMPLE 7 *Graphing with a tangent frame.* Draw a tangent frame for the curve $y = x^3 - 3x + 3$. Then draw the curve.

Solution We first make a slope table that includes the points where the curve has horizontal tangents (Fig. 1.69a). From the slope formula $m = 3x^2 - 3$ in Example 4 we see that $m = 0$ when $x = \pm 1$. We also see that the slope is negative for x between -1 and $+1$. Elsewhere, the slope m is positive.

We then plot the points from the table and draw the tangents by sketching lines that have roughly the right slope (Fig. 1.69b).

The tangents make a frame within which to draw the curve (Fig. 1.69c). □

PROBLEMS

Graph the functions in Problems 1–15 by carrying out the following steps:

a) Use the method of Examples 2 and 4 to find a formula that gives the slope at any point $P(x, y)$ on the given curve.
b) Make a slope table like the one in Example 7 (Fig. 1.69a), including any points where the curve has a horizontal tangent (slope = 0). Use the information in the table to draw a tangent frame for the curve. Sketch the curve.

1. $y = x^2 - 2x - 3$
2. $y = 2x^2 - x - 1$
3. $y = 4 - x^2$
4. $y = x^2 - 4x$
5. $y = x^2 - 4x + 4$
6. $y = x^2 + 4x + 4$
7. $y = 6 + x - x^2$
8. $y = 6 + 5x - x^2$
9. $y = x^2 + 3x + 2$
10. $y = 2 - x - x^2$
11. $y = 2x^3 + 3x^2 - 12x + 7$
12. $y = x^3 - 3x$
13. $y = x^3 - 12x$
14. $y = x^2(4x + 3) + 1$
15. $y = x^3 - 3x^2 + 4$

16. In Example 1, estimate the increase that took place in the fruit fly population during the tenth day by using a ruler to estimate the slope of the population curve in Fig. 1.64 at $t = 10$.

17. a) Sketch the parabola $y = x^2$, showing the points $(-2, 4)$, $(-1, 1)$, $(0, 0)$, $(\frac{1}{2}, \frac{1}{4})$, and $(1, 1)$.
b) Write an equation for the line tangent to the parabola at $P(1, 1)$.
c) Calculate the slopes of the secants through $P(1, 1)$ and the four points $(-2, 4)$, $(-1, 1)$, $(0, 0)$, and $(\frac{1}{2}, \frac{1}{4})$.
d) Let Δx be a small increment in x. Express the slope m_{PQ} of the secant through the points $P(1, 1)$ and $Q(1 - \Delta x, (1 - \Delta x)^2)$ as a function of Δx. Find the limit of m_{PQ} as Δx approaches zero. What is the relation between this limit and the slope of the tangent to the parabola at $P(1, 1)$?

18. Let Q be a point on the curve $y = x^3 - 3x + 3$ whose x-coordinate equals $h \neq 0$.
a) Find the slope of the secant through Q and the point $P(0, 3)$.
b) Find the limit of the slope of the secant in (a) as Q approaches P.
c) Write an equation for the tangent to the curve at P.

19. Find equations for the tangents to the curve $y = x^3 - 3x + 3$ of Example 5 at each of the following points:
a) $(3, 21)$;
b) $(-3, -15)$;
c) $(\sqrt{2}, 3 - \sqrt{2})$.

***Toolkit* programs**	
Function Evaluator	Super * Grapher
Secant Lines	

1.7
The Slope of the Curve $y = f(x)$. Derivatives

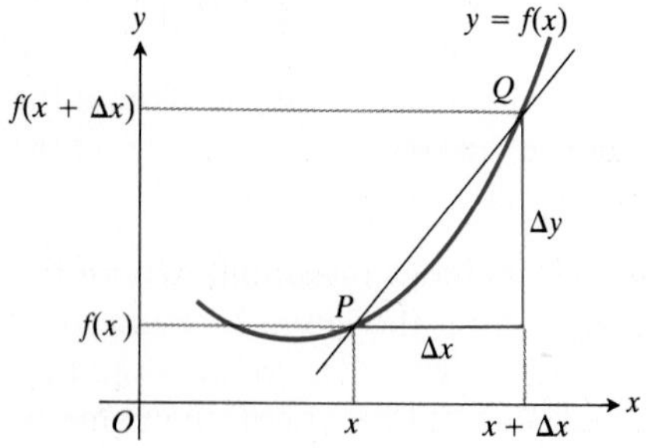

1.70 The slope of the line PQ is

$$\frac{f(x + \Delta x) - f(x)}{\Delta x}.$$

In Article 1.6 we saw how tangents and slopes can be defined for quadratic and cubic curves by what we might call the "delta method." The method works on such a wide range of functions $y = f(x)$ that it is worth exploring in general terms.

Let $P(x, y)$ be a fixed point on the curve $y = f(x)$. If $Q(x + \Delta x, y + \Delta y)$ is another point on the curve, then

$$y + \Delta y = f(x + \Delta x),$$

as in Fig. 1.70. From this we subtract $y = f(x)$ to obtain

$$\Delta y = f(x + \Delta x) - f(x).$$

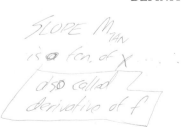

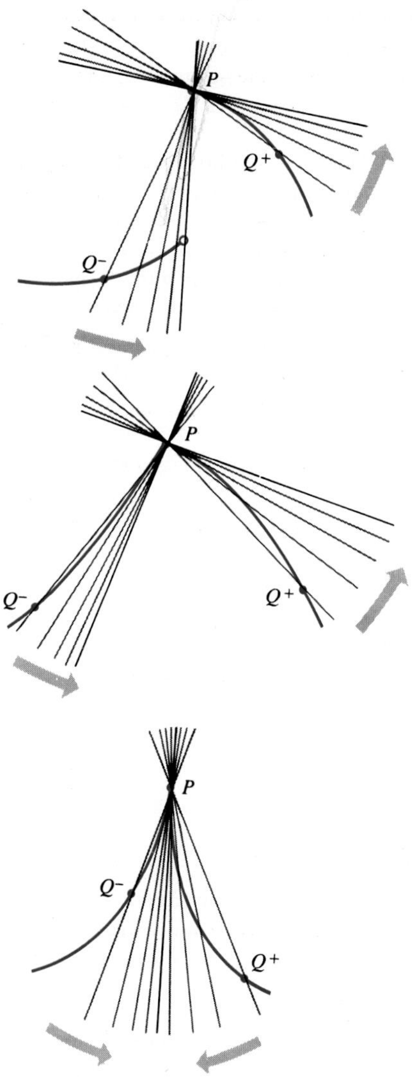

1.71 The derivative of $y = f(x)$ is not defined at a point where the graph of the function has a break, corner, or sharp point.

Then the slope of the secant line PQ is

$$m_{\text{sec}} = \frac{\Delta y}{\Delta x} = \frac{f(x + \Delta x) - f(x)}{\Delta x}. \tag{1}$$

The division in Eq. (1) can only be *indicated* when we talk about a general function $f(x)$, but for any specific function like the function $f(x) = x^3 - 3x + 3$ in Example 5 of Article 1.6 this division must be carried out *before* we do the next operation.

After performing the division in Eq. (1), *we hold x fixed* and let Δx approach zero. If m_{sec} approaches a value that depends only upon x, we define this value to be the *slope* m_{tan} *of the tangent to the curve at* $P(x, y)$.

The mathematical symbols for the definition of the slope m_{tan} are

$$\begin{aligned} m_{\text{tan}} &= \lim_{Q \to P} m_{\text{sec}} \\ &= \lim_{\Delta x \to 0} \frac{\Delta y}{\Delta x} \\ &= \lim_{\Delta x \to 0} \frac{f(x + \Delta x) - f(x)}{\Delta x}. \end{aligned} \tag{2}$$

The notation "lim" with $\Delta x \to 0$ beneath is read "the limit, as Δx approaches 0, of. . . ." The concept of limit, vital to much of modern mathematics, will be studied in detail in Article 1.9. For now, we shall keep our dealings with the notion informal. The fraction on the right in (2) is called the *difference quotient* for f.

The slope m_{tan} is itself a function of x, defined at every point x at which the limit in Eq. (2) exists. We usually denote this slope function by $f'(x)$, and call it the *derivative* of f. Thus the values of the derivative $f'(x)$ of the function $f(x)$ are defined by the rule

$$f'(x) = \lim_{\Delta x \to 0} \frac{f(x + \Delta x) - f(x)}{\Delta x}. \tag{3}$$

The domain of $f'(x)$ is a subset of the domain of f, and for most of the functions considered in this book, it will turn out that $f'(x)$ exists at all or all but a few of the values at which f itself is defined. The exceptional points will generally be points where the graph of $y = f(x)$ has corners, breaks, or peaks like the ones shown in Fig. 1.71.

DEFINITION

Derivative

The *derivative* of a function $y = f(x)$ is the function $f'(x)$ whose value at each x is defined by the rule

$$f'(x) = \lim_{\Delta x \to 0} \frac{f(x + \Delta x) - f(x)}{\Delta x}, \tag{3}$$

whenever the limit on the right exists. The domain of f' is the set of points in the domain of f at which the limit on the right exists.

The branch of mathematics that deals with determining the values of x at which a function $y = f(x)$ has a derivative, and with finding the value of $f'(x)$ at such a point, is called *differential calculus*.

EXAMPLE 1 The derivative of $f(x) = x^2$ is $f'(x) = 2x$, according to Example 3 in Article 1.6. It is defined for all values of x. □

EXAMPLE 2 The derivative of $f(x) = x^3 - 3x + 3$ is $f'(x) = 3x^2 - 3$, according to Example 5 in Article 1.6. It is defined for all values of x. □

The most common notations for the derivative of $y = f(x)$ besides $f'(x)$ are

$$y' \quad \text{(pronounced "y prime")}$$

$$\frac{dy}{dx} \quad \text{(pronounced "d y d x")}$$

$$\frac{df}{dx} \quad \text{(pronounced "d f d x")}$$

The notation dy/dx and df/dx may be interpreted as

$$\frac{dy}{dx} = \frac{d}{dx}(y)$$

and

$$\frac{df}{dx} = \frac{d}{dx}(f),$$

where d/dx stands for the operation of taking the derivative with respect to x. Thus we also read dy/dx as "the derivative of y with respect to x" and df/dx as "the derivative of f with respect to x." See Fig. 1.72.

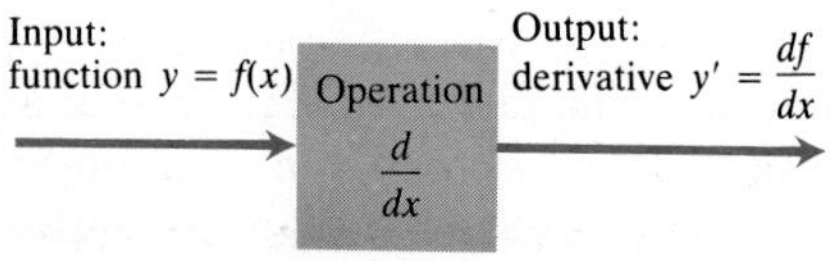

1.72 Flow diagram for the operation of taking a derivative with respect to x.

EXAMPLE 3 Show that the derivative of $f(x) = mx + b$ is $f' = m$. That is, the derivative of f is the slope of the line $y = mx + b$. (See Fig. 1.73)

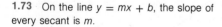

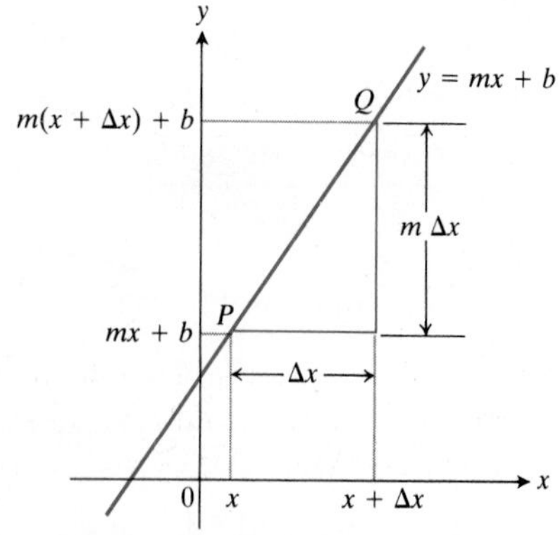

1.73 On the line $y = mx + b$, the slope of every secant is m.

Solution We calculate the limit in Eq. (3) with $f(x) = mx + b$. The calculation has four basic steps.

STEP 1. Write out $f(x + \Delta x)$ and $f(x)$:

$$f(x + \Delta x) = m(x + \Delta x) + b$$
$$= mx + m\,\Delta x + b,$$
$$f(x) = mx + b.$$

STEP 2. Subtract $f(x)$ from $f(x + \Delta x)$:

$$f(x + \Delta x) - f(x) = m\,\Delta x.$$

STEP 3. Divide by Δx:

$$\frac{f(x + \Delta x) - f(x)}{\Delta x} = \frac{m\,\Delta x}{\Delta x} = m.$$

STEP 4. Calculate the limit as Δx approaches zero:

$$f'(x) = \lim_{\Delta x \to 0} \frac{f(x + \Delta x) - f(x)}{\Delta x} = \lim_{\Delta x \to 0} m = m.$$

The derivative $f'(x)$ is defined at every value of x. It has the constant value m, the slope of the line. □

In the following example, we show that the function $y = |x|$ has a derivative at every value of x except $x = 0$.

EXAMPLE 4 Find the derivative of $y = |x|$.

Solution When x is positive, $y = |x| = x$. From Example 3 with $m = 1$ and $b = 0$, we know that the derivative of $y = x$ is $y' = 1$.

When x is negative, $y = |x| = -x$. The derivative of $y = -x$ (by Example 3 with $m = -1$ and $b = 0$) is $y' = -1$.

Therefore,

$$\text{if } y = |x|, \text{ then } y' = \begin{cases} 1 & \text{if } x > 0, \\ -1 & \text{if } x < 0. \end{cases} \tag{4}$$

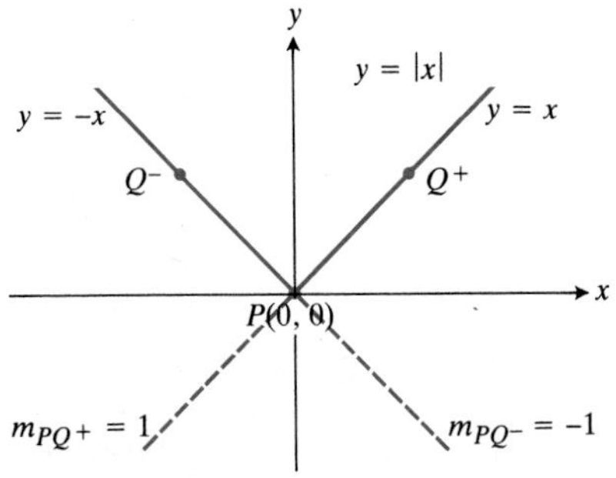

1.74 The lines $y = x$ and $y = -x$ are the only secants to the graph of $y = |x|$ that pass through the origin.

At $x = 0$, the function $y = |x|$ has no derivative. To see this, look at the secant slopes of which the derivative, to exist, would have to be the limit (Fig. 1.74). There are *only two* different secants to the curve $y = |x|$ through the point $P(0, 0)$. These are the lines $y = x$ and $y = -x$. If Q is any point on the graph to the right of P, the secant PQ is the line $y = x$, whose slope is $+1$. If Q is any point on the graph to the left of P, the secant PQ is the line $y = -x$, whose slope is -1.

As Q approaches P along the graph, the secants themselves remain stationary. For the limit in Eq. (3) to exist, the slopes of the right- and left-hand secants would have to come together as Q approaches P. They never do. No matter how close Q comes to P, the slope is -1 on the left, $+1$ on the right. □

The difference quotient

$$\frac{f(x + \Delta x) - f(x)}{\Delta x}$$

in Eq. (3) takes on different forms if we change the way we label the points on the graph of $f(x)$. With the labeling in Fig. 1.75, the derivative of f at x is expressed as

$$f'(x) = \lim_{h \to 0} \frac{f(x + h) - f(x)}{h}. \tag{5}$$

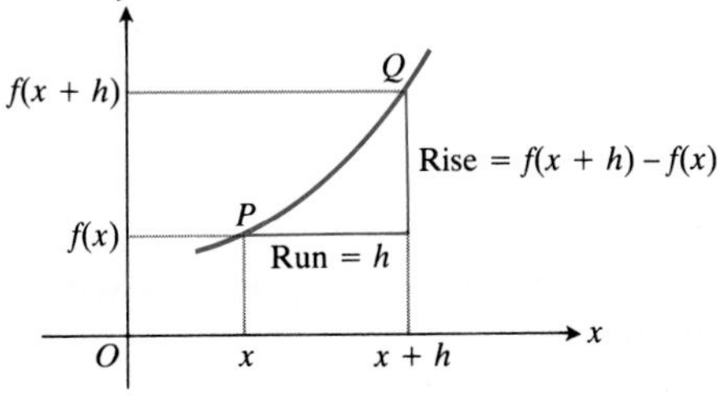

1.75 When the difference between the abscissas of P and Q is called h instead of Δx, the defining equation for the derivative of $y = f(x)$ is

$$f'(x) = \lim_{h \to 0} \frac{f(x + h) - f(x)}{h}.$$

Equation (5) is Eq. (3) with h in place of Δx. The replacement of Δx by a single letter makes the formulas that arise in calculating a derivative easier to write and read. We use Eq. (5) to calculate the derivative in the following example.

EXAMPLE 5 Find dy/dx if $y = \sqrt{x}$ and $x > 0$.

Solution We use Eq. (5) with $f(x + h) = \sqrt{x + h}$ and $f(x) = \sqrt{x}$ to form the quotient

$$\frac{f(x + h) - f(x)}{h} = \frac{\sqrt{x + h} - \sqrt{x}}{h}. \tag{6}$$

Unfortunately, this will involve division by 0 if we replace h by 0. We therefore look for an equivalent expression in which this difficulty does not arise. If we rationalize the numerator in (6), we find

$$\frac{\sqrt{x + h} - \sqrt{x}}{h} = \frac{\sqrt{x + h} - \sqrt{x}}{h} \cdot \frac{\sqrt{x + h} + \sqrt{x}}{\sqrt{x + h} + \sqrt{x}}$$

$$= \frac{(x + h) - x}{h(\sqrt{x + h} + \sqrt{x})} = \frac{1}{\sqrt{x + h} + \sqrt{x}}. \tag{7}$$

Now as h approaches 0, the denominator in the final form approaches $\sqrt{x} + \sqrt{x} = 2\sqrt{x}$, which is positive because $x > 0$. Therefore,

$$\frac{dy}{dx} = \lim_{h \to 0} \frac{\sqrt{x + h} - \sqrt{x}}{h}$$

$$= \lim_{h \to 0} \frac{1}{\sqrt{x + h} + \sqrt{x}} = \frac{1}{2\sqrt{x}}. \ \square \tag{8}$$

EXAMPLE 6 Find the slope of the curve $y = \sqrt{x}$ at $x = 4$.

Solution The slope is the value of the derivative of $y = \sqrt{x}$ at $x = 4$. Equation (8) gives

$$\left.\frac{dy}{dx}\right|_{x=4} = \frac{1}{2\sqrt{4}} = \frac{1}{4}. \ \square$$

If we label the graph of a function $y = f(x)$ as in Fig. 1.76, then we can calculate the values of its derivative $y' = f'(x)$ from the equation

$$f'(x) = \lim_{a \to x} \frac{f(a) - f(x)}{a - x}. \tag{9}$$

Equation (9) is Eq. (3) with $x + \Delta x = a$ and $\Delta x = a - x$. This equation is particularly convenient for calculating the derivatives of some functions (but not others). We use it to calculate the derivative of $f(x) = 1/x$ in the following example.

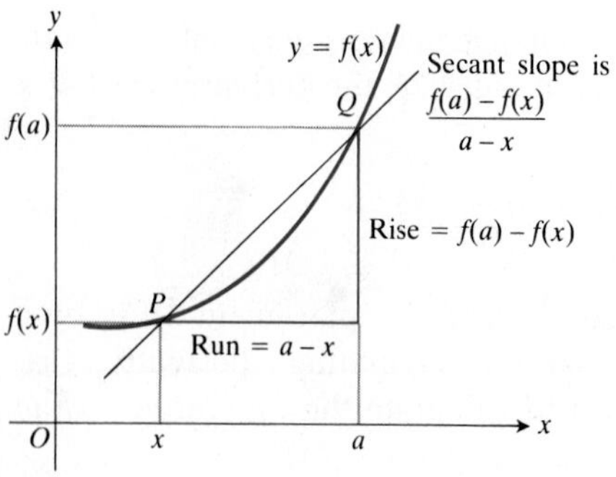

1.76 $f'(x) = \lim_{a \to x} \frac{f(a) - f(x)}{a - x}$

EXAMPLE 7 Calculate the derivative of the function $f(x) = 1/x$.

Solution The domain of f consists of all $x \neq 0$. We use Eq. (9) with $f(x) = 1/x$ and $f(a) = 1/a$ to form the quotient

$$\frac{f(a) - f(x)}{a - x} = \frac{(1/a) - (1/x)}{a - x} = \frac{(x - a)/ax}{a - x} = -\frac{1}{ax}.$$

As a approaches x, the product ax approaches $x \cdot x = x^2$, and we have

$$f'(x) = \lim_{a \to x} \frac{f(a) - f(x)}{a - x} = \lim_{a \to x} -\frac{1}{ax} = -\frac{1}{x^2}. \tag{10}$$

Note that x is held fixed in this calculation, while a approaches x. We conclude from Eq. (10) that the derivative of the function $f(x) = 1/x$ is $f'(x) = -1/x^2$ for every $x \neq 0$. □

Estimating $f'(x)$ from a Graph of $f(x)$

When we record data in a laboratory or in the field, we are often recording the values of a function $y = f(x)$. We might be recording the pressure in a gas as a function of volume at a given temperature or the size of a population as a function of time. To see what the function looks like we usually plot the data points and fit a curve through them.

Although we may have no formula for the function $y = f(x)$ from which to calculate the derivative $y = f'(x)$, it is still possible to graph f' by estimating slopes on the graph of f. The following example shows how this can be done and what we can learn from the graph of f'.

EXAMPLE 8 Graph the derivative of the function $y = f(x)$ whose graph is shown in Fig. 1.77(a).

1.77 We made the graph of $y = f'(x)$ in (b) by plotting slopes from the graph of $y = f(x)$ in (a). The ordinate of B' is the slope at B, and so on. The graph of $y = f'(x)$ is a visual record of how the slope of f changes with x.

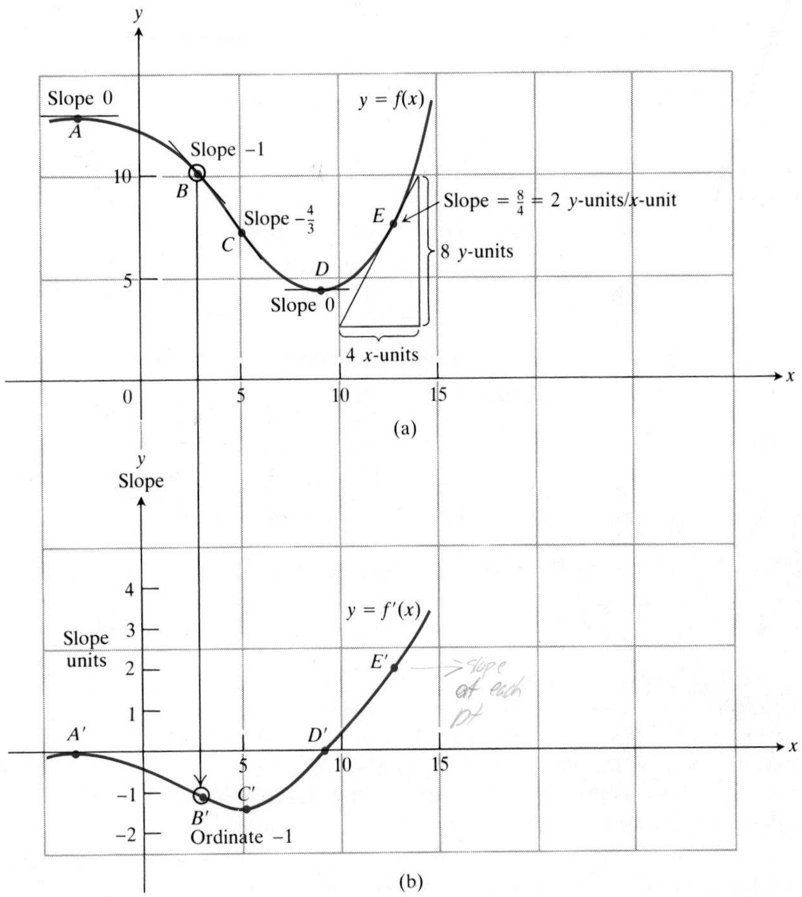

Solution We estimate the slope of the graph of f in y-units per x-unit at frequent intervals. We then plot the estimates in a coordinate plane with the horizontal axis in x-units and the vertical axis in slope units (Fig. 1.77b). We draw a smooth curve through the plotted points.

From the graph of $y = f'(x)$ we can see at a glance:

Where the rate at which f changes (grows) is positive, negative, and zero;

The rough size of the growth rate at any x (remember, this is an estimate) and its relation to the size of $f(x)$;

Where the rate of change itself is increasing or decreasing. (See Problems 22–24.) □

PROBLEMS

For each of the functions f in Problems 1–20, use Eqs. (3), (5), or (9) to find the derivative $f'(x)$. Then find the slope of the tangent to the curve $y = f(x)$ at $x = 3$ and write an equation for the tangent line.

1. $f(x) = x^2$ **2.** $f(x) = x^3$

3. $f(x) = 2x + 3$ **4.** $f(x) = x^2 - x + 1$

5. $f(x) = \frac{1}{x}$ **6.** $f(x) = \frac{1}{x^2}$

7. $f(x) = \frac{1}{2x + 1}$ **8.** $f(x) = \frac{x}{x + 1}$

9. $f(x) = 2x^2 - x + 5$ **10.** $f(x) = x^3 - 12x + 11$

11. $f(x) = x^4$ **12.** $f(x) = ax^2 + bx + c$ (a, b, c constants)

13. $f(x) = x - \frac{1}{x}$ **14.** $f(x) = ax + \frac{b}{x}$ (a, b constants)

15. $f(x) = \sqrt{2x}$ **16.** $f(x) = \sqrt{x + 1}$

17. $f(x) = \sqrt{2x + 3}$ **18.** $f(x) = \frac{1}{\sqrt{x}}$

19. $f(x) = \frac{1}{\sqrt{2x + 3}}$ **20.** $f(x) = \sqrt{x^2 + 1}$

21. Graph the derivative of $f(x) = |x|$ (Example 4). Then graph the function $y = |x|/x$, $x \neq 0$. What can you conclude?

Problems 22–24 are about the graphs shown in Fig. 1.78. The graphs in part (a) show the numbers of rabbits and foxes in a small arctic population. They are plotted as functions of time for 200 days. The number of rabbits increases at first, as the rabbits reproduce. But the foxes prey on the rabbits and as the number of foxes increases, the rabbit population levels off and then drops. Figure 1.78(b) is the graph of the derivative of the rabbit population. It was made by plotting slopes, as in Example 7.

Note that the time when the rabbit population is dropping fastest does not coincide with the time when the fox population is largest. It occurs before the fox population peaks, about when the fox population is growing most rapidly.

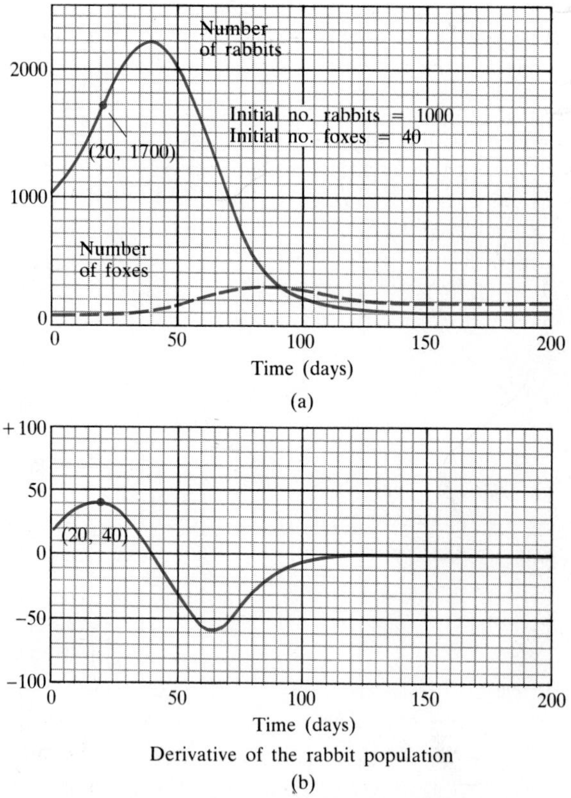

1.78 Rabbits and foxes in an arctic predator–prey food chain. (*Source: Differentiation*, by W. U. Walton *et al.*, Project CALC, Education Development Center, 1975, p. 86.)

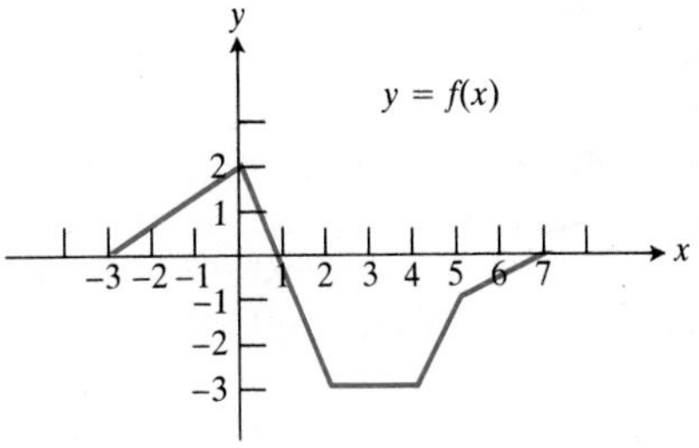

Figure 1.79

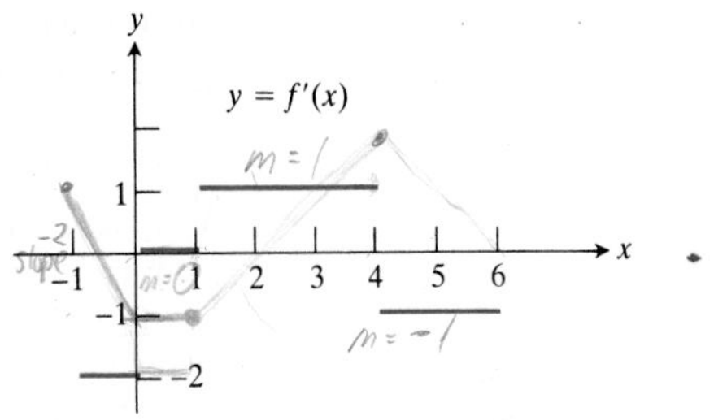

Figure 1.80

22. What is the value of the derivative of the rabbit population in Fig. 1.78 when the number of rabbits is largest? smallest?

23. What is the size of the rabbit population in Fig. 1.78 when its derivative is largest? smallest?

24. In what units should the slope of the fox population curve be measured?

25. a) Use the graphical technique of Example 8 to graph the derivative of the fruit fly population shown in Fig. 1.64. What units should be used on the horizontal and vertical axes?

   b) During what days does the fruit fly population seem to be increasing fastest? slowest?

26. The graph of the function $y = f(x)$ in Fig. 1.79 is made up of line segments joined end to end.

   a) Graph the derivative of the function. Call the vertical axis the y'-axis. The graph should show a step function.

b) At what values of x between -3 and 7 is the derivative not defined?

27. Use the following information about a function $y = f(x)$ to graph the function for $-1 \le x \le 6$.

   i) The graph of f is made of line segments joined end to end.

   ii) The graph of f starts at the point $(-1, 1)$.

   iii) The derivative of f is the step function shown in Fig. 1.80.

28. Use Eq. (9) to calculate the derivative of $y = \sqrt{x}$. (*Hint:* Since $x > 0$, we can assume that the a in Eq. (9) is positive, so that we can write $a - x$ in the denominator of (9) as $a - x = (\sqrt{a} + \sqrt{x})(\sqrt{a} - \sqrt{x})$.)

***Toolkit* programs**

Derivative Grapher   Super ∗ Grapher
Secant Lines

1.8
Velocity and Other Rates of Change

When a body moves in a straight line it is customary to represent the line of motion by a coordinate axis. We choose an origin, a positive direction, and a unit of distance on the line. We describe the motion by an equation that gives the coordinate s of the body as a function of t, the amount of time that has elapsed since the motion began. The coordinate is called the body's *position* at time t. A change in the position coordinate between two times then gives the *displacement* or net distance traveled by the body over the time interval. The displacement divided by the time traveled is, in turn, the body's average velocity for the time interval. As we shall see, this leads in a natural way to defining the body's velocity at any instant t.

EXAMPLE 1 Figure 1.81 shows a ball bearing released at time $t = 0$ seconds to fall straight down. The equation that expresses the distance s

displacement
– *Change in the position coordinate between 2 times*

$$\frac{displacement}{time} = average\ velocity\ for\ time\ interval$$

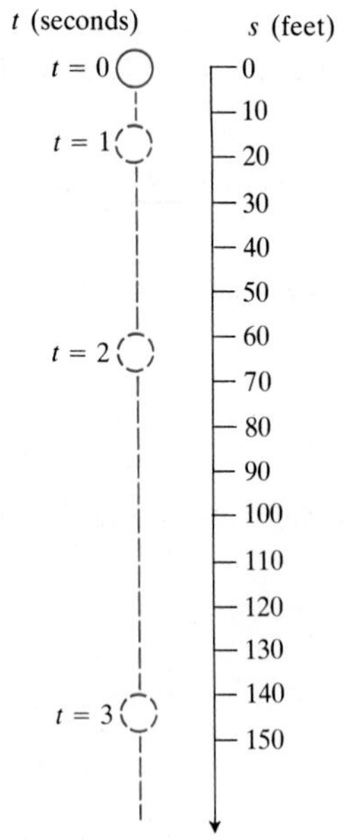

1.81 Distance fallen by a 3-cm ball bearing released at $t = 0$ seconds.

fallen as a function of the time t is

$$s = \frac{1}{2}gt^2. \tag{1}$$

The number g is the acceleration due to gravity at the surface of the earth. It has been determined experimentally to be about 32 ft/s² if s is measured in feet and t in seconds (abbreviated as "s" throughout). Thus,

$$s = \frac{1}{2} \cdot 32t^2 = 16t^2 \text{ ft}. \tag{2}$$

If s is measured in centimeters, then $g = 980$ and

$$s = \frac{1}{2} \cdot 980 = 490t^2 \text{ cm}. \tag{3}$$

During the first two seconds, the ball falls

$$s = 16(2)^2 = 64 \text{ ft},$$

or

$$s = 490(2)^2 = 1960 \text{ cm} = 19.6 \text{ m}. \quad \square$$

Suppose we already know the equation of motion of a body (still along a line) to be

$$s = f(t), \tag{4}$$

and we want to find the *velocity* of the body at some instant t. How shall we *define* the instantaneous velocity of a moving body? If we assume that distance and time are the fundamental physical quantities that we can measure, we might reason as follows:

In the interval from time t to time $t + \Delta t$ the body moves from position $s = f(t)$ to position

$$s + \Delta s = f(t + \Delta t), \tag{5}$$

for a net change in position (displacement) of

$$\Delta s = f(t + \Delta t) - f(t). \tag{6}$$

(All this can be measured by clocks and tapelines, say.)

The *average velocity* of the body over the time interval from t to $t + \Delta t$ is the displacement Δs divided by the time traveled Δt:

DEFINITION

Average velocity

The *average velocity* of a body moving along a line is

$$v_{av} = \frac{\text{displacement}}{\text{time traveled}} = \frac{\Delta s}{\Delta t} = \frac{f(t + \Delta t) - f(t)}{\Delta t}. \tag{7}$$

For example, a sprinter who runs 100 meters in 10 seconds has an average velocity of

$$v_{av} = \frac{\Delta s}{\Delta t} = \frac{100 \text{ m}}{10 \text{ s}} = 10 \text{ m/s}.$$

To obtain the *instantaneous velocity* v of the moving body at time t, or what we simply call the *velocity* at time t, we take the limit of the average velocities as Δt approaches zero:

$$v = \lim_{\Delta t \to 0} v_{av} = \lim_{\Delta t \to 0} \frac{\Delta s}{\Delta t} = \lim_{\Delta t \to 0} \frac{f(t + \Delta t) - f(t)}{\Delta t}. \qquad (8)$$

In other words, we define v to be the derivative of $s = f(t)$ with respect to t, as follows.

DEFINITION

Instantaneous velocity

The *instantaneous velocity* of a body (moving along a line) at any instant t is the derivative of its position coordinate $s = f(t)$ with respect to t:

$$v = \frac{ds}{dt} = f'(t). \qquad (9)$$

EXAMPLE 2 A body falls freely as in Example 1. Find its velocity as a function of t. How fast (ft/s) is the body falling 2 seconds after release?

Solution We use Eq. (8) with $f(t) = \frac{1}{2}gt^2$ to calculate the velocity v as a limit of average velocities. The body's average velocity over a time interval of length Δt is

$$\frac{\Delta s}{\Delta t} = \frac{\frac{1}{2}g(t + \Delta t)^2 - \frac{1}{2}gt^2}{\Delta t}$$

$$= \frac{1}{2}g\frac{t^2 + 2t\,\Delta t + (\Delta t)^2 - t^2}{\Delta t} = \frac{1}{2}g(2t + \Delta t).$$

The velocity at time t is

$$v = \lim_{\Delta t \to 0} \frac{\Delta s}{\Delta t} = \lim_{\Delta t \to 0} \frac{1}{2}g(2t + \Delta t) = \frac{1}{2}g(2t + 0) = gt. \qquad (10)$$

In short, the velocity is

$$v = gt. \qquad (11)$$

With s in feet and t in seconds, $g = 32$ and

$$v = 32\,t \text{ ft/s}. \qquad (12)$$

Two seconds after release, the velocity is

$$v = 32(2) = 64 \text{ ft/s}. \ \square$$

When we graph $s = f(t)$ as a function of t, the average and instantaneous velocities have geometric interpretations (Fig. 1.82). The average velocity is the slope of a secant line. the instantaneous velocity at time t is the slope of the tangent at the point $(t, f(t))$. We can use the interpretation of a tangent slope as a velocity to estimate velocities when our information about a motion is given by a graph rather than an equation. (See Problems 12–14 and 26.)

There are many other applications of the notion of average rate and instantaneous rate.

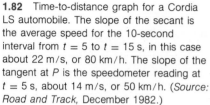

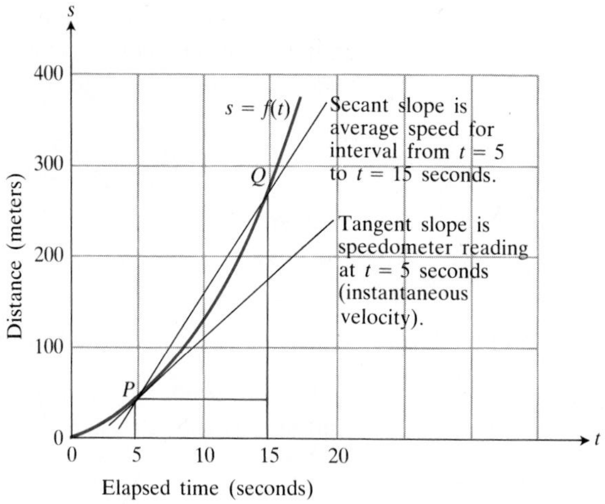

1.82 Time-to-distance graph for a Cordia LS automobile. The slope of the secant is the average speed for the 10-second interval from $t = 5$ to $t = 15$ s, in this case about 22 m/s, or 80 km/h. The slope of the tangent at P is the speedometer reading at $t = 5$ s, about 14 m/s, or 50 km/h. (*Source: Road and Track,* December 1982.)

EXAMPLE 3 The quantity of water Q (gal) in a reservoir at time t (min) is a function of t. Water may flow into or out of the reservoir. As it does so, suppose that Q changes by an amount ΔQ from time t to time $(t + \Delta t)$. Then the average and instantaneous rates of change of Q with respect to t are

Average rate: $\dfrac{\Delta Q}{\Delta t}$ (gal/min);

Instantaneous rate: $\dfrac{dQ}{dt} = \lim_{\Delta t \to 0} \dfrac{\Delta Q}{\Delta t}$ (gal/min). □

Although it is natural to think of rates of change in terms of motion and time, there is no need to be so restrictive. We can define the average rate of change for any function $y = f(x)$ over any interval in its domain and define the instantaneous rate of change as a limit of average rates of change whenever the limit exists.

DEFINITION

Rates of Change

The *average rate of change* of a function $y = f(x)$ over the interval from x to $x + \Delta x$ is

$$\text{Average rate of change} = \frac{f(x + \Delta x) - f(x)}{\Delta x}.$$

The *instantaneous rate of change* of f at x is the derivative

$$f'(x) = \lim_{\Delta x \to 0} \frac{f(x + \Delta x) - f(x)}{\Delta x} = \lim_{\Delta x \to 0} (\text{Average rate of change}),$$

provided the limit exists.

Thus the instantaneous rate of change of f with respect to x is the value of f' at x. Note that we still use the term "instantaneous" even when x does not represent time.

Limits are extremely important in calculus and its applications, and we shall devote the remainder of this chapter to them.

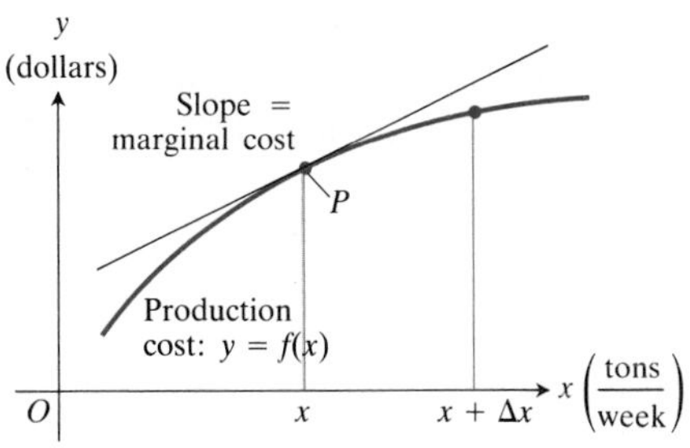

1.83 Weekly steel production: $y = f(x)$ is the cost of producing x tons. The cost of producing an additional Δx tons is given by $\Delta y = f(x + \Delta x) - f(x)$.

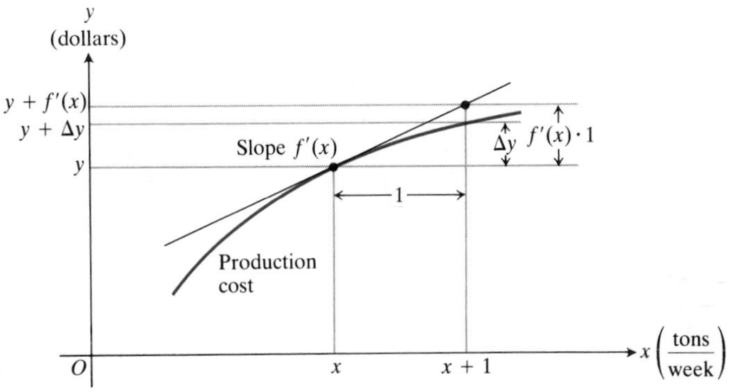

1.84 As weekly steel production increases from x to $x + 1$ tons, the cost curve rises by the amount Δy. The tangent line rises by the amount: slope • run $= f'(x) \cdot 1 = f'(x)$.

EXAMPLE 4 *Marginal cost.* Economists often call the derivative of a function the *marginal value* of the function. Suppose, for example, that it costs a company $y = f(x)$ dollars to produce x tons of steel in a week. It costs more to produce $x + \Delta x$ tons a week, say $y + \Delta y$ dollars. The average increase in cost per additional ton is $\Delta y/\Delta x$. The limit of this ratio as Δx approaches zero is the *marginal cost* of producing x tons of steel in a week. (See Fig. 1.83.)

How are we to interpret marginal cost? First of all, it is the slope of the graph of $y = f(x)$ at the point marked P in Fig. 1.83. But there is more.

Figure 1.84 shows a slightly enlarged view of the curve and its tangent at P. We can see that if the company, currently producing x tons, increases production by one ton, then $f'(x)$ is approximately the incremental cost Δy of producing that one ton. That is,

$$\Delta y \approx f'(x) \quad \text{when } \Delta x = 1. \tag{13}$$

Herein lies the economic importance of marginal cost. It gives an estimate or prediction of the cost of producing one more unit beyond the present production level. It is the approximate cost of producing one more car, one more radio, one more washing machine, whatever.

In fact, if the cost curve stays close to the tangent for Δx units beyond x rather than just one unit beyond x, as in Fig. 1.85, the cost of producing the Δx units is approximately

$$\Delta y \approx f'(x)\, \Delta x. \tag{14}$$

1.85 Sometimes $f'(x)\,\Delta x$ is a good approximation of Δy.

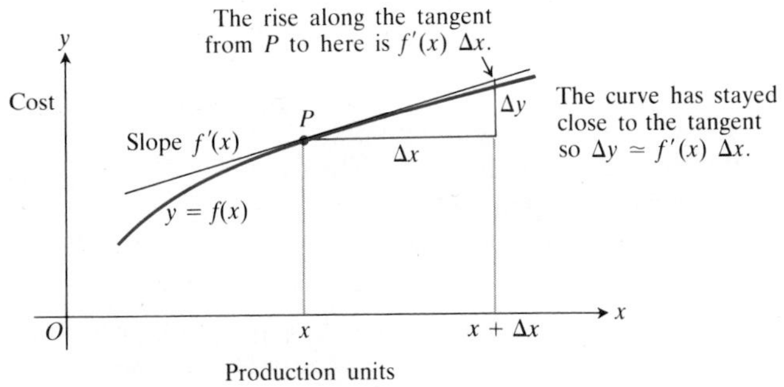

[handwritten margin notes: displacement $\Delta s = f(t + \Delta t) - f(t)$; $v_{av} = \frac{\Delta s}{\Delta t}$; $f'(r) = \frac{f(t+\Delta t) - f(t)}{\Delta t}$; or $2at + b$; plug in t]

Under the right circumstances, therefore, we might use Eq. (14) to predict the added cost Δy of quite an extensive increment Δx in production. Of course, the soundness of the prediction would depend on having a reliable estimate of the error in the approximation $\Delta y \approx f'(x)\,\Delta x$. Predicting change and estimating possible errors in predictions are important in science, and we shall return to them in Chapters 2 and 3. □

Formulas for Free Fall Near the Earth's Surface

1. $s = \frac{1}{2}gt^2$ $s =$ distance, $t =$ time, $g =$ gravitational constant

2. $s = 16t^2$ $s =$ feet, $t =$ seconds, $g = 32$ ft/s^2

3. $s = 490t^2$ $s =$ centimeters, $t =$ seconds, $g = 980$ cm/s^2

4. $s = 4.9t^2$ $s =$ meters, $t =$ seconds, $g = 9.8$ m/s^2

PROBLEMS

1. If a, b, c are constants and
$$f(t) = at^2 + bt + c,$$
show that
$$f'(t) = \lim_{\Delta t \to 0} \frac{f(t + \Delta t) - f(t)}{\Delta t} = 2at + b.$$

The laws of motion in Problems 2–10 give the position $s = f(t)$ of a moving body as a function of t, with s measured in meters and t in seconds.
a) Find the displacement and average velocity for the time interval from $t = 0$ to $t = 2$ seconds.
b) Use the formula $f'(t) = 2at + b$ from Problem 1 to express the velocity $v = ds/dt$ as a function of t, by inspection.
c) Use the formula obtained in (b) to find the body's velocity at $t = 2$ seconds.

2. $s = 2t^2 + 5t - 3$

3. $s = \frac{1}{2}gt^2 + v_0 t + s_0$
(g, v_0, s_0 constants)

4. $s = 4t + 3$

5. $s = t^2 - 3t + 2$

6. $s = 4 - 2t - t^2$

7. $s = (2t + 3)^2$

8. $s = (2 - t)^2$

9. $s = 3 - 2t^2$

10. $s = 64t - 16t^2$

11. Figure 1.86 shows a multiflash photograph of two balls falling after being released from rest. The rulers in the figure are marked in centimeters.
a) Which of Eqs. (2) and (3) is appropriate here?
b) How long did it take the balls to fall the first 180 cm? What was their average velocity for this period?
c) How long was the time between consecutive flashes?

12. a) Use the time-to-speed graph in Fig. 1.82 to estimate the car's average velocity for the first 10 seconds.
b) Estimate the car's speedometer reading at $t = 15$ seconds by estimating the slope of the curve with a straightedge.

13. The following data give the coordinates s of a moving body for various values of t. Plot s versus t on coordinate paper and sketch a smooth curve through the given points. Assuming that this smooth curve represents the motion of the body, estimate the velocity (a) at $t = 1.0$; (b) at $t = 2.5$; (c) at $t = 2.0$.

s (in ft)	10	38	58	70	74	70	58	38	10
t (in sec)	0	0.5	1.0	1.5	2.0	2.5	3.0	3.5	4.0

14. When a chemical reaction was allowed to run for t minutes, it produced the amounts of substance $A(t)$ shown in the following table.†

t (in min)	10	15	20	25	30	35	40
A(t) (in moles)	26.5	36.5	44.8	52.1	57.1	61.3	64.4

---

†Data from *Some Mathematical Models in Biology*, Revised Edition, R. M. Thrall, J. A. Mortimer, K. R. Rebman, R. F. Baum, eds., December 1967, PB-202 364, p. 72; distributed by N.T.I.S., U.S. Department of Commerce.

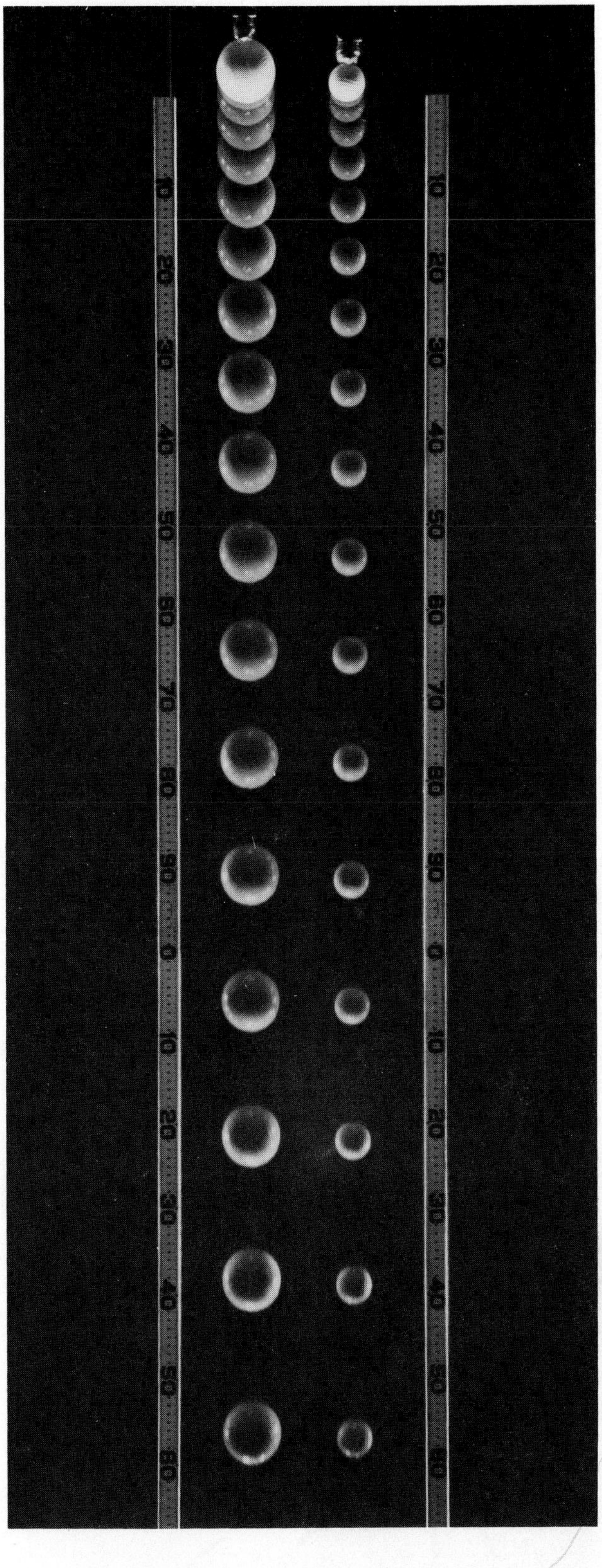

1.86 Two balls falling from rest. (Courtesy of Education Development Center.)

a) Find the average rate of the reaction for the interval from $t = 20$ to $t = 30$.

b) Plot the data points from the table, draw a smooth curve through them, and estimate the instantaneous rate of the reaction at $t = 25$.

When a model rocket is launched, the propellant burns for a few seconds, accelerating the rocket upward. After burnout, the rocket coasts upward for a while, then begins a period of free fall back to the ground. A small explosive charge pops out a parachute shortly after the rocket starts down. The parachute slows the rocket to keep it from breaking when it lands. Figure 1.87 shows velocity data from the flight of a model rocket. Use the data to answer the questions in Problems 15–19.

15. How fast was the rocket climbing when the engine stopped?

16. For how many seconds did the engine burn?

17. When did the rocket reach its highest point? What was its velocity then?

18. When did the parachute pop out? How fast was the rocket falling then?

19. How long did the rocket fall before the parachute popped out?

20. Suppose that the dollar cost of producing x washing machines is $f(x) = 2000 + 100x - 0.1x^2$.

a) Find the average cost of producing 100 washing machines.

b) Find the marginal cost of producing 100 washing machines.

c) Show that the marginal cost for 100 washing machines is approximately the cost of producing one more washing machine after the first 100 have been made, by computing the cost directly.

21. When a bactericide was added to a nutrient broth in which bacteria were growing, the bacteria population continued to grow for a while, but then stopped

1.87 Velocity of a model rocket (Problems 15–19).

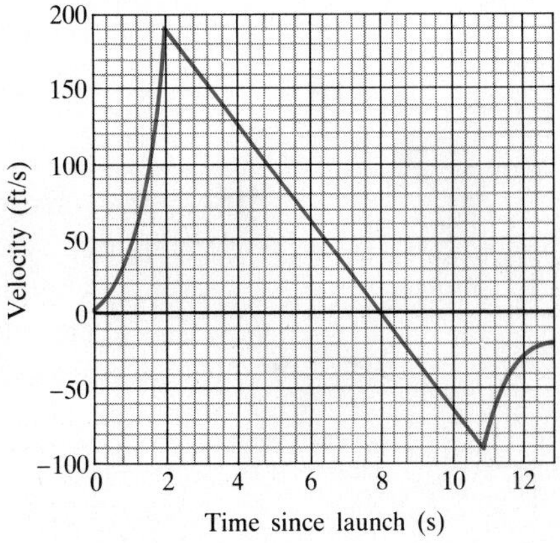

growing and began to decline. The size of the population at time t (hours) was $b(t) = 10^6 + 10^4t - 10^3t^2$. Use the result of Problem 1 to find the growth rates at (a) $t = 0$; (b) $t = 5$; and (c) $t = 10$ hours.

22. A tank is to be drained for cleaning. If Q represents the number of gallons of water in the tank t minutes after the tank has started to drain, and $Q = 200(30 - t)^2$, how fast is the water running out at the end of 10 minutes? What is the *average* rate at which the water flows out during the first 10 minutes?

23. a) If the radius of a circle changes from r to $(r + \Delta r)$, what is the average rate of change of the area of the circle with respect to the radius?
b) Find the instantaneous rate of change of the area with respect to the radius.

24. The volume V (ft^3) of a sphere of radius r (ft) is $V = \frac{4}{3}\pi r^3$. Find the rate of change of V with respect to r.

25. The radius r and altitude h of a certain cone are equal at all times. Find the rate of change of the volume $V = \frac{1}{3}\pi r^2 h$ with respect to h.

26. The three graphs in Fig. 1.88 show the distance traveled (miles), velocity (miles per hour), and acceleration (the derivative of the velocity, measured in miles per hour per second) for each second of a two-minute automobile trip. Which graph shows (a) distance? (b) velocity? (c) acceleration? (d) The vertical axis for the velocity graph is marked in units of 5 mph. Estimate the maximum and minimum values of the acceleration.

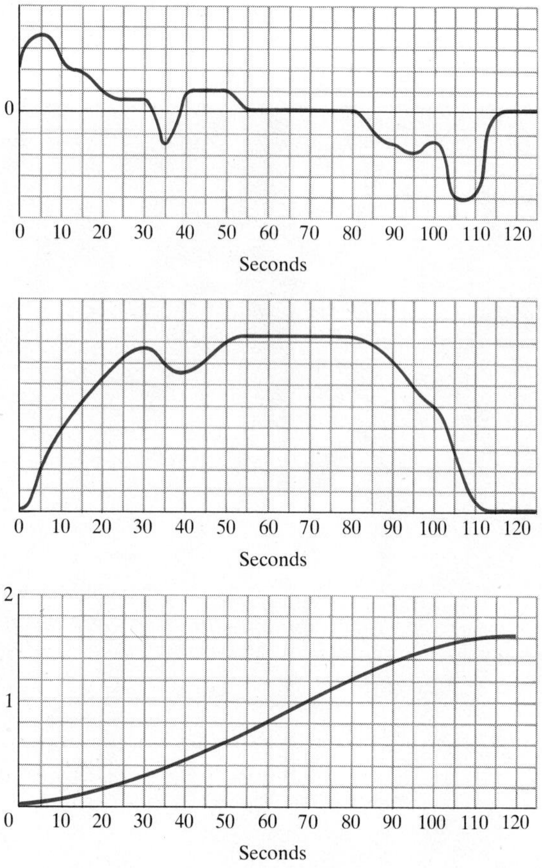

1.88 A two-minute automobile trip. Which of these curves is distance? velocity? acceleration?

1.9 Properties of Limits

The Definition

In the preceding articles we defined tangent slopes to be limits of secant slopes, and instantaneous rates of change (velocities, marginal cost) to be limiting values of average rates of change (average velocity, average cost). We found these limits from such expressions as

$$\text{Average change in } f \text{ over the interval from } x \text{ to } x + \Delta x = \frac{f(x + \Delta x) - f(x)}{\Delta x}. \tag{1}$$

In each case we were able to change the expression on the right into one in which we could set Δx equal to 0 to calculate the limit we wanted.

The problem is that we cannot always change the right-hand side of Eq. (1) into a form in which we can substitute 0 for Δx. For example, suppose we wanted to find the slope of the curve $y = \sin x$ by letting Δx approach 0 in the expression

$$\frac{\sin (x + \Delta x) - \sin x}{\Delta x}. \tag{2}$$

There is no stage at which the limit can be found by substituting $\Delta x = 0$, and rewriting the expression doesn't help.

Besides this strategic problem, there is also a practical one. Even when we can change the right-hand side to an expression that does not involve division by Δx, it may require a great deal of work to do so. We need better ways to calculate limits, and we will develop them in this article.

In the process of taking the limit of

$$\frac{f(x + \Delta x) - f(x)}{\Delta x} \tag{3}$$

as Δx approaches zero, we hold x fixed while Δx varies. That is, we treat the difference quotient (3) as a function of the single variable Δx. Therefore, our first step in developing a practical method of calculating limits of difference quotients will be to define what it means for a function of a single variable to have a limit as the variable approaches a fixed number. Doing this will enable us to treat the limit concept independently of its connection with the calculation of derivatives.

So long as we are discussing functions of a single variable, we might as well use a notation for them that is simpler than the one in Eq. (3). We will denote the variable by t and the function by $F(t)$, and set about defining what it means for F to have a limit L as t approaches a predetermined value c. Once we have the definition out of the way and know how to calculate a variety of limits, we can then go back to our original purpose for limits and use them to calculate the derivative of a function $y = f(x)$ by calculating the limit L of the function

$$F(t) = \frac{f(x + t) - f(x)}{t}, \tag{4}$$

as t approaches $c = 0$. In the meantime, we don't have to think about derivatives, we can think just about limits.

Suppose we are watching the values of a function $F(t)$ as t moves toward c without actually taking on the value c (just as we will let Δx approach 0 without taking on the value 0). What do we have to know about the behavior of the values of $F(t)$ to say that they have L as a limit? What observable pattern in their behavior would guarantee their eventual approach to L?

Certainly we want to be able to say that $F(t)$ stays within one tenth of a unit of L as soon as t stays within a certain radius r_1 of c, as shown here:

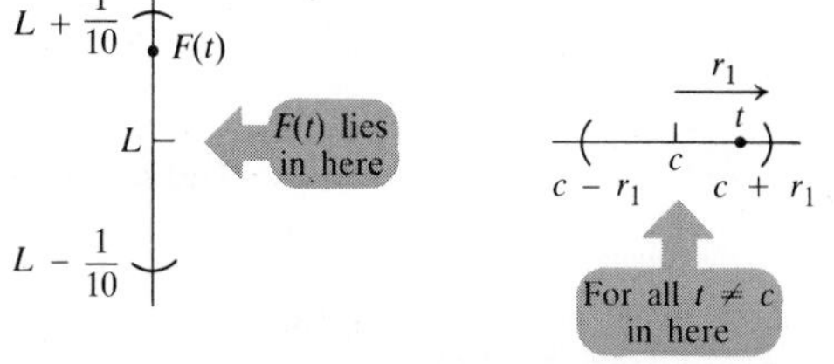

But that in itself is not enough, since as t continues on its course toward c, what is to prevent $F(t)$ from jittering about within the interval $(L - \frac{1}{10}, L + \frac{1}{10})$ instead of tending toward L?

We need to say also that as t continues toward c, $F(t)$ will eventually have to get still closer to L. We might say this by requiring $F(t)$ to lie within $\frac{1}{100}$ of a unit of L for all values of t within some smaller radius r_2 of c:

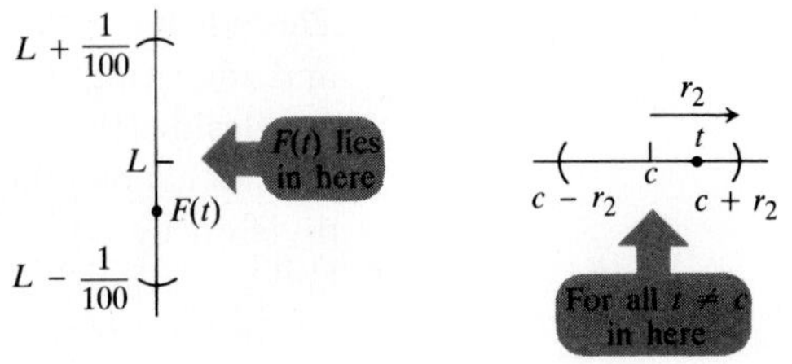

But this is not enough either. What if $F(t)$ skips about within the interval $(L - \frac{1}{100}, L + \frac{1}{100})$ from then on, without heading toward L? We had better require that $F(t)$ also lie within $1/1000$ of a unit of L after a while. That is, for all values of t within some still smaller radius r_3 of c, all the values of $F(t)$ lie in the interval

$$L - \frac{1}{1000} < F(t) < L + \frac{1}{1000},$$

as shown below:

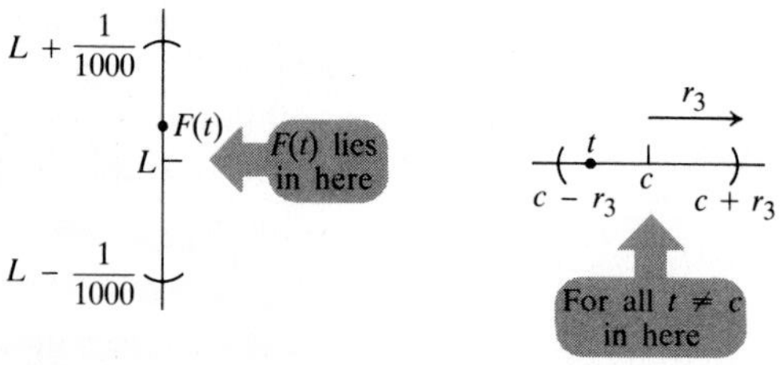

This still does not guarantee that $F(t)$ will now move toward L as t approaches c. Even if $F(t)$ has not skipped about before, it might start now. We need more.

We need to require that for *every* interval about L, no matter how small, we can find an interval of numbers about c all of whose F-values lie within that interval about L. In other words, given *any* positive radius ε about L, there exists some positive radius δ about c such that for all t within δ units of c (except $t = c$ itself) the values of $F(t)$ lie within ε units of L:

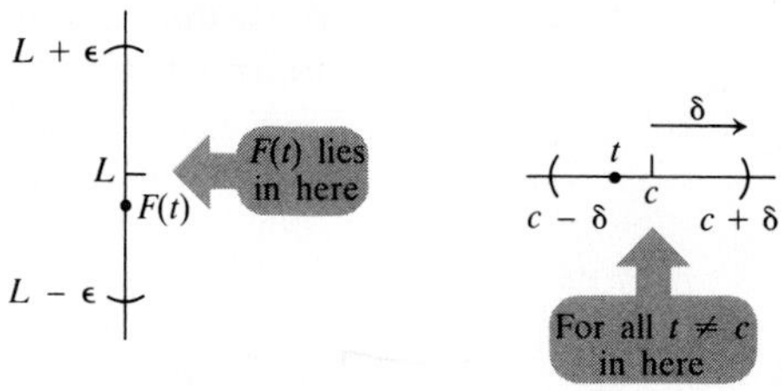

Thus, the closer t stays to c without equaling c, the closer $F(t)$ must stay to L.

DEFINITION

Limit

The *limit* of $F(t)$ as t approaches c is the number L if:

Given any radius $\varepsilon > 0$ about L there exists a radius $\delta > 0$ about c such that for all t

$$0 < |t - c| < \delta \quad \text{implies} \quad |F(t) - L| < \varepsilon. \tag{5}$$

We can write "the limit of $F(t)$ as t approaches c is L" as

$$\lim_{t \to c} F(t) = L.$$

Roughly speaking, to say that $F(t)$ approaches the limit L as t approaches c means that for any tolerance ε there is a small number δ (which depends on ε) such that $F(t)$ will stay within ε units of L if t is restricted to lie within δ units of c.

We might think of machining something like a generator shaft to a close tolerance. We try for a diameter L, but nothing is perfect and we must be satisfied to get the diameter $F(t)$ within $L \pm \varepsilon$. The δ is how accurate our control setting t must be to guarantee this degree of accuracy in the diameter of the shaft.

Examples—Testing the Definition

Whenever we make a new definition, it is a good idea to test it against familiar examples to see if it gives results consistent with past experience. For instance, our experience tells us that as t approaches 1, $5t$ approaches $5(1) = 5$ and $5t - 3$ approaches $5(1) - 3 = 2$. If our definition were to tell us that either of these limits were zero, or anything absurd like that, we would throw it out and look for a new definition. The following three examples are included in part to show that the definition in Eq. (5) gives the kinds of results we want.

EXAMPLE 1 Show that $\lim_{t \to 1}(5t - 3) = 2$ according to the definition of limit.

Solution We apply the definition of limit with $c = 1$, $F(t) = 5t - 3$, and $L = 2$. To satisfy the definition, we need to show that for any $\varepsilon > 0$ (radius about $L = 2$) there exists a $\delta > 0$ (radius about $c = 1$) such that for all t

$$0 < |t - 1| < \delta \quad \Rightarrow \quad |(5t - 3) - 2| < \varepsilon.$$

(The symbol $\Rightarrow$ is read "implies.") We change the ε-inequality from

$$|(5t - 3) - 2| < \varepsilon$$

to

$$|5t - 5| < \varepsilon$$

to

$$5|t - 1| < \varepsilon$$

to

$$|t - 1| < \varepsilon/5.$$

These inequalities are all equivalent. Therefore, the original ε-inequality will hold if $|t - 1| < \varepsilon/5$. We therefore take $\delta = \varepsilon/5$. See Fig. 1.89.

The value $\delta = \varepsilon/5$ is not the only value of δ that will make the ε-inequality hold. Any smaller positive δ will do as well. The definition does not ask for a "best" δ, just one that will work. □

EXAMPLE 2 Confirm that $\lim_{t \to c} t = c$ according to the definition of limit.

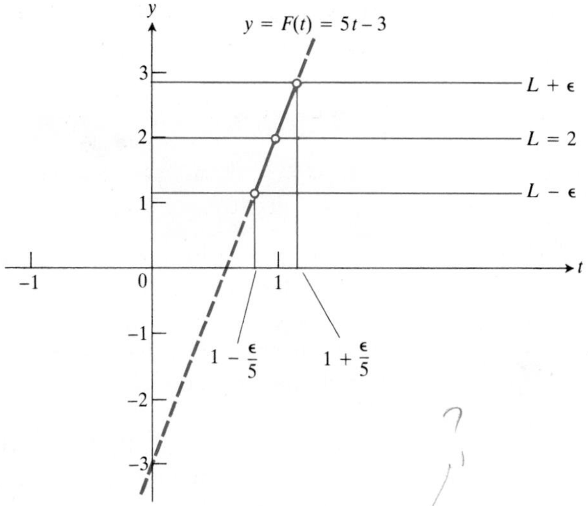

1.89 For every $t \neq 1$,
$1 - \varepsilon/5 < t < 1 + \varepsilon/5 \Rightarrow 2 - \varepsilon < 5t - 3 < 2 + \varepsilon.$

Solution We apply the definition of limit with $F(t) = t$ and $L = c$. To satisfy the definition, we must show that given any $\varepsilon > 0$ (radius about $L = c$) there exists a $\delta > 0$ (radius about c on the t-axis) such that for all t

$$0 < |t - c| < \delta \quad \Rightarrow \quad |t - c| < \varepsilon.$$

The ε-inequality will hold if δ is ε or any smaller positive number. □

When we read

$$F(t) \to L \quad \text{as} \quad t \to c$$

as "$F(t)$ approaches L as t approaches c," the verb "approaches" has a connotation of motion that is not always justified. For example, a constant function that has the value $F(t) = k$ for all values of t certainly has the limit k as t approaches any number c. We see this in the following example.

EXAMPLE 3 Let $F(t) = k$ be the function whose value is k for every t. Show that $\lim_{t\to c} F(t) = k$ for any c.

Solution We apply the definition of limit with $F(t) = k$ and $L = k$. We must show that for any $\varepsilon > 0$ there exists a $\delta > 0$ such that for all t

$$0 < |t - c| < \delta \quad \Rightarrow \quad |k - k| < \varepsilon.$$

Any positive δ will work because $k - k = 0$ is less than ε for all t. □

Right Limits and Left Limits

Sometimes the values of a function $F(t)$ tend to different limits as t approaches a number c from different sides. When this happens, we call the limit of F as t approaches c from the right the *right-hand limit* of F at c, and the limit of F as t approaches c from the left the *left-hand limit* of F at c.

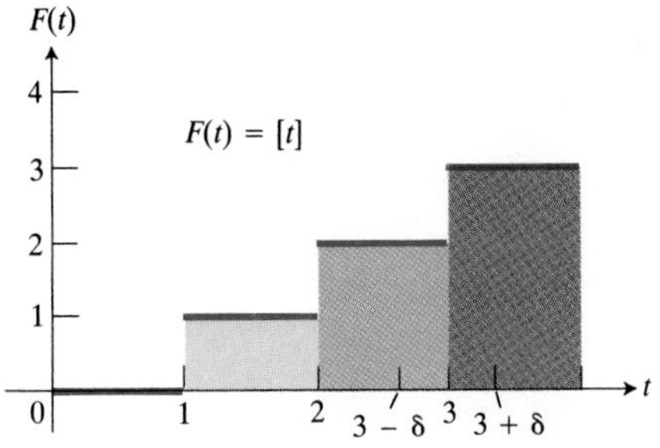

1.90 At each integer the greatest integer function has different right- and left-hand limits.

EXAMPLE 4 Show that $F(t) = [t]$ has no limit as t approaches 3 (Fig. 1.90).

Solution Our first guess might be that there is a limit, $L = 3$, since the values of $F(t) = [t]$ are close to 3 when t is equal to or slightly *greater* than 3. But when t is slightly *less* than 3, say $t = 2.9999$, then $[t] = 2$. That is, if δ is any positive number less than one, then

$$[t] = 2 \quad \text{if} \quad 3 - \delta < t < 3,$$

whereas

$$[t] = 3 \quad \text{if} \quad 3 < t < 3 + \delta.$$

Thus if we are challenged with a small positive ε, for example $\varepsilon = 0.01$, we cannot find a $\delta > 0$ that makes

$$|[t] - 3| < \varepsilon \quad \text{for} \quad 0 < |t - 3| < \delta.$$

In fact, no number L will work as the limit in this case. When t is near 3, *some* of the functional values of $[t]$ are 2 while others are 3. Hence the functional values are not all close to any one number L. Therefore,

$$\lim_{t \to 3} [t] \text{ does not exist.}$$

The right- and left-hand limits do, however, exist at 3. The right-hand limit is

$$L^+ = \lim_{t \to 3^+} [t] = 3,$$

and the left-hand limit is

$$L^- = \lim_{t \to 3^-} [t] = 2.$$

The notation $t \to 3^+$ may be read "t approaches 3 from above" (or "from the right," or "through values larger than 3") with an analogous meaning for $t \to 3^-$. □

In general, the right- and left-hand limits mentioned in Example 4 are defined as follows.

DEFINITION

Right-hand limit: $\lim_{t \to c^+} F(t) = L$

The limit of the function $F(t)$ as t approaches c from the right equals L if:

Given any $\varepsilon > 0$ (radius about L) there exists a $\delta > 0$ (radius to the right of c) such that for all t

$$c < t < c + \delta \quad \Rightarrow \quad |F(t) - L| < \varepsilon. \tag{6}$$

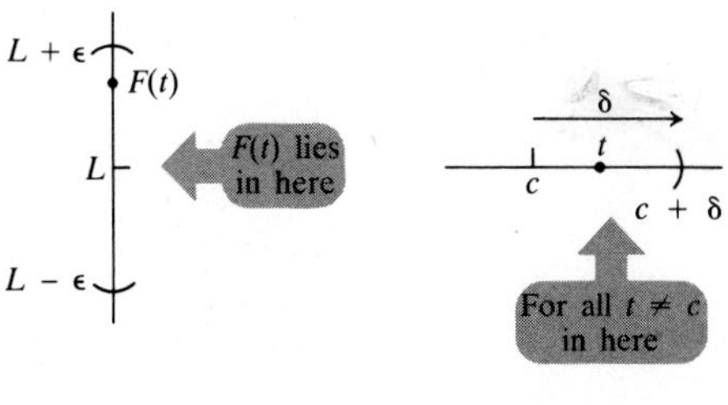

DEFINITION

Left-hand limit: $\lim_{t \to c^-} F(t) = L$

The limit of the function $F(t)$ as t approaches c from the left equals L if:

Given any $\varepsilon > 0$ (radius about L) there exists a $\delta > 0$ (radius to the left of c) such that for all t

$$c - \delta < t < c \quad \Rightarrow \quad |F(t) - L| < \varepsilon. \tag{7}$$

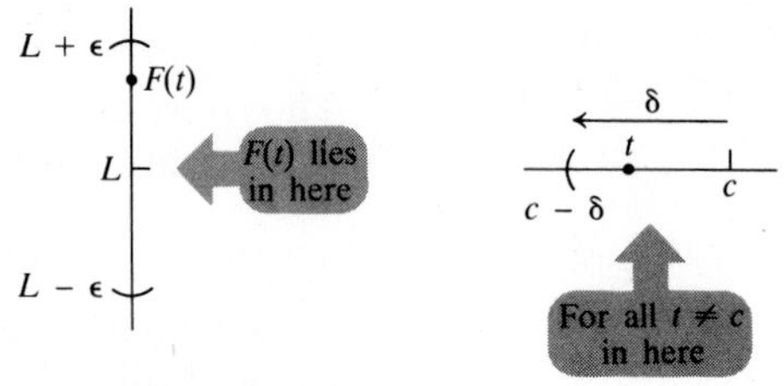

We can see the relation between the right- and left-hand limits of a function at a point and the limit defined earlier by comparing Eqs. (6) and (7) with Eq. (5). If we subtract c from the δ-inequalities in (6) and (7) they become

$$0 < t - c < \delta \quad \Rightarrow \quad |F(t) - L| < \varepsilon \tag{6'}$$

and

$$-\delta < t - c < 0 \quad \Rightarrow \quad |F(t) - L| < \varepsilon. \tag{7'}$$

Together, Eqs. (6′) and (7′) say the same thing as

$$0 < |t - c| < \delta \quad \Rightarrow \quad |F(t) - L| < \varepsilon, \tag{5}$$

which is (5) in the definition of limit. In short, $F(t)$ has a limit at a point if and only if the right-hand and left-hand limits there exist and are equal.

$$\lim_{t \to c} F(t) = L \quad \Leftrightarrow \quad \lim_{t \to c^+} F(t) = L \quad \text{and} \quad \lim_{t \to c^-} F(t) = L \tag{8}$$

We sometimes call $\lim_{t \to c} F(t)$ the *two-sided* limit of F at c to distinguish it from the right-hand and left-hand limits of F at c.

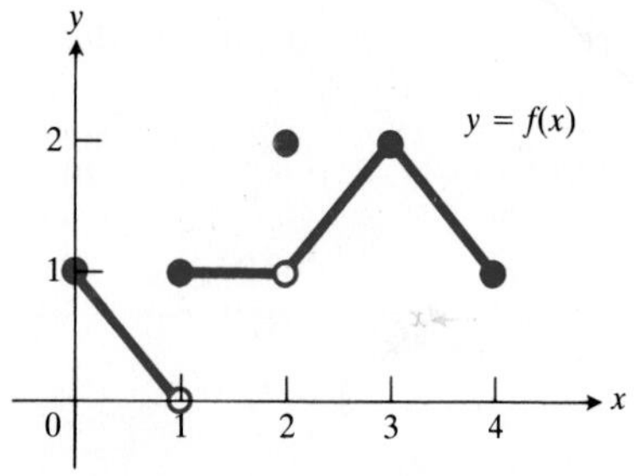

1.91 Example 5 discusses the limit properties of the function $y = f(x)$ graphed above.

EXAMPLE 5 All the following statements about the function $y = f(x)$ graphed in Fig. 1.91 are true.

At $x = 0$: $\lim_{x \to 0^+} f(x) = 1$.

At $x = 1$: $\lim_{x \to 1^-} f(x) = 0$ even though $f(1) = 1$,

$\lim_{x \to 1^+} f(x) = 1$,

$f(x)$ has no limit as $x \to 1$. (The right- and left-hand limits at 1 are not equal.)

At $x = 2$: $\lim_{x\to 2^-} f(x) = 1,$

$\lim_{x\to 2^+} f(x) = 1,$

$\lim_{x\to 2} f(x) = 1$ even though $f(2) = 2.$

At $x = 3$: $\lim_{x\to 3^-} f(x) = \lim_{x\to 3^+} f(x) = \lim_{x\to 3} f(x) = f(3) = 2.$

At $x = 4$: $\lim_{x\to 4^-} f(x) = 1.$

At every other point c between 0 and 4, $f(x)$ has a limit as $x \to c$. □

EXAMPLE 6 Show that $f(x) = |x|$ has no derivative at $x = 0$.

Solution We are asked to show that the difference quotient

$$\frac{|0 + \Delta x| - |0|}{\Delta x} = \frac{|\Delta x|}{\Delta x} \tag{9}$$

has no limit as $\Delta x \to 0$. Since $|\Delta x| = \Delta x$ when $\Delta x > 0$, the right-hand limit at 0 is

$$\lim_{\Delta x\to 0^+} \frac{|\Delta x|}{\Delta x} = \lim_{\Delta x\to 0^+} \frac{\Delta x}{\Delta x} = \lim_{\Delta x\to 0^+} 1 = 1. \tag{10}$$

Since $|\Delta x| = -\Delta x$ when $\Delta x < 0$, the left-hand limit is

$$\lim_{\Delta x\to 0^-} \frac{|\Delta x|}{\Delta x} = \lim_{\Delta x\to 0^-} \frac{-\Delta x}{\Delta x} = \lim_{\Delta x\to 0^-} -1 = -1. \tag{11}$$

The right-hand and left-hand limits at 0 are not equal. Therefore, the difference quotient (9) has no limit as $\Delta x \to 0$. □

Sometimes a function $f(x)$ has neither a right- nor a left-hand limit at a point. The following example shows how this can happen.

EXAMPLE 7 Show that the function $y = \sin(1/x)$ has no limit as $x \to 0$.

Solution See Fig. 1.92. In every interval about $x = 0$, the function takes on all values between -1 and $+1$. Hence there is no single number L that the function values all stay close to when x is close to 0. This is true even

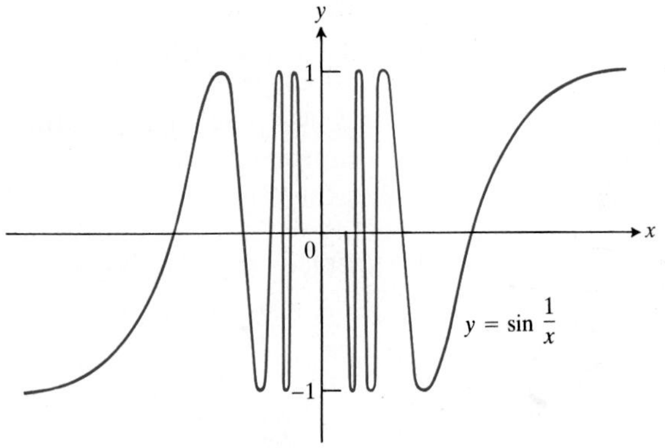

1.92 The function $y = \sin(1/x)$ has neither a right- nor a left-hand limit as x approaches 0.

when we restrict x to positive values or to negative values. In other words, this function has neither a right-hand limit nor a left-hand limit as x approaches 0. □

Calculation Techniques

Now we come to a theorem that tells how to calculate the limits of the sums and products of all the functions whose limits we already know. It also explains how to calculate the limits of many of the ratios of these functions.

THEOREM 1

If $\lim_{t\to c} F_1(t) = L_1$ and $\lim_{t\to c} F_2(t) = L_2$, then

i) $\lim\,[F_1(t) + F_2(t)] = L_1 + L_2$,

ii) $\lim\,[F_1(t) - F_2(t)] = L_1 - L_2$,

iii) $\lim\,[F_1(t)F_2(t)] = L_1L_2$,

iv) $\lim\,[kF_2(t)] = kL_2$ (any number k),

v) $\lim \dfrac{F_1(t)}{F_2(t)} = \dfrac{L_1}{L_2}$ if $L_2 \neq 0$.

The limits are all to be taken as $t \to c$.

In words, the formulas in Theorem 1 say:

i) The limit of the sum of two functions is the sum of their limits.

ii) The limit of the difference of two functions is the difference of their limits.

iii) The limit of a product of two functions is the product of their limits.

iv) The limit of a constant times a function is the constant times the limit.

v) The limit of a quotient of two functions is the quotient of their limits, provided the denominator does not tend to zero.

For future reference, the limits in L_1 and L_2 in Theorem 1 must be finite, that is, real numbers rather than the "infinite" limits discussed in Article 1.10.

A proof of Theorem 1 is given in Appendix 3. Informally, we can paraphrase the theorem in terms that make it highly reasonable: When t is close to c, $F_1(t)$ is close to L_1 and $F_2(t)$ is close to L_2. Then we naturally think that $F_1(t) + F_2(t)$ is close to $L_1 + L_2$; $F_1(t) - F_2(t)$ is close to $L_1 - L_2$; $F_1(t)F_2(t)$ is close to L_1L_2; $kF_2(t)$ is close to kL_2; and that $F_1(t)/F_2(t)$ is close to L_1/L_2 if L_2 is not zero.

What keeps this discussion from being a proof is that the word "close" is not precise. The phrases "arbitrarily close to" and "sufficiently close to" might improve the argument a bit, but the full ε and δ treatment in Appendix 3 is the clincher.

EXAMPLE 8 Since we know from Examples 2 and 3 that $\lim_{t\to c} t = c$ and $\lim_{t\to c} k = k$, we have

$$\lim_{t\to c} t^2 = \lim\,(t)(t) = (c)(c) = c^2 \qquad \text{(from iii)},$$

$$\lim_{t\to c} (t^2 - 5) = c^2 - 5 \qquad \text{(from ii)},$$

$$\lim_{t\to c} 4t^2 = 4c^2 \qquad \text{(from iv)},$$

and, if $c \neq 0$,

$$\lim_{t \to c} \frac{t^2 - 5}{4t^2} = \frac{c^2 - 5}{4c^2} \qquad \text{(from v).} \quad \square$$

As Example 8 suggests, the limit of any polynomial function $f(t)$ as $t \to c$ is $f(c)$. In other words, the limit can be found by evaluating the polynomial at $t = c$. Similarly, the limit of the ratio $f(t)/g(t)$ of two polynomials as $t \to c$ is $f(c)/g(c)$, provided $g(c) \neq 0$.

THEOREM 2

Polynomials

If $f(t) = a_0t^n + a_1t^{n-1} + \cdots + a_n$ is any polynomial function, then

$$\lim_{t \to c} f(t) = f(c) = a_0c^n + a_1c^{n-1} + \cdots + a_n.$$

THEOREM 3

Quotients of polynomials

If $f(t)$ and $g(t)$ are polynomials, then

$$\lim_{t \to c} \frac{f(t)}{g(t)} = \frac{f(c)}{g(c)}, \quad \text{provided } g(c) \neq 0.$$

The proofs of Theorems 2 and 3 are set out in Appendix 3, Problems 1–5. Theorems 2 and 3 are the theorems behind some of the limit calculations in Articles 1.6 and 1.7.

EXAMPLE 9 In calculating the slope of $y = x^3 - 3x + 3$ from Example 5 in Article 1.6, we find from Theorem 2 that

$$\lim_{\Delta x \to 0} (3x^2 + 3x\,\Delta x + (\Delta x)^2 - 3) = 3x^2 + 3x(0) + (0)^2 - 3 = 3x^2 - 3.$$

Note that, with x fixed, $3x^2 + 3x\,\Delta x + (\Delta x)^2 - 3$ is a polynomial function of Δx. $\square$

EXAMPLE 10 Find

$$\lim_{t \to 2} \frac{t^2 + 2t + 4}{t + 2}.$$

Solution The function whose limit we are to find is the quotient of two polynomials. The denominator $t + 2$ is not 0 when $t = 2$. Therefore, the limit is the value of the quotient at $t = 2$, as in Theorem 3:

$$\lim_{t \to 2} \frac{t^2 + 2t + 4}{t + 2} = \frac{(2)^2 + 2(2) + 4}{2 + 2} = \frac{12}{4} = 3. \quad \square$$

EXAMPLE 11 Find

$$\lim_{t \to 2} \frac{t^3 - 8}{t^2 - 4}.$$

Solution The denominator is 0 at $t = 2$ and we cannot apply Theorem 3 directly. However, if we factor the numerator and denominator we find that

$$\frac{t^3 - 8}{t^2 - 4} = \frac{(t - 2)(t^2 + 2t + 4)}{(t - 2)(t + 2)}.$$

For $t \neq 2$,

$$\frac{t - 2}{t - 2} = 1.$$

Therefore, for all values of t different from 2 (the values that really determine the limit as $t \to 2$),

$$\lim_{t \to 2} \frac{t^3 - 8}{t^2 - 4} = \lim_{t \to 2} \frac{t^2 + 2t + 4}{t + 2} = \frac{(2)^2 + 2(2) + 4}{2 + 2} = \frac{12}{4} = 3.$$

To calculate this limit, we divided the numerator and denominator by a common factor and evaluated the result at $t = 2$. □

Example 11 illustrates an important mathematical point about limits: The limit of a function $f(t)$ as $t \to c$ *never* depends on what happens when $t = c$. The limit (if it exists at all) is *entirely determined* by the values f has when $t \neq c$. In Example 11, the quotient $f(t) = (t^3 - 8)/(t^2 - 4)$ is not even defined at $t = 2$. Yet its limit as $t \to 2$ exists and is equal to 3. The limit is entirely determined by the values the quotient takes on when $t \neq 2$.

We conclude this article with a theorem that will be used repeatedly in our later work.

THEOREM 4

The Sandwich Theorem

Suppose that

$$f(t) \leq g(t) \leq h(t)$$

for all $t \neq c$ in some interval about c, and that $f(t)$ and $h(t)$ approach the same limit L as t approaches c. Then,

$$\lim_{t \to c} g(t) = L.$$

See Fig. 1.93. The Sandwich Theorem holds for right-hand and left-hand limits as well as two-sided limits.

The idea behind Theorem 4 is that if $f(t)$ and $h(t)$ head toward L with $g(t)$ "sandwiched" in between, they take $g(t)$ along with them:

$$\begin{array}{ccccc} f(t) & \leq & g(t) & \leq & h(t) \\ \downarrow & & \downarrow & & \downarrow \\ L & \leq & \lim g(t) & \leq & L. \end{array}$$

We might think of $g(t)$ as a ball between two table tennis paddles that move closer together as $t \to c$. A proof of the theorem can be found at the end of Appendix 3.

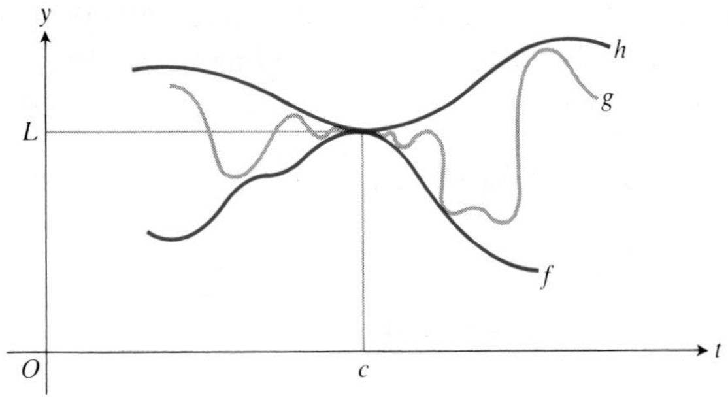

1.93 The Sandwich Theorem. If $\lim_{t\to c} f(t)$ and $\lim_{t\to c} h(t)$ both equal L, then $\lim_{t\to c} g(t) = L$ because $f(t) \le g(t) \le h(t)$ for all values of t near c. The graph of g is caught between the graphs of f and h.

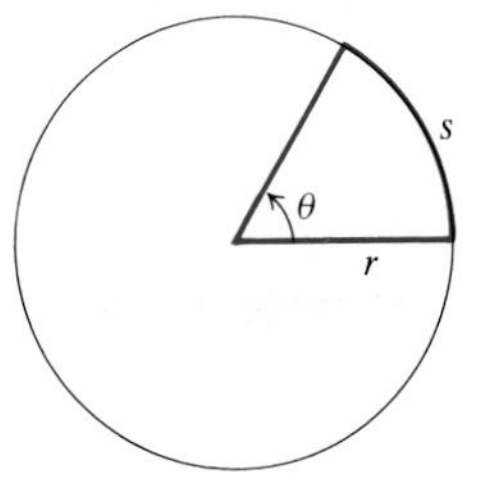

1.94 $\theta = s/r$ radians.

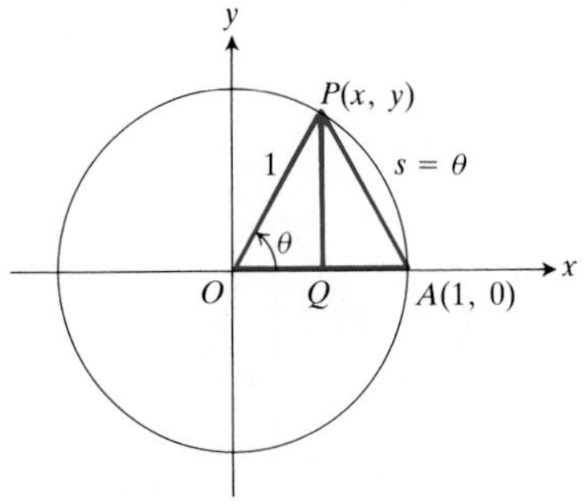

1.95 Radian measure on a unit circle: $r = 1$, $s = \theta$.

EXAMPLE 12 As an application of the Sandwich Theorem, we show that

$$\lim_{\theta\to 0} \sin\theta = 0, \tag{12}$$

$$\lim_{\theta\to 0} \cos\theta = 1, \tag{13}$$

$$\lim_{\theta\to 0} \frac{\sin\theta}{\theta} = 1, \tag{14}$$

provided that θ is measured in radians.

Solution The radian measure of an angle can be defined by

$$\theta = \frac{s}{r}, \tag{15}$$

where s is the length of the arc the angle intercepts on a circle of radius r when the center of the circle is at the angle's vertex. Figure 1.94 illustrates this definition. (There is more about radian measure in the trigonometry review in Article 2.7.)

In Fig. 1.95, O is the center of a unit circle and θ is the radian measure of an acute angle AOP. Note that $s = \theta$ under these conditions (because $s = r\theta$ and $r = 1$). Now $\triangle APQ$ is a right triangle with legs of length

$$QP = \sin\theta, \qquad AQ = 1 - \cos\theta.$$

From the Pythagorean theorem and the fact that $AP < \theta$, we get

$$\sin^2\theta + (1-\cos\theta)^2 = (AP)^2 < \theta^2. \tag{16}$$

Both terms on the left side of Eq. (16) are positive, so each is smaller than their sum, and hence is less than θ^2:

$$\sin^2\theta < \theta^2, \tag{17}$$

$$(1-\cos\theta)^2 < \theta^2, \tag{18}$$

or

$$-\theta < \sin\theta < \theta, \tag{19}$$

$$-\theta < 1 - \cos\theta < \theta. \tag{20}$$

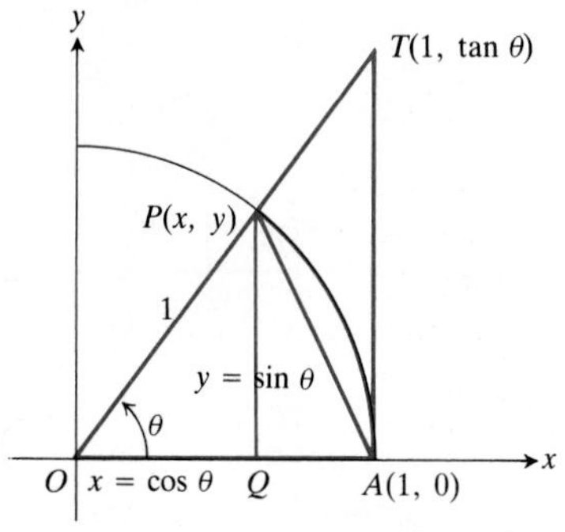

1.96 Area $\triangle OAP <$ area sector $OAP <$ area $\triangle OAT$.

Now we let θ approach 0. Since $\lim_{\theta\to 0} -\theta = 0$ and $\lim_{\theta\to 0} \theta = 0$, the Sandwich Theorem says that in the limit Eqs. (19) and (20) become

$$0 \le \lim_{\theta\to 0} \sin\theta \le 0, \tag{21}$$

$$0 \le \lim_{\theta\to 0} (1 - \cos\theta) \le 0. \tag{22}$$

Therefore, since

$$0 = \lim_{\theta\to 0} (1 - \cos\theta) = 1 - \lim_{\theta\to 0} \cos\theta,$$

we have

$$\lim_{\theta\to 0} \sin\theta = 0, \qquad \lim_{\theta\to 0} \cos\theta = 1.$$

To establish Eq. (14) we show that the right-hand and left-hand limits of $(\sin\theta)/\theta$ at 0 are both equal to 1.

To show that the right-hand limit is equal to 1, we assume that θ is positive and less than $\pi/2$ in Fig. 1.96. We compare the areas of $\triangle OAP$, sector OAP, and $\triangle OAT$, and note that

$$\text{Area } \triangle OAP < \text{Area sector } OAP < \text{Area } \triangle OAT. \tag{23}$$

These areas may be expressed in terms of θ as follows:

$$\text{Area } \triangle OAP = \frac{1}{2}\text{ base} \times \text{height} = \frac{1}{2}(1)(\sin\theta) = \frac{1}{2}\sin\theta, \tag{24}$$

$$\text{Area sector } OAP = \frac{1}{2} r^2\theta = \frac{1}{2}(1)^2\theta = \frac{\theta}{2}, \tag{25}$$

$$\text{Area } \triangle OAT = \frac{1}{2}\text{ base} \times \text{height} = \frac{1}{2}(1)(\tan\theta) = \frac{1}{2}\tan\theta, \tag{26}$$

so that

$$\frac{1}{2}\sin\theta < \frac{1}{2}\theta < \frac{1}{2}\tan\theta. \tag{27}$$

The inequality in (27) will go the same way if we divide all three terms by the positive number $(\frac{1}{2})\sin\theta$:

$$1 < \frac{\theta}{\sin\theta} < \frac{1}{\cos\theta}. \tag{28}$$

We next take reciprocals in (28), which reverses the inequalities:

$$\cos\theta < \frac{\sin\theta}{\theta} < 1. \tag{29}$$

Since $\cos\theta$ approaches 1 as θ approaches 0, the Sandwich Theorem tells us that

$$1 \le \lim_{\theta\to 0^+} \frac{\sin\theta}{\theta} \le 1, \tag{30}$$

or

$$\lim_{\theta\to 0^+} \frac{\sin\theta}{\theta} = 1. \tag{31}$$

The limits in (30) and (31) are right-hand limits because we have been dealing with values of θ between 0 and $\pi/2$.

Table 1.3

Degrees	θ (radians)	$\sin\theta$	$\dfrac{\sin\theta}{\theta}$
0°	0	0	Undefined
1°	0.017453	0.017452	0.99994
2°	0.034907	0.034899	0.9998
5°	0.08727	0.08716	0.9987
10°	0.17453	0.17365	0.995

The geometrical argument leading to Eq. (27) and to the limit in (31) is based on the assumption that θ is positive, but the same limit is obtained for $(\sin\theta)/\theta$ if θ approaches 0 from the left. For if $\theta = -\alpha$ and α is positive, then

$$\frac{\sin\theta}{\theta} = \frac{\sin(-\alpha)}{-\alpha} = \frac{-\sin(\alpha)}{-\alpha} = \frac{\sin\alpha}{\alpha}. \tag{32}$$

Therefore,

$$\lim_{\theta\to 0^-} \frac{\sin\theta}{\theta} = \lim_{\alpha\to 0^+} \frac{\sin\alpha}{\alpha} = 1. \tag{33}$$

This concludes the proof that $\lim_{\theta\to 0} (\sin\theta)/\theta = 1$. □

Table 1.3 shows $\sin\theta$ and $(\sin\theta)/\theta$ for a few selected values of θ near zero. To get the effect of $(\sin\theta)/\theta$ approaching 1, read the last column upward from the bottom.

EXAMPLE 13

$$\lim_{x\to 0} \frac{\sin 3x}{x} = \lim_{3x\to 0} \frac{3\sin 3x}{3x} = 3\lim_{\theta\to 0}\left(\frac{\sin\theta}{\theta}\right) = 3. \ \square$$

EXAMPLE 14

$$\lim_{x\to 0} \frac{\tan x}{x} = \lim_{x\to 0}\left(\frac{\sin x}{x}\frac{1}{\cos x}\right)$$

$$= \left[\lim_{x\to 0}\left(\frac{\sin x}{x}\right)\right]\left[\lim_{x\to 0}\left(\frac{1}{\cos x}\right)\right] = (1)(1) = 1. \ \square$$

PROBLEMS

Find the limits in Problems 1–26.

1. $\lim_{x\to 2} 2x$
2. $\lim_{x\to 0} 2x$
3. $\lim_{x\to 4} 4$
4. $\lim_{x\to -2} 4$
5. $\lim_{x\to 1} (3x - 1)$
6. $\lim_{x\to 1/3} (3x - 1)$
7. $\lim_{x\to 5} x^2$
8. $\lim_{x\to 2} x(2 - x)$
9. $\lim_{x\to 0} (x^2 - 2x + 1)$
10. $\lim_{x\to 5} (x^2 - 3x - 18)$
11. $\lim_{\Delta x\to 0} 2x + \Delta x$
12. $\lim_{t\to 2} t^2$
13. $\lim_{x\to 1} |x - 1|$
14. $\lim_{x\to -3} |x|$
15. $\lim_{x\to 0} x[x]$
16. $\lim_{x\to 1} x(2x - 1)$
17. $\lim_{x\to 2} 3x(2x - 1)$
18. $\lim_{x\to 2} 3(2x - 1)(x + 1)$
19. $\lim_{x\to 2} 3x^2(2x - 1)$
20. $\lim_{x\to 2} 3(2x - 1)(x + 1)^2$
21. $\lim_{x\to -1} (x + 3)$
22. $\lim_{x\to -1} (x + 3)^2$
23. $\lim_{x\to -1} (x^2 + 6x + 9)$
24. $\lim_{x\to -2} (x + 3)^{171}$

25. $\lim_{x\to -4} (x + 3)^{1984}$

26. $\lim_{x\to 1} (x^3 + 3x^2 - 2x - 17)$

27. Suppose $\lim_{x\to c} f(x) = 5$ and $\lim_{x\to c} g(x) = -2$.
a) What is $\lim_{x\to c} f(x)g(x)$?
b) What is $\lim_{x\to c} 2f(x)g(x)$?

28. Suppose $\lim_{x\to b} f(x) = 7$ and $\lim_{x\to b} g(x) = -3$. What is $\lim_{x\to b} [f(x) + g(x)]$?

29. Suppose $\lim_{x\to b} f(x) = 7$ and $\lim_{x\to b} g(x) = -3$. Find
a) $\lim_{x\to b} (f(x) + g(x))$, b) $\lim_{x\to b} f(x) \cdot g(x)$,
c) $\lim_{x\to b} 4g(x)$, d) $\lim_{x\to b} f(x)/g(x)$.

30. Suppose $\lim_{x\to -2} p(x) = 4$, $\lim_{x\to -2} r(x) = 0$, and $\lim_{x\to -2} s(x) = -3$. Find
a) $\lim_{x\to -2} (p(x) + r(x) + s(x))$,
b) $\lim_{x\to -2} p(x) \cdot r(x) \cdot s(x)$.

Find the limits in Problems 31–51.

31. $\lim_{t\to 2} \dfrac{t+3}{t+2}$ **32.** $\lim_{x\to 5} \dfrac{x^2-25}{x-5}$

33. $\lim_{x\to 5} \dfrac{x^2-25}{x+5}$ **34.** $\lim_{x\to 5} \dfrac{x-5}{x^2-25}$, if it exists

35. $\lim_{x\to 5} \dfrac{x+5}{x^2-25}$, if it exists **36.** $\lim_{x\to 1} \dfrac{x^2-x-2}{x^2-1}$

37. $\lim_{x\to 0} \dfrac{5x^3+8x^2}{3x^4-16x^2}$ **38.** $\lim_{y\to 2} \dfrac{y^2+5y+6}{y+2}$

39. $\lim_{y\to 2} \dfrac{y^2-5y+6}{y-2}$ **40.** $\lim_{x\to -5} \dfrac{x^2+3x-10}{x+5}$

41. $\lim_{x\to 4} \dfrac{x-4}{x^2-5x+4}$ **42.** $\lim_{t\to 3} \dfrac{t^2-3t+2}{t^2-1}$

43. $\lim_{t\to 1} \dfrac{t^2-3t+2}{t^2-1}$ **44.** $\lim_{x\to 2} \dfrac{x-2}{x^2-6x+8}$

45. $\lim_{x\to -3} \dfrac{x^2+4x+3}{x-3}$ **46.** $\lim_{x\to -3} \dfrac{x+3}{x^2+4x+3}$

47. $\lim_{x\to -2} \dfrac{x^2+x-2}{x^2-4}$ **48.** $\lim_{x\to 1} \dfrac{x^2+x-2}{x-2}$

49. $\lim_{x\to 2} \dfrac{x^2-7x+10}{x-2}$ **50.** $\lim_{t\to 1} \dfrac{t^3-1}{t-1}$

51. $\lim_{x\to a} \dfrac{x^3-a^3}{x^4-a^4}$

52. Find $\lim_{x\to 1} [(x^n - 1)/(x - 1)]$, n a positive integer. Compare your result with the limits in Problems 50 and 51.

53. Give an example of functions f and g such that $f(x) + g(x)$ approaches a limit as $x \to 0$ even though $f(x)$ and $g(x)$ separately do not approach limits as $x \to 0$.

54. Repeat Problem 53 for $f(x) \cdot g(x)$ in place of $f(x) + g(x)$.

55. Give an example of functions f and g such that $\lim_{x\to 0} f(x)/g(x)$ exists, but at least one of the functions f and g fails to have a limit as $x \to 0$.

Each of the limits in Problems 56 and 57 is the derivative of a function at $x = 0$. Find the function.

56. $\lim_{h\to 0} \dfrac{(1+h)^2 - 1}{h}$ **57.** $\lim_{h\to 0} \dfrac{|-1+h| - |-1|}{h}$

58. Which of the following statements are true of the function f defined for $-1 \le x \le 3$ in Fig. 1.97?
a) $\lim_{x\to -1^+} f(x) = 1$ b) $\lim_{x\to 2} f(x)$ does not exist.
c) $\lim_{x\to 2} f(x) = 2$ d) $\lim_{x\to 1^-} f(x) = 2$
e) $\lim_{x\to 1^+} f(x) = 1$ f) $\lim_{x\to 1} f(x)$ does not exist.
g) $\lim_{x\to 0^+} f(x) = \lim_{x\to 0^-} f(x)$
h) $\lim_{x\to c} f(x)$ exists at every c between -1 and $+1$.
i) $\lim_{x\to c} f(x)$ exists at every c between 1 and 3.

Figure 1.97

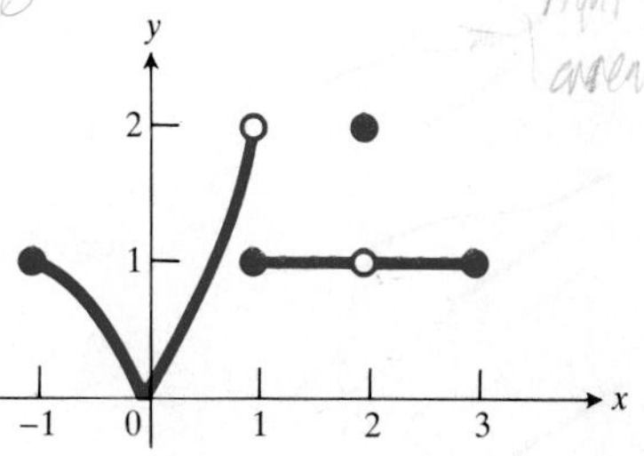

Graph the functions in Problems 59 and 60. Then answer the following questions about them:
a) At what points c in the domain of f does $\lim_{x\to c}$ exist?
b) At what points does only the left-hand limit exist?
c) At what points does only the right-hand limit exist?

59. $f(x) = \begin{cases} \sqrt{1-x^2} & \text{if } 0 \le x < 1, \\ 1 & \text{if } 1 \le x < 2, \\ 2 & \text{if } x = 2. \end{cases}$

60. $f(x) = \begin{cases} x & \text{if } -1 \le x < 0, \text{ or } 0 < x \le 1, \\ 1 & \text{if } x = 0, \\ 0 & \text{if } x < -1, \text{ or } x > 1. \end{cases}$

Find the limits in Problems 61–69. The square brackets in Problems 61–64 denote the greatest integer function.

61. $\lim_{x\to 0^+} [x]$ **62.** $\lim_{x\to 0^-} [x]$

63. $\lim_{x\to 0.5} [x]$ **64.** $\lim_{x\to 2^-} \dfrac{x}{[x]}$

65. $\lim_{x\to 0^+} \dfrac{x}{|x|}$ **66.** $\lim_{x\to 0^-} \dfrac{x}{|x|}$

67. $\lim_{x\to 3^+} \dfrac{x^2-9}{|x-3|}$ **68.** $\lim_{x\to 3^-} \dfrac{x^2-9}{|x-3|}$

69. $\lim_{x\to 2^-} \dfrac{x^2|4x-8|}{x-2}$

70. For what values of c does the greatest integer function $f(x) = [x]$ approach a limit as $x \to c$?

71. For what values of c does $f(x) = x/|x|$ approach a limit as $x \to c$?

Find the limits in Problems 72–88.

72. $\lim_{x\to 0} \dfrac{1+\sin x}{\cos x}$

73. $\lim_{x\to 0^+} \cos x$

74. $\lim_{t\to 0} \dfrac{t}{\sin t}$

75. $\lim_{h\to 0} \dfrac{\sin^2 h}{h}$

76. $\lim_{h\to 0} \dfrac{\sin^2 h}{h^2}$

77. $\lim_{t\to 0} \dfrac{2\sin t\cos t}{t}$

78. $\lim_{x\to 0} \tan x$

79. $\lim_{\theta\to 0} \dfrac{\tan\theta}{\theta}$

80. $\lim_{x\to 0^-} \sin x$

81. $\lim_{x\to 0^+} \dfrac{\sin x}{|x|}$

82. $\lim_{x\to 0^-} \dfrac{\sin x}{|x|}$

83. $\lim_{x\to 0} x\cos x$

84. $\lim_{x\to 0} \dfrac{\sin 2x}{x}$

85. $\lim_{x\to 0} \dfrac{\sin 5x}{\sin 3x}$

86. $\lim_{y\to 0} \dfrac{\tan 2y}{3y}$

87. $\lim_{x\to 0} \dfrac{\sin 2x}{2x^2+x}$

88. a) Show that $-|x| \le x\sin(1/x) \le |x|$ for all $x \ne 0$.
b) Use the Sandwich Theorem and the inequality in (a) to calculate $\lim_{x\to 0} x\sin(1/x)$.

89. The inequality

$$1 - \frac{x^2}{6} < \frac{\sin x}{x} < 1$$

holds when x is measured in radians and $|x| < 1$. Use this inequality and the Sandwich Theorem to find $\lim_{x\to 0} (\sin x)/x$.

90. Does $\lim_{x\to 0} (\sin x)/|x|$ exist? If so, what is it? If not, why not?

91. For each function below, find the limit L of $F(t)$ as $t \to c$. Then show that given $\varepsilon > 0$ there exists a $\delta > 0$ such that for all t

$$0 < |t-c| < \delta \Rightarrow |L - F(t)| < \varepsilon.$$

In each case, draw a graph similar to Fig. 1.89.

a) $F(t) = 5 - 3t, \quad c = 2$

b) $F(t) = 7, \quad c = -1$

c) $F(t) = \dfrac{t^2-4}{t-2}, \quad c = 2$

d) $F(t) = \dfrac{t^2+6t+5}{t+5}, \quad c = 5$

e) $F(t) = \dfrac{3t^2+8t-3}{2t+6}, \quad c = -3$

f) $F(t) = \dfrac{4}{t}, \quad c = 2$

g) $F(t) = \dfrac{(1/t)-(1/3)}{t-3}, \quad c = 3$

92. Find a domain $0 < |t-3| < \delta$ such that, when t is restricted to this domain, the difference between t^2 and 9 will be numerically smaller than (a) $\frac{1}{10}$; (b) $\frac{1}{100}$; and (c) ε, where ε may be any positive number.

93. Repeat Problem 92 using $t^2 + t$ and 12 in place of t^2 and 9.

94. (*Calculator*). It is sometimes easy to guess the value of a limit once the limit is known to exist.

a) Ignoring the question of whether the limit below exists (it does, and is finite), use a calculator to guess its value. First take $\Delta x = 0.1, 0.01, 0.001, \ldots,$ continuing until you are ready to guess the right-hand limit. Then test your guess by using $\Delta x = -0.1, -0.01, \ldots$:

$$\lim_{\Delta x\to 0} \frac{\sqrt{4+\Delta x}-2}{\Delta x}.$$

b) Prove that the limit in (a) exists, and relate it to a derivative.

95. (*Calculator*). To estimate the value of the derivative of $f(x) = \sqrt{9-x^2}$ at $x = 0$, write out the appropriate difference quotient and proceed as in Problem 94.

***Toolkit* programs**

Function Evaluator	Limit Problems
Limit Definition	

1.10 Infinity as a Limit

The function

$$y = \frac{1}{x}, \tag{1}$$

graphed in Fig. 1.98, is defined for all real numbers except $x = 0$. Evidently, the following statements apply.

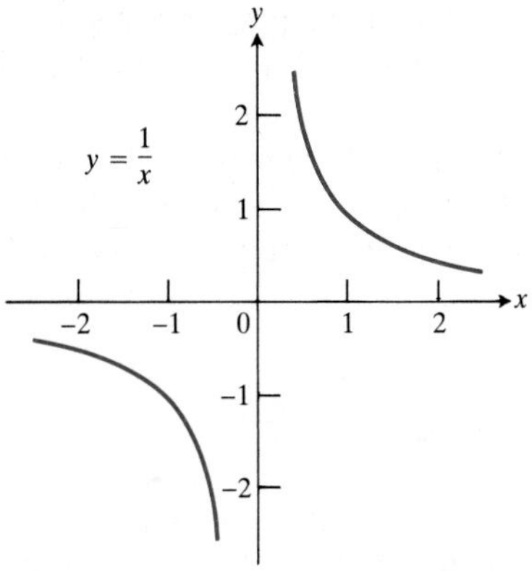

1.98 The graph of $y = 1/x$.

a) When x is small and positive, 1/x is large and positive. For example,

$$\frac{1}{0.001} = 1000.$$

b) When x is small and negative, 1/x is large and negative. For example,

$$\frac{1}{-0.001} = -1000.$$

c) When x is large and positive, 1/x is small and positive. For example,

$$\frac{1}{10{,}000} = 0.0001.$$

d) When x is large and negative, 1/x is small and negative. For example,

$$\frac{1}{-10{,}000} = -0.0001.$$

These facts are sometimes abbreviated by saying:

a) As x approaches 0 from the right, 1/x tends to ∞.
b) As x approaches 0 from the left, 1/x tends to $-\infty$.
c) As x tends to ∞, 1/x approaches 0.
d) As x tends to $-\infty$, 1/x approaches 0.

The symbol ∞, *infinity,* does not represent any real number. We cannot use ∞ in arithmetic in the usual way, but it is convenient to be able to say things like "the limit of 1/x as x approaches infinity is 0," and we can do so according to the following definition.

DEFINITION

Limit as $x \to \infty$ or $x \to -\infty$

1. The limit of the function $f(x)$ as x approaches infinity is the number L,

$$\lim_{x\to\infty} f(x) = L,$$

if: Given any $\varepsilon > 0$ there exists a number M such that for all x

$$M < x \quad \Rightarrow \quad |f(x) - L| < \varepsilon. \tag{2}$$

2. The limit of $f(x)$ as x approaches minus infinity is the number L,

$$\lim_{x\to-\infty} f(x) = L,$$

if: Given any $\varepsilon > 0$ there exists a number N such that for all x

$$x < N \quad \Rightarrow \quad |f(x) - L| < \varepsilon. \tag{3}$$

In less formal language,

$$\lim_{x\to\infty} f(x) = L$$

means that $f(x)$ can be made as close as desired to L by making x large enough and positive. Similarly,

$$\lim_{x\to-\infty} f(x) = L$$

means that $f(x)$ can be made as close as desired to L by making x large enough and negative (that is, by taking x far enough out on the negative x-axis).

Graphically, the inequalities in statements (2) and (3) mean that the curve $y = f(x)$ stays between the lines $L - \varepsilon$ and $L + \varepsilon$ for $|x|$ sufficiently large, as in Figs. 1.99 and 1.100 that accompany Examples 1 and 7.

EXAMPLE 1 Show that

$$\lim_{x\to\infty} \frac{1}{x} = 0 \qquad \text{and} \qquad \lim_{x\to-\infty} \frac{1}{x} = 0 \tag{4}$$

according to the definitions of limit as $x \to \infty$ and $x \to -\infty$.

Solution See Fig. 1.99. We have

$$\left|\frac{1}{x} - 0\right| = \left|\frac{1}{x}\right| < \varepsilon \tag{5}$$

for any $\varepsilon > 0$, provided

$$|x| > \frac{1}{\varepsilon}. \tag{6}$$

Thus $1/x$ lies within ε of 0 for all $x > M = 1/\varepsilon$ and for all $x < N = -1/\varepsilon$. □

EXAMPLE 2 Show that if k is any number, then

$$\lim_{x\to\infty} k = k \qquad \text{and} \qquad \lim_{x\to-\infty} k = k \tag{7}$$

according to the definitions of limit as $x \to \infty$ and $x \to -\infty$.

Solution When we apply the definitions with $f(x) = k$ and $L = k$, we have

$$|k - k| = 0 < \varepsilon \qquad \text{for any } \varepsilon > 0. \;\square$$

1.99 When $|x| > 1/\varepsilon$, the curve $y = 1/x$ lies between the lines $y = \varepsilon$ and $y = -\varepsilon$.

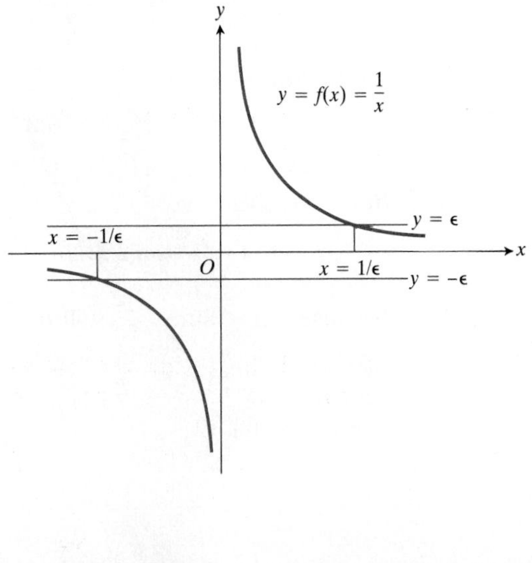

The following theorem about limits of sums, differences, products, and quotients is analogous to the corresponding theorem for limits when $x \to c$. It tells us how to combine results like the ones in Examples 1 and 2 to calculate other limits.

THEOREM 5 If

$$\lim_{x\to\infty} f(x) = L_1 \qquad \text{and} \qquad \lim_{x\to\infty} g(x) = L_2,$$

where L_1 *and* L_2 are (finite) real numbers, then

i) $\lim_{x\to\infty} [f(x) + g(x)] = L_1 + L_2,$

ii) $\lim_{x\to\infty} [f(x) - g(x)] = L_1 - L_2,$

iii) $\lim_{x\to\infty} f(x)g(x) = L_1L_2,$

iv) $\lim_{x\to\infty} kf(x) = kL_1$ (any number k),

v) $\lim_{x\to\infty} \dfrac{f(x)}{g(x)} = \dfrac{L_1}{L_2}$ if $L_2 \neq 0.$

These results hold for $x \to -\infty$ as well as for $x \to \infty$.

EXAMPLE 3

$$\lim_{x\to\infty} \frac{x}{7x + 4} = \lim_{x\to\infty} \frac{1}{7 + (4/x)} = \frac{1}{7 + 0} = \frac{1}{7} \quad \square$$

EXAMPLE 4

$$\lim_{x\to-\infty} \frac{1}{x^2} = \lim_{x\to-\infty} \frac{1}{x}\cdot\frac{1}{x} = 0\cdot 0 = 0 \quad \square$$

EXAMPLE 5

$$\lim_{x\to\infty} \frac{2x^2 - x + 3}{3x^2 + 5} = \lim_{x\to\infty} \frac{2 - (1/x) + (3/x^2)}{3 + (5/x^2)} = \frac{2 - 0 + 0}{3 + 0} = \frac{2}{3} \quad \square$$

EXAMPLE 6

$$\lim_{x\to-\infty} \frac{5x + 3}{2x^2 - 1} = \lim_{x\to-\infty} \frac{(5/x) + (3/x^2)}{2 - (1/x^2)} = \frac{0 + 0}{2 - 0} = 0 \quad \square$$

EXAMPLE 7 Find

$$\lim_{x\to\infty} \left(2 + \frac{\sin x}{x}\right).$$

Solution We have

$$\lim_{x\to\infty} 2 = 2 \qquad \text{and} \qquad \lim_{x\to\infty} \frac{\sin x}{x} = 0$$

because $-1 \leq \sin x \leq 1$ while $x \to \infty$. Therefore,

$$\lim_{x\to\infty} \left(2 + \frac{\sin x}{x}\right) = \lim_{x\to 2} 2 + \lim_{x\to\infty} \frac{\sin x}{x} = 2 + 0 = 2.$$

See Fig. 1.100. $\square$

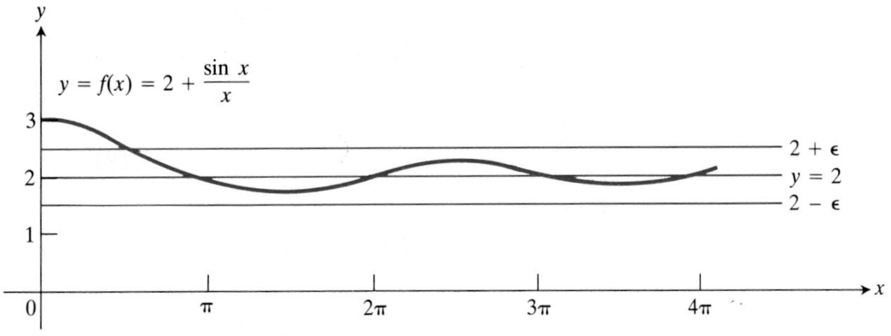

1.100 The graph of $y = 2 + (\sin x)/x$ oscillates about the line $y = 2$. The amplitude of the oscillations decreases toward zero as $x \to \infty$. The curve lies between the lines $y = 2 + \varepsilon$ and $y = 2 - \varepsilon$ when $x > 1/\varepsilon$.

As suggested by the behavior of $1/x$ as $x \to 0$, and other functions, we sometimes want to say such things as

$$\lim_{x \to c} f(x) = \infty, \tag{8}$$

$$\lim_{x \to c^+} f(x) = \infty, \tag{9}$$

$$\lim_{x \to c^-} f(x) = \infty, \tag{10}$$

$$\lim_{x \to \infty} f(x) = \infty, \tag{11}$$

$$\lim_{x \to -\infty} f(x) = \infty. \tag{12}$$

In every instance, we mean that the value of $f(x)$ eventually exceeds any positive real number B. That is, for any real number B no matter how large, the condition

$$f(x) > B \tag{13}$$

is satisfied for values of x in some restricted set, usually depending on B. The restricted set in (8) has the form

$$0 < |x - c| < \delta.$$

In (9), the set is an interval to the right of c:

$$c < x < c + \delta.$$

In (10), the set is an interval to the left of c:

$$c - \delta < x < c.$$

In (11), the set is an infinite interval of the form

$$M < x < \infty.$$

In (12), the set is an infinite interval of the form

$$-\infty < x < M.$$

Replacing the condition $f(x) > B$ in (13) by the condition

$$f(x) < -B, \tag{14}$$

where $-B$ is any negative real number, we can similarly define statements like

$$\lim_{x\to c} f(x) = -\infty, \qquad \lim_{x\to c^+} f(x) = -\infty, \qquad \lim_{x\to c^-} f(x) = -\infty,$$

$$\lim_{x\to \infty} f(x) = -\infty, \qquad \lim_{x\to -\infty} f(x) = -\infty.$$

EXAMPLE 8

a) $\lim_{x\to 0^+} \frac{1}{x} = \infty$

b) $\lim_{x\to 0^-} \frac{1}{x} = -\infty$

c) $\lim_{x\to 0} \frac{1}{x^2} = \infty$

d) $\lim_{x\to \infty} \sqrt{x} = \infty$

e) $\lim_{x\to 1^-} \frac{1}{x-1} = -\infty$

f) $\lim_{x\to 2^+} \frac{1}{x^2-4} = \infty$

g) $\lim_{x\to 2^-} \frac{1}{x^2-4} = -\infty$

h) $\lim_{x\to -\infty} 2x - \frac{3}{x} = -\infty$ □

The graphs of $y = 1/x^2$, $y = 1/(x-1)$, and $y = 1/(x^2-4)$ are shown in Fig. 1.101. In Chapter 3 we shall explore the general subject of graphing functions like these.

EXAMPLE 9

$$\lim_{x\to -\infty} \frac{2x^2-3}{7x+4} = \lim_{x\to -\infty} \frac{2x-(3/x)}{7+(4/x)} = -\infty \quad \square$$

As many of the examples above illustrate, one method for calculating the limit of a quotient of two polynomials as $x \to \infty$ is to divide the numerator and denominator by the largest power of x in the denominator, and then watch the new numerator and denominator as $x \to \infty$. (See Problem 62.)

An alternative method for calculating the limit of a quotient of two polynomials as $x \to \infty$ is to make the substitution $x = 1/h$ and calculate the limit as $h \to 0^+$. We do this in the next two examples.

1.101 Graphs of some of the functions in Example 8.

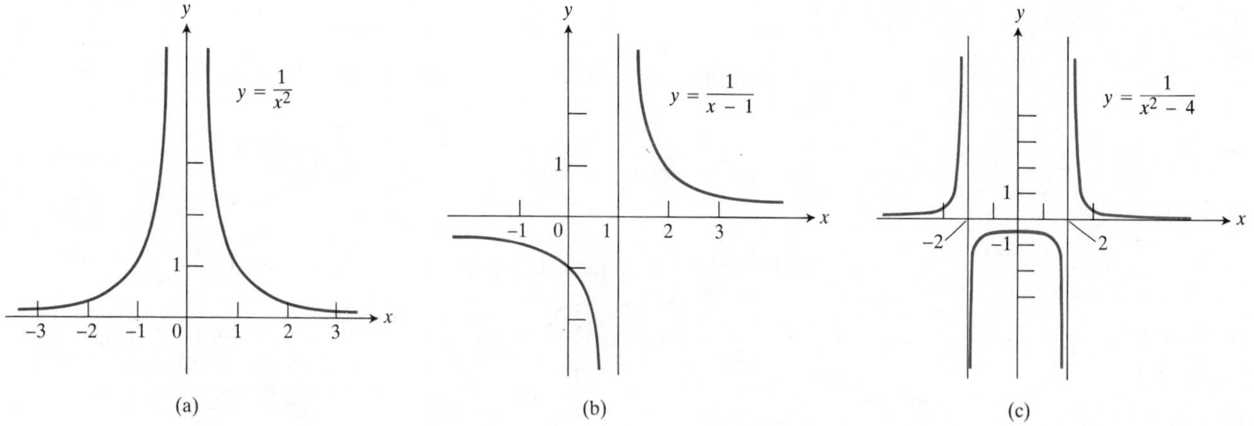

EXAMPLE 10 (Compare with Example 5.)

$$\lim_{x\to\infty}\frac{2x^2 - x + 3}{3x^2 + 5} = \lim_{h\to 0^+}\frac{(2/h^2) - (1/h) + 3}{(3/h^2) + 5}$$
$$= \lim_{h\to 0^+}\frac{2 - h + 3h^2}{3 + 5h^2} = \frac{2 - 0 + 3(0)^2}{3 + 5(0)^2} = \frac{2}{3} \ \square$$

EXAMPLE 11 (Compare with Example 6.)

$$\lim_{x\to\infty}\frac{5x + 3}{2x^2 - 1} = \lim_{h\to 0^+}\frac{(5/h) + 3}{(2/h^2) - 1} = \lim_{h\to 0^+}\frac{5h + 3h^2}{2 - h^2} = \frac{0}{2} = 0 \ \square$$

Similarly, to calculate the limit of a quotient of two polynomials as $x \to -\infty$, we may substitute $h = 1/x$ and calculate the limit as $h \to 0^-$.

EXAMPLE 12 (Compare with Example 9.)

$$\lim_{x\to -\infty}\frac{2x^2 - 3}{7x + 4} = \lim_{h\to 0^-}\frac{(2/h^2) - 3}{(7/h) + 4} = \lim_{h\to 0^-}\frac{2 - 3h^2}{7h + 4h^2} = -\infty \ \square$$

The substitution $x = 1/h$ may help in calculating limits of other of functions as well.

EXAMPLE 13

$$\lim_{x\to\infty} x \sin\frac{1}{x} = \lim_{h\to 0^+}\frac{\sin h}{h} = 1 \ \square$$

PROBLEMS

Find the limits in Problems 1–32.

1. $\lim_{x\to\infty}\frac{2x + 3}{5x + 7}$ **2.** $\lim_{t\to\infty}\frac{t^3 + 7}{t^4}$

3. $\lim_{x\to\infty}\frac{x + 1}{x^2 + 3}$ **4.** $\lim_{x\to\infty}\frac{3x^2 - 6x}{4x - 8}$

5. $\lim_{y\to\infty}\frac{3y + 7}{y^2 - 2}$ **6.** $\lim_{x\to\infty}\frac{7x - 28}{x^3}$

7. $\lim_{t\to\infty}\frac{t^2 - 2t + 3}{2t^2 + 5t - 3}$ **8.** $\lim_{t\to\infty}\frac{t^2 + 1}{t + 1}$

9. $\lim_{x\to\infty}\frac{x}{x - 1}$ **10.** $\lim_{x\to\infty}[x]$

11. $\lim_{x\to -\infty}|x|$ **12.** $\lim_{x\to -\infty}\frac{1}{|x|}$

13. $\lim_{a\to\infty}\frac{|a|}{|a| + 1}$ **14.** $\lim_{t\to -\infty}\frac{t}{t + 1}$

15. $\lim_{x\to\infty}\frac{3x^3 + 5x^2 - 7}{10x^3 - 11x + 5}$ **16.** $\lim_{x\to\infty}\left(\frac{1}{x} + 1\right)\left(\frac{5x^2 - 1}{x^2}\right)$

17. $\lim_{s\to\infty}\left(\frac{s}{s + 1}\right)\left(\frac{s^2}{5 + s^2}\right)$ **18.** $\lim_{x\to\infty}\frac{8x^{23} - 7x^2 + 5}{2x^{23} + x^{22}}$

19. $\lim_{r\to -\infty}\frac{8r^2 + 7r}{4r^2}$ **20.** $\lim_{x\to\infty}\frac{7x^3}{x^3 - 3x^2 + 6x}$

21. $\lim_{y\to\infty}\frac{y^4}{y^4 - 7y^3 + 7y^2 + 9}$ **22.** $\lim_{x\to\infty}\frac{5x^3 - 6x + 2}{10x^3 + 5}$

23. $\lim_{x\to\infty}\frac{x - 3}{x^2 - 5x + 4}$ **24.** $\lim_{x\to\infty}\frac{9x^4 + x}{2x^4 + 4x^2 - x + 6}$

25. $\lim_{x\to\infty}\frac{-2x^3 - 2x + 3}{3x^3 + 3x^2 - 5x}$ **26.** $\lim_{x\to -\infty}\frac{-2x^3 - 2x + 3}{3x^3 + 3x^2 - 5x}$

27. $\lim_{x\to\infty}\frac{x + \sin x}{x + \cos x}$ **28.** $\lim_{x\to -\infty}\frac{1 - x^2}{1 + 2x^2}$

29. $\lim_{x\to\infty}\left(\frac{1}{x^4} + \frac{1}{x}\right)$ **30.** $\lim_{x\to\infty}\frac{\sin 2x}{x}$

31. $\lim_{x\to\infty}\cos\frac{1}{x} + 1$ **32.** $\lim_{y\to\infty}\frac{1}{y^2 + 5}$

Find the limits in Problems 33–50.

33. $\lim_{x\to 0^+}\frac{1}{3x}$ **34.** $\lim_{x\to 0^-}\frac{2}{x}$

35. $\lim_{x\to 0^+}\frac{5}{2x}$ **36.** $\lim_{t\to 2^+}\frac{t^2 + 4}{t - 2}$

37. $\lim_{t\to 2}\frac{t^2 - 4}{t - 2}$ **38.** $\lim_{x\to 2^-}\frac{x}{x - 2}$

39. $\lim_{x\to 1^+}\frac{x}{x - 1}$ **40.** $\lim_{x\to 0}\frac{|x|}{|x| + 1}$

41. $\lim\limits_{x\to-1^-} \dfrac{1}{x+1}$

42. $\lim\limits_{x\to 0} \dfrac{1}{|x|}$

43. $\lim\limits_{x\to-2^+} \dfrac{1}{x+2}$

44. $\lim\limits_{x\to 3^-} \dfrac{x^2}{x-3}$

45. $\lim\limits_{x\to 3} \dfrac{x-3}{x^2}$

46. $\lim\limits_{x\to 1^+} \dfrac{2}{x^2-1}$

47. $\lim\limits_{x\to 2^-} \dfrac{x^2+5}{x-2}$

48. $\lim\limits_{x\to 2} \dfrac{x-2}{x^2+5}$

49. $\lim\limits_{x\to-5} \dfrac{x^2+3x-10}{x+5}$

50. $\lim\limits_{x\to 1^+} \dfrac{x+4}{x^2+2x-3}$

51. Find

$$\lim \frac{x-1}{2x^2-7x+5}$$

as (a) $x \to 0$, (b) $x \to \infty$, and (c) $x \to 1$.

52. Find the domain and range of the function

$$y = \sqrt{\frac{1}{x} - 1}.$$

Sketch the graphs of the functions in Problems 53–59.

53. $\dfrac{1}{x-2}$

54. $\dfrac{x}{x-2}$

55. $1 + \dfrac{1}{x}$

56. $\dfrac{1}{|x|}$

57. $\dfrac{1}{x+1}$

58. $\dfrac{x}{x+1}$

59. $\dfrac{|x|}{|x|+1}$

60. Find $\lim_{x\to\infty} f(x)$ if

$$\frac{2x-3}{x} < f(x) < \frac{2x^2+5x}{x^2}.$$

61. a) As $x \to 0^+$ the functions $1/x$, $1/x^2$, and $1/\sqrt{x}$ all become infinite. Which function increases most rapidly? least rapidly? (*Hint:* Consider ratios of these functions as $x \to 0^+$.)

b) Repeat (a) for $x \to +\infty$. Which function decreases most rapidly? least rapidly?

62. Let $f(x) = a_nx^n + a_{n-1}x^{n-1} + \cdots + a_1x + a_0$ be a polynomial of degree n, $g(x) = b_mx^m + b_{m-1}x^{m-1} + \cdots + b_1x + b_0$ a polynomial of degree m. Show that $\lim_{x\to\infty} f(x)/g(x)$ is a_n/b_m if $m = n$, 0 if $m > n$, infinite if $m < n$. (*Hint:* Divide the numerator and denominator of the fraction by x^m. What happens to x^n/x^m as $x \to \infty$ if $m = n$? if $m > n$? if $m < n$?)

1.11 Continuous Functions

A function $y = f(x)$ that can be graphed over each interval of its domain with one continuous motion of the pen is an example of a *continuous function.* The height of the graph over the interval varies continuously with x. At each interior point of the function's domain, like the point c in Fig. 1.102, the function value $f(c)$ is the limit of the function values on either side; that is,

$$f(c) = \lim_{x\to c} f(x). \tag{1}$$

The function value at each endpoint is also the limit of the nearby function values. At the left endpoint a in Fig. 1.102,

$$f(a) = \lim_{x\to a^+} f(x). \tag{2}$$

1.102 Continuous at a, b, and c.

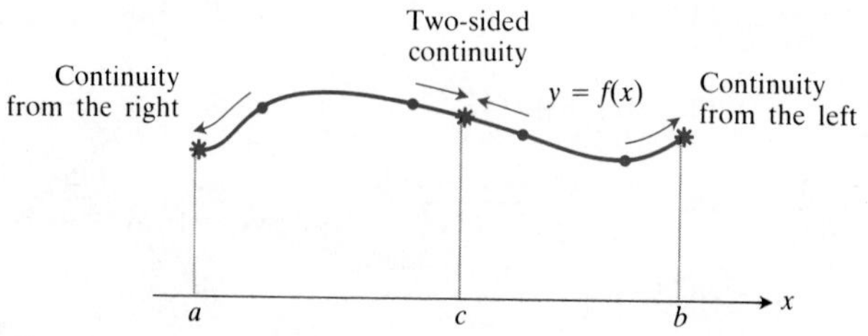

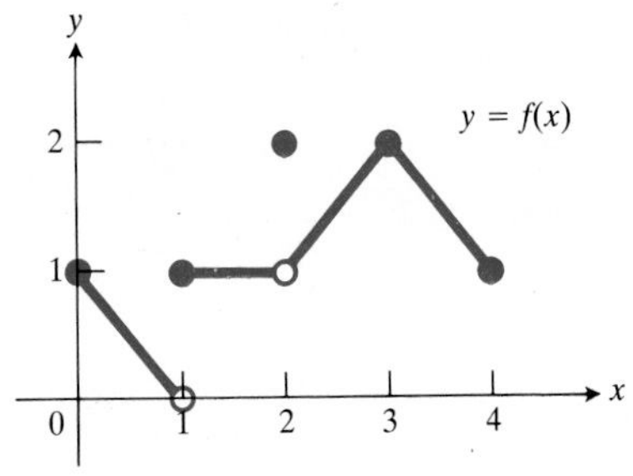

1.103 Discontinuous at $x = 1$ and $x = 2$.

At the right endpoint b,

$$f(b) = \lim_{x \to b^-} f(x). \tag{3}$$

To be specific, let us look again at the function $y = f(x)$ of Example 5 in Article 1.9. Its graph is reproduced here as Fig. 1.103.

EXAMPLE 1 The function in Fig. 1.103 is continuous at every point in its domain except $x = 1$ and $x = 2$. At these points there are breaks in the graph. Note the relation between the limit of f and the value of f at each point of the function's domain.

Points of discontinuity.

At $x = 1$: $\lim_{x \to 1} f(x)$ does not exist.

At $x = 2$: $\lim_{x \to 2} f(x) = 1$, but $1 \neq f(2)$.

Points at which f is continuous.

At $x = 0$: $\lim_{x \to 0^+} f(x) = f(0)$.

At $x = 4$: $\lim_{x \to 4^-} f(x) = f(4)$.

At every point $0 < c < 4$ except $x = 1, 2$: $\lim_{x \to c} f(x) = f(c)$. □

We now come to the formal definition of continuity at a point in a function's domain. In the definition we distinguish between continuity at an endpoint (which involves a one-sided limit) and continuity at an interior point (which involves a two-sided limit).

DEFINITIONS

Continuity at an Interior Point

A function $y = f(x)$ is continuous at an interior point c of its domain if

$$\lim_{x \to c} f(x) = f(c). \tag{1}$$

Continuity at an Endpoint

A function $y = f(x)$ is continuous at a left endpoint a of its domain if

$$\lim_{x \to a^+} f(x) = f(a). \tag{2}$$

A function $y = f(x)$ is continuous at a right endpoint b of its domain if

$$\lim_{x \to b^-} f(x) = f(b). \tag{3}$$

Continuous

A function is continuous if it is continuous at each point of its domain.

Functions are usually tested for continuity by applying the following test.

The Continuity Test

The function $y = f(x)$ is continuous at $x = c$ if and only if *all three* of the following statements are true:

1. $f(c)$ exists (c is in the domain of f),
2. $\lim_{x\to c} f(x)$ exists (f has a limit as $x \to c$),
3. $\lim_{x\to c} f(x) = f(c)$ (the limit equals the function value).

(The limit in the continuity test is to be two-sided if c is an interior point of the domain of f; it is to be the appropriate one-sided limit if c is an endpoint of the domain.)

EXAMPLE 2 When applied to the function $y = f(x)$ of Example 1 (Fig. 1.103) at the points $x = 0, 1, 2, 3$, and 4, the continuity test gives the following results.

a) f is continuous at $x = 0$ because
   i) $f(0)$ exists (it equals 1),
   ii) $\lim_{x\to 0^+} f(x) = 1$ (f has a limit as $x \to 0^+$),
   iii) $\lim_{x\to 0^+} = f(0)$ (the limit equals the function value).

b) f is *not* continuous at $x = 1$ because $\lim_{x\to 1} f(x)$ does not exist. The function fails part (2) of the test. (The right-hand and left-hand limits exist at $x = 1$, but they are not equal.)

c) f is *not* continuous at $x = 2$ because $\lim_{x\to 2} f(x) \neq f(2)$. The function fails part (3) of the test.

d) f is continuous at $x = 3$ because
   i) $f(3)$ exists (it equals 2),
   ii) $\lim_{x\to 3} f(x) = 2$ (f has a limit as $x \to 3$),
   iii) $\lim_{x\to 3} f(x) = f(3)$ (the limit equals the function value).

e) f is continuous at $x = 4$ because
   i) $f(4)$ exists (it equals 1),
   ii) $\lim_{x\to 4^-} f(x) = 1$ (f has a limit as $x \to 4^-$),
   iii) $\lim_{x\to 4^-} f(x) = f(4)$ (the limit equals the function value). □

EXAMPLE 3 The function $y = 1/x$ is continuous at every value of x except $x = 0$. The function is not defined at $x = 0$, and therefore fails part (1) of the continuity test at $x = 0$. See Fig. 1.104. □

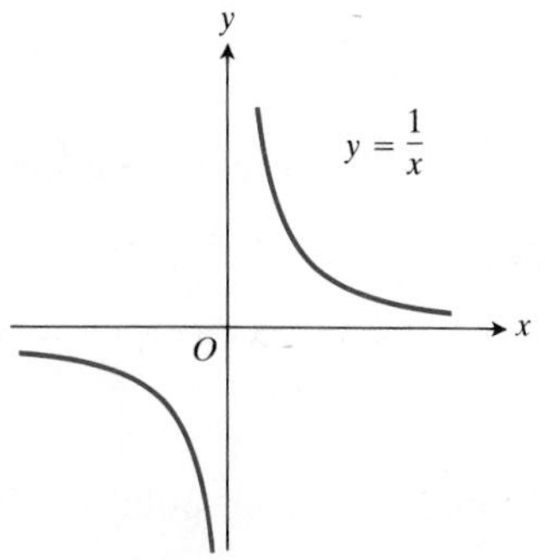

1.104 The function $y = 1/x$ is continuous at every point except $x = 0$.

EXAMPLE 4 The greatest integer function $y = [x]$ is discontinuous (fails to be continuous) at every integer. It does not approach a limit at any integer, and therefore fails part (2) of the continuity test at every integer. □

EXAMPLE 5 The functions $y = \sin x$ and $y = \cos x$ are continuous at $x = 0$. By Example 12 in Article 1.9,

$$\lim_{x\to 0} \sin x = 0 = \sin 0 \quad \text{and} \quad \lim_{x\to 0} \cos x = 1 = \cos 0. \ \square$$

EXAMPLE 6 *Polynomials and quotients of polynomials.*

a) Every polynomial $f(x) = a_n x^n + \cdots + a_1 x + a_0$ is continuous. By Theorem 2 in Article 1.9, $\lim_{x\to c} f(x) = f(c)$ at every point c.

b) Every quotient $f(x)/g(x)$ of polynomials is continuous except where $g(x) = 0$. By Theorem 3 in Article 1.9, $\lim_{x\to c} (f(x)/g(x)) = f(c)/g(c)$ at every point c at which g does not vanish. □

As you might have guessed, there is a theorem that says that sums, differences, products, and quotients of continuous functions are continuous at every point at which the combinations are defined.

THEOREM 6

If the functions $f(x)$ and $g(x)$ are continuous at $x = c$, then all of the following combinations are continuous at $x = c$:

i) $f(x) + g(x)$,

ii) $f(x) - g(x)$,

iii) $f(x)g(x)$,

iv) $kg(x)$ (any number k).

v) Furthermore, if $g(c) \neq 0$, then $f(x)/g(x)$ is continuous at $x = c$.

Proof. Theorem 6 is an immediate consequence of Theorem 1 in Article 1.9. If Theorem 1 were restated for the functions f and g, and it would say that if $\lim_{x\to c} f(x) = f(c)$ and $\lim_{x\to c} g(x) = g(c)$, then

i) $\lim_{x\to c} [f(x) + g(x)] = f(c) + g(c)$,

ii) $\lim_{x\to c} [f(x) - g(x)] = f(c) - g(c)$,

iii) $\lim_{x\to c} f(x)g(x) = f(c)g(c)$,

iv) $\lim_{x\to c} kg(x) = kg(c)$ (any number k),

v) $\lim_{x\to c} \dfrac{f(x)}{g(x)} = \dfrac{f(c)}{g(c)}$, provided $g(c) \neq 0$.

In other words, the limits of the functions in (i)–(v) as $x \to c$ exist and are the values of the functions at $x = c$. Therefore, each function fulfills the three requirements of the continuity test at any interior point $x = c$ of its domain. Similar arguments based on right-hand and left-hand limits establish the theorem for continuity at endpoints. ■

EXAMPLE 7 The following functions are continuous at every value of x:

$$f(x) = x^{14} + 20x^4, \qquad g(x) = 5x(2 - x) + 1/(x^2 + 1).$$

The following function is continuous at every value of x except $x = 5$ and $x = -2$:

$$h(x) = \frac{x + 3}{x^2 - 3x - 10} = \frac{x + 3}{(x - 5)(x + 2)}. \quad □$$

EXAMPLE 8 The function $y = (\sin x)/x$ is continuous at every point except $x = 0$. In this it is like the function $y = 1/x$. But $y = (\sin x)/x$ is

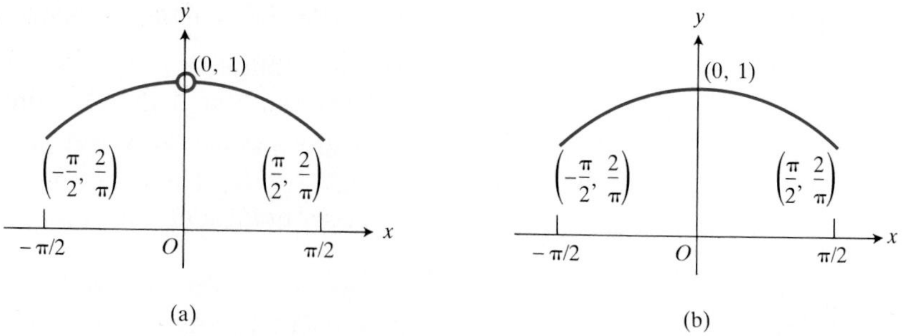

1.105 The graph (a) of $f(x) = (\sin x)/x$ for $-\pi/2 \le x \le \pi/2$ does not include the point (0, 1) because the function is not defined at $x = 0$. But we can remove the discontinuity from the graph by defining $f(0) = 1$. When we fill in the missing point this way, we get a continuous curve as shown in (b).

different from $y = 1/x$ in that it has a finite limit as $x \to 0$. It is therefore possible to extend the function's domain to include the point $x = 0$ in such a way that the extended function is continuous at $x = 0$. We define

$$f(x) = \begin{cases} \dfrac{\sin x}{x} & x \neq 0, \\ 1 & x = 0. \end{cases} \tag{4}$$

See Fig. 1.105. The function $f(x)$ is continuous at $x = 0$ because

$$\lim_{x\to 0} \frac{\sin x}{x} = f(0). \; \square$$

The function f in Example 8 is called the continuous extension of $(\sin x)/x$ to the origin. In extending $(\sin x)/x$ to the origin we "removed" its one point of discontinuity, $x = 0$. We therefore call the point $x = 0$ a removable discontinuity of $(\sin x)/x$, according to the following definition.

DEFINITIONS

Removable Discontinuity and Continuous Extension

If a function $y = f(x)$ is discontinuous at $x = c$, but $\lim_{x\to c} f(x)$ exists, then c is called a *removable discontinuity* of f.

If we define $f(c)$ to be $\lim_{x\to c} f(x)$, then this new f, which is continuous at $x = c$, is called the *continuous extension* of f to c.

(The limit in these definitions is understood to be finite.)

EXAMPLE 9 Is it possible to define $f(2)$ in a way that extends

$$f(x) = \frac{x^2 + x - 6}{x^2 - 4}$$

to be continuous at $x = 2$? If so, what value should $f(2)$ have?

Solution For f to be continuous at $x = 2$ we must have

$$\lim_{x\to 2} f(x) = f(2).$$

Does f have a limit at $x = 2$, and, if so, what is it? To answer this question, we try to factor the numerator and denominator of the expression for $f(x)$ to see if there is a way to rewrite it to avoid division by zero when $x = 2$. We find

$$f(x) = \frac{x^2 + x - 6}{x^2 - 4} = \frac{(x - 2)(x + 3)}{(x - 2)(x + 2)} = \frac{x + 3}{x + 2}.$$

Therefore,

$$\lim_{x\to 2} f(x) = \lim_{x\to 2} \frac{x + 3}{x + 2} = \frac{2 + 3}{2 + 2} = \frac{5}{4},$$

and defining $f(2) = \frac{5}{4}$ will make

$$\lim_{x\to 2} f(x) = f(2).$$

The extended function

$$f(x) = \begin{cases} \dfrac{x^2 + x - 6}{x^2 - 4} & \text{if } x \neq 2, \\[2ex] \dfrac{5}{4} & \text{if } x = 2, \end{cases}$$

is continuous at $x = 2$, because $\lim_{x\to 2} f(x)$ exists and equals $f(2)$. □

REMARK. Some discontinuities cannot be removed. When the right- and left-hand limits of a function $y = f(x)$ are not equal at a point $x = c$, it is not possible to define (or redefine) $f(c)$ in a way that makes f continuous at $x = c$. Nor is it possible to define or redefine $f(c)$ to make f continuous at $x = c$ if f has an infinite limit as $x \to c^+$ or $x \to c^-$. Nevertheless, functions with nonremovable discontinuities play important roles in mathematical modeling. Functions with "jump" discontinuities (like step functions) model electric pulse generators, digital outputs, heart beats, and motion that begins or ceases abruptly. Functions with "poles" (points at which they approach infinity; $1/x^2$ has a pole at $x = 0$) model gravitational and electrical phenomena in physics and engineering.

Differentiable Functions

One of the virtues of differentiable functions (functions that have derivatives) is that they are all continuous.

THEOREM 7

> A function is continuous at every point at which it has a derivative. That is, if $y = f(x)$ has a derivative $f'(c)$ at $x = c$, then f is continuous at $x = c$.

Proof. Our task is to show that

$$\lim_{x\to c} f(x) = f(c).$$

If we label the graph of $f(x)$ as in Fig. 1.106, the derivative $f'(c)$ can be calculated from the equation

$$f'(c) = \lim_{x\to c} \frac{f(x) - f(c)}{x - c}. \tag{5}$$

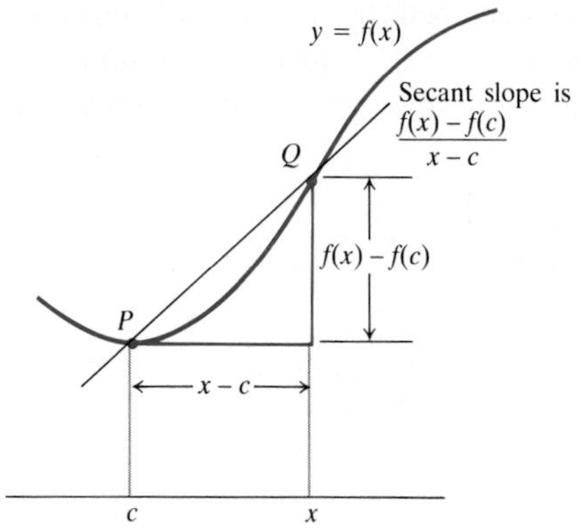

Figure 1.106

The idea of the proof is this: As $x \to c$, the denominator $x - c$ approaches 0. Therefore, if the limit in (5) is to be finite, the numerator $f(x) - f(c)$ must also approach 0. In other words, $f(x)$ must approach $f(c)$ as $x \to c$.

Formally, we use the fact that the limit of a product of functions is the product of their limits to show that

$$\begin{aligned}\lim_{x\to c}[f(x) - f(c)] &= \lim_{x\to c}\left[(x-c)\frac{f(x)-f(c)}{x-c}\right]\\ &= \lim_{x\to c}(x-c)\lim_{x\to c}\frac{f(x)-f(c)}{x-c}\\ &= 0\cdot f'(c) = 0. \qquad (6)\end{aligned}$$

The equation $\lim_{x\to c}[f(x) - f(c)] = 0$ implies that $\lim_{x\to c} f(x) = f(c)$. This is what we set out to show. ■

EXAMPLE 10 The following functions are continuous:

a) $y = \sqrt{x}$ (differentiable if $x > 0$ by Example 5 in Article 1.7 and continuous at $x = 0$ because $\lim_{x\to 0^+}\sqrt{x} = 0$);

b) $y = x^2$ (differentiable by Example 1 in Article 1.7);

c) $y = |x|$ (differentiable if $x \neq 0$ by Example 4 in Article 1.7 and continuous at $x = 0$ because $\lim_{x\to 0}|x| = 0$). □

Although differentiability implies continuity, the converse is not true, as the following example shows.

EXAMPLE 11 The function $y = |x|$ is continuous at $x = 0$ even though it has no derivative at $x = 0$. □

Composites of Continuous Functions

All composites of continuous functions are continuous. This means that composites like

$$y = \sin\sqrt{x}, \qquad y = |\cos x|$$

are continuous at every point at which they are defined. The idea is that if $f(x)$ is continuous at $x = c$ and $g(x)$ is continuous at $x = f(c)$, then $g(f(x))$ is continuous at $x = c$. See Fig. 1.107. (In Article 2.8, we shall establish the differentiability of the sine and cosine at every value of x, hence, by Theorem 7 above, their continuity at every value of x.)

THEOREM 8

> If f is continuous at c and g is continuous at $f(c)$, then the composite $g(f(x))$ is continuous at c.

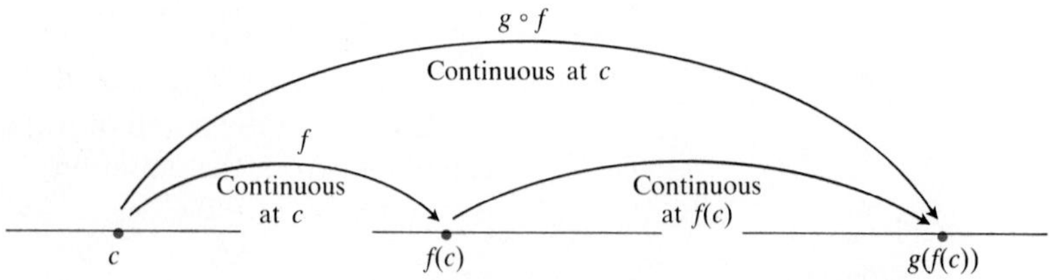

1.107 Composites of continuous functions are continuous.

For an outline of the proof of Theorem 8, see Problem 6 in Appendix 3.

EXAMPLE 12 Show that the function

$$y = \left|\frac{x \sin x}{x^2 + 2}\right|$$

is continuous at every value of x.

Solution The function y is the composite of the continuous functions

$$f(x) = \frac{x \sin x}{x^2 + 2} \qquad \text{and} \qquad g(x) = |x|.$$

The function $f(x)$ is continuous by Theorem 6, the function $g(x)$ by Example 10, and their composite $g(f(x))$ by Theorem 8. □

If a composite function $g(f(x))$ is continuous at a point $x = c$, its limit as $x \to c$ is $g(f(c))$.

EXAMPLE 13

a) $\lim_{x \to 1} \sin \sqrt{x - 1} = \sin \sqrt{1 - 1} = \sin 0 = 0$

b) $\lim_{x \to 0} |1 + \cos x| = |1 + \cos 0| = |1 + 1| = 2$ □

Important Properties of Functions that Are Continuous on Bounded Closed Intervals

We study continuous functions because they have useful properties that may be lacking in other functions.

The following two theorems describe properties that any function will have if it is continuous on a bounded closed interval. Theorem 9 says that a function that is continuous on a closed interval $[a, b]$ takes on a maximum value and a minimum value on the interval. We always look for these values when we graph a function (Fig. 1.108). Theorem 10 says that the graph of a continuous function over a closed interval cannot have any gaps. Theorem 10 has consequences for algebra (root finding) as well as for geometry (graphing), as we shall see in a moment.

The proofs of these theorems depend on a property of the real numbers called *completeness*, and will not be given in this book. They can be found in most texts on advanced calculus.

THEOREM 9

The Existence of Maximum and Minimum Values

Suppose that $f(x)$ is continuous for all x in the closed interval $[a, b]$. Then f has a minimum value m and a maximum value M on $[a, b]$. That is, there are numbers α and β in $[a, b]$ such that $f(\alpha) = m$, $f(\beta) = M$, and

$$m \leq f(x) \leq M \quad \text{for all } x \text{ in } [a, b].$$

(See Fig. 1.108.)

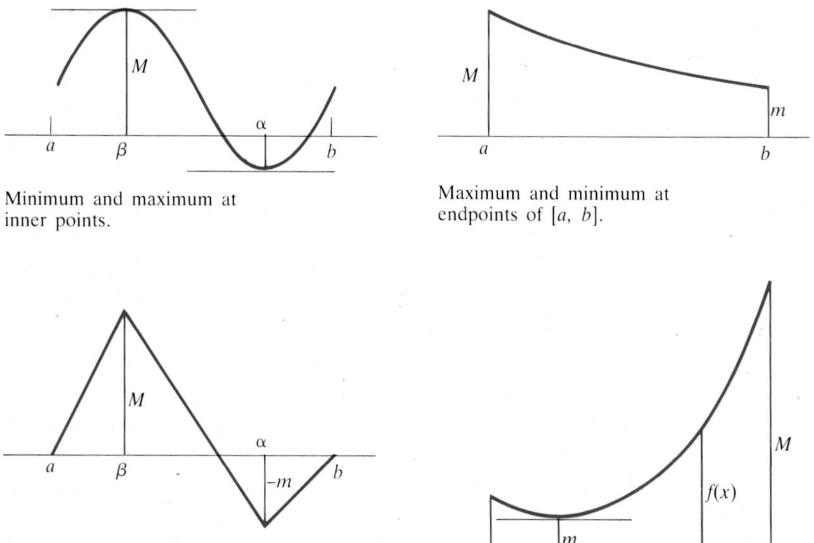

Minimum and maximum at inner points.

Maximum and minimum at endpoints of $[a, b]$.

Minimum and maximum at interior points α and β where the slope is not zero. This function is continuous on $[a, b]$ but is not differentiable at α and β.

Minimum m at inner point α of interval; maximum M at endpoint b.

1.108 A continuous function over the domain $a \leq x \leq b$ has a minimum and a maximum (low point and high point).

THEOREM 10

The Intermediate Value Theorem

If f is continuous on the closed interval $[a, b]$ and N is any number between $f(a)$ and $f(b)$, then there is at least one number c between a and b such that $f(c) = N$. (See Fig. 1.109.)

The Intermediate Value Theorem gets its name from the fact that f takes on all intermediate values between $f(a)$ and $f(b)$ as x varies from a to b. In particular, if $f(a) < 0$ and $f(b) > 0$ (or vice versa), then f takes on the value $N = 0$ at some point c between a and b.

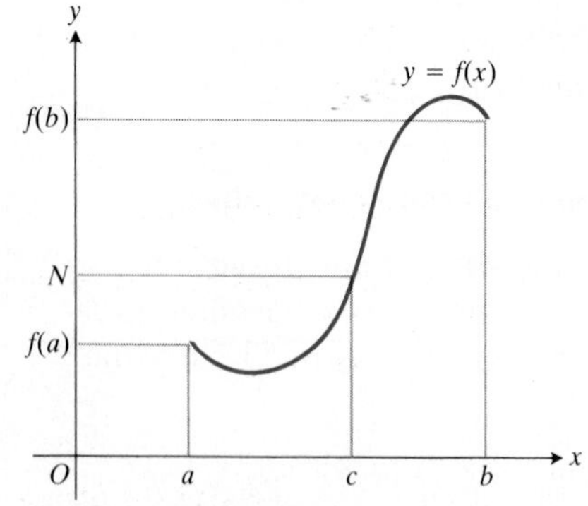

1.109 A continuous function $y = f(x)$ on $[a, b]$ takes on every value N between $f(a)$ and $f(b)$.

Consequence for root finding. If $f(a)$ and $f(b)$ differ in sign, then the equation $f(x) = 0$ has at least one solution in the interval (a, b). This fact is frequently used to help locate roots of equations, as we shall see in Chapter 2. Incidentally, a point at which $f(x) = 0$ is often called a *zero* of f.

Consequence for graphing. If $f(a)$ and $f(b)$ differ in sign, then the graph of f crosses the x-axis at some point between a and b.

EXAMPLE 14 Is there a real number that is one less than its cube?

Solution The number we seek must satisfy the equation $x = x^3 - 1$, or $x^3 - x - 1 = 0$. In other words, we are looking for a zero of the function $f(x) = x^3 - x - 1$. By trial we find that $f(1) = -1$ and $f(2) = 8 - 2 - 1 = 5$. We conclude from the Intermediate Value Theorem with $N = 0$

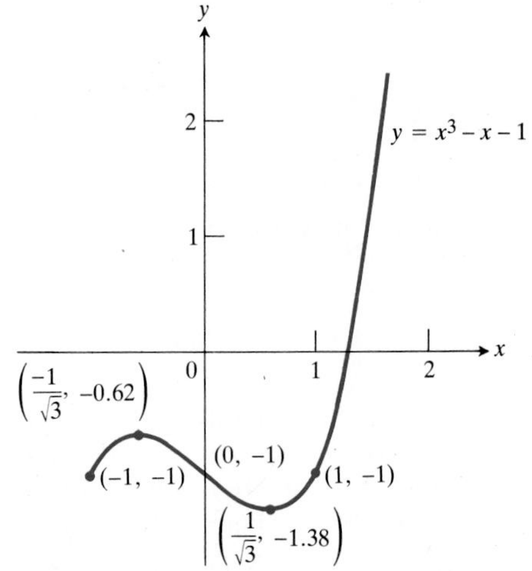

1.110 The graph of $f(x) = x^3 - x - 1$ crosses the x-axis between $x = 1$ and $x = 2$.

that there is a number c between 0 and 2 such that $f(c) = c^3 - c - 1 = 0$. This is the number we seek, for $c = c^3 - 1$. See Fig. 1.110. □

A few concluding remarks. For any function $y = f(x)$ it is important to distinguish between continuity at $x = c$ and having a limit as $x \to c$. The limit, $\lim_{x\to c} f(x)$, is where the function is headed as $x \to c$. Continuity is the property of arriving at the point where $f(x)$ has been headed when x actually gets to c. (Someone is home when you get there, so to speak.) If the limit is what you expect as $x \to c$, and the number $f(c)$ is what you get when $x = c$, then the function is continuous at c if what you get equals what you expect.

Finally, remember the test for continuity:

1. Does $f(c)$ exist?
2. Does $\lim_{x\to c} f(x)$ exist?
3. Does $\lim_{x\to c} f(x) = f(c)$?

For f to be continuous at $x = c$, all three answers must be *yes*.

PROBLEMS

Problems 1–6 are about the function

$$y = f(x) = \begin{cases} x^2 - 1 & -1 \le x < 0, \\ 2x, & 0 \le x < 1, \\ 1, & x = 1 \\ -2x + 4, & 1 < x < 2, \\ 0, & 2 < x \le 3. \end{cases}$$

This function is graphed in Fig. 1.111.

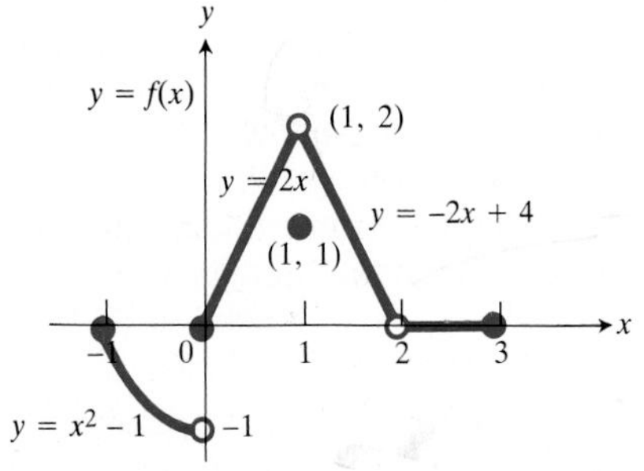

1.111 The function $y = f(x)$ for Problems 1–6.

1. a) Does $f(-1)$ exist?
b) Does $\lim_{x\to -1^+} f(x)$ exist?
c) Does $\lim_{x\to -1^+} f(x) = f(-1)$?
d) Is f continuous at $x = -1$?

2. a) Does $f(1)$ exist?
b) Does $\lim_{x\to 1} f(x)$ exist?
c) Does $\lim_{x\to 1} f(x) = f(1)$?
d) Is f continuous at $x = 1$?

3. a) Is f defined at $x = 2$? (Look at the definition of f.)
b) Is f continuous at $x = 2$?

4. At what values of x is f continuous?

5. a) What is the value of $\lim_{x\to 2} f(x)$?
b) What value should be assigned to $f(2)$ to make f continuous at $x = 2$?

6. To what new value should $f(1)$ be changed to make f continuous at $x = 1$?

7. At what points is the function

$$y = f(x) = \begin{cases} 0, & x < 0, \\ 1, & 0 \le x \le 1, \\ 1, & 0 < x, \end{cases}$$

continuous? (*Hint:* Graph the function.)

8. Let $y = f(x)$ be defined by

$$\begin{aligned} y &= 1 && \text{for } x < 0, \\ y &= \sqrt{1 - x^2} && \text{for } 0 \le x \le 1, \\ y &= x - 1 && \text{for } x > 1. \end{aligned}$$

Is y continuous? (*Hint:* Graph the function.)

At what points are the functions in the following problems in Article 1.9 continuous?

9. Problem 58

10. Problem 59

11. Problem 60

Find the points (if any) at which the functions in Problems 12–21 are *not* continuous.

12. $y = \dfrac{1}{x - 2}$

13. $y = \dfrac{1}{(x + 2)^2}$

14. $y = \dfrac{x}{x+1}$

15. $y = \dfrac{x+1}{x^2 - 4x + 3}$

16. $y = |x - 1|$

17. $y = \dfrac{x+3}{x^2 - 3x - 10}$

18. $y = \dfrac{x^2 - 1}{x^3 - 1}$

19. $y = \dfrac{1}{x^2 + 1}$

20. $y = \dfrac{\cos x}{x}$

21. $y = \dfrac{|x|}{x}$

22. The function $f(x)$ is defined by $f(x) = (x^2 - 1)/(x - 1)$ when $x \neq 1$, and by $f(1) = 2$. Is f continuous, or discontinuous, at $x = 1$? Explain.

23. Define $g(3)$ in a way that extends $g(x) = (x^2 - 9)/(x - 3)$ to be continuous at $x = 3$.

24. Define $h(2)$ in a way that extends $h(x) = (x^2 + 3x - 10)/(x - 2)$ to be continuous at $x = 2$.

25. a) Graph the function

$$f(x) = \begin{cases} x, & 0 \le x \le 1, \\ 2 - x, & 1 < x \le 2. \end{cases}$$

b) Is f continuous at $x = 1$?
c) Does f have a derivative at $x = 1$?

26. How should $f(2)$ be redefined in Fig. 1.103 to make the function continuous?

27. What value should be assigned to a to make the function

$$f(x) = \begin{cases} x^2 - 1, & x < 3, \\ 2ax, & x \ge 3, \end{cases}$$

continuous at $x = 3$?

28. What value should be assigned to b to make the function

$$g(x) = \begin{cases} x^3, & x < \frac{1}{2}, \\ bx^2, & x \ge \frac{1}{2}, \end{cases}$$

continuous at $x = \frac{1}{2}$?

29. The function $f(x) = (x^2 - 16)/|x - 4|$ is not continuous at $x = 4$. Is the discontinuity removable? If so, what should $f(4)$ be? If not, why not? (*Hint:* Calculate the right- and left-hand limits of f at $x = 4$.)

30. The function $f(x) = \sin(1/x)$ is not continuous at $x = 0$. Is the discontinuity removable? If so, what value should $f(0)$ take? If not, why not? (See Fig. 1.92 for the graph.)

31. At what points is the function $y = 1/[x]$ discontinuous?

Find the limits in Problems 32–34.

32. $\lim\limits_{x \to 0} \dfrac{1 + \cos x}{2}$

33. $\lim\limits_{x \to 0} \cos\left(1 - \dfrac{\sin x}{x}\right)$

34. $\lim\limits_{x \to \infty} \sqrt{\dfrac{x+1}{4x-1}}$

35. What is the maximum of $y = |x|$ for $-1 \le x \le 1$? the minimum?

36. At what values of x does the function in Fig. 1.103 take on its maximum value? Does the function take on a minimum value? Explain.

37. Does the function $y = x^2$ have a maximum value on the open interval $-1 < x < 1$? a minimum value? Explain.

38. On the closed interval $0 \le x \le 1$ the function $y = [x]$ takes on a minimum value $m = 0$ and a maximum value $M = 1$ even though the function is discontinuous at $x = 1$. Does this violate Theorem 9? Why?

39. A continuous function $y = f(x)$ is known to be negative at $x = 0$ and positive at $x = 1$. Why does the equation $f(x) = 0$ have at least one solution between $x = 0$ and $x = 1$? Illustrate with a sketch.

40. Assuming $y = \cos x$ to be continuous, show that the equation $\cos x = x$ has at least one solution. (*Hint:* Show that the function $f(x) = \cos x - x$ has at least one real zero.)

41. Show that every polynomial of odd degree has at least one real zero.

42. The function $f(x) = |x|$ is continuous at $x = 0$. Given a positive number ε, how small must δ be in order for $|x - 0| < \delta$ to imply $|f(x) - 0| < \varepsilon$?

43. a) Graph the function $y = f(x)$ defined by

$$f(x) = \begin{cases} x \sin(1/x) & \text{if } x \neq 0, \\ 0 & \text{if } x = 0. \end{cases}$$

(*Hint:* The graph of f lies in the "bow"-shaped portion of the plane bounded by the lines $y = x$ and $y = -x$ and containing the x-axis. See Fig. 1.92 for a graph of $y = \sin(1/x)$.)

b) Show that the function $y = f(x)$ in (a) is continuous at $x = 0$. (*Hint:* First show that $|x \sin(1/x)| \le |x|$ for all x. Then answer the question, How small does $|x - 0|$ need to be to make $|x \sin(1/x) - 0|$ less than ε?)

44. Let f be a continuous function, and suppose that $f(c)$ is positive. Show that there is some interval about c, say $c - \delta < x < c + \delta$, throughout which $f(x)$ remains positive. Illustrate with a sketch. (*Hint:* Take $\varepsilon = f(c)/2$.)

***Toolkit* programs**

Continuity at a Point	Super * Grapher
Limit Problems	

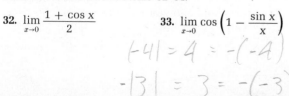

REVIEW QUESTIONS AND EXERCISES

1. Define what is meant by the *slope* of a straight line. How would you find the slope of a straight line from its graph? from its equation?

2. In Fig. 1.112, the lines L_1, L_2, and L_3 have slopes m_1, m_2, m_3, respectively. Which slope is algebraically least? greatest? Write the three slopes in order of increasing size with the symbols for "less than" or "greater than" correctly inserted between them.

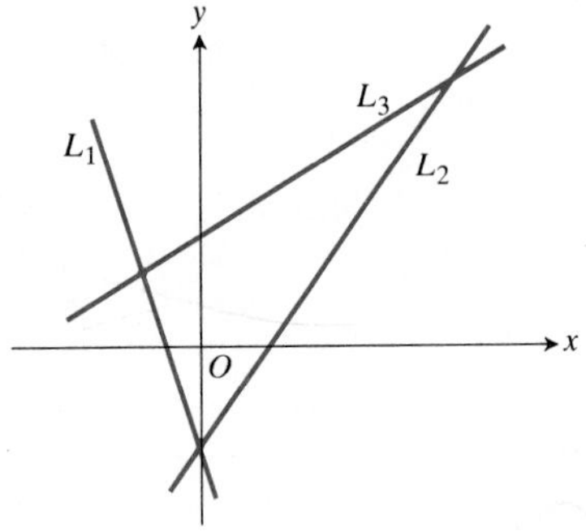

1.112 The lines for Review Question 2.

3. Describe the family of lines $y - y_1 = m(x - x_1)$:
   a) if (x_1, y_1) is fixed and different lines are drawn for different values of m;
   b) if m and x_1 are fixed and different lines are drawn for different values of y_1.

4. Define *function*. What is the *domain* of a function? What is its *range*?

5. The domain of a certain function is the set $0 \le x \le 2$. The range of the function is the single number $y = 1$.
   a) Sketch and describe the graph of the function.
   b) Write an expression for the function.
   c) If we interchange the axes in part (a), does the new graph so obtained represent a function? Explain.

6. If $f(x) = 1/x$ and $g(x) = 1/\sqrt{x}$, what are the natural domains of f, g, $f - g$, $f \cdot g$, f/g, g/f, and the composites $f(g(x))$ and $g(f(x))$? What is the domain of $h(x) = g(x + 4)$?

7. Write a rule of the form $y = f(x)$ for the function f that is the composite of $y = |x - 10|$ followed by $y = \sqrt{1 - x}$. Find the largest possible domain of f and the corresponding range of y.

8. *Definition*. A *rational number* is one that can be expressed in the form p/q, where p and q are integers with no common divisor greater than one, and $q > 0$. A real number that is not rational is called *irrational*.

   With these definitions in mind, plot a few points on the graph of the function that maps the rational number p/q (reduced to lowest terms as above) into $1/q$, and maps each irrational number into zero. This function has been described as the "ruler function" because of the resemblance of its graph to the markings on an ordinary ruler showing inches, half-inches, quarters, eighths, and sixteenths by lines of different lengths. The edge of the ruler corresponds to the x-axis. Do you see why this terminology is rather appropriate?

9. Is it appropriate to define a *tangent line* to a curve C at a point P on C as a line that has just the one point P in common with C? Illustrate your discussion with graphs.

10. Define, carefully, the concept of *slope of a tangent* to a curve at a point on the curve.

11. Define the concept of *average velocity;* of *instantaneous velocity.*

12. What more general concept includes both the concept of slope of the tangent to a curve and the concept of instantaneous velocity?

13. Define the derivative of a function at a point in its domain. Illustrate your definition by applying it to the function f defined by $f(x) = x^2$, at $x = 2$.

14. A function f, whose domain is the set of all real numbers, has the property that $f(x + h) = f(x) \cdot f(h)$ for all x and h; and $f(0) \neq 0$.
    a) Show that $f(0) = 1$. (*Hint:* Let $h = x = 0$.)
    b) If f has a derivative at 0, show that f has a derivative at every real number x, and that
    $$f'(x) = f(x) \cdot f'(0).$$

15. Give an example of a continuous function, defined for all real x, that fails to have a derivative (a) at some point; (b) at several points; (c) at infinitely many points.

16. Suppose F is a function whose values are all less than or equal to some constant M; $F(t) \le M$. Prove: If $\lim_{t \to c} F(t) = L$, then $L \le M$. (*Suggestion:* An indirect proof may be used to show that $L > M$ is false. For, if $L > M$, we may take $\frac{1}{2}(L - M)$ as a positive number ε, apply the definition of limit, and arrive at a contradiction.)

17. Let $f(x) = mx + b$. Prove that, for any positive number ε, the number $\delta = \varepsilon/(1 + |m|)$ is a positive number such that
    $$|f(x) - f(c)| < \varepsilon \quad \text{when } |x - c| < \delta.$$
    What conclusion can you now draw about the limit of $f(x)$ as x approaches c?

18. Suppose that $\lim_{t\to c} F(t) = -7$ and $\lim_{t\to c} G(t) = 0$. Find the limit as t approaches c of each of the following functions.

a) $F^2(t)$ b) $3 \cdot F(t)$ c) $F(t) \cdot G(t)$ d) $\dfrac{F(t)}{G(t) + 7}$

19. Define what it means to say that
a) $\lim_{x\to 2} f(x) = 5$; b) $\lim_{x\to 2} g(x) = k$.

20. Evaluate the following limits.

a) $\lim_{x\to 1} \dfrac{x + 1}{x + 2}$

b) $\lim_{x\to 1} \dfrac{x - 1}{x + 2}$

c) $\lim_{x\to 2} \dfrac{x^2 - x - 2}{x^2 - 4}$

d) $\lim_{x\to 0} \dfrac{\sin 2x}{4x}$

21. Determine which of the following indicated limits exist and evaluate those that do.

a) $\lim_{x\to\infty} \dfrac{2x + 3}{5x + 7}$

b) $\lim_{z\to\infty} \dfrac{4z^3 + 3z^2 + 2}{z^4 + 5z^2 + 1}$

c) $\lim_{x\to\infty} \dfrac{\sin x}{\sqrt{x}}$

d) $\lim_{x\to\infty} \dfrac{2 + 3x}{1 + 5x}$

22. Graph and discuss the continuity of

$$f(x) = \begin{cases} x + 1/x, & \text{for } x < 0, \\ -x^3, & \text{for } 0 \le x \le 1, \\ -1, & \text{for } 1 < x < 2, \\ 1, & \text{for } x = 2, \\ 0, & \text{for } x > 2. \end{cases}$$

23. Give an example of a removable discontinuity.

24. Give an example of a function that is defined on $0 \le x \le 1$, continuous in the open interval $0 < x < 1$, and discontinuous at $x = 0$.

25. State and prove a theorem about the relationship between continuity and differentiability of a function at a point in its domain.

26. True or false? If $y = f(x)$ is continuous, $f(1) = 0$, and $f(2) = 3$, then f takes on the value 2.5 at some point between $x = 1$ and $x = 2$. Explain.

27. The function $y = 1/x$ does not take on either a maximum or a minimum value on the interval $[-1, 1]$. Does this contradict Theorem 9? Why?

28. Read the article "The Lever of Mahomet," by R. Courant and H. Robbins, and the accompanying "Commentary on Continuity," by J. R. Newman, in *World of Mathematics*, Vol. 4 (New York: Simon and Schuster, 1960), pp. 2410–2413.

MISCELLANEOUS PROBLEMS

1. a) Plot the points $A(8, 1)$, $B(2, 10)$, $C(-4, 6)$, $D(2, -3)$, $E(\frac{14}{3}, 6)$.
b) Find the slopes of the lines AB, BC, CD, DA, CE, BD.
c) Do four of the five points A, B, C, D, E form a parallelogram? (Why?)
d) Do three of the five points lie on a common straight line? (Why?)
e) Does the origin (0, 0) lie on a straight line through two of the five points? (Why?)
f) Find equations of the lines AB, CD, AD, CE, BD.
g) Find the coordinates of the points in which the lines AB, CD, AD, CE, BD intersect the x- and y-axes.

2. Given the straight line $2y - 3x = 4$ and the point $(1, -3)$.
a) Find the equation of the straight line through the given point and perpendicular to the given line.
b) Find the distance between the given point and the given line.

3. Plot the three points $A(6, 4)$, $B(4, -3)$, and $C(-2, 3)$.
a) Is triangle ABC a right triangle? Why?
b) Is it isosceles? Why?
c) Does the origin lie inside, outside, or on the boundary of the triangle? Why?
d) If C is replaced by a point $C'(-2, y)$ such that angle $C'BA$ is a right angle, find y, the ordinate of C'.

4. Find the equations of the straight lines passing through the origin that are tangent to the circle of center (2, 1) and radius 2.

5. Let $P_1(x_1, y_1)$ and $P_2(x_2, y_2)$ be any two points. Find the coordinates of the midpoint of the line segment P_1P_2.

6. The x- and y-intercepts of a line L are, respectively, a and b. ($a, b \neq 0$.) Show that an equation of L is $(x/a) + (y/b) = 1$.

7. Given the line L: $Ax + By = C$, find (a) its slope; (b) its y-intercept; (c) its x-intercept; (d) the line through the origin perpendicular to L.

8. Show that the distance between a point $P(x_1, y_1)$ and a line $Ax + By = C$ is equal to

$$\frac{|Ax_1 + By_1 - C|}{\sqrt{A^2 + B^2}}.$$

When $Ax_1 + By_1 = C$: P_1 is on L.

When $Ax_1 + By_1 > C$: P_1 is on one side of L.
When $Ax_1 + By_1 < C$: P_1 is on the opposite side of L.

(There are neat solutions of this problem in Vol. 59, 1952, of the *American Mathematical Monthly,* pp. 242 and 248.)

9. How many circles can you find tangent to the following three lines?

$$L_1\colon\ x + y = 1; \quad L_2\colon\ y = x + 1; \quad L_3\colon\ x - 3y = 1$$

Give the center and radius of at least one such circle. You may use the result of Problem 8.

10. Find, in terms of b, b', and m, the perpendicular distance between the parallel lines $y = mx + b$ and $y = mx + b'$.

11. Given the two lines

$$L_1\colon\ a_1x + b_1y + c_1 = 0;$$
$$L_2\colon\ a_2x + b_2y + c_2 = 0.$$

If k is a constant, describe the set of points whose coordinates satisfy the equation

$$(a_1x + b_1y + c_1) + k(a_2x + b_2y + c_2) = 0.$$

12. Determine the coordinates of the point on the straight line $y = 3x + 1$ that is equidistant from $(0, 0)$ and $(-3, 4)$.

13. Find an equation of a straight line that is perpendicular to $5x - y = 1$ and is such that the area of the triangle formed by the x-axis, the y-axis, and the straight line is equal to 5.

14. Given the equation $y = (x^2 + 2)/(x^2 - 1)$, express x in terms of y and determine the values of y for which x is real.

15. Express the area A and the circumference C of a circle as functions of the radius r. Express A as a function of C.

16. In each of the following functions, what is the largest domain of x and the corresponding range of y?

a) $y = \dfrac{1}{1 + x}$ b) $y = \dfrac{1}{1 + x^2}$

c) $y = \dfrac{1}{1 + \sqrt{x}}$ d) $y = \dfrac{1}{\sqrt{3 - x}}$

17. Without the use of the absolute-value symbol, describe the domain of x for which $|x + 1| < 4$.

18. If $y = 2x + |2 - x|$, express x in terms of y.

19. For what range of values of y does the equation $y = x + |2 - x|$ determine x as a single-valued function of y? Solve for x in terms of y on this set of values.

20. If $y = x + (1/x)$, express x in terms of y and determine the values of y for which x is real.

21. a) If $f(x) = x^2 + 2x - 3$, find $f(-2)$; $f(-1)$; $f(x_1)$; $f(x_1 + \Delta x)$.
b) If $f(x) = x - (1/x)$, show that $f(1/x) = -f(x) = f(-x)$.

22. Sketch the graph of each of the following equations.
a) $y = |x - 2| + 2$ b) $y = x^2 - 1$
c) Find the point on the curve in part (b) where the tangent to the curve makes an angle of 45° with the positive x-axis.

23. Sketch a graph of the function $y = |x + 2| + x$ for the domain $-5 \leq x \leq 2$. What is the range?

24. Show that the expression

$$M(a, b) = \frac{(a + b)}{2} + \frac{|a - b|}{2}$$

is equal to a when $a \geq b$ and is equal to b when $b \geq a$. In other words, $M(a, b)$ gives the larger of the two numbers a and b. Find a similar expression, $m(a, b)$, that gives the smaller of the two numbers.

25. For each of the following expressions $f(x)$, sketch first the graph of $y = f(x)$, then the graph of $y = |f(x)|$, and finally the graph of $y = f(x)/2 + |f(x)|/2$.
a) $f(x) = (x - 2)(x + 1)$ b) $f(x) = x^2$
c) $f(x) = -x^2$ d) $f(x) = 4 - x^2$

26. *Lagrange interpolation formula.* Let (x_1, y_1), (x_2, y_2), . . . , (x_n, y_n) be n points in the plane, no two of them having the same abscissas. Find a polynomial, $f(x)$, of degree $(n - 1)$, that takes the value y_1 at x_1, y_2 at x_2, . . . , y_n at x_n; that is, $f(x_i) = y_i (i = 1, 2, \ldots, n)$. (*Hint:*

$$f(x) = y_1\phi_1(x) + y_2\phi_2(x) + \cdots + y_n\phi_n(x),$$

where $\phi_k(x)$ is a polynomial that is zero at x_i $(i \neq k)$ and $\phi_k(x_k) = 1$.)

27. Let $f(x) = ax + b$ and $g(x) = cx + d$. What condition must be satisfied by the constants a, b, c, and d to make $f(g(x))$ and $g(f(x))$ identical?

28. Let $f(x) = (ax + b)/(cx + d)$. If $d = -a$, show that $f(f(x)) = x$ for all values of x.

29. If $f(x) = x/(x - 1)$, find (a) $f(1/x)$; (b) $f(-x)$; (c) $f(f(x))$; (d) $f(1/f(x))$.

30. Using the definition of the derivative, find $f'(x)$ if $f(x)$ is:
a) $(x - 1)/(x + 1)$; b) $x^{3/2}$; c) $x^{1/3}$.

31. Use the definition of the derivative to find:
a) $f'(x)$ if $f(x) = x^2 - 3x - 4$;
b) $\dfrac{dy}{dx}$ if $y = \dfrac{1}{3x} + 2x$;
c) $f'(t)$ if $f(t) = \sqrt{t - 4}$.

32. a) By the Δ-method, find the slope of the curve $y = 2x^3 + 2$ at the point $(1, 4)$.
b) At which point of the curve in (a) is the tangent to the curve parallel to the x-axis? Sketch the curve.

33. If $f(x) = 2x/(x - 1)$, find (a) $f(0)$, $f(-1)$, $f(1/x)$; (b) $\Delta f(x)/\Delta x$; (c) $f'(x)$, using the result of part (b).

34. Given $y = 180x - 16x^2$. Using the method of Article 1.6, find the slope of the curve at the point (x_1, y_1). Sketch the curve. At what point does the curve have a horizontal tangent?

35. Find the velocity $v = ds/dt$ if the particle's position at time t is $s = 180t - 16t^2$. When does the velocity vanish?

36. If a ball is thrown straight up with a velocity of 32 ft/s, its height after t seconds is given by the equation $s = 32t - 16t^2$. At what instant will the ball be at its highest point, and how high will it rise?

37. If the pressure P and volume V of a gas are related by the formula $P = 1/V$, find (a) the average rate of change of P with respect to V; (b) the rate of change of P with respect to V at the instant when $V = 2$.

38. The volume V (in^3) of water remaining in a leaking pail after t seconds is $V = 2000 - 40t + 0.2t^2$. How fast is the volume decreasing when $t = 30$?

Evaluate the following limits, or show that the limit does not exist.

39. $\lim_{x\to\infty} \dfrac{x + \sin x}{2x + 5}$

40. $\lim_{x\to\infty} \dfrac{1 + \sin x}{x}$

41. $\lim_{x\to 1} \dfrac{x^2 - 4}{x^3 - 8}$

42. $\lim_{x\to 0} \dfrac{x}{\tan 3x}$

43. $\lim_{x\to\infty} \dfrac{x \sin x}{x + \sin x}$

44. $\lim_{x\to a} \dfrac{x^2 - a^2}{x - a}$

45. $\lim_{x\to a} \dfrac{x^2 - a^2}{x + a}$

46. $\lim_{h\to 0} \dfrac{(x + h)^2 - x^2}{h}$

47. $\lim_{h\to 0} \dfrac{\sqrt{x + h} - \sqrt{x}}{h}$

48. $\lim_{\Delta x\to 0} \dfrac{1/(x + \Delta x) - 1/x}{\Delta x}$

49. $\lim_{x\to 0^+} \dfrac{1}{x}$

50. $\lim_{x\to 1} \dfrac{1 - \sqrt{x}}{1 - x}$

51. $\lim_{x\to 1} \dfrac{(2x - 3)(\sqrt{x} - 1)}{2x^2 + x - 3}$

52. $\lim_{x\to\infty} (1 - x \cos x)$

53. $\lim_{x\to 1} \dfrac{\sqrt{x + 1} - \sqrt{2x}}{x^2 - x}$

Find each of the following limits.

54. $\lim_{x\to 0^+} \dfrac{|x|}{x}$

55. $\lim_{x\to 0^-} \dfrac{|x|}{x}$

56. $\lim_{x\to 4^-} ([x] - x)$

57. $\lim_{x\to 4^+} ([x] - x)$

58. $\lim_{x\to 3^+} \dfrac{[x]^2 - 9}{x^2 - 9}$

59. $\lim_{x\to 3^-} \dfrac{[x]^2 - 9}{x^2 - 9}$

60. $\lim_{x\to 4^+} [[x]]$

61. $\lim_{x\to 0^+} \dfrac{\sqrt{x}}{\sqrt{4 + \sqrt{x}} - 2}$

62. Given $(x - 1)/(2x^2 - 7x + 5) = f(x)$, find (a) the limit of $f(x)$ as $x \to \infty$, (b) the limit of $f(x)$ as $x \to 1$, (c) $f(-1/x)$, $f(0)$, $1/f(x)$.

63. Compute the coordinates of the point of intersection of the straight lines

$$3x + 5y = 1, \qquad (2 + c)x + 5c^2y = 1,$$

and determine the limiting position of this point as c tends to 1.

64. Find

a) $\lim_{n\to\infty} (\sqrt{n^2 + 1} - n)$, b) $\lim_{n\to\infty} (\sqrt{n^2 + n} - n)$.

65. Given $\varepsilon > 0$, find $\delta > 0$ such that $\sqrt{t^2 - 1} < \varepsilon$ when $0 < (t - 1) < \delta$.

66. Given $\varepsilon > 0$, find M such that

$$\left|\frac{t^2 + t}{t^2 - 1} - 1\right| < \varepsilon \quad \text{when } t > M.$$

67. Suppose that $f(x) = x^3 - 3x^2 - 4x + 12$ and

$$h(x) = \begin{cases} \dfrac{f(x)}{x - 3} & \text{for } x \neq 3, \\ k & \text{for } x = 3. \end{cases}$$

a) Find all zeros of f.
b) Find the value of k that makes h continuous at $x = 3$.
c) Using the value of k found in (b), determine whether h is an even function.

68. Define $y = [x]$ to be the greatest integer less than x. Graph
a) $y = [x]$,
b) $y = [x] - [x)$.

69. *Properties of inequalities.* If a and b are any two real numbers, we say a is less than b and write $a < b$ if (and only if) $b - a$ is positive. If $a < b$ we also say that b is greater than a ($b > a$). Prove the following properties of inequalities:
a) If $a < b$, then $a + c < b + c$ and $a - c < b - c$ for any real number c.
b) If $a < b$ and $c < d$, then $a + c < b + d$. Is it also true that $a - c < b - d$? If so, prove it; if not, give a counterexample.
c) If a and b are both positive (or both negative) and $a < b$, then $1/b < 1/a$.
d) If $a < 0 < b$, then $1/a < 0 < 1/b$.
e) If $a < b$ and $c > 0$, then $ac < bc$.
f) If $a < b$ and $c < 0$, then $bc < ac$.

70. *Properties of absolute values*
a) Prove that $|a| < |b|$ if and only if $a^2 < b^2$.
b) Prove that $|a + b| \leq |a| + |b|$.
c) Prove that $|a - b| \geq ||a| - |b||$.
d) Prove, by mathematical induction, that

$$|a_1 + a_2 + \cdots + a_n| \leq |a_1| + |a_2| + \cdots + |a_n|.$$

e) Using the result from part (d), prove that

$$|a_1 + a_2 + \cdots + a_n| \geq |a_1| - |a_2| - \cdots - |a_n|.$$

2 Derivatives

2.1 Formal Differentiation

In Chapter 1 we studied limits and used the notion of limit as a tool to define the derivative of a function. The notion of limit is a general notion for functions; the derivative is a limit of a particular kind. We saw how derivatives can be interpreted as slopes of curves (Article 1.7) and as instantaneous rates of change (Article 1.8).

If we want the slope of a curve $y = f(x)$ at just one point, say $x = x_1$, $y = y_1$, we might try to calculate

$$\text{Slope at } P_1(x_1, y_1) = \lim_{h \to 0} \frac{f(x_1 + h) - f(x_1)}{h} \tag{1}$$

numerically. With a calculator and a formula for $f(x)$ that can be evaluated with the calculator, we choose some sequence of values of h, like 0.1, 0.01, 0.001, 0.0001, and so on. With x_1 held fixed, we calculate successive values of the difference quotient on the right-hand side of Eq. (1), and from these values we deduce what we think the limit would be, if there were a limit. But this does not tell us that the limit exists. And even if the limit does exist, we may not be able to find its value with sufficient accuracy by straight numerical work.

In addition to the numerical method of finding a slope, we were able to determine the slope algebraically in some instances. The result of the algebra was a formula that gave useful information about the graph of the function and the function's rate of change over most if not all of the function's domain:

Function	Derivative formula
$y = mx + b$	$y' = m$
$y = x^2$	$y' = 2x$
$y = \|x\|$	$y' = \begin{cases} -1 & \text{if } x < 1, \\ 1 & \text{if } x > 1 \end{cases}$
$y = ax^2 + bx + c$	$y' = 2ax + b$

In the last instance (Article 1.8, Problems 1–10), we have a very simple way of going directly from the equation of the function to its derivative if the equation is any quadratic expression in x.

The obvious question to ask is whether there are also simple rules for general cubic equations, or polynomials of higher degree. The purpose of this chapter is to provide an affirmative answer to this question, and to go beyond this to develop formal rules for calculating the derivatives of other functions as well. We will also see how these rules are used in applications and in the further development of calculus itself.

The use of formal rules to find derivatives is called *formal differentiation*. When we find the derivative of a function, whether by Fermat's Δ-process or a standard formula, we say that we have *differentiated* the function. In calculus, the term "differentiate" means "find the derivative of."

2.2 Polynomial Functions and their Derivatives

In the practice of calculus we often want to find derivatives of polynomial functions like

$$f(x) = x^3 - 2x + 3, \qquad h(x) = 7x^5, \qquad \phi(x) = \tfrac{4}{3}.$$

This means that we must be able to differentiate powers of x, constants times powers of x, and sums and differences of these.

A single term of the form cx^n, where c is a constant and n is zero or a positive integer, is called a *monomial* in x. A sum of a finite number of monomials in x is called a *polynomial* in x. Our approach to differentiating polynomials will be first to find a formula for differentiating monomials, and then to find a rule that will allow us to construct the derivative of any polynomial from the derivatives of the monomials from which it is made.

As we know from Chapter 1, derivatives are defined in the following way:

DEFINITION

Derivative

Let $y = f(x)$ define a function of x. If the limit

$$\frac{dy}{dx} = f'(x) = \lim_{\Delta x \to 0} \frac{f(x + \Delta x) - f(x)}{\Delta x} \tag{1a}$$

exists and is finite, we call this limit the *derivative* of f at x and say that f is *differentiable* at x.

In our work it will save time to write Δy for the increment $f(x + \Delta x) - f(x)$. When we do this, the limit in Eq. (1a) becomes

$$\frac{dy}{dx} = \lim_{\Delta x \to 0} \frac{\Delta y}{\Delta x}. \tag{1b}$$

RULE 1

The derivative of a constant is zero.

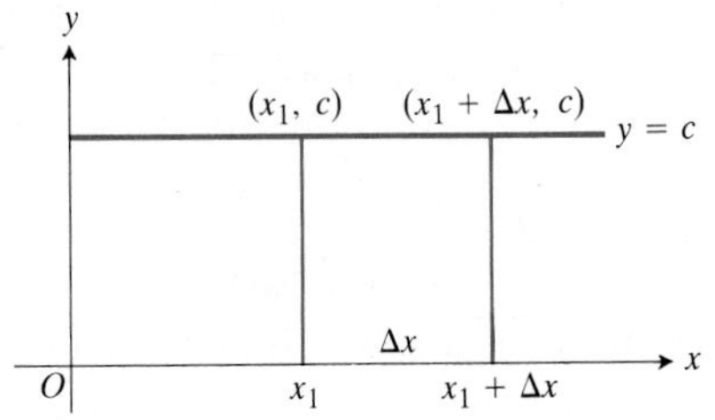

2.1 The slope of the graph of $y =$ constant is zero.

Rule 1 says that if $y = f(x) = c$ is a constant function, then $dy/dx = 0$. The geometric meaning of this fact is that the slope of the line $y = c$ is zero (Fig. 2.1).

To see why $dy/dx = 0$, let Δx be an increment in x. The corresponding increment in y is

$$\Delta y = f(x + \Delta x) - f(x) = c - c = 0.$$

Dividing by Δx gives

$$\frac{\Delta y}{\Delta x} = \frac{0}{\Delta x} = 0,$$

and thereby

$$\frac{dy}{dx} = \lim_{\Delta x \to 0} \frac{\Delta y}{\Delta x} = 0.$$

RULE 2

Positive Integer Powers of x

If n is a positive integer, then

$$\frac{d}{dx}(x^n) = nx^{n-1}. \tag{2}$$

Proof of Rule 2. To prove Rule 2, we let

$$y = f(x) = x^n.$$

Then

$$\frac{\Delta y}{\Delta x} = \frac{f(x + \Delta x) - f(x)}{\Delta x} = \frac{(x + \Delta x)^n - x^n}{\Delta x}. \tag{3}$$

Since n is a positive integer, we can apply the algebraic formula

$$a^n - b^n = (a - b)(a^{n-1} + a^{n-2}b + \cdots + ab^{n-2} + b^{n-1}),$$

with $a = x + \Delta x$, $b = x$, $a - b = \Delta x$, to the expression $(x + \Delta x)^n - x^n$ on the right-hand side of Eq. (3). This gives

$$\begin{aligned}\frac{\Delta y}{\Delta x} &= \frac{(x + \Delta x)^n - x^n}{\Delta x} \\ &= \frac{(\Delta x)((x + \Delta x)^{n-1} + (x + \Delta x)^{n-2}x + \cdots + (x + \Delta x)x^{n-2} + x^{n-1})}{\Delta x} \\ &= \underbrace{((x + \Delta x)^{n-1} + (x + \Delta x)^{n-2}x + \cdots + (x + \Delta x)x^{n-2} + x^{n-1})}_{n \text{ terms, each with limit } x^{n-1} \text{ as } \Delta x \to 0}.\end{aligned} \tag{4}$$

We now let Δx approach zero and find that

$$\begin{aligned}\frac{dy}{dx} &= \lim_{\Delta x \to 0} \frac{\Delta y}{\Delta x} \\ &= \underbrace{((x + 0)^{n-1} + (x + 0)^{n-2}x + \cdots + (x + 0)x^{n-2} + x^{n-1})}_{n \text{ terms}} \\ &= \underbrace{(x^{n-1} + x^{n-1} + \cdots + x^{n-1} + x^{n-1})}_{n \text{ copies of } x^{n-1}} \\ &= nx^{n-1}.\end{aligned} \tag{5}$$

In short,

$$\frac{dy}{dx} = nx^{n-1},$$

which is Eq. (2). ■

To apply the power rule for positive integers, we subtract 1 from the original exponent n and multiply the resulting power of x by the original exponent.

EXAMPLE 1

$$\frac{d}{dx}(x) = \frac{d}{dx}(x^1) = 1 \cdot x^0 = 1,$$

$$\frac{d}{dx}(x^2) = 2x^1 = 2x,$$

$$\frac{d}{dx}(x^3) = 3x^2,$$

$$\frac{d}{dx}(x^4) = 4x^3,$$

$$\frac{d}{dx}(x^5) = 5x^4 \quad \square$$

Although we have discussed only polynomials so far, the next two rules work for *any* differentiable function.

RULE 3

Constant Multiples

If $u = f(x)$ is any differentiable function of x, and c is a constant, then

$$\frac{d}{dx}(cu) = c\frac{du}{dx}. \tag{6}$$

Rule 3 says that the derivative of a constant multiple of a function is the same constant times the derivative of the function.

Proof of Rule 3. This follows immediately from the fact that $u = f(x)$ has a derivative:

$$\begin{aligned}\frac{d}{dx}cu &= \lim_{\Delta x \to 0} \frac{cf(x+\Delta x) - cf(x)}{\Delta x} \\ &= \lim_{\Delta x \to 0} c\frac{f(x+\Delta x) - f(x)}{\Delta x} \\ &= c \lim_{\Delta x \to 0} \frac{f(x+\Delta x) - f(x)}{\Delta x} \\ &= c\frac{du}{dx}. \ ■ \end{aligned} \tag{7}$$

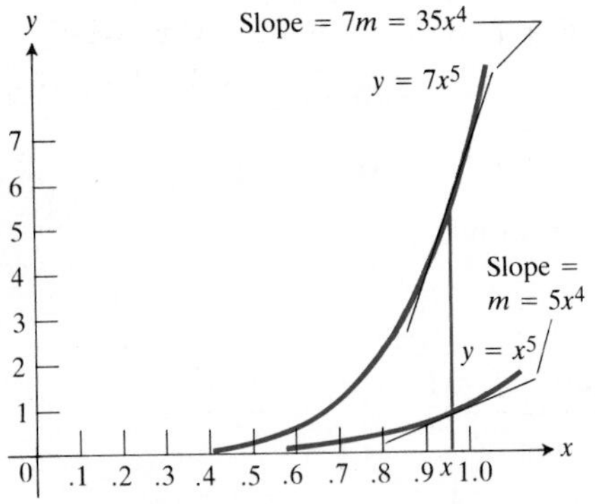

2.2 Graphs of $y = x^5$ and the stretched curve $y = 7x^5$. Multiplying the ordinates by 7 multiplies all the slopes by 7.

EXAMPLE 2

$$\frac{d(7x^5)}{dx} = 7 \cdot 5x^4 = 35x^4$$

Geometrically this says that if we stretch the graph of $y = x^5$ in the y-direction by multiplying each ordinate by 7, then we multiply each slope by 7 also (Fig. 2.2). □

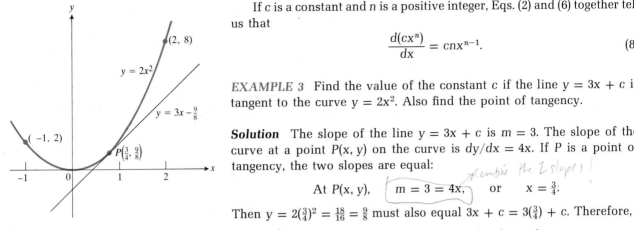

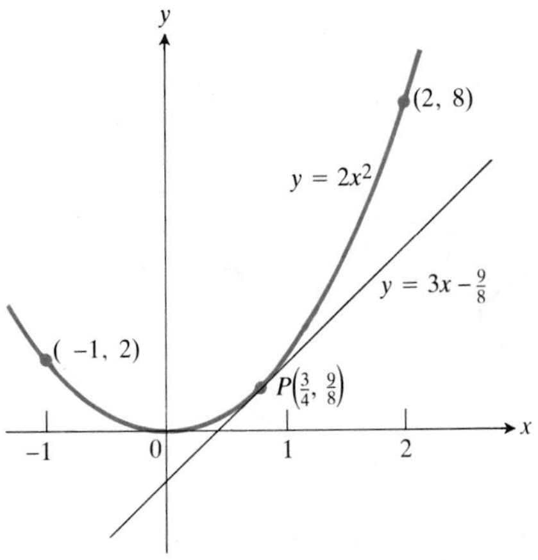

2.3 The line $y = 3x + c$ is tangent to the curve $y = 2x^2$ at $P(0.75, 1.125)$ if $c = -\frac{9}{8} = -1.125$.

If c is a constant and n is a positive integer, Eqs. (2) and (6) together tell us that

$$\frac{d(cx^n)}{dx} = cnx^{n-1}. \tag{8}$$

EXAMPLE 3 Find the value of the constant c if the line $y = 3x + c$ is tangent to the curve $y = 2x^2$. Also find the point of tangency.

Solution The slope of the line $y = 3x + c$ is $m = 3$. The slope of the curve at a point $P(x, y)$ on the curve is $dy/dx = 4x$. If P is a point of tangency, the two slopes are equal:

$$\text{At } P(x, y), \quad m = 3 = 4x, \quad \text{or} \quad x = \tfrac{3}{4}.$$

Then $y = 2(\frac{3}{4})^2 = \frac{18}{16} = \frac{9}{8}$ must also equal $3x + c = 3(\frac{3}{4}) + c$. Therefore,

$$\tfrac{9}{4} + c = \tfrac{9}{8} \quad \text{and} \quad c = \tfrac{9}{8} - \tfrac{9}{4} = -\tfrac{9}{8}.$$

The point of tangency is

$$P(x, y) = P(\tfrac{3}{4}, \tfrac{9}{8}) = P(0.75, 1.125).$$

(See Fig. 2.3.) □

RULE 4

The Sum Rule

If u and v are differentiable functions of x, then their sum $u + v$ is a differential function of x and

$$\frac{d}{dx}(u + v) = \frac{du}{dx} + \frac{dv}{dx} \tag{9}$$

at all values of x at which the derivatives of u and v both exist. Similarly, the derivative of the sum of any finite number of differentiable functions is the sum of their derivatives.

The idea is that if u and v both have derivatives at x, then their sum also has a derivative at x that is the sum of the derivatives of u and v at x.

Proof of Rule 4. To prove the first part of Rule 4, let

$$y = u + v,$$

where we suppose that u and v are differentiable functions of x. When x is replaced by $x + \Delta x$, the new values of the variables will satisfy the equation

$$y + \Delta y = (u + \Delta u) + (v + \Delta v).$$

We subtract $y = u + v$ from this and obtain

$$\Delta y = \Delta u + \Delta v,$$

and hence

$$\frac{\Delta y}{\Delta x} = \frac{\Delta u}{\Delta x} + \frac{\Delta v}{\Delta x}.$$

When Δx approaches zero, we get

$$\frac{dy}{dx} = \lim_{\Delta x \to 0} \frac{\Delta y}{\Delta x} = \lim_{\Delta x \to 0} \left(\frac{\Delta u}{\Delta x} + \frac{\Delta v}{\Delta x}\right) = \lim_{\Delta x \to 0} \frac{\Delta u}{\Delta x} + \lim_{\Delta x \to 0} \frac{\Delta v}{\Delta x} = \frac{du}{dx} + \frac{dv}{dx}.$$

Therefore,

$$\frac{d(u + v)}{dx} = \frac{du}{dx} + \frac{dv}{dx}.$$

This equation says that the derivative of the sum of two terms is the sum of their derivatives.

We may proceed by mathematical induction (Appendix 5) to establish the result for the sum of any finite number of terms. For example, if

$$y = u_1 + u_2 + u_3$$

is a sum of differentiable functions of x, then we may take

$$u = u_1 + u_2, \qquad v = u_3,$$

and apply the result already established for the sum of two terms, namely,

$$\frac{dy}{dx} = \frac{d(u_1 + u_2)}{dx} + \frac{du_3}{dx}.$$

Since the first term is again a sum of two terms, we have

$$\frac{d(u_1 + u_2)}{dx} = \frac{du_1}{dx} + \frac{du_2}{dx},$$

so that

$$\frac{d(u_1 + u_2 + u_3)}{dx} = \frac{du_1}{dx} + \frac{du_2}{dx} + \frac{du_3}{dx}.$$

Finally, if it has been established for some integer n that

$$\frac{d(u_1 + u_2 + \cdots + u_n)}{dx} = \frac{du_1}{dx} + \frac{du_2}{dx} + \cdots + \frac{du_n}{dx},$$

and we let

$$y = u + v,$$

with

$$u = u_1 + u_2 + \cdots + u_n, \qquad v = u_{n+1},$$

then we find in the same way as above that

$$\frac{d(u_1 + u_2 + \cdots u_{n+1})}{dx} = \frac{du_1}{dx} + \frac{du_2}{dx} + \cdots + \frac{du_{n+1}}{dx}.$$

This enables us to conclude that if Rule 4 is true for a sum of n terms it is also true for a sum of $(n + 1)$ terms. Since Rule 4 is already established for the sum of two terms, we conclude from the mathematical induction principle that it is true for the sum of any finite number of terms. ■

EXAMPLE 4 Find dy/dx if $y = x^3 + 7x^2 - 5x + 4$.

Solution We find the derivatives of the separate terms and add the results:

$$\begin{aligned}\frac{dy}{dx} &= \frac{d(x^3)}{dx} + \frac{d(7x^2)}{dx} + \frac{d(-5x)}{dx} + \frac{d(4)}{dx}\\ &= 3x^2 + 14x - 5x^0 + 0\\ &= 3x^2 + 14x - 5. \ \square\end{aligned}$$

Second Derivatives

The derivative

$$y' = \frac{dy}{dx}$$

is the *first derivative* of y with respect to x, and its derivative

$$y'' = \frac{dy'}{dx} = \frac{d}{dx}\left(\frac{dy}{dx}\right)$$

is called the *second derivative of* y *with respect to* x. Thus the second derivative is the derivative of the derivative.

The operation of taking the derivative of a function twice in succession is denoted by

$$\frac{d}{dx}\left(\frac{d}{dx}\cdots\right), \qquad \text{or} \qquad \frac{d^2}{dx^2}(\cdots).$$

In this notation, we write the second derivative of y with respect to x as

$$\frac{d^2y}{dx^2}.$$

In general terms, the result of differentiating a function $y = f(x)$ n times in succession is denoted by $y^{(n)}$, $f^{(n)}(x)$, or d^ny/dx^n.

EXAMPLE 5 For example, if $y = x^3 - 3x^2 + 2$, then

$$y' = \frac{dy}{dx} = 3x^2 - 6x, \qquad y''' = \frac{d^3y}{dx^3} = 6,$$

$$y'' = \frac{d^2y}{dx^2} = 6x - 6, \qquad y^{(iv)} = \frac{d^4y}{dx^4} = 0. \ \square$$

Velocity and Acceleration

In mechanics, if $s = f(t)$ gives the position of a moving body at time t (as discussed in Article 1.8), then

the first derivative ds/dt gives the *velocity;* and

the second derivative d^2s/dt^2 gives the *acceleration* of the body at time t.

Thus velocity is how fast position is changing; acceleration is how fast velocity is changing (that is, how quickly the body picks up or loses speed).

EXAMPLE 6 A heavy rock, blasted vertically upward with a velocity of 160 ft/s (about 109 mph) reaches a height of $s(t) = 160t - 16t^2$ feet after t seconds.

a) How high does the rock go?

b) How fast is the rock traveling when it reaches 256 ft above the ground on the way up? on the way down?

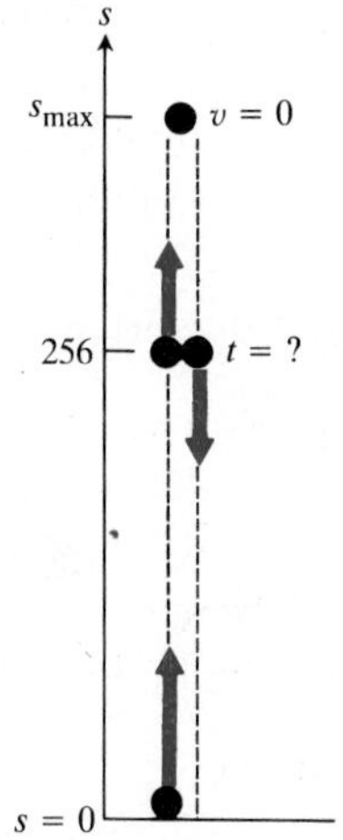

2.4 The flight of the rock in Example 6.

Solution *Part (a).* To find how high the rock goes, we find the height at the time when the rock's velocity is zero (Fig. 2.4). As a function of t, the velocity is

$$v(t) = \frac{ds}{dt} = \frac{d}{dt}(160t - 16t^2) = 160 - (2)(16)t = 160 - 32t \text{ ft/s}.$$

The velocity is zero when

$$160 - 32t = 0 \quad \text{or} \quad t = 5 \text{ s}.$$

The rock's height at $t = 5$ s is

$$s(5) = 160(5) - 16(5)^2 = 800 - 400 = 400 \text{ ft}.$$

Part (b). To find the rock's velocity at 256 ft on the way up and again on the way down, we find the two values of t for which

$$s(t) = 160t - 16t^2 = 256. \tag{10}$$

To solve Eq. (10) we write

$$16t^2 - 160 + 256 = 0$$
$$(t - 2)(t - 8) = 0$$
$$t = 2 \text{ s}, \qquad t = 8 \text{ s}.$$

The rock is 256 feet above the ground 2 seconds after the explosion, and again 8 seconds after. The rock's velocities at these times are

$$v(2) = 160 - 32(2) = 160 - 64 = 96 \text{ ft/s},$$
$$v(8) = 160 - 32(8) = 160 - 256 = -96 \text{ ft/s}.$$

The downward velocity is negative because s is decreasing when $t = 8$. □

PROBLEMS

In Problems 1–5, s represents the position of a moving body at time t. Find the velocity $v = ds/dt$ and the acceleration $a = dv/dt = d^2s/dt^2$.

1. $s = t^2 - 4t + 3$

2. $s = 2t^3 - 5t^2 + 4t - 3$

3. $s = gt^2/2 + v_0t + s_0$, (g, v_0, s_0 constants)

4. $s = 3 + 4t - t^2$

5. $s = (2t + 3)^2$

Find $y' = dy/dx$ and $y'' = dy'/dx$ in Problems 6–15.

6. $y = x^4 - 7x^3 + 2x^2 + 5$

7. $y = 5x^3 - 3x^5$

8. $y = 4x^2 - 8x + 1$

9. $y = \frac{x^4}{4} - \frac{x^3}{3} + \frac{x^2}{2} - x + 3$

10. $y = 2x^4 - 4x^2 - 8$

11. $12y = 6x^4 - 18x^2 - 12x$

12. $y = 3x^7 - 7x^3 + 21x^2$

13. $y = x^2(x^3 - 1)$

14. $y = (x - 2)(x + 3)$

15. $y = (3x - 1)(2x + 5)$

16. Find the tangent to each curve at the given point.
a) $y = x^3$ at (2, 8)
b) $y = 2x^2 + 4x - 3$ at (1, 3)
c) $y = x^3 - 6x^2 + 5x$ at the origin

17. Which of the following is the slope of the line tangent to the curve $y = x^2 + 5x$ at $x = 3$?
a) 24 b) $-\frac{5}{2}$ c) 11 d) 8

18. Which of the following is the slope of the line $3x - 2y + 12 = 0$?
a) 6 b) 3 c) $\frac{3}{2}$ d) $\frac{2}{3}$

19. Find the equation of the line perpendicular to the tangent to the curve $y = x^3 - 3x + 1$ at the point (2, 3).

20. A rock, thrown vertically upward from the surface of the moon at a velocity of 24 m/s (about 86 km/h), reaches a height of $s = 24t - 0.8t^2$ meters after t seconds.
a) Find the acceleration of gravity at the moon's surface.

b) How long did it take the rock to reach its highest point?
c) How high did the rock go?
d) How long did it take the rock to reach half its maximum height?
e) How long was the rock aloft?
f) The rock is dropped into a crevasse and seen to strike the bottom 30 seconds later. How deep is the crevasse? How fast was the rock going when it hit the bottom?

21. On the earth, the rock in Problem 20 would reach a height of $s = 24t - 4.9t^2$ meters after t seconds. How high would the rock go?

22. Here are the equations for free fall at the surfaces of Mars and Jupiter (s in meters, t in seconds). Find the acceleration of gravity on each planet.
a) Mars: $s = 1.86t^2$
b) Jupiter: $s = 11.44t^2$
c) How many seconds would a rock have to fall to reach a velocity of 100 km/h on each planet?

23. A body moves in a straight line according to the law of motion $s = t^3 - 4t^2 - 3t$. Find its acceleration each time the velocity is zero.

24. Find the lines tangent to the curve $y = x^3 + x$, where the slope is equal to 4. What is the smallest value the slope of this curve can ever have and where on the curve does the slope equal this smallest value?

25. Find the points on the curve $y = 2x^3 - 3x^2 - 12x + 20$ where the tangent is parallel to the x-axis.

26. Suppose that you have plotted the point $P(x, x^2)$ on the graph of $y = x^2$. To construct the tangent to the graph at P, locate the point $T(x/2, 0)$ on the x-axis and draw the line through T and P. Show that this construction is correct.

27. The tangent to the curve $y = x^n$ at $P(x_1, y_1)$ intersects the x-axis at $T(t, 0)$. Express t in terms of n and x_1, then show how the result can be used to construct the tangent at P.

28. Find the x- and y-intercepts of the line L that is tangent to the curve $y = x^3$ at $A(-2, -8)$.

29. A line L is drawn tangent to the curve $y = x^3 - x$ at the point $A(-1, 0)$. Where else does this line intersect the curve?

30. *Curvature.* It is shown later in this text that one can define a number that represents the curvature of $y = f(x)$ at $P(x, y)$. The formula is

$$\text{Curvature} = |y''|/(1 + (y')^2)^{3/2},$$

where $y' = dy/dx$ and y'' is the second derivative. Use this formula to find the curvature of $y = x^2$:
a) at the origin; b) at (1, 1); c) as $x \to \infty$.

31. Find the values of the constants a, b, and c if the curve $y = ax^2 + bx + c$ is to pass through the point (1, 2) and is to be tangent to the line $y = x$ at the origin.

32. Find the constants a, b, and c so the the two curves $y = x^2 + ax + b$ and $y = cx - x^2$ shall be tangent to each other at the point (1, 0).

33. Find the constant c if the curve $y = x^2 + c$ is to be tangent to the line $y = x$.

> ***Toolkit* programs**
> Derivative Grapher

2.3 Products, Powers, and Quotients

In the preceding article we learned how to find derivatives of polynomials. We did this by finding the derivative of a monomial in x, then proving that the derivative of the sum of a finite number of differentiable functions of any kind is the sum of their derivatives.

In this article, we derive formulas for finding derivatives with respect to x of

Products: $y = uv$;

Powers: $y = u^n$, any integer n;

Quotients: $y = u/v$;

where u and v are differentiable functions of x. In particular, we will be able to find the derivative of any *rational* function (quotient of two polynomials).

Products

RULE 5

The Product Rule

The product of two differentiable functions u and v is differentiable and

$$\frac{d}{dx}(uv) = u\frac{dv}{dx} + v\frac{du}{dx}. \tag{1}$$

As with the statement of the Sum Rule in Article 2.2, the Product Rule in Eq. (1) is understood to hold only at values of x at which the derivatives of u and v both exist. At such a value of x, the rule says, the derivative of the product uv is u times the derivative of v, plus v times the derivative of u.

Proof of Rule 5. To prove Rule 5 we let

$$y = uv,$$

where u and v are differentiable functions of x. Let Δx be an increment in x, and let the corresponding increments in y, u, v be denoted by Δy, Δu, Δv. These latter increments may be either positive, negative, or zero, but $\Delta x \neq 0$. Then

$$y + \Delta y = (u + \Delta u)(v + \Delta v) = uv + u\,\Delta v + v\,\Delta u + \Delta u\,\Delta v. \tag{2}$$

Figure 2.5 gives a geometric interpretation of the terms on the right-hand side of this equation.

We subtract $y = uv$ from both sides of Eq. (2) to get

$$\Delta y = u\,\Delta v + v\,\Delta u + \Delta u\,\Delta v,$$

and divide by Δx to get

$$\frac{\Delta y}{\Delta x} = u\frac{\Delta v}{\Delta x} + v\frac{\Delta u}{\Delta x} + \Delta u\frac{\Delta v}{\Delta x}.$$

When Δx approaches zero, so will Δu, since

$$\lim \Delta u = \lim\left(\frac{\Delta u}{\Delta x}\Delta x\right) = \lim\frac{\Delta u}{\Delta x}\lim \Delta x = \frac{du}{dx}\cdot 0 = 0.$$

Thus,

$$\begin{aligned}\lim\frac{\Delta y}{\Delta x} &= \lim\left(u\frac{\Delta v}{\Delta x} + v\frac{\Delta u}{\Delta x} + \Delta u\frac{\Delta v}{\Delta x}\right)\\ &= \lim u\frac{\Delta v}{\Delta x} + \lim v\frac{\Delta u}{\Delta x} + \lim \Delta u\frac{\Delta u}{\Delta x}\\ &= \lim u \lim\frac{\Delta v}{\Delta x} + \lim v \lim\frac{\Delta u}{\Delta x} + \lim \Delta u \lim\frac{\Delta v}{\Delta x}\\ &= u\frac{dv}{dx} + v\frac{du}{dx} + 0\cdot\frac{dv}{dx}.\end{aligned}$$

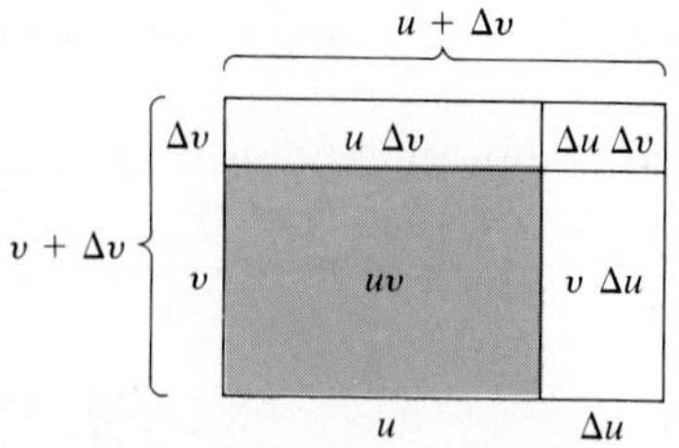

2.5 The area of the shaded rectangle is *uv*. When *u* and *v* change by increments Δu and Δv, their product $y = uv$ changes by the amount $\Delta y = u\,\Delta v + v\,\Delta u + \Delta u\,\Delta v$.

That is,

$$\frac{dy}{dx} = u\frac{dv}{dx} + v\frac{du}{dx},$$

which establishes Eq. (1). ■

EXAMPLE 1 Find the derivative of $y = (x^2 + 1)(x^3 + 3)$.

Solution From the Product Rule with

$$u = x^2 + 1, \qquad v = x^3 + 3,$$

we find

$$\frac{d}{dx}[(x^2 + 1)(x^3 + 3)] = (x^2 + 1)(3x^2) + (2x)(x^3 + 3)$$
$$= 3x^4 + 3x^2 + 2x^4 + 6x$$
$$= 5x^4 + 3x^2 + 6x. \ \square$$

This particular example can be done as well (perhaps better) by multiplying out the original expression for y and differentiating the resulting polynomial. We do that now as a check. From

$$y = (x^2 + 1)(x^3 + 3) = x^5 + x^3 + 3x^2 + 3,$$

we obtain

$$\frac{dy}{dx} = 5x^4 + 3x^2 + 6x,$$

in agreement with our first calculation.

There are, however, times when the Product Rule *must* be used, as the next example shows.

EXAMPLE 2 Let $y = uv$ be the product of the functions u and v, and suppose that

$$u(2) = 3, \qquad u'(2) = -4, \qquad v(2) = 1, \quad \text{and} \quad v'(2) = 2.$$

Find $y'(2)$.

Solution From the Product Rule, in the form

$$y' = (uv)' = uv' + vu',$$

we have

$$y'(2) = u(2)v'(2) + v(2)u'(2) = (3)(2) + (1)(-4) = 6 - 4 = 2. \ \square$$

Note that the derivative of a product is *not* the product of the derivatives. Instead, we add together two terms $u(dv/dx)$ and $v(du/dx)$. In the first of these we leave u untouched and differentiate v, and in the second we differentiate u and leave v alone. In fact, it is possible to extend the formula, by the method of mathematical induction, to show that the derivative of a product

$$y = u_1u_2 \cdots u_n$$

of a finite number of differentiable functions is given by

$$\frac{d}{dx}(u_1u_2 \cdots u_n)$$
$$= \frac{du_1}{dx} \cdot u_2 \cdots u_n + u_1\frac{du_2}{dx} \cdots u_n + \cdots + u_1u_2 \cdots u_{n-1}\frac{du_n}{dx}, \qquad (3)$$

where the right-hand side of the equation consists of the sum of the n terms obtained by multiplying the derivative of each one of the factors by the other $(n - 1)$ factors undifferentiated.

Powers

RULE 6

Positive Integer Powers of a Differentiable Function

If u is a differentiable function of x and n is a positive integer, then u^n is differentiable and

$$\frac{d}{dx}(u^n) = nu^{n-1}\frac{du}{dx}. \tag{4}$$

For $n = 1$ we interpret Eq. (2) as

$$\frac{d}{dx}(u) = u^0\frac{du}{dx} = \frac{du}{dx}, \tag{5}$$

which is certainly true if $u \neq 0$. If $u = 0$, we get 0^0, which is an indeterminate expression but which we interpret here as being 1 to be consistent.

For $n = 2$ we apply the Product Rule to the function $y = u \cdot u$ to obtain

$$\begin{aligned}\frac{d}{dx}(u^2) &= \frac{d}{dx}(u \cdot u)\\ &= u\frac{du}{dx} + u\frac{du}{dx} = 2u\frac{du}{dx}.\end{aligned} \tag{6}$$

Having established the Power Rule for $n = 2$, we may proceed by mathematical induction to establish it for all positive integer values of n.

Suppose that the rule has been established for some positive integer k, so that

$$\frac{d}{dx}(u^k) = ku^{k-1}\frac{du}{dx}. \tag{7}$$

We may then show that the rule holds for the next integer, $k + 1$, by the following argument. We let $y = u^{k+1}$, and rewrite y as the product of the two functions u and $u^k = v$:

$$y = u^{k+1} = u \cdot u^k.$$

We then apply the Product Rule,

$$\frac{dy}{dx} = u\frac{dv}{dx} + v\frac{du}{dx},$$

with $v = u^k$ to find the derivative of y:

$$\begin{aligned}\frac{dy}{dx} &= \frac{d}{dx}(u \cdot u^k)\\ &= u\frac{d}{dx}(u^k) + u^k\frac{du}{dx}\\ &= u\left(ku^{k-1}\frac{du}{dx}\right) + u^k\frac{du}{dx} \qquad \text{(by Eq. 7)}\\ &= ku^k\frac{du}{dx} + u^k\frac{du}{dx}\\ &= (k+1)u^k\frac{du}{dx}.\end{aligned} \tag{8}$$

This enables us to conclude that if Rule 6 holds for the exponent $n = k$, then it also holds for $n = k + 1$. Since Rule 6 is already established for $n = 2$, we conclude that it holds for any positive integer n.

Note that Rule 2,

$$\frac{d}{dx}(x^n) = nx^{n-1}, \tag{9}$$

is a special case of the Power Rule,

$$\frac{d}{dx}(u^n) = nu^{n-1}\frac{du}{dx}, \tag{10}$$

obtained by taking $u = x$:

$$\frac{d}{dx}(x^n) = nx^{n-1}\frac{dx}{dx} = nx^{n-1}\cdot 1 = nx^{n-1}. \tag{11}$$

We do not write the dx/dx as part of Eq. (9) because $dx/dx = 1$. To see why du/dx is a factor in Eq. (10), look for its appearance as the Product Rule is used in Eqs. (6) and (8).

EXAMPLE 3 Find the derivative of

$$y = (x^2 - 3x + 1)^5.$$

Solution We apply Rule 6 with $u = (x^2 - 3x + 1)$ and $n = 5$:

$$\begin{aligned}\frac{dy}{dx} &= \frac{d}{dx}(x^2 - 3x + 1)^5 \\ &= 5(x^2 - 3x + 1)^4 \cdot \frac{d}{dx}(x^2 - 3x + 1) \\ &= 5(x^2 - 3x + 1)^4 \cdot (2x - 3) \\ &= 5(2x - 3)(x^2 - 3x + 1)^4. \ \square\end{aligned}$$

WARNING. Do not forget the du/dx term in Eq. (4). The differentiation is not correct without it. In Example 3, the differentiation would not be correct without the factor

$$\frac{d}{dx}(x^2 - 3x + 1) = (2x - 3).$$

EXAMPLE 4 Find the derivative of $y = (x^2 + 1)^3(x^3 - 1)^2$.

Solution We could, of course, expand everything here and write y as a polynomial in x, but this is not necessary. Instead, we first use the Product Rule:

$$\frac{dy}{dx} = (x^2 + 1)^3\frac{d}{dx}(x^3 - 1)^2 + (x^3 - 1)^2\frac{d}{dx}(x^2 + 1)^3.$$

Then we evaluate the remaining derivatives by the Power Rule:

$$\begin{aligned}\frac{d}{dx}(x^3 - 1)^2 &= 2(x^3 - 1)\frac{d}{dx}(x^3 - 1) \\ &= 2(x^3 - 1)\cdot 3x^2 \\ &= 6x^2(x^3 - 1)\end{aligned}$$

and

$$\frac{d}{dx}(x^2+1)^3 = 3(x^2+1)^2\frac{d}{dx}(x^2+1)$$
$$= 3(x^2+1)^2\cdot 2x$$
$$= 6x(x^2+1)^2.$$

We substitute these into the earlier equation and have

$$\frac{dy}{dx} = (x^2+1)^3\cdot 6x^2(x^3-1) + (x^3-1)^2\cdot 6x(x^2+1)^2$$
$$= 6x(x^2+1)^2(x^3-1)[x(x^2+1)+(x^3-1)]$$
$$= 6x(x^2+1)^2(x^3-1)(2x^3+x-1). \square$$

Quotients

The ratio or quotient u/v of two polynomials in x is generally not a polynomial. Such a ratio is called a *rational function* of x, where the word *ratio* is the key to the use of the word rational.

Rational functions play a key role in computation because they are the most elaborate functions a digital computer can evaluate directly. The next differentiation rule, however, is not restricted to rational functions. It applies to the quotient of any two differentiable functions.

RULE 7

The Quotient Rule

At a point where $v \neq 0$, the quotient $y = u/v$ of two differentiable functions is differentiable and

$$\frac{d}{dx}\left(\frac{u}{v}\right) = \frac{v\dfrac{du}{dx} - u\dfrac{dv}{dx}}{v^2}. \tag{12}$$

As with the rules for differentiating sums and products of differentiable functions, Eq. (12) in the Quotient Rule is understood to hold only at values of x at which both u and v are differentiable.

Proof of Rule 7. To establish Eq. (12), consider a point x where $v \neq 0$ and where u and v are differentiable. Let x be given an increment Δx and let Δy, Δu, Δv be the corresponding increments in y, u, v. Then, as $\Delta x \to 0$,

$$\lim\,(v+\Delta v) = \lim v + \lim \Delta v$$

while

$$\lim \Delta v = \lim \frac{\Delta v}{\Delta x}\cdot \Delta x = \frac{dv}{dx}\cdot 0 = 0.$$

Therefore, the value of $v + \Delta v$ is close to the value of v when $x + \Delta x$ is close to x; that is, when Δx is near zero. In particular, since $v \neq 0$ at x, it follows that $v + \Delta v \neq 0$ when Δx is *near* zero, say when $0 < |\Delta x| < h$. Let Δx be so restricted. Then $v + \Delta v \neq 0$ and

$$y + \Delta y = \frac{u+\Delta u}{v+\Delta v}.$$

From this we subtract

$$y = \frac{u}{v}$$

and obtain

$$\Delta y = \frac{u + \Delta u}{v + \Delta v} - \frac{u}{v} = \frac{(vu + v\,\Delta u) - (uv + u\,\Delta v)}{v(v + \Delta v)} = \frac{v\,\Delta u - u\,\Delta v}{v(v + \Delta v)}.$$

We divide this by Δx and have

$$\frac{\Delta y}{\Delta x} = \frac{v\dfrac{\Delta u}{\Delta x} - u\dfrac{\Delta v}{\Delta x}}{v(v + \Delta v)}.$$

When Δx approaches zero,

$$\lim \frac{\Delta u}{\Delta x} = \frac{du}{dx}, \qquad \lim \frac{\Delta v}{\Delta x} = \frac{dv}{dx},$$

and

$$\lim v(v + \Delta v) = \lim v \lim (v + \Delta v) = v^2 \neq 0.$$

Therefore,

$$\frac{dy}{dx} = \lim \frac{\Delta y}{\Delta x} = \frac{\lim \left(v\dfrac{\Delta u}{\Delta x} - u\dfrac{\Delta v}{\Delta x}\right)}{\lim v(v + \Delta v)} = \frac{v\dfrac{du}{dx} - u\dfrac{dv}{dx}}{v^2},$$

which is Eq. (12). ■

EXAMPLE 5 Find the derivative of

$$y = \frac{x^2 + 1}{x^2 - 1}.$$

Solution We apply the Quotient Rule (Eq. 12):

$$\frac{dy}{dx} = \frac{(x^2 - 1) \cdot 2x - (x^2 + 1) \cdot 2x}{(x^2 - 1)^2} = \frac{-4x}{(x^2 - 1)^2}. \quad \square$$

RULE 8

Negative Integer Powers of a Differentiable Function

At a point where $u = g(x)$ is differentiable and different from zero, the derivative of

$$y = u^n$$

is given by

$$\frac{d(u^n)}{dx} = nu^{n-1}\frac{du}{dx} \tag{13}$$

if n is a negative integer.

Proof of Rule 8. Rule 8 is the extension of Eq. (4) to the case where n is a *negative integer*. To prove it, we combine the results in Eqs. (12) and (4). Let

$$y = u^{-m} = \frac{1}{u^m},$$

where $-m$ is a negative integer, so that m is a positive integer. Then, using (12) for the derivative of a quotient, we have

$$\frac{dy}{dx} = \frac{d\left(\frac{1}{u^m}\right)}{dx} = \frac{u^m \frac{d(1)}{dx} - 1\frac{d(u^m)}{dx}}{(u^m)^2} \tag{14}$$

at any point where u is differentiable and different from zero. Now the various derivatives on the right-hand side of (14) can be evaluated by formulas already proved, namely,

$$\frac{d(1)}{dx} = 0,$$

since 1 is a constant, and

$$\frac{d(u^m)}{dx} = mu^{m-1}\frac{du}{dx},$$

since m is a *positive integer*. Therefore,

$$\frac{dy}{dx} = \frac{u^m \cdot 0 - 1 \cdot mu^{m-1}\frac{du}{dx}}{u^{2m}} = -mu^{-m-1}\frac{du}{dx}.$$

If $-m$ is replaced by its equivalent value n, this equation reduces to Eq. (13). ■

EXAMPLE 6 Find the derivative of

$$y = x^2 + \frac{1}{x^2}.$$

Solution We write $y = x^2 + x^{-2}$. Then

$$\frac{dy}{dx} = 2x^{2-1}\frac{dx}{dx} + (-2)x^{-2-1}\frac{dx}{dx} = 2x - 2x^{-3}. \quad \square$$

As a practical matter in differentiation, it is usually best to treat a reciprocal

$$\frac{1}{[u(x)]^n}$$

as a function raised to a power, and not as a quotient.

EXAMPLE 7 The most effective way to find the derivative of

$$y = \frac{1}{(x^2 - 1)^5}$$

is to write $y = (x^2 - 1)^{-5}$, and calculate

$$y' = -5(x^2 - 1)^{-6} \cdot \frac{d}{dx}(x^2 - 1) = \frac{-5}{(x^2 - 1)^6} \cdot (2x) = \frac{-10x}{(x^2 - 1)^6}.$$

If you treat

$$y = \frac{1}{(x^2 - 1)^5}$$

as a quotient, with $u = 1$ and $v = (x^2 - 1)^5$, the first step in calculating

y' is

$$y' = \frac{(x^2 - 1)^5 \cdot \frac{d}{dx}(1) - 1 \cdot \frac{d}{dx}(x^2 - 1)^5}{((x^2 - 1)^5)^2}.$$

This is correct, but cumbersome. □

As Example 7 suggests, the choice of what rules to use in solving a differentiation problem can make a difference in how much work you have to do. Here are two more examples.

EXAMPLE 8 The easiest way to find y' if

$$y = \left(\frac{2x - 1}{x + 7}\right)^3$$

is to use the Power Rule first and then the Quotient Rule. We start with

$$y' = 3 \cdot \left(\frac{2x - 1}{x + 7}\right)^2 \cdot \frac{d}{dx}\left(\frac{2x - 1}{x + 7}\right).$$

We are in for more work if we begin with

$$y = (2x - 1)^3(x + 7)^{-3},$$

$$y' = (2x - 1)^3 \cdot \frac{d}{dx}(x + 7)^{-3} + (x + 7)^{-3} \cdot \frac{d}{dx}(2x - 1)^3. \quad \square$$

EXAMPLE 9 Do not use the Quotient Rule to find the derivative of

$$y = \frac{(x - 1)(x^2 - 2x)}{x^4}.$$

Instead, expand the numerator, and write

$$y = \frac{(x - 1)(x^2 - 2x)}{x^4} = \frac{x^3 - 3x^2 + 2x}{x^4} = x^{-1} - 3x^{-2} + 2x^{-3}.$$

Then use the Sum and Power Rules:

$$\frac{dy}{dx} = -x^{-2} - 3(-2)x^{-3} + 2(-3)x^{-4} = -\frac{1}{x^2} + \frac{6}{x^3} - \frac{6}{x^4}. \quad \square$$

Using Derivatives to Rewrite Polynomials. Approximations

If we are especially interested in approximating the values of a polynomial function $y = f(x)$ for values of x near, say, $x = 2$, it is better to express f in powers of $(x - 2)$ instead of powers of x. The next example shows how this may be done, and what can be accomplished by writing the polynomial this way.

EXAMPLE 10 Write the polynomial

$$f(x) = x^3 - 2x + 3 \tag{15}$$

in the form

$$f(x) = C_0 + C_1(x - 2) + C_2(x - 2)^2 + C_3(x - 2)^3, \tag{16}$$

where C_0, C_1, C_2, and C_3 are constants.

$C_0 + C_1(x-2) + C_2(x-2)^2 + C_3(x-2)^2$

Solution We differentiate both expressions for $f(x)$ three times, and substitute $x = 2$ to get

$$\begin{aligned}
f(x) &= C_0 + C_1(x-2) + C_2(x-2)^2 + C_3(x-2)^3, & f(2) &= C_0;\\
f'(x) &= C_1 + 2C_2(x-2) + 3C_3(x-2)^2, & f'(2) &= C_1;\\
f''(x) &= 2C_2 + 6C_3(x-2), & f''(2) &= 2C_2;\\
f'''(x) &= 6C_3, & f'''(2) &= 6C_3;
\end{aligned}$$

and

$$\begin{aligned}
f(x) &= x^3 - 2x + 3, & f(2) &= 8 - 4 + 3 = 7;\\
f'(x) &= 3x^2 - 2, & f'(2) &= 12 - 2 = 10;\\
f''(x) &= 6x, & f''(2) &= 12;\\
f'''(x) &= 6, & f'''(2) &= 6.
\end{aligned}$$

This gives two expressions for the value of each derivative of f at $x = 2$. Equating them, we find

$$\begin{aligned}
C_0 &= f(2) = 7,\\
C_1 &= f'(2) = 10,\\
C_2 &= \tfrac{1}{2}f''(2) = 6,\\
C_3 &= \tfrac{1}{6}f'''(2) = 1.
\end{aligned}$$

Therefore,

$$\begin{aligned}
f(x) &= x^3 - 2x + 3\\
&= 7 + 10(x-2) + 6(x-2)^2 + (x-2)^3. \square
\end{aligned}$$

Two things are important in making an approximation: first, the form of the approximation, and second, the size of the error. For the present example we can say that, when x is near 2, the successive powers of $(x - 2)$ are increasingly small, and

$$\begin{aligned}
&f(x) \approx 7, \quad \text{with an error about } 10(x-2);\\
&f(x) \approx 7 + 10(x-2), \quad \text{with an error about } 6(x-2)^2;\\
&f(x) \approx 7 + 10(x-2) + 6(x-2)^2, \quad \text{with an error exactly } (x-2)^3;\\
&f(x) = 7 + 10(x-2) + 6(x-2)^2 + (x-2)^3, \quad \text{with no error.}
\end{aligned}$$

Figure 2.6 shows the first two of these approximations to $f(x)$.

The approximation

$$f(x) \approx 7 + 10(x-2)$$

is the *linear* or *first order* approximation of f obtained by using the tangent line

$$y = 7 + 10(x-2)$$

to approximate f near the point $x = 2$ (Fig. 2.6). Linear approximations are important in science because they let us replace complicated formulas by simple ones over an interval with a tolerable error. We will see this again in Articles 2.5 and 2.9.

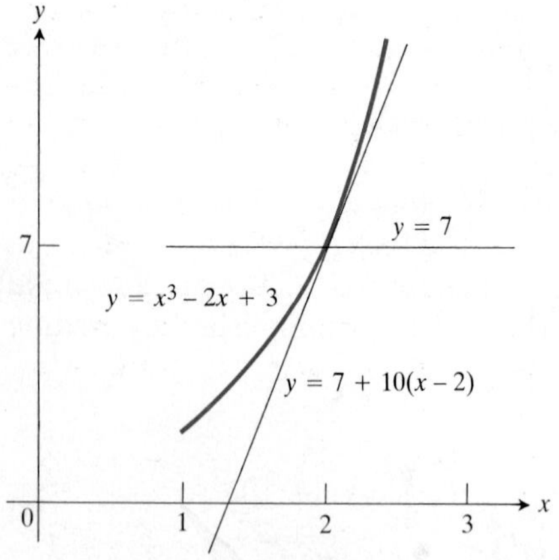

2.6 The graph of $y = x^3 - 2x + 3$ near $x = 2$ with horizontal and tangent line approximations.

PROBLEMS

Find dy/dx in Problems 1–24.

1. $y = \dfrac{x^3}{3} - \dfrac{x^2}{2} + x - 1$ **2.** $y = (x - 1)^3(x + 2)^4$

3. $y = (x^2 + 1)^5$ **4.** $y = (x^3 - 3x)^4$

5. $y = (x + 1)^2(x^2 + 1)^{-3}$ **6.** $y = \dfrac{2x + 1}{x^2 - 1}$

7. $y = \dfrac{2x + 5}{3x - 2}$ **8.** $y = \left(\dfrac{x + 1}{x - 1}\right)^2$

9. $y = (1 - x)(1 + x^2)^{-1}$ **10.** $y = (x + 1)^2(x^2 + 2x)^{-2}$

11. $y = \dfrac{5}{(2x - 3)^4}$ **12.** $y = (x - 1)^3(x + 2)$

13. $y = (5 - x)(4 - 2x)$ **14.** $y = [(5 - x)(4 - 2x)]^2$

15. $y = (2x - 1)^3(x + 7)^{-3}$ **16.** $y = \dfrac{x^3 + 7}{x}$

17. $y = (2x^3 - 3x^2 + 6x)^{-5}$ **18.** $y = \dfrac{x^2}{(x - 1)^2}$

19. $y = \dfrac{(x - 1)^2}{x^2}$ **20.** $y = \dfrac{-1}{15(5x - 1)^3}$

21. $y = \dfrac{12}{x} - \dfrac{4}{x^3} + \dfrac{3}{x^4}$

22. $y = \dfrac{(x - 1)(x^2 + x + 1)}{x^3}$

23. $y = \dfrac{(x^2 - 1)}{x^2 + x - 2}$

24. $y = \dfrac{(x^2 + x)(x^2 - x + 1)}{x^4}$

Find ds/dt in Problems 25–32.

25. $s = \dfrac{t}{t^2 + 1}$ **26.** $s = (2t + 3)^3$

27. $s = (t^2 - t)^{-2}$ **28.** $s = t^2(t + 1)^{-1}$

29. $s = \dfrac{2t}{3t^2 + 1}$ **30.** $s = (t + t^{-1})^2$

31. $s = (t^2 + 3t)^3$

32. $s = (t^2 - 7t)(5 - 2t^3 + t^4)/t^3$

33. Suppose that u and v are functions of x that are differentiable at $x = 0$, and that

$$u(0) = 5, \quad u'(0) = -3, \quad v(0) = -1, \quad v'(0) = 2.$$

(The primes denote differentiation with respect to x.) Find the values of the derivatives below at $x = 0$.

a) $\dfrac{d}{dx}(uv)$ b) $\dfrac{d}{dx}\left(\dfrac{u}{v}\right)$ c) $\dfrac{d}{dx}\left(\dfrac{v}{u}\right)$

d) $\dfrac{d}{dx}(7v - 2u)$ e) $\dfrac{d}{dx}(u^3)$ f) $\dfrac{d}{dx}(5v^{-3})$

34. Find an equation for the tangent to the curve $y = x/(x^2 + 1)$ at the origin.

35. Find an equation of the tangent to the curve $y = x + (1/x)$ at $x = 2$.

36. Find the first and second derivatives of $f(x) = (x^2 + 3x + 1)^3$.

37. Find d^2y/dx^2 if $y = (3 - 2x)^{-1}$.

38. Differentiate $x = 5y/(y + 1)$ with respect to y.

39. In Example 10, $f(x) = x^3 - 2x + 3$ was claimed to be equal to $7 + 10(x - 2) + 6(x - 2)^2 + (x - 2)^3$. Expand the terms in this latter expression, using the binomial theorem, and thus show that the two expressions are identical.

40. Suppose that the function $f(x) = x^3 - 2x + 3$ of Example 10 is put into the form $f(x) = a_0 + a_1(x - 1) + a_2(x - 1)^2 + a_3(x - 1)^3$, where a_0, a_1, a_2, and a_3 are constants. Determine these constants in terms of $f(1)$, $f'(1)$, $f''(1)$, and $f'''(1)$. If the linear expression $L_1(x) = a_0 + a_1(x - 1)$ is used to approximate $f(x)$ for values of x near 1, about how large is the error? What is the limit of

$$\frac{f(x) - L_1(x)}{(x - 1)^2}$$

as x approaches 1?

41. *Production.* Economists often use the expression "rate of growth" in relative rather than absolute terms. For example, in a given industry, let $u = f(t)$ be the number of people in the labor force at time t. (This function will be treated as though it were differentiable even though it is an integer-valued step function. We approximate the step function by a smooth curve.)

Let $v = g(t)$ be the average production per person in the labor force at time t. The total production is then $y = uv$. If the labor force is growing at the rate of 4 percent per year ($du/dt = 0.04u$) and the production per worker is growing at the rate of 5 percent per year ($dv/dt = 0.05v$), find the rate of growth of the total production, y.

42. Suppose that the labor force in Problem 41 is decreasing at the rate of 2 percent per year while the production per person is increasing at the rate of 3 percent per year. Is the total production increasing, or decreasing, and at what rate?

43. *Rate of a chemical reaction.* When two chemicals, A and B, combine to form a product P, the rate dP/dt at which the product is formed is called the *reaction rate*. Suppose that one molecule of P is formed from exactly one molecule of A and one of B, and that the initial molar masses of A and B are equal, both having value a. Under these conditions, the amount of P at any time t after mixing is given by the function $P(t) = a^2kt/(akt + 1)$, where k is a constant of proportionality in the chemical Law of Mass Action, representing the affinity of the two chemicals.

a) Find the reaction rate, dP/dt.

b) Find the time at which the reaction is proceeding at its fastest rate and the value of dP/dt at this time.

44. Let c be a constant, and u a differentiable function of x. Show that Rule 3, which says that

$$\frac{d}{dx}(cu) = c\frac{du}{dx},$$

is a special case of the Product Rule.

45. Show that Rule 6 is a special case of the finite product rule given in Eq. (3).

46. With the book closed, state and prove the formula for the derivative of the product of two differentiable functions u and v.

47. With the book closed, state and prove the formula for the derivative of the quotient u/v of two differentiable functions u and v.

48. With the book closed, state and prove the formula for the derivative with respect to x of u^n, where n is a positive integer and u is a differentiable function of x.

***Toolkit* programs**

Derivative Grapher

2.4 Implicit Differentiation and Fractional Powers

Most of the functions with which we have dealt have been given by equations of the form $y = f(x)$ that express y explicitly in terms of x. Quite often, however, we encounter equations like

$$x^2 + y^2 - 1 = 0, \quad \text{an equation for the unit circle at the origin,}$$
$$y^2 - x = 0, \quad \text{the parabola in Fig. 2.7,} \tag{1}$$

that do not give y explicitly in terms of x. Nevertheless, each of these equations defines a relation between x and y. When a definite number from some domain is substituted for x, the resulting equation determines one or more values for y. The xy-pairs so obtained can be plotted in the plane to graph the equation.

Now, the graph of an arbitrary equation $F(x, y) = 0$ in x and y may fail to be the graph of a function $y = f(x)$ because some vertical lines intersect it more than once. For example, in Fig. 2.8, the numbers y_1, y_2, and y_3 all correspond to the same value $x = x_0$. The pairs (x_0, y_1), (x_0, y_2), and (x_0, y_3) all satisfy the equation $F(x, y) = 0$, and the corresponding points lie on the graph of $F(x, y) = 0$.

However, various *parts* of the curve may be considered as the graphs of functions of x. For example, the segment AB of the curve in Fig. 2.8 is the graph of some function $y = f(x)$ that satisfies $f(x_0) = y_3$ and is defined on an open interval (a, b) containing x_0. If x is any point in (a, b), then the pair $(x, f(x)) = (x, y)$ satisfies the original equation $F(x, y) = 0$. We say that the equation $F(x, y) = 0$ has defined f *implicitly* on (a, b), even though it has not defined f by giving $y = f(x)$ explicitly in terms of x.

EXAMPLE 1 The graph of the equation $F(x, y) = x^2 + y^2 - 4 = 0$ is the circle $x^2 + y^2 = 4$, which is not the graph of any function of x (Fig. 2.9). For each x in the open interval $(-2, 2)$ there are *two* corresponding values

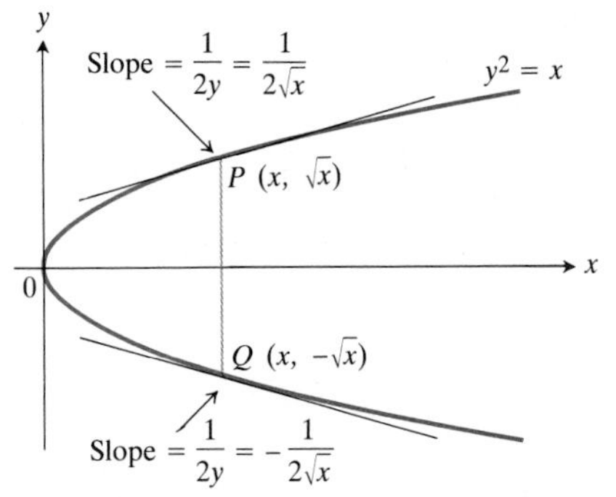

2.7 The slopes of the tangents to the parabola $y^2 = x$ at the points directly above and below x are both given by the formula $m_{\tan} = 1/(2y)$.

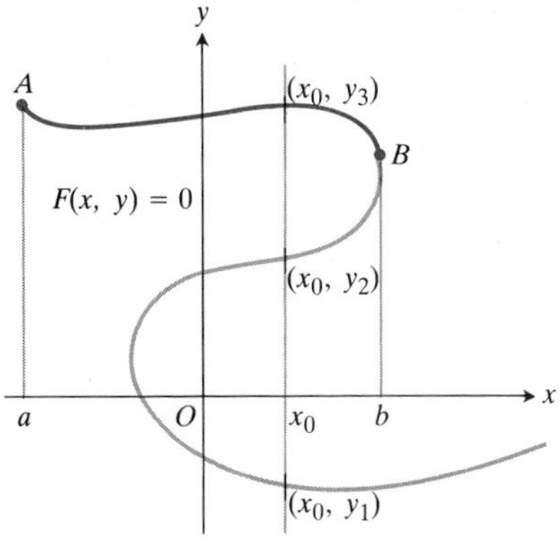

2.8 The graph of an equation in x and y of the form $F(x, y) = 0$ may not be the graph of a function of x. Some vertical lines may intersect it more than once. Portions of the graph, however, like the arc from A to B above, may be considered as the graphs of functions of x.

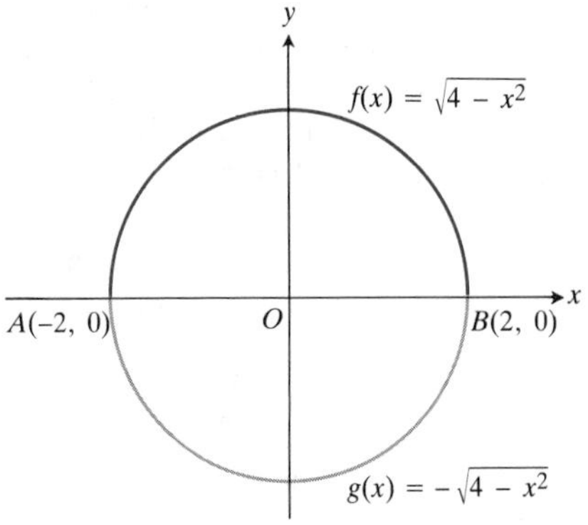

2.9 The graph of the equation $F(x, y) = x^2 + y^2 - 4 = 0$ is the circle $x^2 + y^2 = 4$ shown here. The upper semicircle AB is the graph of $y = f(x) = \sqrt{4 - x^2}$. The lower semicircle AB is the graph of $y = g(x) = -\sqrt{4 - x^2}$.

of y, namely, $y = \sqrt{4 - x^2}$ and $y = -\sqrt{4 - x^2}$. However, the upper and lower semicircles are the graphs of the functions $f(x) = \sqrt{4 - x^2}$ and $g(x) = -\sqrt{4 - x^2}$, respectively. The pairs $(x, y) = (x, f(x)) = (x, \sqrt{4 - x^2})$ and $(x, y) = (x, g(x)) = (x, -\sqrt{4 - x^2})$ satisfy the equation $x^2 + y^2 - 4 = 0$ whenever $-2 \leq x \leq 2$.

The functions f and g are also differentiable for $-2 < x < 2$, but since their graphs have vertical tangents at $x = \pm 2$ they are not differentiable at those points. □

The Method of Implicit Differentiation

If a curve $F(x, y) = 0$ is smooth enough to have a tangent at each point, then various parts of it will be the graphs of *differentiable* functions, as in Example 1. Under favorable circumstances we may be able to calculate the derivatives of these functions directly from the formula for F, without having to solve first for y. Indeed, in an equation like

$$x^5 + 4xy^3 - 3y^7 - 2 = 0, \tag{2}$$

it would generally not be possible to solve for y. Nevertheless, it is often possible to calculate dy/dx from such an equation by the method known as *implicit differentiation*. Using this method we simply treat y as an unknown but differentiable function of x and apply the rules for finding derivatives of u^n, uv, u/v, and so on, that we have already developed. (A discussion of the validity of the assumption that an equation like Eq. 2 does define y as one or more differentiable functions of x can be found in most textbooks on advanced calculus.)

EXAMPLE 2 Find dy/dx if $y^2 - x = 0$.

Solution The relation $y^2 - x = 0$ requires $x \geq 0$, and if we solve the equation for y we get

$$y = \pm\sqrt{x}. \tag{3}$$

For each positive value of x there are two corresponding values of y, not just one as required for a function. But we can look at the two different functions

$$y = \sqrt{x}, \qquad y = -\sqrt{x}. \tag{4}$$

(See Fig. 2.7.)

For $y = \sqrt{x}$, we have

$$\frac{dy}{dx} = \frac{1}{2\sqrt{x}} = \frac{1}{2y} \qquad \text{(from Article 1.7, Example 5).}$$

For $y = -\sqrt{x}$ we have

$$\frac{dy}{dx} = -\frac{1}{2\sqrt{x}} = \frac{1}{2(-\sqrt{x})} = \frac{1}{2y}.$$

In either case,

$$\frac{dy}{dx} = \frac{1}{2y}. \tag{5}$$

Now look at what would have happened had we just *assumed* y to be a differentiable function of x and differentiated both sides of the equation

$$y^2 - x = 0$$

with respect to x. We would have calculated

$$\frac{d}{dx}(y^2 - x) = \frac{d}{dx}(0)$$

$$\frac{d}{dx}(y^2) - \frac{d}{dx}(x) = 0$$

$$2y\frac{dy}{dx} - 1 = 0$$

$$\frac{dy}{dx} = \frac{1}{2y}. \tag{6}$$

This is the same answer we obtained in Eq. (5). □

Another way to look at what is going on in Example 2 is to observe that $y^2 - x = 0$ implicitly defines the functions $y = \sqrt{x}$ and $y = -\sqrt{x}$ in the sense that

$$(\sqrt{x})^2 - x = 0 \quad \text{for all } x \geq 0$$

and

$$(-\sqrt{x})^2 - x = 0 \quad \text{for all } x \geq 0.$$

In each equation, the left-hand side is identical to the constant function 0. Therefore, the derivative of the left-hand side is the derivative of zero and we are led once again to the single equation

$$\frac{d}{dx}(y^2 - x) = \frac{d}{dx}(0)$$

from which we calculated dy/dx.

EXAMPLE 3 Find dy/dx if $xy = 1$.

Solution This is another example in which we can solve for y first, or not, as we choose.

If we assume y to be a differentiable function of x without solving for y explicitly, and differentiate both sides of the equation $xy = 1$ with respect to x, we get

$$\frac{d}{dx}(xy) = \frac{d}{dx}(1)$$

$$x\frac{dy}{dx} + y\frac{dx}{dx} = 0 \qquad \text{(Product Rule applied to } xy\text{)}$$

$$x\frac{dy}{dx} + y = 0$$

$$\frac{dy}{dx} = -\frac{y}{x}. \tag{7}$$

If we solve for y and write $y = x^{-1}$, we get

$$\frac{dy}{dx} = \frac{d}{dx}(x^{-1}) = -x^{-2} = -\frac{1}{x^2}. \tag{8}$$

We see that the answers in (7) and (8) agree if we substitute $y = 1/x$ in Eq. (7). □

The foregoing examples were intended to convince you that it is all right to differentiate implicitly. We shall not be able to justify the method rigorously, however, because it involves more sophisticated ideas than we take up in this text.

EXAMPLE 4 Find dy/dx if $x^5 + 4xy^3 - 3y^5 = 2$.

Solution We differentiate both sides of the given equation with respect to x, treating y as a differentiable function of x. Thus,

$$\frac{d}{dx}(x^5) + \frac{d}{dx}(4xy^3) - \frac{d}{dx}(3y^5) = \frac{d}{dx}(2),$$

or

$$5x^4 + 4\left(x\frac{d(y^3)}{dx} + y^3\frac{dx}{dx}\right) - 15y^4\frac{dy}{dx} = 0,$$

$$5x^4 + 4\left(3xy^2\frac{dy}{dx} + y^3\right) - 15y^4\frac{dy}{dx} = 0,$$

$$(12xy^2 - 15y^4)\frac{dy}{dx} = -(5x^4 + 4y^3).$$

Finally, at all points where $12xy^2 - 15y^4 \neq 0$, we have

$$\frac{dy}{dx} = \frac{5x^4 + 4y^3}{15y^4 - 12xy^2}.$$

Note that the rule

$$\frac{d}{dx}u^n = nu^{n-1}\frac{du}{dx}$$

becomes

$$\frac{d}{dx}y^5 = 5y^4\frac{dy}{dx}$$

when applied to y^5. □

EXAMPLE 5 *Second derivatives.* Find d^2y/dx^2 if $2x^3 - 3y^2 = 7$.

Solution We differentiate both sides of the given equation twice with respect to x, treating both y and $y' = dy/dx$ as differentiable functions of x. First, we have

$$\frac{d}{dx}(2x^3) - \frac{d}{dx}(3y^2) = \frac{d}{dx}(7)$$

$$6x^2 - 6yy' = 0$$

$$x^2 - yy' = 0. \tag{9}$$

Then, differentiating both sides a second time and applying the Product Rule to differentiate yy', we have

$$\frac{d}{dx}(x^2) - \frac{d}{dx}(yy') = 0$$

$$2x - \left(y\frac{dy'}{dx} + y'\frac{dy}{dx}\right) = 0$$

$$2x - yy'' - (y')^2 = 0$$

$$yy'' = 2x - (y')^2. \tag{10}$$

At all points where $y \neq 0$, Eqs. (9) and (10) yield the following:

$$y' = \frac{x^2}{y}, \tag{11a}$$

$$y'' = \frac{2x - (y')^2}{y} = \frac{2x - (x^4/y^2)}{y} = \frac{2xy^2 - x^4}{y^3}. \tag{11b}$$

We can express both y' and y'' in terms of x and y. □

Tangents

Applying the method of implicit differentiation to an equation $F(x, y) = 0$ usually expresses dy/dx in terms of both x and y. This is no real handicap since, if we want the slope of the tangent to a curve at a known point (x_1, y_1), for instance, we need only substitute x_1 for x and y_1 for y in the final expression for dy/dx.

EXAMPLE 6 Find the slope of the tangent to the curve $x^2 + xy + y^2 = 7$ at the point (1, 2). (See Fig. 2.10.)

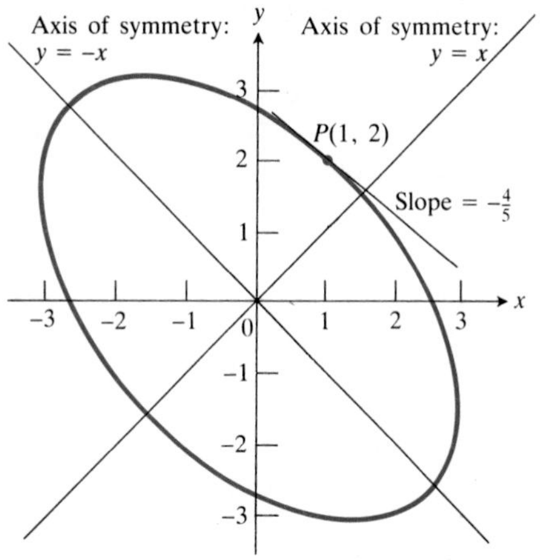

2.10 The graph of $x^2 + xy + y^2 = 7$ with slope of tangent at $P(1, 2)$: $(dy/dx)_{(1,2)} = -\frac{4}{5}$.

Solution We differentiate both sides of the equation with respect to x, noting that xy is a product, uv, and that y^2 has the form u^n, while 7 is a constant. Thus

$$2x\frac{dx}{dx} + \left(x\frac{dy}{dx} + y\frac{dx}{dx}\right) + 2y\frac{dy}{dx} = 0$$

$$(x + 2y)\frac{dy}{dx} = -(2x + y),$$

and wherever $x + 2y \neq 0$, we have

$$\frac{dy}{dx} = -\frac{2x + y}{x + 2y}.$$

In particular, at (1, 2) we have

$$\left(\frac{dy}{dx}\right)_{(1,2)} = -\frac{4}{5}. \tag{12}$$

The expression on the left is read "the value of dy/dx at the point (1, 2)." Another common notation for this value is

$$\left.\frac{dy}{dx}\right|_{(1,2)}. \tag{13}$$

Note that at points where $2x + y = 0$, the slope is zero and the tangent is parallel to the x-axis. Where $x + 2y = 0$, the tangent is parallel to the y-axis: At these points we cannot solve for dy/dx but we can solve for dx/dy, and it is zero. □

Normal Lines

Descartes, Fermat, and Newton, among many others, were interested in the design of lenses for microscopes and telescopes. Some features of Newton's telescope designs are still in use today. In the law that describes the way in which the direction of a light ray is changed by passing through the surface of a lens, the important angles are the angles that the light makes with the line perpendicular to the surface of the lens at the point of entry (the angles A and B in Fig. 2.11). This perpendicular line is called the "normal" to the surface at the point of entry. In a profile view of a lens

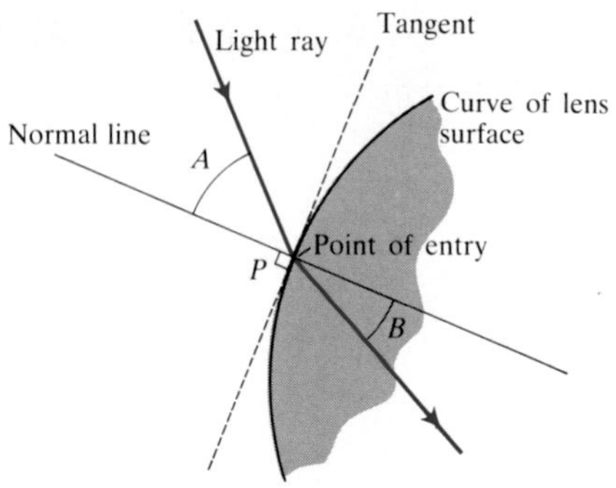

2.11 The profile or cutaway view of a lens, showing the bending (refraction) of a light ray as it passes through the lens surface.

like the one in Fig. 2.11, the normal is the line perpendicular to the tangent to the curve at the point of entry.

In calculus we define the line *normal* to a differentiable curve at a point P, whether the curve represents the surface of a lens or not, to be the line perpendicular to the tangent to the curve at P.

EXAMPLE 7 Find the lines tangent and normal to the curve

$$y^2 - 6x^2 + 4y + 19 = 0$$

at the point (2, 1).

Solution We differentiate both sides of the equation with respect to x and solve for (dy/dx):

$$2y\frac{dy}{dx} - 12x + 4\frac{dy}{dx} = 0$$

$$\frac{dy}{dx}(2y + 4) = 12x$$

$$\frac{dy}{dx} = \frac{6x}{y + 2}.$$

We then evaluate the derivative at $x = 2$, $y = 1$, to obtain

$$\left.\frac{dy}{dx}\right|_{(2,1)} = \left.\frac{6x}{y + 2}\right|_{(2,1)} = \frac{12}{3} = 4.$$

Therefore, the tangent to the curve at the point (2, 1) is

$$y - 1 = 4(x - 2).$$

The slope of the normal is $-1/4$ and its equation is

$$y - 1 = -\frac{1}{4}(x - 2). \ \square$$

Fractional Powers

The method of implicit differentiation enables us to establish the Power Rule for fractional exponents:

RULE 9

Power Rule for Fractional Exponents

If u is a differentiable function of x, and p and q are integers with $q > 0$, then

$$\frac{d}{dx}u^{p/q} = \frac{p}{q}u^{(p/q)-1}\frac{du}{dx}, \qquad (14)$$

provided $u \neq 0$ if $p/q < 1$.

This is the familiar rule for the derivative of u^n, but extended to the case in which $n = p/q$ is any rational number. As with earlier rules, Eq. (14) is understood to hold only at values of x at which $u^{p/q}$, $u^{(p/q)-1}$, and du/dx are all defined as real numbers. Rule 9 then says that we can obtain the value of the derivative of $u^{p/q}$ at x by evaluating the right-hand side of Eq. (14) at x.

The next example illustrates Eq. (14) and the restrictions that may have to be placed on the domains of x-values for the equation to hold.

EXAMPLE 8

a) $\frac{d}{dx}(x^{1/2}) = \frac{1}{2}x^{-1/2} = \frac{1}{2\sqrt{x}}$ for $x > 0$

b) $\frac{d}{dx}(x^{1/5}) = \frac{1}{5}x^{-4/5}$ for $x \neq 0$

c) $\frac{d}{dx}(x^{-4/3}) = -\frac{4}{3}x^{-7/3}$ for $x \neq 0$

d) $\frac{d}{dx}(1 - x^2)^{-1/2} = -\frac{1}{2}(1 - x^2)^{-3/2}(-2x)$

$$= \frac{x}{(1 - x^2)^{3/2}} \qquad \text{for } |x| < 1 \ \square$$

Proof of Rule 9. To establish Rule 9, let

$$y = u^{p/q},$$

which means that

$$y^q = u^p.$$

Then, differentiating both sides of the equation implicitly and using the familiar formulas for the derivatives of y^q and u^p (since p and q are integers, these formulas are valid), we obtain

$$qy^{q-1}\frac{dy}{dx} = pu^{p-1}\frac{du}{dx}.$$

Hence if $y \neq 0$, we have

$$\frac{dy}{dx} = \frac{pu^{p-1}}{qy^{q-1}}\frac{du}{dx}. \tag{15}$$

But

$$y^{q-1} = (u^{p/q})^{q-1} = u^{p-(p/q)},$$

so that

$$\frac{dy}{dx} = \frac{p}{q}\frac{u^{p-1}}{u^{p-(p/q)}}\frac{du}{dx}$$

$$= \frac{p}{q}u^{(p/q)-1}\frac{du}{dx}.$$

This establishes Eq. (14). ■

The restriction $y \neq 0$ in Eq. (15) is the same as the restriction $u \neq 0$ but was made without reference to whether p/q was or was not less than one. The restriction is not needed if $p/q = 1$, since we then have $y = u$ and $dy/dx = du/dx$.

The case $y = x^{3/2}$ is typical of the case

$$y = u^{p/q}, \qquad p/q > 1,$$

and in the following example we shall show that the slope of this curve is 0 when $x = y = 0$.

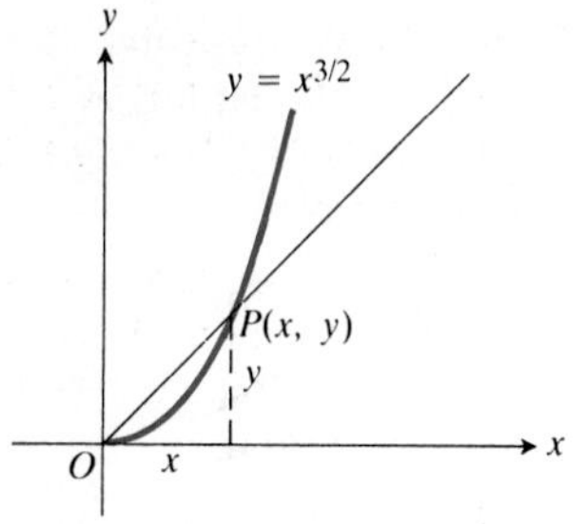

2.12 The graph of $y = x^{3/2}$. The slope at $x = 0$ is $\lim m_{OP} = 0$.

EXAMPLE 9   Investigate the slope of the curve $y = x^{3/2}$ at the origin.

Solution   For $y = x^{3/2}$, Eq. (14) gives

$$\frac{dy}{dx} = \frac{3}{2}x^{1/2} = \frac{3}{2}\sqrt{x},$$

which is zero when $x = 0$. But the graph of $y = x^{3/2}$ stops abruptly at $x = 0$ (Fig. 2.12), while derivatives as we know them are supposed to be calculated as two-sided limits.

We may, however, define the slope at $(0, 0)$ as the limit of slopes of secants drawn through $(0, 0)$ and a point $P(x, y)$ on the curve to the right of $(0, 0)$. (See Fig. 2.12.) The slope of OP is

$$m = \frac{\text{rise}}{\text{run}} = \frac{y - 0}{x - 0} = \frac{x^{3/2}}{x} = x^{1/2}.$$

As $P \to (0, 0)$, we have

$$m_0 = \lim_{x \to 0^+} m = \lim_{x \to 0^+} x^{1/2} = 0,$$

in agreement with the earlier result from Eq. (14).

We shall call this limit the derivative of $y = x^{3/2}$ at $x = 0$ and say that the graph has a horizontal tangent there. $\square$

Differentiability at an Endpoint

The right-hand limit calculated in Example 9 is a special case of the limit

$$\lim_{\Delta x \to 0^+} \frac{f(x + \Delta x) - f(x)}{\Delta x},$$

which is called the *right-hand derivative* of f at x when it exists. Similarly, the limit

$$\lim_{\Delta x \to 0^-} \frac{f(x + \Delta x) - f(x)}{\Delta x},$$

when it exists, is called the *left-hand derivative* of f at x. These two definitions allow us to extend the notion of differentiability to the endpoints of a function's domain. We say that a function f defined on a closed interval $[a, b]$ is differentiable at a if its right-hand derivative exists at a. We say that f is differentiable at b if its left-hand derivative exists at b. For f to be differentiable at any point between a and b, however, the ordinary "two-sided" derivative

$$f'(x) = \lim_{\Delta x \to 0} \frac{f(x + \Delta x) - f(x)}{\Delta x}$$

must exist as usual.

$$2x + 2y\left(\frac{dy}{dx}\right) = \emptyset$$

$$\frac{dy}{dx} = \frac{-2x}{2y} = -\frac{x}{y}$$

PROBLEMS

In Problems 1–29, find dy/dx.

1. $x^2 + y^2 = 1$

2. $y^2 = \dfrac{x - 1}{x + 1}$

3. $x^2 + xy = 2$

4. $x^2 y + xy^2 = 6$

5. $y^2 = x^3$

6. $x^{2/3} + y^{2/3} = 1$

7. $x^{1/2} + y^{1/2} = 1$

8. $x^3 - xy + y^3 = 1$

9. $x^2 = \dfrac{x - y}{x + y}$

10. $y = \dfrac{x}{\sqrt{x^2 + 1}}$

11. $y = x\sqrt{x^2 + 1}$

12. $y^2 = x^2 + \dfrac{1}{x^2}$

13. $2xy + y^2 = x + y$

14. $y = \sqrt{x} + \sqrt[3]{x} + \sqrt[4]{x}$

15. $y^2 = \dfrac{x^2 - 1}{x^2 + 1}$

16. $(x + y)^3 + (x - y)^3 = x^4 + y^4$

17. $(3x + 7)^5 = 2y^3$

18. $y = (x + 5)^4(x^2 - 2)^3$

19. $\dfrac{1}{y} + \dfrac{1}{x} = 1$

20. $y = (x^2 + 5x)^3$

21. $y^2 = x^2 - x$

22. $x^2y^2 = x^2 + y^2$

23. $y = \dfrac{\sqrt[3]{x^2 + 3}}{x}$

24. $y = x^2\sqrt{1 - x^2}$

25. $x^3 + y^3 = 18xy$

26. $y = (3x^2 + 5x + 1)^{3/2}$

27. $y = (2x + 5)^{-1/5}$

28. $y = 3(2x^{-1/2} + 1)^{-1/3}$

29. $y = \sqrt{1 - \sqrt{x}}$

30. Find dT/dD if $T^2 = 1.66 \times 10^{-19}D^3$. (This equation expresses Kepler's third law for the solar system.)

31. Find dT/dL if $T^2 = 4\pi^2L/g$. (This equation gives the period T of a simple pendulum in terms of its length, L, and the acceleration of gravity, g.)

32. Find the x- and y-intercepts of the line tangent to the curve $y = x^{1/2}$ at $x = 4$.

33. Which of the following may be true if $f''(x) = x^{-1/3}$?

a) $f(x) = \dfrac{3}{2}x^{2/3} - 3$ b) $f(x) = \dfrac{9}{10}x^{5/3} - 7$

c) $f'''(x) = -\dfrac{1}{3}x^{-4/3}$ d) $f'(x) = \dfrac{3}{2}x^{2/3} + 6$

34. Find db/da if $b = a^{2/3}$. What restrictions, if any, should be put on the domain of a?

35. a) By differentiating the equation $x^2 - y^2 = 1$ implicitly, show that $dy/dx = x/y$.

b) By differentiating both sides of the equation $dy/dx = x/y$ implicitly, show that

$$\frac{d^2y}{dx^2} = \frac{y - x\left(\dfrac{dy}{dx}\right)}{y^2} = \frac{y - \dfrac{x^2}{y}}{y^2} = \frac{y^2 - x^2}{y^3},$$

or, since $y^2 - x^2 = -1$ from the original equation,

$$\frac{d^2y}{dx^2} = \frac{-1}{y^3}.$$

In Problems 36–39, find dy/dx and then d^2y/dx^2 by implicit differentiation.

36. $x^2 + y^2 = 1$

37. $x^3 + y^3 = 1$

38. $x^{2/3} + y^{2/3} = 1$

39. $xy + y^2 = 1$

40. The position function of a particle moving on a coordinate line is given by $s(t) = \sqrt{1 + 4t}$, with s in meters and t in seconds. Find the particle's velocity and acceleration when $t = 6$ seconds.

41. A particle of mass m moves along the x-axis. The velocity $v = dx/dt$ and position x satisfy the equation

$$m(v^2 - v_0^2) = k(x_0^2 - x^2),$$

where k, v_0, and x_0 are constants. Show, by implicitly differentiating this equation with respect to t, that whenever $v \neq 0$,

$$m\frac{dv}{dt} = -kx.$$

In Problems 42–48, find the lines that are (a) tangent and (b) normal to the curve at the given point.

42. $x^2 + xy - y^2 = 1$, $P(2, 3)$

43. $x^2 + y^2 = 25$, $P(3, -4)$

44. $x^2y^2 = 9$, $P(-1, 3)$

45. $\dfrac{x - y}{x - 2y} = 2$, $P(3, 1)$

46. $(y - x)^2 = 2x + 4$, $P(6, 2)$

47. $y^2 - 2x - 4y - 1 = 0$, $P(-2, 1)$

48. $xy + 2x - 5y = 2$, $P(3, 2)$

49. Find the normals to the curve $xy + 2x - y = 0$ that are parallel to the line $2x + y = 0$.

50. If three normals can be drawn from the point $A(a, 0)$ to the curve $y^2 = x$, show that a must be greater than $\frac{1}{2}$. One normal is always the x-axis. Find the value of a for which the other two are perpendicular.

51. The line that is normal to the curve $y = x^2 + 2x - 3$ at $(1, 0)$ intersects the curve at what other point?

52. Show that the normal to the circle $x^2 + y^2 = a^2$ at any point (x_1, y_1) passes through the origin.

53. Find the two points where the curve $x^2 + xy + y^2 = 7$ crosses the x-axis, and show that the tangents to the curve at these points are parallel. What is the common slope of these tangents?

54. Find points on the curve $x^2 + xy + y^2 = 7$ (a) where the tangent is parallel to the x-axis; (b) where the tangent is parallel to the y-axis. (In the latter case, dy/dx is not defined, but dx/dy is. What value does dx/dy have at these points?)

55. Use the quadratic formula to solve $x^2 + xy + y^2 - 7 = 0$ for y in terms of x. Then find dy/dx at $P(1, 2)$ directly from an expression of the form $y = f(x)$.

35b 49 51

56. The function

$$f(x) = x^{1/2} + x^{-1/2}$$

is to be approximated near $x = 4$ by a quadratic function

$$Q(x) = C_0 + C_1(x - 4) + C_2(x - 4)^2$$

by choosing coefficients C_0, C_1, and C_2 in such a way that $f(4) = Q(4)$, $f'(4) = Q'(4)$, and $f''(4) = Q''(4)$. Find the coefficients. Use the result to estimate the difference between $f(x)$ and the tangent-line approximation $L(x) = C_0 + C_1(x - 4)$, and compare this estimate of error with the actual error at $x = 4.41 = (2.1)^2$. (We use "error" to mean $f(x) - L(x)$.)

> ***Toolkit* Programs**
> Derivative Grapher

2.5
Tangent Line Approximations

In science and engineering we can sometimes approximate complicated functions with simpler functions that still give enough accuracy to carry out the work at hand. It is important to know how this is done, and in this article we study the simplest of the useful approximations, the tangent line approximation. (Approximations are also important in mathematics because many proofs depend on being able to make a manageable approximation that can be shown to come arbitrarily close to the quantity under study.)

As you can see from the enlargements in Fig. 2.13, the tangent to a curve lies close to the curve near the point of tangency P. In fact, the closer you get to P (see the last enlargement), the more the curve looks like the tangent line.

Since a straight line is simpler than other curves, and since the tangent to a differentiable curve always runs close to the curve near the point of tangency, the tangent line can provide a useful approximation to the curve.

To make this important idea applicable, consider Fig. 2.14, which shows a function $y = f(x)$ near a point $x = a$ where the function has a derivative. Suppose we know the value of y that corresponds to $x = a$ (so

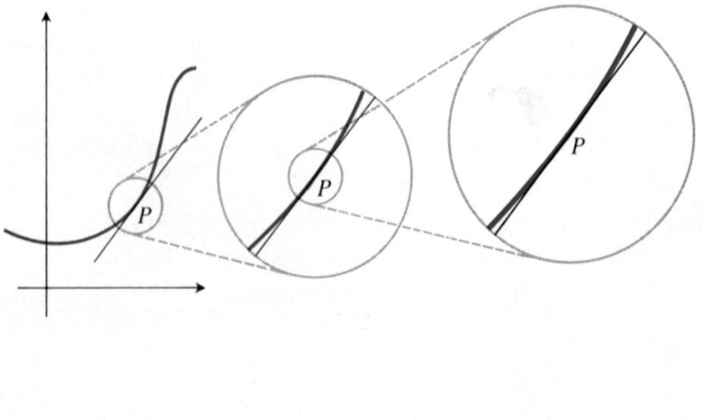

2.13 Successive enlargements of a differentiable curve near a point of tangency.

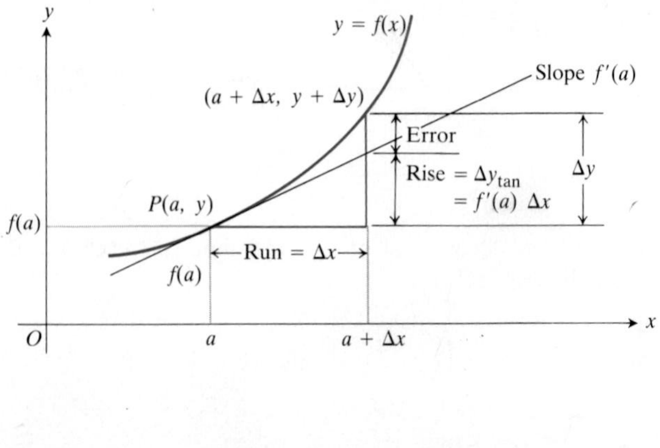

2.14 If we know $f'(a)$, $f'(a)$, and Δx, we can estimate Δy from the tangent line increment $\Delta y_{\tan}$.

we know the point P), and that we change x by an amount Δx. How can we get a simple estimate of the new value of y?

The new x-coordinate is $a + \Delta x$, while the new y-coordinate (the value we wish to approximate) is $f(a + \Delta x) = y + \Delta y$. Since we know y, all we need to do to estimate $f(x + \Delta x)$ is to approximate Δy. As you can see in Fig. 2.14,

$$\Delta y \approx \Delta y_{\tan} \tag{1}$$

and

$$\frac{\Delta y_{\tan}}{\Delta x} = \text{slope of tangent line} = f'(a). \tag{2}$$

Therefore,

$$\Delta y_{\tan} = f'(a)\,\Delta x. \tag{3}$$

Consequently, to approximate the new value of y, we add the tangent line increment $f'(a)\,\Delta x$ to the old value of y:

$$\underbrace{f(a + \Delta x)}_{\text{new } y} \approx \underbrace{f(a)}_{\text{old } y} + \underbrace{f'(a)\,\Delta x.}_{\Delta y_{\tan}} \tag{4}$$

The Tangent Line Increment $\Delta y_{\tan}$ Approximates Δy

If $y = f(x)$ is differentiable at $x = a$, and x changes from a to $a + \Delta x$, then

$$\Delta y \approx \Delta y_{\tan} = f'(a)\,\Delta x \tag{5}$$

and

$$\text{New } y = f(a) + \Delta y \approx f(a) + f'(a)\,\Delta x. \tag{6}$$

The Approximation Error

How good is the approximation? We measure the error by subtracting the approximation from the exact answer:

$$\begin{aligned}\text{Error} = \underbrace{f(a + \Delta x)}_{\text{exact}} - \underbrace{(f(a) + f'(a)\,\Delta x)}_{\text{approximation}} &= f(a + \Delta x) - f(a) - f'(a)\,\Delta x \\ &= \underbrace{\left[\frac{f(a + \Delta x) - f(a)}{\Delta x} - f'(a)\right]}_{\text{Call this } \varepsilon.} \Delta x \\ &= \varepsilon \cdot \Delta x. \end{aligned} \tag{7}$$

Now, as $\Delta x \to 0$, the quantity

$$\frac{f(a + \Delta x) - f(a)}{\Delta x}$$

approaches $f'(a)$ (remember the definition of $f'(a)$) so the quantity in brackets becomes a very small number ε. In fact,

$$\varepsilon \to 0 \quad \text{as} \quad \Delta x \to 0.$$

Thus, when Δx is small, the error $\varepsilon \cdot \Delta x$ is even smaller than Δx.

Rearranging the terms in Eq. (7) gives

$$\Delta y = f'(a)\,\Delta x + \varepsilon \cdot \Delta x \qquad (8)$$

(an equation we shall use later, in Article 2.6) and the following equation, which fits together all the pieces we have been talking about:

$$\underbrace{f(a + \Delta x)}_{\text{new } y} = \underbrace{f(a)}_{\text{old } y} + \overbrace{\underbrace{f'(a)\,\Delta x}_{\Delta y_{\tan}} + \underbrace{\varepsilon \cdot \Delta x.}_{\text{error}}}^{\Delta y} \qquad (9)$$

EXAMPLE 1 If the radius of a circle is increased from $r = a$ to $r = a + \Delta r$, estimate the increase in the area $A = \pi r^2$. Then calculate the increase directly and compare the results with Eq. (8).

Solution To estimate ΔA we use the tangent line increment (Eq. 5):

$$\Delta A \approx \Delta A_{\tan} = \left(\frac{dA}{dr}\bigg|_{r=a}\right) \cdot \Delta r = 2\pi a\,\Delta r.$$

The direct calculation of ΔA is

$$\begin{aligned}\Delta A &= \pi(a + \Delta r)^2 - \pi a^2 \\ &= \pi a^2 + 2\pi a\,\Delta r + \pi(\Delta r)^2 - \pi a^2 \\ &= \underbrace{2\pi a\,\Delta r}_{\frac{dA}{dr}\cdot\Delta r} + \underbrace{(\pi\,\Delta r)(\Delta r).}_{\varepsilon\cdot\Delta r}\end{aligned}$$

If $a = 7$ cm and $\Delta r = 0.1$ cm, then

$$\Delta A = \underbrace{2\pi(7)(0.1)}_{\text{approximation}} + \underbrace{\pi(0.1)(0.1)}_{\text{error } = \varepsilon\cdot\Delta r} = 1.4\pi + 0.01\pi = 1.41\pi.$$

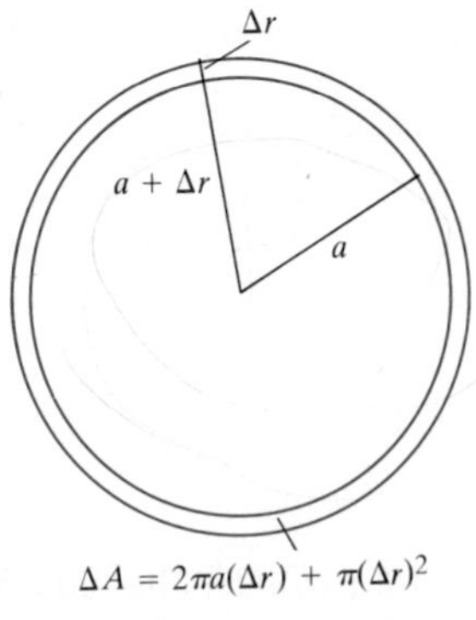

2.15 When Δr is small, the contribution of $\pi(\Delta r)^2$ to ΔA can be neglected and ΔA calculated from the approximation $\Delta A \approx 2\pi a\,\Delta r$.

Geometrically, $2\pi a\,\Delta r$ is the product of the circumference of the original circle with the radius change Δr (Fig. 2.15). When Δr is small, this accounts for nearly all the change in area. The remaining bit, $(\pi\,\Delta r)(\Delta r)$, is the product of two numbers that are both small when Δr is small. □

EXAMPLE 2 An edge of a cube is measured as 6 inches with a possible error of 0.05 inch. The volume of the cube is to be calculated from this measurement. Estimate the possible error in the volume calculation. Express the estimated error as a percent of the calculated volume.

Solution If x denotes the length of the edge, the volume is $V = x^3$. We have a measured value of 6 in. for x, with an error of

$$|\Delta x| \le 0.05 \text{ in.}$$

According to Eq. (5), the variation in volume ΔV caused by an increment Δx in x is about equal to the tangent line approximation:

$$\Delta V \approx \left(\left.\frac{dV}{dx}\right|_{x=6}\right) \cdot \Delta x = (3x^2|_{x=6}) \cdot \Delta x = 108\,\Delta x.$$

If $|\Delta x| \leq 0.05$, then

$$|\Delta V| \approx 108|\Delta x| \leq 108(0.05) = 5.4 \text{ in}^3.$$

The volume calculated might differ from the true volume by as much as 5.4 in^3 either way. As a percent of the calculated volume, the error is

$$\frac{\text{Error}}{V} \times 100 = \frac{5.4 \times 100}{6^3} = \frac{540}{216} = 2.5\%. \;\square$$

EXAMPLE 3 Suppose that the earth were a perfect sphere and that we determined its radius to be 3939 ± 0.1 miles. What effect would the tolerance ± 0.1 have on our estimate of the earth's surface area?

Solution Since $S = 4\pi r^2$, $S' = 8\pi r$ and the tangent line increment in S that corresponds to $\Delta r = 0.1$ is

$$\Delta S_{\tan} = 8\pi r\,\Delta r = 8\pi(3959)(0.1) = 9950 \text{ square miles}$$

(to the nearest square mile).

In geographical terms we see that an error equal to the area of the state of Maryland can result from only a 0.1-mile error (= 528 feet) in the earth's radius. However, 9950 mi^2 is a relatively small error (only 0.005%) when compared to the calculated surface area of the earth,

$$4\pi(3959)^2 = 196{,}961{,}284 \text{ mi}^2. \;\square$$

EXAMPLE 4 About how accurately should we measure the radius r of a sphere to calculate the surface area $S = 4\pi r^2$ within 1% of its true value?

Solution We want any inaccuracy Δr in our measurement to be small enough to make the corresponding increment ΔS in surface area satisfy the inequality

$$|\Delta S| \leq \frac{1}{100} S = \frac{1}{100} 4\pi r^2. \tag{10}$$

The tangent line approximation of ΔS (Eq. 5, with $dS/dr = 8\pi r$) is

$$\Delta S \approx 8\pi r\,\Delta r.$$

Combining this approximation with Eq. (10) gives us

$$|8\pi r\,\Delta r| \leq \frac{4\pi r^2}{100}$$

$$|\Delta r| \leq \frac{r}{200} = \frac{1}{2}\frac{r}{100}.$$

We must measure the radius with an error that is no more than one half of one percent of the true value. $\square$

Linear Approximations of Functions

The tangent to the curve $y = f(x)$ at $x = a$ is the line through the point $(a, f(a))$ whose slope is $f'(a)$. Its point-slope equation is

$$y = f(a) + f'(a)(x - a). \tag{11}$$

If we label the graph of $y = f(x)$ as in Fig. 2.16, substituting x for $a + \Delta x$ and $x - a$ for Δx in Fig. 2.14, Eq. (9) takes the following form:

$$f(x) = \underbrace{f(a) + f'(a)(x - a)}_{\substack{\text{tangent line} \\ \text{approximation} \\ \text{of } f(x)}} + \underbrace{\varepsilon \cdot (x - a)}_{\substack{\text{error} \\ \text{term}}} \tag{12}$$

and $\varepsilon \to 0$ as $x \to a$.

Equation (12) says that if f has a derivative at $x = a$, then we can approximate the values of $f(x)$ over an entire interval about a by the values of $y = f(a) + f'(a)(x - a)$ with an error of $\varepsilon \cdot (x - a)$. We call this approximation the *linear approximation* of f near a.

Linear Approximation of $f(x)$ near $x = a$

If $y = f(x)$ is differentiable at $x = a$, then

$$f(x) \approx f(a) + f'(a)(x - a) \tag{13}$$

for values of x near a.

This approximation is easy to compute from the values of $f(a)$ and $f'(a)$, and its error can be significantly smaller than $\Delta x = x - a$ when x is close to a. In Article 3.10 we shall find out how to estimate this error from the values of the second derivative of f.

EXAMPLE 5 Find the linear approximation to $f(x) = \sqrt{1 + x}$ near $x = 0$. Use it to estimate $\sqrt{1.2}$, $\sqrt{1.05}$, and $\sqrt{1.005}$.

Solution We evaluate Eq. (13) for $f(x) = \sqrt{1 + x}$ and $a = 0$.

The derivative of f is

$$f'(x) = \frac{1}{2}(1 + x)^{-1/2} = \frac{1}{2\sqrt{1 + x}}.$$

Its value at $x = 0$ is

$$f'(0) = \frac{1}{2\sqrt{1 + 0}} = \frac{1}{2}.$$

We substitute $f(a) = 1$, $f'(0) = \frac{1}{2}$, and $a = 0$ into Eq. (13) to get

$$\sqrt{1 + x} \approx 1 + \frac{x}{2}.$$

This is the linear approximation to $\sqrt{1 + x}$ at $x = 0$ (Fig. 2.17).

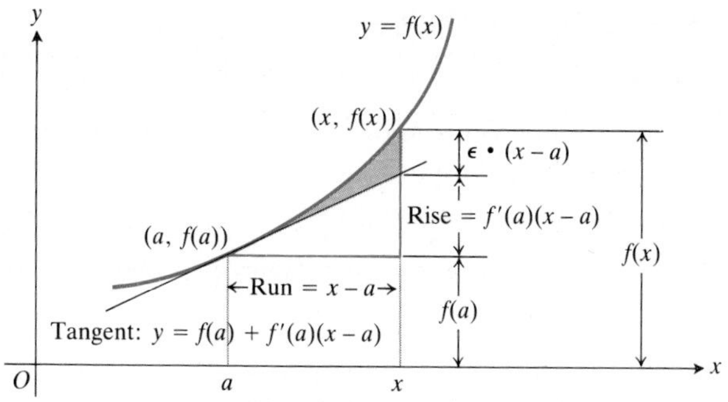

2.16 For values of x close to a, the sum $y = f(a) + f'(a)(x - a)$ accounts for nearly all of $f(x)$.

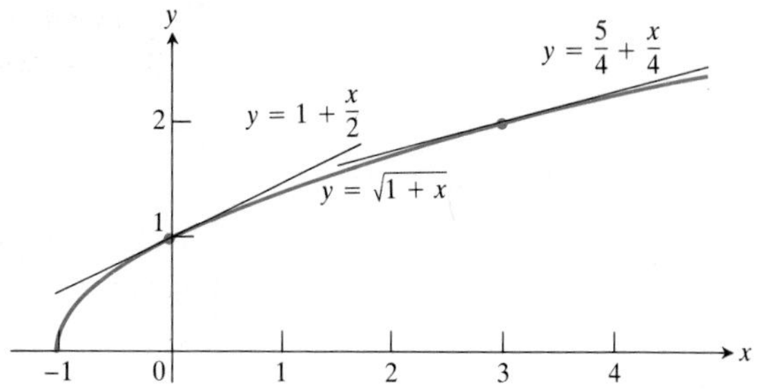

2.17 The graph of $y = \sqrt{1 + x}$ and its tangents at $x = 0$ and $x = 3$.

When $x = 0.2$, 0.05, and 0.005, this approximation gives

$$\sqrt{1.2} \approx 1 + \frac{0.2}{2} = 1.10 \qquad \text{(accurate to 2 decimals)},$$

$$\sqrt{1.05} \approx 1 + \frac{0.05}{2} = 1.025 \qquad \text{(accurate to 3 decimals)},$$

$$\sqrt{1.005} \approx 1 + \frac{0.005}{2} = 1.00250 \quad \text{(accurate to 5 decimals)}. \ \square$$

EXAMPLE 6 Find the linear approximation to $f(x) = \sqrt{1 + x}$ near $x = 3$. Use it to estimate $\sqrt{4.2}$. What value does a calculator give for $\sqrt{4.2}$?

Solution We evaluate Eq. (13) for $f(x) = \sqrt{1 + x}$, $f'(x) = 1/(2\sqrt{1 + x})$, and $a = 3$. With

$$f(3) = 2, \qquad f'(3) = \frac{1}{2\sqrt{1 + 3}} = \frac{1}{4},$$

Eq. (13) gives

$$\sqrt{1 + x} \approx 2 + \frac{1}{4}(x - 3) = 2 + \frac{x}{4} - \frac{3}{4} = \frac{5}{4} + \frac{x}{4}.$$

The linear approximation near $x = 3$ is

$$\sqrt{1 + x} \approx \frac{5}{4} + \frac{x}{4}$$

(Fig. 2.14).

To estimate $\sqrt{4.2}$ we have

$$\sqrt{4.2} = \sqrt{1 + 3.2} \approx \frac{5}{4} + \frac{3.2}{4} = 1.25 + 0.8 = 2.05.$$

A calculator gives $\sqrt{4.2} \approx 2.04939$ to five places, which differs from the estimate of 2.05 by less than 1/1000. $\square$

The utility of the estimates in Examples 5 and 6 is not in calculating particular square roots. We can do that more quickly with a calculator. The utility lies in finding linear formulas that can replace $\sqrt{1 + x}$ over a *whole interval* with enough accuracy to be useful. We are not yet ready to control the accuracy of such replacements, but we shall be by the time we reach Article 3.10.

An important linear approximation, used in physics and engineering, is

$$(1 + x)^k \approx 1 + kx \quad \text{(any number } k\text{).} \tag{14}$$

(See Problem 48.) The approximation is good for values of x close to zero. For example, when x is small,

$$\sqrt{1 + x} \approx 1 + \frac{x}{2},$$

$$\sqrt[3]{1 + 5x^4} \approx 1 + \frac{1}{3}(5x^4) = 1 + \frac{5}{3}x^4,$$

$$\frac{1}{\sqrt{1 - x^2}} = (1 - x^2)^{-1/2}$$

$$\approx 1 + \left(-\frac{1}{2}\right)(-x^2) = 1 + \frac{1}{2}x^2. \tag{15}$$

REMARK. Here is an example of how Eq. (15) is used in an applied problem.

Newton's second law,

$$F = \frac{d}{dt}(mv) = m\frac{dv}{dt} = ma,$$

is stated with the assumption that mass is constant, but we know this is not strictly true because the mass of a body increases with velocity. In Einstein's corrected formula, mass has the value

$$m = \frac{m_0}{\sqrt{1 - v^2/c^2}}, \tag{16}$$

where the "rest mass" m_0 represents the mass of a body that is not moving and c is the speed of light, which is about 300,000 kilometers per second. When v is very small compared to c it is safe to use the approximation

$$\frac{1}{\sqrt{1 - v^2/c^2}} = 1 + \frac{1}{2}\left(\frac{v^2}{c^2}\right)$$

(which is Eq. 15 with $x = v/c$) to write

$$m = \frac{m_0}{\sqrt{1 - v^2/c^2}} = m_0\left(1 + \frac{1}{2}\left(\frac{v^2}{c^2}\right)\right) = m_0 + \frac{1}{2}m_0v^2\left(\frac{1}{c^2}\right),$$

or

$$m = m_0 + \frac{1}{2}m_0v^2\left(\frac{1}{c^2}\right). \tag{17}$$

Equation (17) expresses the increase in mass that results from the added velocity v.

In Newtonian physics, $\frac{1}{2}m_0v^2$ is the kinetic energy (KE) of the body, and if we rewrite (17) in the form

$$(m - m_0)c^2 = \frac{1}{2}m_0v^2,$$

we see that

$$(m - m_0)c^2 = \frac{1}{2}m_0v^2 = \frac{1}{2}m_0v^2 - \frac{1}{2}m_0(0)^2 = \Delta(\text{KE}),$$

or

$$(\Delta m)c^2 = \Delta(\text{KE}).$$

In other words, the change in kinetic energy $\Delta(\text{KE})$ in going from velocity 0 to velocity v equals $(\Delta m)c^2$.

Such energy changes ordinarily represent extremely small changes in mass. The energy released by a 20-kiloton atomic bomb, for instance, is the result of converting only one gram of mass into energy. The products of the explosion weigh only one gram less than the material exploded. It may help to put this in perspective if you remember that a penny weighs about 3 grams.

Table 2.1 lists three standard linear approximations for values of x near zero. The first was derived in Example 5. The remaining two are derived in Problems 47 and 48.

Table 2.1 Standard linear approximations, $x \approx 0$

1. $\sqrt{1 + x} \approx 1 + \frac{x}{2}$
2. $\frac{1}{1 - x} \approx 1 + x$
3. $(1 + x)^k \approx 1 + kx$ (any number k)

PROBLEMS

Each of Problems 1–6 gives a function $y = f(x)$, a point a, and an increment Δx. Find (a) the linear approximation of $f(x)$ near a; (b) $\Delta y_{\tan}$; (c) Δy; and (d) the difference, $\Delta y - \Delta y_{\tan}$. Then (e) graph the function and its linear approximation near $x = a$.

1. $f(x) = x^2 + 2x$, $a = 0$, $\Delta x = 0.1$
2. $f(x) = 2x^2 + 4x - 3$, $a = -1$, $\Delta x = 0.1$
3. $f(x) = x^3 - x$, $a = 1$, $\Delta x = 0.1$
4. $f(x) = x^4$, $a = 1$, $\Delta x = 0.1$
5. $f(x) = x^{-1}$, $a = 0.5$, $\Delta x = 0.1$
6. $f(x) = x^3 - 2x + 3$, $a = 2$, $\Delta x = 0.1$

In Problems 7–14, find the tangent line increments for the given changes in volume and surface area. (The formulas for volume and surface area may be found in Appendix 7.)

7. The volume of a sphere when the radius changes by an amount Δr.
8. The surface area of a sphere when the radius changes by an amount Δr.
9. The volume of a cube when the edge lengths change by an amount Δx.
10. The surface area of a cube when the edge lengths change by an amount Δx.
11. The volume of a right circular cylinder when the radius changes by an amount Δr and the height does not change.
12. The lateral surface area of a right circular cylinder when the height changes by an amount Δh and the radius does not change.
13. The volume of a right circular cone when the radius changes by an amount Δr and the height does not change.
14. The lateral surface area of a right circular cone when the height changes by an amount Δh and the radius does not change.

In Problems 15–20, find the linear approximation to the given function near the given point a. Then use it to estimate the given function value. If you have a calculator, compare your estimate with the calculator's value.

15. $f(x) = \sqrt{1+x}$, $a = 8$, $f(9.1)$
16. $f(x) = \sqrt{x}$, $a = 4$, $f(4.1)$
17. $f(x) = \sqrt[3]{x}$, $a = 8$, $f(8.5)$
18. $f(x) = x^{-1}$, $a = 2$, $f(2.1)$
19. $f(x) = \sqrt{x^2+9}$, $a = -4$, $f(-4.2)$
20. $f(x) = \dfrac{x}{x+1}$, $a = 1$, $f(1.3)$

21. The radius of a circle is increased from 2.00 to 2.02 meters.
 a) Estimate the change in area.
 b) Calculate the error in the estimate in (a) as a percent of the original area.

22. The edge of a cube is measured as 10 cm with a possible error of one percent. The cube's volume is to be calculated from this measurement. About how much error is possible in the volume calculation?

23. The diameter of a sphere is measured as 100 ± 1 cm and the volume is calculated from this measurement. Estimate the possible error in the calculation.

24. About how accurately should you measure the side of a square to be sure of calculating the area within two percent of its true value?

25. a) About how accurately should the edge of a cube be measured to be sure of calculating the cube's surface area with an error of no more than two percent?
 b) How accurately can the cube's volume be calculated from the edge measurement in (a)? That is, what is the relative error (error as a percent of the true volume)?

26. The height and radius of a right circular cone are equal, so that the volume of the cone is $V = (\frac{1}{3})\pi h^3$. The volume is to be calculated from a measurement of h, and must be calculated with an error of no more than one percent. Find approximately the greatest error that can be tolerated in the measurement of h, expressed as a percent of h.

27. The circumference of a great circle of a sphere is measured as 10 cm, with a possible error of 0.4 cm. The measurement is then used to calculate the radius. The radius is then used to calculate the surface area and volume of the sphere. Find the possible errors in the calculated value of (a) the radius, (b) the surface area, and (c) the volume.

28. Find the allowable percentage error in measuring the diameter d of a sphere if the volume is to be calculated correctly to within three percent. (The percentage error is the absolute error $|\Delta d|$ expressed as a percent of d.)

29. a) About how accurately must the interior diameter of a 10-meter-high cylindrical storage tank be measured to calculate the tank's volume to within one percent of its true value?
 b) About how accurately must the tank's exterior diameter be measured to calculate the amount of paint it will take to paint the side of the tank within five percent of the true amount?

30. A manufacturer contracts to mint coins for the federal government. How much variation Δr in the radius of the coins can be tolerated in manufacture if the coins are to weigh within 1/1000 of their ideal weight? Assume that the thickness does not vary.

31. The period T of a clock pendulum (time for one full swing and back) is given by the formula $T^2 = 4\pi^2 L/g$, where T is measured in seconds, $g = 32.2$ ft/s^2, and L, the length of the pendulum, is measured in feet. Find approximately:
 a) the length of a clock pendulum whose period is $T = 1$ second;
 b) the change ΔT in T if the pendulum in (a) is lengthened 0.01 ft; and
 c) the amount the clock gains or loses in a day as a result of the period's changing by the amount ΔT found in (b).

32. The volume $y = x^3$ of a cube of edge x increases by an amount Δy when x increases by an amount Δx. Show that Δy can be represented geometrically as the sum of the volumes of:
 a) three slabs of dimensions x by x by Δx,
 b) three bars of dimensions x by Δx by Δx,
 c) one cube of dimensions Δx by Δx by Δx.
 Illustrate with a sketch.

Use the formulas in Table 2.1 to find linear approximations of the following functions for small values of x.

33. $(1+x)^2$
34. $(1+x)^3$
35. $\dfrac{1}{(1+x)^5}$
36. $\dfrac{4}{(1+x)^2}$
37. $\dfrac{2}{(1-x)^4}$
38. $(1-x)^6$
39. $2\sqrt{1+x}$
40. $3(1+x)^{1/3}$
41. $\dfrac{1}{1+x}$
42. $\dfrac{1}{\sqrt{1+x}}$

43. Use the standard linear approximations from Table 2.1 to calculate
 a) $(1.0002)^{100}$, b) $\sqrt[3]{1.009}$, c) $1/(0.999)$.

44. (*Calculator*) a) Test the linear approximation

$$\sqrt{1+x} \approx 1 + x/2$$

for $x = \pm 1, \pm 0.1, \pm 0.01$, and ± 0.001.
 b) Repeat (a) for the linear approximation

$$(1+x)^4 \approx 1 + 4x.$$

45. Let $g(x) = \sqrt{x} + \sqrt{1+x} - 4$.

a) Find $g(3) < 0$ and $g(4) > 0$ to show (by the Intermediate Value Theorem, Article 1.11) that the equation $g(x) = 0$ has a solution between $x = 3$ and $x = 4$.

b) To estimate the solution of $g(x) = 0$, replace the square roots by their linear approximations near $x = 3$ and solve the resulting linear equation.

c) (*Calculator*) Check your estimate in the original equation.

46. Let

$$f(x) = \frac{2}{1-x} + \sqrt{1+x} - 3.1.$$

a) Find $f(0) < 0$ and $f(\frac{1}{2}) > 0$ to show (by the Intermediate Value Theorem, Article 1.11) that the equation $f(x) = 0$ has a solution between $x = 0$ and $x = \frac{1}{2}$.

b) To estimate the solution of $f(x) = 0$, replace $2/(1-x)$ and $\sqrt{1+x}$ by their standard linear approximations near $x = 0$ and solve the resulting linear equation.

c) (*Calculator*) Check your estimate in the original equation.

47. Show that the first two formulas in Table 2.1 are special cases of the third formula.

48. We know that the Power Rule,

$$\frac{d}{dx}(1+x)^k = k(1+x)^{k-1},$$

holds for any rational number k. In Chapter 6 we shall show that the rule holds for any irrational number k as well. Assuming this result for now, verify the third formula in Table 2.1 by showing that the linear approximation of $f(x) = (1+x)^k$ near $a = 0$ is $(1+x)^k \approx 1 + kx$.

49. To what relative speed should a body at rest be accelerated to increase its mass by 1%?

2.6 The Chain Rule and Parametric Equations

The rule for calculating the derivative of the composite of two differentiable functions is, roughly speaking, that the derivative of their composite is the product of their derivatives. This rule is called the *Chain Rule*. In this article we shall see why the Chain Rule works and study its application to equations of motion.

EXAMPLE 1 In the gear train shown in Fig. 2.18, the ratios of the radii of gears A, B, and C are 3:1:2. If A turns t times, then B turns $x = 3t$ times and C turns $x/2 = (3/2)t$ times. In terms of derivatives,

$$\frac{dx}{dt} = 3$$

(B turns at three times A's rate, three turns for A's one),

$$\frac{dy}{dx} = \frac{1}{2}$$

(C turns at half B's rate, one half turn for each turn of B), and

$$\frac{dy}{dt} = \frac{3}{2} = \frac{1}{2} \cdot 3 = \frac{dy}{dx}\frac{dx}{dt}$$

(C turns at three-halves A's rate, three halves of a turn for A's one). To calculate dy/dt we can multiply dy/dx by dx/dt. □

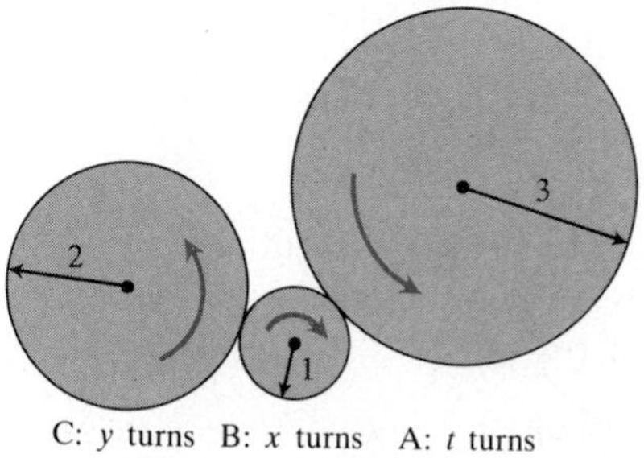

C: y turns B: x turns A: t turns

2.18 If wheel A takes t turns, then wheel B takes x turns and wheel C takes y turns. By comparing circumferences we see that $dy/dx = 1/2$ and $dx/dt = 3$. What is dy/dt?

EXAMPLE 2 A particle moves along the line $y = 5x - 2$ in such a way that its x-coordinate at time t is $x = 3t$. Calculate dy/dt.

Solution As a function of t,

$$y = 5x - 2 = 5(3t) - 2 = 15t - 2.$$

Therefore,

$$\frac{dy}{dt} = \frac{d}{dt}(15t - 2) = 15.$$

Note that

$$\frac{dy}{dx} = 5, \qquad \frac{dx}{dt} = 3,$$

and

$$\frac{dy}{dt} = 3 \cdot 5 = \frac{dy}{dx}\frac{dx}{dt}. \quad \square$$

EXAMPLE 3 The function $y = (5t + 1)^2$ is the composite of $y = x^2$ and $x = 5t + 1$. The rule for differentiating powers gives

$$\frac{dy}{dt} = \frac{d}{dt}(5t + 1)^2 = 2(5t + 1) \cdot \frac{d}{dt}(5t + 1) = 2(5t + 1) \cdot 5.$$

The expression on the right is the product of

$$\frac{dy}{dx} = 2x = 2(5t + 1) \qquad \text{and} \qquad \frac{dx}{dt} = \frac{d}{dt}(5t + 1) = 5.$$

Once again,

$$\frac{dy}{dt} = \frac{dy}{dx} \cdot \frac{dx}{dt}. \quad \square$$

The Chain Rule

The examples above all work because the derivative of a composite $g \circ f$ of two differentiable functions is the product of their derivatives. This is the observation we wish to state formally as the Chain Rule. (As in Article 1.5, the notation $g \circ f$ denotes the composite of the functions f and g with g following f. That is, the value of $g \circ f$ at x is $(g \circ f)(x) = g(f(x))$.)

RULE 10

The Chain Rule

Suppose that $h = g \circ f$ is the composite of the differentiable functions $y = g(x)$ and $x = f(t)$. Then h is a differentiable function of t whose derivative at each value of t is

$$h'(t) = g'(f(t)) \cdot f'(t). \tag{1}$$

Equation (1) says that to calculate $h'(t)$ we multiply the values of f' at t and g' at $f(t)$. See Fig. 2.19.

The virtue of the formulation of Eq. (1) is that it tells how each derivative is to be evaluated. Another formulation of the Chain Rule, one that

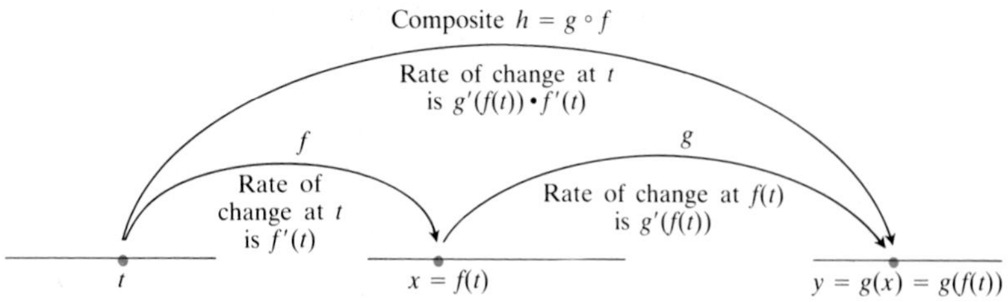

2.19 Rates of change multiply: The derivative of $h = g \circ f$ at t is the derivative of f at t times the derivative of g at $f(t)$.

gives less information but is sometimes easier to remember, is

$$\frac{dy}{dt} = \frac{dy}{dx}\frac{dx}{dt}. \tag{1a}$$

Still another is

$$\left.\frac{dy}{dt}\right|_t = \left.\frac{dy}{dx}\right|_{f(t)} \left.\frac{dx}{dt}\right|_t. \tag{1b}$$

Equation (1a) is particularly memorable because it suggests thinking of the derivatives as fractions with the dx's canceling on the right to produce the fraction on the left. As an aid to the memory this is fine. It makes no sense mathematically because the derivatives are not fractions. (We shall have more to say about this in Article 4.8 when we introduce the notion of differential.)

Proof of the Chain Rule. The idea behind the Chain Rule is this: If $x = f(t)$ is differentiable at t_0, then an increment Δt produces an increment Δx such that

$$\Delta x = f'(t_0)\,\Delta t + \varepsilon_1\,\Delta t = [f'(t_0) + \varepsilon_1]\,\Delta t; \tag{2a}$$

and if $y = g(x)$ is differentiable at $x_0 = f(t_0)$, then an increment Δx produces an increment Δy such that

$$\Delta y = g'(x_0)\,\Delta x + \varepsilon_2\,\Delta x = [g'(x_0) + \varepsilon_2]\,\Delta x. \tag{2b}$$

Equations (2a) and (2b) are versions of Eq. (8) in Article 2.5, relating the increments Δx and Δy to their tangent line approximations. In these equations, $\varepsilon_1 \to 0$ when $\Delta t \to 0$, and $\varepsilon_2 \to 0$ when $\Delta x \to 0$.

Combining (2a) and (2b) gives us

$$\Delta y = [g'(x_0) + \varepsilon_2]\,\Delta x = [g'(x_0) + \varepsilon_2][f'(t_0) + \varepsilon_1]\,\Delta t. \tag{3}$$

Dividing through by Δt gives

$$\begin{aligned}\frac{\Delta y}{\Delta t} &= [g'(x_0) + \varepsilon_2][f'(t_0) + \varepsilon_1] \qquad (4)\\ &= g'(x_0)f'(t_0) + \varepsilon_2 f'(t_0) + \varepsilon_1 g'(x_0) + \varepsilon_1\varepsilon_2.\end{aligned}$$

As Δt approaches zero, so do Δx, ε_1, and ε_2, and we get

$$\lim \frac{\Delta y}{\Delta t} = g'(x_0)f'(t_0),$$

which is the same as

$$\frac{dy}{dt}(t_0) = g'(f(t_0))f'(t_0). \quad ■ \tag{5}$$

EXAMPLE 4 Find dy/dt at $t = -1$ if

$$y = x^3 + 5x - 4 = g(x), \qquad \text{and} \qquad x = t^2 + t = f(t).$$

Solution From Eq. (1c) we have

$$\left.\frac{dy}{dt}\right|_{t=-1} = \left.\frac{dy}{dx}\right|_{x=f(-1)} \left.\frac{dx}{dt}\right|_{t=-1}$$
$$= (3x^2 + 5)_{x=0}(2t + 1)_{t=-1} = (5)(-1) = -5. \ \square$$

EXAMPLE 5 The familiar rule for differentiating powers of functions is a special case of the Chain Rule.

If u is a differentiable function of x, and $y = u^n$, then the Chain Rule in the form

$$\frac{dy}{dx} = \frac{dy}{du} \cdot \frac{du}{dx}$$

gives

$$\frac{dy}{dx} = \frac{d}{du}(u^n) \cdot \frac{du}{dx} = nu^{n-1}\frac{du}{dx}. \ \square$$

When using the Chain Rule, it sometimes helps to think in the following way. If $y = g(f(x))$, then

$$y' = g'(f(x)) \cdot f'(x)$$

says: Differentiate g, leaving everything "inside" (that is, $f(x)$) alone; then multiply by the derivative of the "inside." In practice, we then have

$$y = \underbrace{(x^3 + x^2)}_{\text{"inside" left alone}}{}^{13}$$

$$y' = \overbrace{13(x^3 + x^2)^{12}} \cdot \underbrace{(3x^2 + 2x)}_{\substack{\text{derivative}\\ \text{of the "inside"}}}.$$

EXAMPLE 6 *Snowball melting.* How long does it take a snowball to melt?

Discussion We start with a mathematical model. Let us assume that the snowball is, approximately, a sphere of radius r and volume $V = (\frac{4}{3})\pi r^3$. (Of course, the snowball is not a perfect sphere, but we can apply mathematics only to a mathematical model of the situation, and we choose one that seems reasonable and is not too complex.) In the same way, we choose some hypothesis about the rate at which the volume of the snowball is changing. One model is to assume that the volume decreases at a rate that is proportional to the surface area: In mathematical terms,

$$\frac{dV}{dt} = -k(4\pi r^2).$$

We tacitly assume that k, the proportionality factor, is a constant. (It probably depends on several things, like the relative humidity of the surrounding air, the air temperature, the incidence of sunlight or its absence, to name a few.) Finally, we need at least one more bit of information: How long has it taken for the snowball to melt some specific percent? We have nothing to guide us unless we make one or more observations, but let us now assume a particular set of conditions in which the snowball melted $\frac{1}{4}$ of its volume in two hours. (You could use letters instead of these precise numbers: say n percent in h hours. Then your answer would be in terms of n and h.) Now to work. Mathematically, we have the following problem.

Given:

$$V = \tfrac{4}{3}\pi r^3 \qquad \text{and} \qquad \frac{dV}{dt} = -k(4\pi r^2),$$

and $V = V_0$ when $t = 0$, and $V = \frac{3}{4}V_0$ when $t = 2$ hr.

To find: The value of t when $V = 0$.

We apply the Chain Rule to differentiate $V = (\frac{4}{3})\pi r^3$ with respect to t:

$$\frac{dV}{dt} = \frac{4}{3}\pi(3r^2)\frac{dr}{dt} = 4\pi r^2\frac{dr}{dt}.$$

We set this equal to the given rate, $-k(4\pi r^2)$, and divide by $4\pi r^2$ to get

$$\frac{dr}{dt} = -k.$$

The radius is *decreasing* at the constant rate of k radius units per hour. Thus, in t hours the radius decreases from r_0 to

$$r = r_0 - kt.$$

In two hours, r decreases from r_0 to

$$r_2 = r_0 - 2k,$$

so that

$$k = \frac{r_0 - r_2}{2}.$$

The melting time is the value of t that makes $r = 0$, or $kt = r_0$:

$$t_{\text{melt}} = \frac{r_0}{k} = \frac{2r_0}{r_0 - r_2} = \frac{2}{1 - r_2/r_0}.$$

But

$$\frac{r_2}{r_0} = \frac{\left(\dfrac{3}{4\pi}V_2\right)^{1/3}}{\left(\dfrac{3}{4\pi}V_0\right)^{1/3}} = \left(\frac{V_2}{V_0}\right)^{1/3} = \left(\frac{\dfrac{3}{4}V_0}{V_0}\right)^{1/3} = \left(\frac{3}{4}\right)^{1/3} \approx 0.91.$$

Therefore,

$$t_{\text{melt}} \approx \frac{2}{1 - 0.91} \approx 22 \text{ hr}.$$

If $\frac{1}{4}$ of the snowball melts in two hours, it takes nearly 20 hours for the rest of it to melt. □

REMARK. If we were natural scientists who were really interested in testing our model, we could collect some data and compare it with the results of the mathematics. One practical application may lie in analyzing the proposal to tow large icebergs from polar waters to offshore locations near Southern California to provide fresh water from the melting ice. As a first approximation, we might assume that the iceberg is a large cube, or a pyramid, or a sphere.

Parametric Equations

Instead of describing a curve by expressing the y-coordinate of a point $P(x, y)$ on the curve as a function of x, it is sometimes more convenient to describe the curve by expressing *both* coordinates as functions of a third variable, say,

$$x = f(t), \qquad y = g(t). \tag{6}$$

The equations in (6) are then called *parametric equations* for x and y, and the variable t is called a *parameter*. We might think of t as a variable that "controls" the values of x and y. In many applications, t denotes time, and the point $(x, y) = (f(t), g(t))$ is the position of a moving particle at time t.

Suppose now that x and y are differentiable functions of t, and that dx/dt is never zero on the interval I of t-values with which we are working. Then the equation $x = f(t)$ defines t implicitly as a differentiable function of x, say, $t = h(x)$. (We shall touch on the reason for this briefly in Articles 2.10 and 3.8, but the full explanation is too much to give here. It may be found in a text on advanced calculus or real variables under the heading "inverse function theorem.")

Accordingly, $y = g(t) = g(h(x))$ is also a differentiable function of x, because it is a composite of differentiable functions. The derivatives dy/dx, dy/dt, and dx/dt now all exist and, with $dx/dt \neq 0$, we may solve the equation

$$\frac{dy}{dt} = \frac{dy}{dx} \cdot \frac{dx}{dt}$$

(Eq. 1b) for dy/dx to obtain

$$\frac{dy}{dx} = \frac{dy/dt}{dx/dt}. \tag{7}$$

From this equation we can calculate the derivative dy/dx as a function of t from the derivatives dy/dt and dx/dt.

EXAMPLE 7 If

$$x = 2t + 3, \qquad y = t^2 - 1,$$

find the value of dy/dx at $t = 6$. Also, find dy/dx as a function of x.

Solution Equation (7) gives dy/dx as a function of t:

$$\frac{dy}{dx} = \frac{dy/dt}{dx/dt} = \frac{2t}{2} = t = \frac{x - 3}{2}.$$

When $t = 6$, $dy/dx = 6$. □

EXAMPLE 8 Sketch the path traced by the point $P(x, y)$ if

$$x = t + \frac{1}{t}, \qquad y = t - \frac{1}{t} \tag{8}$$

for each positive real number t.

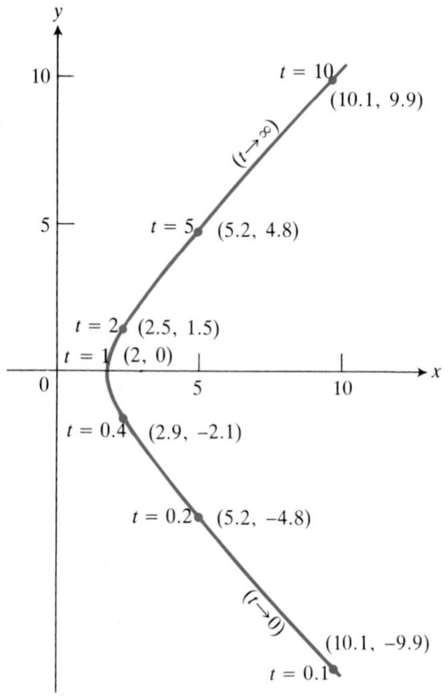

2.20 The graph of $x = t + (1/t)$, $y = t - (1/t)$, $t > 0$. (Part shown is for $0.1 \le t \le 10$.)

Table 2.2 Values of $x = t + (1/t)$ and $y = t - (1/t)$ for selected values of t.

t	$1/t$	x	y
0.1	10.0	10.1	−9.9
0.2	5.0	5.2	−4.8
0.4	2.5	2.9	−2.1
1.0	1.0	2.0	0.0
2.0	0.5	2.5	1.5
5.0	0.2	5.2	4.8
10.0	0.1	10.1	9.9

Solution We make a brief table of values, plot the points, and draw a smooth curve through them (Fig. 2.20). From Table 2.2, or from Eqs. (8), we see that, if t is replaced by $1/t$, the value of x is unchanged and only the sign of y changes. Thus the graph is symmetric with respect to the x-axis. For large values of t, the difference between x and y is small:

$$x - y = \frac{2}{t}, \tag{9a}$$

and if we add the two equations in (8) we get

$$x + y = 2t. \tag{9b}$$

It is now easy to eliminate t between Eqs. (9a) and (9b) by multiplication:

$$(x - y)(x + y) = x^2 - y^2 = 4. \tag{10}$$

The coordinates of all the points described by the parametric equations (8) satisfy Eq. (10), but the converse is not true because Eq. (10) does not require x to be positive.

Figure 2.20 shows the part of the graph that corresponds to $0.1 \le t \le 10$. For smaller positive values of t, the curve extends downward and to the right, approaching the line $y = -x$. As t tends to $+\infty$, the curve extends up and to the right and approaches the line $y = x$.

If we want the slope at any point on the curve, other than at $t = 1$ where the tangent is vertical, we use the Chain Rule in the form

$$\frac{dy}{dx} = \frac{dy/dt}{dx/dt} = \frac{1 + t^{-2}}{1 - t^{-2}} = \frac{t^2 + 1}{t^2 - 1}. \tag{11}$$

For example, when $t = 5$, $x = 5.2$, $y = 4.8$, and $dy/dx = 26/24$. The line tangent to the graph at this point is

$$y - 4.8 = \tfrac{13}{12}(x - 5.2). \ \square$$

REMARK. In the foregoing example, we can also find dy/dx by differentiating Eq. (10) implicitly:

$$2x - 2y\frac{dy}{dx} = 0, \qquad \frac{dy}{dx} = \frac{x}{y} = \frac{t + t^{-1}}{t - t^{-1}} = \frac{t^2 + 1}{t^2 - 1}.$$

We get a result that is equivalent to Eq. (11). We expect the same result, because the Chain Rule works that way. The derivative dy/dx does not exist at $t = 1$, where $dx/dt = 0$ and $y = 0$.

Second Derivatives in Parametric Form

The second derivative of y with respect to x is obtained by differentiating y with respect to x twice:

$$\frac{d^2y}{dx^2} = \frac{d}{dx}\left[\frac{d}{dx}(y)\right].$$

If the parametric equations

$$x = f(t), \qquad y = g(t) = g(h(x))$$

define y as a twice-differentiable function of x, then we may calculate

$$\frac{dy}{dx} = y' = \frac{dy/dt}{dx/dt}$$

from Eq. (7), and calculate d^2y/dx^2 from the equation

$$\frac{d^2y}{dx^2} = \frac{dy'}{dx} = \frac{dy'/dt}{dx/dt}, \tag{12}$$

which can be obtained from Eq. (7) with y' in place of y.

Equation (12) says that to find the second derivative of y with respect to x, we:

1. express $y' = dy/dx$ in terms of t;
2. differentiate y' with respect to t;
3. divide the result by dx/dt.

EXAMPLE 9 Find d^2y/dx^2 if $x = t - t^2$ and $y = t - t^3$.

Solution

$$y' = \frac{dy}{dx} = \frac{dy/dt}{dx/dt} = \frac{1 - 3t^2}{1 - 2t},$$

$$\frac{d^2y}{dx^2} = \frac{dy'/dt}{dx/dt} = \frac{\frac{d}{dt}\left[\frac{1 - 3t^2}{1 - 2t}\right]}{(1 - 2t)}$$

$$= \frac{(1 - 2t)\cdot(-6t) - (1 - 3t^2)\cdot(-2)}{(1 - 2t)^3} = \frac{2 - 6t + 6t^2}{(1 - 2t)^3} \quad \square$$

PROBLEMS

In Problems 1–10, find dy/dt by the Chain Rule, expressing the results in terms of t.

1. $y = x^2, \quad x = 2t - 5$

2. $y = x^4, \quad x = \sqrt[3]{t}$

3. $y = 8 - \frac{x}{3}, \quad x = t^3$

4. $y + 4x^2 = 7, \quad x + \frac{5}{4}t = 3$

5. $2x - 3y = 9, \quad 2x + \frac{t}{3} = 1$

6. $y = x^{-1}, \quad x = t^2 - 3t + 8$

7. $y = \sqrt{x + 2}, \quad x = \frac{2}{t}, \ t > 0$

8. $y = \frac{x^2}{x^2 + 1}, \quad x = \sqrt{2t + 1}$

9. $y = x^2 + 3x - 7, \quad x = 2t + 1$

10. $y = x^{2/3}, \quad x = t^2 + 1$

11. Find dz/dx if $z = w^2 - w^{-1}, \quad w = 3x$.

12. Find dy/dx if $y = 2v^3 + 2v^{-3}, \quad v = (3x + 2)^{2/3}$.

13. Find dr/dt if $r = (s + 1)^{1/2}, \quad s = 16t^2 - 20t$.

14. Find da/db if $a = 7r^3 - 2, \quad r = 1 - 1/b$.

15. Find du/dv if $u = t + 1/t, \quad t = 1 - 1/v$.

In Problems 16–19, each pair of equations represents a curve in parametric form. In each case, find the equation of the curve in the form $y = f(x)$ by eliminating t from the two equations. Then calculate dy/dx, dy/dt, and dx/dt, and verify that they satisfy the Chain Rule (Eq. 7).

16. $x = 3t + 1, \quad y = t^2$

17. $x = t^2, \quad y = t^3$

18. $x = \frac{t}{1 - t}, \quad y = t^2$

19. $x = \frac{t}{1 + t}, \quad y = \frac{t^2}{1 + t}$

In Problems 20–26, graph the curve represented by the given equations.

20. $x = 2t - 5, \quad y = 4t - 7$

21. $x = 1 - t, \quad y = 1 + t$

22. $x = 3t, \quad y = 9t^2$

23. $x = t, \quad y = 1/t$

24. $x = \sqrt{t}, \quad y = t \qquad$ (*Hint:* x cannot be negative.)

25. $x = t, \quad y = \sqrt{t}$

26. $x = \cos t, \quad y = \sin t$

27. If $x = 4t - 5$ and $y = t^2$, which of the following is the value of dy/dx when $t = 2$?
a) 2 b) 4 c) 1 d) 1/2

28. If $x = 3t^2 + 2$ and $y = 2t^4 - 1$, which of the following is the value of dy/dx when $t = 1$?
a) 8 b) 4/3 c) 6 d) 3/4

The parametric equations in Problems 29–33 describe curves in the plane. Find (a) the slope of the curve at the point (x, y) at which $t = 2$, and (b) the tangent to the curve at this point.

29. $x = t + 1/t, \quad y = t - 1/t$

30. $x = \sqrt{2t^2 + 1}, \quad y = (2t + 1)^2$

31. $x = t\sqrt{2t + 5}, \quad y = (4t)^{1/3}$

32. $x = \dfrac{t - 1}{t + 1}, \quad y = \dfrac{t + 1}{t - 1}$

33. $x = t^{-2}, \quad y = \sqrt{t^2 + 12}$

34. Find the equation of the tangent to the curve

$$x = \frac{1}{t} + t^2, \qquad y = t^2 - t + 1$$

at the point $(2, 1)$.

35. Given $x = 80t$ and $y = 64t - 16t^2$, find the value of t for which $dy/dx = 0$.

36. A particle P moves along the curve $x^2y^3 = 27$. At the time when P is at $(1, 3)$, $dy/dt = 10$. Find the value of dx/dt at this time.

37. If a point traces the circle $x^2 + y^2 = 25$, and if $dx/dt = 4$ when the point reaches $(3, 4)$, find dy/dt there.

38. The velocity of a falling body is $k\sqrt{s}$ meters per second (k constant) at the instant the body has fallen a distance of s meters from its starting point. Show that the acceleration of the body is constant.

39. Let $f(x) = x^2$ and $g(x) = |x|$. Show that the composites $f \circ g$ and $g \circ f$ are both differentiable at $x = 0$ even though g has no derivative at $x = 0$. Does this contradict the Chain Rule?

Use Eq. (12) to find d^2y/dx^2 from the parametric equations in each of the following problems.

40. Problem 20 **41.** Problem 21

42. Problem 22 **43.** Problem 23

44. Problem 24

45. Find d^2y/dx^2 from the parametric equations in
a) Problem 29, b) Problem 32.

46. *Simple pendulum and temperature variation.* For oscillations of small amplitude, the relation between the period T and the length L of a simple pendulum may be approximated by the equation

$$T = 2\pi\sqrt{\frac{L}{g}},$$

where $g = 980\ \text{cm/s}^2$ is the acceleration due to gravity. When the temperature θ changes, the length L either increases or decreases at a rate that is proportional to L:

$$\frac{dL}{d\theta} = kL,$$

where k is a proportionality constant. What is the rate of change of the period with respect to temperature?

47. *Measuring the acceleration of gravity.* When the length L of a clock pendulum is held constant by controlling its temperature, the pendulum's period T depends on the acceleration of gravity, g. The period will therefore vary slightly as the clock is moved from place to place on the earth's surface, depending on the change in g. By keeping track of ΔT, one can estimate the variation in g from the equation $T = 2\pi(L/g)^{1/2}$ that relates T, g, and L.

a) With L held constant and g as the independent variable, calculate $\Delta T_{\tan}$. Use the estimate $\Delta T \approx \Delta T_{\tan}$ to relate ΔT to Δg and answer the questions in (b) and (c).
b) If g increases, will T increase, or decrease? Will a pendulum clock speed up, or slow down, if g increases?
c) A clock with a 100-cm pendulum is moved from a location where $g = 980\ \text{cm/s}^2$ to a new location. This increases the period by $\Delta T = 0.001$ s. Estimate Δg and the value of g at the new location.

***Toolkit* Programs**

Parametric Equations   Super * Grapher

2.7

Brief Review of Trigonometry. Angles between Curves

Many natural phenomena are periodic; that is, they repeat after definite periods of time. Such phenomena are most readily studied with trigonometric functions, particularly sines and cosines. Our object in this article and the next is to apply the operations of the calculus to these functions, but before we do, we shall review some of their properties.

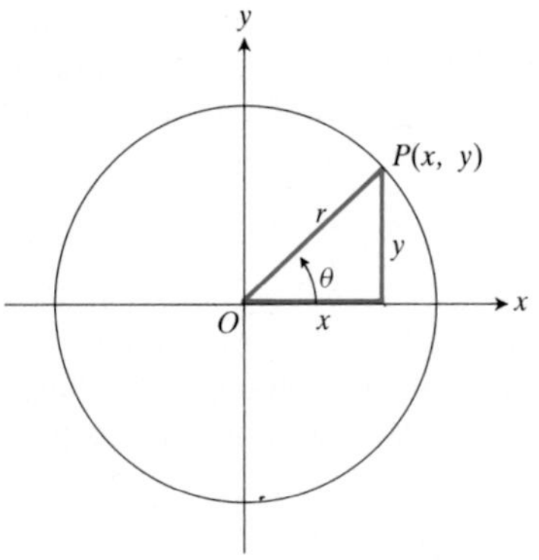

2.21 Angle θ in standard position.

When an angle of measure θ is placed in standard position at the center of a circle of radius r, as in Fig. 2.21, the trigonometric functions of θ are defined by the equations

$$\sin\theta = \frac{y}{r}, \qquad \cos\theta = \frac{x}{r}, \qquad \tan\theta = \frac{y}{x},$$
$$\csc\theta = \frac{r}{y}, \qquad \sec\theta = \frac{r}{x}, \qquad \cot\theta = \frac{x}{y}. \tag{1}$$

Observe that $\tan\theta$ and $\sec\theta$ are not defined for values of θ for which $x = 0$. In radian measure, this means that $\pi/2, 3\pi/2, \ldots, -\pi/2, -3\pi/2, \ldots$ are excluded from the domains of the tangent and the secant functions. Similarly, $\cot\theta$ and $\csc\theta$ are not defined for values of θ corresponding to $y = 0$: that is, for $\theta = 0, \pi, 2\pi, \ldots, -\pi, -2\pi, \ldots$. For those values of θ where the functions are defined, it follows from Eqs. (1) that

$$\tan\theta = \frac{\sin\theta}{\cos\theta}, \qquad \csc\theta = \frac{1}{\sin\theta}, \qquad \sec\theta = \frac{1}{\cos\theta}, \qquad \cot\theta = \frac{1}{\tan\theta}.$$

Since

$$x^2 + y^2 = r^2,$$

by the Pythagorean theorem, we have

$$\cos^2\theta + \sin^2\theta = 1. \tag{2}$$

It is also useful to express the coordinates of $P(x, y)$ in terms of r and θ as follows:

$$x = r\cos\theta, \qquad y = r\sin\theta. \tag{3}$$

When $\theta = 0$ in Fig. 2.21, we have $y = 0$ and $x = r$; hence, from the definitions (1), we obtain

$$\sin 0 = 0, \qquad \cos 0 = 1.$$

Similarly, for a right angle, $\theta = \pi/2$, we have $x = 0$, $y = r$; hence

$$\sin\frac{\pi}{2} = 1, \qquad \cos\frac{\pi}{2} = 0.$$

Calculating Sines and Cosines

If (x, y) is a point on a circle with radius $r = 1$ unit, then Eqs. (3) become

$$x = \cos\theta, \qquad y = \sin\theta.$$

The cosine and sine of any angle can therefore be calculated from an acute reference triangle made by dropping a perpendicular to the x-axis as shown in Fig. 2.22. The ratios are read from the triangle, and the signs determined by the quadrant in which the angle lies.

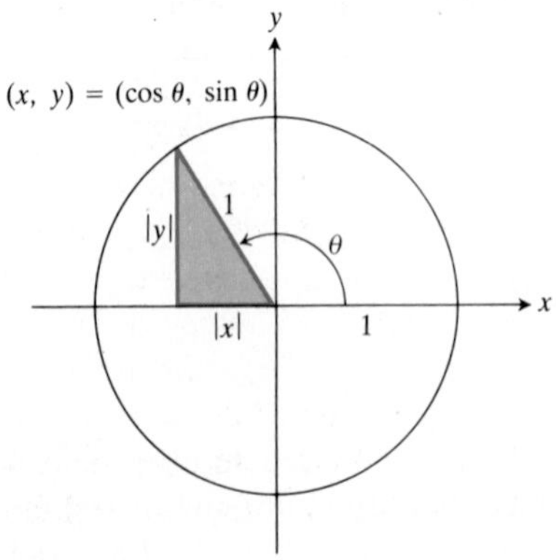

2.22 The acute reference triangle for an angle θ.

EXAMPLE 1

a) From Fig. 2.23(a):

$$\cos\left(-\frac{\pi}{4}\right) = \frac{\sqrt{2}}{2}, \qquad \sin\left(-\frac{\pi}{4}\right) = -\frac{\sqrt{2}}{2}.$$

b) From Fig. 2.23(b):

$$\cos\left(\frac{2\pi}{3}\right) = -\frac{1}{2}, \qquad \sin\left(\frac{2\pi}{3}\right) = \frac{\sqrt{3}}{2}. \ \square$$

$$\left(\cos\frac{2\pi}{3},\ \sin\frac{2\pi}{3}\right)=\left(-\frac{1}{2},\ \frac{\sqrt{3}}{2}\right)$$

$$\left(\frac{\sqrt{2}}{2},\ -\frac{\sqrt{2}}{2}\right)=\left(\cos-\frac{\pi}{4},\ \sin-\frac{\pi}{4}\right)$$

(a)   (b)

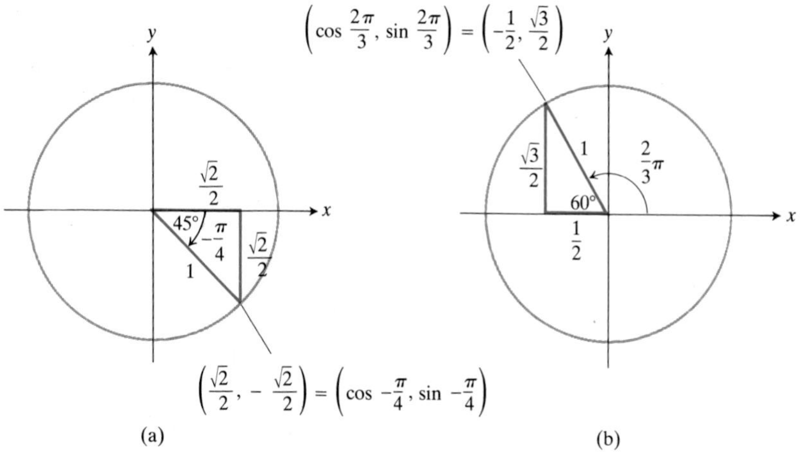

2.23   Calculating the sine and cosine of (a) $-\pi/4$ radians and (b) $2\pi/3$ radians.

Radian Measure

The radian measure of angle ABC at the center B of the unit circle in Fig. 2.24 is defined to be the length θ of the circular arc AC. If $A'C'$ is the arc cut by the angle from a second circle centered at B, then the circular sectors $A'B'C'$ and ABC are similar. In particular, their ratios of arc length to radius are equal. In the notation of Fig. 2.24, this means that

$$\frac{\text{length of arc } A'C'}{r}=\frac{\text{length of arc } AC}{1},$$

or

$$\frac{s}{r}=\frac{\theta}{1}=\theta. \tag{4a}$$

This is true no matter how large or small the radius of the second circle may be. Thus for any circle centered at B, the ratio s/r of the length of the intercepted arc to the circle's radius always gives the angle's radian measure.

Equation (4a) is sometimes written in the form

$$s=r\theta, \tag{4b}$$

from which it is convenient to calculate the arc length s when the radius r and angle's radian measure θ are known.

A useful interpretation of radian measure is easy to get if we take $r=1$ in Eq. (4b). Then the central angle θ, in radians, is just equal to the arc s subtended by θ. We may imagine the circumference of the circle marked off with a scale from which we may read θ. We think of a number scale, like the y-axis shifted one unit to the right, as having been wrapped around the circle. The unit on this number scale is the same as the unit radius. We put the zero of the scale at the place where the initial ray crosses the circle, and then we wrap the positive end of the scale around the circle in the counterclockwise direction, and wrap the negative end around in the opposite direction (see Fig. 2.25). Then θ can be read from this curved s-"axis."

Two points on the s-axis that are exactly 2π units apart will map onto the same point on the unit circle when the wrapping is carried out. For

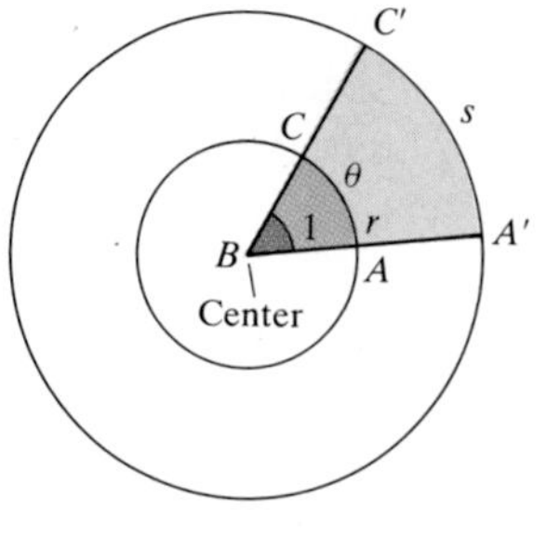

2.24   The radian measure of the angle centered at B is $s/r=\theta/1=\theta$.

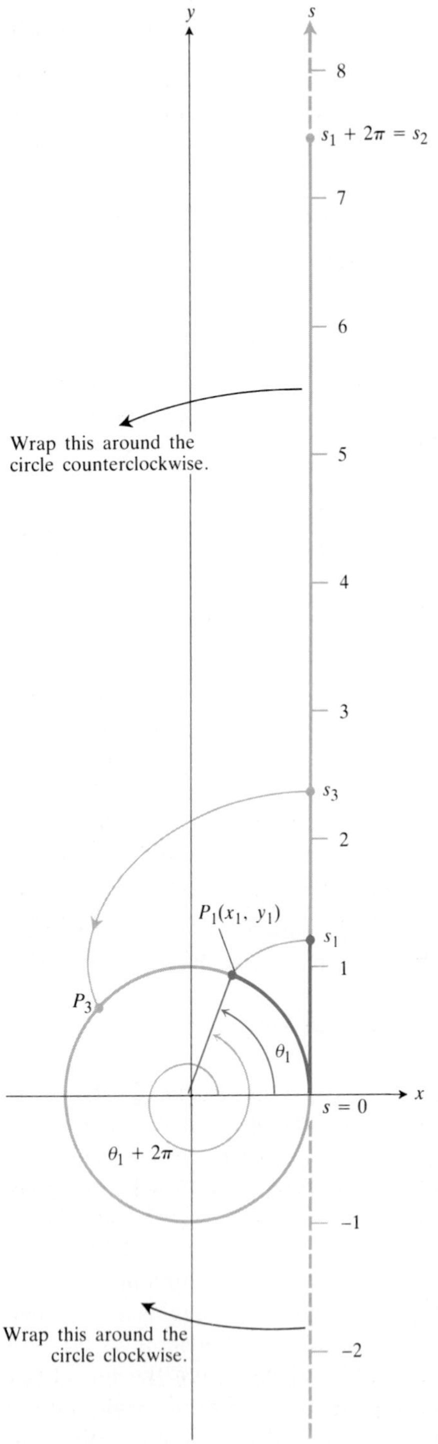

2.25 The s-axis wrapped around the unit circle.

example, if $P_1(x_1, y_1)$ is the point to which an arc of length s_1 reaches, then arcs of length $s_1 + 2\pi$, $s_1 + 4\pi$, and so on, will reach exactly the same point after going completely around the circle one, or two, or more, times. Similarly, P_1 will be the image of points on the negative s-axis at $s_1 - 2\pi$, $s_1 - 4\pi$, and so on. Thus, from the wrapped s-axis, we could read

$$\theta_1 = s_1,$$

or

$$\theta_1 + 2\pi, \quad \theta_1 + 4\pi, \ldots, \quad \theta_1 - 2\pi, \quad \theta_1 - 4\pi, \ldots.$$

A unit of arc length $s = 1$ radius subtends a central angle of $57°18'$ (approximately), so

$$1 \text{ radian} \approx 57°18'. \tag{5}$$

We find this, and other relations between degree measure and radian measure, by using the fact that the full circumference has arc length $s = 2\pi r$ and central angle 360°. Therefore,

$$360° = 2\pi \text{ radians}, \tag{6a}$$

$$180° = \pi \text{ radians} = 3.14159\ldots \text{ radians}, \tag{6b}$$

$$\left(\frac{180}{\pi}\right)^{\circ} = 1 \text{ radian} \approx 57°17'44.8'', \tag{6c}$$

$$1° = \frac{2\pi}{360} = \frac{\pi}{180} \approx 0.01745 \text{ radians}. \tag{6d}$$

It should be emphasized, however, that the radian measure of an angle is dimensionless, since r and s in Eqs. (4a, b) both represent lengths measured in identical units, for instance feet, inches, centimeters, or light-years. Thus $\theta = 2.7$ is to be interpreted as a pure number. The sine and cosine of 2.7 are the ordinate and abscissa, respectively, of the point $P(x, y)$ on a circle of radius r at the end of an arc of length 2.7 radii. For practical purposes we could convert 2.7 radians to $2.7(180/\pi)$ degrees and say

$$\sin 2.7 = \sin\left[2.7\left(\frac{180}{\pi}\right)^{\circ}\right] \approx \sin\,[154°41'55'']$$
$$\approx 0.42738.$$

Table 2.3 gives the values of the sine, cosine, and tangent functions for selected values of θ.

Periodicity

The mapping from the real numbers s onto points $P(x, y)$ on the unit circle by the wrapping process described above and illustrated in Fig. 2.25 defines the coordinates as functions of s because Eqs. (1) apply, with $\theta = s$ and $r = 1$:

$$x = \cos\theta = \cos s, \qquad y = \sin\theta = \sin s.$$

Because $s + 2\pi$ maps onto the same point that s does, it follows that

$$\begin{aligned}\cos(\theta + 2\pi) &= \cos\theta,\\ \sin(\theta + 2\pi) &= \sin\theta.\end{aligned} \tag{7}$$

Table 2.3 Values of sin θ, cos θ, and tan θ for selected values of θ.

Degrees	-180	-135	-90	-45	0	45	90	135	180
θ (radians)	$-\pi$	$-3\pi/4$	$-\pi/2$	$-\pi/4$	0	$\pi/4$	$\pi/2$	$3\pi/4$	π
sin θ	0	$-\sqrt{2}/2$	-1	$-\sqrt{2}/2$	0	$\sqrt{2}/2$	1	$\sqrt{2}/2$	0
cos θ	-1	$-\sqrt{2}/2$	0	$\sqrt{2}/2$	1	$\sqrt{2}/2$	0	$-\sqrt{2}/2$	-1
tan θ	0	1		-1	0	1		-1	0

Equations (7) are *identities;* that is, they are true for all real numbers θ. These identities would be true for $\theta' = \theta + 2\pi$:

$$\cos \theta' = \cos(\theta' - 2\pi) \quad \text{and} \quad \sin \theta' = \sin(\theta' - 2\pi). \tag{8}$$

Equations (7) and (8) say that 2π can be added to or subtracted from the domain variable of the sine and cosine functions with no change in the function values. The same process could be repeated any number of times. Consequently,

$$\begin{aligned} \cos(\theta + 2n\pi) &= \cos \theta, \\ \sin(\theta + 2n\pi) &= \sin \theta, \qquad n = 0, \pm 1, \pm 2, \ldots. \end{aligned} \tag{9}$$

Figure 2.26 shows graphs of the curves

$$y = \sin x \quad \text{and} \quad y = \cos x.$$

The portion of each curve between 0 and 2π is repeated endlessly to the left and to the right. We also note that the cosine curve is the same as the sine curve shifted to the left an amount $\pi/2$.

EXAMPLE 2† The builders of the Trans-Alaska Pipeline used insulated pads to keep the heat from the hot oil in the pipeline from melting the permanently frozen soil beneath. To design the pads, it was necessary to take into account the variation in air temperature throughout the year. The variation was represented in the calculations by a *general sine function* of the form

$$f(x) = A \sin\left[\frac{2\pi}{B}(x - C)\right] + D,$$

where $|A|$ is the *amplitude,* $|B|$ is the *period,* C is the *horizontal shift,* and D is the *vertical shift* (Fig. 2.27). Figure 2.28 shows how such a function can be used to represent temperature data. The data points in the figure are plots of the mean air temperature for Fairbanks, Alaska, based on records of the National Weather Service from 1941 to 1970. The sine function that is used to fit the data is

$$f(x) = 37 \sin\left[\frac{2\pi}{365}(x - 101)\right] + 25,$$

where f is temperature in degrees Fahrenheit, and x is the number of the day counting from the beginning of the year.

The fit is remarkably good. □

† From "Is the curve of temperature variation a sine curve?" by B. M. Lando and C. A. Lando. *The Mathematics Teacher,* September 1977, Vol. 7, No. 6, pp. 534–537.

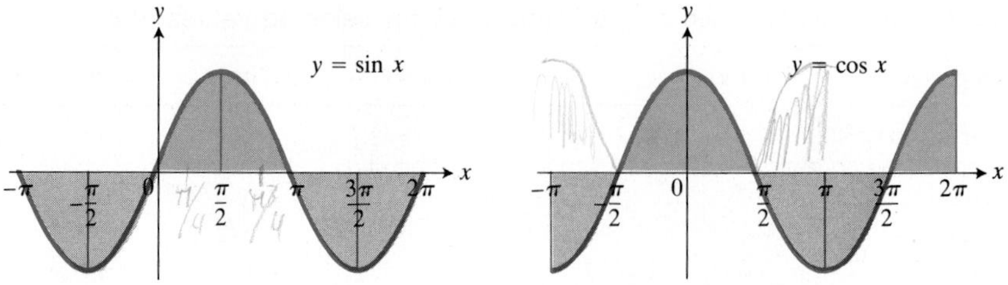

2.26 The value of the sine at x is the value of the cosine at $(x - \pi/2)$. That is, $\sin x = \cos(x - \pi/2)$.

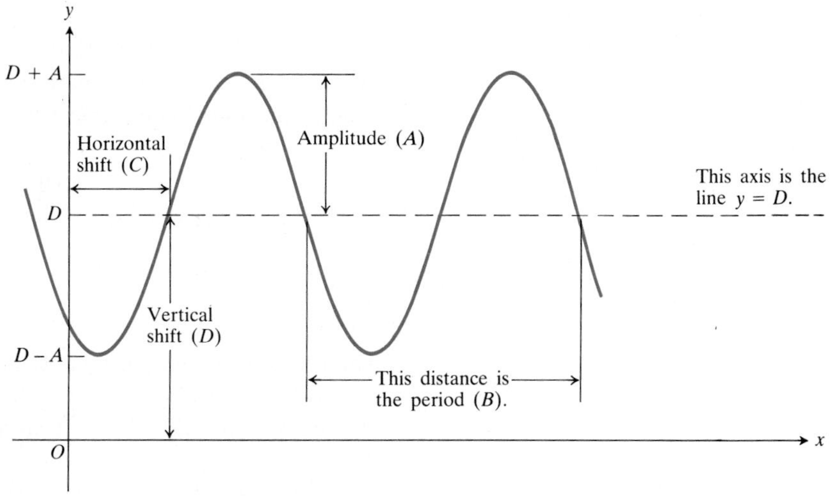

2.27 The general sine curve $y = A \sin[(2\pi/B)(x - C)] + D$, shown for A, B, C, and D positive.

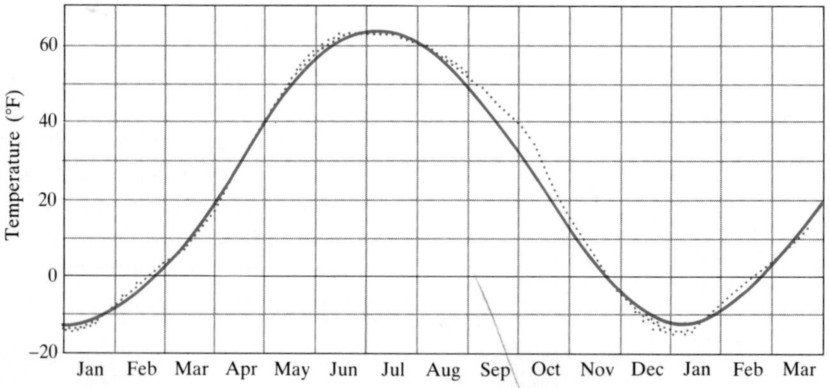

2.28 Normal mean air temperatures at Fairbanks, Alaska, plotted as data points. The approximating sine function is

$$f(x) = 37 \sin\left[\frac{2\pi}{365}(x - 101)\right] + 25.$$

(From "Is the curve of temperature variation a sine curve?" by B. M. Lando and C. A. Lando. *The Mathematics Teacher*, September 1977, **7,** 6; Fig. 2, p. 535.)

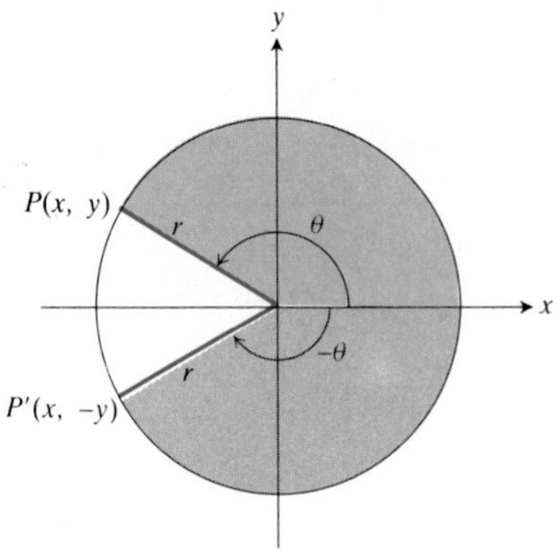

2.29 Angles of opposite sign.

Sum and Difference Formulas

Figure 2.29 shows two angles of opposite sign but of equal magnitude. By symmetry, the points $P(x, y)$ and $P'(x, -y)$, where the rays of the two angles θ and $-\theta$ intersect the circle, have equal abscissas, and ordinates that differ only in sign. Hence we have

$$\sin(-\theta) = -\frac{y}{r} = -\sin\theta, \tag{10a}$$

$$\cos(-\theta) = \frac{x}{r} = \cos\theta. \tag{10b}$$

For example,

$$\sin\left(-\frac{\pi}{2}\right) = -\sin\frac{\pi}{2} = -1, \quad \text{and} \quad \cos\left(-\frac{\pi}{2}\right) = \cos\frac{\pi}{2} = 0.$$

It will be helpful, for reasons that will soon be apparent, to review the formulas

$$\sin(A+B) = \sin A\cos B + \cos A\sin B, \tag{11a}$$

$$\cos(A+B) = \cos A\cos B - \sin A\sin B, \tag{11b}$$

together with two formulas obtained from these by replacing B by $-B$ and recalling that

$$\sin(-B) = -\sin B, \qquad \cos(-B) = \cos B; \tag{11c}$$

namely,

$$\sin(A-B) = \sin A\cos B - \cos A\sin B, \tag{11d}$$

$$\cos(A-B) = \cos A\cos B + \sin A\sin B. \tag{11e}$$

These formulas are derived from the law of cosines in Appendix 6.

The standard formula for the tangent of the difference of two angles is a consequence of Eqs. (11d) and (11e):

$$\tan(A-B) = \frac{\sin(A-B)}{\cos(A-B)} = \frac{\sin A\cos B - \cos A\sin B}{\cos A\cos B + \sin A\sin B}.$$

Dividing the numerator and denominator of the fraction on the right by $\cos A\cos B$ gives

$$\tan(A-B) = \frac{\tan A - \tan B}{1+\tan A\tan B}. \tag{11f}$$

Similarly, for the tangent of the sum of two angles, Eqs. (11a) and (11b) give

$$\tan(A+B) = \frac{\sin(A+B)}{\cos(A+B)} = \frac{\sin A\cos B + \cos A\sin B}{\cos A\cos B - \sin A\sin B},$$

or

$$\tan(A+B) = \frac{\tan A + \tan B}{1-\tan A\tan B}. \tag{11g}$$

Double-Angle Formulas

From Eqs. (11a) and (11b) with $A = B = \theta$ we have

$$\sin 2\theta = 2\sin\theta\cos\theta, \tag{12a}$$

$$\cos 2\theta = \cos^2\theta - \sin^2\theta. \tag{12b}$$

Half-Angle Formulas and a Useful Limit

From Eq. (11*b*) with $A = B = \theta$ we have

$$\cos 2\theta = \cos^2\theta - \sin^2\theta = \cos^2\theta - (1 - \cos^2\theta) = 2\cos^2\theta - 1,$$

so that

$$\cos^2\theta = \frac{1 + \cos 2\theta}{2}. \tag{13a}$$

Similarly,

$$\cos 2\theta = \cos^2\theta - \sin^2\theta = (1 - \sin^2\theta) - \sin^2\theta = 1 - 2\sin^2\theta,$$

so that

$$\sin^2\theta = \frac{1 - \cos 2\theta}{2}. \tag{13b}$$

Equation (13a) is equivalent to

$$\cos\theta = \sqrt{\frac{1 + \cos 2\theta}{2}} \quad \text{when } \cos\theta > 0,$$

and

$$\cos\theta = -\sqrt{\frac{1 + \cos 2\theta}{2}} \quad \text{when } \cos\theta < 0.$$

That is,

$$\cos\theta = \pm\sqrt{\frac{1 + \cos 2\theta}{2}}, \tag{13c}$$

the sign being (+) when θ lies in the first and fourth quadrants and (−) when θ lies in the second and third quadrants.

Similarly, Eq. (13b) leads to

$$\sin\theta = \pm\sqrt{\frac{1 - \cos 2\theta}{2}}, \tag{13d}$$

the sign being (+) when θ lies in the first and second quadrants and (−) otherwise.

When θ is replaced by $\theta/2$ in (13c) and (13d), we get

$$\cos\frac{\theta}{2} = \pm\sqrt{\frac{1 + \cos\theta}{2}}, \tag{13e}$$

$$\sin\frac{\theta}{2} = \pm\sqrt{\frac{1 - \cos\theta}{2}}, \tag{13f}$$

the signs now depending on the location of $\theta/2$.

The Equations 13a–13f are sometimes called the *half-angle formulas* for the sine and cosine.

By dividing Eq. (13b) through by θ and turning the resulting equation around, we get

$$\frac{1 - \cos 2\theta}{2\theta} = \frac{\sin^2\theta}{\theta}, \tag{14}$$

so that

$$\begin{aligned}\lim_{\theta\to 0}\frac{1 - \cos 2\theta}{2\theta} &= \lim_{\theta\to 0}\frac{\sin^2\theta}{\theta}\\ &= \lim_{\theta\to 0}\left(\frac{\sin\theta}{\theta}\cdot\sin\theta\right) = \lim_{\theta\to 0}\left(\frac{\sin\theta}{\theta}\right)\cdot\lim_{\theta\to 0}\sin\theta = 1\cdot 0 = 0.\end{aligned}$$

The substitution $h = 2\theta$ gives the following simpler expression:

$$\lim_{h\to 0} \frac{1 - \cos h}{h} = 0. \tag{15}$$

We shall use this limit in calculating the derivative of $y = \sin x$ in Article 2.8.

Angles between Curves

The angles between two differentiable curves C_1 and C_2 at a point of intersection are the angles between the curves' tangents there, as shown in Fig. 2.30. If the tangents are perpendicular, the curves are said to be *orthogonal* ("right-angled"). If the tangents have slopes

$$m_1 = \tan\theta_1, \qquad m_2 = \tan\theta_2,$$

then the condition for orthogonality is

$$m_2 = -\frac{1}{m_1}.$$

If the tangents are not perpendicular, then

$$\beta = \phi_2 - \phi_1$$

is different from 90°, $m_1 m_2 \neq -1$, and Eq. (11f) gives

$$\tan\beta = \tan(\theta_2 - \theta_1) = \frac{\tan\theta_2 - \tan\theta_1}{1 + \tan\theta_2 \tan\theta_1} = \frac{m_2 - m_1}{1 + m_1 m_2}. \tag{16}$$

In practice, when we use the right-hand side of Eq. (16) to calculate the angle between two nonorthogonal curves, we may calculate either the acute angle β or its supplement β'. Which one we get will depend on which curve we call C_1 and which we call C_2. Changing the numbering interchanges ϕ_1 and ϕ_2 and m_1 and m_2, and Eq. (16) then gives

$$\begin{aligned}\tan(\text{new } \phi_2 - \text{new } \phi_1) &= \tan(\text{old } \phi_1 - \text{old } \phi_2) = \tan(-\beta)\\ &= -\tan\beta = \tan(\pi - \beta).\end{aligned} \tag{17}$$

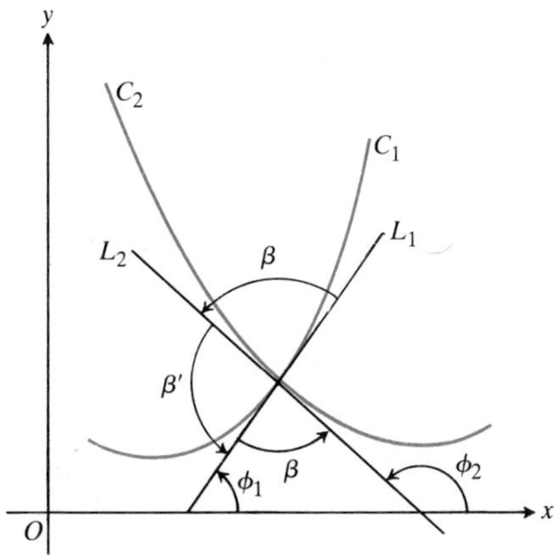

2.30 The angles β and β' between two smooth (differentiable) curves at a point of intersection are the angles between the tangents. They are measured counterclockwise.

(See Problem 51 for an explanation of the last equality.)

We are therefore led to the following working rule for calculating angles between differentiable curves.

Working Rule for Calculating Angles between Differentiable Curves at a Point of Intersection

If the tangents at the point have slopes m_1 and m_2, then the curves are orthogonal if

$$m_2 = -1/m_1. \tag{18}$$

If $m_2 \neq -1/m_1$, then the tangent of the acute angle β between the curves is

$$\tan\beta = \left|\frac{m_2 - m_1}{1 + m_1 m_2}\right|. \tag{19}$$

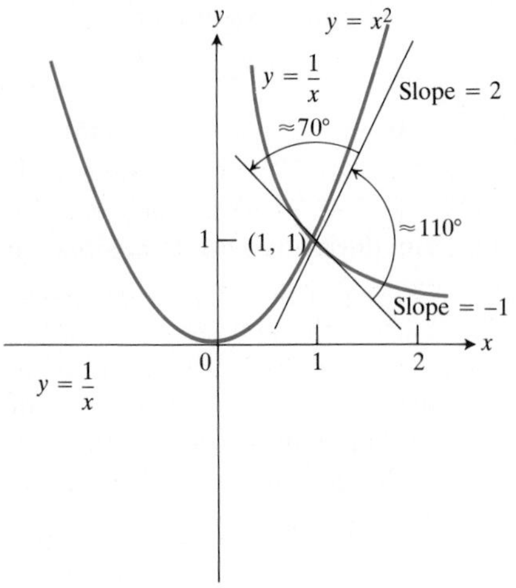

2.31 The angles between the curves $y = x^2$ and $y = 1/x$ at the point (1, 1) are found in Example 3 to be about 70° and 110°.

EXAMPLE 3 Find the angles between the curves

$$y = x^2, \qquad xy = 1,$$

at each point of intersection.

Solution In this case (see Fig. 2.31), the curves intersect at the single point (1, 1), as we discover by solving the given equations simultaneously.

The slopes of the curves at (1, 1) are

$$m_1: \quad y = x^2, \qquad\qquad m_2: \quad y = \frac{1}{x},$$

$$y' = 2x, \qquad\qquad y' = -\frac{1}{x^2},$$

$$y'|_{(1,1)} = 2 \cdot 1 = 2; \qquad\qquad y'|_{(1,1)} = -\frac{1}{(1)^2} = -1.$$

The curves are not orthogonal, because $m_2 \neq -1/m_1$. We therefore use Eq. (19) to calculate

$$\tan \beta = \left|\frac{m_2 - m_1}{1 + m_1 m_2}\right| = \left|\frac{-1 - 2}{1 + 2(-1)}\right| = \left|\frac{-3}{-1}\right| = |3| = 3.$$

The angle β is the angle whose tangent is 3, or about 70° (to the nearest degree, from the tables at the back of the book). The obtuse angle between the curves is therefore about 180° − 70° = 110°, and the problem is solved.

The choice of $y = x^2$ for curve 1 and $y = 1/x$ for curve 2 was arbitrary. We might equally well have numbered the curves the other way around, obtaining

$$m_1 = -1, \qquad m_2 = 2.$$

In this case, $(m_2 - m_1)/(1 + m_1 m_2)$ would have been

$$\frac{m_2 - m_1}{1 + m_1 m_2} = \frac{2 - (-1)}{1 - (-1)(2)} = \frac{3}{-1} = -3,$$

which is $-\tan \beta$. The absolute value formula in Eq. (19), however, gives

$$\tan \beta = |-3| = 3,$$

as before. □

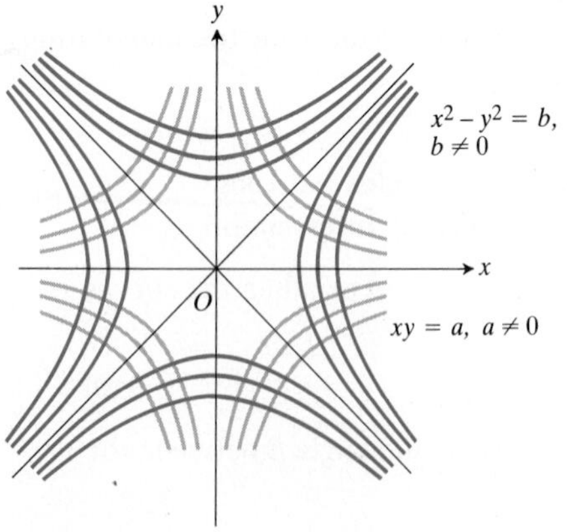

2.32 Each curve is orthogonal to every curve it meets in the other family.

EXAMPLE 4 Show that every curve of the family

$$xy = a, \qquad a \neq 0, \tag{20a}$$

is orthogonal to every curve of the family

$$x^2 - y^2 = b, \qquad b \neq 0. \tag{20b}$$

Solution The two families of curves are sketched in Fig. 2.32. Their slopes can be obtained by implicit differentiation. At a point $P(x, y)$ on any curve of (20a), the slope is

$$\frac{dy}{dx} = -\frac{y}{x}, \tag{21a}$$

and on any curve of (20b) the slope is

$$\frac{dy}{dx} = \frac{x}{y}. \tag{21b}$$

At a point of intersection, the values of x and y in (21b) are the same as in (21a), and the two curves are orthogonal because these slopes are negative reciprocals of each other. The cases $x = 0$ or $y = 0$ cannot occur in Eqs. (21) if (x, y) is a point of intersection of any curve (20a) and a curve (20b), since a and b are restricted to be constants different from zero. That every curve in (20a) does in fact intersect every curve in (20b) follows from the fact that the equation

$$x^2 - \frac{a^2}{x^2} = b,$$

which results from eliminating y between (20a) and (20b), has real roots for every pair of nonzero real constants a and b. □

The curves in (20b) are called *orthogonal trajectories* of the curves in (20a). Such mutually orthogonal systems of curves are of particular importance in physical problems related to electrical potential, where the curves in one family correspond to lines of flow and those in the other family correspond to lines of constant potential. They also occur in hydrodynamics and in heat-flow problems.

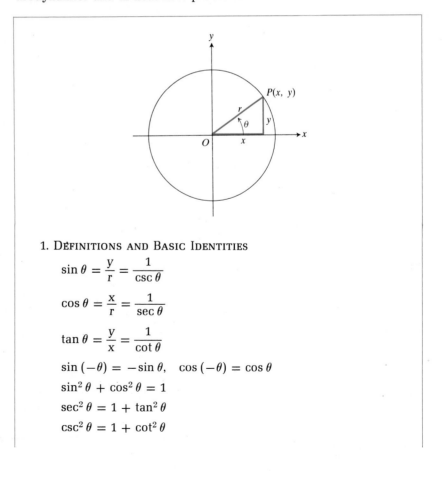

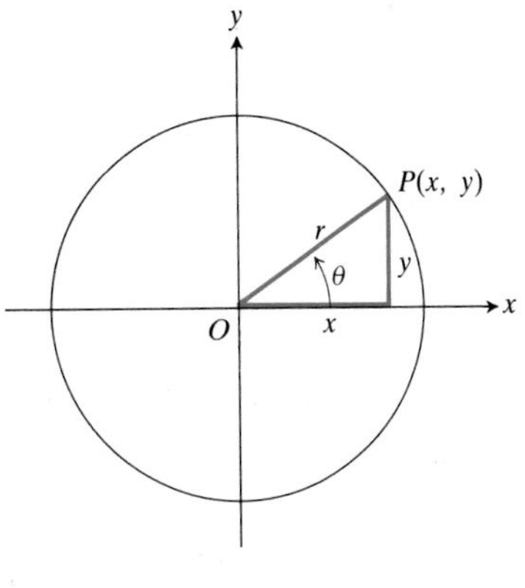

1. DEFINITIONS AND BASIC IDENTITIES

$$\sin\theta = \frac{y}{r} = \frac{1}{\csc\theta}$$

$$\cos\theta = \frac{x}{r} = \frac{1}{\sec\theta}$$

$$\tan\theta = \frac{y}{x} = \frac{1}{\cot\theta}$$

$$\sin(-\theta) = -\sin\theta, \quad \cos(-\theta) = \cos\theta$$

$$\sin^2\theta + \cos^2\theta = 1$$

$$\sec^2\theta = 1 + \tan^2\theta$$

$$\csc^2\theta = 1 + \cot^2\theta$$

$$\sin 2\theta = 2 \sin \theta \cos \theta, \quad \cos 2\theta = \cos^2 \theta - \sin^2 \theta$$

$$\cos^2 \theta = \frac{1 + \cos 2\theta}{2}, \quad \sin^2 \theta = \frac{1 - \cos 2\theta}{2}$$

$$\sin (A + B) = \sin A \cos B + \cos A \sin B$$

$$\sin (A - B) = \sin A \cos B - \cos A \sin B$$

$$\cos (A + B) = \cos A \cos B - \sin A \sin B$$

$$\cos (A - B) = \cos A \cos B + \sin A \sin B$$

$$\tan (A + B) = \frac{\tan A + \tan B}{1 - \tan A \tan B}$$

$$\tan (A - B) = \frac{\tan A - \tan B}{1 + \tan A \tan B}$$

$$\sin \left(A - \frac{\pi}{2}\right) = -\cos A, \quad \cos \left(A - \frac{\pi}{2}\right) = \sin A$$

$$\sin \left(A + \frac{\pi}{2}\right) = \cos A, \quad \cos \left(A + \frac{\pi}{2}\right) = -\sin A$$

2. Angles and Sides of a Triangle†

Law of cosines: $c^2 = a^2 + b^2 - 2ab \cos C.$

Law of sines: $\dfrac{\sin A}{a} = \dfrac{\sin B}{b} = \dfrac{\sin C}{c}.$

Area $= \frac{1}{2}bc \sin A = \frac{1}{2}ac \sin B = \frac{1}{2}ab \sin C.$

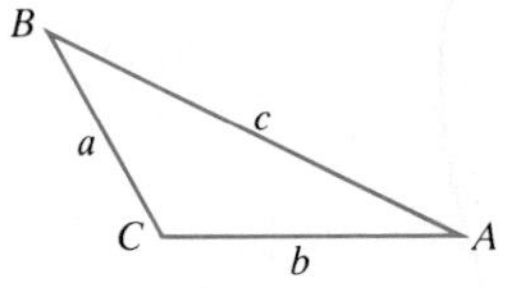

PROBLEMS

Hints for graphing sine and cosine functions:

i) Find the amplitude, period, and shift.
ii) Draw the curve roughly.
iii) Draw and label the axes; then finish the sketch.

Remember: *Angles in radians.*

Graph the equations in Problems 1–20.

1. $y = 2 \sin x$

2. $y = 5 \sin 2x$

3. $y = \sin(-x)$

4. $y = \sin 2\pi x$

5. $y = 2 \cos 3x$

6. $y = \tan (x/3)$

7. $y = \sin (x + (\pi/2))$

8. $y = \cos (x - (\pi/2))$

9. $y = |\cos x|$

10. $y = \frac{1}{2}(|\cos x| + \cos x)$

11. $y = \frac{1}{2}(|\sin x| - \sin x)$

12. $y = \sin^2 x$

13. $y = \cos^2 x$

14. $y = \sin x + \cos x$

15. $y = \sin x - \cos x$

16. $y = \cos 2\pi(x + 1)$

17. $y = 2 \cos (4x - 2\pi), \ -\pi \le x \le \pi$

18. $y = \sin (x - (\pi/4)), \ -\pi \le x \le \pi$

19. $y = \sec x$ and $y = \cos x$ together, $-2\pi \le x \le 2\pi$

†Not derived in this article.

20. $y = \sin x$, $y = \cos x$, $y = \tan x$ together, $-2\pi \leq x \leq 2\pi$

In Problems 21–27, graph the curve given by the parametric equations over the given parameter domain. Then, thinking of the equations as equations of motion, say where the point $P(x, y)$ starts and stops, and in which direction the point moves as t increases. (*Hint:* $\cos^2 t + \sin^2 t = 1$.)

21. $x = \cos t$, $y = \sin t$, $0 \leq t \leq 2\pi$

22. $x = \cos t$, $y = \sin t$, $0 \leq t \leq \pi$

23. $x = \cos 2\pi t$, $y = \sin 2\pi t$, $0 \leq t \leq 1$

24. $x = \cos(\pi - t)$, $y = \sin(\pi - t)$, $0 \leq t \leq \pi$

25. $x = 3\cos t$, $y = 3\sin t$, $0 \leq t \leq 2\pi$

26. $x = \cos t$, $y = -\sin t$, $0 \leq t \leq 2\pi$

27. $x = \cos^2 t$, $y = \sin^2 t$, $0 \leq t \leq \pi/2$

28. Find a Cartesian equation for the circle defined by the parametric equations $x = a\cos t$, $y = a\sin t$, $0 \leq t \leq 2\pi$.

29. Find parametric equations for the motion of a particle that traces the circle $x^2 + y^2 = 4$ once in the (a) clockwise and (b) counterclockwise direction. Use the parameter interval $0 \leq t \leq 2\pi$ in each case.

30. Repeat Problem 29 for the parameter domains
a) $0 \leq t \leq \pi$,
b) $0 \leq t \leq 1$.

31. Find the (a) amplitude, (b) period, (c) horizontal shift, and (d) vertical shift of the general sine function

$$f(x) = 37\sin\left[\frac{2\pi}{365}(x - 101)\right] + 25.$$

32. Use the equation in Problem 31 to approximate the answers to the following questions about the temperatures in Fairbanks, Alaska, shown in Fig. 2.28. Assume that the year has 365 days.
a) What is the highest mean daily temperature shown?
b) What is the lowest mean daily temperature shown?
c) What is the average of the highest and lowest mean daily temperatures shown? Why is this average the vertical shift of the function?

Find the limits in Problems 33–36.

33. $\lim_{h\to 0} \frac{(\sin h)(1 - \cos h)}{h^2}$ **34.** $\lim_{x\to 0} \frac{1 - \cos x}{x^2}$

35. $\lim_{x\to 0} \frac{1 - \cos x}{\sin x}$ (*Hint:* Divide numerator and denominator by x.)

36. $\lim_{x\to 0} x \cot x$

37. In Eq. (11e) take $B = A$. Does the result agree with something else you know?

38. In Eq. (11d) take $B = A$. Does the result agree with something you already know?

Find the angles between the pairs of curves in Problems 39–44.

39. $3x + y = 5$, $2x - y = 4$

40. $y = \sqrt{3}x - 1$, $y = -\sqrt{3}x + 2$

41. $y = \sqrt{3/4 + x}$, $y = \sqrt{3/4 - x}$

42. $y = x^3$, $y = \sqrt{x}$ (two points of intersection)

43. $y = x^2$, $y = \sqrt[3]{x}$ (two points of intersection)

44. $x^2 + y^2 = 16$, $y^2 = 6x$ (two points of intersection)

45. What horizontal line crosses the curve $y = \sqrt{x}$ at a 45° angle?

46. Show that the curves $2x^2 + 3y^2 = 5$ and $y^2 = x^3$ are orthogonal.

47. Find the interior angles of the triangle whose vertices are $A(1, 1)$, $B(3, -1)$, and $C(5, 2)$.

48. Find the slope of the line that bisects the angle ACB in Problem 47.

49. Let A, B, C be the interior angles of a triangle in which no angle is a right angle. By calculating $\tan(A + B + C)$ and observing that

$$A + B + C = 180°,$$

show that

$$\tan A + \tan B + \tan C = \tan A \tan B \tan C.$$

50. Calculate the interior angles of the triangle whose vertices are $A(1, 2)$, $B(2, -1)$, and $C(-1, 1)$, and check your results by showing that they satisfy the equation

$$\tan A + \tan B + \tan C = \tan A \tan B \tan C.$$

(See Problem 49.)

51. Use Eq. (11f) to show that $\tan(\pi - \beta) = \tan(-\beta)$.

52. A function $f(\theta)$ is said to be

an even function of θ if $f(-\theta) = f(\theta)$,
an odd function of θ if $f(-\theta) = -f(\theta)$.

Which of the six trigonometric functions are even, and which are odd?

53. Derive the following formulas from the identities for $\sin(A \pm B)$ and $\cos(A \pm B)$.

a) $\sin\left(x - \frac{\pi}{2}\right) = -\cos x$

b) $\sin\left(x + \frac{\pi}{2}\right) = \cos x$

c) $\cos\left(x - \frac{\pi}{2}\right) = \sin x$

d) $\cos\left(x + \frac{\pi}{2}\right) = -\sin x$

e) $\sin(\pi - x) = \sin x$

f) $\cos(\pi - x) = -\cos x$

***Toolkit* programs**	
Function Evaluator	
Limit Problems	Parametric Equations
Name That Function	Super * Grapher

2.8

Derivatives of Trigonometric Functions

The derivative of $y = \sin x$ with respect to x is the limit

$$\frac{dy}{dx} = \lim_{h \to 0} \frac{\sin(x+h) - \sin x}{h}.$$

To calculate this limit, we use three ideas from our earlier work:

1. $\lim_{h \to 0} \frac{\sin h}{h} = 1$ (Article 1.9, Eq. 14)
2. $\lim_{h \to 0} \frac{\cos h - 1}{h} = 0$ (Article 2.7, Eq. 15)
3. $\sin(x + h) = \sin x \cos h + \cos x \sin h$ (Article 2.7, Eq. 11a)

Then, taking all limits as $h \to 0$, we have

$$\begin{aligned}
\frac{dy}{dx} &= \lim \frac{\sin(x+h) - \sin x}{h} \\
&= \lim \frac{\sin x \cos h + \cos x \sin h - \sin x}{h} \\
&= \lim \frac{\sin x(\cos h - 1) + \cos x \sin h}{h} \\
&= \lim \left(\sin x \frac{\cos h - 1}{h}\right) + \lim \left(\cos x \frac{\sin h}{h}\right) \\
&= \sin x \lim \frac{\cos h - 1}{h} + \cos x \lim \frac{\sin h}{h} \\
&= \sin x \cdot 0 + \cos x \cdot 1 \\
&= \cos x.
\end{aligned}$$

The derivative of $y = \sin x$ with respect to x is

$$\frac{d}{dx} \sin x = \cos x. \tag{1}$$

If u is a differentiable function of x, we can apply the Chain Rule

$$\frac{dy}{dx} = \frac{dy}{du}\frac{du}{dx}$$

to $y = \sin u$ with the following result:

$$\frac{d}{dx} \sin u = \cos u \frac{du}{dx}. \tag{2}$$

EXAMPLE 1

a) $$\frac{d}{dx} \sin 2x = \cos 2x \frac{d}{dx}(2x)$$
$$= 2 \cos 2x$$

b) $$\frac{d}{dx} \sin x^5 = \cos x^5 \frac{d}{dx}(x^5)$$
$$= 5x^4 \cos x^5$$

c) $$\frac{d}{dx} \sin \sqrt{3x} = \cos \sqrt{3x} \frac{d}{dx}(\sqrt{3x})$$
$$= \cos \sqrt{3x} \frac{1}{2\sqrt{3x}} \cdot \frac{d}{dx}(3x)$$
$$= \frac{3 \cos \sqrt{3x}}{2\sqrt{3x}}$$

In (c) we used the Chain Rule twice. □

The answer to the question "Why do we use radian measure in calculus?" is contained in the argument that shows that the derivative of the sine is the cosine. The argument requires that

$$\lim_{h \to 0} \frac{\sin h}{h} = 1.$$

The limit is 1 *only* if h is measured in radians.

EXAMPLE 2 Find dy/dx by implicit differentiation if

$$xy + \sin y = 0.$$

Solution We differentiate both sides of the equation, treating y as a differentiable function of x:

$$x\frac{dy}{dx} + y + \cos y \frac{dy}{dx} = 0$$

$$(x + \cos y)\frac{dy}{dx} + y = 0$$

$$\frac{dy}{dx} = -\frac{y}{x + \cos y}. \quad \square$$

Derivative of cos *u*

To obtain a formula for the derivative of $\cos u$, we use the identities

$$\cos u = \sin\left(\frac{\pi}{2} - u\right), \qquad \sin u = \cos\left(\frac{\pi}{2} - u\right).$$

Thus

$$\begin{aligned}\frac{d}{dx}\cos u &= \frac{d}{dx}\sin\left(\frac{\pi}{2}-u\right)\\ &= \cos\left(\frac{\pi}{2}-u\right)\frac{d}{dx}\left(\frac{\pi}{2}-u\right)\\ &= \sin u\left(-\frac{du}{dx}\right),\end{aligned}$$

or

$$\frac{d}{dx}(\cos u) = -\sin u\frac{du}{dx}. \tag{3}$$

This equation tells us that the derivative of the cosine of a differentiable function is minus the sine of the same function times the derivative of the function.

EXAMPLE 3

a) $\dfrac{d}{dx}\cos 3x = -\sin 3x\dfrac{d}{dx}(3x) = -3\sin 3x$

b) $$\begin{aligned}\frac{d}{dx}\cos^2 3x &= 2\cos 3x\frac{d}{dx}\cos 3x\\ &= 2\cos 3x(-3\sin 3x)\\ &= -6\sin 3x\cos 3x \quad \square\end{aligned}$$

There is more than one correct way to write the answer in Example 3(b). For example, we can use the identity

$$2\sin\theta\cos\theta = \sin 2\theta$$

with $\theta = 3x$ to continue the example another step by writing the answer as

$$-6\sin 3x\cos 3x = -3\sin 6x.$$

If you find that your answer to a differentiation problem in trigonometry differs from someone else's, you may both be right.

EXAMPLE 4 Find the linear approximation to $f(x) = \cos x$ near $x = \pi/2$.

Solution The linear approximation (Article 2.5, Eq. 13), is

$$f(x) \approx f(a) + f'(a)(x-a).$$

With $f(x) = \cos x$ and $a = \pi/2$, we have

$$\begin{aligned}f(\pi/2) &= \cos(\pi/2) = 0,\\ f'(x) &= -\sin x,\\ f'(\pi/2) &= \sin(\pi/2) = -1,\end{aligned}$$

and

$$\cos x \approx 0 + -1\left(x-\frac{\pi}{2}\right) = -x+\frac{\pi}{2}.$$

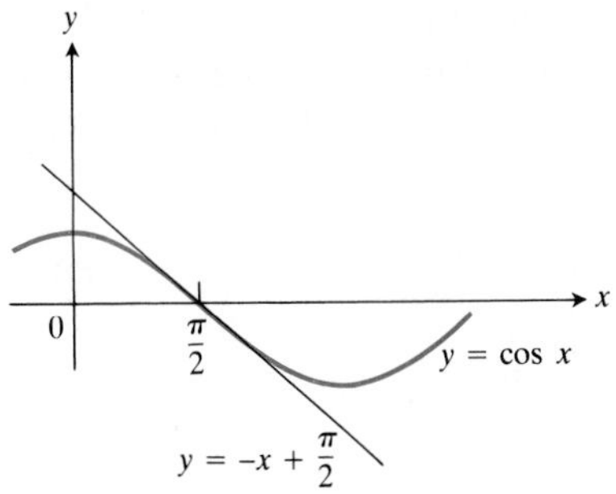

2.33 The linear approximation of $y = \cos x$ at $x = \pi/2$ is $y = -x + (\pi/2)$.

The linear approximation to cos x near $x = \pi/2$ is

$$\cos x \approx -x + \frac{\pi}{2}.$$

The graphs are shown in Fig. 2.33. □

Derivatives of Other Trigonometric Functions

Since sin x and cos x are differentiable functions of x, the functions

$$\tan x = \frac{\sin x}{\cos x}, \qquad \cot x = \frac{\cos x}{\sin x}, \qquad \sec x = \frac{1}{\cos x}, \qquad \csc x = \frac{1}{\sin x}, \tag{4}$$

are differentiable at every value of x at which they are defined. Their derivatives can be calculated from the Quotient Rule.

EXAMPLE 5

a) Find dy/dx if $y = \tan x$.

b) Find dy/dx if $y = \tan u$ and u is a differentiable function of x.

Solution

a)
$$\frac{d}{dx}\tan x = \frac{d}{dx}\left(\frac{\sin x}{\cos x}\right)$$
$$= \frac{\cos x \frac{d}{dx}(\sin x) - \sin x \frac{d}{dx}(\cos x)}{\cos^2 x}$$
$$= \frac{\cos x \cos x - \sin x(-\sin x)}{\cos^2 x}$$
$$= \frac{\cos^2 x + \sin^2 x}{\cos^2 x} = \frac{1}{\cos^2 x}$$
$$= \sec^2 x$$

Thus,

$$\frac{d}{dx}\tan x = \sec^2 x. \tag{5}$$

b) If u is a differentiable function of x, then we can apply the Chain Rule

$$\frac{dy}{dx} = \frac{dy}{du}\frac{du}{dx}$$

to $y = \tan u$ with the following result:

$$\frac{d}{dx}\tan u = \sec^2 u \frac{du}{dx}. \ \square \tag{6}$$

If u is a differentiable function of x, the differentiation rules for the other three functions in (4) can be derived in much the same way that the derivative of the tangent was derived in Example 5. The results are:

$$\frac{d}{dx}\sec u = \sec u \tan u \frac{du}{dx}, \tag{7}$$

$$\frac{d}{dx}\csc u = -\csc u \cot u \frac{du}{dx}, \tag{8}$$

$$\frac{d}{dx}\cot u = -\csc^2 u \frac{du}{dx}. \tag{9}$$

In differentiation problems you can always get along by converting all the trigonometric functions to sines and cosines before differentiating.

EXAMPLE 6 Find dy/dx if

$$y = \sec^2 5x = (\cos 5x)^{-2}.$$

Solution METHOD 1. We apply the formula for the derivative of a function to a power,

$$\frac{d(u^n)}{dx} = nu^{n-1}\frac{du}{dx}.$$

We then have

$$\frac{dy}{dx} = -2(\cos 5x)^{-3}\frac{d(\cos 5x)}{dx}$$

$$= (-2\sec^3 5x)\left(-\sin 5x\frac{d(5x)}{dx}\right)$$

$$= +10\sec^3 5x \sin 5x. \tag{10}$$

METHOD 2. We use Eq. (7) and the Power Rule to get

$$\frac{d}{dx}\sec^2 5x = 2(\sec 5x)^1 \cdot \frac{d}{dx}(\sec 5x) \tag{11}$$

$$= 2\sec 5x \cdot \sec 5x \tan 5x \cdot \frac{d}{dx}(5x) \tag{12}$$

$$= 10\sec^2 5x \tan 5x. \ \square \tag{13}$$

Why do the answers in Eqs. (10) and (13) differ? Here is another example in which answers that look different are both correct. If we change the $\tan 5x$ in Eq. (13) to

$$\tan 5x = \frac{\sin 5x}{\cos 5x},$$

we find that

$$10\sec^2 5x \tan 5x = 10\sec^2 5x\frac{\sin 5x}{\cos 5x}$$

$$= 10\sec^2 5x \cdot \frac{1}{\cos 5x} \cdot \sin 5x$$

$$= 10\sec^2 5x \cdot \sec 5x \cdot \sin 5x$$

$$= 10\sec^3 5x \sin 5x,$$

which is the answer in Eq. (10).

EXAMPLE 7 Find the linear approximation to $f(x) = \tan x$ near $x = 0$.

Solution From Article 2.5, Eq. (13), we know that the linear approximation is

$$f(x) \approx f(a) + f'(a)(x - a).$$

With $f(x) = \tan x$ and $a = 0$, we have

$$f(0) = \tan(0) = 0, \qquad f'(x) = \sec^2 x, \qquad f'(0) = \sec^2(0) = 1,$$

and

$$\tan x \approx 0 + 1(x - 0) = x.$$

The linear approximation to $\tan x$ near $a = 0$ is

$$\tan x \approx x. \quad \square$$

Continuity

Since the trigonometric functions we have been studying are differentiable, they are also continuous (Theorem 7 in Article 1.11). That is,

$$\lim_{x \to a} f(x) = f(a)$$

for all these functions. This means we can calculate various limits of trigonometric functions as $x \to a$ by evaluating the functions at $x = a$.

Table 2.4 Standard linear approximations, $x \approx 0$

1. $\sin x \approx x$
2. $\cos x \approx 1$
3. $\tan x \approx x$

Table 2.4 lists the standard linear approximations for the sine, cosine, and tangent functions near $x = 0$. The approximations for the sine and cosine are derived in Problem 56. The approximation for the tangent was derived in Example 7.

Derivatives of the Basic Trigonometric Functions

If u is a differentiable function of x, then:

1. $\dfrac{d}{dx}\sin u = \cos u \dfrac{du}{dx}$,
2. $\dfrac{d}{dx}\cos u = -\sin u \dfrac{du}{dx}$,
3. $\dfrac{d}{dx}\tan u = \sec^2 u \dfrac{du}{dx}$,
4. $\dfrac{d}{dx}\sec u = \sec u \tan u \dfrac{du}{dx}$,
5. $\dfrac{d}{dx}\csc u = -\csc u \cot u \dfrac{du}{dx}$,
6. $\dfrac{d}{dx}\cot u = -\csc^2 u \dfrac{du}{dx}$.

PROBLEMS

In Problems 1–36, find dy/dx.

1. $y = \sin(x + 1)$
2. $y = -\cos x$
3. $y = \sin(x/2)$
4. $y = \sin(-x)$
5. $y = \cos 5x$
6. $y = \cos(-x)$
7. $y = \cos(-2x)$
8. $y = \sin 7x$
9. $y = \sin(3x + 4)$
10. $y = \cos(2 - x)$
11. $y = x \sin x$
12. $y = \sin 5(x - 1)$
13. $y = x \sin x + \cos x$
14. $y = \dfrac{1}{\sin x}$
15. $y = \dfrac{1}{\cos x}$
16. $y = \dfrac{\sin x}{\cos x}$
17. $y = \sec x$
18. $y = \cot x$
19. $y = \sec(1 - x)$
20. $y = \dfrac{2}{\cos 3x}$
21. $y = \tan 2x$
22. $y = \cos(ax + b)$
23. $y = \sin^2 x$
24. $y = \sin^2 x + \cos^2 x$
25. $y = \cos^2 5x$
26. $y = \cot^2 x$
27. $y = \tan(5x - 1)$
28. $y = \sin x - x \cos x$
29. $y = 2 \sin x \cos x$
30. $y = \sec(x^2 + 1)$
31. $y = \sqrt{2 + \cos 2x}$
32. $y = \sin(1 - x^2)$
33. $y = \cos \sqrt{x}$
34. $y = \sec^2 x - \tan^2 x$
35. $y = \sqrt{\dfrac{1 + \cos 2x}{2}}$ (*Hint:* Use a half-angle formula first.)
36. $y = \sin^2 x^2$

Assume that each of the equations in Problems 37–41 defines y as a differentiable function of x. Find dy/dx by implicit differentiation.

37. $x = \tan y$
38. $x = \sin y$
39. $y^2 = \sin^4 2x + \cos^4 2x$
40. $x + \sin y = xy$
41. $x + \tan(xy) = 0$

42. Assume that the equation $2xy + \pi \sin y = 2\pi$ defines y as a differentiable function of x. Find dy/dx when $x = 1$ and $y = \pi/2$.

43. Find an equation for the tangent to the curve $x \sin 2y = y \cos 2x$ at the point $(\pi/4, \pi/2)$.

Find the limits in Problems 44–51.

44. $\lim_{x \to 2} \sin\left(\frac{1}{x} - \frac{1}{2}\right)$
45. $\lim_{x \to \pi/4} \dfrac{\sin x}{\cos x}$
46. $\lim_{x \to -\pi} \cos^2 x$
47. $\lim_{x \to \pi} \sec(1 + \cos x)$
48. $\lim_{x \to 0} (\sec x + \tan x)$
49. $\lim_{x \to 0} x \csc x$
50. $\lim_{h \to 0} \dfrac{\sin(a + h) - \sin a}{h}$
51. $\lim_{h \to 0} \dfrac{\cos(a + h) - \cos a}{h}$

52. Find an equation for the tangent to the curve $y = \sin mx$ at $x = 0$.

53. Find dy/dx at $t = \pi/4$ if:
a) $x = \cos^2 t, \quad y = \sin^2 t$;
b) $x = \cos^3 t, \quad y = \sin^3 t$;
c) $x = \tan^2 t, \quad y = \sin 2t$.

54. By about how much does the value of $\cos x$ change if x increases a small amount from $\pi/2$ to $\pi/2 + \Delta x$?

55. Graph $y = \tan x$ and its linear approximation $y = x$ together for $-\pi/4 \le x \le \pi/4$.

56. Find the linear approximation for each function $f(x)$ near the given point. Illustrate with a sketch.
a) $f(x) = \sin x$ near $x = \pi/2$
b) $f(x) = \sin x$ near $x = \pi$
c) $f(x) = \cos x$ near $x = 0$
d) $f(x) = \cos x$ near $x = -\pi/2$
e) $f(x) = \tan x$ near $x = \pi/4$
f) $f(x) = \sec x$ near $x = \pi/4$
g) $f(x) = \tan x$ near $x = -\pi/4$
h) $f(x) = \sec x$ near $x = -\pi/4$

57. Is there a value of b that makes

$$f(x) = \begin{cases} x + b & \text{for } x < 0, \\ \cos x & \text{for } x \ge 0, \end{cases}$$

continuous at $x = 0$? If so, what is it? If not, why not?

58. Figure 2.34 shows a boat 1 km off shore, sweeping the shore with a searchlight. The light turns at a constant rate (angular velocity) $d\theta/dt = -3/5$ radians per second.
a) Express x (see Fig. 2.34) in terms of θ.
b) Differentiate both sides of the equation you obtained in (a) with respect to t. Substitute $d\theta/dt = -3/5$. This will express dx/dt (the rate at which the light moves along the shore) as a function of θ.
c) How fast is the light moving along the shore when it reaches point A?
d) How many revolutions per minute is 0.6 radians per second?

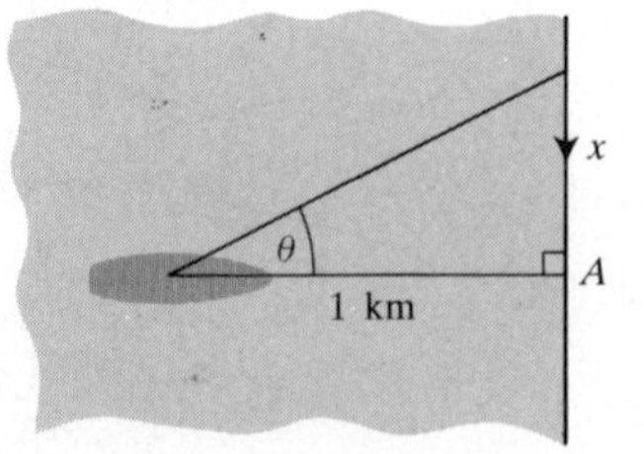

2.34 The boat in Problem 58.

59. *A Kepler equation.* The problem of locating a planet in its orbit at a given time and date leads to solving "Kepler" equations of the form

$$f(x) = x - 1 - \frac{1}{2}\sin x = 0.$$

a) Find $f(0) < 0$ and $f(2) > 0$ to show that $f(x)$ has a zero between $x = 0$ and $x = 2$.
b) Estimate the solution to the equation $f(x) = 0$ by replacing $\sin x$ by its linear approximation near $x = \pi/2$ and solving the resulting linear equation.
c) (*Calculator*) Check your solution in the original equation.

60. A useful linear approximation to

$$\frac{1}{1 + \tan x}$$

can be obtained by combining the approximations

$$\frac{1}{1+x} \approx 1 - x \quad \text{and} \quad \tan x \approx x$$

to get

$$\frac{1}{1 + \tan x} \approx 1 - x.$$

Show that this is the linear approximation of $1/(1 + \tan x)$ obtained from Eq. (13) in Article 2.5.

61. Find the linear approximation to $f(x) = \sqrt{1 + x} + \sin x$ near $x = 0$. How is it related to the individual linear approximations for $\sqrt{1 + x}$ and $\sin x$?

62. Let $f(x) = \sqrt{1 + x} + \sin x - 0.5$.
a) Find $f(-\pi/4) < 0$ and $f(0) > 0$ to show that the equation $f(x) = 0$ has a solution between $-\pi/4$ and 0.
b) To estimate the solution of $f(x) = 0$, replace $\sqrt{1 + x}$ and $\sin x$ by their linear approximations near $x = 0$ and solve the resulting linear equation.
c) (*Calculator*) Check your estimate in the original equation.

63. Carry out the following steps to estimate the solution of

$$2\cos x = \sqrt{1 + x}.$$

a) Let $f(x) = 2\cos x - \sqrt{1 + x}$. Find $f(0) > 0$ and $f(\pi/2) < 0$ to show that $f(x)$ has a zero between 0 and $\pi/2$.
b) Find the linear approximations for $\cos x$ near $x = \pi/4$ and $\sqrt{1 + x}$ near $x = 0.69$.
c) (*Calculator*) To estimate the solution of the original equation, replace $\cos x$ and $\sqrt{1 + x}$ by their linear approximations from (b) and solve the resulting linear equation for x. Check your estimate in the original equation.

64. Derive Eq. (7) by writing $\sec u = 1/\cos u$ and differentiating with respect to x.

65. Derive Eq. (8) by writing $\csc u = 1/\sin u$ and differentiating with respect to x.

66. Derive Eq. (9) by writing $\cot u = \cos u/\sin u$ and differentiating with respect to x.

***Toolkit* programs**

Derivative Grapher	Super * Grapher
Function Evaluator	

2.9 Newton's Method for Approximating Solutions of Equations

We know simple formulas for solving linear and quadratic equations. There are somewhat more complicated formulas for cubic and quartic equations. At one time it was hoped that similar formulas might be found for quintic and higher-degree equations, but a young Norwegian mathematician, Niels Henrik Abel (1802–1829), showed that no formulas like these are possible for equations of degree greater than four.

When exact formulas are not available, we can turn to one of a number of numerical techniques for approximating the solutions we seek. One of these is the Newton or, as it is more accurately named, the Newton-Raphson method, discussed in this article. If you have access to a computer or a programmable calculator, you can easily write a program to do the arithmetic. If not, you can still see how it can be done. The procedure is as follows.

The Procedure for Newton's Method

1. Guess a first approximation to a root of the equation $f(x) = 0$. A graph of $y = f(x)$ will help.
2. Use the first approximation to get a second, the second to get a third, and so on. To go from the nth approximation x_n to the next approximation x_{n+1}, use the formula

$$x_{n+1} = x_n - \frac{f(x_n)}{f'(x_n)}, \tag{1}$$

where $f'(x_n)$ is the derivative of f at x_n.

We first show how the method works, and then go to the theory behind it.

In our first example we find decimal approximations to $\sqrt{2}$ by estimating the positive root of the equation $f(x) = x^2 - 2 = 0$.

EXAMPLE 1 Find the positive root of the equation $f(x) = x^2 - 2 = 0$.

Solution With $f(x) = x^2 - 2$ and $f'(x) = 2x$, Eq. (1) becomes

$$x_{n+1} = x_n - \frac{x_n^2 - 2}{2x_n}. \tag{2}$$

To use our calculator efficiently, we rewrite Eq. (2) in a form that uses fewer arithmetic operations:

$$x_{n+1} = x_n - \frac{x_n^2 - 2}{2x_n} = \frac{x_n^2 + 2}{2x_n} = \frac{x_n}{2} + \frac{1}{x_n}. \tag{3}$$

The equation

$$x_{n+1} = \frac{x_n}{2} + \frac{1}{x_n} \tag{4}$$

permits us to go from each approximation to the next by the following steps:

	Operation	Result	Example
[x]	Enter x	x	1.5
[$\frac{1}{x}$] [STO]	Store the reciprocal	$\frac{1}{x}$	0.66667
[$\frac{1}{x}$]	Take the reciprocal of the display again	x	1.5
[÷] [2]	Divide by 2	$\frac{x}{2}$	0.75
[+] [RCL] [=]	Add memory to display	$\frac{x}{2} + \frac{1}{x}$	1.41667

With the starting value $x_0 = 1$, we get the results in the first column of the following table (to five decimal places, $\sqrt{2} = 1.41421$).

	Error	Number of correct figures
$x_0 = 1$	-0.41421	1
$x_1 = 1.5$	$+0.08579$	1
$x_2 = 1.41667$	0.00245	3
$x_3 = 1.41422$	0.00001	5

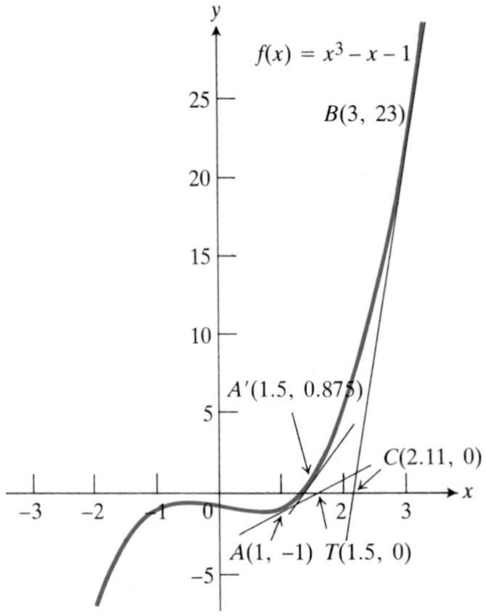

2.35 Newton's method for finding a root of $f(x) = 0$ uses the tangent to approximate the curve.

Newton's method is the method used by most calculators to calculate roots because it converges so fast.† If the arithmetic in the table had been carried to 13 decimal places rather than 5, then going one step further to x_4 would have given $\sqrt{2}$ to more than 10 decimal places. □

EXAMPLE 2 Use the procedure described above to find where the graph of $y = x^3$ intersects the line $y = x + 1$.

Solution The curve and line cross when $x^3 = x + 1$. We put this equation into the form $f(x) = 0$ with

$$f(x) = x^3 - x - 1.$$

Then $f'(x) = 3x^2 - 1$. We used a brief table of values of $f(x)$, $f'(x)$, and $x - f(x)/f'(x)$ to construct the graph shown in Fig. 2.35. The curve crosses the x-axis at a root of the equation $f(x) = 0$. We have taken $x_0 = 1$ as our first approximation. We then used Eq. (1) repeatedly to get x_1, x_2, and so on, as shown in Table 2.5.

Table 2.5 Approximate data for Fig. 2.35.

n	x_n	$f(x_n)$	$f'(x_n)$	$x_{n+1} = x_n - \dfrac{f(x_n)}{f'(x_n)}$
0	1	-1	2	1.5
1	1.5	0.875	5.75	1.347826087
2	1.347826087	0.100682174	4.449905482	1.325200399
3	1.325200399	0.002058363	4.268468293	1.324718174
4	1.324718174	0.000000925	4.264634722	1.324717957
5	1.324717957	-5×10^{-10}	4.264632997	1.324717957

At $n = 5$ we come to the result $x_5 = x_4 = 1.324717957$. When $x_{n+1} = x_n$, Eq. (1) shows that $f(x_n) = 0$. Hence we have found a solution of $f(x) = 0$ to nine decimals, or so it appears. (Our calculator shows only ten figures, and we cannot guarantee the accuracy of the ninth decimal, though we believe it to be correct.) □

†An estimate of the error in Newton's method, calculated from the second derivative of f, may be found in many books on numerical analysis, among them Gerald and Wheatley's *Applied Numerical Analysis*, 3d ed. (Reading, Massachusetts: Addison-Wesley, 1984).

What is the theory behind the method? It is this: We use the tangent to approximate the graph of $y = f(x)$ near the point $P(x_n, y_n)$, where $y_n = f(x_n)$ is small, and we let x_{n+1} be the value of x where that tangent line crosses the x-axis. (We assume that the slope $f'(x_n)$ of the tangent is not zero.) The equation of the tangent is

$$y - y_n = f'(x_n)(x - x_n). \tag{5}$$

We put $y_n = f(x_n)$ and $y = 0$ into Eq. (5) and solve for x to get

$$x = x_n - \frac{f(x_n)}{f'(x_n)}.$$

REMARK 1. The method doesn't work if $f'(x_n) = 0$. In that case, choose a new starting place. Of course, it may happen that $f(x) = 0$ and $f'(x) = 0$ have a common root. To detect whether that is so, we could first find the solutions of $f'(x) = 0$, and then check the value of $f(x)$ at such places.

In Example 2, we started at the point $A(1, -1)$ in Fig. 2.35. The tangent at A crosses the x-axis at $T(1.5, 0)$. We take $x_1 = 1.5$, $y_1 = f(x_1) = 0.875$, and approximate the graph near $A'(1.5, 0.875)$ by the line tangent to the curve at A'. That tangent crosses the x-axis at $x_2 = 1.347\ldots$. Each successive approximation becomes the input for the right side of Eq. (1) to get the next approximation. It is clear from Eq. (1) that $x_{n+1} = x_n$ implies that $f(x_n) = 0$, and conversely. Therefore, if the process stops with $x_{n+1} = x_n$, as in the example, we have a zero of $f(x)$.

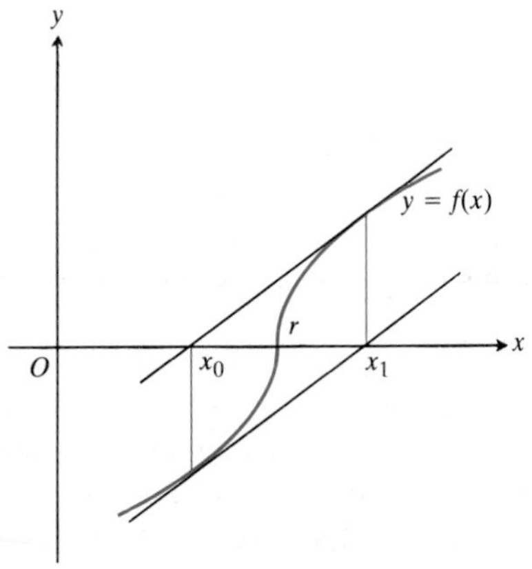

2.36 The graph of a function for which Newton's method fails to converge.

REMARK 2. In Fig. 2.35 we have also indicated that the process might have started at the point $B(3, 23)$ on the curve, with $x_0 = 3$. Point B is quite far from the x-axis, but the tangent at B crosses the x-axis at $C(2.11\ldots, 0)$, so that x_1 is still an improvement over x_0. If we take $x_0 = 3$ and use Eq. (1) repeatedly as before, with $f(x) = x^3 - x - 1$ and $f'(x) = 3x^2 - 1$, we confirm the nine-place solution $x_6 = x_5 = 1.324717957$ in six steps.

The curve in Fig. 2.35 has a high turning point at $x = -1/\sqrt{3}$ and a low turning point at $x = +1/\sqrt{3}$. We would not expect good results from Newton's method if we were to start with x_0 between these points, but we can start any place between A and B and get the answer. It would not be very clever to do so, but we could even begin far to the right of B, for example with $x_0 = 10$. It takes a bit longer, but the process still converges to the same answer as before.

REMARK 3. Newton's method does not always converge. For instance, if

$$f(x) = \begin{cases} \sqrt{x - r} & \text{for } x \geq r, \\ -\sqrt{r - x} & \text{for } x \leq r, \end{cases} \tag{6}$$

the graph will be like that shown in Fig. 2.36. If we begin with $x_0 = r - h$, we get $x_1 = r + h$, and successive approximations go back and forth between these two values. No amount of iteration will bring us any closer to the root r than our first guess.

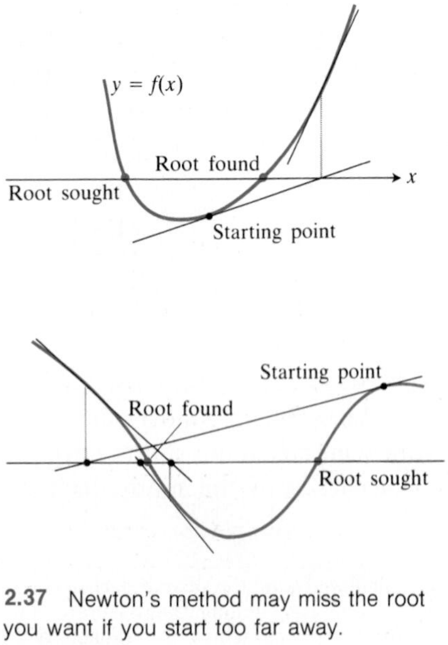

2.37 Newton's method may miss the root you want if you start too far away.

REMARK 4. If Newton's method does converge, it converges to a root of $f(x)$. However, the method may converge to a root different from the expected one if the starting value is not close enough to the root sought. Figure 2.37 shows two of the ways in which this might happen.

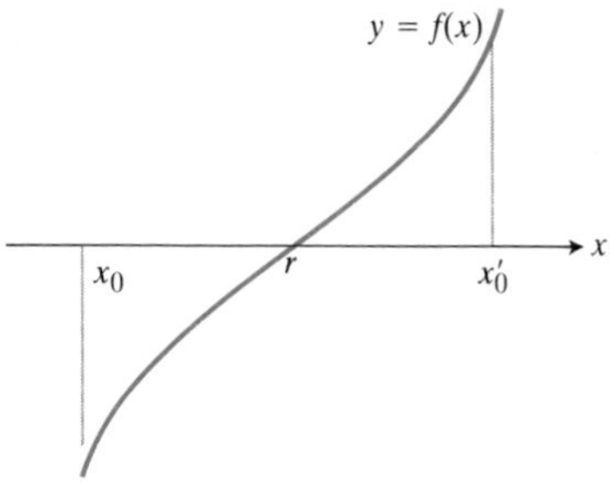

2.38 The curve $y = f(x)$ is convex toward the axis between x_0 and r and between x'_0 and r. Newton's method will converge to r from either starting point.

REMARK 5. When will Newton's method converge? If the inequality

$$\left|\frac{f(x)f''(x)}{[f'(x)]^2}\right| < 1 \tag{7}$$

holds for all values of x in an interval about a root r of f, then the method will converge to r for any starting value x_0 in that interval. This is a *sufficient*, but not a necessary, condition. The method can (and does) converge in some cases when there is no interval about r in which the inequality (7) holds.

J. Raymond Mouraille (1768), and later Joseph Fourier (1768–1830), independently, discovered that Newton's method will always work if the curve $y = f(x)$ is convex ("bulges") toward the axis in the interval between x_0 and the root sought. See Fig. 2.38.

PROBLEMS

In Problems 1–6, sketch the graph of $y = f(x)$ and show that sign $f(a) = -\text{sign } f(b)$. Then use Newton's method to estimate the root of the equation $f(x) = 0$ between a and b to three decimal places.

1. $f(x) = x^2 + x - 1, \quad a = 0, \quad b = 1$

2. $f(x) = x^3 + x - 1, \quad a = 0, \quad b = 1$

3. $f(x) = x^4 + x - 3, \quad a = 1, \quad b = 2$

4. $f(x) = x^4 - 2, \quad a = 1, \quad b = 2$

5. $f(x) = 2 - x^4, \quad a = -1, \quad b = -2$

6. $f(x) = \sqrt{2x+1} - \sqrt{x+4}, \quad a = 2, \quad b = 4$

7. Suppose our first guess is lucky, in the sense that x_0 is a root of $f(x) = 0$. What happens to x_1 and later approximations?

8. (*Calculator*) Show that $f(x) = x^3 + 2x - 4$ has a root between $x = 1$ and $x = 2$. Find the root to five decimal places.

9. (*Calculator*) Show that $f(x) = x^4 - x^3 - 75$ has a root between $x = 3$ and $x = 4$. Find the root to five decimal places.

10. a) Explain why the following four statements ask for the same information:

i) Find the roots of $f(x) = x^3 - 3x - 1$.

ii) Find the x-coordinates of the intersections of the curve $y = x^3$ with the line $y = 3x + 1$.

iii) Find the x-coordinates of the points where the curve $y = x^3 - 3x$ crosses the horizontal line $y = 1$.

iv) Find the values of x where the derivative of $g(x) = \frac{1}{4}x^4 - \frac{3}{2}x^2 - x + 5$ equals zero.

b) Sketch the graph of $f(x) = x^3 - 3x - 1$ over the interval $-2 \le x \le 1$.

c) (*Calculator*) Find the positive root of $f(x) = x^3 - 3x - 1$ to five decimal places.

d) (*Calculator*) Find the two negative roots of $f(x) = x^3 - 3x - 1$ to five decimal places.

11. (*Calculator*) Estimate π to five decimal places by applying Newton's method to the equation $\tan x = 0$ with $x_0 = 3$. Remember to use radians.

12. You plan to estimate $\pi/2$ to five decimal places by solving the equation $\cos x = 0$ by Newton's method. Does it matter what your starting value is? Explain.

13. (*Calculator*) Find the point of intersection of the curve $y = \cos x$ with the line $y = x$ to five decimal places.

14. (*Calculator*) Graphing $f(x) = x - 1 - \frac{1}{2}\sin x$ suggests that the function has a root near $x = 1.5$. Use one application of Newton's method to improve this estimate. That is, start with $x_0 = 1.5$ and find x_1. (The value of the root is 1.49870 to five decimal places.) Remember to use radians.

15. (*Programmable calculator*) Find two real roots of the equation $x^4 - 2x^3 - x^2 - 2x + 2 = 0$ to six decimal places.

16. To find $x = \sqrt[q]{a}$, we apply Newton's method to $f(x) = x^q - a$. Here we assume that a is a positive real number and q is a positive integer. Show that x_1 is a "weighted average" of x_0 and a/x_0^{q-1}, and find the coefficients m_0, m_1 such that

$$x_1 = m_0x_0 + m_1\left(\frac{a}{x_0^{q-1}}\right), \qquad \begin{matrix} m_0 > 0, \; m_1 > 0, \\ m_0 + m_1 = 1. \end{matrix}$$

What conclusion would you reach if x_0 and a/x_0^{q-1} were equal? What would be the value of x_1 in that case? (You may also wish to read the article by J. P. Ballantine, "An Averaging Method of Extracting Roots," *American Mathematical Monthly,* **63,** 1956, pp. 249–252, where more efficient ways of averaging are discussed. Also see J. S. Frame, "The Solution of Equations by Continued Fractions," *American Mathematical Monthly,* **60,** 1953, pp. 293–305.)

17. Show that Newton's method applied to $f(x)$ in Eq. (6) leads to $x_1 = r + h$ if $x_0 = r - h$, and to $x_1 = r - h$ if $x_0 = r + h$, where $h > 0$. Interpret the result geometrically.

18. (See Remark 3.) Is it possible that successive approximations actually get "worse," in that x_{n+1} is farther away from the root r than x_n is? Can you find such a "pathological" example? (*Hint:* Try cube roots in place of square roots in Eq. 6.)

***Toolkit* programs**	
Function Evaluator	Sequences and Series
Root Finder	Super * Grapher

2.10
Inverse Functions and the Picard Method

As you know, a function is a rule that assigns a number in its range to each number in its domain. Some functions, like

$$y = \sin x, \qquad y = x^2, \quad \text{and} \quad y = 3,$$

can give the same output for different inputs. But other functions, like

$$y = \sqrt{x}, \qquad y = x^3, \quad \text{and} \quad y = 4x - 4,$$

always give different outputs for different inputs. Functions that always give different outputs for different inputs are called *one-to-one* functions.

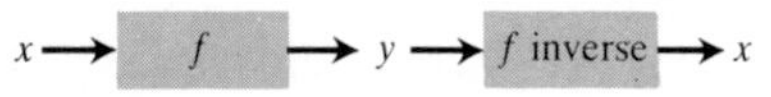

2.39 The inverse of a function f sends every output of f back to the input from which it came.

Since each output of a one-to-one function comes from just one input, any one-to-one function can be reversed to turn the outputs back into the inputs from which they came (Fig. 2.39).

The function defined by reversing a one-to-one function f is called the *inverse of f.* The symbol for the inverse of f is f^{-1}, read "f inverse." The symbol -1 in f^{-1} is *not* an exponent: $f^{-1}(x)$ does not mean $1/f(x)$. As you can see in Fig. 2.40, the result of composing f and f^{-1} in either order is the *identity function,* the function that assigns each number to itself.

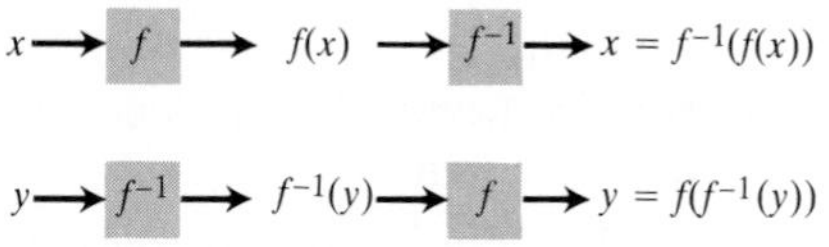

2.40 If $y = f(x)$ is a one-to-one function, then $f^{-1}(f(x)) = x$ and $f(f^{-1}(y)) = y$. Each of the composites $f^{-1} \circ f$ and $f \circ f^{-1}$ is the identity function on its domain.

Figure 2.41 shows the graph of the function $y = \sqrt{x}$, whose domain is $x \geq 0$ and whose range is $y \geq 0$. For each input x_0 the function gives a single output, $y = \sqrt{x_0}$. Since every nonnegative y is the image of just one x under this function, we can reverse the construction. That is, we can start with a $y \geq 0$ and then go over to the curve and down to $x = y^2$ on the x-axis. The construction in reverse defines the function $g(y) = y^2$, the inverse of $f(x) = \sqrt{x}$.

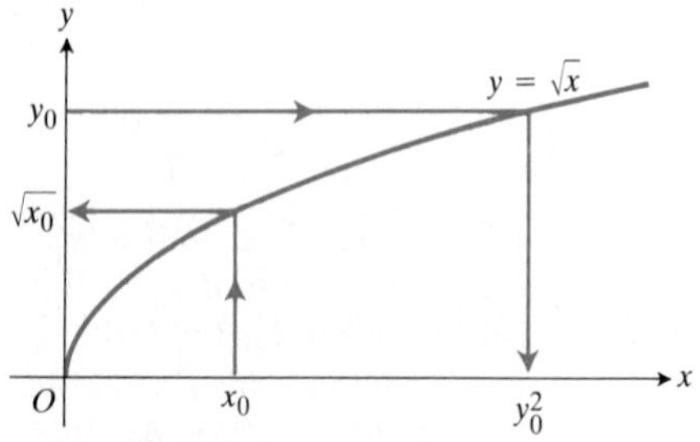

2.41 The inverse of $y = \sqrt{x} = f(x)$ is $x = y^2 = g(y)$.

The algebraic description of what we see in Fig. 2.41 is

$$\begin{aligned} g(f(x)) &= (\sqrt{x})^2 = x, \\ f(g(y)) &= \sqrt{y^2} = y. \end{aligned} \tag{1}$$

(In general, $\sqrt{y^2} = |y|$, but since $y \geq 0$ in this case, we get $\sqrt{y^2} = y$.)

In inverse notation,

$$g = f^{-1}.$$

Note that the equations in (1) also say that f is the inverse of g:

$$f = g^{-1}.$$

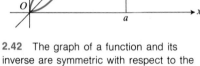

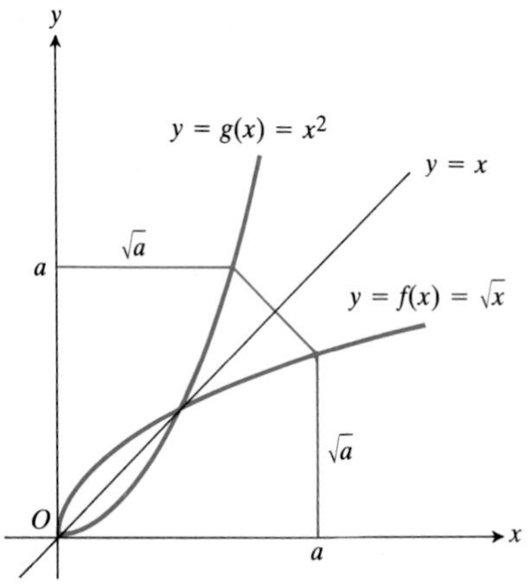

2.42 The graph of a function and its inverse are symmetric with respect to the line $y = x$.

Every one-to-one function is the inverse of its own inverse:

$$(f^{-1})^{-1} = f.$$

If the squaring function g is expressed with x as the independent variable, so that $g(x) = x^2$, the graph of g can be obtained by reflecting the graph of f in the 45° line $y = x$, as in Fig. 2.42. This is because any two points (x, y) and (y, x) whose coordinates are reversed are symmetric with respect to the line $y = x$. The pairs $(a, \sqrt{a})$ and $(\sqrt{a}, a)$ have this symmetry. The first lies on the graph of $f(x) = \sqrt{x}$, the second on the graph of $g(x) = x^2$.

If a one-to-one function is defined by an equation $y = f(x)$ that we can solve for x in terms of y, we get an equation in the form $x = g(y)$ that expresses the inverse of f in terms of y. To express the inverse in a form in which x denotes the independent variable, we interchange the letters x and y in the equation $x = g(y)$. Here are some examples.

EXAMPLE 1 Find the inverse of

$$y = \frac{1}{4}x + 3.$$

Solution We solve the given equation for x in terms of y:

$$x = 4y - 12.$$

We then interchange x and y in the formula $x = 4y - 12$ to get

$$y = 4x - 12.$$

The inverse of $y = \frac{1}{4}x + 3$ is $y = 4x - 12$.

To check, we let

$$f(x) = \frac{1}{4}x + 3, \qquad g(x) = 4x - 12.$$

Then,

$$g(f(x)) = 4\left(\frac{1}{4}x + 3\right) - 12 = x + 12 - 12 = x,$$
$$f(g(x)) = \frac{1}{4}(4x - 12) + 3 = x - 3 + 3 = x. \ \square$$

EXAMPLE 2 Find the inverse of $y = \sqrt{x}$.

Solution We solve the given equation for x in terms of y to get

$$x = y^2.$$

Then we interchange x and y to get

$$y = x^2.$$

This gives an expression for the inverse in which the independent variable is denoted by x. To check that the functions $y = x^2$ and $y = \sqrt{x}$ are inverses, we calculate that their composite in either order is the identity function:

$$\sqrt{x^2} = |x| = x \quad \text{and } (\sqrt{x})^2 = x.$$

(The equality $|x| = x$ holds because $x \geq 0$ in this example.) □

EXAMPLE 3 Find the inverse of the function $y = 8x^3$.

Solution We solve the equation $y = 8x^3$ for x in terms of y, getting

$$x = \frac{\sqrt[3]{y}}{2}.$$

We interchange x and y to get

$$y = \frac{\sqrt[3]{x}}{2}.$$

The inverse of $y = 8x^3$ is $y = \sqrt[3]{x}/2$.

To check the calculation we form the composites of the two functions to see that

$$y = 8\left(\frac{\sqrt[3]{x}}{2}\right)^3 = 8\left(\frac{x}{8}\right) = x, \qquad y = \frac{1}{2}\sqrt[3]{8x^3} = \frac{1}{2}(2x) = x.$$

Each composite is the identity function. □

Derivatives of Inverses of Differentiable Functions

If f and g are inverses of each other, their graphs are mirror images of each other across the line $y = x$, as in Fig. 2.43. Thus if L_1 is a line tangent to the graph of f at $(a, f(a))$ and L_2 is the mirror image of L_1 across the line $y = x$, it is reasonable to expect L_2 to be tangent to the graph of g at $f(a)$. Since rise/run on L_2 corresponds to run/rise on L_1, we see that the slope m_2 of L_2 is the reciprocal of the slope m_1 of L_1:

$$m_2 = \frac{1}{m_1} = \frac{1}{f'(a)}. \tag{2}$$

If L_2 really is tangent to the graph of g at $(f(a), a)$, then the slope of L_2 is $g'(f(a))$ and we have

$$g'(f(a)) = \frac{1}{f'(a)}. \tag{3}$$

The rule for calculating derivatives of inverses of differentiable functions (proved in more advanced texts) is the following.

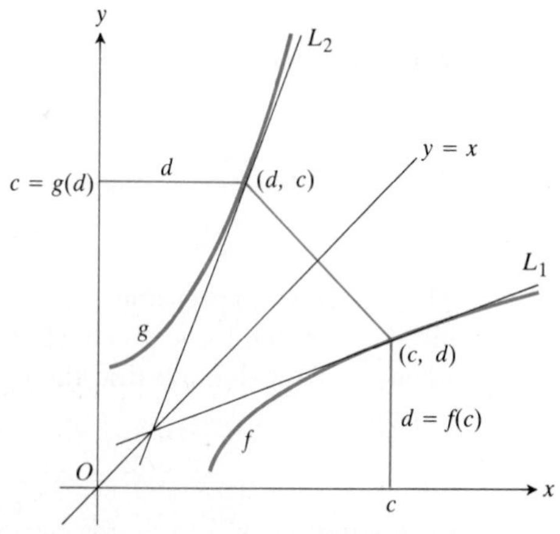

2.43 Graphs of inverse functions f and g and tangents L_1 and L_2 at corresponding points (c, d) and (d, c). Note that $d = f(c)$ and $c = g(d)$.

RULE 11

The Inverse Function Rule

Let f be a one-to-one function defined on an interval I. If f is differentiable, and if f' never takes on the value zero, then $g = f^{-1}$ is differentiable, and the value of its derivative at the point $t = f(a)$ is

$$g'(t) = \frac{1}{f'(a)}, \tag{3a}$$

where a is the unique number in I satisfying $f(a) = t$. In other words, at any point $f(a)$ the value of g' is

$$g'(f(a)) = \frac{1}{f'(a)}. \tag{3b}$$

REMARK. In Article 3.8 we shall show as a corollary of the Mean Value Theorem that if f is differentiable on an open interval I, and if f' is never zero on I, then f is necessarily one-to-one (and therefore has an inverse).

In discussing parametric equations in Article 2.6 we said that if the derivative of a function $x = f(t)$ were never zero on an interval I, then the equation $x = f(t)$ defined t implicitly as a differentiable function of x. What we had in mind, we can now say more precisely, was that f had to be one-to-one because f' was never zero. Therefore, f had to have an inverse, $t = h(x)$. We called the inverse $t = h(x)$ instead of $t = f^{-1}(x)$ because the notion of inverse functions had not yet been introduced.

EXAMPLE 4 If $f(x) = \sqrt{x}$ and $g(x) = x^2$, then $f(9) = \sqrt{9} = 3$. Show that $g'(3) = 1/f'(9)$, as Eq. (3b) predicts.

Solution We have

$$g'(x) = 2x, \qquad f'(x) = \frac{1}{2\sqrt{x}},$$

$$g'(3) = 6, \qquad f'(9) = \frac{1}{2 \cdot 3} = \frac{1}{6}.$$

Therefore,

$$g'(3) = \frac{1}{f'(9)}. \quad \square$$

EXAMPLE 5 Verify Eq. (3b) for the function $f(x) = 4x - 12$ and its inverse

$$g(x) = \frac{1}{4}x + 3.$$

Solution We have

$$f'(x) = 4 \qquad \text{and} \qquad g'(x) = \frac{1}{4} = \frac{1}{f'(x)}$$

for all values of x. $\square$

EXAMPLE 6 Verify Eq. (3b) for the function $f(x) = x^3$ and its inverse $g(x) = \sqrt[3]{x}$ at the point $a = 2$. That is, show that

$$g'(f(2)) = \frac{1}{f'(2)}.$$

Solution We have

$$f(2) = 2^3 = 8, \qquad g'(x) = \frac{1}{3}x^{-2/3},$$

$$f'(x) = 3x^2, \qquad g'(8) = \frac{1}{3}(8)^{-2/3} = \frac{1}{3 \cdot 8^{2/3}} = \frac{1}{3 \cdot 4} = \frac{1}{12},$$

$$f'(2) = 3 \cdot 2^2 = 12.$$

Thus,

$$g'(8) = \frac{1}{f'(2)}. \quad \square$$

The virtue of Eq. (3b) is that it tells exactly how the derivative of $g = f^{-1}$ is to be calculated at $f(a)$: Take the reciprocal of the value of f' at a. As with the Chain Rule, there is a shorter formulation that gives less information but may be easier to remember: If $y = f(x)$ and its inverse $x = g(y)$ are differentiable, then

$$\frac{dx}{dy} = \frac{1}{\dfrac{dy}{dx}}. \tag{4}$$

Sometimes we can use Eq. (4) to calculate a derivative indirectly when it is not convenient to calculate it directly, as in the next example.

EXAMPLE 7 Find dy/dx when

$$x = \sin y - \sqrt{y}.$$

Solution We differentiate both sides of the given equation with respect to y to obtain dx/dy:

$$\frac{dx}{dy} = \cos y - \frac{1}{2\sqrt{y}} = \frac{2\sqrt{y}\cos y - 1}{2\sqrt{y}}.$$

Therefore, from Eq. (4), we have

$$\frac{dy}{dx} = \frac{1}{\dfrac{dx}{dy}} = \frac{2\sqrt{y}}{2\sqrt{y}\cos y - 1}. \quad \square$$

Other Ways to Look at Rule 11

Equation (3b) can be rewritten in the form

$$g'(f(a)) \cdot f'(a) = 1, \tag{5}$$

which may remind you of the Chain Rule. Indeed, there is a connection. If f and g are differentiable functions that are inverses of each other, then

$$(g \circ f)(x) = x \tag{6}$$

and by the Chain Rule (Article 2.6, Eq. 1),

$$(g \circ f)'(a) = g'(f(a)) \cdot f'(a). \tag{7}$$

If g is the inverse of f, then $(g \circ f)(x) = g(f(x)) = x$ and the left-hand side of (7) reduces to 1 so that we have

$$1 = g'(f(a)) \cdot f'(a). \tag{8}$$

If $f'(a) \neq 0$, Eq. (8) can be solved for $g'(f(a))$ to obtain

$$g'(f(a)) = \frac{1}{f'(a)},$$

which is Rule 11.

Still another way to look at Rule 11 is this: If $x = f^{-1}(y)$, and if $y = f(x)$ is differentiable at $x = a$, then

$$\Delta y \approx f'(a)\,\Delta x.$$

This means that y is changing about $f'(a)$ times as fast as x, and x is changing about $1/f'(a)$ times as fast as y.

The Picard Method for Finding Roots

The problem of finding the roots of the equation

$$f(x) = 0 \tag{9}$$

is equivalent to that of finding the solutions of the equation

$$g(x) = f(x) + x = x, \tag{10}$$

obtained by adding x to both sides of $f(x) = 0$. Any value of x that satisfies (10) satisfies (9), and conversely. By this simple change we cast (9) into a form that renders it solvable on a computer by a powerful method that we are about to present. The method is called the *Picard method* (after Charles Emile Picard, 1856–1941), the *$g(x) = x$ method,* or simply the *method of iteration*.

If g is a function whose range is contained in its domain, we can start with a point x_0 in the domain and apply g again and again to get

$$x_1 = g(x_0), \qquad x_2 = g(x_1), \qquad x_3 = g(x_2), \ldots. \tag{11}$$

Under some simple restrictions that we will describe after Example 10, the sequence $x_0, x_1, x_2, \ldots$ will converge to (approach) a point x for which

$$g(x) = x.$$

This point solves the equation $f(x) = 0$, because

$$\begin{aligned} f(x) &= g(x) - x \\ &= x - x = 0. \end{aligned} \tag{12}$$

A point x for which $g(x) = x$ is called a *fixed point* of g, and what we are observing in Eq. (12) is that the fixed points of g are precisely the roots of the equation $f(x) = 0$. We shall study the convergence of sequences in detail in Chapter 11.

We begin with an example whose outcome we know in advance so that we can see how the successive approximations in (11) work.

EXAMPLE 8 Solve the equation

$$\frac{1}{4}x + 3 = x$$

by the Picard method.

Table 2.6

x	$g(x) = \frac{1}{4}x + 3$
$x_0 = 1$	$\frac{1}{4} \cdot 1 + 3 = 3.25 = x_1$
$x_1 = 3.25$	$3.8125 = x_2$
$x_2 = 3.8125$	$3.95313 = x_3$
$x_3 = 3.95313$	$3.98828 = x_4$
x_4	$3.99707 = x_5$
x_5	3.99927 .
x_6	3.99982 .
x_7	3.99995 .
x_8	3.99999 .
x_9	4.00000

Solution By algebra we know that the solution is $x = 4$. To apply the Picard method, we choose a starting point, say $x_0 = 1$, and calculate the successive values of

$$g(x) = \frac{1}{4}x + 3$$

as in Eq. (11). Table 2.6 lists the results. The fixed point of g is found to five decimal places in ten steps.

The geometry of the method is shown in Fig. 2.44. We start with $x_0 = 1$ and calculate the first y-value, $g(x_0)$. This becomes the second x-value, x_1. The second y-value, $g(x_1)$, becomes the third x-value, x_2, and so on. The process can be shown as a path that starts at $x_0 = 1$, moves vertically to $(x_0, g(x_0)) = (x_0, x_1)$, horizontally to (x_1, x_1), vertically to $(x_1, g(x_1))$, and so on. The path converges to the point where the graph of g meets the line $y = x$, the point where $g(x) = x$. □

EXAMPLE 9 Solve the equation $\cos x = x$ by the Picard method.

Solution We choose a starting value, say $x_0 = 1$, and calculate

$$x_0 = 1, \qquad x_1 = \cos 1, \qquad x_2 = \cos x_1, \ldots .$$

On a scientific calculator set for radian measure this amounts to nothing more than entering 1 and pressing the cosine key again and again until the display stops changing! As you press the key you will note that the successive approximations lie alternately above and below the fixed point

$$x = 0.7390851332.$$

Figure 2.45 shows that the values oscillate this way because the path of the procedure spirals around the fixed point. □

EXAMPLE 10 The Picard method cannot be relied on to solve the equation

$$g(x) = 4x - 12 = x.$$

2.44 The geometric interpretation of the Picard solution of the equation $g(x) = (1/4)x + 3 = x$ of Example 8.

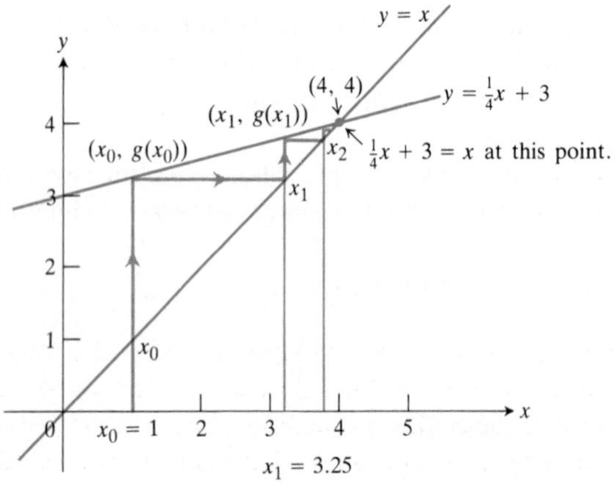

2.45 The solution of $\cos x = x$ by the Picard method starting with $x_0 = 1$.

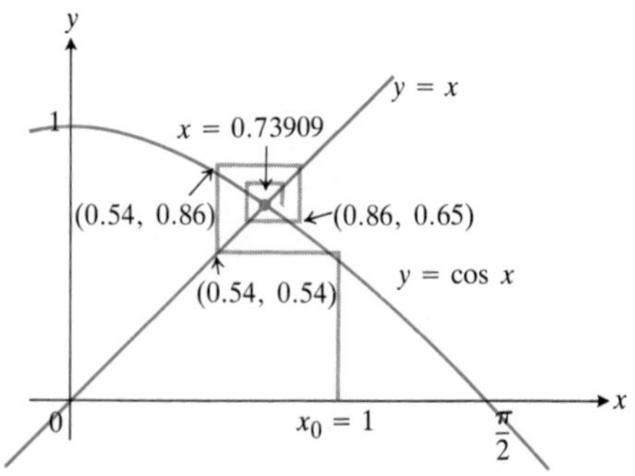

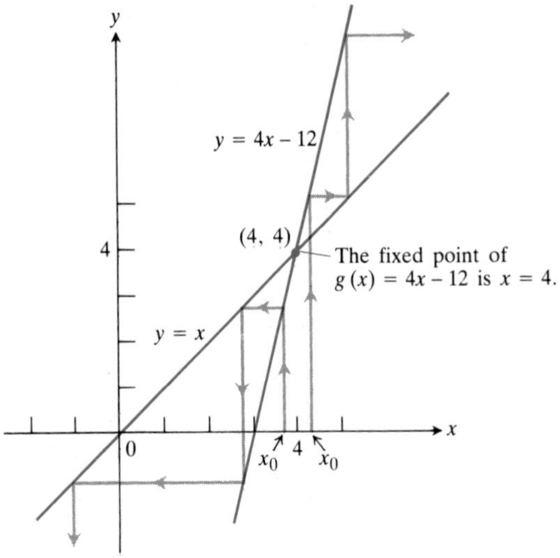

2.46 Applying the Picard method $g(x) = 4x - 12$ will not find the fixed point.

As Fig. 2.46 shows, any choice of x_0 except $x = 4$ (the solution) moves us away from the solution. □

The difficulty in Example 10 can be traced to the fact that the slope of the line $y = 4x - 12$ is greater than 1 (the slope of the line $y = x$). Conversely, the process worked in Example 8 because the slope of the line $y = \frac{1}{4}x + 3$ was less than 1. In general terms, the Picard method works when $g(x)$ and $g'(x)$ are continuous and

$$|g'(x)| < 1 \tag{13}$$

for all x in an interval (a, b) about a root r of the equation $g(x) = x$. Then any choice of x_0 in (a, b) will lead to r. (The conditions on g here are *sufficient* conditions, but not always necessary. In some instances, the method may find r even if one or more of these conditions does not hold.) We shall return to these conditions when we study convergence (Article 11.6).

Example 10 showed that we cannot apply the Picard method to find a fixed point of $g(x) = 4x - 12$. But note that we *can* apply the method successfully to find a fixed point of

$$g^{-1}(x) = \tfrac{1}{4}x + 3$$

because the derivative of g^{-1} is $\frac{1}{4}$, whose absolute value is less than 1 on any interval. In Example 8 we found the fixed point of g^{-1} to be $r = 4$. Furthermore, $r = 4$ is also a fixed point of g, since

$$g(4) = 4(4) - 12 = 4.$$

In finding the fixed point of g^{-1}, we found the fixed point of g.

A function $y = g(x)$ and its inverse always have the same fixed points. The algebraic reason is that if $g(x) = x$, then

$$g^{-1}(x) = g^{-1}(g(x)) = x.$$

The geometric reason is that the graphs of g and g^{-1} are symmetric about the line $y = x$ and cross the line $y = x$ at the same points.

We now see that the application of the Picard method is quite broad. For if $y = g(x)$ is a one-to-one function, with a continuous derivative whose absolute value is greater than 1 on an interval (a, b) about a fixed point r of g, then the derivative of g^{-1}, being the reciprocal of the derivative of g, has absolute value less than 1 on (a, b). The Picard method applied to g^{-1} will find r.

For more about the Picard method read Carroll O. Wilde's *The Contraction Mapping Principle*, UMAP Unit 326 (Lexington, Mass.: COMAP, Inc.).

PROBLEMS

In Problems 1–8, (a) find the inverse $g(x)$ of the function $f(x)$, (b) graph f and g together, and (c) verify that Rule 11 applies to f and g at the points c and $f(c)$.

1. $f(x) = 2x + 3, \quad c = -1$
2. $f(x) = 5 - 4x, \quad c = \frac{1}{2}$
3. $f(x) = \frac{1}{5}x + 7, \quad c = -1$
4. $f(x) = 2x^2, \quad x \geq 0, \quad c = 1$
5. $f(x) = \sqrt{x^2 + 1}, \quad c = \sqrt{3}$
6. $f(x) = (x - 1)^{1/3}, \quad c = 9$
7. $f(x) = x + 1, \quad c = 2$
8. $f(x) = -x + 2, \quad c = 0$

In Problems 9–16, find the inverse $f^{-1}(x)$ of each function and verify that $f(f^{-1}(x)) = f^{-1}(f(x)) = x$.

9. $f(x) = x^5$
10. $f(x) = x^4, \quad x \geq 0$
11. $f(x) = x^{2/3}, \quad x \geq 0$
12. $f(x) = \frac{1}{2}x - \frac{7}{2}$
13. $f(x) = (x - 1)^2$
14. $f(x) = x^3 + 1$
15. $f(x) = x^{-2}$
16. $f(x) = x^{-3}$

In Problems 17–21, find dy/dx.

17. $x = \sqrt{y} + y^2 - 1$
18. $x = \sqrt[3]{3 - y^3 - y^6}$
19. $x = 5\sqrt{1 - y^2}$
20. $x = \cos y + y^2$
21. $x = \sqrt{y} - \sqrt[3]{y}$

22. a) Sketch the graphs of $y = x^3$ and $y = x^{1/3}$ for $-2 \leq x \leq 2$, and sketch the lines tangent to them at $(1, 1)$ and $(-1, -1)$.
    b) Which of these functions fails to have a derivative at one value of x, and what is that x? What is the slope of the other curve at x? What lines are tangent to the curves at that x?
23. a) Sketch the curve $y = 1/x$ and observe that it is symmetric about the line $y = x$.
    b) What is the slope of the curve at $P(a, 1/a)$? at $P'(1/a, a)$? What is the product of these slopes?
    c) Find the inverse of the function $f(x) = 1/x$.
24. Let $f(x) = x^2 - 4x - 3$, $x > 2$, and let g be the inverse of f. Find the values of g' when $f(x) = 2$.
25. Find the point at which the tangent to the curve $x = y^2 + y + 1$ is perpendicular to the x-axis.
26. Find the inverse of the function $f(x) = mx + b$, $m \neq 0$. Is there any value of m for which f is its own inverse? Explain.
27. Find the inverse $g(x)$ of the function $f(x) = 1 + 1/x$. Then show that $f(g(x)) = g(f(x)) = x$, and that $g'(f(x)) = 1/f'(x)$.

(*Calculator*) Solve the equations in Problems 28–33 by the Picard method.

28. $\sin x = x$
29. $\sqrt{x} = x$
30. $x^2 = x$
31. $1.2 \sin x = x$
32. $\cos x = -x$
33. $\cos x = x + 1$
34. Solving the equation $\sqrt{x} = x$ by the Picard method (Problem 29) finds the solution $x = 1$ but not the solution $x = 0$. Why? (*Hint:* Graph $y = x$ and $y = \sqrt{x}$ together.)
35. Solving the equation $x^2 = x$ by the Picard method (Problem 30) finds the solution $x = 0$ but not the solution $x = 1$. Why? (*Hint:* Graph $y = x^2$ and $y = x$ together.)

(*Calculator*) Solve the equations in Problems 36–38 by rewriting them in the form $g(x) = x$ and finding a fixed point of g.

36. $x - \sin x = 0.1$
37. $x - \frac{1}{2}\sin x = 1$
38. $\sqrt{x} = 4 - \sqrt{1 + x}$ (*Hint:* Square both sides first.)
39. (*Calculator. Programmable feature helpful but not necessary*) *The sonobuoy problem*, C. O. Wilde, loc cit. In submarine location problems it is often necessary to find the submarine's closest point of approach (CPA) to a sonobuoy (sound-detector) in the water. Suppose that the submarine travels on a parabolic path $y = x^2$ and that the buoy is located at the point $(2, -\frac{1}{2})$ as shown in Fig. 2.47. As we shall see in Chapter 3 (Article 3.5, Problem 17), the value of x that minimizes the distance between the point (x, x^2) and the point $(2, -\frac{1}{2})$ is a solution of the equation

$$\frac{1}{x^2 + 1} = x.$$

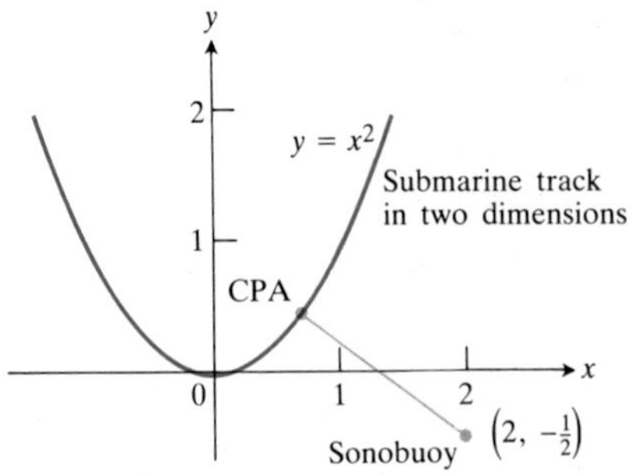

2.47 The diagram for Problem 39. CPA = closest point of approach.

Solve this equation by the Picard method to find the CPA to five decimal places.

40. A function $y = f(x)$ is one-to-one if, for all points x_1 and x_2 in the function's domain, $x_1 \neq x_2 \Rightarrow f(x_1) \neq f(x_2)$. Show that the inverse of a one-to-one function is one-to-one.

***Toolkit* programs**

Function Evaluator	Sequences and Series
Picard's Fixed Point Method	Super * Grapher

2.11 Summary of Derivative Formulas

The derivative formulas developed in this chapter are listed below as Formulas 1–12. We will add to this list as we continue through the text.

The functions u and v that appear in the formulas are assumed to be differentiable functions of x. The letter c denotes an arbitrary constant. It is further assumed that each formula is valid only at values of x at which the formula's individual terms are all defined. The variables and functions are not allowed to take on values that require division by zero or taking square roots or other even roots of negative numbers. For example, Formula 5 calculates the derivative of u/v with respect to x only at values of x where u, v, du/dx, and dv/dx are all defined and $v \neq 0$.

Derivatives

1. $\dfrac{dc}{dx} = 0$ (any constant c)
2. $\dfrac{d}{dx}(cu) = c\dfrac{du}{dx}$ (any constant c)
3. $\dfrac{d}{dx}(u + v) = \dfrac{du}{dx} + \dfrac{dv}{dx}$
4. $\dfrac{d}{dx}(uv) = u\dfrac{dv}{dx} + v\dfrac{du}{dx}$
5. $\dfrac{d}{dx}\left(\dfrac{u}{v}\right) = \dfrac{v\dfrac{du}{dx} - u\dfrac{dv}{dx}}{v^2}$
6. $\dfrac{d}{dx}u^n = nu^{n-1}\dfrac{du}{dx}$

6a. $\dfrac{d}{dx}cx^n = cnx^{n-1}$

7. $\dfrac{d}{dx}\sin u = \cos u\dfrac{du}{dx}$
8. $\dfrac{d}{dx}\cos u = -\sin u\dfrac{du}{dx}$
9. $\dfrac{d}{dx}\tan u = \sec^2 u\dfrac{du}{dx}$
10. $\dfrac{d}{dx}\cot u = -\csc^2 u\dfrac{du}{dx}$
11. $\dfrac{d}{dx}\sec u = \sec u \tan u\dfrac{du}{dx}$
12. $\dfrac{d}{dx}\csc u = -\csc u \cot u\dfrac{du}{dx}$

Chain Rule

If y is a differentiable function of u and u is a differentiable function of x, then y is a differentiable function of x and

$$\frac{dy}{dx} = \frac{dy}{du}\frac{du}{dx}.$$

(Article 2.6 shows how this formula is to be used.)

IN ANOTHER FORM

If f is differentiable at x and g is differentiable at $f(x)$, then the composite $g \circ f$ is differentiable at x and

$$\frac{d}{dx}g(f(x)) = g'(f(x)) \cdot f'(x).$$

REVIEW QUESTIONS AND EXERCISES

1. Using the definition of the derivative, deduce the formula for the derivative of a product uv of two differentiable functions.
2. In the formula for the derivative of uv, let $v = u$, and thereby deduce a formula for the derivative of u^2. Repeat the process, with $v = u^2$, and get a formula for the derivative of u^3. Extend the result, by the method of mathematical induction, to deduce the formula for the derivative of u^n for every positive integer n.
3. Explain how the three formulas

   a) $\dfrac{d(x^n)}{dx} = nx^{n-1}$,

   b) $\dfrac{d(cu)}{dx} = c\dfrac{du}{dx}$,

   c) $\dfrac{d(u+v)}{dx} = \dfrac{du}{dx} + \dfrac{dv}{dx}$

   let us differentiate any polynomial.
4. What formula do we need, in addition to the three listed in Exercise 3 above, to differentiate rational functions?
5. Does the derivative of a polynomial function exist at every point of its domain? What is the largest domain the function can have? Does the derivative of a rational function exist at every point in its domain? What real numbers, if any, must be excluded from the domain of a rational function?
6. *Definition of algebraic function*. Let $y = f(x)$ define a function of x such that every pair of numbers (x, y) belonging to f satisfy an irreducible equation of the form

$$P_0(x)y^n + P_1(x)y^{n-1} + \cdots + P_{n-1}(x)y + P_n(x) = 0 \quad (\alpha)$$

   for some positive integer n, with coefficients $P_0(x)$, . . . , $P_n(x)$ polynomials in x, and $P_0(x)$ not identically zero. Then y is said to be an *algebraic function* of x. What technique of this chapter can be used to find the derivative of an algebraic function if the polynomial coefficients $P_0(x)$, . . . , $P_n(x)$ in its defining equation (α) are given?
7. Show that the function f defined by $f(x) = x^{2/3}$ is an algebraic function by finding an appropriate equation of type (α) in the definition in Exercise 6 above. On what domain of values of x is this function defined? Where is it continuous? Where is it differentiable?
8. Show that every rational function is algebraic. Do you think the converse is also true? Explain.
9. It can be shown (though not very easily) that sums, quotients, products, powers, and roots of algebraic functions are algebraic functions. Thus

$$f(x) = x\sqrt{3x^2+1} + \frac{5x^{4/3}}{3x+2}, \qquad x \neq -\frac{2}{3},$$

   defines an algebraic function. Find its derivative. What formulas of this chapter are used in finding derivatives of functions like this one?
10. Show that $y = |x|$ satisfies the equation $y^2 - x^2 = 0$. Is the absolute value function algebraic? What is its

derivative? Where does the derivative exist? Where is the absolute value function continuous?

11. Suppose that $y = f(x)$ has a derivative at $x = a$, and that we change x by an amount Δx. How can we estimate the resulting change in y?

12. What is meant by the linear approximation of a function $y = f(x)$ near a point where the function has a derivative? Give six examples of linear approximations near $x = 0$.

13. State the Chain Rule for derivatives. Prove it with the book closed.

14. Write expressions for $\sin(A + B)$ and $\cos(A + B)$.

15. Under what assumption is it true that

$$\lim_{h \to 0} \frac{\sin h}{h} = 1?$$

16. State the half-angle formulas. Use one of them to prove that

$$\lim_{h \to 0} \frac{1 - \cos h}{h} = 0.$$

How is this limit used in the proof that

$$\frac{d}{dx} \sin x = \cos x?$$

17. Let A_n be the area bounded by a regular n-sided polygon inscribed in a circle of radius r. Show that $A_n = (n/2)r^2 \sin(2\pi/n)$. Find $\lim A_n$ as $n \to \infty$. Does the result agree with what you know about the area of a circle?

18. Describe Newton's method for solving equations. Give an example. What is the theory behind the method?

19. Give an example of a function that is (a) one-to-one; (b) not one-to-one.

20. Suppose that the domain of a one-to-one function f is $[a, b]$ and that the range of f is $[f(a), f(b)]$. Describe the inverse of f. What are its domain and range?

21. *Continuation of Exercise 20.* If $g(y) = x$ whenever $f(x) = y$, then

$$g(f(x)) = x$$

for all $a \leq x \leq b$. Suppose that f is differentiable. Apply the Chain Rule to the composite $h = g \circ f$ to deduce a general formula for the derivative of an inverse function. What restriction must be placed on the derivative of f?

22. Describe the Picard method for solving the equation $g(x) = x$. How does the method help to solve equations of the form $f(x) = 0$? If g is one-to-one and the Picard method fails to work on g, why might the method work on g^{-1}?

MISCELLANEOUS PROBLEMS

Find dy/dx in Problems 1–58.

1. $y = \dfrac{x}{\sqrt{x^2 - 4}}$

2. $x^2 + xy + y^2 - 5x = 2$

3. $xy + y^2 = 1$

4. $x^3 + 4xy - 3y^3 = 2$

5. $x^2y + xy^2 = 10$

6. $y = (x + 1)^2(x^2 + 2x)^{-2}$

7. $y = \cos(1 - 2x)$

8. $y = \dfrac{\cos x}{\sin x}$

9. $y = \dfrac{x}{x + 1}$

10. $y = \sqrt{2x + 1}$

11. $y = x^2\sqrt{x^2 - a^2}$

12. $y = \dfrac{2x + 1}{2x - 1}$

13. $y = \dfrac{x^2}{1 - x^2}$

14. $y = (x^2 + x + 1)^3$

15. $y = \sec^2(5x)$

16. $y^3 = \sin^3 x + \cos^3 x$

17. $y = \dfrac{(2x^2 + 5x)^{3/2}}{3}$

18. $y = \dfrac{3}{(2x^2 + 5x)^{3/2}}$

19. $xy^2 + \sqrt{xy} = 2$

20. $x^2 - y^2 = xy$

21. $x^{2/3} + y^{2/3} = a^{2/3}$

22. $x^{1/2} + y^{1/2} = a^{1/2}$

23. $xy = 1$

24. $\sqrt{xy} = 1$

25. $(x + 2y)^2 + 2xy^2 = 6$

26. $y = \sqrt{\dfrac{1 - x}{1 + x^2}}$

27. $y^2 = \dfrac{x}{x + 1}$

28. $x^2y + xy^2 = 6(x^2 + y^2)$

29. $xy + 2x + 3y = 1$

30. $x^2 + xy + y^2 + x + y + 1 = 0$

31. $x^3 - xy + y^3 = 1$

32. $xy^3 + 3x^2y^2 = 7$

33. $y = \sqrt{\dfrac{1 + x}{1 - x}}$

34. $y = \sqrt{x} + 1 + \dfrac{1}{\sqrt{x}}$

35. $y = (x^3 + 1)^{1/3}$

36. $y = x^2 \sin^5 2x$

37. $y = \cot 2x$

38. $y = \sin^2(1 + 3x)$

39. $y = \dfrac{\sin x}{\cos^2 x}$

40. $y = \sin^3 2x$

41. $y = x^2 \cos 8x$

42. $y = \sin(\cos^2 x)$

43. $y = \dfrac{\sin x}{1 + \cos x}$

44. $y = \dfrac{\sin^2 x}{\cos x}$

45. $y = \csc x$

46. $y = \cot x^2$

47. $y = \cos(\sin^2 x)$

48. $y = \dfrac{\sin x}{x}$

49. $y = \sec^2 x$

50. $y = \sec x \sin x$

51. $y = \cos(\sin 3x)$

52. $y = u^2 - 1, \quad x = u^2 + 1$

53. $y = \sqrt{2t + t^2}, \quad t = 2x + 3$

54. $x = \dfrac{t}{1 + t^2}, \quad y = 1 + t^2$

55. $t = \dfrac{x}{1 + x^2}, \quad y = x^2 + t^2$

56. $x = t^2 - 1, \quad y = 3t^4 - t^2$

57. $x = t^2 + t, \quad y = t^3 - 1$

58. $x = \cos 3t, \quad y = \sin(t^2 + 1)$

59. Find the slope of $y = x/(x^2 + 1)$ at the origin. Write the equation of the tangent line at the origin.

60. Write the equation of the tangent at (2, 2) to the curve

$$x^2 - 2xy + y^2 + 2x + y - 6 = 0.$$

61. Which of the following statements could be true if $f''(x) = x^{1/3}$?

I. $f(x) = \frac{9}{28}x^{7/3} + 9$ II. $f'(x) = \frac{9}{28}x^{7/3} - 2$

III. $f'(x) = \frac{3}{4}x^{4/3} + 6$ IV. $f(x) = \frac{3}{4}x^{4/3} - 4$

a) I only   b) III only

c) II and IV only   d) I and III only

62. Graph together for $-\pi \le x \le \pi$: $y = \sin x$, $y = \cos x$, $y = \sin 2x$, $y = \sin(x - (\pi/4))$.

63. Find the tangents to the curve $x^2y + xy^2 = 6$ at the points where $x = 1$.

64. Find the tangent to the curve at the given point.

a) $x^2 + 2y^2 = 9$ at (1, 2)

b) $x^3 + y^2 = 2$ at (1, 1)

65. The designer of a 30-ft-diameter spherical hot-air balloon (a cutaway view is shown in Fig. 2.48) wishes to suspend the gondola 8 ft below the bottom of the balloon with suspension cables tangent to the surface of the balloon. Two of the cables are shown running from the top edges of the gondola to their points of tangency, (−12, −9) and (12, −9). How wide must the gondola be?

2.48 The balloon for Problem 65.

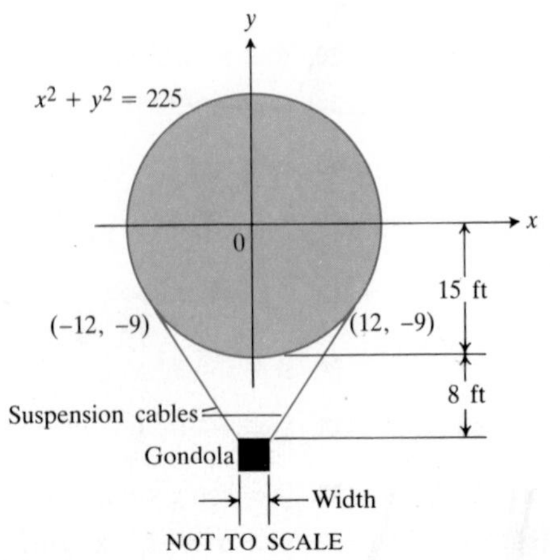

66. Determine the constant c such that the straight line joining the points (0, 3) and (5, −2) is tangent to the curve $y = c/(x + 1)$.

67. What is the slope of the curve $y = 2x^2 - 6x + 3$ at the point on the curve where $x = 2$? What is the equation of the tangent line to the curve at this point?

68. Find the points on the curve $y = 2x^3 - 3x^2 - 12x + 20$ where the tangent is parallel to the x-axis.

69. Find the derivatives of the following functions.

a) $y = (x^2 + 2x)^5$ b) $f(t) = \sqrt{3t^2 - 2t}$

c) $f(r) = \sqrt{r^2 + 5} + \sqrt{r^2 - 5}$

d) $f(x) = \dfrac{x^2 - 1}{x^2 + 1}$

70. Find the equation of the tangent to the curve $y = 2/\sqrt{x - 1}$ at the point on the curve where $x = 10$.

71. Write the equation of the straight line passing through the point (2, 1) and normal to the curve $x^2 = 4y$.

72. Use the definition of the derivative to find dy/dx for $y = \sqrt{2x + 3}$ and then check the result by finding the same derivative by the power formula.

73. Find the value of

$$\lim_{\Delta x \to 0} \frac{[2 - 3(x + \Delta x)]^2 - [2 - 3x]^2}{\Delta x}$$

and specify the function $f(x)$ of which this is the derivative.

74. Find the slope of the curve $x^2y + xy^2 = 6$ at the point (1, 2).

75. A cylindrical can of height 6 in. and radius r in. has volume $V = 6\pi r^2$ in³. What is the difference between ΔV and $\Delta V_{\tan}$ as r varies? What is the geometric significance of $\Delta V_{\tan}$?

76. If a hemispherical bowl of radius 10 in. is filled with water to a depth of x in., the volume of water is given by $v = \pi[10 - (x/3)]x^2$. Find the rate of increase of the volume per inch increase of the depth.

77. A bus will hold 60 people. If the number x of people per trip who use the bus is related to the fare charged (p nickels) by the law $p = [3 - (x/40)]^2$, write an expression for the total revenue per trip received by the bus company. What number x_1 of people per trip will make the marginal revenue equal to zero? What is the corresponding fare?

78. Prove Eq. (3) in Article 2.3 by mathematical induction.

79. Given $y = x - x^2$, find the rate of change of y^2 with respect to x^2 (expressed in terms of x).

80. If $x = 3t + 1$ and $y = t^2 + t$, find dy/dt, dx/dt, and dy/dx. Eliminate t to obtain y as a function of x, and then determine dy/dx directly. Do the results check?

81. A particle projected vertically upward with a speed of a ft/s reaches an elevation $s = at - 16t^2$ ft at the end of t seconds. What must the initial velocity be for the particle to travel 49 ft upward before it starts coming back down?

82. Find the rate of change of $\sqrt{x^2 + 16}$ with respect to $x/(x - 1)$ at $x = 3$.

83. The circle $(x - h)^2 + (y - k)^2 = r^2$ is tangent to the curve $y = x^2 + 1$ at the point (1, 2).
a) Find the possible locations of the point (h, k).
b) If, also, the circle and the curve have the same second derivative at (1, 2), find h, k, and r. Sketch the curve and the circle.

84. If $y = x^2 + 1$ and $u = \sqrt{x^2 + 1}$, find dy/du.

85. If $x = y^2 + y$ and $u = (x^2 + x)^{3/2}$, find dy/du.

86. If $f'(x) = \sqrt{3x^2 - 1}$ and $y = f(x^2)$, find dy/dx.

87. If $f'(x) = \sin(x^2)$ and $y = f((2x - 1)/(x + 1))$, find dy/dx.

88. Given $y = 3 \sin 2x$ and $x = u^2 + \pi$, find the value of dy/du when $u = 0$.

89. If $0 < x < \pi/2$, prove that $x > \sin x > 2x/\pi$.

90. If $y = x\sqrt{2x - 3}$, find d^2y/dx^2.

91. Find the value of d^2y/dx^2 in the equation $y^3 + y = x$ at the point (2, 1).

92. If $x = t - t^2$, $y = t - t^3$, find the values of dy/dx and d^2y/dx^2 at $t = 1$.

93. Prove Leibniz's rule:

a) $$\frac{d^2(uv)}{dx^2} = \frac{d^2u}{dx^2}\cdot v + 2\frac{du}{dx}\frac{dv}{dx} + u\frac{d^2v}{dx^2},$$

b) $$\frac{d^3(uv)}{dx^3} = \frac{d^3u}{dx^3}\cdot v + 3\frac{d^2u}{dx^2}\frac{dv}{dx} + 3\frac{du}{dx}\frac{d^2v}{dx^2} + u\frac{d^3v}{dx^3},$$

c) $$\frac{d^n(uv)}{dx^n} = \frac{d^nu}{dx^n}\cdot v + n\frac{d^{n-1}u}{dx^{n-1}}\frac{dv}{dx} + \cdots + \frac{n(n-1)\cdots(n-k+1)}{k!}\frac{d^{n-k}u}{dx^{n-k}}\frac{d^kv}{dx^k} + \cdots + u\frac{d^nv}{dx^n}.$$

The terms on the right-hand side of this equation may be obtained from the terms in the binomial expansion $(a + b)^n$ by replacing $a^{n-k}b^k$ by $(d^{n-k}u/dx^{n-k})\cdot(d^kv/dx^k)$ for $k = 0, 1, 2, \ldots, n$, and interpreting d^0u/dx^0 as being u itself.

94. Find d^3y/dx^3 in each of the following cases.
a) $y = \sqrt{2x - 1}$
b) $y = \dfrac{1}{3x + 2}$
c) $y = ax^3 + bx^2 + cx + d$

95. If $f(x) = (x - a)^n g(x)$, where $g(x)$ is a polynomial and $g(a) \neq 0$, show that $f(a) = 0 = f'(a) = \cdots = f^{(n-1)}(a)$; but $f^{(n)}(a) = n!\, g(a) \neq 0$.

96. If $y = 2x^2 - 3x + 5$, find Δy for $x = 3$ and $\Delta x = 0.1$. Approximate Δy by finding $\Delta y_{\tan}$.

97. a) Show that the perimeter P_n of an n-sided regular polygon inscribed in a circle of radius r is $P_n = 2nr \sin(\pi/n)$.
b) Find the limit of P_n as $n \to \infty$. Is the answer consistent with what you know about the circumference of a circle?

98. Find dy/dx and d^2y/dx^2 if $x = \cos 3t$ and $y = \sin^2 3t$.

99. To compute the height h of a lamppost, the length a of the shadow of a 6-ft pole is measured. The pole is 20 ft from the lamppost. If $a = 15$ ft, with a possible error of less than one inch, find the height of the lamppost and estimate the possible error in height.

100. Suppose a function f satisfies the following two conditions for all x and y:
i) $f(x + y) = f(x)\cdot f(y)$;
ii) $f(x) = 1 + xg(x)$ where $\lim_{x\to 0} g(x) = 1$.

Prove that (a) the derivative $f'(x)$ exists; and (b) $f'(x) = f(x)$.

101. A particle moves around the circle $x^2 + y^2 = 1$. The velocity component parallel to the x-axis is $dx/dt = y$. Find dy/dt. Does the particle travel clockwise, or counterclockwise?

102. Suppose that an even function f is differentiable for all values of x, and that $f(c) = 1$, $f'(c) = 5$ at some point $c > 0$.
a) Find $f'(-c)$ and $f'(0)$.
b) Let L_1 and L_2 be the tangents to the graph of f at $x = c$ and $x = -c$. Find the coordinates of the point P in which L_1 and L_2 intersect (in terms of c).

103. Find all values of the constants m and b for which the function

$$y = \begin{cases} \sin x & \text{for } x < \pi, \\ mx + b & \text{for } x \geq \pi, \end{cases}$$

is (a) continuous at $x = \pi$; (b) differentiable at $x = \pi$.

104. Does the function

$$f(x) = \begin{cases} \dfrac{1 - \cos x}{x} & \text{for } x \neq 0, \\ 0 & \text{for } x = 0, \end{cases}$$

have a derivative at $x = 0$? Explain.

105. a) Show that the function

$$f(x) = \begin{cases} x^2 \sin\dfrac{1}{x} & \text{for } x \neq 0, \\ 0 & \text{at } x = 0, \end{cases}$$

is differentiable at $x = 0$. (Use the definition of derivative.)
b) Find $f'(x)$ for $x \neq 0$.
c) Is f' continuous at $x = 0$? Explain.

3 Applications of Derivatives

3.1 Curve Sketching. The Sign of the First Derivative

The derivative of a function $y = f(x)$ is the rate at which y changes with respect to x. It defines the slope of the function's graph at x and allows us to estimate how much y changes when we change x by a small amount. The derivative also allows us to construct a systematic procedure (Newton's method) for estimating whatever zeros f may have. In this article and the next two we shall see that the first and second derivatives together tell how the graph of a function is shaped. Toward the end of the chapter we shall also see how to use the second derivative to estimate the error in the linear approximation of a function.

If a function has a derivative over an interval, then we know from the theorems in Article 1.11 that it is continuous over the interval and that its graph over the interval is unbroken. Thus the graphs of $y = \sin x$ and $y = \cos x$ and the graphs of all polynomials remain unbroken no matter how far they are extended from the origin. The graphs of $y = \tan x$ and of rational functions are unbroken over all the intervals on which their derivatives are defined.

We can gain even more information about the graph of a differentiable function if we know where its derivative is positive, negative, and zero. For, as we shall see, this tells us where the graph is rising, is falling, and has a horizontal tangent. Here are some examples.

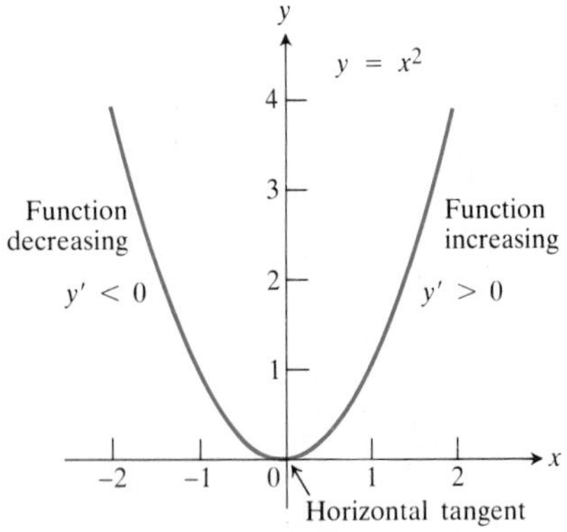

3.1 The graph of $y = x^2$.

EXAMPLE 1 See Fig. 3.1. The function $y = x^2$ decreases on $(-\infty, 0)$, where $y' = 2x < 0$. It increases on $(0, \infty)$, where $y' = 2x > 0$. At $x = 0$, the point of transition between the falling and rising portions of the curve, $y' = 0$ and the curve has a horizontal tangent. □

EXAMPLE 2 See Fig. 3.2. The function $y = \sin x$ increases on $(-\pi/2, \pi/2)$ where its derivative $y' = \cos x$ is positive, and again on $(3\pi/2, 2\pi)$. It decreases on $(-\pi, -\pi/2)$ and on $(\pi/2, 3\pi/2)$, where $y' = \cos x$ is negative. At $x = \pm\pi/2$ and $x = 3\pi/2$, the points of transition between increasing and decreasing, $y' = 0$ and the curve has a horizontal tangent. □

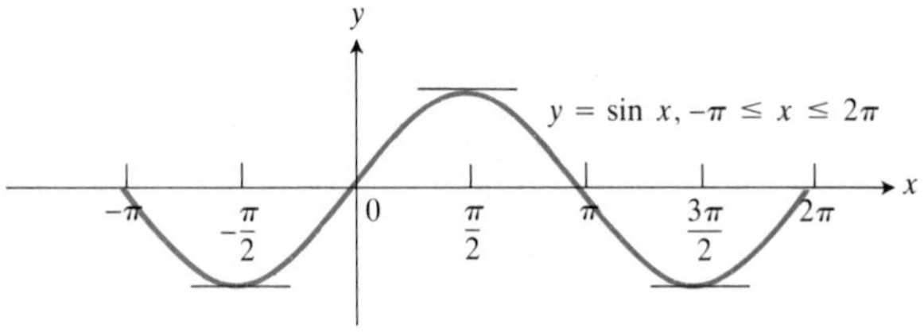

3.2 The graph of $y = \sin x$, $-\pi \le x \le 2\pi$. Horizontal tangents are located at the points of transition between rise and fall.

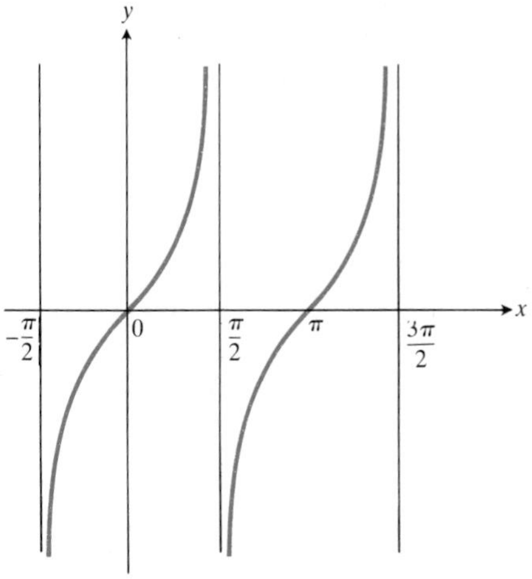

3.3 The graph of $y = \tan x$ increases for $-\pi/2 < x < \pi/2$ and again for $\pi/2 < x < 3\pi/2$.

EXAMPLE 3 See Fig. 3.3. The function $f(x) = \tan x$ increases on $(-\pi/2, \pi/2)$ and $(\pi/2, 3\pi/2)$, where $f'(x) = \sec^2 x = 1/\cos^2 x > 0$. □

EXAMPLE 4 See Fig. 3.4. The function $y = 1/x^2$ increases from left to right on $(-\infty, 0)$ where $y' = -2/x^3 > 0$, and decreases from left to right on $(0, \infty)$ where $y' = -2/x^3 < 0$. The derivative is not defined at $x = 0$, the point of transition from the interval on which y increases to the interval on which y decreases. □

EXAMPLE 5 See Fig. 3.5. The function $f(x) = 1/x$ decreases from left to right on both $(-\infty, 0)$ and $(0, \infty)$. The derivative $f'(x) = -1/x^2$ is negative on both intervals, but is not defined at their common endpoint $x = 0$. □

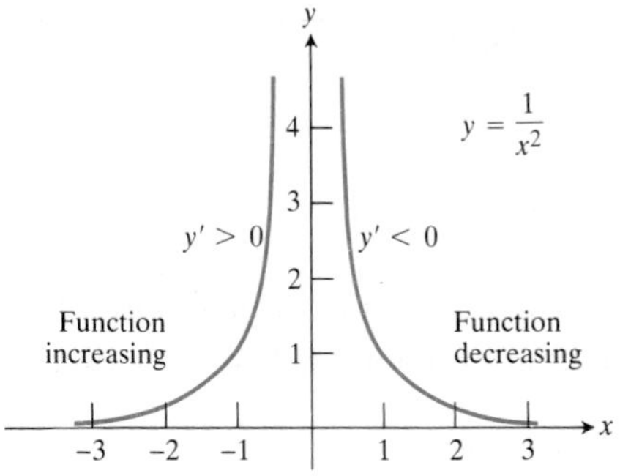

3.4 The graph of $y = 1/x^2$.

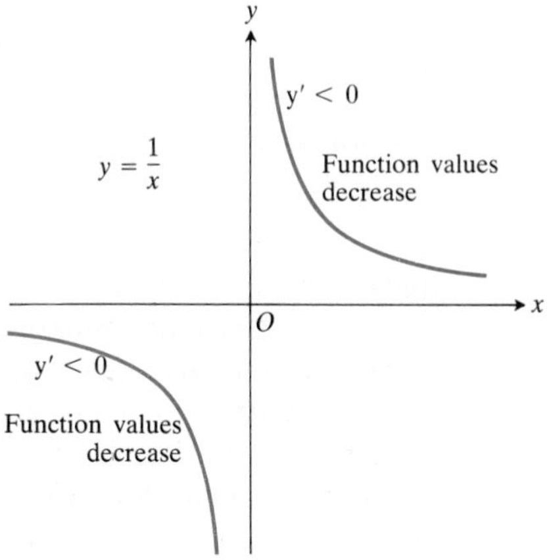

3.5 The graph of $y = 1/x$.

Increasing and Decreasing Functions

The notions of increasing and decreasing are *always* defined in terms of increasing x, that is, from left to right, as in the examples above. A function $y = f(x)$ is an increasing function of x if y increases as x increases. A point tracing the graph of an increasing function rises as it moves from left to right. Similarly, $y = f(x)$ is a decreasing function of x if y decreases as x increases. The graph of a decreasing function falls as x increases. As Examples 1–4 show, a function may increase over one interval and decrease over another.

Increasing and Decreasing Functions

A function $y = f(x)$ is an *increasing function* on an interval I if

$$x_1 < x_2 \quad \Rightarrow \quad f(x_1) < f(x_2)$$

for all x_1 and x_2 in I.

A function $y = f(x)$ is a *decreasing function* on an interval I if

$$x_1 < x_2 \quad \Rightarrow \quad f(x_2) < f(x_1)$$

for all x_1 and x_2 in I.

We will show later (in Article 3.8) that if $y = f(x)$ is differentiable, with $f'(x) > 0$ at every point of an interval I, then f increases on I. Similarly, if $f'(x) < 0$ at every point of I, then f decreases on I. We will assume these facts for now, as the first derivative test for rise and fall.

The First Derivative Test for Rise and Fall

Suppose that $y = f(x)$ has a derivative at every point x of an interval I. Then:

1. f increases on I if $f'(x) > 0$ for all x in I, and
2. f decreases on I if $f'(x) < 0$ for all x in I.

In geometric terms, the first derivative test says that a differentiable function increases where its graph has positive slopes and decreases where its graph has negative slopes.

Horizontal Tangents

If the derivative f' of a function $y = f(x)$ is continuous, then f' can go from negative to positive values only by going through zero. (This is a consequence of the Intermediate Value Theorem for continuous functions, Theorem 10 in Article 1.11.) At such a transition point the graph of f has a horizontal tangent (Fig. 3.6).

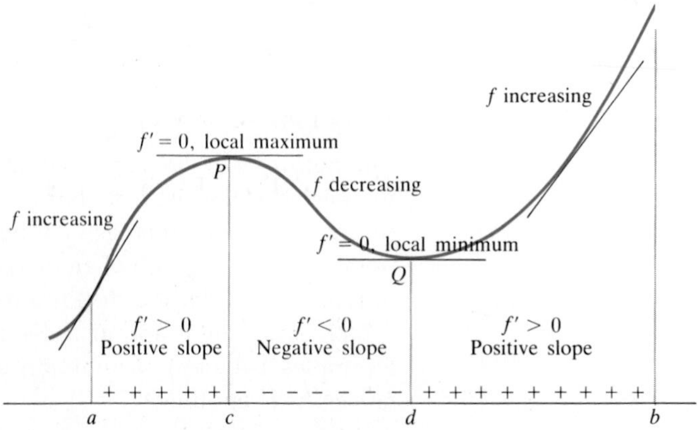

3.6 The function $y = f(x)$ increases on (a, c) where $f' > 0$, decreases on (c, d) where $f' < 0$, and increases again on (d, b). The transitions are marked by horizontal tangents.

x	y	y'
−1	0	9
0	4	0
1	2	−3
2	0	0
3	4	9

(a)

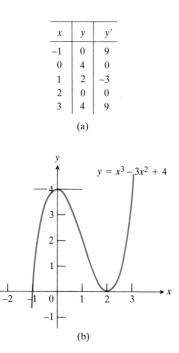

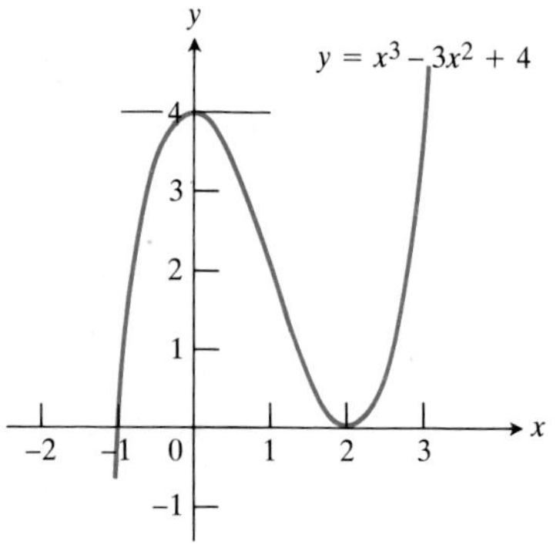

(b)

3.7 (a) Slope table. (b) The graph of $y = x^3 - 3x^2 + 4$ rises on $(-\infty, 0)$ to a local maximum of 4 at $x = 0$, falls to a local minimum of 0 at $x = 2$, then rises again on $(2, \infty)$.

If f' changes continuously from positive to negative values as x passes from left to right through a point c, then the value of f at c is a *local maximum* value of f, as shown in Fig. 3.6. That is, $f(c)$ is the largest value the function takes on in the immediate neighborhood of $x = c$. Similarly, if f' changes from negative to positive values as x passes from left to right through a point d, then the value of f at d is a *local minimum* value of f. That is, $f(d)$ is the smallest value f takes in the immediate neighborhood of d. The graph "bottoms out" at $x = d$. (We will give more formal definitions of local maximum and local minimum when we study the theory of maximum and minimum values of functions in Article 3.4.)

Graphing

EXAMPLE 6 Graph the function $y = x^3 - 3x^2 + 4$.

Solution See Fig. 3.7. We first find the intercepts (points where the graph crosses or touches the axes). The y-intercept is

$$y = (0)^3 - 3(0)^2 + 4 = 4.$$

Factoring the polynomial

$$y = x^3 - 3x^2 + 4 = (x + 1)(x - 2)^2$$

shows the x-intercepts to be $x = -1$ and $x = 2$.

We then find where the derivative of the function is positive, negative, and zero. The derivative is

$$y' = 3x^2 - 6x = 3x(x - 2),$$

which is zero at $x = 0$ and $x = 2$. The curve has horizontal tangents at these values.

The derivative y' is positive to the left of $x = 0$, where x and $(x - 2)$ are both negative, and positive to the right of $x = 2$, where x and $(x - 2)$ are both positive. The function therefore increases on the intervals $(-\infty, 0)$ and $(2, \infty)$.

Between $x = 0$ and $x = 2$, the derivative $y' = 3x(x - 2)$ is negative because $3x > 0$ while $x - 2 < 0$. The function therefore decreases on the interval $(0, 2)$.

We now construct a small table of function values and slopes (see Fig. 3.7a), which includes the intercepts and the points of transition between the rising and falling portions of the curve.

Finally, we plot the points and use the information about how the curve rises and falls to complete the sketch shown in Fig. 3.7(b). At $x = 0$ the function has a local maximum value of $y = 4$. At $x = 2$ the function has a local minimum value of $y = 0$. □

EXAMPLE 7 Sketch the curve

$$y = f(x) = \frac{1}{3}x^3 - 2x^2 + 3x + 2.$$

Solution The y-intercept is $y = 2$, but it is not easy to find even one x-intercept because the polynomial does not factor nicely into linear factors. We might notice that there is a root between $x = -1$ and $x = 0$ because $f(-1) < 0$ and $f(0) = 2$, but the root is not easily estimated. Fortu-

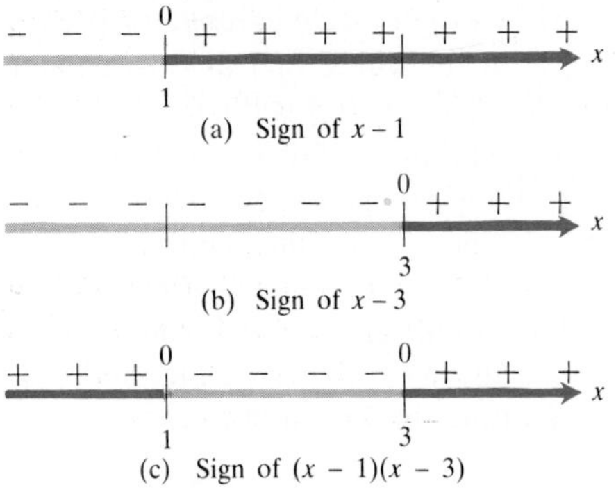

3.8 The sign pattern of the product $(x - 1)(x - 3)$.

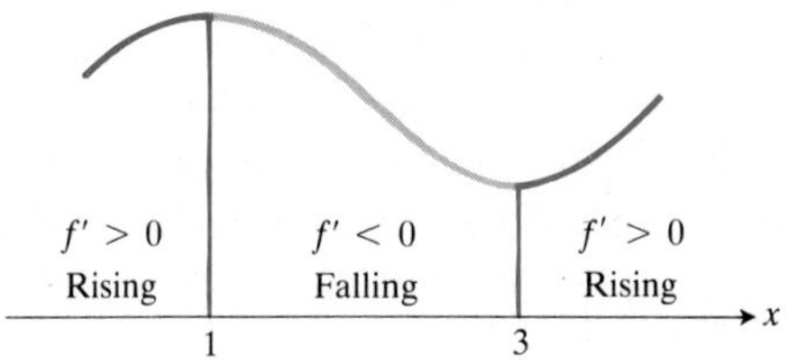

3.9 A function whose derivative has the sign pattern shown in Fig. 3.8(c) must have a graph something like this.

nately, we do not need to know the x-intercepts to sketch the curve. The first derivative of the function will tell us all we need to know about where the curve rises, falls, and has horizontal tangents.

The first derivative,

$$\frac{dy}{dx} = x^2 - 4x + 3 = (x - 1)(x - 3),$$

is zero when $x = 1$ and $x = 3$. The curve therefore has horizontal tangents at $x = 1$ and $x = 3$. Moreover, these values of x mark the transition between positive and negative slopes, as we shall now see.

The sign of dy/dx depends on the signs of the two factors $(x - 1)$ and $(x - 3)$. Since the sign of $(x - 1)$ is negative when x is to the left of 1 and positive to the right, we have the pattern of signs indicated in Fig. 3.8(a). Similarly, the sign of $(x - 3)$ is shown in Fig. 3.8(b). The sign of the product $dy/dx = (x - 1)(x - 3)$ is shown in Fig. 3.8(c). We can get a rough idea of the shape of the curve just from this pattern of signs if we sketch a curve that is rising when $x < 1$, falling when $1 < x < 3$, and rising again when $x > 3$ (Fig. 3.9). The function has a local maximum at $x = 1$ and a local minimum at $x = 3$.

To get a more accurate curve, we would construct a table of values extending from $x = 0$ to $x = 4$, say, that included the transition points between the rising and falling portions of the curve (Fig. 3.10). □

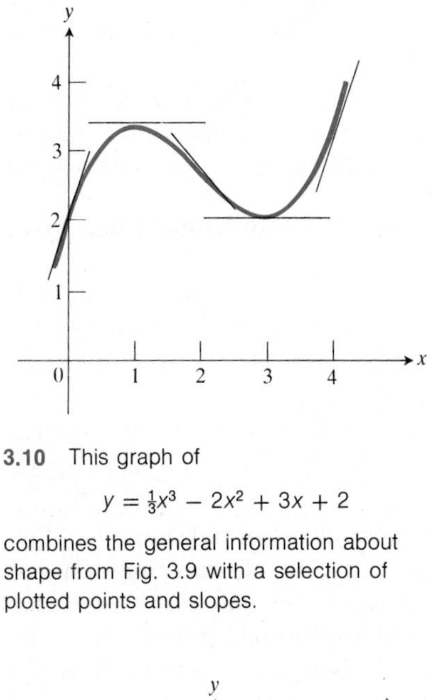

3.10 This graph of

$$y = \tfrac{1}{3}x^3 - 2x^2 + 3x + 2$$

combines the general information about shape from Fig. 3.9 with a selection of plotted points and slopes.

A Horizontal Tangent Without a Maximum or a Minimum

If $y = f(x)$ has a continuous derivative f' that changes sign as x passes through a point c, we know that $f'(c) = 0$. However, a change in sign does not always occur when the derivative is zero. The curve may cross its horizontal tangent and keep on rising as $y = x^3$ does at (0, 0) in Fig. 3.11. Or the curve may cross its horizontal tangent and keep on falling, as $y = -x^3$ does at (0, 0) in Fig. 3.11. Neither function has a local maximum value or a local minimum value at $x = 0$.

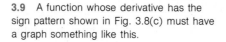

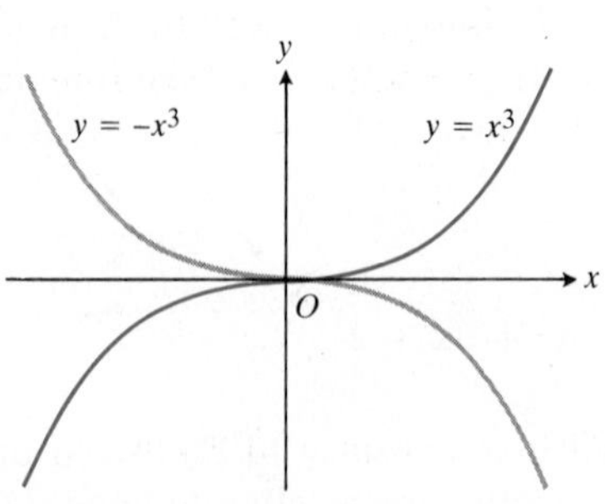

Figure 3.11

In Fig. 3.11 we also see that the conditions $f' > 0$ for increasing and $f' < 0$ for decreasing are not always necessary. The function $y = f(x) =$

x^3 increases on the entire x-axis because $x_1 < x_2 \Rightarrow (x_1)^3 < (x_2)^3$ for all x_1 and x_2. Yet the first derivative $y' = 3x^2$ is zero at $x = 0$. Similarly, $y = -x^3$ decreases for all x even though its derivative is zero at the origin.

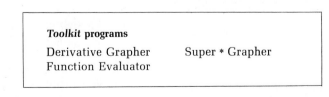

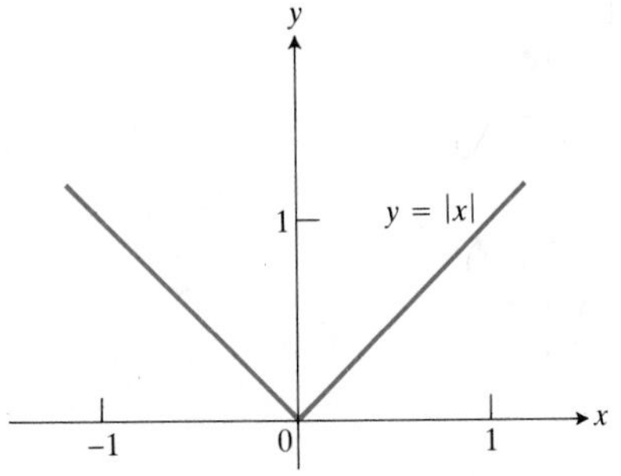

Figure 3.12

A Maximum or Minimum Without a Horizontal Tangent

The intervals on which the function $y = 1/x^2$ rises and falls are separated by the point $x = 0$ at which the function is not continuous (Fig. 3.4). But even if a function is continuous, the intervals on which it rises and falls may be separated by points at which its derivative fails to exist.

EXAMPLE 8 See Fig. 3.12. The function $y = |x|$ decreases on $(-\infty, 0)$ where $y' = -1$, increases on $(0, \infty)$ where $y' = 1$, and has no derivative at $x = 0$. The transition from negative to positive slopes, from falling to rising, takes place at a point where the derivative fails to exist. □

PROBLEMS

In Problems 1–8, find the values of x that satisfy the given inequality.

1. $x^2 - x - 2 < 0$ **2.** $x^2 - 4x - 45 > 0$

3. $x - x^2 > 0$ **4.** $x^3 - x^2 > 0$

5. $x^3 - 4x < 0$ **6.** $-2/(x + 1)^3 < 0$

7. $(x - 1)(x + 1)(x - 5) > 0$ **8.** $x^3 - 2x^2 - 8x < 0$

In Problems 9–25, find dy/dx and the intervals of x-values on which $y = f(x)$ is increasing or decreasing. Sketch each curve, showing the points of transition between the rising and falling portions. Find any local maximum and local minimum values that the function has when $y' = 0$.

9. $y = x^2 - x + 1$ **10.** $y = 12 - 12x + 2x^2$

11. $y = \frac{x^3}{3} - \frac{x^2}{2} - 2x + \frac{1}{3}$ **12.** $y = 2x^3 - 3x^2 + 3$

13. $y = x^3 - 27x + 36$ **14.** $y = x^4 - 8x^2 + 16$

15. $y = 3x^2 - 2x^3$

16. $y = (x - 2)(x - 11)(x + 13)$

17. $y = x^4$ **18.** $y = x^{4/3}$

19. $y = 1/x^3$ **20.** $y = 1/(x - 1)$

21. $y = 1/(x + 1)^2$ **22.** $y = 9x - x^3$

23. $y = \cos x, \quad -3\pi/2 < x < 3\pi/2$

24. $y = \sec x, \quad -\pi/2 < x < \pi/2$

25. $y = x|x|$

26. Graph $y = |\cos x|$, $-\pi \le x \le \pi$. What are the maximum and minimum values of the function, and where are they taken on?

27. Graph

$$y = \frac{1}{2}[|\sin x| + \sin x], \qquad 0 \le x \le 2\pi.$$

What are the maximum and minimum values of the function, and where are they taken on?

28. If $y = f(x)$ has a derivative at $x = c$ and $f'(c) = 0$, must f have a local maximum or local minimum value at $x = c$? Explain.

29. Give an example of a function that is continuous for all x, that decreases on $(-\infty, 0)$ and $(0, \infty)$, but has no derivative at $x = 0$. (*Hint:* Piece two functions together.)

30. Find all values of the constants m and b for which the function

$$f(x) = \begin{cases} mx + b & \text{for } x < 1, \\ x^2 + 1 & \text{for } x \ge 1, \end{cases}$$

is (a) continuous; (b) differentiable.

31. Let $y = x - 2 \sin x$, $0 \le x \le \pi$.

a) Find where $y' < 0$, $y' = 0$, and $y' > 0$ on $[0, \pi]$.

b) Plot the endpoints of the curve and any points where $y' = 0$. Then sketch the curve.

c) The equation $x - 2 \sin x = 0$ has two solutions on $[0, \pi]$, one of them being $x = 0$. Estimate the other solution to two decimal places.

32. Suppose that $y = x^n$, and n is a positive integer. Determine the values of x for which y is increasing and decreasing if (a) n is even; (b) n is odd.

***Toolkit* programs**

Derivative Grapher Super * Grapher
Function Evaluator

3.2

Concavity and Points of Inflection

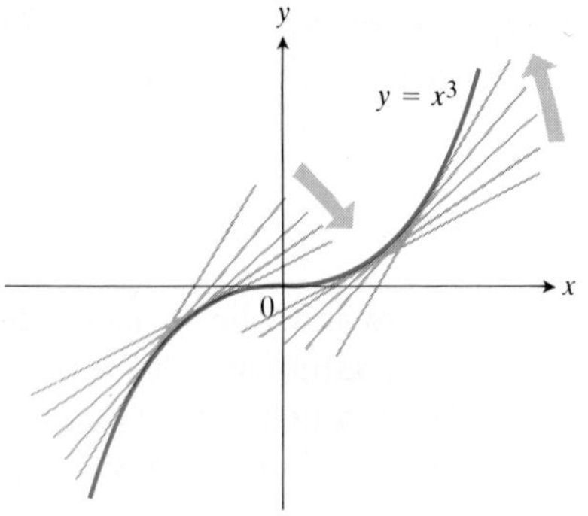

3.13 The graph of $y = x^3$ is concave down on the left, concave up on the right.

As we can see in Fig. 3.13, the function $y = x^3$ increases as x increases, but the portions of the curve that lie over the intervals $(-\infty, 0)$ and $(0, \infty)$ turn in different ways. If we come in from the left along the curve toward the origin, the curve turns to our right. As we leave the origin, the curve turns to our left. The left portion "bends down." The right portion "bends up."

To describe the turning another way, we can watch the tangent to the curve as x increases from left to right. As the point of tangency moves toward (0, 0) from the left, the tangent turns clockwise and the slope $y' = 3x^2$ decreases. The tangent becomes horizontal at (0, 0). As the point of tangency moves into the first quadrant, the tangent rises from its horizontal position to turn counterclockwise, and the slope $y' = 3x^2$ increases.

We say that the curve $y = x^3$ is *concave down* on the interval $(-\infty, 0)$ where y' decreases and *concave up* on the interval $(0, \infty)$ where y' increases. Thus we have the following definition of concavity for graphs of differentiable functions.

DEFINITIONS

Concave Up and Concave Down

The graph of a differentiable function $y = f(x)$ is *concave down* on an interval where y' decreases and *concave up* on an interval where y' increases.

If a function $y = f(x)$ has a second derivative as well as a first (as do most of the functions we deal with in this text), we can apply the first derivative test (discussed in Article 3.1) to the function $f' = y'$ to conclude that y' decreases if $y'' < 0$ and increases if $y'' > 0$. We therefore have a test that we can apply to the formula $y = f(x)$ to determine the concavity of its graph. It is called the *second derivative test for concavity.*

The Second Derivative Test for Concavity

The graph of $y = f(x)$ is

concave down on any interval where $y'' < 0$,

concave up on any interval where $y'' > 0$.

The idea is that if $y'' < 0$, then y' decreases as x increases and the tangent turns clockwise. Conversely, if $y'' > 0$, then y' increases as x increases and the tangent turns counterclockwise.

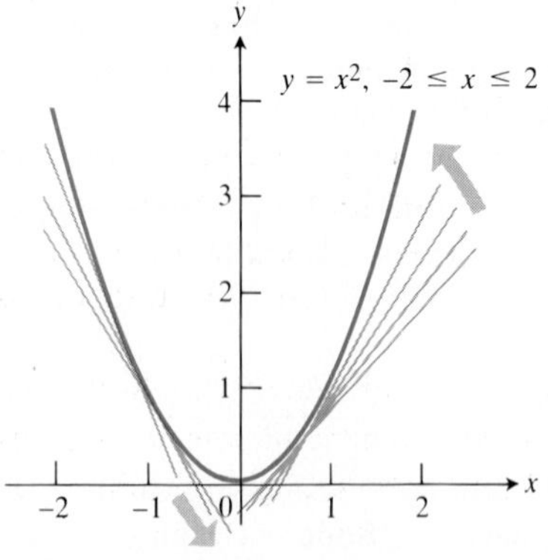

3.14 Concave up: The tangent turns counterclockwise as x increases; y' is increasing.

EXAMPLE 1 See Fig. 3.14. The curve $y = x^2$ is concave up on the entire x-axis because $y'' = 2 > 0$. □

EXAMPLE 2 See Fig. 3.15. The curve $y = \sin x$ is concave down over $0 < x < \pi$ because $y'' = -\sin x < 0$ on this interval. □

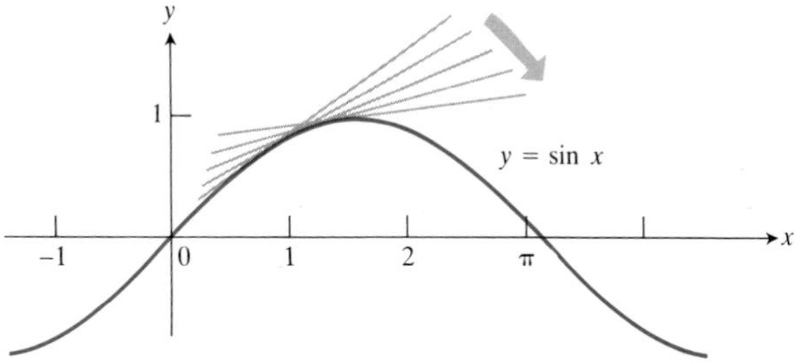

3.15 Concave down: The tangent turns clockwise as x increases; y' decreasing.

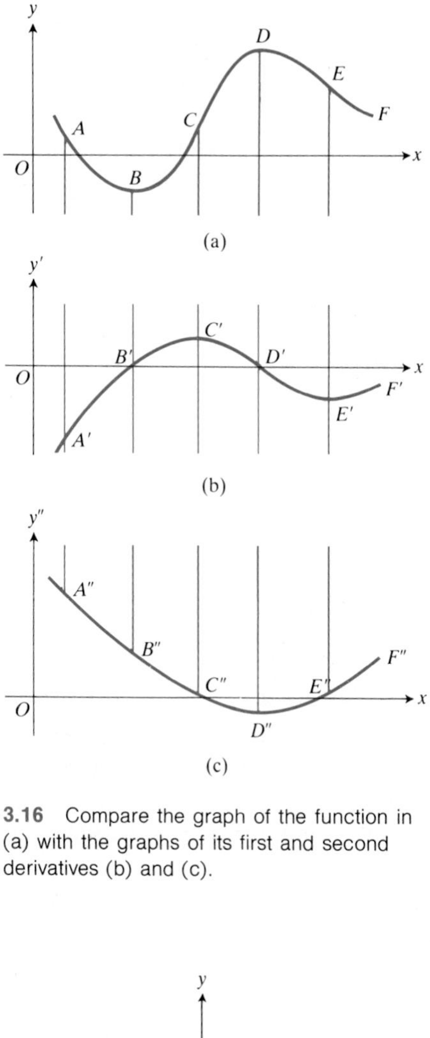

3.16 Compare the graph of the function in (a) with the graphs of its first and second derivatives (b) and (c).

Figure 3.16 shows the interplay between a typical function $y = f(x)$ and its first two derivatives. The arc ABC of the y-curve is concave up, CDE is concave down, and EF is again concave up. To focus attention, let us consider a section near A on arc ABC. Here y' is negative and the y-curve slopes downward to the right. But as we travel through A toward B we find the slope becomes less negative. That is, y' is an increasing function of x. Therefore, the y'-curve slopes upward at A'. Its own slope, which is y'', is positive there. The same kind of argument applies at all points along the arc ABC: y' is an increasing function of x so its derivative y'' is positive. This is indicated by drawing the arc $A''B''C''$ of the y''-curve above the x-axis.

Similarly, where the y-curve is concave down (along CDE) the y'-curve is falling, so its slope y'' is negative. At the point C, where the y-curve changes from concave up to concave down, y'' is zero.

Inflection

A point on a curve $y = f(x)$ where the concavity changes from up to down or vice versa is called a *point of inflection*. Thus a point of inflection on a twice-differentiable curve is a point where y'' is positive on one side, negative on the other. If y'' is continuous, this implies that y'' must be zero at a point of inflection. Unfortunately, it is possible for y'' to equal zero at a point that is *not* a point of inflection, as we shall see in Example 4. Also, a point of inflection on a graph may occur where y'' fails to exist, as in Example 5.

EXAMPLE 3 See Fig. 3.13. The curve $y = x^3$ has a point of inflection at $x = 0$ where $y'' = 6x$ changes sign. □

EXAMPLE 4 See Fig. 3.17. The curve $y = x^4$ has no inflection point at $x = 0$ even though $y'' = 12x^2$ is zero there. The second derivative does not change sign at $x = 0$ (in fact, y'' is never negative). The curve is concave up over the entire x-axis because $y' = 4x^3$ is an increasing function on $(-\infty, \infty)$.

We see in this example that although the condition $y'' > 0$ in the second derivative test for concavity is a sufficient condition, it is not a necessary one. The curve $y = x^4$ is concave up even though y'' fails to be greater than zero at the origin. □

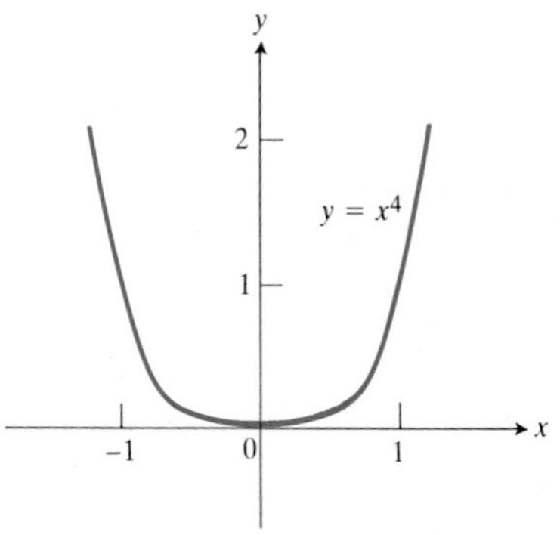

3.17 The graph of $y = x^4$ has no inflection point at the origin even though $y''(0) = 0$.

Most points of inflection occur where $f'' = 0$, but a point of inflection may occur where f'' is undefined, as the next example shows.

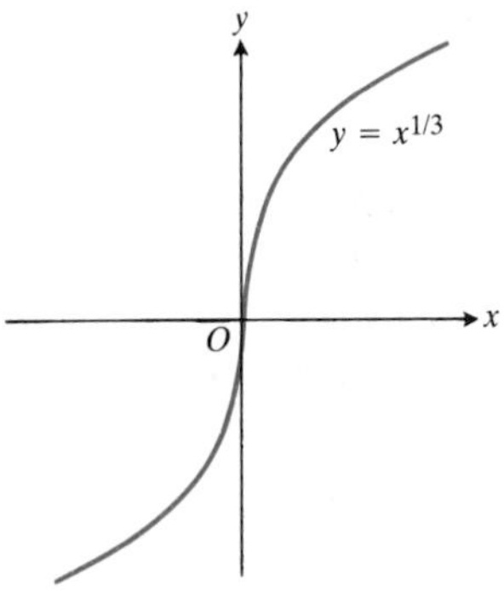

3.18 The graph of $y = x^{1/3}$ shows that a point where y'' fails to exist can be a point of inflection.

EXAMPLE 5 See Fig. 3.18. The curve $y = x^{1/3}$ has a point of inflection at $x = 0$ even though the second derivative does not exist there. To see this, we calculate y'' for $x \neq 0$:

$$y = x^{1/3},$$
$$y' = \tfrac{1}{3}x^{-2/3},$$
$$y'' = -\tfrac{2}{9}x^{-5/3}.$$

As $x \to 0$, y'' becomes infinite. Yet the curve is concave up for $x < 0$ (where $y'' > 0$ and y' is increasing) and concave down for $x > 0$ (where $y'' < 0$ and y' is decreasing). □

We now use what we have learned so far to graph a polynomial of degree three. The steps 1–5 in the solution give a general procedure for graphing.

EXAMPLE 6 See Fig. 3.19. Sketch the curve

$$y = \tfrac{1}{6}(x^3 - 6x^2 + 9x + 6).$$

Solution

1. We calculate y' and y'':

$$y' = \tfrac{1}{6}(3x^2 - 12x + 9) = \tfrac{1}{2}(x^2 - 4x + 3),$$
$$y'' = \tfrac{1}{2}(2x - 4) = x - 2.$$

2. We find the points where $y' = 0$ and determine where y' is positive and where it is negative. This gives information about where the curve rises and falls. The points where $y' = 0$ may give local maximum and minimum values of y.

   When factored,

$$y' = \tfrac{1}{2}(x - 1)(x - 3),$$

   which is zero when $x = 1$ and $x = 3$. As shown in Fig. 3.20, y' is positive when $x < 1$, negative when $1 < x < 3$, and positive again when $x > 3$. The curve has a local maximum at $x = 1$ (where y' changes from + to −) and a local minimum at $x = 3$ (where y' changes from − to +).

3.19 A graph of $y = \frac{1}{6}(x^3 - 6x^2 + 9x + 6)$ drawn after studying the values of y' and y''.

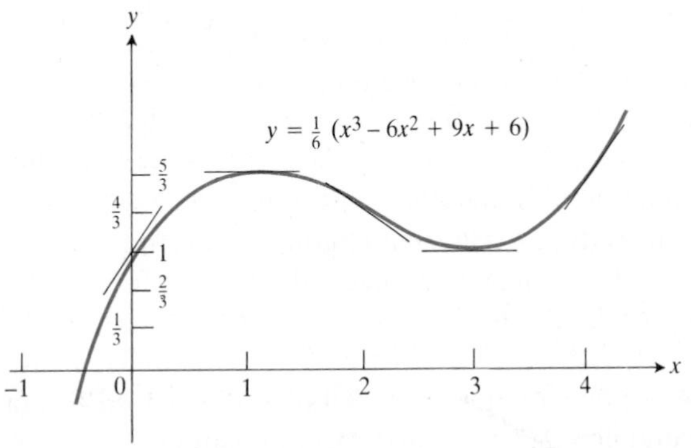

3.20 The sign of $y' = (x - 1)(x - 3)$.

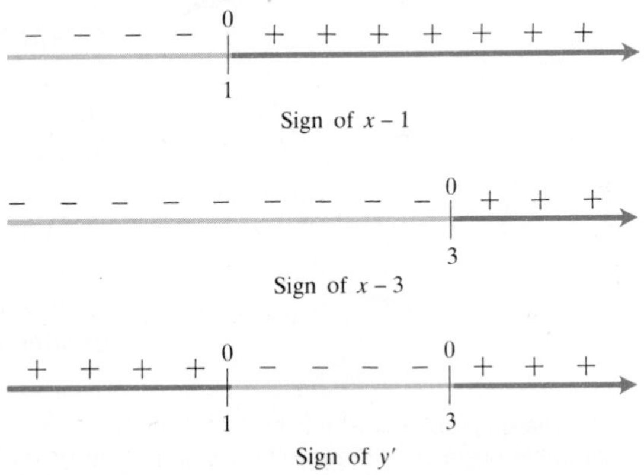

Table 3.1

x	y	y′	y″	Conclusions
-1	$-\frac{5}{3}$	$+4$	$-$	Rising; concave down
0	1	$+\frac{3}{2}$	$-$	Rising; concave down
1	$\frac{5}{3}$	0	$-$	Local maximum
2	$\frac{4}{3}$	$-\frac{1}{2}$	0	Falling; point of inflection
3	1	0	$+$	Local minimum
4	$\frac{5}{3}$	$+\frac{3}{2}$	$+$	Rising; concave up

3. We now find where $y'' = 0$ and determine where y'' is positive and where it is negative. This gives information about concavity. The points where $y'' = 0$ may be points of inflection.

   We have

$$y'' = x - 2,$$

   which is zero when $x = 2$, negative when $x < 2$, and positive when $x > 2$. The curve is concave down on $(-\infty, 2)$ and concave up on $(2, \infty)$. The point $x = 2$ is a point of inflection because y'' changes sign at $x = 2$.

4. We make a short table of values of y, y', and y'' that includes information we have gathered so far. We also enter the y-intercept and the values of y at a few additional points to get general information about the curve (Table 3.1).

5. We plot the points from the table, sketch tangents, and use the information about rise, fall, and convexity to draw the graph (Fig. 3.19). □

PROBLEMS

In Problems 1–12, find the intervals of x-values on which the curve is (a) rising, (b) falling, (c) concave up, (d) concave down. Sketch the curve, showing the points of inflection I and the points where the function has local maximum values M and local minimum values m.

1. $y = x^2 - 4x + 3$

2. $y = 20x - x^2$

3. $y = x^3 - 3x + 3$

4. $y = 4 + 3x - x^3$

5. $y = x^3 - 6x^2 + 9x + 1$

6. $y = \dfrac{x^3}{3} - \dfrac{x^2}{2} - 6x$

7. $y = (x - 2)^3 + 1$

8. $y = x^{2/3}$

9. $y = \tan x, \quad -\pi/2 < x < \pi/2$

10. $y = \cos x, \quad 0 \le x \le 2\pi$

11. $y = -x^4$

12. $y = |x^3|$

13. Sketch a smooth curve $y = f(x)$ illustrating

$$f(1) = 0,$$
$$f'(x) < 0 \quad \text{for } x < 1,$$
$$f'(x) > 0 \quad \text{for } x > 1.$$

14. Sketch a smooth curve $y = f(x)$ illustrating

$$f(1) = 0,$$
$$f''(x) < 0 \quad \text{for } x < 1,$$
$$f''(x) > 0 \quad \text{for } x > 1.$$

Sketch the curves in Problems 15–26, indicating inflection points and local maximum and minimum values.

15. $y = 6 - 2x - x^2$

16. $y = 2x^2 - 4x + 3$

17. $y = x(6 - 2x)^2$

18. $y = (x - 1)^2(x + 2)$

19. $y = 12 - 12x + x^3$

20. $y = x^3 - 3x^2 + 2$

21. $y = x^3 - 6x^2 - 135x$

22. $y = x^3 - 33x^2 + 216x$

23. $y = x^4 - 32x + 48$

24. $y = 3x^4 - 4x^3$

25. $y = x + \sin 2x$

26. $y = \sin x + \cos x$

In Problems 27–30, calculate dx/dy to find the vertical tangents. Then sketch the curves.

27. $x = y^3 + 3y^2 + 3y + 2$

28. $x = y^3 + 3y^2 - 9y - 11$

29. $x = y^4 - 2y^2 + 2$

30. $x = \sin y$

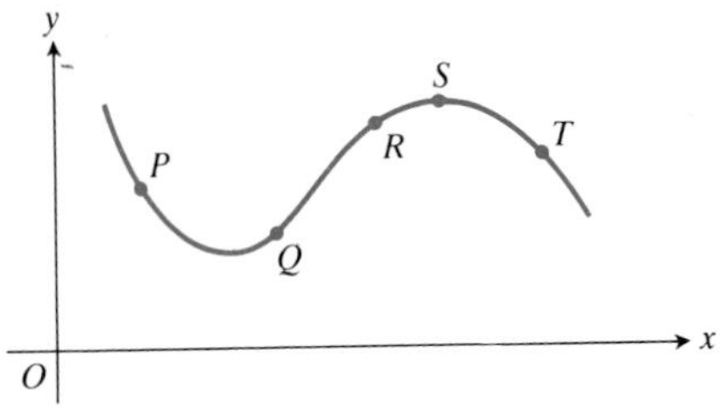

3.21 The curve for Problem 31.

31. Figure 3.21 shows the graph of a function $y = f(x)$. At which of the five points indicated on the graph (a) are $f'(x)$ and $f''(x)$ both negative? (b) is $f'(x)$ negative and $f''(x)$ positive?

32. Sketch a continuous curve $y = f(x)$ having the following characteristics:

$$f(-2) = 8, \quad f'(2) = f'(-2) = 0,$$
$$f(0) = 4, \quad f'(x) < 0 \quad \text{for } |x| < 2,$$
$$f(2) = 0, \quad f''(x) < 0 \quad \text{for } x < 0,$$
$$f'(x) > 0 \quad \text{for } |x| > 2, \quad f''(x) > 0 \quad \text{for } x > 0.$$

33. Sketch a continuous curve $y = f(x)$ with the following properties. Label coordinates where possible.

x	y	Curve
$x < 2$		Falling, concave up
2	1	Horizontal tangent
$2 < x < 4$		Rising, concave up
4	4	Inflection point
$4 < x < 6$		Rising, concave down
6	7	Horizontal tangent
$x > 6$		Falling, concave down

34. Sketch a continuous curve $y = f(x)$ having

$$f'(x) > 0 \quad \text{for } x < 2, \qquad f'(x) < 0 \quad \text{for } x > 2,$$

a) if $f'(x)$ is continuous at $x = 2$;
b) if $f'(x) \to 1$ as $x \to 2^-$ and $f'(x) \to -1$ as $x \to 2^+$;
c) if $f'(x) = 1$ for all $x < 2$ and $f'(x) = -1$ for all $x > 2$.

35. Sketch a continuous curve $y = f(x)$ for $x > 0$ if

$$f(1) = 0 \quad \text{and} \quad f'(x) = 1/x \quad \text{for all } x > 0.$$

Is such a curve necessarily concave upward or concave downward?

36. The curve $y = x^3 + bx^2 + cx + d$ (b, c, and d constant) will have a point of inflection when $x = 1$ if b has which of the following values?
a) 2 b) -2 c) 3 d) -3

37. Sketch the curve

$$y = 2x^3 + 2x^2 - 2x - 1$$

after locating its maximum, minimum, and inflection points. Then answer the following questions from your graph.
a) How many times and approximately where does the curve cross the x-axis?
b) How many times and approximately where would the curve cross the x-axis if $+3$ were added to all the y-values?
c) How many times and approximately where would the curve cross the x-axis if -3 were added to all the y-values?

38. Show that the curve

$$y = x + \sin x$$

has no maximum or minimum points even though it does have points where dy/dx is zero. Sketch the curve.

39. Find all local maximum points, local minimum points, and points of inflection on the curve $y = x^4 + 8x^3 - 270x^2$.

40. The graph of the function $f(x) = x^4 - 2x^2 - 4$ resembles a molar.
a) Graph the function and label the points of inflection.
b) Estimate the positive root from an initial guess of $x_0 = 2$ with one iteration of Newton's method.

41. The slope of a curve $y = f(x)$ is

$$\frac{dy}{dx} = 3(x-1)^2(x-2)^3(x-3)^4(x-4).$$

a) For what value or values of x does y have a local maximum?
b) For what value or values of x does y have a local minimum?

42. Recall that a function $y = f(x)$ is *odd* if $f(-x) = -f(x)$, and *even* if $f(-x) = f(x)$, for all x in its domain.
a) Suppose that an odd function is known to be increasing on the interval $[-1, 0]$. What can be said about its behavior on $[0, 1]$?
b) Suppose that an even function is known to be increasing on $[-1, 0]$. What can be said about its behavior on $[0, 1]$?
c) Suppose that an odd function is concave up on $[0, 1]$. What can be said about its concavity on $[-1, 0]$?
d) Suppose that an even function is concave down on $[0, 1]$. What can be said about its concavity on $[-1, 0]$?

***Toolkit* programs**

Derivative Grapher
Function Evaluator
Super * Grapher

3.3 Asymptotes and Symmetry

Any equation

$$F(x, y) = 0 \tag{1}$$

in x and y determines a graph in the plane that is the set of all points (x, y) whose coordinates satisfy the equation. We discussed equations like this briefly in Article 2.4. Any equation in x and y can be put in the form of Eq. (1) by moving all its terms to the left side. Thus,

$$x^2 + y^2 = 1, \qquad y = x^2, \qquad y \cos x = 3x + 1,$$

can be put in the equivalent forms

$$x^2 + y^2 - 1 = 0,$$
$$y - x^2 = 0,$$
$$y \cos x - 3x - 1 = 0,$$

and so on.

Symmetry

Certain kinds of symmetries in the graph of the equation $F(x, y) = 0$ can easily be detected in the following way.

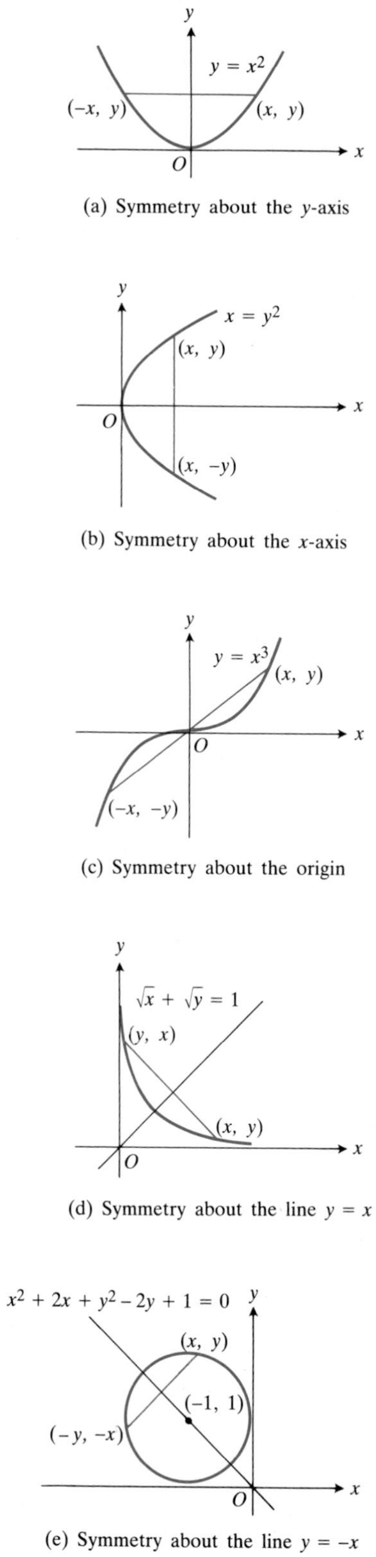

(a) Symmetry about the y-axis

(b) Symmetry about the x-axis

(c) Symmetry about the origin

(d) Symmetry about the line $y = x$

(e) Symmetry about the line $y = -x$

3.22 Coordinate tests for symmetry.

Symmetry about the y-axis The graph (curve) is symmetric about the y-axis if the equation is unaltered when x is replaced by $-x$; that is, if

$$F(x, y) = F(-x, y),$$

then the points (x, y) and $(-x, y)$ both lie on the curve (satisfy the equation $F = 0$) if either one of them does. In particular, an equation that contains only even powers of x represents such a curve (Fig. 3.22a).

Symmetry about the x-axis The graph is symmetric about the x-axis if the equation is unaltered when y is replaced by $-y$; that is, if

$$F(x, y) = F(x, -y).$$

In particular, this will happen if only even powers of y occur in the equation (Fig. 3.22b).

Symmetry about the origin (See Fig. 3.22c.) The equation is unaltered when x and y are replaced by $-x$ and $-y$; that is,

$$F(x, y) = F(-x, -y).$$

Symmetry about the line $y = x$ (See Fig. 3.22d.) The equation is unaltered when x and y are interchanged, that is,

$$F(x, y) = F(y, x).$$

Symmetry about the line $y = -x$ (See Fig. 3.22e.) The equation is unaltered when x and y are replaced by $-y$ and $-x$; that is,

$$F(x, y) = F(-y, -x).$$

EXAMPLE 1 The graph of the equation $x^2 + y^2 = 1$ is symmetric about both axes, the origin, the line $y = x$, and the line $y = -x$. □

EXAMPLE 2 The graph of the equation $x^2 - y^2 = 1$ is symmetric about both axes and the origin and not symmetric about the lines $y = x$ and $y = -x$. □

EXAMPLE 3 The graph of the equation $xy = 1$ is not symmetric about either axis but is symmetric about the origin and about the lines $y = x$ and $y = -x$. □

Horizontal and Vertical Asymptotes

As a point P on the graph of a function $y = f(x)$ moves farther and farther away from the origin, it may happen that the distance between P and some fixed line tends to zero. In other words, the curve "approaches" the line as it gets further from the origin. In such a case, the line is called an *asymptote* of the graph. For instance, the x-axis and y-axis are asymptotes of the curves $y = 1/x$ and $y = 1/x^2$ (Figs. 3.5 and 3.4).

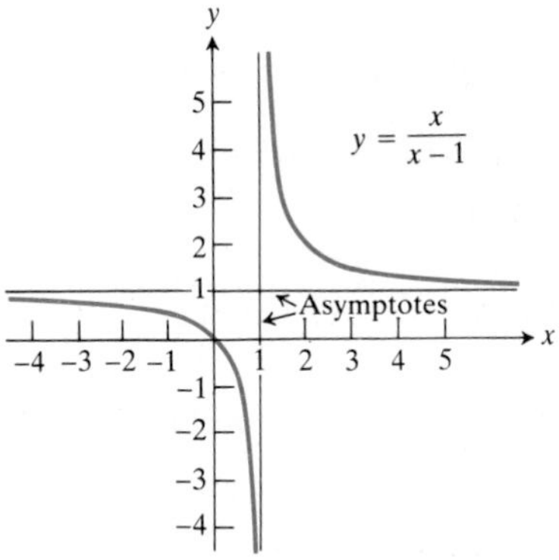

3.23 The graph of $y = x/(x - 1)$.

EXAMPLE 4 Figure 3.23 shows the asymptotes of

$$y = \frac{x}{x - 1}.$$

The line $y = 1$ is an asymptote on the right because

$$\lim_{x \to \infty} \frac{x}{x - 1} = 1.$$

It is also an asymptote on the left because y again approaches 1 as $x \to -\infty$.

The line $x = 1$ is a vertical asymptote of the right-hand branch of the graph because

$$\lim_{x \to 1^+} \frac{x}{x - 1} = \infty,$$

and of the left-hand branch because

$$\lim_{x \to 1^-} \frac{x}{x - 1} = -\infty. \quad \square$$

In general, we have the following description of horizontal and vertical asymptotes.

DEFINITIONS

Horizontal and Vertical Asymptotes

A line $y = b$ is a *horizontal asymptote* of the graph of a function $y = f(x)$ if either

$$\lim_{x \to \infty} f(x) = b \quad \text{or} \quad \lim_{x \to -\infty} f(x) = b.$$

A line $x = a$ is a *vertical asymptote* of the graph if either

$$\lim_{x \to a^+} f(x) = \pm\infty \quad \text{or} \quad \lim_{x \to a^-} f(x) = \pm\infty.$$

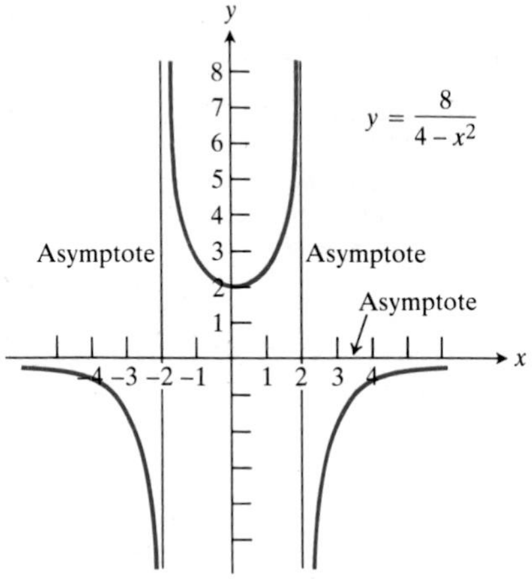

3.24 The graph of $y = 8/(4 - x^2)$.

EXAMPLE 5 Figure 3.24 shows the graph of

$$y = \frac{8}{4 - x^2}.$$

The line $y = 0$ is an asymptote on the right because $y \to 0$ as $x \to \infty$, and an asymptote on the left because $y \to 0$ as $x \to -\infty$.

The line $x = 2$ is an asymptote because

$$\lim_{x \to 2^-} \frac{8}{4 - x^2} = \infty,$$

and again because

$$\lim_{x \to 2^+} \frac{8}{4 - x^2} = -\infty.$$

Similarly, the line $x = -2$ is an asymptote because $y \to \infty$ as $x \to -2^+$ and again because $y \to -\infty$ as $x \to -2^-$. □

Oblique Asymptotes

If a rational function is a quotient of two polynomials that have no common factor, and if the degree of the numerator exceeds the degree of the denominator by 1, as it does in

$$y = \frac{x^2 - 3}{2x - 4},$$

then the graph will have an *oblique asymptote* (Fig. 3.25). To see why this is so, we divide $x^2 - 3$ by $2x - 4$ to obtain

$$\begin{array}{r} \frac{x}{2} + 1 \\ 2x - 4 \overline{)\, x^2 - 3} \\ \underline{x^2 - 2x} \\ 2x - 3 \\ \underline{2x - 4} \\ 1 \end{array}$$

3.25 The graph of $y = (x^2 - 3)/(2x - 4)$.

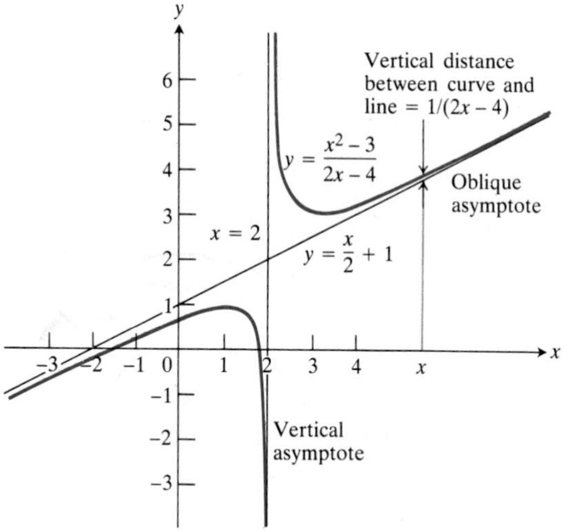

This tells us that

$$y = \frac{x^2 - 3}{2x - 4} = \underbrace{\frac{x}{2} + 1}_{\text{linear}} + \underbrace{\frac{1}{2x - 4}}_{\substack{\text{remainder} \\ \text{goes to 0} \\ \text{as } x \to \infty}}. \tag{2}$$

From this representation we see that

$$y - \left(\frac{x}{2} + 1\right) = \frac{1}{2x - 4}$$

and that

$$\lim_{x\to\infty} \left[y - \left(\frac{x}{2} + 1\right)\right] = \lim_{x\to\infty} \frac{1}{2x - 4} = 0.$$

Thus the vertical distance between the curve $y = (x^2 - 3)/(2x - 4)$ and the line $y = x/2 + 1$ approaches zero as $x \to \infty$. Therefore, the line

$$y = \frac{x}{2} + 1$$

is an asymptote of the curve.

Note too that as $x \to \infty$ the slope

$$y' = \frac{1}{2} - \frac{2}{(2x - 4)^2}$$

of the curve approaches $\frac{1}{2}$, which is the slope of the line.

The curve also has a vertical asymptote at $x = 2$.

EXAMPLE 6 Graph the function

$$y = x + \frac{1}{x}.$$

Solution (See Fig. 3.26.) The graph is symmetric about the origin. Replacing x and y by $-x$ and $-y$ gives

$$-y = -x - \frac{1}{x}, \quad \text{or} \quad y = x + \frac{1}{x}.$$

The equation is not altered by the substitution.

If we let $x \to 0$ from the right, $y \to \infty$. If we let $x \to 0$ from the left, $y \to -\infty$. The line $x = 0$ is a vertical asymptote.

$$\text{As } x \to \infty, \quad \frac{1}{x} \to 0 \quad \text{and} \quad (y - x) = \frac{1}{x} \to 0.$$

Therefore, the line $y = x$ is also an asymptote of the graph.

The first derivative

$$y' = 1 - \frac{1}{x^2}$$

equals zero when $x = -1$ and $x = 1$, and is not defined at $x = 0$. The derivative is positive when $x < -1$, negative when $-1 < x < 0$ and $0 < x < 1$, and positive when $x > 1$ (Fig. 3.26). Accordingly, the graph rises on $(-\infty, -1)$ and $(1, \infty)$, and falls on $(-1, 0)$ and $(0, 1)$. The function has a

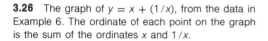

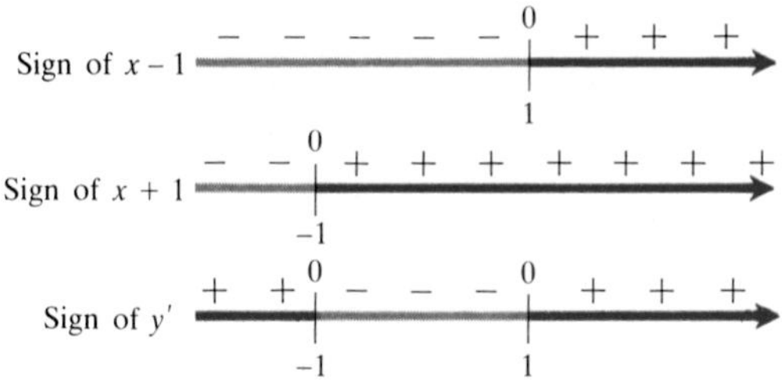

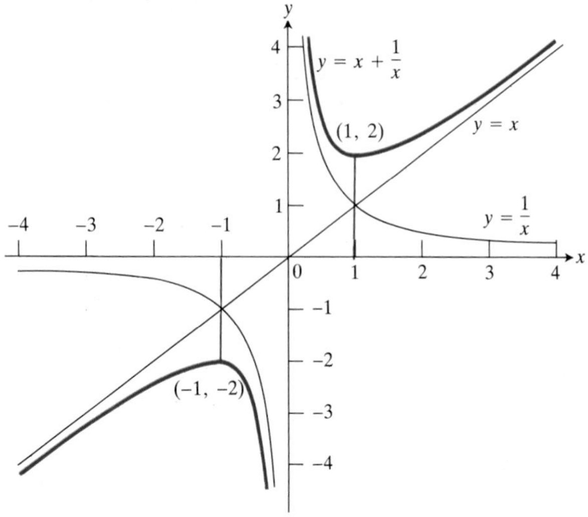

3.26 The graph of $y = x + (1/x)$, from the data in Example 6. The ordinate of each point on the graph is the sum of the ordinates x and $1/x$.

local maximum value at $x = -1$ where y' changes from $(+)$ to $(-)$, and a local minimum value at $x = 1$ where y' changes from $(-)$ to $(+)$. The second derivative

$$y'' = \frac{2}{x^3}$$

is negative when $x < 0$, positive when $x > 0$, and not defined at $x = 0$. The curve is concave down when $x < 0$, concave up when $x > 0$.

When x is small the term x contributes very little to the sum

$$y = x + \frac{1}{x} \tag{3}$$

in comparison to the contribution of $1/x$. The term $1/x$ is the dominant term for small values of x and we can expect the function to behave like $1/x$ when x is small. On the other hand, when x is numerically very large, the term $1/x$ contributes very little to the value of y in comparison to the contribution of x. The term x is the dominant term for large values of x and we can expect the function to behave like x when x is large. (Hence the asymptote $y = x$.)

Therefore, by looking at the formula in (3) we can tell two things right away about the behavior of y:

$$y \approx \frac{1}{x} \quad \text{when } x \text{ is small} \qquad \text{and} \qquad y \approx x \quad \text{when } x \text{ is large.} \tag{4}$$

Of all the observations we can make quickly about y, these are probably the most useful.

To graph $y = x + 1/x$, we draw the asymptote $y = x$, sketch the curve $y = 1/x$, and plot the local maximum and minimum points. We then sketch a curve that fits these and has the symmetry and other properties we have discovered. □

Behavior for Special Values of *x*

Here are some other examples of the behavior of functions for small, large, and other special values of x.

EXAMPLE 7 If

$$y = \frac{2}{x} + 6x^2,$$

then

$$y \approx \frac{2}{x}, \quad x \text{ small},$$

$$y \approx 6x^2, \quad x \text{ large. } \square$$

EXAMPLE 8 From Eq. (2) we have

$$y = \frac{x^2 - 3}{2x - 4} = \frac{x}{2} + 1 + \frac{1}{2x - 4},$$

$$y \approx \frac{x}{2} + 1, \quad |x| \text{ large},$$

$$y \approx 2 + \frac{1}{2x - 4}, \quad x \text{ near } 2.$$

(See Fig. 3.25.) □

EXAMPLE 9 For

$$y = \frac{x + 3}{x + 2} = 1 + \frac{1}{x + 2},$$

we have

$$y \approx 1, \quad |x| \text{ large},$$

$$y \approx 1 + \frac{1}{x + 2}, \quad x \text{ near } -2. \ \square$$

EXAMPLE 10 From Example 2 we have

$$y = \frac{8}{4 - x^2} = \frac{8}{(2 - x)(2 + x)},$$

so that

$$y \approx \frac{8}{(2 - x)(4)} = \frac{2}{2 - x}, \quad x \text{ near } 2,$$

$$y \approx \frac{8}{4(2 + x)} = \frac{2}{2 + x}, \quad x \text{ near } -2.$$

(See Fig. 3.24.) □

EXAMPLE 11 Graph the equation

$$y^2(x^2 - x) = x^2 + 1. \tag{5}$$

Solution We solve for y to obtain

$$y = \pm\sqrt{\frac{x^2 + 1}{x(x - 1)}} \tag{6}$$

(see Fig. 3.27). The expression under the radical must not be negative, so no portion of the curve lies between the lines $x = 0$ and $x = 1$. But all values of $x > 1$ and all negative values of x, that is, $x < 0$, are permissible.

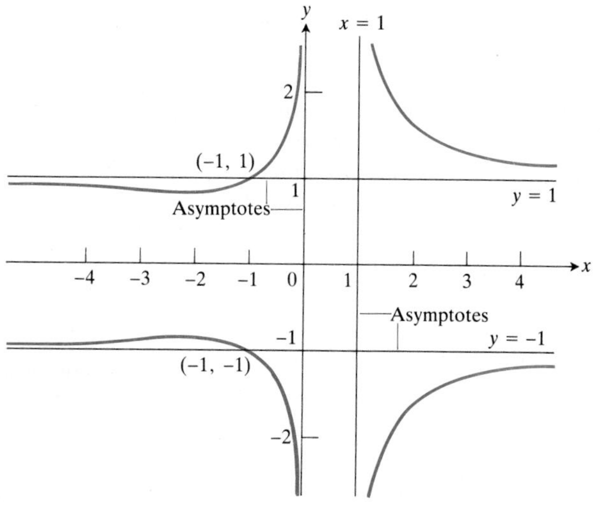

3.27 The graph of $y^2(x^2 - x) = x^2 + 1$ has four asymptotes.

As x approaches zero *from the left*, $y \to \pm\infty$ and as x approaches one *from the right*, $y \to \pm\infty$. The lines $x = 0$ and $x = 1$ are asymptotes of the curve.

Since arbitrarily large values of $|x|$ are permitted, we also investigate the behavior of the curve as $x \to -\infty$ and again as $x \to +\infty$. If we write the equation of the curve in the equivalent form

$$y^2 = \frac{x^2 + 1}{x^2 - x} = \frac{1 + (1/x^2)}{1 - (1/x)}, \tag{7}$$

we see that $y^2 \to 1$ as $x \to \infty$ or as $x \to -\infty$. Thus the lines $y = 1$ and $y = -1$ are also asymptotes of the curve.

We reexamine Eq. (7) to see if y^2 is necessarily always greater than one, or always less than one, or may be equal to one. When x tends to $+\infty$, the numerator in (7) is greater than one, while the denominator is less than one; hence the fraction is larger than one, that is,

$$y^2 > 1, \qquad y^2 \to 1 \qquad \text{as } x \to +\infty. \tag{8}$$

On the other hand, when x tends to $-\infty$, both $1/x^2$ and $-1/x$ are positive, so that both the numerator and denominator in (7) are greater than one. However, $1/x^2$ will be less than $-1/x$, so that the numerator is less than the denominator; that is,

$$y^2 < 1, \qquad y^2 \to 1 \qquad \text{as } x \to -\infty. \tag{9}$$

To determine whether y^2 may ever equal 1, we try it, say in (5), and find

$$y^2 = 1 \qquad \text{when } x = -1. \tag{10}$$

The curve is symmetric about the x-axis, because y may be replaced by $-y$ without changing the equation. There is no x-intercept, because putting $y = 0$ in Eq. (5) requires $0 = x^2 + 1$, which has only imaginary roots.

Using this information, we can sketch the curve with a fair degree of accuracy (see Fig. 3.27). □

PROBLEMS

Analyze each of the equations below to investigate the following properties of the curve: (a) symmetry, (b) restrictions on x and y, (c) intercepts, (d) asymptotes, (e) slope at the intercepts. Draw the asymptotes, locate a few points, and sketch the curve, taking into account the information discovered above.

1. $y = \frac{1}{2x - 3}$

2. $y = \frac{1}{x^2 - 1}$

3. $y = 2 + \frac{4}{4 - x^2}$

4. $y = \frac{x}{2} + \frac{2}{x}$

5. $y = x - \frac{1}{x}$

6. $y = \frac{x + 3}{x + 2}$ (See Example 9.)

7. $y = \frac{x}{x + 1}$ (*Hint:* Divide x by x + 1.)

8. $y = \frac{x - 4}{x - 5}$ (*Hint:* Divide x − 4 by x − 5.)

9. $y = x^2 + 1$

10. $y = \frac{1}{x^2 + 1}$

11. $y = x^2 - 1$

12. $y = \frac{1}{x^2 - 1}$

13. $y = \frac{x^2}{x - 1}$

14. $y = \left(\frac{x}{x - 1}\right)^2$

15. $y = \frac{x^2 + 1}{x - 1}$

16. $y = \frac{x^2 + 1}{1 - x}$

17. $y = \frac{x^2 - 4}{x - 1}$

18. $y = \frac{x^2 - 9}{x - 5}$

19. $y = \frac{x^2 - 1}{2x + 4}$

20. $y = \frac{x^2 + x - 2}{x - 2}$

21. $y = \frac{2}{x} + 6x^2$

22. $y = x^2 + \frac{1}{x^2}$

23. $y = \frac{x^2 - 4}{x^2 - 1}$

24. $y = \frac{x^2 + 8}{x^2 - 4}$

25. $y = \frac{x^2 + 1}{x^2 - 4x + 3}$

26. $y^2 = x(x - 2)$

27. $y^2 = \frac{x}{x - 2}$

28. $x^4 + y^4 = 1$

29. $x^2 = \frac{1 + y^2}{1 - y^2}$

30. $x^2 = \frac{y^2 + 1}{y^2 - 1}$

31. $y = \frac{x - 1}{x^2(x - 2)}$

32. $y = \frac{x^2 + 1}{x^3 - 4x}$

33. The graph of the function $f(x) = 2 + (\sin x)/x$, $x > 0$ (see Fig. 1.100) crosses the line $y = 2$ whenever $\sin x = 0$, and therefore crosses the line infinitely often as $x \to \infty$. Show that the slope of the curve nevertheless approaches the slope of the line as $x \to \infty$.

34. Let

$$f(x) = \frac{x^3 - 4x}{x^3 - x}, \quad x \neq 0.$$

a) Find the limits of f as $x \to 0$ and $x \to \pm\infty$.
b) What value should be assigned to $f(0)$ to make f continuous at $x = 0$?
c) Find equations for the vertical and horizontal asymptotes of f.
d) Describe the symmetry of f.
e) Sketch a graph of the continuous extension of f.

35. What symmetries must the graph of $y = f(x)$ have if (a) f is an even function of x? (b) f is an odd function of x? (c) What symmetries do the graphs of $y = \sin x$ and $y = \tan x$ have?

36. Is it possible for a curve to be symmetric with respect to both the origin and the y-axis without being symmetric to the x-axis? If so, give an example. If not, why not?

***Toolkit* programs**

Derivative Grapher
Function Evaluator
Super * Grapher

3.4 Maxima and Minima: Theory

Figure 3.28 shows a point $x = c$ where a function $y = f(x)$ has a minimum value. If we move away from c to either side, the curve rises and the function value gets bigger. However, if we were to sketch more of the curve (Fig. 3.29) we might find that the function took on smaller values somewhere else. We would then know that $f(c)$ was not the absolute minimum of the function, but it would still be a relative or local minimum.

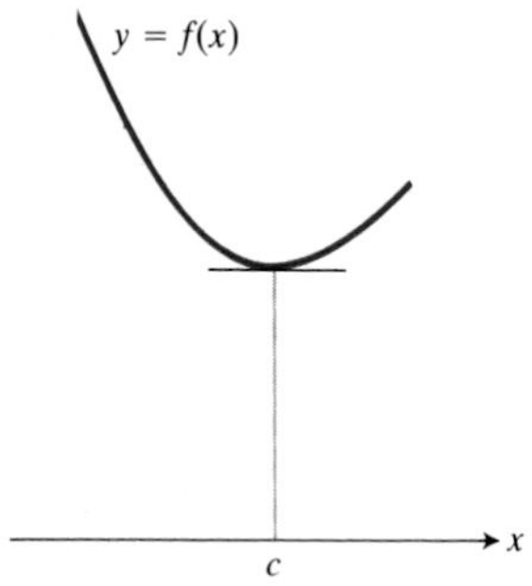

3.28 f has a relative minimum at $x = c$.

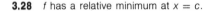

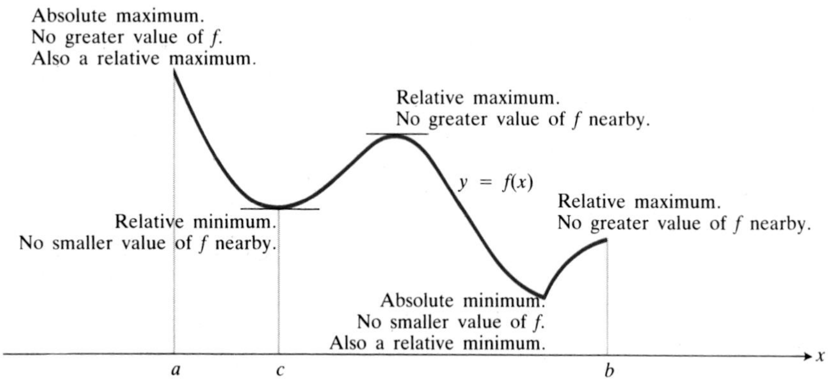

3.29 How maxima and minima are classified.

A function f is said to have a *relative* or *local minimum* at $x = c$ if

$$f(c) \leq f(x)$$

for all values of x in some open interval about c. (If c is an endpoint of the domain of f, the interval is to be half open, containing c as an endpoint.) The interval might be small or it might be large, but no value of the function in the interval is less than $f(c)$. For a *local* or *relative maximum* at $x = c$,

$$f(x) \leq f(c)$$

for all x in some interval about c. No value of the function in this interval is greater than $f(c)$.

The word *relative* or *local* is used to distinguish such a point from an *absolute* maximum or minimum that would occur if either of the inequalities

$$f(x) \leq f(c) \quad \text{or} \quad f(c) \leq f(x)$$

held for all x in the domain of f, not just for all x in an appropriate interval about c.

The function

$$f(x) = x + \frac{1}{x}$$

in Fig. 3.26 has a relative minimum at $x = 1$, because

$$x + \frac{1}{x} \geq 2$$

for all values of x in a small interval about 1. But the inequality is not satisfied when x is negative. In fact, the function has a relative maximum at $x = -1$. It has no absolute maximum or absolute minimum.

We have encountered relative maxima and minima in sketching curves and have observed that they occur at transition points between rising and falling portions of a curve. At such points we have also observed that dy/dx is usually zero, but that in exceptional cases dy/dx may not exist (as with $y = |x|$ at $x = 0$, shown in Fig. 3.12). The following theorem says that if a function $y = f(x)$ is differentiable on an open interval, then all of its relative maxima and minima in the interval are to be found where $dy/dx = 0$.

THEOREM 1

> Suppose that a function f defined on an interval $[a, b]$ has a relative maximum or minimum at an interior point $x = c$ of the interval (that is, $a < c < b$). If f is differentiable at c, then
>
> $$f'(c) = 0. \tag{1}$$

Proof. We give the proof for the case of a relative minimum at $x = c$. Namely, we shall assume that

$$f(c) \leq f(x)$$

for all x in an interval about c, which is equivalent to saying that

$$f(c) \leq f(c + h)$$

for all h in an interval about zero. By hypothesis,

$$f'(c) = \lim_{h \to 0} \frac{f(c + h) - f(c)}{h} \tag{2}$$

exists as a definite number, which we want to prove is zero. When h is small, we have

$$\frac{f(c + h) - f(c)}{h} \geq 0 \quad \text{if } h > 0$$

and

$$\frac{f(c + h) - f(c)}{h} \leq 0 \quad \text{if } h < 0,$$

because in both cases the numerator is positive or zero. Hence if we let $h \to 0$ through positive values, we have

$$f'(c) \geq 0$$

from the first case; but if we let $h \to 0$ through negative values, we also have

$$f'(c) \leq 0$$

from the second case. Since the derivative is assumed to exist, we must have the same limit in both cases, so

$$0 \leq f'(c) \leq 0.$$

The only way this can happen is to have

$$f'(c) = 0.$$

The proof in case of a relative maximum at $x = c$ is similar. ■

CAUTION. Be careful not to read into the theorem more than it says. It says that $f' = 0$ at every interior point where f has a relative maximum or minimum and f' exists. It does not say what happens if a maximum or minimum occurs

a) at a point c where f' fails to exist, or

b) at an endpoint of the domain of f.

Neither does it say that f must have a maximum or minimum at every place where f' is zero.

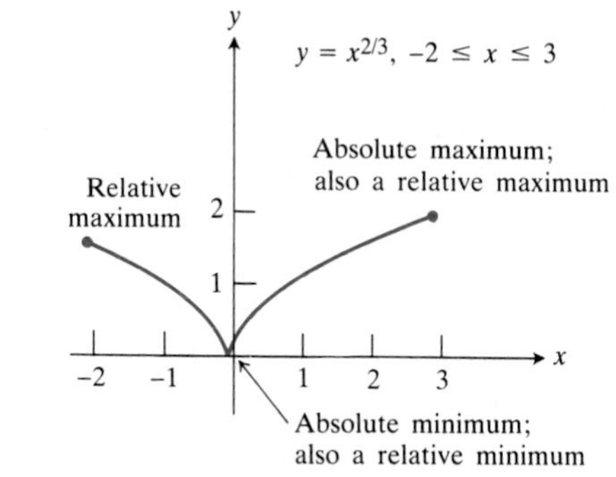

3.30 A function can have a minimum value at a point where its derivative does not exist. One way this can happen is shown here, where the curve has a vertical tangent at $x = 0$. Another way is shown in Fig. 3.12, where $y = |x|$ has no tangent at all at $x = 0$.

EXAMPLE 1 Figure 3.30 shows the graph of $y = x^{2/3}$, on $[-2, 3]$. The derivative

$$y' = \frac{2}{3}x^{-1/3} = \frac{2}{3\sqrt[3]{x}}$$

does not exist at $x = 0$ where y has its minimum value of zero. In fact, the curve has a vertical tangent at (0, 0) because

$$\frac{dx}{dy} = \frac{3}{2}x^{1/3}$$

is zero. Also, $f(x) = x^{2/3}$ has relative maximum values at $x = -2$ and $x = 3$, although neither $f'(-2)$ nor $f'(3)$ is zero. □

Example 1 shows that when a maximum or minimum occurs at the *end* of a curve that exists only over a limited interval, the derivative need not vanish at such a point. This does not contradict Theorem 1 in any way. The argument that establishes the theorem does not apply to endpoints because the limit of the difference quotient in Eq. (2) is required to be a two-sided limit at $x = c$ for the subsequent argument to work.

The point (0, 0) on the curve $y = x^3$ is an example of a point where a curve has zero slope without having either a maximum or a minimum there.

EXAMPLE 2 Find all maxima and minima of the function

$$y = x^3 - 3x + 2, \qquad -\infty < x < \infty.$$

Solution See Fig. 3.31. The function is differentiable at every point of $(-\infty, \infty)$, so Theorem 1 says that all the relative maxima and minima occur where the derivative

$$y' = 3x^2 - 3 = 3 \cdot (x - 1)(x + 1)$$

is zero, that is, at

$$x = 1 \quad \text{and} \quad x = -1.$$

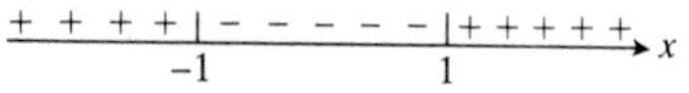

(a) Sign of $y' = 3(x-1)(x+1)$

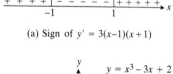

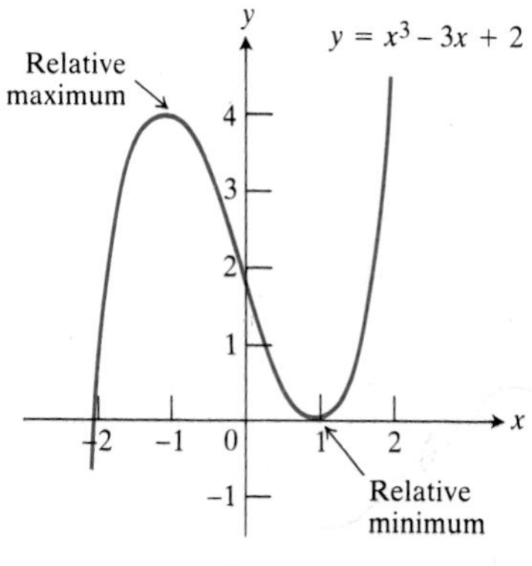

(b)

3.31 $y = x^3 - 3x + 2$

The sign pattern of the derivative is shown in Fig. 3.31(a). There is a relative maximum at $x = -1$ where y' changes from (+) to (−) as x increases, and a relative minimum at $x = 1$ where y' changes from (−) to (+). The corresponding values of y are

$$y(-1) = 4, \qquad \text{relative maximum,}$$
$$y(1) = 0, \qquad \text{relative minimum.}$$

Another way to determine whether the function has extreme values at $x = -1$ and $x = 1$ is to calculate

$$y'' = 6x$$

at these points to determine whether the curve is concave up, is concave down, or has a point of inflection. At $x = -1$,

$$y'' = -6 < 0,$$

which tells that the curve is concave down and y has a relative maximum.

At $x = 1$,

$$y'' = 6 > 0,$$

which tells that the curve is concave up and y has a relative minimum.

When the second derivative of a function is hard to work with or inconvenient to calculate, checking y' for a sign change will generally be the quicker of the two ways to classify the points where $y' = 0$. □

As Example 2 shows, once we have found a point c where the first derivative of a function $y = f(x)$ is zero, it is still necessary to determine whether the function takes on a maximum, a minimum, or neither, at $x = c$. The second derivative test for concavity from Article 3.2 can help us here. For if $y'' < 0$ in an open interval about c, then the curve is concave down and f has a local maximum at $x = c$. Similarly, if $y'' > 0$, then the curve is concave up and f has a local minimum at $x = c$. Thus if y'' is easy to calculate in an interval about a point c where $y' = 0$, we have a convenient way to test for a relative maximum or minimum at c.

THEOREM 2

Second Derivative Test for Relative Maxima and Minima†

If $f'(c) = 0$, and $f'' < 0$ in an open interval containing c, then f has a relative maximum at c. If $f'(c) = 0$ and $f'' > 0$ in an open interval containing c, then f has a relative minimum at c.

When we combine the result of Theorem 1 above with Theorem 9 from Article 1.11, which says that a continuous function on a closed interval $[a, b]$ takes on a maximum value and a minimum value at least once on $[a, b]$, we come up with a useful way to hunt for a differentiable function's *absolute* maximum and minimum on a closed interval. We describe the method in the next theorem.

THEOREM 3

Location of Absolute Maxima and Minima in Closed Intervals

If a function f is continuous on a closed interval $[a, b]$ and differentiable throughout the open interval (a, b), then f takes on an absolute maximum value M and an absolute minimum value m on $[a, b]$. Furthermore, the only points in $[a, b]$ where f may take on these extreme values are

1. the points of (a, b) where $f' = 0$, and
2. the endpoints a and b.

Proof. Theorem 9 in Article 1.11 says that because f is continuous, it takes on an absolute maximum and an absolute minimum at least once in the interval $[a, b]$. The statement of the theorem in Article 1.11 does not use the word "absolute," but the minimum value m and maximum value M described there are absolute because, as stated in the theorem,

$$m \le f(x) \le M$$

for all x in $[a, b]$.

†For a stronger version of the test, see Miscellaneous Problem 95, Chapter 3.

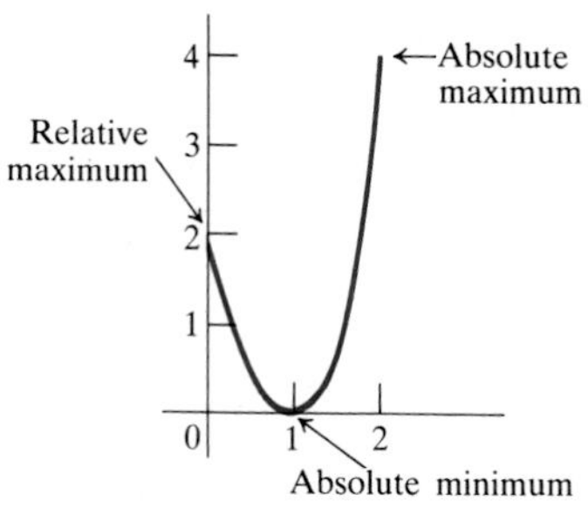

3.32 The graph of $y = x^3 - 3x + 2$ on $[0, 2]$.

Theorem 1, just proved, says that if these occur at an interior point of $[a, b]$, then they occur at a point where $f' = 0$. Thus the extreme values of f occur either where $f' = 0$ or at a or b. ■

EXAMPLE 3 Find the absolute maximum and minimum values of the function

$$y = x^3 - 3x + 2, \qquad 0 \le x \le 2$$

on the closed interval $[0, 2]$.

Solution The function is continuous on $[0, 2]$ and differentiable throughout $(0, 2)$ (see Fig. 3.32). We may therefore apply Theorem 3 with $a = 0$, $b = 2$ to conclude that the absolute maximum and minimum function values occur either (1) at the points in $(0, 2)$ where $y' = 0$, or (2) at the endpoints $a = 0$, $b = 2$.

The derivative $y' = 3x^2 - 3$ is zero on the interval $(0, 2)$ only at $x = 1$. A comparison of the values

$$y(1) = 0, \qquad y(0) = 2, \quad \text{and} \quad y(2) = 4$$

shows an absolute minimum of zero at $x = 1$ and an absolute maximum of four at $x = 2$. □

Summary

To find the maximum and minimum values of a function $y = f(x)$, locate

1. the points where f' is zero or fails to exist, and
2. the endpoints (if any) of the domain of f.

These are the only candidates for the values of x where f takes on an extreme value.

The points where the derivative is zero or fails to exist are called *critical points* of f. By comparing the values of f at the critical points and endpoints with each other, and with the values of f at nearby points, decide which of them, if any, are local (or absolute) maxima and minima. Be sure to look for sign changes in y'. Also, the second derivative may help (if it is easy to calculate).

For a function continuous on a closed interval $[a, b]$ and differentiable on the interior (a, b), one need only examine the values of the function at the endpoints and where $f' = 0$. Nothing else is required.

EXAMPLE 4 Let $f(x) = |4 - x^2|$ on $[-3, 3]$. Find the local (and absolute) maxima and minima.

Solution See Fig. 3.33. We rewrite the formula for f:

$$f(x) = \begin{cases} 4 - x^2 & \text{for } |x| \le 2, \\ x^2 - 4 & \text{for } 2 < |x| \le 3. \end{cases}$$

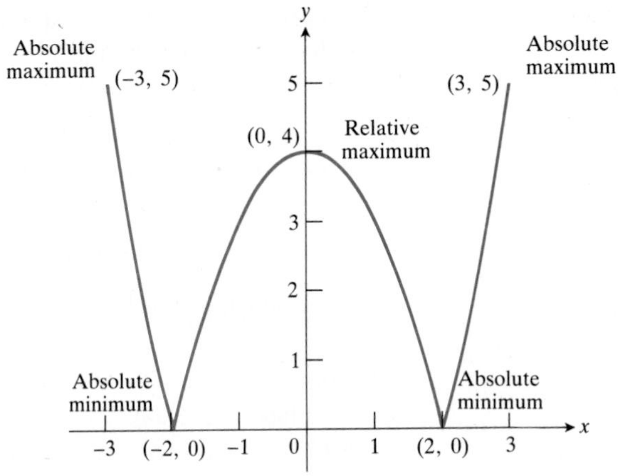

3.33 The graph of $y = |4 - x^2|$, $-3 \leq x \leq 3$.

The derivative is given by

$$f'(x) = \begin{cases} -2x & \text{for } -2 < x < 2, \\ 2x & \text{for } -3 < x < -2 \text{ or } 2 < x < 3. \end{cases}$$

Hence the critical points are

$x = 0$ where the derivative is zero,

$x = -2$ and $x = 2$ where the derivative fails to exist because the slopes do not match where the pieces of the curve are joined.

The endpoints of the domain are $x = -3$ and $x = 3$.

The function values at the critical points and endpoints are

$$f(0) = 4, \qquad f(2) = f(-2) = 0, \qquad f(3) = f(-3) = 5.$$

There are minima (relative and absolute) at $x = \pm 2$, maxima (relative and absolute) at $x = \pm 3$, and a relative maximum at $x = 0$. □

PROBLEMS

Find the critical points for each of the functions in Problems 1–23. For each critical point, determine whether the function has a local maximum or a local minimum there, or neither. If possible, find the absolute maximum and minimum values of the function on the indicated domain.

1. $y = x - x^2$ on $[0, 1]$

2. $y = x - x^3$

3. $y = x - x^3$ on $[0, 1]$

4. $y = x^3 - 3x^2 + 2$

5. $y = x^3 - 147x$

6. $y = x^3 - 2x^2 + x$ on $[-1, 2]$

7. $y = |x - x^2|$ on $[-2, 2]$

8. $y = 2x$ on $[0, 3]$

9. $y = \dfrac{1}{3 - x}$ on $[0, 4]$

10. $y = \sqrt{4 - x^2}$

11. $y = x^4 - 4x$ on $[0, 2]$

12. $y = \tan x$ on $[0, \pi/2)$

13. $y = \sec x$ on $(-\pi/2, \pi/2)$

14. $y = 2 \sin x + \cos 2x$ on $[0, \pi/2]$

15. $y = x^4 - 8x^3 - 270x^2$

16. $y = x^4 - \dfrac{x^3}{3} - 2x^2 + x - 1$

17. $y = \dfrac{x}{1 + |x|}$ on $(-\infty, \infty)$

18. $y = \dfrac{|x|}{1 + |x|}$ on $[0, \infty)$

19. $y = x - [x]$ on $(-\infty, \infty)$, where $[x]$ is the greatest integer in x.

20. $y = (x - x^2)^{-1}$ on $(0, 1)$

21. $y = |x^3|$ on $[-2, 3]$. What would occur if the domain were changed to $[-2, 3)$?

22. $y = \begin{cases} -x & \text{for } x \leq 0, \\ 2x - x^2 & \text{for } x > 0 \end{cases}$

23. $y = \begin{cases} 3 - x \text{ on } [0, 2], \\ \frac{1}{2}x^2 \text{ on } (2, 3] \end{cases}$

24. Show that the sum of any positive number and its reciprocal is at least 2.

25. Find the critical points, asymptotes, points of inflection, and graph the function

$$y = \frac{x}{x^2 + 1}.$$

26. Test the function

$$y = \frac{x^3}{6} + \frac{x^2}{2} - 1 + \cos x$$

for the existence of a relative maximum or minimum at $x = 0$.

27. Suppose that a function $y = f(x)$ is known to be differentiable for all values of x and to have a relative maximum at $x = c$. Which of the following must be true of the graph of f'?
a) It has a point of inflection at $x = c$.
b) It crosses the x-axis at $x = c$.
c) It has a relative maximum or minimum at $x = c$.

28. Find the maximum height of the curve $y = 4 \sin x - 3 \cos x$ above the x-axis.

29. Find the maximum height of the curve $y = 4 \sin^2 x - 3 \cos^2 x$ above the x-axis.

3.5

Maxima and Minima: Problems

The differential calculus is a powerful tool for solving problems that call for minimizing or maximizing a function. We shall give examples, and then summarize the technique in a list of rules.

EXAMPLE 1 Find two positive numbers whose sum is 20 and whose product is as large as possible.

Solution See Fig. 3.34. If one of the numbers is x, then the other is $(20 - x)$, and their product is

$$f(x) = x(20 - x) = 20x - x^2.$$

In this context, the domain of f can be restricted to the interval $0 \leq x \leq 20$. By Theorem 3 in Article 3.4, the function has an absolute maximum on $[0, 20]$, which is located either at $x = 0$, $x = 20$, or an interior point where $f' = 0$. The derivative

$$f'(x) = 20 - 2x = 2(10 - x)$$

equals zero only when $x = 10$. The absolute maximum is therefore one of the three numbers

$$f(0) = 0, \qquad f(10) = 10(20 - 10) = 100, \qquad f(20) = 0.$$

The function takes on the largest of these, 100, when $x = 10$. The two numbers we seek are therefore $x = 10$ and $(20 - x) = 10$. □

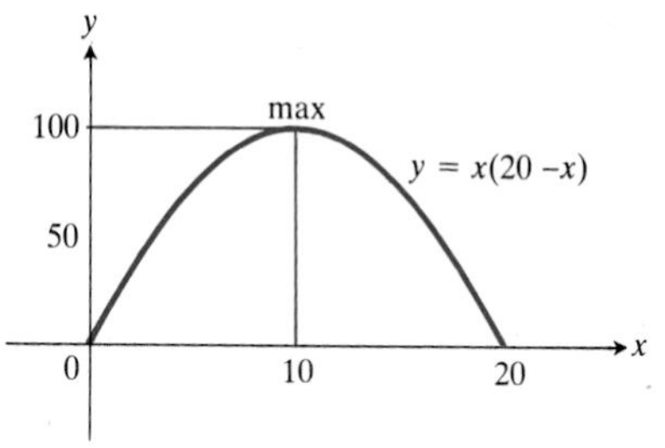

3.34 The product $y = x(20 - x)$ reaches a maximum value of 100 when $x = 10$.

EXAMPLE 2 A square sheet of tin a inches on a side is to be used to make an open-top box by cutting a small square of tin from each corner and bending up the sides. How large a square should be cut from each corner to make the box have as large a volume as possible?

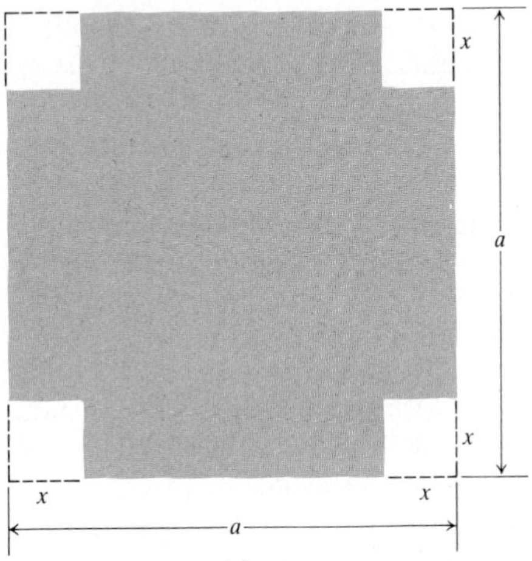

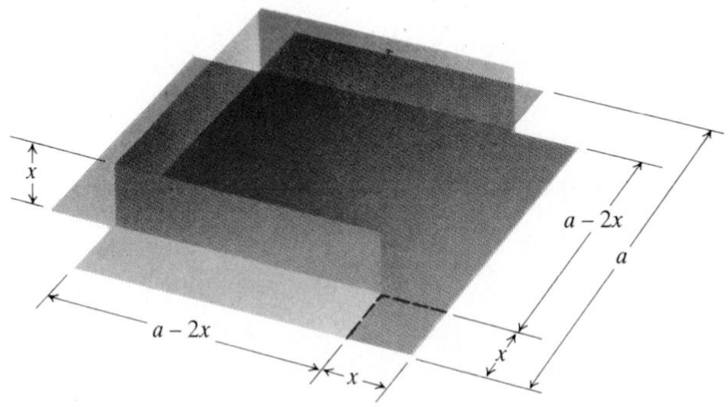

3.35 To make an open box, squares are cut from the corners of a square sheet of tin (a) and the sides are bent up (b). What value of x gives the largest volume?

Solution We first draw a figure to illustrate the problem (Fig. 3.35). In the figure, the side of the square cut from each corner is taken to be x inches and the volume of the box in cubic inches is then given by

$$y = x(a - 2x)^2, \qquad 0 \le x \le a/2. \tag{1}$$

The restrictions placed on x in Eq. (1) are imposed by the fact that one can neither cut a negative amount of material from a corner nor cut away more than the total amount present. It is also evident that $y = 0$ when $x = 0$ or when $x = a/2$, so that the maximum volume y must occur at a value of x between 0 and $a/2$. The function in (1) has a derivative at every such point, and hence, by Theorem 3 in Article 3.4, the maximum occurs at an interior point of $[0, a/2]$ where $y' = 0$. From Eq. (1) we find

$$y = a^2x - 4ax^2 + 4x^3,$$
$$y' = a^2 - 8ax + 12x^2 = (a - 2x)(a - 6x),$$

so that

$$y' = 0 \qquad \text{when } x = \frac{a}{6} \quad \text{or} \quad x = \frac{a}{2}.$$

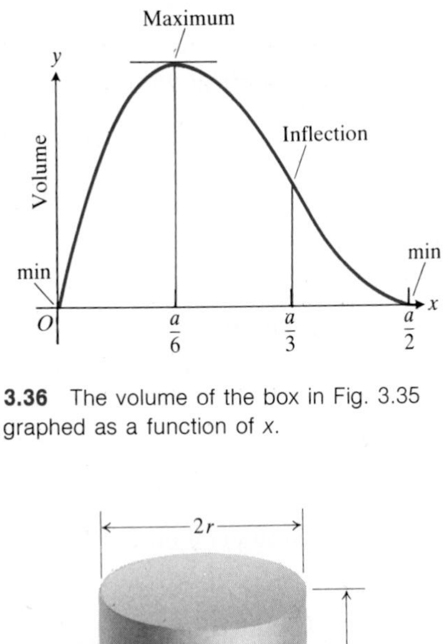

3.36 The volume of the box in Fig. 3.35 graphed as a function of x.

Of these, only $x = a/6$ lies in the interior of $[0, a/2]$. The maximum therefore occurs at $x = a/6$. Each corner square should have dimensions $a/6$ by $a/6$ to produce a box of maximum volume. The graph of the volume is shown in Fig. 3.36. □

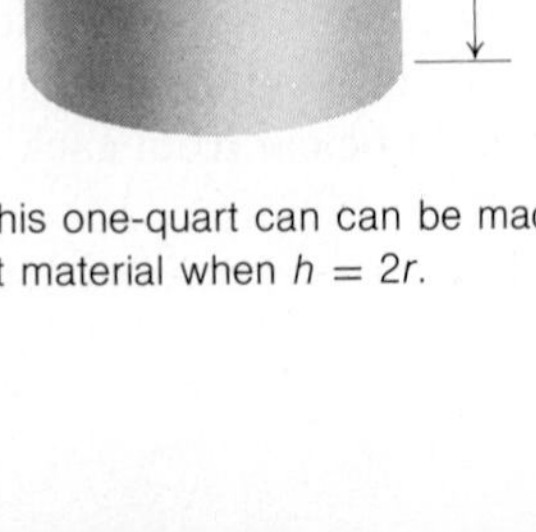

3.37 This one-quart can can be made from the least material when $h = 2r$.

EXAMPLE 3 An oil can is to be made in the form of a right circular cylinder to contain one quart of oil. What dimensions of the can will require the least amount of material?

Solution Again we start with a figure (Fig. 3.37). The requirement that the can hold a quart of oil is the same as

$$V = \pi r^2 h = a^3, \tag{2a}$$

if the radius r and altitude h are in inches and a^3 is the number of cubic inches in a quart ($a^3 = 57.75$).

How shall we interpret the phrase "least material"? One possibility is to ignore the thickness of the material and the waste in manufacturing. Then we ask for dimensions r and h that make the total surface area

$$A = 2\pi r^2 + 2\pi rh \tag{2b}$$

as small as possible while still satisfying the constraint in (2a). Problem 34 shows one of the ways we might take the cost of wasted material into account.

In either case, what kind of oil can do we expect? Not a tall, thin one like a six-foot piece of pipe one inch in diameter, nor a short, wide one like a nine-inch pie pan. These both hold about a quart, but use more metal than the standard oil can we see in stores and gas stations. We expect something in between.

We are not quite ready to apply the methods used in Examples 1 and 2, because Eq. (2b) expresses A as a function of *two* variables, r and h, and our methods call for A to be expressed as a function of just *one* variable. However, Eq. (2a) can be used to express one of the variables r or h in terms of the other; in fact, we find

$$h = \frac{a^3}{\pi r^2} \tag{2c}$$

or

$$r = \sqrt{\frac{a^3}{\pi h}}. \tag{2d}$$

The division in (2c) and (2d) is legitimate because neither r nor h can be zero, and only the positive square root is used in (2d) because the radius r can never be negative. If we substitute from (2c) into (2b), we have

$$A = 2\pi r^2 + \frac{2a^3}{r}, \qquad 0 < r, \tag{2e}$$

and now we can apply our previous methods. A minimum of A can occur only at a point where

$$\frac{dA}{dr} = 4\pi r - 2a^3 r^{-2} \tag{2f}$$

is zero, that is, where

$$4\pi r = \frac{2a^3}{r^2}, \qquad r^3 = \frac{a^3}{2\pi}, \qquad \text{or} \qquad r = \frac{a}{\sqrt[3]{2\pi}}. \tag{2g}$$

At such a value of r,

$$\frac{d^2A}{dr^2} = 4\pi + 4a^3 r^{-3} = 12\pi > 0,$$

which indicates a relative minimum. Since the second derivative is *always positive* for $0 < r$, the curve is concave upward everywhere, and there can be no other relative minimum, so we have also found the absolute minimum. From (2g) and (2c), we find

$$r = \frac{a}{\sqrt[3]{2\pi}} = \sqrt[3]{\frac{V}{2\pi}}, \qquad h = \frac{2a}{\sqrt[3]{2\pi}} = 2\sqrt[3]{\frac{V}{2\pi}}$$

as the dimensions of the can of volume V having minimum surface area.

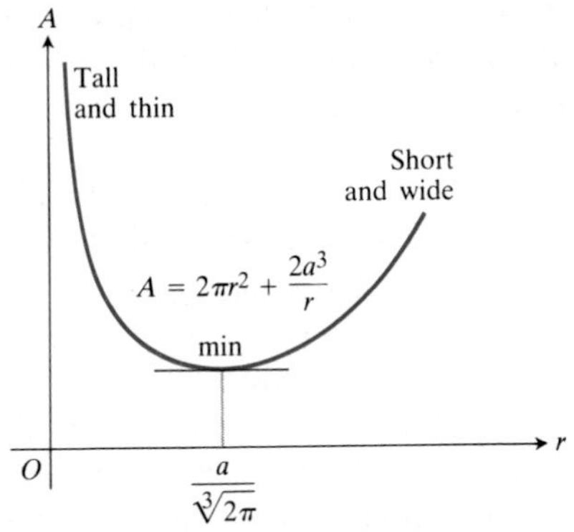

3.38 A graph of $A = 2\pi r^2 + (2a^3/r)$.

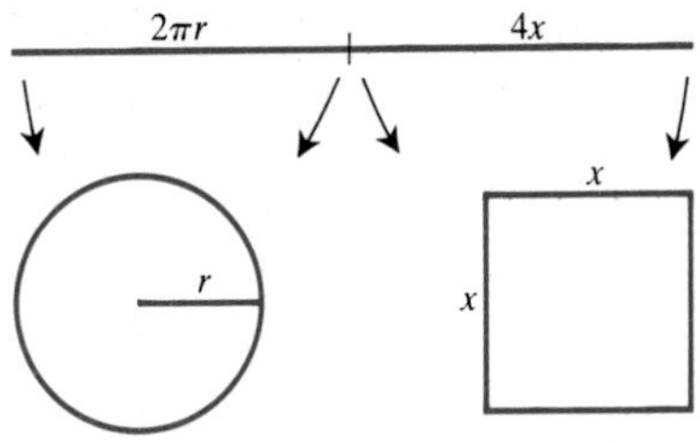

3.39 A circle and a square made from a wire $2\pi r + 4x$ units long. How large an area can they enclose together?

For $V = 1\text{ qt} = 57.75\text{ in}^3$, this comes to

$$r = 2.1\text{ in.}, \qquad h = 4.2\text{ in.}$$

The height of the can is equal to its diameter. Figure 3.38 shows the graph of A as a function of r. □

Some problems about maxima and minima have their solution at an endpoint of the function's domain. The next example discusses one such problem.

EXAMPLE 4 A wire of length L is available for making a circle and a square (see Fig. 3.39). How should the wire be divided between the two shapes to make the sum of the areas enclosed a maximum?

Solution In the notation of Fig. 3.39, the sum of the areas is

$$A = \pi r^2 + x^2, \tag{3a}$$

where r and x must satisfy the equation

$$L = 2\pi r + 4x. \tag{3b}$$

We could solve (3b) for x in terms of r, but instead we shall differentiate both (3a) and (3b), treating A and x as functions of r for $0 \leq 2\pi r \leq L$. Then

$$\frac{dA}{dr} = 2\pi r + 2x\frac{dx}{dr}, \tag{3c}$$

and

$$\frac{dL}{dr} = 2\pi + 4\frac{dx}{dr} = 0, \qquad \frac{dx}{dr} = -\frac{\pi}{2}, \tag{3d}$$

where dL/dr is zero because L is a constant. If we substitute dx/dr from (3d) into (3c), we get

$$\frac{dA}{dr} = \pi(2r - x) \tag{3e}$$

and note that

$$\frac{d^2A}{dr^2} = \pi\left(2 - \frac{dx}{dr}\right) = \pi\left(2 + \frac{\pi}{2}\right) \tag{3f}$$

is a positive constant. This means that the graph of A as a function of r is always concave up. The curve will therefore have a relative maximum at one or both endpoints, and in either case the larger endpoint value (if there is a larger) will be the maximum value of the function for the interval

$$0 \leq r \leq \frac{L}{2\pi}$$

on which A is defined. When

$$r = 0: \qquad x = \frac{L}{4}, \quad A = \frac{1}{16}L^2,$$

and when

$$r = \frac{L}{2\pi}: \qquad x = 0, \quad A = \frac{1}{4\pi}L^2.$$

The maximum value of A occurs when $r = L/2\pi$. Therefore, the wire should not be cut at all, but bent into a circle for maximum total area.

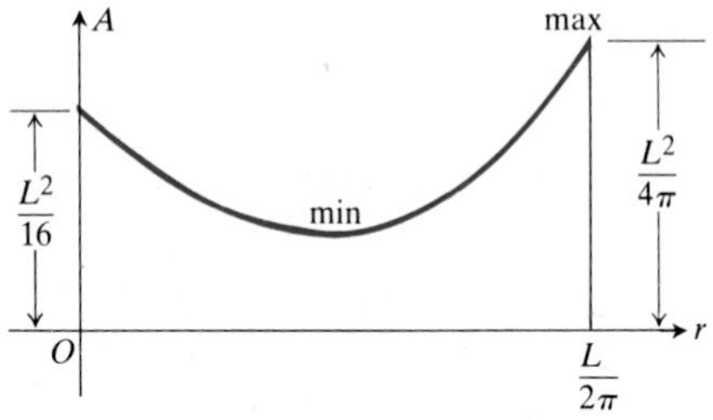

3.40 The sum of the areas in Fig. 3.39 graphed as a function of r.

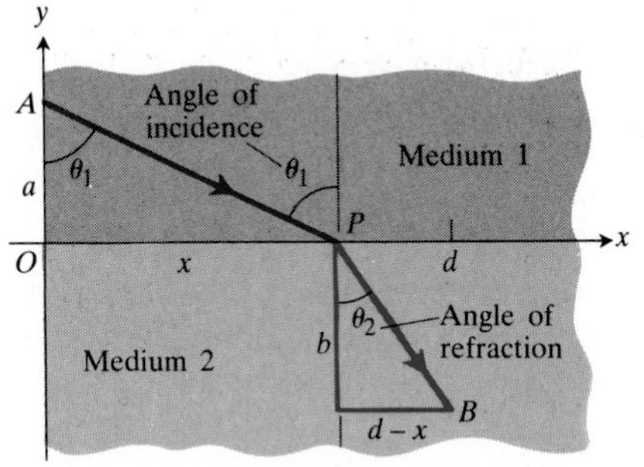

3.41 A light ray is refracted (deflected from its path) when it passes from one medium to another. θ_1 is the angle of incidence, and θ_2 is the angle of refraction.

The graph of A as a function of r looks like the curve in Fig. 3.40 (Problem 36), which shows two relative maxima, but we do not need to know this to find the absolute maximum. □

EXAMPLE 5 Fermat's principle, and Snell's law. The speed of light depends on the medium through which it travels, and is generally slower in denser media. In a vacuum light travels at the famous speed $c = 3 \times 10^8$ m/s, but in the earth's atmosphere it travels slower than that, and in glass slower still.

Fermat's principle in optics states that light travels from one point to another along a path for which the time of travel is a minimum. Let us find the path that a ray of light will follow in going from a point A in a medium where the speed of light is c_1 to a point B in a second medium where its speed is c_2.

Solution We assume that the two points lie in the xy-plane and that the x-axis separates the two media as in Fig. 3.41.

In either medium, where the speed of light remains constant, "shortest time" means "shortest path," and the ray of light will follow a straight line. Hence the path from A to B will consist of a line segment from A to a boundary point P, followed by another line segment from P to B. According to the formula

$$\text{time} = \frac{\text{distance}}{\text{rate}},$$

the time required for light to travel from A to P is

$$t_1 = \frac{\sqrt{a^2 + x^2}}{c_1}.$$

From P to B the time required is

$$t_2 = \frac{\sqrt{b^2 + (d - x)^2}}{c_2}.$$

We therefore seek to minimize

$$t = t_1 + t_2 = \frac{\sqrt{a^2 + x^2}}{c_1} + \frac{\sqrt{b^2 + (d - x)^2}}{c_2}. \tag{4a}$$

We find

$$\frac{dt}{dx} = \frac{x}{c_1\sqrt{a^2 + x^2}} - \frac{(d - x)}{c_2\sqrt{b^2 + (d - x)^2}} \tag{4b}$$

or

$$\frac{dt}{dx} = \frac{\sin\theta_1}{c_1} - \frac{\sin\theta_2}{c_2}, \tag{4c}$$

if we use the angles θ_1 and θ_2 in the figure.

If we restrict x to the interval $0 \le x \le d$, then t has a negative derivative at $x = 0$ and a positive derivative at $x = d$, while at the value of x, say x_c, for which

$$\frac{\sin\theta_1}{c_1} = \frac{\sin\theta_2}{c_2}, \tag{4d}$$

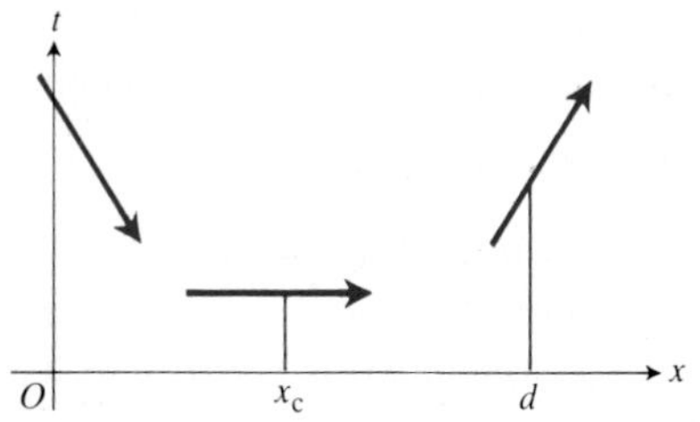

3.42 These tangents to the graph of Eq. (4a) suggest that t has a minimum at $x = x_c$.

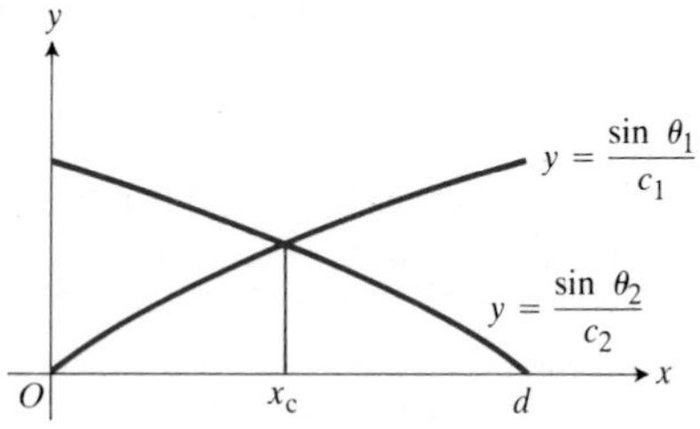

3.43 At $x = x_c$, $(\sin\theta_1)/c_1 = (\sin\theta_2)/c_2$.

dt/dx is zero. Figure 3.42 indicates the directions of the tangents to the curve giving t as a function of x, at these three points, and suggests that t will indeed be at a minimum at $x = x_c$.

If we want further information on this point, by referring to Fig. 3.41 we see that an increase in x will cause P to move to the right, making θ_1 larger, and hence also $\sin\theta_1$ larger, but will have the opposite effect on θ_2 and $\sin\theta_2$. Figure 3.43 shows that since $(\sin\theta_1)/c_1$ is an increasing function of x which is zero at $x = 0$, while $(\sin\theta_2)/c_2$ is a decreasing function of x which is zero at $x = d$, the two curves can cross at only one point, $x = x_c$, between 0 and d. To the right of x_c, the curve for $(\sin\theta_1)/c_1$ is above the curve for $(\sin\theta_2)/c_2$, but these roles are reversed to the left of x_c, so that

$$\frac{dt}{dx} = \frac{\sin\theta_1}{c_1} - \frac{\sin\theta_2}{c_2} \begin{cases} \text{is negative for } x < x_c, \\ \text{is zero for } x = x_c, \\ \text{is positive for } x > x_c, \end{cases}$$

and the minimum of t does indeed occur at $x = x_c$.

Instead of determining x_c explicitly, it is customary to characterize the path followed by the ray of light by leaving the equation for $dt/dx = 0$ in the form (4d), which is known as the *law of refraction* or Snell's law.† □

Strategy for Finding Maxima and Minima

1. *A figure.* When possible, draw a figure to illustrate the problem and label the parts that are important in the problem. Keep track of which letters represent constants and which represent variables.
2. *An equation.* Write an equation for the quantity that is to be a maximum or a minimum. If the quantity is denoted by y, it is desirable to express it in terms of a single independent variable x. This may require some algebraic manipulation and the use of information given in the problem.
3. *Critical points test.* If $y = f(x)$ is the quantity to be a maximum or a minimum, test each value of x for which $y' = 0$ to determine whether y has a maximum, a minimum, or neither. The usual tests are:

   a) Check the sign of y': If

   $$y' \text{ is } \begin{cases} \text{positive for } x < c, \\ \text{zero at } x = c, \\ \text{negative for } x > c, \end{cases}$$

   y has a maximum at $x = c$. But if y' changes from $(-)$ to $(+)$ as x advances through c, then y has a minimum. If y' does not change sign, y may or may not have an extreme value at $x = c$.

†See F. W. Sears, M. W. Zemansky, and H. D. Young, *University Physics*, 6th ed. (Reading, Mass.: Addison-Wesley, 1982), Chapter 38.

b) The second derivative test:

If $y'' < 0$ when $y' = 0$, then y is a maximum.

If $y'' > 0$ when $y' = 0$, then y is a minimum.

If $y'' = 0$ when $y' = 0$, use another test.

If y' fails to exist at some point (as in Figs. 3.30 and 3.33), examine this point for a possible maximum or minimum.

4. *Endpoints test.* Examine every endpoint of the domain of y for a possible maximum or minimum.

Finding Extremes Without Using the Second Derivative

The following argument can be applied to many problems in which we are trying to find an extreme value, say a maximum, of a function. Suppose that we know that:

1. We can restrict our search to a closed interval I;
2. The function is continuous and differentiable everywhere (we might know this from a formula for the function, or from physical considerations); and
3. The function does not attain a maximum at either endpoint of I.

Then we know that the function has at least one maximum at an interior point of I, at which the derivative must be zero. Therefore, if we find that the derivative of the function is zero at only one interior point of I, then this is where the function takes on its maximum. A similar argument applies to a search for a minimum value.

EXAMPLE 6 A producer can sell x items per week at a price $P = 200 - 0.01x$ cents, and it costs $y = 50x + 20{,}000$ cents to make x items. What is the most profitable number to make?

Solution The total weekly revenue from x items is

$$xP = 200x - 0.01x^2.$$

The profit T is revenue minus cost:

$$\begin{aligned} T = xP - y &= (200x - 0.01x^2) - (50x + 20{,}000) \\ &= 150x - 0.01x^2 - 20{,}000. \end{aligned}$$

For very large values of x, say beyond a million, T is negative. Therefore, T takes on its maximum value somewhere in the interval $0 \leq x \leq 10^6$. The formula for T shows that T is differentiable at every x, and obviously T does not take on its maximum at either endpoint, 0 or 10^6. The derivative

$$\frac{dT}{dx} = 150 - 0.02x$$

is zero only when

$$x = 7500.$$

Therefore, $x = 7500$ is the production level for maximum profit. □

To solve the problem in Example 6 it would have been as effective to calculate the second derivative $d^2T/dx^2 = -0.02$ and conclude that T has a relative maximum at $x = 7500$. A quick look at the domain of x or the graph of T is all that would then be needed to show that the maximum was absolute. The virtue of the argument that we gave instead is that it can apply to a function whose second derivative is nonexistent or difficult to compute.

PROBLEMS

1. Show that the rectangle that has maximum area for a given perimeter is a square.

2. Find the dimensions of the rectangle of greatest area that can be inscribed in a semicircle of radius r.

3. Find the area of the largest rectangle with lower base on the x-axis and upper vertices on the curve $y = 12 - x^2$.

4. An open rectangular box is to be made from a piece of cardboard 8 in. wide and 15 in. long by cutting a square from each corner and bending up the sides. Find the dimensions of the box of largest volume.

5. One side of an open field is bounded by a straight river. How would you put a fence around the other three sides of a rectangular plot in order to enclose as great an area as possible with a given length of fence?

6. An open storage bin with square base and vertical sides is to be constructed from a given amount of material. Determine its dimensions if its volume is a maximum. Neglect the thickness of the material and waste in construction.

7. A box with square base and open top is to hold 32 in^3. Find the dimensions that require the least amount of material. Neglect the thickness of the material and waste in construction.

8. A variable line through the point (1, 2) intersects the x-axis at $A(a, 0)$ and the y-axis at $B(0, b)$. Find the area of the triangle AOB of least area if both a and b are positive.

9. A poster is to contain 50 in^2 of printed matter with margins of 4 in. each at top and bottom and 2 in. at each side. Find the overall dimensions if the total area of the poster is a minimum.

10. The height of an object moving vertically is given by

$$s = -16t^2 + 96t + 112,$$

with s in feet and t in seconds. Find (a) its velocity when $t = 0$, (b) its maximum height, and (c) its velocity when $s = 0$.

11. A rectangular plot of land containing 216 m^2 is to be enclosed by a fence and divided into two equal parts by another fence parallel to one of the sides. What dimensions of the outer rectangle require the smallest total length for the two fences? How much fence is needed?

12. Two sides of a triangle are to have lengths a cm and b cm. What is the largest area the triangle can have? (*Hint:* What angle between a and b gives the largest area?)

13. What are the dimensions of the lightest open-top right circular cylindrical can that can be made of material weighing 1 gm per cm^2 if the volume of the can is to be 1000 cm^3?

14. Find the largest possible value of $2x + y$ if x and y are the lengths of the sides of a right triangle whose hypotenuse is $\sqrt{5}$ units long.

15. A container with a rectangular base, rectangular sides, and no top is to have a volume of 2 m^3. The width of the base is to be 1 m. When cut to size, material costs \$10 per square meter for the base and \$5 per square meter for the sides. What is the cost of the least expensive container?

16. The U. S. Post Office will accept a box for shipment only if the sum of the length and girth (distance around) does not exceed 84 in. Find the dimensions of the largest acceptable box with a square end (Fig. 3.44).

17. (Conclusion of the sonobuoy problem, Problem 39 in Article 2.10.) Show that the value of x that minimizes the distance between the points (x, x^2) and $(2, -\frac{1}{2})$ in Fig. 2.47 is a solution of the equation $1/(x^2 + 1) = x$. (*Hint:* Minimize the square of the distance.)

3.44 The box in Problem 16.

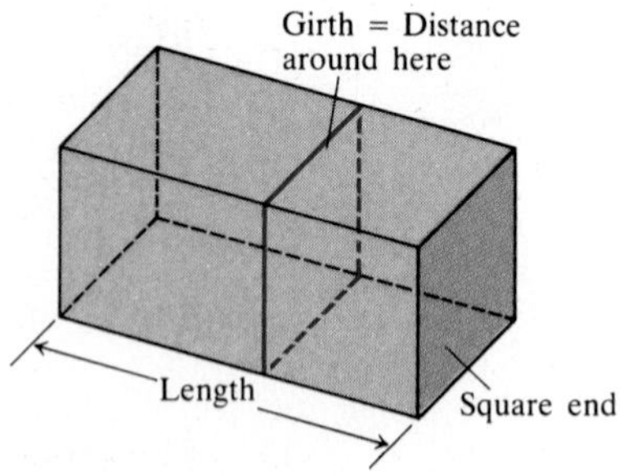

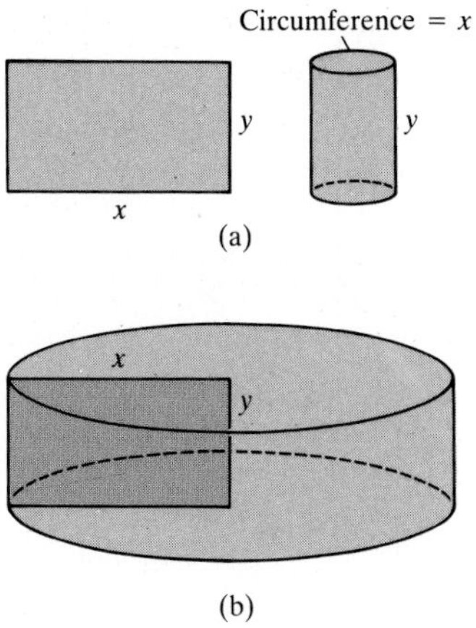

3.45 The rectangle and cylinders in Problem 18.

18. Compare the answers to the following two construction problems.

a) A rectangle of dimensions x cm by y cm is to be rolled into the cylinder shown in Fig. 3.45(a). The perimeter of the sheet is 36 cm. What values of x and y give the largest cylinder volume? What is this volume?

b) The rectangle in (a) is revolved about one of the edges of length y to sweep out the cylinder in Fig. 3.45(b). What values of x and y give the largest cylinder volume now? What is this volume?

19. A right triangle of given hypotenuse is rotated about one of its legs to generate a right circular cone. Find the cone of greatest volume.

20. It costs a manufacturer c dollars each to manufacture and distribute a certain item. If the items sell at x dollars each, the number sold is given by $n = a/(x - c) + b(100 - x)$, where a and b are certain positive constants. What selling price will bring a maximum profit?

21. A cantilever beam of length L has one end built into a wall, while the other end is simply supported. If the beam weighs w lb per unit length, its deflection y at distance x from the built-in end satisfies the equation

$$48EIy = w(2x^4 - 5Lx^3 + 3L^2x^2),$$

where E and I are constants depending on the material of the beam and the shape of its cross section. How far from the built-in end does the maximum deflection occur?

22. Determine the constant a in order that the function

$$f(x) = x^2 + \frac{a}{x}$$

may have (a) a relative minimum at $x = 2$, (b) a relative minimum at $x = -3$, (c) a point of inflection at $x = 1$. (d) Show that the function cannot have a relative maximum for any value of a.

23. Determine the constants a and b in order that the function

$$f(x) = x^3 + ax^2 + bx + c$$

may have (a) a relative maximum at $x = -1$ and a relative minimum at $x = 3$, (b) a relative minimum at $x = 4$ and a point of inflection at $x = 1$.

24. A wire of length L is cut into two pieces, one being bent to form a square and the other to form an equilateral triangle. How should the wire be cut (a) if the sum of the two areas is a minimum? (b) if the sum of the areas is a maximum?

25. Find the points on the curve $5x^2 - 6xy + 5y^2 = 4$ that are nearest the origin.

26. Find the point on the curve $y = \sqrt{x}$ nearest the point $(c, 0)$, (a) if $c \geq \frac{1}{2}$, (b) if $c < \frac{1}{2}$.

27. Find the volume of the largest right circular cone that can be inscribed in a sphere of radius r.

28. Find the volume of the largest right circular cylinder that can be inscribed in a sphere of radius r.

29. Show that the volume of the largest right circular cylinder that can be inscribed in a given right circular cone is $\frac{4}{9}$ the volume of the cone.

30. The strength of a rectangular beam is proportional to the product of its width and the square of its depth. Find the dimensions of the strongest beam that can be cut from a circular cylindrical log of radius r.

31. The stiffness of a rectangular beam is proportional to the product of its breadth and the cube of its depth but is not related to its length. Find the proportions of the stiffest beam that can be cut from a log of given diameter.

32. The intensity of illumination at any point is proportional to the product of the strength of the light source and the inverse of the square of the distance from the source. If two sources of relative strengths a and b are a distance c apart, at what point on the line joining them will the intensity be a minimum? Assume the intensity at any point is the sum of intensities from the two sources.

33. A window is in the form of a rectangle surmounted by a semicircle. If the rectangle is of clear glass while the semicircle is of colored glass, which transmits only half as much light per square foot as clear glass does, and the total perimeter is fixed, find the proportions of the window that will admit the most light.

34. Right circular cylindrical tin cans are to be manufactured to contain a given volume. There is no waste involved in cutting the tin for the vertical sides, but the tops of radius r are cut from squares that measure $2r$ units on a side. The total amount of material

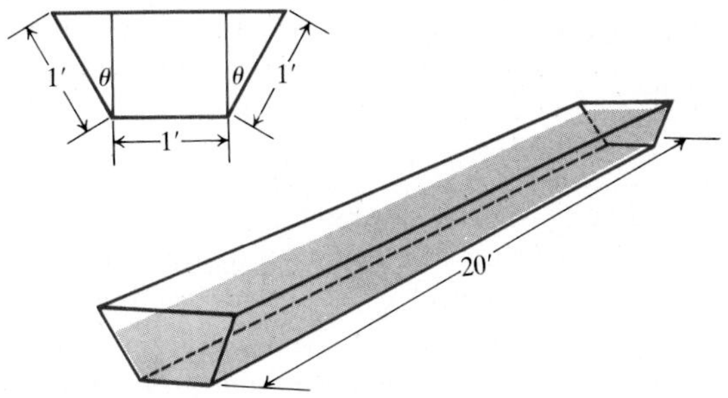

3.46 The trough in Problem 38.

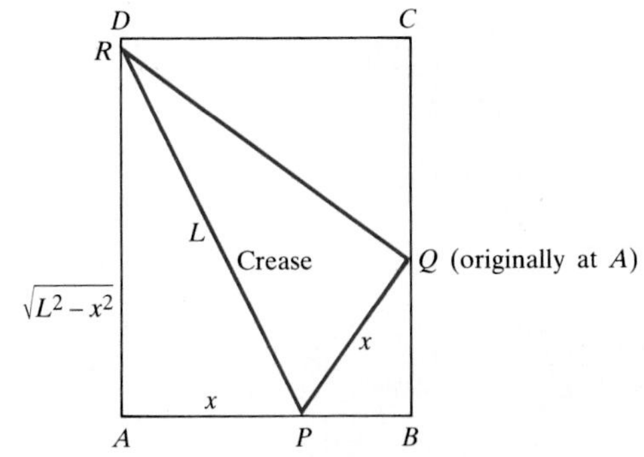

3.47 The paper in Problem 40.

consumed by each can is therefore $A = 8r^2 + 2\pi rh$ (rather than the area $A = 2\pi r^2 + 2\pi rh$ in Example 3). Find the ratio of height to diameter for the most economical cans.

35. A closed container is made from a right circular cylinder of radius r and height h with a hemispherical dome on top. Find the relationship between r and h that maximizes the volume for a given surface area.

36. Show that the area function A in Example 4 has an absolute minimum value at $r = L/(2\pi + 8)$. Then calculate the minimum value of A and show that the graph of A as a function of r is like the one shown in Fig. 3.40. (*Hint:* Express dA/dr in terms of r and L.)

37. Is the function $y = 3 + 4\cos\theta + \cos 2\theta$ ever negative?

38. The trough shown in Fig. 3.46 is to be constructed from a piece of metal 20 ft long and 3 ft wide. The trough is to be made by turning up strips of width 1 ft to make equal angles θ with the vertical.

a) Express the volume of the trough in terms of the angle θ.

b) Find the maximum possible volume of the trough.

39. Find all maxima and minima of $y = \sin x + \cos x$.

40. A rectangular sheet of $8\frac{1}{2} \times 11$ inch paper is placed on a flat surface and one of the vertices is lifted up and placed on the opposite longer edge, while the other three vertices are held in their original positions. With all four vertices now held fixed, the paper is smoothed flat, as shown in Fig. 3.47. The problem is to find the minimum possible length of the crease.

a) Try it with paper.

b) Show that $L^2 = 2x^3/(2x - 8.5)$, $4.25 < x < 8.5$.

c) Minimize L^2.

41. A silo is to be made in the form of a cylinder surmounted by a hemisphere. The cost of construction per square foot of surface area is twice as great for the hemisphere as for the cylinder. Determine the dimensions to be used if the volume is fixed and the cost of construction is to be a minimum. Neglect the thickness of the silo and waste in construction.

42. If the sum of the surface areas of a cube and a sphere is constant, what is the ratio of an edge of the cube to the diameter of the sphere when (a) the sum of their volumes is a minimum? (b) the sum of their volumes is a maximum?

†43. Two towns, located on the same side of a straight river, agree to construct a pumping station and filtering plant at the river's edge, to be used jointly to supply the towns with water. If the distances of the two towns from the river are a and b and the distance between them is c, show that the sum of the lengths of the pipe lines joining them to the pumping station is at least as great as $\sqrt{c^2 + 4ab}$.

†44. Light from a source A is reflected to a point B by a plane mirror. If the time required for the light to travel from A to the mirror and then to B is a minimum, show that the angle of incidence is equal to the angle of reflection.

45. Let $f(x)$ and $g(x)$ be the differentiable functions on $a \le x \le b$ whose graphs are shown in Fig. 3.48. The point c is the point where the vertical distance D between the curves is the greatest. Show that the tangents to the curves at $x = c$ are parallel.

46. In Example 4 in Article 1.8 we introduced the *marginal cost* dy/dx of producing x tons of steel per week as the derivative of the cost y with respect to x. The derivative at any particular x was the approximate cost of producing the next ton of steel.

To sell x tons of steel per week, the company prices its steel at P dollars per ton. The company's *revenue* is then the product xP. Its *marginal revenue* is the derivative of xP with respect to x, or the rate of change of revenue per unit increase in production. The company's profit T is the difference between revenue and cost: $T = xP - y$.

†You may prefer to do these problems without calculus.

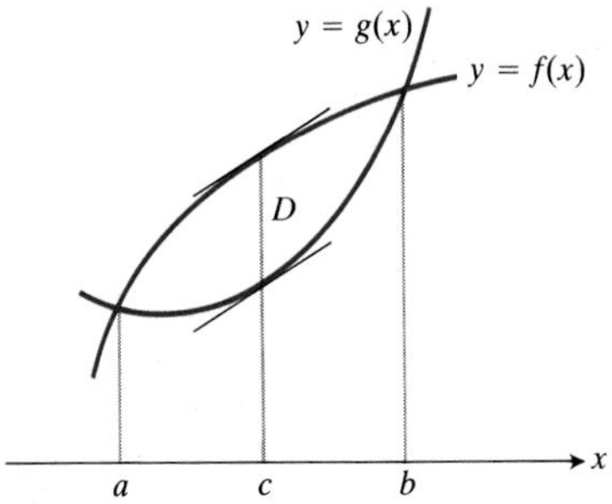

3.48 The graphs for Problem 45.

The company wants to adjust production to achieve maximum profit.

a) Show that if the profit can be maximized then it is maximized at the value of x for which marginal revenue equals marginal cost.

b) What condition involving the second derivatives of P and y would assure that the point of equality in (a) was one of maximum profit (and not, say, one of minimum profit)?

47. The wall shown in Fig. 3.49 is 8 feet high and stands 27 feet from the building. What is the length of the shortest straight beam that will reach to the side of the building from the ground outside the wall?

48. A rope with a ring in one end is looped over two pegs in a horizontal line. The free end, after being passed through the ring, has a weight suspended from it, so that the rope hangs taut. If the rope slips freely over the pegs and through the ring, the weight will descend as far as possible. Assume that the length of the rope is at least four times as great as the distance between the pegs, and that the configuration of the rope is symmetric with respect to the line of the vertical part of the rope. (The symmetry assumption can be justified on the grounds that the rope and weight will take a rest position that minimizes the potential energy of the system. See Problem 39 in Article 8.5.) Find the angle formed at the bottom of the loop.

3.49 The diagram for Problem 47.

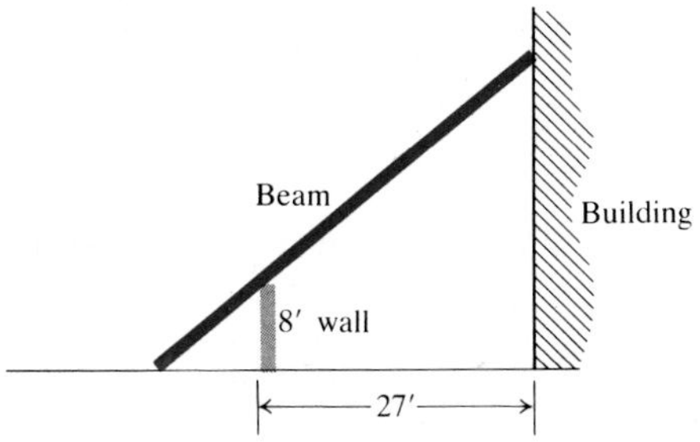

49. *Autocatalytic reactions.* A catalyst for a chemical reaction is a substance that controls the rate of the reaction without undergoing any permanent change in itself. An autocatalytic reaction is one whose product is a catalyst for its own formation. Such a reaction may proceed slowly at first if the amount of catalyst present is small, and slowly again at the end when most of the original substance is used up. But in between, when both the substance and its product are abundant, the reaction may proceed at a faster rate.

In some cases it is reasonable to assume that the rate $v = dx/dt$ of the reaction is proportional both to the amount of the original substance present and to the amount of product. That is, v may be considered to be a function of x alone, and

$$v = kx(a - x) = kax - kx^2,$$

where

x = the amount of product,

a = the amount of substance at the beginning,

k = a positive constant.

At what value of x does the rate v have a maximum? What is the maximum value of v?

***Toolkit* programs**

Derivative Grapher Super * Grapher

3.6 Related Rates

In this article we look at problems that ask for the rate at which some variable changes. In each case the rate is a derivative that has to be computed from the rate at which some other variable is known to change. To find it, we write an equation that relates the two variables. We then differentiate both sides of the equation to express the derivative we seek in terms of the derivative we know.

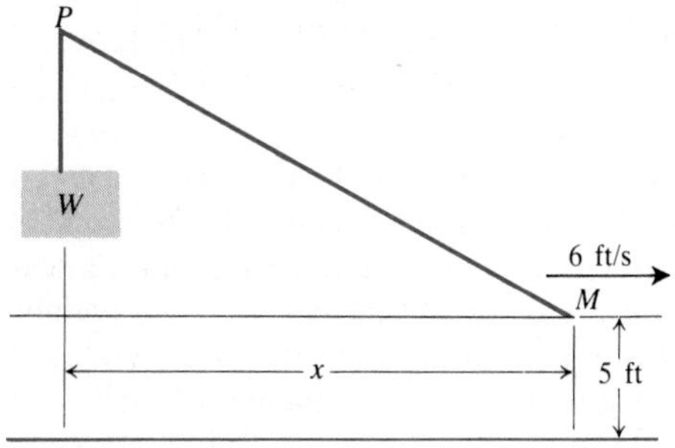

3.50 If the worker at M moves to the right at the speed shown, then $dx/dt = 6$ ft/s.

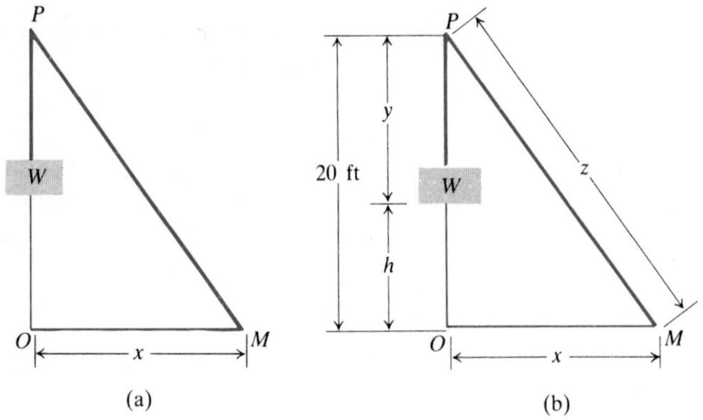

3.51 A typical sketch for a related rate problem models the problem for any time t, not just for the time in question. The reason for this is that certain distances in the figure vary (here x, z, h, and $20 - h$) and it is important to treat them as variables and not as constants.

Very often the key to relating the variables in such a problem is drawing a picture that shows the geometric relations between them.

EXAMPLE 1 Figure 3.50 shows a rope running through a pulley at P and bearing a weight W at one end. The other end is held 5 feet above the ground in a worker's hand M. The worker walks away from the line PW at the rate of 6 ft/s.

How do we use mathematics to describe the motion? If x represents the distance in feet from the vertical line PW to the worker's hand, and t represents time in seconds, then x is an increasing function of t. The statement that the worker walks away from the line PW at 6 ft/s translates into the mathematical statement

$$\frac{dx}{dt} = 6.$$

On the other hand, if the worker were to walk *toward* the line PW at the rate of 6 ft/s, we would have

$$\frac{dx}{dt} = -6,$$

because x would then be a decreasing function of t. □

EXAMPLE 2 Suppose that the pulley in Example 1 is 25 ft above ground, the rope is 45 ft long, and the worker is walking away from the line PW at the rate of 6 ft/s. How fast is the weight being raised when $x = 15$ ft?

Solution To picture the problem we draw and label a preliminary sketch, shown in Fig. 3.51(a). The sketch shows the worker's hand M, the pulley P, and the horizontal line OM from the worker's hand to a point directly below the pulley. The angle at O is a right angle.

The variables are the distance x, the height of the weight above O, the distance from the weight to P, and the length of rope from P to M.

The constants are the height of P above O (20 ft because O is 5 ft above the ground), the total length of rope (45 ft), and the rate at which the worker walks (6 ft/s).

We are asked to find the rate at which the weight is being raised when $x = 15$ ft. If we let h be the height of the weight W above O and denote time in seconds by t, then we can write this rate as dh/dt. In mathematical terms, we are asked to find dh/dt when $x = 15$ ft.

We label the other changing dimensions in the sketch (Fig. 3.51b). All variables are to be differentiable functions of t, and $dx/dt = 6$.

We now write equations to express the relationships that hold between the variables at all values of t under consideration:

$$\begin{aligned} &20 - h + z = 45 \qquad &&\text{(the total length of rope is 45 ft);} \\ &20^2 + x^2 = z^2 \qquad &&\text{(the angle at } O \text{ is a right angle).} \end{aligned} \tag{1}$$

If these equations are differentiated with respect to t, the new equations will tell how the derivatives of the variables are related. From this information we can express dh/dt in terms of x and dx/dt.

Actually, it will save time to obtain first a single equation relating x (whose rate is given) with h (whose rate is wanted) before we take derivatives. We do this by eliminating z from the equations in (1):

$$z = 45 - (20 - h) = 25 + h,$$

$$20^2 + x^2 = z^2 = (25 + h)^2$$

$$20^2 + x^2 = (25 + h)^2. \tag{2}$$

Both sides of Eq. (2) are functions of t, and the equation says they represent the same function of t. Thus when we differentiate both sides of (2) with respect to t we will have another equation:

$$2x\frac{dx}{dt} = 2(25 + h)\frac{dh}{dt}. \tag{3}$$

We can solve Eq. (3) for dh/dt:

$$\frac{dh}{dt} = \frac{x}{25 + h}\frac{dx}{dt}. \tag{4}$$

We want the value of dh/dt when

$$x = 15$$

and

$$\frac{dx}{dt} = 6.$$

It remains only to find h at this instant from Eq. (2):

$$20^2 + 15^2 = (25 + h)^2$$

$$(25 + h)^2 = 625$$

$$h = 0.$$

Equation (4) now gives

$$\frac{dh}{dt} = \frac{15}{25 + 0}\cdot 6 = \frac{18}{5}$$

as the rate at which the weight is being raised when $x = 15$ ft. □

The problem in Example 2 is called a problem in related rates. It is typical of such problems that

a) certain variables are related in a definite way for all values of t under consideration;

b) the values of some or all of these variables and the rates of change of some of them are given at some particular instant; and

c) it is required to find the rate of change of one or more of them at this instant.

The variables are considered to be functions of time. If the equations that relate them for all values of t are differentiated with respect to t, the new equations will tell how their rates of change are related.

With occasional variation, related rate problems can be solved in the following way.

Strategy for Solving Related Rate Problems

1. *Draw and label a picture.* Note what changes and what does not.
2. *Write down what you are asked to find.* Express it in terms of a variable. Show the variable in the picture. (In Example 2, you were asked to find the rate dh/dt.)
3. *Name the remaining variables and constants.* Show them in the sketch. Write down any numerical information about them.
4. *Write equations that relate the variables and constants.*
5. *Substitute (if desirable) and differentiate.* Get a single equation that expresses the rate you want in terms of the rates and other quantities whose values you know.

We followed the strategy explicitly in Example 2, and we shall do so again in Example 3.

EXAMPLE 3 A ladder 26 ft long leans against a vertical wall. The foot of the ladder is drawn away from the wall at a rate of 4 ft/s. How fast is the top of the ladder sliding down the wall when the foot of the ladder is 10 ft from the wall?

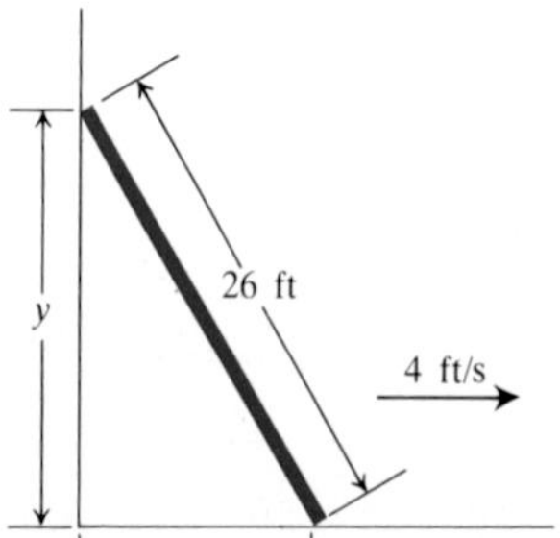

3.52 When the foot of the ladder slides away from the wall, y decreases and x increases.

Solution STEP 1. *Draw a picture.* Figure 3.52 shows the ladder against the wall. The variables are the height of the top of the ladder above the ground and the distance of the base of the ladder from the wall. The constants are the length of the ladder and the rate at which the base moves away from the wall.

STEP 2. *Write down what we are asked to find.* We are asked to find a rate. If y denotes the height of the top of the ladder above the ground in feet and t denotes the time in seconds, then the rate we seek is dy/dt. What is the value of dy/dt when the base of the ladder is 10 ft from the wall?

STEP 3. *Name the remaining variables.* Let x be the distance of the base of the ladder from the wall. Then $dx/dt = 4$ ft/s.

STEP 4. *Write equations that relate the variables.* The variables are related by the equation

$$x^2 + y^2 = 26^2.$$

STEP 5. *Differentiate* (we have only one equation so there is no call for substitution) to express dy/dt in terms of dx/dt:

$$2x\frac{dx}{dt} + 2y\frac{dy}{dt} = 0$$

$$\frac{dy}{dt} = -\frac{x}{y}\frac{dx}{dy}. \tag{5}$$

When $x = 10$,

$$y = \sqrt{26^2 - 10^2} = 24, \qquad \frac{dx}{dt} = 4, \qquad \text{and} \qquad \frac{dy}{dt} = -\frac{10}{24}(4) = -\frac{5}{3}.$$

Therefore, the top of the ladder is moving down (y is decreasing) at the rate of $\frac{5}{3}$ feet per second when the ladder is 10 ft from the wall. □

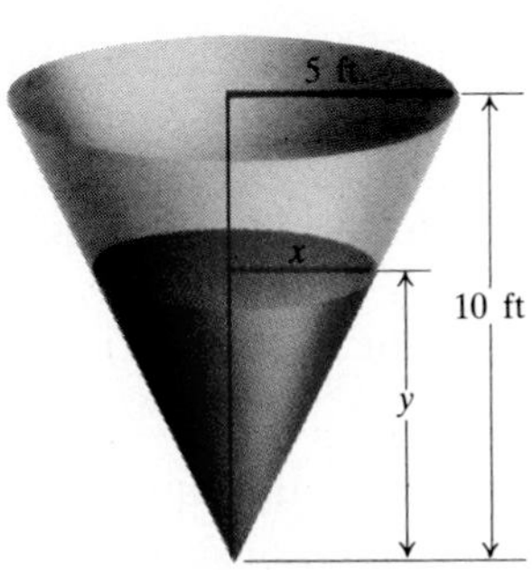

3.53 To show that the water level is changing in this conical tank, the depth of the water is denoted by a variable, y. The rate at which the level changes is then dy/dt.

EXAMPLE 4 Water runs into the conical tank shown in Fig. 3.53 at the constant rate of 2 ft^3 per minute. How fast is the water level rising when the water is 6 ft deep?

Solution The variables in the problem are

v = the volume (ft^3) of water in the tank at time t (min),
x = the radius (ft) of the surface of the water at time t,
y = the depth (ft) of water in the tank at time t.

The constants are the dimensions of the tank and the rate

$$\frac{dv}{dt} = 2\ \text{ft}^3/\text{min}$$

at which the tank fills.

We are asked for dy/dt when $y = 6$.

The relationship between the variables v and y is expressed by the equation

$$v = \frac{1}{3}\pi x^2 y. \tag{6}$$

This equation involves x as well as v and y, but we can eliminate x since, by similar triangles in Fig. 3.53,

$$\frac{x}{y} = \frac{5}{10} \quad \text{or} \quad x = \frac{1}{2}y.$$

Therefore,

$$v = \frac{1}{12}\pi y^3. \tag{7}$$

We now differentiate Eq. (7) to express dy/dt in terms of dv/dt:

$$\frac{dv}{dt} = \frac{1}{4}\pi y^2\frac{dy}{dt} \quad \text{or} \quad \frac{dy}{dt} = \frac{4}{\pi y^2}\frac{dv}{dt}. \tag{8}$$

When $dv/dt = 2$ and $y = 6$,

$$\frac{dy}{dt} = \frac{4 \cdot 2}{\pi \cdot 36} = \frac{2}{9\pi} \approx 0.071\ \text{ft/min}.$$

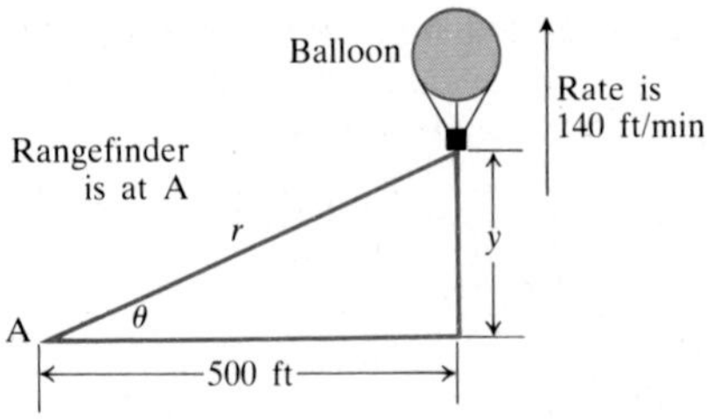

3.54 The rising balloon in Example 5.

When the water is 6 feet deep, the water level is rising at the rate of 0.071 feet per minute. □

EXAMPLE 5 A balloon rising from the ground at 140 ft/min is tracked by a rangefinder at point A, located 500 ft from the point of liftoff (see Fig. 3.54). Find the rate at which the angle at A and the range r are changing when the balloon is 500 ft above the ground.

Solution

The angle at A. The variable θ is related to y by the equation

$$\tan\theta = \frac{y}{500}.$$

The derivatives of θ and y with respect to time t are therefore related by the equation

$$\sec^2\theta\frac{d\theta}{dt} = \frac{1}{500}\frac{dy}{dt}. \tag{9}$$

When $y = 500$ ft, $\theta = \pi/4$, and $\sec^2\theta = (\sqrt{2})^2 = 2$. Also, $dy/dt =$ 140 ft/min. When we substitute these values into Eq. (9), we find

$$2\frac{d\theta}{dt} = \frac{1}{500}(140) \qquad \text{or} \qquad \frac{d\theta}{dt} = \frac{140}{1000} = 0.14 \text{ radians/min.}$$

When $y = 500$ ft, the angle at A is increasing at the rate of 0.14 radians per minute.

The range r. The variables r and y are related by the equation

$$r^2 = 500^2 + y^2,$$

and their time derivatives by the equations

$$2r\frac{dr}{dt} = 2y\frac{dy}{dt}, \qquad \text{or} \qquad \frac{dr}{dt} = \frac{y}{r}\frac{dy}{dt}.$$

When $y = 500$, $r = \sqrt{500^2 + 500^2} = 500\sqrt{2}$, $dy/dt = 140$ ft/min, and

$$\frac{dr}{dt} = \frac{500}{500\sqrt{2}}140 = \frac{140\sqrt{2}}{2} = 70\sqrt{2} \text{ ft/min.}$$

When $y = 500$ ft, the range is increasing at the rate of $70\sqrt{2}$ feet per minute. □

PROBLEMS

1. Let A be the area of a circle of radius r. How is dA/dt related to dr/dt?
2. Let V be the volume of a sphere of radius r. How is dV/dt related to dr/dt?
3. Let V be the volume of a cube of side s. How is dV/dt related to ds/dt?
4. A circular plate of metal is heated in an oven so that its radius increases at the rate of 0.1 mm per minute. At what rate is the plate's area increasing when the radius is 50 cm?
5. Ohm's law for an electrical circuit like the one shown in Fig. 3.55 states that

$$V = IR,$$

where V is the voltage, I is the current in amperes, and R is the resistance in ohms. Suppose V is in-

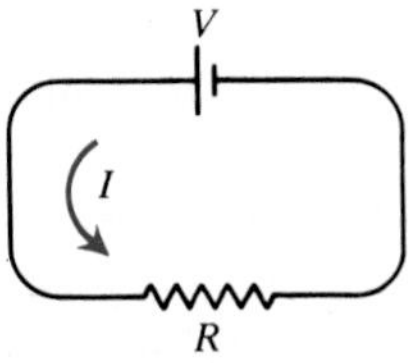

3.55 The current in this circuit obeys Ohm's law (Problem 5).

creasing at the rate of one volt per second while I is decreasing at the rate of one third ampere per second. Let t denote time in seconds.

a) What is the value of dV/dt?
b) What is the value of dI/dt?
c) What equation relates dR/dt to dV/dt and dI/dt?
d) Find the rate at which R is changing when $V = 12$ volts and $I = 2$ amperes. Is R increasing, or decreasing?

6. The length l of a rectangle is decreasing at the rate of 2 cm/s and the width w is increasing at the rate of 2 cm/s. When $l = 12$ cm and $w = 5$ cm, find the rates of change of (a) the area, (b) the perimeter, and (c) the lengths of the diagonals of the rectangle. Which of these quantities are decreasing and which are increasing?

7. A baseball diamond is a square 90 ft on a side (Fig. 3.56). A player runs from first base to second base at a rate of 16 ft/s. At what rate is the player's distance from third base decreasing when the player is 30 ft from first base? Carry out the following steps to answer this question.

a) Assign variables to the distances between the player and second and third bases. What are the values of the variables at the time in question?
b) How are the variables related?
c) How are the derivatives of the variables related?
d) Calculate the rate at which the player's distance from third base is changing.

3.56 The baseball diamond in Problem 7.

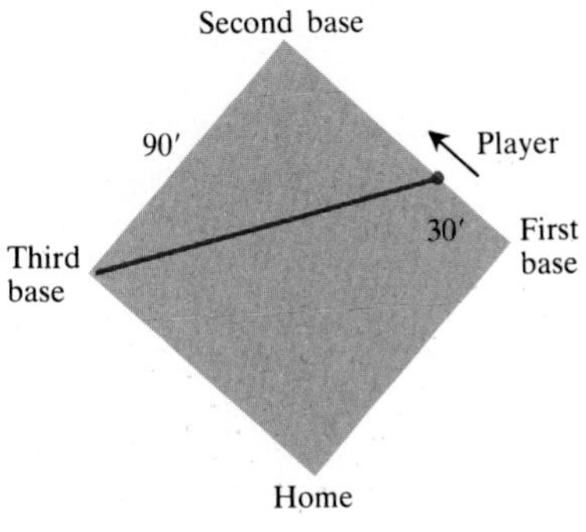

8. Let V be the volume and S the total surface area of a solid right circular cylinder that is 5 ft high and has radius r ft. Find dV/dS when $r = 3$.

9. At what rate is the area of the triangle formed by the ladder, ground, and wall in Example 3 changing when $x = 10$? If the motion began with the ladder flat against the wall at $t = 0$ seconds, and $x = 4t$, at what time was the area of the triangle largest?

10. Sand falls onto a conical pile at the rate of 10 ft^3/min. The radius of the base of the pile is always equal to one half its altitude. How fast is the altitude of the pile increasing when it is 5 ft deep?

11. Suppose that a raindrop is a perfect sphere. Assume that, through condensation, the raindrop accumulates moisture at a rate proportional to its surface area. Show that the radius increases at a constant rate.

12. Point A moves along the x-axis at the constant rate of a ft/s while point B moves along the y-axis at the constant rate of b ft/s. Find how fast the distance between them is changing when A is at the point $(x, 0)$ and B is at the point $(0, y)$.

13. A spherical balloon is inflated with gas at the rate of 100 ft^3/min. Assuming that the gas pressure remains constant, how fast is the radius of the balloon increasing at the instant when the radius is 3 ft?

14. A boat is pulled in to a dock by a rope with one end attached to the bow of the boat, the other end passing through a ring attached to the dock at a point 4 ft higher than the bow of the boat. If the rope is pulled in at the rate of 2 ft/s, how fast is the boat approaching the dock when 10 ft of rope are out?

15. A balloon is 200 ft off the ground and rising vertically at the constant rate of 15 ft/s. An automobile passes beneath it traveling along a straight road at the constant rate of 45 mi/hr = 66 ft/s. How fast is the distance between them changing one second later?

16. Water is withdrawn from a conical reservoir 8 ft in diameter and 10 ft deep (vertex down) at the constant rate of 5 ft^3/min. How fast is the water level falling when the depth of water in the reservoir is 6 ft?

17. A point moves on the curve $3x^2 - y^2 = 12$ so that its y-coordinate increases at the constant rate of 6 meters per second. At what rate is the x-coordinate changing when $x = 4$ m? What is the slope of the curve when $x = 4$ m?

18. A particle moves around the circle $x^2 + y^2 = 1$ with an x-velocity component $dx/dt = y$. Find dy/dt. Does the particle travel in the clockwise, or counterclockwise, direction around the circle?

19. A man 6 ft tall walks at the rate of 5 ft/s toward a street light that is 16 ft above the ground. At what rate is the tip of his shadow moving? At what rate is the length of his shadow changing when he is 10 ft from the base of the light?

20. When air changes volume adiabatically (without any heat being added to it), the pressure p and the volume v satisfy the relationship $pv^{1.4}$ = constant. At a certain instant the pressure is 50 lb/in^2 and the volume is 32 in^3 and is decreasing at the rate of 4 in^3/s. How rapidly is the pressure changing at this instant?

21. A light is at the top of a pole 50 ft high. A ball is dropped from the same height from a point 30 ft away from the light. How fast is the shadow of the ball moving along the ground $\frac{1}{2}$ second later? (Assume the ball falls a distance $s = 16t^2$ ft in t seconds.)

22. A girl flies a kite at a height of 300 ft, the wind carrying the kite horizontally away from her at a rate of 25 ft/s. How fast must she let out the string when the kite is 500 ft away from her?

23. A spherical iron ball 8 in. in diameter is coated with a layer of ice of uniform thickness. If the ice melts at the rate of 10 in^3 per minute, how fast is the thickness of the ice decreasing when it is 2 in. thick? How fast is the outer surface area of ice decreasing?

24. A highway patrol plane flies 1 mile above a straight road at a steady ground speed of 120 miles per hour. The pilot sees an oncoming car and, with radar, determines that the line-of-sight distance from plane to car is 1.5 miles, decreasing at the rate of 136 miles per hour. Find the car's speed along the highway.

25. The shadow of an 80-ft building on level ground is 100 ft long. If the angle made by the sun with the ground is decreasing at the rate of 15° per hour, at what rate is the shadow length increasing?

26. At noon ship A was 12 nautical miles due north of ship B. Ship A was sailing south at 12 knots (nautical miles per hour), and continued to do so all day. Ship B was sailing east at 8 knots, and continued to do so all day.
   a) How rapidly was the distance between the ships changing at noon? one hour later?
   b) The visibility that day was 5 nautical miles. Did the ships ever catch sight of each other?

27. Two ships A and B are sailing straight away from the point O along routes such that the angle $AOB = 120°$. How fast is the distance between them changing if, at a certain instant, $OA = 8$ mi, $OB = 6$ mi, ship A is sailing at the rate of 20 mi/hr, and ship B at the rate of 30 mi/hr? (*Hint:* Use the law of cosines.)

***Toolkit* programs**	
Derivative Grapher	Super * Grapher

3.7
Rolle's Theorem

There is strong geometric evidence that between two points where a smooth curve $y = f(x)$ crosses the x-axis there is at least one point on the curve where the tangent is horizontal (Fig. 3.57a). This would say that if a differentiable function is zero at two values, then its derivative is zero somewhere in between. But this need not be the case if the graph has a corner where f' fails to exist (Fig. 3.57b). More precisely, we have the following theorem.

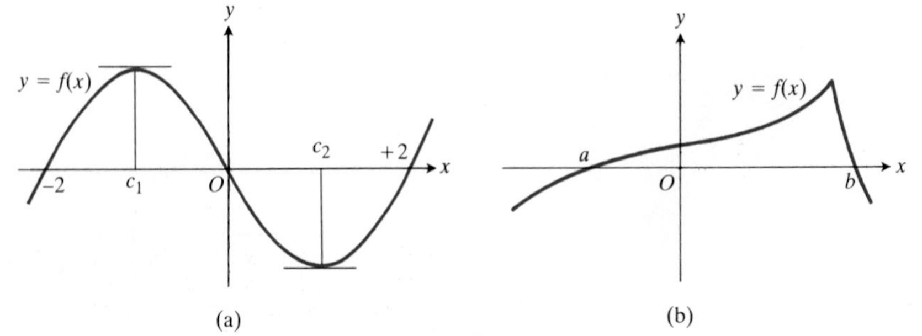

3.57 The functions graphed in (a) and (b) are both continuous. Graph (b), however, has no horizontal tangent between the two points where it crosses the x-axis. This could not have happened if $y = f(x)$ had been differentiable at every point between $x = a$ and $x = b$. The proof of Rolle's Theorem explains why.

THEOREM

Rolle's Theorem†

Suppose that $y = f(x)$ is continuous over the closed interval $a \leq x \leq b$ and differentiable on the open interval $a < x < b$. If

$$f(a) = f(b) = 0,$$

then there is at least one number c between a and b where f' is zero. That is,

$$f'(c) = 0 \quad \text{for some } c, \qquad a < c < b.$$

Proof. If $f(x) = 0$ for all x between a and b, then $f'(x) = 0$ for all x between a and b (the derivative of a constant function is zero) and the theorem is true.

But if $f(x)$ is not zero everywhere between a and b, then either it is positive some place, or negative some place, or both. In any case, the function will then have a maximum positive value or a minimum negative value, or both (by Theorem 9, Article 1.11). That is, it has an extreme value at a point c where $f(c)$ is negative (in the case of a minimum) or $f(c)$ is positive (in the case of a maximum). In either case, c is neither a nor b, since

$$f(a) = f(b) = 0, \qquad f(c) \neq 0.$$

Therefore c lies between a and b and Theorem 1 of Article 3.4 applies, showing that the derivative must be zero at $x = c$:

$$f'(c) = 0 \quad \text{for some } c, \qquad a < c < b. \blacksquare$$

EXAMPLE 1 The polynomial

$$y = x^3 - 4x = f(x)$$

is continuous and differentiable for all x, $-\infty < x < +\infty$. If we take

$$a = -2, \qquad b = +2,$$

the hypotheses of Rolle's Theorem are satisfied, since

$$f(-2) = f(+2) = 0.$$

Thus

$$f'(x) = 3x^2 - 4$$

must be zero at least once between -2 and $+2$. In fact, we find

$$3x^2 - 4 = 0$$

at

$$x = c_1 = -\frac{2\sqrt{3}}{3} \quad \text{and} \quad x = c_2 = +\frac{2\sqrt{3}}{3}. \square$$

† Published in 1691 in *Méthode pour Résoudre les Egalités* by the French mathematician Michel Rolle. See "Rolle's Theorem," in *A Source Book in Mathematics* by D. E. Smith (New York: McGraw-Hill, 1929), p. 253.

Rolle's Theorem may be combined with the Intermediate Value Theorem of Article 1.11 to obtain a criterion for isolating the real solutions of an equation $f(x) = 0$. Suppose that a and b are two real numbers such that

a) $f(x)$ and its first derivative $f'(x)$ are continuous for $a \le x \le b$,
b) $f(a)$ and $f(b)$ have opposite signs,
c) $f'(x)$ is different from zero for all values of x between a and b.

Then there is one and only one solution of the equation $f(x) = 0$ between a and b.

EXAMPLE 2 Show that the equation

$$x^3 + 3x + 1 = 0$$

has exactly one real root.

Solution Let

$$f(x) = x^3 + 3x + 1.$$

Then the derivative

$$f'(x) = 3x^2 + 3$$

is never zero (because it is always positive). Now, if there were even two points $x = a$ and $x = b$ where $f(x)$ was zero, Rolle's Theorem would guarantee the existence of a point $x = c$ where f' was zero. Therefore, f has no more than one zero. It does in fact have one zero, because the Intermediate Value Theorem tells us that the graph of $y = f(x)$ crosses the x-axis somewhere between $x = -1$ (where $y = -3$) and $x = 0$ (where $y = 1$). (See Fig. 3.58.) □

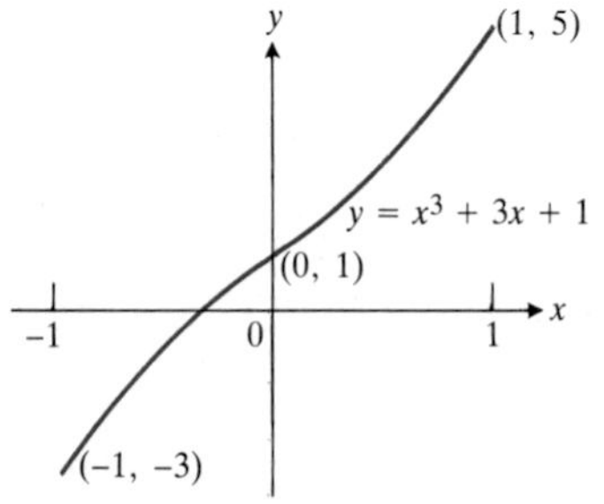

3.58 The only real zero of the polynomial $y = x^3 + 3x + 1$ is the one shown here between -1 and 0.

PROBLEMS

Without trying to solve the equations exactly, show that the equation $f(x) = 0$ has one and only one real root between the given pair of numbers for each of the following functions $f(x)$.

1. $x^4 + 3x + 1$, $\quad p = -2, \; q = -1$

2. $x^4 + 2x^3 - 2$, $\quad p = 0, \; q = 1$

3. $2x^3 - 3x^2 - 12x - 6$, $\quad p = -1, \; q = 0$

4. Let $f(x)$, together with its first two derivatives $f'(x)$ and $f''(x)$, be continuous for $a \le x \le b$. Suppose the curve $y = f(x)$ intersects the x-axis in at least three different places between a and b inclusive. Use Rolle's Theorem to show that the equation $f''(x) = 0$ has at least one real root between a and b. Generalize this result.

5. a) Plot the zeros of each polynomial on a line together with the zeros of its first derivative:

i) $y = x^2 - 4$ $\quad$ ii) $y = x^2 + 8x + 15$

iii) $y = x^3 - 3x^2 + 4 = (x + 1)(x - 2)^2$

iv) $y = x^3 - 33x^2 + 216x = x(x - 9)(x - 24)$

What pattern do you see?

b) Use Rolle's Theorem to prove that between every two zeros of the polynomial $x^n + a_{n-1}x^{n-1} + \cdots + a_1x + a_0$ there lies a zero of the polynomial

$$nx^{n-1} + (n-1)a_{n-1}x^{n-2} + \cdots + a_1.$$

6. The function

$$y = f(x) = \begin{cases} x & \text{if } 0 \le x < 1, \\ 0 & \text{if } x = 1, \end{cases}$$

is zero at $x = 0$ and at $x = 1$. Its derivative, $y' = 1$, is different from zero at every point between 0 and 1. Why doesn't that contradict Rolle's Theorem?

7. Suppose that a function $y = f(x)$ is continuous on $[a, b]$ and differentiable on (a, b). Show that if f' is never zero on (a, b), then $f(a) \ne f(b)$.

Toolkit programs

Derivative Grapher $\quad$ Super * Grapher

3.8

The Mean Value Theorem

Figure 3.59 shows the graph of a differentiable function f over an interval $a \le x \le b$, and there appears to be a point c between a and b at which the tangent is parallel to the chord AB that joins the endpoints of the curve. If we draw another smooth curve from A to B, as in Fig. 3.60, there again appears to be a point (in this case four) where the tangent is parallel to the chord. A generalization of Rolle's Theorem called the Mean Value Theorem says that this will always happen if the curve is continuous on $[a, b]$ and differentiable on (a, b). (It is not necessary for the function to be differentiable at $x = a$ and $x = b$.)

THEOREM

The Mean Value Theorem

If $y = f(x)$ is continuous at each point of $[a, b]$ and differentiable at each point of (a, b), then there is at least one number c between a and b for which

$$\frac{f(b) - f(a)}{b - a} = f'(c). \qquad (1)$$

Proof. To see the idea behind the proof, consider sliding the chord AB in Fig. 3.60 upward while keeping it parallel to its original position. If we slide it far enough it will lose contact with the curve. But to lose contact with the curve it will have to come first to a point where it is tangent to the curve. This is a point where the vertical distance from the curve to the chord has a local maximum. The argument we are about to give is based on the idea of measuring the vertical distance from the chord to the curve. We use Rolle's Theorem to find a point $x = c$ where the derivative of this distance is zero and then show that the slope $f'(c)$ is the slope of the chord AB.

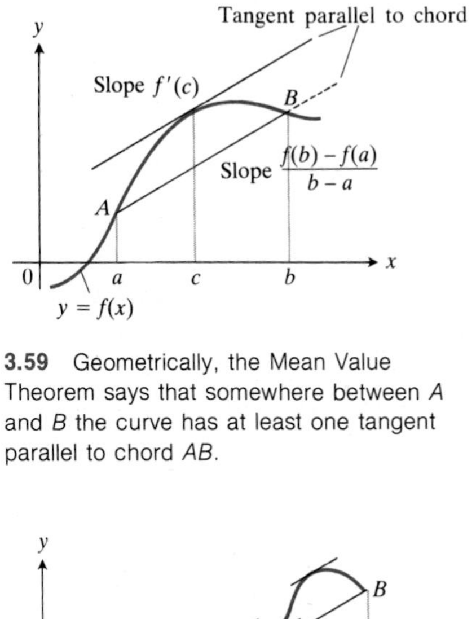

3.59 Geometrically, the Mean Value Theorem says that somewhere between A and B the curve has at least one tangent parallel to chord AB.

If x is any point on $[a, b]$, then the vertical distance between the chord and the curve directly above x in Fig. 3.61 is measured by

$$F(x) = RS = f(x) - PR. \qquad (2)$$

Now, PR is the y-coordinate of the point on the chord directly above x, and we may find it from an equation for the chord. The slope of the chord is

$$m = \frac{f(b) - f(a)}{b - a},$$

and a point on the chord is $(a, f(a))$. From these we get the point-slope equation

$$y - f(a) = \frac{f(b) - f(a)}{b - a}(x - a).$$

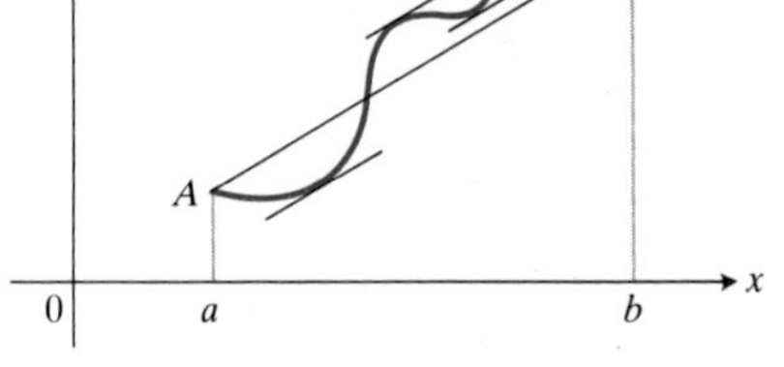

3.60 This smooth curve from A to B has four tangents parallel to chord AB.

The height of the chord above x is therefore

$$PR = y = f(a) + \frac{f(b) - f(a)}{b - a}(x - a).$$

3.61 If f is continuous on $[a, b]$ and differentiable on (a, b), then the function $F(x)$ that measures the distance between R and S is also continuous on $[a, b]$ and differentiable on (a, b). Since $F(a) = 0$ and $F(b) = 0$, F satisfies all the requirements of Rolle's Theorem. Therefore, $F' = 0$ at some point c between a and b. When we translate the equation $F'(c) = 0$ into a statement about f, we obtain one of the most useful theorems of calculus.

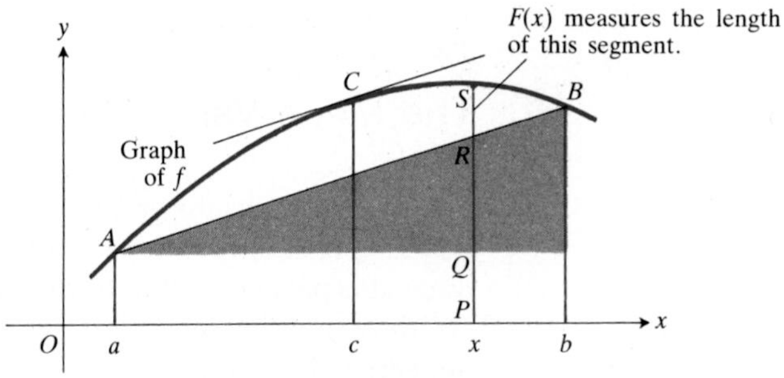

Hence the distance RS in Eq. (2) is

$$F(x) = f(x) - PR = f(x) - \left[f(a) + \frac{f(b) - f(a)}{b - a}(x - a)\right]$$

or

$$F(x) = f(x) - f(a) - \frac{f(b) - f(a)}{b - a}(x - a). \qquad (3)$$

The function $F(x)$ turns out to satisfy all the hypotheses of Rolle's Theorem. It is continuous on $[a, b]$ because $f(x)$ and $(x - a)$ are continuous on $[a, b]$. It is differentiable on (a, b) because $f(x)$ and $(x - a)$ are differentiable on (a, b). In addition,

$$F(a) = f(a) - f(a) - \frac{f(b) - f(a)}{b - a}(a - a) = 0, \qquad (4)$$

$$F(b) = f(b) - f(a) - \frac{f(b) - f(a)}{b - a}(b - a) = 0. \qquad (5)$$

By Rolle's Theorem, therefore,

$$F'(c) = 0 \qquad (6)$$

at some point c between a and b.

To see what Eq. (6) says about the original function f, we differentiate both sides of Eq. (3) with respect to x and substitute $x = c$ in the resulting expression for $F'(x)$:

$$F'(x) = f'(x) - \frac{f(b) - f(a)}{b - a} \cdot \frac{d}{dx}(x - a)$$
$$= f'(x) - \frac{f(b) - f(a)}{b - a}.$$

Since $F'(c) = 0$, this gives

$$0 = F'(c) = f'(c) - \frac{f(b) - f(a)}{b - a},$$
$$0 = f'(c) - \frac{f(b) - f(a)}{b - a},$$
$$\frac{f(b) - f(a)}{b - a} = f'(c),$$

which is the equation we started out to prove. ■

If $f'(x)$ is continuous on $[a, b]$, then by Theorem 9, Article 1.11, it has a maximum value max f' and a minimum value min f' on $[a, b]$. Since the number $f'(c)$ cannot exceed max f' nor be less than min f', the equation

$$\frac{f(b) - f(a)}{b - a} = f'(c), \tag{1}$$

gives us the inequality

$$\min f' \leq \frac{f(b) - f(a)}{b - a} \leq \max f', \tag{7}$$

where max and min refer to the values of f' on the interval $[a, b]$.

If we interpret $y = f(x)$ as distance traveled up to time x, then Eq. (7) says that the average speed from time a to time b is no greater than the maximum speed and no less than the minimum speed. Equation (1) says that at some instant c during the trip the speed was exactly equal to the average speed for the trip.

The importance of the Mean Value Theorem lies in the numerical estimates that sometimes come from Eq. (7) and in the mathematical conclusions that can be drawn from Eq. (1) (a few of which we shall see in a moment). Estimates based on an extended version of the Mean Value Theorem will be the subject of Article 3.10.

We usually do not know any more about the number c than Eq. (1) tells us, which is that c exists. In a few exceptional cases one's curiosity about the identity of c can be satisfied, as in the next two examples. Keep in mind, however, that our ability to identify c is the exception rather than the rule, and that the importance of the Mean Value Theorem lies elsewhere.

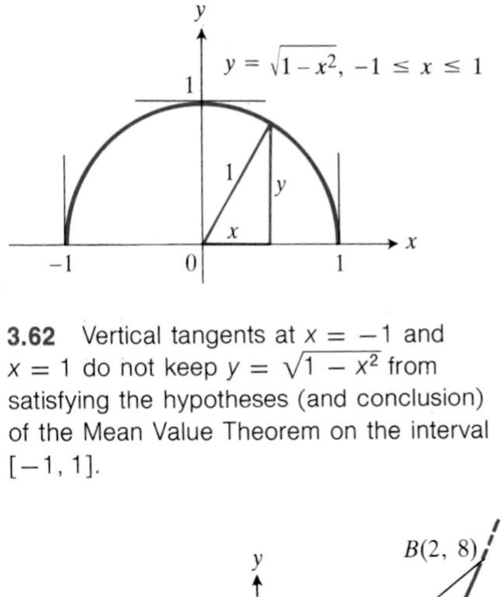

3.62 Vertical tangents at $x = -1$ and $x = 1$ do not keep $y = \sqrt{1 - x^2}$ from satisfying the hypotheses (and conclusion) of the Mean Value Theorem on the interval $[-1, 1]$.

EXAMPLE 1 See Fig. 3.62. The function $y = \sqrt{1 - x^2}$ satisfies the hypotheses (and conclusion) of the Mean Value Theorem on the interval $[-1, 1]$. It is continuous on the closed interval, and the derivative

$$y' = \frac{-x}{\sqrt{1 - x^2}}$$

is defined on the open interval $(-1, 1)$. The graph has a horizontal tangent at $x = 0$. Note that the function is not differentiable at $x = -1$ and $x = 1$. □

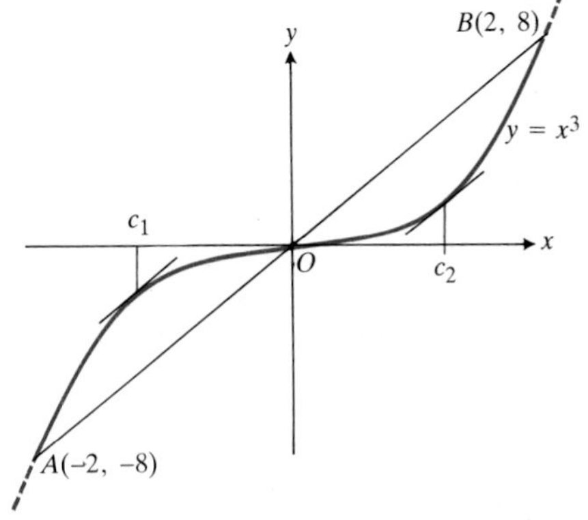

3.63 At $c = \pm\frac{2}{3}\sqrt{3}$, the tangents to the curve are parallel to the chord AB.

EXAMPLE 2 See Fig. 3.63. Let $f(x) = x^3$, $a = -2$, and $b = +2$. Then

$$f'(x) = 3x^2, \qquad f'(c) = 3c^2,$$
$$f(b) = 2^3 = 8, \qquad f(a) = (-2)^3 = -8,$$

so that

$$f'(c) = \frac{f(b) - f(a)}{b - a} = \frac{8 - (-8)}{2 - (-2)} = \frac{16}{4} = 4$$

becomes

$$3c^2 = 4, \qquad c = \pm\tfrac{2}{3}\sqrt{3}.$$

There are thus two values of c between $a = -2$ and $b = +2$, where the tangent to the curve $y = x^3$ is parallel to the chord through $A(-2, -8)$ and $B(2, 8)$. □

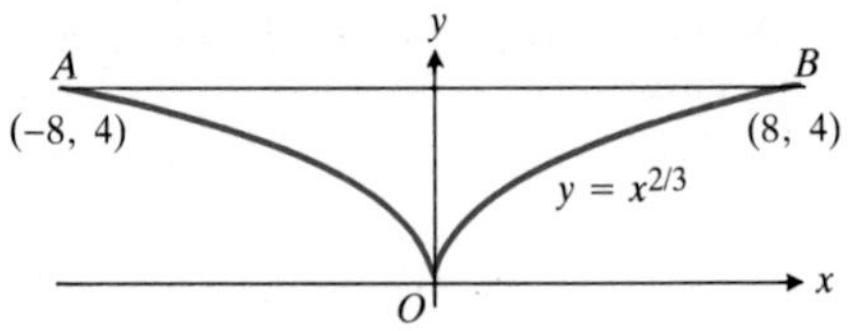

3.64 Having a tangent at each point does not mean having a derivative at each point and does not guarantee the conclusion of the Mean Value Theorem. The graph of $y = x^{2/3}$ has a tangent at every point (the tangent at O is vertical) but none of the tangents is parallel to the chord AB.

EXAMPLE 3 See Fig. 3.64. Let $f(x) = x^{2/3}$, $a = -8$, $b = +8$. Then

$$f'(x) = \frac{2}{3}x^{-1/3} = \frac{2}{3\sqrt[3]{x}}$$

exists everywhere between a and b *except* at $x = 0$. We find that

$$\frac{f(b) - f(a)}{b - a} = \frac{(8)^{2/3} - (-8)^{2/3}}{8 - (-8)} = \frac{4 - 4}{16} = 0,$$

but that

$$f'(c) = \frac{2}{3\sqrt[3]{c}}$$

is not zero for any finite value of c. Equation (1) does not hold in this case due to the failure of the derivative to exist at a point, namely $x = 0$, between $a = -8$ and $b = 8$. Note, however, that the curve $y = x^{2/3}$ does have a tangent at $x = 0$ (it is vertical) and therefore has a tangent everywhere between a and b. □

The equation in the Mean Value Theorem is often stated in the form

$$f(x) = f(a) + f'(c)(x - a), \tag{8}$$

obtained by writing x for b in Eq. (1) and solving for $f(x)$. In this new formulation, we think of x as an independent variable on $[a, b]$, and the number c lies in the interval between a and x. In Article 3.10, we shall explore the resemblance between Eq. (8) and the linear approximation

$$f(x) \approx f(a) + f'(a)(x - a)$$

that we encountered in Article 2.5.

In the next example, we show how Eq. (7) may be used to estimate the value of a function f at a particular value of x when a and f' are known.

EXAMPLE 4 Estimate $f(1)$ if

$$f'(x) = \frac{1}{5 - x^2} \quad \text{and} \quad f(0) = 2.$$

Solution (Using Eq. 7.) With $a = 0$, $b = 1$, and $f'(x) = 1/(5 - x^2)$, Eq. (7) becomes

$$\min \frac{1}{5 - x^2} \le \frac{f(1) - f(0)}{1 - 0} \le \max \frac{1}{5 - x^2},$$

where min and max refer to $[0, 1]$. On $[0, 1]$ the derivative $1/(5 - x^2)$ is smallest when $x = 0$ and largest when $x = 1$, so the inequalities in the equation above give

$$\frac{1}{5 - 0} \le f(1) - 2 \le \frac{1}{5 - 1},$$

$$2 + \frac{1}{5} \le f(1) \le 2 + \frac{1}{4},$$

$$2.2 \le f(1) \le 2.25. \ \square$$

In Article 3.1 we stated, as the first derivative test for rise and fall, the fact that a differentiable function f increases on intervals where $f' > 0$ and decreases on intervals where $f' < 0$. This fact can now be deduced from the Mean Value Theorem, in the following way.

COROLLARY 1

> Suppose that f is continuous on $[a, b]$ and differentiable on (a, b). If $f'(x) > 0$ throughout (a, b), then f is an increasing function on $[a, b]$. If $f'(x) < 0$ throughout (a, b), then f is decreasing on $[a, b]$. In either case, f is one-to-one.

Proof. Let x_1 and x_2 be any two numbers in $[a, b]$ such that $x_1 < x_2$. Apply the Mean Value Theorem to f on $[x_1, x_2]$:

$$f(x_2) - f(x_1) = f'(c)(x_2 - x_1) \tag{9}$$

for some c between x_1 and x_2. The sign of the right-hand side of (9) is the same as the sign of $f'(c)$ because $x_2 - x_1$ is positive. Therefore

$$f(x_2) > f(x_1) \qquad \text{if } f'(x) \text{ is positive on } (a, b)$$

(f is increasing) and

$$f(x_2) < f(x_1) \qquad \text{if } f'(x) \text{ is negative on } (a, b)$$

(f is decreasing). In either case, $x_1 \neq x_2$ implies that $f(x_1) \neq f(x_2)$, so f is one-to-one. ■

COROLLARY 2

> If
>
> $$F'(x) = 0 \qquad \text{for all } x \text{ in } (a, b), \tag{10}$$
>
> then F is constant throughout (a, b): there is a constant C such that
>
> $$F(x) = C \qquad \text{for all } x \text{ in } (a, b). \tag{11}$$

Corollary 2 says that the only functions whose derivatives are equal to zero throughout an interval are the functions that are constant on that interval.

COROLLARY 3

> If each of the functions F_1 and F_2 has its derivative equal to $f(x)$ for $a < x < b$, that is, if
>
> $$F_1'(x) = F_2'(x) = f(x), \qquad a < x < b,$$
>
> then there is a constant C such that
>
> $$F_1(x) - F_2(x) = C, \qquad \text{for all } x \text{ in } (a, b).$$

Corollary 3 says that the only way two functions can have identical rates of change on an interval is for their values to differ by some fixed constant on the interval. For example, we know that

$$\frac{d}{dx}(x^2) = 2x.$$

Same but one is bumped up by a constant

Therefore, if $F(x)$ is any differentiable function whose rate of change with respect to x is 2x, that is, if

$$\frac{dF}{dx} = 2x,$$

then

$$F(x) = x^2 + C \qquad \text{for some constant } C.$$

In determining that $F(x) = x^2 + C$, we say that we have *determined F up to a constant.* Techniques for determining functions from their rates of change are extremely important in science and engineering and we will begin their development in Chapter 4.

The proofs of Corollaries 2 and 3 are left as exercises.

PROBLEMS

In Problems 1–5, a, b, and c refer to the equation $f(b) - f(a) = (b - a)f'(c)$, which expresses the Mean Value Theorem. Given $f(x)$, a, and b, find c.

1. $f(x) = x^2 + 2x - 1, \quad a = 0, \quad b = 1$

2. $f(x) = x^3, \quad a = 0, \quad b = 3$

3. $f(x) = x^{2/3}, \quad a = 0, \quad b = 1$

4. $f(x) = x + \dfrac{1}{x}, \quad a = \frac{1}{2}, \quad b = 2$

5. $f(x) = \sqrt{x - 1}, \quad a = 1, \quad b = 3$

6. Suppose you know that $f(x)$ is differentiable and that $f'(x)$ always has a value between -1 and $+1$. Show that

$$|f(x) - f(a)| \le |x - a|.$$

7. Let $P_1(x_1, y_1)$ and $P_2(x_2, y_2)$ be any two points on the parabola

$$y = Ax^2 + Bx + C,$$

and let $P_3(x_3, y_3)$ be the point on the arc P_1P_2 where the tangent is parallel to the chord P_1P_2. Show that $x_3 = (x_1 + x_2)/2$.

8. Use the Mean Value Theorem to show that

$$|\sin b - \sin a| \le |b - a|.$$

9. Does the function $y = x$, $0 \le x \le 1$, satisfy the hypotheses of the Mean Value Theorem? If not, why not? If so, what value or values could c have?

10. The function $y = |4 - x^2|$, $-3 \le x \le 3$ (Fig. 3.33), has a horizontal tangent at $x = 0$ even though the function is not differentiable at $x = -2$ and $x = 2$. Does this contradict the Mean Value Theorem? Explain.

11. The function $y = 1/x$ is continuous and differentiable on the interval $0 < x < 1$ but its derivative is never zero on this interval. Does this contradict the Mean Value Theorem? Explain.

12. Suppose that $y = f(x)$ is continuous on $[0, 1]$ and differentiable on $(0, 1)$, and that $f(0) = 0$, $f(1) = 1$. Show that the derivative of f must equal 1 at some point between $x = 0$ and $x = 1$.

13. A motorist drove 30 miles during a one-hour trip. Show that the car's speed was equal to 30 miles per hour at least once during the trip.

14. Suppose that $f'(x) = 1/(1 + \cos x)$ for $0 \le x \le \pi/2$ and that $f(0) = 3$. Estimate $f(\pi/2)$.

15. Suppose that f is differentiable for all values of x, $f(-3) = -3$, $f(3) = 3$, and $|f'(x)| \le 1$. Show that $f(0) = 0$.

16. Suppose that f is continuous on $[a, b]$ and differentiable on (a, b), and that $f(a) < f(b)$. Show that f' is positive at some point between a and b.

17. The formula $f(x) = 3x + b$ gives a different function for every value of b. All of these functions, however, have the same derivative with respect to x, namely, $f'(x) = 3$. Are these the only differentiable functions of x whose derivative is 3? Could there be any others? Explain.

18. Show that

$$\frac{d}{dx}\left(\frac{x}{x+1}\right) = \frac{d}{dx}\left(-\frac{1}{x+1}\right),$$

even though

$$\frac{x}{x+1} \ne -\frac{1}{x+1}.$$

Doesn't this contradict Corollary 3 of the Mean Value Theorem? Explain.

19. Prove Corollary 2.

20. Prove Corollary 3.

***Toolkit* programs**

Derivative Grapher
Function Evaluator
Super * Grapher

3.9 Indeterminate Forms and l'Hôpital's Rule

As an elegant application of the Mean Value Theorem we prove a rule that extends our ability to calculate limits. The rule is named after Guillaume François Antoine de l'Hôpital (1661–1704), Marquis de St. Mesme, the French nobleman who wrote the first calculus text.

The Indeterminate Form 0/0

We sometimes want to know the value of a limit that looks like

$$\lim_{x \to a} \frac{f(x)}{g(x)} \qquad \text{where } f(a) = g(a) = 0. \tag{1}$$

It cannot be evaluated by substituting $x = a$, since this produces 0/0, a meaningless expression known as an *indeterminate form*.

As we have seen, the value of the limit in (1) is hard to predict:

$$\lim_{x \to 2} \frac{x^2 - 4}{x - 2} = \lim_{x \to 2} (x + 2) = 4$$

$$\lim_{x \to 0} \frac{\sin x}{x} = 1$$

$$\lim_{x \to 0} \frac{1 - \cos x}{x} = 0.$$

The limit

$$f'(a) = \lim_{x \to a} \frac{f(x) - f(a)}{x - a}$$

from which we calculate derivatives always produces the indeterminate form 0/0. Our success in calculating derivatives suggests that we might turn things around and use derivatives to calculate limits that lead to indeterminate forms. For example,

$$\lim_{x \to 0} \frac{\sin x}{x} = \lim_{x \to 0} \frac{\sin x - \sin 0}{x - 0} = \frac{d}{dx}(\sin x)\bigg|_{x=0} = \cos 0 = 1.$$

L'Hôpital's rule gives an explicit connection between derivatives and limits that lead to the indeterminate form 0/0.

THEOREM 4

L'Hôpital's Rule (first form)

Suppose that $f(a) = g(a) = 0$, that $f'(a)$ and $g'(a)$ exist, and that $g'(a) \neq 0$. Then

$$\lim_{x\to a} \frac{f(x)}{g(x)} = \frac{f'(a)}{g'(a)}.$$

Proof. Working backward from $f'(a)$ and $g'(a)$, which are themselves limits, we have

$$\frac{f'(a)}{g'(a)} = \frac{\displaystyle\lim_{x\to a} \frac{f(x) - f(a)}{x - a}}{\displaystyle\lim_{x\to a} \frac{g(x) - g(a)}{x - a}} = \lim_{x\to a} \frac{\dfrac{f(x) - f(a)}{x - a}}{\dfrac{g(x) - g(a)}{x - a}}$$

$$= \lim_{x\to a} \frac{f(x) - f(a)}{g(x) - g(a)} = \lim_{x\to a} \frac{f(x) - 0}{g(x) - 0} = \lim_{x\to a} \frac{f(x)}{g(x)}. \blacksquare$$

EXAMPLE 1

a) $\displaystyle\lim_{x\to 0} \frac{3x - \sin x}{x} = \frac{3 - \cos x}{1}\bigg|_{x=0} = 2$

b) $\displaystyle\lim_{x\to 0} \frac{\sqrt{1 + x} - 1}{x} = \frac{\dfrac{1}{2\sqrt{1 + x}}}{1}\Bigg|_{x=0} = \frac{1}{2}$

c) $\displaystyle\lim_{x\to 0} \frac{x - \sin x}{x^3} = \frac{1 - \cos x}{3x^2}\bigg|_{x=0} = ?$ □

What can we do about the limit in *c*? The first form of l'Hôpital's rule does not tell us what the limit is because the derivative of $g(x) = x^3$ is zero at $x = 0$. However, there is a stronger form of l'Hôpital's rule that says that whenever the rule gives 0/0 we can apply it again, repeating the process until we get a different result.

Roughly speaking, the stronger form of l'Hôpital's rule says that if $f(a) = g(a) = 0$, then

$$\lim_{x\to a} \frac{f(x)}{g(x)} = \lim_{x\to a} \frac{f'(x)}{g'(x)}$$

provided the derivatives and limit on the right exist.

With this stronger rule we can finish the work begun in Example 1(c):

$$\lim_{x\to 0} \frac{x - \sin x}{x^3} = \lim_{x\to 0} \frac{1 - \cos x}{3x^2} \qquad \left[\text{still } \frac{0}{0}\right]$$

$$= \lim_{x\to 0} \frac{\sin x}{6x} \qquad \left[\text{still } \frac{0}{0}\right]$$

$$= \lim_{x\to 0} \frac{\cos x}{6} = \frac{1}{6}.$$

Notice that to apply l'Hôpital's rule to f/g we divide the derivative of f by the derivative of g. Do not fall into the trap of taking the derivative of f/g. The quotient to use is f'/g', not $(f/g)'$.

The proof of the stronger form of l'Hôpital's rule is based on a mean value theorem discovered by Augustin L. Cauchy that uses two functions instead of one.

THEOREM 5

Cauchy's Mean Value Theorem

Suppose that the functions $f(x)$ and $g(x)$ are continuous for $a \leq x \leq b$ and differentiable for $a < x < b$, and that $g'(x) \neq 0$ for $a < x < b$. Then there exists a number c between a and b such that

$$\frac{f'(c)}{g'(c)} = \frac{f(b) - f(a)}{g(b) - g(a)}. \tag{2}$$

Proof. We apply the Mean Value Theorem of Article 3.8 twice. First we use it to show that $g(b) \neq g(a)$. For if $g(b)$ did equal $g(a)$, then the Mean Value Theorem would apply to g, to give

$$g'(c) = \frac{g(b) - g(a)}{b - a} = 0$$

for some c between a and b. But this cannot happen because $g'(x) \neq 0$ for $a < x < b$.

We next apply the Mean Value Theorem to the function

$$F(x) = f(x) - f(a) - \frac{f(b) - f(a)}{g(b) - g(a)}[g(x) - g(a)]. \tag{3}$$

This function is continuous and differentiable where f and g are, and $F(b) = F(a) = 0$. Therefore there is a number c between a and b for which $F'(c) = 0$. In terms of f and g this says

$$F'(c) = f'(c) - \frac{f(b) - f(a)}{g(b) - g(a)}[g'(c)] = 0, \tag{4}$$

or

$$\frac{f'(c)}{g'(c)} = \frac{f(b) - f(a)}{g(b) - g(a)},$$

which is Eq. (2). ■

THEOREM 6

L'Hôpital's Rule (stronger form)

Suppose that

$$f(x_0) = g(x_0) = 0,$$

and that the functions f and g are both differentiable on an open interval (a, b) that contains the point x_0. Suppose also that $g' \neq 0$ at every point in (a, b) except possibly x_0. Then

$$\lim_{x \to x_0} \frac{f(x)}{g(x)} = \lim_{x \to x_0} \frac{f'(x)}{g'(x)} \tag{5}$$

provided the limit on the right exists.

Proof. We first establish Eq. (5) for the case $x \to x_0^+$. The method needs almost no change to apply to $x \to x_0^-$, and the combination of these two cases establishes the result.

Suppose that x lies to the right of x_0. Then $g'(x) \neq 0$ and we can apply Cauchy's Mean Value Theorem to the closed interval from x_0 to x. This produces a number c between x_0 and x such that

$$\frac{f'(c)}{g'(c)} = \frac{f(x) - f(x_0)}{g(x) - g(x_0)}. \tag{6}$$

But $f(x_0) = g(x_0) = 0$, so that

$$\frac{f'(c)}{g'(c)} = \frac{f(x)}{g(x)}. \tag{7}$$

As x approaches x_0, c approaches x_0, because it lies between x and x_0. Therefore,

$$\lim_{x\to x_0^+} \frac{f(x)}{g(x)} = \lim_{c\to x_0^+} \frac{f'(c)}{g'(c)} = \lim_{x\to x_0^+} \frac{f'(x)}{g'(x)}.$$

This establishes l'Hôpital's rule for the case where x approaches x_0 from above. The case where x approaches x_0 from below is proved by applying Cauchy's Mean Value Theorem to the closed interval $[x, x_0]$, $x < x_0$. ■

In practice, the functions we deal with in this book satisfy the hypotheses of l'Hôpital's rule.

> To find
>
> $$\lim_{x\to x_0} \frac{f(x)}{g(x)}$$
>
> by l'Hôpital's rule, we proceed to differentiate f and g so long as we still get the form 0/0 at $x = x_0$. But as soon as one or the other of these derivatives is different from zero at $x = x_0$ we stop differentiating. L'Hôpital's rule does not apply when either the numerator or denominator has a finite nonzero limit.

EXAMPLE 2

$$\lim_{x\to 0} \frac{\sqrt{1+x} - 1 - (x/2)}{x^2} \qquad \left[\frac{0}{0}\right]$$

$$= \lim_{x\to 0} \frac{\frac{1}{2}(1+x)^{-1/2} - \frac{1}{2}}{2x} \qquad \left[\text{still } \frac{0}{0}\right]$$

$$= \lim_{x\to 0} \frac{-\frac{1}{4}(1+x)^{-3/2}}{2} = -\frac{1}{8}. \ \square$$

EXAMPLE 3

$$\lim_{x\to 0} \frac{1 - \cos x}{x + x^2} \qquad \left[\frac{0}{0}\right]$$

$$= \lim_{x\to 0} \frac{\sin x}{1 + 2x} = 0.$$

If we continue to differentiate in an attempt to apply l'Hôpital's rule once more, we get

$$\lim_{x\to 0} \frac{1 - \cos x}{x + x^2} = \lim_{x\to 0} \frac{\cos x}{2} = \frac{1}{2},$$

which is wrong. □

In applying l'Hôpital's rule we may reach a point where one of the derivatives is zero at $x = x_0$ and the other is not. Then the limit of the fraction is either zero, as in Example 3, or infinity, as in the next example.

EXAMPLE 4

$$\lim_{x\to 0} \frac{\sin x}{x^2} \qquad \left[\frac{0}{0}\right]$$

$$= \lim_{x\to 0} \frac{\cos x}{2x} = \infty. \ \square$$

The Indeterminate Forms ∞/∞, $\infty \cdot 0$, and $\infty - \infty$

Sometimes when we try to evaluate a limit as $x \to a$ by substituting $x = a$ we get an ambiguous expression like ∞/∞, $\infty \cdot 0$, or $\infty - \infty$, instead of 0/0. In more advanced books it is proved that l'Hôpital's rule applies to the indeterminate form ∞/∞ as well as to 0/0. If $f(x) \to \infty$ and $g(x) \to \infty$ as $x \to a$, then

$$\lim_{x\to a} \frac{f(x)}{g(x)} = \lim_{x\to a} \frac{f'(x)}{g'(x)}$$

provided the limit on the right exists.† In the notation $x \to a$, a may be either finite or infinite.

The forms $\infty \cdot 0$ and $\infty - \infty$ can sometimes be handled by changing the expressions algebraically to get 0/0 or ∞/∞ instead.

EXAMPLE 5

$$\lim_{x\to 0} \frac{x - 2x^2}{3x^2 + 5x} = \lim_{x\to 0} \frac{1 - 4x}{6x + 5} = \frac{1}{5},$$

$$\lim_{x\to \infty} \frac{x - 2x^2}{3x^2 + 5x} = \lim_{x\to \infty} \frac{1 - 4x}{6x + 5} = \lim_{x\to \infty} \frac{-4}{6} = -\frac{2}{3}. \ \square$$

EXAMPLE 6 The limit

$$\lim_{x\to \infty} x \sin \frac{1}{x}$$

leads to the form $\infty \cdot 0$, but we can change to the form 0/0 by writing $x = 1/t$ and letting $t \to 0$:

$$\lim_{x\to \infty} x \sin \frac{1}{x} = \lim_{t\to 0} \frac{1}{t} \cdot \sin t = \lim_{t\to 0} \frac{\sin t}{t} = 1. \ \square$$

EXAMPLE 7 Find

$$\lim_{x\to 0} \left(\frac{1}{\sin x} - \frac{1}{x}\right).$$

Solution If $x \to 0^+$, then $\sin x \to 0^+$ and $1/\sin x \to +\infty$, while $1/x \to +\infty$. The expression $(1/\sin x) - (1/x)$ formally becomes $+\infty - (+\infty)$, which is indeterminate. On the other hand, if $x \to 0^-$, then $1/\sin x \to -\infty$

†There is also a proof of l'Hôpital's rule for both forms 0/0 and ∞/∞ in the article "L'Hôpital's Rule," by A. E. Taylor, *American Mathematical Monthly*, 59, 20–24 (1952).

and $1/x \to -\infty$, so that $(1/\sin x) - (1/x)$ becomes $-\infty + \infty$, which is also indeterminate. But we may also write

$$\frac{1}{\sin x} - \frac{1}{x} = \frac{x - \sin x}{x \sin x}$$

and apply l'Hôpital's rule to the expression on the right. Thus,

$$\begin{aligned}\lim_{x\to 0}\left(\frac{1}{\sin x} - \frac{1}{x}\right) &= \lim_{x\to 0}\frac{x - \sin x}{x \sin x} && \left[\frac{0}{0}\right]\\ &= \lim_{x\to 0}\frac{1 - \cos x}{\sin x + x\cos x} && \left[\text{still } \frac{0}{0}\right]\\ &= \lim_{x\to 0}\frac{\sin x}{2\cos x - x\sin x} = 0. \ \square\end{aligned}$$

PROBLEMS

Find the limits in Problems 1–24.

1. $\lim_{x\to 2}\frac{x-2}{x^2-4}$

2. $\lim_{t\to\infty}\frac{6t+5}{3t-8}$

3. $\lim_{x\to\infty}\frac{5x^2-3x}{7x^2+1}$

4. $\lim_{x\to 1}\frac{x^3-1}{4x^3-x-3}$

5. $\lim_{t\to 0}\frac{\sin t^2}{t}$

6. $\lim_{x\to\pi/2}\frac{2x-\pi}{\cos x}$

7. $\lim_{x\to 0}\frac{\sin 5x}{x}$

8. $\lim_{t\to 0}\frac{\cos t - 1}{t^2}$

9. $\lim_{\theta\to\pi}\frac{\sin\theta}{\pi-\theta}$

10. $\lim_{x\to\pi/2}\frac{1-\sin x}{1+\cos 2x}$

11. $\lim_{x\to\pi/4}\frac{\sin x - \cos x}{x-\pi/4}$

12. $\lim_{x\to\pi/3}\frac{\cos x - 0.5}{x-\pi/3}$

13. $\lim_{x\to(\pi/2)} -\left(x-\frac{\pi}{2}\right)\tan x$

14. $\lim_{x\to 0}\frac{2x}{x+7\sqrt{x}}$

15. $\lim_{x\to 1}\frac{2x^2-(3x+1)\sqrt{x}+2}{x-1}$

16. $\lim_{x\to 2}\frac{\sqrt{x^2+5}-3}{x^2-4}$

17. $\lim_{x\to 0}\frac{\sqrt{a(a+x)}-a}{x}, \quad a>0$

18. $\lim_{t\to 0}\frac{10(\sin t - t)}{t^3}$

19. $\lim_{x\to 0}\frac{x(\cos x - 1)}{\sin x - x}$

20. $\lim_{h\to 0}\frac{\sin(a+h)-\sin a}{h}$

21. $\lim_{r\to 1}\frac{a(r^n-1)}{r-1}$, n a positive integer

22. $\lim_{x\to 0^+}\left(\frac{1}{x}-\frac{1}{\sqrt{x}}\right)$

23. $\lim_{x\to\infty}(x-\sqrt{x^2+x})$

24. Which is correct, (a) or (b)? Explain.

a) $\lim_{x\to 3}\frac{x-3}{x^2-3} = \lim_{x\to 3}\frac{1}{2x} = \frac{1}{6}$

b) $\lim_{x\to 3}\frac{x-3}{x^2-3} = \frac{0}{6} = 0$

25. L'Hôpital's rule does not help with

$$\lim_{x\to\infty}\frac{\sqrt{10x+1}}{\sqrt{x+1}}.$$

Find the limit some other way.

26. In Fig. 3.65, the circle has radius OA equal to 1, and AB is tangent to the circle at A. The arc AC has radian measure θ and the segment AB also has length θ. The line through B and C crosses the x-axis at $P(x, 0)$.

a) Show that the length of PA is

$$1 - x = \frac{\theta(1-\cos\theta)}{\theta - \sin\theta}.$$

b) Find $\lim_{\theta\to 0}(1-x)$.

c) Show that $\lim_{\theta\to\infty}[(1-x)-(1-\cos\theta)] = 0$.

Interpret this geometrically.

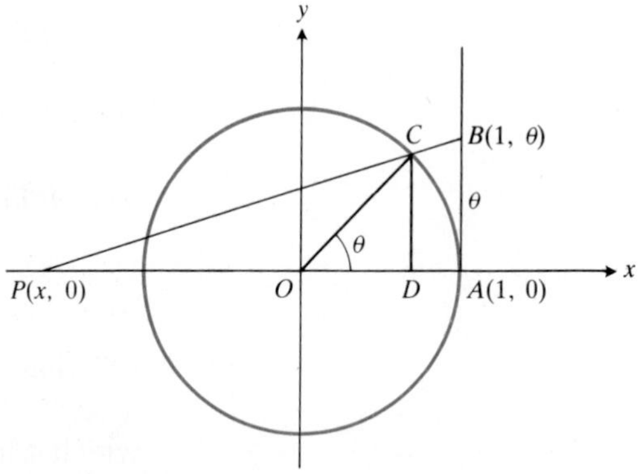

3.65 Diagram for Problem 26.

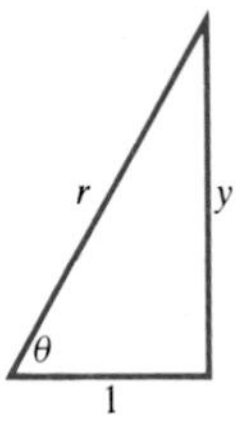

3.66 Diagram for Problem 27.

27. (Fig. 3.66) A right triangle has one leg of length 1, another of length y, and a hypotenuse of length r. The angle opposite y has radian measure θ. Find the limits as $\theta \to \pi/2$ of (a) $r - y$; (b) $r^2 - y^2$; (c) $r^3 - y^3$.

28. Consider the functions

$$f(x) = \begin{cases} x + 2 & \text{for } x \neq 0, \\ 0 & \text{for } x = 0, \end{cases}$$

$$g(x) = \begin{cases} x + 1 & \text{for } x \neq 0, \\ 0 & \text{for } x = 0, \end{cases}$$

and let $x_0 = 0$. Show that

$$\lim_{x \to x_0} \frac{f'(x)}{g'(x)} = 1$$

but

$$\lim_{x \to x_0} \frac{f(x)}{g(x)} = 2.$$

Does this contradict l'Hôpital's rule?

In Problems 29–31, find all values of c that satisfy Eq. (2), the conclusion of Cauchy's Mean Value Theorem, for the given functions and interval.

29. $f(x) = x$, $g(x) = x^2$, $(a, b) = (-2, 0)$

30. $f(x) = x$, $g(x) = x^2$, (a, b) arbitrary

31. $f(x) = x^3/3 - 4x$, $g(x) = x^2$, $(a, b) = (0, 3)$

***Toolkit* programs**

Limit Problems   Super * Grapher

3.10

Extending the Mean Value Theorem to Taylor's Formula. Estimating Approximation Errors

Earlier we introduced the Mean Value Theorem, which says that if a function f is continuous on an interval $[a, b]$ and differentiable on (a, b), then

$$f(b) = f(a) + f'(c)(b - a) \tag{1a}$$

for some point c between a and b. If we think of b as an independent variable we can rewrite (1a) in the form

$$f(x) = f(a) + f'(c)(x - a), \tag{1b}$$

valid for the interval from a to x, with c between a and x.

The right-hand side of (1b) looks like the linearization of f near a. If f' is continuous and c is close to a (as it will have to be if x is close to a), then $f'(c)$ is close to $f'(a)$, and (1b) gives

$$f(x) \approx f(a) + f'(a)(x - a), \tag{1c}$$

which *is* the linear approximation of f near a. In Article 2.5 we produced and used linearizations without knowing exactly how good they were. Now, with an extended version of the Mean Value Theorem, we shall see that the error in (1c) is proportional to $(x - a)^2$. Therefore, if $(x - a)$ is small the error will be very small. Knowing this, we shall be able to look at some of the approximations we have been using, to find out how good they are.

THEOREM

The First Extended Mean Value Theorem

If f and f' are continuous on $[a, b]$ and f' is differentiable on (a, b), then there exists a number c_2 between a and b such that

$$f(b) = f(a) + f'(a)(b - a) + \frac{f''(c_2)}{2}(b - a)^2. \tag{2a}$$

Proof. The significance of Eq. (2a) is not the fact that some number K satisfies the equation

$$f(b) = f(a) + f'(a)(b - a) + K(b - a)^2, \tag{2b}$$

but the fact that the value of K defined by this equation is actually equal to

$$K = \frac{f''(c_2)}{2} \tag{3}$$

for some point c_2 in the interval between a and b. Therefore, given that K is the number that satisfies (2b), we set out to show that K must satisfy (3) for some number c_2 between a and b.

Equation (2b) says that when $x = b$, the function $f(x)$ has the same value as the function

$$f(a) + f'(a)(x - a) + K(x - a)^2.$$

These two functions also have the same value when $x = a$ (namely, $f(a)$) but generally their difference,

$$F(x) = f(x) - f(a) - f'(a)(x - a) - K(x - a)^2, \tag{4}$$

is different from zero.

The function $F(x)$ satisfies all the hypotheses of Rolle's Theorem on the interval $[a, b]$: $F(a) = 0$, $F(b) = 0$, F is continuous on $[a, b]$ because f, $(x - a)$, and $(x - a)^2$ are, and F is differentiable on (a, b) for the same reason. Therefore, $F' = 0$ at some point c_1 between a and b:

$$F'(c_1) = 0, \qquad a < c_1 < b.$$

Because $F'(c_1) = 0$, the derivative

$$F'(x) = f'(x) - f'(a) - 2K(x - a) \tag{5}$$

satisfies all the hypotheses of Rolle's Theorem on the interval $[a, c_1]$: $F'(c_1) = 0$, and substituting $x = a$ in Eq. (5) shows that $F'(a) = 0$. Also, F' is continuous on $[a, c_1]$ and differentiable on (a, c_1) because both f' and $(x - a)$ are. Therefore, the derivative

$$F''(x) = f''(x) - 2K$$

is zero at some point c_2 between a and c_1, which means that

$$0 = f''(c_2) - 2K,$$

or

$$K = \frac{f''(c_2)}{2}.$$

Substituting this value of K in (2b) gives Eq. (2a), which is what we set out to prove. ■

REMARK. The proof and theorem remain valid if $b < a$, provided one refers to "the interval with endpoints a and b" rather than explicitly to $[a, b]$ or (a, b). If one does this, and takes similar care with the other intervals that appear in the proof, then the arguments go through as before.

Linear Approximations

We are now in a position to calculate the error in the linear approximation

$$f(x) \approx f(a) + f'(a)(x - a)$$

that we used in Article 2.5. We begin by regarding b as an independent variable in Eq. (2a) and rewriting this equation in the form

$$f(x) = f(a) + f'(a)(x - a) + \frac{f''(c_2)}{2}(x - a)^2, \tag{6}$$

with the understanding that c_2 lies between a and x. This equation holds for $x < a$ as well as $x > a$, by the remark at the end of the proof of the First Extended Mean Value Theorem. From Eq. (6) we get the linear approximation

$$f(x) \approx f(a) + f'(a)(x - a), \qquad x \approx a, \tag{7}$$

valid on the interval from a to x with an error of

$$e_1(x) = \frac{f''(c_2)}{2}(x - a)^2. \tag{8}$$

If f'' is continuous on the closed interval from a to x, then it has a maximum value on the interval and $e_1(x)$ satisfies the inequality

$$|e_1(x)| \leq \frac{\max|f''|}{2}(x - a)^2, \tag{9}$$

where max refers to the interval joining a and x.

This inequality gives an *upper bound* for the error on the interval from a to x that is of practical value in many cases.

We summarize this important result, as follows:

The Linear Approximation of $f(x)$ Near $x = a$

Suppose that f, f', and f'' are continuous on the closed interval joining a and x. Then,

$$f(x) \approx f(a) + f'(a)(x - a) \tag{10}$$

with an error $e_1(x)$ that satisfies the inequality

$$|e_1(x)| \leq \frac{\max|f''|}{2}(x - a)^2. \tag{11}$$

The "max" in Eq. (11) refers to the values of f'' on the interval joining a and x.

EXAMPLE 1 Verify the approximation

$$\frac{1}{1-x} \approx 1 + x, \qquad x \approx 0,$$

from Article 2.5, and investigate its error for $|x| \leq 0.1$.

Solution With $a = 0$ and

$$f(x) = (1-x)^{-1}, \qquad f(0) = 1$$
$$f'(x) = (1-x)^{-2}, \qquad f'(0) = 1$$
$$f''(x) = 2(1-x)^{-3},$$

Equation (7) gives

$$\frac{1}{1-x} \approx 1 + 1(x-0) = 1 + x.$$

For the error, Eq. (11) gives

$$|e_1(x)| \leq \max \left| \frac{2(1-x)^{-3}}{2} \right| (x-0)^2 = \max \left| \frac{1}{(1-x)^3} \right| x^2.$$

On the interval $|x| \leq 0.1$, the largest the right-hand side of the inequality can be is

$$\frac{1}{(1-0.1)^3}(0.1)^2 = \frac{(0.1)^2}{(0.9)^3} < 0.014.$$

Therefore,

$$|e_1(x)| < 0.014 \qquad \text{if} \qquad |x| \leq 0.1. \; \square$$

Quadratic Approximations

To get more accuracy in a linear approximation we add a quadratic term. Typical quadratic approximations near $x = 0$ are

$$\frac{1}{1-x} \approx 1 + x + x^2$$
$$\sqrt{1+x} \approx 1 + \frac{x}{2} - \frac{x^2}{8}$$
$$\cos x \approx 1 - \frac{x^2}{2}$$
$$\sin x \approx x$$

Others can be obtained from these with algebra. For $x \approx 0$ we have

$$\frac{1}{2-x} = \frac{1}{2}\left(\frac{1}{1-x/2}\right) \approx \frac{1}{2}\left(1 + \frac{x}{2} + \left(\frac{x}{2}\right)^2\right) = \frac{1}{2} + \frac{x}{4} + \frac{x^2}{8}.$$

$$\frac{1+x}{1-x} \approx (1+x)(1+x+x^2) \approx 1 + 2x + 2x^2 \qquad \text{(ignoring the } x^3 \text{ term).}$$

To combine approximations correctly, each approximation must be valid at the time it is applied:

Right: $$\frac{1}{1-\sin x} \approx \frac{1}{1-x} \approx 1 + x + x^2, \qquad x \approx 0.$$

Wrong: $$\sin\left(\frac{1}{1-x}\right) \underset{(1)}{\approx} \frac{1}{1-x} \underset{(2)}{\approx} 1 + x + x^2.$$

The second combination is wrong because (1) requires $1/(1-x) \approx 0$, hence x large, while (2) requires $x \approx 0$.

We can get a good estimate for the error in quadratic approximations by extending the Mean Value Theorem one step further in the following way.

THEOREM

The Second Extended Mean Value Theorem

If f, f', and f'' are continuous on $[a, b]$ and f'' is also differentiable on (a, b), then there exists a number c_3 between a and b for which

$$f(b) = f(a) + f'(a)(b-a) + \frac{f''(a)}{2}(b-a)^2 + \frac{f'''(c_3)}{6}(b-a)^3. \tag{12}$$

The proof of this second extended mean value theorem, which resembles the proof of the first, is a special case of the proof outlined in Miscellaneous Problem 88 at the end of the chapter.

In applications we usually write Eq. (12) with x in place of b, writing

$$f(x) = f(a) + f'(a)(x-a) + \frac{f''(a)}{2}(x-a)^2 + \frac{f'''(c_3)}{6}(x-a)^3, \tag{13}$$

with the understanding that c_3 lies between a and x. From (13) we get the quadratic approximation

$$f(x) \approx f(a) + f'(a)(x-a) + \frac{f''(a)}{2}(x-a)^2, \tag{14}$$

valid on the interval between a and x with an error of

$$e_2(x) = \frac{f'''(c_3)}{6}(x-a)^3. \tag{15}$$

Note that the first two terms on the right-hand side of (14) give the standard linear approximation of f. To get the quadratic approximation we have only to add the quadratic term without changing the linear part.

If $f'''(x)$ is continuous on the closed interval from a to x, then it has a maximum value on the interval, and Eq. (15) tells us that

$$|e_2(x)| \le \max\left|\frac{f'''(x)}{6}\right| |x-a|^3. \tag{16}$$

EXAMPLE 2 Verify the quadratic approximation

$$\frac{1}{1-x} \approx 1 + x + x^2, \qquad x \approx 0$$

and determine its accuracy for $|x| \le 0.1$.

Solution With $a = 0$ and

$$\begin{aligned} f(x) &= (1-x)^{-1}, & f(0) &= 1 \\ f'(x) &= (1-x)^{-2}, & f'(0) &= 1 \\ f''(x) &= 2(1-x)^{-3}, & f''(0) &= 2 \\ f'''(x) &= 6(1-x)^{-4}, \end{aligned}$$

Equation (14) gives

$$\frac{1}{1-x} \approx 1 + (x-0) + \frac{2}{2}(x-0)^2 = 1 + x + x^2.$$

For the error, Eq. (16) gives the upper bound

$$\begin{aligned} |e_2(x)| &\le \max\left|\frac{6(1-x)^{-4}}{6}\right| |x^3| = \max\left|\frac{1}{(1-x)^4}\right| |x^3| \\ &= \frac{1}{\min(1-x)^4}|x^3|. \end{aligned}$$

When $|x| \le 0.1$ the largest the right-hand side can be is

$$\frac{(0.1)^3}{(0.9)^4} < 0.0016.$$

Therefore,

$$|e_2(x)| < 0.0016. \;\square$$

This is an improvement over the upper bound of 0.014 found for the error in the linear approximation of $1/(1-x)$ in Example 1.

EXAMPLE 3 Show that the quadratic approximation of $f(x) = \sin x$ near $x = 0$ is

$$\sin x \approx x, \qquad x \approx 0.$$

Solution With $a = 0$, we have

$$\begin{aligned} f(x) &= \sin x & f(0) &= 0 \\ f'(x) &= \cos x & f'(0) &= 1 \\ f''(x) &= -\sin x & f''(0) &= 0 \\ f'''(x) &= -\cos x. \end{aligned}$$

Equations (14) and (16) give

$$\sin x \approx 0 + 1(x-0) + 0(x-0)^2 = x,$$

$$|e_2(x)| \le \max\frac{|\cos x|}{6}|x^3| \le \frac{|x^3|}{6}.$$

The *quadratic* approximation to $f(x) = \sin x$ is therefore

$$\sin x \approx x, \qquad \text{with an error bound } |e_2(x)| \le \frac{|x^3|}{6}.$$

How can the approximation be quadratic if there is no x^2 term? There is a quadratic term, but its coefficient is zero, the important fact being that no quadratic term is missing and we can estimate the error with $e_2(x)$ instead of $e_1(x)$.

How good is the approximation?

For $|x| \le 0.1$, $\quad |e_2(x)| \le \dfrac{(0.1)^3}{6} < 0.00017.$

For $|x| \le 0.25$, $\quad |e_2(x)| \le \dfrac{(0.25)^3}{6} < 0.0026.$ □

Taylor's Theorem

The extended mean value theorems are special cases of a theorem called *Taylor's Theorem*, which holds for any natural number n. The most convenient statement of the theorem for our purposes is the following:

THEOREM

Taylor's Theorem

If f and its first n derivatives $f', f'', \ldots, f^{(n)}$ are continuous on $[a, b]$ and if $f^{(n)}$ is differentiable on (a, b), then there exists a number c_{n+1} between a and b such that

$$f(b) = f(a) + f'(a)(b-a) + \frac{f''(a)}{2}(b-a)^2 + \cdots + \frac{f^{(n)}(a)}{n!}(b-a)^n + \frac{f^{(n+1)}(c_{n+1})}{(n+1)!}(b-a)^{n+1}. \tag{17}$$

The proof of Taylor's Theorem is outlined in Problem 88 in the Miscellaneous Problems at the end of the chapter.

Equation (17) provides extremely accurate polynomial approximations for a large class of functions that have derivatives of all orders, a class that includes many more functions than we have seen so far. We shall explore Taylor's Theorem and its powerful consequences in Chapter 12.

Table 3.2 Approximation Formulas

Linear	**Quadratic**
Approximation:	
$f(x) \approx f(a) + f'(a)(x-a)$	$f(x) \approx f(a) + f'(a)(x-a) + \dfrac{f''(a)}{2}(x-a)^2$
Error bound:	
$\lvert e_1(x)\rvert \le \dfrac{\max \lvert f''(c)\rvert}{2}(x-a)^2$	$\lvert e_2(x)\rvert \le \dfrac{\max \lvert f'''(c)\rvert}{6}\lvert x-a\rvert^3$
c between a and x	c between a and x
Common approximations for $x \approx 0$:	
$\dfrac{1}{1-x} \approx 1 + x$	$\dfrac{1}{1-x} \approx 1 + x + x^2$
$\sqrt{1+x} \approx 1 + \dfrac{x}{2}$	$\sqrt{1+x} \approx 1 + \dfrac{x}{2} - \dfrac{x^2}{8}$
$\sin x \approx x$	$\sin x \approx x$
$\cos x \approx 1$	$\cos x \approx 1 - \dfrac{x^2}{2}$
$(1+x)^k \approx 1 + kx$ (any number k)	$(1+x)^k \approx 1 + kx + \dfrac{k(k-1)}{2}x^2$ (any number k)

PROBLEMS

Use formulas from Table 3.2 to find quadratic approximations for the functions in Problems 1–5 near the given values of x.

1. $x \sin x, \quad x \approx 0$
2. $\sqrt{1 + \sin x}, \quad x \approx 0$
3. $\cos \sqrt{1 + x}, \quad x \approx 0$
4. $\dfrac{\sec x}{1 - x}, \quad x \approx 0$
5. $\sqrt{x}, \quad x \approx 1$
   (*Hint:* $x = 1 + (x - 1)$, with $(x - 1) \approx 0$.)
6. Find upper bounds for the error $|e_1(x)|$ in the following linearizations.
   a) $\sqrt{1 + x} \approx 1 + (x/2), \quad |x| \le 0.1$
   b) $\cos x \approx 1, \quad |x| \le 0.1$

In Problems 7–9: (a) Derive the quadratic approximation for the function near $x = 0$. (b) Find a numerical upper bound for the error $|e_2(x)|$ of each approximation for the interval $|x| \le 0.1$.

7. $\sqrt{1 + x}$
8. $\tan x$
9. $\cos x$

In Problems 10–13: (a) Find the quadratic approximation of the function near the given point a. (b) Use Eq. (16) to find a numerical upper bound for the error of each approximation for the interval $|x - a| \le 0.1$.

10. $\sin x, \quad a = \pi/2$
11. $\sin x, \quad a = \pi$
12. $\cos x, \quad a = \pi/2$
13. $\cos x, \quad a = \pi$
14. For what values of x will the error $|e_2(x)|$ in the approximation

$$\cos x \approx 1 - \frac{x^2}{2}$$

be (a) less than 1/100? (b) less than 1% of $|x|$?

15. Assuming that the differentiation formula

$$\frac{d}{dx} u^k = k u^{k-1} \frac{du}{dx}$$

holds for any number k, show that the quadratic approximation of $(1 + x)^k$ is

$$(1 + x)^k \approx 1 + kx + \frac{k(k - 1)}{2} x^2.$$

16. Verify Eq. (6) for $f(x) = 3x^2 + 2x + 4, \quad a = 1.$
17. Verify Eq. (13) for $f(x) = x^3 + 5x - 7, \quad a = 1.$
18. Use Eq. (6) to prove the following:

    Let f be continuous and have continuous first and second derivatives. Suppose that $f'(a) = 0$. Then
    a) f has a relative maximum at a if $f'' \le 0$ throughout an interval whose interior contains a;
    b) f has a relative minimum at a if $f'' \ge 0$ throughout an interval whose interior contains a.
19. Using Taylor's Theorem, prove that a polynomial $f(x)$ of degree n may be written, precisely, in the form

$$f(x) = f(a) + f'(a)(x - a) + \frac{f''(a)}{2}(x - a)^2 + \cdots + \frac{f^{(n)}(a)}{n!}(x - a)^n.$$

20. Prove the case $n = 3$ of Taylor's Theorem.

***Toolkit* programs**	
Derivative Grapher	Super * Grapher

REVIEW QUESTIONS AND EXERCISES

1. Discuss the significance of the signs of the first and second derivatives. Sketch a small portion of a curve, illustrating how it looks near a point where
   a) both y' and y'' are positive;
   b) $y' > 0$, $y'' < 0$;
   c) $y' < 0$, $y'' > 0$;
   d) $y' < 0$, $y'' < 0$.
2. Define *point of inflection*. How do you find points of inflection from an equation of a curve?
3. Give an example of a function whose graph has an oblique asymptote; a vertical asymptote; a horizontal asymptote. Sketch the graphs and asymptotes.
4. Discuss criteria for symmetry of a curve with respect to (a) the x-axis; (b) the y-axis; (c) the origin; (d) the line $y = x$; (e) the line $y = -x$.
5. How do you locate local maximum and minimum points of a curve? Discuss exceptional points, such as endpoints and points where the derivative fails to exist, in addition to the nonexceptional type. Illustrate with graphs.
6. Let n be a positive integer. For which values of n does the curve $y = x^n$ have (a) a local minimum at the origin; (b) a point of inflection at the origin?
7. Outline a general method for solving "max-min" problems.
8. Outline a general method for solving "related rates" problems.
9. What are the hypotheses of Rolle's Theorem? What is the conclusion?
10. Is the converse of Rolle's Theorem true?

11. With the book closed, state and prove the Mean Value Theorem. What is its geometrical interpretation?
12. We know that if $F(x) = x^2$, then $F'(x) = 2x$. If someone knows a function G such that $G'(x) = 2x$ but $G(x) \neq x^2$, what can be said about the difference $G(2) - G(1)$? Explain.
13. Describe l'Hôpital's rule. How do you know how many times to use it in a given problem? Give an example.
14. Give the linear and quadratic approximations of three different functions. What are the general formulas for the linear and quadratic approximations of a function? How can we find upper bounds for the errors in these approximations?
15. Read the article "Mathematics in warfare" by F. W. Lanchester, *World of Mathematics*, Vol. 4 (New York: Simon and Schuster, 1960), pp. 2138–2157, as a discussion of a practical problem in "related rates."

MISCELLANEOUS PROBLEMS

In Problems 1–18 find y' and y''. Determine in each case the sets of values of x for which
a) y is increasing (as x increases),
b) y is decreasing (as x increases),
c) the graph is concave up,
d) the graph is concave down.
Also sketch the graph in each case, indicating maxima, minima, asymptotes, and points of inflection.

1. $y = 9x - x^2$
2. $y = x^3 - 5x^2 + 3x$
3. $y = 4x^3 - x^4$
4. $y = 4x + x^{-1}$
5. $y = x^2 + 4x^{-1}$
6. $y = x + 4x^{-2}$
7. $y = 5 - x^{2/3}$
8. $y = \dfrac{x-1}{x+1}$
9. $y = x - \dfrac{4}{x}$
10. $y = x^4 - 2x^2$
11. $y = \dfrac{x^2}{ax+b}$, $a > 0$, $b > 0$
12. $y = 2x^3 - 9x^2 + 12x$
13. $y = (x-1)(x+1)^2$
14. $y = x^2 - \frac{1}{6}x^3$
15. $y = 2x^3 - 9x^2 + 12x - 4$
16. $y = 3x(x-1)(x+1)^2$
17. $xy^2 = 3(1-x)$
18. $y = 2\cos x + \cos^2 x$

In Problems 19–27, determine the following properties of the curves whose equations are given: (A) symmetry, (B) restrictions on x and y, (C) intercepts, (D) asymptotes, (E) slope at intercepts. Use this information in sketching the curves.

19. a) $y^2 = x(4-x)$ b) $y^2 = x(x-4)$ c) $y^2 = \dfrac{x}{4-x}$
20. a) $y = x + \dfrac{1}{x^2}$ b) $y^2 = x + \dfrac{1}{x^2}$ c) $y = x^2 + \dfrac{1}{x}$
21. a) $y = x(x+1)(x-2)$ b) $y^2 = x(x+1)(x-2)$
22. a) $y = \dfrac{8}{4+x^2}$ b) $y = \dfrac{8}{4-x^2}$ c) $y = \dfrac{8x}{4+x^2}$
23. a) $xy = x^2 + 1$ b) $y = \dfrac{x^2}{x-1}$
24. a) $y^2 = x^4 - x^2$ b) $y^2 = \dfrac{x-1}{x-2}$
25. $x^2y - y = 4(x-2)$
26. $y = \dfrac{x^2+1}{x^2-1}$
27. $x^2 + xy + y^2 = 3$
28. A certain graph has an equation of the form

$$ay^2 + by = \frac{cx + d}{ex^2 + fx + g},$$

where a, b, c, d, e, f, and g are constants whose value in each case is either 0 or 1. From the following information about the graph, determine the constants and give a reason for your choice in each case:

Extent. The curve does not exist for $x < -1$. All values of y are permissible.

Symmetry. The curve is symmetric about the x-axis.

Intercepts. No y-intercept; x-intercept at $(-1, 0)$.

Asymptotes. Both axes; no others.

Sketch the graph.

29. The slope of a curve at any point (x, y) is given by the equation

$$\frac{dy}{dx} = 6(x-1)(x-2)^2(x-3)^3(x-4)^4.$$

a) For what value (or values) of x is y a maximum? Why?
b) For what value (or values) of x is y a minimum? Why?

30. A particle moves along the x-axis with velocity $v = dx/dt = f(x)$. Show that its acceleration is $f(x)f'(x)$.

31. A meteorite entering the earth's atmosphere has velocity inversely proportional to $\sqrt{s}$ when at distance s from the center of the earth. Show that its acceleration is inversely proportional to s^2.

32. The volume of a cube is increasing at a rate of 300 in^3/min at the instant when the edge is 20 in. Find the rate at which the edge is changing.

33. Sand falling at the rate of 3 ft^3/min forms a conical pile whose radius always equals twice the height. Find the rate at which the height is changing at the instant when the height is 10 ft.

34. The volume of a sphere is decreasing at the rate of 12π ft^3/min. Find the rates at which the radius and the surface area are changing at the instant when the radius is 20 ft. Also find approximately how much the radius and surface area may be expected to change in the following 6 seconds.

35. At a certain instant airplane A is flying a level course at 500 mi/hr. At the same time, airplane B is straight above airplane A and flying at the rate of 700 mi/hr on a course that intercepts A's course at a point C that is 4 mi from B and 2 mi from A. (a) At the instant in question, how fast is the distance between the airplanes decreasing? (b) What is the minimum distance between the airplanes, if they continue on the present courses at constant speed?

36. A point moves along the curve $y^2 = x^3$ in such a way that its distance from the origin increases at the constant rate of 2 units per second. Find dx/dt at $(2, 2\sqrt{2})$.

37. Refer to the triangle in Fig. 3.52. How fast is its area changing when $x = 17\sqrt{2}$?

38. Suppose the cone in Fig. 3.53 has a small opening at the vertex through which the water escapes at the rate of $0.08\sqrt{y}$ ft^3/min when its depth is y. Water is also running into the cone at a constant rate of c ft^3/min. When the depth is $6\frac{1}{4}$ ft, the depth of the water is observed to be increasing at the rate of 0.02 ft/min. Under these conditions, will the tank fill? Give a reason for your answer.

39. A particle projected vertically upward from the surface of the earth with initial velocity v_0 has velocity $\sqrt{v_0^2 - 2gR[1 - (R/s)]}$ when it reaches a distance $s \geq R$ from the *center* of the earth. Here R is the radius of the earth. Show that the acceleration is inversely proportional to s^2.

40. Given a triangle ABC. Let D and E be points on the sides AB and AC, respectively, such that DE is parallel to BC. Let the distance between BC and DE equal x. Show that the derivative, with respect to x, of the area $BCED$ is equal to the length of DE.

41. Points A and B move along the x- and y-axes, respectively, in such a way that the perpendicular distance r (inches) from the origin to AB remains constant. How fast is OA changing, and is it increasing, or decreasing, when $OB = 2r$ and B is moving toward O at the rate of $0.3r$ in/s?

42. Ships A and B start from O at the same time. Ship A travels due east at a rate of 15 mi/hr. Ship B travels in a straight course making an angle of 60° with the path of ship A at a rate of 20 mi/hr. How fast are they separating at the end of 2 hr?

43. Water is being poured into an inverted conical tank (vertex down) at the rate of 2 ft^3/min. How fast is the water level rising when the depth of the water is 5 ft? The radius of the base of the cone is 3 ft and the altitude is 10 ft.

44. Divide 20 into two parts (not necessarily integers) such that the product of one part with the square of the other shall be a maximum.

45. Find the largest value of $f(x) = 4x^3 - 8x^2 + 5x$ for $0 \leq x \leq 2$. Give reasons for your answer.

46. Find two *positive* numbers whose sum is 36 and whose product is as large as possible. Can the problem be solved if the product is to be as small as possible?

47. Determine the coefficients a, b, c, d so that the curve whose equation is $y = ax^3 + bx^2 + cx + d$ has a maximum at $(-1, 10)$ and an inflection point at $(1, -6)$.

48. Find the number that most exceeds its square.

49. The perimeter p and area A of a circular sector ("piece of pie") of radius r and arc length s are given by $p = 2r + s$; $A = \frac{1}{2}rs$. If the perimeter is 100 ft, what value of r will produce a maximum area?

50. If a ball is thrown vertically upward with a velocity of 32 ft/s, its height after t seconds is given by the equation $s = 32t - 16t^2$. At what instant will the ball be at its highest point, and how high will it rise?

51. A right circular cone has altitude 12 ft and radius of base 6 ft. A cone is inscribed with its vertex at the center of the base of the given cone and its base parallel to the base of the given cone. Find the dimensions of the cone of maximum volume that can be so inscribed.

52. A soup can is to be made in the form of a right circular cylinder to contain 16π in^3. What dimensions of the can will require the least amount of material?

53. An isosceles triangle is drawn with its vertex at the origin, its base parallel to and above the x-axis and the vertices of its base on the curve $12y = 36 - x^2$. Determine the area of the largest such triangle.

54. A tire manufacturer is able to make x (hundred) grade A tires and y (hundred) grade B tires per day, where $y = (40 - 10x)/(5 - x)$, with $0 \leq x \leq 4$. If the profit on each grade A tire is twice the profit on a grade B tire, what is the most profitable number of grade A tires per day to make?

55. Find the points on the curve $x^2 - y^2 = 1$ that are nearest the point $P(a, 0)$ in case (a) $a = 4$, (b) $a = 2$, (c) $a = \sqrt{2}$.

56. A motorist, in a desert 5 mi from a point A, which is the nearest point on a long straight road, wishes to get to a point B on the road. If the car can travel at 15 mi/hr on the desert and 39 mi/hr on the road, find the point where it must meet the road to get to B in the shortest possible time if (a) B is 5 mi from A, (b) B is 10 mi from A, (c) B is 1 mi from A.

57. A drilling rig, a km off shore, is to be connected by pipe to a refinery on shore, b km down the coast from the rig, as shown in Fig. 3.67. If underwater pipe costs w dollars/km and land-based pipe costs $l < w$ dollars/km, what values of x and y give the least expensive connection?

58. Points A and B are ends of a diameter of a circle and C is a point on the circumference. Which of the following statements about triangle ABC is (or are) true?

a) The area is a maximum when the triangle is isosceles.

b) The area is a minimum when the triangle is isosceles.

c) The perimeter is a maximum when the triangle is isosceles.

d) The perimeter is a minimum when the triangle is isosceles.

59. The base and the perimeter of a triangle are fixed. Determine the remaining two sides if the area is to be a maximum.

60. The base b and the area k of a triangle are fixed. Determine the base angles if the angle at the vertex opposite b is to be a maximum.

61. A line is drawn through a fixed point (a, b) to meet the axes Ox, Oy in P and Q. Show that the minimum values of PQ, $OP + OQ$, and $OP \cdot OQ$ are, respectively,

$$(a^{2/3} + b^{2/3})^{3/2}, \quad (\sqrt{a} + \sqrt{b})^2, \quad \text{and} \quad 4ab.$$

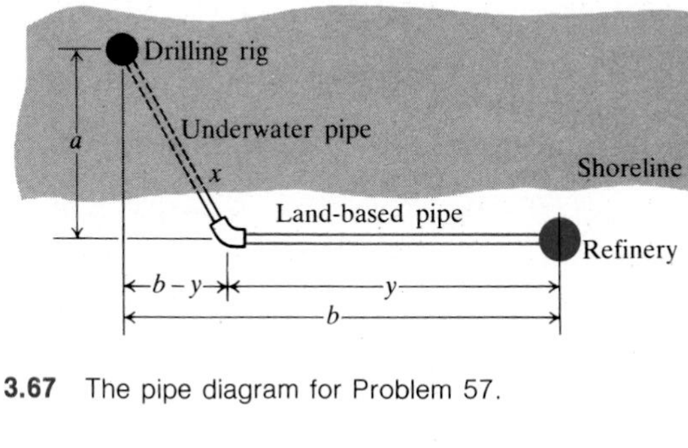

3.67 The pipe diagram for Problem 57.

62. Find the smallest value of the constant m if $mx - 1 + (1/x)$ is to be greater than or equal to zero for all positive values of x.

63. Let s be the distance between the fixed point $P_1(x_1, y_1)$ and a point $P(x, y)$ on the line

$$L: \quad ax + by + c = 0.$$

Using calculus methods, (a) show that s^2 is a minimum when P_1P is perpendicular to L, and (b) show that the minimum distance is

$$|ax_1 + by_1 + c|/\sqrt{a^2 + b^2}.$$

64. A playing field is to be built in the shape of a rectangle plus a semicircular area at each end. A 400-m race track is to form the perimeter of the field. Find the dimensions of the field if the rectangular part is to have as large an area as possible.

65. If $ax + (b/x) \geq c$ for all positive values of x, where a, b, and c are positive constants, show that $ab \geq c^2/4$.

66. Prove that if $ax^2 + (b/x) \geq c$ for all positive values of x, where a, b, and c are positive constants, then $27ab^2 \geq 4c^3$.

67. Given $f(x) = ax^2 + 2bx + c$ with $a > 0$. By considering the minimum, prove that $f(x) \geq 0$ for all real x if, and only if, $b^2 - ac \leq 0$.

68. In Problem 67, take

$$f(x) = (a_1x + b_1)^2 + (a_2x + b_2)^2 + \cdots + (a_nx + b_n)^2,$$

and deduce Schwarz's inequality:

$$(a_1b_1 + a_2b_2 + \cdots + a_nb_n)^2 \leq (a_1^2 + a_2^2 + \cdots + a_n^2)(b_1^2 + b_2^2 + \cdots + b_n^2).$$

69. In Problem 68 prove that equality can hold only in case there is a real number x such that $b_i = -a_ix$ for every $i = 1, 2, \ldots, n$.

70. If x is positive and m is greater than one, prove that $x^m - 1 - m(x - 1)$ is not negative.

71. What are the dimensions of the rectangular plot of greatest area that can be laid out within a triangle of base 36 ft and altitude 12 ft? Assume that one side of the rectangle lies on the base of the triangle.

72. Find the width across the top of an isosceles trapezoid of base 12 in. and slant sides 6 in. if its area is a maximum.

73. A fence h m high runs parallel to and w m from a vertical wall. Find the length of the shortest ladder that will reach from the ground across the top of the fence to the wall.

74. Assuming that the cost per hour of running the *Queen Elizabeth II* is $a + bv^n$, where a, b, and n are positive constants, $n > 1$, and v is the velocity through the water, find the speed for making the run from Liverpool to New York at minimum cost.

75. A flower bed is to be in the shape of a circular sector of radius r and central angle θ (i.e., like a piece of pie). Find r and θ if the area is fixed and the perimeter is a minimum.

76. A reservoir is to be built in the form of a right circular cone and the lateral area waterproofed. If the capacity of the reservoir is to be 72π ft^3 and one gallon of waterproofing material will cover 80 ft^2, how many gallons are required?

77. Given two concentric circles, C_1 of radius r_1 and C_2 of radius r_2, $r_2 > r_1 > 0$. Let A be the area between them.

a) How fast is A increasing (or decreasing) when $r_1 = 4$ cm and is increasing at the rate of 0.02 cm/s while $r_2 = 6$ cm and is increasing at the rate of 0.01 cm/s?

b) Suppose that at time $t = 0$, r_1 is 3 cm and r_2 is 5 cm, and that for $t > 0$, r_1 increases at the constant rate of a cm/s and r_2 increases at the constant rate of b cm/s. If $(\frac{3}{5})a < b < a$, find when the area A will be a maximum.

78. Given two concentric spheres, S_1 of radius r_1 and S_2 of radius r_2, $r_2 > r_1 > 0$. Let V be the volume between them. Suppose that at time $t = 0$, $r_1 = r$ in. and $r_2 = R$ in., and that for $t > 0$, r_1 increases at the constant rate of a in/s and r_2 increases at the constant rate of b in/s. If $a > b > ar^2/R^2$, find when V will be a maximum.

79. The motion of a particle in a straight line is given by $s = \lambda t - (1 + \lambda^4)t^2$. Show that the particle moves forward initially when λ is positive but ultimately retreats. Show also that for different values of λ the maximum possible distance that the particle can move forward is $\frac{1}{8}$.

80. Let $h(x) = f(x)g(x)$ be the product of two functions that have first and second derivatives and are positive; that is, $f(x) > 0$, $g(x) > 0$.

a) Is it true, if f and g both have a relative maximum at $x = a$, that h has a relative maximum at $x = a$?

b) Is it true, if f and g both have a point of inflection at $x = a$, that h has a point of inflection at $x = a$?

For both (a) and (b) either give a proof or construct a numerical example showing that the statement is false.

81. The numbers $c_1, c_2, \ldots, c_n$ are recorded in an experiment. It is desired to determine a number x with the property that

$$(c_1 - x)^2 + (c_2 - x)^2 + (c_3 - x)^2 + \cdots + (c_n - x)^2$$

shall be a minimum. Find x.

82. The four points

$$(-2, -\tfrac{1}{2}),\ (0, 1),\ (1, 2),\ \text{and}\ (3, 3)$$

are observed to lie more or less close to a straight line of equation $y = mx + 1$. Find m if the sum

$$(y_1 - mx_1 - 1)^2 + (y_2 - mx_2 - 1)^2 + (y_3 - mx_3 - 1)^2 + (y_4 - mx_4 - 1)^2$$

is to be a minimum, where $(x_1, y_1), \ldots, (x_4, y_4)$ are the coordinates of the given points.

83. The *geometric mean* of the n positive numbers $a_1, a_2, \ldots, a_n$ is the nth root of $a_1a_2 \cdots a_n$ and the arithmetic mean is $(a_1 + a_2 + \cdots + a_n)/n$. Show that if $a_1, a_2, \ldots, a_{n-1}$ are fixed and $a_n = x$ is permitted to vary over the set of positive real numbers, the ratio of the arithmetic mean to the geometric mean is a minimum when x is the arithmetic mean of $a_1, a_2, \ldots, a_{n-1}$.

84. The curve $(y + 1)^3 = x^2$ passes through the points $(1, 0)$ and $(-1, 0)$. Does Rolle's Theorem justify the conclusion that dy/dx vanishes for some value of x in the interval $-1 \leq x \leq 1$? Give reasons for your answer.

85. If $a < 0 < b$ and $f(x) = x^{-1/3}$, show that there is no c that satisfies Eq. (1), Article 3.8. Illustrate with a sketch of the graph.

86. If $a < 0 < b$ and $f(x) = x^{1/3}$, show that there is a value of c that satisfies Eq. (1), Article 3.8, even though the function fails to have a derivative at $x = 0$. Illustrate with a sketch of the graph.

87. Show that the equation $f(x) = 2x^3 - 3x^2 + 6x + 6 = 0$ has exactly one real root and find its value accurate to two significant figures. (*Hint:* $f(-1) = -5$, $f(0) = +6$, and $f'(x) > 0$ for all real x).

88. *Extended Mean Value Theorem.* Suppose $f(x)$ and its derivatives $f'(x), f''(x), \ldots, f^{(n)}(x)$ of order one through n are continuous on $a \leq x \leq b$, and $f^{(n+1)}(x)$ exists for $a < x < b$. If

$$F(x) = f(x) - f(a) - (x - a)f'(a) - (x - a)^2f''(a)/2! - \cdots - \frac{(x - a)^n f^{(n)}(a)}{n!} - K(x - a)^{n+1},$$

where K is chosen so that $F(b) = 0$, show that

a) $F(a) = F(b) = 0$,

b) $F'(a) = F''(a) = \cdots = F^{(n)}(a) = 0$,

c) there exist numbers $c_1, c_2, c_3, \ldots, c_{n+1}$ such that

$$a < c_{n+1} < c_n < \cdots < c_2 < c_1 < b$$

and such that

$$F'(c_1) = 0 = F''(c_2) = F'''(c_3) = \cdots = F^{(n)}(c_n) = F^{(n+1)}(c_{n+1}).$$

d) Hence, deduce that $K = [f^{(n+1)}(c_{n+1})]/(n+1)!$ for c_{n+1} as above in (c); or, in other words, since $F(b) = 0$,

$$f(b) = f(a) + f'(a)(b-a) + \frac{f''(a)}{2!}(b-a)^2 + \cdots + \frac{f^{(n)}(a)}{n!}(b-a)^n + \frac{f^{(n+1)}(c_{n+1})}{(n+1)!}(b-a)^{(n+1)}$$

for some c_{n+1}, $a < c_{n+1} < b$. (*Amer. Math. Monthly*, **60** (1953), p. 415, James Wolfe.)

89. Suppose that it costs a company $y = a + bx$ dollars to produce x units per week. It can sell x units per week at a price of $P = c - ex$ dollars per unit. (a) What production level maximizes the profit? (b) What is the corresponding price? (c) What is the weekly profit at this level of production? (d) At what price should each item be sold to maximize profits if the government imposes a tax of t dollars per item sold? Comment on the difference between this price and the price before tax.

90. Evaluate the following limits.

a) $\lim_{x\to 0} \frac{2 \sin 5x}{3x}$ b) $\lim_{x\to 0} \sin 5x \cot 3x$

c) $\lim_{x\to 0} x \csc^2 \sqrt{2x}$ d) $\lim_{x\to \pi/2} (\sec x - \tan x)$

e) $\lim_{x\to 0} \frac{x - \sin x}{x - \tan x}$ f) $\lim_{x\to 0} \frac{\sin x^2}{x \sin x}$

g) $\lim_{x\to 0} \frac{\sec x - 1}{x^2}$

h) $\lim_{x\to 2} \frac{x^3 - 8}{x^2 - 4}$

91. L'Hôpital's rule does not help with the following limits. Find them some other way.

a) $\lim_{x\to\infty} \frac{\sqrt{x+5}}{\sqrt{x}+5}$ b) $\lim_{x\to\infty} \frac{2x}{x + 7\sqrt{x}}$

92. If $f'(x) \leq 2$ for all x, what is the most f can increase on the interval [0, 6]?

93. Suppose that f is continuous and differentiable on $[a, b]$. Show that if $f'(x) \leq 0$ on $[a, c)$ and $f'(x) \geq 0$ on $(c, b]$, $a < c < b$, then $f(x)$ is never less than $f(c)$ on $[a, b]$.

94. a) Show that

$$-\frac{1}{2} \leq \frac{x}{1+x^2} \leq \frac{1}{2}$$

for any value of x.

b) Let f be a function whose derivative f' is defined by

$$f'(x) = \frac{x}{1+x^2}.$$

Use the result in (a) to show that

$$|f(b) - f(a)| \leq \frac{1}{2}|b - a|$$

for any a and b with $a \neq b$.

95. In Article 3.4, Theorem 2 gives a second derivative test for relative maxima and minima. The test can be strengthened by replacing the phrase "in an open interval containing c" by the phrase "at c" in both parts of the theorem. For example, if $f'(c) = 0$ and $f''(c) < 0$, then f has a relative maximum at c. Prove that this is so. [Suggestion: Let $\varepsilon = (1/2)|f''(c)|$ and use the fact that

$$f''(c) = \lim_{h\to 0} \frac{f'(c+h) - f'(c)}{h} = \lim_{h\to 0} \frac{f'(c+h)}{h}$$

to conclude that, for some $\delta > 0$,

$$0 < |h| < \delta \Rightarrow \frac{f'(c+h)}{h} < f''(c) + \varepsilon < 0.$$

Thus $f'(c+h)$ is positive for $-\delta < h < 0$ and is negative for $0 < h < \delta$.]

4 Integration

4.1 Introduction

In the preceding chapters we pursued one of the two main branches of the calculus, namely, *differential calculus*. We shall now turn our attention to the other main branch of the subject, *integral calculus*. Today, "to integrate" has two meanings in calculus. The deeper and more fundamental meaning is nearly the same as the nontechnical definition: "to indicate the whole of; to give the sum or total of" (Webster). The mathematical meaning of the word in this sense will be amply illustrated in finding areas bounded by curves, volumes of various solids, lengths of curves, centers of gravity, and other applications.

The second mathematical meaning of the verb "integrate" is "to find a function whose derivative is given." This is the aspect of integration that we shall discuss in the next three articles.

As we shall see, the two kinds of integration are intimately connected.

4.2 Indefinite Integrals

The acceleration due to gravity near the surface of the earth is 9.8 m/s². This means that the velocity v of a body falling freely in a vacuum near the earth's surface changes at the rate of

$$\frac{dv}{dt} = 9.8 \text{ m/s}^2. \tag{1}$$

If the body is dropped from rest, what will its velocity be t seconds after it is released?

If we are to determine v it must be from the only two facts we know about v as a function of t, which are

1. $\dfrac{dv}{dt} = 9.8$ (the acceleration is 9.8 m/s²);
2. $v(0) = 0$ (the initial velocity of the body is zero).

We start with the equation for the acceleration and ask, "What functions of t have derivatives exactly equal to 9.8?" Our experience tells us that one answer is

$$v = 9.8t,$$

but there are other answers as well, since

$$v = 9.8t + 5, \qquad v = 9.8t - \sqrt{2}, \qquad v = 9.8t + 5\pi$$

are also functions whose derivatives are 9.8. Indeed,

$$v = 9.8t + C \tag{2}$$

is an answer for any value of the constant C.

Does Eq. (2) account for all functions whose derivatives are 9.8? The answer is yes, because Corollary 3 of the Mean Value Theorem says that any function whose derivative is 9.8 can differ from the function $9.8t$ only by a constant.

Having established that the velocity of the falling body is

$$v(t) = 9.8t + C \tag{3}$$

for some value of C, we can substitute $t = 0$ into Eq. (3) and use the fact that $v(0) = 0$ to find that

$$0 = 9.8(0) + C,$$

or

$$C = 0.$$

The velocity of the falling body t seconds after release is therefore

$$v(t) = 9.8t \text{ m/s}. \tag{4}$$

Differential Equations

From a mathematical point of view, the problem of determining the velocity of the falling body from its acceleration is a special case of finding a function $y = F(x)$ whose derivative is given by an equation

$$\frac{dy}{dx} = f(x) \tag{5}$$

over some interval of x-values (a, b). An equation like (5) that has a derivative in it is called a *differential equation*. Equation (5) gives the derivative dy/dx as a function of x, but a differential equation may also express the derivative of an unknown function $y = F(x)$ as a function of y as well as x, as in the equation

$$\frac{dy}{dx} = 2xy^2.$$

A differential equation like

$$\frac{d^2y}{dx^2} + 6xy\frac{dy}{dx} + 3x^2y^3 = 0$$

may also involve higher order derivatives. For the time being, however, we shall restrict our attention to equations that contain a single derivative. Differential equations of a more general nature will be considered in Chapter 18.

A function $y = F(x)$ is called a *solution* of the differential equation (5) if F is differentiable throughout the domain (a, b) and

$$\frac{d}{dx}F(x) = f(x).$$

We also say that $F(x)$ is an *antiderivative* or a *primitive* of $f(x)$. To *solve* Eq. (5) means to find all the functions on the interval (a, b) that are antiderivatives of f. In the falling body example, we solved the differential equation $dv/dt = 9.8$ on the interval $t \geq 0$ by finding the *general solution* $v(t) = 9.8t + C$. We then solved the falling body problem by finding the *particular solution* $v = 9.8t$, obtained from the general solution by setting C equal to 0.

If $F(x)$ is an antiderivative of $f(x)$, then $F(x) + C$ is also an antiderivative when C is any constant whatever. For, if $F(x)$ satisfies Eq. (5), then

$$\frac{d}{dx}(F(x) + C) = \frac{dF}{dx} + \frac{dC}{dx} = f(x) + 0 = f(x). \tag{6}$$

What may not be clear, however, is whether there are any other antiderivatives of $f(x)$ besides those given by the formula $F(x) + C$. Again, this question is answered by the third corollary of the Mean Value Theorem in Article 3.8, which says that any function y whose derivative is $f(x)$ can differ from F only by a constant. The idea behind the corollary is that if a function's derivative is zero, then the function is equal to some constant C. Hence, if y and $F(x)$ have the same derivative, then

$$\frac{d}{dx}(y - F(x)) = \frac{dy}{dx} - \frac{dF}{dx} = f(x) - f(x) = 0,$$

so that $(y - F(x)) = C$ and $y = F(x) + C$. Therefore, if $F(x)$ is any solution whatever of the equation $dy/dx = f(x)$, then *all* solutions of the equation are given by the formula

$$y = F(x) + C.$$

Indefinite Integrals

The set of all antiderivatives of $f(x)$ is called the *indefinite integral* of f with respect to x and is denoted by the symbol

$$\int f(x)\,dx.$$

The fact that the formula $F(x) + C$ gives all the antiderivatives of f is indicated by writing

$$\int f(x)\,dx = F(x) + C. \tag{7}$$

The symbol $\int$ is called an "integral sign" and Eq. (7) is read "The integral of $f(x)$ with respect to x is $F(x)$ plus C." The function f is called the *integrand* of the integral, and C is called the *constant of integration*. The dx tells us that the *variable of integration* is x.

In the falling body example, we found that the integral of 9.8 with respect to t is $9.8t + C$:

$$\int 9.8\,dt = 9.8t + C.$$

Here the integrand is the constant function 9.8, and the variable of integration is *t*.

EXAMPLE 1 Solve the differential equation

$$\frac{dy}{dx} = 3x^2.$$

Solution We know from experience that

$$\frac{d}{dx}(x^3) = 3x^2.$$

Therefore,

$$y = \int 3x^2\,dx = x^3 + C. \square$$

Separation of Variables

If both x and y occur in the expression for dy/dx in a differential equation, we have an equation that does not give one variable explicitly in terms of the other. Solving the differential equation

$$\frac{dy}{dx} = x^2\sqrt{y} \tag{8}$$

means finding a function $y = f(x)$ whose derivative dy/dx at any particular x equals x^2 times the value of $\sqrt{y}$ at x. While equations like (8) usually do have solutions, as we shall see in Chapter 18, the solutions are not always easy to find. A method that sometimes works involves combining all the y terms with dy/dx on one side of the equation, putting all the x terms on the other side. This is called *separating the variables*. If we can separate the variables, we may then be able to integrate as in the following example.

EXAMPLE 2 Solve the differential equation

$$\frac{dy}{dx} = x^2\sqrt{y}, \quad y > 0. \tag{8}$$

Solution We divide both sides of the equation by $\sqrt{y}$ to separate the variables and obtain

$$y^{-1/2}\frac{dy}{dx} = x^2. \tag{9}$$

Under the assumption that the original equation defines y as a differentiable function of x we see that the left side of Eq. (9) is the derivative of the function $2y^{1/2}$ with respect to x. The right side is the derivative of $x^3/3$ with respect to x. Therefore, when we integrate both sides of (9) with respect to x we get

$$\int\left(y^{-1/2}\frac{dy}{dx}\right)dx = \int x^2\,dx,$$

$$2y^{1/2} + C_1 = \frac{x^3}{3} + C_2,$$

$$2y^{1/2} = \frac{x^3}{3} + C,$$

where we have combined C_1 and C_2 into a single constant $C = C_2 - C_1$. There is no need to describe the family of solutions with two constants when one will do, nor is any greater generality achieved by doing so. □

Integration Formulas

Integration, as described above, requires the ability to guess the answer. But the following formulas help to reduce the amount of guesswork in many cases. In these formulas, u and v denote differentiable functions of some independent variable (say x), and a, n, and C are constants.

Integration Formulas

INTEGRAL	MATCHING DERIVATIVE
a) $\int \frac{du}{dx}\,dx = u(x) + C$	$\frac{d}{dx}[u(x) + C] = \frac{du}{dx}$
b) $\int au(x)\,dx = a\int u(x)\,dx$ (if a is a constant)	$\frac{d}{dx}(au) = a\frac{du}{dx}$
c) $\int [u(x) + v(x)]\,dx = \int u(x)\,dx + \int v(x)\,dx$	$\frac{d}{dx}(u + v) = \frac{du}{dx} + \frac{dv}{dx}$
d) $\int u^n \frac{du}{dx}\,dx = \frac{u^{n+1}}{n+1} + C$ $(n \neq -1)$	$\frac{d}{dx}\left(\frac{u^{n+1}}{n+1}\right) = u^n \frac{du}{dx}$

In words, these formulas say

a) The integral of the derivative of a differentiable function u is u plus an arbitrary constant.

b) A constant may be moved across the integral sign. (*Caution:* Functions of the variable of integration must not be moved across the integral sign.)

c) The integral of the sum of two functions is the sum of their integrals. This extends to the sum of any finite number of functions:

$$\int [u_1 + u_2 + \cdots + u_n]\,dx = \int u_1\,dx + \int u_2\,dx + \cdots + \int u_n\,dx.$$

d) If $n \neq -1$, the integral of $u^n\,du/dx$ is obtained by adding 1 to the exponent, dividing by the new exponent, and adding an arbitrary constant.

Formula (a) is a restatement of the definition of the indefinite integral as the set of all functions with a given derivative. It says that any function whose derivative is du/dx must be given by the formula $u(x) + C$ for some value of C. The remaining formulas come from "reversing" derivative formulas from Chapter 2. We shall not derive these integration formulas but instead shall show by example how they are used to evaluate indefinite integrals.

EXAMPLE 3

$$\int (5x - x^2 + 2)\,dx = 5\int x\,dx - \int x^2\,dx + \int 2\,dx$$
$$= \frac{5}{2}x^2 - \frac{x^3}{3} + 2x + C.$$

$$\int x^{1/2}\,dx = \frac{x^{3/2}}{3/2} + C = \frac{2}{3}x^{3/2} + C.$$

$$\int (x^2 + 5)^2\,dx = \int (x^4 + 10x^2 + 25)\,dx = \frac{x^5}{5} + \frac{10}{3}x^3 + 25x + C.$$

$$\int (x^2 + 5)^2\,2x\,dx = \frac{(x^2 + 5)^3}{3} + C.$$

In the last equation we applied the formula

$$\int u^n \frac{du}{dx}\,dx = \frac{u^{n+1}}{n+1} + C, \quad (n \neq -1)$$

with $u = x^2 + 5$, $n = 2$, and $du/dx = 2x$. □

The Guess and Check Strategy

The examples we have seen so far succeeded because our experience with derivatives allowed us to guess the answers. But what if we don't know what to guess?

There is a procedure to try if we don't see how to integrate a given function right away. The procedure assumes that we have enough experience to make a reasonable guess at the answer, but it does not require us to guess right the first time. The steps are

1. Write down a guess.
2. Compare its derivative with the integrand.
3. Modify the guess accordingly.
4. Check the result and make further improvements as necessary.
5. Add C.

We use this procedure in the next example.

EXAMPLE 4 Evaluate the integral

$$\int \sqrt{2x + 1}\,dx.$$

Solution We seek a function whose derivative is $(2x + 1)^{1/2}$. We add 1 to the exponent and guess

$$(2x + 1)^{3/2},$$

whose derivative is

$$\frac{3}{2}(2x + 1)^{1/2} \cdot 2 = 3(2x + 1)^{1/2}.$$

This differs from the integrand $(2x + 1)^{1/2}$ by a factor of 3. In other words, our trial function is 3 times too large. We divide it by 3 to obtain the next candidate,

$$\frac{1}{3}(2x + 1)^{3/2}.$$

The derivative of this new function is

$$\frac{3}{2}\cdot\frac{1}{3}(2x+1)^{1/2}\cdot 2 = (2x+1)^{1/2},$$

the function whose integral we were asked to find. We conclude that

$$\int \sqrt{2x+1}\,dx = \frac{1}{3}(2x+1)^{3/2} + C. \ \square$$

PROBLEMS

Solve the differential equations in Problems 1–26. Remember to separate the variables when necessary.

1. $\dfrac{dy}{dx} = 2x - 7$

2. $\dfrac{dy}{dx} = 7 - 2x$

3. $\dfrac{dy}{dx} = x^2 + 1$

4. $\dfrac{dy}{dx} = \dfrac{1}{x^2}, \quad x > 0$

5. $\dfrac{dy}{dx} = \dfrac{-5}{x^2}, \quad x > 0$

6. $\dfrac{dy}{dx} = 3x^2 - 2x + 5$

7. $\dfrac{dy}{dx} = (x-2)^4$

8. $\dfrac{dy}{dx} = (5x-2)^4$

9. $\dfrac{dy}{dx} = \dfrac{1}{x^2} + x, \quad x > 0$

10. $\dfrac{dy}{dx} = x + \sqrt{2x}$

11. $\dfrac{dy}{dx} = \dfrac{x}{y}, \quad y > 0$

12. $\dfrac{dy}{dx} = \sqrt{\dfrac{x}{y}}, \quad y > 0$

13. $\dfrac{dy}{dx} = \dfrac{x+1}{y-1}, \quad y > 1$

14. $\dfrac{dy}{dx} = \dfrac{\sqrt{x+1}}{\sqrt{y-1}}, \quad y > 1$

15. $\dfrac{dy}{dx} = \sqrt{xy}, \quad x > 0, y > 0$

16. $\dfrac{dy}{dx} = x\sqrt{x^2+1}$

17. $\dfrac{dy}{dx} = 2xy^2, \quad y > 0$

18. $\dfrac{dy}{dx} = \sqrt[3]{y/x}, \quad x > 0, \quad y > 0$

19. $x^3\dfrac{dy}{dx} = -2, \quad x > 0$

20. $x^2\dfrac{dy}{dx} = \dfrac{1}{y^2 + \sqrt{y}}, \quad x > 0, \quad y > 0$

21. $\dfrac{ds}{dt} = 3t^2 + 4t - 6$

22. $\dfrac{dr}{dz} = (2z+1)^3$

23. $\dfrac{du}{dv} = 2u^2(4v^3 + 4v^{-3}), \quad v > 0, \quad u > 0$

24. $\dfrac{dx}{dt} = 8\sqrt{x}, \quad x > 0$

25. $\dfrac{dy}{dt} = (2t + t^{-1})^2, \quad t > 0$

26. $\dfrac{dy}{dz} = \sqrt{(z^2 - z^{-2})^2 + 4}, \quad z > 0$

Evaluate the integrals in Problems 27–46.

27. $\int (2x+3)\,dx$

28. $\int (x^2 - \sqrt{x})\,dx$

29. $\int (x-1)^{243}\,dx$

30. $\int (2-7t)^{2/3}\,dt$

31. $\int \sqrt{1-x}\,dx$

32. $\int \dfrac{1}{\sqrt{1-x}}\,dx$

33. $\int (2-t)^{2/3}$

34. $\int (3x-1)^{234}\,dx$

35. $\int (1+x^3)^2\,dx$

36. $\int (1+x^3)^2\,3x^2\,dx$

37. $\int 3x^2\sqrt{1+x^3}\,dx$

38. $\int \dfrac{3x^2}{\sqrt{1+x^3}}\,dx$

39. $\int \sqrt{2+5y}\,dy$

40. $\int \dfrac{dx}{(3x+2)^2}$

41. $\int \dfrac{3r\,dr}{\sqrt{1-r^2}}$

42. $\int x\sqrt{2x^2+1}\,dx$

43. $\int t^2(1+2t^3)^{-(2/3)}\,dt$

44. $\int \dfrac{y\,dy}{\sqrt{2y^2+1}}$

45. $\int \left(\sqrt{x} + \dfrac{1}{\sqrt{x}}\right)dx$

46. $\int \dfrac{(z+1)\,dz}{\sqrt[3]{z^2+2z+2}}$

47. The acceleration due to gravity on the moon is 1.6 m/s^2. If a rock is dropped into a crevasse, how fast will it be going just before it hits the bottom 30 seconds later?

48. A rocket lifts off the surface of the earth with a constant acceleration of 20 m/s^2. How fast will the rocket be going one minute later?

4.3

Applications. Determining Constants of Integration

Differential equations like Eqs. (1) and (9) of Article 4.2 arise in chemistry, physics, mathematics, and all branches of engineering. Some of these applications will be illustrated in the examples that follow. Before proceeding with these, however, let us consider the meaning of the arbitrary constant C, which always enters when we integrate a differential equation $dy/dx = f(x)$ to find the solution $y = F(x) + C$.

If we graph the particular solution $y = F(x)$ (taking $C = 0$), then any other solution curve $y = F(x) + C$ is obtained by shifting this curve through a vertical displacement C. Thus we obtain, as in Fig. 4.1, a family of "parallel" curves. They are parallel in the sense that the slope of the tangent to any one of them at the point of abscissa x is $f(x)$.

This family of parallel curves has the property that, given any point $P(x_0, y_0)$ with x_0 in the domain of F, there is one and only one curve of the family that passes through this particular point. For the curve to pass through P_0, the equation must be satisfied by its coordinates. This uniquely specifies the value of C to be

$$C = y_0 - F(x_0).$$

With C thus determined, we get a definite function expressing y in terms of x.

The condition imposed that $y = y_0$ when $x = x_0$ is called an *initial condition*. The name comes from problems where time is the independent variable and initial velocities or initial positions of moving bodies are specified.

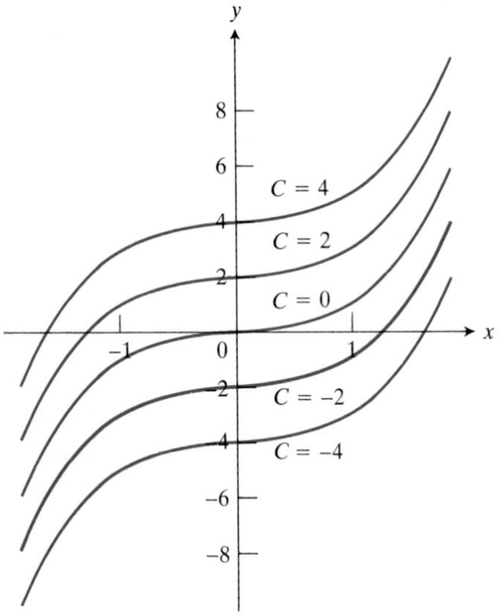

4.1 Selected curves of the family $y = x^3 + C$.

EXAMPLE 1 The velocity, at time t, of a moving body is given by

$$\frac{ds}{dt} = at,$$

where a is a constant and s is the body's position at time t. If $s = s_0$ when $t = 0$, find s as a function of t.

Solution We want to solve the problem that consists of

$$\text{The differential equation:} \quad \frac{ds}{dt} = at \tag{1}$$

and

$$\text{The initial condition:} \quad s = s_0 \quad \text{when } t = 0. \tag{2}$$

We integrate both sides of the differential equation with respect to t:

$$s = \int at\,dt$$

$$= a\frac{t^2}{2} + C.$$

The constant of integration may now be determined from the initial condition, which says that

$$s_0 = a\frac{(0)^2}{2} + C, \qquad \text{or} \qquad C = s_0.$$

Hence,

$$s = a\frac{t^2}{2} + s_0. \ \square$$

EXAMPLE 2 See Fig. 4.1. Find the curve whose slope at the point (x, y) is $3x^2$ if the curve is also required to pass through the point $(1, -1)$.

Solution In mathematical language, we have the following problem:

$$\text{Differential equation:} \qquad \frac{dy}{dx} = 3x^2,$$

$$\text{Initial condition:} \qquad y = -1 \quad \text{when } x = 1.$$

First, we integrate both sides of the differential equation with respect to x:

$$y = \int \frac{dy}{dx}\,dx = \int 3x^2\,dx = x^3 + C.$$

Then we impose the initial condition to evaluate C:

$$-1 = 1^3 + C; \qquad C = -2.$$

We substitute the value of C into the solution of the differential equation, and obtain the particular curve that passes through the given point, namely,

$$y = x^3 - 2. \ \square$$

EXAMPLE 3 A projectile is fired straight up from a platform 10 ft above the ground, with initial velocity of 160 ft/s. The only force affecting the motion of the projectile during its flight is from gravity, which produces a downward acceleration of 32 ft/s². Find an equation for the height of the projectile above the ground as a function of time t if $t = 0$ when the projectile is fired.

Solution If s denotes the height of the projectile above the ground, then the velocity and acceleration of the projectile are

$$v = \frac{ds}{dt} \qquad \text{and} \qquad a = \frac{dv}{dt} = \frac{d^2s}{dt^2}.$$

Since the projectile is fired upward, against the force of gravity, the velocity will be a decreasing function of t. Therefore, the equation to solve is

$$\frac{dv}{dt} = -32 \text{ ft/s}^2,$$

subject to the initial conditions

$$v(0) = 160 \text{ ft/s} \qquad \text{and} \qquad s(0) = 10 \text{ ft}.$$

We integrate once to find

$$v = \frac{ds}{dt} = -32t + v_0,$$

and a second time to find

$$s = -16t^2 + v_0 t + s_0.$$

We determine the values of the constants v_0 and s_0 from the initial conditions:

$$v_0 = v(0) = 160, \qquad s_0 = s(0) = 10.$$

The equation of motion is

$$s = -16t^2 + 160t + 10. \ \square$$

PROBLEMS

In Problems 1–6, find the position s as a function of t from the given velocity $v = ds/dt$. Evaluate the constant of integration so as to have $s = s_0$ when $t = 0$.

1. $v = 3t^2$
2. $v = 2t + 1$
3. $v = (t + 1)^2$
4. $v = (t^2 + 1)^2$
5. $v = (t + 1)^{-2}$
6. $v = \sqrt{2gs}$ (g = constant)

In Problems 7–11, find the velocity v and position s as functions of t from the given acceleration $a = dv/dt$. Evaluate the constants of integration so as to have $v = v_0$ and $s = s_0$ when $t = 0$.

7. $a = g$ (constant)
8. $a = t$
9. $a = \sqrt[3]{2t + 1}$
10. $a = (2t + 1)^{-3}$
11. $a = (t^2 + 1)^2$

In Problems 12–26, solve the differential equations subject to the prescribed initial conditions.

12. $\frac{dy}{dx} = 9x^2 - 4x + 5, \quad x = -1, \ y = 0$
13. $\frac{dy}{dx} = 4(x - 7)^3, \quad x = 8, \ y = 10$
14. $\frac{dy}{dx} = x^{1/2} + x^{1/4}, \quad x = 0, \ y = -2$
15. $\frac{dy}{dx} = \frac{x^2 + 1}{x^2}, \quad x = 1, \ y = 1$
16. $\frac{dy}{dx} = x\sqrt{y}, \quad x = 0, \ y = 1$
17. $\frac{dy}{dx} = 2xy^2, \quad x = 1, \ y = 1$
18. $\frac{dy}{dx} = x^3y^2, \quad x = 0, \ y = 47$
19. $\frac{dy}{dx} = (x + x^{-1})^2, \quad x = 1, \ y = 1$
20. $\frac{dy}{dx} = \frac{1}{x^2\sqrt{y + 1}}, \quad x = 1, \ y = -1$
21. $\frac{dy}{dx} = x\sqrt{1 + x^2}, \quad x = 0, \ y = -3$
22. $\frac{dy}{dx} = \frac{4\sqrt{(1 + y^2)^3}}{y}, \quad x = 0, \ y = 1$
23. $\frac{d^2y}{dx^2} = 2 - 6x, \quad x = 0, \ y = 1, \ \frac{dy}{dx} = 4$
24. $\frac{d^3y}{dx^3} = 6, \quad x = 0, \ y = 5, \ \frac{dy}{dx} = 0, \ \frac{d^2y}{dx^2} = -8$
25. $\frac{d^2y}{dx^2} = \frac{3x}{8}$, the graph of y passes through the point (4, 4) with slope 3.
26. $\frac{d^2y}{dx^2} = 2x - 3x^2 + 1$, the graph of y passes through the origin and the point (1, 1).
27. The graph of $y = f(x)$ passes through the point (9, 4). Also, the line tangent to the graph at any point (x, y) has the slope $3\sqrt{x}$. Find f.
28. If $y = 3$ when $x = 3$, and $dy/dx = 2x/y^2$, find the value of y when $x = 1$.
29. With approximately what velocity does a diver enter the water after diving from a 30-m platform? (Use $g = 9.8 \text{ m/s}^2$.)
30. The acceleration due to gravity on Mars is 3.72 m/s^2. A rock is thrown straight up from the surface with an initial velocity of 23 m/s. How high does it go? (*Hint:* When is the velocity zero?)
31. The gravitational attraction exerted by the earth on a particle of mass m at distance s from the center is given by $F = -mgR^2s^{-2}$, where R is the radius of the

earth. F is negative because the force acts in opposition to increasing s (Fig. 4.2). If a particle is projected vertically upward from the surface of the earth with initial velocity $v_0 = \sqrt{2gR}$, apply Newton's second law $F = ma$ with $a = dv/dt = (dv/ds)(ds/dt) = v(dv/ds)$ to show that $v = v_0\sqrt{R/s}$ and that $s^{3/2} = R^{3/2}[1 + (3v_0t/2R)]$.

The initial velocity $v_0 = \sqrt{2gR}$ (11.2 kilometers per second) is known as the "escape velocity" since the displacement s tends to infinity with increasing t if the initial velocity is this large or greater. Actually, a somewhat larger initial velocity is required to escape the earth's gravitational attraction, due to the retardation effect of air resistance, which we neglected here for the sake of simplicity.

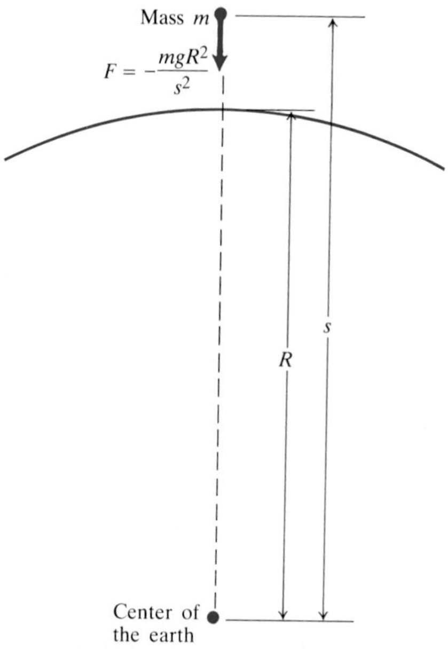

4.2 A mass m that is s km from the earth's center.

Toolkit programs

Antiderivatives and Direction Fields

4.4 Integrals of Trigonometric Functions

From the derivative formulas

$$\cos x = \frac{d}{dx}(\sin x), \qquad \sin x = \frac{d}{dx}(-\cos x)$$

we get the integration formulas

$$\int \cos x\, dx = \sin x + C, \tag{1a}$$

$$\int \sin x\, dx = -\cos x + C. \tag{1b}$$

If u is a differentiable function of x, the Chain Rule applied to $\sin u$ gives

$$\frac{d}{dx}\sin u = \cos u \frac{du}{dx}.$$

When we turn this equation around, to get

$$\cos u \frac{du}{dx} = \frac{d}{dx}\sin u,$$

we see that

$$\int \cos u \frac{du}{dx}\, dx = \sin u + C. \tag{2a}$$

Similarly,

$$\frac{d}{dx} \cos u = -\sin u \frac{du}{dx},$$

$$\sin u \frac{du}{dx} = -\frac{d}{dx}(\cos u) = \frac{d}{dx}(-\cos u),$$

and

$$\int \sin u \frac{du}{dx}\, dx = -\cos u + C. \tag{2b}$$

The formulas in Eqs. (1) and (2) enable us to evaluate a variety of trigonometric integrals.

EXAMPLE 1 Evaluate the integral

$$\int \cos 2x\, dx.$$

Solution METHOD 1. We are looking for a function whose derivative is cos 2x, so we try sin 2x. We find

$$\frac{d}{dx} \sin 2x = 2 \cos 2x,$$

so that

$$\frac{d}{dx}\left(\frac{1}{2} \sin 2x\right) = \cos 2x.$$

Therefore,

$$\int \cos 2x\, dx = \frac{1}{2} \sin 2x + C.$$

METHOD 2. We see the similarity between

$$\int \cos 2x\, dx \tag{3}$$

and

$$\int \cos u \frac{du}{dx}\, dx \quad \text{with } u = 2x,$$

which equals

$$\int \cos 2x \cdot 2\, dx. \tag{4}$$

The only difference between (3) and (4) is that the integrand in (3) lacks the factor $du/dx = 2$ that would make the match complete. However, we can supply the missing factor by moving a constant across the integral sign! We write

$$\begin{aligned}
\int \cos 2x\, dx &= \frac{1}{2} \cdot 2 \int \cos 2x\, dx \\
&= \frac{1}{2} \int \cos 2x \cdot 2\, dx \\
&= \frac{1}{2} \int \cos u \frac{du}{dx}\, dx && (u = 2x) \\
&= \frac{1}{2} \sin u + C && \text{(from Eq. 2a)} \\
&= \frac{1}{2} \sin 2x + C, && (\text{because } u = 2x)
\end{aligned}$$

in agreement with our earlier calculation.

The use of substitutions like $u = 2x$ to change integrals to standard forms is a powerful tool and will be treated formally in Article 4.8 after we have worked more with integrals. □

EXAMPLE 2 Evaluate the integral

$$\int \sin (7x + 5)\, dx.$$

Solution METHOD 1. We are looking for a function whose derivative is $\sin (7x + 5)$. We try

$$\cos (7x + 5),$$

whose derivative is

$$-\sin (7x + 5) \cdot 7.$$

This differs from the integrand by a factor of -7. We therefore divide the trial function by -7, obtaining

$$-\frac{1}{7} \cos (7x + 5).$$

The derivative of this new function is

$$-\frac{1}{7} \cdot -\sin (7x + 5) \cdot 7 = \sin (7x + 5),$$

which is what we want. We conclude that

$$\int \sin (7x + 5)\, dx = -\frac{1}{7} \cos (7x + 5) + C.$$

METHOD 2. We notice the similarity between

$$\int \sin (7x + 5)\, dx \tag{5}$$

and

$$\int \sin u \frac{du}{dx}\, dx \quad \text{with } u = (7x + 5),$$

which equals

$$\int \sin (7x + 5) \cdot 7\, dx. \tag{6}$$

The only difference between (5) and (6) is that the integrand in (5) lacks the factor $du/dx = 7$. We can supply the missing factor, as we did in Example 1, by moving a constant across the integral sign. We write

$$\begin{aligned}\int \sin(7x+5)\,dx &= \frac{1}{7}\cdot 7\int \sin(7x+5)\,dx\\ &= \frac{1}{7}\int \sin(7x+5)\cdot 7\,dx\\ &= \frac{1}{7}\int \sin u\,\frac{du}{dx}\,dx \quad (u = 7x+5)\\ &= \frac{1}{7}(-\cos u) + C\\ &= -\frac{1}{7}\cos(7x+5) + C. \ \square\end{aligned}$$

EXAMPLE 3 Evaluate the integral

$$\int \frac{\cos 2x}{\sin^3 2x}\,dx.$$

Solution METHOD 1. Since

$$\frac{d}{dx}\sin 2x = 2\cos 2x,$$

we see that the numerator of the integrand is

$$\cos 2x = \frac{1}{2}\frac{d}{dx}(\sin 2x).$$

Thus,

$$\begin{aligned}\int \frac{\cos 2x}{\sin^3 2x}\,dx &= \int (\sin 2x)^{-3}\cdot\frac{1}{2}\cdot\frac{d}{dx}(\sin 2x)\,dx\\ &= \frac{1}{2}\int (\sin 2x)^{-3}\cdot\frac{d}{dx}(\sin 2x)\,dx.\end{aligned}$$

The integral on the right now has the form

$$\frac{1}{2}\int u^{-3}\frac{du}{dx}\,dx, \quad u = \sin 2x.$$

Since

$$\frac{1}{2}\int u^{-3}\frac{du}{dx}\,dx = \frac{1}{2}\frac{u^{-2}}{-2} + C = \frac{-1}{4u^2} + C$$

(from formula (d) of Article 4.2 with $n = 3$), our original integral is

$$\int \frac{\cos 2x}{\sin^3 2x}\,dx = \frac{1}{2}\int (\sin 2x)^{-3}\cdot\frac{d}{dx}(\sin 2x)\,dx = \frac{-1}{4\sin^2 2x} + C.$$

METHOD 2. We are looking for a function whose derivative is

$$\frac{\cos 2x}{\sin^3 2x}.$$

We see a similarity between

$$\frac{1}{\sin^3 2x}\cdot\cos 2x \quad \text{and} \quad \frac{1}{u^3}\cdot\frac{du}{dx}.$$

We know that the integral of the second of these is a multiple of $1/u^2$. We therefore guess that the integral of the first is something like

$$\frac{1}{\sin^2 2x}. \tag{2}$$

We compare the derivative

$$\frac{d}{dx}\left(\frac{1}{\sin^2 2x}\right) = \frac{-2}{\sin^3 2x} \cdot \cos 2x \cdot 2 = -4\frac{\cos 2x}{\sin^3 2x}$$

with the integrand and see that we are off by a factor of -4. We therefore divide (2) by -4 to get

$$\frac{-1}{4 \sin^2 2x},$$

whose derivative,

$$-\frac{1}{4} \cdot \frac{-2}{\sin^3 2x} \cdot \cos 2x \cdot 2 = \frac{\cos 2x}{\sin^3 2x},$$

is now the integrand. We conclude that

$$\int \frac{\cos 2x}{\sin^3 2x}\, dx = \frac{-1}{4 \sin^2 2x} + C. \ \square$$

Both methods used in Example 3 took advantage of the similarity between the integrand and the standard form whose integral is

$$\int \frac{1}{u^3} \frac{du}{dx}\, dx = \frac{u^{-2}}{-2} + C.$$

In Article 4.8 we shall develop a systematic way of converting integrals to standard forms that we can usually find in an integral table.

The formulas from Article 2.8 for the derivatives of the secant, cosecant, tangent, and cotangent of an angle in radian measure lead to the following integral formulas.

$$\int \sec^2 u \frac{du}{dx}\, dx = \tan u + C$$

$$\int \csc^2 u \frac{du}{dx}\, dx = -\cot u + C$$

$$\int \sec u \tan u \frac{du}{dx}\, dx = \sec u + C$$

$$\int \csc u \cot u \frac{du}{dx}\, dx = -\csc u + C$$

In each formula, u is assumed to be a differentiable function of x. Each formula can be checked by differentiating the right-hand side with respect to x. In each case, the Chain Rule applies to produce the integrand on the left.

EXAMPLE 4

a) $\int \frac{1}{\cos^2 2x}\, dx = \int \sec^2 2x\, dx = \frac{1}{2} \tan 2x + C$

b) $\int \frac{\sin x}{\cos^2 x}\, dx = \int \frac{1}{\cos x} \cdot \frac{\sin x}{\cos x}\, dx = \int \sec x \tan x\, dx = \sec x + C$

c) $\int \sec^2 x \tan x\, dx = \int \sec x(\sec x \tan x)\, dx$

$$= \int \sec x \frac{d}{dx}(\sec x)\, dx$$

$$= \frac{1}{2}\sec^2 x + C. \ \square$$

The integrals of $\sin^2 x$ and $\cos^2 x$ often arise in applications, and as our last example we show how they may be evaluated with the aid of the half-angle formulas from Article 2.7.

EXAMPLE 5 Evaluate

$$\int \cos^2 x\, dx \quad \text{and} \quad \int \sin^2 x\, dx.$$

Solution With the half-angle formula

$$\cos^2 x = \frac{1 + \cos 2x}{2},$$

we have

$$\int \cos^2 x\, dx = \int \frac{1 + \cos 2x}{2} = \int \frac{1}{2}\, dx + \int \frac{\cos 2x}{2}\, dx$$

$$= \frac{x}{2} + \frac{1}{2}\frac{\sin 2x}{2} + C = \frac{x}{2} + \frac{\sin 2x}{4} + C.$$

Similarly,

$$\int \sin^2 x\, dx = \int \frac{1 - \cos 2x}{2}\, dx = \int \frac{1}{2}\, dx - \int \frac{\cos 2x}{2}\, dx$$

$$= \frac{x}{2} - \frac{1}{2}\frac{\sin 2x}{2} + C = \frac{x}{2} - \frac{\sin 2x}{4} + C. \ \square$$

PROBLEMS

In Problems 1–36, evaluate the integrals.

1. $\int \sin 3x\, dx$
2. $\int \cos(2x + 4)\, dx$
3. $\int \sec^2(x + 2)\, dx$
4. $\int \sec 2x \tan 2x\, dx$
5. $\int \csc(x + \pi/2)\cot(x + \pi/2)\, dx$
6. $\int \csc^2(2x - 3)\, dx$
7. $\int x \sin(2x^2)\, dx$
8. $\int (\cos\sqrt{x})\frac{dx}{\sqrt{x}}$
9. $\int \sin 2t\, dt$
10. $\int \cos(3\theta - 1)\, d\theta$
11. $\int 4 \cos 3y\, dy$
12. $\int 2 \sin z \cos z\, dz$
13. $\int \sin^2 x \cos x\, dx$
14. $\int \cos^2 2y \sin 2y\, dy$
15. $\int \sec^2 2\theta\, d\theta$
16. $\int \sec^3 x \tan x\, dx$
17. $\int \sec\frac{x}{2}\tan\frac{x}{2}\, dx$
18. $\int \frac{d\theta}{\cos^2\theta}$
19. $\int \frac{d\theta}{\sin^2\theta}$
20. $\int \csc^2 5\theta \cot 5\theta\, d\theta$
21. $\int \cos^2 y\, dy$ (*Hint:* Use a half-angle formula.)
22. $\int \sin^2(x/2)\, dx$ (*Hint:* Use a half-angle formula.)
23. $\int (1 - \sin^2 3t)\cos 3t\, dt$
24. $\int \frac{\sin x\, dx}{\cos^2 x}$
25. $\int \frac{\cos x\, dx}{\sin^2 x}$
26. $\int \sqrt{2 + \sin 3t}\cos 3t\, dt$
27. $\int \frac{\sin 2t\, dt}{\sqrt{2 - \cos 2t}}$
28. $\int \sin^3\frac{y}{2}\cos\frac{y}{2}\, dy$

29. $\int \cos^2 \frac{2x}{3} \sin \frac{2x}{3}\, dx$

30. $\int \frac{\sec^2 u}{\tan^2 u}\, du$

31. $\int \sec \theta(\sec \theta + \tan \theta)\, d\theta$

32. $\int (1 + \tan^2 \theta)\, d\theta$

33. $\int (\sec^2 y + \csc^2 y)\, dy$

34. $\int (1 + \sin 2t)^{3/2} \cos 2t\, dt$

35. $\int (3 \sin 2x + 4 \cos 3x)\, dx$

36. $\int \sin t \cos t(\sin t + \cos t)\, dt$

37. Which of the following are antiderivatives of $f(x) = \sec x \tan x$?
a) $-\sec x + \pi/6$
b) $-\tan x + \pi/3$
c) $\sec x(\sec^2 x) + \tan x(\sec x \tan x)$
d) $\sec x - \pi/4$

38. Which of the following are antiderivatives of $f(x) = \csc^2 x$?
a) $-2 \csc x(\csc x \cot x) + C$
b) $-\cot x + \pi/6$
c) $\cot x - \pi/3$
d) $-(\cot x + C)$

In Problems 39–41, solve the differential equations subject to the given initial conditions.

39. $2y \frac{dy}{dx} = 5x - 3 \sin x, \quad x = 0,\ y = 0$

40. $\frac{dy}{dx} = \frac{\pi \cos \pi x}{\sqrt{y}}, \quad x = \frac{1}{2},\ y = 1$

41. $y^{(4)} = \cos x, \quad x = 0,\ y = 3,\ y' = 2,\ y'' = 1,\ y''' = 0$

42. The velocity of a particle moving back and forth on a line is $v = ds/dt = 6 \sin 2t$ m/s for all t. If $s = 0$ when $t = 0$, find the value of s when $t = \pi/2$ s.

43. The acceleration of a particle moving back and forth on a line is $a = d^2s/dt^2 = \pi^2 \cos \pi t$ m/s^2 for all t. If $s = 0$ and $v = 8$ m/s when $t = 0$, find s when $t = 1$ s.

44. We can treat the integral of $2 \sin x \cos x\, dx$ in three different ways:

i) $\int 2 \sin x \cos x\, dx = \int 2 \sin x \frac{d}{dx}(\sin x)\, dx$
$= \sin^2 x + C_1$

ii) $\int 2 \sin x \cos x\, dx = \int -2 \cos x \frac{d}{dx}(\cos x)\, dx$
$= -\cos^2 x + C_2$

iii) $\int 2 \sin x \cos x\, dx = \int \sin 2x\, dx$
$= -\frac{1}{2} \cos 2x + C_3$

Can all three integrations be correct? Explain.

***Toolkit* programs**	
Antiderivatives and Direction Fields	Function Evaluator

4.5

Definite Integrals. The Area under a Curve

The area of a circle is easily computed from the familiar formula $A = \pi r^2$. But the idea behind this simple formula isn't so simple. In fact, it is the subtle concept of a *limit*, the area of the circle being *defined* as the limit of areas of inscribed (or circumscribed) regular polygons as the number of sides *increases without bound*. A similar idea is involved in the definition we now introduce for other plane areas.

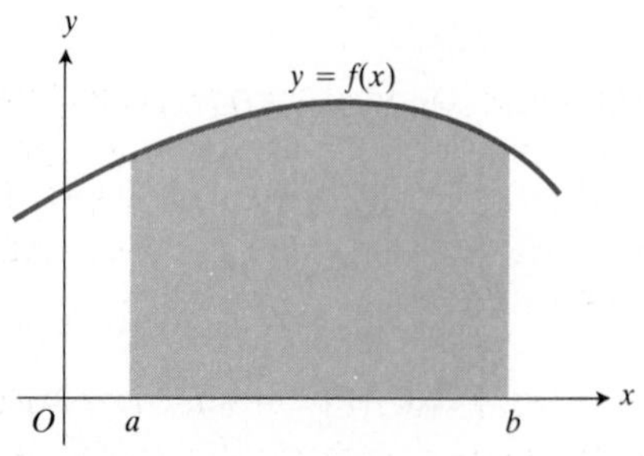

4.3 Area under a curve.

Approximating with Rectangle Areas

Let $y = f(x)$ define a continuous function of x on the closed interval $[a, b]$. For simplicity, we shall also suppose that $f(x)$ is nonnegative over $[a, b]$. We consider the problem of calculating the area bounded above by the graph of f, on the sides by the lines $x = a$ and $x = b$, and below by the x-axis (Fig. 4.3).

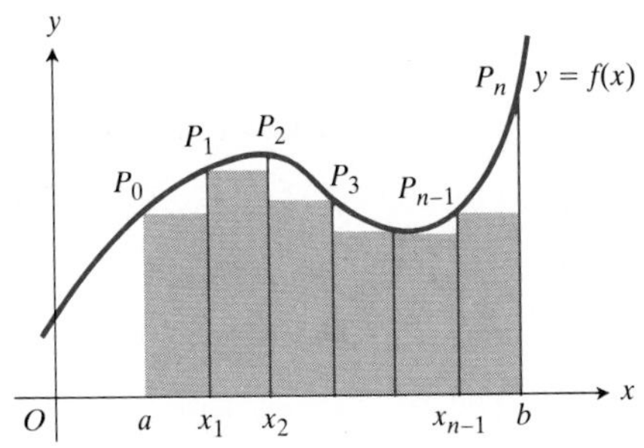

4.4 The area under the curve $y = f(x)$ from $x = a$ to $x = b$ is divided into strips. Each strip is then approximated by a rectangle.

We divide the area into n thin strips of uniform width $\Delta x = (b - a)/n$ by lines perpendicular to the x-axis through the endpoints $x = a$ and $x = b$ and many intermediate points, which we number as $x_1, x_2, \ldots, x_{n-1}$ (Fig. 4.4). We use an inscribed rectangle to approximate the area in each strip. For instance, in the figure, we approximate the area of the strip $aP_0P_1x_1$ by the shaded rectangle of altitude aP_0 and base ax_1. The area of this rectangle is

$$f(a) \cdot (x_1 - a) = f(a) \cdot \Delta x,$$

since the height aP_0 is the value of f at $x = a$, and the length of the base is $x_1 - a = \Delta x$. Similarly, the inscribed rectangle in the second strip has area

$$f(x_1) \cdot \Delta x.$$

Continuing in this fashion, we inscribe a rectangle in each strip.

Increasing Functions

If the function increases with x as in Fig. 4.5, then the altitude of each rectangle is the length of its left edge, and we have

$$\begin{aligned}
&\text{Area of first rectangle} &&= f(a) \cdot \Delta x,\\
&\text{Area of second rectangle} &&= f(x_1) \cdot \Delta x,\\
&\text{Area of third rectangle} &&= f(x_2) \cdot \Delta x,\\
&\qquad\vdots\\
&\text{Area of } n\text{th and last rectangle} &&= f(x_{n-1}) \cdot \Delta x.
\end{aligned}$$

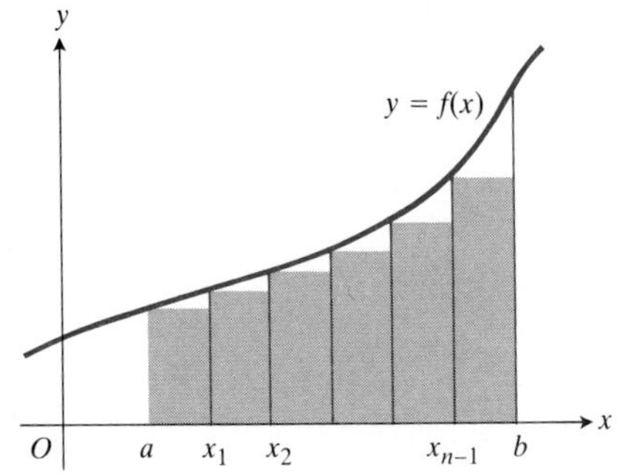

4.5 Rectangles under the graph of an increasing function.

EXAMPLE 1 Estimate the area under the curve $f(x) = 1 + x^2$ with $a = 0$, $b = 1$, and $n = 4$.

Solution See Fig. 4.6. There are $n - 1 = 3$ intermediate points $x_1 = \frac{1}{4}$, $x_2 = \frac{1}{2}$, and $x_3 = \frac{3}{4}$, which divide the interval $0 \le x \le 1$ into $n = 4$ subintervals, each of length $\Delta x = \frac{1}{4}$. The inscribed rectangles have the following areas.

$$\begin{aligned}
f(0) \cdot \Delta x &= 1 \cdot \tfrac{1}{4} = \tfrac{16}{64}\\
f(\tfrac{1}{4}) \cdot \Delta x &= \tfrac{17}{16} \cdot \tfrac{1}{4} = \tfrac{17}{64}\\
f(\tfrac{1}{2}) \cdot \Delta x &= \tfrac{5}{4} \cdot \tfrac{1}{4} = \tfrac{20}{64}\\
f(\tfrac{3}{4}) \cdot \Delta x &= \tfrac{25}{16} \cdot \tfrac{1}{4} = \tfrac{25}{64}\\
\text{Sum} &= \tfrac{78}{64} = 1.21875
\end{aligned}$$

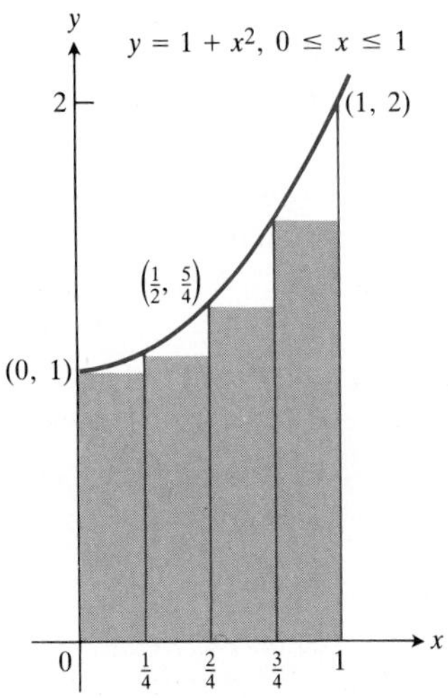

4.6 Rectangles for estimating the area under the graph of $y = 1 + x^2$, $0 \le x \le 1$.

This sum is the estimate of the area under the curve. Since the rectangles in Fig. 4.6 do not cover the entire region under the curve, we may expect 1.21875 to be an underestimate. In fact, by using methods we shall soon develop, we shall find that the area is exactly 4/3. Thus our estimate of 1.21875 is about 8 percent too small. □

One way to improve the accuracy of the estimate would be to use more rectangles. Another would be to use other geometric shapes such as trapezoids, as we shall see in Article 4.9.

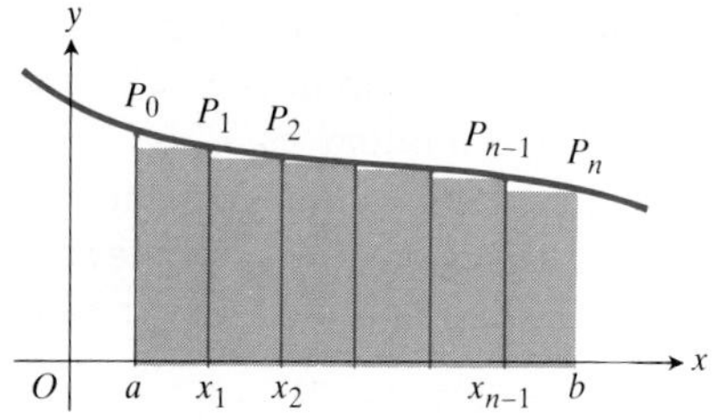

4.7 Rectangles under the graph of a decreasing function.

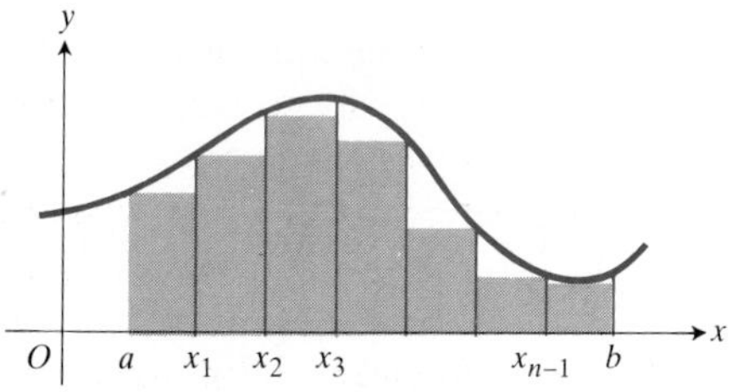

4.8 Rectangles under a curve that rises and falls between *a* and *b*.

Decreasing Functions

If the curve slopes downward as in Fig. 4.7, then the altitude of each inscribed rectangle is the length of its right edge, and we have areas as follows:

First rectangle	$f(x_1) \cdot \Delta x$,
Second rectangle	$f(x_2) \cdot \Delta x$,
Third rectangle	$f(x_3) \cdot \Delta x$,
⋮	
*n*th and last	$f(b) \cdot \Delta x$.

More generally, the curve may rise and fall between $x = a$ and $x = b$, as in Fig. 4.8. But there is a number c_1, between a and x_1 inclusive, such that the first inscribed rectangle has area $f(c_1) \cdot \Delta x$; and a number c_2 in the second closed subinterval such that the area of the second inscribed rectangle is $f(c_2) \cdot \Delta x$; and so on. The number c_1 is the place between a and x_1, inclusive, where f is minimized for the first subinterval. Similarly, the minimum value of f for x in the second subinterval is attained at c_2, and so on.

The sum of the areas of these inscribed rectangles is

$$S_n = f(c_1) \cdot \Delta x + f(c_2) \cdot \Delta x + \cdots + f(c_n) \cdot \Delta x. \tag{1}$$

We often abbreviate sums like these by using the capital Greek letter Σ (sigma) to denote the word "sum";

$$S_n = \sum_{k=1}^{n} f(c_k)\, \Delta x \tag{2}$$

("S_n is the sum from $k = 1$ to n of f of c_k delta x"). This is called writing the sum in sigma notation. Note that each term of the sum in (1) has the form $f(c_k) \cdot \Delta x$, with only the subscript on c changing from one term to another. We have indicated the subscript by k, but we could equally well have used i or j or any other symbol not currently in use for something else. In the first term in the sum on the right-hand side of Eq. (1) the subscript is $k = 1$; in the second, $k = 2$; and so on to the last, or nth, in which $k = n$. We indicate this by writing $k = 1$ below the Σ in (2), to say that the sum is to *start* with the term we get by replacing k by 1 in the expression that follows. The n above the sigma tells us where to *stop*. For instance, if $n = 4$, we have

$$\sum_{k=1}^{4} f(c_k)\, \Delta x = f(c_1)\, \Delta x + f(c_2)\, \Delta x + f(c_3)\, \Delta x + f(c_4)\, \Delta x.$$

The only thing that changes from one summand to the next is the number k. We replace k by 1, then by 2, then 3, then 4. Then we add.

EXAMPLE 2

a) $\sum_{k=1}^{5} k^2 = 1^2 + 2^2 + 3^2 + 4^2 + 5^2$

b) $\sum_{k=1}^{3} \frac{k}{k+1} = \frac{1}{1+1} + \frac{2}{2+1} + \frac{3}{3+1}$

c) $\sum_{j=0}^{2} \frac{j+1}{j+2} = \frac{0+1}{0+2} + \frac{1+1}{1+2} + \frac{2+1}{2+2}$

d) $\sum_{i=1}^{4} x_i = x_1 + x_2 + x_3 + x_4$

e) $\sum_{k=1}^{4} x^k = x + x^2 + x^3 + x^4$ □

The Area under a Graph

We turn our attention once more to the area under a continuous curve. We know general formulas for areas of regions enclosed by triangles, trapezoids, and circles, which are all shapes from classical Greek geometry. But there are no general formulas from precalculus mathematics for the more arbitrary shapes that arise as regions under the graphs of continuous nonnegative functions. We have reached the point therefore where we have to *define* these areas, and we have just discussed the mechanism for doing so. We shall define them as limits of the areas of inscribed rectangles. That these limits always exist is a consequence of a theorem we shall state later in this article.

DEFINITION

Area

The *area* under the graph of a nonnegative continuous function f over an interval $[a, b]$ is the limit of the sums of the areas of inscribed rectangles of equal base length as their number n increases without bound. In symbols,

$$A = \lim_{n\to\infty} [f(c_1)\,\Delta x + f(c_2)\,\Delta x + \cdots + f(c_n)\,\Delta x]$$

$$= \lim_{n\to\infty} \sum_{k=1}^{n} f(c_k)\,\Delta x, \tag{3}$$

where $f(c_k)$ is the smallest value of f on the interval $[x_{k-1}, x_k]$.

The limit in Eq. (3) always exists, as we shall explain shortly, and is usually not hard to compute by techniques developed in Article 4.7.

Riemann Integrals

The existence of the limit in (3) is a consequence of a more general limit theorem that applies to any continuous function on the interval $[a, b]$. In the more general theorem the function may have negative values. We shall introduce the theorem first, then discuss why it works, but we shall not go through the rigors of a complete proof. For convenience the pictures we draw will show positive functions, but the general processes illustrated by the pictures are valid for arbitrary continuous functions as well.

Given a function f, continuous on $[a, b]$, we begin by inserting points

$$x_1, x_2, \ldots, x_{k-1}, x_k, \ldots, x_{n-1}$$

between a and b as shown in Fig. 4.9. These points subdivide $[a, b]$ into n subintervals of lengths

$$\Delta x_1 = x_1 - a, \Delta x_2 = x_2 - x_1, \ldots, \Delta x_n = b - x_{n-1},$$

which need not be equal.

Since f is continuous, it has a minimum value $\min_k$ and a maximum value $\max_k$ on each subinterval. The areas of the shaded rectangles in Fig. 4.9(a) add up to what we call the *lower sum:*

$$L = \min_1 \Delta x_1 + \min_2 \Delta x_2 + \cdots + \min_n \Delta x_n. \tag{4}$$

The areas of the shaded rectangles in Fig. 4.9(b) add up to the *upper sum:*

$$U = \max_1 \Delta x_1 + \max_2 \Delta x_2 + \cdots + \max_n \Delta x_n. \tag{5}$$

The difference $U - L$ between the upper and lower sums is the sum of the areas of the shaded blocks in Fig. 4.9(c).

The idea conveyed by Fig. 4.9(c), and the idea we want to pursue, is that the more finely we subdivide $[a, b]$ the less area there will be in $U - L$. To make this important idea useful we have to say more precisely what we mean by subdividing $[a, b]$ more finely. Our goal is to improve the way the tops of the rectangles fit the curve and thereby decrease the difference between U and L. At the very least, we want the number of x's to increase in a way that makes the bases of the rectangles smaller. In other words, we want "subdivide more finely" to mean "subdivide $[a, b]$ in a way that makes the longest of the subintervals smaller than before." We therefore define the *norm* of the subdivision

$$a, x_1, x_2, \ldots, x_{n-1}, b$$

to be the longest subinterval length. Then, as norm $\to 0$, the subintervals become more numerous and shorter at the same time. In Fig. 4.9(c) this means that as norm $\to 0$, the blocks increase in number and get less wide (although their total length remains $b - a$). As they get less wide they also get less tall. As Fig. 4.9(d) suggests, we can make the difference $U - L$ less than any prescribed positive ε by taking the norm of the subdivision of $[a, b]$ close enough to zero. In other words

$$\lim_{\text{norm}\to 0} (U - L) = 0, \tag{6}$$

and, as shown in more advanced texts,

$$\lim_{\text{norm}\to 0} L = \lim_{\text{norm}\to 0} U. \tag{7}$$

The fact that Eqs. (6) and (7) hold for any continuous function, and not just for the one shown in Fig. 4.9, is the consequence of a special property continuous functions have on a bounded closed interval called *uniform continuity.* This is the property that guarantees that as the norm approaches zero the blocks shown in Fig. 4.9(c) that make up the difference between U and L get less tall as well as less wide, and we can make them as short as we like by making them narrow enough. The fact that we are passing over the $\varepsilon - \delta$ arguments associated with uniform continuity here

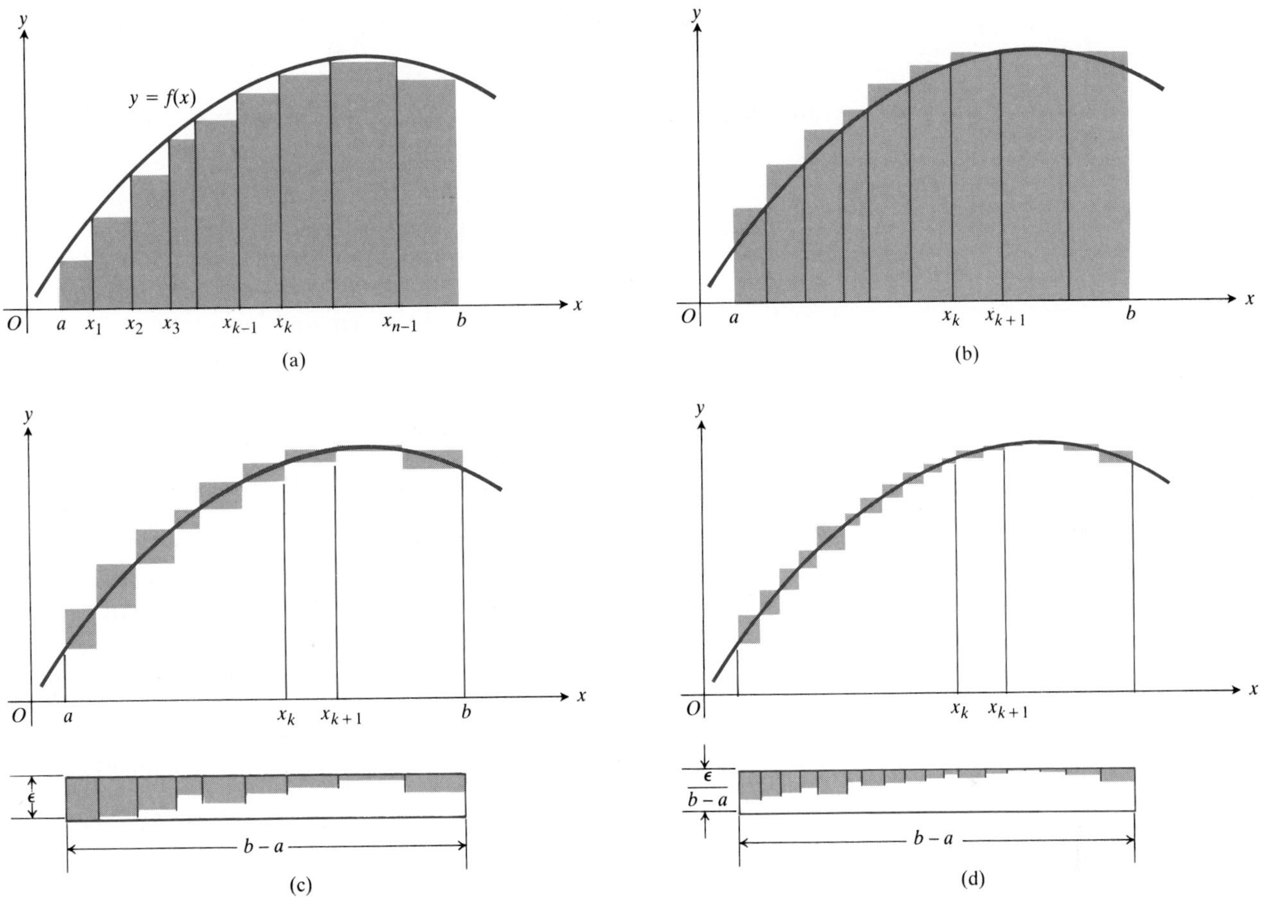

4.9 Given $\epsilon > 0$, there is a corresponding $\delta > 0$ such that all the blocks in part (d) have height less than $\epsilon/(b - a)$ if their maximum width is less than δ. This makes $0 \leq U - L =$ sum of blocks $< (\epsilon/(b - a)) \cdot (b - a) = \epsilon$.

is one of the things that keeps the derivation of Eq. (7) from being a proof. But the argument is right in spirit and gives a faithful portrait of the real proof.

Now, assuming that we know (7) for any continuous function on $[a, b]$, suppose that in each interval $[x_{k-1}, x_k]$ of a subdivision of $[a, b]$ we choose a point c_k to form the sum

$$S = \sum f(c_k)\, \Delta x_k. \tag{8}$$

Like the sums

$$L = \sum \min_k \Delta x_k, \qquad \text{and} \qquad U = \sum \max_k \Delta x_k,$$

S is a sum of function values times interval lengths. But c_k is randomly chosen and all we know about the number $f(c_k)$ is that

$$\min_k \leq f(c_k) \leq \max_k.$$

This is enough, however, to tell us that

$$L \leq S \leq U, \tag{9}$$

and therefore that

$$\lim_{\text{norm}\to 0} L = \lim_{\text{norm}\to 0} S = \lim_{\text{norm}\to 0} U \tag{10}$$

(by a slightly modified version of the Sandwich Theorem in Article 1.9). In other words, the sum S has the same limit that L and U have.

Pause for a moment to see how remarkable the conclusion in (10) really is. It says that *no matter how* we choose the point c_k to form the sum

$$S = \sum f(c_k)\,\Delta x_k$$

for a continuous function on an interval $[a, b]$, we always get the same limit as the norm of the subdivision approaches zero. We can choose every c_k so that $f(c_k)$ is the maximum value of f on $[x_{k-1}, x_k]$. The limit is the same. We can choose c_k so that $f(c_k)$ is the minimum value of f on $[x_{k-1}, x_k]$. The limit is the same. We can choose every c_k at random. No matter, the limit is still the same.

This fact was first discovered (without uniform continuity) by Cauchy (1823) and later put on a solid logical foundation (with uniform continuity) by other mathematicians of the nineteenth century. The limit, called the *Riemann integral* of f over the interval $[a, b]$, and denoted by

$$\int_a^b f(x)\,dx,$$

bears the name of Georg Friedrich Bernhard Riemann (1826–1866), whose idea it was to trap the limit between upper and lower sums. The sum

$$S = \sum f(c_k)\,\Delta x_k$$

is called an *approximating sum* for the integral. The numbers a and b are called the *limits* of integration, a being the *lower limit* and b the *upper limit.*

THEOREM

The Integral Existence Theorem

EXISTENCE OF THE RIEMANN INTEGRAL

If f is continuous on $[a, b]$, then

$$\int_a^b f(x)\,dx = \lim_{\text{norm}\to 0} \sum f(c_k)\,\Delta x_k$$

exists and is the same number for any choice of the numbers c_k.

The full proof of the theorem can be found in most texts on mathematical analysis.

The area we defined earlier in Eq. (3) for the region under the graph of a nonnegative function over an interval $[a, b]$ was the Riemann integral of the function:

$$\text{Area} = \int_a^b f(x)\,dx, \quad \text{when } f(x) \geq 0.$$

As we shall see in Article 4.7, and again in Chapter 5, Riemann integrals have many other interpretations as well.

Definite Integrals

The integral

$$\int_a^b f(x)\,dx$$

is called the *definite integral* of f over $[a, b]$ and may also be denoted by

$$\int_a^b f(t)\,dt, \qquad \int_a^b f(u)\,du,$$

and so on. The variable of integration can be any letter like x, t, or u that is not currently in use for something else.

The point to remember is that $\int_a^b f(x)\,dx$ is a number defined in a certain way as a limit of approximating sums over the interval from a to b on the x-axis. If we use another name for the axis, say the t-axis, then the appropriate symbol for the integral is $\int_a^b f(t)\,dt$, but the integral is still the *same number*.

Incidently, Leibniz chose the notation $\int$ for the integral because it suggested the S in Sum.

If f is continuous on $[a, b]$, then $\int_a^b f(x)\,dx$ exists and we define $\int_b^a f(x)\,dx$ to be its negative:

$$\int_b^a f(x)\,dx = -\int_a^b f(x)\,dx.$$

If $a = b$, we define $\int_a^b f(x)\,dx$ to be zero:

$$\int_a^a f(x)\,dx = 0.$$

Definite integrals of continuous functions have useful algebraic properties that come from defining the integrals as limits of finite sums, because these sums have the corresponding properties.

Properties of Definite Integrals

I1. $\int_a^b k\,f(x)\,dx = k\int_a^b f(x)\,dx$ (any number k)

I2. $\int_a^b [f(x) + g(x)]\,dx = \int_a^b f(x)\,dx + \int_a^b g(x)\,dx$

I3. $\int_a^b [f(x) - g(x)]\,dx = \int_a^b f(x)\,dx - \int_a^b g(x)\,dx$

I4. $\int_a^b f(x)\,dx \geq 0$ if $f(x) \geq 0$ on $[a, b]$

I5. $\int_a^b f(x)\,dx \leq \int_a^b g(x)\,dx$ if $f(x) \leq g(x)$ on $[a, b]$

I6. $\min f \cdot (b - a) \leq \int_a^b f(x)\,dx \leq \max f \cdot (b - a)$, where min and max refer to the minimum and maximum value of f on $[a, b]$.

I7. $\int_b^a f(x)\,dx = -\int_a^b f(x)\,dx$ (a definition)

I8. $\int_a^a f(x)\,dx = 0$ (a definition)

I9. $\int_a^b f(x)\,dx + \int_b^c f(x)\,dx = \int_a^c f(x)\,dx$

The equality in I9 holds for any a, b, and c provided f is continuous on the intervals joining them. Figure 4.10 illustrates I9 for areas.

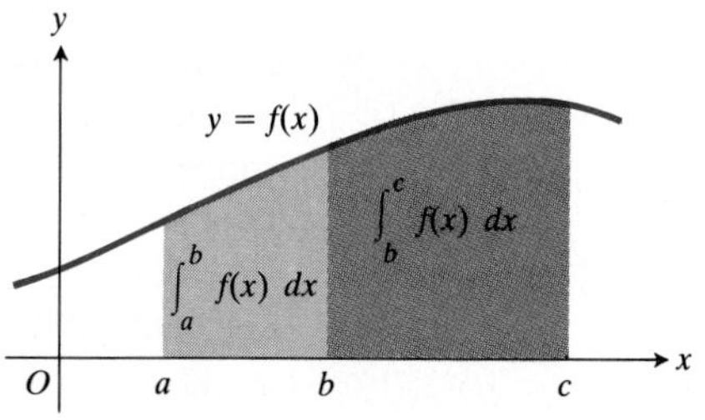

4.10 $\int_a^b f(x)\,dx + \int_b^c f(x)\,dx = \int_a^c f(x)\,dx.$

EXAMPLE 3 Suppose

$$\int_{-1}^{1} f(x)\,dx = 5, \quad \int_{1}^{4} f(x)\,dx = -2, \quad \int_{-1}^{1} g(x)\,dx = 7,$$

and

$$h(x) \geq f(x) \quad \text{on } [-1, 1].$$

Then

1. $\int_{-1}^{1} 3f(x)\,dx = 3\int_{-1}^{1} f(x)\,dx = 15$
2. $\int_{-1}^{1} [2f(x) + 3g(x)]\,dx = 2(5) + 3(7) = 31$
3. $\int_{-1}^{1} [f(x) - g(x)]\,dx = 5 - 7 = -2$
4. $\int_{-1}^{4} f(x)\,dx = \int_{-1}^{1} f(x)\,dx + \int_{1}^{4} f(x)\,dx = 5 - 2 = 3$
5. $\int_{-1}^{1} h(x)\,dx \geq \int_{-1}^{1} f(x)\,dx = 5.$ □

The notation for the definite integral suggests the indefinite integrals we worked with in Articles 4.2 through 4.4. The connection between the two kinds of integrals is one of the most important relationships in calculus, and it will be developed in Article 4.7. But first, in Article 4.6, we shall calculate a few areas without the benefit of the more modern mathematical tools of Article 4.7. The contrast between the methods of Articles 4.6 and 4.7 is striking and typifies the difference between the mathematics of the world into which Newton and Leibniz were born and the mathematics of the world that Newton and Leibniz left behind.

PROBLEMS

In Problems 1–5, sketch the graph of the given equation over the interval $a \leq x \leq b$. Divide the interval into $n = 4$ subintervals each of length $\Delta x = (b - a)/4$. (a) Sketch the inscribed rectangles and compute the sum of their areas. (b) Do the same using the circumscribed in place of the inscribed rectangle in each subinterval (Fig. 4.11).

1. $y = 2x + 1, \quad a = 0, \quad b = 1$

2. $y = x^2, \quad a = -1, \quad b = 1$

3. $y = \sin x, \quad a = 0, \quad b = \pi$

4. $y = 1/x, \quad a = 1, \quad b = 2$

5. $y = \sqrt{x}, \quad a = 0, \quad b = 4$

Write out the following sums, as in Example 2.

6. $\sum_{k=1}^{5} \frac{1}{k}$ **7.** $\sum_{i=-1}^{3} 2^i$ **8.** $\sum_{n=1}^{4} \cos n\pi x$

Find the value of each sum.

9. $\sum_{n=0}^{4} \frac{n}{4}$ **10.** $\sum_{k=1}^{3} \frac{k-1}{k}$ **11.** $\sum_{m=0}^{5} \sin \frac{m\pi}{2}$

12. $\sum_{i=1}^{4} (i^2 - 1)$ **13.** $\sum_{i=0}^{3} (i^2 + 5)$ **14.** $\sum_{k=0}^{5} \frac{1}{2^k}$

15. Which of the following express $1 + 2 + 4 + 8 + 16 + 32$ in summation notation?

a) $\sum_{j=2}^{7} 2^{j-2}$ b) $\sum_{k=0}^{5} 2^k$ c) $\sum_{j=0}^{5} 2^j$ d) $\sum_{j=1}^{6} 2^{j-1}$

16. Suppose f and g are continuous and that

$$\int_1^2 f(x)\,dx = -4, \quad \int_2^5 f(x)\,dx = 6, \quad \int_1^5 g(x)\,dx = 8.$$

Find

a) $\int_1^5 f(x)\,dx$, b) $\int_5^1 -4f(x)\,dx$,

c) $\int_1^5 [4f(x) - 2g(x)]\,dx$.

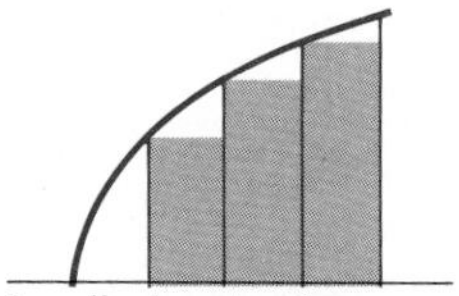

Inscribed rectangles

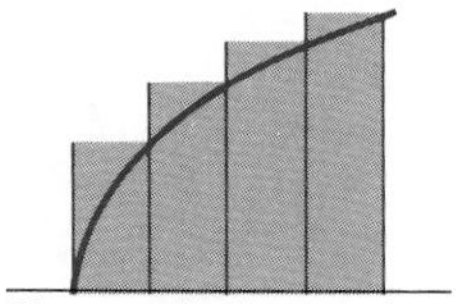

Circumscribed rectangles

Figure 4.11

17. Suppose f and h are continuous and that

$$\int_1^7 f(x)\,dx = -1, \quad \int_7^9 f(x)\,dx = 5, \quad \int_7^9 h(x) = 4.$$

Find

a) $\int_1^9 -2f(x)\,dx$, b) $\int_7^9 [2f(x) - h(x)]\,dx$,

c) $\int_9^7 f(x)\,dx$, d) $\int_7^7 [f(x) + h(x)]\,dx$.

18. Suppose f is continuous on the interval $[0, 4]$, and that

$$\int_0^3 f(x)\,dx = 3, \quad \int_0^4 f(z)\,dz = 7.$$

Find

$$\int_3^4 f(y)\,dy.$$

19. Suppose that the graph of f is always rising toward the right between $x = a$ and $x = b$, as in Fig. 4.12. Take $\Delta x = (b - a)/n$. Show, by reference to Fig. 4.12, that the difference between the upper and lower sums is representable graphically as the area $[f(b) - f(a)]\,\Delta x$ of the rectangle R. (*Hint:* The difference $U - L$ is the sum of areas of rectangles with diagonals P_0P_1, P_1P_2, and so forth along the curve, and there is no overlapping when these are all displaced horizontally into the rectangle R.)

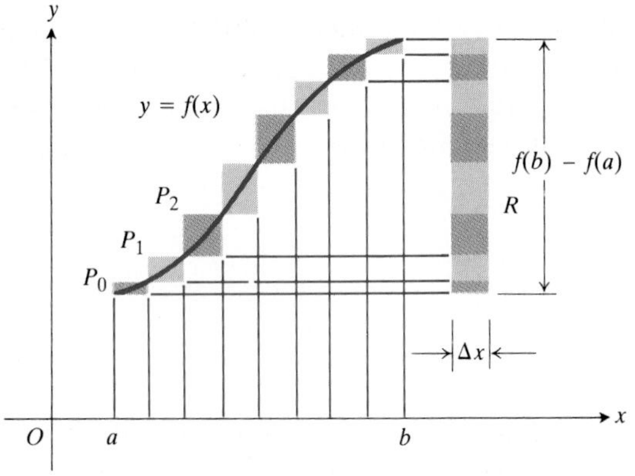

4.12 If f is an increasing function, the blocks in $U - L$ fill the rectangle on the right without overlapping. See Problems 19–21.

20. Draw a figure to represent a continuous curve $y = f(x)$ that is always falling to the right between $x = a$ and $x = b$. Suppose the subdivisions Δx_k are again equal, $\Delta x_k = \Delta x = (b - a)/n$. Obtain an expression for the difference $U - L$ analogous to the expression in Problem 19.

21. In Problem 19 or 20, if the Δx_k's are not all equal, show that

$$U - L \leq |f(b) - f(a)|(\Delta x_{\max}),$$

where $\Delta x_{\max}$ is the largest of the Δx_k's for

$$k = 1, 2, \ldots, n.$$

***Toolkit* programs**

Integration

4.6
Calculating Areas as Limits

In Article 4.5 we defined the area under the graph of a nonnegative function $y = f(x)$ over an interval $[a, b]$ to be a limit of sums of inscribed rectangles. We also saw that this limit was a special case of a limit called the definite integral of f, which could be defined for any continuous function on $[a, b]$. We computed a few sums, but no limits.

In the present article we shall compute limits with the assistance of algebraic formulas (which we first develop). Our purpose is to show how area as we have defined it might be calculated with the mathematics of the late Renaissance, before the seventeenth-century development of the fundamental theorems of calculus, which will be the subject of the next article. As you will see, each area calculation in the present article requires its own bit of ingenuity, and the formulas developed to solve a given problem do not seem to apply to any other problem. This lack of generality is one of the features that distinguishes the early calculus of the Renaissance from the later methods of Newton and Leibniz.

To begin, we need the following formulas:

$$\sum_{k=1}^{n} k = 1 + 2 + 3 + \cdots + n = \frac{n(n+1)}{2},$$

$$\sum_{k=1}^{n} k^2 = 1^2 + 2^2 + 3^2 + \cdots + n^2 = \frac{n(n+1)(2n+1)}{6},$$

which we shall prove by the method of mathematical induction. This consists of showing that each formula is true when $n = 1$, and that if the formula is true for any integer n, then it is also true for the next integer, $n + 1$. We shall also show how such formulas might be discovered.

First, consider the sum of first powers:

$$F(n) = 1 + 2 + 3 + \cdots + n.$$

Table 4.1 shows briefly how $F(n)$ increases with n. The last column exhibits $F(n)/n$, the ratio of $F(n)$ to n.

The last column suggests that the ratio $F(n)/n$ is equal to $(n + 1)/2$. At least this is the case for all the entries in the table ($n = 1, 2, 3, 4, 5, 6$). In other words, the formula

$$\frac{F(n)}{n} = \frac{n+1}{2},$$

or

$$1 + 2 + 3 + \cdots + n = \frac{n(n+1)}{2}, \tag{1}$$

is true for $n = 1, 2, 3, 4, 5, 6$. Suppose now that n is any integer for which (1) is known to be true (at the moment, n could be any integer from 1 through 6). Then if $(n + 1)$ were added to both sides of the equation, the new equation

$$1 + 2 + 3 + \cdots + n + (n+1) = \frac{n(n+1)}{2} + (n+1) \tag{2}$$

would also be true for that same n. But the right side of (2) is

$$\frac{n(n+1)}{2} + (n+1) = \frac{(n+1)}{2}(n+2) = \frac{(n+1)(n+2)}{2},$$

so that (2) becomes

$$1 + 2 + 3 + \cdots + n + (n+1) = \frac{(n+1)((n+1)+1)}{2},$$

which is just like Eq. (1) except that n is replaced by $n + 1$. Thus if Eq. (1) is true for an integer n, it is also true for the next integer $n + 1$. Hence we now know that it is true for $n + 1 = 7$, since it was true for $n = 6$. Then we can say it is true for $n + 1 = 8$, since it is true for $n = 7$. By the principle of mathematical induction, then, it is true for every positive integer n.

Now let's consider the squares. Let

$$Q(n) = 1^2 + 2^2 + 3^2 + \cdots + n^2$$

be the sum of the squares of the first n positive integers. Obviously this grows faster than $F(n)$, the sum of first powers, but let us look at the ratio of $Q(n)$ to $F(n)$ to compare them. (See Table 4.2.)

Table 4.1

n	$F(n) = 1 + 2 + 3 + \cdots + n$	$F(n)/n$
1	1	$1 = \frac{2}{2}$
2	$1 + 2 = 3$	$\frac{3}{2} = \frac{3}{2}$
3	$1 + 2 + 3 = 6$	$\frac{6}{3} = \frac{4}{2}$
4	$1 + 2 + 3 + 4 = 10$	$\frac{10}{4} = \frac{5}{2}$
5	$1 + 2 + 3 + 4 + 5 = 15$	$\frac{15}{5} = \frac{6}{2}$
6	$1 + 2 + 3 + 4 + 5 + 6 = 21$	$\frac{21}{6} = \frac{7}{2}$

Table 4.2

n	$F(n)$	$Q(n) = 1^2 + 2^2 + 3^2 + \cdots + n^2$	$Q(n)/F(n)$
1	1	$1^2 = 1$	$\frac{1}{1} = \frac{3}{3}$
2	3	$1^2 + 2^2 = 5$	$\frac{5}{3} = \frac{5}{3}$
3	6	$1^2 + 2^2 + 3^2 = 14$	$\frac{14}{6} = \frac{7}{3}$
4	10	$1^2 + 2^2 + 3^2 + 4^2 = 30$	$\frac{30}{10} = \frac{9}{3}$
5	15	$1^2 + 2^2 + 3^2 + 4^2 + 5^2 = 55$	$\frac{55}{15} = \frac{11}{3}$
6	21	$1^2 + 2^2 + 3^2 + 4^2 + 5^2 + 6^2 = 91$	$\frac{91}{21} = \frac{13}{3}$

We note how regular the last column is: $\frac{3}{3}$, $\frac{5}{3}$, $\frac{7}{3}$, and so on. In fact it is just $(2n + 1)/3$ for $n = 1, 2, 3, 4, 5, 6$; that is,

$$Q(n) = F(n) \cdot \frac{2n + 1}{3}.$$

But from Eq. (1), $F(n) = n(n + 1)/2$, and hence

$$Q(n) = 1^2 + 2^2 + 3^2 + \cdots + n^2 = \frac{n(n + 1)(2n + 1)}{6} \tag{3}$$

is true for the integers n from 1 through 6. To establish it for all other positive integers, we proceed as before. Start with any n for which (3) is true and add $(n + 1)^2$. Then

$$\begin{aligned} 1^2 + 2^2 + 3^2 + \cdots + n^2 + (n + 1)^2 &= \frac{n(n + 1)(2n + 1)}{6} + (n + 1)^2 \\ &= \frac{(n + 1)}{6}[n(2n + 1) + 6(n + 1)] \\ &= \frac{(n + 1)}{6}(2n^2 + 7n + 6) = \frac{(n + 1)(n + 2)(2n + 3)}{6}. \end{aligned} \tag{4}$$

We note that the last expression in (4) is the same as the last expression in (3) with n replaced by $n + 1$. In other words, if the formula in (3) is true for any integer n, we have just shown that it is true for $n + 1$. Since we know it is true for $n = 6$, it is also true for $n + 1 = 7$. And now that we know it is true for $n = 7$, it follows that it is true for $n + 1 = 8$, and so on. It is true for every positive integer n by the principle of mathematical induction.

We now apply these formulas to find areas under two graphs.

EXAMPLE 1 Let a, b, and m be positive numbers, with $a < b$. Find the area under the graph $y = mx$, $a \leq x \leq b$ (Fig. 4.13).

Solution We calculate

$$\int_a^b mx\,dx$$

as a limit. Let n be a positive integer and divide the interval $[a, b]$ into n subintervals each of length $\Delta x = (b - a)/n$, by inserting the points

$$\begin{aligned} x_1 &= a + \Delta x, \\ x_2 &= a + 2\,\Delta x, \\ x_3 &= a + 3\,\Delta x, \\ &\vdots \\ x_{n-1} &= a + (n - 1)\,\Delta x. \end{aligned}$$

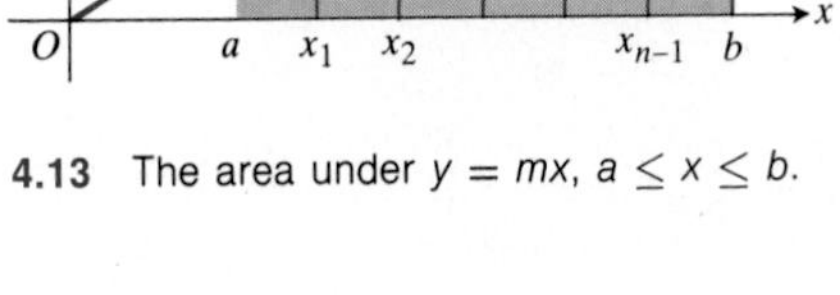

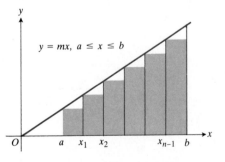

4.13 The area under $y = mx$, $a \leq x \leq b$.

The inscribed rectangles have areas

$$\begin{aligned} f(a)\,\Delta x &= ma \cdot \Delta x, \\ f(x_1)\,\Delta x &= m(a + \Delta x) \cdot \Delta x, \\ f(x_2)\,\Delta x &= m(a + 2\,\Delta x) \cdot \Delta x, \\ &\vdots \\ f(x_{n-1})\,\Delta x &= m[a + (n - 1)\,\Delta x] \cdot \Delta x, \end{aligned}$$

whose sum is

$$\begin{aligned} S_n &= m[a + (a + \Delta x) + (a + 2\,\Delta x) + \cdots + (a + (n-1)\,\Delta x)] \cdot \Delta x \\ &= m[na + (1 + 2 + \cdots + (n-1))\,\Delta x]\,\Delta x \\ &= m\left[na + \frac{(n-1)n}{2}\,\Delta x\right]\Delta x \\ &= m\left[a + \frac{n-1}{2}\,\Delta x\right] n\,\Delta x \qquad \left(\Delta x = \frac{b-a}{n}\right) \\ &= m\left[a + \frac{b-a}{2} \cdot \frac{n-1}{n}\right] \cdot (b-a). \end{aligned}$$

The area under the graph is defined to be the limit of S_n as $n \to \infty$. In the final form, the only place n appears is in the fraction

$$\frac{n-1}{n} = 1 - \frac{1}{n},$$

and $1/n \to 0$ as $n \to \infty$, so

$$\lim \frac{n-1}{n} = 1 - 0 = 1.$$

Therefore,

$$\begin{aligned} \int_a^b mx\,dx = \lim S_n &= m\left(a + \frac{b-a}{2}\right) \cdot (b-a) \\ &= \frac{ma + mb}{2} \cdot (b-a) = m\left(\frac{b^2}{2} - \frac{a^2}{2}\right). \end{aligned}$$

This is the area of a trapezoid, with "bases" (in this case vertical) ma and mb and with altitude $(b - a)$. □

EXAMPLE 2 Find the area under the graph of $y = x^2$, $0 \le x \le b$ (Fig. 4.14).

Solution We calculate

$$\int_0^b x^2\,dx$$

as a limit. Divide the interval $0 \le x \le b$ into n (> 0) subintervals each of length $\Delta x = b/n$, by inserting the points

$$x_1 = \Delta x, \quad x_2 = 2\,\Delta x, \quad x_3 = 3\,\Delta x, \quad \ldots, \quad x_{n-1} = (n-1)\,\Delta x.$$

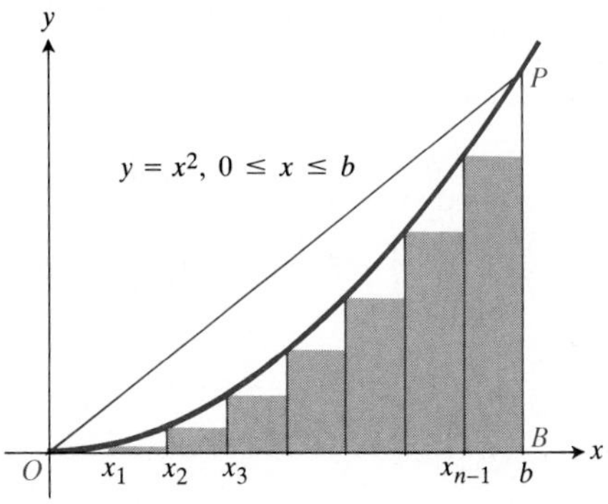

4.14 The area under $y = x^2$, $0 \le x \le b$.

The inscribed rectangles have areas

$$\begin{aligned} f(0)\,\Delta x &= 0, \\ f(x_1)\,\Delta x &= (\Delta x)^2\,\Delta x, \\ f(x_2)\,\Delta x &= (2\,\Delta x)^2\,\Delta x, \\ f(x_3)\,\Delta x &= (3\,\Delta x)^2\,\Delta x, \\ &\;\vdots \\ f(x_{n-1})\,\Delta x &= ((n-1)\,\Delta x)^2\,\Delta x. \end{aligned}$$

The sum of these areas is

$$\begin{aligned} S_n &= (1^2 + 2^2 + 3^2 + \cdots + (n-1)^2)(\Delta x)^3 \\ &= \frac{(n-1)n(2n-1)}{6} \cdot \left(\frac{b}{n}\right)^3 \\ &= \frac{b^3}{6} \cdot \frac{n-1}{n} \cdot \frac{n}{n} \cdot \frac{2n-1}{n} \\ &= \frac{b^3}{6} \cdot \left(1 - \frac{1}{n}\right) \cdot \left(2 - \frac{1}{n}\right). \end{aligned}$$

To find the area under the graph, we let n increase without bound and get

$$A = \int_0^b x^2\,dx = \lim S_n = \frac{b^3}{6} \cdot 1 \cdot 2 = \frac{b^3}{3}.$$

Therefore the area under the curve is $\frac{1}{3}$ the base b times the "altitude" b^2. The triangle OBP in Fig. 4.14 has area $\frac{1}{2}b \cdot b^2 = b^3/2$, and the area under the curve turns out to be somewhat smaller, as we should expect.

The formula

$$\int_0^b x^2\,dx = \frac{b^3}{3}$$

gives

$$\int_0^1 x^2\,dx = \frac{1}{3}, \quad \int_0^2 x^2\,dx = \frac{8}{3}, \quad \int_0^3 x^2\,dx = \frac{27}{3} = 9,$$

and so on. □

PROBLEMS

1. Verify the formula

$$\sum_{k=1}^{n} k^3 = 1^3 + 2^3 + \cdots + n^3 = \left(\frac{n(n+1)}{2}\right)^2$$

for $n = 1, 2, 3$. Then add $(n+1)^3$ and thereby prove by mathematical induction (as in the text) that the formula is true for all positive integers n.

2. Using the result of Problem 1 and the method of Example 2 in the text, show that the area under the graph of $y = x^3$ over the interval $0 \le x \le b$ is $b^4/4$.

3. Find the area under the graph of $y = mx$ over the interval $a \le x \le b$ by using *circumscribed* rectangles in place of the inscribed rectangles of Example 1 in the text.

4. Find the area under the curve $y = x^2$ over the interval $0 \le x \le b$ by using circumscribed rectangles in place of the inscribed rectangles of Example 2 in the text.

5. Do Problem 2 above by using circumscribed rectangles instead of inscribed rectangles.

Establish the formulas in Problems 6 and 7 for every positive integer n, by showing (a) that the formula is correct for $n = 1$, and (b) if true for n, the formula is also true for $n + 1$.

6. $\displaystyle\sum_{k=1}^{n} (2k - 1) = 1 + 3 + 5 + \cdots + (2n - 1) = n^2$

7. $\displaystyle\sum_{k=1}^{n} \frac{1}{k(k+1)} = \frac{1}{1 \cdot 2} + \frac{1}{2 \cdot 3} + \cdots + \frac{1}{n \cdot (n+1)} = \frac{n}{n+1}$

8. Use the results of Examples 1 and 2 and the properties of definite integrals from Article 4.5 to calculate the following integrals.

a) $\displaystyle\int_0^2 3x\,dx$ b) $\displaystyle\int_2^3 4x\,dx$ c) $\displaystyle\int_0^2 x^2\,dx$

d) $\displaystyle\int_0^2 (x^2 - 5x)\,dx$ e) $\displaystyle\int_0^1 x^2\,dx$ f) $\displaystyle\int_1^2 x^2\,dx$

g) $\displaystyle\int_1^3 x^2\,dx$ h) $\displaystyle\int_2^3 x^2\,dx$

9. Interpret

$$\int_0^1 (1 + x^2)\,dx$$

as an area and calculate the integral from the result of Example 2. (The graph of $y = 1 + x^2$ is shown in Fig. 4.6.)

10. Let

$$S_n = \frac{1}{n}\left[\frac{1}{n} + \frac{2}{n} + \frac{3}{n} + \cdots + \frac{n-1}{n}\right].$$

Calculate

$$\lim_{n\to\infty} S_n$$

by showing that S_n is an approximating sum of the integral

$$\int_0^1 x\,dx$$

(whose value we know from Example 1). (*Hint:* Subdivide [0, 1] into n intervals of equal length and write out the approximating sum for inscribed rectangles.)

11. Let

$$S_n = \frac{1^2}{n^3} + \frac{2^2}{n^3} + \cdots + \frac{(n-1)^2}{n^3}.$$

To calculate $\lim_{n\to\infty} S_n$, show that

$$S_n = \frac{1}{n}\left[\left(\frac{1}{n}\right)^2 + \left(\frac{2}{n}\right)^2 + \cdots + \left(\frac{n-1}{n}\right)^2\right]$$

and interpret S_n as an approximating sum of the integral

$$\int_0^1 x^2\,dx$$

(whose value we know from Example 2). (*Hint:* Subdivide [0, 1] into n intervals of equal length and write out the approximating sum for inscribed rectangles.)

12. Let

$$S_n = 1 + 2 + \cdots + n = \frac{n(n+1)}{2}$$

be the sum of the first n integers.

a) Use mathematical induction to show that

$$S_n^{(2)} = S_1 + S_2 + \cdots + S_n = \frac{n(n+1)(n+2)}{2\cdot 3}.$$

b) Use mathematical induction to show that

$$S_n^{(3)} = S_1^{(2)} + S_2^{(2)} + \cdots S_n^{(2)} = \frac{n(n+1)(n+2)(n+3)}{2\cdot 3\cdot 4}.$$

c) For any $k > 1$, let

$$S_n^{(k)} = S_1^{(k-1)} + \cdots + S_n^{(k-1)}.$$

Guess a formula for $S_n^{(k)}$.

13. The formula

$$\sin h + \sin 2h + \sin 3h + \cdots + \sin mh = \frac{\cos(h/2) - \cos(m + \frac{1}{2})h}{2\sin(h/2)}$$

may be assumed for this problem. Use it to find the area under the graph of $f(x) = \sin x$ over $[0, \pi/2]$, in two steps:

a) Divide the interval $[0, \pi/2]$ into n equal subintervals and calculate the corresponding upper sum U; then

b) find the limit of U as $n \to \infty$ and

$$\Delta x = (b - a)/n \to 0.$$

***Toolkit* programs**	
Integration	Sequences and Series

4.7 The Fundamental Theorems of Integral Calculus

In Article 4.5 we defined the area under a curve as a definite integral and showed how to estimate it by adding areas of inscribed rectangles. Nothing more than arithmetic was involved in these calculations, but we paid the price of getting only an estimate of the true area. In Article 4.6 we used the definition of the definite integral as a limit to compute areas exactly, but at the cost of extensive algebraic preliminaries. In this article we follow the trail blazed by Newton and Leibniz to show how definite integrals can be computed with calculus.

The First Fundamental Theorem

The first theorem we present tells how to calculate the definite integral of a continuous function f from any antiderivative of f.

THEOREM 1

The First Fundamental Theorem of Integral Calculus

If f is continuous on $[a, b]$ and F is any antiderivative of f on $[a, b]$, then

$$\int_a^b f(x)\,dx = F(b) - F(a). \tag{1}$$

The theorem says that to calculate the definite integral of f over $[a, b]$ all we need to do is

1. find an antiderivative F of f, and
2. calculate the number $\int_a^b f(x)\,dx = F(x)]_a^b = F(b) - F(a)$.

Note that Theorem 1 does not say that f *has* an antiderivative F. It says that *if* f has an antiderivative F, then the integral of f from a to b may be calculated as $F(b)–F(a)$. It will be shown later in this article that f actually has an antiderivative, and Theorem 1 itself will be proven at that time.

EXAMPLE 1 The area under the line $y = mx$ in Example 1, Article 4.6, is

$$\int_a^b mx\,dx = \frac{mx^2}{2}\Big]_a^b = \frac{mb^2}{2} - \frac{ma^2}{2}. \ \square$$

EXAMPLE 2 The area under the curve $y = x^2$ in Example 2, Article 4.6, is

$$\int_0^b x^2\,dx = \frac{x^3}{3}\Big]_0^b = \frac{b^3}{3} - 0 = \frac{b^3}{3}. \ \square$$

EXAMPLE 3 Calculate the area bounded by the x-axis and the parabola $y = 6 - x - x^2$.

Solution We find where the curve crosses the x-axis by setting

$$y = 0 = 6 - x - x^2 = (3 + x)(2 - x),$$

which gives

$$x = -3 \qquad \text{or} \qquad x = 2.$$

The curve is sketched in Fig. 4.15.

The area is

$$\int_{-3}^2 (6 - x - x^2)\,dx = 6x - \frac{x^2}{2} - \frac{x^3}{3}\Big]_{-3}^2$$

$$= (12 - 2 - \tfrac{8}{3}) - (-18 - \tfrac{9}{2} + \tfrac{27}{3}) = 20\tfrac{5}{6}.$$

The curve in Fig. 4.15 is an arch of a parabola, and it is interesting to note that the area is exactly equal to two-thirds the base times the altitude:

$$\tfrac{2}{3}(5)(\tfrac{25}{4}) = \tfrac{125}{6} = 20\tfrac{5}{6}. \ \square$$

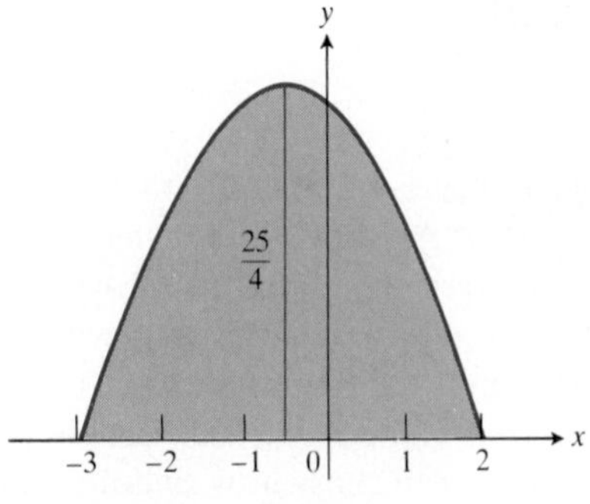

4.15 The parabolic arch $y = 6 - x - x^2$, $-3 \le x \le 2$.

EXAMPLE 4 Show that the area under one arch of the curve $y = \sin x$ is 2.

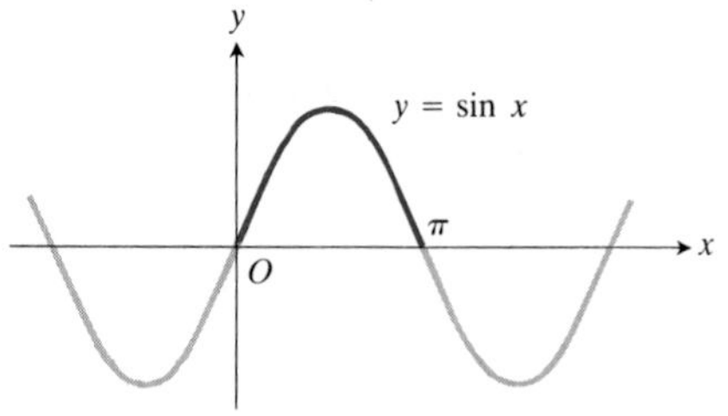

4.16 One arch of the curve $y = \sin x$.

Solution One arch of the sine curve extends from $x = 0$ to $x = \pi$ (Fig. 4.16). Therefore the area is

$$\int_0^{\pi} \sin x\, dx = -\cos x\Big]_0^{\pi} = -\cos \pi - (-\cos 0) = -(-1) + 1 = 2. \quad \square$$

EXAMPLE 5 This is an illustration of some of the algebraic properties of definite integrals presented at the end of Article 4.5.

a) $\int_0^{\pi} 5 \sin x\, dx = 5 \int_0^{\pi} \sin x = 5 \cdot 2 = 10 \qquad \begin{pmatrix}\text{using the result}\\ \text{of Example 4}\end{pmatrix}$

b) $$\int_1^2 \left[3 - \frac{6}{x^2}\right] dx = \int_1^2 3\, dx - \int_1^2 \frac{6}{x^2}\, dx$$
$$= \Big[3x\Big]_1^2 - \Big[-\frac{6}{x}\Big]_1^2 = 6 - 3 + 3 - 6 = 0 \quad \square$$

If a continuous function f has no negative values over $[a, b]$, then its integral over $[a, b]$ is the area under its graph. If f is negative (or nonpositive) over the interval, then the integral is the negative of the area, as in the next example.

EXAMPLE 6 Find the area between the curve $y = x^2 - 4$ and the x-axis from $x = -2$ to $x = 2$. See Fig. 4.17(a).

Solution If we integrate $f(x) = x^2 - 4$ over the interval from -2 to 2, we find

$$\int_{-2}^{2} (x^2 - 4)\, dx = \frac{x^3}{3} - 4x\Big]_{-2}^{2} = \left(\frac{8}{3} - 8\right) - \left(-\frac{8}{3} + 8\right) = -\frac{32}{3}.$$

The area between the curve and the y-axis from $x = -2$ to $x = 2$ contains $\frac{32}{3}$ units of area.

4.17 These graphs enclose the same amount of area with the x-axis, but the definite integrals of the two functions from -2 to 2 differ in sign.

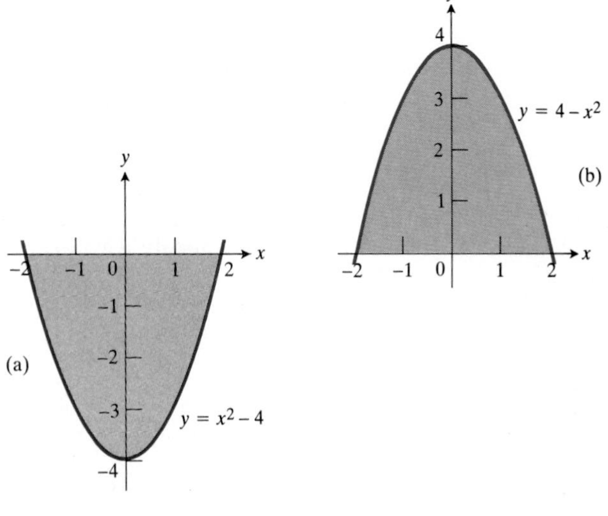

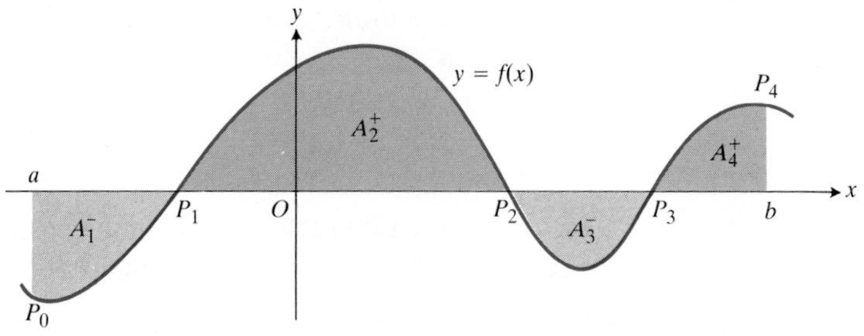

4.18 The integral $\int_a^b f(x)\,dx$ is the algebraic sum of signed areas.

The graph of

$$y = g(x) = -f(x) = 4 - x^2, \qquad -2 \le x \le 2$$

in Fig. 4.17(b) is the mirror image of the graph of f across the x-axis. The area between the graph of g and the x-axis is

$$\int_{-2}^{2} (4 - x^2)\,dx = 4x - \frac{x^3}{3}\bigg]_{-2}^{2} = \frac{32}{3}. \ \square$$

If the graph of $y = f(x)$, $a \le x \le b$, is partly below and partly above the x-axis, as in Fig. 4.18, then

$$\lim \sum f(c_k)\,\Delta x = \int_a^b f(x)\,dx = F(x)\bigg]_a^b = F(b) - F(a)$$

is the algebraic sum of *signed* areas, positive areas above the x-axis, negative areas below. For example, if the absolute values of the areas between the curve and the x-axis in Fig. 4.18 are A_1, A_2, A_3, A_4, then the definite integral of f from a to b is equal to

$$\int_a^b f(x)\,dx = -A_1 + A_2 - A_3 + A_4.$$

Thus if we wanted the sum of the absolute values of these signed areas, that is,

$$A = |-A_1| + A_2 + |-A_3| + A_4,$$

we should need to find the abscissas s_1, s_2, s_3 of the points P_1, P_2, P_3 where the curve crosses the x-axis. We would then compute, separately,

$$-A_1 = \int_a^{s_1} f(x)\,dx, \qquad A_2 = \int_{s_1}^{s_2} f(x)\,dx,$$

$$-A_3 = \int_{s_2}^{s_3} f(x)\,dx, \qquad A_4 = \int_{s_3}^{b} f(x)\,dx,$$

and add their absolute values.

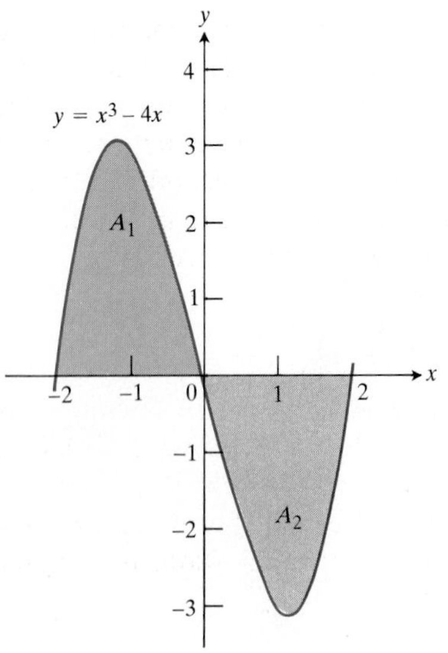

4.19 Graph of $y = x^3 - 4x$, $-2 \le x \le 2$.

EXAMPLE 7 Find the total area bounded by the curve $y = x^3 - 4x$ and the x-axis.

Solution The graph in Fig. 4.19 lies above the x-axis from -2 to 0, below from 0 to $+2$. (The polynomial $x^3 - 4x$ factors as

$$x^3 - 4x = x(x - 2)(x + 2),$$

and we determine the sign of the product from the signs of the three factors.)

$$A_1 = \int_{-2}^{0} (x^3 - 4x)\,dx = \frac{x^4}{4} - 2x^2\bigg]_{-2}^{0} = 0 - (4 - 8) = +4,$$

$$-A_2 = \int_{0}^{2} (x^3 - 4x)\,dx = \frac{x^4}{4} - 2x^2\bigg]_{0}^{2} = (4 - 8) - 0 = -4,$$

$$A_1 + |-A_2| = 4 + |-4| = 8. \square$$

Functions Defined by Integrals. The Second Fundamental Theorem

The integral of any continuous function $f(t)$ from $t = a$ to $t = x$ defines a number

$$F(x) = \int_a^x f(t)\,dt \tag{2}$$

that we can treat as a function of x. For every value of x in the domain of f the integral gives the output $F(x)$.

This gives an important way of defining new functions. For example, in Chapter 6 we shall define the natural logarithm of a number $x > 0$ by the formula

$$\ln x = \int_1^x \frac{1}{t}\,dt, \quad x > 0. \tag{3}$$

The error *function*

$$\text{erf}\,(x) = \frac{2}{\sqrt{\pi}} \int_0^x e^{-t^2}\,dt \tag{4}$$

is important in probability as well as in the theory of heat flow and signal transmission in physics and engineering. Its value may be found in standard mathematical tables.

People who first encounter functions like ln x and erf (x), defined by integrals, sometimes feel that the definition serves only to give a complicated description of a function that must have a simple formula somewhere. But there are no simpler formulas for ln x and erf (x) than the ones given in (3) and (4). Although they may seem unfamiliar, the integral formulas for ln x and erf (x) give a way of calculating the functions to any degree of accuracy we please by standard numerical techniques for estimating integrals. (We shall study two of the easier ones in Article 4.9.)

There is a very practical difference between the formulas

$$\int \frac{1}{t}\,dt \qquad \text{and} \qquad \int_1^x \frac{1}{t}\,dt, \qquad x > 0.$$

The indefinite integral (on the left) is a collection of functions that the formula for the integral does not seem to tell much about. The definite integral (on the right) is a specific function of x whose values we can compute to any degree of accuracy we please. It is as specific as $\sqrt{2}$ and π, numbers that we might otherwise represent in computations by the decimal approximations 1.41421 and 3.14159.

The formula

$$F(x) = \int_a^x f(t)\,dt$$

is important in mathematics for another reason, which is that $F(x)$ is differentiable (if f is continuous) and

$$\frac{dF}{dx} = f(x). \tag{5}$$

This means, for instance, that the differential equation (5) has a solution for any continuous function f. To put it another way, every continuous function is the derivative of some other function. In other words still, every continuous function has an antiderivative.

THEOREM 2

The Second Fundamental Theorem of Integral Calculus

If f is continuous on $[a, b]$ then

$$F(x) = \int_a^x f(t)\, dt$$

is differentiable at every point x in $[a, b]$ and

$$\frac{dF}{dx} = \frac{d}{dx}\int_a^x f(t)\, dt = f(x). \tag{6}$$

COROLLARY

Existence of Antiderivatives of Continuous Functions

If $y = f(x)$ is continuous on $[a, b]$, then there exists a function $F(x)$ whose derivative on $[a, b]$ is f.

Proof of the corollary. Take $F(x) = \int_a^x f(t)\, dt$. The integral exists by the Integral Existence Theorem, Article 4.5, and $dF/dx = f(x)$ by the Second Fundamental Theorem. ∎

In a moment we shall prove the Second Fundamental Theorem by calculating the derivative from its definition as the limit as $\Delta x \to 0$ of the difference quotient

$$\frac{F(x + \Delta x) - F(x)}{\Delta x} = \frac{\int_a^{x+\Delta x} f(t)\, dt - \int_a^x f(t)\, dt}{\Delta x} = \frac{\int_x^{x+\Delta x} f(t)\, dt}{\Delta x}.$$

That is, we shall calculate

$$\lim_{\Delta x \to 0} \frac{\int_x^{x+\Delta x} f(t)\, dt}{\Delta x}. \tag{7}$$

But first, let's look at some of the geometry behind the theorem.

When $f \geq 0$, the limit in (7) has the geometric interpretation shown in Fig. 4.20. The integral from x to $x + \Delta x$ is the area of the strip under the graph of f from x to $x + \Delta x$. The strip is about $f(x)$ high by Δx wide, so that

$$\text{Area} = \int_x^{x+\Delta x} f(t)\, dt \approx f(x)\, \Delta x.$$

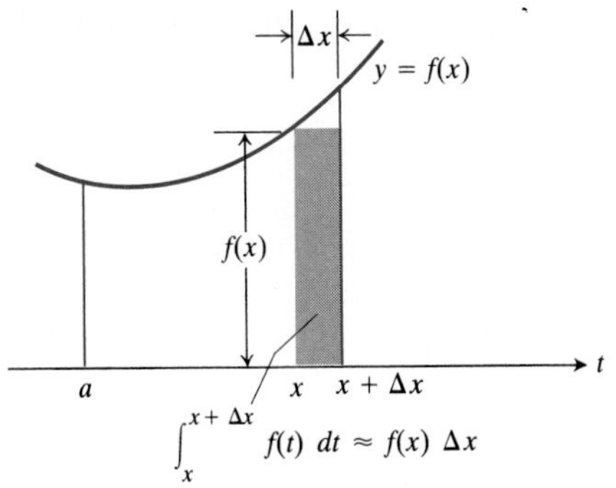

4.20
$$\frac{1}{\Delta x}\int_x^{x+\Delta x} f(t)\,dt \approx \frac{1}{\Delta x} f(x)\,\Delta x = f(x).$$

Dividing through by Δx gives

$$\frac{\text{Area}}{\Delta x} = \frac{\displaystyle\int_x^{x+\Delta x} f(t)\,dt}{\Delta x} \approx f(x).$$

The approximation improves as $\Delta x \to 0$, and we expect (and find) equality in the limit.

Another way that helps in understanding Theorem 2 is to think of Eq. (6) dynamically. Think of t as moving to the right along the t-axis and sweeping out area with a vertical pole $f(t)$ units high. As t passes through x, the instantaneous rate at which area is being added to the shaded region shown in Fig. 4.21 is $f(x)$. Or, as in Fig. 4.22, think of covering the region under the curve $y = f(t)$ from left to right by unrolling a carpet of variable width $f(t)$. The rate at which the floor is being covered as the carpet rolls past x is $f(x)$.

Now to the proof for an arbitrary continuous function.

Proof of the Second Fundamental Theorem. We prove the theorem for arbitrary continuous functions by showing that

$$\lim \frac{\displaystyle\int_a^{x+\Delta x} f(t)\,dt}{\Delta x} = f(x)$$

holds for $\Delta x \to 0^+$ and for $\Delta x \to 0^-$. This will show that the two-sided limit as $\Delta x \to 0$ exists and equals $f(x)$.

For $\Delta x \to 0^+$ and $a < x < b$ the calculation goes as follows: Since $a < x < b$, we can start with values of Δx small enough to make $x + \Delta x$ lie between a and b. Since f is continuous on the closed interval from x to $x + \Delta x$, it has a minimum value min f and a maximum value max f on the interval. The integral inequality I6 in Article 4.5 applied to the interval $[x, x + \Delta x]$ therefore gives

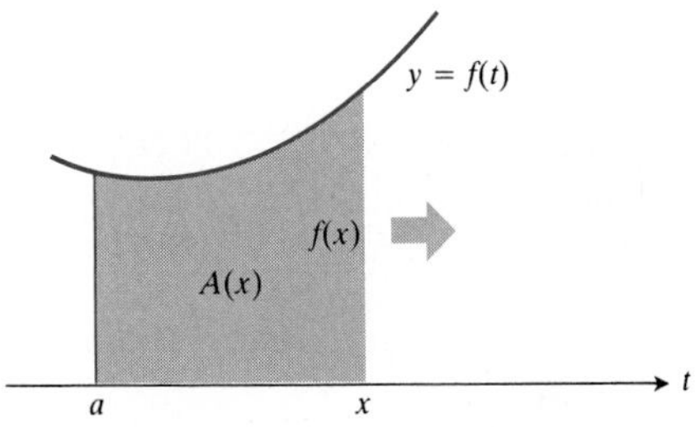

4.21 $\dfrac{dA}{dx} = f(x)$.

$$(\min f)\,\Delta x \le \int_x^{x+\Delta x} f(t)\,dt \le (\max f)\,\Delta x,$$

$$\min f \le \frac{\displaystyle\int_x^{x+\Delta x} f(t)\,dt}{\Delta x} \le \max f. \tag{8}$$

Now, f, being continuous, takes on the value max f at some point c in $[x, x + \Delta x]$ and the value min f at some point c' in $[x, x + \Delta x]$. That is,

$$\min f = f(c')$$

and

$$\max f = f(c).$$

With these substitutions, (8) becomes

$$f(c') \le \frac{\displaystyle\int_x^{x+\Delta x} f(t)\,dt}{\Delta x} \le f(c). \tag{9}$$

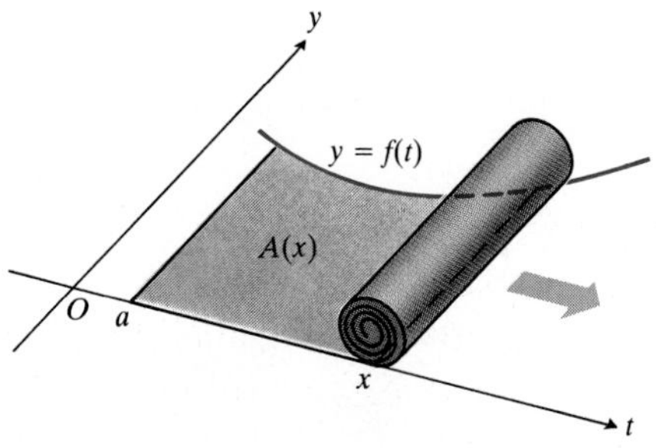

4.22 $\dfrac{dA}{dx} = f(x)$ is the width of the carpet's leading edge.

As $\Delta x \to 0^+$, c' and c both approach x (they are trapped between x and $x + \Delta x$), and $f(c')$ and $f(c)$ both approach $f(x)$ (because f is continuous at x). Therefore, the right-hand and left-hand sides of (9) both approach $f(x)$ and

$$\lim_{\Delta x \to 0^+} \frac{\int_x^{x+\Delta x} f(t)\,dt}{\Delta x} = f(x) \tag{10}$$

by the Sandwich Theorem.

A similar argument shows that

$$\lim_{\Delta x \to 0^-} \frac{\int_x^{x+\Delta x} f(t)\,dt}{\Delta x} = f(x). \tag{11}$$

Together (10) and (11) prove that

$$\frac{d}{dx}\int_a^x f(t)\,dt = f(x), \tag{6}$$

which is Eq. (6).

If $x = a$, we interpret (6) as the right-hand derivative obtained by taking $\Delta x \to 0^+$. If $x = b$, we interpret (6) as the left-hand derivative obtained by taking $\Delta x \to 0^-$.

This concludes the proof of the Second Fundamental Theorem. ■

Proof of the First Fundamental Theorem. Let us now see how the First Fundamental Theorem follows from the Second. The First Fundamental Theorem says:

If f is continuous and $F' = f$, then

$$\int_a^b f(x)\,dx = F(b) - F(a). \tag{1}$$

To prove the theorem, we use Corollary 3 of the Mean Value Theorem in Article 3.8, which says that any two functions that have f as a derivative on $[a, b]$ must differ by some fixed constant throughout $[a, b]$. We already know one function whose derivative equals f, namely,

$$G(x) = \int_a^x f(t)\,dt.$$

Therefore, if F is any other such function, then

$$F(x) = G(x) + C \tag{12}$$

throughout $[a, b]$ for some constant C. When we use (12) to calculate $F(b) - F(a)$ we find that

$$\begin{aligned} F(b) - F(a) &= [G(b) + C] - [G(a) + C] \\ &= G(b) - G(a) \\ &= \int_a^b f(t)\,dt - \int_a^a f(t)\,dt \\ &= \int_a^b f(t)\,dt - 0 \\ &= \int_a^b f(t)\,dt, \end{aligned}$$

which establishes Eq. (1) and the First Fundamental Theorem. ■

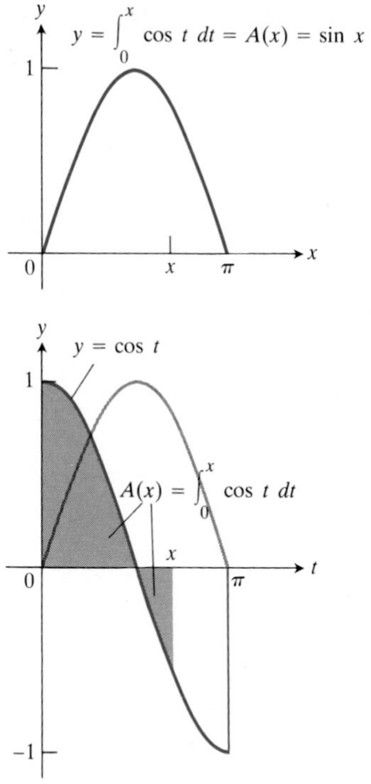

4.23 The upper graph ($y = \sin x$) gives the signed area between the lower graph ($y = \cos t$) and the x-axis.

EXAMPLE 8 See Fig. 4.23. The integral of $y = \cos t$ from 0 to x is

$$F(x) = \int_0^x \cos t\, dt = \sin(x) - \sin(0) = \sin x - 0 = \sin x.$$

By the Second Fundamental Theorem,

$$\frac{dF}{dx} = \cos x,$$

which agrees with the direct calculation

$$\frac{d}{dx}(\sin x) = \cos x.$$

(Nothing new is being discovered here; we're just checking Theorem 2.) □

EXAMPLE 9 Calculate dy/dx if

$$y = \int_0^{x^2} \cos t\, dt.$$

Solution The Chain Rule is the key to doing this problem. We treat y as the composite of

$$f(u) = \int_0^u \cos t\, dt \qquad \text{and} \qquad u = x^2.$$

From the Chain Rule, in the form

$$\frac{dy}{dx} = \frac{df}{du}\frac{du}{dx},$$

we get

$$\frac{dy}{dx} = \cos u \cdot \frac{d}{dx}(x^2) = \cos x^2 \cdot 2x = 2x \cos x^2.$$

Therefore,

$$\frac{d}{dx}\int_0^{x^2} \cos t\, dt = 2x \cos x^2. \ \square$$

EXAMPLE 10 Calculate dy/dx if

$$y = \int_{x^2}^0 \cos t\, dt.$$

Solution We write

$$y = \int_{x^2}^0 \cos t\, dt = -\int_0^{x^2} \cos t\, dt,$$

and apply the result of Example 9 to the integral on the right:

$$\frac{dy}{dx} = -\frac{d}{dx}\int_0^{x^2} \cos t\, dt = -2x \cos x^2. \ \square$$

Leibniz's Rule

In applications we sometimes encounter functions defined by integrals that have variable upper limits and variable lower limits at the same time:

$$y = \int_{\sin x}^{x^2} (1 + t)\, dt, \qquad y = \int_{\sqrt{x}}^{2\sqrt{x}} \sin t^2\, dt.$$

The first integral can be evaluated directly, but the second cannot. We may calculate the derivative of either integral, however, by a formula called Leibniz's Rule:

THEOREM 3

Leibniz's Rule

If f is continuous on $[a, b]$, and $u(x)$ and $v(x)$ are differentiable functions of x whose values lie in $[a, b]$, then

$$\frac{d}{dx}\int_{u(x)}^{v(x)} f(t)\,dt = f(v(x))\frac{dv}{dx} - f(u(x))\frac{du}{dx}. \qquad (13)$$

EXAMPLE 11 Find dy/dx by Leibniz's rule if

$$y = \int_{\sin x}^{x^2} (1 + t)\,dt.$$

Solution From Eq. (13) we have

$$\begin{aligned}\frac{dy}{dx} &= (1 + x^2)\frac{d}{dx}(x^2) - (1 + \sin x)\frac{d}{dx}(\sin x)\\ &= (1 + x^2)(2x) - (1 + \sin x)(\cos x)\\ &= 2x + 2x^3 - \cos x - \sin x \cos x.\end{aligned}$$

In this case we could also find dy/dx by first integrating with respect to t, and then differentiating with respect to x, but Leibniz's rule is quicker. □

Before we prove Leibniz's rule, there is a geometric interpretation in Fig. 4.24 worth looking at. Think of a carpet of variable height $f(t)$ that is being rolled up at the left at the same time x as it is being unrolled at the right. (In this interpretation x is time, not t.) At time x, the floor is covered from $u(x)$ to $v(x)$. The rate du/dx at which carpet is being rolled up need not be the same as the rate dv/dx at which the carpet is being laid down. At any given time x, the area covered by carpet is

$$A(x) = \int_{u(x)}^{v(x)} f(t)\,dt.$$

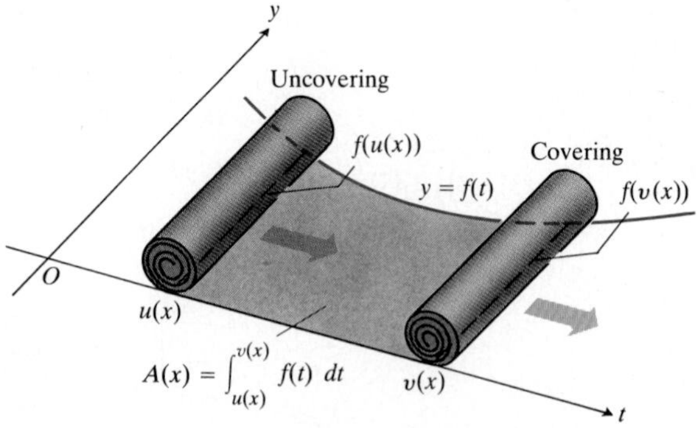

4.24 Maya's carpet.

$$\frac{dA}{dx} = f(v(x))\frac{dv}{dx} - f(u(x))\frac{du}{dx}.$$

Now, at what rate is the covered area changing? At the instant x, $A(x)$ is increasing by the width $f(v(x))$ of the unrolling carpet times the rate dv/dx at which the carpet is being unrolled. That is, $A(x)$ is being increased at the rate

$$f(v(x))\,\frac{dv}{dx}.$$

At the same time, A is being decreased at the rate

$$f(u(x))\,\frac{du}{dx},$$

the width at the end that is being rolled up times the rate du/dx. The net change in A is therefore

$$\frac{dA}{dx} = f(v(x))\,\frac{dv}{dx} - f(u(x))\,\frac{du}{dx},$$

which is precisely Leibniz's rule.

Proof of Leibniz's Rule. To prove Leibniz's rule, let F be an antiderivative of f on the interval $[a, b]$. Then

$$\int_{u(x)}^{v(x)} f(t)\,dt = F(v(x)) - F(u(x)).$$

We differentiate both sides of this equation, using the Chain Rule on the right, to get

$$\begin{aligned}\frac{d}{dx}\int_{u(x)}^{v(x)} f(t)\,dt &= \frac{d}{dx}[F(v(x)) - F(u(x))]\\ &= F'(v(x))\,\frac{dv}{dx} - F'(u(x))\,\frac{du}{dx}\\ &= f(v(x))\,\frac{dv}{dx} - f(u(x))\,\frac{du}{dx},\end{aligned}$$

which is Leibniz's rule. ■

Approximating Finite Sums with Integrals

In applications of calculus, integrals are sometimes used to approximate finite sums, the reverse of the usual procedure of using sums to approximate integrals.

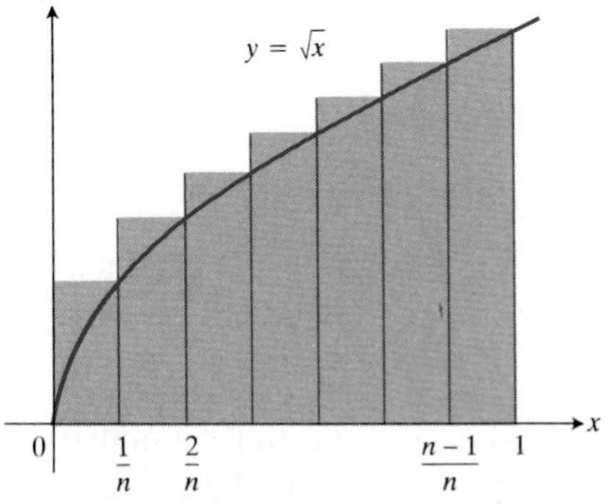

4.25 The relation of the circumscribed rectangles to the integral $\int_0^1 \sqrt{x}\,dx$ leads to an estimate of the sum $1 + \sqrt{2} + \sqrt{3} + \cdots + \sqrt{n}$. See Example 12.

EXAMPLE 12 Estimate the sum of the square roots of the first n positive integers,

$$\sqrt{1} + \sqrt{2} + \cdots + \sqrt{n}.$$

Solution See Fig. 4.25. The integral

$$\int_0^1 \sqrt{x}\,dx = \frac{2}{3}x^{3/2}\Big]_0^1 = \frac{2}{3}$$

is the limit of the sums

$$\begin{aligned}S_n &= \sqrt{\frac{1}{n}}\cdot\frac{1}{n} + \sqrt{\frac{2}{n}}\cdot\frac{1}{n} + \cdots + \sqrt{\frac{n}{n}}\cdot\frac{1}{n}\\ &= \frac{\sqrt{1} + \sqrt{2} + \cdots + \sqrt{n}}{n^{3/2}}.\end{aligned}$$

When n is large, S_n will be close to its limit, 2/3. Therefore, when n is large, the sum of the roots will be about $(2/3)n^{3/2}$:

$$\text{Root sum} = \sqrt{1} + \sqrt{2} + \cdots + \sqrt{n} \approx \frac{2}{3}n^{3/2}, \quad n \text{ large.}$$

For $n = 10$, Root sum ≈ 22.5, while $(2/3)n^{3/2} \approx 21.1$. The relative error is about 6%. When $n = 50$, Root sum ≈ 239.0, while $(2/3)n^{3/2} \approx 235.7$. The relative error is about 1.4%. □

PROBLEMS

In Problems 1–10, find the area bounded by the x-axis, the given curve $y = f(x)$, and the given vertical lines.

1. $y = x^2 + 1, \quad x = 0, \quad x = 3$
2. $y = 2x + 3, \quad x = 0, \quad x = 1$
3. $y = \sqrt{2x + 1}, \quad x = 0, \quad x = 4$
4. $y = \dfrac{1}{\sqrt{2x + 1}}, \quad x = 0, \quad x = 4$
5. $y = \dfrac{1}{(2x + 1)^2}, \quad x = 1, \quad x = 2$
6. $y = (2x + 1)^2, \quad x = -1, \quad x = 3$
7. $y = x^3 + 2x + 1, \quad x = 0, \quad x = 2$
8. $y = x\sqrt{2x^2 + 1}, \quad x = 0, \quad x = 2$
9. $y = \dfrac{x}{\sqrt{2x^2 + 1}}, \quad x = 0, \quad x = 2$
10. $y = \dfrac{x}{(2x^2 + 1)^2}, \quad x = 0, \quad x = 2$
11. Find the area bounded by the coordinate axes and the line $x + y = 1$.
12. Find the area between the curve $y = 4 - x^2$ and the x-axis.
13. Find the area between the curve $y = 1/\sqrt{x}$, the x-axis, and the lines $x = 1$, $x = 4$.
14. Find the area between the curve $y = \sqrt{1 - x}$ and the coordinate axes.
15. Find the area between the curve $x = 1 - y^2$ and the y-axis.
16. Find the area contained between the x-axis and one arch of the curve $y = \cos 3x$.
17. Find the area between the x-axis and one arch of the curve $y = \sin^2 3x$. (*Hint:* $\sin^2 A = (\frac{1}{2})(1 - \cos 2A)$.)
18. The graph of $y = \sqrt{a^2 - x^2}$ over $-a \le x \le a$ is a semicircle of radius a.
    a) Using this fact, explain why it is true that
    $$\int_{-a}^{a} \sqrt{a^2 - x^2}\, dx = \frac{1}{2}\pi a^2.$$
    b) Evaluate
    $$\int_{0}^{a} \sqrt{a^2 - x^2}\, dx.$$

Evaluate the following integrals.

19. $\displaystyle\int_1^2 (2x + 5)\, dx$
20. $\displaystyle\int_0^1 (x^2 - 2x + 3)\, dx$
21. $\displaystyle\int_{-1}^1 (x + 1)^2\, dx$
22. $\displaystyle\int_0^2 \sqrt{4x + 1}\, dx$
23. $\displaystyle\int_0^\pi \sin x\, dx$
24. $\displaystyle\int_0^\pi \cos x\, dx$
25. $\displaystyle\int_{\pi/4}^{\pi/2} \frac{\cos x\, dx}{\sin^2 x}$
26. $\displaystyle\int_0^{\pi/6} \frac{\sin 2x}{\cos^2 2x}\, dx$
27. $\displaystyle\int_0^\pi \sin^2 x\, dx$
28. $\displaystyle\int_0^{2\pi/\omega} \cos^2 (\omega t)\, dt \quad (\omega \text{ constant})$
29. $\displaystyle\int_0^1 \frac{dx}{(2x + 1)^3}$
30. $\displaystyle\int_{-1}^0 x\sqrt{1 - x^2}\, dx$
31. $\displaystyle\int_0^1 \sqrt{5x + 4}\, dx$
32. $\displaystyle\int_{-2}^0 (4 - w)^2\, dw$
33. $\displaystyle\int_{-1}^0 \left(\frac{x^7}{2} - x^{15}\right) dx$
34. $\displaystyle\int_0^2 (t + 1)(t^2 + 4)\, dt$
35. $\displaystyle\int_1^2 \frac{x^2 + 1}{x^2}\, dx$
36. $\displaystyle\int_9^4 \frac{1 - \sqrt{u}}{\sqrt{u}}\, du$
37. $\displaystyle\int_{\pi/6}^{\pi/2} \cos^2 \theta \sin \theta\, d\theta$
38. $\displaystyle\int_0^\pi x \cos\left(2x - \frac{\pi}{2}\right) dx$
39. $\displaystyle\int_{-4}^4 |x|\, dx$
40. $\displaystyle\int_0^\pi \frac{1}{2}[\cos x + |\cos x|]\, dx$
41. Find $\displaystyle\int_{-1}^3 f(x)\, dx$ if $f(x) = \begin{cases} 3 - x, & x \le 2 \\ x/2, & x > 2. \end{cases}$
42. In Article 4.5 we introduced a definition that extended the notion of area from the class of triangles and polygons to a larger class of regions bounded by continuous curves. Whenever we bring in a new definition like that it is a good idea to be sure that the new and old definitions agree on objects to which they both apply. For example, does the integral definition of area give $A = (\frac{1}{2})bh$ for the area of a right triangle with base b and height h? To find out, use an integral to calculate the area of the triangle shown in Fig. 4.26.

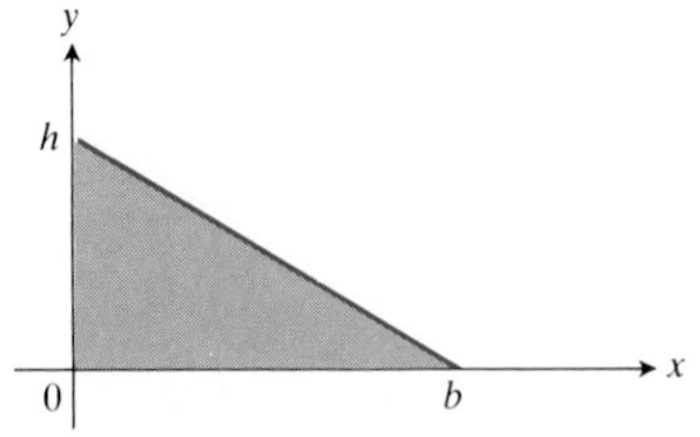

4.26 Is the area of this triangle still $(1/2)bh$? See Problem 42.

43. Find the area between the x-axis and the curve $y = x \sin x$ for (a) $0 \le x \le \pi$, (b) $\pi \le x \le 2\pi$.

Find dF/dx in Problems 44–53.

44. $F(x) = \int_0^x \sqrt{1 + t^2}\, dt$ **45.** $F(x) = \int_1^x \frac{dt}{t}$

46. $F(x) = \int_x^1 \sqrt{1 - t^2}\, dt$ **47.** $F(x) = \int_0^x \frac{dt}{1 + t^2}$

48. $F(x) = \int_1^{2x} \cos(t^2)\, dt$ **49.** $F(x) = \int_1^{x^2} \frac{dt}{1 + \sqrt{1 - t}}$

50. $F(x) = \int_{\sin x}^0 \frac{dt}{2 + t}$ **51.** $F(x) = \int_{1/x}^x \frac{1}{t}\, dt$

52. $F(x) = \int_{\cos x}^{\sin x} \frac{1}{1 - t^2}\, dt$ **53.** $F(x) = \int_{\sqrt{x}}^{2\sqrt{x}} \sin t^2\, dt$

54. Find the value of x that maximizes the integral

$$\int_x^{x+3} t(5 - t)\, dt.$$

(Problems like this arise in the mathematical theory of political elections. See the article "The Entry Problem in a Political Race," by Steven J. Brams and Philip D. Straffin, Jr., in *Political Equilibrium,* Peter Ordeshook and Kenneth Shepfle, Editors, Kluwer-Nijhoff, Boston, 1982, pp. 181–195.)

55. Use l'Hôpital's rule to find

$$\lim_{x \to 0} \frac{1}{x^3} \int_0^x \frac{t^2}{t^4 + 1}\, dt.$$

56. Find (a) the linear approximation and (b) the quadratic approximation of the function

$$y = 2 + \int_0^x \frac{10}{1 + t}\, dt$$

near $x = 0$.

57. Show that

$$y(x) = \frac{1}{a} \int_0^x f(t) \sin a(x - t)\, dt$$

is a solution of the differential equation

$$y'' + a^2 y = f(x), \qquad y(0) = y'(0) = 0.$$

58. Suppose that f has a positive derivative for all values of x and that $f(1) = 0$. Which of the following statements about the function

$$y = \int_0^x f(t)\, dt$$

must be true?

a) y is a differentiable function of x.
b) y is a continuous function of x.
c) The graph of y has a horizontal tangent at $x = 1$.
d) y has a local maximum at $x = 1$.
e) y has a local minimum at $x = 1$.
f) The graph of y has an inflection point at $x = 1$.
g) The graph of dy/dx crosses the x-axis at $x = 1$.

59. Find $f(4)$ if

a) $\int_0^x f(t)\, dt = x \cos \pi x$,

b) $\int_0^{x^2} f(t)\, dt = x \cos \pi x$,

c) $\int_0^{f(x)} t^2\, dt = x \cos \pi x$. (*Hint:* Integrate.)

60. Find $f(\pi/2)$ from the following two pieces of information: (i) $f(x)$ is continuous, and (ii) the area under the graph of f and over the interval $[0, a]$ is

$$\frac{a^2}{2} + \frac{a}{2} \sin a + \frac{\pi}{2} \cos a.$$

61. Let p be a fixed real number and let f be a continuous function that satisfies the equation

$$f(x + p) = f(x)$$

for all x. (For example, $\sin(x + 2\pi) = \sin x$ for all x.)

a) Show that the integral

$$\int_a^{a+p} f(t)\, dt$$

has the same value for every number a. That is, the integral is a constant function of a. (*Hint:* What do you know about the derivative of a constant function?)

b) Find the value of the integral in part (a) if $f(t) = \sin t$ and $p = 2\pi$.

62. What is wrong with the following "proof" of the Second Fundamental Theorem?

STEP 1. Let $G(x) = \int_a^x f(t)\, dt$.
STEP 2. Let F be an antiderivative of f.
STEP 3. Then,

$$G(x) = \int_a^x f(t)\, dt = F(x) - F(a).$$

STEP 4. Differentiate both sides of the equation in Step 3 to get

$$\frac{dG}{dx} = \frac{d}{dx}[F(x) - F(a)] = \frac{d}{dx} F(x) = f(x).$$

63. Evaluate

$$\lim_{n\to\infty} \frac{1^5 + 2^5 + 3^5 + \cdots + n^5}{n^6}$$

by showing that the limit is

$$\int_0^1 x^5\, dx$$

and evaluating the integral.

64. (See Problem 63.) Evaluate

$$\lim_{n\to\infty} \frac{1}{n^4}[1^3 + 2^3 + 3^3 + \cdots + n^3].$$

Toolkit programs	
Integration	Derivative Grapher

4.8 Integration by Substitution. Differentials

The integral

$$\int \sin^2 x \cos x\, dx$$

can be evaluated by setting $u = \sin x$, $du = \cos x\, dx$, and evaluating

$$\int u^2\, du$$

to get

$$\int u^2\, du = \frac{u^3}{3} + C = \frac{\sin^3 x}{3} + C. \tag{1a}$$

Substitution
$u = \sin x$

It is the presence of the factor

$$\cos x = \frac{d}{dx} \sin x$$

that makes this work. By the Chain Rule,

$$\frac{d}{dx}\left(\frac{\sin^3 x}{3}\right) = \sin^2 x \cos x,$$

and therefore

$$\int \sin^2 x \cos x\, dx = \int \sin^2 x \frac{d}{dx}(\sin x)\, dx = \frac{\sin^3 x}{3} + C. \tag{1b}$$

By the
Chain Rule

The integral

$$\int \sqrt{1 + x^2} \cdot 2x\, dx$$

can be evaluated by setting $u = 1 + x^2$, $du = 2x\, dx$, and evaluating

$$\int \sqrt{u}\, du$$

to get

$$\int \sqrt{u}\, du = \int u^{1/2}\, du = \frac{2}{3} u^{3/2} + C = \frac{2}{3}(1 + x^2)^{3/2} + C. \tag{2a}$$

Substitution
$u = 1 + x^2$

It is the presence of the factor

$$\frac{d}{dx}(1 + x^2) = 2x$$

that makes this work. By the Chain Rule

$$\frac{d}{dx}\left(\frac{2}{3}(1 + x^2)^{3/2}\right) = (1 + x^2)^{1/2} \cdot 2x,$$

so that

$$\int \sqrt{1 + x^2}\, 2x\, dx = \int \sqrt{1 + x^2}\, \frac{d}{dx}(1 + x^2)\, dx = \frac{2}{3}(1 + x^2)^{3/2} + C. \tag{2b}$$

Chain Rule

In general, if f and g' are continuous functions, the integral

$$\int f(g(x))g'(x)\, dx$$

can be evaluated by setting $u = g(x)$, calculating

$$\int f(u)\, du,$$

and substituting $u = g(x)$ in the result. The transformation between

$$\int f(g(x))g'(x)\, dx \qquad \text{and} \qquad \int f(u)\, du$$

is accomplished by the formal substitutions

$$u = g(x), \qquad du = g'(x)\, dx. \tag{3}$$

The evaluation of integrals by the substitutions in (3) is called the *substitution method* of integration. This method (when it works) reduces integrals to standard forms we know how to integrate. We shall look at some examples, then go briefly into the theory behind the method.

EXAMPLE 1 Evaluate

$$\int \frac{x\, dx}{\sqrt{4 - x^2}}.$$

Solution With

$$u = 4 - x^2, \qquad du = -2x\, dx, \qquad -\frac{1}{2}\, du = x\, dx,$$

the integral becomes

$$\int \frac{x\,dx}{\sqrt{4-x^2}} = \int \frac{(-\frac{1}{2})\,du}{\sqrt{u}}$$
$$= -\frac{1}{2}\int \frac{du}{\sqrt{u}}$$
$$= -\frac{1}{2}(2u^{1/2}) + C$$
$$= -u^{1/2} + C$$
$$= -\sqrt{4-x^2} + C. \quad \square$$

EXAMPLE 2 Evaluate

$$\int \frac{(z+1)\,dz}{\sqrt[3]{3z^2+6z+5}}.$$

Solution With

$$u = 3z^2 + 6z + 5, \qquad du = (6z+6)\,dz = 6(z+1)\,dz,$$
$$\frac{1}{6}\,du = (z+1)\,dz,$$

the integral becomes

$$\int \frac{(z+1)\,dz}{\sqrt[3]{3z^2+6z+5}} = \int \frac{(\frac{1}{6})\,du}{u^{1/3}}$$
$$= \frac{1}{6}\int u^{-1/3}\,du = \frac{1}{6}\cdot\frac{3}{2}u^{2/3} + C$$
$$= \frac{1}{4}(3z^2+6z+5)^{2/3} + C. \quad \square$$

There is often more than one way to make a successful substitution, as the next example shows.

EXAMPLE 3 Evaluate

$$\int 16x \sin^3(2x^2+1)\cos(2x^2+1)\,dx.$$

Solution METHOD 1. With

$$u = 2x^2 + 1, \qquad du = 4x\,dx,$$

the integral becomes

$$\int 16x \sin^3(2x^2+1)\cos(2x^2+1)\,dx$$
$$= \int 4\sin^3 u \cos u\,du \tag{4}$$
$$= \sin^4 u + C$$
$$= \sin^4(2x^2+1) + C.$$

If we do not see how to integrate (4) we can make another substitution.

We set

$$v = \sin u, \qquad dv = \cos u \, du,$$

to get

$$\begin{aligned}\int 4 \sin^3 u \cos u \, du &= \int 4v^3 \, dv = v^4 + C \\ &= \sin^4 u + C = \sin^4 (2x^2 + 1) + C.\end{aligned}$$

We arrive at the answer either way.

METHOD 2. With

$$\begin{aligned} u &= \sin (2x^2 + 1), \\ du &= \cos (2x^2 + 1) \cdot 4x \, dx, \\ 4 \, du &= 16x \cos (2x^2 + 1) \, dx, \end{aligned}$$

the integral becomes

$$\begin{aligned}\int 16x \sin^3 (2x^2 + 1) \cos (2x^2 + 1) \, dx &= \int 4u^3 \, du \\ &= u^4 + C = \sin^4 (2x^2 + 1) + C. \ \square\end{aligned}$$

The following statements summarize the substitution method of integration.

The Substitution Method of Integration

To evaluate the integral

$$\int f(g(x))g'(x) \, dx,$$

carry out the following steps:

1. Substitute $u = g(x)$ and $du = g'(x) \, dx$ to obtain the integral

$$\int f(u) \, du.$$

2. Evaluate $\int f(u) \, du$ by integrating with respect to u.
3. Replace u by $g(x)$ in the resulting expression.

Why the Substitution Method Works

If F is an antiderivative of f, then, by the Chain Rule,

$$\frac{d}{dx} F(g(x)) = F'(g(x))g'(x).$$

Therefore,

$$\int f(g(x))g'(x) \, dx = \int F'(g(x))g'(x) \, dx = F(g(x)) + C. \tag{5a}$$

Chain Rule

Setting $u = g(x)$ and $du = g'(x) \, dx$ in the integral on the left gives the same result because

$$\int f(u) \, du = F(u) + C = F(g(x)) + C. \tag{5b}$$

Because $u = g(x)$

Evaluating Definite Integrals by Substitution: How to Change the Limits of Integration

If we apply the substitution method to *definite* integrals we have the following formula, which first appeared in print in a book written by Isaac Barrow (1630–1677), Newton's teacher and predecessor at Cambridge University.

$$\int_a^b f(g(x))g'(x)\,dx = \int_{u=g(a)}^{u=g(b)} f(u)\,du. \quad \begin{pmatrix} u = g(x), \\ du = g'(x)\,dx \end{pmatrix} \qquad (6)$$

This formula will hold if g' is continuous on the interval that joins a and b, and f is continuous on the set of values taken on by g.

EXAMPLE 4 Evaluate

$$\int_0^{\pi/2} \sin^2 x \cos x\,dx.$$

Solution We use Eq. (6) with the substitutions

$$u = g(x) = \sin x, \qquad g(0) = \sin(0) = 0,$$
$$du = g'(x)\,dx = \cos x\,dx, \qquad g(\pi/2) = \sin(\pi/2) = 1,$$

to get

$$\int_0^{\pi/2} \sin^2 x \cos x\,dx = \int_{u=g(0)}^{u=g(\pi/2)} u^2\,du = \int_0^1 u^2\,du = \frac{u^3}{3}\Big]_0^1 = \frac{1}{3}. \quad \square$$

To prove Eq. (6), let F be an antiderivative of f, so that $F' = f$. Then,

$$\int_a^b f(g(x))g'(x)\,dx = \int_a^b F'(g(x))g'(x)\,dx$$
$$= F(g(x))\Big]_{x=a}^{x=b} = F(g(b)) - F(g(a)) = \int_{g(a)}^{g(b)} f(u)\,du.$$

In evaluating definite integrals by substitution we now have a choice of methods for dealing with limits, as shown in the next example.

EXAMPLE 5 Evaluate

$$\int_0^2 \frac{6x^2\,dx}{\sqrt{2x^3+9}}.$$

Solution METHOD 1. With

$$u = g(x) = 2x^3 + 9, \qquad du = g'(x)\,dx = 6x^2\,dx,$$

the integral becomes

$$\int_0^2 \frac{6x^2\,dx}{\sqrt{2x^3+9}} = \int_{u=g(0)}^{u=g(2)} \frac{du}{\sqrt{u}} \qquad \text{(by Eq. (6))}$$
$$= \int_9^{25} u^{-1/2}\,du$$
$$= 2u^{1/2}\Big]_{u=9}^{u=25} = 10 - 6 = 4.$$

METHOD 2. With $u = 2x^3 + 9$, $du = 6x^2\,dx$ as before, we get

$$\int \frac{6x^2\,dx}{\sqrt{2x^3+9}} = \int \frac{du}{\sqrt{u}} = 2u^{1/2} = 2(2x^3+9)^{1/2} + C,$$

so that

$$\int_0^2 \frac{6x^2\,dx}{\sqrt{2x^3+9}} = 2(2x^3+9)^{1/2}\Big]_0^2 = 2(25)^{1/2} - 2(9)^{1/2} = 4.$$

In Method 1, we expressed the limits of integration in terms of u, and applied formula (6). In other words, we evaluated the transformed integral with transformed limits.

In Method 2, we used a substitution to find an antiderivative of $6x^2/\sqrt{2x^3+9}$ that could be evaluated at the original limits of integration $x = 0$ and $x = 2$.

Which method is better? Here they seem equally good. At times Method 1 seems easier than Method 2, but at other times Method 2 is easier. The moral is to know both methods. Use whichever one seems better at the time. □

Differentials

You may have noticed in the algebra of the examples of this article that we have treated the symbols du and dx like real variables. These symbols are called *differentials*, du being the differential of u and dx the differential of x.

If $u = g(x)$ and g is differentiable, then when we formally divide both sides of

$$du = g'(x)\,dx$$

by dx, we obtain

$$\frac{du}{dx} = g'(x). \tag{7}$$

Equation (7) says that we may regard the symbol for the derivative du/dx as a quotient of differentials. In many calculations it is convenient to do so. For example, the Chain Rule

$$\frac{dy}{dt} = \frac{dy}{dx}\frac{dx}{dt}$$

may be remembered as a formula in which the dx's cancel in the fractions on the right to produce the fraction on the left.

Although differentials like du and dx are only defined formally, to make the symbols in calculus easier to use, they may be given a meaning as real variables by using them to represent increments in x and u. The equation

$$du = g'(x)\,dx$$

could then be interpreted as the equation

$$\Delta u_{\tan} = g'(x)\,\Delta x \tag{8}$$

that gives the tangent line increment in u (Article 2.5). In this article, however, we give them no meaning outside their formal meaning in the context of differentiation and integration.

Derivatives in Differential Notation The derivative formulas we derived in Chapter 2 can all be expressed in terms of differentials. The differential formulas have the advantage of being shorter and easier to remember.

For example, after multiplying both sides by dx, the formula

$$\frac{d}{dx}(u + v) = \frac{du}{dx} + \frac{dv}{dx} \tag{9a}$$

becomes

$$d(u + v) = du + dv, \tag{9b}$$

which says that the differential of the function $(u + v)$ is the differential of u plus the differential of v. It is still assumed that u and v are differentiable functions, but the name of the independent variable no longer appears in the formula. We do not need to mention it as long as we understand that (9b) is an abbreviation for (9a).

By multiplying each of the derivative formulas from Chapter 2 by dx, we can obtain corresponding formulas for differentials. These are listed in Table 4.3.

Table 4.3 Formulas for derivatives repeated in the notation of differentials

Derivatives	Differentials
1. $\frac{dc}{dx} = 0$	1′. $dc = 0$
2. $\frac{d(cu)}{dx} = c\frac{du}{dx}$	2′. $d(cu) = c\,du$
3. $\frac{d(u+v)}{dx} = \frac{du}{dx} + \frac{dv}{dx}$	3′. $d(u + v) = du + dv$
4. $\frac{d(uv)}{dx} = u\frac{dv}{dx} + v\frac{du}{dx}$	4′. $d(uv) = u\,dv + v\,du$
5. $\frac{d\left(\frac{u}{v}\right)}{dx} = \frac{v\frac{du}{dx} - u\frac{dv}{dx}}{v^2}$	5′. $d\left(\frac{u}{v}\right) = \frac{v\,du - u\,dv}{v^2}$
6. $\frac{du^n}{dx} = nu^{n-1}\frac{du}{dx}$	6′. $d(u^n) = nu^{n-1}\,du$
6a. $\frac{dcx^n}{dx} = cnx^{n-1}$	6a′. $d(cx^n) = cnx^{n-1}\,dx$
7. $\frac{d \sin u}{dx} = \cos u\frac{du}{dx}$	7′. $d(\sin u) = \cos u\,du$
8. $\frac{d \cos u}{dx} = -\sin u\frac{du}{dx}$	8′. $d(\cos u) = -\sin u\,du$
9. $\frac{d \tan u}{dx} = \sec^2 u\frac{du}{dx}$	9′. $d(\tan u) = \sec^2 u\,du$
10. $\frac{d \cot u}{dx} = -\csc^2 u\frac{du}{dx}$	10′. $d(\cot u) = -\csc^2 u\,du$
11. $\frac{d \sec u}{dx} = \sec u \tan u\frac{du}{dx}$	11′. $d(\sec u) = \sec u \tan u\,du$
12. $\frac{d \csc u}{dx} = -\csc u \cot u\frac{du}{dx}$	12′. $d(\csc u) = -\csc u \cot u\,du$

Any problem involving differentials, say that of finding dy when y is given as a function of x, may be handled either

a) by finding dy/dx and multiplying by dx, or

b) by direct use of Formulas 1′–12′.

EXAMPLE 6 Find the differentials of

a) $\sin(2x)$, b) $\cos^2(3x)$, c) $\dfrac{x}{x^2+1}$.

Solutions

a) $d\sin(2x) = \cos(2x)d(2x) = 2\cos(2x)\,dx$.

b)
$$\begin{aligned} d\cos^2(3x) &= 2\cos(3x)d\cos(3x) \\ &= 2\cos(3x)\cdot(-\sin 3x)d(3x) \\ &= -6\sin 3x\cos 3x\,dx \\ &= -3\sin(6x)\,dx. \end{aligned}$$

c)
$$\begin{aligned} d\frac{x}{x^2+1} &= \frac{(x^2+1)\,dx - x d(x^2+1)}{(x^2+1)^2} \\ &= \frac{(x^2+1)\,dx - x(2x\,dx)}{(x^2+1)^2} = \frac{(1-x^2)\,dx}{(x^2+1)^2}. \ \square \end{aligned}$$

It should be noted that a *differential* on the left side of an equation, say dy, also calls for a *differential*, usually dx, on the right side of the equation. Thus we never have $dy = 3x^2$, but instead $dy = 3x^2\,dx$.

Integrals in Differential Notation The notation of differentials allows us to express integrals in a shorthand that often proves useful. We will not do much of it here, but this is a good place to mention the notation for future reference. If u is a differentiable function of x, for example, then the integral of du/dx with respect to x is sometimes written simply as the integral of du:

$$\int \frac{du}{dx}\,dx = \int du. \tag{10}$$

Thus, the integral of du is to be evaluated as

$$\int du = \int \frac{du}{dx}\,dx = u + C, \tag{11}$$

or simply

$$\int du = u + C. \tag{12}$$

For instance, if $u = \sin x$, then $d(\sin x) = \cos x\,dx$, and

$$\int d(\sin x) = \sin x + C \tag{13}$$

is short for

$$\int \cos x\,dx = \int \frac{d}{dx}(\sin x)\,dx = \sin x + C. \tag{14}$$

Similarly, if u and v are both differentiable functions of x, then the equation

$$\int \frac{d(uv)}{dx}\,dx = \int\left[u\frac{dv}{dx} + v\frac{du}{dx}\right]dx = \int u\frac{dv}{dx}\,dx + \int v\frac{du}{dx}\,dx \tag{15}$$

can be written in differential form as

$$\int d(uv) = \int u\,dv + \int v\,du. \tag{16}$$

Abbreviations like these can come in handy as a way of remembering more complicated formulas, as we shall see in Chapters 5 and 7.

PROBLEMS

Evaluate the integrals in Problems 1–30.

1. $\int 2x\sqrt{x^2-1}\,dx$
2. $\int x\sqrt{2x^2-1}\,dx$
3. $\int_0^2 x^2\sqrt{1+x^3}\,dx$
4. $\int_0^{\pi/4} \tan x \sec^2 x\,dx$
5. $\int_0^{\pi} 3\cos^2 x \sin x\,dx$
6. $\int_0^{\sqrt{7}} x(x^2+1)^{1/3}\,dx$
7. $\int x(x^2+1)^{10}\,dx$
8. $\int (1-\cos 3x)\sin 3x\,dx$
9. $\int \frac{dt}{2\sqrt{1+t}}$
10. $\int_0^{\sqrt{3}} \frac{4x}{\sqrt{x^2+1}}\,dx$
11. $\int \frac{x^{1/3}}{(1+x^{4/3})^2}\,dx$
12. $\int 5r\sqrt{1-r^2}\,dr$
13. $\int_0^{\pi} \frac{\sin x}{(3+\cos x)^2}\,dx$
14. $\int \frac{\cos x}{\sqrt{1+\sin x}}\,dx$
15. $\int_0^1 \sqrt{t^5+2t}\,(5t^4+2)\,dt$
16. $\int x\cos(2x^2)\,dx$
17. $\int \cos^2 2x \sin 2x\,dx$
18. $\int x^2\cos(x^3+1)\,dx$
19. $\int_{\pi/4}^{\pi/2} \frac{\cos x}{\sin^2 x}\,dx$
20. $\int x^4(7-x^5)^3\,dx$
21. $\int \frac{x^3}{\sqrt[4]{1+x^4}}\,dx$
22. $\int_{-\pi/4}^{\pi/4} \tan^2 x \sec^2 x\,dx$
23. $\int_{-\pi/4}^{\pi/4} \tan^3 x \sec^2 x\,dx$
24. $\int \frac{x\cos\sqrt{3x^2-6}}{\sqrt{3x^2-6}}$
25. $\int \frac{1}{x^2+4x+4}\,dx$
26. $\int \frac{1}{y^2-2y+1}\,dy$
27. $\int_0^a x\sqrt{a^2-x^2}\,dx$
28. $\int \frac{\sin[(z-1)/3]}{\cos^2[(z-1)/3]}\,dz$
29. $\int \frac{1}{\sqrt{x}(1+\sqrt{x})^2}\,dx$
30. $\int_0^1 (y^4+4y^2+1)^2(y^3+2y)\,dy$

In Problems 31–38, find dy.

31. $y = x^3 - 3x^2 + 5x - 7$
32. $y^2 = (3x^2+1)^{3/2}$
33. $xy^2 + x^2y = 4$
34. $y = \frac{2x}{1+x^2}$
35. $y = x\sqrt{1-x^2}$
36. $y = \frac{x+1}{x^2-2x+4}$
37. $y = \frac{(1-x)^3}{2-3x}$
38. $y = \frac{1+x-x^2}{1-x}$

In Problems 39–41, solve the differential equations subject to the given initial conditions.

39. $\frac{dy}{dx} = \frac{5\cos x}{\sin^2 x}, \quad x = \pi/2,\ y = 10$
40. $\frac{dy}{dx} = 3x^2\sqrt{1+x^3}, \quad x = 0,\ y = 1$
41. $\frac{dy}{dx} = \frac{\sqrt{y^2+1}}{y}\cos x, \quad x = \pi,\ y = \sqrt{3}$
42. Which of the following methods could be used to evaluate

$$\int x^2(x^3-1)^5\,dx?$$

a) Expand $(x^3-1)^5$ and then multiply by x^2 to get a polynomial to integrate term by term.
b) Factor x^2 out to get an integral of the form $x^2\int u^n\,du$.
c) Use the substitution $u = x^3 - 1$ to get an integral of the form $\int u^n\,du$.

43. The substitution $u = \tan x$ gives

$$\int \sec^2 x \tan x\,dx = \int \tan x \cdot \sec^2 x\,dx = \int u\,du = \frac{u^2}{2} + C = \frac{\tan^2 x}{2} + C.$$

The substitution $u = \sec x$ gives

$$\int \sec^2 x \tan x\,dx = \int \sec x \cdot \sec x \tan x\,dx = \int u\,du = \frac{u^2}{2} + C = \frac{\sec^2 x}{2} + C.$$

Can both integrations be correct? Explain.

44. To evaluate

$$\int \sqrt{1+\sin^2(x-1)}\,\sin(x-1)\cos(x-1)\,dx$$

try these sets of substitutions:
a) first $u = x - 1$, then $v = \sin u$, then $w = 1 + v^2$;
b) first $v = \sin(x-1)$, then $w = 1 + v^2$.

45. Show that if f is continuous, then

$$\int_0^1 f(x)\,dx = \int_0^1 f(1-t)\,dt.$$

46. a) Graph the curve $y = x\sqrt{3 - x^2}$.
b) Find the area between the curve and the x-axis.

47. Evaluate

a) $\int_{0.5}^{1} \frac{\pi}{x^2} \sin(\pi/x)\, dx$, b) $\int_{0.01}^{1} \frac{\pi}{x^2} \sin(\pi/x)\, dx$.

48. A basic property of the definite integral is its invariance under translation, as expressed by the equation

$$\int_a^b f(x)\, dx = \int_{a-c}^{b-c} f(x + c)\, dx. \tag{17}$$

This equation will hold if f is continuous and defined for the necessary values of x. For each of the functions below, graph $f(x)$ and $f(x + c)$ together and convince yourself that Eq. (17) is reasonable. Then evaluate both sides of Eq. (17).
a) $f(x) = x^2$, $a = 0$, $b = 1$, $c = 1$
b) $f(x) = \sin x$, $a = 0$, $b = \pi$, $c = \pi/2$
c) $f(x) = \sqrt{x - 4}$, $a = 4$, $b = 8$, $c = 5$

49. Use the method of substitution for definite integrals (Eq. (6)) to verify Eq. (17) in Problem 48.

50. Suppose $F(x)$ is an antiderivative of

$$f(x) = \frac{\sin x}{x}, \quad x \neq 0,$$

Express

$$\int_1^3 \frac{\sin 2x}{x}\, dx$$

in terms of F.

51. Suppose that

$$\int_0^1 f(x)\, dx = 3.$$

Find

$$\int_{-1}^0 f(x)\, dx$$

if a) f is odd, b) f is even.

52. Suppose that the function $h(x)$ is even and continuous for all x.
a) Show that the product $h(x) \sin x$ is odd.
b) Show that, for any number a,

$$\int_{-a}^0 h(x) \sin x\, dx = -\int_0^a h(x) \sin x\, dx.$$

(*Hint:* Use the substitution $u = -x$.)
c) Use the result in (b) to show that

$$\int_{-a}^a h(x) \sin x\, dx = 0.$$

d) Show that

$$\int_{-\pi/4}^{\pi/4} \sec x \sin x\, dx = 0.$$

53. Show that

$$\int_{-a}^a h(x)\, dx = \begin{cases} 0 & \text{if } h \text{ is odd} \\ 2\int_0^a h(x)\, dx & \text{if } h \text{ is even.} \end{cases}$$

54. The integral

$$\int_{-a}^a \sqrt{a^2 - x^2}\, dx$$

can be evaluated by using the substitution

$$x = a \cos \theta, \quad \pi \geq \theta \geq 0$$

to replace $\sqrt{a^2 - x^2}\, dx$ by $-a^2 \sin^2 \theta\, d\theta$. Show that

$$\int_{-a}^a \sqrt{a^2 - x^2}\, dx = \int_\pi^0 -a^2 \sin^2 \theta\, d\theta$$
$$= -\frac{a^2}{2} \int_\pi^0 (1 - \cos 2\theta)\, d\theta,$$

and evaluate this last integral.

4.9 Rules for Approximating Definite Integrals

We know how to evaluate a definite integral when we know an antiderivative of the integrand, but there are simple functions like

$$\frac{\sin x}{x}$$

whose antiderivatives have no elementary formula. There are also functions important in physics and engineering, like

$$\operatorname{erf}(x) = \frac{2}{\sqrt{\pi}} \int_0^x e^{-t^2}\, dt,$$

whose values can be found only by estimating an integral by numerical methods.

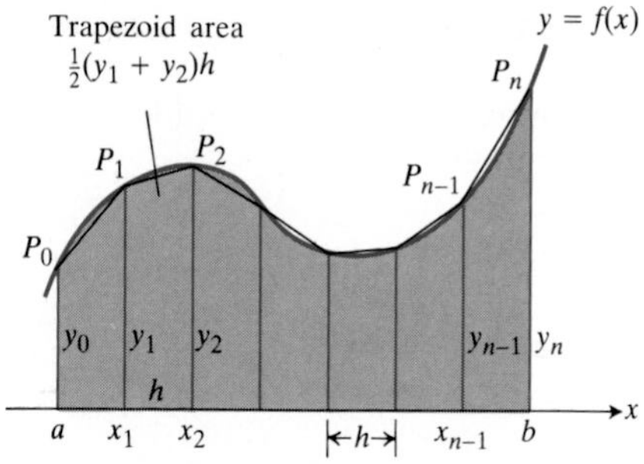

4.27 To estimate the shaded area, add the areas of the trapezoids.

Trapezoidal Approximation

One of the simplest methods of numerical integration is based on estimating the area under a curve with trapezoids instead of rectangles, as in Fig. 4.27. To approximate

$$\int_a^b f(x)\,dx,$$

we divide $[a, b]$ into n subintervals by inserting the points

$$x_1, x_2, \ldots, x_{n-1}$$

between $x_0 = a$ and $x_n = b$. It is not necessary for the points to be evenly spaced, but the resulting formula is simpler if they are. We suppose therefore that each subinterval has the length

$$h = \Delta x = \frac{b-a}{n}.$$

The *trapezoidal rule* says: To estimate $\int_a^b f(x)\,dx$, use the *trapezoidal approximation* T given by

$$\begin{aligned} T &= \frac{1}{2}(y_0 + y_1)h + \frac{1}{2}(y_1 + y_2)h + \cdots \\ &\quad + \frac{1}{2}(y_{n-2} + y_{n-1})h + \frac{1}{2}(y_{n-1} + y_n)h \\ &= h\left(\frac{1}{2}y_0 + y_1 + y_2 + \cdots + y_{n-1} + \frac{1}{2}y_n\right) \\ &= \frac{h}{2}(y_0 + 2y_1 + 2y_2 + \cdots + 2y_{n-1} + y_n), \end{aligned} \tag{1a}$$

where

$$y_0 = f(x_0),\ y_1 = f(x_1),\ \ldots,\ y_n = f(x_n).$$

Since $h = (b-a)/n$, we may also write the approximation in the following way.

RULE

Trapezoidal Rule

$$\int_a^b f(x)\,dx \approx T$$

$$= \frac{b-a}{2n}(y_0 + 2y_1 + 2y_2 + \cdots + 2y_{n-1} + y_n) \tag{1b}$$

(n subintervals of equal length)

EXAMPLE 1 Use the trapezoidal rule with $n = 4$ to estimate $\int_1^2 x^2\,dx$ and compare this approximation with the exact value of the integral.

Solution The exact value of this integral is

$$\int_1^2 x^2\,dx = \frac{x^3}{3}\bigg]_1^2 = \frac{7}{3} \approx 2.33333.$$

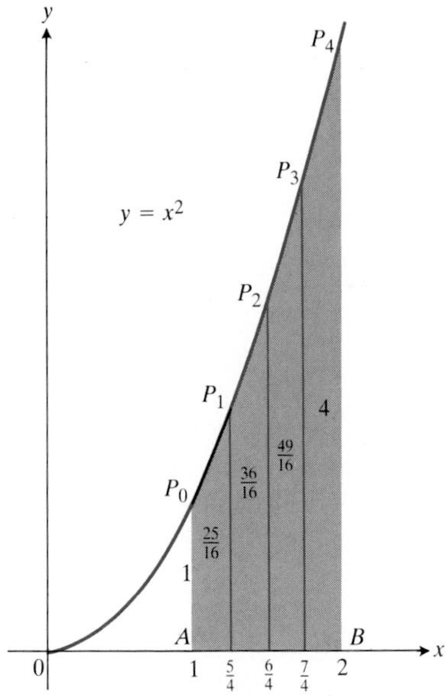

4.28 The trapezoidal approximation to the area under the graph of $y = x^2$, $1 \leq x \leq 2$, is a slight overestimate.

For the trapezoidal approximation we have

$$a = 1, \qquad b = 2, \qquad n = 4, \qquad h = \tfrac{1}{4},$$

so that

$$\begin{aligned} x_0 &= a &&= 1, & y_0 &= f(x_0) = 1^2 = \tfrac{16}{16} \\ x_1 &= a + h &&= \tfrac{5}{4}, & y_1 &= f(x_1) = (\tfrac{5}{4})^2 = \tfrac{25}{16} \\ x_2 &= a + 2h &&= \tfrac{6}{4}, & y_2 &= f(x_2) = (\tfrac{6}{4})^2 = \tfrac{36}{16} \\ x_3 &= a + 3h &&= \tfrac{7}{4}, & y_3 &= f(x_3) = (\tfrac{7}{4})^2 = \tfrac{49}{16} \\ x_4 &= b &&= 2, & y_4 &= f(x_4) = (\tfrac{8}{4})^2 = \tfrac{64}{16} \end{aligned}$$

and

$$\begin{aligned} T &= \frac{b-a}{2n}(y_0 + 2y_1 + 2y_2 + 2y_3 + y_4) \\ &= \frac{1}{8}\left(\frac{75}{4}\right) = \frac{75}{32} = 2.34375. \end{aligned}$$

Thus the approximation is too large by about half a percent. As the graph in Fig. 4.28 shows, each approximating trapezoid contains slightly more area than the corresponding strip of area under the curve. □

Accuracy of the Trapezoidal Approximation

As n increases and $h = \Delta x$ approaches zero, T approaches the exact value of $\int_a^b f(x)\,dx$ as limit. To see why, write

$$\begin{aligned} T &= h\left(\frac{1}{2}y_0 + y_1 + y_2 + \cdots + y_{n-1} + \frac{1}{2}y_n\right) \\ &= (y_1 + y_2 + \cdots + y_n)\,\Delta x + \frac{1}{2}(y_0 - y_n)\,\Delta x \\ &= \sum f(x_k)\,\Delta x + \frac{1}{2}[f(a) - f(b)]\,\Delta x. \end{aligned} \tag{2}$$

As $n \to \infty$ and $\Delta x \to 0$,

$$\sum f(x_k)\,\Delta x \to \int_a^b f(x)\,dx \qquad \text{and} \qquad \frac{1}{2}[f(a) - f(b)]\,\Delta x \to 0.$$

Therefore,

$$\lim_{n\to\infty} T = \int_a^b f(x)\,dx + 0 = \int_a^b f(x)\,dx.$$

This means that in theory we can make the difference between T and the integral as small as we want by taking n large enough. In practice, though, how do we tell how large n should be for a given tolerance?

By an extension of the Mean Value Theorem it is possible to prove that if f and f' are continuous on $[a, b]$ and f' is differentiable on (a, b), then there is a number c in (a, b) such that

$$\int_a^b f(x)\,dx = T - \frac{b-a}{12}h^2 f''(c). \tag{3}$$

(We will not prove (3) here, but proofs can be found in a number of texts on numerical analysis or advanced calculus.) Thus, as $h \to 0$, the error

$$E_T = \frac{b-a}{12} h^2 f''(c) \tag{4}$$

approaches zero as the *square* of h.

The inequality

$$|E_T| \leq \frac{b-a}{12} h^2 \max |f''(x)|, \tag{5}$$

where max refers to $[a, b]$, gives an upper bound for the absolute value of the error in the trapezoidal approximation in case f'' is continuous on $[a, b]$. In practice, we usually cannot find the exact value of $\max|f''(x)|$ in Eq. (5) and have to estimate an upper bound or "worst case" for it instead. If M is *any* upper bound for $\max|f''(x)|$, then

$$|E_T| \leq \frac{b-a}{12} h^2 M, \tag{5a}$$

and this is the inequality we normally use in estimating $|E_T|$. We find the best M we can and go from there to an estimate for $|E_T|$. This may sound careless, but it works. To make $|E_T|$ small for a given M, we just make h small.

Error Bounds for the Trapezoidal Rule

$$|E_T| \leq \frac{b-a}{12} h^2 \max |f''(x)|, \tag{5}$$

$$|E_T| \leq \frac{b-a}{12} h^2 M. \tag{5a}$$

In these formulas, $\max|f''(x)|$ refers to the values $|f''(x)|$ takes on $[a, b]$, and M is any upper bound for $\max|f''(x)|$.

EXAMPLE 2 Obtain an upper bound for the error in the approximation of $\int_1^2 x^2\,dx$ in Example 1.

Solution With $b - a = 1$, $h = 1/4$, and

$$f(x) = x^2, \qquad f'(x) = 2x, \qquad f''(x) = 2,$$

Equation (5) gives

$$|E_T| \leq \frac{1}{12}\left(\frac{1}{4}\right)^2(2) = \frac{1}{96}.$$

This is precisely what we find when we subtract $T = \frac{75}{32}$ from $\int_1^2 x^2\,dx = \frac{7}{3}$, since $\frac{7}{3} - \frac{75}{32} = -\frac{1}{96}$. Here we are able to give the error *exactly*, since the second derivative of $f(x) = x^2$ is a constant and we have no uncertainty in the term $f''(c)$in Eq. (4). We are not always this lucky, and in most cases the best we can do is to *estimate* the difference between the integral and T. □

EXAMPLE 3 How many subdivisions should be used in the trapezoidal rule to approximate

$$\ln 2 = \int_1^2 \frac{1}{x}\,dx$$

with an error of less than 10^{-4}?

Solution To determine n, the number of subdivisions, we use Eq. (5) with

$$b - a = 2 - 1 = 1, \qquad h = \frac{b-a}{n} = \frac{1}{n},$$

$$f''(x) = \frac{d^2}{dx^2}(x^{-1}) = 2x^{-3} = \frac{2}{x^3}.$$

Then,

$$|E_T| \le \frac{b-a}{12}h^2 \max|f''(x)| = \frac{1}{12}\left(\frac{1}{n}\right)^2 \max\left|\frac{2}{x^3}\right|,$$

where max refers to [1, 2].

This is one of the rare cases in which we actually can find max $|f''|$, as opposed to having to settle for an estimate. On [1, 2], $y = 2/x^3$ decreases steadily from a maximum of $y = 2$ to a minimum of $y = 1/4$ (Fig. 4.29). Therefore,

$$|E_T| \le \frac{1}{12}\left(\frac{1}{n}\right)^2 \cdot 2 = \frac{1}{6n^2}.$$

The error's absolute value will therefore be less than 10^{-4} if

$$\frac{1}{6n^2} < 10^{-4}, \qquad \frac{10^4}{6} < n^2, \qquad \frac{100}{\sqrt{6}} < n, \qquad \text{or} \qquad 40.98 < n.$$

The first integer beyond 40.98 is $n = 41$. With $n = 41$ subdivisions we can guarantee calculating ln 2 with an error of less than 10^{-4}. Of course, any larger n will work, too. □

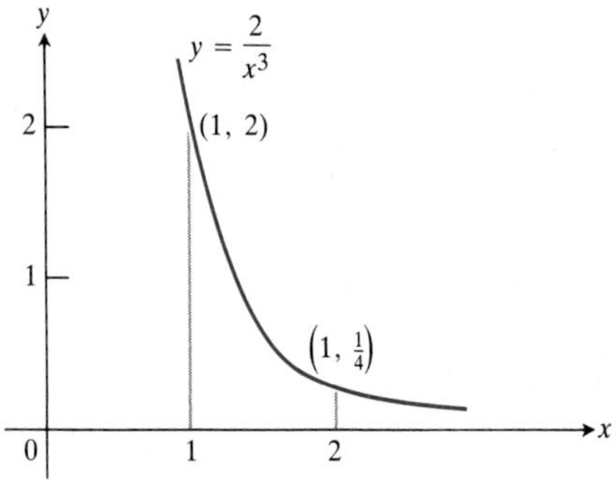

4.29 The continuous function $y = 2/x^3$ has its maximum value on [1, 2] at $x = 1$.

Simpson's Rule

Any three noncollinear points in the plane can be fitted with a parabola, and Simpson's rule is based on approximating curves with parabolas, as shown in Fig. 4.30, instead of trapezoids. The shaded area under the parabola in Fig. 4.30 is

$$A_p = \frac{h}{3}[y_0 + 4y_1 + y_2],$$

and applying this formula successively along a continuous curve $y = f(x)$ from $x = a$ to $x = b$ leads to an estimate of $\int_a^b f(x)\,dx$ that is generally more accurate than T for a given step size h.

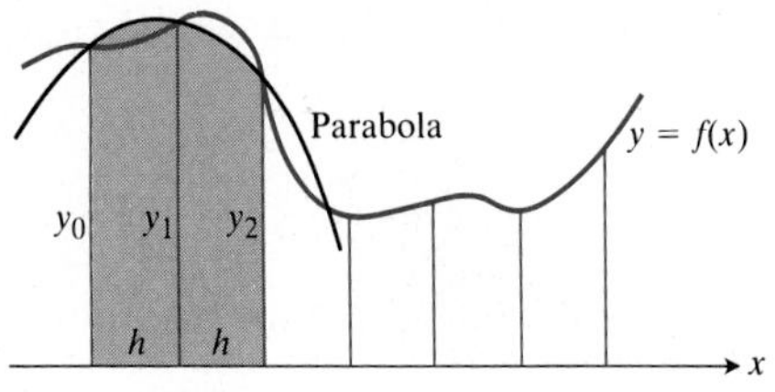

4.30 Simpson's rule approximates short stretches of curves with parabolas.

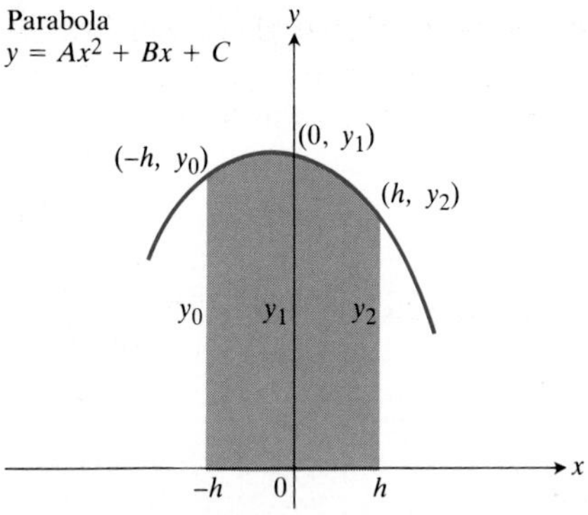

4.31 By integrating from $-h$ to h, the shaded area is found to be

$$A_p = \frac{h}{3}[y_0 + 4y_1 + y_2].$$

We can derive the formula for A_p in the following way. To simplify the algebra, we use the coordinate system shown in Fig. 4.31. The area under the parabola will be the same no matter where the y-axis is, as long as we preserve the vertical scale. The parabola has an equation of the form

$$y = Ax^2 + Bx + C,$$

so the area under it from $x = -h$ to $x = h$ is

$$\begin{aligned} A_p &= \int_{-h}^{h} (Ax^2 + Bx + C)\, dx \\ &= \frac{Ax^3}{3} + \frac{Bx^2}{2} + Cx\Big]_{-h}^{h} \\ &= \frac{2Ah^3}{3} + 2Ch = \frac{h}{3}[2Ah^2 + 6C]. \end{aligned}$$

Since the curve passes through the three points $(-h, y_0)$, $(0, y_1)$, and (h, y_2), we also have

$$y_0 = Ah^2 - Bh + C, \qquad y_1 = C, \qquad y_2 = Ah^2 + Bh + C,$$

from which there follows

$$\begin{aligned} C &= y_1, \\ Ah^2 - Bh &= y_0 - y_1, \\ Ah^2 + Bh &= y_2 - y_1, \\ 2Ah^2 &= y_0 + y_2 - 2y_1. \end{aligned}$$

Hence, expressing the area A_p in terms of the ordinates y_0, y_1, and y_2, we have

$$A_p = \frac{h}{3}[2Ah^2 + 6C] = \frac{h}{3}[(y_0 + y_2 - 2y_1) + 6y_1],$$

or

$$A_p = \frac{h}{3}[y_0 + 4y_1 + y_2].$$

Simpson's rule follows from applying the formula for A_p to successive pieces of the curve $y = f(x)$ between $x = a$ and $x = b$. Each separate piece of the curve, covering an x-subinterval of width $2h$, is approximated by an arc of a parabola through its ends and its midpoint. The areas under the parabolic arcs are then added to give Simpson's rule.

RULE

Simpson's Rule

$$\int_a^b f(x)\, dx \simeq S$$

$$= \frac{h}{3}[y_0 + 4y_1 + 2y_2 + 4y_3 + 2y_4 + \cdots + 2y_{n-2} + 4y_{n-1} + y_n]$$

$$(n \text{ even}, \quad h = (b - a)/n) \qquad (6)$$

The y's in Eq. (6) are the values of $y = f(x)$ at the points

$$x_0 = a, \quad x_1 = a + h, \quad x_2 = a + 2h, \quad \ldots, \quad x_n = a + nh = b,$$

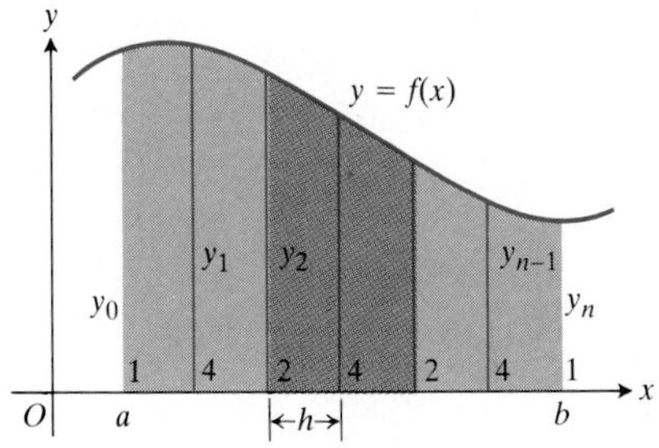

4.32 The number of subdivisions must be even for Simpson's rule to work.

that subdivide the interval $[a, b]$ into n equal subintervals of width $h = (b - a)/n$. (See Fig. 4.32.) The number n must be even to apply the method.

To estimate the error in Simpson's rule, we have the formula

$$\int_a^b f(x)\,dx = S - \frac{b-a}{180}h^4 f^{(4)}(c), \tag{7}$$

where c is some point in (a, b). Formula (7) can be derived from the Extended Mean Value Theorem (see almost any text on numerical analysis or advanced calculus). The formula assumes that f and its first three derivatives are continuous on $[a, b]$ and that $f^{(3)}$ is differentiable throughout (a, b).

From Eq. (7) we see that as $h \to 0$ the error

$$E_S = \frac{b-a}{180}h^4 f^{(4)}(c) \tag{8}$$

in the Simpson approximation approaches zero as the *fourth power* of h. This helps to explain why S often gives a better approximation of $\int_a^b f(x)\,dx$ for a given h than T does, when $h < 1$. The inequality

$$|E_S| \le \frac{b-a}{180}h^4 \max|f^{(4)}(x)|, \tag{9}$$

where max refers to $[a, b]$, gives a useful bound for the absolute value of the error in Simpson's rule when $f^{(4)}$ is continuous on $[a, b]$.

As with $\max|f''|$ in the error formula for the trapezoidal rule, we usually cannot find the exact value of $\max|f^{(4)}(x)|$ on the interval of integration. In practice, we replace it with a reasonable upper bound M. If M is any upper bound for $\max|f^{(4)}x|$ on $[a, b]$, then

$$|E_S| \le \frac{b-a}{180}h^4 M. \tag{9a}$$

This is the formula we usually use in estimating error when we apply Simpson's rule. We find a reasonable value for M and go to an estimate of $|E_S|$ from there.

Error Bounds for Simpson's Rule

$$|E_S| \le \frac{b-a}{180}h^4 \max|f^{(4)}(x)|, \tag{9}$$

$$|E_S| \le \frac{b-a}{180}h^4 M. \tag{9a}$$

In these formulas, $\max|f^{(4)}|$ refers to the values $|f^{(4)}(x)|$ takes on $[a, b]$, and M is any upper bound for $\max|f^{(4)}(x)|$.

If $f(x)$ is a polynomial of degree less than four, then $f^{(4)}(c) = 0$ in Eq. (7) and the approximation S gives the exact value of the integral. We illustrate this in the next example, in which, for comparison, we approximate the value of an integral whose value we already know.

EXAMPLE 4 Approximate $\int_0^1 4x^3\,dx$ by the trapezoidal rule and by Simpson's rule with $n = 2$.

Solution The exact value of the integral is

$$\int_0^1 4x^3\,dx = x^4\Big]_0^1 = 1.$$

The trapezoidal rule with

$$h = \frac{b-a}{n} = \frac{1-0}{2} = \frac{1}{2}$$

gives

$$T = \frac{1}{2\cdot 2}[0 + 2\cdot 4(\tfrac{1}{2})^3 + 4(1)^3] = \tfrac{1}{4}[1+4] = \tfrac{5}{4}.$$

Simpson's rule gives

$$S = \frac{1}{2\cdot 3}[0 + 4\cdot 4(\tfrac{1}{2})^3 + 4(1)^3] = \tfrac{1}{6}[6] = 1.$$

Simpson's rule gives the exact value in this case because $f(x) = 4x^3$ is a polynomial of degree less than four. Since $f^{(4)}(x) = 0$, $E_S = 0$. □

Both of the approximation rules of this article can give good estimates of the integral of a continuous function from a table of values of the function when h is small enough. Thus we can sometimes expect to obtain useful estimates of the integral of a function even when we do not know a formula for the function. This is the case in many practical applications in which the source of our information about a function is a set of specific values of the function measured in the laboratory or in the field.

EXAMPLE 5 A town wants to drain and fill the small swamp shown in Fig. 4.33. The swamp averages 5 ft deep. About how many cubic yards of dirt will be needed to fill the area after the swamp is drained?

Solution To calculate the volume of the swamp we estimate the surface area A and multiply by 5. To estimate A we use Simpson's rule with $h = 20$ ft and the y's equal to the distances measured across the swamp shown in Fig. 4.33:

$$A \approx \frac{20}{3}(146 + 488 + 152 + 216 + 80 + 120 + 14) = \frac{20}{3}(1216) \approx 8107 \text{ ft}^2.$$

The volume of the swamp is about

$$\text{Volume} \approx 8107 \times 5 \text{ ft}^3 = 40{,}535 \text{ ft}^3 \approx 1500 \text{ yd}^3.$$ □

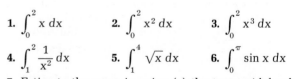

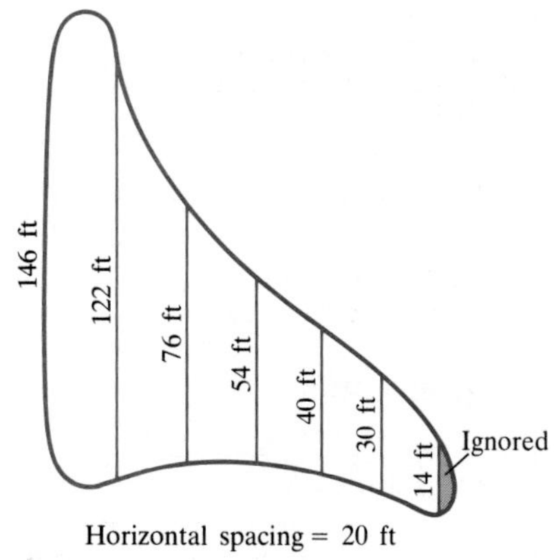

4.33 The swamp in Example 5.

PROBLEMS

Approximate each of the integrals in Problems 1–6 with $n = 4$ by (a) the trapezoidal rule and (b) Simpson's rule. Compare your answers with (c) the exact value in each case. Use the error terms in (d) Eq. (5) and (e) Eq. (9) to estimate the minimum number of subdivisions needed to approximate the integral with an error of less than 10^{-5} by each of the two rules.

1. $\int_0^2 x\,dx$ **2.** $\int_0^2 x^2\,dx$ **3.** $\int_0^2 x^3\,dx$

4. $\int_1^2 \frac{1}{x^2}\,dx$ **5.** $\int_1^4 \sqrt{x}\,dx$ **6.** $\int_0^\pi \sin x\,dx$

7. Estimate the error in using (a) the trapezoidal rule and (b) Simpson's rule to approximate $\int_1^2 (1/x)\,dx$ with $n = 10$.

8. Repeat Example 3 with Simpson's rule in place of the trapezoidal rule.

9. Interpret the meaning of the sign of the error term in Eq. (4) in case the graph of $y = f(x)$ is (a) concave upward over $a < x < b$, (b) concave downward. Illustrate both cases with sketches.

10. (*Calculator*) To estimate $\pi/2$, use the trapezoidal rule with $n = 6$ to estimate the area enclosed by the x-axis and the semicircle $y = \sqrt{1 - x^2}$. Is the estimate too high or too low? Repeat the calculation using Simpson's rule with $n = 6$ in place of the Trapezoidal rule. Is the resulting estimate too high, or too low? ($\pi/2 \approx 1.570796327$.)

11. If the integral

$$\ln 2 = \int_1^2 \frac{1}{t}\, dt$$

were estimated with $n = 4$, how accurate an estimate could you expect from T? from S?

12. (*Calculator*) The design of a new airplane requires a gasoline tank of constant cross-sectional area in each wing. A scale drawing of a cross section is shown in Fig. 4.34. The tank must hold 5000 lb of gasoline that has a density of 50 lb/ft^3. Estimate the length of the tank.

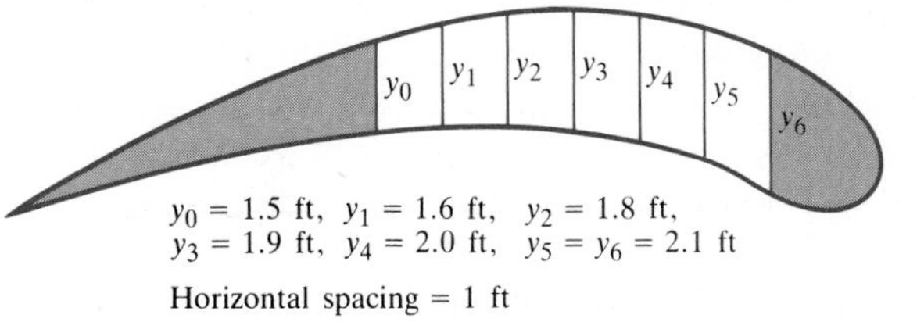

4.34 The airplane wing and tank cross section for Problem 12.

13. (*Calculator*) The rate at which flashbulbs give off light varies during the flash. For some bulbs, the light output, measured in *lumens*, reaches a peak and fades quickly, as shown in Fig. 4.35(a). For other bulbs, the light, instead of reaching a peak, stays at a moderate level for a relatively longer period of time, as shown in Fig. 4.35(b). To calculate how much light reaches the film in a camera, we must know when the shutter opens and closes. A typical focal-plane shutter opens 20 milliseconds and closes 70 milliseconds after the button is pressed. The amount A of light emitted by the flashbulb in this interval is

$$A = \int_{20}^{70} L(t)\, dt \quad \text{lumen-milliseconds},$$

where $L(t)$ is the lumen output of the bulb as a function of time. Use the trapezoidal rule and the numerical data from Tables 4.4 and 4.5 to estimate A for each of the given bulbs, and find out which bulb gets more light to the film.†

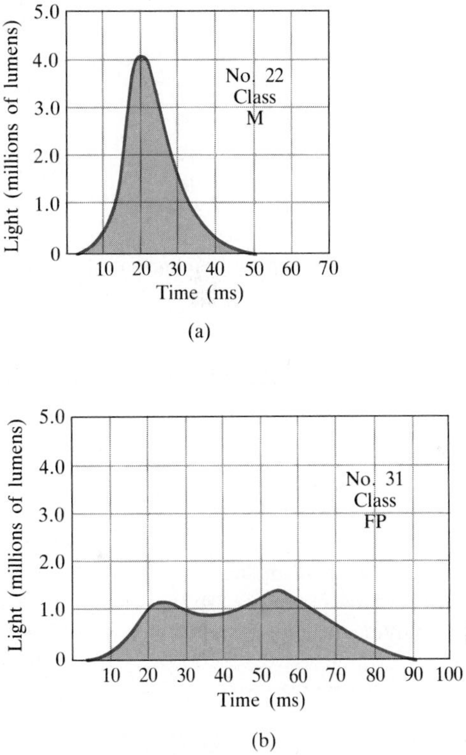

4.35 Flashbulb output data from Tables 4.4 and 4.5 plotted and connected by smooth curves.

Table 4.4 Light output (in millions of lumens) vs. time (in milliseconds) for No. 22 flashbulb

Time after Ignition	Light Output	Time after Ignition	Light Output
0	0	30	1.7
5	0.2	35	0.7
10	0.5	40	0.35
15	2.6	45	0.2
20	4.2	50	0
25	3.0		

Data from *Photographic Lamp and Equipment Guide*, P4–15P, Gen. Elec. Company, Cleveland, Ohio.

† From *Integration*, by W. U. Walton, *et al.*, Project CALC, Education Development Center, Inc., Newton, MA (1975), p. 83.

Table 4.5 Light output (in millions of lumens) vs. time (in milliseconds) for No. 31 flashbulb

Time after Ignition	Light Output	Time after Ignition	Light Output
0	0	50	1.3
5	0.1	55	1.4
10	0.3	60	1.3
15	0.7	65	1.0
20	1.0	70	0.8
25	1.2	75	0.6
30	1.0	80	0.3
35	0.9	85	0.2
40	1.0	90	0
45	1.1		

Data from *Photographic Lamp and Equipment Guide*, P4–15P, Gen. Elec. Company, Cleveland, Ohio.

14. (*Calculator*) As we shall see in Chapter 12, the function

$$f(x) = \begin{cases} \dfrac{\sin x}{x}, & x \neq 0 \\ 1, & x = 0 \end{cases}$$

has derivatives of all orders at every value of x. It can therefore be integrated numerically by Simpson's rule over any closed interval.

a) Use the fact that $|f^{(4)}| \leq 1$ on $[-\pi/2, \pi/2]$ to give an upper bound for the error that would occur if

$$\int_{-\pi/2}^{\pi/2} f(x)\,dx$$

were estimated by Simpson's rule with $n = 4$.

b) Estimate the integral in (a) by Simpson's rule with $n = 4$.

c) Express the upper bound in (a) as a percent of the estimate in (b).

15. (*Calculator*) The error function

$$\operatorname{erf}(x) = \frac{2}{\sqrt{\pi}} \int_0^x e^{-t^2}\,dt$$

must be evaluated numerically because there is no elementary expression for the antiderivative of e^{-t^2}.

a) Use Simpson's rule with $n = 10$ to estimate erf(1).

b) On [0, 1].

$$\left| \frac{d^4}{dt^4} e^{-t^2} \right| \leq 12.$$

Give an upper bound for the absolute value of error of the estimate in (a).

***Toolkit* programs**	
Integration	Integral Evaluator

REVIEW QUESTIONS AND EXERCISES

1. Can there be more than one indefinite integral of a given function? If more than one exist, how are they related? What theorem of Chapter 3 is the key to this relationship?

2. If the acceleration of a moving body is given as a function of time (t), what further information do you need to find the law of motion, $s = f(t)$? How is s found?

3. Write a formula for the area bounded above by the semicircle

$$y = \sqrt{r^2 - t^2},$$

below by the t-axis, on the left by the y-axis, and on the right by the line $t = x$. Without making any calculations, what do you know must be the derivative, with respect to x, of this expression? Explain.

4. What numerical methods do you know for approximating a definite integral? How can an approximation usually be improved? What upper bounds can be calculated for the error?

5. What are the First and Second Fundamental Theorems of Integral Calculus? How can the Second be used to prove the First?

6. Must a continuous function have an antiderivative? Explain.

7. Give an example of integration by substitution. Does the method work for both indefinite integrals and definite integrals?

8. True, or false?
   a) If $\int_a^b f(x)\,dx$ exists, then f is differentiable.
   b) If f is differentiable, then $\int_a^b f(x)\,dx$ exists.
   c) If f is continuous on $[a, b]$ then $\int_a^b f(x)\,dx$ exists.
   d) If f is continuous on $[a, b]$ then $F(x) = \int_a^x f(t)\,dt$ is continuous on $[a, b]$.
   e) If f is continuous on $[a, b]$ then $F(x) = \int_a^x f(t)\,dt$ is differentiable on $[a, b]$.

9. Refer to the Euler-Maclaurin Summation Formula in a textbook on numerical analysis, or in *Handbook of Mathematical Functions* (Dover Publications). How does it relate to the trapezoidal rule?

MISCELLANEOUS PROBLEMS

Solve the differential equations in Problems 1–5.

1. $\dfrac{dy}{dx} = xy^2$ **2.** $\dfrac{dy}{dx} = \sqrt{1 + x + y + xy}$

3. $\dfrac{dy}{dx} = \dfrac{x^2 - 1}{y^2 + 1}$ **4.** $\dfrac{dx}{dy} = \dfrac{y - \sqrt{y}}{x\sqrt{x}}$

5. $\dfrac{dx}{dy} = \left(\dfrac{2 + x}{3 - y}\right)^2$

6. Solve each of the following differential equations subject to the prescribed initial conditions.

a) $\dfrac{dy}{dx} = x\sqrt{x^2 - 4}, \quad x = 2, \quad y = 3$

b) $\dfrac{dy}{dx} = xy^3, \quad x = 0, \quad y = 1$

7. Can there be a curve satisfying the following conditions? d^2y/dx^2 is everywhere equal to zero and, when $x = 0$, $y = 0$ and $dy/dx = 1$. Give a reason for your answer.

8. Find an equation of the curve whose slope at the point (x, y) is $3x^2 + 2$, if the curve is required to pass through the point $(1, -1)$.

9. A particle moves along the x-axis. Its acceleration is $a = -t^2$. At $t = 0$, the particle is at the origin. In the course of its motion, it reaches the point $x = b$, where $b > 0$, but no point beyond b. Determine its velocity at $t = 0$.

10. A particle moves with acceleration $a = \sqrt{t} - (1/\sqrt{t})$. Assuming that the velocity $v = 4/3$ and the position $s = -4/15$ when $t = 0$, find

a) the velocity v in terms of t,

b) the position s in terms of t.

11. A particle is accelerated with acceleration $3 + 2t$, where t is the time. At $t = 0$, the velocity is 4. Find the velocity as a function of time and the distance between the position of the particle at time zero and at time 4.

12. The acceleration of a particle moving along the x-axis is given by $d^2x/dt^2 = -4x$. If the particle starts from rest at $x = 5$, find the velocity when it first reaches $x = 3$.

13. Let $f(x)$, $g(x)$ be two continuously differentiable functions satisfying the relationships $f'(x) = g(x)$ and $f''(x) = -f(x)$. Let $h(x) = f^2(x) + g^2(x)$. If $h(0) = 5$, find $h(10)$.

14. The family of straight lines $y = ax + b$ (a, b arbitrary constants) can be characterized by the relation $y'' = 0$. Find a similar relation satisfied by the family of all circles

$$(x - h)^2 + (y - h)^2 = r^2,$$

where h and r are arbitrary constants. (*Hint:* Eliminate h and r from the set of three equations including the given one and two obtained by successive differentiation.)

15. Assume that the brakes of an automobile produce a constant deceleration of k ft/s^2. (a) Determine what k must be to bring an automobile traveling 60 mi/hr (88 ft/s) to rest in a distance of 100 ft from the point where the brakes are applied. (b) With the same k, how far would a car traveling 30 mi/hr travel before being brought to a stop?

16. Solve the differential equation $dy/dx = x\sqrt{1 + x^2}$ subject to the condition that $y = -2$ when $x = 0$.

17. The acceleration due to gravity is 32 ft/s^2. A stone is thrown upward from the ground with a speed of 96 ft/s. Find the height to which the stone rises in t seconds. What is the maximum height reached by the stone?

18. (*Amer. Math. Monthly* (1955), M. S. Klamkin.) Show that the following procedure will produce a continuous polygonal "curve" whose slope at the point (x_k, y_k) will be $f(x_k)$. First, sketch the auxiliary curve C: $y = xf(x)$. Then through the point $P_0(x_0, y_0)$, construct the line $x = x_0$ intersecting C in $Q_0(x_0, x_0 f(x_0))$. Through P_0 draw a line segment P_0P_1 parallel to OQ_0. Then the slope of P_0P_1 is $f(x_0)$. Now take $P_1(x_1, y_1)$ on this segment to lie close to P_0. For example, take $x_1 = x_0 + h$, where h is small. Then find $Q_1(x_1, x_1 f(x_1))$ on C and through P_1 draw a line segment P_1P_2 parallel to OQ_1. Continue the process by taking $P_2(x_2, y_2)$ close to P_1, with $x_2 = x_1 + h$; then find $Q_2(x_2, x_2 f(x_2))$ on C and through P_2 draw a line segment P_2P_3 parallel to OQ_2; and so on.

19. (a) Apply the procedure of Problem 18 to the case $f(x) = 1/x$ with $x_0 = 1$, $y_0 = 1$, and $h = \frac{1}{4}$. Continue the process until you reach the point $P_4(x_4, y_4)$. What is your value of y_4? (b) Repeat the construction of part (a), but with $h = \frac{1}{8}$, and continue until you reach $x_8 = 2$. What is your value of y_8?

20. A body is moving with velocity 16 ft/s when it is suddenly subjected to a deceleration. If the deceleration is proportional to the square root of the velocity, and the body comes to rest in 4 seconds,

a) how fast is the body moving 2 s after it begins decelerating, and

b) how far does the body travel before coming to rest?

Evaluate the integrals in Problems 21–33.

21. $\displaystyle\int \frac{x^3 + 1}{x^2}\,dx$ **22.** $\displaystyle\int y\sqrt{1 + y^2}\,dy$

23. $\int t^{1/3}(1 + t^{4/3})^{-7}\, dt$

24. $\int \frac{(1 + \sqrt{u})^{1/2}\, du}{\sqrt{u}}$

25. $\int \frac{dr}{\sqrt[3]{(7 - 5r)^2}}$

26. $\int \cos 4x\, dx$

27. $\int \sin^2 3x \cos 3x\, dx$

28. $\int \frac{\cos x\, dx}{\sqrt{\sin x}}$

29. $\int \cos(2x - 1)\, dx$

30. $\int \frac{y\, dy}{\sqrt{25 - 4y^2}}$

31. $\int \frac{dt}{t\sqrt{2t}}$

32. $\int (x^2 - \sqrt{x})\, dx$

33. $\int \frac{dx}{(2 - 3x)^2}$

34. If one side and the opposite angle of a triangle are fixed, prove that the area is a maximum when the triangle is isosceles.

35. A light hangs above the center of a table of radius r ft. The illumination at any point on the table is directly proportional to the cosine of the angle of incidence (i.e., the angle a ray of light makes with the normal) and is inversely proportional to the square of the distance from the light. How far should the light be above the table to give the strongest illumination at the edge of the table?

36. If A, B, C are constants, $AB \neq 0$, prove that the graph of the curve $y = A \sin(Bx + C)$ is always concave toward the x-axis except at its points of inflection, which are its points of intersection with the x-axis.

37. Two particles move on the same straight line so that their distances from a fixed point O, at any time t, are

$$x_1 = a \sin bt \quad \text{and} \quad x_2 = a \sin[bt + (\pi/3)],$$

where a and b are constants, $ab \neq 0$. Find the greatest distance between them.

38. If the identity $\sin(x + a) = \sin x \cos a + \cos x \sin a$ is differentiated with respect to x, is the resulting equation also an identity? Does this principle apply to the equation $x^2 - 2x - 8 = 0$? Explain.

39. A revolving beacon light in a lighthouse $\frac{1}{2}$ mile offshore makes two revolutions per minute. If the shoreline is a straight line, how fast is the ray of light moving along the shore when it passes a point one mile from the lighthouse?

40. The coordinates of a moving particle are $x = a \cos^3 \theta$ and $y = a \sin^3 \theta$. If a is a positive constant and θ increases at the constant rate of ω rad/s, find the magnitude of the velocity vector.

41. The area bounded by the x-axis, the curve $y = f(x)$, and the lines $x = 1$, $x = b$, is equal to $\sqrt{b^2 + 1} - \sqrt{2}$ for all $b > 1$. Find $f(x)$.

42. Let $f(x)$ be a continuous function. Express

$$\lim_{n\to\infty} \frac{1}{n}\left[f\left(\frac{1}{n}\right) + f\left(\frac{2}{n}\right) + \cdots + f\left(\frac{n}{n}\right)\right]$$

as a definite integral.

43. Use the result of Problem 42 to evaluate

a) $\lim_{n\to\infty} \frac{1}{n^{16}}[1^{15} + 2^{15} + 3^{15} + \cdots + n^{15}]$,

b) $\lim_{n\to\infty} \frac{\sqrt{1} + \sqrt{2} + \sqrt{3} + \cdots + \sqrt{n}}{n^{3/2}}$,

c) $\lim_{n\to\infty} \frac{1}{n}\left[\sin\frac{\pi}{n} + \sin\frac{2\pi}{n} + \sin\frac{3\pi}{n} + \cdots + \sin\frac{n\pi}{n}\right]$.

What can be said about the following limits?

d) $\lim_{n\to\infty} \frac{1}{n^{17}}[1^{15} + 2^{15} + 3^{15} + \cdots + n^{15}]$

e) $\lim_{n\to\infty} \frac{1}{n^{15}}[1^{15} + 2^{15} + 3^{15} + \cdots + n^{15}]$

44. Find

a) $\lim_{h\to 0} \frac{1}{h}\int_x^{x+h} \frac{du}{u + \sqrt{u^2 + 1}}$,

b) $\lim_{x\to x_1} \left[\frac{x}{x - x_1}\int_{x_1}^{x} f(t)\, dt\right]$.

45. Variables x and y are related by the equation

$$x = \int_0^y \frac{1}{\sqrt{1 + 4t^2}}\, dt.$$

Show that d^2y/dx^2 is proportional to y and find the constant of proportionality.

46. a) Show that the area A_n of an n-sided regular polygon in a circle of radius r is

$$A_n = \frac{nr^2}{2} \sin\frac{2\pi}{n}.$$

b) Find the limit of A_n as $n \to \infty$. Is this answer consistent with what you know already about the area of a circle?

47. Prove that

$$\int_0^x \left(\int_0^u f(t)\, dt\right) du = \int_0^x f(u)(x - u)\, du.$$

(*Hint:* Express the integral on the right-hand side as the difference of two integrals. Then show that both sides of the original equation have the same derivative with respect to x.)

5 Applications of Definite Integrals

5.1 Introduction

In Chapter 4 we discovered the connection between sums of the form

$$S_a^b = \sum_a^b f(x)\,\Delta x \tag{1}$$

and integration as the inverse of differentiation. The abbreviated notation here is read "the sum from a to b of $f(x)\,\Delta x$." When f is continuous on $a \le x \le b$, we found that the *limit* of S_a^b as Δx approaches zero is the number

$$\int_a^b f(x)\,dx = F(b) - F(a),$$

where F is any antiderivative of f. We applied this to the problem of computing the area between the x-axis and the graph of $y = f(x)$, $a \le x \le b$.

In this chapter we shall extend the applications to the following topics: area between two curves, distance, volumes, lengths of curves, areas of surfaces of revolution, average value of a function, center of mass, centroid, hydrostatic force, and work.

5.2 Area between Two Curves

Suppose that

$$y_1 = f_1(x) \qquad \text{and} \qquad y_2 = f_2(x)$$

are continuous for $a \le x \le b$, and that

$$f_1(x) \ge f_2(x) \qquad \text{for} \quad a \le x \le b.$$

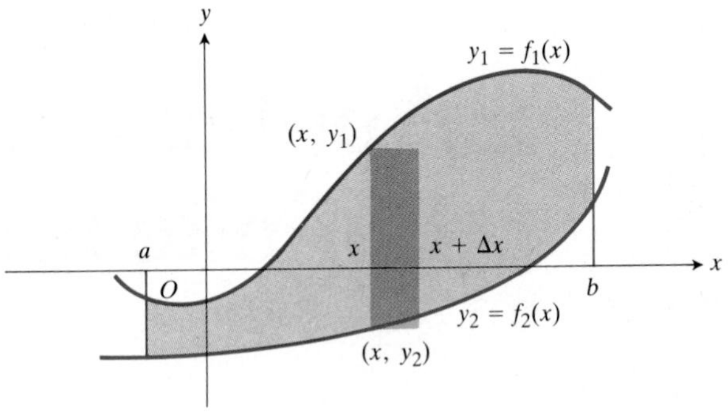

5.1 The area between two curves can be approximated by adding the areas of rectangular strips that reach from one curve to the other.

Then the y_1 curve lies above the y_2 curve from a to b (Fig. 5.1) and we consider the problem of finding the area bounded above by the y_1 curve, below by the y_2 curve, and on the sides by the vertical lines $x = a$, $x = b$.

To define the area, we first cover the region between the curves with vertical rectangles that stretch from curve to curve, like the one shown in Fig. 5.1 whose area is

$$[y_1 - y_2]\,\Delta x = [f_1(x) - f_2(x)]\,\Delta x.$$

The rectangles are defined as usual, by dividing the interval $a \leq x \leq b$ into n subintervals of width $\Delta x = (b - a)/n$ by the points

$$a = x_0, x_1, x_2, \ldots, x_{n-1}, x_n = b.$$

The area of the typical rectangle is then

$$[f_1(x_k) - f_2(x_k)]\,\Delta x.$$

The sum of the n rectangle areas is

$$S_n = \sum_{k=1}^{n} [f_1(x_k) - f_2(x_k)]\,\Delta x = \sum_{k=1}^{n} f_1(x_k)\,\Delta x - \sum_{k=1}^{n} f_2(x_k)\,\Delta x.$$

The limit of S_n as $n \to \infty$ is the number we define to be the area between the curves.

When the functions f_1 and f_2 are integrable (as they are when they are continuous) the limit of the sums S_n exists and may be calculated as the difference of two integrals:

DEFINITION

Area Between Two Curves

$$\text{Area} = \lim S_n = \int_a^b f_1(x)\,dx - \int_a^b f_2(x)\,dx$$

$$= \int_a^b [f_1(x) - f_2(x)]\,dx. \qquad (1)$$

One way to remember the formula for the area in (1) is to think of adding, by an integral, vertical strips of "area" $[f_1(x) - f_2(x)]\,dx$ along the x-axis from $x = a$ to $x = b$. The differential dx plays a dual role in the notation here, in representing both the width of the rectangular strip and the variable of integration in the integral.

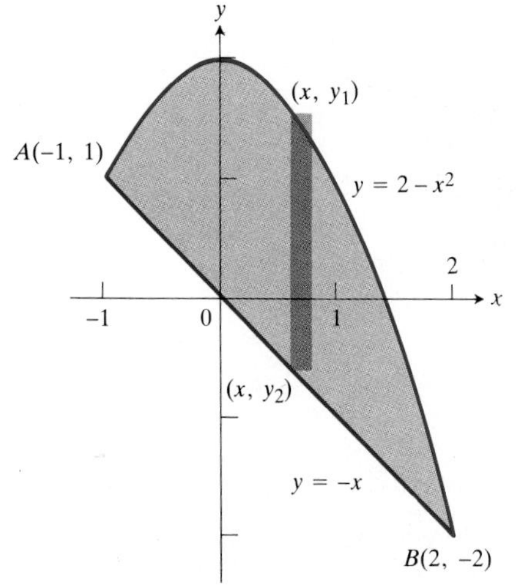

5.2 Calculating the area between $y = 2 - x^2$ and $y = -x$.

EXAMPLE 1 Find the area (Fig. 5.2) bounded by the parabola

$$y = 2 - x^2$$

and the straight line

$$y = -x.$$

Solution We first find where the curves intersect by finding points that satisfy both equations simultaneously. That is, we solve

$$2 - x^2 = -x$$

or

$$x^2 - x - 2 = 0, \qquad (x - 2)(x + 1) = 0, \qquad x = -1, 2.$$

The points of intersection are thus $A(-1, 1)$ and $B(2, -2)$. For all values of x between -1 and $+2$, the curve

$$y_1 = 2 - x^2$$

lies above the line

$$y_2 = -x$$

by an amount

$$y_1 - y_2 = (2 - x^2) - (-x) = 2 - x^2 + x.$$

This is the altitude of a typical rectangle used to approximate that portion of the area lying between x and $x + \Delta x$. The total area A is approximated by

$$A \approx \sum_{-1}^{2} (y_1 - y_2)\,\Delta x = \sum_{-1}^{2} (2 - x^2 + x)\,\Delta x$$

and is given exactly by

$$A = \lim_{\Delta x \to 0} \sum_{-1}^{2} (2 - x^2 + x)\,\Delta x = \int_{-1}^{2} (2 - x^2 + x)\,dx = \frac{9}{2}. \quad \square$$

It is sometimes convenient to find an area by integrating a horizontal strip over an interval on the y-axis, as in the first part of the next example.

EXAMPLE 2 Find the area bounded on the right by the line $y = x - 2$, on the left by the parobola $x = y^2$, and below by the x-axis.

Solution METHOD 1. Integration with respect to y (Fig. 5.3). The points of intersection of the parabola and the line can be found by solving the equations $y = x - 2$ and $x = y^2$ simultaneously. We substitute $x = y^2$ into $y = x - 2$ to find

$$y = y^2 - 2,$$
$$y^2 - y - 2 = 0,$$
$$(y - 2)(y + 1) = 0,$$
$$y = 2, \qquad y = -1.$$

Of the two possibilities, only $y = 2$ gives a point of intersection in the first quadrant where the region lies, namely the point (4, 2).

The length of the horizontal strip from the point (y^2, y) on the left curve to the point $(y + 2, y)$ on the right curve is

$$y + 2 - y^2,$$

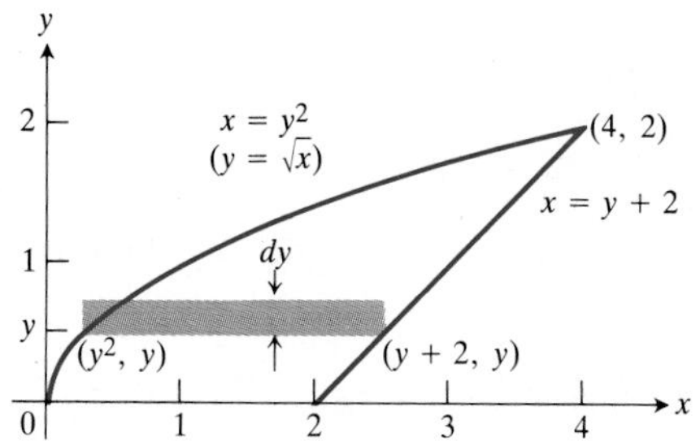

5.3 The area of the horizontal strip is length × width = $(y + 2 - y^2)\,dy$.

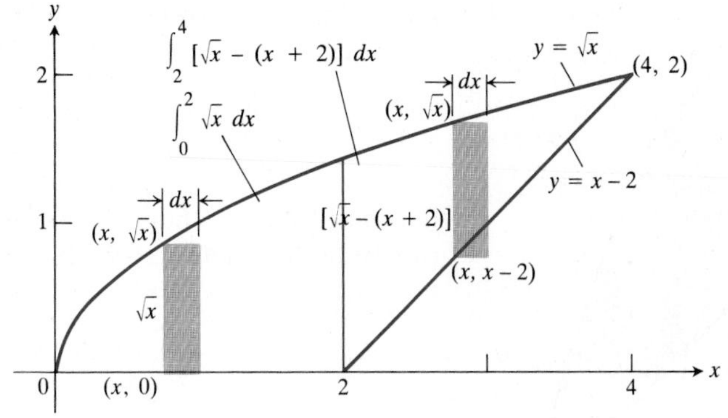

5.4 The area of this region can be expressed as the sum of the two integrals shown.

its width is dy, and we find the area between the curves by integrating the strip of area $(y + 2 - y^2)\, dy$ from $y = 0$ to $y = 2$:

$$\int_0^2 (y + 2 - y^2)\, dy = \left[\frac{y^2}{2} + 2y - \frac{y^3}{3}\right]_0^2 = 2 + 4 - \frac{8}{3} = \frac{10}{3}.$$

METHOD 2. Integration with respect to x (Fig. 5.4). (This way is not as easy as integrating with respect to y.) We divide the area into two portions by the vertical line $x = 2$ and calculate each portion as an integral with respect to x:

$$\begin{aligned} \text{Area} &= \int_0^2 \sqrt{x}\, dx + \int_2^4 [\sqrt{x} - (x - 2)]\, dx \\ &= \left[\frac{2}{3} x^{3/2}\right]_0^2 + \left[\frac{2}{3} x^{3/2}\right]_2^4 - \left[\frac{x^2}{2} - 2x\right]_2^4 \\ &= \frac{2}{3}(2)^{3/2} + \frac{2}{3}(4)^{3/2} - \frac{2}{3}(2)^{3/2} - \left(\frac{16}{2} - 8\right) + \left(\frac{4}{2} - 4\right) \\ &= \frac{2}{3}(8) - 2 \\ &= \frac{10}{3}. \end{aligned}$$

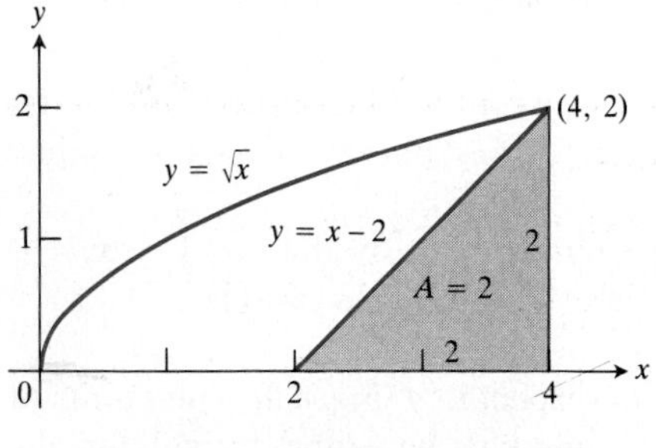

5.5 The area of the unshaded region is also the area under the curve $y = \sqrt{x}$, $0 \le x \le 4$, minus the area of the triangle.

METHOD 3. Subtracting areas (Fig. 5.5). (This is the fastest method because of the geometry of this particular problem.) The area we wish to calculate is the area between the x-axis, $0 \le x \le 4$, and the curve $y = \sqrt{x}$, *minus* the area of a triangle with base 2 and the height 2:

$$\int_0^4 \sqrt{x}\, dx - \frac{1}{2}(2)(2) = \frac{2}{3} x^{3/2}\Big]_0^4 - 2 = \frac{2}{3}(8) - 2 = \frac{10}{3}. \quad \square$$

PROBLEMS

1. Make a sketch to represent a region that is bounded on the right by a continuous curve $x = f(y)$, on the left by a continuous curve $x = g(y)$, below by the line $y = a$, and above by the line $y = b$. Divide the region into n horizontal strips each of altitude $\Delta y = (b - a)/n$ and express the area of the region (a) as a limit of a sum of areas of rectangles, and (b) as an appropriate definite integral.

In Problems 2–20, find the areas bounded by the given curves and lines.

2. The x-axis and the curve $y = 2x - x^2$.
3. The y-axis and the curve $x = y^2 - y^3$.
4. The curve $y^2 = x$ and the line $x = 4$.
5. The curve $y = 2x - x^2$ and the line $y = -3$.

6. The curve $y = x^2$ and the line $y = x$.

7. The curve $x = 3y - y^2$ and the line $x + y = 3$.

8. The curves $y = x^4 - 2x^2$ and $y = 2x^2$.

9. The curve $x = y^2$ and the line $x = y + 2$.

10. The curve $y = x^4$ and the line $y = 8x$.

11. The curves $x = y^3$, $x = y^2$.

12. The curve $y^3 = x$ and the line $y = x$.

13. The curve $y = \sin(\pi x/2)$ and the line $y = x$.

14. The curves $y = \sec^2 x$, $y = \tan^2 x$ and the lines $x = -\pi/4$, $x = \pi/4$.

15. The curve $y = x^2 - 2x$ and the line $y = x$.

16. The curve $x = 10 - y^2$ and the line $x = 1$.

17. The curves $x = y^2$, $x = -2y^2 + 3$.

18. The curves $y = x^2$, $y = -x^2 + 4x$.

19. The line $y = x$ and the curve $y = 2 - (x - 2)^2$.

20. The curves $y = \cos(\pi x/2)$, $y = 1 - x^2$.

21. Find the area of the "triangular" region in the first quadrant bounded by the y-axis and the curves $y = \sin x$, $y = \cos x$.

22. Find the area in the first quadrant bounded by $y = \sqrt{4 - x}$, $x = 0$, $y = 0$.

23. Find the area bounded on the right by $x + y = 2$, on the left by $y = x^2$, and below by the x-axis.

24. Find the area bounded on the right by $y = 6 - x$, on the left by $y = \sqrt{x}$, and below by $y = 1$.

25. Find the area between $y = 3 - x^2$ and $y = -1$, (a) by integration with respect to x, (b) by integration with respect to y.

26. Find the area between $y = \cos x$ and $y = \sin x$, $\pi/4 \le x \le 5\pi/4$.

27. Find the area between $y = x$ and $y = x^3$ from $x = -1$ to $x = 1$.

28. The area bounded by the curve $y = x^2$ and the line $y = 4$ is divided into two equal portions by the line $y = c$.

a) Find c by integrating with respect to y. (This puts c into the limits of integration.)

b) Find c by integrating with respect to x. (This puts c into the integrand.)

29. Figure 5.6 shows a triangle AOC and a shaded region cut from the parabola $y = x^2$ by a horizontal line. Find the limit as $a \to 0$ of the ratio of the area of the triangle to the area of the shaded region.

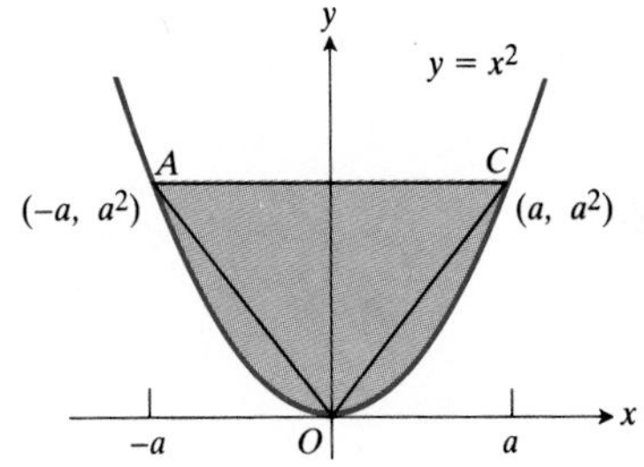

5.6 The figure for Problem 29.

30. Find the area bounded by the curve $\sqrt{x} + \sqrt{y} = 1$ and the coordinate axes.

31. Use Simpson's rule with $n = 10$ to approximate the area between the curves $y = 1/(1 + x^2)$ and $y = -1/(1 + x^2)$, $-1 \le x \le 1$.

***Toolkit* programs**	
Integral Evaluator	Super * Grapher

5.3 Distance

As a second application of the basic principles involved in the use of the fundamental theorems of integral calculus we shall calculate the distance traveled by a body moving along a straight line with velocity

$$v = \frac{ds}{dt} = f(t). \qquad (1)$$

For the moment we shall assume that f is nonnegative as well as continuous for $a \le t \le b$. This means that the body moves in only one direction and, although it may pause, it does not back up.

Now, there are two approaches to calculating the total distance traveled by the body from $t = a$ to $t = b$. The methods are different but yield the same result.

First Method If we can solve the differential equation

$$\frac{ds}{dt} = f(t), \tag{2}$$

then we can determine the position s of the body as a function of t, say

$$s = F(t) + C. \tag{3}$$

The total distance traveled between $t = a$ and $t = b$ by a body that never backs up is the distance between its position when $t = a$ and its position when $t = b$. Thus,

$$\begin{aligned} \text{Distance traveled} &= s(b) - s(a) \\ &= (F(b) + C) - (F(a) + C) \\ &= F(b) - F(a). \end{aligned}$$

Since

$$\int_a^b f(t)\,dt = F(t)\Big]_a^b = F(b) - F(a),$$

we see that

$$\text{Distance traveled} = \int_a^b f(t)\,dt. \tag{4}$$

This is the formula we use for calculating the distance when $ds/dt = f(t) \geq 0$. If the velocity $f(t)$ changes sign during the trip, we have to use a different formula, which we shall come to later in the article.

Second Method In this method we show how the simple formula

$$\text{Distance} = \text{velocity} \times \text{time},$$

which applies only to *constant* velocities, can be extended to a variable velocity, provided we apply it to short subintervals Δt. We imagine the total time interval $[a, b]$ divided into n subintervals, each of duration $\Delta t = (b - a)/n$, by the points

$$a = t_0, t_1, \ldots, t_{n-1}, t_n = b.$$

The velocity at the beginning of the interval is

$$v_0 = f(t_0) = f(a).$$

If Δt is small, the velocity remains nearly constant throughout the time from $t_0 = a$ to $t_1 = a + \Delta t$. Hence, during the first subinterval of time, the body travels a distance Δs_0 that is approximately equal to $v_0\,\Delta t$:

$$\Delta s_0 \approx v_0\,\Delta t = f(t_0)\,\Delta t. \tag{5}$$

By reasoning in the same manner for the second, third, . . . , nth subintervals, we conclude that

$$\begin{aligned} \Delta s_1 &\approx f(t_1)\,\Delta t \\ \Delta s_2 &\approx f(t_2)\,\Delta t \\ &\vdots \\ \Delta s_{n-1} &\approx f(t_{n-1})\,\Delta t \end{aligned}$$

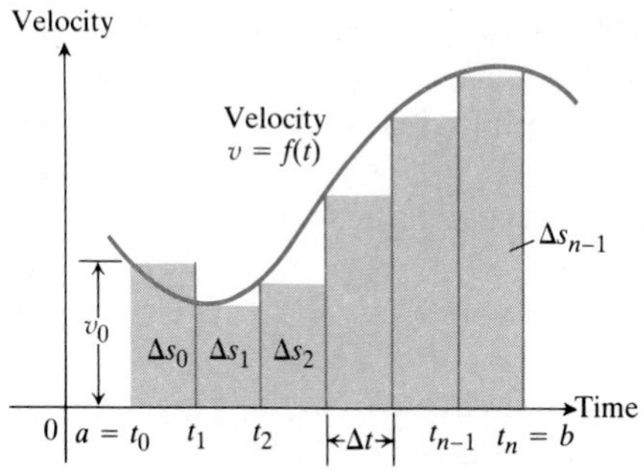

5.7 The area under the velocity curve $v = f(t)$ is approximated by a sum of rectangle areas. Each rectangle area represents a distance traveled at a constant speed.

(Fig. 5.7), where $\Delta s_1, \Delta s_2, \ldots, \Delta s_{n-1}$ represent the distance traveled during these time intervals. Therefore the total distance traveled between $t = a$ and $t = b$ is

$$\begin{aligned} \text{Distance traveled} &= \Delta s_0 + \Delta s_1 + \cdots + \Delta s_{n-1} \\ &\approx f(t_0)\,\Delta t + f(t_1)\,\Delta t + \cdots + f(t_{n-1})\,\Delta t \\ &\approx \sum f(t_k)\,\Delta t. \end{aligned} \tag{6}$$

As $n \to \infty$ the sums tend to the limit

$$\lim \sum f(t_k)\,\Delta t = \int_a^b f(t)\,dt, \tag{7}$$

by virtue of the Integral Existence Theorem, Article 4.5. Once again we are led to Eq. (4).

EXAMPLE 1 A rock thrown straight down from the Golden Gate Bridge at time $t = 0$ has a velocity of

$$v = 9.8t + 8 \text{ m/s}.$$

How far does it fall in the first 4 seconds? (Neglect air friction.)

Solution Since the velocity is nonnegative, we may calculate the distance the rock falls during the first four seconds from Eq. (4):

$$\begin{aligned} \text{Distance fallen} &= \int_0^4 (9.8t + 8)\,dt \\ &= 4.9t^2 + 8t\Big]_0^4 = 78.4 + 32 = 110.4 \text{ meters. } \square \end{aligned}$$

If the velocity changes sign during the interval (a, b), and the body changes direction one or more times, then the integral in Eq. (4) *gives only the net change* in the position coordinate from $t = a$ to $t = b$. This net change is called the body's *displacement* for the time interval. It tells how much the body is displaced by the trip from its original position. If the velocity changes sign, the displacement will not be the same as the total distance traveled. For example, if the body moves 5 feet forward and then 5 feet backward to its original position, it travels 10 feet but the resulting displacement is zero. If the body moves 5 feet forward then 7 feet backward, it travels 12 feet but the resulting displacement is -2 feet. The value of the position coordinate s shows a net decrease of 2 feet.

The integral of the velocity,

$$\int_a^b f(t)\,dt,$$

permits distance traveled forward and backward to cancel. If we want to determine the total distance traveled, we calculate the integral of the *absolute value* of the velocity,

$$\text{Distance traveled} = \int_a^b |f(t)|\,dt. \tag{8}$$

This can be done in practice by integrating separately over the intervals where v is positive and where v is negative and adding the absolute values of the results. The absolute value of the velocity, $|f(t)|$, is called the body's *speed*. Thus, the total distance traveled is the integral of the body's speed (Eq. 8), as opposed to the integral of the body's velocity (Eq. 4). Only if $f(t) \geq 0$ for the time traveled are the two equal.

EXAMPLE 2 Find the position $s(t)$ of a particle moving on a line if

$$\frac{ds}{dt} = v = 5 \cos \pi t \text{ m/s} \tag{9}$$

and $s(0) = 2$. Also, find the total distance traveled by the particle from $t = 0$ to $t = \frac{3}{2}$ seconds, and the particle's displacement for this time period. See Fig. 5.8.

Solution To express s as a function of t we solve Eq. (9) subject to the initial condition $s(0) = 2$. The general solution of

$$\frac{ds}{dt} = 5 \cos \pi t$$

is

$$s = \frac{5}{\pi} \sin \pi t + C.$$

When $t = 0$,

$$s(0) = \frac{5}{\pi} \sin \pi(0) + C, \qquad 2 = 0 + C, \qquad C = 2.$$

Therefore the position of the particle at time t is

$$s(t) = \frac{5}{\pi} \sin \pi t + 2.$$

The total distance traveled from time $t = 0$ seconds to $t = \frac{3}{2}$ seconds is

$$\text{Distance traveled} = \int_{t=0}^{t=3/2} |5 \cos \pi t| \, dt,$$

from Eq. (8). The velocity $v = 5 \cos \pi t$ is positive on $(0, \frac{1}{2})$ and negative on $(\frac{1}{2}, \frac{3}{2})$ (see Fig. 5.8a). Therefore the total distance traveled by the particle from $t = 0$ to $t = \frac{3}{2}$ is

$$\begin{aligned}\int_0^{3/2} |5 \cos \pi| \, dt &= \int_0^{1/2} 5 \cos \pi t \, dt + \int_{1/2}^{3/2} -5 \cos \pi t \, dt \\ &= \frac{5}{\pi} \sin \pi t\Big]_0^{1/2} - \frac{5}{\pi} \sin \pi t\Big]_{1/2}^{3/2} \\ &= \frac{5}{\pi}(1 - 0) - \frac{5}{\pi}(-1 - 1) \\ &= \frac{15}{\pi} \text{ m.}\end{aligned}$$

The displacement, or net change in the position $s(t)$ from $t = 0$ to $t = \frac{3}{2}$, is

$$\text{Displacement} = \int_0^{3/2} 5 \cos \pi t = \frac{5}{\pi} \sin \pi t\Big|_0^{3/2} = \frac{5}{\pi}(-1) = -\frac{5}{\pi} \text{ m.}$$

In other words, the value of s showed a net decrease of $5/\pi$ meters over the time interval from $t = 0$ to $t = \frac{3}{2}$ seconds (Fig. 5.8b).

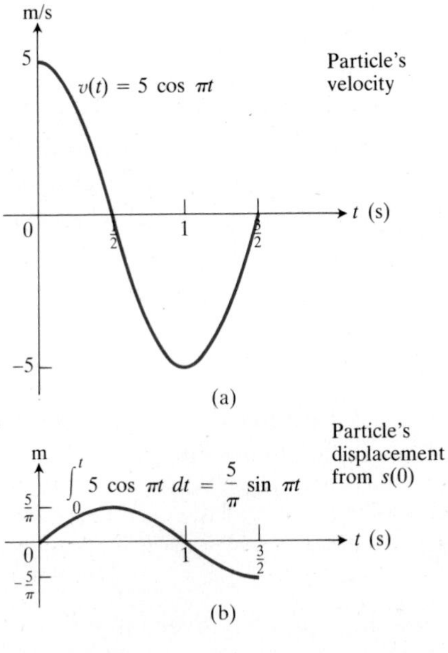

5.8 The velocity (a) and displacement (b) of the particle in Example 2.

Since we have a formula for s(t) in this example, we can also calculate the displacement directly:

$$s(0) = \frac{5}{\pi} \sin(0) + 2 = 2 \text{ m},$$

$$s\left(\frac{3}{2}\right) = \frac{5}{\pi} \sin\left(\frac{3\pi}{2}\right) + 2 = \left(-\frac{5}{\pi} + 2\right) \text{m},$$

and

$$\text{Displacement} = s\left(\frac{3}{2}\right) - s(0) = -\frac{5}{\pi} \text{ m},$$

in agreement with our earlier calculation. □

For linear motion in which the velocity is never negative, the diagram in Fig. 5.9 shows the relationship between

$$\text{Distance traveled} = \int_{t_0}^{t} v(\tau)\, d\tau \qquad (v \geq 0),$$

and acceleration

$$a = \frac{dv}{dt}$$

to the graph of the velocity v(t). The distance traveled is the area under the velocity curve, and the acceleration is the slope of the velocity curve. Incidentally, Galileo Galilei (1564–1642), who dropped iron balls from the leaning tower in his hometown, Pisa, to study their motion, knew that the area under the time–velocity curve represented distance.

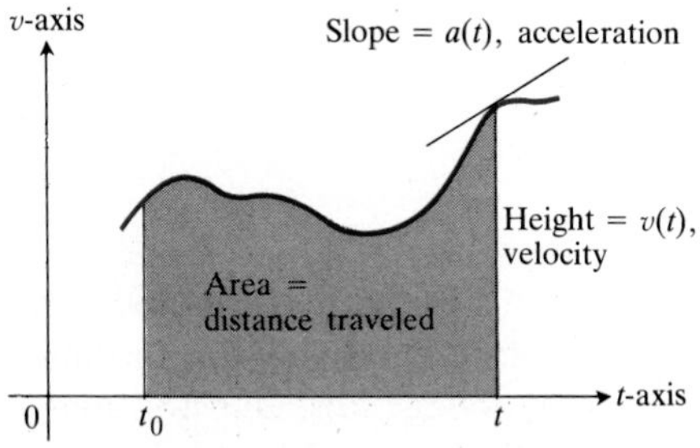

5.9 The graph of velocity versus time for a moving body.

PROBLEMS

In Problems 1–8 the function $v = f(t)$ represents the velocity v (m/s) of a moving body as a function of the time t (s). Sketch enough of the graph of v versus t to find when the velocity is (a) positive, and (b) negative. Then find (c) the displacement, and (d) the total distance traveled by the body between $t = a$ and $t = b$.

1. $v = 2t + 1, \quad 0 \leq t \leq 2$

2. $v = t^2 - t - 2, \quad 0 \leq t \leq 3$

3. $v = t - \frac{8}{t^2}, \quad 1 \leq t \leq 3$

4. $v = |t - 1|, \quad 0 \leq t \leq 2$

5. $v = 6 \sin 3t, \quad 0 \leq t \leq \frac{\pi}{2}$

6. $v = 4 \cos 2t, \quad 0 \leq t \leq \pi$

7. $v = \sin t + \cos t, \quad 0 \leq t \leq \pi$

8. $v = \sin t \sqrt{2 + 2 \cos t}, \quad 0 \leq t \leq \pi$

In Problems 9–13 the function $a = f(t)$ represents the acceleration (m/s^2) of a moving body and v_0 is its velocity at time $t = 0$. Find the *distance* traveled by the body between time $t = 0$ and $t = 2$.

9. $a = \sin t, \quad v_0 = 2$ **10.** $a = 1 - \cos t, \quad v_0 = 0$

11. $a = g(\text{const.}), \quad v_0 = 0$

12. $a = \sqrt{4t + 1}, \quad v_0 = -4\tfrac{1}{3}$

13. $a = \frac{1}{\sqrt{4t + 1}}, \quad v_0 = 1$

Problems 14–16 give the velocity $v = f(t)$ m/s of a body for a time interval $[a, b]$ and the initial position $s(a)$ at the time $t = a$ seconds. Find (a) the position s(t) as a function of t, (b) the total distance traveled from $t = a$ to $t = b$, and (c) the body's displacement for the time period. Calculate the displacement in part (c) both as an integral and directly from the formula for s(t) obtained in part (a).

14. $v = 5 \cos \pi t, \quad [0, \tfrac{3}{2}], \quad s(0) = -1$

15. $v = -\sin t, \quad [0, 2\pi], \quad s(0) = 1$

16. $v = \cos t + |\cos t|, \quad [0, \pi], \quad s(0) = 0$

17. Suppose water flows into a tank at the rate of f(t) gal/min, where f is a given, positive, continuous function of t. Let the amount of water in the tank at time $t = 0$ be Q_0 gal. Apply theorems from Chapter 4 to show that the amount of water in the tank at any later time $t = b$ is

$$Q = Q_0 + \int_0^b f(t)\, dt.$$

18. (*Calculator*) Table 5.1 shows the speed of an automobile every 10 seconds from start to stop on a two-minute trip. Use Simpson's rule to estimate how far the car went. Compare your results with the graphs in Fig. 1.88.

Table 5.1 Selected Speeds of a Car on a Two-minute Trip

Time (s)	Velocity (mph)	Time (s)	Velocity (mph)
0	0	70	66
10	32	80	66
20	51	90	58
30	57	100	40
40	54	110	6
50	64	120	0
60	66		

***Toolkit* programs**

Derivative Grapher
Integral Evaluator
Super * Grapher

5.4 Calculating Volumes by Slicing

Now that we have defined the areas of regions in the plane bounded by the graphs of continuous functions, we are in a good position to define volumes of solids whose cross sections are plane regions. Our first step is to define the volume of a cylinder of known base area A and height h to be Ah. This extends the formula

$$\text{Volume} = \text{base} \times \text{altitude}$$

from classical geometry to cylinders with more arbitrary bases, like the one shown in Fig. 5.10.

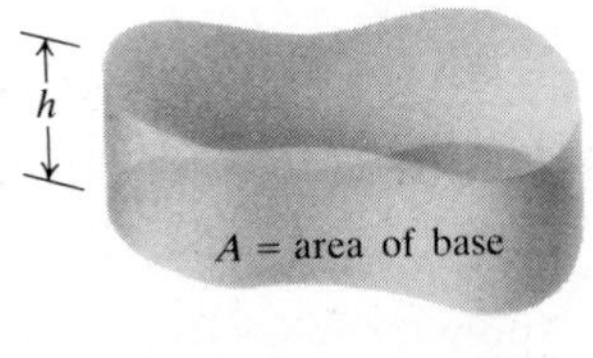

5.10 The volume of a cylinder with an arbitrary base and height h can now be defined to be Ah.

With this as a starting point, the volumes of many solids can be defined by the "method of slicing." Suppose, for example, that the solid is bounded by two parallel planes perpendicular to the x-axis at $x = a$ and $x = b$. Imagine the solid to be cut into thin slices of thickness Δx by planes perpendicular to the x-axis. Then the total volume V of the solid can be defined as the sum of the volumes of these slices (see Fig. 5.11).

However we define the volume of a representative slice between x and $x + \Delta x$, we want it to be a number that lies somewhere between the cylinder volume $A_{\min}\,\Delta x$ and the cylinder volume $A_{\max}\,\Delta x$, where $A_{\min}$ and $A_{\max}$ are the smallest and largest cross-sectional areas in the interval $[x, x + \Delta x]$. (See Fig. 5.12.) The portion of the solid sliced by the planes at x and $x + \Delta x$ contains the smaller cylinder and is itself contained in the larger. In symbols, then, we want

$$A_{\min}\,\Delta x \le \Delta V \le A_{\max}\,\Delta x \tag{1}$$

Slice volume-to-be

If the cross-sectional area of the solid is a continuous function $A(x)$, then A takes on its minimum value at some point c in $[x, x + \Delta x]$ and takes on its maximum value at some point c' in $[x, x + \Delta x]$. That is,

$$A_{\min}\,\Delta x = A(c)\,\Delta x, \qquad A_{\max}\,\Delta x = A(c')\,\Delta x. \tag{2}$$

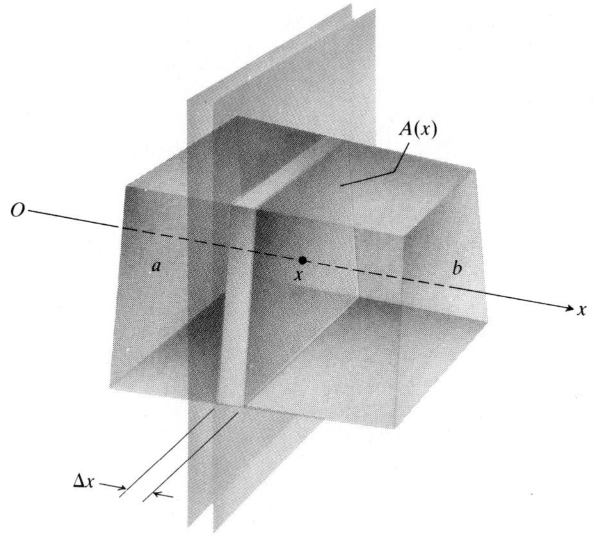

5.11 The volume of the solid is the sum of the volumes of slices of thickness Δx.

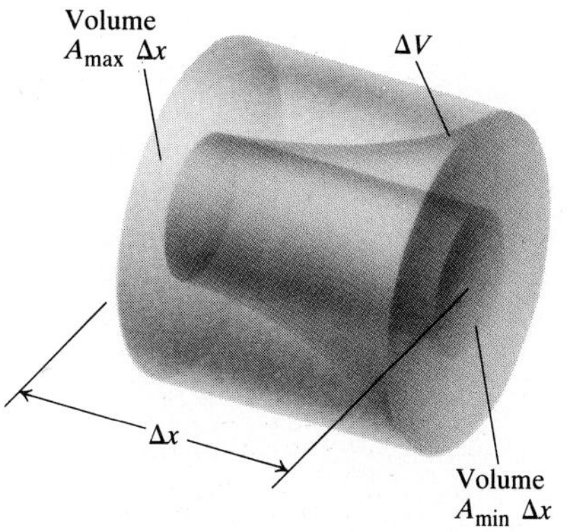

5.12 The trumpet-shaped solid slice lies between two cylinders whose volumes we can calculate.

With these substitutions, (1) becomes

$$A(c)\,\Delta x \leq \Delta V \leq A(c')\,\Delta x. \tag{3}$$

Adding these volumes for all slices from $x = a$ to $x = b$ gives

$$\sum_a^b A(c)\,\Delta x \leq \sum_a^b \Delta V \leq \sum_a^b A(c')\,\Delta x. \tag{4}$$

As $\Delta x \to 0$ the sums on the left and right in (4) both approach $\int_a^b A(x)\,dx$. We therefore define the volume V of the solid to be this integral.

DEFINITION

> The *volume* of a solid of known cross-sectional area $A(x)$ from $x = a$ to $x = b$ is
>
> $$\text{Volume} = \int_a^b A(x)\,dx. \tag{5}$$

To use this definition we do not need $A(x)$ to be continuous as long as the integral exists.

The practical steps involved in using Eq. (5) to calculate volumes of solids are as follows:

1. Draw the region whose volume is to be found.
2. Find $A(x)$.
3. Find the limits of integration.
4. Integrate.

Volume of a Solid of Revolution

The solid generated by rotating a plane region about an axis in its plane is called a solid of revolution. To find the volume of a solid like the one shown in Fig. 5.13, we need only observe that the cross-sectional area $A(x)$ in Eq. (5) is the area of a circle of radius $r = y = f(x)$, so that

$$A(x) = \pi r^2 = \pi[f(x)]^2. \tag{6}$$

EXAMPLE 1 Suppose the curve in Fig. 5.13 represents the graph of

$$y = \sqrt{x}$$

from (0, 0) to (4, 2). Then the cross-sectional area of the representative slice is the continuous function

$$A(x) = \pi y^2 = \pi(\sqrt{x})^2 = \pi x$$

and the volume is

$$V = \lim_{\Delta x \to 0} \sum_0^4 \pi x \, \Delta x = \int_0^4 \pi x \, dx = \pi \frac{x^2}{2}\Big]_0^4 = 8\pi. \; \square$$

EXAMPLE 2 The circle

$$x^2 + y^2 = a^2$$

is rotated about the x-axis to generate a sphere. Find its volume.

Solution We imagine the sphere cut into thin slices by planes perpendicular to the x-axis (Fig. 5.14). The cross-sectional area at a typical point x between $-a$ and a is

$$A(x) = \pi y^2 = \pi(a^2 - x^2).$$

Therefore, the volume is

$$V = \int_{-a}^{a} A(x)\, dx = \int_{-a}^{a} \pi(a^2 - x^2)\, dx = \pi\left[a^2x - \frac{x^3}{3}\right]_{-a}^{a} = \frac{4}{3}\pi a^3.$$

5.13 A slice perpendicular to the axis of a solid of revolution.

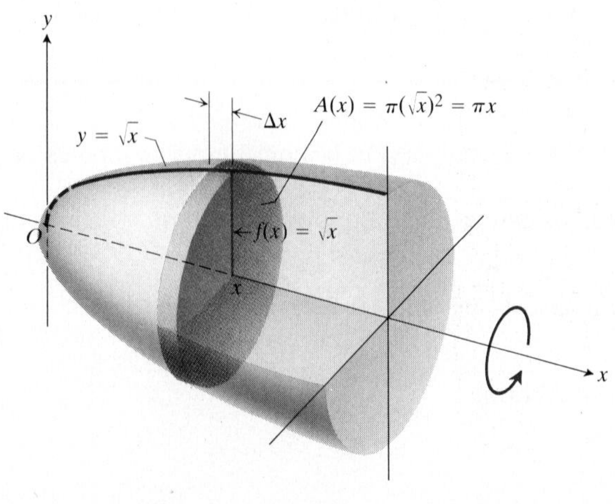

5.14 The sphere generated by rotating the circle $x^2 + y^2 = a^2$ about the x-axis.

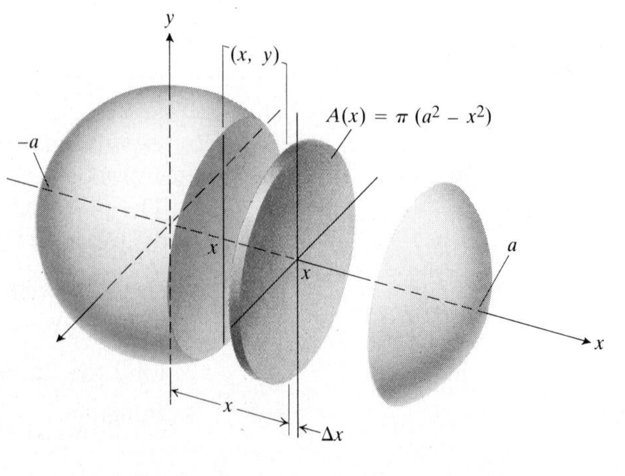

To see how the calculation here corresponds to the modeling that led to the definition of volume, note that the volume of a typical slice between the two planes at x and x + Δx is approximately

$$\pi y^2 \Delta x = \pi(a^2 - x^2)\,\Delta x,$$

and the sum of all slices is approximately

$$V \approx \sum_{-a}^{a} \pi(a^2 - x^2)\,\Delta x.$$

The exact volume is

$$V = \lim_{\Delta x \to 0} \sum_{-a}^{a} \pi(a^2 - x^2)\,\Delta x = \int_{-a}^{a} \pi(a^2 - x^2)\,dx = \frac{4}{3}\pi a^3. \quad \square$$

Volumes of Other Solids

The method of slicing and the formula

$$V = \int_a^b A(x)\,dx$$

can be used to calculate volumes of solids that are not solids of revolution. We slice the solids by planes perpendicular to a coordinate axis as before, but instead of being discs the cross sections whose areas appear in the formula are some other shape.

EXAMPLE 3 A wedge is cut from a right circular cylinder of radius r by two planes. One plane is perpendicular to the axis of the cylinder while the second makes an angle α with the first and intersects it at the center of the cylinder. Find the volume of the wedge.

Solution Again, we shall discuss the modeling process as we go through the calculation.

The volume ΔV of the slice between y and y + Δy in Fig. 5.15 is approximately

$$\Delta V \approx A(y)\,\Delta y$$

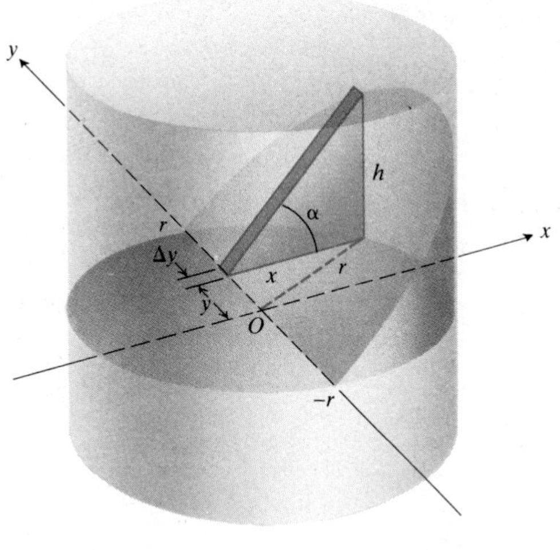

5.15 The curved wedge can be cut into triangular slices.

where

$$A(y) = \frac{1}{2}xh$$

is the area of the triangle that forms one face of the slice and is to be expressed as a function of y. By trigonometry,

$$h = x \tan \alpha,$$

and by the Pythagorean theorem,

$$x^2 = r^2 - y^2.$$

Hence

$$A(y) = \frac{1}{2}x^2 \tan \alpha = \frac{1}{2}\tan \alpha(r^2 - y^2).$$

The total volume is given by

$$V = \lim_{\Delta y \to 0} \sum_{-r}^{r} A(y)\,\Delta y = \int_{-r}^{r} \frac{1}{2}\tan \alpha(r^2 - y^2)\,dy$$

$$= \frac{1}{2}\tan \alpha \int_{-r}^{r} (r^2 - y^2)\,dy = \frac{1}{2}\tan \alpha \left[r^2 y - \frac{y^3}{3}\right]_{y=-r}^{y=r}$$

$$= \frac{1}{2}\tan \alpha \left[\left(r^3 - \frac{r^3}{3}\right) - \left(-r^3 + \frac{r^3}{3}\right)\right]$$

$$= \frac{1}{2}\tan \alpha \left[\frac{2}{3}r^3 + \frac{2}{3}r^3\right] = \frac{2}{3}r^3 \tan \alpha.$$

The volume of the wedge is

$$V = \frac{2}{3}r^3 \tan \alpha. \;\square$$

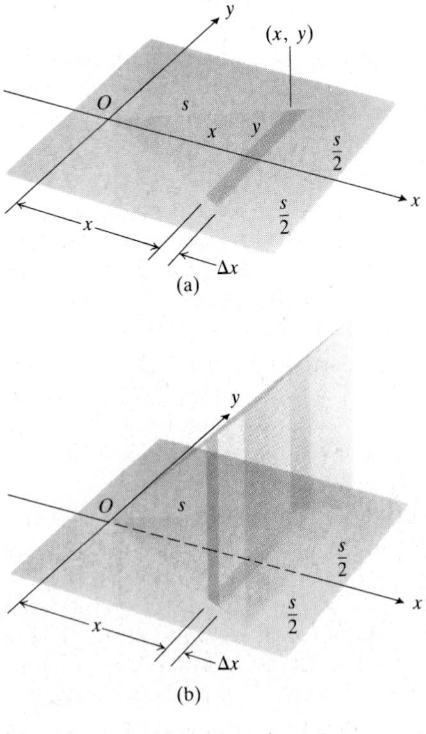

5.16 A solid with a triangular base (a), and square sections (b).

EXAMPLE 4 The base of the solid in Fig. 5.16 is an equilateral triangle of side s, with one vertex at the origin; one altitude of the triangle lies along the x-axis. Each plane section perpendicular to the x-axis is a square, one side of which lies in the base of the solid. Find the volume of the solid.

Solution Figure 5.16(a) shows the base of the solid with a strip of width Δx, which corresponds to a slice of the solid of volume

$$\Delta V \approx A(x)\,\Delta x,$$

where

$$A(x) = (2y)^2 = 4y^2$$

is the area of a face of the slice. The slice extends upward from the xy-plane of Fig. 5.16(a). It is possible to find the required volume from the statement of the problem by reference to this figure and without visualizing the actual solid. A perspective view of the solid is, however, given in Fig. 5.16(b). It is required to find y in terms of x so that the area $A(x)$ will be given as a function of x. Since the base triangle is equilateral, its altitude h is given by

$$h^2 = s^2 - \left(\frac{s}{2}\right)^2 = \frac{3}{4}s^2$$

or

$$h = \frac{s}{2}\sqrt{3}.$$

Then, by similar triangles, we find

$$\frac{2y}{x} = \frac{s}{h} = \frac{2}{\sqrt{3}} \qquad \text{or} \qquad y = \frac{x}{\sqrt{3}}.$$

This gives

$$A(x) = 4y^2 = \frac{4}{3}x^2,$$

and

$$V = \int_0^h A(x)\,dx = \int_0^{s\sqrt{3}/2} \frac{4}{3}x^2\,dx = \frac{s^3}{2\sqrt{3}} = \frac{1}{3}hs^2. \quad \square$$

PROBLEMS

In Problems 1–10, find the volumes generated when the plane regions bounded by the given curves and lines are rotated about the x-axis. (*Note:* $x = 0$ is the y-axis and $y = 0$ is the x-axis.)

1. $x + y = 2, \quad x = 0, \quad y = 0$

2. $y = \sin x, \quad y = 0, \quad 0 \le x \le \pi$

3. $y = x - x^2, \quad y = 0$

4. $y = -3x - x^2, \quad y = 0$

5. $y = x^2 - 2x, \quad y = 0$

6. $y = x^3, \quad x = 2, \quad y = 0$

7. $y = x^4, \quad x = 1, \quad y = 0$

8. $y = \sqrt{\cos x}, \quad 0 \le x \le \pi/2, \quad x = 0, \quad y = 0$

9. $y = \sec x, \quad x = -\pi/4, \quad x = \pi/3, \quad y = 0$

10. $y = x^3 + 1, \quad x = 2, \quad y = 0$

In Problems 11–15, find the volumes generated when the plane regions bounded by the given curves and lines are rotated about the y-axis.

11. $y = x/2, \quad x = 0, \quad y = 2$

12. $x = \sqrt{4 - y}, \quad x = 0, \quad y = 0$

13. $x = 1 - y^2, \quad x = 0$

14. $x = y^{3/2}, \quad x = 0, \quad y = 3$

15. $xy = 1, \quad x = 0, \quad y = 1, \quad y = 2$

16. Find the volume generated when the region bounded by $y = \sin x \cos x$, $0 \le x \le \pi/2$, is revolved about the x-axis.

17. Find the volume generated when the region in the first quadrant bounded by $y = \tan x$ and $x = \pi/3$ is revolved about the x-axis.

18. Find the volume generated when the region bounded by $y = \sqrt{x}$, $y = 2$, and $x = 0$ is rotated
a) about the y-axis,
b) about the line $y = 2$.

19. Find the volume generated when the region bounded by $y = 3 - x^2$ and $y = -1$ is resolved about the line $y = -1$.

20. By integration find the volume generated by rotating the triangle with vertices at $(0, 0)$, $(h, 0)$, (h, r)
a) about the x-axis, b) about the y-axis.

21. (a) A hemispherical bowl of radius a contains water to a depth h. Find the volume of water in the bowl. (b) (Review problem on related rates.) Water runs into a hemispherical bowl of radius 5 ft at the rate of 0.2 ft³/s. How fast is the water level in the bowl rising when the water is 4 ft deep?

22. A football has a volume that is approximately the same as the volume generated by rotating the region inside the ellipse $b^2x^2 + a^2y^2 = a^2b^2$ (where a and b are constants) about the x-axis. Find the volume so generated.

23. The cross sections of a certain solid by planes perpendicular to the x-axis are circles with diameters extending from the curve $y = x^2$ to the curve $y = 8 - x^2$. The solid lies between the points of intersection of these two curves. Find its volume.

24. The base of a certain solid is the circle $x^2 + y^2 = a^2$. Each plane section of the solid cut out by a plane perpendicular to the x-axis is a square with one edge of the square in the base of the solid. Find the volume of the solid.

25. Two great circles, lying in planes that are perpendicular to each other, are marked on a sphere of radius a. A portion of the sphere is then shaved off in such a manner that any plane section of the remaining solid, perpendicular to the common diameter of the two great circles, is a square with vertices on these circles. Find the volume of the solid that remains.

26. The base of a certain solid is the circle $x^2 + y^2 = a^2$. Each plane section of the solid cut out by a plane perpendicular to the y-axis is an isosceles right triangle with one leg in the base of the solid. Find the volume.

27. The base of a certain solid is the region between the x-axis and the curve $y = \sin x$ between $x = 0$ and $x = \pi/2$. Each plane section of the solid perpendicular to the x-axis is an equilateral triangle with one side in the base of the solid. Find the volume.

28. A rectangular swimming pool is 30 ft wide and 50 ft long. The depth of water h ft at distance x ft from one end of the pool is measured at 5-ft intervals and found to be as follows:

x(ft)	h(ft)	x(ft)	h(ft)
0	6.0	30	11.5
5	8.2	35	11.9
10	9.1	40	12.3
15	9.9	45	12.7
20	10.5	50	13.0
25	11.0		

Use the trapezoidal rule to estimate the volume of water in the pool.

5.5 Volumes Modeled with Shells and Washers

When we revolve a region in the plane about a line in the plane to sweep out a volume, we may choose to use rectangular strips that are parallel to the axis of rotation and therefore sweep out cylindrical shells instead of disks. The calculation of the volume in these cases is straightforward but involves modeling that differs enough from the examples of the preceding article to warrant separate attention.

Suppose the tinted region $PQRS$ in Fig. 5.17 is revolved around the y-axis. We can compute the volume generated in the following way. Consider a strip of area between x and $x + \Delta x$. When this strip is revolved around the y-axis, it generates a hollow, thin-walled shell of inner radius x, outer radius $x + \Delta x$, and volume ΔV.

5.17 The volume swept out when $PQRS$ is revolved about the y-axis is a union of cylindrical shells like the one shown here.

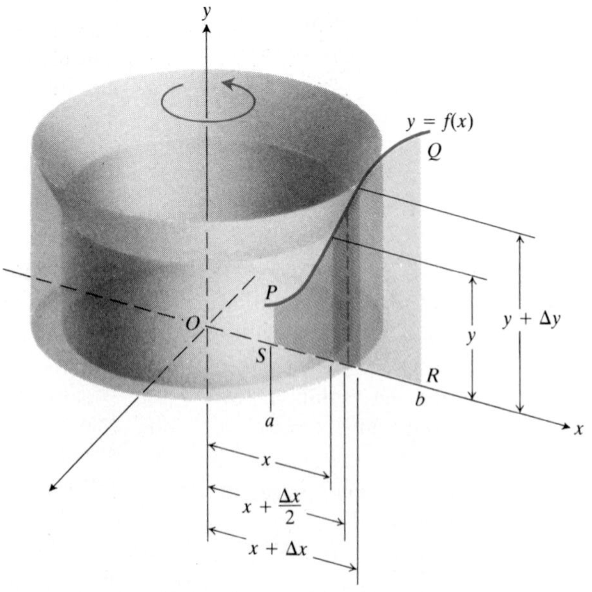

17. The region bounded by $y = 2x - x^2$ and $y = x$, about the y-axis.

18. The region bounded by $y = \sqrt{x}$, $y = 2$, $x = 0$, (a) about the x-axis; (b) about the line $x = 4$.

19. The region bounded by the y-axis, and by the curves $y = \cos x$ and $y = \sin x$ for $0 \leq x \leq \pi/4$, about the x-axis.

20. The region bounded by $y = 0$ and the curve $y = 8x^2 - 8x^3$, $0 \leq x \leq 1$, about the y-axis.

21. The region between the curves $y = 2x^2$ and $y = x^4 - 2x^2$ about the y-axis.

22. The region in the first quadrant bounded by $y = x^2$, $x + y = 2$, and the x-axis, about the x-axis.

23. Use cylindrical shells and the formula $\int x \sin x \, dx = \sin x - x \cos x + C$ to find the volume generated by revolving about the y-axis the area bounded by $y = 0$ and the curve $y = \sin x$, $0 \leq x \leq \pi$.

24. The area bounded by the curve $y = x^2$ and the line $y = 4$ generates various solids of revolution when rotated as follows:
a) about the y-axis,
b) about the line $y = 4$,
c) about the x-axis,
d) about the line $y = -1$,
e) about the line $x = 2$.
Find the volume generated in each case.

25. The circle $x^2 + y^2 = a^2$ is rotated about the line $x = b$ $(b > a)$ to generate a torus. Find the volume generated. (*Hint:* $\int_{-a}^{a} \sqrt{a^2 - x^2}\, dx = \pi a^2/2$, since it is the area of a semicircle of radius a.)

5.6 Length of a Plane Curve

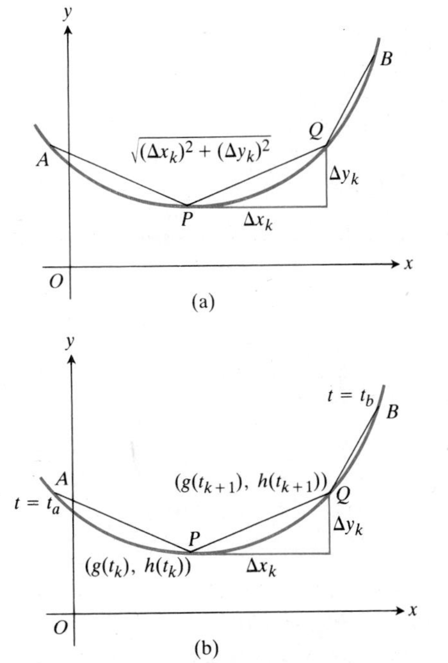

5.24 (a) The arc AB is approximated by the polygonal path $APQB$. The length of the arc is defined to be the limit (when it exists) of the lengths of successively finer polygonal approximations. (b) If the curve is given parametrically by the equations $x = g(t)$, $y = h(t)$, $t_a \leq t \leq t_b$, then $\Delta x_k = g(t_{k+1}) - g(t_k)$, $\Delta y_k = h(t_{k+1}) - h(t_k)$, and the length of PQ is

$$\sqrt{(\Delta x_k)^2 + (\Delta y_k)^2} = \sqrt{(g(t_{k+1}) - g(t_k))^2 + (h(t_{k+1}) - h(t_k))^2}.$$

Divide the arc AB (Fig. 5.24) into n pieces and join the successive points of division by straight lines. A representative line, such as PQ, will have length

$$PQ = \sqrt{(\Delta x_k)^2 + (\Delta y_k)^2}.$$

The length of the curve AB is approximately

$$L_A^B \approx \sum_{k=1}^{n} \sqrt{(\Delta x_k)^2 + (\Delta y_k)^2}.$$

When the number of division points is increased indefinitely while the lengths of the individual segments tend to zero, we obtain

$$L_A^B = \lim_{n\to\infty} \sum_{k=1}^{n} \sqrt{(\Delta x_k)^2 + (\Delta y_k)^2}, \tag{1}$$

provided the limit exists. The sum on the right side of (1) is not in the standard form to which we can apply the fundamental theorems of integral calculus, but it can be put into such a form, as follows:

Suppose that the function $y = f(x)$ is continuous and has a continuous derivative at each point of $[a, b]$. Then, by the Mean Value Theorem, there is some point $P(c_k, d_k)$ between P and Q on the curve where the tangent to the curve is parallel to the chord PQ. That is,

$$f'(c_k) = \frac{\Delta y_k}{\Delta x_k} \quad \text{or} \quad \Delta y_k = f'(c_k)\, \Delta x_k.$$

Hence (1) may also be written in the form

$$L_A^B = \lim_{n\to\infty} \sum_{k=1}^{n} \sqrt{(\Delta x_k)^2 + (f'(c_k)\,\Delta x_k)^2} = \lim_{\Delta x\to 0} \left(\sum_{a}^{b} \sqrt{1 + (f'(c_k))^2}\, \Delta x \right)$$

or

$$L_A^B = \int_a^b \sqrt{1 + (f'(x))^2}\,dx = \int_a^b \sqrt{1 + \left(\frac{dy}{dx}\right)^2}\,dx$$
$$= \int_a^b \sqrt{1 + (y')^2}\,dx, \tag{2a}$$

where we have written dy/dx and y' for $f'(x)$.

It is sometimes convenient, if x can be expressed as a function of y, to interchange the roles of x and y. The analogue of Eq. (2a) in this case is

$$L_A^B = \int_c^d \sqrt{1 + \left(\frac{dx}{dy}\right)^2}\,dy. \tag{2b}$$

EXAMPLE 1 Find the length of the curve

$$y = \frac{4\sqrt{2}}{3}x^{3/2} - 1$$

from $x = 0$ to $x = 1$.

Solution We calculate the length from formula (2a) as follows:

$$y = \frac{4\sqrt{2}}{3}x^{3/2} - 1$$
$$y' = \frac{4\sqrt{2}}{3}\cdot\frac{3}{2}x^{1/2} = 2\sqrt{2}x^{1/2}$$
$$1 + (y')^2 = 1 + (2\sqrt{2}x^{1/2})^2 = 1 + 8x$$
$$\int_0^1 \sqrt{1 + (y')^2}\,dx = \int_0^1 \sqrt{1 + 8x}\,dx = \frac{2}{3}\cdot\frac{1}{8}(1 + 8x)^{3/2}\Big]_0^1 = \frac{13}{6}. \quad \square$$

There is a particularly useful formula for calculating the length of a curve that is given parametrically. Let the equations be

$$x = g(t), \qquad y = h(t), \tag{3}$$

and let t_k, t_{k+1} be the values of t at P and Q, respectively. Suppose that arc AB (Fig. 5.24b) is traced just once by $P(x, y)$ as t goes from t_a at A to t_b at B. If the functions $g(t)$ and $h(t)$ are continuously differentiable for t between t_a and t_b inclusive, the Mean Value Theorem may be applied to Eq. (3) to give

$$\Delta x_k = x_{k+1} - x_k = g(t_{k+1}) - g(t_k) = g'(t_k')\,\Delta t_k,$$
$$\Delta y_k = y_{k+1} - y_k = h(t_{k+1}) - h(t_k) = h'(t_k'')\,\Delta t_k,$$

where t_k' and t_k'' are two suitably chosen values of t between t_k and t_{k+1}. Then (1) becomes

$$L_A^B = \lim_{n\to\infty} \sum_{k=1}^{n} \sqrt{(g'(t_k'))^2 + (h'(t_k''))^2}\,\Delta t_k. \tag{4}$$

The sum on the right-hand side of Eq. (4) is not the Riemann sum of any function because the points t_k' and t_k'' need not be the same. However, a theorem called Bliss's Theorem (proved in more advanced texts) assures us that the sums converge to the integral we would like them to converge to, namely, the integral

$$L_A^B = \int_{t_a}^{t_b} \sqrt{\left(\frac{dx}{dt}\right)^2 + \left(\frac{dy}{dt}\right)^2}\, dt. \tag{5}$$

There is nothing inherently dependent upon the choice of t for the parameter in Eqs. (3) and (5). Any other variable could serve as well, and if θ is used instead of t in (3) in representing the curve, then we need only replace t by θ in (4) and (5) as well.

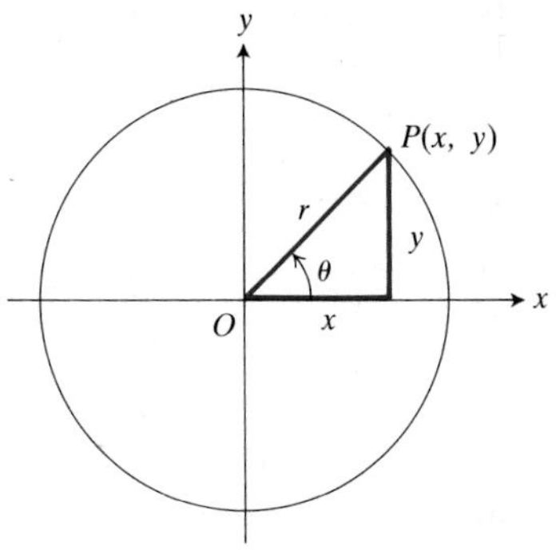

5.25 As θ grows from $\theta = 0$ to $\theta = 2\pi$, the point $P(x, y)$ travels around the circle exactly once.

EXAMPLE 2 The coordinates (x, y) of a point on a circle of radius r can be expressed in terms of the central angle θ (Fig. 5.25) as

$$x = r\cos\theta, \qquad y = r\sin\theta.$$

The point P(x, y) moves once around the circle as θ varies from 0 to 2π, so that the circumference of the circle is given by

$$C = \int_0^{2\pi} \sqrt{\left(\frac{dx}{d\theta}\right)^2 + \left(\frac{dy}{d\theta}\right)^2}\, d\theta.$$

We find

$$\frac{dx}{d\theta} = -r\sin\theta, \qquad \frac{dy}{d\theta} = r\cos\theta,$$

so that

$$\left(\frac{dx}{d\theta}\right)^2 + \left(\frac{dy}{d\theta}\right)^2 = r^2(\sin^2\theta + \cos^2\theta) = r^2,$$

and hence

$$C = \int_0^{2\pi} r\, d\theta = r[\theta]_0^{2\pi} = 2\pi r. \ \square$$

EXAMPLE 3 Find the distance traveled between $t = 0$ and $t = \pi/2$ by a particle P(x, y) whose position at time t is given by

$$x = \sin^2 t, \qquad y = \cos^2 t.$$

Solution We calculate the length of the path traced by the particle from $t = 0$ to $t = \pi/2$ from Eq. (5) as follows:

$$x = \sin^2 t \qquad\qquad y = \cos^2 t$$

$$\frac{dx}{dt} = 2\sin t\cos t \qquad \frac{dy}{dt} = -2\cos t\sin t$$

$$\begin{aligned}\int_0^{\pi/2} \sqrt{\left(\frac{dx}{dt}\right)^2 + \left(\frac{dy}{dt}\right)^2}\, dt &= \int_0^{\pi/2} \sqrt{8\sin^2 t\cos^2 t}\, dt \\ &= \int_0^{\pi/2} \sqrt{2}\cdot 2\sin t\cos t\, dt \\ &= \sqrt{2}\cdot\sin^2 t\Big]_0^{\pi/2} \\ &= \sqrt{2}.\end{aligned}$$

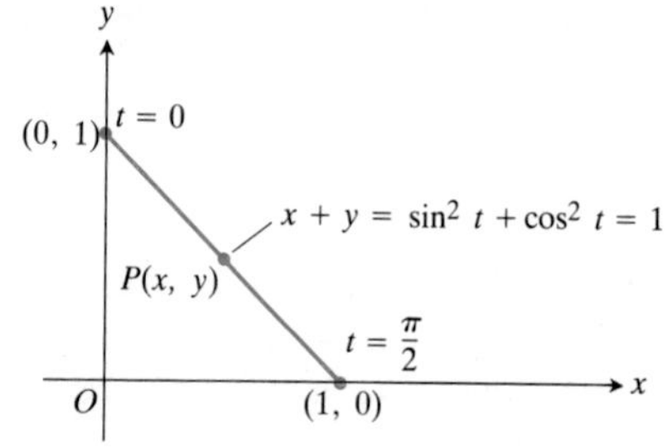

5.26 The path traced by the particle $P(x, y)$ whose position at time t is given by $x = \sin^2 t$, $y = \cos^2 t$, $0 \leq t \leq \pi/2$.

In this case we can check our result geometrically, since

$$x + y = \sin^2 t + \cos^2 t = 1$$

for all values of t. The path of the particle is the segment of the line

$$x + y = 1$$

that runs from (0, 1), where $t = 0$, to (1, 0), where $t = \pi/2$, as shown in Fig. 5.26. The length of this segment is

$$\sqrt{(0-1)^2 + (1-0)^2} = \sqrt{2}. \ \square$$

Equation (5) is frequently written in terms of differentials in place of derivatives. This is done formally by writing $(dt)^2$ under the radical in place of the dt outside the radical, and then writing

$$\left(\frac{dx}{dt}\right)^2 (dt)^2 = \left(\frac{dx}{dt}\,dt\right)^2 = (dx)^2$$

and

$$\left(\frac{dy}{dt}\right)^2 (dt)^2 = \left(\frac{dy}{dt}\,dt\right)^2 = (dy)^2.$$

It is also customary to eliminate the parentheses in $(dx)^2$ and write dx^2 instead, so that Eq. (5) is written

$$L = \int \sqrt{dx^2 + dy^2}. \tag{6}$$

Of course, dx and dy must both be expressed in terms of one and the same variable, and appropriate limits must be supplied in (6) before the integration can be performed.

A useful way to remember Eq. (6) is to write

$$ds = \sqrt{dx^2 + dy^2} \tag{7}$$

and treat ds as the differential of arc length, which can be integrated between appropriate limits to give the total length of a curve. Figure 5.27(a) gives the exact interpretation of ds corresponding to Eq. (7). Figure 5.27(b) is not strictly accurate but is to be thought of as a simplified version of Fig. 5.27(a).

With Eq. (7) in mind, the quickest way to recall the formulas for arc length is to remember the equation

$$\text{Arc length} = \int ds. \tag{8}$$

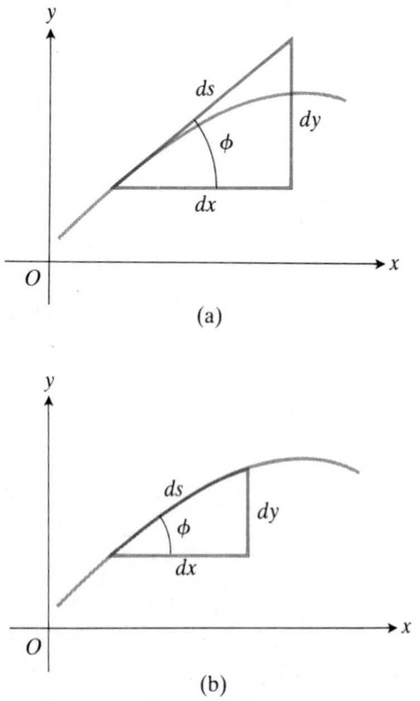

5.27 Diagrams for remembering the equation $ds = \sqrt{dx^2 + dy^2}$.

EXAMPLE 4 Find the length of the curve $y = x^{2/3}$ between $x = -1$ and $x = 8$.

Solution From the equation of the curve, we find

$$\frac{dy}{dx} = \frac{2}{3}x^{-1/3}.$$

Since this becomes infinite at the origin (see Fig. 5.28), we use Eq. (2b) instead of (2a) to find the length of the curve. Then we need to find dx/dy. Since the equation of the curve,

$$y = x^{2/3},$$

can also be written as

$$x = \pm y^{3/2},$$

we find

$$\frac{dx}{dy} = \pm\frac{3}{2}y^{1/2},$$

$$dx = \pm\frac{3}{2}y^{1/2}\,dy,$$

$$ds^2 = dx^2 + dy^2 = \left(\frac{9}{4}y + 1\right)dy^2,$$

$$ds = \sqrt{\frac{9}{4}y + 1}\,dy.$$

The portion of the curve between $A(-1, 1)$ and the origin has length

$$L_1 = \int_0^1 \sqrt{\frac{9}{4}y + 1}\,dy,$$

and the rest of the curve from the origin to $B(8, 4)$ has length

$$L_2 = \int_0^4 \sqrt{\frac{9}{4}y + 1}\,dy,$$

and the total length is

$$L = L_1 + L_2.$$

It is necessary to calculate L with two separate integrals because $x = \pm y^{3/2}$ needs to be separated into two distinct functions of y. On AO, we have $x = -y^{3/2}$, $0 \le y \le 1$; on OB we have $x = +y^{3/2}$, $0 \le y \le 4$.

To evaluate the given integrals, we substitute

$$u = \frac{9}{4}y + 1, \qquad du = \frac{9}{4}dy, \qquad dy = \frac{4}{9}du,$$

to get

$$\int\left(\frac{9}{4}y + 1\right)^{1/2} dy = \frac{4}{9}\int u^{1/2}\,du = \frac{8}{27}u^{3/2} + C$$

and

$$L = \frac{8}{27}\left(\frac{9}{4}y + 1\right)^{3/2}\Bigg]_0^1 + \frac{8}{27}\left(\frac{9}{4}y + 1\right)^{3/2}\Bigg]_0^4$$

$$= \frac{1}{27}(13\sqrt{13} + 80\sqrt{10} - 16) = 10.5.$$

To check against gross errors, we calculate the sum of the lengths of the two inscribed chords:

$$AO + OB = \sqrt{2} + \sqrt{80} \approx 10.4,$$

which appears to be satisfactory.

The curve in Fig. 5.28 has a cusp at (0, 0) where the slope becomes infinite. If we were to reconstruct the derivation of Eq. (2) for this particular curve, we would see that the crucial step that required an application of the Mean Value Theorem could not have been taken for the case of a chord PQ from a point P to the left of the cusp to a point Q to its right. For this reason, if for no other, when one or more cusps occur between the ends of a portion of a curve whose length is to be calculated, it is best to calculate the lengths of portions of the curve between cusps and add the results. We recall that the Mean Value Theorem of Article 3.8 is still valid

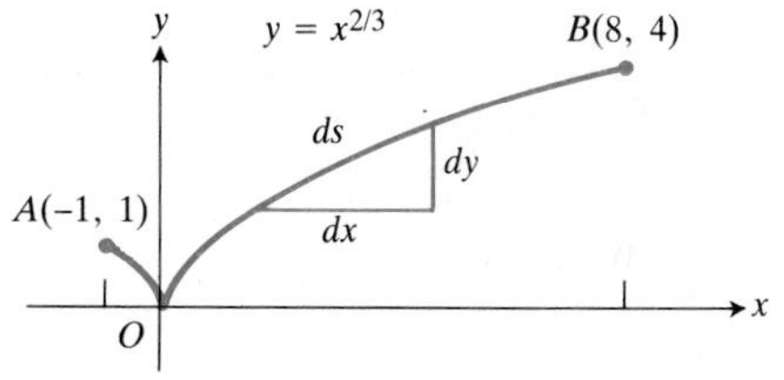

5.28 To compute the length of $y = x^{2/3}$ between A and B one writes $x = -y^{3/2}$ for the part from A to O, and $x = y^{3/2}$ for the part from O to B, and uses Eq. (2b) twice. The discussion at the end of Example 4 explains why.

even when the derivative becomes infinite at an extremity of the interval where it is to be applied, so that the derivation of Eq. (2) would be valid for the separate portions of a curve lying *between* cusps (or other discontinuities of dy/dx, such as occur at corners). Thus in the example just worked, we find the lengths L_1 from $A(-1, 1)$ to O (up to the cusp) and L_2 from O to $B(8, 4)$, then add the results to obtain $L = L_1 + L_2$. This was done by the sum of the two integrals. Had there been no point of discontinuity of dy/dx, it would not have been necessary to take two separate integrals, and Eq. (2a) could have been used. □

Arc Length Formulas

PATH EQUATION	ARC LENGTH
$y = f(x)$, $a \leq x \leq b$	$L = \int_a^b \sqrt{1 + (f'(x))^2}\, dx = \int_a^b \sqrt{1 + \left(\frac{dy}{dx}\right)^2}\, dx$
$x = g(y)$, $c \leq y \leq d$	$L = \int_c^d \sqrt{1 + (g'(y))^2}\, dy = \int_c^d \sqrt{1 + \left(\frac{dx}{dy}\right)^2}\, dy$
$x = x(t)$, $y = y(t)$, $t_a \leq t \leq t_b$	$L = \int_{t_a}^{t_b} \sqrt{\left(\frac{dx}{dt}\right)^2 + \left(\frac{dy}{dt}\right)^2}\, dt$
Brief formula:	$L = \int ds$, where $ds = \sqrt{dx^2 + dy^2}$

PROBLEMS

Find the lengths of the curves in Problems 1–6.

1. $y = \frac{1}{3}(x^2 + 2)^{3/2}$ from $x = 0$ to $x = 3$.

2. $y = x^{3/2}$ from $(0, 0)$ to $(4, 8)$.

3. $9x^2 = 4y^3$ from $(0, 0)$ to $(2\sqrt{3}, 3)$.

4. $y = (x^3/3) + 1/(4x)$ from $x = 1$ to $x = 3$.

5. $x = (y^4/4) + 1/(8y^2)$ from $y = 1$ to $y = 2$.

6. $(y + 1)^2 = 4x^3$ from $x = 0$ to $x = 1$.

7. The coordinates of a point $P(x, y)$ on the four-cusped hypocycloid are given by $x = a \cos \theta$, $y = a \sin \theta$. Find the total length of the curve (Fig. 5.29).

8. Find the length of the curve

$$y = \int_0^x \sqrt{\cos 2t}\, dt$$

from $x = 0$ to $x = \pi/4$.

9. Find the distance traveled between $t = 0$ and $t = \pi$ by the particle $P(x, y)$ whose position at time t is

$$x = \cos t, \qquad y = t + \sin t.$$

(*Hint:* Use a half-angle formula to simplify the integration.)

10. Find the distance traveled between $t = 0$ and $t = \pi/2$ by a particle $P(x, y)$ whose position at time t is given by

$$x = a \cos t + at \sin t, \qquad y = a \sin t - at \cos t,$$

where a is a positive constant.

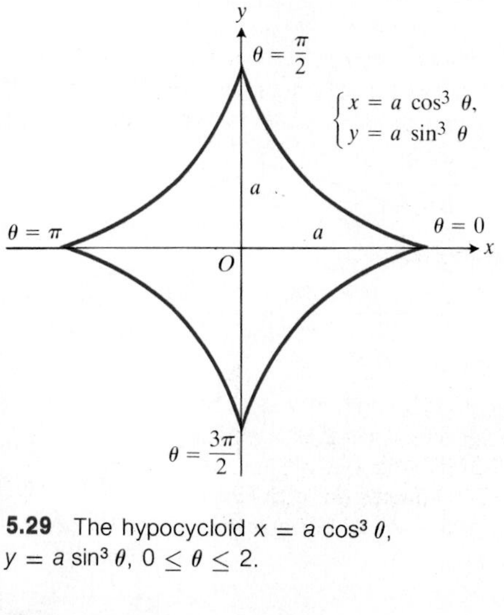

5.29 The hypocycloid $x = a \cos^3 \theta$, $y = a \sin^3 \theta$, $0 \leq \theta \leq 2$.

11. Find the length of the curve

$$x = t - \sin t, \qquad y = 1 - \cos t, \qquad 0 \le t \le 2\pi.$$

(*Hint:* $\sqrt{2 - 2\cos t} = 2\sqrt{(1 - \cos t)/2}$.)

12. Find the distance traveled by the particle $P(x, y)$ between $t = 0$ and $t = 4$ if the position at time t is given by

$$x = \frac{t^2}{2}, \qquad y = \frac{1}{3}(2t + 1)^{3/2}.$$

13. The position of a particle $P(x, y)$ at time t is given by

$$x = \frac{1}{3}(2t + 3)^{3/2}, \qquad y = \frac{t^2}{2} + t.$$

Find the distance it travels between $t = 0$ and $t = 3$.

14. (*Calculator*) The length of one arch of the curve $y = \sin x$ is given by

$$L = \int_0^{\pi} \sqrt{1 + \cos^2 x}\, dx.$$

Estimate L by Simpson's rule with $n = 8$.

15. (*Calculator*) A company wants to make sheets of corrugated iron roofing like the one shown in Fig. 5.30. The cross sections of the corrugated sheets conform to the curve

$$y = \sin \frac{3\pi}{20} x, \qquad 0 \le x \le 20 \text{ in.}$$

If the roofing is to be stamped from flat sheets by a process that does not stretch the material, how wide should the original material be? To find out, use Simpson's rule with $n = 10$ to approximate the length of the sine curve.

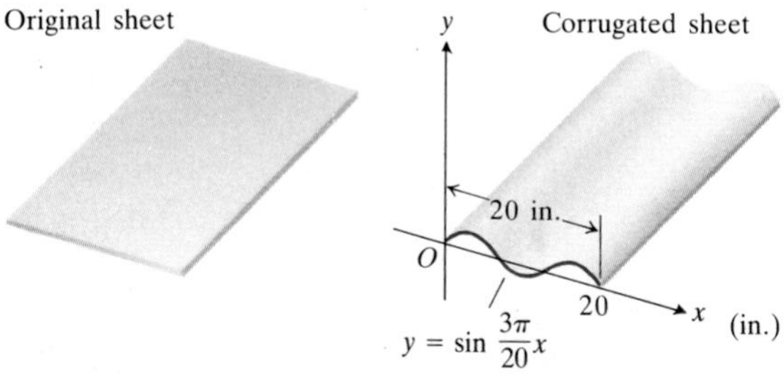

5.30 How wide does the original sheet have to be? See Problem 15.

***Toolkit* programs**	
Integral Evaluator	Super * Grapher
Parametric Equations	

5.7 Area of a Surface of Revolution

When you jump rope, the rope sweeps out a surface in the space around you. As you might imagine, the area of the surface depends on the length of the rope. In this article we shall explore the relation between the area of a surface of revolution and the length of the curve that generates it.

The Definition of Surface Area

Suppose that a curve AB in the xy-plane like the one shown in Fig. 5.31(a) is revolved about the x-axis to generate a surface. If AB is approximated

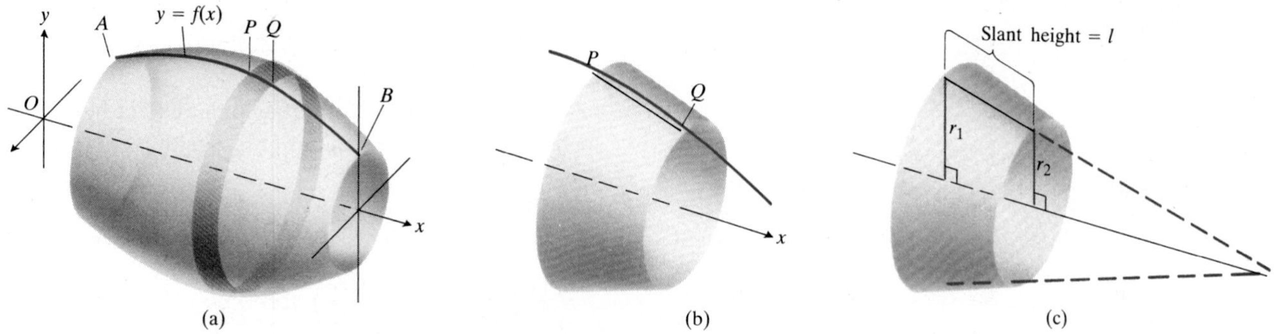

5.31 The surface swept out by revolving the curve AB about the x-axis is a union of bands like the one swept out by the arc PQ shown in (a). The line segment joining P to Q sweeps out a frustum of a cone, shown in (b). The important dimensions of the frustum are shown in (c).

by an inscribed polygon like the ones used to define arc length in Article 5.6, then each segment PQ of the polygon will sweep out part of a cone whose axis lies along the x-axis (magnified view in Fig. 5.31b). This part is called a *frustum* of the cone (frustum is Latin for "piece"). If its base radii are r_1 and r_2, as in Fig. 5.31(c), and its slant height is l, then its lateral surface area A is

$$A = \pi(r_1 + r_2)l. \tag{1}$$

The total of the frustum areas swept out by the segments of the inscribed polygon from A to B will give an approximation to the area S of the surface swept out by the curve AB. This approximation leads to an integral formula for S in the following way.

If we let the coordinates of P be (x, y) and of Q be $(x + \Delta x, y + \Delta y)$, then the dimensions of the frustum swept out by the line segment PQ are

$$r_1 = y, \qquad r_2 = y + \Delta y, \qquad l = \sqrt{(\Delta x)^2 + (\Delta y)^2}. \tag{2}$$

The lateral surface area A of the frustum, from Eq. (1), is

$$A = \pi(r_1 + r_2)l = \pi(2y + \Delta y)\sqrt{(\Delta x)^2 + (\Delta y)^2}. \tag{3}$$

Adding the individual frustum areas over the interval $[a, b]$ from left to right gives

$$\text{Cone frustum sum} = \sum_{x=a}^{b} \pi(2y + \Delta y)\sqrt{(\Delta x)^2 + (\Delta y)^2}$$

$$= \sum_{a}^{b} 2\pi\left(y + \frac{1}{2}\Delta y\right)\sqrt{1 + \left(\frac{\Delta y}{\Delta x}\right)^2}\,\Delta x.$$

If y and dy/dx are continuous functions of x it can be shown (although we shall not do so here) that these sums approach the limit

$$S = \int_a^b 2\pi y\sqrt{1 + \left(\frac{dy}{dx}\right)^2}\,dx. \tag{4}$$

We therefore define the area of the surface to be the value of this integral.

Equation (4) is easily remembered if we write

$$\sqrt{1 + \left(\frac{dy}{dx}\right)^2}\,dx = ds$$

and take

$$S = \int 2\pi y\,ds. \tag{5}$$

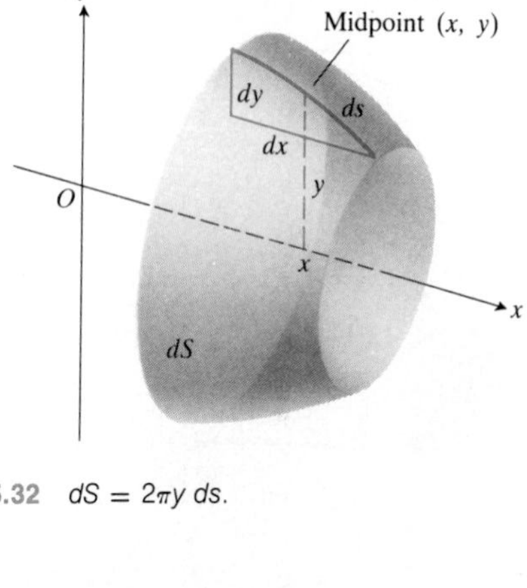

5.32 $dS = 2\pi y\,ds$.

We may then interpret $2\pi y\,ds$ as the product of a

$$\text{Circumference} = 2\pi y$$

and a

$$\text{Slant height} = ds.$$

Thus $2\pi y\,ds$ gives the lateral area of a frustum of a cone of slant height ds if the point (x, y) is the midpoint of the element of arc length ds (Fig. 5.32).

Other Useful Formulas

If the axis of revolution is the y-axis, the formula that replaces (4) is

$$S = \int_c^d 2\pi x\sqrt{1 + \left(\frac{dx}{dy}\right)^2}\,dy. \tag{6}$$

If the curve that sweeps out the surface is given in parametric form with x and y as functions of a third variable t that varies from t_a to t_b, then we may compute S from the formula

$$S = \int_{t_a}^{t_b} 2\pi\rho \sqrt{\left(\frac{dx}{dt}\right)^2 + \left(\frac{dy}{dt}\right)^2}\, dt, \tag{7}$$

where ρ is the distance from the axis of revolution to the element of arc length, and is expressed as a function of t.

Generally (Fig. 5.33) we may write

$$S = \int 2\pi\rho\, ds, \tag{8}$$

where, again, ρ is the distance from the axis of revolution to the element of arc length ds. Both ρ and ds need to be expressed in terms of some one variable and limits must be supplied to (8) in any particular problem.

EXAMPLE 1 Find the area of the surface obtained by revolving the curve

$$y = \sqrt{x}, \qquad 0 \le x \le 2$$

about the x-axis (Fig. 5.34).

Solution To find the surface area we use Eq. (4) with $y = \sqrt{x}$:

$$y = \sqrt{x},$$

$$y' = \frac{1}{2\sqrt{x}},$$

$$\sqrt{1 + (y')^2} = \sqrt{1 + \frac{1}{4x}} = \sqrt{\frac{4x+1}{4x}} = \frac{\sqrt{4x+1}}{2\sqrt{x}},$$

$$\int_0^2 2\pi y\sqrt{1+(y')^2}\, dx = \int_0^2 2\pi\sqrt{x}\, \frac{\sqrt{4x+1}}{2\sqrt{x}}\, dx$$

$$= \pi \int_0^2 \sqrt{4x+1}\, dx$$

$$= \pi \cdot \frac{2}{3} \cdot \frac{1}{4}(4x+1)^{3/2}\Big]_0^2 = \frac{\pi}{6}(27 - 1) = \frac{13\pi}{3}. \quad \square$$

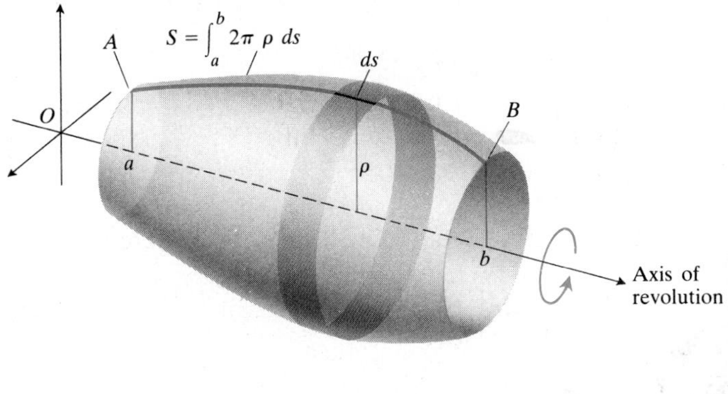

5.33 The area of the surface swept out by revolving arc *AB* about the axis shown here is $\int_a^b 2\pi\rho\, ds$. The specific formulation depends on the formulas for ρ and *ds*.

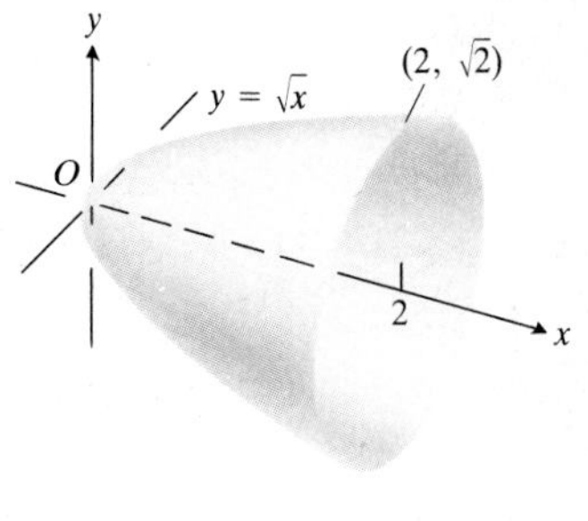

5.34 The area of this surface is calculated in Example 1.

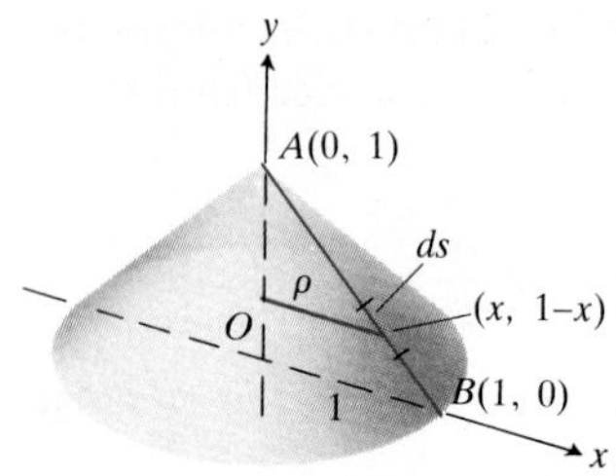

5.35 Revolving the line segment *AB* about the *y*-axis generates a cone.

EXAMPLE 2 The line segment

$$x = \sin^2 t, \qquad y = \cos^2 t, \qquad 0 \le t \le \frac{\pi}{2}$$

of Example 3, Article 5.6, is revolved about the y-axis to generate a cone. Find its surface area.

Solution See Fig. 5.35. We use Eq. (7). The distance from the axis of revolution to a typical element of arc ds is

$$\rho = x = \sin^2 t.$$

With this expression in place of ρ and with

$$\frac{dx}{dt} = 2 \sin t \cos t, \qquad \frac{dy}{dt} = -2 \cos t \sin t,$$

Equation 7 gives

$$\begin{aligned}\int_{t_a}^{t_b} 2\pi\rho\sqrt{\left(\frac{dx}{dt}\right)^2 + \left(\frac{dy}{dt}\right)^2}\, dt &= \int_0^{\pi/2} 2\pi \sin^2 t\sqrt{8 \sin^2 t \cos^2 t}\, dt \\ &= 4\pi\sqrt{2}\int_0^{\pi/2} \sin^3 t \cos t\, dt \\ &= 4\pi\sqrt{2}\,\frac{1}{4}\sin^4 t\Big]_0^{\pi/2} \\ &= \pi\sqrt{2}(1 - 0) \\ &= \pi\sqrt{2}.\end{aligned}$$

In this case we can also calculate the surface area from the formula for the lateral surface area of a cone:

$$\text{Lateral surface area} = \frac{\text{base circumference}}{2} \times \text{slant height} = \pi\sqrt{2}. \ \square$$

EXAMPLE 3 Find the area of the sphere generated by revolving the circle

$$x^2 + y^2 = r^2$$

about the x-axis (Fig. 5.36).

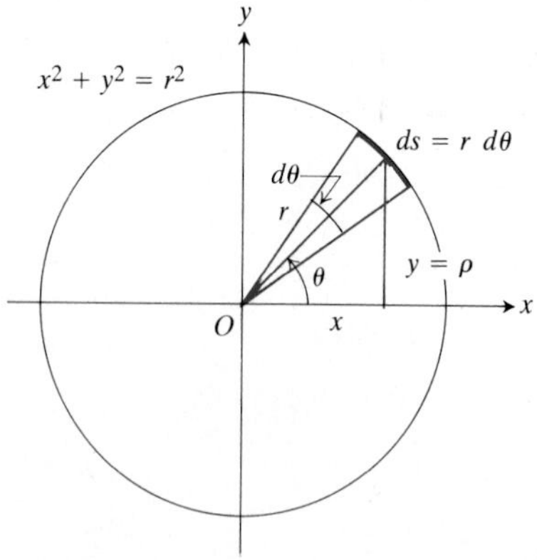

5.36 To compute the surface area of the sphere obtained by revolving the circle $x^2 + y^2 = r^2$ about the x-axis, let θ vary from $\theta = 0$ to $\theta = \pi$. The top half of the circle generates the entire sphere.

Solution We represent the top half of the circle, which generates the entire sphere, by

$$x = r\cos\theta, \qquad y = r\sin\theta, \qquad 0 \le \theta \le \pi.$$

Then

$$dx = -r\sin\theta\, d\theta, \qquad dy = r\cos\theta\, d\theta$$

and

$$ds = \sqrt{dx^2 + dy^2} = \sqrt{r^2\sin^2\theta + r^2\cos^2\theta}\, d\theta = r\, d\theta.$$

With

$$\rho = y = r\sin\theta,$$

Eq. (8) now gives

$$\begin{aligned} S = \int_A^B 2\pi\rho\, ds &= \int_0^{\pi} 2\pi r\sin\theta\, r\, d\theta \\ &= 2\pi r^2\int_0^{\pi}\sin\theta\, d\theta \\ &= 2\pi r^2\,[-\cos\theta]_0^{\pi} = 2\pi r^2\,[1 + 1] \\ &= 4\pi r^2. \ \square\end{aligned}$$

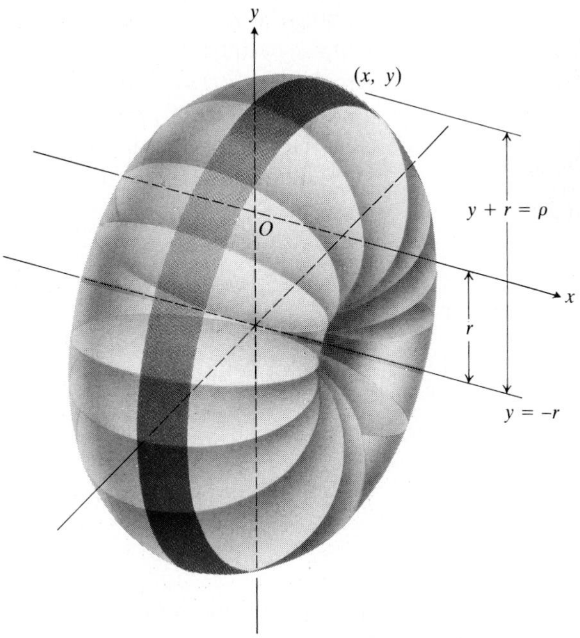

5.37 The circle of Fig. 5.36 revolved about the line $y = -r$.

EXAMPLE 4 The circle of Example 3 is revolved about the line $y = -r$, which is tangent to the circle at the point $(0, -r)$. Find the area of the surface generated. See Fig. 5.37.

Solution Here it takes the whole circle to generate the surface. This means that we must let θ vary from 0 to 2π. The radius of rotation is

$$\rho = y + r,$$

and we have

$$\begin{aligned} 2\pi\,\rho\,ds &= 2\pi(y + r)r\,d\theta \\ &= 2\pi(r\sin\theta + r)r\,d\theta \\ &= 2\pi r^2(\sin\theta + 1)\,d\theta. \end{aligned}$$

Hence, Eq. (8) gives

$$\begin{aligned} S = \int_A^B 2\pi\,\rho\,ds &= \int_0^{2\pi} 2\pi r^2(\sin\theta + 1)\,d\theta \\ &= 2\pi r^2[-\cos\theta + \theta]_0^{2\pi} \\ &= 2\pi r^2[(-1 + 2\pi) - (-1 + 0)] \\ &= 4\pi^2 r^2. \ \square \end{aligned}$$

Cylindrical vs. Conical Bands

Why not approximate the surface area S by inscribed *cylinders* and arrive at a result

$$S = \int 2\pi y\,dx,$$

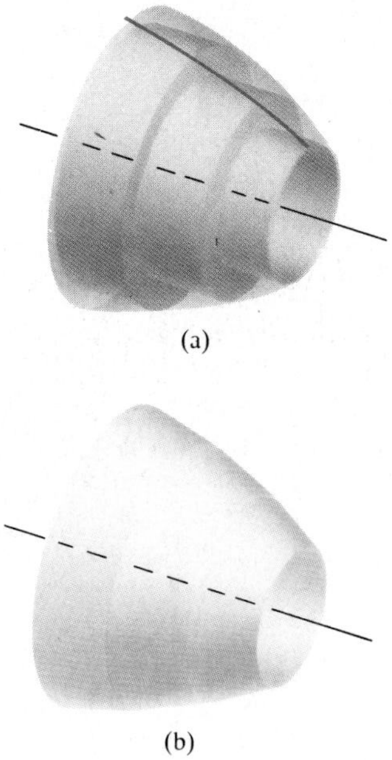

5.38 Why not use cylindrical bands (a) instead of conical bands (b) to approximate surface area?

having dx instead of ds, to replace Eq. (5)? In other words (Fig. 5.38), why not use cylindrical bands instead of conical bands to approximate surface area? Since we know from our discussion on volume that inscribed cylinders work perfectly well for *volumes* of revolution, why not use them for *surfaces* of revolution also?

The answer hinges on the fact that the approximations

$$\Delta S \approx 2\pi y\,\Delta x \qquad \text{(inscribed cylindrical band)} \qquad \text{(a)}$$

and

$$\Delta S \approx 2\pi y\sqrt{(\Delta x)^2 + (\Delta y)^2} \qquad \text{(inscribed conical band)} \qquad \text{(b)}$$

lead to *different* answers for the surface area when we pass to the corresponding definite integrals. This is because their ratio is

$$\frac{2\pi y\sqrt{(\Delta x)^2 + (\Delta y)^2}}{2\pi y\,\Delta x} = \sqrt{1 + \left(\frac{\Delta y}{\Delta x}\right)^2},$$

which has the limiting value

$$\lim_{\Delta x \to 0}\sqrt{1 + \left(\frac{\Delta y}{\Delta x}\right)^2} = \sqrt{1 + \left(\frac{dy}{dx}\right)^2},$$

and this is different from 1, unless $dy/dx = 0$, which is generally not true. We must therefore choose between (a) and (b). We choose (b) because the surface area formula it leads to is consistent with the formulas from classical geometry (Examples 2 and 3, and Problem 8). That is, the integral in Eq. (4) and the classical Greek definitions agree on any objects to which they both apply.

Area Formulas for Surfaces of Revolution

EQUATION OF GENERATING CURVE	SURFACE AREA
$y = f(x)$, $a \le x \le b$	$S = \int_a^b 2\pi y\sqrt{1 + \left(\frac{dy}{dx}\right)^2}\,dx$
$x = g(y)$, $c \le y \le d$	$S = \int_c^d 2\pi x\sqrt{1 + \left(\frac{dx}{dy}\right)^2}\,dy$
$x = x(t)$, $y = y(t)$, $t_a \le t \le t_b$	$S = \int_{t_a}^{t_b} 2\pi\rho\sqrt{\left(\frac{dx}{dt}\right)^2 + \left(\frac{dy}{dt}\right)^2}\,dt$ (ρ is the distance from the axis of revolution to the generating curve, expressed as a function of t.)

Brief formula: $S = \int 2\pi\rho\,ds$

PROBLEMS

1. Find the area of the surface generated by rotating about the x-axis the arc of the curve $y = x^3$ between $x = 0$ and $x = 1$.

2. Find the area of the surface generated by rotating about the y-axis the arc of the curve $y = x^2$ between $(0, 0)$ and $(2, 4)$.

3. The arc of the curve $y = (x^3/3) + (1/4x)$ from $x = 1$ to $x = 3$ is rotated about the line $y = -1$. Find the surface area generated.
4. The arc of the curve $x = (y^4/4) + (1/8y^2)$ from $y = 1$ to $y = 2$ is rotated about the x-axis. Find the surface area generated.
5. Find the area of the surface generated by rotating about the y-axis the curve $y = (x^2/2) + \frac{1}{2}, 0 \leq x \leq 1$.
6. Find the surface area generated by revolving the ellipse $x^2 + 4y^2 = 4$
   a) about the x-axis,
   b) about the y-axis.
7. Find the area of the surface generated by rotating the portion of the curve $y = \frac{1}{3}(x^2 + 2)^{3/2}$ between $x = 0$ and $x = 3$ about the y-axis.
8. Use Eq. (4) to calculate the lateral surface area of the cone frustum shown in Fig. 5.31(c).
9. a) Find the area of the surface generated by rotating the curve $x = \cos t$, $y = 1 + \sin t$, $0 \leq t \leq 2\pi$, about the x-axis.
   b) Graph the curve and compare the result in (a) with the result in Example 4.
10. Find the area of the surface generated by rotating the curve $x = t^2$, $y = t$, $0 \leq t \leq 1$, about the x-axis.
11. Find the area of the surface generated by rotating the hypocycloid $x = a \cos^3 \theta$, $y = a \sin^3 \theta$ shown in Fig. 5.29 about the x-axis.
12. The curve described by the particle $P(x, y)$,

$$x = t + 1, \qquad y = \frac{t^2}{2} + t,$$

from $t = 0$ to $t = 4$, is rotated about the y-axis. Find the surface area that is generated.
13. The loop of the curve $9x^2 = y(3 - y)^2$ is rotated about the x-axis. Find the surface area generated.
14. If you have a spherical loaf of bread and cut it into slices of equal width, then each slice will have the same amount of crust. To see why, suppose that the semicircle $y = \sqrt{r^2 - x^2}$ (Fig. 5.39) is rotated about the x-axis to generate a sphere. Show that the area of the band swept out by an arc AB that stands above an interval of width h does not depend on the location of the interval.

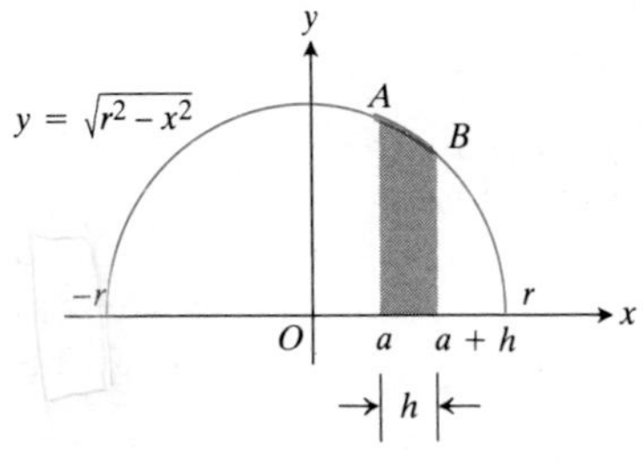

5.39 The semicircle of Problem 14.

5.8 Average Value of a Function

The process of finding the average value of a finite number of data is familiar to all students. For example, if $y_1, y_2, \ldots, y_n$ are the grades of a class of n students on a calculus quiz, then the class average on the quiz is

$$y_{av} = \frac{y_1 + y_2 + \cdots + y_n}{n}. \tag{1}$$

When the number of data is infinite, it is not feasible to use Eq. (1) (since it is likely to take on the meaningless form ∞/∞). This situation arises, in particular, when the data y are given by a continuous function

$$y = f(x), \qquad a \leq x \leq b.$$

In this case, the *average value of y, with respect to x,* is defined to be

$$(y_{av})_x = \frac{1}{b - a} \int_a^b f(x)\, dx. \tag{2}$$

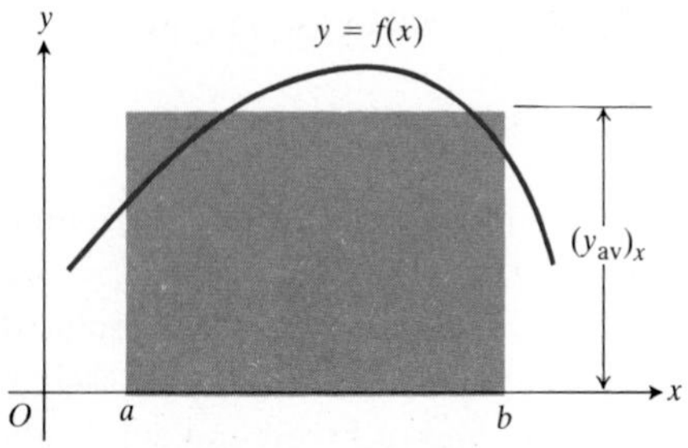

5.40 One interpretation of $(y_{av})_x = [1/(b-a)] \int_a^b f(x)\,dx$ comes from interpreting the integral as an area.

The curve in Fig. 5.40, for example, might represent temperature as a function of time over a twenty-four hour period. Equation (2) would then give the "average temperature" for the day.

Equation (2) is a *definition*, and hence is not subject to proof. Nevertheless, some discussion may help to explain why this particular formula is used to define the average. One might arrive at it as follows. From the total "population" of x-values, $a \le x \le b$, we select a representative "sample," $x_1, x_2, \ldots, x_n$, uniformly distributed between a and b. Then, using Eq. (1), we calculate the average of the functional values

$$y_1 = f(x_1), \qquad y_2 = f(x_2), \qquad \ldots, \qquad y_n = f(x_n)$$

associated with these representative x's. This gives us

$$\frac{y_1 + y_2 + \cdots + y_n}{n} = \frac{f(x_1) + f(x_2) + \cdots + f(x_n)}{n}. \tag{3}$$

Since we require that the x's be uniformly distributed between a and b, let us take the spacing to be Δx, with

$$x_2 - x_1 = x_3 - x_2 = \cdots = x_n - x_{n-1} = \Delta x$$

and

$$\Delta x = \frac{b-a}{n}.$$

Then, in (3), let us replace the n in the denominator by $(b-a)/\Delta x$, thus obtaining

$$\frac{f(x_1) + f(x_2) + \cdots + f(x_n)}{(b-a)/\Delta x} = \frac{f(x_1)\,\Delta x + f(x_2)\,\Delta x + \cdots + f(x_n)\,\Delta x}{b-a}.$$

If, now, n is very large and Δx small, the expression

$$f(x_1)\,\Delta x + f(x_2)\,\Delta x + \cdots + f(x_n)\,\Delta x = \sum_{k=1}^{n} f(x_k)\,\Delta x$$

is very nearly equal to $\int_a^b f(x)\,dx$. In fact, if we take limits, letting $n \to \infty$, we obtain precisely

$$\lim_{n\to\infty} \frac{f(x_1)\,\Delta x + f(x_2)\,\Delta x + \cdots + f(x_n)\,\Delta x}{b-a} = \frac{1}{b-a}\int_a^b f(x)\,dx.$$

This is the expression that defines the average value of y in Eq. (2).

EXAMPLE 1 The average value of the function $y = \sqrt{x}$ with respect to x from $x = 0$ to $x = 4$ is

$$(y_{av})_x = \frac{1}{4}\int_0^4 \sqrt{x}\,dx = \frac{1}{4}\cdot\frac{2}{3}x^{3/2}\Big]_0^4$$

$$= \frac{1}{6}\cdot 8 = \frac{4}{3}. \qquad \square$$

When it is possible to express y as a function of more than one variable, say as a function of u as well as a function of x, then $(y_{av})_u$ need not equal $(y_{av})_x$. In the next example we calculate the average velocity of a body falling in a vacuum first as an average over time, then as an average over distance.

EXAMPLE 2 For a body falling from rest in a vacuum,

$$s = \frac{1}{2}gt^2, \qquad v = gt, \qquad v = \sqrt{2gs}.$$

Suppose we calculate the average velocity, first with respect to t and second with respect to s, from $t_1 = 0$, $s_1 = 0$, to $t_2 > 0$, $s_2 = \frac{1}{2}g(t_2)^2$. Then, by definition,

$$(v_{av})_t = \frac{1}{t_2 - 0}\int_0^{t_2} gt\,dt = \frac{1}{2}gt_2 = \frac{1}{2}v_2,$$

$$(v_{av})_s = \frac{1}{s_2 - 0}\int_0^{s_2} \sqrt{2gs}\,ds = \frac{2}{3}\sqrt{2gs_2} = \frac{2}{3}v_2. \quad \square$$

EXAMPLE 3 (This is a geometric interpretation of Eq. (2) for $f \geq 0$.) If both sides of Eq. (2) are multiplied by $b - a$, we have

$$(y_{av})_x \cdot (b - a) = \int_a^b f(x)\,dx. \tag{4}$$

The right-hand side of Eq. (4) is the area between the curve $y = f(x)$ and the x-axis, from $x = a$ to $x = b$. The left-hand side of the equation can be interpreted as the area of a rectangle of altitude $(y_{av})_x$ and of base $b - a$. Hence, Eq. (4) provides a geometric interpretation of $(y_{av})_x$ as the ordinate of the curve $y = f(x)$ that should be used as altitude if one wishes to construct a rectangle whose base is the interval $a \leq x \leq b$ and whose area is equal to the area under the curve (Fig. 5.40). $\square$

EXAMPLE 4 The distance traveled by a body moving with speed $|v(t)|$ along a straight line from time $t = a$ to time $t = b$ is

$$\text{Distance traveled} = \int_a^b |v(t)|\,dt.$$

The average speed for the trip is therefore

$$\text{Average speed} = \frac{\text{distance traveled}}{b - a} = \frac{1}{b - a}\int_a^b |v(t)|\,dt.$$

Thus, the average speed is the average value of $|v(t)|$ over $[a, b]$. $\square$

We know from the discussion following Eq. (7) in Article 3.8 that at some instant during the motion in Example 4 the speed $|v|$ is exactly equal to the average speed for the trip. This is generally true for the average value of a function f on an interval $[a, b]$. That is, if f is continuous on $[a, b]$, then at some point in $[a, b]$ the value of f is exactly equal to the average value of f over $[a, b]$. We call this fact the Mean Value Theorem for Integrals.

THEOREM

The Mean Value Theorem for Integrals

If f is continuous on $[a, b]$, then there exists a number c in $[a, b]$ at which

$$f(c) = \frac{1}{b - a}\int_a^b f(x)\,dx. \tag{5}$$

Proof. The proof is a consequence of the Intermediate Value Theorem of Article 1.11, which says that the continuous function f takes on every value between its minimum value $\min f$ and its maximum value $\max f$ on $[a, b]$. We know from Property I6 in Article 4.5 that

$$\min f \cdot (b - a) \le \int_a^b f(x)\,dx \le \max f \cdot (b - a).$$

This means that

$$\min f \le \frac{1}{b-a}\int_a^b f(x)\,dx \le \max f. \tag{6}$$

The significance of Eq. (6) is that the average value of f is a number between $\min f$ and $\max f$. It must therefore be taken on by f at some point c in $[a, b]$. That is,

$$f(c) = \frac{1}{b-a}\int_a^b f(x)\,dx,$$

for some point c in $[a, b]$, which is Eq. (5). ■

Average values of functions play a significant role in calculating the effective voltage and current in an electric circuit.

EXAMPLE 5 The electric current in our household power supply is an alternating current whose flow can be modeled by a sine function

$$i = I \sin \omega t \tag{7}$$

like the one graphed in Fig. 5.41. Equation (7) expresses the current i in amperes as a function of time t in seconds. The amplitude I is the current's peak value, and $2\pi/\omega$ is the period.

The average value of i over a half-cycle is

$$i_{av} = \frac{1}{\pi/\omega - 0}\int_0^{\pi/\omega} I \sin \omega t\,dt = \frac{2}{\pi} I,$$

and the average value over a whole cycle is

$$i_{av} = \frac{\omega}{2\pi}\int_0^{2\pi/\omega} I \sin \omega t\,dt = 0.$$

If the current were measured with a standard moving coil galvanometer, the meter would read zero!

To measure the current effectively one uses an instrument that measures the square root of the average value of the square of the current, namely

$$I_{rms} = \sqrt{(i^2)_{av}}. \tag{8}$$

The subscript rms stands for "root-mean-square." Since the average value of $i^2 = I^2 \sin^2 \omega t$ over a whole cycle is

$$(i^2)_{av} = \frac{\omega}{2\pi}\int_0^{2\pi/\omega} I^2 \sin^2 \omega t\,dt = \frac{I^2}{2}, \tag{9}$$

the rms current is

$$I_{rms} = \sqrt{\frac{I^2}{2}} = \frac{I}{\sqrt{2}}. \tag{10}$$

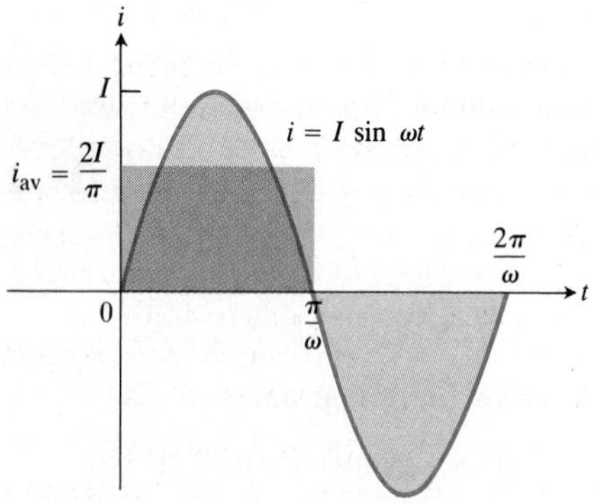

5.41 The average value of a sinusoidal current over a half-cycle is $2I/\pi$. The average over a complete cycle is zero.

In the same way, the root-mean-square value of a sinusoidal voltage $v = V \sin \omega t$ is

$$V_{\text{rms}} = \frac{V}{\sqrt{2}}. \tag{11}$$

Household voltages and currents are always referred to in terms of their rms values. Thus, "115 volts ac" means that the rms voltage is 115 volts. The voltage amplitude V, from Eq. (11), is

$$V = \sqrt{2}\, V_{\text{rms}} \approx 163 \text{ volts. } \square$$

PROBLEMS

In Problems 1–5 find the average value with respect to x, over the given domain, of the given function $f(x)$. In each case, draw a graph of the curve $y = f(x)$, and sketch a rectangle whose altitude is the average ordinate.

1. a) $\sin x,\quad 0 \le x \le \pi/2$ b) $\sin x,\quad 0 \le x \le 2\pi$

2. a) $\sin^2 x,\quad 0 \le x \le \pi/2$ b) $\sin^2 x,\quad \pi \le x \le 2\pi$

3. $\sqrt{2x + 1},\quad 4 \le x \le 12$

4. $\frac{1}{2} + \frac{1}{2} \cos 2x,\quad 0 \le x \le \pi$

5. $\alpha x + \beta;\quad a \le x \le b\quad [\alpha, \beta, a, b, \text{ constants}]$

6. Carry out the integration in Eq. (9), Example 5.

7. Show that the root-mean-square value of the voltage $v = V \sin \omega t$ in Example 5 is $V_{\text{rms}} = V/\sqrt{2}$.

8. If a household electric circuit is rated "20 amps" (rms), what is the peak value (amplitude) of the allowable current?

9. The costs of warehouse space, utilities, insurance, and security can be a large part of the cost of doing business, and a firm's average daily inventory can play a significant role in determining these costs.

For example, suppose that a wholesale grocer receives a shipment of 1200 cases of chocolate bars every 30 days. The chocolate is sold to retailers at a steady rate; and x days after the shipment arrives, the inventory $I(x)$ of cases still on hand is $I(x) = 1200 - 40x$. Compute the average daily inventory.

10. Solon Container receives 450 drums of plastic pellets every 30 days. The inventory function (drums on hand as a function of days) is $I(x) = 450 - x^2/2$. Find the average daily inventory. If the holding cost for one drum is 2¢ per day, find the total daily holding cost.

11. Mitchell Mailorder receives a shipment of 600 cases of athletic socks every 60 days. The number of cases on hand t days after the shipment arrives is $I(t) = 600 - 20\sqrt{15t}$. Find the average daily inventory. If the holding cost for one case is $\frac{1}{2}$¢ per day, find the total daily holding cost.

12. (*Calculator*) Compute the average value of the temperature function

$$f(x) = 37 \sin\left[\frac{2\pi}{365}(x - 101)\right] + 25$$

for a 365-day year (see Example 2 in Article 2.7). This is one way to estimate the annual mean air temperature in Fairbanks, Alaska. The National Weather Service's official figure, a numerical average of the daily normal mean air temperatures for the year, is 25.7°F, which is slightly higher than the average value of $f(x)$. Figure 2.28 shows why.

13. (*Calculator*) (a) Use the trapezoidal rule to estimate the average lumen output of the No. 22 flashbulb of Problem 13, Article 4.9, for the time interval from $t = 0$ to $t = 60$ milliseconds. Take the data from Table 4.4. (b) How many seconds would a 60-watt incandescent light bulb rated at 765 lumens have to burn to put out as many lumen-milliseconds as the No. 22 flashbulb did in (a)?

14. For a body falling freely on the moon, where the acceleration of gravity is 1.62 m/s²,

$$s = 0.81t^2, \qquad v = 1.62t, \qquad v = \sqrt{3.24s},$$

s in meters, t in seconds, v in meters per second. During the first two seconds after being released from rest a body will fall $s = 0.81(2)^2 = 3.24$ meters.

a) Graph $v = 1.62t$ for $0 \le t \le 2$, and $v = \sqrt{3.24s}$ for $0 \le s \le 3.24$.

b) Find the average velocity of the body with respect to time for the first two seconds of fall.

c) Find the average velocity of the body with respect to distance for the first 3.24 meters of fall.

15. Let f be a function that is differentiable on $[a, b]$. In Chapter 1 we defined the average rate of change of f on $[a, b]$ to be

$$\frac{f(b) - f(a)}{b - a}$$

and the instantaneous rate of change of f at x to be $f'(x)$. In this article, we defined the average value of a function. For the new definition of average to be consistent with the old one, we should have

$$\frac{f(b) - f(a)}{b - a} = \text{average value of } f' \text{ on } [a, b].$$

Show that this is the case.

16. Given a circle C of radius a, and a diameter AB of C. Chords are drawn perpendicular to AB, intercepting equal segments along AB. Find the limit of the average of the lengths of these chords, as the number of chords tends to infinity. (*Hint:* $\int_{-a}^{a} \sqrt{a^2 - x^2}\, dx$ is $\frac{1}{2}\pi a^2$, since it is the area of a semicircle of radius a.)

17. Solve Problem 16 under the modified assumption that the chords intercept equal arcs along the circumference of C.

18. Solve Problem 16 using the *squares* of the lengths, in place of the lengths of the chords.

19. Solve Problem 17 using the *squares* of the lengths, in place of the lengths of the chords.

5.9

Moments and Center of Mass

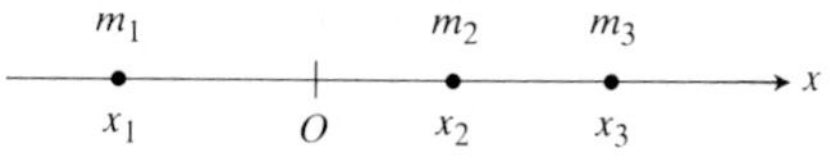

5.42 Masses on the x-axis.

If masses $m_1, m_2, \ldots, m_n$ are placed along the x-axis at distances $x_1, x_2, \ldots, x_n$ from the origin (Fig. 5.42), then their *moment* about the origin is defined to be

$$x_1m_1 + x_2m_2 + \cdots + x_nm_n = \sum_{k=1}^{n} x_km_k. \tag{1}$$

If all the mass

$$m_1 + m_2 + \cdots + m_n = \sum_{k=1}^{n} m_k \tag{2}$$

is concentrated at one point of abscissa $\bar{x}$ ("x bar"), the total moment is

$$\bar{x}\left(\sum_{k=1}^{n} m_k\right).$$

The position of $\bar{x}$ for which this is the same as the total moment in (1) is called the *center of mass*. The condition

$$\bar{x}\sum_{k=1}^{n} m_k = \sum_{k=1}^{n} x_km_k$$

thus determines

$$\bar{x} = \frac{\sum_{k=1}^{n} x_km_k}{\sum_{k=1}^{n} m_k}. \tag{3}$$

EXAMPLE 1 The principle of a moment underlies the simple seesaw. For instance (Fig. 5.43), suppose one child weighs 80 lb and sits 5 ft from the point O, while the other child weighs 100 lb and is 4 ft from O. Then the child at the left end of the seesaw produces a moment of $5 \times 80 = 400$ ft-lb, which tends to rotate the plank counterclockwise about O. The child at the right end of the plank produces a moment of $4 \times 100 = 400$ ft-lb, which tends to rotate the plank clockwise around O. If we introduce coordinates $x_1 = -5$ and $x_2 = +4$, we find

$$x_1m_1 + x_2m_2 = (-5)(80) + (4)(100) = -400 + 400 = 0$$

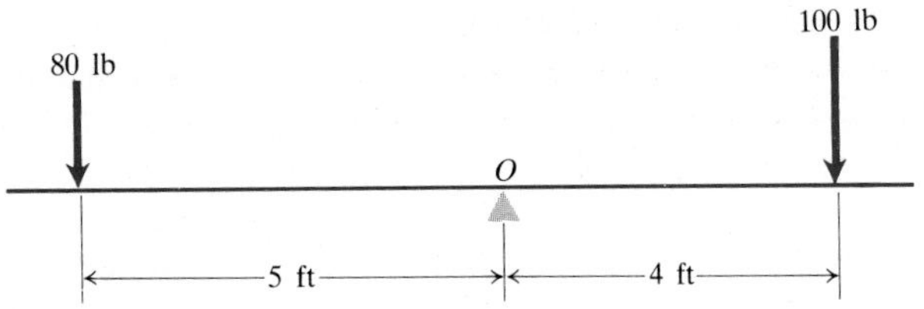

5.43 A 100-lb weight four feet to the right of O balances an 80-lb weight five feet to the left of O.

as the resultant moment about O. The same moment, zero, would be obtained if both children were at O, which is the center of their mass. □

If, instead of being placed on the x-axis, the masses are located in the xy-plane at points

$$(x_1, y_1), \quad (x_2, y_2), \quad \ldots, \quad (x_n, y_n),$$

in that order, then we define their moments with respect to the y-axis and with respect to the x-axis as

Moment with respect to the y-axis:

$$M_y = x_1m_1 + x_2m_2 + \cdots + x_nm_n = \sum_{k=1}^{n} x_km_k,$$

Moment with respect to the x-axis:

$$M_x = y_1m_1 + y_2m_2 + \cdots + y_nm_n = \sum_{k=1}^{n} y_km_k.$$

The center of mass is the point $(\bar{x}, \bar{y})$,

$$\bar{x} = \frac{M_y}{M} = \frac{\sum x_km_k}{\sum m_k}, \qquad \bar{y} = \frac{M_x}{M} = \frac{\sum y_km_k}{\sum m_k}, \tag{4}$$

where the total mass could be concentrated and still give the same total moments M_y and M_x.

In space, three coordinates are needed to specify the position of a point. If the masses are located at the points

$$(x_1, y_1, z_1), \quad (x_2, y_2, z_2), \quad \ldots, \quad (x_n, y_n, z_n),$$

we define their moments with respect to the three coordinate planes as

$$M_{yz} = x_1m_1 + x_2m_2 + \cdots + x_nm_n = \sum_{k=1}^{n} x_km_k,$$

$$M_{zx} = y_1m_1 + y_2m_2 + \cdots + y_nm_n = \sum_{k=1}^{n} y_km_k,$$

$$M_{xy} = z_1m_1 + z_2m_2 + \cdots + z_nm_n = \sum_{k=1}^{n} z_km_k.$$

(M_{yz} = Moment with respect to the yz-plane, etc.)

The center of mass $(\overline{x}, \overline{y}, \overline{z})$ is the point where the total mass could be concentrated without altering these moments. Its coordinates therefore are given by

$$\overline{x} = \frac{M_{yz}}{\sum m_k}, \qquad \overline{y} = \frac{M_{zx}}{\sum m_k}, \qquad \overline{z} = \frac{M_{xy}}{\sum m_k}. \tag{5}$$

Now most physical objects with which we deal are composed of enormously large numbers of molecules. It would be extremely difficult, and in most cases unnecessary, for us to concern ourselves with the molecular structure of a physical object, such as a pendulum, whose motion is to be studied as a whole. Instead, we make simplifying assumptions that we recognize as being only approximately correct. One such assumption is that the matter in a given solid is continuously distributed throughout the solid. Furthermore, if P is a point in the solid and ΔV is an element of volume that contains P, and if Δm is the mass of ΔV, then we assume that the ratio $\Delta m/\Delta V$ tends to a definite limit

$$\delta = \lim_{\Delta V \to 0} \frac{\Delta m}{\Delta V}, \tag{6}$$

as the largest dimension of ΔV approaches zero. The limit δ is called the *density* of the solid at the point P. Its dimensions are mass per unit volume. It is customary to write Eq. (6) in the alternative forms

$$\delta = dm/dV, \qquad dm = \delta\, dV. \tag{7}$$

If a solid is divided into small pieces ΔV of mass Δm and if $P(\tilde{x}, \tilde{y}, \tilde{z})$ is a point in ΔV and δ is the density at P, then

$$\Delta m \approx \delta\, \Delta V.$$

(The symbol ~ over the x, y, and z in $P(\tilde{x}, \tilde{y}, \tilde{z})$ is a *tilde* (rhymes with "Hilda"). Thus, $\tilde{x}$ is read "x tilde," and so on.) The moments of Δm with respect to the coordinate planes are not defined by what we have done thus far. But now we may think of replacing the Δm that fills the volume ΔV by an equal mass all concentrated at the point P. The moments of this concentrated mass with respect to the coordinate planes are

$$\tilde{x}\, \Delta m, \qquad \tilde{y}\, \Delta m, \qquad \tilde{z}\, \Delta m.$$

Now we add the moments of all the concentrated masses in all the volume elements ΔV and take the limit as the ΔV's approach zero. This leads us to the following *definitions* of the moments M_{yz}, etc., for the mass as a whole:

$$M_{yz} = \lim_{\Delta m \to 0} \sum \tilde{x}\, \Delta m = \int \tilde{x}\, dm,$$

$$M_{zx} = \lim_{\Delta m \to 0} \sum \tilde{y}\, \Delta m = \int \tilde{y}\, dm,$$

$$M_{xy} = \lim_{\Delta m \to 0} \sum \tilde{z}\, \Delta m = \int \tilde{z}\, dm.$$

From these, we then deduce the equations

$$\overline{x} = \frac{\int \tilde{x}\, dm}{\int dm}, \qquad \overline{y} = \frac{\int \tilde{y}\, dm}{\int dm}, \qquad \overline{z} = \frac{\int \tilde{z}\, dm}{\int dm} \tag{8}$$

for the center of mass of the solid as a whole.

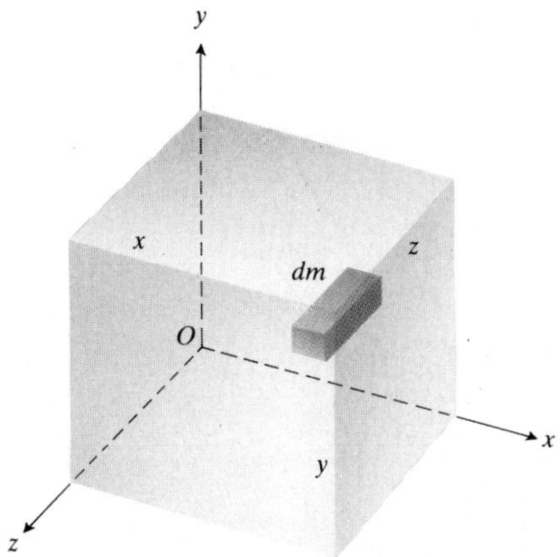

5.44 Mass dm = density × dV.

In theory, the element dm is approximately the mass in a volume element dV which has three small dimensions as in Fig. 5.44. In a later chapter we shall see how to evaluate the integrals (8) that arise when this is done. In practice, however, we can usually take

$$dm = \delta\, dV$$

with a volume element dV which has only one small dimension. Then we must interpret $\tilde{x}$, $\tilde{y}$, $\tilde{z}$ in the integrands in (8) as the coordinates of the *center of mass of the element dm*.

In many problems of practical importance the density δ is constant and the solid has a plane of symmetry. Then it is easy to see that the center of mass lies in this plane of symmetry. For we may, with no loss of generality, choose our coordinate reference frame in such a way that the yz-plane is the plane of symmetry. Then for every element of mass Δm with a positive $\tilde{x}$ there is a symmetrically located element of mass with a corresponding negative $\tilde{x}$. These two elements have moments that are equal in magnitude and of opposite signs. The whole mass is made up of such symmetric pairs of elements and the sum of their moments about the yz-plane is zero. Therefore $\bar{x} = 0$; that is, the center of mass lies in the plane of symmetry. If there are two planes of symmetry, their intersection is an axis of symmetry and the center of mass must lie on this axis, since it lies in both planes of symmetry.

The most frequently encountered distributions of mass are

a) along a thin wire or filament: $dm = \delta_1\, ds$,

b) in a thin plate or shell: $dm = \delta_2\, dA$ or $\delta_2\, dS$,

c) in a solid: $dm = \delta_3\, dV$,

where

$$\begin{aligned} ds &= \text{element of arc length},\\ \delta_1 &= \text{mass per unit length of the wire},\\ dA \text{ or } dS &= \text{element of area},\\ \delta_2 &= \text{mass per unit area of the plate or shell},\\ dV &= \text{element of volume},\\ \delta_3 &= \text{mass per unit volume of the solid}. \end{aligned}$$

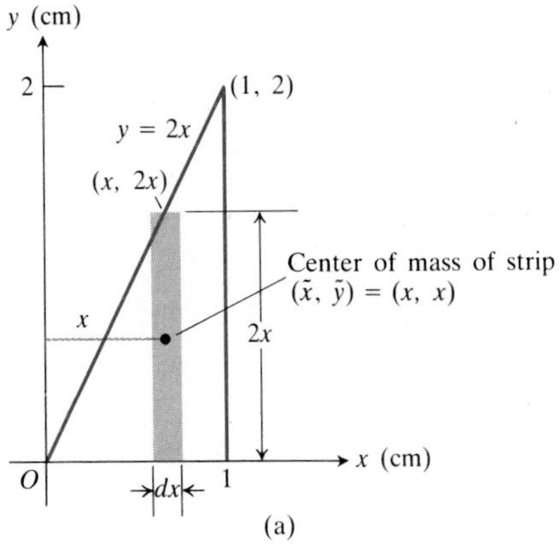

(a)

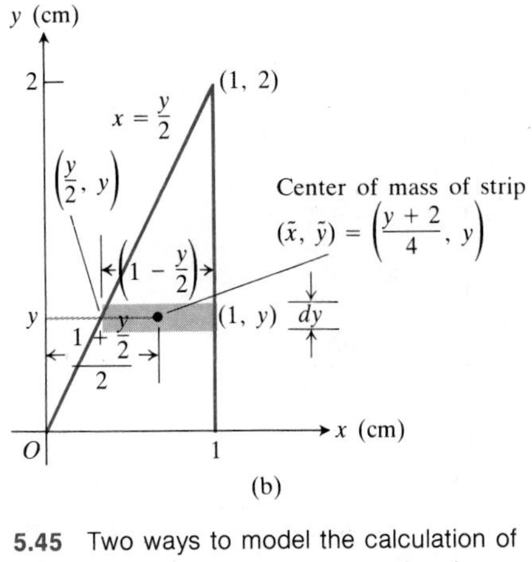

(b)

5.45 Two ways to model the calculation of the moment M_y of the triangular plate in Example 2.

EXAMPLE 2 The triangular plate shown in Fig. 5.45, bounded by the lines $y = 0$, $y = 2x$, and $x = 1$, has a uniform density of $\delta = 3\text{g/cm}^2$. Find (a) the plate's moment M_y about the y-axis, (b) the plate's mass M, and (c) the x-coordinate of the plate's center of mass.

Solution METHOD 1. *Vertical strips* (Fig. 5.45a).

a) The moment M_y: The typical vertical strip has

center of mass (c.m.): $(\tilde{x}, \tilde{y}) = (x, x)$

length: $2x$

width: dx

area: $dA = 2x\, dx$

mass: $dm = \delta\, dA = 3 \cdot 2x\, dx = 6x\, dx$

distance of c.m. from y-axis: $\tilde{x} = x$.

The moment of the strip about the y-axis is

$$\tilde{x}\,dm = x \cdot 6x\,dx = 6x^2\,dx.$$

The moment of the plate about the y-axis is therefore

$$M_y = \int \tilde{x}\,dm = \int_0^1 6x^2\,dx = 2x^3\Big]_0^1 = 2\text{ gm}\cdot\text{cm}.$$

b) The plate's mass:

$$M = \int dm = \int_0^1 6x\,dx = 3x^2\Big]_0^1 = 3\text{ gm}.$$

c) The x-coordinate of the plate's center of mass:

$$\bar{x} = \frac{M_y}{M} = \frac{2\text{ gm}\cdot\text{cm}}{3\text{ gm}} = \frac{2}{3}\text{ cm}.$$

METHOD 2. *Horizontal strips* (Fig. 5.45b).

a) The moment M_y: The typical horizontal strip has

c.m.: $(\tilde{x}, \tilde{y}) = \left(\frac{1}{2}\left(1 + \frac{y}{2}\right), y\right) = \left(\frac{y+2}{4}, y\right)$

length: $1 - \frac{y}{2} = \frac{2-y}{2}$

width: dy

area: $dA = \frac{2-y}{2}\,dy$

mass: $dm = \delta dA = 3 \cdot \frac{2-y}{2}\,dy$

distance of c.m. to y-axis: $\tilde{x} = \frac{y+2}{4}$.

The moment of the strip about the y-axis is

$$\tilde{x}\,dm = \frac{y+2}{4} \cdot 3 \cdot \frac{2-y}{2}\,dy = \frac{3}{8}(4 - y^2)\,dy.$$

The moment of the plate about the y-axis is

$$M_y = \int \tilde{x}\,dm = \int_0^2 \frac{3}{8}(4 - y^2)\,dy = \frac{3}{8}\left[4y - \frac{y^3}{3}\right]_0^2 = \frac{3}{8}\left(\frac{16}{3}\right) = 2\text{ gm}\cdot\text{cm}.$$

b) The plate's mass:

$$M = \int dm = \int_0^2 \frac{3}{2}(2 - y)\,dy = \frac{3}{2}\left[2y - \frac{y^2}{2}\right]_0^2 = \frac{3}{2}[4 - 2] = 3\text{ gm}.$$

c) The x-coordinate of the plate's center of gravity:

$$\bar{x} = \frac{M_y}{M} = \frac{2\text{ gm}\cdot\text{cm}}{3\text{ gm}} = \frac{2}{3}\text{ cm}. \quad \square$$

EXAMPLE 3 Find the center of mass of a thin homogeneous (constant density) plate covering the region bounded above by the parabola $y = 4 - x^2$ and below by the x-axis. See Fig. 5.46.

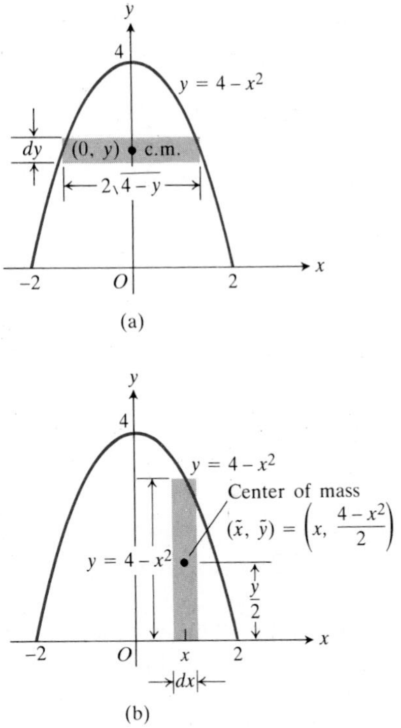

5.46 Modeling the problem in Example 3 with horizontal strips (a) leads to a difficult integration, so we model with vertical strips (b) instead.

Solution Since the plate is symmetric about the y-axis and its density is constant, the center of mass lies on the y-axis. This means that

$$\bar{x} = 0.$$

It remains to find $\bar{y} = M_x/M$.

A trial calculation with horizontal strips (Fig. 5.46a) leads to a difficult integration,

$$M_x = \int_0^4 2\delta y\sqrt{4 - y}\, dy.$$

We therefore model the distribution of mass with vertical strips instead (Fig. 5.46b). The typical vertical strip has

center of mass (c.m.): $(\tilde{x}, \tilde{y}) = \left(x, \dfrac{4 - x^2}{2}\right)$

length: $4 - x^2$

width: dx

area: $dA = (4 - x^2)\, dx$

mass: $dm = \delta\, dA = \delta(4 - x^2)\, dx$

distance from c.m. to x-axis: $\tilde{y} = \dfrac{4 - x^2}{2}$.

The moment of the strip about the x-axis is

$$\tilde{y}\, dm = \frac{4 - x^2}{2} \cdot \delta(4 - x^2)\, dx = \frac{\delta}{2}(4 - x^2)^2\, dx.$$

The moment of the plate about the x-axis is

$$M_x = \int \tilde{y}\, dm = \int_{-2}^{2} \frac{\delta}{2}(4 - x^2)^2\, dx = \frac{256}{15}\delta.$$

The mass of the plate is

$$M = \int dm = \int_{-2}^{2} \delta(4 - x^2)\, dx = \frac{32}{3}\delta.$$

Therefore,

$$\bar{y} = \frac{M_x}{M} = \frac{\frac{256}{15}\delta}{\frac{32}{3}\delta} = \frac{8}{5}.$$

The plate's center of mass is the point

$$(\bar{x}, \bar{y}) = \left(0, \frac{8}{5}\right). \quad \square$$

EXAMPLE 4 Show that the center of mass of a thin homogeneous triangular plate of base b and altitude h lies at the intersection of the medians. See Fig. 5.47.

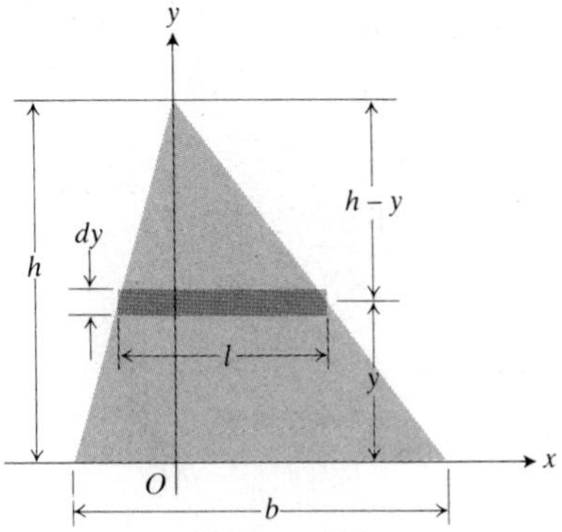

5.47 Calculating the mass of a thin triangular plate.

Solution We base one side of the triangle on the x-axis with the opposite vertex on the positive y-axis. The mass of a typical horizontal strip is

$$dm = \delta_2\, dA = \delta_2\, l\, dy,$$

where δ is the density and l is the width of the triangle at distance y above its base. By similar triangles,

$$\frac{l}{b} = \frac{h - y}{h} \quad \text{or} \quad l = \frac{b}{h}(h - y),$$

so that

$$dm = \delta \frac{b}{h}(h - y)\, dy.$$

For the y-coordinate of the center of mass of the strip, we have $\tilde{y} = y$. For the entire plate,

$$\bar{y} = \frac{\int y\, dm}{\int dm} = \frac{\int_0^h \delta \frac{b}{h} y(h - y)\, dy}{\int_0^h \delta \frac{b}{h}(h - y)\, dy} = \frac{1}{3}h.$$

Thus the center of mass lies above the base of the triangle at a distance one third of the way toward the opposite vertex. By considering each side in turn as a base of the triangle, this result shows that the center of mass lies at the intersection of the medians. (The medians intersect at a point two thirds of the way from any vertex to the opposite side.) □

EXAMPLE 5 A thin homogeneous wire is bent to form a semicircle of radius r (Fig. 5.48). Find its center of mass.

Solution Here we take

$$dm = \delta\, ds,$$

where ds is an element of arc length of the wire and

$$\delta = \frac{M}{L} = \frac{M}{\pi r}$$

is the mass per unit length of the wire. In terms of the central angle θ measured in radians (as usual), we have

$$ds = r\, d\theta$$

and

$$\tilde{x} = r \cos \theta, \qquad \tilde{y} = r \sin \theta.$$

Hence

$$\bar{x} = \frac{\int \tilde{x}\, dm}{\int dm} = \frac{\int_0^\pi r \cos \theta\, \delta\, r\, d\theta}{\int_0^\pi \delta\, r\, d\theta} = \frac{\delta\, r^2 [\sin \theta]_0^\pi}{\delta\, r[\theta]_0^\pi} = 0,$$

$$\bar{y} = \frac{\int \tilde{y}\, dm}{\int dm} = \frac{\int_0^\pi r \sin \theta\, \delta\, r\, d\theta}{\int_0^\pi \delta\, r\, d\theta} = \frac{\delta\, r^2[-\cos \theta]_0^\pi}{\delta\, r[\theta]_0^\pi} = \frac{2}{\pi} r.$$

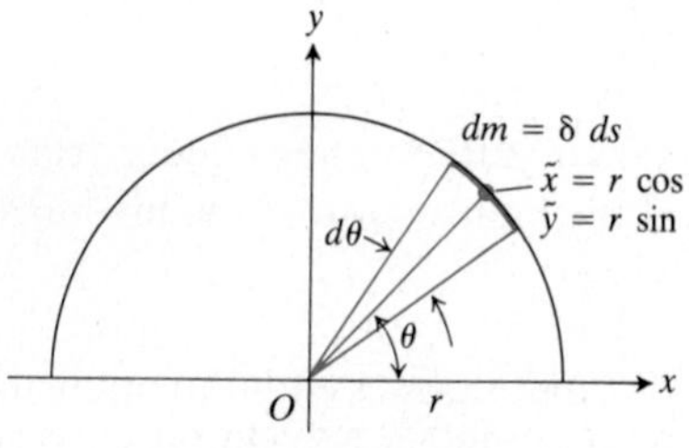

5.48 Calculating the mass of a semicircular wire.

The center of mass is therefore on the y-axis at a distance $2/\pi$ (roughly two thirds) of the way up from the origin toward the intercept (0, r). □

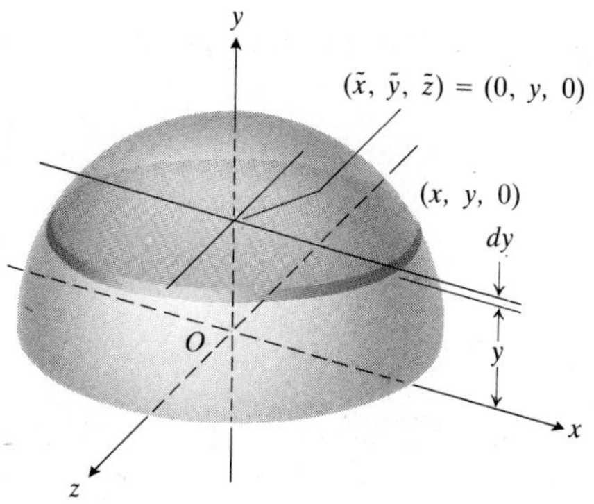

5.49 Finding the mass of a hemispherical solid by slicing parallel to its base.

EXAMPLE 6 Find the center of mass of a solid hemisphere its density at any point P is proportional to the distance betw base of the hemisphere.

Solution Imagine the solid cut into slices of thickness dy by planes perpendicular to the y-axis (Fig. 5.49), and take

$$dm = \delta\, dV,$$

where

$$dV = A(y)\, dy$$

is the volume of the representative slice at distance y above the base of the hemisphere, which is in the xz-plane, and where

$$\delta = ky \qquad (k = \text{constant})$$

is the density of the solid in this slice. The area of a face of the slice dV is

$$A(y) = \pi x^2,$$

where $x^2 + y^2 = r^2$; that is,

$$dV = A(y)\, dy = \pi(r^2 - y^2)\, dy;$$

so that

$$dm = k\pi(r^2 - y^2)y\, dy.$$

The center of mass of the slice may be taken at its geometric center,

$$(\tilde{x}, \tilde{y}, \tilde{z}) = (0, y, 0),$$

so that

$$\bar{x} = \bar{z} = 0$$

and

$$\bar{y} = \frac{\int y\, dm}{\int dm} = \frac{\int_0^r k\pi(r^2 - y^2)y^2\, dy}{\int_0^r k\pi(r^2 - y^2)y\, dy} = \frac{k\pi\left[\frac{r^2y^3}{3} - \frac{y^5}{5}\right]_0^r}{k\pi\left[\frac{r^2y^2}{2} - \frac{y^4}{4}\right]_0^r} = \frac{8}{15}r. \quad \square$$

Moment Formulas

MASS DISTRIBUTED OVER A PLANE REGION

$dm = \delta_2\, dA$ = mass of element with area dA, density δ_2

$(\tilde{x}, \tilde{y})$ = c.m. of mass element dm

M = mass = $\int dm$

M_x = moment about x-axis = $\int \tilde{y}\, dm$

M_y = moment about y-axis = $\int \tilde{x}\, dm$

$(\bar{x}, \bar{y})$ = overall center of mass

$$\bar{x} = \frac{M_x}{M} = \frac{\int \tilde{x}\, dm}{\int dm}, \qquad \bar{y} = \frac{M_y}{M} = \frac{\int \tilde{y}\, dm}{\int dm}$$

MASS IN SPACE

$dm = \delta_3\, dV$ = mass of element with volume dV, density δ_3

$(\tilde{x}, \tilde{y}, \tilde{z})$ = c.m. of mass element dm

M = mass = $\int dm$

M_{yz} = moment about yz-plane = $\int \tilde{x}\, dm$

M_{xz} = moment about xz-plane = $\int \tilde{y}\, dm$

M_{xy} = moment about xy-plane = $\int \tilde{z}\, dm$

$(\bar{x}, \bar{y}, \bar{z})$ = overall center of mass

$$\bar{x} = \frac{M_{yz}}{M} = \frac{\int \tilde{x}\, dm}{\int dm}, \qquad \bar{y} = \frac{M_{xz}}{M} = \frac{\int \tilde{y}\, dm}{\int dm},$$
$$\bar{z} = \frac{M_{xy}}{M} = \frac{\int \tilde{z}\, dm}{\int dm}.$$

PROBLEMS

In Problems 1–5, find the center of mass of a thin homogeneous plate covering the given portion of the xy-plane.

1. The first quadrant of the circle $x^2 + y^2 = a^2$.

2. The region bounded by the parabola $y = h^2 - x^2$ and the x-axis.

3. The "triangular" region in the first quadrant between the circle $x^2 + y^2 = a^2$ and the lines $x = a$, $y = a$.

4. The region between the x-axis and the curve $y = \sin x$ between $x = 0$ and $x = \pi$. (*Hint:* Take $dA = y\, dx$ and $\tilde{y} = \frac{1}{2}y$.)

5. The region between the y-axis and the curve $x = 2y - y^2$.

6. Find the distance, from the base, of the center of mass of a thin triangular plate of base b and altitude h if its density varies as the square root of the distance from the base.

7. In Problem 6, suppose that the density varies as the square of the distance.

8. The density of a thin plate bounded by $y = x^2$ and $y = 2x + 3$ is $\delta(x) = x + 2$. Find the coordinates of the center of mass.

9. A thin rod of length L lies along the x-axis from $x = 0$ to $x = L$. Its density is $\delta(x) = x^4$. Find the x-coordinate of the center of mass.

10. Find the center of mass of a homogeneous solid right circular cone.

11. Find the center of mass of a solid right circular cone if the density varies as the distance from the base.

12. In Problem 11 suppose that the density varies as the square of the distance.

13. In Example 5 (see Fig. 5.48) suppose that the density is $\delta = k \sin \theta$, k being constant. Find the center of mass.

5.10

Centroids and Centers of Gravity

For a mass of uniform density, Eqs. (8), Article 5.9, reduce to

$$\bar{x} = \frac{\int \tilde{x}\, \delta_3\, dV}{\int \delta_3\, dV} = \frac{\delta_3 \int \tilde{x}\, dV}{\delta_3 \int dV} = \frac{\int \tilde{x}\, dV}{\int dV},$$
$$\bar{y} = \frac{\int \tilde{y}\, dV}{\int dV}, \qquad \bar{z} = \frac{\int \tilde{z}\, dV}{\int dV} \tag{1}$$

for a solid, with similar equations having dA or ds in place of dV in the case of a plate or a wire. Since these expressions involve only the geometric objects, namely, solids, plane regions, and curves, we speak of the point $(\bar{x}, \bar{y}, \bar{z})$ in such cases as the *centroid* of the object. The term *center of gravity* is also used, since the line of action of the forces due to gravity acting on the elements of the object passes through this point. (See Sears, Zemansky, Young, *University Physics*, Sixth edition (1982), Chapters 7 and 8, for a discussion of physical concepts involved.)

EXAMPLE 1 Find the center of gravity of a solid hemisphere of radius r.

Solution As in Example 6, Article 5.9, imagine the solid (Fig. 5.49) cut into slices of thickness dy by planes perpendicular to the y-axis. The centroid of a slice is its geometric center, on the y-axis, at (0, y, 0). Its moment with respect to the xz-plane is

$$dM_{xz} = y\,dV = y\pi(r^2 - y^2)\,dy.$$

Hence

$$M_{xz} = \pi \int_0^r (r^2y - y^3)\,dy = \frac{\pi r^4}{4}.$$

Since the volume of the hemisphere is

$$V = \frac{2}{3}\pi r^3,$$

we have

$$\bar{y} = \frac{M_{xz}}{V} = \frac{3}{8}r. \ \square$$

EXAMPLE 2 Find the center of gravity of a thin hemispherical shell of inner radius r and thickness t.

Solution We shall solve this problem, (a) exactly, by using the results of the preceding example; then (b) we shall see how the position of the center of gravity changes as we hold r fixed and let $t \to 0$. Finally, (c) we shall solve the problem approximately without using the results of Example 1, by considering the center of mass of an imaginary distribution of mass over the surface of the hemisphere of radius r, assuming that the thickness t is negligible in comparison with r (as, for example, the gold leaf covering the dome of the State House in Boston).

a) Let M_1, M_2, and M; V_1, V_2, and V denote, respectively, the moments with respect to the xz-plane and the volumes of the solid hemisphere of radius r, the solid hemisphere of radius (r + t), and the hemispherical shell of thickness t and inner radius r. Since the moment of the sum of two masses is the sum of their moments and

$$V_1 + V = V_2,$$

we also have

$$M_1 + M = M_2 \qquad \text{or} \qquad M = M_2 - M_1.$$

But, by Example 1,

$$M_2 = \frac{\pi}{4}(r + t)^4, \qquad M_1 = \frac{\pi}{4}r^4,$$

as a suitable modification of Eq. (1) when the force is variable, where $\bar{F}$ represents the *average* value of the variable force throughout the total displacement. However, the determination of $\bar{F}$ itself involves an integration in the general case, so that it is usually as easy to apply the principles illustrated in Example 1 as it is to apply Eq. (3). In fact, Eq. (3) may be interpreted as *defining* $\bar{F}$. □

In a manner entirely analogous to that in the example above, we are led to define

$$W = \int_a^b F\,ds \tag{4}$$

to be the work done by a variable force (which, however, always acts along a given direction) as the point of application undergoes a displacement from $s = a$ to $s = b$.

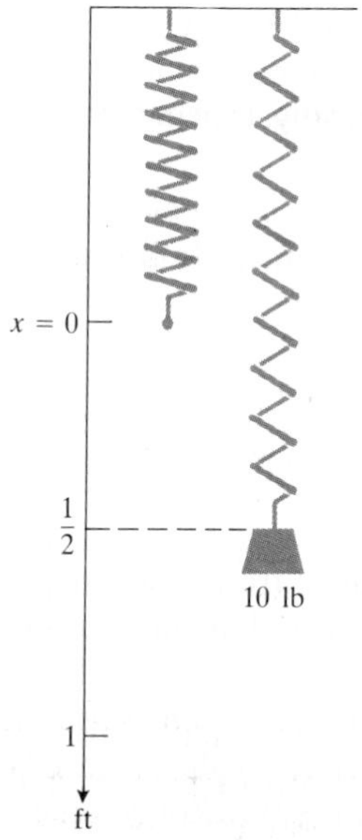

5.62 A 10-lb weight stretches the spring half a foot beyond its unstressed length.

EXAMPLE 2 A spring has a natural length of $L = 3$ ft. A force of 10 lb stretches the spring to a length of 3.5 ft. Find the spring constant. Then calculate the amount of work done in stretching the spring from its natural length to a length of 5 ft. How much work is done in stretching the spring from 4 ft to 5 ft? How far beyond its natural length will a 15-lb force stretch the spring?

Solution We find the spring constant from Eq. (2). A force of 10 lb stretches the spring 0.5 ft, so that

$$10 = c(0.5), \qquad c = 20 \text{ lb/ft}.$$

To calculate the work done in stretching the spring 2 ft beyond its natural length, imagine the spring hanging parallel to the x-axis, as shown in Fig. 5.62. Then, by Eq. (4), with $F = cx = 20x$,

$$W = \int_0^2 20x\,dx = 10x^2\Big]_0^2 = 40 \text{ foot-pounds.}$$

To find the work done in stretching the spring from a length of 4 ft to a length of 5 ft, we calculate

$$W = \int_1^2 20x\,dx = 10x^2\Big]_1^2 = 30 \text{ foot-pounds.}$$

To learn how far a 15-lb force will stretch the spring, we solve Eq. (2) with $F = 15$ and $c = 20$:

$$15 = 20x, \qquad x = \frac{15}{20} \text{ ft} = 9 \text{ in.}$$

No calculus is involved in this last computation. □

Equation (4) leads to an interesting theorem of mechanics.

THEOREM

Let F denote the resultant of all forces acting on a particle of mass m. Let the direction of F remain constant. Then, whether the magnitude of F is constant or variable, the work done on the particle by the force F is equal to the change in the kinetic energy of the particle.

The ingredients required for a proof of this theorem are

a) Eq. (4), which gives the work,

b) the definition of kinetic energy, K.E. $= \frac{1}{2}mv^2$, and

c) Newton's second law: $F = m(dv/dt)$.

The fact that

$$v = \frac{ds}{dt}$$

and

$$\frac{dv}{dt} = \frac{dv}{ds}\frac{ds}{dt} = \frac{dv}{ds}v$$

and Newton's second law enable us to write Eq. (4) in the form

$$W = \int_{s=a}^{s=b} mv\frac{dv}{ds}\,ds = \int_{v=v_a}^{v=v_b} mv\,dv,$$

which leads to

$$W = \frac{1}{2}mv^2\Big]_{v_a}^{v_b} = \frac{1}{2}mv_b^2 - \frac{1}{2}mv_a^2.$$

That is, the work done by the force F is the kinetic energy at b minus the kinetic energy at a or, more simply, the change in kinetic energy:

$$W = \Delta(\text{K.E.}).$$

We now consider another situation where the simple Eq. (1) cannot be applied to the *total* but can be applied to a *small piece.*

EXAMPLE 3 Calculate the amount of work required to pump all the water from a full hemispherical bowl of radius r ft to a distance h ft above the top of the bowl (Fig. 5.63).

Solution We introduce coordinate axes as shown in Fig. 5.63 and imagine the bowl to be divided into thin slices by planes perpendicular to the x-axis between $x = 0$ and $x = r$. The representative slice between the planes at x and $x + \Delta x$ has a volume ΔV that is approximately

$$\Delta V \approx \pi y^2\,\Delta x = \pi(r^2 - x^2)\,\Delta x \text{ ft}^3.$$

The force F required to lift *this slice* is equal to its weight:

$$w\,\Delta V \approx \pi w(r^2 - x^2)\,\Delta x \text{ pounds},$$

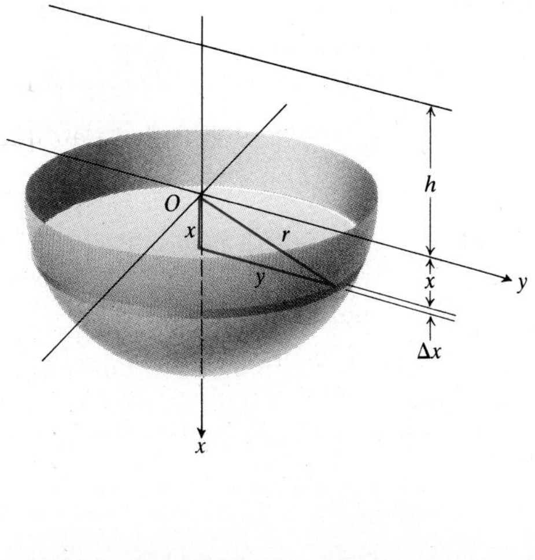

5.63 To calculate the work required to pump the water from a bowl, think of lifting the water out one slab at a time.

where w is the weight of a cubic foot of water. Finally, the *distance* through which this force must act lies between

$$h + x \quad \text{and} \quad h + x + \Delta x \text{ ft}$$

so that the work ΔW done in *lifting this one slice* is approximately

$$\Delta W \approx \pi w(r^2 - x^2)(h + x)\,\Delta x \text{ foot-pounds,}$$

where we again have suppressed all higher powers of Δx. This is justified in the *limit* of the sum, and the total work is

$$\begin{aligned} W &= \lim_{\Delta x \to 0} \sum_0^r \pi w(r^2 - x^2)(h + x)\,\Delta x \\ &= \int_0^r \pi w(h + x)(r^2 - x^2)\,dx \\ &= hw \int_0^r \pi(r^2 - x^2)\,dx + w \int_0^r \pi x(r^2 - x^2)\,dx \\ &= hwV + \bar{x}wV \text{ foot-pounds.} \end{aligned}$$

Here wV is the weight of the whole bowlful of water of volume V, and the second integral may be interpreted physically as giving the work required in pumping *all* the water from the depth of the center of gravity of the bowl to the level $x = 0$, while the first integral gives the work done in pumping the whole bowlful of water from the level $x = 0$ up a distance of h feet. (See Problem 19.) The actual evaluation of the integrals leads to the formula

$$W = \frac{2}{3}\pi r^3 w\left(h + \frac{3}{8}r\right). \quad \square$$

PROBLEMS

1. If the spring in Fig. 5.61 has natural length $L = 18$ in., and a force of 10 lb is sufficient to compress it to a length of 16 in., what is the value of the "spring constant" c for the particular spring in question? How much work is done in compressing it from a length of 16 in. to a length of 12 in.?
2. A spring has a natural length of 10 in. An 800-lb force stretches the spring to 14 in. (a) Find the spring constant. (b) How much work is done in stretching the spring from 10 in. to 12 in.? (c) How far beyond its natural length will a 1600-lb force stretch the spring?
3. A 10,000-lb force compresses a spring from its natural length of 12 in. to a length of 11 in. How much work is done in compressing the spring
   a) from 12 to 11.5 in.?
   b) from 11.5 in. to 11 in.?
4. A spring has a natural length of 2 ft. A 1-lb force stretched the spring 5 ft (from 2 ft to 7 ft). How much work did the 1-lb force do? If the spring is stretched by a 2-lb force, what will its total length be?
5. A bathroom scale is depressed $\frac{1}{16}$ in. when a 150-lb person stands on it. Assuming that the scale behaves like a spring, find how much work is required to depress the scale $\frac{1}{8}$ in. from its natural height. How much weight is required to compress the scale this much?
6. Answer the questions of Problem 1 if the law of force is $F = c \sin(\pi x/2L)$ in place of Eq. (2).
7. Two electrons repel each other with a force inversely proportional to the square of the distance between them. Suppose one electron is held fixed at the point $(1, 0)$ on the x-axis. Find the work required to move a second electron along the x-axis from the point $(-1, 0)$ to the origin.
8. If two electrons are held stationary at the points $(-1, 0)$ and $(1, 0)$ on the x-axis, find the work done in moving a third electron from $(5, 0)$ to $(3, 0)$ along the x-axis.
9. If a straight hole could be bored through the center of the earth, a particle of mass m falling in this hole would be attracted toward the center of the earth with a force $mg(r/R)$ when it is at distance r from the center. (R is the radius of earth; g is the acceleration due to gravity at the surface of the earth.) How much work is done on the particle as it falls from the surface to the center of the earth?

10. A bag of sand originally weighing 144 lb is lifted at a constant rate of 3 ft/min. The sand leaks out uniformly at such a rate that half of the sand is lost when the bag has been lifted 18 ft. Find the work done in lifting the bag this distance.

11. Gas in a cylinder of constant cross-sectional area A expands or is compressed by the motion of a piston. If p is the pressure of the gas in pounds per square inch and v is its volume in cubic inches, show that the work done by the gas as it goes from an initial state (p_1, v_1) to a second state (p_2, v_2) is

$$W = \int_{(p_1,v_1)}^{(p_2,v_2)} p\,dv \quad \text{inch-pounds.}$$

(*Hint:* Take the x-axis perpendicular to the face of the piston. Then $dv = A\,dx$ and $F = pA$.)

12. Use the result of Problem 11 to find the work done on the gas in compressing it from $v_1 = 243$ in^3 to $v_2 = 32$ in^3 if the initial pressure $p_1 = 50$ lb/in^2 and the pressure and volume satisfy the law for adiabatic change of state, $pv^{1.4} = \text{constant}$.

13. Find the work done in pumping all the water out of a conical reservoir of radius 10 ft at the top, altitude 8 ft, to a height of 6 ft above the top of the reservoir.

14. Find the work done in Problem 13 if, at the beginning, the reservoir is filled to a depth of 5 ft and the water is pumped just to the top of the reservoir.

15. The driver of a leaky 800 gal tank truck carrying water from the base to the summit of Mt. Washington discovered upon arrival that the tank was only half full. The truck started out full, climbed at a steady rate, and took 50 minutes to accomplish an elevation change of 4750 ft. The water leaked out at a steady rate. How much work was done in carrying the water to the top? (Do not include the work done on the truck and driver. Water weighs 8 lb/U.S. gal.)

16. You are in charge of the evacuation and repair of a liquid gas storage tank shown in Fig. 5.64. The tank is a hemisphere of radius 10 ft and is full of liquid gas with a density of 56 lb/ft^3. A firm states that it can empty the tank at a cost of $\frac{1}{2}$ cent per foot-pound of work. Find the work required to empty the tank by pumping the liquid to an outlet 2 ft above the top of the tank. If you have \$5000 budgeted for the pumping job, can you afford to hire the firm?

17. Your town has decided to drill a well to increase its water supply. As the town engineer, you have determined that a water tower will be necessary to provide the pressure needed for distribution, and you have designed the system shown in Fig. 5.65. The well is to be 300 ft deep. The water will be lifted through a 4-in. pipe into the base of a cylindrical tank 20 ft in diameter and 25 ft high. The bottom of the tank will be 60 ft above the ground. The pump will be submerged in the well beneath the surface of the water. The pump performs work at the rate of 1650 foot-pounds per second. How long will it take to fill the tank the first time? (Include the work to fill the pipe.)

18. Water, of density 62.5 lb/ft^3, pours into the tank shown in Fig. 5.66 at a rate of 4 ft^3 per minute. The cross sections of the tank are semicircles 4 ft in diam-

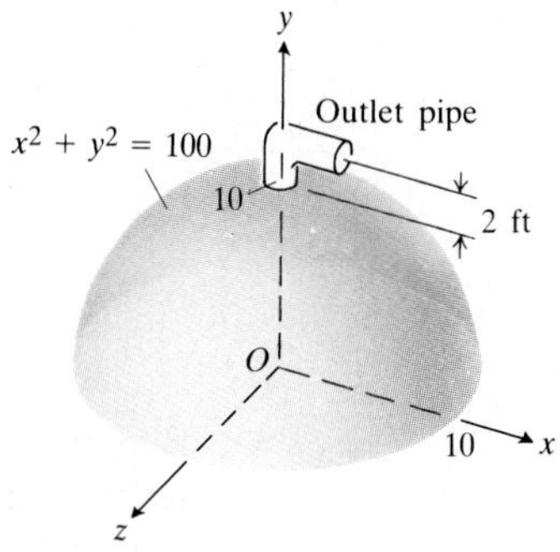

5.64 The tank in Problem 16.

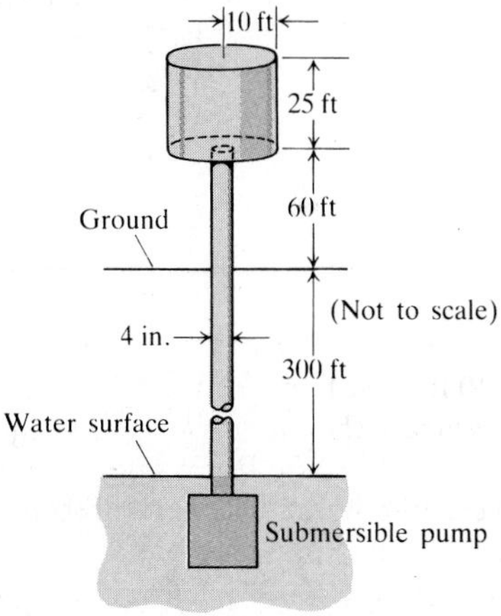

5.65 The water tower in Problem 17.

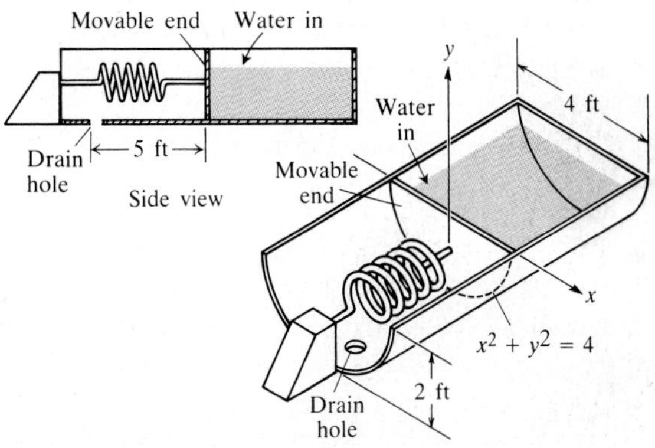

5.66 The tank in Problem 18.

eter. One end of the tank is movable, but moving it to increase the volume of the tank compresses a spring. The spring constant is $k = 100$ lb/ft. If the end of the tank moves 5 ft against the spring, water will flow out of a hole in the bottom at a rate of 5 ft^3/min. Will the tank overflow?

19. Prove that, no matter what the shape of the container in Fig. 5.63, the total work is the sum of two terms, one of which is

$$W_1 = hw \int dV$$

and represents the total work done in lifting a bowlful of water a distance h, while the other is

$$W_2 = w \int x\, dV$$

and represents the work done in lifting a bowlful of water a distance equal to the depth of the center of gravity of the bowl.

REVIEW QUESTIONS AND EXERCISES

1. List nine applications of the definite integral presented in this chapter.
2. How do you define the area bounded above by a curve $y = f_1(x)$, below by a curve $y = f_2(x)$, and on the sides by $x = a$ and $x = b$, $a < b$?
3. How do you define the volume generated by rotating a plane region about an axis in its plane and not intersecting the region?
4. How do you define the length of a plane curve? Can you extend this to a curve in three-dimensional space?
5. How do you define the surface area of a sphere? Of other surfaces of revolution?
6. How do you define the average value of a function over an interval?
7. How do you define center of mass?
8. How do you define the hydrostatic force on the face of a dam?
9. How do you define work done by a variable force?

MISCELLANEOUS PROBLEMS

1. Sketch the graphs of the equations $y = 2 - x^2$ and $x + y = 0$ in one diagram, and find the area bounded by them.
2. Find the maximum and minimum points of the curve $y = x^3 - 3x^2$ and find the total area bounded by this curve and the x-axis. Sketch.

In Problems 3–15, find the area bounded by the given curves and lines. A sketch is usually helpful.

3. $y = x,\quad y = 1/x^2,\quad x = 2$
4. $y = x,\quad y = 1/\sqrt{x},\quad x = 2$
5. $y = x + 1,\quad y = 3 - x^2$
6. $y = 2x^2,\quad y = x^2 + 2x + 3$
7. $x = 2y^2,\quad x = 0,\quad y = 3$
8. $4x = y^2 - 4,\quad 4x = y + 16$
9. $x^{1/2} + y^{1/2} = a^{1/2},\quad x = 0,\quad y = 0$
10. $y = \sin x,\quad y = \sqrt{2}x/2$
11. $y^2 = 9x,\quad y = 3x^2/8$
12. $y = \sin x,\quad y = x,\quad 0 \le x \le \pi/4$
13. $y = x\sqrt{2x^2 + 1},\quad x = 0,\quad x = 2$
14. $y^2 = 4x$ and $y = 4x - 2$
15. $y = 2 - x^2$ and $y = x^2 - 6$
16. The function $v = 3t^2 - 15t + 18$ represents the velocity v (ft/s) of a moving body as a function of the time t (s).
    a) Find that portion of the time interval $0 \le t \le 3$ in which the velocity is positive.
    b) Find that portion where v is negative.
    c) Find the total distance traveled by the body from $t = 0$ to $t = 3$.
17. The area from 0 to x under a certain graph is given to be
    $$A = (1 + 3x)^{1/2} - 1,\quad x \ge 0.$$
    a) Find the *average* rate of change of A with respect to x as x increases from 1 to 8.
    b) Find the *instantaneous* rate of change of A with respect to x at $x = 5$.
    c) Find the ordinate (height) y of the graph as a function of x.
    d) Find the average value of the ordinate (height) y, with respect to x, as x increases from 1 to 8.

18. A solid is generated by rotating, about the x-axis, the region bounded by the curve $y = f(x)$, the x-axis, and the lines $x = a$, $x = b$. Its volume, for all $b > a$, is $b^2 - ab$. Find $f(x)$.

19. Find the volume generated by rotating the region bounded by the given curves and lines about the line indicated:

a) $y = x^2$, $y = 0$, $x = 3$, about the x-axis.
b) $y = x^2$, $y = 0$, $x = 3$, about the line $x = -3$.
c) $y = x^2$, $y = 0$, $x = 3$, about the y-axis, first integrating with respect to x and then integrating with respect to y.
d) $x = 4y - y^2$, $x = 0$, about the y-axis.
e) $x = 4y - y^2$, $x = 0$, about the x-axis.

20. A solid is generated by rotating the curve $y = f(x)$, $0 \leq x \leq a$, about the x-axis. Its volume for all a is $a^2 + a$. Find $f(x)$.

21. The region bounded by the curve $y^2 = 4x$ and the straight line $y = x$ is rotated about the x-axis. Find the volume generated.

22. Sketch the region bounded by the curve $y^2 = 4ax$, the line $x = a$, and the x-axis. Find the volumes generated by rotating this region in the following ways: (a) about the x-axis, (b) about the line $x = a$, (c) about the y-axis.

23. The region bounded by the curve $y = x/\sqrt{x^3 + 8}$, x-axis, and the line $x = 2$ is rotated about the y-axis, generating a volume. Set up the integral that should be used to evaluate the volume and then evaluate the integral.

24. Find the volume of the solid generated by rotating the larger region bounded by $y^2 = x - 1$, $x = 3$, and $y = 1$ about the y-axis.

25. The region bounded by the curve $y^2 = 4ax$ and the line $x = a$ is rotated about the line $x = 2a$. Find the volume generated.

26. A twisted solid is generated as follows: We are given a fixed line L in space, and a square of side s in a plane perpendicular to L. One vertex of the square is on L. As this vertex moves a distance h along L, the square turns through a full revolution, with L as the axis. Find the volume generated by this motion. What would the volume be if the square had turned through two full revolutions in moving the same distance along L?

27. Find the center of gravity of a solid right circular cone of altitude h and base radius r.

28. Find the volume generated by rotating about the x-axis the region bounded by the x-axis and one arch of the curve $y = \sin 2x$.

29. A round hole of radius $\sqrt{3}$ ft is bored through the center of a solid sphere of radius 2 ft. Find the volume cut out.

30. The cross section of a certain solid in any plane perpendicular to the x-axis is a circle having diameter AB with A on the curve $y^2 = 4x$ and B on the curve $x^2 = 4y$. Find the volume of the solid lying between the points of intersection of the curves.

31. The base of a solid is the area bounded by $y^2 = 4ax$ and $x = a$. Each cross section perpendicular to the x-axis is an equilateral triangle. Find the volume of the solid.

32. Find the length of the curve $y = (\frac{2}{3})x^{3/2} - (\frac{1}{2})x^{1/2}$ from $x = 0$ to $x = 4$.

33. Find the surface area generated when the curve of Problem 32 is rotated about the y-axis.

34. Find the length of the curve $x = (\frac{3}{5})y^{5/3} - (\frac{3}{4})y^{1/3}$ from $y = 0$ to $y = 1$.

35. Find the surface area generated when the curve of Problem 34 is rotated about the line $y = -1$.

36. Find the average value of y with respect to x for that part of curve $y = \sqrt{ax}$ between $x = a$ and $x = 3a$.

37. Find the average value of y^2 with respect to x for the curve $ay = b\sqrt{a^2 - x^2}$ between $x = 0$ and $x = a$. Also find the average value of y with respect to x^2 for $0 \leq x \leq a$.

38. Consider the curve $y = f(x)$, $x \geq 0$ such that $f(0) = a$. Let $s(x)$ denote the arc length along the curve from $(0, a)$ to $(x, f(x))$.

a) Find $f(x)$ if $s(x) = Ax$. (What are the permissible values of A?)
b) Is it possible for $s(x)$ to equal x^n, $n > 1$? Give a reason for your answer.

39. A point moves in a straight line during the time from $t = 0$ to $t = 3$ according to the law $s = 120t - 16t^2$.

a) Find the average value of the velocity, with respect to time, for these three seconds. (Compare with the "average velocity," Article 1.8, Eq. (7).)
b) Find the average value of the velocity with respect to the distance s during the three seconds.

40. Sketch a smooth curve through the points

$$A(1, 3),\quad B(3, 5),\quad C(5, 6),\quad D(7, 6),\quad E(9, 7),\quad F(11, 10).$$

The region bounded by this curve, the x-axis, and the lines $x = 1$, $x = 11$, is rotated about the x-axis to generate a solid. Use the trapezoidal rule to approximate the volume generated. Also determine approximately the average value of the circular cross-sectional area (with respect to x.)

41. Determine the center of mass of a thin homogeneous plate covering the region enclosed by the curves $y^2 = 8x$ and $y = x^2$.

42. Find the center of mass of a homogeneous plate covering the region in the first quadrant bounded by the curve $4y = x^2$, the y-axis, and the line $y = 4$.

Here the polynomials are $P_2(x) = x + 1$, $P_1(x) = 0$, and $P_0(x) = -1$. Polynomials and rational functions are algebraic, and all sums, products, quotients, powers, and roots of algebraic functions are algebraic. We could have defined the transcendental functions to be the functions that are not algebraic.

The six basic trigonometric functions are transcendental, as are the inverse trigonometric functions and the exponential and logarithmic functions that are the main subject of the present chapter. The name "transcendental" was coined by Euler in talking about numbers that are not roots of polynomial equations. Such numbers "transcend the power of algebraic methods," as he put it. Transcendental functions are important in solving problems in engineering and physics, in addition to being important in mathematics itself. We shall see a number of applications of these functions as the chapter goes on.

At this stage it might be well to recall the derivative formulas we have encountered so far. The functions u and v in these formulas are assumed to be differentiable functions of x.

Derivatives

1. $\dfrac{dc}{dx} = 0$ (any constant c)
2. $\dfrac{d}{dx}(cu) = c\dfrac{du}{dx}$
3. $\dfrac{d}{dx}(u + v) = \dfrac{du}{dx} + \dfrac{dv}{dx}$
4. $\dfrac{d}{dx}(uv) = u\dfrac{dv}{dx} + v\dfrac{du}{dx}$
5. $\dfrac{d}{dx}\left(\dfrac{u}{v}\right) = \dfrac{v\dfrac{du}{dx} - u\dfrac{dv}{dx}}{v^2}$
6. $\dfrac{d}{dx}u^n = nu^{n-1}\dfrac{du}{dx}$
7. $\dfrac{d}{dx}\sin u = \cos u\dfrac{du}{dx}$
8. $\dfrac{d}{dx}\cos u = -\sin u\dfrac{du}{dx}$
9. $\dfrac{d}{dx}\tan u = \sec^2 u\dfrac{du}{dx}$
10. $\dfrac{d}{dx}\cot u = -\csc^2 u\dfrac{du}{dx}$
11. $\dfrac{d}{dx}\sec u = \sec u \tan u\dfrac{du}{dx}$
12. $\dfrac{d}{dx}\csc u = -\csc u \cot u\dfrac{du}{dx}$

Chain Rule: If y is a differentiable function of u and u is a differentiable function of x, then y is a differentiable function of x and

$$\frac{dy}{dx} = \frac{dy}{du}\frac{du}{dx}.$$

(Article 2.6 shows how this formula is to be used.)

In another form: If f is differentiable at x and g is differentiable at $f(x)$, then the composite $g \circ f$ is differentiable at x and

$$\frac{d}{dx}g(f(x)) = g'(f(x)) \cdot f'(x).$$

6.2 The Inverse Trigonometric Functions

The function y = sin x is not one-to-one; it runs through its full range of values from −1 to 1 twice on every interval of length 2π. If we restrict the domain of the sine to the interval from $-\pi/2$ to $\pi/2$, however, we find that the restricted function

$$y = \sin x, \quad -\pi/2 \le x \le \pi/2, \tag{1a}$$

is one-to-one. It therefore has an inverse (Article 2.10), which we denote by

$$y = \sin^{-1} x. \tag{1b}$$

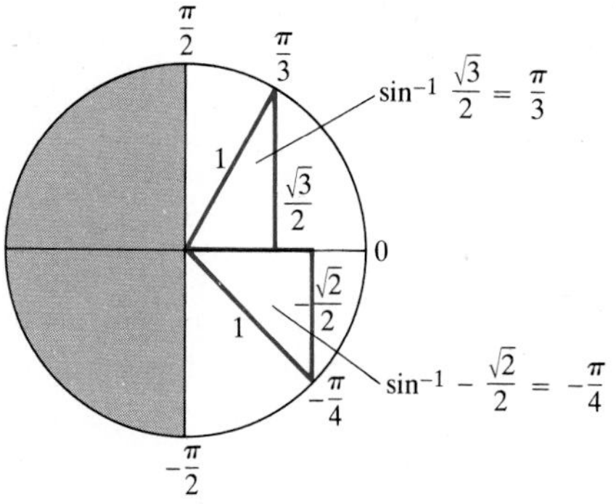

6.1 The angle whose measure is $y = \sin^{-1} x$ ranges from $-\pi/2$ to $\pi/2$.

This equation is read "y equals the arc sine of x," or "y equals the inverse sine of x." For every value of x in the interval $[-1, 1]$, $y = \sin^{-1} x$ is the number in the interval $[-\pi/2, \pi/2]$ whose sine is x. For instance (Fig. 6.1),

$$\begin{aligned}
&\sin^{-1} 0 = 0 && \text{because} \quad \sin 0 = 0,\\
&\sin^{-1} \sqrt{3}/2 = \pi/3 && \text{because} \quad \sin \pi/3 = \sqrt{3}/2,\\
&\sin^{-1} 1 = \pi/2 && \text{because} \quad \sin \pi/2 = 1,\\
&\sin^{-1} (-\sqrt{2}/2) = -\pi/4 && \text{because} \quad \sin(-\pi/4) = -\sqrt{2}/2.
\end{aligned}$$

The graph of $y = \sin^{-1} x$ is shown in Fig. 6.2. The gray curve in the figure is the reflection of the graph of y = sin x across the line y = x, and so it is the graph of x = sin y. The graph of $y = \sin^{-1} x$ is the portion of this curve between $y = -\pi/2$ and $y = \pi/2$.

The minus one in $y = \sin^{-1} x$ is not an exponent; it means "inverse," not "reciprocal." The *reciprocal* of sin x is

$$(\sin x)^{-1} = \frac{1}{\sin x} = \csc x.$$

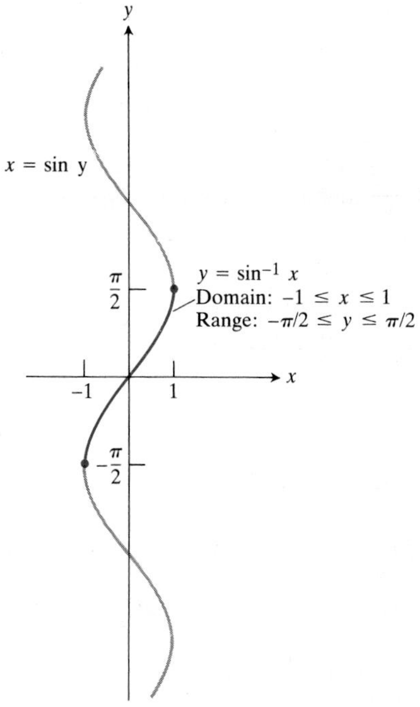

6.2 The graph of $y = \sin^{-1} x$.

The graph of the arc sine in Fig. 6.2 is symmetric about the origin because the graph of x = sin y is symmetric about the origin. Algebraically, this means that

$$\sin^{-1}(-x) = -\sin^{-1} x \tag{2}$$

for every x in the domain of the arc sine, which is another way to say that the function

$$y = \sin^{-1} x \text{ is odd.}$$

Like the sine function, the cosine function y = cos x is not one-to-one, but its restriction to the interval $[0, \pi]$,

$$y = \cos x, \quad 0 \le x \le \pi, \tag{3a}$$

is one-to-one. The restricted function therefore has an inverse,

$$y = \cos^{-1} x, \tag{3b}$$

which we call the *arc cosine* of x, or the *inverse cosine* of x. For each

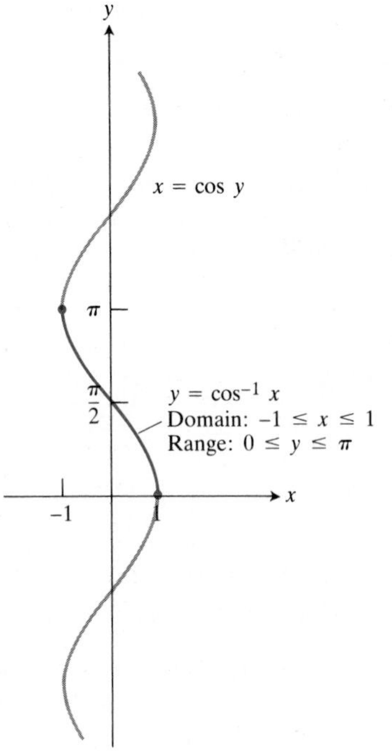

6.3 The graph of $y = \cos^{-1} x = \frac{\pi}{2} - \sin^{-1} x$.

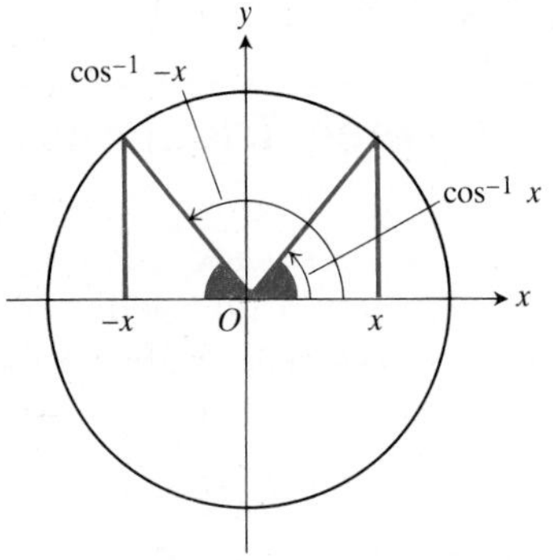

6.4 In this figure, $\cos^{-1} x + \cos^{-1}(-x) = \pi$.

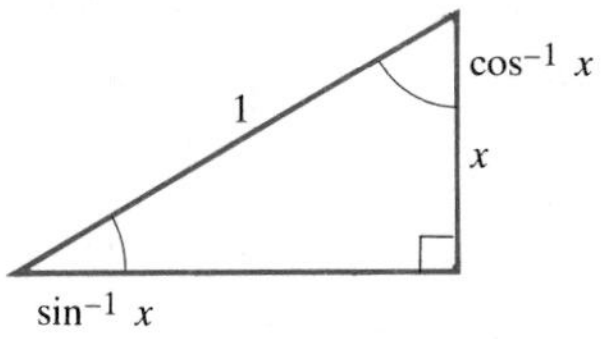

6.5 In this figure, $\sin^{-1} x + \cos^{-1} x = \pi/2$.

value of x in the interval $[-1, 1]$, $y = \cos^{-1} x$ is the number in the interval $[0, \pi]$ whose cosine is x. The graph of $y = \cos^{-1} x$ is shown in Fig. 6.3.

As we can see in Fig. 6.4, the arc cosine of x satisfies the identity

$$\cos^{-1} x + \cos^{-1}(-x) = \pi \tag{4a}$$

or

$$\cos^{-1}(-x) = \pi - \cos^{-1} x. \tag{4b}$$

We can also see in the triangle in Fig. 6.5 that for $x > 0$

$$\sin^{-1} x + \cos^{-1} x = \pi/2 \tag{5}$$

because $\sin^{-1} x$ and $\cos^{-1} x$ are then complementary angles in a right triangle whose hypotenuse is 1 unit long and one of whose legs is x units long. Equation (5) holds for the other values of x in $[-1, 1]$ as well, but we cannot draw this conclusion from the geometry of the triangle in Fig. 6.5. It is, however, a consequence of Eqs. (2) and (4b) (Problem 35).

6.6 The branch chosen for $y = \tan^{-1} x$ is the one through the origin.

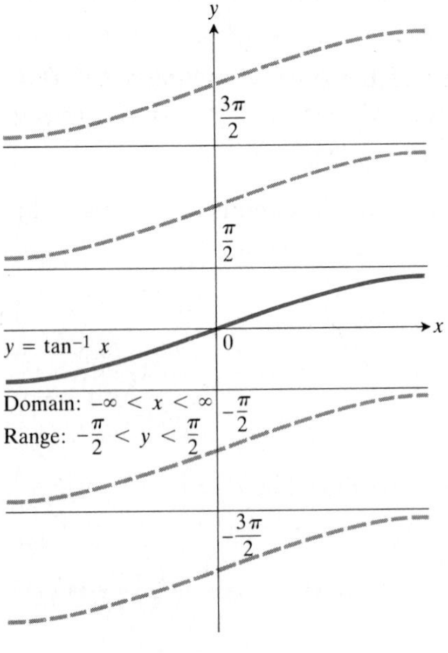

The Inverses of tan x, sec x, csc x, cot x

The other four basic trigonometric functions, $y = \tan x$, $y = \sec x$, $y = \csc x$, and $y = \cot x$, also have inverses when suitably restricted. The inverse of

$$y = \tan x, \quad -\pi/2 < x < \pi/2, \tag{6a}$$

is denoted by

$$y = \tan^{-1} x. \tag{6b}$$

The domain of the arc tangent is the entire real line, the range is the open interval $(-\pi/2, \pi/2)$. For every value of x, $y = \tan^{-1} x$ is the angle between $-\pi/2$ and $\pi/2$ whose tangent is x. The graph of $y = \tan^{-1} x$ is shown in Fig. 6.6.

The graph of $y = \tan^{-1} x$ is symmetric about the origin because it is a branch of the graph of $x = \tan y$ that is symmetric about the origin. Alge-

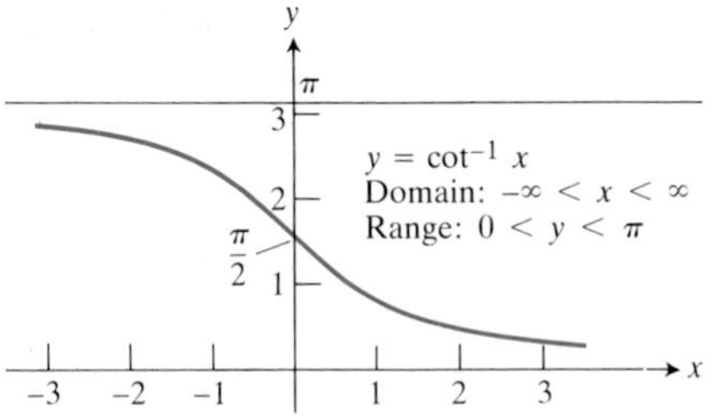

6.7 The graph of $y = \cot^{-1} x = \pi/2 - \tan^{-1} x$.

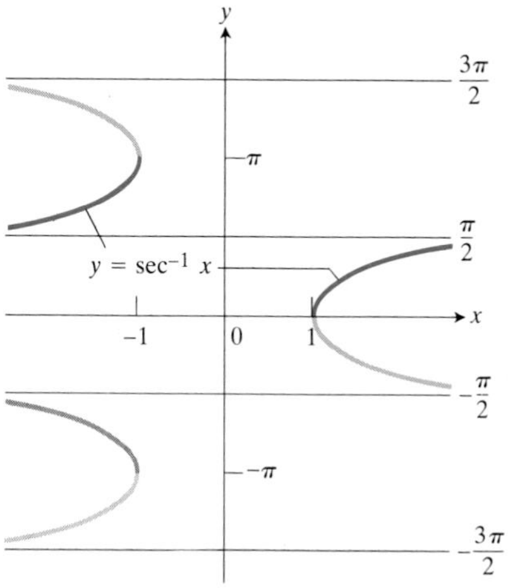

6.8 $y = \sec^{-1} x = \cos^{-1}(1/x)$ is defined for $|x| \geq 1$.

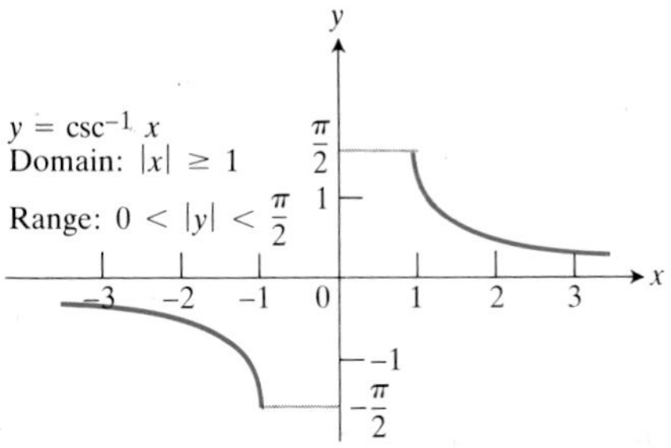

6.9 The graph of $y = \csc^{-1} = \sin^{-1}(1/x)$.

braically, this means that

$$\tan^{-1}(-x) = -\tan^{-1} x. \tag{7}$$

Like the arc sine, the arc tangent is an odd function of x.

The inverses of the (restricted) functions

$$y = \cot x, \quad 0 < x < \pi, \tag{8a}$$

$$y = \sec x, \quad 0 \leq x \leq \pi, \quad x \neq \pi/2, \tag{8b}$$

$$y = \csc x, \quad -\pi/2 \leq x \leq \pi/2, \quad x \neq 0, \tag{8c}$$

are chosen to be the functions graphed in Figs. 6.7, 6.8, and 6.9. They are chosen this way to satisfy the relationships

$$\cot^{-1} x = \pi/2 - \tan^{-1} x, \tag{9a}$$

$$\sec^{-1} x = \cos^{-1}(1/x), \tag{9b}$$

$$\csc^{-1} x = \sin^{-1}(1/x). \tag{9c}$$

We shall not dwell on these relationships here, but they come in handy when we want to find the values of $\cot^{-1} x$, $\sec^{-1} x$, and $\csc^{-1} x$ on a calculator that gives only $\tan^{-1} x$, $\cos^{-1} x$, and $\sin^{-1} x$.

In Problem 34 you will be asked to establish one further identity:

$$\sec^{-1}(-x) = \pi - \sec^{-1} x. \tag{10}$$

It follows from Eqs. (4b) and (9b).

WARNING. Some writers choose $\sec^{-1} x$ to lie between 0 and $\pi/2$ when x is positive, and between $-\pi$ and $-\pi/2$ when x is negative (hence as a negative angle in the third quadrant, as shown by the grey curve in Fig. 6.8). This has the advantage of simplifying the formula for the derivative of $\sec^{-1} x$ but has the disadvantage of failing to satisfy Eq. (9b) when x is negative. Also, some mathematical tables give third-quadrant values for $\sec^{-1} x$ instead of the second-quadrant values used in this book. Watch out for this when you use tables.

The right-triangle interpretations of the inverse trigonometric functions shown in Fig. 6.10 can be useful in integration problems that require substitutions. We will use some of them in Chapter 7.

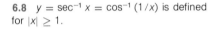

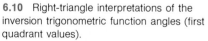

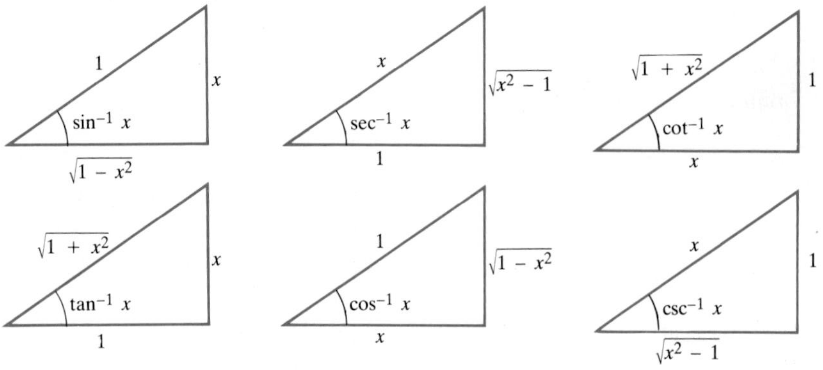

6.10 Right-triangle interpretations of the inversion trigonometric function angles (first quadrant values).

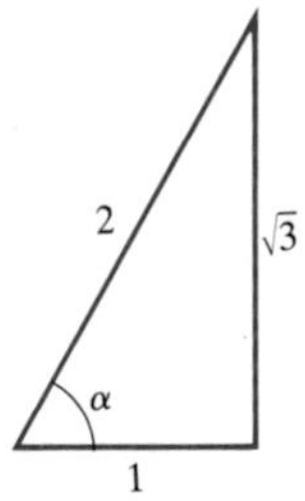

6.11 If $\alpha = \sin^{-1}(\sqrt{3}/2)$, then the values of the trigonometric functions of α can be read from the triangle above.

EXAMPLE 1 Given that

$$\alpha = \sin^{-1}\frac{\sqrt{3}}{2},$$

find $\csc\alpha$, $\cos\alpha$, $\sec\alpha$, $\tan\alpha$, and $\cot\alpha$.

Solution We draw a reference triangle like the one shown in Fig. 6.11, with hypotenuse 2 and vertical side $\sqrt{3}$. We then calculate the length of the remaining side to be $\sqrt{(2)^2 - (\sqrt{3})^2} = 1$. The values of the trigonometric functions of α are then read from the triangle as ratios of side lengths:

$$\csc\alpha = \frac{2}{\sqrt{3}} = \frac{2\sqrt{3}}{3},$$

$$\cos\alpha = \frac{1}{2},$$

$$\sec\alpha = 2$$

$$\tan\alpha = \frac{\sqrt{3}}{1} = \sqrt{3},$$

$$\cot\alpha = \frac{1}{\sqrt{3}} = \frac{\sqrt{3}}{3}. \quad \square$$

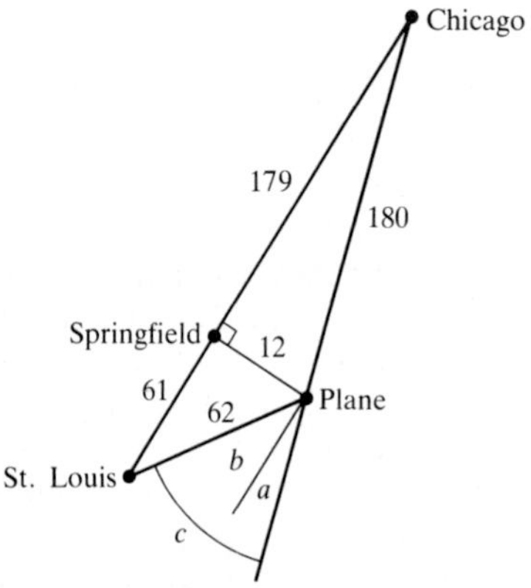

6.12 Diagram for drift correction (Example 2). Distances in miles. Drawing not to scale.

EXAMPLE 2 ***Drift correction.*** During an airplane flight from Chicago to St. Louis the navigator determines that the plane is 12 miles off course, as shown in Fig. 6.12. Find the angle a for a parallel course, the angle b, and the correction angle $c = a + b$.

Solution

$$a = \sin^{-1}\frac{12}{180} \approx 0.067 \text{ radians} \approx 3.8°,$$

$$b = \sin^{-1}\frac{12}{62} \approx 0.195 \text{ radians} \approx 11.2°,$$

$$c = a + b \approx 15°. \quad \square$$

PROBLEMS

1. Given that $\alpha = \sin^{-1}\frac{1}{2}$, find $\cos\alpha$, $\tan\alpha$, $\sec\alpha$, $\csc\alpha$.

2. Given that $\alpha = \cos^{-1}(-\frac{1}{2})$, find $\sin\alpha$, $\tan\alpha$, $\sec\alpha$, $\csc\alpha$.

Evaluate the given expression in Problems 3–17.

3. $\sin\left(\cos^{-1}\frac{\sqrt{2}}{2}\right)$

4. $\tan\left(\sin^{-1}\left(-\frac{1}{2}\right)\right)$

5. $\sec\left(\cos^{-1}\frac{1}{2}\right)$

6. $\cot\left(\sin^{-1}\left(-\frac{1}{2}\right)\right)$

7. $\csc(\sec^{-1} 2)$

8. $\cos(\tan^{-1}(-\sqrt{3}))$

9. $\cos(\cot^{-1} 1)$

10. $\csc\left(\sin^{-1}\left(-\frac{\sqrt{2}}{2}\right)\right)$

11. $\cot(\cos^{-1} 0)$

12. $\sec\left(\tan^{-1}\left(-\frac{1}{2}\right)\right)$

13. $\tan(\sec^{-1} 1)$

14. $\sin(\csc^{-1}(-1))$

15. $\sin^{-1}(1) - \sin^{-1}(-1)$

16. $\tan^{-1}(1) - \tan^{-1}(-1)$

17. $\sec^{-1}(2) - \sec^{-1}(-2)$

Simplify the expressions in Problems 18–23.

18. $\sin(\sin^{-1} 0.735)$

19. $\cos(\sin^{-1} 0.8)$

20. $\sin(2\sin^{-1} 0.8)$

21. $\tan^{-1}(\tan\pi/3)$

22. $\cos^{-1}(-\sin\pi/6)$

23. $\sec^{-1}(\sec(-30°))$ [The answer is *not* $-30°$.]

Find the limits in Problems 24–31. (*Hint:* If in doubt, look at the graph of the function.)

24. $\lim_{x \to 1^-} \sin^{-1} x$ **25.** $\lim_{x \to -1^+} \cos^{-1} x$

26. $\lim_{x \to \infty} \tan^{-1} x$ **27.** $\lim_{x \to -\infty} \tan^{-1} x$

28. $\lim_{x \to \infty} \sec^{-1} x$ **29.** $\lim_{x \to -\infty} \sec^{-1} x$

30. $\lim_{x \to \infty} \csc^{-1} x$ **31.** $\lim_{x \to -\infty} \csc^{-1} x$

32. (*Calculator*) Find the angle α in Fig. 6.13.

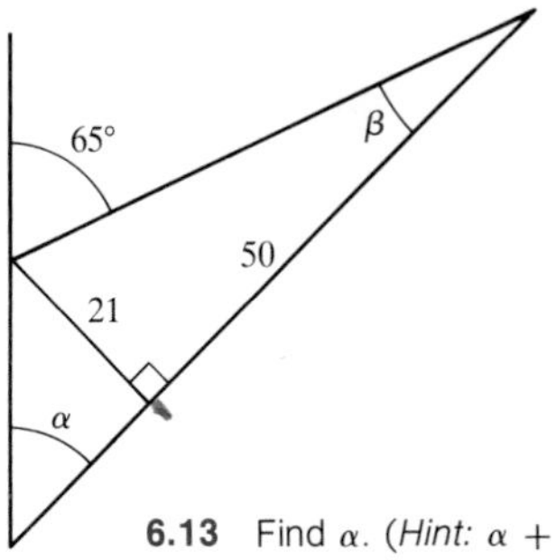

6.13 Find α. (*Hint:* $\alpha + \beta = 65°$.)

33. A picture a feet high is placed on a wall with its base b feet above the level of an observer's eye. If the observer stands x feet from the wall, show that the angle of vision α subtended by the picture is given by

$$\alpha = \cot^{-1} \frac{x}{a+b} - \cot^{-1} \frac{x}{b}.$$

34. Combine Eqs. (4b) and (9b) to show that

$$\sec^{-1}(-x) = \pi - \sec^{-1} x.$$

35. Figure 6.5 shows that the equation

$$\sin^{-1} x + \cos^{-1} x = \pi/2 \tag{5}$$

holds for $0 < x < 1$. To show that the equation holds for all x in $[-1, 1]$, carry out the following steps:

a) Show by direct calculation that

$$\sin^{-1}(1) + \cos^{-1}(1) = \pi/2,$$
$$\sin^{-1}(0) + \cos^{-1}(0) = \pi/2,$$
$$\sin^{-1}(-1) + \cos^{-1}(-1) = \pi/2.$$

b) Then, for values of x in $(-1, 0)$, let $x = -a, a > 0$, and apply Eqs. (2) and (4b) to the sum $\sin^{-1}(-a) + \cos^{-1}(-a)$.

36. When a ray of light passes from one medium (say air) where it travels with a velocity c_1 into a second medium (for example, water) where it travels with velocity c_2, the angle of incidence θ_1 and angle of refraction θ_2 are related by Snell's law:

$$\frac{\sin \theta_1}{c_1} = \frac{\sin \theta_2}{c_2}$$

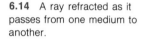

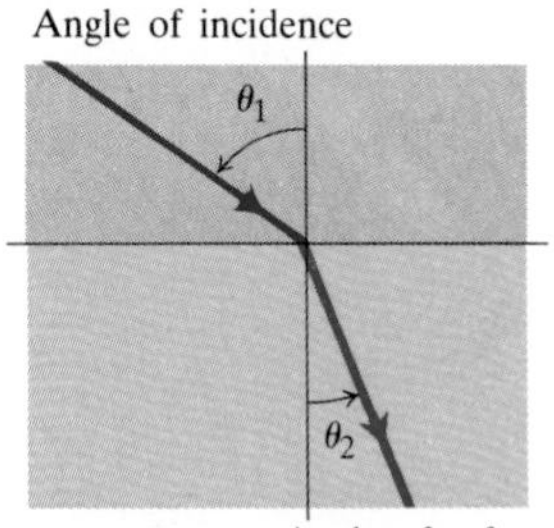

6.14 A ray refracted as it passes from one medium to another.

(Fig. 6.14). The quotient $c_1/c_2 = n_{12}$ is called the *index of refraction* of medium 2 with respect to medium 1.

a) Express θ_2 as a function of θ_1.

b) Find the largest value of θ_1 for which the expression for θ_2 in part (a) is defined. (For values of θ_1 larger than this, the incoming light will be reflected.) (*Remark.* Snell's law also applies in acoustics. See, for example, R. G. Lindsay and K. D. Kryter, "Acoustics," *American Institute of Physics Handbook,* 3rd Edition, 1972.)

37. Find all values of c that satisfy the conclusion of Cauchy's Mean Value Theorem (Article 3.9) for $f(x) = \sin x$, $g(x) = \cos x$, on the interval $0 \le x \le \pi/2$.

38. Find the volume generated by revolving the region in the first quadrant bounded by $y = \tan^{-1} x$, $y = 0$, and $x = 1$ about the y-axis.

39. *Optimal branching angle for pipes.* When a smaller pipe branches off from a larger one in a flow system, we may want it to run off at an angle that is "best" from some energy-saving point of view. We might require, for instance, that energy loss due to friction be minimized along the section AOB shown in Fig. 6.15. In this diagram, B is a given point to be reached by the smaller pipe, A is a point in the larger pipe upstream from B, and O is the point where the branching occurs. A law due to Poiseuille states that the loss of energy due to friction in nonturbulent flow is proportional to the length of the path and inversely proportional to the fourth power of the radius. Thus, the loss along AO is $(kd_1)/R^4$ and along OB is $(kd_2)/r^4$, where k is a constant, d_1 is the length of AO, d_2 is the length of OB, R is the radius of the larger pipe, and r is the radius of the smaller pipe. The angle θ is to be such as to minimize the sum of these two losses:

$$L = k\frac{d_1}{R^4} + k\frac{d_2}{r^4}.$$

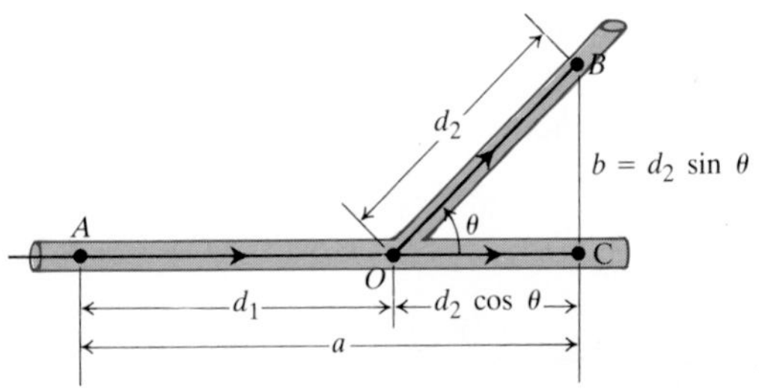

6.15 The smaller pipe *OB* branches away from the larger *AOC* at an angle θ that minimizes the friction loss along *AO* and *OB*. The optimum angle is found to be $\theta_c = \cos^{-1}(r^4/R^4)$, where *r* is the radius of the smaller pipe and *R* is the radius of the larger (Problem 39).

In our model, we assume that $AC = a$ and $BC = b$ are fixed. Thus we have the relations

$$d_1 + d_2 \cos\theta = a, \qquad d_2 \sin\theta = b,$$

so that

$$d_2 = b \csc\theta$$

and

$$d_1 = a - d_2 \cos\theta = a - b \cot\theta.$$

We can express the total loss L as a function of θ:

$$L = k\left(\frac{a - b\cot\theta}{R^4} + \frac{b\csc\theta}{r^4}\right). \tag{11}$$

a) Show that the critical value of θ for which $dL/d\theta$ equals zero is

$$\theta_c = \cos^{-1}\frac{r^4}{R^4}.$$

b) (*Calculator or tables*) If the ratio of the pipe radii is $r/R = 5/6$, estimate to the nearest degree the optimal branching angle given in part (a).

Toolkit programs

Function Evaluator   Super * Grapher

6.3 Derivatives of the Inverse Trigonometric Functions and Related Integrals

Derivative Formulas

The derivatives and differentials of the inverse trigonometric functions are given by the following formulas, whose derivation will be discussed in this article. In each of these formulas, u is assumed to be a differentiable function of x.

DERIVATIVES		DIFFERENTIALS	
13. $\dfrac{d(\sin^{-1} u)}{dx} = \dfrac{du/dx}{\sqrt{1-u^2}},$	$-1 < u < 1$	13′. $d(\sin^{-1} u) = \dfrac{du}{\sqrt{1-u^2}},$	$-1 < u < 1$
14. $\dfrac{d(\cos^{-1} u)}{dx} = -\dfrac{du/dx}{\sqrt{1-u^2}},$	$-1 < u < 1$	14′. $d(\cos^{-1} u) = -\dfrac{du}{\sqrt{1-u^2}},$	$-1 < u < 1$
15. $\dfrac{d(\tan^{-1} u)}{dx} = \dfrac{du/dx}{1+u^2}$		15′. $d(\tan^{-1} u) = \dfrac{du}{1+u^2}$	
16. $\dfrac{d(\cot^{-1} u)}{dx} = -\dfrac{du/dx}{1+u^2}$		16′. $d(\cot^{-1} u) = -\dfrac{du}{1+u^2}$	
17. $\dfrac{d(\sec^{-1} u)}{dx} = \dfrac{du/dx}{\lvert u\rvert\sqrt{u^2-1}},$	$\lvert u\rvert > 1$	17′. $d(\sec^{-1} u) = \dfrac{du}{\lvert u\rvert\sqrt{u^2-1}},$	$\lvert u\rvert > 1$
18. $\dfrac{d(\csc^{-1} u)}{dx} = \dfrac{-du/dx}{\lvert u\rvert\sqrt{u^2-1}},$	$\lvert u\rvert > 1$	18′. $d(\csc^{-1} u) = \dfrac{-du}{\lvert u\rvert\sqrt{u^2-1}},$	$\lvert u\rvert > 1$

EXAMPLE 1

a) $$\frac{d}{dx}\sin^{-1} x^2 = \frac{1}{\sqrt{1-(x^2)^2}}\cdot\frac{d}{dx}(x^2)$$
$$= \frac{2x}{\sqrt{1-x^4}}$$

b) $$\frac{d}{dx}\tan^{-1}\sqrt{x+1} = \frac{1}{1+(\sqrt{x+1})^2}\cdot\frac{d}{dx}(\sqrt{x+1})$$
$$= \frac{1}{x+2}\cdot\frac{1}{2\sqrt{x+1}}$$
$$= \frac{1}{2\sqrt{x+1}(x+2)}$$

c) $$\frac{d}{dx}\sec^{-1}(3x) = \frac{1}{|3x|\sqrt{(3x)^2-1}}\cdot\frac{d}{dx}(3x)$$
$$= \frac{3}{|3x|\sqrt{9x^2-1}}$$
$$= \frac{1}{|x|\sqrt{9x^2-1}}. \quad \square$$

To show how the derivative formulas listed above may be derived, we shall prove formulas 13 and 17.

The Derivative of $y = \sin^{-1} u$ We know from Article 2.10 that the arc sine is differentiable, being the inverse of a differentiable function. If u is a differentiable function of x, with $-1 < u < 1$, then the composite

$$y = \sin^{-1} u \tag{1}$$

is defined, and differentiable by the Chain Rule (Article 2.6). To calculate dy/dx, we rewrite Eq. (1) as

$$\sin y = u, \tag{2}$$

and differentiate both sides with respect to x:

$$\frac{d}{dx}\sin y = \frac{du}{dx},$$

$$\cos y\frac{dy}{dx} = \frac{du}{dx}. \tag{3}$$

Since $-\pi/2 < y < \pi/2$ (because $-1 < u < 1$), $\cos y \neq 0$ and we may divide both sides of (3) by $\cos y$ to obtain

$$\frac{dy}{dx} = \frac{1}{\cos y}\frac{du}{dx}. \tag{4}$$

Since also

$$\cos y = \sqrt{1-\sin^2 y}$$
$$= \sqrt{1-u^2}$$

for $-\pi/2 < y < \pi/2$, the cosine being positive for these values of y, Eq. (4) can be written as

$$\frac{dy}{dx} = \frac{1}{\sqrt{1 - u^2}} \frac{du}{dx}. \tag{5}$$

This is formula 13, which we set out to prove. Referring to Fig. 6.2, note that the graph of $y = \sin^{-1} x$ has vertical tangents at $x = \pm 1$, corresponding to the fact that

$$\frac{d}{dx} \sin^{-1} x = \frac{1}{\sqrt{1 - x^2}}$$

is undefined for $x = \pm 1$.

The Derivative of $y = \sec^{-1} u$ Let $u > 1$ or $u < -1$, and let

$$y = \sec^{-1} u.$$

Then, $0 < y < \pi/2$ or $\pi/2 < y < \pi$, and

$$\sec y = u,$$

$$\sec y \tan y \frac{dy}{dx} = \frac{du}{dx}. \tag{6}$$

Since $|\sec y| \geq 1$ and $\tan y \neq 0$ for the values of y in question, we may divide both sides of (6) by sec y tan y to obtain

$$\frac{dy}{dx} = \frac{1}{\sec y \tan y} \frac{du}{dx}. \tag{7}$$

Substituting

$$\sec y = u, \qquad \tan y = \pm\sqrt{\sec^2 y - 1} = \pm\sqrt{u^2 - 1},$$

we have

$$\frac{dy}{dx} = \frac{1}{u(\pm\sqrt{u^2 - 1})} \frac{du}{dx}.$$

The ambiguous sign is determined by the sign of tan y and hence is

$$+ \quad \text{if } 0 < y < \frac{\pi}{2}, \qquad \text{that is, if u is } +;$$

$$- \quad \text{if } \frac{\pi}{2} < y < \pi, \qquad \text{that is, if u is } -.$$

That is,

$$\frac{dy}{dx} = \begin{cases} \dfrac{1}{u\sqrt{u^2 - 1}} \dfrac{du}{dx} & \text{if } u > 1, \\ \dfrac{1}{-u\sqrt{u^2 - 1}} \dfrac{du}{dx} & \text{if } u < -1, \end{cases}$$

and both of these are summarized in the single formula

$$\frac{dy}{dx} = \frac{1}{|u|\sqrt{u^2 - 1}} \frac{du}{dx}.$$

Integration Formulas

The derivative formulas 13 to 18 are used directly as given, but even more important are the three basic integration formulas obtained from them, listed in the following table.

Integrals Leading to Inverse Trigonometric Functions

13. $\displaystyle\int \frac{du}{\sqrt{1-u^2}} = \sin^{-1} u + C$

15. $\displaystyle\int \frac{du}{1+u^2} = \tan^{-1} u + C$

17. $\displaystyle\int \frac{du}{u\sqrt{u^2-1}} = \int \frac{d(-u)}{(-u)\sqrt{u^2-1}} = \sec^{-1} |u| + C$

Note that the integral formula 13 is valid only if $u^2 < 1$, and that formula 17 is valid only if $u^2 > 1$. Formula 15 holds for all values of u.

Something more should be said about formula 17. In the first place, if we are evaluating a *definite* integral of the form

$$\int_a^b \frac{dx}{x\sqrt{x^2-1}},$$

both a and b should be greater than 1 or both should be less than -1, for otherwise we would be attempting to integrate an expression that becomes infinite, or imaginary, or both, within the interval of integration. On the other hand, if we are looking for all curves of the form $y = f(x)$ such that $f'(x) = 1/(x\sqrt{x^2-1})$, then our answer is

$$y = \begin{cases} \sec^{-1} x + C_1 & \text{if } x > 1, \\ \sec^{-1}(-x) + C_2 & \text{if } x < -1; \end{cases}$$

and the two constants C_1 and C_2 need not be the same.

We can also replace $\sec^{-1}(-x)$ by $\pi - \sec^{-1} x$ by Eq. (10) of Article 6.2 and then let $C_2 + \pi = C'$, and write

$$y = \begin{cases} \sec^{-1} x + C & \text{if } x > 1, \\ -\sec^{-1} x + C' & \text{if } x < -1. \end{cases}$$

EXAMPLE 2 Evaluate

a) $\displaystyle\int_0^1 \frac{dx}{1+x^2}$ and b) $\displaystyle\int_{2/\sqrt{3}}^{2/\sqrt{2}} \frac{dx}{x\sqrt{x^2-1}}.$

Solution

a) $\displaystyle\int_0^1 \frac{dx}{1+x^2} = \tan^{-1} x\Big]_0^1 = \tan^{-1} 1 - \tan^{-1} 0 = \frac{\pi}{4} - 0 = \frac{\pi}{4}.$

b) $\displaystyle\int_{2/\sqrt{3}}^{2/\sqrt{2}} \frac{dx}{x\sqrt{x^2-1}} = \sec^{-1} x\Big]_{2/\sqrt{3}}^{2/\sqrt{2}} = \cos^{-1}(\sqrt{2}/2) - \cos^{-1}(\sqrt{3}/2)$

$$= \frac{\pi}{4} - \frac{\pi}{6} = \frac{\pi}{12}. \quad \square$$

EXAMPLE 3 Evaluate

$$\int \frac{x^2\,dx}{\sqrt{1-x^6}}.$$

Solution The resemblance between the given integral and the standard form

$$\int \frac{du}{\sqrt{1-u^2}} = \sin^{-1} u + C$$

suggests the formal substitution

$$u = x^3, \qquad du = 3x^2\,dx.$$

Indeed,

$$\begin{aligned}\int \frac{x^2\,dx}{\sqrt{1-x^6}} &= \frac{1}{3}\int \frac{3x^2\,dx}{\sqrt{1-(x^3)^2}}\\ &= \frac{1}{3}\int \frac{du}{\sqrt{1-u^2}}\\ &= \frac{1}{3}\sin^{-1} u + C\\ &= \frac{1}{3}\sin^{-1}(x^3) + C. \ \square\end{aligned}$$

EXAMPLE 4 Find a curve whose slope at the point (x, y) is $1/(x\sqrt{x^2-1})$ and that passes through the points $(-2, -2)$ and $(\sqrt{2}, 0)$.

Solution From the discussion before Example 2, we can write

$$y = \begin{cases} \sec^{-1} x + C & \text{if } x > 1, \quad (8a)\\ -\sec^{-1} x + C' & \text{if } x < -1. \quad (8b)\end{cases}$$

Using Eq. (8a) with $x = \sqrt{2}$ and $y = 0$, we have

$$0 = \sec^{-1}\sqrt{2} + C, \qquad C = -\sec^{-1}\sqrt{2} = -\frac{\pi}{4}.$$

To find the value of C' we use Eq. (8b) and the point $(-2, -2)$ to get

$$-2 = -\sec^{-1}(-2) + C',$$

$$C' = \sec^{-1}(-2) - 2 = \frac{2\pi}{3} - 2.$$

Therefore the curve is described by the equations

$$y = \sec^{-1} x - \frac{\pi}{4} \qquad \text{if } x > 1,$$

$$y = -\sec^{-1} x + \frac{2\pi}{3} - 2 \qquad \text{if } x < -1.$$

The graph is shown in Fig. 6.16. $\square$

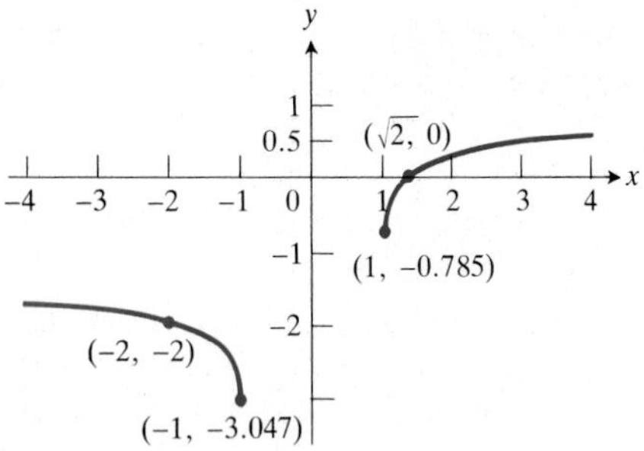

6.16 Graph of curve satisfying $dy/dx = 1/(x\sqrt{x^2-1})$ and passing through the points $(-2, -2)$ and $(\sqrt{2}, 0)$. The two branches have the same shape. Either branch can be moved up or down any number of units without changing the slope.

PROBLEMS

1. Derive formula 14.

2. Derive formula 15.

3. Derive formula 16.

4. Derive formula 18.

In Problems 5–26, find dy/dx.

5. $y = \cos^{-1} x^2$

6. $y = \cos^{-1}(1/x)$

7. $y = \tan^{-1}\sqrt{x}$

8. $y = \cot^{-1}\sqrt{x}$

9. $y = 5\tan^{-1} 3x$

10. $y = \sin^{-1}(1 - x)$

11. $y = \sin^{-1}(x/2)$

12. $y = \frac{1}{3}\tan^{-1}(x/3)$

13. $y = \sec^{-1} 5x$

14. $y = \cos^{-1} 2x$

15. $y = \csc^{-1}(x^2 + 1)$

16. $y = \csc^{-1}\sqrt{x} + \sec^{-1}\sqrt{x}$

17. $y = \csc^{-1}\sqrt{x + 1}$

18. $y = \cot^{-1}\sqrt{x - 1}$

19. $y = x\sqrt{1 - x^2} - \cos^{-1} x$

20. $y = \sqrt{x^2 - 4} - 2\sec^{-1}(x/2)$

21. $y = \cot^{-1}\dfrac{2}{x} + \tan^{-1}\dfrac{x}{2}$

22. $y = \sin^{-1}\dfrac{x - 1}{x + 1}$

23. $y = \tan^{-1}\dfrac{x - 1}{x + 1}$

24. $y = x\sin^{-1} x + \sqrt{1 - x^2}$

25. $y = x(\sin^{-1} x)^2 - 2x + 2\sqrt{1 - x^2}\sin^{-1} x$

26. $y = x\cos^{-1} 2x - \frac{1}{2}\sqrt{1 - 4x^2}$

Evaluate the integrals in Problems 27–39.

27. $\displaystyle\int_0^{1/2} \frac{dx}{\sqrt{1 - x^2}}$

28. $\displaystyle\int_{-1}^{1} \frac{dx}{1 + x^2}$

29. $\displaystyle\int_{\sqrt{2}}^{2} \frac{dx}{x\sqrt{x^2 - 1}}$

30. $\displaystyle\int_{-2}^{-\sqrt{2}} \frac{dx}{x\sqrt{x^2 - 1}}$

31. $\displaystyle\int \frac{dx}{\sqrt{1 - 4x^2}}$

32. $\displaystyle\int_{1/\sqrt{3}}^{1} \frac{dx}{x\sqrt{4x^2 - 1}}$

33. $\displaystyle\int_0^1 \frac{x}{1 + x^4}\,dx$

34. $\displaystyle\int_{-\pi/4}^{\pi/4} \frac{\cos x}{\sqrt{1 - \sin^2 x}}\,dx$

35. $\displaystyle\int_0^1 \frac{4\,dx}{1 + x^2}$

36. $\displaystyle\int_0^{1/6} \frac{dx}{\sqrt{1 - 9x^2}}$

37. $\displaystyle\int \frac{dx}{\sqrt{x}\sqrt{1 - x}}$

38. $\displaystyle\int \frac{dx}{\sqrt{1 - (x - 1)^2}}$

39. $\displaystyle\int \frac{2(x - 1)}{\sqrt{1 - (x - 1)^4}}\,dx$

40. The integral

$$\int \frac{x\,dx}{\sqrt{1 - x^2}}$$

does not involve inverse trigonometric functions. Evaluate it some other way.

Evaluate the limits in Problems 41–46.

41. $\displaystyle\lim_{x\to 0} \frac{\sin^{-1} 2x}{x}$

42. $\displaystyle\lim_{x\to 0} \frac{2\tan^{-1} 3x}{5x}$

43. $\displaystyle\lim_{x\to 0} x^{-3}(\sin^{-1} x - x)$

44. $\displaystyle\lim_{x\to 0} x^{-3}(\tan^{-1} x - x)$

45. $\displaystyle\lim_{b\to 1^-} \int_0^b \frac{dx}{\sqrt{1 - x^2}}$

46. $\displaystyle\lim_{x\to\infty} \frac{\int_0^x \tan^{-1} t\,dt}{x}$

47. How far from the wall should the observer stand to maximize the angle of vision α in Problem 33, Article 6.2?

48. Find a curve whose slope at $P(x, y)$ is $1/(x\sqrt{4x^2 - 1})$ if it contains the points $A(-1, -1)$ and $B(1, 2)$.

49. Let $f(x) = \sin^{-1} x + \cos^{-1} x$. Find

a) $f'(x)$, b) $f(0.32)$.

50. A beachcomber, walking 4 kilometers per hour along a straight shore is tracked by a rotating light 1 kilometer offshore. Find the rate (radians per hour) at which the light rotates when the beachcomber is 2 kilometers from the point on the shore closest to the light. (The beachcomber is walking toward this point.)

51. Can the following integrations, (a) and (b), both be correct? Explain.

a) $\displaystyle\int \frac{dx}{\sqrt{1 - x^2}} = \sin^{-1} x + C$

b) $\displaystyle\int \frac{dx}{\sqrt{1 - x^2}} = -\int -\frac{dx}{\sqrt{1 - x^2}} = -\cos^{-1} x + C$

52. a) Show that the functions

$$f(x) = \sin^{-1}\frac{x - 1}{x + 1} \quad \text{and} \quad g(x) = 2\tan^{-1}\sqrt{x}$$

(which are both defined for $x \geq 0$) have the same derivative, and therefore that

$$f(x) = g(x) + C. \tag{9}$$

b) Find C. (*Hint:* Evaluate both sides of Eq. (9) for a particular value of x.)

53. Find the volume swept out by revolving about the x-axis the "triangular" region in the first quadrant bounded by the curve $y = 1/\sqrt{1 + x^2}$ and the lines $x = 0$ and $y = x/\sqrt{2}$.

***Toolkit* programs**

Derivative Grapher   Super * Grapher

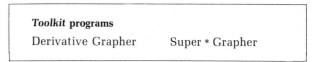

6.4 The Natural Logarithm and its Derivative

Introduction

In algebra we usually define the common logarithm or logarithm to the base 10 of a number b to be the power to which we raise 10 to get b:

$$10^{\log b} = b. \tag{1}$$

We assume there is such a number, and we use logarithms to simplify the work of multiplying numbers that have many decimal places. The simplification comes from the equation

$$\log ax = \log a + \log x. \tag{2}$$

For instance, we find the product of two numbers M and N by finding the number whose logarithm is $\log M + \log N$, as in following scheme:

$$\begin{array}{ccc} M \times N & \xrightarrow{\log} & \log(M \times N) \\ & & \| \\ (M \times N) & \xleftarrow{\text{antilog}} & \log M + \log N. \end{array} \tag{3}$$

But where does Eq. (2) come from, and how do we calculate the logarithms and arbitrary powers that appear in Eq. (1)? The answers to these questions lie in the articles ahead, where we shall also study exponential functions and see why the differentiation rule

$$\frac{d}{dx}x^n = nx^{n-1}$$

holds not just for rational exponents but for any real number n. In Article 6.9 we shall also see some of the applications of logarithmic and exponential functions that account for the importance of these functions in modern science and engineering.

The invention of logarithms was the biggest improvement in arithmetic in the sixteenth and seventeenth centuries. Logarithms made possible the arithmetic of navigation (among other things). Today they are even more important than they first became when John Napier developed them around 1594.†

The Natural Logarithm and Its Derivative

We begin by defining the *natural logarithm* of a positive number x. This number, denoted by $\ln x$, is defined as the value of the integral of the function $y = 1/t$ from $t = 1$ to $t = x$.

$$\ln x = \int_1^x \frac{1}{t}\,dt, \qquad x > 0 \tag{4}$$

†The first discovery of logarithms is credited to a Scottish nobleman, John Napier (1550–1617). For a biographical sketch of Napier, see the *World of Mathematics*, Vol. 1, "The Great Mathematicians," by H. W. Turnbull (New York: Simon and Schuster, 1962), pp. 121–125.

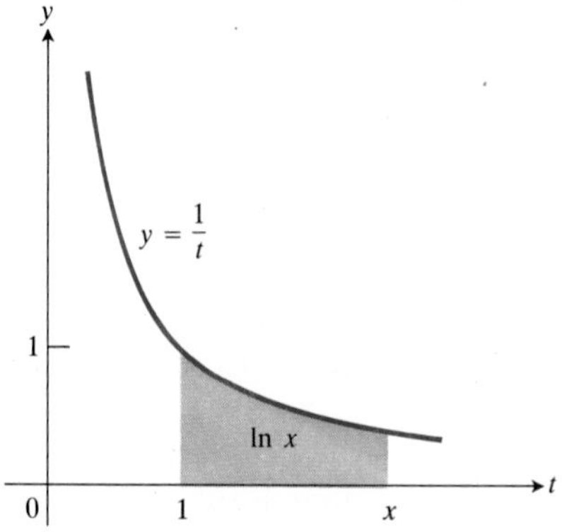

6.17 $\ln x = \int_1^x \frac{1}{t}\,dt, \qquad \frac{d(\ln x)}{dx} = \frac{1}{x}.$

For any x greater than 1, this integral represents the area bounded above by the curve $y = 1/t$, below by the t-axis, on the left by the line $t = 1$, and on the right by the line $t = x$. (See Fig. 6.17.)

If $x = 1$, the left and right boundaries of the area are identical and the area is zero,

$$\ln 1 = \int_1^1 \frac{1}{t}\,dt = 0. \tag{5}$$

If x is less than 1, then the left boundary is the line $t = x$ and the right boundary is $t = 1$. In this case,

$$\ln x = \int_1^x \frac{1}{t}\,dt = -\int_x^1 \frac{1}{t}\,dt \tag{6}$$

is the negative of the area under the curve between x and 1.

In all cases, if x is any positive number, the value of the definite integral in Eq. (4) can be calculated to as many decimal places as may be desired by using inscribed rectangles, circumscribed rectangles, or trapezoids to approximate the appropriate area. (See Article 4.9. Another method for computing natural logarithms, by means of series, will be discussed in Chapter 12. Also see Problem 57 below.) In any event, Eq. (4) defines a computable function of x over the domain $0 < x < +\infty$. We study its range in Article 6.5.

Since the function $y = \ln x$ is defined by the integral

$$\ln x = \int_1^x \frac{1}{t}\,dt, \qquad x > 0,$$

it follows at once, from the Second Fundamental Theorem of Calculus, Article 4.7, that

$$\frac{d}{dx}\ln x = \frac{1}{x}. \tag{7}$$

If u is a differentiable function of x, then the Chain Rule

$$\frac{dy}{dx} = \frac{dy}{du}\frac{du}{dx}$$

gives the more general formula

$$\frac{d}{dx}\ln u = \frac{1}{u}\frac{du}{dx}. \tag{8}$$

EXAMPLE 1 Find dy/dx if $y = \ln(3x^2 + 4)$.

Solution

$$\frac{dy}{dx} = \frac{1}{3x^2 + 4}\,\frac{d(3x^2 + 4)}{dx}$$

$$= \frac{6x}{3x^2 + 4}. \square$$

The Integral $\int \frac{1}{u}\,du = \ln|u| + C$

The integral formula

$$\int u^n\,du = \frac{u^{n+1}}{n+1} + C, \quad n \neq -1,$$

that we derived in Chapter 2 failed to cover the case $n = -1$, because there is no power of u whose derivative is $1/u$.

We are now in a position to treat this exceptional case, however, because Eq. (8) leads at once to the equation

$$\int u^{-1}\,du = \int \frac{1}{u}\,du = \int \frac{1}{u}\frac{du}{dx}\,dx = \ln u + C, \tag{9a}$$

provided u is positive. But if u is negative then $-u$ is positive and

$$\int \frac{du}{u} = \int \frac{d(-u)}{-u} = \ln(-u) + C'. \tag{9b}$$

The two results (9a, b) can be combined into a single result, namely,

$$\int \frac{du}{u} = \begin{cases} \ln u + C & \text{if } u > 0 \\ \ln(-u) + C' & \text{if } u < 0, \end{cases} \tag{10}$$

or, if u does not change sign on the given domain,

$$\int \frac{du}{u} = \ln|u| + C. \tag{11}$$

In applications, it is important to remember that the function u here can be any differentiable function $u = f(x)$. Equation (11) says that integrals of a certain *form* lead to logarithms. That is,

$$\int \frac{f'(x)}{f(x)}\,dx = \ln|f(x)| + C \tag{12}$$

whenever $f(x)$ is a differentiable function that does not change sign on the given domain.

EXAMPLE 2 Evaluate

$$\int \frac{6x}{3x^2 + 4}\,dx.$$

Solution This integral has the form

$$\int \frac{f'(x)}{f(x)}\,dx$$

with $f(x) = 3x^2 + 4$ and $f'(x) = 6x$. The function f does not change sign (it is always positive). Therefore Eq. (12) applies, and

$$\int \frac{6x}{3x^2 + 4}\,dx = \ln|3x^2 + 4| + C = \ln(3x^2 + 4) + C.$$

We can do without the absolute value bars because $3x^2 + 4$ is positive. □

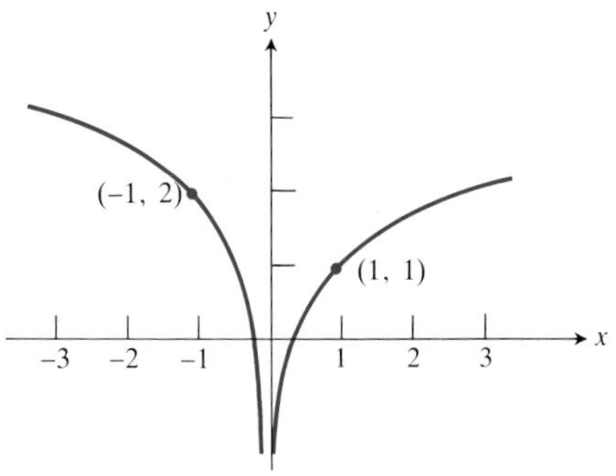

6.18 Graph of function $f(x)$ whose derivative is $1/x$ with $f(-1) = 2$ and $f(1) = 1$:

$$f(x) = \begin{cases} 2 + \ln(-x) & \text{when } x \text{ is negative,} \\ 1 + \ln x & \text{when } x \text{ is positive,} \end{cases}$$

The two constants of integration C and C' are not equal.

EXAMPLE 3 Find a function $y = f(x)$ such that $dy/dx = 1/x$, and $f(1) = 1$ and $f(-1) = 2$. Sketch the solution.

Solution There are constants C and C' such that

$$y = \begin{cases} \ln x + C & \text{if } x \text{ is positive,} \\ \ln(-x) + C' & \text{if } x \text{ is negative.} \end{cases}$$

Substituting $y = 1$ when $x = 1$, we get

$$1 = \ln 1 + C = 0 + C, \qquad \text{so} \quad C = 1.$$

Likewise, putting $y = 2$ when $x = -1$, we get

$$2 = \ln(-(-1)) + C' = \ln 1 + C' = 0 + C', \qquad \text{so} \quad C' = 2.$$

The complete solution is therefore

$$y = f(x) = \begin{cases} \ln x + 1 & \text{if } x \text{ is positive,} \\ \ln(-x) + 2 & \text{if } x \text{ is negative.} \end{cases}$$

A graph of the solution is shown in Fig. 6.18. (For a general analysis of the graph of $y = \ln x$, see Article 6.5.) □

EXAMPLE 4 Evaluate

$$\int \frac{\cos\theta\, d\theta}{2 + \sin\theta}.$$

Solution In Eq. (11), let

$$u = 2 + \sin\theta.$$

Then

$$du = \frac{du}{d\theta}\, d\theta = \cos\theta\, d\theta$$

so that

$$\int \frac{\cos\theta\, d\theta}{2 + \sin\theta} = \int \frac{du}{u} = \ln|u| + C$$
$$= \ln|2 + \sin\theta| + C$$
$$= \ln(2 + \sin\theta) + C. \ \square$$

EXAMPLE 5 Evaluate

$$\int \tan x\, dx.$$

Solution We express the integrand in terms of the sine and cosine (often a good tactic when integrating trigonometric functions),

$$\int \tan x\, dx = \int \frac{\sin x}{\cos x}\, dx,$$

and let

$$u = \cos x, \qquad du = -\sin x\, dx.$$

Then,

$$\int \tan x = \int \frac{\sin x}{\cos x}\, dx = \int \frac{-du}{u} = -\ln|u| + C = -\ln|\cos x| + C. \ \square$$

The Linear and Quadratic Approximations of ln (1 + x)

The standard linear and quadratic approximations of the logarithm function come from approximating $y = \ln(1 + x)$ near $x = 0$. The approximations (shown below) may be obtained directly from the formulas in Table 3.2, Article 3.10. An alternative to approximating $y = \ln(1 + x)$ is to approximate $y = \ln x$ near $x = 1$, but this leads to polynomials in powers of $(x - 1)$. See Problems 55–57.

Standard Approximations of $y = \ln(1 + x)$ Near $x = 0$

$$\ln(1 + x) \approx x \tag{13}$$

$$\ln(1 + x) \approx x - \frac{x^2}{2} \tag{14}$$

EXAMPLE 6 Linear approximations from Eq. (13):

$$\ln(1.2) = \ln(1 + 0.2) \approx 0.2$$
$$\ln(0.8) = \ln(1 - 0.2) \approx -0.2$$

Quadratic approximations from Eq. (14):

$$\ln(1.2) = \ln(1 + 0.2) \approx 0.2 - \frac{(0.2)^2}{2} = 0.2 - \frac{0.04}{2} = 0.18$$

$$\ln(0.8) = \ln(1 - 0.2) \approx -0.2 - \frac{(-0.2)^2}{2} = -0.2 - \frac{0.04}{2} = -0.22$$

The values of these logarithms to five decimal places are

$$\ln(1.2) \approx 0.18232, \qquad \ln(0.8) \approx -0.22314. \ \square$$

Comparison with Simpson's Rule While the approximations in Eqs. (13) and (14) are good for replacing $\ln(1 + x)$ by x and $(x - x^2/2)$ over short intervals about $x = 0$, when it comes to estimating the value of a *particular* logarithm Simpson's rule usually gives better results. For example, with $n = 2$, Simpson's rule gives

$$\ln 1.2 = \int_1^{1.2} \frac{1}{t}\,dt \approx \frac{0.1}{3}\left[1 + 4\left(\frac{1}{1.1}\right) + \frac{1}{1.2}\right] = \frac{1}{30}[5.46970] \approx 0.18232,$$

which is ln 1.2 to five decimal places. (Compare with Example 6.) Similarly, and again with $n = 2$, Simpson's rule gives

$$\ln 0.8 \approx -0.22314,$$

accurate to five places.

REMARK. Common logarithms, or logarithms to the base 10, are related to the natural logarithms by a rule that we shall come to in Article 6.8. Any function $f(x)$ that is defined on the domain of positive numbers x and that satisfies the condition

$$f(ax) = f(a) + f(x)$$

may be called a logarithmic function. The function that we have introduced as ln x has these properties: that is,

$$\ln x = \int_1^x \frac{1}{t}\,dt$$

is defined for all positive values of x, and, as we shall see in Article 6.5

$$\ln ax = \ln a + \ln x. \tag{15}$$

(One method of establishing Eq. (15) is outlined in Problem 58. We shall prove it by a different method in Article 6.5.) It is the property of areas under the curve $y = 1/t$ expressed by Eq. (15) that motivates the definition of ln x that we have given.

Important Formulas

1. $\ln x = \int_1^x \frac{1}{t}\,dt, \qquad x > 0$ (definition of ln x)
2. $\frac{d}{dx}\ln x = \frac{1}{x}$
3. $\frac{d}{dx}\ln u = \frac{1}{u}\frac{du}{dx}$
4. $\int \frac{du}{u} = \begin{cases} \ln u + C & \text{if } u > 0 \\ \ln(-u) + C & \text{if } u < 0 \end{cases}$
5. $\int \frac{du}{u} = \ln|u| + C$ (if u does not change sign)

PROBLEMS

In Problems 1–22, find dy/dx.

1. $\ln 2x$
2. $\ln 5x$
3. $\ln kx$ (k constant)
4. $(\ln x)^2$
5. $\ln(10/x)$
6. $y = \ln(x^2 + 2x)$
7. $y = (\ln x)^3$
8. $y = \ln(\cos x)$
9. $y = \ln(\tan x + \sec x)$
10. $y = x \ln x - x$
11. $y = x^3 \ln(2x)$
12. $\ln(\csc x)$
13. $\tan^{-1}(\ln x)$
14. $\ln(\ln x)$
15. $x^2 \ln(x^2)$
16. $\ln(x - \sqrt{x^2 + 1})$
17. $y = \ln(x^2 + 4) - x\tan^{-1}\frac{x}{2}$
18. $y = \ln x - \frac{1}{2}\ln(1 + x^2) - \frac{\tan^{-1} x}{x}$
19. $y = x(\ln x)^3$
20. $y = x[\sin(\ln x) + \cos(\ln x)]$
21. $y = x\sec^{-1} x - \ln(x + \sqrt{x^2 - 1}), \quad x > 1$
22. $y = x\ln(a^2 + x^2) - 2x + 2a\tan^{-1}\frac{x}{a}$

Evaluate the integrals in Problems 23–45.

23. $\int \frac{dx}{x}$
24. $\int \frac{2dx}{x}$
25. $\int \frac{dx}{2x}$
26. $\int \frac{dx}{x + 2}$
27. $\int_0^1 \frac{dx}{x + 1}$
28. $\int_{-1}^0 \frac{dx}{1 - x}$
29. $\int \frac{dx}{2x + 3}$
30. $\int \frac{dx}{2 - 3x}$
31. $\int \frac{x\,dx}{4x^2 + 1}$
32. $\int \frac{\sin x\,dx}{2 - \cos x}$
33. $\int \frac{\cos x\,dx}{\sin x}$
34. $\int \frac{2x - 5}{x}\,dx$
35. $\int \frac{x\,dx}{x + 1}$
36. $\int \frac{x^2\,dx}{4 - x^3}$
37. $\int \frac{x\,dx}{1 - x^2}$
38. $\int \frac{dx}{\sqrt{x}(1 + \sqrt{x})}$
39. $\int (\ln x)^2 \frac{dx}{x}$
40. $\int \frac{dx}{(2x + 3)^2}$

41. $\int \tan x \, dx$

42. $\int \frac{dx}{x \ln x}$

43. $\int \frac{\ln x}{x} dx$

44. $\int \frac{\sin 3x}{5 - 2\cos 3x} dx$

45. $\int x \ln dx$ (*Hint:* Take $x^2 \ln x$ as your first guess.)

Evaluate the limits in Problems 46–50.

46. $\lim_{t\to 0} \frac{\ln(1+2t) - 2t}{t^2}$

47. $\lim_{x\to\infty} \frac{\ln x}{x}$

48. $\lim_{h\to 0^+} h \ln h$

49. $\lim_{\theta\to 0^+} \frac{\ln(\sin\theta)}{\cot\theta}$

50. $\lim_{x\to\infty} \frac{\ln(\ln x)}{\ln x}$

51. Find the volume generated by rotating about the x-axis the area in the first quadrant bounded by $y = 1/\sqrt{x}$, the x-axis, the line $x = 1$, and the line $x = 2$.

52. Find the centroid (center of mass) of the plane region bounded by $y = (1 + x^2)^{-1}$, $y = -(1 + x^2)^{-1}$, $x = 0$, and $x = 1$.

53. Solve the differential equation

$$\frac{dy}{dx} = \frac{1 + (1/x)}{1 + (1/y)}, \qquad x > 0, \quad y > 0,$$

subject to the condition that $x = 1$ when $y = 1$.

54. a) Use Simpson's rule with $n = 8$ to estimate ln 5.
b) (*Calculator*) Compare the result in (a) with the value of ln 5 given by a calculator.

55. a) Derive the linear and quadratic approximations of $y = \ln(1 + x)$ in Eqs. (13) and (14).
b) Use Eqs. (13) and (14) to estimate ln 1.16 and ln 0.84.
c) Use Simpson's rule with $n = 2$ to estimate ln 1.16 and ln 0.84.
d) (*Calculator*) Compare the results in (b) and (c) with the values a calculator gives for the logarithms.

56. Another way to derive the quadratic approximation of $\ln(1 + h)$ near $h = 0$ is by using the line L tangent to the curve $y = 1/t$ at $A(1, 1)$, as the upper boundary of a trapezoid of altitude $|h|$. Sketch. Show that this leads to the approximation

$$\ln(1 + h) \approx h - \frac{h^2}{2}.$$

57. Another approach to approximating the logarithm.
a) Show, by long division or otherwise, that

$$\frac{1}{1 + u} = 1 - u + u^2 - \frac{u^3}{1 + u}.$$

(The division could be continued. We stop here only for the sake of illustration.)
b) In Eq. (4), make the substitution

$$t = 1 + u, \qquad dt = du$$

and make the corresponding change in the limits of integration, thus obtaining

$$\ln x = \int_0^{x-1} \frac{du}{1 + u}.$$

c) Combine the results of (a) and (b) to obtain

$$\ln x = \int_0^{x-1} \left(1 - u + u^2 - \frac{u^3}{1 + u}\right) du,$$

or

$$\ln x = (x - 1) - \frac{1}{2}(x - 1)^2 + \frac{1}{3}(x - 1)^3 - R,$$

where

$$R = \int_0^{x-1} \frac{u^3}{1 + u} du.$$

d) Show that, if $x > 1$ and $0 \le u \le x - 1$, then

$$\frac{u^3}{1 + u} \le u^3.$$

Hence, deduce that

$$R \le \int_0^{x-1} u^3 \, du = \frac{(x - 1)^4}{4}.$$

e) Combining the results of (c) and (d), show that the approximation

$$\ln x \approx (x - 1) - \frac{1}{2}(x - 1)^2 + \frac{1}{3}(x - 1)^3$$

tends to overestimate the value of ln x, but with an error not greater than $(x - 1)^4/4$.
f) Use the result of (e) to estimate ln 1.2.

58. *A proof of Eq. (15).* a) Prove that the area under the curve $y = 1/t$ over $[1, x]$ is equal to the area under the same curve over $[a, ax]$ where a is any positive constant. *Suggestion.* Take any partition of $[1, x]$ and multiply each coordinate by a to get the corresponding partition of $[a, ax]$. Sketch. If $f(t_i)\,\Delta t_i$ is the area of a rectangle over one of the subdivisions in $[1, x]$, what is the rectangle whose area is represented by $f(at_i)(a\,\Delta t_i)$? Are these two rectangles equal in area?
b) In part (a) you have shown that the area under the curve $y = 1/t$ over $[a, ax]$ is equal to ln x. By Eq. (4), the area over $[1, a]$ is ln a. Now combine these results to show that $\ln ax = \ln a + \ln x$. (*Remark.* It is easier to see what goes on if you assume that both a and x are greater than 1 in the foregoing. The language has to differ if either a or x is less than 1, but the conclusion $\ln ax = \ln a + \ln x$ is still true. Our proof in Article 6.5 is more straightforward and does not require special cases.)

***Toolkit* programs**

Derivative Grapher
Integral Evaluator
Super * Grapher

6.5 Properties of Natural Logarithms. The Graph of $y = \ln x$

In this article we shall establish the following important properties of natural logarithms and use them to graph the function $y = \ln x$:

$$\ln ax = \ln a + \ln x \tag{1}$$

$$\ln \frac{x}{a} = \ln x - \ln a \tag{2}$$

$$\ln x^n = n \ln x \tag{3}$$

These properties hold provided x and a are positive. For the moment there is also an added restriction that the exponent n be a rational number, but we shall remove this restriction in Article 6.7.

EXAMPLE 1

$$\ln\left(\frac{1}{8}\right) = \ln 1 - \ln 8 = 0 - \ln 2^3 = -3 \ln 2$$

$$\ln 4 - \ln 5 = \ln\left(\frac{4}{5}\right) = \ln 0.8$$

$$\ln \sqrt[3]{25} = \ln (25)^{1/3} = \left(\frac{1}{3}\right) \ln 25 = \left(\frac{1}{3}\right) \ln 5^2 = \left(\frac{2}{3}\right) \ln 5 \; \square$$

The proofs of Eqs. (1)–(3) are based on the fact that

$$y = \ln x$$

satisfies the differential equation

$$\frac{dy}{dx} = \frac{1}{x} \quad \text{for all} \quad x > 0,$$

plus the fact that if two functions have the same derivative for all $x > 0$, then the two functions can differ only by a constant. That is, if

$$\frac{dy_1}{dx} = \frac{dy_2}{dx} \quad \text{for all} \quad x > 0,$$

then

$$y_1 = y_2 + \text{constant} \qquad \text{for all} \quad x > 0.$$

$\ln ax = \ln a + \ln x$

To prove Eq. (1), let

$$y_1 = \ln ax, \qquad y_2 = \ln x.$$

Then

$$\frac{dy_1}{dx} = \frac{1}{ax}\frac{d(ax)}{dx} = \frac{a}{ax} = \frac{1}{x} = \frac{dy_2}{dx}. \tag{4}$$

Therefore we have

$$\ln ax = \ln x + C. \tag{5}$$

To evaluate C it suffices to substitute $x = 1$:

$$\ln a = \ln 1 + C = 0 + C,$$

which gives $C = \ln a$. Hence, by (5),

$$\ln ax = \ln x + \ln a,$$

which is Eq. (1).

$\ln (x/a) = \ln x - \ln a$

To prove Eq. (2), we first put $x = 1/a$ in Eq. (1) and recall that $\ln 1 = 0$:

$$0 = \ln 1 = \ln\left(a \cdot \frac{1}{a}\right) = \ln a + \ln\left(\frac{1}{a}\right)$$

so that

$$\ln \frac{1}{a} = -\ln a. \tag{6}$$

Now apply Eq. (1) with a replaced by $1/a$ and $\ln a$ replaced by $\ln (1/a) = -\ln a$:

$$\ln \frac{x}{a} = \ln\left(x \cdot \frac{1}{a}\right) = \ln x + \ln \frac{1}{a} = \ln x - \ln a.$$

The result is Eq. (2).

$\ln x^n = n \ln x$

To prove Eq. (3), let

$$y_1 = \ln x^n.$$

Then

$$\frac{dy_1}{dx} = \frac{1}{x^n} \cdot \frac{d}{dx}(x^n) = \frac{1}{x^n} \cdot nx^{n-1} = \frac{n}{x} = \frac{d}{dx}(n \ln x).$$

Hence $y_1 = \ln x^n$ and $y_2 = n \ln x$ have equal derivatives, so they differ at most by a constant:

$$\ln x^n = n \ln x + C. \tag{7}$$

But by taking $x = 1$, and remembering that $\ln 1 = 0$, we find $C = 0$, which gives Eq. (3). In particular, if $n = 1/m$, where m is a positive integer, we have

$$\ln \sqrt[m]{x} = \ln x^{1/m} = \frac{1}{m} \ln x. \tag{8}$$

Simplifying Derivative Calculations

The numerical properties in Eqs. (1)–(3) allow us to simplify the calculation of the derivatives of logarithms of products and quotients of functions.

EXAMPLE 2 Find dy/dx if

$$y = \ln \frac{x\sqrt{x+5}}{(x-1)^3}.$$

Solution By applying Eqs. (1)–(3) we find that

$$\begin{aligned} y &= \ln \frac{x\sqrt{x+5}}{(x-1)^3} \\ &= \ln x\sqrt{x+5} - \ln (x-1)^3 && \text{(from Eq. 2)} \\ &= \ln x + \ln \sqrt{x+5} - \ln (x-1)^3 && \text{(from Eq. 1)} \\ &= \ln x + \frac{1}{2} \ln (x+5) - 3 \ln (x-1). && \text{(from Eq. 3)} \end{aligned}$$

Therefore,

$$\frac{dy}{dx} = \frac{1}{x} + \frac{1}{2(x+5)} - \frac{3}{x-1}. \quad \square$$

Logarithmic Differentiation

The derivative of a function given by a complicated equation can sometimes be calculated more quickly if we take the logarithm of both sides of the equation before differentiating. The process, called *logarithmic differentiation,* is illustrated in the next example.

EXAMPLE 3 Find dy/dx if

$$y^{2/3} = \frac{(x^2+1)(3x+4)^{1/2}}{\sqrt[5]{(2x-3)(x^2-4)}}, \qquad x > 2. \tag{9}$$

Solution First, we take the logarithm of both sides of Eq. (9). Since $\ln y^{2/3} = (2/3) \ln y$, this gives

$$\frac{2}{3} \ln y = \ln (x^2+1) + \frac{1}{2} \ln (3x+4) - \frac{1}{5} \ln (2x-3) - \frac{1}{5} \ln (x^2-4). \tag{10}$$

We then take the derivative of both sides, using implicit differentiation on the left:

$$\frac{2}{3} \cdot \frac{1}{y} \frac{dy}{dx} = \frac{2x}{x^2+1} + \frac{1}{2} \cdot \frac{3}{3x+4} - \frac{1}{5} \cdot \frac{2}{2x-3} - \frac{1}{5} \cdot \frac{2x}{x^2-4}.$$

Finally, we solve for dy/dx:

$$\frac{dy}{dx} = \frac{3y}{2} \left[\frac{2x}{x^2+1} + \frac{3/2}{3x+4} - \frac{2/5}{2x-3} - \frac{2x/5}{x^2-4} \right].$$

(The restriction $x > 2$ ensures that all the quantities whose logarithms are indicated in (10) above are positive. If we want dy/dx at a point where y is negative, we may multiply our equation by -1 before taking logarithms; for if y is negative, then $-y$ is positive and $\ln(-y)$ is defined.) $\square$

The Graph of $y = \ln x$

The slope of the curve

$$y = \ln x \tag{11}$$

is given by

$$\frac{dy}{dx} = \frac{1}{x}, \tag{12}$$

which is positive for all $x > 0$. Hence the graph of $y = \ln x$ steadily rises from left to right. Since the derivative is continuous, the function $\ln x$ is itself continuous, and the curve has a continuously turning tangent.

The second derivative,

$$\frac{d^2y}{dx^2} = -\frac{1}{x^2}, \tag{13}$$

is always negative, so the curve (11) is always concave downward.

The curve passes through the point $(1, 0)$, since $\ln 1 = 0$. At this point its slope is $+1$, so the tangent line at this point makes an angle of $45°$ with the x-axis (if we use equal units on the x- and y-axes).

If we refer to the definition of ln 2 as an integral,

$$\ln 2 = \int_1^2 \frac{1}{t}\,dt,$$

we see that it may be interpreted as the area in Fig. 6.17 with $x = 2$. By considering the areas of rectangles of base 1 and altitudes 1 or $\frac{1}{2}$, respectively circumscribed over and inscribed under the given area, we see that

$$0.5 < \ln 2 < 1.0.$$

In fact, by more extensive calculations, the value of ln 2 is found to be

$$\ln 2 \approx 0.69315,$$

to five decimal places. By Eq. (3) we then have

$$\ln 4 = \ln 2^2 = 2 \ln 2 \approx 1.38630,$$
$$\ln 8 = \ln 2^3 = 3 \ln 2 \approx 2.07944,$$
$$\ln \frac{1}{2} = \ln 2^{-1} = -\ln 2 \approx -0.69315,$$
$$\ln \frac{1}{4} = \ln 2^{-2} = -2 \ln 2 \approx -1.38630,$$

and so on.

We now plot the points that correspond to $x = \frac{1}{4}, \frac{1}{2}, 1, 2, 4, 8$ on the curve $y = \ln x$ and connect them with a smooth curve. The curve we draw should have slope $1/x$ at the point of abscissa x and should be concave downward. The curve is shown in Fig. 6.19.

Since ln 2 is greater than 0.5 and $\ln 2^n = n \ln 2$, we have

$$\ln 2 > 0.5$$
$$\ln 4 > 2(0.5) = 1$$
$$\ln 8 > 3(0.5) = 1.5$$
$$\ln 16 > 4(0.5) = 2$$
$$\vdots$$
$$\ln 2^n > n(0.5) = \frac{n}{2},$$

6.19 Graph of $y = \ln x$.

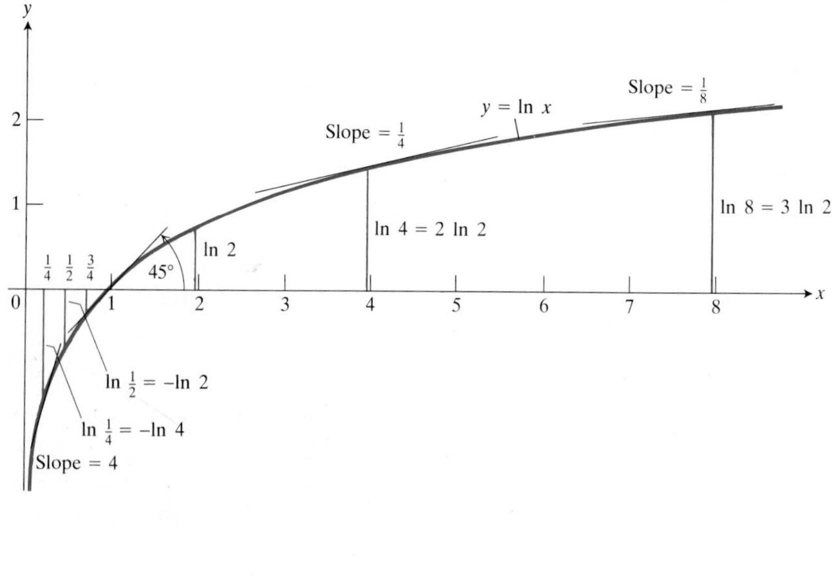

and hence ln x increases without limit as x does. That is,

$$\ln x \to +\infty \quad \text{as} \quad x \to +\infty. \tag{14}$$

On the other hand, as x approaches zero through positive values, $1/x$ tends to plus infinity. Hence, from Eq. (6) we have

$$\ln x = -\ln \frac{1}{x} \to -\infty \quad \text{as} \quad x \to 0^+. \tag{15}$$

The y-axis is a vertical asymptote of the graph of $y = \ln x$.

Properties of $y = \ln x$

1. Domain: The set of all positive real numbers, $x > 0$.
2. Range: The set of all real numbers $-\infty < y < \infty$.
3. It is a continuous, increasing function everywhere on its domain. If $x_1 > x_2 > 0$, then $\ln x_1 > \ln x_2$. It is a one-to-one function from its domain to its range. (It therefore has an inverse, which will be the subject of the next article.)
4. Products, quotients, and powers: If a and x are any two positive numbers, then

$$\ln ax = \ln a + \ln x, \tag{1}$$

$$\ln \frac{x}{a} = \ln x - \ln a, \tag{2}$$

$$\ln x^n = n \ln x. \tag{3}$$

PROBLEMS

Express the logarithms in Problems 1–10 in terms of the two given logarithms $a = \ln 2$, $b = \ln 3$. (For example: $\ln 1.5 = \ln \frac{3}{2} = \ln 3 - \ln 2 = b - a$.)

1. $\ln 16$ **2.** $\ln \sqrt[3]{9}$ **3.** $\ln 2\sqrt{2}$

4. $\ln 0.25$ **5.** $\ln \frac{4}{9}$ **6.** $\ln 12$

7. $\ln \frac{9}{8}$ **8.** $\ln 36$ **9.** $\ln 4.5$

10. $\ln \sqrt{13.5}$

In Problems 11–22, find dy/dx.

11. $y = \ln \sqrt{x^2 + 5}$ **12.** $y = \ln x^{3/2}$

13. $y = \ln \dfrac{1}{x\sqrt{x + 1}}$ **14.** $y = \ln \sqrt[3]{\cos x}$

15. $y = \ln (\sin x \sin 2x)$ **16.** $y = \ln (x\sqrt{x^2 + 1})$

17. $y = \ln (3x\sqrt{x + 2})$ **18.** $y = \dfrac{1}{2} \ln \dfrac{1 + x}{1 - x}$

19. $y = \dfrac{1}{3} \ln \dfrac{x^3}{1 + x^3}$ **20.** $y = \ln \dfrac{x}{2 + 3x}$

21. $y = \ln \dfrac{(x^2 + 1)^5}{\sqrt{1 - x}}$ **22.** $y = \displaystyle\int_{\sqrt{x}}^{\sqrt[3]{x}} \ln t \, dt$

In Problems 23–31, find dy/dx by logarithmic differentiation.

23. $y^2 = x(x + 1) \quad (x > 0)$

24. $y = \sqrt[3]{\dfrac{x + 1}{x - 1}} \quad (x > 1)$

25. $y = \sqrt{x + 3} \sin x \cos x, \quad 0 < x < \pi/2$

26. $y = \dfrac{x\sqrt{x^2 + 1}}{(x + 1)^{2/3}}, \quad x > 0$

27. $y = \sqrt[3]{\dfrac{x(x - 2)}{x^2 + 1}}, \quad x > 2$

28. $y^5 = \sqrt{\dfrac{(x + 1)^5}{(x + 2)^{10}}}$

29. $y = \sqrt[3]{\dfrac{x(x + 1)(x - 2)}{(x^2 + 1)(2x + 3)}}, \quad x > 2$

30. $y^{4/5} = \dfrac{\sqrt{\sin x \cos x}}{1 + 2 \ln x}$

31. $\sqrt{y} = \dfrac{x^5 \tan^{-1} x}{(3 - 2x)\sqrt[3]{x}}$

32. a) What is the largest possible domain of

$$y = \sqrt{\frac{(x+1)(x+2)}{(3-x)(4-x)}}?$$

b) Find dy/dx.

Evaluate the integrals in Problems 33–36.

33. $\int_{-1}^{1} \frac{dx}{x+3}$ **34.** $\int_{0}^{6} \frac{dx}{x+2}$

35. $\int_{\pi/4}^{\pi/2} \cot x\, dx$

36. a) $\int_{0}^{\sqrt{3}} \frac{dx}{1+x^2}$ b) $\int_{0}^{\sqrt{3}} \frac{x\, dx}{1+x^2}$

37. Graph (a) $y = \ln |x|$, (b) $y = |\ln x|$.

38. Graph $y = \ln x$ and $y = \ln 2x$ together. (*Hint:* Before you start, apply Eq. (1) to $\ln 2x$.)

39. Find $\lim_{x\to\infty} \int_{x}^{2x} (1/t)\, dt$.

40. Find the area of the "triangular" region in the first quadrant bounded by $x = 2$, $y = 2$, and the hyperbola $xy = 2$.

41. a) Find the area of the region in the first quadrant bounded by the x-axis, the curve $y = \tan x$, and the line $x = \pi/3$.

b) Find the volume of the solid generated by revolving the region in (a) about the x-axis.

42. Repeat Problem 41 for the region bounded by the curves $y = 4/x$ and $y = (x-3)^2$.

43. Use the result of Problem 45, Article 6.4, to find the center of gravity of the plane region bounded by the x-axis, the curve $y = \ln x$, and the line $x = 2$.

44. Solve the differential equation

$$y'' = \sec^2 x$$

subject to the condition that $y = 0$ and $y' = 1$ when $x = 0$.

45. Show that the curve

$$y = \ln x - (x-1) + \tfrac{1}{2}(x-1)^2$$

has a point of inflection and a horizontal tangent at (1, 0). Sketch the curve for $0 < x < 2$. (Compare Problem 57, Article 6.4.)

46. a) Use Eq. (9), Article 3.10, to obtain an upper bound for the error in the linear approximation

$$\ln(1+x) \approx x \tag{16}$$

over the interval $0 \le x \le 0.1$.

b) Sketch the graphs of $y = x$ and $y = \ln(1+x)$ together to show that the maximum error in the linear approximation (16) on the interval [0, 0.1] occurs when $x = 0.1$.

c) (*Calculator*) Find the difference $\ln(1.1) - 0.1$ with a calculator (this, too, is an approximation), and compare with the upper bound obtained in part (a).

47. Let x_0 be the abscissa of the point $(x_0, 1)$ in which the line $y = 1$ intersects the curve $y = \ln x$. Show that the curve $y = x \ln x$ has a minimum at $[1/x_0, -(1/x_0)]$ and sketch the curve for $x > 0$. Does this curve possess a point of inflection? If so, find it (or them), and if not, state why not.

48. Sketch the curve $x = \ln y$ for $y > 0$. What is its slope at the point (x, y)? Is it always concave upward, or downward, or does it have inflection points? Give reasons for your answers.

49. a) Using one circumscribed rectangle to approximate the area under the curve $y = 1/t$ from $t = 1$ to $t = 2$, show that $\ln 2 < 1$.

b) Using three inscribed rectangles to approximate the area under the curve $y = 1/t$ from $t = 1$ to $t = 4$, show that $\ln 4 > 1$.

c) Then apply a theorem about continuous functions from Article 1.11 to show that $\ln(c) = 1$ for some number c between 1 and 4. This shows that the number whose natural logarithm is 1 lies between 1 and 4. This number, named e by Euler, is the subject of the next article.

50. Use mathematical induction (Appendix 5) to show that at any value of x where the functions $u_1(x), \ldots, u_n(x)$ are all positive

$$\ln(u_1 u_2 \cdots u_n) = \ln u_1 + \ln u_2 + \cdots + \ln u_n.$$

6.6 The Exponential Function e^x

The Definition of $y = e^x$

Since $\ln 2 < 1$ and $\ln 4 > 1$, the Intermediate Value Theorem for continuous functions assures us that $\ln x = 1$ for some value of x between $x = 2$ and $x = 4$. This value of x is denoted by e (which stands for Euler, who named

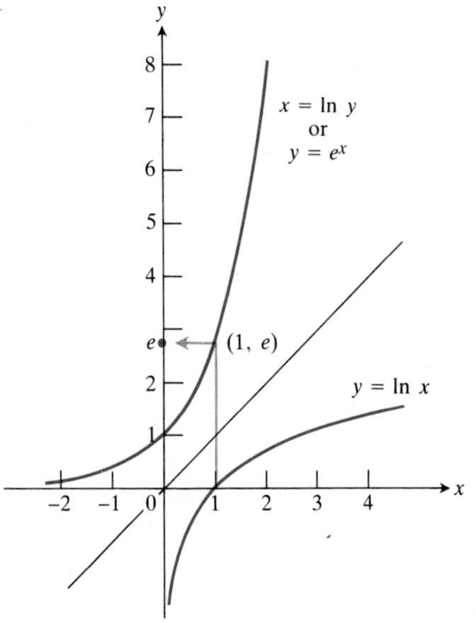

6.20 The graph of $y = \ln x$ and its inverse $y = e^x$.

the number after himself when he studied it in the early seventeen hundreds). Thus the number e satisfies the equation

$$\ln e = 1. \tag{1}$$

It is one of the most important numbers in mathematics.

Since

$$\ln 2 < \ln e < \ln 4,$$

the value of e lies between 2 and 4. If you hold a ruler along the line $y = 1$ in Fig. 6.19, you will see that the value of e is greater than 2 but somewhat less than 3. In Chapter 12 we shall see how the value of e may be computed to any desired number of decimal places by using series. Its value to 15 decimal places is

$$e = 2.7\ 1828\ 1828\ 45\ 90\ 45\ \ldots \tag{2}$$

By Eq. (3), Article 6.5, and Eq. (1) above, we have

$$\ln e^n = n \ln e = n \tag{3}$$

for any rational number n. Thus, for example,

$$\ln e^2 = 2, \qquad \ln e^3 = 3, \qquad \ln e^{-1} = -1, \qquad \ln \sqrt{e} = \tfrac{1}{2}.$$

Since only one number can have its natural logarithm equal to n, for any given n, we may restate Eq. (3) by saying that the number whose natural logarithm is n is e^n; that is, the anti-natural-logarithm of n is e^n.

Since $y = \ln x$ is one-to-one and differentiable, it has a differentiable inverse,

$$y = \ln^{-1} x.$$

The graph of $y = \ln^{-1} x$ (Fig. 6.20) is the reflection of the graph of $y = \ln x$ across the line $y = x$. (It is also the graph of the function $x = \ln y$). We know from Eq. (3) that when x takes on a rational value n, the value of $y = \ln^{-1} x$ is

$$\ln^{-1} n = \ln^{-1} (\ln e^n) = e^n, \tag{4}$$

or

$$e^n = \ln^{-1} n. \tag{5}$$

Since $\ln^{-1} x$ is defined for all values of x, the equation

$$e^x = \ln^{-1} x.$$

gives a way to define e^x for irrational values of x that is consistent with the values that e^x has when x is rational. In other words, we can define e^x for all values of x by the following rule:

DEFINITION

The Function $y = e^x$

For every real number x, the number e^x is defined to be $\ln^{-1} x$:

$$e^x = \ln^{-1} x.$$

That is,

$$y = e^x \qquad \text{if and only if} \qquad x = \ln y.$$

The resulting function is a differentiable function of x, defined for all real x, $-\infty < x < +\infty$. It is called the *exponential function*, with e as base and exponent x. The graph of $y = e^x$ in Fig. 6.20 may be obtained by reflecting the graph of $y = \ln x$ across the line $y = x$. Note, in particular, that $e^0 = 1$ since $\ln 1 = 0$. Note also (Fig. 6.20) that

$$\lim_{x\to\infty} e^x = \infty, \tag{7a}$$

$$\lim_{x\to-\infty} e^x = 0. \tag{7b}$$

Equations Involving ln *x* and e^x

Because $y = e^x$ and $y = \ln x$ are inverses of one another, we have

$$e^{\ln x} = x \tag{8a}$$

for all $x > 0$ and

$$\ln e^x = x \tag{8b}$$

for all values of x.

EXAMPLE 1

a) $e^{\ln 2} = 2$

b) $e^{\ln(x^2+1)} = x^2 + 1$

c) $\ln e^{-1.3} = -1.3$

d) $\ln e^{\sin x} = \sin x$

e) $\ln \dfrac{e^{2x}}{5} = \ln e^{2x} - \ln 5 = 2x - \ln 5$ □

EXAMPLE 2 Solve for y.

a) $\ln y = x^2$

b) $e^{3y} = 2 + \cos x$

c) $\ln (y - 1) - \ln y = 3x$

Solution

a) $\ln y = x^2$

Exponentiate: $e^{\ln y} = e^{x^2}$

$$y = e^{x^2}.$$

b) $e^{3y} = 2 + \cos x$

Take the logarithm of both sides: $\ln e^{3y} = \ln (2 + \cos x)$

$$3y = \ln (2 + \cos x)$$

$$y = \frac{1}{3} \ln (2 + \cos x).$$

c) $\ln (y - 1) - \ln y = 3x$

Combine the logarithms: $\ln \dfrac{y-1}{y} = 3x$

Exponentiate: $\dfrac{y-1}{y} = e^{3x}$

Solve as usual:

$$y - 1 = ye^{3x}$$
$$y - ye^{3x} = 1$$
$$y(1 - e^{3x}) = 1$$
$$y = \frac{1}{1 - e^{3x}}. \quad \square$$

Two Useful Operating Rules

To remove logarithms from an equation, exponentiate both sides.
To remove exponentials, take the logarithm of both sides.

The Derivative of $y = e^x$

To find the derivative of

$$y = e^x, \tag{9}$$

we take the logarithm of both sides,

$$\ln y = x,$$

and differentiate implicitly with respect to x. Then

$$\frac{1}{y}\frac{dy}{dx} = 1 \quad \text{or} \quad \frac{dy}{dx} = y. \tag{10}$$

Since $y = e^x$, this gives

$$\frac{de^x}{dx} = e^x.$$

Here is a function that is not changed by differentiation! It is indestructible. It can be differentiated again and again without changing. In this, the exponential function is like the story of a student who asked a guru what was holding up the earth. The answer was that the earth was upheld by an elephant, and the student naturally wanted to know what held up that elephant. The guru paused a moment, then replied "It's elephants all the way down."

If we didn't know it already, we could determine that $y = e^x$ is an increasing function of x from the fact that its derivative is positive.

We obtain a formula for the derivative of e^u, where u is a differentiable function of x, by applying the Chain Rule:

$$\frac{de^u}{dx} = \frac{de^u}{du}\frac{du}{dx} = e^u\frac{du}{dx}. \tag{11}$$

This in turn leads to the integration formula

$$\int e^u\,du = e^u + C. \tag{12}$$

EXAMPLE 3 Find dy/dx if $y = e^{\tan^{-1} x}$.

Solution

$$\begin{aligned}\frac{dy}{dx} &= e^{\tan^{-1} x} \frac{d \tan^{-1} x}{dx} \\ &= e^{\tan^{-1} x} \cdot \frac{1}{1 + x^2} \cdot \frac{dx}{dx} \\ &= \frac{e^{\tan^{-1} x}}{1 + x^2}. \quad \square\end{aligned}$$

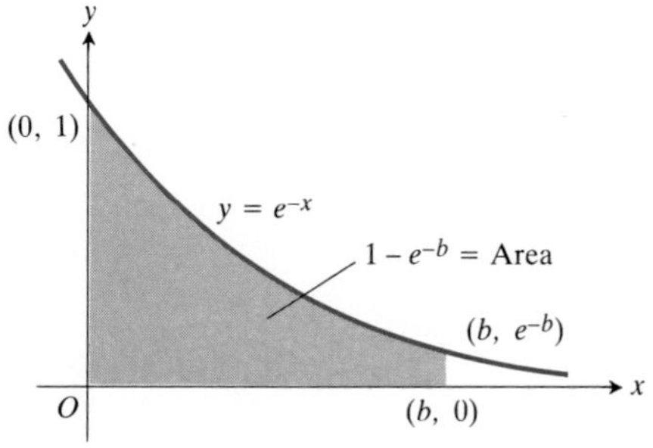

6.21 The shaded area is $1 - e^{-b}$.

EXAMPLE 4 Find the area under the curve $y = e^{-x}$ from $x = 0$ to $x = b(> 0)$ and show that this area *remains finite* as $b \to +\infty$. (See Fig. 6.21).

Solution The area from $x = 0$ to $x = b$ is

$$A_0^b = \int_0^b y\, dx = \int_0^b e^{-x}\, dx.$$

To evaluate this integral, we compare it with our standard forms and see that it is almost, but not exactly, like

$$\int e^u\, du = e^u + C.$$

To bring it into precisely this form, we let

$$u = -x, \qquad du = -dx, \qquad dx = -du,$$

so that

$$\int e^{-x}\, dx = \int e^u(-du) = -\int e^u\, du = -e^u + C = -e^{-x} + C.$$

Therefore

$$A_0^b = -e^{-x}\Big]_0^b = -e^{-b} + e^0 = 1 - e^{-b}.$$

From this we see that the number of square units of area in the shaded region of Fig. 6.21 is somewhat less than unity, for b large and positive. Moreover

$$\lim_{b \to +\infty} A_0^b = \lim_{b \to +\infty} (1 - e^{-b}) = 1,$$

since (see Eq. (14) below)

$$\lim_{b \to +\infty} e^{-b} = \lim_{b \to +\infty} \frac{1}{e^b} = 0. \quad \square$$

The exponential function obeys the familiar rules of exponents from algebra:

$$e^{x_1} \cdot e^{x_2} = e^{x_1 + x_2} \tag{13}$$

$$e^{-x} = \frac{1}{e^x} \tag{14}$$

These equations hold for all real numbers and can be derived from the relationship that e^x has with $\ln x$. To establish Eq. (13), let

$$y_1 = e^{x_1} \qquad \text{and} \qquad y_2 = e^{x_2},$$

then, by definition,

$$x_1 = \ln y_1 \quad \text{and} \quad x_2 = \ln y_2,$$

so that

$$x_1 + x_2 = \ln y_1 + \ln y_2 = \ln y_1 y_2$$

by Eq. (1), Article 6.6. Therefore, by definition (6) again,

$$y_1 y_2 = e^{x_1 + x_2},$$

which establishes Eq. (13).

To establish Eq. (14), let

$$y = e^{-x}.$$

Then, by definition,

$$-x = \ln y$$

and, from Eq. (6), Article 6.5,

$$x = -\ln y = \ln\left(\frac{1}{y}\right).$$

Therefore

$$\frac{1}{y} = e^x, \quad \text{or} \quad y = \frac{1}{e^x},$$

which establishes Eq. (14).

EXAMPLE 5 Simplify

a) $e^{\ln 2 + 3\ln x}$

b) $e^{2x - \ln x}$

Solution

a) $e^{\ln 2 + 3\ln x} = e^{\ln 2} \cdot e^{3\ln x}$ (from Eq. 13)

$= 2 \cdot e^{\ln x^3}$ $\left(\begin{array}{l}\text{definition of } e^x \text{ and a} \\ \text{property of logarithms}\end{array}\right)$

$= 2x^3.$

b) $e^{2x - \ln x} = e^{2x} \cdot e^{-\ln x}$ (from Eq. 13)

$= \dfrac{e^{2x}}{e^{\ln x}}$ (from Eq. 14)

$= \dfrac{e^{2x}}{x}.$ (definition of e^x) □

EXAMPLE 6 Solve the differential equation

$$\frac{dy}{dx} = 2xe^{-y}, \quad x > \sqrt{3} \tag{15}$$

subject to the condition that $y = 0$ when $x = 2$.

Solution We can separate the variables in Eq. (15) if we multiply both sides by e^y. We get

$$e^y \cdot \frac{dy}{dx} = e^y \cdot 2xe^{-y} = 2xe^{y-y} = 2xe^0 = 2x,$$

or

$$e^y \frac{dy}{dx} = 2x.$$

We now integrate both sides with respect to x to obtain

$$e^y = x^2 + C$$

and use the condition $y = 0$ when $x = 2$ to find

$$C = e^0 - 4 = 1 - 4 = -3.$$

Therefore,

$$e^y = x^2 - 3. \tag{17}$$

To solve this equation for y we take the logarithm of both sides:

$$\ln e^y = \ln (x^2 - 3),$$

$$y = \ln (x^2 - 3). \tag{18}$$

Note that the solution is valid for $x > \sqrt{3}$.

It is always a good idea to check the solution of a differential equation in the original equation. From Eq. (18) and then Eq. (17) we have

$$\frac{dy}{dx} = \frac{d}{dx} \ln (x^2 - 3) = \frac{2x}{x^2 - 3} = \frac{2x}{e^y} = 2xe^{-y}.$$

↑ Eq. (18) ↑ Eq. (17)

The function y and its derivative dy/dx therefore satisfy Eq. (15). □

Properties of $y = e^x$

1. $y = e^x$ is the inverse of $y = \ln x$.

   Domain: The set of all real numbers, $-\infty < x < \infty$.

   Range: The set of all positive numbers, $y > 0$.

2. Its derivative is

$$\frac{d}{dx}(e^x) = e^x.$$

3. It is continuous (because it is differentiable) and is an increasing function of x.

4. If u is any differentiable function of x, then

$$\frac{d}{dx} e^u = e^u \frac{du}{dx} \quad \text{and} \quad \int e^u \, du = e^u + C.$$

5. $e^{x_1} \cdot e^{x_2} = e^{x_1 + x_2}$ and $e^{-x} = 1/e^x$

Standard approximations of e^x, $x \approx 0$

Linear: $e^x \approx 1 + x$

Quadratic: $e^x \approx 1 + x + \frac{x^2}{2}$

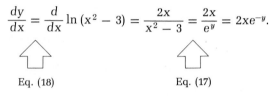

NOTE. The linear and quadratic approximations of e^x near $x = 0$ are the subject of Problem 92. The computation of values of e^x will be treated in more detail in Chapter 12.

A table of e^x and e^{-x} is included at the end of the book for convenience and to give an appreciation of the rapid rate of increase of the exponential function for positive values of x. Scientific calculators give ln x and e^x, as well as other elementary functions.

PROBLEMS

Simplify the expressions in Problems 1–14.

1. $e^{\ln x}$ **2.** $\ln (e^x)$ **3.** $e^{-\ln (x^2)}$

4. $\ln (e^{-x^2})$ **5.** $\ln (e^{1/x})$ **6.** $\ln (1/e^x)$

7. $e^{\ln (1/x)}$ **8.** $e^{-\ln (1/x)}$ **9.** $e^{\ln 2+\ln x}$

10. $e^{2\ln x}$ **11.** $\ln (e^{x-x^2})$ **12.** $\ln (x^2e^{-2x})$

13. $e^{x+\ln x}$ **14.** $e^{\ln x-2\ln y}$

In Problems 15–20, solve for y.

15. $e^{\sqrt{y}} = x^2$

16. $e^{2y} = x^2$

17. $e^{x^2} \cdot e^{(2x+1)} = e^y$

18. $\ln (y - 1) = x + \ln x$

19. $\ln (y - 2) = \ln (\sin x) - x$

20. $\ln (y^2 - 1) - \ln (y + 1) = \sin x$

Find dy/dx in Problems 21–49.

21. $y = e^{3x}$ **22.** $y = e^{x+1}$

23. $y = e^{5-7x}$ **24.** $y = \cos e^x$

25. $y = x^2e^x$ **26.** $y = \sin e^{-x}$

27. $y = e^{\sin x}$ **28.** $y = e^{x^2} \cdot e^{-x}$

29. $y = \ln (3xe^{-x})$

30. $y = e^{2x}(2 \cos 3x + 3 \sin 3x)$

31. $y = \ln \dfrac{e^x}{1 + e^x}$ **32.** $y = \dfrac{1}{2}(e^x - e^{-x})$

33. $y = \dfrac{1}{2}(e^x + e^{-x})$ **34.** $y = \dfrac{e^x - e^{-x}}{e^x + e^{-x}}$

35. $y = e^{\sin^{-1} x}$ **36.** $y = (1 + 2x)e^{-2x}$

37. $y = (9x^2 - 6x + 2)e^{3x}$ **38.** $y = \dfrac{ax - 1}{a^2}e^{ax}$

39. $y = x^2e^{-x^2}$ **40.** $y = e^x \ln x$

41. $y = \tan^{-1} (e^x)$ **42.** $y = \sec^{-1} (e^{2x})$

43. $y = e^{1/x}$

44. $y = x^3e^{-2x} \cos 5x$
(*Hint:* Use logarithmic differentiation.)

45. $y = \displaystyle\int_0^{\ln x} \sin e^t\, dt, \quad x > 0$

46. $\ln y = x \sin x$

47. $\ln xy = e^{x+y}$

48. $e^{2x} = \sin (x + 3y)$

49. $\tan y = e^x + \ln x$

Evaluate the integrals in Problems 50–64.

50. $\displaystyle\int e^{2x}\, dx$ **51.** $\displaystyle\int xe^{x^2}\, dx$

52. $\displaystyle\int e^{\sin x} \cos x\, dx$ **53.** $\displaystyle\int e^{x/3}\, dx$

54. $\displaystyle\int e^{-x}\, dx$ **55.** $\displaystyle\int_0^2 e^{x/2}\, dx$

56. $\displaystyle\int_0^1 e^{\ln \sqrt{x}}\, dx$ **57.** $\displaystyle\int_0^1 \frac{dx}{e^x}$

58. $\displaystyle\int \frac{4\, dx}{e^{3x}}$ **59.** $\displaystyle\int_0^1 \frac{e^x\, dx}{1 + e^x}$

60. $\displaystyle\int \frac{e^x\, dx}{1 + 2e^x}$ **61.** $\displaystyle\int_e^{e^2} \frac{dx}{x \ln x}$

62. $\displaystyle\int \frac{(e^x - e^{-x})\, dx}{(e^x + e^{-x})}$

63. $\displaystyle\int_0^{\ln 2} \frac{e^x\, dx}{1 + e^{2x}}$ (*Hint:* Let $u = e^x$.)

64. $\displaystyle\int_1^4 \frac{e^{\sqrt{x}}\, dx}{\sqrt{x}}$ (*Hint:* Let $u = \sqrt{x}$.)

Use l'Hôpital's rule to evaluate the following limits.

65. $\displaystyle\lim_{h\to 0} \frac{e^h - (1 + h)}{h^2}$ **66.** $\displaystyle\lim_{x\to \infty} \frac{e^x}{x^5}$

67. $\displaystyle\lim_{x\to 0} \frac{\sin x}{e^x - 1}$ **68.** $\displaystyle\lim_{x\to \infty} \frac{x^2 + e^x}{x + e^x}$

69. $\displaystyle\lim_{x\to \infty} xe^{-x}$ **70.** $\displaystyle\lim_{x\to \infty} x^2e^x$

71. The limit

$$\lim_{x\to 0} \frac{e^x - 1}{x}$$

is the derivative of $y = e^x$ at some point. What point? Evaluate the limit.

Graph the functions in Problems 72–76, taking into account concavity, extreme values, and inflection points.

72. $y = x$, $y = e^x$, and $y = \ln x$ (in a common graph)

73. $y = (\ln x)/x$, $x > 0$ **74.** $y = (1 + t)e^{-t}$

75. $y = e^{-t} \cos t$, $0 \leq t \leq 2\pi$

76. $y = e^{-x^2}$

77. (*Calculator*) To estimate the value of e, apply Newton's method (Article 2.9) to find a root of the equation $f(x) = \ln x - 1 = 0$. Start with $x_0 = 3$, and calculate x_3. Compare your result with $e = 2.718281828$.

78. Show that the tangent to the curve $y = e^x$ at the point $P(x, y)$ and the perpendicular from P to the x-axis always intersect the axis at points that are one unit apart (Fig. 6.22). This is another way to show how rapidly the graph of $y = e^x$ climbs.

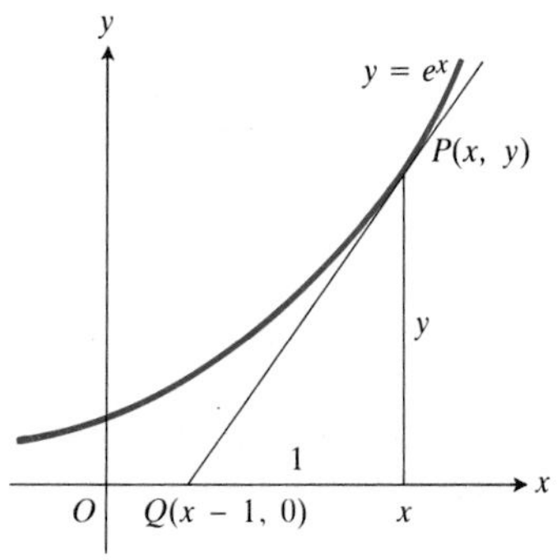

6.22 Line *QP* is tangent to the curve at *P* (Problem 78).

79. a) Find all maxima and minima (relative and absolute) of the function

$$y = e^{\sin x}, \qquad -\pi \leq x \leq 2\pi.$$

b) Graph the function.

80. Find the absolute maximum and minimum values of

$$f(x) = e^x - 2x$$

on the interval [0, 1].

81. Find the maximum value of $f(x) = x^2 \ln (1/x)$.

82. The graph of $y = (x - 3)^2 e^x$ has a horizontal tangent at the point $P(3, 0)$. Does y have a relative extreme value at $x = 3$, or is P a point of inflection?

83. The region between the curve $y = e^{-x}$ and the x-axis from $x = 1$ to $x = \ln 10$ is revolved about the x-axis. Find the volume thus swept out.

84. Find the length of the curve $x = e^t \sin t$, $y = e^t \cos t$, $0 \leq t \leq \pi$.

85. a) Show that $y = Ce^{ax}$ is a solution of the differential equation $dy/dx = ay$ for any choice of the constant C.

b) Using the result of (a), find a solution of the differential equation $dy/dt = -2y$ satisfying the initial condition $y = 3$ when $t = 0$.

86. Find the value of the constant r for which $y = e^{rx}$ is a solution of the differential equation

$$y'' - 4y' + 4y = 0.$$

In Problems 87–90, solve the differential equations subject to the given initial conditions.

87. $\dfrac{dy}{dx} = e^{-x}$, $\quad y = 0$ when $x = 4$

88. $e^y \dfrac{dy}{dx} = 2xe^{x^2-1}$, $\quad y = 0$ when $x = 1$

89. $\dfrac{1}{y+1}\dfrac{dy}{dx} = \dfrac{1}{2x}$, $\quad x > 0$, $\quad y = 1$ when $x = 2$

90. $\dfrac{1}{y+1}\dfrac{dy}{dx} = \dfrac{1}{x^2}$, $\quad x > 0$, $\quad y = 0$ when $x = 1$

91. A point $P(x, y)$ moves in the plane in such a way that

$$\frac{dx}{dt} = \frac{1}{t+2} \qquad \text{and} \qquad \frac{dy}{dt} = 2t$$

for $t \geq 0$.

a) Express x and y as functions of t if $x = \ln 2$ and $y = 1$ when $t = 0$.

b) Express y in terms of x.

c) Express x in terms of y.

d) Find the average rate of change in y with respect to x as t varies from 0 to 2.

e) Find dy/dx when $t = 1$.

92. Find (a) the linear and (b) the quadratic approximation to $f(x) = e^x$ near $x = 0$.

93. Find the quadratic approximation of

$$\text{erf}(x) = \frac{2}{\sqrt{\pi}} \int_0^x e^{-t^2}\, dt$$

near $x = 0$.

94. If a particle moves along the x-axis so that its position at time t is given by $x = ae^{\omega t} + be^{-\omega t}$, where a, b, ω, are constants, show that it is repelled from the origin with a force proportional to the displacement. (Assume that force = mass × acceleration.)

Problems 95–98 deal with hyperbolic functions. These are treated in greater detail in Chapter 9.

Certain combinations of e^x and e^{-x} behave in some respects like the trigonometric functions (and quite unlike the trigonometric functions in other ways). The hyperbolic cosine of x, written cosh x, and hyperbolic sine of x, written sinh x, are defined by the relations:

$$\cosh x = \frac{1}{2}(e^x + e^{-x}), \qquad \sinh x = \frac{1}{2}(e^x - e^{-x}).$$

Using these definitions, prove the following:

95. $\frac{d}{dx}(\cosh x) = \sinh x, \quad \frac{d}{dx}(\sinh x) = \cosh x$

96. $\cosh^2 x - \sinh^2 x = 1, \ \cosh^2 x + \sinh^2 x = \cosh(2x)$

97. $\cosh(-x) = \cosh x, \quad \sinh(-x) = -\sinh x$

98. Using the tables of e^x and e^{-x} in the back of the book, sketch graphs of $y = \cosh x$ and $y = \sinh x$.

***Toolkit* programs**

Integral Evaluator	Root Finder
Name That Function	Super * Grapher
Parametric Equations	

6.7

The Functions a^x and a^u

The functions e^x and ln x give us the mathematics we need to define arbitrary powers of positive numbers.

If a is any positive number, then its natural logarithm ln a is defined, and we know that

$$a = e^{\ln a}, \tag{1}$$

(from the definition of e^x). In other words, ln a is the power to which the base e must be raised to give the number a.

Suppose now that we want to raise a to an arbitrary power x. How can we do so? This is equivalent to raising $e^{\ln a}$ to the power x, which we can do if we define

$$(e^{\ln a})^x = e^{x \ln a}. \tag{2}$$

The number $e^{x \ln a}$ is defined because e^u is defined for any number u. So, to define a^x, we think of a as $e^{\ln a}$ and apply the definition in Eq. (2):

$$a^x = (e^{\ln a})^x = e^{x \ln a}. \tag{3}$$

DEFINITION

Definition of a^x

If $a > 0$ and x is any number, then

$$a^x = e^{x \ln a}. \tag{4}$$

The definition in Eq. (4) is equivalent to saying that

$$\ln a^x = x \ln a \tag{5a}$$

(take the logarithm of both sides of Eq. 4). If x is a rational number n, then Eq. (5a) says that

$$\ln a^n = n \ln a, \tag{5b}$$

which is the same as Eq. (3) in Article 6.5. What we are considering in the present article is the case in which x may be any real number.

From Eqs. (4) and (5a) we can obtain the law of exponents that for any positive number a,

$$a^{xy} = (a^x)^y = (a^y)^x. \tag{6}$$

Thus,

$$e^{3 \ln 2} = (e^3)^{\ln 2} = (e^{\ln 2})^3 = (2)^3 = 8. \tag{7}$$

This law generalizes the integer exponent law

$$a^{mn} = (a^m)^n,$$

which says, for instance, that

$$x^6 = (x^2)^3 = (x^3)^2.$$

The steps in the derivation of Eq. (6) are laid out in Problem 48.

EXAMPLE 1 From Eq. (6) we have

$$e^{3 \ln x} = (e^{\ln x})^3 = x^3,$$

in agreement with the calculation

$$e^{3 \ln x} = e^{\ln x^3} = x^3$$

that was part of Example 5(a) in Article 6.6. □

Equation (4) is the basis for the algorithm used by some small calculators to compute y^x. Humans have no trouble computing $(-2)^3$, because we know that it is just $-(2^3)$, or -8. But suppose that we thought we should calculate the result using Eq. (4):

$$(-2)^3 = e^{3 \ln (-2)},$$

and we have learned that -2 is not in the domain of the function $\ln x$. We would do something similar to what the calculator does: flash a signal that signifies "error." The number a in Eq. (4) *must* be positive.

The Derivatives of a^x and a^u

To calculate the derivative of a^x with respect to x we differentiate both sides of Eq. (4), using the Chain Rule on the right. This gives

$$\begin{aligned}\frac{d}{dx} a^x &= \frac{d}{dx} e^{x \ln a} \\ &= e^{x \ln a} \cdot \frac{d}{dx} x \ln a \quad \text{(Chain Rule)} \\ &= e^{x \ln a} \cdot \ln a \\ &= a^x \ln a,\end{aligned}$$

because $\ln a$ is a constant in this calculation. Thus, we have the following formula:

Derivative of a^x

If $a > 0$, then

$$\frac{d}{dx} a^x = a^x \ln a. \tag{8}$$

Formula (8) shows why e is the most desirable base to use in calculus. When $a = e$, $\ln a = \ln e = 1$ and Eq. (8) above reduces to

$$\frac{d}{dx} e^x = e^x.$$

EXAMPLE 2 From Eq. (8) we have

a) $\dfrac{d}{dx} 3^x = 3^x \ln 3,$

b) $\dfrac{d}{dx} e^x = e^x \ln e = e^x \cdot 1 = e^x.$ □

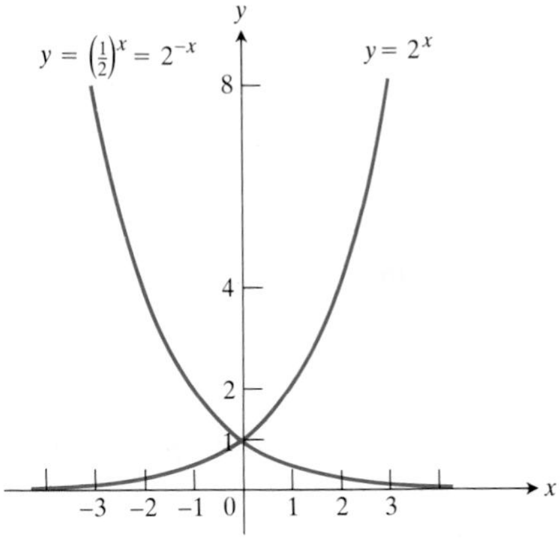

6.23 The functions $y = 2^x$ and $y = (\frac{1}{2})^x$ are both one-to-one.

From the formula

$$\frac{d}{dx} a^x = a^x \ln a, \qquad a > 0,$$

we can see that the derivative of $y = a^x$ is positive if $a > 1$ and negative if $0 < a < 1$. Thus, $y = a^x$ is an increasing function of x if $a > 1$, and a decreasing function of x if $0 < a < 1$. In either case, $y = a^x$ is one-to-one. It therefore has an inverse (which will be the subject of the next article). Figure 6.23 shows the graphs of $y = 2^x$ (increasing and one-to-one) and $y = (\frac{1}{2})^x$ (decreasing and one-to-one).

Combining Eq. (8) with the Chain Rule,

$$\frac{dy}{dx} = \frac{dy}{du}\frac{du}{dx},$$

gives a formula for the derivative of $y = a^u$ when u is any differentiable function of x:

$$\frac{d}{dx} a^u = \frac{d}{du} a^u \cdot \frac{du}{dx} = a^u \ln a \frac{du}{dx}.$$

Thus we have the following formula for the derivative of a^u.

Derivative of a^u

If $a > 0$ and u is any differentiable function of x, then

$$\frac{d}{dx} a^u = a^u \ln a \frac{du}{dx}. \qquad (9a)$$

EXAMPLE 3 From Eq. (9a) we have:

$$\frac{d}{dx} 3^{\sin x} = 3^{\sin x} \ln 3 \cdot \frac{d}{dx} \sin x = 3^{\sin x} \ln 3 \cos x. \ \square$$

In practice, there is little reason to memorize formula (9a) because the derivative of a^u can always be calculated by the method of logarithmic differentiation. The main reason for producing the formula here is to derive from it the companion integration formula. We shall come to that shortly.

EXAMPLE 4 Calculate the derivative of

$$y = 3^{\sin x} \tag{10}$$

by logarithmic differentiation.

Solution We take the natural logarithm of both sides of (10), differentiate, and solve for dy/dx:

$$\begin{aligned} y &= 3^{\sin x} \\ \ln y &= \ln 3^{\sin x} = \sin x \ln 3 \\ \frac{1}{y}\frac{dy}{dx} &= \ln 3 \cos x \\ \frac{dy}{dx} &= y \ln 3 \cos x = 3^{\sin x} \ln 3 \cos x. \end{aligned}$$

This agrees with the result in Example 3. □

When $a \neq 1$, so that $\ln a \neq 0$, the formula in (9a) can be rewritten in the form

$$a^u \frac{du}{dx} = \frac{1}{\ln a} \frac{d}{dx} a^u. \tag{9b}$$

Integrating both sides of this equation with respect to x then yields

$$\int a^u \frac{du}{dx}\, dx = \int \frac{1}{\ln a}\left(\frac{d}{dx} a^u\right) dx = \frac{1}{\ln a} \int \left(\frac{d}{dx} a^u\right) dx = \frac{1}{\ln a} a^u + C. \tag{9c}$$

The integral on the left in (9c) is usually written as

$$\int a^u \frac{du}{dx}\, dx = \int a^u\, du,$$

so that (9c) takes the following form:

$$\int a^u\, du = \frac{1}{\ln a} a^u + C, \quad a \neq 1,\ a > 0. \tag{9d}$$

EXAMPLE 5 From Eq. (9d) we have

a) $\int 2^x\, dx = \frac{1}{\ln 2} 2^x + C,$

b) $\int 2^{\sin x} \cos x\, dx = \frac{1}{\ln 2} 2^{\sin x} + C.$

To evaluate the first integral we apply Eq. (9d) with

$$a = 2, \qquad u = x, \qquad du = dx.$$

To evaluate the second integral we apply Eq. (9d) with

$$a = 2, \qquad u = \sin x, \qquad du = \cos x\, dx.$$

This gives

$$\int 2^{\sin x} \cos x\, dx = \int 2^u\, du = \frac{1}{\ln 2} 2^u + C = \frac{1}{\ln 2} 2^{\sin x} + C. \ \square$$

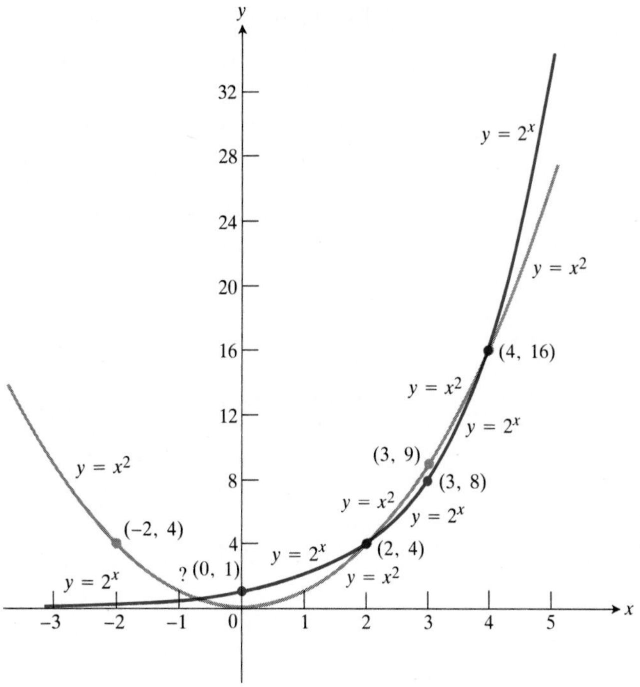

6.24 Where do the curves $y = x^2$ and $y = 2^x$ intersect? (At $x = 2$, $x = 4$, and $x = ?$)

EXAMPLE 6 Find where the curves $y = x^2$ and $y = 2^x$ intersect.

Solution It is easy to plot points on the two curves for $x = -3, -2, -1, 0, 1, 2, 3, 4, 5$, and then sketch a smooth curve for each of the two equations (Fig. 6.24). The curves obviously intersect at $x = 2$ and 4. There is a third intersection between -1 and 0. To find it, we put $f(x) = 2^x - x^2$ and use Newton's method to locate a root of $f(x) = 0$. Recall that the general procedure (Article 2.9) is to make an initial guess (we shall put $x_0 = -1$), and then use the formula

$$x_{n+1} = x_n - \frac{f(x_n)}{f'(x_n)}$$

to get other approximations. For $f(x) = 2^x - x^2$ we get

$$x' = x - \frac{2^x - x^2}{2^x \ln 2 - 2x}$$

where, for simplicity of notation, we have written x' in place of x_{n+1} on the left and x in place of x_n on the right. The successive approximations are:

$x_0 = -1$ $\quad x_4 = -0.766665$

$x_1 = -0.7869$ $\quad x_5 = -0.76666470$

$x_2 = -0.7668$ $\quad x_6 = -0.766664696$

$x_3 = -0.7667$ $\quad x_7 = -0.766664696$

$x_6 = x_7 \qquad (x_6)^2 = 0.587774756$

$2^{x_6} = 0.587774756$

We continued the iterations until we got nine-decimal agreement. For accuracy to three decimals, $x = -0.767$, and we have the answer after two iterations. □

Derivatives of Powers

Now that we can raise all positive numbers to arbitrary powers, we can extend the power rule for differentiation to its final form:

RULE

Power Rule

If n is any real constant and u is a positive differentiable function of x, then

$$\frac{d}{dx} u^n = n u^{n-1} \frac{du}{dx} \qquad \text{(any real number } n\text{).} \tag{11}$$

Proof of the power rule. Let $y = u^n$. Then

$$\ln y = \ln u^n = n \ln u,$$

$$\frac{1}{y}\frac{dy}{dx} = n\frac{1}{u}\frac{du}{dx},$$

$$\frac{dy}{dx} = n\frac{y}{u}\frac{du}{dx} = n\frac{u^n}{u}\frac{du}{dx}$$

$$= n u^{n-1}\frac{du}{dx}. \blacksquare$$

The ability to raise positive numbers to arbitrary powers also makes it possible to define functions like

$$x^x, \quad x^{\sin x}, \quad \text{and} \quad (1 + x)^{1/x}$$

for $x > 0$. The derivatives of functions like these can be found by logarithmic differentiation.

EXAMPLE 7 Find dy/dx if $y = x^x$, $x > 0$.

Solution

$$\ln y = \ln x^x = x \ln x,$$

$$\frac{1}{y}\frac{dy}{dx} = x \cdot \frac{1}{x} + \ln x = 1 + \ln x,$$

$$\frac{dy}{dx} = y(1 + \ln x) = x^x(1 + \ln x). \square$$

Other Functions

Logarithms can be used in calculating limits like the one in the next example. The idea, as you will see, is this: To calculate the limit of a positive function we calculate the limit of the logarithm of the function and then exponentiate that limit.

EXAMPLE 8 Show that

$$\lim_{n\to\infty}\left(1 + \frac{1}{n}\right)^n = \lim_{x\to 0^+}(1 + x)^{1/x} = e. \tag{12}$$

Solution We let

$$f(x) = (1 + x)^{1/x}$$

and work with the expression

$$\ln f(x) = \ln (1 + x)^{1/x} = \frac{1}{x} \ln (1 + x) = \frac{\ln (1 + x)}{x}.$$

We find from l'Hôpital's rule that

$$\lim_{x \to 0^+} \ln f(x) = \lim_{x \to 0^+} \frac{\ln (1 + x)}{x} = \lim_{x \to 0^+} \frac{\dfrac{1}{1 + x}}{1} = 1.$$

Therefore,

$$\ln f(x) \to 1 \quad \text{as} \quad x \to 0^+.$$

We can now exponentiate to see that as $x \to 0^+$,

$$(1 + x)^{1/x} = f(x) = e^{\ln f(x)} \to e^1 = e$$

(because the exponential function is continuous). That is,

$$\lim_{x \to 0^+} (1 + x)^{1/x} = e,$$

which is Eq. (12). □

REMARK. The limit

$$\lim_{n \to \infty} \left(1 + \frac{1}{n}\right)^n = \lim_{x \to 0^+} (1 + x)^{1/x} = e$$

in Eq. (12) gives a way to define e independently of the definition of the natural logarithm. The proof that this limit exists must of course be different from the proof in Example 8, which uses the logarithm.

Properties of $y = a^x$, $a > 0$, $a \neq 1$

If a is a positive real number and $a \neq 1$, then the function $y = a^x$ has the following properties.

1. It is defined by the equation

$$a^x = e^{x \ln a}.$$

Domain: the set of all real numbers, $-\infty < x < \infty$.

Range: the set of all positive real numbers, $y > 0$.

2. Its derivative is

$$\frac{d}{dx} a^x = a^x \ln a.$$

3. It is continuous (because it is differentiable), increasing if $a > 1$, decreasing if $0 < a < 1$, and one-to-one in either case.

4. If u is any differentiable function of x, then

$$\frac{d}{dx} a^u = a^u \ln a \frac{du}{dx} \quad \text{and} \quad \int a^u \, du = \frac{1}{\ln a} a^u + C.$$

PROBLEMS

In Problems 1–15, find dy/dx.

1. $y = 2^x$
2. $y = 2^{3x}$
3. $y = 8^x$
4. $y = 3^{2x}$
5. $y = 9^x$
6. $y = 2^x 3^x$
7. $y = x^{\sin x}, \quad x > 0$
8. $y = (\sin x)^{\tan x}, \quad \sin x > 0$
9. $y = 2^{\sec x}$
10. $y = x^{\ln x}, \quad x > 0$
11. $y = (\cos x)^x, \quad \cos x > 0$
12. $y = (1 - x)^x, \quad x < 1$
13. $y = x2^{(x^2)}$
14. $y = 2^x \ln x$
15. $y = (\cos x)^{\sqrt{x}}, \quad x > 0, \quad \cos x > 0$

Evaluate the integrals in Problems 16–29.

16. $\int 5^x\,dx$
17. $\int 3^{2x}\,dx$
18. $\int 2^{(x+1)}\,dx$
19. $\int 5^{-x}\,dx$
20. $\int e^{\sin x}\cos x\,dx$
21. $\int_0^1 \frac{1}{2^x}\,dx$
22. $\int 2^{\sin 3t}\cos 3t\,dt$
23. $\int_0^{\ln 2} e^{-2x}\,dx$
24. $\int_0^{1.2} 3^x\,dx$
25. $\int_1^{\sqrt{2}} x2^{-x^2}\,dx$
26. $\int_0^1 5^{2t-2}\,dt$
27. $\int_1^{e^2} \frac{2\ln x}{x}\,dx$
28. $\int 4^{(e^x)}e^x\,dx$
29. $\int_0^{\pi/6} (\cos\theta)4^{-\sin\theta}\,d\theta$

30. Which of the following integrals is larger?
    a) $\int_0^1 2^{(3x)}\,dx$
    b) $\int_0^1 3^{(2x)}\,dx$

31. Solve for x:
    a) $4^x = 2^x$
    b) $x^x = 2^x,\ x > 0$
    c) $3^x = 2^{x+1}$
    d) $4^{-x} = 3^{x+2}$

32. Calculate the derivative with respect to x of the following functions.
    a) $y = 2^{\ln x}$
    b) $y = \ln 2^x$
    c) $y = \ln x^2$
    d) $y = (\ln x)^2$

33. a) The function $y = 3^x$ is continuous because it is defined as a composite of continuous functions. What functions?
    b) Find $\lim_{x\to 3} 3^x$.

Find the limits in Problems 34–36.

34. $\lim_{x\to\infty} 2^{-x}$
35. $\lim_{x\to-\infty} 3^x$
36. $\lim_{x\to 0} \frac{3^{\sin x} - 1}{x}$

Use logarithms to find the limits in Problems 37–39.

37. $\lim_{x\to 1} x^{1/(x-1)}$
38. $\lim_{x\to 0} (e^x + x)^{1/x}$
39. $\lim_{x\to\infty} x^{1/x}$

40. Logarithms are no help in finding the following limits. Find them some other way.
    a) $\lim_{x\to\infty} \frac{3^x - 5}{4(3^x + 2)}$
    b) $\lim_{x\to-\infty} \frac{3^x - 5}{4(3^x + 2)}$

41. Let a be a number greater than one. Prove that the graph of $y = a^x$ has the following characteristics:
    a) If $x_1 > x_2$, then $a^{x_1} > a^{x_2}$.
    b) The graph is everywhere concave upward.
    c) The graph lies entirely above the x-axis.
    d) The slope at any point is proportional to the ordinate there, and the proportionality factor is the slope at the y-intercept of the graph.
    e) The curve approaches the negative x-axis as $x \to -\infty$.

42. Suppose that a and b are positive numbers such that $a^b = b^a$. (For example, $2^4 = 4^2$.)
    a) Show that $(\ln a)/a = (\ln b)/b$.
    b) Show that the graph of the function $f(x) = (\ln x)/x$ has a maximum at $x = e$, a point of inflection at $x = e^{3/2}$, approaches minus infinity as x approaches zero through positive values, and approaches zero as x approaches plus infinity. Sketch its graph.
    c) Use the results of parts (a) and (b) to show that
       i) if $0 < a \le 1$ and $a^b = b^a$, then $b = a$, and
       ii) if $1 < a < e$, or if $a > e$, then there is exactly one number $b \ne a$ such that $a^b = b^a$.

43. In Example 6, we found three zeros for the function $f(x) = 2^x - x^2$. Might there be others? One way to find out is to study $f'(x)$ to see how many zeros it has.
    a) Show that each zero of $f'(x)$ corresponds to an intersection of the graphs of $y = 2^x \ln 2$ and $y = 2x$. Further, show that the first of these is everywhere concave upward while the second is a straight line. Sketch their graphs.
    b) From the analysis of part (a), what can you conclude about the number of zeros of $f'(x)$? Of $f(x)$?
    c) (*Calculator*) If you have a calculator that can easily evaluate $2^x \ln 2$ and $2^x (\ln 2)^2$, find the zeros of $f'(x)$ to three-decimal accuracy. Compare your answers with what you might expect from Fig. 6.24.

44. (*Calculator*) Find out how close you can come to the limit

$$e = \lim_{n\to\infty} (1 + (1/n))^n \approx 2.718281828459045$$

with your calculator by taking $n = 10, 10^2, 10^3, \ldots$. You can expect the approximations to approach e at first, but after a while they will move away as round-off errors take their toll.

45. Find the maximum value of
a) $x^{1/x}$ for $x > 0$,
b) x^{1/x^2} for $x > 0$,
c) x^{1/x^n} for $x > 0$ and n a positive integer.

46. Show that

$$\lim_{x\to\infty} x^{1/x^n} = 1$$

if n is a positive integer.

47. Show that $(ab)^u = a^u b^u$ if a and b are any positive numbers and u is any real number.

48. The steps in the derivation of Eq. (6) are the following:

$$a^{xy} = e^{xy \ln a} \qquad \text{(a)}$$
$$= e^{y \cdot x \ln a} \qquad \text{(b)}$$
$$= e^{y \cdot \ln (a^x)} \qquad \text{(c)}$$
$$= (a^x)^y. \qquad \text{(d)}$$

Explain the roles that Eqs. (4) and (5a) play in these steps.

49. For what positive values of x does $x^{(x^x)} = (x^x)^x$?

***Toolkit* programs**

Function Evaluator
Root Finder
Super * Grapher

6.8 The Function $y = \log_a u$. Relative Growth Rates of Functions

From the previous article we know that when a is a positive number different from 1, the function $y = a^x$ is differentiable and one-to-one. It therefore has a differentiable inverse, which we call the *logarithm of x to the base a* and denote by

$$y = \log_a x. \qquad (1)$$

Since $y = a^x$ and $y = \log_a x$ are inverses of one another, their composite in either order is the identity function. Thus we obtain the equations

$$\log_a (a^x) = x \qquad \text{(for all } x\text{)}, \qquad (2a)$$

and

$$a^{(\log_a x)} = x \qquad \text{(for each positive } x\text{)}. \qquad (2b)$$

In words, Eq. (2b) says that the logarithm of x to the base a is the exponent to which we raise a to get x. For example,

$$\log_a (1) = 0 \quad \text{since} \quad a^0 = 1,$$
$$\log_a (a) = 1 \quad \text{since} \quad a^1 = a,$$
$$\log_5 25 = 2 \quad \text{since} \quad 5^2 = 25,$$
$$\log_2 (1/4) = -2 \quad \text{since} \quad 2^{-2} = 1/4.$$

In every case, the logarithm of x is the exponent to which the base is raised to give x. In the equation

$$2^{-2} = 1/4,$$

2 is the base, and -2, the logarithm of 1/4 to the base 2, is the power to which we raise 2 to get 1/4.

Figure 6.25 shows the graphs of the functions $y = 2^x$ and $y = \log_2 x$.

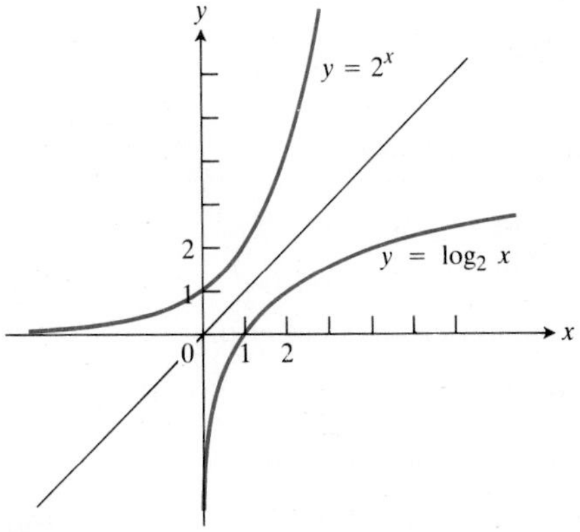

6.25 The graph of $y = 2^x$ and its inverse $y = \log_2 x$.

Calculation of $\log_a x$

The number $y = \log_a x$ can always be calculated from the natural logarithms of a and x by the following formula:

$$\log_a x = \frac{\ln x}{\ln a} \tag{3}$$

This formula can be derived in the following way. If

$$y = \log_a x,$$

then

$$a^y = x.$$

Therefore,

$$\ln a^y = \ln x,$$
$$y \ln a = \ln x,$$

and

$$y = \frac{\ln x}{\ln a},$$

which is Eq. (3).

EXAMPLE 1 From Eq. (3) we find

$$\log_2 (10) = \frac{\ln 10}{\ln 2} \approx \frac{2.30259}{0.69315} \approx 3.32193.$$

In other words,

$$10 \approx 2^{3.32193}. \ \square$$

EXAMPLE 2 Calculate

$$y = \log_4 8.$$

Solution *Method* 1. From Eq. (3) we have

$$y = \log_4 8 = \frac{\ln 8}{\ln 4} = \frac{\ln 2^3}{\ln 2^2} = \frac{3 \ln 2}{2 \ln 2} = \frac{3}{2}.$$

METHOD 2. In this case we can also calculate the value of

$$y = \log_4 8$$

directly from Eq. (2b) by finding the value of y for which

$$4^y = 2^{2y} = 8 = 2^3. \tag{4}$$

From Eq. (4) we read $2y = 3$, or $y = 3/2$. In other words,

$$\log_4 8 = 3/2 \qquad \text{because} \qquad 4^{3/2} = (\sqrt{4})^3 = 8. \ \square$$

Properties of $\log_a u$

The following properties follow readily from Eq. (3) above and the analogous properties of ln u.

$$\log_a uv = \log_a u + \log_a v \tag{5a}$$

$$\log_a \frac{u}{v} = \log_a u - \log_a v \tag{5b}$$

$$\log_a u^v = v \log_a u \tag{5c}$$

Equation (5c), for example, is derived as follows:

$$\log_a u^v = \frac{\ln u^v}{\ln a} = \frac{v \ln u}{\ln a} = v \log_a u,$$

where we make use of Eq. (3), then Eq. (5a) of Article 6.7, and then Eq. (3) again.

If u is a differentiable positive function of x, we may differentiate both sides of

$$\log_a u = \frac{\ln u}{\ln a}$$

to obtain

$$\frac{d}{dx} \log_a u = \frac{d}{dx} \frac{\ln u}{\ln a} = \frac{1}{\ln a} \frac{d}{dx} \ln u = \frac{1}{\ln a} \frac{1}{u} \frac{du}{dx} = \frac{1}{u \ln a} \frac{du}{dx}.$$

$$\frac{d}{dx} \log_a u = \frac{1}{u \ln a} \frac{du}{dx} \tag{6}$$

EXAMPLE 3 Calculate the derivative of

$$y = \log_{10} (3x + 1).$$

Solution From Eq. (6) with $a = 10$ and $u = 3x + 1$ we have

$$\frac{d}{dx} \log_{10} (3x + 1) = \frac{1}{(3x + 1) \ln 10} \frac{d}{dx} (3x + 1) = \frac{3}{(3x + 1) \ln 10}. \quad \square$$

REMARK ON NOTATION. Many advanced texts and research papers in mathematics use $\log x$, with no base specified, to represent the natural logarithm, $\ln x$. Most texts in the physical sciences use $\log x$ to represent $\log_{10} x$.

Base-10 logarithms are often called *common logarithms*.

Applications

Common logarithms are often used in scientific formulas. For example, earthquake intensity is reported on the Richter scale. Here the formula is

$$\text{Magnitude } R = \log_{10}\left(\frac{a}{T}\right) + B, \tag{7}$$

where a is the amplitude of the ground motion in microns at the receiving station, T is the period of the seismic wave in seconds, and B is an empirical factor that allows for weakening of the seismic wave with increasing distance from the epicenter of the earthquake. For an earthquake 10,000 km from the receiving station, $B = 6.8$. If the recorded vertical ground motion is $a = 10$ microns, and the period is $T = 1$ second, then the magnitude of the earthquake is

$$R = \log_{10}\left(\frac{10}{1}\right) + 6.8 = 7.8 \tag{8}$$

on the Richter scale. An earthquake of this magnitude does great damage near its epicenter. Damage begins at a magnitude of 5 and destruction is almost total at a magnitude of 8. The largest earthquake ever recorded had a magnitude of 8.5. The San Fernando, California, earthquake of 1971 (with a damage bill of billions of dollars) was only of magnitude 6.6.

Other examples of the use of common logarithms are the (logarithmic) decibel scale for measuring loudness, and the pH scale for measuring acidity. The pH value (hydrogen potential) of a solution is the common logarithm of the reciprocal of the solution's hydronium ion concentration, $[H_3O^+]$:

$$\text{pH} = \log_{10}\frac{1}{[H_3O^+]} = -\log_{10}[H_3O^+]. \tag{9}$$

The hydronium ion concentration is measured in moles per liter. Vinegar has a pH of 3, distilled water a pH of 7, sea water a pH of 8.15, and household ammonia a pH of 12. The total scale ranges from 1 to 14.

Relative Rates of Growth

Although it has not been discussed, you may have noticed that exponential functions like

$$y = 2^x \qquad \text{and} \qquad y = e^x$$

seem to grow much more rapidly as x increases than the polynomials and rational functions we graphed in Chapter 3. These exponentials certainly grow more rapidly than the function $y = x$, as Figs. 6.20 and 6.23 show, and you can see $y = 2^x$ beginning to outgrow $y = x^2$ as x increases in Fig. 6.24. In fact, as $x \to \infty$ the functions $y = 2^x$ and $y = e^x$ grow faster than any positive power of x, even $x^{1,000,000}$ (Problem 25).

To get a feeling for how rapidly the values of $y = e^x$ grow with increasing x, think of graphing the function on a large blackboard, with the axes scaled in centimeters. At $x = 1$ cm, the graph is $e^1 \approx 3$ cm above the x-axis. At $x = 6$ cm, the graph is $e^6 \approx 403$ cm ≈ 4 m high (it is about to go through the ceiling if it hasn't done so already). At $x = 10$ cm, the graph is $e^{10} \approx 22{,}026$ cm ≈ 220 m high, higher than most buildings. At $x = 24$ cm, the graph is more than halfway to the moon, and at $x = 43$ cm from the origin, the graph is high enough to reach past the nearest neighboring star, Proxima Centauri:

$$\begin{aligned} e^{43} &\approx 4.7 \times 10^{18}\ \text{cm} \\ &= 4.7 \times 10^{16}\ \text{m} \\ &= 4.7 \times 10^{13}\ \text{km} \\ &\approx 1.57 \times 10^{8}\ \text{light-seconds} \\ &\quad \text{(light travels at 300,000 km/s in a vacuum)} \\ &\approx 5.0\ \text{light years.} \end{aligned} \tag{10}$$

(The distance to Proxima Centauri is about 4.3 light years.) Yet, with $x = 43$ cm from the origin, the graph is still less than two feet to the right of the y-axis.

In contrast, logarithmic functions like

$$y = \log_2 x \qquad \text{and} \qquad y = \ln x$$

grow more slowly as $x \to \infty$ than any positive power of x (see Problem 27). With axes scaled in centimeters, you have to go more than 4 light years out on the x-axis to find a point where the graph of $y = \ln x$ is even $y = 43$ cm high.

These important comparisons of exponential, polynomial, and logarithmic functions can be made precise by defining what it means for a function $y = f(x)$ to grow faster than another function $y = g(x)$ as $x \to \infty$. We say that

f(x) grows faster than g(x) as $x \to \infty$ if

$$\lim_{x\to\infty} \frac{f(x)}{g(x)} = \infty, \tag{11a}$$

and

f(x) grows slower than g(x) as $x \to \infty$ if

$$\lim_{x\to\infty} \frac{f(x)}{g(x)} = 0. \tag{11b}$$

If

$$\lim_{x\to\infty} \frac{f(x)}{g(x)} = L \neq 0 \qquad (L \text{ a finite nonzero limit}), \tag{12}$$

then f and g are said to *grow at the same rate as* $x \to \infty$.

According to these definitions, $y = 2x$ does not grow faster than $y = x$. The two functions grow at the same rate because

$$\lim_{x\to\infty} \frac{2x}{x} = \lim_{x\to\infty} 2 = 2,$$

which is a finite nonzero limit. The reason for this apparent disregard of common sense is that we want "f grows faster than g" to mean that for large x-values, g is negligible when compared to f.

EXAMPLE 4 $y = e^x$ grows faster than $y = x^2$ because

$$\lim_{x\to\infty} \frac{e^x}{x^2} = \infty,$$

as we can see by two applications of l'Hôpital's rule:

$$\lim_{x\to\infty} \frac{e^x}{x^2} = \lim_{x\to\infty} \frac{e^x}{2x} = \lim_{x\to\infty} \frac{e^x}{2} = \infty. \ \square$$

EXAMPLE 5 $y = 3^x$ grows faster than $y = 2^x$ as $x \to \infty$ because

$$\lim_{x\to\infty} \frac{3^x}{2^x} = \lim_{x\to\infty} \left(\frac{3}{2}\right)^x = \infty.$$

Conversely, $y = 2^x$ grows slower than $y = 3^x$ because

$$\lim_{x\to\infty} \frac{2^x}{3^x} = \lim_{x\to\infty} \left(\frac{2}{3}\right)^x = 0. \ \square$$

If $f(x)$ grows faster than $g(x)$ as $x \to \infty$, then $g(x)$ grows slower than $f(x)$ as $x \to \infty$.

EXAMPLE 6 $y = x^2$ grows faster than $y = \ln x$ as $x \to \infty$ because

$$\lim_{x\to\infty} \frac{x^2}{\ln x} = \lim_{x\to\infty} \frac{2x}{1/x} = \lim_{x\to\infty} 2x^2 = \infty.$$

l'Hôpital's rule

Again, $y = \ln x$ grows slower than $y = x^2$ because

$$\lim_{x\to\infty} \frac{\ln x}{x^2} = \lim_{x\to\infty} \frac{1/x}{2x} = \lim_{x\to\infty} \frac{1}{2x^2} = 0. \quad \square$$

l'Hôpital's rule

EXAMPLE 7 Compare the rates of growth of

$$y = \log_2 x \qquad \text{and} \qquad y = \ln x$$

as $x \to \infty$.

Solution We take the ratio of the functions in either order (it does not matter which), and calculate the limit of the ratio as $x \to \infty$:

$$\lim_{x\to\infty} \frac{\log_2 x}{\ln x} = \lim_{x\to\infty} \frac{\ln x/\ln 2}{\ln x} = \frac{1}{\ln 2}.$$

Eq. (3)

The limit is finite and different from zero. These two logarithmic functions therefore grow at the same rate even though their bases are different. $\square$

As Example 7 may have suggested, any two logarithmic functions

$$y = \log_a x \qquad \text{and} \qquad y = \log_b x$$

grow at the same rate as $x \to \infty$. To see why, we have only to calculate the limit

$$\lim_{x\to\infty} \frac{\log_a x}{\log_b x} = \lim_{x\to\infty} \frac{\ln x/\ln a}{\ln x/\ln b} = \frac{\ln b}{\ln a},$$

to see that the limit is finite and different from zero.

In contrast to the way logarithmic functions behave, two different exponential functions

$$y = a^x \qquad \text{and} \qquad y = b^x$$

will grow at different rates as $x \to \infty$, as we can see from the calculation

$$\lim_{x\to\infty} \frac{a^x}{b^x} = \lim_{x\to\infty} \left(\frac{a}{b}\right)^x = \begin{cases} \infty & \text{if } a > b \\ 0 & \text{if } a < b. \end{cases}$$

If $a > b$, then a^x grows faster than b^x.

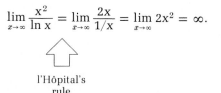

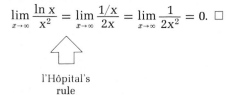

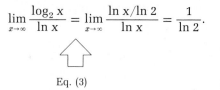

Brief Summary of the Definitions in This Chapter

$$\ln x = \int_1^x \frac{1}{t}\,dt, \qquad x > 0$$

⇩

e^x = inverse of $\ln x$

⇩

$a^x = e^{(\ln a)x}$ (any positive number a)

⇩

$\log_a x$ = inverse of a^x

PROBLEMS

In Problems 1 and 2, determine the given logarithms.

1. a) $\log_4 16$ b) $\log_8 32$ c) $\log_5 0.04$ d) $\log_{0.5} 4$

2. a) $\log_2 4$ b) $\log_4 2$ c) $\log_8 16$ d) $\log_{32} 4$

3. Find x if $3^{\log_3 7} + 2^{\log_2 5} = 5^{\log_5 x}$.

Find dy/dx in Problems 4–15.

4. $y = \log_4 x$

5. $y = \log_4 x^2$

6. $y = \log_{10} e^x$

7. $y = \log_5 \sqrt{x}$

8. $y = \ln 2 \cdot \log_2 x$

9. $y = \log_2 (1/x)$

10. $y = 1/\log_2 x$

11. $y = \ln 10^x$

12. $y = \log_5 (x + 1)$

13. $y = \log_2 (\ln x)$

14. $y = \log_7 (\sin x)$, $\sin x > 0$

15. $y = e^{\log_{10} x}$

16. Which function is changing faster at $x = 10$,

$$f_1(x) = \ln x \quad \text{or} \quad f_2(x) = \log_2 x?$$

17. Given $\ln 2 = 0.69315$, $\ln 10 = 2.30259$, and $\log_{10} 2 = 0.30103$. Find
a) $\log_{10} 20$, $\log_{10} 200$, $\log_{10} 0.2$, $\log_{10} 0.02$;
b) $\ln 20$, $\ln 200$, $\ln 0.2$, $\ln 0.02$.

18. Show that $\log_b u = \log_a u \cdot \log_b a$ if a, b, and u are positive numbers, $a \neq 1$, $b \neq 1$.

19. Find $\lim_{x\to\infty} \dfrac{\log_2 x}{\log_3 x}$.

20. Which of the following functions grow slower than $y = e^x$ as $x \to \infty$?
a) $y = x + 3$ b) $y = x^3 - 3x + 1$
c) $y = \sqrt{x}$ d) $y = 4^x$
e) $y = (5/2)^x$ f) $y = \ln x$
g) $y = \log_{10} x$ h) $y = e^{-x}$
i) $y = e^{x+1}$ j) $y = (\frac{1}{2})e^x$

21. Which of the following functions grow faster than $y = x^2 - 1$ as $x \to \infty$? Which grow at the same rate as $y = x^2 - 1$? Which grow slower?
a) $y = x^2 + 4x$ b) $y = x^3 + 3$
c) $y = x^5$ d) $y = 15x + 3$
e) $y = \sqrt{x^4 + 5x}$ f) $y = (x + 1)^2$
g) $y = \ln x$ h) $y = \ln (x^2)$
i) $y = \ln (10^x)$ j) $y = 2^x$

22. Which of the following functions grow at the same rate as $y = \ln x$ as $x \to \infty$?
a) $y = \log_3 x$ b) $y = \log_2 x^2$
c) $y = \log_{10} \sqrt{x}$ d) $y = 1/x$
e) $y = 1/\sqrt{x}$ f) $y = e^{-x}$
g) $y = x$ h) $y = 5 \ln x$
i) $y = 2$ j) $y = \sin x$

23. Order the following functions from fastest growing to slowest growing as $x \to \infty$:
a) e^x, b) x^x, c) $(\ln x)^x$, d) $e^{x/2}$.

24. Compare the growth rates as $x \to \infty$ of $y = \ln x$ and $y = \ln (\ln x)$.

25. Show that $y = e^x$ grows faster as $x \to \infty$ than $y = x^n$ for any positive integer n (even $x^{1,000,000}$). (*Hint:* What is the nth derivative of x^n?)

26. What do the conclusions about the limits in Article 1.10, Problem 62, imply about the relative growth rates of polynomials as $x \to \infty$?

27. Show that $y = \ln x$ grows slower as $x \to \infty$ than $y = x^{1/n}$ for any positive number n.

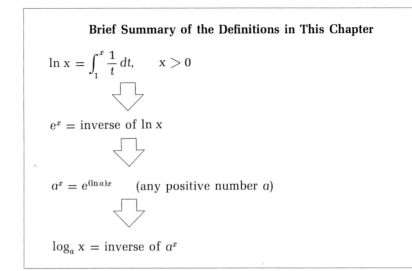

where k is a constant that is positive if the population is increasing and negative if it is decreasing. To solve Eq. (8) we divide through by y and integrate with respect to t, to obtain

$$\int \frac{1}{y}\frac{dy}{dt}\,dt = \int k\,dt,$$

$$\ln y = kt + C. \tag{9}$$

We have omitted the usual absolute value bars on the left because y is positive.

It follows from (9) that

$$y = e^{kt+C} = e^{kt}\cdot e^{C}$$
$$= Ae^{kt},$$

where $A = e^C$. If y_0 denotes the population when $t = 0$, then $A = y_0$ and

$$\boxed{y = y_0 e^{kt}.} \tag{10}$$

This equation is called the *law of exponential growth*, although "law of exponential change" might be more appropriate.

EXAMPLE 2 *The half-life of a radioactive element.* The decay of a radioactive element can be described by Eq. (10) because the decay rate is proportional to the number of radioactive nuclei present. The *half-life* of a radioactive element is the time required for half of the radioactive nuclei originally present in any sample to decay. Show that the half-life of a radioactive element is a constant that does not depend on the number of radioactive nuclei initially present in the sample.

Solution Equation (10) says that after time t has passed the number of radioactive nuclei present is

$$y = y_0 e^{kt},$$

where y_0 is the number originally present. We seek the value of t at which

$$y_0 e^{kt} = \frac{1}{2}y_0,$$

for this will be the time when the number of radioactive nuclei present equals half the original number. The y_0's cancel in this equation to leave

$$e^{kt} = \frac{1}{2},$$

$$kt = \ln\frac{1}{2} = -\ln 2,$$

$$t = -\frac{\ln 2}{k}. \tag{11}$$

This value of t is the half-life of the radioactive element. As Eq. (11) shows, it depends only on the value of k (the so-called decay constant of the element), and not on the number y_0 of nuclei originally present. □

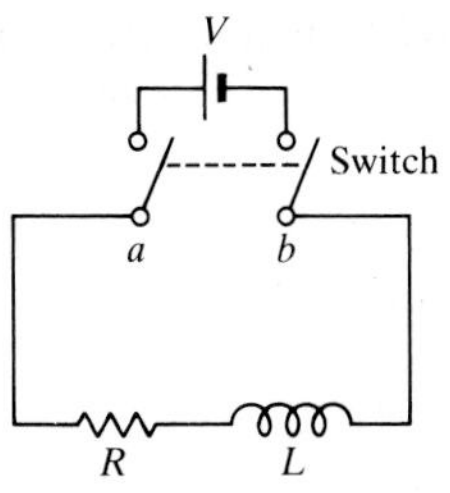

6.26 The *R–L* circuit.

The *R–L* Circuit

The diagram in Fig. 6.26 represents an electrical series circuit whose total resistance is R ohms and whose self-inductance, shown schematically as a coil, is L Henries (hence the name "R–L circuit"). There is also a switch whose terminals at a and b can be closed to connect a constant electrical source of V volts.

Ohms law, $V = RI$, has to be modified for such a circuit. The modified form is

$$L\frac{di}{dt} + Ri = V, \tag{12}$$

where i is the current in amperes, and t is the time in seconds. (See Sears, Zemansky, and Young, *University Physics*, Sixth edition (1982), Chapter 34.) From this equation it is possible to predict how the current will flow after the switch is closed.

EXAMPLE 3 Find an equation that shows how the current in the R–L circuit (Fig. 6.26) will flow as a function of time.

Solution We assume that the switch in Fig. 6.26 is closed at time $t = 0$ and solve Eq. (12) under the assumption that

$$i = 0 \quad \text{when} \quad t = 0. \tag{13}$$

The first step is to rewrite Eq. (12) as

$$L\frac{di}{dt} = V - Ri.$$

We then divide through by $V - Ri$ to obtain

$$\frac{L}{V - Ri}\frac{di}{dt} = 1$$

and integrate with respect to t:

$$\int \frac{L}{V - Ri}\frac{di}{dt}\,dt = \int dt = t + C. \tag{14}$$

To evaluate the integral on the left we substitute

$$u = V - Ri,$$

so that

$$du = -R\frac{di}{dt}\,dt,$$

$$-\frac{1}{R}\,du = \frac{di}{dt}dt, \tag{15}$$

and (14) becomes

$$\int \frac{L}{u} \cdot -\frac{1}{R}\,du = t + C$$

$$-\frac{L}{R}\int \frac{du}{u} = t + C. \tag{16}$$

From the initial conditions (13) we see that $u = V - Ri$ is positive when

$t = 0$. We therefore assume it remains positive, so that $|u| = u$ and the integral in (16) gives

$$-\frac{L}{R}\ln u = t + C, \tag{17}$$

or

$$-\frac{L}{R}\ln (V - Ri) = t + C. \tag{18}$$

The condition $i = 0$ when $t = 0$ enables us to evaluate the constant of integration:

$$-\frac{L}{R}\ln (V - 0) = C, \qquad \text{or} \qquad C = -\frac{L}{R}\ln V.$$

When we substitute this back into (18) and transpose it to the left side, we obtain

$$\frac{L}{R}[\ln V - \ln (V - Ri)] = t,$$

$$\ln \frac{V}{V - Ri} = \frac{Rt}{L},$$

$$\frac{V}{V - Ri} = e^{Rt/L},$$

$$\frac{V - Ri}{V} = e^{-Rt/L},$$

$$1 - \frac{R}{V}i = e^{-Rt/L},$$

or

$$i = \frac{V}{R}(1 - e^{-Rt/L}). \tag{19}$$

We see from this that the current (for $t > 0$) is always less than V/R but that it approaches the *steady state value*

$$I = \lim_{t\to\infty} \frac{V}{R}(1 - e^{-Rt/L}) = \frac{V}{R}(1 - 0) = \frac{V}{R}. \tag{20}$$

The current $I = V/R$ is the current that will flow in the circuit if either $L = 0$ (no inductance) or $di/dt = 0$ (steady current, i = constant). The graph of the current-versus-time relation (19) is shown in Fig. 6.27(a). □

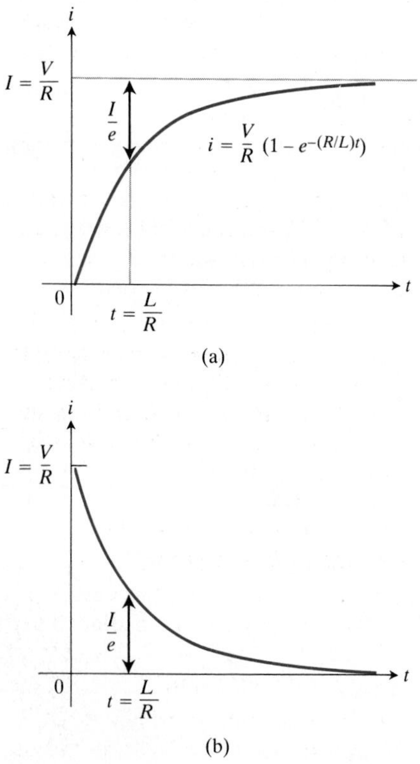

6.27 (a) (R–L circuit, solution of Example 3.) Growth of current in a circuit containing inductance and resistance. I is the current's steady-state value.
(b) (Problem 15.) Decay of current in a circuit containing inductance and resistance.

Newton's law of cooling:	$T - T_s = Ae^{kt}$
Law of exponential growth:	$y = y_0e^{kt}$
R–L circuit equations:	$L\frac{di}{dt} + Ri = V$
	$i = \frac{V}{R}(1 - e^{-Rt/L})$

PROBLEMS

1. *Cooling cocoa.* Suppose that a cup of cocoa cooled from 90 °C to 60 °C after 10 minutes in a room whose temperature was 20 °C.
   a) How much longer would it take the cocoa to cool to 35 °C?
   b) Instead of being left to stand in the room, the cup of 90 °C cocoa is put in a freezer whose temperature is −15 °C. How long will it take the cocoa to cool from 90 °C to 35 °C?
2. *Body of unknown temperature.* A body at an unknown temperature is placed in a room that is held at 30 °F. After 10 minutes the temperature of the body is 0 °F, and after 20 minutes the temperature of the body is 15 °F. What was the body's initial temperature?
3. *Surrounding medium of unknown temperature.* A pan of warm water (46 °C) was put in a refrigerator. Ten minutes later the water temperature was 39 °C, and ten minutes after that it was 33 °C. Assuming Newton's law of cooling, how cold was the refrigerator?
4. *Atmospheric pressure.* The earth's atmospheric pressure p is often modeled by assuming that the rate dp/dh at which p changes with the altitude h above sea level is proportional to h. Suppose that the pressure at sea level is 1013 millibars (a millibar is 1000 dynes per square centimeter, so 1013 millibars is about 14.7 pounds per square inch) and that the pressure at an altitude of 20 km is 50 millibars.
   a) Solve the equation $dp/dh = kh$ (k a constant) to express p in terms of h. Determine the values of k and the constant of integration from the given initial conditions.
   b) What is the atmospheric pressure at $h = 50$ km?
   c) At what altitude is the pressure equal to 900 millibars?
5. *Growth of bacteria.* Suppose that the bacteria in a colony can grow unchecked, by the law of exponential growth. The colony starts with one bacterium, and doubles every half hour. How many bacteria will the colony contain at the end of 24 hours? (Under favorable laboratory conditions, the number of cholera bacteria can double every 30 minutes. Of course, in an infected person, many of the bacteria are destroyed, but this example helps to explain why a person who feels well in the morning may be dangerously ill by evening.)
6. *Discharging capacitor.* As a result of leakage, an electrical capacitor discharges at a rate proportional to the charge. If the charge Q has the value Q_0 at the time $t = 0$, find Q as a function of t.
7. *Carbon-14 dating.* The half-lives of radioactive elements can sometimes be used to date events from the earth's past. The ages of rocks more than 2 billion years old have been measured by the extent of the radioactive decay of uranium (half-life 4.5 billion years!). In a living organism, the ratio of radioactive carbon, carbon-14, to ordinary carbon stays fairly constant during the lifetime of the organism, being approximately equal to the ratio in the organism's surroundings at the time. After the organism's death, however, no new carbon is ingested, and the proportion of carbon-14 in the organism's remains decreases as the carbon-14 decays. Since the half-life of carbon-14 is known to be about 5700 years, it is possible to estimate the age of organic remains by comparing the proportion of carbon-14 they contain with the proportion assumed to have been in the organism's environment at the time it lived. Archeologists have dated shells (which contain $CaCO_3$), seeds, and wooden artifacts this way. The estimate of 15,500 years for the age of the cave paintings at Lascaux, France, is based on carbon-14 dating.
   a) Find k in Eq. (10) for carbon-14.
   b) What is the age of a sample of charcoal in which 90 percent of the carbon-14 has decayed?
   c) The charcoal from a tree killed in the volcanic eruption that formed Crater Lake in Oregon contained $44\frac{1}{2}$ percent of the carbon-14 found in living matter. About how old is Crater Lake?
8. To see the effect of a relatively small error in the estimate of the amount of carbon-14 in a sample being dated, consider this hypothetical situation:
   a) A fossilized bone found in central Illinois in the year 2000 AD is found to contain 17% of its original carbon-14 content. Estimate the year the animal died.
   b) Repeat (a) assuming 18% instead of 17%.
   c) Repeat (a) assuming 16% instead of 17%.
9. The half-life of Polonium is 140 days, but your sample will not be useful to you after its radioactivity has fallen below 90% of its original level. About how many days can you use the Polonium?
10. *Blood sugar.* If glucose is fed intravenously at a constant rate, the change in the overall concentration $c(t)$ of glucose in the blood with respect to time may be described by the differential equation

$$\frac{dc}{dt} = \frac{G}{100V} - kc.$$

In this equation, G, V, and k are positive constants, G being the rate at which glucose is admitted, in milli-

grams per minute, and V the volume of blood in the body, in liters (around 5 liters for an adult). The concentration $c(t)$ is measured in milligrams per centiliter. The term $-kc$ is included because the glucose is assumed to be changing continually into other molecules at a rate proportional to its concentration.
a) Solve the equation for $c(t)$, using c_0 to denote $c(0)$.
b) Find the steady state concentration, $\lim_{t\to\infty} c(t)$.

11. *The incidence of a disease.* In the course of any given year, the number y of cases of a disease is reduced by 10%. If there are 10,000 cases today, about how many years will it take to reduce the number of cases to less than a thousand?

12. *Resistance proportional to velocity.* Suppose a body of mass m moving in a straight line with velocity v encounters a resistance proportional to the velocity and this is the only force acting on the body. If the body starts with velocity v_0, how far does it travel in time t? Assume $F = d(mv)/dt$.

13. *The R–L circuit.* How many seconds after the switch in the circuit in Example 3 is closed will it take the current to reach half of its steady-state value? Note that this time does not depend on V. Compare your answer with Eq. (11).

14. What will be the current in the R–L circuit when $t = L/R$ seconds? (The number L/R is called the *time constant* of the circuit.)

15. If there is a steady current in an R–L circuit and the switch in Fig. 6.26 is thrown open, the decaying current obeys the equation

$$L\frac{di}{dt} + Ri = 0$$

(Eq. 12 with $V = 0$).
a) Solve this equation for i.
b) How long after the switch is thrown will it take the current to decay to half its original value?
c) What is the value of the current when $t = L/R$? (This value of t is called the *time constant* for the open circuit.) See Fig. 6.27(b).

***Toolkit* programs**

Antiderivatives and Direction Fields	Super * Grapher

6.10
Compound Interest and Benjamin Franklin's Will

We are all familiar with arithmetic progressions and geometric progressions. In an arithmetic progression

$$a, \quad a + b, \quad a + 2b, \quad a + 3b, \quad \ldots$$

each term is a fixed amount more (or less) than the term before it. In a geometric progression

$$a, \quad ar, \quad ar^2, \quad ar^3, \quad \ldots$$

each term is a fixed *multiple* of the term before it.

The way money on deposit in a savings account grows is like a geometric progression. For example, if a bank pays 6 percent annual interest and compounds interest quarterly (every three months), an amount A dollars at the start of a three-month interest period receives an additional

$$\frac{0.06}{4}A = \frac{r}{4}A \qquad (r = 0.06)$$

at the end of the interest period. Because the interest is added to the original amount A, the amount at the end of the period is

$$A + \frac{r}{4}A = \left(1 + \frac{r}{4}\right)A.$$

The amounts at the ends of n successive interest periods form a geometric progression

$$A, \quad \left(1 + \frac{r}{4}\right)A, \quad \left(1 + \frac{r}{4}\right)^2 A, \quad \ldots, \quad \left(1 + \frac{r}{4}\right)^n A.$$

In particular, at the end of t years the number of interest periods is $n = 4t$ and the amount is

$$A_t = \left(1 + \frac{r}{4}\right)^{4t} A_0 \qquad (4t = \text{an integer}), \tag{1}$$

where A_0 is the original amount.

EXAMPLE 1 A bank agrees to pay 6 percent interest compounded quarterly for the next eight years. How much should you deposit now to have \$1000 in the account eight years from now?

Solution We put $A_8 = 1000$ in Eq. (1), with $r = 0.06$ and $t = 8$:

$$1000 = A_0(1.015)^{32}$$

and solve for A_0:

$$A_0 = 1000(1.015)^{-32} = 620.9929 \quad \text{(to four decimals).}$$

Thus a deposit of \$621 would yield \$1000 in eight years if the interest is 6 percent compounded quarterly. □

More generally, suppose the rate r per year is fixed, but the number of compounding periods per year is k. Then Eq. (1) becomes

$$A_t = A_0\left(1 + \frac{r}{k}\right)^{kt} = A_0\left[\left(1 + \frac{r}{k}\right)^k\right]^t. \tag{2}$$

The expression

$$f(k) = \left(1 + \frac{r}{k}\right)^k \tag{3}$$

has a limit as k increases without bound. To find what that limit is, we substitute

$$x = \frac{r}{k}, \qquad k = \frac{r}{x}$$

in Eq. (3) to obtain

$$f(k) = \left(1 + \frac{r}{k}\right)^k = (1 + x)^{r/x} = [(1 + x)^{1/x}]^r. \tag{4}$$

Now, as $k \to \infty$, $x = r/k \to 0^+$. Therefore,

$$\lim_{k\to\infty} f(k) = \lim_{x\to 0^+} [(1 + x)^{1/x}]^r = e^r, \tag{4'}$$

as calculated in Example 8 of Article 6.7. Thus, from Eq. (2), as k increases without bound, we get

$$\lim A_t = A_0\, e^{rt}. \tag{5}$$

When we took the limit as k increases without bound in Eq. (2) above, we went to the extreme of *compounding the interest continuously*. The result is the function

$$A(t) = A_0 e^{rt}. \tag{6}$$

In this relation, we need not restrict t to be an integer—it can be any real number greater than or equal to zero. The derivative of this function is

$$\frac{dA}{dt} = rA_0 e^{rt} = rA. \tag{7}$$

If we let Δt be a small time increment and estimate the increase that takes place in A from t to $t + \Delta t$ by calculating the tangent line increment $\Delta A_{\tan} = rA\,\Delta t$, we obtain

$$\Delta A \approx rA\,\Delta t = A(r\,\Delta t). \tag{8}$$

The increase is jointly proportional to the amount A at time t and to the annual interest rate multiplied by Δt.

EXAMPLE 2 Suppose \$621 is deposited in an account with interest at the rate of 6 percent and the interest is compounded continuously for eight years. Find the amount at the time $t = 8$.

Solution The phrase "compounded continuously" corresponds mathematically to the limit we got in Eq. (5) as $k \to \infty$. Thus, we put $r = 0.06$, $t = 8$, and $A_0 = 621$ in Eq. (6), to obtain the answer

$$A(8) = 621e^{0.48} = 1003.5822 \quad \text{(to four decimals).}$$

The effect of continuous compounding, as compared with quarterly, has been an addition of \$3.58. A bank might decide it would be worth this additional amount to be able to advertise "we compound your money every second, night and day—better than that, we compound the interest continuously." □

Benjamin Franklin's Will The Franklin Technical Institute of Boston owes its existence to a provision in a codicil to the will of Benjamin Franklin. In part, it reads:

> I was born in Boston, New England and owe my first instruction in Literature to the free Grammar Schools established there: I have therefore already considered those schools in my Will. . . . I have considered that among Artisans good Apprentices are most likely to make good citizens . . . I wish to be useful even after my Death, if possible, in forming and advancing other young men that may be serviceable to their Country in both Boston and Philadelphia. To this end I devote Two thousand Pounds Sterling, which I give, one thousand thereof to the Inhabitants of the Town of Boston in Massachusetts, and the other thousand to the inhabitants of the City of Philadelphia, in Trust and for the Uses, Interests and Purposes hereinafter mentioned and declared.

Franklin's plan was to lend money to young apprentices at 5 percent interest with the provision that each borrower should pay each year

> . . . with the yearly Interest, one tenth part of the Principal, which sums of Principal and Interest shall be again let to fresh Borrowers.
>
> If this plan is executed and succeeds as projected without interruption for one hundred Years, the Sum will then be one hundred and thirty-one thousand Pounds of which I would have the Managers of the Donation to the Inhabitants of the Town of Boston, then lay out at their discretion one hundred thousand Pounds in Public Works. . . . The remaining thirty-one thousand Pounds, I would have continued to be let out on Interest in the manner above directed for another hundred Years. . . . At the end of this second term if no unfortunate accident has prevented the operation the sum will be Four Millions and Sixty-one Thousand Pounds.

It was not always possible to find as many borrowers as Franklin had planned, but the managers of the trust did the best they could; they lent money to medical students as well as others. At the end of one hundred years from the reception of the Franklin gift, January, 1894, the fund had grown from one thousand pounds to almost exactly ninety thousand pounds. In one hundred years the original capital had multiplied about ninety times instead of the one hundred and thirty-one times Franklin had anticipated.

EXAMPLE 3 What rate of interest, compounded continuously for one hundred years, would have multiplied Benjamin Franklin's original capital by 90?

Solution The mathematical model we assume is the one expressed by the equation

$$A = A_0 e^{rt},$$

where A_0 is the original capital, r is the interest rate per year (but compounded continuously), and t is the time in years. If we put $t = 100$ and $A = 90A_0$, we get

$$90A_0 = A_0 e^{100r},$$

which leads to

$$100r = \ln 90 = 4.50$$

or

$$r = 0.045 \quad \text{or} \quad 4.5 \text{ percent. } \square$$

PROBLEMS

1. Suppose that the GNP in a country is increasing at an annual rate of 4 percent. How many years, at that rate of growth, are required to double the GNP?

2. In Benjamin Franklin's estimate that the original one thousand pounds would grow to 131 thousand in 100 years, he was using an annual rate of 5 percent and compounding once each year. What rate of interest per year when compounded continuously for 100 years would multiply the original amount by 131?

3. *Rule of 70.* If you use the approximation $\ln 2 \approx 0.70$ (in place of 0.69315...), you can derive a rule of thumb that says "If a quantity grows at the rate of 5 percent per year compounded continuously over the next 70 years, then it will double in about $\frac{70}{5} = 14$ years, so it will be multiplied by $2^5 = 32$ during the 70-year span. If the annual rate is r percent, it will be multiplied by about 2^r during a life span of 70 years." Show how this is derived. (A similar "rule of 72" uses 72 instead of 70, probably because it has more integer factors.)

4. In an article about the rise of the U.S. Consumer Price Index on Friday, February 23, 1979, the *Batavia*

Daily News said "A one percent monthly increase would translate into a twelve percent annual increase." On the same day, an article about the Consumer Price Index in the *Buffalo Evening News* said "A one percent monthly increase would translate to a nearly thirteen percent annual increase." Which (if either) is correct?

5. John Napier, who invented natural logarithms, was the first person to answer the question, "What happens if you invest an amount of money at 100 percent interest, compounded continuously?"
a) What does happen?
b) How long does it take to triple your money?
c) How much can you earn in a year?

Toolkit programs
Sequences and Series

REVIEW QUESTIONS AND EXERCISES

1. Review the formulas for derivatives of sums, products, quotients, powers, sines, and cosines. Using these, develop formulas for derivatives of the other trigonometric functions.

2. Define the inverse sine, inverse cosine, inverse tangent, and inverse secant functions. What is the *domain* of each? What is its *range*? What is the derivative of each? Sketch the graph of each.

3. Define the natural logarithm function. What is its domain? What is its range? What is its derivative? Sketch its graph.

4. What is the inverse function of the natural logarithm called? What is the domain of this function? What is its range? What is its derivative? Sketch its graph.

5. Define the term "algebraic function." Define "transcendental function." To which class (algebraic or transcendental) do you think the greatest integer function (Fig. 1.58) belongs?

[It is often difficult to prove that a particular function belongs to one class or the other. Numbers are also classified as algebraic and transcendental. *References:* (1) W. J. Le Veque, *Topics in Number Theory* (Reading, Mass.: Addison-Wesley, 1956), Vol. II, Chap. 5, pp. 161–200; (2) I. M. Niven, *Irrational Numbers*, Carus monograph number 11 of the Mathematical Association of America, (1956). Both π and e are transcendental, and proofs may be found in the references cited. That π is transcendental was first proved by Ferdinand Lindemann in 1882. One result that follows from the transcendence of π is that it is impossible to construct, with unmarked straightedge and compass alone, a square whose area is equal to the area of a given circle. The first proof that e is transcendental was published by Charles Hermite in 1873.]

6. Let n be an integer greater than 1. Divide the interval $1 \le x \le n$ into n equal subintervals and write an expression for the trapezoidal approximation to the area under the graph of $y = 1/x$, $1 \le x \le n$, based on this subdivision. Is this approximation less than, equal to, or greater than $\ln n$? Give a reason for your answer.

7. Prove that $\lim_{x\to\infty} (\ln x)/x = 0$, starting with the definition of $\ln x$ as an integral.

8. How would you calculate $(8.73)^{2.75}$? (If you have a calculator with a y^x key, of course, you just put $y = 8.73$ and $x = 2.75$.) How would you calculate $(-8)^{1/3}$? (When this was tried on a calculator, the 'error' symbol came on. Why?)

9. What does it mean for a function $f(x)$ to grow faster than a function $g(x)$ as x approaches infinity? What does it mean for f to grow slower than g, or at the same rate as g? Give examples.

10. What is the meaning of the differential equation $dy/dt = ky$? What solution of this equation satisfies the initial condition $y = y_0$ when $t = 0$? Give examples of applications of this equation.

MISCELLANEOUS EXERCISES

In Problems 1–25, (a) find dy/dx, and (b) sketch the curve.

1. $y = \sin 2x$

2. $y = 4 \cos (2x + \pi/4)$

3. $y = 2 \sin x + \sin 2x$

4. $y = \dfrac{\sin x}{1 + \cos x}$

5. $x = \cos y$

6. $x = \tan y$

7. $y = 4 \sin\left(\dfrac{x}{2} + \pi\right)$

8. $y = 1 - \sin x$

9. $y = x - \sin x$

10. $y = \dfrac{\sin x}{x}$ $\left(\begin{array}{l}\text{What happens as } x \to 0?\\ \text{As } x \to \infty?\end{array}\right)$

11. $y = \dfrac{\sin x}{x^2}$ (What happens as $x \to 0$? As $x \to \infty$?)

12. $y = x \sin \dfrac{1}{x}$ (What happens as $x \to \infty$? As $x \to 0$?)

13. $y = x^2 \sin \dfrac{1}{x}$ (What happens as $x \to \infty$? As $x \to 0$?)

14. $x = \tan \dfrac{\pi y}{2}$

15. $y = \dfrac{e^x - e^{-x}}{e^x + e^{-x}}$

16. $y = xe^{-x}$

17. a) $y = x \ln x$, b) $y = \sqrt{x} \ln x$

18. a) $y = \dfrac{\ln x}{x}$ b) $y = \dfrac{\ln x}{\sqrt{x}}$

c) $y = \dfrac{\ln x}{x^2}$

19. $y = e^{-x} \sin 2x$

20. a) $y = 2x - \frac{1}{2}e^{2x}$ b) $y = e^{[2x-(1/2)e^{2x}]}$

21. a) $y = x - e^x$ b) $y = e^{(x-e^x)}$

22. $y = \ln (x + \sqrt{x^2 + 1})$

23. $y = \frac{1}{2} \ln \dfrac{1 + x}{1 - x}$

24. $y = \ln (x^2 + 4)$

25. $y = x \tan^{-1} \dfrac{x}{2}$

Find dy/dx in Problems 26–50.

26. $y = \dfrac{e^{2x} - e^{-2x}}{e^{2x} + e^{-2x}}$

27. $y = \ln \dfrac{\sec x + \tan x}{\sec x - \tan x}$

28. $y = x^2e^{2x} \sin 3x$

29. $y = \sin^{-1} (x^2) - xe^{x^2}$

30. $y = \ln \left(\dfrac{x^4}{1 + x^3}\right) + 7^{x^{2/3}}$

31. $y = (x^2 + 2)^{2-x}$

32. $y = \dfrac{\ln x}{e^x}$

33. $y = \ln \dfrac{x}{\sqrt{x^2 + 1}}$

34. $y = \ln (4 - 3x)$

35. $y = \ln (3x^2 + 4x)$

36. $y = x (\ln x)^3$

37. $y = x \ln (x^3)$

38. $y = x^3 \ln x$

39. $y = \ln e^x$

40. $y = x^2e^x$

41. $x = \ln y$

42. $y = \ln (\ln x)$

43. $y = \ln (2xe^{2x})$

44. $y = (\frac{1}{5})e^x (\sin 2x - 2 \cos 2x)$

45. $y = x^{\tan 3x}$

46. $y = e^{\sin^2 x}$

47. $y = e^{\sec x}$

48. $y = e^{\ln (\sin (e^x))}$

49. $\ln (x - y) = e^{xy}$, $x > y$

50. $y = \ln \dfrac{e^x}{1 + e^x}$

51. Find dy/dx by logarithmic differentiation for

a) $y = \dfrac{x}{x^2 + 1}$, $x > 0$; b) $y = \sqrt[3]{\dfrac{x(x - 2)}{x^2 + 1}}$, $x > 2$;

c) $y = \dfrac{(2x - 5)\sqrt{8x^2 + 1}}{(x^3 + 2)^2}$.

52. If $f(x) = x + e^{4x}$, find $f(0)$ and $f'(0)$, and find an approximation for $f(0.01)$.

53. Find dy/dx for each of the following.

a) $y = a^{x^2-x}$ b) $y = \ln \dfrac{e^x}{1 + e^x}$

c) $y = x \ln x - x$ d) $y = x^{1/x}$

54. Sketch the graphs of $y = \ln (1 - x)$ and $y = \ln (1/x)$.

55. Sketch (in a common figure) the graphs of $y = e^{-x}$ and $y = -e^{-x}$.

56. Solve for x: $\tan^{-1} x - \cot^{-1} x = \pi/4$.

57. If $\dfrac{dy}{dx} = \dfrac{e^x - e^{-x}}{e^x + e^{-x}}$, find $\dfrac{d^2y}{dx^2}$ and y.

58. If $dy/dx = 2/e^y$ and $y = 0$ when $x = 5$, find y as a function of x.

59. $P(x_1, y_1)$ and $Q(x_2, y_2)$ are any two points (in the first quadrant) lying on the hyperbola $xy = R$ (R positive). Show that the area bounded by the arc PQ, the lines $x = x_1$, $x = x_2$, and the x-axis is equal to the area bounded by the arc PQ, the lines $y = y_1$, $y = y_2$, and the y-axis.

60. By the trapezoidal rule, find $\int_0^2 e^{-x^2}\, dx$, using $n = 4$.

61. In a capacitor discharging electricity, the rate of change of the voltage in volts per second is proportional to the voltage, being numerically equal to minus one-fortieth of the voltage. Express the voltage as a function of the time. In how many seconds will the voltage decrease to 10 percent of its original value?

62. A curve $y = f(x)$ goes through the points (0, 0) and (x_1, y_1). It divides the rectangle $0 \le x \le x_1$, $0 \le y \le y_1$ into two regions: A above the curve and B below. Find the curve if the area of A is twice the area of B for all choices of $x_1 > 0$ and $y_1 > 0$.

63. Find a curve passing through the origin and such that the length s of the curve between the origin and any point (x, y) of the curve is given by

$$s = e^x + y - 1.$$

64. A particle starts at the origin and moves along the x-axis in such a way that its velocity at the point (x, 0) is given by the formula $dx/dt = \cos^2 \pi x$. How long will it take to cover the distance from the origin to the point $x = \frac{1}{4}$? Will it ever reach the point $x = \frac{1}{2}$? Why?

65. A particle moves in a straight line with acceleration $a = 4/(4 - t)^2$. If when $t = 0$ the velocity is equal to 2, find how far the particle moves between $t = 1$ and $t = 2$.

66. The velocity of a certain particle moving along the x-axis is proportional to x. At time $t = 0$ the particle is located at $x = 2$ and at time $t = 10$ it is at $x = 4$. Find its position at $t = 5$.

67. Solve the differential equation $dy/dx = y^2e^{-x}$ if $y = 2$ when $x = 0$.

68. It is estimated that the population of a certain country is now increasing at a rate of 2 percent per year. Assuming that this (instantaneous) rate will continue indefinitely, estimate what the population N will be t years from now, if the population now is N_0. How many years will it take for the population to double?

69. If $y = (e^{2x} - 1)/(e^{2x} + 1)$, show that $dy/dx = 1 - y^2$.

70. Find the area of the surface generated by revolving the curve

$$y = \frac{1}{2}(e^x + e^{-x}), \qquad 0 \le x \le \ln\sqrt{2},$$

about the x-axis.

71. Find the volume generated when the area bounded by $x = \sec y$, $x = 0$, $y = 0$, $y = \pi/3$, is rotated about the y-axis.

72. Find the length of the curve

$$\frac{x}{a} = \left(\frac{y}{b}\right)^2 - \frac{1}{8}\left(\frac{b^2}{a^2}\right)\ln\left(\frac{y}{b}\right)$$

from $y = b$ to $y = 3b$, assuming a and b to be positive constants.

73. Find the volume generated by rotating about the x-axis the area bounded by $y = e^x$, $y = 0$, $x = 0$, $x = 2$.

74. Find the area bounded by the curve

$$y = \left(\frac{a}{2}\right)(e^{x/a} + e^{-x/a}),$$

the x-axis, and the lines $x = -a$ and $x = +a$.

75. The portion of a tangent to a curve included between the x-axis and the point of tangency is bisected by the y-axis. If the curve passes through (1, 2), find its equation.

Evaluate the integrals in Problems 76–87.

76. $\displaystyle\int \frac{dx}{4 - 3x}$

77. $\displaystyle\int \frac{5\,dx}{x - 3}$

78. $\displaystyle\int_0^2 \frac{x\,dx}{x^2 + 2}$

79. $\displaystyle\int_0^2 \frac{x\,dx}{(x^2 + 2)^2}$

80. $\displaystyle\int \frac{x + 1}{x}\,dx$

81. $\displaystyle\int \frac{x}{2x + 1}\,dx$

82. $\displaystyle\int_0^1 \frac{x^2\,dx}{2 - x^3}$

83. $\displaystyle\int_0^3 x(e^{x^2-1})\,dx$

84. $\displaystyle\int_1^3 \frac{dx}{x}$

85. $\displaystyle\int_0^5 \frac{x\,dx}{x^2 + 1}$

86. $\displaystyle\int_0^1 (e^x + 1)\,dx$

87. $\displaystyle\int \frac{\sec^3 x + e^{\sin x}}{\sec x}\,dx$

88. Calculate $f'(2)$ if $f(x) = e^{g(x)}$ and

$$g(x) = \int_2^x \frac{t}{1 + t^4}\,dt.$$

Find the limits in Problems 89–91.

89. $\displaystyle\lim_{t\to 0} \frac{te^t}{4 - 4e^t}$

90. $\displaystyle\lim_{x\to 4} \frac{e^{x-4} + 4 - x}{\cos^2(\pi x)}$

91. $\displaystyle\lim_{x\to 0} (e^x + x)^{1/x}$

92. Prove by mathematical induction that

$$\frac{d^n}{dx^n}(xe^x) = (x + n)e^x$$

for any positive integer n.

93. Prove by l'Hôpital's rule and mathematical induction that

$$\lim_{x\to\infty} \frac{(\ln x)^n}{x} = 0$$

for any positive integer n.

94. A solar station is to be located at a point between two tall buildings, one of height h, the other of height $2h$. The buildings are a distance d apart. How far from the taller building should the station be placed to maximize the number of hours it will be in the sun?

95. Find the area under the curve $y = t\sin(t - 1)$, $x = \ln(t)$ from $x = 0$ to $x = \ln(\pi + 1)$.

96. The processing of raw sugar involves a step called "inversion," which changes the structure of the raw sugar molecules. After the process is begun, the time rate of change of the amount of raw sugar present varies as the amount of raw sugar remaining. If 1000 lb of raw sugar has been reduced to 800 lb after 10 hr, how much raw sugar will remain after 24 hr?

97. A cylindrical tank of radius 10 ft and height 20 ft, with its axis vertical, is full of water but has a leak at the bottom. Assuming that water escapes at a rate proportional to the depth of water in the tank and that 10 percent escapes during the first hour, find a formula for the volume of water left in the tank after t hr.

98. Let p be a positive integer ≥ 2. Show that

$$\lim_{n\to\infty}\left(\frac{1}{n + 1} + \frac{1}{n + 2} + \cdots + \frac{1}{p \cdot n}\right) = \ln p.$$

99. a) If $(\ln x)/x = (\ln 2)/2$, does it necessarily follow that $x = 2$?

b) If $(\ln x)/x = -2\ln 2$, does it necessarily follow that $x = \frac{1}{2}$? Give reasons for your answers.

100. Given that $\lim_{x\to\infty} (\ln x)/x = 0$ (see Article 6.4, Problem 47). Prove that

a) $\lim_{x\to\infty} (\ln x)/x^h = 0$ if h is any positive constant,

b) $\lim_{x\to+\infty} x^n/e^x = 0$ if n is any constant.

101. Show that $\lim_{h\to 0} (e^h - 1)/h = 1$, by considering the definition of the derivative of e^x at $x = 0$.

102. (*Calculator*) Prove that if x is any positive number, $\lim_{n\to\infty} n(\sqrt[n]{x} - 1) = \ln x$. (*Hint:* Take $x = e^{nh}$ and apply Problem 84.) *Remark.* This result provides a method for finding ln x (to any desired finite number of decimal places) using nothing fancier than repeated use of the operation of extracting square roots. For we may take $n = 2^k$, and then $\sqrt[n]{x}$ is obtained from x by taking k successive square roots.

103. Show that $(x^2/4) < x - \ln(1 + x) < (x^2/2)$ if

$$0 < x < 1.$$

(*Hint:* Let $f(x) = x - \ln(1 + x)$ and show that $(x/2) < f'(x) < x$.)

104. Prove that the area under the graph of $y = 1/x$ over the interval $a \le x \le b$ $(a > 0)$ is the same as the area over the interval $ka \le x \le kb$, for any $k > 0$.

105. Find the limit, as $n \to \infty$, of

$$\frac{e^{1/n} + e^{2/n} + \cdots + e^{(n-1)/n} + e^{n/n}}{n}.$$

106. Compare the growth rates of $y = e^{x^2}$ and $y = x^x$ and $x \to \infty$.

107. During World War II it was necessary to administer blood tests to large numbers of recruits. There are two standard ways to administer the test to N people: (1) Each person can be tested separately. (2) The blood samples of x people can be pooled and tested as one large sample. If the test is negative, this one test is sufficient for the x people. If the test is positive, then each of the x people must be tested separately, and a total of x + 1 tests are then required for the x people. Using the second method and some probability theory it can be shown that, on the average, the total number of tests y will be

$$y = N\left(1 - q^x + \frac{1}{x}\right).$$

With $q = 0.99$ and $N = 1000$, show that the integer value of x that minimizes y is $x = 11$ and that the integer value of x that maximizes y is $x = 895$. (The second result is not important to the real-life situation.) The group testing method (2) was used in World War II with a saving of 80 percent over the individual testing method (1); not with the given q, however.

7 Integration Methods

7.1 Basic Integration Formulas

Since indefinite integration is defined as the inverse of differentiation, evaluating an integral

$$\int f(x)\,dx \tag{1}$$

is equivalent to finding a function F such that $F'(x) = f(x)$ or, in differential notation, such that

$$dF(x) = f(x)\,dx. \tag{2}$$

At first this may seem like a hopeless task, especially if our first approach is by arbitrary trial and error. We would quickly realize that it is impossible to try *all functions* as F in Eq. (2) hoping to find one that does the job.

To reduce the amount of trial and error involved in integration, it is useful to build up a table of standard types of integral formulas by reversing formulas for differentials, as we have done in the previous chapters. Then we try to match any integral that confronts us against one of the standard types. This usually involves a certain amount of algebraic manipulation.

A table of integrals, more extensive than the list of basic formulas given here, may be found inside the covers of this book. To use such a table intelligently, it is necessary to become familiar with techniques for reducing a given integral to a form that matches an entry in the table. The examples and problems in this book should serve to develop skill with these techniques. To concentrate on the techniques without becoming entangled in a mass of algebra, these problems and examples have been kept fairly simple. In fact, *these particular problems* can frequently be solved immediately by consulting an integral table. But solving them this way right now would defeat the purpose of developing the skill likely to be needed for later work. And it is this skill that is important, rather than the specific answer to any given problem.

Success in integration often hinges on the ability to spot what part of the integrand should be called u in order that one will also have du, so that a known formula can be applied. This means that the *first require-*

ment for skill in integration is a thorough mastery of the formulas for differentiation. We therefore list certain formulas for differentials that are based on the derivative formulas in previous chapters, together with their integral counterparts. (See page 446.)

At present, let us just consider this list a handy reference and *not* a challenge to our memories! A bit of examination will probably convince us that we are already rather familiar with the first twelve formulas for *differentials* and, so far as *integration* is concerned, our training in technique will show us how to get the *integrals* in the last six cases without memorizing the formulas for the differentials of the inverse trigonometric functions.

Just what types of functions can we integrate directly with this short table of integrals?

Powers

$$\int u^n\,du, \qquad \int \frac{du}{u}.$$

Exponentials

$$\int e^u\,du, \qquad \int a^u\,du.$$

Trigonometric functions

$$\int \sin u\,du, \qquad \int \cos u\,du.$$

Algebraic functions

$$\int \frac{du}{\sqrt{1-u^2}}, \qquad \int \frac{du}{1+u^2}, \qquad \int \frac{du}{u\sqrt{u^2-1}}.$$

Of course, there are additional combinations of trigonometric functions such as $\int \sec^2 u\,du$, etc., which are integrable accidentally, so to speak.

What common types of functions are *not* included in the table?

Logarithms

$$\int \ln u\,du, \qquad \int \log_a u\,du.$$

Trigonometric functions

$$\int \tan u\,du, \qquad \int \cot u\,du, \qquad \int \sec u\,du, \qquad \int \csc u\,du.$$

Algebraic functions

$$\int \frac{du}{a^2+u^2}, \qquad \text{etc., with } a^2 \neq 1.$$

$$\int \sqrt{a^2 \pm u^2}\,du, \qquad \int \sqrt{u^2-a^2}\,du, \qquad \text{etc.}$$

Inverse Functions

$$\int \sin^{-1} u\,du, \qquad \int \tan^{-1} u\,du, \qquad \text{etc.}$$

Summary of Differential Formulas and Corresponding Integrals

1. $du = \frac{du}{dx}dx$	1. $\int du = u + C$
2. $d(au) = a\,du$	2. $\int a\,du = a\int du$
3. $d(u + v) = du + dv$	3. $\int (du + dv) = \int du + \int dv$
4. $d(u^n) = nu^{n-1}\,du$	4. $\int u^n\,du = \frac{u^{n+1}}{n+1} + C, \quad n \neq -1$
5. $d(\ln u) = \frac{du}{u}$	5. $\int \frac{du}{u} = \ln \|u\| + C$
6. a) $d(e^u) = e^u\,du$	6. a) $\int e^u\,du = e^u + C$
b) $d(a^u) = a^u \ln a\,du$	b) $\int a^u\,du = \frac{a^u}{\ln a} + C$
7. $d(\sin u) = \cos u\,du$	7. $\int \cos u\,du = \sin u + C$
8. $d(\cos u) = -\sin u\,du$	8. $\int \sin u\,du = -\cos u + C$
9. $d(\tan u) = \sec^2 u\,du$	9. $\int \sec^2 u\,du = \tan u + C$
10. $d(\cot u) = -\csc^2 u\,du$	10. $\int \csc^2 u\,du = -\cot u + C$
11. $d(\sec u) = \sec u \tan u\,du$	11. $\int \sec u \tan u\,du = \sec u + C$
12. $d(\csc u) = -\csc u \cot u\,du$	12. $\int \csc u \cot u\,du = -\csc u + C$
13. $d(\sin^{-1} u) = \frac{du}{\sqrt{1-u^2}}$ 14. $d(\cos^{-1} u) = \frac{-du}{\sqrt{1-u^2}}$	13. and 14. $\int \frac{du}{\sqrt{1-u^2}} = \begin{cases} \sin^{-1} u + C \\ -\cos^{-1} u + C' \end{cases}$
15. $d(\tan^{-1} u) = \frac{du}{1+u^2}$ 16. $d(\cot^{-1} u) = \frac{-du}{1+u^2}$	15. and 16. $\int \frac{du}{1+u^2} = \begin{cases} \tan^{-1} u + C \\ -\cot^{-1} u + C' \end{cases}$
17. $d(\sec^{-1} u) = \frac{du}{\|u\|\sqrt{u^2-1}}$ 18. $d(\csc^{-1} u) = \frac{-du}{\|u\|\sqrt{u^2-1}}$	17. and 18. $\int \frac{du}{u\sqrt{u^2-1}} = \begin{cases} \sec^{-1} \|u\| + C \\ -\csc^{-1} \|u\| + C' \end{cases}$

We shall eventually see how to evaluate all of these and some others, but there are no methods that will solve *all* integration problems in terms of elementary functions.

When we encounter an integral that does not match one of the standard forms in our table we can try to apply one of the two basic methods from Chapter 4: the method of substitution, and what we might call the "trial" method. The goal of the substitution method is to transform the integral into a standard form. In the trial method we guess a trial solution (it does not have to be exactly right) and modify it after comparing its derivative with the integrand. This is illustrated in Example 5.

EXAMPLE 1 Evaluate

$$\int \frac{dx}{1+4x^2}.$$

Solution The given integral resembles the standard form

$$\int \frac{du}{1+u^2} = \tan^{-1} u + C$$

(Formula 15 in the list) with $2x$ in place of u. With the substitution

$$u = 2x,$$

which leads to

$$du = 2\,dx \qquad \text{and} \qquad dx = \frac{1}{2}\,du,$$

we have

$$\int \frac{dx}{1+4x^2} = \int \frac{(\frac{1}{2})\,du}{1+u^2} = \frac{1}{2}\int \frac{du}{1+u^2}$$
$$= \frac{1}{2}\tan^{-1} u + C = \frac{1}{2}\tan^{-1} 2x + C. \;\Box$$

EXAMPLE 2 Evaluate

$$\int \sqrt{1-5x}\,dx.$$

Solution We let

$$u = 1 - 5x,$$

so that

$$du = -5\,dx, \qquad dx = -\frac{1}{5}\,du,$$

and use Formula 4 with $n = 1/2$ to obtain

$$\int \sqrt{1-5x}\,dx = \int \sqrt{u}\cdot -\frac{1}{5}\,du = -\frac{1}{5}\int \sqrt{u}\,du$$
$$= -\frac{1}{5}\cdot\frac{2}{3}u^{3/2} + C = -\frac{2}{15}(1-5x)^{3/2} + C. \;\Box$$

EXAMPLE 3 Evaluate

$$\int_0^{\pi/2} \sqrt{1 + \sin x} \cos x \, dx.$$

Solution With

$$u = 1 + \sin x, \qquad du = \cos x \, dx,$$

we get

$$\int \sqrt{1 + \sin x} \cos x \, dx = \int \sqrt{u} \, du = \int u^{1/2} \, du$$

$$= \frac{2}{3} u^{3/2} + C = \frac{2}{3}(1 + \sin x)^{3/2} + C. \quad \square$$

What about the limits of integration? Our options (Article 4.8) are (1) to evaluate the transformed integral with transformed limits, or (2) to evaluate the original integral with the original limits.

1. *With transformed limits.* When

$$x = 0, \qquad u = 1 + \sin(0) = 1,$$

and when

$$x = \pi/2, \qquad u = 1 + \sin(\pi/2) = 2.$$

Therefore,

$$\int_0^{\pi/2} \sqrt{1 + \sin x} \cos x \, dx = \int_1^2 \sqrt{u} \, du = \frac{2}{3} u^{3/2} \Big]_1^2 = \frac{2}{3}(2\sqrt{2} - 1).$$

2. *With the original limits.*

$$\int_0^{\pi/2} \sqrt{1 + \sin x} \cos x \, dx = \frac{2}{3}(1 + \sin x)^{3/2} \Big]_0^{\pi/2}$$

$$= \frac{2}{3}(1 + 1)^{3/2} - \frac{2}{3}(1 + 0)^{3/2}$$

$$= \frac{2}{3}(2\sqrt{2} - 1). \quad \square$$

EXAMPLE 4 Evaluate

$$\int \frac{\sin(\ln x)}{x} \, dx.$$

Solution We substitute

$$u = \ln x$$

so that

$$du = \frac{1}{x} \, dx,$$

and

$$\int \frac{\sin(\ln x)}{x} \, dx = \int \sin u \, du = -\cos u + C = -\cos(\ln x) + C. \quad \square$$

EXAMPLE 5 Evaluate

$$\int xe^x\,dx.$$

Solution There is no convenient substitution, so we hunt for a function whose derivative is xe^x. Starting with

$$xe^x$$

we differentiate with respect to x to find

$$\frac{d}{dx}xe^x = xe^x + e^x.$$

The result contains the xe^x we want, but also an e^x we do not want. To eliminate the unwanted term e^x we subtract e^x from the trial function. This produces the function

$$xe^x - e^x,$$

whose derivative is now

$$xe^x + e^x - e^x = xe^x,$$

which is the integrand in the original integral. We conclude that

$$\int xe^x\,dx = xe^x - e^x + C. \square$$

REMARK. The work of Joel Moses and others at MIT in developing the computer program MACSYMA, which evaluates indefinite integrals by symbolic manipulation, has led to a renewed interest in determining which integrals can be expressed as finite algebraic combinations of elementary functions (like the integrals on the endpapers of this book), and which require infinite series (Chapter 12) or some other means for their evaluation. Examples of the latter include the error function

$$\operatorname{erf}(x) = \frac{2}{\sqrt{\pi}}\int_0^x e^{-t^2}\,dt$$

and integrals like

$$\int \sin x^2\,dx \qquad \text{and} \qquad \int \sqrt{1+x^4}\,dx$$

that arise in engineering and physics. These, and a number of others, like

$$\int \frac{e^x}{x}\,dx, \qquad \int e^{e^x}\,dx, \qquad \int \frac{1}{\ln x}\,dx,$$

$$\int \ln(\ln x)\,dx, \qquad \int \frac{\sin x}{x}\,dx,$$

$$\int \sqrt{1-k^2\sin^2 x}\,dx, \qquad 0<k<1,$$

may seem so simple that one may be tempted to spend time trying to integrate them just to see how they turn out. It can be proved, however, that no simple expressions for these integrals exist. That is, there is no way to express these integrals as finite combinations of the so-called elementary functions we have studied so far. The same statement applies to any integral that can be changed into one of these by substitution. None of the integrals you will be asked to evaluate in the present chapter falls into this category, but you may encounter nonelementary integrals in your other work from time to time. The infinite series techniques of Chapter 12 will then be the ones to apply.

PROBLEMS

Evaluate the integrals in Problems 1–14 by using the given substitution to reduce the integral to a standard form.

1. $\int 6x\sqrt{3x^2+5}\,dx, \quad u = 3x^2+5$

2. $\int \frac{16x\,dx}{8x^2+2}, \quad u = 8x^2+2$

3. $\int_0^{\sqrt{\ln 2}} xe^{x^2}\,dx, \quad u = x^2$

4. $\int \frac{\cos x\,dx}{\sqrt{1+\sin x}}, \quad u = 1+\sin x$

5. $\int_0^1 \frac{x\,dx}{\sqrt{8x^2+1}}, \quad u = 8x^2+1$

6. $\int \frac{\sin x}{3+4\cos x}\,dx, \quad u = 3+4\cos x$

7. $\int e^x \sec^2(e^x)\,dx, \quad u = e^x$

8. $\int \frac{dy}{\sqrt{y}(1+y)}, \quad u = \sqrt{y}$

9. $\int e^x\sqrt{3+4e^x}\,dx, \quad u = 3+4e^x$

10. $\int_4^9 \frac{dx}{x-\sqrt{x}}, \quad u = \sqrt{x}$

11. $\int \frac{dx}{x\ln x}, \quad u = \ln x$

12. $\int_0^2 \frac{e^x\,dx}{1+e^x}, \quad u = 1+e^x$

13. $\int_0^1 e^{\sqrt{x}}\,dx, \quad u = \sqrt{x}$

14. $\int \frac{\tan\sqrt{x}}{\sqrt{x}}\,dx, \quad u = \sqrt{x}$

Evaluate the integrals in Problems 15–68 by trial or by reducing the integrand to a standard form. If you use a substitution, indicate what you have called u and refer by number to the standard formula used.

15. $\int \sqrt{2x+3}\,dx$

16. $\int \frac{dx}{3x+5}$

17. $\int \frac{dx}{(2x-7)^2}$

18. $\int \frac{(x+1)\,dx}{x^2+2x+3}$

19. $\int \frac{\sin x\,dx}{2+\cos x}$

20. $\int \tan^3 2x \sec^2 2x\,dx$

21. $\int \frac{x\,dx}{\sqrt{1-4x^2}}$

22. $\int x^{1/3}\sqrt{x^{4/3}-1}\,dx$

23. $\int_0^{\sqrt{2}/2} \frac{dx}{\sqrt{1-x^2}}$

24. $\int_0^{\pi/2} \sin 2x\,dx$

25. $\int_0^1 \frac{3x}{1+x^2}\,dx$

26. $\int_0^{\pi} x\sin x\,dx$

27. $\int_0^{\pi/4} \frac{\sin^2 2x}{1+\cos 2x}\,dx$ (*Hint:* $\sin^2 2x = 1-\cos^2 2x$.)

28. $\int_{\sqrt{2}}^2 \frac{dy}{y\sqrt{y^2-1}}$

29. $\int \frac{2\,dx}{\sqrt{1-4x^2}}$

30. $\int \frac{2v\,dv}{\sqrt{1-v^4}}$

31. $\int \frac{x\,dx}{(3x^2+4)^3}$

32. $\int x^2\sqrt{x^3+5}\,dx$

33. $\int \frac{x^2\,dx}{\sqrt{x^3+5}}$

34. $\int \frac{x\,dx}{4x^2+1}$

35. $\int e^{2x}\,dx$

36. $\int e^{\cos x}\sin x\,dx$

37. $\int \frac{dx}{e^{3x}}$

38. $\int \frac{e^{\sqrt{x+1}}}{\sqrt{x+1}}\,dx$

39. $\int \frac{e^x\,dx}{1+e^{2x}}$

40. $\int \frac{dt}{1+9t^2}$

41. $\int \cos^2 x\sin x\,dx$

42. $\int \frac{\cos x\,dx}{\sin^3 x}$

43. $\int \cot^3 x\csc^2 x\,dx$

44. $\int \tan 3x\sec^2 3x\,dx$

45. $\int \frac{e^{2x}+e^{-2x}}{e^{2x}-e^{-2x}}\,dx$

46. $\int \sin 2x\cos^2 2x\,dx$

47. $\int (1+\cos\theta)^3\sin\theta\,d\theta$

48. $\int te^{-t^2}\,dt$

49. $\int \frac{dt}{t\sqrt{4t^2-1}}$

50. $\int \frac{dx}{\sqrt{e^{2x}-1}}$
(*Hint:* Multiply numerator and denominator by e^x.)

51. $\int \frac{\cos x\,dx}{\sin x}$

52. $\int \frac{\cos x\,dx}{1+\sin x}$

53. $\int \sec^3 x\tan x\,dx$

54. $\int \frac{\sin\theta\,d\theta}{\sqrt{1+\cos\theta}}$

55. $\int e^{\tan 3x}\sec^2 3x\,dx$

56. $\int \cos 2t\sqrt{4-\sin 2t}\,dt$

57. $\int \frac{1 + \cos 2x}{\sin^2 2x}\,dx$

58. $\int \frac{\sin^2 2x}{1 + \cos 2x}\,dx$

59. $\int \frac{\csc^2 2t}{\sqrt{1 + \cot 2t}}\,dt$

60. $\int e^{3x}\,dx$

61. $\int \frac{e^{\tan^{-1} 2t}}{1 + 4t^2}\,dt$

62. $\int xe^{-x^2}\,dx$

63. $\int 3^x\,dx$

64. $\int 10^{2x}\,dx$

65. $\int \frac{\ln x}{x}\,dx$

66. $\int \frac{\cos(\ln x)}{x}\,dx$

67. $\int \frac{dx}{\sqrt{x}(1 + \sqrt{x})}$

68. $\int \ln(x^2 + 1) \cdot 2x(x^2 + 1)^{-1}\,dx$

69. Evaluate

$$\int xe^{x^2}\,dx$$

with the substitutions (a) $u = x^2$ and (b) $u = e^{x^2}$.

70. Can both of the following integrations be correct? Explain.

a) $\int \frac{dx}{\sqrt{1 - x^2}} = \sin^{-1} x + C$

b) $\int \frac{dx}{\sqrt{1 - x^2}} = -\cos^{-1} x + C$

71. Solve for C in each of the following equations.

a) $\tan^{-1} u + \cot^{-1} u = C$

b) $\sec^{-1} |u| + \csc^{-1} |u| = C$

72. Each of the integrals in (a) through (c) may be evaluated easily for a particular numerical value of n. Choose this value and integrate. For example,

$$\int x^n \cos(x^2)\,dx$$

is evaluated easily for $n = 1$:

$$\int x \cos(x^2)\,dx = \frac{1}{2}\sin(x^2) + C.$$

a) $\int x^n \ln x\,dx$, b) $\int x^n e^{x^3}\,dx$, c) $\int x^n \sin \sqrt{x}\,dx$.

7.2 Integration by Parts

There are really just two systematic methods of integration. One of these is the method of substitution. The other, called *integration by parts*, depends on the product rule

$$\frac{d}{dx}(uv) = u\frac{dv}{dx} + v\frac{du}{dx}$$

whose differential form is

$$d(uv) = u\,dv + v\,du$$

or

$$u\,dv = d(uv) - v\,du.$$

When this is integrated we have

$$\int u\,dv = uv - \int v\,du. \qquad (1)$$

Formula (1) expresses one integral, $\int u\,dv$, in terms of a second integral, $\int v\,du$. By proper choice of u and dv, the second integral may be simpler than the first.

In a definite integral, appropriate limits must be supplied. We may then interpret the formula for integration by parts,

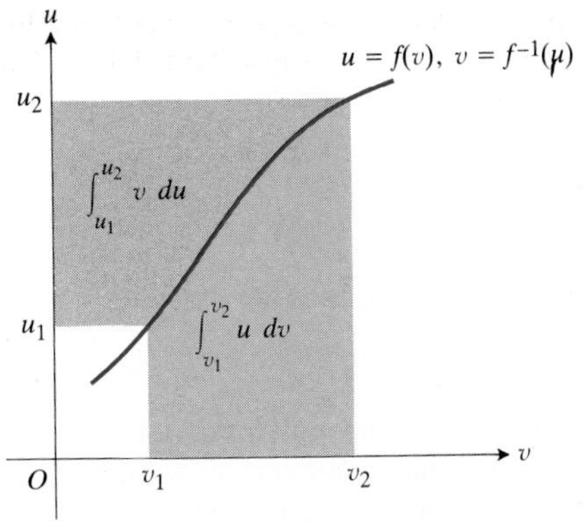

7.1 The area of the blue region, $\int_{v_1}^{v_2} u\,dv$, is equal to the area of the large rectangle, u_2v_2, minus the areas of the small rectangle, u_1v_1, and the grey region, $\int_{u_1}^{u_2} v\,du$. In symbols,

$$\int_{v_1}^{v_2} u\,dv = (u_2v_2 - u_1v_1) - \int_{u_1}^{u_2} v\,du.$$

$$\int_{v_1}^{v_2} u\,dv = (u_2v_2 - u_1v_1) - \int_{u_1}^{u_2} v\,du, \tag{2}$$

geometrically in terms of areas (see Fig. 7.1).

EXAMPLE 1 Evaluate

$$\int \ln x\,dx.$$

Solution If we try to match $\ln x\,dx$ with $u\,dv$, we may take $u = \ln x$ and $dv = dx$. Then, to use (1), we see that we also require

$$du = d(\ln x) = \frac{dx}{x}$$

and

$$v = \int dv = \int dx = x + C_1,$$

so that

$$\begin{aligned}\int \ln x\,dx &= (\ln x)(x + C_1) - \int (x + C_1)\frac{dx}{x} + C_2\\ &= x\ln x + C_1\ln x - \int dx - \int C_1\frac{dx}{x} + C_2\\ &= x\ln x + C_1 \ln x - x - C_1\ln x + C_2\\ &= x\ln x - x + C_2.\end{aligned}$$

Note that the first constant of integration C_1 does not appear in the final answer. This is generally true, for if we write $v + C_1$ in place of v in the right side of (1), we obtain

$$\begin{aligned}u(v + C_1) - \int (v + C_1)\,du &= uv + C_1u - \int v\,du - \int C_1\,du\\ &= uv - \int v\,du,\end{aligned}$$

where it is understood that a constant of integration must still be added in the final result. It is therefore customary to drop the first constant of integration when determining v as $\int dv$. (See Problem 42.) □

EXAMPLE 2 Evaluate

$$\int \tan^{-1} x\,dx.$$

Solution This is typical of the inverse trigonometric functions. Let

$$u = \tan^{-1} x \quad \text{and} \quad dv = dx;$$

then

$$du = \frac{dx}{1 + x^2} \quad \text{and} \quad v = x,$$

so that

$$\int \tan^{-1} x\,dx = x\tan^{-1} x - \int \frac{x\,dx}{1 + x^2} = x\tan^{-1} x - \frac{1}{2}\ln(1 + x^2) + C. \;\square$$

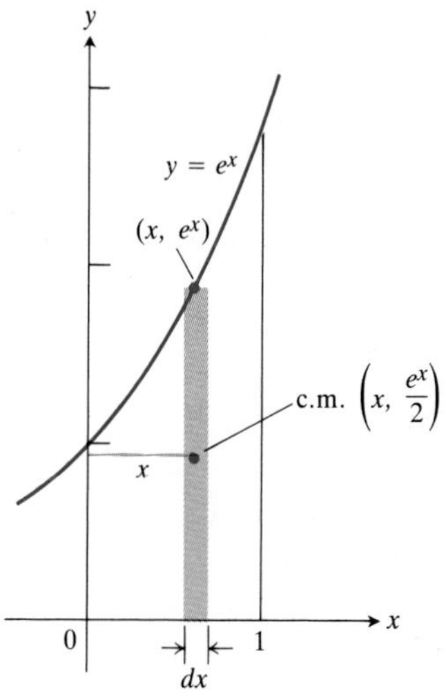

7.2 The moment of the strip about the y-axis is $x\,\delta\,dA = \delta\,xe^x\,dx$.

EXAMPLE 3 Find the moment about the y-axis of a thin homogeneous plate of density δ covering the region in the first quadrant bounded by the curve $y = e^x$ and the line $x = 1$. See Fig. 7.2.

Solution A typical vertical strip has

center of mass (c.m.): $(\tilde{x}, \tilde{y}) = \left(x, \frac{e^x}{2}\right)$
length: e^x
width: dx
area: $dA = e^x\,dx$
mass: $dm = \delta\,dA = \delta\,e^x\,dx.$

The moment of the strip about the y-axis is therefore

$$x\,dm = x \cdot \delta\,e^x\,dx = \delta\,x\,e^x\,dx.$$

The moment of the plate about the y-axis is the integral

$$M_y = \int x\,dm = \delta \int_0^1 x\,e^x\,dx.$$

To evaluate this integral we let

$$u = x \qquad \text{and} \qquad v = \int e^x\,dx = e^x.$$

Integration by parts then gives

$$\int_0^1 xe^x\,dx = xe^x\Big]_0^1 - \int_0^1 e^x\,dx = e - [e^x]_0^1 = e - [e - 1] = 1.$$

The moment of the plate about the y-axis is therefore

$$M_y = \delta \int_0^1 xe^x\,dx = \delta \cdot 1 = \delta. \;\square$$

The integration in Example 3 illustrates a general rule for choosing the u and the dv in integration by parts: If you can, choose u to be something that becomes simpler when differentiated, and choose dv to be something whose integral v is not complicated. This way, the integral on the right side of the equation

$$\int u\,dv = uv - \int v\,du$$

will be simpler than the one on the left.

Right: With

$$u = x, \qquad v = \int dv = \int e^x\,dx = e^x,$$

we obtain

$$\int xe^x\,dx = xe^x - \int e^x\,dx = xe^x - e^x + C. \tag{3}$$

Wrong: With

$$u = e^x, \qquad v = \int dv = \int x\,dx = \frac{x^2}{2},$$

we obtain

$$\int xe^x\,dx = e^x \cdot \frac{x^2}{2} - \int \frac{x^2}{2} e^x\,dx. \tag{4}$$

Equation (4) is correct, but the integral on the right is more complicated than the original integral. It leaves us worse off than before.

Sometimes an integration by parts must be repeated to obtain an answer, as in the next example.

EXAMPLE 4 Evaluate

$$\int x^2 e^x \, dx.$$

Solution Let

$$u = x^2 \qquad \text{and} \qquad dv = e^x \, dx.$$

Then

$$du = 2x \, dx \qquad \text{and} \qquad v = e^x,$$

so that

$$\int x^2 e^x \, dx = x^2 e^x - 2 \int x e^x \, dx.$$

The integral on the right is similar to the original integral, except that we have reduced the power of x from 2 to 1. If we could now reduce it from 1 to 0, we could see success ahead. In $\int x e^x \, dx$, we therefore let

$$U = x \qquad \text{and} \qquad dV = e^x \, dx,$$

so that

$$dU = dx \qquad \text{and} \qquad V = e^x.$$

Then

$$\int x e^x \, dx = x e^x - \int e^x \, dx = x e^x - e^x + C',$$

and

$$\int x^2 e^x \, dx = x^2 e^x - 2x e^x + 2e^x + C. \quad \square$$

EXAMPLE 5 Obtain a "reduction" formula that expresses the integral

$$\int \cos^n x \, dx$$

in terms of an integral of a lower power of cos x.

Solution We may think of $\cos^n x$ as $\cos^{n-1} x \cdot \cos x$. Then we let

$$u = \cos^{n-1} x \qquad \text{and} \qquad dv = \cos x \, dx,$$

so that

$$du = (n-1) \cos^{n-2} x (-\sin x \, dx) \qquad \text{and} \qquad v = \sin x.$$

Hence

$$\begin{aligned} \int \cos^n x \, dx &= \cos^{n-1} x \sin x + (n-1) \int \sin^2 x \cos^{n-2} x \, dx \\ &= \cos^{n-1} x \sin x + (n-1) \int (1 - \cos^2 x) \cos^{n-2} x \, dx, \\ &= \cos^{n-1} x \sin x + (n-1) \int \cos^{n-2} x \, dx - (n-1) \int \cos^n x \, dx. \end{aligned}$$

If we add

$$(n-1)\int \cos^n x\,dx$$

to both sides of this equation, we obtain

$$n\int \cos^n x\,dx = \cos^{n-1} x \sin x + (n-1)\int \cos^{n-2} x\,dx.$$

We then divide through by n, and the final result is

$$\int \cos^n x\,dx = \frac{\cos^{n-1} x \sin x}{n} + \frac{n-1}{n}\int \cos^{n-2} x\,dx. \qquad (5)$$

This allows us to reduce the exponent on cos x by 2 and is a very useful formula. When n is a positive integer, we may apply the formula repeatedly until the remaining integral is either

$$\int \cos x\,dx = \sin x + C \qquad \text{or} \qquad \int \cos^0 x\,dx = \int dx = x + C. \ \square$$

The companion formula for sines (Problem 44) is

$$\int \sin^n x\,dx = -\frac{\sin^{n-1} x \cos x}{n} + \frac{n-1}{n}\int \sin^{n-2} x\,dx. \qquad (6)$$

EXAMPLE 6 Evaluate

$$\int \cos^4 x\,dx.$$

Solution With $n = 4$ in Eq. (5) we get

$$\int \cos^4 x\,dx = \frac{\cos^3 x \sin x}{4} + \frac{3}{4}\int \cos^2 x\,dx,$$

and with $n = 2$,

$$\int \cos^2 x\,dx = \frac{\cos x \sin x}{2} + \frac{1}{2}\int dx.$$

Therefore,

$$\int \cos^4 x\,dx = \frac{\cos^3 x \sin x}{4} + \frac{3}{4}\left(\frac{\cos x \sin x}{2} + \frac{1}{2}x\right) + C. \ \square$$

The integral in the next example occurs in electrical engineering problems. Its evaluation requires two integrations by parts, followed by solving for the unknown integral as in Example 5.

EXAMPLE 7 Evaluate

$$\int e^{ax} \cos bx\,dx.$$

Solution Let

$$u = e^{ax} \qquad \text{and} \qquad dv = \cos bx\, dx.$$

Then

$$du = ae^{ax}\, dx \qquad \text{and} \qquad v = \frac{1}{b} \sin bx,$$

so that

$$\int e^{ax} \cos bx\, dx = \frac{e^{ax} \sin bx}{b} - \frac{a}{b} \int e^{ax} \sin bx\, dx.$$

The second integral is like the first except that it has sin bx in place of cos bx. If we apply integration by parts to it, letting

$$U = e^{ax} \qquad \text{and} \qquad dV = \sin bx\, dx,$$

then

$$dU = ae^{ax}\, dx \qquad \text{and} \qquad V = -\frac{1}{b} \cos bx,$$

so that

$$\int e^{ax} \cos bx\, dx = \frac{e^{ax} \sin bx}{b} - \frac{a}{b}\left[-\frac{e^{ax} \cos bx}{b} + \frac{a}{b} \int e^{ax} \cos bx\, dx\right].$$

Now the unknown integral appears on the left with a coefficient of unity and on the right with a coefficient of $-a^2/b^2$. Transposing this term to the left and dividing by the new coefficient

$$1 + \frac{a^2}{b^2} = \frac{a^2 + b^2}{b^2},$$

we have

$$\int e^{ax} \cos bx\, dx = e^{ax}\left(\frac{b \sin bx + a \cos bx}{a^2 + b^2}\right) + C. \;\square \qquad (7)$$

Tabular Integration

Integrals of the form

$$\int f(x)\, g(x)\, dx$$

in which f can be differentiated repeatedly to become zero and g can be integrated repeatedly without difficulty are natural candidates for integration by parts. If many repetitions are required, however, the calculations can be cumbersome. There is a way to organize the calculations in situations like this, called *tabular integration,* that saves a great deal of work. The method is illustrated in the following examples.

EXAMPLE 8 Evaluate

$$\int x^2 e^x\, dx$$

by tabular integration.

Solution With $f(x) = x^2$ and $g(x) = e^x$ we have

$f(x)$ and its derivatives		$g(x)$ and its integrals
x^2	$(+)$	e^x
$2x$	$(-)$	e^x
2	$(+)$	e^x
0		e^x.

The products of the functions connected by the arrows are added, with the middle sign changed, to obtain

$$\int x^2e^x\,dx = x^2e^x - 2xe^x + 2e^x + C,$$

in agreement with Example 4. □

EXAMPLE 9 Evaluate

$$\int x^3 \sin x\,dx$$

by tabular integration.

Solution With $f(x) = x^3$ and $g(x) = \sin x$ we have

$f(x)$ and its derivatives		$g(x)$ and its integrals
x^3	$(+)$	$\sin x$
$3x^2$	$(-)$	$-\cos x$
$6x$	$(+)$	$-\sin x$
6	$(-)$	$\cos x$
0		$\sin x$.

The products of the functions connected by the arrows are added, with every other sign changed, to obtain

$$\int x^3 \sin x\,dx = -x^3\cos x + 3x^2 \sin x + 6x\cos x - 6\sin x + C. \ \square$$

REMARK. In addition to being a powerful technique for evaluating indefinite integrals, integration by parts plays an important role in preparing definite integrals for computer evaluation.

PROBLEMS

In Problems 1–34, evaluate the integrals.

1. $\int x \sin x\,dx$

2. $\int x \cos x\,dx$

3. $\int x^2 \sin x\,dx$

4. $\int x^2 \cos x\,dx$

5. $\int x \ln x\,dx$

6. $\int x^2 \ln x\,dx$

7. $\int x^n \ln ax\,dx, \quad n \neq -1$

8. $\int \ln (x + 1)\,dx$

9. $\int x^3e^x\,dx$

10. $\int x^4e^{-x}\,dx$

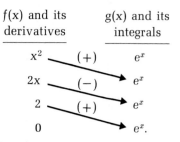

11. $\int x^5 e^x \, dx$

12. $\int x^2 e^{4x} \, dx$

13. $\int x^2 2^x \, dx$

14. $\int_0^{\pi/2} x^2 \sin 2x \, dx$

15. $\int_0^{\pi/2} x^3 \cos 2x \, dx$

16. $\int x^4 \cos x \, dx$

17. $\int x^5 \sin x \, dx$

18. $\int \cos^6 x \, dx$

19. $\int \sin^2 x \, dx$

20. $\int \sin^4 x \, dx$

21. $\int \sin^{-1} ax \, dx$

22. $\int x \sin ax \, dx$

23. $\int x^2 \cos ax \, dx$

24. $\int x \cos (2x + 1) \, dx$

25. $\int_1^2 x \sec^{-1} x \, dx$

26. $\int_1^4 \sec^{-1} \sqrt{x} \, dx$

27. $\int \frac{\ln x}{x} \, dx$

28. $\int_0^1 x\sqrt{1 - x} \, dx$

29. $\int e^{2x} \cos 3x \, dx$

30. $\int x \sin^{-1} (a/x) \, dx$

31. $\int e^{ax} \sin bx \, dx$

32. $\int \sin (\ln x) \, dx$

33. $\int \cos (\ln x) \, dx$

34. $\int x (\ln x)^2 \, dx$

35. Use cylindrical shells to find the volume of the solid obtained by revolving the area bounded by $x = 0$, $y = 0$, and $y = \cos x$, $0 \le x \le \pi/2$, about the y-axis.

36. Find the centroid of the area bounded by $y = x^2 e^x$, $y = 0$, $x = 1$.

37. Find the centroid of the area bounded by $y = \ln x$, $x = 1$, $y = 1$.

38. Find the volume swept out when the region in the first quadrant bounded by the x-axis and the curve $y = x \sin x$, $0 \le x \le \pi$, is revolved about the y-axis.

39. Find the moment about the y-axis of a thin plate of density $\delta = (1 + x)$ covering the region bounded by the x-axis and the curve $y = \sin x$, $0 \le x \le \pi$.

40. Find the volume of the solid generated by revolving about the y-axis the region bounded by the x-axis and the curve $y = x\sqrt{1 - x}$, $0 \le x \le 1$.

41. Evaluate $\int_0^1 e^{\sqrt{x}} \, dx$ by integration by parts. (*Hint:* First let $u^2 = x$.)

42. While one usually drops the constant of integration in determining v as $\int dv$ in integration by parts, choosing the constant to be different from zero can occasionally be helpful. As a case in point, evaluate

$$\int x \tan^{-1} x \, dx$$

with $u = \tan^{-1} x$ and $v = (x^2/2) + 1/2$.

In Problems 43–45, derive the reduction formula in (a) and use it to evaluate the integral in (b).

43. a) $\int x^m (\ln x)^n \, dx = \frac{x^{m+1} (\ln x)^n}{m + 1} - \frac{n}{m + 1} \int x^m (\ln x)^{n-1} \, dx$

b) $\int x^3 (\ln x)^2 \, dx$

44. a) $\int \sin^n x \, dx = -\frac{\sin^{n-1} x \cos x}{n} + \frac{n - 1}{n} \int \sin^{n-2} x \, dx$

b) $\int_0^{\pi/6} \sin^4 3x \, dx$

45. a) $\int x^n e^x \, dx = x^n e^x - n \int x^{n-1} e^x \, dx$

b) $\int x^3 e^x \, dx$

7.3 Products and Powers of Trigonometric Functions

The formula

$$\int u^n \, du = \begin{cases} \dfrac{u^{n+1}}{n + 1} + C, & n \neq -1, \\ \ln |u| + C, & n = -1, \end{cases}$$

may be used to evaluate certain integrals involving powers of the trigonometric functions, as illustrated by the examples that follow. The same methods work for powers of other functions. Pay attention to the *methods* rather than trying to remember specific results.

EXAMPLE 1 Evaluate

$$\int \sin^n ax \cos ax \, dx.$$

Solution If we let $u = \sin ax$, then

$$\begin{aligned} du &= \cos ax \, d(ax) \\ &= a \cos ax \, dx, \end{aligned}$$

so that we need only multiply the integral by unity in the form of a times $1/a$. Since a and $1/a$ are constants, we may write a inside the integral sign and $1/a$ in front of the integral sign to have

$$\begin{aligned} \int \sin^n ax \cos ax \, dx &= \frac{1}{a} \int (\sin ax)^n (a \cos ax \, dx) \\ &= \frac{1}{a} \int u^n \, du \\ &= \begin{cases} \dfrac{1}{a} \dfrac{u^{n+1}}{n+1} + C, & n \neq -1, \\ \dfrac{1}{a} \ln |u| + C, & n = -1; \end{cases} \end{aligned}$$

that is,

$$\int \sin^n ax \cos ax \, dx = \frac{\sin^{n+1} ax}{(n+1)a} + C, \qquad n \neq -1, \tag{1a}$$

and if $n = -1$, we get

$$\int \cot ax \, dx = \int \frac{\cos ax \, dx}{\sin ax} = \frac{1}{a} \ln |\sin ax| + C, \tag{1b}$$

provided $\sin ax$ does not change sign on the domain of ax. (The sign restriction comes from the derivation of the formula

$$\int \frac{du}{u} = \ln |u| + C$$

in Article 6.4.) Note that the success of the method depends upon having $\cos ax$ to go with the dx as part of du. □

By a similar method we can obtain

$$\int \cos^n ax \sin ax \, dx = \frac{-\cos^{n+1} ax}{(n+1)a} + C, \qquad n \neq -1, \tag{2a}$$

and, when $n = -1$,

$$\int \tan ax \, dx = \int \frac{\sin ax}{\cos ax} \, dx = -\frac{1}{a} \ln |\cos ax| + C. \tag{2b}$$

Again, Eq. (2b) holds provided $\cos ax$ does not change sign on the domain of ax.

EXAMPLE 2 Evaluate

$$\int \sin^3 x\, dx.$$

Solution The method of the previous example does not work because there is no cos x to go with dx to give du if we try letting $u = \sin x$. But if we write

$$\sin^3 x = \sin^2 x \cdot \sin x = (1 - \cos^2 x) \cdot \sin x$$

and let

$$u = \cos x, \qquad du = -\sin x\, dx,$$

we have

$$\begin{aligned}\int \sin^3 x\, dx &= \int (1 - \cos^2 x) \cdot \sin x\, dx = \int (1 - u^2) \cdot (-du) \\ &= \int (u^2 - 1)\, du = \frac{1}{3}u^3 - u + C = \frac{1}{3}\cos^3 x - \cos x + C. \quad \square\end{aligned}$$

The method of Example 2 may be applied whenever an *odd* power of sin x or cos x is to be integrated. For example, any positive odd power of cos x has the form

$$\cos^{2n+1} x = \cos^{2n} x \cdot \cos x = (\cos^2 x)^n \cdot \cos x = (1 - \sin^2 x)^n \cdot \cos x,$$

with n an integer ≥ 0. Then if we let $u = \sin x$, $du = \cos x\, dx$, we have

$$\begin{aligned}\int \cos^{2n+1} x\, dx &= \int (1 - \sin^2 x)^n \cdot \cos x\, dx \\ &= \int (1 - u^2)^n \cdot du. \qquad (3a)\end{aligned}$$

The expression $(1 - u^2)^n$ may now be expanded by the binomial theorem and the result evaluated as a sum of individual integrals of the type $\int u^m\, du$.

Similarly, each positive odd power of sin x has the form

$$\begin{aligned}\sin^{2n+1} x &= \sin^{2n} x \cdot \sin x \\ &= (\sin^2 x)^n \cdot \sin x \\ &= (1 - \cos^2 x)^n \cdot \sin x.\end{aligned}$$

If we let $u = \cos x$, $du = -\sin x\, dx$, then

$$\begin{aligned}\int \sin^{2n+1} x\, dx &= \int (1 - \cos^2 x)^n \cdot \sin x\, dx \\ &= -\int (1 - u^2)^n\, du. \qquad (3b)\end{aligned}$$

EXAMPLE 3 Evaluate

$$\int \sec x \tan x\, dx.$$

Solution Of course this is a standard form already, so there is no real problem in finding an answer. But in trigonometry we often express all trigonometric functions in terms of sines and cosines, and we now investigate what this does to the integral in question:

$$\int \sec x \tan x \, dx = \int \frac{1}{\cos x} \frac{\sin x}{\cos x} \, dx = \int \frac{\sin x \, dx}{\cos^2 x}.$$

Taking a cue from the previous examples (keeping in mind du as well as u), we let

$$u = \cos x, \qquad du = -\sin x \, dx,$$

and then

$$\int \frac{\sin x \, dx}{\cos^2 x} = \int \frac{-du}{u^2} = -\int u^{-2} \, du = \frac{-u^{-1}}{-1} + C = \frac{1}{u} + C$$

$$= \frac{1}{\cos x} + C = \sec x + C. \ \square$$

EXAMPLE 4 Evaluate

$$\int \tan^4 x \, dx.$$

Solution This does not lend itself readily to the use of sines and cosines, since both of them occur to even powers. We say: $u = \tan x$ would require $du = \sec^2 x \, dx$. Is there some way to include $\sec^2 x$? Yes; there is an identity involving tangents and secants. How does it go? Since

$$\sin^2 x + \cos^2 x = 1,$$

if we divide through by $\cos^2 x$ we get

$$\tan^2 x + 1 = \sec^2 x \qquad \text{or} \qquad \tan^2 x = \sec^2 x - 1.$$

Then

$$\int \tan^4 x \, dx = \int \tan^2 x \cdot \tan^2 x \, dx = \int \tan^2 x \cdot (\sec^2 x - 1) \, dx$$

$$= \int \tan^2 x \sec^2 x \, dx - \int \tan^2 x \, dx$$

$$= \int \tan^2 x \sec^2 x \, dx - \int (\sec^2 x - 1) \, dx$$

$$= \int \tan^2 x \sec^2 x \, dx - \int \sec^2 x \, dx + \int dx.$$

In the first two of these, let

$$u = \tan x, \qquad du = \sec^2 x \, dx$$

and have

$$\int u^2 \, du - \int du = \frac{1}{3} u^3 - u + C'.$$

The other is a standard form, so

$$\int \tan^4 x \, dx = \frac{1}{3} \tan^3 x - \tan x + x + C. \ \square$$

The method of Example 4 works for any *even* power of tan x, but what is still better is a *reduction formula* derived as follows:

$$\int \tan^n x\, dx = \int \tan^{n-2} x(\sec^2 x - 1)\, dx$$

$$= \int \tan^{n-2} x \sec^2 x\, dx - \int \tan^{n-2} x\, dx$$

$$= \frac{\tan^{n-1} x}{n-1} - \int \tan^{n-2} x\, dx. \qquad (4)$$

Equation (4) reduces the problem of integrating $\tan^n x\, dx$ to the problem of integrating $\tan^{n-2} x\, dx$. Since this decreases the exponent on tan x by 2, a repetition with the same formula will reduce the exponent by 2 again, and so on. Applying this to the problem above, we have

$$n = 4: \quad \int \tan^4 x\, dx = \frac{\tan^3 x}{3} - \int \tan^2 x\, dx,$$

$$n = 2: \quad \int \tan^2 x\, dx = \frac{\tan x}{1} - \int \tan^0 x\, dx,$$

$$n = 0: \quad \int \tan^0 x\, dx = \int 1\, dx = x + C.$$

Therefore,

$$\int \tan^4 x\, dx = \frac{1}{3}\tan^3 x - [\tan x - x + C]$$

$$= \frac{1}{3}\tan^3 x - \tan x + x + C'.$$

This reduction formula works whether the original exponent n is even or odd, but if the exponent is odd, say $2m + 1$, then after m steps it will be reduced by $2m$, leaving

$$\int \tan x = -\ln|\cos x| + C$$

as the final integral to be evaluated.

These examples illustrate how it is possible to use the trigonometric identities

$$\sin^2 x + \cos^2 x = 1, \qquad \tan^2 x + 1 = \sec^2 x,$$

and others readily derived from these, to evaluate the integrals of

a) *odd* powers of sin x or cos x,

b) *any* integral powers of tan x (or cot x), and

c) *even* powers of sec x (or csc x).

The even powers of sec x, say $\sec^{2n} x$, can all be reduced to powers of tan x by employing the substitution $\sec^2 x = 1 + \tan^2 x$, and then using the reduction formula for integrating powers of tan x after expanding $\sec^{2n} x = (1 + \tan^2 x)^n$ by the binomial theorem. But it is even simpler to use the following method.

$$\int \sec^{2n} x\, dx = \int \sec^{2n-2} x \sec^2 x\, dx = \int (\sec^2 x)^{n-1} \sec^2 x\, dx$$
$$= \int (1 + \tan^2 x)^{n-1} \sec^2 x\, dx$$
$$= \int (1 + u^2)^{n-1}\, du \qquad (u = \tan x). \tag{5}$$

When $(1 + u^2)^{n-1}$ is expanded by the binomial theorem, the resulting polynomial in u may be integrated term by term.

EXAMPLE 5

$$\int \sec^6 x\, dx = \int \sec^4 x \cdot \sec^2 x\, dx = \int (1 + \tan^2 x)^2 \cdot \sec^2 x\, dx$$
$$= \int (1 + 2u^2 + u^4)\, du \qquad (u = \tan x)$$
$$= u + \frac{2u^3}{3} + \frac{u^5}{5} + C = \tan x + 2\frac{\tan^3 x}{3} + \frac{\tan^5 x}{5} + C. \ \square$$

EXAMPLE 6 Evaluate

$$\int \sec x\, dx.$$

Solution This is hard to evaluate unless one has seen the following trick!

$$\sec x = \frac{\sec x(\tan x + \sec x)}{\sec x + \tan x} = \frac{\sec x \tan x + \sec^2 x}{\sec x + \tan x}.$$

In this form the numerator is the derivative of the denominator. Therefore

$$\int \sec x\, dx = \int \frac{\sec x \tan x + \sec^2 x}{\sec x + \tan x}\, dx$$
$$= \int \frac{du}{u} \qquad (u = \sec x + \tan x)$$
$$= \ln |u| + C.$$

That is,

$$\int \sec x\, dx = \ln |\sec x + \tan x| + C. \ \square \tag{6}$$

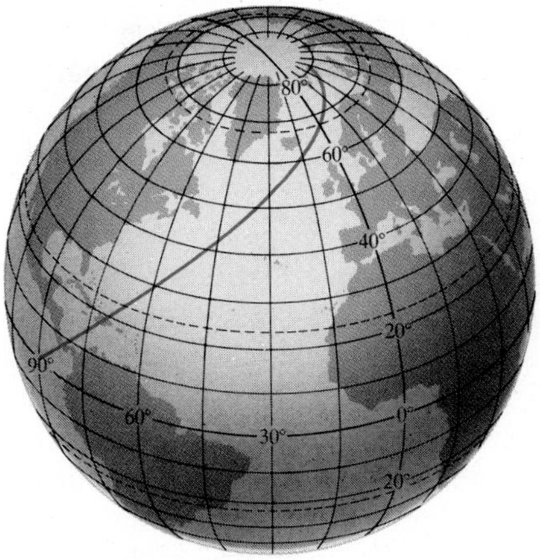

7.3 A flight with a constant bearing of 45° E of N from the Galapagos Islands in the Pacific to Franz Josef Land in the Arctic Ocean.

EXAMPLE 7 *Mercator's world map.* The integral of the secant plays an important role in making maps for compass navigation. The easiest course for a sailor to steer is a course whose compass heading is constant. This might be a course of 45° (northeast), for example, or a course of 225° (southwest), or whatever. Such a course will lie along a spiral that winds around the globe toward one of the poles (Fig. 7.3), unless the course runs due north or south, or lies parallel to the equator.

In 1569, Gerhard Krämer, a Flemish surveyor and geographer known to us by his Latinized last name, Mercator, made a world map on which

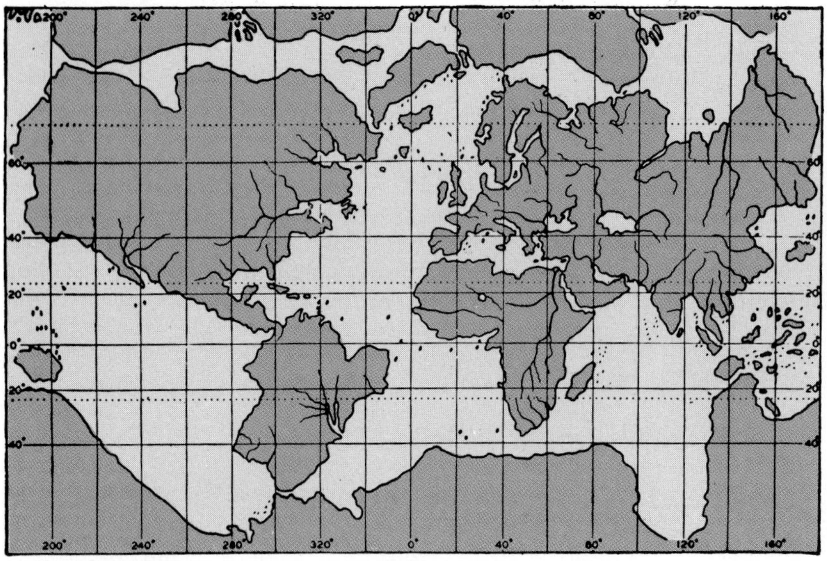

7.4 Sketch of Mercator's map of 1569.

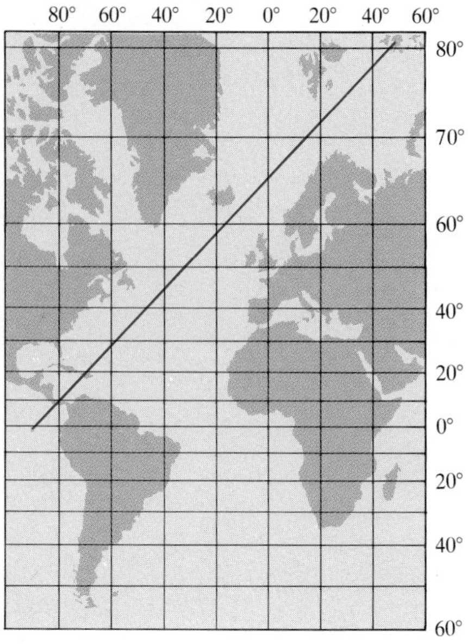

7.5 The flight of Fig. 7.3 traced on a Mercator map.

all spirals of constant compass heading appeared as straight lines (Fig. 7.4). This fantastic achievement met what must have been one of the most pressing navigational needs of all time. For a sailor could then read the compass heading for a voyage between any two points from the direction of a straight line connecting them on Mercator's map (Fig. 7.5).

If you look closely at Fig. 7.5, you will see that the vertical lines of longitude that meet at the poles on a globe have been spread apart to lie parallel on the map. The horizontal lines of latitude that are shown every 10° are parallel also, as they are on the globe, but they are not evenly spaced. The spacing between them increases toward the poles.

The secant function plays a role in determining the correct spacing of all these lines. The scaling factor by which horizontal distances are increased at a fixed latitude $\theta°$ to spread the lines of longitude apart is precisely $\sec\theta$. There is no spread at the equator, where $\sec 0° = 1$. At latitude 30° north or south, the spreading is accomplished by multiplying all horizontal distances by the factor $\sec 30° = 2/\sqrt{3} \approx 1.15$. At 60°, the factor is $\sec 60° = 2$. The closer you move toward the poles, the more the longitudes have to be spread to be parallel. The lines of latitude are spread apart toward the poles to match the spreading of the longitudes, but the formula for the spreading is complicated by the fact that the scaling factor $\sec\theta$ increases with the latitude θ. Thus, the factor to be used for stretching an interval of latitude is not a constant on the interval. This difficulty is overcome by integration. If R is the radius of the globe being modeled (Fig. 7.6), then the distance D between the lines that are drawn on the map to show the equator and the latitude $\alpha°$ is R times the integral of the secant from zero to α:

$$D = R\int_0^{\alpha} \sec x \, dx. \qquad (7)$$

Therefore, the map distance between two lines of north latitude, say, at $\alpha°$

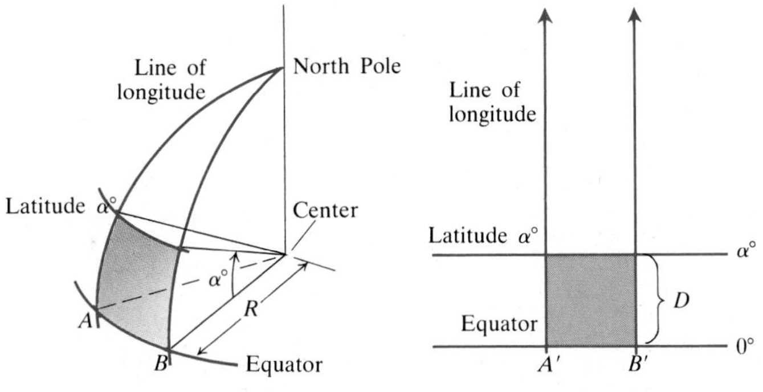

7.6 Lines of latitude and longitude.

and $\beta°$ ($\alpha < \beta$) is

$$R\int_0^\beta \sec x\,dx - R\int_0^\alpha \sec x\,dx = R\int_\alpha^\beta \sec x\,dx$$

$$= R\ln|\sec x + \tan x|\Big]_\alpha^\beta. \quad \square \qquad (8)$$

EXAMPLE 8 Suppose that the equatorial length of a Mercator map just matches the equator of a globe of radius 25 cm. Then Eq. (7) gives the spacing on the map between the equator and the latitude 20° north as

$$25\int_0^{20°} \sec x\,dx = 25\ln|\sec x + \tan x|\Big]_0^{20°} \approx 9 \text{ cm}.$$

The spacing between 60° and 80° north is given by Eq. (8) as

$$25\int_{60°}^{80°} \sec x\,dx = 25\ln|\sec x + \tan x|\Big]_{60°}^{80°} \approx 28 \text{ cm}.$$

The navigational properties of a Mercator map are achieved at the expense of a considerable distortion of distance. (For a very readable derivation and discussion of the formula in Eq. (7), see Philip M. Tuchinsky's *Mercator's World Map and the Calculus,* UMAP Unit No. 206, COMAP, Inc., Lexington, MA, 1979.) □

Products of Sines and Cosines

The integrals,

$$\int \sin mx \sin nx\,dx, \quad \int \sin mx \cos nx\,dx, \quad \int \cos mx \cos nx\,dx, \qquad (9)$$

arise in alternating-current theory, heat transfer problems, bending of beams, cable stress analysis in suspension bridges, and many other places where trigonometric series are applied to problems in mathematics, science, and engineering. They can be evaluated by integration by parts, but two such integrations are required in each case. A simpler way to evaluate them is to exploit the identities

$$\sin mx \sin nx = \tfrac{1}{2}[\cos(m-n)x - \cos(m+n)x], \qquad (10a)$$

$$\sin mx \cos nx = \tfrac{1}{2}[\sin(m-n)x + \sin(m+n)x], \qquad (10b)$$

$$\cos mx \cos nx = \tfrac{1}{2}[\cos(m-n)x + \cos(m+n)x]. \qquad (10c)$$

These identities follow at once from

$$\begin{aligned}\cos(A+B) &= \cos A\cos B - \sin A\sin B,\\ \cos(A-B) &= \cos A\cos B + \sin A\sin B,\end{aligned} \tag{11a}$$

and

$$\begin{aligned}\sin(A+B) &= \sin A\cos B + \cos A\sin B,\\ \sin(A-B) &= \sin A\cos B - \cos A\sin B.\end{aligned} \tag{11b}$$

For example, if we add the two equations in (11a) and then divide by 2, we obtain (10c) by taking $A = mx$ and $B = nx$. The identity in (10a) is obtained in a similar fashion by subtracting the first equation in (11a) from the second equation. Finally, if we add the two equations in (11b) we are led to the identity in (10b).

EXAMPLE 9 By applying (10b) with $m = 3$ and $n = 5$ we get

$$\begin{aligned}\int \sin 3x\cos 5x\,dx &= \frac{1}{2}\int [\sin(-2x) + \sin 8x]\,dx\\ &= \frac{1}{2}\int (\sin 8x - \sin 2x)\,dx\\ &= -\frac{\cos 8x}{16} + \frac{\cos 2x}{4} + C. \;\square\end{aligned}$$

EXAMPLE 10 Show that if m and n are integers, and $m^2 \neq n^2$, then

$$\int_0^{2\pi} \cos mx\cos nx\,dx = 0.$$

Solution From formula (10c) we have

$$\begin{aligned}\int_0^{2\pi} \cos mx\cos nx\,dx &= \int_0^{2\pi} \frac{1}{2}[\cos(m-n)x + \cos(m+n)x]\,dx\\ &= \frac{1}{2}\left[\frac{\sin(m-n)x}{m-n} + \frac{\sin(m+n)x}{m+n}\right]_0^{2\pi}.\end{aligned} \tag{12}$$

(The integration requires $m - n \neq 0$ and $m + n \neq 0$, both of which are true if $m^2 \neq n^2$.) Now, $(m - n)$ and $(m + n)$ are integers because m and n are integers. Furthermore, the sine of any integer multiple of 2π is zero. The terms in brackets in (12) are therefore zero at both the upper and lower limits of integration, and

$$\int_0^{2\pi} \cos mx\cos nx\,dx = 0. \;\square$$

Formulas Developed in This Article

1a) $\displaystyle\int \sin^n ax\cos ax\,dx = \frac{\sin^{n+1} ax}{(n+1)a} + C, \quad n \neq -1$

1b) $\displaystyle\int \cot ax\,dx = \int \frac{\cos ax\,dx}{\sin ax} = \frac{1}{a}\ln|\sin ax| + C$

2a) $\displaystyle\int \cos^n ax\sin ax\,dx = \frac{-\cos^{n+1} ax}{(n+1)a} + C, \quad n \neq -1$

2b) $\int \tan ax\, dx = \int \frac{\sin ax}{\cos ax} dx = -\frac{1}{a} \ln |\cos ax| + C$

3a) $\int \cos^{2n+1} x\, dx = \int (1 - \sin^2 x)^n \cdot \cos x\, dx$

$= \int (1 - u^2)^n \cdot du, \qquad u = \sin x$

3b) $\int \sin^{2n+1} x\, dx = \int (1 - \cos^2 x)^n \cdot \sin x\, dx$

$= -\int (1 - u^2)^n \cdot du, \qquad u = \cos x$

4) $\int \tan^n x\, dx = \frac{\tan^{n-1} x}{n-1} - \int \tan^{n-2} x\, dx$

5) $\int \sec^{2n} x\, dx = \int (1 + \tan^2 x)^{n-1} \sec^2 x\, dx$

$= \int (1 + u^2)^{n-1} \cdot du, \qquad u = \tan x$

6) $\int \sec x\, dx = \ln |\sec x + \tan x| + C$

The expressions $(1 - u^2)^n$ and $(1 + u^2)^{n-1}$ on the right-hand sides of (3a), (3b), and (5) are expanded by the binomial theorem before integration.

PROBLEMS

Evaluate the following integrals. In evaluating the definite integrals you may wish to refer to the triangles in Fig. 7.7.

1. $\int_0^{\pi/4} \sin^5 x \cos x\, dx$

2. $\int \sin^{3/2} x \cos x\, dx$

3. $\int \sqrt{\cos x} \sin x\, dx$

4. $\int \sin t \sqrt{1 + \cos t}\, dt$

5. $\int \frac{\sin \theta\, d\theta}{2 - \cos \theta}$

6. $\int \frac{\sec^2 2x\, dx}{1 + \tan 2x}$

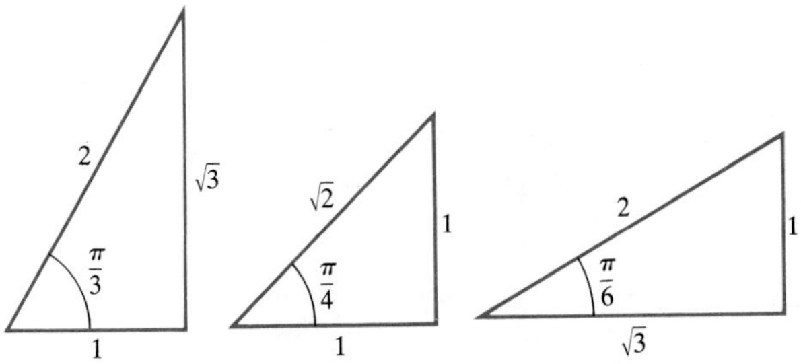

7.7 Reference triangles for evaluating trigonometric functions at $\pi/6$, $\pi/4$, and $\pi/3$ radians.

7. $\int \tan 3x\, dx$

8. $\int \cot 5x\, dx$

9. $\int \cos^3 x\, dx$

10. $\int \tan^3 x\, dx$

11. $\int \tan^2 4\theta\, d\theta$

12. $\int_0^{\pi/4} \tan^4 x\, dx$

13. $\int_{\pi/6}^{\pi/3} \tan^5 t\, dt$

14. $\int_{\pi/3}^{\pi/2} \csc (x/2) \cot (x/2)\, dx$

15. a) $\int \sin^3 x \cos^2 x\, dx$ b) $\int \frac{\sin^3 x\, dx}{\cos^2 x}$

16. $\int \sec^n x \tan x\, dx$

17. $\int \tan^n x \sec^2 x\, dx$

18. $\int \sin^n x \cos x\, dx$

19. $\int \cos^n x \sin x\, dx$

20. $\int \sin^2 3x \cos 3x\, dx$

21. $\int \cos^3 2x \sin 2x\, dx$

22. $\int \sec^4 3x \tan 3x\, dx$

23. $\int_{-\pi/2}^{\pi/2} \sin^2 y \cos^3 y \, dy$

24. $\int_{-\pi/2}^{\pi/2} \sin^2 2x \cos^3 2x \, dx$

25. $\int_{\pi/6}^{\pi/3} \tan^3 2x \sec^2 2x \, dx$

26. $\int_{-\pi/4}^{\pi/4} \tan^3 x \sec^3 x \, dx$

27. $\int \cot^3 x \csc^2 x \, dx$

28. $\int \cos^4 x \sin^3 x \, dx$

29. $\int \sec^4 3x \, dx$

30. $\int \cos^3 2x \, dx$

31. $\int \tan^3 2x \, dx$

32. $\int \tan^3 x \sec x \, dx$

33. $\int \sin^3 x \, dx$

34. $\int \frac{\cos x \, dx}{(1 + \sin x)^2}$

35. $\int \frac{\sec^2 x \, dx}{2 + \tan x}$

36. $\int \frac{\cos^3 t \, dt}{\sin^2 t}$

37. $\int \cot^3 x \, dx$

38. $\int \csc^3 2t \cot 2t \, dt$

39. $\int \csc^4 x \, dx$

40. $\int \frac{dx}{\cos x}$

41. $\int_0^{\pi/4} \frac{dx}{\sqrt{1 - \sin^2 x}}$

42. $\int_0^{\pi/4} x \sec^2 x \, dx$

43. $\int_0^{\pi/3} \frac{dx}{1 + \sin x}$

(*Hint:* Multiply numerator and denominator by $1 - \sin x$.)

44. $\int \csc \theta \, d\theta$

(*Hint:* Repeat Example 6 with cofunctions, using the relation $\csc^2 \theta = \cot^2 \theta + 1$.)

45. $\int_{-\pi}^{\pi} \sin 3x \cos 2x \, dx$

46. $\int_0^{\pi/2} \cos 3x \sin 2x \, dx$

47. $\int_{-\pi}^{\pi} \sin^2 3x \, dx$

48. $\int \sin x \cos x \, dx$

49. Use the result of Problem 44 and the identity $2 \sin A \cos A = \sin 2A$ to evaluate

$$\int \frac{\sec 2x \csc 2x}{2} \, dx.$$

50. Find the area of the surface generated by revolving the arc

$$x = t^{2/3}, \quad y = t^2/2, \quad 0 \le t \le 2,$$

about the x-axis.

51. Find the length of the curve

$$y = \ln(\cos x), \quad 0 \le x \le \pi/3.$$

52. Find the length of the curve

$$y = \ln(\sec x), \quad 0 \le x \le \pi/4.$$

53. Find the center of gravity of the region bounded by the x-axis, the curve $y = \sec x$, and the lines $x = -\pi/4$, $x = \pi/4$.

54. (*Calculator, or tables*) How far apart should the following lines of latitude be on the Mercator map of Example 8?

a) 30° and 45° north (about the latitudes of New Orleans, La., and Minneapolis, Minn.);
b) 45° and 60° north (about the latitudes of Salem, Ore., and Seward, Al.).

55. Show that

a) $\int_0^{2\pi} \sin mx \sin nx \, dx = 0,$

b) $\int_0^{2\pi} \sin px \cos qx \, dx = 0,$

when m, n, p, and q are integers and $m^2 \neq n^2$.

56. Two functions f and g are said to be *orthogonal* on an interval $a \le x \le b$ if $\int_a^b f(x)g(x) \, dx = 0$.

a) Prove that $\sin mx$ and $\sin nx$ are orthogonal on any interval of length 2π provided m and n are integers such that $m^2 \neq n^2$.
b) Prove the same for $\cos mx$ and $\cos nx$.
c) Prove the same for $\sin mx$ and $\cos nx$ even if $m = n$.

(Example 10 and Problem 55 prove these statements for the interval $[0, 2\pi]$.)

57. Derive a reduction formula for $\int \cot^n ax \, dx$ and use it to evaluate $\int \cot^4 3x \, dx$.

7.4 Even Powers of Sines and Cosines

In Article 7.3 we saw how to evaluate integrals of odd powers of sines and cosines. To set the stage for our discussion of even powers, we begin by looking at the technique we would normally use for odd powers and then examine what goes wrong when we try to use the same technique for even powers.

EXAMPLE 1 Evaluate

$$\int \cos^{2/3} x \sin^5 x \, dx.$$

Solution Here we have sin x to an *odd* power. So we put one factor of sin x with dx and the remaining sine factors, namely $\sin^4 x$, can be expressed in terms of cos x without introducing any square roots, as follows:

$$\sin^4 x = (\sin^2 x)^2 = (1 - \cos^2 x)^2.$$

The sin x goes well with dx when we take

$$u = \cos x, \qquad du = -\sin x \, dx$$

and evaluate the integral as follows:

$$\begin{aligned}\int \cos^{2/3} x \sin^5 x \, dx &= \int \cos^{2/3} x (1 - \cos^2 x)^2 \sin x \, dx \\ &= \int u^{2/3}(1 - u^2)^2(-du) \\ &= -\int (u^{2/3} - 2u^{8/3} + u^{14/3}) \, du \\ &= -\left[\frac{3}{5}u^{5/3} - \frac{6}{11}u^{11/3} + \frac{3}{17}u^{17/3}\right] + C \\ &= -\cos^{5/3} x \left[\frac{3}{5} - \frac{6}{11}\cos^2 x + \frac{3}{17}\cos^4 x\right] + C. \ \square\end{aligned}$$

DISCUSSION. If sin x had appeared to an *even* power in the integral, as it does, say, in the integral

$$\int \cos^{2/3} x \sin^4 x \, dx,$$

then the method above would not work. The substitution $\sin^4 x = (1 - \cos^2 x)^2$ would reduce the integral to

$$\int \cos^{2/3} x \sin^4 x \, dx = \int \cos^{2/3} x (1 - \cos^2 x)^2 \, dx,$$

and there would be no $-\sin x \, dx = du$ to go with the $\cos x = u$.

When the exponents m and n in the integral

$$\int \sin^m x \cos^n x \, dx \tag{1}$$

are even integers, we turn to one or both of the identities

$$\sin^2 A = \frac{1}{2}(1 - \cos 2A), \tag{2a}$$

$$\cos^2 A = \frac{1}{2}(1 + \cos 2A). \tag{2b}$$

These identities may be derived by adding or subtracting the equations

$$\cos^2 A + \sin^2 A = 1, \qquad \cos^2 A - \sin^2 A = \cos 2A,$$

and dividing by two.

EXAMPLE 2

$$\int \cos^4 x\, dx = \int (\cos^2 x)^2\, dx = \int \frac{1}{4}(1 + \cos 2x)^2\, dx \qquad (\text{2b with } A = x)$$

$$= \frac{1}{4}\int (1 + 2\cos 2x + \cos^2 2x)\, dx$$

$$= \frac{1}{4}\int \left[1 + 2\cos 2x + \frac{1}{2}(1 + \cos 4x)\right] dx \qquad \begin{pmatrix}\text{2b with}\\ A = 2x\end{pmatrix}$$

$$= \frac{3}{8}x + \frac{1}{4}\sin 2x + \frac{1}{32}\sin 4x + C. \ \square$$

An integral like

$$\int \sin^2 x \cos^4 x\, dx,$$

which involves even powers of both sin x and cos x, can be changed to a sum of integrals that involve only powers of one of them. Then these may be handled by the method of Example 2.

EXAMPLE 3

$$\int \sin^2 x \cos^4 x\, dx = \int (1 - \cos^2 x)\cos^4 x\, dx$$

$$= \int \cos^4 x\, dx - \int \cos^6 x\, dx.$$

We evaluated $\int \cos^4 x\, dx$ above, and

$$\int \cos^6 x\, dx = \int (\cos^2 x)^3\, dx = \frac{1}{8}\int (1 + \cos 2x)^3\, dx$$

$$= \frac{1}{8}\int (1 + 3\cos 2x + 3\cos^2 2x + \cos^3 2x)\, dx.$$

We now know how to handle each term of the integrand. The result is

$$\int \cos^6 x\, dx = \frac{5}{16}x + \frac{1}{4}\sin 2x + \frac{3}{64}\sin 4x - \frac{1}{48}\sin^3 2x + C.$$

Combining this with the result of Example 2 gives

$$\int \sin^2 x \cos^4 x\, dx = \int \cos^4 x\, dx - \int \cos^6 x\, dx$$

$$= \frac{1}{16}x - \frac{1}{64}\sin 4x + \frac{1}{48}\sin^3 2x + C. \ \square$$

EXAMPLE 4 Evaluate

$$\int_{-\pi/2}^{\pi/2} \sqrt{1 - \cos^2 t}\, dt.$$

Solution 1 We can use the substitution

$$\sin^2 t = 1 - \cos^2 t$$

to write

$$\int_{-\pi/2}^{\pi/2} \sqrt{1 - \cos^2 t}\, dt = \int_{-\pi/2}^{\pi/2} \sqrt{\sin^2 t}\, dt = \int_{-\pi/2}^{\pi/2} |\sin t|\, dt,$$

but we must not remove the absolute value signs without first checking the sign of sin t on the domain $-\pi/2 \le t \le \pi/2$. Indeed, sin t is negative to the left of 0 on this domain so that

$$|\sin t| = \begin{cases} -\sin t, & -\pi/2 \le t \le 0, \\ \sin t, & 0 \le t \le \pi/2. \end{cases}$$

To integrate correctly we must therefore write

$$\begin{aligned} \int_{-\pi/2}^{\pi/2} |\sin t|\, dt &= \int_{-\pi/2}^{0} -\sin t\, dt + \int_{0}^{\pi/2} \sin t\, dt \\ &= \cos t]_{-\pi/2}^{0} - \cos t]_{0}^{\pi/2} \\ &= (1 - 0) - (0 - 1) \\ &= 2. \end{aligned}$$

Solution 2 Another way to evaluate the integral is to observe that the function $|\sin t|$ is even, so that

$$\int_{-\pi/2}^{\pi/2} |\sin t|\, dt = 2\int_{0}^{\pi/2} |\sin t|\, dt = 2\int_{0}^{\pi/2} \sin t\, dt = -2\cos t\Big]_{0}^{\pi/2} = 2.$$

If we disregard the absolute value bars and mistakenly write $\sqrt{1 - \cos^2 t} = \sin t$, we obtain the following result,

$$\int_{-\pi/2}^{\pi/2} \sin t\, dt = -\cos t\Big]_{-\pi/2}^{\pi/2} = -0 + 0 = 0,$$

which is not the value of the original integral. □

EXAMPLE 5 Evaluate

$$\int_{0}^{\pi/4} \sqrt{1 + \cos 4x}\, dx.$$

Solution To eliminate the square root we use the identity

$$\cos^2 A = \frac{1 + \cos 2A}{2},$$

or

$$1 + \cos 2A = 2\cos^2 A,$$

from Eq. (2b). With $A = 2x$, this becomes

$$1 + \cos 4x = 2\cos^2 2x.$$

Therefore,

$$\begin{aligned} \int_{0}^{\pi/4} \sqrt{1 + \cos 4x}\, dx &= \int_{0}^{\pi/4} \sqrt{2\cos^2 2x}\, dx \\ &= \int_{0}^{\pi/4} \sqrt{2}\sqrt{\cos^2 2x}\, dx \\ &= \sqrt{2}\int_{0}^{\pi/4} |\cos 2x|\, dx \\ &= \sqrt{2}\int_{0}^{\pi/4} \cos 2x\, dx \qquad \left(\begin{matrix}\text{because } \cos 2x \ge 0 \\ \text{on } [0, \pi/4]\end{matrix}\right) \\ &= \sqrt{2}\left[\frac{\sin 2x}{2}\right]_{0}^{\pi/4} \\ &= \frac{\sqrt{2}}{2}[1 - 0] \\ &= \frac{\sqrt{2}}{2}. \end{aligned}$$ □

PROBLEMS

Evaluate the integrals in Problems 1–24.

1. $\int \sin^2 x \cos^2 x\, dx$

2. $\int \sin^4 y \cos^2 y\, dy$

3. $\int \sin^2 2t\, dt$

4. $\int \cos^2 3\theta\, d\theta$

5. $\int \sin^4 ax\, dx$

6. $\int \frac{\sin^4 x\, dx}{\cos^2 x}$

7. $\int \frac{dx}{\cos^2 x}$

8. $\int \frac{dx}{\sin^4 x}$

9. $\int \frac{\sin^2 z}{\cos z}\, dz$

10. $\int \frac{\sin^2 x}{\cos^2 x}\, dx$

11. $\int \frac{\sin^3 x}{\cos^2 x}\, dx$

12. $\int_0^{\pi} \sin^2 x\, dx$

13. $\int \frac{\cos 2t\, dt}{\sin^4 2t}$

14. $\int \sin^6 x\, dx$

15. $\int_0^{2\pi} \sqrt{\frac{1 - \cos t}{2}}\, dt$

16. $\int_0^{\pi} \sqrt{1 - \cos 2x}\, dx$

17. $\int_0^{\pi/10} \sqrt{1 + \cos 5\theta}\, d\theta$

18. $\int_0^{2\pi} \sqrt{1 + \cos (y/4)}\, dy$

19. $\int \theta \sqrt{1 - \cos \theta}\, d\theta$

(*Hint:* Substitute $\sqrt{1 - \cos \theta} = \sqrt{2} \sin (\theta/2)$.)

20. $\int_0^{\pi} \sqrt{1 - \sin^2 t}\, dt$

21. $\int_{-\pi/4}^{\pi/4} \sqrt{1 + \tan^2 x}\, dx$

22. $\int_{-\pi/4}^{\pi/4} \sqrt{\sec^2 x - 1}\, dx$

23. $\int_0^{\pi} \sqrt{1 - \cos^2 \theta}\, d\theta$

24. $\int_0^{\pi} \sqrt{1 - \cos^2 2x}\, dx$

25. The graph of $x = t - \sin t$, $y = 1 - \cos t$, $0 \le t \le 2\pi$, is an arch standing on the x-axis. Find the surface area generated by rotating the arch about the x-axis.

26. Find the volume generated by revolving one arch of the curve $y = \sin x$ about the x-axis.

27. Find the area between the x-axis and the curve $y = \sqrt{1 + \cos 4x}$, $0 \le x \le \pi$.

7.5

Trigonometric Substitutions in Integrals Involving $a^2 + u^2$, $\sqrt{a^2 - u^2}$, $\sqrt{a^2 + u^2}$, $\sqrt{u^2 - a^2}$

Some of these follow directly from the integral formulas in Article 7.1. For example, we have

$$\int \frac{du}{1 + u^2} = \tan^{-1} u + C, \tag{1}$$

and, to evaluate

$$\int \frac{du}{a^2 + u^2}, \tag{2}$$

we substitute

$$u = az, \quad du = a\, dz$$

and proceed as follows:

$$\int \frac{du}{a^2 + u^2} = \int \frac{a\, dz}{a^2 + a^2 z^2} = \frac{1}{a} \int \frac{dz}{1 + z^2}$$

$$= \frac{1}{a} \tan^{-1} z + C = \frac{1}{a} \tan^{-1} \frac{u}{a} + C.$$

That is,

$$\int \frac{du}{a^2 + u^2} = \frac{1}{a} \tan^{-1} \frac{u}{a} + C. \tag{3}$$

The essential feature here was the introduction of a new variable by the substitution

$$z = \frac{u}{a} \qquad \text{or} \qquad u = az.$$

Such a substitution allows the a terms to be brought outside the integral sign, and the resulting integral may then match one of the inverse trigonometric formulas.

The following alternative substitutions, however, allow us to bypass the job of learning these formulas. In addition, these substitutions permit the evaluation of many other integrals. To arrive at these substitutions we use the identities

$$\begin{aligned} 1 - \sin^2 \theta &= \cos^2 \theta, \\ 1 + \tan^2 \theta &= \sec^2 \theta, \\ \sec^2 \theta - 1 &= \tan^2 \theta. \end{aligned} \tag{4a}$$

These may be multiplied by a^2 to obtain

$$\begin{aligned} a^2 - a^2 \sin^2 \theta &= a^2 \cos^2 \theta, \\ a^2 + a^2 \tan^2 \theta &= a^2 \sec^2 \theta, \\ a^2 \sec^2 \theta - a^2 &= a^2 \tan^2 \theta, \end{aligned} \tag{4b}$$

with the result that the substitutions listed below have the following effects:

$u = a \sin \theta$	replaces	$a^2 - u^2$	by	$a^2 \cos^2 \theta$,	(5a)
$u = a \tan \theta$	replaces	$a^2 + u^2$	by	$a^2 \sec^2 \theta$,	(5b)
$u = a \sec \theta$	replaces	$u^2 - a^2$	by	$a^2 \tan^2 \theta$.	(5c)

It is thus seen that corresponding to each of the *binomial* expressions

$$a^2 - u^2, \qquad a^2 + u^2, \qquad \text{and} \qquad u^2 - a^2,$$

we have a substitution that replaces the binomial by a single squared term. The particular substitution to use will depend upon the form of the integrand. In the examples that follow, a is a positive constant.

EXAMPLE 1 Evaluate

$$\int \frac{du}{\sqrt{a^2 - u^2}}, \qquad |a| > |u|.$$

Solution We try the substitution (5a):

$$u = a \sin \theta, \qquad du = a \cos \theta \, d\theta,$$
$$a^2 - u^2 = a^2(1 - \sin^2 \theta) = a^2 \cos^2 \theta.$$

Then

$$\int \frac{du}{\sqrt{a^2 - u^2}} = \int \frac{a \cos \theta \, d\theta}{\sqrt{a^2 \cos^2 \theta}}$$

$$= \int \frac{a \cos \theta \, d\theta}{\pm a \cos \theta} \quad (\pm \text{ depends on sign of } \cos \theta)$$

$$= \pm \int d\theta = \pm(\theta + C).$$

Since $\sin \theta = u/a$,

$$\theta = \sin^{-1} \frac{u}{a}$$

and

$$\int \frac{du}{\sqrt{a^2 - u^2}} = \pm\left(\sin^{-1} \frac{u}{a} + C\right).$$

Since $\theta = \sin^{-1} u/a$ lies between $-\pi/2$ and $\pi/2$, where the values of $\cos \theta$ are nonnegative, the ambiguous sign is $+$; that is,

$$\int \frac{du}{\sqrt{a^2 - u^2}} = \sin^{-1} \frac{u}{a} + C. \; \square \qquad (6)$$

EXAMPLE 2 Evaluate

$$\int \frac{du}{\sqrt{a^2 + u^2}}, \qquad a > 0.$$

Solution This time we try

$$u = a \tan \theta,$$
$$du = a \sec^2 \theta \, d\theta,$$
$$a^2 + u^2 = a^2(1 + \tan^2 \theta) = a^2 \sec^2 \theta.$$

Then

$$\int \frac{du}{\sqrt{a^2 + u^2}} = \int \frac{a \sec^2 \theta \, d\theta}{\sqrt{a^2 \sec^2 \theta}}$$

$$= \pm \int \sec \theta \, d\theta \quad (\pm \text{ depends on sign of } \sec \theta).$$

By Eq. (6), Article 7.3, we know that

$$\int \sec \theta \, d\theta = \ln |\sec \theta + \tan \theta| + C.$$

If we take

$$\theta = \tan^{-1} \frac{u}{a}, \qquad -\frac{\pi}{2} < \theta < \frac{\pi}{2},$$

then $\sec \theta$ is positive, and

$$\int \frac{du}{\sqrt{a^2 + u^2}} = \int \sec \theta \, d\theta$$

$$= \ln |\sec \theta + \tan \theta| + C = \ln \left| \frac{\sqrt{a^2 + u^2}}{a} + \frac{u}{a} \right| + C$$

$$= \ln |\sqrt{a^2 + u^2} + u| + C',$$

where

$$C' = C - \ln a.$$

That is,

$$\int \frac{du}{\sqrt{a^2 + u^2}} = \ln|\sqrt{a^2 + u^2} + u| + C'. \quad \square \tag{7}$$

EXAMPLE 3 Evaluate

$$\int \frac{du}{\sqrt{u^2 - a^2}}, \qquad |u| > a > 0.$$

Solution We try the substitution

$$\begin{aligned} u &= a \sec\theta, \\ du &= a \sec\theta \tan\theta \, d\theta, \\ u^2 - a^2 &= a^2(\sec^2\theta - 1) = a^2 \tan^2\theta. \end{aligned}$$

Then

$$\begin{aligned} \int \frac{du}{\sqrt{u^2 - a^2}} &= \int \frac{a \sec\theta \tan\theta \, d\theta}{\sqrt{a^2 \tan^2\theta}} \\ &= \pm \int \sec\theta \, d\theta \quad (\pm \text{ depends on sign of } \tan\theta). \end{aligned}$$

If we take

$$\theta = \sec^{-1}\frac{u}{a}, \quad 0 \le \theta \le \pi,$$

then

$$\begin{aligned} &\tan\theta \text{ is positive} \quad \text{if } 0 < \theta < \pi/2, \\ &\tan\theta \text{ is negative} \quad \text{if } \pi/2 < \theta < \pi, \end{aligned}$$

and, from Eq. (6) of Article 7.3,

$$\pm \int \sec\theta \, d\theta = \pm\ln|\sec\theta + \tan\theta| + C.$$

When $\tan\theta$ is positive, we must use the plus sign; when $\tan\theta$ is negative, the minus sign. Moreover,

$$\sec\theta = \frac{u}{a}, \qquad \tan\theta = \pm\frac{\sqrt{u^2 - a^2}}{a}.$$

So we have

$$\begin{aligned} \int \frac{du}{\sqrt{u^2 - a^2}} &= \pm\ln\left|\frac{u}{a} \pm \frac{\sqrt{u^2 - a^2}}{a}\right| + C \\ &= \begin{cases} \ln\left|\dfrac{u}{a} + \dfrac{\sqrt{u^2 - a^2}}{a}\right| + C \\ \text{or} \\ -\ln\left|\dfrac{u}{a} - \dfrac{\sqrt{u^2 - a^2}}{a}\right| + C. \end{cases} \end{aligned}$$

But the two forms are actually equal, because

$$\begin{aligned}
-\ln\left|\frac{u}{a}-\frac{\sqrt{u^2-a^2}}{a}\right| &= \ln\left|\frac{a}{u-\sqrt{u^2-a^2}}\right| \\
&= \ln\left|\frac{a(u+\sqrt{u^2-a^2})}{(u-\sqrt{u^2-a^2})(u+\sqrt{u^2-a^2})}\right| \\
&= \ln\left|\frac{a(u+\sqrt{u^2-a^2})}{a^2}\right| \\
&= \ln\left|\frac{u+\sqrt{u^2-a^2}}{a}\right|.
\end{aligned}$$

Therefore

$$\int \frac{du}{\sqrt{u^2-a^2}} = \ln|u+\sqrt{u^2-a^2}| + C', \tag{8}$$

where we have replaced $C - \ln a$ by C'. □

EXAMPLE 4 Evaluate

$$\int \frac{du}{u\sqrt{u^2-a^2}}.$$

Solution This follows directly from Formula 17 in Article 7.1. We let

$$u = az, \quad du = a\,dz.$$

Then

$$\begin{aligned}
\int \frac{du}{u\sqrt{u^2-a^2}} &= \int \frac{a\,dz}{az\sqrt{a^2z^2-a^2}} \\
&= \frac{1}{a}\int \frac{dz}{z\sqrt{z^2-1}} \\
&= \frac{1}{a}\sec^{-1}|z| + C \\
&= \frac{1}{a}\sec^{-1}\left|\frac{u}{a}\right| + C.
\end{aligned}$$

That is,

$$\int \frac{du}{u\sqrt{u^2-a^2}} = \frac{1}{a}\sec^{-1}\left|\frac{u}{a}\right| + C. \;\square \tag{9}$$

EXAMPLE 5 Evaluate

$$\int \frac{x^2\,dx}{\sqrt{9-x^2}}.$$

Solution If we substitute

$$x = 3\sin\theta, \quad \frac{-\pi}{2} < \theta < \frac{\pi}{2},$$

$$dx = 3\cos\theta\,d\theta,$$

$$9 - x^2 = 9(1-\sin^2\theta) = 9\cos^2\theta,$$

then we find

$$\int \frac{x^2\,dx}{\sqrt{9 - x^2}} = \int \frac{9\sin^2\theta \cdot 3\cos\theta\,d\theta}{3\cos\theta} = 9\int \sin^2\theta\,d\theta.$$

This is considerably simpler than the integral we started with. To evaluate it, we use the identity

$$\sin^2\theta = \frac{1}{2}(1 - \cos 2\theta).$$

Then

$$\int \sin^2\theta\,d\theta = \frac{1}{2}\int (1 - \cos 2\theta)\,d\theta = \frac{1}{2}\left[\theta - \frac{1}{2}\sin 2\theta\right] + C$$
$$= \frac{1}{2}[\theta - \sin\theta\cos\theta] + C.$$

Substituting this above, we get

$$\int \frac{x^2\,dx}{\sqrt{9 - x^2}} = \frac{9}{2}[\theta - \sin\theta\cos\theta] + C.$$

To express $\sin\theta$ in terms of x we use the original substitution $x = 3\sin\theta$, which gives

$$\sin\theta = \frac{x}{3}.$$

To express $\cos\theta$ in terms of x we may take one of two approaches. One is to use the algebraic relation

$$\cos\theta = \sqrt{1 - \sin^2\theta} = \sqrt{1 - \left(\frac{x}{3}\right)^2} = \frac{\sqrt{9 - x^2}}{3}.$$

The other is to draw right triangles (Fig. 7.8) with θ in one corner and whose sides express the relation

$$\sin\theta = \frac{x}{3}.$$

With hypotenuse 3 and opposite side x, the side adjacent to θ has length $\sqrt{9 - x^2}$ (from the Pythagorean theorem). The triangles therefore give

$$\cos\theta = \frac{\sqrt{9 - x^2}}{3}.$$

The geometric approach is sometimes preferred to the algebraic approach in problems like this because it gives the values of all trigonometric functions of the angle in question. You do not need to make a separate computation for each function.

To continue the calculation, then,

$$\int \frac{x^2\,dx}{\sqrt{9 - x^2}} = \frac{9}{2}[\theta - \sin\theta\cos\theta] + C$$
$$= \frac{9}{2}\left[\sin^{-1}\frac{x}{3} - \frac{x}{3}\cdot\frac{\sqrt{9 - x^2}}{3}\right] + C$$
$$= \frac{9}{2}\left[\sin^{-1}\frac{x}{3} - \frac{x\sqrt{9 - x^2}}{9}\right] + C. \quad \square$$

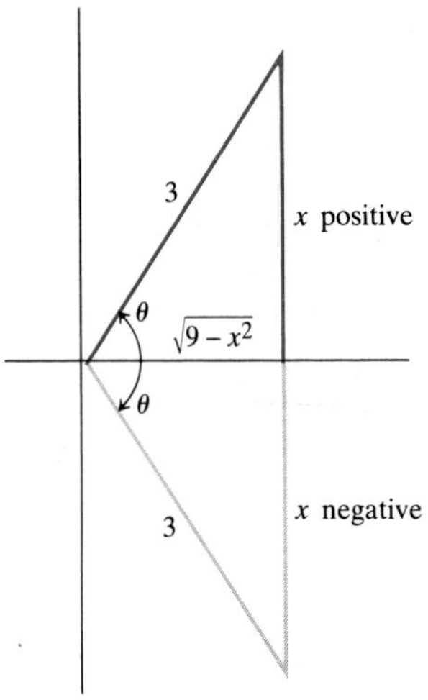

7.8 If $-\pi/2 < \theta < \pi/2$ and $\sin\theta = x/3$, then $\cos\theta = \sqrt{9 - x^2}/3$ in either triangle.

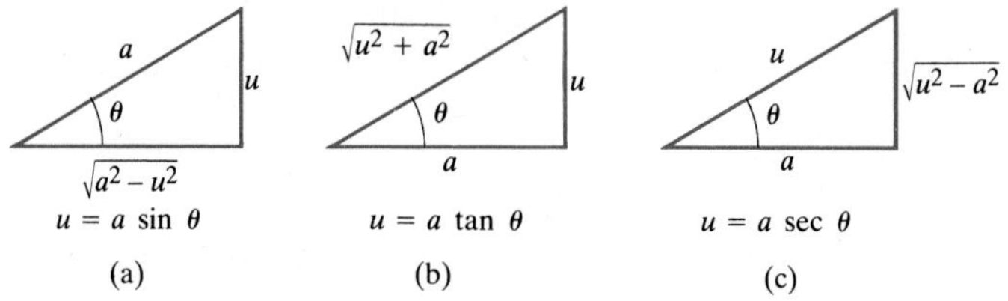

7.9 If $u = a \sin\theta$, then the values of the trigonometric functions of θ can be read from triangle (a). Triangles (b) and (c) give the values of the trigonometric functions of θ for the substitutions $u = a \tan\theta$ and $u = a \sec\theta$.

The trigonometric substitutions (a), (b), and (c) of Eq. (5) can be remembered by thinking of the theorem of Pythagoras and taking a and u as two of the sides and θ as an angle in a right triangle. Thus $\sqrt{a^2 - u^2}$ suggests a as hypotenuse and u as a leg, $\sqrt{a^2 + u^2}$ suggests a and u as the legs, and $\sqrt{u^2 - a^2}$ suggests u as hypotenuse and a as a leg. These situations are shown in Fig. 7.9.

The trigonometric identities, Eqs. (4a), are simply equivalent expressions of the theorem of Pythagoras applied to the right triangles in Fig. 7.10.

Note that whenever we evaluate an indefinite integral by changing variables, i.e., by substitution, we express the final result in terms of the original variable of integration. When we have a definite integral to evaluate, however, another course of action is open to us. We can transform the limits of integration by the same formula we use to change the variable of integration, and then evaluate the transformed integral with the transformed limits.

EXAMPLE 6 Find the area of the quarter circle bounded by the curve $y = \sqrt{1 - x^2}$ in the first quadrant.

Solution The area of the quarter circle is

$$\int_0^1 \sqrt{1 - x^2}\, dx.$$

If we substitute $x = \sin\theta$, $dx = \cos\theta\, d\theta$, then $\theta = 0$ when $x = 0$, and $\theta = \pi/2$ when $x = 1$. Accordingly,

$$\int_0^1 \sqrt{1 - x^2}\, dx = \int_0^{\pi/2} \sqrt{1 - \sin^2\theta}\, \cos\theta\, d\theta = \int_0^{\pi/2} \cos^2\theta\, d\theta$$

$$= \int_0^{\pi/2} \frac{1 + \cos 2\theta}{2}\, d\theta = \frac{\theta}{2} + \frac{\sin 2\theta}{4}\Big]_0^{\pi/2} = \frac{\pi}{4} + 0 = \frac{\pi}{4}. \quad \square$$

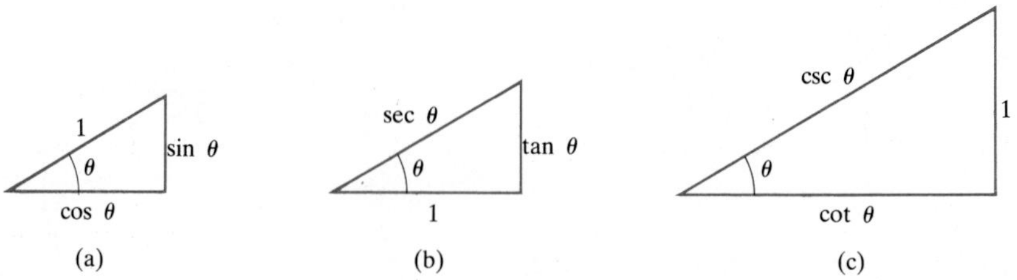

7.10 The triangles in (b) and (c) are obtained from the basic triangle in (a) by dividing all its sides by $\cos\theta$ or $\sin\theta$, respectively, to obtain similar triangles.

Integration Formulas from Article 7.5

1. $\int \frac{du}{a^2 + u^2} = \frac{1}{a} \tan^{-1} \frac{u}{a} + C$ (3)

2. $\int \frac{du}{\sqrt{a^2 - u^2}} = \sin^{-1} \frac{u}{a} + C$ (6)

3. $\int \frac{du}{\sqrt{a^2 + u^2}} = \ln |\sqrt{a^2 + u^2} + u| + C$ (7)

4. $\int \frac{du}{\sqrt{u^2 - a^2}} = \ln |u + \sqrt{u^2 - a^2}| + C$ (8)

5. $\int \frac{du}{u\sqrt{u^2 - a^2}} = \frac{1}{a} \sec^{-1} \left|\frac{u}{a}\right| + C$ (9)

PROBLEMS

In Problems 1–15, evaluate the integrals.

1. $\int_{-2}^{2} \frac{dx}{4 + x^2}$

2. $\int \frac{dx}{8 + 2x^2}$

3. $\int \frac{dx}{1 + 4x^2}$

4. $\int_0^{3/2} \frac{dx}{\sqrt{9 - x^2}}$

5. $\int \frac{2\,dx}{\sqrt{1 - 4x^2}}$

6. $\int \frac{dx}{\sqrt{9 - 4x^2}}$

7. $\int \frac{dy}{\sqrt{25 + y^2}}$

8. $\int \frac{3\,dy}{\sqrt{1 + 9y^2}}$

9. $\int \frac{dy}{\sqrt{25 + 9y^2}}$

10. $\int \frac{dz}{\sqrt{z^2 - 4}}$

11. $\int \frac{3\,dz}{\sqrt{9z^2 - 1}}$

12. $\int \frac{dz}{\sqrt{25z^2 - 9}}$

13. $\int \frac{dx}{x\sqrt{x^2 - 16}}$

14. $\int \frac{dx}{x\sqrt{x^2 - 3}}$

15. $\int \frac{dx}{2x\sqrt{4x^2 - 1}}$

In Problems 16–23, evaluate the integrals with the substitutions given.

16. $\int \frac{dy}{y^2\sqrt{y^2 - 16}}$, $y = 4 \sec u$

17. $\int_2^4 \sqrt{x^2 - 4}\,dx$, $x = 2 \sec u$

18. $\int_{1/2}^1 \frac{\sqrt{1 - x}}{x}\,dx$, $x = \cos^2 u$

19. $\int \frac{x^3\,dx}{\sqrt{1 - x^2}}$, $x = \cos u$

20. $\int \frac{dx}{x^2\sqrt{9 - x^2}}$, $x = 3 \sin u$

21. $\int_0^{1/2} \frac{dx}{\sqrt{1 + x^2}}$, $x = \tan u$

22. $\int_0^1 x^4\sqrt{1 - x^2}\,dx$, $x = \sin u$

23. $\int \frac{dx}{x^2\sqrt{x^2 - 1}}$, $x = \csc u$

In Problems 24–46, evaluate the integrals.

24. $\int_0^{0.6a} \frac{x\,dx}{\sqrt{a^2 - x^2}}$

25. $\int \frac{dx}{\sqrt{1 - 4x^2}}$

26. $\int_0^a \sqrt{a^2 - x^2}\,dx$

27. $\int \frac{dx}{\sqrt{4 - (x - 1)^2}}$

28. $\int \sec 2t\,dt$

29. $\int_0^2 \frac{dx}{\sqrt{4 + x^2}}$

30. $\int_0^1 \frac{dx}{\sqrt{4 - x^2}}$

31. $\int \frac{x\,dx}{\sqrt{4 + x^2}}$

32. $\int_0^1 \frac{dx}{4 - x^2}$

33. $\int_0^1 \frac{x^3\,dx}{\sqrt{x^2 + 1}}$

34. $\int \frac{dx}{x\sqrt{a^2 + x^2}}$

35. $\int \frac{x + 1}{\sqrt{4 - x^2}}\,dx$

36. $\int \frac{dx}{\sqrt{2 - 5x^2}}$

37. $\int \frac{\sin\theta\,d\theta}{\sqrt{2 - \cos^2\theta}}$

38. $\int \frac{dx}{x\sqrt{x^2 - a^2}}$

39. $\int \frac{dx}{x\sqrt{a^2 - x^2}}$

40. $\int \frac{dx}{(a^2 - x^2)^{3/2}}$

41. $\int \frac{dx}{(a^2 + x^2)^2}$

42. $\int \frac{\sqrt{1 - x^2}}{x^2} dx$

43. $\int x \sin^{-1} x \, dx$

44. $\int_0^{(1/2)\ln 3} \frac{e^x}{1 + e^{2x}} dx$

45. $\int \frac{dx}{x^2\sqrt{1 - x^2}}$

46. $\int \frac{4x^2 \, dx}{(1 - x^2)^{3/2}}$

47. Evaluate

$$\int \frac{y \, dy}{\sqrt{16 - y^2}}$$

a) without a trigonometric substitution,
b) with a trigonometric substitution.

48. Evaluate

a) $\int_0^2 \frac{x \, dx}{4 + x^2}$, b) $\int_0^2 \frac{dx}{4 + x^2}$.

49. Evaluate

$$\int \frac{x \, dx}{(x^2 - 1)^{3/2}}$$

a) without a trigonometric substitution,
b) with a trigonometric substitution.

50. Evaluate

a) $\int_0^{\sqrt{3}/2} \frac{x \, dx}{\sqrt{1 - x^2}}$, b) $\int_0^{\sqrt{3}/2} \frac{dx}{\sqrt{1 - x^2}}$.

51. Find the area bounded by the curve $y = \sqrt{1 - (x^2/9)}$ in the first quadrant.

52. Suppose $\theta = \sin^{-1}(u/2)$. Express $\cos\theta$ and $\tan\theta$ in terms of u. (See Fig. 7.9.)

53. Find the solution of the differential equation

$$\frac{dy}{dx} = y^2\left(1 - \frac{1}{\sqrt{4 - x^2}}\right)$$

that passes through the point $(\sqrt{2}, 4/\pi)$.

54. Derive the reduction formula in part (a) and apply it to evaluate the integrals in parts (b) and (c).

a) $\int \sec^n x \, dx = \frac{\sec^{n-2} x \tan x}{(n - 1)} + \frac{n - 2}{n - 1}\int \sec^{n-2} x \, dx$

b) $\int \sec^3 x \, dx$ (See Eq. (6), Article 7.3.)

c) $\int \sqrt{a^2 + x^2} \, dx$ (*Hint:* Let $x = a \tan\theta$.)

55. Express in terms of u and a:

a) $\sin\left(\tan^{-1}\frac{u}{a}\right)$, b) $\cos\left(\sec^{-1}\frac{u}{a}\right)$.

7.6

Integrals Involving $ax^2 + bx + c$

The general quadratic

$$f(x) = ax^2 + bx + c, \quad a \neq 0 \tag{1}$$

can be reduced to the form $a(u^2 + B)$ by completing the square, as follows:

$$ax^2 + bx + c = a\left(x^2 + \frac{b}{a}x\right) + c = a\left(x^2 + \frac{b}{a}x + \frac{b^2}{4a^2}\right) + c - \frac{b^2}{4a}$$
$$= a\left(x + \frac{b}{2a}\right)^2 + \frac{4ac - b^2}{4a},$$

and then substituting

$$u = x + \frac{b}{2a}, \quad B = \frac{4ac - b^2}{4a^2}, \tag{2}$$

which gives us

$$f(x) = ax^2 + bx + c = a(u^2 + B).$$

When the integrand involves the square root of $f(x) = ax^2 + bx + c$, we restrict attention to the case where $f(x)$ is not negative. If a is negative and

B is positive, the square root is imaginary. We disregard this case, and consider

$$\sqrt{a(u^2 + B)}$$

1. when a is positive, and
2. when a and B are both negative.

In the first case, the relation

$$\sqrt{a(u^2 + B)} = \sqrt{a}\sqrt{u^2 + B}$$

allows us to reduce the problem to a consideration of $\sqrt{u^2 + B}$, which we handle by the methods of Article 7.5. In the second case, when a and B are both negative, then $-a$ and $-B$ are positive and

$$\begin{aligned}\sqrt{a(u^2 + B)} &= \sqrt{-a(-B - u^2)} \\ &= \sqrt{-a}\sqrt{-B - u^2} \qquad (-a > 0,\ -B > 0),\end{aligned}$$

and, taking $-B = A^2$, say, we consider $\sqrt{A^2 - u^2}$ as in Article 7.5.

When the integrand does not involve an even root of $f(x)$, there need be no restriction on the signs of a and B. In these cases also, we try to use the trigonometric substitutions of the previous section, or to apply the integration formulas derived there.

EXAMPLE 1 Evaluate

$$\int \frac{dx}{\sqrt{2x - x^2}}.$$

Solution With $u = x - 1$,

$$\sqrt{2x - x^2} = \sqrt{-(x^2 - 2x)} = \sqrt{-(x^2 - 2x + 1) + 1} = \sqrt{1 - u^2}$$

and

$$\int \frac{dx}{\sqrt{2x - x^2}} = \int \frac{du}{\sqrt{1 - u^2}} = \sin^{-1} u + C.$$

That is,

$$\int \frac{dx}{\sqrt{2x - x^2}} = \sin^{-1}(x - 1) + C. \ \square$$

EXAMPLE 2 Evaluate

$$\int \frac{dx}{4x^2 + 4x + 2}.$$

Solution Again we start with algebraic transformations:

$$\begin{aligned}4x^2 + 4x + 2 &= 4(x^2 + x) + 2 \\ &= 4(x^2 + x + \tfrac{1}{4}) + (2 - \tfrac{4}{4}) \\ &= 4(x + \tfrac{1}{2})^2 + 1.\end{aligned}$$

Then we let

$$u = x + \tfrac{1}{2}, \qquad du = dx,$$

and

$$\int \frac{dx}{4x^2 + 4x + 2} = \int \frac{du}{4u^2 + 1} = \frac{1}{4} \int \frac{du}{u^2 + \frac{1}{4}}$$
$$= \frac{1}{4} \int \frac{du}{u^2 + a^2} \qquad (a = \tfrac{1}{2})$$
$$= \frac{1}{4} \frac{1}{a} \tan^{-1} \frac{u}{a} + C$$
$$= \frac{1}{2} \tan^{-1} (2x + 1) + C. \ \square$$

EXAMPLE 3 Evaluate

$$\int \frac{(x + 1)\,dx}{\sqrt{2x^2 - 6x + 4}}.$$

Solution The quadratic part may be reduced as follows:

$$2x^2 - 6x + 4 = 2(x^2 - 3x) + 4$$
$$= 2(x^2 - 3x + \tfrac{9}{4}) + 4 - \tfrac{9}{2} = 2(u^2 - a^2)$$

with

$$u = x - \tfrac{3}{2}, \qquad a = \tfrac{1}{2}.$$

Then

$$x = u + \tfrac{3}{2}, \qquad dx = du, \qquad x + 1 = u + \tfrac{5}{2},$$

and

$$\int \frac{(x + 1)\,dx}{\sqrt{2x^2 - 6x + 4}} = \int \frac{(u + \frac{5}{2})\,du}{\sqrt{2(u^2 - a^2)}}$$
$$= \frac{1}{\sqrt{2}} \int \frac{u\,du}{\sqrt{u^2 - a^2}} + \frac{5}{2\sqrt{2}} \int \frac{du}{\sqrt{u^2 - a^2}}.$$

In the former of these, we let

$$z = u^2 - a^2, \qquad dz = 2u\,du, \qquad u\,du = \tfrac{1}{2}\,dz,$$

and to the latter we apply Eq. (8), Article 7.5. That is,

$$\frac{1}{\sqrt{2}} \int \frac{u\,du}{\sqrt{u^2 - a^2}} = \frac{1}{2\sqrt{2}} \int \frac{dz}{\sqrt{z}} = \frac{1}{2\sqrt{2}} \int z^{-(1/2)}\,dz$$
$$= \frac{1}{2\sqrt{2}} \frac{z^{1/2}}{\frac{1}{2}} + C_1 = \sqrt{\frac{u^2 - a^2}{2}} + C_1$$

and

$$\frac{5}{2\sqrt{2}} \int \frac{du}{\sqrt{u^2 - a^2}} = \frac{5}{2\sqrt{2}} \ln |u + \sqrt{u^2 - a^2}| + C_2,$$

so that

$$\int \frac{(x + 1)\,dx}{\sqrt{2x^2 - 6x + 4}}$$
$$= \sqrt{\frac{u^2 - a^2}{2}} + \frac{5}{2\sqrt{2}} \ln |u + \sqrt{u^2 - a^2}| + C$$
$$= \sqrt{\frac{x^2 - 3x + 2}{2}} + \frac{5}{2\sqrt{2}} \ln \left| x - \frac{3}{2} + \sqrt{x^2 - 3x + 2} \right| + C. \ \square$$

PROBLEMS

In Problems 1–24, evaluate the integrals.

1. $\int_1^3 \frac{dx}{x^2 - 2x + 5}$
2. $\int \frac{dx}{\sqrt{x^2 - 2x + 5}}$
3. $\int \frac{x\,dx}{x^2 - 2x + 5}$
4. $\int \frac{x\,dx}{\sqrt{x^2 - 2x + 5}}$
5. $\int_1^2 \frac{dx}{x^2 - 2x + 4}$
6. $\int_1^2 \frac{3\,dx}{9x^2 - 6x + 5}$
7. $\int \frac{dx}{\sqrt{9x^2 - 6x + 5}}$
8. $\int \frac{3x\,dx}{9x^2 - 6x + 5}$
9. $\int \frac{x\,dx}{\sqrt{9x^2 - 6x + 5}}$
10. $\int \frac{dx}{\sqrt{x^2 - 2x}}$
11. $\int \frac{dx}{\sqrt{x^2 + 2x}}$
12. $\int_{-2}^1 \frac{dx}{\sqrt{x^2 + 4x + 13}}$
13. $\int \frac{x\,dx}{\sqrt{x^2 + 4x + 13}}$
14. $\int_{-1}^0 \frac{dx}{\sqrt{3 - 2x - x^2}}$
15. $\int \frac{dx}{\sqrt{x^2 - 2x - 3}}$
16. $\int \frac{dx}{(x + 1)\sqrt{x^2 + 2x}}$
17. $\int \frac{(x + 1)\,dx}{\sqrt{2x - x^2}}$
18. $\int \frac{(x - 1)\,dx}{\sqrt{x^2 - 4x + 3}}$
19. $\int \frac{x\,dx}{\sqrt{5 + 4x - x^2}}$
20. $\int \frac{dx}{\sqrt{x^2 - 2x - 8}}$
21. $\int \frac{(1 - x)\,dx}{\sqrt{8 + 2x - x^2}}$
22. $\int \frac{x\,dx}{\sqrt{x^2 + 4x + 5}}$
23. $\int \frac{x\,dx}{x^2 + 4x + 5}$
24. $\int \frac{(2x + 3)\,dx}{4x^2 + 4x + 5}$

Find the limits in Problems 25–26.

25. $\lim_{a \to 1^+} \int_a^3 \frac{dx}{(x + 1)\sqrt{x^2 + 2x - 3}}$
26. $\lim_{a \to -5^+} \int_a^{-4} \frac{dx}{\sqrt{-x^2 - 8x - 15}}$

7.7 Partial Fractions

In algebra we learned how to combine fractions over a common denominator. In integration, it is desirable to reverse the process and split a fraction into a sum of fractions with simpler denominators, which are easier to integrate. The technique for doing so is called the *method of partial fractions*. We shall look at some examples, present the method in general terms, and then describe a shortcut called the *Heaviside "cover-up" method* (invented by Oliver Heaviside (1850–1925), a pioneer in electrical engineering and vector analysis).

EXAMPLE 1

$$\frac{2}{x + 1} + \frac{3}{x - 3} = \frac{2(x - 3) + 3(x + 1)}{(x + 1)(x - 3)} = \frac{5x - 3}{(x + 1)(x - 3)}. \tag{1}$$

Adding the fractions on the left produces the fraction on the right. The reverse process consists in finding constants A and B such that

$$\frac{5x - 3}{(x + 1)(x - 3)} = \frac{A}{x + 1} + \frac{B}{x - 3}.$$

(Pretend, for a moment, we don't know that $A = 2$, $B = 3$ will work.) We call A and B *undetermined coefficients*. Clearing of fractions, we have

$$\begin{aligned} 5x - 3 &= A(x - 3) + B(x + 1) \\ &= (A + B)x - 3A + B. \end{aligned}$$

This will be an identity in x if and only if coefficients of like powers of x on the two sides of the equation are equal:

$$A + B = 5, \qquad -3A + B = -3.$$

These two equations in two unknowns determine

$$A = 2, \qquad B = 3. \ \square$$

EXAMPLE 2 *Two linear factors in the denominator.* Evaluate

$$\int \frac{5x - 3}{(x + 1)(x - 3)}\, dx.$$

Solution From Example 1,

$$\begin{aligned} \int \frac{5x - 3}{(x + 1)(x - 3)}\, dx &= \int \frac{2}{x + 1}\, dx + \int \frac{3}{x - 3}\, dx \\ &= 2 \ln |x + 1| + 3 \ln |x - 3| + C. \ \square \end{aligned}$$

EXAMPLE 3 *A repeated linear factor in the denominator.* Express

$$\frac{6x + 7}{(x + 2)^2}$$

as a sum of partial fractions.

Solution Since the denominator has a repeated linear factor, $(x + 2)^2$, we must express the fraction in the form

$$\frac{6x + 7}{(x + 2)^2} = \frac{A}{x + 2} + \frac{B}{(x + 2)^2}. \tag{2}$$

Clearing (2) of fractions gives

$$6x + 7 = A(x + 2) + B = Ax + (2A + B).$$

Matching coefficients of like terms gives $A = 6$, $B = -5$. Hence,

$$\frac{6x + 7}{(x + 2)^2} = \frac{6}{x + 2} - \frac{5}{(x + 2)^2}. \ \square \tag{3}$$

EXAMPLE 4 *A quadratic factor in the denominator.* Express

$$\frac{-2x + 4}{(x^2 + 1)(x - 1)^2}$$

as a sum of partial fractions.

Solution Since the denominator has a quadratic factor as well as a repeated linear factor, we write

$$\frac{-2x + 4}{(x^2 + 1)(x - 1)^2} = \frac{Ax + B}{x^2 + 1} + \frac{C}{x - 1} + \frac{D}{(x - 1)^2}. \tag{4}$$

Note that the numerator over $x^2 + 1$ is linear and not constant. Then

$$\begin{aligned} -2x + 4 &= (Ax + B)(x - 1)^2 + C(x - 1)(x^2 + 1) + D(x^2 + 1) \\ &= (A + C)x^3 + (-2A + B - C + D)x^2 \\ &\quad + (A - 2B + C)x + (B - C + D). \end{aligned}$$

In order for this to be an identity in x, it is both necessary and sufficient that the coefficient of each power of x be the same on the left side of the equation as it is on the right side. Equating these coefficients leads to the following equations:

$$\begin{aligned} &\text{Coefficient of } x^3: && 0 = A + C, \\ &\text{Coefficient of } x^2: && 0 = -2A + B - C + D \\ &\text{Coefficient of } x^1: && -2 = A - 2B + C, \\ &\text{Coefficient of } x^0: && 4 = B - C + D. \end{aligned}$$

If we subtract the fourth equation from the second, we obtain

$$-4 = -2A, \qquad A = 2.$$

Then from the first equation, we have

$$C = -A = -2.$$

Knowing A and C, we find B from the third equation,

$$B = 1.$$

Finally, from the fourth equation, we have

$$D = 4 - B + C = 1.$$

Hence

$$\frac{-2x + 4}{(x^2 + 1)(x - 1)^2} = \frac{2x + 1}{x^2 + 1} - \frac{2}{x - 1} + \frac{1}{(x - 1)^2}. \quad \square$$

EXAMPLE 5 Evaluate

$$\int \frac{x^5 - x^4 - 3x + 5}{x^4 - 2x^3 + 2x^2 - 2x + 1}\, dx.$$

Solution The integrand is a fraction, but the degree of the numerator is greater than the degree of the denominator. Hence we divide first, obtaining

$$\frac{x^5 - x^4 - 3x + 5}{x^4 - 2x^3 + 2x^2 - 2x + 1} = x + 1 + \frac{-2x + 4}{x^4 - 2x^3 + 2x^2 - 2x + 1}. \tag{5}$$

The denominator factors as follows:

$$x^4 - 2x^3 + 2x^2 - 2x + 1 = (x^2 + 1)(x - 1)^2.$$

By the result of Example 4, we have, for the remainder term,

$$\frac{-2x + 4}{(x^2 + 1)(x - 1)^2} = \frac{2x + 1}{x^2 + 1} - \frac{2}{x - 1} + \frac{1}{(x - 1)^2}. \tag{6}$$

Hence, substituting from (6) into (5), multiplying by dx, and integrating, we have

$$\begin{aligned} &\int \frac{x^5 - x^4 - 3x + 5}{x^4 - 2x^3 + 2x^2 - 2x + 1}\, dx \\ &\quad = \int \left[x + 1 + \frac{2x + 1}{x^2 + 1} - \frac{2}{x - 1} + \frac{1}{(x - 1)^2}\right] dx \\ &\quad = \int \left[x + 1 + \frac{2x}{x^2 + 1} + \frac{1}{x^2 + 1} - \frac{2}{x - 1} + \frac{1}{(x - 1)^2}\right] dx \\ &\quad = \frac{x^2}{2} + x + \ln(x^2 + 1) + \tan^{-1} x - 2 \ln |x - 1| - \frac{1}{x - 1} + C. \quad \square \end{aligned}$$

General Description of the Method

Success in separating

$$\frac{f(x)}{g(x)} \tag{7}$$

into a sum of partial fractions hinges on two things:

1. *The degree of* $f(x)$ *should be less than the degree of* $g(x)$. (If this is not the case, we first perform a long division, then work with the remainder term. This remainder can always be put into the required form.)
2. *The factors of* $g(x)$ *should be known.* (Theoretically, any polynomial $g(x)$ with real coefficients can be expressed as a product of real linear and quadratic factors. In practice, g may be hard to factor.)

If these two conditions are met we can carry out the following steps.

FIRST. Let $x - r$ be a linear factor of $g(x)$. Suppose $(x - r)^m$ is the highest power of $x - r$ that divides $g(x)$. Then assign the sum of m partial fractions to this factor, as follows:

$$\frac{A_1}{x - r} + \frac{A_2}{(x - r)^2} + \cdots + \frac{A_m}{(x - r)^m}.$$

Do this for each distinct linear factor of $g(x)$.

SECOND. Let $x^2 + px + q$ be a quadratic factor of $g(x)$. Suppose

$$(x^2 + px + q)^n$$

is the highest power of this factor that divides $g(x)$. Then, to this factor, assign the sum of the n partial fractions:

$$\frac{B_1x + C_1}{x^2 + px + q} + \frac{B_2x + C_2}{(x^2 + px + q)^2} + \cdots + \frac{B_nx + C_n}{(x^2 + px + q)^n}.$$

Do this for each distinct quadratic factor of $g(x)$.

THIRD. Set the original fraction $f(x)/g(x)$ equal to the sum of all these partial fractions. Clear the resulting equation of fractions and arrange the terms in decreasing powers of x.

FOURTH. Equate the coefficients of corresponding powers of x, and solve the resulting equations for the undetermined coefficients.

Proofs that $f(x)/g(x)$ can be written as a sum of partial fractions as described here are given in advanced algebra texts.

The Heaviside "Cover-up" Method

When the degree of the polynomial $f(x)$ is less than the degree of $g(x)$, and

$$g(x) = (x - r_1)(x - r_2) \cdots (x - r_n) \tag{9}$$

is a product of n different linear factors, each to the first power, there is a quick way to expand $f(x)/g(x)$ by partial fractions.

EXAMPLE 6 Find A, B, and C in the partial-fractions expansion

$$\frac{x^2 + 1}{(x - 1)(x - 2)(x - 3)} = \frac{A}{x - 1} + \frac{B}{x - 2} + \frac{C}{x - 3}. \tag{10}$$

Solution If we multiply both sides of Eq. (10) by $(x - 1)$ to get

$$\frac{x^2 + 1}{(x-2)(x-3)} = A + \frac{B(x-1)}{x-2} + \frac{C(x-1)}{x-3}, \tag{11}$$

and set $x = 1$, the resulting equation gives the value of A:

$$\frac{(1)^2 + 1}{(1-2)(1-3)} = A + 0 + 0,$$
$$A = 1.$$

Thus, the value of A is the number we would have obtained if we had covered the factor $(x - 1)$ in the denominator of the original fraction

$$\frac{x^2 + 1}{(x-1)(x-2)(x-3)} \tag{12}$$

and evaluated the rest at $x = 1$:

$$A = \frac{(1)^2 + 1}{\boxed{(x-1)}\,(1-2)(1-3)} = \frac{2}{(-1)(-2)} = 1. \tag{13}$$

Cover

Similarly, we can find the value of B in Eq. (10) by covering the factor $(x - 2)$ in (12) and evaluating the rest at $x = 2$:

$$B = \frac{(2)^2 + 1}{(2-1)\,\boxed{(x-2)}\,(2-3)} = \frac{5}{(1)(-1)} = -5. \tag{14}$$

Cover

Finally, C can be found by covering the $(x - 3)$ in (12) and evaluating the rest at $x = 3$:

$$C = \frac{(3)^2 + 1}{(3-1)(3-2)\,\boxed{(x-3)}} = \frac{10}{(2)(1)} = 5. \ \square \tag{15}$$

Cover

In general, to expand a quotient of polynomials $f(x)/g(x)$ by partial fractions when the degree of $f(x)$ is less than the degree of $g(x)$, and $g(x)$ is the product of n linear factors, each to the first power, one may proceed in the following way. First, write the quotient with $g(x)$ factored:

$$\frac{f(x)}{g(x)} = \frac{f(x)}{(x-r_1)(x-r_2)\cdots(x-r_n)}. \tag{16}$$

Then, cover the factors $(x - r_i)$ of $g(x)$ in (16) one at a time, each time replacing all the uncovered x's by the number r_i. This gives a number A_i

for each root r_i:

$$
\begin{aligned}
A_1 &= \frac{f(r_1)}{(r_1 - r_2)\cdots(r_1 - r_n)},\\
A_2 &= \frac{f(r_2)}{(r_2 - r_1)(r_2 - r_3)\cdots(r_2 - r_n)},\\
&\vdots\\
A_n &= \frac{f(r_n)}{(r_n - r_1)(r_n - r_2)\cdots(r_n - r_{n-1})}.
\end{aligned}
\tag{17}
$$

The partial-fraction expansion of $f(x)/g(x)$ is then

$$\frac{f(x)}{g(x)} = \frac{A_1}{(x - r_1)} + \frac{A_2}{(x - r_2)} + \cdots + \frac{A_n}{(x - r_n)}. \tag{18}$$

EXAMPLE 7 Evaluate

$$\int \frac{x + 4}{x^3 + 3x^2 - 10x}\,dx.$$

Solution The degree of $f(x) = x + 4$ is less than the degree of $g(x) = x^3 + 3x^2 - 10x$, and, with $g(x)$ factored,

$$\frac{x + 4}{x^3 + 3x^2 - 10x} = \frac{x + 4}{x(x - 2)(x + 5)}.$$

The roots of $g(x)$ are $r_1 = 0$, $r_2 = 2$, and $r_3 = -5$. We find

$$A_1 = \frac{0 + 4}{\boxed{x}\,(0 - 2)(0 + 5)} = \frac{4}{(-2)(5)} = -\frac{2}{5},$$

Cover

$$A_2 = \frac{2 + 4}{2\,\boxed{(x - 2)}\,(2 + 5)} = \frac{6}{(2)(7)} = \frac{3}{7},$$

Cover

$$A_3 = \frac{-5 + 4}{(-5)(-5 - 2)\,\boxed{(x + 5)}} = \frac{-1}{(-5)(-7)} = -\frac{1}{35}.$$

Cover

Therefore,

$$\frac{x + 4}{x(x - 2)(x + 5)} = -\frac{2}{5x} + \frac{3}{7(x - 2)} - \frac{1}{35(x + 5)},$$

and

$$\int \frac{x + 4}{x(x - 2)(x + 5)}\,dx = -\frac{2}{5}\ln|x| + \frac{3}{7}\ln|x - 2| - \frac{1}{35}\ln|x + 5| + C. \ \square$$

Other Ways to Determine the Constants

Another way to determine the constants that appear in partial fractions is to differentiate, as in the next example.

EXAMPLE 8 Find A, B, and C in the equation

$$\frac{x-1}{(x+1)^3} = \frac{A}{x+1} + \frac{B}{(x+1)^2} + \frac{C}{(x+1)^3}.$$

Solution We first clear of fractions to get

$$x - 1 = A(x+1)^2 + B(x+1) + C.$$

Substituting $x = -1$ shows $C = -2$. We then differentiate both sides with respect to x to get

$$1 = 2A(x+1) + B.$$

Substituting $x = -1$ shows $B = 1$. We differentiate again to get

$$0 = 2A,$$

which shows $A = 0$. Hence

$$\frac{x-1}{(x+1)^3} = \frac{1}{(x+1)^2} - \frac{2}{(x+1)^3}. \quad \square$$

In many problems it is easiest to assign small values to x, like $x = 0$, ± 1, ± 2, to get equations in A, B, C, D, and so on. This often provides a fast alternative to the Heaviside method.

EXAMPLE 9 Find A, B, and C in

$$\frac{x^2+1}{(x-1)(x-2)(x-3)} = \frac{A}{x-1} + \frac{B}{x-2} + \frac{C}{x-3}.$$

Solution Clear of fractions to get

$$x^2 + 1 = A(x-2)(x-3) + B(x-1)(x-3) + C(x-1)(x-2).$$

Then let $x = 1, 2, 3$ successively to find A, B, and C:

$$x = 1: \quad (1)^2 + 1 = A(-1)(-2) + B(0) + C(0)$$
$$2 = 2A$$
$$A = 1,$$

$$x = 2: \quad (2)^2 + 1 = A(0) + B(1)(-1) + C(0)$$
$$5 = -B$$
$$B = -5,$$

$$x = 3: \quad (3)^2 + 1 = A(0) + B(0) + C(2)(1)$$
$$10 = 2C$$
$$C = 5.$$

Conclusion:

$$\frac{x^2+1}{(x-1)(x-2)(x-3)} = \frac{1}{x-1} - \frac{5}{x-2} + \frac{5}{x-3}. \quad \square$$

PROBLEMS

In Problems 1–10, expand by partial fractions.

1. $\frac{5x-13}{(x-3)(x-2)}$

2. $\frac{5x-7}{x^2-3x+2}$

3. $\frac{x+4}{(x+1)^2}$

4. $\frac{2x+2}{x^2-2x+1}$

5. $\frac{x+1}{x^2(x-1)}$

6. $\frac{z}{z^3-z^2-6z}$

7. $\frac{x^2+8}{x^2-5x+6}$ (Remember to divide first.)

8. $\frac{4}{x^3+4x}$

9. $\frac{3}{x^2(x^2+9)}$

10. $\frac{x^3-1}{(x^2+x+1)^2}$

In Problems 11–49, evaluate the integrals.

11. $\int_0^{1/2} \frac{dx}{1-x^2}$

12. $\int_1^2 \frac{dx}{x^2+2x}$

13. $\int_0^{2\sqrt{2}} \frac{x^3}{x^2+1}\,dx$

14. $\int_1^2 \frac{dx}{x^3+x}$

15. $\int_{1/4}^{3/4} \frac{dx}{x-x^2}$

16. $\int_{-1}^0 \frac{x\,dx}{x^2-3x+2}$

17. $\int \frac{x+4}{x^2+5x-6}\,dx$

18. $\int \frac{2x+1}{x^2-7x+12}\,dx$

19. $\int_0^1 \frac{3x^2}{x^2+2x+1}\,dx$

20. $\int \frac{d\theta}{\theta^3+\theta^2-2\theta}$

21. $\int \frac{x\,dx}{x^2+4x-5}$

22. $\int_4^8 \frac{x\,dx}{x^2-2x-3}$

23. $\int \frac{(x+1)\,dx}{x^2+4x-5}$

24. $\int \frac{x^3\,dx}{x^2+2x+1}$

25. $\int_1^3 \frac{dx}{x(x+1)^2}$

26. $\int_0^1 \frac{dx}{(x+1)(x^2+1)}$

27. $\int \frac{(x+3)\,dx}{2x^3-8x}$

28. $\int \frac{\cos x\,dx}{\sin^2 x+\sin x-6}$

29. $\int_0^{\sqrt{3}} \frac{5x^2\,dx}{x^2+1}$

30. $\int \frac{x^3\,dx}{x^2-2x+1}$

31. $\int \frac{dx}{x(x^2+x+1)}$

32. $\int_{\pi/3}^{\pi/2} \frac{\sin\theta\,d\theta}{\cos^2\theta+\cos\theta-2}$

33. $\int \frac{3x^2+x+4}{x^3+x}\,dx$

34. $\int \frac{dx}{(x^2-1)^2}$

35. $\int \frac{x^3+4x^2}{x^2+4x+3}\,dx$

36. $\int \frac{4x+4}{x^2(x^2+2)}\,dx$

37. $\int \frac{x^2+2x+1}{(x^2+1)^2}\,dx$

38. $\int \frac{x^3-x}{(x^2+1)(x-1)^2}\,dx$

39. $\int \frac{2x}{(x^2+1)(x-1)^2}\,dx$

40. $\int \frac{x^2}{(x-1)(x^2+2x+1)}\,dx$

41. $\int_0^{\ln 2} \frac{e^t\,dt}{e^{2t}+3e^t+2}$

42. $\int_0^1 \frac{dx}{(x^2+1)^2}$

43. $\int \frac{x^4\,dx}{(x^2+1)^2}$

44. $\int \frac{(4x^3-20x)\,dx}{x^4-10x^2+9}$

45. $\int \frac{1-\sqrt{x}}{1+\sqrt{x}}\,dx$ (*Hint:* Substitute $u=\sqrt{x}$.)

46. $\int \frac{dx}{1+\sqrt{x}}$ (*Hint:* Substitute $u=\sqrt{x}$.)

47. $\int \frac{dx}{\sqrt{x}+\sqrt[3]{x}}$ (*Hint:* Substitute $x=u^6$.)

48. $\int x\ln(x+5)\,dx$

49. $\int \ln(x^2+1)\,dx$

50. To integrate $\int x^2\,dx/(x^2-1)$ by partial fractions, we would first divide x^2 by x^2-1 to get

$$\frac{x^2}{x^2-1}=1+\frac{1}{x^2-1}.$$

Suppose that we ignore the fact that x^2 and x^2-1 have the same degree, and try to write

$$\frac{x^2}{x^2-1}=\frac{x^2}{(x-1)(x+1)}=\frac{A}{x-1}+\frac{B}{x+1}.$$

What goes wrong? Find out by trying to solve for A and B.

51. Many chemical reactions are the result of an interaction of two molecules that undergo a change to produce a new product. The rate of the reaction typically depends on the concentration of the two kinds of molecules. If a is the amount of substance A and b is the amount of substance B at time $t=0$, and if x is the amount of product at time t, then the rate of formation of x may be given by the differential equation

$$\frac{dx}{dt}=k(a-x)(b-x),$$

or

$$\frac{1}{(a-x)(b-x)}\frac{dx}{dt}=k$$

(k a constant for the reaction). Integrate both sides of this equation with respect to t to obtain a relation between x and t; (a) if $a=b$, and (b) if $a\neq b$. Assume in each case that $x=0$ when $t=0$.

52. The equation that describes the autocatalytic reaction of Article 3.5, Problem 49, can be written as

$$\frac{dx}{dt} = kx(a - x).$$

Read Problem 49 for background (there is no need to solve it). Then solve the equation above to find x as a function of t. Assume that $x = x_0$ when $t = 0$.

***Toolkit* programs**

Partial Fractions

7.8 $z = \tan(x/2)$

It has been discovered that the substitution

$$z = \tan\frac{x}{2} \tag{1}$$

will reduce the problem of integrating any rational function of sin x and cos x to a problem involving a rational function of z. This in turn can be integrated by partial fractions. Thus the substitution (1) is a very powerful tool. It is cumbersome, however, and used only when the simpler methods outlined previously have failed.

To see the effect of the substitution, we calculate

$$\begin{aligned}\cos x = 2\cos^2\frac{x}{2} - 1 &= \frac{2}{\sec^2(x/2)} - 1 \\ &= \frac{2}{1 + \tan^2(x/2)} - 1 = \frac{2}{1 + z^2} - 1\end{aligned}$$

or

$$\cos x = \frac{1 - z^2}{1 + z^2}, \tag{2a}$$

and

$$\begin{aligned}\sin x = 2\sin\frac{x}{2}\cos\frac{x}{2} &= 2\frac{\sin(x/2)}{\cos(x/2)}\cdot\cos^2\frac{x}{2} \\ &= 2\tan\frac{x}{2}\cdot\frac{1}{\sec^2(x/2)} = \frac{2\tan(x/2)}{1 + \tan^2(x/2)}\end{aligned}$$

or

$$\sin x = \frac{2z}{1 + z^2}. \tag{2b}$$

Finally,

$$x = 2\tan^{-1} z,$$

so that

$$dx = \frac{2\,dz}{1 + z^2}. \tag{2c}$$

EXAMPLE 1

$$\int \sec x\,dx = \int \frac{dx}{\cos x} \qquad \text{becomes} \qquad \int \frac{2\,dz}{1+z^2}\cdot\frac{1+z^2}{1-z^2} = \int \frac{2\,dz}{1-z^2}.$$

To this we apply the method of partial fractions:

$$\frac{2}{1-z^2} = \frac{A}{1-z} + \frac{B}{1+z},$$

$$2 = A(1+z) + B(1-z) = (A+B) + (A-B)z,$$

which requires

$$A + B = 2, \qquad A - B = 0.$$

Hence

$$A = B = 1$$

and

$$\begin{aligned}\int \frac{2\,dz}{1-z^2} = \int \frac{dz}{1-z} + \int \frac{dz}{1+z} &= -\ln|1-z| + \ln|1+z| + C\\ &= \ln\left|\frac{1+z}{1-z}\right| + C\\ &= \ln\left|\frac{1+\tan(x/2)}{1-\tan(x/2)}\right| + C\\ &= \ln\left|\frac{\tan(\pi/4)+\tan(x/2)}{1-\tan(\pi/4)\tan(x/2)}\right| + C\\ &= \ln\left|\tan\left(\frac{\pi}{4}+\frac{x}{2}\right)\right| + C.\end{aligned}$$

That is,

$$\int \sec x\,dx = \ln\left|\tan\left(\frac{\pi}{4}+\frac{x}{2}\right)\right| + C \tag{3}$$

is an alternative to Eq. (6), Article 7.3. □

EXAMPLE 2

$$\int \frac{dx}{1+\cos x} = \int \frac{2\,dz}{1+z^2}\,\frac{1+z^2}{2} = \int dz = z + C = \tan\frac{x}{2} + C. \;\square$$

EXAMPLE 3

$$\begin{aligned}\int \frac{dx}{2+\sin x} &= \int \frac{2\,dz}{1+z^2}\left[\frac{1+z^2}{2+2z+2z^2}\right] = \int \frac{dz}{z^2+z+1} = \int \frac{dz}{(z+\frac{1}{2})^2+\frac{3}{4}}\\ &= \int \frac{du}{u^2+a^2} \qquad \left[u = z + \frac{1}{2},\quad a = \frac{\sqrt{3}}{2}\right]\\ &= \frac{1}{a}\tan^{-1}\frac{u}{a} + C = \frac{2}{\sqrt{3}}\tan^{-1}\frac{2z+1}{\sqrt{3}} + C\\ &= \frac{2}{\sqrt{3}}\tan^{-1}\frac{1+2\tan(x/2)}{\sqrt{3}} + C. \;\square\end{aligned}$$

PROBLEMS

Evaluate the following integrals. (Integrals like the ones in Problems 1 and 3 arise in calculating the average angular velocity of the output shaft of a universal joint when the input and output shafts are not aligned.)

1. $\int_0^{\pi} \frac{dx}{1+\sin x}$

2. $\int_{\pi/2}^{\pi} \frac{dx}{1-\cos x}$

3. $\int \frac{dx}{1-\sin x}$

4. $\int_0^{\pi/2} \frac{dx}{2+\cos x}$

5. $\int \frac{\cos x\, dx}{2-\cos x}$

6. $\int \frac{dx}{1+\sin x+\cos x}$

7. $\int \frac{dx}{\sin x-\cos x}$

8. $\int_{\pi/2}^{2\pi/3} \frac{dx}{\sin x+\tan x}$

7.9 Improper Integrals

The integral

$$\int_0^1 \sqrt{\frac{1+x}{1-x}}\, dx,$$

which arises in lifting line theory in aerodynamics, is called an *improper* integral because the integrand

$$f(x) = \sqrt{\frac{1+x}{1-x}}$$

becomes infinite at one of the limits of integration, in this case at $x = 1$.

If we interpret the integral in terms of area under the curve $y = f(x)$, we see that the curve extends to infinity as x approaches 1^- (Fig. 7.11), and the area from $x = 0$ to $x = 1$ is not well defined. Nevertheless, we can define the area from $x = 0$ to $x = b$, where b is any positive number less

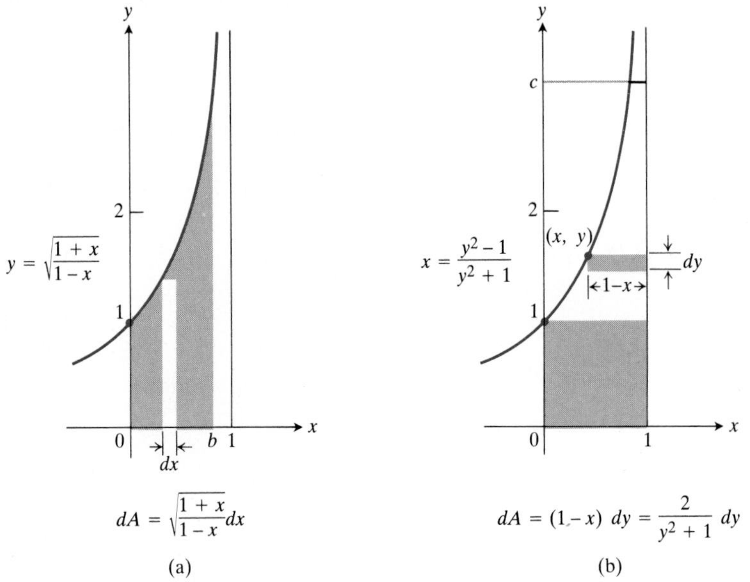

7.11 Two schemes for calculating $\int_0^1 \sqrt{(1+x)/(1-x)}\, dx$.

than 1, and we can take b as close to 1 as we please. The integral

$$\int_0^b f(x)\,dx$$

may or may not have a finite limit as b approaches 1 from the left. If it does, we *define* the value of the integral (and the area under the curve) from 0 to 1 to be this limit:

$$\int_0^1 f(x)\,dx = \lim_{b\to 1^-} \int_0^b f(x)\,dx.$$

In this case we also say that the improper integral *converges*. On the other hand, if the integral has no definite finite limit as $b \to 1^-$ we say the integral *diverges* and that the area under the curve is infinite.

EXAMPLE 1 Determine whether

$$\int_0^1 \sqrt{\frac{1+x}{1-x}}\,dx$$

converges or diverges.

Solution Multiplying numerator and denominator by $\sqrt{1+x}$ gives

$$\int \sqrt{\frac{1+x}{1-x}}\,dx = \int \frac{1+x}{\sqrt{1-x^2}}\,dx = \sin^{-1} x - \sqrt{1-x^2} + C.$$

Therefore,

$$\begin{aligned}
\lim_{b\to 1^-} \int_0^b \sqrt{\frac{1+x}{1-x}}\,dx &= \lim_{b\to 1^-} [\sin^{-1} x - \sqrt{1-x^2}]_0^b \\
&= \lim_{b\to 1^-} [\sin^{-1} b - \sqrt{1-b^2} + 1] \\
&= \sin^{-1} 1 - 0 + 1 = (\pi/2) + 1.
\end{aligned}$$

The same result is obtained if we find the area by using horizontal strips and integrating with respect to y (Fig. 7.11b). In this case we define the portion of the area above the shaded square to be

$$\lim_{c\to\infty} \int_1^c (1-x)\,dy.$$

After expressing x in terms of y we find the limit to be

$$\begin{aligned}
\lim_{c\to\infty} \int_1^c (1-x)\,dy &= \lim_{c\to\infty} \int_1^c \frac{2}{y^2+1}\,dy \\
&= \lim_{c\to\infty} 2\tan^{-1} y]_1^c \\
&= \lim_{c\to\infty} 2\tan^{-1} c - 2\cdot\frac{\pi}{4} = 2\cdot\frac{\pi}{2} - \frac{\pi}{2} = \frac{\pi}{2}.
\end{aligned}$$

Including the shaded square, the area is $(\pi/2) + 1$, in agreement with our first calculation. □

In place of

$$\lim_{c\to\infty} \int_1^c \frac{2}{y^2+1}\,dy$$

we normally write

$$\int_1^\infty \frac{2}{y^2+1}\,dy.$$

This is a second kind of improper integral, improper because one of its limits of integration is infinite.

DEFINITION

Improper Integral

An integral $\int_a^b f(x)\,dx$ is *improper* if

1. f becomes infinite at one or more points in the interval of integration, or
2. one or both of the limits of integration is infinite.

EXAMPLE 2 In the integral

$$\int_0^1 \frac{dx}{x},$$

the function

$$f(x) = \frac{1}{x}$$

becomes infinite at $x = 0$. We cut off the point $x = 0$ and start our integration at some positive number $b < 1$. (See Fig. 7.12.) That is, we consider the integral

$$\int_b^1 \frac{dx}{x} = \ln x\Big]_b^1 = \ln 1 - \ln b = \ln\frac{1}{b},$$

and investigate its behavior as b approaches zero from the right. Since

$$\lim_{b\to 0^+}\int_b^1 \frac{dx}{x} = \lim_{b\to 0^+}\left(\ln\frac{1}{b}\right) = +\infty,$$

we say that the integral from $x = 0$ to $x = 1$ *diverges*. □

The method to be used when the function f becomes infinite at an interior point of the interval of integration is illustrated in the following example.

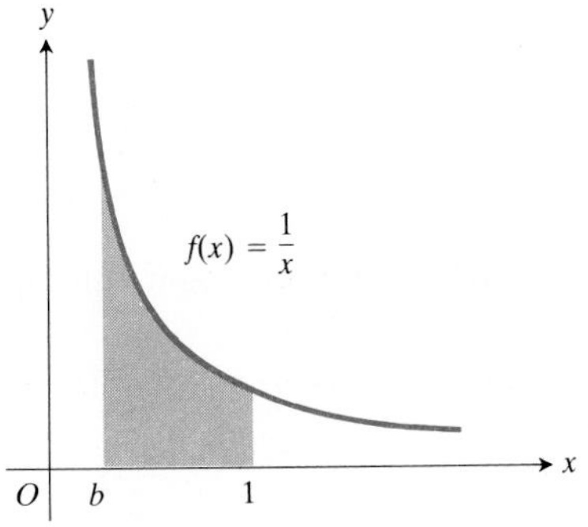

7.12 $\int_0^1 \frac{dx}{x} = \lim_{b\to 0^+}\int_b^1 \frac{dx}{x}.$

EXAMPLE 3 In the integral

$$\int_0^3 \frac{dx}{(x-1)^{2/3}},$$

the function

$$f(x) = \frac{1}{(x-1)^{2/3}}$$

becomes infinite at $x = 1$, which lies between the limits of integration 0 and 3. In such a case, we again cut out the point where $f(x)$ becomes infinite. This time we integrate from 0 to b, where b is slightly less than 1,

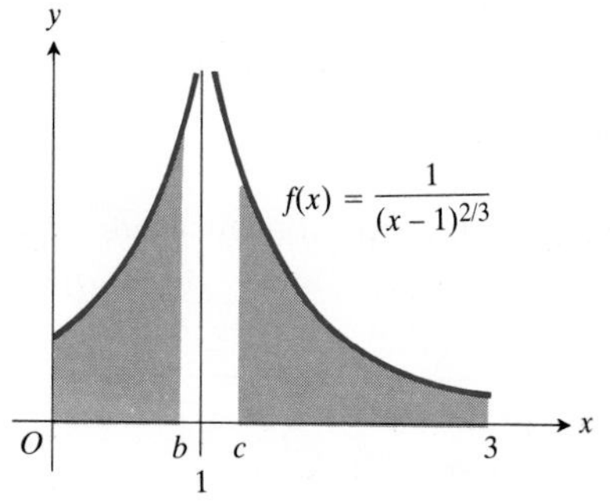

7.13 $\int_0^3 f(x)\,dx = \lim_{b\to 1^-}\int_0^b f(x)\,dx + \lim_{c\to 1^+}\int_c^3 f(x)\,dx.$

and start again on the other side of 1 at c and integrate from c to 3 (Fig. 7.13). This gives two integrals to investigate:

$$\int_0^b \frac{dx}{(x-1)^{2/3}} \quad\text{and}\quad \int_c^3 \frac{dx}{(x-1)^{2/3}}.$$

If the first of these has a definite limit as $b \to 1^-$ and if the second also has a definite limit as $c \to 1^+$, then we say that the improper integral converges and that its value is given by

$$\int_0^3 \frac{dx}{(x-1)^{2/3}} = \lim_{b\to 1^-}\int_0^b \frac{dx}{(x-1)^{2/3}} + \lim_{c\to 1^+}\int_c^3 \frac{dx}{(x-1)^{2/3}}.$$

If either limit fails to exist, we say that the given improper integral *diverges*. For this example,

$$\lim_{b\to 1^-}\int_0^b (x-1)^{-2/3}\,dx = \lim_{b\to 1^-}[3(b-1)^{1/3} - 3(0-1)^{1/3}] = +3$$

and

$$\lim_{c\to 1^+}\int_c^3 (x-1)^{-2/3}\,dx = \lim_{c\to 1^+}[3(3-1)^{1/3} - 3(c-1)^{1/3}] = 3\sqrt[3]{2}.$$

Since both limits exist and are finite, the original integral is said to converge and its value is $3 + 3\sqrt[3]{2}$. □

7.14 (a) The shaded area, ln b, does not approach a finite limit as $b \to \infty$. (b) The shaded area, $1 - (1/b)$, approaches 1 as $b \to \infty$.

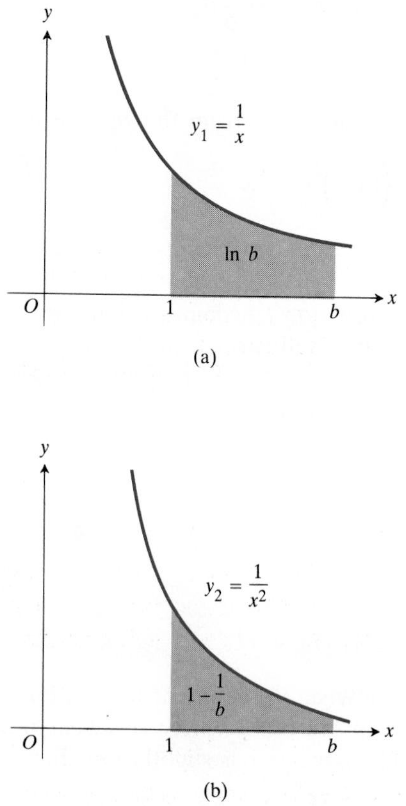

The convergence of the integral

$$\int_1^\infty \frac{dx}{x^p}$$

depends on the value of the exponent p. The next example illustrates this with $p = 1$ and $p = 2$.

EXAMPLE 4 Determine whether the following improper integrals converge or diverge:

$$\int_1^\infty \frac{dx}{x} \quad\text{and}\quad \int_1^\infty \frac{dx}{x^2}.$$

Solution The two curves

$$y_1 = \frac{1}{x} \quad\text{and}\quad y_2 = \frac{1}{x^2}$$

both approach the x-axis as $x \to \infty$ (Fig. 7.14). In the first case,

$$\int_1^b \frac{dx}{x} = \ln x\Big]_1^b = \ln b.$$

If we now let b take on larger and larger positive values, the logarithm of b increases indefinitely and

$$\lim_{b\to\infty}\int_1^b \frac{dx}{x} = \infty.$$

We therefore say that

$$\int_1^\infty \frac{dx}{x} = \infty$$

and that the integral *diverges*.

In the second case,

$$\int_1^b \frac{dx}{x^2} = -\frac{1}{x}\Big]_1^b = 1 - \frac{1}{b}.$$

This does have a definite limit, namely 1, as b increases indefinitely. We therefore say that the integral from 1 to ∞ converges and that its value is 1. That is,

$$\int_1^\infty \frac{dx}{x^2} = \lim_{b\to\infty} \int_1^b \frac{dx}{x^2} = \lim_{b\to\infty}\left(1 - \frac{1}{b}\right) = 1. \square$$

Generally, the integral

$$\int_1^\infty \frac{dx}{x^p}$$

converges when $p > 1$ but diverges when $p \le 1$ (Problem 44).

Sometimes we can determine whether an improper integral converges without having to evaluate it. Instead, we compare it to an integral whose convergence or divergence we already know. This is the case with the next example, an integral important in probability theory.

EXAMPLE 5 Determine whether the integral

$$\int_1^\infty e^{-x^2}\,dx$$

converges or diverges.

Solution Even though we cannot find any simpler expression for

$$I(b) = \int_1^b e^{-x^2}\,dx,$$

we can show that

$$\lim_{b\to\infty} I(b)$$

exists and is finite.

The function $I(b)$ represents the area between the x-axis and the curve

$$y = e^{-x^2}$$

between $x = 1$ and $x = b$. Clearly, this is an increasing function of b, so that there are two alternatives: either

a) $I(b)$ becomes infinite as $b \to \infty$,

or

b) $I(b)$ has a finite limit as $b \to \infty$.

We show that the first alternative cannot be the true one. We do this by comparing the area under the given curve $y = e^{-x^2}$ with the area under the curve $y = e^{-x}$ (Fig. 7.15). The latter area, from $x = 1$ to $x = b$, is given by the integral

$$\int_1^b e^{-x}\,dx = -e^{-x}\Big]_1^b = e^{-1} - e^{-b},$$

which approaches the finite limit e^{-1} as $b \to \infty$. Since e^{-x^2} is less than e^{-x}

7.15 The graphs of $y = e^{-x^2}$ and $y = e^{-x}$.

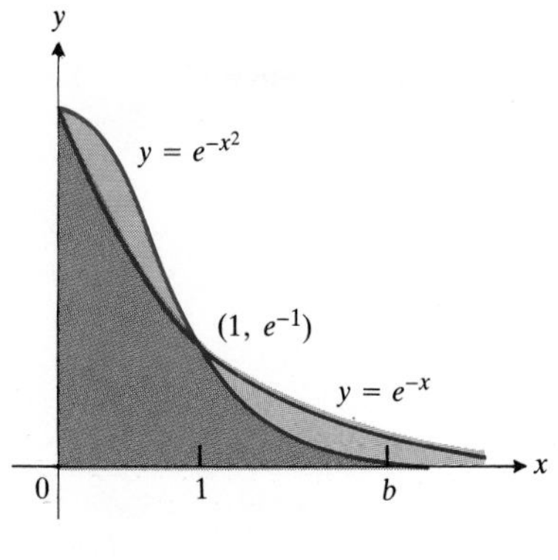

for all x greater than one, the area under the given curve is certainly no greater than e^{-1} no matter how large b is.

The discussion above is summarized in the following inequalities:

$$I(b) = \int_1^b e^{-x^2}\,dx \le \int_1^b e^{-x}\,dx = e^{-1} - e^{-b} \le e^{-1};$$

that is,

$$I(b) \le e^{-1} < 0.37.$$

Therefore $I(b)$ does not become infinite as $b \to \infty$, so alternative (a) is ruled out and alternative (b) must hold; that is,

$$\int_1^\infty e^{-x^2}\,dx = \lim_{b\to\infty} \int_1^b e^{-x^2}\,dx$$

converges to a definite finite value. We have not calculated what this limit is, but we know that it exists and is less than 0.37. □

We say that a positive function f *dominates* a positive function g as $x \to \infty$ if

$$g(x) \le f(x)$$

for all values of x beyond some point a. For instance, $f(x) = e^{-x}$ dominates $g(x) = e^{-x^2}$ as $x \to \infty$ because

$$e^{-x^2} \le e^{-x}$$

for all $x > a = 1$.

If f dominates g as $x \to \infty$, then

$$\int_a^b g(x)\,dx \le \int_a^b f(x)\,dx, \qquad b > a,$$

and from this it can be argued as in Example 5 that

$$\int_a^\infty g(x)\,dx \text{ converges if } \int_a^\infty f(x) \text{ converges.}$$

Turning this around says that

$$\int_a^\infty f(x)\,dx \text{ diverges if } \int_a^\infty g(x) \text{ diverges.}$$

We state these results as a theorem and then give examples.

THEOREM 1

Domination Test for Convergence and Divergence of Improper Integrals

If $0 \le g(x) \le f(x)$ for all $x > a$, then

1. $\int_a^\infty g(x)\,dx$ converges if $\int_a^\infty f(x)\,dx$ converges,
2. $\int_a^\infty f(x)\,dx$ diverges if $\int_a^\infty g(x)\,dx$ diverges.

(Theorem 1 assumes that f and g are integrable over every finite interval $[a, b]$, which will be true, for instance, if the functions are continuous.)

EXAMPLE 6

a) $\int_1^\infty \frac{1}{x^2+1}\,dx$ converges because $\frac{1}{x^2+1} < \frac{1}{x^2}$ and $\int_1^\infty \frac{1}{x^2}\,dx$ converges.

b) $\int_1^\infty \frac{1}{e^{2x}}\,dx$ converges because $\frac{1}{e^{2x}} < \frac{1}{e^x}$ and $\int_1^\infty \frac{1}{e^x}\,dx$ converges.

c) $\int_1^\infty \frac{1}{\sqrt{x}}\,dx$ diverges because $\frac{1}{x} < \frac{1}{\sqrt{x}}$ for $x > 1$ and $\int_1^\infty \frac{1}{x}\,dx$ diverges.

d) $\int_1^\infty \left(\frac{1}{x} + \frac{1}{x^2}\right) dx$ diverges because $\frac{1}{x} < \frac{1}{x} + \frac{1}{x^2}$ and $\int_1^\infty \frac{1}{x}\,dx$ diverges. □

Another useful result, which we shall state here but not prove, is called the limit comparison test for improper integrals. (You will find a similar result for infinite series in Chapter 11.) It goes like this:

THEOREM 2

Limit Comparison Test for Convergence and Divergence of Improper Integrals

Suppose that $f(x)$ and $g(x)$ are positive functions and that

$$\lim_{x\to\infty} \frac{f(x)}{g(x)} = L, \quad 0 < L < \infty.$$

Then, $\int_a^\infty f(x)\,dx$ and $\int_a^\infty g(x)\,dx$ both converge or both diverge.

Like Theorem 1, Theorem 2 assumes that f and g are integrable over every finite interval $[a, b]$.

In the language of Article 6.8, Theorem 2 says that *if two positive functions grow at the same rate as $x \to \infty$, then their integrals from a to ∞ behave alike: they both converge, or both diverge.* This does not mean, however, that their integrals must have the same value, as the next example shows.

EXAMPLE 7 Compare

$$\int_1^\infty \frac{dx}{x^2} \quad \text{and} \quad \int_1^\infty \frac{dx}{1+x^2}$$

with the limit comparison test.

Solution With

$$f(x) = \frac{1}{x^2}, \qquad g(x) = \frac{1}{1+x^2},$$

we have

$$\lim_{x\to\infty} \frac{1}{x^2}\,\frac{1+x^2}{1} = \lim_{x\to\infty} \frac{1+x^2}{x^2} = \lim_{x\to\infty}\left(\frac{1}{x^2} + 1\right) = 0 + 1 = 1,$$

as a positive finite limit. Therefore

$$\int_1^\infty \frac{dx}{x^2} \text{ converges} \Rightarrow \int_1^\infty \frac{dx}{1+x^2} \text{ converges.}$$

Note, however, that the two integrals converge to different values:

$$\int_1^\infty \frac{dx}{x^2} = 1,$$

from Example 4, and

$$\begin{aligned}\int_1^\infty \frac{dx}{1+x^2} &= \lim_{b\to\infty}\int_1^b \frac{dx}{1+x^2} \qquad \text{(the definition)}\\ &= \lim_{b\to\infty}[\tan^{-1} b - \tan^{-1} 1]\\ &= \frac{\pi}{2} - \frac{\pi}{4} = \frac{\pi}{4}. \ \square\end{aligned}$$

EXAMPLE 8

$$\int_1^\infty \frac{3}{e^x+5}\,dx \text{ converges because } \int_1^\infty \frac{1}{e^x}\,dx \text{ converges}$$

and

$$\lim_{x\to 0} \frac{1}{e^x}\cdot\frac{e^x+5}{3} = \lim_{x\to\infty}\frac{e^x+5}{3e^x} = \lim_{x\to\infty}\left(\frac{1}{3}+\frac{5}{3e^x}\right) = \frac{1}{3}+0 = \frac{1}{3},$$

a positive finite limit. As far as the convergence of the improper integral is concerned, $3/(e^x+5)$ behaves like $1/e^x$. □

EXAMPLE 9

$$\int_3^\infty \frac{1}{e^{2x}-10e^x}\,dx \text{ converges because } \int_3^\infty \frac{1}{e^{2x}}\,dx \text{ converges}$$

and

$$\lim_{x\to\infty}\frac{1}{e^{2x}}\cdot\frac{e^{2x}-10e^x}{1} = \lim_{x\to\infty}\left(1-\frac{10}{e^x}\right) = 1+0 = 1,$$

a positive finite limit. As far as the convergence of the improper integral is concerned, $1/(e^{2x}-10e^x)$ behaves like $1/e^{2x}$. □

We know that

$$\int_1^\infty \frac{1}{x^2}\,dx$$

converges, but what about integrals like

$$\int_2^\infty \frac{1}{x^2}\,dx \qquad \text{and} \qquad \int_{100}^\infty \frac{1}{x^2}\,dx?$$

The answer is that they converge, too. The existence of the limits

$$\lim_{b\to\infty}\int_2^b \frac{1}{x^2}\,dx \qquad \text{and} \qquad \lim_{b\to\infty}\int_{100}^b \frac{1}{x^2}\,dx$$

does not depend on the starting points $a = 2$ and $a = 100$, but only on the values of $1/x^2$ as x approaches infinity. Indeed, we find for any positive a that

$$\lim_{b\to\infty} \int_a^b \frac{1}{x^2}\,dx = \lim_{b\to\infty}\left(-\frac{1}{b} + \frac{1}{a}\right) = \frac{1}{a}.$$

The *value* of the limit depends on the value of a, but the *existence* of the limit does not. The limit exists for any positive a.

Similarly, the integrals

$$\int_2^\infty \frac{1}{x}\,dx \quad \text{and} \quad \int_{100}^\infty \frac{1}{x}\,dx$$

both diverge. We find for any positive a that

$$\lim_{b\to\infty} \int_a^b \frac{1}{x}\,dx = \lim_{b\to\infty} (\ln b - \ln a) = \infty.$$

The convergence and divergence of integrals that are improper only because the upper limit is infinite does not depend on their lower limits of integration. This gives us a great deal of freedom in comparing integrals to decide the question of convergence.

EXAMPLE 10

a) $\int_6^\infty \frac{1}{\sqrt{x}}\,dx$ diverges because $\int_1^\infty \frac{1}{x}\,dx$ diverges and $\frac{1}{\sqrt{x}} > \frac{1}{x}$ for $x > 1$.

b) $\int_6^\infty \frac{1}{\sqrt{x-5}}\,dx$ diverges because $\int_6^\infty \frac{1}{\sqrt{x}}$ diverges and

$$\begin{aligned}\lim_{x\to\infty} \frac{1}{\sqrt{x}} \cdot \frac{\sqrt{x-5}}{1} &= \lim_{x\to\infty} \frac{\sqrt{x-5}}{\sqrt{x}} \\ &= \lim_{x\to\infty} \sqrt{1 - \frac{5}{x}} \\ &= \sqrt{1-0} = 1,\end{aligned}$$

a positive finite limit. □

An improper integral may diverge without becoming infinite.

EXAMPLE 11 The integral

$$\int_0^b \cos x\,dx = \sin b$$

takes all values between -1 and $+1$ as b varies between $2n\pi - \pi/2$ and $2n\pi + \pi/2$, where n is any integer. Hence,

$$\lim_{b\to\infty} \int_0^b \cos x\,dx$$

does not exist. We might say that this integral "diverges by oscillation." □

PROBLEMS

In Problems 1–12, evaluate the integrals.

1. $\int_0^\infty \frac{dx}{x^2+1}$

2. $\int_0^1 \frac{dx}{\sqrt{x}}$

3. $\int_{-1}^1 \frac{dx}{x^{2/3}}$

4. $\int_1^\infty \frac{dx}{x^{1.001}}$

5. $\int_0^4 \frac{dx}{\sqrt{4-x}}$

6. $\int_0^1 \frac{dx}{\sqrt{1-x^2}}$

7. $\int_0^\infty e^{-x} \cos x \, dx$

8. $\int_0^1 \frac{dx}{x^{0.999}}$

9. $\int_{-\infty}^2 \frac{dx}{4-x}$

10. $\int_{-1}^0 \sqrt{\frac{1+x}{1-x}}\, dx$

11. $\int_2^\infty \frac{1}{x^2-x}\, dx$

12. $\int_0^\infty \frac{dx}{(1+x)\sqrt{x}}$

In Problems 13–41, determine whether the integrals converge or diverge. (In some cases, you may not need to evaluate the integral to decide. Name any tests you use.)

13. $\int_1^\infty \frac{dx}{\sqrt{x}}$

14. $\int_1^\infty \frac{dx}{x^3}$

15. $\int_1^\infty \frac{dx}{x^3+1}$

16. $\int_0^\infty \frac{dx}{x^3}$

17. $\int_0^\infty \frac{dx}{x^3+1}$

18. $\int_0^\infty \frac{dx}{1+e^x}$

19. $\int_0^{\pi/2} \tan x \, dx$

20. $\int_{-1}^1 \frac{dx}{x^2}$

21. $\int_{-1}^1 \frac{dx}{x^{2/5}}$

22. $\int_0^\infty \frac{dx}{\sqrt{x}}$

23. $\int_2^\infty \frac{dx}{\sqrt{x-1}}$

24. $\int_1^\infty \frac{5}{x}\, dx$

25. $\int_0^2 \frac{dx}{1-x^2}$

26. $\int_2^\infty \frac{dx}{(x+1)^2}$

27. $\int_0^4 \frac{dx}{\sqrt{4-x}}$

28. $\int_{-1}^1 \frac{dx}{\sqrt[3]{x}}$

29. $\int_0^\infty x^2 e^{-x}\, dx$

30. $\int_1^\infty \frac{\sqrt{x+1}}{x^2}\, dx$

31. $\int_\pi^\infty \frac{2+\cos x}{x}\, dx$

32. $\int_1^\infty \frac{\ln x}{x}\, dx$

33. $\int_6^\infty \frac{1}{\sqrt{x+5}}\, dx$

34. $\int_1^\infty \frac{dx}{\sqrt{2x+10}}$

35. $\int_2^\infty \frac{1}{x^2-1}\, dx$

36. $\int_1^\infty \frac{1}{e^{\ln x}}\, dx$

37. $\int_2^\infty \frac{1}{\ln x}\, dx$

38. $\int_1^\infty \frac{1}{\sqrt{e^x - x}}\, dx$

39. $\int_1^\infty \frac{1}{e^x - 2^x}\, dx$

40. $\int_2^\infty \frac{1}{x^3-5}\, dx$

41. $\int_0^\infty \frac{dx}{\sqrt{x+x^4}}$

(*Hint:* Compare the integral with $\int dx/\sqrt{x}$ for x near zero and with $\int dx/x^2$ for large x.)

42. Show that

$$\int_3^\infty e^{-3x}\, dx = \tfrac{1}{3}e^{-9} = 0.000041,$$

and hence that $\int_3^\infty e^{-x^2} dx < 0.000041$. Therefore, $\int_0^\infty e^{-x^2}\, dx$ can be replaced by $\int_0^3 e^{-x^2}\, dx$ without introducing any errors in the first three decimal places of the answer. Evaluate this last integral by Simpson's rule with $n = 6$. (This illustrates one method by which a convergent improper integral may be approximated numerically.)

43. As Example 4 shows, the integral $\int_1^\infty (dx/x)$ diverges. This means that the integral

$$\int_1^\infty 2\pi \frac{1}{x}\sqrt{1+\frac{1}{x^4}}\, dx,$$

which measures the *surface area* of the solid of revolution traced out by revolving the curve $y = 1/x$, $1 \le x$, about the x-axis, diverges also. For, by comparing the two integrals, we see that, for every finite value $b > 1$,

$$\int_1^b 2\pi \frac{1}{x}\sqrt{1+\frac{1}{x^4}}\, dx > \int_1^b \frac{dx}{x}.$$

However, the integral

$$\int_1^\infty \pi\left(\frac{1}{x}\right)^2 dx$$

for the *volume* of the solid converges. Calculate it. This solid of revolution is sometimes described as a can that does not hold enough paint to cover its outside surface.

44. Show that

$$\int_1^\infty \frac{dx}{x^p} = \frac{1}{p-1} \quad \text{when } p > 1,$$

but that the integral is infinite when $p < 1$. Example 4 shows what happens when $p = 1$.

45. Find the values of p for which each integral converges:

a) $\int_1^2 \frac{dx}{x(\ln x)^p}$, b) $\int_2^\infty \frac{dx}{x(\ln x)^p}$.

46. Find the coordinates of the centroid of the region in the first quadrant bounded by the x-axis and the curve $y = e^{-x}$.

47. Find the volume obtained by revolving the region in Problem 46
a) about the x-axis, b) about the y-axis.

48. Find the center of gravity of the region bounded by the curves $y = \pm(1 - x^2)^{-1/2}$ and the lines $x = 0$, $x = 1$.

49. Find the area between the curves $y = \sec x$ and $y = \tan x$ for $0 \le x \le \pi/2$.

7.10 Using Integral Tables

The numbered integration formulas on the endpapers at the front and back of this book are stated in terms of constants a, b, c, m, n, and so on. These constants can usually assume any real value and need not be integers. Occasional limitations on their values are stated along with the formulas. Formula 5 requires $n \neq -1$, and Formula 11 requires $n \neq -2$.

The formulas also assume that the constants do not take on values that require dividing by zero or taking even roots of negative numbers. For example, Formula 8 assumes $a \neq 0$, and Formula 13(a) cannot be used when b is positive.

The following examples show how the integration formulas are commonly used.

EXAMPLE 1 Evaluate

$$\int x(2x + 5)^{-1}\,dx.$$

Solution We use Formula 8 (not 7),

$$\int x(ax + b)^{-1}\,dx = \frac{x}{a} - \frac{b}{a^2}\ln|ax + b| + C,$$

with $a = 2$ and $b = 5$:

$$\int x(2x + 5)^{-1}\,dx = \frac{x}{2} - \frac{5}{4}\ln|2x + 5| + C. \quad \square$$

EXAMPLE 2 Evaluate

$$\int \frac{dx}{x\sqrt{2x + 4}}.$$

Solution We use Formula 13(b),

$$\int \frac{dx}{x\sqrt{ax + b}} = \frac{1}{\sqrt{b}}\ln\left|\frac{\sqrt{ax + b} - \sqrt{b}}{\sqrt{ax + b} + \sqrt{b}}\right| + C, \qquad \text{if } b > 0,$$

with $a = 2$ and $b = 4$:

$$\int \frac{dx}{x\sqrt{2x+4}} = \frac{1}{\sqrt{4}} \ln \left| \frac{\sqrt{2x+4} - \sqrt{4}}{\sqrt{2x+4} + \sqrt{4}} \right| + C$$

$$= \frac{1}{2} \ln \left| \frac{\sqrt{2x+4} - 2}{\sqrt{2x+4} + 2} \right| + C. \square$$

EXAMPLE 3 Evaluate

$$\int \frac{dx}{x\sqrt{2x-4}}.$$

Solution We use Formula 13(a),

$$\int \frac{dx}{x\sqrt{ax+b}} = \frac{2}{\sqrt{-b}} \tan^{-1} \sqrt{\frac{ax+b}{-b}} + C, \qquad \text{if } b < 0,$$

with $a = 2$ and $b = -4$:

$$\int \frac{dx}{x\sqrt{2x-4}} = \frac{2}{\sqrt{-(-4)}} \tan^{-1} \sqrt{\frac{2x-4}{-(-4)}} + C = \tan^{-1} \sqrt{\frac{x-2}{2}} + C. \square$$

EXAMPLE 4 Evaluate

$$\int \frac{dx}{x^2\sqrt{2x-4}}.$$

Solution We begin with Formula 15,

$$\int \frac{dx}{x^2\sqrt{ax+b}} = -\frac{\sqrt{ax+b}}{bx} - \frac{a}{2b} \int \frac{dx}{x\sqrt{ax+b}} + C,$$

with $a = 2$ and $b = -4$:

$$\int \frac{dx}{x^2\sqrt{2x-4}} = -\frac{\sqrt{2x-4}}{-4x} + \frac{2}{2 \cdot 4} \int \frac{dx}{x\sqrt{2x-4}} + C.$$

We then use Formula 13(a) to evaluate the integral on the right (Example 3) to obtain

$$\int \frac{dx}{x^2\sqrt{2x+4}} = \frac{\sqrt{2x-4}}{4x} + \frac{1}{4} \tan^{-1} \sqrt{\frac{x-2}{2}} + C. \square$$

EXAMPLE 5 Evaluate

$$\int \sec^5 x \, dx.$$

Solution We use Formula 92,

$$\int \sec^n ax \, dx = \frac{\sec^{n-2} ax \tan ax}{a(n-1)} + \frac{n-2}{n-1} \int \sec^{n-2} ax \, dx, \qquad n \neq 1,$$

with $n = 5$ and $a = 1$:

$$\int \sec^5 x \, dx = \frac{\sec^3 x \tan x}{4} + \frac{3}{4} \int \sec^3 x \, dx.$$

To evaluate the integral on the right, we use Formula 92 again, with $n = 3$:

$$\int \sec^3 x\,dx = \frac{\sec x \tan x}{2} + \frac{1}{2}\int \sec x\,dx$$

$$= \frac{\sec x \tan x}{2} + \frac{1}{2}\ln|\sec x + \tan x| + C.$$

The combined result is

$$\int \sec^5 x\,dx = \frac{\sec^3 x \tan x}{4} + \frac{3 \sec x \tan x}{8} + \frac{3}{8}\ln|\sec x + \tan x| + C. \quad \square$$

EXAMPLE 6 Evaluate

$$\int x \sin^{-1} x\,dx.$$

Solution We use Formula 99,

$$\int x^n \sin^{-1} ax\,dx = \frac{x^{n+1}}{n+1}\sin^{-1} ax - \frac{a}{n+1}\int \frac{x^{n+1}\,dx}{\sqrt{1-a^2x^2}}, \qquad n \neq -1,$$

with $n = 1$ and $a = 1$:

$$\int x \sin^{-1} x\,dx = \frac{x^2}{2}\sin^{-1} x - \frac{1}{2}\int \frac{x^2\,dx}{\sqrt{1-x^2}}.$$

The integral on the right is found in the table as Formula 33,

$$\int \frac{x^2}{\sqrt{a^2-x^2}}\,dx = \frac{a^2}{2}\sin^{-1}\frac{x}{a} - \frac{1}{2}x\sqrt{a^2-x^2} + C,$$

with $a = 1$:

$$\int \frac{x^2\,dx}{\sqrt{1-x^2}} = \frac{1}{2}\sin^{-1} x - \frac{1}{2}x\sqrt{1-x^2} + C.$$

The combined result is

$$\int x \sin^{-1} x\,dx = \frac{x^2}{2}\sin^{-1} x - \frac{1}{2}\left(\frac{1}{2}\sin^{-1} x - \frac{1}{2}x\sqrt{1-x^2}\right) + C$$

$$= \left(\frac{x^2}{2} - \frac{1}{4}\right)\sin^{-1} x + \frac{1}{4}x\sqrt{1-x^2} + C. \quad \square$$

PROBLEMS

The formulas referred to by number in Problems 1–5 are to be found in the table of integrals on the endpapers.

1. Verify Formula 55 by differentiating the right side.

2. Verify Formula 76 by differentiating the right side.

3. Verify Formula 9 by integrating

$$\int \frac{x}{(ax+b)^2}\,dx$$

with the substitution $u = ax + b$.

4. Verify Formula 46 by integrating

$$\int \frac{dx}{x^2\sqrt{x^2-a^2}}$$

with the substitution $x = a \sec u$.

5. To derive Formula 104, integrate $\int xe^{ax}$ by parts.

Use formulas from the table of integrals (endpapers) to evaluate the integrals in Problems 6–29.

6. $\displaystyle\int_0^{\pi/2} \sin^{10} x\,dx$

7. $\displaystyle\int_0^{\pi/2} \cos^9 x\,dx$

8. $\int_0^{\infty} e^{-x^2} dx$
9. $\int x \cos^{-1} x\, dx$
10. $\int x^3 5^x\, dx$
11. $\int \sec^4 x\, dx$
12. $\int \csc^5 x\, dx$
13. $\int \frac{dx}{x\sqrt{x-3}}$
14. $\int x \tan^{-1} 2x\, dx$
15. $\int \frac{dx}{(9-x^2)^2}$
16. $\int \frac{\sqrt{4x+9}}{x^2} dx$
17. $\int \frac{dx}{x^2\sqrt{7+x^2}}$
18. $\int \frac{dx}{x^2\sqrt{7-x^2}}$
19. $\int \frac{\sqrt{x^2-2}}{x} dx$
20. $\int \frac{dx}{5+4\sin 2x}$
21. $\int \frac{dx}{4+5\sin 2x}$
22. $\int \tan^4 5x\, dx$
23. $\int \tan^3 \frac{x}{2}\, dx$
24. $\int \frac{x}{\sqrt{x-2}} dx$
(*Hint:* The n in Formula 7 need not be an integer.)
25. $\int x\sqrt{ax+b}\, dx$
26. $\int \frac{\sqrt{3x-4}}{x} dx$
27. $\int 3^{ex} dx$
28. $\int_0^{\infty} x^{10}e^{-x} dx$
29. $\int_0^1 x^6 e^x\, dx$

REVIEW QUESTIONS AND EXERCISES

1. What are some of the general methods presented in this chapter for finding an indefinite integral?
2. What substitution(s) would you consider trying if the integrand contained the following?
   a) $\sqrt{x^2+9}$ b) $\sqrt{x^2-9}$
   c) $\sqrt{9-x^2}$ d) $\sin^3 x \cos^2 x$
   e) $\sin^2 x \cos^2 x$ f) $\frac{1+\sin\theta}{2+\cos\theta}$
3. What method(s) would you try if the integrand contained the following?
   a) $\sin^{-1} x$ b) $\ln x$
   c) $\sqrt{1+2x-x^2}$ d) $x \sin x$
   e) $\frac{2x+3}{x^2-5x+6}$ f) $\sin 5x \cos 3x$
   g) $\frac{1-\sqrt{x}}{1+\sqrt[4]{x}}$ h) $x\sqrt{2x+3}$
4. Discuss two types of improper integral. Define convergence and divergence of each type. Give examples of convergent and divergent integrals of each type. What tests are there for convergence and divergence of improper integrals?

MISCELLANEOUS PROBLEMS

Evaluate the following integrals:

1. $\int \frac{\cos x\, dx}{\sqrt{1+\sin x}}$
2. $\int \frac{\sin^{-1} x\, dx}{\sqrt{1-x^2}}$
3. $\int \frac{\tan x\, dx}{\cos^2 x}$
4. $\int \frac{y}{y^4+1} dy$
5. $\int e^{\ln\sqrt{x}} dx$
6. $\int \frac{\cos\sqrt{x}}{\sqrt{x}} dx$
7. $\int \frac{dx}{\sqrt{x^2+2x+2}}$
8. $\int \frac{(3x-7)\, dx}{(x-1)(x-2)(x-3)}$
9. $\int x^2 e^x\, dx$
10. $\int \sqrt{x^2+1}\, dx$
11. $\int \frac{e^t\, dt}{1+e^{2t}}$
12. $\int \frac{dx}{e^x+e^{-x}}$
13. $\int \frac{dx}{(x+1)\sqrt{x}}$
14. $\int \frac{dx}{\sqrt{1+\sqrt{x}}}$
15. $\int t^{2/3}(t^{5/3}+1)^{2/3}\, dt$
16. $\int \frac{\cot x\, dx}{\ln(\sin x)}$
17. $\int \frac{dt}{\sqrt{e^t+1}}$
18. $\int \frac{dt}{\sqrt{1-e^{-t}}}$
19. $\int \frac{\sin x e^{\sec x}}{\cos^2 x} dx$
20. $\int \frac{\cos x\, dx}{1+\sin^2 x}$
21. $\int \frac{dx}{\sqrt{2x-x^2}}$
22. $\int \frac{\sin x\, dx}{1+\cos^2 x}$
23. $\int \frac{\cos 2t}{1+\sin 2t} dt$
24. $\int \frac{dx}{\sin x \cos x}$
25. $\int \sqrt{1+\sin x}\, dx$
26. $\int \sqrt{1-\sin x}\, dx$
27. $\int \frac{dx}{\sqrt{(a^2-x^2)^3}}$
28. $\int \frac{dx}{\sqrt{(a^2+x^2)^3}}$
29. $\int \frac{\sin x\, dx}{\cos^2 x - 5\cos x + 4}$
30. $\int \frac{e^{2x}\, dx}{\sqrt[3]{1+e^x}}$
31. $\int \frac{dx}{x(x+1)(x+2)\cdots(x+m)}$

32. $\int \frac{dx}{x^6 - 1}$

33. $\int \frac{dy}{y(2y^3 + 1)^2}$

34. $\int \frac{x\,dx}{1 + \sqrt{x}}$

35. $\int \frac{dx}{x(x^2 + 1)^2}$

36. $\int \ln \sqrt{x - 1}\,dx$

37. $\int \frac{dx}{e^x - 1}$

38. $\int \frac{d\theta}{1 - \tan^2 \theta}$

39. $\int \frac{(x + 1)\,dx}{x^2(x - 1)}$

40. $\int \frac{x\,dx}{x^2 + 4x + 3}$

41. $\int \frac{du}{(e^u - e^{-u})^2}$

42. $\int \frac{4\,dx}{x^3 + 4x}$

43. $\int \frac{dx}{5x^2 + 8x + 5}$

44. $\int \frac{\sqrt{x^2 - a^2}}{x}\,dx$

45. $\int e^x \cos 2x\,dx$

46. $\int \frac{dx}{x(3\sqrt{x} + 1)}$

47. $\int \frac{dx}{x(1 + \sqrt[3]{x})}$

48. $\int \frac{\cot \theta\,d\theta}{1 + \sin^2 \theta}$

49. $\int \frac{z^5\,dz}{\sqrt{1 + z^2}}$

50. $\int \frac{e^{4t}\,dt}{(1 + e^{2t})^{2/3}}$

51. $\int \frac{dx}{x^{1/5}\sqrt{1 + x^{4/5}}}$

52. $\int x \sec^2 x\,dx$

53. $\int x \sin^{-1} x\,dx$

54. $\int \frac{(x^3 + x^2)\,dx}{x^2 + x - 2}$

55. $\int \frac{x^3 + 1}{x^3 - x}\,dx$

56. $\int \frac{x\,dx}{(x - 1)^2}$

57. $\int \frac{(2e^{2x} - e^x)\,dx}{\sqrt{3e^{2x} - 6e^x - 1}}$

58. $\int \frac{(x + 1)\,dx}{(x^2 + 2x - 3)^{2/3}}$

59. $\int \frac{dy}{(2y + 1)\sqrt{y^2 + y}}$

60. $\int \frac{dx}{x^2\sqrt{a^2 - x^2}}$

61. $\int (1 - x^2)^{3/2}\,dx$

62. $\int \ln (x + \sqrt{1 + x^2})\,dx$

63. $\int x \tan^2 x\,dx$

64. $\int \frac{\tan^{-1} x}{x^2}\,dx$

65. $\int x \cos^2 x\,dx$

66. $\int x^2 \sin x\,dx$

67. $\int x \sin^2 x\,dx$

68. $\int \frac{dt}{t^4 + 4t^2 + 3}$

69. $\int \frac{du}{e^{4u} + 4e^{2u} + 3}$

70. $\int x \ln \sqrt{x + 2}\,dx$

71. $\int (x + 1)^2 e^x\,dx$

72. $\int \sec^{-1} x\,dx$

73. $\int \frac{8\,dx}{x^4 + 2x^3}$

74. $\int \frac{x\,dx}{x^4 - 16}$

75. $\int_0^{\pi/2} \frac{\cos x\,dx}{\sqrt{1 + \cos x}}$

76. $\int \frac{\cos x\,dx}{\sin^3 x - \sin x}$

77. $\int \frac{du}{(e^u + e^{-u})^2}$

78. $\int \frac{x\,dx}{1 + \sqrt{x} + x}$

79. $\int \frac{\sec^2 t\,dt}{\sec^2 t - 3 \tan t + 1}$

80. $\int \frac{dt}{\sec^2 t + \tan^2 t}$

81. $\int \frac{dx}{1 + \cos^2 x}$

82. $\int e^{2t} \cos (e^t)\,dt$

83. $\int \ln \sqrt{x^2 + 1}\,dx$

84. $\int x \ln (x^3 + x)\,dx$

85. $\int x^3 e^{x^2}\,dx$

86. $\int \frac{\cos x\,dx}{\sqrt{4 - \cos^2 x}}$

87. $\int \frac{\sec^2 x\,dx}{\sqrt{4 - \sec^2 x}}$

88. $\int x^2 \sin (1 - x)\,dx$

89. $\int \frac{dx}{(x^2 + 1)(2 + \tan^{-1} x)}$

90. $\int \frac{dx}{1 + 2 \sin x}$

91. $\int \frac{dx}{\sin^3 x}$

92. $\int \frac{dx}{\cot^3 x}$

93. $\int (\sin^{-1} x)^2\,dx$

94. $\int x \ln \sqrt[3]{3x + 1}\,dx$

95. $\int \frac{x^3\,dx}{(x^2 + 1)^2}$

96. $\int \frac{x\,dx}{\sqrt{1 - x}}$

97. $\int x\sqrt{2x + 1}\,dx$

98. $\int \ln (x + \sqrt{x^2 - 1})\,dx$

99. $\int \ln (x - \sqrt{x^2 - 1})\,dx$

100. $\int \frac{dt}{t - \sqrt{1 - t^2}}$

101. $\int e^{-x} \tan^{-1} (e^x)\,dx$

102. $\int \sin^{-1} \sqrt{x}\,dx$

103. $\int \ln (x + \sqrt{x})\,dx$

104. $\int \tan^{-1} \sqrt{x}\,dx$

105. $\int \ln (x^2 + x)\,dx$

106. $\int \ln (\sqrt{x} + \sqrt{1 + x})\,dx$

107. $\int \cos \sqrt{x}\,dx$

108. $\int \sin \sqrt{x}\,dx$

109. $\int \tan^{-1} \sqrt{x + 1}\,dx$

110. $\int \sqrt{1 - x^2} \sin^{-1} x\,dx$

111. $\int x \sin^2 (2x)\,dx$

112. $\int \frac{\tan x\,dx}{\tan x + \sec x}$

113. $\int \frac{dt}{\sqrt{e^{2t} + 1}}$

114. $\int \frac{dx}{(\cos^2 x + 4 \sin x - 5) \cos x}$

115. $\int \frac{dt}{a + be^{ct}}$, $a, b, c \neq 0$

116. $\int \sqrt{\frac{1 - \cos x}{\cos \alpha - \cos x}}\, dx$, α constant, $0 < \alpha < x < \pi$

117. $\int \frac{dx}{9x^4 + x^2}$

118. $\int \ln \sqrt{1 + x^2}\, dx$

119. $\int \ln (2x^2 + 4)\, dx$

120. $\int \frac{x^3}{\sqrt{1 - x^2}}\, dx$

121. $\int \frac{dx}{x(2 + \ln x)}$

122. $\int \frac{\cos 2x - 1}{\cos 2x + 1}\, dx$

123. $\int \frac{dx}{x^3 + 1}$

124. $\int \frac{e^{2x}\, dx}{\sqrt[4]{e^x + 1}}$

125. $\int \frac{2 \sin \sqrt{x}}{\sqrt{x} \sec \sqrt{x}}\, dx$

126. $\int (16 + x^2)^{-3/2}\, dx$

127. $\int \sin \sqrt{x + 1}\, dx$

128. $\int \cos \sqrt{1 - x}\, dx$

In Problems 129–135, evaluate each of the limits by identifying it with an appropriate definite integral and evaluating the latter.

129. $\lim_{n\to\infty} \left(\frac{1}{n + 1} + \frac{1}{n + 2} + \cdots + \frac{1}{2n}\right)$

130. $\lim_{n\to\infty} \left(\frac{1}{\sqrt{n^2}} + \frac{1}{\sqrt{n^2 + n}} + \frac{1}{\sqrt{n^2 + 2n}} + \cdots + \frac{1}{\sqrt{n^2 + (n - 1)n}}\right)$

131. $\lim_{n\to\infty} (\sin 0 + \sin (\pi/n) + \sin (2\pi/n) + \cdots + \sin [(n - 1)\pi/n])/n$

132. $\lim_{n\to\infty} \left(\frac{1 + \sqrt[n]{e} + \sqrt[n]{e^2} + \sqrt[n]{e^3} + \cdots + \sqrt[n]{e^{n-1}}}{n}\right)$

133. $\lim_{n\to\infty} \left(\frac{n}{n^2 + 0^2} + \frac{n}{n^2 + 1^2} + \frac{n}{n^2 + 2^2} + \cdots + \frac{n}{n^2 + (n + 1)^2}\right)$

134. $\lim_{n\to\infty} \sum_{k=1}^{n} \ln \sqrt[n]{1 + \frac{k}{n}}$

135. $\lim_{n\to\infty} \sum_{k=0}^{n-1} \frac{1}{\sqrt{n^2 - k^2}}$

136. Show that $\int_0^\infty x^3 e^{-x^2}\, dx$ is a convergent integral, and evaluate it.

137. Show that $\int_0^1 \ln x\, dx$ is a convergent integral and find its value. Sketch $y = \ln x$ for $0 < x \leq 1$.

138. Assuming that $|\alpha| \neq |\beta|$, prove that

$$\lim_{T\to\infty} \frac{1}{T} \int_0^T \sin \alpha x \sin \beta x\, dx = 0.$$

139. Evaluate

$$\lim_{h\to 0} \frac{1}{h} \int_2^{2+h} e^{-x^2}\, dx.$$

140. At points of the curve $y^2 = 4px$, lines of length $h = y$ are drawn perpendicular to its plane. Find the area of the surface formed by these lines at points of the curve between $(0, 0)$ and $(p, 2p)$.

141. A plane figure is bounded by a 90° arc of a circle of radius r and a straight line. Find its area and its centroid.

142. Find the coordinates of the center of gravity of the area bounded by the curves $y = e^x$, $x = 1$, and $y = 1$.

143. At points on a circle of radius r, perpendiculars to its plane are erected, the perpendicular at each point P being of length ks, where s is the arc of the circle from a fixed point A to P. Find the area of the surface formed by the perpendiculars along the arc beginning at A and extending once around the circle.

144. A plate in the first quadrant, bounded by the curves $y = e^x$, $y = 1$, and $x = 4$, is submerged vertically in water with its upper corner on the surface. The surface of the water is given by the line $y = e^4$. Find the total force on one side of the plate if the weight of water is 62.5 lb/ft^3, and if the units of x and y are also measured in feet.

145. The area under the curve $y = e^{-x}$, for $x \geq 0$, is rotated about the x-axis. If A is the area under the curve, V is the volume generated, and S is the surface area of this volume: (a) is A finite? (b) is V finite? (c) is S finite? Give reasons for your answers.

146. Find the length of arc of $y = \ln x$ from $x = 1$ to $x = e$.

147. The arc of Problem 146 is rotated about the y-axis. Find the surface area generated. Check your answer by comparing it with the area of a frustum of a suitably related cone.

148. Find the length of arc of the curve $y = e^x$ from $x = 0$ to $x = 2$.

149. The arc of Problem 148 is rotated about the x-axis. Find the surface area generated. Compare this area with the area of a suitably related frustum of a cone to check your answer.

150. One arch of the curve $y = \cos x$ is rotated about the x-axis. Find the surface area generated.

151. Use Simpson's rule with $n = 8$ to approximate the length of arc of $y = \cos x$ from $x = 0$ to $x = \pi/2$. Check your answer by consulting a table of elliptic integrals if one is available.

152. A thin wire is bent into the shape of one arch of the curve $y = \cos x$. Find its center of mass (see Problems 150, 151).

153. The area between the graph of $y = \ln(1/x)$, the x-axis, and the y-axis is revolved around the x-axis. Find the volume of the solid it generates.

154. Find the total perimeter of the curve $x^{2/3} + y^{2/3} = a^{2/3}$.

155. Find the area of the surface generated by rotating the curve $x^{2/3} + y^{2/3} = a^{2/3}$ about the x-axis.

156. A thin homogeneous wire is bent into the shape of one arch of the curve $x^{2/3} + y^{2/3} = a^{2/3}$. Find its center of mass (see Problems 154, 155).

157. Determine whether the integral $\int_1^\infty \ln x(dx/x^2)$ converges or diverges.

158. Sketch the curves

a) $y = x - e^x$; show behavior for large $|x|$,

b) $y = e^{(x-e^x)}$; show behavior for large $|x|$.

Show that the following integrals converge and compute their values:

c) $\int_{-\infty}^{b} e^{(x-e^x)}\,dx$, d) $\int_{-\infty}^{\infty} e^{(x-e^x)}\,dx$.

159. The gamma function $\Gamma(x)$ is defined, for $x > 0$, by the definite integral

$$\Gamma(x) = \int_0^\infty t^{x-1}e^{-t}\,dt.$$

a) Sketch graphs of the integrand $y = t^{x-1}e^{-t}$ vs. t, $t > 0$, for the three typical cases $x = \frac{1}{2}$, 1, 3. Find maxima, minima, and points of inflection (if they exist).

b) Show that the integral converges if $x > 0$.

c) Show that the integral diverges if $x \le 0$.

d) Using integration by parts, show that $\Gamma(x + 1) = x\Gamma(x)$, for $x > 0$.

e) Using the result of part (d), show that, if n is a positive integer, $\Gamma(n) = (n - 1)!$, where $0! = 1$ (by definition) and if m is a positive integer, $m!$ is the product of the positive integers from 1 through m inclusive.

f) Discuss how one might compute a table of values of $\Gamma(x)$ for the domain $1 \le x \le 2$, say. Consult such a table, if available, and sketch the graph of $y = \Gamma(x)$ for $0 < x \le 3$.

160. Determine whether the integral $\int_0^\infty t^{x-1}(\ln t)e^{-t}\,dt$ converges or diverges for fixed $x > 0$. Sketch the integrand $y = t^{x-1}(\ln t)e^{-t}$ vs. t, $t > 0$, for the two cases $x = \frac{1}{2}$ and $x = 3$.

161. Solve the differential equation $d^2x/dt^2 = -k^2x$, subject to the conditions $x = a$, $dx/dt = 0$ when $t = 0$. (*Hint:* Let $dx/dt = v$, $d^2x/dt^2 = dv/dt = v\,dv/dx$.)

162. Find $\lim_{n\to\infty} \int_0^1 \frac{ny^{n-1}}{1 + y}\,dy$.

163. Prove that if n is a positive integer, or zero, then

$$\int_{-1}^{1} (1 - x^2)^n\,dx = \frac{2^{2n+1}(n!)^2}{(2n + 1)!},$$

with $0! = 1$; $n! = 1 \cdot 2 \cdots n$ for $n \ge 1$.

164. Show that if $p(x)$ is a polynomial of degree n, then

$$\int e^x p(x)\,dx = e^x[p(x) - p'(x) + p''(x) - \cdots + (-1)^n p^{(n)}(x)].$$

8 Plane Analytic Geometry

8.1 Conic Sections

This chapter is about how the conic sections that originated in Greek geometry are described today by quadratic equations as curves in the coordinate plane. The Greeks of Plato's time described these curves as the curves in which a plane could intersect a cone (hence the name "conic section"). There are a number of possibilities, as you can see in Fig. 8.1.

Kepler's use of the conic sections in his *Commentaries on the motions of Mars* in 1609, the work in which he announced his first two laws (elliptical orbits, equal areas in equal times), led to an energetic reexamination of conic sections for properties that might be useful in astronomy. Optics, an interest of mathematicians since Greek times (Claudius Ptolemy was interested in refraction, but when Snell's law eluded him he fitted his data with a parabola), received greatly increased attention after the invention of the telescope and microscope in the beginning of the seventeenth century. The resulting interest in the design of lenses and mirrors led to an interest in the shapes of their surfaces and, since these surfaces are surfaces of revolution, to an interest in the shapes of their generating curves. For many of the curved mirrors used in reflecting telescopes, these curves are conic sections. Geographical exploration created a need for maps and for studying the courses of ships as represented on globes and maps, as we saw in connection with our study of the integral of the secant in Chapter 7. The introduction of the notion of a moving earth called for new principles of mechanics to account for the paths of moving objects. This, too, meant a study of curves. Among moving objects, projectiles became important because cannons could by then fire at targets thousands of meters away. To a first approximation, projectiles move along parabolas, as we shall see in Chapter 13.

As the mathematicians of the sixteenth and seventeenth centuries studied Greek works, they began to realize that the Greek methods of proof lacked generality. A special method had to be devised for nearly every theorem (as we saw in the calculations of Article 4.6). The approach to the conic sections was eventually altered from a purely geometric one (sections of a cone) to one that involved the notions of coordinates and distance. Guidobaldo del Monte, for example, in 1579 defined the ellipse

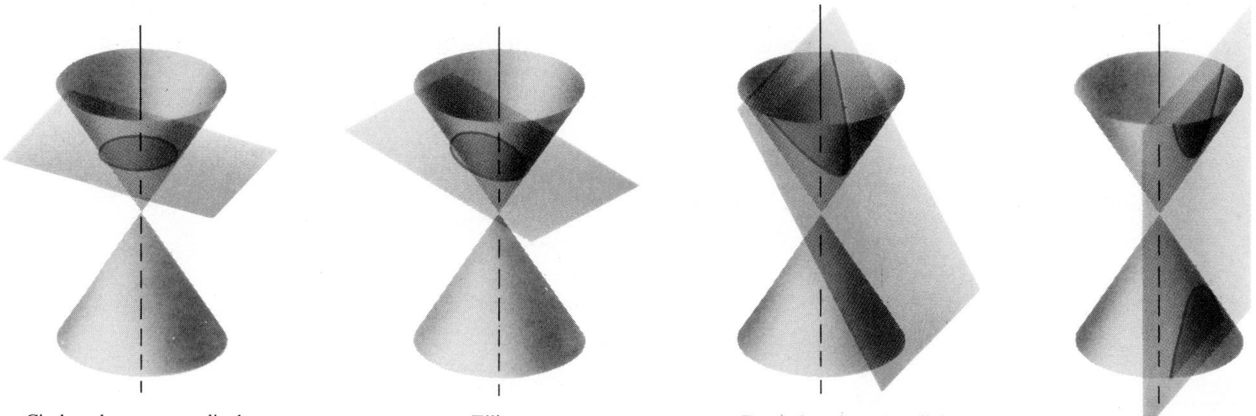

Circle: plane perpendicular to cone axis

Ellipse

Parabola: plane parallel to side of cone

Hyperbola: plane parallel to cone axis

(a)

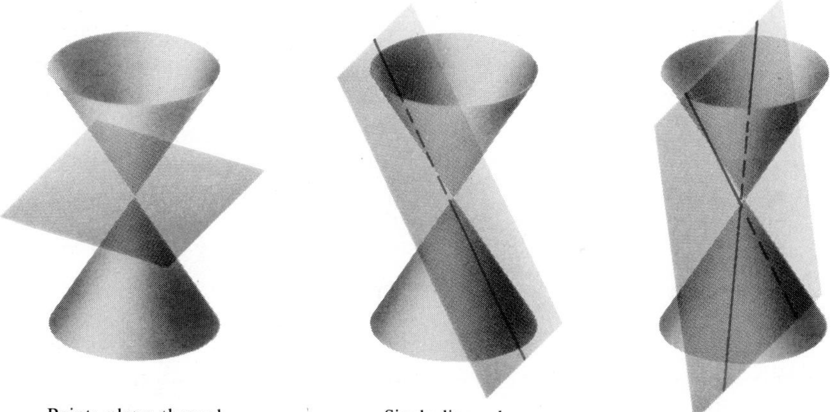

Point: plane through cone vertex only

Single line: plane tangent to cone

Pair of lines

(b)

8.1 (a) The "standard" conic sections are the curves in which a plane cuts a cone. In the case of a hyperbola, the intersection consists of two curves, not one, each called a 'branch' of the hyperbola. The "degenerate" conic sections in (b) are obtained by passing the plane of intersection through the cone's vertex.

as the set of points in a plane the sum of whose distances from the foci is a constant, the definition we shall use in Article 8.5.

From an algebraic point of view, the present chapter sets the stage for the amazing result that the graph of a quadratic equation in x and y in the plane (if any) is always a conic section or two parallel lines.

From the point of view of applications, the study of conic sections in the seventeenth century provided the mathematics needed to describe the paths of comets, planets, and asteroids that move through space under the influence of gravitational forces. The same mathematics describes the orbits of the satellites we launch today. Once we know that the path of a moving body, whether it be a planet or an electron, is a conic section, we know many other properties of the body's motion as well.

If you like history, you will find a great deal to interest you about conics in Morris Kline's *Mathematical Thought from Ancient to Modern Times* (Oxford University Press, 1972).

8.2

Equations from the Distance Formula

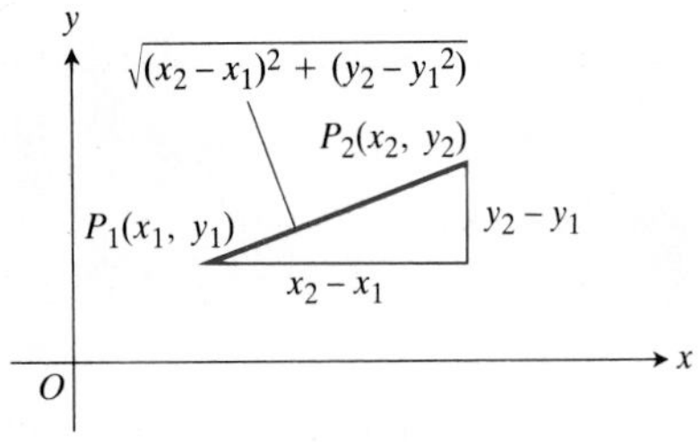

8.2 The distance between two points P_1 and P_2 is computed from their coordinates.

The distance d between two points (x_1, y_1) and (x_2, y_2) in the plane is calculated from their coordinates by the formula

$$d = \sqrt{(x_2 - x_1)^2 + (y_2 - y_1)^2}. \tag{1}$$

This formula comes from applying the Pythagorean theorem to the triangle in Fig. 8.2. The formula is particularly useful in finding equations for curves whose geometric character depends on one or more distances, as in the following example.

EXAMPLE 1 Find an equation for the set of points $P(x, y)$ that are equidistant from the origin O and the line L: $x = 4$.

Solution The distance between P and L is the perpendicular distance PQ between $P(x, y)$ and the point $Q(4, y)$ on L that has the same y-coordinate as P. (See Fig. 8.3.) Thus,

$$PQ = \sqrt{(4 - x)^2 + (y - y)^2} = |4 - x|.$$

The distance OP is

$$OP = \sqrt{x^2 + y^2}.$$

The condition to be satisfied by P is

$$OP = PQ, \qquad \text{or} \qquad \sqrt{x^2 + y^2} = |4 - x|. \tag{2}$$

If (2) holds, so does the equation we get by squaring:

$$x^2 + y^2 = 16 - 8x + x^2$$

or

$$y^2 = 16 - 8x. \tag{3}$$

That is, if a point is equidistant from O and L, then its coordinates must satisfy Eq. (3).

It is also true that any point whose coordinates satisfy Eq. (3) lies equidistant from O and L. For, if

$$y^2 = 16 - 8x,$$

then

$$\sqrt{x^2 + y^2} = \sqrt{x^2 + (16 - 8x)} = \sqrt{(x - 4)^2} = |x - 4| = |4 - x|,$$

and hence

$$OP = PQ.$$

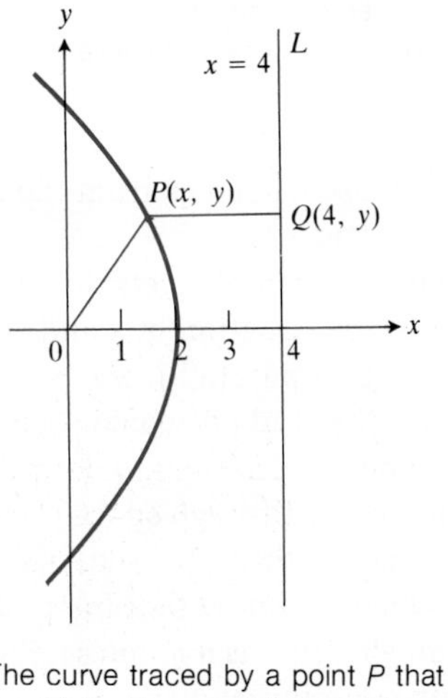

8.3 The curve traced by a point P that stays equidistant from point O and from line L.

Therefore Eq. (3) expresses both the necessary and sufficient condition for $P(x, y)$ to be equidistant from O and L. In other words, Eq. (3) is an equation for these equidistant points. □

PROBLEMS

Use the distance formula to find equations for the sets of points $P(x, y)$ that satisfy the conditions in Problems 1-11. Graph each equation.

1. P is equidistant from the origin and the line $y = -4$.
2. P is equidistant from the point $(0, 1)$ and the line $y = -1$.
3. P is equidistant from the two points $A(-2, 1)$ and $B(2, -3)$.
4. The distance from P to $F_1(-1, 0)$ is twice its distance to $F_2(2, 0)$.
5. The product of its distances from $F_1(-2, 0)$ and $F_2(2, 0)$ is 4.
6. The sum of the distances from P to $F_1(1, 0)$ and $F_2(0, 1)$ is constant and the curve passes through the origin.
7. The distance of P from the line $x = -2$ is 2 times its distance from the point $(2, 0)$.
8. The distance from P to $(-\sqrt{2}, -\sqrt{2})$ is 2 plus the distance from P to $(\sqrt{2}, \sqrt{2})$.
9. The distance of P from the point $(-3, 0)$ is 4 more than its distance from the point $(3, 0)$.
10. The distance of P from the line $y = 1$ is 3 less than its distance from the origin.
11. P is 3 units from the point $(2, 3)$.
12. Find the points on the line $x - y = 1$ that are 2 units from the point $(3, 0)$.
13. Find a point that is equidistant from the three points $A(0, 1)$, $B(1, 0)$, $C(4, 3)$. What is the radius of the circle through A, B, and C?
14. Find the distance between the point $P_1(x_1, y_1)$ and the straight line $Ax + By = C$.

8.3 Circles

DEFINITION

> A *circle* is the set of points in a plane whose distance from a given fixed point in the plane is a constant.

Equation of a Circle

Let $C(h, k)$ be the given fixed point, the *center* of the circle. Let r be the constant distance, the *radius* of the circle. Let $P(x, y)$ be a point on the circle (Fig. 8.4a). Then

$$CP = r, \tag{1}$$

or

$$\sqrt{(x - h)^2 + (y - k)^2} = r,$$

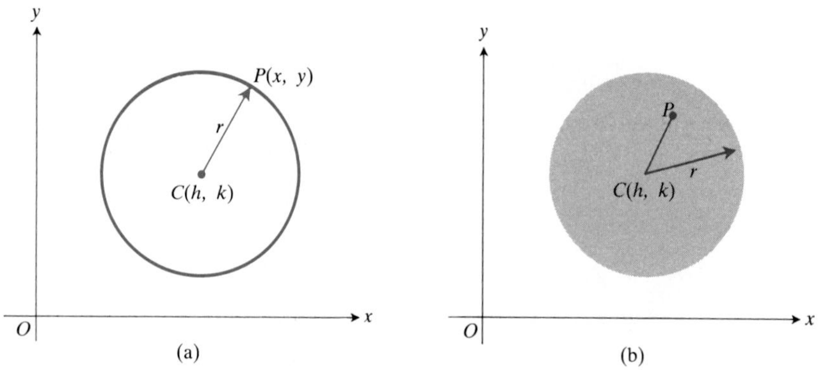

8.4 (a) $(x - h)^2 + (y - k)^2 = r^2$. (b) The region $(x - h)^2 + (y - k)^2 < r^2$ is the interior of the circle with center $C(h, k)$ and radius r.

or

$$(x - h)^2 + (y - k)^2 = r^2. \tag{2}$$

If (1) is satisfied, so is (2), and if (2) is satisfied, so is (1). Therefore (2) is an equation for the points of the circle.

EXAMPLE 1 Find an equation for the circle with center at the origin and with radius r.

Solution If $h = k = 0$, Eq. (2) becomes

$$x^2 + y^2 = r^2. \ \square \tag{3}$$

EXAMPLE 2 Find the circle through the origin with center at $C(2, -1)$.

Solution With $(h, k) = (2, -1)$, Eq. (2) takes the form

$$(x - 2)^2 + (y + 1)^2 = r^2.$$

Since the circle goes through the origin, $x = y = 0$ must satisfy the equation. Hence

$$(0 - 2)^2 + (0 + 1)^2 = r^2 \quad \text{or} \quad r^2 = 5.$$

The equation is

$$(x - 2)^2 + (y + 1)^2 = 5. \ \square$$

EXAMPLE 3 What points $P(x, y)$ satisfy the inequality

$$(x - h)^2 + (y - k)^2 < r^2? \tag{4}$$

Solution The left side of (4) is the square of the distance CP from $C(h, k)$ to $P(x, y)$. The inequality is satisfied if and only if

$$CP < r;$$

that is, if and only if P lies inside the circle of radius r with center at $C(h, k)$, Fig. 8.4(b). $\square$

EXAMPLE 4 Find the center and radius of the circle

$$x^2 + y^2 + 4x - 6y = 12.$$

Solution We complete the squares in the x terms and y terms and get

$$(x^2 + 4x + 4) + (y^2 - 6y + 9) = 12 + 4 + 9,$$

or

$$(x + 2)^2 + (y - 3)^2 = 25.$$

This is Eq. (2) with $(h, k) = (-2, 3)$ and $r^2 = 25$. It therefore represents a circle with

Center: $C(-2, 3)$,
Radius: $r = 5$.

See Fig. 8.5. $\square$

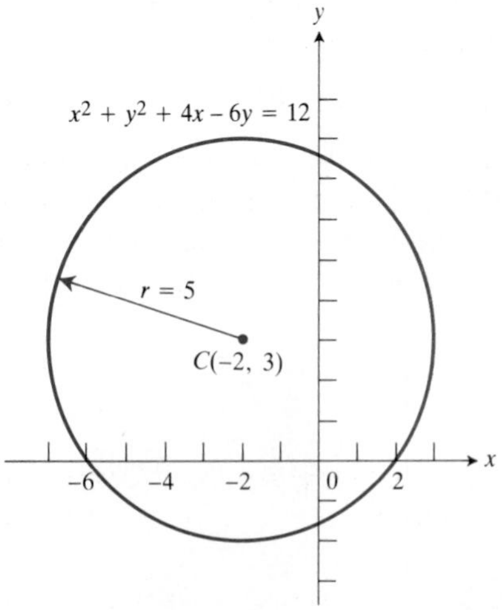

8.5 The graph of $x^2 + y^2 + 4x - 6y = 12$.

An equation of the form

$$Ax^2 + Ay^2 + Dx + Ey + F = 0, \qquad A \neq 0, \tag{5}$$

can often be reduced to the form of Eq. (2) by completing the squares as we did in Example 4. More specifically, we may divide (5) by A and write

$$\left(x^2 + \frac{D}{A}x\right) + \left(y^2 + \frac{E}{A}y\right) = -\frac{F}{A}. \tag{6}$$

To complete the squares for x, add $(D/2A)^2 = D^2/(4A^2)$; and for y add $(E/2A)^2 = E^2/(4A^2)$. Of course we must add to both sides of Eq. (6), thus obtaining

$$\left(x + \frac{D}{2A}\right)^2 + \left(y + \frac{E}{2A}\right)^2 = -\frac{F}{A} + \frac{D^2 + E^2}{4A^2} = \frac{D^2 + E^2 - 4AF}{4A^2}. \tag{7}$$

Equation (7) is like Eq. (2), with

$$r^2 = \frac{D^2 + E^2 - 4AF}{4A^2}, \tag{8}$$

provided the expression on the right is positive. Then Eq. (5) represents a circle with

Center: $\left(-\frac{D}{2A}, -\frac{E}{2A}\right)$

Radius: $r = \sqrt{(D^2 + E^2 - 4AF)/4A^2}$.

If the right-hand side of Eq. (8) is zero, the curve reduces to a single point.

If the right-hand side of Eq. (8) is negative, no points in the plane satisfy either (7) or (5).

In analyzing the equation

$$Ax^2 + Ay^2 + Dx + Ey + F = 0, \qquad A \neq 0, \tag{5}$$

we recommend that you use the method of Example 4 instead of memorizing the formulas (6), (7), and (8). It is enough to remember that *a quadratic equation like* (5), *in which* x^2 *and* y^2 *have equal coefficients, and with no* xy *term, represents a circle* (or a single point, or no point at all).

Dividing Eq. (5) by A gives an equation of the form

$$x^2 + y^2 + ax + by + c = 0, \tag{9}$$

where a, b, and c are constants. These constants can often be determined from geometric conditions. For instance, the circle might be known to pass through three specific (noncollinear) points; or to be tangent to three specific nonconcurrent lines; or to be tangent to two lines and to pass through a point not on either line.

PROBLEMS

In Problems 1–4, find the circle with center $C(h, k)$ and radius r.

1. $C(0, 2)$, $r = 2$

2. $C(-2, 0)$, $r = 3$

3. $C(3, -4)$, $r = 5$

4. $C(1, 1)$, $r = \sqrt{2}$

5. Write an inequality that describes the points that lie inside the circle with center $C(-2, -1)$ and radius $r = \sqrt{6}$.

6. Write an inequality that describes the points that lie outside the circle with center $C(-4, 2)$ and radius $r = 4$.

In Problems 7–16, find the center and radius of the given circle. Then draw the circle.

7. $x^2 + y^2 = 16$ **8.** $x^2 + y^2 + 6y = 0$

9. $x^2 + y^2 - 2y = 3$ **10.** $x^2 + y^2 + 2x = 8$

11. $x^2 + 4x + y^2 = 12$ **12.** $3x^2 + 3y^2 + 6x = 1$

13. $x^2 + y^2 + 2x + 2y = -1$

14. $x^2 + y^2 - 6x + 2y + 1 = 0$

15. $2x^2 + 2y^2 + x + y = 0$

16. $2x^2 + 2y^2 - 28x + 12y + 114 = 0$

What points satisfy the inequalities in Problems 17 and 18?

17. $x^2 + y^2 + 2x - 4y + 5 \leq 0$

18. $x^2 + y^2 + 4x + 4y + 9 \geq 0$

19. The center of a circle is $C(2, 2)$. The circle goes through the point $A(4, 5)$ Find its equation.

20. The center of a circle is $C(-1, 1)$. The circle is tangent to the line $x + 2y = 4$. Find its equation.

21. A circle passes through the points $A(2, -2)$ and $B(3, 4)$. Its center is on the line $x + y = 2$. Find its equation.

22. Show geometrically that the lines that are drawn from the exterior point $P_1(x_1, y_1)$ tangent to the circle $(x - h)^2 + (y - k)^2 = r^2$ have length l given by

$$l^2 = (x_1 - h)^2 + (y_1 - k)^2 - r^2.$$

23. Find an equation for the circle that passes through the three points, $A(2, 3)$, $B(3, 2)$, and $C(-4, 3)$.

24. Find an equation for the coordinates of the point $P(x, y)$ if the sum of the squares of its distances from the two points $(-5, 2)$ and $(1, 4)$ is always 52. Identify and sketch the curve.

25. Is the point (0.1, 3.1) inside, outside, or on the circle

$$x^2 + y^2 - 2x - 4y + 3 = 0?$$

Why?

26. If the distance from $P(x, y)$ to the point (6, 0) is twice its distance from the point (0, 3), show that P lies on a circle and find the center and radius.

27. Show that the line normal (perpendicular) to the circle $x^2 + y^2 = a^2$ at any point (x_1, y_1) on the circle passes through the origin.

28. Find the circle inscribed in the triangle whose sides are the lines $4x + 3y = 24$, $3x - 4y = 18$, $4x - 3y + 32 = 0$. (*Hint:* The distance between the point (h, k) and the line $Ax + By = C$ is

$$\frac{|Ah + Bk - C|}{\sqrt{A^2 + B^2}}$$

by the answer to Problem 14, Article 8.2. Establish on which side of each line the center (h, k) of the circle lies, and use that information to determine how to handle the absolute values.)

29. Let P be a point outside a given circle C. Let PT be tangent to C at T. Let the line PN from P through the center of C intersect C at M and N. Prove that $PM \cdot PN = (PT)^2$.

30. It is known that any angle inscribed in a semicircle is a right angle. Prove the converse: i.e., if for every choice of the point $P(x, y)$ on a curve C joining O and A, the angle OPA is a right angle, then the curve is a circle or a semicircle having OA as diameter.

***Toolkit* programs**

Conic Sections   Super * Grapher

8.4 Parabolas

DEFINITION

A *parabola* is the set of points in a plane that are equidistant from a given fixed point and fixed line in the plane.

The fixed point is called the *focus* of the parabola and the fixed line the *directrix*.

If the focus F lies on the directrix L, then the parabola is nothing more than the line through F perpendicular to L. We consider this to be a degenerate case.

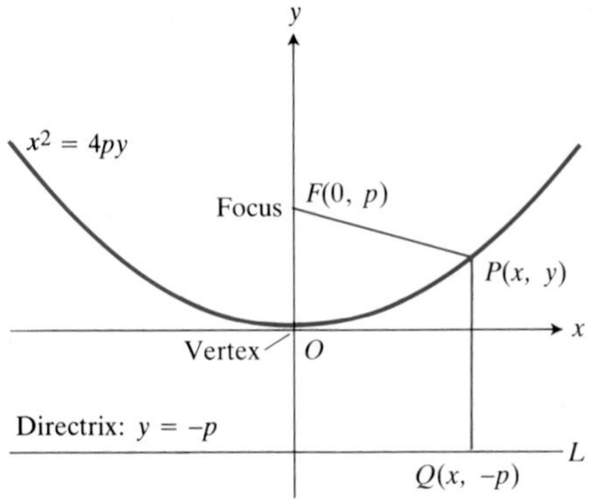

8.6 The parabola $x^2 = 4py$.

Parabolas That Open Upward

If F does not lie on L, then we may choose a coordinate system that results in a simple equation for the parabola by taking the y-axis through F perpendicular to L, and taking the origin halfway between F and L. If the distance between F and L is $2p$, we may assign F coordinates $(0, p)$ and the equation of L is $y = -p$, as in Fig. 8.6.

With reference to Fig. 8.6, a point $P(x, y)$ lies on the parabola if and only if the distances PF and PQ are equal:

$$PF = PQ, \tag{1}$$

where $Q(x, -p)$ is the foot of the perpendicular from P to L. From the distance formula,

$$PF = \sqrt{x^2 + (y - p)^2} \quad \text{and} \quad PQ = \sqrt{(y + p)^2}.$$

When we equate these two expressions, square, and simplify, we get

$$x^2 = 4py. \tag{2}$$

This equation must be satisfied by any point on the parabola. Conversely, if (2) is satisfied, then we have

$$\begin{aligned} PF &= \sqrt{x^2 + (y - p)^2} = \sqrt{4py + (y^2 - 2py + p^2)} \\ &= \sqrt{(y + p)^2} = PQ, \end{aligned}$$

and $P(x, y)$ is on the parabola. In other words, the parabola is the graph of the equation $x^2 = 4py$.

Axis and Vertex

Since p is positive in Eq. (2), y cannot be negative (if x is to be real) and the curve lies on or above the x-axis.

The curve is symmetric about the y-axis since x appears only to an even power.

The axis of symmetry of the parabola is called the *axis* of the parabola. The point on this axis midway between the focus and the directrix is on the parabola, since it is equidistant from the focus and the directrix. It is called the *vertex* of the parabola. The origin is the vertex of the parabola in Fig. 8.6. The tangent to a parabola at its vertex is parallel to the directrix. From Eq. (2) we find the slope of the tangent at any point is $dy/dx = x/2p$, and this is zero at the origin.

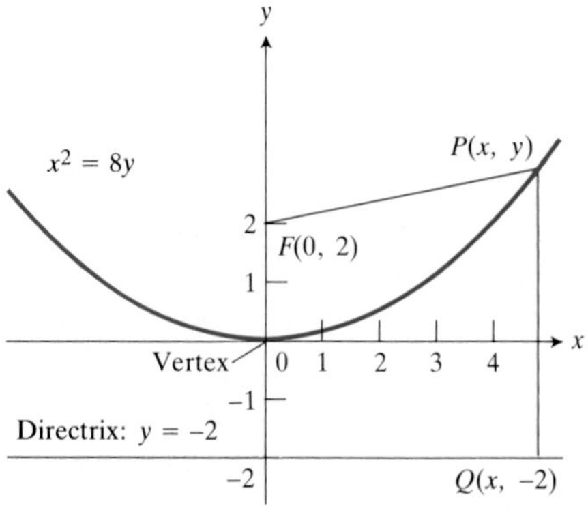

8.7 The parabola $x^2 = 8y$.

EXAMPLE 1 Find the focus and directrix of the parabola

$$x^2 = 8y. \tag{3}$$

Solution See Fig. 8.7. Equation (3) is Eq. (2) with

$$4p = 8, \qquad p = 2.$$

The focus is on the y-axis $p = 2$ units from the vertex, that is, at

Focus: $F(0, 2)$.

The directrix $y = -p$ is the line $y = -2$:

Directrix: $y = -2$. □

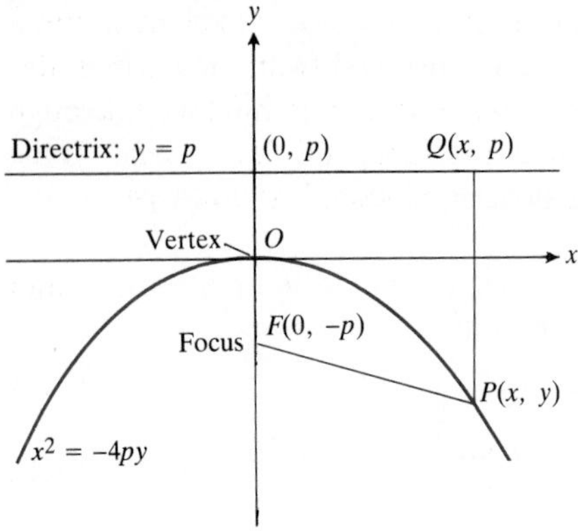

8.8 The parabola $x^2 = -4py$.

Parabolas That Open Downward

If the parabola opens downward, as in Fig. 8.8, with its focus at $F(0, -p)$ and its directrix the line $y = p$, then Eq. (2) becomes

$$x^2 = -4py. \tag{4}$$

Parabolas That Open to the Right Or Left

We may also interchange the roles of x and y in Eqs. (2) and (4). The resulting equations

$$y^2 = 4px \tag{5a}$$

and

$$y^2 = -4px \tag{5b}$$

also represent parabolas (Fig. 8.9), but now they are symmetric about the x-axis because y appears only to an even power. The vertex is still at the origin. The directrix is perpendicular to the axis of symmetry, and p units from the vertex. The focus is on the axis of symmetry, also p units from the vertex, and "inside" the curve. The parabola in (5a) opens toward the right, because x must be greater than or equal to zero, while the parabola in (5b) opens toward the left.

Translation of Axes

If the vertex of the parabola is at the point $V(h, k)$, Eqs. (2), (4), and (5) no longer apply. However, it is easy to determine what the appropriate equation is, by introducing a new coordinate system, with its origin O' at V, and its axes parallel to the original axes. Every point P in the plane then has two sets of coordinates, say (x, y) in the original system, and (x', y') in

8.9 (a) The parabola $y^2 = 4px$.
(b) The parabola $y^2 = -4px$.

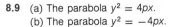

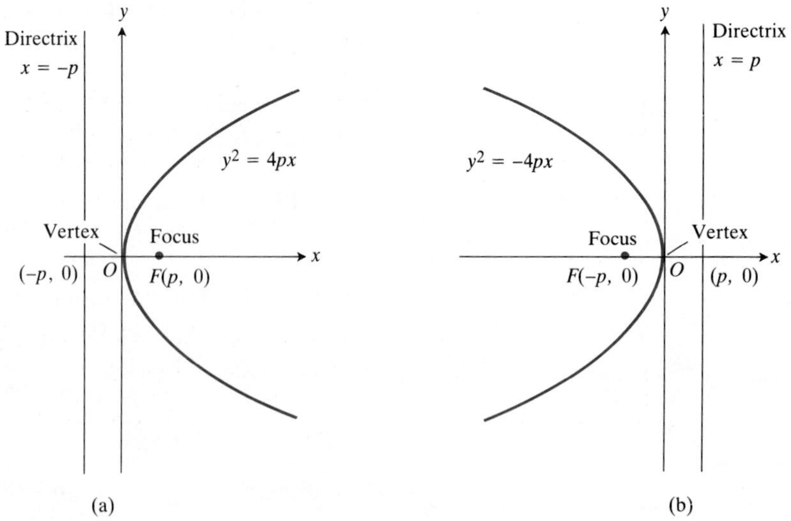

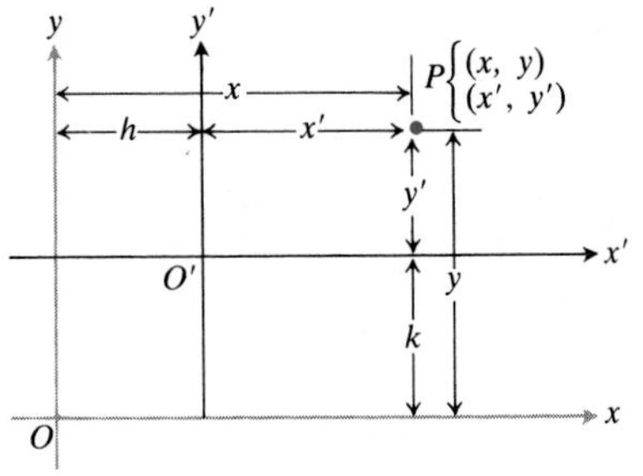

8.10 Translation of axes. Note that $h + x' = x$ and $k + y' = y$. Solving for x and y gives $x' = x - h$, $y' = y - k$.

the new. To move from O to P, we have a horizontal displacement x and a vertical displacement y. The value of x is obtained from two horizontal displacements: h from O to O′ and x′ from O′ to P (Fig. 8.10). Similarly, y is obtained from two vertical displacements: k from O to O′ and y′ from O′ to P. Thus the two sets of coordinates are related as follows:

$$\begin{cases} x = x' + h \\ y = y' + k \end{cases} \tag{6a}$$

or

$$\begin{cases} x' = x - h \\ y' = y - k. \end{cases} \tag{6b}$$

Equations (6) are called the equations for *translation of axes,* because the new coordinate axes may be obtained by translating the old ones.

Suppose, now, we consider a parabola, with vertex $V(h, k)$ and opening upward as in Fig. 8.11. In terms of x′y′-coordinates, Eq. (2) provides an equation of the parabola in the form

$$(x')^2 = 4py'. \tag{7}$$

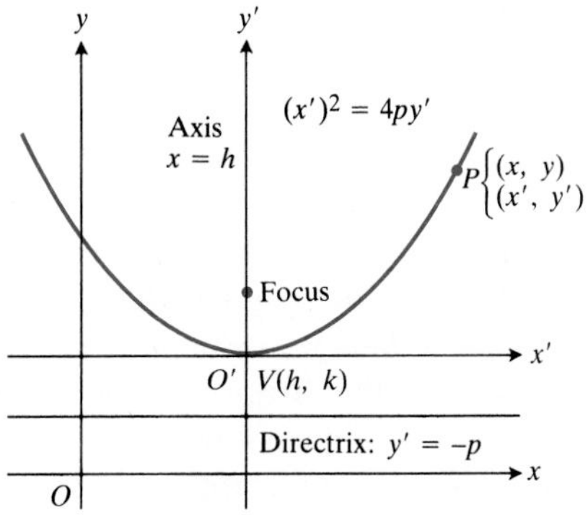

8.11 Parabola with vertex at $V(h, k)$.

By using Eqs. (6a), we may express this in xy-coordinates by the equation

$$(x - h)^2 = 4p(y - k). \tag{8a}$$

The axis of the parabola in (8a) is the line $x = h$. The equation of this line is obtained by setting the quadratic term $(x - h)^2$ in (8a) equal to zero. The focus is on the axis of symmetry, p units above the vertex at $x = h$, $y = k + p$. The directrix is p units below the vertex and perpendicular to the axis of symmetry, thus having equation $y = k - p$.

Other forms of equations of parabolas are

$$(x - h)^2 = -4p(y - k), \tag{8b}$$
$$(y - k)^2 = 4p(x - h), \tag{8c}$$
$$(y - k)^2 = -4p(x - h). \tag{8d}$$

Equation (8b) has a graph symmetric about $x = h$, opening downward; (8c) symmetric about $y = k$ and opening to the right ($x \geq h$); (8d) symmetric about $y = k$ and opening to the left ($x \leq h$).

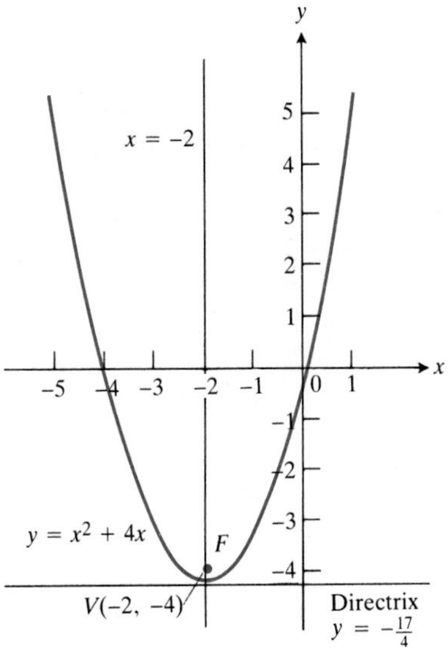

8.12 The parabola $y = x^2 + 4x$.

EXAMPLE 2 Discuss the parabola

$$y = x^2 + 4x. \tag{9}$$

Solution See Fig. 8.12. We complete the square in the x terms by adding 4 to both sides of Eq. (9):

$$y + 4 = x^2 + 4x + 4 = (x + 2)^2,$$

or

$$(x + 2)^2 = y + 4.$$

This has the form

$$(x - h)^2 = 4p(y - k)$$

with

$$h = -2, \quad k = -4, \quad 4p = 1, \quad p = 1/4.$$

The parabola's vertex is $V(h, k) = V(-2, -4)$:

Vertex: $V(-2, -4)$.

Its axis of symmetry is the line $(x + 2)^2 = 0$, or $x = -2$:

Axis: $x = -2$.

The parabola opens upward because $y \geq -4$ for real x. The focus is on the line $x = -2$ one fourth of a unit above the vertex, at $F(-2, -\frac{15}{4})$:

Focus: $F(-2, -15/4)$.

The directrix is parallel to the x-axis, one fourth of a unit below the vertex:

Directrix: $y = -17/4$. □

The Key Feature of the Equation

The key feature of an equation that represents a parabola that lies in the xy-plane with its axis parallel to one of the coordinate axes is that it is quadratic in that coordinate and linear in the other. Whenever we have such an equation we may reduce it to one of the standard forms (8a, b, c, d) by completing the square in the coordinate that appears quadratically. We then put the linear terms in the form $\pm 4p(x - h)$ or $\pm 4p(y - k)$ with p positive. Then the information about vertex, distance from vertex to focus, axis of symmetry, and direction the curve opens can all be read from the equation in this standard form.

EXAMPLE 3

Analyze the equation

$$2x^2 + 5y - 3x + 4 = 0.$$

Solution This equation is quadratic in x, linear in y. We divide by 2, the coefficient of x^2, and collect all the x terms on one side of the equation:

$$x^2 - \tfrac{3}{2}x = -\tfrac{5}{2}y - 2.$$

Now we complete the square by adding $(-\frac{3}{4})^2 = \frac{9}{16}$ to both sides:

$$(x - \tfrac{3}{4})^2 = -\tfrac{5}{2}y - 2 + \tfrac{9}{16} = -\tfrac{5}{2}y - \tfrac{23}{16}.$$

To get the y terms in the form $-4p(y - k)$, we factor out $-\frac{5}{2}$, and write

$$(x - \tfrac{3}{4})^2 = -\tfrac{5}{2}(y + \tfrac{23}{40}).$$

This has the form

$$(x - h)^2 = -4p(y - k)$$

with

$$h = \tfrac{3}{4}, \qquad k = -\tfrac{23}{40}, \qquad 4p = \tfrac{5}{2}.$$

Hence the vertex is $V(\frac{3}{4}, -\frac{23}{40})$. The axis of symmetry is $(x - \frac{3}{4})^2 = 0$, or

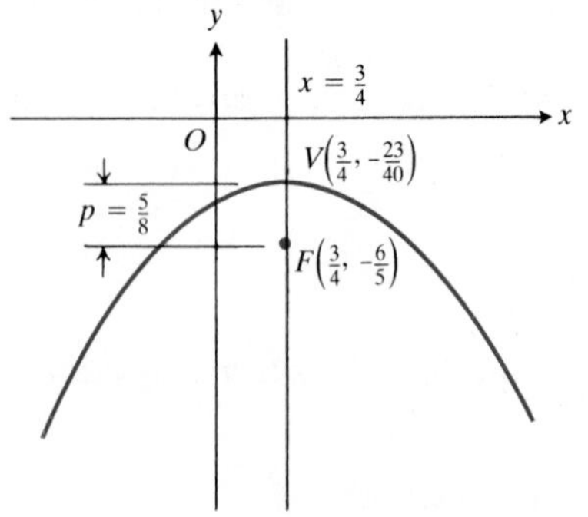

8.13 The parabola $2x^2 + 5y - 3x + 4 = 0$.

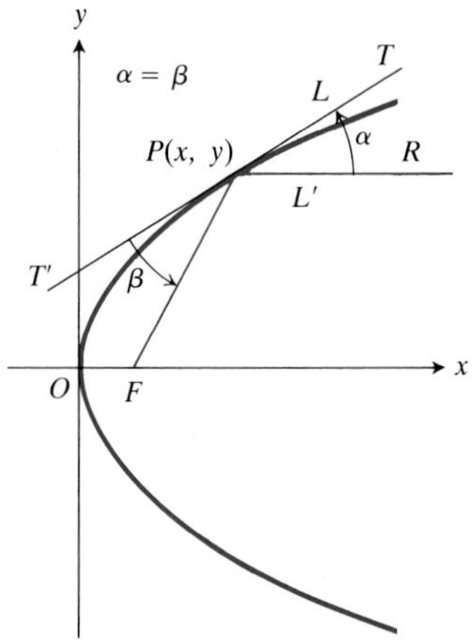

8.14 In a parabolic reflector, the angles α and β are equal.

$x = \frac{3}{4}$; and the distance from the vertex to the focus is

$$p = \tfrac{5}{8}.$$

Since $y - k$ must here be ≤ 0 for real x, the curve opens downward, and the focus is p units below the vertex at $F(\frac{3}{4}, -\frac{6}{5})$. The graph is shown in Fig. 8.13. □

The Reflective Property of Parabolas

In Fig. 8.14, $T'PT$ is tangent to the parabola, F is the focus, and PL' is parallel to the axis of the parabola. The angles $\alpha = \angle L'PT$ and $\beta = \angle T'PF$ are equal (see Problem 35). This accounts for the property of a parabolic reflector that rays originating from the focus are reflected parallel to the axis, or rays coming into the reflector parallel to the axis are reflected to the focus. This property is used in parabolic mirrors of telescopes and in parabolic radar antennas.

Applications

For a very readable account of the many roles played by parabolas in astronomy, radio communications, radar, wind tunnel photography, submarine tracking, and bridge construction, see Lee Whitt's "The Standup Conic Presents: The Parabola and Applications," *The UMAP Journal*, Volume 3, Number 3, 1982, pp. 285–313.

PROBLEMS

Each of Problems 1–8 gives the vertex V and focus F of a parabola. Find equations for the parabola and its directrix. Then sketch the parabola, showing the focus, vertex, and directrix.

1. $V(0, 0)$, $F(0, 2)$ **2.** $V(0, 0)$, $F(-2, 0)$

3. $V(0, 0)$, $F(0, -4)$ **4.** $V(0, 0)$, $F(4, 0)$

5. $V(-2, 3)$, $F(-2, 4)$ **6.** $V(0, 3)$, $F(-1, 3)$

7. $V(-3, 1)$, $F(0, 1)$ **8.** $V(1, -3)$, $F(1, 0)$

Each of Problems 9–14 gives the vertex V and directrix L of a parabola. Find an equation for the parabola and the coordinates of its focus. Then graph the parabola, showing the focus, vertex, and directrix.

9. $V(2, 0)$, L is the y-axis.

10. $V(1, -2)$, L is the x-axis.

11. $V(-3, 1)$, L is the line $x = 1$.

12. $V(-2, -2)$, L is the line $y = -3$.

13. $V(0, 1)$, L is the line $x = -1$.

14. $V(0, 1)$, L is the line $y = 2$.

In Problems 15–18, find the vertex, axis, focus, and directrix of the given parabola. Then sketch the parabola, showing these features.

15. $y^2 = 8x$ **16.** $y^2 + 36x = 0$

17. $x^2 = 100y$ **18.** $x^2 - 9y = 0$

In Problems 19–28, find the vertex, axis of symmetry, focus, and directrix of the given parabola. Sketch the curve, showing these features.

19. $x^2 + 8y - 2x = 7$ **20.** $x^2 - 2y + 8x + 10 = 0$

21. $y^2 + 4x = 8$ **22.** $x^2 - 8y = 4$

23. $x^2 + 2x - 4y - 3 = 0$ **24.** $y^2 + x + y = 0$

25. $4y^2 - 8y + 3x - 2 = 0$ **26.** $y^2 + 6y + 2x + 5 = 0$

27. $3x^2 - 8y - 12x = 4$ **28.** $3x - 2y^2 - 4y + 7 = 0$

29. What points satisfy the inequality $x^2 < 8y$? Sketch.

30. Find the length of the chord perpendicular to the axis of the parabola $y^2 = 4px$ and passing through the focus. (This chord is called the "latus rectum" of the parabola.)

31. Find the equation of the parabolic arch of base b and altitude h in Fig. 8.15.

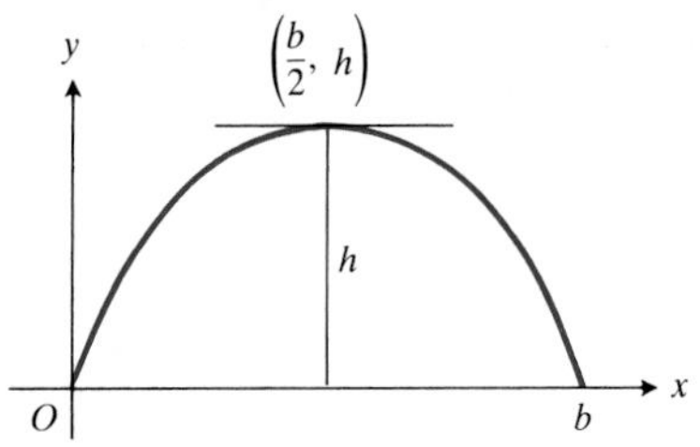

8.15 The parabolic arch of Problems 31 and 37.

32. Given the three points $(-1, 2)$, $(1, -1)$, and $(2, 1)$.
a) Find a parabola passing through the given points and having its axis parallel to the x-axis.
b) Find a parabola passing through the given points and having its axis parallel to the y-axis.

33. Suppose a and b are positive numbers. Sketch the parabolas

$$y^2 = 4a^2 - 4ax \quad \text{and} \quad y^2 = 4b^2 + 4bx$$

in the same diagram. Show that they have a common focus, the same for any a and b. Show that they intersect at $(a - b, \pm 2\sqrt{ab})$, and that each "a-parabola" is orthogonal to every "b-parabola." (Using different values of a and b, we obtain families of confocal parabolas. Each family is a set of orthogonal trajectories of the other family. See Article 2.7.)

34. What points satisfy the equation

$$(2x + y - 3)(x^2 + y^2 - 4)(x^2 - 8y) = 0?$$

Give a reason for your answer.

35. Prove that the angles α and β in Fig. 8.14 are equal.

36. Prove that the tangent to the parabola $y^2 = 4px$ at $P_1(x_1, y_1)$ intersects the axis of symmetry x_1 units to the left of the vertex. (This provides a simple method for constructing the tangent to a parabola at any point on it.)

37. Show that the area of a parabolic sector of altitude h and base b is $\frac{2}{3}bh$ (see Problem 31).

38. Show that the volume generated by rotating the region bounded by the parabola $y = (4h/b^2)x^2$ and the line $y = h$ about the y-axis is equal to one and one-half times the volume of the corresponding inscribed cone.

39. The condition for equilibrium of the section OP of a cable that supports a weight of w pounds per foot measured along the horizontal (Fig. 8.16) is

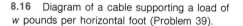

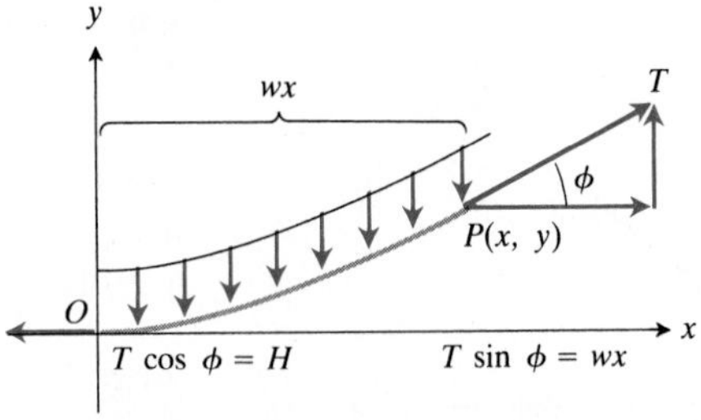

8.16 Diagram of a cable supporting a load of w pounds per horizontal foot (Problem 39).

$$\frac{dy}{dx} = \frac{wx}{H}\left(= \frac{T \sin \phi}{T \cos \phi}\right),$$

where the origin O is taken at the low point of the cable and H is the horizontal tension at O. Show that the curve in which the cable hangs is a parabola. (For this reason, parabolas are used to model the main cables in suspension bridges. Cables and chains that hang "freely" when suspended from their two ends, with no additional load, hang in curves called *catenaries*, studied in Article 9.5. The word catenary comes from *catena*, the Latin word for chain.)

40. *Reflective property of parabolas.* Assume, from optics, that when a ray of light is reflected by a mirror the angle of incidence is equal to the angle of reflection. If a mirror is formed by rotating a parabola about its axis and silvering the resulting surface, show that a ray of light emanating from the focus of the parabola is reflected parallel to the axis.

***Toolkit* programs**

Conic Sections   Super * Grapher

8.5 Ellipses

DEFINITION

An *ellipse* is the set of points in a plane whose distances from two fixed points in the plane have a constant sum.

Equations for Ellipses

If the two fixed points, called *foci*, are taken at $F_1(-c, 0)$ and $F_2(c, 0)$ as in Fig. 8.17, and the sum of the distances $PF_1 + PF_2$ is denoted by $2a$,

$$PF_1 + PF_2 = 2a,$$

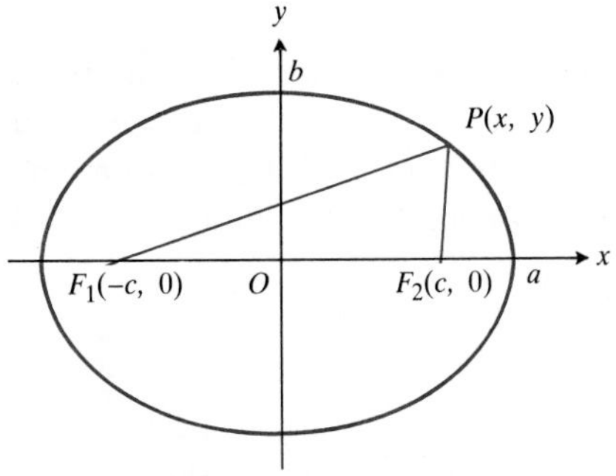

8.17 The ellipse $(x^2/a^2) + (y^2/b^2) = 1$, $a > b$.

then the coordinates of a point $P(x, y)$ on the ellipse satisfy the equation

$$\sqrt{(x+c)^2+y^2} + \sqrt{(x-c)^2+y^2} = 2a.$$

To simplify this expression, we transpose the second radical to the right side of the equation, square and simplify twice to obtain

$$\frac{x^2}{a^2} + \frac{y^2}{a^2-c^2} = 1. \tag{1}$$

Since the sum $PF_1 + PF_2 = 2a$ of two sides of the triangle F_1F_2P is greater than the third side $F_1F_2 = 2c$, the term $(a^2 - c^2)$ in (1) is positive and has a real positive square root, which we denote by b:

$$b = \sqrt{a^2 - c^2}. \tag{2}$$

Then (1) takes the more compact form

$$\frac{x^2}{a^2} + \frac{y^2}{b^2} = 1. \tag{3}$$

Equation (3) reveals that the curve is symmetric about both axes and lies inside the rectangle bounded by the lines $x = a$, $x = -a$, $y = b$, $y = -b$. The intercepts of the curve are at $(\pm a, 0)$ and $(0, \pm b)$. The curve intersects each axis at an angle of 90°, since the slope

$$\frac{dy}{dx} = \frac{-b^2x}{a^2y}$$

is zero at $x = 0$, $y = \pm b$ and is infinite at $y = 0$, $x = \pm a$.

We have shown that the coordinates of P satisfy (1) if P satisfies the geometric condition $PF_1 + PF_2 = 2a$. Suppose now that x and y satisfy (1) with $0 < c < a$. Then

$$y^2 = (a^2 - c^2)\frac{a^2 - x^2}{a^2},$$

and substituting this in the radicals below, we find that

$$PF_1 = \sqrt{(x+c)^2 + y^2} = \left|a + \frac{c}{a}x\right| \tag{4a}$$

and

$$PF_2 = \sqrt{(x-c)^2 + y^2} = \left|a - \frac{c}{a}x\right|. \tag{4b}$$

Since x is restricted to the interval $-a \le x \le a$, the value of $(c/a)x$ lies between $-c$ and c, so that both $a + (c/a)x$ and $a - (c/a)x$ are positive, both being between $a + c$ and $a - c$. Hence the absolute values in (4a) and (4b) yield

$$PF_1 = a + \frac{c}{a}x, \qquad PF_2 = a - \frac{c}{a}x. \tag{5}$$

Adding these, we see that $PF_1 + PF_2$ has the value $2a$ independent of the position of P on the curve. Thus the *geometric property* and *algebraic equation* are equivalent.

Axes

In Eq. (3), $b^2 = a^2 - c^2$ is less than a^2. The *major axis* of the ellipse is the segment of length $2a$ between the x-intercepts $(\pm a, 0)$. The *minor axis* is the segment of length $2b$ between the y-intercepts $(0, \pm b)$. The number a is called the *semimajor axis,* and the number b the *semiminor axis.* If $a = 4$, $b = 3$, then Eq. (3) is

$$\frac{x^2}{16} + \frac{y^2}{9} = 1. \tag{6a}$$

If we interchange the roles of x and y in (6a), we get the equation

$$\frac{x^2}{9} + \frac{y^2}{16} = 1, \tag{6b}$$

which represents an ellipse with its major axis vertical rather than horizontal. Graphs of Eqs. (6a) and (6b) are given in Fig. 8.18 (a) and (b).

There is never any cause for confusion in analyzing equations like (6a) and (6b). We simply find the intercepts on the axes of symmetry; then we know which way the major axis runs, because it is the longer of the two axes. The foci are always on the major axis.

If we use the letters a, b, and c to represent the lengths of semimajor axis, semiminor axis, and *half-distance* between foci, then Eq. (2) tells us that

$$b^2 = a^2 - c^2 \qquad \text{or} \qquad a^2 = b^2 + c^2. \tag{7}$$

Hence a is the hypotenuse of a right triangle of sides b and c, as in Fig. 8.18. When we start with an equation like (6a) or (6b), we can read off a^2 and b^2 from it at once. Then Eq. (7) determines c^2 as their difference. So in either of Eqs. (6a) and (6b) we have

$$c^2 = 16 - 9 = 7.$$

Therefore the foci are $\sqrt{7}$ units from the center of the ellipse as shown.

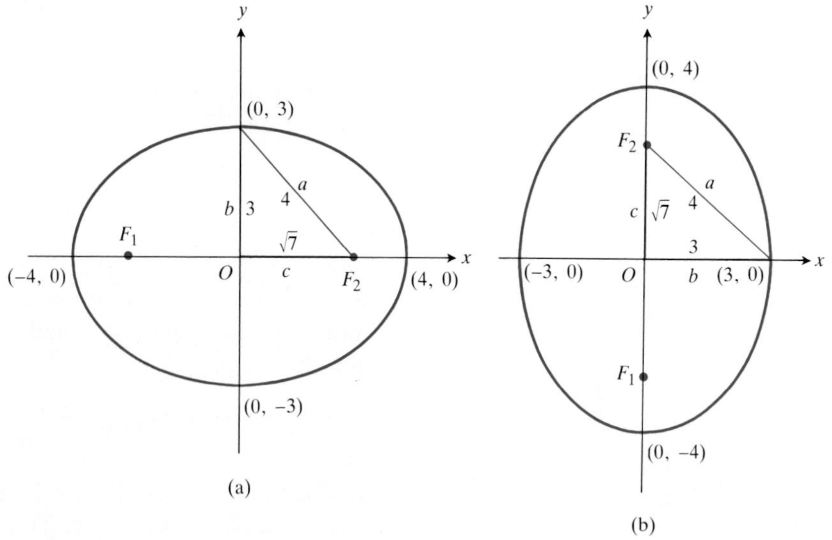

8.18 (a) The major axis of $(x^2/16) + (y^2/9) = 1$ is horizontal. (b) The major axis of $(x^2/9) + (y^2/16) = 1$ is vertical.

Center not at the Origin

The *center* of an ellipse is defined as the point of intersection of its axes of symmetry. If the center is at $C(h, k)$, and the axes of the ellipse are parallel to the x- and y-axes, then we may introduce new coordinates

$$x' = x - h, \qquad y' = y - k, \tag{8}$$

using C as origin O' of $x'y'$-coordinates. The equation of the ellipse in the new coordinates is either

$$\frac{x'^2}{a^2} + \frac{y'^2}{b^2} = 1 \tag{9a}$$

or

$$\frac{x'^2}{b^2} + \frac{y'^2}{a^2} = 1, \tag{9b}$$

depending upon which way the major axis runs.

EXAMPLE 1 Analyze the equation $9x^2 + 4y^2 + 36x - 8y + 4 = 0$.

Solution We collect the x and y terms separately to get

$$9(x^2 + 4x) + 4(y^2 - 2y) = -4,$$

and then complete the square in each set of parentheses to obtain

$$9(x^2 + 4x + 4) + 4(y^2 - 2y + 1) = -4 + 36 + 4.$$

We now divide both sides by 36 and write

$$\frac{(x + 2)^2}{4} + \frac{(y - 1)^2}{9} = 1.$$

Setting

$$x' = x + 2, \qquad y' = y - 1,$$

we see that the new origin $x' = 0$, $y' = 0$, is the same as the point $x = -2$, $y = 1$. In terms of the new coordinates, the equation is

$$\frac{x'^2}{4} + \frac{y'^2}{9} = 1.$$

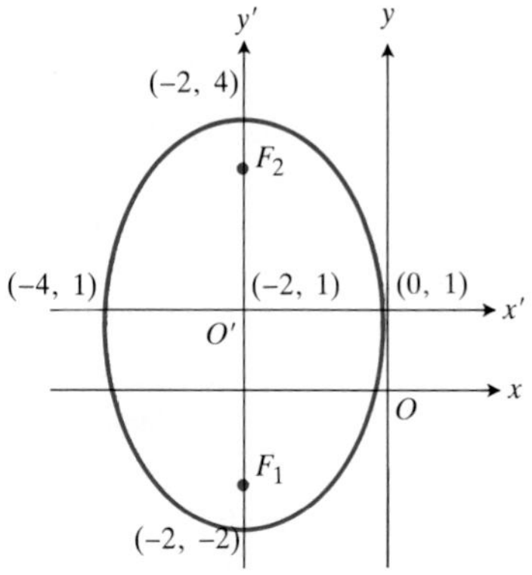

8.19 The ellipse $9x^2 + 4y^2 + 36x - 8y + 4 = 0$.

This equation represents an ellipse with intercepts at $(0, \pm 3)$ on the y'-axis and $(\pm 2, 0)$ on the x'-axis (Fig. 8.19). To locate the foci, we use the relation $c^2 + b^2 = a^2$ to find

$$c = \sqrt{a^2 - b^2} = \sqrt{9 - 4} = \sqrt{5}.$$

The foci are at the points $(0, \pm\sqrt{5})$ on the y'-axis or at $(-2, 1 \pm \sqrt{5})$ in terms of the original coordinates. □

Two Key Properties

The essential *geometric* property of an ellipse (Fig. 8.17) is that the sum of the distances from any point on it to the two foci is a constant, namely,

$$PF_1 + PF_2 = 2a.$$

The essential *algebraic* property of its equation, when written in the form of a quadratic without a cross-product term, is that the x^2 and y^2 terms have the same sign.

To discuss the properties of the ellipse in more detail, we shall assume that its equation has been reduced to the form

$$\frac{x^2}{a^2} + \frac{y^2}{b^2} = 1, \qquad a > b > 0. \tag{10}$$

Although the distance c from the center of the ellipse to either focus does not appear in its equation, we may still determine c as in the examples above from the equation

$$c^2 = a^2 - b^2, \tag{11}$$

(which comes from rewriting Eq. (7)).

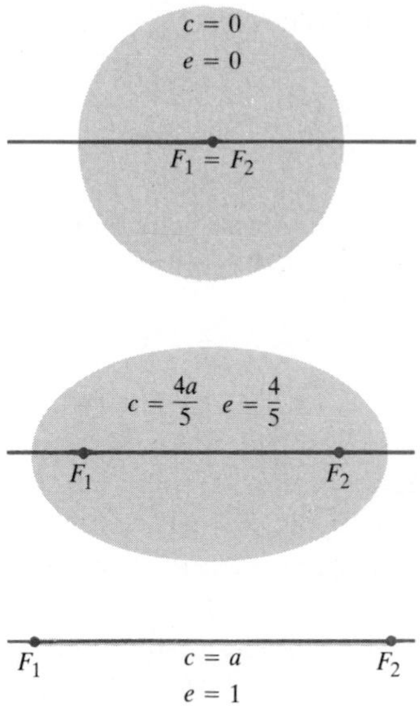

8.20 The eccentricity of an ellipse describes its shape.

Eccentricity

If we keep a fixed and vary the focal distance c over the interval $0 \le c \le a$, the resulting ellipses will vary in shape. They are circular when $c = 0$ (so that $a = b$), and become flatter as c increases, until in the extreme case $c = a$, the "ellipse" reduces to the line segment F_1F_2 joining the two foci (Fig. 8.20). The ratio

$$e = \frac{c}{a}, \tag{12}$$

called the *eccentricity* of the ellipse, varies from 0 to 1 and indicates the degree of departure from circularity.

Table 8.1 Eccentricities of planetary orbits

Mercury	0.21	Saturn	0.06
Venus	0.01	Uranus	0.05
Earth	0.02	Neptune	0.01
Mars	0.09	Pluto	0.25
Jupiter	0.05		

The planets in the solar system revolve around the sun in elliptical orbits with the sun at one focus. Most of the planets, including the earth, have orbits that are nearly circular, as can be seen from the eccentricities in Table 8.1. Pluto, however, has a fairly eccentric orbit, with $e = 0.25$, as does Mercury, with $e = 0.21$. Other members of the solar system have orbits that are even more eccentric. Icarus, an asteroid about one mile wide that revolves around the sun every 409 earth days, has an orbital eccentricity of 0.83 (Fig. 8.21).

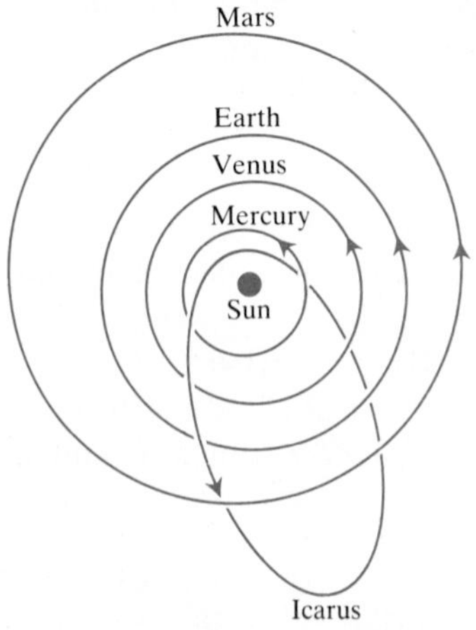

8.21 The orbit of the asteroid Icarus.

EXAMPLE 2 The orbit of the comet Kohoutek (Fig. 8.22) is about 44 astronomical units wide by 3,600 astronomical units long. (One *astronomical unit* (AU) is the earth's mean distance from the sun, about 92,600,000 miles.) Find the eccentricity of the orbit.

Solution

$$e = \frac{\sqrt{a^2 - b^2}}{a} = \frac{\sqrt{(1800)^2 - (22)^2}}{1800} \approx 0.9999. \square$$

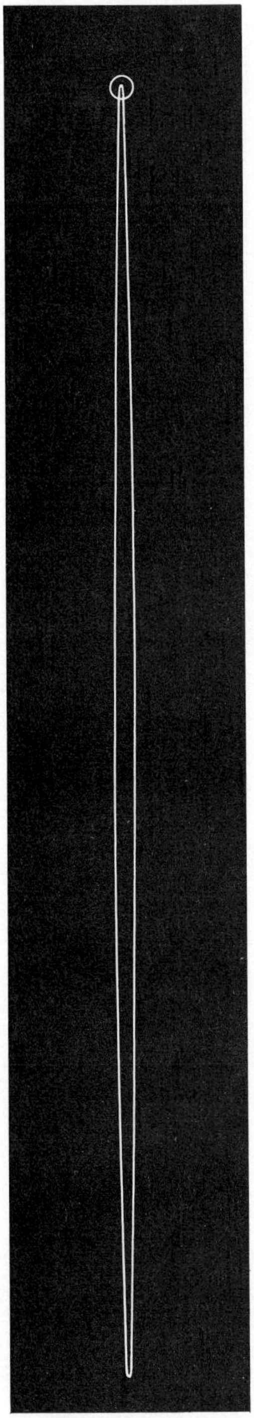

8.22 The orbit of Comet Kohoutek, shown approximately to scale. The small circle shows the outer limit of the orbit of Pluto.

Halley's Comet

Edmund Halley (1656–1742), the British mathematician and astronomer who financed the publication of Newton's *Principia,* used Newton's theory to calculate the orbit of the great comet of 1682, since known as Halley's comet. He predicted that it would reappear in 1758, which it did. Halley's comet returns every 76 years or so, and records of its appearance have now been identified as far back as 2200 years ago.

Last seen in 1911, Halley's comet is scheduled (at this writing) to round the sun again during the fall, winter, and spring of 1985–1986. It was already within view of the great 200-inch telescope on Mount Palomar, California, on October 16, 1982.

The orbit of Halley's comet is an ellipse 36.18 AU long by 9.12 AU wide. Its eccentricity is

$$e = \frac{\sqrt{a^2 - b^2}}{a} = \frac{\sqrt{(18.09)^2 - (4.56)^2}}{18.09} \approx 0.97.$$

The best times to view the comet will be in November 1985 and April 1986. The comet will come within 57.7 million miles of the earth on November 27, 1985, on its way in toward the sun, and within 39 million miles of the earth on April 11, 1986, on its way out.

Classifying Conic Sections by Eccentricity

While a parabola has one focus and one directrix, each ellipse has two foci and two directrices. The directrices are the lines perpendicular to the major axis of the ellipse at distances $\pm a/e$ from its center. The *parabola* has the property that

$$PF = 1 \cdot PD \tag{13}$$

for any point P on it, where F is the focus and D is the point nearest P on the directrix. For an *ellipse,* it is not difficult to show that the equations that replace (13) are

$$PF_1 = e \cdot PD_1, \qquad PF_2 = e \cdot PD_2. \tag{14}$$

Here e is the eccentricity, P is any point on the ellipse, F_1 and F_2 are the foci, and D_1 and D_2 are the points nearest P on the two directrices. In Eq. (14), the corresponding directrix and focus must be used; that is, if one uses the distance from P to the focus F_1, one must also use the distance from P to the directrix at the same end of the ellipse (see Fig. 8.23). We thus associate the directrix $x = -a/e$ with the focus $F_1(-c, 0)$, and the directrix $x = a/e$ with the focus $F_2(c, 0)$. In terms of the semimajor axis a and eccentricity $e < 1$, as one goes away from the center along the major axis,

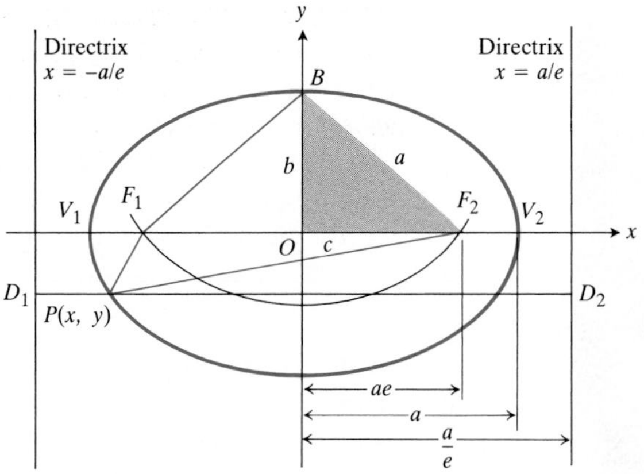

8.23 The ellipse $(x^2/a^2) + (y^2/b^2) = c^2$.

one finds successively

a *focus* at distance ae from the center,

a *vertex* at distance a from the center,

a *directrix* at distance a/e from the center.

The "focus–directrix" property unites the parabola, ellipse, and hyperbola (next article) in the following way. Suppose that the distance PF of a point P from a fixed point F (the focus) is a constant multiple of its distance from a fixed line (the directrix). That is, suppose

$$PF = e \cdot PD \tag{15}$$

where e is the constant of proportionality. Then the path traced by P is

a) a *parabola* if $e = 1$,

b) an *ellipse* of eccentricity e if $e < 1$, and

c) a *hyperbola* of eccentricity e if $e > 1$.

Thus the parabola is a very special case.

Construction

The quickest way to construct an ellipse uses the definition directly. A loop of string around two tacks F_1 and F_2 is pulled tight with a pencil point P and held taut as the pencil traces a curve (Fig. 8.24). The curve is an ellipse because $PF_1 + PF_2$ is constant (being equal to the length of the string loop minus the distance between the tacks).

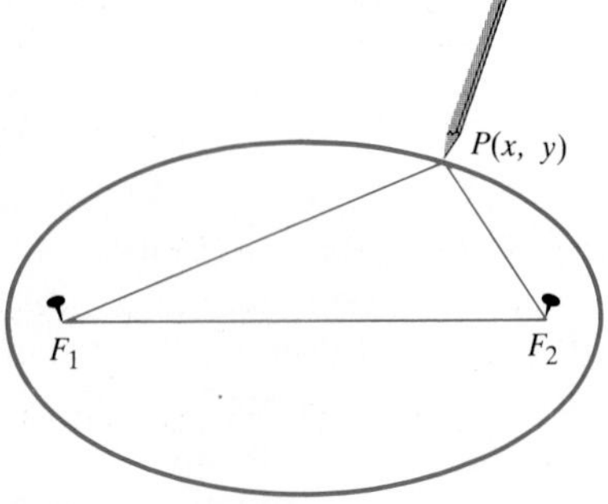

8.24 $PF_1 + PF_2$ is constant.

Applications

Ellipses appear in airplane wings (British Spitfire), and sometimes in gears designed for racing bicycles. Stereo systems often have elliptical styli, and water pipes are sometimes designed with elliptical cross sections to allow the pipe to expand without breaking when the water freezes. Ellipses also appear in instruments used to study aircraft noise in wind tunnels (sound at one focus can be received at the other with relatively little noise from other sources). The triggering mechanisms in some lasers are elliptical, and stones on a beach become more and more elliptical as they are ground down by waves. There are also applications of ellipses to fossil formation. The Ellipsolith, once interpreted as a separate species, is now known to be an elliptically deformed Nautilus. For details of these applications and many more, see Lee Whitt's "The Standup Conic Presents: The Ellipse and Applications," *The UMAP Journal,* Volume 4, Number 2, 1983.

PROBLEMS

In Problems 1–5, find an equation for the ellipse that has given center C, focus F, and semimajor axis a. Then calculate the eccentricity, and graph the ellipse.

1. $C(0, 0)$, $F(0, 2)$, $a = 4$

2. $C(0, 0)$, $F(-3, 0)$, $a = 5$

3. $C(0, 2)$, $F(0, 0)$, $a = 3$

4. $C(-3, 0)$, $F(-3, -2)$, $a = 4$

5. $C(2, 2)$, $F(-1, 2)$, $a = \sqrt{10}$

6. The endpoints of the major and minor axes of an ellipse are $(1, 1)$, $(3, 4)$, $(1, 7)$, and $(-1, 4)$. Sketch the ellipse, give its equation, and find its foci.

Find the centers, vertices, and foci of the ellipses in Problems 7–20.

7. $\dfrac{(x-7)^2}{4} + \dfrac{(y-5)^2}{25} = 1$

8. $\dfrac{(x+3)^2}{16} + \dfrac{(y-1)^2}{4} = 1$

9. $\dfrac{(x+1)^2}{9} + \dfrac{(y+4)^2}{25} = 1$

10. $\dfrac{(x-8)^2}{25} + \dfrac{y^2}{81} = 1$

11. $25(x-3)^2 + 4(y-1)^2 = 100$

12. $9(x-4)^2 + 16(y-3)^2 = 144$

13. $x^2 + 10x + 25y^2 = 0$

14. $x^2 + 16y^2 + 96y + 128 = 0$

15. $x^2 + 9y^2 - 4x + 18y + 4 = 0$

16. $4x^2 + y^2 - 32x + 16y + 124 = 0$

17. $4x^2 + y^2 - 16x + 4y + 16 = 0$

18. $x^2 + 4y^2 + 2x + 8y + 1 = 0$

19. $9x^2 + 16y^2 + 18x - 96y + 9 = 0$

20. $25x^2 + 9y^2 - 100x + 54y - 44 = 0$

21. Sketch each of the following ellipses:

a) $9x^2 + 4y^2 = 36$

b) $4x^2 + 9y^2 = 144$

c) $\dfrac{(x-1)^2}{16} + \dfrac{(y+2)^2}{4} = 1$

d) $4x^2 + y^2 = 1$

e) $16(x-2)^2 + 9(y+3)^2 = 144$

22. Find an equation for the ellipse that passes through the origin and has foci at $(-1, 1)$ and $(1, 1)$.

23. Find the eccentricity and the directrices of the ellipse $(x^2/7) + (y^2/16) = 1$.

24. For what values of r is the circle $x^2 + y^2 = r^2$ tangent to the ellipse $x^2 + 4y^2 = 36$?

25. Find the volume generated by rotating an ellipse of semiaxes a and b $(a > b)$ about its major axis.

26. Set up the integrals that give (a) the area of a quadrant of the circle $x^2 + y^2 = a^2$, (b) the area of a quadrant of the ellipse $b^2x^2 + a^2y^2 = a^2b^2$. Show that the integral in (b) is b/a times the integral in (a), and deduce the area of the ellipse from the known area of the circle.

27. *Reflective property of ellipses.* An ellipsoid is generated by rotating an ellipse about its major axis. The inside surface of the ellipsoid is silvered to produce a mirror. Show that a ray of light emanating from one focus will be reflected to the other focus. (Sound waves also follow such paths, and this property of ellipsoids accounts for phenomena in certain "whispering galleries.")

28. Find the length of the chord perpendicular to the major axis of the ellipse $b^2x^2 + a^2y^2 = a^2b^2$ and passing through a focus. (This chord is called the "latus rectum" of the ellipse.)

29. Find the equation of an ellipse of eccentricity $\frac{2}{3}$ if the line $x = 9$ is one directrix and the corresponding focus is at $(4, 0)$.

30. Find the values of the constants A, B, and C if the ellipse

$$4x^2 + y^2 + Ax + By + C = 0$$

is to be tangent to the x-axis at the origin and to pass through the point $(-1, 2)$.

31. Show that the line tangent to the ellipse $(x^2/a^2) + (y^2/b^2) = 1$ at the point $P_1(x_1, y_1)$ on it is

$$\frac{xx_1}{a^2} + \frac{yy_1}{b^2} = 1.$$

In Problems 32–34, graph the sets of points whose coordinates satisfy the given inequality or equation.

32. $9x^2 + 16y^2 < 144$

33. $(x^2 + 4y)(2x - y - 3)(x^2 + y^2 - 25) \times (x^2 + 4y^2 - 4) = 0$

34. $(x^2 + y^2 - 1)(9x^2 + 4y^2 - 36) < 0$

35. Draw an ellipse of eccentricity $\frac{4}{5}$.

36. Draw the orbit of Pluto to scale.

37. a) Write an equation for the orbit of Halley's comet in a coordinate system in which the sun lies at the origin and the other focus lies on the positive x-axis, scaled in astronomical units.

b) About how close does the comet come to the sun in astronomical units? in miles?

c) What is the farthest the comet gets from the sun in astronomical units? in miles? When was the comet last that far away?

d) Use Kepler's third law to estimate the comet's mean distance from the sun, given that its period is about 76 years.

38. (*Calculator*) The equations $x = a\cos\theta$, $y = b\sin\theta$, $0 \le \theta \le 2\pi$, are parametric equations for an ellipse.

a) Show that $(x^2/a^2) + (y^2/b^2) = 1$.

b) Show that the total length (perimeter) of the ellipse is

$$4a\int_0^{\pi/2} \sqrt{1 - e^2\cos^2\theta}\, d\theta,$$

where e is the eccentricity of the ellipse.

c) The integral in (b) is called an *elliptic integral*. It has no elementary antiderivative. Estimate the length of the ellipse when $a = 1$ and $e = \frac{1}{2}$ by the trapezoidal rule with $n = 10$.

d) The absolute value of the second derivative of $f(\theta) = \sqrt{1 - e^2\cos^2\theta}$ is less than one. Based on this, what estimate does Eq. (5), Article 4.9, give of the error in the approximation in (c)?

39. Problem 48, Article 3.5, considered a rope with a ring at one end looped over two pegs in a horizontal line. The free end of the rope passed through the ring to a weight that pulled the rope taut. It was assumed that the rope, which slipped freely over the pegs and through the ring, would come to rest in a configuration symmetric with respect to the line of the vertical part of the rope.

a) Show that for each fixed position of the ring on the rope the possible locations of the ring in space lie on an ellipse with foci at the pegs.

b) Justify the original symmetry assumption by combining the result in (a) with the assumption that the rope and weight will take a rest position of minimal potential energy.

***Toolkit* programs**

Conic Sections   Super * Grapher

8.6 Hyperbolas

DEFINITION

A *hyperbola* is the set of points in a plane whose distances from two fixed points in the plane have a constant difference.

The Equation of a Hyperbola

If we take the two fixed points, called *foci*, at $F_1(-c, 0)$ and $F_2(c, 0)$ and the constant equal to $2a$ (see Fig. 8.25), then a point $P(x, y)$ lies on the hyperbola if and only if

$$\sqrt{(x + c)^2 + y^2} - \sqrt{(x - c)^2 + y^2} = 2a$$

or

$$\sqrt{(x - c)^2 + y^2} - \sqrt{(x + c)^2 + y^2} = 2a.$$

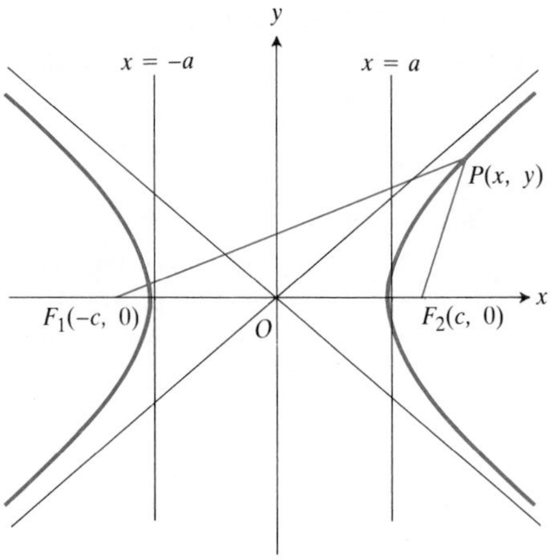

8.25 $PF_1 - PF_2 = 2a$.

The second equation is like the first, with $2a$ replaced by $-2a$. Hence we write the first one with $\pm 2a$, transpose one radical to the right side of the equation, square, and simplify. One radical still remains. We isolate it and square again. We then obtain the equation

$$\frac{x^2}{a^2} + \frac{y^2}{a^2 - c^2} = 1. \tag{1}$$

So far, this is just like the equation for an ellipse. But now $a^2 - c^2$ is negative, because the *difference* in two sides of the triangle F_1F_2P is less than the third side:

$$2a < 2c.$$

So in this case $c^2 - a^2$ is positive and has a real positive square root, which we call b:

$$b = \sqrt{c^2 - a^2}, \qquad \text{or} \qquad a^2 - c^2 = -b^2. \tag{2}$$

The equation of the hyperbola now becomes

$$\boxed{\frac{x^2}{a^2} - \frac{y^2}{b^2} = 1,} \tag{3}$$

which is analogous to the equation of an ellipse. The only differences are the minus sign in the equation of the hyperbola, and the new relation among a, b, and c given by Eq. (2).

The hyperbola, like the ellipse, is symmetric with respect to both axes and the origin, but it has no real y-intercepts and in fact no portion of the curve lies between the lines $x = a$ and $x = -a$.

If we start with a point $P(x, y)$ whose coordinates satisfy Eq. (3), the distances PF_1 and PF_2 will be given by

$$PF_1 = \sqrt{(x + c)^2 + y^2} = \left|a + \frac{c}{a}x\right|, \tag{4a}$$

$$PF_2 = \sqrt{(x - c)^2 + y^2} = \left|a - \frac{c}{a}x\right|, \tag{4b}$$

as for the ellipse. But now c is greater than a, and P is either to the right of the line $x = a$, that is $x > a$, or else P is to the left of the line $x = -a$, and then $x < -a$. The absolute values in Eqs. (4) work out to be

$$\left.\begin{aligned} PF_1 &= a + \frac{c}{a}x \\ PF_2 &= \frac{c}{a}x - a \end{aligned}\right\} \text{ if } x > a \tag{5a}$$

and

$$\left.\begin{aligned} PF_1 &= -\left(a + \frac{c}{a}x\right) \\ PF_2 &= a - \frac{c}{a}x \end{aligned}\right\} \text{ if } x < -a. \tag{5b}$$

Thus, when P is to the right of the line $x = a$, the condition $PF_1 - PF_2 = 2a$ is satisfied, while if P is to the left of $x = -a$, the condition $PF_2 - PF_1 = 2a$

is fulfilled (Fig. 8.25). In either case, *any point P that satisfies the geometric conditions satisfies Eq. (3). Conversely, any point that satisfies Eq. (3) also satisfies the geometric conditions.*

Asymptotes

As we saw in Chapter 3, it may happen that as a point P on a curve moves farther and farther away from the origin, the distance between P and some fixed line tends to zero. In other words, the curve gets closer and closer to the line as it moves away from the origin. We studied such lines in Chapter 3 and called them *asymptotes*.

As we shall now see, the hyperbola

$$\frac{x^2}{a^2} - \frac{y^2}{b^2} = 1 \tag{3}$$

has two asymptotes, the lines

$$y = \pm\frac{b}{a}.$$

The left-hand side of (3) can be factored and the equation written in the form

$$\left(\frac{x}{a} - \frac{y}{b}\right)\left(\frac{x}{a} + \frac{y}{b}\right) = 1$$

or

$$\frac{x}{a} - \frac{y}{b} = \frac{ab}{bx + ay}. \tag{6a}$$

Analysis of (3) shows that one branch of the curve lies in the first quadrant and has infinite extent. If the point P moves along this branch so that x and y both become infinite, then the right-hand side of (6a) tends to zero; hence the left-hand side must do likewise. That is

$$\lim_{\substack{x\to\infty\\ y\to\infty}} \left(\frac{x}{a} - \frac{y}{b}\right) = 0, \tag{6b}$$

which leads us to speculate that the straight line

$$\frac{x}{a} - \frac{y}{b} = 0 \tag{7}$$

may be an asymptote of the curve.

To see that it is, we first investigate the vertical distance between the curve

$$y = \frac{b}{a}\sqrt{x^2 - a^2}$$

and the line

$$y = \frac{b}{a}x$$

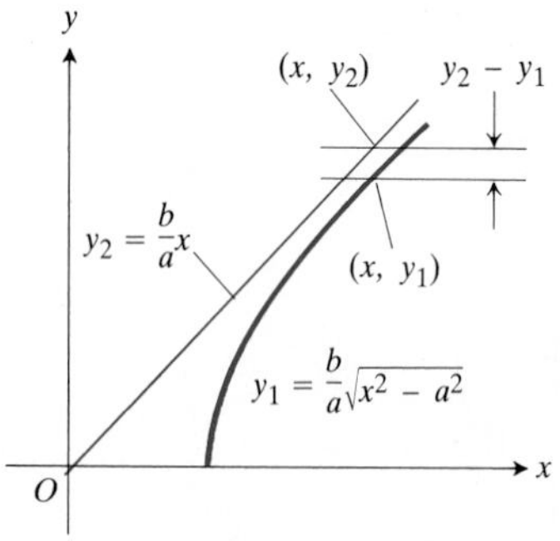

8.26 As $x \to \infty$, the point (x, y_1) rises on the hyperbola toward the asymptote $y_2 = (b/a)x$.

(Fig. 8.26). As $x \to \infty$, we find

$$\begin{aligned}\lim_{x\to\infty}\left(\frac{b}{a}x - \frac{b}{a}\sqrt{x^2 - a^2}\right) &= b\lim_{x\to\infty}\left(\frac{x}{a} - \frac{\sqrt{x^2 - a^2}}{a}\right)\\ &= b\lim_{\substack{x\to\infty\\ y\to\infty}}\left(\frac{x}{a} - \frac{y}{b}\right)\\ &= b\cdot 0 \qquad \text{(by Eq. 6b)}\\ &= 0.\end{aligned} \tag{8}$$

Since the vertical distance between the line and the hyperbola approaches zero as $x \to 0$, the perpendicular distance from points on the hyperbola to the line $y = (b/a)x$ also approaches zero. Therefore, the line $y = (b/a)x$ is an asymptote of the hyperbola.

By symmetry, the line

$$\frac{x}{a} + \frac{y}{b} = 0 \tag{9}$$

is also an asymptote of the hyperbola.

Both asymptotes may be obtained by replacing the "one" on the right-hand side of Eq. (3) by zero and then factoring.

REMARK. Some definitions of asymptote would also require the slope of the hyperbola to approach the slope of the line (7) as $x \to \infty$. To show that this is the case here, we calculate the slope of the curve:

$$y = \frac{b}{a}\sqrt{x^2 - a^2}, \qquad \frac{dy}{dx} = \frac{b}{a}\frac{x}{\sqrt{x^2 - a^2}},$$

from which we see that

$$\lim_{x\to\infty} \frac{dy}{dx} = \frac{b}{a} \lim_{x\to\infty} \frac{x}{\sqrt{x^2 - a^2}} = \frac{b}{a}, \tag{10}$$

which is the slope of the asymptote $y = (b/a)x$.

Graphing Hyperbolas

In graphing a hyperbola (Fig. 8.27), it is helpful to mark off distances a to the right and to the left of the origin along the x-axis and distances b above and below the origin along the y-axis, and to construct a rectangle whose sides pass through these points, parallel to the coordinate axes. The diagonals of this rectangle, extended, are the asymptotes of the hyperbola. The semidiagonal $c = \sqrt{a^2 + b^2}$ can also be used as the radius of a circle that will cut the x-axis in two points, $F_1(-c, 0)$ and $F_2(c, 0)$, which are the foci of the hyperbola.

If we interchange x and y in Eq. (3), the new equation

$$\frac{y^2}{a^2} - \frac{x^2}{b^2} = 1 \tag{11}$$

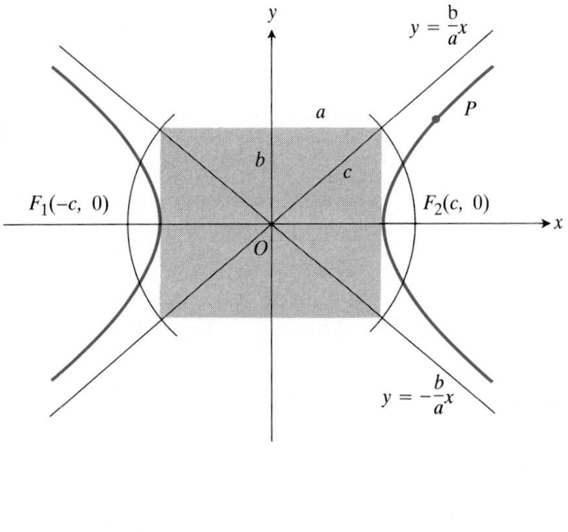

8.27 The hyperbola $(x^2/a^2) - (y^2/b^2) = 1$.

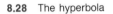

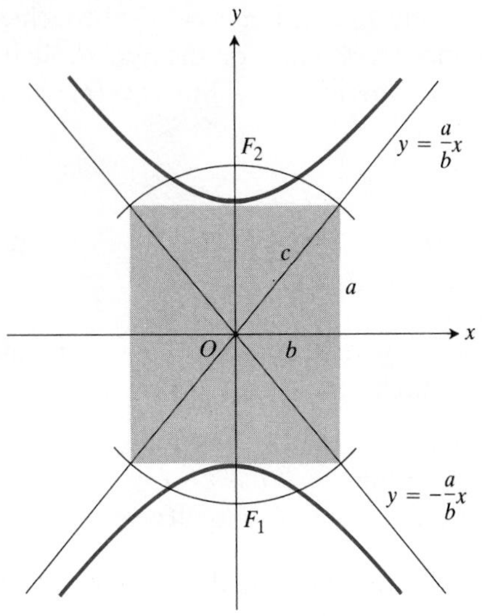

8.28 The hyperbola $(y^2/a^2) - (x^2/b^2) = 1$.

represents a hyperbola with foci on the y-axis. Its graph is shown in Fig. 8.28.

Center not at the Origin

The *center* of a hyperbola is the point of intersection of its axes of symmetry. If the center is $C(h, k)$, we may introduce a translation to new coordinates

$$x' = x - h, \qquad y' = y - k \tag{12}$$

with origin O' at the center. In terms of the new coordinates, the equation of the hyperbola is either

$$\frac{x'^2}{a^2} - \frac{y'^2}{b^2} = 1 \tag{13}$$

or

$$\frac{y'^2}{a^2} - \frac{x'^2}{b^2} = 1. \tag{14}$$

EXAMPLE 1 Analyze the equation $x^2 - 4y^2 + 2x + 8y - 7 = 0$.

Solution We complete the squares in the x and y terms separately and reduce to standard form:

$$(x^2 + 2x) - 4(y^2 - 2y) = 7,$$
$$(x^2 + 2x + 1) - 4(y^2 - 2y + 1) = 7 + 1 - 4,$$
$$\frac{(x+1)^2}{4} - (y-1)^2 = 1.$$

The translation

$$x' = x + 1, \qquad y' = y - 1$$

reduces the equation to

$$\frac{x'^2}{4} - \frac{y'^2}{1} = 1,$$

which represents a hyperbola with center at $x' = 0$, $y' = 0$, or $x = -1$, $y = 1$, having

$$a^2 = 4, \qquad b^2 = 1, \qquad c^2 = a^2 + b^2 = 5$$

and asymptotes

$$\frac{x'}{2} - y' = 0, \qquad \frac{x'}{2} + y' = 0.$$

The foci have coordinates $(\pm\sqrt{5}, 0)$ relative to the new axes or, since

$$x = x' - 1, \qquad y = y' + 1,$$

the coordinates relative to the original axes are $(-1 \pm \sqrt{5}, 1)$. The curve is sketched in Fig. 8.29. □

EXAMPLE 2 Analyze the equation $x^2 - 4y^2 - 2x + 8y - 2 = 0$.

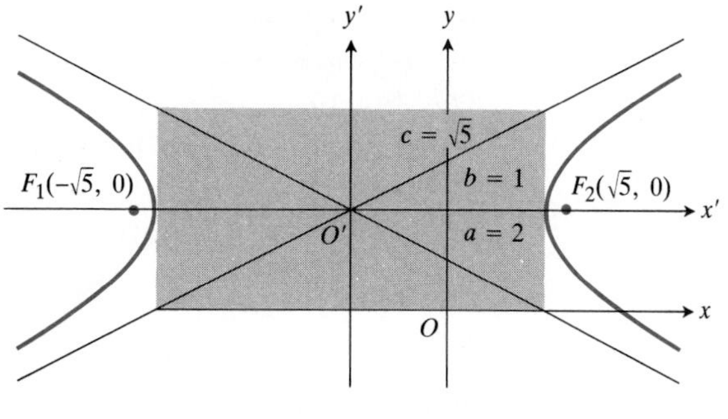

8.29 The xy-equation of the hyperbola shown to the right is $x^2 - 4y^2 + 2x + 8y - 7 = 0$. The $x'y'$-equation is $(x'^2/4) - (y'^2/1) = 1$.

Solution Proceeding as before, we obtain

$$(x - 1)^2 - 4(y - 1)^2 = 2 + 1 - 4 = -1.$$

The standard form requires a plus one on the right-hand side of the equation, so we change signs and have

$$4(y - 1)^2 - (x - 1)^2 = 1.$$

To compare this equation with (4), we write the first term as $(y - 1)^2$ divided by 0.25:

$$\frac{(y - 1)^2}{0.25} - \frac{(x - 1)^2}{1} = 1.$$

The translation $x' = x - 1$, $y' = y - 1$ replaces this by

$$\frac{y'^2}{0.25} - \frac{x'^2}{1} = 1,$$

which represents a hyperbola with center at $x' = y' = 0$, or $x = y = 1$. The curve has intercepts at $(0, \pm 0.5)$ on the y'-axis but does not cross the x'-axis. Here (14) applies, with

$$a^2 = 0.25, \qquad b^2 = 1, \qquad c^2 = a^2 + b^2 = 1.25.$$

The lines (Fig. 8.30)

$$\frac{y'}{0.5} - x' = 0, \qquad \frac{y'}{0.5} + x' = 0$$

are the asymptotes, while the foci are at $(0, \pm\sqrt{1.25})$ on the y'-axis. □

8.30 The hyperbola

$(y'^2/0.25) - (x'^2/1) = 1$.

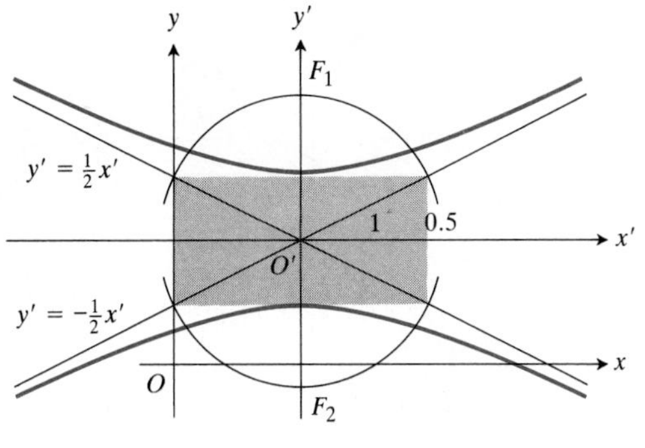

Eccentricity

There is no restriction $a > b$ for the hyperbola as there is for the ellipse. The direction in which the hyperbola opens is controlled by the *signs* rather than by the relative *sizes* of the coefficients of the quadratic terms.

In our further discussion of the hyperbola, we shall assume that it has been referred to axes through its center and that its equation has the form

$$\frac{x^2}{a^2} - \frac{y^2}{b^2} = 1. \tag{15}$$

Then

$$b = \sqrt{c^2 - a^2}$$

and

$$c^2 = a^2 + b^2. \tag{16}$$

As for the ellipse, we define the *eccentricity* e of the hyperbola to be

$$e = \frac{c}{a}.$$

Since $c > a$, the eccentricity of a hyperbola is greater than one. The lines

$$x = \frac{a}{e}, \qquad x = -\frac{a}{e}$$

are the *directrices*.

The Focus-Directrix Property

We shall now verify that a point $P(x, y)$ whose coordinates satisfy Eq. (15) also has the property that

$$PF_1 = e \cdot PD_1 \tag{17a}$$

and

$$PF_2 = e \cdot PD_2, \tag{17b}$$

where $F_1(-c, 0)$ and $F_2(c, 0)$ are the foci while $D_1(-a/e, y)$ and $D_2(a/e, y)$ are the points nearest P on the directrices.

8.31 $PF_1 = e \cdot PD_1$ and $PF_2 = e \cdot PD_2$.

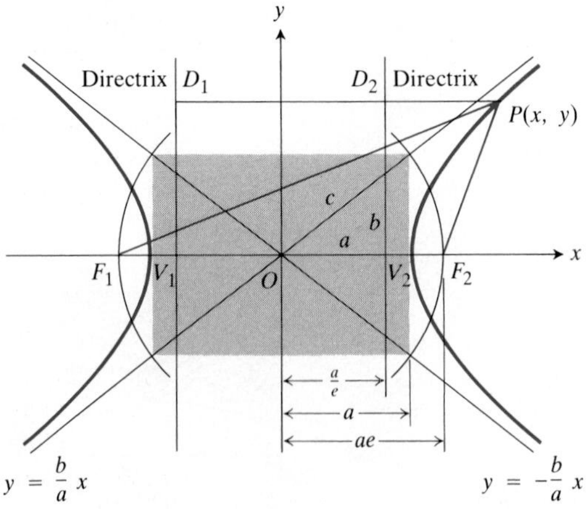

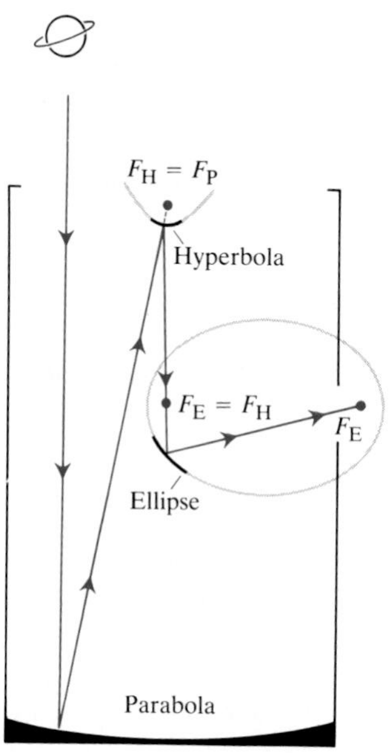

8.32 In this schematic drawing of a reflecting telescope, light from a planet reflects off a primary parabolic mirror toward the mirror's focus F_P. It is then reflected by a small hyperbolic mirror, whose focus is $F_H = F_P$, toward the second focus of the hyperbola, $F_E = F_H$. Since this focus is shared by an ellipse, the light is reflected by the elliptical mirror to the ellipse's second focus to be seen by an observer. See Problem 16 for more about the reflective property of the hyperbola used here.

We shall content ourselves with establishing the results (17a, b) for any point P on the right branch of the hyperbola; the method is the same when P is on the left branch. Reference to Eqs. (5a) then shows that

$$PF_1 = \frac{c}{a}x + a = e\left(x + \frac{a}{e}\right), \qquad PF_2 = \frac{c}{a}x - a = e\left(x - \frac{a}{e}\right), \tag{18a}$$

while we see from Fig. 8.31 that

$$PD_1 = x + \frac{a}{e}, \qquad PD_2 = x - \frac{a}{e}. \tag{18b}$$

These results combine to establish the "focus-and-directrix" properties of the hyperbola expressed in Eqs. (17a, b). Conversely, if Eqs. (18a) are satisfied, it is also true that

$$PF_1 - PF_2 = 2a;$$

that is, P satisfies the requirement that the difference of its distances from the two foci is constant.

Applications

Hyperbolic paths arise in Einstein's theory of special relativity and form the basis for the LORAN radio navigation system (LORAN is short for Long Range Navigation). A comet that does not return to its sun follows a hyperbolic path (the probability of its being parabolic is zero). Reflecting telescopes like the 200-inch Hale telescope on Mount Palomar in California, and the space telescope NASA plans to launch in early 1985, use small hyperbolic mirrors in conjunction with their larger parabolic ones (schematic drawing in Fig. 8.32). For more about applications of hyperbolas, see Lee Whitt's "The Standup Conic Presents: The Hyperbola and Applications," *The UMAP Journal,* Volume 5, Number 1, 1984.

PROBLEMS

1. Sketch each of the following hyperbolas:

a) $\dfrac{x^2}{9} - \dfrac{y^2}{16} = 1$, b) $\dfrac{x^2}{16} - \dfrac{y^2}{9} = 1$,

c) $\dfrac{y^2}{9} - \dfrac{x^2}{16} = 1$, d) $\dfrac{x^2}{9} - \dfrac{y^2}{16} = -1$.

In Problems 2–8, find the center, vertices, foci, eccentricity, and asymptotes of the given hyperbola. Sketch the curve.

2. $9(x - 2)^2 - 4(y + 3)^2 = 36$

3. $4(x - 2)^2 - 9(y + 3)^2 = 36$

4. $4(y + 3)^2 - 9(x - 2)^2 = 1$

5. $5x^2 - 4y^2 + 20x + 8y = 4$

6. $4x^2 = y^2 - 4y + 8$

7. $4y^2 = x^2 - 4x$

8. $4x^2 - 5y^2 - 16x + 10y + 31 = 0$

9. Show that the line tangent to the hyperbola $b^2x^2 - a^2y^2 = a^2b^2$ at a point $P(x_1, y_1)$ on it has an equation that may be written in the form $b^2xx_1 - a^2yy_1 = a^2b^2$.

10. Find the volume generated when the area bounded by the hyperbola $b^2x^2 - a^2y^2 = a^2b^2$ and the line $x = c$, through its focus $(c, 0)$, is rotated about the y-axis.

11. Show that the equation

$$\frac{x^2}{9 - C} + \frac{y^2}{5 - C} = 1$$

represents (a) an ellipse if C is any constant less than 5, (b) a hyperbola if C is any constant between 5 and 9, (c) no graph at all if C is greater than 9. Show that each ellipse in (a) and each hyperbola in (b) has foci at the two points $(\pm 2, 0)$, independent of the value of C.

12. Find the equation of the hyperbola with foci at (0, 0) and (0, 4) if it is required to pass through the point (12, 9).

13. One focus of a hyperbola is located at the point $(1, -3)$ and the corresponding directrix is the line $y = 2$. Find the equation of the hyperbola if its eccentricity is 3/2.

14. Find out what you can about the DECCA system of air navigation. (See *Time*, Feb. 23, 1959, p. 87, or the reference in Problem 15.)

15. A radio signal was sent simultaneously from towers *A* and *B* located several hundred miles apart on the California coast. A ship offshore received the signal from *A* 1400 microseconds before it received the signal from *B*.

a) Assume that the radio signals traveled at 980 ft per microsecond. What can be said about the approximate location of the ship relative to the two towers?

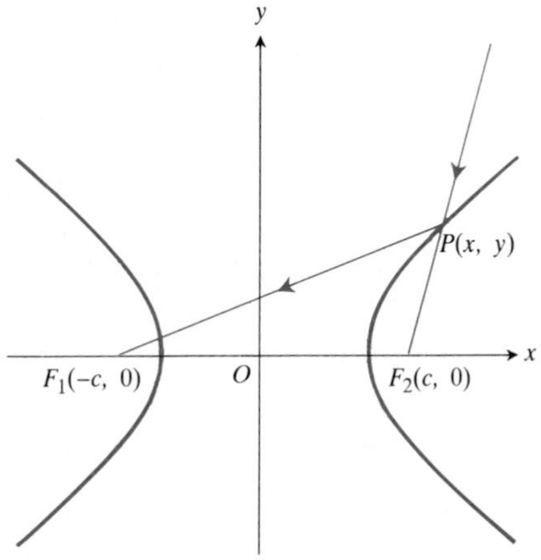

8.33 Reflective property of hyperbolas.

b) Find out what you can about LORAN and other hyperbolic radio-navigation systems. (See, for example, Nathaniel Bowditch's *American Practical Navigator*, Vol. I, U.S. Defense Mapping Agency Hydrographic Center, Publication No. 9, 1977, Chapter XLIII.)

16. *Reflective property of hyperbolas.* Show that a ray of light directed toward one focus of a hyperbolic mirror, as in Fig. 8.33, is reflected toward the other focus. (*Hint:* Show that the tangent to the hyperbola at the point *P* in Fig. 8.33 bisects the angle made by the segments PF_1 and PF_2.)

17. Show that an ellipse and a hyperbola that have the same foci, *A* and *B*, as in Fig. 8.34, cross at right angles at their points of intersection. (*Hint:* A ray of light emanating from the focus *A* that met the hyperbola at *P* would be reflected from the hyperbola as if it came directly from *B* (Problem 16). The same light ray would be reflected off the ellipse to pass through *B*. Thus, *BPC* is a straight line.)

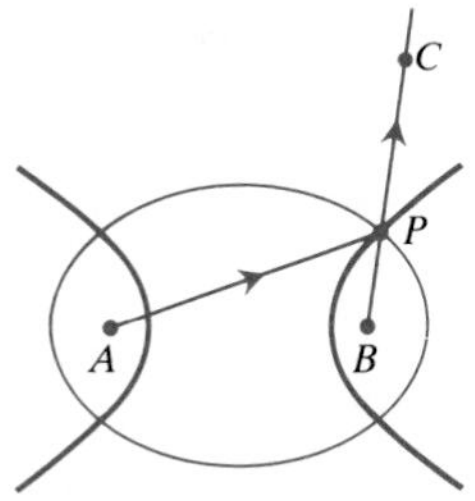

8.34 The confocal ellipse and hyperbola of Problem 17.

***Toolkit* programs**	
Conic Sections	Super * Grapher

8.7
Quadratic Curves

The equations of the circle, parabola, ellipse, and hyperbola are special cases of the general quadratic equation

$$Ax^2 + Bxy + Cy^2 + Dx + Ey + F = 0, \tag{1}$$

in which the coefficients *A*, *B*, and *C* are not all zero. For example, the circle

$$(x - h)^2 + (y - k)^2 = r^2$$

may be obtained from Eq. (1) by taking

$$A = C = 1, \quad B = 0, \quad D = -2h,$$
$$E = -2k, \quad F = h^2 + k^2 - r^2,$$

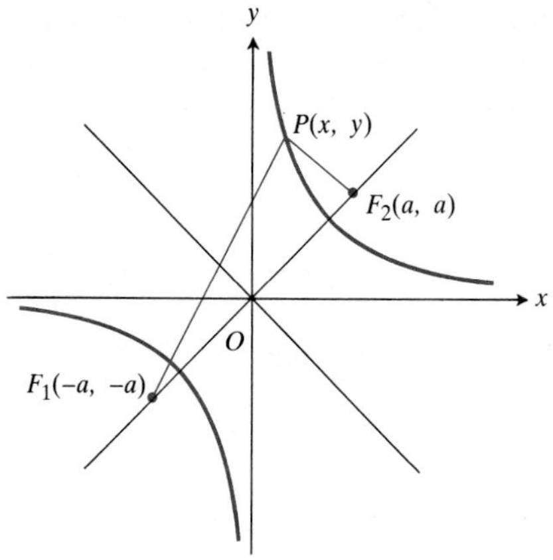

8.35 The hyperbola $2xy = a^2$.

and the parabola

$$x^2 = 4py$$

by taking

$$A = 1, \quad E = -4p, \quad B = C = D = F = 0.$$

Even the straight line is a special case of (1), with $A = B = C = 0$, but this reduces (1) to a linear equation.

The terms Ax^2, Bxy, and Cy^2 are the quadratic terms in (1), and in this article we shall investigate the nature of the curve represented by (1) when at least one of these terms is present. It turns out that with only a few exceptions the graph of (1) is a conic section.

The "cross-product" term, Bxy, has not appeared in the equations we found in Articles 8.3 to 8.6. This is a consequence of our having chosen the coordinate axes so that at least one of them lies parallel to an axis of symmetry of the curve in question. However, suppose we seek the equation of a hyperbola with foci at $F_1(-a, -a)$ and $F_2(a, a)$, and with $|PF_1 - PF_2| = 2a$ as in Fig. 8.35. Then

$$\sqrt{(x + a)^2 + (y + a)^2} - \sqrt{(x - a)^2 + (y - a)^2} = \pm 2a,$$

and when we transpose one radical, square, solve for the radical that still appears, and square again, this reduces to

$$2xy = a^2, \tag{2}$$

which is a special case of Eq. (1) in which the cross-product term is present. The asymptotes of the hyperbola in Eq. (2) are the x- and y-axes, and the transverse axis of the hyperbola (the axis of symmetry on which the foci lie) makes an angle of 45° with the coordinate axes. As in this example, the cross-product term is present only in some similar circumstance, where the axes of the conic are tilted.

Rotation of Axes

The equation of a curve can be modified by referring it to x′y′-axes by a rotation through an angle α in the counterclockwise direction. In the notation of Fig. 8.36, we have

$$\begin{aligned} x &= OM = OP\cos(\theta + \alpha), \\ y &= MP = OP\sin(\theta + \alpha), \end{aligned} \tag{3a}$$

while

$$\begin{aligned} x' &= OM' = OP\cos\theta, \\ y' &= M'P = OP\sin\theta. \end{aligned} \tag{3b}$$

After substituting

$$\cos(\theta + \alpha) = \cos\theta\cos\alpha - \sin\theta\sin\alpha,$$
$$\sin(\theta + \alpha) = \sin\theta\cos\alpha + \cos\theta\sin\alpha,$$

in (3a) and taking account of (3b), we find the following equations:

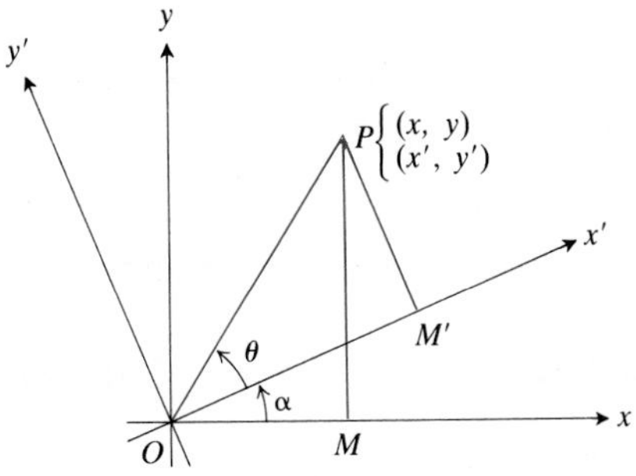

8.36 A counterclockwise rotation through angle α about the origin.

Equations for Rotation of Axes

$$\begin{aligned} x &= x'\cos\alpha - y'\sin\alpha, \\ y &= x'\sin\alpha + y'\cos\alpha. \end{aligned} \tag{4}$$

These equations are used in calculators to calculate sines and cosines of arbitrary angles from a small "wired-in" library of small angles that are used as building blocks for larger angles. (You will find more about this application in the remark at the end of this article.)

EXAMPLE 1 The x- and y-axes are rotated $\alpha = 45°$ about the origin. Find an equation for the hyperbola of Eq. (2) in the new coordinates.

Solution Since $\cos 45° = \sin 45° = \sqrt{\frac{1}{2}}$, we substitute

$$x = \frac{x' - y'}{\sqrt{2}}, \qquad y = \frac{x' + y'}{\sqrt{2}}$$

from Eqs. (4) into Eq. (2) and obtain

$$(x')^2 - (y')^2 = a^2,$$

which is equivalent to Eq. (3), Article 8.6, with $b = a$. □

Eliminating the Cross-product Term

If we apply the equations for rotation of axes (4) to the general quadratic equation (1), we obtain a new quadratic equation of the form

$$A'x'^2 + B'x'y' + C'y'^2 + D'x' + E'y' + F' = 0. \tag{5}$$

The new coefficients are related to the old ones by the equations

$$\begin{aligned}
A' &= A\cos^2\alpha + B\cos\alpha\sin\alpha + C\sin^2\alpha,\\
B' &= B(\cos^2\alpha - \sin^2\alpha) + 2(C - A)\sin\alpha\cos\alpha,\\
C' &= A\sin^2\alpha - B\sin\alpha\cos\alpha + C\cos^2\alpha,\\
D' &= D\cos\alpha + E\sin\alpha,\\
E' &= -D\sin\alpha + E\cos\alpha,\\
F' &= F.
\end{aligned} \tag{6}$$

If we start with an equation for a curve in which the cross-product term is present ($B \neq 0$), we can always find a rotation angle α that produces an equation in which no cross-product term appears ($B' = 0$). To find α, we put $B' = 0$ in the second equation in (6) and solve for α. It is easier to do this if we note that

$$\cos^2\alpha - \sin^2\alpha = \cos 2\alpha,$$

$$2\sin\alpha\cos\alpha = \sin 2\alpha,$$

so that

$$B' = B\cos 2\alpha + (C - A)\sin 2\alpha.$$

Hence B' will vanish if we choose α so that

$$B\cos 2\alpha + (C - A)\sin 2\alpha = 0.$$

In practice, this means finding a suitable value of α from one of the two equations

$$\cot 2\alpha = \frac{A - C}{B}, \quad \text{or} \quad \tan 2\alpha = \frac{B}{A - C}. \tag{7}$$

EXAMPLE 2 The coordinate axes are to be rotated through an angle α to produce an equation for the curve

$$x^2 + xy + y^2 = 3 \tag{8}$$

that has no cross-product term. Find α and the new equation.

Solution The equation (8) has $A = B = C = 1$. Choose α according to Eq. (7):

$$\cot 2\alpha = 0, \qquad 2\alpha = 90°, \qquad \alpha = 45°.$$

Substituting $\alpha = 45°$, $A = B = C = 1$, $D = E = 0$, and $F = -3$ into Eqs. (6) gives

$$A' = \frac{3}{2}, \qquad B' = 0, \qquad C' = \frac{1}{2}, \qquad D' = E' = 0, \qquad F' = -3.$$

Equation (5) then gives

$$\frac{3}{2}x'^2 + \frac{1}{2}y'^2 - 3 = 0 \qquad \text{or} \qquad 3x'^2 + y'^2 = 6.$$

This is the equation of an ellipse with foci on the new y'-axis. □

The Classification of Quadratic Curves

We can now return to our analysis of the graph of the general quadratic equation.

Since axes may always be rotated to eliminate the cross-product term, there is no loss of generality in assuming that this has been done. Then the quadratic equation in (5), with $B' = 0$, will look like Eq. (1) with $B = 0$:

$$Ax^2 + Cy^2 + Dx + Ey + F = 0. \tag{9}$$

Equation 9 represents

a) A straight line if $A = C = 0$, and not both D and E vanish. (If the left-hand side of the equation can be factored into a product of two linear factors, then the equation represents one or two straight lines.)

b) A circle if $A = C \neq 0$. (Special cases: the graph is a point, or there is no graph at all (if no x-y pairs satisfy the equation)).

c) A parabola if (9) is quadratic in one variable, linear in the other.

d) An ellipse if A and C are both positive or both negative. (Special cases: a single point, or no graph at all.)

e) A hyperbola if A and C have opposite signs, both different from zero. (Special case: a pair of straight lines. Example: $x^2 - y^2 = 0$.)

Summary

Any quadratic equation in x and y represents a circle, parabola, ellipse, or hyperbola (except for special cases in which the graph is a point, a line, a pair of lines, or fails to exist).

Every circle, parabola, ellipse, and hyperbola has a quadratic equation. To identify the curve from its equation, we

1. rotate axes (if necessary) to eliminate the cross-product term, and
2. translate axes (as desired) to reduce the equation to a standard form we recognize.

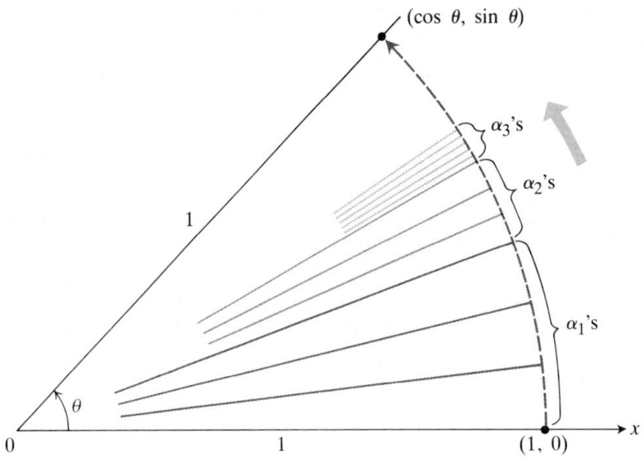

8.37 To calculate the sine and cosine of an angle θ between 0 and 2π, the calculator rotates the point (1, 0) to an appropriate location on the unit circle and displays the resulting coordinates.

REMARK. Some calculators use rotations of axes to calculate sines and cosines of arbitrary angles. The procedure goes something like this: The calculator has, stored,

1. ten angles, or so, say

$$\alpha_1 = \sin^{-1}(10^{-1}),\ \alpha_2 = \sin^{-1}(10^{-2}),\ \ldots,\ \alpha_{10} = \sin^{-1}(10^{-10}),$$

and

2. twenty numbers, the sines and cosines of the angles $\alpha_1, \alpha_2, \ldots, \alpha_{10}$.

To calculate the sine and cosine of an arbitrary angle θ, one enters θ (in radian measure) into the calculator. The calculator subtracts or adds multiples of 2π to θ to replace θ by the angle between 0 and 2π that has the same sine and cosine as θ (we shall continue to call the angle "θ"). The calculator then "writes" θ as a sum of multiples of α_1 (as many as possible, without overshooting) plus multiples of α_2 (again, as many as possible), and so on, working its way to α_{10}. This gives

$$\theta \approx m_1\alpha_1 + m_2\alpha_2 + \cdots + m_{10}\alpha_{10}.$$

The calculator then rotates the point (1, 0) through m_1 copies of α_1 (through α_1, m_1 times in succession), plus m_2 copies of α_2, and so on, finishing off with m_{10} copies of α_{10}. See Fig. 8.37. The coordinates of the final position of (1, 0) on the unit circle are the values the calculator gives for $(\cos\theta, \sin\theta)$.

PROBLEMS

Transform the equations in Problems 1–6 by a rotation of axes into an equation with no cross-product term. (The answers to these problems will vary, depending on the size and sense of the rotation.)

1. $xy = 2$

2. $x^2 + xy + y^2 = 1$

3. $2y^2 + \sqrt{3}xy - x^2 = 2$

4. $3x^2 + 2xy + 3y^2 = 19$

5. $x^2 - 3xy + y^2 = 5$

6. $3x^2 + 4\sqrt{3}xy - y^2 = 7$

7. Find the sine and cosine of an angle through which the coordinate axes may be rotated to eliminate the cross-product term from the equation

$$14x^2 + 16xy + 2y^2 - 10x + 26370y - 17 = 0.$$

Do not carry out the rotation.

8. Use the definition of an ellipse to write an equation for the ellipse with foci at $F_1(-1, 0)$ and $F_2(0, \sqrt{3})$ if it passes through the point (1, 0). Through what angle α should the axes be rotated to eliminate the cross-product term from the equation found? (Do not carry out the rotation.)

9. Show that the equation $x^2 + y^2 = r^2$ becomes $x'^2 + y'^2 = r^2$ for every choice of the angle α in the equations for rotation of axes.

10. Show that $A' + C' = A + C$ for every choice of the angle α in Eqs. (6).

11. Show that $B'^2 - 4A'C' = B^2 - 4AC$ for every choice of the angle α in Eqs. (6).

12. Show that a rotation of the axes through 45° will eliminate the cross-product term from Eq. (1) whenever $A = C$.

13. Find the equation of the curve $x^2 + 2xy + y^2 = 1$ after a rotation of axes that makes $A' = 0$ in Eq. (6).

14. Find the angles of intersection between the ellipse $x^2 + xy + y^2 = 7$ and the line $y = 2x$. (Angles between curves are treated in Article 2.7.)

***Toolkit* programs**	
Conic Sections	Super * Grapher

8.8

Parabola, Ellipse, or Hyperbola? The Discriminant Tells

In this article we show how to tell quickly whether a quadratic equation

$$Ax^2 + Bxy + Cy^2 + Dx + Ey + F = 0 \tag{1}$$

represents a parabola, ellipse, or hyperbola. The test does not require the elimination of the xy-term.

If $B \neq 0$, then a rotation of axes through an angle α determined by

$$\cot 2\alpha = \frac{A - C}{B} \tag{2}$$

will transform the equation to the equivalent form

$$A'x'^2 + B'x'y' + C'y'^2 + D'x' + E'y' + F' = 0 \tag{3}$$

with new coefficients $A', \ldots, F'$ related to the old as in Eq. (6), Article 8.7, and with $B' = 0$ for the particular choice of α satisfying Eq. (2) above.

Now the nature of the curve whose equation is (3) with cross-product term removed is determined as follows:

a) a *parabola* if A' or $C' = 0$, that is, if $A'C' = 0$;

b) an *ellipse* if A' and C' have the same sign, that is, if $A'C' > 0$; and

c) a *hyperbola* if A' and C' have opposite signs, that is, if $A'C' < 0$.

But it can be verified by use of Eq. (6), Article 8.7, that the coefficients A, B, C, and A', B', C' satisfy the following condition:

$$B^2 - 4AC = B'^2 - 4A'C' \tag{4}$$

for *any* rotation of axes. This means that the quantity $B^2 - 4AC$ *is not changed* by a rotation of axes. But when the particular rotation is performed that makes $B' = 0$, the right-hand side of (4) becomes simply $-4A'C'$. The criteria above, expressed in terms of A' and C', can now be expressed in terms of the *discriminant:*

$$\text{Discriminant} = B^2 - 4AC. \tag{5}$$

Namely, the curve is

a) a *parabola* if $B^2 - 4AC = 0$,
b) an *ellipse* if $B^2 - 4AC < 0$,
c) a *hyperbola* if $B^2 - 4AC > 0$,

with the understanding that certain degenerate cases may arise.

EXAMPLE 1

a) $3x^2 - 6xy + 3y^2 + 2x - 7 = 0$ represents a parabola because

$$B^2 - 4AC = (-6)^2 - 4 \cdot 3 \cdot 3 = 36 - 36 = 0.$$

b) $x^2 + xy + y^2 - 1 = 0$ represents an ellipse because

$$B^2 - 4AC = (1)^2 - 4 \cdot 1 \cdot 1 = -3 < 0.$$

c) $xy - y^2 - 5y + 1 = 0$ represents a hyperbola because

$$B^2 - 4AC = (1)^2 - 4(0)(-1) = 1 > 0. \square$$

Another invariant associated with Eqs. (1) and (3) is the sum of the coefficients of the squared terms. For it is evident from Eqs. (6), Article 8.7, that

$$A' + C' = A(\cos^2 \alpha + \sin^2 \alpha) + C(\sin^2 \alpha + \cos^2 \alpha)$$

or

$$A' + C' = A + C, \tag{6}$$

since

$$\sin^2 \alpha + \cos^2 \alpha = 1 \qquad \text{for any angle } \alpha.$$

The two invariants (4) and (6) may be used to check against numerical errors in performing a rotation of axes of a quadratic equation. They may also be used to find the coefficients A' and C' in Eq. (3) when the axes are rotated to make $B' = 0$.

EXAMPLE 2 Find the equation to which

$$x^2 + xy + y^2 = 1$$

reduces when the axes are rotated to eliminate the cross-product term.

Solution From the original equation we find

$$B^2 - 4AC = -3, \qquad A + C = 2,$$

Then, taking $B' = 0$, we have from (4) and (6),

$$-4A'C' = -3, \qquad A' + C' = 2,$$

from which we know the curve to be an ellipse. Substituting $C' = 2 - A'$ from the second of these into the first, we obtain

$$-4A'(2 - A') = -3,$$
$$4A'^2 - 8A' + 3 = 0,$$
$$(2A' - 3)(2A' - 1) = 0,$$

and

$$A' = \tfrac{3}{2} \quad \text{or} \quad A' = \tfrac{1}{2}.$$

The corresponding values of C' are

$$C' = \tfrac{1}{2} \quad \text{or} \quad C' = \tfrac{3}{2}.$$

The equation in the new coordinates is therefore

$$\tfrac{3}{2}x'^2 + \tfrac{1}{2}y'^2 = 1 \tag{7a}$$

or

$$\tfrac{1}{2}x'^2 + \tfrac{3}{2}y'^2 = 1. \tag{7b}$$

The advantage of Eqs. (7a and b) over the one originally given is that they give ready information about the shape of the ellipse: the lengths of the axes, the distance from center to foci, and the eccentricity. □

PROBLEMS

Use the discriminant to classify the quadratic curves in Problems 1–17 as representing a circle, a parabola, an ellipse, or a hyperbola.

1. $x^2 - y^2 - 1 = 0$

2. $25x^2 + 9y^2 - 225 = 0$

3. $y^2 - 4x - 4 = 0$

4. $x^2 + y^2 - 10 = 0$

5. $x^2 + 4y^2 - 4x - 8y + 4 = 0$

6. $x^2 + y^2 + xy + x - y = 3$

7. $2x^2 - y^2 + 4xy - 2x + 3y = 6$

8. $x^2 + 4xy + 4y^2 - 3x = 6$

9. $x^2 + y^2 + 3x - 2y = 10$

10. $xy + y^2 - 3x = 5$

11. $3x^2 + 6xy + 3y^2 - 4x + 5y = 12$

12. $x^2 - y^2 = 1$

13. $2x^2 + 3y^2 - 4x = 7$

14. $x^2 - 3xy + 3y^2 + 6y = 7$

15. $25x^2 - 4y^2 - 350x = 0$

16. $6x^2 + 3xy + 2y^2 + 17y + 2 = 0$

17. $3x^2 + 12xy + 12y^2 + 435x - 9y + 72 = 0$

18. When $B^2 - 4AC$ is negative, the equation

$$Ax^2 + Bxy + Cy^2 = 1$$

represents an ellipse. If the semiaxes have lengths a and b, the area of the ellipse is πab. Show that the area of the ellipse given above is $2\pi/\sqrt{4AC - B^2}$.

19. Show, by reference to Eq. (6), Article 8.7, that

$$D'^2 + E'^2 = D^2 + E^2$$

for every angle of rotation α.

20. If $C = -A$ in Eq. (1), show that there is a rotation of axes for which $A' = C' = 0$ in the resulting Eq. (3). Find the angle α that makes $A' = C' = 0$ in this case. (*Hint:* Since $A' + C' = 0$, one need only make the further requirement that $A' = 0$ in Eq. (3).)

8.9

Sections of a Cone

The circle, parabola, ellipse, and hyperbola are called *conic* sections because each may be obtained by cutting a cone by a plane. If the cutting plane is perpendicular to the axis of the cone, the section is a circle.

In general, suppose the cutting plane makes an acute angle α with the axis of the cone and let the acute angle between the side and axis of the

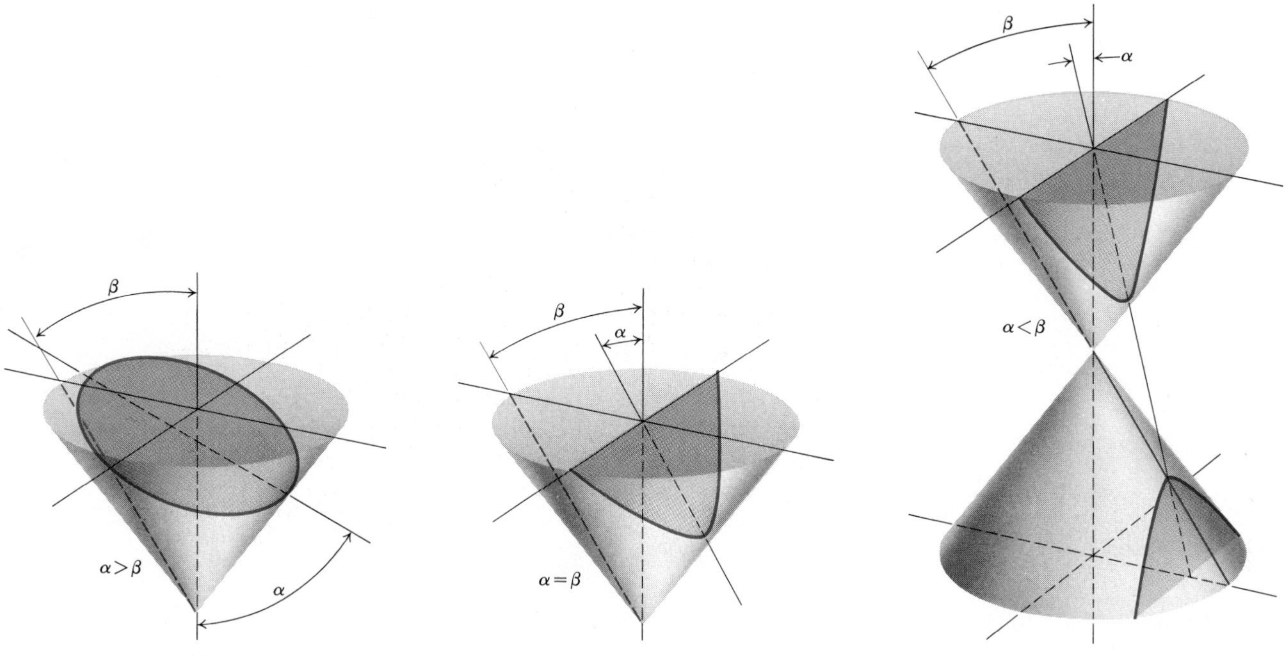

(a) (b) (c)

8.38 The intersection of a plane and cone in (a) an ellipse, (b) a parabola, (c) a hyperbola. The angle β is the angle between the side and axis of the cone. The angle α is the acute angle between the plane and the cone's axis.

cone be β (see Fig. 8.38). Then the section is

i) a circle, if $\alpha = 90°$;
ii) an ellipse, if $\beta < \alpha < 90°$;
iii) a parabola, if $\alpha = \beta$;
iv) a hyperbola, if $0 \leq \alpha < \beta$.

The connection between these curves as we have defined them and the sections of a cone can be seen in Fig. 8.39. The figure shows an ellipse, but the argument works for the other cases as well. The general construction goes as follows:

A sphere is inscribed tangent to the cone along a circle C, and tangent to the cutting plane at a point F. Point P is any point on the conic section. We shall see that F is a focus, and that the line L, in which the cutting plane and the plane of the circle C intersect, is a directrix of the curve. To this end let Q be the point where the line through P parallel to the axis of the cone intersects the plane of C, let A be the point where the line joining P to the vertex of the cone touches C, and let PD be perpendicular to line L at D. Then PA and PF are two lines tangent to the same sphere from a common point P and hence have the same length:

$$PA = PF.$$

Also, from the right triangle PQA, we have

$$PQ = PA \cos \beta;$$

and from the right triangle PQD, we find that

$$PQ = PD \cos \alpha.$$

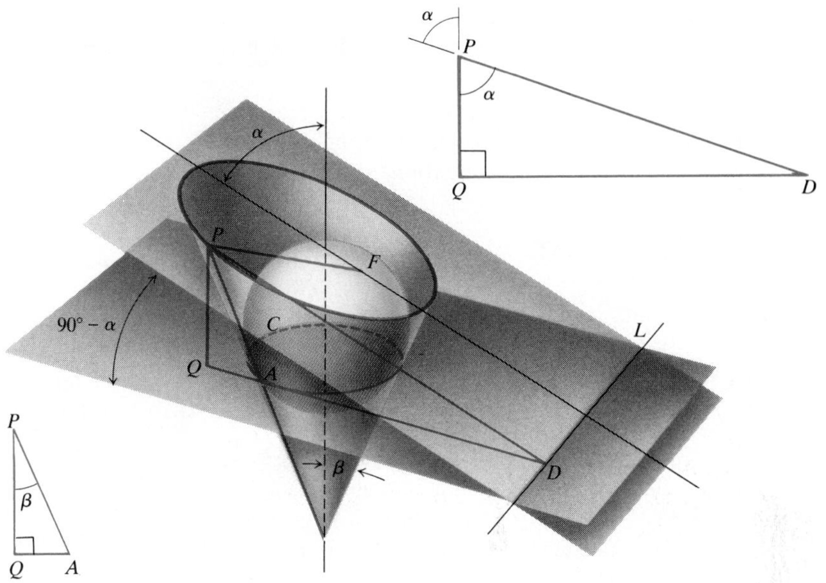

8.39 The line L is the directrix that corresponds to the focus F of the ellipse.

Hence

$$PA \cos \beta = PD \cos \alpha,$$

or

$$\frac{PA}{PD} = \frac{\cos \alpha}{\cos \beta}. \tag{1}$$

But since $PA = PF$, this means that

$$\frac{PF}{PD} = \frac{\cos \alpha}{\cos \beta}, \tag{2}$$

or

$$PF = \frac{\cos \alpha}{\cos \beta} PD. \tag{3}$$

Since α and β are constant for a given cone and cutting plane, Eq. (3) has the form

$$PF = e \cdot PD. \tag{4}$$

This characterizes P as belonging to a parabola, an ellipse, or a hyperbola, with focus at F and directrix L, according as $e = 1$, $e < 1$, or $e > 1$, where

$$e = \frac{\cos \alpha}{\cos \beta}$$

is thus identified with the eccentricity.

PROBLEMS

1. Sketch a figure similar to Fig. 8.39 when the conic section is a parabola, and carry through the argument of this article on the basis of your figure.

2. Sketch a figure similar to Fig. 8.39 when the conic section is a hyperbola and carry through the argument of this article on the basis of your figure.

3. Which parts of the construction described in this article become impossible when the conic section is a circle?

4. Let one directrix be the line $x = -p$ and take the corresponding focus at the origin. Use the equation $PF = e \cdot PD$ to derive the equation of the general conic section of eccentricity e. If e is neither 0 nor 1, show that the center of the conic section has coordinates

$$\left(\frac{pe^2}{1 - e^2}, 0\right).$$

REVIEW QUESTIONS AND EXERCISES

1. Name the conic sections. Where does the name come from?

2. What kind of equation characterizes the conic sections?

3. If the equation of a conic section is given, and it contains no xy-term, how can you tell by inspection whether it is a parabola, circle, ellipse, or hyperbola? How can you tell what the curve is if there is an xy-term in the equation?

4. What are the coordinate equations (a) for a translation of axes? (b) for a rotation of axes? Illustrate with diagrams.

5. What two quantities that are associated with the equation of a conic section remain invariant under a rotation of axes?

6. Sketch a parabola and label its vertex, focus, axis, and directrix. What is the definition of a parabola? What is the equation of your parabola?

7. Sketch an ellipse and label its vertices, foci, axes, and directrices. What is the definition of an ellipse? What is the equation of your ellipse?

8. Sketch a hyperbola and label its vertices, foci, axes, asymptotes, and directrices. What is the definition of a hyperbola? What is the equation of your hyperbola?

9. A ripple tank is made by bending a strip of tin around the perimeter of an ellipse for the wall of the tank and soldering a flat bottom onto this. An inch or two of water is put in the tank and the experimenter pokes a finger into it, right at one focus of the ellipse. Ripples radiate outward through the water, reflect from the strip around the edge of the tank, and in a short time a drop of water spurts up at the second focus. Why?

MISCELLANEOUS PROBLEMS

1. Let $P(x, y)$ be a point on the curve

$$x^2 + xy + y^2 = 3,$$

and $P'(kx, ky)$ a point on the line OP from the origin to P. If k is held constant, find an equation for the curve traced by P' as P traces the curve C.

2. Sketch the graph whose equation is

$$(y - x + 2)(2y + x - 4) = 0.$$

3. Each of the following inequalities describes one or more regions of the xy-plane. Sketch first the curve obtained by replacing the inequality sign by an equal sign; then indicate the region that contains the points whose coordinates satisfy the given inequality

a) $x < 3$ b) $x < y$ c) $x^2 < y$
d) $x^2 + y^2 > 4$ e) $x^2 + xy + y^2 < 3$
f) $x^2 + xy + y^2 > 3$

4. Write an equation of the tangent, at (2, 2), to the curve

$$x^2 - 2xy + y^2 + 2x + y - 6 = 0.$$

5. Sketch the curves $xy = 2$ and $x^2 - y^2 = 3$ in one diagram and show that they intersect at right angles.

6. Find equations of the tangents to the curve

$$y = x^3 - 6x + 2$$

that are parallel to the line $y = 6x - 2$.

7. Prove that if a line is drawn tangent to the curve $y^2 = kx$ at a point $P(x, y)$ not at the origin, then the portion of the tangent that lies between the x-axis and P is bisected by the y-axis.

8. Through the point $P(x, y)$ on the curve $y^2 = kx$, lines are drawn parallel to the axes. The rectangular region bounded by these two lines and the axes is divided into two portions by the given curve.
   a) If these two are rotated about the y-axis, show that they generate two solids whose volumes are in the ratio of four to one.
   b) What is the ratio of the volumes of the solids generated when these regions are rotated about the x-axis?

9. Show that the curves $2x^2 + 3y^2 = a^2$ and $ky^2 = x^3$ are orthogonal for all values of the constants a and k

($a \neq 0$, $k \neq 0$). Sketch the four curves corresponding to $a = 2$, $a = 4$, $k = \frac{1}{2}$, $k = -2$ in one diagram.

10. Show, analytically, that an angle inscribed in a semicircle is a right angle.

11. Two points P, Q are called symmetric with respect to a circle if P and Q lie on the same ray through the center and if the product of their distances from the center is equal to the square of the radius. Given that Q traverses the straight line $x + 2y - 5 = 0$, find the path of the point P that is symmetric to Q with respect to the circle $x^2 + y^2 = 4$.

12. A point $P(x, y)$ moves so that the ratio of its distances from two fixed points is a constant k. Show that the point traces a circle if $k \neq 1$, and a straight line if $k = 1$.

13. Show that the centers of all chords of the parabola $x^2 = 4py$ with slope m lie on a straight line, and find its equation.

14. The line through the focus F and the point $P(x_1, y_1)$ on the parabola $y^2 = 4px$ intersects the parabola in a second point $Q(x_2, y_2)$. Find the coordinates of Q in terms of y_1 and p. If O is the vertex and PO cuts the directrix at R, prove that QR is parallel to the axis of the parabola.

15. Find the point (or points) on the curve $x^2 = y^3$ nearest the point $P(0, 4)$. Sketch the curve and the shortest line from P to the curve.

16. Prove that every line through the center of the circle

$$(x - h)^2 + (y - k)^2 = r^2$$

is orthogonal to the circle.

17. Find all points on the curve $x^2 + 2xy + 3y^2 = 3$, where the tangent line is perpendicular to the line $x + y = 1$.

18. A line PT is drawn tangent to the curve $xy = x + y$ at the point $P(-2, \frac{2}{3})$. Find equations of two lines that are normal to the curve and perpendicular to PT.

19. Graph each of the following equations:
a) $(x + y)(x^2 + y^2 - 1) = 0$;
b) $(x + y)(x^2 + y^2 - 1) = 1$.
(*Hint:* In part (b), consider intersections of the curve with the line $x + y = k$ for different values of the constant k.)

20. Find the center and radius of the circle through the two points $A(2, 0)$ and $B(6, 0)$ and tangent to the curve $y = x^2$.

21. Find the center of the circle that passes through the point $(0, 1)$ and is tangent to the curve $y = x^2$ at $(2, 4)$.

22. Let L_1, L_2, L_3 be three straight lines, no two of which are parallel. Let $E_i = a_i x + b_i y + c_i = 0$, $i = 1, 2, 3$, be the equation of the line L_i.
a) Describe the graph of the equation $E_1E_2 + hE_2E_3 + kE_1E_3 = 0$, assuming h and k are constants.
b) Use the method of part (a) and determine h and k so that the equation represents a circle through the points of intersection of the lines

$$x + y - 2 = 0, \qquad x - y + 2 = 0, \qquad y - 2x = 0.$$

c) Find a parabola, axis vertical, through the points of intersection of the lines in (b).

23. A comet moves in a parabolic orbit with the sun at the focus. When the comet is 4×10^7 miles from the sun, the line from the sun to it makes an angle of $60°$ with the axis of the orbit (drawn in the direction in which the orbit opens). How near does the comet come to the sun?

24. Sketch the curves $y^2 = 4x + 4$ and $y^2 = 64 - 16x$ together, and find the angles at which they intersect.

25. Find an equation of the curve such that the distance from any point $P(x, y)$ on the curve to the line $x = 3$ is the same as its distance to the point $(4, 0)$. Sketch the curve.

26. Two radar stations lying along an east–west line are separated by 20 mi. Choose a coordinate system in which their positions are $(-10, 0)$ and $(10, 0)$. A low-flying plane traveling from west to east is known to have a speed of v_0 mi/s. At $t = 0$ a signal is sent from the station at $(-10, 0)$, bounces off the plane, and is received at $(10, 0)$ $30/c$ seconds later (c is the velocity of the signal). When $t = 10/v_0$, another signal is sent out from the station at $(-10, 0)$, reflects off the plane, and is once again received $30/c$ seconds later by the other station. Find the position of the plane when it reflects the second signal under the assumption that v_0 is much less than c.

27. A line is drawn tangent to the parabola $y^2 = 4px$ at a point $P(x, y)$ on the curve. Let A be the point where this tangent line crosses the axis of the parabola, let F be the focus, and let PD be the line parallel to the axis of the parabola and intersecting the directrix at D. Prove that $AFPD$ is a rhombus.

28. Find an equation for the curve traced by point $P(x, y)$ if the distance from P to the vertex of the parabola $x^2 = 8y$ is twice the distance from P to the focus. Identify the curve.

29. Prove that the tangent to a parabola at a point P cuts the axis of the parabola at a point whose distance from the vertex equals the distance from P to the tangent at the vertex.

30. Discuss and sketch the graph of the equation $x^4 - (y^2 - 9)^2 = 0$.

31. Show that the curve C: $x^4 - (y^2 - 9)^2 = 1$ approaches part of the curve $x^4 - (y^2 - 9)^2 = 0$ as the point $P(x, y)$ moves farther and farther away from the origin. Sketch. Do any points of C lie inside the circle $x^2 + y^2 = 9$? Give a reason for your answer.

32. The ellipse $(x^2/a^2) + (y^2/b^2) = 1$ divides the plane into two regions; one inside the ellipse, the other outside. Show that points in one of these regions have coordinates that satisfy the inequality $(x^2/a^2) + (y^2/b^2) < 1$, while in the other, $(x^2/a^2) + (y^2/b^2) > 1$. (Consider the effect of replacing x, y in the given equation by $x' = kx$, $y' = ky$, with $k < 1$ in one case and $k > 1$ in the other.)

33. Find an equation of an ellipse with foci at (1, 0) and (5, 0), and one vertex at the origin.

34. Let $F_1 = (3, 0)$, $F_2 = (0, 5)$, $P = (-1, 3)$.
a) Find the distances F_1P and F_2P.
b) Does the origin O lie inside or outside the ellipse that has F_1 and F_2 as its foci and that passes through the point P? Why?

35. Find the greatest possible area of a rectangle inscribed in the ellipse $(x^2/a^2) + (y^2/b^2) = 1$, with sides parallel to the coordinate axes.

36. Show that the line $y = mx + c$ is tangent to the conic section $Ax^2 + y^2 = 1$ if and only if the constants A, m, and c satisfy the condition $A(c^2 - 1) = m^2$.

37. Starting from the general equation for the conic, find the equation of the conic with the following properties: (a) it is symmetric with respect to the origin; (b) it passes through the point (1, 0); (c) the line $y = 1$ is tangent to it at the point $(-2, 1)$.

38. Find an ellipse with one vertex at the point (3, 1), the nearer focus at the point (1, 1), and eccentricity $\frac{2}{3}$.

39. By a suitable rotation of axes, show that the equation $xy - x - y = 1$ represents a hyperbola. Sketch.

40. Find an equation of a hyperbola with eccentricity equal to $\sqrt{2}$ and with vertices at the points (2, 0) and (−2, 0).

41. Sketch the conic $\sqrt{2}y - 2xy = 3$. Locate its center and find its eccentricity.

42. If c is a fixed positive constant, then

$$\frac{x^2}{t^2} + \frac{y^2}{t^2 - c^2} = 1, \qquad c^2 < t^2,$$

defines a family of ellipses, any member of which is characterized by a particular value of t. Show that every member of the family

$$\frac{x^2}{t^2} - \frac{y^2}{c^2 - t^2} = 1, \qquad t^2 < c^2,$$

intersects any member of the first family at right angles.

43. Graph the equation $|x| + |y| = 1$. Find the area it encloses.

44. Show that if the tangent to a curve at a point P(x, y) passes through the origin, then $dy/dx = y/x$ at P. Hence show that no tangent can be drawn from the origin to the hyperbola $x^2 - y^2 = 1$.

45. a) Find the coordinates of the center and the foci, the lengths of the axes, and the eccentricity of the ellipse $x^2 + 4y^2 - 4x + 8y - 1 = 0$.
b) Do likewise with the hyperbola $3x^2 - y^2 + 12x - 3y = 0$, and in addition find equations of its asymptotes.

46. If the ends of a line segment of constant length move along perpendicular lines, show that a point P on the segment, at distances a and b from the ends, describes an ellipse.

47. Sketch the curves:
a) $(9x^2 + 4y^2 - 36)(4x^2 + 9y^2 - 36) = 0$,
b) $(9x^2 + 4y^2 - 36)(4x^2 + 9y^2 - 36) = 1$.
Is the curve in (b) bounded or does it extend to points arbitrarily far from the origin? Give a reason for your answer.

48. Let p, q be positive numbers with $q < p$. If r is a third number, prove that the equation $[x^2/(p - r)] + [y^2/(q - r)] = 1$ represents (a) an ellipse if $r < q$, (b) a hyperbola if $q < r < p$, (c) nothing if $p < r$. Prove that all these ellipses and hyperbolas have the same foci, and find these foci.

49. Find the eccentricity of the hyperbola $xy = 1$.

50. On a level plane the sound of a rifle and that of the bullet striking the target are heard at the same instant. What is the location of the hearer?

51. Show that any tangent to the hyperbola $xy = a^2$ determines with its asymptotes a triangle of area $2a^2$.

52. Given the hyperbola $9x^2 - 4y^2 - 18x - 16y + 29 = 0$. Find the coordinates of the center and foci, and the equations of the asymptotes. Sketch.

53. By an appropriate rotation, eliminate the xy-term from the equation,

$$7x^2 - 8xy + y^2 = 9.$$

54. Show that the tangent to the conic section

$$Ax^2 + Bxy + Cy^2 + Dx + Ey + F = 0$$

at a point (x_1, y_1) on it has an equation that may be written in the form

$$Axx_1 + B\left(\frac{x_1y + xy_1}{2}\right) + Cyy_1 + D\left(\frac{x + x_1}{2}\right) + E\left(\frac{y + y_1}{2}\right) + F = 0.$$

55. a) Find the eccentricity and center of the conic

$$x^2 + 12y^2 - 6x - 48y + 9 = 0.$$

b) Find the vertex of the conic

$$x^2 - 6x - 12y + 9 = 0.$$

c) Sketch the conics together.

56. Find an equation of the circle passing through the three points common to the conics of Problem 55.

57. Two vertices A, B of a triangle are fixed. The third vertex $C(x, y)$ moves in such a way that $\angle A = 2(\angle B)$. Find the path traced by C.

58. Find the equation into which $x^{1/2} + y^{1/2} = a^{1/2}$ is transformed by a rotation of axes through 45° and elimination of radicals.

59. Show that $dx^2 + dy^2$ is invariant under any rotation of axes about the origin.

60. Show that $x\,dy - y\,dx$ is invariant under any rotation of axes about the origin.

61. Graph the equation $x^{2n} + y^{2n} = a^{2n}$ for the following values of n: (a) 1, (b) 2, (c) 100. In each instance find where the curve cuts the line $y = x$.

9 Hyperbolic Functions

9.1 Introduction

In this chapter we shall consider the special combinations of e^x and e^{-x} called *hyperbolic functions*. We study these functions for two reasons. One is that they are used to solve engineering problems. The tension at any point in a hanging cable, like an electric transmission line suspended from its two ends, is calculated with hyperbolic functions. (We shall investigate hanging cables in Article 9.5.) The designers of the St. Louis arch used hyperbolic functions to predict the structure's internal forces. Hyperbolic functions are also used to describe wave motion in liquids and elastic solids.†

A second reason for studying hyperbolic functions is that they are useful in solving differential equations that arise in physics and engineering.

9.2 Definitions and Identities

The combinations

$$\cosh u = \frac{1}{2}(e^u + e^{-u}) \qquad \text{(\textit{hyperbolic cosine} of } u)$$
$$\sinh u = \frac{1}{2}(e^u - e^{-u}) \qquad \text{(\textit{hyperbolic sine} of } u) \tag{1}$$

occur so frequently in applications that it has been found convenient to give special names to them. It may not be clear at the moment why the names about to be introduced are appropriate, but it will become apparent as we proceed. These functions are related to each other by rules very

†F. W. Sears, M. W. Zemansky, and H. D. Young, *University Physics*, 6th ed. (Reading, Mass.: Addison-Wesley, 1982), Chapter 21.

as a fraction, we get

$$\begin{aligned}\frac{d(\tanh u)}{dx} &= \frac{\cosh u[d(\sinh u)/dx] - \sinh u[d(\cosh u)/dx]}{\cosh^2 u} \\ &= \frac{\cosh^2 u(du/dx) - \sinh^2 u(du/dx)}{\cosh^2 u} \\ &= \frac{(\cosh^2 u - \sinh^2 u)(du/dx)}{\cosh^2 u} = \frac{1}{\cosh^2 u}\frac{du}{dx} \\ &= \operatorname{sech}^2 u \frac{du}{dx}.\end{aligned}$$

In a similar manner, we may establish the rest of the formulas in the following list:

$$\begin{aligned}\frac{d(\tanh u)}{dx} &= \operatorname{sech}^2 u \frac{du}{dx}, \\ \frac{d(\coth u)}{dx} &= -\operatorname{csch}^2 u \frac{du}{dx}, \\ \frac{d(\operatorname{sech} u)}{dx} &= -\operatorname{sech} u \tanh u \frac{du}{dx}, \\ \frac{d(\operatorname{csch} u)}{dx} &= -\operatorname{csch} u \coth u \frac{du}{dx}.\end{aligned} \tag{3}$$

Each of these derivative formulas has a matching differential formula. These in turn may be integrated at once to produce the following integration formulas:

$$\begin{aligned}\int \sinh u\, du &= \cosh u + C. \\ \int \cosh u\, du &= \sinh u + C. \\ \int \operatorname{sech}^2 u\, du &= \tanh u + C. \\ \int \operatorname{csch}^2 u\, du &= -\coth u + C. \\ \int \operatorname{sech} u \tanh u\, du &= -\operatorname{sech} u + C. \\ \int \operatorname{csch} u \coth u\, du &= -\operatorname{csch} u + C.\end{aligned} \tag{4}$$

EXAMPLE 1 Show that $y = a \cosh (x/a)$ satisfies the differential equation

$$\frac{d^2y}{dx^2} = \frac{w}{H}\sqrt{1 + \left(\frac{dy}{dx}\right)^2}, \tag{5}$$

provided $a = H/w$, where H and w are constants.

Solution By differentiating $y = a \cosh (x/a)$, we find

$$\frac{dy}{dx} = a\left(\sinh \frac{x}{a}\right)\frac{1}{a} = \sinh \frac{x}{a}, \qquad \frac{d^2y}{dx^2} = \left(\cosh \frac{x}{a}\right)\frac{1}{a}.$$

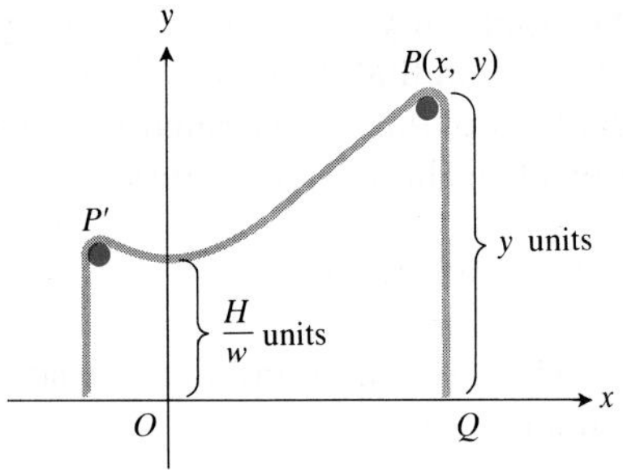

9.4 If the axes are positioned so that the lowest point of the uniform cable is H/w units above the x-axis, then the tension at each point $P(x, y)$ on the cable is exactly equal to the weight of a piece of cable y units long.

over a smooth peg while the cable is held at P so that it does not slip, and if the cable is then cut off at the point Q where it crosses the x-axis (Fig. 9.4), it may then be released at P and the weight wy of the section of cable PQ will be just sufficient to prevent the cable from slipping. If this is carried out at two different points P and P', the cable may be draped over two smooth pegs without slipping, provided the free ends reach just to the x-axis. The curve $y = a \cosh (x/a)$ is called a *catenary* from the Latin word *catena,* meaning chain. The x-axis is called the *directrix* of the catenary. □

EXAMPLE 4 Evaluate

$$\int \coth 5x \, dx.$$

Solution

$$\begin{aligned}\int \coth 5x \, dx &= \int \frac{\cosh 5x}{\sinh 5x} dx \\ &= \frac{1}{5}\int \frac{du}{u} \qquad (u = \sinh 5x) \\ &= \frac{1}{5} \ln |u| + C = \frac{1}{5} \ln |\sinh x| + C. \ \square\end{aligned}$$

EXAMPLE 5 Evaluate

$$\int_0^1 \sinh^2 x \, dx.$$

Solution From the identity (12b) in Article 9.2,

$$\sinh^2 x = \frac{\cosh 2x - 1}{2}.$$

Therefore,

$$\begin{aligned}\int_0^1 \sinh^2 x \, dx &= \int_0^1 \frac{\cosh 2x - 1}{2} dx \\ &= \frac{\sinh 2x}{4} - \frac{x}{2}\bigg]_0^1 = \frac{\sinh 2}{4} - \frac{1}{2} \approx 0.40672. \ \square\end{aligned}$$

EXAMPLE 6 Evaluate

$$\int x \sinh x \, dx.$$

Solution *Integration by parts.* We let

$$u = x, \qquad dv = \sinh x \, dx, \qquad v = \cosh x,$$

so that

$$\int x \sinh x \, dx = x \cosh x - \int \cosh x \, dx = x \cosh x - \sinh x + C. \ \square$$

PROBLEMS

1. Verify the formulas given in the text for the derivatives of
a) $y = \coth u$ b) $y = \operatorname{sech} u$ c) $y = \operatorname{csch} u$.

Find dy/dx in Problems 2–17.

2. $y = \sinh 3x$
3. $y = \cosh^2 5x$
4. $y = \cosh^2 5x - \sinh^2 5x$
5. $y = \tanh 2x$
6. $y = \coth(\tan x)$
7. $y = \operatorname{sech}^3 x$
8. $y = 4 \operatorname{csch}(x/4)$
9. $\sinh y = \tan x$
10. $y = \operatorname{sech}^2 x + \tanh^2 x$
11. $y = \operatorname{csch}^2 x$
12. $y = \sin^{-1}(\tanh x)$
13. $y = x - (\frac{1}{4}) \coth 4x$
14. $y = \ln |\tanh(x/2)|$
15. $y = x^4 \sinh x$
16. $y = x \sinh 2x - (\frac{1}{2}) \cosh 2x$
17. $y = x \sinh x - \cosh x$

Evaluate the integrals in Problems 18–32.

18. $\int \cosh(2x + 1)\, dx$
19. $\int \tanh x\, dx$
20. $\int \frac{\sinh x}{\cosh^4 x}\, dx$
21. $\int \frac{4\, dx}{(e^x + e^{-x})^2}$
22. $\int \frac{e^x - e^{-x}}{e^x + e^{-x}}\, dx$
23. $\int \tanh^2 x\, dx$
24. $\int \frac{\sinh \sqrt{x}}{\sqrt{x}}\, dx$
25. $\int \cosh^2 3x\, dx$
26. $\int \sqrt{\cosh x - 1}\, dx$
27. $\int \cosh^2 5x\, dx$
28. $\int 2 \cosh x \sinh x\, dx$
29. $\int \cosh^3 x\, dx$
30. $\int \frac{\sinh x}{1 + \cosh x}\, dx$
31. $\int \operatorname{csch} x\, dx$
32. $\int \operatorname{sech}^3 5x \tanh 5x\, dx$

Use tables to evaluate the integrals in Problems 33–36.

33. $\int \sinh^4 3x\, dx$
34. $\int \operatorname{sech} 7x\, dx$
35. $\int e^{3x} \cosh 2x\, dx$
36. $\int \tanh^3 x\, dx$

37. Find the area of the hyperbolic sector AOP bounded by the arc AP and the lines OA, OP through the origin, in Fig. 9.1.
38. Find the length of $y = \cosh x$, $0 \le x \le 1$.
39. (*Calculator*) The line $y = (x/2) + 1$ crosses the catenary $y = \cosh x$ at the point (0, 1). Experiment with a calculator to collect evidence for the fact that the line also crosses the catenary approximately at the point (0.9308, 1.4654).
40. (*Calculator*) Two successive poles supporting an electric power line are 100 feet apart, the supporting members being at the same level. If the wire dips 25 feet at the center, estimate (a) the length of the wire between supports, and (b) the tension in the wire at its lowest point if its weight is $w = 0.3$ lb/ft. (*Hint:* First approximate a from the equation $(25/a) + 1 = \cosh(50/a)$, which can be related to Problem 39 with $x = 50/a$.)

***Toolkit* programs**	
Root Finder	Super * Grapher

9.4 The Inverse Hyperbolic Functions

Since

$$\frac{d}{dx} \sinh x = \cosh x \tag{1}$$

is always positive, the hyperbolic sine is one-to-one over its entire domain, $-\infty < x < \infty$. It therefore has an inverse, which we denote in the usual way by

$$y = \sinh^{-1} x. \tag{2}$$

The minus one in $\sinh^{-1} x$ does not mean "reciprocal"; it means "inverse." For every value of x in the interval $-\infty < x < \infty$, the value of $y = \sinh^{-1} x$ is the number whose hyperbolic sine is x.

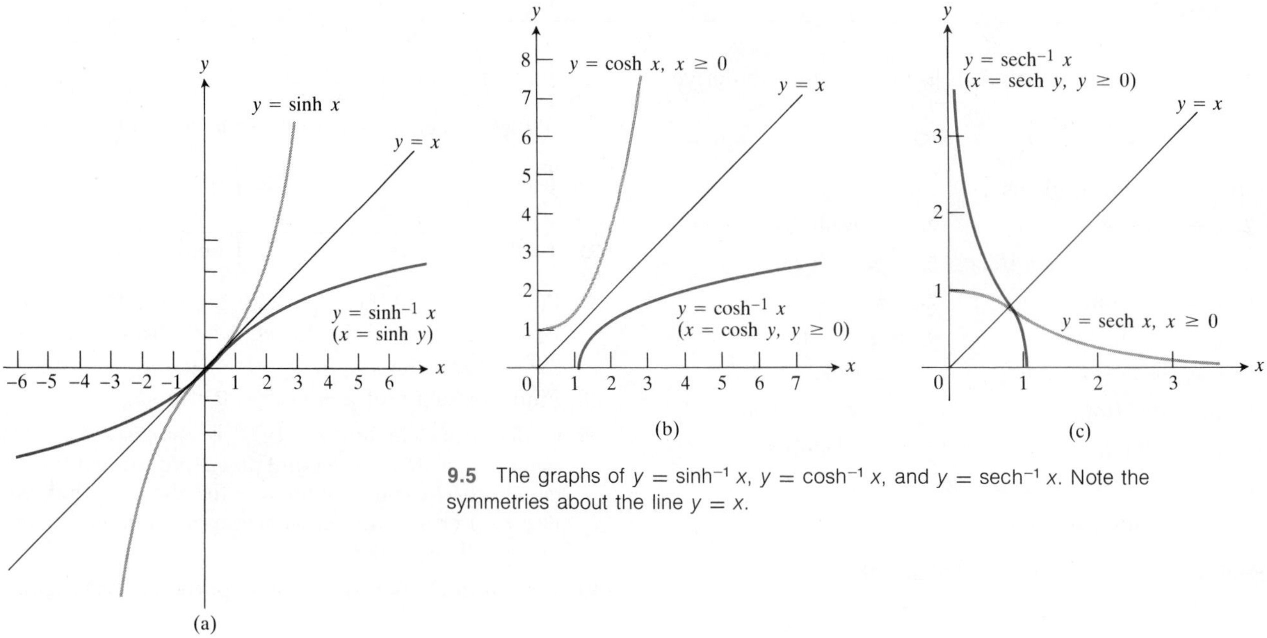

9.5 The graphs of $y = \sinh^{-1} x$, $y = \cosh^{-1} x$, and $y = \text{sech}^{-1} x$. Note the symmetries about the line $y = x$.

We call $\sinh^{-1} x$ the *inverse hyperbolic sine* of x. The graphs of $y = \sinh x$ and $y = \sinh^{-1} x$ are shown in Fig. 9.5(a).

The function $y = \cosh x$ is not one-to-one, as we can see quickly enough from its graph in Fig. 9.2(a). But the restricted function

$$y = \cosh x, \qquad x \geq 0, \tag{3}$$

is one-to-one, and therefore has an inverse, denoted by

$$y = \cosh^{-1} x. \tag{4}$$

For every value of $x \geq 1$, $y = \cosh^{-1} x$ is the number in the interval $0 \leq y < \infty$ whose hyperbolic cosine is x. The graphs of $y = \cosh x$, $x \geq 0$, and $y = \cosh^{-1} x$ are shown in Fig. 9.5(b).

Like $y = \cosh x$, the function

$$y = \text{sech}\, x = \frac{1}{\cosh x}$$

fails to be one-to-one, but its restriction

$$y = \text{sech}\, x, \qquad x \geq 0, \tag{5}$$

to nonnegative values of x does have an inverse, denoted by

$$y = \text{sech}^{-1} x. \tag{6}$$

For every value of x in the interval (0, 1], $y = \text{sech}^{-1} x$ is the nonnegative number whose hyperbolic secant is x. The graphs of $y = \text{sech}\, x$, $x \geq 0$, and $y = \text{sech}^{-1} x$ are shown in Fig. 9.5(c).

The hyperbolic tangent, cotangent, and cosecant are one-to-one on their domains and therefore have inverses, denoted by

$$y = \tanh^{-1} x, \qquad y = \coth^{-1} x, \qquad y = \text{csch}^{-1} x. \tag{7}$$

These are graphed in Fig. 9.6.

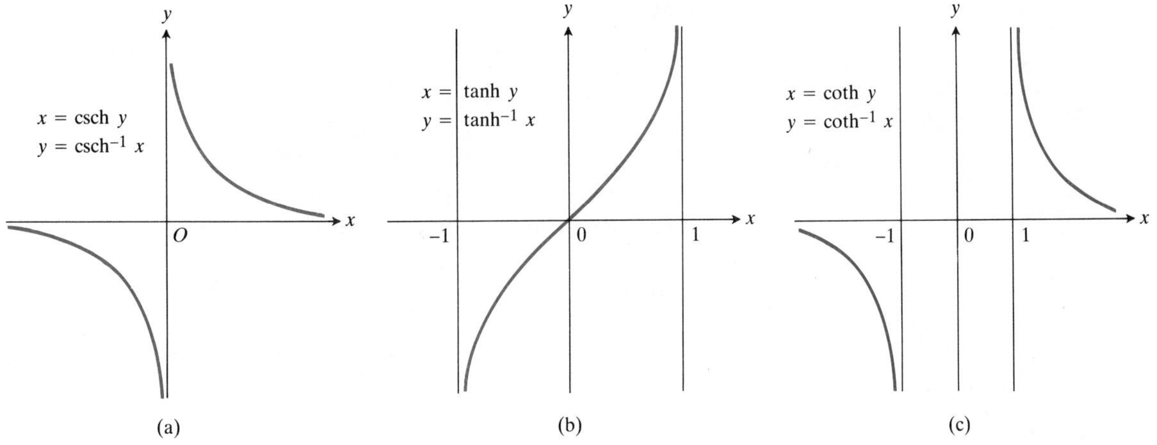

9.6 The graphs of $y = \text{csch}^{-1} x$, $y = \tanh^{-1} x$, and $y = \coth^{-1} x$.

The inverse hyperbolic secant, cosecant, and cotangent satisfy the identities

$$\text{sech}^{-1} x = \cosh^{-1} \frac{1}{x}, \tag{8}$$

$$\text{csch}^{-1} x = \sinh^{-1} \frac{1}{x}, \tag{9}$$

$$\coth^{-1} x = \tanh^{-1} \frac{1}{x}. \tag{10}$$

These identities come in handy when we have to calculate the values of $\text{sech}^{-1} x$, $\text{csch}^{-1} x$, and $\coth^{-1} x$ on a calculator that gives only $\cosh^{-1} x$, $\sinh^{-1} x$, and $\tanh^{-1} x$.

Numerical values of inverse hyperbolic functions are given in tables such as *Handbook of Mathematical Functions* (formerly published by the National Bureau of Standards, now printed by Dover Publications), which gives their values to ten decimal places.† They are also given by many scientific calculators to eight or ten places.

When hyperbolic function keys are not available on a calculator, it is still possible to evaluate the inverse hyperbolic functions by expressing them in terms of logarithms, as illustrated below for $\tanh^{-1} x$. Let $y = \tanh^{-1} x$; then $\tanh y = x$ or

$$x = \frac{\sinh y}{\cosh y} = \frac{\frac{1}{2}(e^y - e^{-y})}{\frac{1}{2}(e^y + e^{-y})} = \frac{e^y - (1/e^y)}{e^y + (1/e^y)} = \frac{e^{2y} - 1}{e^{2y} + 1}.$$

We now solve this equation for e^{2y}:

$$xe^{2y} + x = e^{2y} - 1,$$

$$1 + x = e^{2y}(1 - x),$$

$$e^{2y} = \frac{1 + x}{1 - x}.$$

†The notations arc sinh x, etc., are often used for inverse hyperbolic sine, etc.

Hence

$$y = \tanh^{-1} x = \frac{1}{2} \ln \frac{1 + x}{1 - x}, \qquad |x| < 1. \tag{11}$$

The variable x in Eq. (11) is restricted to the domain $|x| < 1$, since $x = \tanh y$ lies in this interval for all real values of y, $-\infty < y < +\infty$.

EXAMPLE 1

$$\tanh^{-1} 0.25 = \frac{1}{2} \ln \frac{1.25}{0.75} = \frac{1}{2} \ln \frac{5}{3}$$
$$= \frac{1}{2}(\ln 5 - \ln 3) \approx 0.25541. \ \square$$

The expressions for the other inverse hyperbolic functions in terms of logarithms are found in a similar manner. They are

$$\begin{aligned}
\sinh^{-1} x &= \ln (x + \sqrt{x^2 + 1}), \quad -\infty < x < \infty,\\
\cosh^{-1} x &= \ln (x + \sqrt{x^2 - 1}), \quad x \geq 1,\\
\operatorname{sech}^{-1} x &= \ln \left(\frac{1 + \sqrt{1 - x^2}}{x}\right) = \cosh^{-1}\left(\frac{1}{x}\right), \quad 0 < x \leq 1,\\
\operatorname{csch}^{-1} x &= \ln \left(\frac{1}{x} + \frac{\sqrt{1 + x^2}}{|x|}\right) = \sinh^{-1}\left(\frac{1}{x}\right), \quad x \neq 0,\\
\coth^{-1} x &= \frac{1}{2} \ln \frac{x + 1}{x - 1} = \tanh^{-1}\left(\frac{1}{x}\right), \quad |x| > 1.
\end{aligned} \tag{12}$$

These logarithmic expressions are, on the whole, rather cumbersome, and the inverse hyperbolic functions provide a useful shorthand wherever these expressions arise.

The chief merit of the inverse hyperbolic functions lies in their usefulness in integration. This will easily be understood after we have derived the following formulas for their derivatives:

$$\frac{d(\sinh^{-1} u)}{dx} = \frac{1}{\sqrt{1 + u^2}} \frac{du}{dx}, \tag{13}$$

$$\frac{d(\cosh^{-1} u)}{dx} = \frac{1}{\sqrt{u^2 - 1}} \frac{du}{dx}, \tag{14}$$

$$\frac{d(\tanh^{-1} u)}{dx} = \frac{1}{1 - u^2} \frac{du}{dx}, \quad |u| < 1, \tag{15}$$

$$\frac{d(\coth^{-1} u)}{dx} = \frac{1}{1 - u^2} \frac{du}{dx}, \quad |u| > 1, \tag{16}$$

$$\frac{d(\operatorname{sech}^{-1} u)}{dx} = \frac{-du/dx}{u\sqrt{1 - u^2}}, \tag{17}$$

$$\frac{d(\operatorname{csch}^{-1} u)}{dx} = \frac{-du/dx}{|u|\sqrt{1 + u^2}}. \tag{18}$$

The proofs of these all follow the same method. We illustrate the case of $\cosh^{-1} u$. To this end, let

$$y = \cosh^{-1} u.$$

Then

$$\cosh y = u,$$

$$\sinh y \frac{dy}{dx} = \frac{du}{dx},$$

$$\frac{dy}{dx} = \frac{1}{\sinh y} \frac{du}{dx}.$$

But

$$\cosh^2 y - \sinh^2 y = 1,$$

and

$$\cosh y = u,$$

so that

$$\begin{aligned}\sinh y &= \pm\sqrt{\cosh^2 y - 1} \\ &= \pm\sqrt{u^2 - 1}\end{aligned}$$

and

$$\frac{dy}{dx} = \frac{1}{\pm\sqrt{u^2 - 1}} \frac{du}{dx}.$$

The ambiguous sign will be + if we restrict attention to the principal value, $y = \cosh^{-1} u$, $y \geq 0$, for then $\sinh y \geq 0$ and the ambiguous sign is the same as the sign of $\sinh y$. Thus Eq. (14) is established. The identities (5a) and (5b) of Article 9.2, with y in place of u, will be found to be useful in proving Eqs. (15) to (18). The derivation is straightforward.

The restrictions $|u| < 1$ and $|u| > 1$ in (15) and (16) are due to the fact that

$$\text{if} \quad y = \tanh^{-1} u, \qquad \text{then} \qquad u = \tanh y,$$

and since

$$-1 < \tanh y < 1,$$

this means $|u| < 1$. Similarly,

$$y = \coth^{-1} u, \qquad u = \coth y,$$

requires $|u| > 1$. The distinction becomes important when we invert the formulas to get integration formulas, since otherwise we would be unable to tell whether we should write $\tanh^{-1} u$ or $\coth^{-1} u$ for

$$\int \frac{du}{1 - u^2}.$$

The integration formulas at the top of the next page follow at once from Eqs. (13)–(18). The integrals in Formulas 1, 2, 4, and 5 can all be evaluated by trigonometric substitutions as well, and the integral in Formula 3 can be evaluated by partial fractions. (See Problems 29–32.)

1. $\int \frac{du}{\sqrt{1+u^2}} = \sinh^{-1} u + C$

2. $\int \frac{du}{\sqrt{u^2-1}} = \cosh^{-1} u + C$

3. $\int \frac{du}{1-u^2} = \begin{Bmatrix} \tanh^{-1} u + C & \text{if } |u| < 1 \\ \coth^{-1} u + C & \text{if } |u| > 1 \end{Bmatrix} = \frac{1}{2} \ln \left| \frac{1+u}{1-u} \right| + C$

4. $\int \frac{du}{u\sqrt{1-u^2}} = -\text{sech}^{-1} |u| + C = -\cosh^{-1}\left(\frac{1}{|u|}\right) + C$

5. $\int \frac{du}{u\sqrt{1+u^2}} = -\text{csch}^{-1} |u| + C = -\sinh^{-1}\left(\frac{1}{|u|}\right) + C$

PROBLEMS

1. Solve the equation $x = \sinh y = \frac{1}{2}(e^y - e^{-y})$ for e^y in terms of x, and thus show that $y = \ln(x + \sqrt{1+x^2})$. (This equation expresses $\sinh^{-1} x$ as a logarithm.)

2. Express $\cosh^{-1} x$ in terms of logarithms by using the method of Problem 1.

3. Establish Eq. (13).

4. Establish Eq. (15).

5. Establish Eq. (17).

In Problems 6–20, find dy/dx.

6. $y = \sinh^{-1}(2x)$

7. $y = \tanh^{-1}(\cos x)$

8. $y = \cosh^{-1}(\sec x)$

9. $y = \coth^{-1}(\sec x)$

10. $y = \text{sech}^{-1}(\sin 2x)$

11. $y = \cosh^{-1} x^2$

12. $y = \sinh^{-1} \sqrt{x-1}$

13. $y = \text{csch}^{-1}(\tan x)$

14. $y = \sinh^{-1}(1/x)$

15. $y = \cosh^{-1} \sqrt{x+1}$

16. $y = \sinh^{-1}(\tan x)$

17. $y = \coth^{-1}(\sin x)$

18. $y = \sqrt{1+x^2} - \sinh^{-1} \frac{1}{x}$

19. $y = \frac{x}{2}\sqrt{x^2-1} - \frac{1}{2}\cosh^{-1} x$

20. $y = 2\cosh^{-1}\left(\frac{x}{2}\right) + \frac{x}{2}\sqrt{x^2-4}$

Evaluate the integrals in Problems 21–28.

21. $\int \frac{dx}{\sqrt{1+4x^2}}$

22. $\int \frac{dx}{\sqrt{4+x^2}}$

23. $\int_0^{0.5} \frac{dx}{1-x^2}$

24. $\int_{5/4}^{2} \frac{dx}{1-x^2}$

25. $\int \frac{dx}{x\sqrt{4+x^2}}$

26. $\int \frac{\cos x}{\sqrt{1+\sin^2 x}} dx$

27. $\int \sinh^{-1} x \, dx$

28. $\int \tanh^{-1} x \, dx$

Use trigonometric substitutions to evaluate the integrals in Problems 29–31.

29. $\int \frac{du}{\sqrt{u^2-1}}, \quad u \geq 1$

30. $\int \frac{du}{u\sqrt{u^2-1}}$

31. $\int \frac{du}{u\sqrt{u^2+1}}$

32. Evaluate the following integral by partial fractions:

$$\int \frac{du}{u^2-1}.$$

Use tables to evaluate the integrals in Problems 33–38.

33. $\int \frac{\sqrt{x^2-25}}{x} dx$

34. $\int \frac{\sqrt{x^2-25}}{x^2} dx$

35. $\int \frac{x^2}{\sqrt{x^2-25}} dx$

36. $\int x^2\sqrt{x^2+3}\, dx$

37. $\sqrt{x^2-4}\, dx$

38. $\int x^2\sqrt{x^2-4}\, dx$

39. *Retarded free fall.* If a body of mass m falling from rest under the action of gravity encounters an air resistance proportional to the square of the velocity, then the velocity v at time t satisfies the differential equation

$$m\left(\frac{dv}{dt}\right) = mg - kv^2,$$

where k is a constant of proportionality and $v = 0$ when $t = 0$. Show that

$$v = \sqrt{\frac{mg}{k}} \tanh\left(\sqrt{\frac{gk}{m}}\, t\right),$$

and hence deduce that the body approaches a "limiting velocity" equal to $\sqrt{mg/k}$ as $t \to \infty$.

***Toolkit* programs**

Root Finder

Super * Grapher

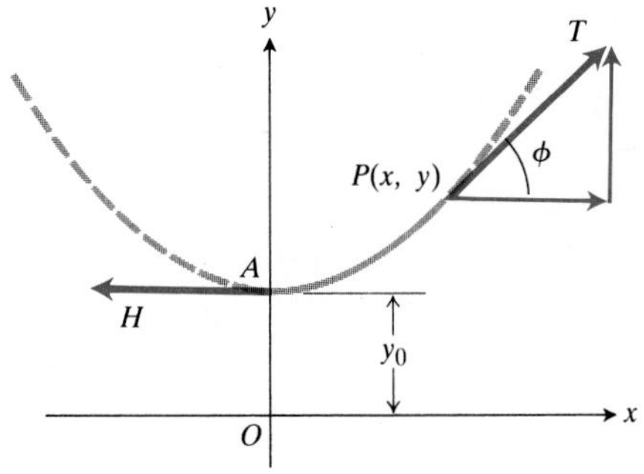

9.7 A section of hanging cable.

9.5 Hanging Cables

We conclude this chapter by deriving the solution of the differential equation

$$\frac{d^2y}{dx^2} = \frac{w}{H}\sqrt{1 + \left(\frac{dy}{dx}\right)^2}, \tag{1}$$

which is the equation of equilibrium of forces on a hanging cable discussed in Article 9.3, Example 2. Since Eq. (1) involves the second derivative, two conditions are required to determine the constants of integration. By choosing the y-axis to be the vertical line through the lowest point of the cable (Fig. 9.7), one condition becomes

$$\frac{dy}{dx} = 0 \quad \text{when} \quad x = 0. \tag{2a}$$

Then we may still move the x-axis up or down to suit our convenience. That is, we let

$$y = y_0 \quad \text{when} \quad x = 0, \tag{2b}$$

and we may choose y_0 later to give us the simplest form in our final answer.

When solving an equation such as (1), we may effectively introduce a single letter to represent dy/dx. The letter p is most often used. Thus we let

$$\frac{dy}{dx} = p, \qquad \frac{d^2y}{dx^2} = \frac{dp}{dx}, \tag{3}$$

so that Eq. (1) takes the form

$$\frac{dp}{dx} = \frac{w}{H}\sqrt{1 + p^2} \qquad \text{or} \qquad \frac{1}{\sqrt{1 + p^2}} \cdot \frac{dp}{dx} = \frac{w}{H}.$$

When we integrate both sides of the second equation with respect to x, we obtain

$$\int \frac{dp}{\sqrt{1 + p^2}} = \int \frac{w}{h}\,dx \tag{4a}$$

$$\sinh^{-1} p = \frac{w}{H}x + C_1. \tag{4b}$$

Since $p = dy/dx = 0$ when $x = 0$, we determine the constant of integration

$$\sinh^{-1} 0 = C_1,$$

from which (see Fig. 9.7)

$$C_1 = 0.$$

Hence

$$\sinh^{-1} p = \frac{w}{H}x \qquad \text{or} \qquad p = \sinh\left(\frac{w}{H}x\right). \tag{5a}$$

We substitute $p = dy/dx$, multiply by dx, and have

$$dy = \sinh\left(\frac{w}{H}x\right) dx$$

or

$$y = \int \sinh\left(\frac{w}{H}x\right) dx = \frac{H}{w}\int \sinh\left(\frac{w}{H}x\right) d\left(\frac{w}{H}x\right)$$

$$= \frac{H}{w}\cosh\left(\frac{w}{H}x\right) + C_2. \tag{5b}$$

The condition $y = y_0$ when $x = 0$ determines C_2:

$$y_0 = \frac{H}{w}\cosh 0 + C_2, \qquad C_2 = y_0 - \frac{H}{w}, \tag{5c}$$

and hence

$$y = \frac{H}{w}\cosh\left(\frac{w}{H}x\right) + y_0 - \frac{H}{w}. \tag{5d}$$

Clearly, this equation will have a simpler form if we choose y_0 so that

$$y_0 = \frac{H}{w}.$$

We do so. The answer then takes the form

$$y = \frac{H}{w}\cosh\left(\frac{w}{H}x\right)$$

or

$$y = a\cosh\frac{x}{a} \quad \text{with} \quad a = \frac{H}{w}. \tag{6}$$

PROBLEMS

1. Find the length of arc of the catenary $y = a\cosh x/a$ from $A(0, a)$ to $P_1(x_1, y_1)$, $x_1 > 0$.

2. Show that the area bounded by the x-axis, the catenary $y = a\cosh x/a$, the y-axis, and the vertical line through $P_1(x_1, y_1)$, $x_1 > 0$, is the same as the area of a rectangle of altitude a and base s, where s is the length of the arc from $A(0, a)$ to P_1. (See Problem 1.)

3. The catenary $y = a\cosh x/a$ is revolved about the x-axis. Find the surface area generated by the portion of the curve between the points $A(0, a)$ and $P_1(x_1, y_1)$, $x_1 > 0$. (Incidentally, of all continuously differentiable curves $y = f(x)$, $f(x) > 0$, from $A(0, a)$ to $P(x_1, y_1)$, the catenary generates the surface of revolution of least area.)

4. Find the center of gravity of the arc of the catenary $y = a\cosh x/a$ between two symmetrically located points $P_0(-x_1, y_1)$ and $P_1(x_1, y_1)$.

5. Find the volume generated when the area of Problem 2 is revolved about the x-axis.

6. The length of the arc AP (Fig. 9.7) is $s = a\sinh x/a$. (See Problem 1.)
   a) Show that the coordinates of $P(x, y)$ may be expressed as functions of the arc length s, as follows:
   $$x = a\sinh^{-1}\frac{s}{a}, \qquad y = \sqrt{s^2 + a^2}.$$
   b) Calculate dx/ds and dy/ds from part (a) above and verify that $(dx/ds)^2 + (dy/ds)^2 = 1$.

7. (*Calculator or tables*) A cable 32 feet long and weighing 2 pounds per foot has its ends fastened at the same level to two posts 30 feet apart.
   a) Show that the constant a in Eq. (6) must satisfy the equation
   $$\sinh u = \frac{16}{15}u, \qquad u = \frac{15}{a}. \qquad \text{(See Problem 6a.)}$$
   b) Sketch graphs of the curves $y_1 = \sinh u$, $y_2 = (16/15)u$ and show (with the aid of a calculator or tables) that they intersect at $u = 0$ and $u = \pm 0.6$ (approximately).

c) Using the results of part (b), find the sag in the cable at its center.

d) Using the results of part (b), find the tension in the cable at its lowest point.

REVIEW QUESTIONS AND EXERCISES

1. Define each of the hyperbolic functions.
2. State three trigonometric identities (such as formulas for $\sin(A + B)$, $\cos(A - B)$, $\cos^2 A + \sin^2 A = 1$, etc.). What are the corresponding hyperbolic identities? Verify them.
3. Develop formulas for derivatives of the six hyperbolic functions.
4. What are the domain and range of $y = \sinh x$? $y = \cosh x$? $y = \tanh x$?
5. State some differences between the graphs of the trigonometric functions and their hyperbolic counterparts (for example, sine and sinh, cosine and cosh, tangent and tanh).
6. If $y = A \sin(at) + B \cos(at)$, then $y'' = -a^2 y$. What is the corresponding differential equation satisfied by $y = A \sinh(at) + B \cosh(at)$?
7. Define $y = \sinh^{-1} x$ and $y = \cosh^{-1} x$. What are their domains? ranges? derivatives?

MISCELLANEOUS PROBLEMS

1. Prove the hyperbolic identity $\cosh 2x = \cosh^2 x + \sinh^2 x$.
2. Verify that $\tanh x = \sinh 2x/(1 + \cosh 2x)$.
3. Sketch the curves $y = \cosh x$ and $y = \sinh x$ together. To each positive value of x there corresponds a point P on $y = \sinh x$ and a point Q on $y = \cosh x$. Calculate the limit of the distance PQ as x becomes infinitely large.
4. If $\cosh x = \frac{5}{4}$, find sinh x and tanh x.
5. If $\operatorname{csch} x = -\frac{9}{40}$, find cosh x and tanh x.
6. If $\tanh x > \frac{5}{13}$, show that $\sinh x > 0.4$ and $\operatorname{sech} x < 0.95$.
7. Let $P(x, y)$ be a point on the curve $y = \tanh x$ (Fig. 9.2b). Let AB be the vertical line segment through P with A and B on the asymptotes of the curve. Let C be a semicircle with AB as diameter. Let L be a line through P perpendicular to AB and cutting C in a point Q. Show that $PQ = \operatorname{sech} x$.
8. Prove that $\sinh 3u = 3 \sinh u + 4 \sinh^3 u$.
9. Find equations of the asymptotes of the hyperbola represented by the equation $y = \tanh(\frac{1}{2} \ln x)$.
10. A particle moves along the x-axis according to one of the following laws: (a) $x = a \cos kt + b \sin kt$, (b) $x = a \cosh kt + b \sinh kt$. In both cases, show that the acceleration is proportional to x, but that in the first case it is always directed toward the origin while in the second case it is directed away from the origin.
11. Show that $y = \cosh x$, $\sinh x$, $\cos x$, and $\sin x$ all satisfy the relationship $d^4 y/dx^4 = y$.

In Problems 12–21, find dy/dx.

12. $y = \sinh^2 3x$
13. $\tan x = \tanh^2 y$
14. $\sin^{-1} x = \operatorname{sech} y$
15. $\sinh y = \sec x$
16. $\tan^{-1} y = \tanh^{-1} x$
17. $y = \tanh(\ln x)$
18. $x = \cosh(\ln y)$
19. $y = \sinh(\tan^{-1} e^{3x})$
20. $y = \sinh^{-1}(\tan x)$
21. $y^2 + x \cosh y + \sinh^2 x = 50$

Evaluate the integrals in Problems 22–31.

22. $\displaystyle\int \frac{d\theta}{\sinh \theta + \cosh \theta}$
23. $\displaystyle\int \frac{\cosh \theta \, d\theta}{\sinh \theta + \cosh \theta}$
24. $\displaystyle\int \sinh^3 x \, dx$
25. $\displaystyle\int e^x \sinh 2x \, dx$
26. $\displaystyle\int \frac{e^{2x} - 1}{e^{2x} + 1} dx$
27. $\displaystyle\int_0^1 \frac{dx}{4 - x^2}$
28. $\displaystyle\int_3^5 \frac{dx}{4 - x^2}$
29. $\displaystyle\int \frac{e^t \, dt}{\sqrt{1 + e^{2t}}}$
30. $\displaystyle\int \frac{\sin x \, dx}{1 - \cos^2 x}$
31. $\displaystyle\int \frac{\sec^2 \theta \, d\theta}{\sqrt{\tan^2 \theta - 1}}$

Sketch the curves in Problems 32–34.

32. $y = \dfrac{1}{2} \ln \dfrac{1 + \tanh x}{1 - \tanh x}$
33. $y = \tan\left(\dfrac{\pi}{2} \tanh x\right)$
34. $\cosh y = 1 + \dfrac{x^2}{2}$
35. If the arc s of the catenary $y = a \cosh(x/a)$ is measured from the lowest point, show that $dy/dx = s/a$.
36. Evaluate the limit, as $x \to \infty$, of $\cosh^{-1} x - \ln x$.
37. Evaluate

$$\lim_{x\to\infty} \int_1^x \left(\frac{1}{\sqrt{1 + t^2}} - \frac{1}{t}\right) dt.$$

10 Polar Coordinates

10.1 Polar Coordinates

One of the original ideas offered by Newton's *The Method of Fluxions and Infinite Series* (written about 1671 but not published until 1736) was the use of new coordinate systems, among them the system of polar coordinates that locates points by reference to what we might call their distance and compass direction from a fixed point. Polar coordinates are important because in polar coordinates the conic sections are all given by a single equation. We shall explore this amazing fact in Article 10.3. The polar equation for conic sections is important in physics and astronomy, where it arises in the study of planetary motion and the derivation of Kepler's three laws.

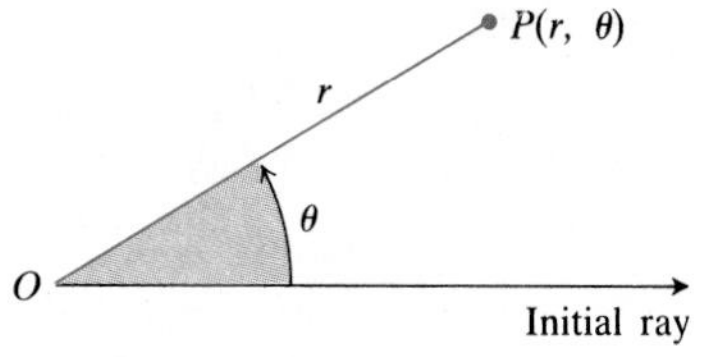

10.1 Polar coordinates.

To define polar coordinates we first fix an *origin* O and an *initial ray* from O, as shown in Fig. 10.1. Then each point P can be assigned *polar coordinates* (r, θ) in which the first number, r, gives the directed distance from O to P and the second number, θ, gives the directed angle from the initial ray to the segment OP:

$$P(r, \theta) \qquad (1)$$

Directed distance from O to P — Directed angle from initial ray to OP.

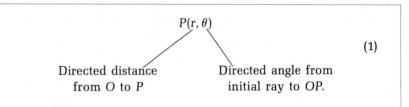

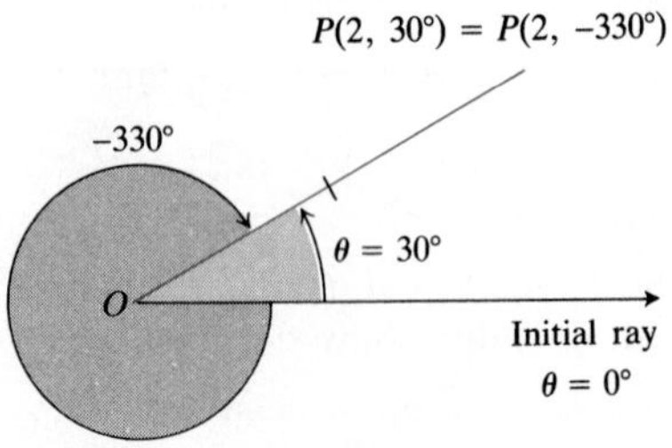

10.2 The ray $\theta = 30°$ is the same as the ray $\theta = -330°$.

As we shall see, each point in the plane has many different coordinate pairs.

As in trigonometry, the angle θ is *positive* when measured counterclockwise and negative when measured clockwise (Fig. 10.1). But the angle associated with a given point is not unique (Fig. 10.2). For instance, the point 2 units from the origin, along the ray $\theta = 30°$, has polar coordinates $r = 2$, $\theta = 30°$. It also has coordinates $r = 2$, $\theta = -330°$, and $r = 2$, $\theta = 390°$.

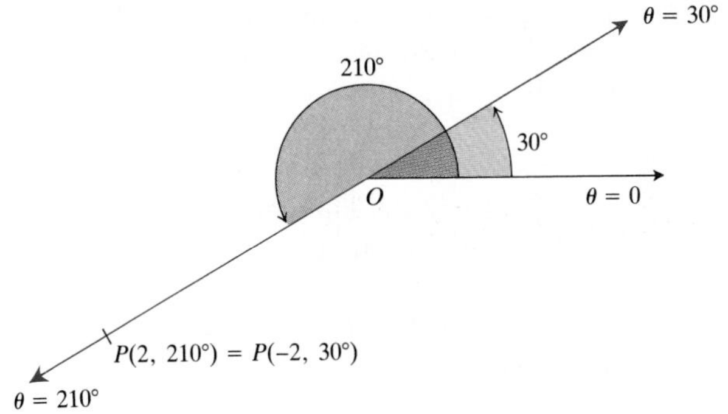

10.3 The rays $\theta = 30°$ and $\theta = 210°$ make a line.

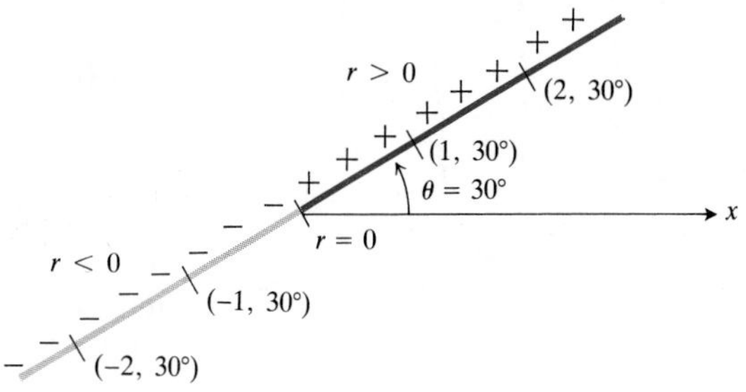

10.4 The terminal ray $\theta = \pi/6$ and its negative.

There are occasions when we wish to allow r to be negative. That is why we say "directed distance" in (1). The ray $\theta = 30°$ and the ray $\theta = 210°$ together make up a complete line through O (Fig. 10.3). The point $P(2, 210°)$ two units from O on the ray $\theta = 210°$ has polar coordinates $r = 2$, $\theta = 210°$. It can be reached by standing at O and facing out along the initial ray, if you first turn 210° counterclockwise, and then go forward two units. You would reach the same point by turning only 30° counterclockwise from the initial ray and then going *backward* two units. So we say that the point also has polar coordinates $r = -2$, $\theta = 30°$.

Whenever the angle between two rays is 180°, the rays make a straight line. We then say that each ray is the *negative* or *opposite* of the other. Points on the ray $\theta = \alpha$ have polar coordinates (r, α) with $r \geq 0$. Points on the negative ray, $\theta = \alpha + 180°$, have coordinates (r, α) with $r \leq 0$. The origin is $r = 0$. (See Fig. 10.4 for the ray $\theta = 30°$ and its negative. *Caution.* The "negative" of the ray $\theta = 30°$ is the ray $\theta = 30° + 180° = 210°$ and *not* the ray $\theta = -30°$. "Negative" refers to the directed distance r, not to the angle θ.)

There is a great advantage in being able to use polar and cartesian coordinates simultaneously. To do this, we use a common origin and take the initial ray as the positive x-axis. The ray $\theta = 90°$ is the positive y-axis. The coordinate sets (Fig. 10.5) are then related by the equations

$$x = r\cos\theta, \quad y = r\sin\theta, \qquad \text{or} \qquad x^2 + y^2 = r^2, \quad \frac{y}{x} = \tan\theta. \tag{2}$$

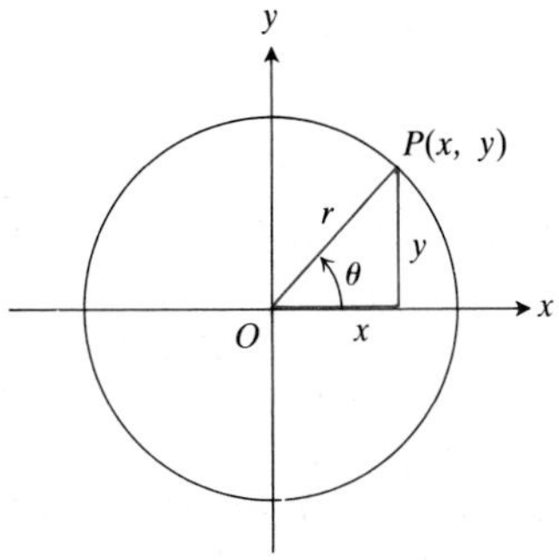

10.5 Polar and cartesian coordinates.

These are the equations that define $\sin\theta$ and $\cos\theta$ when r is positive. They are also valid if r is negative, because $\cos(\theta + 180°) = -\cos\theta$, $\sin(\theta + 180°) = -\sin\theta$, so positive r's on the ray $\theta + 180°$ correspond to negative r's on the ray θ. When $r = 0$, then $x = y = 0$, and P is the origin. That is, the equation

$$r = 0, \qquad \text{any value } \theta, \tag{3}$$

denotes the origin.

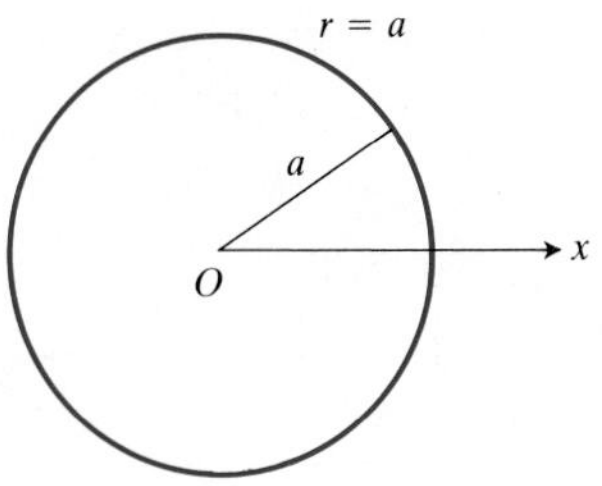

10.6 The polar equation for the circle shown above is $r = a$.

If we hold r fixed at a constant nonzero value $r = a$, then P moves along the circle with center O and radius a, tracing the circle once as θ varies from 0 to 360° (Fig. 10.6). Thus,

$$r = a \tag{4}$$

is the equation of a circle (provided $a \neq 0$). The equation $r = 1$ is an equation for the unit circle in the plane. The equation $r = -1$ is another equation for the same circle.

If we hold θ fixed and let r take on all real values, the point P will trace a straight line through the origin. Thus

$$\theta = \alpha \tag{5}$$

is an equation for the line through the origin that makes an angle α with (measured from) the initial ray. The equation $\theta = 30°$ gives the line in Fig. 10.4. With θ fixed, positive values of r give points on the terminal side of the angle; *negative* values of r give points on the opposite ray.

A given point may be represented in more than one way in polar coordinates. For example, the point (2, 30°), or (2, $\pi/6$), has the representations

$$(2, 30), \qquad (2, -330°), \qquad (-2, 210°), \qquad (-2, -150°), \tag{6}$$

among many others.

EXAMPLE 1 Find all the polar coordinates of the point (2, 30°).

Solution See Fig. 10.7.

10.7 The point P (2, 30°) has many different polar coordinates.

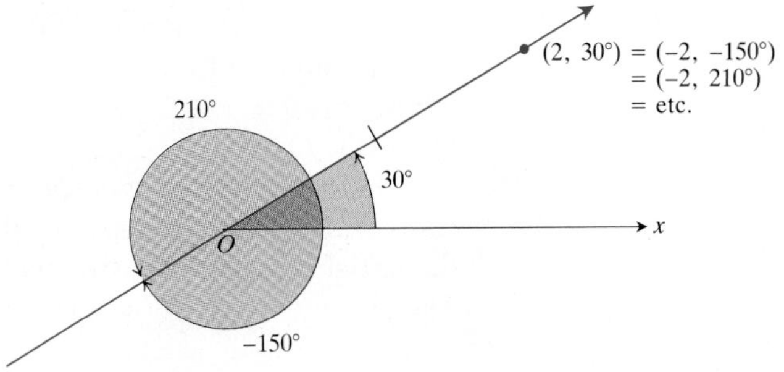

For $r = 2$: The angles

$$\begin{aligned} &30° \pm 1 \cdot 360° \\ &30° \pm 2 \cdot 360° \\ &30° \pm 3 \cdot 360° \\ &\quad\vdots \end{aligned} \tag{7}$$

all have the same terminal ray as the angle 30°. Therefore, the polar coordinates

$$(2, 30° + n \cdot 360°), \qquad n = 0, \pm 1, \pm 2, \ldots \tag{8}$$

all identify the point (2, 30°).

For $r = -2$: See Fig. 10.7. Numerous as they are, the coordinates in (8) are not the only polar coordinates of the point (2, 30°). The coordinates

$$\begin{aligned}&(-2, -150°)\\&(-2, -150° \pm 1 \cdot 360°)\\&(-2, -150° \pm 2 \cdot 360°)\\&(-2, -150° \pm 3 \cdot 360°)\\&\quad\vdots\end{aligned} \tag{9}$$

also represent the point (2, 30°). These are summarized in the formula

$$(-2, -150° + n \cdot 360°), \qquad n = 0, \pm 1, \pm 2, \ldots . \tag{10}$$

Radian measure: If we measure angles in radians, the formulas that correspond to (8) and (10) are

$$\left(2, \frac{\pi}{6} + 2n\pi\right), \qquad n = 0, \pm 1, \pm 2, \ldots \tag{11}$$

and

$$\left(-2, -\frac{5\pi}{6} + 2n\pi\right), \qquad n = 0, \pm 1, \pm 2, \ldots . \tag{12}$$

When $n = 0$, these formulas give

$$(2, \pi/6) \qquad \text{and} \qquad (-2, -5\pi/6).$$

When $n = 1$, they give

$$(2, 13\pi/6) \qquad \text{and} \qquad (-2, 7\pi/6),$$

and so on. □

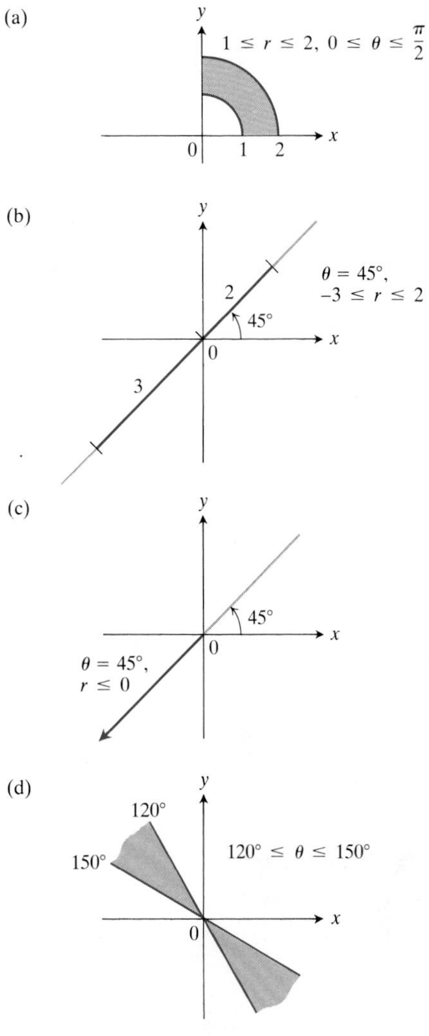

Figure 10.8

EXAMPLE 2 The graphs in Figs. 10.8 (a) through (d) show the sets of points whose polar coordinates satisfy the given equations and inequalities. □

To find cartesian equations equivalent to polar equations, and vice versa, we use one or more of the substitutions

$$r\cos\theta = x, \qquad r\sin\theta = y, \qquad r^2 = x^2 + y^2, \qquad \frac{y}{x} = \tan\theta, \tag{13}$$

as in the next example.

EXAMPLE 3

Polar equation	**Cartesian equivalent**
$r^2\cos^2\theta - r^2\sin^2\theta = 1$	$x^2 - y^2 = 1$
$r^2\cos\theta\sin\theta = 4$	$xy = 4$
$r\cos\theta = 2$	$x = 2$
$r^2 = 3r\sin\theta$	$x^2 + y^2 = 3y$ □

Trigonometric identities like

$$\cos(A - B) = \cos A\cos B + \sin A\sin B \tag{14}$$

allow the conversion of still other polar equations into cartesian form, as in Example 4.

EXAMPLE 4 Find a cartesian equivalent of the polar equation

$$r \cos\left(\theta - \frac{\pi}{3}\right) = 3.$$

Solution Using Eq. (14), with $A = \theta$ and $B = \pi/3$, we find

$$r \cos\left(\theta - \frac{\pi}{3}\right) = 3,$$

$$r\left[\cos\theta \cos\frac{\pi}{3} + \sin\theta \sin\frac{\pi}{3}\right] = 3,$$

$$r\cos\theta \cdot \frac{1}{2} + r\sin\theta \cdot \frac{\sqrt{3}}{2} = 3,$$

$$\frac{1}{2}x + \frac{\sqrt{3}}{2}y = 3,$$

$$x + \sqrt{3}y = 6. \square$$

EXAMPLE 5 Replace the following polar equations by equivalent cartesian equations, and identify their graphs:

a) $r\cos\theta = -4$

b) $r^2 = 4r\cos\theta$

c) $r = \dfrac{4}{2\cos\theta - \sin\theta}$.

Solution We use the substitutions $r\cos\theta = x$, $r\sin\theta = y$, $r^2 = x^2 + y^2$.

a) $r\cos\theta = -4$

The cartesian equation: $r\cos\theta = -4,$

$$x = -4.$$

The graph: Vertical line through $x = -4$ on the x-axis.

b) $r^2 = 4r\cos\theta$

The cartesian equation: $r^2 = 4r\cos\theta,$

$$x^2 + y^2 = 4x,$$

$$x^2 - 4x + y^2 = 0,$$

$$x^2 - 4x + 4 + y^2 = 4, \quad \begin{pmatrix}\text{completing}\\ \text{the square}\end{pmatrix}$$

$$(x - 2)^2 + y^2 = 4.$$

The graph: Circle, radius 2, center $(h, k) = (2, 0)$.

c) $r = \dfrac{4}{2\cos\theta - \sin\theta}$

The cartesian equation: $r(2\cos\theta - \sin\theta) = 4,$

$$2r\cos\theta - r\sin\theta = 4,$$

$$2x - y = 4,$$

$$y = 2x - 4.$$

The graph: Line, slope $m = 2$, y-intercept $b = -4$. $\square$

EXAMPLE 6 Show that the point $(2, \pi/2)$ lies on the curve $r = 2 \cos 2\theta$.

Solution It may seem at first that the point $(2, \pi/2)$ does not lie on the curve, because substituting the given coordinates into the equation gives

$$2 = 2 \cos 2\left(\frac{\pi}{2}\right) = 2 \cos \pi = -2,$$

which is not a true equality. The magnitude is right, but the sign is wrong. This suggests looking for a pair of coordinates for the given points in which r is negative, for example,

$$\left(-2, -\frac{\pi}{2}\right).$$

When we try these in the equation $r = 2 \cos 2\theta$, we find

$$-2 = 2 \cos 2\left(-\frac{\pi}{2}\right) = 2(-1) = -2,$$

and the equation is satisfied. The point $(2, \pi/2)$ is seen to lie on the curve after all. □

Solving Polar Equations Simultaneously

The fact that a point may be represented in different ways in polar coordinates makes added care necessary in determining the points in which the graphs of polar equations intersect. For example, the point $(2a, \pi)$ lies on the curve

$$r^2 = 4a^2 \cos \theta \tag{15}$$

even though its coordinates as given do not satisfy the equation, because the same point is represented by $(-2a, 0)$ and these coordinates do satisfy the equation. The same point $(2a, \pi)$ lies on the curve

$$r = a(1 - \cos \theta), \tag{16}$$

and hence this point should be included among the points of intersection of the two curves represented by Eqs. (15) and (16). But if we solve the equations simultaneously by substituting $\cos \theta = r^2/4a^2$ from (15) into (16) and then solving the resulting quadratic equation

$$\left(\frac{r}{a}\right)^2 + 4\left(\frac{r}{a}\right) - 4 = 0$$

for

$$\frac{r}{a} = -2 \pm 2\sqrt{2}, \tag{17}$$

we do *not* obtain the point $(2a, \pi)$ as a point of intersection. The reason is simple enough; namely, the point is not on the curves "simultaneously" in the sense of being reached at the same value of θ, since it is reached in the one case when $\theta = 0$ and in the other case when $\theta = \pi$. It is as though two ships describe paths that intersect at a point, but the ships do not collide because they reach the point of intersection at different times! The curves represented by Eqs. (15) and (16) are shown in Fig. 10.11(c), and they are seen to intersect at the four points

$$(0, 0), \qquad (2a, \pi), \qquad (r_1, \theta_1), \qquad (r_1, -\theta_1), \tag{18a}$$

where

$$r_1 = (-2 + 2\sqrt{2})a,$$
$$\cos\theta_1 = 1 - \frac{r_1}{a} = 3 - 2\sqrt{2}. \tag{18b}$$

The last two of these points are found from the simultaneous solution. The first two are disclosed by graphing the curves. (See Problem 53.)

We will take a closer look at graphing polar equations in the next article.

PROBLEMS

1. Plot the following points, given in polar form with angles in degrees. Then find *all* polar coordinates of each point.
a) $(2, 90°)$ b) $(2, 0°)$ c) $(-2, 90°)$ d) $(-2, 0°)$

2. Plot the following points, given in polar form with angles in radians. Then find *all* polar coordinates of each point.
a) $(3, \pi/4)$ b) $(-3, \pi/4)$ c) $(3, -\pi/4)$ d) $(-3, -\pi/4)$

In Problems 3–10, graph the set of points $P(r, \theta)$ whose polar coordinates satisfy the given equation, inequality or inequalities.

3. $r = 2$ **4.** $0 \le r \le 2$

5. $r > 1$ **6.** $1 < r < 2$

7. $0° \le \theta \le 30°, \quad r \ge 0$ **8.** $\theta = 120°, \quad r \le -2$

9. $\theta = 60°, \quad -1 \le r \le 3$ **10.** $\theta = 495°, \quad r \ge -1$

Graph the regions given by the inequalities in Problems 11–14 (angles in radians).

11. $\pi/4 \le \theta \le 3\pi/4, \quad 0 \le r \le 1$

12. $1 \le r \le 2$

13. $-\pi/2 \le \theta \le \pi/2, \quad 1 \le r \le 2$

14. $0 \le \theta \le \pi/2, \quad 1 \le |r| \le 2$

Replace the polar equations in Problems 15–30 by equivalent cartesian equations. Then graph the equations.

15. $r\cos\theta = 2$ **16.** $r\sin\theta = -1$

17. $r\sin\theta = 4$ **18.** $r\cos\theta = 0$

19. $r\sin\theta = 0$ **20.** $r\cos\theta = -3$

21. $r\cos\theta + r\sin\theta = 1$ **22.** $r\sin\theta = r\cos\theta$

23. $r^2 = 1$ **24.** $r^2 = 4r\sin\theta$

25. $r\sin\theta = e^{r\cos\theta}$ **26.** $r^2 + 2r^2\cos\theta\sin\theta = 1$

27. $r = \dfrac{5}{\sin\theta - 2\cos\theta}$ **28.** $r = \cos\theta + \sin\theta$

29. $r = 4\tan\theta\sec\theta$ **30.** $r + \sin\theta = 2\cos\theta$

Replace the cartesian equations in Problems 31–40 by equivalent polar equations.

31. $x = 7$ **32.** $y = 1$

33. $x = y$ **34.** $x - y = 3$

35. $x^2 + y^2 = 4$ **36.** $x^2 - y^2 = 1$

37. $\dfrac{x^2}{9} + \dfrac{y^2}{4} = 1$ **38.** $xy = 2$

39. $y^2 = 4x$ **40.** $x^2 - y^2 = 25\sqrt{x^2 + y^2}$

In Problems 41–44, use the trigonometric formulas in Eqs. (11a–e) of Article 2.7 to expand the left-hand side of each equation. Then replace the resulting polar equations by an equivalent cartesian equation, and sketch the graph.

41. $r\cos(\theta - 60°) = 3$ **42.** $r\sin(\theta + 45°) = 4$

43. $r\sin(45° - \theta) = \sqrt{2}$ **44.** $r\cos(30° - \theta) = 0$

45. Show that $(2, \frac{3}{4}\pi)$ is on the curve $r = 2\sin 2\theta$.

46. Show that $(\frac{1}{2}, \frac{3}{2}\pi)$ is on the curve $r = -\sin(\theta/3)$.

47. Show that the equations $r = \cos\theta + 1$, $r = \cos\theta - 1$, represent the same curve.

In Problems 48–52, find some intersections of the pairs of curves (a = constant).

48. $r^2 = 2a^2\sin 2\theta, \quad r = a$

49. $r = a\sin\theta, \quad r = a\cos\theta$

50. $r = a(1 + \cos\theta), \quad r = a(1 - \sin\theta)$

51. $r = a(1 + \sin\theta), \quad r = 2a\cos\theta$

52. $r = a\cos 2\theta, \quad r = a(1 + \cos\theta)$

53. The simultaneous solution of the equations

$$r^2 = 4a^2\cos\theta, \tag{15}$$
$$r = a(1 - \cos\theta), \tag{16}$$

in the text did not reveal the points $(0, 0)$ and $(2a, \pi)$ in which their graphs intersected.

a) We could have found the point $(2a, \pi)$, however, by replacing the (r, θ) in Eq. (15) by the equivalent $(-r, \theta + \pi)$, to obtain

$$r^2 = 4a^2 \cos \theta,$$
$$(-r)^2 = 4a^2 \cos (\theta + \pi),$$
$$r^2 = -4a^2 \cos \theta. \tag{15'}$$

Solve Eqs. (15′) and (16) simultaneously to show that $(2a, \pi)$ is a common solution. (This will still not reveal that the graphs intersect at $(0, 0)$.)

b) The origin is still a special case. (It often is.) One way to handle it is the following: Set $r = 0$ in Eqs. (15) and (16) and solve each equation for a corresponding value of θ. Since $(0, \theta)$ is the origin for *any* θ, this will show that both curves pass through the origin even if they do so for different θ-values.

10.2 Graphs of Polar Equations

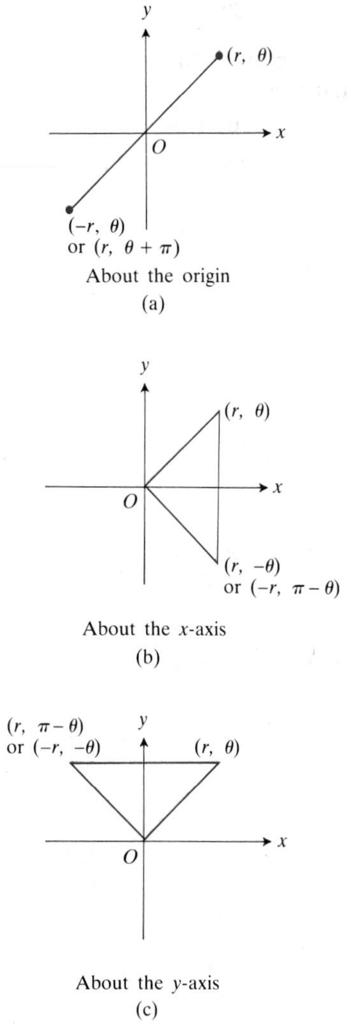

10.9 Some of the polar coordinate tests for symmetry.

Symmetry

The graph of an equation

$$F(r, \theta) = 0$$

consists of all those points whose coordinates (in some form) satisfy the equation. Frequently the equation gives r explicitly in terms of θ, as

$$r = f(\theta).$$

As many points as desired may then be obtained by substituting values of θ and calculating the corresponding values of r. In particular, it is desirable to plot the points where r is a maximum or a minimum and to find the values of θ at which the curve passes through the origin, if that occurs.

Certain types of *symmetry* are readily detected. For example, the curve is

a) symmetric about the origin if the equation is unchanged when r is replaced by $-r$, or when θ is replaced by $\theta + \pi$;

b) symmetric about the x-axis if the equation is unchanged when θ is replaced by $-\theta$, or the pair (r, θ) by the pair $(-r, \pi - \theta)$;

c) symmetric about the y-axis if the equation is unchanged when θ is replaced by $\pi - \theta$, or the pair (r, θ) by the pair $(-r, -\theta)$.

These are readily verified by considering the symmetrically located points in Fig. 10.9.

Particular Curves

EXAMPLE 1 *Cardioids.* Discuss and sketch the curve $r = a(1 - \cos \theta)$, where a is a positive constant.

Discussion Since $\cos(-\theta) = \cos \theta$, the equation is unaltered when θ is replaced by $-\theta$; hence the curve is symmetric about the x-axis (Fig. 10.9b).

θ	r
0	0
$\frac{\pi}{3}$	$\frac{a}{2}$
$\frac{\pi}{2}$	a
$\frac{2\pi}{3}$	$\frac{3a}{2}$
π	$2a$

(a)

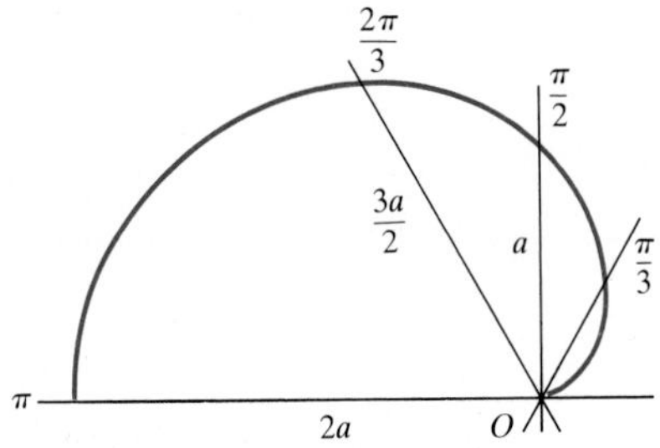

(b)

10.10 (a) Values of $r = a(1 - \cos\theta)$ for selected values of θ. (b) A smooth curve sketched through the points from part (a).

Also, since

$$-1 \le \cos\theta \le 1,$$

the values of r vary between 0 and $2a$. The minimum value, $r = 0$, occurs at $\theta = 0$, and the maximum value, $r = 2a$, occurs at $\theta = \pi$. Moreover, as θ varies from 0 to π, $\cos\theta$ decreases from 1 to -1; hence $1 - \cos\theta$ increases from 0 to 2. That is, r increases from 0 to $2a$ as the radius vector OP swings from $\theta = 0$ to $\theta = \pi$. We make a table of values (Fig. 10.10a) and plot the corresponding points. Now we sketch a smooth curve through them (Fig. 10.10b) in such a way that r increases as θ increases, since

$$\frac{dr}{d\theta} = a\sin\theta$$

is positive for $0 < \theta < \pi$. Then we exploit the symmetry of the curve and reflect this portion across the x-axis. The result is the curve shown in Fig. 10.11(a), which is called a *cardioid* because of its heart-shaped appearance.

We may investigate the slope of the cardioid at the origin as follows: Let P be a point on the curve in the first quadrant, where P is destined to approach O along the curve (see Fig. 10.12). Then, as $P \to O$, the slope of OP ($= \tan\theta$) approaches the slope of the tangent at O. Since $P \to O$ as $\theta \to 0$,

$$\lim_{\theta\to 0} (\text{slope of } OP) = \lim_{\theta\to 0} \tan\theta = 0.$$

That is, the slope of the tangent to the curve at the origin is zero. □

Slope at the Origin

The process by which we found the tangent to the cardioid at the origin at the end of Example 1 works for any smooth curve through the origin. If the curve passes through the origin when $\theta = \theta_0$, then the discussion in Example 1 would be modified only to the extent of saying that $P \to O$

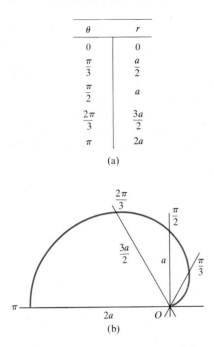

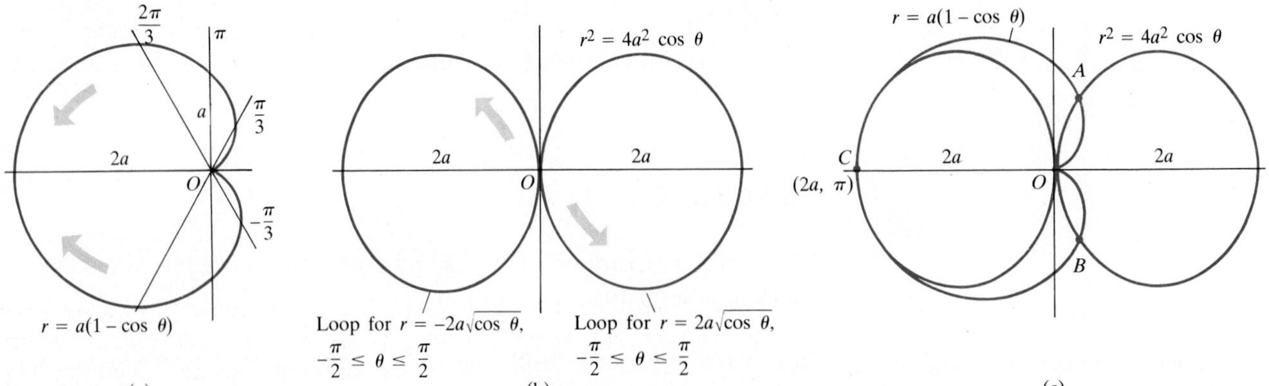

(a) (b) (c)

10.11 (a) $r = a(1 - \cos\theta)$. (b) $r^2 = 4a^2\cos\theta$. (c) The four points of intersection of the curves in (a) and (b). Only two of the four (A and B) can be found by simultaneous solution. The intersections at O and C are disclosed here by graphing. The arrows show the directions in which the curves are traced as θ increases. In (b), the values of θ for which $\cos\theta$ is negative do not yield real values of r.

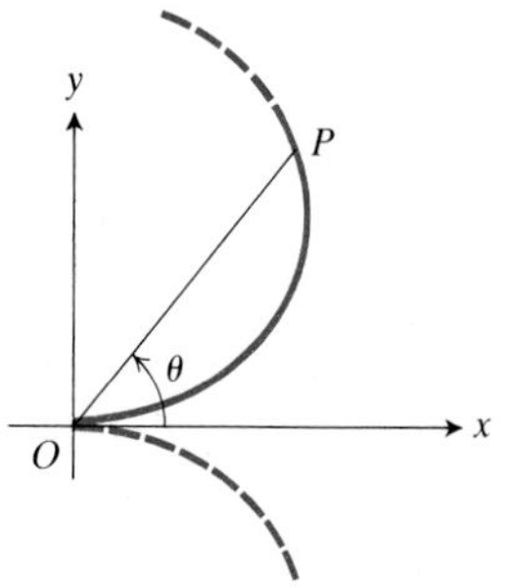

10.12 As $P \to O$ along the cardioid, $\theta \to 0$. At the origin, the cardioid has a horizontal tangent.

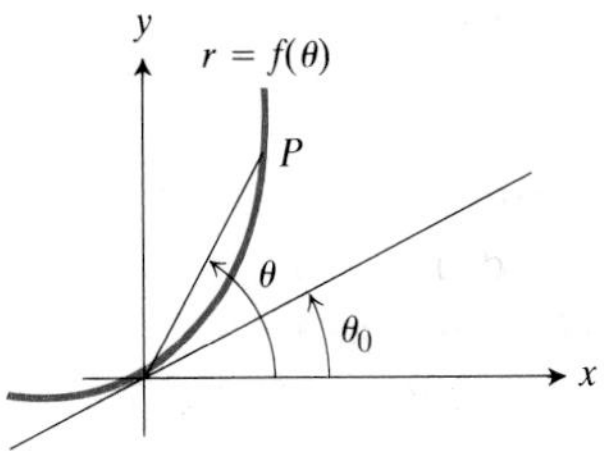

10.13 If the curve $r = f(\theta)$ passes through the origin at $\theta = \theta_0$, and if f has a derivative at $\theta = \theta_0$, then the line $\theta = \theta_0$ is tangent to the curve at the origin.

along the curve as $\theta \to \theta_0$, and hence

$$\left(\frac{dy}{dx}\right)_{\theta=\theta_0} = \lim_{\theta\to\theta_0} (\tan\theta) = \tan\theta_0.$$

But $(dy/dx)_{\theta=\theta_0}$ is also the tangent of the angle between the x-axis and the curve at this point. Hence the line $\theta = \theta_0$ is tangent to the curve at the origin. In other words, whenever a curve passes through the origin for a value θ_0 of θ and the derivative $dr/d\theta$ exists at that point, it does so *tangent* to the line $\theta = \theta_0$. See Fig. 10.13.

EXAMPLE 2 $r^2 = 4a^2 \cos\theta$. This curve is symmetric about the origin. Two values,

$$r = \pm 2a\sqrt{\cos\theta},$$

correspond to each value of θ for which $\cos\theta > 0$, namely,

$$-\frac{\pi}{2} < \theta < \frac{\pi}{2}.$$

Furthermore, the curve is symmetric about the x-axis, since θ may be replaced by $-\theta$ without altering the value of $\cos\theta$. The curve passes through the origin at $\theta = \pi/2$ and is tangent to the y-axis at this point. Since $\cos\theta$ never exceeds unity, the maximum value of r is $2a$, which occurs at $\theta = 0$. As θ increases from 0 to $\pi/2$, $|r|$ decreases from $2a$ to 0. The curve is sketched in Fig. 10.11(b). □

A Technique for Graphing

One way to graph a polar equation $r = f(\theta)$ is to make a table of $r\theta$-values, plot the corresponding points, and connect them in order of increasing θ. This can work well if enough points have been plotted to reveal all the loops and dimples in the graph. In this section we describe another method of graphing that is usually quicker and more reliable. The steps in the new method are

1. first graph $r = f(\theta)$ in the *cartesian* $r\theta$-plane,
2. then use the cartesian graph as a "table" and guide to sketch the *polar* coordinate graph.

This method is better than simple point plotting because the first cartesian graph, even when hastily drawn, shows at a glance where r is positive, negative, and nonexistent, as well as where r is increasing and decreasing. As examples, we graph the functions $r = 1 + \cos(\theta/2)$ and $r^2 = \sin 2\theta$.

EXAMPLE 3 Graph the curve

$$r = 1 + \cos\frac{\theta}{2}.$$

Solution We first graph $r = 1 + \cos(\theta/2)$ in the cartesian $r\theta$-plane. See Fig. 10.14. The arrows from the θ-axis to the curve give the radius vectors for graphing $r = 1 + \cos(\theta/2)$ in the polar plane, as shown in Fig. 10.15. □

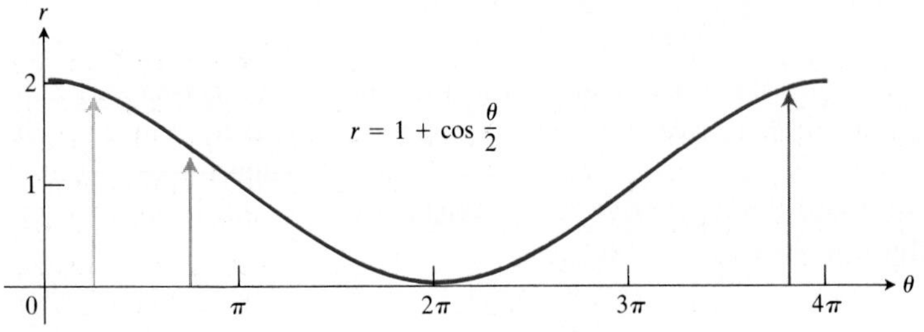

Figure 10.14

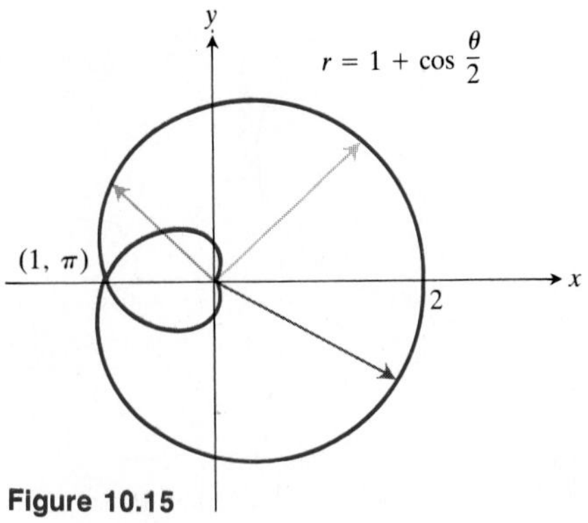

Figure 10.15

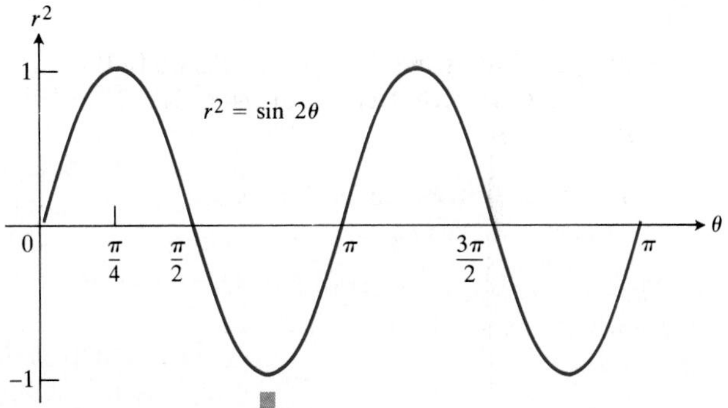

Figure 10.16

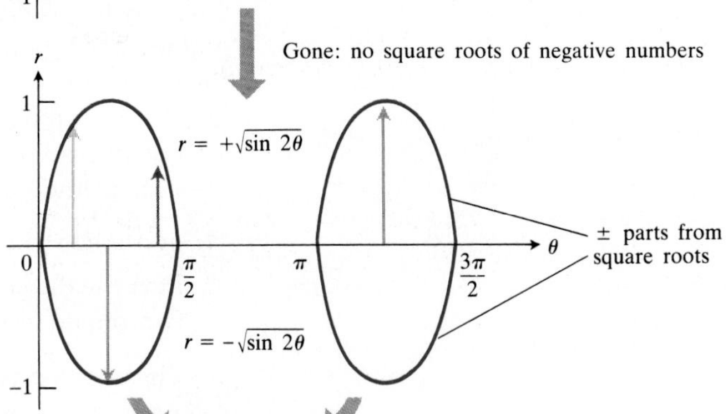

Figure 10.17

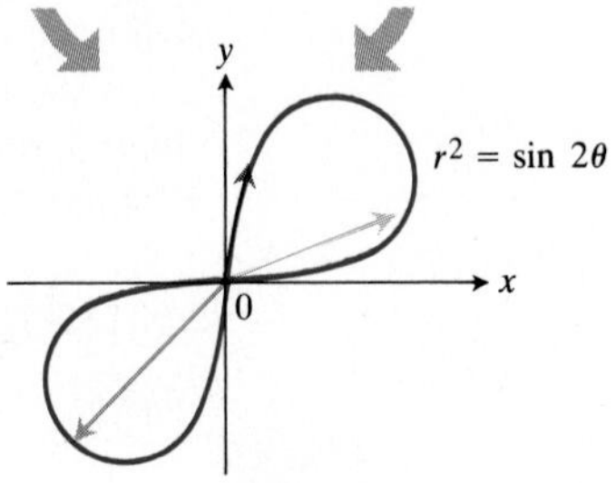

Figure 10.18

EXAMPLE 4 Graph the curve

$$r^2 = \sin 2\theta.$$

Solution Here we begin by plotting r^2 (not r) as a function of θ in the cartesian $r^2\theta$-plane. See Fig. 10.16. We pass from there to the graph of $r = \pm\sqrt{\sin 2\theta}$ in the $r\theta$-plane (Fig. 10.17), and then draw the polar graph (Fig. 10.18). The graph in Fig. 10.17 "covers" the final polar graph in Fig. 10.18 twice. We could have managed with either loop alone, with the two upper halves, or with the two lower halves. The double covering does no harm, however, and we actually learn a little more about the behavior of the function this way. □

PROBLEMS

1. Find the polar form of the equation of the line $3x + 4y = 5$.
2. If the polar coordinates (r, θ) of a point P satisfy the equation $r = 2a \cos \theta$, what equation is satisfied by the cartesian coordinates (x, y) of P?

Discuss and sketch the curves in Problems 3–10.

3. $r = a(1 + \cos \theta)$
4. $r = a(1 - \sin \theta)$
5. $r = a \sin 2\theta$
6. $r^2 = 2a^2 \cos 2\theta$
7. $r = a(2 + \sin \theta)$
8. $r = a(1 + 2 \sin \theta)$
9. $r = \theta$
10. $r = a \sin (\theta/2)$
11. Graph the following limaçons (limaçon, "leema-sahn," is an old French word for snail).
    a) $r = \cos \theta + 2$
    b) $r = \frac{3}{2} \cos \theta + \frac{3}{2}$
    c) $r = 2 \cos \theta + 1$
12. In how many distinct points do the curves $r = 1 - \cos \theta$ and $r = 1 + \cos \theta$ intersect?
13. In how many distinct points do the curves $r = 2 \sin 2\theta$ and $r = 1$ intersect?
14. Find the points on the curve $r = a(1 + \cos \theta)$, where the tangent is (a) parallel to the x-axis, (b) parallel to the y-axis. (*Hint:* Use the given equation to express $x = r \cos \theta$ and $y = r \sin \theta$ in terms of θ. Then calculate dy/dx from the chain rule

$$dy/dx = (dy/d\theta)/(dx/d\theta).$$

15. Sketch the curves $r = a(1 + \cos \theta)$ and $r = 3a \cos \theta$ in one diagram and find the angle between their tangents at the point of intersection that lies in the first quadrant. (See Problem 14.)

> ***Toolkit* programs**
>
> Parametric Equations   Super * Grapher

10.3 Polar Equations of Conics and Other Curves

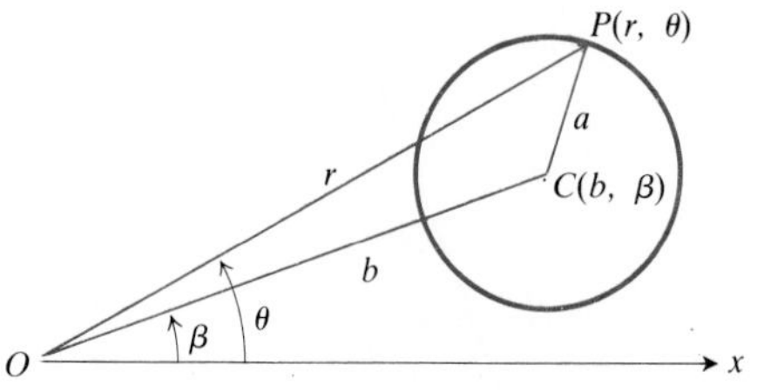

10.19 To find an equation for the circle shown above, apply the law of cosines to triangle *OCP*.

EXAMPLE 1 *Circles.* Find the polar equation of the circle of radius a with center at (b, β).

Solution We let $P(r, \theta)$ be a representative point on the circle and apply the law of cosines to the triangle OCP (Fig. 10.19) to obtain

$$a^2 = b^2 + r^2 - 2br \cos (\theta - \beta). \tag{1}$$

If the circle passes through the origin, then $b = a$ and the equation takes

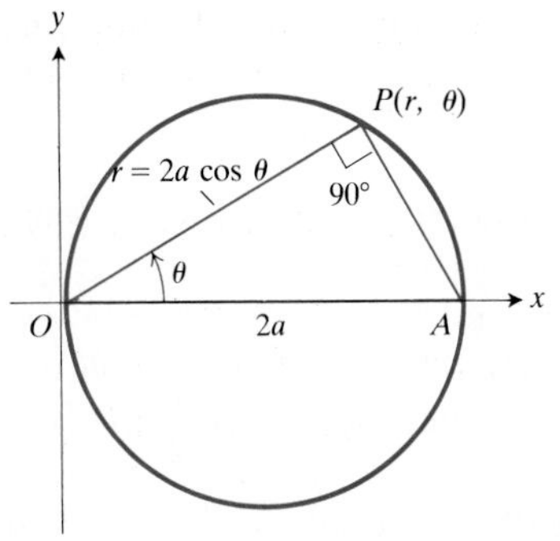

10.20 The circle $r = 2a \cos \theta$.

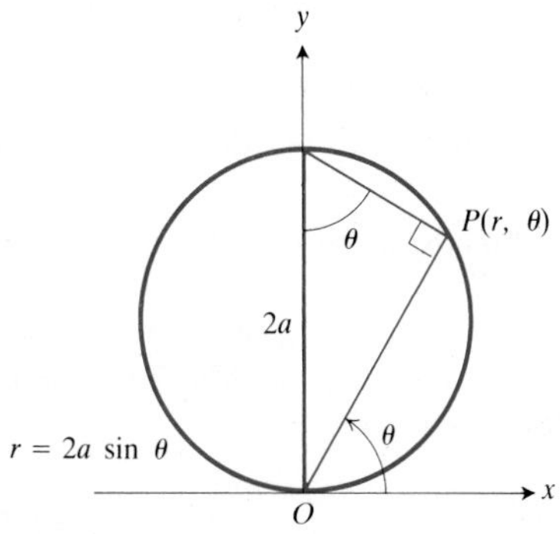

10.21 $r = 2a \sin \theta$ is the equation of a circle through the origin with its center on the positive y-axis.

the simpler form

$$r[r - 2a \cos (\theta - \beta)] = 0$$

or

$$r = 2a \cos (\theta - \beta). \tag{2}$$

In particular, if $\beta = 0$, Eq. (2) reduces to

$$r = 2a \cos \theta \tag{3}$$

(Fig. 10.20) while if $\beta = 90°$, so that the center of the circle lies on the y-axis (Fig. 10.21), Eq. (2) reduces to

$$r = 2a \sin \theta. \ \square \tag{4}$$

EXAMPLE 2 *Lines.* Suppose that the perpendicular from the origin to the line L meets L at the point $N(p, \beta)$. Find an equation for L.

Solution We let $P(r, \theta)$ be a typical point on L (Fig. 10.22) and read the result

$$\cos (\theta - \beta) = \frac{p}{r} \tag{5}$$

or

$$r \cos (\theta - \beta) = p \tag{6}$$

from the right triangle ONP.

If L were perpendicular to the x-axis, β would be 0 and Eq. (6) would reduce to

$$r \cos \theta = p \qquad \text{or} \qquad x = p. \tag{7}$$

In fact, if we rotate the axes through the angle β, as shown in Fig. 10.23, the new polar coordinates (r', θ') are related to the old polar coordinates by the equations

$$r' = r, \qquad \theta' = \theta - \beta. \tag{8}$$

When these are substituted for r and $(\theta - \beta)$ in Eq. (6), we get

$$r' \cos \theta' = p, \qquad \text{or} \qquad x' = p. \ \square$$

10.22 The relation $p/r = \cos (\theta - \beta)$, or $r \cos (\theta - \beta) = p$, can be read from triangle ONP in this picture.

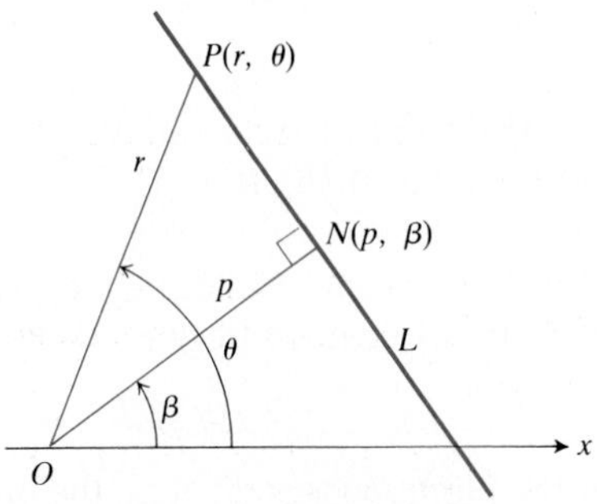

10.23 In the $x'y'$-coordinates shown here, the line L in Fig. 10.22 has the equation $x' = p$.

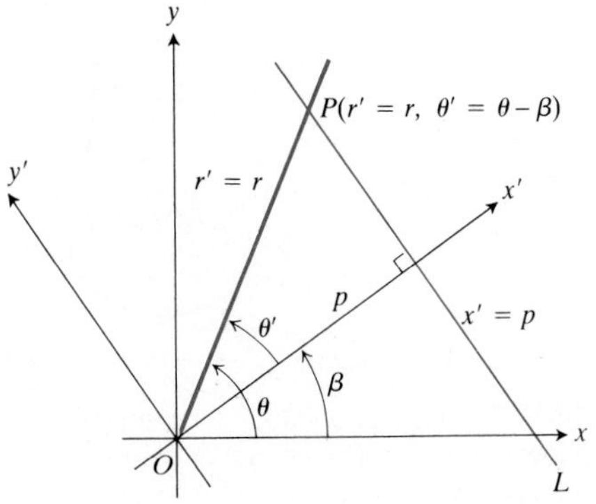

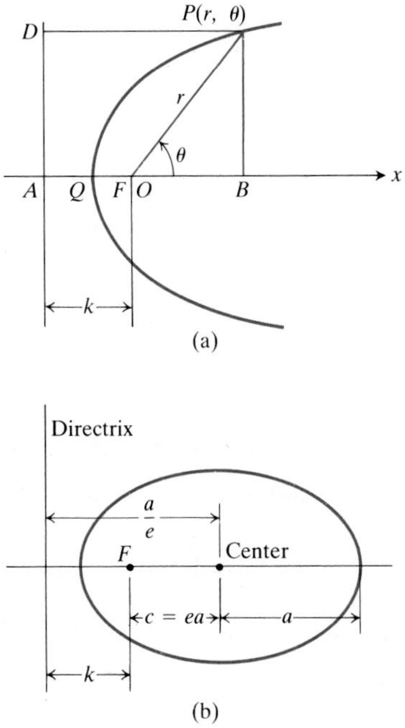

10.24 (a) If $PF = e \cdot PD$, then $r = e(k + r\cos\theta)$. When this is solved for r we get $r = ke/(1 - e\cos\theta)$. (b) In an ellipse with semimajor axis a, the distance from focus to directrix is $k = (a/e) - ae$, so that $ke = a(1 - e^2)$.

EXAMPLE 3 Parabolas, ellipses, and hyperbolas. Find the polar equation of the conic section of eccentricity e if the focus is at the origin and the associated directrix is the line $x = -k$.

Solution We adopt the notation of Fig. 10.24(a) and use the focus-and-directrix property

$$PF = e \cdot PD, \tag{9}$$

which allows us to handle the parabola, ellipse, and hyperbola all at the same time.

By taking the origin at the focus F, we have

$$PF = r,$$

while

$$PD = AB = AF + FB = k + r\cos\theta.$$

Then Eq. (9) is the same as

$$r = e(k + r\cos\theta). \tag{10}$$

If we solve this equation for r, we get

The Polar Equation for Conic Sections

$$r = \frac{ke}{1 - e\cos\theta}. \tag{11}$$

Equation (11) is the standard equation for the conic sections in polar coordinates. Typical special cases are obtained by taking

$$e = 1, \qquad r = \frac{k}{1 - \cos\theta} = \frac{k}{2}\csc^2\frac{\theta}{2}, \tag{12a}$$

which represents a *parabola;*

$$e = \frac{1}{2}, \qquad r = \frac{k}{2 - \cos\theta}, \tag{12b}$$

which represents an *ellipse;* and

$$e = 2, \qquad r = \frac{2k}{1 - 2\cos\theta}, \tag{12c}$$

which represents a *hyperbola.*

The denominator in Eq. (12b) for the ellipse can never vanish, so that r remains finite for all values of θ. But r becomes infinite as θ approaches 0 in (12a) and as θ approaches $\pi/3$ in (12c).

In terms of semimajor axis and eccentricity for an ellipse (Fig. 10.24b),

$$k = \frac{a}{e} - ae \qquad \text{or} \qquad ke = a(1 - e^2). \tag{13}$$

Therefore the equation for an ellipse with semimajor axis a and eccentricity e is

$$r = \frac{a(1 - e^2)}{1 - e\cos\theta}. \tag{14}$$

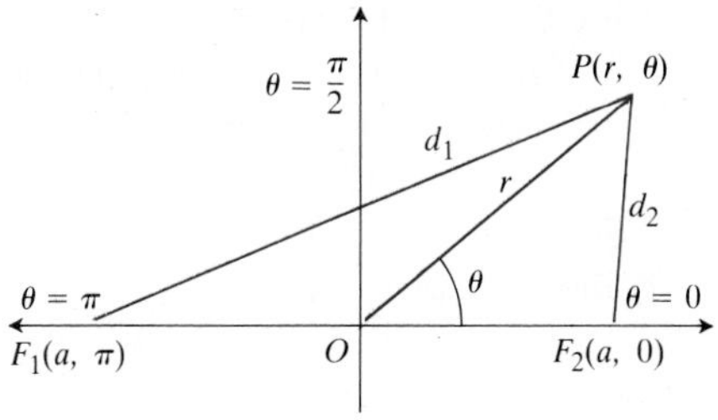

10.25 Example 4 shows that if P is constrained to move so that $PF_1 \cdot PF_2 = a^2$, then P traces out the curve $r^2 = 2a^2 \cos 2\theta$. The curve is shown in Fig. 10.26.

If we let e approach 0 in Eq. (14) we get $r = a$. That is, the circle of radius a, centered at the origin, is a limiting case of the ellipses represented by Eq. (14). □

EXAMPLE 4 *Lemniscates.* Find a polar equation for the path of a point P that moves in such a way that the product of its distances from the two points $F_1(a, \pi)$ and $F_2(a, 0)$ is equal to a^2. (Angles in radian measure.)

Solution We let $P(r, \theta)$ be a typical point on the curve. Then, with reference to Fig. 10.25, P must move in such a way that

$$d_1 d_2 = a^2, \tag{15}$$

where $d_1 = PF_1$, and $d_2 = PF_2$. We apply the law of cosines twice: once to triangle OPF_2,

$$d_2^2 = r^2 + a^2 - 2ar \cos \theta, \tag{16}$$

and again to triangle OPF_1,

$$\begin{aligned} d_1^2 &= r^2 + a^2 - 2ar \cos (\pi - \theta) \\ &= r^2 + a^2 + 2ar \cos \theta. \end{aligned} \tag{17}$$

Hence,

$$\begin{aligned} d_1^2 d_2^2 &= (r^2 + a^2 - 2ar \cos \theta)(r^2 + a^2 + 2ar \cos \theta) \\ &= (r^2 + a^2)^2 - 4a^2 r^2 \cos^2 \theta \\ &= r^4 + 2a^2 r^2 + a^4 - 4a^2 r^2 \cos^2 \theta \\ &= a^4 + r^4 + 2a^2 r^2 (1 - 2 \cos^2 \theta) \\ &= a^4 + r^4 - 2a^2 r^2 \cos 2\theta. \end{aligned} \tag{18}$$

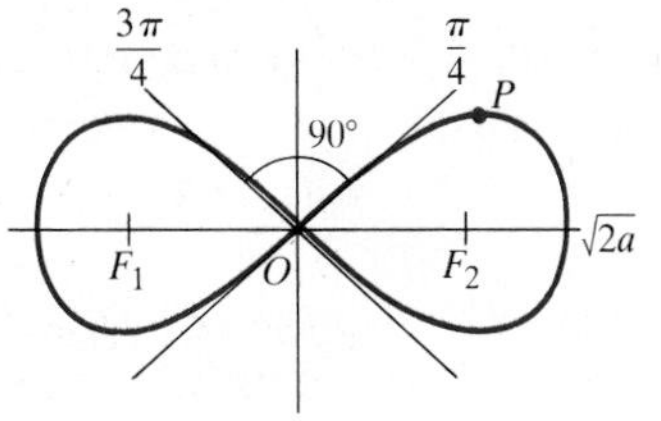

10.26 The lemniscate $r^2 = 2a^2 \cos 2\theta$.

Since $d_1^2 d_2^2 = a^4$, Eq. (18) reduces to

$$a^4 = a^4 + r^4 - 2a^2 r^2 \cos 2\theta.$$

After subtracting a^4 from both sides and dividing by r^2 (and noting that $r = 0$ is still a solution of the remaining equation), we find

$$r^2 = 2a^2 \cos 2\theta. \tag{19}$$

The graph of Eq. (19), called a *lemniscate*, is shown in Fig. 10.26.

The equivalent equation for a lemniscate with foci F_1 and F_2 on the y-axis is

$$r^2 = -2a^2 \cos 2\theta, \tag{20}$$

(Fig. 10.27). □

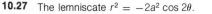

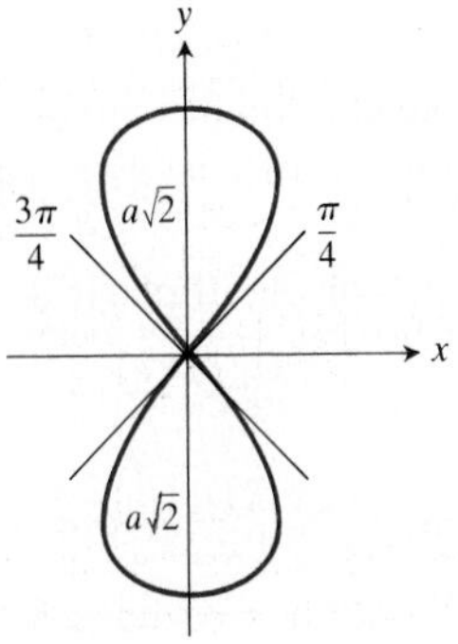

10.27 The lemniscate $r^2 = -2a^2 \cos 2\theta$.

PROBLEMS

In Problems 1–14, determine the cartesian equation and sketch the given curve.

1. $r = 4 \cos \theta$ **2.** $r = 6 \sin \theta$

3. $r = -2 \cos \theta$ **4.** $r = -2 \sin \theta$

5. $r = \sin 2\theta$ **6.** $r = \sin 3\theta$

7. $r^2 = 8 \cos 2\theta$ **8.** $r^2 = 4 \sin 2\theta$

9. $r = 8(1 - 2 \cos \theta)$ **10.** $r = 1/(2 - \cos \theta)$

11. $r = 4(1 - \cos \theta)$ **12.** $r(2 - 2 \cos \theta) = 4$

13. $r(3 - 6 \cos \theta) = 12$ **14.** $r(10 - 5 \cos \theta) = 25$

15. Sketch the following curves:
a) $r = 2 \cos (\theta + 45°)$, b) $r = 4 \csc (\theta - 30°)$,
c) $r = 5 \sec (60° - \theta)$, d) $r = 3 \sin (\theta + 30°)$,
e) $r = a + a \cos (\theta - 30°)$, f) $0 \le r \le 2 - 2 \cos \theta$.

Each of the graphs in Problems 16–23 is the graph of exactly one of the equations (a)–(l) in the list below. Find the equation for each graph.

a) $r = \cos 2\theta$ b) $r \cos \theta = 1$

c) $r = \dfrac{6}{1 - 2 \cos \theta}$ d) $r = \sin 2\theta$

e) $r = \theta$ f) $r^2 = \cos 2\theta$

g) $r = 1 + \cos \theta$ h) $r = 1 - \sin \theta$

i) $r = \dfrac{2}{1 - \cos \theta}$ j) $r^2 = \sin 2\theta$

k) $r = -\sin \theta$ l) $r = 2 \cos \theta + 1$

16. Four-leaved rose.

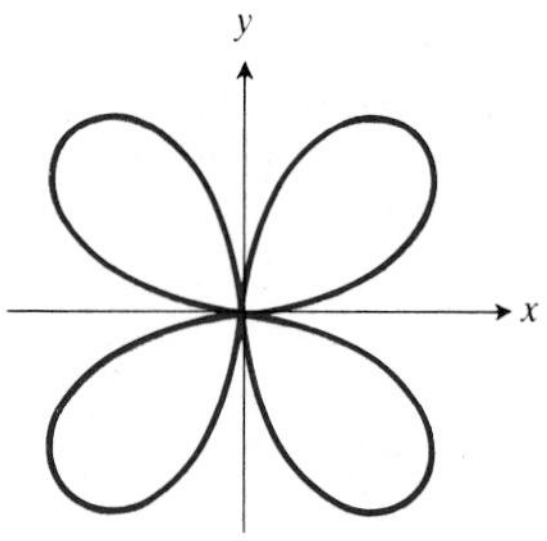

17. Spiral.

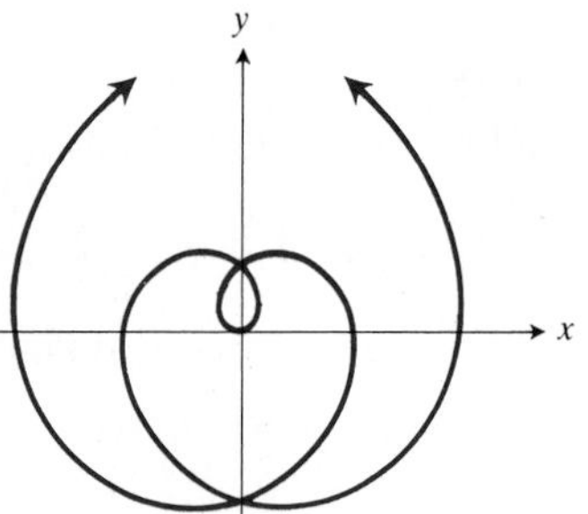

18. Limaçon.

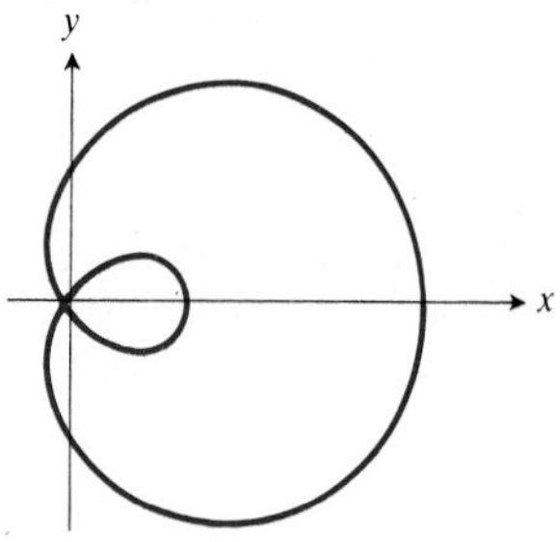

19. Lemniscate.

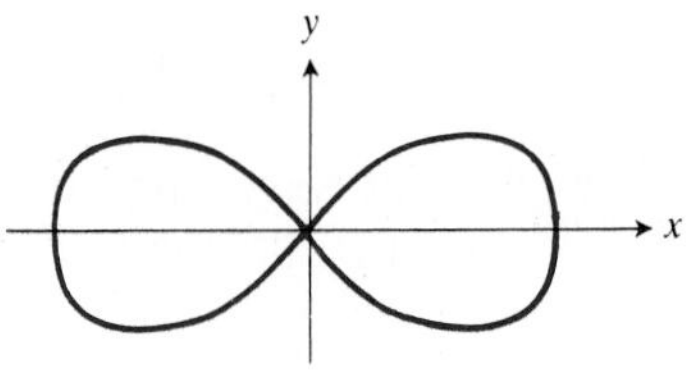

20. Circle.

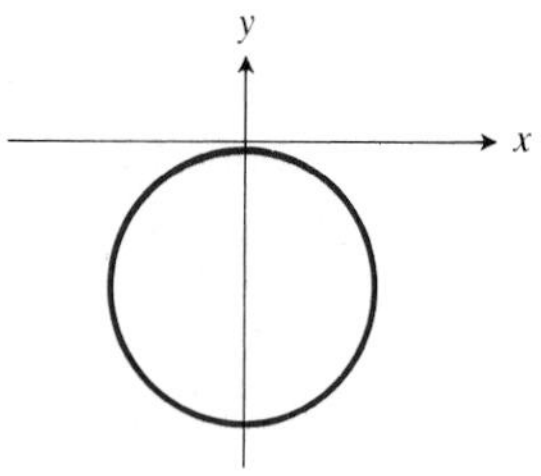

21. Cardioid.

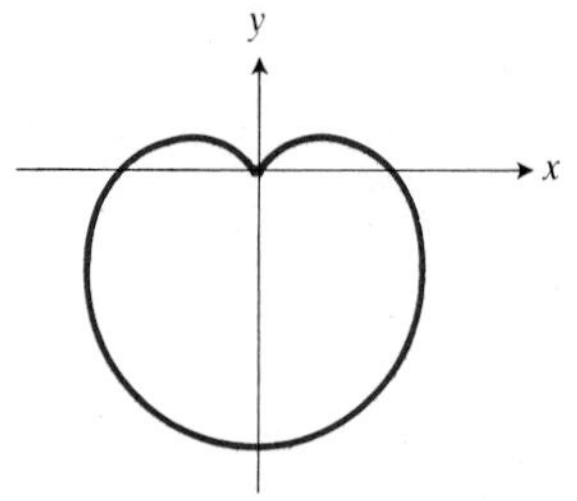

22. Parabola.

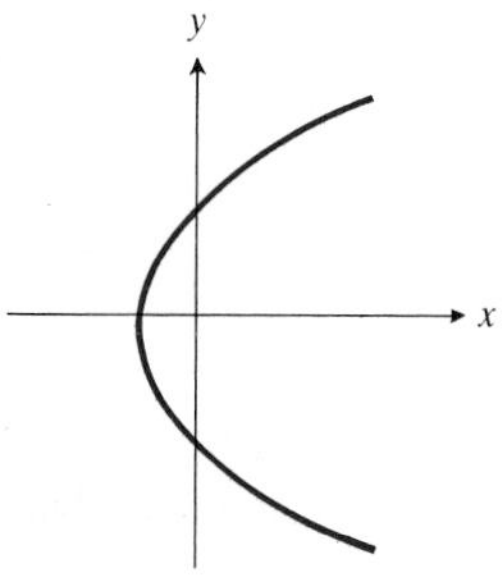

23. Lemniscate.

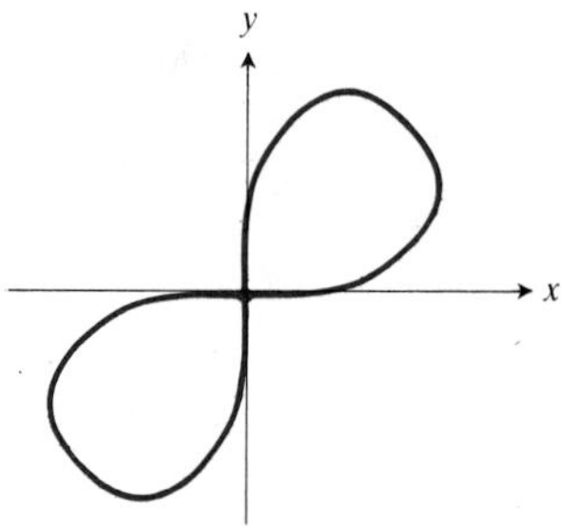

24. a) Find rectangular coordinates for the equations $r = 2 \sin \theta$ and $r = \csc \theta$.
b) Graph the equations in part (a) together, and label the points of intersection in both cartesian and polar coordinates.

25. Repeat Problem 24 for the curves $r = 2 \cos \theta$ and $r = \sec \theta$.

26. a) How can the angle β and the distance p (see Fig. 10.22 and Eq. 6) be determined from the cartesian equation $ax + by = c$ for a straight line?
b) Specifically, find p, β, and the polar equation for the straight line $\sqrt{3}x + y = 6$.

27. The focus of a parabola is at the origin and its directrix is the line $r \cos \theta = -4$. Find the polar equation of the parabola.

28. One focus of a hyperbola of eccentricity $\frac{5}{4}$ is at the origin and the corresponding directrix is the line $r \cos \theta = 9$. Find the polar coordinates of the second focus. Also determine the polar equation of the hyperbola and sketch.

29. One focus of an ellipse of eccentricity $\frac{1}{4}$ is at the origin. The corresponding directrix is the line $r \cos \theta = 8$. Find the coordinates of the other focus. Write a polar equation for the ellipse, and sketch.

30. Find the major axis (the length $2a$, in the notation of Article 8.6) of the hyperbola

$$r = \frac{6}{1 + 2\cos\theta}.$$

31. Determine the slopes of the lines that are tangent to the lemniscate $r^2 = 2a^2 \cos 2\theta$ at the origin.

32. Equation (12b), $r = k/(2 - \cos\theta)$, represents an ellipse with one focus at the origin. Sketch the curve and its directrices for the case $k = 2$ and locate the center of the ellipse.

33. In the case $e > 1$, Eq. (11) represents a hyperbola.
a) Show that the equation can also be written

$$r = \frac{a(e^2 - 1)}{1 - e\cos\theta}$$

where $2a$ is the distance between the vertices of the hyperbola.

b) From the polar form of the equation, determine the slopes of the asymptotes of the hyperbola of eccentricity e.

34. Write a polar equation for the parabola with focus at the origin and directrix the line $r\cos(\theta - \pi/2) = 2$.

In Problems 35–39, find a polar equation and sketch the given curve.

35. $x^2 + y^2 - 2ay = 0$

36. $(x^2 + y^2)^2 + 2ax(x^2 + y^2) - a^2y^2 = 0$

37. $x\cos\alpha + y\sin\alpha = p \qquad (\alpha, p \text{ constants})$

38. $y^2 = 4ax + 4a^2$

39. $(x^2 + y^2)^2 = x^2 - y^2$

***Toolkit* programs**

Parametric Equations   Super * Grapher

10.4 Integrals

Area in the Plane

The area AOB in Fig. 10.28 is bounded by the rays $\theta = \alpha$, $\theta = \beta$, and the curve $r = f(\theta)$. We imagine the angle AOB as being divided into n parts and we approximate the area in a typical sector POQ by the area of a *circular* sector of radius r and central angle $\Delta\theta$ (Fig. 10.29). That is,

$$\text{Area of } POQ \approx \tfrac{1}{2}r^2\,\Delta\theta,$$

and hence the entire area AOB is approximately

$$\sum_{\theta=\alpha}^{\beta} \tfrac{1}{2}r^2\,\Delta\theta.$$

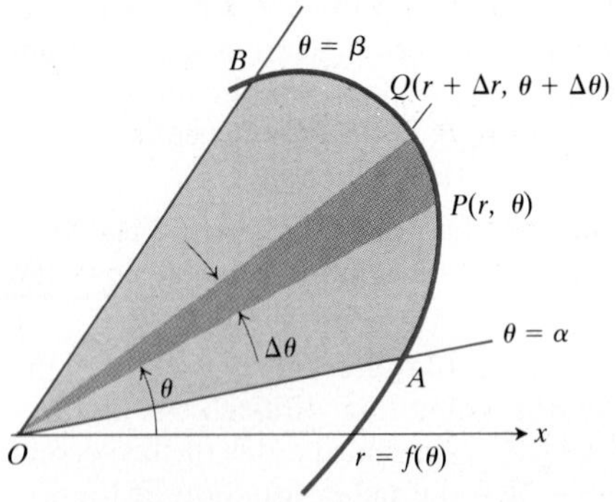

10.28 To derive a formula for the area swept out by the radius vector OP as P moves from A to B along the curve, the area is divided into sectors.

In fact, if the function $r = f(\theta)$, which represents the polar curve, is a continuous function of θ for $\alpha \le \theta \le \beta$, then there is a θ_k between θ and $\theta + \Delta\theta$ such that the circular sector of radius

$$r_k = f(\theta_k)$$

and central angle $\Delta\theta$ gives the *exact* area of POQ (Fig. 10.29). Then the entire area is given exactly by

$$A = \sum \tfrac{1}{2}r_k^2\,\Delta\theta = \sum \tfrac{1}{2}[f(\theta_k)]^2\,\Delta\theta.$$

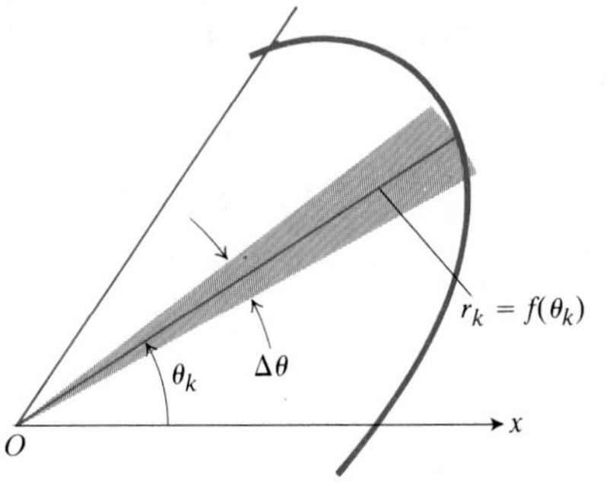

10.29 For some θ_k between θ and $\theta + \Delta\theta$, the area of the shaded circular sector just equals the area of the sector *POQ* bounded by the curve shown in Fig. 10.28.

If we let $\Delta\theta \to 0$, we see that

$$A = \lim_{\Delta\theta \to 0} \sum \tfrac{1}{2}[f(\theta_k)]^2\, \Delta\theta = \tfrac{1}{2}\int_\alpha^\beta [f(\theta)]^2\, d\theta,$$

or

$$A = \int_\alpha^\beta \tfrac{1}{2} r^2\, d\theta. \qquad (1)$$

This result may also be remembered as the integral of the differential element of area (Fig. 10.30):

$$dA = \tfrac{1}{2} r^2\, d\theta,$$

taken between the appropriate limits on θ.

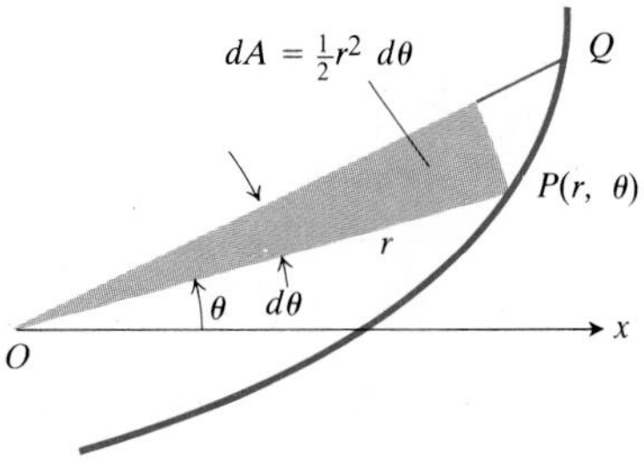

10.30 The area of the sector *POQ* is approximated by the area of the shaded circular sector.

EXAMPLE 1 Find the area that is inside the circle $r = a$ and outside the cardioid $r = a(1 - \cos\theta)$. (See Fig. 10.31.)

Solution We take a representative element of area

$$dA = dA_1 - dA_2 = \tfrac{1}{2} r_1^2\, d\theta - \tfrac{1}{2} r_2^2\, d\theta,$$

where

$$r_1 = a, \qquad r_2 = a(1 - \cos\theta).$$

Such elements of area belong to the region inside the circle and outside the cardioid provided θ lies between

$$-\frac{\pi}{2} \quad \text{and} \quad +\frac{\pi}{2},$$

where the curves intersect. Hence

$$A = \int_{-\pi/2}^{\pi/2} \frac{1}{2}(r_1^2 - r_2^2)\, d\theta = \int_{-\pi/2}^{\pi/2} \frac{a^2}{2}(2\cos\theta - \cos^2\theta)\, d\theta$$

$$= a^2 \int_{-\pi/2}^{\pi/2} \cos\theta\, d\theta - \frac{a^2}{2}\int_{-\pi/2}^{\pi/2} \frac{1 + \cos 2\theta}{2}\, d\theta = a^2\left(2 - \frac{\pi}{4}\right).$$

As a check against gross errors, we observe that this is roughly 80 percent of the area of a semicircle of radius a, and this seems reasonable when we look at the area in Fig. 10.31. □

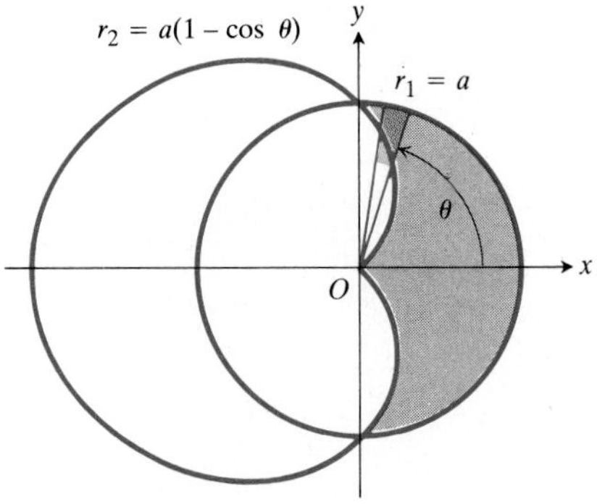

10.31 The shaded area is obtained from the integral $\int_{-\pi/2}^{\pi/2} \frac{1}{2}(r_1^2 - r_2^2)\, d\theta$.

EXAMPLE 2 Find the area inside the smaller loop of the limaçon

$$r = 2\cos\theta + 1.$$

Solution After sketching the curve (Fig. 10.32), we see that the smaller loop is traced out by the point (r, θ) as θ increases from $\theta = 2\pi/3$ to $\theta = 4\pi/3$. Since the curve is symmetric about the x-axis (the equation is

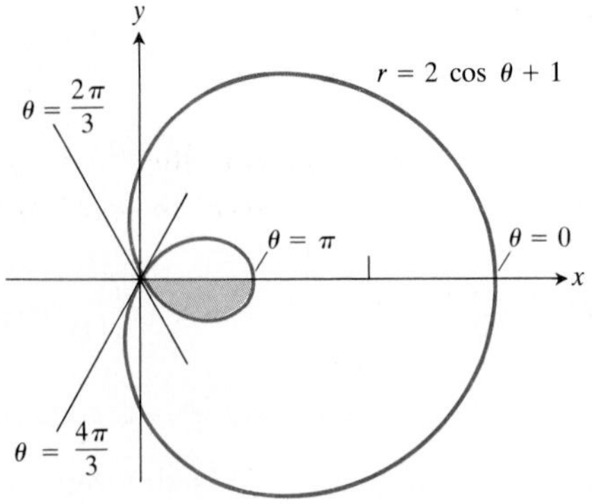

10.32 The limaçon in Example 2.

unaltered when we replace θ by $-\theta$), we may calculate the area of the shaded half of the inner loop by integrating from $\theta = 2\pi/3$ to $\theta = \pi$. The area A we seek will be twice the value of the resulting integral:

$$A = 2\int_{2\pi/3}^{\pi} \frac{1}{2} r^2\, d\theta = \int_{2\pi/3}^{\pi} r^2\, d\theta.$$

Since

$$\begin{aligned} r^2 &= (2\cos\theta + 1)^2 \\ &= 4\cos^2\theta + 4\cos\theta + 1 \\ &= 4\cdot\frac{1+\cos 2\theta}{2} + 4\cos\theta + 1 \qquad \text{(using a half-angle formula)} \\ &= 2 + 2\cos 2\theta + 4\cos\theta + 1 \\ &= 3 + 2\cos 2\theta + 4\cos\theta, \end{aligned}$$

we have

$$\begin{aligned} A &= \int_{2\pi/3}^{\pi} (3 + 2\cos 2\theta + 4\cos\theta)\, d\theta \\ &= [3\theta + \sin 2\theta + 4\sin\theta]_{2\pi/3}^{\pi} \\ &= [3\pi] - \left[2\pi - \frac{\sqrt{3}}{2} + 4\cdot\frac{\sqrt{3}}{2}\right] \\ &= \pi - \frac{3\sqrt{3}}{2}. \ \square \end{aligned}$$

Arc Length and Surface Area

One may obtain an expression for the arc length differential ds by squaring and adding the differentials

$$\begin{aligned} dx &= d(r\cos\theta) = -r\sin\theta\, d\theta + \cos\theta\, dr, \\ dy &= d(r\sin\theta) = r\cos\theta\, d\theta + \sin\theta\, dr. \end{aligned}$$

After the arithmetic is done we find that

$$ds^2 = dx^2 + dy^2 = r^2\, d\theta^2 + dr^2. \tag{2}$$

The result is easily remembered in the form

$$ds^2 = r^2\, d\theta^2 + dr^2 \tag{3}$$

if we refer to the "differential triangle" shown in Fig. 10.33. We simply think of dr and $r\, d\theta$ as the two legs and ds as the hypotenuse of an ordinary right triangle with the angle ψ opposite the side $r\, d\theta$.

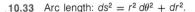

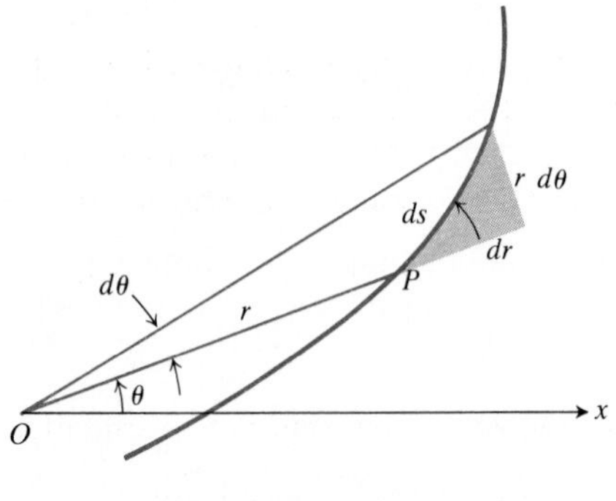

10.33 Arc length: $ds^2 = r^2\, d\theta^2 + dr^2$.

TO FIND THE LENGTH OF A POLAR CURVE:

1. find ds from the equation $ds^2 = r^2\, d\theta^2 + dr^2$, and
2. integrate between appropriate limits.

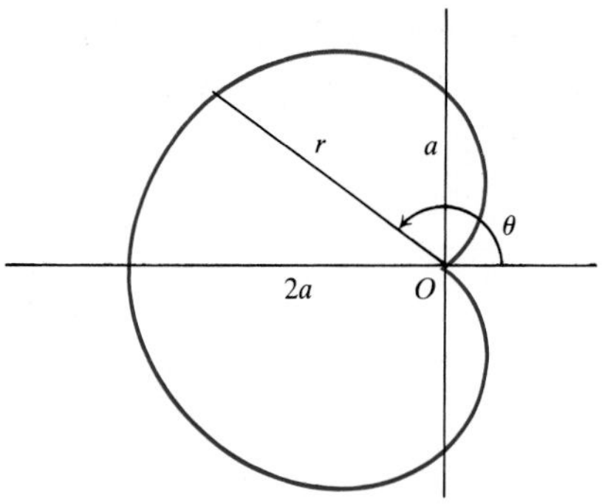

10.34 Cardioid: $r = a(1 - \cos\theta)$.

EXAMPLE 3 Find the length of the cardioid

$$r = a(1 - \cos\theta).$$

Solution We graph the cardioid (Fig. 10.34) to determine the limits of integration. As θ goes from 0 to 2π, we start at the origin, trace the cardioid once counterclockwise, and return to the origin.

To calculate the length of the curve we substitute

$$r = a(1 - \cos\theta), \qquad dr = a\sin\theta\, d\theta,$$

into Eq. (3) to obtain

$$\begin{aligned} ds^2 &= r^2\, d\theta^2 + dr^2 \\ &= a^2(1 - \cos\theta)^2\, d\theta^2 + a^2\sin^2\theta\, d\theta^2 \\ &= 2a^2\, d\theta^2[1 - \cos\theta] \end{aligned}$$

and

$$ds = a\sqrt{2}\sqrt{1 - \cos\theta}\, d\theta.$$

To integrate the expression on the right, we substitute

$$1 - \cos\theta = 2\sin^2\frac{\theta}{2}$$

so that

$$ds = 2a\left|\sin\frac{\theta}{2}\right|\, d\theta.$$

Since $\sin\theta/2$ is not negative when θ varies from 0 to 2π, we obtain

$$s = \int_0^{2\pi} 2a\sin\frac{\theta}{2}\, d\theta = -4a\cos\frac{\theta}{2}\Big]_0^{2\pi} = 8a.$$

If we wish to take advantage of the symmetry of the cardioid, we can calculate the length of the upper portion from $\theta = 0$ to $\theta = \pi$ and double the result:

Length of upper half: $\frac{s}{2} = \int_0^{\pi} 2a\sin\frac{\theta}{2} = -4a\cos\frac{\theta}{2}\Big]_0^{\pi} = 4a$

Length of cardioid: $s = 2(4a) = 8a.$

In letting θ range from 0 to 2π, we start at the cusp at the origin, go once around the smooth portion of the cardioid and return to the cusp. In doing this we do not pass *across* the cusp (Fig. 10.34). If, on the other hand, we take the cardioid $r = a(1 + \cos\theta)$, the appropriate procedure for avoiding the cusp (Fig. 10.35) is to let θ increase from $-\pi$ to π, or we may use the symmetry of the curve and calculate half of the total length by letting θ vary from 0 to π. □

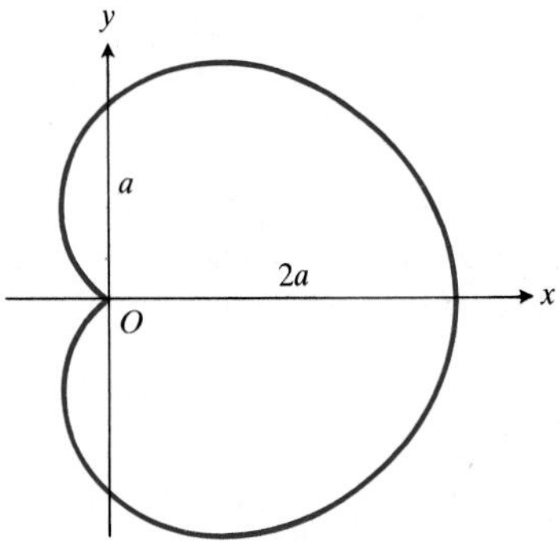

10.35 Cardioid: $r = a(1 + \cos\theta)$.

EXAMPLE 4 The lemniscate

$$r^2 = 2a^2\cos 2\theta$$

is revolved about the y-axis. Find the area of the surface generated.

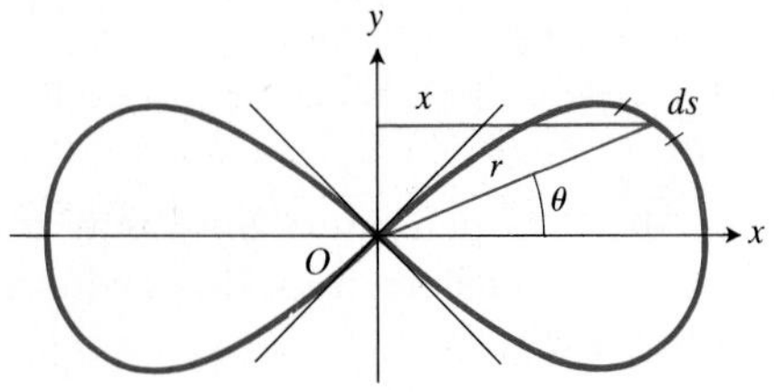

10.36 Lemniscate: $r^2 = 2a^2 \cos 2\theta$.

Solution An element of arc length ds (Fig. 10.36) generates a portion of surface area

$$dS = 2\pi x\, ds,$$

where

$$x = r\cos\theta, \qquad ds = \sqrt{dr^2 + r^2\, d\theta^2}.$$

That is,

$$dS = 2\pi r\cos\theta\sqrt{dr^2 + r^2\, d\theta^2} = 2\pi\cos\theta\sqrt{r^2\, dr^2 + r^4\, d\theta^2}.$$

From the equation of the curve, we get

$$r\, dr = -2a^2 \sin 2\theta\, d\theta.$$

Then

$$(r^2\, dr^2 + r^4\, d\theta^2) = (2a^2\, d\theta)^2(\sin^2 2\theta + \cos^2 2\theta)$$

and

$$dS = 4\pi a^2 \cos\theta\, d\theta.$$

The total surface area is generated by the loop of the lemniscate to the right of the y-axis between $\theta = -\pi/4$ and $\theta = +\pi/4$, so that

$$S = \int_{-\pi/4}^{\pi/4} 4\pi a^2 \cos\theta\, d\theta = 4\pi a^2\sqrt{2}. \ \square$$

PROBLEMS

NOTE. The identities

$$\sin^2\theta = \tfrac{1}{2}(1 - \cos 2\theta), \qquad \cos^2\theta = \tfrac{1}{2}(1 + \cos 2\theta)$$

may be used to evaluate $\int \sin^2\theta\, d\theta$ and $\int \sin^2\theta\, d\theta$ in some of the problems that follow.

Find the *total area* of the regions described in Problems 1–15. Sketch each region.

1. Inside the cardioid $r = a(1 + \cos\theta)$.
2. Inside the circle $r = 2a\sin\theta$.
3. Inside the lemniscate $r^2 = 2a^2\cos 2\theta$.
4. Inside the lemniscate $r^2 = 2a^2\cos 2\theta$ but not included in the circle $r = a$.
5. Inside the curve $r = a(2 + \cos\theta)$.
6. Common to the circles $r = 2a\cos\theta$ and $r = 2a\sin\theta$.
7. Inside the circle $r = 3a\cos\theta$ but outside the cardioid $r = a(1 + \cos\theta)$.
8. Inside the circle $r = -2\cos\theta$ and outside the circle $r = 1$.
9. Shared by the circles $r = a$ and $r = 2a\sin\theta$.
10. Shared by the cardioids $r = a(1 + \cos\theta)$ and $r = a(1 - \cos\theta)$.
11. Inside one leaf of the rose $r = \cos 2\theta$.
12. Inside one loop of the lemniscate $r^2 = 4\sin 2\theta$.
13. Inside the curve $r = 4 + 2\cos\theta$.
14. Inside the six-leaved rose $r^2 = 2a^2\sin 3\theta$.
15. a) Inside the large loop of the limaçon of Example 2.
    b) Inside the large loop of the limaçon but outside the small loop.
16. Since the center of gravity of a triangle is located on a median at a distance $\frac{2}{3}$ of the way from a vertex to the opposite base, the lever arm for the moment about the x-axis of the area of triangle POQ in Fig. 10.28 is $\frac{2}{3}r\sin\theta + \epsilon$, where $\epsilon \to 0$ as $\Delta\theta \to 0$. Deduce that the center of gravity of the region AOB in the figure is given by

$$\bar{y} = \frac{\int \frac{2}{3}r\sin\theta \cdot \frac{1}{2}r^2\, d\theta}{\int \frac{1}{2}r^2\, d\theta}, \qquad \bar{x} = \frac{\int \frac{2}{3}r\cos\theta \cdot \frac{1}{2}r^2\, d\theta}{\int \frac{1}{2}r^2\, d\theta},$$

with limits $\theta = \alpha$ to $\theta = \beta$ on all integrals.

17. Use the results of Problem 16 to find the center of gravity of the area bounded by the cardioid $r = a(1 + \cos\theta)$.
18. Use the results of Problem 16 to find the center of gravity of the area of a semicircle of radius a.

19. As usual, when we have a new formula, it is a good idea to try it out on familiar objects to be sure it calculates the results we want. Use the formula $ds^2 = r^2\,d\theta^2 + dr^2$ from Eq. (3) to calculate the circumferences of the following circles:
a) $r = a$, b) $r = a\cos\theta$, c) $r = a\sin\theta$.

20. Find the length of the cardioid $r = a(1 + \cos\theta)$. (*Hint:* $\int \sqrt{1 + \cos\theta}\,d\theta = \int \sqrt{2}\,|\cos(\theta/2)|\,d\theta$.)

21. Find the length of the curve $r = a\sin^2(\theta/2)$ from $\theta = 0$ to $\theta = \pi$. Sketch the curve.

22. Find the length of the parabolic spiral $r = a\theta^2$ between $\theta = 0$ and $\theta = \pi$. Sketch the curve.

23. Find the length of the curve $r = a\sin^3(\theta/3)$ between $\theta = 0$ and $\theta = \pi$. Sketch the curve.

24. A thin, uniform wire is bent into the shape of the cardioid $r = a(1 + \cos\theta)$. Find its center of gravity $(\bar{x}, \bar{y})$. (*Hint:* $\int \cos\theta\cos(\theta/2)\,d\theta$ can be evaluated by substituting $\cos\theta = 1 - 2\sin^2(\theta/2)$ and then letting $u = \sin(\theta/2)$.)

25. The lemniscate $r^2 = 2a^2\cos 2\theta$ is rotated about the x-axis. Find the area of the surface generated.

26. Find the surface area generated by rotating the curve $r = 2a\cos\theta$ about the line $\theta = \pi/2$. Sketch.

27. Find the surface area generated by rotating the first-quadrant portion of the curve $r = 1 + \cos\theta$ about the x-axis.

REVIEW QUESTIONS AND EXERCISES

1. Make a diagram to show the standard relations between cartesian coordinates (x, y) and polar coordinates (r, θ). Express each set of coordinates in terms of the other kind.

2. If a point has polar coordinates (r_1, θ_1), what other polar coordinates represent the point?

3. What is the expression for area between curves in polar coordinates?

4. What is the expression for length of curve in polar coordinates? For area of a surface of revolution?

5. What are some criteria for symmetry of a curve if its polar coordinates satisfy the equation $r = f(\theta)$? Illustrate your discussion with specific examples.

6. A satellite is in an orbit that passes over the North and South Poles of the earth. When it is over the North Pole it is at the highest point of its orbit, 1000 miles above the earth's surface. Above the South Pole it is at the lowest point of its orbit, 300 miles above the earth's surface.

a) Assuming that the orbit (with reference to the earth) is an ellipse with one focus at the center of the earth, find its eccentricity. (Take the diameter of the earth to be 8000 miles.)

b) Using the north–south axis of the earth as the x-axis, and the center of the earth as origin, find a polar equation of the orbit.

MISCELLANEOUS PROBLEMS

Discuss and sketch each of the curves in Problems 1–14 (where a is a positive constant).

1. $r = a\theta$

2. $r = a(1 + \cos 2\theta)$

3. a) $r = a\sec\theta$ b) $r = a\csc\theta$
c) $r = a\sec\theta + a\csc\theta$

4. $r = a\sin\left(\theta + \dfrac{\pi}{3}\right)$

5. $r^2 + 2r(\cos\theta + \sin\theta) = 7$

6. $r = a\cos\theta - a\sin\theta$

7. $r\cos\dfrac{\theta}{2} = a$

8. $r^2 = a^2\sin\theta$

9. $r^2 = 2a^2\sin 2\theta$

10. $r = a(1 - 2\sin 3\theta)$

11. a) $r = \cos 2\theta$ b) $r^2 = \cos 2\theta$

12. a) $r = 1 + \cos\theta$ b) $r = \dfrac{1}{1 + \cos\theta}$

13. a) $r = \dfrac{2}{1 - \cos\theta}$ b) $r = \dfrac{2}{1 + \sin\theta}$

14. a) $r = \dfrac{1}{2 + \cos\theta}$ b) $r = \dfrac{1}{2 + \sin\theta}$

Sketch each of the pairs of curves in Problems 15–20, and find all points of intersection.

15. $r = a$, $r = 2a\sin\theta$

16. $r = a$, $r = a(1 - \sin\theta)$

17. $r = a\sec\theta$, $r = 2a\sin\theta$

18. $r = a\cos\theta$, $r = a(1 + \cos\theta)$

19. $r = a(1 + \cos 2\theta)$, $r = a \cos 2\theta$

20. $r^2 = 4 \cos 2\theta$, $r^2 = \sec 2\theta$

21. Find the equation, in polar coordinates, of a parabola whose focus is at $r = 0$ and whose vertex is at $r = 1$, $\theta = 0$.

22. Find the polar equation of the straight line with intercepts a and b on the lines $\theta = 0$, $\theta = \pi/2$.

23. Find the equation of a circle with center on the line $\theta = \pi$, of radius a, and passing through the origin.

24. Find the polar equation of a parabola with focus at the origin and vertex at $(a, \pi/4)$.

25. Find the polar equation of an ellipse with one focus at the origin, the other at (2, 0), and a vertex at (4, 0).

26. Find the polar equation of a hyperbola with one focus at the origin, center at $(2, \pi/2)$, and vertex at $(1, \pi/2)$.

27. Three LORAN stations are located (in polar coordinates) at $(a, 0)$, $(0, 0)$, and $(a, \pi/4)$. Radio signals are sent out from the three stations simultaneously. A ship receiving the signals notes that the signals from the second and third stations arrive $a/2v$ seconds later than that from the first. If v is the velocity of a radio signal, what is the location of the ship in polar coordinates? (Also see Article 8.6, Problem 15.)

28. The cardoid $r = a(1 - \cos \theta)$ is rotated about the x-axis.

a) Find the area of the surface generated. (*Hint:* You may use the fact that $\sin \theta = 2 \sin (\theta/2) \cos (\theta/2)$, and $1 - \cos \theta = 2 \sin^2 (\theta/2)$ to evaluate your integral.)

b) Set up the definite integral (or integrals) that would be used to find the centroid of the area in part (a).

c) Find the centroid of the area in part (a).

29. If $r = a \cos^3 (\theta/3)$, show that $ds = a \cos^2 (\theta/3)\, d\theta$ and determine the perimeter of the curve.

30. Find the area that lies inside the curve $r = 2a \cos 2\theta$ and outside the curve $r = a\sqrt{2}$.

31. Sketch the curves

$$r = 2a \cos^2 (\theta/2) \qquad \text{and} \qquad r = 2a \sin^2 (\theta/2),$$

and find the area they have in common.

Find the total area enclosed by each of the curves in Problems 32–38.

32. $r^2 = a^2 \cos 2\theta$

33. $r = a(2 - \cos \theta)$

34. $r = a(1 + \cos 2\theta)$

35. $r = 2a \cos \theta$

36. $r = 2a \sin 3\theta$

37. $r^2 = 2a^2 \sin 3\theta$

38. $r^2 = 2a^2 \cos^2 (\theta/2)$

39. Find the area that is inside the cardioid $r = a(1 + \sin \theta)$ and outside the circle $r = a \sin \theta$.

40. Find the area bounded by one loop of the lemniscate $r^2 = 9 \cos 2\theta$.

The angle between the radius vector and the tangent line

In cartesian coordinates, when we want to discuss the direction of a curve at a point, we use the angle ϕ from the positive x-axis to the tangent line. In polar coordinates, it is more convenient to make use of the angle ψ (psi) from the *radius vector* to the tangent line. Then the relationship

$$\phi = \theta + \psi, \tag{1}$$

which can be read from Fig. 10.37, makes it a simple matter to find ϕ if that angle is desired instead of ψ.

Suppose the equation of the curve is given in the form $r = f(\theta)$, where $f(\theta)$ is a differentiable function of θ. Then from

$$x = r \cos \theta, \qquad y = r \sin \theta, \tag{2}$$

we see that x and y are differentiable functions of θ with

$$\frac{dx}{d\theta} = -r \sin \theta + \cos \theta \frac{dr}{d\theta}, \qquad \frac{dy}{d\theta} = r \cos \theta + \sin \theta \frac{dr}{d\theta}. \tag{3}$$

Since $\psi = \phi - \theta$ from (1),

$$\tan \psi = \tan (\phi - \theta) = \frac{\tan \phi - \tan \theta}{1 + \tan \phi \tan \theta},$$

while

$$\tan \phi = \frac{dy/d\theta}{dx/d\theta}, \qquad \tan \theta = \frac{y}{x}.$$

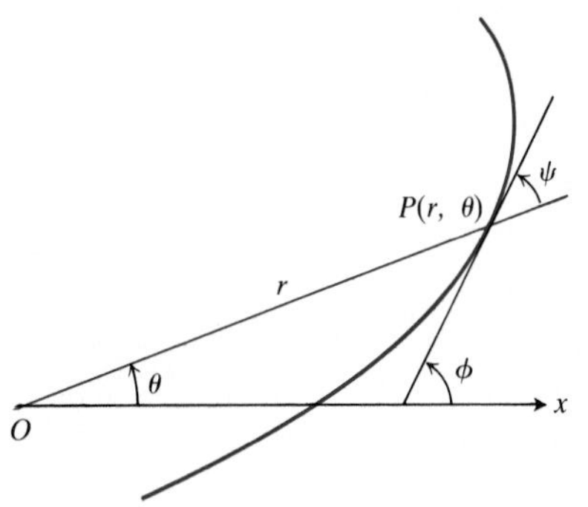

10.37 The angle ψ between the tangent line and the radius vector.

Hence

$$\tan\psi = \frac{\dfrac{dy/d\theta}{dx/d\theta} - \dfrac{y}{x}}{1 + \dfrac{y}{x}\dfrac{dy/d\theta}{dx/d\theta}} = \frac{x\dfrac{dy}{d\theta} - y\dfrac{dx}{d\theta}}{x\dfrac{dx}{d\theta} + y\dfrac{dy}{d\theta}}. \tag{4}$$

The numerator in the last expression in Eq. (4) is found by substitution from Eqs. (2) and (3) to be

$$x\frac{dy}{d\theta} - y\frac{dx}{d\theta} = r^2.$$

Similarly, the denominator is

$$x\frac{dx}{d\theta} + y\frac{dy}{d\theta} = r\frac{dr}{d\theta}.$$

When we substitute these into Eq. (4), we obtain the very simple final result

$$\tan\psi = \frac{r}{dr/d\theta}. \tag{5}$$

41. Find the value of $\tan\psi$ for the curve $r = \sin^4(\theta/4)$.

42. Find the angle between the curve $r = 2a\sin 3\theta$ and its tangent when $\theta = \pi/3$.

43. For the hyperbolic spiral $r\theta = a$ show that $\psi = 3\pi/4$ when $\theta = 1$ radian, and that $\psi \to \pi/2$ as the spiral winds around the origin. Sketch the curve and indicate ψ for $\theta = 1$ radian.

44. Show, by reference to a figure, that the angle β between the tangents to two curves at a point of intersection may be found from the formula

$$\tan\beta = \frac{\tan\psi_2 - \tan\psi_1}{1 + \tan\psi_2\tan\psi_1}. \tag{6}$$

When will the two curves intersect orthogonally?

45. Find a point of intersection of the parabolas

$$r = \frac{1}{1 - \cos\theta} \quad \text{and} \quad r = \frac{3}{1 + \cos\theta}$$

and find the angle between the tangents to these curves at this point.

46. Find points on the cardioid $r = a(1 + \cos\theta)$ where the tangent line is horizontal.

47. Show that the parabolas $r = a/(1 + \cos\theta)$, $r = b/(1 - \cos\theta)$ are orthogonal at each point of intersection ($ab \neq 0$).

48. Find the angle between the line $\theta = \pi/2$ and the cardioid $r = a(1 - \cos\theta)$ at their intersection.

49. Find the angle between the line $r = 3\sec\theta$ and the curve $r = 4(1 + \cos\theta)$ at one of their intersections.

50. Find the slope of the tangent line to the curve $r = a\tan(\theta/2)$ at $\theta = \pi/2$.

51. Check that the two curves $r = 1/(1 - \cos\theta)$ and $r = 3/(1 + \cos\theta)$ intersect at the point $(2, \pi/3)$. Find the angle between the tangents to these curves at this point.

52. The equation $r^2 = 2\csc 2\theta$ represents a curve in polar coordinates.
a) Sketch the curve.
b) Find the equation of the curve in rectangular coordinates.
c) Find the angle at which the curve intersects the line $\theta = \pi/4$.

53. A given curve cuts all rays θ = constant at the constant angle α.
a) Show that the area bounded by the curve and two rays $\theta = \theta_1$, $\theta = \theta_2$, is proportional to $r_2^2 - r_1^2$, where (r_1, θ_1) and (r_2, θ_2) are polar coordinates of the ends of the arc of the curve between these rays. Find the factor of proportionality.
b) Show that the length of the arc of the curve in part (a) is proportional to $r_2 - r_1$ and find the proportionality constant.

54. Let P be a point on the hyperbola $r^2\sin 2\theta = 2a^2$. Show that the triangle formed by OP, the tangent at P, and the initial line is isosceles.

11 Sequences and Infinite Series

11.1 Introduction

In this chapter and the next we deal with two related topics: *sequences* and *series*. We begin with an example that shows how a sequence of numbers can be generated by an iterative process like Picard's method for solving an equation.

EXAMPLE 1 Use Picard's method to find approximations to a solution of the equation $\sin x - x^2 = 0$.

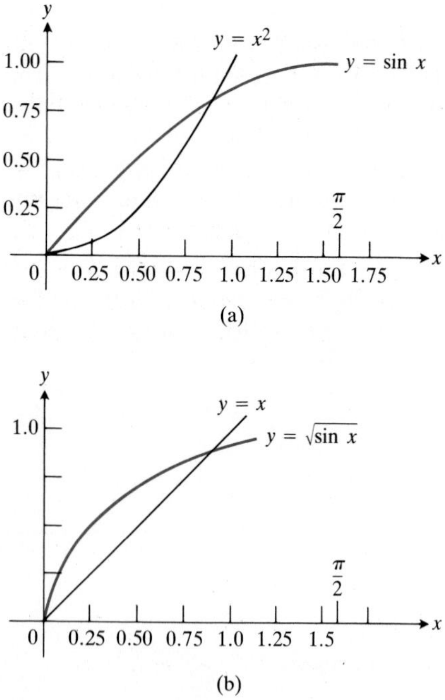

11.1 (a) The graphs of $y = x^2$ and $y = \sin x$ intersect near $x = 1$. (b) We solve $x = \sqrt{\sin x}$ by iteration with $x_0 = 1$ and $x_{n+1} = \sqrt{\sin x_n}$.

Solution (See Art. 2.10 for a review of Picard's method.) Figure 11.1 shows graphs of portions of the curves

$$y = \sin x \qquad \text{and} \qquad y = x^2.$$

The curves intersect at the origin and between $x = 0$ and $x = 1$. To estimate the latter intersection, we rewrite the equation in the form

$$x = \sqrt{\sin x}.$$

Let $x_0 = 1$, and define $x_1, x_2, x_3, \ldots$ by the iterative formula

$$x_{n+1} = \sqrt{\sin x_n} \qquad \text{for } n = 0, 1, 2, \ldots. \tag{1}$$

A computer or programmable calculator can easily be programmed to grind out

$$x_1 = 0.917\ldots$$
$$x_2 = 0.891\ldots$$
$$x_3 = 0.881\ldots$$

and so on. After each x_n there is a next term, x_{n+1}. In principle, the process could be continued indefinitely. In practice, we would probably stop with x_{19}, say, in this particular problem when we see that to nine decimal places, $x_{19} = x_{20} = x_{21}$. The numerical value that we got is

$$x_{19} = 0.876726216. \tag{2}$$

If we could carry out the calculations with no limit on the number of decimal places, we could get an unending sequence of numbers $x_1, x_2, x_3,$

$\ldots, x_n, \ldots$ that approach a limit that is the desired solution of $\sin x - x^2 = 0$. □

There are many examples of infinite sequences in mathematics—for example, the sequence of prime numbers:

$$2, 3, 5, 7, 11, 13, 17, 19, 23, 29, 31, \ldots.$$

In this case, we do not have a formula for the nth prime. Still simpler is the sequence of even numbers:

$$2, 4, 6, 8, \ldots, 2n, \ldots$$

where we do have a formula $x_n = 2n$ for the nth even number.

Sometimes the terms of a sequence are, or can be, formed by simple addition. For example, the decimal approximations to 1/3 are

$$0.3, \qquad 0.33 = 0.3 + 0.03, \qquad 0.333 = 0.3 + 0.03 + 0.003,$$

and so on. Here we could also define the terms of the sequence by saying that $x_1 = 0.3$ and $x_{n+1} = x_n + 3/10^{n+1}$.

This chapter deals primarily with sequences of constants. The next chapter will deal with sequences (and series, which are special kinds of sequences) of *functions*.

PROBLEMS

In Problems 1–5, write out in explicit form each of the terms $x_1, x_2, x_3, x_4, x_5, x_6$.

1. $x_0 = 1$, and $x_{n+1} = x_n + (\frac{1}{2})^{n+1}$

2. $x_0 = 1$, and $x_{n+1} = \dfrac{x_n}{n+1}$

3. $x_0 = 2$, and $x_{n+1} = \dfrac{x_n}{2}$

4. $x_0 = -2$, and $x_{n+1} = \dfrac{n}{n+1}x_n$

5. $x_0 = x_1 = 1$, and $x_{n+2} = x_n + x_{n+1}$. (This is called the Fibonacci sequence. The ratio $r_n = x_n/x_{n+1}$ gives a related sequence $r_1, r_2, \ldots$.)

6. If we try to solve the equation $\sin x = x^2$ by taking $x_0 = 1$ and $x_{n+1} = \sin^{-1}(x_n^2)$, what is x_1? x_2? Do you think this process will converge to an x that satisfies the given equation $\sin x = x^2$?

7. (*Calculator*) Put $x_0 = 1$ and let $x_{n+1} = x_n + \sin x_n - x_n^2$. When you get $x_{n+1} = x_n$, you have solved $\sin x_n - x_n^2 = 0$. Compare your answer with Eq. (2).

8. (*Calculator*) Use Picard's method to generate sufficiently many terms of the sequence $x_1, x_2, \ldots$, defined by $x_0 = 1$, $x_{n+1} = (\sin x_n)^{1/3}$, to estimate a root of the equation $\sin x = x^3$.

11.2 Sequences of Numbers

We continue our study of sequences of numbers with a definition.

DEFINITION 1

A *sequence* of numbers is a function whose domain is the set of positive integers.

of simplicity. A potential disadvantage, however, is the fact that several a_n's may turn out to be the same for different values of n, as they are in the sequence defined by the rule $a_n = 3$, in which every term is 3. On the other hand, the points (n, a_n) are distinct for different values of n.

As Fig. 11.2 shows, the sequences of Example 1 exhibit different kinds of behavior. The sequences $\{1/n\}$, $\{(-1)^{n+1}(1/n)\}$, and $\{1 - 1/n\}$ seem to approach single limiting values as n increases, and the sequence $\{3\}$ is already at a limiting value from the very first. On the other hand, terms of the sequence $\{(-1)^{n+1}(1 - 1/n)\}$ seem to accumulate near two different values, -1 and 1, while the terms of $\{n - 1\}$ get larger and larger and do not accumulate anywhere.

To distinguish sequences that approach a unique limiting value L, as n increases, from those that do not, we say that they *converge*, according to the following definition.

DEFINITION 2

The sequence $\{a_n\}$ converges to the number L if to every positive number ε there corresponds an index N such that

$$|a_n - L| < \varepsilon \qquad \text{for all} \quad n > N. \tag{4}$$

In other words, $\{a_n\}$ converges to L if, for every positive ε, there is an index N such that all terms after the Nth lie within a radius ε of L. Or, all but finitely many (namely the first N) terms of the sequence lie within a radius ε of L. (See Fig. 11.3 and look once more at the sequences in Fig. 11.2.) We indicate the fact that $\{a_n\}$ converges to L by writing

$$\lim_{n\to\infty} a_n = L \qquad \text{or} \qquad a_n \to L \quad \text{as } n \to \infty,$$

and we call L the *limit* of the sequence $\{a_n\}$. If no such limit exists, we say that $\{a_n\}$ *diverges.*

EXAMPLE 2 $\{1/n\}$ converges to 0.

Solution Let $\varepsilon > 0$ be given. We begin by writing down the inequality (4), with $a_n = 1/n$ and $L = 0$. This gives

$$|a_n - L| = \left|\frac{1}{n} - 0\right| = \frac{1}{n} < \varepsilon, \tag{5}$$

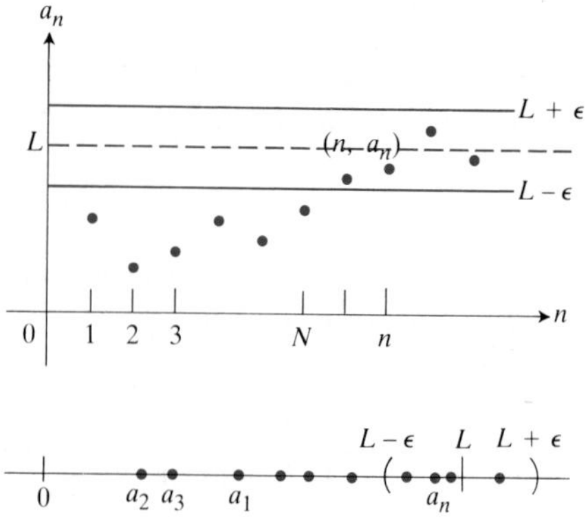

11.3 $a_n \to L$ if L is a horizontal asymptote of $\{(n, a_n)\}$. In this figure, all the a_n's after a_N lie within ϵ of L.

and therefore we seek an integer N such that

$$\frac{1}{n} < \varepsilon \qquad \text{for all} \quad n > N. \tag{6}$$

Certainly

$$\frac{1}{n} < \varepsilon \qquad \text{for all} \quad n > \frac{1}{\varepsilon}, \tag{7}$$

but there is no reason to expect $1/\varepsilon$ to be an integer. This minor difficulty is easily overcome: We just choose any integer $N > 1/\varepsilon$. Then every index n greater than N will automatically be greater than $1/\varepsilon$. In short, for this choice of N we can guarantee (6). The criterion set forth in Definition 2 for convergence to 0 is satisfied. □

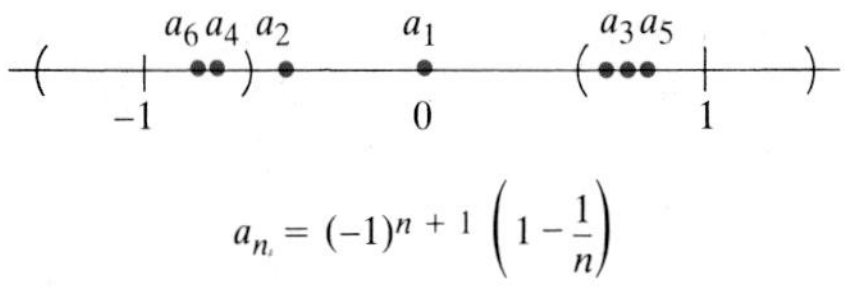

Neither the ε-interval about 1 nor the ε-interval about −1 contains a complete tail of the sequence.

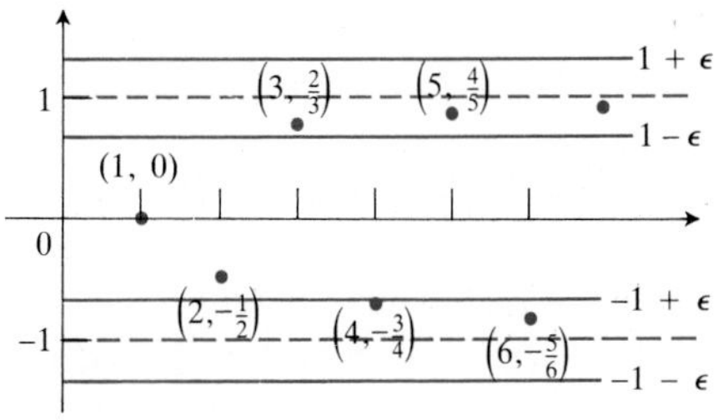

Neither of the ε-bands shown here contains all the points (n, a_n) from some index onward.

11.4 The sequence $\{(-1)^{n+1}[1 - (1/n)]\}$ diverges.

EXAMPLE 3 If k is any number, then the constant sequence $\{k\}$, defined by $a_n = k$ for all n, converges to k.

Solution Let $\varepsilon > 0$ be given. When we take both $a_n = k$ and $L = k$ on the left of the inequality in (4), we find

$$|a_n - L| = |k - k| = 0, \tag{8}$$

which is less than any positive ε for every $n \geq 1$. Hence $N = 1$ will work. □

EXAMPLE 4 The sequence $\{(-1)^{n+1}(1 - 1/n)\}$ diverges. To see why, pick a positive ε smaller than 1 so that the bands shown in Fig. 11.4 about the lines $y = 1$ and $y = -1$ do not overlap. Any $\varepsilon < 1$ will do. Convergence to 1 would require every point of the graph from some index on to lie inside the upper band, but this will never happen. As soon as a point (n, a_n) lies within the upper band, every alternate point starting with $(n + 1, a_{n+1})$ will lie within the lower band. Likewise, the sequence cannot converge to -1. On the other hand, because the terms of the sequence get increasingly close to 1 and -1 alternately, they never accumulate near any other value. □

A *tail* of a sequence $\{a_n\}$ is the collection of all the terms whose indices are greater than some index N; in other words, one of the sets $\{a_n | n > N\}$. Another way to say $a_n \to L$ is to say that every ε-interval about L contains a tail. The convergence (or divergence) of a sequence, as well as the limit of a convergent sequence, depends only on the tail behavior of the sequence.

As Example 4 suggests, a sequence cannot have more than one limit: there cannot be two different numbers L and L' such that every ε-interval about each one contains a complete tail. (See Problem 62.)

The behavior of the sequence $\{(-1)^{n+1}(1 - 1/n)\}$ is qualitatively different from that of $\{n - 1\}$, which diverges because it outgrows every real number L. We describe the behavior of $\{n - 1\}$ by writing

$$\lim_{n\to\infty} (n - 1) = \infty.$$

In speaking of infinity as a limit of a sequence $\{a_n\}$, we do not mean that the difference between a_n and infinity becomes small as n increases. We mean that a_n becomes numerically large as n increases.

When a sequence is defined iteratively, we may not have a formula that expresses the nth term as a function of n. For example, the sequence of numbers $\{x_n\}$ of Example 1 of Art. 11.1 appears to converge to a solution of the equation $\sin x - x^2 = 0$. But we have no explicit formula for x_n in terms of n and we don't yet have enough expertise to be sure that the sequence converges. We shall return to this point in Art. 11.6.

REMARK. Picard's method and Newton's method are familiar examples of schemes that define sequences iteratively. To return to the example of

$$\sin x - x^2 = 0,$$

we could base an iteration on the alternate form

$$x = \sin^{-1}(x^2)$$

leading to

$$x_{n+1} = \sin^{-1}(x_n^2),$$

which won't work if we start with $x_0 = 1$. (It leads to $x_1 = 1.57\ldots$, and there is no $\sin^{-1}(x_1^2)$ for such x_1.) But, when we use the inverse function $\sqrt{\sin x}$, the formula

$$x_{n+1} = \sqrt{\sin x_n}, \qquad x_0 = 1,$$

does produce a convergent sequence. (See Art. 2.10 for a discussion of criteria for convergence to a solution of $x = g(x)$.)

The study of limits would be a cumbersome business if every question about convergence had to be answered by applying Definition 2 directly, as we have done so far. Fortunately, there are three theorems that will make this process largely unnecessary from now on. The first two are practically the same as Theorems 1 and 4 in Article 1.9.

THEOREM 1

If $A = \lim_{n\to\infty} a_n$ and $B = \lim_{n\to\infty} b_n$ both exist and are finite, then

i) $\lim \{a_n + b_n\} = A + B$,

ii) $\lim \{ka_n\} = kA$ (k any number),

iii) $\lim \{a_n \cdot b_n\} = A \cdot B$,

iv) $\lim \left\{\dfrac{a_n}{b_n}\right\} = \dfrac{A}{B}$, provided $B \neq 0$ and b_n is never 0,

it being understood that all of the limits are to be taken as $n \to \infty$.

By combining Theorem 1 with Examples 2 and 3, we can proceed immediately to

$$\lim_{n\to\infty} -\frac{1}{n} = -1 \cdot \lim_{n\to\infty} \frac{1}{n} = -1 \cdot 0 = 0,$$

$$\lim_{n\to\infty} \left(1 - \frac{1}{n}\right) = \lim_{n\to\infty} 1 - \lim_{n\to\infty} \frac{1}{n} = 1 - 0 = 1,$$

$$\lim_{n\to\infty} \frac{5}{n^2} = 5 \cdot \lim_{n\to\infty} \frac{1}{n} \cdot \lim_{n\to\infty} \frac{1}{n} = 5 \cdot 0 \cdot 0 = 0,$$

$$\lim_{n\to\infty} \frac{4 - 7n^6}{n^6 + 3} = \lim_{n\to\infty} \frac{(4/n^6) - 7}{1 + (3/n^6)} = \frac{0 - 7}{1 + 0} = -7.$$

A corollary of Theorem 1 that will be useful later on is that every nonzero multiple of a divergent sequence is divergent.

COROLLARY

If the sequence $\{a_n\}$ diverges, and if c is any number different from 0, then the sequence $\{ca_n\}$ diverges.

Proof of the corollary. Suppose, on the contrary, that $\{ca_n\}$ converges. Then, by taking $k = 1/c$ in part (ii) of Theorem 1, we see that the sequence

$$\left\{\frac{1}{c} \cdot ca_n\right\} = \{a_n\}$$

converges. Thus $\{ca_n\}$ cannot converge unless $\{a_n\}$ converges. If $\{a_n\}$ does not converge, then $\{ca_n\}$ does not converge. ■

The next theorem is the sequence version of the Sandwich Theorem of Article 1.9.

THEOREM 2 If $a_n \le b_n \le c_n$ for all n beyond some index N, and if $\lim a_n = \lim c_n = L$, then $\lim b_n = L$ also.

An immediate consequence of Theorem 2 is that, if $|b_n| \le c_n$ and $c_n \to 0$, then $b_n \to 0$ because $-c_n \le b_n \le c_n$. We use this fact in the next example.

EXAMPLE 5

$$\frac{\cos n}{n} \to 0 \qquad \text{because} \quad 0 \le \left|\frac{\cos n}{n}\right| = \frac{|\cos n|}{n} \le \frac{1}{n}. \ \square$$

EXAMPLE 6

$$\frac{1}{2^n} \to 0 \qquad \text{because} \quad 0 \le \frac{1}{2^n} \le \frac{1}{n}. \ \square$$

EXAMPLE 7

$$(-1)^n\frac{1}{n} \to 0 \qquad \text{because} \quad 0 \le \left|(-1)^n\frac{1}{n}\right| \le \frac{1}{n}. \ \square$$

The application of Theorems 1 and 2 is broadened by a theorem that says that the result of applying a continuous function to a convergent sequence is again a convergent sequence. We state the theorem without proof. (See Problem 65 for an outline of the proof.)

THEOREM 3 If $a_n \to L$ and if f is a function that is continuous at L and defined at all the a_n's, then $f(a_n) \to f(L)$.

EXAMPLE 8 Find $\lim_{n\to\infty} \sqrt{(n+1)/n}$.

Solution Let $f(x) = \sqrt{x}$ and $a_n = (n+1)/n$. Then

$$a_n \to 1 \qquad \text{and} \qquad f(a_n) = \sqrt{a_n} \to f(1) = \sqrt{1} = 1$$

because $f(x)$ is continuous at $x = 1$. □

EXAMPLE 9 Find $\lim_{n\to\infty} 2^{1/n}$.

Solution Let $f(x) = 2^x$ and $a_n = 1/n$. Then

$$a_n \to 0 \qquad \text{and} \qquad f(a_n) = 2^{1/n} \to 2^0 = 1$$

because 2^x is continuous at $x = 0$. □

THEOREM 4

Suppose that $f(x)$ is a function defined for all $x \geq n_0$ and $\{a_n\}$ is a sequence such that $a_n = f(n)$ when $n \geq n_0$. If

$$\lim_{x\to\infty} f(x) = L, \qquad \text{then} \qquad \lim_{n\to\infty} a_n = L.$$

Proof. Let $\varepsilon > 0$. Suppose that L is a finite limit such that

$$\lim_{x\to\infty} f(x) = L.$$

Then there is a number M such that

$$x > M \Rightarrow |f(x) - L| < \varepsilon.$$

Let N be an integer such that

$$N \geq n_0 \qquad \text{and} \qquad N > M.$$

Then

$$n > N \Rightarrow a_n = f(n) \qquad \text{and} \qquad |a_n - L| = |f(n) - L| < \varepsilon. \blacksquare$$

The proof would require modification in case $f(x) \to +\infty$ or $f(x) \to -\infty$, but in either of these cases the sequence $\{a_n\}$ would diverge.

L'Hôpital's rule can be used to determine the limits of some sequences. The next example shows how.

EXAMPLE 10 Find $\lim_{n\to\infty} (\ln n)/n$.

Solution The function $(\ln x)/x$ is defined for all $x \geq 1$ and agrees with the given sequence on the positive integers. Therefore $\lim_{n\to\infty} (\ln n)/n$ will equal $\lim_{x\to\infty} (\ln x)/x$ if the latter exists. A single application of l'Hôpital's rule shows that

$$\lim_{x\to\infty} \frac{\ln x}{x} = \lim_{x\to\infty} \frac{1/x}{1} = \frac{0}{1} = 0.$$

We conclude that $\lim_{n\to\infty} (\ln n)/n = 0$. □

When we use l'Hôpital's rule to find the limit of a sequence, we often treat n as a continuous real variable, and differentiate directly with respect to n. This saves us from having to rewrite the formula for a_n as we did in Example 10.

EXAMPLE 11 Find $\lim_{n\to\infty} (2^n/5n)$.

Solution By l'Hôpital's rule,

$$\lim_{n\to\infty} \frac{2^n}{5n} = \lim_{n\to\infty} \frac{2^n \cdot \ln 2}{5} = \infty. \ \square$$

When the terms of the sequence are ratios of polynomials in n, we have a choice of methods for finding the limit: either use l'Hôpital's rule or divide the numerator and denominator by an appropriate power of n. (For example, we could divide both numerator and denominator by the highest power of n in one or the other.) The following shows what happens when the degree of the numerator is equal to, less than, or greater than, the degree of the denominator.

EXAMPLE 12 Find the following limits as $n \to \infty$.

$$\lim \frac{n^2 - 2n + 1}{2n^2 + 5} = \lim \frac{1 - (2/n) + (1/n^2)}{2 + (5/n^2)} = \frac{1}{2},$$

$$\lim \frac{n^3 + 5n}{n^4 - 6} = \lim \frac{(1/n) + (5/n^3)}{1 - (6/n^4)} = 0,$$

$$\lim \frac{n^2 - 5}{n + 1} = \lim \frac{2n}{1} = \infty \qquad \text{(l'Hôpital's rule).} \ \square$$

PROBLEMS

In Problems 1–10, write a_1, a_2, a_3, and a_4 for each sequence $\{a_n\}$. Determine which of the sequences converge and which diverge. Find the limit of each sequence that converges.

1. $a_n = \dfrac{1-n}{n^2}$

2. $a_n = \dfrac{n}{2^n}$

3. $a_n = \left(\dfrac{1}{3}\right)^n$

4. $a_n = \dfrac{1}{n!}$

5. $a_n = \dfrac{(-1)^{n+1}}{2n-1}$

6. $a_n = 2 + (-1)^n$

7. $a_n = \cos \dfrac{n\pi}{2}$

8. $a_n = 8^{1/n}$

9. $a_n = \dfrac{(-1)^{n-1}}{\sqrt{n}}$

10. $a_n = \sin^2 \dfrac{1}{n} + \cos^2 \dfrac{1}{n}$

In Problems 11–54, determine which of the sequences $\{a_n\}$ converge and which diverge. Find the limit of each sequence that converges.

11. $a_n = \dfrac{1}{10n}$

12. $a_n = \dfrac{n}{10}$

13. $a_n = 1 + \dfrac{(-1)^n}{n}$

14. $a_n = \dfrac{1 + (-1)^n}{n}$

15. $a_n = (-1)^n\left(1 - \dfrac{1}{n}\right)$

16. $a_n = 1 + (-1)^n$

17. $a_n = \dfrac{2n+1}{1-3n}$

18. $a_n = \dfrac{n^2 - n}{2n^2 + n}$

19. $a_n = \sqrt{\dfrac{2n}{n+1}}$

20. $a_n = \dfrac{\sin n}{n}$

21. $a_n = \sin \pi n$

22. $a_n = \sin\left(\dfrac{\pi}{2} + \dfrac{1}{n}\right)$

23. $a_n = n\pi \cos n\pi$

24. $a_n = \dfrac{\sin^2 n}{2^n}$

25. $a_n = \dfrac{n^2}{(n+1)^2}$

26. $a_n = \dfrac{\sqrt{n} - 1}{\sqrt{n}}$

27. $a_n = \dfrac{1 - 5n^4}{n^4 + 8n^3}$

28. $a_n = \sqrt[n]{3^{2n+1}}$

29. $a_n = \tanh n$

30. $a_n = \dfrac{\ln n}{\sqrt{n}}$

31. $a_n = \dfrac{2(n+1)+1}{2n+1}$

32. $a_n = \dfrac{(n+1)!}{n!}$

33. $a_n = 5$

34. $a_n = 5^n$

35. $a_n = (0.5)^n$

36. $a_n = \dfrac{10^{n+1}}{10^n}$

37. $a_n = \dfrac{n^n}{(n+1)^{n+1}}$

38. $a_n = (0.03)^{1/n}$

39. $a_n = \sqrt{2 - \dfrac{1}{n}}$

40. $a_n = 2 + (0.1)^n$

41. $a_n = \dfrac{3^n}{n^3}$

42. $a_n = \dfrac{\ln(n+1)}{n+1}$

43. $a_n = \ln n - \ln(n+1)$

44. $a_n = \dfrac{1 - 2^n}{2^n}$

45. $a_n = \dfrac{n^2 - 2n + 1}{n - 1}$

46. $a_n = \dfrac{n + (-1)^n}{n}$

47. $a_n = \left(-\dfrac{1}{2}\right)^n$

48. $a_n = \dfrac{\ln n}{\ln 2n}$

49. $a_n = \tan^{-1} n$

50. $a_n = \sinh(\ln n)$

51. $a_n = n \sin \dfrac{1}{n}$

52. $a_n = \dfrac{2n + \sin n}{n + \cos 5n}$

53. $a_n = \dfrac{n^2}{2n-1} \sin \dfrac{1}{n}$

54. $a_n = n\left(1 - \cos \dfrac{1}{n}\right)$

55. Show that $\lim_{n\to\infty} (n!/n^n) = 0$. (*Hint:* Expand the numerator and denominator and compare the quotient with $1/n$.)

56. (*Calculator*) The formula $x_{n+1} = (x_n + a/x_n)/2$ is the one produced by Newton's method to generate a sequence of approximations to the positive solution of $x^2 - a = 0$, $a > 0$. Starting with $x_1 = 1$ and $a = 3$, use the formula to calculate successive terms of the sequence until you have approximated $\sqrt{3}$ as accurately as your calculator permits.

57. (*Calculator*) If your calculator has a square-root key, enter $x = 10$ and take successive square roots to approximate the terms of the sequence $10^{1/2}$, $10^{1/4}$, $10^{1/8}$, . . . , continuing as far as your calculator permits. Repeat, with $x = 0.1$. Try other positive numbers above and below 1. When you have enough evidence, guess the answers to these questions: Does $\lim_{n\to\infty} x^{1/n}$ exist when $x > 0$? Does it matter what x is?

58. (*Calculator*) If you start with a reasonable value of x_1, then the rule $x_{n+1} = x_n + \cos x_n$ will generate a sequence that converges to $\pi/2$. Figure 11.5 shows why. The convergence is rapid. With $x_1 = 1$, calculate x_2, x_3, and x_4. Find out what happens when you start with $x_1 = 5$. Remember to use radians.

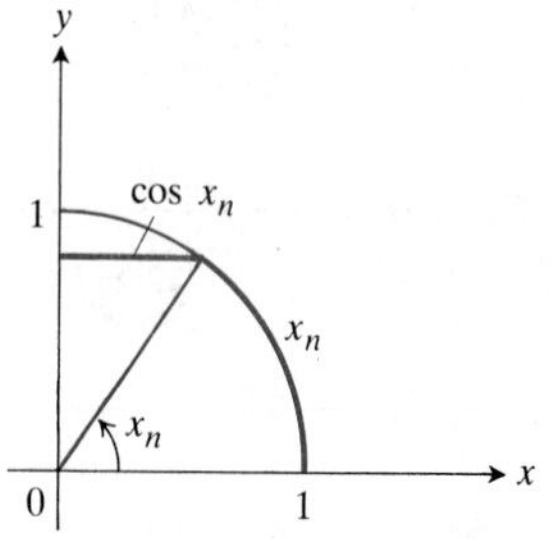

11.5 The length $\pi/2$ of the circular arc is approximated by $x_n + \cos x_n$.

59. Suppose that $f(x)$ is defined for all $0 \le x \le 1$, that f is differentiable at $x = 0$, and that $f(0) = 0$. Define a sequence $\{a_n\}$ by the rule $a_n = nf(1/n)$. Show that $\lim a_n = f'(0)$.

Use the result of Problem 59 to find the limits of the sequences in Problems 60 and 61.

60. $a_n = n \tan^{-1}\dfrac{1}{n}$

61. $a_n = n(e^{1/n} - 1)$

62. Prove that a sequence $\{a_n\}$ cannot have two different limits L and L'. (*Hint:* Take $\epsilon = \frac{1}{2}|L - L'|$ in Eq. 4.)

63. Under the hypotheses of Theorem 4, prove that if $\lim_{x\to\infty} f(x) = +\infty$, then the sequence $\{a_n\}$ diverges.

64. Prove Theorem 2.

65. Prove Theorem 3. (Outline of proof. Assume the hypotheses of the theorem and let ϵ be any positive number. For this ϵ there is a $\delta > 0$ such that

$$|f(x) - f(L)| < \epsilon \quad \text{when} \quad |x - L| < \delta.$$

For such a $\delta > 0$, there is an index N such that

$$|a_n - L| < \delta \quad \text{when } n > N.$$

What is the conclusion?)

Toolkit programs

Sequences and Series

11.3 Limits That Arise Frequently

Some limits arise so frequently that they are worth special attention. In this article we investigate these limits and look at examples in which they occur.

1. $\lim\limits_{n\to\infty} \dfrac{\ln n}{n} = 0$
2. $\lim\limits_{n\to\infty} \sqrt[n]{n} = 1$
3. $\lim\limits_{n\to\infty} x^{1/n} = 1 \quad (x > 0)$
4. $\lim\limits_{n\to\infty} x^n = 0 \quad (|x| < 1)$
5. $\lim\limits_{n\to\infty} \left(1 + \dfrac{x}{n}\right)^n = e^x \quad (\text{any } x)$
6. $\lim\limits_{n\to\infty} \dfrac{x^n}{n!} = 0 \quad (\text{any } x)$

1. $\lim\limits_{n\to\infty} \dfrac{\ln n}{n} = 0$

This limit was calculated in Example 10 of Article 11.2.

2. $\lim_{n\to\infty} \sqrt[n]{n} = 1$

Let $a_n = n^{1/n}$. Then

$$\ln a_n = \ln n^{1/n} = \frac{1}{n}\ln n \to 0, \tag{1}$$

so that, by applying Theorem 3 of Article 11.2 to $f(x) = e^x$, we have

$$n^{1/n} = a_n = e^{\ln a_n} \to e^0 = 1. \tag{2}$$

3. $\lim_{n\to\infty} x^{1/n} = 1$, **if** $x > 0$

Let $a_n = x^{1/n}$. Then

$$\ln a_n = \ln x^{1/n} = \frac{1}{n}\ln x \to 0, \tag{3}$$

because x remains fixed while n gets large. Thus, again by Theorem 3, with $f(x) = e^x$,

$$x^{1/n}\, a_n = e^{\ln a_n} \to e^0 = 1. \tag{4}$$

4. $\lim_{n\to\infty} x^n = 0$, **if** $|x| < 1$

Our scheme here is to show that the criteria of Definition 2 of Article 11.2 are satisfied with $L = 0$. That is, we show that to each $\varepsilon > 0$ there corresponds an index N so large that

$$|x^n| < \varepsilon \qquad \text{for} \quad n > N. \tag{5}$$

Since $\varepsilon^{1/n} \to 1$, while $|x| < 1$, there is an index N for which

$$|x| < \varepsilon^{1/N}. \tag{6}$$

In other words,

$$|x^N| = |x|^N < \varepsilon. \tag{7}$$

This is the index we seek, because

$$|x^n| < |x^N| \qquad \text{for} \quad n > N. \tag{8}$$

Combining (7) and (8) produces

$$|x^n| < |x^N| < \varepsilon \qquad \text{for} \quad n > N, \tag{9}$$

which is just what we needed to show.

5. $\lim_{n\to\infty} \left(1 + \frac{x}{n}\right)^n = e^x$ **(any** x**)**

Let

$$a_n = \left(1 + \frac{x}{n}\right)^n.$$

Then

$$\ln a_n = \ln\left(1 + \frac{x}{n}\right)^n = n \ln\left(1 + \frac{x}{n}\right) \to x,$$

as we can see by the following application of l'Hôpital's rule, in which we differentiate with respect to n:

$$\lim_{n\to\infty} n \ln\left(1 + \frac{x}{n}\right) = \lim_{n\to\infty} \frac{\ln(1 + x/n)}{1/n}$$

$$= \lim_{n\to\infty} \frac{\left(\dfrac{1}{1 + x/n}\right)\cdot\left(-\dfrac{x}{n^2}\right)}{-1/n^2} = \lim_{n\to\infty} \frac{x}{1 + x/n} = x.$$

Thus, by Theorem 3 of Art. 11.2, with $f(x) = e^x$,

$$\left(1 + \frac{x}{n}\right)^n = a_n = e^{\ln a_n} \to e^x.$$

6. $\displaystyle\lim_{n\to\infty} \frac{x^n}{n!} = 0$ **(any x)**

Since

$$-\frac{|x|^n}{n!} \le \frac{x^n}{n!} \le \frac{|x|^n}{n!},$$

all we really need to show is that $|x|^n/n! \to 0$. The first step is to choose an integer $M > |x|$, so that

$$\frac{|x|}{M} < 1 \qquad \text{and} \qquad \left(\frac{|x|}{M}\right)^n \to 0.$$

We then restrict our attention to values of $n > M$. For these values of n, we can write

$$\frac{|x|^n}{n!} = \frac{|x|^n}{1\cdot 2 \cdots M \cdot \underbrace{(M+1)(M+2)\cdots n}_{(n-M)\text{ factors}}}$$

$$\le \frac{|x|^n}{M!\,M^{n-M}} = \frac{|x|^n M^M}{M!\,M^n} = \frac{M^M}{M!}\left(\frac{|x|}{M}\right)^n.$$

Thus,

$$0 \le \frac{|x|^n}{n!} \le \frac{M^M}{M!}\left(\frac{|x|}{M}\right)^n.$$

Now, the constant $M^M/M!$ does not change with n. Thus the Sandwich Theorem tells us that

$$\frac{|x|^n}{n!} \to 0 \qquad \text{because} \qquad \left(\frac{|x|}{M}\right)^n \to 0.$$

REMARK. It is important to note that x is fixed and only n varies in the limits in Formulas 3 through 6, above.

A large number of limits can be found directly from the six limits we have just calculated.

EXAMPLES

1. If $|x| < 1$, then $x^{n+4} = x^4 \cdot x^n \to x^4 \cdot 0 = 0$. □
2. $\sqrt[n]{2n} = \sqrt[n]{2}\sqrt[n]{n} \to 1 \cdot 1 = 1$. □
3. $\left(1 + \dfrac{1}{n}\right)^{2n} = \left[\left(1 + \dfrac{1}{n}\right)^n\right]^2 \to e^2$. □
4. $\dfrac{100^n}{n!} \to 0$. □

5. $\dfrac{x^{n+1}}{(n+1)!} = \dfrac{x}{(n+1)} \cdot \dfrac{x^n}{n!} \to 0 \cdot 0 = 0.$ □

Still other limits can be calculated by using logarithms or l'Hôpital's rule, as in the calculations of limits (2), (3), and (5) at the beginning of this article.

EXAMPLE 6 Find $\lim_{n\to\infty} (\ln(3n+5))/n$.

Solution By l'Hôpital's rule,

$$\lim_{n\to\infty} \frac{\ln(3n+5)}{n} = \lim_{n\to\infty} \frac{3/(3n+5)}{1} = 0. \ \square$$

EXAMPLE 7 Find $\lim_{n\to\infty} \sqrt[n]{3n+5}$.

Solution Let

$$a_n = \sqrt[n]{3n+5} = (3n+5)^{1/n}.$$

Then,

$$\ln a_n = \ln(3n+5)^{1/n} = \frac{\ln(3n+5)}{n} \to 0,$$

as in Example 6. Therefore,

$$a_n = e^{\ln a_n} \to e^0 = 1,$$

by Theorem 3 of Article 11.2. □

We know that $\ln n$ increases more slowly than n does as $n \to \infty$, because $(\ln n)/n \to 0$. But we can say much more: $\ln n$ increases more slowly than $\sqrt{n}$, $\sqrt[3]{n}$, or even $n^{0.00001}$. In fact, if c is any positive constant $(\ln n)/n^c \to 0$ as $n \to \infty$. The next example establishes this fact.

EXAMPLE 8 Show that

$$\lim_{n\to\infty} \frac{\ln n}{n^c} = 0$$

if c is any positive constant.

Solution Apply l'Hôpital's rule and get

$$\lim \frac{\ln n}{n^c} = \lim \frac{1/n}{cn^{c-1}} = \lim \frac{1}{cn^c} = 0. \ \square$$

PROBLEMS

Determine which of the following sequences $\{a_n\}$ converge and which diverge. Find the limit of each sequence that converges.

1. $a_n = \dfrac{1+\ln n}{n}$

2. $a_n = \dfrac{\ln n}{3n}$

3. $a_n = \dfrac{(-4)^n}{n!}$

4. $a_n = \sqrt[n]{10n}$

5. $a_n = (0.5)^n$

6. $a_n = \dfrac{1}{(0.9)^n}$

7. $a_n = \left(1 + \dfrac{7}{n}\right)^n$

8. $a_n = \left(\dfrac{n+5}{n}\right)^n$

9. $a_n = \dfrac{\ln(n+1)}{n}$

10. $a_n = \sqrt[n]{n+1}$

11. $a_n = \dfrac{n!}{10^{6n}}$

12. $a_n = \dfrac{1}{\sqrt{2^n}}$

13. $a_n = \sqrt[2n]{n}$

14. $a_n = (n + 4)^{1/(n+4)}$

15. $a_n = \dfrac{1}{3^{2n-1}}$

16. $a_n = \ln\left(1 + \dfrac{1}{n}\right)^n$

17. $a_n = \left(\dfrac{n}{n+1}\right)^n$

18. $a_n = \left(1 + \dfrac{1}{n}\right)^{-n}$

19. $a_n = \dfrac{\ln(2n + 1)}{n}$

20. $a_n = \sqrt[n]{2n + 1}$

21. $a_n = \sqrt[n]{\dfrac{x^n}{2n + 1}}, \quad x > 0$

22. $a_n = \sqrt[n]{n^2}$

23. $a_n = \sqrt[n]{n^2 + n}$

24. $a_n = \dfrac{3^n \cdot 6^n}{2^{-n} \cdot n!}$

25. $a_n = \left(\dfrac{3}{n}\right)^{1/n}$

26. $a_n = \sqrt[n]{4^n n}$

27. $a_n = \left(1 - \dfrac{1}{n}\right)^n$

28. $a_n = \left(1 - \dfrac{1}{n^2}\right)^n$

29. $a_n = \dfrac{\ln(n^2)}{n}$

30. $a_n = \dfrac{(\ln n)^{200}}{n}$

31. $a_n = \dfrac{\ln n}{n^{1/n}}$

32. $a_n = \dfrac{1}{n}\displaystyle\int_1^n \frac{1}{x}\,dx$

33. $a_n = \displaystyle\int_1^n \frac{1}{x^p}\,dx, \quad p > 1$

(*Calculator*) In Problems 34–36, use a calculator to find a value of N such that the given inequality is satisfied for $n \geq N$.

34. $|\sqrt[n]{0.5} - 1| < 10^{-3}$

35. $|\sqrt[n]{n} - 1| < 10^{-3}$

36. $\dfrac{2^n}{n!} < 10^{-9}$

(*Hint:* If you do not have a factorial key, then write

$$\frac{2^n}{n!} = \left(\frac{2}{1}\right)\left(\frac{2}{2}\right)\cdots\left(\frac{2}{n}\right).$$

That is, calculate successive terms by multiplying by 2 and dividing by the next value of n.)

37. In Example 8, we assumed that obviously if $c > 0$, then $1/n^c \to 0$ as $n \to \infty$. Write out a formal proof of this fact. (*Hint:* If $\varepsilon = 0.001$ and $c = 0.04$, how large should N be in order to be sure that $|1/n^c - 0| < \varepsilon$ when $n > N$?)

Toolkit programs
Sequences and Series

11.4 Infinite Series

Infinite series are sequences of a special kind: those in which the nth term is the sum of the first n terms of a related sequence.

EXAMPLE 1 Suppose that we start with the sequence

$$1, \frac{1}{2}, \frac{1}{4}, \frac{1}{8}, \frac{1}{16}, \frac{1}{32}, \frac{1}{64}, \ldots$$

and then form a new sequence by successively adding 1, 2, 3, . . . terms of the original sequence. We denote the terms of the first sequence by a_n and the terms of the second sequence by s_n. For this example, we have

$$s_1 = a_1 = 1,$$

$$s_2 = a_1 + a_2 = 1 + \frac{1}{2} = \frac{3}{2},$$

$$s_3 = a_1 + a_2 + a_3 = 1 + \frac{1}{2} + \frac{1}{4} = \frac{7}{4},$$

as the first three terms of the sequence $\{s_n\}$. □

When a sequence $\{s_n\}$ is formed in this way from a given sequence $\{a_n\}$ by the rule

$$s_n = a_1 + a_2 + \cdots + a_n = \sum_{k=1}^{n} a_k,$$

the result is called an *infinite series*. The series $\{s_n\}$ is usually denoted by

$$\sum_{k=1}^{\infty} a_k \quad \text{or} \quad \sum_{n=1}^{\infty} a_n \quad \text{or simply} \sum a_n,$$

to show how the numbers s_n are obtained from the terms a_n.

We formalize these remarks in the following definition.

DEFINITION

1. If $\{a_n\}$ is a given sequence and s_n is defined by

$$s_n = a_1 + a_2 + \cdots + a_n = \sum_{k=1}^{n} a_k,$$

then the sequence $\{s_n\}$ is called an *infinite series*.

2. Instead of $\{s_n\}$, we usually write $\Sigma_{n=1}^{\infty} a_n$ or simply $\Sigma\, a_n$ (to show how the sums s_n are related to the terms $a_1, a_2, \ldots$).
3. The number $s_n = \Sigma_{k=1}^{n} a_k$ is called the nth *partial sum* of the series $\Sigma\, a_k$.
4. The number a_n is called the nth *term* of the series (and it is also the nth term of the original sequence $\{a_n\}$).
5. The series $\Sigma\, a_n$ is said to *converge* to a number L if and only if

$$L = \lim_{n\to\infty} s_n = \lim_{n\to\infty} \sum_{k=1}^{n} a_k,$$

in which case we call L the *sum* of the series and write

$$\sum_{n=1}^{\infty} a_n = L \quad \text{or} \quad a_1 + a_2 + \cdots + a_n + \cdots = L.$$

If no such limit exists (i.e., if $\{s_n\}$ diverges), the series is said to *diverge*.

We shall illustrate one method of finding the sum of an infinite series with the repeating decimal

$$0.3333\ldots = \frac{3}{10} + \frac{3}{100} + \frac{3}{1000} + \frac{3}{10{,}000} + \cdots$$

$$s_1 = \frac{3}{10},$$

$$s_2 = \frac{3}{10} + \frac{3}{10^2},$$

$$s_n = \frac{3}{10} + \frac{3}{10^2} + \cdots + \frac{3}{10^n}.$$

We can obtain a simple expression for s_n in closed form as follows: We multiply both sides of the equation for s_n by $\frac{1}{10}$ and obtain

$$\frac{1}{10}s_n = \frac{3}{10^2} + \frac{3}{10^3} + \cdots + \frac{3}{10^n} + \frac{3}{10^{n+1}}.$$

When we subtract this from s_n, we have

$$s_n - \frac{1}{10}s_n = \frac{3}{10} - \frac{3}{10^{n+1}} = \frac{3}{10}\left(1 - \frac{1}{10^n}\right).$$

Therefore,

$$\frac{9}{10}s_n = \frac{3}{10}\left(1 - \frac{1}{10^n}\right) \quad \text{and} \quad s_n = \frac{3}{9}\left(1 - \frac{1}{10^n}\right).$$

As $n \to \infty$, $(\frac{1}{10})^n \to 0$ and

$$\lim_{n\to\infty} s_n = \frac{3}{9} = \frac{1}{3}.$$

We therefore say that the sum of the infinite series

$$\frac{3}{10^1} + \frac{3}{10^2} + \frac{3}{10^3} + \cdots + \frac{3}{10^n} + \cdots$$

is $\frac{1}{3}$, and we write

$$\sum_{n=1}^{\infty} \frac{3}{10^n} = \frac{1}{3}.$$

The decimal 0.333 . . . is a special kind of *geometric series.*

DEFINITION

A series of the form

$$a + ar + ar^2 + ar^3 + \cdots + ar^{n-1} + \cdots \tag{1}$$

is called a *geometric series.* The ratio of any term to the one before it is r.

The ratio r can be positive, as in

$$1 + \frac{1}{2} + \frac{1}{4} + \cdots + \frac{1}{2^{n-1}} + \cdots, \tag{2}$$

or negative, as in

$$1 - \frac{1}{3} + \frac{1}{9} - \cdots + (-1)^n \frac{1}{3^{n-1}} + \cdots. \tag{3}$$

The sum of the first n terms of (1) is

$$s_n = a + ar + ar^2 + \cdots + ar^{n-1}. \tag{4}$$

Multiplying both sides of (4) by r gives

$$rs_n = ar + ar^2 + \cdots + ar^{n-1} + ar^n. \tag{5}$$

When we subtract (5) from (4), nearly all the terms cancel on the right side, leaving

$$s_n - rs_n = a - ar^n,$$

or

$$(1 - r)s_n = a(1 - r^n). \tag{6}$$

If $r \neq 1$, we may divide (6) by $(1 - r)$ to obtain

$$s_n = \frac{a(1 - r^n)}{1 - r}, \qquad r \neq 1. \tag{7a}$$

On the other hand, if $r = 1$ in (4), we get

$$s_n = na, \qquad r = 1. \tag{7b}$$

We are interested in the limit as $n \to \infty$ in Eqs. (7a) and (7b). Clearly, (7b) has no finite limit if $a \neq 0$. If $a = 0$, the series (1) is just

$$0 + 0 + 0 + \cdots,$$

which converges to the sum zero.

If $r \neq 1$, we use (7a). In the right side of (7a), n appears only in the expression r^n. This approaches zero as $n \to \infty$ if $|r| < 1$. Therefore,

$$\lim_{n\to\infty} s_n = \lim_{n\to\infty} \frac{a(1 - r^n)}{1 - r} = \frac{a}{1 - r} \quad \text{if} \quad |r| < 1. \tag{8a}$$

If we recall that $r^0 = 1$ when $r \neq 0$ we can summarize by writing

$$a + ar + ar^2 + \cdots + ar^{n-1} + \cdots = a \sum_{n=1}^{\infty} r^{n-1} = \frac{a}{1 - r}, \tag{8b}$$

or

$$a + ar + ar^2 + \cdots + ar^{n-1} + \cdots = a \sum_{n=0}^{\infty} r^n = \frac{a}{1 - r} \tag{8c}$$

if $0 < |r| < 1$. If $r = 0$, the series still converges, to $a/(1 - r) = a$. If $|r| > 1$, then $|r^n| \to \infty$, and (1) diverges.

The remaining case is where $r = -1$. Then $s_1 = a$, $s_2 = a - a = 0$, $s_3 = a$, $s_4 = 0$, and so on. If $a \neq 0$, this sequence of partial sums has no limit as $n \to \infty$, and the series (1) diverges.

We have thus proved the following theorem.

THEOREM 1

Geometric Series Theorem

If $|r| < 1$, the geometric series

$$a + ar + ar^2 + \cdots + ar^{n-1} + \cdots$$

converges to $a/(1 - r)$. If $|r| \geq 1$, the series diverges unless $a = 0$. If $a = 0$, the series converges to 0.

EXAMPLE 1 Geometric series with $a = \frac{1}{9}$ and $r = \frac{1}{3}$.

$$\frac{1}{9} + \frac{1}{27} + \frac{1}{81} + \cdots = \frac{1}{9}\left(1 + \frac{1}{3} + \frac{1}{3^2} + \cdots\right) = \frac{1/9}{1-(1/3)} = \frac{1}{6}. \quad \square$$

EXAMPLE 2 Geometric series with $a = 4$ and $r = -\frac{1}{2}$.

$$4 - 2 + 1 - \frac{1}{2} + \frac{1}{4} - \cdots = 4\left(1 - \frac{1}{2} + \frac{1}{4} - \frac{1}{8} + \frac{1}{16} - \cdots\right)$$
$$= \frac{4}{1 + (1/2)} = \frac{8}{3}. \quad \square$$

EXAMPLE 3 A ball is dropped from a meters above a flat surface. Each time the ball hits after falling a distance h, it rebounds a distance rh, where r is a positive number less than one. Find the total distance the ball travels up and down. See Fig. 11.6.

Solution The distance is given by the series

$$s = a + 2ar + 2ar^2 + 2ar^3 + \cdots.$$

The terms following the first term form a geometric series of sum $2ar/(1 - r)$. Hence the distance is

$$s = a + \frac{2ar}{1-r} = a\frac{1+r}{1-r}.$$

For instance, if a is 6 meters and $r = \frac{2}{3}$, the distance is

$$s = 6\frac{1+\frac{2}{3}}{1-\frac{2}{3}} = 30 \text{ m.} \quad \square$$

EXAMPLE 4 If we take $a = 1$ and $r = x$ in the geometric series theorem, we obtain

$$1 + x + x^2 + \cdots + x^{n-1} + \cdots = \frac{1}{1-x}, \qquad |x| < 1. \quad \square \tag{9}$$

REMARK 1. We were fortunate in the case of the geometric series to have found the closed-form expressions

$$s_n = \begin{cases} a\dfrac{1-r^n}{1-r} & \text{when } r \neq 1, \\ na & \text{when } r = 1, \end{cases}$$

from which we could get the precise results given by Theorem 1. There are not many other types of series where such closed-form expressions are available. (The next example is one of those rare cases.) Most of the remainder of this chapter will be devoted to tests that we can apply to the individual terms a_n of the series $\Sigma\, a_n$ to determine whether the series converges or diverges, without having to calculate the partial sums s_n. It will turn out that we can do so for a great many series. If a series does converge, we still have the question of determining its sum. Chapter 12 will help to some extent in doing that, but for a great many series we will still be left with no alternative but to compute numerical values of the partial sums and use those to estimate the true sum.

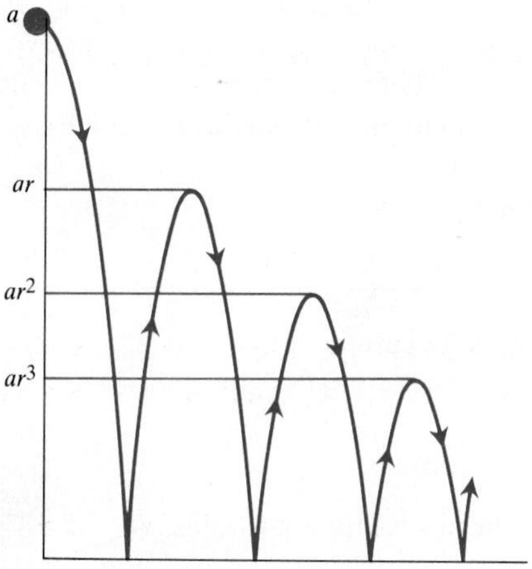

11.6 The height of each rebound is reduced by the factor r.

As noted above, the next example is another series whose sum can be found exactly by first finding a closed expression, or formula, for the kth partial sum s_k.

EXAMPLE 5 Determine whether $\Sigma_{n=1}^{\infty} [1/n(n+1)]$ converges. If it does, find the sum.

Solution We begin by looking for a pattern in the sequence of partial sums that might lead us to a closed expression for s_k. The key to success here, as in the integration

$$\int \frac{dx}{x(x+1)} = \int \frac{dx}{x} - \int \frac{dx}{x+1},$$

is the use of partial fractions:

$$\frac{1}{k(k+1)} = \frac{1}{k} - \frac{1}{k+1}. \tag{10}$$

This permits us to write the partial sum

$$\sum_{n=1}^{k} \frac{1}{n(n+1)} = \frac{1}{1 \cdot 2} + \frac{1}{2 \cdot 3} + \cdots + \frac{1}{k \cdot (k+1)}$$

as

$$s_k = \left(\frac{1}{1} - \frac{1}{2}\right) + \left(\frac{1}{2} - \frac{1}{3}\right) + \cdots + \left(\frac{1}{k} - \frac{1}{k+1}\right). \tag{11}$$

By removing parentheses on the right, and combining terms, we find that

$$s_k = 1 - \frac{1}{k+1} = \frac{k}{k+1}. \tag{12}$$

From this expression for s_k, we see immediately that $s_k \to 1$. Therefore the series does converge, and

$$\sum_{n=1}^{\infty} \frac{1}{n(n+1)} = 1. \ \square \tag{13}$$

There are, of course, other series that diverge besides geometric series with $|r| \geq 1$.

EXAMPLE 6

$$\sum_{n=1}^{\infty} n^2 = 1 + 4 + 9 + \cdots + n^2 + \cdots$$

diverges because the partial sums grow beyond every number L. The number $s_n = 1 + 4 + 9 + \cdots + n^2$ is greater than or equal to n^2 at each stage. $\square$

EXAMPLE 7

$$\sum_{n=1}^{\infty} \frac{n+1}{n} = \frac{2}{1} + \frac{3}{2} + \frac{4}{3} + \cdots + \frac{n+1}{n} + \cdots$$

diverges because the sequence of partial sums eventually outgrows every preassigned number: each term is greater than 1, so the sum of n terms is greater than n. □

A series can diverge without having its partial sums become large. For instance, the partial sums may oscillate between two extremes, as they do in the next example.

EXAMPLE 8 $\Sigma_{n=1}^{\infty} (-1)^{n+1}$ diverges because its partial sums alternate between 1 and 0:

$$s_1 = (-1)^2 = 1,$$
$$s_2 = (-1)^2 + (-1)^3 = 1 - 1 = 0,$$
$$s_3 = (-1)^2 + (-1)^3 + (-1)^4 = 1 - 1 + 1 = 1,$$

and so on. □

The next theorem provides a quick way to detect the kind of divergence that occurred in Examples 6, 7, and 8.

THEOREM 2

The *n*th-term Test for Divergence

If $\lim_{n\to\infty} a_n \neq 0$, or if $\lim_{n\to\infty} a_n$ fails to exist, then $\Sigma_{n=1}^{\infty} a_n$ diverges.

When we apply Theorem 2 to the series in Examples 6, 7, and 8, we find that

$$\sum_{n=1}^{\infty} n^2 \qquad \text{diverges because } n^2 \to \infty,$$

$$\sum_{n=1}^{\infty} \frac{n+1}{n} \qquad \text{diverges because } \frac{n+1}{n} \to 1 \neq 0,$$

$$\sum_{n=1}^{\infty} (-1)^{n+1} \qquad \text{diverges because } \lim_{n\to\infty} (-1)^{n+1} \text{ does not exist.}$$

Proof of the theorem. We prove Theorem 2 by showing that if $\Sigma\, a_n$ converges, then $\lim_{n\to\infty} a_n = 0$. Let

$$s_n = a_1 + a_2 + \cdots + a_n,$$

and suppose that $\Sigma\, a_n$ converges to S; that is

$$s_n \to S.$$

Then, corresponding to any preassigned number $\epsilon > 0$, there is an index N such that all the terms of the sequence $\{s_n\}$ after the Nth one lie between $S - (\epsilon/2)$ and $S + (\epsilon/2)$. Hence, no two of them may differ by as much as ϵ. That is, if m and n are both greater than N, then

$$|s_n - s_m| < \epsilon.$$

In particular, this inequality holds if $m = n - 1$ and $n > N + 1$, so that

$$|a_n| = |s_n - s_{n-1}| < \epsilon \qquad \text{when} \quad n > N + 1.$$

Since ϵ was any positive number whatsoever, this means that

$$\lim_{n\to\infty} a_n = 0. \blacksquare$$

In the next example, we use both Theorems 1 and 2.

EXAMPLE 9 Determine whether each series converges or diverges. If it converges, find its sum.

a) $\displaystyle\sum_{n=1}^{\infty} 2\left(\cos\frac{\pi}{3}\right)^n$ b) $\displaystyle\sum_{n=0}^{\infty} \left(\tan\frac{\pi}{4}\right)^n$

c) $\displaystyle\sum_{n=1}^{\infty} \frac{n}{2n+5}$ d) $\displaystyle\sum_{n=1}^{\infty} \frac{5(-1)^n}{4^n}$

Solution

a) $\cos \pi/3 = \frac{1}{2}$. This is a geometric series with first term $a_1 = 2(\cos \pi/3) = 1$ and ratio $r = \cos \pi/3 = \frac{1}{2}$; so the series *converges* and its *sum* is $a_1/(1 - r) = 1/(1 - \frac{1}{2}) = 2$.

b) $\tan \pi/4 = 1$. The nth term does not have 0 as its limit, so the series *diverges*.

c) $a_n = \dfrac{n}{2n+5}$ and $\displaystyle\lim_{n\to\infty} \frac{n}{2n+5} = \frac{1}{2} \neq 0$. Series (c) *diverges*.

d) This is a geometric series with first term $a_1 = -5/4$ and ratio $r = -1/4$. Series (d) *converges* and its *sum* is

$$\frac{a_1}{1-r} = \frac{-5/4}{1+\frac{1}{4}} = -1. \qquad \square$$

Because of how it is proved, Theorem 2 is often stated in the following shorter way.

THEOREM 3 If $\Sigma_{n=1}^{\infty} a_n$ converges, then $a_n \to 0$.

A Word of Caution Theorem 3 does *not* say that if $a_n \to 0$ then $\Sigma\, a_n$ converges. The series $\Sigma\, a_n$ may diverge even though $a_n \to 0$. Thus, $\lim a_n = 0$ is a necessary but not a sufficient condition for the series $\Sigma\, a_n$ to converge.

EXAMPLE 10 The series

$$1 + \underbrace{\frac{1}{2} + \frac{1}{2}}_{2 \text{ terms}} + \underbrace{\frac{1}{4} + \frac{1}{4} + \frac{1}{4} + \frac{1}{4}}_{4 \text{ terms}} + \cdots + \underbrace{\frac{1}{2^n} + \frac{1}{2^n} + \cdots + \frac{1}{2^n}}_{2^n \text{ terms}} + \cdots$$

diverges even though its terms form a sequence that converges to 0. $\square$

Whenever we have two convergent series we can add them, subtract them, and multiply them by constants, to make other convergent series. The next theorem gives the details.

THEOREM 4

> If $A = \Sigma_{n=1}^{\infty} a_n$ and $B = \Sigma_{n=1}^{\infty} b_n$ both exist and are finite, then
>
> i) $\sum_{n=1}^{\infty} (a_n + b_n) = A + B$,
>
> ii) $\sum_{n=1}^{\infty} ka_n = k \sum_{n=1}^{\infty} a_n = kA$ (*k* any number).

Proof of the theorem. Let

$$A_n = a_1 + a_2 + \cdots + a_n, \qquad B_n = b_1 + b_2 + \cdots + b_n.$$

Then the partial sums of $\Sigma_{n=1}^{\infty} (a_n + b_n)$ are

$$\begin{aligned} S_n &= (a_1 + b_1) + (a_2 + b_2) + \cdots + (a_n + b_n) \\ &= (a_1 + \cdots + a_n) + (b_1 + \cdots + b_n) \\ &= A_n + B_n. \end{aligned}$$

Since $A_n \to A$ and $B_n \to B$, we have $S_n \to A + B$. The partial sums of $\Sigma_{n=1}^{\infty} (ka_n)$ are

$$S_n = ka_1 + ka_2 + \cdots + ka_n = k(a_1 + a_2 + \cdots + a_n) = kA_n,$$

which converge to kA. ■

REMARK 2. If you think there should be two more parts of Theorem 4 to match those of Theorem 1, Article 11.2, see Problems 44–46. Also, see Problem 41, Article 11.5, and Problem 39, Article 11.7.

Part (ii) of Theorem 4 says that every multiple of a convergent series converges. A companion to this is the next corollary, which says that every *nonzero* multiple of a divergent series diverges.

COROLLARY

> If $\Sigma_{n=1}^{\infty} a_n$ diverges, and if c is any number different from 0, then the series of multiples $\Sigma_{n=1}^{\infty} ca_n$ diverges.

Proof of the corollary. Suppose, to the contrary, that $\Sigma_{n=1}^{\infty} ca_n$ actually converges. Then, when we take $k = 1/c$ in part (ii) of Theorem 4 we find that

$$\frac{1}{c} \cdot \sum_{n=1}^{\infty} ca_n = \sum_{n=1}^{\infty} \frac{1}{c} \cdot ca_n = \sum_{n=1}^{\infty} a_n$$

converges. That is, $\Sigma_{n=1}^{\infty} ca_n$ cannot converge unless $\Sigma_{n=1}^{\infty} a_n$ also converges. Thus, if $\Sigma_{n=1}^{\infty} a_n$ diverges, then $\Sigma_{n=1}^{\infty} ca_n$ must diverge. ■

REMARK 3. An immediate consequence of Theorem 4 is that if $A = \Sigma_{n=1}^{\infty} a_n$ and $B = \Sigma_{n=1}^{\infty} b_n$, then

$$\sum (a_n - b_n) = \sum a_n + \sum (-1)b_n = \sum a_n - \sum b_n = A - B. \tag{14}$$

The series $\Sigma_{n=1}^{\infty} (a_n - b_n)$ is called the *difference* of $\Sigma_{n=1}^{\infty} a_n$ and $\Sigma_{n=1}^{\infty} b_n$, while $\Sigma_{n=1}^{\infty} (a_n + b_n)$ is called their *sum*.

EXAMPLE 11

a) $$\sum_{n=1}^{\infty} \frac{4}{2^{n-1}} = 4 \sum_{n=1}^{\infty} \frac{1}{2^{n-1}} = 4 \frac{1}{1 - \frac{1}{2}} = 8,$$

b) $$\sum_{n=0}^{\infty} \frac{3^n - 2^n}{6^n} = \sum_{n=0}^{\infty} \left(\frac{1}{2^n} - \frac{1}{3^n}\right) = \sum_{n=0}^{\infty} \frac{1}{2^n} - \sum_{n=0}^{\infty} \frac{1}{3^n}$$

$$= \frac{1}{1 - \frac{1}{2}} - \frac{1}{1 - \frac{1}{3}} = 2 - \frac{3}{2} = \frac{1}{2}. \quad \square$$

REMARK 4. A finite number of terms can always be deleted from or added to a series without altering its convergence or divergence. If $\Sigma_{n=1}^{\infty} a_n$ converges and k is an index greater than 1, then $\Sigma_{n=k}^{\infty} a_n$ converges, and

$$\sum_{n=1}^{\infty} a_n = a_1 + a_2 + \cdots + a_{k-1} + \sum_{n=k}^{\infty} a_n. \tag{15}$$

Conversely, if $\Sigma_{n=k}^{\infty} a_n$ converges for any $k > 1$, then $\Sigma_{n=1}^{\infty} a_n$ converges and the sums continue to be related as in Eq. (15). Thus, for example,

$$\sum_{n=1}^{\infty} \frac{1}{5^n} = \frac{1}{5} + \frac{1}{25} + \frac{1}{125} + \sum_{n=4}^{\infty} \frac{1}{5^n} \tag{16}$$

and

$$\sum_{n=4}^{\infty} \frac{1}{5^n} = \sum_{n=1}^{\infty} \frac{1}{5^n} - \frac{1}{5} - \frac{1}{25} - \frac{1}{125}. \tag{17}$$

Note that while the addition or removal of a finite number of terms from a series has no effect on the convergence or divergence of the series, these operations can change the *sum* of a convergent series.

REMARK 5. The indexing of the terms of a series can be changed without altering convergence of the series. For example, the geometric series that starts with

$$1 + \frac{1}{2} + \frac{1}{4} + \cdots$$

can be described as

$$\sum_{n=0}^{\infty} \frac{1}{2^n} \quad \text{or} \quad \sum_{n=-4}^{\infty} \frac{1}{2^{n+4}} \quad \text{or} \quad \sum_{n=5}^{\infty} \frac{1}{2^{n-5}}. \tag{18}$$

The partial sums remain the same no matter what indexing is chosen, so that we are free to start indexing with whatever integer we want. Preference is usually given to an indexing that leads to a simple expression. In Example 11(b) we chose to start with $n = 0$ instead of $n = 1$, because this allowed us to describe the series we had in mind as:

$$\sum_{n=0}^{\infty} \frac{3^n - 2^n}{6^n} \quad \text{instead of} \quad \sum_{n=1}^{\infty} \frac{3^{n-1} - 2^{n-1}}{6^{n-1}}. \tag{19}$$

PROBLEMS

In Problems 1–8, find a closed expression for the sum s_n of the first n terms of each series. Then compute the sum of the series if the series converges.

1. $2 + \frac{2}{3} + \frac{2}{9} + \frac{2}{27} + \cdots + \frac{2}{3^{n-1}} + \cdots$

2. $\frac{9}{100} + \frac{9}{100^2} + \frac{9}{100^3} + \cdots + \frac{9}{100^n} + \cdots$

3. $1 + e^{-1} + e^{-2} + \cdots + e^{-(n-1)} + \cdots$

4. $1 - \frac{1}{2} + \frac{1}{4} - \frac{1}{8} + \cdots + (-1)^{n-1}\frac{1}{2^{n-1}} + \cdots$

5. $1 - 2 + 4 - 8 + \cdots + (-1)^{n-1}2^{n-1} + \cdots$

6. $\frac{1}{2 \cdot 3} + \frac{1}{3 \cdot 4} + \frac{1}{4 \cdot 5} + \cdots + \frac{1}{(n+1)(n+2)} + \cdots$

7. $\ln\frac{1}{2} + \ln\frac{2}{3} + \ln\frac{3}{4} + \cdots + \ln\frac{n}{n+1} + \cdots$

8. $1 + 2 + 3 + \cdots + n + \cdots$

9. The series in Problem 6 can be described as

$$\sum_{n=1}^{\infty} \frac{1}{(n+1)(n+2)}.$$

It can also be described as a summation beginning with $n = -1$:

$$\sum_{n=-1}^{\infty} \frac{1}{(n+3)(n+4)}.$$

Describe the series as a summation beginning with
a) $n = -2$, b) $n = 0$, c) $n = 5$.

10. a) A ball is dropped from a height of 4 m. Each time it strikes the pavement after falling from a height of h meters, it rebounds to a height of $0.75h$ meters. Find the total distance traveled up and down by the ball.
b) Calculate the total time the ball is traveling. (*Hint:* $y = \frac{1}{2}gt^2 = \frac{1}{2}(9.8)t^2$, t in seconds, y in meters.)

In Problems 11–18, find the sum of the series.

11. $\sum_{n=0}^{\infty} \frac{1}{4^n}$

12. $\sum_{n=2}^{\infty} \frac{1}{4^n}$

13. $\sum_{n=1}^{\infty} \frac{7}{4^n}$

14. $\sum_{n=0}^{\infty} (-1)^n \frac{5}{4^n}$

15. $\sum_{n=0}^{\infty} \left(\frac{5}{2^n} + \frac{1}{3^n}\right)$

16. $\sum_{n=0}^{\infty} \left(\frac{5}{2^n} - \frac{1}{3^n}\right)$

17. $\sum_{n=0}^{\infty} \left(\frac{2^n}{5^n}\right)$

18. $\sum_{n=0}^{\infty} \left(\frac{2^{n+1}}{5^n}\right)$

Use partial fractions to find the sum of the series in Problems 19–22.

19. $\sum_{n=1}^{\infty} \frac{4}{(4n-3)(4n+1)}$

20. $\sum_{n=1}^{\infty} \frac{1}{(4n-3)(4n+1)}$

21. $\sum_{n=3}^{\infty} \frac{4}{(4n-3)(4n+1)}$

22. $\sum_{n=1}^{\infty} \frac{2n+1}{n^2(n+1)^2}$

23. a) Express the repeating decimal

$$0.234\,234\,234\ldots$$

as an infinite series, and give the sum as a ratio p/q of two integers.
b) Is it true that *every* repeating decimal is a rational number p/q? Give a reason for your answer.

24. Express the decimal number

$$1.24\,123\,123\,123\ldots,$$

which begins to repeat after the first three figures, as a rational number p/q.

In Problems 25–38, determine whether each series converges or diverges. If it converges, find the sum.

25. $\sum_{n=0}^{\infty} \left(\frac{1}{\sqrt{2}}\right)^n$

26. $\sum_{n=1}^{\infty} \ln\frac{1}{n}$

27. $\sum_{n=1}^{\infty} (-1)^{n+1}\frac{3}{2^n}$

28. $\sum_{n=1}^{\infty} (\sqrt{2})^n$

29. $\sum_{n=0}^{\infty} \cos n\pi$

30. $\sum_{n=0}^{\infty} \frac{\cos n\pi}{5^n}$

31. $\sum_{n=0}^{\infty} e^{-2n}$

32. $\sum_{n=1}^{\infty} \frac{n^2+1}{n}$

33. $\sum_{n=1}^{\infty} (-1)^{n+1}n$

34. $\sum_{n=1}^{\infty} \frac{2}{10^n}$

35. $\sum_{n=0}^{\infty} \frac{2^n - 1}{3^n}$

36. $\sum_{n=1}^{\infty} \left(1 - \frac{1}{n}\right)^n$

37. $\sum_{n=0}^{\infty} \frac{n!}{1000^n}$

38. $\sum_{n=0}^{\infty} \frac{1}{x^n}$, $|x| > 1$

In Problems 39 and 40, the equalities are instances of Theorem 1. Give the value of a and of r in each case.

39. $\frac{1}{1+x} = \sum_{n=0}^{\infty} (-1)^n x^n$, $|x| < 1$

40. $\frac{1}{1+x^2} = \sum_{n=0}^{\infty} (-1)^n x^{2n}$, $|x| < 1$

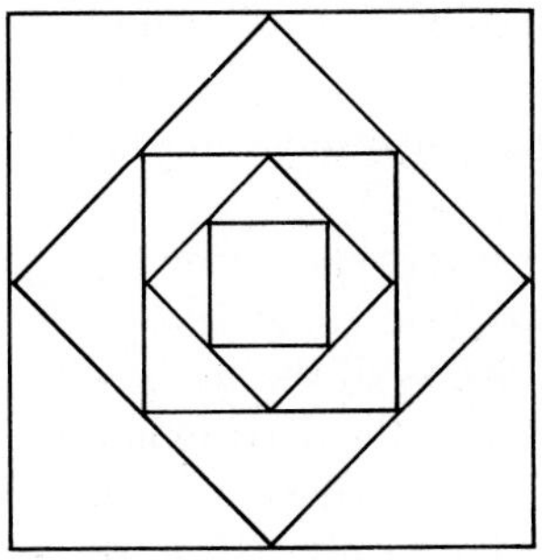

Figure 11.7

41. Figure 11.7 shows the first five of an infinite series of squares. The outermost square has an area of 4, and each of the other squares is obtained by joining the midpoints of the sides of the square before it. Find the sum of the areas of all the squares.

42. Find a closed-form expression for the nth partial sum of the series $\Sigma_{n=1}^{\infty}(-1)^{n+1}$.

43. Show by example that the term-by-term sum of two divergent series may converge.

44. Find convergent geometric series $A = \Sigma_{n=1}^{\infty} a_n$ and $B = \Sigma_{n=1}^{\infty} b_n$ that illustrate the fact that $\Sigma_{n=1}^{\infty} a_n \cdot b_n$ may converge without being equal to $A \cdot B$.

45. Show by example that $\Sigma_{n=1}^{\infty}(a_n/b_n)$ may diverge even though $\Sigma_{n=1}^{\infty} a_n$ and $\Sigma_{n=1}^{\infty} b_n$ converge and no $b_n = 0$.

46. Show by example that $\Sigma_{n=1}^{\infty}(a_n/b_n)$ may converge to something other than A/B even when $A = \Sigma_{n=1}^{\infty} a_n$, $B = \Sigma_{n=1}^{\infty} b_n \neq 0$, and no $b_n = 0$.

47. Show that if $\Sigma_{n=1}^{\infty} a_n$ converges, and $a_n \neq 0$ for all n, then $\Sigma_{n=1}^{\infty}(1/a_n)$ diverges.

48. a) Verify by long division that

$$\frac{1}{1+t} = 1 - t + t^2 - t^3 + \cdots + (-1)^n t^n + \frac{(-1)^{n+1}t^{n+1}}{1+t}.$$

b) By integrating both sides of the equation in part (a) with respect to t, from 0 to x, show that

$$\ln(1+x) = x - \frac{x^2}{2} + \frac{x^3}{3} - \frac{x^4}{4} + \cdots + (-1)^n \frac{x^{n+1}}{n+1} + R,$$

where

$$R = (-1)^{n+1}\int_0^x \frac{t^{n+1}}{1+t}\,dt.$$

c) If $x > 0$, show that

$$|R| \leq \int_0^x t^{n+1}\,dt = \frac{x^{n+2}}{n+2}.$$

(*Hint:* As t varies from 0 to x, $1 + t \geq 1$.)

d) If $x = \frac{1}{2}$, how large should n be in part (c) above if we want to be able to guarantee that $|R| < 0.001$? Write a polynomial that approximates $\ln(1 + x)$ to this degree of accuracy for $0 \leq x \leq \frac{1}{2}$.

e) If $x = 1$, how large should n be in part (c) above if we want to be able to guarantee that $|R| < 0.001$?

Toolkit programs

Sequences and Series

11.5

Tests for Convergence of Series with Nonnegative Terms

In this article we shall study series that do not have negative terms. The reason for this restriction is that the partial sums of these series always form increasing sequences, and increasing sequences *that are bounded from above* always converge, as we shall see. Thus, to show that a series of nonnegative terms converges, we need only show that there is some number beyond which the partial sums never go.

It may at first seem to be a drawback that this approach establishes the fact of convergence without actually producing the sum of the series in question. Surely it would be better to compute sums of series directly from nice formulas for their partial sums. But in most cases such formulas are not available, and in their absence we have to turn instead to a two-step procedure of first establishing convergence and then approximating the sum. In this article and the next, we focus on the first of these two steps.

Surprisingly enough it is not a severe restriction to begin our study of convergence with the temporary exclusion of series that have one or more negative terms, for it is a fact, as we shall see in the next article, that a series $\Sigma_{n=1}^{\infty} a_n$ will converge whenever the corresponding series of absolute values $\Sigma_{n=1}^{\infty} |a_n|$ converges. (The converse is not true. Articles 11.6 and 11.7 will deal with these matters more fully.) Thus, once we know that

$$\sum_{n=1}^{\infty} \frac{1}{n^2} = 1 + \frac{1}{4} + \frac{1}{9} + \frac{1}{16} + \frac{1}{25} + \cdots \tag{1}$$

converges, we will know that *all* of the series like

$$1 - \frac{1}{4} + \frac{1}{9} - \frac{1}{16} + \frac{1}{25} + \cdots \tag{2}$$

and

$$-1 - \frac{1}{4} + \frac{1}{9} + \frac{1}{16} - \frac{1}{25} - \cdots, \tag{3}$$

that can be obtained from (1) by changing the sign of one or more terms, also converge! We might not know at first what they converge to, but at least we know they converge, and that is a first and necessary step toward estimating their sums.

Increasing Sequences

Suppose now that $\Sigma_{n=1}^{\infty} a_n$ is an infinite series that has no negative terms. That is, $a_n \geq 0$ for every n. Then, when we calculate the partial sums s_1, s_2, s_3, and so on, we see that each one is greater than or equal to its predecessor because $s_{n+1} = s_n + a_n$. That is,

$$s_1 \leq s_2 \leq s_3 \leq \cdots \leq s_n \leq s_{n+1} \leq \cdots. \tag{4}$$

A sequence $\{s_n\}$ like the one in (4), with the property that $s_n \leq s_{n+1}$ for every n, is called an *increasing* sequence.

There are two types of increasing sequences—those that increase beyond any finite bound and those that don't. The former diverge to infinity. We turn our attention to the other kind: those that do not grow beyond all bounds. Such a sequence is said to be *bounded from above,* and any number M such that

$$s_n \leq M \qquad \text{for all } n$$

is called *an upper bound* of the sequence.

EXAMPLE 1 If $s_n = n/(n + 1)$, then 1 is an upper bound and so is any larger number, like $\sqrt{2}$, 5, 17. No number smaller than 1 is an upper bound, so that for this sequence 1 is the *least upper bound.* □

When an increasing sequence is bounded from above, we may ask, "Must it have a *least* upper bound?". The answer is yes, but we shall not prove this fact. We shall prove that *if L is the least upper bound, then the sequence converges to L.* The following argument shows why L is the limit.

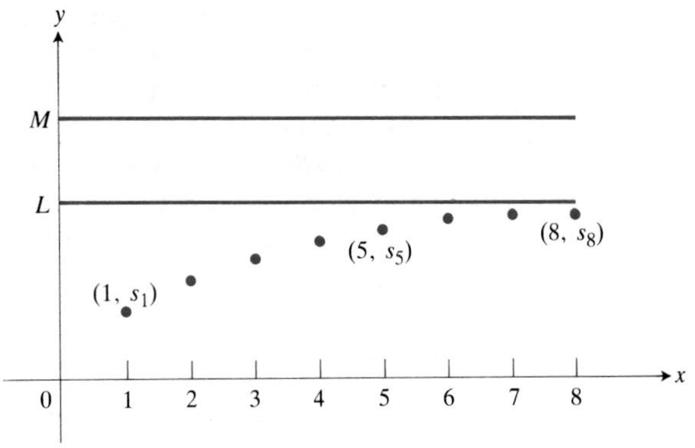

11.8 When the terms of an increasing sequence have an upper bound M, they have a limit $L \leq M$.

When we plot the points $(1, s_1), (2, s_2), \ldots, (n, s_n)$ in the xy-plane, if M is an upper bound of the sequence, all these plotted points will lie on or below the line $y = M$ (Fig. 11.8). It seems intuitively clear that there ought to be a lowest such line $y = L$. That would mean that none of the points (n, s_n) is above $y = L$, but that some do lie above any lower line $y = L - \epsilon$, if ϵ is a positive number. Then the sequence converges to L as limit, because the number L has the properties (a) $s_n \leq L$ for *all* values of n, (b) given any $\epsilon > 0$, there exists at least one integer N such that

$$s_N > L - \epsilon.$$

Then the fact that $\{s_n\}$ is an increasing sequence tells us further that

$$s_n \geq s_N > L - \epsilon \qquad \text{for all} \qquad n \geq N.$$

This means that *all* the numbers s_n, beyond the Nth one in the sequence, lie within ϵ distance of L. This is precisely the condition for L to be the limit of the sequence s_n,

$$L = \lim_{n\to\infty} s_n.$$

The facts for increasing sequences are summarized in the following theorem.

THEOREM 1

The Increasing Sequence Theorem

Let $\{s_n\}$ be an increasing sequence of real numbers. Then one or other of the following alternatives must hold:

A. The terms of the sequence are all less than or equal to some finite constant M. In this case, the sequence has a finite limit L that is also less than or equal to M. (In fact, L is the least upper bound of the sequence $\{s_n\}$.)

B. The sequence diverges to plus infinity: that is, the numbers in the sequence ultimately exceed every preassigned number.

Alternative *B* of the theorem is what happens when there are points (n, s_n) above any given line $y = M$, no matter how large M may be.

Let us now apply Theorem 1 to the convergence of infinite series of nonnegative numbers. If $\Sigma\, a_n$ is such a series, its sequence of partial sums $\{s_n\}$ is an increasing sequence. Therefore, $\{s_n\}$, and hence $\Sigma\, a_n$, will converge if and only if the numbers s_n have an upper bound. The question is how to find out in any particular instance whether the s_n's do have an upper bound.

Sometimes we can show that the s_n's are bounded above by showing that each one is less than or equal to the corresponding partial sum of a series that is already known to converge. The next example shows how this can happen.

EXAMPLE 2

$$\sum_{n=0}^{\infty} \frac{1}{n!} = 1 + \frac{1}{1!} + \frac{1}{2!} + \frac{1}{3!} + \cdots \tag{5}$$

converges because its terms are all positive and less than or equal to the corresponding terms of

$$1 + \sum_{n=0}^{\infty} \frac{1}{2^n} = 1 + 1 + \frac{1}{2} + \frac{1}{2^2} + \cdots. \tag{6}$$

To see how this relationship between these two series leads to an upper bound for the partial sums of $\Sigma_{n=0}^{\infty}\,(1/n!)$, let

$$s_n = 1 + \frac{1}{1!} + \frac{1}{2!} + \cdots + \frac{1}{n!},$$

and observe that, for each n,

$$s_n \leq 1 + 1 + \frac{1}{2} + \frac{1}{2^2} + \cdots + \frac{1}{2^n} < 1 + \sum_{n=0}^{\infty} \frac{1}{2^n} = 1 + \frac{1}{1 - \frac{1}{2}} = 3.$$

Thus the partial sums of $\Sigma_{n=0}^{\infty}\,(1/n!)$ are all less than 3. Therefore, $\Sigma_{n=0}^{\infty}\,(1/n!)$ converges.

Just because 3 is an upper bound for the partial sums of $\Sigma_{n=0}^{\infty}\,(1/n!)$ we cannot conclude that the series converges to 3. The series actually converges to $e = 2.71828\ldots$. □

We established the convergence of the series in Example 2 by comparing it with a series that was already known to converge. This kind of comparison is typical of a procedure called the *comparison test* for convergence of series of nonnegative terms.

Comparison Test for Series of Nonnegative Terms

Let $\Sigma_{n=1}^{\infty}\, a_n$ be a series that has no negative terms.

A. Test for *convergence* of $\Sigma\, a_n$. The series $\Sigma\, a_n$ converges if there is a convergent series of nonnegative terms $\Sigma\, c_n$ with $a_n \leq c_n$ for all $n > n_0$.

B. Test for *divergence* of $\Sigma\, a_n$. The series $\Sigma\, a_n$ diverges if there is a divergent series of nonnegative terms $\Sigma\, d_n$ with $a_n \geq d_n$ for all $n \geq n_0$.

In part (A), the partial sums of the series $\Sigma\, a_n$ are bounded from above by

$$M = a_1 + a_2 + \cdots + a_{n_0} + \sum_{n=n_0+1}^{\infty} c_n.$$

Therefore, they form an increasing sequence that has a limit L that is less than or equal to M.

In part (B), the partial sums for $\Sigma\, a_n$ are not bounded from above: if they were, the partial sums for $\Sigma\, d_n$ would be bounded by

$$M' = d_1 + d_2 + \cdots + d_{n_0} + \sum_{n=n_0+1}^{\infty} a_n.$$

That would imply convergence of $\Sigma\, d_n$. Therefore, divergence of $\Sigma\, d_n$ implies divergence of $\Sigma\, a_n$.

To apply the comparison test to a series, we do not have to include the early terms of the series. We can start the test with any index N, provided we include all the terms of the series being tested from there on.

EXAMPLE 3 The convergence of the series

$$5 + \frac{2}{3} + 1 + \frac{1}{7} + \frac{1}{2} + \frac{1}{3!} + \frac{1}{4!} + \cdots + \frac{1}{k!} + \cdots$$

can be established by ignoring the first four terms and comparing the remainder of the series from the fifth term on (the fifth term is $\frac{1}{2}$) with the convergent geometric series

$$\sum_{n=1}^{\infty} \frac{1}{2^n} = \frac{1}{2} + \frac{1}{4} + \frac{1}{8} + \cdots. \ \square$$

To apply the comparison test we need to have on hand a list of series that are known to converge and a list of series that are known to diverge. Our next example adds a divergent series to the list.

EXAMPLE 4 The *harmonic series*

$$\sum_{n=1}^{\infty} \frac{1}{n} = 1 + \frac{1}{2} + \frac{1}{3} + \frac{1}{4} + \cdots$$

diverges.

To see why, we represent the terms of the series as the areas of rectangles each of base unity and having altitudes $1, \frac{1}{2}, \frac{1}{3}, \ldots$, as in Fig. 11.9. The sum of the first n terms of the series,

$$s_n = 1 + \frac{1}{2} + \frac{1}{3} + \cdots + \frac{1}{n},$$

could be thought to represent the sum of the areas of n rectangles each of which is greater than the area underneath the corresponding portion of

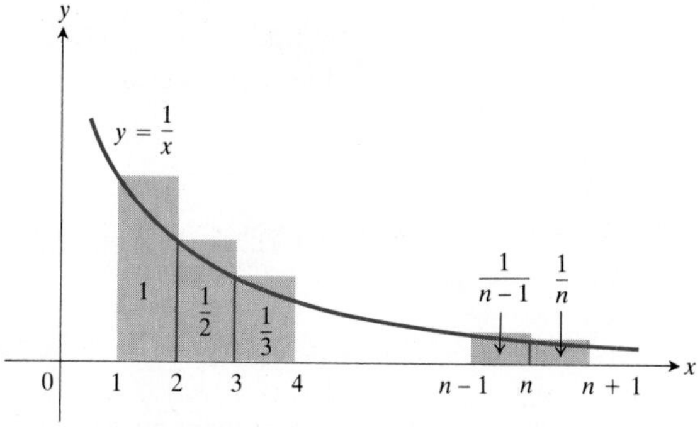

11.9 $1 + \frac{1}{2} + \frac{1}{3} + \cdots + \frac{1}{n} > \int_1^{n+1} \frac{1}{x}\,dx = \ln(n+1)$

the curve $y = 1/x$. Thus s_n is greater than the area under this curve between $x = 1$ and $x = n + 1$:

$$s_n > \int_1^{n+1} \frac{dx}{x} = \ln(n+1).$$

Therefore $s_n \to +\infty$ because $\ln(n+1) \to +\infty$. The series

$$1 + \frac{1}{2} + \frac{1}{3} + \cdots + \frac{1}{n} + \cdots$$

diverges to plus infinity. □

The harmonic series $\Sigma_{n=1}^{\infty}(1/n)$ is another series whose divergence cannot be detected by the nth-term test for divergence. The series diverges in spite of the fact that $1/n \to 0$.

We know that every nonzero multiple of a divergent series diverges (Corollary of Theorem 4 in the preceding article). Therefore, the divergence of the harmonic series implies the divergence of series like

$$\sum_{n=1}^{\infty} \frac{1}{2n} = \frac{1}{2} + \frac{1}{4} + \frac{1}{6} + \frac{1}{8} + \cdots$$

and

$$\sum_{n=1}^{\infty} \frac{1}{100n} = \frac{1}{100} + \frac{1}{200} + \frac{1}{300} + \frac{1}{400} + \cdots.$$

The Integral Test

In Example 4 we deduced the divergence of the harmonic series by comparing its sequence of partial sums with a divergent sequence of integrals. This comparison is a special case of a general comparison process called the *integral test*, a test that gives criteria for convergence as well as for divergence of series whose terms are positive.

Integral Test

Let the function $y = f(x)$, obtained by introducing the continuous variable x in place of the discrete variable n in the nth term of the positive series

$$\sum_{n=1}^{\infty} a_n,$$

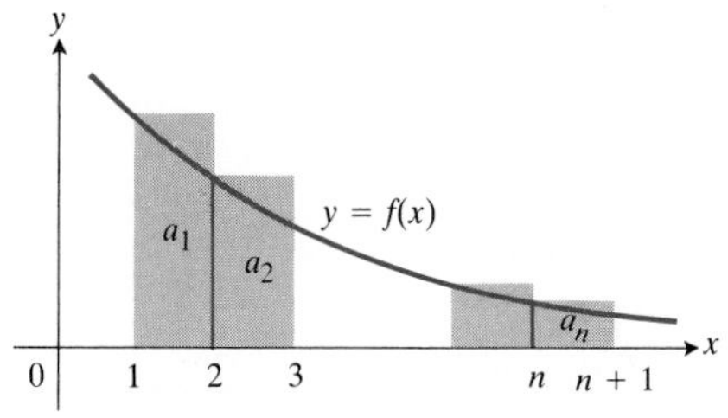

(a)

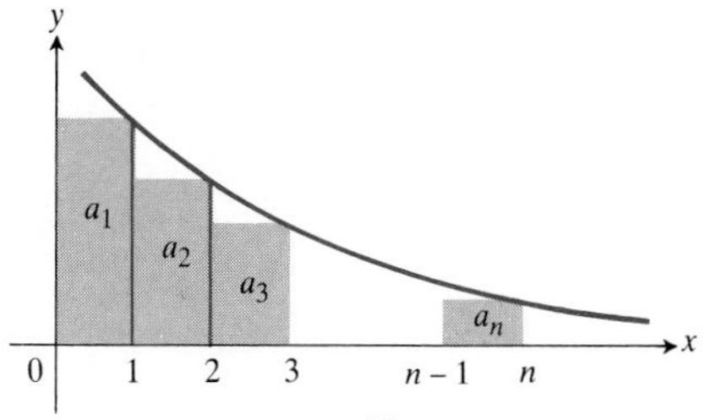

(b)

11.10 $\int_1^{n+1} f(x)\,dx \le a_1 + a_2 + \cdots + a_n$

$\le a_1 + \int_1^n f(x)\,dx.$

be a decreasing function of x for $x \ge 1$. Then the series and the integral

$$\int_1^\infty f(x)\,dx$$

both converge or both diverge.

Proof. We start with the assumption that f is a decreasing function with $f(n) = a_n$ for every n. This leads us to observe that the rectangles in Fig. 11.10(a), which have areas $a_1, a_2, \ldots, a_n$, collectively enclose more area than that under the curve $y = f(x)$ from $x = 1$ to $x = n + 1$. That is,

$$\int_1^{n+1} f(x)\,dx \le a_1 + a_2 + \cdots + a_n.$$

In Fig. 11.10(b) the rectangles have been faced to the left instead of to the right. If we momentarily disregard the first rectangle, of area a_1, we see that

$$a_2 + a_3 + \cdots + a_n \le \int_1^n f(x)\,dx.$$

If we include a_1, we have

$$a_1 + a_2 + \cdots + a_n \le a_1 + \int_1^n f(x)\,dx.$$

Combining these results gives

$$\int_1^{n+1} f(x)\,dx \le a_1 + a_2 + \cdots + a_n \le a_1 + \int_1^n f(x)\,dx. \qquad (7)$$

If the integral $\int_1^\infty f(x)\,dx$ is finite the right-hand inequality shows that $\sum_{n=1}^\infty a_n$ is also finite. But if $\int_1^\infty f(x)\,dx$ is infinite, then the left-hand inequality shows that the series is also infinite.

Hence the series and the integral are both finite or both infinite. ■

EXAMPLE 5 *The p-series.* If p is a real constant, the series

$$\sum_{n=1}^\infty \frac{1}{n^p} = \frac{1}{1^p} + \frac{1}{2^p} + \frac{1}{3^p} + \cdots + \frac{1}{n^p} + \cdots \qquad (8)$$

converges if $p > 1$ and diverges if $p \le 1$. To prove this, let

$$f(x) = \frac{1}{x^p}.$$

Then, if $p > 1$, we have

$$\int_1^\infty x^{-p}\,dx = \lim_{b\to\infty} \frac{x^{-p+1}}{-p+1}\bigg|_1^b$$

$$= \frac{1}{p-1},$$

which is finite. Hence the p-series converges if p is greater than one.

If $p = 1$, we have

$$1 + \frac{1}{2} + \frac{1}{3} + \cdots + \frac{1}{n} + \cdots,$$

which we already know diverges. Or, by the integral test,

$$\int_1^{\infty} x^{-1}\, dx = \lim_{b\to\infty} \ln x \Big|_1^b = +\infty,$$

and, since the integral diverges, the series does likewise.

Finally, if $p < 1$, then the terms of the p-series are greater than the corresponding terms of the divergent harmonic series. Hence the p-series diverges, by the comparison test.

Thus, we have convergence for $p > 1$, but divergence for every other value of p. □

Given a series $\Sigma\, a_n$ we have two questions:

1. Does the series converge?
2. If it converges, what is its sum?

Most of this chapter is devoted to the first question. But, as a practical matter, the second question is just as important for a scientist or engineer. We digress briefly to address that question now.

Estimation of Remainders by Integrals

The difference $R_n = L - s_n$ between the sum of a convergent series and its nth partial sum is called a *remainder* or a *truncation error*. Since R_n itself is given as an infinite series, which, in principle, is as difficult to evaluate as the original series, you might think that there would be no advantage in singling out R_n for attention. But sometimes even a crude estimate for R_n can lead to an estimate of L that is closer to L than s_n is.

Suppose, for example, that we are interested in learning the numerical value of the series

$$\sum_{k=1}^{\infty} \frac{1}{k^2} = \frac{1}{1^2} + \frac{1}{2^2} + \frac{1}{3^2} + \cdots.$$

This is a p-series with $p = 2$, and hence is known to converge. This means that the sequence of partial sums

$$s_n = \frac{1}{1^2} + \frac{1}{2^2} + \cdots + \frac{1}{n^2}$$

has a limit L. If we want to know L to a couple of decimal places, we might try to find an integer n such that the corresponding *finite* sum s_n differs from L by less, say, than 0.005. Then we would use this s_n in place of L, to two decimals. If we write

$$L = \sum_{k=1}^{\infty} \frac{1}{k^2} = \frac{1}{1^2} + \frac{1}{2^2} + \cdots + \frac{1}{n^2} + \frac{1}{(n+1)^2} + \cdots,$$

we see that

$$R_n = L - s_n = \frac{1}{(n+1)^2} + \cdots.$$

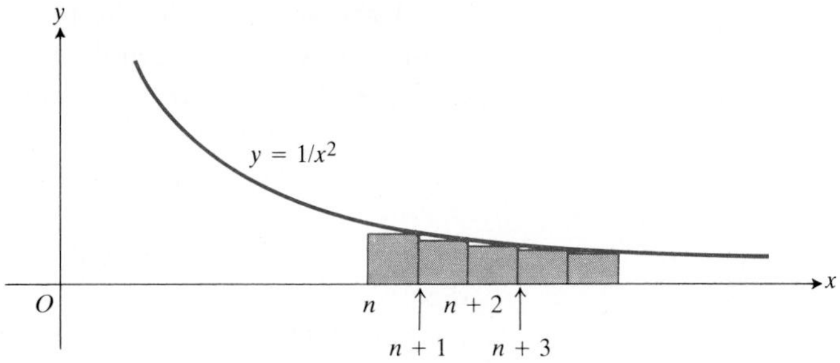

11.11 The rectilinear area R_n is
$$\frac{1}{(n+1)^2}+\frac{1}{(n+2)^2}+\frac{1}{(n+3)^2}+\cdots < \int_n^\infty \frac{dx}{x^2}.$$

We estimate the error R_n by comparing it with the area under the curve

$$y = \frac{1}{x^2}$$

from $x = n$ to ∞.

From Fig. 11.11 we see that

$$R_n < \int_n^\infty \frac{1}{x^2}\,dx = \frac{1}{n},$$

which tells us that, by taking 200 terms of the series, we can be sure that the difference between the sum L of the entire series and the sum s_{200} of these 200 terms will be less than $1/200 = 0.005$.

A somewhat closer estimate of R_n results from using the trapezoidal rule to approximate the area under the curve in Fig. 11.11. Let us write u_k for $1/k^2$ and consider the trapezoidal approximation

$$\begin{aligned} T_n &= \sum_{k=n}^{\infty} \frac{1}{2}(u_k + u_{k+1}) = \frac{1}{2}(u_n + u_{n+1}) + \frac{1}{2}(u_{n+1} + u_{n+2}) + \cdots \\ &= \frac{1}{2}u_n + u_{n+1} + u_{n+2} + \cdots = \frac{1}{2}u_n + R_n. \end{aligned}$$

Now since the curve $y = 1/x^2$ is concave upward.

$$T_n > \int_n^\infty \frac{1}{x^2}\,dx = \frac{1}{n},$$

and we have

$$R_n = T_n - \frac{1}{2}u_n > \frac{1}{n} - \frac{1}{2n^2}.$$

We now know that

$$\frac{1}{n} > R_n > \frac{1}{n} - \frac{1}{2n^2},$$

and $L = s_n + R_n$ may be estimated as follows:

$$s_n + \frac{1}{n} > L > s_n + \frac{1}{n} - \frac{1}{2n^2}. \tag{9}$$

Thus, by using $s_n + 1/n$ in place of s_n to estimate L, we shall be making an error which is numerically less than $1/(2n^2)$. By taking $n \geq 10$, this error is then made less than 0.005. The difference in time required to compute the sum of 10 terms versus 200 terms is sufficiently great to make this sharper analysis of practical importance.

What we have done in the case of this specific example may be done in any case where the graph of the function $y = f(x)$ is concave upward as in Fig. 11.11. We find that when $\int_n^\infty f(x)\,dx$ exists,

$$u_1 + u_2 + \cdots + u_n + \int_n^\infty f(x)\,dx \tag{10}$$

tends to overestimate the value of the series, but by an amount that is less than $u_n/2$.

For excellent articles on the subject of estimating remainders, see "Estimating Remainders," by R. P. Boas, *Mathematics Magazine,* Volume 51, Number 2, March 1978, pp. 83–89, and "Partial Sums of Infinite Series and How They Grow," R. P. Boas, *American Mathematical Monthly,* Volume 84, Number 4, April 1977, pp. 237–258.

The Limit Comparison Test

We resume discussion of criteria for determining the convergence or divergence of a given series $\Sigma\, a_n$. Observe that we can combine the comparison tests with the information we have about the geometric series and the various p-series. There is a more powerful form of the comparison test that is known as the *limit comparison test,* which we take up next. It is particularly handy in dealing with series in which a_n is a rational function of n. The next example will illustrate what we mean.

EXAMPLE 6 Do you think $\Sigma\, a_n$ converges? or diverges? Why?

a) $a_n = \dfrac{2n}{n^2 - n + 1}$ b) $a_n = \dfrac{2n^3 + 100n^2 + 1000}{\frac{1}{8}n^6 - n + 2}$

Discussion In determining convergence or divergence, only the tails count. And, when n is very large, the highest powers of n in both numerator and denominator are what count the most. So, in (a), we reason this way:

$$a_n = \frac{2n}{n^2 - n + 1} \quad \text{behaves about like} \quad \frac{2n}{n^2} = \frac{2}{n},$$

and we guess that $\Sigma\, a_n$ diverges by comparison with $\Sigma\, 1/n$. In (b), we reason that a_n will behave about like $2n^3/(1/8)n^6 = 16/n^3$ and we guess that the series will converge, by comparison with $\Sigma\, 1/n^3$, a p-series with $p = 3$.

To be more precise, in part (a) we can take

$$a_n = \frac{2n}{n^2 - n + 1} \quad \text{and} \quad d_n = \frac{1}{n}$$

and look at the ratio

$$\frac{a_n}{d_n} = \frac{2n^2}{n^2 - n + 1} = \frac{2}{1 - \left(\frac{1}{n}\right) + \left(\frac{1}{n^2}\right)}.$$

Clearly, as $n \to \infty$ the limit is 2:

$$\lim \frac{a_n}{d_n} = 2.$$

This means that, in particular, if we take $\varepsilon = 1$ in the definition of limit, we know there is an index N such that a_n/d_n is within 1 unit of this limit for all $n \geq N$:

$$2 - 1 \leq a_n/d_n \leq 2 + 1 \qquad \text{for} \quad n \geq N.$$

Thus $a_n \geq d_n$ for $n \geq N$. Therefore, by the comparison test, $\Sigma\, a_n$ diverges because $\Sigma\, d_n$ diverges.

In part (b), if we let $c_n = 1/n^3$, we can easily show that

$$\lim \frac{a_n}{c_n} = 16.$$

Taking $\varepsilon = 1$ in the definition of limit, we can conclude that there is an index N' such that a_n/c_n is between 15 and 17 when $n \geq N'$. Since $\Sigma\, c_n$ converges, so also does $\Sigma\, 17c_n$ and thus $\Sigma\, a_n$. □

Our rather rough guesswork paved the way for successful choices of comparison series. We make all of this more precise in the following *limit comparison test.*

Limit Comparison Test

A. Test for convergence. If $a_n \geq 0$ for $n \geq n_0$ and there is a convergent series $\Sigma\, c_n$ such that $c_n > 0$ and

$$\lim \frac{a_n}{c_n} < \infty, \tag{11}$$

then $\Sigma\, a_n$ converges.

B. Test for divergence. If $a_n \geq 0$ for $n \geq n_0$ and there is a divergent series $\Sigma\, d_n$ such that $d_n > 0$ and

$$\lim \frac{a_n}{d_n} > 0, \tag{12}$$

then $\Sigma\, a_n$ diverges.

We shall not formally prove these results. Part (A) follows easily from the fact that if (11) holds, then there is an index $N \geq n_0$ and a constant M such that $a_n < Mc_n$ when $n > N$, and $\Sigma\, Mc_n$ converges. Similarly, if (12) holds, there is an index $N' \geq n_0$ and a constant $k > 0$ such that $a_n > kd_n$ for $n \geq N'$, and $\Sigma\, kd_n$ diverges.

EXAMPLE 7 Do the following series converge, or diverge?

a) $\displaystyle \frac{3}{4} + \frac{5}{9} + \frac{7}{16} + \frac{9}{25} + \cdots = \sum_{n=1}^{\infty} \frac{2n+1}{(n+1)^2}$

b) $\displaystyle \frac{1}{2} + \frac{2}{3} + \frac{3}{4} + \frac{4}{5} + \frac{5}{6} + \cdots = \sum_{n=1}^{\infty} \frac{n}{n+1}$

c) $\displaystyle \frac{101}{3} + \frac{102}{10} + \frac{103}{29} + \cdots = \sum_{n=1}^{\infty} \frac{100+n}{n^3+2}$

d) $\frac{1}{1} + \frac{1}{3} + \frac{1}{7} + \cdots = \sum_{n=1}^{\infty} \frac{1}{2^n - 1}$

Solution a) Let $a_n = (2n + 1)/(n^2 + 2n + 1)$ and $d_n = 1/n$. Then

$$\sum d_n \text{ diverges} \qquad \text{and} \qquad \lim \frac{a_n}{d_n} = \lim \frac{2n^2 + n}{n^2 + 2n + 1} = 2,$$

so

$$\sum a_n \text{ diverges.}$$

b) Let $b_n = n/(n + 1)$. Then $\lim b_n = 1 \neq 0$, so $\Sigma\, b_n$ diverges, by the nth term test.

c) Let $a_n = (100 + n)/(n^3 + 2)$. When n is large, this ought to compare with $n/n^3 = 1/n^2$, so we let $c_n = 1/n^2$. Then we apply the limit comparison theorem:

$$\sum c_n \text{ converges} \qquad \text{and} \qquad \lim \frac{a_n}{c_n} = \lim \frac{n^3 + 100n^2}{n^3 + 2} = 1,$$

so

$$\sum a_n \text{ converges.}$$

d) Let $a_n = 1/(2^n - 1)$ and $c_n = 1/2^n$. (We reason that $2^n - 1$ behaves about like 2^n when n is large.) Then

$$\frac{a_n}{c_n} = \frac{2^n}{2^n - 1} = \frac{1}{1 - (1/2)^n} \to 1 \quad \text{as} \quad n \to \infty.$$

Because $\Sigma\, c_n$ converges, we conclude that $\Sigma\, a_n$ does also. □

When we use the p-series for comparison, it is essential that p be a constant, as shown by this example.

EXAMPLE 8 Does the series

$$1^{-2} + 2^{-3/2} + 3^{-4/3} + 4^{-5/4} + \cdots + n^{-(n+1)/n} + \cdots$$

converge, or diverge?

Solution Let

$$a_n = n^{-(n+1)/n} = \frac{1}{n^{1+(1/n)}}.$$

This looks a bit like $1/n^p$ with $p = 1 + (1/n)$, which is greater than 1. But $1 + (1/n)$ isn't constant, so we need to be careful about drawing a conclusion. In fact

$$n^{1+(1/n)} = n(n)^{1/n} \qquad \text{and} \qquad (n)^{1/n} \to 1 \quad \text{as} \quad n \to \infty.$$

Let's use the limit comparison test with $d_n = 1/n$. Then

$$\frac{a_n}{d_n} = \frac{n}{n(n)^{1/n}} = \frac{1}{(n)^{1/n}} \to 1 \quad \text{as} \quad n \to \infty.$$

We know that $\Sigma\, d_n$ diverges, and when we apply the limit comparison test we conclude that $\Sigma\, a_n$ also diverges. □

EXAMPLE 9 Does the series $\Sigma_{n=1}^{\infty} \ln n/n^{3/2}$ converge?

Solution Let $a_n = \ln n/n^{3/2}$ and $c_n = n^c/n^{3/2}$, where c is a positive constant that is less than $\frac{1}{2}$. For example, we might choose $c = \frac{1}{4}$. Then $c_n = 1/n^{3/2-c} = 1/n^p$ with $p = \frac{3}{2} - c > 1$. (When $c = \frac{1}{4}$, $p = \frac{5}{4}$.) Hence $\Sigma\, c_n$ converges. Because $\ln n$ goes to infinity more slowly than n^c, for any positive constant c (Article 11.3, Example 8), we have reason to believe that $\Sigma\, a_n$ also converges. To verify this hunch, we apply the limit comparison test:

$$\lim \frac{a_n}{c_n} = \lim \frac{\ln n}{n^c} = \lim \frac{1/n}{cn^{c-1}} \qquad \text{(by l'Hôpital's rule)}$$

$$= \lim \frac{1}{cn^c} = 0.$$

The given series converges. □

The Ratio Test

It is not always possible to tell whether a particular series converges by using the comparison test or the limit comparison test. We might not be able to find a series to compare it with. Nor is it always possible to use the integral test to answer whatever questions of convergence then remain unanswered. The terms of the series might not decrease as n increases, or we might not find a formula for the nth term that we can integrate. What we really need is an intrinsic test that can be applied directly to the terms of the given series $\Sigma\, a_n$. The next two tests are intrinsic and they are easy to apply. The first of these, the *ratio test,* measures the rate of growth (or decline) by examining the ratio a_{n+1}/a_n. For a geometric series, this rate of growth is a constant, and the series converges if and only if it is less than 1 in absolute value. But even if the ratio is not constant, we may be able to find a geometric series for comparison, as illustrated in the next example.

EXAMPLE 10 Let $a_1 = 1$ and define a_{n+1} to be

$$a_{n+1} = \frac{n}{2n+1} a_n.$$

Does the series $\Sigma\, a_n$ converge, or diverge?

Solution We begin by writing out a few terms of the series:

$$a_1 = 1, \qquad a_2 = \frac{1}{3}, \qquad a_3 = \frac{2}{5} a_2 = \frac{1 \cdot 2}{3 \cdot 5}, \qquad a_4 = \frac{3}{7} a_3 = \frac{1 \cdot 2 \cdot 3}{3 \cdot 5 \cdot 7}.$$

We observe that each term is somewhat less than $\frac{1}{2}$ the term before it, because $n/(2n + 1)$ is less than $\frac{1}{2}$. Therefore the terms of the given series are less than or equal to the terms of the geometric series

$$1 + \left(\frac{1}{2}\right) + \left(\frac{1}{4}\right) + \cdots + \left(\frac{1}{2}\right)^{n-1} + \cdots$$

that converges to 2. So our series also converges, and we now know that its sum is less than 2. □

In *proving* the ratio test, we shall make a comparison with appropriate geometric series as in the example above, but when we *apply* it we don't actually make a direct comparison.

The Ratio Test

Let $\Sigma\, a_n$ be a series with positive terms, and suppose that

$$\lim_{n\to\infty} \frac{a_{n+1}}{a_n} = \rho \qquad \text{(Greek letter rho).}$$

Then,

a) the series *converges* if $\rho < 1$,

b) the series *diverges* if $\rho > 1$,

c) the series *may converge or it may diverge* if $\rho = 1$. (The test provides no information.)

Proof. a) Assume first that $\rho < 1$ and let r be a number between ρ and 1. Then the number

$$\varepsilon = r - \rho$$

is positive. Since

$$\frac{a_{n+1}}{a_n} \to \rho,$$

a_{n+1}/a_n must lie within ε of ρ when n is large enough, say, for all $n \geq N$. In particular,

$$\frac{a_{n+1}}{a_n} < \rho + \varepsilon = r, \qquad \text{when } n > N.$$

That is,

$$\begin{aligned} &a_{N+1} < ra_N, \\ &a_{N+2} < ra_{N+1} < r^2 a_N, \\ &a_{N+3} < ra_{N+2} < r^3 a_N, \\ &\vdots \\ &a_{N+m} < ra_{N+m-1} < r^m a_N. \end{aligned}$$

These inequalities show that the terms of our series, after the Nth term, approach zero more rapidly than the terms in a geometric series with ratio $r < 1$. More precisely, consider the series $\Sigma\, c_n$, where $c_n = a_n$ for $n = 1, 2, \ldots, N$ and $c_{N+1} = ra_N$, $c_{N+2} = r^2 a_N, \ldots, c_{N+m} = r^m a_N, \ldots$. Now $a_n \leq c_n$ for all n, and

$$\begin{aligned} \sum_{n=1}^{\infty} c_n &= a_1 + a_2 + \cdots + a_{N-1} + a_N + ra_N + r^2 a_N + \cdots \\ &= a_1 + a_2 + \cdots + a_{N-1} + a_N(1 + r + r^2 + \cdots). \end{aligned}$$

Because $|r| < 1$, the geometric series $1 + r + r^2 + \cdots$ converges, and hence so does $\Sigma\, c_n$. By the comparison test, $\Sigma\, a_n$ also converges.

b) Next, suppose $\rho > 1$. Then, from some index M on, we have

$$\frac{a_{n+1}}{a_n} > 1 \qquad \text{or} \qquad a_M < a_{M+1} < a_{M+2} < \cdots.$$

Hence, the terms of the series do not approach 0 as n becomes infinite, and the series diverges, by the nth-term test.

c) Finally, the two series

$$\sum_{n=1}^{\infty} \frac{1}{n} \quad \text{and} \quad \sum_{n=1}^{\infty} \frac{1}{n^2}$$

show that, when $\rho = 1$, some other test for convergence must be used.

$$\text{For } \sum_{n=1}^{\infty} \frac{1}{n}: \qquad \frac{a_{n+1}}{a_n} = \frac{1/(n+1)}{1/n} = \frac{n}{n+1} \to 1.$$

$$\text{For } \sum_{n=1}^{\infty} \frac{1}{n^2}: \qquad \frac{a_{n+1}}{a_n} = \frac{1/(n+1)^2}{1/n^2} = \left(\frac{n}{n+1}\right)^2 \to 1^2 = 1.$$

In both cases $\rho = 1$, yet the first series diverges while the second converges. ■

The ratio test is often effective when the terms of the series contain factorials of expressions involving n, or expressions raised to the nth power, or combinations, as in the next example. We recall that

$$n! = 1 \cdot 2 \cdot 3 \cdot \cdots \cdot n$$

implies that

$$(n+1)! = (n+1)n!$$

and that $(2(n+1))! = (2n+2)! = (2n+2)(2n+1)(2n)!$. These facts are used in parts (a) and (b) of the example. Part (b) also uses the fact that $4^{n+1}/4^n = 4$. Making use of appropriate cancellation properties of factorials and powers leads to simplified expressions for the ratio a_{n+1}/a_n.

EXAMPLE 11 Test the following series for convergence or divergence, using the ratio test.

a) $\displaystyle\sum_{n=1}^{\infty} \frac{n!n!}{(2n)!}$ b) $\displaystyle\sum_{n=1}^{\infty} \frac{4^n n!n!}{(2n)!}$ c) $\displaystyle\sum_{n=0}^{\infty} \frac{2^n + 5}{3^n}$

Solution a) If $a_n = n!n!/(2n)!$, then $a_{n+1} = (n+1)!(n+1)!/(2n+2)!$, and

$$\frac{a_{n+1}}{a_n} = \frac{(n+1)!(n+1)!(2n)!}{n!n!(2n+2)(2n+1)(2n)!}$$
$$= \frac{(n+1)(n+1)}{(2n+2)(2n+1)} = \frac{n+1}{4n+2} \to \frac{1}{4}.$$

The series converges because $\rho = \frac{1}{4}$ is less than 1.

b) If $a_n = 4^n n!n!/(2n)!$, then

$$\frac{a_{n+1}}{a_n} = \frac{4^{n+1}(n+1)!(n+1)!}{(2n+2)(2n+1)(2n)!} \times \frac{(2n)!}{4^n n!n!}$$
$$= \frac{4(n+1)(n+1)}{(2n+2)(2n+1)} = \frac{2(n+1)}{2n+1} \to 1.$$

Because the limit is $\rho = 1$, we cannot decide on the basis of the ratio test alone whether the series converges or diverges. However, when we note that $a_{n+1}/a_n = (2n + 2)/(2n + 1)$, we conclude that a_{n+1} is always greater than a_n because $(2n + 2)/(2n + 1)$ is always greater than 1. Therefore, all terms are greater than or equal to $a_1 = 2$ and the nth term does not go to zero as n tends to infinity. Hence, by the nth term test, the series diverges.

c) For the series $\Sigma_{n=0}^{\infty} (2^n + 5)/3^n$,

$$\frac{a_{n+1}}{a_n} = \frac{(2^{n+1} + 5)/3^{n+1}}{(2^n + 5)/3^n} = \frac{1}{3} \cdot \frac{2^{n+1} + 5}{2^n + 5}$$
$$= \frac{1}{3} \cdot \left(\frac{2 + 5 \cdot 2^{-n}}{1 + 5 \cdot 2^{-n}}\right) \to \frac{1}{3} \cdot \frac{2}{1} = \frac{2}{3}.$$

The series converges because $\rho = \frac{2}{3}$ is less than 1.

This does *not* mean that $\frac{2}{3}$ is the sum of the series. In fact,

$$\sum_{n=0}^{\infty} \frac{2^n + 5}{3^n} = \sum_{n=0}^{\infty} \left(\frac{2}{3}\right)^n + \sum_{n=0}^{\infty} \frac{5}{3^n} = \frac{1}{1 - \frac{2}{3}} + \frac{5}{1 - \frac{1}{3}} = 10\tfrac{1}{2}. \quad \square$$

REMARK 1. While the ratio test is useful in the types of series just discussed, it is not very useful for series like p-series.

In the next example, the series is expressed in terms of powers of x. By applying the ratio test, we can learn for what values of x the series converges. For those values, the series defines a function of x. (In Chapter 12, we discuss such power series in more detail.)

EXAMPLE 12 For what values of x does the series

$$x + \frac{x^3}{3} + \frac{x^5}{5} + \frac{x^7}{7} + \cdots + \frac{x^{2n-1}}{2n - 1} + \cdots$$

converge?

Solution The nth term of the series is

$$a_n = \frac{x^{2n-1}}{2n - 1}.$$

We consider first the case where x is positive. Then the series is a positive series and

$$\frac{a_{n+1}}{a_n} = \frac{(2n - 1)x^2}{(2n + 1)} \to x^2.$$

The ratio test therefore tells us that the series converges if x is positive and less than one and diverges if x is greater than one.

Since only odd powers of x occur in the series, we see that the series simply changes sign when x is replaced by $-x$. Therefore the series also converges for $-1 < x \leq 0$ and diverges for $x < -1$. The series converges to zero when $x = 0$.

We know, thus far, that the series

converges for $|x| < 1$,
diverges for $|x| > 1$,

but we don't know what happens when $|x| = 1$. To test at $x = 1$, we apply the integral test to the series

$$1+\frac{1}{3}+\frac{1}{5}+\frac{1}{7}+\cdots+\frac{1}{2n-1}+\cdots,$$

which we get by taking $x = 1$ in the original series. The companion integral is

$$\int_1^{\infty}\frac{dx}{2x-1}=\frac{1}{2}\ln(2x-1)\bigg|_1^{\infty}=\infty.$$

Hence the series diverges to $+\infty$ when $x = 1$. It diverges to $-\infty$ when $x = -1$. Therefore the only values of x for which the given series converges are $-1 < x < 1$. □

REMARK 2. When $\rho < 1$, the ratio test is also useful in estimating the truncation error that results from using

$$s_N = a_1 + a_2 + \cdots + a_N$$

as an approximation to the sum of a convergent series of positive terms

$$S = a_1 + a_2 + \cdots + a_N + (a_{N+1} + \cdots);$$

for, if we know that

$$r_1 \le \frac{a_{n+1}}{a_n} \le r_2 \qquad \text{for} \qquad n \ge N, \tag{13}$$

where r_1 and r_2 are constants that are both less than one, then the inequalities

$$r_1 a_n \le a_{n+1} \le r_2 a_n \qquad (n = N, N+1, N+2, \ldots)$$

enable us to deduce that

$$a_N(r_1 + r_1^2 + r_1^3 + \cdots) \le \sum_{n=N+1}^{\infty} a_n \le a_N(r_2 + r_2^2 + r_2^3 + \cdots). \tag{14}$$

The two geometric series have sums

$$r_1 + r_1^2 + r_1^3 + \cdots = \frac{r_1}{1-r_1}, \qquad r_2 + r_2^2 + r_2^3 + \cdots = \frac{r_2}{1-r_2}.$$

Hence, the error

$$R_N = \sum_{n=N+1}^{\infty} a_n$$

lies between

$$a_N\frac{r_1}{1-r_1} \quad \text{and} \quad a_N\frac{r_2}{1-r_2}.$$

That is,

$$a_N\frac{r_1}{1-r_1} \le S - s_N \le a_N\frac{r_2}{1-r_2} \tag{15}$$

if

$$0 \le r_1 \le \frac{a_{n+1}}{a_n} \le r_2 < 1 \qquad \text{for} \quad n \ge N.$$

converges, from which it follows that

$$\sum_{n=1}^{\infty} a_n = a_1 + \cdots + a_{N-1} + \sum_{n=N}^{\infty} a_n$$

converges.

b) Suppose that $\rho > 1$. Then, for all indices beyond some index M, we have

$$\sqrt[n]{a_n} > 1,$$

so that

$$a_n > 1 \quad \text{for} \quad n > M,$$

and the terms of the series do not converge to 0. The series therefore diverges by the nth-term test.

c) The series $\Sigma_{n=1}^{\infty} (1/n)$ and $\Sigma_{n=1}^{\infty} (1/n^2)$ show that the test is not conclusive when $\rho = 1$. The first series diverges and the second converges, but in both cases $\sqrt[n]{a_n} \to 1$. ■

EXAMPLE 14 For the series $\Sigma_{n=1}^{\infty} (1/n^n)$,

$$\sqrt[n]{\frac{1}{n^n}} = \frac{1}{n} \to 0.$$

The series converges. □

EXAMPLE 15 For the series $\Sigma_{n=1}^{\infty} (2^n/n^2)$,

$$\sqrt[n]{\frac{2^n}{n^2}} = \frac{2}{\sqrt[n]{n^2}} = \frac{2}{(\sqrt[n]{n})^2} \to \frac{2}{1^2} = 2.$$

The series diverges. □

EXAMPLE 16 For the series $\Sigma_{n=1}^{\infty} (1 - 1/n)^n = 0 + \frac{1}{4} + \frac{8}{27} + \cdots$,

$$\sqrt[n]{\left(1 - \frac{1}{n}\right)^n} = \left(1 - \frac{1}{n}\right) \to 1.$$

Because $\rho = 1$, the root test is not conclusive. However, if we apply the nth-term test for divergence, we find that

$$\left(1 - \frac{1}{n}\right)^n = \left(1 + \frac{-1}{n}\right)^n \to e^{-1} = \frac{1}{e}.$$

The series diverges. □

List of Tests

We now have seven tests for divergence and convergence of infinite series:

1. The nth-term test for divergence (applies to all series).
2. The "bounded from above" test (applies to partial sums of nonnegative series).
3. A. The comparison test for convergence (nonnegative series for $n \geq n_0$).

B. The comparison test for divergence (nonnegative series for $n \geq n_0$).

4. A. The limit comparison test for convergence (as for 3A).
   B. The limit comparison test for divergence (as for 3B).
5. The integral test (positive decreasing series).
6. The ratio test (positive series).
7. The nth root test (nonnegative series for $n \geq n_0$).

NOTE. These tests can also be applied to settle questions about the convergence or divergence of series of nonpositive or negative terms. Just factor -1 from the series in question, and test the resulting series of nonnegative or positive terms.

EXAMPLE 17

$$\sum_{n=1}^{\infty} -\frac{1}{n} = -1 \cdot \sum_{n=1}^{\infty} \frac{1}{n} \quad \text{diverges.} \ \Box$$

EXAMPLE 18

$$\sum_{n=0}^{\infty} -\frac{1}{2^n} = -1 \cdot \sum_{n=0}^{\infty} \frac{1}{2^n} = -1 \cdot 2 = -2. \ \Box$$

PROBLEMS

In Problems 1–30, determine whether the given series converges or diverges. In each case, give a reason for your answer.

1. $\sum_{n=1}^{\infty} \frac{1}{10^n}$

2. $\sum_{n=1}^{\infty} \frac{n}{n+2}$

3. $\sum_{n=1}^{\infty} \frac{\sin^2 n}{2^n}$

4. $\sum_{n=1}^{\infty} \frac{5}{n}$

5. $\sum_{n=1}^{\infty} \frac{n^3}{2^n}$

6. $\sum_{n=1}^{\infty} -\frac{1}{8^n}$

7. $\sum_{n=1}^{\infty} \frac{\ln n}{n}$

8. $\sum_{n=1}^{\infty} \frac{1}{n\sqrt{n}}$

9. $\sum_{n=1}^{\infty} \frac{2^n}{3^n}$

10. $\sum_{n=0}^{\infty} \frac{-2}{n+1}$

11. $\sum_{n=1}^{\infty} \frac{1}{1+\ln n}$

12. $\sum_{n=1}^{\infty} \frac{\ln n}{\sqrt{n+1}\, n}$

13. $\sum_{n=1}^{\infty} \frac{2^n}{n+1}$

14. $\sum_{n=1}^{\infty} \left(\frac{n}{3n+1}\right)^n$

15. $\sum_{n=1}^{\infty} -\frac{n^2}{2^n}$

16. $\sum_{n=1}^{\infty} \frac{1}{\sqrt{n}} \frac{(\ln n)^{10}}{n^{2/3}}$

17. $\sum_{n=1}^{\infty} \frac{1}{\sqrt{n^3+2}}$

18. $\sum_{n=1}^{\infty} \frac{1}{\sqrt[n]{2}}$

19. $\sum_{n=1}^{\infty} \frac{(n+1)(n+2)}{n!}$

20. $\sum_{n=1}^{\infty} \frac{\sqrt{n}}{n^2+1}$

21. $\sum_{n=1}^{\infty} \frac{n}{n^2+1}$

22. $\sum_{n=1}^{\infty} n^2 e^{-n}$

23. $\sum_{n=1}^{\infty} \left(1+\frac{1}{n}\right)^n$

24. $\sum_{n=1}^{\infty} \frac{1}{3^{n-1}+1}$

25. $\sum_{n=1}^{\infty} \frac{(n+3)!}{3!n!3^n}$

26. $\sum_{n=2}^{\infty} \frac{1}{n \ln n}$

27. $\sum_{n=1}^{\infty} \frac{1}{(2n+1)!}$

28. $\sum_{n=1}^{\infty} \frac{1}{(\ln 2)^n}$

29. $\sum_{n=1}^{\infty} \frac{n!}{n^n}$

30. $\sum_{n=1}^{\infty} \frac{1-n}{n \cdot 2^n}$

31. Show that the series

$$\sum_{n=1}^{\infty} \frac{1}{2n-1} = 1 + \frac{1}{3} + \frac{1}{5} + \cdots$$

diverges. (*Hint:* Compare the series with a multiple of the harmonic series.)

In Problems 32–34, find all values of x for which the given series converge. Begin with the ratio test or the root test, and then apply other tests as needed.

32. $\sum_{n=1}^{\infty} \left(\frac{x^2+1}{3}\right)^n$ **33.** $\sum_{n=1}^{\infty} \frac{x^{2n+1}}{n^2}$

34. $\sum_{n=1}^{\infty} \left(\frac{1}{|x|}\right)^n$

35. (*Calculator*) Use Inequality (9) of this article to estimate $\Sigma_{n=1}^{\infty} (1/n^2)$ with an error less than 0.005. Compare your result with the value given in Problem 36.

36. (*Calculator*) Euler discovered that

$$\sum_{n=1}^{\infty} \frac{1}{n^2} = \sum_{n=1}^{\infty} \frac{3[(n-1)!]^2}{(2n)!} = \frac{\pi^2}{6}.$$

Compute s_6 for each series. To 10 decimal places, $\pi^2/6 = 1.6449340668$.

37. (*Calculator*) Use the expression in (10) of this article to find the value of $\Sigma_{n=1}^{\infty} (1/n^4) = (\pi^4/90)$ with an error less than 10^{-6}.

38. (*Calculator*) To estimate partial sums of the divergent harmonic series, Inequality (7) with $f(x) = 1/x$ tells us that

$$\ln n < 1 + \frac{1}{2} + \cdots + \frac{1}{n} < 1 + \ln n.$$

Suppose that the summation started with $s_1 = 1$ thirteen billion years ago (one estimate of the age of the universe) and that a new term has been added every *second* since then. How large would you expect s_n to be today?

39. There are no values of x for which $\Sigma_{n=1}^{\infty} (1/nx)$ converges. Why?

40. Show that if $\Sigma_{n=1}^{\infty} a_n$ is a convergent series of nonnegative numbers then the series $\Sigma_{n=1}^{\infty} (a_n/n)$ converges.

41. Show that if $\Sigma\, a_n$ and $\Sigma\, b_n$ are convergent series with $a_n \geq 0$ and $b_n \geq 0$, then $\Sigma\, a_n b_n$ converges. (*Hint:* From some index on, $a_n b_n < a_n + b_n$.)

42. A sequence of numbers

$$s_1 \geq s_2 \geq \cdots \geq s_n \geq s_{n+1} \geq \cdots,$$

in which $s_n \geq s_{n+1}$ for every n, is called a *decreasing sequence*. A sequence $\{s_n\}$ is *bounded from below* if there is a finite constant M with $M \leq s_n$ for every n. Such a number M is called a *lower bound* for the sequence. Deduce from Theorem 1 that a decreasing sequence that is bounded from below converges, and that a decreasing sequence that is not bounded from below diverges.

43. The *Cauchy condensation test* says: Let $\{a_n\}$ be a decreasing sequence ($a_n \geq a_{n+1}$, all n) of positive terms that converges to 0. Then,

$$\sum a_n \text{ converges} \quad \text{if and only if} \quad \sum 2^n a_{2^n} \text{ converges}.$$

For example, $\Sigma\,(1/n)$ diverges because $\Sigma\, 2^n \cdot (1/2^n) = \Sigma\, 1$. Show why the test works.

44. Use the Cauchy condensation test of Problem 43 to show that

a) $\sum_{n=2}^{\infty} \frac{1}{n \ln n}$ diverges.

b) $\sum_{n=1}^{\infty} \frac{1}{n^p}$ converges if $p > 1$ and diverges if $p \leq 1$.

45. Pictures like the one in Fig. 11.9 suggest that, as n increases, there is very little change in the difference between the sum

$$1 + \frac{1}{2} + \cdots + \frac{1}{n}$$

and the integral

$$\ln n = \int_1^n \frac{1}{x}\,dx.$$

To explore this idea, carry out the following steps.

a) By taking $f(x) = 1/x$ in inequality (7), show that

$$\ln(n+1) \leq 1 + \frac{1}{2} + \cdots + \frac{1}{n} \leq 1 + \ln n$$

or

$$0 < \ln(n+1) - \ln n \leq 1 + \frac{1}{2} + \cdots + \frac{1}{n} - \ln n \leq 1.$$

Thus, the sequence

$$a_n = 1 + \frac{1}{2} + \cdots + \frac{1}{n} - \ln n$$

is bounded from below and from above.

b) Show that

$$\frac{1}{n+1} < \int_n^{n+1} \frac{1}{x}\,dx = \ln(n+1) - \ln n,$$

so that the sequence $\{a_n\}$ in part (a) is decreasing.

Since a decreasing sequence that is bounded from below converges (Problem 42) the numbers a_n defined in (a) converge:

$$1 + \frac{1}{2} + \cdots + \frac{1}{n} - \ln n \to \gamma.$$

The number γ, whose value is 0.5772 . . . , is called *Euler's constant*. In contrast to other special numbers like π and e, no other expression with a simple law of formulation has ever been found for γ.

46. *Logarithmic p-series.* Let p be a positive constant. Show that

$$\int_2^{\infty} \frac{dx}{x(\ln x)^p}$$

converges if and only if $p > 1$. (The integral does not start at 1, but at 2, because $\ln 1 = 0$.) What can you deduce about convergence or divergence of the following series?

a) $\sum_{n=2}^{\infty} \frac{1}{n \ln n}$

b) $\sum_{n=2}^{\infty} \frac{1}{n(\ln n)^{1.01}}$

c) $\sum_{n=5}^{\infty} \frac{n^{1/2}}{(\ln n)^3}$

d) $\sum_{n=3}^{\infty} \frac{1}{n \ln (n^3)}$

e) $\sum_{n=2}^{\infty} \frac{1}{n(\ln n)^{(n+1)/n}}$

***Tookit* programs**

Sequences and Series

11.6 Absolute Convergence

We now extend to series that have both positive and negative terms the techniques that we have developed for answering questions about the convergence of series of nonnegative numbers. The extension is made possible by a theorem that says that, if a series converges after all its negative terms have been made positive, then the unaltered series converges also.

THEOREM 1

If $\Sigma_{n=1}^{\infty} |a_n|$ converges, then $\Sigma_{n=1}^{\infty} a_n$ converges.

Proof of the theorem. For each n,

$$-|a_n| \leq a_n \leq |a_n|,$$

so that

$$0 \leq a_n + |a_n| \leq 2|a_n|.$$

If $\Sigma_{n=1}^{\infty} |a_n|$ converges, then $\Sigma_{n=1}^{\infty} 2|a_n|$ converges and, by the comparison test, the nonnegative series

$$\sum_{n=1}^{\infty} (a_n + |a_n|)$$

converges. The equality $a_n = (a_n + |a_n|) - |a_n|$ now lets us express $\Sigma_{n=1}^{\infty} a_n$ as the difference of two convergent series:

$$\sum_{n=1}^{\infty} a_n = \sum_{n=1}^{\infty} (a_n + |a_n| - |a_n|) = \sum_{n=1}^{\infty} (a_n + |a_n|) - \sum_{n=1}^{\infty} |a_n|.$$

Therefore, $\Sigma_{n=1}^{\infty} a_n$ converges. ■

An obvious corollary of Theorem 1 (which is also called the *contrapositive* form of the theorem) is

COROLLARY

If $\Sigma\, a_n$ diverges, then $\Sigma\, |a_n|$ diverges.

(See Problem 19.)

DEFINITION

A series $\Sigma_{n=1}^{\infty} a_n$ is said to *converge absolutely* if $\Sigma_{n=1}^{\infty} |a_n|$ converges.

Our theorem can now be rephrased to say that *every absolutely convergent series converges.* We shall see in the next article, however, that the converse of this statement is false. Many convergent series do not converge absolutely. That is, there are many series whose convergence depends on the presence of infinitely many positive and negative terms arranged in a particular order.

Here are some examples of how the theorem can and cannot be used to determine convergence.

EXAMPLE 1 For $\sum_{n=1}^{\infty} (-1)^{n+1} \frac{1}{n^2} = 1 - \frac{1}{4} + \frac{1}{9} - \frac{1}{16} + \cdots,$

the corresponding series of absolute values is

$$\sum_{n=1}^{\infty} \frac{1}{n^2} = 1 + \frac{1}{4} + \frac{1}{9} + \frac{1}{16} + \cdots,$$

which converges because it is a p-series with $p = 2 > 1$ (Article 11.5). Therefore

$$\sum_{n=1}^{\infty} (-1)^{n+1} \frac{1}{n^2}$$

converges absolutely. Therefore

$$\sum_{n=1}^{\infty} (-1)^{n+1} \frac{1}{n^2}$$

converges. □

EXAMPLE 2 For $\sum_{n=1}^{\infty} \frac{\sin n}{n^2} = \frac{\sin 1}{1} + \frac{\sin 2}{4} + \frac{\sin 3}{9} + \cdots,$

the corresponding series of absolute values is

$$\sum_{n=1}^{\infty} \left| \frac{\sin n}{n^2} \right| = \frac{|\sin 1|}{1} + \frac{|\sin 2|}{4} + \cdots,$$

which converges by comparison with $\Sigma_{n=1}^{\infty} (1/n^2)$, because $|\sin n| \le 1$ for every n. The original series converges absolutely; therefore it converges. □

EXAMPLE 3 For $\sum_{n=1}^{\infty} (-1)^{n+1} \frac{1}{n} = 1 - \frac{1}{2} + \frac{1}{3} - \frac{1}{4} + \cdots,$

the corresponding series of absolute values is

$$\sum_{n=1}^{\infty} \frac{1}{n} = 1 + \frac{1}{2} + \frac{1}{3} + \frac{1}{4} + \cdots,$$

which diverges. *We can draw no conclusion from this about the convergence or divergence of the original series.* Some other test must be found. In fact, the original series converges, as we shall see in the next article. □

EXAMPLE 4 The series

$$\sum_{n=1}^{\infty} (-1)^n \frac{n}{5n+1} = -\frac{1}{6} + \frac{2}{11} - \frac{3}{16} + \frac{4}{21} - \cdots$$

does not converge, by the nth-term test. Therefore, by the Corollary of Theorem 1, the series does not converge absolutely. □

REMARK. We know that $\Sigma\, a_n$ converges if $\Sigma\, |a_n|$ converges, but the two series will generally not converge to the same sum. For example,

$$\sum_{n=0}^{\infty} \left|\frac{(-1)^n}{2^n}\right| = \sum_{n=0}^{\infty} \frac{1}{2^n} = \frac{1}{1-\frac{1}{2}} = 2,$$

while

$$\sum_{n=0}^{\infty} \frac{(-1)^n}{2^n} = \frac{1}{1+\frac{1}{2}} = \frac{2}{3}.$$

In fact, when a series $\Sigma\, a_n$ converges absolutely, we can expect $\Sigma\, a_n$ to equal $\Sigma\, |a_n|$ only if none of the numbers a_n is negative.

One other important fact about absolutely convergent series is

THEOREM 2

The Rearrangement Theorem for Absolutely Convergent Series

If $\Sigma\, a_n$ converges absolutely, and $b_1, b_2, \ldots, b_n, \ldots$ is any arrangement of the sequence $\{a_n\}$, then

$$\sum b_n \text{ converges}$$

and

$$\sum_{n=1}^{\infty} b_n = \sum_{n=1}^{\infty} a_n.$$

(For an outline of the proof, see Problem 22.)

EXAMPLE 5 As we saw in Example 1, the series

$$1 - \left(\frac{1}{4}\right) + \left(\frac{1}{9}\right) - \left(\frac{1}{16}\right) + \cdots + (-1)^{n-1}\left(\frac{1}{n^2}\right) + \cdots \tag{1}$$

converges absolutely. A possible rearrangement of the terms of the series might start with a positive term, then 2 negative terms, then 3 positive terms, then 4 negative terms, and so on: after k terms of one sign, take $k + 1$ terms of the opposite sign. The first ten terms of this series look like this:

$$1 - \frac{1}{4} - \frac{1}{16} + \frac{1}{9} + \frac{1}{25} + \frac{1}{49} - \frac{1}{36} - \frac{1}{64} - \frac{1}{100} - \frac{1}{144} + \cdots. \tag{2}$$

The rearrangement theorem says that both series converge to the same value. In this problem, if we had the second series to begin with, we would probably be glad to exchange it for the first, if we knew that we could. We can do even better: the sum of either series is also equal to

$$\sum_{n=1}^{\infty} \frac{1}{(2n-1)^2} - \sum_{n=1}^{\infty} \frac{1}{(2n)^2}. \tag{3}$$

(See Problem 23.) □

Multiplication of Series

Don't panic: it is possible to multiply two series, by applying the distributive law. For convenience of notation, we shall write the first series as

$$a_0 + a_1 + a_2 + \cdots + a_n + \cdots \tag{1}$$

and the second as

$$b_0 + b_1 + b_2 + \cdots + b_n + \cdots \tag{2}$$

and the product as

$$a_0b_0 + (a_0b_1 + a_1b_0) + (a_0b_2 + a_1b_1 + a_2b_0) + \cdots \\ + (a_0b_n + a_1b_{n-1} + \cdots + a_kb_{n-k} + \cdots + a_nb_0) + \cdots. \tag{3}$$

Thus every term in the first series appears as a factor with every term in the second series, in the form a_kb_{n-k}.

EXAMPLE 6 Let

$$a_0 = 1, \quad a_1 = \frac{1}{2}, \quad \ldots, \quad a_n = \frac{1}{2^n} \tag{4}$$

and

$$b_0 = 1, \quad b_1 = \frac{1}{3}, \quad \ldots, \quad b_n = \frac{1}{3^n}. \tag{5}$$

The products in (3) are

$$a_0b_0 = 1, \quad a_0b_1 = \frac{1}{3}, \quad a_1b_0 = \frac{1}{2}, \quad a_0b_2 = \frac{1}{9}, \quad a_1b_1 = \frac{1}{6}, \\ a_2b_0 = \frac{1}{4}$$

and, in general, all terms of the form

$$a_kb_{n-k} = \frac{1}{2^k3^{n-k}}, \qquad n = 0, 1, 2, \ldots \quad \text{and} \quad k = 0, 1, 2, \ldots, n. \ \square$$

The following theorem tells us that the sum of all these products is the sum of the first series, which is 2, times the sum of the second series, which is $\frac{3}{2}$ (both are geometric series: $r_1 = \frac{1}{2}$ and $r_2 = \frac{1}{3}$). By the way in which the terms of the product series are formed, we see that every number that is not divisible by a prime greater than 3 appears as a denominator, precisely once. By rearranging the terms (using Theorem 2) we can indicate the sum as

$$\frac{1}{1} + \frac{1}{2} + \frac{1}{3} + \frac{1}{4} + \frac{1}{6} + \frac{1}{8} + \frac{1}{9} + \frac{1}{12} + \frac{1}{16} + \frac{1}{18} + \frac{1}{24} + \frac{1}{32} + \cdots. \tag{6}$$

The sum of the series in (6) is 3.

We now state Theorem 3 (without proof).

THEOREM 3

Let $\Sigma\, a_n$ and $\Sigma\, b_n$ be absolutely convergent series. Define $\Sigma\, c_n$ by the equations

$$c_0 = a_0 b_0, \qquad c_1 = a_0 b_1 + a_1 b_0, \ldots, c_n = \sum_{k=0}^{n} a_k b_{n-k}. \tag{7}$$

Then $\Sigma\, c_n$ converges absolutely and

$$\sum_{n=0}^{\infty} c_n = \left(\sum_{n=0}^{\infty} a_n\right) \times \left(\sum_{n=0}^{\infty} b_n\right). \tag{8}$$

EXAMPLE 7 The sum of the reciprocals of those integers that can be expressed as products of the form $3^a 7^b$, where a and b independently take all nonnegative integer values, is

$$\frac{1}{1-\frac{1}{3}} \times \frac{1}{1-\frac{1}{7}} =$$
$$\left(1 + \frac{1}{3} + \frac{1}{9} + \cdots + \frac{1}{3^n} + \cdots\right) \times \left(1 + \frac{1}{7} + \frac{1}{49} + \cdots + \frac{1}{7^n} + \cdots\right),$$

or

$$\frac{3}{2} \times \frac{7}{6} = \frac{7}{4}. \quad \square$$

We shall have more to do with multiplication of series in Chapter 12.

Picard's Method

We conclude this section with a proof that the Picard iteration procedure based upon the iterative scheme $x_{n+1} = g(x_n)$ produces a sequence $\{x_n\}$ that converges to a solution of the equation $x = g(x)$ under suitable conditions. We shall convert the sequence $\{x_n\}$ into an infinite series, prove the absolute convergence of that series, conclude the argument by claiming the convergence of the series, and thus the convergence of $\{x_n\}$. We do this as follows. Let

$$a_0 = x_0, \qquad a_1 = x_1 - x_0, \qquad a_2 = x_2 - x_1, \ldots, a_{n+1} = x_{n+1} - x_n. \tag{9}$$

Then we also have

$$\begin{aligned} x_0 &= a_0, \\ x_1 &= a_0 + a_1, \\ x_2 &= a_0 + a_1 + a_2, \\ &\vdots \\ x_{n+1} &= a_0 + a_1 + a_2 + \cdots + a_{n+1}. \end{aligned} \tag{10}$$

Equations (10) show that if $\Sigma\, a_n$ converges, then the sequence $\{x_{n+1}\}$ also converges. Then, of course, the sequence $\{x_n\}$ converges to the same limit and this common limit satisfies the equation $x = g(x)$. Under what conditions? Those stated in the next theorem.

THEOREM 4

Picard Convergence Theorem

Let $g(x)$ be continuous on a closed interval $[a, b]$ and differentiable in the open interval (a, b). Let $x_0, x_1, x_2, \ldots$ be in (a, b) and satisfy

$$x_{n+1} = g(x_n). \tag{11}$$

If there is a constant M that is less than 1, such that

$$|g'(x)| \leq M \qquad \text{for all } x \text{ in } (a, b), \tag{12}$$

then $\{x_n\}$ converges to a solution of the equation

$$x = g(x). \tag{13}$$

Proof. Using the notation of Eqs. (9) and (10) above and the Mean Value Theorem, we see that

$$a_2 = x_2 - x_1 = g(x_1) - g(x_0) = (x_1 - x_0)g'(c_1)$$

for some c_1 between x_1 and x_0. Therefore, c_1 is in (a, b) and if condition (12) is satisfied, we can conclude that

$$|a_2| \leq M|x_1 - x_0| = M|a_1|.$$

In the same way,

$$a_3 = g(x_2) - g(x_1) = (x_2 - x_1)g'(c_2) = a_2 g'(c_2)$$

for some c_2 in (a, b). Thus, if (12) is satisfied,

$$|a_3| \leq |a_2|M \leq |a_1|M^2.$$

Continue in this manner. The conclusion will be that if the hypotheses of Theorem 4 are satisfied, then

$$\begin{aligned} |a_2| &\leq M|a_1|, \\ |a_3| &\leq M^2|a_1| \\ &\vdots \\ |a_n| &\leq M^{n-1}|a_1|. \end{aligned}$$

The series

$$a_0 + a_1 + a_2 + \cdots + a_n + \cdots$$

converges absolutely because the series

$$|a_0| + |a_1|(1 + M + M^2 + \cdots + M^n + \cdots)$$

converges when M is a positive constant less than 1.

We conclude that the sequence $\{x_{n+1}\}$ converges to the following limit:

$$x = \lim_{N\to\infty} x_{N+1} = \lim_{N\to\infty} \sum_{k=0}^{N} a_k.$$

Because $g(x)$ is continuous at $x = \lim x_n$, it is also true that $\lim g(x_n) = g(x)$ and

$$x = \lim x_{n+1} = \lim g(x_n) = g(x). \quad \blacksquare$$

EXAMPLE 8 In Article 11.1, we investigated the sequence generated by $x_0 = 1$, $x_{n+1} = \sqrt{\sin x_n}$. In this example,

$$g(x) = \sqrt{\sin x}$$

and we can take $a = \pi/4$, $b = \pi/2$ because all the terms of the sequence lie in this interval. The graph of g is concave downward over (a, b) because $g''(x)$ is negative there. We can therefore take M to be any positive constant less than 1 and greater than $g'(\pi/4) \approx 0.42045$. (We can do still better when we see that all terms of the sequence $\{x_n\}$ are in the interval $(a, b]$ with $a = 0.87672621$ and $b = 1$. Then we can take $M = 0.365$, which is slightly greater than $g'(a)$.)

In general, the error estimate is given by

$$|x - x_N| \leq \left(\sum_{n=N+1}^{\infty} M^{n-1} \right) \times |x_1 - x_0| = \frac{M^N}{1 - M} \times |x_1 - x_0|. \qquad (14)$$

With $x_0 = 1$, $x_1 = \sqrt{\sin 1} \doteq 0.917$, $N = 20$, and $M = 0.365$, we find that x_{20} differs from the limit by less than 3×10^{-10} for the example in Article 11.1. (The numbers that we calculated involved round-off errors, so we cannot claim 9-place accuracy for our answer.) □

PROBLEMS

In Problems 1–18, determine whether the series converge absolutely. In each case give a reason for the convergence or divergence of the corresponding series of absolute values.

1. $\sum_{n=1}^{\infty} \frac{1}{n^2}$

2. $\sum_{n=1}^{\infty} \frac{1}{(-n)^3}$

3. $\sum_{n=1}^{\infty} \frac{1-n}{n^2}$

4. $\sum_{n=1}^{\infty} \left(-\frac{1}{5}\right)^n$

5. $\sum_{n=1}^{\infty} \frac{-1}{n^2 + 2n + 1}$

6. $\sum_{n=1}^{\infty} \frac{(-1)^n}{2n}$

7. $\sum_{n=1}^{\infty} \frac{\cos n\pi}{n\sqrt{n}}$

8. $\sum_{n=1}^{\infty} \frac{-10}{n}$

9. $\sum_{n=0}^{\infty} \frac{(-1)^n}{(2n)!}$

10. $\sum_{n=0}^{\infty} \frac{(-1)^n}{(2n+1)!}$

11. $\sum_{n=2}^{\infty} (-1)^n \frac{n}{n+1}$

12. $\sum_{n=1}^{\infty} \frac{-n}{2^n}$

13. $\sum_{n=1}^{\infty} (5)^{-n}$

14. $\sum_{n=1}^{\infty} \left(\frac{1}{2^n} - 1\right)$

15. $\sum_{n=1}^{\infty} \frac{(-100)^n}{n!}$

16. $\sum_{n=2}^{\infty} (-1)^n \frac{\ln n}{\ln n^2}$

17. $\sum_{n=1}^{\infty} \frac{2-n}{n^3}$

18. $\sum_{n=1}^{\infty} \left(\frac{1}{2^n} - \frac{1}{3^n}\right)$

19. Show that if $\Sigma_{n=1}^{\infty} a_n$ diverges, then $\Sigma_{n=1}^{\infty} |a_n|$ diverges.

20. Show that if $\Sigma_{n=1}^{\infty} a_n$ converges absolutely, then

$$\left| \sum_{n=1}^{\infty} a_n \right| \leq \sum_{n=1}^{\infty} |a_n|.$$

21. Show that if $\Sigma_{n=1}^{\infty} a_n$ and $\Sigma_{n=1}^{\infty} b_n$ both converge absolutely, then so does

a) $\Sigma (a_n + b_n)$ b) $\Sigma (a_n - b_n)$
c) Σka_n (k any number)

22. Prove Theorem 2. (Outline of proof. Assume the hypotheses. Let $\varepsilon > 0$. Show there is an index N_1 such that

$$\sum_{n=N_1}^{\infty} |a_n| < \frac{\varepsilon}{2}$$

and an index $N_2 \geq N_1$ such that $|s_{N_2} - L| < \varepsilon/2$, where $s_n = \Sigma_{k=1}^{n} a_k$. Because all of the terms $a_1, a_2, \ldots, a_{N_2}$ appear somewhere in the sequence $\{b_n\}$, there is an index $N_3 \geq N_2$ such that if $n \geq N_3$, then $\Sigma_{k=1}^{n} b_k - s_{N_2}$ is, at most, a sum of terms of the form a_m with $m \geq N_1$. Therefore, if $n \geq N_3$,

$$\left| \sum_{k=1}^{n} b_k - L \right| \leq \left| \sum_{k=1}^{n} b_k - s_{N_2} \right| + |s_{N_2} - L|$$

$$\leq \sum_{k=N_1}^{\infty} |a_k| + |s_{N_2} - L| < \varepsilon. \ \blacksquare$$

23. Establish the following: If $\Sigma\,|a_n|$ converges and $b_n = a_n$ when $a_n \geq 0$, while $b_n = 0$ when $a_n < 0$, then $\Sigma\, b_n$ converges. Likewise $\Sigma\, c_n$ converges, where $c_n = 0$ when $a_n \geq 0$ and $c_n = -a_n$ when $a_n < 0$. In other words, when the original series converges absolutely, its positive terms by themselves form a convergent series, and so do the negative terms. And

$$\sum_{n=1}^{\infty} a_n = \sum_{n=1}^{\infty} b_n - \sum_{n=1}^{\infty} c_n$$

because $b_n = \frac{1}{2}(a_n + |a_n|)$ and $c_n = \frac{1}{2}(-a_n + |a_n|)$.

25. In the Picard convergence theorem, if $M = 1/2$, show that the error (as estimated by inequality 14) is cut by a factor of 1/2 at each successive iteration.

24. In Example 6, where $a_n = 1/2^n$ and $b_n = 1/3^n$, let $c_n = a_0b_n + a_1b_{n-1} + \cdots + a_nb_0$, as in Eq. (3). Since the number a_kb_{n-k} is less than or equal to $(1/2)^n$, show that $c_n \leq n + 1/2^n$ and then prove that $\Sigma\, c_n$ converges. Does it converge absolutely?

26. Let $x_0 = 1$ and define the sequence $\{x_n\}$ by the formula $x_{n+1} = \cos x_n$. Show that one could take $M = \sin 1 \approx 0.84$ in the Picard convergence theorem. If we make a fresh start with $x_0 = 0.8$, we get $x_1 = 0.6967$ and all the rest of the sequence $\{x_n\}$ will be between these numbers. What value of M can you now use?

***Toolkit* programs**

Sequences and Series

11.7 Alternating Series. Conditional Convergence

When some of the terms of a series $\Sigma\, a_n$ are positive and others are negative, the series converges if $\Sigma\,|a_n|$ converges. Thus we may apply any of our tests for convergence of nonnegative series, provided we apply them to the series of absolute values. But we do not know, when the series of absolute values *diverges,* whether the *original* series diverges or converges. If it converges, in this case, we say that it *converges conditionally.*

We shall discuss one simple case of series with mixed signs, namely, series that take the form

$$a_1 - a_2 + a_3 - a_4 + \cdots (-1)^{n+1}a_n + \cdots, \tag{1}$$

with all a's > 0. Such series are called *alternating series* because successive terms have alternate signs. Examples of alternating series are

$$1 - \frac{1}{2} + \frac{1}{3} - \frac{1}{4} + \frac{1}{5} - \frac{1}{6} + \cdots, \tag{2}$$

$$\frac{1}{\ln 2} - \frac{1}{\ln 3} + \frac{1}{\ln 4} - \frac{1}{\ln 5} + \cdots, \tag{3}$$

$$1 - \sqrt{2} + \sqrt{3} - \sqrt{4} + \cdots. \tag{4}$$

The series

$$1 - \frac{1}{2} - \frac{1}{4} + \frac{1}{6} - \frac{1}{8} - \frac{1}{10} + \frac{1}{12} \cdots \tag{5}$$

is *not* an alternating series. The signs of its terms do not alternate.

DEFINITION A sequence $\{a_n\}$ is called a *decreasing sequence* if $a_n \geq a_{n+1}$ for every n.

Examples of decreasing sequences are

$$1, \frac{1}{2}, \frac{1}{3}, \frac{1}{4}, \ldots, \quad \text{and} \quad 1, 1, 1, 1, \ldots.$$

The sequence

$$\frac{1}{3}, \frac{1}{2}, \frac{1}{6}, \frac{1}{4}, \ldots, \frac{1}{3n}, \frac{1}{2n}, \ldots$$

is *not* a decreasing sequence even though it converges to 0 from above.

One reason for selecting alternating series for study is that every alternating series whose numbers a_n form a decreasing sequence with limit 0 converges. Another reason is that whenever an alternating series converges, it is easy to estimate its sum. This fortunate combination of assured convergence and easy estimation gives us an opportunity to see how a wide variety of series behave. We look first at the convergence.

THEOREM 1

Leibniz's Theorem

$\Sigma_{n=1}^{\infty} (-1)^{n+1} a_n$ converges if all three of the following conditions are satisfied:

1. The a_n's are all positive,
2. $a_n \geq a_{n+1}$ for every n,
3. $a_n \to 0$.

Proof. If n is an even integer, say $n = 2m$, then the sum of the first n terms is

$$\begin{aligned} s_{2m} &= (a_1 - a_2) + (a_3 - a_4) + \cdots + (a_{2m-1} - a_{2m}) \\ &= a_1 - (a_2 - a_3) - (a_4 - a_5) - \cdots - (a_{2m-2} - a_{2m-1}) - a_{2m}. \end{aligned}$$

The first equality exhibits s_{2m} as the sum of m nonnegative terms, since each expression in parentheses is positive or zero. Hence $s_{2m+2} \geq s_{2m}$, and the sequence $\{s_{2m}\}$ is increasing. The second equality shows that $s_{2m} \leq a_1$. Since $\{s_{2m}\}$ is increasing and bounded from above it has a limit, say

$$\lim_{n\to\infty} s_{2m} = L. \tag{6}$$

If n is an odd integer, say $n = 2m + 1$, then the sum of the first n terms is

$$s_{2m+1} = s_{2m} + a_{2m+1}.$$

Since $a_n \to 0$,

$$\lim_{m\to\infty} a_{2m+1} = 0.$$

Hence, as $m \to \infty$,

$$s_{2m+1} = s_{2m} + a_{2m+1} \to L + 0 = L. \tag{7}$$

Finally, we may combine (6) and (7) and say simply

$$\lim_{n\to\infty} s_n = L. \blacksquare$$

Here are some examples of what Theorem 1 can do.

EXAMPLE 1 The *alternating harmonic* series

$$\sum_{n=1}^{\infty} (-1)^{n+1}\frac{1}{n} = 1 - \frac{1}{2} + \frac{1}{3} - \frac{1}{4} + \cdots$$

satisfies the three requirements of the theorem; therefore it converges. It converges conditionally because the corresponding series of absolute values is the harmonic series, which diverges. □

EXAMPLE 2

$$\sum_{n=1}^{\infty} (-1)^{n+1}\sqrt{n} = 1 - \sqrt{2} + \sqrt{3} - \sqrt{4} + \cdots$$

diverges by the nth-term test. □

EXAMPLE 3 Theorem 1 gives no information about

$$\frac{2}{1} - \frac{1}{1} + \frac{2}{2} - \frac{1}{2} + \frac{2}{3} - \frac{1}{3} + \cdots + \frac{2}{n} - \frac{1}{n} + \cdots.$$

The sequence $\frac{2}{1}, \frac{1}{1}, \frac{2}{2}, \frac{1}{2}, \frac{2}{3}, \frac{1}{3}, \ldots$ is not a decreasing sequence. Some other test must be found. When we group the terms of the series in consecutive pairs

$$\left(\frac{2}{1} - \frac{1}{1}\right) + \left(\frac{2}{2} - \frac{1}{2}\right) + \left(\frac{2}{3} - \frac{1}{3}\right) + \cdots + \left(\frac{2}{n} - \frac{1}{n}\right) + \cdots,$$

we see that the $2n$th partial sum of the given series is the same number as the nth partial sum of the harmonic series. Thus the sequence of partial sums, and hence the series, diverges. □

EXAMPLE 4 Does the series $\sum_{n=2}^{\infty} (-1)^n(\ln n/(n+1))$ converge? While it is clear that this is an alternating series and $a_n = \ln n/(n+1)$ approaches 0 as $n \to \infty$, it isn't clear at first glance that the sequence $\{a_n\}$ is a decreasing sequence. We might consider the corresponding function

$$f(x) = \frac{\ln x}{x+1}$$

whose derivative is

$$f'(x) = \frac{(x+1)/x - \ln x}{(x+1)^2}.$$

The derivative is negative, and the function f decreasing, if $\ln x$ is greater than $(x+1)/x = 1 + 1/x$. For $x = 1, 2, \ldots, 1 + (1/x) \leq 2$, so the sequence decreases when $\ln n > 2$. Since $e^2 \approx 7.4$, we conclude that the original sequence decreases for $n \geq 8$. That is, we can apply Leibniz's theorem to conclude that $\sum_{n=8}^{\infty} (-1)^n(\ln n/(n+1))$ converges. Therefore the original series also converges. □

We use the following graphical interpretation of the partial sums to gain added insight into the way in which an alternating series converges

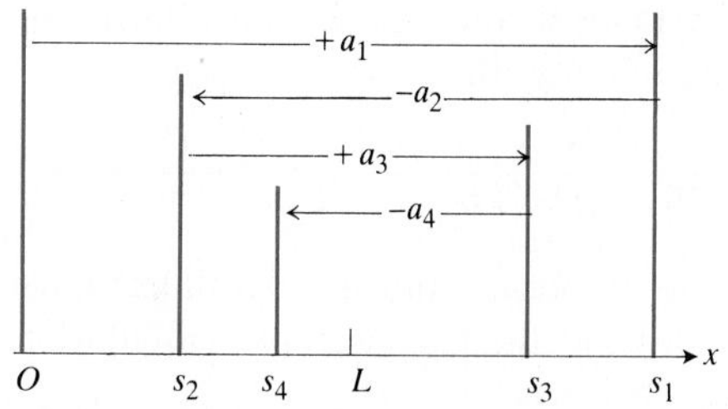

11.12 The partial sums of an alternating series that satisfies the hypotheses of Leibniz's theorem straddle their limit.

to its limit L when the three conditions of the theorem are satisfied. Starting from the origin O on a scale of real numbers (Fig. 11.12), we lay off the positive distance

$$s_1 = a_1.$$

To find the point corresponding to

$$s_2 = a_1 - a_2$$

we must back up a distance equal to a_2. Since $a_2 \leq a_1$, we do not back up any farther than O at most. Next we go forward a distance a_3 and mark the point corresponding to

$$s_3 = a_1 - a_2 + a_3.$$

Since $a_3 \leq a_2$, we go forward by an amount that is no greater than the previous backward step; that is, s_3 is less than or equal to s_1. We continue in this seesaw fashion, backing up or going forward as the signs in the series demand. But each forward or backward step is shorter than (or at most the same size as) the preceding step, because $a_{n+1} \leq a_n$. And since the nth term approaches zero as n increases, the size of step we take forward or backward gets smaller and smaller. We thus oscillate across the limit L, but the amplitude of oscillation continually decreases and approaches zero as its limit. The even-numbered partial sums $s_2, s_4, s_6, \ldots, s_{2m}$ continually increase toward L, while the odd-numbered sums $s_1, s_3, s_5, \ldots, s_{2m+1}$ continually decrease toward L. The limit L is between any two successive sums s_n and s_{n+1} and hence differs from s_n by an amount less than a_{n+1}.

It is because

$$|L - s_n| < a_{n+1} \qquad \text{for every } n \tag{8}$$

that we can make useful estimates of the sums of convergent alternating series.

THEOREM 2

The Alternating Series Estimation Theorem

If

$$\sum_{n=1}^{\infty} (-1)^{n+1} a_n$$

is an alternating series that satisfies the three conditions of Theorem 1, then

$$s_n = a_1 - a_2 + \cdots + (-1)^{n+1} a_n$$

approximates the sum L of the series with an error whose absolute value is less than a_{n+1}, the numerical value of the first unused term. Furthermore, the remainder, $L - s_n$, has the same sign as the first unused term.

We will leave the determination of the sign of the remainder as an exercise (see Problem 40).

EXAMPLE 5 Let us first try the estimation theorem on an alternating series whose sum we already know, namely, the geometric series

$$\sum_{n=0}^{\infty} (-1)^n \frac{1}{2^n} = 1 - \frac{1}{2} + \frac{1}{4} - \frac{1}{8} + \frac{1}{16} - \frac{1}{32} + \frac{1}{64} - \frac{1}{128} \;\vdots\; + \frac{1}{256} - \cdots.$$

Theorem 2 says that when we truncate the series after the eighth term, we throw away a total that is positive and less than $\frac{1}{256}$. A rapid calculation shows that the sum of the first eight terms is

$$0.6640625.$$

The sum of the series is

$$\frac{1}{1-(-\frac{1}{2})} = \frac{1}{\frac{3}{2}} = \frac{2}{3}. \tag{9}$$

The difference,

$$\frac{2}{3} - 0.6640625 = 0.0026041666\ldots,$$

is positive and less than

$$\frac{1}{256} = 0.00390625. \;\square$$

We shall show in Chapter 12 that a series for computing $\ln(1+x)$ when $|x| < 1$ is

$$\ln(1+x) = x - \frac{x^2}{2} + \frac{x^3}{3} - \cdots + (-1)^{n+1}\frac{x^n}{n} \cdots. \tag{10}$$

For $0 < x < 1$, this series satisfies all three conditions of Theorem 1, and we may use the estimation theorem to see how good an approximation of $\ln(1+x)$ we get from the first few terms of the series.

EXAMPLE 6 Calculate $\ln(1.1)$ with the approximation

$$\ln(1+x) \approx x - \frac{x^2}{2}, \tag{11}$$

and estimate the error involved. Is $x - (x^2/2)$ too large, or too small in this case?

Solution

$$\ln(1.1) \approx (0.1) - \frac{(0.1)^2}{2} = 0.095.$$

This approximation differs from the exact value of $\ln 1.1$ by less than

$$\frac{(0.1)^3}{3} = 0.000333\ldots.$$

Since the sign of this, the first unused term, is positive, the remainder is positive. That is, 0.095 underestimates $\ln(1.1)$. $\square$

EXAMPLE 7 How many terms of the series (10) do we need to use in order to be sure of calculating $\ln(1.2)$ with an error of less than 10^{-6}?

Solution

$$\ln(1.2) = (0.2) - \frac{(0.2)^2}{2} + \frac{(0.2)^3}{3} - \cdots$$

We find by trial that the eighth term

$$-\frac{(0.2)^8}{8} = -3.2 \times 10^{-7}$$

is the first term in the series whose absolute value is less than 10^{-6}. Therefore the sum of the first *seven* terms will give ln (1.2) with an error of less than 10^{-6}. The use of more terms would give an approximation that is even better, but seven terms are enough to guarantee the accuracy we wanted. Note also that we have not shown that six terms would *not* provide that accuracy. □

CAUTION ABOUT REARRANGEMENTS. If we rearrange an infinite number of terms of a conditionally convergent series, we can get results that are far different from the sum of the original series. We illustrate some of the things that can happen in the next example.

EXAMPLE 8 Consider the alternating harmonic series

$$\frac{1}{1} - \frac{1}{2} + \frac{1}{3} - \frac{1}{4} + \frac{1}{5} - \frac{1}{6} + \frac{1}{7} - \frac{1}{8} + \frac{1}{9} - \frac{1}{10} + \frac{1}{11} - \cdots.$$

Here, the series of terms $\Sigma\, 1/(2n - 1)$ diverges to $+\infty$ and the series of terms $\Sigma\, -1/2n$ diverges to $-\infty$. We can always add enough positive terms (no matter how far out in the sequence of odd-numbered terms we begin) to get an arbitrarily large sum. Similarly, with the negative terms, no matter how far out we start, we can add enough consecutive even numbered terms to get a negative sum of arbitrarily large absolute value. If we wished to do so, we could start adding odd numbered terms until we had a sum equal to $+3$, say; then follow that with enough consecutive negative terms to make the new total less than -4. We could then add enough positive terms to make the total greater than $+5$ and follow with consecutive unused negative terms to make a new total less than -6, and so on. In this way, we could make the swings arbitrarily large in both the positive and negative directions.

Another possibility, with the same series, is to focus on a particular limit. Suppose we try to get sums that converge to 1. We start with the first term, $\frac{1}{1}$, and then subtract $\frac{1}{2}$. Next we add $\frac{1}{3}$ and $\frac{1}{5}$, which brings the total back to 1 or above. Then we subtract consecutive negative terms until the total is less than 1. Continue in this manner: when the sum is below 1, add positive terms until the total is 1 or more; then subtract (i.e., add negative) terms until the total is once more less than 1. This process can be continued indefinitely. Because both the odd-numbered terms and the even-numbered terms of the original series approach zero as $n \to \infty$, the amount by which our partial sums exceed 1 or fall below it approaches zero. So the new series converges to 1. The rearranged series starts like this:

$$\frac{1}{1} - \frac{1}{2} + \frac{1}{3} + \frac{1}{5} - \frac{1}{4} + \frac{1}{7} + \frac{1}{9} - \frac{1}{6} + \frac{1}{11} + \frac{1}{13} - \frac{1}{8} + \frac{1}{15} + \frac{1}{17} - \frac{1}{10}$$
$$+ \frac{1}{19} + \frac{1}{21} - \frac{1}{12} + \frac{1}{23} + \frac{1}{25} - \frac{1}{14} + \frac{1}{27} - \frac{1}{16} + \cdots.$$ □

The kind of behavior illustrated by this example is typical of what can happen with any conditionally convergent series. Moral: add the terms of such a series in the order given.

PROBLEMS

In Problems 1–10, determine which of the alternating series converge and which diverge.

1. $\sum_{n=1}^{\infty} (-1)^{n+1} \frac{1}{n^2}$

2. $\sum_{n=2}^{\infty} (-1)^{n+1} \frac{1}{\ln n}$

3. $\sum_{n=1}^{\infty} (-1)^{n+1}$

4. $\sum_{n=1}^{\infty} (-1)^{n+1} \frac{10^n}{n^{10}}$

5. $\sum_{n=1}^{\infty} (-1)^{n+1} \frac{\sqrt{n}+1}{n+1}$

6. $\sum_{n=1}^{\infty} (-1)^{n+1} \frac{\ln n}{n}$

7. $\sum_{n=1}^{\infty} (-1)^{n+1} \frac{1}{n^{3/2}}$

8. $\sum_{n=1}^{\infty} (-1)^{n+1} \frac{\ln n}{\ln n^2}$

9. $\sum_{n=1}^{\infty} (-1)^{n} \ln \left(1 + \frac{1}{n}\right)$

10. $\sum_{n=1}^{\infty} (-1)^{n+1} \frac{3\sqrt{n+1}}{\sqrt{n}+1}$

In Problems 11–28, determine whether the series are absolutely convergent, conditionally convergent, or divergent.

11. $\sum_{n=1}^{\infty} (-1)^{n+1}(0.1)^n$

12. $\sum_{n=1}^{\infty} (-1)^{n+1} \frac{1}{\sqrt{n}}$

13. $\sum_{n=1}^{\infty} (-1)^{n+1} \frac{n}{n^3+1}$

14. $\sum_{n=1}^{\infty} \frac{n!}{2^n}$

15. $\sum_{n=1}^{\infty} (-1)^{n} \frac{1}{n+3}$

16. $\sum_{n=1}^{\infty} (-1)^{n} \frac{\sin n}{n^2}$

17. $\sum_{n=1}^{\infty} (-1)^{n+1} \frac{3+n}{5+n}$

18. $\sum_{n=2}^{\infty} (-1)^{n} \frac{1}{\ln n^3}$

19. $\sum_{n=1}^{\infty} (-1)^{n+1} \frac{1+n}{n^2}$

20. $\sum_{n=1}^{\infty} \frac{(-2)^{n+1}}{n+5^n}$

21. $\sum_{n=1}^{\infty} n^2 (\frac{2}{3})^n$

22. $\sum_{n=1}^{\infty} (-1)^{n+1} (\sqrt[n]{10})$

23. $\sum_{n=1}^{\infty} (-1)^{n} \frac{\tan^{-1} n}{n^2+1}$

24. $\sum_{n=2}^{\infty} (-1)^{n+1} \frac{1}{n \ln n}$

25. $\sum_{n=1}^{\infty} \left(\frac{1}{n} - \frac{1}{2n}\right)$

26. $\sum_{n=1}^{\infty} (-1)^{n+1} \frac{(0.1)^n}{n}$

27. $\sum_{n=1}^{\infty} (-1)^{n+1} (\sqrt{n+1} - \sqrt{n})$

28. $\sum_{n=1}^{\infty} \frac{(-1)^{n+1}(n!)^2}{(2n)!}$

In Problems 29–32, estimate the magnitude of the error if the first four terms are used to approximate the series.

29. $\sum_{n=1}^{\infty} (-1)^{n+1} \frac{1}{n}$

30. $\sum_{n=1}^{\infty} (-1)^{n+1} \frac{1}{10^n}$

31. $\ln(1.01) = \sum_{n=1}^{\infty} (-1)^{n+1} \frac{(0.01)^n}{n}$

32. $\frac{1}{1+t} = \sum_{n=0}^{\infty} (-1)^n t^n, \quad 0 < t < 1$

Approximate the sums in Problems 33 and 34 to five decimal places (magnitude of the error less than 5×10^{-6}).

33. $\sum_{n=0}^{\infty} (-1)^n \frac{1}{(2n)!}$ (This is cos 1, the cosine of one radian.)

34. $\sum_{n=0}^{\infty} (-1)^n \frac{1}{n!}$ (This is $1/e$.)

35. a) The series

$$\frac{1}{3} - \frac{1}{2} + \frac{1}{9} - \frac{1}{4} + \frac{1}{27} - \frac{1}{8} + \cdots + \frac{1}{3^n} - \frac{1}{2^n} + \cdots$$

does not meet one of the conditions of Theorem 1. Which one?

b) Find the sum of the series in (a).

36. The limit L of an alternating series that satisfies the conditions of Theorem 1 lies between the values of any two consecutive partial sums. This suggests using the average

$$\frac{s_n + s_{n+1}}{2} = s_n + \frac{1}{2} a_{n+1}$$

to estimate L. Compute

$$s_{20} + \frac{1}{2} \cdot \frac{1}{21}$$

as an approximation to the sum of the alternating harmonic series. The exact sum is $\ln 2 \approx 0.6931\ldots$.

37. Show that whenever an alternating series is approximated by one of its partial sums, if the three conditions of Leibniz's theorem are satisfied, then the remainder (sum of the unused terms) has the same sign as the first unused term. (*Hint:* Group the terms of the remainder in consecutive pairs.)

38. Prove the "zipper" theorem for sequences: If $\{a_n\}$ and $\{b_n\}$ both converge to L, then the sequence

$$a_1, \ b_1, \ a_2, \ b_2, \ \ldots, \ a_n, \ b_n, \ \ldots$$

also converges to L.

39. Show by example that $\Sigma_{n=1}^{\infty} a_n b_n$ may diverge even though $\Sigma_{n=1}^{\infty} a_n$ and $\Sigma_{n=1}^{\infty} b_n$ both converge.

40. Prove that the remainder, $L - s_n$, has the sign specified in Theorem 2.

41. (*Calculator*) In Example 8, suppose the goal is to arrange the terms to get a new series that converges to $-\frac{1}{2}$. Start the new arrangement with the first negative term, which is $-\frac{1}{2}$. Whenever you have a sum that is less than or equal to $-\frac{1}{2}$, start introducing positive terms, taken in order, until the new total is greater than $-\frac{1}{2}$. Then add negative terms until the total is less than or equal to $-\frac{1}{2}$ again. Continue this process until your partial sums have been above the target at least three times and finish at or below it. If s_n is the sum of the first n terms of your new series, plot the points (n, s_n) to illustrate how the sums are behaving.

***Toolkit* programs**

Sequences and Series

REVIEW QUESTIONS AND EXERCISES

1. Define "sequence," "series," "sequence of partial sums of a series."
2. Define "convergence" (a) of a sequence, (b) of an infinite series.
3. Which of the following statements are true, and which are false?
   a) If a sequence does not converge, then it diverges.
   b) If a sequence $\{n, f(n)\}$ does not converge, then $f(n)$ tends to infinity as n does.
   c) If a series does not converge, then its nth term does not approach zero as n tends to infinity.
   d) If the nth term of a series does not approach zero as n tends to infinity, then the series diverges.
   e) If a sequence $\{n, f(n)\}$ converges, then there is a number L such that $f(n)$ lies within 1 unit of L (i) for all values of n, (ii) for all but a finite number of values of n.
   f) If all partial sums of a series are less than some constant L, then the series converges.
   g) If a series converges, then its partial sums s_n are bounded (that is, $m \le s_n \le M$ for some constants m and M).
4. List three tests for convergence (or divergence) of an infinite series.
5. Under what circumstances do you know that a bounded sequence converges?
6. Define "absolute convergence" and "conditional convergence." Give examples of series that are (a) absolutely convergent, (b) conditionally convergent.
7. What test is usually used to decide whether a given alternating series converges? Give examples of convergent and divergent alternating series.

MISCELLANEOUS PROBLEMS

1. Find explicitly the nth partial sum of the series $\Sigma_{n=2}^{\infty} \ln(1 - 1/n^2)$, and thereby determine whether the series converges.
2. Evaluate $\Sigma_{k=2}^{\infty} 1/(k^2 - 1)$ by finding the nth partial sum and taking the limit as n becomes infinite.
3. Prove that the sequence $\{x_n\}$ and the series $\Sigma_{k=1}^{\infty}(x_{k+1} - x_k)$ both converge or both diverge.
4. In an attempt to find a root of the equation $x = f(x)$, a first approximation x_1 is estimated from the graphs of $y = x$ and $y = f(x)$. Then $x_2, x_3, \ldots, x_n, \ldots$ are computed successively from the formula $x_n = f(x_{n-1})$. If the points $x_1, x_2, \ldots, x_n, \ldots$ all lie on an interval $a \le x \le b$ on which $f(x)$ has a derivative such that $|f'(x)| < M < 1$, show that the sequence $\{x_n\}$ converges to a root of the given equation.
5. Assuming $|x| > 1$, show that
$$\frac{1}{1-x} = -\frac{1}{x} - \frac{1}{x^2} - \frac{1}{x^3} - \cdots.$$
6. Does the series $\Sigma_{n=1}^{\infty} \operatorname{sech} n$ converge? Why?
7. Does $\Sigma_{n=1}^{\infty} (-1)^n \tanh n$ converge? Why?

Establish the convergence or divergence of the series whose nth terms are given in Problems 8–19.

8. $\dfrac{1}{\ln(n+1)}$
9. $\dfrac{n}{2(n+1)(n+2)}$
10. $\dfrac{\sqrt{n+1} - \sqrt{n}}{\sqrt{n}}$
11. $\dfrac{1}{n(\ln n)^2}, \quad n \ge 2$
12. $\dfrac{1 + (-2)^{n-1}}{2^n}$
13. $\dfrac{n}{1000n^2 + 1}$

14. $e^n/n!$

15. $\dfrac{1}{n\sqrt{n^2+1}}$

16. $\dfrac{1}{n^{1+1/n}}$

17. $\dfrac{1\cdot 3\cdot 5\cdots(2n-1)}{2\cdot 4\cdot 6\cdots(2n)}$

18. $\dfrac{n^2}{n^3+1}$

19. $\dfrac{n+1}{n!}$

20. If the following series converges, find the sum

$$\sum_{n=1}^{\infty}\frac{1}{(n+1)(n+2)}.$$

21. a) Suppose $a_1, a_2, a_3, \ldots, a_n$ are positive numbers satisfying the following conditions:
i) $a_1 \geq a_2 \geq a_3 \geq \cdots$,
ii) the series $a_2 + a_4 + a_8 + a_{16} + \cdots$ diverges.

Show that the series

$$\frac{a_1}{1}+\frac{a_2}{2}+\frac{a_3}{3}+\cdots$$

diverges.

b) Use the result above to show that

$$\sum_{n=2}^{\infty}\frac{1}{n\ln n}$$

diverges.

22. Given $a_n \neq 1$, $a_n > 0$, $\Sigma\, a_n$ converges.
a) Show that $\Sigma\, a_n^2$ converges.
b) Does $\Sigma\, a_n/(1-a_n)$ converge? Explain.

23. Show that $\Sigma_{n=2}^{\infty}\, 1/[n(\ln n)^k]$ converges for $k > 1$.

12 Power Series

12.1 Power Series for Functions

The rational operations of arithmetic are addition, subtraction, multiplication, and division. Using only these simple operations, we can evaluate any rational function of x. But other functions, such as $\sqrt{x}$, ln x, cos x, and so on, cannot be evaluated so simply. These functions occur so frequently, however, that their values have been printed in mathematical tables, and many calculators and computers have been programmed to produce them on demand. One may wonder where the values in the tables came from. By and large, these numbers came from calculating partial sums of power series.

DEFINITION

A *power series* is a series of the form

$$\sum_{n=0}^{\infty} a_n x^n = a_0 + a_1 x + a_2 x^2 + \cdots .$$

In this article we shall show how a power series can arise when we seek to approximate a function

$$y = f(x) \tag{1}$$

by a sequence of polynomials $f_n(x)$ of the form

$$f_n(x) = a_0 + a_1 x + a_2 x^2 + \cdots + a_n x^n. \tag{2}$$

We shall be interested, at least at first, in making the approximation for values of x near 0, because we want the term $a_n x^n$ to decrease as n increases. Hence we focus our attention on a portion of the curve $y = f(x)$ near the point $A(0, f(0))$, as shown in Fig. 12.1.

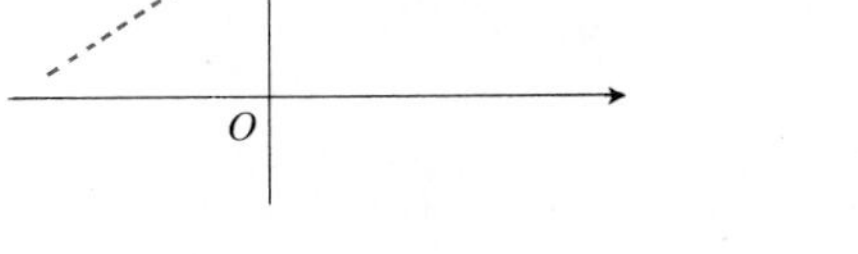

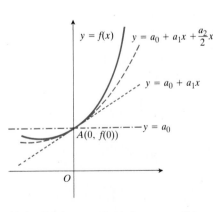

12.1 $f(x)$ is approximated near $x = 0$ by polynomials whose derivatives at $x = 0$ match the derivatives of f.

1. The graph of the polynomial $f_0(x) = a_0$ of degree zero will pass through $(0, f(0))$ if we take

$$a_0 = f(0).$$

2. The graph of the polynomial $f_1(x) = a_0 + a_1 x$ will pass through

$(0, f(0))$ and have the same slope as the given curve at that point if we choose

$$a_0 = f(0) \quad \text{and} \quad a_1 = f'(0).$$

3. The graph of the polynomial $f_2(x) = a_0 + a_1x + a_2x^2$ will pass through $(0, f(0))$ and have the same first and second derivative as the given curve at that point if

$$a_0 = f(0), \quad a_1 = f'(0), \quad \text{and } a_2 = \frac{f''(0)}{2}.$$

4. In general, the polynomial $f_n(x) = a_0 + a_1x + a_2x^2 + \cdots + a_nx^n$, which we choose to approximate $y = f(x)$ near $x = 0$, is the one whose graph passes through $(0, f(0))$ and whose first n derivatives match the derivatives of $f(x)$ at $x = 0$. To match the derivatives of f_n to those of f at $x = 0$, we merely have to choose the coefficients a_0 through a_n properly. To see how this may be done, we write down the polynomial and its derivatives as follows:

$$\begin{aligned} f_n(x) &= a_0 + a_1x + a_2x^2 + a_3x^3 + \cdots + a_nx^n \\ f_n'(x) &= a_1 + 2a_2x + 3a_3x^2 + \cdots + na_nx^{n-1} \\ f_n''(x) &= 2a_2 + 3 \cdot 2a_3x + \cdots + n(n-1)a_nx^{n-2} \\ &\vdots \\ f_n^{(n)}(x) &= (n!)a_n. \end{aligned}$$

When we substitute 0 for x in the array above, we find that

$$a_0 = f(0), \quad a_1 = f'(0), \quad a_2 = \frac{f''(0)}{2!}, \quad \ldots, \quad a_n = \frac{f^{(n)}(0)}{n!}.$$

Thus,

$$f_n(x) = f(0) + f'(0)x + \frac{f''(0)}{2!}x^2 + \cdots + \frac{f^{(n)}(0)}{n!}x^n \tag{3}$$

is the polynomial we seek. Its graph passes through the point $(0, f(0))$, and its first n derivatives match the first n derivatives of $y = f(x)$ at $x = 0$. It is called the nth-degree *Taylor polynomial of f at $x = 0$.*

EXAMPLE 1 Find the Taylor polynomials $f_n(x)$ for the function $f(x) = e^x$ at $x = 0$.

Solution Expressed in terms of x, the given function and its derivatives are

$$f(x) = e^x, \quad f'(x) = e^x, \quad \ldots, \quad f^{(n)}(x) = e^x,$$

so that

$$f(0) = e^0 = 1, \quad f'(0) = 1, \quad \ldots, \quad f^{(n)}(0) = 1,$$

and

$$f_n(x) = 1 + x + \frac{x^2}{2!} + \frac{x^3}{3!} + \cdots + \frac{x^n}{n!}.$$

See Fig. 12.2. □

EXAMPLE 2 Find the Taylor polynomials $f_n(x)$ for $f(x) = \cos x$ at $x = 0$.

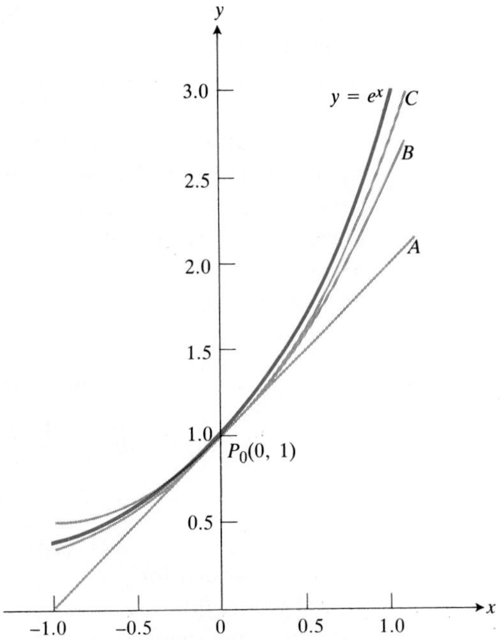

12.2 The graph of the function $y = e^x$, and graphs of three approximating polynomials, (*A*) a straight line, (*B*) a parabola, and (*C*) a cubic curve.

Solution The cosine and its derivatives are

$$\begin{aligned} f(x) &= \cos x, & f'(x) &= -\sin x, \\ f''(x) &= -\cos x, & f^{(3)}(x) &= \sin x, \\ &\vdots & &\vdots \\ f^{(2k)}(x) &= (-1)^k \cos x, & f^{(2k+1)}(x) &= (-1)^{k+1} \sin x. \end{aligned}$$

When $x = 0$, the cosines are 1 and the sines are 0, so that

$$f^{(2k)}(0) = (-1)^k, \qquad f^{(2k+1)}(0) = 0.$$

The Taylor polynomials have only even-powered terms, and for $n = 2k$ we have

$$f_{2k}(x) = 1 - \frac{x^2}{2!} + \frac{x^4}{4!} - \cdots + (-1)^k \frac{x^{2k}}{(2k)!}. \tag{4}$$

Figure 12.3 shows how well these polynomials can be expected to approx-

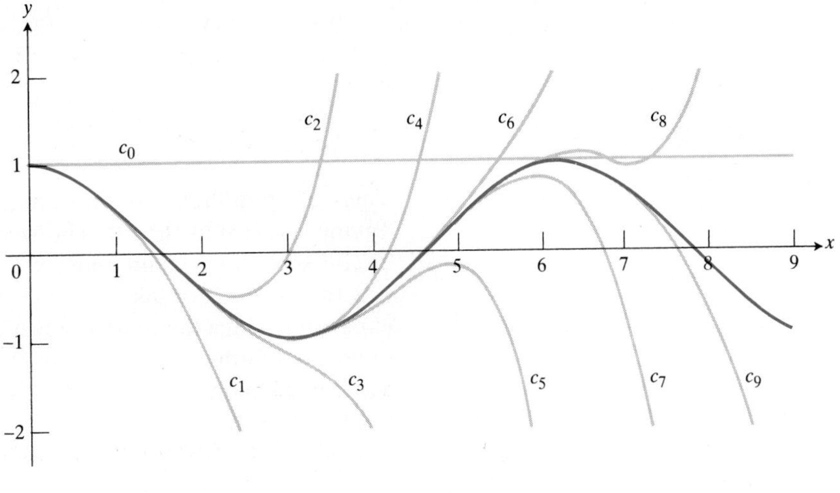

12.3 The polynomials

$$c_n(x) = \sum_{k=0}^{n} [(-1)^k x^{2k}/(2k)!]$$

converge to $\cos x$ as $n \to \infty$. (Adapted from Helen M. Kammerer, *American Mathematical Monthly,* 43(1936), 293–294.)

imate $y = \cos x$ near $x = 0$. Only the right-hand portions of the graphs are shown because the graphs are symmetric about the y-axis. □

The degrees of the Taylor polynomials of a given function are limited by the degree of differentiability of the function at $x = 0$. But if $f(x)$ has derivatives of all orders at the origin, it is natural to ask whether, for a fixed value of x, the values of these approximating polynomials converge to $f(x)$ as $n \to \infty$. Now, these polynomials are precisely the partial sums of a series known as the *Maclaurin series* for f:

The Maclaurin Series for f

$$f(0) + f'(0)x + \frac{f''(0)}{2!}x^2 + \cdots + \frac{f^{(n)}(0)}{n!}x^n + \cdots. \tag{5}$$

Thus, the question just posed is equivalent to asking whether the Maclaurin series for f converges to $f(x)$ as a sum. It certainly has the correct value, $f(0)$, at $x = 0$, but how far away from $x = 0$ may we go and still have convergence? And if the series does converge away from $x = 0$, does it still converge to $f(x)$? The graphs in Figs. 12.2 and 12.3 are encouraging, and the next few articles will confirm that we can normally expect a Maclaurin series to converge to its function in an interval about the origin. For many functions, this interval is the entire x-axis.

If, instead of approximating the values of f near zero, we are concerned with values of x near some other point a, we write our approximating polynomials in powers of $(x - a)$:

$$f_n(x) = a_0 + a_1(x - a) + a_2(x - a)^2 + \cdots + a_n(x - a)^n. \tag{6}$$

When we now determine the coefficients $a_0, a_1, \ldots, a_n$, so that the polynomial and its first n derivatives agree with the given function and its derivatives at $x = a$, we are led to a series that is called the *Taylor series expansion of f about x = a*, or simply the *Taylor series for f at x = a*.

The Taylor Series for f at $x = a$

$$f(a) + f'(a)(x - a) + \frac{f''(a)}{2!}(x - a)^2 + \cdots + \frac{f^{(n)}(a)}{n!}(x - a)^n + \cdots. \tag{7}$$

There are two things to notice here. The first is that Maclaurin series are Taylor series with $a = 0$. The second is that a function cannot have a Taylor series expansion about $x = a$ unless it has finite derivatives of all orders at $x = a$. For instance, $f(x) = \ln x$ does not have a Maclaurin series expansion, since the function itself, to say nothing of its derivatives, does not have a finite value at $x = 0$. On the other hand, it does have a Taylor series expansion in powers of $(x - 1)$, since $\ln x$ and all its derivatives are finite at $x = 1$.

Here are some examples of Taylor series.

EXAMPLE 3 From the formula derived for the Taylor polynomials of cos x in Example 2, it follows immediately that

$$\sum_{k=0}^{\infty} (-1)^k \frac{x^{2k}}{(2k)!} = 1 - \frac{x^2}{2!} + \frac{x^4}{4!} - \frac{x^6}{6!} + \cdots$$

is the Maclaurin series for cos x. □

EXAMPLE 4 Find the Taylor series expansion of cos x about the point $a = 2\pi$.

Solution The values of cos x and its derivatives at $a = 2\pi$ are the same as their values at $a = 0$. Therefore,

$$f^{(2k)}(2\pi) = f^{(2k)}(0) = (-1)^k \qquad \text{and} \qquad f^{(2k+1)}(2\pi) = f^{(2k+1)}(0) = 0,$$

as in Example 2. The required series is

$$\sum_{k=0}^{\infty} (-1)^{2k} \frac{(x - 2\pi)^{2k}}{(2k)!} = 1 - \frac{(x - 2\pi)^2}{2!} + \frac{(x - 2\pi)^4}{4!} - \cdots. \; \square$$

REMARK. There is a convention about how formulas like

$$\frac{x^{2k}}{(2k)!} \qquad \text{and} \qquad \frac{(x - 2\pi)^{2k}}{(2k)!},$$

which arise in the power series of Examples 3 and 4, are to be evaluated when $k = 0$. Besides the usual agreement that $0! = 1$, we also assume that

$$\frac{x^0}{0!} = \frac{1}{1} = 1 \qquad \text{and} \qquad \frac{(x - 2\pi)^0}{0!} = \frac{1}{1} = 1,$$

even when $x = 0$ or 2π.

The Binomial Series

One of the most celebrated series of all times, the *binomial series,* is the Maclaurin series for the function $f(x) = (1 + x)^m$. Newton used it to estimate integrals (we will, in Article 12.5) and it can be used to give accurate estimates of roots. To derive the series, we first list the function and its derivatives:

$$\begin{aligned}
f(x) &= (1 + x)^m, \\
f'(x) &= m(1 + x)^{m-1}, \\
f''(x) &= m(m - 1)(1 + x)^{m-2}, \\
f'''(x) &= m(m - 1)(m - 2)(1 + x)^{m-3}, \\
&\vdots \\
f^{(k)}(x) &= m(m - 1)(m - 2) \cdots (m - k + 1)(1 + x)^{m-k}.
\end{aligned}$$

We then substitute the values of these at $x = 0$ in the basic Maclaurin series (5) to obtain

$$1 + mx + \frac{m(m-1)}{2!}x^2 + \cdots + \frac{m(m-1)(m-2)\cdots(m-k+1)}{k!}x^k + \cdots. \quad (8)$$

If m is an integer ≥ 0, the series terminates after $(m + 1)$ terms, because the coefficients from $k = m + 1$ on are 0. But when k is not an integer, the series is infinite. (For a proof that the series converges to $(1 + x)^m$ when $|x| < 1$, see Courant and John's *Introduction to Calculus and Analysis,* Wiley-Interscience, 1974.)

EXAMPLE 5 Use the binomial series to estimate $\sqrt{1.25}$ with an error of less than 0.001.

Solution We take $x = \frac{1}{4}$ and $m = \frac{1}{2}$ in (8) to obtain

$$\left(1+\frac{1}{4}\right)^{1/2} = 1 + \frac{1}{2}\left(\frac{1}{4}\right) + \frac{(\frac{1}{2})(-\frac{1}{2})}{2!}\left(\frac{1}{4}\right)^2 + \frac{(\frac{1}{2})(-\frac{1}{2})(-\frac{3}{2})}{3!}\left(\frac{1}{4}\right)^3 + \cdots$$

$$= 1 + \frac{1}{8} - \frac{1}{128} + \frac{1}{1024} - \frac{1}{32768} + \cdots.$$

The series alternates after the first term, so that the approximation

$$\sqrt{1.25} \approx 1 + \frac{1}{8} - \frac{1}{128} \approx 1.117$$

is within $\frac{1}{1024}$ of the exact value and thus has the required accuracy. □

PROBLEMS

In Problems 1–9, use Eq. (3) to write the Taylor polynomials $f_3(x)$ and $f_4(x)$ for each of the following functions $f(x)$ at $x = 0$. In each case, your first step should be to complete a table like the one shown below.

n	$f^{(n)}(x)$	$f^{(n)}(0)$
0		
1		
2		
3		
4		

1. e^{-x} **2.** $\sin x$ **3.** $\cos x$

4. $\sin\left(x + \frac{\pi}{2}\right)$ **5.** $\sinh x$ **6.** $\cosh x$

7. $x^4 - 2x + 1$ **8.** $x^3 - 2x + 1$ **9.** $x^2 - 2x + 1$

In Problems 10–13, find the Maclaurin series for each function.

10. $\dfrac{1}{1+x}$ **11.** x^2

12. $(1 + x)^2$ **13.** $(1 + x)^{3/2}$

14. Find the Maclaurin series for $f(x) = 1/(1 - x)$. Show that the series diverges when $|x| \geq 1$ and converges when $|x| < 1$.

In Problems 15–20, use Eq. (7) to write the Taylor series expansion of the given function about the given point a.

15. $f(x) = e^x$, $a = 10$ **16.** $f(x) = x^2$, $a = \frac{1}{2}$

17. $f(x) = \ln x$, $a = 1$ **18.** $f(x) = \sqrt{x}$, $a = 4$

19. $f(x) = \dfrac{1}{x}$, $a = -1$ **20.** $f(x) = \cos x$, $a = -\dfrac{\pi}{4}$

In Problems 21 and 22, write the sum of the first three terms of the Taylor series for the given function about the given point a.

21. $f(x) = \tan x$, $a = \dfrac{\pi}{4}$ **22.** $f(x) = \ln \cos x$, $a = \dfrac{\pi}{3}$

23. Use the binomial theorem to estimate $\sqrt{1.02}$ with an error of less than 0.001.

12.2

Taylor's Theorem with Remainder: Sines, Cosines, and e^x

In the previous article, we asked when a Taylor series for a function can be expected to converge to the function. In this article, we answer the question with a theorem named after the English mathematician Brook Taylor (1685–1731). In Article 3.10, we stated Taylor's Theorem and referred to an outline of the proof. We restate the theorem here and fill in some of the details of the proof.

THEOREM 1

Taylor's Theorem

If f and its first n derivatives $f', f'', \ldots, f^{(n)}$ are continuous on $[a, b]$, or on $[b, a]$, and $f^{(n)}$ is differentiable on (a, b), or on (b, a), then there exists a number c between a and b such that

$$f(b) = f(a) + f'(a)(b-a) + \frac{f''(a)}{2!}(b-a)^2 + \cdots + \frac{f^{(n)}(a)}{n!}(b-a)^n + \frac{f^{(n+1)}(c)}{(n+1)!}(b-a)^{n+1}.$$

Proof. We assume that f satisfies the hypotheses of the theorem and define the Taylor polynomial about a of degree n:

$$f_n(x) = f(a) + f'(a)(x-a) + \frac{f''(a)}{2!}(x-a)^2 + \cdots + \frac{f^{(n)}(a)}{n!}(x-a)^n. \tag{1}$$

This polynomial and its first n derivatives match the function f and its first n derivatives at $x = a$. We do not disturb that matching by adding another term of the form $K(x-a)^{n+1}$, where K is any constant, because such a function and its first n derivatives are all equal to zero at $x = a$. Therefore, the new function

$$\phi_n(x) = f_n(x) + K(x-a)^{n+1} \tag{2}$$

and its first n derivatives still agree with f and its first n derivatives at $x = a$.

We now choose that particular value of K that makes the curve $y = \phi_n(x)$ agree with the original curve $y = f(x)$ at $x = b$. This can be done: we need only satisfy

$$f(b) = f_n(b) + K(b-a)^{n+1} \tag{3a}$$

or

$$K = \frac{f(b) - f_n(b)}{(b-a)^{n+1}}. \tag{3b}$$

With K defined by Eq. (3b), let $F(x) = f(x) - \phi_n(x)$, so that $F(x)$ measures the difference between the original function f and approximating function

ϕ_n, for each x in $[a, b]$, or in $[b, a]$ if $b < a$. To simplify the notation, we assume $a < b$, so that a is the left endpoint of all intervals mentioned. The same proof is valid if a is the right endpoint, instead of the left endpoint (for example, $[b, a]$, (b, a), $(c_1, a), \ldots, (c_n, a)$).

The remainder of the proof makes repeated use of Rolle's Theorem. First, because $F(a) = F(b) = 0$ and both F and F' are continuous on $[a, b]$, we know that

$$F'(c_1) = 0 \qquad \text{for some } c_1 \text{ in } (a, b).$$

Next, because $F'(a) = F'(c_1) = 0$ and both F' and F'' are continuous on $[a, c_1]$, we know that

$$F''(c_2) = 0 \qquad \text{for some } c_2 \text{ in } (a, c_1).$$

Rolle's Theorem, applied successively to $F'', F''', \ldots, F^{(n-1)}$ implies the existence of

$$\begin{aligned}
&c_3 \text{ in } (a, c_2) && \text{such that } F'''(c_3) = 0,\\
&c_4 \text{ in } (a, c_3) && \text{such that } F^{(iv)}(c_4) = 0,\\
&\vdots\\
&c_n \text{ in } (a, c_{n-1}) && \text{such that } F^{(n)}(c_n) = 0.
\end{aligned}$$

Finally, because $F^{(n)}$ is continuous on $[a, c_n]$ and differentiable on (a, c_n) and $F^{(n)}(a) = F^{(n)}(c_n) = 0$, Rolle's Theorem implies that there is a number c_{n+1} in (a, c_n) such that

$$F^{(n+1)}(c_{n+1}) = 0. \tag{4}$$

When we differentiate

$$F(x) = f(x) - f_n(x) - K(x - a)^{n+1}$$

$n + 1$ times, we get

$$F^{(n+1)}(x) = f^{(n+1)}(x) - 0 - (n + 1)!\,K. \tag{5}$$

Eqs. (4) and (5) together lead to the result

$$K = \frac{f^{(n+1)}(c)}{(n + 1)!} \qquad \text{for some number } c = c_{n+1} \text{ in } (a, b). \tag{6}$$

Combining Eqs. (3b) and (6), we have

$$\frac{f(b) - f_n(b)}{(b - a)^{n+1}} = \frac{f^{(n+1)}(c)}{(n + 1)!}$$

or

$$f(b) = f_n(b) + \frac{f^{(n+1)}(c)}{(n + 1)!}(b - a)^{n+1} \qquad \text{for some } c \text{ between } a \text{ and } b. \blacksquare$$

COROLLARY 1

If f has derivatives of all orders in an open interval I containing a, then for each positive integer n and for each x in I,

$$f(x) = f(a) + f'(a)(x - a) + \frac{f''(a)}{2!}(x - a)^2 + \cdots + \frac{f^{(n)}(a)}{n!}(x - a)^n + R_n(x, a) \tag{7a}$$

where

$$R_n(x, a) = \frac{f^{(n+1)}(c)}{(n+1)!}(x-a)^{n+1} \qquad \text{for some } c \text{ between } a \text{ and } x. \tag{7b}$$

The corollary follows at once from Taylor's Theorem because the existence of derivatives of all orders in an interval I implies the continuity of those derivatives and we have merely replaced b by x in the final formula.

The function $R_n(x, a)$ is called the remainder of order n: it's the difference $f(x) - f_n(x)$ where $f_n(x)$ is the Taylor polynomial of degree n used to approximate $f(x)$ near $x = a$. This difference, also called the "error" in the approximation $f_n(x)$, can often be estimated by using Eq. (7b) as in the next example.

When $R_n(x, a) \to 0$ as $n \to \infty$, for all x in some interval around $x = a$, we say that the Taylor-series expansion for $f(x)$ converges to $f(x)$ on that interval and write

$$f(x) = \sum_{k=0}^{\infty} \frac{f^{(k)}(a)}{k!}(x-a)^k. \tag{8}$$

EXAMPLE 1 Show that the Taylor series of $f(x) = e^x$ about $a = 0$ converges to $f(x)$ for every real value of x.

Solution Let $f(x) = e^x$. This function and all its derivatives are continuous at every point, so Taylor's Theorem may be applied with any convenient value of a. We take $a = 0$, since the values of f and its derivatives are easy to compute there. Taylor's Theorem leads to

$$e^x = 1 + x + \frac{x^2}{2!} + \frac{x^3}{3!} + \cdots + \frac{x^n}{n!} + R_n(x, 0) \tag{9a}$$

where

$$R_n(x, 0) = \frac{e^c}{(n+1)!}x^{n+1} \qquad \text{for some } c \text{ between } 0 \text{ and } x. \tag{9b}$$

Because e^x is an increasing function of x, and c is between 0 and x, the value of e^c is between 1 and e^x. Therefore, if x is negative, so is c and $e^c < 1$; if x is positive, so is c and $e^c < e^x$. Thus we can write

$$R_n(x, 0) < \frac{|x|^{n+1}}{(n+1)!} \qquad \text{when} \quad x < 0, \tag{9c}$$

and

$$R_n(x, 0) < e^x \frac{x^{n+1}}{(n+1)!} \qquad \text{when} \quad x > 0. \tag{9d}$$

When $x = 0$, the first term of the series in (9a) is $1 = e^0$, so the "error" is zero. Finally, because

$$\lim_{n\to\infty} \frac{x^{n+1}}{(n+1)!} = 0 \qquad \text{for every } x,$$

it is also true that

$$\lim_{n\to\infty} R_n(x, 0) = 0 \qquad \text{for every value of } x:$$

$$e^x = \sum_{k=0}^{\infty} \frac{x^k}{k!} = 1 + x + \frac{x^2}{2!} + \cdots + \frac{x^k}{k!} + \cdots. \square \tag{10}$$

Estimating the Remainder

It is often possible to estimate $R_n(x, a)$ as we did in Example 1. This method of estimation is so convenient that we state it as a theorem for future reference.

THEOREM 2

The Remainder Estimation Theorem

If there are positive constants M and r such that $|f^{(n+1)}(t)| \le Mr^{n+1}$ for all t between a and x, inclusive, then the remainder term $R_n(x, a)$ in Taylor's Theorem satisfies the inequality

$$|R_n(x, a)| \le M\frac{r^{n+1}|x - a|^{n+1}}{(n + 1)!}.$$

Furthermore, if the conditions above hold and all the other conditions of Taylor's Theorem are satisfied by $f(x)$, then the series converges to $f(x)$.

We are now ready to look at some examples of how the Remainder Estimation Theorem and Taylor's Theorem can be used together to settle questions of convergence. As you will see, they can also be used to determine the accuracy with which a function is approximated by one of its Taylor polynomials.

EXAMPLE 2 The Maclaurin series for sin x converges to sin x for all x.

Expressed in terms of x, the function and its derivatives are

$$\begin{aligned} f(x) &= \sin x, & f'(x) &= \cos x, \\ f''(x) &= -\sin x, & f'''(x) &= -\cos x, \\ &\vdots & &\vdots \\ f^{(2k)}(x) &= (-1)^k \sin x, & f^{(2k+1)}(x) &= (-1)^k \cos x, \end{aligned}$$

so that

$$f^{(2k)}(0) = 0 \qquad \text{and} \qquad f^{(2k+1)}(0) = (-1)^k.$$

The series has only odd-powered terms and, for $n = 2k + 1$, Taylor's formula gives

$$\sin x = x - \frac{x^3}{3!} + \frac{x^5}{5!} - \cdots + \frac{(-1)^k x^{2k+1}}{(2k + 1)!} + R_{2k+1}(x, 0).$$

Now, since all the derivatives of sin x have absolute values less than or equal to 1, we can apply the Remainder Estimation Theorem with $M = 1$ and $r = 1$ to obtain

$$|R_{2k+1}(x, 0)| \le 1 \cdot \frac{|x|^{2k+2}}{(2k + 2)!}.$$

Since $[|x|^{2k+2}/(2k+2)!] \to 0$ as $k \to \infty$, whatever the value of x,

$$R_{2k+1}(x, 0) \to 0,$$

and the Maclaurin series for sin x converges to sin x for every x:

$$\sin x = \sum_{k=0}^{\infty} \frac{(-1)^k x^{2k+1}}{(2k+1)!} = x - \frac{x^3}{3!} + \frac{x^5}{5!} - \frac{x^7}{7!} + \cdots. \quad \square \tag{11}$$

EXAMPLE 3 The Maclaurin series for cos x converges to cos x for every value of x.

We begin by adding the remainder term to the Taylor polynomial for cos x in Eq. (4) of the previous article, to obtain Taylor's formula for cos x with $n = 2k$:

$$\cos x = 1 - \frac{x^2}{2!} + \frac{x^4}{4!} - \cdots + (-1)^k \frac{x^{2k}}{(2k)!} + R_{2k}(x, 0).$$

Since the derivatives of the cosine have absolute value less than or equal to 1, we apply the Remainder Estimation Theorem with $M = 1$ and $r = 1$ to obtain

$$|R_{2k}(x, 0)| \le 1 \cdot \frac{|x|^{2k+1}}{(2k+1)!}.$$

For every value of x, $R_{2k} \to 0$ as $k \to \infty$. Therefore, the series converges to cos x for every value of x:

$$\cos x = \sum_{k=0}^{\infty} \frac{(-1)^k x^{2k}}{(2k)!} = 1 - \frac{x^2}{2!} + \frac{x^4}{4!} - \frac{x^6}{6!} + \cdots. \quad \square \tag{12}$$

EXAMPLE 4 Find the Maclaurin series for cos 2x and show that it converges to cos 2x for every value of x.

Solution The Maclaurin series for cos x converges to cos x for every value of x, and therefore converges for every value of 2x:

$$\cos 2x = \sum_{k=0}^{\infty} \frac{(-1)^k (2x)^{2k}}{(2k)!} = 1 - \frac{(2x)^2}{2!} + \frac{(2x)^4}{4!} - \frac{(2x)^6}{6!} + \cdots. \quad \square$$

Taylor series can be added, subtracted, and multiplied by constants, just as other series can, and the results are once again Taylor series. The Taylor series for $f(x) + g(x)$ is the sum of the Taylor series for $f(x)$ and $g(x)$, because the nth derivative of $f(x) + g(x)$ is $f^{(n)}(x) + g^{(n)}(x)$, and so on. In the next example, we add the series for e^x and e^{-x} and divide by 2, to obtain the Taylor series for cosh x.

EXAMPLE 5 Find the Taylor series for cosh x.

Solution

$$e^x = 1 + x + \frac{x^2}{2!} + \frac{x^3}{3!} + \frac{x^4}{4!} + \frac{x^5}{5!} + \cdots,$$

$$e^{-x} = 1 - x + \frac{x^2}{2!} - \frac{x^3}{3!} + \frac{x^4}{4!} - \frac{x^5}{5!} + \cdots;$$

$$\cosh x = \frac{e^x + e^{-x}}{2} = 1 + \frac{x^2}{2!} + \frac{x^4}{4!} + \cdots = \sum_{k=0}^{\infty} \frac{x^{2k}}{(2k)!}. \quad \square$$

EXAMPLE 6 *The identity* $e^{i\theta} = \cos\theta + i\sin\theta$. Up to this point we have not used imaginary numbers in our study of series. We recall that complex numbers occur in solving quadratic equations. The formula for the roots of the quadratic equation

$$ax^2 + bx + c = 0$$

is

$$x = \frac{-b \pm \sqrt{b^2 - 4ac}}{2a},$$

when a, b, and c are real numbers and $a \neq 0$. When the discriminant $D = b^2 - 4ac$ is negative, the two roots are complex numbers

$$u + iv \quad \text{and} \quad u - iv,$$

where

$$u = -\frac{b}{2a}, \qquad v = \frac{\sqrt{4ac - b^2}}{2a}.$$

We review these facts mainly to recall the symbol i:

$$i = \sqrt{-1},$$

and to remind ourselves that

$$i^2 = -1, \qquad i^3 = i^2 i = -i, \qquad i^4 = i^2 i^2 = 1, \qquad i^5 = i^4 i = i,$$

and so on.

With these facts in mind, we replace x by $i\theta$ in the Maclaurin series for e^x and simplify to get

$$\begin{aligned} e^{i\theta} &= 1 + \frac{i\theta}{1!} + \frac{i^2\theta^2}{2!} + \frac{i^3\theta^3}{3!} + \frac{i^4\theta^4}{4!} + \frac{i^5\theta^5}{5!} + \frac{i^6\theta^6}{6!} + \cdots \\ &= \left(1 - \frac{\theta^2}{2!} + \frac{\theta^4}{4!} - \frac{\theta^6}{6!} + \cdots\right) + i\left(\theta - \frac{\theta^3}{3!} + \frac{\theta^5}{5!} - \cdots\right) \\ &= \cos\theta + i\sin\theta. \end{aligned}$$

It would not be accurate to say that the calculations just completed have proved that

$$e^{i\theta} = \cos\theta + i\sin\theta. \tag{13}$$

Rather, we shall adopt the point of view that Eq. (13) is the *definition* of $e^{i\theta}$. This definition, which is standard, is motivated by the series expansions for $\cos\theta$, $\sin\theta$, and e^x with $x = i\theta$.

Once we have accepted Eq. (13) as the definition of $e^{i\theta}$, we quickly verify

The Law of Addition of Imaginary Exponents

If θ_1 and θ_2 are any real numbers then

$$e^{i\theta_1}e^{i\theta_2} = e^{i(\theta_1+\theta_2)}. \tag{14}$$

Proof. By definition,

$$e^{i\theta_1} = \cos\theta_1 + i\sin\theta_1, \qquad e^{i\theta_2} = \cos\theta_2 + i\sin\theta_2.$$

Multiplying and simplifying, we get

$$e^{i\theta_1}e^{i\theta_2} = (\cos\theta_1\cos\theta_2 - \sin\theta_1\sin\theta_2) + i(\sin\theta_1\cos\theta_2 + \cos\theta_1\sin\theta_2) = e^{i(\theta_1+\theta_2)}. \blacksquare$$

Notice also that when $\theta = 0$, $i\theta = 0$ and Eq. (13) yields

$$e^0 = \cos 0 + i \sin 0 = 1,$$

which is consistent with $e^x = 1$ when $x = 0$.

If $\theta_1 = \theta$ and $\theta_2 = -\theta$, then $\theta_1 + \theta_2 = 0$, and Eq. (14) yields the result

$$e^{i\theta}e^{-i\theta} = e^0 = 1,$$

so that

$$e^{-i\theta} = 1/e^{i\theta}. \tag{15}$$

Thus the usual laws for the exponential function continue to apply to the function $e^{i\theta}$ defined by Eq. (13). □

Truncation Error

Here are some examples of how to use the Remainder Estimation Theorem to estimate truncation error.

EXAMPLE 7 Calculate e with an error of less than 10^{-6}.

Solution We can use the result of Example 1, Eq. (9a), with $x = 1$ to write

$$e = 1 + 1 + \frac{1}{2!} + \cdots + \frac{1}{n!} + R_n(1, 0)$$

with

$$R_n(1, 0) = e^c\frac{1}{(n+1)!} \qquad \text{for some } c \text{ between 0 and 1.}$$

For the purposes of this example, we do not assume that we already know that $e = 2.71828\ldots$, but we have earlier shown that $e < 3$. Hence, we are certain that

$$\frac{1}{(n+1)!} < R_n(1, 0) < \frac{3}{(n+1)!}$$

because $1 < e^c < 3$.

By trial we find that $1/9! > 10^{-6}$, while $3/10! < 10^{-6}$. Thus we should take $(n + 1)$ to be at least 10, or n to be at least 9. With an error of less than 10^{-6},

$$e = 1 + 1 + \frac{1}{2} + \frac{1}{3!} + \cdots + \frac{1}{9!} \approx 2.718282. \ \square$$

EXAMPLE 8 For what values of x can sin x be replaced by $x - (x^3/3!)$ with an error of magnitude no greater than 3×10^{-4}?

Solution Here we can take advantage of the fact that the Maclaurin series for sin x is an alternating series for every nonzero value of x. According to the Alternating Series Estimation Theorem in Article 11.7, the error in truncating

$$\sin x = x - \frac{x^3}{3!} \Big| + \frac{x^5}{5!} - \cdots$$

after $(x^3/3!)$ is no greater than

$$\left|\frac{x^5}{5!}\right| = \frac{|x|^5}{120}.$$

Therefore the error will be less than or equal to 3×10^{-4} if

$$\frac{|x|^5}{120} < 3 \times 10^{-4} \qquad \text{or} \qquad |x| < \sqrt[5]{360 \times 10^{-4}} \approx 0.514.$$

The Alternating Series Estimation Theorem tells us something that the Remainder Estimation Theorem does not: namely, that the estimate $x - (x^3/3!)$ for $\sin x$ is an underestimate when x is positive, because then $x^5/120$ is positive.

Figure 12.4 shows the graph of $\sin x$, along with the graphs of a number of its approximating Taylor polynomials. Note that the graph of $s_1 = x - (x^3/3!)$ is almost indistinguishable from the sine curve when $-1 \le x \le 1$. However, it crosses the x-axis at $\pm\sqrt{6} \approx \pm 2.45$, whereas the sine curve crosses the axis at $\pm\pi \approx \pm 3.14$.

One might wonder how the estimate given by the Remainder Estimation Theorem would compare with the one we just obtained from the Alternating Series Estimation Theorem. If we write

$$\sin x = x - \frac{x^3}{3!} + R_3,$$

then the Remainder Estimation Theorem gives

$$|R_3| \le 1 \cdot \frac{|x|^4}{4!} = \frac{|x|^4}{24},$$

which is not very good. But, when we recognize that $x - x^3/3! = 0 + x + 0x^2 - x^3/3! + 0x^4$ is the Taylor polynomial of degree 4 as well as of

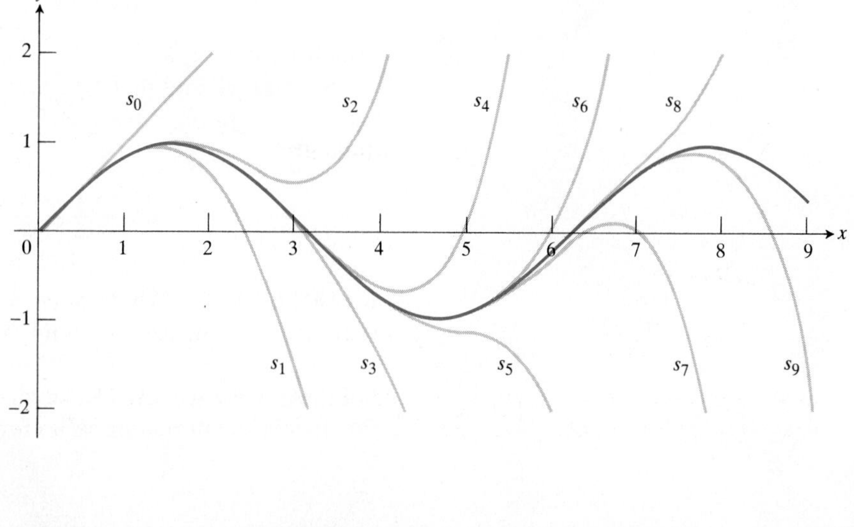

12.4 The polynomials

$$s_n(x) = \sum_{k=0}^{n} [(-1)^k x^{2k+1}/(2k+1)!]$$

converge to $\sin x$ as $n \to \infty$. (Adapted from Helen M. Kammerer. *American Mathematical Monthly,* 43(1936), 293–294.)

degree 3, then we have

$$\sin x = x - \frac{x^3}{3!} + 0 + R_4,$$

and the Remainder Estimation Theorem with $M = r = 1$ gives

$$|R_4| \le 1 \cdot \frac{|x|^5}{5!} = \frac{|x|^5}{120}.$$

This is what we had from the Alternating Series Estimation Theorem. □

In the preceding examples of application of the Remainder Estimation Theorem, we have been able to take $r = 1$. In the next example, $f(x) = \sin 2x$, a factor of 2 enters each time we differentiate, so that we have $r = 2$.

EXAMPLE 9 Let $f(x) = \sin 2x$. For what values of x is the approximation $\sin 2x \approx (2x) - (2x)^3/3! + (2x)^5/5!$ not in error by more than 5×10^{-6}?

Solution Because the Taylor polynomials of degree 5 and of degree 6 for sin 2x are identical, in that they differ only in a term that has a zero coefficient, we are certain that the error is no greater than $2^7|x|^7/7!$ (by comparison with Example 8). We therefore solve the inequality

$$\frac{|2x|^7}{7!} < 5 \times 10^{-6}.$$

The result is

$$|2x| < \sqrt[7]{7! \times 5 \times 10^{-6}} = \sqrt[7]{0.025200} \approx 0.59106$$

or

$$|x| < 0.29553 \text{ (radians)}. \ \square$$

PROBLEMS

In Problems 1–6, write the Maclaurin series for each function.

1. $e^{x/2}$
2. $\sin 3x$
3. $5 \cos \dfrac{x}{\pi}$
4. $\sinh x$
5. $\dfrac{x^2}{2} - 1 + \cos x$
6. $\cos^2 x = \dfrac{1 + \cos 2x}{2}$
7. Use series to verify that
   a) $\cos(-x) = \cos x$
   b) $\sin(-x) = -\sin x$
8. Show that
$$e^x = e^a\left[1 + (x - a) + \frac{(x-a)^2}{2!} + \cdots\right].$$

In Problems 9–11, write Taylor's formula (Eq. 7a), with $n = 2$ and $a = 0$, for the given function.

9. $\dfrac{1}{1 + x}$
10. $\ln(1 + x)$
11. $\sqrt{1 + x}$
12. Find the Taylor series for e^x at $a = 1$. Compare your series with the result in Problem 8.
13. For approximately what values of x can one replace sin x by $x - (x^3/6)$ with an error of magnitude no greater than 5×10^{-4}?
14. If cos x is replaced by $1 - (x^2/2)$ and $|x| < 0.5$, what estimate can be made of the error? Does $1 - (x^2/2)$ tend to be too large, or too small?
15. How close is the approximation $\sin x = x$ when $|x| < 10^{-3}$? For which of these values of x is $x < \sin x$?
16. The estimate $\sqrt{1 + x} = 1 + (x/2)$ is used when $|x|$ is small. Estimate the error when $|x| < 0.01$.

17. The approximation $e^x = 1 + x + (x^2/2)$ is used when x is small. Use the Remainder Estimation Theorem to estimate the error when $|x| < 0.1$.

18. When $x < 0$, the series for e^x is an alternating series. Use the Alternating Series Estimation Theorem to estimate the error that results from replacing e^x by $1 + x + (x^2/2)$ when $-0.1 < x < 0$. Compare with Problem 17.

19. Estimate the error in the approximation $\sinh x = x + (x^3/3!)$ when $|x| < 0.5$. (*Hint:* Use R_4, not R_3.)

20. When $0 \le h \le 0.01$, show that e^h may be replaced by $1 + h$ with an error of magnitude no greater than six-tenths of one percent of h. Use $e^{0.1} = 1.105$.

21. Let $f(x)$ and $g(x)$ have derivatives of all orders at $a = 0$. Show that the Maclaurin series for $f + g$ is

$$\sum_{n=0}^{\infty} \frac{f^{(n)}(0) + g^{(n)}(0)}{n!} x^n.$$

22. Each of the following sums is the value of an elementary function at some point. Find the function and the point.

a) $(0.1) - \dfrac{(0.1)^3}{3!} + \dfrac{(0.1)^5}{5!} - \cdots + \dfrac{(-1)^k (0.1)^{2k+1}}{(2k+1)!} + \cdots$

b) $1 - \dfrac{\pi^2}{4^2 \cdot 2!} + \dfrac{\pi^4}{4^4 \cdot 4!} - \cdots + \dfrac{(-1)^k (\pi)^{2k}}{4^{2k} \cdot (2k)!} + \cdots$

c) $1 + \dfrac{1}{2!} + \dfrac{1}{4!} + \cdots + \dfrac{1}{(2k)!} + \cdots$

23. Express each of the following in the form $u + iv$ with u and v real:

a) $e^{i\pi}$ b) $e^{i\pi/4}$

c) $e^{-i\pi/2}$ d) $e^{i\pi} \cdot e^{-i\pi/2}$

24. Using Eq. (13), show that

$$\cos\theta = \frac{e^{i\theta} + e^{-i\theta}}{2} \quad \text{and} \quad \sin\theta = \frac{e^{i\theta} - e^{-i\theta}}{2i}.$$

These are sometimes called Euler's identities.

25. Use the results of Problem 24 to show that

$$\cos^3\theta = \frac{1}{4}\cos 3\theta + \frac{3}{4}\cos\theta$$

and

$$\sin^3\theta = -\frac{1}{4}\sin 3\theta + \frac{3}{4}\sin\theta.$$

26. When a and b are real, we define $e^{(a+ib)x}$ to be $e^{ax}(\cos bx + i \sin bx)$. From this definition, show that

$$\frac{d}{dx} e^{(a+ib)x} = (a + ib)e^{(a+ib)x}.$$

27. Two complex numbers, $a + ib$ and $c + id$, are equal if and only if $a = c$ and $b = d$. Use this fact to evaluate

$$\int e^{ax} \cos bx\, dx \quad \text{and} \quad \int e^{ax} \sin bx\, dx$$

from

$$\int e^{(a+ib)x}\, dx = \frac{a - ib}{a^2 + b^2} e^{(a+ib)x} + C,$$

where

$$C = C_1 + iC_2$$

is a complex constant of integration.

***Toolkit* programs**

Taylor Series

12.3 Further Computations, Logarithms, Arctangents, and π

The Taylor-series expansion

$$f(x) = f(a) + f'(a)(x - a) + \frac{f''(a)}{2!}(x - a)^2 + \cdots + \frac{f^{(n)}(a)}{n!}(x - a)^n + R_n(x, a) \tag{1}$$

expresses the value of the function at x in terms of its value and the values of its derivatives at a, plus a remainder term, which we hope is so small that it may safely be omitted. In applying series to numerical computations, it is therefore *necessary* that a be chosen so that $f(a), f'(a), f''(a), \ldots$ are known. In dealing with the trigonometric functions, for example, one might take $a = 0, \pm\pi/6, \pm\pi/4, \pm\pi/3, \pm\pi/2$, and so on. It is also clear that it is *desirable* to choose the value of a near to the value of x for which the function is to be computed, in order to make $(x - a)$ small, so that the terms of the series decrease rapidly as n increases.

EXAMPLE 1 What value of a might one choose in the Taylor series (1) to compute sin 35°?

Solution We could choose $a = 0$ and use the series

$$\sin x = x - \frac{x^3}{3!} + \frac{x^5}{5!} - \cdots + (-1)^n \frac{x^{2n+1}}{(2n+1)!} + 0 \cdot x^{2n+2} + R_{2n+2}(x, 0), \tag{2}$$

or we could choose $a = \pi/6$ (which corresponds to 30°) and use the series

$$\sin x = \sin\frac{\pi}{6} + \cos\frac{\pi}{6}\left(x - \frac{\pi}{6}\right) - \sin\frac{\pi}{6}\frac{(x - \pi/6)^2}{2!} - \cos\frac{\pi}{6}\frac{(x - \pi/6)^3}{3!} + \cdots + \sin\left(\frac{\pi}{6} + n\frac{\pi}{2}\right)\frac{(x - \pi/6)^n}{n!} + R_n\left(x, \frac{\pi}{6}\right).$$

The remainder in the series (2) satisfies the inequality

$$|R_{2n+2}(x, 0)| \le \frac{|x|^{2n+3}}{(2n+3)!}, \tag{3}$$

which tends to zero as n becomes infinite, no matter how large $|x|$ may be. We could therefore calculate sin 35° by placing

$$x = \frac{35\pi}{180} = 0.6108652$$

in the approximation

$$\sin x \approx x - \frac{x^3}{6} + \frac{x^5}{120} - \frac{x^7}{5040},$$

with an error of magnitude no greater than 3.3×10^{-8}, since

$$\left|R_8\left(\frac{35\pi}{180}, 0\right)\right| < \frac{(0.611)^9}{9!} < 3.3 \times 10^{-8}.$$

By using the series with $a = \pi/6$, we could obtain equal accuracy with a smaller exponent n, but at the expense of introducing $\cos \pi/6 = \sqrt{3}/2$ as one of the coefficients. In this series, with $a = \pi/6$, we would take

$$x = \frac{35\pi}{180},$$

but the quantity that appears raised to the various powers is

$$x - \frac{\pi}{6} = \frac{5\pi}{180} = 0.0872665,$$

which decreases rapidly when raised to high powers.

As a matter of fact, various trigonometric identities may be used, such as

$$\sin\left(\frac{\pi}{2} - x\right) = \cos x,$$

to facilitate the calculation of the sine or cosine of any angle with the Maclaurin series of the two functions. This method of finding the sine or cosine of an angle is used in computers. □

Computation of Logarithms

Natural logarithms may be computed from series. The starting point is the series for $\ln(1 + x)$ in powers of x:

$$\ln(1 + x) = x - \frac{x^2}{2} + \frac{x^3}{3} - \cdots + (-1)^{n-1}\frac{x^n}{n} + \cdots.$$

This series may be found directly from the Taylor-series expansion, Eq. (1), with $a = 0$. It may also be obtained by integrating the geometric series for $1/(1 + t)$ from $t = 0$ to $t = x$:

$$\int_0^x \frac{dt}{1+t} = \int_0^x (1 - t + t^2 - t^3 + \cdots)\,dt,$$

$$\ln(1 + t)\Big]_0^x = t - \frac{t^2}{2} + \frac{t^3}{3} - \frac{t^4}{4} + \cdots\Big]_0^x,$$

$$\ln(1 + x) = x - \frac{x^2}{2} + \frac{x^3}{3} - \frac{x^4}{4} + \cdots. \tag{4}$$

The expansion (4) is valid for $|x| < 1$, since then the remainder, $R_n(x, 0)$, approaches zero as $n \to \infty$, as we shall now see. The remainder is given by the integral of the remainder in the geometric series, that is,

$$R_n(x, 0) = \int_0^x \frac{(-1)^n t^n}{1+t}\,dt. \tag{5}$$

We now suppose that $|x| < 1$. For every t between 0 and x inclusive we have

$$|1 + t| \geq 1 - |x|$$

and

$$|(-1)^n t^n| = |t|^n,$$

so that

$$\left|\frac{(-1)^n t^n}{1+t}\right| \leq \frac{|t|^n}{1 - |x|}.$$

Therefore,

$$|R_n(x, 0)| \leq \int_0^{|x|} \frac{t^n}{1 - |x|}\,dt = \frac{1}{n+1} \cdot \frac{|x|^{n+1}}{1 - |x|}. \tag{6}$$

When $n \to \infty$, the right-hand side of the inequality (6) approaches zero, and so must the left-hand side. Thus (4) holds for $|x| < 1$.

Computation of π

Archimedes (287–212 B.C.) gave the approximation

$$3\frac{1}{7} > \pi > 3\frac{10}{71},$$

in the third century B.C. A French mathematician, Viéta (1540–1603), gave the formula

$$\frac{2}{\pi} = \sqrt{\tfrac{1}{2}} \times \sqrt{\left(\tfrac{1}{2} + \tfrac{1}{2}\sqrt{\tfrac{1}{2}}\right)} \times \sqrt{\left(\tfrac{1}{2} + \tfrac{1}{2}\sqrt{\left(\tfrac{1}{2} + \tfrac{1}{2}\sqrt{\tfrac{1}{2}}\right)}\right)} \times \cdots,$$

which Turnbull† calls "the first actual formula for the time-honoured number π." Other interesting formulas for π include‡

$$\frac{\pi}{4} = \cfrac{1}{1 + \cfrac{1^2}{2 + \cfrac{3^2}{2 + \cfrac{5^2}{2 + \cdots}}}},$$

which is credited to Lord Brouncker, an Irish peer;

$$\frac{\pi}{4} = \frac{2 \times 4 \times 4 \times 6 \times 6 \times 8 \times \cdots}{3 \times 3 \times 5 \times 5 \times 7 \times 7 \times \cdots},$$

discovered by the English mathematician Wallis; and

$$\frac{\pi}{4} = 1 - \frac{1}{3} + \frac{1}{5} - \frac{1}{7} + \cdots,$$

known as Leibniz's formula.

We now turn our attention to the series for $\tan^{-1} x$, since it leads to the Leibniz formula and others from which π has been computed to a great many decimal places.

Since

$$\tan^{-1} x = \int_0^x \frac{dt}{1 + t^2},$$

we integrate the geometric series, with remainder,

$$\frac{1}{1 + t^2} = 1 - t^2 + t^4 - t^6 + \cdots + (-1)^n t^{2n} + \frac{(-1)^{n+1} t^{2n+2}}{1 + t^2}. \qquad (7)$$

Thus

$$\tan^{-1} x = x - \frac{x^3}{3} + \frac{x^5}{5} - \frac{x^7}{7} + \cdots + (-1)^n \frac{x^{2n+1}}{2n + 1} + R,$$

where

$$R = \int_0^x \frac{(-1)^{n+1} t^{2n+2}}{1 + t^2}\, dt.$$

The denominator of the integrand is greater than or equal to 1; hence

$$|R| \le \int_0^{|x|} t^{2n+2}\, dt = \frac{|x|^{2n+3}}{2n + 3}.$$

If $|x| \le 1$, the right side of this inequality approaches zero as $n \to \infty$.

† *World of Mathematics*, Vol. 1 (New York: Simon and Schuster, 1956), p. 121.
‡ *Ibid.*, p. 138.

Therefore R also approaches zero and we have

$$\tan^{-1} x = \sum_{n=0}^{\infty} \frac{(-1)^n x^{2n+1}}{2n+1},$$

or

$$\tan^{-1} x = x - \frac{x^3}{3} + \frac{x^5}{5} - \frac{x^7}{7} + \cdots, \quad |x| \le 1. \tag{8}$$

When we put $x = 1$, $\tan^{-1} 1 = \pi/4$, in Eq. (8) we get Leibniz's formula:

$$\frac{\pi}{4} = 1 - \frac{1}{3} + \frac{1}{5} - \frac{1}{7} + \frac{1}{9} - \cdots + \frac{(-1)^{n-1}}{2n-1} + \cdots.$$

Because this series converges very slowly, it is not used in approximating π to many decimal places. The series for $\tan^{-1}x$ converges most rapidly when x is near zero. For that reason, people who use the series for $\tan^{-1}x$ to compute π use various trigonometric identities.

For example, if

$$\alpha = \tan^{-1}\frac{1}{2} \quad \text{and} \quad \beta = \tan^{-1}\frac{1}{3},$$

then

$$\tan(\alpha + \beta) = \frac{\tan\alpha + \tan\beta}{1 - \tan\alpha\tan\beta} = \frac{\frac{1}{2} + \frac{1}{3}}{1 - \frac{1}{6}} = 1 = \tan\frac{\pi}{4}$$

and

$$\frac{\pi}{4} = \alpha + \beta = \tan^{-1}\frac{1}{2} + \tan^{-1}\frac{1}{3}. \tag{9}$$

Now Eq. (8) may be used with $x = \frac{1}{2}$ to evaluate $\tan^{-1}\frac{1}{2}$ and with $x = \frac{1}{3}$ to give $\tan^{-1}\frac{1}{3}$. The sum of these results, multiplied by 4, gives π.

In 1961, π was computed to more than 100,000 decimal places on an IBM 7090 computer. More recently, in 1973, Jean Guilloud and Martine Bouyer computed π to 1,000,000 decimal places on a CDC 7600 computer, by applying the arctangent series (8) to the formula

$$\pi = 48\tan^{-1}\frac{1}{18} + 32\tan^{-1}\frac{1}{57} - 20\tan^{-1}\frac{1}{239}. \tag{10}$$

They checked their work with the formula

$$\pi = 24\tan^{-1}\frac{1}{8} + 8\tan^{-1}\frac{1}{57} + 4\tan^{-1}\frac{1}{239}. \tag{11}$$

A number of current computations of π are being carried out with an algorithm discovered by Eugene Salamin (Problem 17). The algorithm produces sequences that converge to π even more rapidly than the sequence of partial sums of the arctangent series in Eqs. (10) and (11).†

REMARK. Two types of numerical error tend to occur in computing with series. The first is the *truncation error,* which is the remainder $R_n(x, a)$. This is the only error we have discussed so far.

---

†For a delightful account of attempts to compute, and even to *legislate*(!) the value of π, see Chapter 12 of David A. Smith's *Interface: Calculus and the Computer* (Boston, Massachusetts: Houghton Mifflin, 1976).

FREQUENTLY-USED MACLAURIN SERIES

$$\frac{1}{1-x} = 1 + x + x^2 + \cdots + x^n + \cdots = \sum_{n=0}^{\infty} x^n, \qquad |x| < 1$$

$$\frac{1}{1+x} = 1 - x + x^2 - \cdots + (-x)^n + \cdots = \sum_{n=0}^{\infty} (-1)^n x^n, \qquad |x| < 1$$

$$e^x = 1 + x + \frac{x^2}{2!} + \cdots + \frac{x^n}{n!} + \cdots = \sum_{n=0}^{\infty} \frac{x^n}{n!}, \qquad |x| < \infty$$

$$\sin x = x - \frac{x^3}{3!} + \frac{x^5}{5!} - \cdots + (-1)^n \frac{x^{2n+1}}{(2n+1)!} + \cdots = \sum_{n=0}^{\infty} \frac{(-1)^n x^{2n+1}}{(2n+1)!}, \qquad |x| < \infty$$

$$\cos x = 1 - \frac{x^2}{2!} + \frac{x^4}{4!} - \cdots + (-1)^n \frac{x^{2n}}{(2n)!} + \cdots = \sum_{n=0}^{\infty} \frac{(-1)^n x^{2n}}{(2n)!}, \qquad |x| < \infty$$

$$\ln(1+x) = x - \frac{x^2}{2} + \frac{x^3}{3} - \cdots + (-1)^{n-1} \frac{x^n}{n} + \cdots = \sum_{n=1}^{\infty} \frac{(-1)^{n-1} x^n}{n}, \qquad -1 < x \le 1$$

$$\ln \frac{1+x}{1-x} = 2 \tanh^{-1} x = 2\left(x + \frac{x^3}{3} + \frac{x^5}{5} + \cdots + \frac{x^{2n+1}}{2n+1} + \cdots\right) = 2 \sum_{n=0}^{\infty} \frac{x^{2n+1}}{2n+1}, \qquad |x| < 1$$

$$\tan^{-1} x = x - \frac{x^3}{3} + \frac{x^5}{5} - \cdots + (-1)^{n-1} \frac{x^{2n-1}}{2n-1} + \cdots = \sum_{n=1}^{\infty} \frac{(-1)^{n-1} x^{2n-1}}{2n-1}, \qquad |x| \le 1$$

BINOMIAL SERIES

$$(1+x)^m = 1 + mx + \frac{m(m-1)\,x^2}{2!} + \frac{m(m-1)(m-2)\,x^3}{3!} + \cdots$$
$$+ \frac{m(m-1)(m-2)\cdots(m-k+1)\,x^k}{k!} + \cdots$$
$$= 1 + \sum_{k=1}^{\infty} \binom{m}{k} x^k, \qquad |x| < 1$$

where

$$\binom{m}{1} = m,$$
$$\binom{m}{2} = \frac{m(m-1)}{2!},$$
$$\binom{m}{k} = \frac{m(m-1)\cdots(m-k+1)}{k!} \qquad \text{for } k \ge 3.$$

NOTE. It is customary to define $\binom{m}{0}$ to be 1 and to take $x^0 = 1$ (even in the usually excluded case where $x = 0$) in order to write the binomial series compactly as

$$(1+x)^m = \sum_{k=0}^{\infty} \binom{m}{k} x^k, \qquad |x| < 1.$$

If m is a *positive integer*, the series terminates at x^m and the result converges for all x.

The second is the *round-off error* that enters in calculating the sum of the finite number of terms

$$f(a) + f'(a)(x - a) + \cdots + \frac{f^{(n)}(a)(x - a)^n}{n!}$$

when we approximate each of these terms by a decimal number with only a finite number of decimal places. For example, taking 0.3333 in place of $\frac{1}{3}$ introduces a round-off error equal to $10^{-4}/3$. There are likely to be round-off errors associated with each term, some of these being positive and some negative. When the need for accuracy is paramount, it is important to control both the truncation error and the round-off errors. *Truncation* errors can be reduced by taking more terms of the series; *round-off* errors can be reduced by taking more decimal places.

PROBLEMS

In Problems 1–6, use a suitable series to calculate the indicated quantity to three decimal places. In each case, show that the remainder term does not exceed 5×10^{-4}.

1. cos 31° **2.** tan 46° **3.** sin 6.3

4. cos 69 **5.** ln 1.25 **6.** $\tan^{-1} 1.02$

7. Find the Maclaurin series for ln (1 + 2x). For what values of x does the series converge?

8. For what values of x can one replace ln (1 + x) by x with an error of magnitude no greater than one percent of the absolute value of x?

Use series to evaluate the integrals in Problems 9 and 10 to three decimals.

9. $\displaystyle\int_0^{0.1} \frac{\sin x}{x}\,dx$ **10.** $\displaystyle\int_0^{0.1} e^{-x^2}\,dx$

11. Show that the ordinate of the catenary $y = a \cosh x/a$ deviates from the ordinate of the parabola $x^2 = 2a(y - a)$ by less than $0.003|a|$ over the range $|x/a| \le \frac{1}{3}$.

12. a) Replace x by −x in Eq. (4) to obtain a series for ln (1 − x). Combine this with the series for ln (1 + x) to show that

$$\ln\frac{1 + x}{1 - x} = 2\left(x + \frac{x^3}{3} + \frac{x^5}{5} + \cdots\right) \quad \text{for } |x| < 1.$$

b) For what value of x is (1 + x)/(1 − x) = 2? Use that value of x in the series of part (a) to estimate ln 2 to 3 decimals.

13. Find the sum of the series

$$\frac{1}{2} - \frac{1}{2}\left(\frac{1}{2}\right)^2 + \frac{1}{3}\left(\frac{1}{2}\right)^3 - \frac{1}{4}\left(\frac{1}{2}\right)^4 + \cdots.$$

14. How many terms of the series for $\tan^{-1} 1$ would you have to add for the Alternating Series Estimation Theorem to guarantee a calculation of $\pi/4$ to two decimals?

15. (*Calculator*) Equations (8) and (10) yield a series that converges to $\pi/4$ fairly rapidly. Estimate π to three decimal places with this series. In contrast, the convergence of $\sum_{n=1}^{\infty} (1/n^2)$ to $\pi^2/6$ is so slow that even fifty terms will not yield two-place accuracy.

16. (*Calculator*) a) Find π to two decimals with the formulas of Lord Brouncker and Wallis.

b) If your calculator is programmable, use Viéta's formula to calculate π to five decimal places.

17. (*Calculator*) A special case of Salamin's algorithm for estimating π begins with defining sequences $\{a_n\}$ and $\{b_n\}$ by the rules

$$a_0 = 1, \qquad b_0 = \frac{1}{\sqrt{2}},$$

$$a_{n+1} = \frac{(a_n + b_n)}{2}, \qquad b_{n+1} = \sqrt{a_n b_n},$$

Then the sequence $\{c_n\}$ defined for $n \ge 1$ by

$$c_n = \frac{4a_n b_n}{1 - \sum_{j=1}^{n} 2^{j+1}(a_j^2 - b_j^2)}$$

converges to π. Calculate c_3. (E. Salamin, "Computation of π using arithmetic-geometric mean," *Mathematics of Computation*, **30**, July, 1976, pp. 565–570.)

18. Show that the series in Eq. (8) for $\tan^{-1} x$ diverges for $|x| > 1$.

19. Show that

$$\int_0^x \frac{dt}{1 - t^2} = \int_0^x \left(1 + t^2 + t^4 + \cdots + t^{2n} + \frac{t^{2n+2}}{1 - t^2}\right) dt$$

or, in other words, that

$$\tanh^{-1} x = x + \frac{x^3}{3} + \frac{x^5}{5} + \cdots + \frac{x^{2n+1}}{2n + 1} + R,$$

where

$$R = \int_0^x \frac{t^{2n+2}}{1-t^2}\,dt.$$

20. Show that R in Problem 19 is no greater than

$$\frac{1}{1-x^2}\cdot\frac{|x|^{2n+3}}{2n+3}, \quad \text{if } x^2 < 1.$$

21. a) Differentiate the identity

$$\frac{1}{1-x} = 1 + x + x^2 + \cdots + x^n + \frac{x^{n+1}}{1-x}$$

to obtain the expansion

$$\frac{1}{(1-x)^2} = 1 + 2x + 3x^2 + \cdots + nx^{n-1} + R.$$

b) Prove that, if $|x| < 1$, then $R \to 0$ as $n \to \infty$.

c) In one throw of two dice, the probability of getting a score of 7 is $p = \frac{1}{6}$. If the dice are thrown repeatedly, the probability that a 7 will appear for the first time at the nth throw is $q^{n-1}p$, where $q = 1 - p = \frac{5}{6}$. The expected number of throws until a 7 first appears is $\Sigma_{n=1}^{\infty} nq^{n-1}p$. Evaluate this series numerically.

d) In applying statistical quality control to an industrial operation, an engineer inspects items taken at random from the assembly line. Each item sampled is classified as "good' or "bad." If the probability of a good item is p and of a bad item is $q = 1 - p$, the probability that the first bad item found is the nth inspected is $p^{n-1}q$. The average number inspected up to and including the first bad item found is $\Sigma_{n=1}^{\infty} np^{n-1}q$. Evaluate this series, assuming $0 < p < 1$.

22. In probability theory, a random variable X may assume the values 1, 2, 3, . . . , with probabilities $p_1, p_2, p_3, \ldots$, where p_k is the probability that X is equal to k ($k = 1, 2, \ldots$). It is customary to assume $p_k \geq 0$ and $\Sigma_{k=1}^{\infty} p_k = 1$. The *expected value* of X denoted by E(X) is defined as $\Sigma_{k=1}^{\infty} kp_k$, provided this series converges. In each of the following cases, show that $\Sigma p_k = 1$ and find E(X), if it exists. (*Hint:* See Problem 21.)

a) $p_k = 2^{-k}$

b) $p_k = \dfrac{5^{k-1}}{6^k}$

c) $p_k = \dfrac{1}{k(k+1)} = \dfrac{1}{k} - \dfrac{1}{k+1}$

***Toolkit* programs**

Sequences and Series   Taylor Series

12.4 Indeterminate Forms

In considering the ratio of two functions $f(x)$ and $g(x)$, we sometimes wish to know the value

$$\lim_{x\to a} \frac{f(x)}{g(x)} \tag{1}$$

at a point a where $f(x)$ and $g(x)$ are both zero. L'Hôpital's rule is often a help, but the differentiation involved can be time-consuming, especially if the rule has to be applied several times to reach a determinate form. In many instances, the limit in (1) can be calculated more quickly if the functions involved have power series expansions about $x = a$. In fact, the ease and reliability of the kind of calculation we are about to illustrate contributed to the early popularity of power series. The theoretical justification of the technique is too long to discuss here, but the formal manipulations are worth learning by themselves.

Suppose, then, that the functions f and g both have series expansions in powers of $x - a$,

$$f(x) = f(a) + f'(a)\cdot(x-a) + \frac{f''(a)}{2!}(x-a)^2 + \cdots, \tag{2a}$$

$$g(x) = g(a) + g'(a)\cdot(x-a) + \frac{g''(a)}{2!}(x-a)^2 + \cdots, \tag{2b}$$

that are known to us and that converge in some interval $|x - a| < \delta$. We then proceed to calculate the limit (1), provided the limit exists, in the manner shown by the following examples.

EXAMPLE 1 Evaluate $\lim_{x\to 1} [(\ln x)/(x - 1)]$.

Solution Let $f(x) = \ln x$, $g(x) = x - 1$. The Taylor series for $f(x)$, with $a = 1$, is found as follows:

$$\begin{aligned} f(x) &= \ln x, & f(1) &= \ln 1 = 0,\\ f'(x) &= 1/x, & f'(1) &= 1,\\ f''(x) &= -1/x^2, & f''(1) &= -1, \end{aligned}$$

so that

$$\ln x = 0 + (x - 1) - \frac{1}{2}(x - 1)^2 + \cdots.$$

Hence

$$\frac{\ln x}{x - 1} = 1 - \frac{1}{2}(x - 1) + \cdots$$

and

$$\lim_{x\to 1} \frac{\ln x}{x - 1} = \lim_{x\to 1} [1 - \frac{1}{2}(x - 1) + \cdots] = 1. \ \square$$

EXAMPLE 2 Evaluate $\lim_{x\to 0} [(\sin x - \tan x)/x^3]$.

Solution The Maclaurin series for $\sin x$ and $\tan x$, to terms in x^5, are

$$\sin x = x - \frac{x^3}{3!} + \frac{x^5}{5!} - \cdots,$$

$$\tan x = x + \frac{x^3}{3} + \frac{2x^5}{15} + \cdots.$$

Hence

$$\sin x - \tan x = -\frac{x^3}{2} - \frac{x^5}{8} - \cdots = x^3\left(-\frac{1}{2} - \frac{x^2}{8} - \cdots\right),$$

and

$$\lim_{x\to 0} \frac{\sin x - \tan x}{x^3} = \lim_{x\to 0} \left(-\frac{1}{2} - \frac{x^2}{8} - \cdots\right) = -\frac{1}{2}. \ \square$$

When we apply series to compute the limit $\lim_{x\to 0} (1/\sin x - 1/x)$ of Example 7, Article 3.9, we not only compute the limit successfully but also discover a nice approximation formula for $\csc x$.

EXAMPLE 3 Find

$$\lim_{x\to 0} \left(\frac{1}{\sin x} - \frac{1}{x}\right).$$

Solution

$$\frac{1}{\sin x} - \frac{1}{x} = \frac{x - \sin x}{x \sin x} = \frac{x - \left[x - \frac{x^3}{3!} + \frac{x^5}{5!} - \cdots\right]}{x \cdot \left[x - \frac{x^3}{3!} + \frac{x^5}{5!} - \cdots\right]}$$

$$= \frac{x^3\left[\frac{1}{3!} - \frac{x^2}{5!} + \cdots\right]}{x^2\left[1 - \frac{x^2}{3!} + \cdots\right]}$$

$$= x\,\frac{\frac{1}{3!} - \frac{x^2}{5!} + \cdots}{1 - \frac{x^2}{3!} + \cdots}.$$

Therefore,

$$\lim_{x\to 0}\left(\frac{1}{\sin x} - \frac{1}{x}\right) = \lim_{x\to 0}\left[x\,\frac{\frac{1}{3!} - \frac{x^2}{5!} + \cdots}{1 - \frac{x^2}{3!} + \cdots}\right] = 0.$$

In fact, from the series expressions above we can see that if $|x|$ is small, then

$$\frac{1}{\sin x} - \frac{1}{x} \approx x \cdot \frac{1}{3!} = \frac{x}{6} \qquad \text{or} \qquad \csc x \approx \frac{1}{x} + \frac{x}{6}. \ \square$$

PROBLEMS

Use series to evaluate the limits in Problems 1–20.

1. $\lim_{h\to 0} \frac{\sin h}{h}$

2. $\lim_{x\to 0} \frac{e^x - (1 + x)}{x^2}$

3. $\lim_{t\to 0} \frac{1 - \cos t - \frac{1}{2}t^2}{t^4}$

4. $\lim_{x\to\infty} x \sin \frac{1}{x}$

5. $\lim_{x\to 0} \frac{x^2}{1 - \cosh x}$

6. $\lim_{h\to 0} \frac{(\sin h)/h - \cos h}{h^2}$

7. $\lim_{x\to 0} \frac{1 - \cos x}{\sin x}$

8. $\lim_{x\to 0} \frac{\sin x}{e^x - 1}$

9. $\lim_{z\to 0} \frac{\sin (z^2) - \sinh (z^2)}{z^6}$

10. $\lim_{t\to 0} \frac{\cos t - \cosh t}{t^2}$

11. $\lim_{x\to 0} \frac{\sin x - x + \frac{x^3}{6}}{x^5}$

12. $\lim_{x\to 0} \frac{e^x - e^{-x} - 2x}{x - \sin x}$

13. $\lim_{x\to 0} \frac{x - \tan^{-1} x}{x^3}$

14. $\lim_{x\to 0} \frac{\tan x - \sin x}{x^3 \cos x}$

15. $\lim_{x\to\infty} x^2(e^{-1/x^2} - 1)$

16. $\lim_{x\to 0} \frac{\ln (1 + x^2)}{1 - \cos x}$

17. $\lim_{x\to 0} \frac{\tan 3x}{x}$

18. $\lim_{x\to 1} \frac{\ln x^2}{x - 1}$

19. $\lim_{x\to\infty} \frac{x^{100}}{e^x}$

20. $\lim_{x\to 0} \left(\frac{1}{2 - 2\cos x} - \frac{1}{x^2}\right)$

21. a) Prove that $\int_0^x e^{t^2}\,dt \to +\infty$ as $x \to +\infty$.
b) Find $\lim_{x\to\infty} x \int_0^x e^{t^2 - x^2}\,dt$.

22. Find values of r and s such that

$$\lim_{x\to 0} (x^{-3} \sin 3x + rx^{-2} + s) = 0.$$

23. (*Calculator*) The approximation for $\csc x$ in Example 3 leads to the approximation $\sin x \approx 6x/(6 + x^2)$. Evaluate both sides of this approximation for $x = \pm 1.0$, ± 0.1, and ± 0.01 radians. Try these values of x in the approximation $\sin x \approx x$. Which approximation appears to give better results?

***Toolkit* programs**

Taylor Series

12.5

Convergence of Power Series. Integration, Differentiation, Multiplication, and Division

We now know that some power series, like the Maclaurin series for sin x, cos x, and e^x, converge for all values of x, while others, like the series we derived for ln (1 + x) and $\tan^{-1} x$, converge only on finite intervals. But we learned all this by analyzing remainder formulas, and we have yet to face the question of how to investigate the convergence of a power series when there is no remainder formula to analyze. Moreover, all of the power series we have worked with have been Taylor series of functions for which we already had expressions in closed forms. What about other power series? Are they Taylor series, too, of functions otherwise unknown?

The first step in answering these questions is to note that a power series $\sum_{n=0}^{\infty} a_n x^n$ defines a function whenever it converges, namely, the function f whose value at each x is the number

$$f(x) = \sum_{n=0}^{\infty} a_n x^n. \tag{1}$$

We can then ask what kind of domain f has, how f is to be differentiated and integrated (if at all), whether f has a Taylor series, and, if it has, how its Taylor series is related to the defining series $\sum_{n=0}^{\infty} a_n x^n$.

The questions of what domain f has, and for what values the series (1) may be expected to converge, are answered by Theorem 1 and the discussion that follows it. We will prove Theorem 1, and then, after looking at examples, will proceed to Theorems 2 and 3, which answer the questions of whether f *can* be differentiated and integrated and *how* to do so when it can be. Theorem 3 also solves a problem that arose many chapters ago but that has remained unsolved until now: that of finding convenient expressions for evaluating integrals like

$$\int_0^1 \sin x^2\, dx \qquad \text{and} \qquad \int_0^{0.5} \sqrt{1 + x^4}\, dx,$$

which frequently arise in applications. Finally, we will see that, in the interior of its domain of definition, the function f does have a Maclaurin series, and that this is none other than the defining series $\sum_{n=0}^{\infty} a_n x^n$.

THEOREM 1

The Convergence Theorem for Power Series

If a power series

$$\sum_{n=0}^{\infty} a_n x^n = a_0 + a_1 x + a_2 x^2 + \cdots \tag{2}$$

converges for $x = c$ $(c \neq 0)$, then it converges absolutely for all $|x| < |c|$. If the series diverges for $x = d$, then it diverges for all $|x| > |d|$.

Proof. Suppose the series

$$\sum_{n=0}^{\infty} a_n c^n \tag{3}$$

converges. Then

$$\lim_{n\to\infty} a_n c^n = 0.$$

Hence, there is an index N such that

$$|a_n c^n| < 1 \qquad \text{for all} \quad n \geq N.$$

That is,

$$|a_n| < \frac{1}{|c|^n} \qquad \text{for} \quad n \geq N. \tag{4}$$

Now take any x such that $|x| < |c|$ and consider

$$|a_0| + |a_1 x| + \cdots + |a_{N-1}x^{N-1}| + |a_N x^N| + |a_{N+1}x^{N+1}| + \cdots.$$

There is only a finite number of terms prior to $|a_N x^N|$ and their sum is finite. Starting with $|a_N x^N|$ and beyond, the terms are less than

$$\left|\frac{x}{c}\right|^N + \left|\frac{x}{c}\right|^{N+1} + \left|\frac{x}{c}\right|^{N+2} + \cdots \tag{5}$$

by virtue of the inequality (4). But the series in (5) is a geometric series with ratio $r = |x/c|$, which is less than one, since $|x| < |c|$. Hence the series (5) converges, so that the original series (3) converges absolutely. This proves the first half of the theorem.

The second half of the theorem involves nothing new. For if the series diverges at $x = d$ and converges at a value x_0 with $|x_0| > |d|$, we may take $c = x_0$ in the first half of the theorem and conclude that the series converges absolutely at d. But the series cannot both converge absolutely and diverge at one and the same time. Hence, if it diverges at d, it diverges for all $|x| > |d|$. ■

The significance of Theorem 1 is that a power series always behaves in exactly *one* of the following three ways (Fig. 12.5).

1. It converges at $x = 0$ and diverges everywhere else.
2. There is a positive number c such that the series diverges for $|x| > c$ but converges absolutely for $|x| < c$. It may or may not converge at either of the endpoints $x = c$ and $x = -c$.
3. It converges absolutely for every x.

In Case 2, the set of points at which the series converges is a finite interval. We know from past examples that this interval may be open, half open, or closed, depending on the series in question. But no matter which kind of interval it is, c is called the *radius of convergence* of the series, and the convergence is absolute at every point in the interior of the interval. The interval is called the *interval of convergence*. If a power series converges absolutely for all values of x, we say that its radius of convergence is infinite. If it converges only at $x = 0$, we say that the radius of convergence is 0.

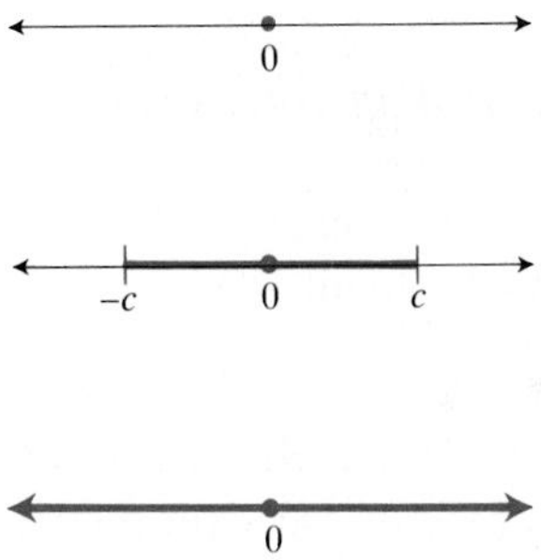

Figure 12.5

As examples of power series whose radii of convergence are infinite, we have the Taylor series of sin x, cos x, and e^x. These series converge for every value of $c = 2x$, and therefore converge absolutely for every value of x.

As examples of series whose radii of convergence are finite we have

Series	Interval of convergence
$\frac{1}{1-x} = 1 + x + x^2 + \cdots$	$-1 < x < 1$
$\ln(1+x) = x - \frac{x^2}{2} + \frac{x^3}{3} - \frac{x^4}{4} + \cdots$	$-1 < x \leq 1$
$\tan^{-1} x = x - \frac{x^3}{3} + \frac{x^5}{5} - \cdots$	$-1 \leq x \leq 1$

The interval of convergence of a power series $\Sigma_{n=0}^{\infty} a_n x^n$ can often be found by applying the ratio test or the root test to the series of absolute values,

$$\sum_{n=0}^{\infty} |a_n x^n|.$$

Thus, if

$$\rho = \lim_{n\to\infty} \left| \frac{a_{n+1}x^{n+1}}{a_n x^n} \right| \quad \text{or if} \quad \rho = \lim_{n\to\infty} \sqrt[n]{|a_n x^n|},$$

then,

a) $\Sigma\, |a_n x^n|$ converges at all values of x for which $\rho < 1$,

b) $\Sigma\, |a_n x^n|$ diverges at all values of x for which $\rho > 1$,

c) $\Sigma\, |a_n x^n|$ may either converge or diverge at a value of x for which $\rho = 1$.

How do these three alternatives translate into statements about the series $\Sigma\, a_n x^n$? Case (a) says that $\Sigma\, a_n x^n$ converges absolutely at all values of x for which $\rho < 1$. Case (c) does not tell us anything more about the series $\Sigma\, a_n x^n$ than it does about the series $\Sigma\, |a_n x^n|$. Either series might converge or diverge at a value of x for which $\rho = 1$. In Case (b), we can actually conclude that $\Sigma\, a_n x^n$ diverges at all values of x for which $\rho > 1$. The argument goes like this: As you may recall from the discussions in Article 11.5, the fact that ρ is greater than 1 means that either

$$0 < |a_n x^n| < |a_{n+1}x^{n+1}| < |a_{n+2}x^{n+2}| < \cdots$$

or

$$\sqrt[n]{|a_n x^n|} > 1$$

for n sufficiently large. Thus the terms of the series do not approach 0 as n becomes infinite, and the series diverges *with or without absolute values*, by the nth-term test.

Therefore, the ratio and root tests, when successfully applied to $\Sigma\, |a_n x^n|$, lead us to the following conclusions about $\Sigma\, a_n x^n$:

A) $\Sigma\, a_n x^n$ converges absolutely for all values of x for which $\rho < 1$,

B) $\Sigma\, a_n x^n$ diverges at all values of x for which $\rho > 1$,

C) $\Sigma\, a_n x^n$ may either converge or diverge at a value of x for which $\rho = 1$.

In Case (C), another test is needed.

EXAMPLE 1 Find the interval of convergence of

$$\sum_{n=1}^{\infty} \frac{x^n}{n}. \tag{6}$$

Solution We apply the ratio test to the series of absolute values, and find

$$\rho = \lim_{n\to\infty} \left| \frac{x^{n+1}}{n+1} \cdot \frac{n}{x^n} \right| = |x|.$$

Therefore the original series converges absolutely if $|x| < 1$ and diverges if $|x| > 1$. When $x = +1$, the series becomes

$$1 + \frac{1}{2} + \frac{1}{3} + \frac{1}{4} + \cdots,$$

which diverges. When $x = -1$, the series becomes

$$-\left(1 - \frac{1}{2} + \frac{1}{3} - \frac{1}{4} + \cdots\right),$$

which converges, by Leibnitz's Theorem. Therefore the series (6) converges for $-1 \le x < 1$ and diverges for all other values of x. □

EXAMPLE 2 For what values of x does the series

$$\sum_{n=1}^{\infty} \frac{(2x-5)^n}{n^2}$$

converge?

Solution We treat the series as a power series in the variable $2x - 5$. An application of the root test to the series of absolute values yields

$$\rho = \lim_{n\to\infty} \sqrt[n]{\left|\frac{(2x-5)^n}{n^2}\right|} = \lim_{n\to\infty} \frac{|2x-5|}{\sqrt[n]{n^2}} = \frac{|2x-5|}{1} = |2x-5|.$$

The series converges absolutely for

$$|2x-5| < 1 \quad \text{or} \quad -1 < 2x - 5 < 1$$

or

$$4 < 2x < 6 \quad \text{or} \quad 2 < x < 3.$$

When $x = 2$, the series is $\Sigma_{n=1}^{\infty} [(-1)^n/n^2]$, which converges.

When $x = 3$, the series is $\Sigma_{n=1}^{\infty} [(1)^n/n^2]$, which converges. Therefore, the interval of convergence is $2 \le x \le 3$. □

Sometimes the comparison test does as well as any.

EXAMPLE 3 For what values of x does

$$\sum_{n=1}^{\infty} \frac{\cos^n x}{n!}$$

converge?

Solution For every value of x,

$$\left|\frac{\cos^n x}{n!}\right| \le \frac{1}{n!}.$$

The series converges for every value of x. □

The next theorem says that a function defined by a power series has derivatives of all orders at every point in the interior of its interval of convergence. The derivatives can be obtained as power series by differentiating the terms of the original series. The first derivative is obtained by differentiating the terms of the original series once:

$$\frac{d}{dx}\sum_{n=0}^{\infty}(a_n x^n) = \sum_{n=0}^{\infty}\frac{d}{dx}(a_n x^n) = \sum_{n=0}^{\infty} n a_n x^{n-1}.$$

For the second derivative, the terms are differentiated again, and so on. We state the theorem without proof, and go directly to the examples.

THEOREM 2

The Term-by-Term Differentiation Theorem

If $f(x) = \Sigma_{n=0}^{\infty} a_n x^n$ has radius of convergence c, then,

1. $\sum_{n=0}^{\infty} n a_n x^{n-1}$ also has radius of convergence c,
2. $f(x)$ is differentiable on $(-c, c)$, and
3. $f'(x) = \sum_{n=0}^{\infty} n a_n x^{n-1}$ on $(-c, c)$.

Ostensibly, Theorem 2 mentions only f and f'. But because f' has the same radius of convergence that f has, the theorem applies equally well to f', saying that it has a derivative f'' on $(-c, c)$. This in turn implies that f'' is differentiable on $(-c, c)$, and so on. Thus, if $f(x) = \Sigma_{n=0}^{\infty} a_n x^n$ converges on $(-c, c)$, it has derivatives of all orders at every point of $(-c, c)$.

EXAMPLE 4 The relation $(d/dx)(\sin x) = \cos x$ is easily checked by differentiating the series for sin x term by term:

$$\sin x = x - \frac{x^3}{3!} + \frac{x^5}{5!} - \frac{x^7}{7!} + \cdots$$

$$\frac{d}{dx}(\sin x) = 1 - \frac{x^2}{2!} + \frac{x^4}{4!} - \frac{x^6}{6!} + \cdots = \cos x. \ \square$$

Convergence at one or both endpoints of the interval of convergence of a power series may be lost in the process of differentiation. That is why Theorem 2 mentions only the *open* interval $(-c, c)$.

EXAMPLE 5 The series $f(x) = \Sigma_{n=1}^{\infty} (x^n/n)$ of Example 1 converges for $-1 \le x < 1$. The series of derivatives

$$f'(x) = \sum_{n=1}^{\infty} x^{n-1} = 1 + x + x^2 + x^3 + \cdots$$

is a geometric series that converges only for $-1 < x < 1$. The series diverges at the endpoint $x = -1$, as well as at the endpoint $x = 1$. □

Example 5 shows, however, that when the terms of a series are integrated, the resulting series may converge at an endpoint that was not a point of convergence before. The justification for term-by-term integration of a series is the following theorem, which we also state without proof.

THEOREM 3

The Term-by-Term Integration Theorem

If $f(x) = \sum_{n=0}^{\infty} a_n x^n$ has radius of convergence c, then,

1. $\displaystyle\sum_{n=0}^{\infty} \frac{a_n x^{n+1}}{n+1}$ also has radius of convergence c,

2. $\int f(x)\,dx$ exists for x in $(-c, c)$,

3. $\displaystyle\int f(x)\,dx = \sum_{n=0}^{\infty} \frac{a_n x^{n+1}}{n+1} + C$ on $(-c, c)$.

EXAMPLE 6 The series

$$\frac{1}{1+t} = 1 - t + t^2 - t^3 \cdots$$

converges on the open interval $-1 < t < 1$. Therefore,

$$\begin{aligned} \ln(1+x) &= \int_0^x \frac{1}{1+t}\,dt \\ &= t - \frac{t^2}{2} + \frac{t^3}{3} - \frac{t^4}{4} + \cdots \Big]_0^x, \\ &= x - \frac{x^2}{2} + \frac{x^3}{3} - \frac{x^4}{4} + \cdots, \qquad -1 < x < 1. \end{aligned}$$

As you know, the latter series also converges at $x = 1$, but that was not guaranteed by the theorem. □

EXAMPLE 7 By replacing t by t^2 in the series of Example 6, we obtain

$$\frac{1}{1+t^2} = 1 - t^2 + t^4 - t^6 + \cdots, \qquad -1 < t < 1.$$

Therefore

$$\tan^{-1} x = \int_0^x \frac{1}{1+t^2}\,dt = t - \frac{t^3}{3} + \frac{t^5}{5} - \frac{t^7}{7} + \cdots \Big]_0^x$$

$$= x - \frac{x^3}{3} + \frac{x^5}{5} - \frac{x^7}{7} + \cdots, \qquad -1 < x < 1.$$

This is not as refined a result as the one we obtained in Article 12.3, where we were able to show that the interval of convergence was $-1 \leq x \leq 1$ by analyzing a remainder. But the result here is obtained more quickly. □

EXAMPLE 8 Express

$$\int \sin x^2\,dx$$

as a power series.

Solution From the series for $\sin x$ we obtain

$$\sin x^2 = x^2 - \frac{x^6}{3!} + \frac{x^{10}}{5!} - \frac{x^{14}}{7!} + \cdots, \qquad -\infty < x < \infty.$$

Therefore,

$$\int \sin x^2\,dx = C + \frac{x^3}{3} - \frac{x^7}{7\cdot 3!} + \frac{x^{11}}{11\cdot 5!} - \frac{x^{15}}{15\cdot 7!} + \cdots,$$

$$-\infty < x < \infty. \ \square$$

EXAMPLE 9 Estimate $\int_0^1 \sin x^2\,dx$ with an error of less than 0.001.

Solution From the indefinite integral in Example 8,

$$\int_0^1 \sin x^2\,dx = \frac{1}{3} - \frac{1}{7\cdot 3!} + \frac{1}{11\cdot 5!} - \frac{1}{15\cdot 7!} + \frac{1}{19\cdot 9!} - \cdots.$$

The series alternates, and we find by trial that

$$\frac{1}{11\cdot 5!} \approx 0.00076$$

is the first term to be numerically less than 0.001. The sum of the preceding two terms gives

$$\int_0^1 \sin x^2\,dx \approx \frac{1}{3} - \frac{1}{42}$$
$$\approx 0.310.$$

With two more terms we could estimate

$$\int_0^1 \sin x^2\,dx \approx 0.310268$$

with an error of less than 10^{-6}; and with only one term beyond that we have

$$\int_0^1 \sin x^2\,dx \approx \frac{1}{3} - \frac{1}{42} + \frac{1}{1320} - \frac{1}{75600} + \frac{1}{6894720}$$
$$\approx 0.310268303,$$

with an error of less than 10^{-9}. To guarantee this accuracy with the error formula for the trapezoid rule would require using about 13,000 subintervals. $\square$

EXAMPLE 10 Estimate $\int_0^{0.5} \sqrt{1 + x^4}\,dx$ with an error of less than 10^{-4}.

Solution The binomial expansion of $(1 + x^4)^{1/2}$ is

$$(1 + x^4)^{1/2} = 1 + \frac{1}{2}x^4 - \frac{1}{8}x^8 + \cdots,$$

a series whose terms alternate in sign after the second term. Therefore,

$$\int_0^{0.5} \sqrt{1 + x^4}\, dx = x + \frac{1}{2 \cdot 5} x^5 - \frac{1}{8 \cdot 9} x^9 + \cdots \Big]_0^{0.5}$$
$$= 0.5 + 0.0031 - 0.00003 + \cdots$$
$$\approx 0.5031,$$

with an error of magnitude less than 0.00003. □

The Maclaurin Series for $\sum_{n=0}^{\infty} a_n x^n$

At the beginning of this article we asked whether a function

$$f(x) = \sum_{n=0}^{\infty} a_n x^n$$

defined by a convergent power series has a Taylor series. We can now answer that a function defined by a power series with a radius of convergence $c > 0$ has a Maclaurin series that converges to the function at every point of $(-c, c)$. Why? Because the Maclaurin series for the function $f(x) = \Sigma_{n=0}^{\infty} a_n x^n$ is the series $\Sigma_{n=0}^{\infty} a_n x^n$ itself. To see this, we differentiate

$$f(x) = a_0 + a_1 x + a_2 x^2 + \cdots + a_n x^n + \cdots$$

term by term and substitute $x = 0$ in each derivative $f^{(n)}(x)$. This produces

$$f^{(n)}(0) = n!\, a_n \qquad \text{or} \qquad a_n = \frac{f^{(n)}(0)}{n!}$$

for every n. Thus,

$$f(x) = \sum_{n=0}^{\infty} a_n x^n$$
$$= \sum_{n=0}^{\infty} \frac{f^{(n)}(0)}{n!} x^n, \qquad -c < x < c. \tag{7}$$

An immediate consequence of this is that series like

$$x \sin x = x^2 - \frac{x^4}{3!} + \frac{x^6}{5!} - \frac{x^8}{7!} + \cdots,$$

and

$$x^2 e^x = x^2 + x^3 + \frac{x^4}{2!} + \frac{x^5}{3!} + \cdots,$$

which are obtained by multiplying Maclaurin series by powers of x, as well as series obtained by integration and differentiation of power series, are themselves the Maclaurin series of the functions they represent.

Another consequence of (7) is that if two power series $\Sigma_{n=0}^{\infty} a_n x^n$ and $\Sigma_{n=0}^{\infty} b_n x^n$ are equal for all values of x in an open interval that contains the origin $x = 0$, then $a_n = b_n$ for every n. For if

$$f(x) = \sum_{n=0}^{\infty} a_n x^n$$
$$= \sum_{n=0}^{\infty} b_n x^n, \qquad -c < x < c,$$

then a_n and b_n are both equal to $f^{(n)}(0)/n!$.

Multiplication of Power Series

We illustrate with an example.

EXAMPLE 11 Find the first five terms in the Maclaurin series expansion for $e^x \cos x$ by multiplying together the series for e^x and for $\cos x$.

Solution From Article 12.2, Eqs. (10) and (12), we have the series expansions

$$e^x = 1 + x + \frac{x^2}{2!} + \frac{x^3}{3!} + \frac{x^4}{4!} + \cdots,$$

$$\cos x = 1 - \frac{x^2}{2!} + \frac{x^4}{4!} - \cdots.$$

Obviously, if we need only terms involving x^n for $n \le 4$, we can truncate both series at their x^4-terms and multiply the resulting polynomials, discarding everything involving higher powers like $x^5, \ldots, x^8$. The result is

$$\begin{aligned} e^x \cos x &= 1 + x + x^2\left(\frac{1}{2!} - \frac{1}{2!}\right) + x^3\left(\frac{1}{3!} - \frac{1}{2!}\right) \\ &\quad + x^4\left(\frac{1}{4!} - \frac{1}{2!2!} + \frac{1}{4!}\right) + \cdots \\ &= 1 + x - \frac{x^3}{3} - \frac{x^4}{6} + \cdots. \ \square \end{aligned} \tag{8}$$

We shall not prove it here, but using $\cos x = (e^{ix} + e^{-ix})/2$ we could establish the result

$$e^x \cos x = \frac{1}{2}\sum_{n=0}^{\infty} \frac{(1+i)^n + (1-i)^n}{n!} x^n = \sum_{n=0}^{\infty} \frac{(\sqrt{2})^n \cos(n\pi/4)}{n!} x^n, \tag{9}$$

because

$$1 + i = \sqrt{2}\, e^{i\pi/4}, \qquad 1 - i = \sqrt{2}\, e^{-i\pi/4}.$$

REMARK. If $f(x) = u(x) + i\, v(x)$, where $u(x)$ and $v(x)$ are real valued functions of x, we call $u(x)$ the *real part* of $f(x)$ and write

$$u(x) = \operatorname{Re} f(x).$$

Similarly, $v(x)$ is called the *imaginary part* of $f(x)$ (that's right, even though v is itself real!) and write

$$v(x) = \operatorname{Im} f(x).$$

With this notation, and recalling that $e^{ix} = \cos x + i \sin x$, we see that

$$e^x \cos x = \operatorname{Re}(e^x e^{ix}) = \operatorname{Re}(e^{(1+i)x})$$

and

$$e^x \sin x = \operatorname{Im}(e^x e^{ix}) = \operatorname{Im}(e^{(1+i)x}).$$

From Eq. (9), it would be easy to show that the series for $e^x \cos x$ converges absolutely for all real values of x. This is also guaranteed by the following theorem, which we shall not prove.

THEOREM 4

The Series Multiplication Theorem for Power Series

If both $\Sigma\, a_n x^n$ and $\Sigma\, b_n x^n$ converge absolutely for $|x| < R$, and

$$c_n = a_0 b_n + a_1 b_{n-1} + a_2 b_{n-2} + \cdots + a_{n-1} b_1 + a_n b_0$$

$$= \sum_{k=0}^{n} a_k b_{n-k}, \tag{10a}$$

then the series $\Sigma\, c_n x^n$ also converges absolutely for $|x| < R$, and

$$(a_0 + a_1 x + a_2 x^2 + \cdots)\cdot(b_0 + b_1 x + b_2 x^2 + \cdots)$$
$$= c_0 + c_1 x + c_2 x^2 + \cdots. \tag{10b}$$

Division of Series

Again, we illustrate with an example.

EXAMPLE 12 Find some of the terms in the Maclaurin series for tan x by dividing the series for sin x by the series for cos x.

Solution We proceed as in ordinary algebraic long division, keeping track of terms up to x^5 and disregarding all higher powers of x:

$$\begin{array}{rl}
 & x + \dfrac{x^3}{3} + \dfrac{2}{15}x^5 + \cdots = \tan x \\
\cos x = 1 - \dfrac{x^2}{2!} + \dfrac{x^4}{4!} - \cdots \Big) & x - \dfrac{x^3}{3!} + \dfrac{x^5}{5!} - \cdots = \sin x \\
 & x - \dfrac{x^3}{2!} + \dfrac{x^5}{4!} \cdots \\
 & \dfrac{x^3}{3} - \dfrac{x^5}{30} \cdots \\
 & \dfrac{x^3}{3} - \dfrac{x^5}{6} \cdots \\
 & \dfrac{2}{15}x^5 \cdots \\
 & \dfrac{2}{15}x^5
\end{array}$$

To terms in x^5, we thus have

$$\tan x = \frac{\sin x}{\cos x} = x + \frac{x^3}{3} + \frac{2}{15}x^5 + \cdots. \tag{11}$$

We used these first few terms of the series for tan x in Article 12.4, Example 2. □

REMARK. Because $\cos(\pi/2) = 0$, we certainly cannot expect the Maclaurin series for tan x to converge outside the interval $|x| < \pi/2$. For the full Maclaurin series, which is not easy to obtain either by long division or by direct application of Taylor's formula, the reader is referred to page 75, Eq. 4.3.67, of the Handbook of Mathematical Functions, Applied Mathematics Series, of the U.S. Department of Commerce, National Bureau of Standards, AMS 55, edited by M. Abramowitz and I. A. Stegun, Dover Publications, Inc., New York.

There is another way to calculate the coefficients in the Maclaurin series for the quotient $f(x)/g(x)$, a way that is readily adaptable to machine computation. The facts are as follows:

1. If $f(x) = \Sigma\, a_n x^n$ for $|x| < R_1$, $g(x) = \Sigma\, b_n x^n$ for $|x| < R_2$, and $b_0 = g(0) \neq 0$, then $f(x)/g(x)$ has a power series representation $\Sigma\, c_n x^n$ on some interval $(-h, h)$.
2. Within that interval,

$$\begin{aligned}\Sigma\, a_n x^n &= f(x)\\ &= g(x)\cdot\frac{f(x)}{g(x)}\\ &= \Sigma\, b_n x^n \cdot \Sigma\, c_n x^n\\ &= \sum_n \left(\sum_{k=0}^{n} b_k c_{n-k}\right) x^n,\end{aligned}$$

and hence

$$a_n = \sum_{k=0}^{n} b_k c_{n-k}.$$

In other words,

$$\begin{aligned}a_0 &= b_0 c_0\\ a_1 &= b_0 c_1 + b_1 c_0, \qquad \text{etc.,}\end{aligned}$$

so that the coefficients c_n can be found one after the other in this way:

$$c_0 = \frac{a_0}{b_0}, \qquad c_1 = \frac{a_1 - b_1 c_0}{b_0}, \tag{12a}$$

and, for all $n \geq 1$,

$$\begin{aligned}c_n &= \frac{a_n - (b_1 c_{n-1} + b_2 c_{n-2} + \cdots + b_n c_0)}{b_0}\\ &= \frac{a_n - \Sigma_{k=1}^{n} b_k c_{n-k}}{b_0}.\end{aligned} \tag{12b}$$

EXAMPLE 13 Let

$$f(x) = \sin x = x - \frac{x^3}{3!} + \frac{x^5}{5!} - \cdots$$

so that

$$a_0 = 0,\ a_1 = 1,\ a_2 = 0,\ a_3 = -\frac{1}{3!},\ a_4 = 0, \ldots$$

and let

$$g(x) = \cos x = 1 - \frac{x^2}{2!} + \frac{x^4}{4!} - \frac{x^6}{6!} + \cdots$$

so that

$$b_0 = 1,\ b_1 = 0,\ b_2 = -\frac{1}{2!},\ b_3 = 0,\ b_4 = \frac{1}{4!},\ \text{etc.}$$

Then

$$\tan x = \frac{\sin x}{\cos x} = c_0 + c_1 x + c_2 x^2 + \cdots$$

with

$$c_0 = \frac{a_0}{b_0} = 0,\ c_1 = \frac{a_1 - b_1 c_0}{b_0} = \frac{1 - 0}{1} = 1,$$

and so on. When we know the values of $c_0, c_1, \ldots, c_{n-1}$, the value of c_n is given by Eq. (12b) in terms of known coefficients. □

PROBLEMS

In Problems 1–20, find the interval of absolute convergence. If the interval is finite, determine whether the series converges at each endpoint.

1. $\sum_{n=0}^{\infty} x^n$
2. $\sum_{n=0}^{\infty} n^2 x^n$
3. $\sum_{n=1}^{\infty} \frac{nx^n}{2^n}$
4. $\sum_{n=0}^{\infty} \frac{(2x)^n}{n!}$
5. $\sum_{n=0}^{\infty} \frac{(-1)^n x^{2n+1}}{(2n+1)!}$
6. $\sum_{n=1}^{\infty} (-1)^{n-1} \frac{(x-1)^n}{n}$
7. $\sum_{n=0}^{\infty} \frac{n^2}{2^n}(x+2)^n$
8. $\sum_{n=0}^{\infty} \frac{x^{2n+1}}{2n+1}$
9. $\sum_{n=0}^{\infty} (-1)^n \frac{x^{2n+1}}{2n+1}$
10. $\sum_{n=1}^{\infty} \frac{(x-2)^n}{n^2}$
11. $\sum_{n=0}^{\infty} \frac{\cos nx}{2^n}$
12. $\sum_{n=1}^{\infty} \frac{2^n x^n}{n^5}$
13. $\sum_{n=0}^{\infty} \frac{x^n e^n}{n+1}$
14. $\sum_{n=1}^{\infty} \frac{(\cos x)^n}{n^n}$
15. $\sum_{n=0}^{\infty} n^n x^n$
16. $\sum_{n=0}^{\infty} \frac{(3x+6)^n}{n!}$
17. $\sum_{n=1}^{\infty} (-2)^n (n+1)(x-1)^n$
18. $\sum_{n=1}^{\infty} \frac{(-1)^{n+1}(x-2)^n}{n \cdot 2^n}$
19. $\sum_{n=0}^{\infty} \left(\frac{x^2-1}{2}\right)^n$
20. $\sum_{n=1}^{\infty} \frac{(x+3)^{n-1}}{n}$

21. Find the sum of the series in Problem 16.

22. When the series of Problem 19 converges, to what does it converge?

23. Use series to verify that

a) $\frac{d}{dx}(\cos x) = -\sin x$, b) $\int_0^x \cos t\, dt = \sin x$,

c) $y = e^x$ is a solution of the equation $y' = y$.

24. Obtain the Maclaurin series for $1/(1+x)^2$ from the series for $-1/(1+x)$.

25. Use the Maclaurin series for $1/(1-x^2)$ to obtain a series for $2x/(1-x^2)^2$.

26. Use the identity $\sin^2 x = (1 - \cos 2x)/2$ to obtain a series for $\sin^2 x$. Then differentiate this series to obtain a series for $2 \sin x \cos x$. Check that this is the series for $\sin 2x$.

(*Calculator*) In Problems 27–34, use series and a calculator to estimate each integral with an error of magnitude less than 0.001.

27. $\int_0^{0.2} \sin x^2\, dx$
28. $\int_0^{0.1} \tan^{-1} x\, dx$
29. $\int_0^{0.1} x^2 e^{-x^2}\, dx$
30. $\int_0^{0.1} \frac{\tan^{-1} x}{x}\, dx$
31. $\int_0^{0.4} \frac{1 - e^{-x}}{x}\, dx$
32. $\int_0^{0.1} \frac{\ln(1+x)}{x}\, dx$
33. $\int_0^{0.1} \frac{1}{\sqrt{1+x^4}}\, dx$
34. $\int_0^{0.25} \sqrt[3]{1+x^2}\, dx$

35. (*Calculator*) a) Obtain a power series for

$$\sinh^{-1} x = \int_0^x \frac{dt}{\sqrt{1+t^2}}.$$

b) Use the result of (a) to estimate $\sinh^{-1} 0.25$ to three decimal places.

36. (*Calculator*) Estimate $\int_0^1 \cos x^2\, dx$ with an error of less than one millionth.

37. Show by example that there are power series that converge only at $x = 0$.

38. Show by examples that the convergence of a series at an endpoint of its interval of convergence may be either conditional or absolute.

39. Let r be any positive number. Use Theorem 1 to show that if $\sum_{n=0}^{\infty} a_n x^n$ converges for $-r < x < r$, then it converges absolutely for $-r < x < r$.

40. Use the ratio test to show that the binomial series converges for $|x| < 1$. (This still does not show that the series converges to $(1 + x)^m$.)

41. Find terms through x^5 of the Maclaurin series for $e^x \sin x$ by appropriate multiplication. The series is the imaginary part of the series for

$$e^x \cdot e^{ix} = e^{(1+i)x}.$$

Use this fact to check your answer. For what values of x should the series for $e^x \sin x$ converge? Why?

42. Divide 1 by a sufficient number of terms of the Maclaurin series for $\cos x$ to obtain terms through x^4 in the Maclaurin series for $\sec x$. Where do you think the resulting complete Maclaurin series should converge?

43. Integrate the first three nonzero terms of the Maclaurin series for $\tan t$ from 0 to x to get the first three nonzero terms in the Maclaurin series for $\ln \sec x$.

44. (Continuation) Another way to get some of the terms in the Maclaurin series for $\ln \sec x$ is as follows: Let $\sec x = 1 + y$, so that $y = x^2/2 + 5x^4/24 + \cdots$ (from Problem 42). Then, for $|y| < 1$,

$$\ln \sec x = \ln (1 + y) = y - \frac{y^2}{2} + \cdots.$$

Neglecting powers of x higher than x^4, show that this also leads to

$$\ln \sec x = \frac{x^2}{2} + \frac{x^4}{12} + \cdots.$$

45. A circle of radius $r_1 = 1$ is inscribed in an equilateral triangle. A circle of radius r_2 passes through the vertices of that triangle and is inscribed in a square. A circle of radius r_3 passes through the vertices of that square and is inscribed in a regular pentagon. Continue in this fashion: a circle of radius r_{n-1} passes through the vertices of a regular polygon of n sides and is inscribed in a regular polygon of $n + 1$ sides.

a) Show that $r_2 = r_1 \sec (\pi/3)$, $r_3 = r_2 \sec (\pi/4)$, and, in general, that

$$r_n = r_{n-1} \sec \frac{\pi}{n+1}.$$

b) Next, show that

$$\ln r_n = \ln r_1 + \ln \sec \frac{\pi}{3} + \ln \sec \frac{\pi}{4} + \cdots + \ln \sec \frac{\pi}{n+1}.$$

c) Does r_n have a finite limit as n tends to infinity? (Suggestion: Compare the series $\sum_{n=3}^{\infty} \ln \sec (\pi/n)$ with the series $\sum_{n=3}^{\infty} 1/n^2$ using the limit comparison test.) You may wish to use the answer to Problem 44.

***Toolkit* programs**

Taylor Series

REVIEW QUESTIONS AND EXERCISES

1. State Taylor's Theorem, with remainder.
2. It can be shown (though not very simply) that the function f defined by

$$f(x) = \begin{cases} 0 & \text{when } x = 0, \\ e^{-1/x^2} & \text{when } x \neq 0 \end{cases}$$

is everywhere continuous, together with its derivatives of all orders. At 0, the derivatives are all equal to 0.

a) Write the Taylor series expansion of f in powers of x.

b) What is the remainder $R_n(x, 0)$ for this function?

Does the Taylor series for f converge to $f(x)$ at some value of x different from zero? Give a reason for your answer.

3. If a Taylor series in powers of $x - a$ is to be used for the numerical evaluation of a function, what is necessary or desirable in the choice of a?
4. Describe a method that may be useful in finding $\lim_{x \to a} f(x)/g(x)$ if $f(a) = g(a) = 0$. Illustrate.
5. What tests may be used to find the interval of convergence of a power series? Do they also work at the endpoints of the interval? Illustrate with examples.

MISCELLANEOUS PROBLEMS

1. a) Find the expansion, in powers of x, of $x^2/(1 + x)$.

b) Does the series expansion of $x^2/(1 + x)$ in powers of x converge when $x = 2$? (Give a brief reason.)

2. Obtain the Maclaurin series expansion for $\sin^{-1} x$ by integrating the series for $(1 - t^2)^{-1/2}$ from 0 to x. Find the intervals of convergence of these series.

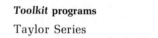

3. Obtain the first four terms in the Maclaurin series for $e^{\sin x}$ by substituting the series for $y = \sin x$ in the series for e^y.

4. Assuming $|x| > 1$, obtain the expansions

$$\tan^{-1} x = \frac{\pi}{2} - \frac{1}{x} + \frac{1}{3x^3} - \frac{1}{5x^5} + \cdots, \qquad x > 1,$$

$$\tan^{-1} x = -\frac{\pi}{2} - \frac{1}{x} + \frac{1}{3x^3} - \frac{1}{5x^5} + \cdots, \qquad x < -1,$$

by integrating the series

$$\frac{1}{1+t^2} = \frac{1}{t^2} \cdot \frac{1}{1+(1/t^2)} = \frac{1}{t^2} - \frac{1}{t^4} + \frac{1}{t^6} - \frac{1}{t^8} + \cdots$$

from $x (>1)$ to $+\infty$ or from $-\infty$ to $x\ (< -1)$.

5. a) Obtain the Maclaurin series, through the term in x^6, for $\ln(\cos x)$ by substituting the series for $y = 1 - \cos x$ in the series for $\ln(1 - y)$.
b) Use the result of part (a) to estimate

$$\int_0^{0.1} \ln(\cos x)\, dx$$

to five decimal places.

6. Compute $\int_0^1 [(\sin x)/x]\, dx$ to three decimal places.

7. Compute $\int_0^1 e^{-x^2}\, dx$ to three decimal places.

8. Expand the function $f(x) = \sqrt{1 + x^2}$ in powers of $(x - 1)$, obtaining three nonvanishing terms.

9. Expand the function $f(x) = 1/(1 - x)$ in powers of $(x - 2)$, and find the interval of convergence.

10. Expand $f(x) = 1/(x + 1)$ in powers of $(x - 3)$.

11. Expand $\cos x$ in powers of $(x - \pi/3)$.

12. Find the first three terms of the Taylor series expansion of the function $1/x$ about the point π.

13. Let f and g be functions satisfying the following conditions: (a) $f(0) = 1$, (b) $f'(x) = g(x)$, $g'(x) = f(x)$, (c) $g(0) = 0$. Estimate $f(1)$ to three decimal places.

14. Suppose $f(x) = \sum_{n=0}^{\infty} a_n x^n$. Prove that (a) if $f(x)$ is an even function, then $a_1 = a_3 = a_5 = \cdots = 0$; (b) if $f(x)$ is an odd function, then $a_0 = a_2 = a_4 = \cdots = 0$.

15. Find the first four terms (up to x^3) of the Maclaurin series of $f(x) = e^{(e^x)}$.

16. Estimate the error involved in using $x - x^2/2$ as an approximation to $\ln(1 + x)$ for values of x between 0 and 0.2, inclusive.

17. If $(1 + x)^{1/3}$ is replaced by $1 + x/3$ and $0 \le x \le \frac{1}{10}$, what estimate can be given for the error?

18. Use series to find

$$\lim_{x \to 0} \frac{\ln(1 - x) - \sin x}{1 - \cos^2 x}.$$

19. Find $\lim_{x\to 0} [(\sin x)/x]^{1/x^2}$.

20. Given a series of positive numbers a_n such that Σa_n converges. Does $\Sigma \ln(1 + a_n)$ converge? Explain.

In Problems 21–28, find the interval of convergence of each series. Test for convergence at the endpoints if the interval is finite.

21. $1 + \dfrac{x+2}{3 \cdot 1} + \dfrac{(x+2)^2}{3^2 \cdot 2} + \cdots + \dfrac{(x+2)^n}{3^n \cdot n} + \cdots$

22. $1 + \dfrac{(x-1)^2}{2!} + \dfrac{(x-1)^4}{4!} + \cdots + \dfrac{(x-1)^{2n-2}}{(2n-2)!} + \cdots$

23. $\displaystyle\sum_{n=1}^{\infty} \frac{x^n}{n^n}$

24. $\displaystyle\sum_{n=1}^{\infty} \frac{n!x^n}{n^n}$

25. $\displaystyle\sum_{n=0}^{\infty} \frac{n+1}{2n+1} \frac{(x-3)^n}{2^n}$

26. $\displaystyle\sum_{n=0}^{\infty} \frac{n+1}{2n+1} \frac{(x-2)^n}{3^n}$

27. $\displaystyle\sum_{n=1}^{\infty} \frac{(-1)^{n-1}(x-1)^n}{n^2}$

28. $\displaystyle\sum_{n=1}^{\infty} \frac{x^n}{\sqrt{n}}$

In Problems 29–31, determine *all* the values of x for which the series converge.

29. $\displaystyle\sum_{n=1}^{\infty} \frac{(x-2)^{3n}}{n!}$

30. $\displaystyle\sum_{n=1}^{\infty} \frac{2^n(\sin x)^n}{n^2}$

31. $\displaystyle\sum_{n=1}^{\infty} \frac{1}{n}\left(\frac{x-1}{x}\right)^n$

32. A function is defined by the power series

$$y = 1 + \frac{1}{6}x^3 + \frac{1}{180}x^6 + \cdots + \frac{1 \cdot 4 \cdot 7 \cdots (3n-2)}{(3n)!}x^{3n} + \cdots.$$

a) Find the interval of convergence of the series.
b) Show that there exist two constants a and b such that the function so defined satisfies a differential equation of the form $y'' = x^a y + b$.

33. If $a_n > 0$ and the series $\sum_{n=1}^{\infty} a_n$ converges, prove that $\sum_{n=1}^{\infty} a_n/(1 + a_n)$ converges.

34. If $1 > a_n > 0$ and $\sum_{n=1}^{\infty} a_n$ converges, prove that $\sum_{n=1}^{\infty} \ln(1 - a_n)$ converges. (*Hint:* First show that $|\ln(1 - a_n)| \le a_n/(1 - a_n)$.)

35. An infinite product, indicated by $\prod_{n=1}^{\infty} (1 + a_n)$, is said to converge if the series $\sum_{n=1}^{\infty} \ln(1 + a_n)$ converges. (The series is the natural logarithm of the product.) Prove that the product converges if every $a_n > -1$ and $\sum_{n=1}^{\infty} |a_n|$ converges. (*Hint:* Show that

$$|\ln(1 + a_n)| \le \frac{|a_n|}{1 - |a_n|} < 2|a_n|$$

when $|a_n| < \frac{1}{2}$.)

36. By multiplying the appropriate terms of the Maclaurin series for $\tan^{-1} x$ and $\ln(1 + x)$, find the terms through x^5 in the Maclaurin series for the product $(\tan^{-1} x) \cdot \ln(1 + x)$.

37. By appropriate division or multiplication of series, find the terms through x^5 in the Maclaurin series for $\tan^{-1} x/(1 - x)$.

38. When $f(0)$ and $g(0)$ are both zero, it may be possible to redefine the quotient $f(x)/g(x)$ as having the value $f'(0)/g'(0)$ provided $g'(0) \neq 0$. The result is equivalent to dividing both the numerator and the denominator by x in the series that represent f and g. For example, for all $x \neq 0$, we can write

$$\frac{x}{e^x - 1} = \frac{1}{1 + \dfrac{x}{2} + \dfrac{x^2}{6} + \dfrac{x^3}{24} + \cdots + \dfrac{x^n}{(n+1)!} + \cdots}.$$

By appropriate division or multiplication of series, find the coefficients c_0, c_1, and c_2 for the series representation $\Sigma\, c_n x^n$ of the function on the right.

13 Vectors

13.1 Vector Components and the Unit Vectors i and j.

Some physical quantities, like length and mass, are completely determined when their magnitudes are given in terms of specific units. Such quantities are called *scalars*. Other quantities, like forces and velocities, in which the direction as well as the magnitude is important, are called *vectors*. It is customary to represent a vector by a directed line segment whose direction represents the direction of the vector and whose length, in terms of some chosen unit, represents its magnitude.

We shall ordinarily deal with "free vectors," meaning that a vector is free to move about under parallel displacements. We say that two vectors are *equal* if they have the same direction and the same magnitude. See Fig. 13.1.

Addition

Two vectors $\mathbf{v}_1$ and $\mathbf{v}_2$, may be added geometrically by drawing a representative of $\mathbf{v}_1$, say from A to B as in Fig. 13.2(a), and then a vector equal to $\mathbf{v}_2$ starting from the terminal point of $\mathbf{v}_1$.† In Fig. 13.2(a), $\mathbf{v}_2 = \overrightarrow{BC}$. The sum $\mathbf{v}_1 + \mathbf{v}_2$ is then the vector from the starting point A of $\mathbf{v}_1$ to the terminal point C of $\mathbf{v}_2$. That is, if

$$\mathbf{v}_1 = \overrightarrow{AB}, \qquad \text{and} \qquad \mathbf{v}_2 = \overrightarrow{BC},$$

then

$$\mathbf{v}_1 + \mathbf{v}_2 = \overrightarrow{AB} + \overrightarrow{BC} = \overrightarrow{AC}.$$

Whenever a vector $\mathbf{v}$ can be written as a sum

$$\mathbf{v} = \mathbf{v}_1 + \mathbf{v}_2,$$

the vectors $\mathbf{v}_1$ and $\mathbf{v}_2$ are called *components of* $\mathbf{v}$.

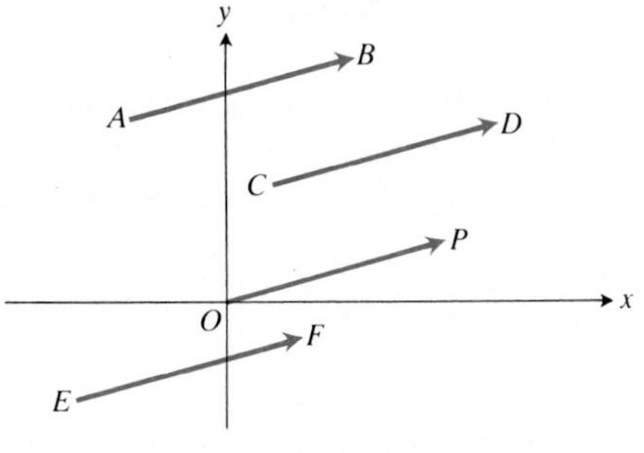

13.1 The arrows $\overrightarrow{AB}$, $\overrightarrow{CD}$, $\overrightarrow{OP}$, and $\overrightarrow{EF}$ shown here have the same direction and the same length. They all represent the same vector, and we write $\overrightarrow{AB} = \overrightarrow{CD} = \overrightarrow{OP} = \overrightarrow{EF}$.

† In print, vectors are usually indicated by bold-faced Roman letters. In handwritten work it is customary to draw small arrows above letters that represent vectors, so: $\vec{v}$.

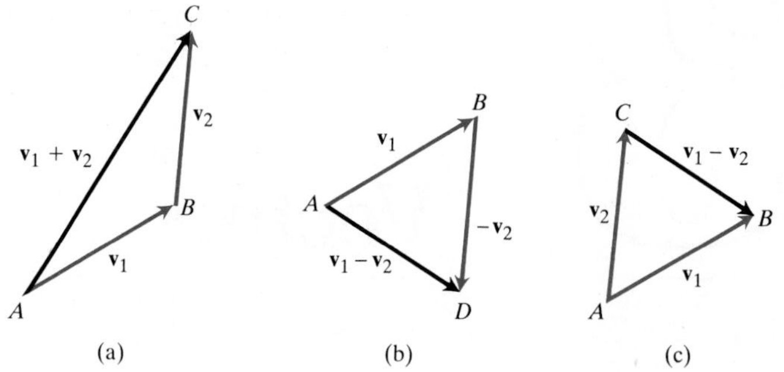

13.2 The sum (a) and difference (b and c) of two vectors $\mathbf{v}_1$ and $\mathbf{v}_2$.

The most satisfactory algebra of vectors is based on a representation of each vector in terms of components parallel to the axes of a cartesian coordinate system. This is accomplished by using the same unit of length on the two axes, with vectors of unit length along the axes used as *basic vectors* in terms of which the other vectors in the plane may be expressed.

Along the x-axis we choose the vector $\mathbf{i}$ from (0, 0) to (1, 0), as in Fig. 13.3. Along the y-axis we choose the vector $\mathbf{j}$ from (0, 0) to (0, 1). Then $a\mathbf{i}$, a being a scalar, represents a vector parallel to the x-axis, having magnitude $|a|$ and pointing to the right if a is positive, to the left if a is negative. Similarly, $b\mathbf{j}$ is a vector parallel to the y-axis and having the same direction as $\mathbf{j}$ if b is positive, or the opposite direction if b is negative. Figure 13.3 shows how the vector $\mathbf{v} = \overrightarrow{AC}$ is "resolved" into its x- and y- components as the sum

$$\mathbf{v} = a\mathbf{i} + b\mathbf{j}.$$

In this context it will also be convenient to call the numbers a and b *components* of $\mathbf{v}$.

Components give us a way to define the equality of vectors algebraically:

DEFINITION

Equality of Vectors (Algebraic Definition)

$$a\mathbf{i} + b\mathbf{j} = a'\mathbf{i} + b'\mathbf{j} \Leftrightarrow a = a' \text{ and } b = b' \qquad (1)$$

That is, two vectors are equal if and only if their corresponding components are equal. Thus, in Fig. 13.3, the vector $\overrightarrow{AB}$ and the vector $\overrightarrow{OP}$ from (0, 0) to $(a, 0)$ are both equal to $a\mathbf{i}$.

Vectors may be added algebraically by adding their components in the following way. In Fig. 13.3, we see that $a\mathbf{i} + b\mathbf{j}$ is the vector hypotenuse of a right triangle whose vector sides are $a\mathbf{i}$ and $b\mathbf{j}$. If two vectors $\mathbf{v}_1$ and $\mathbf{v}_2$ are given in terms of components,

$$\mathbf{v}_1 = a_1\mathbf{i} + b_1\mathbf{j}, \qquad \mathbf{v}_2 = a_2\mathbf{i} + b_2\mathbf{j},$$

then

$$\mathbf{v}_1 + \mathbf{v}_2 = (a_1 + a_2)\mathbf{i} + (b_1 + b_2)\mathbf{j} \qquad (2)$$

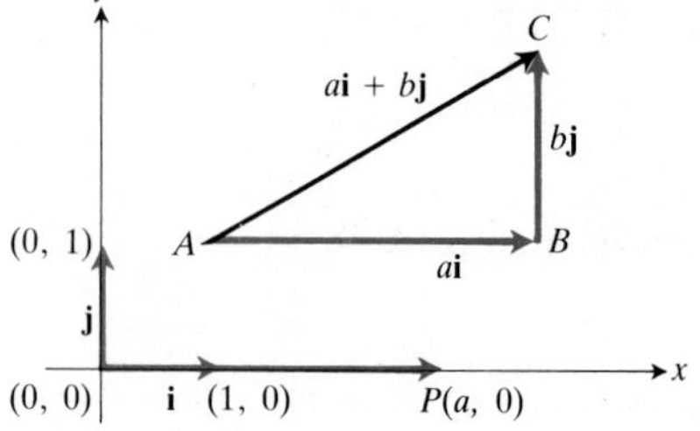

13.3 The vector $\overrightarrow{AC}$ expressed as a multiple of $\mathbf{i}$ plus a multiple of $\mathbf{j}$.

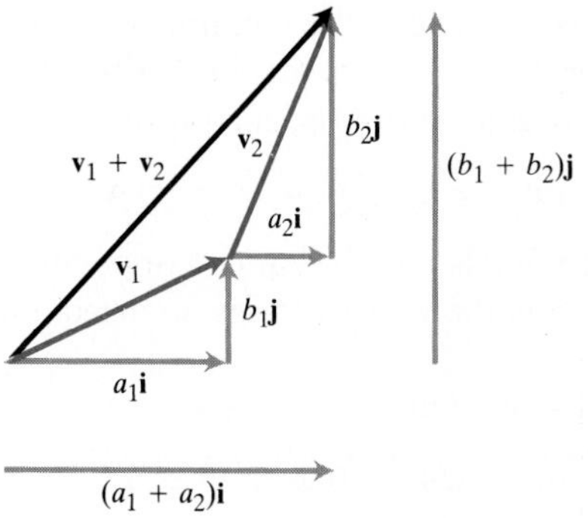

13.4 If $\mathbf{v}_1 = a_1\mathbf{i} + b_1\mathbf{j}$ and $\mathbf{v}_2 = a_2\mathbf{i} + b_2\mathbf{j}$, then $\mathbf{v}_1 + \mathbf{v}_2 = (a_1 + a_2)\mathbf{i} + (b_1 + b_2)\mathbf{j}$.

has x- and y-components obtained by adding the x- and y-components of $\mathbf{v}_1$ and $\mathbf{v}_2$. See Fig. 13.4.

EXAMPLE 1

$$(2\mathbf{i} - 4\mathbf{j}) + (5\mathbf{i} + 3\mathbf{j}) = (2 + 5)\mathbf{i} + (-4 + 3)\mathbf{j} = 7\mathbf{i} - \mathbf{j}. \square$$

Subtraction

The negative of a vector $\mathbf{v}$ is the vector $-\mathbf{v}$ that has the same length as $\mathbf{v}$ but points in the opposite direction. To subtract a vector $\mathbf{v}_2$ from a vector $\mathbf{v}_1$, we add $-\mathbf{v}_2$ to $\mathbf{v}_1$. This may be done geometrically by drawing $-\mathbf{v}_2$ from the tip of $\mathbf{v}_1$ and then drawing the vector from the initial point of $\mathbf{v}_1$ to the tip of $-\mathbf{v}_2$, as shown in Fig. 13.2(b), where

$$\overrightarrow{AD} = \overrightarrow{AB} + \overrightarrow{BD} = \mathbf{v}_1 + (-\mathbf{v}_2) = \mathbf{v}_1 - \mathbf{v}_2.$$

Another way to subtract $\mathbf{v}_2$ from $\mathbf{v}_1$ is to draw them both with a common initial point and then draw the vector from the tip of $\mathbf{v}_2$ to the tip of $\mathbf{v}_1$. This is illustrated in Fig. 13.2(c), where

$$\overrightarrow{CB} = \overrightarrow{CA} + \overrightarrow{AB} = -\mathbf{v}_2 + \mathbf{v}_1 = \mathbf{v}_1 - \mathbf{v}_2.$$

Thus, $\overrightarrow{CB}$ is the vector that when added to $\mathbf{v}_2$ gives $\mathbf{v}_1$:

$$\overrightarrow{CB} + \mathbf{v}_2 = (\mathbf{v}_1 - \mathbf{v}_2) + \mathbf{v}_2 = \mathbf{v}_1.$$

In terms of components, vector subtraction follows the simple algebraic law

$$\mathbf{v}_1 - \mathbf{v}_2 = (a_1 - a_2)\mathbf{i} + (b_1 - b_2)\mathbf{j}, \qquad (3)$$

which says that corresponding components are subtracted.

EXAMPLE 2

$$(6\mathbf{i} + 2\mathbf{j}) - (3\mathbf{i} - 5\mathbf{j}) = (6 - 3)\mathbf{i} + (2 - (-5))\mathbf{j} = 3\mathbf{i} + 7\mathbf{j}. \square$$

Length of a Vector

The length of the vector $\mathbf{v} = a\mathbf{i} + b\mathbf{j}$ is usually denoted by $|\mathbf{v}|$, which may be read "the magnitude of v." Figure 13.3 shows that $\mathbf{v}$ is the hypotenuse of a right triangle whose legs have lengths $|a|$ and $|b|$. Hence we may apply the theorem of Pythagoras to obtain

$$|\mathbf{v}| = |a\mathbf{i} + b\mathbf{j}| = \sqrt{a^2 + b^2}. \qquad (4)$$

EXAMPLE 3

$$|3\mathbf{i} - 5\mathbf{j}| = \sqrt{(3)^2 + (-5)^2} = \sqrt{9 + 25} = \sqrt{34}. \square$$

Multiplication by Scalars

The algebraic operation of multiplying a vector $\mathbf{v} = a\mathbf{i} + b\mathbf{j}$ by a scalar c is also simple, namely

$$c(a\mathbf{i} + b\mathbf{j}) = (ca)\mathbf{i} + (cb)\mathbf{j}. \qquad (5)$$

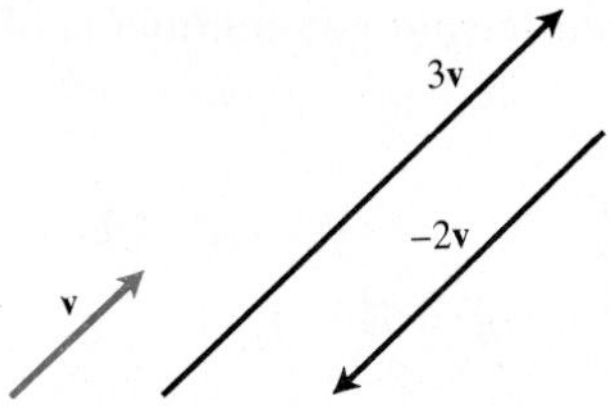

13.5 Scalar multiples of **v**.

Note how the unit vectors **i** and **j** allow us to keep the components separated from one another when we operate on vectors algebraically. Geometrically, $c\mathbf{v}$ is a vector whose length is $|c|$ times the length of **v**:

$$|c\mathbf{v}| = |(ca)\mathbf{i} + (cb)\mathbf{j}| = \sqrt{(ca)^2 + (cb)^2} = |c|\sqrt{a^2 + b^2} = |c|\,|\mathbf{v}|. \quad (6)$$

The direction of $c\mathbf{v}$ agrees with that of **v** if c is positive, and is opposite to that of **v** if c is negative (Fig. 13.5). If $c = 0$, the vector $c\mathbf{v}$ has no direction.

EXAMPLE 4 Let $c = 2$ and $\mathbf{v} = -3\mathbf{i} + 4\mathbf{j}$. Then,

$$|\mathbf{v}| = |-3\mathbf{i} + 4\mathbf{j}| = \sqrt{(-3)^2 + (4)^2} = \sqrt{9 + 16} = \sqrt{25} = 5,$$

and

$$|2\mathbf{v}| = |2(-3\mathbf{i} + 4\mathbf{j})| = |-6\mathbf{i} + 8\mathbf{j}| = \sqrt{(-6)^2 + (8)^2} = \sqrt{36 + 64}$$
$$= \sqrt{100} = 10 = 2|\mathbf{v}|.$$

If c had been -2 instead of 2, we would have found

$$|-2\mathbf{v}| = |-2(-3\mathbf{i} + 4\mathbf{j}) = |6\mathbf{i} - 8\mathbf{j}| = \sqrt{(6)^2 + (-8)^2} = \sqrt{100}$$
$$= 10 = |-2|\,|\mathbf{v}|. \;\square$$

Zero Vector

The vector

$$\mathbf{0} = 0\mathbf{i} + 0\mathbf{j} \quad (7)$$

is called the *zero vector*. It is the only vector whose length is zero, as we can see from the fact that

$$|a\mathbf{i} + b\mathbf{j}| = \sqrt{a^2 + b^2} = 0 \quad \Leftrightarrow \quad a = b = 0.$$

Unit Vector

Any vector **u** whose length is equal to the unit of length used along the coordinate axes is called a *unit vector*. If **u** is the unit vector obtained by rotating **i** through an angle θ in the positive direction, then (Fig. 13.6) **u** has a horizontal component

$$u_x = \cos\theta$$

and a vertical component

$$u_y = \sin\theta,$$

so that

$$\mathbf{u} = \mathbf{i}\cos\theta + \mathbf{j}\sin\theta. \quad (8)$$

If we allow the angle θ in Eq. (8) to vary from 0 to 2π, then the point P in Fig. 13.6 traces the unit circle $x^2 + y^2 = 1$ once in the counterclockwise direction. Every unit vector in the plane is given by Eq. (8) for some value of θ.

In physics it is common to denote unit vectors with small "hats," as in $\hat{u}$ (pronounced "u hat"). In hat notation, **i** and **j** become $\hat{\imath}$ and $\hat{\jmath}$.

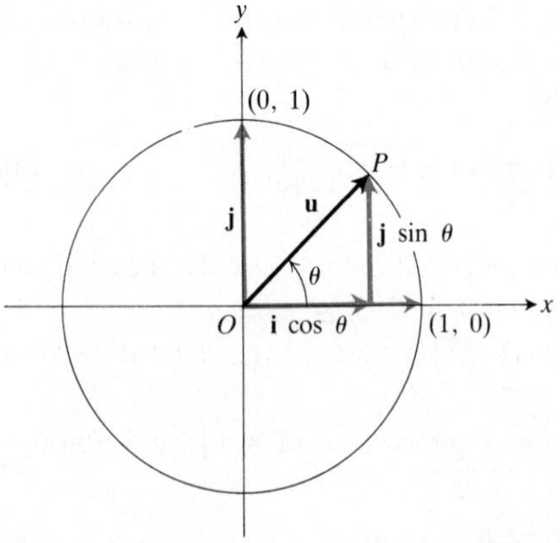

13.6 $\mathbf{u} = \mathbf{i}\cos\theta + \mathbf{j}\sin\theta$.

Direction

It is common in fields like classical electricity and magnetism, which use vectors a great deal, to define *direction* of a nonzero vector **A** to be the unit vector obtained by dividing **A** by its own length:

$$\text{Direction of } \mathbf{A} = \frac{\mathbf{A}}{|\mathbf{A}|}. \tag{9}$$

Instead of just saying that the unit vector $\mathbf{A}/|\mathbf{A}|$ *represents* the direction of **A**, we say that it *is* the direction of **A**.

To see that $\mathbf{A}/|\mathbf{A}|$ is indeed a unit vector, we can calculate its length directly:

$$\text{Length of } \frac{\mathbf{A}}{|\mathbf{A}|} = \left|\frac{\mathbf{A}}{|\mathbf{A}|}\right| = \left|\frac{1}{|\mathbf{A}|}\mathbf{A}\right| = \frac{1}{|\mathbf{A}|}|\mathbf{A}| = 1. \tag{10}$$

The zero vector **0** has no defined direction.

EXAMPLE 5 The direction of $\mathbf{A} = 3\mathbf{i} - 4\mathbf{j}$ is

$$\text{Direction of } \mathbf{A} = \frac{\mathbf{A}}{|\mathbf{A}|} = \frac{3\mathbf{i} - 4\mathbf{j}}{\sqrt{(3)^2 + (-4)^2}} = \frac{3\mathbf{i} - 4\mathbf{j}}{\sqrt{25}} = \frac{3}{5}\mathbf{i} - \frac{4}{5}\mathbf{j}.$$

To check the length, we can calculate

$$\left|\frac{3}{5}\mathbf{i} - \frac{4}{5}\mathbf{j}\right| = \sqrt{\left(\frac{3}{5}\right)^2 + \left(\frac{4}{3}\right)^2} = \sqrt{\frac{9}{25} + \frac{16}{25}} = \sqrt{\frac{25}{25}} = 1. \ \square$$

It follows from the definition of direction that two nonzero vectors **A** and **B** have the same direction if and only if

$$\frac{\mathbf{A}}{|\mathbf{A}|} = \frac{\mathbf{B}}{|\mathbf{B}|}, \tag{11}$$

or

$$\mathbf{A} = \frac{|\mathbf{A}|}{|\mathbf{B}|}\mathbf{B}. \tag{12}$$

Thus, if **A** and **B** have the same direction, then **A** is a positive scalar multiple of **B**. Conversely, if

$$\mathbf{A} = k\mathbf{B}, \qquad k > 0,$$

then

$$\frac{\mathbf{A}}{|\mathbf{A}|} = \frac{k\mathbf{B}}{|k\mathbf{B}|} = \frac{k}{|k|}\frac{\mathbf{B}}{|\mathbf{B}|} = \frac{k}{k}\frac{\mathbf{B}}{|\mathbf{B}|} = \frac{\mathbf{B}}{|\mathbf{B}|}. \tag{13}$$

Therefore, two nonzero vectors **A** and **B** point in the same direction if and only if **A** is a positive scalar multiple of **B**.

We say that two nonzero vectors **A** and **B** point in *opposite* directions if their directions are opposite in sign:

$$\frac{\mathbf{A}}{|\mathbf{A}|} = -\frac{\mathbf{B}}{|\mathbf{B}|}. \tag{14}$$

From this it follows (Problem 22) that **A** and **B** have opposite directions if and only if **A** is a negative scalar multiple of **B**.

EXAMPLE 6

a) Same direction: $\mathbf{A} = 3\mathbf{i} - 4\mathbf{j}$ and $\mathbf{B} = \frac{3}{2}\mathbf{i} - 2\mathbf{j} = \frac{1}{2}\mathbf{A}$.

b) Opposite directions: $\mathbf{A} = 3\mathbf{i} - 4\mathbf{j}$ and $\mathbf{B} = -\mathbf{i} + \frac{4}{3}\mathbf{j} = -\frac{1}{3}\mathbf{A}$. □

Vectors Tangent and Normal to a Curve

Two vectors are said to be *parallel* if they are scalar multiples of each other, or the line segments representing them are parallel. Similarly, a vector is parallel to a line if the segments that represent the vector are parallel to the line. When we talk of a vector's being *tangent* or *normal* to a curve at a point, we mean that the vector is parallel to the line that is tangent or normal (perpendicular) to the curve at the point. The next example shows how such a vector may be found.

EXAMPLE 7 Find unit vectors tangent and normal to the curve $y = (x^3/2) + \frac{1}{2}$ at the point (1, 1).

Solution The slope of the line tangent to the curve at the point (1, 1) is

$$y' = \left.\frac{3x^2}{2}\right|_{x=1} = \frac{3}{2}.$$

We find a unit vector with this slope. The vector $\mathbf{v} = 2\mathbf{i} + 3\mathbf{j}$ has slope $\frac{3}{2}$ (Fig. 13.7), as does every nonzero multiple of $\mathbf{v}$. To find a multiple of $\mathbf{v}$ that is a unit vector, we divide $\mathbf{v}$ by its length,

$$|\mathbf{v}| = \sqrt{2^2 + 3^2} = \sqrt{13}.$$

This produces the unit vector

$$\mathbf{u} = \frac{\mathbf{v}}{|\mathbf{v}|} = \frac{2}{\sqrt{13}}\mathbf{i} + \frac{3}{\sqrt{13}}\mathbf{j}.$$

The vector $\mathbf{u}$ is tangent to the curve at (1, 1) because it has the same direction as $\mathbf{v}$. Of course, the vector

$$-\mathbf{u} = -\frac{2}{\sqrt{13}}\mathbf{i} - \frac{3}{\sqrt{13}}\mathbf{j},$$

which points in the direction opposite to $\mathbf{u}$, is also tangent to the curve at (1, 1). Without some additional requirement, there is no reason to prefer one of these vectors to the other.

To find unit vectors normal to the curve at (1, 1), we look for unit vectors whose slopes are the negative reciprocal of the slope of $\mathbf{u}$. This is quickly done by interchanging the components of $\mathbf{u}$ and changing the sign of one of them. We obtain

$$\mathbf{n} = \frac{3}{\sqrt{13}}\mathbf{i} - \frac{2}{\sqrt{13}}\mathbf{j}, \quad \text{and} \quad -\mathbf{n} = -\frac{3}{\sqrt{13}}\mathbf{i} + \frac{2}{\sqrt{13}}\mathbf{j}.$$

Again, either one will do. The vectors have opposite directions, but both are normal to the curve at the point (1, 1). □

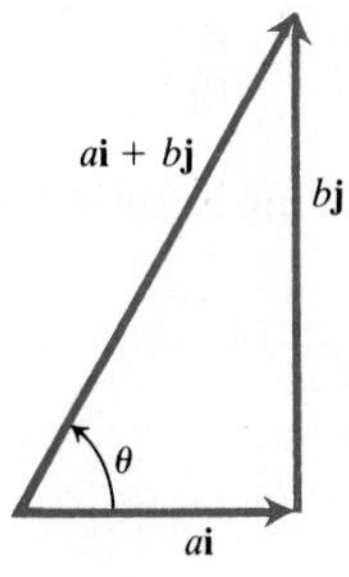

13.7 If $a \neq 0$, the vector $a\mathbf{i} + b\mathbf{j}$ has slope $b/a = \tan\theta$.

Vectors that are tangent and normal to curves in space will be discussed in the next chapter.

PROBLEMS

1. The three vectors **A, B,** and **C** shown below lie in a plane. Copy them on a sheet of paper.

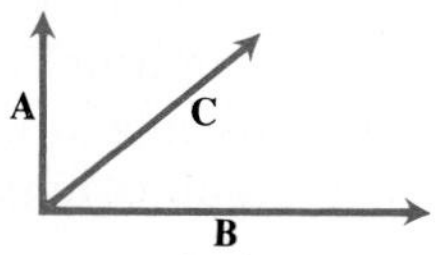

Then, by arranging the vectors in head-to-tail fashion, sketch
   a) **A** + **B**, to which then add **C**
   b) **A** + **C**, to which then add **B**
   c) **B** + **C**, to which then add **A**
   d) **A** − **B**, to which then add **C**
   e) **A** + **C**, from which then subtract **B**

In Problems 2–11 express each of the vectors in the form $a\mathbf{i} + b\mathbf{j}$. Indicate all quantities graphically.

2. $\overrightarrow{P_1P_2}$ if P_1 is the point (1, 3) and P_2 is the point (2, −1).
3. $\overrightarrow{OP_3}$ if O is the origin and P_3 is the midpoint of the vector $\overrightarrow{P_1P_2}$ joining $P_1(2, -1)$ and $P_2(-4, 3)$.
4. The vector from the point $A(2, 3)$ to the origin.
5. The sum of the vectors $\overrightarrow{AB}$ and $\overrightarrow{CD}$, given the four points $A(1, -1)$, $B(2, 0)$, $C(-1, 3)$, and $D(-2, 2)$.
6. A unit vector making an angle of 30° with the positive x-axis.
7. The unit vector obtained by rotating **j** through 120° in the clockwise direction.
8. A unit vector having the same direction as the vector $3\mathbf{i} - 4\mathbf{j}$.
9. A unit vector tangent to the curve $y = x^2$ at the point (2, 4).
10. A unit vector normal to the curve $y = x^2$ at the point $P(2, 4)$ and pointing from P toward the concave side of the curve (that is, an "inner" normal).
11. a) The unit vector tangent to the curve $y = x^2 + 2x$ at the point (1, 3) that points toward the x-axis.
    b) The unit vector normal to the curve at the point (1, 3) obtained by rotating the unit tangent vector found in part (a) 90° clockwise.
12. Show the unit vectors given by Eq. (8) for $\theta = 0, \pi/4, \pi/2, 2\pi/3, 5\pi/4$, and $5\pi/3$, together with a graph of the circle $x^2 + y^2 = 1$.

In Problems 13–18, find the length and direction of the vectors, and the angle that each makes with the positive x-axis.

13. $\mathbf{i} + \mathbf{j}$	**14.** $2\mathbf{i} - 3\mathbf{j}$
15. $\sqrt{3}\mathbf{i} + \mathbf{j}$	**16.** $-2\mathbf{i} + 3\mathbf{j}$
17. $5\mathbf{i} + 12\mathbf{j}$	**18.** $-5\mathbf{i} - 12\mathbf{j}$

19. The *speed* of a particle moving with velocity **v** is defined to be $|\mathbf{v}|$, the length (or magnitude) of **v**. Find the particle's speed if its velocity is $\mathbf{v} = -4\mathbf{i} + 2\mathbf{j}$.
20. Show that $\mathbf{A} = 3\mathbf{i} + 6\mathbf{j}$ and $\mathbf{B} = -\mathbf{i} - 2\mathbf{j}$ have opposite directions. Sketch them together.
21. Show that $\mathbf{C} = 3\mathbf{i} + 6\mathbf{j}$ and $\mathbf{D} = \frac{1}{2}\mathbf{i} + \mathbf{j}$ have the same direction.
22. Show that any two nonzero vectors **A** and **B** have opposite directions if and only if **A** is a negative scalar multiple of **B**. (*Hint:* See Eq. 14.)
23. Let A, B, C, D be the vertices, in order, of a quadrilateral. Let A', B', C', D' be the midpoints of the sides AB, BC, CD, and DA, in order. Prove that $A'B'C'D'$ is a parallelogram. (*Hint:* First show that $\overrightarrow{A'B'} = \overrightarrow{D'C'} = \frac{1}{2}\overrightarrow{AC}$.)
24. Show that the diagonals of a parallelogram bisect each other. (*Method:* Let A be one vertex and let M and N be the midpoints of the diagonals. Then show that $\overrightarrow{AM} = \overrightarrow{AN}$.)

13.2 Modeling Projectile Motion

In newtonian mechanics, the motion of a particle in a plane is usually described by a pair of differential equations

$$F_x = m\frac{dv_x}{dt}, \qquad F_y = m\frac{dv_y}{dt} \tag{1}$$

that express Newton's second law of motion

$$\mathbf{F} = m\mathbf{a} = m\frac{d\mathbf{v}}{dt} = m\frac{dv_x}{dt}\mathbf{i} + m\frac{dv_y}{dt}\mathbf{j} \tag{2}$$

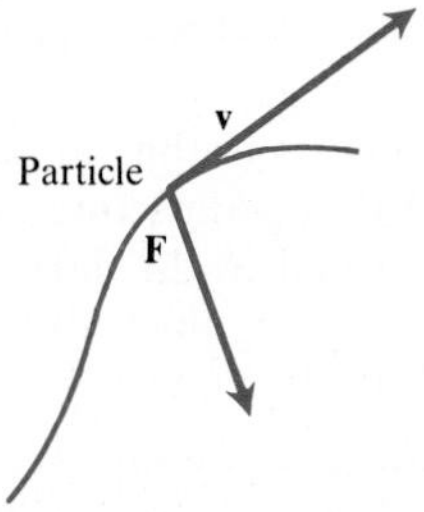

The force and velocity vectors of the motion of a particle in a plane might look like this at a particular time *t*.

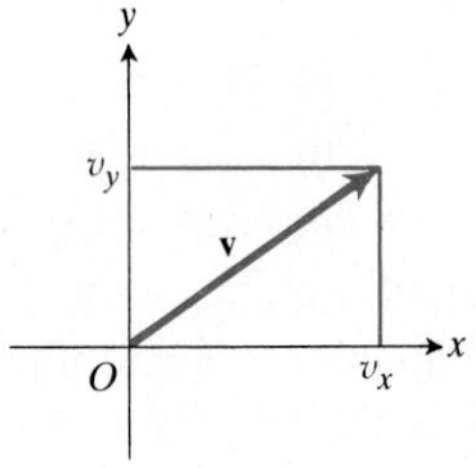

The components v_x and v_y of **v**.

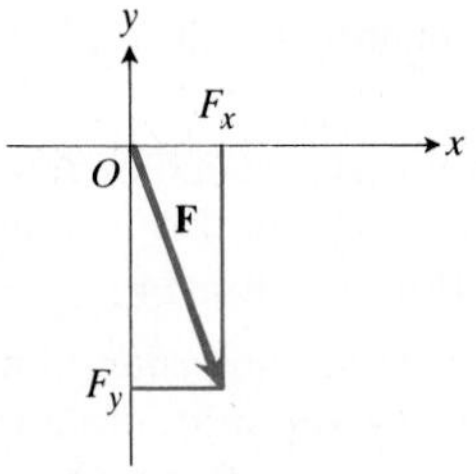

The components F_x and F_y of **F**.

13.8 Motion of a particle in a plane.

in parametric form. Here, **F** is a vector that represents a force acting on a particle of mass m at time t. The vector $\mathbf{v} = v_x\mathbf{i} + v_y\mathbf{j}$ is the velocity vector of the particle at time t (Fig. 13.8). The quantities F_x and F_y are the x- and y-components of **F**, while v_x and v_y are the components of **v**.

If we know the position and velocity of the particle at some given instant, then the position of the particle at any later time can usually be found by integrating Eqs. (1) with respect to time. The constants of integration are determined from given initial conditions. The result is another pair of parametric equations

$$x = f(t), \qquad y = g(t) \tag{3}$$

that give the coordinates x and y of the particle as functions of t.

The equations in (3) contain more information about the motion of the particle than the cartesian equation

$$y = F(x) \tag{4}$$

that we get from (3) by eliminating t. The parametric equations tell where the particle goes and *when* it gets to any given place, whereas the cartesian equation only tells the curve along which the particle travels. (Sometimes, too, a parametric representation of a curve is all that is possible; that is, a parameter cannot always be eliminated in practice.)

EXAMPLE 1 A projectile is fired with an initial velocity v_0 ft/s at an angle of elevation α. Assuming that gravity is the only force acting on the projectile, find parametric equations that give the coordinates of the projectile's position at any time t.

Solution We introduce coordinate axes with the origin at the point where the projectile begins its motion (Fig. 13.9). The distance traveled by the projectile over the ground is measured along the x-axis, and the height of the projectile above the ground is measured along the y-axis. At any time t, the projectile's position is given by a coordinate pair $x(t)$, $y(t)$, that we assume to be differentiable functions of t. If we measure distance in feet and time in seconds, with $t = 0$ at the instant the projectile is fired, then

13.9 Ideal projectiles move along parabolas. The schematic figure here shows the lengths of the velocity vector and its components, first at time $t = 0$, and then later in the flight.

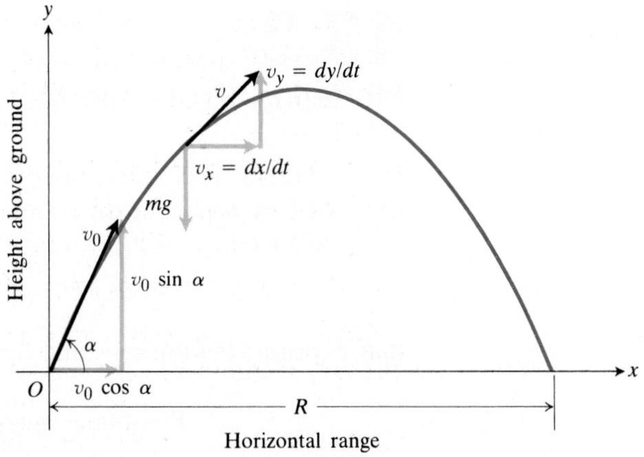

the initial conditions for the projectile's motion are

$$t = 0 \text{ sec}, \qquad x = 0 \text{ ft}, \qquad y = 0 \text{ ft},$$
$$v_x(0) = \left.\frac{dx}{dt}\right|_{t=0} = v_0 \cos\alpha \quad \text{ft/s}, \qquad v_y(0) = \left.\frac{dy}{dt}\right|_{t=0} = v_0 \sin\alpha \quad \text{ft/s}. \tag{5}$$

If the projectile is to travel only a few miles, and not go very high, it will cause no serious error to model the force of gravity with a constant vector **F** that points straight down. Its x- and y-components are

$$F_x = 0 \text{ lb}, \qquad F_y = -mg \text{ lb},$$

where m is the mass of the projectile and $g = 32$ ft/s is the acceleration of gravity. With these values for F_x and F_y, the equations in (1) become

$$0 = m\frac{d^2x}{dt^2}, \qquad -mg = m\frac{d^2y}{dt^2}. \tag{6}$$

To solve these equations for x and y, we integrate each one twice. This introduces four constants of integration, which may be evaluated by using the initial conditions (5). From the first equation in (6), we get

$$\frac{d^2x}{dt^2} = 0 \text{ ft/s}^2, \qquad \frac{dx}{dt} = c_1 \text{ ft/s}, \qquad x = c_1t + c_2 \text{ ft}, \tag{7a}$$

and from the second equation in (6),

$$\frac{d^2y}{dt^2} = -g \text{ ft/s}^2, \qquad \frac{dy}{dt} = -gt + c_3 \text{ ft/s},$$
$$y = -\frac{1}{2}gt^2 + c_3t + c_4 \text{ ft}. \tag{7b}$$

From the initial conditions, we find

$$c_1 = v_0 \cos\alpha \text{ ft/s}, \qquad c_2 = 0 \text{ ft}, \qquad c_3 = v_0 \sin\alpha \text{ ft/s}, \qquad c_4 = 0 \text{ ft}. \tag{7c}$$

The position of the projectile t seconds after firing is

$$x = (v_0 \cos\alpha)t \text{ ft}, \qquad y = -\frac{1}{2}gt^2 + (v_0 \sin\alpha)t \text{ ft}. \;\square \tag{8}$$

Height, Range, and Angle of Elevation

For a given angle of elevation α and a given muzzle velocity v_0, the position of the projectile in Example 1 at any time may be determined from the equations in (8). These equations may also be used to answer such questions as

1. How high does the projectile rise? When is it at its highest?
2. How far away does the projectile land, and how does the horizontal range R vary with the angle of elevation?
3. What angle of elevation gives the maximum range?

First, the projectile will reach its highest point when its y-(vertical) velocity component is zero, that is, when

$$\frac{dy}{dt} = -gt + v_0 \sin\alpha = 0 \text{ ft/s}$$

or

$$t = t_m = \frac{v_0 \sin \alpha}{g} \text{ sec.}$$

For this value of t, the value of y is

$$y_{\max} = -\frac{1}{2}g(t_m)^2 + (v_0 \sin \alpha)t_m = \frac{(v_0 \sin \alpha)^2}{2g} \text{ ft.}$$

Second, to find R we first find the time when the projectile strikes the ground. That is, we find the value of t for which $y = 0$. We then find the value of x for this value of t:

$$y = t\left(-\frac{1}{2}gt + v_0 \sin \alpha\right) = 0 \text{ ft}$$

when

$$t = 0 \qquad \text{or} \qquad t = \frac{2v_0 \sin \alpha}{g} = 2t_m \text{ sec.}$$

Since $t = 0$ is the instant when the projectile is fired, $t = 2t_m$ is the time when the projectile hits the ground. The corresponding value of x is

$$R = (v_0 \cos \alpha)(2t_m) = v_0 \cos \alpha \frac{2v_0 \sin \alpha}{g} = \frac{v_0^2}{g} \sin 2\alpha \text{ ft.}$$

Finally, the formula just given for R shows that the maximum range for a given muzzle velocity is obtained when $\sin 2\alpha = 1$, or $\alpha = 45°$.

A cartesian equation for the path of the projectile is readily obtained from (8). We need only substitute

$$t = \frac{x}{v_0 \cos \alpha}$$

from the first into the second equation of (8) to eliminate t and obtain

$$y = -\left(\frac{g}{2v_0^2 \cos^2 \alpha}\right)x^2 + (\tan \alpha)x. \tag{9}$$

Since this equation is linear in y and quadratic in x, it represents a *parabola*. Thus the path of a projectile (neglecting air resistance) is a parabola.

Differential equations of motion that take air resistance into account are usually too complicated for straightforward integration. The M.I.T. differential analyzers (early computers) were used to solve such equations during World War II to build up "range tables." In following moving targets, keeping track of *time* is of great importance, so that equations or tables that give x and y in terms of t are preferred over other forms.

Position Vectors

The vector $\mathbf{R}(t)$ from the origin to the position $(x(t), y(t))$ that the projectile in Example 1 has at time t is called the *position vector* of the projectile. It tells where the projectile is at time t. We can describe the motion in terms of this vector by writing the single vector equation

$$\mathbf{R}(t) = \mathbf{i}(v_0 \cos \alpha)t + \mathbf{j}\left(-\frac{1}{2}gt^2 + (v_0 \sin \alpha)t\right). \quad (10)$$

This formulation has the advantage that we can immediately write down the velocity and acceleration vectors of the motion by differentiating $\mathbf{R}(t)$ component by component with respect to t. We shall show how this is done in Chapter 14, where our study of vector functions and their derivatives will lead to Kepler's laws of planetary motion.

Other Motion

Not all motion can be considered to be projectile motion, and it is therefore a good idea to consider parametric equations in a less specific setting, as we do in the next example.

EXAMPLE 2 Find parametric equations for the plane curve traced by the point $P(x, y)$ if

$$\frac{dx}{dt} = 1 + x^2, \qquad \frac{dy}{dt} = \cos t, \qquad 0 \le t \le \frac{\pi}{4} \qquad (11)$$

and

$$x = 1, \qquad y = \sqrt{2} \qquad \text{when } t = \frac{\pi}{4}.$$

Solution The general rule for solving pairs of differential equations like these is to solve first for one coordinate in terms of t, then for the other. In this case we may begin with either equation, say with

$$\frac{dx}{dt} = 1 + x^2.$$

We separate the variables, as in Article 4.2, and integrate both sides of the resulting equation with respect to t to get

$$\frac{1}{1 + x^2}\frac{dx}{dt} = 1, \qquad \int \frac{1}{1 + x^2}\frac{dx}{dt}\,dt = \int dt, \qquad \tan^{-1} x = t + C.$$

We then determine C from the condition that $x = 1$ when $t = \pi/4$:

$$\tan^{-1} 1 = \frac{\pi}{4} + C, \qquad C = 0.$$

This gives $\tan^{-1} x = t$ or $x = \tan t$.

To express y in terms of t we integrate the second equation in (11):

$$\frac{dy}{dt} = \cos t, \qquad y = \sin t + C_1.$$

From the condition that $y = \sqrt{2}$ when $t = \pi/4$ we find

$$\sqrt{2} = \sin\frac{\pi}{4} + C_1, \qquad C_1 = \frac{\sqrt{2}}{2}.$$

Thus,

$$y = \sin t + \frac{\sqrt{2}}{2}.$$

The equations we seek are

$$x = \tan t, \quad \text{and} \quad y = \sin t + \frac{\sqrt{2}}{2}.$$

To check our solution (always a good idea) we differentiate both equations with respect to t,

$$\frac{dx}{dt} = \frac{d}{dt}(\tan t) = \sec^2 t = 1 + \tan^2 t = 1 + x^2,$$

$$\frac{dy}{dt} = \frac{d}{dt}\left(\sin t + \frac{\sqrt{2}}{2}\right) = \cos t,$$

to see that the original equations in (11) are satisfied. □

PROBLEMS

In Problems 1–4, the projectile is assumed to obey the laws of motion discussed above, in which air resistance is neglected.

1. Find two values of the angle of elevation that will enable a projectile to reach a target on the same level as the gun and 25,000 feet distance from it if the initial velocity is 1000 ft/s. Determine the times of flight corresponding to these two angles.
2. Show that doubling the initial velocity of a projectile multiplies both the maximum height and the range by a factor of four.
3. Show that a projectile attains three quarters of its maximum height in one half the time required to reach that maximum.
4. Suppose a target moving at the constant rate of a ft/s is level with and b ft away from a gun at the instant the gun is fired. If the target moves in a horizontal line directly away from the gun, show that the muzzle velocity v_0 and angle of elevation α must satisfy the equation.

$$v_0^2 \sin 2\alpha - 2av_0 \sin \alpha - bg = 0$$

if the projectile is to strike the target.

5. A human cannonball is to be fired with an initial speed of $v_0 = 80\sqrt{10/3}$ ft/s. The circus performer (of the right caliber, naturally) hopes to land on a special cushion located 200 ft down range. The circus is being held in a large room with a flat ceiling 75 ft high. Can the performer be fired to the cushion without striking the ceiling? If so, what should the cannon's angle of elevation be?

In Problems 6–10, find parametric equations for the curve traced by the point $P(x, y)$ if its coordinates satisfy the given differential equations and conditions. Sketch the curve. It may help to find a cartesian equation for the curve before you sketch.

6. $\dfrac{dx}{dt} = x, \quad \dfrac{dy}{dt} = y; \quad x = 1$, and $y = m$ (a constant) when $t = 0$
7. $\dfrac{dx}{dt} = x, \quad \dfrac{dy}{dt} = -y; \quad x > 0,\ y > 0$, and $x = 1,\ y = 2$ when $t = 0$
8. $\dfrac{dx}{dt} = x, \quad \dfrac{dy}{dt} = -x^2; \quad x > 0$, and $x = 1,\ y = -4$ when $t = 0$
   (*Hint:* First solve for x in terms of t. Then use this result in finding y.)
9. $\dfrac{dx}{dt} = y, \quad \dfrac{dy}{dt} = y^2; \quad t < 1$, and $x = 0,\ y = 1$ when $t = 0$
10. $\dfrac{dx}{dt} = e^{-x}, \quad \dfrac{dy}{dt} = tx; \quad t > 0$, and $x = 0,\ y = 0$ when $t = 1$
11. a) Find parametric equations for the curve traced by the point $P(x, y)$ if

$$\frac{dx}{dt} = \sqrt{1 - x^2}, \qquad \frac{dy}{dt} = x,$$

and $x = 0,\ y = -1$ when $t = 0$.

   b) Find a cartesian equation for the curve.
   c) What are the coordinates of P when $t = \pi/2$? $t = \pi$?
12. Find parametric equations for the curve traced by the point $P(x, y)$ if

$$\frac{dx}{dt} = \sqrt{1 - x^2}, \qquad \frac{dy}{dt} = x^2,$$

and $x = 0,\ y = 1$ when $t = 0$. (*Hint:* Express x in terms of t before attempting to solve for y.)

***Toolkit* programs**

Parametric Equations   Super * Grapher

13.3 Parametric Equations in Analytic Geometry

The solutions of differential equations of motion are not the only ways in which parametric equations arise.

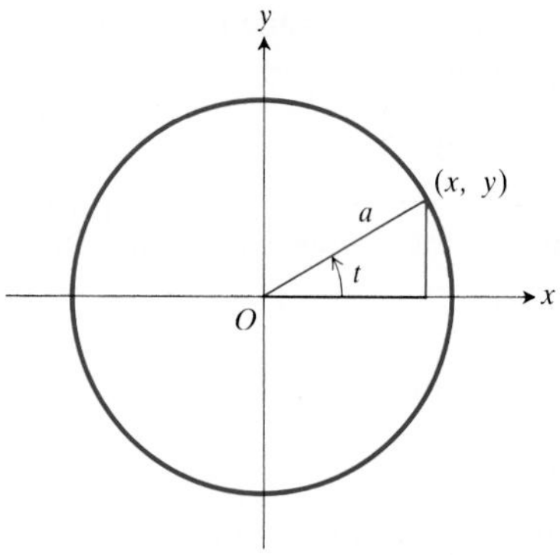

13.10 Circle: $x = a \cos t$, $y = a \sin t$.

EXAMPLE 1 *Parametric equations for the circle* $x^2 + y^2 = a^2$ (Fig. 13.10):

$$x = a \cos t, \qquad y = a \sin t, \qquad 0 \le t \le 2\pi. \ \square \tag{1}$$

EXAMPLE 2 *Parametric equations for the ellipse* $x^2/a^2 + y^2/b^2 = 1$ (Fig. 13.11):

$$x = a \cos t, \qquad y = b \sin t, \qquad 0 \le t \le 2\pi. \tag{2}$$

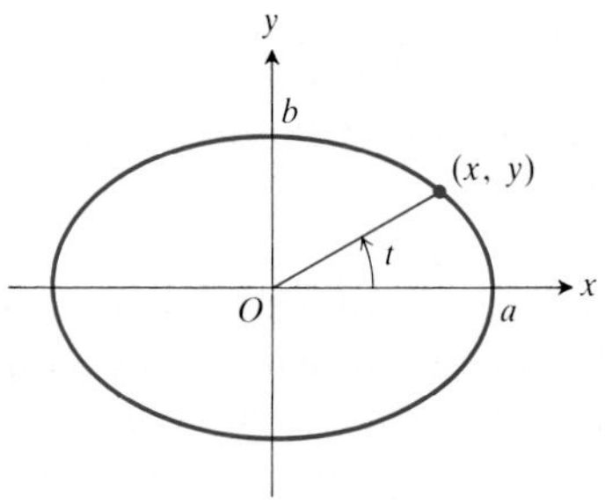

13.11 Ellipse: $x = a \cos t$, $y = b \sin t$.

To see that these are parametric equations for the ellipse we first substitute $x = a \cos t$, $y = b \sin t$ into the cartesian equation:

$$\frac{x^2}{a^2} + \frac{y^2}{b^2} = \frac{a^2 \cos^2 t}{a^2} + \frac{b^2 \sin^2 t}{b^2} = \cos^2 t + \sin^2 t = 1. \tag{3}$$

This shows that the point $(a \cos t, b \sin t)$ lies on the ellipse.

We then observe that as t increases from 0 to 2π, x varies continuously from a to $-a$ to a, and y varies continuously from 0 to b, to 0, to $-b$, and back to 0. Thus, the point $(x, y) = (a \cos t, b \sin t)$ goes "once around the clock," tracing the entire ellipse. $\square$

Sometimes the parametric equations of a curve and the cartesian equation are not coextensive.

EXAMPLE 3 *One branch of an hyperbola.* Suppose the parametric equations of a curve are

$$x = \cosh \theta, \qquad y = \sinh \theta. \tag{4}$$

Then the hyperbolic identity

$$\cosh^2 \theta - \sinh^2 \theta = 1$$

enables us to eliminate θ and write

$$x^2 - y^2 = 1 \tag{5}$$

as a cartesian equation of the curve. Closer scrutiny, however, shows that Eq. (5) *includes too much,* for $x = \cosh \theta$ is always positive, so the parametric equations represent a curve lying wholly to the right of the y-axis, whereas the cartesian equation (5) represents both the right- and left-hand branches of the hyperbola (Fig. 13.12). The left-hand branch could be excluded by taking only positive values of x. That is,

$$x = \sqrt{1 + y^2} \tag{6}$$

does represent the curve given by (4). $\square$

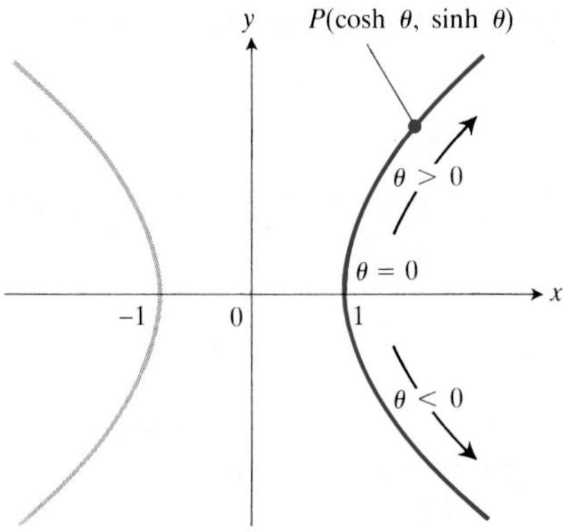

13.12 The parametric equations $x = \cosh \theta$, $y = \sinh \theta$, $-\infty < \theta < \infty$, give only the right branch of the hyperbola $x^2 - y^2 = 1$, because $\cosh \theta \ge 1$.

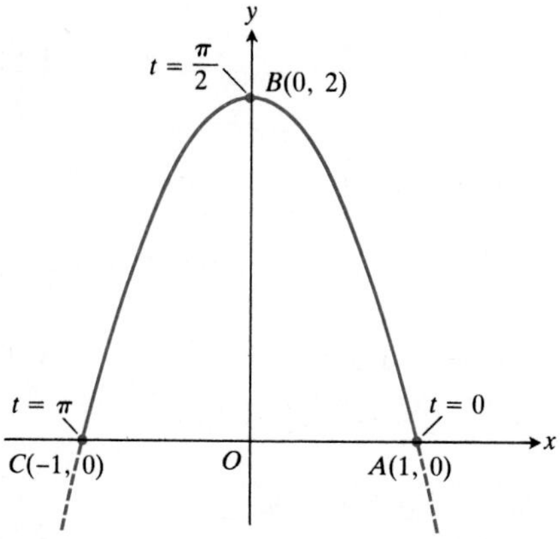

13.13 $x = \cos t$, $y = 1 - \cos 2t$.

EXAMPLE 4 A parabolic arch. Sketch the curve traced by the point $P(x, y)$ whose coordinates satisfy the equations

$$x = \cos t, \qquad y = 1 - \cos 2t. \tag{7}$$

Solution We find a cartesian equation for the curve by eliminating t:

$$y = 1 - \cos 2t = 1 - 2\cos^2 t + 1 = 2 - 2x^2.$$

Thus every point of the graph of (7) lies on the parabola

$$y = 2 - 2x^2. \tag{8}$$

The parametric equations in (7), however, describe only the portion of the parabola (Fig. 13.13) for which

$$-1 \le x = \cos t \le 1 \qquad \text{and} \qquad 0 \le y = 1 - \cos 2t \le 2.$$

From (7) we see that the point $P(x, y)$ starts at $A(1, 0)$ when $t = 0$. It then moves up and to the left as t increases, arriving at $B(0, 2)$ when $t = \pi/2$. It continues on to $C(-1, 0)$ as t increases to π. As t varies from π to 2π, the point retraces the arch CBA back to A. Since x and y are periodic, x with period 2π and y with period π, any further variation in t results in retracing a portion of the arch. □

EXAMPLE 5 Trochoids and cycloids. A wheel of radius a rolls along a horizontal straight line without slipping. Find the curve traced by a point P on a spoke of the wheel b units from its center. Such a curve is called a *trochoid* (one Greek word for wheel is *trochos*). When $b = a$, P is on the circumference and the curve is called a *cycloid*. This is like the path traveled by the head of a nail in a tire.

Solution In Fig. 13.14 we take the x-axis to be the line the wheel rolls along, with the y-axis through a low point of the trochoid. It is customary to use the angle ϕ through which CP has rotated as the parameter. Since the circle rolls without slipping, the distance OM that the wheel has moved horizontally is just equal to the circular arc $MN = a\phi$. (Roll the wheel back. Then N will fall at the origin O.) The xy-coordinates of C are therefore

$$h = a\phi, \qquad k = a. \tag{9}$$

We now introduce $x'y'$-axes parallel to the xy-axes and having their origin at C (Fig. 13.15). The xy- and $x'y'$-coordinates of P are related by the

13.14 Trochoid: $x = a\phi - b \sin \phi$, $y = a - b \cos \phi$.

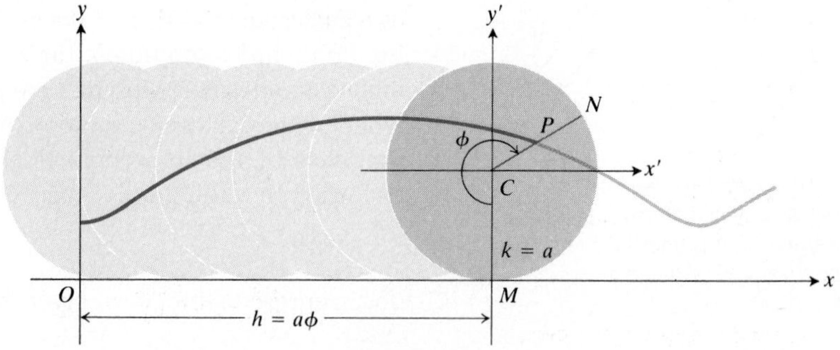

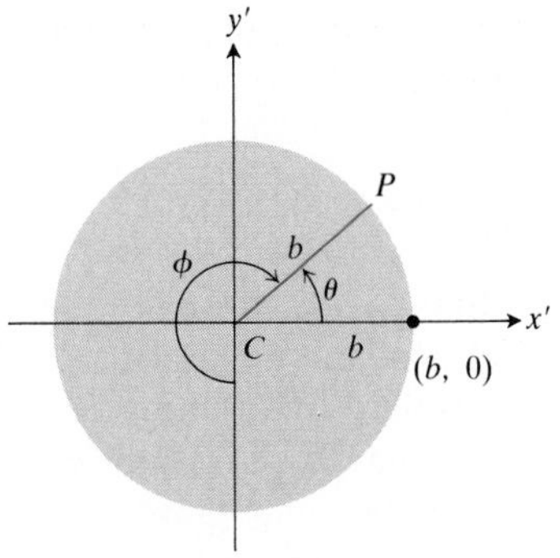

13.15 The $x'y'$-coordinates of P are $x' = b\cos\theta$, $y' = b\sin\theta$.

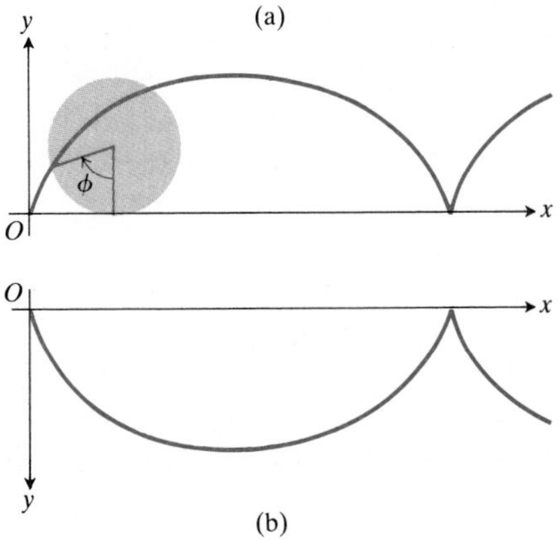

13.16 Cycloid: $x = a(\phi - \sin\phi)$, $y = a(1 - \cos\phi)$.

equations

$$x = h + x', \qquad y = k + y'. \tag{10}$$

From Fig. 13.15 we may immediately read

$$x' = b\cos\theta, \qquad y' = b\sin\theta,$$

or, since

$$\theta = \frac{3\pi}{2} - \phi,$$

$$x' = -b\sin\phi, \qquad y' = -b\cos\phi. \tag{11}$$

We substitute these results and Eqs. (10) into (9) and obtain

$$x = a\phi - b\sin\phi,$$
$$y = a - b\cos\phi \tag{12}$$

as parametric equations of the trochoid.

The cycloid (Fig. 13.16a),

$$x = a(\phi - \sin\phi), \qquad y = a(1 - \cos\phi), \tag{13}$$

obtained from (12) by taking $b = a$, is the most important special case. □

Brachistochrones and Tautochrones (Optional)

If we reflect both the cycloid and the y-axis across the x-axis, Eqs. (13) still apply, and the resulting curve (Fig. 13.16b) has several interesting properties, one of which we shall now discuss without proof. The proofs belong to a branch of mathematics known as the calculus of variations. Much of the fundamental theory of this subject is attributed to the Bernoulli brothers, John and James, who were friendly rivals and stimulated each other with mathematical problems in the form of challenges. One of these, the brachistochrone problem, was: Among all smooth curves joining two given points, to find that one along which a bead, subject only to the force of gravity, might slide *in the shortest time.*

The two points, labeled P_0 and P_1 in Fig. 13.17, may be taken to lie in a vertical plane at the origin and at (x_1, y_1), respectively. We can formulate the problem in mathematical terms as follows. The kinetic energy of the bead at the start is zero, since its velocity is zero. The work done by gravity in moving the particle from (0, 0) to any point (x, y) is mgy and this must be equal to the change in kinetic energy; that is,

$$mgy = \frac{1}{2}mv^2 - \frac{1}{2}m(0)^2.$$

Thus the velocity

$$v = ds/dt$$

that the particle has when it reaches $P(x, y)$ is

$$v = \sqrt{2gy}.$$

That is,

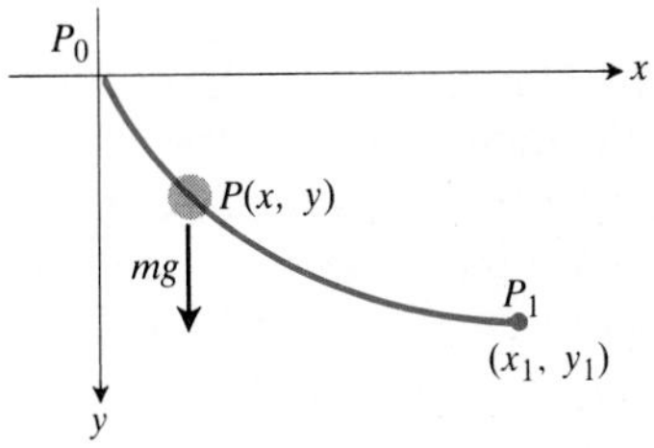

13.17 A bead sliding down a cycloid.

$$\frac{ds}{dt} = \sqrt{2gy} \qquad \text{or} \qquad dt = \frac{ds}{\sqrt{2gy}} = \frac{\sqrt{1 + \left(\frac{dy}{dx}\right)^2}\,dx}{\sqrt{2gy}}.$$

The time t_1 required for the bead to slide from P_0 to P_1 depends upon the particular curve $y = f(x)$ along which it moves and is given by

$$t_1 = \int_0^{x_1} \sqrt{\frac{1 + (f'(x))^2}{2gf(x)}}\, dx. \tag{14}$$

The problem is *to find the curve* $y = f(x)$ that passes through the points $P_0(0, 0)$ and $P_1(x_1, y_1)$ and minimizes the value of the integral in Eq. (14).

At first sight, one might guess that the straight line joining P_0 and P_1 would also yield the shortest time, but a moment's reflection will cast some doubt on this conjecture, for there may be some gain in time by having the particle start to fall vertically at first, thereby building up its velocity more quickly than if it were to slide along an inclined path. With this increased velocity, one may be able to afford to travel over a longer path and still reach P_1 in a shorter time. The solution of the problem is beyond the present book, but the brachistochrone curve is actually an arc of a cycloid through P_0 and P_1, having a cusp at the origin.

If we write Eq. (14) in the equivalent form

$$t_1 = \int \sqrt{\frac{dx^2 + dy^2}{2gy}}$$

and then substitute Eqs. (15) into this, we obtain

$$t_1 = \int_0^{\phi_1} \sqrt{\frac{a^2(2 - 2\cos\phi)}{2ga(1 - \cos\phi)}}\, d\phi = \phi_1 \sqrt{\frac{a}{g}}$$

as the time required for the particle to slide from P_0 to P_1. The time required to reach the bottom of the arc is obtained by taking $\phi_1 = \pi$. Now it is a remarkable fact, which we shall soon demonstrate, that the time required to slide along the cycloid from $(0, 0)$ to the lowest point $(a\pi, 2a)$ is the same as the time required for the particle, starting from rest, to slide from *any intermediate point* of the arc, say (x_0, y_0), to $(a\pi, 2a)$. For the latter case, one has

$$v = \sqrt{2g(y - y_0)}$$

as the velocity at $P(x, y)$, and the time required is

$$T = \int_{\phi_0}^{\pi} \sqrt{\frac{a^2(2 - 2\cos\phi)}{2ag(\cos\phi_0 - \cos\phi)}}\, d\phi = \sqrt{\frac{a}{g}} \int_{\phi_0}^{\pi} \sqrt{\frac{1 - \cos\phi}{\cos\phi_0 - \cos\phi}}\, d\phi$$

$$= \sqrt{\frac{a}{g}} \int_{\phi_0}^{\pi} \sqrt{\frac{2\sin^2(\phi/2)}{[2\cos^2(\phi_0/2) - 1] - [2\cos^2(\phi/2) - 1]}}\, d\phi$$

$$= 2\sqrt{\frac{a}{g}} \left[-\sin^{-1} \frac{\cos(\phi/2)}{\cos(\phi_0/2)} \right]_{\phi_0}^{\pi} = 2\sqrt{\frac{a}{g}} (-\sin^{-1} 0 + \sin^{-1} 1) = \pi \sqrt{\frac{a}{g}}.$$

Since this answer is independent of the value of ϕ_0, it follows that the same length of time is required to reach the lowest point on the cycloid no matter where on the arc the particle is released from rest. Thus, in Fig. 13.18, three particles that start at the same time from O, A, and B will reach C simultaneously. In this sense, the cycloid is also a *tautochrone* (meaning "the same time") as well as being a brachistochrone (meaning "shortest time").

13.18 Beads released on the cycloid at *O*, *A*, and *B* will all take the same amount of time to reach *C*.

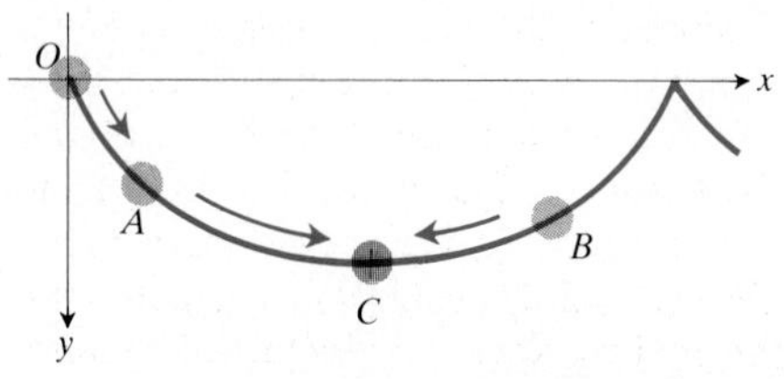

Standard Parametric Equations

For the circle $x^2 + y^2 = a^2$: $\begin{cases} x = a\cos t, \\ y = a\sin t, \end{cases}$ $0 \le t \le 2\pi$

For the ellipse $\dfrac{x^2}{a^2} + \dfrac{y^2}{b^2} = 1$: $\begin{cases} x = a\cos t, \\ y = b\sin t, \end{cases}$ $0 \le t \le 2\pi$

Cycloid: $x = a(\phi - \sin\phi)$, $y = a(1 - \cos\phi)$

PROBLEMS

In Problems 1–17, sketch the curve traced by the point $P(x, y)$ as the parameter t varies over the given domain. Also find a cartesian equation for each curve.

1. $x = \cos t$, $y = \sin t$, $0 \le t \le 2\pi$

2. $x = \cos 2t$, $y = \sin 2t$, $0 \le t \le \pi$

3. $x = 4\cos t$, $y = 2\sin t$, $0 \le t \le 2\pi$

4. $x = 4\cos t$, $y = 5\sin t$, $0 \le t \le 2\pi$

5. $x = \cos 2t$, $y = \sin t$, $0 \le t \le 2\pi$

6. $x = \cos t$, $y = \sin 2t$, $0 \le t \le 2\pi$

7. $x = \sec t$, $y = \tan t$, $-\pi/2 < t < \pi/2$

8. $x = \csc t$, $y = \cot t$, $0 < t < \pi$

9. $x = t - \sin t$, $y = 1 - \cos t$, $0 \le t \le 2\pi$

10. $x = 2 + 4\sin t$, $y = 3 - 2\cos t$, $0 \le t \le 2\pi$

11. $x = t^3$, $y = t^2$, $-\infty < t < \infty$

12. $x = 2t + 3$, $y = 4t^2 - 9$, $-\infty < t < \infty$

13. $x = \cosh t$, $y = 2\sinh t$, $0 \le t < \infty$

14. $x = 2 + 1/t$, $y = 2 - t$, $0 < t < \infty$

15. $x = t + 1$, $y = t^2 + 4$, $0 \le t < \infty$

16. $x = t^2 + t$, $y = t^2 - t$, $-\infty < t < \infty$

17. $x = 3 + 2\,\text{sech}\, t$, $y = 4 - 3\tanh t$, $-\infty < t < \infty$

18. Find parametric equations of the semicircle

$$x^2 + y^2 = a^2, \qquad y > 0,$$

using as parameter the slope $t = dy/dx$ of the tangent to the curve at (x, y).

19. Find parametric equations of the semicircle

$$x^2 + y^2 = a^2, \qquad y > 0,$$

using as parameter the variable θ defined by the equation $x = a\tanh\theta$.

20. Find parametric equations of the circle

$$x^2 + y^2 = a^2,$$

using as parameter the arc length s measured counterclockwise from the point $(a, 0)$ to the point (x, y).

21. Find parametric equations of the catenary $y = a\cosh x/a$, using as parameter the length of arc s from the point $(0, a)$ to the point (x, y), with the sign of s taken to be the same as the sign of x.

22. If a string wound around a fixed circle is unwound while held taut in the plane of the circle, its end traces an *involute* of the circle (Fig. 13.19). Let the fixed circle be located with its center at the origin O and have radius a. Let the initial position of the tracing point P be $A(a, 0)$ and let the unwound portion of the string PT be tangent to the circle at T. Derive parametric equations of the involute, using the angle AOT as the parameter t.

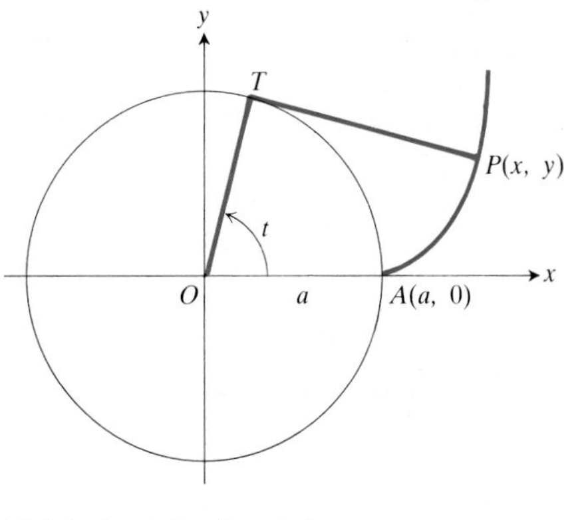

13.19 Involute of a circle.

23. When a circle rolls externally on the circumference of a second, fixed circle, any point P on the circumference of the rolling circle describes an *epicycloid* (Fig. 13.20). Let the fixed circle have its center at the origin O and have radius a. Let the radius of the rolling circle be b and let the initial position of the tracing point P be $A(a, 0)$. Determine parametric equations of the epicycloid, using as parameter the angle θ from the positive x-axis to the line of centers.

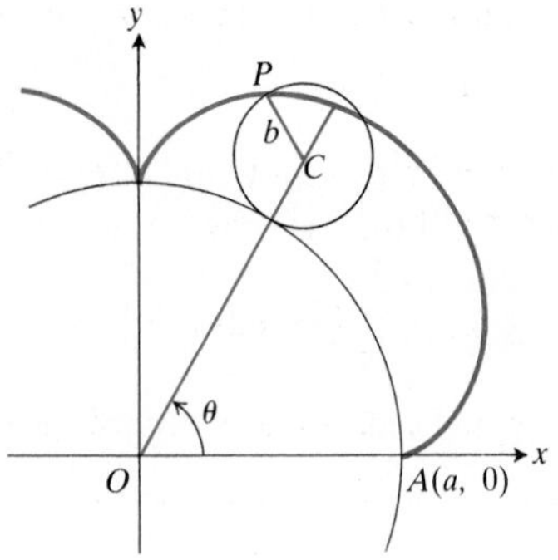

13.20 Epicycloid, with $b = a/4$.

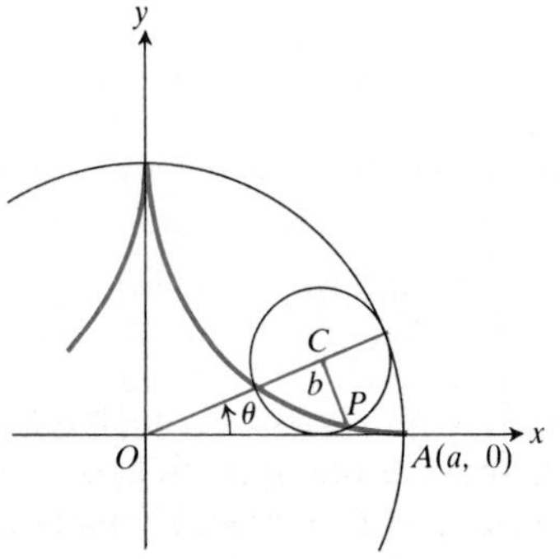

13.21 Hypocycloid, with $b = a/4$.

24. When a circle rolls on the inside of a fixed circle any point P on the circumference of the rolling circle describes a *hypocycloid*. Let the fixed circle be $x^2 + y^2 = a^2$, let the radius of the rolling circle be b, and let the initial position of the tracing point P be $A(a, 0)$. Use the angle θ from the positive x-axis to the line of centers as parameter and determine parametric equations of the hypocycloid. In particular, if $b = a/4$, as in Fig. 13.21, show that

$$x = a \cos^3 \theta, \qquad y = a \sin^3 \theta.$$

25. Find the length of one arch of the cycloid

$$x = a(\phi - \sin \phi), \qquad y = a(1 - \cos \phi).$$

26. Show that the slope of the cycloid

$$x = a(\phi - \sin \phi), \qquad y = a(1 - \cos \phi)$$

is $dy/dx = \cot \phi/2$. In particular, the tangent to the cycloid is vertical when ϕ is 0 or 2π.

27. Show that the slope of the trochoid

$$x = a\phi - b \sin \phi, \qquad y = a - b \cos \phi$$

is always finite if $b < a$.

28. The *witch of Maria Agnesi* is a bell-shaped curve that may be constructed as follows: Let C be a circle of radius a having its center at $(0, a)$ on the y-axis (Fig. 13.22). The variable line OA through the origin

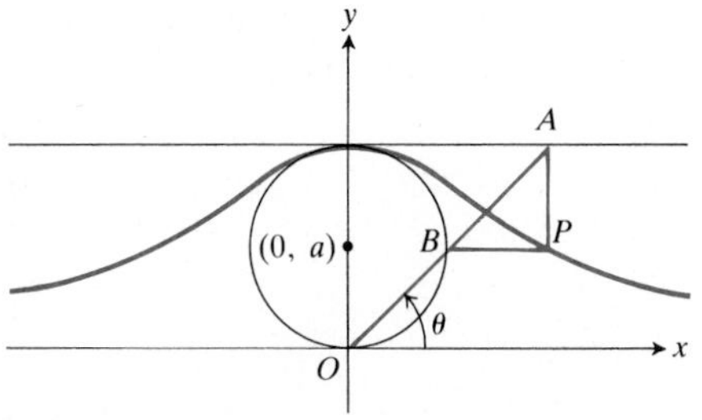

13.22 The witch of Maria Agnesi.

O intersects the line $y = 2a$ in the point A and intersects the circle in the point B. A point P on the witch is now located by taking the intersection of lines through A and B parallel to the y- and x-axes, respectively.

a) Find parametric equations of the witch, using as parameter the angle θ from the x-axis to the line OA.

b) Also find a cartesian equation for the witch.

Historical note: Maria Gaetana Agnesi (1718–1799), the daughter of a mathematics professor at the University of Bologna, wrote the first comprehensive calculus text. In four books, the text treated algebra and geometry, differential calculus, integral calculus, and differential equations. The text was translated into French and English, and it is a mistranslation that is responsible for our calling Agnesi's bell-shaped curve "the witch" today. This name, in fact, is found only in texts written in English. Agnesi's own name for the curve was "versiera," from the Latin verb *vertere,* to turn. The translator, a Cambridge scholar who had learned Italian expressly for the purpose of translating Agnesi's text, probably confused the Latin *versiera* with the Italian *avversiera,* "wife of the devil," carefully translating the latter as "the witch."

29. The following question appeared on a college entrance examination a few years ago.

Figure 13.23

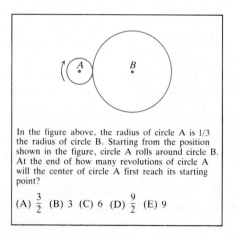

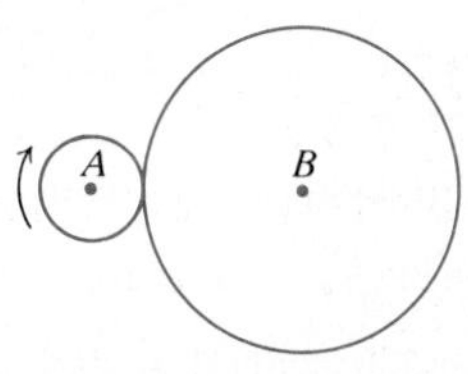

In the figure above, the radius of circle A is 1/3 the radius of circle B. Starting from the position shown in the figure, circle A rolls around circle B. At the end of how many revolutions of circle A will the center of circle A first reach its starting point?

(A) $\frac{3}{2}$ (B) 3 (C) 6 (D) $\frac{9}{2}$ (E) 9

None of the choices offered with the question was correct. What is the correct answer?

30. (*Calculator*) An automobile tire of radius 1 ft has a pebble stuck in the tread. Estimate to the nearest foot the length of the arched path traced by the pebble when the car goes one mile. Start by finding the ratio of the length of one arch of a cycloid to its base length.

31. The region bounded by the ellipse $x = 2\cos t$, $y = \sin t$, $0 \le t \le 2\pi$, is revolved about the y-axis. Find the volume swept out.

32. The equations $r = e^{2t}$, $\theta = 3t$, $0 \le t \le \pi/6$ give a curve in the polar coordinate plane.

a) Find the area bounded by the curve in the first quadrant.
b) Find the length of the curve.

33. Find parametric equations and a cartesian equation for the figure traced by the point $P(x, y)$ if

$$\frac{dx}{dt} = -2y, \qquad \frac{dy}{dt} = \cos t,$$

and $x = 3$, $y = 0$ when $t = 0$. Identify the figure.

***Toolkit* programs**

Parametric Equations   Super * Grapher

13.4
Space Coordinates

The extension of coordinate geometry to three dimensions was begun in the seventeenth century, mainly by Descartes, Fermat, and Philippe de la Hire (1640–1718), a painter who later turned to geometry and astronomy. The full development, however, was the work of the eighteenth century. John Bernoulli, in a letter to Leibniz in 1715, introduced the three cartesian coordinate planes we use today, and the subsequent work of Euler, Lagrange, and Gaspard Monge (1746–1818) brought the analytic geometry of three dimensional space to the point we know today.

In this article, we look briefly at three coordinate systems for space. The first, cartesian coordinates, is the system we shall use the most. Cylindrical and spherical coordinates, however, will come in handy when we study integration in Chapter 16, because surfaces that have complicated cartesian equations sometimes have simpler equations in one of these other systems.

Cartesian Coordinates

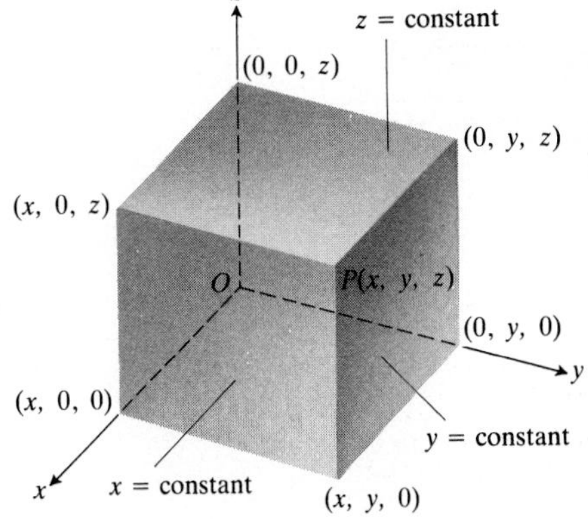

13.24 Right-handed cartesian coordinate system.

Figure 13.24 shows a system of mutually orthogonal coordinate axes, Ox, Oy, and Oz. The system is called *right-handed* because a right-threaded screw pointing along Oz will advance when turned from Ox to Oy through an angle, say, of 90°. The cartesian coordinates of a point $P(x, y, z)$ in space may be read from the coordinate axes by passing planes through P perpendicular to each axis. The points on the x-axis have their y- and z-coordinates both zero. That is, they have coordinates of the form $(x, 0, 0)$. Points in a plane perpendicular to the z-axis, say, all have the same z-coordinate. Thus the points in the plane perpendicular to the z-axis and 5 units above the xy-plane all have coordinates of the form $(x, y, 5)$. We can write $z = 5$ as an equation for this plane. The three planes

$$x = 2, \qquad y = 3, \qquad z = 5$$

intersect in the point $P(2, 3, 5)$. The points of the yz-plane are obtained by setting $x = 0$. The three coordinate planes $x = 0$, $y = 0$, $z = 0$ divide the

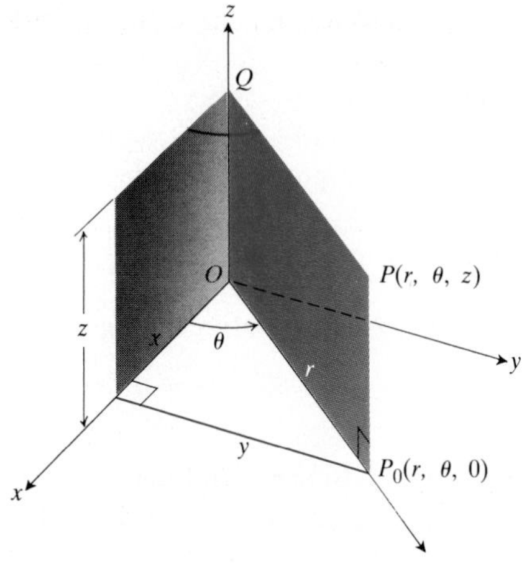

13.25 Cylindrical coordinates.

space into eight cells, called octants. The octant in which all three coordinates are positive is called the *first octant,* but there is no conventional numbering of the remaining seven octants.

EXAMPLE 1 Describe the set of points $P(x, y, z)$ whose cartesian coordinates satisfy the simultaneous equations

$$x^2 + y^2 = 4, \qquad z = 3.$$

Solution The points all lie in the horizontal plane $z = 3$, and in this plane they lie on the circle $x^2 + y^2 = 4$. Thus we may describe the set as the circle $x^2 + y^2 = 4$ in the plane $z = 3$. □

Cylindrical Coordinates

It is frequently convenient to use cylindrical coordinates (r, θ, z) to locate a point in space. These are just the polar coordinates (r, θ) used instead of (x, y) in the plane $z = 0$, coupled with the z-coordinate. Cylindrical and cartesian coordinates are therefore related by the following equations (Fig. 13.25):

Equations Relating Cartesian and Cylindrical Coordinates

$$\begin{aligned} x &= r\cos\theta, & r^2 &= x^2 + y^2, \\ y &= r\sin\theta, & \tan\theta &= y/x, \\ & & z &= z. \end{aligned} \tag{1}$$

The equation $r =$ constant describes a circular cylinder of radius r whose axis is the z-axis, $r = 0$ being an equation for the z-axis itself. The equation $\theta =$ constant describes a plane containing the z-axis and making an angle θ with the positive x-axis (Fig. 13.26). (Some authors require the

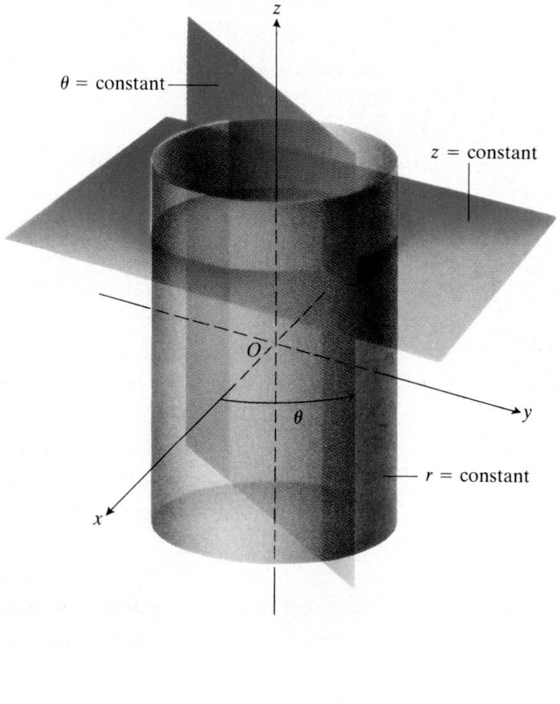

13.26 Some planes and cylinders have simple equations in cylindrical coordinates.

values of r in cylindrical coordinates to be nonnegative. In this case, the equation θ = constant describes a half-plane fanning out from the z-axis.)

Cylindrical coordinates are convenient when there is an axis of symmetry in a physical problem.

EXAMPLE 2 Describe the set of points $P(r, \theta, z)$ whose cylindrical coordinates satisfy the simultaneous equations

$$r = 2, \qquad \theta = \frac{\pi}{4}.$$

Solution These points make up the line of intersection of the cylinder $r = 2$ (cylinder of radius 2 about the z-axis) and the half-plane $\theta = \pi/4$, $r > 0$ (half-plane containing the z-axis and making an angle of $\pi/4$ radians with the positive x-axis). Thus we have the line that passes through the point $(2, \pi/4, 0)$ parallel to the z-axis. □

Spherical Coordinates

Spherical coordinates are useful when there is a center of symmetry that we can take as the origin. The *spherical coordinates* (ρ, ϕ, θ) are shown in Fig. 13.27.

The first coordinate $\rho = |OP|$ is the distance from the origin to the point P. It is never negative. The equation ρ = constant describes the surface of a sphere of radius ρ with center at O (Fig. 13.28).

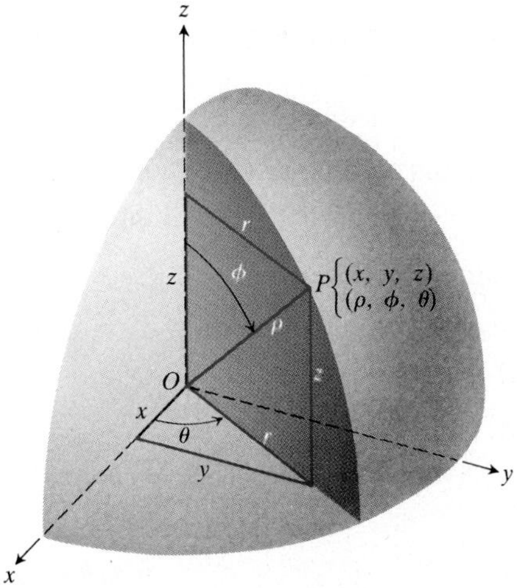

13.27 Spherical coordinates.

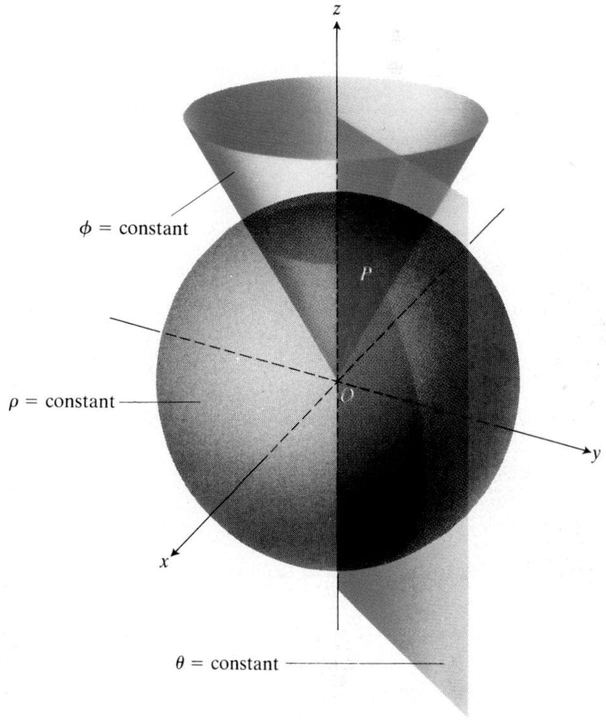

13.28 Spheres and cones whose centers are at the origin have simple equations in spherical coordinates.

The second spherical coordinate, ϕ, is the angle measured down from the z-axis to the line OP. The equation ϕ = constant describes a cone with vertex at O, axis Oz, and generating angle ϕ, provided we broaden our interpretation of the word "cone" to include the xy-plane for which $\phi = \pi/2$ and cones with generating angles greater than $\pi/2$.

The third spherical coordinate θ is the same as the angle θ in cylindrical coordinates, namely, the angle from the xz-plane to the plane through P and the z-axis. But, in contrast with cylindrical coordinates, the equation θ = constant in spherical coordinates defines a half-plane (Fig. 13.28) (because $\rho \geq 0$ and $0 \leq \phi \leq \pi$).

Incidentally, some books give spherical coordinates in the order (ρ, θ, ϕ) with the θ and ϕ reversed, and you should watch out for this when you read elsewhere.

EXAMPLE 3 Describe the set of points $P(\rho, \phi, \theta)$ whose spherical coordinates satisfy the simultaneous equations

$$\rho = 1, \qquad \phi = \frac{\pi}{3}.$$

Solution The equation $\rho = 1$ is an equation for the sphere of radius one unit centered at the origin. The equation $\phi = \pi/3$ is an equation for the cone that stands on its vertex at the origin and opens upward to make an angle of $\phi = \pi/3$ radians with the positive z-axis. The points in question make up the horizontal circle in which this cone intersects the sphere. □

From Fig. 13.27 we may read the following relationships between the cartesian (x, y, z), cylindrical (r, θ, z), and spherical (ρ, ϕ, θ) coordinate systems:

Equations Relating Cartesian and Cylindrical Coordinates to Spherical Coordinates

$$\begin{aligned} r &= \rho \sin \phi, & x &= r \cos \theta, & x &= \rho \sin \phi \cos \theta, \\ z &= \rho \cos \phi, & y &= r \sin \theta, & y &= \rho \sin \phi \sin \theta, \\ \theta &= \theta, & z &= z, & z &= \rho \cos \phi. \end{aligned} \tag{2}$$

Every point in space can be given spherical coordinates restricted to the ranges

$$\rho \geq 0, \qquad 0 \leq \phi \leq \pi, \qquad 0 \leq \theta < 2\pi. \tag{3}$$

PROBLEMS

In Problems 1–10, describe the set of points P(x, y, z) whose cartesian coordinates satisfy the given pairs of simultaneous equations.

1. $x = 1, \quad y = 1$
2. $x = a, \quad y = b \qquad$ (a and b constant)
3. $y = x, \quad z = 5$
4. $x^2 + y^2 = 4, \quad z = 0$
5. $x^2 + y^2 = 4, \quad z = -2$
6. $x^2 + z^2 = 4, \quad y = 0$
7. $y^2 + z^2 = 1, \quad x = 0$

8. $z = 4y^2, \quad x = 0$

9. $z = y, \quad x = 0$

10. $y^2/a^2 + z^2/b^2 = 1, \quad x = 0$

The following table gives the coordinates of specific points in space in one of three coordinate systems. In Problems 11–19, find coordinates for each point in the other two systems. There may be more than one right answer, because points in cylindrical and spherical coordinates may have many coordinate triples.

	Cartesian (x, y, z)	Cylindrical (r, θ, z)	Spherical (ρ, ϕ, θ)
11.	(1, 0, 0)		
12.	(0, 1, 0)		
13.	(0, 0, 1)		
14.		(1, 0, 0)	
15.		$(\sqrt{2}, 0, 1)$	
16.		$(1, \pi/2, 1)$	
17.			$(\sqrt{3}, \pi/3, -\pi/2)$
18.			$(2\sqrt{2}, \pi/2, 3\pi/2)$
19.			$(\sqrt{2}, \pi, 3\pi/2)$

In Problems 20–23, describe the set of points $P(r, \theta, z)$ whose cylindrical coordinates satisfy the given pairs of simultaneous equations. Sketch.

20. $r = 2, \quad z = 3$

21. $\theta = \pi/6, \quad z = r$

22. $r = 3, \quad z = 2\theta$

23. $r = 2\theta, \quad z = 3\theta$

In Problems 24–28, describe the set of points $P(\rho, \phi, \theta)$ whose spherical coordinates satisfy the given pairs of simultaneous equations. Sketch.

24. $\rho = 5, \quad \theta = \pi/4$

25. $\rho = 5, \quad \phi = \pi/4$

26. $\theta = \pi/4, \quad \phi = \pi/4$

27. $\theta = \pi/2, \quad \rho = 4 \cos \phi$

28. $\rho = 1, \quad \theta = \phi, \quad 0 \leq \theta \leq \pi/2$

In Problems 29–40, translate the equations from the given coordinate system (cartesian, cylindrical, or spherical) into equations in the other two systems.

29. $x^2 + y^2 + z^2 = 4$

30. $x^2 + y^2 + z^2 = 4z$

31. $z^2 = r^2$

32. $\rho = 6 \cos \phi$

33. $x = y$

34. $\phi = 0$

35. $\theta = 0$

36. $z = -2$

37. $\rho = 1$

38. $\rho \cos \phi = 3$

39. $\rho \sin \phi \cos \theta = 2$

40. $x^2 + y^2 = 5$

Describe the sets in Problems 40–44.

40. $x \geq 0$

41. $3 \leq \rho \leq 5$

42. $r \geq 2, \quad \rho \leq 5$

43. $0 \leq \theta \leq \pi/4, \quad 0 \leq \phi \leq \pi/4, \quad \rho \geq 0$

44. $4x^2 + 9y^2 \leq 36$

13.5 Vectors and Distance in Space

Vectors in space are the three-dimensional analog of vectors in the plane and are subject to the same rules of addition, subtraction, and scalar multiplication that govern vectors in the plane. The vectors from the origin to the points whose cartesian coordinates are (1, 0, 0), (0, 1, 0), and (0, 0, 1) are the basic unit vectors. We denote them by **i**, **j**, and **k**, and write the vector from the origin O to the point $P(x, y, z)$ as

$$\mathbf{R} = \overrightarrow{OP} = \mathbf{i}x + \mathbf{j}y + \mathbf{k}z. \tag{1}$$

Vector Between Two Points

If $P_1(x_1, y_1, z_1)$ and $P_2(x_2, y_2, z_2)$ are two points in space (Fig. 13.29), then the vector from P_1 to P_2 is the vector sum

$$\overrightarrow{P_1P_2} = \overrightarrow{P_1O} + \overrightarrow{OP_2}.$$

Since

$$\overrightarrow{P_1O} = -\overrightarrow{OP_1},$$

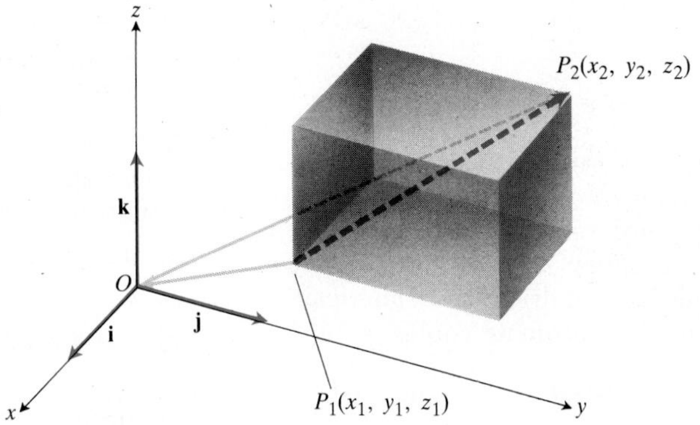

13.29 $\overrightarrow{P_1P_2} = \overrightarrow{P_1O} + \overrightarrow{OP_2}$.

this is the same as

$$\overrightarrow{P_1P_2} = \overrightarrow{OP_2} - \overrightarrow{OP_1}$$

or

$$\overrightarrow{P_1P_2} = \mathbf{i}(x_2 - x_1) + \mathbf{j}(y_2 - y_1) + \mathbf{k}(z_2 - z_1). \tag{2}$$

The vector from $P_1(x_1, y_1, z_1)$ to $P_2(x_2, y_2, z_2)$ is

$$\overrightarrow{P_1P_2} = (x_2 - x_1)\mathbf{i} + (y_2 - y_1)\mathbf{j} + (z_2 - z_1)\mathbf{k}.$$

The length of any vector

$$\mathbf{A} = a\mathbf{i} + b\mathbf{j} + c\mathbf{k}$$

is readily determined by applying the theorem of Pythagoras twice. In the right triangle ABC (Fig. 13.30),

$$|\overrightarrow{AC}| = |a\mathbf{i} + b\mathbf{j}| = \sqrt{a^2 + b^2},$$

and in the right triangle ACD,

$$|\overrightarrow{AD}| = \sqrt{|\overrightarrow{AC}|^2 + |\overrightarrow{CD}|^2} = \sqrt{(a^2 + b^2) + c^2}.$$

That is,

$$|a\mathbf{i} + b\mathbf{j} + c\mathbf{k}| = \sqrt{a^2 + b^2 + c^2}. \tag{3}$$

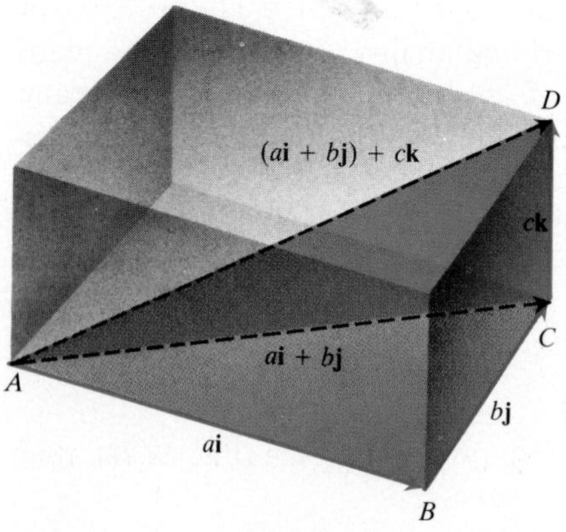

13.30 The length $|\overrightarrow{AD}|$ of the vector $\overrightarrow{AD}$ can be determined by applying the Pythagorean theorem to the right triangles ABC and ACD.

Distance

If we apply this result to the vector $\overrightarrow{P_1P_2}$ of Eq. (2), we obtain a formula for the distance between two points:

$$|\overrightarrow{P_1P_2}| = \sqrt{(x_2 - x_1)^2 + (y_2 - y_1)^2 + (z_2 - z_1)^2}. \tag{4}$$

EXAMPLE 1 The distance between

$$P_1(-2, 3, 0) \quad \text{and} \quad P_2(2, 1, 5)$$

is

$$|\overrightarrow{P_1P_2}| = \sqrt{(-2 - 2)^2 + (3 - 1)^2 + (0 - 5)^2} = \sqrt{16 + 4 + 25}$$
$$= \sqrt{45} = 3\sqrt{5}. \ \square$$

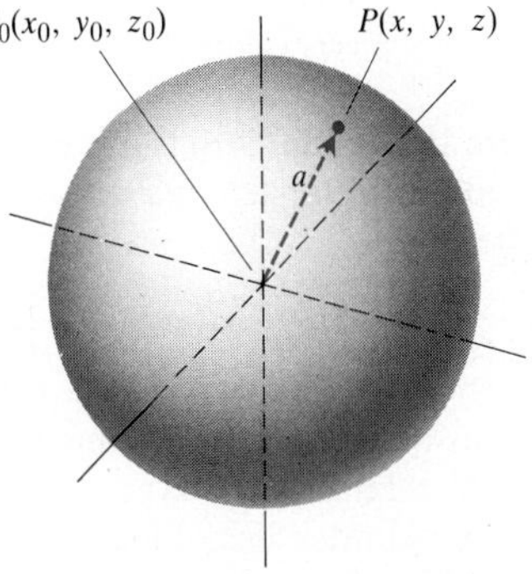

13.31 The equation of the sphere of radius a centered at the point (x_0, y_0, z_0) is $(x - x_0)^2 + (y - y_0)^2 + (z - z_0)^2 = a^2$.

Spheres

Equation (4) may be used to determine an equation for the sphere of radius a with center at $P_0(x_0, y_0, z_0)$ (Fig. 13.31). The point P is on the sphere if and only if

$$|\overrightarrow{P_0P}| = a,$$

or

$$(x - x_0)^2 + (y - y_0)^2 + (z - z_0)^2 = a^2. \tag{5}$$

EXAMPLE 2 Find the center and radius of the sphere

$$x^2 + y^2 + z^2 + 2x - 4y = 0.$$

Solution Complete the squares in the given equation to obtain

$$x^2 + 2x + 1 + y^2 - 4y + 4 + z^2 = 0 + 1 + 4$$
$$(x + 1)^2 + (y - 2)^2 + z^2 = 5.$$

Comparison with Eq. (5) shows that $x_0 = -1$, $y_0 = 2$, $z_0 = 0$, and $a = \sqrt{5}$. The center is $(-1, 2, 0)$ and the radius is $\sqrt{5}$. □

EXAMPLE 3 *Midpoints.* Find the coordinates of the midpoint M of the line segment that joins the points $P_1(x_1, y_1, z_1)$ and $P_2(x_2, y_2, z_2)$.

Solution Since

$$\overrightarrow{OM} = \overrightarrow{OP_1} + \frac{1}{2}(\overrightarrow{P_1P_2}) = \overrightarrow{OP_1} + \frac{1}{2}(\overrightarrow{OP_2} - \overrightarrow{OP_1}) = \frac{1}{2}(\overrightarrow{OP_1} + \overrightarrow{OP_2})$$
$$= \frac{x_1 + x_2}{2}\mathbf{i} + \frac{y_1 + y_2}{2}\mathbf{j} + \frac{z_1 + z_2}{2}\mathbf{k},$$

we see that the midpoint M is the point

$$M = \left(\frac{x_1 + x_2}{2}, \frac{y_1 + y_2}{2}, \frac{z_1 + z_2}{2}\right)$$

whose coordinates are obtained by averaging the coordinates of the points P_1 and P_2.

For instance, the midpoint of the segment joining $P_1(3, -2, 0)$ and $P_2(7, 4, 4)$ is

$$\left(\frac{3 + 7}{2}, \frac{-2 + 4}{2}, \frac{0 + 4}{2}\right) = (5, 1, 2). \ \square$$

Direction

We define the *direction* of a nonzero vector $\mathbf{A}$ in space to be the unit vector obtained by dividing $\mathbf{A}$ by its own length:

$$\text{Direction of } \mathbf{A} = \frac{\mathbf{A}}{|\mathbf{A}|}. \tag{6}$$

EXAMPLE 4 If $\mathbf{A} = 2\mathbf{i} - 3\mathbf{j} + 7\mathbf{k}$, then its length is $\sqrt{4 + 9 + 49} = \sqrt{62}$, and

$$\text{Direction of } (2\mathbf{i} - 3\mathbf{j} + 7\mathbf{k}) = \frac{2\mathbf{i} - 3\mathbf{j} + 7\mathbf{k}}{\sqrt{62}}.$$

In Problem 11 you will be asked to show that the vector found here really is a unit vector. □

PROBLEMS

Find the centers and radii of the spheres in Problems 1–4.

1. $x^2 + y^2 + z^2 + 4x - 4z = 0$

2. $2x^2 + 2y^2 + 2z^2 + x + y + z = 9$

3. $x^2 + y^2 + z^2 - 2az = 0$

4. $3x^2 + 3y^2 + 3z^2 + 2y - 2z = 9$

5. Find the distance between the point $P(x, y, z)$ and (a) the x-axis, (b) the y-axis, (c) the z-axis, (d) the xy-plane.

6. The distance from $P(x, y, z)$ to the origin is d_1 and the distance from P to $A(0, 0, 3)$ is d_2. Write an equation for the coordinates of P if

a) $d_1 = 2d_2$, b) $d_1 + d_2 = 6$, c) $|d_1 - d_2| = 2$.

Find the lengths of the following vectors.

7. $2\mathbf{i} + \mathbf{j} - 2\mathbf{k}$ **8.** $3\mathbf{i} - 6\mathbf{j} + 2\mathbf{k}$

9. $\mathbf{i} + 4\mathbf{j} - 8\mathbf{k}$ **10.** $9\mathbf{i} - 2\mathbf{j} + 6\mathbf{k}$

11. Show that the vector $(2\mathbf{i} - 3\mathbf{j} + 7\mathbf{k})/\sqrt{62}$ found in Example 4 is a unit vector.

12. Find the directions of the following vectors:

a) $4\mathbf{i} + 3\mathbf{j} + 5\mathbf{k}$ b) $\dfrac{\mathbf{i}}{\sqrt{3}} + \dfrac{\mathbf{j}}{\sqrt{3}} + \dfrac{\mathbf{k}}{\sqrt{3}}$

c) $\dfrac{3}{5}\mathbf{i} + \dfrac{4}{5}\mathbf{k}$ d) $6\mathbf{i}$

e) $-4\mathbf{j}$ f) $\mathbf{i} + \mathbf{j}$

g) $\mathbf{i} + \mathbf{j} + \mathbf{k}$

13. Find the vector from the origin O to the point of intersection of the medians of the triangle whose vertices are the three points

$$A(1, -1, 2), \quad B(2, 1, 3), \quad C(-1, 2, -1).$$

14. A bug is crawling straight up the side of a rotating right circular cylinder of radius 2 ft. At time $t = 0$, it is at the point $(2, 0, 0)$ relative to a fixed set of xyz-axes. The axis of the cylinder lies along the z-axis. Assume that the bug travels on the cylinder along a line parallel to the z-axis at the rate of c ft/s, and that the cylinder rotates (counterclockwise as viewed from above) at the rate of b radians/second. If $P(x, y, z)$ is the bug's position at the end of t seconds, show that

$$\overrightarrow{OP} = \mathbf{i}(2\cos bt) + \mathbf{j}(2\sin bt) + \mathbf{k}(ct).$$

15. Let $ABCD$ be a general (not necessarily planar) quadrilateral in space. Show that the two segments joining the midpoints of opposite sides of $ABCD$ bisect each other (*Hint:* Show that the segments have the same midpoint.)

16. Use similar triangles to find the coordinates of the point Q that divides the segment from $P_1(x_1, y_1, z_1)$ to $P_2(x_2, y_2, z_2)$ into two lengths whose ratio is $p/q = r$ (that is, (distance P_1Q)/(distance QP_2) $= p/q = r$).

17. Show that $|c\mathbf{v}| = |c||\mathbf{v}|$ for any real number c and any vector $\mathbf{v} = \overrightarrow{P_1P_2}$.

13.6 The Scalar Product of Two Vectors

Definition

In mechanics, the work done by a constant force $\mathbf{F}$ when the point of application of $\mathbf{F}$ undergoes a displacement $\overrightarrow{PQ}$ (Fig. 13.32) is defined to be

$$\text{Work} = (|\mathbf{F}|\cos\theta)|\overrightarrow{PQ}| = |\mathbf{F}||\overrightarrow{PQ}|\cos\theta. \tag{1}$$

For instance, if the magnitude of $\mathbf{F}$ is $|\mathbf{F}| = 40$ newtons (about 9 pounds), $|\overrightarrow{PQ}| = 3$ m, and $\theta = 60°$ so that $\cos\theta = \frac{1}{2}$, then the work done is

$$\text{Work} = (40)(3)\left(\frac{1}{2}\right) = 60 \text{ newton meters.}$$

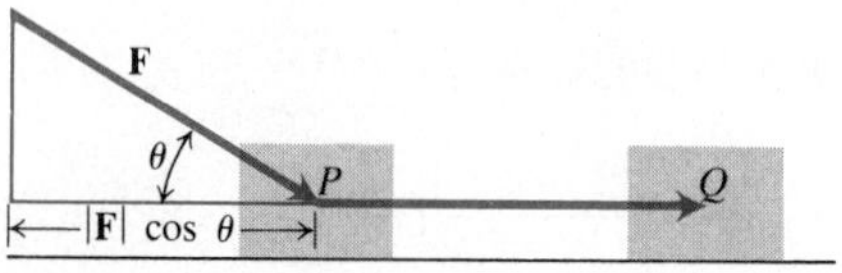

13.32 The work done by **F** during a displacement $\overrightarrow{PQ}$ is $(|\mathbf{F}|\cos\theta)|\overrightarrow{PQ}|$.

In mathematics, we call the quantity

$$|\mathbf{F}||\overrightarrow{PQ}|\cos\theta$$

the *scalar product* of $\mathbf{F}$ and $\overrightarrow{PQ}$.

DEFINITION

Scalar Product

The *scalar product* $\mathbf{A} \cdot \mathbf{B}$ ("$\mathbf{A}$ dot $\mathbf{B}$") of two vectors $\mathbf{A}$ and $\mathbf{B}$ is the number

$$\mathbf{A} \cdot \mathbf{B} = |\mathbf{A}||\mathbf{B}| \cos \theta, \qquad (2)$$

where θ measures the smaller angle determined by $\mathbf{A}$ and $\mathbf{B}$ when their initial points coincide (as in Fig. 13.33).

In words, the scalar product of $\mathbf{A}$ and $\mathbf{B}$ is the length of $\mathbf{A}$ times the length of $\mathbf{B}$ times the cosine of the angle between them. The scalar product is a scalar, not a vector. It is sometimes called the *dot product,* because of the dot in the notation $\mathbf{A} \cdot \mathbf{B}$. In more advanced settings in mathematics, it is also called the *inner product* of $\mathbf{A}$ and $\mathbf{B}$.

Geometric and Algebraic Properties

If the dot product is negative, then $\cos \theta$ is negative and the angle between $\mathbf{A}$ and $\mathbf{B}$ is greater than 90°.

From Eq. (2) we can see that interchanging the two factors $\mathbf{A}$ and $\mathbf{B}$ does not change the dot product. That is,

$$\mathbf{A} \cdot \mathbf{B} = \mathbf{B} \cdot \mathbf{A} \qquad (3)$$

Scalar multiplication is commutative.

If c is any number, then

$$(c\mathbf{A}) \cdot \mathbf{B} = c(\mathbf{A} \cdot \mathbf{B}) = \mathbf{A} \cdot (c\mathbf{B}) \qquad (4)$$

(Problem 43).

The angle a vector $\mathbf{A}$ makes with itself is $\theta = 0$, and $\cos 0 = 1$. Therefore,

$$\mathbf{A} \cdot \mathbf{A} = |\mathbf{A}||\mathbf{A}|(1) = |\mathbf{A}|^2 \qquad \text{or} \qquad |\mathbf{A}| = \sqrt{\mathbf{A} \cdot \mathbf{A}} \qquad (5)$$

This gives a convenient way to calculate a vector's length, as we shall see when we work Example 1.

The vector we get by projecting $\mathbf{B}$ onto the line through $\mathbf{A}$ is called the *vector projection of* $\mathbf{B}$ *onto* $\mathbf{A}$. We denote it by $\text{proj}_{\mathbf{A}}\,\mathbf{B}$ (Fig. 13.33).

The *component of* $\mathbf{B}$ *in the direction of* $\mathbf{A}$ is a number that is plus or minus the length of the vector projection of $\mathbf{B}$ onto $\mathbf{A}$. The sign is plus if $\text{proj}_{\mathbf{A}}\,\mathbf{B}$ has the same direction as $+\mathbf{A}$, and is minus if it has the same

13.33 Vector projections of **B** onto **A**. In (a), the component of **B** in the direction of **A** is the length of the vector projection. In (b), it is *minus* the length of the vector projection.

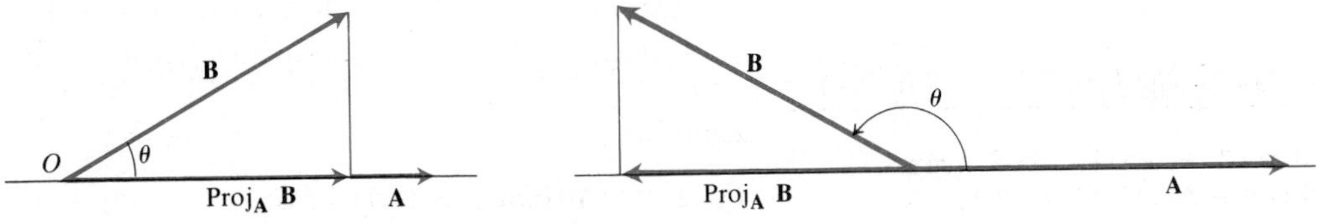

direction as $-\mathbf{A}$. In either case, the component of $\mathbf{B}$ in the direction of $\mathbf{A}$ is equal to $|\mathbf{B}|\cos\theta$ (see Fig. 13.33 again).

The dot product gives a convenient way to calculate the component of $\mathbf{B}$ in the direction of $\mathbf{A}$. We solve Eq. (2) for $|\mathbf{B}|\cos\theta$ to get

$$\mathbf{B}\text{-component in }\mathbf{A}\text{-direction} = |\mathbf{B}|\cos\theta = \frac{\mathbf{A}\cdot\mathbf{B}}{|\mathbf{A}|} = \mathbf{B}\cdot\frac{\mathbf{A}}{|\mathbf{A}|}. \tag{6}$$

Thus the component of $\mathbf{B}$ in the direction of $\mathbf{A}$ is the dot product of $\mathbf{B}$ with the direction of $\mathbf{A}$.

Multiplying both sides of Eq. (6) by $|\mathbf{A}|$ leads to a geometric interpretation of $\mathbf{A}\cdot\mathbf{B}$:

$$\begin{aligned}\mathbf{A}\cdot\mathbf{B} &= |\mathbf{A}|(|\mathbf{B}|\cos\theta)\\ &= (\text{length of }\mathbf{A})\text{ times }(\mathbf{B}\text{-component in }\mathbf{A}\text{-direction}).\end{aligned} \tag{7}$$

Of course we may interchange the roles of $|\mathbf{A}|$ and $|\mathbf{B}|$ and write the dot product in the alternative form of

$$\begin{aligned}\mathbf{A}\cdot\mathbf{B} &= |\mathbf{B}|(|\mathbf{A}|\cos\theta)\\ &= (\text{length of }\mathbf{B})\text{ times }(\mathbf{A}\text{-component in }\mathbf{B}\text{-direction}).\end{aligned} \tag{8}$$

Calculation

To calculate $\mathbf{A}\cdot\mathbf{B}$ from the components of $\mathbf{A}$ and $\mathbf{B}$, we let

$$\mathbf{A} = a_1\mathbf{i} + a_2\mathbf{j} + a_3\mathbf{k}, \qquad \mathbf{B} = b_1\mathbf{i} + b_2\mathbf{j} + b_3\mathbf{k},$$

and

$$\mathbf{C} = \mathbf{B} - \mathbf{A} = (b_1 - a_1)\mathbf{i} + (b_2 - a_2)\mathbf{j} + (b_3 - a_3)\mathbf{k}.$$

Then we apply the law of cosines to a triangle whose sides represent the vectors $\mathbf{A}$, $\mathbf{B}$, and $\mathbf{C}$ (Fig. 13.34) and obtain

$$\begin{aligned}|\mathbf{C}|^2 &= |\mathbf{A}|^2 + |\mathbf{B}|^2 - 2|\mathbf{A}||\mathbf{B}|\cos\theta,\\ |\mathbf{A}||\mathbf{B}|\cos\theta &= \frac{|\mathbf{A}|^2 + |\mathbf{B}|^2 - |\mathbf{C}|^2}{2}.\end{aligned} \tag{9}$$

The left side of this equation is $\mathbf{A}\cdot\mathbf{B}$, and we may calculate all terms on the right side of (9) by applying Eq. (3) of Article 13.5 to find the lengths of $\mathbf{A}$, $\mathbf{B}$, and $\mathbf{C}$. The result of this algebra is the formula

$$\boxed{\mathbf{A}\cdot\mathbf{B} = a_1b_1 + a_2b_2 + a_3b_3.} \tag{10}$$

Thus, to find the scalar product of two given vectors we multiply their *corresponding* $\mathbf{i}$, $\mathbf{j}$, and $\mathbf{k}$ components together and add the results. In particular, from Eq. (5) we have

$$|\mathbf{A}| = \sqrt{\mathbf{A}\cdot\mathbf{A}} = \sqrt{a_1^2 + a_2^2 + a_3^2}. \tag{11}$$

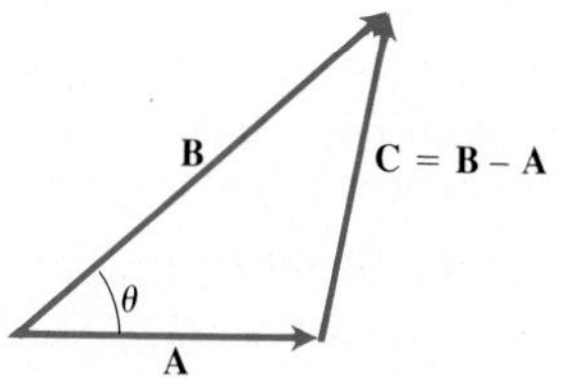

13.34 Equation (10) is obtained by applying the law of cosines to a triangle whose sides represent **A**, **B**, and **C** = **B** − **A**.

EXAMPLE 1 Find the angle θ between $\mathbf{A} = \mathbf{i} - 2\mathbf{j} - 2\mathbf{k}$ and $\mathbf{B} = 6\mathbf{i} + 3\mathbf{j} + 2\mathbf{k}$. Also, find the component of $\mathbf{B}$ in the direction of $\mathbf{A}$.

Solution

$$\mathbf{A}\cdot\mathbf{B} = (1)(6) + (-2)(3) + (-2)(2) = 6 - 6 - 4 = -4$$

from Eq. (10), while

$$\mathbf{A} \cdot \mathbf{B} = |\mathbf{A}||\mathbf{B}| \cos \theta$$

from Eq. (2). Since $|\mathbf{A}| = \sqrt{1 + 4 + 4} = 3$ and $|\mathbf{B}| = \sqrt{36 + 9 + 4} = 7$, we have

$$\cos \theta = \frac{\mathbf{A} \cdot \mathbf{B}}{|\mathbf{A}||\mathbf{B}|} = \frac{-4}{21}, \qquad \text{or} \qquad \theta = \cos^{-1} \frac{-4}{21} \approx 101°.$$

The component of **B** in the direction of **A** is

$$\mathbf{B} \cdot \frac{\mathbf{A}}{|\mathbf{A}|} = -\frac{4}{3}.$$

This is the negative of the length of the vector projection of **B** onto **A.** □

From Eq. (10), it is readily seen that if

$$\mathbf{C} = c_1\mathbf{i} + c_2\mathbf{j} + c_3\mathbf{k}$$

is any third vector, then

$$\begin{aligned} \mathbf{A} \cdot (\mathbf{B} + \mathbf{C}) &= a_1(b_1 + c_1) + a_2(b_2 + c_2) + a_3(b_3 + c_3) \\ &= (a_1b_1 + a_2b_2 + a_3b_3) + (a_1c_1 + a_2c_2 + a_3c_3) \\ &= \mathbf{A} \cdot \mathbf{B} + \mathbf{A} \cdot \mathbf{C}. \end{aligned}$$

Hence scalar multiplication obeys the *distributive* law:

$$\mathbf{A} \cdot (\mathbf{B} + \mathbf{C}) = \mathbf{A} \cdot \mathbf{B} + \mathbf{A} \cdot \mathbf{C}. \tag{12}$$

If we combine this with the commutative law, Eq. (3), it is also evident that

$$(\mathbf{A} + \mathbf{B}) \cdot \mathbf{C} = \mathbf{A} \cdot \mathbf{C} + \mathbf{B} \cdot \mathbf{C}. \tag{13}$$

Equations (12) and (13) together permit us to multiply sums of vectors by the familiar laws of algebra. For example,

$$(\mathbf{A} + \mathbf{B}) \cdot (\mathbf{C} + \mathbf{D}) = \mathbf{A} \cdot \mathbf{C} + \mathbf{A} \cdot \mathbf{D} + \mathbf{B} \cdot \mathbf{C} + \mathbf{B} \cdot \mathbf{D}. \tag{14}$$

Orthogonal Vectors

It is clear from Eq. (2) that the dot product of two vectors is zero when the vectors are perpendicular, since $\cos 90° = 0$. Conversely, if $\mathbf{A} \cdot \mathbf{B} = 0$ then one of the vectors is zero or else the vectors are perpendicular. The zero vector has no definite direction, and we can adopt the convention that it is perpendicular to any vector. Then we can say that $\mathbf{A} \cdot \mathbf{B} = 0$ if and only if the vectors **A** and **B** are perpendicular. Perpendicular vectors are also said to be *orthogonal.*

The zero vector $\mathbf{0} = 0\mathbf{i} + 0\mathbf{j} + 0\mathbf{k}$ is orthogonal to *every* vector, because its dot product with every vector is zero.

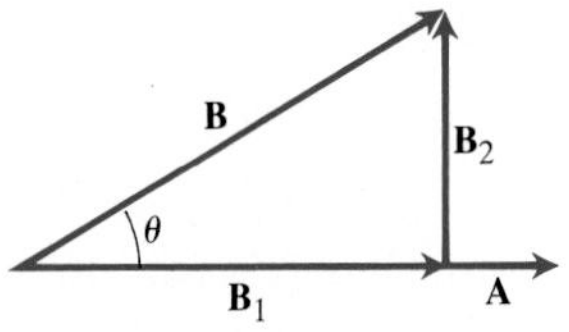

13.35 The vector **B** as the sum of vectors parallel and perpendicular to **A**.

EXAMPLE 2 Let **A** and **B** be nonzero vectors. Express the vector **B** as the sum of a vector $\mathbf{B}_1$ parallel to **A** and a vector $\mathbf{B}_2$ perpendicular to **A** (Fig. 13.35).

Solution We wish to write

$$\mathbf{B} = \mathbf{B}_1 + \mathbf{B}_2 \quad \text{with} \quad \mathbf{B}_1 = c\mathbf{A} \quad \text{and} \quad \mathbf{B}_2 \cdot \mathbf{A} = 0. \tag{15}$$

To determine an appropriate value for the number c, we have

$$\mathbf{B}_2 = \mathbf{B} - \mathbf{B}_1 = \mathbf{B} - c\mathbf{A} \tag{16}$$

and

$$\begin{aligned} 0 &= \mathbf{B}_2 \cdot \mathbf{A} = (\mathbf{B} - c\mathbf{A}) \cdot \mathbf{A} = \mathbf{B} \cdot \mathbf{A} - c(\mathbf{A} \cdot \mathbf{A}) \\ c(\mathbf{A} \cdot \mathbf{A}) &= \mathbf{B} \cdot \mathbf{A} \\ c &= \frac{\mathbf{B} \cdot \mathbf{A}}{\mathbf{A} \cdot \mathbf{A}}. \end{aligned} \tag{17}$$

From (17) and (15) we now get the equations for calculating $\mathbf{B}_1$ and $\mathbf{B}_2$:

$$\mathbf{B}_1 = \frac{\mathbf{B} \cdot \mathbf{A}}{\mathbf{A} \cdot \mathbf{A}}\mathbf{A}, \qquad \mathbf{B}_2 = \mathbf{B} - \mathbf{B}_1. \tag{18}$$

For example, if

$$\mathbf{B} = 2\mathbf{i} + \mathbf{j} - 3\mathbf{k} \qquad \text{and} \qquad \mathbf{A} = 3\mathbf{i} - \mathbf{j},$$

then

$$\frac{\mathbf{B} \cdot \mathbf{A}}{\mathbf{A} \cdot \mathbf{A}} = \frac{6 - 1}{9 + 1} = \frac{1}{2},$$

and

$$\mathbf{B}_1 = \frac{1}{2}\mathbf{A} = \frac{3}{2}\mathbf{i} - \frac{1}{2}\mathbf{j}$$

is parallel to **A**, while

$$\mathbf{B}_2 = \mathbf{B} - \mathbf{B}_1 = \frac{1}{2}\mathbf{i} + \frac{3}{2}\mathbf{j} - 3\mathbf{k}$$

is perpendicular to **A**. □

As Fig. 13.35 suggests, the vector $\mathbf{B}_1$ in Eq. (18) is the vector projection of **B** onto **A** (Problem 44).

EXAMPLE 3 Show that the vector $\mathbf{N} = a\mathbf{i} + b\mathbf{j}$ is perpendicular to the line $ax + by = c$ in the xy-plane (Fig. 13.36).

Solution Let $P_1(x_1, y_1)$ and $P_2(x_2, y_2)$ be any two points on the line; that is,

$$ax_1 + by_1 = c, \qquad ax_2 + by_2 = c.$$

By subtraction, we eliminate c and obtain

$$a(x_2 - x_1) + b(y_2 - y_1) = 0,$$

or

$$(a\mathbf{i} + b\mathbf{j}) \cdot [(x_2 - x_1)\mathbf{i} + (y_2 - y_1)\mathbf{j}] = 0. \tag{19}$$

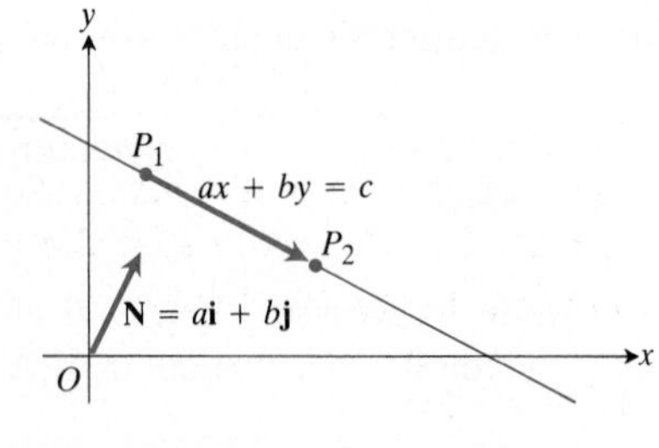

13.36 The vector $\mathbf{N} = a\mathbf{i} + b\mathbf{j}$ is normal to the line $ax + by = c$. Example 3 explains why.

Now $(x_2 - x_1)\mathbf{i} + (y_2 - y_1)\mathbf{j} = \overrightarrow{P_1P_2}$ is a vector joining two points on the line, while $\mathbf{N} = a\mathbf{i} + b\mathbf{j}$ is the given vector. Equation (19) says that either $\mathbf{N} = \mathbf{0}$, or $\overrightarrow{P_1P_2} = \mathbf{0}$, or else $\mathbf{N} \perp \overrightarrow{P_1P_2}$. But $ax + by = c$ is assumed to be an honest equation of a straight line, so that a and b are not both zero and

$\mathbf{N} \neq \mathbf{0}$. Furthermore, we may surely choose P_2 different from P_1 on the line to make $\overrightarrow{P_1P_2} \neq \mathbf{0}$. Hence $\mathbf{N} \perp \overrightarrow{P_1P_2}$.

For example, $\mathbf{N} = 2\mathbf{i} - 3\mathbf{j}$ is normal to the line $2x - 3y = 5$. □

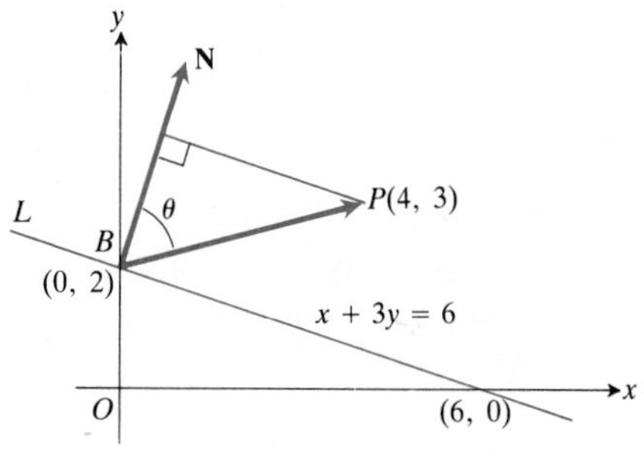

13.37 The distance from P to line L is the length of the vector projection of $\overrightarrow{BP}$ onto **N**.

EXAMPLE 4 Using vector methods, find the distance d between the point $P(4, 3)$ and the line L: $x + 3y = 6$ (Fig. 13.37).

Solution The line cuts the y-axis at $B(0, 2)$. At B, draw the vector

$$\mathbf{N} = \mathbf{i} + 3\mathbf{j}$$

normal to L (see Example 3). Then the distance between P and L is the component of $\overrightarrow{BP}$ in the direction of **N**. Since

$$\overrightarrow{BP} = (4 - 0)\mathbf{i} + (3 - 2)\mathbf{j} = 4\mathbf{i} + \mathbf{j},$$

we have

$$d = |\text{proj}_{\mathbf{N}}\, \overrightarrow{BP}| = \left|\overrightarrow{BP} \cdot \frac{\mathbf{N}}{|\mathbf{N}|}\right| = \frac{4 + 3}{\sqrt{10}} = \frac{7\sqrt{10}}{10}. \ \square$$

Unit Vectors and Direction Cosines

Suppose that **u** is a unit vector that makes angles α, β, and γ with the coordinate axes when we represent it by an arrow with its initial point at the origin. Figure 13.38 shows the case in which the three angles are acute. Then

$$\begin{aligned} \mathbf{u} \cdot \mathbf{i} &= |\mathbf{u}||\mathbf{i}| \cos \alpha = \cos \alpha, \\ \mathbf{u} \cdot \mathbf{j} &= |\mathbf{u}||\mathbf{j}| \cos \beta = \cos \beta, \\ \mathbf{u} \cdot \mathbf{k} &= |\mathbf{u}||\mathbf{k}| \cos \gamma = \cos \gamma. \end{aligned} \tag{20}$$

If we write

$$\mathbf{u} = u_1\mathbf{i} + u_2\mathbf{j} + u_3\mathbf{k},$$

then we also see that

$$\mathbf{u} \cdot \mathbf{i} = u_1, \qquad \mathbf{u} \cdot \mathbf{j} = u_2, \qquad \mathbf{u} \cdot \mathbf{k} = u_3. \tag{21}$$

Thus,

$$u_1 = \cos \alpha, \qquad u_2 = \cos \beta, \qquad u_3 = \cos \gamma. \tag{22}$$

That is, the components of any unit vector **u** are the cosines of the angles it makes with the coordinate axes:

$$\mathbf{u} = u_1\mathbf{i} + u_2\mathbf{j} + u_3\mathbf{k} = \mathbf{i} \cos \alpha + \mathbf{j} \cos \beta + \mathbf{k} \cos \gamma. \tag{23}$$

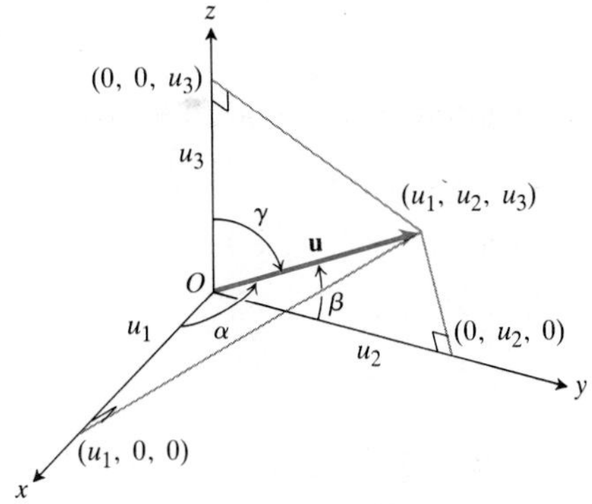

13.38 If $\mathbf{u} = u_1\mathbf{i} + u_2\mathbf{j} + u_3\mathbf{k}$ is a unit vector, then $u_1 = \cos \alpha$, $u_2 = \cos \beta$, $u_3 = \cos \gamma$ from the three right triangles shown here.

If **A** is any nonzero vector, then **A** and its direction $\mathbf{u} = \mathbf{A}/|\mathbf{A}|$ make the same angles with the coordinate axes. To find the cosines of these angles we need only calculate the components of the unit vector

$$\frac{\mathbf{A}}{|\mathbf{A}|} = \mathbf{i} \cos \alpha + \mathbf{j} \cos \beta + \mathbf{k} \cos \gamma.$$

These cosines are called the *direction cosines* of **A**. That is, the direction cosines of **A** are the components of the direction of **A**.

PROBLEMS

In Problems 1–10, calculate $\mathbf{A}\cdot\mathbf{B}$, $|\mathbf{A}|$, and $|\mathbf{B}|$, and find the cosine of the angle between **A** and **B**.

1. $\mathbf{A} = 3\mathbf{i} + 2\mathbf{j}$, $\mathbf{B} = 5\mathbf{j} + \mathbf{k}$

2. $\mathbf{A} = \mathbf{i}$, $\mathbf{B} = 5\mathbf{j} - 3\mathbf{k}$

3. $\mathbf{A} = 3\mathbf{i} - 2\mathbf{j} - \mathbf{k}$, $\mathbf{B} = -2\mathbf{j}$

4. $\mathbf{A} = -2\mathbf{i} + 7\mathbf{j}$, $\mathbf{B} = \mathbf{k}$

5. $\mathbf{A} = 5\mathbf{j} - 3\mathbf{k}$, $\mathbf{B} = \mathbf{i} + \mathbf{j} + \mathbf{k}$

6. $\mathbf{A} = \frac{1}{\sqrt{2}}\mathbf{i} + \frac{1}{\sqrt{3}}\mathbf{j} + \frac{1}{\sqrt{6}}\mathbf{k}$, $\mathbf{B} = \frac{1}{\sqrt{2}}\mathbf{j} - \mathbf{k}$

7. $\mathbf{A} = -\mathbf{i} + \mathbf{j}$, $\mathbf{B} = \sqrt{2}\mathbf{i} + \sqrt{3}\mathbf{j} + 2\mathbf{k}$

8. $\mathbf{A} = \mathbf{i} + \mathbf{k}$, $\mathbf{B} = \mathbf{i} + \mathbf{j} + \mathbf{k}$

9. $\mathbf{A} = 2\mathbf{i} - 4\mathbf{j} + \sqrt{5}\mathbf{k}$, $\mathbf{B} = -2\mathbf{i} + 4\mathbf{j} - \sqrt{5}\mathbf{k}$

10. $\mathbf{A} = -5\mathbf{i} + \mathbf{j}$, $\mathbf{B} = 2\mathbf{i} + \sqrt{17}\mathbf{j} + 10\mathbf{k}$

11. Find the interior angles of the triangle ABC whose vertices are $A(-1, 0, 2)$, $B(2, 1, -1)$, and $C(1, -2, 2)$.

12. Find the angle between $\mathbf{A} = 2\mathbf{i} + 2\mathbf{j} + \mathbf{k}$ and $\mathbf{B} = 2\mathbf{i} + 10\mathbf{j} - 11\mathbf{k}$.

13. Find the angle between the diagonal of a cube and the diagonal of one of its faces. (*Hint:* Use a cube three of whose edges represent the vectors **i**, **j**, and **k**.)

14. Find the angle between the diagonal of a cube and one of the edges it meets at a vertex.

15. Find (a) the vector projection of $\mathbf{B} = \mathbf{i} + 3\mathbf{j} + 4\mathbf{k}$ onto the vector $\mathbf{A} = 10\mathbf{i} + 11\mathbf{j} - 2\mathbf{k}$ and (b) the component of **B** in the direction of **A**.

16. Find the component of $\mathbf{A} = 2\mathbf{i} + 2\mathbf{j} + \mathbf{k}$ in the direction of $\mathbf{B} = 2\mathbf{i} + 10\mathbf{j} - 11\mathbf{k}$.

17. Find the point $A(a, a, 0)$ at the foot of the perpendicular from the point $B(2, 4, -3)$ to the line $y = x$ in the xy-plane.

18. How many lines through the origin make angles of 60° with both the y- and z-axes? What angles do these lines make with the positive x-axis?

19. Write the vector $\mathbf{B} = 3\mathbf{j} + 4\mathbf{k}$ as the sum of a vector $\mathbf{B}_1$ parallel to $\mathbf{A} = \mathbf{i} + \mathbf{j}$ and a vector $\mathbf{B}_2$ perpendicular to **A**.

20. Repeat Problem 19 for $\mathbf{B} = \mathbf{i} + \mathbf{j} + \mathbf{k}$.

21. Find the distance in the xy-plane between the line $x + 3y = 6$ and
a) the point (2, 8), b) the origin.

22. Show that each of the vectors

$$\mathbf{A} = \frac{1}{\sqrt{3}}(\mathbf{i} - \mathbf{j} + \mathbf{k}),$$

$$\mathbf{B} = \frac{1}{\sqrt{2}}(\mathbf{j} + \mathbf{k}),$$

$$\mathbf{C} = \frac{1}{\sqrt{6}}(-2\mathbf{i} - \mathbf{j} + \mathbf{k})$$

is orthogonal to the other two.

23. Find the vector projections of $\mathbf{D} = \mathbf{i} + \mathbf{j} + \mathbf{k}$ on the vectors (a) **A**, (b) **B**, and (c) **C** of Problem 22. Then (d) show that **D** is the sum of these vector projections.

In Problems 24–32, find the direction cosines of the vector **A** in the following problems.

24. Problem 1	**25.** Problem 2	**26.** Problem 3
27. Problem 4	**28.** Problem 5	**29.** Problem 6
30. Problem 7	**31.** Problem 8	**32.** Problem 9

33. Find the work done by a force $\mathbf{F} = -5\mathbf{k}$ (magnitude 5 newtons) as its point of application moves from the origin to the point (1, 1, 1).

34. A locomotive exerted a constant force of 60,000 newtons on a freight train while drawing it for 1 km along a straight track. How much work did the locomotive do?

35. How much work is performed in sliding a crate 20 m along a loading dock by pushing on it with a constant force of 200 newtons at an angle of 30° from the horizontal?

36. Suppose it is known that $\mathbf{A}\cdot\mathbf{B}_1 = \mathbf{A}\cdot\mathbf{B}_2$, and **A** is not zero, but nothing more is known about the vectors $\mathbf{B}_1$ and $\mathbf{B}_2$. Is it permissible to cancel **A** from both sides of the equation? Give a reason for your answer.

37. In Fig. 13.2 it looks as if $\mathbf{v}_1 + \mathbf{v}_2$ and $\mathbf{v}_1 - \mathbf{v}_2$ are orthogonal. Is this mere coincidence, or are there circumstances under which we may expect the sum of two vectors to be perpendicular to the difference of the same two vectors? Find out by expanding the left-hand side of the equation

$$(\mathbf{v}_1 + \mathbf{v}_2)\cdot(\mathbf{v}_1 - \mathbf{v}_2) = 0.$$

38. Show that scalar multiplication is *positive definite;* that is, $\mathbf{A}\cdot\mathbf{A} \geq 0$ for every vector **A**, and $\mathbf{A}\cdot\mathbf{A} = 0$ if and only if **A** is the zero vector.

39. If **R** is the vector from the origin O to $P(x, y, z)$ and **k** is the unit vector along the z-axis, show geometrically that the equation

$$\frac{\mathbf{R}\cdot\mathbf{k}}{|\mathbf{R}|} = \cos 45°$$

represents a cone with vertex at the origin and generating angle of 45°. Express the equation in cartesian form.

40. If $a = |\mathbf{A}|$ and $b = |\mathbf{B}|$, show that the vector

$$\mathbf{C} = a\mathbf{B} + b\mathbf{A}$$

bisects the angle between **A** and **B**.

41. With the same notation as in Problem 40, show that the vectors $a\mathbf{B} + b\mathbf{A}$ and $\mathbf{A}b - \mathbf{B}a$ are perpendicular.

42. Using vector methods, show that the distance d between the point (x_1, y_1) and the line $ax + by = c$ is

$$d = \frac{|ax_1 + by_1 - c|}{\sqrt{a^2 + b^2}}.$$

43. Show that if c is any number, and $\mathbf{A}$ and $\mathbf{B}$ any vectors, then

$$(c\mathbf{A}) \cdot \mathbf{B} = c(\mathbf{A} \cdot \mathbf{B}) = \mathbf{A} \cdot (c\mathbf{B}).$$

44. Show that the vector $\mathbf{B}_1$ in Example 2 is the vector projection of $\mathbf{B}$ onto $\mathbf{A}$.

45. Let α, β, and γ be the angles that a line through the origin makes with the positive x-, y-, and z-axes. Show that $\cos^2\alpha + \cos^2\beta + \cos^2\gamma = 1$.

13.7 The Vector Product of Two Vectors in Space

Cross Products

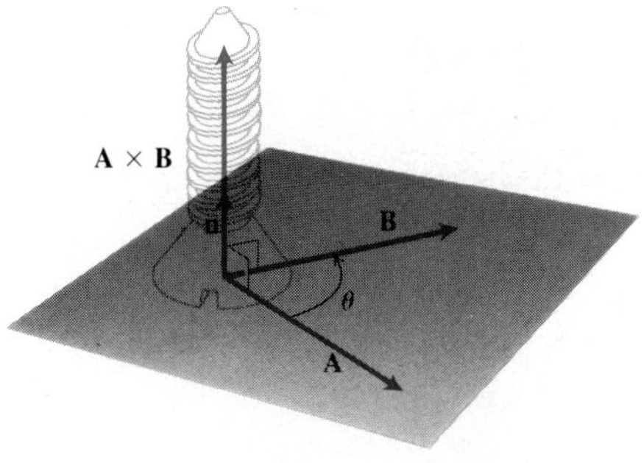

13.39 $\mathbf{A} \times \mathbf{B}$ and $\mathbf{B} \times \mathbf{A}$ have the same magnitude but point in opposite directions from the plane of $\mathbf{A}$ and $\mathbf{B}$.

Two nonzero vectors $\mathbf{A}$ and $\mathbf{B}$ in space may be moved without changing their magnitude or direction so that their initial points coincide. Suppose that this has been done and let θ be the angle from $\mathbf{A}$ to $\mathbf{B}$, with $0 \le \theta \le \pi$. Unless $\mathbf{A}$ and $\mathbf{B}$ are parallel, they now determine a plane. Let $\mathbf{n}$ be a unit vector perpendicular to this plane and pointing in the direction a right-threaded screw advances when its head is rotated from $\mathbf{A}$ to $\mathbf{B}$ through the angle θ. The *vector product*, or *cross product*, of $\mathbf{A}$ and $\mathbf{B}$, in that order, is then defined by the equation (Fig. 13.39)

$$\mathbf{A} \times \mathbf{B} = \mathbf{n}|\mathbf{A}||\mathbf{B}| \sin\theta. \tag{1}$$

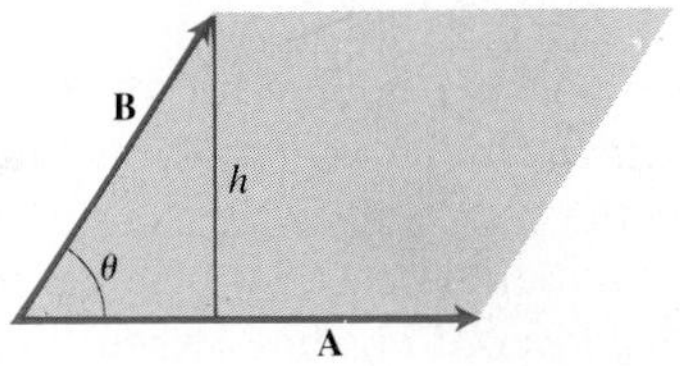

13.40 The area of the parallelogram is $|\mathbf{A} \times \mathbf{B}|$.

Like the definition of the scalar product of two vectors, given in Article 13.6, the definition of the vector product given here is coordinate-free. We emphasize, however, that the vector product $\mathbf{A} \times \mathbf{B}$ is a *vector*, while the scalar product $\mathbf{A} \cdot \mathbf{B}$ is a *scalar*. Cross products play an important role in the study of electricity and magnetism.

If $\mathbf{A}$ and $\mathbf{B}$ are parallel, then $\theta = 0$ or $180°$ and $\sin\theta = 0$, so that $\mathbf{A} \times \mathbf{B} = \mathbf{0}$. In this case, the direction of $\mathbf{n}$ is not determined, but this is immaterial, since the zero vector has no specific direction. In all other cases, however, $\mathbf{n}$ is determined and the cross product is a vector having the same direction as $\mathbf{n}$ and having magnitude equal to the area, $|\mathbf{A}||\mathbf{B}| \sin\theta$, of the parallelogram determined by the vectors $\mathbf{A}$ and $\mathbf{B}$ (Fig. 13.40).

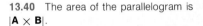

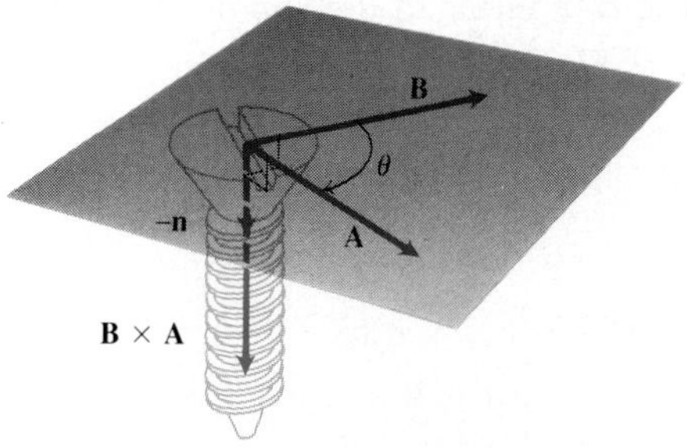

13.41 $\mathbf{A} \times \mathbf{B}$ and $\mathbf{B} \times \mathbf{A}$ have the same magnitude but point in opposite directions from the plane of $\mathbf{A}$ and $\mathbf{B}$.

If the order of the factors $\mathbf{A}$ and $\mathbf{B}$ is reversed in the construction of the cross product, the direction of the unit vector perpendicular to their plane is reversed (Fig. 13.41). The right-handed screw that turns through θ from $\mathbf{B}$ to $\mathbf{A}$ points the other way. The original unit vector $\mathbf{n}$ is now replaced by $-\mathbf{n}$, with the result that

$$\mathbf{B} \times \mathbf{A} = -\mathbf{A} \times \mathbf{B}. \tag{2}$$

Thus, cross-product multiplication is not commutative. Reversing the order of the factors changes the sign of the product.

When the definition of the cross product is applied to the unit vectors, **i**, **j**, and **k**, one readily finds that

$$\begin{aligned} \mathbf{i} \times \mathbf{j} &= -\mathbf{j} \times \mathbf{i} = \mathbf{k}, \\ \mathbf{j} \times \mathbf{k} &= -\mathbf{k} \times \mathbf{j} = \mathbf{i}, \\ \mathbf{k} \times \mathbf{i} &= -\mathbf{i} \times \mathbf{k} = \mathbf{j}, \end{aligned} \tag{3}$$

while

$$\mathbf{i} \times \mathbf{i} = \mathbf{j} \times \mathbf{j} = \mathbf{k} \times \mathbf{k} = \mathbf{0}.$$

The Associative and Distributive Laws

The associative law

$$(r\mathbf{A}) \times (s\mathbf{B}) = (rs)\mathbf{A} \times \mathbf{B}, \tag{4}$$

follows from the geometric meaning of the cross product, but the distributive law

$$\mathbf{A} \times (\mathbf{B} + \mathbf{C}) = \mathbf{A} \times \mathbf{B} + \mathbf{A} \times \mathbf{C} \tag{5}$$

is not so easy to establish. We shall assume it here and leave its proof to Appendix 10.

The companion law

$$(\mathbf{B} + \mathbf{C}) \times \mathbf{A} = \mathbf{B} \times \mathbf{A} + \mathbf{C} \times \mathbf{A} \tag{6}$$

now follows at once from Eq. (5) if we multiply both sides of Eq. (5) by minus one and take account of the fact that interchanging the two factors in a cross product changes the sign of the result.

From Eqs. (4), (5), and (6), we may conclude that cross-product multiplication of vectors follows the ordinary laws of algebra, *except that the order of the factors is not reversible.*

The Determinant Formula

Our next objective is to express $\mathbf{A} \times \mathbf{B}$ in terms of the components of **A** and **B**. If we apply the results we have obtained so far to calculate $\mathbf{A} \times \mathbf{B}$ with

$$\mathbf{A} = a_1\mathbf{i} + a_2\mathbf{j} + a_3\mathbf{k}, \qquad \mathbf{B} = b_1\mathbf{i} + b_2\mathbf{j} + b_3\mathbf{k},$$

we obtain

$$\begin{aligned} \mathbf{A} \times \mathbf{B} &= (a_1\mathbf{i} + a_2\mathbf{j} + a_3\mathbf{k}) \times (b_1\mathbf{i} + b_2\mathbf{j} + b_3\mathbf{k}) \\ &= a_1b_1\mathbf{i} \times \mathbf{i} + a_1b_2\mathbf{i} \times \mathbf{j} + a_1b_3\mathbf{i} \times \mathbf{k} + a_2b_1\mathbf{j} \times \mathbf{i} \\ &\quad + a_2b_2\mathbf{j} \times \mathbf{j} + a_2b_3\mathbf{j} \times \mathbf{k} + a_3b_1\mathbf{k} \times \mathbf{i} \\ &\quad + a_3b_2\mathbf{k} \times \mathbf{j} + a_3b_3\mathbf{k} \times \mathbf{k} \\ &= \mathbf{i}(a_2b_3 - a_3b_2) + \mathbf{j}(a_3b_1 - a_1b_3) + \mathbf{k}(a_1b_2 - a_2b_1), \end{aligned} \tag{7}$$

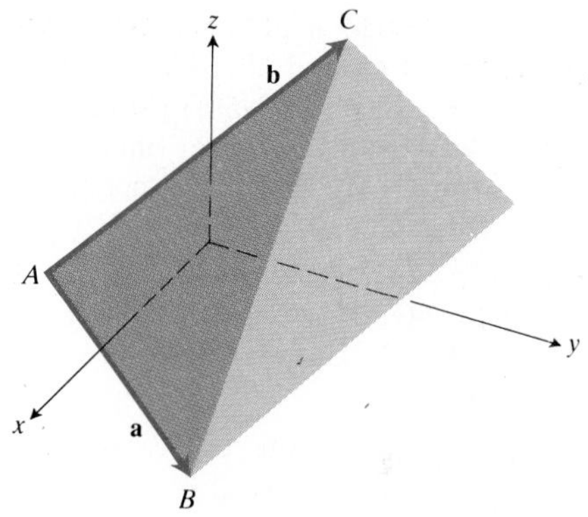

13.42 The area of ΔABC is half of $|\mathbf{a} \times \mathbf{b}|$.

where Eqs. (3) have been used to evaluate the products $\mathbf{i} \times \mathbf{i} = \mathbf{0}$, $\mathbf{i} \times \mathbf{j} = \mathbf{k}$, etc. The terms on the right-hand side of Eq. (7) are the same as the terms in the expansion of the third-order determinant below, so that the cross product may conveniently be calculated from the equation

$$\mathbf{A} \times \mathbf{B} = \begin{vmatrix} \mathbf{i} & \mathbf{j} & \mathbf{k} \\ a_1 & a_2 & a_3 \\ b_1 & b_2 & b_3 \end{vmatrix}. \tag{8}$$

(Determinants are reviewed in Appendix 1.)

EXAMPLE 1 Find the area of the triangle whose vertices are $A(1, -1, 0)$, $B(2, 1, -1)$, and $C(-1, 1, 2)$ (Fig. 13.42).

Solution Two sides of the given triangle are represented by the vectors

$$\mathbf{a} = \overrightarrow{AB} = (2 - 1)\mathbf{i} + (1 + 1)\mathbf{j} + (-1 - 0)\mathbf{k} = \mathbf{i} + 2\mathbf{j} - \mathbf{k},$$
$$\mathbf{b} = \overrightarrow{AC} = (-1 - 1)\mathbf{i} + (1 + 1)\mathbf{j} + (2 - 0)\mathbf{k} = -2\mathbf{i} + 2\mathbf{j} + 2\mathbf{k}.$$

The area of the triangle is one half the area of the parallelogram represented by these vectors. The area of the parallelogram is the magnitude of the vector

$$\mathbf{c} = \mathbf{a} \times \mathbf{b} = \begin{vmatrix} \mathbf{i} & \mathbf{j} & \mathbf{k} \\ 1 & 2 & -1 \\ -2 & 2 & 2 \end{vmatrix}$$

$$= \mathbf{i}\begin{vmatrix} 2 & -1 \\ 2 & 2 \end{vmatrix} - \mathbf{j}\begin{vmatrix} 1 & -1 \\ -2 & 2 \end{vmatrix} + \mathbf{k}\begin{vmatrix} 1 & 2 \\ -2 & 2 \end{vmatrix} = 6\mathbf{i} + 6\mathbf{k},$$

which is $|\mathbf{c}| = \sqrt{6^2 + 6^2} = 6\sqrt{2}$. Therefore, the area of the triangle is $\frac{1}{2}|\mathbf{a} \times \mathbf{b}| = 3\sqrt{2}$. □

EXAMPLE 2 Find a unit vector perpendicular to both $\mathbf{A} = 2\mathbf{i} + \mathbf{j} - \mathbf{k}$ and $\mathbf{B} = \mathbf{i} - \mathbf{j} + 2\mathbf{k}$.

Solution The vector $\mathbf{N} = \mathbf{A} \times \mathbf{B}$ is perpendicular to both $\mathbf{A}$ and $\mathbf{B}$. We divide $\mathbf{N}$ by $|\mathbf{N}|$ to obtain a unit vector $\mathbf{u}$ that has the same direction as $\mathbf{N}$:

$$\mathbf{u} = \frac{\mathbf{N}}{|\mathbf{N}|} = \frac{\mathbf{A} \times \mathbf{B}}{|\mathbf{A} \times \mathbf{B}|} = \frac{\mathbf{i} - 5\mathbf{j} - 3\mathbf{k}}{\sqrt{1^2 + (-5)^2 + (-3)^2}} = \frac{\mathbf{i} - 5\mathbf{j} - 3\mathbf{k}}{\sqrt{35}}.$$

Either $\mathbf{u}$ or its negative will do. □

PROBLEMS

1. Find $\mathbf{A} \times \mathbf{B}$ if $\mathbf{A} = 2\mathbf{i} - 2\mathbf{j} - \mathbf{k}$, $\mathbf{B} = \mathbf{i} + \mathbf{j} + \mathbf{k}$.
2. Find the direction of $\mathbf{A} \times \mathbf{B}$ if
   a) $\mathbf{A} = \mathbf{i}$, $\mathbf{B} = \mathbf{j}$
   b) $\mathbf{A} = \mathbf{i} + \mathbf{k}$, $\mathbf{B} = \mathbf{j}$
   c) $\mathbf{A} = \mathbf{i} + \mathbf{k}$, $\mathbf{B} = \mathbf{j} + \mathbf{k}$
   d) $\mathbf{A} = \sqrt{5}\mathbf{i} - 2\mathbf{j} - 3\mathbf{k}$, $\mathbf{B} = 2\mathbf{i} + 2\mathbf{j} + \mathbf{k}$
3. Find a vector $\mathbf{N}$ perpendicular to the plane determined by the points $A(1, -1, 2)$, $B(2, 0, -1)$, and $C(0, 2, 1)$.
4. Find a vector that is perpendicular to both of the vectors $\mathbf{A} = \mathbf{i} + \mathbf{j} + \mathbf{k}$ and $\mathbf{B} = \mathbf{i} + \mathbf{j}$.
5. Find the distance between the origin and the plane

ABC of Problem 3 by projecting $\overrightarrow{OA}$ onto the normal vector $\mathbf{N}$.

6. Find the area of the triangle ABC in Problem 3.
7. If $\overrightarrow{AB} \times \overrightarrow{AC} = 2\mathbf{i} - 4\mathbf{j} + 4\mathbf{k}$, find the area of triangle ABC.
8. a) Find a vector normal to the plane through the three points

$$A(0, 2, 1), \quad B(2, 1, 2), \quad C(1, 1, 3).$$

   b) Find the area of the triangle formed by the three points in part (a).
9. If $\mathbf{A} = 2\mathbf{i} - \mathbf{j}$, $\mathbf{B} = \mathbf{i} + 3\mathbf{j} - 2\mathbf{k}$, find $\mathbf{A} \times \mathbf{B}$. Then calculate $(\mathbf{A} \times \mathbf{B}) \cdot \mathbf{A}$ and $(\mathbf{A} \times \mathbf{B}) \cdot \mathbf{B}$.
10. Is $(\mathbf{A} \times \mathbf{B}) \cdot \mathbf{A}$ always zero? Explain. What about $(\mathbf{A} \times \mathbf{B}) \cdot \mathbf{B}$?
11. Let $\mathbf{A} = 5\mathbf{i} - \mathbf{j} + \mathbf{k}$, $\mathbf{B} = \mathbf{j} - 5\mathbf{k}$, $\mathbf{C} = -15\mathbf{i} + 3\mathbf{j} - 3\mathbf{k}$. Which pairs of vectors are (a) perpendicular? (b) parallel?
12. The vector $\mathbf{A}$ is 4 units long and its direction is $\mathbf{i}$. The vector $\mathbf{B}$ is 6 units long, and its direction is $\mathbf{k}$.
    a) What is the direction of $\mathbf{A} \times \mathbf{B}$? of $\mathbf{B} \times \mathbf{A}$?
    b) What is the magnitude of $\mathbf{A} \times \mathbf{B}$? of $\mathbf{B} \times \mathbf{A}$?
13. $\mathbf{A} = 3\mathbf{i} + \mathbf{j} - \mathbf{k}$ is normal to a plane M_1 and $\mathbf{B} = 2\mathbf{i} - \mathbf{j} + \mathbf{k}$ is normal to a second plane M_2. (a) Find the angle between the two normals. (b) Do the two planes intersect? Give a reason for your answer. (c) If the two planes do intersect, find a vector parallel to their line of intersection.
14. Let $\mathbf{A}$ be a nonzero vector. Show that (a) $\mathbf{A} \times \mathbf{B} = \mathbf{A} \times \mathbf{C}$ does not guarantee $\mathbf{B} = \mathbf{C}$ (see Problem 36, Article 13.6); (b) $\mathbf{A} \cdot \mathbf{B} = \mathbf{A} \cdot \mathbf{C}$ and $\mathbf{A} \times \mathbf{B} = \mathbf{A} \times \mathbf{C}$ together imply $\mathbf{B} = \mathbf{C}$.
15. Vectors from the origin to the points A, B, C are given by $\mathbf{A} = \mathbf{i} - \mathbf{j} + \mathbf{k}$, $\mathbf{B} = 2\mathbf{i} + 3\mathbf{j} - \mathbf{k}$, $\mathbf{C} = -\mathbf{i} + 2\mathbf{j} + 2\mathbf{k}$. Find all points $P(x, y, z)$ that satisfy the following requirements: $\overrightarrow{OP}$ is a unit vector perpendicular to $\mathbf{C}$ and P lies in the plane determined by $\mathbf{A}$ and $\mathbf{B}$.
16. Find the distance between the line L_1 through the two points $A(1, 0, -1)$, $B(-1, 1, 0)$ and the line L_2 through the two points $C(3, 1, -1)$, $D(4, 5, -2)$. (The distance is to be measured along a line perpendicular to L_1 and L_2. First find a vector $\mathbf{N}$ perpendicular to both lines, and then project $\overrightarrow{AC}$ onto $\mathbf{N}$.)
17. Repeat Problem 16 for the points $A(4, 0, 2)$, $B(2, 4, 1)$, $C(1, 3, 2)$, $D(2, 2, 4)$.

13.8 Equations of Lines, Line Segments, and Planes

Lines

Suppose L is a line in space that passes through a point $P_0(x_0, y_0, z_0)$ and is parallel to a given nonzero vector

$$\mathbf{v} = A\mathbf{i} + B\mathbf{j} + C\mathbf{k}.$$

Then L is the set of all points $P(x, y, z)$ for which the vector $\overrightarrow{P_0P}$ is parallel to $\mathbf{v}$ (Fig. 13.43).

That is, P is on the line L if and only if there is a scalar t such that

$$\overrightarrow{P_0P} = t\mathbf{v} \tag{1a}$$

or

$$(x - x_0)\mathbf{i} + (y - y_0)\mathbf{j} + (z - z_0)\mathbf{k} = t(A\mathbf{i} + B\mathbf{j} + C\mathbf{k}). \tag{1b}$$

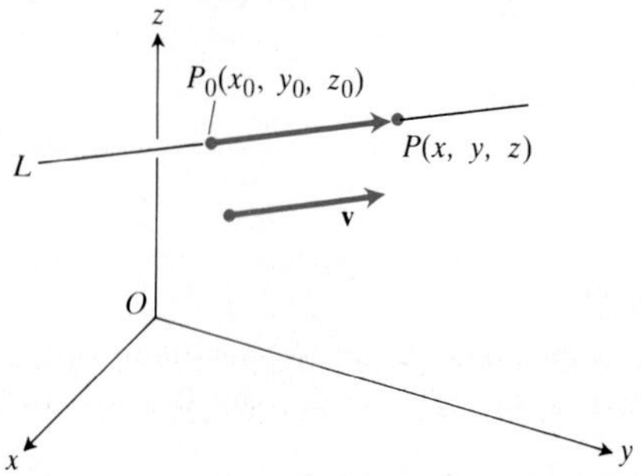

13.43 P is on the line through P_0 parallel to $\mathbf{v}$ if and only if $\overrightarrow{P_0P}$ is a scalar multiple of $\mathbf{v}$.

When we match the components in Eq. (1b) we have

$$x - x_0 = tA, \qquad y - y_0 = tB, \qquad z - z_0 = tC \tag{2a}$$

or

$$x = x_0 + tA, \qquad y = y_0 + tB, \qquad z = z_0 + tC. \tag{2b}$$

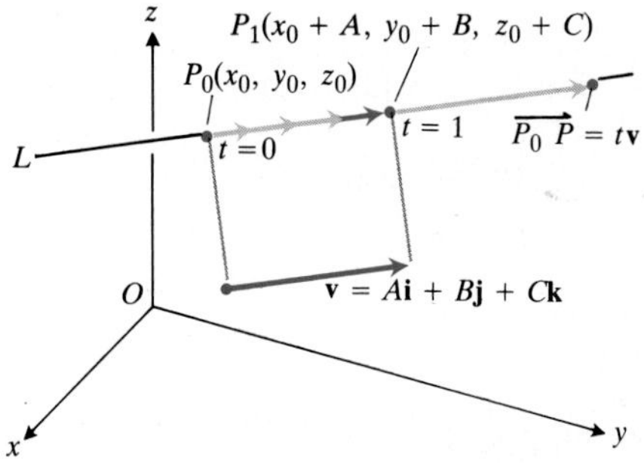

13.44 As t increases from 0 to 1, the point P moves from $P_0(x_0, y_0, z_0)$ to $P_1(x_0 + A, y_0 + B, z_0 + C)$.

The equations in (2a) and (2b) are *parametric equations* for the line, as opposed to Eqs. (1a) and (1b), which are *vector equations* for the line. If we let t vary between $-\infty$ and $+\infty$, the point $P(x, y, z)$ given by Eq. (2b) will traverse the entire line L through P_0.

If we let t vary through the closed interval $[a, b]$ from $t = a$ to $t = b$, the point $P(x, y, z)$ given by Eq. (2b) will trace out the segment from the point where $t = a$ to the point where $t = b$.

For example, when t varies from 0 to 1, the point

$$P(x, y, z) = P(x_0 + tA, x_0 + tB, x_0 + tC)$$

in Fig. 13.44 moves from $P_0(x_0, y_0, z_0)$ to $P_1(x_0 + A, y_0 + B, z_0 + C)$.

We may eliminate t from the equations in (2) to obtain the following *cartesian equations* for the line:

$$\frac{x - x_0}{A} = \frac{y - y_0}{B} = \frac{z - z_0}{C}, \tag{3}$$

provided A, B, and C are different from zero.

EXAMPLE 1 The line through $P(2, -9, 5)$ parallel to $\mathbf{v} = 3\mathbf{i} - \mathbf{j} + 4\mathbf{k}$ is

$$\frac{x - 2}{3} = \frac{y + 9}{-1} = \frac{z - 5}{4}.$$

Remember: To say that these are cartesian equations for the line is to say that (x, y, z) lies on the line if and only if these equations hold. □

If any one of the constants A, B, or C is zero in Eqs. (3), the corresponding numerator is also zero. This follows at once from the parametric equations (2), which show, for example, that

$$x - x_0 = tA \qquad \text{and} \qquad A = 0$$

together imply that

$$x - x_0 = 0.$$

Thus, when one of the denominators in (3) is zero, we interpret the equations to say that the corresponding numerator is zero. With this interpretation, Eqs. (3) may always be used.

EXAMPLE 2 Find parametric and cartesian equations for the line through $P(2, -9, 5)$ parallel to $\mathbf{v} = 2\mathbf{j} + 3\mathbf{k}$.

Solution The parametric equations given by Eq. (2b) are

$$x = 2 + 0 \cdot t, \qquad y = -9 + 2t, \qquad z = 5 + 3t$$

or

$$x = 2, \qquad y = -9 + 2t, \qquad z = 5 + 3t. \tag{4}$$

As t increases from $t = 0$ to $t = 1$, the point (x, y, z) in (4) traces the segment from

$$t = 0: \qquad (2, -9 + 2 \cdot 0, 5 + 3 \cdot 0) = (2, -9, 5)$$

to

$$t = 1: \qquad (2, -9 + 2 \cdot 1, 5 + 3 \cdot 1) = (2, -7, 8).$$

The cartesian equations found by eliminating t in Eq. (4) are

$$x = 2, \qquad \frac{y + 9}{2} = \frac{z - 5}{3}. \ \square \tag{5}$$

Planes

To obtain an equation for a *plane,* we suppose that a point $P_0(x_0, y_0, z_0)$ on the plane and a nonzero vector

$$\mathbf{N} = A\mathbf{i} + B\mathbf{j} + C\mathbf{k} \tag{6}$$

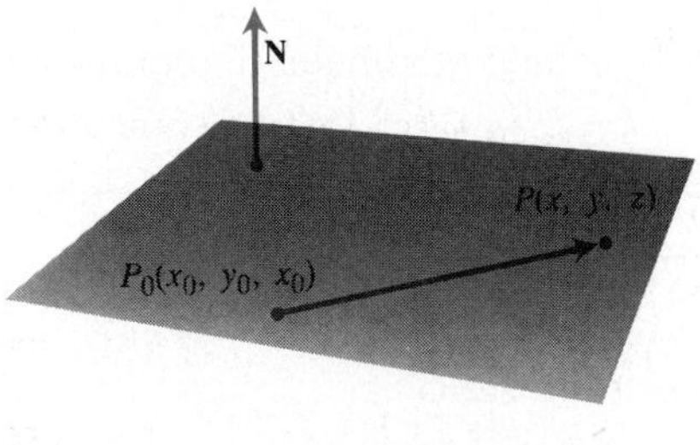

13.45 *P* lies in the plane through P_0 perpendicular to **N** if and only if $\overrightarrow{P_0P} \cdot \mathbf{N} = 0$.

perpendicular to the plane are given (Fig. 13.45). Then the point $P(x, y, z)$ will lie in the plane if and only if the vector $\overrightarrow{P_0P}$ is perpendicular to **N**; that is, if and only if

$$\mathbf{N} \cdot \overrightarrow{P_0P} = 0$$

or

$$A(x - x_0) + B(y - y_0) + C(z - z_0) = 0. \tag{7}$$

This equation may also be put in the form

$$Ax + By + Cz = Ax_0 + By_0 + Cz_0 \tag{8}$$

or

$$Ax + By + Cz = D, \qquad \text{where} \qquad D = Ax_0 + By_0 + Cz_0. \tag{9}$$

Conversely, if we start from any linear equation such as (9), we may find a point $P_0(x_0, y_0, z_0)$ whose coordinates do satisfy it; that is, such that

$$Ax_0 + By_0 + Cz_0 = D.$$

Then, by subtraction, we may put the given equation (9) into the form of Eq. (7) and factor it into the dot product

$$\mathbf{N} \cdot \overrightarrow{P_0P} = 0,$$

with **N** as in Eq. (6). This says that the constant vector **N** is perpendicular to the vector $\overrightarrow{P_0P}$ for every pair of points P_0 and P whose coordinates satisfy the equation. Hence the set of points $P(x, y, z)$ whose coordinates satisfy such a linear equation is a plane and the vector $A\mathbf{i} + B\mathbf{j} + C\mathbf{k}$, with the same coefficients that x, y, and z have in the given equation, is normal to the plane.

Equation (7) is called a *vector equation* for the plane through P, perpendicular to **N**. Equation (9) is a *cartesian equation* for the plane.

EXAMPLE 3 Find a cartesian equation for the plane through $P_0(-3, 0, 7)$ perpendicular to the vector $\mathbf{N} = 5\mathbf{i} + 2\mathbf{j} - \mathbf{k}$.

Solution From Eq. (9),

$$D = 5(-3) + 2(0) - 1(7) = -15 - 7 = -22,$$

and the corresponding cartesian equation for this line is

$$5x + 2x - y = -22. \square$$

EXAMPLE 4 Find the distance d between the point $P(2, -3, 4)$ and the plane $x + 2y + 2z = 13$.

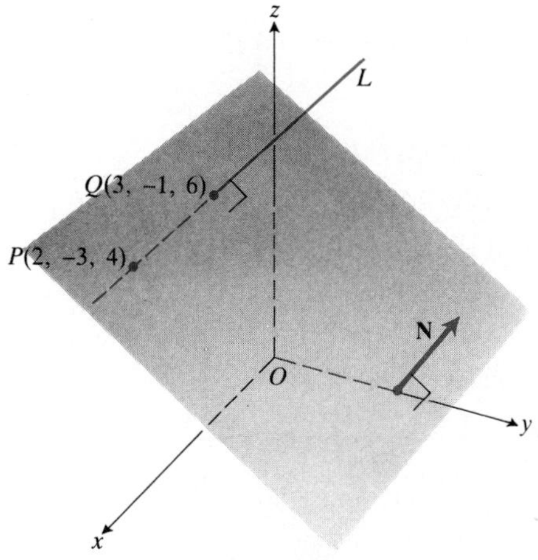

13.46 The distance between P and the plane is the distance between P and Q.

Solution 1 Carry out the following steps:

STEP 1. Find a line L through P normal to the plane.

STEP 2. Find the coordinates of the point Q in which the line meets the plane.

STEP 3. Compute the distance between P and Q.

The vector $\mathbf{N} = \mathbf{i} + 2\mathbf{j} + 2\mathbf{k}$ is normal to the given plane, and the line

$$x = 2 + t, \qquad y = -3 + 2t, \qquad z = 4 + 2t \tag{10}$$

goes through P, and, being parallel to **N**, is normal to the plane. The point (x, y, z) from (10) will lie in the plane

$$x + 2y + 2z = 13$$

if

$$\begin{aligned} (2 + t) + 2(-3 + 2t) + 2(4 + 2t) &= 13, \\ 9t + 4 &= 13, \\ t &= 1. \end{aligned}$$

Thus, the point Q in which the line meets the plane can be obtained by setting $t = 1$ in Eq. (11):

$$Q(x, y, z) = (2 + 1, -3 + 2, 4 + 2) = (3, -1, 6).$$

The distance between $P(2, -3, 4)$ and $Q(3, -1, 6)$ (Fig. 13.46) is

$$\sqrt{(2-3)^2 + (-3+1)^2 + (4-6)^2} = \sqrt{1 + 4 + 4} = \sqrt{9} = 3.$$

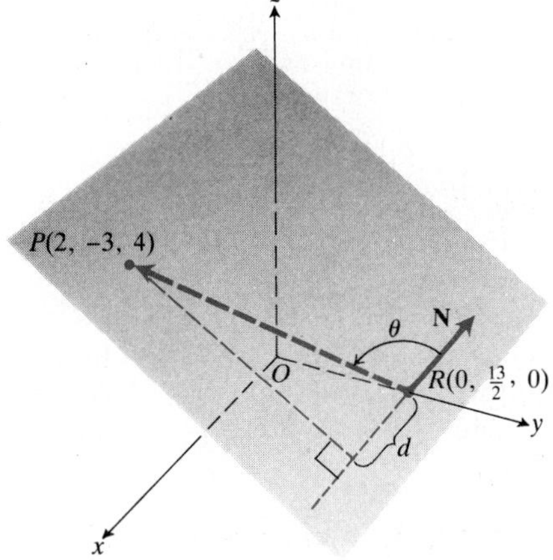

13.47 d is the length of the $\text{proj}_{\mathbf{N}}\overrightarrow{RP}$.

Solution 2 Let R be any point in the plane, and find the component of $\overrightarrow{RP}$ in the direction of **N**. This will be plus d or minus d, the sign depending on the direction of the vector projection of $\overrightarrow{RP}$ onto **N**. Figure 13.47 shows that the component is negative in this case, but we do not need to know this to find d.

Since R can be any point in the given plane, we might as well choose R to be a point whose coordinates are simple, say the point $R(0, \frac{13}{2}, 0)$ where the plane meets the y-axis. Then,

$$\overrightarrow{RP} = (2 - 0)\mathbf{i} + (-3 - \tfrac{13}{2})\mathbf{j} + (4 - 0)\mathbf{k} = 2\mathbf{i} - \tfrac{19}{2}\mathbf{j} + 4\mathbf{k}.$$

The component of $\overrightarrow{RP}$ in the direction of $\mathbf{N} = \mathbf{i} + 2\mathbf{j} + 2\mathbf{k}$ is

$$\overrightarrow{RP} \cdot \frac{\mathbf{N}}{|\mathbf{N}|} = \frac{\mathbf{N} \cdot \overrightarrow{RP}}{|\mathbf{N}|} = \frac{2 - 19 + 8}{\sqrt{(1)^2 + (2)^2 + (2)^2}} = \frac{-9}{\sqrt{9}} = -3.$$

Therefore, $d = 3$. □

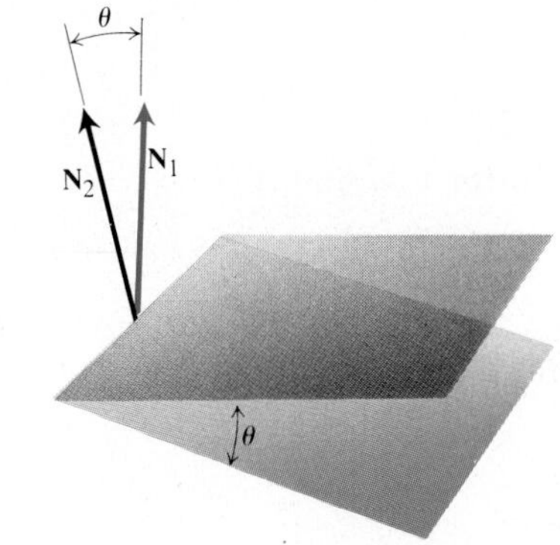

13.48 The angle between two planes can be obtained from their normals.

EXAMPLE 5 Find the angle between the two planes $3x - 6y - 2z = 7$ and $2x + y - 2z = 5$.

Solution Clearly the angle between two planes (Fig. 13.48) is the same as the angle between their normals. (Actually there are two angles in each case, namely θ and $180° - \theta$.) From the equations of the planes we may read off their normal vectors:

$$\mathbf{N}_1 = 3\mathbf{i} - 6\mathbf{j} - 2\mathbf{k}, \qquad \mathbf{N}_2 = 2\mathbf{i} + \mathbf{j} - 2\mathbf{k}.$$

Then

$$\cos\theta = \frac{\mathbf{N}_1 \cdot \mathbf{N}_2}{|\mathbf{N}_1||\mathbf{N}_2|} = \frac{4}{21}, \qquad \theta = \cos^{-1}\left(\frac{4}{21}\right) \approx 79°. \ \square$$

EXAMPLE 6 Find a vector parallel to the line of intersection of the two planes in Example 5.

Solution The requirements are met by the vector

$$\mathbf{v} = \mathbf{N}_1 \times \mathbf{N}_2 = \begin{vmatrix} \mathbf{i} & \mathbf{j} & \mathbf{k} \\ 3 & -6 & -2 \\ 2 & 1 & -2 \end{vmatrix} = 14\mathbf{i} + 2\mathbf{j} + 15\mathbf{k}.$$

The vector $\mathbf{v}$ is perpendicular to both of the normals $\mathbf{N}_1$ and $\mathbf{N}_2$, and is therefore parallel to both planes. □

Intersecting Planes

EXAMPLE 7 Find cartesian equations for the line in which the planes of Example 5 intersect.

Solution We find a vector parallel to the line, a point on the line, and use Eq. (3). Example 6 gives

$$\mathbf{v} = 14\mathbf{i} + 2\mathbf{j} + 15\mathbf{k}$$

parallel to the line. To find a point on the line we find a point common to the two planes. Substituting $z = 0$ into the plane equations from Example 5 and solving for x and y simultaneously gives the point $(3, -1, 0)$. We therefore obtain

$$\frac{(x - 3)}{14} = \frac{y - (-1)}{2} = \frac{z - 0}{15}$$

or

$$\frac{x-3}{14} = \frac{y+1}{2} = \frac{z}{15}$$

as cartesian equations for the line. □

EXAMPLE 8 Find an equation of the plane that passes through the two points $P_1(1, 0, -1)$ and $P_2(-1, 2, 1)$ and is parallel to the line of intersection of the planes $3x + y - 2z = 6$ and $4x - y + 3z = 0$.

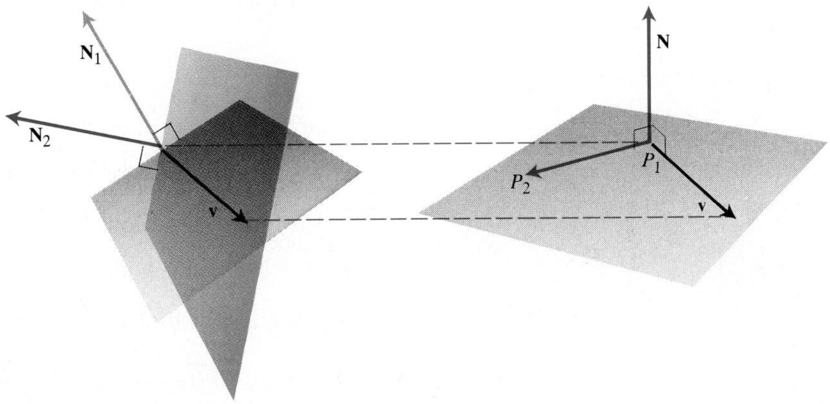

13.49 Constructing a plane through P_1 and P_2 that is parallel to the line of intersection of two other planes.

Solution The coordinates of either one of the points P_1 or P_2 will do for the x_0, y_0, and z_0 in Eq. (9). What remains, then, is to find a vector $\mathbf{N}$ normal to the plane in question to furnish the coefficients A, B, and C of Eq. (9) (Fig. 13.49).

The line of intersection of the two given planes is parallel to the vector

$$\mathbf{v} = \mathbf{N}_1 \times \mathbf{N}_2 = \begin{vmatrix} \mathbf{i} & \mathbf{j} & \mathbf{k} \\ 3 & 1 & -2 \\ 4 & -1 & 3 \end{vmatrix} = \mathbf{i} - 17\mathbf{j} - 7\mathbf{k},$$

where $\mathbf{N}_1$ and $\mathbf{N}_2$ are normals to the two given planes. The vector

$$\overrightarrow{P_1P_2} = -2\mathbf{i} + 2\mathbf{j} + 2\mathbf{k}$$

is to lie in the required plane. Now we may slide $\mathbf{v}$ parallel to itself until it also lies in the required plane (since the plane is to be parallel to $\mathbf{v}$). Hence we may take

$$\mathbf{N} = \overrightarrow{P_1P_2} \times \mathbf{v} = 20\mathbf{i} - 12\mathbf{j} + 32\mathbf{k}$$

as a vector normal to the plane. Actually,

$$\frac{1}{4}\mathbf{N} = 5\mathbf{i} - 3\mathbf{j} + 8\mathbf{k}$$

serves just as well. From this normal vector, we may substitute

$$A = 5, \qquad B = -3, \qquad C = 8$$

in Eq. (9), together with $x_0 = 1$, $y_0 = 0$, $z_0 = -1$, since the point $(1, 0, -1)$ is to lie in the plane. The required plane is therefore

$$5x - 3y + 8z = -3. \ \square$$

Equations for Lines and Planes in Space

Line through $P_0(x_0, y_0, z_0)$ parallel to $\mathbf{v} = A\mathbf{i} + B\mathbf{j} + C\mathbf{k}$:

Vector equation: $\overrightarrow{P_0P} = t\mathbf{v}$

Parametric equations:

$$x = x_0 + tA, \qquad y = y_0 + tB, \qquad z = z_0 + tC$$

Cartesian equations:

$$\frac{x - x_0}{A} = \frac{y - y_0}{B} = \frac{z - z_0}{C}$$

Plane through $P_0(x_0, y_0, z_0)$ perpendicular to $\mathbf{N} = A\mathbf{i} + B\mathbf{j} + C\mathbf{k}$:

Vector equation: $\mathbf{N} \cdot \overrightarrow{P_0P} = 0$

Cartesian equation:

$$Ax + By + Cz = D, \qquad D = Ax_0 + By_0 + Cz_0$$

Distance from point P to a plane:

$$\text{Distance} = \left|\overrightarrow{PR} \cdot \frac{\mathbf{N}}{|\mathbf{N}|}\right| = \left|\frac{\mathbf{N} \cdot \overrightarrow{PR}}{|\mathbf{N}|}\right| \qquad \begin{pmatrix}\mathbf{N} \text{ any normal to the plane,} \\ R \text{ any point in the plane}\end{pmatrix}$$

PROBLEMS

1. Find the coordinates of the point P in which the line

$$\frac{x - 1}{2} = \frac{y + 1}{-1} = \frac{z}{3}$$

intersects the plane $3x + 2y - z = 5$.

2. Find parametric and cartesian equations of the line joining the points $A(1, 2, -1)$ and $B(-1, 0, 1)$.
3. Find parametric equations for the line through the points $(1, 2, 0)$ and $(1, 1, -1)$.
4. Find equations for the line through $P(2, 4, 5)$ perpendicular to the plane $3x + 7y - 5z = 21$.
5. Find the point of intersection of the lines

$$\frac{x}{1} = \frac{y}{3} = \frac{z}{2} \quad \text{and} \quad \frac{x - 2}{1} = \frac{y - 3}{4} = \frac{z - 4}{2}.$$

6. Find an equation for the plane through the point $P_0(0, 2, -1)$ perpendicular to $\mathbf{N} = 3\mathbf{i} - 2\mathbf{j} - \mathbf{k}$.
7. Find an equation for the plane through $(1, -1, 3)$ parallel to the plane $3x + y + z = 7$.
8. Find an equation for the plane through the points $(1, 1, -1)$, $(2, 0, 2)$, and $(0, -2, 1)$.
9. Find an equation for the plane through the points $(2, 4, 5)$, $(1, 5, 7)$, and $(-1, 6, 8)$.
10. Find an equation for the plane through the points $(1, 0, 0)$, $(0, 1, 0)$, and $(0, 0, 1)$.
11. Find an equation for the plane through $P_0(2, 4, 5)$ perpendicular to the line

$$\frac{x - 5}{1} = \frac{y - 1}{3} = \frac{z}{4}.$$

12. Find an equation for the plane through the origin that contains the line

$$\frac{x - 1}{2} = \frac{y + 1}{3} = \frac{z}{4}.$$

13. Find a plane through $A(1, -2, 1)$ perpendicular to the vector from the origin to A.

14. Find a plane through $P_0(2, 1, -1)$ perpendicular to the line of intersection of the planes $2x + y - z = 3$, $x + 2y + z = 2$.

15. Find a plane through the points $P_0(1, 2, 3)$ and $P_2(3, 2, 1)$ perpendicular to the plane $4x - y + 2z = 7$.

16. Find the distance from the point $(1, 1, 1)$ to the plane $2x - 3y + 7z = 11$.

17. a) Find an equation for the plane through $P_0(4, 2, 1)$ perpendicular to $\mathbf{N} = 6\mathbf{i} - 2\mathbf{j} + 3\mathbf{k}$.
b) Find the distance from the origin to the plane in part (a).
c) Find the distance from the point $(6, 6, 8)$ to the plane in part (a).
d) Find equations for the line through $P_0(4, 2, -1)$ perpendicular to the plane in part (a).

18. Find the angle between the planes
a) $x + y + z = 5$, $x + 2y + z = 7$,
b) $x + y = 1$, $x + z = 1$,
c) $5x + y - z = 10$, $x - 2y + 3 = -1$.

19. Find the line of intersection of the planes
a) $x + y + z = 1$, $x + y = 2$,
b) $3x - 6y - 2z = 3$, $2x + y - 2z = 2$,
c) $x - 2y + 4z = 2$, $x + y - 2z = 5$.

20. The line L through the origin is normal to the plane $2x - y - z = 4$. Find the point in which L meets the plane $x + y - 2z = 2$.

21. a) What is meant by the angle between a line and a plane?
b) Find the acute angle between the line

$$x = -1 + 2t, \qquad y = 3t, \qquad z = 3 + 6t$$

and the plane $10x + 2y - 11z = 3$.

22. Let $P_i(x_i, y_i, z_i)$, $i = 1, 2, 3$, be three points. What set is described by the equation

$$\begin{vmatrix} x & y & z & 1 \\ x_1 & y_1 & z_1 & 1 \\ x_2 & y_2 & z_2 & 1 \\ x_3 & y_3 & z_3 & 1 \end{vmatrix} = 0?$$

23. Find the distance between the origin and the line

$$\frac{x-5}{3} = \frac{y-5}{4} = \frac{z+3}{-5}.$$

24. The equation $\mathbf{N} \cdot \overrightarrow{P_0P} = 0$ represents a plane through P_0 perpendicular to $\mathbf{N}$. What set does the inequality $\mathbf{N} \cdot \overrightarrow{P_0P} > 0$ represent? Give a reason for your answer.

25. The unit vector $\mathbf{u}$ makes angles α, β, γ, respectively, with the positive x-, y-, z-axes. Find the plane normal to $\mathbf{u}$ through $P_0(x_0, y_0, z_0)$.

26. Show that the planes obtained by substituting different values for the constant D in the equation

$$2x + 3y - 6z = D$$

are parallel. Find the distance between two of these planes, one corresponding, say, to $D = D_1$ and the other to $D = D_2$.

27. Prove that the line

$$x = 1 + 2t, \qquad y = -1 + 3t, \qquad z = 2 + 4t$$

is parallel to the plane $x - 2y + z = 6$.

28. a) Prove that three points A, B, C are collinear if and only if $\overrightarrow{AC} \times \overrightarrow{AB} = \mathbf{0}$.
b) Are the points $A(1, 2, -3)$, $B(3, 1, 0)$, $C(-3, 4, -9)$ collinear?

29. Prove that four points A, B, C, D are coplanar if and only if $\overrightarrow{AD} \cdot (\overrightarrow{AB} \times \overrightarrow{BC}) = 0$.

30. Show that the line of intersection of the planes

$$x + 2y - 2z = 5 \quad \text{and} \quad 5x - 2y - z = 0$$

is parallel to the line

$$x = -3 + 2t, \qquad y = 3t, \qquad z = 1 + 4t.$$

Find the plane determined by these two lines.

31. Show that the lines

$$\frac{x+1}{3} = \frac{y-6}{1} = \frac{z-3}{2}$$

and

$$\frac{x-6}{2} = \frac{y-11}{2} = \frac{z-3}{-1}$$

intersect. Find the plane determined by these two lines.

32. Show, by vector methods, that the distance from the point $P_1(x_1, y_1, z_1)$ to the plane $Ax + By + Cz - D = 0$ is

$$\frac{|Ax_1 + By_1 + Cz_1 - D|}{\sqrt{A^2 + B^2 + C^2}}.$$

***Toolkit* programs**

3D Grapher

13.9

Products of Three Vectors or More

Products of three vectors or more often arise in physical and engineering problems. For example, the electromotive force $d\mathcal{E}$ induced in an element of a conducting wire $d\mathbf{l}$ moving with velocity $\mathbf{v}$ through a magnetic field $\mathbf{B}$ is given by $d\mathcal{E} = (\mathbf{v} \times \mathbf{B}) \cdot d\mathbf{l}$.† Here the factor in parentheses is a vector, and the result of forming the scalar product of this vector and $d\mathbf{l}$ is a scalar. It is a real economy in thinking to represent the result in the compact vector form that removes the necessity of carrying factors such as the sine of the angle between $\mathbf{B}$ and $\mathbf{v}$ and the cosine of the angle between the normal to their plane and the vector $d\mathbf{l}$. These are all taken into account by the given product of the three vectors.

Triple Scalar Product

The product $(\mathbf{A} \times \mathbf{B}) \cdot \mathbf{C}$, called the *triple scalar product,* has the following geometrical significance. The vector $\mathbf{N} = \mathbf{A} \times \mathbf{B}$ is normal to the base of the parallelepiped determined by the vectors $\mathbf{A}$, $\mathbf{B}$, and $\mathbf{C}$ in Fig. 13.50. The magnitude of $\mathbf{N}$ equals the area of the base determined by $\mathbf{A}$ and $\mathbf{B}$. Thus

$$(\mathbf{A} \times \mathbf{B}) \cdot \mathbf{C} = \mathbf{N} \cdot \mathbf{C} = |\mathbf{N}|\,|\mathbf{C}| \cos\theta$$

is, except perhaps for sign, the *volume of a box* of edges $\mathbf{A}$, $\mathbf{B}$, and $\mathbf{C}$, since

$$|\mathbf{N}| = |\mathbf{A} \times \mathbf{B}| = \text{area of base}$$

and

$$|\mathbf{C}| \cos\theta = \pm h = \pm\text{altitude of box.}$$

If $\mathbf{C}$ and $\mathbf{A} \times \mathbf{B}$ lie on the same side of the plane determined by $\mathbf{A}$ and $\mathbf{B}$, the triple scalar product will be positive. But if the vectors $\mathbf{A}$, $\mathbf{B}$, and $\mathbf{C}$ form a left-handed system, then $(\mathbf{A} \times \mathbf{B}) \cdot \mathbf{C}$ is negative. By successively considering the plane of $\mathbf{B}$ and $\mathbf{C}$, then the plane of $\mathbf{C}$ and $\mathbf{A}$, as the base of the parallelepiped, we can readily see that

$$(\mathbf{A} \times \mathbf{B}) \cdot \mathbf{C} = (\mathbf{B} \times \mathbf{C}) \cdot \mathbf{A} = (\mathbf{C} \times \mathbf{A}) \cdot \mathbf{B}. \tag{1}$$

Since the dot product is commutative, we also have

$$(\mathbf{B} \times \mathbf{C}) \cdot \mathbf{A} = \mathbf{A} \cdot (\mathbf{B} \times \mathbf{C}),$$

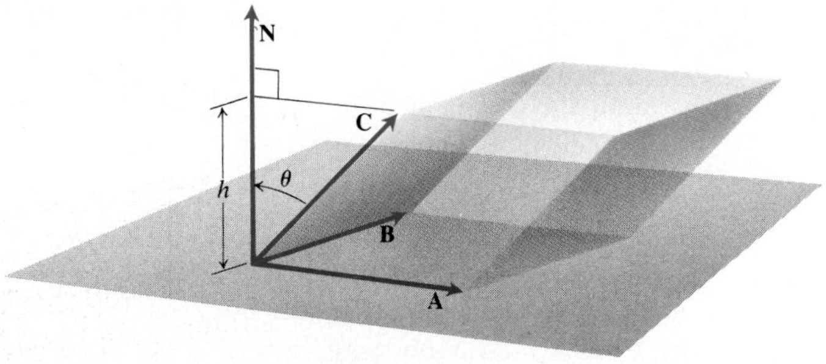

13.50 Except perhaps for sign, the number $(\mathbf{A} \times \mathbf{B}) \cdot \mathbf{C}$ is the volume of the parallelepiped (parallelogram-sided box) shown here.

† See F. W. Sears, M. W. Zemansky, H. D. Young, *University Physics,* 6th ed. (Reading, Mass.: Addison-Wesley, 1982), Chapter 33.

so that Eq. (1) gives the result

$$(\mathbf{A} \times \mathbf{B}) \cdot \mathbf{C} = \mathbf{A} \cdot (\mathbf{B} \times \mathbf{C}). \tag{2}$$

Equation (2) says that the dot and the cross may be interchanged in the triple scalar product, provided only that the multiplications are performed in a way that "makes sense." Thus $(\mathbf{A} \cdot \mathbf{B}) \times \mathbf{C}$ is excluded on the ground that $(\mathbf{A} \cdot \mathbf{B})$ is a scalar and we never "cross" a scalar and a vector.

The triple scalar product in Eq. (2) is conveniently expressed in determinant form as follows:

$$\begin{aligned}\mathbf{A} \cdot (\mathbf{B} \times \mathbf{C}) &= A \cdot \left[\begin{vmatrix} b_2 & b_3 \\ c_2 & c_3 \end{vmatrix} \mathbf{i} - \begin{vmatrix} b_1 & b_3 \\ c_1 & c_3 \end{vmatrix} \mathbf{j} + \begin{vmatrix} b_1 & b_2 \\ c_1 & c_2 \end{vmatrix} \mathbf{k} \right] \\ &= a_1 \begin{vmatrix} b_2 & b_3 \\ c_2 & c_3 \end{vmatrix} - a_2 \begin{vmatrix} b_1 & b_3 \\ c_1 & c_3 \end{vmatrix} + a_3 \begin{vmatrix} b_1 & b_2 \\ c_1 & c_2 \end{vmatrix} \qquad (3) \\ &= \begin{vmatrix} a_1 & a_2 & a_3 \\ b_1 & b_2 & b_3 \\ c_1 & c_2 & c_3 \end{vmatrix}.\end{aligned}$$

$$\mathbf{A} \cdot (\mathbf{B} \times \mathbf{C}) = \begin{vmatrix} a_1 & a_2 & a_3 \\ b_1 & b_2 & b_3 \\ c_1 & c_2 & c_3 \end{vmatrix}.$$

EXAMPLE 1 See Fig. 13.51. Let

$$\mathbf{A} = \overrightarrow{PQ}, \qquad \mathbf{B} = \overrightarrow{PS}$$
$$\mathbf{A}' = \overrightarrow{P'Q'}, \qquad \mathbf{B}' = \overrightarrow{P'S'}$$

be sides of parallelograms $PQRS$ and $P'Q'R'S'$ that are related in such a way that PP', QQ', and SS' are parallel to one another and to the unit vector **n.** Show that

$$(\mathbf{A} \times \mathbf{B}) \cdot \mathbf{n} = (\mathbf{A}' \times \mathbf{B}') \cdot \mathbf{n} \tag{4}$$

and discuss the geometrical meaning of this identity.

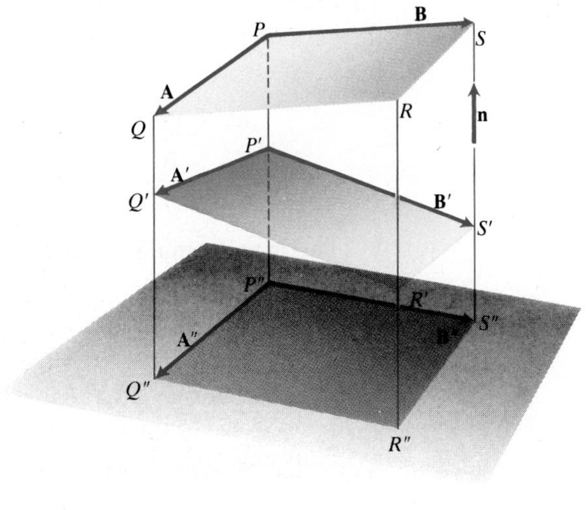

13.51 If $|\mathbf{n}| = 1$, then $(\mathbf{A} \times \mathbf{B}) \cdot \mathbf{n}$ is the area of the projection of the parallelogram determined by **A** and **B** on a plane perpendicular to **n**.

Verification of Eq. (4) From the way the parallelograms are related, it follows that

$$\begin{aligned}\mathbf{A} &= \overrightarrow{PQ} = \overrightarrow{PP'} + \overrightarrow{P'Q'} + \overrightarrow{Q'Q} \\ &= \overrightarrow{P'Q'} + (\overrightarrow{PP'} - \overrightarrow{QQ'}) \\ &= \mathbf{A}' + s\mathbf{n}\end{aligned}$$

for some scalar s, since both $\overrightarrow{PP'}$ and QQ' are parallel to $\mathbf{n}$. Similarly,

$$\mathbf{B} = \mathbf{B}' + t\mathbf{n}$$

for some scalar t. Hence

$$\begin{aligned}\mathbf{A} \times \mathbf{B} &= (\mathbf{A}' + s\mathbf{n}) \times (\mathbf{B}' + t\mathbf{n}) \\ &= \mathbf{A}' \times \mathbf{B}' + t(\mathbf{A}' \times \mathbf{n}) + s(\mathbf{n} \times \mathbf{B}') + st(\mathbf{n} \times \mathbf{n}). \qquad (5)\end{aligned}$$

But $\mathbf{n} \times \mathbf{n} = \mathbf{0}$, while $\mathbf{A}' \times \mathbf{n}$ and $\mathbf{n} \times \mathbf{B}'$ are both perpendicular to $\mathbf{n}$. Therefore when we dot both sides of (5) with $\mathbf{n}$ we get Eq. (4).

Geometrical meaning of Eq. (4) The result (4) says that when the parallelograms $PQRS$ and $P'Q'R'S'$ are any two plane sections of a prism with sides parallel to $\mathbf{n}$, then the box determined by $\mathbf{A}$, $\mathbf{B}$, and $\mathbf{n}$ has the same volume as the box determined by $\mathbf{A}'$, $\mathbf{B}'$, and $\mathbf{n}$. Thus, in particular, we may replace the right-hand side of (4) by $(\mathbf{A}'' \times \mathbf{B}'') \cdot \mathbf{n}$, where $\mathbf{A}''$ and $\mathbf{B}''$ are sides of a *right* section $P''Q''R''S''$ as in Fig. 13.51. Then $\mathbf{A}'' \times \mathbf{B}''$ is parallel to $\mathbf{n}$, and

$$\mathbf{A}'' \times \mathbf{B}'' = (\text{Area right section}) \text{ times } \mathbf{n}$$

and

$$(\mathbf{A}'' \times \mathbf{B}'') \cdot \mathbf{n} = \text{Area right section.}$$

Therefore, by Eq. (4), we have the following interpretation:

> $(\mathbf{A} \times \mathbf{B}) \cdot \mathbf{n}$ is the area of the orthogonal projection of the parallelogram determined by $\mathbf{A}$ and $\mathbf{B}$ onto a plane whose unit normal is $\mathbf{n}$. (6)

This assumes that $\mathbf{A} \times \mathbf{B}$ and $\mathbf{n}$ lie on the same side of the plane $PQRS$. If they are on opposite sides, take the absolute value to get the area. Except possibly for sign, then,

$$(\mathbf{A} \times \mathbf{B}) \cdot \mathbf{k} = \text{Area of projection in the } xy\text{-plane}, \qquad (7a)$$

$$(\mathbf{A} \times \mathbf{B}) \cdot \mathbf{j} = \text{Area of projection in the } xz\text{-plane}, \qquad (7b)$$

$$(\mathbf{A} \times \mathbf{B}) \cdot \mathbf{i} = \text{Area of projection in the } yz\text{-plane}. \qquad (7c)$$

The area formula in (6) will be used in calculating surface areas in Article 16.9. □

Triple Vector Product

The triple vector products $(\mathbf{A} \times \mathbf{B}) \times \mathbf{C}$ and $\mathbf{A} \times (\mathbf{B} \times \mathbf{C})$ are usually not equal, but each of them can be evaluated rather simply by formulas that we shall now derive.

We start by showing that the vector product $(\mathbf{A} \times \mathbf{B}) \times \mathbf{C}$ is given by

$$(\mathbf{A} \times \mathbf{B}) \times \mathbf{C} = (\mathbf{A} \cdot \mathbf{C})\mathbf{B} - (\mathbf{B} \cdot \mathbf{C})\mathbf{A}. \tag{8}$$

In this formula, $\mathbf{A} \cdot \mathbf{C}$ is a scalar multiplying $\mathbf{B}$, and $\mathbf{B} \cdot \mathbf{C}$ is a scalar multiplying $\mathbf{A}$.

CASE 1. If one of the vectors is the zero vector, Eq. (8) is true because both sides of it are zero.

CASE 2. If none of the vectors is zero, but if $\mathbf{B} = s\mathbf{A}$ for some scalar s, then both sides of Eq. (8) are zero again.

CASE 3. Suppose that none of the vectors is zero and that $\mathbf{A}$ and $\mathbf{B}$ are not parallel. The vector on the left in Eq. (8) is parallel to the plane determined by $\mathbf{A}$ and $\mathbf{B}$, so that it is possible to find scalars m and n such that

$$(\mathbf{A} \times \mathbf{B}) \times \mathbf{C} = m\mathbf{A} + n\mathbf{B}. \tag{9}$$

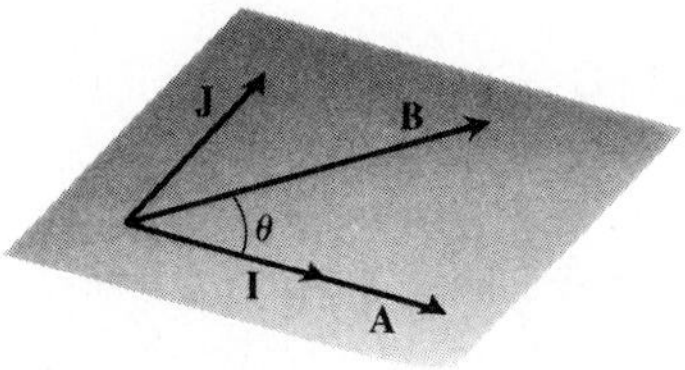

13.52 Orthogonal vectors **I** and **J** in the plane of **A** and **B**.

To calculate m and n, we introduce orthogonal unit vectors $\mathbf{I}$ and $\mathbf{J}$ in the plane of $\mathbf{A}$ and $\mathbf{B}$ with $\mathbf{I} = \mathbf{A}/|\mathbf{A}|$ (Fig. 13.52). We also introduce a third unit vector $\mathbf{K} = \mathbf{I} \times \mathbf{J}$, and write all our vectors in terms of these unit vectors $\mathbf{I}$, $\mathbf{J}$, and $\mathbf{K}$:

$$\begin{aligned} \mathbf{A} &= a_1\mathbf{I}, \\ \mathbf{B} &= b_1\mathbf{I} + b_2\mathbf{J}, \\ \mathbf{C} &= c_1\mathbf{I} + c_2\mathbf{J} + c_3\mathbf{K}. \end{aligned} \tag{10}$$

Then

$$\mathbf{A} \times \mathbf{B} = a_1 b_2 \mathbf{K}$$

and

$$(\mathbf{A} \times \mathbf{B}) \times \mathbf{C} = a_1 b_2 c_1 \mathbf{J} - a_1 b_2 c_2 \mathbf{I}. \tag{11}$$

Comparing this with the right-hand side of Eq. (9), we have

$$m(a_1\mathbf{I}) + n(b_1\mathbf{I} + b_2\mathbf{J}) = a_1 b_2 c_1 \mathbf{J} - a_1 b_2 c_2 \mathbf{I}.$$

This is equivalent to the pair of scalar equations

$$\begin{aligned} ma_1 + nb_1 &= -a_1 b_2 c_2, \\ nb_2 &= a_1 b_2 c_1. \end{aligned}$$

If b_2 were equal to zero, $\mathbf{A}$ and $\mathbf{B}$ would be parallel, contrary to hypothesis. Hence b_2 is not zero and we may solve the last equation for n. We find

$$n = a_1 c_1 = \mathbf{A} \cdot \mathbf{C}.$$

Then, by substitution,

$$\begin{aligned} ma_1 &= -nb_1 - a_1 b_2 c_2 \\ &= -a_1 c_1 b_1 - a_1 b_2 c_2, \end{aligned}$$

and since $|\mathbf{A}| = a_1 \neq 0$, we may divide by a_1 and have

$$m = -(b_1 c_1 + b_2 c_2) = -(\mathbf{B} \cdot \mathbf{C}).$$

When these values are substituted for m and n in Eq. (9), we obtain the result given in Eq. (8).

The identity

$$(\mathbf{B} \times \mathbf{C}) \times \mathbf{A} = (\mathbf{B} \cdot \mathbf{A})\mathbf{C} - (\mathbf{C} \cdot \mathbf{A})\mathbf{B} \tag{12a}$$

follows from Eq. (8) by a simple interchange of the letters **A, B,** and **C.** If we now interchange the factors $\mathbf{B} \times \mathbf{C}$ and **A** we must change the sign on the right-hand side of the equation. This gives the following identity, which is a companion of Eq. (8):

$$\mathbf{A} \times (\mathbf{B} \times \mathbf{C}) = (\mathbf{A} \cdot \mathbf{C})\mathbf{B} - (\mathbf{A} \cdot \mathbf{B})\mathbf{C}. \tag{12b}$$

EXAMPLE 2 Verify Eq. (8) for the vectors

$$\mathbf{A} = \mathbf{i} - \mathbf{j} + 2\mathbf{k}, \qquad \mathbf{B} = 2\mathbf{i} + \mathbf{j} + \mathbf{k}, \qquad \mathbf{C} = \mathbf{i} + 2\mathbf{j} - \mathbf{k}.$$

Solution Since

$$\mathbf{A} \cdot \mathbf{C} = -3, \qquad \mathbf{B} \cdot \mathbf{C} = 3,$$

the right-hand side of Eq. (8) is

$$(\mathbf{A} \cdot \mathbf{C})\mathbf{B} - (\mathbf{B} \cdot \mathbf{C})\mathbf{A} = -3\mathbf{B} - 3\mathbf{A} = -3(3\mathbf{i} + 3\mathbf{k}) = -9\mathbf{i} - 9\mathbf{k}.$$

To calculate the left-hand side of Eq. (8) we have

$$\mathbf{A} \times \mathbf{B} = \begin{vmatrix} \mathbf{i} & \mathbf{j} & \mathbf{k} \\ 1 & -1 & 2 \\ 2 & 1 & 1 \end{vmatrix} = -3\mathbf{i} + 3\mathbf{j} + 3\mathbf{k},$$

so that

$$(\mathbf{A} \times \mathbf{B}) \times \mathbf{C} = \begin{vmatrix} \mathbf{i} & \mathbf{j} & \mathbf{k} \\ -3 & 3 & 3 \\ 1 & 2 & -1 \end{vmatrix} = -9\mathbf{i} - 9\mathbf{k}. \ \square$$

EXAMPLE 3 Use Eqs. (8) and (12b) to express

$$(\mathbf{A} \times \mathbf{B}) \times (\mathbf{C} \times \mathbf{D})$$

in terms of scalar multiplication and cross products involving no more than two factors.

Solution Write, for convenience,

$$\mathbf{C} \times \mathbf{D} = \mathbf{V}.$$

Then use Eq. (8) to evaluate

$$(\mathbf{A} \times \mathbf{B}) \times \mathbf{V} = (\mathbf{A} \cdot \mathbf{V})\mathbf{B} - (\mathbf{B} \cdot \mathbf{V})\mathbf{A}$$

or

$$(\mathbf{A} \times \mathbf{B}) \times (\mathbf{C} \times \mathbf{D}) = (\mathbf{A} \cdot \mathbf{C} \times \mathbf{D})\mathbf{B} - (\mathbf{B} \cdot \mathbf{C} \times \mathbf{D})\mathbf{A}.$$

The result, as written, expresses the answer as a scalar times **B** minus a scalar times **A.** One could also represent the answer as a scalar times **C** minus a scalar times **D.** Geometrically, the vector is parallel to the line of intersection of the **A, B**-plane and the **C, D**-plane. □

PROBLEMS

In Problems 1–3, take

$$\mathbf{A} = 4\mathbf{i} - 8\mathbf{j} + \mathbf{k},$$
$$\mathbf{B} = 2\mathbf{i} + \mathbf{j} - 2\mathbf{k},$$
$$\mathbf{C} = 3\mathbf{i} - 4\mathbf{j} + 12\mathbf{k}.$$

1. Find $(\mathbf{A} \cdot \mathbf{B})\mathbf{C}$ and $\mathbf{A}(\mathbf{B} \cdot \mathbf{C})$.

2. Find the volume of the box having $\mathbf{A}$, $\mathbf{B}$, $\mathbf{C}$ as three co-terminous edges.

3. a) Find $\mathbf{A} \times \mathbf{B}$ and use the result to find $(\mathbf{A} \times \mathbf{B}) \times \mathbf{C}$.
b) Find $(\mathbf{A} \times \mathbf{B}) \times \mathbf{C}$ by another method.

4. Find the volume of the parallelepiped whose edges are determined by the vectors $\mathbf{A} = 3\mathbf{j}$, $\mathbf{B} = -2\mathbf{i} + \mathbf{j}$, and $\mathbf{C} = -\mathbf{i} + \mathbf{j} + \mathbf{k}$.

In Problems 5–7, find the volume of the tetrahedron whose vertices are the given points.

5. $(0, 0, 0)$, $(1, -1, 1)$, $(2, 1, -2)$, $(-1, 2, -1)$

6. $(0, 0, 0)$, $(2, 1, 0)$, $(2, -1, 1)$, $(1, 0, 2)$

7. $(1, 0, 3)$, $(2, 1, 1)$, $(0, 0, 2)$, $(3, 4, 1)$

8. Which of the following are *not always* true?
a) $\mathbf{A} \times \mathbf{B} = \mathbf{B} \times \mathbf{A}$
b) $\mathbf{A} \times (\mathbf{B} + \mathbf{C}) = \mathbf{A} \times \mathbf{B} + \mathbf{A} \times \mathbf{C}$
c) $\mathbf{A} \times (-\mathbf{A}) = \mathbf{0}$
d) $\mathbf{A} \cdot (\mathbf{B} \times \mathbf{C}) = (\mathbf{A} \cdot \mathbf{B}) \times \mathbf{C}$
e) $(\mathbf{A} \times \mathbf{B}) \cdot \mathbf{A} = \mathbf{B} \cdot (\mathbf{A} \times \mathbf{B})$
f) $(\mathbf{A} \times \mathbf{A}) \cdot \mathbf{A} = 0$
g) $(\mathbf{A} \times \mathbf{B}) \cdot \mathbf{C} = (\mathbf{B} \times \mathbf{C}) \cdot \mathbf{A}$
h) $(\mathbf{A} \times \mathbf{B}) \times \mathbf{C} = \mathbf{A} \times (\mathbf{B} \times \mathbf{C})$

9. Suppose that

$$\mathbf{A} \cdot \mathbf{A} = 4, \quad \mathbf{B} \cdot \mathbf{B} = 4, \quad \mathbf{A} \cdot \mathbf{B} = 0,$$
$$(\mathbf{A} \times \mathbf{B}) \times \mathbf{C} = \mathbf{0}, \quad (\mathbf{A} \times \mathbf{B}) \cdot \mathbf{C} = 8.$$

Find each of the following:
a) $\mathbf{A} \cdot \mathbf{C}$, b) $|\mathbf{C}|$, c) $\mathbf{B} \times \mathbf{C}$.

(*Hint:* Picture the vectors, and think geometrically. Use basic, coordinate-free definitions. Avoid long calculations.)

10. Prove that any vector $\mathbf{A}$ satisfies the identity

$$\mathbf{A} = \tfrac{1}{2}[\mathbf{i} \times (\mathbf{A} \times \mathbf{i}) + \mathbf{j} \times (\mathbf{A} \times \mathbf{j}) + \mathbf{k} \times (\mathbf{A} \times \mathbf{k})].$$

11. Express the product $\mathbf{R} = (\mathbf{A} \times \mathbf{B}) \times (\mathbf{C} \times \mathbf{D})$ in the form $a\mathbf{C} + b\mathbf{D}$ with scalars a and b.

12. Use Eq. (3) to show that
a) $\mathbf{A} \cdot (\mathbf{C} \times \mathbf{B}) = -\mathbf{A} \cdot (\mathbf{B} \times \mathbf{C})$,
b) $\mathbf{A} \cdot (\mathbf{A} \times \mathbf{B}) = 0$,
c) $(\mathbf{A} + \mathbf{D}) \cdot (\mathbf{B} \times \mathbf{C}) = \mathbf{A} \cdot (\mathbf{B} \times \mathbf{C}) + \mathbf{D} \cdot (\mathbf{B} \times \mathbf{C})$.
Interpret the results geometrically.

13. Explain the statement in the text that $(\mathbf{A} \times \mathbf{B}) \times \mathbf{C}$ is parallel to the plane determined by $\mathbf{A}$ and $\mathbf{B}$. Illustrate with a sketch.

14. Explain the statement, at the end of Example 3, that $(\mathbf{A} \times \mathbf{B}) \times (\mathbf{C} \times \mathbf{D})$ is parallel to the line of intersection of the **A**, **B**-plane and the **C**, **D**-plane. Illustrate with a sketch.

15. Find a line in the plane of $P_0(0, 0, 0)$, $P_1(2, 2, 0)$, $P_2(0, 1, -2)$, and perpendicular to the line

$$\frac{x+1}{3} = \frac{y-1}{2} = 2z.$$

16. Let $P(1, 2, -1)$, $Q(3, -1, 4)$, and $R(2, 6, 2)$ be three vertices of a parallelogram $PQRS$.
a) Find the coordinates of S.
b) Find the area of $PQRS$.
c) Find the area of the projection of $PQRS$ in the xy-plane; in the yz-plane; in the xz-plane.

17. Show that the area of a parallelogram in space is the square root of the sum of the squares of the areas of its projections on any three mutually orthogonal planes.

13.10 Cylinders

In this article and the next, we shall consider some extensions of analytic geometry to space. We begin with the notion of a surface.

The set of points $P(x, y, z)$ that satisfy an equation

$$F(x, y, z) = 0 \tag{1}$$

may be interpreted in a broad sense as being a surface. The simplest

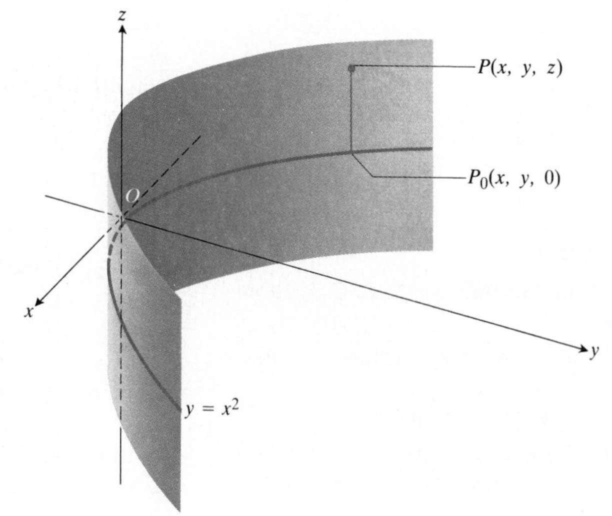

13.53 Parabolic cylinder.

examples of surfaces are planes, which have equations of the form $Ax + By + Cz - D = 0$. Almost as simple as planes are the surfaces called *cylinders*.

In general, a cylinder is a surface that is generated by moving a straight line along a given curve while holding the line parallel to a given fixed line.

EXAMPLE 1 The *parabolic cylinder* of Fig. 13.53 is generated by a line parallel to the z-axis that moves along the curve $y = x^2$ in the xy-plane. If a point $P_0(x, y, 0)$ lies on the parabola, then every point $P(x, y, z)$ with the same x- and y-coordinates lies on the line through P_0 parallel to the z-axis, and hence belongs to the surface. Conversely, if $P(x, y, z)$ lies on the surface, its projection $P_0(x, y, 0)$ on the xy-plane lies on the parabola $y = x^2$, so that its coordinates satisfy the equation $y = x^2$. Regardless of the value of z, the points of the surface are the points whose coordinates satisfy this equation. Thus, the equation $y = x^2$ is an equation for the cylinder as well as for the generating parabola. The cross sections of the cylinder perpendicular to the z-axis are parabolas, too, all of them congruent to the parabola in the xy-plane. □

In general, any curve

$$f(x, y) = 0 \tag{2}$$

in the xy-plane defines a cylinder in space whose equation is also $f(x, y) = 0$, and which is made up of the points of the lines through the curve that are parallel to the z-axis. The lines are sometimes called *elements* of the cylinder.

The discussion above can be carried through for cylinders with elements parallel to the other coordinate axes, and the result is summarized by saying that *an equation in cartesian coordinates, from which one coordinate variable is missing, represents a cylinder with elements parallel to the axis associated with the missing variable.*

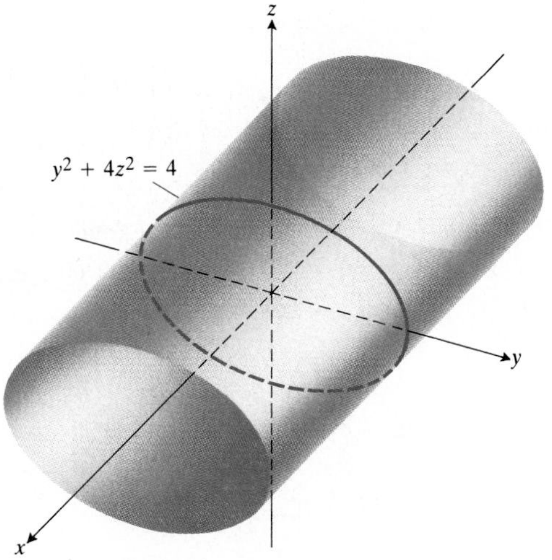

13.54 Elliptic cylinder.

EXAMPLE 2 The surface

$$y^2 + 4z^2 = 4$$

is an *elliptic cylinder* with elements parallel to the x-axis. It extends indefinitely in both the negative and positive directions along the x-axis, which is the axis of the cylinder, since it passes through the centers of the elliptical cross sections of the cylinder (Fig. 13.54). □

EXAMPLE 3 The surface

$$r^2 = 2a \cos 2\theta$$

in cylindrical coordinates is a cylinder with elements parallel to the z-axis. Each section perpendicular to the z-axis is a lemniscate. The cylinder extends indefinitely in both the positive and negative directions along the z-axis (Fig. 13.55). □

13.55 A cylinder whose cross sections are lemniscates.

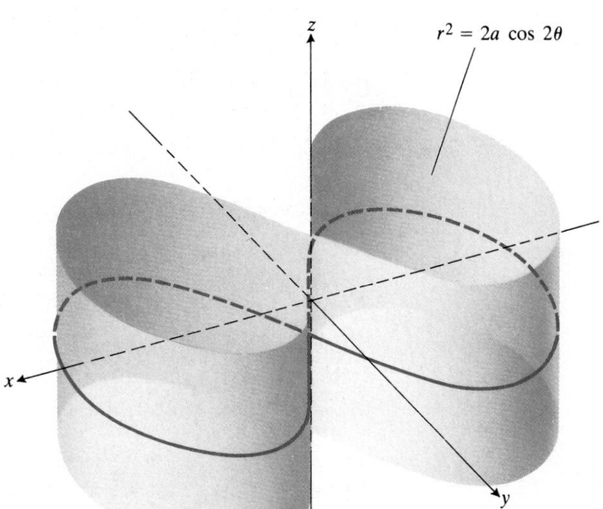

PROBLEMS

Describe and sketch each of the following surfaces [(r, θ, z) are cylindrical coordinates].

1. $x^2 + z^2 = 1$ **2.** $z = x^2$ **3.** $x = -y^2$
4. $4x^2 + y^2 = 4$ **5.** $z = -y$ **6.** $y^2 - x^2 = 1$
7. $x^2 - z^2 = 1$ **8.** $z^2 - y^2 = 1$ **9.** $r = 4$
10. $r = \sin\theta$ **11.** $r = \cos\theta$ **12.** $r = 1 + \cos\theta$
13. $x^2 + y^2 = a^2$ **14.** $y^2 + z^2 - 4z = 0$
15. $x^2 + 4z^2 - 4z = 0$

***Toolkit* programs**
3D Grapher

13.11 Quadric Surfaces

A surface whose equation is a quadratic in the variables x, y, and z is called a *quadric* surface. We indicate briefly how some of the simpler ones may be recognized from their equations.

The *sphere*

$$(x - x_0)^2 + (y - y_0)^2 + (z - z_0)^2 = a^2 \tag{1}$$

with center at (x_0, y_0, z_0) and radius a has already been mentioned in Article 13.5. Likewise, the various *cylinders*

$$Ax^2 + Bxy + Cy^2 + Dx + Ey + F = 0 \tag{2}$$

with elements parallel to the z-axis, and others with elements parallel to the other coordinate axes, are familiar and will not be further discussed. In the examples that follow, we choose coordinate axes that yield simple forms of the equations. For example, we take the origin to be at the center of the ellipsoid in Example 1 below. If the center were at (x_0, y_0, z_0) instead, the equation would simply have $x - x_0$, $y - y_0$, and $z - z_0$, in place of x, y, z, respectively. We take a, b, and c to be positive constants in every case.

EXAMPLE 1 *The ellipsoid* (Fig. 13.56)

$$\frac{x^2}{a^2} + \frac{y^2}{b^2} + \frac{z^2}{c^2} = 1 \tag{3}$$

cuts the coordinate axes at $(\pm a, 0, 0)$, $(0, \pm b, 0)$, and $(0, 0, \pm c)$. It lies inside the rectangular box

$$|x| \le a, \qquad |y| \le b, \qquad |z| \le c.$$

Since only even powers of x, y, and z occur in the equation, this surface is symmetric with respect to each coordinate plane. The sections cut out by the coordinate planes are ellipses. For example,

$$\frac{x^2}{a^2} + \frac{y^2}{b^2} = 1 \qquad \text{when} \quad z = 0.$$

Each section cut out by a plane

$$z = z_1, \qquad |z_1| < c,$$

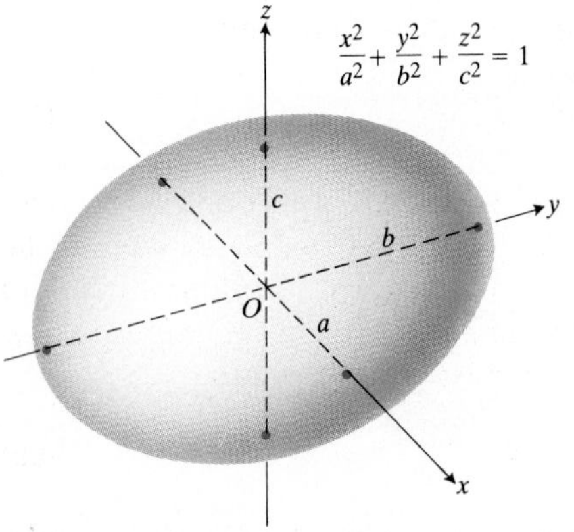

13.56 Ellipsoid.

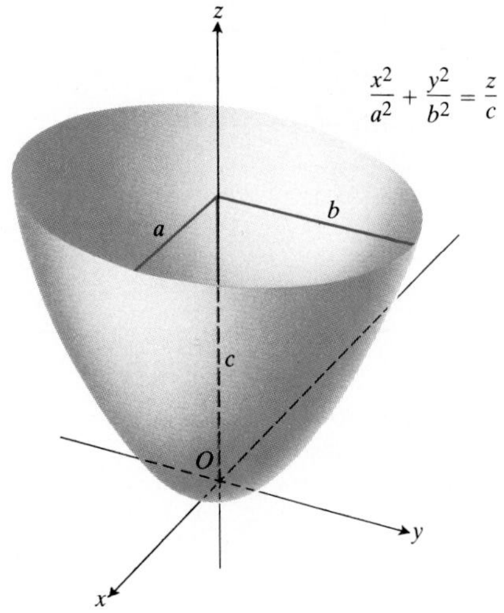

13.57 Elliptic paraboloid.

is an ellipse

$$\frac{x^2}{a^2[1 - (z_1^2/c^2)]} + \frac{y^2}{b^2[1 - (z_1^2/c^2)]} = 1$$

with center on the z-axis and having semiaxes

$$\frac{a}{c}\sqrt{c^2 - z_1^2} \quad \text{and} \quad \frac{b}{c}\sqrt{c^2 - z_1^2}.$$

The surface is sketched in Fig. 13.56. When two of the three semiaxes a, b, and c are equal, the surface is an ellipsoid of revolution, and when all three are equal, it is a sphere. □

EXAMPLE 2 *The elliptic paraboloid* (Fig. 13.57)

$$\frac{x^2}{a^2} + \frac{y^2}{b^2} = \frac{z}{c} \tag{4}$$

is symmetric with respect to the planes $x = 0$ and $y = 0$. The only intercept on the axes is at the origin. Since the left-hand side of the equation is nonnegative, the surface is limited to the region $z \geq 0$. That is, away from the origin it lies above the xy-plane. The section cut out from the surface by the yz-plane is

$$x = 0, \qquad y^2 = \frac{b^2}{c}z,$$

which is a parabola with vertex at the origin and opening upward. Similarly, one finds that when $y = 0$,

$$x^2 = \frac{a^2}{c}z,$$

which is also such a parabola. When $z = 0$, the cut reduces to the single point (0, 0, 0). Each plane $z = z_1 > 0$ perpendicular to the z-axis cuts the surface in an ellipse of semiaxes

$$a\sqrt{z_1/c} \quad \text{and} \quad b\sqrt{z_1/c}.$$

These semiaxes increase in magnitude as z_1 increases. The paraboloid extends indefinitely upward. □

EXAMPLE 3 *Circular paraboloid, or paraboloid of revolution:*

$$\frac{x^2}{a^2} + \frac{y^2}{a^2} = \frac{z}{c}. \tag{5a}$$

The equation is obtained by taking $b = a$ in Eq. (4) for the elliptic paraboloid. The cross sections of the surface by planes perpendicular to the z-axis are circles centered on the z-axis. The cross sections by planes containing the z-axis are congruent parabolas with a common focus at the point $(0, 0, a^2/4c)$. In cylindrical coordinates, (5a) becomes

$$\frac{r^2}{a^2} = \frac{z}{c}. \tag{5b}$$

Shapes cut from circular paraboloids are used for antennas in radio telescopes, satellite trackers, and microwave radio links (Fig. 13.58). □

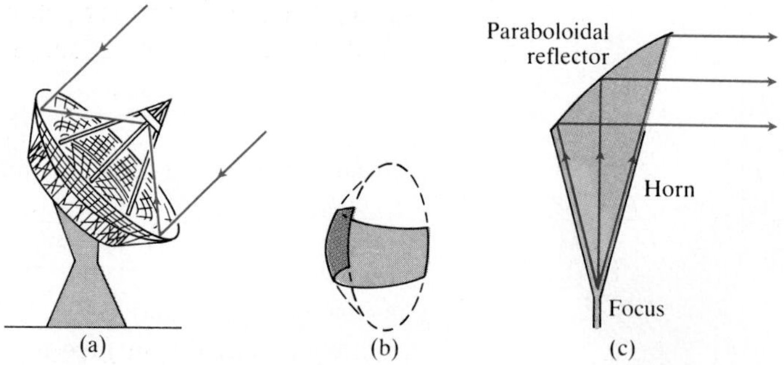

13.58 Many antennas are shaped like pieces of paraboloids of revolution. (a) Radio telescopes use the same principles as optical telescopes. (b) A "rectangular-cut" radar reflector. (c) Horn antenna in a microwave radio link.

EXAMPLE 4 *The elliptic cone* (Fig. 13.59)

$$\frac{x^2}{a^2} + \frac{y^2}{b^2} = \frac{z^2}{c^2} \tag{6}$$

is symmetric with respect to all three coordinate planes. The plane $z = 0$ cuts the surface in the single point (0, 0, 0). The plane $x = 0$ cuts it in the two intersecting straight lines

$$x = 0, \qquad \frac{y}{b} = \pm\frac{z}{c}, \tag{7}$$

and when

$$y = 0, \qquad \frac{x}{a} = \pm\frac{z}{c}. \tag{8}$$

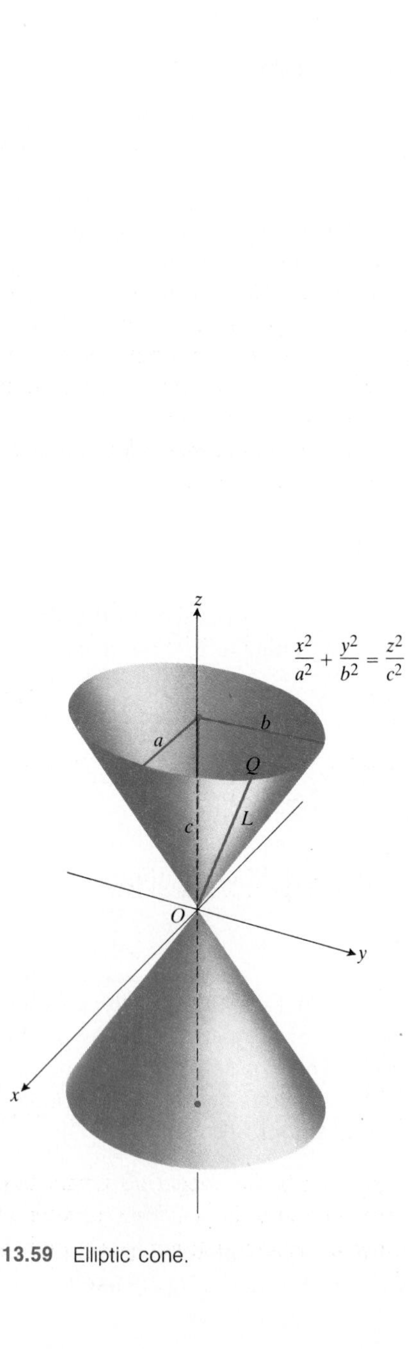

13.59 Elliptic cone.

The section cut out by a plane $z = z_1 > 0$ is an ellipse with center on the z-axis and vertices lying on the straight lines (7) and (8). In fact, the whole surface is generated by a straight line L that passes through the origin and a point Q on the ellipse

$$z = c, \qquad \frac{x^2}{a^2} + \frac{y^2}{b^2} = 1.$$

As the point Q traces out the ellipse, the infinite line L generates the surface, which is a cone with elliptic cross sections. To see why, suppose that $Q(x_1, y_1, z_1)$ is a point on the surface and t is any scalar. Then the vector from O to the point $P(tx_1, ty_1, tz_1)$ is simply t times $\overrightarrow{OQ}$, so that as t varies from $-\infty$ to $+\infty$ the point P traces out the infinite line L. But since Q is assumed to be on the surface, the equation

$$\frac{x_1^2}{a^2} + \frac{y_1^2}{b^2} = \frac{z_1^2}{c^2}$$

is satisfied. Multiplying both sides of this equation by t^2 shows that the point $P(tx_1, ty_1, tz_1)$ is also on the surface. This establishes the validity of the remark that the surface is a cone generated by the line L through O and the point Q on the ellipse.

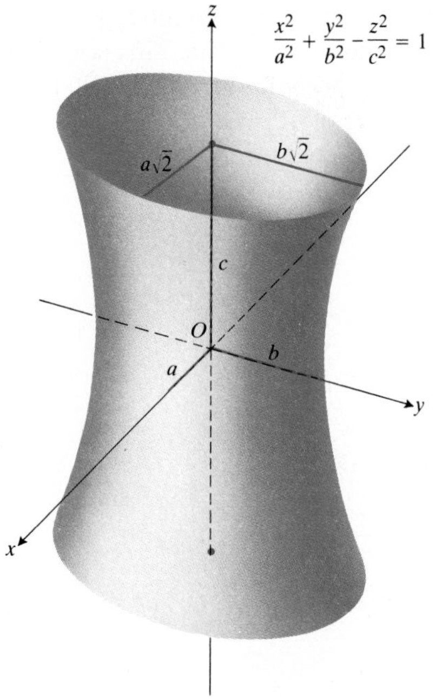

13.60 Hyperboloid of one sheet.

In case $a = b$, the cone is a right circular cone and its equation in cylindrical coordinates is simply

$$\frac{r}{a} = \frac{z}{c}. \square \tag{9}$$

EXAMPLE 5 The hyperboloid of one sheet (Fig. 13.60)

$$\frac{x^2}{a^2} + \frac{y^2}{b^2} - \frac{z^2}{c^2} = 1 \tag{10}$$

is symmetric with respect to each of the three coordinate planes. The sections cut out by the coordinate planes are

$$\begin{aligned} x = 0:& \quad \text{the hyperbola} \quad \frac{y^2}{b^2} - \frac{z^2}{c^2} = 1, \\ y = 0:& \quad \text{the hyperbola} \quad \frac{x^2}{a^2} - \frac{z^2}{c^2} = 1, \\ z = 0:& \quad \text{the ellipse} \quad \frac{x^2}{a^2} + \frac{y^2}{b^2} = 1. \end{aligned} \tag{11}$$

The plane $z = z_1$ cuts the surface in an ellipse with center on the z-axis and vertices on the hyperbolas in (11). The surface is connected, meaning that it is possible to travel from any point on it to any other point on it without leaving the surface. For this reason, it is said to have *one* sheet, in contrast to the next example, which consists of *two* sheets.

In the special case where $a = b$, the surface is a hyperboloid of revolution with equation given in cylindrical coordinates by

$$\frac{r^2}{a^2} - \frac{z^2}{c^2} = 1. \square \tag{12}$$

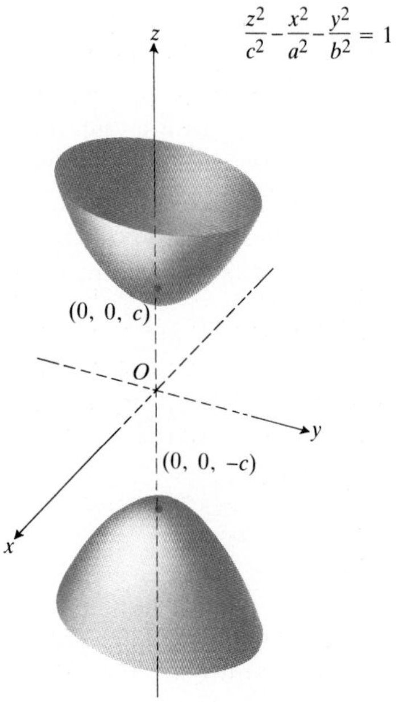

13.61 Hyperboloid of two sheets.

EXAMPLE 6 The hyperboloid of two sheets (Fig. 13.61)

$$\frac{z^2}{c^2} - \frac{x^2}{a^2} - \frac{y^2}{b^2} = 1 \tag{13}$$

is symmetric with respect to the three coordinate planes. The plane $z = 0$ does not intersect the surface; in fact, one must have

$$|z| \geq c$$

for real values of x and y in Eq. (13). The hyperbolic sections

$$x = 0: \quad \frac{z^2}{c^2} - \frac{y^2}{b^2} = 1,$$

$$y = 0: \quad \frac{z^2}{c^2} - \frac{x^2}{a^2} = 1$$

have their vertices and foci on the z-axis. The surface is separated into two portions, one above the plane $z = c$ and the other below the plane $z = -c$. This accounts for its name.

Equations (10) and (13) differ in the number of negative terms that each contains on the left side when the right side is $+1$. The number of negative signs is the same as the number of sheets of the hyperboloid. If

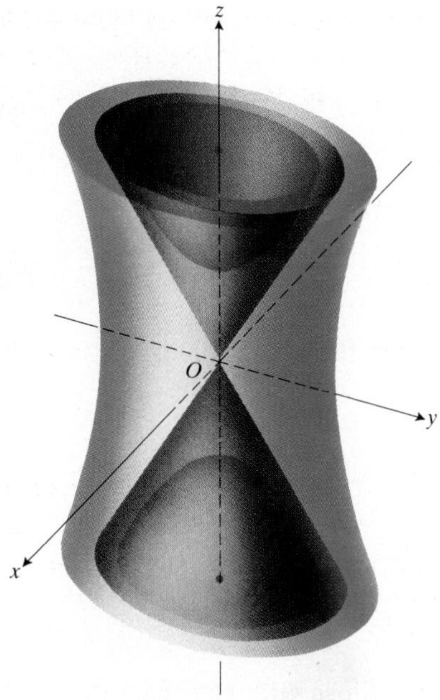

13.62 Cone asymptotic to hyperboloid of one sheet and hyperboloid of two sheets.

we compare with Eq. (6), we see that replacing the one on the right side of either Eq. (10) or (13) by zero gives the equation of a cone. This cone (Fig. 13.62) is, in fact, asymptotic to both of the hyperboloids (10) and (13) in the same way that the lines

$$\frac{x^2}{a^2} - \frac{y^2}{b^2} = 0$$

are asymptotic to the two hyperbolas

$$\frac{x^2}{a^2} - \frac{y^2}{b^2} = \pm 1$$

in the xy-plane. □

EXAMPLE 7 *The hyperbolic paraboloid* (Fig. 13.63)

$$\frac{y^2}{b^2} - \frac{x^2}{a^2} = \frac{z}{c} \tag{14}$$

has symmetry with respect to the planes $x = 0$ and $y = 0$. The sections in these planes are

$$x = 0: \quad y^2 = b^2\frac{z}{c}, \tag{15a}$$

$$y = 0: \quad x^2 = -a^2\frac{z}{c}, \tag{15b}$$

which are parabolas. In the plane $x = 0$, the parabola opens upward and has vertex at the origin. The parabola in the plane $y = 0$ has the same vertex, but it opens downward.

If we cut the surface by a plane $z = z_1 > 0$, the section is a hyperbola,

$$\frac{y^2}{b^2} - \frac{x^2}{a^2} = \frac{z_1}{c}, \tag{16}$$

whose focal axis is parallel to the y-axis and that has its vertices on the parabola in (15a). If, on the other hand, z_1 is negative in Eq. (16), then the focal axis of the hyperbola is parallel to the x-axis, and its vertices lie on the parabola in (15b).

Near the origin the surface is shaped like a saddle. To a person traveling along the surface in the yz-plane, the origin looks like a minimum. To

13.63 Hyperbolic paraboloid.

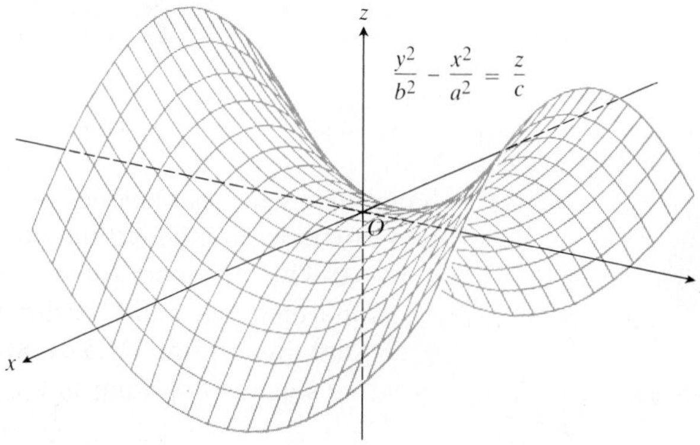

a person traveling in the xz-plane, on the other hand, the origin looks like a maximum. Such a point is called a *minimax* or *saddle point* of a surface (Fig. 13.63). We shall discuss maximum and minimum points on surfaces in Chapter 15.

If $a = b$ in Eq. (14), the surface is not a surface of revolution, but it is possible to express the equation in the alternative form

$$\frac{2x'y'}{a^2} = \frac{z}{c} \tag{17}$$

if we refer it to $x'y'$-axes obtained by rotating the xy-axes through 45°. □

PROBLEMS

In Problems 1–24, describe and sketch each of the surfaces [(r, θ, z) are cylindrical coordinates]. Complete the square, when necessary, to put the equation into one of the standard forms shown in the examples.

1. $x^2 + y^2 = z - 1$

2. $x^2 + y^2 + z^2 + 4x - 6y = 3$

3. $x^2 + 4y^2 + z^2 = 4$

4. $x^2 + 4y^2 + z^2 - 8y = 0$

5. $4x^2 + 4y^2 + 4z^2 - 8y = 0$

6. $x^2 - y^2 + z^2 + 4x - 6y = 9$

7. $x^2 - y^2 - z^2 + 4x - 6y = 9$

8. $z^2 = 4x$

9. $z^2 = 4xy$

10. $z = 4xy$

11. $z = r^2$

12. $z = r$

13. $z^2 = r$

14. $z^2 = x^2 + 4y^2$

15. $z^2 = x^2 - 4y^2$

16. $z^2 = 4y^2 - x^2$

17. $z^2 = x^2 + 4y^2 - 2x + 8y + 4z$

18. $z^2 = x^2 + 4y^2 - 2x + 8y + 4z + 1$

19. $x^2 + 4z^2 = 4$

20. $x = y^2 + 4z^2 + 1$

21. $z = r\cos\theta$

22. $z = r\sin\theta$

23. $z = \sin\theta \ (0 \le \theta \le \pi/2)$

24. $z = \cosh\theta \ (0 \le \theta \le \pi/2)$

25. a) Express the area, $A(z_1)$, of the cross section cut from the ellipsoid

$$\frac{x^2}{a^2} + \frac{y^2}{b^2} + \frac{z^2}{c^2} = 1$$

by the plane $z = z_1$, as a function of z_1. (The area of an ellipse of semiaxes A and B is πAB.)

b) By integration, find the volume of the ellipsoid of part (a). Consider slices made by planes perpendicular to the z-axis. Does your answer give the correct volume of a sphere in case $a = b = c$?

26. By integration, prove that the volume of the segment of the elliptic paraboloid

$$\frac{x^2}{a^2} + \frac{y^2}{b^2} = \frac{z}{c}$$

cut off by the plane $z = h$ is equal to one half the area of its base times its altitude.

27. Given the hyperboloid of one sheet of Eq. (10).

a) By integration, find the volume between the plane $z = 0$ and the plane $z = h$, enclosed by the hyperboloid.

b) Express your answer to part (a) in terms of the altitude h and the areas A_0 and A_h of the plane ends of the segment of the hyperboloid.

c) Verify that the volume of part (a) is also given exactly by the prismoid formula

$$V = h(A_0 + 4A_m + A_h)/6,$$

where A_0 and A_h are the areas of the plane ends of the segment of the hyperboloid and A_m is the area of its midsection cut out by the plane $z = h/2$.

28. If the hyperbolic paraboloid

$$\frac{y^2}{b^2} - \frac{x^2}{a^2} = \frac{z}{c}$$

is cut by the plane $y = y_1$, the resulting curve is a parabola. Find its vertex and focus.

29. What is the nature, in general, of a surface whose equation in spherical coordinates has the form $\rho = F(\phi)$? Give reasons for your answer.

Describe and sketch the following surfaces, which are special cases of Problem 29.

30. $\rho = a\cos\phi$

31. $\rho = a(1 + \cos\phi)$

32. A surface has the equation $z^2 = 3(x^2 + y^2)$ in rectangular coordinates. Find equations for the surface in (a) cylindrical and (b) spherical coordinates. Identify and sketch the surface.

***Toolkit* programs**

3D Grapher

REVIEW QUESTIONS AND EXERCISES

1. When are two vectors equal?
2. How are two vectors added? Subtracted?
3. If a vector is multiplied by a scalar, how is the result related to the original vector? In your discussion include all possible values of the scalar: positive, negative, and zero.
4. In a single diagram, show the cartesian, cylindrical, and spherical coordinates of an arbitrary point P, and write the expressions for each set of coordinates in terms of the other two kinds.
5. What set in space is described by
   a) $x = \text{constant}$, b) $r = \text{constant}$,
   c) $\theta = \text{constant}$, d) $\rho = \text{constant}$,
   e) $\phi = \text{constant}$, f) $ax + by + cz = d$,
   g) $ax^2 + by^2 + cz^2 = d$?
6. What is the length of the vector $a\mathbf{i} + b\mathbf{j} + c\mathbf{k}$? On what theorem of plane geometry does this result depend?
7. Define *scalar product* of two vectors. Which algebraic laws (commutative, associative, distributive) are satisfied by the operations of addition and scalar multiplication of vectors? Which of these laws is (are) not satisfied? Explain. When is the scalar product equal to zero?
8. Suppose that $\mathbf{i}, \mathbf{j}, \mathbf{k}$ is one set of mutually orthogonal unit vectors and that $\mathbf{i}', \mathbf{j}', \mathbf{k}'$ is another set of such vectors. Suppose that all the scalar products of a unit vector from one set with a unit vector from the other set are known. Let
$$\mathbf{A} = a\mathbf{i} + b\mathbf{j} + c\mathbf{k} = a'\mathbf{i}' + b'\mathbf{j}' + c'\mathbf{k}'$$
and express a, b, c in terms of a', b', c'; and conversely. (Expressions involve $\mathbf{i}\cdot\mathbf{i}', \mathbf{i}\cdot\mathbf{j}', \mathbf{i}\cdot\mathbf{k}'$, and so forth.)
9. List four applications of the scalar product.
10. Define *vector product* of two vectors. Which algebraic laws (commutative, associative, distributive) are satisfied by the vector product operation (combined with addition), and which are not? Explain. When is the vector product equal to zero?
11. Derive the formula for expressing the vector product of two vectors as a determinant. What is the effect of interchanging the order of the two vectors and the corresponding rows of the determinant?
12. How may vector and scalar products be used to find the equation of a plane through three given points?
13. With the book closed, develop equations for a line
    a) through two given points,
    b) through one point and parallel to a given line.
14. With the book closed, develop the equation of a plane
    a) through a given point and normal to a given vector,
    b) through one point and parallel to a given plane,
    c) through a point and perpendicular to each of two given planes.
15. What is the geometrical interpretation of
$$\mathbf{A}\cdot(\mathbf{B}\times\mathbf{C})?$$
When is this triple scalar product equal to zero?
16. Given a parallelogram $PQRS$ in space, how could you find a vector normal to its plane and with length equal to its area?
17. What set in space is described by an equation of the form
    a) $f(x, y) = 0$, b) $f(z, r) = 0$,
    c) $z = f(\theta)$, $0 \le \theta \le 2\pi$?
18. Define *quadric surface*. Name and sketch six different quadric surfaces and indicate their equations.

MISCELLANEOUS PROBLEMS

In Problems 1 through 10, find parametric equations of the path traced by the point $P(x, y)$ for the data given.

1. $\dfrac{dx}{dt} = x^2, \quad \dfrac{dy}{dt} = x; \quad t = 0, x = 1, y = 1$
2. $\dfrac{dx}{dt} = \cos^2 x, \quad \dfrac{dy}{dt} = x; \quad t = 0, x = \dfrac{\pi}{4}, y = 0$
3. $\dfrac{dx}{dt} = e^t, \quad \dfrac{dy}{dt} = xe^x; \quad t = 0, x = 1, y = 0$
4. $\dfrac{dx}{dt} = 6\sin 2t, \quad \dfrac{dy}{dt} = 4\cos 2t; \quad t = 0, x = 0, y = 4$
5. $\dfrac{dx}{dt} = 1 - \cos t, \quad \dfrac{dy}{dt} = \sin t; \quad t = 0, x = 0, y = 0$
6. $\dfrac{dx}{dt} = \sqrt{1 + y}, \quad \dfrac{dy}{dt} = y; \quad t = 0, x = 0, y = 1$
7. $\dfrac{dx}{dt} = \operatorname{sech} x, \quad \dfrac{dy}{dt} = x; \quad t = 0, x = 0, y = 0$
8. $\dfrac{dx}{dt} = \cosh\dfrac{t}{2}, \quad \dfrac{dy}{dt} = x; \quad t = 0, x = 2, y = 0$
9. $\dfrac{dx}{dt} = y, \quad \dfrac{dy}{dt} = -x; \quad t = 0, x = 0, y = 4$

10. $\dfrac{d^2x}{dt^2} = -\dfrac{dx}{dt}, \quad \dfrac{dy}{dt} = x; \quad t = 0, x = 1, y = 1, \dfrac{dx}{dt} = 1$

11. A particle is projected with velocity v at an angle α to the horizontal from a point that is at the foot of a hill inclined at an angle ϕ to the horizontal, where

$$0 < \phi < \alpha < (\pi/2).$$

Show that it reaches the ground at a distance

$$\frac{2v^2 \cos \alpha}{g \cos^2 \phi} \sin (\alpha - \phi)$$

measured up the face of the hill. Hence show that the greatest range achieved for a given v is when $\alpha = (\phi/2) + (\pi/4)$.

12. A wheel of radius 4 in. rolls along the x-axis with angular velocity 2 rad/s. Find the curve described by a point on a spoke and 2 in. from the center of the wheel if it starts from the point (0, 2) at time $t = 0$.

13. OA is the diameter of a circle of radius a. AN is tangent to the circle at A. A line through O making angle θ with diameter OA intersects the circle at M and tangent line at N. On ON a point P is located so that $OP = MN$. Taking O as origin, OA along the y-axis, and angle θ as parameter, find parametric equations of the locus described by P.

14. Let a line AB be the x-axis of a system of rectangular coordinates. Let the point C be the point (0, 1). Let the line DE through C intersect AB at F. Let P and P' be the points on DE such that $PF = P'F = a$. Express the coordinates of P and P' in terms of the angle $\theta = \angle CFB$.

15. For the curve

$$x = a(t - \sin t), \qquad y = a(1 - \cos t),$$

find the following quantities.

a) The area bounded by the x-axis and one loop of the curve
b) The length of one loop
c) The area of the surface of revolution obtained by rotating one loop about the x-axis
d) The coordinates of the centroid of the area in (a).

16. In Fig. 13.64, D is the midpoint of side AB and E is one third of the way between C and B. Using vectors, prove that F is midpoint of the line CD.

Figure 13.64

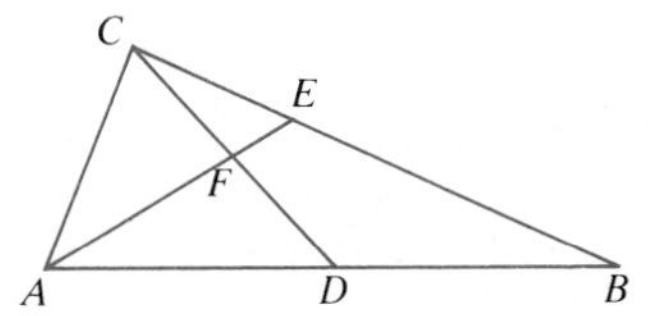

17. The vectors $2\mathbf{i} + 3\mathbf{j}$, $4\mathbf{i} + \mathbf{j}$, and $5\mathbf{i} + y\mathbf{j}$ have their initial points at the origin. Find the value of y so that the vectors terminate on one straight line.

18. $\mathbf{A}$ and $\mathbf{B}$ are vectors from the origin to the two points A and B. The point P is determined by the vector $\overrightarrow{OP} = x\mathbf{A} + y\mathbf{B}$, where x and y are positive numbers whose sum is equal to one. Prove that P lies on the line segment AB.

19. Using vector methods, prove that the segment joining the midpoints of two sides of a triangle is parallel to, and half the length of, the third side.

20. Let ABC be a triangle and let M be the midpoint of AB. Let P be the point on CM that is two thirds of the way from C to M. Let O be any point in space.

a) Show that

$$\overrightarrow{OP} = \left(\frac{\overrightarrow{OA} + \overrightarrow{OB} + \overrightarrow{OC}}{3}\right).$$

b) Show how the result in part (a) leads to the conclusion that the medians of a triangle meet in a point.

21. A, B, C are the vertices of a triangle and a, b, c are the midpoints of the opposite sides. Show that

$$\overrightarrow{Aa} + \overrightarrow{Bb} + \overrightarrow{Cc} = \mathbf{0}.$$

Interpret the result geometrically.

22. Vectors are drawn from the center of a regular polygon to its vertices. Show that their sum is zero.

23. Let $\mathbf{A}$, $\mathbf{B}$, $\mathbf{C}$ be vectors from a common point O to points A, B, C.

a) If A, B, C are collinear, show that three constants x, y, z (not all zero) exist such that

$$x + y + z = 0 \quad \text{and} \quad x\mathbf{A} + y\mathbf{B} + z\mathbf{C} = \mathbf{0}.$$

b) Conversely, if three constants x, y, z (not all zero) exist such that

$$x + y + z = 0 \qquad \text{and} \qquad x\mathbf{A} + y\mathbf{B} + z\mathbf{C} = \mathbf{0},$$

show that A, B, C are collinear.

24. Find the vector projection of $\mathbf{B}$ onto $\mathbf{A}$ if

$$\mathbf{A} = 3\mathbf{i} - \mathbf{j} + \mathbf{k}, \qquad \mathbf{B} = 2\mathbf{i} + \mathbf{j} - 2\mathbf{k}.$$

25. Find the cosine of the angle between the line

$$\frac{1 - x}{4} = \frac{y}{3} = -\frac{z}{5}$$

and the vector $\mathbf{i} + \mathbf{j}$.

26. Given two noncollinear vectors $\mathbf{A}$ and $\mathbf{B}$. Given also that $\mathbf{A}$ can be expressed in the form $\mathbf{A} = \mathbf{C} + \mathbf{D}$, where $\mathbf{C}$ is a vector parallel to $\mathbf{B}$, and $\mathbf{D}$ is a vector perpendicular to $\mathbf{B}$. Express $\mathbf{C}$ and $\mathbf{D}$ in terms of $\mathbf{A}$ and $\mathbf{B}$.

27. The curve whose vector equation is

$$\mathbf{r} = (t^4 + 2t^2 + 1)\mathbf{i} + (1 + 4t - t^4)\mathbf{j}$$

intersects the line $x + y = 0$, $z = 0$. Find the cosine of the angle which the acceleration vector makes with the radius vector at the point of intersection.

28. *Using vectors,* prove that for any four numbers a, b, c, d we have the inequality

$$(a^2 + b^2)(c^2 + d^2) \geq (ac + bd)^2.$$

(*Hint:* Consider $\mathbf{A} = a\mathbf{i} + b\mathbf{j}$ and $\mathbf{B} = c\mathbf{i} + d\mathbf{j}$.)

29. Find a vector parallel to the plane $2x - y - z = 4$ and perpendicular to the vector $\mathbf{i} + \mathbf{j} + \mathbf{k}$.

30. Find a vector which is normal to the plane determined by the points

$$A(1, 0, -1), \quad B(2, -1, 1), \quad C(-1, 1, 2).$$

31. Given vectors

$$\mathbf{A} = 2\mathbf{i} - \mathbf{j} + \mathbf{k},$$
$$\mathbf{B} = \mathbf{i} + 2\mathbf{j} - \mathbf{k},$$
$$\mathbf{C} = \mathbf{i} + \mathbf{j} - 2\mathbf{k},$$

find a *unit* vector in the plane of **B** and **C** that is $\perp\mathbf{A}$.

32. By forming the cross product of two appropriate vectors, derive the trigonometric identity

$$\sin(\alpha - \beta) = \sin\alpha\cos\beta - \cos\alpha\sin\beta.$$

33. Find a vector of *length two* parallel to the line

$$x + 2y + z - 1 = 0, \qquad x - y + 2z + 7 = 0.$$

34. Given a tetrahedron with vertices O, A, B, C. A vector is constructed normal to each face, pointing outward, and having length equal to the area of the face. Using cross products, prove that the sum of these four outward normals is the zero vector.

35. What angle does the line of intersection of the two planes

$$2x + y - z = 0, \qquad x + y + 2z = 0$$

make with the x-axis?

36. Let **A** and **C** be given vectors in space, with $\mathbf{A} \neq \mathbf{0}$ and $\mathbf{A} \cdot \mathbf{C} = 0$, and let d be a given scalar. Find a vector **B** that satisfies both equations $\mathbf{A} \times \mathbf{B} = \mathbf{C}$ and $\mathbf{A} \cdot \mathbf{B} = d$ simultaneously. The answer should be given as a formula involving **A**, **C**, and d.

37. Given any two vectors

$$\mathbf{A} = a_1\mathbf{i} + a_2\mathbf{j}, \qquad \mathbf{B} = b_1\mathbf{i} + b_2\mathbf{j}$$

in the plane, define a new vector, $\mathbf{A} \otimes \mathbf{B}$, called their "circle product," as follows:

$$\mathbf{A} \otimes \mathbf{B} = (a_1b_1 - a_2b_2)\mathbf{i} + (a_1b_2 + a_2b_1)\mathbf{j}.$$

This product satisfies the following algebraic laws.

a) $\mathbf{A} \otimes \mathbf{B} = \mathbf{B} \otimes \mathbf{A}$
b) $\mathbf{A} \otimes (\mathbf{B} \otimes \mathbf{C}) = (\mathbf{A} \otimes \mathbf{B}) \otimes \mathbf{C}$
c) $\mathbf{A} \otimes (\mathbf{B} + \mathbf{C}) = (\mathbf{A} \otimes \mathbf{B}) + (\mathbf{A} \otimes \mathbf{C})$
d) $|\mathbf{A} \otimes \mathbf{B}| = |\mathbf{A}|\,|\mathbf{B}|$

Prove (a) and (d).

38. Find the equations of the straight line that passes through the point (1, 2, 3) and makes an angle of 30° with the positive x-axis and an angle of 60° with the positive y-axis.

39. The line L, whose equations are

$$x - 2z - 3 = 0, \qquad y - 2z = 0,$$

intersects the plane

$$x + 3y - z + 4 = 0.$$

Find the point of intersection P and find the equation of that line in this plane that passes through P and is perpendicular to L.

40. Find the distance between the point (2, 2, 3) and the plane $2x + 3y + 5z = 0$.

41. Given the two parallel planes

$$Ax + By + Cz = D_1, \qquad Ax + By + Cz = D_2,$$

show that the distance between them is given by the formula

$$\frac{|D_1 - D_2|}{|A\mathbf{i} + B\mathbf{j} + C\mathbf{k}|}.$$

42. Consider the straight line through the point (3, 2, 1) and perpendicular to the plane

$$2x - y + 2z + 2 = 0.$$

Compute the coordinates of the point of intersection of that line and that plane.

43. Find the distance between the point (2, 2, 0) and the line

$$x + y = 0, \qquad y - z = 1.$$

44. Find an equation of the plane parallel to the plane $2x - y + 2z + 4 = 0$ if the point (3, 2, −1) is equidistant from both planes.

45. Given the four points

$$A(-2, 0, -3), \qquad B(1, -2, 1),$$
$$C(-2, -\tfrac{13}{5}, \tfrac{26}{5}), \qquad D(\tfrac{16}{5}, -\tfrac{13}{5}, 0).$$

a) Find the equation of the plane through AB that is parallel to CD.
b) Compute the shortest distance between the lines AB and CD.

46. The three vectors

$$\mathbf{A} = 3\mathbf{i} - \mathbf{j} + \mathbf{k},$$
$$\mathbf{B} = \mathbf{i} + 2\mathbf{j} - \mathbf{k},$$
$$\mathbf{C} = \mathbf{i} + \mathbf{j} + \mathbf{k}$$

are all drawn from the origin. Find an equation for the plane through their endpoints.

47. Show that the plane through the three points

$$(x_1, y_1, z_1), \qquad (x_2, y_2, z_2), \qquad (x_3, y_3, z_3)$$

is given by

$$\begin{vmatrix} x_1 - x & y_1 - y & z_1 - z \\ x_2 - x & y_2 - y & z_2 - z \\ x_3 - x & y_3 - y & z_3 - z \end{vmatrix} = 0.$$

48. Given the two straight lines

$$x = a_1t + b_1, \qquad y = a_2t + b_2, \qquad z = a_3t + b_3,$$
$$x = c_1\tau + d_1, \qquad y = c_2\tau + d_2, \qquad z = c_3\tau + d_3,$$

where t and τ are parameters. Show that the necessary and sufficient condition that the two lines either intersect or are parallel is

$$\begin{vmatrix} a_1 & c_1 & b_1 - d_1 \\ a_2 & c_2 & b_2 - d_2 \\ a_3 & c_3 & b_3 - d_3 \end{vmatrix} = 0.$$

49. Given the vectors

$$\mathbf{A} = \mathbf{i} + \mathbf{j} - \mathbf{k},$$
$$\mathbf{B} = 2\mathbf{i} + \mathbf{j} + \mathbf{k},$$
$$\mathbf{C} = -\mathbf{i} - 2\mathbf{j} + 3\mathbf{k},$$

evaluate

a) $\mathbf{A} \cdot (\mathbf{B} \times \mathbf{C})$, b) $\mathbf{A} \times (\mathbf{B} \times \mathbf{C})$.

50. Given four points

$$A\ (1, 1, 1), \qquad B\ (0, 0, 2),$$
$$C\ (0, 3, 0), \qquad D\ (4, 0, 0),$$

find the volume of the tetrahedron with vertices at A, B, C, D, and find the angle between the edges AB and AC.

51. Prove or disprove the formula

$$\mathbf{A} \times [\mathbf{A} \times (\mathbf{A} \times \mathbf{B})] \cdot \mathbf{C} = -|\mathbf{A}|^2\mathbf{A} \cdot \mathbf{B} \times \mathbf{C}.$$

52. If the four vectors **A, B, C, D** are coplanar, show that

$$(\mathbf{A} \times \mathbf{B}) \times (\mathbf{C} \times \mathbf{D}) = \mathbf{0}.$$

53. Prove the following identities, in which **i, j, k** are three mutually perpendicular unit vectors, and **A, B, C** are any vectors.

a) $\mathbf{A} \times (\mathbf{B} \times \mathbf{C}) + \mathbf{B} \times (\mathbf{C} \times \mathbf{A}) + \mathbf{C} \times (\mathbf{A} \times \mathbf{B}) = \mathbf{0}$.

b) $\mathbf{A} \times \mathbf{B} = [\mathbf{A} \cdot (\mathbf{B} \times \mathbf{i})]\mathbf{i} + [\mathbf{A} \cdot (\mathbf{B} \times \mathbf{j})]\mathbf{j} + [\mathbf{A} \cdot (\mathbf{B} \times \mathbf{k})]\mathbf{k}$.

54. Show that

$$(\mathbf{a} \times \mathbf{b}) \cdot (\mathbf{c} \times \mathbf{d}) = \begin{vmatrix} \mathbf{a} \cdot \mathbf{c} & \mathbf{b} \cdot \mathbf{c} \\ \mathbf{a} \cdot \mathbf{d} & \mathbf{b} \cdot \mathbf{d} \end{vmatrix}.$$

55. Sketch the surfaces

a) $(x - 1)^2 + 4(y^2 + z^2) = 16$,

b) $z = r^2$ (cylindrical coordinates),

c) $\rho = a \sin \phi$ (spherical coordinates).

56. Find an equation for the set of points in space whose distance from the point $(2, -1, 3)$ is twice their distance from the xy-plane. Name the surface and find its center of symmetry.

57. Find an equation of the sphere that has the two planes

$$x + y + z - 3 = 0, \qquad x + y + z - 9 = 0$$

as tangent planes, if the two planes

$$2x - y = 0, \qquad 3x - z = 0$$

pass through the center of the sphere.

58. The two cylinders $z^3 - x = 0$ and $x^2 - y = 0$ intersect in a curve C. Find an equation of a cylinder parallel to the x-axis which passes through C. This cylinder traces out a curve C' in the yz-plane. Rotate C' about an y-axis and obtain the equation for the surface so generated.

14 Vector Functions and Their Derivatives

14.1 Derivatives of Vector Functions

Vector Functions

Let **i**, **j**, and **k** be the unit vectors along the x-, y-, and z-axes in space. Then a function

$$\mathbf{F}(t) = \mathbf{i}x(t) + \mathbf{j}y(t) + \mathbf{k}z(t), \tag{1}$$

where $x(t)$, $y(t)$, and $z(t)$ are real-valued functions of the real variable t, is a vector-valued function or *vector function* of t. We shall use such functions primarily to describe the motion of a particle in space or in the plane. When the motion is in the plane, we shall ordinarily assume that coordinates are chosen to make the plane of motion the xy-plane. This allows us to consider motion in the xy-plane as a special case of motion in space, the special condition being that the z-coordinate of the particle is zero.

EXAMPLE 1 The position of a particle in the xy-plane at time t is given by

$$\mathbf{R}(t) = x\mathbf{i} + y\mathbf{j} = e^t\mathbf{i} + te^t\mathbf{j}$$

(Fig. 14.1). Where is the particle at time $t = 0$, and how fast is it moving and in what direction?

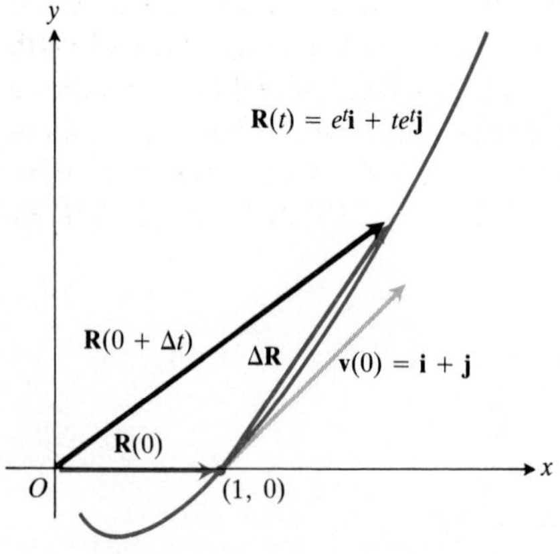

14.1 At $t = 0$, the position and velocity vectors are $\mathbf{R}(0) = \mathbf{i}$ and $\mathbf{v}(0) = \mathbf{i} + \mathbf{j}$.

Solution At $t = 0$, we have $x = 1$ and $y = 0$, so

$$\mathbf{R}(0) = \mathbf{i}.$$

This is the vector from the origin to the position $P(0, 1)$ of the particle at time $t = 0$. Next, how can we find the speed and direction of motion? If you have studied particle mechanics from a vector viewpoint, then you probably know the answer. If not, think where the particle might be a short time Δt past $t = 0$. Both x and y will have increased somewhat, and **R** will have changed by the amount

$$\Delta\mathbf{R} = \mathbf{i}\,\Delta x + \mathbf{j}\,\Delta y.$$

The particle's speed will therefore be approximately

$$\frac{|\Delta \mathbf{R}|}{\Delta t} = \left|\frac{\Delta \mathbf{R}}{\Delta t}\right|.$$

If we let $\Delta t \to 0$, the right-hand side of

$$\frac{\Delta \mathbf{R}}{\Delta t} = \mathbf{i}\frac{\Delta x}{\Delta t} + \mathbf{j}\frac{\Delta y}{\Delta t} \tag{2}$$

will approach the vector

$$\mathbf{v} = \mathbf{i}\frac{dx}{dt} + \mathbf{j}\frac{dy}{dt} = \mathbf{i}e^t + \mathbf{j}(te^t + e^t),$$

whose value at $t = 0$ is

$$\mathbf{v}(0) = \mathbf{i} + \mathbf{j}.$$

The left-hand side of Eq. (2) therefore has the limit

$$\lim_{\Delta t \to 0} \frac{\Delta \mathbf{R}}{\Delta t} = \mathbf{i} + \mathbf{j},$$

and it is natural to call this limit the velocity of the motion at $t = 0$. Accordingly, the particle's speed and direction at $t = 0$ are

$$\text{Speed} = |\mathbf{i} + \mathbf{j}| = \sqrt{2},$$

$$\text{Direction} = \frac{\mathbf{i} + \mathbf{j}}{\sqrt{2}}. \quad \square$$

We now pause to define more precisely the notions of limit and derivative for vector functions. We shall return to velocity in the next article as one of several vector derivatives that are important in studying motion.

Limits

Roughly speaking, we want to say that the limit of the vector $\mathbf{F}(t)$ as $t \to a$ is the vector $\mathbf{L}$ if $\mathbf{F}(t)$ moves into the "position" occupied by $\mathbf{L}$ as $t \to a$. In the limit, the length and direction of $\mathbf{F}$ should match the length and direction of $\mathbf{L}$. If we assume that $\mathbf{L}$ and $\mathbf{F}(t)$ are located so that their initial points coincide, then we may picture the convergence of $\mathbf{F}(t)$ to $\mathbf{L}$ as in Fig. 14.2. This picture shows that for $\mathbf{F}(t)$ to approach $\mathbf{L}$, the difference vector $\mathbf{F}(t) - \mathbf{L}$ must shrink to zero, which is equivalent to saying that its length $|\mathbf{F}(t) - \mathbf{L}|$ goes to zero as $t \to a$. This length is a real-valued function of the single variable t, and we know what it means for such a function to have zero as a limit as $t \to a$. Thus we may define limits of vector-valued functions in terms of the familiar limits of real-valued functions in the following way.

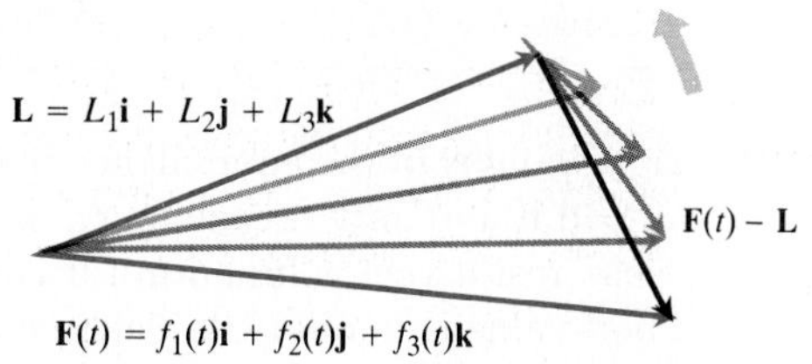

14.2 We say that $\lim_{t\to a} \mathbf{F}(t) = \mathbf{L}$ if $\lim_{t\to a} |\mathbf{F}(t) - \mathbf{L}| = 0$.

DEFINITION

Limit of a Vector-Valued Function

Let $\mathbf{F}(t) = f_1(t)\mathbf{i} + f_2(t)\mathbf{j} + f_3(t)\mathbf{k}$ be a vector valued-function of t. The limit of $\mathbf{F}(t)$ as t approaches the number a is the vector $\mathbf{L}$ if and only if the limit of $|\mathbf{F}(t) - \mathbf{L}|$ as t approaches a is zero. In symbols,

$$\mathbf{L} = \lim_{t\to a} \mathbf{F}(t) \quad \Leftrightarrow \quad \lim_{t\to a} |\mathbf{F}(t) - \mathbf{L}| = 0. \tag{3}$$

In practice, the most useful observation to make is that $\mathbf{F}(t)$ having $\mathbf{L}$ as a limit means that the components of $\mathbf{F}$ have the corresponding components of $\mathbf{L}$ as limits. To be precise, if

$$\mathbf{F}(t) = f_1(t)\mathbf{i} + f_2(t)\mathbf{j} + f_3(t)\mathbf{k} \quad \text{and} \quad \mathbf{L} = L_1\mathbf{i} + L_2\mathbf{j} + L_3\mathbf{k},$$

then it can be shown that

$$\lim \mathbf{F}(t) = \mathbf{L} \quad \Leftrightarrow \quad \lim f_1 = L_1,\ \lim f_2 = L_2, \quad \text{and} \quad \lim f_3 = L_3. \tag{4}$$

This equivalence says that we may calculate limits of vector-valued functions one component at a time.

EXAMPLE 2 Find $\lim_{t\to 0} \mathbf{F}(t)$ if

$$\mathbf{F}(t) = e^t\mathbf{i} + \frac{\sin t}{t}\mathbf{j} + \frac{t^2 + 6}{3 + \ln(1 + t)}\mathbf{k}.$$

Solution The limits of the components of $\mathbf{F}(t)$ as $t \to 0$ are

$$\lim_{t\to 0} e^t = 1, \qquad \lim_{t\to 0} \frac{\sin t}{t} = 1, \qquad \lim_{t\to 0} \frac{t^2 + 6}{3 + \ln(1 + t)} = 2.$$

Therefore,

$$\lim_{t\to 0} \mathbf{F}(t) = \mathbf{i} + \mathbf{j} + 2\mathbf{k}. \ \square$$

Continuity

The definition of continuity of a vector-valued function $\mathbf{F}(t)$ is the same as the definition for continuity of a real-valued function.

DEFINITION

Continuity

A vector-valued function $\mathbf{F}(t)$ is *continuous* at $t = a$ if $\mathbf{F}$ is defined at a, $\lim_{t\to a} \mathbf{F}(t)$ exists, and

$$\lim_{t\to a} \mathbf{F}(t) = \mathbf{F}(a).$$

In view of the equivalence in (4) we may say that $\mathbf{F}(t)$ is continuous at $t = a$ if and only if each component of $\mathbf{F}$ is continuous at $t = a$. Thus we may test a vector function for continuity by applying our knowledge of real-valued functions to each component of $\mathbf{F}$.

EXAMPLE 3 The function

$$\mathbf{F}(t) = \frac{1}{\sqrt{t}}\mathbf{i} + (\sin t)\mathbf{j} + \ln(1 + t^2)\mathbf{k}$$

is continuous at every value of $t > 0$ because each component is continuous for $t > 0$. However, $\mathbf{F}$ is discontinuous for $t \leq 0$ because $1/\sqrt{t}$, the first component of $\mathbf{F}$, is not defined for $t \leq 0$. □

Derivatives

We define the derivative of a vector-valued function $\mathbf{F}(t)$ at a point $t = a$ by the same type of limit equation we use for scalar functions:

$$\mathbf{F}'(a) = \lim_{h \to 0} \frac{\mathbf{F}(a + h) - \mathbf{F}(a)}{h}, \tag{5}$$

provided the limit exists. As we might expect, $\mathbf{F}$ is differentiable at $t = a$ if and only if each of its components is differentiable at $t = a$.

THEOREM

A vector function

$$\mathbf{F}(t) = f_1(t)\mathbf{i} + f_2(t)\mathbf{j} + f_3(t)\mathbf{k}$$

is differentiable at $t = a$ if and only if each of its component functions is differentiable at $t = a$. If this condition is met, then

$$\mathbf{F}'(a) = f_1'(a)\mathbf{i} + f_2'(a)\mathbf{j} + f_3'(a)\mathbf{k}. \tag{6}$$

Proof. The difference quotient we need to consider is

$$\frac{\mathbf{F}(a + h) - \mathbf{F}(a)}{h} = \frac{f_1(a + h) - f_1(a)}{h}\mathbf{i} + \frac{f_2(a + h) - f_2(a)}{h}\mathbf{j} + \frac{f_3(a + h) - f_3(a)}{h}\mathbf{k}. \tag{7}$$

The left-hand side of this equation has a limit as $h \to 0$ if and only if each component on the right-hand side has a limit as $h \to 0$. The ith component on the right has a limit if and only if f_i is differentiable at a. Finally, if each component is differentiable at a, then as $h \to 0$ and we pass to the limit in Eq. (7), we get Eq. (6). This concludes the proof. ■

EXAMPLE 4 The derivative of

$$\mathbf{F}(t) = (\sin^2 t)\mathbf{i} + (\ln t)\mathbf{j} + \tan^{-1}(3t)\mathbf{k}$$

is

$$\mathbf{F}'(t) = (2 \sin t \cos t)\mathbf{i} + \frac{1}{t}\mathbf{j} + \frac{3}{1 + 9t^2}\mathbf{k}.$$

The function and its derivative are defined at every positive value of t. □

The Chain Rule

If $\mathbf{F}(t) = f_1(t)\mathbf{i} + f_2(t)\mathbf{j} + f_3(t)\mathbf{k}$ is a differentiable function of t, and $t = g(s)$ is a differentiable function of s, then the composite $\mathbf{F}(g(s))$ is a differentiable function of s and

$$\frac{d\mathbf{F}}{ds} = \mathbf{F}'(g(s))g'(s). \tag{8}$$

We often write Eq. (8) in the abbreviated form

$$\frac{d\mathbf{F}}{ds} = \frac{d\mathbf{F}}{dt}\frac{dt}{ds}. \tag{9}$$

The Chain Rule given for vector functions by Eq. (9) is an immediate consequence of the Chain Rule for scalar functions that applies to the components f_1, f_2, and f_3. According to this rule, f_1, f_2, and f_3 are differentiable functions of s and

$$\frac{df_1}{ds} = \frac{df_1}{dt}\frac{dt}{ds}, \qquad \frac{df_2}{ds} = \frac{df_2}{dt}\frac{dt}{ds}, \qquad \frac{df_3}{ds} = \frac{df_3}{dt}\frac{dt}{ds}.$$

Therefore,

$$\begin{aligned}\frac{d\mathbf{F}}{ds} &= \left(\frac{df_1}{ds}\right)\mathbf{i} + \left(\frac{df_2}{ds}\right)\mathbf{j} + \left(\frac{df_3}{ds}\right)\mathbf{k}\\ &= \left(\frac{df_1}{dt}\frac{dt}{ds}\right)\mathbf{i} + \left(\frac{df_2}{dt}\frac{dt}{ds}\right)\mathbf{j} + \left(\frac{df_3}{dt}\frac{dt}{ds}\right)\mathbf{k}\\ &= \left(\frac{df_1}{dt}\mathbf{i} + \frac{df_2}{dt}\mathbf{j} + \frac{df_3}{dt}\mathbf{k}\right)\frac{dt}{ds}\\ &= \left(\frac{d\mathbf{F}}{dt}\right)\frac{dt}{ds}.\end{aligned}$$

EXAMPLE 5 Express $d\mathbf{F}/ds$ in terms of s if $\mathbf{F}(t) = \mathbf{i} + \sin(t+1)\mathbf{j} + e^{t+1}\mathbf{k}$ and $t = g(s) = s^2 - 1$.

Solution From the Chain Rule we have

$$\frac{d\mathbf{F}}{ds} = \frac{d\mathbf{F}}{dt}\frac{dt}{ds} = (\cos(t+1)\mathbf{j} + e^{t+1}\mathbf{k})(2s) = 2s\cos(s^2)\mathbf{j} + 2se^{s^2}\mathbf{k}.$$

This is what we would have obtained had we first substituted $t = g(s) = s^2 - 1$ in the formula for $\mathbf{F}(t)$ and then differentiated with respect to s:

$$\mathbf{F}(g(s)) = \mathbf{i} + \sin(s^2)\mathbf{j} + e^{s^2}\mathbf{k},$$

$$\frac{d}{ds}\mathbf{F}(g(s)) = 2s\cos(s^2)\mathbf{j} + 2se^{s^2}\mathbf{k}. \ \square$$

PROBLEMS

1. Find $\lim_{t\to 0}\mathbf{F}(t)$ if $\mathbf{F}(t) = e^t\mathbf{i} + te^t\mathbf{j}$.

2. Find $\lim_{t\to 1}\mathbf{F}(t)$ if $\mathbf{F}(t) = (t-1)\mathbf{i} + 3\mathbf{j} + 7t\mathbf{k}$.

3. Find $\lim_{t\to 0}\mathbf{F}(t)$ if $\mathbf{F}(t) = (e^t\sin t)\mathbf{i} + (e^t\cos t)\mathbf{j} - e^t\mathbf{k}$.

4. Find $\lim_{t\to -3}\mathbf{F}(t)$ if

$$\mathbf{F}(t) = \frac{t^2+4t+3}{t+3}\mathbf{j} + \ln(4+t)\mathbf{k}.$$

5. Find $\lim_{t\to 0}\mathbf{F}(t)$ if

$$\mathbf{F}(t) = \frac{t}{\sin t}\mathbf{i} + \frac{1-\cos t}{t}\mathbf{j} + \mathbf{k}.$$

6. Find $\lim_{t\to 0}\mathbf{F}(t)$ if

$$\mathbf{F}(t) = \frac{t}{\cos t}\mathbf{i} + t\left(1+\frac{1}{t}\right)\mathbf{j} + (2+\sin t)\mathbf{k}.$$

At what values of t are the vector functions in Problems 7–12 continuous?

7. $\mathbf{F}(t) = (\cos t)\mathbf{i} + (\sin t)\mathbf{j} + \mathbf{k}$

8. $\mathbf{F}(t) = (t^2 - 1)\mathbf{i} + 3\mathbf{k}$

9. $\mathbf{F}(t) = e^t\mathbf{i} + \cos\dfrac{1}{t+1}\mathbf{j} + \ln|1+t|\mathbf{k}$

10. $\mathbf{F}(t) = \dfrac{1}{t-1}\mathbf{i} + \dfrac{1}{t-2}\mathbf{j} + \dfrac{1}{t-3}\mathbf{k}$

11. $\mathbf{F}(t) = (\tan^{-1} t)\mathbf{i} + t^2\mathbf{k}$

12. $\mathbf{F}(t) = \sqrt{1-t^2}\,\mathbf{i} + 3t\mathbf{j} - 7\mathbf{k}$

Find the derivatives of the vector functions in Problems 13–18, and give the domain of each derivative.

13. $\mathbf{F}(t) = e^{2t}\mathbf{i} + te^{-t}\mathbf{j}$

14. $\mathbf{F}(t) = (\ln\sqrt{1+t}\,)\mathbf{i} + \sqrt{1-t^2}\,\mathbf{j}$

15. $\mathbf{F}(t) = \mathbf{i} + 3\mathbf{j} - \mathbf{k}$

16. $\mathbf{F}(t) = \mathbf{i}\sin^{-1} 2t + \mathbf{j}\tan^{-1} 3t + \mathbf{k}(1/t)$

17. $\mathbf{F}(x) = \mathbf{i}\sec^{-1} 3x + \mathbf{j}\cosh 2x + \mathbf{k}\tanh 4x$

18. $\mathbf{F}(t) = \mathbf{i}\left(\dfrac{2t-1}{2t+1}\right) + \mathbf{j}\ln(1-4t^2)$

19. Find $\mathbf{F}'(s)$ if $\mathbf{F}(t) = (1/\sqrt{t})\mathbf{i} + (\sin t)\mathbf{j} + \ln(1+t^2)\mathbf{k}$ and $t = \sqrt{s-1}$.

20. Find $\mathbf{F}'(t)$ if $\mathbf{F}(x) = \mathbf{i}xe^x + \mathbf{j}\ln 3x$ and

a) $x = \ln t$, b) $x = e^t$.

21. Find $\mathbf{F}'(x)$ if

$$\mathbf{F}(u) = \mathbf{i}\cos\sqrt{u} + \mathbf{j}\tan^{-1} u - \mathbf{k}(1/(1+\sqrt{u})),$$

$u = x^2 + 2x + 1$, and $x + 1 > 0$.

22. Find $\mathbf{F}'(x)$ if

$$\mathbf{F}(u) = \mathbf{i}\cos\sqrt{u} + \mathbf{j}\tan^{-1} u - \mathbf{k}(1/(1+\sqrt{u})),$$

$u = x^2 + 2x + 1$, and $x + 1 > 0$.

23. Find $\mathbf{F}'(s)$ if

$$\mathbf{F}(t) = -\mathbf{i}\sin t + \mathbf{j}\cos t + \mathbf{k}$$

and $dt/ds = \sqrt{2}$.

24. Find $\mathbf{F}'(s)$ as a function of t if

$$\mathbf{F}(t) = (\cos t + t\sin t)\mathbf{i} + (\sin t - t\cos t)\mathbf{j}$$

and $dt/ds = t$.

25. Find $\mathbf{F}'(t)$ if $\mathbf{F}(\theta) = \mathbf{i}\sin 2\theta + \mathbf{j}\cos 2\theta$ and $d\theta/dt = 2t$.

***Toolkit* programs**	
Parametric Equations	Super * Grapher

14.2 Tangent Vectors, Velocity, and Acceleration

When we model the motion of a particle along a curve in the plane, we want the particle's velocity vector $\mathbf{v}$ to be tangent to the curve and to point in the direction of motion. We also want its length $|\mathbf{v}|$ to be the particle's speed $|ds/dt|$ (where s is arc length measured along the curve from some preselected base point P_0). The remarkable fact is that if

$$\mathbf{R} = x(t)\mathbf{i} + y(t)\mathbf{j}$$

gives the particle's position on the curve at time t, then the vector

$$\frac{d\mathbf{R}}{dt} = \frac{dx}{dt}\mathbf{i} + \frac{dy}{dt}\mathbf{j}$$

has these properties. The slope of the curve at any point $P(x, y)$ where the tangent is not vertical is

$$\frac{dy}{dx} = \frac{dy/dt}{dx/dt} = \text{the slope of } \frac{d\mathbf{R}}{dt}$$

(see Fig. 14.3). Furthermore, if s is arc length measured along the curve from the point $P_0(x(t_0), y(t_0))$, then the value of s at any value of t is

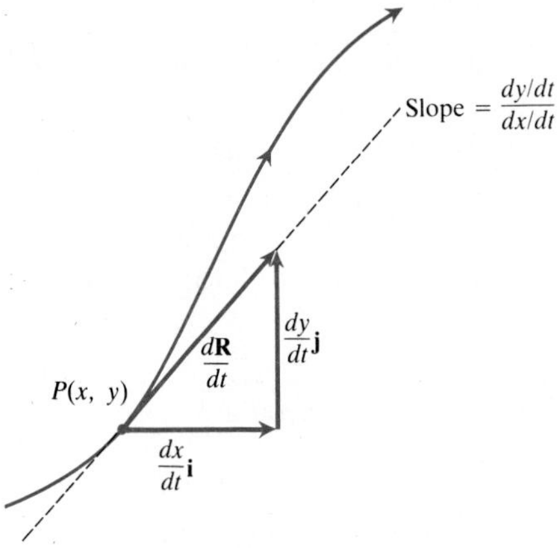

14.3 At each point where $dx/dt \neq 0$ on the curve $\mathbf{R} = x(t)\mathbf{i} + y(t)\mathbf{j}$, the slope of $d\mathbf{R}/dt$ is the slope of the tangent to the curve.

given by

$$s(t) = \begin{cases} \int_{t_0}^{t} \sqrt{x'(\tau)^2 + y'(\tau)^2}\, d\tau & \text{if } t \geq t_0, \\ \\ \int_{t}^{t_0} \sqrt{x'(\tau)^2 + y'(\tau)^2}\, d\tau & \text{if } t < t_0, \end{cases} \tag{1}$$

as we saw in Article 5.6. The Second Fundamental Theorem of Calculus (Article 4.7) applied to these integrals gives

$$\left|\frac{ds}{dt}\right| = \sqrt{x'(t)^2 + y'(t)^2} = \left|\frac{d\mathbf{R}}{dt}\right|. \tag{2}$$

Because $d\mathbf{R}/dt$ has the properties we want a velocity vector to have, we model velocity in mathematics by defining it to be the vector

$$\text{Velocity:} \qquad \mathbf{v} = \frac{d\mathbf{R}}{dt}. \tag{3}$$

At a point where $dx/dt = 0$ and $dy/dt \neq 0$, $d\mathbf{R}/dt = \mathbf{j}(dy/dt)$ is still the appropriate vector. For example, $\mathbf{R}(t) = (\cos t)\mathbf{i} + (\sin t)\mathbf{j}$ describes motion in a counterclockwise direction on a unit circle with center at the origin. Here,

$$\frac{d\mathbf{R}}{dt} = -(\sin t)\mathbf{i} + (\cos t)\mathbf{j}$$

points vertically upward at $t = 0$.

If both dx/dt and dy/dt are zero, then the speed is zero and a vector of zero length has no direction. This happens, for example, at each cusp on the four-cusped hypocycloid

$$x = \cos^3 t, \qquad y = \sin^3 t.$$

The acceleration of the motion is defined to be

$$\text{Acceleration:} \qquad \mathbf{a} = \frac{d\mathbf{v}}{dt} = \frac{d^2\mathbf{R}}{dt^2} = \frac{d^2x}{dt^2}\mathbf{i} + \frac{d^2y}{dt^2}\mathbf{j}. \tag{4}$$

For a particle of constant mass m moving under the action of an applied force $\mathbf{F}$, Newton's second law of motion states that

$$\mathbf{F} = m\mathbf{a}. \tag{5}$$

Since one ordinarily visualizes the force vector as being *applied at P*, it is customary to adopt the same viewpoint about the acceleration vector $\mathbf{a}$ (Fig. 14.4).

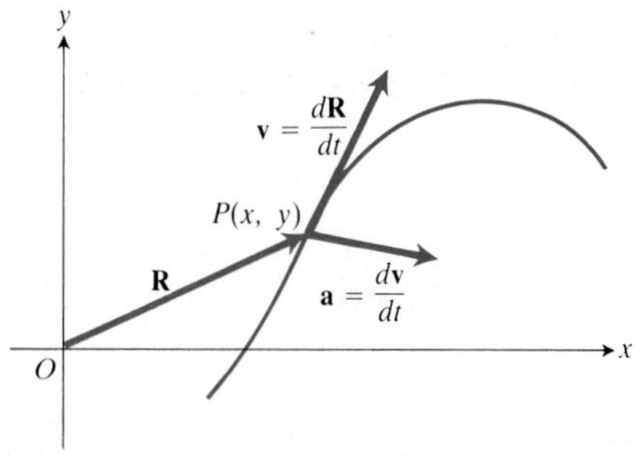

14.4 Typical position, velocity, and acceleration vectors of a particle moving in the plane.

EXAMPLE 1 A particle $P(x, y)$ moves on the hyperbola

$$x = r \cosh \omega t, \qquad y = r \sinh \omega t, \tag{6}$$

where r and ω are positive constants. Then

$$\mathbf{R} = \mathbf{i}(r \cosh \omega t) + \mathbf{j}(r \sinh \omega t),$$

$$\mathbf{v} = \frac{d\mathbf{R}}{dt} = \mathbf{i}(\omega r \sinh \omega t) + \mathbf{j}(\omega r \cosh \omega t),$$

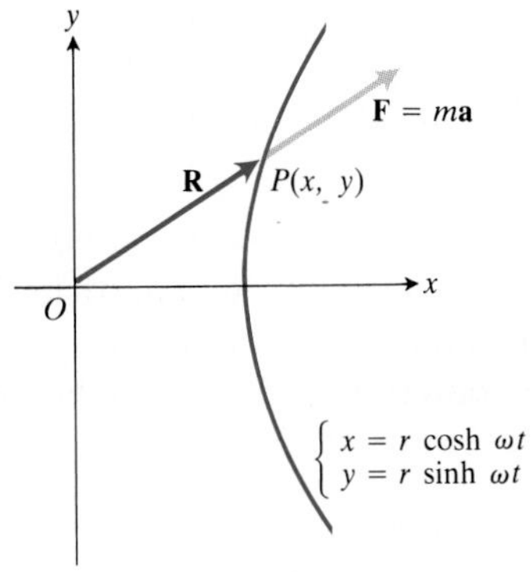

14.5 The force acting on the particle of Example 1 points directly away from O at all times.

and

$$\mathbf{a} = \mathbf{i}(\omega^2 r \cosh \omega t) + \mathbf{j}(\omega^2 r \sinh \omega t) = \omega^2 \mathbf{R}.$$

This means that the force $\mathbf{F} = m\mathbf{a} = m\omega^2\mathbf{R}$ has a magnitude $m\omega^2|\mathbf{R}| = m\omega^2|\overrightarrow{OP}|$ which is directly proportional to the distance OP, and that its direction is the same as the direction of $\mathbf{R}$. Thus the force is directed away from O (Fig. 14.5). □

The next example illustrates how we obtain the path of motion by integrating Eq. (5) when the force $\mathbf{F}$ is a given function of time, and the initial position and initial velocity of the particle are given. In general, the force $\mathbf{F}$ may depend upon the position of P as well as upon the time, and the problem of integrating the differential equations so obtained is usually discussed in textbooks on that subject.

EXAMPLE 2 The force acting on a particle P of mass m in the plane is given as a function of time t by

$$\mathbf{F} = \mathbf{i} \cos t + \mathbf{j} \sin t.$$

If the particle starts at the point (2, 0) with the initial velocity $v_0\mathbf{j}$ (Fig. 14.6), find its position at time t.

Solution If we denote the position vector by

$$\mathbf{R}(t) = \mathbf{i}x(t) + \mathbf{j}y(t),$$

we may restate the problem as follows: Find $\mathbf{R}(t)$ if

$$\mathbf{F} = m\frac{d^2\mathbf{R}}{dt^2} = \mathbf{i} \cos t + \mathbf{j} \sin t \tag{7}$$

and

$$\mathbf{R}(0) = 2\mathbf{i}, \qquad \mathbf{v}(0) = v_0\mathbf{j}. \tag{8}$$

Since $\mathbf{v} = d\mathbf{R}/dt$, we may rewrite (7) in the form

$$m\frac{d\mathbf{v}}{dt} = \mathbf{i} \cos t + \mathbf{j} \sin t.$$

Integrating both sides of this equation with respect to t gives

$$m\mathbf{v} = \mathbf{i} \sin t - \mathbf{j} \cos t + \mathbf{C}_1, \tag{9}$$

where the constant of integration $\mathbf{C}_1$ is a *vector*. The value of $\mathbf{C}_1$ may be found by using the initial condition $\mathbf{v}(0) = v_0\mathbf{j}$ in Eq. (9) with $t = 0$:

$$mv_0\mathbf{j} = -\mathbf{j} + \mathbf{C}_1, \qquad \mathbf{C}_1 = (mv_0 + 1)\mathbf{j}.$$

Thus, Eq. (9) gives

$$m\mathbf{v} = \mathbf{i} \sin t + (mv_0 + 1 - \cos t)\mathbf{j}.$$

Another integration gives

$$m\mathbf{R} = -\mathbf{i} \cos t + \mathbf{j}(mv_0t + t - \sin t) + \mathbf{C}_2. \tag{10}$$

The initial condition $\mathbf{R}(0) = 2\mathbf{i}$ enables us to evaluate $\mathbf{C}_2$:

$$2m\mathbf{i} = -\mathbf{i} + \mathbf{C}_2, \qquad \mathbf{C}_2 = (2m + 1)\mathbf{i}.$$

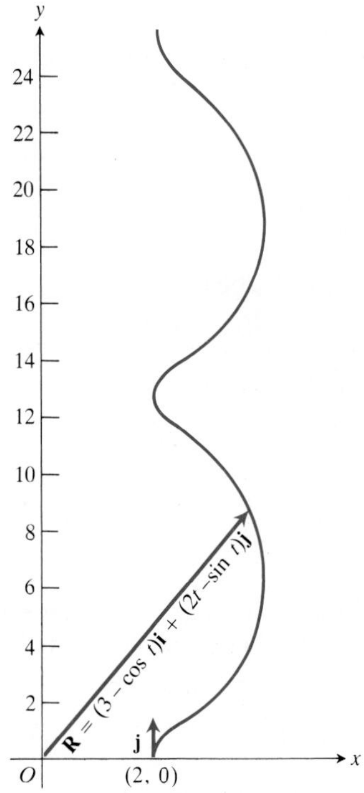

14.6 When $m = v_0 = 1$, the particle P in Example 2 follows the path shown here: $\mathbf{R} = (3 - \cos t)\mathbf{i} + (2t - \sin t)\mathbf{j}$.

Therefore,

$$\mathbf{R}(t) = \frac{1}{m}[(2m + 1 - \cos t)\mathbf{i} + (mv_0 t + t - \sin t)\mathbf{j}]. \quad \square$$

The foregoing equations for velocity and acceleration in two dimensions would be appropriate for describing the motion of a particle moving on a flat surface, such as a water bug skimming over the surface of a pond or a hockey puck sliding on ice. But to describe the flight of a bumblebee or a rocket, we need three coordinates. Thus, if

$$\mathbf{R}(t) = \mathbf{i}x + \mathbf{j}y + \mathbf{k}z, \tag{11}$$

where x, y, z are functions of t that are twice-differentiable, then the velocity of $P(x, y, z)$ is defined to be

$$\mathbf{v} = \frac{d\mathbf{R}}{dt} = \mathbf{i}\frac{dx}{dt} + \mathbf{j}\frac{dy}{dt} + \mathbf{k}\frac{dz}{dt}, \tag{12}$$

and the acceleration is defined to be

$$\mathbf{a} = \mathbf{i}\frac{d^2x}{dt^2} + \mathbf{j}\frac{d^2y}{dt^2} + \mathbf{k}\frac{d^2z}{dt^2}. \tag{13}$$

As in the plane, the length or magnitude of the velocity vector is the *speed* with which the object moves along its path:

$$\text{speed} = |\mathbf{v}|. \tag{14}$$

We shall see why this is so in the next article.

PROBLEMS

In Problems 1–12, $\mathbf{R}$ is the position vector of a moving particle at time t. Find the velocity and acceleration at time t. Then find the velocity, acceleration, and speed at the particular instant given.

1. $\mathbf{R} = (a \cos \omega t)\mathbf{i} + (a \sin \omega t)\mathbf{j}$, a and ω being positive constants: $t = \pi/(3\omega)$.

2. $\mathbf{R} = (2 \cos t)\mathbf{i} + (3 \sin t)\mathbf{j}$, $t = \pi/4$

3. $\mathbf{R} = (t + 1)\mathbf{i} + (t^2 - 1)\mathbf{j}$, $t = 2$

4. $\mathbf{R} = (\cos 2t)\mathbf{i} + (2 \sin t)\mathbf{j}$, $t = 0$

5. $\mathbf{R} = e^t\mathbf{i} + e^{-2t}\mathbf{j}$, $t = \ln 3$

6. $\mathbf{R} = (\sec t)\mathbf{i} + (\tan t)\mathbf{j}$, $t = \pi/6$

7. $\mathbf{R} = (\cosh 3t)\mathbf{i} + (2 \sinh t)\mathbf{j}$, $t = 0$

8. $\mathbf{R} = [\ln (t + 1)]\mathbf{i} + t^2\mathbf{j}$, $t = 1$

9. $\mathbf{R} = (e^{-t})\mathbf{i} + (2 \cos 3t)\mathbf{j} + (2 \sin 3t)\mathbf{k}$, $t = 0$

10. $\mathbf{R} = (1 + t)\mathbf{i} + (t^2/\sqrt{2})\mathbf{j} + (t^3/3)\mathbf{k}$, $t = 1$

11. $\mathbf{R} = (4 \cos t)\mathbf{i} - (3 \sin t)\mathbf{j} + 2t\mathbf{k}$, $t = \pi/3$

12. $\mathbf{R} = (\sqrt{2}/(t + 1))\mathbf{i} + (t^3/3)\mathbf{j} + (\sqrt{2} \ln t)\mathbf{k}$, $t = 1$

In Problems 13–15, $\mathbf{R}$ is the position vector of a particle in space for time $t > 0$. Find the time or times at which the velocity and acceleration vectors are perpendicular.

13. $\mathbf{R} = (2t^3 + 3)\mathbf{i} + (\ln t)\mathbf{j} + 3\mathbf{k}$

14. $\mathbf{R} = (t^4 + 3)\mathbf{i} + (4 \ln t)\mathbf{j} - 5t\mathbf{k}$

15. $\mathbf{R} = (\sin t)\mathbf{i} + (\cos t)\mathbf{j} + t\mathbf{k}$

In Problems 16–18, $\mathbf{R}$ is the position vector of a particle in space. Find the angle between the velocity and acceleration vectors in each case.

16. $\mathbf{R} = (e^t)\mathbf{i} + (e^t \sin t)\mathbf{j} + (e^t \cos t)\mathbf{k}$

17. $\mathbf{R} = (\tan t)\mathbf{i} + (\sinh 2t)\mathbf{j} + (\text{sech } 3t)\mathbf{k}$

18. $\mathbf{R} = (\ln (t^2 + 1))\mathbf{i} + (\tan^{-1} t)\mathbf{j} + (\sqrt{t^2 + 1})\mathbf{k}$

19. Let $\mathbf{R} = t\mathbf{i} + t^2\mathbf{j}$ be the position vector of a particle moving in the xy-plane.
a) Find $\mathbf{v}$ and $\mathbf{a}$.
b) Sketch the path of the motion, and draw $\mathbf{v}$ and $\mathbf{a}$ for $t = 2$ (as vectors starting at the point (2, 4)).

20. In Article 13.2 we derived the equation

$$\mathbf{R} = \mathbf{i}(v_0 \cos \alpha)t + \mathbf{j}\left(-\frac{1}{2}gt^2 + (v_0 \sin \alpha)t\right)$$

for the position vector of a projectile fired from the origin into the first quadrant with an initial speed v_0 at an angle α with the x-axis. Find the velocity and acceleration.

21. A Howitzer at the point (1726, 0) in the xy-plane fires a practice round whose trajectory $\mathbf{R} = x(t)\mathbf{i} + y(t)\mathbf{j}$ is given by

$$\mathbf{R} = (256t + 1726)\mathbf{i} + (-4.9t^2 + 153t)\mathbf{j}, \qquad t \geq 0,$$

with t in seconds, and x and y in meters. Down range, 9900 m from the Howitzer, is a hill that rises 1100 m above the x-axis. Assuming no other obstructions, will the round clear the top of the hill? If so, by how many meters?

22. The acceleration of a particle in the plane as a function of time t is $\mathbf{a}(t) = -3t\mathbf{i}$. When $t = 0$, the particle's velocity and position are $\mathbf{v}(0) = 2\mathbf{j}$ and $\mathbf{R}(0) = 4\mathbf{i}$. Find the velocity $\mathbf{v}$ and position $\mathbf{R}$ as functions of time.

23. The acceleration of a particle in space as a function of time t is $\mathbf{a}(t) = 3t\mathbf{i} + 4\mathbf{j} + \mathbf{k}$. When $t = 0$, the particle's velocity and position are $\mathbf{v}(0) = 4\mathbf{i}$ and $\mathbf{R}(0) = 5\mathbf{j}$. Find the velocity $\mathbf{v}$ and position $\mathbf{R}$ as functions of time.

24. The force acting on a particle of mass m in the plane is given as a function of time t by the equation $\mathbf{F} = (1 + t)^{-1/2}\mathbf{i} - e^{-t}\mathbf{j}$. When $t = 0$, the particle's position and velocity are $\mathbf{R}(0) = (1/3)\mathbf{i} - \mathbf{j}$ and $\mathbf{v}(0) = -\mathbf{i} + 0.5\mathbf{j}$. Find the particle's velocity and position as functions of t.

25. If the projectile encounters a resistance proportional to the velocity, the force is

$$\mathbf{F} = -mg\mathbf{j} - k\frac{d\mathbf{R}}{dt}.$$

Show that one integration of $\mathbf{F} = md^2\mathbf{R}/dt^2$ leads to the differential equation

$$\frac{d\mathbf{R}}{dt} + \frac{k}{m}\mathbf{R} = \mathbf{v}_0 - gt\mathbf{j}.$$

(To solve this equation, one can multiply both sides of the equation by $e^{(k/m)t}$. Then the left side is the derivative of the product $\mathbf{R}e^{(k/m)t}$ and both sides can be integrated.)

26. The plane $z = 2x + 3y$ intersects the cylinder $x^2 + y^2 = 9$ in an ellipse.

a) Express the position of a point $P(x, y, z)$ on this ellipse as a vector function $\mathbf{R} = \overrightarrow{OP} = \mathbf{R}(\theta)$, where θ is a measure of the dihedral angle between the xz-plane and the plane containing the z-axis and P.

b) Using the equations of part (a), find the velocity and acceleration of P, assuming that $d\theta/dt = \omega$ is constant.

***Toolkit* programs**

Parametric Equations Super * Grapher

14.3 Arc Length for Space Curves. The Unit Tangent Vector **T**

Space Curves

Everything we have done so far for two-dimensional motion in the plane can be extended to three-dimensional motion in space. To this end, let $P(x, y, z)$ be a point whose position in space is given by the vector

$$\mathbf{R}(t) = x(t)\mathbf{i} + y(t)\mathbf{j} + z(t)\mathbf{k},$$

where x, y, and z are differentiable functions of t. As t varies continuously, P traces a curve in space.

EXAMPLE 1 The equation

$$\mathbf{R} = (a \cos \omega t)\mathbf{i} + (a \sin \omega t)\mathbf{j} + (bt)\mathbf{k},$$

where a, b, and ω are positive constants, represents a circular helix (Fig. 14.7). The projection of the point $P(x, y, z)$ onto the xy-plane moves around the circle $x^2 + y^2 = a^2$, $z = 0$, as t varies, while the distance $|bt|$ between P and the xy-plane changes steadily with t. □

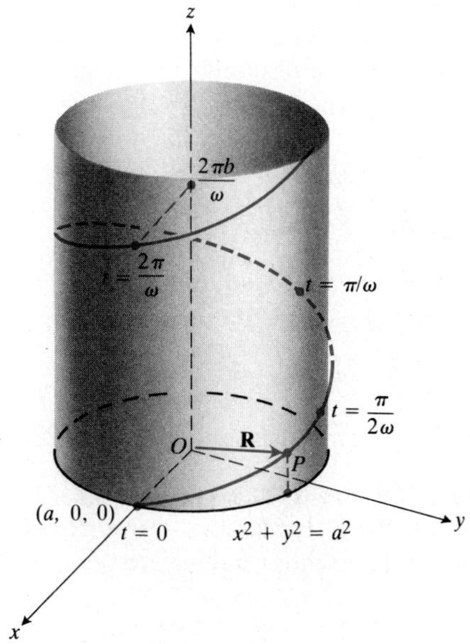

14.7 The helix traced by $\mathbf{R} = (a \cos \omega t)\mathbf{i} + (a \sin \omega t)\mathbf{j} + (bt)\mathbf{k}$, $b > 0$, spirals up from the xy-plane as t increases from zero.

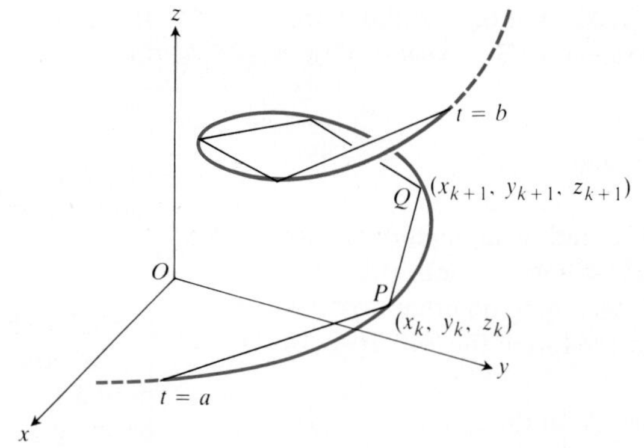

14.8 The length of the arc from $t = a$ to $t = b$ is a limit of lengths of polygonal paths.

Arc Length

As Fig. 14.8 suggests, we define the length of a space curve traced by a moving particle as a limit of lengths

$$\sum_{a}^{b} \sqrt{(x_{k+1} - x_k)^2 + (y_{k+1} - y_k)^2 + (z_{k+1} - z_k)^2} \tag{1}$$

of approximating polygonal paths in much the same way that we defined the length of a plane curve in Article 5.6. The resulting formula for the length of a space curve

$$\mathbf{R}(t) = x(t)\mathbf{i} + y(t)\mathbf{j} + z(t)\mathbf{k}$$

from $t = a$ to $t = b$ is

$$L_a^b = \int_a^b \sqrt{x'(t)^2 + y'(t)^2 + z'(t)^2}\, dt. \tag{2}$$

Thus the length of curve from $t = a$ to $t = b$ is obtained by integrating $|\mathbf{v}| = |d\mathbf{R}/dt|$ from a to b. Dropping the z-term in Eq. (2) gives the arc length formula for curves in the plane.

DEFINITION

Arc Length

The length of the curve $\mathbf{R}(t)$ from $t = a$ to $t = b$ is the integral of the length of the velocity vector from $t = a$ to $t = b$:

$$L_a^b = \int_a^b |\mathbf{v}|\, dt. \tag{3}$$

If we choose a base point $P_0(t_0)$ on the curve, as shown in Fig. 14.9, then the length $s(t)$ along the curve from P_0 to the point $P(t) = (x(t), y(t), z(t))$ for any $t \geq t_0$ is a function of t whose values are given by the integral

$$s(t) = \int_{t_0}^{t} \sqrt{x'(\tau)^2 + y'(\tau)^2 + z'(\tau)^2}\, d\tau. \tag{4}$$

If the derivatives beneath the radical are continuous, then the Second Fundamental Theorem of Calculus tells us that s is a differentiable func-

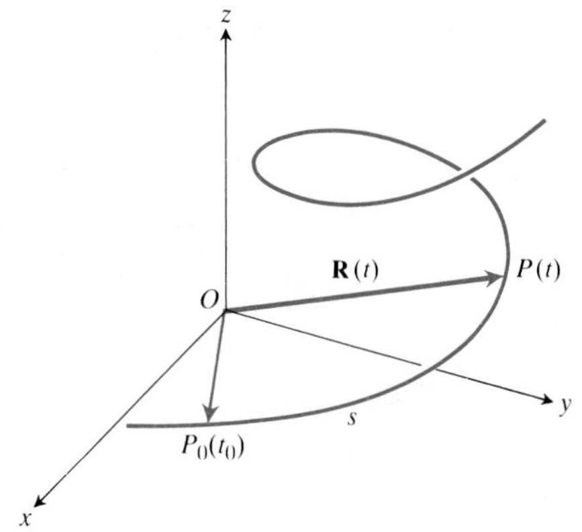

14.9 The distance s along the curve from $P_0(t_0)$ to any point $P(t)$, $t \geq t_0$, is

$$s = \int_{t_0}^{t} |\mathbf{v}|\, d\tau.$$

tion of t whose derivative is

$$\frac{ds}{dt} = \sqrt{x'(t)^2 + y'(t)^2 + z'(t)^2} = \left|\frac{d\mathbf{R}}{dt}\right| = |\mathbf{v}|. \tag{5}$$

Once again we find that the length of the particle's velocity vector is the particle's speed along the curve:

$$\text{Speed} = \left|\frac{ds}{dt}\right| = |\mathbf{v}|. \tag{6}$$

EXAMPLE 2 Find the length of one turn of the helix

$$\mathbf{R} = (\cos t)\mathbf{i} + (\sin t)\mathbf{j} + t\mathbf{k}.$$

Solution This is the helix shown in Fig. 14.7, with $a = b = \omega = 1$. Since $\cos t$ and $\sin t$ have period 2π, the helix makes one full turn as t runs from 0 to 2π. According to Eq. (3), therefore, the length we seek is

$$\int_0^{2\pi} |\mathbf{v}|\, dt = \int_0^{2\pi} \sqrt{(-\sin t)^2 + (\cos t)^2 + (1)^2}\, dt = \int_0^{2\pi} \sqrt{2}\, dt = 2\pi\sqrt{2}.$$

This is $\sqrt{2}$ times the length of the unit circle in the xy-plane over which the helix stands. (Also, see Problem 21.) □

The Unit Tangent Vector T

Suppose that $\mathbf{R}(t)$ is a curve in the plane or in space whose component functions are differentiable, and that $s(t)$ gives arc length along the curve measured from some preselected base point P_0. Then we know from Eq. (4) and the Second Fundamental Theorem that $s(t)$ is a differentiable function of t. If, further, $ds/dt = |\mathbf{v}|$ is never zero (as we shall assume from now on), then $s(t)$ is one-to-one and has an inverse that gives t as a differentiable function of s whose derivative is

$$\frac{dt}{ds} = \frac{1}{|\mathbf{v}|}. \tag{7}$$

(See Articles 2.10 and 3.8.) This makes $\mathbf{R}(t)$ a differentiable function of s, whose derivative can be calculated from the Chain Rule in Article 14.1 to be

$$\frac{d\mathbf{R}}{ds} = \frac{d\mathbf{R}}{dt}\frac{dt}{ds} = \mathbf{v}\frac{dt}{ds}. \tag{8}$$

From this equation we see that

$$\left|\frac{d\mathbf{R}}{ds}\right| = |\mathbf{v}|\left|\frac{dt}{ds}\right| = |\mathbf{v}|\frac{1}{|\mathbf{v}|} = 1. \tag{9}$$

Thus the vector $d\mathbf{R}/ds$ is *always* a unit vector. Furthermore, it points in the direction of $\mathbf{v}$ because it is a positive scalar multiple $dt/ds = 1/|\mathbf{v}|$ of $\mathbf{v}$. We call $d\mathbf{R}/ds$ the unit tangent vector of the curve and denote it by $\mathbf{T}$. See Figs. 14.10 and 14.11.

From a geometric point of view, $\mathbf{T}$ arises in the following way: As $Q \to P$ and $\Delta s \to 0$, the direction of the secant line PQ approaches the direction of the tangent to the curve at P, while the ratio of chord PQ to

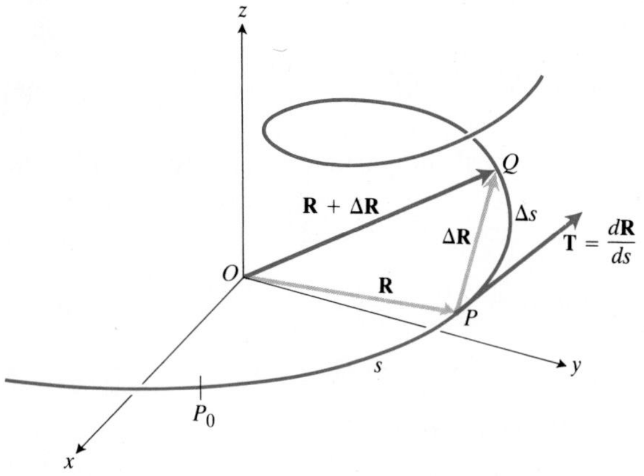

14.10 If **R** is differentiable, then $\mathbf{T} = d\mathbf{R}/ds = \lim_{\Delta s \to 0} \Delta\mathbf{R}/\Delta s$ is a unit vector tangent to the curve traced out by P.

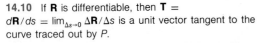

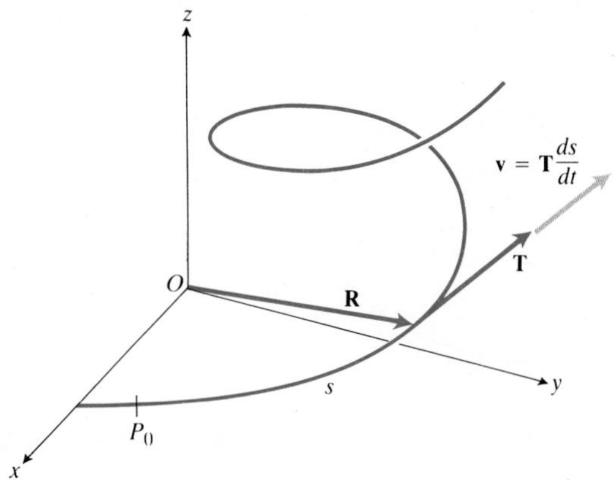

14.11 Divide **v** by ds/dt to get **T**.

arc PQ approaches one (for a "smooth" curve). Therefore the limit of $\Delta\mathbf{R}/\Delta s$ is a unit vector tangent to the curve at P and pointing in the direction in which arc length increases along the curve.

DEFINITION

Unit Tangent Vector

The unit tangent vector to the curve $\mathbf{R}(t)$ is

$$\mathbf{T} = \frac{d\mathbf{R}}{ds}. \tag{10}$$

This definition assumes that $|\mathbf{v}| = |d\mathbf{R}/dt|$ is never zero; we shall require this of all our parameterized curves.

While **T** is defined to be $d\mathbf{R}/ds$, the natural parameterization of a motion in many cases is likely to be time and not arc length. The best way to find **T** in such cases is not from the equation $\mathbf{T} = d\mathbf{R}/ds$ but from the equation

$$\mathbf{T} = \frac{\mathbf{v}}{|\mathbf{v}|}. \tag{11}$$

This equation follows from observing that since **T** is a unit vector in the direction of **v**, it must be $\mathbf{v}/|\mathbf{v}|$.

EXAMPLE 3 Find **T** for the helix of Example 1.

Solution The velocity vector is

$$\mathbf{v} = \mathbf{i}(-a\omega \sin \omega t) + \mathbf{j}(a\omega \cos \omega t) + \mathbf{k}(b),$$

whose length is

$$|\mathbf{v}| = \sqrt{a^2\omega^2 \sin^2 \omega t + a^2\omega^2 \cos^2 \omega t + b^2} = \sqrt{a^2\omega^2 + b^2}.$$

Therefore,

$$\mathbf{T} = \frac{\mathbf{v}}{|\mathbf{v}|} = \frac{a\omega(-\mathbf{i} \sin \omega t + \mathbf{j} \cos \omega t) + b\mathbf{k}}{\sqrt{a^2\omega^2 + b^2}}. \quad \square$$

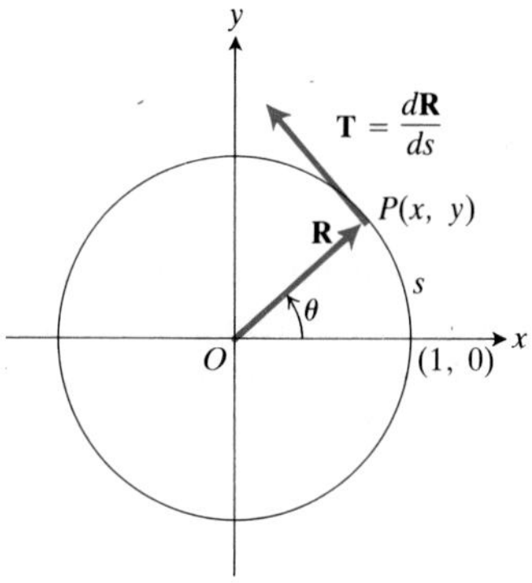

14.12 $\mathbf{R} = \mathbf{i}\cos\theta + \mathbf{j}\sin\theta$.

EXAMPLE 4 Find **T** for the motion

$$\mathbf{R} = (\cos t + t\sin t)\mathbf{i} + (\sin t - t\cos t)\mathbf{j}, \qquad t \geq 0.$$

Solution

$$\mathbf{v} = \frac{d\mathbf{R}}{dt} = (-\sin t + t\cos t + \sin t)\mathbf{i} + (\cos t + t\sin t - \cos t)\mathbf{j}$$

$$= (t\cos t)\mathbf{i} + (t\sin t)\mathbf{j},$$

$$|\mathbf{v}| = \sqrt{t^2\cos^2 t + t^2\sin^2 t} = |t| = t,$$

$$\mathbf{T} = \frac{\mathbf{v}}{|\mathbf{v}|} = (\cos t)\mathbf{i} + (\sin t)\mathbf{j}. \ \square$$

EXAMPLE 5 For the counterclockwise motion

$$\mathbf{R} = \mathbf{i}\cos\theta + \mathbf{j}\sin\theta$$

around the unit circle, we find that

$$\mathbf{v} = (-\sin\theta)\mathbf{i} + (\cos\theta)\mathbf{j}$$

is already a unit vector, so that **v** = **T**. In fact, **T** is **R** rotated 90° counterclockwise (Fig. 14.12). □

PROBLEMS

Each of Problems 1–8 gives the position vector **R** of a particle moving in the plane. Find **T** in each case.

1. $\mathbf{R} = 2\mathbf{i}\cos t + 2\mathbf{j}\sin t$
2. $\mathbf{R} = e^t\mathbf{i} + t^2\mathbf{j}$
3. $\mathbf{R} = (\cos^3 t)\mathbf{i} + (\sin^3 t)\mathbf{j}$
4. $\mathbf{R} = \mathbf{i}x + \mathbf{j}x^2$
5. $\mathbf{R} = (\cos 2t)\mathbf{i} + (2\cos t)\mathbf{j}$
6. $\mathbf{R} = \dfrac{t^3}{3}\mathbf{i} + \dfrac{t^2}{2}\mathbf{j}$
7. $\mathbf{R} = (e^t\cos t)\mathbf{i} + (e^t\sin t)\mathbf{j}$
8. $\mathbf{R} = \cosh t\mathbf{i} + t\mathbf{j}$
9. Find the length of the curve $\mathbf{R} = t\mathbf{i} + (2/3)t^{3/2}\mathbf{j}$ from $t = 0$ to $t = 8$.
10. Find the length of the curve

$$\mathbf{R} = (t\sin t + \cos t)\mathbf{i} + (t\cos t - \sin t)\mathbf{j}$$

from $t = 0$ to $t = 2$.

11. Find **T** at $t = 0$ if $\mathbf{R} = (2t^3 + 3t^2)\mathbf{i} + (t^2 + 2t)\mathbf{j}$

In Problems 12–15, find **T**, the speed $|ds/dt| = |\mathbf{v}|$, and the length of the curve from $t = 0$ to $t = \pi$.

12. $\mathbf{R} = (6\sin 2t)\mathbf{i} + (6\cos 2t)\mathbf{j} + (5t)\mathbf{k}$
13. $\mathbf{R} = (e^t\cos t)\mathbf{i} + (e^t\sin t)\mathbf{j} + e^t\mathbf{k}$
14. $\mathbf{R} = (3\cosh 2t)\mathbf{i} + (3\sinh 2t)\mathbf{j} + (6t)\mathbf{k}$
15. $\mathbf{R} = (3t\cos t)\mathbf{i} + (3t\sin t)\mathbf{j} + (4t)\mathbf{k}$
16. Find **T** when $t = 1$ if $\mathbf{R} = t\mathbf{i} + t^2\mathbf{j} - \dfrac{1}{t}\mathbf{k}$.
17. Find **v**, **T**, the speed $ds/dt = |\mathbf{v}|$, and **a**, all at $t = 0$, if

$$\mathbf{R} = 3t^2\mathbf{i} + (6t + 7)\mathbf{j} + 2t^4\mathbf{k}.$$

18. Find the length of the curve $\mathbf{R} = \mathbf{i} + t\mathbf{j} + t^2\mathbf{k}$ from the point (1, 0, 0) to the point (1, 1, 1).
19. Find the length of the curve

$$\mathbf{R} = (\cos t)\mathbf{i} + (\sin t)\mathbf{j} + t^2\mathbf{k}, \qquad 0 \leq t \leq 1.$$

20. Find the length of the curve

$$\mathbf{R} = t\mathbf{i} + t\mathbf{j} + (4 - t^2)\mathbf{k}$$

from the point (0, 0, 4) to the point (1, 1, 3).

21. The length $2\pi\sqrt{2}$ of one turn of the helix in Example 2 is also the length of the diagonal of a square 2π units on a side. Show how to obtain this square by cutting away and flattening a portion of the cylinder around which the helix winds.

***Tookit* programs**

Parametric Equations Super * Grapher

significant role in the geometry of space curves (see any current text on *differential geometry*). Moving reference frames are important in physics and in the determination of the flight paths of space vehicles.

Curvature and Vector Formulas

In the plane:

$$\kappa = \frac{|d^2y/dx^2|}{[1 + (dy/dx)^2]^{3/2}} \quad (\text{if } y = f(x))$$

$$\kappa = \frac{|d^2x/dy^2|}{[1 + (dx/dy)^2]^{3/2}} \quad (\text{if } x = g(y))$$

$$\kappa = \frac{|\dot{x}\ddot{y} - \dot{y}\ddot{x}|}{[\dot{x}^2 + \dot{y}^2]^{3/2}} \quad (\text{if } x = f(t),\ y = g(t))$$

In plane and space:

$$\kappa = \left|\frac{d\mathbf{T}}{ds}\right| = \frac{|\mathbf{v} \times \mathbf{a}|}{|\mathbf{v}|^3}$$

Radius of curvature: $\rho = 1/\kappa \quad (\text{if } \kappa \neq 0)$

Unit tangent vector: $\mathbf{T} = \dfrac{\mathbf{v}}{|\mathbf{v}|}$

Principal normal vector: $\mathbf{N} = \dfrac{d\mathbf{T}/ds}{|d\mathbf{T}/ds|} = \dfrac{1}{\kappa}\dfrac{d\mathbf{T}}{ds} \quad (\kappa \neq 0)$

Binormal vector: $\mathbf{B} = \mathbf{T} \times \mathbf{N}$

PROBLEMS

Find the curvature of the plane curves in Problems 1–8.

1. $y = \ln(\cos x)$ **2.** $y = e^{2x}$

3. $x = \ln \sec y$ **4.** $x = \frac{1}{3}(y^2 + 2)^{3/2}$

5. $x = \dfrac{y^4}{4} + \dfrac{1}{8y^2}$ **6.** $y = a \cosh(x/a)$

7. $x = 2t + 3, \quad y = 5 - t^2$

8. $x = t^3/3, \quad y = t^2/2$

Find **T**, κ, and **N** for the plane curves in Problems 9–12.

9. $\mathbf{R} = (\cos t + t \sin t)\mathbf{i} + (\sin t - t \cos t)\mathbf{j}$

10. $\mathbf{R} = a(t - \sin t)\mathbf{i} + a(1 - \cos t)\mathbf{j}$

11. $\mathbf{R} = (e^t \cos t)\mathbf{i} + (e^t \sin t)\mathbf{j}$

12. $\mathbf{R} = (\cos^3 t)\mathbf{i} + (\sin^3 t)\mathbf{j}, \quad 0 < t < \pi/2$

Find **T**, κ, **N**, and **B** for the space curves in Problems 13–16.

13. $\mathbf{R} = (6 \sin 2t)\mathbf{i} + (6 \cos 2t)\mathbf{j} + (5t)\mathbf{k}$

14. $\mathbf{R} = (e^t \cos t)\mathbf{i} + (e^t \sin t)\mathbf{j} + (e^t)\mathbf{k}$

15. $\mathbf{R} = \dfrac{1}{3}(1 + t)^{3/2}\mathbf{i} + \dfrac{1}{3}(1 - t)^{3/2}\mathbf{j} + \dfrac{t}{2}\mathbf{k}, \quad -1 < t < 1$

16. $\mathbf{R} = (3 \cosh 2t)\mathbf{i} + (3 \sinh 2t)\mathbf{j} + (6t)\mathbf{k}$

17. Let $\mathbf{R} = (\cos t + t \sin t)\mathbf{i} + (\sin t - t \cos t)\mathbf{j} + 3\mathbf{k}$ describe a space curve for $t \geq 0$. Find **T**, κ, **N**, and **B**. Also, find equations for the line tangent to the curve at the point $(\pi/2, 1, 3)$.

18. Let $\mathbf{R} = (3 \sin t)\mathbf{i} + (3 \cos t)\mathbf{j} + (4t)\mathbf{k}$. Find **T**, κ, **N**, and **B**. Also, find equations for the line tangent to the curve at the point $(0, 3, 8\pi)$.

19. Let $\mathbf{R} = (2 \cos t)\mathbf{i} + (2 \sin t)\mathbf{j} + t\mathbf{k}$. Find **T**, κ, **N**, and **B**. Also, find the arc length s as a function of t. Assume $s = 0$ when $t = 0$ and that s increases with t. (*Hint*: Remember that $ds/dt = |\mathbf{v}|$.)

20. Repeat Problem 19 for

$$\mathbf{R} = (e^t \cos t)\mathbf{i} + (e^t \sin t)\mathbf{j} + 2\mathbf{k}.$$

21. A rocket leaves the point $(1, -2, 3)$ at time $t = 0$ and travels with constant speed 1 unit in a straight line toward the point $(3, 0, 0)$. Find, as functions of t, the
a) position vector **R**,
b) velocity **v**,
c) unit tangent vector **T**,
d) acceleration **a**,
e) curvature κ.

22. Find the radius of curvature of the plane curve $y = \ln(\cos x)$ as a function of x.

23. Find an equation for the circle of curvature of the plane curve

$$\mathbf{R} = (2 \ln t)\mathbf{i} - (t + (1/t))\mathbf{j}.$$

24. Find an equation for the circle of curvature of the curve $y = \sin x$ at the point $(\pi/2, 1)$.

25. Find the equation of the osculating circle associated with the curve $y = e^x$ at the point (0, 1). By calculating dy/dx and d^2y/dx^2 at the point (0, 1) from the equation of this circle, verify that these derivatives have the same values there as do the corresponding derivatives for the curve $y = e^x$. Sketch the curve and the osculating circle.

26. The figure below shows arc length s on the circle $x^2 + y^2 = a^2$, measured counterclockwise from the positive x-axis to a point P. It also shows the angle ϕ that the tangent to the circle at P makes with the x-axis. Use the equations $s = a\theta$, $\phi = \theta + \pi/2$ to calculate the circle's curvature directly from the definition $\kappa = |d\phi/ds|$.

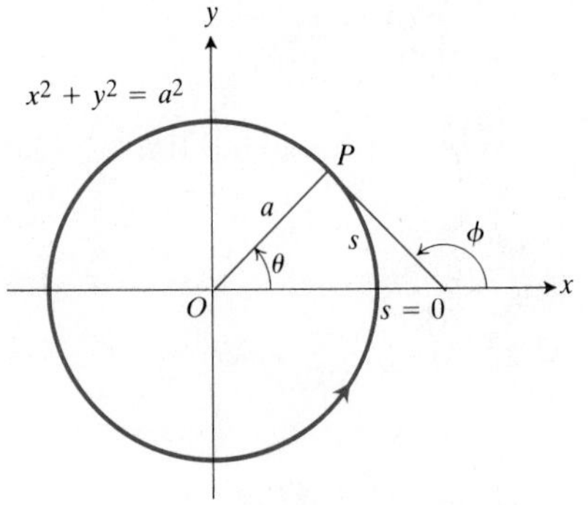

27. The curve $\mathbf{R}(t) = x(t)\mathbf{i} + y(t)\mathbf{j}$ in the plane may be thought of as a curve $\mathbf{R}(t) = x(t)\mathbf{i} + y(t)\mathbf{j} + 0 \cdot \mathbf{k}$ in space whose **k** component is zero. Show that for such a curve the formula $\kappa = |\mathbf{v} \times \mathbf{a}|/|\mathbf{v}|^3$ reduces to

$$\kappa = |\dot{x}\ddot{y} - \dot{y}\ddot{x}|/[\dot{x}^2 + \dot{y}^2]^{3/2}.$$

28. Show that when x and y are considered as functions of arc length s, the unit vectors **T** and **N** may be expressed as follows:

$$\mathbf{T} = \mathbf{i}\frac{dx}{ds} + \mathbf{j}\frac{dy}{ds}, \qquad \mathbf{N} = -\mathbf{i}\frac{dy}{ds} + \mathbf{j}\frac{dx}{ds},$$

where $dx/ds = \cos\phi$, $dy/ds = \sin\phi$, and ϕ is the angle from the positive x-axis to the tangent line.

29. Let $\mathbf{R} = \overrightarrow{OP}$ be the vector from the origin to a moving point P. Let **T** and **N** be the unit tangent and principal normal vectors, respectively, for the curve described by P. Express the velocity and acceleration vectors $d\mathbf{R}/dt$ and $d^2\mathbf{R}/dt^2$ in terms of their **T**- and **N**-components.

***Toolkit* programs**

Parametric Equations   Super * Grapher

14.5 Derivatives of Vector Products. The Tangential and Normal Components of **v** and **a**

If $\mathbf{U}(t)$, $\mathbf{V}(t)$ are differentiable vector functions of t, then the products $\mathbf{U} \cdot \mathbf{V}$ and $\mathbf{U} \times \mathbf{V}$ are differentiable, and their derivatives may be calculated by applying rules like the one we know for products of scalar functions:

$$\frac{d}{dt}(\mathbf{U} \cdot \mathbf{V}) = \frac{d\mathbf{U}}{dt} \cdot \mathbf{V} + \mathbf{U} \cdot \frac{d\mathbf{V}}{dt}, \tag{1}$$

$$\frac{d}{dt}(\mathbf{U} \times \mathbf{V}) = \frac{d\mathbf{U}}{dt} \times \mathbf{V} + \mathbf{U} \times \frac{d\mathbf{V}}{dt}. \tag{2}$$

One way to verify Eqs. (1) and (2) would be to write

$$\begin{aligned} \mathbf{U} &= \mathbf{i}f_1(t) + \mathbf{j}f_2(t) + \mathbf{k}f_3(t), \\ \mathbf{V} &= \mathbf{i}g_1(t) + \mathbf{j}g_2(t) + \mathbf{k}g_3(t), \end{aligned} \tag{3}$$

with the component f's and g's being differentiable functions of t. We would then substitute these expressions for **U** and **V** into Eqs. (1) and (2), multiply out, and apply the ordinary rules for differentiating products of scalar functions. However, instead of appealing to the component-wise verification of the identities in Eqs. (1) and (2), it is instructive to establish these equations by the Δ-process. For example, let

$$\mathbf{W} = \mathbf{U} \times \mathbf{V},$$

where t has some specific value. Then give t an increment Δt and denote the new values of the vectors by $\mathbf{U} + \Delta\mathbf{U}$, etc., so that

$$\begin{aligned} \mathbf{W} + \Delta\mathbf{W} &= (\mathbf{U} + \Delta\mathbf{U}) \times (\mathbf{V} + \Delta\mathbf{V}) \\ &= \mathbf{U} \times \mathbf{V} + \mathbf{U} \times \Delta\mathbf{V} + \Delta\mathbf{U} \times \mathbf{V} + \Delta\mathbf{U} \times \Delta\mathbf{V}, \end{aligned}$$

and

$$\frac{\Delta\mathbf{W}}{\Delta t} = \mathbf{U} \times \frac{\Delta\mathbf{V}}{\Delta t} + \frac{\Delta\mathbf{U}}{\Delta t} \times \mathbf{V} + \frac{\Delta\mathbf{U}}{\Delta t} \times \Delta\mathbf{V}.$$

Now take limits as $\Delta t \to 0$, noting that

$$\lim \frac{\Delta\mathbf{W}}{\Delta t} = \frac{d\mathbf{W}}{dt}, \qquad \lim \frac{\Delta\mathbf{U}}{\Delta t} = \frac{d\mathbf{U}}{dt}, \qquad \lim \Delta\mathbf{V} = \lim \frac{\Delta\mathbf{V}}{\Delta t} \cdot \lim \Delta t = \mathbf{0},$$

so that

$$\frac{d\mathbf{W}}{dt} = \mathbf{U} \times \frac{d\mathbf{V}}{dt} + \frac{d\mathbf{U}}{dt} \times \mathbf{V},$$

which is equivalent to Eq. (2).

EXAMPLE 1 The formula for the derivative of the triple scalar product leads to an interesting identity for the derivative of a determinant of order three. Let

$$\begin{aligned} \mathbf{U} &= u_1\mathbf{i} + u_2\mathbf{j} + u_3\mathbf{k}, \\ \mathbf{V} &= v_1\mathbf{i} + v_2\mathbf{j} + v_3\mathbf{k}, \\ \mathbf{W} &= w_1\mathbf{i} + w_2\mathbf{j} + w_3\mathbf{k}, \end{aligned} \tag{4}$$

where the components are differentiable functions of a scalar t. Then the identity

$$\frac{d}{dt}(\mathbf{U} \cdot \mathbf{V} \times \mathbf{W}) = \frac{d\mathbf{U}}{dt} \cdot \mathbf{V} \times \mathbf{W} + \mathbf{U} \cdot \frac{d\mathbf{V}}{dt} \times \mathbf{W} + \mathbf{U} \cdot \mathbf{V} \times \frac{d\mathbf{W}}{dt} \tag{5}$$

(Problem 21) is equivalent to

$$\frac{d}{dt}\begin{vmatrix} u_1 & u_2 & u_3 \\ v_1 & v_2 & v_3 \\ w_1 & w_2 & w_3 \end{vmatrix} = \begin{vmatrix} \frac{du_1}{dt} & \frac{du_2}{dt} & \frac{du_3}{dt} \\ v_1 & v_2 & v_3 \\ w_1 & w_2 & w_3 \end{vmatrix} + \begin{vmatrix} u_1 & u_2 & u_3 \\ \frac{dv_1}{dt} & \frac{dv_2}{dt} & \frac{dv_3}{dt} \\ w_1 & w_2 & w_3 \end{vmatrix} + \begin{vmatrix} u_1 & u_2 & u_3 \\ v_1 & v_2 & v_3 \\ \frac{dw_1}{dt} & \frac{dw_2}{dt} & \frac{dw_3}{dt} \end{vmatrix} \tag{6}$$

This says that the derivative of a determinant of order three is the sum of three determinants obtained from the original determinant by differentiating one row at a time. The result may be extended to determinants of any order n. □

Derivatives of Vectors of Constant Length

An important geometrical result is obtained by differentiating the identity

$$\mathbf{V}\cdot\mathbf{V} = |\mathbf{V}|^2 \tag{7a}$$

when $\mathbf{V}$ is a vector of constant magnitude. For then $|\mathbf{V}|^2$ is a constant, so its derivative is zero, and one has

$$\mathbf{V}\cdot\frac{d\mathbf{V}}{dt} + \frac{d\mathbf{V}}{dt}\cdot\mathbf{V} = 0$$

or, since the scalar product is commutative,

$$2\mathbf{V}\cdot\frac{d\mathbf{V}}{dt} = 0 \qquad \text{or} \qquad \mathbf{V}\cdot\frac{d\mathbf{V}}{dt} = 0. \tag{7b}$$

This means that either $\mathbf{V}$ is zero, $d\mathbf{V}/dt$ is zero (and hence $\mathbf{V}$ is constant in direction as well as magnitude), or else that $d\mathbf{V}/dt$ is perpendicular to $\mathbf{V}$.

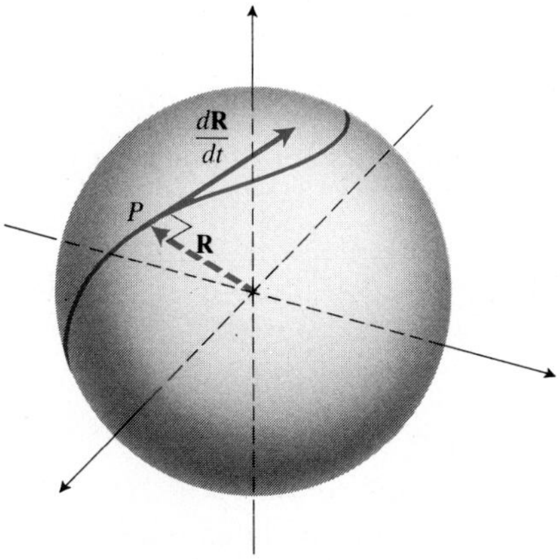

14.18 The velocity vector of a particle P that moves on the surface of a sphere is tangent to the sphere.

EXAMPLE 2 Suppose that a point P moves about on the surface of a sphere. Then the magnitude of the vector $\mathbf{R}$ from the center to P is a constant equal to the radius of the sphere. Therefore, the velocity vector $d\mathbf{R}/dt$ is always perpendicular to $\mathbf{R}$ (Fig. 14.18). □

N · T Is Zero

We can now show that the principal normal vector

$$\mathbf{N} = \frac{1}{\kappa}\frac{d\mathbf{T}}{ds} \tag{8}$$

defined in Article 14.4 is orthogonal to the unit tangent vector $\mathbf{T}$. The vector $\mathbf{T}$ has a constant length of $|\mathbf{T}| = 1$. We may therefore apply Eq. (7b) with $\mathbf{V} = \mathbf{T}$ and $t = s$ to conclude that

$$\mathbf{T}\cdot\frac{d\mathbf{T}}{ds} = 0 \qquad \text{or} \qquad \mathbf{T}\cdot\mathbf{N} = 0. \tag{9}$$

The Tangential and Normal Components of a

In mechanics it is often useful to describe the acceleration of a moving object in terms of its components tangent and normal to the path of motion; that is, in terms of its components in the directions of $\mathbf{T}$ and $\mathbf{N}$. This may be done for curves in either space or plane in the following way.

From the Chain Rule we see that

$$\mathbf{v} = \frac{d\mathbf{R}}{dt} = \frac{d\mathbf{R}}{ds}\frac{ds}{dt} = \mathbf{T}\frac{ds}{dt}, \tag{10}$$

$$\mathbf{a} = \frac{d\mathbf{v}}{dt} = \frac{d}{dt}\left(\mathbf{T}\frac{ds}{dt}\right) = \mathbf{T}\frac{d^2s}{dt^2} + \frac{d\mathbf{T}}{dt}\frac{ds}{dt}. \tag{11}$$

By the Chain Rule and Eq. (8) we also have

$$\frac{d\mathbf{T}}{dt} = \frac{d\mathbf{T}}{ds}\frac{ds}{dt} = \mathbf{N}\kappa\frac{ds}{dt},$$

so that

$$\mathbf{a} = \mathbf{T}\frac{d^2s}{dt^2} + \mathbf{N}\kappa\left(\frac{ds}{dt}\right)^2 = a_T\mathbf{T} + a_N\mathbf{N}. \tag{12}$$

Equation (12) expresses **a** in terms of its tangential and normal components. The *tangential component,*

$$a_T = \frac{d^2s}{dt^2} \tag{13}$$

is simply the derivative of $v = ds/dt$, the particle's speed along its path. The *normal component*

$$a_N = \kappa\left(\frac{ds}{dt}\right)^2 \tag{14}$$

can be written as

$$a_N = \kappa\left(\frac{ds}{dt}\right)^2 = \kappa v^2,$$

the curvature times the square of the speed. This explains why a large normal force must be supplied by friction between the tires and the roadway to hold an automobile on a level road when it makes a sharp turn (large κ) or a moderate turn at high speed (large v^2).

If the particle moves in a circle with constant speed $v = ds/dt$, then $d^2s/dt^2 = 0$, and the only acceleration is the normal acceleration

$$\mathbf{N}\kappa\left(\frac{ds}{dt}\right)^2 = \mathbf{N}\kappa v^2 = \mathbf{N}\frac{v^2}{\rho},$$

which is toward the center of the circle. If the speed around the circle is not constant, then **a** has a nonzero tangential component as well (Fig. 14.19).

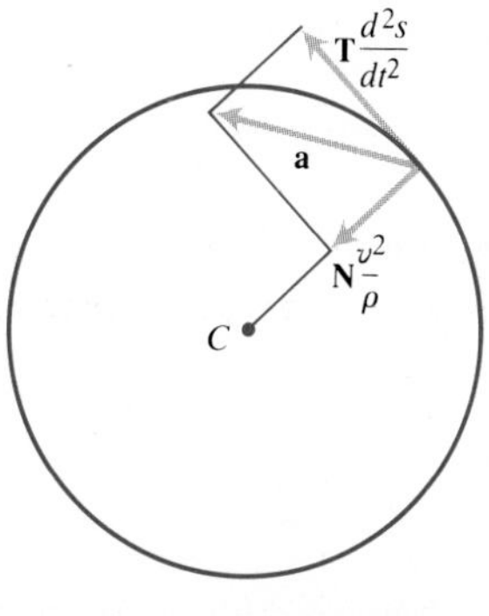

14.19 Tangential and normal components of an acceleration vector.

To calculate a_T we can use Eq. (13) directly, but instead of using Eq. (14) to calculate a_N it is often better to use the equation

$$a_N = \sqrt{|\mathbf{a}|^2 - a_T^2} \tag{15}$$

(which comes from solving the equation

$$|\mathbf{a}|^2 = a_T^2 + a_N^2 \tag{16}$$

for a_N). With Eq. (15) we can find a_N without having to find κ first.

EXAMPLE 3 The position of a moving particle at time $t \geq 0$ is

$$\mathbf{R} = (\cos t + t\sin t)\mathbf{i} + (\sin t - t\cos t)\mathbf{j}.$$

Find **v**, **a**, $|ds/dt|$, a_T, and a_N. Then express **a** in terms of a_T and a_N.

Solution We differentiate **R** with respect to t to find

$$\begin{aligned}\mathbf{v} &= (-\sin t + \sin t + t\cos t)\mathbf{i} + (\cos t - \cos t + t\sin t)\mathbf{j}\\ &= (t\cos t)\mathbf{i} + (t\sin t)\mathbf{j}\end{aligned}$$

and

$$\mathbf{a} = \frac{d\mathbf{v}}{dt} = (\cos t - t \sin t)\mathbf{i} + (\sin t + t \cos t)\mathbf{j}.$$

The speed is

$$\left|\frac{ds}{dt}\right| = |\mathbf{v}| = \sqrt{(t \cos t)^2 + (t \sin t)^2} = t.$$

From Eq. (13),

$$a_T = \frac{d^2s}{dt^2} = \frac{d}{dt}\left(\frac{ds}{dt}\right) = \frac{d}{dt}(t) = 1.$$

From Eq. (15),

$$a_N = \sqrt{|\mathbf{a}|^2 - a_T^2} = \sqrt{(\cos t - t \sin t)^2 + (\sin t + t \cos t)^2 - 1} = t.$$

Finally,

$$\mathbf{a} = a_T\mathbf{T} + a_N\mathbf{N} = (1)\mathbf{T} + t\mathbf{N} = \mathbf{T} + t\mathbf{N}.$$

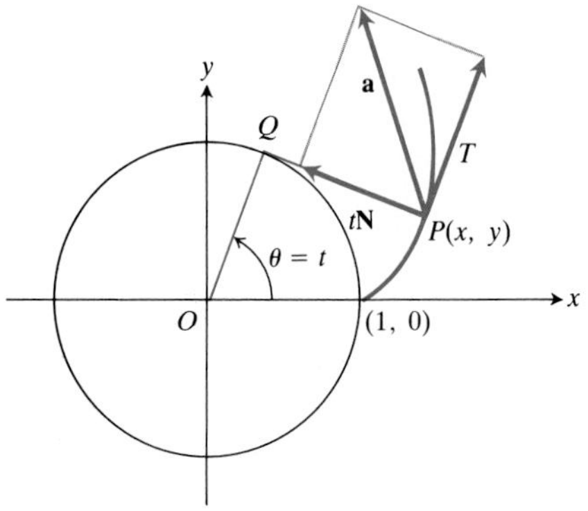

14.20 Motion on the involute of a circle when θ increases steadily with time.

Here the tangential acceleration has constant magnitude and the normal acceleration starts with zero magnitude at $t = 0$ and increases with time. The equations of motion are the same as the parametric equations for the *involute* of a circle of unit radius. This is the path of the endpoint P of a string that is held taut as it is unwound from the circle. To get the parameterization, the origin is taken as the center of the circle (Fig. 14.20), with (1, 0) being the position where P starts. The angle t is measured from the positively directed x-axis counterclockwise to the ray from O to the point of tangency Q.

Incidentally, we can find the radius of curvature from the equation $a_N = v^2/\rho$, since $v^2 = |\mathbf{v}|^2 = t^2$ and $a_N = t$ and hence

$$\rho = \frac{v^2}{a_N} = \frac{t^2}{t} = t. \quad \square$$

The Vector Formula for Curvature

Equations (10) and (12) for **v** and **a** can be used to derive the vector curvature formula

$$\kappa = \frac{|\mathbf{v} \times \mathbf{a}|}{|\mathbf{v}|^3} \tag{17}$$

that we introduced in Article 14.4.

First we compute $\mathbf{v} \times \mathbf{a}$:

$$\mathbf{v} \times \mathbf{a} = \mathbf{T}\frac{ds}{dt} \times \left[\mathbf{T}\frac{d^2s}{dt^2} + \mathbf{N}\kappa\left(\frac{ds}{dt}\right)^2\right] = \mathbf{T} \times \mathbf{N}\kappa\left(\frac{ds}{dt}\right)^3 \tag{18}$$

because we can apply the distributive law for the cross product, and $\mathbf{T} \times \mathbf{T} = \mathbf{0}$. Moreover, $\mathbf{T} \times \mathbf{N}$ is the unit binormal vector **B**. Therefore

$$\mathbf{v} \times \mathbf{a} = \mathbf{B}\kappa\left(\frac{ds}{dt}\right)^3. \tag{19}$$

Since **B** is a *unit* vector, the magnitude of $\mathbf{v} \times \mathbf{a}$ is

$$|\mathbf{v} \times \mathbf{a}| = \kappa\left|\frac{ds}{dt}\right|^3 = \kappa|\mathbf{v}|^3.$$

Finally, if $|\mathbf{v}| \neq 0$, we get (by division)

$$\kappa = \frac{|\mathbf{v} \times \mathbf{a}|}{|\mathbf{v}|^3},$$

which is Eq. (17).

Useful Formulas for Curves in Plane and Space

Position vector in the plane: $\mathbf{R} = x(t)\mathbf{i} + y(t)\mathbf{j}$

Position vector in space: $\mathbf{R} = x(t)\mathbf{i} + y(t)\mathbf{j} + z(t)\mathbf{k}$

Velocity: $\mathbf{v} = \dfrac{d\mathbf{R}}{dt}$

Acceleration: $\mathbf{a} = \dfrac{d\mathbf{v}}{dt} = \dfrac{d^2\mathbf{R}}{dt^2}$

$$\mathbf{T} = \frac{\mathbf{v}}{|\mathbf{v}|} \text{ if } |\mathbf{v}| \neq 0, \qquad \mathbf{v} = \mathbf{T}\frac{ds}{dt}, \qquad \left|\frac{ds}{dt}\right| = |\mathbf{v}| = \text{speed}$$

$$\kappa = \frac{|\mathbf{v} \times \mathbf{a}|}{|\mathbf{v}|^3} \text{ if } |\mathbf{v}| \neq 0, \qquad \kappa\mathbf{N} = \frac{d\mathbf{T}}{ds}$$

$$a_T = \frac{d^2s}{dt^2} = \frac{d}{dt}|\mathbf{v}|, \qquad a_N = \sqrt{|\mathbf{a}|^2 - a_T^2}$$

$$\mathbf{a} = a_T\mathbf{T} + a_N\mathbf{N}$$

PROBLEMS

In Problems 1–7, find $\mathbf{v}$, $\mathbf{a}$, the speed $|ds/dt|$, a_T, and a_N. Then express $\mathbf{a}$ in terms of a_T and a_N (*without* finding $\mathbf{T}$ and $\mathbf{N}$).

1. $\mathbf{R} = (2t + 3)\mathbf{i} + (t^2 - 1)\mathbf{j}$

2. $\mathbf{R} = \ln(t^2 + 1)\mathbf{i} + (t - 2\tan^{-1} t)\mathbf{j}$

3. $\mathbf{R} = (2\cos t)\mathbf{i} + (2\sin t)\mathbf{j}$

4. $\mathbf{R} = (a\cos \omega t)\mathbf{i} + (a\sin \omega t)\mathbf{j}$

5. $\mathbf{R} = (e^t\cos t)\mathbf{i} + (e^t\sin t)\mathbf{j}$

6. $\mathbf{R} = (\cosh 2t)\mathbf{i} + (\sinh 2t)\mathbf{j}$

7. a) $\mathbf{R} = (\cos t)\mathbf{i} + (\sin t)\mathbf{j} + (bt)\mathbf{k}$.
b) Calculate the curvature of the helix in part (a). What effect does increasing $|b|$ have on the curvature? (This explains mathematically why stretching a spring tends to straighten it.)

In Problems 8–14, find $\mathbf{v}$, $\mathbf{a}$, the speed $|ds/dt|$, a_T, and a_N *at the given value of t.*

8. $\mathbf{R} = (1 - \sin t)\mathbf{i} + (1 - \sqrt{2}\cos t)\mathbf{j}$, at $t = \pi/6$

9. $\mathbf{R} = (t + 1)\mathbf{i} + 2t\mathbf{j} + t^2\mathbf{k}$, at $t = 1$

10. $\mathbf{R} = (t\cos t)\mathbf{i} + (t\sin t)\mathbf{j} + t\mathbf{k}$, at $t = 0$

11. $\mathbf{R} = t^2\mathbf{i} + (t + \frac{1}{3}t^3)\mathbf{j} + (t - \frac{1}{3}t^3)\mathbf{k}$, at $t = 0$

12. $\mathbf{R} = (e^t\cos t)\mathbf{i} + (e^t\sin t)\mathbf{j} + \sqrt{2}e^t\mathbf{k}$, at $t = 0$

13. $\mathbf{R} = (2 + 3t + 3t^2)\mathbf{i} + (4t + 4t^2)\mathbf{j} - (6\cos t)\mathbf{k}$, at $t = 0$

14. $\mathbf{R} = (2 + t)\mathbf{i} + (t + 2t^2)\mathbf{j} + (1 + t^2)\mathbf{k}$, at $t = 0$

15. For the curve in Problem 14, find (a) the curvature at $t = 0$, (b) equations for the line tangent to the curve at $t = 0$, and (c) an equation for the plane through the point (2, 0, 1) that contains $v(0)$ and $a(0)$.

16. Deduce from Eq. (12) that a particle will move in a straight line if the normal component of acceleration is identically zero.

17. If a particle moves in a curve with constant speed, show that the force is always directed along the normal.

18. If the force acting on a particle is at all times perpendicular to the direction of motion, show that the speed remains constant.

19. Show that the radius of curvature of a plane curve $\mathbf{R} = x(t)\mathbf{i} + y(t)\mathbf{j}$ is given by

$$\rho = \frac{(\dot{x}^2 + \dot{y}^2)}{\sqrt{\ddot{x}^2 + \ddot{y}^2 - \ddot{s}^2}},$$

where the dots mean derivatives with respect to t ($\dot{x} = dx/dt$, $\ddot{x} = d^2x/dt^2$, and so on) and

$$\ddot{s} = \frac{d}{dt}\sqrt{\dot{x}^2 + \dot{y}^2}.$$

20. Derive Eq. (1) by the Δ-process.

21. Apply Eqs. (1) and (2) to $\mathbf{U} \cdot \mathbf{V}_1$ with $\mathbf{V}_1 = \mathbf{V} \times \mathbf{W}$ and thereby derive Eq. (5) for

$$\frac{d}{dt}[\mathbf{U} \cdot (\mathbf{V} \times \mathbf{W})].$$

22. If $\mathbf{F}(t) = \mathbf{i}f(t) + \mathbf{j}g(t) + \mathbf{k}h(t)$, where f, g, and h are functions of t that have derivatives of orders one, two, and three, show that

$$\frac{d}{dt}\left[\mathbf{F} \cdot \left(\frac{d\mathbf{F}}{dt} \times \frac{d^2\mathbf{F}}{dt^2}\right)\right] = \mathbf{F} \cdot \left(\frac{d\mathbf{F}}{dt} \times \frac{d^3\mathbf{F}}{dt^3}\right).$$

Explain why the answer contains just this one term rather than the three terms that one might expect.

23. With the book closed, derive vector expressions for the velocity and acceleration in terms of tangential and normal components. Check your derivations with those given in the text.

14.6 Planetary Motion and Satellites

In this article we show that the motion of a planet or particle under the influence of a central force field takes place in a plane. We then derive Kepler's second law (equal areas in equal times) and formulate Kepler's first and third laws.

Polar and Cylindrical Coordinates

If a particle P moves on a plane curve whose equation is given in polar coordinates, it is convenient to express the velocity and acceleration vectors in terms of still a third set of unit vectors. We introduce unit vectors

$$\begin{aligned} \mathbf{u}_r &= \mathbf{i}\cos\theta + \mathbf{j}\sin\theta, \\ \mathbf{u}_\theta &= -\mathbf{i}\sin\theta + \mathbf{j}\cos\theta, \end{aligned} \qquad (1)$$

which point respectively along the radius vector $\overrightarrow{OP}$, and at right angles to $\overrightarrow{OP}$ and in the direction of increasing θ, as shown in Fig. 14.21. Then we find from (1) that

$$\begin{aligned} \frac{d\mathbf{u}_r}{d\theta} &= -\mathbf{i}\sin\theta + \mathbf{j}\cos\theta = \mathbf{u}_\theta, \\ \frac{d\mathbf{u}_\theta}{d\theta} &= -\mathbf{i}\cos\theta - \mathbf{j}\sin\theta = -\mathbf{u}_r. \end{aligned} \qquad (2)$$

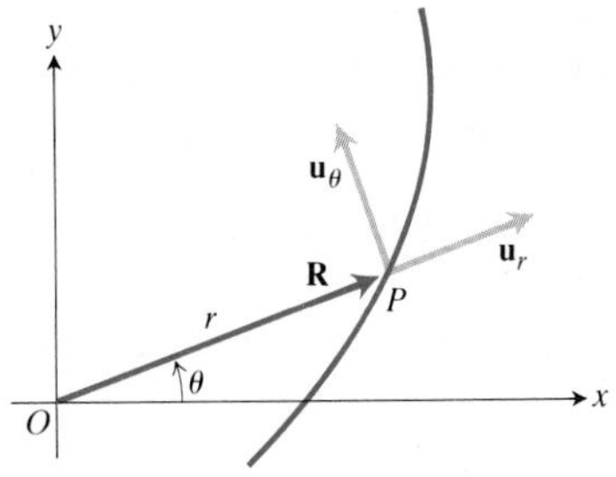

14.21 The unit vectors $\mathbf{u}_r$ and $\mathbf{u}_\theta$.

This says that the result of differentiating either one of the unit vectors $\mathbf{u}_r$ and $\mathbf{u}_\theta$ with respect to θ is equivalent to rotating that vector through 90° in the positive (counterclockwise) direction.

Since the vectors $\mathbf{R} = \overrightarrow{OP}$ and $r\mathbf{u}_r$ have the same direction, and the length of $\mathbf{R}$ is the absolute value of the polar coordinate r of $P(r, \theta)$, we have

$$\mathbf{R} = r\mathbf{u}_r. \qquad (3)$$

To obtain the velocity, we must differentiate this with respect to t, remembering that both r and $\mathbf{u}_r$ may be variables. From (2) and the Chain Rule we get

$$\frac{d\mathbf{u}_r}{dt} = \frac{d\mathbf{u}_r}{d\theta}\frac{d\theta}{dt} = \mathbf{u}_\theta \frac{d\theta}{dt},$$
$$\frac{d\mathbf{u}_\theta}{dt} = \frac{d\mathbf{u}_\theta}{d\theta}\frac{d\theta}{dt} = -\mathbf{u}_r \frac{d\theta}{dt}. \tag{4}$$

Hence

$$\mathbf{v} = \frac{d\mathbf{R}}{dt} = \mathbf{u}_r \frac{dr}{dt} + r\frac{d\mathbf{u}_r}{dt}$$

becomes

$$\mathbf{v} = \mathbf{u}_r \frac{dr}{dt} + \mathbf{u}_\theta r \frac{d\theta}{dt}. \tag{5}$$

Of course this velocity vector is tangent to the curve at P and has magnitude

$$|\mathbf{v}| = \sqrt{(dr/dt)^2 + r^2(d\theta/dt)^2} = |ds/dt|.$$

In fact, if the three sides of the "differential triangle" of sides dr, $r\,d\theta$, and ds are all divided by dt, the result will be a similar triangle having sides dr/dt, $r\,d\theta/dt$, and ds/dt, which illustrates the vector equation

$$\mathbf{v} = \mathbf{T}\frac{ds}{dt} = \mathbf{u}_r \frac{dr}{dt} + \mathbf{u}_\theta r \frac{d\theta}{dt}.$$

(See Fig. 14.22.)

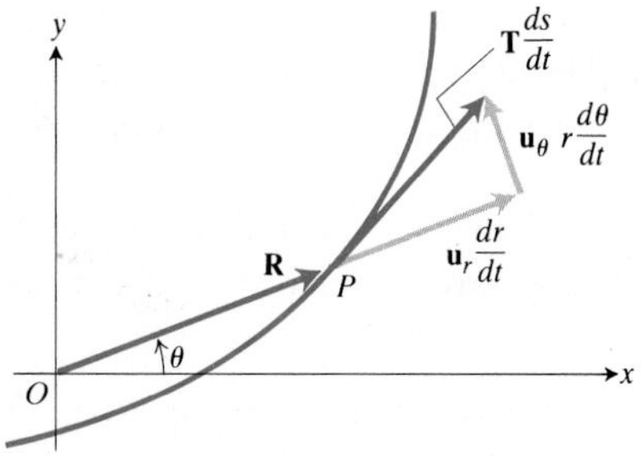

14.22 $\mathbf{u}_r \dfrac{dr}{dt} + \mathbf{u}_\theta r \dfrac{d\theta}{dt}$ is the velocity vector.

The acceleration vector is found by differentiating the velocity vector in (5) as follows:

$$\mathbf{a} = \frac{d\mathbf{v}}{dt} = \left(\mathbf{u}_r \frac{d^2r}{dt^2} + \frac{dr}{dt}\frac{d\mathbf{u}_r}{dt}\right) + \left(\mathbf{u}_\theta r \frac{d^2\theta}{dt^2} + \mathbf{u}_\theta \frac{dr}{dt}\frac{d\theta}{dt} + \frac{d\mathbf{u}_\theta}{dt} r \frac{d\theta}{dt}\right).$$

When Eqs. (4) are used to evaluate the derivatives of $\mathbf{u}_r$ and $\mathbf{u}_\theta$ and the components are separated, the result becomes

$$\mathbf{a} = \mathbf{u}_r\left[\frac{d^2r}{dt^2} - r\left(\frac{d\theta}{dt}\right)^2\right] + \mathbf{u}_\theta\left[r\frac{d^2\theta}{dt^2} + 2\frac{dr}{dt}\frac{d\theta}{dt}\right]. \tag{6}$$

Equations (5) and (6) apply to motion in the xy-plane. It is easy to modify them to apply to motion in space. First, we need to add a term

$\mathbf{k}z$ to the right-hand side of Eq. (3),

$\mathbf{k}\dfrac{dz}{dt}$ to the velocity vector,

$\mathbf{k}\dfrac{d^2z}{dt^2}$ to the acceleration vector.

Thus we get

$$\mathbf{R} = r\mathbf{u}_r + \mathbf{k}z, \tag{7a}$$

$$\mathbf{v} = \mathbf{u}_r \frac{dr}{dt} + \mathbf{u}_\theta r \frac{d\theta}{dt} + \mathbf{k}\frac{dz}{dt}, \tag{7b}$$

$$\mathbf{a} = \mathbf{u}_r\left[\frac{d^2r}{dt^2} - r\left(\frac{d\theta}{dt}\right)^2\right] + \mathbf{u}_\theta\left[r\frac{d^2\theta}{dt^2} + 2\frac{dr}{dt}\frac{d\theta}{dt}\right] + \mathbf{k}\frac{d^2z}{dt^2}. \tag{7c}$$

Equations (7) are particularly useful in connection with cylindrical coordinates. The three vectors $\mathbf{u}_r$, $\mathbf{u}_\theta$, and $\mathbf{k}$ are mutually orthogonal unit vectors that form a right-handed frame:

$$\begin{aligned} \mathbf{u}_r \times \mathbf{u}_\theta &= \mathbf{k}, \\ \mathbf{k} \times \mathbf{u}_r &= \mathbf{u}_\theta, \\ \mathbf{u}_\theta \times \mathbf{k} &= \mathbf{u}_r. \end{aligned} \tag{8}$$

Planets Move in Planes

If we assume that the motion obeys Newton's second law, $\mathbf{F} = m\mathbf{a}$, and that the only force acting on the planet is a gravitational attraction directed toward a fixed point O, of magnitude inversely proportional to the square of the distance of the planet from O, Kepler's laws may be deduced by integrating differential equations of motion. We get these equations from the physical assumptions

$$\mathbf{F} = m\mathbf{a} = m\frac{d^2\mathbf{R}}{dt^2} = -\frac{GmM}{|\mathbf{R}|^2}\frac{\mathbf{R}}{|\mathbf{R}|}, \tag{9}$$

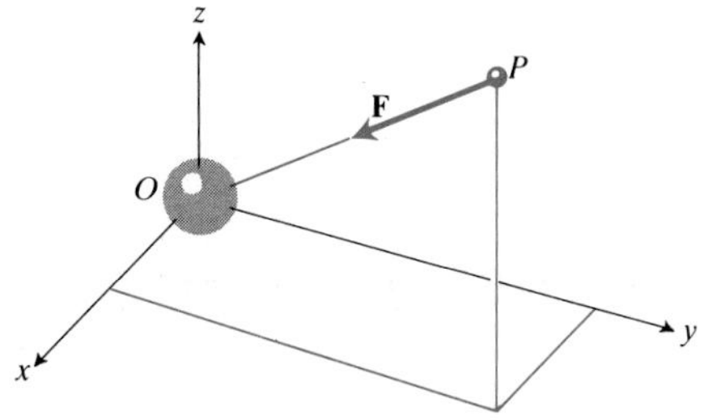

14.23 **F** is called a *central force* if it points toward a fixed point (here the origin) no matter how *P* moves.

where G is the gravitational constant, m the mass of the moving particle (or planet), and M the mass of the attracting object (or sun) at O. Equation (9) tells us that the acceleration vector $d^2\mathbf{R}/dt^2$ is proportional to the position vector $\mathbf{R}$. This is typical of a *central force field* (see Fig. 14.23), and it leads to the result

$$\mathbf{R} \times \frac{d^2\mathbf{R}}{dt^2} = \mathbf{0}, \tag{10}$$

because $\mathbf{A} \times \mathbf{B}$ is always equal to the zero vector if $\mathbf{B}$ is a scalar times $\mathbf{A}$. Now we can deduce that the left-hand side of Eq. (10) is the derivative of the cross product of $\mathbf{R}$ and $d\mathbf{R}/dt$. First we have

$$\frac{d}{dt}\left(\mathbf{R} \times \frac{d\mathbf{R}}{dt}\right) = \frac{d\mathbf{R}}{dt} \times \frac{d\mathbf{R}}{dt} + \mathbf{R} \times \frac{d^2\mathbf{R}}{dt^2} = \mathbf{0} + \mathbf{R} \times \frac{d^2\mathbf{R}}{dt^2},$$

because the cross product of any vector with itself is the zero vector. When the derivative of a vector is zero, each of its components (relative to a fixed coordinate system) is a constant, so the vector is a constant vector. Thus, Eq. (10) implies that

$$\mathbf{R} \times \frac{d\mathbf{R}}{dt} = \mathbf{C}. \tag{11}$$

From the geometric interpretation of the cross product, we conclude from Eq. (11) that both **R** and $\mathbf{v} = d\mathbf{R}/dt$ lie in a plane perpendicular to the fixed vector **C**. We choose a reference frame so that this plane, which includes the path of the particle, is the same as the xy-plane, and so that the unit vector in the direction of **C** is **k**. We introduce polar coordinates in this plane, choosing as initial line $\theta = 0$, the direction of **R** when $|\mathbf{R}|$ is a *minimum*. In planetary motion, this corresponds to *perihelion* position of the planet (the position nearest to the sun). If we also measure the time t from the instant of passage through perihelion, we get the following initial values: when $t = 0$,

$$\begin{aligned} r &= r_0 = \text{minimum value of } r \Rightarrow \left(\frac{dr}{dt}\right)_0 = 0, \\ \theta &= 0, \\ |\mathbf{v}| &= v_0 = \left(r\frac{d\theta}{dt}\right)_0. \end{aligned} \tag{12}$$

Kepler's Second Law

Because the motion takes place in a plane, we can use Eqs. (3), (5), and (6) for **R**, **v**, and **a**. Also, since $\mathbf{R} \times \mathbf{v}$ is constant (Eq. 11), we can evaluate that constant by forming the cross product at time $t = 0$:

$$\mathbf{C} = (\mathbf{R}_0) \times (\mathbf{v}_0) = \mathbf{k}\left(r^2\frac{d\theta}{dt}\right)_0 = \mathbf{k}(r_0 v_0). \tag{13}$$

To get Eq. (13), cross the right-hand side of Eq. (3) and the right-hand side of Eq. (5), and recall that

$$\mathbf{u}_r \times \mathbf{u}_r = \mathbf{0} \quad \text{and} \quad \mathbf{u}_r \times \mathbf{u}_\theta = \mathbf{k}.$$

If we substitute this result for **C** in Eq. (11), we get

$$\mathbf{k}\left(r^2\frac{d\theta}{dt}\right) = \mathbf{k}(r_0 v_0), \qquad \text{or} \qquad r^2\frac{d\theta}{dt} = r_0 v_0. \tag{14}$$

Kepler's second law of planetary motion follows from this, because the element of area in polar coordinates is

$$dA = \frac{1}{2}r^2\,d\theta$$

and Eq. (14) implies that

$$\frac{dA}{dt} = \frac{1}{2}r^2\frac{d\theta}{dt} = \frac{1}{2}r_0 v_0 = \text{constant}. \tag{15}$$

This is the result: *The radius vector sweeps over area at a constant rate in a central force field* (Fig. 14.24).

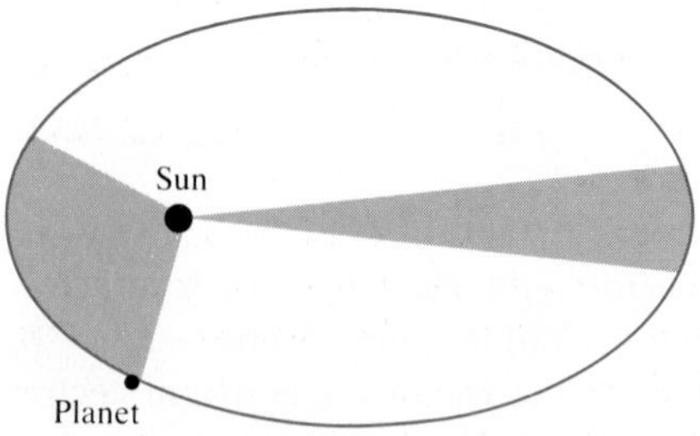

14.24 The line joining a planet and its sun sweeps over equal areas in equal times.

Kepler's First and Third Laws

In deriving Kepler's second law we introduced the constants that appear in Kepler's other two laws, so we can at least state these laws explicitly even though we shall not derive them here.

Table 14.1 Numerical Data

Gravitational constant:	$G = 6.673 \times 10^{-8} \dfrac{\text{cm}^3}{\text{gm} \cdot \text{s}^2}$
Sun's mass:	2×10^{33} gm
Earth's mass:	5.976×10^{27} gm
One year = 365.256 days = 3.156×10^7 seconds	
One mile = 1.609×10^5 cm	

Kepler's first law says that the path of a planet's motion, which we already know lies in a plane, is actually a conic section whose equation is

$$r = \frac{(1/h)}{1 + e\cos\theta}, \tag{16}$$

with

$$\frac{1}{h} = \frac{r_0^2 v_0^2}{GM} \quad \text{and} \quad e = \frac{1}{r_0 h} - 1 = \frac{r_0 v_0^2}{GM} - 1. \tag{17}$$

In the coordinate system we have been using, Eq. (16) represents a conic section of eccentricity e with one focus at the origin (sun's center). For given values of r_0, M, and G, Eq. (17) shows that the eccentricity depends on the v_0, velocity at perihelion. For a circular orbit, $e = 0$ and $v_0 = \sqrt{GM/r_0}$. For $0 < e < 1$, the orbit is an ellipse with one focus at O; for $e = 1$, the orbit is a parabola; and for $e > 1$, the orbit is a hyperbola.

Kepler's third law says that when the orbit is an ellipse, the time T required to go around once (the *period*) is given by the equation

$$\frac{T^2}{a^3} = \frac{4\pi^2}{GM}, \tag{18}$$

where a is the length of the ellipse's semimajor axis. Thus, for every planet in a given solar system, the ratio T^2/a^3 has the constant value $4\pi^2/GM$, where M is the sun's mass and G is Newton's gravitational constant. If a is measured in centimeters, M in grams, and G in $\text{cm}^3/\text{gm} \cdot \text{s}^2$, then Eq. (18) calculates T in seconds.

PROBLEMS

In Problems 1–5, express **v** and **a** in terms of $\mathbf{u}_r$ and $\mathbf{u}_\theta$.

1. $r = a(1 - \cos\theta)$ and $\dfrac{d\theta}{dt} = 3$

2. $r = a\sin 2\theta$ and $\dfrac{d\theta}{dt} = 2t$

3. $r = e^{a\theta}$ and $\dfrac{d\theta}{dt} = 2$

4. $r = a(1 + \sin t)$ and $\theta = 1 - e^{-t}$

5. $r = 2\cos 4t$ and $\theta = 2t$

6. Let $r = \sin\theta$ and $\theta = \pi t$. Express **v** and **a** in terms of $\mathbf{u}_r$ and $\mathbf{u}_\theta$ when $t = 1/4$.

7. Let $\mathbf{R} = (t\cos t)\,\mathbf{i} + (t\sin t)\,\mathbf{j} + t\mathbf{k}$, $t \geq 0$. Express **v** and **a** (a) in the frame **i, j, k**; (b) in the frame $\mathbf{u}_r$, $\mathbf{u}_\theta$, **k**.

8. Let $r = a(1 + \sin t)$, $\theta = 1 - e^{-t}$, $z = \cos t$. Express **v** and **a** in terms of $\mathbf{u}_r$, $\mathbf{u}_\theta$, **k** at $t = 0$.

9. If a particle moves in an ellipse whose polar equation is $r = c/(1 - e\cos\theta)$ and the force is directed toward the origin, show that the magnitude of the force is proportional to $1/r^2$.

10. Since the orbit of the Vanguard satellite had major axis $2a = 10{,}784$ mi (approximately), Eq. (18), with M equal to Earth's mass, should give the period. Compute it.

11. In May 1965, the U.S.S.R. launched Proton I, weighing 26,900 lb (at launch), with a perigee of 118 miles, an apogee of 390 miles, and a period of 92 minutes. Using the period $T = 5520$ s and the relevant data for the mass of the earth and the gravitational constant G, compute the semimajor axis a of the orbit from Eq. (18). Use 1 lb = 2204.62 gm.

12. Read the article *Orbit* in the Encyclopaedia Britannica. What dates does it give for Kepler's announcements of his first two laws? Of his third law?

13. Read the article *Space Exploration* in the most recent Yearbook of the Encyclopaedia Britannica (or other source). Report the perigee, apogee, and orbital period for at least one earth satellite as given in the article you read.

14. In the Encyclopaedia Britannica (or elsewhere), read an article on the scientific work of Kepler. In what way was Kepler's work dependent on earlier work of Tycho Brahe? On contemporary work of Galileo?

15. Without introducing coordinates, give a geometric argument for the validity of the equation

$$\frac{dA}{dt} = \frac{1}{2}\left|\mathbf{R} \times \frac{d\mathbf{R}}{dt}\right|,$$

where $\mathbf{R}$ is the position vector of a particle moving in a plane curve, and dA/dt is the rate at which that vector sweeps out area.

16. For what values of v_0 is the orbit of Eq. (16) a parabola? A circle? An ellipse? A hyperbola?

17. Assuming that the earth's distance from the sun at perihelion is approximately 93,000,000 miles, and that the eccentricity of the earth's orbit about the sun is 0.0167, compute the velocity of the earth in its orbit at perihelion. (Use Eq. 17.)

REVIEW QUESTIONS AND EXERCISES

1. Define the derivative of a vector function.
2. Develop formulas for the derivatives, with respect to θ, of the unit vectors $\mathbf{u}_r$ and $\mathbf{u}_\theta$.
3. Develop vector formulas for velocity and acceleration of a particle moving in a plane curve:
   a) in terms of cartesian coordinates,
   b) in terms of polar coordinates,
   c) in terms of distance traveled along the curve and unit vectors tangent and normal to the curve.
4. a) Define curvature of a plane curve.
   b) Define radius of curvature.
   c) Define center of curvature.
   d) Define osculating circle.
5. Develop a formula for the curvature of a curve whose parametric equations are $x = f(t)$, $y = g(t)$.
6. In what way does the curvature of a curve affect the acceleration of a particle moving along the curve? In particular, discuss the case of constant-speed motion along a curve.
7. State and derive Kepler's second law concerning motion in a central force field.
8. If a vector $\mathbf{V}$ is a differentiable function of t and $|\mathbf{V}| =$ constant, what do you know about $d\mathbf{V}/dt$?
9. Define arc length and curvature of a space curve.
10. Explain how to $\mathbf{T}$, $\mathbf{N}$, and $\mathbf{B}$ for a space curve.
11. Express the vector $\mathbf{R} = \overrightarrow{OP}$ in terms of cylindrical coordinates and the unit vectors $\mathbf{u}_r$, $\mathbf{u}_\theta$, and $\mathbf{k}$.
12. Derive formulas for the velocity $\mathbf{v} = d\mathbf{R}/dt$ and acceleration $\mathbf{a} = d\mathbf{v}/dt$ in terms of cylindrical coordinates and the unit vectors $\mathbf{u}_r$, $\mathbf{u}_\theta$, and $\mathbf{k}$.

MISCELLANEOUS PROBLEMS

1. A particle moves in the xy-plane according to the time law

$$x = 1/\sqrt{1+t^2}, \quad y = t/\sqrt{1+t^2}.$$

   a) Compute the velocity vector and acceleration vector when $t = 1$.
   b) At what time is the speed of the particle a maximum?

2. A circular wheel with unit radius rolls along the x-axis uniformly, rotating one half-turn per second (Fig. 14.25). The position of a point P on the circumference is given by the formula

$$\overrightarrow{OP} = \mathbf{R} = \mathbf{i}(\pi t - \sin \pi t) + \mathbf{j}(1 - \cos \pi t).$$

   a) Determine the velocity (*vector*) $\mathbf{v}$ and the acceleration (*vector*) $\mathbf{a}$ at time t.
   b) Determine the slopes (as functions of t) of the two straight lines PC and PQ joining P to the center C of the wheel and to the point Q that is topmost at the instant.

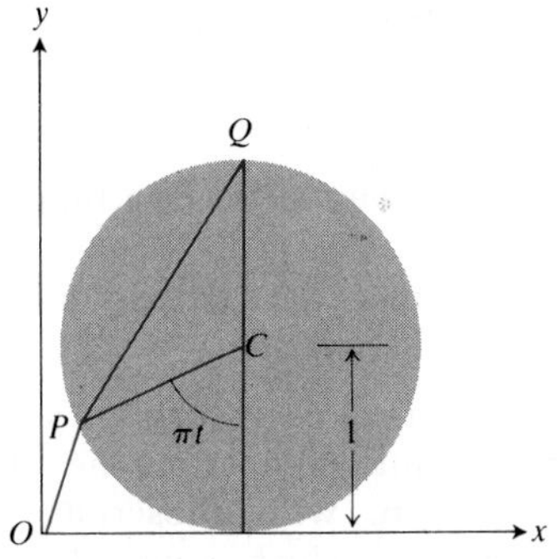

Figure 14.25

c) Show that the directions of the vectors **v** and **a** can be expressed in terms of the straight lines described in part (b).

3. The motion of a particle in the xy-plane is given by

$$\mathbf{R} = \mathbf{i}at \cos t + \mathbf{j}at \sin t.$$

Find the speed, and the tangential and normal components of the acceleration.

4. A particle moves in such a manner that the derivative of the position vector is always perpendicular to the position vector. Show that the particle moves on a circle with center at the origin.

5. The position of a point at time t is given by the formulas $x = e^t \cos t$, $y = e^t \sin t$.
a) Show that $\mathbf{a} = 2\mathbf{v} - 2\mathbf{r}$.
b) Show that the angle between the radius vector $\mathbf{r}$ and the acceleration vector $\mathbf{a}$ is constant, and find this angle.

6. Given the instantaneous velocity $\mathbf{v} = a\mathbf{i} + b\mathbf{j}$ and acceleration $\mathbf{a} = c\mathbf{i} + d\mathbf{j}$ of a particle at a point P on its path of motion, determine the curvature of the path at P.

7. Find the parametric equations, in terms of the parameter θ, of the locus of the center of curvature of the cycloid

$$x = a(\theta - \sin\theta), \quad y = a(1 - \cos\theta).$$

8. Find the point on the curve $y = e^x$ for which the radius of curvature is a minimum.

9. a) Given a closed curve having the property that every line parallel to the x-axis or the y-axis has at most two points in common with the curve. Let

$$x = x(t), \quad y = y(t), \quad \alpha \le t \le \beta,$$

be equations of the curve. Prove that if dx/dt and dy/dt are continuous, then the area bounded by the curve is

$$\frac{1}{2}\left|\int_{\alpha}^{\beta}\left[x(t)\frac{dy}{dt} - y(t)\frac{dx}{dt}\right]dt\right|.$$

b) Use the result of part (a) to find the area inside the ellipse

$$x = a\cos\phi, \quad y = b\sin\phi, \quad 0 \le \phi \le 2\pi.$$

What does the answer become when $a = b$?

10. For the curve defined by the equations

$$x = \int_0^{\theta} \cos\left(\frac{1}{2}\pi t^2\right) dt, \quad y = \int_0^{\theta} \sin\left(\frac{1}{2}\pi t^2\right) dt,$$

calculate the curvature κ as a function of the length of arc s, where s is measured from $(0, 0)$.

11. The curve for which the length of the tangent intercepted between the point of contact and the y-axis is always equal to 1 is called the *tractrix*. Find its equation. Show that the radius of curvature at each point of the curve is inversely proportional to the length of the normal intercepted between the point on the curve and the y-axis. Calculate the length of arc of the tractrix, and find the parametric equations in terms of the length of arc.

12. Let $x = x(t)$, $y = y(t)$ be a closed curve. A constant length p is measured off along the normal to the curve. The extremity of this segment describes a curve that is called a *parallel curve* to the original curve. Find the length of arc, the radius of curvature, and the area enclosed by the parallel curve.

13. Given the curve represented by the parametric equations

$$x = 32t, \quad y = 16t^2 - 4.$$

a) Calculate the radius of curvature of the curve at the point where $t = 3$.
b) Find the length of the curve between the points where $t = 0$ and $t = 1$.

14. Find the velocity, acceleration, and speed of a particle whose position at time t is

$$x = 3\sin t, \quad y = 2\cos t.$$

Also find the tangential and normal components of the acceleration.

15. The position of a particle at time t is given by the equations

$$x = 1 + \cos 2t, \quad y = \sin 2t.$$

Find
a) the normal and tangential components of acceleration at time t,
b) the radius of curvature of the path,
c) the equation of the path in polar coordinates, using the x-axis as the line $\theta = 0$ and the y-axis as the line $\theta = \pi/2$.

16. A particle moves so that its position at time t has the polar coordinates $r = t$, $\theta = t$. Find the velocity $\mathbf{v}$, the acceleration $\mathbf{a}$, and the curvature κ at any time t.

15 Partial Derivatives

15.1 Functions of Two or More Variables

In many applications, the values of the functions under study are determined by the values of more than one independent variable. The function may be as simple as the rule $V = \pi r^2 h$ for calculating the volume of a circular cylinder from its radius and height. Or it might be as complicated as the rule

$$w(x, t) = \cos(1.7 \times 10^{-2} t - 0.2x)e^{-0.2x},$$

which calculates the temperature x feet below the surface of the earth during the tth day of the year as a fraction of the average surface temperature on that day (Example 5 below).

Like functions of a single variable, these functions have *domains* (the sets of allowable input pairs, triples, or whatever, of real numbers) and *ranges* (their sets of output values).

DEFINITION

> Suppose D is a collection of n-tuples of real numbers
>
> $$(x_1, x_2, \ldots, x_n).$$
>
> A *function* f with domain D is a rule that assigns a number
>
> $$w = f(x_1, x_2, \ldots, x_n)$$
>
> to each n-tuple in D. The function's range is the set of w-values the function takes on. The symbol w is called the *dependent variable* of f, and f is said to be a function of the n *independent variables* $x_1, x_2, \ldots, x_n$.

In the function $V = \pi r^2 h$, the dependent variable is V. The independent variables are r and h.

Functions given by formulas are evaluated in the usual way by substituting values of the independent variables and calculating the corresponding values of the dependent variable. Thus the value of the function $w = \sqrt{x^2 + y^2 + z^2}$ at the point $(3, 0, 4)$ is $\sqrt{(3)^2 + (0)^2 + (4)^2} = \sqrt{25} = 5$.

The Convention About Domains

In defining functions of more than one variable we observe the same convention that we do for functions of a single variable (Article 1.6). We never let the independent variables take on values that require division by zero. Also, we require the outputs to be real numbers unless we specifically say otherwise. Points that give imaginary numbers as outputs are to be excluded from the domain. Thus, if $f(x, y) = 1/(x^2 - y^2)$, we require $x^2 \neq y^2$, and if $f(x, y) = \sqrt{y - x^2}$, we require $y \geq x^2$.

Except for these restrictions, the domains of functions unless otherwise stated are assumed to be the largest possible sets for which their defining rules assign real numbers. These sets are sometimes called the "natural" domains of the functions.

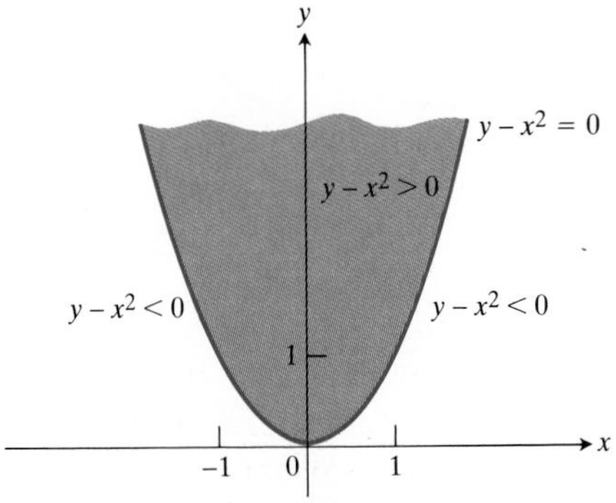

15.1 The domain of $f(x, y) = \sqrt{y - x^2}$ consists of the shaded region and its bounding parabola $y = x^2$. Elsewhere in the plane, $y - x^2 < 0$ and f is not defined.

EXAMPLE 1 Sketch the domain of

$$f(x, y) = \sqrt{y - x^2}.$$

What is the function's range?

Solution According to our convention, the domain D is the set of all number pairs (x, y) in the plane for which $\sqrt{y - x^2}$ is real (Fig. 15.1). This means the points for which

$$y - x^2 \geq 0 \quad \text{or} \quad y \geq x^2.$$

Therefore, D is the set of points that lie above and on the parabola $y = x^2$. The range of f is the set of all nonnegative numbers, $z \geq 0$. □

EXAMPLE 2 What are the domain and range of

$$f(x, y) = \frac{xy}{x^2 - y^2}?$$

Solution By our convention, x and y may take on any values except the ones for which $x^2 - y^2 = 0$, or $x^2 = y^2$. The domain therefore consists of all points in the plane except the points on the lines $y = x$ and $y = -x$. The range of f is the entire set of real numbers $-\infty < z < \infty$, as we may see by letting the point $P(x, y)$ traverse the hyperbola $x^2 - y^2 = 1$. On the right-hand branch of the hyperbola, $x = \sqrt{1 + y^2}$ and the values of f are given by the formula

$$f(x, y) = \frac{(\sqrt{1 + y^2})\, y}{1} = y\sqrt{1 + y^2}. \;\square$$

Graphs, Level Curves, Contours

There are two standard ways to picture a function $z = f(x, y)$. One is to draw a number of its *level curves*, the curves in the domain along which f has some constant value $f(x, y) = c$. The other is to draw the graph of f as a *surface* $z = f(x, y)$ in space, the surface being the set of points $(x, y, f(x, y))$. In the next example we do both.

EXAMPLE 3 Graph

$$z = f(x, y) = 100 - x^2 - y^2,$$

and plot the level curves $f(x, y) = 0$, 51, and 75.

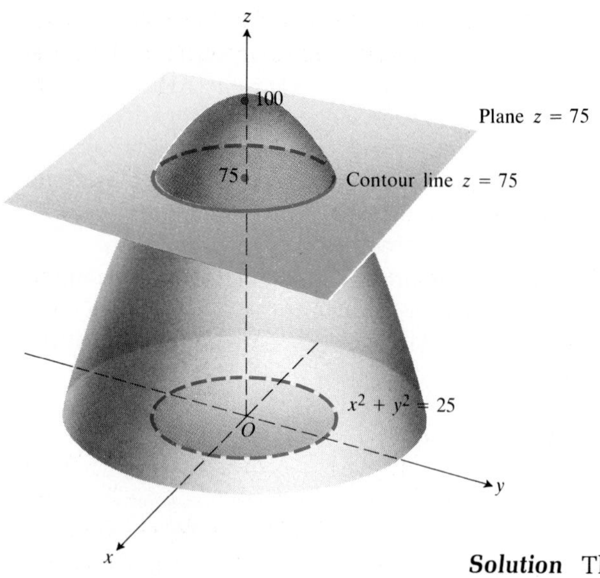

15.2 The contour line $z = 75$ on the surface $z = f(x, y) = 100 - x^2 - y^2$.

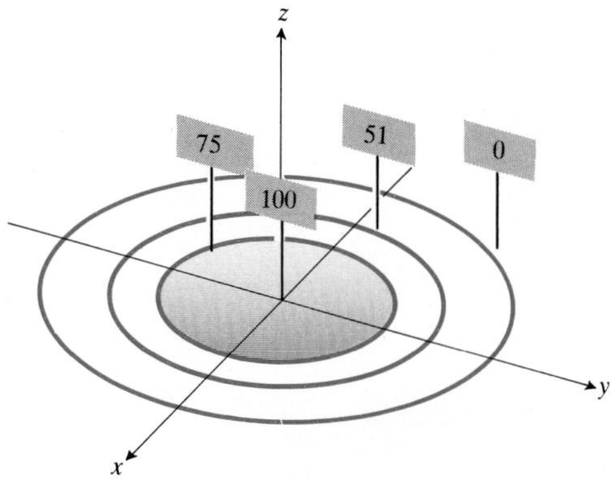

15.3 Level curves in the domain of f.

Solution The graph of f is a paraboloid, shown in Fig. 15.2. Its domain is the entire xy-plane, and its range is the set of real numbers $z \leq 100$. (The figure does not show the graph for negative values of z.)

The level curve $f(x, y) = 0$ is the set of points in the xy-plane at which

$$f(x, y) = 100 - x^2 - y^2 = 0, \qquad \text{or} \qquad x^2 + y^2 = 100,$$

the circle of radius 10 centered at the origin (Fig. 15.3). Similarly, the level curves $f(x, y) = 51$ and $f(x, y) = 75$ are the circles

$$f(x, y) = 100 - x^2 - y^2 = 75 \qquad \text{or} \qquad x^2 + y^2 = 25,$$
$$f(x, y) = 100 - x^2 - y^2 = 51 \qquad \text{or} \qquad x^2 + y^2 = 49. \ \square$$

Contour Lines

The points on a surface $z = f(x, y)$ that represent a fixed function value $f(x, y) = c$ make a *contour line* on the surface. These points all have the same distance $|c|$ above (or below) the xy-plane. In other words, the contour line is the curve in which the surface is cut by the plane $z = c$. Figure 15.2 shows the contour line $z = f(x, y) = 75$ lying directly above the circle $x^2 + y^2 = 25$, which is the level curve $f(x, y) = 75$ in the xy-plane.

Because of the close association of contour lines with level curves there is no firm agreement about which word to use for which kind of curve. On most maps, for example, curves that represent constant elevation are called contours, not level curves. The convention in this book is that level curves lie in the domain of f (in the plane $z = 0$) while contour lines lie on the surface defined by f (in the appropriate plane $z = c$).

Level Surfaces

The points (x, y, z) in space at which a function $w = f(x, y, z)$ of three independent variables takes on a constant value $f(x, y, z) = c$ usually make up a surface called a *level surface* of the function.

EXAMPLE 4 Describe the level surfaces of the function

$$w = \sqrt{x^2 + y^2 + z^2}.$$

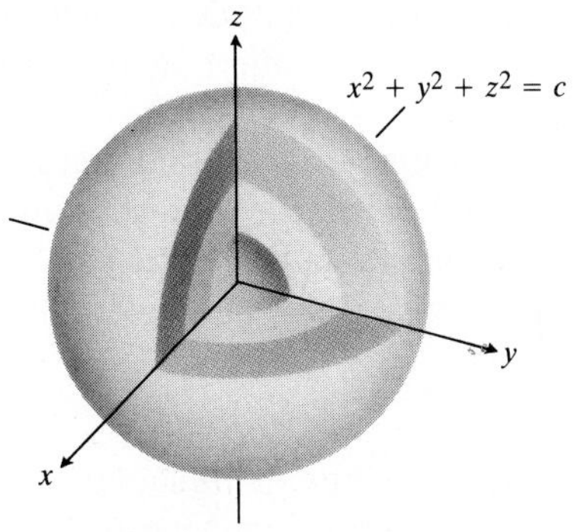

15.4 The level surfaces of $w = \sqrt{x^2 + y^2 + z^2}$ are concentric spheres.

Solution The value of function w can be interpreted as the distance from the origin to the point (x, y, z) in rectangular coordinates. Each level surface

$$\sqrt{x^2 + y^2 + z^2} = a, \qquad a > 0$$

is a sphere of radius a centered at the origin. Figure 15.4 shows a cutaway view of three of these spheres. The "level surface" $\sqrt{x^2 + y^2 + z^2} = 0$ consists of the origin alone.

It is important to keep in mind that we are *not* graphing the function $w = \sqrt{x^2 + y^2 + z^2}$ here. The graph of the function lies in a four-variable space. Instead, we are looking at surfaces in the function's domain. These surfaces show how the function values change as we move about in the domain. If we stay on the sphere of radius a centered at the origin, the function maintains the constant value a. If we move from sphere to sphere, the function values change. They increase as we move away from the origin, and decrease as we move toward it. Thus, the way the function values change as we move away from any given point depends on the direction we take. This dependence of change on direction is an important idea, and we shall return to it in Article 15.6. □

Three-dimensional graphing programs now available for computers make it possible to graph many functions of two variables with only a few keystrokes. We can often get information about a function from one of these graphs more quickly than we can from a formula.

EXAMPLE 5 Figure 15.5 shows a computer-generated graph of the function

$$w(x, t) = \cos(1.7 \times 10^{-2}t - 0.2x)e^{-0.2x} \qquad (t \text{ in days, } x \text{ in feet}).$$

The graph shows how the temperature beneath the earth's surface varies with time. The variation is given as a fraction of the variation at the

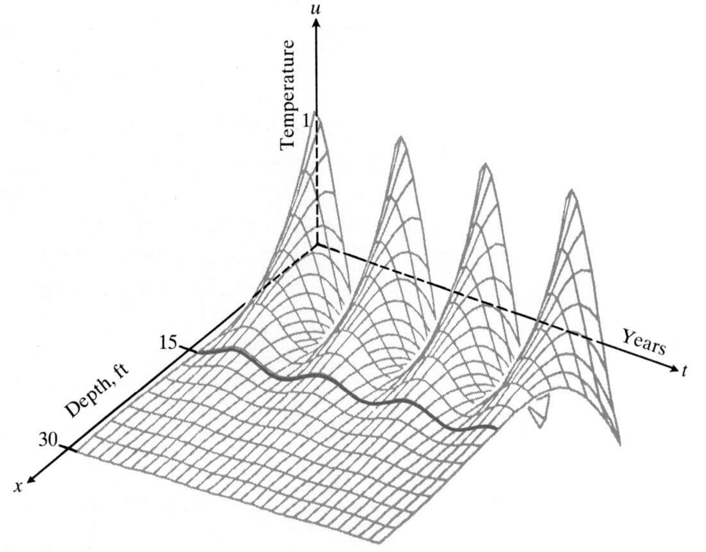

15.5 This computer-generated graph of $w(x, t) = \cos(1.7 \times 10^{-2}t - 0.2x)e^{-0.2x}$ shows the seasonal variation of the temperature below ground as a fraction of surface temperature. $\Delta x = 0.375$ ft, $\Delta t = 15.625$ days. At $x = 15$ ft the variation is only 5% of the variation at the surface. At $x = 30$ ft the variation is less than 0.25% of the surface variation. (Adapted by John G. Aspinall from art provided by Norton Starr for G. C. Berresford's "Differential Equations and Root Cellars," *The UMAP Journal,* Volume 2, Number 3, (1981), pp. 53–75.)

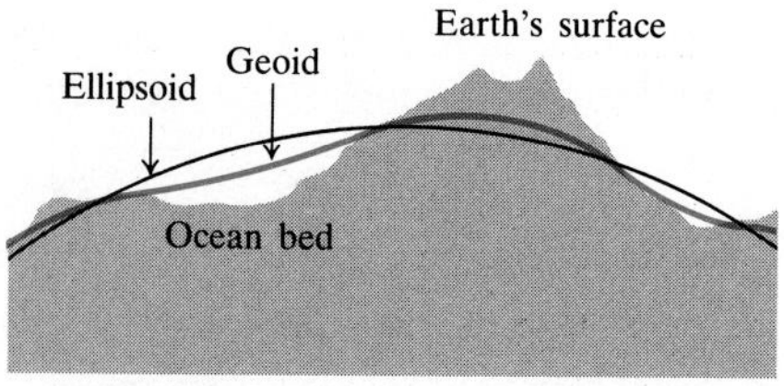

15.6 A profile of the earth, geoid, and ellipsoid.

surface. At a depth of 15 ft, the variation (vertical amplitude in the figure) is about 5 percent of the surface variation. At 30 ft, there is almost no variation in temperature throughout the year.

Another thing the graph shows that is not immediately apparent from the equation for f is that the temperature 15 ft below the surface is about half a year out of phase with the surface temperature. When the temperature is lowest on the surface (late January, say) it is at its highest 15 ft below. The seasons are reversed. □

EXAMPLE 6 Along with the launching and positioning of satellites has come an increased interest in mapping the variations in the earth's gravitational field. The strength of the field may vary significantly from place to place on the surface of the earth. In fact, changes in the performance of pendulum clocks taken by voyagers from Europe to other continents are said to have provided early supporting evidence of Newton's law of gravitation. If one takes the strength of the field at mean sea level as a standard, then the points in the earth's gravitational field that have the same gravitational potential as the standard value constitute a potato-shaped surface that geophysicists call the *geoid*. The geoid differs from the surface of the earth itself in most places, and it also differs from the ellipsoid that is normally used to approximate the surface of the earth (Fig. 15.6). The height of the geoid above or below the ellipsoid is called the *geoidal height*. Geoidal heights are counted as positive when the geoid rises above the ellipsoid, and negative when it dips below the ellipsoid. Figure 15.7 shows contours of geoidal height on a Mercator map of the earth. These contours are the level curves of the geoidal height function. □

15.7 A contour map of geoidal height measured in meters above (+) and below (−) an ideal earth shaped like an ellipsoid.

PROBLEMS

In Problems 1–12, give the domain and range, and describe the level curves.

1. $f(x, y) = 1/xy$
2. $f(x, y) = \sqrt{x + y}$
3. $f(x, y) = 1/\sqrt{x + y}$
4. $f(x, y) = \ln (x + y)$
5. $f(x, y) = \ln (x^2 + y^2)$
6. $f(x, y) = \sqrt{4 - x^2 - y^2}$
7. $f(x, y) = \sqrt{y/x}$
8. $f(x, y) = y/x^2$
9. $f(x, y) = \tan^{-1} (y/x)$
10. $f(x, y) = e^{x-y}$
11. $f(x, y) = \cos xy$
12. $f(x, y) = \sinh (x^2 + y^2)$.

In Problems 13–18, find the domain and range.

13. $f(x, y) = \dfrac{xy}{y^2 - x^2}$
14. $f(x, y, z) = \sqrt{1 - (x^2 + y^2 + z^2)}$
15. $f(x, y, z) = \sqrt{z}$
16. $f(x, y, z) = 5$
17. $f(x, y, z) = \dfrac{x^2}{x^2 + y^2 + z^2}$
18. $f(x, y, z) = \dfrac{1}{1 - (x^2 + y^2 + z^2)}$

Describe the level surfaces of the functions in Problems 19–32. You may wish to refer to Articles 13.8, 13.10, and 13.11.

19. $f(x, y, z) = x^2 + y^2 + z^2$
20. $f(x, y, z) = \ln (x^2 + y^2 + z^2)$
21. $f(x, y, z) = 1/(x^2 + y^2 + z^2)$
22. $f(x, y, z) = (x^2/25) + (y^2/16) + (z^2/9)$
23. $f(x, y, z) = x$
24. $f(x, y, z) = x + y$
25. $f(x, y, z) = x + y + z$
26. $f(x, y, z) = x^2 + y^2$
27. $f(x, y, z) = y^2 + z^2$
28. $f(x, y, z) = x^2 + z^2$
29. $f(x, y, z) = x^2 + y^2 - z$
30. $f(x, y, z) = \sqrt{x^2 + y^2 - z}$
31. $f(x, y, z) = x^2 - y$
32. $f(x, y, z) = \sin y$.

In Problems 33–38, show two ways of representing the function $z = f(x, y)$: (a) by sketching the surface $z = f(x, y)$ in space, and (b) by drawing a family of level curves in the plane.

33. $f(x, y) = x$
34. $f(x, y) = y$
35. $f(x, y) = x^2 + y^2$
36. $f(x, y) = x^2 - y^2$
37. $f(x, y) = \sqrt{y - x^2}$ (Example 1)
38. $f(x, y) = x \sin y$
39. Sketch the level curves of
$$f(x, y) = -(x - 1)^2 - y^2 + 1$$
for $f(x, y) = 1, 0, -3, -8$.
40. Sketch the level curves of $f(x, y) = 2x^2 - y + 1$ for $f(x, y) = 0, 1, 2$.
41. Find an equation for the level curve of $f(x, y) = 16 - x^2 - y^2$ that passes through the point $(2\sqrt{2}, \sqrt{2})$.
42. Repeat Problem 41 for $f(x, y) = \sqrt{x^2 - 1}$ and $(1, 0)$.
43. Sketch in the xy-plane the domain of the function $z = \sqrt{\sin (\pi xy)}$, $xy \geq 0$, and shade the points that lie in the domain.
44. Find the maximum value of the function $w = xyz$ on the line $x = 20 - t$, $y = t$, $z = 20$. At what point or points on the line does the maximum occur? (*Hint:* Along the line, w can be expressed as a function of the single variable t.)

***Toolkit* programs**
3D Grapher

15.2 Limits and Continuity

If the values of a function $z = f(x, y)$ can be made as close as we like to a fixed number L by taking the point (x, y) close to the point (x_0, y_0), but not equal to (x_0, y_0), we say that L is the limit of f as (x, y) approaches (x_0, y_0).

In symbols, we write

$$\lim_{(x,\, y) \to (x_0,\, y_0)} f(x, y) = L, \tag{1}$$

and we say "the limit of f as (x, y) approaches (x_0, y_0) equals L." This is like the limit of a function of one variable, except that there are two independent variables involved instead of one.

For (x, y) to be "close" to (x_0, y_0) means that the cartesian distance $\sqrt{(x - x_0)^2 + (y - y_0)^2}$ is small in some sense. Since

$$|x - x_0| = \sqrt{(x - x_0)^2} \le \sqrt{(x - x_0)^2 + (y - y_0)^2} \tag{2}$$

and

$$|y - y_0| = \sqrt{(y - y_0)^2} \le \sqrt{(x - x_0)^2 + (y - y_0)^2}, \tag{3}$$

the inequality

$$\sqrt{(x - x_0)^2 + (y - y_0)^2} < \delta$$

implies

$$|x - x_0| < \delta \quad \text{and} \quad |y - y_0| < \delta.$$

Conversely, if for some $\delta > 0$ both $|x - x_0| < \delta$ and $|y - y_0| < \delta$, then

$$\sqrt{(x - x_0)^2 + (y - y_0)^2} < \sqrt{\delta^2 + \delta^2} = \sqrt{2}\delta,$$

which is small if δ is small. Thus, in calculating limits we may think either in terms of the distance in the plane or in terms of differences in coordinates. See Fig. 15.8. We therefore have two equivalent definitions of limit.

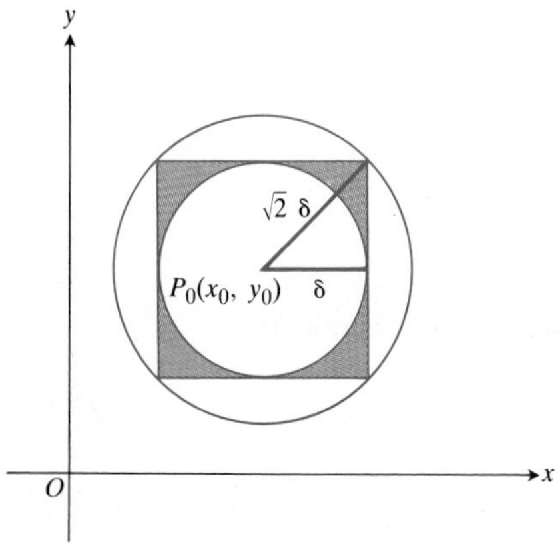

15.8 The open square $|x - x_0| < \delta$, $|y - y_0| < \delta$ lies inside the open disc $\sqrt{(x - x_0)^2 + (y - y_0)^2} < \sqrt{2}\delta$ and contains the open disc $\sqrt{(x - x_0)^2 + (y - y_0)^2} < \delta$.

Two Equivalent Definitions of Limit

The limit of $f(x, y)$ as $(x, y) \to (x_0, y_0)$ is the number L if for any $\varepsilon > 0$ there exists a $\delta > 0$ such that for all points (x, y) either

1. $0 < \sqrt{(x - x_0)^2 + (y - y_0)^2} < \delta \Rightarrow |f(x, y) - L| < \varepsilon$,

or

2. $0 < |x - x_0| < \delta$ and $0 < |y - y_0| < \delta \Rightarrow |f(x, y) - L| < \varepsilon$.

As we did with functions of a single variable in Article 1.10, we may use the definition of limit to prove directly that

$$\lim_{(x,\, y) \to (x_0,\, y_0)} x = x_0, \qquad \lim_{(x,\, y) \to (x_0,\, y_0)} y = y_0, \qquad \lim_{(x,\, y) \to (x_0,\, y_0)} k = k. \tag{4}$$

The proofs are similar and we shall not repeat them here.

There is also a theorem like Theorem 1 of Article 1.10 that says that the limit of a sum $f(x, y) + g(x, y)$ is the sum of the limits of f and g (when they exist), with similar results for differences, products, constant multiples, and quotients.

THEOREM 1

If

$$\lim_{(x, y)\to(x_0, y_0)} f(x, y) = L_1 \quad \text{and} \quad \lim_{(x, y)\to(x_0, y_0)} g(x, y) = L_2,$$

then

i) $\lim[f(x, y) + g(x, y)] = L_1 + L_2$,
ii) $\lim[f(x, y) - g(x, y)] = L_1 - L_2$,
iii) $\lim[f(x, y)g(x, y)] = L_1L_2$,
iv) $\lim[kf(x, y)] = kL_1$ (any number k),
v) $\lim \dfrac{f(x, y)}{g(x, y)} = \dfrac{L_1}{L_2}$ if $L_2 \neq 0$.

The limits are all to be taken as $(x, y) \to (x_0, y_0)$.

When we apply Theorem 1 to the functions and limits in (4) we obtain the useful result that the limits of polynomials and rational functions like

$$x^2 + y^2, \qquad \frac{x - xy + 3}{x^2y + 5xy - y^3}$$

as $(x, y) \to (x_0, y_0)$ may be calculated by evaluating the functions at the point (x_0, y_0). The only requirement is that the functions be defined there. In other words, polynomials and rational functions in two variables are *continuous* wherever they are defined, as we shall see in a moment.

EXAMPLE 1

a) $\displaystyle\lim_{(x, y)\to(3, -4)} (x^2 + y^2) = (3)^2 + (-4)^2 = 9 + 16 = 25.$

b) $\displaystyle\lim_{(x, y)\to(0, 1)} \frac{x - xy + 3}{x^2y + 5xy - y^3} = \frac{0 - 0(1) + 3}{(0)^2(1) + 5(0)(1) - (1)^3} = -3.$ □

Continuity

A function $f(x, y)$ is said to be *continuous* at the point (x_0, y_0) if

1. f is defined at (x_0, y_0),
2. $\displaystyle\lim_{(x, y)\to(x_0, y_0)} f(x, y)$ exists, and (5)
3. $\displaystyle\lim_{(x, y)\to(x_0, y_0)} f(x, y) = f(x_0, y_0)$.

It is a consequence of Theorem 1, although we shall not prove it, that if $f(x, y)$ and $g(x, y)$ are both continuous at a point, then their sum $f(x, y) + g(x, y)$ is continuous at that point. Similar results hold for differences, products, and multiples of continuous functions. Also, the quotient of two continuous functions is continuous wherever it is defined. It follows that polynomials and rational functions in two variables are continuous at any point at which they are defined.

If $z = f(x, y)$ is a continuous function of x and y, and $w = g(z)$ is a continuous function of z, then the composite

$$w = g(f(x, y))$$

is continuous. Thus, composites like

$$e^{x-y}, \qquad \cos\frac{xy}{x^2+1}, \qquad \ln(1 + x^2y^2)$$

are continuous at every point (x, y).

As with functions of a single variable, the general rule is that composites of continuous functions are continuous. The only requirement is that each function be continuous where it is applied.

EXAMPLE 2 Given $\varepsilon > 0$, how close to $(0, 0)$ should we take (x, y) to make $|f(x, y) - f(0, 0)| < \varepsilon$ if

$$f(x, y) = \frac{x + y}{x^2 + y^2 + 1}?$$

Solution Since $f(0, 0) = 0$,

$$|f(x, y) - f(0, 0)| = \left|\frac{x + y}{x^2 + y^2 + 1}\right|,$$

and we are being asked how close to $(0, 0)$ we should take (x, y) to make

$$\left|\frac{x + y}{x^2 + y^2 + 1}\right| < \varepsilon.$$

Since the denominator of the fraction is never less than one, we have

$$\begin{aligned}\left|\frac{x + y}{x^2 + y^2 + 1}\right| &\le |x + y| \le |x| + |y| \\ &\le \sqrt{x^2 + y^2} + \sqrt{x^2 + y^2} = 2\sqrt{x^2 + y^2},\end{aligned}$$

which will be less than ε if

$$\sqrt{x^2 + y^2} < \frac{\varepsilon}{2}.$$

Therefore, the original ε inequality will hold if the distance from (x, y) to $(0, 0)$ is less than $\varepsilon/2$. That is, we may choose our δ in the definition of limit to be the number $\varepsilon/2$. □

EXAMPLE 3 Show that the function

$$f(x, y) = \begin{cases} \dfrac{xy}{x^2 + y^2}, & (x, y) \ne (0, 0) \\ 0, & (x, y) = (0, 0) \end{cases}$$

is continuous at every point except the origin. See Fig. 15.9.

Solution The function f is continuous at any point $(x, y) \ne (0, 0)$ because its values are then given by a rational function of x and y.

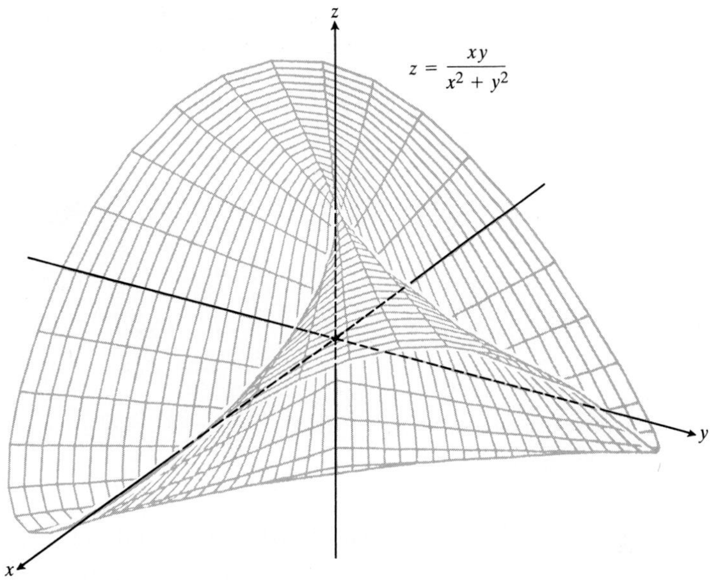

15.9 The graph of

$$f(x, y) = \begin{cases} \dfrac{xy}{x^2 + y^2}, & (x, y) \neq (0, 0) \\ 0, & (x, y) = (0, 0). \end{cases}$$

The only point of the z-axis that belongs to this surface is the point (0, 0, 0).

At (0, 0) the value of f is defined, but f, we claim, has no limit as $(x, y) \to (0, 0)$. The reason is that different paths of approach to the origin may give different limits, as we shall now see.

If $(x, y) \to (0, 0)$ along the line $y = mx$, but $(x, y) \neq (0, 0)$, then

$$f(x, y) = \frac{xy}{x^2 + y^2} = \frac{x(mx)}{x^2 + m^2x^2} = \frac{m}{1 + m^2},$$

$$\lim_{\substack{(x, y) \to (0, 0) \\ \text{along } y = mx}} f(x, y) = \frac{m}{1 + m^2}.$$

This limit changes as m changes. In fact, the substitution $m = \tan \theta$ produces

$$f(x, y) = \frac{m}{1 + m^2} = \frac{\tan \theta}{\sec^2 \theta} = \frac{1}{2} \sin 2\theta,$$

which shows that the limit varies from $-\frac{1}{2}$ to $\frac{1}{2}$ depending on the angle of approach. There is no single number L that we may call the limit of f as $(x, y) \to (0, 0)$. Therefore, the limit fails to exist and the function cannot be continuous at (0, 0). □

EXAMPLE 4 Examine the limits of

$$f(x, y) = \frac{x^2y}{x^4 + y^2}, \qquad (x, y) \neq (0, 0)$$

as $(x, y) \to (0, 0)$ along the line $y = mx$ and along the parabola $y = x^2$. Does f have a limit as $(x, y) \to (0, 0)$?

Solution (Fig. 15.10) Along the line $y = mx$

$$f(x, y) = f(x, mx) = \frac{x^2(mx)}{x^4 + m^2x^2} = \frac{mx}{x^2 + m^2}.$$

Therefore,

$$\lim_{\substack{(x, y) \to (0, 0) \\ \text{along } y = mx}} f(x, y) = \lim_{x \to 0} \frac{mx}{x^2 + m^2} = 0.$$

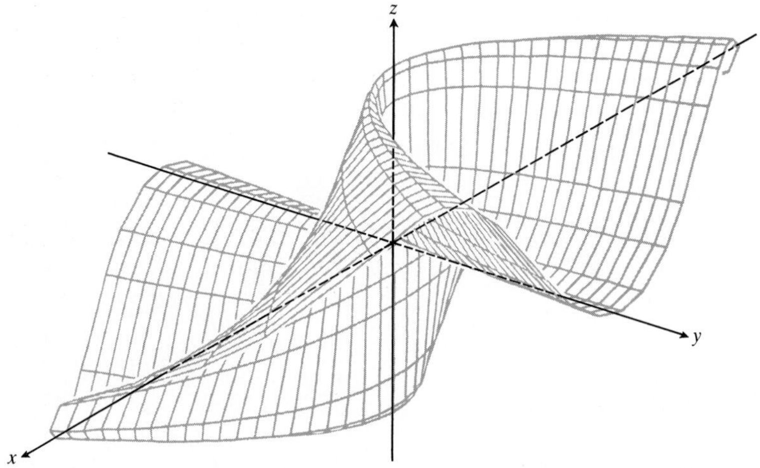

15.10 The graph of

$$f(x, y) = \begin{cases} \dfrac{x^2y}{x^4 + y^2}, & (x, y) \neq (0, 0) \\ 0, & (x, y) = (0, 0). \end{cases}$$

The only point of the z-axis that belongs to this surface is the point (0, 0, 0).

The fact that the limit, taken along all *linear* paths to (0, 0), exists and equals zero may lead you to suspect that $\lim_{(x, y)\to(0, 0)} f(x, y)$ exists. However, along the parabola $y = x^2$,

$$f(x, y) = f(x, x^2) = \frac{x^2(x^2)}{x^4 + x^4} = \frac{1}{2}.$$

Therefore,

$$\lim_{\substack{(x, y)\to(0, 0) \\ \text{along } y=x^2}} f(x, y) = \frac{1}{2}.$$

For f to have a limit as $(x, y) \to (0, 0)$, the limits along all paths of approach would have to agree. Since we have reached different limits on different paths, we conclude that f has no limit as $(x, y) \to (0, 0)$. □

Functions of Three Variables or More

The definitions of limit and continuity for functions of two variables and the conclusions about limits and continuity for sums, products, quotients, and composites all extend to functions of three variables or more. Thus, functions like

$$\ln \sqrt{x^2 + y^2 + z^2} \qquad \text{and} \qquad \frac{e^{x-y+z}}{z^2 + \cos \sqrt{xy}}$$

are continuous at every point at which they are defined, and limits like

$$\lim_{P\to(1, 0, 1)} \frac{e^{x-y+z}}{z^2 + \cos(\sqrt{xy})} = \frac{e^{1-0+1}}{(1)^2 + \cos(0)} = \frac{e^2}{2},$$

where P denotes the point (x, y, z), may be evaluated by direct substitution.

Continuous Functions Defined in Closed, Bounded Regions

As we know, a function of a single variable that is continuous throughout a closed, bounded interval $[a, b]$ takes on an absolute maximum value and an absolute minimum value at least once in $[a, b]$. The same is true of a function $z = f(x, y)$ that is continuous on a closed, bounded region R in the plane (like a line segment, a disc, or a filled-in triangle, square, or rectangle). It takes on an absolute maximum value at some point in R, and an absolute minimum value at some point in R.

Similar results are true for functions of three or more variables. A continuous function $w = f(x, y, z)$, for example, must take on absolute maximum and minimum values on any closed, bounded region (solid ball or cube, spherical shell, rectangular plate) on which it is defined.

Techniques for finding these maximum and minimum values will be the subject of Articles 15.9 and 15.10.

PROBLEMS

Find the limits (if they exist) in Problems 1–18.

1. $\displaystyle\lim_{(x,y)\to(0,0)} \frac{3x^2 - y^2 + 5}{x^2 + y^2 + 2}$

2. $\displaystyle\lim_{(x,y)\to(1,1)} \ln|1 + x^2y^2|$

3. $\displaystyle\lim_{(x,y)\to(0,\ln 2)} e^{x-y}$

4. $\displaystyle\lim_{(x,y)\to(0,4)} \frac{x}{\sqrt{y}}$

5. $\displaystyle\lim_{P\to(1,3,4)} \sqrt{x^2 + y^2 + z^2 - 1}$

6. $\displaystyle\lim_{P\to(1,2,6)} \left(\frac{1}{x} + \frac{1}{y} + \frac{1}{z}\right)$

7. $\displaystyle\lim_{(x,y)\to(0,0)} \frac{e^y \sin x}{x}$

8. $\displaystyle\lim_{(x,y)\to(0,\pi/2)} \sec x \tan y$

9. $\displaystyle\lim_{(x,y)\to(0,0)} \tan^{-1}\frac{1}{\sqrt{x^2 + y^2}}$

10. $\displaystyle\lim_{(x,y)\to(0,0)} \cos\frac{x^2 + y^2}{x + y + 1}$

11. $\displaystyle\lim_{(x,y)\to(1,1)} \cos\sqrt[3]{|xy| - 1}$

12. $\displaystyle\lim_{(x,y)\to(1,0)} \frac{x \sin y}{x^2 + 1}$

13. $\displaystyle\lim_{(x,y)\to(0,0)} y \sin(1/x)$

14. $\displaystyle\lim_{(x,y)\to(0,2)} \left(\frac{\cos x - 1}{x^2}\right)\left(\frac{y - 2}{y^2 - 4}\right)$

15. $\displaystyle\lim_{(x,y)\to(1,1)} \frac{x^2 - 2xy + y^2}{x - y}$

16. $\displaystyle\lim_{(x,y)\to(-2,2)} \frac{xy + y - 2x - 2}{x + 1}$

17. $\displaystyle\lim_{(x,y)\to(1,1)} \frac{x^2 - y^2}{x - y}$

18. $\displaystyle\lim_{(x,y)\to(1,1)} \frac{x^3y^3 - 1}{xy - 1}$

By considering different lines of approach, show that the functions in Problems 19–24 have no limit as $(x, y) \to (0, 0)$.

19. $\dfrac{x + y}{x - y}$

20. $\dfrac{x - y}{x + y}$

21. $\dfrac{x^2 - y^2}{x^2 + y^2}$

22. $\dfrac{x}{\sqrt{x^2 + y^2}}$

23. $\dfrac{x^2 + y^2}{xy}$

24. $\dfrac{xy}{|xy|}$

In Problems 25–28, find how close to the origin one should take the point (x, y) to make $|f(x, y) - f(0, 0)| < 0.01$.

25. $f(x, y) = x^2 + y^2$

26. $f(x, y) = y/(x^2 + 1)$

27. $f(x, y) = (x + y)/(x^2 + 1)$

28. $f(x, y) = (x + y)/(2 + \cos x)$

In Problems 29–32, find how close to the origin one should take the point (x, y, z) to make

$$|f(x, y, z) - f(0, 0, 0)| < \varepsilon$$

for the given ε.

29. $f(x, y, z) = x^2 + y^2 + z^2$, $\quad \varepsilon = 0.01$

30. $f(x, y, z) = xyz$, $\quad \varepsilon = 0.008$

31. $f(x, y, z) = \dfrac{x + y + z}{x^2 + y^2 + z^2 + 1}$, $\quad \varepsilon = 0.015$

32. $f(x, y, z) = \tan^2 x + \tan^2 y + \tan^2 z$, $\quad \varepsilon = 0.03$

33. Is $f(x, y, z) = \sqrt{x^2 + y^2 + z^2}$ continuous at $(0, 0, 0)$? Explain.

34. Show that the function $f(x, y) = x^2/(x^2 + y)$ has no limit as $(x, y) \to (0, 0)$ by examining the limits of f as $(x, y) \to (0, 0)$ along the parabola $y = kx^2$ for selected values of k.

15.3 Partial Derivatives

Partial derivatives are the derivatives we get when we hold constant all but one of the independent variables in a function and differentiate with respect to that one. In this article, we show how partial derivatives arise geometrically and how they may be calculated by applying the rules we already know for differentiating functions of a single variable. We begin our discussion with functions of two independent variables.

If (x_0, y_0) is a point in the domain of a function $z = f(x, y)$, the plane $y = y_0$ will cut the surface $z = f(x, y)$ in the curve $z = f(x, y_0)$, as shown in Fig. 15.11. This curve is the graph of the function $z = f(x, y_0)$ in the plane $y = y_0$. The vertical coordinate in this plane is z, the height above (below) the xy-plane. The horizontal coordinate is x.

The derivative of $z = f(x, y_0)$ with respect to x at $x = x_0$ is defined in the usual way as the limit

$$\left.\frac{d}{dx}f(x, y_0)\right|_{\text{at } x = x_0} = \lim_{\Delta x \to 0} \frac{f(x_0 + \Delta x, y_0) - f(x_0, y_0)}{\Delta x}, \tag{1}$$

provided that the limit exists. The limit is called the *partial derivative* of *f with respect to* x at the point (x_0, y_0). The *slope* of the curve $z = f(x, y_0)$ at $x = x_0$ is defined to be the value of the partial derivative with respect to x there. The *tangent* to the curve $z = f(x, y_0)$ at $x = x_0$ is defined to be the line in the plane $y = y_0$ that passes through the point $(x_0, y_0, f(x_0, y_0))$ with this slope.

The usual notations for the partial derivative of $z = f(x, y)$ with respect to x at (x_0, y_0) are as follows:

$\dfrac{\partial f}{\partial x}(x_0, y_0)$ or $f_x(x_0, y_0)$ "Partial derivative of f with respect to x at (x_0, y_0)" or "f sub x at (x_0, y_0)." Convenient for stressing the point (x_0, y_0).

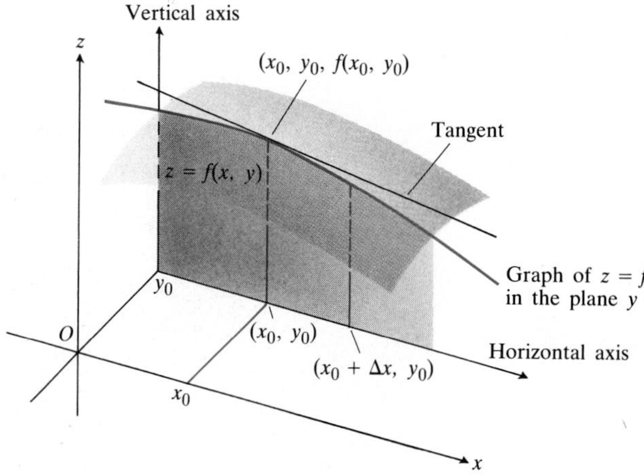

15.11 The intersection of the plane $y = y_0$ with the surface $z = f(x, y)$, viewed from a point above the third quadrant of the xy-plane.

$\left.\dfrac{\partial z}{\partial x}\right\vert_{(x_0, y_0)}$	"Partial derivative of z with respect to x at (x_0, y_0)." Common in science and engineering when you are dealing with variables and do not mention the function explicitly.
$f_x, \dfrac{\partial f}{\partial x}, z_x$, or $\dfrac{\partial z}{\partial x}$	"Partial derivative of f (or z) with respect to x." Convenient when you regard the partial derivative as a function in its own right.

As Eq. (1) shows, $\partial f/\partial x$ is the ordinary derivative of f with respect to x with y held fixed. Thus, to calculate $\partial f/\partial x$ we may differentiate f with respect to x in the usual way while treating y as a constant.

EXAMPLE 1 Find $\partial f/\partial x$ if

$$f(x, y) = 100 - x^2 - y^2.$$

Solution We regard y as a constant and calculate

$$\frac{\partial f}{\partial x} = \frac{d}{dx}(100 - x^2 - y^2) = 0 - 2x - 0 = -2x. \square$$

The definition of the partial derivative $\partial f/\partial y$ of $z = f(x, y)$ with respect to y at (x_0, y_0) is like the definition of $\partial f/\partial x$. We regard $z = f(x_0, y)$ as a function of the single independent variable y (Fig. 15.12) and define

$$\frac{\partial f}{\partial y}(x_0, y_0) = \left.\frac{d}{dy}f(x_0, y)\right\vert_{y=y_0} = \lim_{\Delta y \to 0} \frac{f(x_0, y_0 + \Delta y) - f(x_0, y_0)}{\Delta y}, \quad (2)$$

provided the limit exists. This limit is also taken to be the slope of the curve $z = f(x_0, y)$ in the plane $x = x_0$. The tangent to the curve is the line in the plane $x = x_0$ that passes through $(x_0, y_0, f(x_0, y_0))$ with this slope.

We now have two tangent lines associated with $z = f(x, y)$ at (x_0, y_0). Figure 15.13 shows them together. If $\partial f/\partial x$ and $\partial f/\partial y$ are both continuous at (x_0, y_0) the plane determined by these two lines will be tangent to the

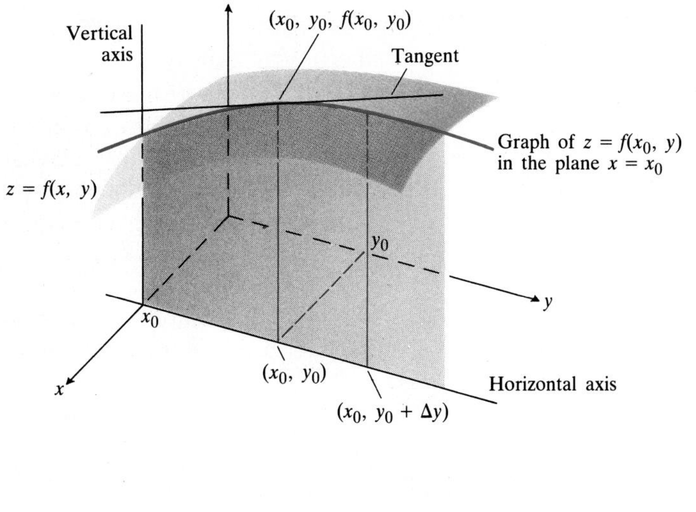

15.12 The intersection of the plane $x = x_0$ with the surface $z = f(x, y)$, viewed from the first quadrant of the xy-plane.

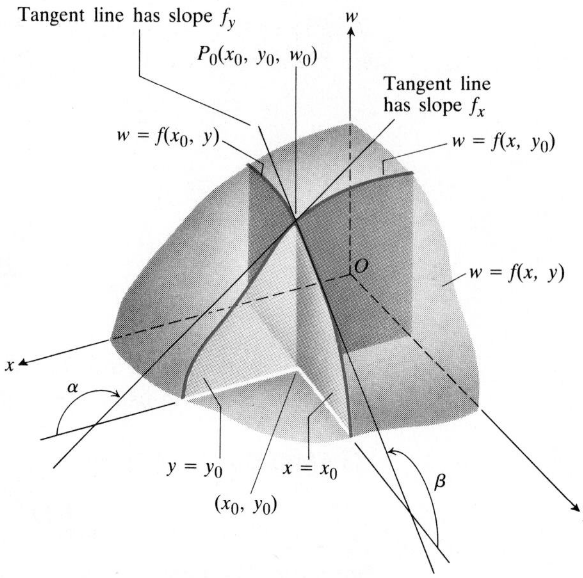

15.13 The plane $y = y_0$ cuts the surface $w = f(x, y)$ in the curve $w = f(x, y_0)$. At each x, the slope of this curve is $f_x(x, y_0)$. Similarly, the plane $x = x_0$ cuts the surface in a curve whose slope is $f_y(x_0, y)$.

surface $z = f(x, y)$ at (x_0, y_0). Tangency is a more complicated phenomenon for surfaces than it is for curves, however, and we shall have to wait until Article 15.6 to deal with it.

EXAMPLE 2 Find $\partial f/\partial y$ if

$$f(x, y) = e^x \ln (x^2 + y^2 + 1).$$

Solution Treating x, e^x, and x^2 as constants we apply the chain rule for functions of a single variable (in this case y) to find

$$\frac{\partial f}{\partial y} = \frac{\partial}{\partial y}(e^x \ln (x^2 + y^2 + 1)) = e^x \cdot \frac{\partial}{\partial x} \ln (x^2 + y^2 + 1)$$

$$= e^x \cdot \frac{1}{x^2 + y^2 + 1} \cdot \frac{\partial}{\partial y}(x^2 + y^2 + 1) = \frac{e^x}{x^2 + y^2 + 1} \cdot 2y$$

$$= \frac{2ye^x}{x^2 + y^2 + 1}. \quad \square$$

Functions with More Than Two Variables

The definitions of the partial derivatives of functions of more than two independent variables are like the definitions for functions of two variables. They are ordinary derivatives with respect to one variable, taken while all the other independent variables are regarded as constants. Thus, if $w = f(x, y, z, u, v)$, we may have as many as five partial derivatives, w_x, w_y, w_z, w_u, and w_v.

EXAMPLE 3 Three resistors of resistances R_1, R_2, and R_3 connected in parallel produce a resistance R given by

$$\frac{1}{R} = \frac{1}{R_1} + \frac{1}{R_2} + \frac{1}{R_3}.$$

Find $\partial R/\partial R_2$.

Solution Treat R_1 and R_3 as constants and differentiate both sides of the equation with respect to R_2. Then

$$-\frac{1}{R^2}\frac{\partial R}{\partial R_2} = -\frac{1}{R_2^2}, \qquad \text{or} \qquad \frac{\partial R}{\partial R_2} = \left(\frac{R}{R_2}\right)^2. \ \square$$

EXAMPLE 4 Find

$$\frac{\partial}{\partial z}(xy)^{\sin z}.$$

Solution We treat xy as a constant and apply the single variable rule

$$\frac{d}{dz}(a)^u = a^u \ln a \frac{du}{dz}$$

with $a = xy$ and $u = \sin z$. The result is

$$\frac{\partial}{\partial z}(xy)^{\sin z} = (xy)^{\sin z} \ln(xy) \frac{d}{dz} \sin z = (xy)^{\sin z} \ln(xy) \cos z. \ \square$$

If $z = f(x, y)$ is continuous we can expect any change in z created by small changes in x and y to be small. If, in addition, f has continuous partial derivatives, we may estimate Δz by the approximation

$$\Delta z \approx f_x(x_0, y_0)\,\Delta x + f_y(x_0, y_0)\,\Delta y. \tag{3}$$

The theorem below establishes this fact and gives some information about the error involved. We call it the increment theorem for functions of two variables. As we shall see, it provides a basis for much of the later work in this chapter. Applications of the approximation in Eq. (3), and a useful estimate of the error involved, will be presented in Article 15.8.

THEOREM 2

Increment Theorem for Functions of Two Variables

Suppose that $z = f(x, y)$ is continuous and has partial derivatives throughout a region

$$R: \quad |x - x_0| < h, \qquad |y - y_0| < k$$

in the xy-plane. Suppose also that Δx and Δy are small enough for the point $(x_0 + \Delta x, y_0 + \Delta y)$ to lie in R. If f_x and f_y are continuous at (x_0, y_0), then the increment

$$\Delta z = f(x_0 + \Delta x, y_0 + \Delta y) - f(x_0, y_0) \tag{4}$$

can be written as

$$\Delta z = f_x(x_0, y_0)\,\Delta x + f_y(x_0, y_0)\,\Delta y + \varepsilon_1\,\Delta x + \varepsilon_2\,\Delta y, \tag{5}$$

where

$$\varepsilon_1, \varepsilon_2 \to 0 \quad \text{as} \quad \Delta x, \Delta y \to 0.$$

Theorem 2 is analagous to the result in Article 2.5 that the increment in a differentiable function $y = g(x)$ of a single variable that takes place when x changes from x_0 to $x_0 + \Delta x$ can be written as

$$\Delta y = g'(x_0)\,\Delta x + \varepsilon \cdot \Delta x, \tag{6}$$

where $\varepsilon \to 0$ as $\Delta x \to 0$. A proof of Theorem 2 may be found in Appendix 4.

Our first application of the increment theorem will be to develop chain rules for differentiating functions of two variables in the next article.

Functions of Three or More Variables

Analogous results hold for functions of any finite number of independent variables. For a function of three variables

$$w = f(x, y, z),$$

that is continuous and has partial derivatives f_x, f_y, f_z at and in some neighborhood of the point (x_0, y_0, z_0), and whose derivatives are continuous at the point, we have

$$\begin{aligned}\Delta w &= f(x_0 + \Delta x, y_0 + \Delta y, z_0 + \Delta z) - f(x_0, y_0, z_0)\\ &= f_x\,\Delta x + f_y\,\Delta y + f_z\,\Delta z + \varepsilon_1\,\Delta x + \varepsilon_2\,\Delta y + \varepsilon_3\,\Delta z, \qquad (7)\end{aligned}$$

where

$$\varepsilon_1, \varepsilon_2, \varepsilon_3 \to 0 \quad \text{when} \quad \Delta x, \Delta y, \text{ and } \Delta z \to 0.$$

The partial derivatives f_x, f_y, f_z in this formula are to be evaluated at the point (x_0, y_0, z_0).

PROBLEMS

Find $\partial f/\partial x$ and $\partial f/\partial y$ in Problems 1–24.

1. $f(x, y) = x + y$
2. $f(x, y) = x^2 + y^2$
3. $f(x, y) = ye^x$
4. $f(x, y) = e^x \cos y$
5. $f(x, y) = e^x \sin y$
6. $f(x, y) = \tan^{-1}(y/x)$
7. $f(x, y) = \ln\sqrt{x^2 + y^2}$
8. $f(x, y) = \cosh(y/x)$
9. $f(x, y) = x$
10. $f(x, y) = y$
11. $f(x, y) = 4$
12. $f(x, y) = \sqrt{9 - x^2 - y^2}$
13. $f(x, y) = x^2 - xy + y^2$
14. $f(x, y) = x/(x^2 + y^2)$
15. $f(x, y) = 1/(x + y)$
16. $f(x, y) = (x + 2)(y + 3)$
17. $f(x, y) = (x + 2)/(y + 3)$
18. $f(x, y) = e^{x \ln y}$
19. $f(x, y) = (x + y)/(xy - 1)$
20. $f(x, y) = 5xy - 7x^2 - y^2 + 3x - 6y + 2$
21. $f(x, y) = \sec(x + y)$
22. $f(x, y) = \tanh(2x + 5y)$
23. $f(x, y) = (xy)^e$
24. $f(x, y) = \ln|\sec xy + \tan xy|$

In Problems 25–36, find the partial derivatives of the given function with respect to each variable.

25. $f(x, y, z, w) = x^2e^{2y+3z}\cos(4w)$
26. $f(x, y, z) = z\sin^{-1}(y/x)$
27. $f(u, v, w) = \dfrac{u^2 - v^2}{v^2 + w^2}$
28. $f(r, \theta, z) = \dfrac{r(2 - \cos 2\theta)}{r^2 + z^2}$
29. $f(x, y, u, v) = \dfrac{x^2 + y^2}{u^2 + v^2}$
30. $f(x, y, r, s) = \sin 2x \cosh 3r + \sinh 3y \cos 4s$
31. $f(x, y, z) = xy + yz + zx$
32. $f(u, v, w) = (u^2 + v^2 + w^2)^{-1/2}$
33. $f(x, y, z) = 1 + y^2 + 2z^2$
34. $f(x, y, z) = (xy)^z$
35. $f(P, Q, R) = PQR$
36. $f(u, v, w, x) = \ln(uvwx)$

In Problems 37 and 38, A, B, C are the angles of a triangle and a, b, c are the respective opposite sides (Fig. 15.14).

37. Express A (explicitly or implicitly) as a function of a, b, c and calculate $\partial A/\partial a$ and $\partial A/\partial b$.

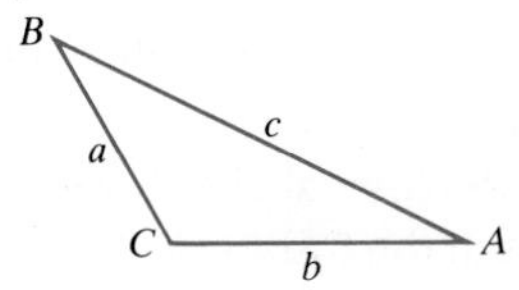

Figure 15.14

38. Express a (explicitly or implicitly) as a function of A, b, B and calculate $\partial a/\partial A$ and $\partial a/\partial B$.

In Problems 39–44, express the spherical coordinates ρ, ϕ, θ as functions of the cartesian coordinates x, y, z and calculate:

39. $\partial\rho/\partial x$ **40.** $\partial\phi/\partial z$ **41.** $\partial\theta/\partial y$

42. $\partial\theta/\partial z$ **43.** $\partial\phi/\partial x$ **44.** $\partial\theta/\partial x$

In Problems 45–47, let $\mathbf{R} = \mathbf{i}x + \mathbf{j}y + \mathbf{k}z$ be the vector from the origin to (x, y, z). Express x, y, z as functions of the spherical coordinates ρ, ϕ, θ, and calculate:

45. $\partial\mathbf{R}/\partial\rho$ **46.** $\partial\mathbf{R}/\partial\phi$ **47.** $\partial\mathbf{R}/\partial\theta$

48. Express the answers to Problems 45–47 in terms of the unit vectors $\mathbf{u}_\rho$, $\mathbf{u}_\phi$, $\mathbf{u}_\theta$ discussed in Miscellaneous Problem 25, Chapter 14.

15.4 Chain Rules

Chain Rules for Functions Defined Along Paths

When we are interested in the temperature $f(x, y, z)$ at points on a path

$$x = x(t), \qquad y = y(t), \qquad z = z(t)$$

in space (or in the pressure or density at points along a path through a gas or fluid) we may think of f as a function of the single variable t. At each t the temperature at the point $(x(t), y(t), z(t))$ on the path is the value of the composite $f(x(t), y(t), z(t))$. If we then wish to know the rate at which f changes with respect to t as we travel along the path, we have only to differentiate this composite with respect to t (provided, of course, df/dt exists).

Many times df/dt can be found by substituting $x(t)$, $y(t)$, and $z(t)$ into the formula for f and differentiating directly with respect to t. But often we work with functions whose formulas are complicated or for which formulas are not readily available. To find their derivatives we need chain rules that express them in terms of the derivatives of the functions being composed. These rules and their application are the subject of this article. As we shall see, they are like the chain rule from Chapter 2, with an additional term for each additional variable.

In Chapter 2, when we had a differentiable function $y = f(x)$ of a single variable x that was itself a differentiable function of t, we found that y was a differentiable function of t with

$$\frac{dy}{dt} = \frac{dy}{dx}\frac{dx}{dt}. \tag{1}$$

If we write $x = g(t)$, then $y = f(x) = f(g(t))$ and Eq. (1) takes the form

$$f'(t) = f'(x)g'(t)$$

or

$$f'(t) = f'(g(t))g'(t). \tag{2}$$

This shows how the derivatives in Eq. (1) are to be evaluated.

Chain Rule for Functions of Two Variables

The analagous formula for a function $z = f(x, y)$ when $x = x(t)$ and $y = y(t)$ are both differentiable functions of t is given in the following theorem.

THEOREM 3

> If $w = f(x, y)$ has continuous partial derivatives f_x and f_y and if $x = x(t)$, $y = y(t)$ are differentiable functions of t, then the composite $w = f(x(t), y(t))$ is a differentiable function of t and
>
> $$\frac{df}{dt} = f_x(x(t), y(t)) \cdot x'(t) + f_y(x(t), y(t)) \cdot y'(t), \tag{3a}$$
>
> or
>
> $$\frac{dw}{dt} = \frac{\partial f}{\partial x}\frac{dx}{dt} + \frac{\partial f}{\partial y}\frac{dy}{dt}. \tag{3b}$$

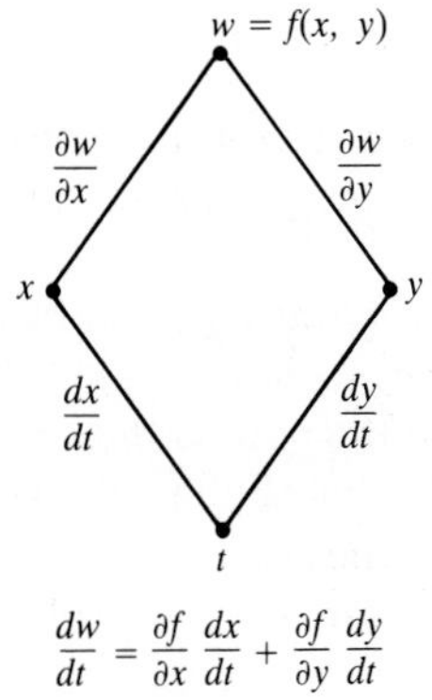

Figure 15.15

The tree diagram in Fig. 15.15 may help in remembering Eq. (3b).

Proof. Let t_0 be any allowable value of t, and let $x_0 = x(t_0)$, $y_0 = y(t_0)$. Suppose that Δx and Δy are the changes that occur in x and y when t is changed from t_0 to $t_0 + \Delta t$. From Theorem 2 in the previous article, the change that occurs in $w = f(x, y)$ is

$$\Delta w = f_x(x_0, y_0)\,\Delta x + f_y(x_0, y_0)\,\Delta y + \varepsilon_1\,\Delta x + \varepsilon_2\,\Delta y, \tag{4}$$

where ε_1 and ε_2 are quantities that approach zero as Δx and $\Delta y \to 0$.

Divide both sides of Eq. (4) by Δt:

$$\frac{\Delta w}{\Delta t} = f_x(x_0, y_0)\frac{\Delta x}{\Delta t} + f_y(x_0, y_0)\frac{\Delta y}{\Delta t} + \varepsilon_1 \cdot \frac{\Delta x}{\Delta t} + \varepsilon_2 \cdot \frac{\Delta y}{\Delta t}. \tag{5}$$

As $\Delta t \to 0$,

$$\frac{\Delta x}{\Delta t} \to x'(t_0) \qquad (x \text{ is differentiable}),$$

$$\frac{\Delta y}{\Delta t} \to y'(t_0) \qquad (y \text{ is differentiable}).$$

Therefore,

$$\lim_{\Delta t \to 0} \frac{\Delta w}{\Delta t} = f_x(x_0, y_0)x'(t_0) + f_y(x_0, y_0)y'(t_0) + 0 \cdot x'(t_0) + 0 \cdot y'(t_0),$$

or

$$\frac{dw}{dt} = f_x(x_0, y_0)x'(t_0) + f_y(x_0, y_0)y'(t_0). \tag{6}$$

The equations we wished to prove are simply statements that Eq. (6) here holds at every allowable value of t, which it does. ■

EXAMPLE 1 Use the chain rule to find the derivative of

$$f(x, y) = xy$$

with respect to t along the path $x = \cos t$, $y = \sin t$.

Solution

$$f_x = y = \sin t, \qquad f_y = x = \cos t,$$

$$\frac{dx}{dt} = -\sin t, \qquad \frac{dy}{dt} = \cos t,$$

$$\begin{aligned}\frac{df}{dt} &= f_x \frac{dx}{dt} + f_y \frac{dy}{dt}\\ &= (\sin t)(-\sin t) + (\cos t)(\cos t)\\ &= -\sin^2 t + \cos^2 t\\ &= \cos 2t.\end{aligned}$$

In this case we can check our result by a more direct calculation. As a function of t,

$$f(x, y) = xy = \cos t \sin t = \frac{1}{2}\sin 2t.$$

Therefore,

$$\frac{df}{dt} = \frac{d}{dt}\left(\frac{1}{2}\sin 2t\right) = \cos 2t. \ \square$$

Chain Rule for Functions of Three Variables

If $w = f(x, y, z)$ has continuous partial derivatives, and $x = x(t)$, $y = y(t)$, $z = z(t)$ are differentiable functions of t, then the appropriate formulas for df/dt are the following:

$$\frac{df}{dt} = f_x \frac{dx}{dt} + f_y \frac{dy}{dt} + f_z \frac{dz}{dt}, \tag{7a}$$

$$\frac{dw}{dt} = \frac{\partial w}{\partial x}\frac{dx}{dt} + \frac{\partial w}{\partial y}\frac{dy}{dt} + \frac{\partial w}{\partial z}\frac{dz}{dt}. \tag{7b}$$

Except for the fact that three intermediate variables (x, y, and z) are involved instead of two, the derivations of Eqs. (7a) and (7b) are identical with the derivation of Eqs. (3a) and (3b). The diagrams for remembering the equations are similar as well (Fig. 15.16).

Chain Rule

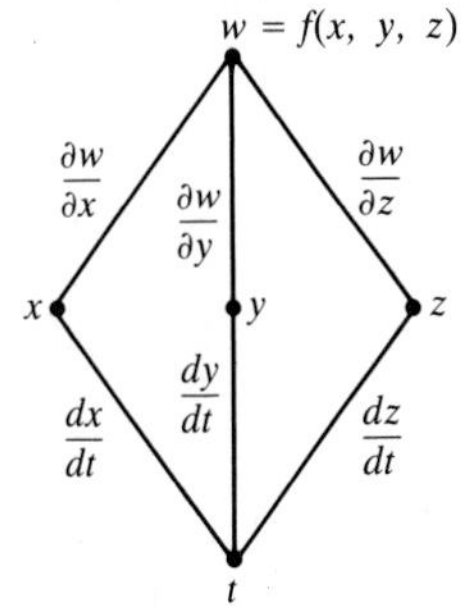

$$\frac{dw}{dt} = \frac{\partial f}{\partial x}\frac{dx}{dt} + \frac{\partial f}{\partial y}\frac{dy}{dt} + \frac{\partial f}{\partial z}\frac{dz}{dt}$$

Figure 15.16

EXAMPLE 2 Express dw/dt as a function of t if $w = xy + z$ and $x = \cos t$, $y = \sin t$, $z = t$. (This derivative shows how w varies along a helix in its domain.)

Solution Starting with Eq. (7b) we have

$$\begin{aligned} \frac{dw}{dt} &= \frac{\partial w}{\partial x}\frac{dx}{dt} + \frac{\partial w}{\partial y}\frac{dy}{dt} + \frac{\partial w}{\partial z}\frac{dz}{dt} \\ &= y(-\sin t) + x(\cos t) + 1(1) \\ &= -\sin^2 t + \cos^2 t + 1 \\ &= 1 + \cos 2t. \ \square \end{aligned}$$

If the parameter t for a curve

$$x = x(t), \qquad y = y(t), \qquad z = z(t)$$

in the domain of a function $f(x, y, z)$ happens to be arc length, then the derivative df/dt may be interpreted as the rate at which f changes with respect to distance along the curve. We shall explore this extremely important interpretation in Article 15.6.

Chain Rules for Functions Defined on Surfaces

No essential complication is introduced by considering a surface (instead of a curve) passing through the domain of a function $w = f(x, y, z)$. For example, we might be interested in how the temperature $w = f(x, y, z)$ varies over the surface of some sphere in space. If the points of the sphere are given in terms of their latitude r and longitude s, then we can think of the temperature on the sphere as a function of these two independent variables. More generally, if $w = f(x, y, z)$, and if

$$x = x(r, s), \qquad y = y(r, s), \qquad z = z(r, s) \tag{8}$$

are any functions whatsoever, then the composite

$$w = f(x(r, s), y(r, s), z(r, s)), \tag{9}$$

when defined, is a function of r and s. If x, y, z, and f also have continuous partial derivatives, then the partial derivatives of w with respect to r and s exist and are given by the following equations:

$$\frac{\partial w}{\partial r} = \frac{\partial f}{\partial x}\frac{\partial x}{\partial r} + \frac{\partial f}{\partial y}\frac{\partial y}{\partial r} + \frac{\partial f}{\partial z}\frac{\partial z}{\partial r}, \tag{10a}$$

$$\frac{\partial w}{\partial s} = \frac{\partial f}{\partial x}\frac{\partial x}{\partial s} + \frac{\partial f}{\partial y}\frac{\partial y}{\partial s} + \frac{\partial f}{\partial z}\frac{\partial z}{\partial s}. \tag{10b}$$

Equation (10a) can be derived from Eq. (7b) by holding s fixed and taking $r = t$. Similarly, Eq. (10b) can be obtained by holding r fixed and taking $s = t$. Or we can go back to increments in each case and apply Eq. (7) of Article 15.3.

The tree diagrams for Eqs. (10a) and (10b) are shown in Fig. 15.17.

in its domain may be calculated from the formula

$$\frac{df}{dt} = \frac{\partial f}{\partial x}\frac{dx}{dt} + \frac{\partial f}{\partial y}\frac{dy}{dt} + \frac{\partial f}{\partial z}\frac{dz}{dt}. \tag{1}$$

To use this formula at any particular point

$$P_0(x_0, y_0, z_0) = P_0(x(t_0), y(t_0), z(t_0)),$$

we evaluate the partial derivatives of f at P_0 and the derivatives of $x(t)$, $y(t)$, and $z(t)$ at t_0.

The derivative df/dt in Eq. (1) is interpreted as the rate of change of f at P_0 with respect to increasing t and therefore depends, among other things, on the direction of motion along the curve. This observation is particularly important when the curve is a straight line through P_0 and the parameter t is arc length (distance along the line) measured from P_0 in the direction of a given unit vector **u**. For then df/dt is the rate of change of f with respect to distance in its domain in the direction of **u.** Thus, by varying **u,** we can find the rates at which f changes with respect to distance as we move through P_0 in different directions. These "directional derivatives" are extremely useful in science and engineering (as well as in mathematics), and one of the goals of this article is to develop a simple formula for calculating them.

Calculating Directional Derivatives

Let

$$\mathbf{u} = u_1\mathbf{i} + u_2\mathbf{j} + u_3\mathbf{k}$$

be a unit vector, and let L be the line through a point P_0 whose equations are

$$x = x_0 + tu_1, \qquad y = y_0 + tu_2, \qquad z = z_0 + tu_3. \tag{2}$$

Then, motion along L in the direction of increasing t is motion in the direction of **u.** Also, $|t|$ measures distance along the line from $P_0(x_0, y_0, z_0)$, as we can see from the following calculation:

$$\begin{aligned}
&\text{Distance from } P_0 \text{ to any point } P(x, y, z) \text{ on this line}\\
&\qquad = \sqrt{(x - x_0)^2 + (y - y_0)^2 + (z - z_0)^2}\\
&\qquad = \sqrt{(tu_1)^2 + (tu_2)^2 + (tu_3)^2}\\
&\qquad = |t|\sqrt{u_1^2 + u_2^2 + u_3^2}\\
&\qquad = |t| \cdot 1 \qquad (\mathbf{u} \text{ is a unit vector})\\
&\qquad = |t|.
\end{aligned}$$

Substituting the derivatives

$$\frac{dx}{dt} = \frac{d}{dt}(x_0 + tu_1) = u_1,$$

$$\frac{dy}{dt} = \frac{d}{dt}(y_0 + tu_2) = u_2,$$

$$\frac{dz}{dt} = \frac{d}{dt}(z_0 + tu_3) = u_3,$$

in Eq. (1) gives

$$\frac{df}{dt} = \frac{\partial f}{\partial x}u_1 + \frac{\partial f}{\partial y}u_2 + \frac{\partial f}{\partial z}u_3. \tag{3}$$

The expression on the right in this equation is the dot product of **u** and the vector

$$\nabla f = \frac{\partial f}{\partial x}\mathbf{i} + \frac{\partial f}{\partial y}\mathbf{j} + \frac{\partial f}{\partial z}\mathbf{k}.$$

This vector is called the *gradient vector* of f at the point P_0. It is customary to picture it as a vector in the domain of f. Its components are calculated by evaluating the three partial derivatives of f at (x_0, y_0, z_0). The derivative on the left-hand side of Eq. (3) is called "the derivative of f at the point P_0 in the direction of **u**." It is often denoted by

$$(D_{\mathbf{u}} f)_{P_0}.$$

DEFINITION

Definitions of Directional Derivative and Gradient

1. If the partial derivatives of $f(x, y, z)$ are defined at $P_0(x_0, y_0, z_0)$, then the *gradient* of f at P_0 is the vector

$$\nabla f = \frac{\partial f}{\partial x}\mathbf{i} + \frac{\partial f}{\partial y}\mathbf{j} + \frac{\partial f}{\partial z}\mathbf{k} \qquad (4)$$

obtained by evaluating the partial derivatives of f at P_0.

2. If $f(x, y, z)$ has continuous partial derivatives at $P_0(x_0, y_0, z_0)$, and **u** is a unit vector, then the *derivative of f at P_0 in the direction of* **u** is the number

$$(D_{\mathbf{u}} f)_{P_0} = (\nabla f)_{P_0} \cdot \mathbf{u}, \qquad (5)$$

which is the scalar product of **u** and the gradient of f at P_0.

Another notation in use for the gradient of f is

$$\text{grad } f,$$

read the way it is written. The symbol ∇f may be read "grad f" as well as "gradient of f" or "del f."

Another common notation for the directional derivative is

$$\left(\frac{df}{ds}\right)_{\mathbf{u},P_0}.$$

EXAMPLE 1 Find the derivative of

$$f(x, y, z) = x^3 - xy^2 - z$$

at $P_0(1, 1, 0)$ in the direction of the vector $\mathbf{A} = 2\mathbf{i} - 3\mathbf{j} + 6\mathbf{k}$.

Solution The direction of **A** is obtained by dividing **A** by its length:

$$|\mathbf{A}| = \sqrt{(2)^2 + (-3)^2 + (6)^2} = \sqrt{49} = 7,$$

$$\mathbf{u} = \frac{\mathbf{A}}{|\mathbf{A}|} = \frac{2}{7}\mathbf{i} - \frac{3}{7}\mathbf{j} + \frac{6}{7}\mathbf{k}.$$

The partial derivatives of f at P_0 are

$$f_x = 3x^2 - y^2|_{(1,1,0)} = 2,$$
$$f_y = -2xy|_{(1,1,0)} = -2,$$
$$f_z = -1|_{(1,1,0)} = -1.$$

The gradient of f at P_0 is

$$\nabla f|_{(1,1,0)} = 2\mathbf{i} - 2\mathbf{j} - \mathbf{k}.$$

The derivative of f at P_0 in the direction $\mathbf{A}$ is therefore

$$\begin{aligned}(D_{\mathbf{u}}f)|_{(1,1,0)} &= \nabla f|_{(1,1,0)} \cdot \mathbf{u} \\ &= (2\mathbf{i} - 2\mathbf{j} - \mathbf{k}) \cdot \left(\frac{2}{7}\mathbf{i} - \frac{3}{7}\mathbf{j} + \frac{6}{7}\mathbf{k}\right) \\ &= \frac{4}{7} + \frac{6}{7} - \frac{6}{7} \\ &= \frac{4}{7}. \quad \square\end{aligned}$$

EXAMPLE 2 Estimate how much

$$f(x, y, z) = xe^y + yz$$

will change if the point $P(x, y, z)$ is moved from $P_0(2, 0, 0)$ straight toward $P_1(4, 1, -2)$ a distance of $\Delta s = 0.1$ units.

Solution We first find the derivative of f at P_0 in the direction of the vector

$$\overrightarrow{P_0P_1} = 2\mathbf{i} + \mathbf{j} - 2\mathbf{k}.$$

The direction of this vector is

$$\mathbf{u} = \frac{\overrightarrow{P_0P_1}}{|\overrightarrow{P_0P_1}|} = \frac{\overrightarrow{P_0P_1}}{3} = \frac{2}{3}\mathbf{i} + \frac{1}{3}\mathbf{j} - \frac{2}{3}\mathbf{k}.$$

The gradient of f at P_0 is

$$\begin{aligned}\nabla f|_{(2,0,0)} &= (e^y\mathbf{i} + (xe^y + z)\mathbf{j} + y\mathbf{k})|_{(2,0,0)} \\ &= \mathbf{i} + 2\mathbf{j}.\end{aligned}$$

Therefore,

$$(D_{\mathbf{u}}f)_{P_0} = (\mathbf{i} + 2\mathbf{j}) \cdot \left(\frac{2}{3}\mathbf{i} + \frac{1}{3}\mathbf{j} - \frac{2}{3}\mathbf{k}\right) = \frac{2}{3} + \frac{2}{3} = \frac{4}{3}.$$

The change Δf in f that results from moving $\Delta s = 0.1$ units away from P_0 in the direction of $\mathbf{u}$ is approximately the derivative of f in this direction times Δs:

$$\Delta f \approx (D_{\mathbf{u}}f)\,\Delta s = \left(\frac{4}{3}\right)(0.1) \approx 0.13. \quad \square$$

If we write the directional derivative in the form

$$D_{\mathbf{u}}f = \nabla f \cdot \mathbf{u} = |\nabla f||\mathbf{u}| \cos\theta = |\nabla f| \cos\theta, \tag{6}$$

the following facts come to light.

Properties of the Directional Derivative

$$D_{\mathbf{u}}f = \nabla f \cdot \mathbf{u} = |\nabla f| \cos \theta$$

1. The directional derivative has its largest positive value when $\cos \theta = 1$, or when $\mathbf{u}$ is the direction of the gradient. That is, f increases most rapidly in its domain in the direction of ∇f. The derivative in this direction is

$$D_{\mathbf{u}}f = |\nabla f| \cos (0) = |\nabla f|.$$

2. Similarly, f decreases most rapidly in the direction of $-\nabla f$. The derivative in this direction is

$$D_{\mathbf{u}}f = |\nabla f| \cos (\pi) = -|\nabla f|.$$

3. $D_{-\mathbf{u}}f = \nabla f \cdot (-\mathbf{u}) = -\nabla f \cdot \mathbf{u} = -D_{\mathbf{u}}f.$ (7)

4. The relationships of the partial derivatives of f to the directional derivative are

$$D_{\mathbf{i}}f = \nabla f \cdot \mathbf{i} = f_x, \qquad D_{\mathbf{j}}f = \nabla f \cdot \mathbf{j} = f_y, \qquad D_{\mathbf{k}}f = \nabla f \cdot \mathbf{k} = f_z.$$

Thus,

f_x = derivative of f in the $\mathbf{i}$ direction,
f_y = derivative of f in the $\mathbf{j}$ direction,
f_z = derivative of f in the $\mathbf{k}$ direction.

Combining these results with Eq. (7) gives

$$D_{-\mathbf{i}}f = -f_x, \qquad D_{-\mathbf{j}}f = -f_y, \qquad D_{-\mathbf{k}}f = -f_z.$$

5. Any direction $\mathbf{u}$ normal (perpendicular) to the gradient is a direction of zero change in f because

$$D_{\mathbf{u}}f = |\nabla f| \cos (\pi/2) = |\nabla f| \cdot 0 = 0.$$

For functions of two variables, we get results much like the ones for functions of three variables. The two-variable formulas are obtained by dropping the z-terms from the three-variable formulas. Thus, for a function $f(x, y)$, and a unit vector $\mathbf{u} = u_1\mathbf{i} + u_2\mathbf{j}$, we have the following formulas:

$$\nabla f = \frac{\partial f}{\partial x}\mathbf{i} + \frac{\partial f}{\partial y}\mathbf{j}, \tag{8}$$

$$D_{\mathbf{u}}f = \nabla f \cdot \mathbf{u} = \frac{\partial f}{\partial x}u_1 + \frac{\partial f}{\partial y}u_2. \tag{9}$$

EXAMPLE 3 a) Find the derivative of

$$f(x, y) = 100 - x^2 - y^2$$

at the point $P_0(3, 4)$ in the direction of the unit vector $\mathbf{u} = u_1\mathbf{i} + u_2\mathbf{j}$. (b) In what direction in its domain (the xy-plane) is f increasing most rapidly at P_0? What is the derivative of f in this direction? (c) Identify the directions in which the derivative of f is zero.

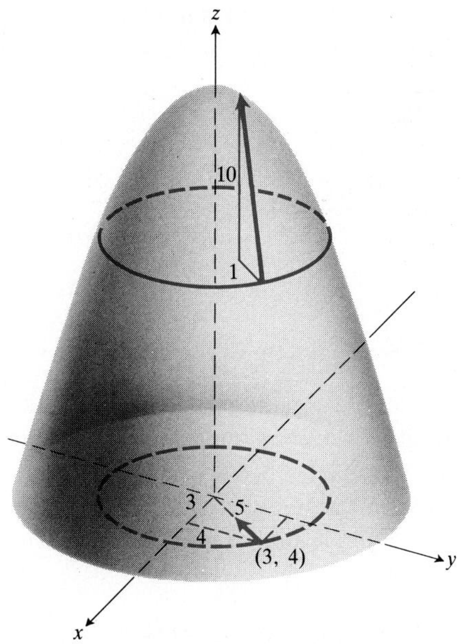

15.21 At (3, 4) the change in $f(x, y) = 100 - x^2 - y^2$ is greatest in the direction toward the origin. This corresponds to the direction of steepest ascent on the surface $z = 100 - x^2 - y^2$.

Solution a) We have

$$f(x, y) = 100 - x^2 - y^2,$$
$$f_x(3, 4) = -2x|_{(3,4)} = -6,$$
$$f_y(3, 4) = -2y|_{(3,4)} = -8,$$

and

$$D_{\mathbf{u}}f = \nabla f \cdot \mathbf{u} = (-6\mathbf{i} - 8\mathbf{j}) \cdot \mathbf{u} = -6u_1 - 8u_2.$$

b) The function increases most rapidly in the direction of the gradient. Since

$$|\nabla f| = \sqrt{(-6)^2 + (-8)^2} = \sqrt{36 + 64} = 10,$$

this direction is

$$\frac{\nabla f}{|\nabla f|} = -\frac{6}{10}\mathbf{i} - \frac{8}{10}\mathbf{j} = -\frac{3}{5}\mathbf{i} - \frac{4}{5}\mathbf{j}.$$

See Fig. 15.21. The derivative in this direction is $|\nabla f| = 10$.

c) The derivative of f is zero in the directions perpendicular to ∇f. We can obtain one of these directions by interchanging the components of

$$\frac{\nabla f}{|\nabla f|} = -\frac{3}{5}\mathbf{i} - \frac{4}{5}\mathbf{j}$$

and changing the sign of the new first component (see Example 7, Article 13.1). The result is

$$\mathbf{n} = \frac{4}{5}\mathbf{i} - \frac{3}{5}\mathbf{j}.$$

As a check, we can calculate $\nabla f \cdot \mathbf{n}$ to see that it is zero:

$$\nabla f \cdot \mathbf{n} = (-6\mathbf{i} - 8\mathbf{j})\left(\frac{4}{5}\mathbf{i} - \frac{3}{5}\mathbf{j}\right)$$
$$= -\frac{24}{5} + \frac{24}{5} = 0.$$

The other direction of zero change in f in the xy-plane is

$$-\mathbf{n} = -\frac{4}{5}\mathbf{i} + \frac{3}{5}\mathbf{j}. \square$$

The Geometry of Gradients and Level Surfaces

Suppose that $f(x, y, z)$ has continuous partial derivatives, that the surface

$$S: \quad f(x, y, z) = c$$

is one of the level surfaces of f, and that

$$x = x(t), \qquad y = y(t), \qquad z = z(t)$$

is a differentiable curve on S through a point P_0 on S. Then

$$f(x(t), y(t), z(t)) = c$$

for every value of t. If we take the derivative of both sides of this equation with respect to t and apply the chain rule on the left-hand side, we get

$$\frac{d}{dt}f(x(t), y(t), z(t)) = \frac{d}{dt}(c) = 0,$$

$$\frac{\partial f}{\partial x}\frac{dx}{dt} + \frac{\partial f}{\partial y}\frac{dy}{dt} + \frac{\partial f}{\partial z}\frac{dz}{dt} = 0,$$

or

$$\nabla f \cdot \mathbf{v} = 0, \tag{10}$$

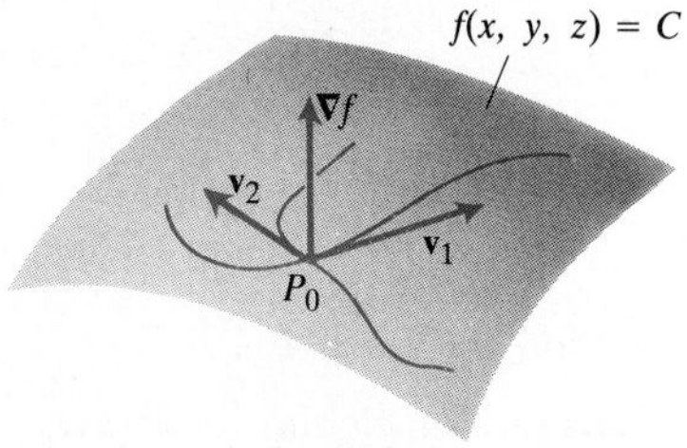

15.22 ∇f is perpendicular to the velocity vector of every differentiable curve in the surface through P_0. The velocity vectors at P_0 therefore lie in a common plane.

where $\mathbf{v}$ is the velocity vector of the curve. We conclude from this that the gradient is perpendicular to the velocity vector of every differentiable curve on S that passes through P_0 (see Fig. 15.22). Therefore, ∇f is perpendicular to the tangent line of every differentiable curve on S that passes through P_0. These lines therefore lie in a single plane, namely the plane through P_0 normal to ∇f, whose equation is

$$f_x(P_0)(x - x_0) + f_y(P_0)(y - y_0) + f_z(P_0)(z - z_0) = 0.$$

We define this plane to be the plane tangent to S at P_0. The line through P_0 perpendicular to this plane and parallel to ∇f at P_0 is called the normal line to the surface at P_0.

Tangent Plane and Normal Line to a Level Surface $f(x, y, z) = c$

If $f(x, y, z)$ has continuous partial derivatives at a point $P_0(x_0, y_0, z_0)$ on the level surface

$$S: \quad f(x, y, z) = c,$$

the *tangent plane* to S at P_0 is the plane through P_0 normal to ∇f at P_0. Its equation is

$$f_x(P_0)(x - x_0) + f_y(P_0)(y - y_0) + f_z(P_0)(z - z_0) = 0. \tag{11}$$

The partial derivatives in this equation are to be evaluated at P_0.

The *normal line* to S at P_0 is the line perpendicular to the tangent plane and parallel to ∇f at P_0, given by the equations

$$\begin{aligned} x &= x_0 + f_x(P_0)t, \\ y &= y_0 + f_y(P_0)t, \\ z &= z_0 + f_z(P_0)t. \end{aligned} \tag{12}$$

If none of the partial derivatives of f is zero at P_0, the normal line is also given by the equations

$$\frac{x - x_0}{f_x(P_0)} = \frac{y - y_0}{f_y(P_0)} = \frac{z - z_0}{f_z(P_0)}. \tag{13}$$

EXAMPLE 4 Find the tangent plane and normal line to the surface

$$x^2 + xyz - z^3 = 1$$

at the point $P_0(1, 1, 1)$.

Solution We first find the gradient vector at P_0:

$$\begin{aligned} f_x(1, 1, 1) &= (2x + yz)|_{(1,1,1)} = 3, \\ f_y(1, 1, 1) &= xz|_{(1,1,1)} = 1, \\ f_z(1, 1, 1) &= (xy - 3z^2)|_{(1,1,1)} = -2, \\ \nabla f|_{(1,1,1)} &= 3\mathbf{i} + \mathbf{j} - 2\mathbf{k}. \end{aligned}$$

From Eq. (11) the plane tangent to the surface at $P_0(1, 1, 1)$ is

$$3(x - 1) + (y - 1) - 2(z - 1) = 0,$$

or

$$3x + y - 2z = 2.$$

From Eq. (12), the normal to the surface at $P_0(1, 1, 1)$ is the line

$$x = 1 + 3t, \qquad y = 1 + t, \qquad z = 1 - 2t. \ \square$$

EXAMPLE 5 The surfaces

$$f(x, y, z) = x^2 + y^2 - z^2 = 1, \qquad g(x, y, z) = x + y + z = 5$$

intersect in a curve C. Find the line tangent to C at the point $P_0(1, 2, 2)$.

Solution The tangent line is normal to both ∇f and ∇g at P_0. Therefore,

$$\mathbf{v} = \nabla f \times \nabla g$$

will be a vector parallel to the line. We use the components of **v** and the coordinates of P_0 to write the equations for the line.

We have

$$\begin{aligned} \nabla f|_{(1,2,2)} &= (2x\mathbf{i} + 2y\mathbf{j} - 2z\mathbf{k})|_{(1,2,2)} \\ &= 2\mathbf{i} + 4\mathbf{j} - 4\mathbf{k}, \end{aligned}$$

$$\nabla g = \mathbf{i} + \mathbf{j} + \mathbf{k}.$$

Therefore,

$$\mathbf{v} = \nabla f \times \nabla g = \begin{vmatrix} \mathbf{i} & \mathbf{j} & \mathbf{k} \\ 2 & 4 & -4 \\ 1 & 1 & 1 \end{vmatrix} = 8\mathbf{i} - 6\mathbf{j} - 2\mathbf{k}.$$

The line is

$$x = 1 + 8t, \qquad y = 2 - 6t, \qquad z = 2 - 2t. \ \square$$

Just as the gradient of $f(x, y, z)$ is perpendicular to the level surfaces of f, the gradient of a function $g(x, y)$ of two variables is perpendicular to the function's level curves (Problems 81 and 82).

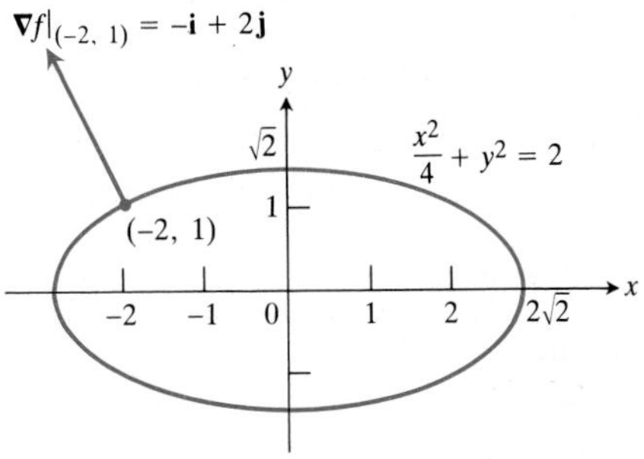

15.23 The level curve and gradient of $f(x, y) = (x^2/4) + y^2$ at $P_0\ (-2, 1)$.

EXAMPLE 6 Sketch the level curve of

$$f(x, y) = \frac{x^2}{4} + y^2$$

that passes through the point $P_0(-2, 1)$. Find a vector in the plane of the curve and normal to it at P_0, and include it in the sketch.

Solution The value of f at P_0 is

$$\frac{(-2)^2}{4} + (1)^2 = 2.$$

See Fig. 15.23. Therefore the level curve of f through P_0 is the ellipse

$$\frac{x^2}{4} + y^2 = 2.$$

For a vector normal to the ellipse at $(-2, 1)$ we may take

$$\nabla f\Big|_{(-2,\,1)} = \left(\frac{x}{2}\mathbf{i} + 2y\mathbf{j}\right)\Big|_{(-2,\,1)} = -\mathbf{i} + 2\mathbf{j}. \ \square$$

The Tangent Plane for a Surface $z = f(x, y)$

The equation for a surface $z = f(x, y)$ can be written in the form

$$\underbrace{f(x, y) - z}_{F(x,\,y,\,z)} = 0. \tag{14}$$

From this we see that the surface is also the level surface

$$F(x, y, z) = 0$$

of the function $w = F(x, y, z)$. Thus if F has continuous partial derivatives, we may define the tangent plane to the surface $z = f(x, y)$ at a point $P_0(x_0, y_0, z_0)$ on the surface to be the plane perpendicular to ∇F at P_0. The equation for this plane is

$$F_x(P_0)(x - x_0) + F_y(P_0)(y - y_0) + F_z(P_0)(z - z_0) = 0.$$

But now

$$\begin{aligned} F_x(P_0) &= \frac{\partial}{\partial x}(f(x, y) - z)\big|_{(x_0,\,y_0,\,z_0)} = f_x(x_0, y_0), \\ F_y(P_0) &= \frac{\partial}{\partial y}(f(x, y) - z)\big|_{(x_0,\,y_0,\,z_0)} = f_y(x_0, y_0), \\ F_z(P_0) &= \frac{\partial}{\partial z}(f(x, y) - z)\big|_{(x_0,\,y_0,\,z_0)} = -1, \end{aligned} \tag{15}$$

and the equation of the tangent plane reduces to

$$f_x(x_0, y_0)(x - x_0) + f_y(x_0, y_0)(y - y_0) - (z - z_0) = 0.$$

We can see from Eqs. (15) that F_x, F_y, and F_z will be continuous at P_0 if f_x and f_y are continuous at (x_0, y_0).

Tangent Plane and Normal Line for a Surface $z = f(x, y)$

If $P_0(x_0, y_0, z_0)$ is a point on the surface $z = f(x, y)$ and f_x and f_y are continuous at (x_0, y_0), then the tangent plane to the surface at P_0 is the plane

$$f_x(x_0, y_0)(x - x_0) + f_y(x_0, y_0)(y - y_0) - (z - z_0) = 0. \quad (16)$$

The normal line to the surface at P_0 is the line

$$\begin{aligned} x &= x_0 + tf_x(x_0, y_0),\\ y &= y_0 + tf_y(x_0, y_0),\\ z &= z_0 - t, \end{aligned} \quad (17)$$

or

$$\frac{x - x_0}{f_x(x_0, y_0)} = \frac{y - y_0}{f_y(x_0, y_0)} = \frac{z - z_0}{-1} \quad (18)$$

if $f_x(x_0, y_0) \neq 0$ and $f_y(x_0, y_0) \neq 0$.

EXAMPLE 7 Find equations for the tangent plane and normal line to the surface

$$z = f(x, y) = 9 - x^2 - y^2$$

at the point $P_0(1, 2, 4)$. See Fig. 15.24.

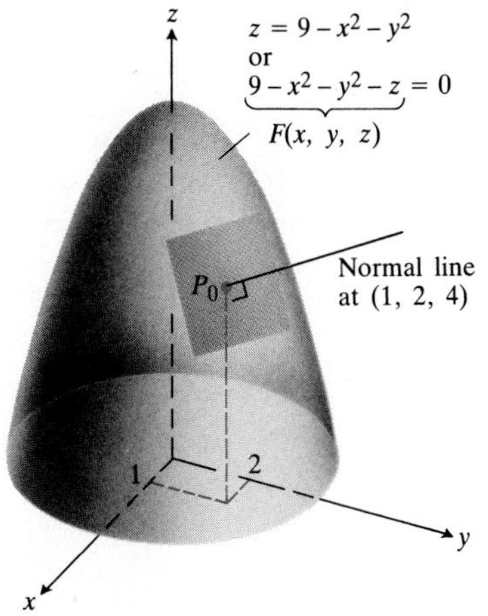

15.24 The tangent plane and normal line to the surface $z = 9 - x^2 - y^2$ at $P_0(1, 2, 4)$.

Solution 1 With

$$f_x(1, 2) = -2x|_{(1,2)} = -2, \qquad f_y(1, 2) = -2y|_{(1,2)} = -4,$$

Equations (16) and (17) give

Tangent plane: $(-2)(x - 1) + (-4)(y - 2) - (z - 4) = 0$ or $2x + 4y + z = 14$,

Normal line: $x = 1 - 2t, \quad y = 2 - 4t, \quad z = 4 - t.$

Solution 2 If we rewrite the equation

$$z = 9 - x^2 - y^2$$

in the form

$$9 - x^2 - y^2 - z = 0,$$

we see that the surface $z = f(x, y)$ is the same as the level surface

$$F(x, y, z) = 0$$

of the function

$$F(x, y, z) = 9 - x^2 - y^2 - z.$$

The tangent plane we seek is therefore in the plane normal to ∇F at $P_0(1, 2, 5)$. With

$$\begin{aligned} F_x &= -2x|_{(1,2,4)} = -2, \\ F_y &= -2y|_{(1,2,4)} = -4, \\ F_z &= -1|_{(1,2,4)} = -1, \end{aligned}$$

Equations (11) and (12) give

Tangent plane: $(-2)(x-1) + (-4)(y-2) + (-1)(z-4) = 0$

or $2x + 4y + z = 14$,

Normal line: $x = 1 - 2t, \quad y = 2 - 4t, \quad z = 4 - t,$

in agreement with Solution 1. □

The Continuity Requirement for Partial Derivatives

Why do we require functions defining surfaces to have continuous first partial derivatives before we define tangent planes? The answer is that surfaces defined by functions with discontinuous partial derivatives may be too uneven to have satisfactory tangent planes. Here are two examples of such surfaces.

EXAMPLE 8 The "ridge" surface

$$z = \frac{1}{2}(||x| - |y|| - |x| - |y|) \tag{19}$$

shown in Fig. 15.25 consists of two upside-down troughs whose "backbones" lie along the x- and y-axes and which are cut to fit together as they come in to the origin.

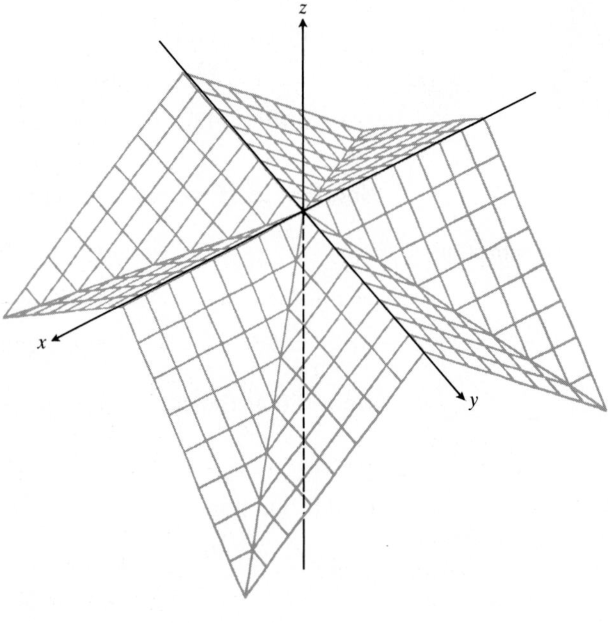

15.25 The "ridge" surface

$z = \frac{1}{2}(||x| - |y|| - |x| - |y|)$

viewed from the point (10, 15, 20).

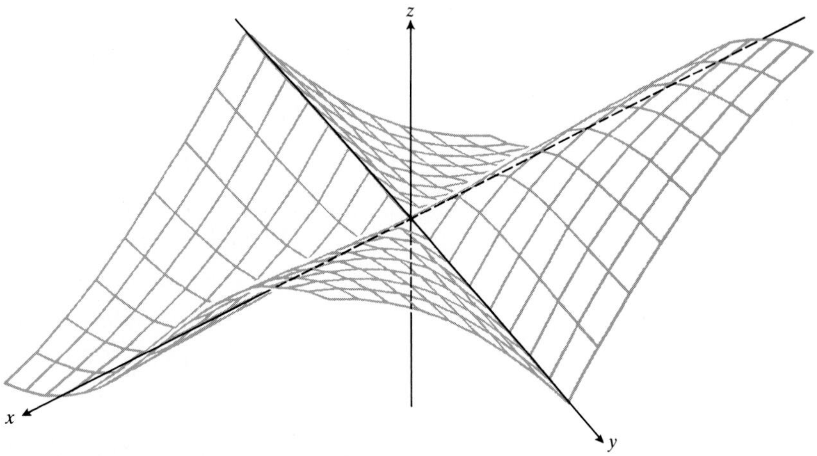

15.26 The "twisted butterfly" surface

$$z = \begin{cases} \dfrac{2|x|y}{\sqrt{x^2 + y^2}}, & (x, y) \neq (0, 0) \\ 0, & (x, y) = (0,0) \end{cases}$$

viewed from the point (10, 15, 10).

There is no unique plane tangent to the surface at the origin or at points where the two troughs meet.

EXAMPLE 9 The "twisted butterfly" surface

$$z = \begin{cases} \dfrac{2|x|y}{\sqrt{x^2 + y^2}}, & (x, y) \neq (0, 0) \\ 0, & (x, y) = (0, 0) \end{cases} \tag{20}$$

shown in Fig. 15.26 can be generated by pivoting the x-axis at the origin and swinging it around like a wobbly compass needle. As the axis swings from $\theta = -\pi/2$ to $\theta = \pi/2$ (using cylindrical coordinates) the point r units from the origin rises, or falls, according to the law

$$z = r \sin 2\theta. \tag{21}$$

For every position the pivoted x-axis takes, there is a line in the surface through the origin perpendicular to it. But the plane determined by these two lines is not tangent to the surface. Otherwise, every such plane would have an equal claim to that honor. □

Algebraic Properties of the Gradient Vector

If $f(x, y, z)$ and $g(x, y, z)$ have partial derivatives, then the constant multiple kf, the sum $f + g$, and the product fg have gradient vectors, and

$$\begin{aligned} &1. \quad \nabla(kf) = k\nabla f \quad \text{(any number } k\text{)}, \\ &2. \quad \nabla(f + g) = \nabla f + \nabla g, \\ &3. \quad \nabla(fg) = f\nabla g + g\nabla f. \end{aligned} \tag{22}$$

The proofs are left as exercises.

EXAMPLE 10 If

$$f(x, y, z) = e^x, \qquad g(x, y, z) = y - z,$$

then

$$\nabla f = e^x\mathbf{i}, \qquad \nabla g = \mathbf{j} - \mathbf{k}.$$

Therefore,

1. $\nabla(2f) = \nabla(2e^x) = 2e^x\mathbf{i} = 2\,\nabla f,$
2. $\nabla(f + g) = \nabla(e^x + y - z) = e^x\mathbf{i} + \mathbf{j} - \mathbf{k} = \nabla f + \nabla g,$
3. $\nabla(fg) = \nabla(ye^x - ze^x)$
   $= (ye^x - ze^x)\mathbf{i} + e^x\mathbf{j} - e^x\mathbf{k}$
   $= (y - z)e^x\mathbf{i} + e^x(\mathbf{j} - \mathbf{k})$
   $= g\,\nabla f + f\,\nabla g.$ □

Formulas for Functions with Continuous First Partial Derivatives

Gradient vector:

$\nabla f = f_x\mathbf{i} + f_y\mathbf{j}$ (two-dimensional)

$\nabla f = f_x\mathbf{i} + f_y\mathbf{j} + f_z\mathbf{k}$ (three-dimensional)

$\nabla f(x, y)$ is normal to the level curves of f.

$\nabla f(x, y, z)$ is normal to the level surfaces of f.

Directional derivative: The derivative of f in the direction $\mathbf{u}$ at the point P_0 is

$$(D_{\mathbf{u}}f)_{P_0} = (\nabla f)_{P_0} \cdot \mathbf{u} = (\text{grad } f)_{P_0} \cdot \mathbf{u}$$

Tangent plane and normal line to the surface $f(x, y, z) = 0$ at the point $P_0(x_0, y_0, z_0)$:

Tangent plane:
$f_x(P_0)(x - x_0) + f_y(P_0)(y - y_0) + f_z(P_0)(z - z_0) = 0$

Normal line:
$x = x_0 + f_x(P_0)t, \quad y = y_0 + f_y(P_0)t, \quad z = z_0 + f_z(P_0)t$

Tangent plane and normal line to a surface $z = f(x, y)$ at the point $P_0(x_0, y_0, f(x_0, y_0))$:

Tangent plane:
$f_x(x_0, y_0)(x - x_0) + f_y(x_0, y_0)(y - y_0) - (z - z_0) = 0$

Normal line:
$x = x_0 + tf_x(x_0, y_0), \quad y = y_0 + tf_y(x_0, y_0), \quad z = z_0 - t$

PROBLEMS

In Problems 1–7, find the electric intensity vector $\mathbf{E} = -\nabla V$ for each potential function V at the given point.

1. $V = x^2 + y^2 - 2z^2$, $(1, 1, 1)$

2. $V = 2z^3 - 3(x^2 + y^2)z$, $(1, 1, 1)$

3. $V = e^{-2y} \cos 2x$, $(\pi/4, 0, 0)$

4. $V = \ln \sqrt{x^2 + y^2}$, $(3, 4, 0)$

5. $V = (x^2 + y^2 + z^2)^{-(1/2)}$, $(1, 2, -2)$

6. $V = e^{3x+4y} \cos 5z$, $(0, 0, \pi/6)$

7. $V = \cos 3x \cos 4y \sinh 5z$, $(0, \pi/4, 0)$

In Problems 8–18, find the derivative of f at the point P_0 in the direction of the vector **A**.

8. $f = x^2 + y^2$, $P_0(1, 0)$, $\mathbf{A} = \mathbf{i} - \mathbf{j}$

9. $f = e^x \sin \pi y$, $P_0(0, 1)$, $\mathbf{A} = 4\mathbf{i} + 4\mathbf{j}$

10. $f = \cos xy$, $P_0(2, \pi/4)$, $\mathbf{A} = 4\mathbf{i} - \mathbf{j}$

11. $f = \dfrac{x - y^2}{x}$, $P_0(1, 1)$, $\mathbf{A} = 12\mathbf{i} + 5\mathbf{j}$

12. $f = x^2 + 2xy - 3y^2$, $P_0(1/2, 1/2)$, $\mathbf{A} = \sqrt{3}\,\mathbf{i} + \mathbf{j}$

13. $f = x \tan^{-1}(y/x)$, $P_0(1, 1)$, $\mathbf{A} = 2\mathbf{i} - \mathbf{j}$

14. $f = xy + yz + zx$, $P_0(1, -1, 2)$, $\mathbf{A} = 10\mathbf{i} + 11\mathbf{j} - 2\mathbf{k}$

15. $f = x^2 + 2y^2 + 3z^2$, $P_0(1, 1, 1)$, $\mathbf{A} = \mathbf{i} + \mathbf{j} + \mathbf{k}$

16. $f = \ln \sqrt{x^2 + y^2 + z^2}$, $P_0(3, 4, 12)$, $\mathbf{A} = 3\mathbf{i} + 6\mathbf{j} - 2\mathbf{k}$

17. $f = e^x \cos yz$, $P_0(0, 0, 0)$, $\mathbf{A} = 2\mathbf{i} + \mathbf{j} - 2\mathbf{k}$

18. $f = \cos xy + e^{yz} + \ln zx$, $P_0(1, 0, 1/2)$, $\mathbf{A} = \mathbf{i} + 2\mathbf{j} + 2\mathbf{k}$

In Problems 19–21, find (a) the direction in which f increases most rapidly at P_0, and (b) the rate at which f changes in that direction.

19. $f(x, y) = x^2 + \cos xy$, $P_0(1, \pi/2)$

20. $f(x, y, z) = e^{xy} + z^2$, $P_0(0, 2, 3)$

21. $f(x, y, z) = (x + y - 2)^2 + (3x - y - 6)^2$, $P_0(1, 1, 0)$

In Problems 22–24, find (a) the direction in which f decreases most rapidly at P_0, and (b) the rate at which f changes in that direction.

22. $f(x, y) = x^2 + xy + y^2$, $P_0(-1, 1)$

23. $f(x, y, z) = (x + y)^2 + (y + z)^2 + (z + x)^2$, $P_0(2, -1, 2)$

24. $f(x, y, z) = z \ln (x^2 + y^2)$, $P_0(1, 1, 1)$

25. Use the directional derivative to estimate how much $f(x, y) = \cos \pi xy + xy^2$ will change if the point $P(x, y)$ is moved from $P(-1, -1)$ a distance of $\Delta s = 0.1$ unit along the vector $\mathbf{A} = \mathbf{i} + \mathbf{j}$.

26. By about how much will

$$f(x, y, z) = \ln \sqrt{x^2 + y^2 + z^2}$$

change if the point $P(x, y, z)$ is moved from $P(3, 4, 12)$ a distance of $\Delta s = 0.1$ unit along the vector $3\mathbf{i} + 6\mathbf{j} - 2\mathbf{k}$?

27. By about how much will

$$f(x, y, z) = e^x \cos yz$$

change as the point $P(x, y, z)$ moves from the origin at a distance of $\Delta s = 0.1$ unit in the direction of $\mathbf{A} = 2\mathbf{i} + \mathbf{j} - 2\mathbf{k}$?

28. By about how much will

$$f(x, y, z) = x + x \cos z - y \sin z + y$$

change if the point $P(x, y, z)$ moves from $P_0(2, -1, 0)$ a distance of $\Delta s = 0.2$ units toward the point $P_1(0, 1, 2)$?

29. In which two directions is the derivative of $f(x, y) = xy + y^2$ at the point $P_0(2, 5)$ equal to zero?

30. In which two directions is the derivative of $f(x, y) = (x^2 - y^2)/(x^2 + y^2)$ at $P_0(1, 1)$ equal to zero?

In Problems 31–35, sketch the level curve of $f(x, y)$ through the point P_0. Find a vector normal to the curve at P_0 and include it in your sketch.

31. $f(x, y) = x^2 - y^2$, $P_0(2, -1)$

32. $f(x, y) = x^2/3 + 3y^2/4$, $P_0(2, 2\sqrt{5}/3)$

33. $f(x, y) = x^2 - y$, $P_0(\sqrt{2}, 1)$

34. $f(x, y) = \sqrt{x^2 - y}$, $P_0(1, 0)$

35. $f(x, y) = 12 + 2x^2 - y^2$, $P_0(1, 2)$

In Problems 36–43, find equations for (a) the tangent plane and (b) the normal line to the given level surface at the point P_0.

36. $x^2 + y^2 + z^2 = 3$, $P_0(1, 1, 1)$

37. $x^2 + y^2 - z^2 = 18$, $P_0(3, 5, -4)$

38. $z^2 - x^2 - y^2 = 0$, $P_0(3, 4, -5)$

39. $2z - x^2 = 0$, $P_0(2, 0, 2)$

40. $z - \ln (x^2 + y^2) = 0$, $P_0(1, 0, 0)$

41. $(x + y)^2 + z^2 = 25$, $P_0(1, 2, 4)$

42. $x^2 + 2xy - y^2 + z^2 = 7$, $P_0(1, -1, 3)$

43. $\cos \pi x - x^2 y + e^{xz} + yz = 4$, $P_0(0, 1, 2)$

In Problems 44–57, find (a) the tangent plane and (b) the normal line to the given surface at the given point.

44. $z = x^2 + y^2$, $(3, 4, 25)$

45. $z = \sqrt{9 - x^2 - y^2}$, $(1, -2, 2)$

46. $z = x^2 - xy - y^2$, $(1, 1, -1)$

47. $z = \tan^{-1} \dfrac{y}{x}$, $(1, 1, \pi/4)$

48. $z = x/\sqrt{x^2 + y^2}$, $(3, -4, \frac{3}{5})$

49. $z = 9x^2 + y^2$, $(0, 0, 0)$

50. $z = \cos (\pi x/2)$, $(1, 0, 0)$

51. $z = 1 - x - y$, $(0, 1, 0)$

52. $z = (x + y)/(xy - 1)$, $(1, 2, 3)$

53. $z = x^2 + y^2 - 2xy + 3y - x + 4$, $(2, -3, 18)$

54. $x = 2z^2 - zy + y^3$, $(-4, -2, 1)$ (as a surface over the yz-plane)

55. $x = 1 - y^2 - z^2$, $(0, 0, 1)$

56. $y = 4 - x^2 - 4z^2$, $(0, 0, 1)$ (as a surface over the xz-plane)

57. $y = \sin x$, $(0, 0, 0)$

In Problems 58–65, sketch the portion of the given level surface that lies in the first octant (x, y, and z nonnegative). Then find a vector perpendicular to the surface at the given point P_0. Include the vector in your sketch.

58. $x^2 + y^2 + z^2 = 9$, $P_0(1, 2, 2)$

59. $x + y^2 + z = 2$, $P_0(\frac{1}{2}, 1, \frac{1}{2})$

60. $2x + y^2 + z = 4$, $P_0(1, 1, 1)$

61. $x^2 + y^2 + z = 4$, $P_0(0, 1, 3)$

62. $2x + 3y + 6z = 18$, $P_0(3, 2, 1)$

63. $z - \sqrt{9 - x^2 - y^2} = 0$, $P_0(1, 2, 2)$

64. $1/\sqrt{x^2 + y^2 + z^2} = 1$, $P_0(\frac{1}{2}, \frac{1}{2}, \sqrt{2}/2)$

65. $x^2 + 2y^2 - z = 2$, $P_0(1, 1, 1)$

66. The derivative of $f(x, y)$ at $P_0(1, 2)$ in the direction of the vector $\mathbf{i} + \mathbf{j}$ is $2\sqrt{2}$, and in the direction of the vector $-2\mathbf{j}$ is -3. What is the derivative of f in the direction of the vector $-\mathbf{i} - 2\mathbf{j}$?

67. The derivative of $f(x, y, z)$ at a given point P is greatest in the direction of the vector $\mathbf{A} = \mathbf{i} + \mathbf{j} - \mathbf{k}$. In this direction the value of the derivative is $2\sqrt{3}$. (a) Find the gradient vector of f at P. (b) Find the derivative of f at P in the direction of the vector $\mathbf{i} + \mathbf{j}$.

68. a) Find equations for all lines through the origin normal to the surface $xy + z = 2$.
b) Find the points in which these lines meet the surface.

69. Find the points on the surface

$$(y + z)^2 + (z - x)^2 = 16$$

where the normal is parallel to the yz-plane.

70. Find the points on the surface

$$xy + yz + zx - x - z^2 = 0$$

where the tangent plane is parallel to the xy-plane.

71. A curve is said to be *tangent to a surface at a point P* if the line tangent to the curve at P lies in the plane that is tangent to the surface at P.
a) Show that the curve $x = \ln t$, $y = t \ln t$, $z = t$ is tangent to the surface

$$xz^2 - yz + \cos xy = 1$$

at $P(0, 0, 1)$.
b) Show that the curve $x = (t^3/4) - 2$, $y = (4/t) - 3$, $z = \cos(t - 2)$ is tangent to the surface

$$x^3 + y^3 + z^3 - xyz = 0$$

at $t = 2$.

72. Find the derivative of $f(x, y, z) = xyz$ in the direction of the velocity vector of the helix

$$\mathbf{R}(t) = (\cos 3t)\mathbf{i} + (\sin 3t)\mathbf{j} + 3t\mathbf{k}$$

at time $t = \pi/3$.

73. For the function $f(x, y) = x^2y + 2y^2x$, at the point $P_0(1, 3)$, find
a) the direction of greatest increase in f,
b) the derivative of f in the direction of greatest increase in f,
c) the direction of greatest decrease in f,
d) the directions in which the derivative of f is zero,
e) an equation for the plane tangent to the surface $z = f(x, y)$ at $(1, 3, 21)$.

74. If $z = f(x, y)$ has continuous partial derivatives at the point $P_0(x_0, y_0)$, which of the following statements are true?
a) If $\mathbf{u}$ is a unit vector, then the derivative of f in the direction of $\mathbf{u}$ is $(f_x(x_0, y_0)\mathbf{i} + f_y(x_0, y_0)\mathbf{j}) \cdot \mathbf{u}$.
b) The derivative of f in the direction of $\mathbf{u}$ is a vector.
c) The directional derivative of f at P_0 has its greatest value in the direction of ∇f.
d) At (x_0, y_0), ∇f is normal to the level curve $f(x, y) = f(x_0, y_0)$.

In Problems 75–80, find equations for the line tangent to the curve of intersection of the two surfaces at the given point.

75. Surfaces: $x^2 + y^2 = 4$, $z = x^2 + y^2$
Point: $(\sqrt{2}, \sqrt{2}, 4)$

76. Surfaces: $xyz = 1$, $x^2 + 2y^2 + 3z^2 = 6$
Point: $(1, 1, 1)$

77. Surfaces: $x^2 + 2y + 2z = 4$, $y = 1$
Point: $(1, 1, \frac{1}{2})$

78. Surfaces: $x + y^2 + z = 2$, $y = 1$
Point: $(\frac{1}{2}, 1, \frac{1}{2})$

79. Surfaces: $x + y^2 + 2z = 4$, $x = 1$
Point: $(1, 1, 1)$

80. Surfaces: $x^3 + 3x^2y^2 + y^3 + 4xy - z^2 = 0$,
$x^2 + y^2 + z^2 = 11$
Point: $(1, 1, 3)$

81. Suppose that $g(x, y)$ has continuous partial derivatives and that g has the constant value c on the differentiable curve $x = x(t)$, $y = y(t)$. That is,

$$g(x(t), y(t)) = c$$

for all values of t. Differentiate both sides of this equation with respect to t to show that ∇g is normal to the tangent vector at every point of the curve.

82. *Tangent lines to level curves.* A quick way to find the tangent line to a level curve $f(x, y) = c$ at a point $P_0(x_0, y_0)$ on the curve is as follows: Since ∇f is normal to the tangent line, the points $P(x, y)$ on the line satisfy the equation $\nabla f \cdot \overrightarrow{P_0P} = 0$, or

$$f_x(x_0, y_0)(x - x_0) + f_y(x_0, y_0)(y - y_0) = 0.$$

Use this equation to find the lines tangent to the following curves at the given points.

a) $x^2 + y^2 = 4$, $P_0(\sqrt{2}, \sqrt{2})$
b) $x^2 + xy + y^2 = 7$, $P_0(1, 2)$
(This was Example 6 in Article 2.4.)
c) $x^5 + 4xy^3 - 3y^5 = 2$, $P_0(1, 1)$

83. Show that the curve

$$R = -t\mathbf{i} + \sqrt{t}\,\mathbf{j} + (\ln t)\mathbf{k}$$

intersects the surface

$$z = \ln\left(\frac{y - 2x^2 - y^2}{4}\right)$$

at a right angle when $t = 1$ (i.e., that the curve's velocity vector is normal to the surface's tangent plane).

84. Suppose cylindrical coordinates r, θ, z are introduced into a function $w = f(x, y, z)$ to yield $w = F(r, \theta, z)$. Show that the gradient may be expressed in terms of cylindrical coordinates and the unit vectors $\mathbf{u}_r$, $\mathbf{u}_\theta$, $\mathbf{k}$ as follows:

$$\nabla w = \mathbf{u}_r \frac{\partial w}{\partial r} + \frac{1}{r}\mathbf{u}_\theta \frac{\partial w}{\partial \theta} + \mathbf{k}\frac{\partial w}{\partial z}.$$

(*Hint:* The component of ∇w in the direction of $\mathbf{u}_r$ is equal to the directional derivative in that direction. But this is precisely $\partial w/\partial r$. Reason similarly for the components of ∇w in the directions of $\mathbf{u}_\theta$ and $\mathbf{k}$.)

85. Verify the result given in Problem 84 by transforming the given expression on the right-hand side of the equation into $\mathbf{i}$, $\mathbf{j}$, $\mathbf{k}$ components and replacing the cylindrical coordinates r, θ by cartesian coordinates x, y, and making use of the chain rule for partial derivatives.

86. Express the gradient in terms of spherical coordinates and the appropriate unit vectors $\mathbf{u}_\rho$, $\mathbf{u}_\phi$, $\mathbf{u}_\theta$. Use a geometrical argument to determine the component of ∇w in each of these directions. (See the hint for Problem 84.)

87. Verify the answer obtained in Problem 86 by transforming the expression you obtained in spherical coordinates back into cartesian coordinates. Make use of a chain rule.

88. In Fig. 15.13 let

$$\mathbf{R} = \mathbf{i}x + \mathbf{j}y + \mathbf{k}f(x, y)$$

be the vector from the origin to (x, y, w). What can you say about the direction of the vector (a) $\partial\mathbf{R}/\partial x$, (b) $\partial\mathbf{R}/\partial y$? (c) Calculate the vector product

$$\mathbf{v} = \left(\frac{\partial \mathbf{R}}{\partial x}\right) \times \left(\frac{\partial \mathbf{R}}{\partial y}\right).$$

What can you say about the direction of this vector $\mathbf{v}$ with respect to the surface $w = f(x, y)$?

89. Verify the gradient formulas in Eq. (22).

15.7 Higher Order Derivatives. Partial Differential Equations from Physics

This article is devoted mainly to second order partial derivatives, which are denoted by

$$\frac{\partial^2 f}{\partial x^2}, \quad \frac{\partial^2 y}{\partial y^2}, \quad \frac{\partial^2 f}{\partial x\, \partial y}, \quad \frac{\partial^2 f}{\partial y\, \partial x}$$

or

$$f_{xx}, \quad f_{yy}, \quad f_{yx}, \quad f_{xy},$$

and defined by the equations

$$\frac{\partial^2 f}{\partial x^2} = \frac{\partial}{\partial x}\left(\frac{\partial f}{\partial x}\right), \qquad \frac{\partial^2 f}{\partial x\, \partial y} = \frac{\partial}{\partial x}\left(\frac{\partial f}{\partial y}\right),$$

and so on. Note the order in which the derivatives are taken:

$\dfrac{\partial^2 f}{\partial x\, \partial y}$ Differentiate first with respect to y, then with respect to x,

f_{yx} Means the same thing.

Second order partial derivatives appear in equations that express important physical laws for wave motion, heat flow, and gravitational potential. (A gravitational potential function measures the work done in moving a unit of mass against a gravitational field from a central reference point, say the Sun, to a new position, say Mars.) Second order partial derivatives are also used to test for maxima and minima of functions of two variables, as we shall see in the next article.

EXAMPLE 1 If

$$f(x, y) = x \cos y + ye^x,$$

then

$$\frac{\partial f}{\partial x} = \cos y + ye^x,$$

$$\frac{\partial}{\partial y}\left(\frac{\partial f}{\partial x}\right) = -\sin y + e^x = \frac{\partial^2 f}{\partial y\,\partial x},$$

$$\frac{\partial}{\partial x}\left(\frac{\partial f}{\partial x}\right) = ye^x = \frac{\partial^2 f}{\partial x^2},$$

$$\frac{\partial}{\partial x}\left(\frac{\partial^2 f}{\partial x^2}\right) = ye^x = \frac{\partial^3 f}{\partial x^3},$$

$$\frac{\partial}{\partial y}\left(\frac{\partial^2 f}{\partial x^2}\right) = e^x = \frac{\partial^3 f}{\partial y\,\partial x^2},$$

while

$$\frac{\partial f}{\partial y} = -x \sin y + e^x,$$

$$\frac{\partial}{\partial x}\left(\frac{\partial f}{\partial y}\right) = -\sin y + e^x = \frac{\partial^2 f}{\partial x\,\partial y},$$

$$\frac{\partial}{\partial y}\left(\frac{\partial f}{\partial y}\right) = -x \cos y = \frac{\partial^2 f}{\partial y^2},$$

$$\frac{\partial}{\partial x}\left(\frac{\partial^2 f}{\partial x\,\partial y}\right) = e^x = \frac{\partial^3 f}{\partial x^2\,\partial y}. \quad \square$$

You may have noticed in Example 1 that the "mixed" second order partial derivatives

$$\frac{\partial^2 f}{\partial y\,\partial x} \quad \text{and} \quad \frac{\partial^2 f}{\partial x\,\partial y}$$

were equal. This was not a mere coincidence. Whenever $f, f_x, f_y, f_{xy}, f_{yx}$ are all continuous at a point, the mixed partial derivatives f_{xy} and f_{yx} will be equal at that point.

THEOREM 4

The Mixed-Derivative Theorem

If $f(x, y)$ and its partial derivatives f_x, f_y, f_{xy}, f_{yx} are defined in a region containing a point (a, b) and are all continuous at (a, b), then

$$f_{xy}(a, b) = f_{yx}(a, b). \tag{1}$$

A proof of Theorem 4 is given in Appendix 4.

EXAMPLE 2 Find z_{xy} if

$$z = f(u, v), \qquad u = x^2 - y^2, \qquad v = 2xy,$$

and f and its partial derivatives are all continuous.

Solution From the chain rule,

$$z_x = \frac{\partial}{\partial x} f(u, v) = f_u u_x + f_v v_x = f_u 2x + f_v 2y = 2xf_u + 2yf_v.$$

Therefore,

$$\begin{aligned} z_{xy} &= \frac{\partial}{\partial y}(2xf_u + 2yf_v) \\ &= 2x\frac{\partial}{\partial y}(f_u) + f_u\frac{\partial}{\partial y}(2x) + 2y\frac{\partial}{\partial y}(f_v) + f_v\frac{\partial}{\partial y}(2y) \\ &= 2x\frac{\partial}{\partial y}(f_u) + 2y\frac{\partial}{\partial y}(f_v) + 2f_v. \end{aligned} \tag{2}$$

We use the chain rule again to calculate the two remaining derivatives:

$$\frac{\partial}{\partial y}(f_u) = f_{uu}\frac{\partial u}{\partial y} + f_{uv}\frac{\partial v}{\partial y} = -2yf_{uu} + 2xf_{uv},$$

$$\frac{\partial}{\partial y}(f_v) = f_{vu}\frac{\partial u}{\partial y} + f_{vv}\frac{\partial v}{\partial y} = -2yf_{vu} + 2xf_{vv}.$$

Thus we may continue from Eq. (2) to get

$$\begin{aligned} z_{xy} &= 2x(-2yf_{uu} + 2xf_{uv}) + 2y(-2yf_{vu} + 2xf_{vv}) + 2f_v \\ &= -4xyf_{uu} + 4x^2f_{uv} - 4y^2f_{vu} + 4xyf_{vv} + 2f_v && (3) \\ &= 4xy(f_{vv} - f_{uu}) + 4(x^2 - y^2)f_{uv} + 2f_v, && (4) \end{aligned}$$

or, if we want the answer entirely in terms of u and v,

$$z_{xy} = 2v(f_{vv} - f_{uu}) + 4uf_{uv} + 2f_v.$$

We used the equality $f_{uv} = f_{vu}$ to get from (3) to (4). □

EXAMPLE 3 Find d^2w/dt^2 if

$$w = f(x, y), \qquad x = e^t, \qquad y = 2t - 1,$$

and w and its partial derivatives are all continuous.

Solution We find dw/dt, and then differentiate it with respect to t. From the chain rule,

$$\frac{dw}{dt} = f_x\frac{dx}{dt} + f_y\frac{dy}{dt} = f_x \cdot e^t + f_y \cdot 2 = e^t f_x + 2f_y.$$

Therefore,

$$\begin{aligned} \frac{d^2w}{dt^2} &= \frac{d}{dt}(e^t f_x) + \frac{d}{dt}(2f_y) \\ &= e^t f_x + e^t\frac{d}{dt}(f_x) + 2\frac{d}{dt}(f_y). \end{aligned} \tag{5}$$

We use the chain rule again to calculate the two remaining derivatives:

$$\frac{d}{dt}(f_x) = f_{xx}\frac{dx}{dt} + f_{xy}\frac{dy}{dt} = f_{xx}e^t + f_{xy}\cdot 2 = e^t f_{xx} + 2f_{xy}.$$

$$\frac{d}{dt}(f_y) = f_{yx}\frac{dx}{dt} + f_{yy}\frac{dy}{dt} = f_{yx}e^t + f_{yy}\cdot 2 = e^t f_{yx} + 2f_{yy}.$$

Substituting these in Eq. (5) gives

$$\begin{aligned}\frac{d^2w}{dt^2} &= e^t f_x + e^t(e^t f_{xx} + 2f_{xy}) + 2(e^t f_{yx} + 2f_{yy})\\ &= e^t f_x + e^{2t} f_{xx} + 2e^t f_{xy} + 2e^t f_{yx} + 4f_{yy}\\ &= e^t f_x + e^{2t} f_{xx} + 4e^t f_{xy} + 4f_{yy}. \quad \square\end{aligned}$$

Partial Derivatives of Still Higher Order

If we refer once more to Example 1 we note not only that

$$\frac{\partial^2 f}{\partial x\,\partial y} = \frac{\partial^2 f}{\partial y\,\partial x}$$

but also that

$$\frac{\partial^3 f}{\partial x^2\,\partial y} = \frac{\partial^3 f}{\partial y\,\partial x^2}.$$

This equality may be derived from Theorem 4 as follows:

$$\begin{aligned}\frac{\partial^3 f}{\partial x^2\,\partial y} &= \frac{\partial}{\partial x}\left(\frac{\partial^2 f}{\partial x\,\partial y}\right) = \frac{\partial}{\partial x}\left(\frac{\partial^2 f}{\partial y\,\partial x}\right)\\ &= \frac{\partial}{\partial x}\left(\frac{\partial}{\partial y} f_x\right) = \frac{\partial}{\partial y}\left(\frac{\partial}{\partial x} f_x\right)\\ &= \frac{\partial}{\partial y}\left(\frac{\partial^2 f}{\partial x^2}\right) = \frac{\partial^3 f}{\partial y\,\partial x^2}.\end{aligned}$$

In fact, if all the partial derivatives that appear are continuous, the notation

$$\frac{\partial^{m+n} f}{\partial x^m\,\partial y^n}$$

may be used to denote the result of differentiating the function $f(x, y)$ m times with respect to x and n times with respect to y. The final result will be the same no matter what the order of differentiation is.

Being able to control the order of differentiation can work to our advantage, as we see in the next example.

EXAMPLE 4 To calculate

$$\frac{\partial^5}{\partial x^2\,\partial y^3}(x \sin y + e^y) \tag{6}$$

we can differentiate with respect to x first to show without any further work that the final result is zero:

$$\frac{\partial^5}{\partial x^2\,\partial y^3}(x \sin y + e^y) = \frac{\partial^3}{\partial y^3}\frac{\partial^2}{\partial x^2}(x \sin y + e^y)$$

$$= \frac{\partial^3}{\partial y^3}\frac{\partial}{\partial x}(\sin y) = \frac{\partial^3}{\partial y^3}(0) = 0. \ \square$$

The One-dimensional Heat Equation (= Diffusion Equation = Telegraph Equation)

If $w(x, t)$ represents the temperature at position x at time t in a uniform conducting rod with perfectly insulated sides (no heat flow through the sides—see Fig. 15.27), then the partial derivatives w_{xx} and w_t satisfy a differential equation of the following form:

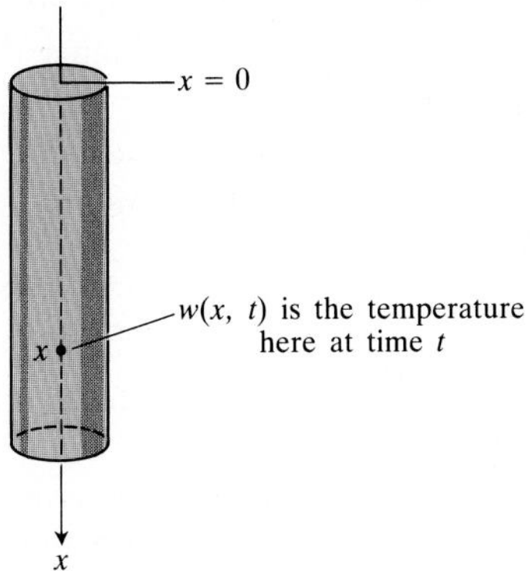

15.27 The temperature distribution in a heat-conducting rod satisfies the equation

$$w_{xx} = \frac{1}{c^2}w_t.$$

$$w_{xx} = \frac{1}{c^2}w_t. \tag{7}$$

This equation is called the *one-dimensional heat equation*. The value of c^2, which is determined by the material from which the rod is made, has been determined experimentally for a broad range of materials, and for a given application one finds the appropriate value in a table.

Equation (7) can be used effectively to model the seasonal temperature variation beneath the earth's surface, $w(x, t)$ then being the temperature on the tth day of the year x ft below the surface. (One thinks of a vertical core of the earth's surface as an insulated rod.) For dry soil,

$$c^2 = 0.19 \text{ ft}^2/\text{day},$$

and Eq. (7) becomes

$$w_{xx} = \frac{1}{0.19}w_t, \qquad \text{or} \qquad w_t = 0.19w_{xx}. \tag{8}$$

The solution that matches the seasonal temperature variation at the earth's surface is

$$w(x, t) = \cos(1.7 \times 10^{-2}t - 0.2x)e^{-0.2x}. \tag{9}$$

This can be verified by calculating w_{xx} and w_t, and seeing that $w_t = 0.19w_{xx}$. The calculation is tedious, however, and we shall omit it. The solution in (9) was graphed in Fig. 15.5.

In chemistry and biochemistry, the heat equation is known as the *diffusion equation*. In this context, $w(x, t)$ represents the concentration of a dissolved substance, a salt, for instance, diffusing along a tube filled with liquid. The value of $w(x, t)$ is the concentration at point x at time t. In other applications, $w(x, t)$ represents the diffusion of a gas down a long, thin pipe.

In electrical engineering, the heat equation appears in the forms

$$v_{xx} = RCv_t \tag{10}$$

and

$$i_{xx} = RCi_t, \tag{11}$$

which are known as the *telegraph equations*. These equations describe the voltage v and the flow of current i in a coaxial cable, or in any other cable in which leakage and inductance are negligible. The functions and constants in these equations are

$$\begin{aligned} v(x, t) &= \text{voltage at point } x \text{ at time } t, \\ R &= \text{resistance per unit length}, \\ C &= \text{capacitance to ground per unit of cable length}, \\ i(x, t) &= \text{current at point } x \text{ at time } t. \end{aligned}$$

Partial differential equations are generally hard to solve, in part because their solutions can take so many different forms. You will not be asked to solve any of the equations that appear in the problems that follow, but only to verify that given functions are solutions of given equations. There are still benefits to be gained from this: a chance to become acquainted with important equations, and an opportunity to practice the chain rule in another professional context.

PROBLEMS

In Problems 1–6, find $\partial^2 f/\partial x^2$, $\partial^2 f/\partial y^2$, and $\partial^2 f/\partial y\, \partial x$.

1. $f(x, y) = \ln(x^2 + y^2)$

2. $f(x, y) = e^x \ln(3 - y^2)$

3. $f(x, y) = x^2 y + \cos y + y \sin x$

4. $f(x, y) = (x - y)/xy$

5. $f(x, y, z) = xyz$

6. $f(x, y, z) = xy + yz + zx$

In Problems 7–10, verify that $w_{xy} = w_{yx}$.

7. $w = \ln(2x + 3y)$

8. $w = \tan^{-1}(y/x)$

9. $w = xy^2 + x^2y^3 + x^3y^4$

10. $w = e^x \sinh y + \cos(2x - 3y)$

11. Which order of differentiation will calculate f_{xy} faster: x first, or y first? Try to answer without writing anything down.

a) $f(x, y) = x \sin y + e^y$
b) $f(x, y) = 1/x$
c) $f(x, y) = y + (x/y)$
d) $f(x, y) = y + x^2y + 4y^3 - \ln(y^2 + 1)$
e) $f(x, y) = x^2 + 5xy + \sin x + 7e^x$
f) $f(x, y) = x \ln xy$

12. The derivative $\partial^5 f/\partial x^2\, \partial y^3$ is zero for each of the following functions. To show this as quickly as possible, which variable should one differentiate with respect to first: x, or y? Try to answer without writing anything down.

a) $f(x, y) = y^2x^4e^x + 2$
b) $f(x, y) = y + y^2[\ln(x^2 + 5) + \sin x - 7x^3]$
c) $f(x, y) = xe^y + x \sin y - x \cos y$
d) $f(x, y) = x \int_1^y e^{t^2/2}\, dt$

In Problems 13–18, assume that the necessary derivatives exist, and that the functions and derivatives are all continuous.

13. Find w_{xy} if $w = f(u, v)$, $u = x + y$, and $v = xy$. Express the answer in terms of u, v, f_u, and f_v.

14. Find z_{xx} if $z = f(u, v)$, $u = x^2 - y^2$, and $v = 2xy$. Express the answer in terms of x, y, f_u, and f_v.

15. Find u_{ss} if $u = f(x, y)$, $x = r^2 + s^2$ and $y = 2rs$. Express the answer in terms of r, s, f_x, and f_y.

16. Find $\partial^2 w/\partial x\, \partial y$ if $w = f(u, v)$, $u = x + y$, and $v = y^2$. Express the answer in terms of x, y, f_u, and f_v.

17. Find d^2w/dt^2 if $w = f(x, y)$, $x = \sin t$, and $y = t^2$.

18. Let $w = f(u)$, where $u = xg(y)$. Show that $w_{xx} = f''(xg(y)) \cdot g^2(y)$.

19. Find the value of $\partial^2 w/\partial\theta^2$ at $r = 2$, $\theta = \pi/2$ if $w = f(x, y)$, $x = r\cos\theta$, $y = r\sin\theta$, and $f_x = f_y = f_{xx} = f_{yy} = 1$ when $r = 2$, $\theta = \pi/2$. (Be sure to work out all the derivatives before substituting numerical values.)

Laplace equations

The *three-dimensional Laplace equation*

$$\frac{\partial^2 f}{\partial x^2} + \frac{\partial^2 f}{\partial y^2} + \frac{\partial^2 f}{\partial z^2} = 0 \tag{12}$$

is satisfied by steady-state heat distributions $T = f(x, y, z)$ in space, by gravitational potentials, and by electrostatic potentials. The *two-dimensional Laplace equation*,

$$\frac{\partial^2 f}{\partial x^2} + \frac{\partial^2 f}{\partial y^2} = 0, \tag{13}$$

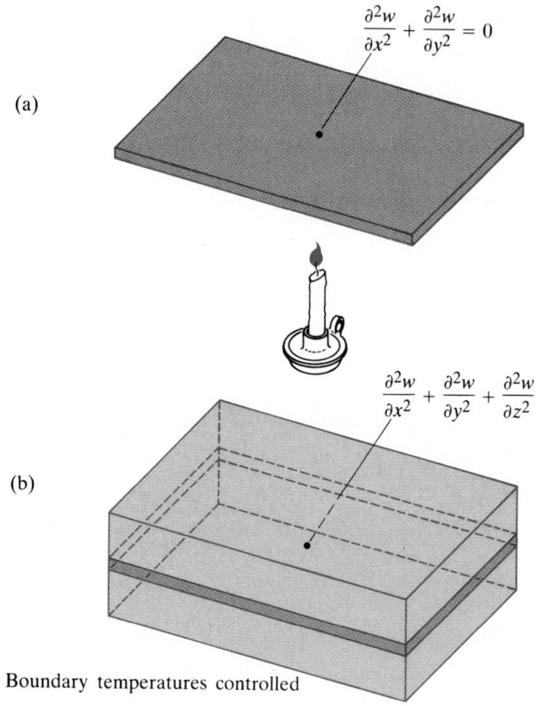

15.28 Steady-state heat distributions in planes and solids satisfy Laplace equations. The plane (a) may be treated as a thin slice of the solid (b) perpendicular to the z-axis.

obtained by dropping the $\partial^2 f/\partial z^2$ term from (12), describes potentials and steady-state heat distributions in a plane. See Fig. 15.28.

Show that the functions in Problems 20–27 satisfy a Laplace equation.

20. $f = x^2 + y^2 - 2z^2$

21. $f = 2z^3 - 3(x^2 + y^2)z$

22. $f = e^{-2y} \cos 2x$

23. $f = \ln \sqrt{x^2 + y^2}$

24. $f = (x^2 + y^2 + z^2)^{-1/2}$

25. $f = e^{3x+4y} \cos 5z$

26. $f = \tan^{-1}(y/x)$

27. $f = \cos 3x \cos 4y \sinh 5z$

28. Show that if $w = f(u, v)$ satisfies the Laplace equation $f_{uu} + f_{vv} = 0$, and if $u = (x^2 - y^2)/2$, $v = xy$, then $w_{xx} + w_{yy} = 0$.

29. Show that $u = f(x - iy) + g(x + iy)$ is a solution of the equation

$$\frac{\partial^2 u}{\partial x^2} + \frac{\partial^2 u}{\partial y^2} = 0,$$

if all the necessary partial derivatives exist. (Here, $i = \sqrt{-1}$.)

30. For what values of n does

$$f(x, y, z) = (x^2 + y^2 + z^2)^n$$

satisfy the three-dimensional Laplace equation (12)?

31. Find all solutions of the two-dimensional Laplace equation (13) of the form
a) $f(x, y) = ax^2 + bxy + c^2$,
b) $f(x, y) = ax^3 + bx^2y + cxy^2 + dy^3$.

The wave equation

If we stand on an ocean shore and take a snapshot of the waves, the picture shows a regular pattern of peaks and valleys at an instant in time (Fig. 15.29). We see periodic vertical motion in space, with respect to distance. If we stand in the water we can feel the rise and fall of the water as the waves go by. We see periodic vertical motion in time. In physics, this beautiful symmetry is expressed by the *one-dimensional wave equation*

$$\frac{\partial^2 w}{\partial t^2} = c^2 \frac{\partial^2 w}{\partial x^2}, \tag{14}$$

where w is the wave height, x is the distance variable, t is the time variable, and c is the velocity with which the waves are propagated.

In our example, x is the distance across the ocean's surface, but in other applications x might be the distance along a vibrating string, distance through air (sound waves), or distance through space (light waves). The number c varies with the medium and type of wave.

Show that the functions in Problems 32–37 are solutions of the wave equation, Eq. (14).

32. $w = \sin(x + ct)$

33. $w = \cos(2x + 2ct)$

34. $w = \sin(x + ct) + \cos(2x + 2ct)$

35. $w = \ln(2x + 2ct)$

36. $w = \tan(2x - 2ct)$

37. $w = 5\cos(3x + 3ct) - 7\sinh(4x - 4ct)$

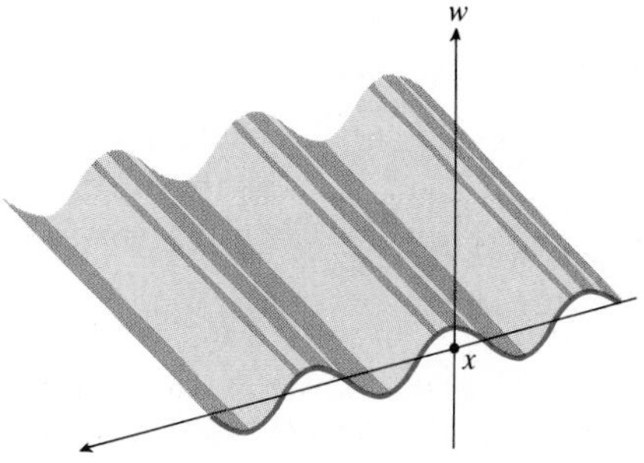

15.29 Waves in water at an instant in time. As time passes,

$$\frac{\partial^2 w}{\partial t^2} = c^2 \frac{\partial^2 w}{\partial x^2}.$$

constant and $w = f(u) + g(v)$, where ...nd $v = x - ct$, show that

$$\ldots = c^2 \frac{\partial^2 w}{\partial x^2} = c^2(f''(u) + g''(v)),$$

assuming that all the necessary derivatives exist.

The heat equation

39. Find all solutions of the one-dimensional heat equation $w_t = c^2 w_{xx}$ of the form $w = e^{rt} \sin \pi x$, r a constant.

40. Find all solutions of the one-dimensional heat equation $w_t = c^2 w_{xx}$ that have the form $w = e^{rt} \sin kx$ and that satisfy the conditions that $w(0, t) = 0$ and $w(L, t) = 0$. What happens to the solution as $t \to \infty$?

41. Let $u = f(y)$ be a differentiable function of y, and let $y = x - tg(u)$, where g is any function of u. Show that u satisfies the equation

$$\frac{\partial u}{\partial t} + g(u) \frac{\partial u}{\partial x} = 0.$$

15.8

Linear Approximation and Increment Estimation

Linearization

If a function $f(x, y)$ is continuous and has continuous first partial derivatives at a point (x_0, y_0), then the surface $z = f(x, y)$ has a tangent plane at the point $P_0(x_0, y_0, z_0) = (x_0, y_0, f(x_0, y_0))$ given by the equation

$$z = T(x, y) = f(x_0, y_0) + f_x(x_0, y_0)(x - x_0) + f_y(x_0, y_0)(y - y_0). \quad (1)$$

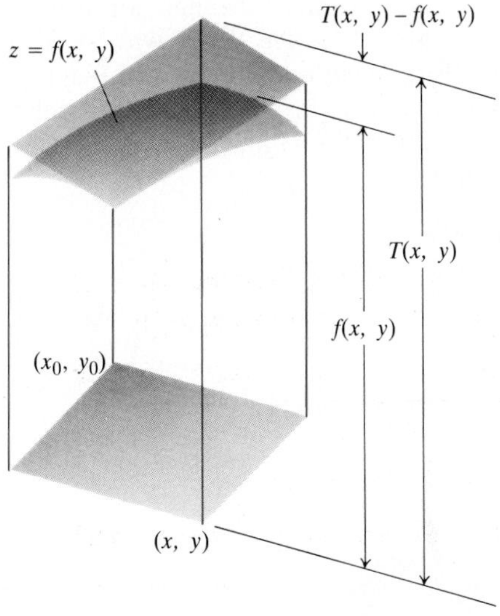

15.30 The tangent plane and surface are close together near the point of tangency.

Since the surface and plane will be close together near P_0, we can use the values of T to approximate the values of f (Fig. 15.30). The function $T(x, y)$ is called the *linearization* or *linear approximation* of f at (x_0, y_0).

The approximation

$$f(x, y) \approx T(x, y) \quad (2)$$

plays the same role for functions of two variables that the tangent line approximation plays for functions of a single variable. It provides a useful, simple way to estimate the values of functions that arise in science, engineering, and mathematics.

How good is the approximation in (2)? From the increment theorem for functions of two variables (Theorem 2, Article 15.3), we know that if $(x_0 + \Delta x, y_0 + \Delta y)$ is a point near (x_0, y_0), then

$$f(x_0 + \Delta x, y_0 + \Delta y) = f(x_0, y_0) + f_x(x_0, y_0)\,\Delta x + f_y(x_0, y_0)\,\Delta y + \varepsilon_1 \Delta x + \varepsilon_2 \Delta y, \quad (3)$$

where $\varepsilon_1, \varepsilon_2 \to 0$ as $\Delta x, \Delta y \to 0$.

To compare the values of $T(x, y)$ with those of $f(x, y)$ we change the notation in Eq. (3), writing

$$\Delta x = (x - x_0), \qquad \Delta y = (y - y_0)$$
$$x = x_0 + \Delta x, \qquad y = y_0 + \Delta y.$$

With these changes, Eq. (3) becomes

$$f(x, y) = \underbrace{f(x_0, y_0) + f_x(x_0, y_0)(x - x_0) + f_y(x_0, y_0)(y - y_0)}_{\text{This part is } T(x, y).} + \varepsilon_1(x - x_0) + \varepsilon_2(y - y_0). \quad (4)$$

Thus,

$$\begin{aligned}|f(x, y) - T(x, y)| &= |\varepsilon_1(x - x_0) + \varepsilon_2(y - y_0)| \\ &\leq |\varepsilon_1|\,|x - x_0| + |\varepsilon_2|\,|y - y_0|.\end{aligned} \tag{5}$$

This inequality, combined with the fact that $\varepsilon_1, \varepsilon_2 \to 0$ as $\Delta x, \Delta y \to 0$, assures us that the approximation $f(x, y) \approx T(x, y)$ will be close in some rectangle

$$R\colon |x - x_0| \leq \delta_1, \qquad |y - y_0| \leq \delta_2,$$

centered at (x_0, y_0).

To see how close the approximation is, we need the second order partial derivatives of f. If f_{xx}, f_{yy}, f_{xy} are continuous throughout R, then their absolute values are all less than or equal to

$$B = \max \text{ on } R \text{ of } |f_{xx}|, |f_{yy}|, |f_{xy}|, \tag{6}$$

and it turns out (for reasons explained in Article 15.9) that

$$|f(x, y) - T(x, y)| \leq \tfrac{1}{2}B(|x - x_0| + |y - y_0|)^2 \tag{7}$$

throughout R. (In practice the exact value of B is hard to find, and we replace it with any reasonable upper bound.) If we set $x - x_0 = \Delta x$, $y - y_0 = \Delta y$, Eq. (7) takes the form

$$|f(x, y) - T(x, y)| \leq \tfrac{1}{2}B(|\Delta x| + |\Delta y|)^2. \tag{8}$$

The following statements summarize the discussion so far:

Linear Approximation of $f(x, y)$ Near (x_0, y_0)

Suppose $f(x, y)$ and its first and second order partial derivatives are continuous throughout a rectangle R

$$R\colon |x - x_0| \leq \delta_1, \qquad |y - y_0| \leq \delta_2,$$

centered at (x_0, y_0), and that $|f_{xx}|$, $|f_{yy}|$, and $|f_{xy}|$ are less than or equal to some number B throughout R. Then throughout R

$$f(x, y) \approx \underbrace{f(x_0, y_0) + f_x(x_0, y_0)(x - x_0) + f_y(x_0, y_0)(y - y_0)}_{T(x,\,y)} \tag{9}$$

with an error E that is bounded by the inequality

$$|E| = |f(x, y) - T(x, y)| \leq \tfrac{1}{2}B(|x - x_0| + |y - y_0|)^2. \tag{10}$$

EXAMPLE 1 Find the linearization $T(x, y)$ of

$$f(x, y) = x^2 - xy + \tfrac{1}{2}y^2 + 3$$

at the point (3, 2). Then use Eq. (10) to give an upper bound for the error in the approximation $f(x, y) \approx T(x, y)$ over the rectangle

$$R\colon |x - 3| \leq 0.1, \qquad |y - 2| \leq 0.1.$$

Express the possible error as a percent of the value of f at (3, 2), the center of the rectangle.

Solution

$$f(3, 2) = 8,$$
$$f_x(3, 2) = 2x - y|_{(3,2)} = 4,$$
$$f_y(3, 2) = -x + y|_{(3,2)} = -1.$$

Therefore,

$$\begin{aligned} T(x, y) &= f(3, 2) + f_x(3, 2)(x - 3) + f_y(3, 2)(y - 2) \\ &= 8 + 4(x - 3) - (y - 2) \\ &= 4x - y - 2. \end{aligned}$$

To get an upper bound for the error in the approximation

$$x^2 - xy + \tfrac{1}{2}y^2 + 3 \approx 4x - y - 2 \tag{11}$$

over the rectangle R, we calculate

$$|f_{xx}| = |2| = 2,$$
$$|f_{yy}| = |1| = 1,$$
$$|f_{xy}| = |-1| = 1.$$

The largest of these is $B = 2$, and Eq. (10) gives

$$|E| = |f(x, y) - T(x, y)| \le \tfrac{1}{2}(2)(|x - 3| + |y - 2|)^2$$

throughout R. Since

$$|x - 3| \le 0.1 \qquad \text{and} \qquad |y - 2| \le 0.1,$$

on R, we have

$$|E| \le (0.1 + 0.1)^2 = 0.04.$$

Since $f(3, 2) = 8$, the estimate in (11) will be in error by no more than

$$\frac{0.04}{8} = 0.005 = 0.5\%$$

of the value of f at (3, 2) as long as (x, y) lies in R. □

EXAMPLE 2 Show that

$$f(x, y) = \frac{1}{1 + x - y} \approx 1 - x + y \tag{12}$$

for $(x, y) \approx (0, 0)$.

Solution The function f is continuous for $|x - y| < 1$, as are the derivatives

$$f_x(x, y) = \frac{-1}{(1 + x - y)^2}, \qquad f_y(x, y) = \frac{1}{(1 + x - y)^2}. \tag{13}$$

At (0, 0) Eq. (9) takes the form

$$f(x, y) \approx f(0, 0) + xf_x(0, 0) + yf_y(0, 0),$$

or

$$f(x, y) = \frac{1}{1 + x - y} \approx 1 + x(-1) + y(1) = 1 - x + y,$$

which is Eq. (12). □

EXAMPLE 3 Find the linear approximation of

$$f(x, y) = \frac{1}{1 + x - y}$$

near the point (2, 1).

Solution Using Eq. (9) and the formulas for f_x and f_y from Example 2 gives

$$f(x, y) \approx f(2, 1) + f_x(2, 1)(x - 2) + f_y(2, 1)(y - 1),$$

$$\frac{1}{1 + x - y} \approx \frac{1}{2} + \left(-\frac{1}{4}\right)(x - 2) + \left(\frac{1}{4}\right)(y - 1) = \frac{3}{4} - \frac{x}{4} + \frac{y}{4}. \quad \square$$

Increment Estimation

To the extent that a function $f(x, y)$ is approximated by its linearization $T(x, y)$, changes in f are approximated by changes in T. To be precise, suppose we start at a point $P_0(x_0, y_0)$ and move to a nearby point $Q(x, y) = (x_0 + \Delta x, y_0 + \Delta y)$. Then

$$\begin{aligned} T(x, y) &= f(x_0, y_0) + f_x(x_0, y_0)(x - x_0) + f_y(x_0, y_0)(y - y_0) \\ &= f(x_0, y_0) + f_x(x_0, y_0)\,\Delta x + f_y(x_0, y_0)\,\Delta y, \end{aligned}$$

and the change in the value of T resulting from the move from P to Q is

$$\begin{aligned} \Delta T &= T(x, y) - T(x_0, y_0) \\ &= T(x, y) - f(x_0, y_0) = f_x(x_0, y_0)\,\Delta x + f_y(x_0, y_0)\,\Delta y. \end{aligned} \tag{14}$$

See Fig. 15.31.

The change in the value of f that results from the move from P to Q is

$$\Delta f = f(x, y) - f(x_0, y_0),$$

which, according to Eq. (9), can be approximated as

$$\begin{aligned} \Delta f = f(x, y) - f(x_0, y_0) &\approx f_x(x_0, y_0)(x - x_0) + f_y(x_0, y_0)(y - y_0) \\ &\approx \underbrace{f_x(x_0, y_0)\,\Delta x + f_y(x_0, y_0)\,\Delta y}_{\Delta T}. \end{aligned}$$

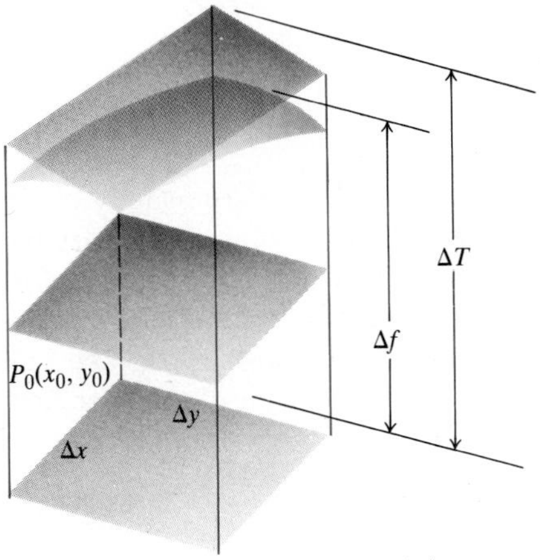

15.31 $\Delta T = T(x, y) - f(x_0, y_0)$.

The right-hand side of this approximation is the formula we derived for ΔT in Eq. (14).

Increment Estimation Formula for f(x, y)

$$\Delta f = f(x, y) - f(x_0, y_0) \approx \underbrace{f_x(x_0, y_0)\,\Delta x + f_y(x_0, y_0)\,\Delta y}_{\Delta T} \tag{15}$$

The increment approximation in (15) gives a way to see how *sensitive* a function $f(x, y)$ is to small changes in x and y near the point (x_0, y_0). This is illustrated in the following example.

EXAMPLE 4 How sensitive is the volume

$$V = \pi r^2 h$$

of a right circular cylinder to small changes in its radius and height near the point $(r_0, h_0) = (1, 5)$?

Solution We use the increment approximation in Eq. (15) to obtain

$$\begin{aligned}\Delta V &\approx V_r(r_0, h_0)\,\Delta r + V_h(r_0, h_0)\,\Delta h \\ &\approx V_r(1, 5)\,\Delta r + V_h(1, 5)\,\Delta h \\ &\approx 2\pi r h|_{(1,5)}\,\Delta r + \pi r^2|_{(1,5)}\,\Delta h = 10\pi\,\Delta r + \pi\,\Delta h,\end{aligned}$$

or

$$\Delta V \approx 10\pi\,\Delta r + \pi\,\Delta h. \tag{16}$$

This shows that a one-unit change in r will change V by about 10π units. A one-unit change in h will change V by about π units. Therefore, the volume of a cylinder with radius $r = 1$ and height $h = 5$ is nearly ten times as sensitive to small changes in r as it is to small changes in h.

In contrast, if the values of r and h are reversed, so that $r = 5$ and $h = 1$, then

$$\Delta V \approx 2\pi r h|_{(5,1)}\,\Delta r + \pi r^2|_{(5,1)}\,\Delta h = 10\pi r\,\Delta r + 25\pi\,\Delta h.$$

The volume is now more sensitive to small changes in h than it is to small changes in r. Thus, the sensitivity to change depends not only on the increments but also on the relative sizes of r and h (Fig. 15.32). □

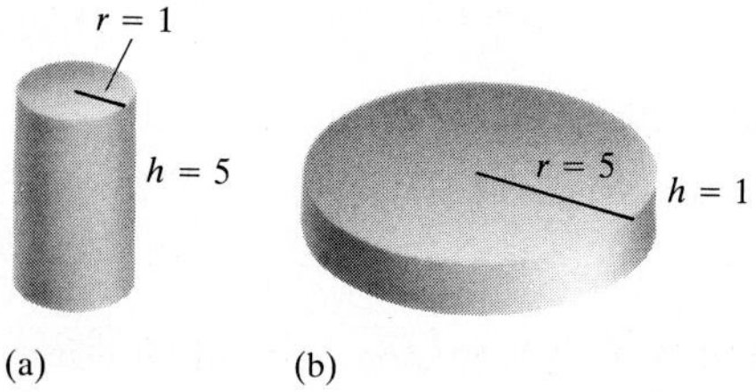

15.32 The volume of cylinder (a) is more sensitive to a small change in r than it is to an equally small change in h. The volume of cylinder (b) is more sensitive to small changes in h than it is to small changes in r.

There is a general rule to be learned from Example 4: A function is most sensitive to small changes in the variables that give the largest partial derivatives.

If the value of a function $z = f(x, y)$ changes from z_0 to $z_0 + \Delta z$, the change Δz may be looked at from three points of view:

Absolute change:	Δz	
Relative change:	$\dfrac{\Delta z}{z_0}$	(17)
Percentage change:	$\dfrac{\Delta z}{z_0} \times 100$	

In practice, percentage change and relative change are usually more important than absolute change. Knowing that a line voltage may vary by ±5 volts is useful information only if we know what the line voltage is supposed to be in the first place. If it is supposed to be 220,000 volts, then 5 volts more or less won't matter. If it is supposed to be 10 volts, then ±5 volts is likely to be significant because it is ±50 percent of the specified value.

EXAMPLE 5 How is the relative change in

$$V = \pi r^2 h$$

related to relative change in r and h? How are the percentage changes related?

Solution Starting with

$$\Delta V \approx V_r(r, h)\,\Delta r + V_h(r, h)\,\Delta h = 2\pi rh\,\Delta r + \pi r^2\,\Delta h,$$

we divide both sides by $V = \pi r^2 h$ to obtain

$$\frac{\Delta V}{V} \approx \frac{2\pi rh}{\pi r^2 h}\,\Delta r + \frac{\pi r^2}{\pi r^2 h}\,\Delta h$$

or

$$\frac{\Delta V}{V} \approx 2\frac{\Delta r}{r} + \frac{\Delta h}{h}. \tag{18}$$

The relative change in V is the relative change in h plus twice the relative change in r. Multiplying through by 100 gives

$$\frac{\Delta V}{V} \times 100 \approx 2\frac{\Delta r}{r} \times 100 + \frac{\Delta h}{h} \times 100, \tag{19}$$

which shows that the percentage change in V is about equal to the percentage change in h plus twice the percentage change in r.

A 3 percent increase in r and a 2 percent decrease in h give

$$\frac{r}{\Delta r} \times 100 = 3, \qquad \frac{h}{\Delta h} \times 100 = -2,$$

respectively. The resulting change in V is an increase of about four percent:

$$\frac{\Delta V}{V} \times 100 \approx 2(3) + (-2) = 6 - 2 = 4\%. \ \square$$

EXAMPLE 6 If r is measured with an accuracy of ± 2 percent and h with an accuracy of ± 0.5 percent, about how accurately can we calculate V from the formula

$$V = \pi r^2 h?$$

Solution From Eq. (18) in Example 5 we have

$$\frac{\Delta V}{V} \approx 2\frac{\Delta r}{r} + \frac{\Delta h}{h}.$$

We are told that

$$\Delta r = \pm 2\frac{r}{100}, \qquad \Delta h = \pm 0.5\frac{h}{100} = \pm\frac{h}{200},$$

so that

$$|\Delta r| \le \frac{r}{50}, \qquad |\Delta h| \le \frac{h}{200}.$$

Therefore,

$$\left|\frac{\Delta V}{V}\right| \approx \left|2\frac{\Delta r}{r} + \frac{\Delta h}{h}\right| \le \frac{2}{r}|\Delta r| + \frac{1}{h}|\Delta h|$$

$$\le \frac{2}{r}\frac{r}{50} + \frac{1}{h}\frac{h}{200} = \frac{1}{25} + \frac{1}{200} = 0.045.$$

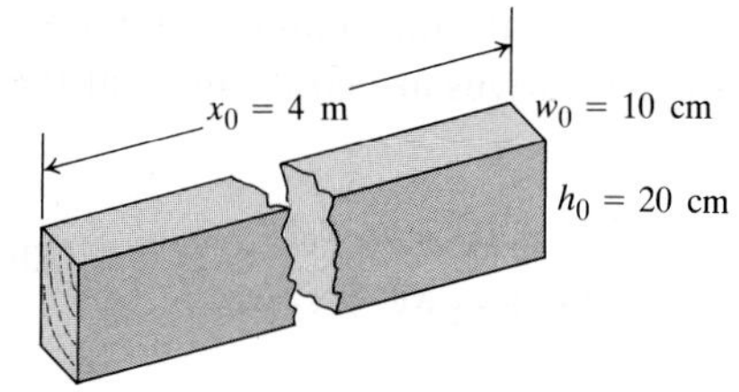

15.34 The dimensions of the beam in Example 8.

where

$p =$ the load (kg/meter of beam length),
$x =$ the length between supports (m),
$w =$ the width of the beam (m),
$h =$ the height of the beam (m),
$C =$ a constant that depends on the units of measurement and on the material from which the beam is made.

When

$$\Delta S \approx S_p\,\Delta p + S_x\,\Delta x + S_w\,\Delta w + S_h\,\Delta h$$

is written out for a particular set of values p_0, x_0, w_0, h_0, and simplified, the resulting approximation is

$$\Delta S \approx S_0\left[\frac{\Delta p}{p_0} + \frac{4\,\Delta x}{x_0} - \frac{\Delta w}{w_0} - \frac{3\,\Delta h}{h_0}\right], \tag{25}$$

where $S_0 = S(p_0, x_0, h_0, w_0) = Cp_0x_0^4/w_0h_0^3$.

At $p_0 = 100$ kg/m, $x_0 = 4$ m, $w_0 = 0.1$ m, and $h_0 = 0.2$ m,

$$\Delta S \approx S_0\left[\frac{\Delta p}{100} + \Delta x - 10\,\Delta w - 15\,\Delta h\right]. \tag{26}$$

(See Fig. 15.34.) □

Conclusions about this beam from Eq. (26) Since Δp and Δx appear with positive coefficients in Eq. (26), increases in p and in x will increase the sag. But Δw and Δh appear with negative coefficients, so that increases in w and h will *decrease* the sag (make the beam stiffer). The sag is not very sensitive to small changes in load, because the coefficient of Δp is 1/100. The coefficient of Δh is a negative number of greater magnitude than the coefficient of Δw. Therefore, making the beam $\Delta h = 1$ cm higher will decrease the sag more than making the beam $\Delta w = 1$ cm wider.

PROBLEMS

In Problems 1–6, find the linearization $T(x, y)$ of the function $f(x, y)$ at each point.

1. $f(x, y) = x^2 + y^2 + 1$ at (a) $(0, 0)$, (b) $(1, 1)$.

2. $f(x, y) = x^3y^4$ at (a) $(1, 1)$, (b) $(0, 0)$.

3. $f(x, y) = e^x \cos y$ at (a) $(0, 0)$, (b) $(0, \pi/2)$.

4. $f(x, y) = (x + y + 2)^2$ at (a) $(0, 0)$, (b) $(1, 2)$.

5. $f(x, y) = 3x - 4y + 5$ at (a) $(0, 0)$, (b) $(1, 1)$.

6. $f(x, y) = e^{2y-x}$ at (a) $(0, 0)$, (b) $(1, 2)$.

In Problems 7–10, find the linearization $T(x, y, z)$ of the function $f(x, y, z)$ at each point.

7. $f(x, y, z) = \sqrt{x^2 + y^2 + z^2}$ at
(a) $(1, 0, 0)$, (b) $(1, 1, 0)$, (c) $(1, 1, 1)$.

8. $f(x, y, z) = (\sin xy)/z$ at
(a) $(\pi/2, 1, 1)$, (b) $(2, 0, 1)$.

9. $f(x, y, z) = e^x \cos(y + z)$ at
(a) $(0, 0, 0)$, (b) $(0, \pi/4, \pi/4)$.

10. $f(x, y, z) = \tan^{-1}(xyz)$ at
(a) $(1, 0, 0)$, (b) $(1, 1, 0)$, (c) $(1, 1, 1)$.

In Problems 11–15, find the linearization $T(x, y)$ of the function $f(x, y)$ at the point $P_0(x_0, y_0)$. Then use Eq. (10) to give an upper bound for the error in the approximation $f(x, y) \approx T(x, y)$ over the given rectangle R.

11. $f(x, y) = \frac{1}{2}x^2 + xy + \frac{1}{4}y^2 + 3x - 3y + 4$ at $P_0(2, 2)$.

R: $|x - 2| \le 0.1$, $|y - 2| \le 0.1$

12. $f(x, y) \times 1 + y + x \cos y$ at $P_0(0, 0)$.

R: $|x| \le 0.2$, $|y| \le 0.2$

(Use $|\cos y| \le 1$ and $|\sin y| \le 1$ in estimating E.)

13. $f(x, y) = \ln x + \ln y$ at $P_0(1, 1)$.

R: $|x - 1| \leq 0.2, \quad |y - 1| \leq 0.2$

14. $f(x, y) = e^x \cos y$ at $P_0(0, 0)$.

R: $|x| \leq 0.1, \quad |y| \leq 0.1$

(Use $e^x \leq 3$ and $|\cos y| \leq 1$ in estimating E.)

15. $f(x, y) = xy^2 + y\cos(x - 1)$ at $P_0(1, 2)$.

R: $|x - 1| \leq 0.1, \quad |y - 2| \leq 0.1$

16. Estimate how much simultaneous errors of 2 percent in a, b, and c may affect the product abc.

17. Find the linearization $T(x, y, z)$ of the function $f(x, y, z)$ at the point $P_0(x_0, y_0, z_0)$. Then use Eq. (23) to give a bound for the error in the approximation $f(x, y, z) \approx T(x, y, z)$ over the given rectangular region R.

a) $f(x, y, z) = x^2 + xy + yz + (1/4)z^2$ at $P_0(1, 1, 2)$.
R: $|x - 1| \leq 0.01, |y - 1| \leq 0.01, |z - 2| \leq 0.08$

b) $f(x, y, z) = xy + 2yz - 3xz$ at $P_0(1, 1, 0)$.
R: $|x - 1| \leq 0.01, |y - 1| \leq 0.01, |z| \leq 0.01$

c) $f(x, y, z) = \sqrt{2}\cos x \sin(y + z)$ at $P_0(0, 0, \pi/4)$.
R: $|x| \leq 0.01, |y| \leq 0.01, |z - \pi/4| \leq 0.01$.

18. What relationship must hold between the r and h in Example 4 if the volume of the cylinder is to be equally sensitive to small changes in the two variables?

19. The beam of Example 8 is tipped on its side, so that $h = 0.1$ m and $w = 0.2$ m.

a) What is the approximation for ΔS now?

b) Compare the sensitivity of the beam to a small change in height with its sensitivity to a change of the same amount in width.

20. Estimate the amount of material in a hollow rectangular box whose inside measurements are 5 ft long, 3 ft wide, 2 ft deep, if the box is made of lumber that is $\frac{1}{2}$ in. thick and the box has no top.

21. The area of a triangle is $A = \frac{1}{2}ab \sin C$, where a and b are two sides of the triangle and C is the included angle. In surveying a particular triangular plot of land, a and b are measured to be 150 ft and 200 ft, respectively, and C is read to be 60°. By how much (approximately) is the computed area in error if a and b are in error by $\frac{1}{2}$ ft each and C is in error by 2°?

22. Suppose that $u = xe^y + y\sin z$, and that x, y, and z can be measured with maximum possible errors of ± 0.2, ± 0.6, and $\pm\pi/180$, respectively. Estimate the maximum possible error in calculating u from the measured values $x = 2$, $y = \ln 3$, $z = \pi/2$.

23. Suppose T is to be found from the formula $T = x\cosh y$, where x and y are found to be 2 and $\ln(2)$ with maximum possible errors of $|\Delta x| = 0.1$ and $|\Delta y| = 0.02$. Estimate the maximum possible error in the computed value of T.

24. About how accurately may $V = \pi r^2 h$ be calculated from measurements of r and h that are in error by 1%?

25. If $r = 5$ cm and $h = 12$ cm to the nearest millimeter, what should we expect the maximum percentage error in calculating $V = \pi r^2 h$ to be?

26. To estimate the volume of a cylinder of radius about 2 m and height about 3 m, about how accurately should the radius and height be measured so that the error in the volume estimate will not exceed 0.1 m³? Assume that the possible error Δr in measuring r is equal to the possible error Δh in measuring h.

27. Give a reasonable square centered at (1, 1) over which the value of $f(x, y) = x^3y^4$ will not vary by more than ± 0.1.

28. When an x ohm and y ohm resistor are in parallel, the resistance R they produce may be calculated from the formula $1/R = 1/x + 1/y$. By about what percentage will R change if x increases from 20 to 20.1 ohms and y decreases from 25 to 24.9 ohms?

29. a) If $x = 3 \pm 0.01$, $y = 4 \pm 0.01$, with what accuracy can the polar coordinates r and θ of the point (x, y) be calculated from the formulas $r^2 = x^2 + y^2$, $\theta = \tan^{-1}(y/x)$?

b) At the point $(x, y) = (3, 4)$, are r and θ more sensitive to changes in x, or to changes in y? Draw a figure showing x, y, r, θ and confirm your results by appealing to the geometry of the figure.

30. A function $w = f(x, y)$ is said to be *differentiable* at $P(a, b)$ if there are constants M and N (possibly depending on f and P) such that

$$\Delta w = M\,\Delta x + N\,\Delta y + \alpha[|\Delta x| + |\Delta y|]$$

and $\alpha \to 0$ as $|\Delta x| + |\Delta y| \to 0$. Here $\Delta w = f(a + \Delta x, b + \Delta y) - f(a, b)$. Prove that if f is differentiable at (a, b), then $M = f_x(a, b)$ and $N = f_y(a, b)$.

31. If the function $w = f(x, y)$ is differentiable at $P(a, b)$ (see Problem 30), prove it is continuous there.

32. To which of the variables K, M, and h is the function $Q = \sqrt{2KM/h}$ most sensitive near the point $(K_0, M_0, h_0) = (2, 20, 0.05)$? least sensitive?

33. a) Around the point (1, 0) is $f(x, y) = x^2(y + 1)$ more sensitive to changes in x, or to changes in y?

b) What ratio of Δx to Δy will make ΔT in Eq. (15) equal to zero at (1, 0)?

34. If $|a|$ is much greater than $|b|$, $|c|$, and $|d|$, to which entry is the value of the determinant

$$\begin{vmatrix} a & b \\ c & d \end{vmatrix}$$

most sensitive?

35. *Method of steepest descent.* Suppose it is desired to find a solution of the equation $f(x, y, z) = 0$. Let

$P_0(x_0, y_0, z_0)$ be a first guess, and suppose $f(x_0, y_0, z_0) = f_0$ is not zero. Let $(\nabla f)_0$ be the gradient vector normal to the surface $f(x, y, z) = f_0$ at P_0. If f_0 is positive, we want to decrease the value of f. The gradient points in the direction of most rapid increase, its negative in the direction of "steepest descent." We therefore take as next approximation

$$x_1 = x_0 - hf_x(x_0, y_0, z_0),$$
$$y_1 = y_0 - hf_y(x_0, y_0, z_0),$$
$$z_1 = z_0 - hf_z(x_0, y_0, z_0).$$

What value of h corresponds to making $\Delta T = -f_0$? What change is suggested if f_0 is negative? (The method could be applied to the problem of solving the simultaneous equations

$$x - y + z = 3,$$
$$x^2 + y^2 + z^2 = 20,$$
$$xyz = 8,$$

by writing

$$f(x, y, z) = (x - y + z - 3)^2 + (x^2 + y^2 + z^2 - 20)^2 + (xyz - 8)^2.)$$

15.9 Maxima, Minima, and Saddle Points

As we can see in Figs. 15.35 and 15.36, functions of two variables can have relative and absolute maximum and minimum values just the way functions of a single variable can. The way we go about finding these extreme values is much the same, except that we now have more derivatives to work with because we have more variables.

Testing for Extreme Values

1. *Look for critical points and boundary points.*

As we know, a continuous function $z = f(x, y)$ takes on an absolute maximum value and an absolute minimum value on any closed, bounded region R on which it is defined. Moreover (as we shall show later), these and any other relative maxima and minima can occur only at

i) boundary points of R,

ii) interior points of R where $f_x = f_y = 0$, or points where f_x or f_y fail to exist. (We call these the *critical points* of f.)

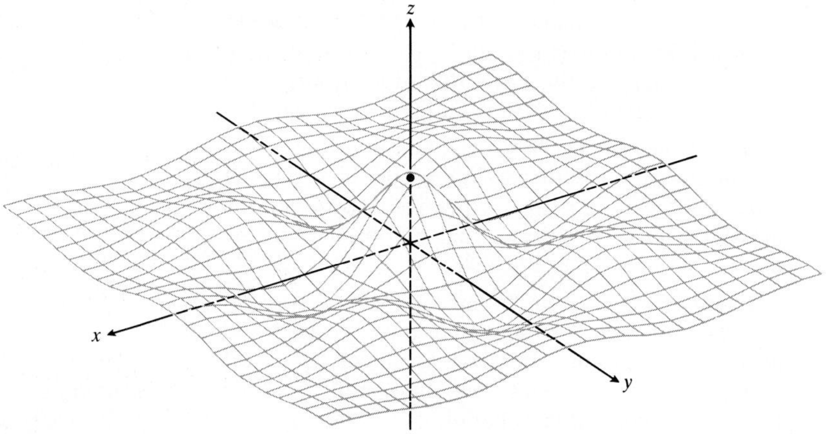

15.35 The function $z = (\cos x)(\cos y)e^{-\sqrt{x^2+y^2}}$ has many relative maxima and minima.

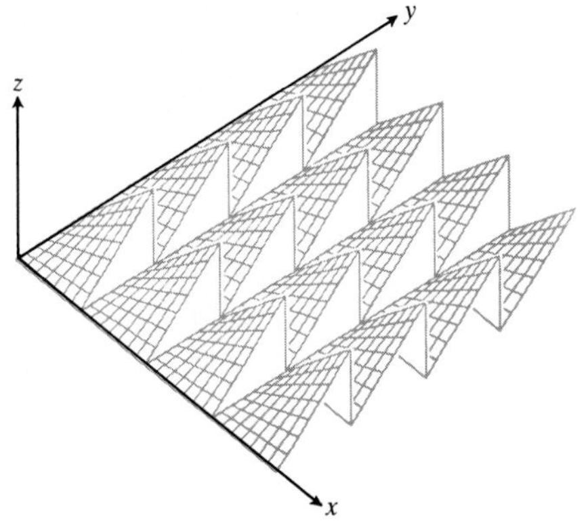

15.36 The surface $z = (x - [x])(y - [y])$ for $x \geq 0$, $y \geq 0$, where $[x]$ is the greatest integer function.

Usually there are only a few such points, so we can evaluate f at all of them and choose the largest and smallest values. No further tests are required if all we wish to find is the absolute maximum and minimum.

2. *If there are no boundary points, look for critical points.*

A function $z = f(x, y)$ defined on a region R without boundary points (for instance, the interior of a disc, triangle, or rectangle, a quadrant minus the axes, or the entire plane) may have no relative maxima or minima on R. However, if it does they must occur at points of R where

$$f_x = f_y = 0.$$

or where one or both of these derivatives fails to exist (as with $z = \sqrt{x^2 + y^2}$ at (0, 0)).

3. *A second derivative test may be applied at interior points where $f_x = f_y = 0$ and the first and second order partial derivatives of f are continuous.*

The fact that $f_x = f_y = 0$ at an interior point (a, b) does not guarantee that f will have an extreme value there. There is, however, a second derivative test that may help to identify the behavior of f at (a, b). It goes like this:

If $f_x(a, b) = f_y(a, b) = 0$, then

i) f has a *relative maximum* at (a, b) if $f_{xx} < 0$ and $f_{xx}f_{yy} - f_{xy}^2 > 0$ at (a, b).
ii) f has a *relative minimum* at (a, b) if $f_{xx} > 0$ and $f_{xx}f_{yy} - f_{xy}^2 > 0$ at (a, b).
iii) f has a *saddle point* at (a, b) if $f_{xx}f_{yy} - f_{xy}^2 < 0$ at (a, b).
iv) The test is *inconclusive* at (a, b) if $f_{xx}f_{yy} - f_{xy}^2 = 0$ at (a, b). We must find some other way to determine the behavior of f at (a, b).

The expression $f_{xx}f_{yy} - f_{xy}^2$ is called the *discriminant* of f. It is sometimes easier to remember in the determinant form

$$f_{xx}f_{yy} - f_{xy}^2 = \begin{vmatrix} f_{xx} & f_{xy} \\ f_{yx} & f_{yy} \end{vmatrix}.$$

We shall now look at examples that show these tests at work. After that, we shall show why the condition $f_x = f_y = 0$ is a necessary condition for having an extreme value at an interior point of the domain of a differentiable function and look into the mathematics behind the second derivative test.

The Tests at Work

In the first example we look at the function $f(x, y) = x^2 + y^2$, whose behavior we already know from looking at the formula: Its value is zero at the origin and increases steadily as (x, y) moves away from the origin. The point of Example 1 is to show how the tests above reveal this.

EXAMPLE 1 Find the extreme value of

$$f(x, y) = x^2 + y^2.$$

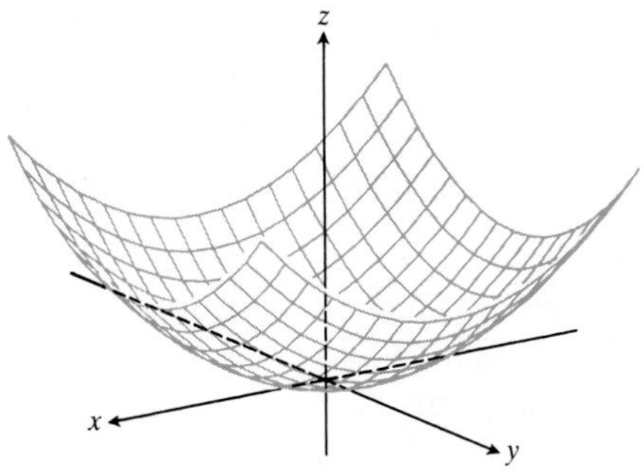

15.37 The function $f(x, y) = x^2 + y^2$ has an absolute minimum at the origin.

Solution The domain of f has no boundary points, for it is the entire plane (Fig. 15.37). The derivatives $f_x = 2x$ and $f_y = 2y$ exist everywhere. Therefore, relative maxima and minima can occur only where

$$f_x = 2x = 0 \qquad \text{and} \qquad f_y = 2y = 0.$$

The only possibility is the origin, where the value of f is zero. Since f is never negative, we see that zero is an absolute minimum.

We have not needed the second derivative test at all. Had we used it, we would have found

$$f_{xx} = 2, \qquad f_{yy} = 2, \qquad f_{xy} = 0$$

and

$$f_{xx}f_{yy} - f_{xy}^2 = (2)(2) - (0)^2 = 4 > 0,$$

identifying $(0, 0)$ as a relative minimum. This in itself does not identify $(0, 0)$ as an absolute minimum. It takes more information to do that. □

EXAMPLE 2 Find the extreme values of the function

$$f(x, y) = xy - x^2 - y^2 - 2x - 2y + 4.$$

Solution The function is defined and differentiable for all x and y and therefore has extreme values only at the points where f_x and f_y are simultaneously zero. This leads to

$$f_x = y - 2x - 2 = 0, \qquad f_y = x - 2y - 2 = 0,$$

or

$$x = y = -2.$$

Therefore, the point $(-2, -2)$ is the only point where f may take on an extreme value. To see if it does so, we calculate

$$f_{xx} = -2, \qquad f_{yy} = -2, \qquad f_{xy} = 1.$$

The discriminant of f at $(a, b) = (-2, -2)$ is

$$f_{xx}f_{yy} - f_{xy}^2 = (-2)(-2) - (1)^2 = 4 - 1 = 3.$$

The combination

$$f_{xx} < 0 \qquad \text{and} \qquad f_{xx}f_{xy} - f_{xy}^2 > 0$$

tells us that f has a relative maximum at $(-2, -2)$. The value of f at this point is $f(-2, -2) = 8$. □

EXAMPLE 3 Find the extreme values of

$$f(x, y) = xy.$$

Solution Since the function is differentiable everywhere and its domain has no boundary points (Fig. 15.38), the function can assume extreme values only where

$$f_x = y = 0 \qquad \text{and} \qquad f_y = x = 0.$$

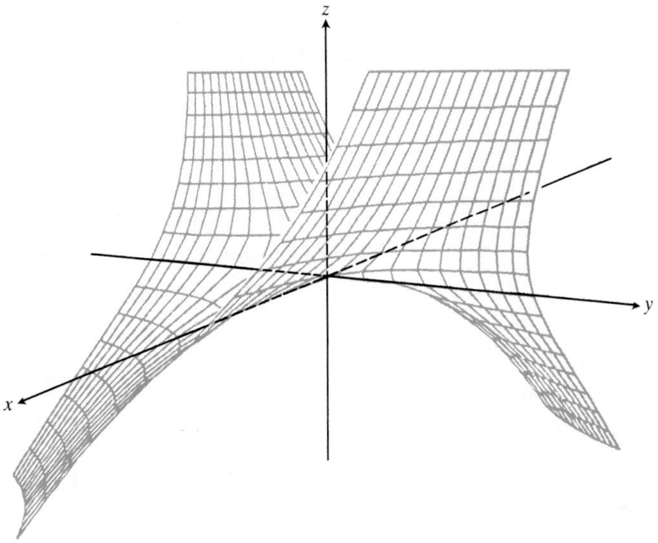

15.38 The surface $z = xy$ has a saddle point at the origin.

Thus, the origin is the only point where f might have an extreme value. To see what happens there, we calculate

$$f_{xx} = 0, \qquad f_{yy} = 0, \qquad f_{xy} = 1.$$

The discriminant,

$$f_{xx}f_{yy} - f_{xy}^2 = -1,$$

is negative. Therefore the function has a saddle point at (0, 0). We conclude that $f(x, y) = xy$ assumes no extreme values at all.

If we restrict the domain of f to the disc $x^2 + y^2 \leq 1$, then the maximum value of f is $+\frac{1}{2}$ and the minimum is $-\frac{1}{2}$. (Change to polar coordinates: $xy = r^2 \sin\theta \cos\theta = \frac{1}{2}r^2 \sin 2\theta$.) □

EXAMPLE 4 Find the absolute maximum and minimum values of

$$f(x, y) = 2 + 2x + 2y - x^2 - y^2$$

on the triangular plate in the first quadrant bounded by the lines $x = 0$, $y = 0$, $y = 9 - x$.

Solution The only places where f can assume these values are points on the boundary of the triangle (Fig. 15.39) and points inside it where $f_x = f_y = 0$.

Interior points. For these we have

$$f_x = 2 - 2x = 0,$$
$$f_y = 2 - 2y = 0,$$

or

$$(x, y) = (1, 1),$$

where

$$f(1, 1) = 4.$$

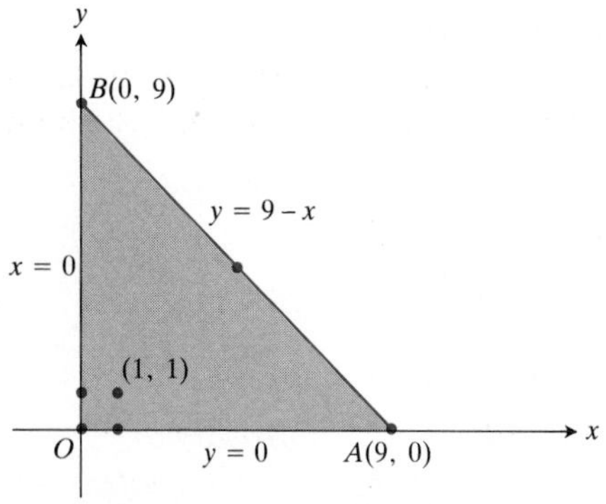

15.39 This triangular plate is the domain of the function in Example 4.

Boundary points. We take the triangle one side at a time:

1. On the segment OA, $y = 0$. The function

$$f(x, y) = f(x, 0) = 2 + 2x - x^2$$

may now be regarded as a function of x defined on the closed interval $0 \le x \le 9$.

Its extreme values (we know from Chapter 3) may occur at the endpoints

$$x = 0 \quad \text{where } f(0, 0) = 2,$$
$$x = 9 \quad \text{where } f(9, 0) = 2 + 18 - 81 = -61,$$

and at the interior points where

$$f'(x, 0) = 2 - 2x = 0.$$

The only interior point where $f'(x, 0) = 0$ is $x = 1$, where

$$f(x, 0) = f(1, 0) = 3.$$

2. On the segment OB, $x = 0$ and

$$f(x, y) = f(0, y) = 2 + 2y - y^2.$$

We know from the symmetry of f in x and y, and from the analysis we just carried out, that the candidates on this segment are

$$f(0, 0) = 2, \qquad f(0, 9) = -61, \qquad f(0, 1) = 3.$$

3. We have already accounted for the values of f at the endpoints of AB, so we have only to look at the interior points of AB. With

$$y = 9 - x,$$

we have

$$f(x, y) = 2 + 2x + 2(9 - x) - x^2 - (9 - x)^2 = -61 + 18x - 2x^2.$$

Setting

$$f' = 18 - 4x = 0$$

gives

$$x = \frac{18}{4} = \frac{9}{2}.$$

At this value of x,

$$y = 9 - \frac{9}{2} = \frac{9}{2},$$

and

$$f(x, y) = -\frac{41}{2}.$$

Summary. We look at all the candidates:

$$4, \quad 2, \quad -61, \quad 3, \quad -\frac{41}{2}.$$

The maximum is 4, which f assumes at (1, 1). The minimum is −61, assumed at (0, 9) and (9, 0). □

The Condition $f_x(a, b) = f_y(a, b) = 0$

Suppose that a relative maximum value of the function $f(x, y)$ occurs at an interior point (a, b) of the function's domain and that $\partial f/\partial x$ and $\partial f/\partial y$ both exist at (a, b). (See Fig. 15.40.) Then,

1. $x = a$ is an interior point of the domain of the curve

   $$C_1\colon z = f(x, b)$$

   in which the plane $y = b$ cuts the surface $z = f(x, y)$.
2. The function $z = f(x, b)$ has a relative maximum value at $x = a$.
3. The value of the derivative of $z = f(x, b)$ at $x = a$ is therefore zero.

Since this derivative is precisely $\partial f/\partial x\,|_{(a,b)}$, we conclude that

$$\left.\frac{\partial f}{\partial x}\right|_{(a,b)} = 0.$$

A similar argument with the function $z = f(a, y)$ shows that

$$\left.\frac{\partial f}{\partial y}\right|_{(a,b)} = 0.$$

Thus,

$$f_x(a, b) = f_y(a, b) = 0$$

is a necessary condition for f to have an extreme value at an interior point (a, b) where both f_x and f_y exist.

Remember, though, that these derivatives are also zero at a saddle point, and that a function can assume an extreme value at a point where one or both of the derivatives fail to exist.

Derivation of the Second Derivative Test

The second derivative test stated at the beginning of this article is one that we may sometimes apply to determine whether a function $z = f(x, y)$ has a relative maximum or minimum value at a point $P(a, b)$ where f_x and f_y both vanish. The derivation of the test involves a clever application of Taylor's theorem from Article 12.2.

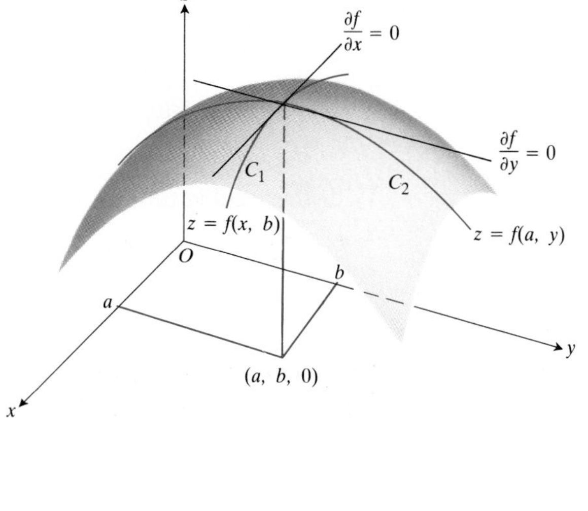

15.40 The maximum of f occurs at $x = a$, $y = b$.

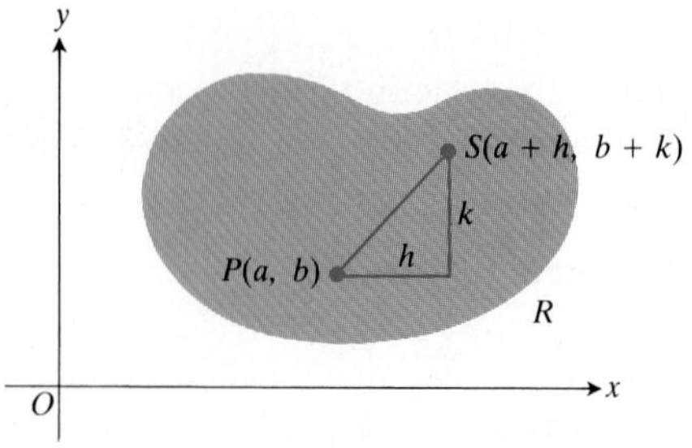

15.41 A region R whose interior contains the point $P(a, b)$.

We assume that f and its first and second order partial derivatives are continuous in some region R about $P(a, b)$, as shown in Fig. 15.41. We let $S(a + h, b + k)$ be a point close enough to P for the segment PS to lie in R, and we describe PS with the parametric equations

$$x = a + th, \qquad y = b + tk, \qquad 0 \le t \le 1. \tag{1}$$

We may now study the values of $f(x, y)$ along PS by considering the function

$$F(t) = f(a + ht, b + kt).$$

First of all, F is a differentiable function of t because the first partial derivatives of f are continuous and $x = a + th$, $y = b + tk$ are differentiable functions of t. The chain rule gives

$$F'(t) = \frac{\partial f}{\partial x}\frac{dx}{dt} + \frac{\partial f}{\partial y}\frac{dy}{dt} = h\frac{\partial f}{\partial x} + k\frac{\partial f}{\partial y}. \tag{2}$$

Also, F' is continuous on the closed interval $0 \le t \le 1$ because $\partial f/\partial x$ and $\partial f/\partial y$ are continuous throughout R.

Next, $F'(t)$ is a differentiable function of t because $\partial f/\partial x$ and $\partial f/\partial y$ have continuous partial derivatives. The chain rule gives

$$F''(t) = h\frac{\partial F'(t)}{\partial x} + k\frac{\partial F'(t)}{\partial y} = h\left(h\frac{\partial^2 f}{\partial x^2} + k\frac{\partial^2 f}{\partial y\,\partial x}\right) + k\left(h\frac{\partial^2 f}{\partial x\,\partial y} + k\frac{\partial^2 f}{\partial y^2}\right)$$

$$= h^2\frac{\partial^2 f}{\partial x^2} + 2hk\frac{\partial^2 f}{\partial x\,\partial y} + k^2\frac{\partial^2 f}{\partial y^2}. \tag{3}$$

Thus, $F(t)$ satisfies the hypotheses of Taylor's theorem for $n = 2$ on the interval $[0, 1]$. Namely, F and F' are continuous on $[0, 1]$ and F' is differentiable on $(0, 1)$. Therefore,

$$F(1) = F(0) + F'(0)(1 - 0) + F''(c)\frac{(1 - 0)^2}{2},$$

or

$$F(1) = F(0) + F'(0) + \frac{1}{2}F''(c), \tag{4}$$

for some number c between 0 and 1.

If we express F and its derivatives in terms of f, we find

$$\begin{aligned} F(1) &= f(a + h, b + k),\\ F(0) &= f(a, b),\\ F'(0) &= hf_x(a, b) + kf_y(a, b),\\ F''(c) &= (h^2 f_{xx} + 2hkf_{xy} + k^2 f_{yy})|_{(a+ch,\, b+ck)}. \end{aligned} \tag{5}$$

With these substitutions, Eq. (4) becomes

$$\begin{aligned} f(a + h, b + k) = f(a, b) &+ hf_x(a, b) + kf_y(a, b)\\ &+ \frac{1}{2}(h^2 f_{xx} + 2hkf_{xy} + k^2 f_{yy})|_{(a+ch,\, b+ck)}. \end{aligned} \tag{6}$$

This is the equation from which we now derive the second derivative test.

Suppose now that f_x and f_y are both zero at (a, b) and we wish to determine whether f has a maximum or minimum at (a, b). We may then rewrite Eq. (6) in the form

$$f(a + h, b + k) - f(a, b) = \frac{1}{2}(h^2 f_{xx} + 2hk f_{xy} + k^2 f_{yy})|_{(a+ch,\, b+ck)}. \quad (7)$$

Since a maximum or minimum value of f at (a, b) is reflected in the sign of $f(a + h, b + k) - f(a, b)$, we are led to considering the sign of

$$Q(c) = (h^2 f_{xx} + 2hk f_{xy} + k^2 f_{yy})|_{(a+ch,\, b+ck)}, \quad (8)$$

the quadratic expression on the right-hand side of Eq. (7) (without the one-half).

The second order derivatives in $Q(c)$ are to be evaluated at some point $(a + ch, b + ck)$ on the segment PS. This is not convenient for our purposes, since we know nothing more about c than $0 < c < 1$. However, since f_{xx}, f_{xy}, and f_{yy} are continuous throughout R, the values of these derivatives at $(a + ch, b + ck)$ will be nearly the same as their values at (a, b) when h and k are small. In particular, if $Q(0) \neq 0$, the sign of $Q(c)$ will be the same as the sign of $Q(0)$ for sufficiently small values of h and k.

Now, we wish to predict the sign of

$$Q(0) = h^2 f_{xx}(a, b) + 2hk f_{xy}(a, b) + k^2 f_{yy}(a, b) \quad (9)$$

from the signs of f_{xx} and $f_{xx}f_{yy} - f_{xy}^2$ at (a, b). We multiply both sides of Eq. (9) by f_{xx} and rearrange the right-hand side to get

$$f_{xx}Q(0) = (hf_{xx} + kf_{xy})^2 + (f_{xx}f_{yy} - f_{xy}^2)k^2. \quad (10)$$

The sign of $Q(0)$ can be determined from this equation, and we are led to the following criteria for the behavior of $f(x, y)$ at (a, b).

1. If $f_{xx} < 0$ and $f_{xx}f_{yy} - f_{xy}^2 > 0$ at (a, b), then $Q(0) < 0$ for all sufficiently small nonzero values of h and k, and f has a *relative maximum* value at (a, b).
2. If $f_{xx} > 0$ and $f_{xx}f_{yy} - f_{xy}^2 > 0$ at (a, b), then $Q(0) > 0$ for all sufficiently small nonzero values of h and k, and f has a *relative minimum* value at (a, b).
3. If $f_{xx}f_{yy} - f_{xy}^2 < 0$ at (a, b), then it can be shown that there are combinations of arbitrarily small nonzero values of h and k for which $Q(0) > 0$, and others for which $Q(0) < 0$. Thus, arbitrarily close to the point $P_0(a, b, f(a, b))$ on the surface $z = f(x, y)$ there are points above P_0, and points below P_0. The function f therefore has a *saddle point* at (a, b).
4. Finally, if $f_{xx}f_{yy} - f_{xy}^2 = 0$, we can draw *no conclusion* about the sign of $Q(c)$, and some other test is needed. It follows from (10) that $f_{xx}Q(0) \geq 0$ for all choices of h and k, but if $Q(0) = 0$ no conclusion can be drawn about the sign of $Q(c)$.

This completes the derivation of the second derivative test.

Derivation of the Error Estimate for Linear Approximations

We now derive the formula

$$E|f(x, y) - T(x, y)| \leq \frac{1}{2}B(|\Delta x| + |\Delta y|)^2, \quad (11)$$

which we used in Article 15.7 to estimate the difference between the values of a function $f(x, y)$ and its linearization $T(x, y)$ near a point (x_0, y_0). We begin by rewriting Eq. (6) in the following way:

$$f(x, y) = \underbrace{f(x_0, y_0) + f_x(x_0, y_0)(x - x_0) + f_y(x_0, y_0)(y - y_0)}_{T(x,\,y)} \tag{12}$$
$$+ \underbrace{\frac{1}{2}(\Delta x^2 f_{xx} + 2\,\Delta x\,\Delta y f_{xy} + \Delta y^2 f_{yy})\big|_{(x_0+c\,\Delta x,\; y_0+c\,\Delta y)}}_{\text{Error } E}$$

The substitutions that produce this equation from Eq. (6) are $(a, b) = (x_0, y_0)$, $h = \Delta x = (x - x_0)$, $k = \Delta y = (y - y_0)$, $x = x_0 + \Delta x$, $y = y_0 + \Delta y$. For Eq. (12) to hold, we assume f and its first and second order partial derivatives to be continuous in some region R whose interior contains (x_0, y_0). The equation calculates the values of f at the points (x, y) in R.

From Eq. (12) we see that the error in using $T(x, y)$ to approximate $f(x, y)$ is precisely

$$f(x, y) - T(x, y) = E.$$

If $|f_{xx}|$, $|f_{xy}|$, and $|f_{yy}|$ are all less than or equal to some number B throughout R (as they will be if they are continuous and R is closed and bounded), then we obtain

$$\begin{aligned}|f(x, y) - T(x, y)| &= |E| \\ &= \frac{1}{2}\left|\,\Delta x^2 f_{xx} + 2\,\Delta x\,\Delta y f_{xy} + \Delta y^2 f_{yy}\right|_{(x_0+c\,\Delta x,\; y_0+c\,\Delta y)} \\ &\le \frac{1}{2}B(|\Delta x| + |\Delta y|)^2,\end{aligned} \tag{13}$$

which is Eq. (11).

Taylor's Theorem for Functions of Two Variables

Equation (2) may be interpreted as saying that applying d/dt to $F(t) = f(a + ht, b + kt)$ gives the same result as applying

$$\left(h\frac{\partial}{\partial x} + k\frac{\partial}{\partial y}\right)$$

to $f(x, y)$. Similarly, Eq. (3) says that applying d^2/dt^2 to $F(t)$ gives the same result as applying

$$\left(h\frac{\partial}{\partial x} + k\frac{\partial}{\partial y}\right)^2 = h^2\frac{\partial^2}{\partial x^2} + 2hk\frac{\partial^2}{\partial x\,\partial y} + k^2\frac{\partial^2}{\partial y^2}$$

to $f(x, y)$. These are the first two instances of a more general formula that says that

$$F^{(n)}(t) = \frac{d^n}{dt^n}F(t) = \left(h\frac{\partial}{\partial x} + k\frac{\partial}{\partial y}\right)^n f(x, y), \tag{14}$$

where the expression in parentheses on the right is to be expanded by the binomial theorem and then applied term by term to $f(x, y)$.

Suppose now that $f(x, y)$ is continuous and has partial derivatives of all orders throughout the region R shown in Fig. 15.41. Then we may extend the Maclaurin series for $F(t)$ to more terms,

$$F(t) = F(0) + F'(0) \cdot t + \frac{F''(0)}{2!}t^2 + \cdots + \frac{F^{(n)}(0)}{n!}t^n + \cdots,$$

and take $t = 1$ to obtain

$$F(1) = F(0) + F'(0) + \frac{F''(0)}{2!} + \cdots + \frac{F^{(n)}(0)}{n!} + \cdots.$$

Finally, when we replace each derivative on the right of this last series by its equivalent expression from Eq. (14) evaluated at $t = 0$, and expand each term by the binomial theorem, we arrive at the following formula:

THEOREM 5

The Taylor Series Expansion of the Function $f(x, y)$ About a Point (a, b).

If $f(x, y)$ is continuous and has partial derivatives of all orders throughout a rectangular region R centered at a point (a, b), then throughout R

$$\begin{aligned} f(a+h, b+k) &= f(a, b) + (hf_x + kf_y)_{(a,b)} \\ &+ \frac{1}{2!}(h^2 f_{xx} + 2hkf_{xy} + k^2 f_{yy})_{(a,b)} \\ &+ \frac{1}{3!}(h^3 f_{xxx} + 3h^2 k f_{xxy} + 3hk^2 f_{xyy} + k^3 f_{yyy})_{(a,b)} \\ &+ \cdots \\ &+ \frac{1}{n!}\left[\left(h\frac{\partial}{\partial x} + k\frac{\partial}{\partial y}\right)^n f\right]_{(a,b)} + \cdots. \end{aligned} \tag{15}$$

This expresses the value of the function $f(x, y)$ at $x = a + h$, $y = b + k$, in terms of the values of the function and its partial derivatives at (a, b), and powers of $h = x - a$ and $k = y - b$. This is the Taylor series expansion, about the point (a, b), of the function $f(x, y)$. Analogous formulas hold for functions of more independent variables.

Summary of Max-Min Tests

If $z = f(x, y)$ is continuous, then the extreme values of f may occur only at

i) boundary points of the domain of f,
ii) interior points where $f_x = f_y = 0$,
iii) points where f_x or f_y fail to exist.

If f has continuous first and second order partial derivatives on some open disc containing (a, b), and if $f_x(a, b) = f_y(a, b) = 0$, then

i) $f_{xx} < 0$ and $f_{xx}f_{yy} - f_{xy}^2 > 0$ at (a, b) $\Rightarrow$ *local maximum,*
ii) $f_{xx} > 0$ and $f_{xx}f_{yy} - f_{xy}^2 > 0$ at (a, b) $\Rightarrow$ *local minimum,*
iii) $f_{xx}f_{yy} - f_{xy}^2 < 0$ at (a, b) $\Rightarrow$ *saddle point,*
iv) $f_{xx}f_{yy} - f_{xy}^2 = 0$ at (a, b) $\Rightarrow$ test is *inconclusive.*

PROBLEMS

Test the following surfaces for maxima, minima, and saddle points. Find the function values at these points.

1. $z = x^2 + xy + y^2 + 3x - 3y + 4$
2. $z = x^2 + 3xy + 3y^2 - 6x + 3y - 6$
3. $z = 5xy - 7x^2 + 3x - 6y + 2$
4. $z = 2xy - 5x^2 - 2y^2 + 4x + 4y - 4$
5. $z = x^2 + xy + 3x + 2y + 5$
6. $z = y^2 + xy - 2x - 2y + 2$
7. $z = 2xy - 5x^2 - 2y^2 + 4x - 4$
8. $z = 2xy - x^2 - 2y^2 + 3x + 4$
9. $z = x^2 + xy + y^2 + 3y + 3$
10. $z = 3x^2 + 6xy + 7y^2 - 2x + 4y$
11. $z = 2x^2 + 3xy + 4y^2 - 5x + 2y$
12. $z = 4x^2 - 6xy + 5y^2 - 20x + 26y$
13. $z = x^2 - 4xy + y^2 + 5x - 2y$
14. $z = x^2 + y^2 - 2x + 4y + 6$
15. $z = x^2 - y^2 - 2x + 4y + 6$
16. $z = x^2 - 2xy + 2y^2 - 2x + 2y + 1$
17. $z = x^2 + 2xy$
18. $z = 3 + 2x + 2y - 2x^2 - 2xy - y^2$
19. $z = x^2 + xy + y^2 + x - 4y + 5$
20. $z = x^2 - xy + y^2 + 2x + 2y - 4$
21. $z = 3x^2 - xy + 2y^2 - 8x + 9y + 10$
22. $z = 5x^2 - 4xy + 2y^2 + 4x - 4y + 10$
23. $z = x^3 - y^3 - 2xy + 6$
24. $z = x^3 + y^3 + 3x^2 - 3y^2 - 8$
25. $z = 6x^2 - 2x^3 + 3y^2 + 6xy$
26. $z = 9x^3 + y^3/3 - 4xy$
27. $z = x^3 + y^3 - 3xy + 15$
28. $z = x^3 + 3xy + y^3$
29. $z = 4xy - x^4 - y^4$
30. $z = x \sin y$

In Problems 31–38, find the absolute maxima and minima of the functions on the given domains.

31. $f(x, y) = 2x^2 - 4x + y^2 - 4y + 1$ on the closed triangular plate bounded by the lines $x = 0$, $y = 2$, $y = 2x$ in the first quadrant.
32. $D(x, y) = x^2 - xy + y^2 + 1$ on the closed triangular plate in the first quadrant bounded by the lines $x = 0$, $y = 4$, $y = x$.
33. $f(x, y) = x^2 + y^2$ on the closed triangular plate bounded by the lines $x = 0$, $y = 0$, $y + 2x = 2$ in the first quadrant.
34. $T(x, y) = x^2 + xy + y^2 - 6x$ on the rectangular plate $0 \le x \le 5$, $-3 \le y \le 3$.
35. $T(x, y) = x^2 + xy + y^2 - 6x + 2$ on the rectangular plate $0 \le x \le 5$, $-3 \le y \le 0$.
36. $f(x, y) = 48xy - 32x^3 - 24y^2$ on the rectangular plate $0 \le x \le 1$, $0 \le y \le 1$.
37. $f(x, y) = (x^2 - 4x) \cos y$ over the region $1 \le x \le 3$, $-\pi/4 \le y \le \pi/4$.
38. $f(x, y) = x - (x - \frac{1}{4})(y + \frac{1}{2})$ on the closed triangular region bounded by the lines $x = 0$, $y = 0$, $x + y = 1$ in the first quadrant.
39. A flat circular plate has the shape of the region $x^2 + y^2 \le 1$. The plate, including the boundary where $x^2 + y^2 = 1$, is heated so that the temperature at any point (x, y) is

$$T(x, y) = x^2 + 2y^2 - x.$$

Find the hottest and coldest points on the plate, and the temperature at each of these points.

40. Find the critical point of

$$f(x, y) = xy + 2x - \ln x^2y$$

in the open first quadrant ($x > 0, y > 0$) and show that f takes on a minimum there.

41. Find all relative and absolute maxima and minima of

$$f(x, y) = -x^2 - y^2 + 2x + 2y + 2$$

on the closed first quadrant $x \ge 0$, $y \ge 0$.

42. Determine the maxima, minima, and saddle points of $f(x, y)$, if any, given that
    a) $f_x = 2x - 4y$, $f_y = 2y - 4x$,
    b) $f_x = 2(x - 1)$, $f_y = 2(y - 2)$,
    c) $f_x = 9x^2 - 9$, $f_y = 2y + 4$.
43. Sketch the surface

$$z = \sqrt{x^2 + y^2}$$

over the region R: $|x| \le 1$, $|y| \le 1$. Find the high and low points of the surface over R. Discuss the existence, and the values, of $\partial z/\partial x$ and $\partial z/\partial y$ at these points.

44. The discriminant $f_{xx}f_{yy} - f_{xy}^2$ of each of the following functions is zero at the origin. Determine whether the function has a maximum, a minimum, or neither at the origin by imagining in each case what the surface $z = f(x, y)$ looks like.
    a) $f(x, y) = x^2y^2$
    b) $f(x, y) = 1 - x^2y^2$
    c) $f(x, y) = xy^2$
    d) $f(x, y) = x^3y^2$
    e) $f(x, y) = x^3y^3$
    f) $f(x, y) = x^4y^4$

45. *The error bound for linear approximations of functions of three variables* (Article 15.8, Eq. (23)).

a) Suppose that $f(x, y, z)$ and its first and second order partial derivatives are continuous throughout a region R whose interior contains the point $P(a, b, d)$. Suppose also that the increments h, k, and m are small enough for the line segment PS joining P to the point $S(a + h, b + k, d + m)$ to lie in the interior of R. Parameterize PS by the equations

$$x = a + th,$$
$$y = b + tk,$$
$$z = d + tm, \qquad 0 \le t \le 1.$$

Apply Taylor's theorem to the function

$$F(t) = f(a + th, b + tk, d + tm)$$

to show that

$$\begin{aligned} f(a + h, b + k, d + m) &= f(a, b, c) + hf_x(a, b, c) + kf_y(a, b, c) + mf_z(a, b, c) \\ &\quad + \frac{1}{2}\Big[h^2f_{xx} + k^2f_{yy} + m^2f_{zz} + 2hkf_{xy} \\ &\quad + 2hmf_{xz} + 2kmf_{yz}\Big]_{(a+ch,\, b+ck,\, d+cm)} \end{aligned} \qquad (6')$$

for some c between 0 and 1. This is the three-variable version of Eq. (6) of the present article.

b) Use Eq. (6′) from part (a) to derive the inequality in Article 15.8, Eq. (23).

***Toolkit* programs**
3D Grapher

15.10 Lagrange Multipliers

Constrained Maxima and Minima

As we saw in Article 15.9, it is sometimes necessary to find the extreme values of a function $f(x, y)$ when its domain is subject to some kind of constraint, for example, that it be a particular triangular plate or quadrant of the plane. But, as Fig. 15.42 suggests, functions may be subject to other kinds of constraints as well. In this article, we explore an important method of finding the maxima and minima of constrained functions—the method of *Lagrange multipliers*. We set the stage with two examples, and then discuss the method in general terms and look at more examples.

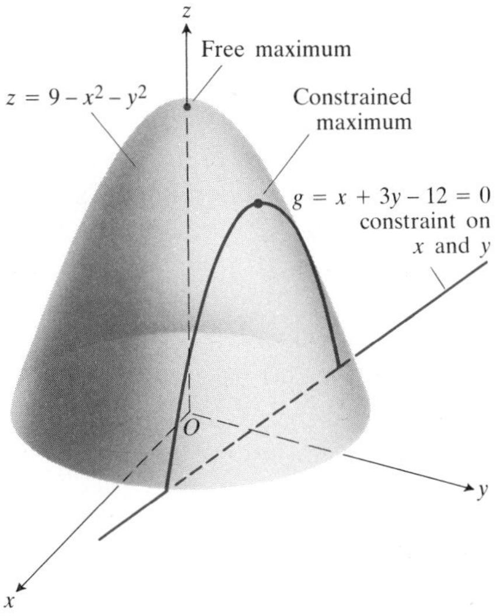

15.42 $f(x, y) = 9 - x^2 - y^2$, subject to the constraint $g(x, y) = x + 3y - 12 = 0$.

EXAMPLE 1 Find the point $P(x, y, z)$ on the plane

$$2x + y - z - 5 = 0$$

that lies closest to the origin.

Solution The problem asks us to find the minimum value of the function

$$|\overrightarrow{OP}| = \sqrt{(x - 0)^2 + (y - 0)^2 + (z - 0)^2} = \sqrt{x^2 + y^2 + z^2}$$

subject to the constraint that

$$2x + y - z - 5 = 0.$$

Since $|\overrightarrow{OP}|$ has a minimum value wherever the function

$$f(x, y, z) = x^2 + y^2 + z^2$$

has a minimum value, we may solve the problem by finding the minimum value of $f(x, y, z)$ subject to the constraint

$$z = 2x + y - 5.$$

This leads to finding the points (x, y) at which the function

$$h(x, y) = f(x, y, 2x + y - 5) = x^2 + y^2 + (2x + y - 5)^2$$

has its minimum value or values. Since the domain of h is the entire xy-plane, the first derivative test of the preceding article tells us that any minima h might have must occur at points where

$$h_x = h_y = 0.$$

This leads to

$$10x + 4y = 20,$$
$$4x + 4y = 10,$$

and the solution.

$$x = \frac{5}{3}, \qquad y = \frac{5}{6}.$$

We may apply either geometric arguments or the second derivative test to show that these values minimize h. The z-coordinate of the corresponding point on the plane $z = 2x + y - 5$ is

$$z = 2\left(\frac{5}{3}\right) + \frac{5}{6} - 5 = -\frac{5}{6}.$$

Therefore, the point we seek is

$$\text{Closest point: } P\left(\frac{5}{3}, \frac{5}{6}, -\frac{5}{6}\right). \ \square$$

Attempts to solve a constrained maximum or minimum problem by substitution, as we might call the method of Example 1, do not always go smoothly, as the next example shows. This is one of the reasons for learning the new method of this article, which does not require us to decide in advance which of the constrained variables to regard as independent.

EXAMPLE 2 Find the minimum distance from the origin to the surface

$$x^2 - z^2 - 1 = 0.$$

Solution 1 We begin by looking for the points $P(x, y, z)$ on the surface that are closest to the origin. To find them, we look for the points $P(x, y, z)$ in space that minimize the square of the distance

$$|\overrightarrow{OP}|^2 = f(x, y, z) = x^2 + y^2 + z^2$$

subject to the constraint that

$$x^2 - z^2 - 1 = 0 \qquad \text{or} \qquad z^2 = x^2 - 1.$$

The values of f at points on this surface are given by the function

$$h(x, y) = x^2 + y^2 + (x^2 - 1) = 2x^2 + y^2 - 1.$$

The only extreme value of h occurs at (0, 0), where

$$h_x = 4x = 0 \qquad \text{and} \qquad h_y = 2y = 0.$$

But here we run into trouble because $z^2 = x^2 - 1 = -1$ means that z is imaginary when $x = 0$. A point $P(x, y, z)$ can lie on the surface $z^2 = x^2 - 1$ only if $|x| \geq 1$. What went wrong?

What happened was that the first derivative test found (as it should have) the point *in the domain of* h where h has a minimum value, but we are looking for points *on the surface* where h has a minimum value.

One way to go from here is to substitute $x^2 = z^2 + 1$ into the formula $f(x, y, z) = x^2 + y^2 + z^2$ and look for points where

$$k(y, z) = (z^2 + 1) + y^2 + z^2 = 1 + y^2 + 2z^2$$

takes on its smallest values. These occur where

$$k_y = 2y = 0 \qquad \text{and} \qquad k_z = 4z = 0,$$

or where

$$y = z = 0.$$

This leads to

$$x^2 = z^2 + 1 = 1, \qquad x = \pm 1.$$

The resulting points, $(\pm 1, 0, 0)$, are now on the surface. Moreover, it is clear from the inequality

$$k(y, z) = 1 + y^2 + 2z^2 \geq 1$$

that the points $(\pm 1, 0, 0)$ give a minimum of k. The minimum distance from the origin to the surface is therefore 1 unit. The surface is shown in Fig. 15.43.

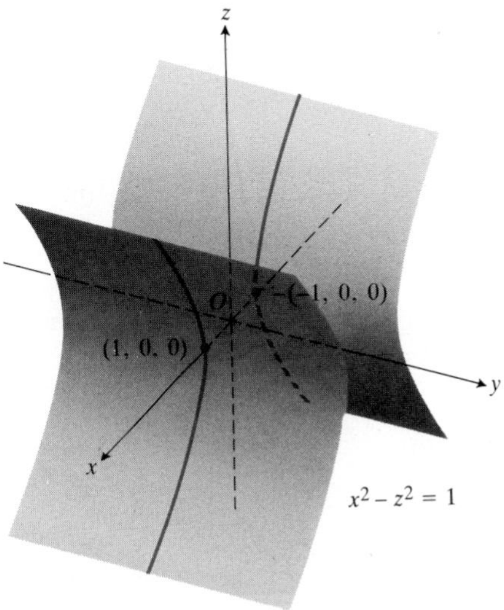

15.43 The surface $x^2 - z^2 = 1$.

Solution 2 Another way to find the minimum distance from the origin to the surface is to start with a small sphere centered at the origin and let it expand like a soap bubble until it just touches the surface. At the two points of contact, the surface and sphere have the same tangent planes and normal lines. Therefore, if the surface and sphere are represented as level surfaces

$$g(x, y, z) = x^2 - z^2 - 1 = 0,$$
$$f(x, y, z) = x^2 + y^2 + z^2 - 1 = 0$$

of the functions g and f, then the gradients ∇g and ∇f will be parallel at the two points of contact. In other words, at each point of contact we will be able to find a scalar λ such that

$$\nabla f = \lambda \nabla g,$$

or

$$2x\mathbf{i} + 2y\mathbf{j} + 2z\mathbf{k} = \lambda(2x\mathbf{i} - 2z\mathbf{k}).$$

Thus, the coordinates of either point of tangency will have to satisfy the three scalar equations

$$2x = \lambda 2x, \tag{1a}$$
$$2y = \lambda \cdot 0 \tag{1b}$$
$$2z = -2z\lambda, \tag{1c}$$

or

$$x(2 - 2\lambda) = 0, \tag{2a}$$
$$y = 0, \tag{2b}$$
$$z(2 + 2\lambda) = 0. \tag{2c}$$

For what values of λ will a point $P(x, y, z)$ whose coordinates satisfy Eqs. (2a)–(2c) lie on the surface $x^2 - z^2 - 1 = 0$? Since every point on the surface has $x \neq 0$, we may divide both sides of Eq. (2a) by x to obtain

$$2 - 2\lambda = 0, \qquad \lambda = 1.$$

For $\lambda = 1$, Eq. (2c) gives

$$z(2 + 2 \cdot 1) = 0, \qquad z = 0.$$

Since $y = 0$ also (from Eq. 2b) the points we seek on the surface all have coordinates of the form

$$(x, 0, 0),$$

which means that $x = \pm 1$. As in Solution 1, we find that the points on the surface closest to the origin are $(\pm 1, 0, 0)$, and the minimum distance is one unit. □

The Method of Lagrange Multipliers

In Solution 2 above we solved the problem presented in Example 2 by the *method of Lagrange multipliers*. In general terms, the method says that the extreme values of a function $f(x, y, z)$ whose variables are subject to a constraint of the form $g(x, y, z) = 0$ are to be found on the surface $g = 0$ at the points where

$$\nabla f = \lambda \nabla g$$

for some scalar λ (called a *Lagrange multiplier*).

To explore the method further and see why it works, we first make the following observation, which we state as a theorem.

THEOREM 6

> Suppose that $f(x, y, z)$ has continuous first partial derivatives in a region that contains the differentiable curve
>
> $$C: \quad \mathbf{R}(t) = x(t)\mathbf{i} + y(t)\mathbf{j} + z(t)\mathbf{k}.$$
>
> If P_0 is a point on C where f has a local maximum or minimum relative to its values on C, then ∇f is perpendicular to C at P_0.

Proof. We show that $\nabla f \cdot \mathbf{v} = 0$ at the points in question, where $\mathbf{v} = d\mathbf{R}/dt$ is the tangent vector to C. The values of f on C are given by the composite function

$$f(x(t), y(t), z(t)),$$

whose derivative with respect to t is

$$\frac{df}{dt} = \frac{\partial f}{\partial x}\frac{dx}{dt} + \frac{\partial f}{\partial y}\frac{dy}{dt} + \frac{\partial f}{\partial z}\frac{dz}{dt} = \nabla f \cdot \mathbf{v}.$$

At any point P_0 where f has a local maximum or minimum relative to its values on the curve, $df/dt = 0$, so that

$$\nabla f \cdot \mathbf{v} = 0. \blacksquare$$

By dropping the z-terms in the statement and proof of Theorem 6 we obtain a similar result for functions of two variables: The points on a differentiable curve

$$\mathbf{R}(t) = x(t)\mathbf{i} + y(t)\mathbf{j}$$

where a function $f(x, y)$ takes on its local maxima and minima relative to its values on the curve are the points where $\nabla f \cdot d\mathbf{R}/dt = 0$.

Theorem 6 is the key to why the method of Lagrange multipliers works, as we shall now see. Suppose that $f(x, y, z)$ and $g(x, y, z)$ have continuous first partial derivatives and that P_0 is a point on the surface $g(x, y, z) = 0$ where f has a local maximum or minimum value relative to its other values on the surface. Then f takes on a local maximum or minimum at P_0 relative to its values on every differentiable curve through P_0 on the surface $g(x, y, z) = 0$. Therefore, ∇f is perpendicular to the velocity vector of every such differentiable curve through P_0. But so is ∇g (because ∇g is perpendicular to the level surface $g = 0$, as we saw in Article 15.6). Therefore, at P_0, ∇f is some scalar multiple λ of ∇g.

The argument for the two-variable case is left as an exercise (Problem 36).

The Method of Lagrange Multipliers

Suppose that $f(x, y, z)$ and $g(x, y, z)$ have continuous partial derivatives. To find the local maximum and minimum values of f subject to the constraint $g(x, y, z) = 0$, find the values of x, y, z, and λ that satisfy the equations

$$\nabla f = \lambda \nabla g \quad \text{and} \quad g(x, y, z) = 0 \tag{3}$$

simultaneously.

Comment on notation. Some books describe the method of Lagrange multipliers without vector notation in the following equivalent way:

To maximize or minimize a function $f(x, y, z)$ subject to the constraint $g(x, y, z) = 0$, construct the auxiliary function

$$H(x, y, z, \lambda) = f(x, y, z) - \lambda g(x, y, z). \tag{4}$$

Then find the values of x, y, z, and λ for which the partial derivatives of H are all zero:

$$H_x = 0, \quad H_y = 0, \quad H_z = 0, \quad H_\lambda = 0. \tag{5}$$

These requirements are equivalent to the requirements in (3), as we can see by calculating

$$\begin{aligned} H_x &= f_x - \lambda g_x = 0 &\text{or}\quad & f_x = \lambda g_x, \\ H_y &= f_y - \lambda g_y = 0 &\text{or}\quad & f_y = \lambda g_y, \\ H_z &= f_z - \lambda g_z = 0 &\text{or}\quad & f_z = \lambda g_z, \\ H_\lambda &= -g(x, y, z) = 0 &\text{or}\quad & g(x, y, z) = 0. \end{aligned} \tag{6}$$

The first three equations give $\nabla f = \lambda \nabla g$, and the last is $g(x, y, z) = 0$.

EXAMPLE 3 Find the greatest and smallest values that the function

$$f(x, y) = xy$$

takes on the ellipse

$$\frac{x^2}{8} + \frac{y^2}{2} = 1.$$

Solution We are asked to find the extreme values of $f(x, y) = xy$ subject to the constraint

$$g(x, y) = \frac{x^2}{8} + \frac{y^2}{2} - 1 = 0.$$

To do so, we first find the values of x, y, and λ for which

$$\nabla f = \lambda \nabla g \qquad \text{and} \qquad g(x, y) = 0.$$

The gradient equation gives

$$y\mathbf{i} + x\mathbf{j} = \frac{\lambda}{4}x\mathbf{i} + \lambda y\mathbf{j},$$

from which we find

$$y = \frac{\lambda}{4}x, \qquad x = \lambda y,$$

and

$$y = \frac{\lambda}{4}(\lambda y) = \frac{\lambda^2}{4}y, \quad \text{so that} \quad \lambda = \pm 2 \quad \text{or} \quad y = 0.$$

We now consider two cases.

CASE 1. If $y = 0$, then $x = y = 0$, but the point (0, 0) is not on the ellipse. Hence, $y \neq 0$.

CASE 2. If $y \neq 0$, then

$$x = \pm 2y.$$

Substituting this in the equation $g(x, y) = 0$ gives

$$\frac{(+2y)^2}{8} + \frac{y^2}{2} = 1$$

$$4y^2 + 4y^2 = 8$$

$$y = \pm 1.$$

The function $f(x, y) = xy$ therefore takes on its extreme values on the ellipse at the four points $(\pm 2, 1)$, $(\pm 2, -1)$, and the extreme values are $xy = 2$ and $xy = -2$. □

Geometric interpretation of the solution (see Fig. 15.44). The level curves of the function $f(x, y) = xy$ are the hyperbolas $xy = c$. The farther they are from the origin, the larger the absolute value of f. We want to find the extreme values of $f(x, y)$, given that the point (x, y) also lies on the ellipse $x^2 + 4y^2 = 8$. Which hyperbolas intersecting the ellipse are farthest from

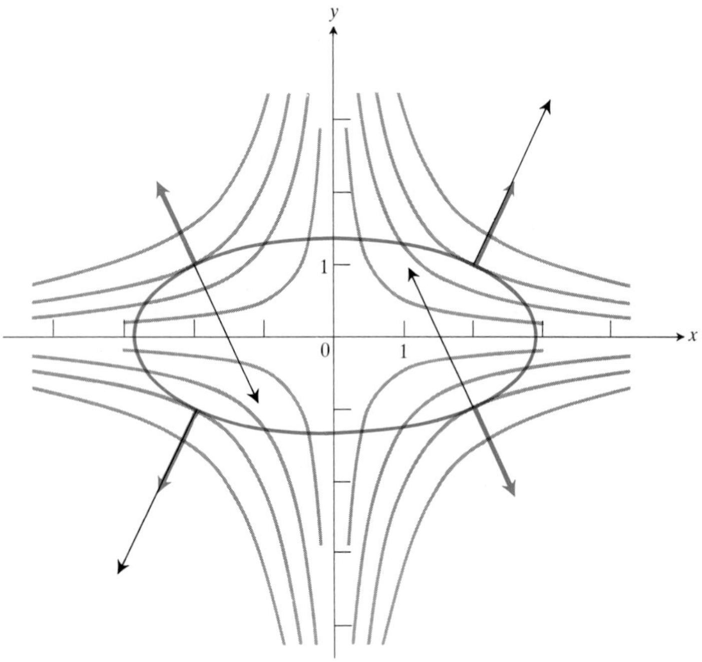

15.44 When subjected to the constraint $g(x, y) = x^2/8 + y^2/2 - 1 = 0$, the function $f(x, y) = xy$ takes on extreme values at the four points $(\pm 2, \pm 1)$.

the origin? The hyperbolas that just graze the ellipse, the ones that are tangent to it. At these points,

a) the normal to the hyperbola is normal to the ellipse;
b) the gradient

$$\nabla f = y\mathbf{i} + x\mathbf{j}$$

is a multiple ($\lambda = \pm 2$) of the gradient

$$\nabla g = \frac{x}{4}\mathbf{i} + y\mathbf{j}.$$

For example, at the point (2, 1),

$$\nabla f = \mathbf{i} + 2\mathbf{j} \quad \text{and} \quad \nabla g = \tfrac{1}{2}\mathbf{i} + \mathbf{j},$$

so that $\nabla f = 2\nabla g$. At the point $(-2, 1)$,

$$\nabla f = \mathbf{i} - 2\mathbf{j} \quad \text{and} \quad \nabla g = -\tfrac{1}{2}\mathbf{i} + \mathbf{j},$$

so that $\nabla f = -2\nabla g$.

EXAMPLE 4 Let a and b be positive constants. Find the maximum and minimum values of the function

$$ax + by$$

subject to the constraint

$$x^2 + y^2 = 1.$$

Solution We model this as a Lagrange multiplier problem with

$$f(x, y) = ax + by, \qquad g(x, y) = x^2 + y^2 - 1$$

and find the values of x, y, and λ that satisfy the equations

$$\nabla f = \lambda \nabla g: \qquad a\mathbf{i} + b\mathbf{j} = 2x\lambda\,\mathbf{i} + 2y\lambda\,\mathbf{j},$$
$$g(x, y) = 0: \qquad x^2 + y^2 - 1 = 0.$$

The gradient equation implies that $\lambda \neq 0$, and gives

$$x = \frac{a}{2\lambda}, \qquad y = \frac{b}{2\lambda}.$$

Note that x and y have the same sign because $a > 0$ and $b > 0$. With these values, the equation $g(x, y) = 0$ gives

$$\frac{a^2}{4\lambda^2} + \frac{b^2}{4\lambda^2} = 1$$
$$a^2 + b^2 = 4\lambda^2$$
$$\lambda = \pm\frac{\sqrt{a^2 + b^2}}{2}.$$

Thus,

$$x = \frac{a}{2\lambda} = \pm\frac{a}{\sqrt{a^2 + b^2}}, \qquad y = \frac{b}{2\lambda} = \pm\frac{b}{\sqrt{a^2 + b^2}},$$

and $f(x, y) = ax + by$ has its extreme values at the two points

$$(x, y) = \pm\left(\frac{a}{\sqrt{a^2 + b^2}}, \frac{b}{\sqrt{a^2 + b^2}}\right).$$

By calculating the values of f at these points we see that its maximum value on the circle $x^2 + y^2 = 1$ is

$$\frac{a^2}{\sqrt{a^2 + b^2}} + \frac{b^2}{\sqrt{a^2 + b^2}} = \sqrt{a^2 + b^2}, \tag{7a}$$

and its minimum value is

$$-\frac{a^2}{\sqrt{a^2 + b^2}} - \frac{b^2}{\sqrt{a^2 + b^2}} = -\sqrt{a^2 + b^2}. \ \Box \tag{7b}$$

Lagrange Multipliers with Two Constraints

If there are two constraints on the variables of $f(x, y, z)$, say

$$g(x, y, z) = 0 \qquad \text{and} \qquad h(x, y, z) = 0,$$

we may find the constrained relative maxima and minima of f by introducing two Lagrange multipliers λ and μ. That is, we locate the points $P(x, y, z)$ where f takes on its constrained extreme values by finding the values of x, y, z, λ, and μ that satisfy the equations

$$\nabla f = \lambda \nabla g + \mu \nabla h, \tag{8a}$$
$$g(x, y, z) = 0, \tag{8b}$$
$$h(x, y, z) = 0 \tag{8c}$$

simultaneously. The functions f, g, and h are required to have continuous first partial derivatives.

The equations in (8) have a nice geometric interpretation. The surfaces $g = 0$ and $h = 0$ (usually) intersect in a differentiable curve, say C (Fig.

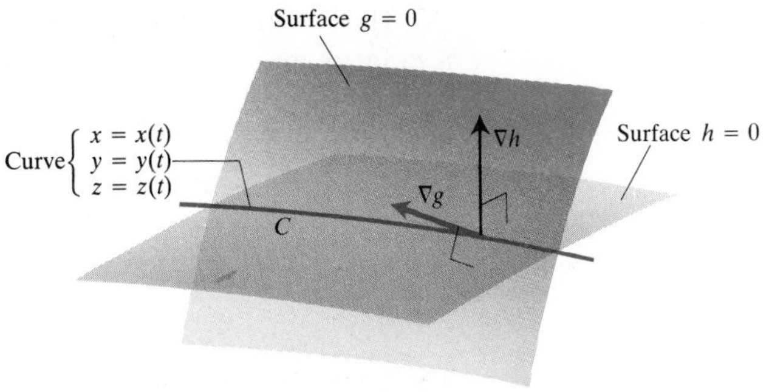

15.45 Vectors ∇g and ∇h are in a plane perpendicular to curve C because ∇g is normal to the surface $g = 0$, and ∇h is normal to the surface $h = 0$.

15.45), and along this curve we seek the points where f has local maximum and minimum values relative to its other values on the curve. These are the points where ∇f is normal to C, as we saw in Theorem 6. But ∇g and ∇h are also normal to C at these points because C lies in the surfaces $g = 0$ and $h = 0$. Therefore ∇f lies in the plane determined by ∇g and ∇h, which means that

$$\nabla f = \lambda \nabla g + \mu \nabla h$$

for some λ and μ. Since the points we seek also lie in both surfaces, their coordinates must satisfy the equations

$$g(x, y, z) = 0 \qquad \text{and} \qquad h(x, y, z) = 0,$$

which are the remaining requirements in (8).

EXAMPLE 5 The cone $z^2 = x^2 + y^2$ is cut by the plane $z = 1 + x + y$ in a conic section C. Find the points on C that are nearest to, and farthest from, the origin.

Solution We model this as a Lagrange multiplier problem in which we find the extreme values of

$$f(x, y, z) = x^2 + y^2 + z^2$$

(the square of the function that measures distance from the origin) subject to the two constraints

$$g(x, y, z) = x^2 + y^2 - z^2 = 0, \tag{9a}$$
$$h(x, y, z) = 1 + x + y - z = 0. \tag{9b}$$

The gradient equation (8a) gives

$$2x\mathbf{i} + 2y\mathbf{j} + 2z\mathbf{k} = \lambda(2x\mathbf{i} + 2y\mathbf{j} - 2z\mathbf{k}) + \mu(\mathbf{i} + \mathbf{j} - \mathbf{k}),$$

which leads to the scalar equations

$$\left.\begin{aligned} 2x &= 2x\lambda + \mu \\ 2y &= 2y\lambda + \mu \end{aligned}\right\} \Rightarrow x - y = (x - y)\lambda, \tag{10a}$$
$$\left.\begin{aligned} 2y &= 2y\lambda + \mu \\ 2z &= -2z\lambda - \mu \end{aligned}\right\} \Rightarrow y + z = (y - z)\lambda. \tag{10b}$$

The equation $x - y = (x - y)\lambda$ is satisfied if $x = y$ or if $x \neq y$ and $\lambda = 1$.

The case $\lambda = 1$ does not lead to a point on the cutting plane (that is, to a point whose coordinates satisfy Eq. 9b). For $\lambda = 1$ implies

$$y + z = y - z \qquad \text{(from Eq. 10b)}$$
$$2z = 0$$
$$z = 0$$

and Eq. (9a) then gives

$$x^2 + y^2 = 0, \qquad x = 0, \qquad y = 0.$$

The point (0, 0, 0) does not satisfy the constraint (9b).

Therefore $\lambda \neq 1$ and we have $x = y$. The plane $x = y$ meets the plane $z = 1 + x + y$ in a line that cuts the cone in just two points, as we can see by substituting $y = x$ and $z = 1 + 2x$ into the cone equations:

$$z^2 = x^2 + y^2$$
$$(1 + 2x)^2 = x^2 + x^2$$
$$2x^2 + 4x + 1 = 0$$
$$x = -1 \pm \frac{\sqrt{2}}{2}.$$

The points are

$$A = (-1 - \sqrt{\tfrac{1}{2}}, -1 - \sqrt{\tfrac{1}{2}}, -1 - \sqrt{2}), \qquad (11a)$$
$$B = (-1 + \sqrt{\tfrac{1}{2}}, -1 + \sqrt{\tfrac{1}{2}}, -1 + \sqrt{2}). \qquad (11b)$$

Now, we know that C is either an ellipse or a hyperbola. If it is an ellipse, we would conclude that B is the point on it nearest the origin, and A the point farthest from the origin. But if it is a hyperbola, then there is no point on it that is farthest away from the origin, and the points A and B are the points on the two branches that are nearest the origin. Problem 33 asks you to think about these possibilities and decide between them. It seems obvious that the critical points should satisfy the condition $x = y$ because all three of the functions f, g, and h treat x and y alike. □

PROBLEMS

1. Find the points on the ellipse $x^2 + 2y^2 = 1$ where $f(x, y) = xy$ has its extreme values.
2. Find the extreme values of $f(x, y) = xy$ subject to the constraint $g(x, y) = x^2 + y^2 - 10 = 0$.
3. Find the maximum value of $f(x, y) = 9 - x^2 - y^2$ on the line $x + 3y = 12$ (Fig. 15.42).
4. Find the minimum distance between the line $y = x + 1$ and the parabola $y^2 = x$.
5. Find the extreme values of $f(x, y) = x^2y$ on the line $x + y = 3$.
6. Find the points on the curve $x^2y = 2$ nearest the origin.
7. Use the method of Lagrange multipliers to find (a) the minimum value of $x + y$, subject to the constraint $xy = 16$; (b) the maximum value of xy, subject to the constraint $x + y = 16$. Comment on the geometry of each solution.
8. Find the points on the curve $x^2 + xy + y^2 = 1$ that are nearest to and farthest from the origin.
9. Find the dimensions of the closed circular can of smallest surface area whose volume is 16π cm^3.
10. Use the method of Lagrange multipliers to find the dimensions of the rectangle of greatest area that can be inscribed in the ellipse $x^2/16 + y^2/9 = 1$ with sides parallel to the coordinate axes.
11. The temperature at a point (x, y) on a metal plate is $T(x, y) = 4x^2 - 4xy + y^2$. An ant on the plate walks around the circle of radius 5 centered at the origin.

What are the highest and lowest temperatures encountered by the ant?

12. A horizontal water tank is to be constructed in the form of a cylinder with hemispherical ends. Find the diameter and the length of the cylindrical portion of the tank if the tank is to hold 8000 cubic feet of water and the least amount of material is to be used in constructing the tank.

13. Find the maximum and minimum values of $x^2 + y^2$ subject to the constraint $x^2 - 2x + y^2 - 4y = 0$.

14. A pentagon is made by mounting an isosceles triangle on top of a rectangle. What dimensions minimize the perimeter for a given area A?

15. Find the point on the plane $x + 2y + 3z = 13$ closest to the point $(1, 1, 1)$.

16. Find the maximum and minimum values of

$$f(x, y, z) = x - 2y + 5z$$

on the sphere

$$x^2 + y^2 + z^2 = 30.$$

17. Find the minimum distance from the surface $x^2 + y^2 - z^2 = 1$ to the origin.

18. Find the point on the surface $z = xy + 1$ nearest the origin.

19. Find the points on the surface $z^2 = xy + 4$ closest to the origin.

20. Find the points on the sphere $x^2 + y^2 + z^2 = 25$ where $f(x, y, z) = x + 2y + 3z$ has its maximum and minimum values.

21. The temperature T at any point (x, y, z) in space is $T = 400xyz^2$. Find the highest temperature on the unit sphere

$$x^2 + y^2 + z^2 = 1.$$

22. Find three real numbers whose sum is 9 and the sum of whose squares is as small as possible.

23. Find the largest product the numbers x, y, and z can have if $x + y + z^2 = 16$.

24. If a, b, and c are positive numbers, find the maximum value that $f(x, y, z) = ax + by + cz$ can take on the sphere $x^2 + y^2 + z^2 = 1$.

25. Find the maximum and minimum values of $f(x, y) = x^2 + 3y^2 + 2y$ on the unit disc $x^2 + y^2 \leq 1$.

26. Find the extreme values of $f(x, y, z) = x^2yz + 1$ on the intersection of the plane $z = 1$ with the sphere $x^2 + y^2 + z^2 = 2$.

27. A space probe in the shape of the ellipsoid

$$4x^2 + y^2 + 4z^2 = 16$$

enters the earth's atmosphere and its surface begins to heat. After one hour, the temperature at the point (x, y, z) on the probe's surface is

$$T(x, y, z) = 8x^2 + 4yz - 16z + 600.$$

Find the hottest point on the probe's surface.

28. You are in charge of erecting a radio telescope on a newly discovered planet. To minimize interference, you want to place it where the magnetic field of the planet is weakest. The planet is spherical with a radius of 6 units. The strength of the magnetic field is given by $M(x, y, z) = 6x - y^2 + xz + 60$ based on a coordinate system whose origin is at the center of the planet. Where should you locate the radio telescope?

29. Given n positive numbers $a_1, a_2, \ldots, a_n$, find the maximum value of the expression

$$w = a_1x_1 + a_2x_2 + \cdots + a_nx_n$$

if the variables $x_1, x_2, \ldots, x_n$ are restricted so that the sum of their squares is 1.

30. A plane of the form

$$z = Ax + By + C$$

is to be "fitted" to the following points (x_i, y_i, z_i):

$$(0, 0, 0), \quad (0, 1, 1), \quad (1, 1, 1), \quad (1, 0, -1).$$

Find the plane that minimizes the sum of squares of the deviations

$$\sum_{i=1}^{4} (Ax_i + By_i + C - z_i)^2.$$

31. Find the maximum value of $f(x, y, z) = x^2 + 2y - z^2$ subject to the constraints $2x - y = 0$ and $y + z = 0$.

32. a) Find the maximum value of $w = xyz$ among all points lying on the intersection of the two planes $x + y + z = 40$ and $z = x + y$.
b) Give a geometric argument supporting the fact that you have found a maximum (and not a minimum) value of xyz subject to the constraints.

33. In the solution of Example 5, two points A and B were located as candidates for maximum or minimum distances from the origin. Use cylindrical coordinates r, θ, z to express the equations of the cone, the plane $z = 1 + x + y$, and the cylinder that contains their curve of intersection and has elements parallel to the z-axis. Is this cylinder circular, elliptical, parabolic, or hyperbolic? (Consider its intersection with the xy-plane.) Express the distance from the origin to a point on the curve of intersection of the cone and the plane as a function of θ. Does it have a minimum? A maximum? What can you now say about the points A and B of Eqs. (11a, b) as solutions in Example 5?

34. In Example 5, the extrema for $f(x, y, z)$ on the cone $z^2 = x^2 + y^2$ and the plane $z = 1 + x + y$ were

found to satisfy $y = x$ as well. Thus $z = 1 + 2x$, and the function to be made a maximum or minimum is $x^2 + y^2 + z^2 = x^2 + x^2 + (1 + 2x)^2$, which can also be written as $6(x + \frac{1}{3})^2 + \frac{1}{3}$. This is obviously a minimum when $x = -\frac{1}{3}$. But the point we get with

$$y = x, \qquad z = 1 + 2x, \qquad x = -\frac{1}{3},$$

is $(-\frac{1}{3}, -\frac{1}{3}, \frac{1}{3})$, which is not on the cone. What's wrong? (A sketch of the situation in the plane $y = x$ may throw some light on the question.)

35. In Example 5, we can determine that for all points on the cone $z^2 = x^2 + y^2$, the square of the distance from the origin to $P(x, y, z)$ is $w = 2(x^2 + y^2)$, which is a function of two independent variables, x and y. But if P is also to be on the plane

$$z = 1 + x + y$$

as well as the cone, then

$$(1 + x + y)^2 = x^2 + y^2.$$

Show that these points have coordinates that satisfy the equation

$$2xy + 2x + 2y + 1 = 0.$$

Interpret this equation in two ways:

a) as a curve in the xy-plane, and

b) as a set of points on a cylinder in 3-space.

Sketch the curve of (a) and find the point or points on it for which w is a minimum. Are there points on this curve for which w is a maximum? Use the information you now have to complete the discussion of Example 5.

36. *An argument for the two-variable version of the method of Lagrange multipliers.* Suppose that $f(x, y)$ and $g(x, y)$ have continuous first partial derivatives and that we wish to find the relative maximum and minimum values of f subject to the constraint that $g(x, y) = 0$. To find these values of f we find the values of x, y, and λ that satisfy the equations

$$\nabla f = \lambda \nabla g \qquad \text{and} \qquad g(x, y) = 0$$

simultaneously.

To prove this, let P_0 be a point on the curve $g(x, y) = 0$ at which f has a local maximum or minimum value relative to its values on the curve. Then carry out the following steps:

STEP 1. Let $x = x(t)$, $y = y(t)$ be a parameterization of the curve $g(x, y) = 0$ in an interval about P_0. Show that $df/dt = 0$ at P_0 and therefore that ∇f is perpendicular to the velocity vector of the curve at P_0. (Assume that neither vector vanishes at P_0.)

STEP 2. Combine the result in Step 1 above with the result of Problem 81 in Article 15.6 to show that ∇f and ∇g are parallel at P_0, and therefore that $\nabla f = \lambda \nabla g$ at P_0 for some number λ.

15.11 Exact Differentials

The Differential of a Function

If $f(x, y)$ and its first order partial derivatives are continuous, and $x = x(t)$, $y = y(t)$ are differentiable functions of t, then we know from the chain rule that f is a differentiable function of t and that

$$\frac{df}{dt} = \frac{\partial f}{\partial x}\frac{dx}{dt} + \frac{\partial f}{\partial y}\frac{dy}{dt}. \tag{1}$$

This equation is sometimes written in differential form as

$$df = \frac{\partial f}{\partial x}\,dx + \frac{\partial f}{\partial y}\,dy, \tag{2}$$

which is like the familiar

$$du = g'(x)\,dx \tag{3}$$

for the differential of a function $u = g(x)$ of a single variable.

In many problems we must recover u from $g'(x)$ by integrating both sides of Eq. (3) to get

$$u = \int g'(x)\,dx, \tag{4}$$

which determines $u = g(x)$ up to a constant. In like manner, we can determine $f(x, y)$ up to a constant from the equation

$$f = \int df = \int \frac{\partial f}{\partial x}\,dx + \int \frac{\partial f}{\partial y}\,dy, \tag{5}$$

in which the integration

$$\int \frac{\partial f}{\partial x}\,dx$$

reverses the partial differentiation with respect to x, and

$$\int \frac{\partial f}{\partial y}\,dy$$

reverses the partial differentiation with respect to y. The following example shows how this is done.

EXAMPLE 1 Find $f(x, y)$ if

$$df = (x^2 + y^2)\,dx + (2xy + 1)\,dy.$$

Discussion. We must try to recover f from the information that

$$\frac{\partial f}{\partial x} = x^2 + y^2 \qquad \text{and} \qquad \frac{\partial f}{\partial y} = 2xy + 1, \tag{6}$$

for this is all we know about f. The first two solutions we give reconstruct f by a method we shall use again in discussing Theorem 7 below. Once you have read them you will be ready for the third solution, which contains a shortcut that makes it the fastest method of all.

Solution 1 (Preview of the construction in Theorem 7.) The partial derivative

$$\frac{\partial f}{\partial x} = x^2 + y^2$$

was calculated by holding y constant and differentiating with respect to x. We may therefore find f by holding y at a constant value and integrating with respect to x. When we do so we get

$$f(x, y) = \int_{y\ \text{const.}} (x^2 + y^2)\,dx = \frac{x^3}{3} + y^2x + k(y), \tag{7}$$

where the constant of integration $k(y)$ is written as a function of y because its value may change with each new y.

To determine $k(y)$, we calculate $\partial f/\partial y$ from (7) and set the result equal to the given partial derivative, $2xy + 1$:

$$\begin{aligned} \frac{\partial}{\partial y}\left(\frac{x^3}{3} + y^2x + k(y)\right) &= 2xy + 1 \\ 2xy + k'(y) &= 2xy + 1 \\ k'(y) &= 1. \end{aligned} \tag{8}$$

This shows that

$$k(y) = y + C$$

and

$$f(x, y) = \frac{x^3}{3} + xy^2 + y + C.$$

We have recovered f up to a constant.

Solution 2 (Like Solution 1, but we integrate first with respect to y.) We begin our recovery of f by integrating $\partial f/\partial y = 2xy + 1$ with respect to y, holding x constant. This gives

$$f(x, y) = \int_{x \text{ const.}} (2xy + 1)\, dy = xy^2 + y + h(x), \tag{9}$$

where the constant of integration $h(x)$ is now to be regarded as a function of x.

To determine $h(x)$, we set

$$\frac{\partial f}{\partial x} = x^2 + y^2$$

to obtain the equations

$$\frac{\partial}{\partial x}(xy^2 + y + h(x)) = x^2 + y^2$$

$$\begin{aligned} y^2 + h'(x) &= x^2 + y^2 \\ h'(x) &= x^2 \\ h(x) &= \frac{x^3}{3} + C. \end{aligned} \tag{10}$$

We conclude, as in Solution 1, that

$$f(x, y) = \frac{x^3}{3} + xy^2 + y + C.$$

Solution 3 *Fastest way:* Do both integrations first.

$(x^2 + y^2)dx$ + $(2xy + 1)dy$

⇓ Integrate with respect to x — ⇓ Integrate with respect to y

$\frac{x^3}{3} + xy^2 + y\text{-terms} + \text{Const.}$ — $xy^2 + y + x\text{-terms} + \text{Const.}$

⇓ ⇓ Terms needed to account for both intermediate expressions.

$$\frac{x^3}{3} + xy^2 + y + C.$$

If you use this method, do not fall into the trap of doubling terms that appear in both intermediate expressions.

Right: $f(x, y) = \frac{x^3}{3} + xy^2 + y + C$

Wrong: $f(x, y) = \frac{x^3}{3} + 2xy^2 + y + C.$ □

Exactness of Differential Forms

An expression like

$$(x^2 + y^2)\,dx + (2xy + 1)\,dy$$

that has the form

$$M(x, y)\,dx + N(x, y)\,dy \tag{11}$$

is called a *differential form* in x and y. Such a form is said to be *exact* over a region R if throughout R it is the differential df of some function f. That is, (11) is exact on R if there exists a function $f(x, y)$ such that

$$\frac{\partial f}{\partial x} = M(x, y) \quad \text{and} \quad \frac{\partial f}{\partial y} = N(x, y)$$

for all (x, y) in R.

In terms of gradients, we may say that $M(x, y)\,dx + N(x, y)\,dy$ is exact over a region R if there exists a function $f(x, y)$ such that

$$\nabla f = M(x, y)\mathbf{i} + N(x, y)\mathbf{j}$$

for all (x, y) in R.

Differential Equations

Our interest in exact differential forms comes partly from a desire to solve differential equations like

$$(x^2 + y^2) + (2xy + 1)\frac{dy}{dx} = 0 \tag{12}$$

that have the form

$$M(x, y) + N(x, y)\frac{dy}{dx} = 0. \tag{13}$$

In the simplest cases we may be able to separate variables and rewrite (13) in the form

$$g(y)\frac{dy}{dx} = h(x),$$

which we can then solve if we are able to integrate both sides with respect to x. If we cannot separate the variables (for example, we cannot do so in Eq. 12), we may still be able to solve the equation by rewriting it in the form

$$M(x, y)\,dx + N(x, y)\,dy = 0. \tag{14}$$

For if the left-hand side turns out to be the differential of a function $f(x, y)$ when we do this, then

$$\frac{\partial f}{\partial x} = M(x, y), \qquad \frac{\partial f}{\partial y} = N(x, y),$$

and Eq. (14) becomes

$$df = 0,$$

and its general solution is

$$f(x, y) = C. \tag{15}$$

Equation (15) may be considered to be a solution of the original equation

$$M(x, y) + N(x, y)\frac{dy}{dx} = 0 \tag{16}$$

in the sense that the equation $f(x, y) = C$ defines y implicitly as one or more differentiable functions of x that solve Eq. (16). To see that this is so we differentiate both sides of $f(x, y) = C$ with respect to x, treating y as a differentiable function of x and applying the chain rule. This gives

$$\begin{aligned} f(x, y) &= C \\ \frac{d}{dx}f(x, y) &= \frac{d}{dx}C \\ \frac{\partial f}{\partial x}\frac{dx}{dx} + \frac{\partial f}{\partial y}\frac{dy}{dx} &= 0 \\ \frac{\partial f}{\partial x} + \frac{\partial f}{\partial y}\frac{dy}{dx} &= 0, \end{aligned} \tag{17}$$

or, returning to the original notation,

$$M(x, y) + N(x, y)\frac{dy}{dx} = 0,$$

which is Eq. (16). Thus, any differentiable function $y = y(x)$ defined implicitly by Eq. (15) is seen to satisfy Eq. (16).

EXAMPLE 2 Find the general solution $f(x, y) = C$ of the differential equation

$$(x^2 + y^2) + (2xy + 1)\frac{dy}{dx} = 0.$$

Then find the particular solution whose graph passes through the point (0, 1).

Solution It is not possible to separate the variables in the given equation, so we rewrite it in the form

$$(x^2 + y^2)\,dx + (2xy + 1)\,dy = 0.$$

We then seek a function $f(x, y)$ whose differential df gives the left-hand side of this equation:

$$df = (x^2 + y^2)\,dx + (2xy + 1)\,dy.$$

We know from Example 1 that

$$f(x, y) = \frac{x^3}{3} + xy^2 + y$$

is such a function. Therefore, the general solution of the given equation is

$$f(x, y) = C, \qquad \text{or} \qquad \frac{x^3}{3} + xy^2 + y = C.$$

The particular solution whose graph passes through the point (0, 1) is found from substitution to be

$$\frac{x^3}{3} + xy^2 + y = 1.$$

See Fig. 15.46. □

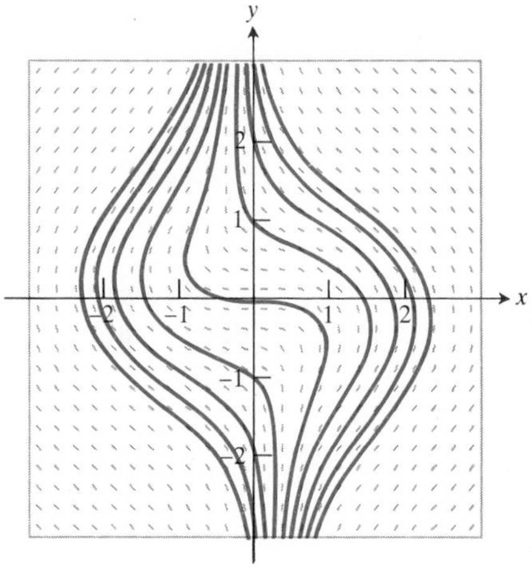

15.46 Solutions: $(x^3/3) + xy^2 + y = C$ for $C = 4, 3, 2, 1, 0, -1, -2, -3, -4$.

The Test for Exactness

Had the differential form

$$(x^2 + y^2)\,dx + (2xy + 1)\,dy$$

not been exact, we could not have proceeded the way we did to solve the equation

$$(x^2 + y^2)\,dx + (2xy + 1)\,dy = 0$$

in Example 2. How can we tell when a form

$$M(x, y)\,dx + N(x, y)\,dy \tag{18}$$

is exact?

The answer lies in the observation that if

$$M(x, y)\,dx + N(x, y)\,dy = f_x\,dx + f_y\,dy = df$$

is the differential of a function $f(x, y)$ that has continuous first and second order partial derivatives, then

$$M_y = f_{xy} = f_{yx} = N_x.$$

These mixed partial derivatives have to be equal by Theorem 4 in Article 15.7. Thus,

$$M_y = N_x$$

is a *necessary* condition for (18) to be exact. We may therefore exclude from consideration any candidates that fail this test.

The amazing fact is that the condition $M_x = N_y$ is also a *sufficient* condition for exactness. That is, if $M_x = N_y$, then (18) is exact. Theorem 7 below gives the details.

Theorem 7 introduces a mild restriction on the geometric nature of the set R on which the functions $M(x, y)$ and $N(x, y)$ are defined. Specifically, R is required to be a simply connected region. The word *region* has the technical meaning here of "connected open set." A subset of the plane is *open* if each of its points is the center of a circle whose interior lies entirely in the set. An *open* subset of the plane is *connected* if every two of its points can be joined by a path that lies entirely in the set. A subset R of the plane is *simply connected* if no simple closed curve in R surrounds points not in R. When you read "simply connected," think "no holes." (See Fig. 15.47.) You need not be concerned with these technicalities now. The plane domains on which most of the functions in this book are defined really are regions or unions of regions, and most of them are simply

Connected and simply connected.

Connected but not simply connected.

Connected and simply connected.

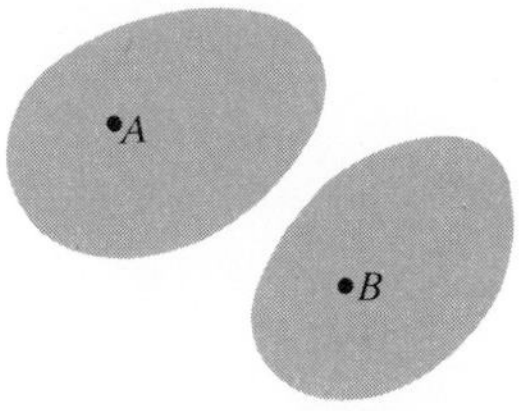

Simply connected but not connected.
No path from A to B lies entirely in the region.

Figure 15.47

connected. It will not be necessary to test every function that comes along. But without these conditions on R (which are usually met in practice), or others like them, some of the most useful theorems of calculus would not be true.

THEOREM 7

Test for Exactness

Suppose that the functions $M(x, y)$ and $N(x, y)$ and their partial derivatives M_x, M_y, N_x, and N_y are continuous for all real values of x and y in a simply connected region R. Then a necessary and sufficient condition for

$$M(x, y)\,dx + N(x, y)\,dy$$

to be an exact differential in R is

$$\frac{\partial M}{\partial y} = \frac{\partial N}{\partial x}.$$

Discussion. We saw before we stated Theorem 7 that $M_y = N_x$ is a necessary condition for exactness, and the argument did not require anything special of the region R. What we have yet to show, however, is that $M_y = N_x$ is sufficient to guarantee exactness. It is here that the simple connectivity comes in, but it is also here that the argument becomes too technical to present in full detail. What we can do here, however, is show how to use the equation

$$M_x = N_y$$

to construct a function $f(x, y)$ whose differential satisfies the equation

$$df = M\,dx + N\,dy. \tag{19}$$

The construction repeats the steps we took in Example 1.

For Eq. (19) to hold we must have

$$\frac{\partial f}{\partial x} = M(x, y), \qquad \frac{\partial f}{\partial y} = N(x, y).$$

Integrating the first of these with respect ot x, we find that for each value of y we have

$$f(x, y) = \int M(x, y)\,dx + k(y), \tag{20}$$

where $k(y)$ is a constant that may vary with the value of y. To see how f may satisfy the second condition, $\partial f/\partial y = N(x, y)$, we differentiate both sides of Eq. (20) with respect to y, obtaining

$$\frac{\partial f}{\partial y} = \frac{\partial}{\partial y}\int M(x, y)\,dx + k'(y). \tag{21}$$

We then set $\partial f/\partial y$ equal to $N(x, y)$ and have

$$\frac{\partial}{\partial y}\int M(x, y)\,dx + k'(y) = N(x, y)$$

$$k'(y) = N(x, y) - \frac{\partial}{\partial y}\int M(x, y)\,dx. \tag{22}$$

We use this differential equation to determine $k(y)$ and substitute the result back into Eq. (20).

The success of this construction of $f(x, y)$ depends on the fact that the right-hand side of Eq. (22) is a function of y alone, and it is exactly this independence of x that is guaranteed by the condition $M_y = N_x$. (If the right-hand side of Eq. (22) depended on x as well as y it could not be equal to $k'(y)$, which involves only y.)

We prove that the right-hand side of Eq. (22) is independent of x by showing that its derivative with respect to x is identical to zero. We calculate

$$\begin{aligned}\frac{\partial}{\partial x}\left(N(x, y) - \frac{\partial}{\partial y}\int M(x, y)\,dx\right) &= \frac{\partial N}{\partial x} - \frac{\partial^2}{\partial x\,\partial y}\int M(x, y)\,dx \\ &= \frac{\partial N}{\partial x} - \frac{\partial}{\partial y}\left(\frac{\partial}{\partial x}\int M(x, y)\,dx\right) \\ &= \frac{\partial N}{\partial x} - \frac{\partial}{\partial y}(M),\end{aligned}$$

and this last expression vanishes exactly when $M_y = N_x$ as we have assumed. ■

EXAMPLE 3 The following differential forms are defined over the region R that consists of the entire xy-plane. Test the forms for exactness on R.

a) $(x^2 + y^2)\,dx - 2xy\,dy$.

b) $(e^x \cos y)\,dx + (1 - e^x \sin y)\,dy$.

Solution We check to see whether $M_y = N_x$ at all points (x, y) in R.

a) Not exact on R because $M_y \neq N_x$ except when $y = 0$:

$$\frac{\partial}{\partial y}(x^2 + y^2) = 2y,$$

$$\frac{\partial}{\partial x}(-2xy) = -2y.$$

b) Exact on R because $M_y = N_x$ for all x and y:

$$\frac{\partial}{\partial y}(e^x \cos y) = -e^x \sin y,$$

$$\frac{\partial}{\partial x}(1 - e^x \sin y) = -e^x \sin y. \;\square$$

PROBLEMS

In Problems 1–14, determine whether the given expression is or is not the differential of some function $f(x, y)$ over the region R that consists of the entire xy-plane. If it is, find f.

1. $2x(x^3 + y^3)\,dx + 3y^2(x^2 + y^2)\,dy$

2. $e^y\,dx + x(e^y + 1)\,dy$

3. $(2x + y)\,dx + (x + 2y)\,dy$

4. $(\cosh y + y \cosh x)\,dx + (\sinh x + x \sinh y)\,dy$

5. $(\sin y + y \sin x)\,dx + (\cos x + x \cos y)\,dy$

6. $(1 + e^x)\,dy + e^x(y - x)\,dx$

7. $e^{x+y}\,dx - e^{x-y}\,dy$

8. $ye^{xy}\,dx + xe^{xy}\,dy$

9. $(y \cos x + \sin y)\,dx + (\sin x + x \cos y - \sin y)\,dy$

10. $(y\cos x + y^2)\,dx + (\sin x)\,dy$

11. $(6xy^5 + 6y)\,dx + (15x^2y^4 + 6x - 10y)\,dy$

12. $(ye^x + y)\,dx + (e^x + 1)\,dy$

13. $(ye^{xy} + 10x)\,dx + xe^{xy}\,dy$

14. $(-2x + e^{2y}\cos 3x)\,dx + (e^y + e^{2y}\sin 3x)\,dy$

15. Find the general solution $f(x, y) = C$ of the differential equation

$$(12xy + 2y^2) + (6x^2 + 4xy)\frac{dy}{dx} = 0.$$

Then find the particular solution whose graph passes through the point $(1, -3)$.

16. Find the general solution $f(x, y) = C$ of the differential equation

$$(e^x \sin y + \cos x) + (e^x \cos y + 3e^{3y})\frac{dy}{dx} = 0.$$

Then find the particular solution whose graph passes through the point $(\pi/2, 0)$.

17. Find the general solution $f(x, y) = C$ of the differential equation

$$\left(\frac{y}{x} + e^y\right) + (\ln x + 2y + xe^y)\frac{dy}{dx} = 0.$$

Then find the particular solution whose graph passes through the point $(1, 0)$.

18. Find the general solution $f(x, y) = C$ of the differential equation

$$3x^2y^2 - (10y^4 - 2yx^3)\frac{dy}{dx} = 0.$$

Then find the particular solution that satisfies the condition $f(1, 1) = 0$.

19. Find the solution of the differential equation

$$(2x + e^{-y}\sin x) + (e^y + e^{-y}\cos x)\frac{dy}{dx} = 0$$

whose graph passes through the origin.

20. a) Find the value of the constant b that makes

$$(y\cos x + b\cos y)\,dx + (x\sin y + \sin x + y)\,dy$$

the differential of a function $f(x, y)$.

b) Find the function $f(x, y)$ that corresponds to the value of b you found in part (a) and satisfies the condition $f(0, 1) = 0$.

21. a) Find the value of α that makes

$$(y^5 + 3x^{1/2}y^3)\,dx + (5xy^4 + 3\alpha x^{3/2}y^2 + 2y)\,dy$$

the differential of a function $g(x, y)$.

b) Find the function $g(x, y)$ that corresponds to the value of α you found in part (a) and satisfies the condition $g(1, -1) = 0$.

22. Let $u = u(x, y)$, $v = v(x, y)$. Assuming that all the necessary partial derivatives exist, establish the following differential formulas:

a) $d(cu) = c\,du$ (any constant c),

b) $d(u^n) = n\,u^{n-1}$ (any number n),

c) $d(uv) = u\,dv + v\,du$,

d) $d\left(\frac{u}{v}\right) = \frac{v\,du - u\,dv}{v^2}$.

15.12

Least Squares

An important application of minimizing a function of two variables is the *method of least squares* for fitting a straight line

$$y = mx + b \tag{1}$$

to a set of experimentally observed points $(x_1, y_1), (x_2, y_2), \ldots, (x_n, y_n)$ (Fig. 15.48). Corresponding to each observed value of x there are two values of y, namely, the observed value y_{obs} and the value predicted by the straight line $y = mx_{obs} + b$. We call the difference

$$\text{Predicted value} - \text{Observed value} = (mx_{obs} + b) - y_{obs} \tag{2}$$

a *deviation*. Each deviation measures the amount by which the predicted value of y differs from the observed value. The set of deviations

$$d_1 = (mx_1 + b) - y_1, \ldots, d_n = (mx_n + b) - y_n \tag{3}$$

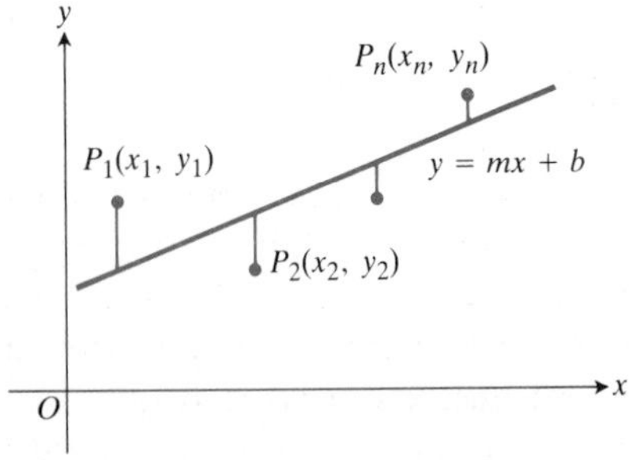

15.48 To fit a line to noncollinear points, we may choose a line that minimizes the sum of the squares of the deviations.

gives a picture of how closely the line of Eq. (1) fits the observed data. The line is a perfect fit if and only if all of the deviations are zero. But in general no straight line will give a perfect fit. Then we are confronted with the problem of finding a line that fits *best* in some sense or other. Here is where the method of least squares comes in.

For a straight line that comes *close* to fitting all of the observed points, some of the deviations will probably be positive and some will be negative. But their squares will all be positive, and the expression

$$f(m, b) = (mx_1 + b - y_1)^2 + (mx_2 + b - y_2)^2 + \cdots + (mx_n + b - y_n)^2 \quad (4)$$

counts a positive deviation d and a negative deviation $-d$ equally. This sum of squares of the deviations depends upon the choice of m and b. It is never negative and it can be zero only if m and b have values that produce a straight line that is a perfect fit.

Whether such a perfectly fitting line can be found or not, the method of least squares says, "take as the line $y = mx + b$ of best fit that one for which the sum of squares of the deviations

$$f(m, b) = d_1^2 + d_2^2 + \cdots + d_n^2$$

is a minimum." Thus we try to find the values of m and b where the surface

$$w = f(m, b)$$

in mbw-space has a low point (Fig. 15.49). To do this, we solve the equations

$$\frac{\partial f}{\partial m} = 0 \quad \text{and} \quad \frac{\partial f}{\partial b} = 0 \quad (5)$$

simultaneously.

These equations are equivalent (Problem 13) to

$$\begin{aligned} \left(\sum x_i^2\right)m + \left(\sum x_i\right)b &= \sum x_i y_i \quad \left(\frac{\partial f}{\partial m} = 0\right), \\ \left(\sum x_i\right)m + nb &= \sum y_i \quad \left(\frac{\partial f}{\partial b} = 0\right). \end{aligned} \quad (6)$$

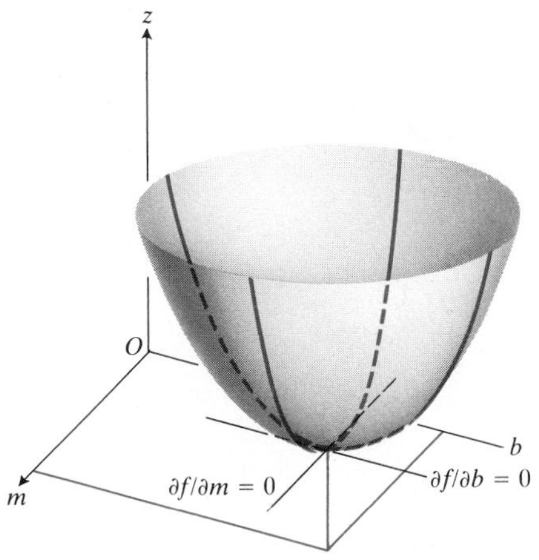

15.49 The sum $f(m, b)$ of the squares of the deviations has a minimum when $\partial f/\partial m$ and $\partial f/\partial b$ are both zero.

where the sums run from $i = 1$ to $i = n$. Note that the variables whose values we wish to determine from Eqs. (6) are m and b. The x's and y's are the coordinates of the data points, which are known, and n is the number of data points. Thus (6) is a system of two equations in the two unknowns m and b, which can be solved for m and b in the familiar way.

It can be proved that the function $f(m, b)$ has an absolute minimum value at the point (m, b) obtained by solving the equations in (6), but the algebra in showing that $f_{mm}f_{bb} - f_{mb}^2 > 0$ is complicated and we shall not go into it here.

Method of Least Squares

To find the *least squares line*

$$y = mx + b$$

for n data points

$$(x_1, y_1), (x_2, y_2), \ldots, (x_n, y_n),$$

solve the equations

$$\left(\sum x_i^2\right)m + \left(\sum x_i\right)b = \sum x_i y_i, \qquad (7)$$

$$\left(\sum x_i\right)m + nb = \sum y_i \qquad (8)$$

simultaneously for m and b.

EXAMPLE Find the least squares line for the points (0, 1), (1, 3), (2, 2), (3, 4), (4, 5).

Solution It is useful to organize the computations in a table, like this:

i	x_i	y_i	x_i^2	$x_i y_i$
1	0	1	0	0
2	1	3	1	3
3	2	2	4	4
4	3	4	9	12
5	4	5	16	20
Σ	10	15	30	39

We then solve Eqs. (7) and (8) with

$$\sum x_i = 10, \quad \sum y_i = 15, \quad \sum x_i^2 = 30, \quad \sum x_i y_i = 39, \quad \text{and } n = 5$$

to get

$$30m + 10b = 39$$
$$10m + 5b = 15$$

and the solution

$$m = 0.9, \qquad b = 1.2.$$

The least squares line is therefore

$$y = 0.9x + 1.2.$$

See Fig. 15.50. □

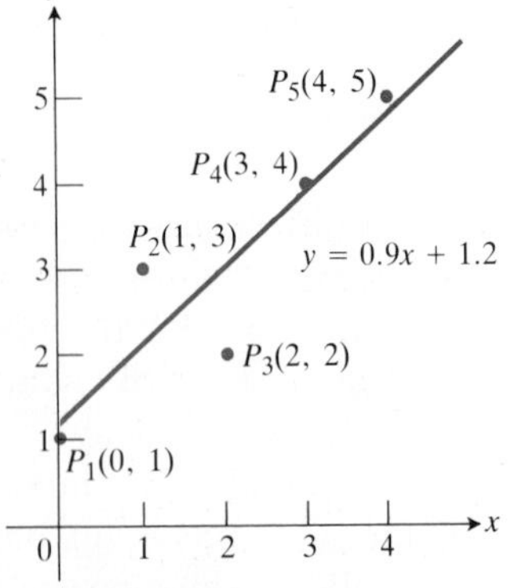

15.50 The least squares line for the data in the Example.

Finding a least squares line $y = mx + b$ allows us to

1. summarize data with a simple expression,
2. predict values of y for other, experimentally untried values of x,
3. handle data analytically.

The method of least squares is so convenient that when data are nonlinear it is common practice to transform one or both of the variables x and y to linearize the data, then fit a line and transform back. (See Problem 7.)

In many fields, the least squares line is called a *regression line* or *line of regression.*

One may also use least squares techniques to fit planes to three-dimensional data but we shall not treat this subject here. (However, see Problem 30 in Article 15.10.)

PROBLEMS

1. In practice, the values of m and b determined by Eqs. (7) and (8) are often calculated from the equations

$$m = \frac{\left(\sum x_i\right)\left(\sum y_i\right) - n\sum x_i y_i}{\left(\sum x_i\right)^2 - n\sum x_i^2}, \tag{9}$$

$$b = \frac{1}{n}\left(\sum y_i - m\sum x_i\right). \tag{10}$$

One finds m first, then uses the value of m in finding b. Show that these values of m and b solve Eqs. (7) and (8).

In Problems 2–6, find the least squares line for each set of data points. Use the linear equation you obtain to predict the value of y that would correspond to $x = 4$.

2. $(-1, 2)$, $(0, 1)$, $(3, -4)$

3. $(-2, 0)$, $(0, 2)$, $(2, 3)$

4. $(0, 0)$, $(1, 2)$, $(2, 3)$

5. $(0, 1)$, $(2, 2)$, $(3, 2)$

6. $(1, 1)$, $(2, 1)$, $(-3, 0)$

7. (*Calculator*) To determine the intermolecular potential between potassium ions and xenon gas, Budenholzer, Gislason, and Jorgensen (June 1, 1977, *Journal of Chemical Physics,* **66,** No. 11; p. 4832) accelerated a beam of potassium ions toward a cell containing xenon and measured the current I of ions leaving the cell as a percentage of the current I_0 entering the cell. This fraction, which is a function of xenon gas pressure, was recorded at five different pressures, with the results shown in Table 13.1.

Table 13.1 Scattering of potassium ions by xenon

x (pressure in millitorr)	$\frac{I}{I_0}$
0.165	0.940
0.399	0.862
0.573	0.810
0.930	0.712
1.281	0.622

As a step in their determination, the authors used the method of least squares to find the slope m and y-intercept b of a line $y = mx + b$, where $y = \ln(I/I_0)$. In particular, they hoped to find that, within the limits of experimental error, b was zero. (a) Write down the value of y for each x in the table and fit a least squares line to the (x, y) data points. Round m and b to three decimal places. (b) Express I/I_0 as a function of x.

8. (*Calculator*) Write a linear equation for the effect of irrigation on the yield of alfalfa by fitting a least squares line to the data in Table 13.2 (from the University of California Experimental Station, *Bulletin* No. 450, p. 8). Plot the data and draw the line.

Table 13.2 Growth of alfalfa

x (total seasonal depth of water applied (in.))	y (average alfalfa yield (tons/acre))
12	5.27
18	5.68
24	6.25
30	7.21
36	8.20
42	8.71

9. (*Calculator*) *Hubble's law* for the expansion of the universe is the linear equation

$$\text{Velocity} = \text{the Hubble constant} \cdot \text{distance}$$

or

$$v = Hx.$$

It says that the velocity with which a galaxy appears to move away from us is proportional to how far away the galaxy lies. The farther away it lies, the faster it recedes. If the velocity is measured in kilometers per second and the distance in millions of light-years, then Hubble's constant is given in kilometers per second per million light-years. H is the rate at which the universe appears to be expanding. For every extra million light-years of distance, the galaxies we can observe recede faster by H kilometers per second.

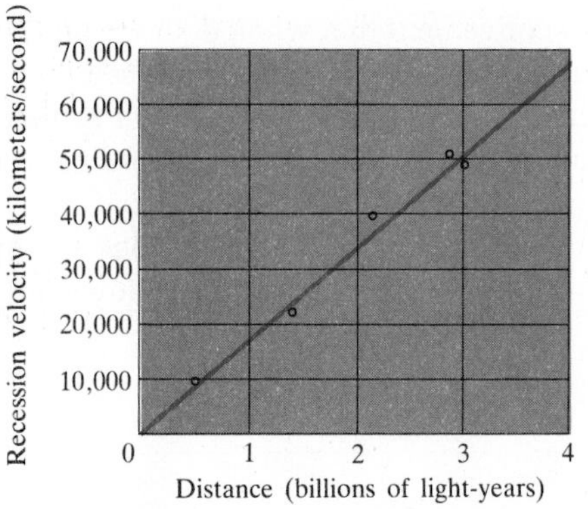

15.51 Velocity vs. distance observed for the galaxies in Table 13–3.

Table 13.3 lists the observed distances and velocities for five galaxies. Discover Hubble's constant H by fitting a least squares line to the data (Fig. 15.51). Round your answer to the nearest integer. You will find that the y-intercept of the line is 240 km/s when rounded to the nearest integer, and not zero. A discrepancy in the intercept is to be expected, given the uncertainties in measurement and, here, the small size of the sample. Note that the discrepancy is a small percentage of the observed recession velocities. (For more information, see Jastrow and Thompson's *Astronomy: Fundamentals and Frontiers*, Third Edition, John Wiley and Sons, Inc., 1977, Chapter 11.)

Table 13.3 Observed velocities and distances of five receding galaxies

Galaxy	Observed distance (10^6 l-yr)	Recession velocity (km/s)
A	500	9,000
B	1,400	22,000
C	2,100	39,000
D	2,900	51,000
E	3,000	49,000

10. (*Calculator*) *Craters of Mars.* One theory of crater formation suggests that the frequency of large craters should fall off as the square of the diameter (Marcus, *Science,* June 21, 1968, p. 1334). Pictures from Mariner IV show the frequencies listed in Table 13.4.

Table 13.4 Crater sizes on Mars

Diameter in km, D	$1/D^2$ (for left value of class interval)	Frequency, F
32–45	0.001	53
45–64	0.0005	22
64–90	0.00025	14
90–128	0.000125	3

Fit a line of the form $F = m(1/D^2) + b$ to the data. Plot the data and draw the line.

11. If $y = mx + b$ is the best-fitting straight line, in the sense of least squares, show that the sum of deviations

$$\sum_{i=1}^{n} (y_i - mx_i - b)$$

is zero. (This means that positive and negative deviations cancel.)

12. Show that the point

$$(\bar{x}, \bar{y}) = \left[\frac{1}{n}\left(\sum_{i=1}^{n} x_i\right), \frac{1}{n}\left(\sum_{i=1}^{n} y_i\right)\right]$$

lies on the straight line $y = mx + b$ that is determined by the method of least squares. (This means that the "best-fitting" line passes through the center of gravity of the n points.)

13. Expand the equations in (4) and (5) to show that they lead to the equations in (6).

REVIEW QUESTIONS AND EXERCISES

1. Let $z = f(x, y)$ be a function of two independent variables defined on a region R in the xy-plane. Describe two graphical ways to represent f.
2. Define level surface and contour line.
3. Give two equivalent definitions of

$$\lim_{(x, y)\to(x_0, y_0)} f(x, y) = L.$$

What is the basic theorem for calculating limits of sums, differences, products, and quotients of functions of two variables?

4. When is a function of two variables continuous at a point of its domain? Give an example of a function that is discontinuous at some point(s) of its domain.
5. When is a function of three variables continuous at a point of its domain? Give an example of a function of three independent variables that is continuous at some points of its domain and discontinuous at at least one place in its domain. Give an example that is discontinuous at all points of a surface $F(x, y, z) = 0$; at all points of a line.

6. Let $z = f(x, y)$. Define $\partial z/\partial x$ and $\partial z/\partial y$ at a point (x_0, y_0) in the domain of f.
7. What does the Increment Theorem for functions of two variables say?
8. State several chain rules for partial derivatives of functions of two and three variables, and give their tree diagrams.
9. Define the gradient of a function $f(x, y, z)$ and describe the role the gradient plays in defining directional derivatives, tangent planes, and normal lines. What is the relation between ∇f and the directions in which f changes most rapidly? What are the analogous results for functions of two variables?
10. Suppose $\mathbf{R}(t) = x(t)\mathbf{i} + y(t)\mathbf{j} + z(t)\mathbf{k}$ is a curve in the domain of a function $f(x, y, z)$. Describe the relation between df/dt, ∇f, and $\mathbf{v} = d\mathbf{R}/dt$. What can be said about ∇f and $\mathbf{v}$ at points of the curve where f has a local maximum or minimum relative to its values on the curve?
11. Why do we require functions defining surfaces to have continuous first partial derivatives before we define tangent planes?
12. What is the basic theorem about mixed second order partial derivatives of functions of two variables? Give an example. What about mixed partial derivatives of higher order?
13. Give three important partial differential equations from physics, along with sample solutions.
14. Give the basic formula for the linear approximation of a function $f(x, y)$ near a point (x_0, y_0), and give an upper bound for the error in the approximation. What formula may be used to estimate increments in f?
15. Describe a way to find the extreme values of a function $f(x, y)$ that is continuous and has continuous first and second order partial derivatives (a) at an interior point of its domain, and (b) at a boundary point. Give an example.
16. Describe the method of Lagrange multipliers and its geometric interpretation as it applies to the problem of maximizing or minimizing a function $f(x, y, z)$
    a) subject to one constraint,
    $$g(x, y, z) = 0;$$
    b) subject to two constraints,
    $$g(x, y, z) = 0 \quad \text{and} \quad h(x, y, z) = 0.$$
17. Sketch some of the level curves of $f(x, y) = 2x + 3y$ and find points on the curve $g(x, y) = x^2 + xy + y^2 - 3 = 0$ that (a) maximize f, (b) minimize f. Does it seem reasonable that the level curve of f at each of these points should be tangent to the curve $g(x, y) = 0$ at each of these points? Is it true?
18. Explain the test for the exactness of the differential form
    $$M(x, y)\,dx + N(x, y)\,dy.$$
    Give examples of forms that pass, and fail, the test. Find the solution of the differential equation
    $$(2x + y)\,dx + (2y + x)\,dy = 0$$
    whose graph passes through the point (1, 1).
19. Describe the method of least squares for fitting a line to a finite set of data points. To what use may we put the least squares line?

MISCELLANEOUS PROBLEMS

1. Let $f(x, y) = (x^2 - y^2)/(x^2 + y^2)$ for $x^2 + y^2 \neq 0$. Is it possible to define the value of f at $x = 0$, $y = 0$, in such a way that the function would be continuous at $x = 0$, $y = 0$? Why?
2. Let the function $f(x, y)$ be defined by the relations
   $$f(x, y) = \frac{\sin^2(x - y)}{|x| + |y|} \quad \text{for } |x| + |y| \neq 0, \quad f(0, 0) = 0.$$
   Is f continuous at $x = 0$, $y = 0$?
3. Prove that if $f(x, y)$ is defined for all x, y, by
   $$f(x, y) = \begin{cases} \dfrac{2xy}{x^2 + y^2} & \text{if } (x, y) \neq (0, 0), \\ 0 & \text{if } x = y = 0, \end{cases}$$
   then
   a) for any fixed x, $f(x, y)$ is a continuous function of y;
   b) for any fixed y, $f(x, y)$ is a continuous function of x;
   c) $f(x, y)$ is not continuous at (0, 0);
   d) $\partial f/\partial x$ and $\partial f/\partial y$ exist at (0, 0) but are not continuous there.
   (This example shows that a function may possess partial derivatives at all points of a region, yet not be continuous in the region.) Contrast the case of a function of one variable, where the existence of a derivative implies continuity.
4. Let $f(x, y)$ be defined and continuous for all x, y (differentiability not assumed). Show that it is always

possible to find arbitrarily many points (x_1, y_1), (x_2, y_2), ..., (x_n, y_n) such that the function has the same value at each of them.

5. Find the first partial derivatives of the following functions:
a) $(\sin xy)^2$, b) $\sin[(xy)^2]$.

6. Let α, β, γ be the angles that a line passing through the origin into the first octant makes with the positive coordinate axes. Consider γ as a function of α and β. Find the value of $\partial\gamma/\partial\alpha$ when $\alpha = \pi/4$, $\beta = \pi/3$, $\gamma = \pi/3$.

7. Let (r, θ) and (x, y) be polar coordinates and cartesian coordinates in the plane. Show geometrically why $\partial r/\partial x$ is not equal to $(\partial x/\partial r)^{-1}$, by appealing to the definitions of these derivatives.

8. Consider the surface whose equation is $x^3z + y^2x^2 + \sin(yz) + 54 = 0$. Give an equation of the tangent plane to the surface at the point $P(3, 0, -2)$, and give equations of the line through P normal to the surface.

9. a) Sketch and name the surface $x^2 - y^2 + z^2 = 4$.
b) Find a vector that is normal to this surface at $(2, -3, 3)$.
c) Find equations of the surface's tangent plane and normal line at $(2, -3, 3)$.

10. a) Find an equation of the plane tangent to the surface

$$x^3 + xy^2 + y^3 + z^3 + 1 = 0$$

at the point $(-2, 1, 2)$.
b) Find equations of the straight line perpendicular to the plane above at the point $(-2, 1, 2)$.

11. The directional derivative of a given function $f(x, y)$ at the point $P_0(1, 2)$ in the direction toward $P_1(2, 3)$ is $2\sqrt{2}$, and in the direction toward $P_2(1, 0)$ it is -3. Compute $\partial f/\partial x$ and $\partial f/\partial y$ at $P_0(1, 2)$, and compute the derivative of f at $P_0(1, 2)$ in the direction toward $P_3(4, 6)$.

12. Let $z = f(x, y)$ have continuous first partial derivatives. Let C be any curve lying on the surface and passing through (x_0, y_0, z_0). Prove that the tangent line to C at (x_0, y_0, z_0) must lie wholly in the plane determined by the tangent lines to the curves C_x and C_y, where C_x is the curve of intersection of $y = y_0$ and $z = f(x, y)$, and C_y is the curve of intersection of $x = x_0$ and the surface.

13. Let $u = xyz$. Show that if x and y are the independent variables (so u and z are functions of x and y), then

$$\partial u/\partial x = xy\,(\partial z/\partial x) + yz;$$

but that if x, y, and z are the independent variables,

$$\partial u/\partial x = yz.$$

14. Let

$$\mathbf{u} = u_1\mathbf{i} + u_2\mathbf{j} + u_3\mathbf{k} \quad \text{and} \quad \mathbf{v} = v_1\mathbf{i} + v_2\mathbf{j} + v_3\mathbf{k}$$

be given constant unit vectors and let $f(x, y, z)$ be a given scalar function. Compute
a) the directional derivative $D_{\mathbf{u}}f$, and
b) the directional derivative $D_{\mathbf{v}}(D_{\mathbf{u}}f)$,
in terms of derivatives of f and the components of $\mathbf{u}$ and $\mathbf{v}$.

15. Consider the function $w = xyz$.
a) Compute the directional derivative of w at the point $(1, 1, 1)$ in the direction of the vector $\mathbf{i} + \mathbf{j} + \mathbf{k}$.
b) Compute the largest value of the directional derivative of w at the point $(1, 1, 1)$.

16. The function $w = f(x, y)$ has, at the point $(1, 2)$, directional derivatives that are equal to $+2$ in the direction toward $(2, 2)$, and -2 in the direction toward $(1, 1)$. What is its directional derivative at $(1, 2)$ in the direction toward $(4, 6)$?

17. Given the function $f(x, y, z) = x^2 + y^2 - 3z$, what is the maximum value of the directional derivative of f at the point $(1, 3, 5)$?

18. Given the function

$$(x - 1)^2 + 2(y + 1)^2 + 3(z - 2)^2 - 6.$$

Find the derivative of the function at the point $(2, 0, 1)$ in the direction of the vector $\mathbf{i} - \mathbf{j} + 2\mathbf{k}$.

19. Find the derivative of the function

$$f(x, y, z) = x^2 - 2y^2 + z^2$$

at the point $(3, 3, 1)$ in the direction of the vector $2\mathbf{i} + \mathbf{j} - \mathbf{k}$.

20. The two equations

$$e^u \cos v - x = 0 \quad \text{and} \quad e^u \sin v - y = 0$$

define u and v as functions of x and y, say $u = u(x, y)$ and $v = v(x, y)$. Show that the angle between the two vectors

$$\left(\frac{\partial u}{\partial x}\right)\mathbf{i} + \left(\frac{\partial u}{\partial y}\right)\mathbf{j} \quad \text{and} \quad \left(\frac{\partial v}{\partial x}\right)\mathbf{i} + \left(\frac{\partial v}{\partial y}\right)\mathbf{j}$$

is constant.

21. a) Find a vector $\mathbf{N}(x, y, z)$ normal to the surface $z = \sqrt{x^2 + y^2} + (x^2 + y^2)^{3/2}$ at the point (x, y, z) of the surface.
b) Find the cosine of the angle γ between $\mathbf{N}(x, y, z)$ and the z-axis. Find the limit of $\cos\gamma$ as $(x, y, z) \to (0, 0, 0)$.

22. Find all points (a, b, c) in space for which the spheres

$$(x - a)^2 + (y - b)^2 + (z - c)^2 = 1$$

and

$$x^2 + y^2 + z^2 = 1$$

will intersect orthogonally. (Their tangents are to be perpendicular at each point of intersection.)

23. a) Find the gradient of the function

$$f(x, y, z) = x^2 + 2xy - y^2 + z^2$$

at the point $P_0(1, -1, 3)$.

b) Find the plane that is tangent to the surface $x^2 + 2xy - y^2 + z^2 = 7$ at $P_0(1, -1, 3)$.

24. Find a unit vector normal to the surface $x^2 + y^2 = 3z$ at the point $(1, 3, \frac{10}{3})$.

25. In a flowing fluid, the density $\rho(x, y, z, t)$ depends on position and time. If

$$\mathbf{V} = \mathbf{V}(x, y, z, t)$$

is the velocity of the fluid particle at the point (x, y, z) at time t, then

$$\frac{d\rho}{dt} = \mathbf{V} \cdot \nabla\rho + \frac{\partial\rho}{\partial t} = V_1\frac{\partial\rho}{\partial x} + V_2\frac{\partial\rho}{\partial y} + V_3\frac{\partial\rho}{\partial z} + \frac{\partial\rho}{\partial t},$$

where $\mathbf{V} = V_1\mathbf{i} + V_2\mathbf{j} + V_3\mathbf{k}$. Explain the physical and geometrical meaning of this relation.

26. Find a constant a such that, at any point of intersection of the two spheres

$$(x - a)^2 + y^2 + z^2 = 3 \quad \text{and} \quad x^2 + (y - 1)^2 + z^2 = 1,$$

their tangent planes will be perpendicular to each other.

27. If the gradient of a function $f(x, y, z)$ is always parallel to the vector $x\mathbf{i} + y\mathbf{j} + z\mathbf{k}$, show that the function must assume the same value at the points $(0, 0, a)$ and $(0, 0, -a)$.

28. Let $f(P)$ denote a function defined for points P in the plane; i.e., to each point P there is attached a real number $f(P)$. Explain how one could introduce the notions of continuity and differentiability of the function and define the vector ∇f *without* introducing a coordinate system. If one introduces a polar coordinate system r, θ, $\mathbf{U}_r$, $\mathbf{U}_\theta$, what form does the vector $\nabla f(r, \theta)$ take?

29. Show that the directional derivative of

$$r = \sqrt{x^2 + y^2 + z^2}$$

equals 1 in any direction at the origin, but that r does not have a gradient vector at the origin.

30. Let $\mathbf{R} = x\mathbf{i} + y\mathbf{j} + z\mathbf{k}$ and $r = |\mathbf{R}|$.

a) From its geometrical interpretation, show that $\nabla r = \mathbf{R}/r$.

b) Show that $\nabla(r^n) = nr^{n-2}\mathbf{R}$.

c) Find a function with gradient equal to $\mathbf{R}$.

d) Show that $\mathbf{R} \cdot d\mathbf{R} = r\,dr$.

e) If $\mathbf{A}$ is a constant vector, show that $\nabla(\mathbf{A} \cdot \mathbf{R}) = \mathbf{A}$.

31. If θ is the polar coordinate in the xy-plane, find the direction and magnitude of $\nabla\theta$.

32. If r_1, r_2 are the distances from the point $P(x, y)$ on an ellipse to its foci, show that the equation $r_1 + r_2 =$ const., satisfied by these distances, requires $\mathbf{U} \cdot \nabla(r_1 + r_2) = 0$, where $\mathbf{U}$ is a unit tangent to the curve. By geometrical interpretation, show that the tangent makes equal angles with the lines to the foci.

33. If A, B are fixed points and θ is the angle at $P(x, y, z)$ subtended by the line segment AB, show that $\nabla\theta$ is normal to the circle through A, B, P.

34. Find the general solution of the partial differential equations:

a) $af_x + bf_y = 0$, a, b constants, b) $yf_x - xf_y = 0$.

(*Hint:* Consider the geometrical meaning of the equations.)

35. When y is eliminated from the two equations $z = f(x, y)$ and $g(x, y) = 0$, the result is expressible in the form $z = h(x)$. Express the derivative $h'(x)$ in terms of $\partial f/\partial x$, $\partial f/\partial y$, $\partial g/\partial x$, $\partial g/\partial y$. Check your formula by computing $h(x)$ and $h'(x)$ explicitly in the example where $f(x, y) = x^2 + y^2$ and $g(x, y) = x^3 + y^2 - x$.

36. Suppose the equation $F(x, y, z) = 0$ defines z as a function of x and y, say $z = f(x, y)$, with derivatives $\partial f/\partial x$ and $\partial f/\partial y$. Suppose also that the same equation $F(x, y, z) = 0$ defines x as a function of y and z, say $x = g(y, z)$, with derivatives $\partial g/\partial y$ and $\partial g/\partial z$. Prove that

$$\frac{\partial g}{\partial y} = -\frac{\partial f/\partial y}{\partial f/\partial x},$$

and also express $\partial g/\partial z$ in terms of $\partial f/\partial x$ and $\partial f/\partial y$.

37. Given $z = x \sin x - y^2$, $\cos y = y \sin z$, find dx/dz.

38. If

$$z = f\left(\frac{x - y}{y}\right),$$

show that $x(\partial z/\partial x) + y(\partial z/\partial y) = 0$.

39. If the substitution $u = (x - y)/2$, $v = (x + y)/2$, changes $f(u, v)$ into $F(x, y)$, express $\partial F/\partial x$ and $\partial F/\partial y$ in terms of the derivatives of $f(u, v)$ with respect to u and v.

40. Given $w = f(x, y)$ with $x = u + v$, $y = u - v$, show that

$$\frac{\partial^2 w}{\partial u\,\partial v} = \frac{\partial^2 w}{\partial x^2} - \frac{\partial^2 w}{\partial y^2}.$$

41. Suppose $f(x, y, z)$ is a function that has continuous partial derivatives and satisfies $f(tx, ty, tz) = t^n f(x, y, z)$ for every quadruple of numbers x, y, z, t (where n is a fixed integer). Show the identity

$$\frac{\partial f}{\partial x}x + \frac{\partial f}{\partial y}y + \frac{\partial f}{\partial z}z = nf.$$

(*Hint:* Differentiate with respect to t; then set $t = 1$.)

42. The substitution $u = x + y$, $v = xy^2$ changes the function $f(u, v)$ into $F(x, y)$. Express the partial derivative $\partial^2 F/\partial x\, \partial y$ in terms of x, y, and the partial derivatives of $f(u, v)$ with respect to u, v.

43. Given $z = u(x, y) \cdot e^{ax+by}$, where $u(x, y)$ is a function of x and y such that $\partial^2 u/\partial x\, \partial y = 0$, ($a$, b constants). Find values of a and b that will make the expression $\partial^2 z/\partial x\, \partial y - \partial z/\partial x - \partial z/\partial y + z$ identically zero.

44. Introducing polar coordinates, $x = r\cos\theta$, $y = r\sin\theta$, changes $f(x, y)$ into $g(r, \theta)$. Compute the value of the second derivative $\partial^2 g/\partial\theta^2$ at the point where $r = 2$ and $\theta = \pi/2$, given that

$$\frac{\partial f}{\partial x} = \frac{\partial f}{\partial y} = \frac{\partial^2 f}{\partial x^2} = \frac{\partial^2 f}{\partial y^2} = 1$$

at that point.

45. Let $w = f(u, v)$ be a function of u, v with continuous partial derivatives, where u, v in turn are functions of independent variables, x, y, z, with continuous partial derivatives. Show that if w is regarded as a function of x, y, z, its gradient at any point (x_0, y_0, z_0) lies in a common plane with the gradients of $u = u(x, y, z)$ and $v = v(x, y, z)$.

46. Show that if a function u has first derivatives that satisfy a relation of the form $F(u_x, u_y) = 0$, and if $\partial F/\partial u_x$ and $\partial F/\partial u_y$ are not both zero, then u also satisfies $u_{xx}u_{yy} - u_{xy}^2 = 0$. (*Hint:* Differentiate both sides of the equation $F = 0$ with respect to x and y.)

47. If $f(x, y) = 0$, find d^2y/dx^2.

48. If $f(x, y, z) = 0$ and $z = x + y$, find dz/dx.

49. The function $v(x, t)$ is defined for $0 \le x \le 1$, $0 \le t$ and satisfies the partial differential equation

$$v_t = v_x(v - x) + a v_{xx}$$

(a = constant > 0) and the boundary conditions $v(0, t) = 0$, $v(1, t) = 1$. Suppose that for each fixed t, $v(x, t)$ is a strictly increasing function of x; that is, $v_x(x, t) > 0$. Show that v and t may be introduced as independent variables and x as dependent variable, and find the partial differential equation satisfied by the function $x(v, t)$. Also find the region of definition of $x(v, t)$ and boundary values that it satisfies. By considering level curves, show geometrically why the assumption $v_x(x, t) > 0$ is necessary for the success of this transformation.

50. Let $f(x, y, z)$ be a function depending only on $r = \sqrt{x^2 + y^2 + z^2}$; that is, $f(x, y, z) = g(r)$. Prove that if $f_{xx} + f_{yy} + f_{zz} = 0$, it follows that

$$f = \left(\frac{a}{r}\right) + b,$$

where a and b are constants.

51. A function $f(x, y)$, defined and differentiable for all x, y, is said to be homogeneous of degree n (a non-negative integer) if $f(tx, ty) = t^n f(x, y)$ for all t, x, and y. For such a function prove:

a) $x(\partial f/\partial x) + y(\partial f/\partial y) = nf(x, y)$ and express this in vector form;

b) $$x^2\left(\frac{\partial^2 f}{\partial x^2}\right) + 2xy\left(\frac{\partial^2 f}{\partial x \partial y}\right) + y^2\left(\frac{\partial^2 f}{\partial y^2}\right) = n(n-1)$$

if f has continuous second partial derivatives;

c) a homogeneous function of degree zero is a constant.

52. Prove the Mean Value Theorem for functions of two variables

$$f(x + h, y + k) - f(x, y) = f_x(x + \theta h, y + \theta k)h + f_y(x + \theta h, y + \theta k)k, \quad 0 < \theta < 1,$$

with suitable assumptions about f. What assumptions? (Apply the Mean Value Theorem for functions of one variable to $F(t) = f(x + ht, y + kt)$.)

53. Prove the theorem: If $f(x, y)$ is defined in a region R, and f_x, f_y exist and are bounded in R, then $f(x, y)$ is continuous in R. (The assumption of boundedness is essential.)

54. For each of the following three surfaces, find all the values of x and y for which z is a maximum or minimum (if there are any). Give complete reasonings.

a) $x^2 + y^2 + z^2 = 3$ b) $x^2 + y^2 = 2z$

c) $x^2 - y^2 = 2z$

55. Find the point(s) on the surface $xyz = 1$ whose distance from the origin is a minimum.

56. A closed rectangular box is to be made to hold a given volume, V in^3. The cost of the material used in the box is a cents/in^2 for top and bottom, b cents/in^2 for front and back, c cents/in^2 for the remaining two sides. What dimensions make the total cost of materials a minimum?

57. Find the maximum value of the function $xye^{-(2x+3y)}$ in the first quadrant.

58. A surface is defined by $z = x^3 + y^3 - 9xy + 27$. Prove that the only possible maxima and minima of z occur at $(0, 0)$ or $(3, 3)$. Prove that $(0, 0)$ is neither a maximum nor a minimum. Determine whether $(3, 3)$ is a maximum or a minimum.

59. Find the minimum volume bounded by the planes $x = 0$, $y = 0$, $z = 0$, and a plane that is tangent to the ellipsoid

$$\frac{x^2}{a^2} + \frac{y^2}{b^2} + \frac{z^2}{c^2} = 1$$

at a point in the octant $x > 0$, $y > 0$, $z > 0$.

60. Let z be defined implicitly as a function of x and y by the equation $\sin(x + y) + \sin(y + z) = 1$. Compute $\partial^2 z/\partial x\, \partial y$ in terms of x, y, and z.

61. Given $z = xy^2 - y \sin x$, calculate the value of $y(\partial^2 z/\partial y\, \partial x) - \partial z/\partial x$.

62. Let $w = z \tan^{-1}(x/y)$. Compute

$$\frac{\partial^2 w}{\partial x^2} + \frac{\partial^2 w}{\partial y^2} + \frac{\partial^2 w}{\partial z^2}.$$

63. Show that the function satisfies the equation.

a) $\int_0^{x/2\sqrt{kt}} e^{-\sigma^2}\, d\sigma, \qquad kf_{xx} - f_t = 0 \qquad (k \text{ const.})$

b) $\phi(x + at) + \psi(x - at), \qquad f_{tt} = a^2 f_{xx}$

64. Let

$$f(x, y) = \begin{cases} xy\dfrac{x^2 - y^2}{x^2 + y^2}, & (x, y) \neq (0, 0), \\ 0, & (x, y) = (0, 0). \end{cases}$$

Find $f_{yx}(0, 0)$ and $f_{xy}(0, 0)$. See Fig. 15.52.

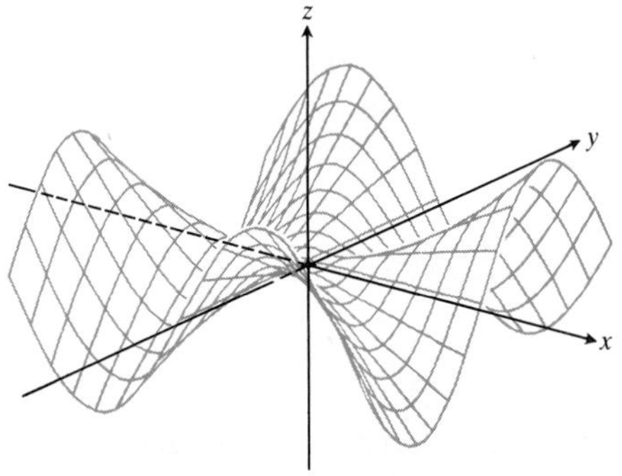

15.52 The surface of Problem 64.

65. Is $2x(x^3 + y^3)\, dx + 3y^2(x^2 + y^2)\, dy$ the differential df of a function $f(x, y)$? If so, find the function.

66. Find a function $w = f(x, y)$ such that $\partial w/\partial x = 1 + e^x \cos y$ and $\partial w/\partial y = 2y - e^x \sin y$, or else explain why no such function exists.

67. In thermodynamics the five quantities S, T, u, p, v are such that any two of them may be considered independent variables, the others then being determined. They are connected by the differential relation $T\, dS = du + p\, dv$. Show that

$$\left(\frac{\partial S}{\partial v}\right)_T = \left(\frac{\partial p}{\partial T}\right)_v \quad \text{and} \quad \left(\frac{\partial v}{\partial S}\right)_p = \left(\frac{\partial T}{\partial p}\right)_S.$$

68. Let

$$f(r, \theta) = \begin{cases} \dfrac{\sin 6r}{6r}, & r \neq 0 \\ 1, & r = 0. \end{cases}$$

(See Fig. 15.53.) Find (a) $\lim_{r \to 0} f(r, \theta)$, (b) $f_r(0, 0)$, (c) $f_\theta(r, \theta)$, $r \neq 0$.

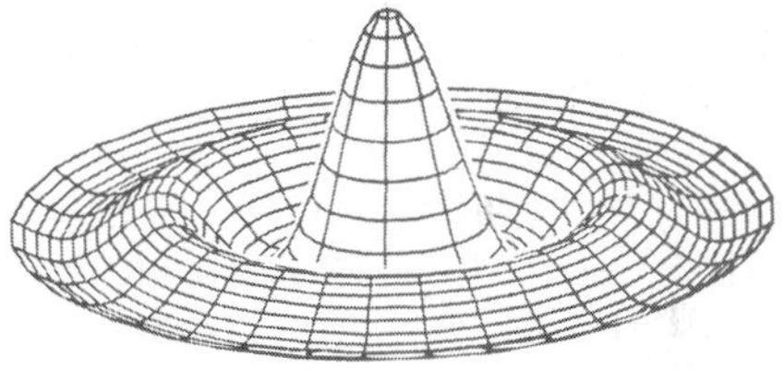

15.53 The surface of Problem 68.

16 Multiple Integrals

16.1 Introduction

Up to now in applications of calculus we have worked with integrals of functions of a single variable. As we know from Chapter 4, the integral of a function $y = f(x)$ over an interval $[a, b]$ is a limit of approximating sums

$$\int_a^b f(x)\,dx = \lim \sum_{k=1}^{n} f(c_k)\,\Delta x_k, \tag{1}$$

where $a = x_0 < x_1 < \cdots < x_n = b$, $\Delta x_k = x_k - x_{k-1}$, and c_k is any point from the interval $[x_{k-1}, x_k]$. The limit in (1) is taken as the norm of the subdivision (length of the longest subinterval) approaches zero. The limit is guaranteed to exist if f is continuous and also exists when f is bounded and has only finitely many points of discontinuity in $[a, b]$. There is no loss in assuming the intervals $[x_k, x_{k+1}]$ to have a common length $\Delta x = (b - a)/n$, and the limit may thus be obtained by letting $\Delta x \to 0$ or $n \to \infty$.

If $f(x) > 0$, then $\int_a^b f(x)\,dx$ represents the area under the graph of f from $x = a$ to $x = b$, but in general the integral has many other important interpretations (distance, volume, arc length, surface area, moment of inertia, mass, hydrostatic pressure, work) depending on the nature and interpretation of f.

In this chapter we shall see that integrals of functions of two or more variables, which are called *multiple integrals* and defined in much the same way as integrals of functions of a single variable, have equally far-reaching interpretations.

16.2 Double Integrals

In this article we define the integral of a function $f(x, y)$ of two variables over a rectangular region in the xy-plane. We then show how such an integral is evaluated and generalize the definition to include bounded regions of a more general nature.

Double Integrals Over Rectangles

Suppose that $f(x, y)$ is defined on a rectangular region R defined by

$$R: a \leq x \leq b, \qquad c \leq y \leq d.$$

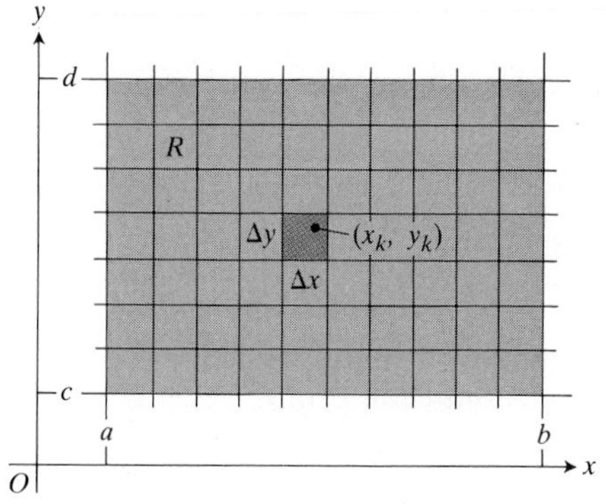

16.1 Rectangular grid subdividing the region R into small rectangles of area $\Delta A_k = \Delta x_k \Delta y_k$.

We imagine R to be covered by a network of lines parallel to the x- and y-axes, as shown in Fig. 16.1. These lines divide R into small pieces of area

$$\Delta A = \Delta x \, \Delta y.$$

We number these in some order

$$\Delta A_1, \Delta A_2, \ldots, \Delta A_n,$$

choose a point (x_k, y_k) in each piece ΔA_k, and form the sum

$$S_n = \sum_{k=1}^{n} f(x_k, y_k) \, \Delta A_k. \tag{2}$$

If f is continuous throughout R, then as we refine the mesh width to make both Δx and Δy go to zero the sums in (2) approach a limit called the *double integral* of f over R that is denoted by

$$\iint_R f(x, y) \, dA \qquad \text{or} \qquad \iint_R f(x, y) \, dx \, dy.$$

Thus,

$$\iint_R f(x, y) \, dA = \lim_{\Delta A \to 0} \sum_{1}^{n} f(x_k, y_k) \, \Delta A_k. \tag{3}$$

As with functions of a single variable, the sums approach this limit no matter how the intervals $[a, b]$ and $[c, d]$ that determine R are subdivided, as long as the norms of the subdivisions both go to zero. The limit in (3) is also independent of the order in which the areas ΔA_k are numbered, and independent of the choice of the point (x_k, y_k) within each ΔA_k. The values of the individual approximating sums S_n depend on these choices, of course, but the sums approach the same limit in the end. The proof of the existence and uniqueness of this limit for a continuous function f is given in more advanced texts. The continuity of f is a sufficient condition for the existence of the double integral, but not a necessary one, and the limit in question exists for many discontinuous functions as well.

Properties of the Double Integral

Like "single" integrals, as we may now call integrals of functions of a single variable, double integrals of continuous functions have algebraic properties that are useful in computations and applications. These properties hold because they hold for the sums from which the integrals are defined. Among them are the following.

I1. $\displaystyle\iint_R kf(x, y) \, dA = k \iint_R f(x, y) \, dA$ (any number k)

I2. $\displaystyle\iint_R [f(x, y) + g(x, y)] \, dA = \iint_R f(x, y) \, dA + \iint_R g(x, y) \, dA$

I3. $\displaystyle\iint_R [f(x, y) - g(x, y)] \, dA = \iint_R f(x, y) \, dA - \iint_R g(x, y) \, dA$

I4. $\iint_R f(x, y)\, dA \geq 0$ if $f(x, y) \geq 0$ on R

I5. $\iint_R f(x, y)\, dA \geq \iint_R g(x, y)\, dA$ if $f(x, y) \geq g(x, y)$ on R

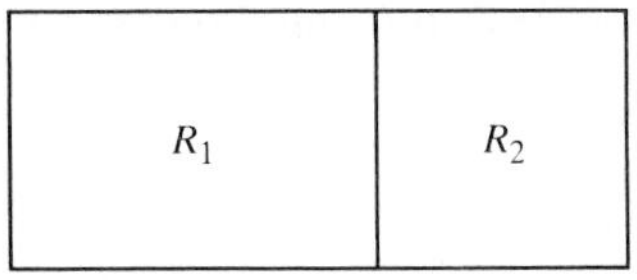

16.2 $\iint_{R_1 \cup R_2} f(x, y)\, dA = \int_{R_1} f(x, y)\, dA + \int_{R_2} f(x, y)\, dA.$

These are like the properties I1–I5 in Article 4.5, and the proofs (which we omit) are similar.

There is also a "domain additivity" property:

I6. $\iint_R f(x, y)\, dA = \iint_{R_1} f(x, y)\, dA + \iint_{R_2} f(x, y)\, dA,$

which holds when R is the union of two nonoverlapping rectangles R_1 and R_2 as shown in Fig. 16.2. Again, we shall omit the proof.

Volume

When $f(x, y) > 0$, we may interpret $\iint_R f(x, y)\, dA$ as the volume of the solid enclosed by R, the planes $x = a$, $x = b$, $y = c$, $y = d$, and the surface $z = f(x, y)$, as shown in Fig. 16.3. Each term $f(x_k, y_k)\, \Delta A_k$ in the sum

$$S_n = \Sigma f(x_k, y_k)\, \Delta A_k$$

is the volume of a vertical rectangular prism that approximates the volume of the portion of the solid that stands directly above the base ΔA_k. The sum S_n thus approximates what we want to call the total volume of the solid, and we *define* this volume to be

$$\text{Volume} = \lim S_n = \iint_R f(x, y)\, dA. \tag{4}$$

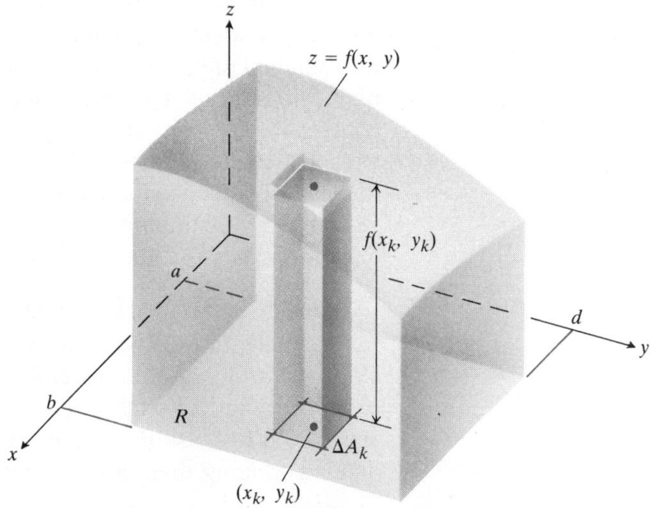

16.3 Approximating a solid with rectangular prisms leads to a definition of volume consistent with past definitions.

Fubini's Theorem for Calculating Double Integrals

We are now ready to calculate our first double integral.

Suppose we wish to calculate the volume under the plane $z = 4 - x - y$ over the region R: $0 \leq x \leq 2$, $0 \leq y \leq 1$ in the xy-plane. If

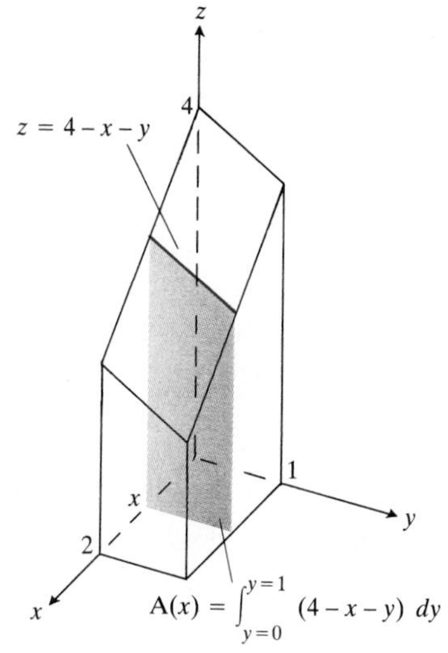

16.4 The cross-sectional area $A(x)$ is obtained by holding x fixed and integrating with respect to y.

we apply the method of slicing from Article 5.4, with slices perpendicular to the x-axis (Fig. 16.4), then the volume is

$$\int_{x=0}^{x=2} A(x)\, dx, \tag{5}$$

where $A(x)$ is the cross-sectional area at x. For each value of x we may calculate $A(x)$ as the integral

$$A(x) = \int_{y=0}^{y=1} (4 - x - y)\, dy, \tag{6}$$

which is the area under the curve $z = 4 - x - y$ in the plane of the cross section at x. In calculating $A(x)$, x is held fixed and the integration takes place with respect to y. Combining (5) and (6) we see that the volume of the entire solid is

$$\begin{aligned} \text{Volume} &= \int_{x=0}^{x=2} A(x)\, dx = \int_{x=0}^{x=2} \left(\int_{x=0}^{x=1} (4 - x - y)\, dy \right) dx \\ &= \int_{x=0}^{x=2} \left[4y - xy - \frac{y^2}{2} \right]_{y=0}^{y=1} dx \\ &= \int_{x=0}^{x=2} \left[\frac{7}{2} - x \right] dx = \frac{7}{2}x - \frac{x^2}{2} \bigg]_0^2 = 5. \end{aligned} \tag{7}$$

If we had just wanted to write instructions for calculating the volume, without carrying out any of the integrations, we could have written

$$\text{Volume} = \int_0^2 \int_0^1 (4 - x - y)\, dy\, dx.$$

The expression on the right, called an *iterated* or *repeated* integral, says that the volume is to be obtained by integrating $4 - x - y$ with respect to y from $y = 0$ to $y = 1$ holding x fixed, and then by integrating the resulting expression in x with respect to x from $x = 0$ to $x = 2$.

What would have happened if we had calculated the volume by slicing with planes perpendicular to the y-axis, as shown in Fig. 16.5? As a function of y the typical cross-sectional area is now

$$A(y) = \int_{x=0}^{x=2} (4 - x - y)\, dx = 4x - \frac{x^2}{2} - xy \bigg]_{x=0}^{x=2} = 6 - 2y. \tag{8}$$

The volume of the entire solid is therefore

$$\text{Volume} = \int_{y=0}^{y=1} A(y)\, dy = \int_{y=0}^{y=1} (6 - 2y)\, dy = 6y - y^2 \bigg]_0^1 = 5,$$

in agreement with our earlier calculation.

Again, we may give instructions for calculating the volume as an iterated integral by writing

$$\text{Volume} = \int_0^1 \int_0^2 (4 - x - y)\, dx\, dy.$$

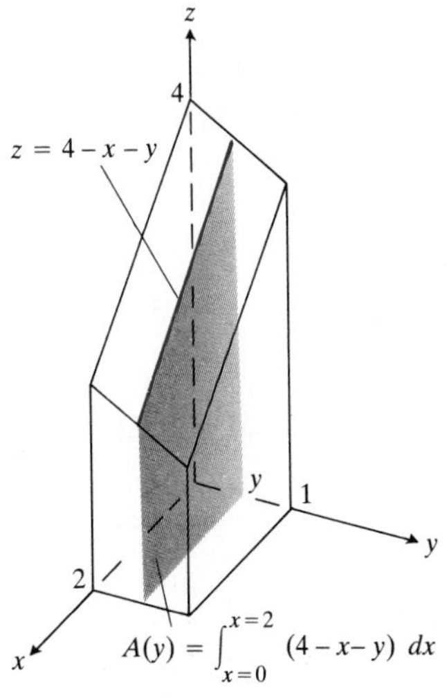

16.5 The cross-sectional area $A(y)$ is obtained by holding y fixed and integrating with respect to x.

The expression on the right says that the volume may be obtained by integrating $4 - x - y$ with respect to x from $x = 0$ to $x = 2$ (as we did in Eq. 8) and by integrating the result with respect to y from $y = 0$ to $y = 1$.

In this iterated integral the order of integration is first x and then y, the reverse of the order we used in (7).

What do these two volume calculations with iterated integrals have to do with the double integral

$$\iint_R (4 - x - y)\, dA$$

over the rectangle R: $0 \le x \le 2$, $0 \le y \le 1$? The answer is that they both give the value of the double integral. A theorem proved by Guido Fubini (1879–1943) and published in 1907 says that the double integral of any continuous function over a rectangle can always be calculated as an iterated integral in either order of integration. (Fubini proved his theorem in much greater generality, but this is how it translates into what we're doing now.)

THEOREM 1

Fubini's Theorem (First Form)

If $f(x, y)$ is continuous on the rectangular region R: $a \le x \le b$, $c \le y \le d$, then

$$\iint_R f(x, y)\, dA = \int_c^d \int_a^b f(x, y)\, dx\, dy = \int_a^b \int_c^d f(x, y)\, dy\, dx.$$

Fubini's theorem says that double integrals over rectangles can always be calculated as iterated integrals. This means that we can evaluate a double integral by integrating one variable at a time, using the integration techniques we already know for functions of a single variable.

Fubini's theorem also says that we may calculate the double integral by integrating in *either* order (a genuine convenience, as we shall see in Example 3). In particular, when we calculate a volume by slicing, we may use either planes perpendicular to the x-axis or planes perpendicular to the y-axis. We get the same answer either way.

Even more important is the fact that Fubini's theorem holds for *any* continuous function $f(x, y)$. In particular, f may have negative values as well as positive values on R, and the integrals we calculate with Fubini's theorem may represent other things besides volumes (as we shall see later on).

EXAMPLE 1 Calculate $\iint_R f(x, y)\, dA$ for

$$f(x, y) = 1 - 6x^2y \quad \text{and} \quad R: 0 \le x \le 2,\ -1 \le y \le 1.$$

Solution By Fubini's theorem

$$\iint_R f(x, y)\, dA = \int_{-1}^{1} \int_0^2 (1 - 6x^2y)\, dx\, dy = \int_{-1}^{1} [x - 2x^3y]_{x=0}^{x=2}\, dy$$

$$= \int_{-1}^{1} [2 - 16y]\, dy = 2y - 8y^2]_{-1}^{1}$$

$$= (2 - 8) - (-2 - 8) = 4.$$

Reversing the order of integration gives the same answer:

$$\int_0^2 \int_{-1}^1 (1 - 6x^2y)\, dy\, dx = \int_0^2 [y - 3x^2y^2]_{y=-1}^{y=1}\, dx$$

$$= \int_0^2 [(1 - 3x^2 - (-1 - 3x^2)]\, dx$$

$$= \int_0^2 2\, dx = 4. \square$$

Double Integrals for Bounded Nonrectangular Regions

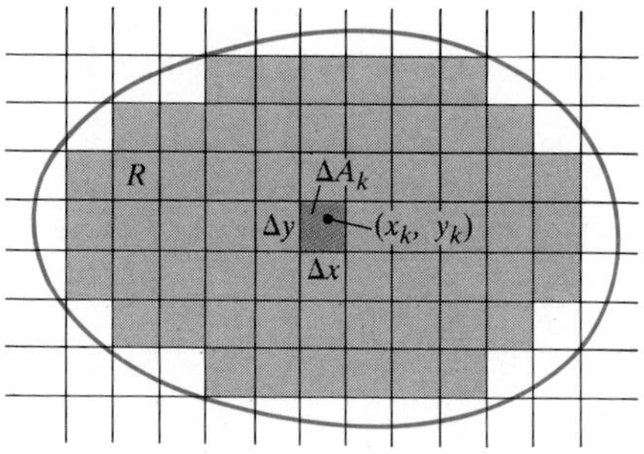

16.6 A rectangular grid subdividing a bounded nonrectangular region into cells.

To define the double integral of a function $f(x, y)$ over a bounded nonrectangular region like the one shown in Fig. 16.6, we again imagine R to be covered by a rectangular grid, but we include in the partial sum only the small pieces of area $\Delta A = \Delta x\, \Delta y$ that lie entirely within the region (shaded in the figure). We number the pieces in some order, choose an arbitrary point (x_k, y_k) in each ΔA_k, and form the sum

$$S_n = \sum_{k=1}^{n} f(x_k, y_k)\, \Delta A_k.$$

The only difference between this sum and the one in Eq. (2) is that now the areas ΔA_k may not cover the entire region R. But, as the mesh becomes increasingly fine and the number of terms in S_n increases, more and more of R is included. If f is continuous, and the boundary of R is made up of a finite number of line segments or smooth curves pieced together end to end, then the sums S_n will have a limit as Δx and Δy approach zero. We call the limit the double integral of f over R:

$$\iint_R f(x, y)\, dA = \lim_{\Delta A \to 0} \sum f(x_k, y_k)\, \Delta A_k.$$

This limit may also exist under less restrictive circumstances, but we shall not pursue this point here.

Double integrals of continuous functions over nonrectangular regions have the algebraic properties I1–I5 listed earlier for integrals over rectangular regions. The domain additivity property corresponding to I6 says that if R is decomposed into nonoverlapping regions R_1 and R_2 with boundaries that are again made of line segments or smooth curves (see Fig. 16.7 for an example), then

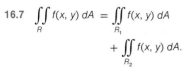

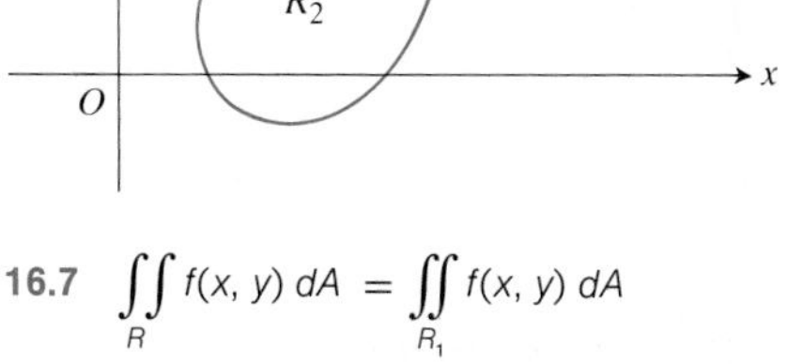

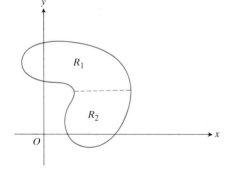

16.7 $\iint_R f(x, y)\, dA = \iint_{R_1} f(x, y)\, dA + \iint_{R_2} f(x, y)\, dA.$

I6′. $\iint_R f(x, y)\, dA = \iint_{R_1} f(x, y)\, dA + \iint_{R_2} f(x, y)\, dA.$

If $f(x, y)$ is positive and continuous over R (Fig. 16.8), we define the volume of the solid region between R and the surface $z = f(x, y)$ to be $\iint_R f(x, y)\, dA$, as before.

If R is a region like the one shown in the xy-plane in Fig. 16.9, bounded "above" and "below" by the curves $y = f_2(x)$ and $y = f_1(x)$, and on the sides by the lines $x = a$, $x = b$, we may again calculate the volume by the method of slicing. We first calculate the cross-sectional area

$$A(x) = \int_{y=f_1(x)}^{y=f_2(x)} f(x, y)\, dy$$

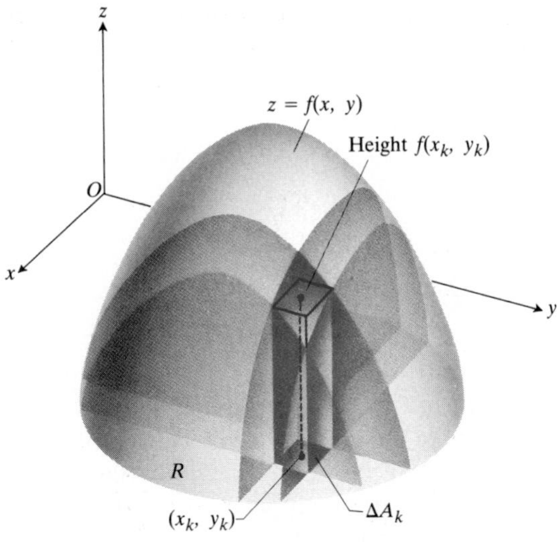

16.8 Volume $= \lim \Sigma f(x_k, y_k)\Delta A_k = \iint_R f(x, y)\, dA$.

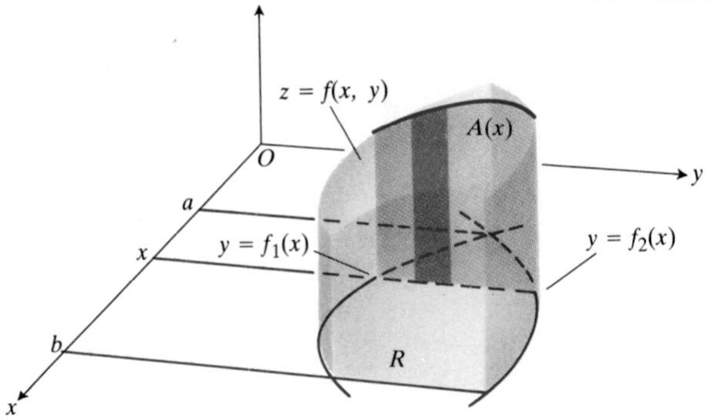

16.9 The area of the vertical slice shown here is

$$A(x) = \int_{f_1(x)}^{f_2(x)} f(x, y)\, dy.$$

This area is integrated from a to b to calculate the volume.

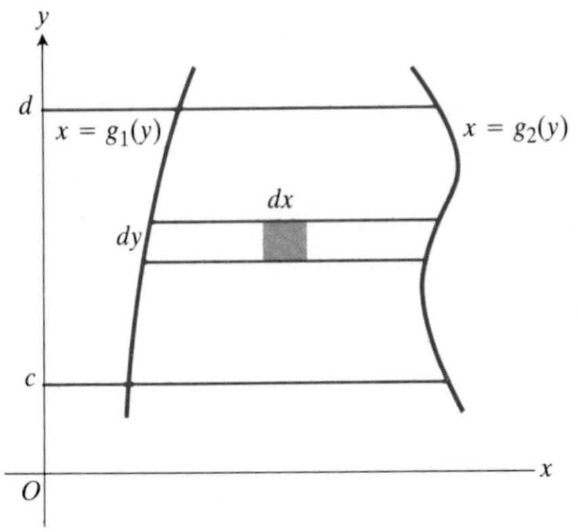

16.10 Area: $\int_c^d \int_{g_1(y)}^{g_2(y)} dx\, dy$.

and then integrate $A(x)$ from $x = a$ to $x = b$ to get the volume as an iterated integral:

$$V = \int_a^b A(x)\, dx = \int_a^b \int_{f_1(x)}^{f_2(x)} f(x, y)\, dy\, dx. \tag{9}$$

Similarly, if R is a region like the one shown in Fig. 16.10, bounded on the right by $x = g_2(y)$, on the left by $x = g_1(y)$, and below and above by the lines $y = c$ and $y = d$, then the volume calculated by slicing is given by the iterated integral

$$\text{Volume} = \int_c^d \int_{g_1(y)}^{g_2(y)} f(x, y)\, dx\, dy. \tag{10}$$

The fact that the iterated integrals in Eqs. (9) and (10) both give the volume that we defined to be the double integral of f over R is a consequence of the following stronger form of Fubini's theorem.

THEOREM 2

Fubini's Theorem (Stronger Form)

Let $f(x, y)$ be continuous on a region R.

1. If R is defined by $a \le x \le b$, $f_1(x) \le y \le f_2(x)$, with f_1 and f_2 continuous on $[a, b]$, then

$$\iint_R f(x, y)\, dA = \int_a^b \int_{f_1(x)}^{f_2(x)} f(x, y)\, dy\, dx.$$

2. If R is defined by $c \le y \le d$, $g_1(y) \le x \le g_2(y)$, with g_1 and g_2 continuous on $[c, d]$, then

$$\iint_R f(x, y)\, dA = \int_c^d \int_{g_1(y)}^{g_2(y)} f(x, y)\, dx\, dy.$$

EXAMPLE 2 Find the volume of the prism whose base is the triangle in the xy-plane bounded by the x-axis and the lines $y = x$ and $x = 1$, and whose top lies in the plane

$$z = f(x, y) = 3 - x - y.$$

Solution For any x between 0 and 1, y may vary from $y = 0$ to $y = x$ (Fig. 16.11). Hence,

$$V = \int_0^1 \int_0^x (3 - x - y)\, dy\, dx = \int_0^1 \left[3y - xy - \frac{y^2}{2}\right]_{y=0}^{y=x} dx$$

$$= \int_0^1 \left(3x - \frac{3x^2}{2}\right) dx = \frac{3x^2}{2} - \frac{x^3}{2}\bigg]_{x=0}^{x=1} = 1.$$

When the order of integration is reversed, the integral for the volume is

$$V = \int_0^1 \int_y^1 (3 - x - y)\, dx\, dy = \int_0^1 \left[3x - \frac{x^2}{2} - xy\right]_{x=y}^{x=1} dy$$

$$= \int_0^1 \left(3 - \frac{1}{2} - y - 3y + \frac{y^2}{2} + y^2\right) dy$$

$$= \int_0^1 \left(\frac{5}{2} - 4y + \frac{3}{2}y^2\right) dy = \frac{5}{2}y - 2y^2 + \frac{y^3}{2}\bigg]_{y=0}^{y=1} = 1.$$

The two integrals are equal, as they should be. □

While Fubini's theorem assures us that a double integral may be calculated as an iterated integral in either order of integration, the value of one integral may be easier to find than the value of the other. The next example shows how this can happen.

Figure 16.11

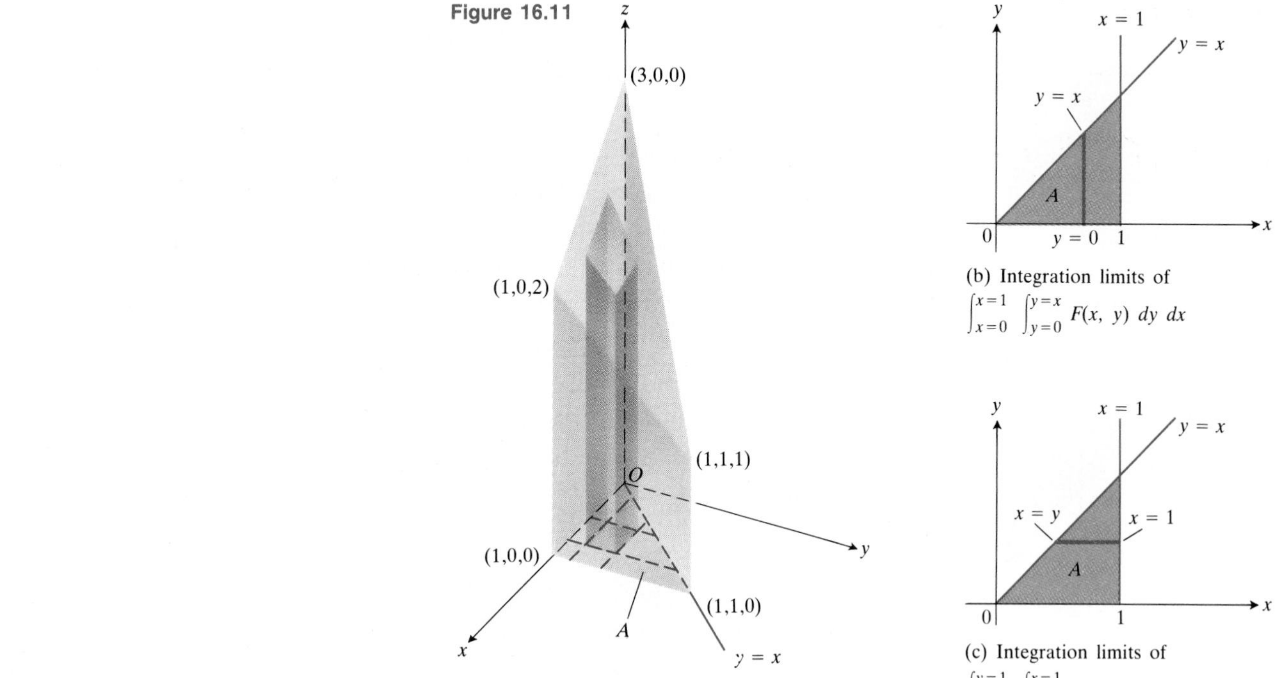

(a) Prism with a triangular base in the xy-plane.

(b) Integration limits of $\int_{x=0}^{x=1} \int_{y=0}^{y=x} F(x, y)\, dy\, dx$

(c) Integration limits of $\int_{y=0}^{y=1} \int_{x=y}^{x=1} F(x, y)\, dx\, dy$

EXAMPLE 3 Calculate

$$\iint_A \frac{\sin x}{x}\, dA,$$

where A is the triangle in the xy-plane bounded by the x-axis, the line $y = x$, and the line $x = 1$.

Solution The region of integration is the same as the one in Example 2. If we integrate first with respect to y and then with respect to x, we find

$$\int_0^1 \left(\int_0^x \frac{\sin x}{x}\, dy \right) dx = \int_0^1 \left(y \frac{\sin x}{x} \right]_{y=0}^{y=x} \Big) dy$$

$$= \int_0^1 \sin x\, dx = -\cos(1) + 1 \approx 0.46.$$

If we reverse the order of integration, and attempt to calculate

$$\int_0^1 \int_y^1 \frac{\sin x}{x}\, dx\, dy,$$

we are stopped by the fact that $\int (\sin x/x)\, dx$ cannot be expressed in terms of elementary functions. □

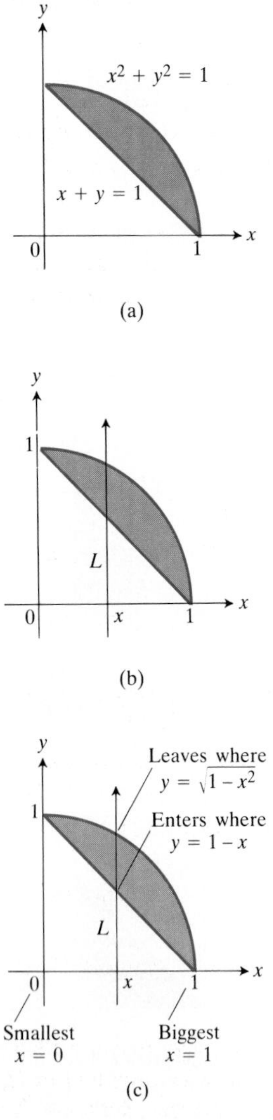

16.12 Finding limits of integration.

Determining the Limits of Integration

The hardest part of evaluating a double integral can be finding the limits of integration. Fortunately, there is a good procedure to follow.

If we want to evaluate

$$\iint_R f(x, y)\, dA$$

over the region R shown in Fig. 16.12(a), integrating first with respect to y and then with respect to x, we take the following steps:

1. We imagine a vertical line L cutting through R in the direction of increasing y (Fig. 16.12b).
2. We integrate from the y-value where L enters R to the y-value where L leaves R (Fig. 16.12c).
3. We choose x-limits that include all the vertical lines that pass through R (Fig. 16.12c). The integral is

$$\int_{x=0}^{x=1} \int_{y=1-x}^{y=\sqrt{1-x^2}} f(x, y)\, dy\, dx.$$

To calculate the same double integral as an iterated integral with the order of integration reversed, the procedure uses horizontal lines (Fig. 16.13) to give

$$\int_{y=0}^{y=1} \int_{x=1-y}^{x=\sqrt{1-y^2}} f(x, y)\, dx\, dy.$$

EXAMPLE 4 Sketch the region of integration of

$$\int_0^2 \int_{x^2}^{2x} f(x, y)\, dy\, dx$$

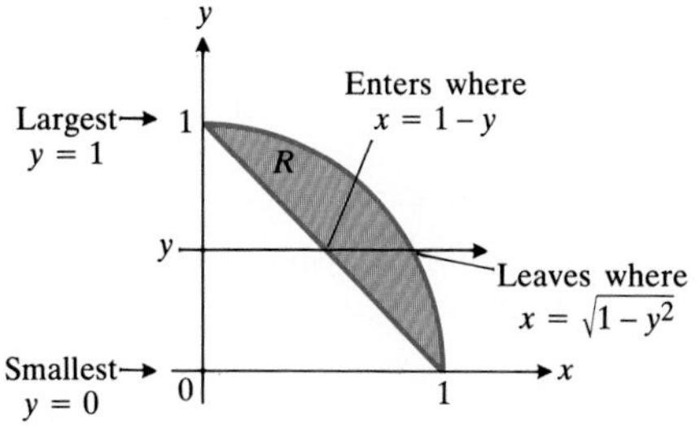

16.13 $\iint_R f(x, y)\, dA = \int_0^1 \int_{1-y}^{\sqrt{1-y^2}} f(x, y)\, dx\, dy.$

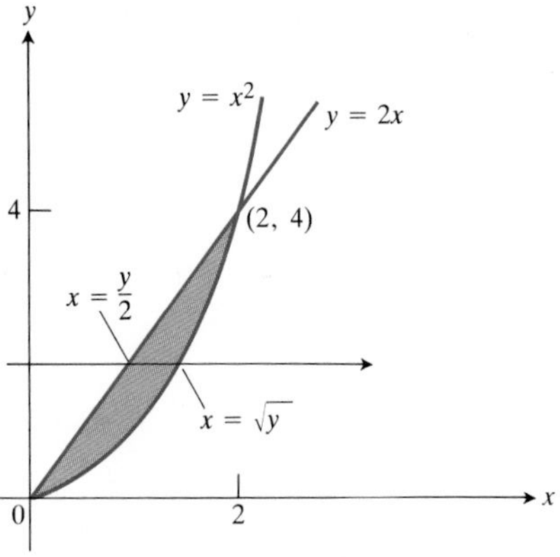

16.14 $\int_0^2 \int_{x^2}^{2x} f(x, y)\, dy\, dx = \int_0^4 \int_{y/2}^{\sqrt{y}} f(x, y)\, dx\, dy.$

and express the integral as an equivalent double integral with the order of integration reversed.

Solution The region of integration is given by the inequalities $x^2 \le y \le 2x$, $0 \le x \le 2$. It is therefore the region bounded by the curves $y = x^2$ and $y = 2x$ between the vertical lines $x = 0$ and $x = 2$, as shown in Fig. 16.14.

To find the limits for the integration in the reverse order we imagine a horizontal line passing from left to right through the region. It enters at $x = y/2$ and leaves at $x = \sqrt{y}$. To include all such lines we let y run from $y = 0$ to $y = 4$. The integral is

$$\int_0^4 \int_{y/2}^{\sqrt{y}} f(x, y)\, dx\, dy. \quad \square$$

PROBLEMS

Evaluate the following integrals and sketch the region over which each integration takes place.

1. $\int_0^3 \int_0^2 (4 - y^2)\, dy\, dx$
2. $\int_{-1}^0 \int_{-1}^1 (x + y + 1)\, dx\, dy$
3. $\int_0^3 \int_{-2}^0 (x^2y - 2xy)\, dy\, dx$
4. $\int_\pi^{2\pi} \int_0^\pi (\sin x + \cos y)\, dx\, dy$
5. $\int_0^\pi \int_0^x x \sin y\, dy\, dx$
6. $\int_1^{\ln 8} \int_0^{\ln y} e^{x+y}\, dx\, dy$
7. $\int_0^\pi \int_0^{\sin x} y\, dy\, dx$
8. $\int_1^2 \int_y^{y^2} dx\, dy$

Evaluate the following integrals.

9. $\int_{10}^1 \int_0^{1/y} ye^{xy}\, dx\, dy$
10. $\int_0^1 \int_0^{x^3} e^{y/x}\, dy\, dx$

In Problems 11–16, integrate the function $f(x, y)$ over the given region.

11. $f(x, y) = x/y$ over the region in the first quadrant bounded by the lines $y = x$, $y = 2x$, $x = 1$, $x = 2$.
12. $f(x, y) = x^2 + y^2$ over the triangular region whose vertices are (0, 0), (1, 0) and (0, 1).
13. $f(x, y) = y - \sqrt{x}$ over the triangular region cut from the first quadrant by the line $x + y = 1$.
14. $f(x, y) = x^2 + 3xy$ over the rectangle R: $0 \le x \le 1$, $0 \le y \le 1$.
15. $f(x, y) = 1/xy$ over the rectangle R: $1 \le x \le 2$, $1 \le y \le 2$.
16. $f(x, y) = y \cos xy$ over the rectangle R: $0 \le x \le \pi$, $0 \le y \le 1$.

In Problems 17–20, sketch the region over which the integration takes place and write an equivalent integral with the order of integration reversed. Evaluate both integrals.

17. $\int_0^2 \int_1^{e^x} dy\,dx$ **18.** $\int_0^1 \int_{\sqrt{y}}^1 dx\,dy$

19. $\int_0^{\sqrt{2}} \int_{-\sqrt{4-2y^2}}^{\sqrt{4-2y^2}} y\,dx\,dy$ **20.** $\int_{-2}^1 \int_{x^2+4x}^{3x+2} dy\,dx$

In Problems 21–26, write an equivalent iterated integral with the order of integration reversed. *Do not integrate.* It will help to sketch the region over which the integration takes place.

21. $\int_0^1 \int_{x^2}^{x} f(x, y)\,dy\,dx$ **22.** $\int_0^1 \int_x^{2x} f(x, y)\,dy\,dx$

23. $\int_0^1 \int_1^{e^x} dy\,dx$ **24.** $\int_0^1 \int_{\sqrt{x}}^1 \cos(x + y)\,dy\,dx$

25. $\int_0^2 \int_0^{x^3} f(x, y)\,dy\,dx$ **26.** $\int_0^1 \int_{-\sqrt{y}}^{\sqrt{y}} f(x, y)\,dx\,dy$

Evaluate the integrals in Problems 27–32 by integrating the equivalent integral obtained by reversing the order of integration.

27. $\int_0^{\pi} \int_x^{\pi} \frac{\sin y}{y}\,dy\,dx$ **28.** $\int_0^1 \int_{2y}^2 \cos(x^2)\,dx\,dy$

29. $\int_0^1 \int_y^1 x^2 e^{xy}\,dx\,dy$ **30.** $\int_0^2 \int_x^2 y^2 \sin xy\,dy\,dx$

31. $\int_0^8 \int_{\sqrt[3]{x}}^2 \frac{dy\,dx}{y^4 + 1}$ **32.** $\int_0^2 \int_0^{4-x^2} \frac{xe^{2y}}{4 - y}\,dy\,dx$

33. Find the volume of the solid whose base is the region in the xy-plane that is bounded by the parabola $y = 4 - x^2$ and the line $y = 3x$, while the top of the solid is bounded by the plane $z = x + 4$.

34. The base of a solid is the region in the xy-plane that is bounded by the circle $x^2 + y^2 = a^2$, while the top of the solid is bounded by the paraboloid $az = x^2 + y^2$. Find the volume.

35. Find the volume in the first octant bounded by the coordinate planes, the cylinder $x^2 + y^2 = 4$, and the plane $z + y = 3$.

36. Find the volume of the solid in the first octant bounded by the plane, the cylinder $y = x^2$, the surface $z = xy$, and the planes $x = 2$, $y = 0$, $z = 0$.

37. Find the volume of the solid in the first octant bounded by the coordinate planes, the plane $x = 3$, and the parabolic cylinder $z = 4 - y^2$.

38. Find the volume of the solid cut from the first octant by the surface $z = 4 - x^2 - y$.

39. Evaluate the integral

$$\int_0^2 \int_{y/2}^1 e^{x^2}\,dx\,dy.$$

40. The volume under the paraboloid $z = x^2 + y^2$ and above a certain region R in the xy-plane is

$$V = \int_0^1 \int_0^y (x^2 + y^2)\,dx\,dy + \int_1^2 \int_0^{2-y} (x^2 + y^2)\,dx\,dy.$$

Sketch the region and express the volume as an iterated integral with the order of integration reversed.

41. Evaluate the integral

$$\int_0^2 (\tan^{-1} \pi x - \tan^{-1} x)\,dx.$$

(*Hint:* Write the integrand as an integral.)

***Toolkit* programs**

Double Integral

16.3 Area

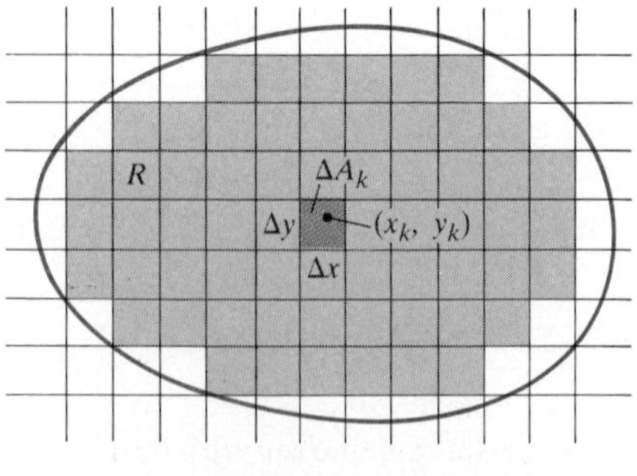

16.15 A rectangular grid subdividing a bounded nonrectangular region into cells.

If we take $f(x, y) = 1$ in the definition of the double integral over a region R in the preceding article, the partial sums reduce to

$$S_n = \sum_{k=1}^{n} f(x_k, y_k)\,\Delta A_k = \sum_{k=1}^{n} \Delta A_k \tag{1}$$

and give an approximation to what we would like to call the area of the region R. As Δx and Δy approach zero, the coverage of R by the ΔA_k's (Fig. 16.15) becomes increasingly complete, and we *define* the area of R to be the limit

$$\text{Area} = \lim \sum \Delta A_k = \iint_R dA. \tag{2}$$

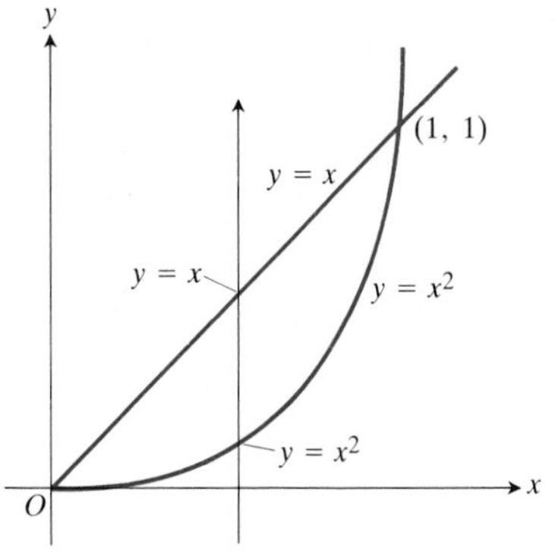

16.16 Area: $\int_0^1 \int_{x^2}^x dy\,dx$.

To evaluate the area integral in (2), we integrate the constant function $f(x, y) \equiv 1$ over R.

EXAMPLE 1 Find the area of the region R bounded by $y = x$ and $y = x^2$ in the first quadrant.

Solution We sketch the region (Fig. 16.16) and calculate the area as

$$A = \int_0^1 \int_{x^2}^x dy\,dx = \int_0^1 (x - x^2)\,dx = \frac{x^2}{2} - \frac{x^3}{3}\bigg]_0^1 = \frac{1}{6}. \quad \square$$

EXAMPLE 2 Find the area of the region R enclosed by the parabola $y = x^2$ and the line $y = x + 2$.

Solution If we divide R into the regions R_1 and R_2 shown in Fig. 16.17, we may calculate the area as

$$A = \iint_{R_1} dA + \iint_{R_2} = \int_0^1 \int_{-\sqrt{y}}^{\sqrt{y}} dx\,dy + \int_1^4 \int_{y-2}^{\sqrt{y}} dx\,dy.$$

On the other hand, reversing the order of integration gives

$$A = \int_{-1}^2 \int_{x^2}^{x+2} dy\,dx.$$

Clearly, this result is simpler and is the only one we would bother to write down in practice. Evaluation of this integral leads to the result

$$A = \int_{-1}^2 y\bigg]_{x^2}^{x+2} dx = \int_{-1}^2 (x + 2 - x^2)\,dx = \frac{9}{2}. \quad \square$$

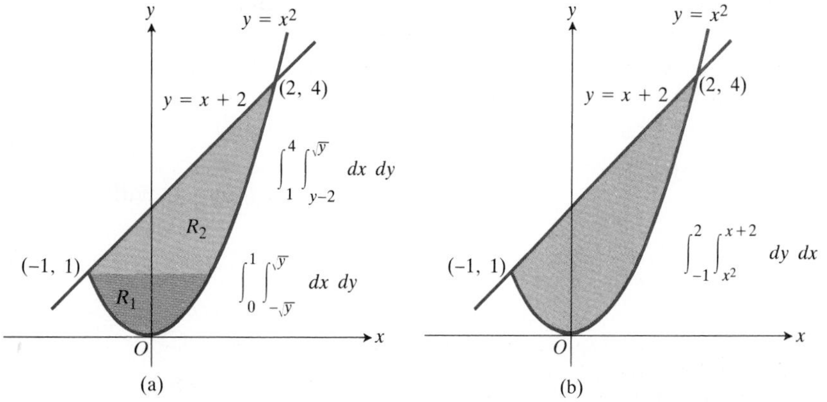

16.17 Calculating the area shown here takes (a) two integrals if the first integration is with respect to x, but (b) only one if the first integration is with respect to y.

PROBLEMS

In Problems 1–8, find the area of the region bounded by the given curves and lines by means of double integration.

1. The coordinate axes and the line $x + y = a$.
2. The x-axis, the curve $y = e^x$, and the lines $x = 0$, $x = 1$.
3. The y-axis, the line $y = 2x$, and the line $y = 4$.
4. The curve $y^2 + x = 0$, and the line $y = x + 2$.
5. The curves $x = y^2$, $x = 2y - y^2$.
6. The semicircle $y = \sqrt{a^2 - x^2}$, the lines $x = \pm a$, and the line $y = -a$.
7. The parabola $x = y - y^2$ and the line $x + y = 0$.

8. Above by $y = x^2$, below by $y = -1$, on the left by $x = -2$, and on the right by $y = 2x - 1$.

The integrals in Problems 9–14 give areas of regions in the xy-plane. Sketch each region. Label each bounding curve with its equation, and give the coordinates of the boundary points where the curves intersect.

9. $\int_0^1 \int_y^{\sqrt{y}} dx\, dy$

10. $\int_0^3 \int_{-x}^{x(2-x)} dy\, dx$

11. $\int_0^{\pi/4} \int_{\sin x}^{\cos x} dy\, dx$

12. $\int_{-1}^2 \int_{y^2}^{y+2} dx\, dy$

13. $\int_{-1}^0 \int_{-2x}^{1-x} dy\, dx + \int_0^2 \int_{-x/2}^{1-x} dy\, dx$

14. $\int_0^2 \int_{x^2-4}^0 dy\, dx + \int_0^4 \int_0^{\sqrt{x}} dy\, dx$

***Toolkit* programs**

Double Integral

16.4 Physical Applications

First and Second Moments

If the representative element of mass dm in a mass that is continuously distributed over some region R of the xy-plane is taken to be

$$\begin{aligned} dm &= \delta(x, y)\, dy\, dx \\ &= \delta(x, y)\, dA, \end{aligned} \tag{1}$$

where $\delta = \delta(x, y)$ is the density at the point (x, y) of R (Fig. 16.18), then double integration may be used to calculate

a) the mass, $$m = \iint \delta(x, y)\, dA, \tag{2}$$

b) the first moment of the mass with respect to the x-axis,

$$M_x = \iint y\, \delta(x, y)\, dA, \tag{3a}$$

c) its first moment with respect to the y-axis,

$$M_y = \iint x\, \delta(x, y)\, dA. \tag{3b}$$

From (2) and (3) we get the coordinates of the center of mass,

$$\bar{x} = \frac{M_y}{m}, \qquad \bar{y} = \frac{M_x}{m}.$$

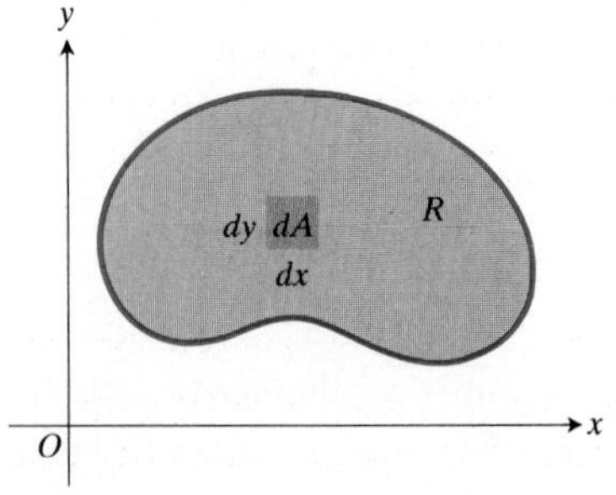

16.18 Area element $dA = dx\, dy$.

Other moments of importance in physical application are the *moments of inertia* of the mass. These are the *second* moments that we get by using the squares instead of the first powers of the "lever-arm" distances x and y. Thus the moment of inertia about the x-axis, denoted by I_x, is defined by

$$I_x = \iint y^2\, \delta(x, y)\, dA. \tag{4}$$

The moment of inertia about the y-axis is

$$I_y = \iint x^2\, \delta(x, y)\, dA. \tag{5}$$

Also of interest is the *polar moment of inertia* about the origin, I_0, given by

$$I_0 = \iint r^2\,\delta(x, y)\,dA = \iint (x^2 + y^2)\,\delta(x, y)\,dA = I_x + I_y. \tag{6}$$

Here $r^2 = x^2 + y^2$ is the square of the distance from the origin to the representative point (x, y) in the element of mass dm.

In all of these integrals, the same limits of integration are to be supplied as would be called for if one were calculating only the area of R.

EXAMPLE 1 A thin plate of uniform (constant) thickness and density δ covers the region in the xy-plane shown in Fig. 16.19. Find the center of mass and the inertial moments I_x, I_y, and I_0.

Solution Using the area computation in Example 1, Article 16.3, we find

$$m = \int_0^1 \int_{x^2}^{x} \delta\,dy\,dx = \delta \int_0^1 \int_{x^2}^{x} dy\,dx = \frac{\delta}{6}.$$

$$M_x = \int_0^1 \int_{x^2}^{x} y\,\delta\,dy\,dx = \delta \int_0^1 \left[\frac{y^2}{2}\right]_{y=x^2}^{y=x} dx$$

$$= \delta \int_0^1 \left(\frac{x^2}{2} - \frac{x^4}{2}\right) dx = \delta\left[\frac{x^3}{6} - \frac{x^5}{10}\right]_0^1 = \frac{\delta}{15}.$$

$$M_y = \int_0^1 \int_{x^2}^{x} x\,\delta\,dy\,dx = \delta \int_0^1 [xy]_{y=x^2}^{y=x}\,dx$$

$$= \delta \int_0^1 (x^2 - x^3)\,dx = \delta\left[\frac{x^3}{3} - \frac{x^4}{4}\right]_0^1 = \frac{\delta}{12}.$$

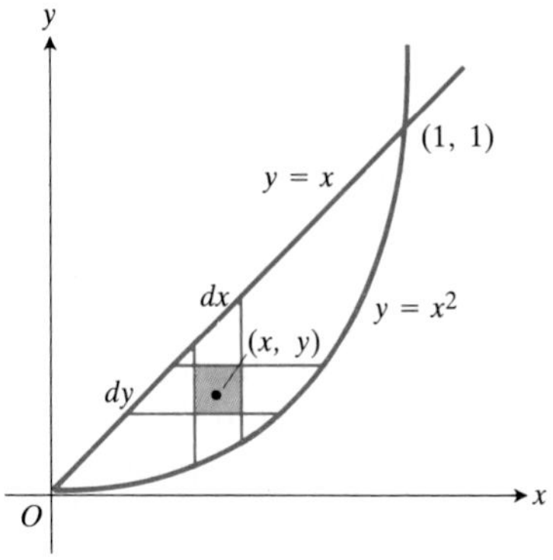

16.19 $I_x = \int_0^1 \int_{x^2}^{x} y^2\,dy\,dx.$

Center of mass $(\bar{x}, \bar{y})$: $\quad \bar{x} = \dfrac{M_y}{m} = \dfrac{1}{2}, \ \bar{y} = \dfrac{M_x}{m} = \dfrac{2}{5}$

$$I_x = \int_0^1 \int_{x^2}^{x} y^2\,\delta\,dy\,dx = \delta \int_0^1 \left[\frac{y^3}{3}\right]_{y=x^2}^{y=x} dx$$

$$= \delta \int_0^1 \left(\frac{x^3}{3} - \frac{x^6}{3}\right) dx = \delta\left[\frac{x^4}{12} - \frac{x^7}{21}\right]_0^1 = \frac{\delta}{28}.$$

$$I_y = \int_0^1 \int_{x^2}^{x} x^2\,\delta\,dy\,dx = \delta \int_0^1 (x^3 - x^4)\,dx = \delta\left[\frac{x^4}{4} - \frac{x^5}{5}\right]_0^1 = \frac{\delta}{20}.$$

We therefore have

$$I_0 = I_x + I_y = \frac{\delta}{28} + \frac{\delta}{20} = \frac{3\delta}{35}.$$

Note that since the density δ in this problem is a constant we are able to move it outside the integral signs. If the density had been given instead as a variable function of x and y, we would have taken this into account by substituting this function for δ before integrating. □

Geometric Figures

Unless otherwise specified, geometric figures in the plane will be treated as objects with constant density $\delta = 1$. The moments of such a figure are then called *moments of area*, and the center of mass is called the figure's *centroid*. Thus, physical bodies have centers of mass (or centers of gravity), while geometric figures have centroids.

the semicircle $y = \sqrt{a^2 - x^2}$, the lines $x = \pm a$, and the line $y = -a$ if $\delta(x, y) = y + a$.

6. The area of the region in the first quadrant bounded by $y = 6x - x^2$ and $y = x$ is $125/6$ square units. Find $\bar{y}$.

7. The area of the region in the first quadrant bounded by $y = 4x - x^2$ and $y = x$ is $9/2$. Find $\bar{x}$.

8. Find the center of mass of a thin plate bounded by $y = x$, $y = 2 - x$, and the x-axis if $\delta(x, y) = 1 + 2x + y$.

9. Find the centroid of the region cut from the first quadrant by the circle $x^2 + y^2 = a^2$. (*Hint:* $\bar{x} = \bar{y}$ and the area is $\pi a^2/4$.)

10. A thin plate bounded by $x^2 + 4y^2 = 12$ and $x = 4y^2$ has a variable density given by $\delta(x, y) = kx$ (k a constant). Find the plate's mass.

11. Find the moment of inertia about the x-axis of the region bounded by the x-axis, the curve $y = e^x$, and the lines $x = 0$, $x = 1$. (For a region, we take $\delta = 1$.)

12. Find the moment of inertia about the x-axis of a thin plate bounded by the curves $x = y^2$, $x = 2y - y^2$ if its density at the point (x, y) is $\delta(x, y) = y + 1$.

13. Find the moment of inertia about the x-axis of a thin plate bounded by the parabola $x = y - y^2$ and the line $x + y = 0$ if $\delta(x, y) = x + y$.

14. Find the moment of inertia with respect to the y-axis of the area bounded by the curve $y = (\sin^2 x)/x^2$ and the interval $\pi \le x \le 2\pi$ on the x-axis.

15. Find the polar moment of inertia about the origin of the triangular region bounded by the y-axis, the line $y = 2x$, and the line $y = 4$.

16. Find the radius of gyration of a uniform slender rod with constant density δ, and of length L, with respect to an axis

a) perpendicular to the axis of the rod through the rod's center of mass,
b) perpendicular to the axis of the rod at one end,
c) parallel to the rod at a distance d from the axis of the rod. Assume that d is very large compared to the radius of the rod.

17. Find the moment of inertia and radius of gyration about the x-axis of each of the following figures.

a) The rectangular region $0 \le x \le b$, $0 \le y \le h$.
b) Any triangular region with base the interval $0 \le x \le b$ on the x-axis and opposite vertex somewhere on the line $y = x$ above the x-axis. (They all have the same moment and radius.)
c) The disc enclosed by the circle $x^2 + y^2 = a^2$.
d) The region cut from the first quadrant by the circle $x^2 + y^2 = a^2$.
e) The region bounded by the ellipse $\dfrac{x^2}{a^2} + \dfrac{y^2}{b^2} = 1$.

18. Find the centroid of the region in the second quadrant bounded by the two axes and the curve $y = e^x$.

19. Find the moment with respect to the y-axis of the region in the first quadrant under the curve $y = e^{-x^2/2}$.

20. Find the centroid of the region in the xy-plane bounded by the curves $y = 1/\sqrt{1 - x^2}$, $y = -1/\sqrt{1 - x^2}$, and the lines $x = 0$, $x = 1$. (*Hint:* Note that $\bar{y} = 0$, by symmetry.)

21. A horizontal cylindrical tank 10 ft in diameter is half full of oil weighing 50 lb/ft³. Find the pressure exerted by the oil on one end of the tank.

22. The average value of a function $f(x, y)$ over a region R is defined to be

$$\frac{1}{\text{area } R} \iint_R f(x, y)\, dA.$$

a) Calculate the average value of the derivative of the function $w = \frac{1}{2}(x^2 + y^2)$ in the direction of the unit vector $\mathbf{u} = u_1\mathbf{i} + u_2\mathbf{j}$ over the region enclosed by the triangle whose vertices are $(0, 0)$, $(0, 1)$ and $(1, 0)$.
b) Show in general that if $w = \frac{1}{2}(x^2 + y^2)$, then the average value of $D_{\mathbf{u}}w$ over a region R is the value of $D_{\mathbf{u}}w$ at the centroid of R.

***Toolkit* programs**

Double Integral

16.5 Changing to Polar Coordinates

When we defined the integral of a function $f(x, y)$ over a region R we divided R with rectangles. Rectangles are easy to describe in rectangular coordinates, and their areas easy to compute. When we work in polar coordinates, however, it is more natural to subdivide R into "polar rectangles," in the way we now describe.

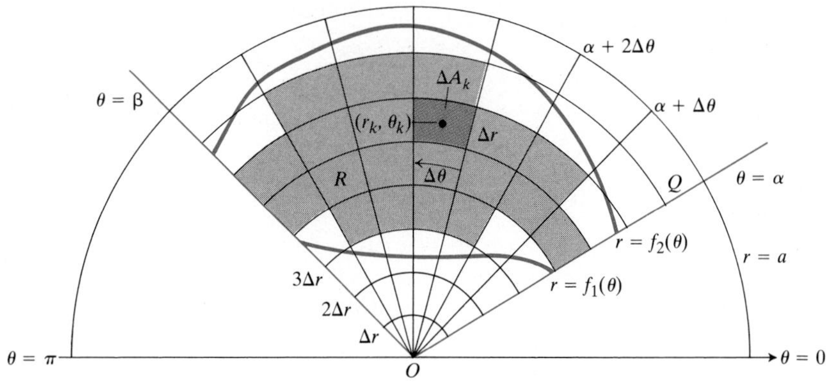

16.20 The region R: $f_1(\theta) \leq r \leq f_2(\theta)$, $\alpha \leq \theta \leq \beta$ is contained in the fan-shaped region Q: $0 \leq r \leq a$, $\alpha \leq \theta \leq \beta$. In the subdivision of Q in polar coordinates, $\Delta A_k = r_k \,\Delta\theta\, \Delta r$.

Suppose that the function $F(r, \theta)$ is defined over a region R bounded by the rays $\theta = \alpha$, $\theta = \beta$, and the continuous curve $r = f_1(\theta)$, $r = f_2(\theta)$, as shown in Fig. 16.20. Suppose that $0 \leq f_1(\theta) \leq f_2(\theta) \leq a$ for all θ between α and β. Then R is contained in the fan-shaped region Q: $0 \leq r \leq a$, $\alpha \leq \theta \leq \beta$. We may cover Q by a grid of circular arcs with centers at 0 and radii

$$\Delta r, 2\,\Delta r, \ldots, m\,\Delta r,$$

where $\Delta r = a/m$, and rays through 0 along

$$\theta = \alpha, \alpha + \Delta\theta, \alpha + 2\,\Delta\theta, \ldots, \alpha + n\,\Delta\theta = \beta,$$

where $\Delta\theta = (\beta - \alpha)/n$. These arcs and rays divide the fan-shaped region Q into small patches called polar rectangles. We number the rectangles that lie inside R, calling their areas

$$\Delta A_1, \Delta A_2, \ldots, \Delta A_n.$$

We let (r_k, θ_k) be the center of the rectangle with area ΔA_k (see Fig. 16.20). By "center" we mean the point that lies halfway between the circular arcs on the ray that bisects them. We then form the sum

$$S_n = \sum_{k=1}^{n} F(r_k, \theta_k)\,\Delta A_k. \tag{1}$$

We may express ΔA_k in terms of Δr and $\Delta\theta$ in the following way. The radius of the inner arc bounding ΔA_k is $r_k - \frac{1}{2}\Delta r$. The area of the circular sector subtended by this arc at the origin is therefore

$$\frac{\Delta\theta}{2\pi} \cdot \pi\left(r_k - \frac{1}{2}\Delta r\right)^2 = \frac{1}{2}\left(r_k - \frac{1}{2}\Delta r\right)^2 \Delta\theta.$$

Similarly, the radius of the outer boundary of ΔA_k is $r_k + \frac{1}{2}\Delta r$, and the area of the sector it subtends is

$$\frac{\Delta\theta}{2\pi} \cdot \pi\left(r_k + \frac{1}{2}\Delta r\right)^2 = \frac{1}{2}\left(r_k + \frac{1}{2}\Delta r\right)^2 \Delta\theta. \tag{2}$$

Therefore,

$$\begin{aligned}\Delta A_k &= \text{area of larger sector} - \text{area of smaller sector}\\ &= \frac{\Delta\theta}{2}\left[\left(r_k + \frac{\Delta r}{2}\right)^2 - \left(r_k - \frac{\Delta r}{2}\right)^2\right]\\ &= \frac{\Delta\theta}{2}[2r_k\,\Delta r]\\ &= r_k\,\Delta r\,\Delta\theta.\end{aligned}$$

Combining this result with Eq. (1) gives

$$S_n = \sum_{k=1}^{n} F(r_k, \theta_k)\,\Delta A_k = \sum_{k=1}^{n} F(r_k, \theta_k) r_k\,\Delta r\,\Delta\theta. \tag{3}$$

If F is continuous on R, then these sums approach a limit

$$\lim S_n = \int_R F(r, \theta)\, dA$$

as Δr and $\Delta\theta$ approach zero, and a version of Fubini's theorem says that this limit may be evaluated by repeated single integrations with respect to r and θ as

$$\iint F(r, \theta)\, dA = \int_{\theta=\alpha}^{\theta=\beta} \int_{r=f_1(\theta)}^{r=f_2(\theta)} F(r, \theta)\, r\, dr\, d\theta. \tag{4}$$

If $F(r, \theta) \equiv 1$ is the constant function whose value is one, then the value of the integral of F over R is the area of R (in agreement with our earlier definition, although we shall not prove this fact). Thus,

$$\text{Area of } R = \iint_R r\, dr\, d\theta. \tag{5}$$

Finding the Limits of Integration

The procedure we used for finding limits of integration for integrals in rectangular coordinates also works for polar coordinates.

EXAMPLE 1 Find the limits of integration for integrating a function $F(r, \theta)$ over the region R that lies inside the cardioid $r = a(1 + \cos\theta)$ and outside the circle $r = a$.

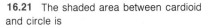

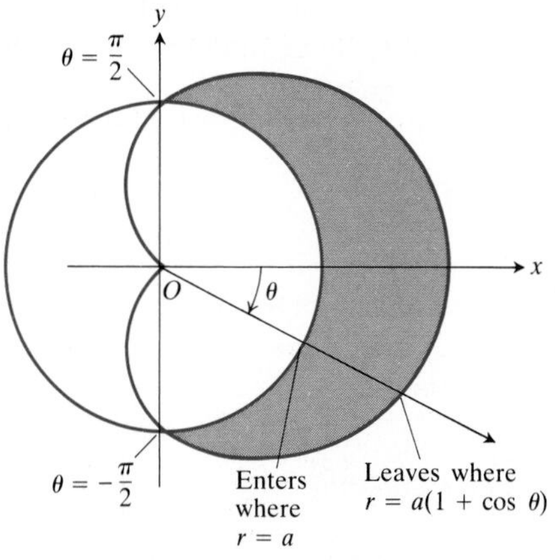

16.21 The shaded area between cardioid and circle is

$$\int_{-\pi/2}^{\pi/2}\int_{a}^{a(1+\cos\theta)} r\, dr\, d\theta.$$

Solution We graph the cardioid and circle (Fig. 16.21) and carry out the following steps:

1. Hold θ fixed, and let r increase to trace a ray out from the origin.
2. Integrate from the r-value where the ray enters R to the r-value where it leaves R.
3. Choose θ-limits to include all the rays from the origin that intersect R.

The result is the integral

$$\int_{-\pi/2}^{\pi/2}\int_{r=a}^{r=a(1+\cos\theta)} F(r, \theta)\, r\, dr\, d\theta. \;\square$$

EXAMPLE 2 Find the area enclosed by the lemniscate $r^2 = 2a^2 \cos 2\theta$.

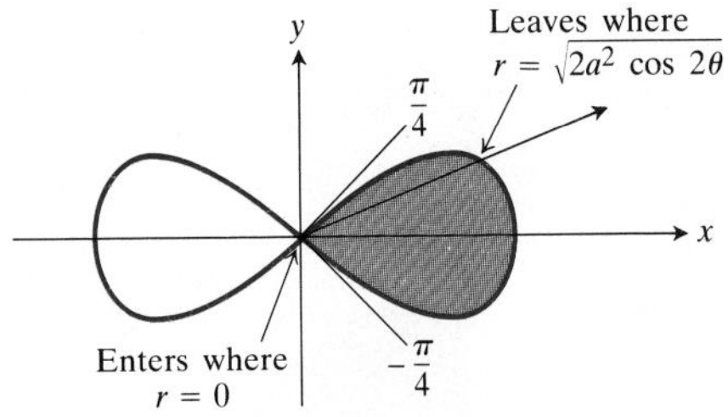

16.22 Limits for integrating over the shaded region bounded by the lemniscate $r^2 = 2a^2 \cos 2\theta$.

Solution We graph the lemniscate (Fig. 16.22) and calculate the area of the right-hand half to be

$$\int_{-\pi/4}^{\pi/4}\int_0^{\sqrt{2a^2\cos 2\theta}} r\,dr\,d\theta = \int_{-\pi/4}^{\pi/4}\left[\frac{r^2}{2}\right]_{r=0}^{r=\sqrt{2a^2\cos 2\theta}} d\theta$$

$$= \int_{-\pi/4}^{\pi/4} a^2\cos 2\theta d\theta = \frac{a^2}{2}\sin 2\theta\Big]_{-\pi/4}^{\pi/4}$$

$$= \frac{a^2}{2}[1-(-1)] = a^2.$$

The total area is therefore $2a^2$. □

Changing Coordinates

If a region G in the uv-plane is transformed into the region R in the xy-plane by differentiable equations of the form

$$x = f(u, v), \qquad y = g(u, v)$$

(see Fig. 16.23), then a function $\phi(x, y)$ defined on R can be thought of as a function

$$\phi(f(u, v), g(u, v))$$

defined on G. It is a theorem from advanced calculus that if all the functions involved are continuous and have continuous first derivatives, then the integral of $\phi(x, y)$ over R and the integral of $\phi(f(u, v), g(u, v))$ over G are related by the equation

$$\iint_R \phi(x, y)\,dx\,dy = \iint_G \phi[f(u, v), g(u, v)]\frac{\partial(x, y)}{\partial(u, v)}\,du\,dv, \tag{6}$$

where $\partial(x, y)/\partial(u, v)$ denotes the determinant of partial derivatives

$$\frac{\partial(x, y)}{\partial(u, v)} = \begin{vmatrix} \dfrac{\partial x}{\partial u} & \dfrac{\partial x}{\partial v} \\ \dfrac{\partial y}{\partial u} & \dfrac{\partial y}{\partial v} \end{vmatrix}. \tag{7}$$

This determinant is called the *Jacobian* of the coordinate transformation $x = f(u, v)$, $y = g(u, v)$.

In the case of polar coordinates, we have r and θ in place of u and v,

$$x = r\cos\theta, \qquad y = r\sin\theta,$$

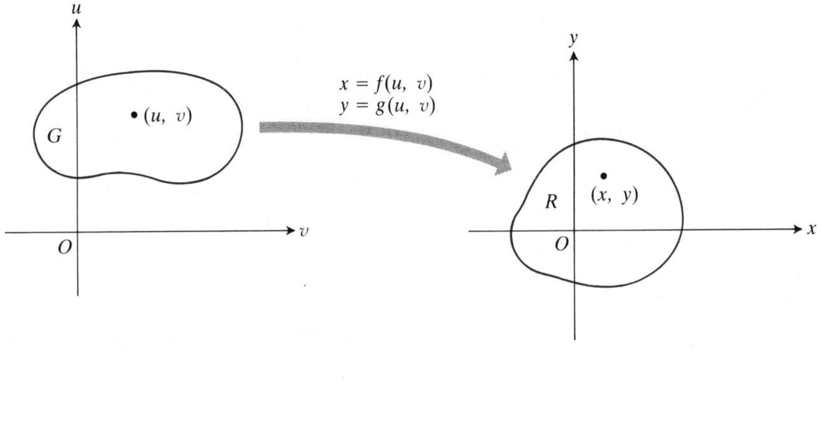

16.23 The equations $x = f(u, v)$, $y = g(u, v)$ allow us to rewrite an integral over R as an equivalent integral over G.

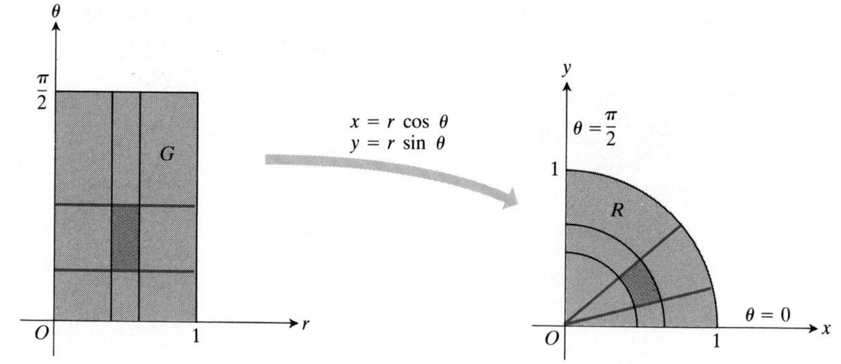

16.24 The equations $x = r\cos\theta$, $y = r\sin\theta$ transform G into R.

and

$$\frac{\partial(x, y)}{\partial(r, \theta)} = \begin{vmatrix} \cos\theta & -r\sin\theta \\ \sin\theta & r\cos\theta \end{vmatrix} = r(\cos^2\theta + \sin^2\theta) = r.$$

Hence, Eq. (6) becomes

$$\iint_R \phi(x, y)\, dx\, dy = \iint_G \phi(r\cos\theta, r\sin\theta) r\, dr\, d\theta, \tag{8}$$

which corresponds to Eq. (4).

Figure 16.24 shows how the equations $x = r\cos\theta$, $y = r\sin\theta$ transform the rectangle G: $0 \le r \le 1$, $0 \le 0 \le \pi/2$ into the quarter circle R bounded by $x^2 + y^2 = 1$ in the first quadrant of the xy-plane.

The object of using the equality of the integrals in Eq. (6) is to simplify integration.

EXAMPLE 3 Find the polar moment of inertia about the origin of a thin plate of density $\delta = 1$ bounded by the quarter circle $x^2 + y^2 = 1$ in the first quadrant.

Solution With reference to Fig. 16.24, Eq. (8) gives

$$\iint_{\substack{\text{quarter} \\ \text{circle } R}} (x^2 + y^2)\, dx\, dy = \iint_{\substack{\text{rectangle} \\ G}} (r^2)\, r\, dr\, d\theta,$$

or

$$\int_0^1 \int_0^{\sqrt{1-x^2}} (x^2 + y^2)\, dy\, dx = \int_0^{\pi/2} \int_0^1 (r^2)\, r\, dr\, d\theta$$

$$= \int_0^{\pi/2} \left[\frac{r^4}{4}\right]_{r=0}^{r=1} d\theta = \int_0^{\pi/2} \frac{1}{4}\, d\theta = \frac{\pi}{8}.$$

To convert the cartesian integral to the polar integral, we took $\phi(x, y) = x^2 + y^2$ in Eq. (8) and used the polar limits that described the quarter circle. □

The general rule for converting a cartesian integral to a polar integral is

1. substitute: $x = r\cos\theta$, $y = r\sin\theta$, $dy\, dx = r\, dr\, d\theta$,
2. supply polar limits of integration.

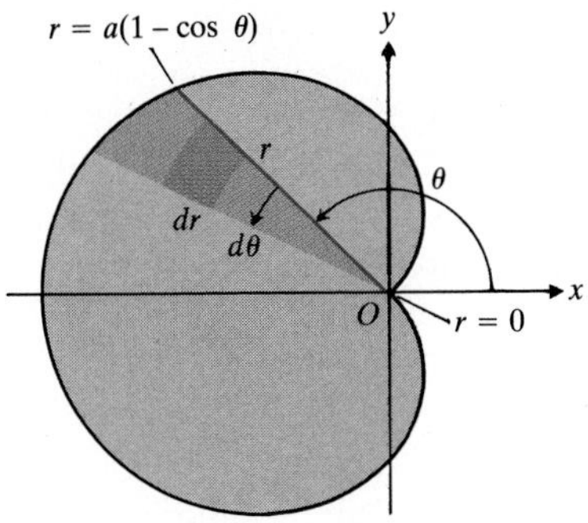

16.25 The cardioid $r = a(1 - \cos\theta)$.

EXAMPLE 4 Find the moment of inertia about the y-axis of the region enclosed by the cardioid

$$r = a(1 - \cos\theta).$$

Solution We take $\delta(x, y) = 1$ when working with a geometric figure. Thus

$$I_y = \iint x^2\delta(x, y)\, dA = \iint x^2\, dA.$$

With

$$x = r\cos\theta, \qquad dA = r\, dr\, d\theta,$$

and limits of integration determined from Fig. 16.25, we have

$$I_y = \int_0^{2\pi}\int_0^{a(1-\cos\theta)} r^3\cos^2\theta\, dr\, d\theta$$

$$= \int_0^{2\pi}\frac{a^4}{4}\cos^2\theta(1 - \cos\theta)^4\, d\theta.$$

The evaluation of the integrals

$$\int_0^{2\pi}\cos^n\theta\, d\theta \qquad (n = 2, 3, 4, 5, 6)$$

that now arise is made easier by use of the reduction formula

$$\int_0^{2\pi}\cos^n\theta\, d\theta = \frac{\cos^{n-1}\theta\sin\theta}{n}\Big]_0^{2\pi} + \frac{n-1}{n}\int_0^{2\pi}\cos^{n-2}\theta\, d\theta$$

or, since $\sin\theta$ vanishes at both limits,

$$\int_0^{2\pi}\cos^n\theta\, d\theta = \frac{n-1}{n}\int_0^{2\pi}\cos^{n-2}\theta\, d\theta.$$

Thus

$$\int_0^{2\pi}\cos^2\theta\, d\theta = \frac{1}{2}\int_0^{2\pi} d\theta = \pi,$$

$$\int_0^{2\pi}\cos^3\theta\, d\theta = \frac{2}{3}\int_0^{2\pi}\cos\theta\, d\theta = \frac{2}{3}\sin\theta\Big]_0^{2\pi} = 0,$$

$$\int_0^{2\pi}\cos^4\theta\, d\theta = \frac{3}{4}\int_0^{2\pi}\cos^2\theta\, d\theta = \frac{3\pi}{4},$$

$$\int_0^{2\pi}\cos^5\theta\, d\theta = \frac{4}{5}\int_0^{2\pi}\cos^3\theta\, d\theta = 0,$$

$$\int_0^{2\pi}\cos^6\theta\, d\theta = \frac{5}{6}\int_0^{2\pi}\cos^4\theta\, d\theta = \frac{5\pi}{8},$$

Therefore

$$I_y = \frac{a^4}{4}\int_0^{2\pi}(\cos^2\theta - 4\cos^3\theta + 6\cos^4\theta - 4\cos^5\theta + \cos^6\theta)\, d\theta$$

$$= \frac{a^4}{4}\left[1 + \frac{18}{4} + \frac{5}{8}\right]\pi = \frac{49\pi a^4}{32}. \; \square$$

PROBLEMS

In Problems 1–8, change the cartesian integral into an equivalent polar integral and evaluate the polar integral.

1. $\int_{-a}^{a}\int_{-\sqrt{a^2-x^2}}^{\sqrt{a^2-x^2}} dy\,dx$ 2. $\int_{0}^{a}\int_{0}^{\sqrt{a^2-y^2}} (x^2+y^2)\,dx\,dy$

3. $\int_{0}^{a/\sqrt{2}}\int_{y}^{\sqrt{a^2-y^2}} x\,dx\,dy$ 4. $\int_{0}^{\infty}\int_{0}^{\infty} e^{-(x^2+y^2)}\,dx\,dy$

5. $\int_{0}^{2}\int_{0}^{x} y\,dy\,dx$ 6. $\int_{0}^{2a}\int_{0}^{\sqrt{2ax-x^2}} x^2\,dy\,dx$

7. $\int_{0}^{3}\int_{0}^{\sqrt{3}x} \frac{dy\,dx}{\sqrt{x^2+y^2}}$

8. $\int_{0}^{2}\int_{0}^{\sqrt{4-x^2}} \frac{xy}{\sqrt{x^2+y^2}}\,dy\,dx$

9. Find the area of the region cut from the first quadrant by the cardioid $r = 1 + \sin\theta$.

10. Find the area of the region common to the cardioids $r = 1 + \cos\theta$ and $r = 1 - \cos\theta$.

11. Find the area cut from the first quadrant by the curve $r = (2 - \sin 2\theta)^{1/2}$.

12. Integrate the function $f(x, y) = 1/(1 - x^2 - y^2)$ over the disc $x^2 + y^2 \leq 3/4$.

13. Find $M_x = \iint y\,dA$ for the region bounded below by the x-axis and above by the cardioid $r = 1 - \cos\theta$.

14. A thin plate in the first quadrant is bounded by the coordinate axes and the circle $x^2 + y^2 = 1$. Find the moment of inertia of the plate with respect to the x-axis if the density varies as the square of the distance from the origin.

15. Find the centroid of the region that lies inside the cardioid $r = a(1 + \cos\theta)$ and outside the circle $r = a$.

16. Find the polar moment of inertia with respect to the origin for the region in Problem 15.

17. The region in Problem 15 is the base of a solid right cylinder whose top lies in the plane $z = x$. Find the volume of the cylinder.

18. Find the area enclosed by one leaf of the rose $r = \cos 3\theta$.

19. The lemniscate $r^2 = 2a^2\cos 2\theta$ is the base of a solid right cylinder whose top is bounded by the sphere $z = \sqrt{2a^2 - r^2}$. Find the volume of the cylinder.

20. a) Find the Jacobian (Eq. 7) of the transformation $x = u$, $y = uv$, and sketch the region G: $1 \leq u \leq 2$, $1 \leq uv \leq 2$ in the uv-plane.
 b) Then use Eq. (6) to transform the integral
$$\int_1^2\int_1^2 \frac{y}{x}\,dy\,dx$$
into an integral over G, and evaluate both integrals.

21. Let R be the region in the first quadrant of the xy-plane bounded by the hyperbolas $xy = 1$, $xy = 9$ and the lines $y = x$, $y = 4x$. Use the transformation $x = u/v$, $y = uv$ with $u > 0$ and $v > 0$ to rewrite $\iint_R dx\,dy$ as an integral over an appropriate region G in the uv-plane. Then evaluate the uv-integral over G.

22. The area πab of the ellipse $x^2/a^2 + y^2/b^2 = 1$ can be found by integrating the function $f(x, y) \equiv 1$ over the region bounded by the ellipse in the xy-plane. Evaluating the integral directly requires a trigonometric substitution. An easier way to evaluate the integral is to use the transformation $x = au$, $y = bv$ and evaluate the transformed integral over the disc G: $u^2 + v^2 \leq 1$ in the uv-plane. Find the area this way.

23. A thin plate of uniform thickness and density covers the region bounded by the ellipse $x^2/a^2 + y^2/b^2 = 1$ in the xy-plane. Find I_0, the polar moment of the plane about the origin. (*Hint:* Use the transformation $x = ar\cos\theta$, $y = br\sin\theta$.)

24. Use the transformation $x = u + (1/2)v$, $y = v$ to evaluate the integral
$$\int_0^2\int_{y/2}^{(y+4)/2} y^3(2x - y)e^{(2x-y)^2}\,dx\,dy$$
by first writing it as an equivalent iterated integral over a region G in the uv-plane.

25. In Article 10.4 we derived the formula $A = \int_\alpha^\beta \frac{1}{2}r^2 d\theta$ for the area of the region swept out by the radius vector $\overrightarrow{OP}$ as P moves along a curve $r = f(\theta)$ from $\theta = \alpha$ to $\theta = \beta$. Show that this formula is also a consequence of Eq. (5) in the present article.

16.6

Triple Integrals in Rectangular Coordinates

If $F(x, y, z)$ is a function defined on a bounded region D in space (a solid ball or truncated cone, for example, or something resembling a swiss cheese, or a finite union of such objects), then the integral of F over D may

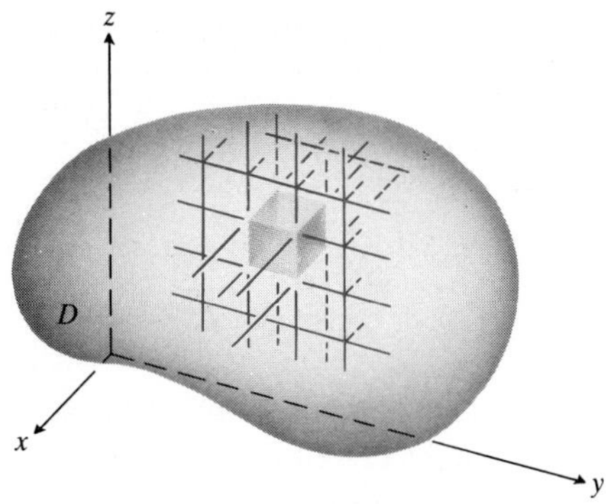

16.26 Partitioning a solid with rectangular cells of volume ΔV.

be defined in the following way. We partition a rectangular region about D into rectangular cells by planes parallel to the coordinate planes, as shown in Fig. 16.26. The cells have dimensions Δx by Δy by Δz. We number the cells that lie inside D in some order

$$\Delta V_1, \Delta V_2, \ldots, \Delta V_n,$$

choose a point (x_k, y_k, z_k) in each ΔV_k, and form the sum

$$S_n = \sum_{k=1}^{n} F(x_k, y_k, z_k)\, \Delta V_k. \tag{1}$$

If F is continuous and the bounding surface of D is made of smooth surfaces joined along continuous curves, then as Δx, Δy, and Δz all approach zero the sums S_n will approach a limit

$$\lim S_n = \iiint_D F(x, y, z)\, dV. \tag{2}$$

We call this limit the *triple integral of F over D*. The limit also exists for some discontinuous functions.

Triple integrals share many algebraic properties with double and single integrals. Writing F for $F(x, y, z)$ and G for $G(x, y, z)$ we have the following.

I1. $\displaystyle\iiint_D kF\, dV = k\iiint_D F\, dV$ (any number k)

I2. $\displaystyle\iiint_D [F + G]\, dV = \iiint_D F\, dV + \iiint_D G\, dV$

I3. $\displaystyle\iiint_D [F - G]\, dV = \iiint_D F\, dV - \iiint_D G\, dV$

I4. $\displaystyle\iiint_D F\, dV \geq 0$ if $F \geq 0$ on D

I5. $\displaystyle\iiint_D F\, dV \geq \iiint_D G\, dV$ if $F \geq G$ on D

The integrals have these properties because the sums that approximate them have these properties.

Triple integrals also have a domain additivity property that proves useful in physics and engineering as well as in mathematics. If the domain D of a continuous function F is partitioned by smooth surfaces into a finite number of cells $D_1, D_2, \ldots, D_n$, then

I6. $\displaystyle\iiint_D F\, dV = \iiint_{D_1} F\, dV + \iiint_{D_2} F\, dV + \cdots + \iiint_{D_n} F\, dV.$

Volume

If $F(x, y, z) \equiv 1$ is the constant function whose value is one, then the sums in Eq. (1) reduce to

$$S_n = \sum_{k=1}^{n} 1 \cdot \Delta V_k = \sum_{k=1}^{n} \Delta V_k. \tag{3}$$

As Δx, Δy, and Δz all approach zero, the cells ΔV_k become smaller and more numerous and fill up more and more of D. We therefore define the volume of D to be the triple integral of the constant function $F(x, y, z) \equiv 1$ over D:

$$\text{Volume of } D = \lim \sum_{k=1}^{n} \Delta V_k = \iiint_D dV. \tag{4}$$

Evaluation

The triple integral is seldom evaluated directly from its definition as a limit. Instead, one applies a three-dimensional version of Fubini's theorem to evaluate the integral by repeated single integrations.

For example, suppose we want to integrate a continuous function $F(x, y, z)$ over a region D that is bounded below by a surface $z = f_1(x, y)$, above by the surface $z = f_2(x, y)$, and on the side by a cylinder C parallel to the z-axis (Fig. 16.27). Let R denote the vertical projection of D onto the xy-plane, which is the region in the xy-plane enclosed by C. The integral of F over D is then evaluated as

$$\iiint_D F(x, y, z)\, dV = \iint_R \left(\int_{f_1(x,y)}^{f_2(x,y)} F(x, y, z)\, dz \right) dy\, dx,$$

or

$$\iiint_D F(x, y, z)\, dV = \iint_R \int_{f_1(x,y)}^{f_2(x,y)} F(x, y, z)\, dz\, dx\, dy, \tag{5}$$

if we omit the parentheses. The z-limits of integration indicate that for every (x, y) in the region R, z may extend from the lower surface $z = f_1(x, y)$ to the upper surface $z = f_2(x, y)$. The y- and x-limits of integration have not been given explicitly in Eq. (5) but are to be determined in the usual way from the boundaries of R.

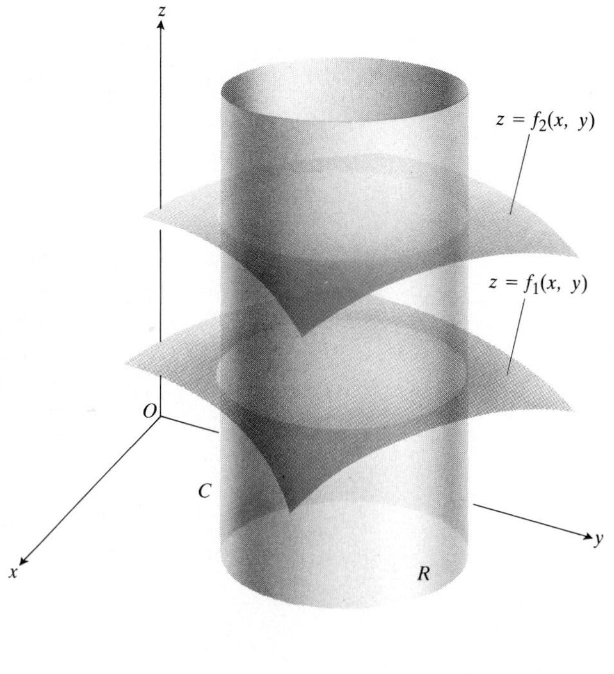

16.27 The enclosed volume can be found by evaluating

$$V = \iint_R \int_{f_1(x,y)}^{f_2(x,y)} dz\, dy\, dx.$$

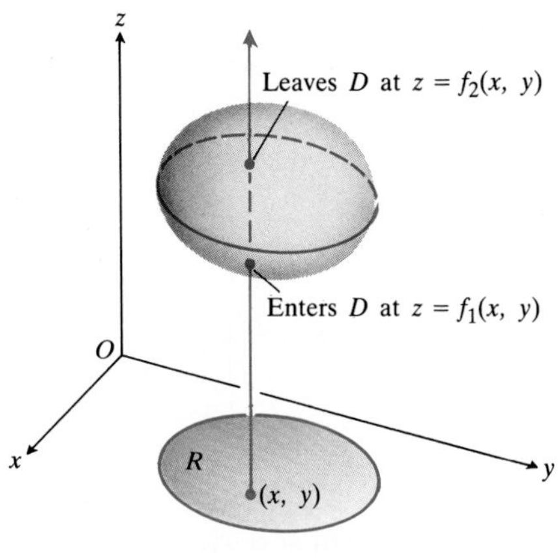

16.28 Finding limits of integration. The boundary of R is defined by the equation $f_1(x, y) = f_2(x, y)$.

In case the lateral surface of the cylinder reduces to zero, as in Fig. 16.28 and the example that follows, we may find the equation of the boundary of R by eliminating z between the two equations $z = f_1(x, y)$ and $z = f_2(x, y)$. This gives

$$f_1(x, y) = f_2(x, y),$$

an equation that contains no z and that defines the boundary of R in the xy-plane.

To supply the z-limits of integration in any particular instance we may use a procedure like the one for double integrals. We imagine a line L through a point (x, y) in R and parallel to the z-axis. As z increases, the line enters D at $z = f_1(x, y)$ and leaves D at $z = f_2(x, y)$. These give the lower and upper limits of the integration with respect to z. The result of this integration is now a function of x and y alone, which we integrate over R, supplying limits in the familiar way.

EXAMPLE 1 Find the volume enclosed between the two surfaces

$$z = x^2 + 3y^2 \quad \text{and} \quad z = 8 - x^2 - y^2.$$

Solution The two surfaces (Fig. 16.29) intersect on the elliptic cylinder

$$x^2 + 3y^2 = 8 - x^2 - y^2,$$

or

$$x^2 + 2y^2 = 4.$$

The volume projects into the region R (in the xy-plane) that is enclosed by the ellipse having this same equation. In the double integral with respect to y and x over R, if we integrate first with respect to y, holding x fixed, y

16.29 The volume between two paraboloids.

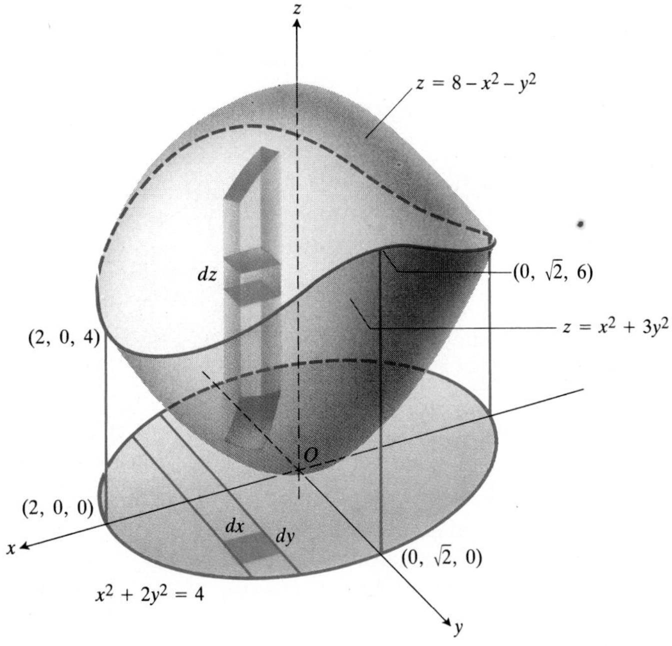

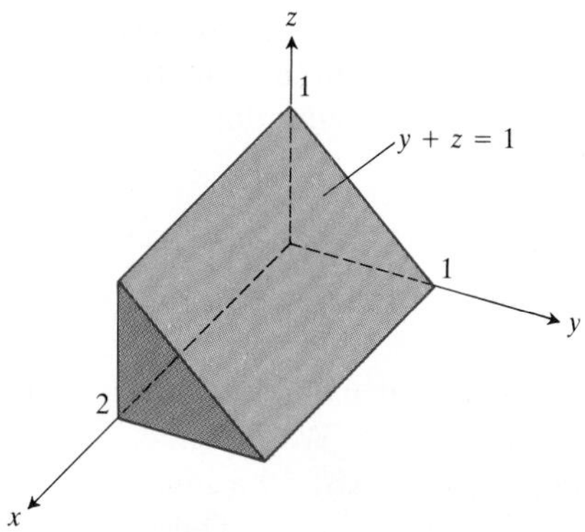

16.30 Example 2 shows how to calculate the volume of this prism with six different iterated triple integrals.

varies from $-\sqrt{(4-x^2)/2}$ to $+\sqrt{(4-x^2)/2}$. Then x varies from -2 to $+2$. Thus we have

$$\begin{aligned}V &= \int_{-2}^{2}\int_{-\sqrt{(4-x^2)/2}}^{\sqrt{(4-x^2)/2}}\int_{x^2+3y^2}^{8-x^2-y^2} dz\,dy\,dx \\ &= \int_{-2}^{2}\int_{-\sqrt{(4-x^2)/2}}^{\sqrt{(4-x^2)/2}} (8-2x^2-4y^2)\,dy\,dx \\ &= \int_{-2}^{2}\left[2(8-2x^2)\sqrt{\frac{4-x^2}{2}} - \frac{8}{3}\left(\frac{4-x^2}{2}\right)^{3/2}\right]dx \\ &= \frac{4\sqrt{2}}{3}\int_{-2}^{2}(4-x^2)^{3/2}\,dx = 8\pi\sqrt{2}. \;\square\end{aligned}$$

As we know, there are sometimes two different orders in which the single integrations that evaluate a double integral may be worked (but not always). For triple integrals there are sometimes (but not always) as many as *six* workable orders of integration. The next example shows an extreme case in which all six are possible.

EXAMPLE 2 Each of the following integrals gives the volume of the solid shown in Fig. 16.30.

a) $\int_0^1\int_0^{1-z}\int_0^2 dx\,dy\,dz$ b) $\int_0^2\int_0^{1-y}\int_0^2 dx\,dz\,dy$

c) $\int_0^1\int_0^2\int_0^{1-z} dy\,dx\,dz$ d) $\int_0^2\int_0^1\int_0^{1-z} dy\,dz\,dx$

e) $\int_0^1\int_0^2\int_0^{1-y} dz\,dx\,dy$ f) $\int_0^2\int_0^1\int_0^{1-y} dz\,dy\,dx$ □

PROBLEMS

1. Write six different iterated triple integrals for the volume of the rectangular solid in the first octant bounded by the coordinate planes and the planes $x = 1$, $y = 2$, and $z = 3$. Evaluate one of the integrals.
2. Write six different iterated triple integrals for the volume of the tetrahedron cut from the first octant by the plane $6x + 3y + 2z = 6$. Evaluate one of the integrals.
3. Write six different iterated triple integrals for the volume in the first octant enclosed by the cylinder $x^2 + z^2 = 4$ and the plane $y = 3$. Evaluate one of the integrals.
4. Write an iterated triple integral in the order *dz dx dy* for the volume of the region bounded by the xy-plane and the paraboloid $z = 4 - x^2 - y^2$.
5. Write an iterated triple integral in the order *dz dy dx* for the volume of the region bounded below by the xy-plane and above by the paraboloid $z = x^2 + y^2$, and lying inside the cylinder $x^2 + y^2 = 4$.
6. Write an iterated integral in the order *dz dy dx* for the volume of the region that lies between the planes $x = 0$ and $x = 1$ in the first octant and is bounded above by the plane $x + y + z = 2$.

Find the volumes in Problems 7–20.

7. The volume of the tetrahedron bounded by the plane $x/a + y/b + z/c = 1$ and the coordinate planes.
8. The volume between the cylinder $z = y^2$ and the xy-plane that is bounded by the four vertical planes $x = 0$, $x = 1$, $y = -1$, $y = 1$.
9. The volume in the first octant bounded by the planes $x + z = 1$, $y + 2z = 2$.
10. The volume in the first octant bounded by the cylinder $x = 4 - y^2$ and the planes $z = y$, $x = 0$, $z = 0$.
11. The volume of the wedge cut from the cylinder $x^2 + y^2 = 1$ by the plane $z = y$ above and the plane $z = 0$ below.
12. The volume of the region in the first octant bounded by the coordinate planes, the cylinder $x^2 + y^2 = 4$, and the plane $x + z = 3$.

13. The volume of the region in the first octant bounded by the coordinate planes, above by the cylinder $x^2 + z = 1$, and on the right by the paraboloid $y = x^2 + z^2$. (*Hint:* Integrate first with respect to y.)
14. The volume enclosed by the cylinders $z = 5 - x^2$, $z = 4x^2$, and the planes $y = 0$, $x + y = 1$.
15. The volume enclosed by the cylinder $y^2 + 4z^2 = 16$ and the planes $x = 0$, $x + y = 4$.
16. The volume bounded below by the plane $z = 0$, laterally by the elliptic cylinder $x^2 + 4y^2 = 4$, and above by the plane $z = x + 2$.
17. The volume common to the two cylinders $x^2 + y^2 = a^2$ and $x^2 + z^2 = a^2$. (See Fig. 16.31.)
18. The volume bounded by the elliptic paraboloids $z = x^2 + 9y^2$ and $z = 18 - x^2 - 9y^2$.
19. The volume of an ellipsoid of semiaxes a, b, c.
20. The volume of the region in the first octant bounded by the coordinate planes, by the plane $z = 3$ below, and by the surface $z = 4 - x^2 - y$ above.
21. Sketch the domain of integration of the integral

$$\int_{-1}^{1}\int_{x^2}^{1}\int_{0}^{1-y} dz\, dy\, dx.$$

Then rewrite the integral as an equivalent iterated integral in the order
a) dy dz dx
b) dy dx dz
c) dx dy dz
d) dx dz dy
e) dz dx dy

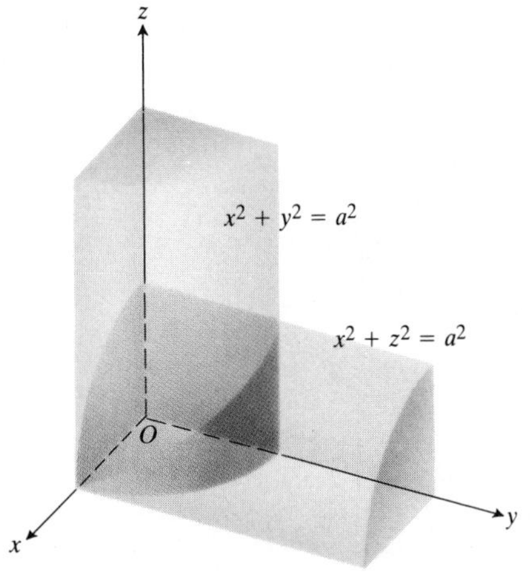

16.31 One eighth of the volume common to the cylinders $x^2 + y^2 = a^2$, $x^2 + z^2 = a^2$.

22. Repeat Problem 21 for the integral

$$\int_{0}^{1}\int_{0}^{1}\int_{0}^{y^2} dz\, dy\, dx.$$

***Toolkit* programs**
Double Integral

16.7
Physical Applications in Three Dimensions

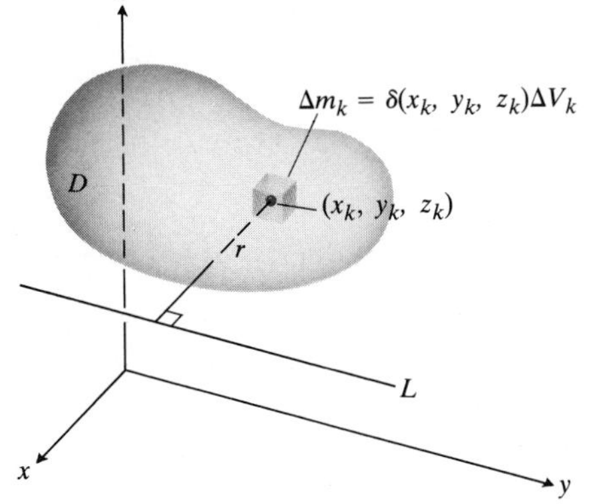

16.32 To define I_L we first imagine D to be subdivided into a finite number of mass elements Δm_k.

If $F(x, y, z) = \delta(x, y, z)$ is the density of an object occupying a region D in space, and we imagine D to be subdivided as in Fig. 16.26 in the preceding article, then the integral of the density,

$$m = \lim \sum_k \delta(x_k, y_k, z_k)\Delta V_k = \iiint_D \delta(x, y, z)\, dV$$

gives the mass of the object.

If $r(x, y, z)$ is the distance from the point (x, y, z) in D to a line L, then the moment of inertia about L of the mass element

$$\Delta m_k = \delta(x_k, y_k, z_k)\Delta V_k$$

shown in Fig. 16.32 is approximately

$$\Delta I_k = r^2(x_k, y_k, z_k)\Delta m_k$$

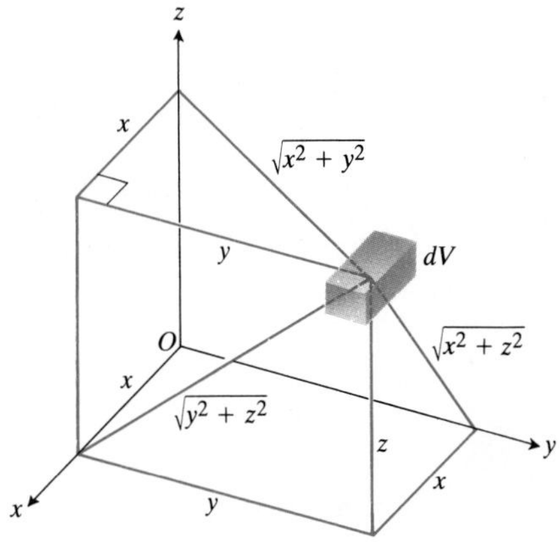

16.33 Distances of dV from the coordinate planes and axes.

and the moment of inertia I_L of the object about L is

$$I_L = \lim \sum_k \Delta I_k$$

$$= \lim \sum_k r^2(x_k, y_k, z_k)\delta(x_k, y_k, z_k)\Delta V_k = \iiint_D r^2\delta\, dV.$$

If L is the x-axis, then $r^2 = y^2 + z^2$ (Fig. 16.33) and

$$I_x = \iiint_D (y^2 + z^2)\delta\, dV.$$

Similarly,

$$I_y = \iiint_D (x^2 + z^2)\delta\, dV \qquad \text{and} \qquad I_z = \iiint_D (x^2 + y^2)\delta\, dV.$$

These and other useful formulas are summarized in the following list.

Mass: $m = \iiint_D \delta\, dV \qquad (\delta = \text{density})$

First moments about the coordinate planes:

$$M_{yz} = \iiint_D x\,\delta\, dV, \qquad M_{xz} = \iiint_D y\,\delta\, dV, \qquad M_{xy} = \iiint_D z\,\delta\, dV$$

Center of mass:

$$\bar{x} = \frac{\iiint x\,\delta\, dV}{m}, \qquad \bar{y} = \frac{\iiint y\,\delta\, dV}{m}, \qquad \bar{z} = \frac{\iiint z\,\delta\, dV}{m}$$

Moments of inertia (second moments):

$$I_x = \iiint (y^2 + z^2)\delta\, dV, \qquad I_y = \iiint (x^2 + z^2)\delta\, dV,$$
$$I_z = \iiint (x^2 + y^2)\delta\, dV \qquad I_L = \iiint r^2\delta\, dV,$$
$$r(x, y, z) = \text{distance from point } (x, y, z) \text{ to line } L$$

Radius of gyration about a line L: $R = \sqrt{I_L/m}$

EXAMPLE 1 Find I_x, I_y, I_z for the rectangular solid of uniform density δ shown in Fig. 16.34.

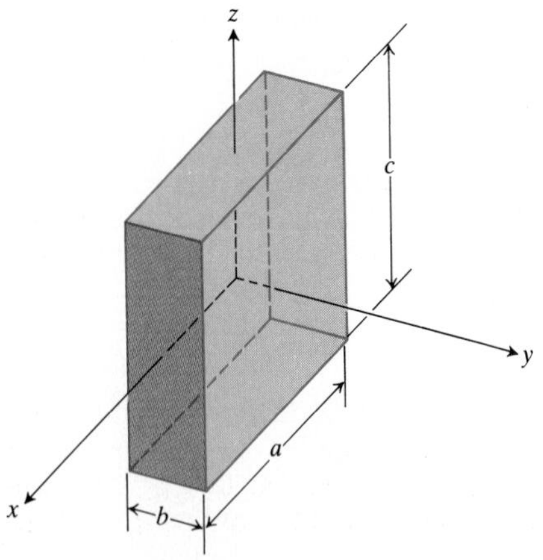

16.34 Example 1 calculates I_x, I_y, and I_z for the block shown here.

Solution

$$I_x = \int_{-c/2}^{c/2}\int_{-b/2}^{b/2}\int_{-a/2}^{a/2} (y^2 + z^2)\delta\, dx\, dy\, dz. \tag{1}$$

We can avoid some of the work of integration by observing that $(y^2 + z^2)\delta$ is an even function of x, y, and z and therefore

$$I_x = 8\int_0^{c/2}\int_0^{b/2}\int_0^{a/2} (y^2 + z^2)\delta\, dx\, dy\, dz = 4a\delta\int_0^{c/2}\int_0^{b/2} (y^2 + z^2)\, dy\, dz$$

$$= 4a\delta\int_0^{c/2}\left[\frac{y^3}{3} + z^2y\right]_{y=0}^{y=b/2} dz = 4a\delta\int_0^{c/2}\left(\frac{b^3}{24} + \frac{z^2b}{2}\right) dz$$

$$= 4a\delta\left(\frac{b^3c}{48} + \frac{c^3b}{48}\right) = \frac{abc\delta}{12}(b^2 + c^2) = \frac{m}{12}(b^2 + c^2).$$

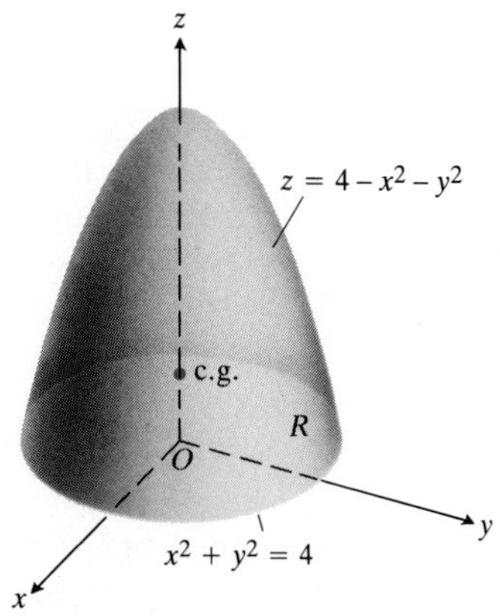

16.35 Example 2 calculates the coordinates of the center of gravity of this solid.

Similarly,

$$I_y = \frac{m}{12}(a^2 + c^2) \qquad \text{and} \qquad I_z = \frac{m}{12}(a^2 + b^2). \ \square$$

EXAMPLE 2 Find the center of gravity of a solid of uniform density δ bounded below by the disc R: $x^2 + y^2 \le 4$ in the plane $z = 0$ and above by the paraboloid $z = 4 - x^2 - y^2$ (Fig. 16.35).

Solution By symmetry, $\bar{x} = \bar{y} = 0$. To find $\bar{z}$ we first calculate

$$\begin{aligned} M_{xy} &= \iint_R \int_{z=0}^{z=4-x^2-y^2} z\,\delta\,dz\,dy\,dx \\ &= \iint_R \left[\frac{z^2}{2}\right]_{z=0}^{z=4-x^2-y^2} \delta\,dy\,dx \\ &= \frac{\delta}{2}\iint_R (4 - x^2 - y^2)^2\,dy\,dx \\ &= \frac{\delta}{2}\int_0^{2\pi}\int_0^2 (4 - r^2)^2\,r\,dr\,d\theta \qquad \text{(polar coordinates)} \\ &= \frac{\delta}{2}\int_0^{2\pi}\left[-\frac{1}{6}(4 - r^2)^3\right]_{r=0}^{r=2} d\theta \\ &= \frac{16\delta}{3}\int_0^{2\pi} d\theta \\ &= \frac{32\pi\delta}{3}. \end{aligned}$$

A similar calculation gives

$$m = \iint_R \int_0^{4-x^2-y^2} \delta\,dz\,dy\,dx = 8\pi\delta\,.$$

Therefore,

$$\bar{z} = \frac{M_{xy}}{m} = \frac{4}{3},$$

and the center of gravity is $(\bar{x}, \bar{y}, \bar{z}) = (0, 0, \frac{4}{3})$. $\square$

Geometric Figures

In moment calculations, geometric figures in space are treated as objects with constant density $\delta = 1$, and the resulting centers of mass are called *centroids*.

PROBLEMS

1. Evaluate the integral for I_x in Eq. (1) directly to show that the shortcut in Example 1 gives the same answer. Use the results in Example 1 to find the radius of gyration of the rectangular solid about each coordinate axis.

2. The axes shown in Fig. 16.36 run through the centroid of the solid wedge parallel to its edges. Using the dimensions shown, find I_x, I_y, and I_z.

3. Find the moments of inertia of the rectangular solid

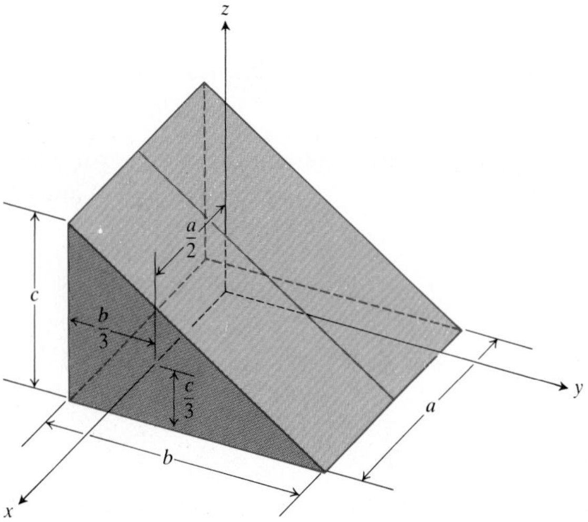

16.36 Figure for Problem 2.

shown in Fig. 16.37 with respect to its edges by calculating I_x, I_y, and I_z.

4. a) Find the centroid and the moments of inertia I_x, I_y, and I_z of the tetrahedron whose vertices are the points (0, 0, 0), (1, 0, 0), (0, 1, 0) and (0, 0, 1).
b) Find the radius of gyration of the tetrahedron about the x-axis. Compare it with the distance from the centroid to the x-axis.

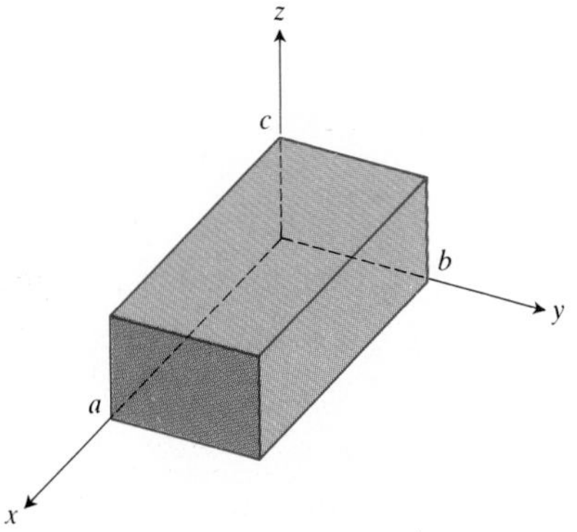

16.37 Figure for Problem 3.

5. A solid "trough" of uniform density δ is bounded below by the surface $z = 4y^2$, above by the plane $z = 4$, and on the ends by the planes $x = 1$ and $x = -1$. Find the center of mass, and the moments of inertia with respect to the three axes.

6. A solid region in the first octant is bounded by the coordinate planes and the plane $x + y + z = 2$. The density of the solid is $\delta(x, y, z) = 2x$. Find the center of mass.

7. A solid in the first octant is bounded below by the plane $z = 0$, on the sides by the plane $y = 0$ and the surface $x = y^2$, and above by the surface $z = 4 - x^2$. The density is $\delta(x, y, z) = kxy$, where k is a constant. Find the mass.

8. Write a triple integral for the mass of a solid hemisphere that is bounded below by the plane $z = 0$, and above by the sphere $x^2 + y^2 + z^2 = 4$, if the density at any point is proportional to the distance of the point from the z-axis. (Do not evaluate the integral.)

9. The density of a solid enclosed by the ellipsoid $9x^2 + 4y^2 + 36z^2 = 36$ is $\delta(x, y, z) = kx$, where k is a constant. Write, but do not evaluate, integral expressions for the solid's mass m and for the moments M_{yz} and I_y.

10. Find the x-coordinate of the centroid of the region bounded below by the plane $z = 0$, laterally by the elliptic cylinder $x^2 + 4y^2 = 4$, and above by the plane $z = x + 2$.

11. A torus of mass m is generated by rotating a circle of radius a about an axis in its plane at distance b from the center (b greater than a). Find its moment of inertia about the axis of revolution.

***Toolkit* programs**
Double Integral

16.8 Integrals in Cylindrical and Spherical Coordinates

Cylindrical Coordinates

Cylindrical coordinates are useful in applications that involve cylinders along the z-axis and planes that contain or are perpendicular to the z-axis, because these surfaces have simple equations of constant coordinate

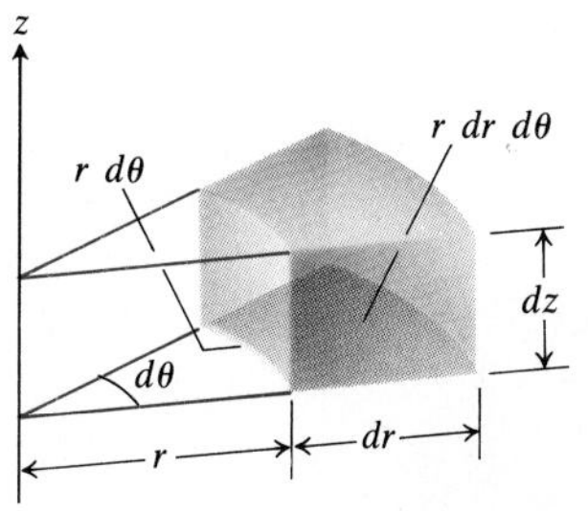

16.38 The volume element $dV = dz\, r\, dr\, d\theta$.

value like

$$r = 4, \qquad \theta = \frac{\pi}{3}, \qquad z = 2.$$

If we are working with a solid that has an axis of symmetry we can often simplify our calculations by taking the axis of symmetry to be the z-axis.

The volume element for subdividing a region in space with cylindrical coordinates is taken to be

$$dV = dz\, r\, dr\, d\theta, \tag{1}$$

as shown in Fig. 16.38. Triple integrals in cylindrical coordinates are then evaluated as iterated integrals, as in the examples that follow.

EXAMPLE 1 Find the centroid of a solid hemisphere of radius a.

Solution We may choose the origin at the center of the sphere and consider the hemisphere that lies above the xy-plane. (See Fig. 16.39.) The equation of the hemispherical surface is

$$z = \sqrt{a^2 - x^2 - y^2}$$

or, in terms of cylindrical coordinates,

$$z = \sqrt{a^2 - r^2}.$$

By symmetry we have

$$\bar{x} = \bar{y} = 0.$$

We calculate $\bar{z}$:

$$\bar{z} = \frac{\iiint z\, dV}{\iiint dV} = \frac{\int_0^{2\pi}\int_0^{a}\int_0^{\sqrt{a^2-r^2}} z\, dz\, r\, dr\, d\theta}{\frac{2}{3}\pi a^3} = \frac{3a}{8}. \quad \square$$

EXAMPLE 2 Find the moment of inertia I_z of the solid that is bounded below by the xy-plane, above by the sphere $x^2 + y^2 + z^2 = 4a^2$, and laterally by the cylinder $r = 2a\cos\theta$.

16.39 The volume element in cylindrical coordinates is $dV = dz\, r\, dr\, d\theta$.

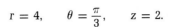

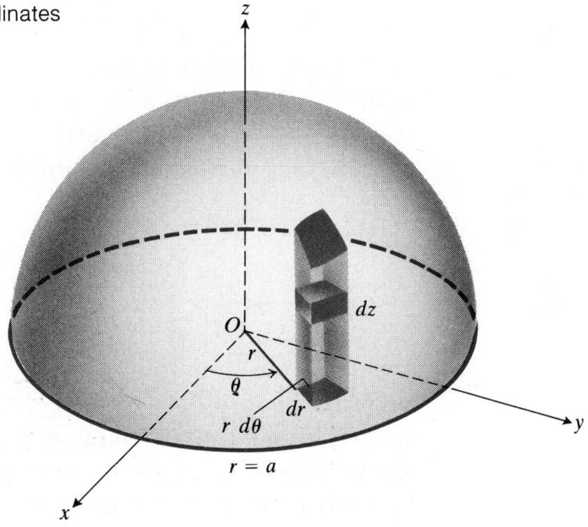

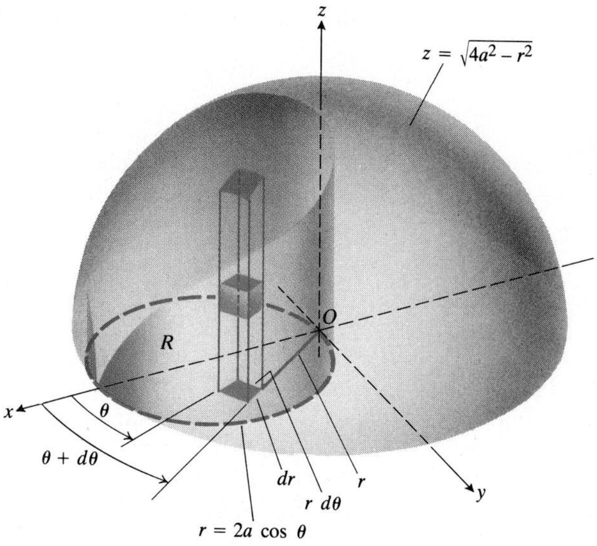

16.40 The solid bounded by the xy-plane, the sphere $x^2 + y^2 + z^2 = 4a^2$, and the cylinder $r = 2a\cos\theta$.

Solution The solid lies in front of the yz-plane, as shown in Fig. 16.40. It projects vertically onto the region R bounded by the circle $r = 2a\cos\theta$ in the xy-plane. Since

$$I_z = \iiint (x^2 + y^2)\,dV$$

(from the list of formulas in Article 16.7), we have

$$I_z = \iiint r^2\,dz\,r\,dr\,d\theta.$$

Integrating first from $z = 0$ to $z = \sqrt{4a^2 - x^2 - y^2} = \sqrt{4a^2 - r^2}$, we find

$$I_z = \iint_R \int_{z=0}^{z=\sqrt{4a^2-r^2}} dz\,r^3\,dr\,d\theta = \int_{-\pi/2}^{\pi/2}\int_0^{2a\cos\theta}\int_0^{\sqrt{4a^2-r^2}} dz\,r^3\,dr\,d\theta$$

$$= \int_{-\pi/2}^{\pi/2}\int_0^{2a\cos\theta} r^3\sqrt{4a^2 - r^2}\,dr\,d\theta = \frac{64a^5}{15}\left(\pi - \frac{26}{15}\right). \square$$

Spherical Coordinates

Spherical coordinates (Fig. 16.41) are useful in applications that involve shapes bounded by spheres centered at the origin, planes through the z-axis, and cones with vertices at the origin whose axes lie along the z-axis. Such surfaces have simple equations of constant coordinate value like

$$\rho = r, \qquad \phi = \frac{\pi}{3}, \qquad \theta = \frac{\pi}{3}.$$

If we are working with a shape that is symmetric with respect to a point, we can often simplify our work by choosing that point as the origin of a spherical coordinate system. Such shapes arise less frequently than shapes with an axis of symmetry, but it is good to be able to handle them when they arise.

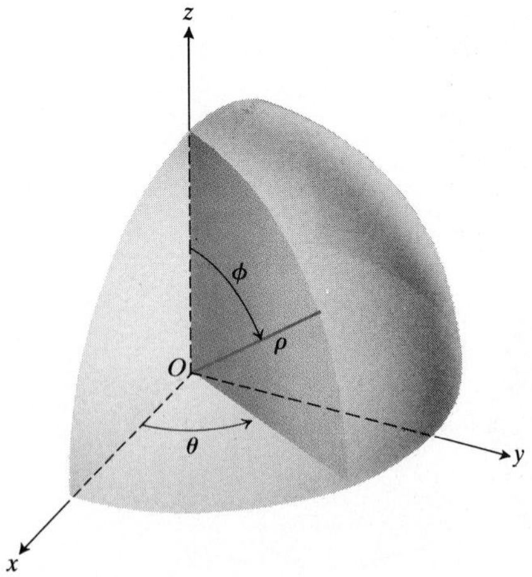

16.41 Spherical coordinates.

The volume element in spherical coordinates is

$$dV = \rho^2 \sin\phi \, d\rho \, d\phi \, d\theta \tag{2}$$

(as shown in Fig. 16.42) and triple integrals take the form

$$\iiint F(\rho, \phi, \theta) dV = \iiint F(\rho, \phi, \theta)\rho^2 \sin\phi \, d\rho \, d\phi \, d\theta. \tag{3}$$

To evaluate these integrals we first integrate with respect to ρ. The procedure for finding the limits of integration for a region D in space is therefore the following:

1. Hold ϕ and θ fixed and let ρ increase. This gives a ray out from the origin.
2. Integrate from the ρ-value where the ray first enters D to the ρ-value where the ray leaves D. This gives the limits for ρ.
3. Hold θ fixed and let ϕ increase. (This gives a family of rays that make a "fan.") Integrate over the ϕ-values for which the rays pass through D.
4. Choose θ-limits that include all the fans that intersect D.

EXAMPLE 3 Find the volume cut from the sphere $\rho = a$ by the cone $\phi = \alpha$. (See Fig. 16.43.)

Solution The volume is given by

$$V = \int_0^{2\pi} \int_0^{\alpha} \int_0^{a} \rho^2 \sin\phi \, d\rho \, d\phi \, d\theta = \frac{2\pi a^3}{3}(1 - \cos\alpha).$$

As a check, we note that the special cases $\alpha = \pi/2$ and $\alpha = \pi$ correspond to the cases of a hemisphere and a sphere, of volumes $2\pi a^3/3$ and $4\pi a^3/3$, respectively. □

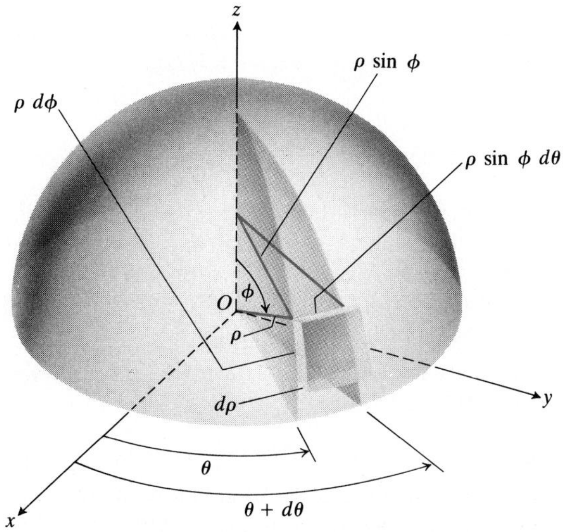

16.42 The volume element in spherical coordinates is $dV = d\rho \cdot \rho \, d\phi \cdot \rho \sin\phi \, d\theta$.

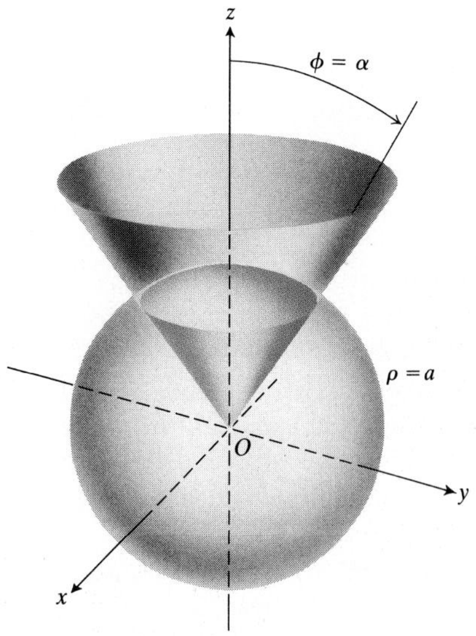

16.43 The volume cut from the sphere $\rho = a$ by the cone $\phi = \alpha$.

EXAMPLE 4 Find I_z for the region in Example 3 if the equation for the cone is $\phi = \pi/3$.

Solution With

$$I_z = \iiint (x^2 + y^2)\, dV$$

and

$$x^2 + y^2 = (\rho \sin \phi \cos \theta)^2 + (\rho \sin \phi \sin \theta)^2 = \rho^2 \sin^2 \phi,$$

we have

$$\begin{aligned} I_z &= \int_0^{2\pi} \int_0^{\pi/3} \int_0^a (\rho^2 \sin^2 \phi)\rho^2 \sin \phi \, d\rho \, d\phi \, d\theta \\ &= \int_0^{2\pi} \int_0^{\pi/3} \int_0^a \rho^4 \sin^3 \phi \, d\rho \, d\phi \, d\theta \\ &= \frac{a^5}{5} \int_0^{2\pi} \int_0^{\pi/3} \sin^3 \phi \, d\phi \, d\theta = \frac{a^5}{5} \int_0^{2\pi} \int_0^{\pi/3} (1 - \cos^2 \phi) \sin \phi \, d\phi \, d\theta \\ &= \frac{a^5}{5} \int_0^{2\pi} \left[-\cos \phi + \frac{\cos^3 \phi}{3} \right]_{\phi=0}^{\phi=\pi/3} d\theta = \frac{a^5}{5} \int_0^{2\pi} \left[\frac{5}{24} \right] d\theta = \frac{\pi a^5}{12}. \quad \square \end{aligned}$$

Coordinate Conversion Formulas

Polar to rectangular	Spherical to cylindrical	Spherical to rectangular
$x = r \cos \theta$	$r = \rho \sin \phi$	$x = \rho \sin \phi \cos \theta$
$y = r \sin \theta$	$z = \rho \cos \phi$	$y = \rho \sin \phi \sin \theta$
$z = z$	$\theta = \theta$	$z = \rho \cos \phi$

(Details in Article 13.4.)

Volume: $\iiint dx\, dy\, dz = \iiint dz\, r\, dr\, d\theta = \iiint \rho^2 \sin \phi \, d\rho \, d\phi \, d\theta$

PROBLEMS

1. Set up an iterated triple integral for the volume of the sphere $x^2 + y^2 + z^2 = 4$ in (a) spherical, (b) cylindrical, and (c) rectangular coordinates.

2. Let D be the smaller spherical cap cut from a solid ball of radius two units by a plane one unit from the center of the sphere. Express the volume of D as an iterated triple integral in (a) spherical, (b) cylindrical, and (c) rectangular coordinates.

3. Convert the following integral (a) to rectangular coordinates, (b) to spherical coordinates:

$$\int_0^{2\pi} \int_0^1 \int_0^{\sqrt{4-r^2}} r\, dz\, r\, dr\, d\theta.$$

4. Let D denote the region in the first octant enclosed below by the cone $z^2 = x^2 + y^2$ and above by the sphere $x^2 + y^2 + z^2 = 8$. Express the volume V of D as an iterated triple integral in (a) cylindrical, and (b) spherical coordinates. Then (c) find V.

5. Give the limits of integration for evaluating the iterated integral

$$\iiint f(r, \theta, z)\, dz\, r\, dr\, d\theta$$

over the region D that is bounded below by the plane $z = 0$, laterally by the circular cylinder $r = \cos \theta$, and above by the paraboloid $z = 3r^2$.

6. Express $I_z = \iiint (x^2 + y^2)\delta dV$ as an iterated triple integral over an arbitrary region D in (a) cylindrical and (b) spherical coordinates.

7. Express the moment of inertia I_z of the solid bounded below by the plane $z = 0$ and above by the sphere $x^2 + y^2 + z^2 = 1$ as an iterated triple integral in (a) cylindrical and (b) spherical coordinates.

8. Express $M_{xy} = \iiint z\delta dV$ as an iterated triple integral over an arbitrary region D in (a) cylindrical and (b) spherical coordinates.

Cylindrical coordinate problems

9. Find the volume bounded below by the plane $z = 0$, laterally by the cylinder $x^2 + y^2 = 1$, and above by the paraboloid $z = x^2 + y^2$.

10. Find the volume bounded below by the paraboloid $z = x^2 + y^2$, laterally by the cylinder $x^2 + y^2 = 1$, and above by the paraboloid $z = x^2 + y^2 + 1$.

11. Find the volume bounded below by the plane $z = 0$ and above by the paraboloid $z = 4 - x^2 - y^2$.

12. Find the volume enclosed by the cylinder $x^2 + y^2 = 4$ and the planes $z = 0$ and $y + z = 4$. (*Hint:* In cylindrical coordinates, $z = 4 - y$ becomes $z = 4 - r\sin\theta$.)

13. Find the volume bounded above by the paraboloid $z = 5 - x^2 - y^2$ and below by the paraboloid $z = 4x^2 + 4y^2$.

14. Find the volume that is bounded above by the paraboloid $z = 9 - x^2 - y^2$, below by the xy-plane, and that lies *outside* the cylinder $x^2 + y^2 = 1$.

15. Find the volume cut from the sphere $x^2 + y^2 + z^2 = 4a^2$ by the cylinder $x^2 + y^2 = a^2$.

16. Find the volume bounded below by the paraboloid $z = x^2 + y^2$ and above by the plane $z = 2y$.

17. Find the volume bounded above by the sphere $x^2 + y^2 + z^2 = 2a^2$ and below by the paraboloid $az = x^2 + y^2$.

18. Find the volume in the first octant bounded by the cylinder $x^2 + y^2 = a^2$ and the planes $x = a$, $y = a$, $z = 0$, $z = x + y$.

19. A hemispherical bowl of radius 5 cm is filled with water to within 3 cm of the top. Find the volume of water in the bowl.

20. A solid of uniform density δ in the first octant is bounded above by the cone $z^2 = x^2 + y^2$, below by the plane $z = 0$, and on the sides by the cylinder $x^2 + y^2 = 4$ and the planes $x = 0$ and $y = 0$. Find the center of gravity of the solid. (*Hint:* $\bar{x} = \bar{y}$.)

21. Find I_x for the region cut from the sphere $x^2 + y^2 + z^2 = 4a^2$ by the cylinder $x^2 + y^2 = a^2$.

22. Find I_z for the region bounded below by the paraboloid $z = x^2 + y^2$ and above by the plane $z = 2y$.

23. Find the centroid of the region bounded above by the sphere $x^2 + y^2 + z^2 = 2a^2$ and below by the paraboloid $az = x^2 + y^2$.

24. Find the moment of inertia of a solid circular cylinder of radius r and height h (a) about the axis of the cylinder, (b) about a line through the centroid perpendicular to the axis of the cylinder.

25. Use cylindrical coordinates to find the moment of inertia of a sphere of radius a and mass m about a diameter.

26. Find the volume generated by rotating the cardioid $r = a(1 - \cos\theta)$ about the x-axis. [*Hint:* Use *double* integration. Rotate an area element dA around the x-axis to generate a volume element dV.]

27. Find the moment of inertia, about the x-axis, of the volume of Problem 26.

28. Find the moment of inertia of a right circular cone of base radius a, altitude h, and mass m about an axis through the vertex and parallel to the base.

29. Find the moment of inertia of a sphere of radius a and mass m with respect to a tangent line.

30. Find the centroid of that portion of the volume of the sphere $r^2 + z^2 = a^2$ that lies between the planes $\theta = -(\pi/4)$ and $\theta = \pi/4$.

31. Find the volume in the first octant that lies between the cylinders $r = 1$ and $r = 2$ and that is bounded below by the xy-plane and above by the surface $z = xy$.

32. Convert the integral

$$\int_{-1}^{1}\int_{0}^{\sqrt{1-y^2}}\int_{0}^{x} (x^2 + y^2)\,dz\,dx\,dy$$

to an equivalent integral in cylindrical coordinates and evaluate the result.

Spherical coordinates

33. Find the volume of the solid bounded above by the sphere $\rho = a$ and below by the cone $\phi = \pi/3$. (The solid resembles a filled ice cream cone.)

34. Find the volume inside the sphere $\rho = a$ that lies between the cones $\phi = \pi/3$ and $\phi = 2\pi/3$.

35. Find the volume cut from the sphere $\rho = a$ by the planes $\theta = 0$ and $\theta = \pi/6$ in the first octant.

36. Find the centroid of the solid in Problem 33.

37. Find the smaller volume cut from the sphere $\rho = 2$ by the plane $z = \sqrt{2}$.

38. Find the volume enclosed by the surface $\rho = a(1 - \cos\phi)$. Compare with Problem 26.

39. Find the volume of the region bounded inside by the surface $\rho = 1 + \cos\phi$ and outside by the sphere $\rho = 2$.

40. The region bounded by the circle $\rho = 2 \sin \phi, \theta = \pi/2$ (circle of radius 1 in the yz-plane tangent to the z-axis at the origin and lying to the right of the z-axis) is revolved about the z-axis to sweep out a solid. Find the volume of the solid. (The first theorem of Pappus in Article 5.11 gives a quick way to check your result.)
41. Find the center of gravity of a homogeneous solid hemisphere of radius a and constant density δ.
42. a) Find the moment of inertia of a solid sphere of radius a and uniform density δ about a diameter of the sphere.
 b) Find the radius of gyration of the sphere about the diameter.
43. Find the radius of gyration, with respect to the diameter, of a spherical shell of mass m bounded by the spheres $\rho = a$ and $\rho = 2a$ if the density is $\delta = \rho^2$.
44. Show that the centroid of a solid circular cone lies on the axis of the cone and one fourth of the way from the base to the vertex.
45. Find the moment of inertia of a solid circular cone of base radius r and height h about the axis of the cone.

16.9 Surface Area

Figure 16.44 shows a piece S of the surface $F(x, y, z) = c$ lying above its "shadow" R on a ground plane directly beneath it. In this article we shall show how to define and calculate the area of such a surface from the formula for F as an integral over R when F and its first partial derivatives are continuous. We shall also investigate the special case in which the surface is given by an equation of the form $z = f(x, y)$.

The first step in defining the area of S is to divide the region R into small rectangles ΔA_k of the kind we would use if we were defining an integral over R. Directly above each ΔA_k there lies a patch of surface $\Delta\sigma_k$ that we may approximate with a portion of ΔP_k of the tangent plane. To be specific, we suppose that ΔP_k is a portion of the plane that is tangent to the surface at the point (x_k, y_k, z_k) directly above the back corner C_k of ΔA_k. If the tangent plane is parallel to R, then ΔP_k will be congruent to ΔA_k. Otherwise it will be a parallelogram whose area is somewhat larger than that of ΔA_k.

Figure 16.45 gives a magnified view of $\Delta\sigma_k$ and ΔP_k, showing the vector $\nabla F(x_k, y_k, z_k)$ and a vector $\mathbf{n}$ that is a unit vector normal to the ground plane. The vector $\mathbf{n}$ is included in the figure because the angle γ it makes with ∇F will prove to be important in later calculations. The other vectors in this picture, $\mathbf{u}$ and $\mathbf{v}$, are the vectors that lie along the edges of the patch ΔP_k in the tangent plane. Both $\mathbf{u} \times \mathbf{v}$ and $\nabla F(x_k, y_k, z_k)$ are normal to ΔP_k.

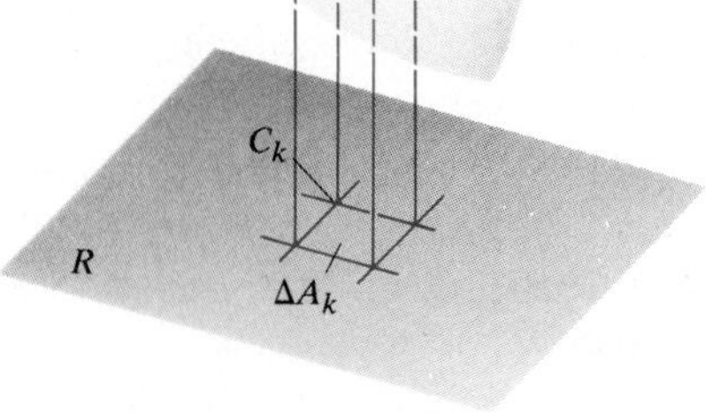

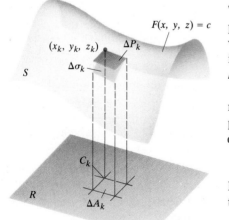

16.44 The tangent plate ΔP_k approximates the surface patch $\Delta\sigma_k$ on the level surface $F(x, y, z) = c$, lying above ΔA_k.

At this point we need a fact that we haven't used since Article 13.9, namely that $|(\mathbf{u} \times \mathbf{v}) \cdot \mathbf{n}|$ is the area of the orthogonal projection of the parallelogram determined by $\mathbf{u}$ and $\mathbf{v}$ onto a plane whose normal is $\mathbf{n}$. In our case, this translates into the statement

$$|(\mathbf{u} \times \mathbf{v}) \cdot \mathbf{n}| = \Delta A_k. \tag{1}$$

Now, $|\mathbf{u} \times \mathbf{v}|$ itself is the area ΔP_k (standard fact about cross products) so that Eq. (1) becomes

$$\underbrace{|\mathbf{u} \times \mathbf{v}|}_{\Delta P_k} \underbrace{|\mathbf{n}|}_{1} \underbrace{|\cos (\text{angle between } \mathbf{u} \times \mathbf{v} \text{ and } \mathbf{n})|}_{\substack{\text{Same as } |\cos\gamma| \text{ because} \\ \nabla F \text{ and } \mathbf{u}\times\mathbf{v} \text{ are both} \\ \text{normal to the tangent plane}}} = \Delta A_k \tag{2}$$

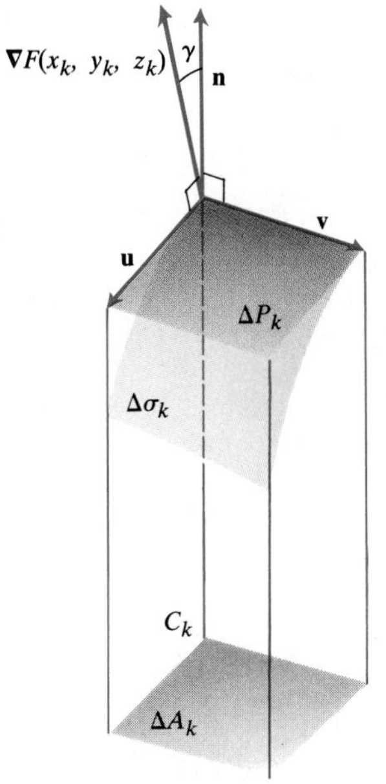

16.45 Magnified view from Fig. 16.44. The vector $\mathbf{u} \times \mathbf{v}$ (not shown) is parallel to ∇F because both vectors are normal to ΔP_k.

or

$$\Delta P_k |\cos \gamma| = \Delta A_k$$

or

$$\Delta P_k = \frac{\Delta A_k}{|\cos \gamma|},$$

provided $\cos \gamma \neq 0$. We will have $\cos \gamma \neq 0$ as long as ∇F is not parallel to the ground plane, or as long as $\nabla F \cdot \mathbf{n} \neq 0$.

Since the patches ΔP_k approximate the surface patches $\Delta\sigma_k$, which fit together to make S, the sum

$$\sum \Delta P_k = \sum \frac{\Delta A_k}{|\cos \gamma|} \tag{3}$$

looks like an approximation of what we might like to call the surface area of S. It also looks as if the approximation would improve if we refined the subdivision of R. In fact, the sums on the right-hand side of Eq. (3) are approximating sums for the double integral

$$\iint_R \frac{1}{|\cos \gamma|}\, dA. \tag{4}$$

We therefore define the area of S to be the value of this integral whenever the integral exists.

For any particular surface $F(x, y, z) = c$ we have

$$|\nabla F \cdot \mathbf{n}| = |\nabla F|\,|\mathbf{n}|\,|\cos \gamma|, \qquad \text{or} \qquad \frac{1}{|\cos \gamma|} = \frac{|\nabla F|\,|\mathbf{n}|}{|\nabla F \cdot \mathbf{n}|} = \frac{|\nabla F|}{|\nabla F \cdot \mathbf{n}|},$$

which combines with Eq. (4) to give the following formula.

$$\text{Surface area} = \iint_R \frac{|\nabla F|}{|\nabla F \cdot \mathbf{n}|}\, dA \tag{5}$$

We reached Eq. (5) under the assumptions that $\nabla F \cdot \mathbf{n} \neq 0$ and that F and its first partial derivatives were continuous over R (so that ∇F would be defined and continuous over R). However, whenever the quotient $|\nabla F|/|\nabla F \cdot \mathbf{n}|$ is integrable over R we may define the value of the integral in Eq. (5) to be the surface area of the portion of the surface $F(x, y, z) = c$ that lies over R.

The surface area defined in Eq. (5) agrees with our earlier definitions of surface area. We shall not prove this, but see Problems 12, 13, 15, and 21.

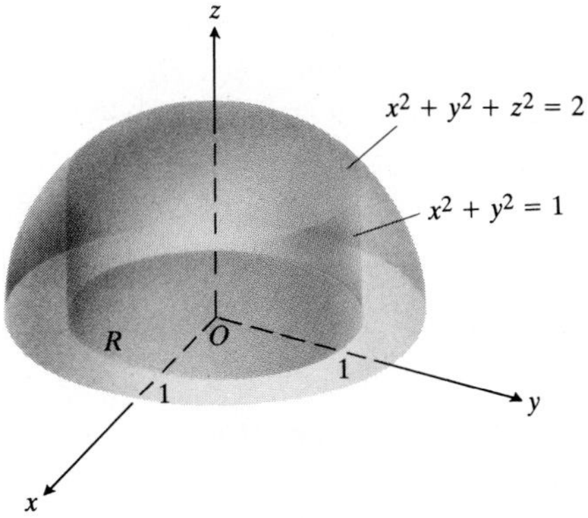

16.46 The cap cut from the hemisphere by the cylinder projects one to one onto the disc R: $x^2 + y^2 \leq 1$.

EXAMPLE 1 Find the area of the upper cap cut from the sphere $x^2 + y^2 + z^2 = 2$ by the cylinder $x^2 + y^2 = 1$ (Fig. 16.46).

Solution The cap, part of the surface $F(x, y, z) = x^2 + y^2 + z^2 = 2$, projects onto the disc R: $x^2 + y^2 \leq 1$ in the xy-plane. At any point (x, y, z) in space we have

$$F = x^2 + y^2 + z^2,$$
$$\nabla F = 2x\mathbf{i} + 2y\mathbf{j} + 2z\mathbf{k},$$
$$|\nabla F| = \sqrt{(2x)^2 + (2y)^2 + (2z)^2} = 2\sqrt{x^2 + y^2 + z^2}.$$

At points on the level surface $F(x, y, z) = x^2 + y^2 + z^2 = 2$, however,

$$|\nabla F| = 2\sqrt{x^2 + y^2 + z^2} = 2\sqrt{2}.$$

Taking $\mathbf{n} = \mathbf{k}$ as the unit vector normal to R gives

$$\nabla F \cdot \mathbf{n} = (2x\mathbf{i} + 2y\mathbf{j} + 2z\mathbf{k}) \cdot \mathbf{k} = 2z.$$

The surface area S is therefore

$$S = \iint_R \frac{|\nabla F|}{|\nabla F \cdot \mathbf{n}|}\, dA = \iint_R \frac{2\sqrt{2}}{2z}\, dA = \sqrt{2} \iint_R \frac{dA}{z}.$$

What do we do about the z?

Since z is the coordinate of a point on the surface $x^2 + y^2 + z^2 = 2$, we may express z in terms of x and y as

$$z = \sqrt{2 - x^2 - y^2}.$$

Thus,

$$\begin{aligned}
\text{Surface area} &= \sqrt{2} \iint_{x^2 + y^2 \le 1} \frac{dA}{\sqrt{2 - x^2 - y^2}} \\
&= \sqrt{2} \int_0^{2\pi} \int_0^1 \frac{r\, dr\, d\theta}{\sqrt{2 - r^2}} \qquad \text{(polar coordinates)} \\
&= \sqrt{2} \int_0^{2\pi} [-(2 - r^2)^{1/2}]_{r=0}^{r=1}\, d\theta \\
&= \sqrt{2} \int_0^{2\pi} (\sqrt{2} - 1)\, d\theta \\
&= 2\pi(2 - \sqrt{2}). \quad \square
\end{aligned}$$

Special Cases

In the case of a smooth surface S given by an equation $z = f(x, y)$ defined over a region R_{xy} in the xy-plane, we may write

$$F(x, y, z) = f(x, y) - z$$

and regard S as the level surface

$$F(x, y, z) = 0$$

of the function F. Taking the normal to R_{xy} to be $\mathbf{n} = \mathbf{k}$ then gives

$$|\nabla F| = |f_x\mathbf{i} + f_y\mathbf{j} - \mathbf{k}| = \sqrt{f_x^2 + f_y^2 + 1},$$

$$|\nabla F \cdot \mathbf{n}| = |-\mathbf{k} \cdot \mathbf{k}| = 1,$$

$$\text{Area of } S = \iint_{R_{xy}} \frac{|\nabla F|}{|\nabla F \cdot \mathbf{n}|}\, dA = \iint_{R_{xy}} \sqrt{f_x^2 + f_y^2 + 1}\, dy\, dx. \tag{6a}$$

Similarly, for the area of a smooth surface $x = f(y, z)$ over a region R_{yz} in the yz-plane we get

$$\text{Area of } S = \iint_{R_{yz}} \sqrt{f_y^2 + f_z^2 + 1}\, dy\, dz \tag{6b}$$

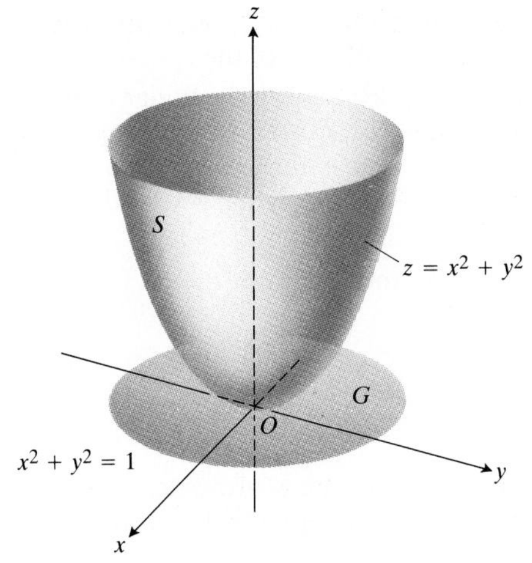

16.47 The area of the parabolic surface above is calculated in Example 2.

and for the area of a smooth surface $y = f(x, z)$ over a region R_{xz} in the xz-plane we get

$$\text{Area of } S = \iint_{R_{xz}} \sqrt{f_x^2 + f_z^2 + 1}\, dx\, dz. \tag{6c}$$

EXAMPLE 2 Find the area of the surface cut from the bottom of the paraboloid $z = x^2 + y^2$ by the plane $z = 1$.

Solution The surface projects onto the disk R: $x^2 + y^2 \le 1$ in the xy-plane (Fig. 16.47). We apply Eq. (6a) with

$$z = f(x, y) = x^2 + y^2$$

to get

$$\text{Area} = \iint_{R_{xy}} \sqrt{f_x^2 + f_y^2 + 1}\, dx\, dy = \iint_{x^2 + y^2 \le 1} \sqrt{4x^2 + 4y^2 + 1}\, dx\, dy$$

$$= \int_0^{2\pi} \int_0^1 \sqrt{4r^2 + 1}\, r\, dr\, d\theta \qquad \text{(polar coordinates)}$$

$$= \frac{\pi}{6}(5\sqrt{5} - 1). \quad \square$$

PROBLEMS

1. Find the area of the surface cut from the paraboloid $z = x^2 + y^2$ by the plane $z = 4$.
2. Find the area of the surface cut from the paraboloid $z = 4 - x^2 - y^2$ by the xy-plane.
3. Find the area cut from the paraboloid $z = 1 - x^2 - y^2$ by the plane $z = -3$. (*Hint:* Integrate over the disc $x^2 + y^2 \le 4$ in the xy-plane.)
4. Find the area of the portion of the surface $z = 9 - x^2 - y^2$ above the plane $z = 5$.
5. Find the area cut from the plane $x + y + z = 5$ by the cylinder whose walls are $x = y^2$ and $x = 2 - y^2$.
6. Find the area cut from the paraboloid $y = 1 - x^2 - z^2$ (a) by the plane $y = 0$, (b) by the plane $y = 3$. (*Hint:* Project onto the xz-plane.)
7. Find the area of the portion of the surface $2z = x^2$ that lies above the triangle bounded by the lines $x = 1$, $y = 0$, and $y = x$ in the xy-plane.
8. Find the area of the surface $y + z = x^2/2$ that lies above the triangle bounded by the lines $x = 2$, $y = 0$, and $y = x$ in the xy-plane.
9. Find the area of the surface $z = 9 - x^2 - y^2$ that lies above the ring R: $1 \le x^2 + y^2 \le 9$ in the xy-plane.
10. Find the area of the surface $z = x^2 + \sqrt{15}\, y - 2 \ln x$ above the square R: $0 \le y \le 1$, $1 \le x \le 2$ in the xy-plane.
11. Write a double iterated integral for the area cut from the upper half of the sphere $x^2 + y^2 + z^2 = 9$ by the elliptic cylinder $x^2 + 4y^2 = 9$.
12. Whenever we replace an old definition with a new one it is a good idea to try out the new one on familiar objects to see that it gives the answers we expect. For instance, is the surface area of a circular cylinder of base radius r and height h still $2\pi rh$? Find out by using Eq. (5) with $\mathbf{n} = \mathbf{k}$ to calculate the area of the surface cut from the cylinder $y^2 + z^2 = r^2$ by the planes $x = 0$ and $x = h > 0$. (Find the area above the xy-plane, and double.)
13. Show that the surface area of the sphere $x^2 + y^2 + z^2 = a^2$ is $4\pi a^2$.
14. Find the area cut from the paraboloid $z = 9 - x^2 - y^2$ by the planes $z = 0$ and $z = 8$.
15. Find the area of the triangle cut from the plane $x/a + y/b + z/c = 1$ (a, b, and $c > 0$) by the coordinate planes. Check your answer by vector methods.
16. Find the area of that portion of the sphere $x^2 + y^2 + z^2 = 2a^2$ that is cut out by the upper nappe of the cone $x^2 + y^2 = z^2$.
17. Find the area cut from the plane $z = cx$ by the cylinder $x^2 + y^2 = a^2$.
18. Find the area of that portion of the cylinder $x^2 + z^2 = a^2$ that lies between the planes $y = \pm a/2$, $x = \pm a/2$.

19. Find the area cut from the surface $az = y^2 - x^2$ by the cylinder $x^2 + y^2 = a^2$.

20. Find the area of that portion of the sphere $x^2 + y^2 + z^2 = 4a^2$ that lies inside the cylinder $x^2 + y^2 = 2ax$. (Figure 16.40 shows the top half.)

21. Find the area of the portion of the sphere $x^2 + y^2 + z^2 = a^2$ that lies in the first octant (a) by integrating over a region in the xy-plane, (b) by integrating over a region in the yz-plane.

22. Find the area of that portion of the cylinder in Problem 20 that lies inside the sphere. (*Hint:* Project the area into the xz-plane. Or use single integration, $\int h\, ds$, where h is the altitude of the cylinder and ds is the element of arc length in the xy-plane.)

23. Derive formulas (6b) and (6c) in the text.

REVIEW QUESTIONS AND EXERCISES

1. Define the double integral of a function of two variables. What geometric interpretation may be given to the integral?
2. List four applications of multiple integration.
3. Define *moment of inertia* and *radius of gyration.*
4. How does a double integral in polar coordinates differ from a double integral in cartesian coordinates? In what way are they alike?
5. What are the fundamental volume elements for triple integrals (a) in cartesian coordinates, (b) in cylindrical coordinates, (c) in spherical coordinates?
6. Describe the general procedures for determining limits of integration in iterated double and triple integrals.
7. How is surface area defined? Which formula in Article 16.9 is the most general one for computing surface area, in the sense that it includes the others as special cases?

MISCELLANEOUS PROBLEMS

1. Reverse the order of integration and evaluate

$$\int_0^4 \int_{-\sqrt{4-y}}^{(y-4)/2} dx\, dy.$$

2. Sketch the region over which the integral

$$\int_0^1 \int_{\sqrt{y}}^{2-\sqrt{y}} xy\, dx\, dy$$

is to be evaluated and find its value.

3. The integral

$$\int_{-1}^1 \int_{x^2}^1 dy\, dx$$

represents the area of a region of the xy-plane. Sketch the region and express the same area as a double integral with the order of integration reversed.

4. The base of a pile of sand covers the region in the xy-plane that is bounded by the parabola $x^2 + y = 6$ and the line $y = x$. The depth of the sand above the point (x, y) is x^2. Sketch the base of the sand pile and a representative element of volume dV, and find the volume of sand in the pile by double integration.

5. In setting up a double integral for the volume V under the paraboloid $z = x^2 + y^2$ and above a certain region R of the xy-plane, the following sum of iterated integrals was obtained:

$$V = \int_0^1 \left(\int_0^y (x^2 + y^2)\, dx\right) dy + \int_1^2 \left(\int_1^{2-y} (x^2 + y^2)\, dx\right) dy.$$

Sketch the region R in the xy-plane and express V as an iterated integral in which the order of integration is reversed.

6. By change of order of integration, show that the following double integral can be reduced to a single integral

$$\int_0^x \int_0^u e^{m(x-t)} f(t)\, dt\, du = \int_0^x (x - t) e^{m(x-t)} f(t)\, dt.$$

Similarly, it can be shown that

$$\int_0^x \int_0^v \int_0^u e^{m(x-t)} f(t)\, dt\, du\, dv = \int_0^x \frac{(x - t)^2}{2!} e^{m(x-t)} f(t)\, dt.$$

Evaluate integrals for the case $f(t) = \cos at$. (This example illustrates that such reductions usually make calculation easier.)

7. Sometimes a multiple integral with variable limits may be changed into one with constant limits. By

changing the order of integration, show that

$$\int_0^1 f(x)\left(\int_0^x \log(x-y)f(y)\,dy\right)dx$$
$$=\int_0^1 f(y)\left(\int_y^1 \log(x-y)\,f(x)\,dx\right)dy$$
$$=\frac{1}{2}\int_0^1\int_0^1 \log|x-y|\,f(x)\,f(y)\,dx\,dy.$$

8. Evaluate the integral

$$\int_0^\infty \frac{e^{-ax}-e^{-bx}}{x}\,dx.$$

(*Hint:* Use the relation

$$\frac{e^{-ax}-e^{-bx}}{x}=\int_a^b e^{-xy}\,dy$$

to form a double integral, and evaluate it by changing the order of integration.)

9. By double integration, find the centroid of that part of the area of the circle $x^2+y^2=a^2$ contained in the first quadrant.

10. Determine the centroid of the plane region that is given in polar coordinates by $0\le r\le a$, $-\alpha\le\theta\le\alpha$.

11. Find the centroid of the region bounded by the lines $\theta=0°$ and $\theta=45°$, and the circles $r=1$ and $r=2$.

12. By double integration, find the centroid of the area between the parabola $x+y^2-2y=0$ and the line $x+2y=0$.

13. For a solid body of constant density, having its center of gravity at the origin, show that the moment of inertia about an axis through (x_0, y_0) parallel to the z-axis is equal to the moment of inertia about the z-axis plus $m(x_0^2+y_0^2)$, where m is the mass of the body.

14. Find the moment of inertia of the angle section shown in Fig. 16.48 (a) with respect to the horizontal base, (b) with respect to a horizontal line through its centroid.

15. Show that, for a uniform elliptic lamina of semiaxes a, b, the moment of inertia about an axis in its plane through the center of the ellipse making an angle α with the axis of length $2a$ is $\frac{1}{4}m(a^2\sin^2\alpha+b^2\cos^2\alpha)$, where m is the mass of the lamina.

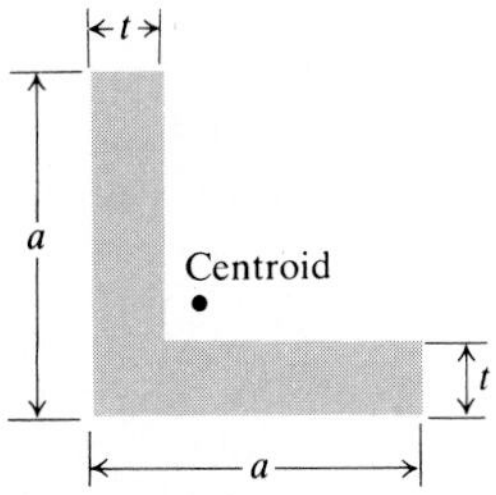

Figure 16.48

16. A counterweight of a flywheel has the form of the smaller segment cut from a circle of radius a by a chord at a distance b from the center $(b<a)$. Find the area of this counterweight and its polar moment of inertia about the center of the circle.

17. Consider an ellipse $(x^2/a^2)+(y^2/b^2)=1$ revolving about the x-axis to generate an ellipsoid. Find the radius of gyration of the ellipsoid with respect to the x-axis.

18. Find the radii of gyration about $\theta=0$ and $\theta=\pi/2$ for the area of a loop of the curve $r^2=a^2\cos 2\theta$, $(a>0)$.

19. The hydrostatic pressure at a depth y in a fluid is wy. Taking the x-axis in the surface of the fluid and the y-axis vertically downward, consider a semicircular lamina, radius a, completely immersed with its bounding diameter horizontal, uppermost, and at a depth c. Show that the depth of the center of pressure is

$$\frac{3\pi a^2+32ac+12\pi c^2}{4(4a+3\pi c)}.$$

The center of pressure is defined as the point where the entire hydrostatic force could be concentrated so as to produce the same first moment of force.

20. Show that

$$\iint \frac{\partial^2 F(x,y)}{\partial x\,\partial y}\,dx\,dy$$

over the rectangle $x_0\le x\le x_1$, $y_0\le y\le y_1$, is

$$F(x_1,y_1)-F(x_0,y_1)-F(x_1,y_0)+F(x_0,y_0).$$

21. Change the following double integral to an equivalent double integral in polar coordinates, and sketch the region of integration.

$$\int_{-a}^{a}\int_0^{\sqrt{a^2-y^2}} x\,dx\,dy.$$

22. A customary method of evaluating the improper integral $I=\int_0^\infty e^{-x^2}\,dx$ is to calculate its square,

$$I^2=\left(\int_0^\infty e^{-x^2}\,dx\right)\left(\int_0^\infty e^{-y^2}\,dy\right)$$
$$=\int_0^\infty\int_0^\infty e^{-(x^2+y^2)}\,dx\,dy.$$

Introduce polar coordinates in the last expression and show that

$$I=\int_0^\infty e^{-x^2}\,dx=\frac{\sqrt{\pi}}{2}.$$

23. By transformation of variables $u = x - y$, $v = y$, show that

$$\int_0^\infty e^{-sx}\,dx \int_0^x f(x-y, y)\,dy = \int_0^\infty \int_0^\infty e^{-s(u+v)} f(u, v)\,du\,dv.$$

24. How must a, b, c be chosen in order that $\int_{-\infty}^{\infty}\int_{-\infty}^{\infty} e^{-(ax^2+2bxy+cy^2)}\,dx\,dy = 1$? (*Hint:* Introduce the transformation

$$s = \alpha x + \beta y, \qquad t = \gamma x + \delta y$$

where $(\alpha\delta - \beta\gamma)^2 = ac - b^2$; then

$$ax^2 + 2bxy + cy^2 = s^2 + t^2.)$$

25. Find the area enclosed by the lemniscate $r^2 = 2a^2 \cos 2\theta$. Also find the moment of inertia of this area about the y-axis.

26. Evaluate the integral

$$\iint \frac{dx\,dy}{(1 + x^2 + y^2)^2}$$

taken (a) over one loop of the lemniscate $(x^2 + y^2)^2 - (x^2 - y^2) = 0$, (b) over the triangle with vertices $(0, 0)$, $(2, 0)$, $(1, \sqrt{3})$. (*Hint:* Transform to polar coordinates.)

27. Show, by transforming to polar coordinates, that

$$K(a) = \int_0^{a \sin\beta} \int_{y \cot\beta}^{\sqrt{a^2-y^2}} \ln(x^2 + y^2)\,dx\,dy = a^2\beta\left(\ln a - \frac{1}{2}\right),$$

where $0 < \beta < \pi/2$. Changing the order of integration, what expression do you obtain?

28. Find the volume bounded by the cylinder $y = \cos x$ and the planes

$$z = y, \qquad x = 0, \qquad x = \pi/2, \qquad \text{and} \qquad z = 0.$$

29. Find the center of mass of the homogeneous pyramid whose base is the square enclosed by the lines $x = 1$, $x = -1$, $y = 1$, $y = -1$, in the plane $z = 0$, and whose vertex is at the point $(0, 0, 1)$.

30. Find the volume bounded above by the sphere $x^2 + y^2 + z^2 = 2a^2$ and below by the paraboloid $az = x^2 + y^2$.

31. Find the volume bounded by the surfaces

$$z = x^2 + y^2 \qquad \text{and} \qquad z = \frac{1}{2}(x^2 + y^2 + 1).$$

32. Determine by triple integration the volume enclosed by the two surfaces $x = y^2 + z^2$ and $x = 1 - y^2$.

33. Find the moment of inertia, with respect to the z-axis, of a solid that is bounded below by the paraboloid $3az = x^2 + y^2$ and above by the sphere $x^2 + y^2 + z^2 = 4a^2$, if its density δ is constant.

34. Find by integration the volume of the ellipsoid

$$\frac{x^2}{a^2} + \frac{y^2}{b^2} + \frac{z^2}{c^2} = 1.$$

35. Evaluate the integral $\iiint |xyz|\,dx\,dy\,dz$ taken throughout the ellipsoid $x^2/a^2 + y^2/b^2 + z^2/c^2 \le 1$. (*Hint:* Introduce new coordinates:

$$x = au, \qquad y = bv, \qquad z = cw.)$$

36. The volume of a certain solid is given by the triple integral

$$\int_0^2 \left[\int_0^{\sqrt{2x-x^2}} \left(\int_{-\sqrt{4-x^2-y^2}}^{\sqrt{4-x^2-y^2}} dz\right) dy\right] dx.$$

a) Describe the solid by giving the equations of all the surfaces that form its boundary.
b) Express the volume as a triple integral in cylindrical coordinates. Give the limits of integration explicitly, but do not evaluate the integral.

37. WARNING: Hard problem. Setting up the integral is straightforward, but integrating the result takes hours. (It took MACSYMA 20 minutes.)

A square hole of side $2b$ is cut symmetrically through a sphere of radius a $(a > b\sqrt{2})$. Find the volume removed.

38. A hole is bored through a sphere, the axis of the hole being a diameter of the sphere. The volume of the solid remaining is given by the integral

$$V = 2\int_0^{2\pi} \int_0^{\sqrt{3}} \int_1^{\sqrt{4-z^2}} r\,dr\,dz\,d\theta.$$

a) By inspecting the given integral, determine the radius of the hole and the radius of the sphere.
b) Calculate the numerical value of the integral.

39. Set up an equivalent triple integral in rectangular coordinates. (Arrange the order so that the first integration is with respect to z, the second with respect to y, and the last with respect to x.)

$$\int_0^{\pi/2} \int_1^{\sqrt{3}} \int_1^{\sqrt{4-r^2}} r^3 \sin\theta \cos\theta\, z^2\,dz\,dr\,d\theta.$$

40. Find the volume bounded by the plane $z = 0$, the cylinder $x^2 + y^2 = a^2$, and the cylinder $az = a^2 - x^2$.

41. Find the volume of that portion of the sphere $r^2 + z^2 = a^2$ that is inside the cylinder $r = a \sin\theta$. (Here r, θ, z are cylindrical coordinates.)

42. Find the moment of inertia, about the z-axis, of the volume that is bounded above by the sphere $\rho = a$ and below by the cone $\phi = \pi/3$. (ρ, ϕ, θ are spherical coordinates.)

43. Find the volume enclosed by the surface $\rho = a \sin \phi$, in spherical coordinates.

44. Find the moment of inertia of the solid of constant density δ bounded by two concentric spheres of radii a and b ($a < b$), about a diameter.

45. Let S be a solid homogeneous sphere of radius a, constant density δ, mass $M = \frac{4}{3}\pi a^3 \delta$. Let P be a particle of mass m situated at distance b ($b > a$) from the center of S. According to Newton, the force of gravitational attraction of the sphere for P is given by the equation

$$\mathbf{F} = \gamma m \iiint \frac{\mathbf{u}\,\delta\, dV}{r^2},$$

where γ is the gravitational constant, $\mathbf{u}$ is a unit vector in the direction from P toward the volume element dV in S, r^2 is the square of the distance from P to dV, and the integration is extended throughout S. Take the origin at the center of the sphere and P at $(0, 0, b)$ on the z-axis, and show that $\mathbf{F} = -(\gamma M m/b^2)\mathbf{k}$. (*Remark.* This result shows that the force is the same as it would be if all the mass of the sphere were concentrated at its center.)

46. The density at P, a point of a solid sphere of radius a and center O, is given to be

$$\rho_0\{1 + \varepsilon \cos\theta + \frac{1}{2}\varepsilon^2(3\cos\theta - 1)\},$$

where θ is the angle OP makes with a fixed radius OQ, and ρ_0 and ε are constants. Find the average density of the sphere.

47. Find the area of the surface $y^2 + z^2 = 2x$ cut off by the plane $x = 1$.

48. Find the area cut from the plane $x + y + z = 1$ by the cylinder $x^2 + y^2 = 1$.

49. Find the area above the xy-plane cut from the cone $x^2 + y^2 = z^2$ by the cylinder $x^2 + y^2 = 2ax$.

50. Find the surface area of that portion of the sphere $r^2 + z^2 = a^2$ that is inside the cylinder $r = a \sin\theta$. ((r, θ, z) are cylindrical coordinates.)

51. Obtain a double integral expressing the surface area cut from the cylinder $z = a^2 - y^2$ by the cylinder $x^2 + y^2 = a^2$, and reduce this double integral to a definite single integral with respect to the variable y.

52. A square hole of side $2\sqrt{2}$ is cut symmetrically through a sphere of radius 2. Show that the area of the surface removed is $16\pi(\sqrt{2} - 1)$.

53. A torus surface is generated by moving a sphere of unit radius whose center travels on a closed plane circle of radius 2. Calculate the area of this surface.

54. Calculate the area of the surface $(x^2 + y^2 + z^2)^2 = x^2 - y^2$. (*Hint:* Use polar coordinates.)

55. Calculate the area of the spherical part of the boundary of the region

$$x^2 + y^2 + z^2 = r^2,$$
$$x^2 + y^2 - rx \geq 0,$$
$$x^2 + y^2 + rx \geq 0.$$

(*Hint:* Integrate first with respect to x and y.)

56. Prove that the potential of a circular disc of mass m per unit area, and of radius a, at a point distant h from the center and on the normal to the disc through the center, is $2\pi m(\sqrt{(a^2 + h^2)} - h)$. (The potential at a point P due to a mass Δm at Q is $\Delta m/r$, where r is the distance from P to Q.)

57. Find the attraction, at the vertex, of a solid right circular cone of mass M, height h, and radius of base a. (The attraction at P due to a mass Δm at Q is $(\Delta m/r^2)\mathbf{u}$, where r is the distance from P to Q, and $\mathbf{u}$ is a unit vector in the direction of $\overrightarrow{PQ}$.)

17 Vector Analysis

17.1 Vector Fields

Suppose that a certain region G of 3-space is occupied by a moving fluid: air, for example, or water. We may imagine that the fluid is made up of an infinite number of particles, and that at time t, the particle that is in position P at that instant has a velocity $\mathbf{v}$. If we stay at P and observe new particles that pass through it, we shall probably see that they have different velocities. This would surely be true, for example, in turbulent motions caused by high winds or stormy seas. Again, if we could take a picture of the velocities of particles at different places at the same instant, we would expect to find that these velocities vary from place to place. Thus the velocity at position P at time t is, in general, a function of both position and time:

$$\mathbf{v} = \mathbf{F}(x, y, z, t). \tag{1}$$

Equation (1) indicates that the velocity $\mathbf{v}$ is a vector function $\mathbf{F}$ of the four independent variables x, y, z, and t. Such functions have many applications, particularly in treatments of flows of material. In hydrodynamics, for example, if $\delta = \delta(x, y, z, t)$ is the *density* of the fluid at (x, y, z) at time t, and we take $\mathbf{F} = \mathbf{i}u + \mathbf{j}v + \mathbf{k}w$ to be the velocity expressed in terms of components, then we are able to derive the Euler partial differential equation of continuity of motion:

$$\frac{\partial \delta}{\partial t} + \frac{\partial(\delta u)}{\partial x} + \frac{\partial(\delta v)}{\partial y} + \frac{\partial(\delta w)}{\partial z} = 0$$

(Article 17.6). Such functions are also applied in physics and electrical engineering; for example, in the study of propagation of electromagnetic waves. Also, much current research activity in applied mathematics has to do with such functions.

Steady-State Flows

In this chapter, we shall deal mainly with flows for which the velocity function, Eq. (1), does not depend on the time t. Such flows are called steady-state flows. They exemplify *vector fields*.

DEFINITION

> If, to each point P in some region G, a vector $\mathbf{F}(P)$ is assigned, the collection of all such vectors is called a *vector field* on G.

In addition to the vector fields that are associated with fluid flows, there are vector *force* fields that are associated with gravitational attraction, magnetic force fields, electric fields, and purely mathematical fields.

EXAMPLE 1 Imagine an ideal fluid flowing with steady-state flow in a long cylindrical pipe of radius a, so that particles at distance r from the central axis are moving parallel to the axis with speed $|\mathbf{v}| = a^2 - r^2$ (Fig. 17.1). Describe this field by a formula for $\mathbf{v}$.

Solution Let the z-axis lie along the axis of the pipe, with positive direction in the direction of the flow. Then, in the usual way, introduce a right-handed cartesian coordinate system with unit vectors along the axes. Since all particles move parallel to the z-axis, the **k**-component of the flow is the only one different from zero. Therefore

$$\mathbf{v} = (a^2 - r^2)\mathbf{k} = (a^2 - x^2 - y^2)\mathbf{k}$$

for points inside the pipe. This vector field is not defined outside the cylinder $x^2 + y^2 = a^2$. If we were to draw the velocity vectors at all points in the disc

$$x^2 + y^2 \le a^2, \qquad z = 0,$$

their tips would describe the surface

$$z = a^2 - r^2$$

(cylindrical coordinates) for $z \ge 0$. Since this field does not depend on z, a similar figure would illustrate the flow field across any cross section of the pipe made by a plane perpendicular to its axis. □

EXAMPLE 2 A fluid is rotating about the z-axis with constant angular velocity ω. Every particle at a distance r from the z-axis and in a plane

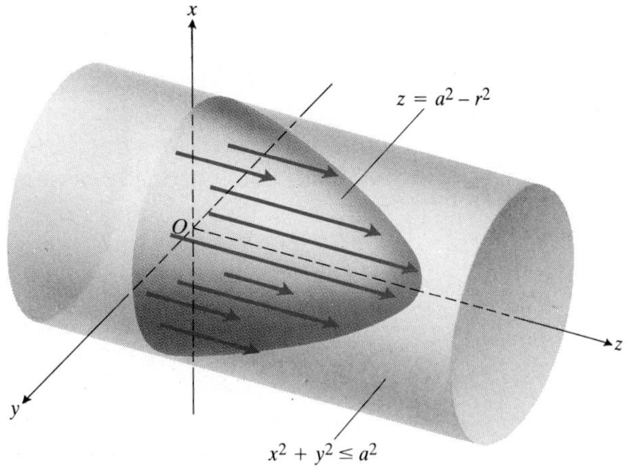

17.1 The flow of fluid in a long cylindrical pipe. The vectors $\mathbf{v} = (a^2 - r^2)\mathbf{k}$ inside the cylinder that have their bases in the xy-plane have their tips on the paraboloid $z = a^2 - r^2$.

perpendicular to the z-axis traces a circle of radius r, and each such particle has constant speed $|\mathbf{v}| = \omega r$. Describe this field by writing an equation for the velocity at $P(x, y, z)$.

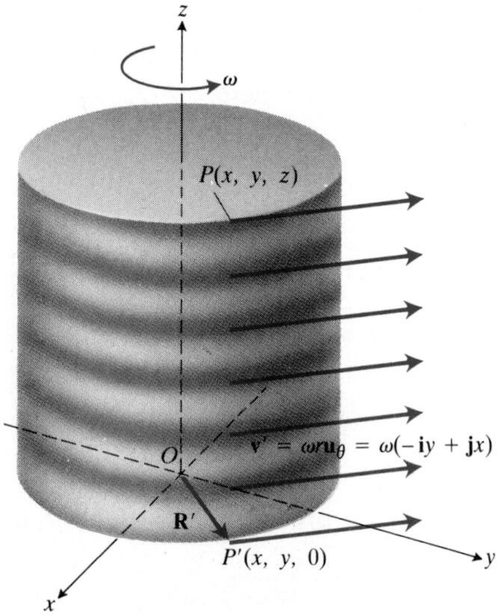

17.2 A steady rotational flow parallel to the *xy*-plane, with constant angular velocity ω in the positive (counterclockwise) direction.

Solution (See Fig. 17.2.) Each particle travels in a circle parallel to the xy-plane. Therefore it is convenient to begin by looking at the projection of such a circle onto this plane. The point $P(x, y, z)$ in space projects onto the image point $P'(x, y, 0)$, and the velocity vector $\mathbf{v}$ of a particle at P projects onto the velocity vector $\mathbf{v}'$ of a particle at P'. We assume that the motion is in the positive, or counterclockwise, direction, as indicated in the figure. The position vector of P' is $\mathbf{R}' = \mathbf{i}x + \mathbf{j}y$, and the vectors $-\mathbf{i}y + \mathbf{j}x$ and $\mathbf{i}y - \mathbf{j}x$ are both perpendicular to $\mathbf{R}'$. All three of these vectors have magnitude

$$\sqrt{x^2 + y^2} = r.$$

The velocity vector we want has magnitude ωr, is perpendicular to $\mathbf{R}'$, and points in the direction of motion. When x and y are both positive (that is, when the particle is in the first quadrant), the velocity has a negative **i**-component and a positive **j**-component. The vector that has these properties is

$$\mathbf{v}' = \omega r\mathbf{u}_\theta = \omega(-\mathbf{i}y + \mathbf{j}x). \tag{2a}$$

This formula can be verified for P' in the other three quadrants as well. For example, in the third quadrant both x and y are negative, so Eq. (2a) gives a vector with a positive **i**-component and a negative **j**-component, which is correct. Also, because the motion of P is in a circle parallel to the circle traced by P', and has the same velocity, we have

$$\mathbf{v} = \omega(-\mathbf{i}y + \mathbf{j}x) \tag{2b}$$

for any point in the fluid. □

EXAMPLE 3 At every point in space, a fluid has a velocity vector that is the sum of a constant velocity vector parallel to the z-axis and a rotational velocity vector given by Eq. (2b). Describe the field.

Solution Let the constant component parallel to the z-axis be $c\mathbf{k}$. Then the resultant field is

$$\mathbf{v} = \omega(-\mathbf{i}y + \mathbf{j}x) + c\mathbf{k}. \ \square \tag{3}$$

EXAMPLE 4 The gravitational force field induced at the point $P(x, y, z)$ in space by a mass M that is taken to lie at an origin is defined to be the force with which M would attract a particle of *unit* mass at P. Describe this field mathematically, assuming the inverse-square law

$$\mathbf{F} = -\frac{GMm}{r^2}\mathbf{r}.$$

Solution Because M and the unit mass at P are assumed to be *point* masses, we don't have to integrate anything; we just write down the force:

$$\mathbf{F} = -\frac{GM(1)}{|\overrightarrow{OP}|^2}\mathbf{r} \tag{4a}$$

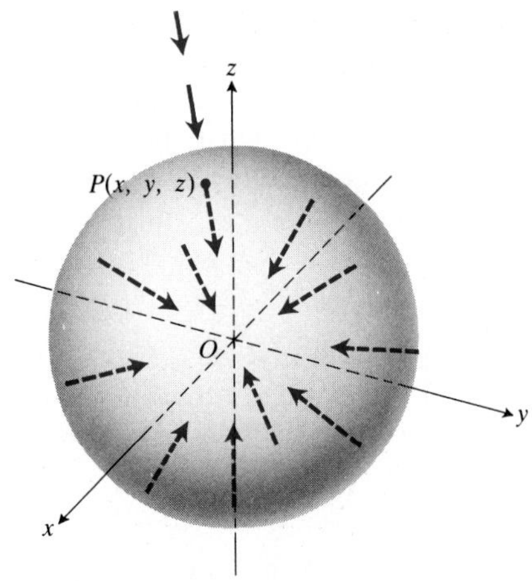

17.3 Some of the vectors of the gravitational field of Example 4.

where G is the gravitational constant, and

$$\mathbf{r} = \frac{\overrightarrow{OP}}{|\overrightarrow{OP}|}.$$

The position vector of P is $\overrightarrow{OP} = \mathbf{i}x + \mathbf{j}y + \mathbf{k}z$, so we find that

$$\mathbf{F} = \frac{-GM(\mathbf{i}x + \mathbf{j}y + \mathbf{k}z)}{(x^2 + y^2 + z^2)^{3/2}} \tag{4b}$$

gives the gravitational force field in question. Its graph would consist of infinitely many vectors, one starting from each point P (except the origin), and pointing straight toward the origin. If P is near the origin, the associated vector is longer than for points farther away from O. For points P on a ray through the origin, the $\mathbf{F}$-vectors would lie along that same ray, and decrease in length in proportion to the square of the distance from O. Figure 17.3 is a partial representation of this field. As you look at the figure, however, you should also imagine that an $\mathbf{F}$-vector is attached to *every* point $P \neq O$, and not just to those shown. At points on the surface of the sphere $|\overrightarrow{OP}| = a$, the vectors all have the same length, and all point toward the center of the sphere. □

Vector fields arise naturally in mathematics when we apply the gradient operator ∇ to a scalar function.

EXAMPLE 5 Suppose that the temperature T at each point $P(x, y, z)$ in some region of space is

$$T = 100 - x^2 - y^2 - z^2, \tag{5}$$

and that

$$\mathbf{F} = \nabla T.$$

Describe this vector field and discuss some of its properties.

Solution We have

$$\begin{aligned}\mathbf{F} = \nabla T = \text{grad } T &= \mathbf{i}\frac{\partial T}{\partial x} + \mathbf{j}\frac{\partial T}{\partial y} + \mathbf{k}\frac{\partial T}{\partial z} \\ &= -2x\mathbf{i} - 2y\mathbf{j} - 2z\mathbf{k} \\ &= -2\mathbf{R},\end{aligned}$$

where

$$\mathbf{R} = \overrightarrow{OP} = \mathbf{i}x + \mathbf{j}y + \mathbf{k}z$$

is the position vector of $P(x, y, z)$. This field is like a central force field, all vectors $\mathbf{F}$ being directed toward the origin. At points on a sphere with $|\overrightarrow{OP}|$ equal to a constant, the magnitude of the field vectors is a constant equal to twice the radius of the sphere. So, to represent the field, we could construct any sphere with center at O and draw a vector from any point P on the surface straight through O to the other side of the sphere. The collection of all such vectors, for points in the domain of the function T of Eq. (5), constitutes the *gradient field* of this particular scalar function.

Any surface on which T is constant is an *isothermal* surface. The isothermal surfaces here are spheres with center at the origin and radius $\sqrt{100 - T}$. Our calculation of $\mathbf{F} = \nabla T = -2\mathbf{R}$ has verified that the gradient of T at P is normal to the isothermal surface through P, because the diameter of such a spherical surface is always normal to the surface. □

PROBLEMS

1. In Example 1, where is a particle's speed (a) the greatest? (b) the least?

2. A vector field $\mathbf{F}$ in space is the field of a force directed toward the origin whose magnitude at each point $P(x, y, z)$ is a constant k times the inverse of the fourth power of the distance from P to the origin. Express $\mathbf{F}$ in terms of $\mathbf{i}$, $\mathbf{j}$, and $\mathbf{k}$ and the coordinates of P.

3. In Example 2, the position vector of P is
$$\mathbf{R} = \mathbf{R}' + z\mathbf{k}.$$
Show that for the motion described, it is correct to say that
$$\frac{d\mathbf{R}}{dt} = \mathbf{v} = \frac{d\mathbf{R}'}{dt} = \mathbf{v}'.$$

4. Describe, in words, the motion of the fluid discussed in Example 3. What path in space is described by a particle of the fluid that goes through the point $A(a, 0, 0)$ at time $t = 0$? Prove your result by taking $x = \cos \omega t$, $y = \sin \omega t$ and integrating the vector equation
$$\frac{d\mathbf{R}}{dt} = \omega(-\mathbf{i}y + \mathbf{j}x) + c\mathbf{k}.$$
with respect to t from $t = 0$ to $t = \tau$.

5. In Example 4, suppose that the mass M is at the point (x_0, y_0, z_0) rather than at the origin. How should Eq. (4b) be modified to describe this new gravitational force field?

In Problems 6–10, find the gradient fields $\mathbf{F}(x, y, z) = \nabla f$ for the given functions f.

6. $f(x, y, z) = x^2 \exp(2y + 3z)$
7. $f(x, y, z) = \ln(x^2 + y^2 + z^2)$
8. $f(x, y, z) = \tan^{-1}(xy/z)$
9. $f(x, y, z) = 2x - 3y + 5z$
10. $f(x, y, z) = (x^2 + y^2 + z^2)^{n/2}$
11. Suppose the density of the fluid in Example 1 is δ = constant at $P(x, y, z)$. Explain why the double integral
$$\int_0^a \int_0^{2\pi} \delta(a^2 - r^2)r\, d\theta\, dr$$
represents the *mass transport* (amount of mass per unit of time) flowing across the disc
$$x^2 + y^2 \le a^2, \qquad z = 0.$$
Evaluate the integral.

17.2 Surface Integrals

In this article we show how to integrate a continuous function over a piecewise smooth surface in space. This generalizes the notion of an integral over a region in a plane, which is a flat surface. As we shall see, these "surface integrals" are closely related to the concept of surface area. Surface integrals have important applications in engineering and physics as well as in mathematics.

Suppose, for example, that we have an electrical charge distributed over a surface $F(x, y, z) = c$ like the one shown in Fig. 17.4, and that the function $g(x, y, z)$ gives the charge density (charge per unit area) at each point on S. Then we may calculate the total charge on S as an integral in the following way.

We subdivide the shadow region R on the ground plane beneath the surface into small rectangles ΔA_k of the kind we would use if we were

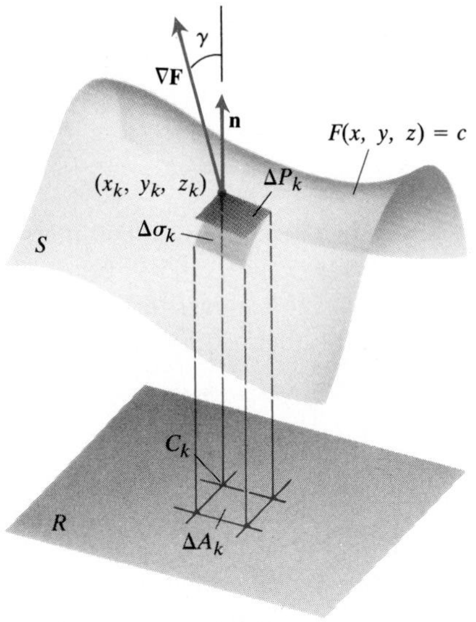

17.4 The integral of a function $g(x, y, z)$ over S is the limit of the sums

$$\sum g(x_k, y_k, z_k)\,\Delta P_k = \sum g(x_k, y_k, z_k)\frac{\Delta A_k}{|\cos\gamma|}.$$

defining the surface area of S. Then directly above each ΔA_k there lies a patch of surface $\Delta\sigma_k$ that we may approximate with a parallelogram-shaped portion of tangent plane, ΔP_k. To be specific we may suppose that ΔP_k is tangent to the surface at the point (x_k, y_k, z_k) directly above the back corner C_k of ΔA_k.

Up to this point the construction proceeds as in the definition of surface area in Article 16.9, but now we take one additional step: We evaluate g at (x_k, y_k, z_k) and approximate the total charge on the surface patch $\Delta\sigma_k$ by the product

$$g(x_k, y_k, z_k)\cdot\Delta P_k.$$

The rationale is that when the subdivision of R is sufficiently fine the value of g throughout $\Delta\sigma_k$ is nearly constant and ΔP_k is nearly the same as $\Delta\sigma_k$. The total charge over S is then approximated by the sum

$$\begin{aligned}\text{Total charge} &\approx \sum g(x_k, y_k, z_k)\,\Delta P_k\\ &\approx \sum g(x_k, y_k, z_k)\frac{\Delta A_k}{|\cos\gamma|},\end{aligned} \tag{1}$$

where γ is the angle between the surface normal ∇F and a unit normal $\mathbf{n}$ to R, as in Article 16.9.

If F, the function defining the surface S, and its first partial derivatives are continuous, and if g is continuous over S, then the sums on the right-hand side of Eq. (1) will approach the limit

$$\iint_R g(x, y, z)\frac{dA}{|\cos\gamma|} = \iint_R g(x, y, z)\frac{|\nabla F|}{|\nabla F\cdot\mathbf{n}|}\,dA \tag{2}$$

as the rectangular subdivision of R is refined in the usual way. This limit is called the *integral of* g *over the surface* S and is often written as

$$\iint_R g(x, y, z)\frac{|\nabla F|}{|\nabla F\cdot\mathbf{n}|}\,dA = \iint_S g\,d\sigma, \tag{3}$$

where the *surface area differential* $d\sigma$ is short for

$$d\sigma = \frac{|\nabla F|}{|\nabla F\cdot\mathbf{n}|}\,dA.$$

As we might expect, the formula in Eq. (3) is used to define the integral of *any* function g over the surface S (as long as the integral exists) and the integral's value takes on different meanings in different applications. For instance, if $g(x, y, z)\equiv 1$ is the constant function whose value is one, then the integral of g over S in Eq. (3) is the surface area of S.

Surface integrals have the same algebraic properties as other kinds of integrals. Writing g for $g(x, y, z)$ and h for $h(x, y, z)$ we have

I1. $\displaystyle\iint_S kg\,d\sigma = k\iint_S g\,d\sigma$ (any number k)

I2. $\displaystyle\iint_S (g+h)\,d\sigma = \iint_S g\,d\sigma + \iint_S h\,d\sigma$

I3. $\displaystyle\iint_S (g-h)\,d\sigma = \iint_S g\,d\sigma - \iint_S h\,d\sigma$

I4. $\iint_S h\,d\sigma \geq 0 \qquad$ if $h \geq 0$ on S

I5. $\iint_S h\,d\sigma \geq \iint_S g\,d\sigma \qquad$ if $h \geq g$ on S.

As usual, the integrals have these properties because their approximating sums have these properties.

There is also a "surface additivity" property. If S is partitioned by line segments and smooth curves into a finite number of nonoverlapping smooth patches $S_1, S_2, \ldots, S_n$, then

I6. $\iint_S h\,d\sigma = \iint_{S_1} h\,d\sigma + \iint_{S_2} h\,d\sigma + \cdots + \iint_{S_n} h\,d\sigma.$

Thus we can integrate a function over the surface of a cube by integrating it over each face and adding the results. We can integrate over any "turtle shell" of welded plates one plate at a time and add the results when we are through.

Special Cases

If a smooth surface S is given by an equation in the form $z = f(x, y)$ defined over a region R_{xy} in the xy-plane, then the formula in Eq. (3) specializes to

$$\iint_S g\,d\sigma = \iint_{R_{xy}} g(x, y, z)\sqrt{f_x^2 + f_y^2 + 1}\,dx\,dy, \tag{4}$$

as we may deduce from the derivation of Eq. (6a) in Article 16.9. Similarly, the surface integral of g over a smooth surface $x = f(y, z)$ that is defined over a region R_{yz} in the yz-plane is

$$\iint_S g\,d\sigma = \iint_{R_{yz}} g(x, y, z)\sqrt{f_y^2 + f_z^2 + 1}\,dy\,dz, \tag{5}$$

and the surface integral of g over a smooth surface $y = f(x, z)$ that is defined over a region R_{xz} in the xz-plane is

$$\iint_S g\,d\sigma = \iint_{R_{xz}} g(x, y, z)\sqrt{f_x^2 + f_z^2 + 1}\,dx\,dz. \tag{6}$$

EXAMPLE 1 Evaluate $\iint z\,d\sigma$ over the hemisphere

$$F(x, y, z) = x^2 + y^2 + z^2 = a^2, \qquad z \geq 0.$$

(This integral gives the first moment of the hemisphere with respect to the xy-plane.)

Solution With $\mathbf{n} = \mathbf{k}$ and base region the disc R: $x^2 + y^2 \leq a^2$, we have

$$|\nabla F| = |2x\mathbf{i} + 2y\mathbf{j} + 2z\mathbf{k}| = 2\sqrt{x^2 + y^2 + z^2} = 2a,$$

$$|\nabla F \cdot \mathbf{n}| = |2z| = 2z \qquad \text{(because } z \geq 0\text{)},$$

$$d\sigma = \frac{|\nabla F|}{|\nabla F \cdot \mathbf{n}|}\,dA = \frac{a}{z}\,dA,$$

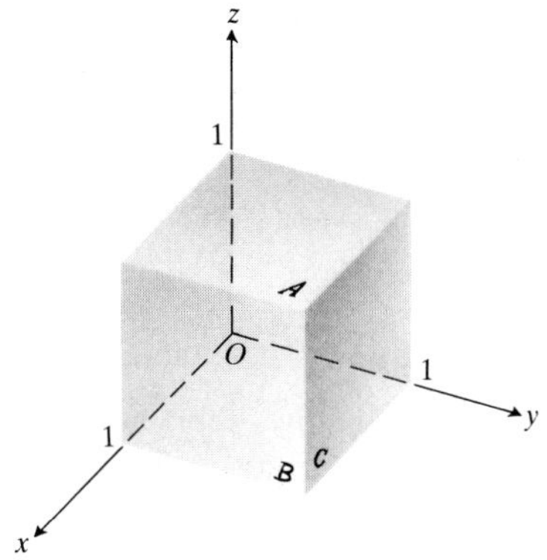

17.5 To integrate a function over the surface of a cube we integrate over each face and add the results.

and

$$\iint_S g\,d\sigma = \iint_{x^2+y^2\le a^2} z\frac{a}{z}\,dA = a\cdot\pi a^2 = \pi a^3. \quad \square$$

EXAMPLE 2 Integrate $g(x, y, z) = xyz$ over the cube in the first octant bounded by the coordinate planes and the planes $x = 1$, $y = 1$, $z = 1$.

Solution The cube is shown in Fig. 17.5. We integrate $g = xyz$ over each face and add the results. Since $g = 0$ on the three faces that lie in the coordinate planes, the integral of g over each of these faces is zero and the integral of g over the cube reduces to

$$\iint_{\text{cube}} xyz\,d\sigma = \iint_{\text{face } A} xyz\,d\sigma + \iint_{\text{face } B} xyz\,d\sigma = \iint_{\text{face } C} xyz\,d\sigma.$$

As a surface, face A is given by the equation

$$z = f(x, y) = 1$$

over the square R_{xy}: $0 \le x \le 1$, $0 \le y \le 1$. Therefore

$$d\sigma = \sqrt{f_x^2 + f_y^2 + 1}\,dx\,dy = \sqrt{0+0+1}\,dx\,dy = dx\,dy.$$

Also,

$$g(x, y, z) = xyz = xy\cdot 1 = xy$$

on face A, so that

$$\iint_{\text{face } A} g\,d\sigma = \iint_{R_{xy}} xy\,dx\,dy = \int_0^1\int_0^1 xy\,dx\,dy = \int_0^1 \frac{y}{2}\,dy = \frac{1}{4}.$$

Similar calculations show

$$\iint_{\text{face } B} g\,d\sigma = \int_0^1\int_0^1 yz\,dy\,dz = \frac{1}{4}, \qquad \iint_{\text{face } C} g\,d\sigma = \int_0^1\int_0^1 xz\,dx\,dz = \frac{1}{4}.$$

Hence,

$$\iint_{\text{cube}} g\,d\sigma = \frac{1}{4} + \frac{1}{4} + \frac{1}{4} = \frac{3}{4}. \quad \square$$

The Surface Integral for Flux

If we can choose a unit normal vector **n** on a surface S in such a way that **n** varies continuously as its initial point moves about the surface, we call the surface *orientable*. Spheres and other closed surfaces in space (surfaces that enclose a solid) are orientable, and by convention we choose **n** on a closed surface to point outward. In any case, once **n** is chosen we call the surface an *oriented surface*. The direction of **n** at any point is called the *positive* direction. Any patch or subportion of an orientable surface is also orientable. The Möbius band shown in Fig. 17.27 in Article 17.7 is a nonorientable surface.

If **F** is a continuous vector field and S is an oriented surface we call the integral over S of $\mathbf{F} \cdot \mathbf{n}$, the component of **F** in the direction of **n**, the flux of **F** across S in the positive direction:

DEFINITION

The *flux* of **F** across S in the direction of **n** is $\iint_S \mathbf{F} \cdot \mathbf{n}\, d\sigma$. (7)

For instance, if **F** is the velocity field in a fluid flow, then $\mathbf{F} \cdot \mathbf{n}$ is the component of the velocity perpendicular to the surface in the positive direction and the flux of **F** is the total flow rate (flow per unit time) across S in the positive direction. We shall discuss this interpretation in detail later in the chapter.

If S is part of a level surface $G(x, y, z) = c$, then **n** may be taken to be one of the two vectors

$$\mathbf{n} = +\frac{\nabla G}{|\nabla G|} \qquad \text{or} \qquad \mathbf{n} = -\frac{\nabla G}{|\nabla G|},$$

depending on which vector gives the preferred direction.

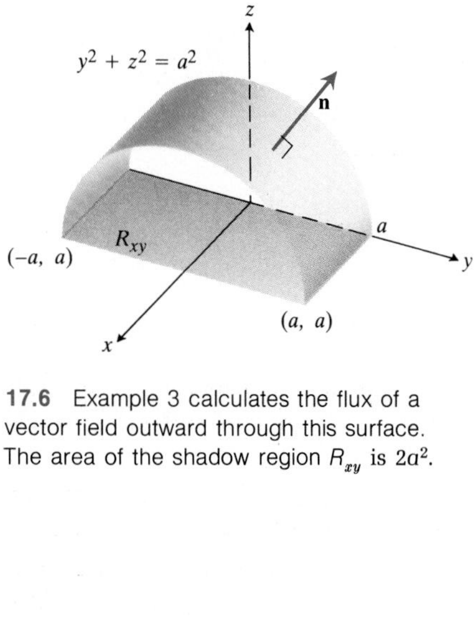

17.6 Example 3 calculates the flux of a vector field outward through this surface. The area of the shadow region R_{xy} is $2a^2$.

EXAMPLE 3 Find the flux of $\mathbf{F} = yz\mathbf{j} + z^2\mathbf{k}$ outward through the surface S cut from the cylinder $y^2 + z^2 = a^2$, $z \geq 0$ by the planes $x = 0$ and $x = a$ (Fig. 17.6).

Solution The outward unit normal to S may be calculated from the gradient of $G(x, y, z) = y^2 + z^2$ to be

$$\mathbf{n} = +\frac{\nabla G}{|\nabla G|} = \frac{2y\mathbf{j} + 2z\mathbf{k}}{\sqrt{4y^2 + 4z^2}} = \frac{2y\mathbf{j} + 2z\mathbf{k}}{2a} = \frac{y}{a}\mathbf{j} + \frac{z}{a}\mathbf{k}.$$

We also have

$$d\sigma = \frac{|\nabla G|}{|\nabla G \cdot \mathbf{k}|}\, dA = \frac{2a}{|2z|}\, dA = \frac{a}{z}\, dA,$$

where we drop the absolute value signs because $z \geq 0$ on S.

On the surface S, the value $\mathbf{F} \cdot \mathbf{n}$ is given by the formula

$$\begin{aligned} \mathbf{F} \cdot \mathbf{n} &= (yz\mathbf{j} + z^2\mathbf{k}) \cdot \left(\frac{y}{a}\mathbf{j} + \frac{z}{a}\mathbf{k}\right) \\ &= \frac{y^2z + z^3}{a} = \frac{z}{a}(y^2 + z^2) \\ &= \frac{z}{a}(a^2) \qquad (y^2 + z^2 = a^2 \text{ on } S) \\ &= az. \end{aligned}$$

Therefore, the flux of **F** outward through S is

$$\iint_S \mathbf{F} \cdot \mathbf{n}\, d\sigma = \iint_S (az)\frac{a}{z}\, dA = \iint_{R_{xy}} a^2\, dx\, dy = a^2 \cdot \text{area}\,(R_{xy}) = 2a^4. \ \square$$

Other Applications

The mass of a thin shell S of material whose density is given by $\delta(x, y, z)$ is

$$m = \text{mass}(S) = \iint_S \delta(x, y, z)\, d\sigma.$$

The first moments of S about the coordinate planes are

$$M_{yz} = \iint_S x\delta\, d\sigma, \qquad M_{xz} = \iint_S y\delta\, d\sigma, \qquad M_{xy} = \iint_S z\delta\, d\sigma.$$

The coordinates of the shell's center of mass are

$$\bar{x} = \frac{\iint_S x\delta\, d\sigma}{m}, \qquad \bar{y} = \frac{\iint_S y\delta\, d\sigma}{m}, \qquad \bar{z} = \frac{\iint_S z\delta\, d\sigma}{m}.$$

The moments of inertia of the shell about the coordinate axes are

$$I_x = \frac{\iint_S (y^2 + z^2)\delta\, d\sigma}{m},$$

$$I_y = \frac{\iint_S (x^2 + z^2)\delta\, d\sigma}{m},$$

$$I_z = \frac{\iint_S (x^2 + y^2)\delta\, d\sigma}{m}.$$

Radii of gyration are defined in the usual way.

The *average value* of a function $g(x, y, z)$ over a surface S is

$$\text{Average value} = \frac{1}{\text{Area}(S)} \iint_S g\, d\sigma = \frac{\iint_S g\, d\sigma}{\iint_S d\sigma}.$$

PROBLEMS

1. Integrate $g(x, y, z) = x + y + z$ over the surface of the unit cube shown in Fig. 17.5.

2. Integrate $h(x, y, z) = y + z$ over the surface of the wedge shown in Fig. 16.30.

3. Integrate $g(x, y, z) = xyz$ over the surface of the rectangular solid shown in Fig. 16.37.

4. Integrate $g(x, y, z) = xyz$ over the surface of the rectangular solid shown in Fig. 16.34.

5. Evaluate $\iint z\, d\sigma$ over the hemisphere

$$x^2 + y^2 + z^2 = a^2, \quad z \le 0.$$

6. Evaluate $\iint z\, d\sigma$ over the sphere $x^2 + y^2 + z^2 = a^2$.

In Problems 7 and 8, let $h(x, y, z) = x + y + z$ and let S be the portion of the plane $z = 2x + 3y$ for which $x \ge 0$, $y \ge 0$, $x + y \le 2$.

7. Evaluate $\iint_S h\, d\sigma$ by projecting S into the xy-plane. Sketch the projection.

8. Evaluate $\iint_S h\, d\sigma$ by projecting S into the yz-plane. Sketch the projection.

9. Find the center of mass of a thin hemispherical shell of radius a and uniform density δ. (Place the shell with its base on the xy-plane.)

10. Find the average height of the hemisphere $x^2 + y^2 + z^2 = a^2$, $z \ge 0$ above the xy-plane.

11. Find the moment of inertia about the z-axis of the surface cut from the cone $z^2 = x^2 + y^2$ by the cylinder $x^2 + y^2 = 2x$.

12. Find the moment of inertia about the z-axis of the hemisphere $x^2 + y^2 + z^2 = a^2$, $z \ge 0$.

In Problems 13–18, find the flux of the given vector field $\mathbf{F}$ through the portion of the sphere $x^2 + y^2 + z^2 = a^2$ that lies in the first octant in the outward direction away from the origin.

13. $\mathbf{F} = \mathbf{n}$

14. $\mathbf{F} = -\mathbf{i}y + \mathbf{j}x$

15. $\mathbf{F} = z\mathbf{k}$

16. $\mathbf{F} = \mathbf{i}x + \mathbf{j}y$

17. $\mathbf{F} = y\mathbf{i} - x\mathbf{j} + \mathbf{k}$

18. $\mathbf{F} = zx\mathbf{i} + zy\mathbf{j} + z^2\mathbf{k}$.

19. Let S be the portion of the cylinder $y = e^x$ in the first octant whose projection parallel to the x-axis onto the yz-plane is the rectangle R_{yz}: $1 \leq y \leq e$, $0 \leq z \leq 1$. Let $\mathbf{n}$ be the unit vector normal to S that points away from the yz-plane. Find the flux of $\mathbf{F} = -2\mathbf{i} + 2y\mathbf{j} + z\mathbf{k}$ through S in the direction of $\mathbf{n}$.

20. Let S be the portion of the cylinder $y = \ln x$ in the first octant whose projection parallel to the y-axis onto the xz-plane is the rectangle R_{xz}: $1 \leq x \leq e$, $0 \leq z \leq 1$. Let $\mathbf{n}$ be the unit vector normal to S that points away from the xz-plane. Find the flux of $\mathbf{F} = 2y\mathbf{j} + z\mathbf{k}$ through S in the direction of $\mathbf{n}$.

21. Find the moment of inertia about the z-axis of the surface of the cube shown in Fig. 17.5. ($\delta = 1$ for a geometric figure.)

22. The sphere $x^2 + y^2 + z^2 = 25$ is cut by the plane $z = 3$, the smaller portion cut off forming a solid that is bounded by a closed surface S made up of a spherical cap S_1 and a flat disc S_2. Find

$$\iint_S \mathbf{F} \cdot \mathbf{n}\, d\sigma = \iint_{S_1} \mathbf{F} \cdot \mathbf{n}\, d\sigma + \iint_{S_2} \mathbf{F} \cdot \mathbf{n}\, d\sigma$$

if $\mathbf{F} = xz\mathbf{i} + yz\mathbf{j} + \mathbf{k}$ and in each integral on the right $\mathbf{n}$ is taken to be the outward-pointing normal to the surface.

23. (*Spherical coordinates.*) Suppose that the surface of the hemisphere in Example 1 is subdivided by arcs of great circles on which the spherical coordinate θ remains constant (meridians of longitude), and by circles parallel to the xy-plane on which ϕ remains constant (parallels of latitude). Let the angular spacings be $\Delta\theta$ and $\Delta\phi$, respectively. Express the integral $\iint_S z\, d\sigma$ of Example 1 in the form

$$\lim_{\substack{\Delta\theta \to 0 \\ \Delta\phi \to 0}} \sum F(\theta, \phi)\, \Delta\theta\, \Delta\phi = \iint F(\theta, \phi)\, d\theta\, d\phi,$$

with appropriate limits of integration, and evaluate. (*Hint:* You should get

$$d\sigma = (r\, d\theta) \cdot (\rho\, d\phi) = \rho^2 \sin\phi\, d\theta\, d\phi = a^2 \sin\phi\, d\theta\, d\phi,$$

where ρ, ϕ, θ are spherical coordinates.)

24. Use the result $d\sigma = a^2 \sin\phi\, d\theta\, d\phi$ from Problem 23 to find the average great-circle distance of a point on the hemisphere $x^2 + y^2 + z^2 = a^2$, $z \geq 0$ from the north pole $N(0, 0, a)$.

***Toolkit* Programs**

Double Integral

17.3

Line Integrals and Work

In mechanics, the work done by a constant force $\mathbf{F}$ when the point of application undergoes a displacement $\Delta\mathbf{R}$ is defined to be $\mathbf{F} \cdot \Delta\mathbf{R}$. When the force $\mathbf{F}$ varies with position, however, and the point of application moves along a curve

$$\mathbf{R}(t) = x(t)\mathbf{i} + y(t)\mathbf{j} + z(t)\mathbf{k}$$

in space from point A to point B, the work done along the curve is defined as the integral

$$\int_A^B \mathbf{F} \cdot d\mathbf{R} = \int_A^B \mathbf{F} \cdot \frac{d\mathbf{R}}{ds}\, ds. \tag{1}$$

This integral is one of the so-called line integrals that are the subject of this article. After defining these integrals and calculating some of them, we shall return to the notion of work and show that in some fields the work done by the force along a curve depends only on the points at which the curve begins and ends, and not on the particular route taken between the points.

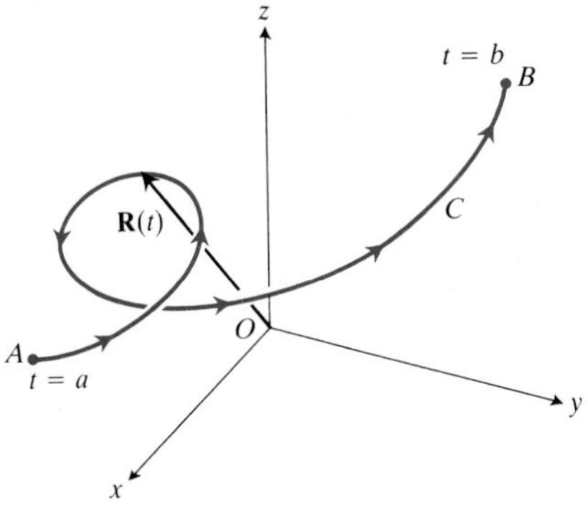

17.7 The directed curve C from A to B.

Line Integrals

Suppose that $w(x, y, z)$ is a function whose domain contains a piecewise smooth space curve C whose position vector

$$\mathbf{R}(t) = x(t)\mathbf{i} + y(t)\mathbf{j} + z(t)\mathbf{k}, \quad a \le t \le b$$

moves from point A to point B as t increases from a to b (Fig. 17.7). We can then define the integral of w over C in much the same way as we would define the integral of w over a piecewise smooth surface in its domain. We subdivide the curve into small pieces of length

$$\Delta s_1, \Delta s_2, \ldots, \Delta s_n,$$

evaluate w at a point (x_k, y_k, z_k) in each piece, form the sums

$$\sum_{k=1}^{n} w(x_k, y_k, z_k)\, \Delta s_k,$$

and define the integral of w over C from A to B to be the limit of these sums as the subdivision of C is refined so that the largest Δs_k approaches zero:

$$\int_C w\, ds = \lim_{n\to\infty} \sum_{k=1}^{n} w(x_k, y_k, z_k)\, \Delta s_k. \tag{2}$$

This limit will exist if w is continuous and the derivatives $x'(t)$, $y'(t)$, and $z'(t)$ are bounded and have only a finite number of discontinuities on the interval $a \le t \le b$. The limit may exist in other cases as well, and whenever it exists we call it the *line integral of w along C from A to B*. Note that if $w(x, y, z) \equiv 1$ is the constant function whose value is one, then the line integral gives the length of C from A to B.

It is a theorem from advanced calculus that the value of the line integral in (2) may be calculated from any parameterization of the curve for which $ds/dt = \sqrt{x'(t)^2 + y'(t)^2 + z'(t)^2}$ is positive from the formula

$$\int_C w\, ds = \int_{t=a}^{t=b} w(x(t), y(t), z(t))\sqrt{x'(t)^2 + y'(t)^2 + z'(t)^2}\, dt, \tag{3}$$

provided the derivatives of $x(t)$, $y(t)$, and $z(t)$ are bounded and piecewise continuous over the interval $a \le t \le b$. Thus the value of the line integral depends only on the nature of w and the geometry of the curve and may be calculated from any convenient parameterization.

EXAMPLE 1 Let C be the line segment from $A(0, 0)$ to $B(1, 1)$, and let $w = x + y^2$. Evaluate $\int_C w\, ds$ for two different parameterizations of C.

Solution 1 If we let

$$x = t \quad \text{and} \quad y = t, \quad 0 \le t \le 1,$$

then we get

$$\int_C w\, ds = \int_0^1 (t + t^2)\sqrt{1 + 1}\, dt = \sqrt{2}\left[\frac{t^2}{2} + \frac{t^3}{3}\right]_0^1 = \frac{5\sqrt{2}}{6}.$$

Solution 2 As a second parameterization of the given segment of the line $y = x$, we let

$$x = \sin t \quad \text{and} \quad y = \sin t, \quad 0 \le t \le \pi/2,$$

and get

$$\begin{aligned}\int_C w\,ds &= \int_0^{\pi/2} (\sin t + \sin^2 t)\sqrt{2\cos^2 t}\,dt \\ &= \sqrt{2}\left[\frac{\sin^2 t}{2} + \frac{\sin^3 t}{3}\right]_0^{\pi/2} = \frac{5\sqrt{2}}{6}. \quad \square\end{aligned}$$

Algebraic Properties

Line integrals inherit the usual algebraic properties from their approximating sums (properties analogous to I1–I6 in the preceding article). The additivity property, as we shall see, is particularly useful. If the curve C is made by joining a finite number of piecewise smooth curves $C_1, C_2, \ldots, C_n$ together end to end, then

I6. $$\int_C w\,ds = \int_{C_1} w\,ds + \int_{C_2} w\,ds + \cdots + \int_{C_n} w\,ds.$$

Thus the integral of w along a polygon is the sum of the integrals over the polygon's sides, and so on.

Line integrals also have the property that changing the direction of integration reverses the sign of the integral:

I7. $$\int_B^A w\,ds = -\int_A^B w\,ds.$$

Thus if we integrate along a curve from A to B and then integrate from B back to A again, the net result is zero:

$$\int_{ABA} w\,ds = \int_A^B w\,ds + \int_B^A w\,ds = \int_A^B w\,ds - \int_A^B w\,ds = 0.$$

Work

To calculate the work done by a force

$$\mathbf{F} = M(x, y, z)\mathbf{i} + N(x, y, z)\mathbf{j} + P(x, y, z)\mathbf{k}$$

whose point of application moves along a curve

$$\mathbf{R}(t) = x(t)\mathbf{i} + y(t)\mathbf{j} + z(t)\mathbf{k}, \qquad a \le t \le b$$

from a point A to a point B in space, we integrate the scalar product

$$w = \mathbf{F} \cdot \frac{d\mathbf{R}}{ds}$$

along the curve from $t = a$ to $t = b$. Thus,

$$\begin{aligned}\text{Work} &= \int \left(\mathbf{F} \cdot \frac{d\mathbf{R}}{ds}\right) ds \\ &= \int_{t=a}^{t=b} (\mathbf{F} \cdot \mathbf{T})\sqrt{x'(t)^2 + y'(t)^2 + z'(t)^2}\,dt, \end{aligned} \tag{4}$$

where the second equality comes from the fact that $d\mathbf{R}/ds$ is the unit tangent vector $\mathbf{T}$.

Equation (4) emphasizes the fact that the work is the value of the line integral along the curve of the tangential component of the force field **F**. If we write

$$d\mathbf{R} = \mathbf{i}\, dx + \mathbf{j}\, dy + \mathbf{k}\, dz,$$

we obtain Eq. (4) in the alternative forms

$$\text{Work} = \int_C \mathbf{F} \cdot \frac{d\mathbf{R}}{ds}\, ds = \int_C \mathbf{F} \cdot d\mathbf{R} = \int_C M\, dx + N\, dy + P\, dz, \tag{5}$$

which are all in common use.

In general the amount of work done by a force **F** along a curve from point A to point B depends on the path as well as on the endpoints A and B.

EXAMPLE 2 The point of application of the force

$$\mathbf{F} = \mathbf{i}(x^2 - y) + \mathbf{j}(y^2 - z) + \mathbf{k}(z^2 - x)$$

moves from the origin O to the point $A(1, 1, 1)$,

a) along the straight line OA, and
b) along the curve

$$x = t, \qquad y = t^2, \qquad z = t^3, \qquad 0 \le t \le 1.$$

Find the work done in each case.

Solution a) Equations for the line OA are

$$x = y = z.$$

The integral to be evaluated is

$$W = \int_C (x^2 - y)\, dx + (y^2 - z)\, dy + (z^2 - x)\, dz$$

which, for the path OA, becomes

$$W = \int_0^1 3(x^2 - x)\, dx = -\frac{1}{2}.$$

b) Along the curve, we get

$$W = \int_0^1 2(t^4 - t^3)t\, dt + 3(t^6 - t)t^2\, dt = -\frac{29}{60}. \quad \square$$

Path Independence

Under certain conditions, the line integral between two points A and B is independent of the path C joining them. That is, the integral in Eq. (5) has the same value for any two paths C_1 and C_2 joining A and B. This happens when the force field **F** is a *gradient field*, that is, when

$$\mathbf{F}(x, y, z) = \nabla f = \mathbf{i}\frac{\partial f}{\partial x} + \mathbf{j}\frac{\partial f}{\partial y} + \mathbf{k}\frac{\partial f}{\partial z},$$

for some differentiable function f. We state this as a formal theorem and prove the sufficiency and necessity of the conditions, with some interpolated remarks.

THEOREM 1 Let **F** be a vector field with components M, N, P, that are continuous throughout some connected region D. Then a necessary and sufficient condition for the integral

$$\int_A^B \mathbf{F} \cdot d\mathbf{R}$$

to be independent of the path joining the points A and B in D is that there exist a differentiable function f such that

$$\mathbf{F} = \nabla f = \mathbf{i}\frac{\partial f}{\partial x} + \mathbf{j}\frac{\partial f}{\partial y} + \mathbf{k}\frac{\partial f}{\partial z} \tag{6}$$

throughout D. Furthermore, if the integral is independent of the path from A to B, then its value is

$$\int_A^B \mathbf{F} \cdot d\mathbf{R} = f(B) - f(A). \tag{7}$$

Proof of sufficiency. First, we suppose that Eq. (6) is satisfied, and then consider A and B to be two points in D (see Fig. 17.8). Suppose that C is any piecewise smooth curve joining A and B:

$$x = x(t), \qquad y = y(t), \qquad z = z(t), \qquad t_1 \le t \le t_2.$$

Along C, $f = f[x(t), y(t), z(t)]$ is a function of t to which we may apply the chain rule to differentiate with respect to t:

$$\begin{aligned}\frac{df}{dt} &= \frac{\partial f}{\partial x}\frac{dx}{dt} + \frac{\partial f}{\partial y}\frac{dy}{dt} + \frac{\partial f}{\partial z}\frac{dz}{dt}\\ &= \nabla f \cdot \left(\mathbf{i}\frac{dx}{dt} + \mathbf{j}\frac{dy}{dt} + \mathbf{k}\frac{dz}{dt}\right)\\ &= \nabla f \cdot \frac{d\mathbf{R}}{dt}.\end{aligned} \tag{8}$$

Because Eq. (6) holds, we also have

$$\mathbf{F} \cdot d\mathbf{R} = \nabla f \cdot d\mathbf{R} = \nabla f \cdot \frac{d\mathbf{R}}{dt}\,dt = \frac{df}{dt}\,dt. \tag{9}$$

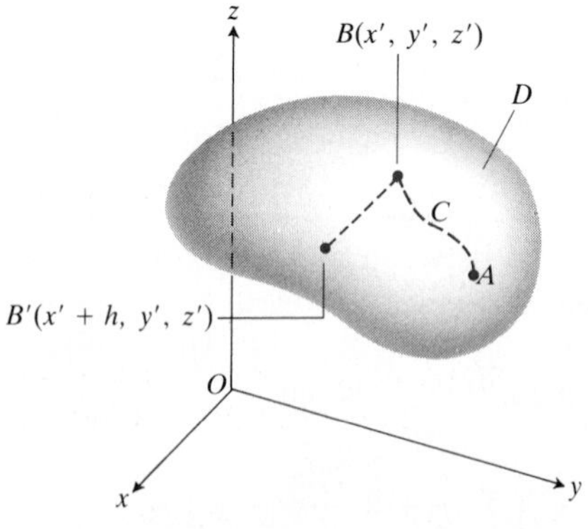

17.8 A piecewise smooth curve C joining points A and B in the region D of Theorem 1.

We now use this result to integrate $\mathbf{F} \cdot d\mathbf{R}$ along C from A to B:

$$\begin{aligned}\int_C \mathbf{F} \cdot d\mathbf{R} &= \int_{t_1}^{t_2} \frac{df}{dt}\,dt\\ &= \int_{t_1}^{t_2} \frac{d}{dt} f(x(t), y(t), z(t))\,dt\\ &= f(x(t), y(t), z(t))\Big]_{t_1}^{t_2}\\ &= f(x(t_2), y(t_2), z(t_2)) - f(x(t_1), y(t_1), z(t_1))\\ &= f(B) - f(A).\end{aligned}$$

Therefore, if $\mathbf{F} = \nabla f$ we have

$$\int_A^B \mathbf{F} \cdot d\mathbf{R} = \int_A^B \nabla f \cdot d\mathbf{R} = f(B) - f(A). \tag{10}$$

The value of the integral $f(B) - f(A)$ does not depend on the path C at all. Equation (10) is the space analog of the First Fundamental Theorem of Integral Calculus (see Article 4.7):

$$\int_a^b f'(x)\,dx = f(b) - f(a).$$

The only difference is that we have $\nabla f \cdot d\mathbf{R}$ in place of $f'(x)\,dx$. This analogy suggests that perhaps there is also a space analog of the fact that any continuous function of a single real variable is the derivative with respect to x of its integral from a to x (again, see Article 4.7). In other words, if we define a function f by the rule

$$f(x', y', z') = \int_A^{(x', y', z')} \mathbf{F} \cdot d\mathbf{R}, \tag{11}$$

perhaps it will be true that

$$\nabla f = \mathbf{F}. \tag{12}$$

Equation (12) is indeed true when the right-hand side of Eq. (11) is path-independent, and the proof of this fact will complete our theorem.

Proof of necessity. We now assume that the line integral in (11) is path-independent, and prove that $\mathbf{F} = \nabla f$ for the function f defined by Eq. (11). We first write $\mathbf{F}$ in terms of its **i**-, **j**-, and **k**-components:

$$\mathbf{F}(x, y, z) = \mathbf{i}M(x, y, z) + \mathbf{j}N(x, y, z) + \mathbf{k}P(x, y, z), \tag{13}$$

and fix the points A and $B(x', y', z')$ in D. To establish Eq. (12), we need to show that the equalities

$$\frac{\partial f}{\partial x} = M, \qquad \frac{\partial f}{\partial y} = N, \qquad \frac{\partial f}{\partial z} = P \tag{14}$$

hold at each point of D. In what follows, we either assume that D is an open set, so that all of its points are interior points, or we restrict our attention to interior points.

The point $B(x', y', z')$ is the center of some small sphere whose interior lies entirely inside D. We let $h \neq 0$ be small enough so that all points on the ray from B to $B'(x' + h, y', z')$ lie in D (Fig. 17.8), and consider the difference quotient

$$\frac{f(x' + h, y', z') - f(x', y', z')}{h} = \frac{1}{h}\int_B^{B'} \mathbf{F} \cdot d\mathbf{R}. \tag{15}$$

Since the integral does not depend on a particular path, we choose one convenient for our purpose:

$$x = x' + th, \qquad y = y', \qquad z = z', \qquad 0 \le t \le 1,$$

along which neither y nor z varies, and along which $d\mathbf{R} = \mathbf{i}\,dx = \mathbf{i}h\,dt$. When this is substituted into Eq. (15), along with $\mathbf{F}$ from Eq. (13), we get

$$\frac{f(x' + h, y', z') - f(x', y', z')}{h} = \frac{1}{h}\int_0^1 M(x' + ht, y', z')h\,dt$$

$$= \int_0^1 M(x' + ht, y', z')\,dt. \tag{16}$$

By hypothesis, $\mathbf{F}$ is continuous, so each component is a continuous function. Thus, given any $\varepsilon > 0$, there is a $\delta > 0$ such that

$$|M(x' + ht, y', z') - M(x', y', z')| < \varepsilon \quad \text{when} \quad |ht| < \delta.$$

This implies that when $|h| < \delta$, the integral in Eq. (16) also differs from

$$\int_0^1 M(x', y', z')\, dt = M(x', y', z') \tag{17}$$

by less than ε. The equality in (17) follows from the fact that the integrand is a constant, and

$$\int_0^1 dt = 1.$$

Therefore, as $h \to 0$ in Eq. (16), the right-hand side has as limit $M(x', y', z')$, so the left-hand side must have the same limit. That, however, is just the partial derivative of f with respect to x at $B(x', y', z')$. Therefore

$$\frac{\partial f}{\partial x} = M \tag{18}$$

holds at each interior point of D.

Equation (18) is the first of the three equalities, Eqs. (14), that are needed to establish Eq. (12). Proofs of the remaining two equalities in (14) are very similar, and you are asked to prove one of them in Problem 42. One sets up a difference quotient like Eq. (15), but takes

$$B' = (x', y' + h, z') \quad \text{or} \quad B' = (x', y', z' + h).$$

In other words, from B, one integrates along a path parallel to the y-axis or parallel to the z-axis. On the former path, $d\mathbf{R} = \mathbf{j}h\, dt$, and on the latter, $d\mathbf{R} = \mathbf{k}h\, dt$. This concludes the proof. ■

EXAMPLE 3 Find a function f such that if $\mathbf{F} = 2x\mathbf{i} + 2y\mathbf{j} + 2z\mathbf{k}$, then $\mathbf{F} = \nabla f$.

Solution We might be lucky and guess

$$f(x, y, z) = x^2 + y^2 + z^2, \tag{19}$$

because $2x$, $2y$, $2z$ are its partial derivatives with respect to x, y, and z. But if we weren't so inspired, then we might try something like Eq. (11). In the first place, the functions $2x$, $2y$, $2z$ are everywhere continuous, so the region D can be all of space. Of course we don't know, until *after* we find that $\mathbf{F}$ is a gradient, that the integral in (11) is path-independent, but we proceed on faith (or at least with hope). The choice of A is up to us, so we make life easy for ourselves by taking $A = (0, 0, 0)$. For the path of integration from A to $B(x', y', z')$, we take the line segment

$$x = x't, \quad y = y't, \quad z = z't, \quad 0 \le t \le 1,$$

along which

$$d\mathbf{R} = (x'\mathbf{i} + y'\mathbf{j} + z'\mathbf{k})\, dt$$

and

$$\begin{aligned} \mathbf{F} \cdot d\mathbf{R} &= (2xx' + 2yy' + 2zz')\, dt \\ &= (2x'^2 + 2y'^2 + 2z'^2)t\, dt. \end{aligned}$$

Therefore, when we substitute into Eq. (11), we get

$$\begin{aligned} f(x', y', z') &= \int_{(0,0,0)}^{(x',y',z')} \mathbf{F} \cdot d\mathbf{R} \\ &= [x'^2 + y'^2 + z'^2] \int_0^1 2t\,dt \\ &= x'^2 + y'^2 + z'^2. \end{aligned}$$

If we delete the primes, this equation is identical with Eq. (19). □

A comment on notation. The upper limit of integration in Eq. (11) is an arbitrary point in the domain of **F**, but we use (x', y', z') to designate it, rather than (x, y, z), because the latter is used for the running point that covers the arc C from A to B during the integration. After we have completed the computation of $f(x', y', z')$, we then delete the primes to express the result as $f(x, y, z)$. The analog in one dimension would be

$$\ln x' = \int_1^{x'} \frac{1}{x}\,dx.$$

We must be careful not to confuse the variable of integration x with the limit of integration x'. We distinguished between these two things in a slightly different manner in Article 6.4, Eq. (4), where we wrote

$$\ln x = \int_1^{x} \frac{1}{t}\,dt.$$

Our purpose there was the same as it is here, however: to maintain a notational difference between the variable *upper limit* and the *variable of integration*.

So far we have only the criterion

$$\mathbf{F} = \nabla f$$

for deciding whether

$$\int_A^B \mathbf{F} \cdot d\mathbf{R}$$

is path-independent. We shall discover another criterion in Eqs. (24) below. If we follow the method indicated by Eq. (11) and illustrated in Example 3 for a field **F** that is *not* path-independent, then we should discover, on trying to verify that $\mathbf{F} = \nabla f$, that it isn't so. The next example illustrates exactly this situation.

EXAMPLE 4 Show that there is no function f such that

$$\mathbf{F} = \nabla f \quad \text{if} \quad \mathbf{F} = y\mathbf{i} - x\mathbf{j}.$$

Solution Here's one way to show it: If there were such a function f, then

$$\frac{\partial f}{\partial x} = y \quad \text{and} \quad \frac{\partial f}{\partial y} = -x,$$

from which we would get

$$\frac{\partial^2 f}{\partial y\,\partial x} = \frac{\partial(y)}{\partial y} = 1 \neq \frac{\partial^2 f}{\partial x\,\partial y} = \frac{\partial(-x)}{\partial x} = -1.$$

But we should have $f_{xy} = f_{yx}$, because

$$f_x = y \qquad \text{and} \qquad f_y = -x$$

are everywhere continuously differentiable. This contradiction shows that no such f exists.

Another method would be to compute $\int \mathbf{F} \cdot d\mathbf{R}$ between two points, say $A(0, 0, 0)$ and $B(1, 1, 0)$, along two different paths. If the answers turn out to be the same, we haven't proved a thing. But if they turn out to be different, then we know that $\mathbf{F}$ is not a gradient field. Problem 5 asks you to do this for two specific paths.

A third method is to proceed blithely with Eq. (11), and get a function f that satisfies Eq. (11) for a particular path, but that fails to satisfy $\mathbf{F} = \nabla f$. Once again we would choose an origin $A = (0, 0, 0)$ and let $B = (x', y', z')$, and then integrate along the segment

$$x = x't, \qquad y = y't, \qquad z = z't, \qquad 0 \leq t \leq 1.$$

We would get

$$\begin{aligned} \mathbf{F} \cdot d\mathbf{R} &= (y\mathbf{i} - x\mathbf{j}) \cdot (\mathbf{i}\, dx + \mathbf{j}\, dy + \mathbf{k}\, dz) \\ &= y\, dx - x\, dy \\ &= (y't)(x'\, dt) - (x't)(y'\, dt) \\ &= t(x'y' - x'y')\, dt = 0\, dt. \end{aligned}$$

Therefore Eq. (11) produces

$$f(x', y', z') = 0 \qquad \text{for all } (x', y', z').$$

This constant function obviously won't have a gradient equal to $y\mathbf{i} - x\mathbf{j}$. In Problem 41 you are asked to explain why this also means that no other function exists whose gradient is the given $\mathbf{F}$. □

Conservative Fields

When $\mathbf{F}$ is a force field such that the work integral from A to B is the same for all paths joining them, the field is said to be *conservative*. Theorem 1 therefore shows that a force field is *conservative* if and only if it is a *gradient* field:

$$\mathbf{F} \text{ is conservative} \quad \Leftrightarrow \quad \mathbf{F} = \nabla f. \tag{20}$$

If the field $\mathbf{F}$ is conservative, the integrand in the work integral,

$$\mathbf{F} \cdot d\mathbf{R} = M\, dx + N\, dy + P\, dz, \tag{21}$$

is an *exact differential*. By this we mean that there is a function f whose differential is equal to the given integrand:

$$df = M\, dx + N\, dy + P\, dz, \tag{22}$$

which holds if and only if

$$M = \frac{\partial f}{\partial x}, \qquad N = \frac{\partial f}{\partial y}, \qquad P = \frac{\partial f}{\partial z}. \tag{23}$$

In Article 15.11, we discussed exact differentials of functions $f(x, y)$ of two variables. The criterion for an exact differential stated in the theorem of that article is easily extended to functions of three or more variables. For functions of three variables, it goes as follows.

THEOREM 2

Let $M(x, y, z)$, $N(x, y, z)$, and $P(x, y, z)$ be continuous, together with their first order partial derivatives. Then a necessary condition for the expression

$$M\,dx + N\,dy + P\,dz$$

to be an exact differential is that the following equations all be satisfied:

$$\frac{\partial M}{\partial y} = \frac{\partial N}{\partial x}, \qquad \frac{\partial M}{\partial z} = \frac{\partial P}{\partial x}, \qquad \frac{\partial N}{\partial z} = \frac{\partial P}{\partial y}. \tag{24}$$

This theorem is a straightforward extension of the theorem of Article 15.11 from the two-dimensional to the three-dimensional case. We shall omit the proof, since it is similar to the proof of the earlier theorem.

EXAMPLE 5 Suppose

$$\mathbf{F} = \mathbf{i}(e^x \cos y + yz) + \mathbf{j}(xz - e^x \sin y) + \mathbf{k}(xy + z).$$

Is $\mathbf{F}$ conservative? If so, find f such that $\mathbf{F} = \nabla f$.

Solution We apply the test of Eqs. (24) to the expression

$$\mathbf{F} \cdot d\mathbf{R} = (e^x \cos y + yz)\,dx + (xz - e^x \sin y)\,dy + (xy + z)\,dz.$$

We let

$$M = e^x \cos y + yz, \qquad N = xz - e^x \sin y, \qquad P = xy + z,$$

and calculate,

$$\frac{\partial M}{\partial z} = y = \frac{\partial P}{\partial x}, \qquad \frac{\partial N}{\partial z} = x = \frac{\partial P}{\partial y}, \qquad \frac{\partial M}{\partial y} = -e^x \sin y + z = \frac{\partial N}{\partial x}.$$

The theorem tells us that there may be a function $f(x, y, z)$ such that

$$\mathbf{F} \cdot d\mathbf{R} = df.$$

We would find f by integrating the system of equations

$$\frac{\partial f}{\partial x} = e^x \cos y + yz, \qquad \frac{\partial f}{\partial y} = xz - e^x \sin y, \qquad \frac{\partial f}{\partial z} = xy + z. \tag{25}$$

We integrate the first of these with respect to x, holding y and z constant, and add an arbitrary function $g(y, z)$ as the "constant of integration"; we thus obtain

$$f(x, y, z) = e^x \cos y + xyz + g(y, z). \tag{26}$$

Next we differentiate this with respect to y and set it equal to $\partial f/\partial y$ as given by the second of Eqs. (25):

$$xz - e^x \sin y = -e^x \sin y + xz + \frac{\partial g}{\partial y},$$

or

$$\frac{\partial g(y, z)}{\partial y} = 0. \tag{27}$$

Integrating Eq. (27) with respect to y, holding z constant, and adding an arbitrary function $h(z)$ as constant of integration, we obtain

$$g(y, z) = h(z). \tag{28}$$

We substitute this into Eq. (26) and then calculate $\partial f/\partial z$, which we compare with the third of Eqs. (25). We find that

$$xy + z = xy + \frac{dh(z)}{dz} \quad \text{or} \quad \frac{dh(z)}{dz} = z,$$

so that

$$h(z) = \frac{z^2}{2} + C.$$

Hence we may write Eq. (26) as

$$f(x, y, z) = e^x \cos y + xyz + (z^2/2) + C.$$

Then, for this function, it is easy to see that

$$\mathbf{F} = \nabla f. \ \square$$

A function $f(x, y, z)$ that has the property that its gradient gives the force vector **F** is called a "potential" function. (Sometimes a minus sign is introduced. For example, the electric intensity of a field is the negative of the potential gradient in the field.

PROBLEMS

1. In Example 1, let C be given by

$$x = t^2, \quad y = t^2, \quad 0 \le t \le 1,$$

and evaluate $\int_C w\, ds$ for $w = x + y^2$.

2. In Example 1, let C be given by $x = f(t) = y$, where $f(0) = 0$ and $f(1) = 1$. Show that if $f'(t)$ is continuous on [0, 1], then $\int_C w\, ds = (5\sqrt{2})/6$, no matter what the particular function f may be.

3. Evaluate $\int \mathbf{F} \cdot d\mathbf{R}$ around the circle

$$x = \cos t, \quad y = \sin t, \quad z = 0, \quad 0 \le t \le 2\pi$$

for the force given in Example 2.

4. In Example 3, evaluate $\int \mathbf{F} \cdot d\mathbf{R}$ along a curve C lying on the sphere $x^2 + y^2 + z^2 = a^2$. Do you need to know anything more about C? Why?

5. Assume $\mathbf{F} = y\mathbf{i} - x\mathbf{j}$, as in Example 4, and take $A = (0, 0, 0)$, $B = (1, 1, 0)$. Evaluate $\int \mathbf{F} \cdot d\mathbf{R}$ for:
a) $x = y = t$, $0 \le t \le 1$;
b) $x = t$, $y = t^2$, $0 \le t \le 1$.
Comment on the meaning of your answers.

6. Evaluate $\int_C \mathbf{F} \cdot d\mathbf{R}$, where the point of application of the force $\mathbf{F} = xy\mathbf{i} - x^2\mathbf{j}$ follows the path C in the xy-plane that consists of
a) the line segment on the x-axis from (1, 0) to (−1, 0),
b) the upper half of the circle $x^2 + y^2 = 1$ from (1, 0) to (−1, 0),
c) the line segment from (1, 0) to (0, 1) followed by the line segment from (0, 1) to (−1, 0).

In Problems 7–16, find the work done by the given force **F** as the point of application moves from (0, 0, 0) to (1, 1, 1)
a) along the straight line $x = y = z$,
b) along the curve $x = t$, $y = t^2$, $z = t^4$, and
c) along the x-axis to (1, 0, 0), then in a straight line to (1, 1, 0), and from there in a straight line to (1, 1, 1).

7. $\mathbf{F} = 2x\mathbf{i} + 3y\mathbf{j} + 4z\mathbf{k}$

8. $\mathbf{F} = 3y\mathbf{i} + 2x\mathbf{j} + 4z\mathbf{k}$

9. $\mathbf{F} = \dfrac{1}{x^2 + 1}\mathbf{j}$

10. $\mathbf{F} = \sqrt{z}\,\mathbf{i} - 2x\mathbf{j} + \sqrt{y}\,\mathbf{k}$

11. $\mathbf{F} = xy\mathbf{i} + yz\mathbf{j} + xz\mathbf{k}$

12. $\mathbf{F} = 3x(x-1)\mathbf{i} + 3z\mathbf{j} + \mathbf{k}$

13. $\mathbf{F} = \mathbf{i}x \sin y + \mathbf{j} \cos y + \mathbf{k}(x + y)$

14. $\mathbf{F} = \mathbf{i}(y + z) + \mathbf{j}(z + x) + \mathbf{k}(x + y)$

15. $\mathbf{F} = e^{y+2z}(\mathbf{i} + \mathbf{j}x + 2\mathbf{k}x)$

16. $\mathbf{F} = \mathbf{i}y \sin z + \mathbf{j}x \sin z + \mathbf{k}xy \cos z$

Evaluate $\int_C \mathbf{F} \cdot d\mathbf{R}$ for the fields and curves in Problems 17–20.

17. $\mathbf{F} = xy\mathbf{i} + y\mathbf{j} - yz\mathbf{k}$
$\mathbf{R} = t\mathbf{i} + t^2\mathbf{j} + t\mathbf{k}, \quad 0 \le t \le 1$

18. $\mathbf{F} = 2y\mathbf{i} + 3x\mathbf{j} + (x + y)\mathbf{k}$
$\mathbf{R} = \mathbf{i} \cos t + \mathbf{j} \sin t + \frac{t}{6}\mathbf{k}, \quad 0 \le t \le 2\pi$

19. $\mathbf{F} = z\mathbf{i} + x\mathbf{j} + y\mathbf{k}$
$\mathbf{R} = \mathbf{i} \sin t + \mathbf{j} \cos t + t\mathbf{k}, \quad 0 \le t \le 2\pi$

20. $\mathbf{F} = 6z\mathbf{i} + y^2\mathbf{j} + 12x\mathbf{k}$
$\mathbf{R} = \mathbf{i} \sin t + \mathbf{j} \cos t + \frac{t}{6}\mathbf{k}, \quad 0 \le t \le 2\pi$

21. Find the work done by the force $\mathbf{F} = -4xy\mathbf{i} + 8y\mathbf{j} + 2\mathbf{k}$ as the point of application moves along the parabola $y = x^2$, $z = 1$ from $A(0, 0, 1)$ to $B(2, 4, 1)$.

22. Evaluate $\int_C y\, ds$ along the curve $y = 2\sqrt{x}$ from (1, 2) to (4, 4).

In Problems 23–26, find a function $f(x, y, z)$ such that $\mathbf{F} = \nabla f$.

23. $\mathbf{F} = 2x\mathbf{i} + 3y\mathbf{j} + 4z\mathbf{k}$

24. $\mathbf{F} = \mathbf{i}(y + z) + \mathbf{j}(z + x) + \mathbf{k}(x + y)$

25. $\mathbf{F} = e^{y+2z}(\mathbf{i} + x\mathbf{j} + 2x\mathbf{k})$

26. $\mathbf{F} = \mathbf{i}y \sin z + \mathbf{j}x \sin z + \mathbf{k}xy \cos z$

27. If A and B are given, prove that the line integral

$$\int_A^B (z^2\, dx + 2y\, dy + 2xz\, dz)$$

is independent of the path of integration.

28. If $\mathbf{F} = y\mathbf{i} + x\mathbf{j}$, evaluate the line integral $\int_A^B \mathbf{F} \cdot d\mathbf{R}$ along the straight line from $A(1, 1, 1)$ to $B(3, 3, 3)$.

29. If $\mathbf{F} = \mathbf{i}x^2 + \mathbf{j}yz + \mathbf{k}y^2$, compute $\int_A^B \mathbf{F} \cdot d\mathbf{R}$, where $A = (0, 0, 0)$, $B = (0, 3, 4)$, along the straight line connecting these points.

30. Let C denote the plane curve whose vector equation is

$$\mathbf{R}(t) = (e^t \cos t)\mathbf{i} + (e^t \sin t)\mathbf{j}.$$

Evaluate the line integral

$$\int \frac{x\, dx + y\, dy}{(x^2 + y^2)^{3/2}}$$

along that arc of C from the point (1, 0) to the point $(e^{2\pi}, 0)$.

Evaluate the line integrals in Problems 31–38.

31. $\int_C 2x \cos y\, dx - x^2 \sin y\, dy$, where C is the path from (1, 0) to (0, 1) on the curve $\mathbf{R} = \mathbf{i} \cos^3 t + \mathbf{j} \sin^3 t$.

32. $\int_C 2x \sin y\, dx + x^2 \cos y\, dy$, where C is the path from (1, 0) to (0, 1) on the curve $\mathbf{R} = t\mathbf{i} + (t - 1)^2\mathbf{j}$.

33. $\int_C (x^2 + y)\, dx + (y^2 + x)\, dy$, where C is the line segment from (1, 1) to (2, 3).

34. $\int_C (y^2x + y)\, dx + (x^2y + x)\, dy$, where C is the line segment from (1, 1) to (2, 3).

35. $\int_C yz\, dx + xz\, dy + xy\, dz$, where C is the line segment from (1, 1, 2) to (3, 5, 0).

36. $$\int_{(0,0,0)}^{(1,2,3)} 2xy\, dx + (x^2 + z^2)\, dy + 2yz\, dz.$$

37. $$\int_{(0,1,1)}^{(2,2,1)} 3x^2\, dx + \frac{z^2}{y}\, dy + 2z \ln y\, dz.$$

38. $$\int_{(1,0,1)}^{(0,1,1)} \sin y \cos x\, dx + \cos y \sin x\, dy + dz$$

39. If the density $\rho(x, y, z)$ of a fluid is a function of the pressure $p(x, y, z)$, and

$$\phi(x, y, z) = \int_{p_0}^{p} (dp/\rho),$$

where p_0 is constant, show that $\nabla\phi = \nabla p/\rho$.

40. If $\mathbf{F} = y\mathbf{i}$, show that the line integral $\int_A^B \mathbf{F} \cdot d\mathbf{R}$ along an arc AB in the xy-plane is equal to an area bounded by the x-axis, the arc, and the ordinates at A and B.

REMARK Despite similarity of appearance and identity of value, the integral of this problem and the integral of earlier calculus are conceptually distinct. The latter is a line integral for which the path lies along the x-axis.

41. In Example 4, when we considered $\mathbf{F} = y\mathbf{i} - x\mathbf{j}$, we found a function $f(x', y', z') = 0$ which expresses the value of the integral

$$\int_{(0,0,0)}^{(x',y',z')} \mathbf{F} \cdot d\mathbf{R}$$

along the line segment from (0, 0, 0) to an arbitrary point (x', y', z'). Using this result, and the first half of the proof of Theorem 1, prove that if $\mathbf{F}$ were a gradient field, say $\mathbf{F} = \nabla g$, then $g - f$ = constant. From this, show that no such g exists for the given $\mathbf{F}$.

42. Using the notations of Eqs. (11) and (13), show that $\partial f/\partial y = N$ holds at each point of D if $\mathbf{F}$ is continuous and if the integral in (11) is path-independent in D.

43. Let $\rho = (x^2 + y^2 + z^2)^{1/2}$. Show that

$$\nabla(\rho^n) = n\rho^{n-2}\mathbf{R},$$

where $\mathbf{R} = \mathbf{i}x + \mathbf{j}y + \mathbf{k}z$. Is there a value of n for which $\mathbf{F} = \nabla(\rho^n)$ represents the "inverse-square law" field? If so, what is this value of n?

44. The "curl" of a vector field

$$\mathbf{F} = \mathbf{i}f(x, y, z) + \mathbf{j}g(x, y, z) + \mathbf{k}h(x, y, z)$$

is defined to be del cross $\mathbf{F}$; that is,

$$\text{curl } \mathbf{F} \equiv \nabla \times \mathbf{F} \equiv \begin{vmatrix} \mathbf{i} & \mathbf{j} & \mathbf{k} \\ \dfrac{\partial}{\partial x} & \dfrac{\partial}{\partial y} & \dfrac{\partial}{\partial z} \\ f & g & h \end{vmatrix},$$

or

$$\text{curl } \mathbf{F} \equiv \mathbf{i}\left(\frac{\partial h}{\partial y} - \frac{\partial g}{\partial z}\right) + \mathbf{j}\left(\frac{\partial f}{\partial z} - \frac{\partial h}{\partial x}\right) + \mathbf{k}\left(\frac{\partial g}{\partial x} - \frac{\partial f}{\partial y}\right),$$

and the "divergence" of a vector field

$$\mathbf{V} = \mathbf{i}u(x, y, z) + \mathbf{j}v(x, y, z) + \mathbf{k}w(x, y, z)$$

is defined to be del dot $\mathbf{V}$; that is,

$$\text{div } \mathbf{V} \equiv \nabla \cdot \mathbf{V} \equiv \frac{\partial u}{\partial x} + \frac{\partial v}{\partial y} + \frac{\partial w}{\partial z}.$$

If the components f, g, h of $\mathbf{F}$ are functions that possess continuous mixed partial derivatives

$$\frac{\partial^2 h}{\partial x\, \partial y}, \quad \ldots,$$

show that

$$\text{div } (\text{curl } \mathbf{F}) = 0.$$

***Toolkit* programs**	
Scalar Fields	Vector Fields

17.4 Two-dimensional Fields. Flux Across a Plane Curve

In this article, we turn our attention to two-dimensional vector fields of the form

$$\mathbf{F} = \mathbf{i}M(x, y) + \mathbf{j}N(x, y). \tag{1}$$

Figure 17.9 shows how such a two-dimensional vector field might look in space. In the figure, for example, $\mathbf{F}$ might represent a fluid flow in which each particle travels in a circle parallel to the xy-plane in such a way that all particles on a given line perpendicular to the xy-plane travel with the same velocity. Example 1 provides another instance, that of an electric field with field strength

$$\mathbf{E} = \frac{\mathbf{i}x + \mathbf{j}y}{x^2 + y^2}. \tag{2}$$

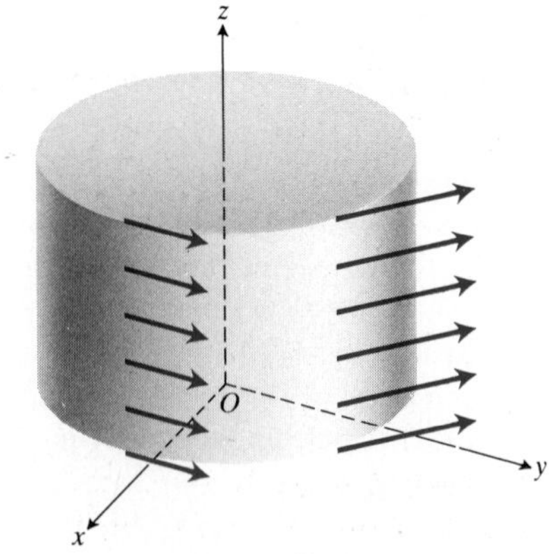

17.9 A two-dimensional vector field in space.

Note that this formulation is like the right-hand side of Eq. (1), with

$$M(x, y) = \frac{x}{x^2 + y^2}, \qquad N(x, y) = \frac{y}{x^2 + y^2}.$$

The essential features of a two-dimensional field are (1) the vectors in $\mathbf{F}$ are all parallel to one plane, which we have taken to be the xy-plane, and (2) in every plane parallel to the xy-plane, the field is the same as it is in that plane. In Eq. (1), the field has a zero $\mathbf{k}$-component everywhere, which makes the vectors parallel to the xy-plane. The $\mathbf{i}$- and $\mathbf{j}$-components do not depend on z, so they are the same in all planes parallel to the xy-plane.

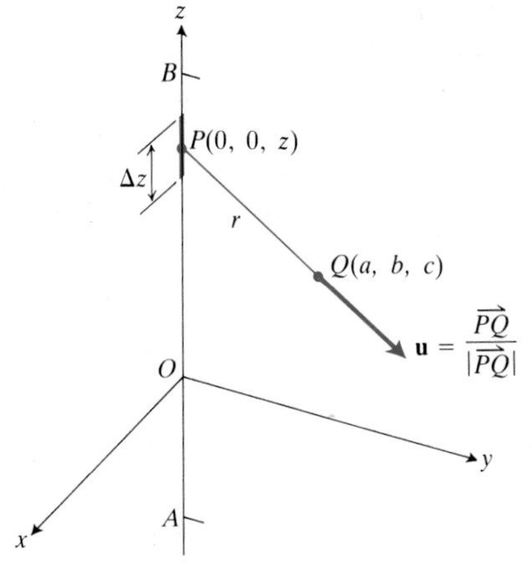

17.10 According to Coulomb's law, a charge $\delta_0\,\Delta z$ at $P(0, 0, z)$ produces a field $\Delta\mathbf{F}$ on a test charge q_0 at $Q(a, b, c)$.

EXAMPLE 1 An infinitely long, thin, straight wire has a uniform electric charge density δ_0. Using Coulomb's law, find the electric field intensity around the wire due to this charge.

Solution From a physics textbook, we find that *Coulomb's law* is an inverse-square law. It says that the force acting on a positive test charge q_0 placed at a distance r from a positive *point charge* q is directed away from q and has magnitude $(4\pi\varepsilon_0)^{-1}(qq_0)/r^2$, where ε_0 is a constant called the *permittivity*. For the charged wire, we have a distributed charge instead of a point charge, but we handle it in the familiar way: We replace this distributed charge by a large number of tiny elements, add, and take limits.

To be specific, suppose that the wire runs along the z-axis from $-\infty$ to $+\infty$. Take a long but finite piece of the wire and divide it into a lot of small segments. One of these is indicated in Fig. 17.10, with its center at P(0, 0, z), and its length equal to Δz. We assume that Δz is so small that we can treat the charge $\delta_0\,\Delta z$ on this segment of the wire as a point charge at P. Now let Q(a, b, c) be any point not on the z-axis, and let Δz denote the force on the test charge at Q due to the point charge at P:

$$\Delta\mathbf{F} = \frac{\delta_0 q_0}{4\pi\varepsilon_0}\frac{\Delta z\mathbf{u}}{a^2 + b^2 + (c - z)^2}, \tag{3a}$$

where $\mathbf{u}$ is a unit vector in the direction of $\overrightarrow{PQ}$:

$$\mathbf{u} = \frac{\mathbf{i}a + \mathbf{j}b + \mathbf{k}(c - z)}{[a^2 + b^2 + (c - z)^2]^{1/2}}. \tag{3b}$$

When we add the vector forces $\Delta\mathbf{F}$ for pieces of wire between $z = A$ and $z = B$ and take the limit as $\Delta z \to 0$, we get an integral of the form

$$\frac{\delta_0 q_0}{4\pi\varepsilon_0}\int_A^B \frac{\mathbf{i}a + \mathbf{j}b + \mathbf{k}(c - z)}{[a^2 + b^2 + (c - z)^2]^{3/2}}\,dz. \tag{3c}$$

The denominator of the integrand behaves about like $|z|^3$ as $|z| \to \infty$, so the three component integrals in (3c) converge as $A \to -\infty$ and $B \to \infty$. As our final integral representation of the field, therefore, we get

$$\mathbf{F} = \frac{\delta_0 q_0}{4\pi\varepsilon_0}\int_{-\infty}^{+\infty} \frac{\mathbf{i}a + \mathbf{j}b + \mathbf{k}(c - z)}{[a^2 + b^2 + (c - z)^2]^{3/2}}\,dz. \tag{4}$$

There are three separate integrals, but two are essentially the same except for the constant coefficients a and b, and the third is zero. It is easy to do the integration by making the substitutions

$$\sqrt{a^2 + b^2} = m,$$

$$z - c = m\tan\theta, \qquad \theta = \tan^{-1}\left(\frac{z - c}{m}\right),$$

$$dz = m\sec^2\theta\,d\theta,$$

and observing that the limits for θ are from $-\pi/2$ to $\pi/2$. In Problem 1 you are asked to finish these calculations and to show that the result is

$$\mathbf{F} = \frac{\delta_0 q_0}{2\pi\varepsilon_0}\left(\frac{\mathbf{i}a + \mathbf{j}b}{a^2 + b^2}\right). \tag{5}$$

Note that $\mathbf{F}$ is a two-dimensional field that does not depend on $\mathbf{k}$ or on c. If we write the resultant field strength $\mathbf{E} = \mathbf{F}/q_0$ as a function of position (x, y, z) instead of (a, b, c), we have

$$\mathbf{E} = k_0\left(\frac{\mathbf{i}x + \mathbf{j}y}{x^2 + y^2}\right), \qquad \text{where } k_0 = \frac{\delta_0}{2\pi\varepsilon_0}. \ \square \tag{6}$$

Mass Transport

There is also a fluid-flow interpretation for the field in Eq. (6) (or Eq. 2). To arrive at that interpretation, imagine a long, thin pipe running along the z-axis, and perforated with a very large number of little holes through which it supplies fluid at a constant rate. (It helps to be a bit vague about the actual physics of this: Don't try to be too literal-minded.) In other words, the z-axis is to be thought of as a *source* from which water flows radially outward to produce a velocity field

$$\mathbf{v} = f(r)\mathbf{u}_r, \tag{7a}$$

where $\mathbf{u}_r$ is the usual unit vector associated with the cylindrical coordinates r, θ, z, and $f(r)$ is a function of r alone. Thus the velocity is all perpendicular to the z-axis, and it is independent of both θ and z. (These are our present interpretations of the phrases "at a constant rate" and "radially outward.")

Now consider the amount of fluid that flows out through the cylinder $r = a$ between the planes $z = 0$ and $z = 1$ in a short interval of time, from t to $t + \Delta t$. According to the law expressed by Eq. (7a), every particle of water that is on the surface $r = a$ moves radially outward a distance Δr, which is approximately $f(a)\,\Delta t$. Thus the volume of fluid that crosses the boundary $r = a$ between $z = 0$ and $z = 1$ in this time interval is approximately the volume between the cylinders $r = a$ and $r = a + f(a)\,\Delta t$, or $2\pi a f(a)\,\Delta t$. If we multiply this by the density δ, we get the *mass* transported through the wall of a unit length of the cylinder $r = a$ in the interval from t to $t + \Delta t$:

$$\Delta m \approx \delta 2\pi a f(a)\,\Delta t. \tag{7b}$$

If we divide both sides of Eq. (7b) by Δt and take the limit as $\Delta t \to 0$, we get the *rate* at which fluid is flowing across the unit length of the cylinder $r = a$:

$$\frac{dm}{dt} = \delta 2\pi a f(a). \tag{7c}$$

For an incompressible fluid such as water, all fluid that flows across the cylinder $r = a$ flows across the cylinder $r = b$ as well (there are no other sources or sinks between the two cylinders). Therefore, for the model under discussion, the rate of mass transport given by Eq. (7c) is independent of a, and its value for any radius $r = a$ is the same as for $r = 1$:

$$\delta 2\pi f(1) = \delta 2\pi a f(a). \tag{7d}$$

From Eq. (7d), we get

$$f(a) = f(1)/a \qquad \text{for any } a > 0.$$

Writing r in place of a and substituting C for f(1), we can rewrite the velocity field (7a) in the form

$$\mathbf{v} = \frac{C}{r}\mathbf{u}_r. \tag{8}$$

If we recall that the position vector in cylindrical coordinates is

$$\mathbf{R} = \overrightarrow{OP} = r\mathbf{u}_r + \mathbf{k}z$$

and that this must also be equal to $\mathbf{R} = \mathbf{i}x + \mathbf{j}y + \mathbf{k}z$, then we conclude that

$$r\mathbf{u}_r = \mathbf{i}x + \mathbf{j}y, \qquad \text{or} \qquad \mathbf{u}_r = \frac{\mathbf{i}x + \mathbf{j}y}{r},$$

Thus,

$$\mathbf{v} = \frac{C}{r}\mathbf{u}_r = \frac{C}{r}\frac{\mathbf{i}x + \mathbf{j}y}{r} = C\frac{\mathbf{i}x + \mathbf{j}y}{x^2 + y^2}, \tag{9}$$

and Eqs. (6) and (8) describe the same field if $C = k_0$.

Plane versus Space

Instead of interpreting the two-dimensional vector fields as we have done in 3-space, we can interpret them simply as fields in the xy-plane itself. Then r in Eqs. (8) and (9) is just the distance from the origin to the point P(x, y) in the plane, and the unit vector is

$$\mathbf{u}_r = \frac{(\mathbf{i}x + \mathbf{j}y)}{r} = \mathbf{i}\cos\theta + \mathbf{j}\sin\theta,$$

where $r > 0$ and θ measures the angle from the positive x-axis to the position vector $\overrightarrow{OP}$. Equation (8) then describes a vector field in the plane that is directed radially outward and whose strength decreases like $1/r$ as r increases. We still use the language of flow across boundaries, but in this interpretation the boundary would be a *curve* in the plane, rather than a unit length of a cylinder. Equation (7d) would be interpreted as saying that the amount of fluid flowing across the unit circle $r = 1$ per unit time is equal to the amount of fluid flowing across the circle $r = a$ in the same time. (This describes conditions after the flow has reached steady state, not during the transient phase.)

We can easily go back and forth between the two interpretations of two-dimensional fields, but henceforth we shall usually treat them as existing just in the xy-plane and ignore the fact that we can project the field onto any plane parallel to the xy-plane and thereby go to the 3-space view.

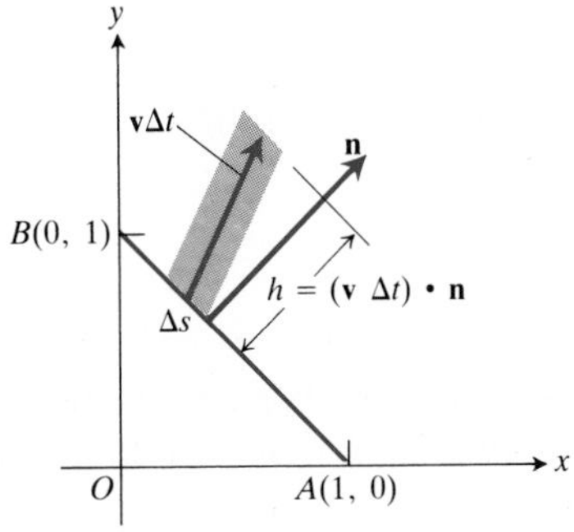

17.11 The fluid that flows across Δs in short time Δt fills a "parallelogram" whose altitude is $h = (\mathbf{v}\,\Delta t)\cdot\mathbf{n}$. The area of the parallelogram is therefore $\mathbf{v}\cdot\mathbf{n}\,\Delta t\,\Delta s$ and the mass of the fluid covering it is $\delta\mathbf{v}\cdot\mathbf{n}\,\Delta t\,\Delta s$.

EXAMPLE 2 Given the velocity field $\mathbf{v} = (\mathbf{i}x + \mathbf{j}y)/r^2$, calculate the mass-transport rate across the line segment AB joining the points A(1, 0) and B(0, 1).

Solution Let δ denote mass per unit area, the density factor by which we multiply the area of a region to get the mass of fluid in that region. (We assume the density to be constant.) Consider a segment of AB having length Δs, with its center at P(x, y) on AB. From Fig. 17.11 we see that the mass of fluid Δm that flows across the segment in time Δt is given, approximately, by

$$\Delta m \approx \delta(\mathbf{v} \cdot \mathbf{n})(\Delta t\, \Delta s), \tag{10}$$

where $\mathbf{n}$ is a unit vector normal to the line AB at P, and pointing away from the origin:

$$\mathbf{n} = \frac{\mathbf{i} + \mathbf{j}}{\sqrt{2}}.$$

If we divide both sides of Eq. (10) by Δt, then sum for all the pieces Δs of the segment AB, and take the limit as $\Delta s \to 0$, we get the *average rate* of mass transport across AB:

$$\frac{\Delta M}{\Delta t} \approx \int_{AB} \delta(\mathbf{v} \cdot \mathbf{n})\, ds.$$

Finally, letting $\Delta t \to 0$ and substituting for $\mathbf{v}$, $\mathbf{n}$, and ds, with

$$x = t, \qquad y = 1 - t, \qquad 0 \le t \le 1$$

as parameterization of the segment AB, we get as the *instantaneous* mass-transport rate

$$\frac{dM}{dt} = \delta \int_0^1 \frac{x + y}{r^2\sqrt{2}} (\sqrt{2}\, dt) = \delta \int_0^1 \frac{dt}{t^2 + (1 - t)^2} = \delta(\pi/2). \tag{11}$$

(In Problem 2 you are asked to verify this integration.) □

Flux

As Fig. 17.12 shows, the right-hand side of Eq. (10) also describes a flow of mass due to a velocity field $\mathbf{v}$ across any smooth curve in the plane in a unit of time Δt. The quantity $\mathbf{v}\, \Delta t$ is, very nearly, the vector displacement of all particles of fluid that were on the segment Δs at time t; hence those particles have swept over a "parallelogram" with dimensions Δs and $|\mathbf{v}\, \Delta t|$. The $\mathbf{n}$-component of $\mathbf{v}\, \Delta t$ is the altitude h of this parallelogram. Its area is therefore approximately

$$(\mathbf{v}\, \Delta t) \cdot \mathbf{n} \text{ times } \Delta s,$$

and the mass of fluid that fills this parallelogram is what flows across the tiny segment Δs between t and $t + \Delta t$.

This same line of reasoning would apply to flows in the xy-plane in general. It leads to the result

$$\frac{dM}{dt} = \int_C \delta(\mathbf{v} \cdot \mathbf{n})\, ds, \tag{12}$$

where dM/dt is the *rate* at which mass is being transported across the curve C, in the direction of the unit normal vector $\mathbf{n}$. One can interpret M as the amount of mass that has crossed C up to time t.

If the oppositely directed normal $\mathbf{n}' = -\mathbf{n}$ is substituted in place of $\mathbf{n}$ in the integral in Eq. (12), the sign of the answer changes. This just means that if flow in one direction across C is considered to be in the positive sense, then flow in the opposite direction is then considered to be negative. If C is a simple closed curve, we usually choose $\mathbf{n}$ to point outward. In Eq. (11), we chose the normal to point away from the origin. We got

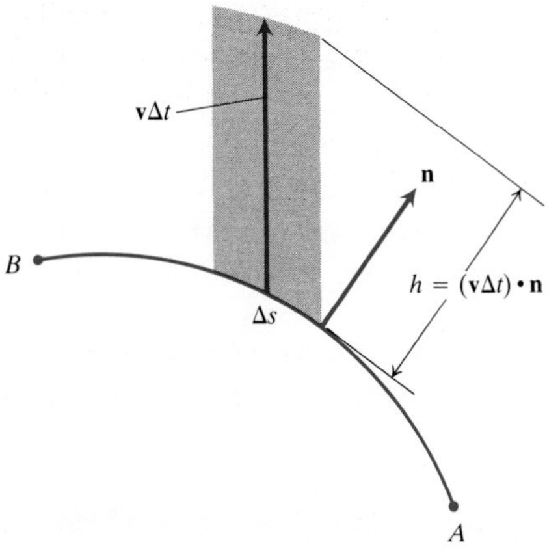

17.12 The fluid that flows across Δs in a short time Δt fills a "parallelogram" whose area is approximately base × height = $\mathbf{v} \cdot \mathbf{n}\, \Delta t\, \Delta s$.

a positive answer because the flow in the first quadrant is generally upward and to the right for the given **v**.

To simplify further discussion, we take

$$\delta \mathbf{v} = \mathbf{F}(x, y)$$

in Eq. (12), and call the resulting integral the *flux* of **F** across *C*:

$$\text{Flux of } \mathbf{F} \text{ across } C = \int_C \mathbf{F} \cdot \mathbf{n}\, ds. \tag{13}$$

We shall use this terminology even when the field **F** has nothing to do with a fluid flow, but you may wish to keep the fluid-flow interpretation in mind too.

The curve *C* in Eq. (13) is to have a direction along it called the *positive* direction, and arc length *s*, measured from some arbitrary starting point, is to increase in this direction. We also assume that *C* is smooth enough to have a tangent vector except at a finite number of points where there may be corners or cusps. Thus, $d\mathbf{R}/ds = \mathbf{T}$ exists almost everywhere and points in the positive direction along *C*.

Because we often want the flux integral to represent flow outward from a region *R* bounded by a simple closed curve *C*, we choose the counterclockwise direction on *C* as positive, and choose the *outward-pointing* unit normal vector as **n.** From Fig. 17.13, we can see that, because

$$\mathbf{T} = \frac{dx}{ds}\mathbf{i} + \frac{dy}{ds}\mathbf{j}, \tag{14a}$$

for the indicated choice of **n** we should choose $\mathbf{n} = \mathbf{T} \times \mathbf{k}$, so that

$$\mathbf{n} = \frac{dy}{ds}\mathbf{i} - \frac{dx}{ds}\mathbf{j}. \tag{14b}$$

As a check, it is easy to see that the dot product of the vectors in Eqs. (14a) and (14b) is zero, that both have unit length, and that when **T** points upward and to the right, **n** has a positive **i**-component and a negative **j**-component. If we proceed around the curve *C* in the direction in which **T** points, with the interior toward our left, then **n**, as given by Eq. (14b), points to our right, as it should.

We now use Eq. (14b) to write the flux of

$$\mathbf{F}(x, y) = \mathbf{i}M(x, y) + \mathbf{j}N(x, y)$$

across *C*:

$$\begin{aligned} \text{Flux across } C &= \int_C \mathbf{F} \cdot \mathbf{n}\, ds \\ &= \int_C \left(M\frac{dy}{ds} - N\frac{dx}{ds}\right) ds \\ &= \int_C (M\, dy - N\, dx). \end{aligned} \tag{15}$$

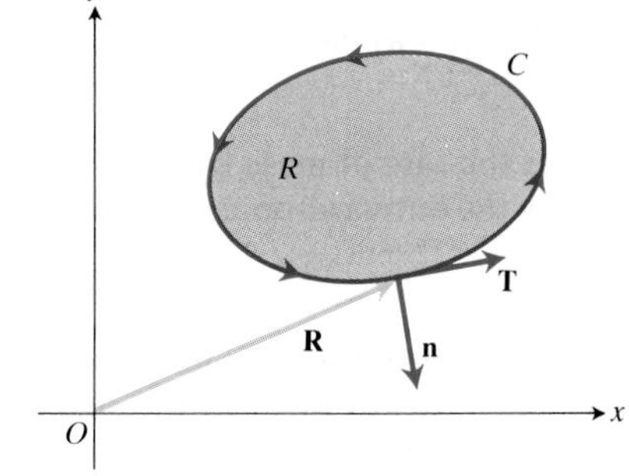

17.13 The position vector $\mathbf{R} = \mathbf{i}x + \mathbf{j}y$, the unit tangent vector $\mathbf{T} = d\mathbf{R}/ds$, and the unit normal vector $\mathbf{n} = \mathbf{T} \times \mathbf{k}$.

In Example 3, the functions M and N are discontinuous at (0, 0), so we cannot immediately apply Green's theorem to C_1: $x^2 + y^2 = 1$, and all of the region inside it. We must delete the origin, which we did by excluding points inside C_h.

In Example 3, we could replace the outer circle C_1 by an ellipse or any other simple closed curve Γ that lies outside C_h (for some positive h). The result would be

$$\oint_\Gamma (M\,dx + N\,dy) + \oint_{C_h} (M\,dx + N\,dy) = 0,$$

which leads to the conclusion

$$\oint_\Gamma \frac{x\,dy - y\,dx}{x^2 + y^2} = 2\pi.$$

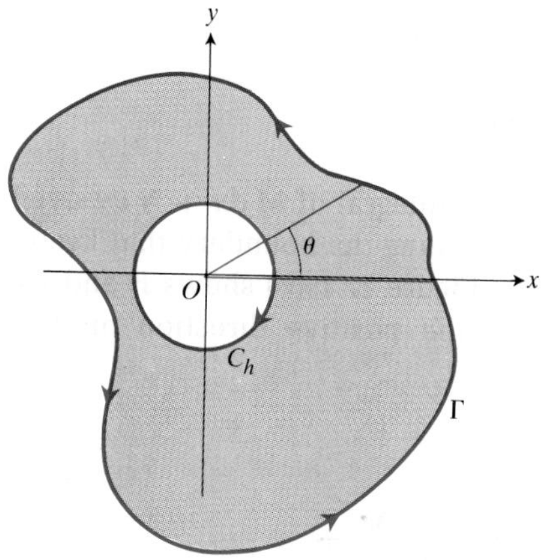

17.21 The region bounded by the circle C_h and the curve Γ.

This result is easily accounted for if we change to polar coordinates for Γ:

$$x = r\cos\theta, \qquad y = r\sin\theta,$$
$$dx = -r\sin\theta\,d\theta + \cos\theta\,dr,$$
$$dy = r\cos\theta\,d\theta + \sin\theta\,dr.$$

For then

$$\frac{x\,dy - y\,dx}{x^2 + y^2} = \frac{r^2(\cos^2\theta + \sin^2\theta)\,d\theta}{r^2} = d\theta;$$

and θ increases by 2π as we progress once around Γ counterclockwise (see Fig. 17.21).

Green's Theorem in Vector Form

Let

$$\mathbf{F} = M\mathbf{i} + N\mathbf{j} + P\mathbf{k} \qquad \text{and} \qquad \mathbf{R} = x\mathbf{i} + y\mathbf{j}.$$

Then the left-hand side of Eq. (3) is given by

$$\oint_C \mathbf{F}\cdot d\mathbf{R} = \oint_C (M\,dx + N\,dy).$$

To express the right-hand side of Eq. (3) in vector form, we use the symbolic vector operator

$$\nabla = \mathbf{i}\frac{\partial}{\partial x} + \mathbf{j}\frac{\partial}{\partial y} + \mathbf{k}\frac{\partial}{\partial z}.$$

We met the del operator in Article 15.6, where we saw that if

$$w = f(x, y, z)$$

is a differentiable scalar function, then ∇w is the gradient of w:

$$\text{grad}\, w = \nabla w = \mathbf{i}\frac{\partial w}{\partial x} + \mathbf{j}\frac{\partial w}{\partial y} + \mathbf{k}\frac{\partial w}{\partial z}.$$

Other uses of the del operator are given in Problem 44, Article 17.3, where the curl of a vector $\mathbf{F}$ is defined as del cross $\mathbf{F}$. Thus, if $\mathbf{F} = M\mathbf{i} + N\mathbf{j} + P\mathbf{k}$, then

$$\operatorname{curl} \mathbf{F} = \nabla \times \mathbf{F} = \begin{vmatrix} \mathbf{i} & \mathbf{j} & \mathbf{k} \\ \dfrac{\partial}{\partial x} & \dfrac{\partial}{\partial y} & \dfrac{\partial}{\partial z} \\ M & N & P \end{vmatrix}$$

$$= \mathbf{i}\left(\frac{\partial P}{\partial y} - \frac{\partial N}{\partial z}\right) + \mathbf{j}\left(\frac{\partial M}{\partial z} - \frac{\partial P}{\partial x}\right) + \mathbf{k}\left(\frac{\partial N}{\partial x} - \frac{\partial M}{\partial y}\right).$$

The component of curl **F** that is normal to a region R in the xy-plane is

$$(\nabla \times \mathbf{F}) \cdot \mathbf{k} = \frac{\partial N}{\partial x} - \frac{\partial M}{\partial y}.$$

Hence Green's theorem can be written in vector form as

$$\oint_C \mathbf{F} \cdot d\mathbf{R} = \iint_R \operatorname{curl} \mathbf{F} \cdot \mathbf{k}\, dx\, dy = \iint_R (\nabla \times \mathbf{F}) \cdot d\mathbf{A}, \qquad (13)$$

where $d\mathbf{A} = \mathbf{k}\, dx\, dy$ is a vector normal to the region R and of magnitude $|d\mathbf{A}| = dx\, dy$. In words, Green's theorem states that the integral around C of the tangential component of **F** is equal to the integral, over the region R bounded by C, of the component of curl **F** that is normal to R. This integral is the flux of curl **F** through R. We shall later extend this result to more general curves and surfaces in a formulation known as Stokes's theorem.

There is a second, *normal,* vector form for Green's theorem. It involves the gradient operator ∇ in another form, one that produces the *divergence.* Now the integrand of the line integral of Eq. (13) is the *tangential* component of the field **F** because

$$\mathbf{F} \cdot d\mathbf{R} = \left(\mathbf{F} \cdot \frac{d\mathbf{R}}{ds}\right) ds = (\mathbf{F} \cdot \mathbf{T})\, ds.$$

As in Article 17.4, if we let

$$\mathbf{F} = \mathbf{i}M(x, y) + \mathbf{j}N(x, y),$$

and let **G** be the orthogonal field given by

$$\mathbf{G} = \mathbf{i}N(x, y) - \mathbf{j}M(x, y),$$

then it follows that

$$\mathbf{F} \cdot \mathbf{T} = \mathbf{G} \cdot \mathbf{n} = M\frac{dx}{ds} + N\frac{dy}{ds}$$

because

$$\mathbf{T} = \mathbf{i}\frac{dx}{ds} + \mathbf{j}\frac{dy}{ds}, \qquad \mathbf{n} = \mathbf{i}\frac{dy}{ds} - \mathbf{j}\frac{dx}{ds}.$$

Therefore Green's theorem, which says that

$$\oint_C (M\, dx + N\, dy) = \iint_R \left[\frac{\partial N}{\partial x} - \frac{\partial M}{\partial y}\right] dx\, dy,$$

also says that

$$\oint_C \mathbf{G}\cdot\mathbf{n}\,ds = \iint_R \nabla\cdot\mathbf{G}\,dx\,dy, \tag{14}$$

where

$$\nabla\cdot\mathbf{G} = \operatorname{div}\mathbf{G} = \frac{\partial N}{\partial x} + \frac{\partial(-M)}{\partial y}.$$

In words, Eq. (14) says that the line integral of the *normal* component of any vector field **G** around the boundary of a region R in which **G** is continuous and has continuous partial derivatives is equal to the double integral of the *divergence* of **G** over R. In the next article, we shall extend this result to three-dimensional vector fields and discuss the physical interpretation of the divergence. For a three-dimensional vector field

$$\mathbf{F}(x, y, z) = \mathbf{i}M(x, y, z) + \mathbf{j}N(x, y, z) + \mathbf{k}P(x, y, z),$$

the *divergence* is defined to be

$$\operatorname{div}\mathbf{F} = \nabla\cdot\mathbf{F} = \frac{\partial M}{\partial x} + \frac{\partial N}{\partial y} + \frac{\partial P}{\partial z}. \tag{15}$$

PROBLEMS

Verify Green's formula, Eq. (3), with C the circle $x^2 + y^2 = a^2$ and R the disc $x^2 + y^2 \le a^2$, for the functions M and N given in Problems 1–6.

1. $M = x, \quad N = y$

2. $M = N = xy$

3. $M = -x^2y, \quad N = xy^2$

4. $M = N = e^{x+y}$

5. $M = y, \quad N = 0$

6. $M = -y, \quad N = x$

7. Verify Green's theorem for the vector field $\mathbf{F} = xy\mathbf{i} + y^2\mathbf{j}$ and the region **R** enclosed between the parabola $y = x^2$ and the line $y = x$ in the first quadrant.

8. Use Green's theorem to find the area between the ellipse $x = 3\cos t$, $y = 2\sin t$ and the circle $x = \cos t$, $y = \sin t$.

9. Find the area enclosed by the hypocycloid $x = a\cos^3 t$, $y = a\sin^3 t$. (Figure 13.21 shows a portion of this curve.)

10. Let C be the boundary of a region on which Green's theorem holds. Use Green's theorem to calculate

a) $\oint_C f(x)\,dx + g(y)\,dy$,

b) $\oint_C k\,y\,dx + h\,x\,dy$ (k and h constants).

Apply Green's theorem to evaluate the line integrals in Problems 11–16.

11. C is the triangle bounded by $x = 0$, $x + y = 1$, $y = 0$:

$$\oint_C (y^2\,dx + x^2\,dy).$$

12. C is the boundary of $0 \le x \le \pi$, $0 \le y \le \sin x$:

$$\oint_C (3y\,dx + 2x\,dy).$$

13. C is the circle $(x - 2)^2 + (y - 3)^2 = 4$:

$$\oint_C (6y + x)\,dx + (y + 2x)\,dy.$$

14. C is any simple closed curve in the plane for which Green's theorem holds:

$$\oint_C (2x + y^2)\,dx + (2xy + 3y)\,dy.$$

15. C is the boundary of the "triangular" region in the first quadrant enclosed by the x-axis, the line $x = 1$, and the curve $y = x^3$:

$$\oint_C 2xy^3\,dx + 4x^2y^2\,dy.$$

16. C is the circle $(x - 2)^2 + (y - 2)^2 = 4$:

$$\oint_C (4x - 2y)\,dx + (2x - 4y)\,dy.$$

17. Show that $\oint_C 4x^3y\,dx + x^4\,dy = 0$ for any closed curve C to which Green's theorem applies.

18. Show that $\oint_C -y^3\,dx + x^3\,dy$ is positive for any closed curve C to which Green's theorem applies.

19. Show that the value of $\oint_C xy^2\,dx + (x^2y + 2x)\,dy$ around any square depends only on the size of the square and not on its location in the plane.

20. Assuming that all the necessary derivatives exist and are continuous, show that if $f(x, y)$ satisfies the Laplace equation

$$\frac{\partial^2 f}{\partial x^2} + \frac{\partial^2 f}{\partial y^2} = 0,$$

then

$$\oint_C \frac{\partial f}{\partial y}\,dx - \frac{\partial f}{dx}\,dy = 0$$

for all closed curves C to which Green's theorem applies. (The converse is also true: if the line integral is always zero, then f satisfies the Laplace equation.)

21. Let

$$\mathbf{F} = \left(\frac{1}{4}x^2y + \frac{1}{3}y^3\right)\mathbf{i} + x\mathbf{j}.$$

Among all smooth simple closed curves in the plane, oriented counterclockwise, find the curve around which the work done by $\mathbf{F}$ is the greatest. (*Hint:* Where is $(\text{curl } \mathbf{F}) \cdot \mathbf{k}$ positive?)

22. Supply the details necessary to establish Eq. (6).

23. Supply the steps necessary to establish Eq. (10).

24. Suppose that

$$R = \{(x, y): 0 \le y \le \sqrt{a^2 - x^2}, \quad -a \le x \le a\},$$

and that C is the boundary of R.

a) Sketch R and C.

b) Write out the proof of Green's theorem for this region.

Definition. A region R is said to be simply connected if every simple closed curve lying in R can be continuously contracted to a point without its touching any part of the boundary of R. Examples are the interiors of circles, ellipses, cardioids, and rectangles; and, in three dimensions, the region between two concentric spheres. (The annular ring in Fig. 17.20 is not simply connected. Also, see Fig. 15.47.)

25. Show, by a geometric argument, that Green's formula, Eq. (3), holds for any simply connected region R whose boundary is a simple closed curve C, provided R can be decomposed into a finite number of nonoverlapping regions $R_1, R_2, \ldots, R_n$ with boundaries $C_1, C_2, \ldots, C_n$ of a type for which the formula (3) is true for each R_i and C_i, $i = 1, \ldots, n$.

26. Suppose R is a region in the xy-plane, C is its boundary, and the area of R is given by

$$A(R) = \oint_C \frac{1}{2}(x\,dy - y\,dx).$$

Suppose the equations $x = f(u, v)$, $y = g(u, v)$ map R and C in a continuous and one-to-one manner onto a region R' and curve C', respectively, in the uv-plane. Use Green's formula to show that

$$\iint_R dx\,dy = \iint_{R'} \begin{vmatrix} f_u & f_v \\ g_u & g_v \end{vmatrix} du\,dv$$

$$= \iint_{R'} \left(\frac{\partial f}{\partial u}\frac{\partial g}{\partial v} - \frac{\partial f}{\partial v}\frac{\partial g}{\partial u}\right) du\,dv.$$

(*Hint:* Note that

$$\iint_R dx\,dy = \frac{1}{2}\int_C (x\,dy - y\,dx)$$

$$= \frac{1}{2}\int_{C'} \left[f(u, v)\left(\frac{\partial g}{\partial u}du + \frac{\partial g}{\partial v}dv\right)\right.$$

$$\left. - g(u, v)\left(\frac{\partial f}{\partial u}du + \frac{\partial f}{\partial v}dv\right)\right],$$

and apply Green's formula to C' and R'.)

27. Rewrite Eq. (14) in nonvector notation for a vector field $\mathbf{F} = \mathbf{i}M(x, y) + \mathbf{j}N(x, y)$ in place of $\mathbf{G}$. (In other words, first write it in vector form with $\mathbf{F}$ in place of $\mathbf{G}$, and then translate the result into nonvector notation.)

***Toolkit* programs**

Scalar Fields   Vector Fields

17.6 The Divergence Theorem

The divergence theorem states that under appropriate conditions the triple integral

$$\iiint_D \text{div } \mathbf{F}\,dV \tag{1}$$

is equal to the double integral

$$\iint_S \mathbf{F} \cdot \mathbf{n}\, d\sigma. \tag{2}$$

Here $\mathbf{F} = \mathbf{i}M + \mathbf{j}N + \mathbf{k}P$, with M, N, and P continuous functions of (x, y, z) that have continuous first-order partial derivatives;

$$\operatorname{div} \mathbf{F} = \frac{\partial M}{\partial x} + \frac{\partial N}{\partial y} + \frac{\partial P}{\partial z};$$

$\mathbf{n}\, d\sigma$ is a vector element of surface area directed along the unit outer normal vector $\mathbf{n}$; and S is the surface enclosing the region D. We shall first show that (1) and (2) are equal if D is some convex region with no holes, such as the interior of a sphere, or a cube, or an ellipsoid, and if S is a piecewise smooth surface. In addition, we assume that the projection of D into the xy-plane is a simply connected region R_{xy} and that any line perpendicular to the xy-plane at an interior point of R_{xy} intersects the surface S in at most two points, producing surfaces S_1 and S_2:

$$\begin{aligned} S_1&: \quad z_1 = f_1(x, y), \quad (x, y) \text{ in } R_{xy}, \\ S_2&: \quad z_2 = f_2(x, y), \quad (x, y) \text{ in } R_{xy}, \end{aligned}$$

with $z_1 \leq z_2$. Similarly for the projection of D onto the other coordinate planes.

If we write the unit normal vector $\mathbf{n}$ in terms of its direction cosines, as

$$\mathbf{n} = \mathbf{i} \cos \alpha + \mathbf{j} \cos \beta + \mathbf{k} \cos \gamma,$$

then

$$\mathbf{F} \cdot (\mathbf{n}\, d\sigma) = (\mathbf{F} \cdot \mathbf{n})\, d\sigma = (M \cos \alpha + N \cos \beta + P \cos \gamma)\, d\sigma; \tag{3}$$

and the divergence theorem states that

$$\iiint_D \left(\frac{\partial M}{\partial x} + \frac{\partial N}{\partial y} + \frac{\partial P}{\partial z} \right) dx\, dy\, dz = \iint_S (M \cos \alpha + N \cos \beta + P \cos \gamma)\, d\sigma. \tag{4}$$

We see that both sides of Eq. (4) are additive with respect to M, N, and P, and that our task is accomplished if we prove

$$\iiint \frac{\partial M}{\partial x}\, dx\, dy\, dz = \iint M \cos \alpha\, d\sigma, \tag{5a}$$

$$\iiint \frac{\partial N}{\partial y}\, dx\, dy\, dz = \iint N \cos \beta\, d\sigma, \tag{5b}$$

$$\iiint \frac{\partial P}{\partial z}\, dx\, dy\, dz = \iint P \cos \gamma\, d\sigma. \tag{5c}$$

We shall establish (5c) in detail.

Figure 17.22 illustrates the projection of D into the xy-plane. The surface S consists of the *upper part*

$$S_2: \quad z = f_2(x, y), \quad (x, y) \text{ in } R_{xy},$$

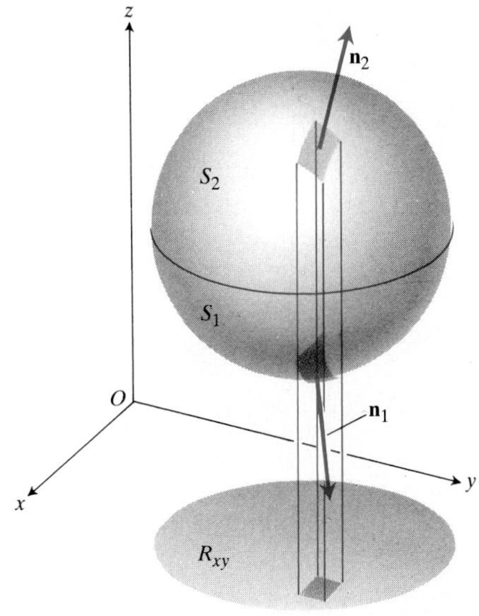

17.22 Regions S_1 and S_2 project onto R_{xy}.

and the *lower part:*

$$S_1: \quad z = f_1(x, y), \quad (x, y) \text{ in } R_{xy}.$$

On the surface S_2, the outer normal has a positive **k**-component, and

$$\cos \gamma_2 \, d\sigma_2 = dx\, dy \tag{6a}$$

is the projection of $d\sigma$ into R_{xy}. On the surface S_1, the outer normal has a negative **k**-component, and

$$\cos \gamma_1 \, d\sigma_1 = -dx\, dy. \tag{6b}$$

Therefore we can evaluate the surface integral on the right-hand side of Eq. (5c) in the following way:

$$\begin{aligned}
\iint_S P \cos \gamma \, d\sigma &= \iint_{S_2} P_2 \cos \gamma_2 \, d\sigma_2 + \iint_{S_1} P_1 \cos \gamma_1 \, d\sigma_1 \\
&= \iint_{R_{xy}} P(x, y, z_2)\, dx\, dy - \iint_{R_{xy}} P(x, y, z_1)\, dx\, dy \\
&= \iint_{R_{xy}} [P(x, y, z_2) - P(x, y, z_1)]\, dx\, dy \\
&= \iint_{R_{xy}} \left[\int_{z_1}^{z_2} \frac{\partial P}{\partial z}\, dz \right] dx\, dy \\
&= \iiint_D \frac{\partial P}{\partial z}\, dz\, dx\, dy.
\end{aligned} \tag{7}$$

Thus we have established Eq. (5c). Proofs for (5a) and (5b) follow the same pattern; or just permute x, y, z; M, N, P; α, β, γ, in order, and get those results from (5c) by renaming the axes. Finally, by addition of (5a, b, c), we get Eq. (4):

THEOREM

Divergence Theorem

$$\iiint_D \operatorname{div} \mathbf{F}\, dV = \iint_S \mathbf{F} \cdot \mathbf{n}\, d\sigma. \tag{8}$$

EXAMPLE 1 Verify Eq. (8) for the sphere

$$x^2 + y^2 + z^2 = a^2$$

if

$$\mathbf{F} = \mathbf{i}x + \mathbf{j}y + \mathbf{k}z.$$

Solution

$$\operatorname{div} \mathbf{F} = \frac{\partial x}{\partial x} + \frac{\partial y}{\partial y} + \frac{\partial z}{\partial z} = 3,$$

so

$$\iiint_D \operatorname{div} \mathbf{F}\, dV = \iiint_D 3\, dV = 3\left(\frac{4}{3}\pi a^3\right) = 4\pi a^3.$$

The outer unit normal to S, calculated from the gradient of $f(x, y, z) = x^2 + y^2 + z^2 - a^2$, is

$$\mathbf{n} = \frac{2(x\mathbf{i} + y\mathbf{j} + z\mathbf{k})}{\sqrt{4(x^2 + y^2 + z^2)}} = \frac{x\mathbf{i} + y\mathbf{j} + z\mathbf{k}}{a},$$

and

$$\mathbf{F} \cdot \mathbf{n}\, d\sigma = \frac{x^2 + y^2 + z^2}{a}\, d\sigma = \frac{a^2}{a}\, d\sigma = a\, d\sigma,$$

because $x^2 + y^2 + z^2 = a^2$ on the surface. Therefore

$$\iint_S \mathbf{F} \cdot d\sigma = \iint_S a\, d\sigma = a(4\pi a^2) = 4\pi a^3. \quad \square$$

EXAMPLE 2 Show that the divergence theorem holds for the cube with faces in the planes

$$x = x_0, \qquad x = x_0 + h,$$
$$y = y_0, \qquad y = y_0 + h,$$
$$z = z_0, \qquad z = z_0 + h,$$

where h is a positive constant.

Solution We compute $\iint \mathbf{F} \cdot \mathbf{n}\, d\sigma$ as the sum of the integrals over the six faces separately. We begin with the two faces perpendicular to the x-axis. For the face $x = x_0$ and the face $x = x_0 + h$, respectively, we have the first and second lines of the following table.

Range of integration	Outward unit normal	$\iint(\mathbf{F} \cdot \mathbf{n})\, d\sigma$
$y_0 \le y \le y_0 + h,\ z_0 \le z \le z_0 + h$	$-\mathbf{i}$	$-\iint M(x_0, y, z)\, dy\, dz$
$y_0 \le y \le y_0 + h,\ z_0 \le z \le z_0 + h$	$\mathbf{i}$	$+\iint M(x_0 + h, y, z)\, dy\, dz$

The sum of the surface integrals over these two faces is

$$\iint (\mathbf{F} \cdot \mathbf{n})\, d\sigma = \iint [M(x_0 + h, y, z) - M(x_0, y, z)]\, dy\, dz$$
$$= \iint \left(\int_{x_0}^{x_0+h} \frac{\partial M}{\partial x}\, dx \right) dy\, dz$$
$$= \iiint_D \frac{\partial M}{\partial x}\, dV.$$

Similarly the sum of the surface integrals over the two faces perpendicular to the y-axis is equal to

$$\iiint_D (\partial N/\partial y)\, dV,$$

and the sum of the surface integrals over the other two faces is equal to

$$\iiint_D \left(\frac{\partial P}{\partial z}\right) dV.$$

Hence the surface integral over the six faces is equal to the sum of three volume integrals, and Eq. (8) holds for the cube:

$$\iint_S \mathbf{F} \cdot \mathbf{n}\, d\sigma = \iiint_D \left(\frac{\partial M}{\partial x} + \frac{\partial N}{\partial y} + \frac{\partial P}{\partial z}\right) dV$$

$$= \iiint_D \operatorname{div} \mathbf{F}\, dV. \ \square$$

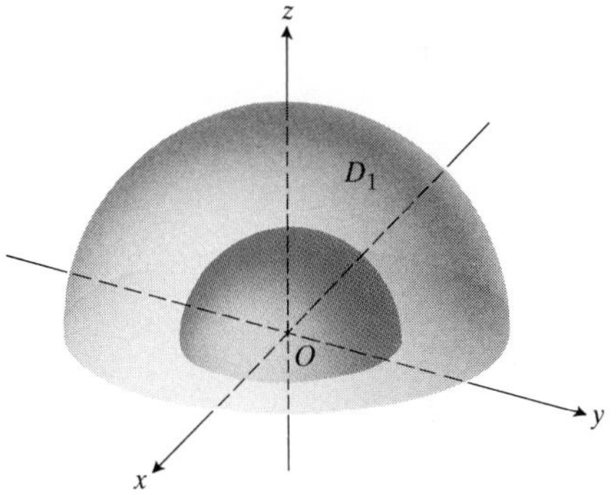

17.23 Upper half of the region between two spheres.

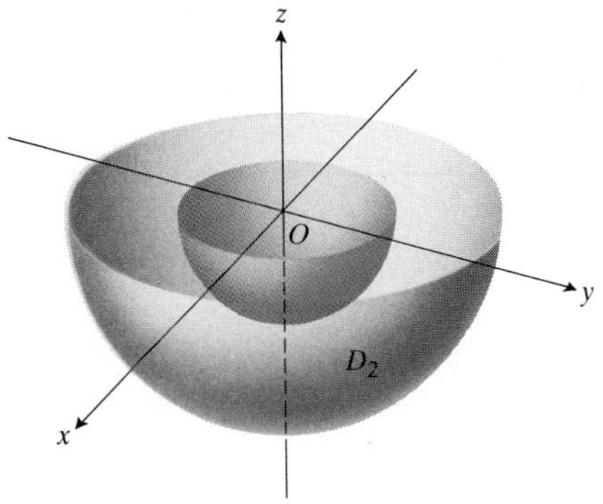

17.24 Lower half of the region between two spheres.

The Divergence Theorem for Other Regions

The divergence theorem can be extended to more complex regions that can be split up into a finite number of simple regions of the type discussed, and to regions that can be defined as limits of simpler regions in certain ways. For example, suppose D is the region between two concentric spheres, and $\mathbf{F}$ has continuously differentiable components throughout D and on the bounding surfaces. Split D by an equatorial plane and apply the divergence theorem to each half separately. The top half, D_1, is shown in Fig. 17.23. The surface that bounds D_1 consists of an outer hemisphere, a plane washer-shaped base, and an inner hemisphere. The divergence theorem says that

$$\iiint_{D_1} \operatorname{div} \mathbf{F}\, dV_1 = \iint_{S_1} \mathbf{F} \cdot \mathbf{n}_1\, d\sigma_1. \tag{9a}$$

The unit normal $\mathbf{n}_1$ that points outward from D_1 points away from the origin along the outer surface, points down along the flat base, and points toward the origin along the inner surface. Next apply the divergence theorem to D_2, as shown in Fig. 17.24:

$$\iiint_{D_2} \operatorname{div} \mathbf{F}\, dV_2 = \iint_{S_2} \mathbf{F} \cdot \mathbf{n}_2\, d\sigma_2. \tag{9b}$$

As we follow $\mathbf{n}_2$ over S_2, pointing outward from D_2, we see that $\mathbf{n}_2$ points upward along the flat surface in the xy-plane, points away from the origin on the outer sphere, and points toward the origin on the inner sphere. When we add (9a) and (9b), the surface integrals over the flat base cancel because of the opposite signs of $\mathbf{n}_1$ and $\mathbf{n}_2$. We thus arrive at the result

$$\iiint_D \operatorname{div} \mathbf{F}\, dV = \iint_S \mathbf{F} \cdot \mathbf{n}\, d\sigma, \tag{10}$$

with D the region between the spheres, S the boundary of D consisting of two spheres, and $\mathbf{n}$ the unit normal to S directed outward from D.

EXAMPLE 3 Verify Eq. (10) for the region

$$1 \le x^2 + y^2 + z^2 \le 4$$

if

$$\mathbf{F} = -\frac{\mathbf{i}x + \mathbf{j}y + \mathbf{k}z}{\rho^3}, \qquad \rho = \sqrt{x^2 + y^2 + z^2}.$$

Solution Observe that

$$\frac{\partial\rho}{\partial x} = \frac{x}{\rho}$$

and

$$\frac{\partial}{\partial x}(x\rho^{-3}) = \rho^{-3} - 3x\rho^{-4}\frac{\partial\rho}{\partial x} = \frac{1}{\rho^3} - \frac{3x^2}{\rho^5}.$$

Thus, throughout the region $1 \le \rho \le 2$, all functions considered are continuous, and

$$\operatorname{div} \mathbf{F} = \frac{-3}{\rho^3} + \frac{3}{\rho^5}(x^2 + y^2 + z^2) = -\frac{3}{\rho^3} + \frac{3\rho^2}{\rho^5} = 0.$$

Therefore

$$\iiint_D \operatorname{div} \mathbf{F}\, dV = 0. \tag{11}$$

On the outer sphere ($\rho = 2$), the positive unit normal is

$$\mathbf{n} = \frac{\mathbf{i}x + \mathbf{j}y + \mathbf{k}z}{\rho},$$

and

$$\mathbf{F} \cdot \mathbf{n}\, d\sigma = -\frac{x^2 + y^2 + z^2}{\rho^4}\, d\sigma = -\frac{1}{\rho^2}\, d\sigma.$$

Hence

$$\iint_{\rho=2} \mathbf{F} \cdot \mathbf{n}\, d\sigma = -\frac{1}{4}\iint_{\rho=2} d\sigma = -\frac{1}{4} \cdot 4\pi\rho^2 = -\pi\rho^2 = -4\pi. \tag{12a}$$

On the inner sphere ($\rho = 1$), the positive unit normal points toward the origin; its equation is

$$\mathbf{n} = \frac{-(\mathbf{i}x + \mathbf{j}y + \mathbf{k}z)}{\rho}.$$

Hence

$$\mathbf{F} \cdot \mathbf{n}\, d\sigma = +\frac{x^2 + y^2 + z^2}{\rho^4} = \frac{1}{\rho^2}\, d\sigma.$$

Thus

$$\iint_{\rho=1} \mathbf{F} \cdot \mathbf{n}\, d\sigma = \iint_{\rho=1} \frac{1}{\rho^2}\, d\sigma = \frac{1}{\rho^2} \cdot 4\pi\rho^2 = 4\pi. \tag{12b}$$

The sum of (12a) and (12b) is the surface integral over the complete boundary of D:

$$-4\pi + 4\pi = 0,$$

which agrees with (11), as it should. □

The Continuity Equation of Hydrodynamics

If $\mathbf{v}(x, y, z)$ is the velocity vector of a differentiable fluid flow through a region D in space, $\delta = \delta(x, y, z, t)$ is the density of the fluid at each point (x, y, z) at time t, and $\mathbf{F} = \delta\mathbf{v}$, then the *continuity equation* is the statement that

$$\operatorname{div} \mathbf{F} + \frac{\partial \delta}{\partial t} = 0. \tag{13}$$

The continuity equation evolves naturally from the divergence theorem

$$\iiint_D \operatorname{div} \mathbf{F}\, dV = \iint_S \mathbf{F} \cdot \mathbf{n}\, d\sigma$$

if the functions involved are continuous, as we shall now show.

First of all, the integral

$$\iint_S \mathbf{F} \cdot \mathbf{n}\, d\sigma$$

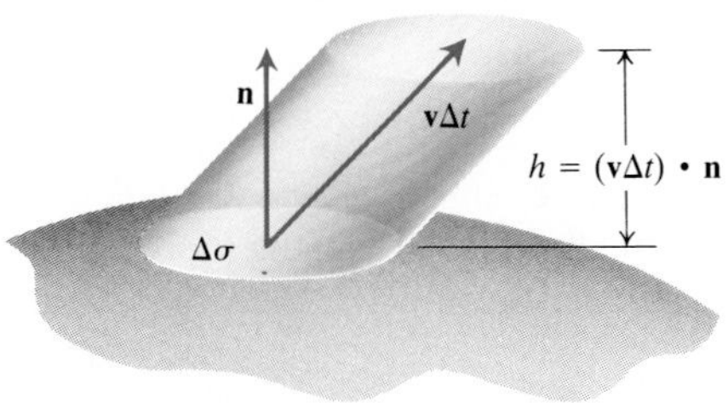

17.25 The fluid that flows upward through the patch $\Delta\sigma$ in a short time Δt fills a "cylinder" whose volume is approximately base $\times$ height $= \mathbf{v} \cdot \mathbf{n}\, \Delta t\, \Delta\sigma$.

is the rate at which mass leaves D across S (leaves, because $\mathbf{n}$ is the outer normal). To see why, consider a patch of area $\Delta\sigma$ on the surface (Fig. 17.25). In a short time interval Δt, the volume ΔV of fluid that flows across the patch is approximately equal to the volume of a cylinder with base area $\Delta\sigma$ and height $(\mathbf{v}\, \Delta t) \cdot \mathbf{n}$, where $\mathbf{v}$ is a velocity vector rooted at a point of the patch:

$$\Delta V \approx \mathbf{v} \cdot \mathbf{n}\, \Delta\sigma\, \Delta t.$$

The mass of this volume of fluid is about

$$\Delta m \approx \delta\mathbf{v} \cdot \mathbf{n}\, \Delta\sigma\, \Delta t,$$

so that the rate at which mass is flowing out of D across the patch is about

$$\frac{\Delta m}{\Delta t} \approx \delta\mathbf{v} \cdot \mathbf{n}\, \Delta\sigma.$$

This leads to the approximation

$$\frac{\Sigma\, \Delta m}{\Delta t} \approx \sum \delta\mathbf{v} \cdot \mathbf{n}\, \Delta\sigma \tag{14}$$

as an estimate of the average rate at which mass flows across S. Finally, letting $\Delta\sigma \to 0$ and $\Delta t \to 0$ gives the instantaneous rate at which mass leaves D across S as

$$\frac{dm}{dt} = \iint_S \delta\mathbf{v} \cdot \mathbf{n}\, d\sigma, \tag{15}$$

which, for our particular flow, is

$$\frac{dm}{dt} = \iint_S \mathbf{F} \cdot \mathbf{n}\, d\sigma. \tag{16}$$

Now let B be a ball centered at a point Q in the flow. The average value of div $\mathbf{F}$ over the ball is

$$\frac{1}{\text{volume}} \iiint_B \operatorname{div} \mathbf{F}\, dV. \tag{17}$$

It is a consequence of the continuity of div $\mathbf{F}$ that div $\mathbf{F}$ actually takes on this value at some point P in B. Thus,

$$\begin{aligned}(\operatorname{div} \mathbf{F})_P &= \frac{1}{\text{volume}} \iiint_B \operatorname{div} \mathbf{F}\, dV \\ &= \frac{\iint_S \mathbf{F} \cdot \mathbf{n}\, d\sigma}{\text{volume}} \\ &= \frac{\text{rate at which mass leaves } B \text{ across its surface } S}{\text{volume of } B}. \end{aligned} \tag{18}$$

The fraction on the right describes decrease in mass per unit volume.

Now let the radius of the ball B approach 0 while the center Q stays fixed. The left-hand side of Eq. (18) converges to $(\operatorname{div} \mathbf{F})_Q$, the right side to $(-\partial\delta/\partial t)_Q$. The equality of these two limits is the continuity equation

$$\operatorname{div} \mathbf{F} = -\frac{\partial \delta}{\partial t}. \tag{19}$$

The continuity equation "explains" div $\mathbf{F}$: the divergence of $\mathbf{F}$ at a point is the rate at which the density of the fluid is decreasing there. If the fluid is incompressible, its density is constant and div $\mathbf{F} = 0$.

The divergence theorem

$$\iiint_D \operatorname{div} \mathbf{F}\, dV = \iint_S \mathbf{F} \cdot \mathbf{n}\, d\sigma$$

now says that the net decrease in density of the region D is accounted for by the mass transported across the surface S. In a way, this is a statement about conservation of mass.

PROBLEMS

In Problems 1–5, verify the divergence theorem for the cube with center at the origin and faces in the planes $x = \pm 1$, $y = \pm 1$, $z = \pm 1$, and $\mathbf{F}$ as given.

1. $\mathbf{F} = 2\mathbf{i} + 3\mathbf{j} + 4\mathbf{k}$
2. $\mathbf{F} = \mathbf{i}x + \mathbf{j}y + \mathbf{k}z$
3. $\mathbf{F} = \mathbf{i}yz + \mathbf{j}xz + \mathbf{k}xy$
4. $\mathbf{F} = \mathbf{i}(x - y) + \mathbf{j}(y - z) + \mathbf{k}(x - y)$
5. $\mathbf{F} = \mathbf{i}x^2 + \mathbf{j}y^2 + \mathbf{k}z^2$

In Problems 6–10, compute both

$$\iiint_D \operatorname{div} \mathbf{F}\, dV \quad \text{and} \quad \iint_S \mathbf{F} \cdot \mathbf{n}\, d\sigma$$

directly. Compare the results with the divergence theorem expressed by Eq. (8), given that

$$\mathbf{F} = \mathbf{i}(x + y) + \mathbf{j}(y + z) + \mathbf{k}(z + x),$$

and given that S bounds the region D given in the problem.

6. $0 \le z \le 4 - x^2 - y^2,\quad 0 \le x^2 + y^2 \le 4$
7. $-4 + x^2 + y^2 \le z \le 4 - x^2 - y^2,\quad 0 \le x^2 + y^2 \le 4$
8. $0 \le x^2 + y^2 \le 9,\quad 0 \le z \le 5$
9. $0 \le x^2 + y^2 + z^2 \le a^2$
10. $|x| \le 1,\quad |y| \le 1,\quad |z| \le 1$

Use the divergence theorem to evaluate the surface integral

$$\iint_S \mathbf{F} \cdot \mathbf{n}\, d\sigma$$

for the surfaces and fields given in Problems 11–14. Take $\mathbf{n}$ to be the outer unit normal in each case.

11. S is the sphere $x^2 + y^2 + z^2 = 4$, and $\mathbf{F} = 2x\mathbf{i} + xz\mathbf{j} + z\mathbf{k}$.

12. S is the surface of the solid in the first octant bounded by the coordinate planes and the sphere $x^2 + y^2 + z^2 = 4$, and $\mathbf{F} = x^2\mathbf{i} - 2xy\mathbf{j} + 3xz\mathbf{k}$.

13. S is the surface of the solid in the first octant bounded by the coordinate planes, the cylinder $x^2 + y^2 = 4$, and the plane $z = 4$, and $\mathbf{F} = (6x^2 + 2xy)\mathbf{i} + (2y + x^2z)\mathbf{j} + 4x^2y^3\mathbf{k}$.

14. S is the surface of the wedge in the first octant bounded by the coordinate planes $x = 0$ and $z = 0$, the plane $z = y$, and the elliptical cylinder $x^2 + 4y^2 = 16$, and $\mathbf{F} = 2xz\mathbf{i} + y\mathbf{j} - z^2\mathbf{k}$.

15. Find the rate of mass transport dm/dt outward through the surface of the closed cylinder $x^2 + y^2 \le 4$, $-1 \le z \le 1$, if the flow vector $\mathbf{F} = \delta\mathbf{v}$ is given by $\mathbf{F} = -y\mathbf{i} + x\mathbf{j} + z\mathbf{k}$.

16. Let S be the spherical cap $x^2 + y^2 + z^2 = 2a^2$, $z \ge a$, together with its base $x^2 + y^2 \le a^2$, $z = a$. Find the flux of $\mathbf{F} = xz\mathbf{i} - yz\mathbf{j} + y^2\mathbf{k}$ outward through S (a) by evaluating $\iint_S \mathbf{F} \cdot \mathbf{n}\, d\sigma$ directly, (b) by applying the divergence theorem.

17. Let S be the closed cube-like surface shown in Fig. 17.26, with its base the unit square in the xy-plane, its four sides lying in the planes $x = 0$, $x = 1$, $y = 0$, $y = 1$, and its top an arbitrary smooth surface whose identity is unknown. Let $\mathbf{F} = x\mathbf{i} - 2y\mathbf{j} + (z + 3)\mathbf{k}$. Suppose that the outward flux of $\mathbf{F}$ through side A is 1 and through side B is -3. Calculate the outward flux of $\mathbf{F}$ through the top.

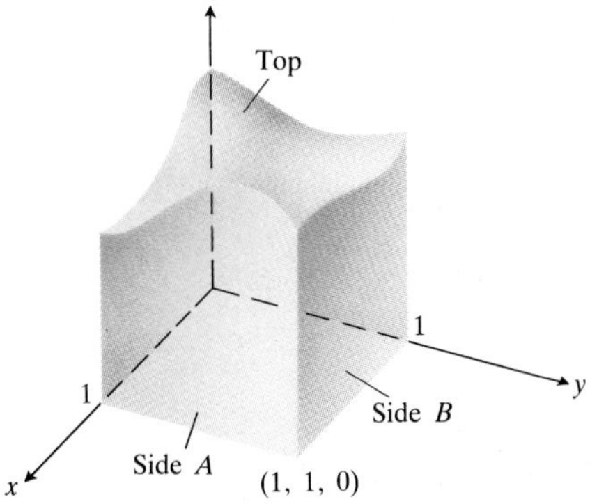

17.26 Problem 17 calculates the flux of a vector field through the top of this solid.

18. Let $\mathbf{F}(x, y, z)$ be a vector field with components that are continuous and differentiable on a portion of space containing a (finite) region D and its bounding surface S. Suppose that the length of the vector $\mathbf{F}(x, y, z)$ never exceeds 1 when (x, y, z) is a point of S. What bound can be placed on the numerical size of the integral

$$\iiint_D \operatorname{div} \mathbf{F}\, dV?$$

Explain.

19. a) Show that the flux of the position vector $\mathbf{F} = x\mathbf{i} + y\mathbf{j} + z\mathbf{k}$ outward through a piecewise smooth closed surface is three times the volume contained by the surface.

b) Verify part (a) for the cube bounded by the planes $x = \pm 1$, $y = \pm 1$, $z = \pm 1$.

c) Verify part (a) for the sphere $x^2 + y^2 + z^2 = a^2$.

d) Let $\mathbf{n}$ be the outward unit normal to a piecewise smooth closed surface S. Show that it is impossible for the position vector $\mathbf{F} = x\mathbf{i} + y\mathbf{j} + z\mathbf{k}$ to be orthogonal to $\mathbf{n}$ at every point of S.

20. Let $\mathbf{F} = M(x, y, z)\mathbf{i} + N(x, y, z)\mathbf{j} + P(x, y, z)\mathbf{k}$ be a vector field whose components M, N, and P are continuous and have continuous second partial derivatives of all kinds.

a) Show by direct computation that div (curl $\mathbf{F}$) = 0.

b) Use the result of part (a) to show that

$$\iint_S (\operatorname{curl} \mathbf{F}) \cdot \mathbf{n}\, d\sigma = 0$$

for any surface to which the divergence theorem applies.

21. Verify the divergence theorem, Eq. (8), for $\mathbf{F} = x\mathbf{i} + y\mathbf{j} + z\mathbf{k}$ over the region D: $4 \le x^2 + y^2 + z^2 \le 9$.

22. *Identities involving div, grad, and curl.*

a) Prove that if ϕ is a scalar function of x, y, z, then

$$\operatorname{curl}(\operatorname{grad} \phi) = \nabla \times (\nabla \phi) = \mathbf{0}.$$

b) State in terms of the vector field $\nabla \times \mathbf{F}$ how would you express the condition that $\mathbf{F} \cdot d\mathbf{R}$ be an exact differential.

Prove the following results: If

$$\mathbf{r} = x\mathbf{i} + y\mathbf{j} + z\mathbf{k},$$

c) $\operatorname{div}(\phi\mathbf{F}) \equiv \nabla \cdot (\phi\mathbf{F}) = \phi\nabla \cdot \mathbf{F} + \mathbf{F} \cdot \nabla\phi$;

d) $\nabla \times (\phi\mathbf{F}) = \phi\nabla \times \mathbf{F} + (\nabla\phi) \times \mathbf{F}$;

e) $\nabla \cdot (\mathbf{F}_1 \times \mathbf{F}_2) = \mathbf{F}_2 \cdot \nabla \times \mathbf{F}_1 - \mathbf{F}_1 \cdot \nabla \times \mathbf{F}_2$;

f) $\nabla \cdot \mathbf{R} = 3$ and $\nabla \times \mathbf{R} = \mathbf{0}$.

23. A function f is said to be *harmonic* in a region D if throughout D it satisfies the Laplace equation

$$\frac{\partial^2 f}{\partial x^2} + \frac{\partial^2 f}{\partial y^2} + \frac{d^2 f}{\partial z^2} = 0.$$

Suppose f is harmonic throughout D, S is the boundary of D, $\mathbf{n}$ is the positive unit normal on S, and $\partial f/\partial n$ is the directional derivative of f in the direction of $\mathbf{n}$. This derivative is called the normal derivative of f. Prove that

$$\iint_S \frac{\partial f}{\partial n}\, d\sigma = 0.$$

(*Hint:* Let $\mathbf{F} = \text{grad } f$.)

24. Prove that if f is harmonic in D (see Problem 23), then

$$\iint_S f\frac{\partial f}{\partial n}\, d\sigma = \iiint_D |\text{grad } f|^2\, dV.$$

(*Hint:* Let $\mathbf{F} = f \text{ grad } f$.)

25. Let S be the eighth of the sphere $x^2 + y^2 + z^2 = a^2$ lying in the first octant, and let $f(x, y, z) = \ln \sqrt{x^2 + y^2 + z^2}$. Calculate

$$\iint_S \frac{\partial f}{\partial n}\, d\sigma.$$

(See Problem 23.)

***Toolkit* programs**
Vector Fields

17.7 Stokes's Theorem

Stokes's theorem is an extension of Green's theorem in vector form to surfaces and curves in three dimensions. It says that the line integral

$$\oint \mathbf{F} \cdot d\mathbf{R} \tag{1}$$

is equal to the surface integral

$$\iint_S (\text{curl } \mathbf{F}) \cdot \mathbf{n}\, d\sigma, \tag{2}$$

under suitable restrictions (i) on the vector

$$\mathbf{F} = \mathbf{i}M + \mathbf{j}N + \mathbf{k}P,$$

(ii) on the simple closed curve C:

$$x = f(t), \qquad y = g(t), \qquad z = h(t), \qquad 0 \le t \le 1,$$

(iii) on the surface

$$S:\ \ \phi(x, y, z) = 0$$

bounded by C.

EXAMPLE 1 Let S be the hemisphere

$$z = \sqrt{4 - x^2 - y^2}, \qquad 0 \le x^2 + y^2 \le 4,$$

lying above the xy-plane, with center at the origin. The boundary of this hemisphere is the circle C:

$$z = 0, \qquad x^2 + y^2 = 4.$$

Show that the integrals in Eqs. (1) and (2) are equal for S, C, and

$$\mathbf{F} = \mathbf{i}y - \mathbf{j}x.$$

Solution The integrand in the line integral (1) is

$$\begin{aligned}\mathbf{F}\cdot d\mathbf{R} &= \mathbf{F}\cdot(\mathbf{i}\,dx + \mathbf{j}\,dy + \mathbf{k}\,dz)\\ &= y\,dx - x\,dy.\end{aligned}$$

By Green's theorem for *plane* curves and surfaces, we have

$$\oint \mathbf{F}\cdot d\mathbf{R} = \oint_C (y\,dx - x\,dy) = \iint\limits_{x^2+y^2\le 4} -2\,dx\,dy = -8\pi. \tag{3}$$

To evaluate the surface integral (2), we compute

$$\text{curl}\,\mathbf{F} = \mathbf{i}\left(\frac{\partial P}{\partial y} - \frac{\partial N}{\partial z}\right) + \mathbf{j}\left(\frac{\partial M}{\partial z} - \frac{\partial P}{\partial x}\right) + \mathbf{k}\left(\frac{\partial N}{\partial x} + \frac{\partial M}{\partial y}\right),$$

with

$$M = y, \qquad N = -x, \qquad P = 0,$$

to get

$$\text{curl}\,\mathbf{F} = -2\mathbf{k}.$$

The unit outer normal to the hemisphere is

$$\mathbf{n} = \frac{\mathbf{i}x + \mathbf{j}y + \mathbf{k}z}{\sqrt{x^2+y^2+z^2}} = \frac{\mathbf{i}x + \mathbf{j}y + \mathbf{k}z}{2}.$$

Therefore,

$$\text{curl}\,\mathbf{F}\cdot\mathbf{n}\,d\sigma = -z\,d\sigma. \tag{4}$$

For element of surface area $d\sigma$ (Article 17.2), we use

$$d\sigma = \sqrt{1 + \left(\frac{\partial z}{\partial x}\right)^2 + \left(\frac{\partial z}{\partial y}\right)^2}\,dx\,dy, \tag{5}$$

with

$$\frac{\partial z}{\partial x} = \frac{-x}{\sqrt{4-x^2-y^2}} \quad\text{and}\quad \frac{\partial z}{\partial y} = \frac{-y}{\sqrt{4-x^2-y^2}},$$

or

$$\frac{\partial z}{\partial x} = \frac{-x}{z} \quad\text{and}\quad \frac{\partial z}{\partial y} = \frac{-y}{z}. \tag{6}$$

From (4), (5), and (6), we get

$$\begin{aligned}\text{curl}\,\mathbf{F}\cdot\mathbf{n}\,d\sigma &= -z\,d\sigma\\ &= -z\sqrt{1 + \frac{x^2}{z^2} + \frac{y^2}{z^2}}\,dx\,dy\\ &= -\sqrt{x^2+y^2+z^2}\,dx\,dy\\ &= -\sqrt{4}\,dx\,dy = -2\,dx\,dy.\end{aligned} \tag{7}$$

Therefore

$$\iint\limits_S \text{curl}\,\mathbf{F}\cdot\mathbf{n}\,d\sigma = \iint\limits_{x^2+y^2\le 4} -2\,dx\,dy = -8\pi, \tag{8}$$

which agrees with the value of the line integral in Eq. (3). □

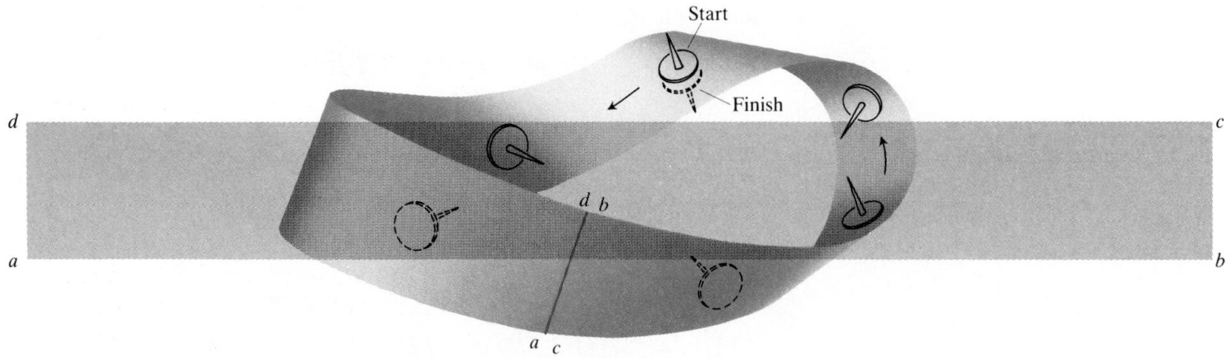

17.27 The Möbius strip (colored band) can be constructed by taking a rectangular strip of paper *abcd*, giving the end *bc* a single twist to interchange the positions of the vertices *b* and *c*, and then pasting the ends of the strip together so as to bring vertices *a* and *c* together, and also *b* and *d*. The Möbius strip is a nonorientable surface.

In Example 1, the surface integral (8) taken over the *hemisphere* turns out to have the same value as a surface integral taken over the *plane base* of that hemisphere. The underlying reason for this equality is that both surface integrals are equal to the line integral around the circle that is their common boundary. (See Problem 14.)

In Stokes's theorem, we require that the surface be *orientable*. By "orientable," we mean that it is possible to consistently assign a unique direction, called *positive*, at each point of S, and that there exists a unit normal $\mathbf{n}$ pointing in this direction. As we move about over the surface S without touching its boundary, the direction cosines of the unit vector $\mathbf{n}$ should vary continuously. Also, when we return to the starting position, $\mathbf{n}$ should return to its initial direction. This rules out a surface like the Möbius strip shown in Fig. 17.27. This surface is nonorientable because a unit normal vector (think of the shaft of a thumbtack) can be continuously moved around the surface without its touching the boundary of the surface, and in such a way that when it is returned to its initial position it will point in a direction *exactly opposite* to its initial direction.

We also want C to have a positive direction that is related to the positive direction on S. We imagine a simple closed curve Γ on S, near the boundary C (see Fig. 17.28), and let $\mathbf{n}$ be normal to S at some point inside Γ. We then assign to Γ a positive direction, the counterclockwise direction as viewed by an observer who is at the end of $\mathbf{n}$ and looking down. (Note that choosing this direction keeps the interior of Γ on our left as we progress around Γ. We could equally well have specified $\mathbf{n}$'s direction by this condition.) Now we move Γ about on S until it touches and is tangent to C. The direction of the positive tangent to Γ at this point of common tangency we shall take to be the positive direction along C. It is a consequence of the orientability of S that a consistent assignment of positive direction along C is induced by this process. The same positive direction is assigned all the way around C, no matter where on S the process is begun. This would not be true of the (nonorientable) Möbius strip.

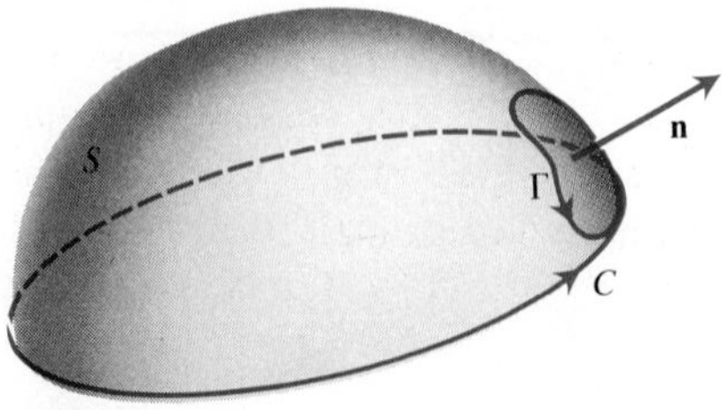

17.28 Orientation of the boundary of an oriented surface.

THEOREM

Stokes's Theorem

Let S be a smooth orientable surface bounded by a closed curve C. Let

$$\mathbf{F} = \mathbf{i}M + \mathbf{j}N + \mathbf{k}P,$$

where M, N, and P and their first order partial derivatives are continuous throughout a region D containing S and C in its interior. Let $\mathbf{n}$ be a positive unit vector normal to S, and let the positive direction around C be the one induced by the positive orientation of S. Then

$$\oint_C \mathbf{F} \cdot d\mathbf{R} = \iint_S \text{curl } \mathbf{F} \cdot \mathbf{n} \, d\sigma, \tag{9}$$

where

$$d\mathbf{R} = \mathbf{i}\,dx + \mathbf{j}\,dy + \mathbf{k}\,dz = \mathbf{T}\,ds$$

and

$$\mathbf{n}\,d\sigma = (\mathbf{i}\cos\alpha + \mathbf{j}\cos\beta + \mathbf{k}\cos\gamma)\,d\sigma.$$

Proof for a polyhedral surface S. Let the surface S be a polyhedral surface consisting of a finite number of plane regions. (Think of one of Buckminster Fuller's geodesic domes.) We apply Green's theorem to each separate panel of S. There are two types of panels:

1. those that are surrounded on all sides by other panels, and
2. those that have one or more edges that are not adjacent to other panels.

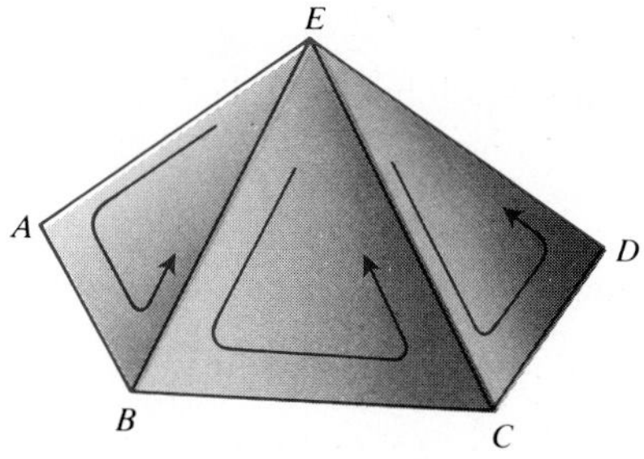

17.29 Part of a polyhedral surface.

Let Δ be part of the boundary of S that consists of those edges of the type 2 panels that are not adjacent to other panels. In Fig. 17.29, the triangles ABE, BCE, and CDE represent a part of S, with $ABCD$ part of the boundary Δ. Applying Green's theorem to the three triangles in turn and adding the results, we get

$$\left(\oint_{ABE} + \oint_{BCE} + \oint_{CDE}\right) \mathbf{F} \cdot d\mathbf{R} = \left(\iint_{ABE} + \iint_{BCE} + \iint_{CDE}\right) \text{curl } \mathbf{F} \cdot \mathbf{n} \, d\sigma. \tag{10}$$

The three line integrals on the left-hand side of Eq. (10) combine into a single line integral taken around the periphery $ABCDE$, because the integrals along interior segments cancel in pairs. For example, the integral along the segment BE in triangle ABE is opposite in sign to the integral along the same segment in triangle EBC. Similarly for the segment CE. Hence (10) reduces to

$$\oint_{ABCDE} \mathbf{F} \cdot d\mathbf{R} = \iint_{ABCDE} \text{curl } \mathbf{F} \cdot \mathbf{n} \, d\sigma.$$

In general, when we apply Green's theorem to all the panels and add the results, we get

$$\oint_{\Delta} \mathbf{F} \cdot d\mathbf{R} = \iint_S \operatorname{curl} \mathbf{F} \cdot \mathbf{n}\, d\sigma. \tag{11}$$

This is Stokes's theorem for a polyhedral surface S. ■

A rigorous proof of Stokes's theorem for more general surfaces is beyond the level of a beginning calculus course.† However, the following intuitive argument shows why one would expect Eq. (9) to be true. Imagine a sequence of polyhedral surfaces

$$S_1, \quad S_2, \quad \ldots,$$

and their corresponding boundaries $\Delta_1, \Delta_2, \ldots$. The surface S_n is constructed in such a way that its boundary Δ_n is inscribed in or tangent to C, the boundary of S, and so that the length of Δ_n approaches the length of C as $n \to \infty$. (C needs to be smooth enough to have a length.) The faces of S_n might be polygonal regions, approximating pieces of S, and such that the area of S_n approaches the area of S as $n \to \infty$. S also needs to have finite area. Assuming that M, N, P, and their partial derivatives are continuous in a region D containing S and C, it is plausible to expect that

$$\oint_{\Delta_n} \mathbf{F} \cdot d\mathbf{R} \qquad \text{approaches} \qquad \oint_C \mathbf{F} \cdot d\mathbf{R}$$

and that

$$\iint_{S_n} \operatorname{curl} \mathbf{F} \cdot \mathbf{n}\, d\sigma_n \qquad \text{approaches} \qquad \iint_S \operatorname{curl} \mathbf{F} \cdot \mathbf{n}\, d\sigma$$

as $n \to \infty$. But if

$$\oint_{\Delta_n} \mathbf{F} \cdot d\mathbf{R} \to \oint_C \mathbf{F} \cdot d\mathbf{R} \tag{12}$$

and

$$\iint_{S_n} \operatorname{curl} \mathbf{F} \cdot \mathbf{n}\, d\sigma \to \iint_S \operatorname{curl} \mathbf{F} \cdot \mathbf{n}\, d\sigma, \tag{13}$$

and if the left-hand sides of (12) and (13) are equal by Stokes's theorem for polyhedra, we then have equality of their limits.

EXAMPLE 2 Let S be the portion of the paraboloid $z = 4 - x^2 - y^2$ that lies above the plane $z = 0$. Let C be their curve of intersection, and let

$$\mathbf{F} = \mathbf{i}(z - y) + \mathbf{j}(z + x) - \mathbf{k}(x + y).$$

Show that

$$\oint_C \mathbf{F} \cdot d\mathbf{R} = \iint_S \operatorname{curl} \mathbf{F} \cdot \mathbf{n}\, d\sigma.$$

Solution The curve C is the circle $x^2 + y^2 = 4$ in the xy-plane. Along C, where $z = 0$ and $d\mathbf{R} = \mathbf{i}\, dx + \mathbf{j}\, dy$, we have

$$\begin{aligned} \mathbf{F} \cdot d\mathbf{R} &= (z - y)\, dx + (z + x)\, dy - (x + y) \cdot 0 \\ &= -y\, dx + x\, dy, \end{aligned}$$

---

† See, for example, Buck, *Advanced Calculus*, 3d ed. (New York: McGraw-Hill, 1978), or Apostol, *Mathematical Analysis*, 2d ed. (Reading, Mass.: Addison-Wesley, 1974).

whose integral around C is twice the area of the circle:

$$\oint_C \mathbf{F} \cdot d\mathbf{R} = 8\pi$$

(see Article 17.5).

For the surface integral, we compute

$$\text{curl } \mathbf{F} = \begin{vmatrix} \mathbf{i} & \mathbf{j} & \mathbf{k} \\ \dfrac{\partial}{\partial x} & \dfrac{\partial}{\partial y} & \dfrac{\partial}{\partial z} \\ z - y & z + x & -x - y \end{vmatrix} = -2\mathbf{i} + 2\mathbf{j} + 2\mathbf{k}.$$

For a positive unit normal (positive for the chosen orientation of C) on the surface

$$S: \quad f(x, y, z) = z - 4 + x^2 + y^2 = 0,$$

we take

$$\mathbf{n} = \frac{\text{grad } f}{|\text{grad } f|} = \frac{2x\mathbf{i} + 2y\mathbf{j} + \mathbf{k}}{\sqrt{4x^2 + 4y^2 + 1}}.$$

The projection of S onto the xy-plane is the region $x^2 + y^2 \leq 4$, and for the element of surface area $d\sigma$, we take

$$d\sigma = \sqrt{\left(\frac{\partial z}{\partial x}\right)^2 + \left(\frac{\partial z}{\partial y}\right)^2 + 1}\, dx\, dy = \sqrt{4x^2 + 4y^2 + 1}\, dx\, dy.$$

Thus

$$\iint_S \text{curl } \mathbf{F} \cdot \mathbf{n}\, d\sigma = \iint_{x^2 + y^2 \leq 4} (-4x + 4y + 2)\, dx\, dy \tag{a}$$

$$= \iint_{x^2 + y^2 \leq 4} 2\, dx\, dy = 8\pi, \tag{b}$$

where (b) follows from (a) because odd powers of x or y integrate to zero over the interior of the circle. □

Stokes's Theorem for Surfaces with Holes

Stokes's theorem can also be extended to a surface S that has one or more holes in it (like a curved slice of Swiss cheese), in a way exactly analogous to Green's theorem: The surface integral over S of the *normal component* of curl **F** is equal to the sum of the line integrals around all the boundaries of S (including boundaries of the holes) of the *tangential component* of **F**, where the boundary curves are to be traced in the positive direction induced by the positive orientation of S.

Circulation

Stokes's theorem provides the following vector interpretation for curl **F**. As in the discussion of divergence, let **v** be the velocity field of a moving fluid, δ the density, and $\mathbf{F} = \delta\mathbf{v}$. Then

$$\oint_C \mathbf{F} \cdot \mathbf{T}\, ds$$

is a measure of the *circulation* of fluid around the closed curve C. By Stokes's theorem, this circulation is also equal to the flux of curl **F** through a surface S spanning C:

$$\oint_C \mathbf{F}\cdot d\mathbf{R} = \iint_S \text{curl } \mathbf{F}\cdot\mathbf{n}\, d\sigma.$$

Suppose we fix a point Q and a direction **u** at Q. Let C be a circle of radius ρ, with center at Q, whose plane is normal to **u.** If curl **F** is continuous at Q, then the average value of the **u**-component of curl **F** over the circular disc bounded by C approaches the **u**-component of curl **F** at Q as $\rho \to 0$:

$$(\text{curl } \mathbf{F}\cdot\mathbf{u})_Q = \lim_{\rho\to 0} \frac{1}{\pi\rho^2} \iint_S \text{curl } \mathbf{F}\cdot\mathbf{u}\, d\sigma. \tag{14}$$

If we replace the double integral on the right-hand side of Eq. (14) by the circulation, we get

$$(\text{curl } \mathbf{F}\cdot\mathbf{u})_Q = \lim_{\rho\to 0} \frac{1}{\pi\rho^2} \oint_C \mathbf{F}\cdot d\mathbf{R}. \tag{15}$$

The left-hand side of Eq. (15) is a maximum at Q when **u** has the same direction as curl **F.** When ρ is small, the right-hand side of Eq. (15) is approximately equal to

$$\frac{1}{\pi\rho^2} \oint_C \mathbf{F}\cdot d\mathbf{R},$$

which is the circulation around C divided by the area of the disc (*circulation density*). Suppose that a small paddle wheel, of radius ρ, is introduced into the fluid at Q, with its axle directed along **u.** The circulation of the fluid around C will affect the rate of spin of the paddle wheel. The wheel will spin fastest when the circulation integral is maximized; therefore it will spin fastest when the axle of the paddle wheel points in the direction of curl **F.** (See Fig. 17.30.)

EXAMPLE 3 A fluid of constant density δ rotates around the z-axis with velocity $\mathbf{v} = \omega(\mathbf{j}x - \mathbf{i}y)$, where ω is a positive constant. If $\mathbf{F} = \delta\mathbf{v}$, find curl **F**, and show its relation to the circulation density.

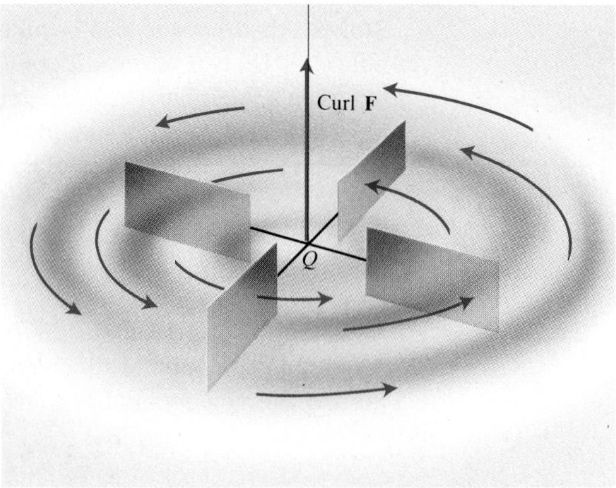

17.30 The paddle-wheel interpretation of curl **F**.

Solution

$$\mathbf{F} = \delta\omega(\mathbf{j}x - \mathbf{i}y),$$

and

$$\text{curl } \mathbf{F} = \begin{vmatrix} \mathbf{i} & \mathbf{j} & \mathbf{k} \\ \dfrac{\partial}{\partial x} & \dfrac{\partial}{\partial y} & \dfrac{\partial}{\partial z} \\ -\delta\omega y & \delta\omega x & 0 \end{vmatrix} = 2\,\delta\omega\mathbf{k}.$$

The work done by a force equal to **F**, as the point of application moves around a circle C of radius ρ, is

$$\oint_C \mathbf{F}\cdot d\mathbf{R}.$$

If C lies in a plane parallel to the xy-plane, Stokes's theorem gives

$$\oint_C \mathbf{F}\cdot d\mathbf{R} = \iint_S \text{curl } \mathbf{F}\cdot\mathbf{n}\, d\sigma = \iint 2\,\delta\omega\mathbf{k}\cdot\mathbf{k}\, dx\, dy = (2\,\delta\omega)(\pi\rho^2).$$

Thus,

$$(\text{curl } \mathbf{F})\cdot\mathbf{k} = 2\,\delta\omega = \frac{1}{\pi\rho^2}\oint_C \mathbf{F}\cdot d\mathbf{R},$$

in agreement with Eq. (15) with $\mathbf{u} = \mathbf{k}$. □

PROBLEMS

1. Verify Stokes's theorem, Eq. (9), when S is the hemisphere $x^2 + y^2 + z^2 = 1$, $z \geq 0$, and C is its boundary, for
a) $\mathbf{F} = x\mathbf{i} + y\mathbf{j} + z\mathbf{k}$,
b) $\mathbf{F} = y\mathbf{i} + z\mathbf{j} + x\mathbf{k}$.

2. Let S be the cylinder $x^2 + y^2 = a^2$, $0 \leq z \leq h$, together with its top, $x^2 + y^2 \leq a^2$, $z = h$. Let $\mathbf{F} = -y\mathbf{i} + x\mathbf{j} + x^2\mathbf{k}$. Use Stokes's theorem to calculate the flux of curl **F** outward through S.

3. Verify Stokes's theorem, Eq. (9), for $\mathbf{F} = y\mathbf{i} + xz\mathbf{j} + x^2\mathbf{k}$ and S the triangular plate whose corners are $(1, 0, 0)$, $(0, 1, 0)$, $(0, 0, 1)$. That is, choose one of the two possible unit normals **n** and a corresponding orientation for the boundary of S. One equation for the plane containing the plate is $x + y + z = 1$.

4. Verify Stokes's theorem, Eq. (9), for $\mathbf{F} = 2y\mathbf{i} + 3x\mathbf{j} - z^2\mathbf{k}$ and S the hemisphere $x^2 + y^2 + z^2 = 9$, $z \geq 0$. Use $\mathbf{n} = (x/3)\mathbf{i} + (y/3)\mathbf{j} + (z/3)\mathbf{k}$.

In Problems 5–8, verify the result of Stokes's theorem for the vector

$$\mathbf{F} = \mathbf{i}(y^2 + z^2) + \mathbf{j}(x^2 + z^2) + \mathbf{k}(x^2 + y^2)$$

for the given surface S and boundary C.

5. S: $z = \sqrt{1 - x^2}$, $-1 \leq x \leq 1$, $-2 \leq y \leq 2$,
$y = 2$, $0 \leq z \leq \sqrt{1 - x^2}$, $-1 \leq x \leq 1$,
$y = -2$, $0 \leq z \leq \sqrt{1 - x^2}$, $-1 \leq x \leq 1$;
C: $z = 0$,
$x = \pm 1$, $-2 \leq y \leq 2$,
$y = \pm 2$, $-1 \leq x \leq 1$

6. The surface S is the surface of the upper half of the cube with one vertex at $(1, 1, 1)$, center at the origin, and edges parallel to the axes; the curve C is the intersection of S with the xy-plane.

7. The surface S is as in Problem 6, with a hole cut out of the top face by the circular disc whose cylindrical coordinates satisfy

$$z = 1, \qquad 0 \leq r \leq \cos\theta, \qquad -\frac{1}{2}\pi \leq \theta \leq \frac{1}{2}\pi.$$

(The circle $z = 1$, $r = \cos\theta$ becomes part of the boundary of S.)

8. The surface S is the surface (excluding the face in the yz-plane) of a pyramid with vertices at the origin O and at $A(1, 0, 0)$, $B(0, 1, 0)$, and $D(0, 0, 1)$; the boundary curve C is the triangle OBD in the yz-plane.

9. Let S be the region bounded by the ellipse C: $4x^2 + y^2 = 4$ in the plane $z = 1$, let $\mathbf{n} = \mathbf{k}$, and let $\mathbf{F} = x^2\mathbf{i} + 2x\mathbf{j} + z^2\mathbf{k}$. Find the value of

$$\oint_C \mathbf{F} \cdot d\mathbf{R}.$$

10. Let $\mathbf{n}$ be the outer normal of the elliptical shell

$$S: \quad 4x^2 + 9y^2 + 36z^2 = 36, \qquad z \geq 0,$$

and let

$$\mathbf{F} = y\mathbf{i} + x^2\mathbf{j} + (x^2 + y^4)^{3/2} \sin e^{\sqrt{xyz}}\mathbf{k}.$$

Use Stokes's theorem to find the value of

$$\iint_S \operatorname{curl} \mathbf{F} \cdot \mathbf{n}\, d\sigma.$$

(*Hint:* One parameterization of the ellipse at the base of the shell is $x = 3\cos t$, $y = 2\sin t$, $0 \leq t \leq 2\pi$.)

11. Suppose $\mathbf{F} = \operatorname{grad} \phi$ is the gradient of a scalar function ϕ having continuous second order partial derivatives

$$\frac{\partial^2 \phi}{\partial x^2}, \quad \frac{\partial^2 \phi}{\partial x\, \partial y}, \quad \ldots$$

throughout a simply connected region D that contains the surface S and its boundary C in the interior of D. What constant value does

$$\oint_C \mathbf{F} \cdot d\mathbf{R}$$

have in such circumstances? Explain.

12. Let $\phi = (x^2 + y^2 + z^2)^{-1/2}$, let $\mathbf{F} = \operatorname{grad} \phi$, and let C be the circle $x^2 + y^2 = a^2$, $z = 0$ in the xy-plane. Show that

$$\oint_C \mathbf{F} \cdot d\mathbf{R} = 0$$

a) by direct evaluation of the integral, and

b) by applying Stokes's theorem with S the hemisphere

$$z = \sqrt{a^2 - x^2 - y^2}, \qquad x^2 + y^2 \leq a^2.$$

13. Suppose that the components of $\mathbf{F}$ are continuous and have continuous second order partial derivatives of all types. Use the divergence theorem and the fact that div (curl $\mathbf{F}$) = 0 (Problem 20, Article 17.6) to show that

$$\iint_S \operatorname{curl} \mathbf{F} \cdot \mathbf{n}\, d\sigma = 0$$

if S is a closed surface like a sphere, an ellipsoid, or a cube.

14. By Stokes's theorem, if S_1 and S_2 are two oriented surfaces having the same positively oriented curve C as boundary, then

$$\iint_{S_1} \operatorname{curl} \mathbf{F} \cdot \mathbf{n}_1\, d\sigma_1 = \oint_C \mathbf{F} \cdot d\mathbf{R} = \iint_{S_2} \operatorname{curl} \mathbf{F} \cdot \mathbf{n}_2\, d\sigma_2.$$

Deduce that

$$\iint_S \operatorname{curl} \mathbf{F} \cdot \mathbf{n}\, d\sigma$$

has the same value for all oriented surfaces S that span C and that induce the same positive direction on C.

15. Use Stokes's theorem to deduce that if curl $\mathbf{F} = \mathbf{0}$ throughout a simply connected region D, then

$$\int_{P_1}^{P_2} \mathbf{F} \cdot d\mathbf{R}$$

has the same value for all simple paths lying in D and joining P_1 and P_2. In other words, $\mathbf{F}$ is conservative.

***Toolkit* programs**

Vector Fields

REVIEW QUESTIONS AND EXERCISES

1. What is a vector field? Give an example of a two-dimensional vector field; of a three-dimensional field.

2. What is the velocity vector field for a fluid rotating about the x-axis if the angular velocity is a constant ω and the flow is counterclockwise as viewed by an observer at (1, 0, 0) looking toward (0, 0, 0)?

3. Give examples of gradient fields
   a) in the plane, b) in space.
   State a property that gradient fields have and other fields do not.

4. If $\mathbf{F} = \nabla f$ is a gradient field, S is a level surface for f, and C is a curve on S, why is it true (or not true) that $\int_C \mathbf{F} \cdot d\mathbf{R} = 0$?

5. Suppose that S is a portion of a level surface of a function $f(x, y, z)$. How could you select an orientation (if S is orientable) on S such that

$$\iint_S \nabla f \cdot \mathbf{n}\, d\sigma = \iint_S |\nabla f|\, d\sigma?$$

Why is the hypothesis that S is a level surface of f important?

6. If $f(x, y, z) = 2x - 3y + e^z$, and C is any smooth curve from $A(0, 0, 0)$ to $(1, 2, \ln 3)$, what is the value of $\int_C \nabla f \cdot d\mathbf{R}$? Why doesn't the answer depend on C?

7. Write a formula for a vector field $\mathbf{F}(x, y, z)$ such that $\mathbf{F} = f(\rho)\mathbf{R}$, where $\rho = |\mathbf{R}|$ and $\mathbf{R} = \mathbf{i}x + \mathbf{j}y + \mathbf{k}z$, if it is true that
   a) $\mathbf{F}$ is directed radially outward from the origin and $|\mathbf{F}| = \rho^{-n}$,
   b) $\mathbf{F}$ is directed toward the origin and $|\mathbf{F}| = 2$,
   c) $\mathbf{F}$ is a gravitational attraction field for a mass M at the origin in which an inverse-*cube* law applies. (Don't worry about the possible nonexistence of such a field.)

8. Give one physical and one geometrical interpretation for a surface integral

$$\iint_S h \, d\sigma.$$

9. State both the normal form and the tangential form of Green's theorem in the plane.

10. State the divergence theorem and show that it applies to the region D described by $1 \le |\mathbf{R}| \le 2$, assuming that S is the total boundary of this region, $\mathbf{n}$ is directed away from D at each point, and $\mathbf{F} = \nabla(1/|\mathbf{R}|)$, $\mathbf{R} = \mathbf{i}x + \mathbf{j}y + \mathbf{k}z$.

MISCELLANEOUS PROBLEMS

In Problems 1–4, describe the vector fields in words and with graphs.

1. $\mathbf{F} = x\mathbf{i} + y\mathbf{j} + z\mathbf{k}$
2. $\mathbf{F} = -x\mathbf{i} - y\mathbf{j} - z\mathbf{k}$
3. $\mathbf{F} = (x - y)\mathbf{i} + (x + y)\mathbf{j}$
4. $\mathbf{F} = (x\mathbf{i} + y\mathbf{j})/(x^2 + y^2)$

In Problems 5–8, evaluate the surface integrals

$$\iint_S h \, d\sigma$$

for the given functions and surfaces.

5. The surface S is the hemisphere $x^2 + y^2 + z^2 = a^2$, $z \ge 0$, and $h(x, y, z) = x + y$.

6. The surface S is the portion of the plane $z = x + y$ for which $x \ge 0$, $y \ge 0$, $z \le \pi$, and

$$h(x, y, z) = \sin z.$$

7. The surface S is $z = 4 - x^2 - y^2$, $z \ge 0$, and $h = z$.

8. The surface S is the sphere $\rho = a$, and $h = z^2$. (You might do the upper and lower hemispheres separately and add; or use spherical coordinates and not project the surface.)

For the functions and surfaces given in Problems 9–12, evaluate

$$\iint_S \frac{\partial f}{\partial n} \, d\sigma,$$

where $\partial f/\partial n$ is the directional derivative of f in the direction of the normal $\mathbf{n}$ in the sense specified.

9. The surface S is the sphere $x^2 + y^2 + z^2 = a^2$, $\mathbf{n}$ is directed outward on S, and $f = x^2 + y^2 + z^2$.

10. The surface and normal are as in Problem 9, and $f = (x^2 + y^2 + z^2)^{-1}$.

11. The surface S is the portion of the plane

$$z = 2x + 3y$$

for which $x \ge 0$, $y \ge 0$, $z \le 5$, $\mathbf{n}$ has a positive $\mathbf{k}$-component, and $f = x + y + z$.

12. The surface S is the one-eighth of the sphere $x^2 + y^2 + z^2 = a^2$ that lies in the first octant, $\mathbf{n}$ is directed inward with respect to the sphere, and

$$f = \ln (x^2 + y^2 + z^2)^{1/2}.$$

In Problems 13–20, evaluate the line integrals

$$\int_C \mathbf{F} \cdot d\mathbf{R}$$

for the given fields $\mathbf{F}$ and paths C.

13. The field $\mathbf{F} = x\mathbf{i} + y\mathbf{j}$ and the circle C:

$$x = \cos t, \qquad y = \sin t, \qquad 0 \le t \le 2\pi.$$

14. The field $\mathbf{F} = -y\mathbf{i} + x\mathbf{j}$ and C as in Problem 13.

15. The field $\mathbf{F} = (x - y)\mathbf{i} + (x + y)\mathbf{j}$ and the circle C in Problem 13.

16. The field $\mathbf{F} = x\mathbf{i} + y\mathbf{j} + z\mathbf{k}$ and the ellipse C, in which the plane $z = 2x + 3y$ cuts the cylinder $x^2 + y^2 = 12$, counterclockwise as viewed from the positive end of the z-axis looking toward the origin.

17. The field $\mathbf{F} = \nabla(xy^2z^3)$ and C as in Problem 16.

18. The field $\mathbf{F} = \nabla \times (x\mathbf{i} + y\mathbf{j} + z\mathbf{k})$ and C as in Problem 13.

19. The field $\mathbf{F}$ as in Problem 18 and C the line segment from the origin to the point $(1, 2, 3)$.

20. The field $\mathbf{F}$ as in Problem 17 and C the line segment from $(1, 1, 1)$ to $(2, 1, -1)$.

21. Heat flows from a hotter body to a cooler body. In three-dimensional heat flow, the fundamental equa-

tion for the rate at which heat flows out of D is

$$\iint_S K \frac{\partial u}{\partial n}\, d\sigma = \iiint_D c\delta \frac{\partial u}{\partial t}\, dV. \tag{1}$$

The symbolism in this equation is as follows:

$u = u(x, y, z, t)$	the temperature at the point (x, y, z) at time t,
K	the thermal conductivity coefficient,
δ	the mass density,
c	the specific heat coefficient. This is the amount of heat required to raise one unit of mass of the material of the body one degree,
S	the boundary surface of the region D,
$\dfrac{\partial u}{\partial n}$	the directional derivative in the direction of the outward normal to S.

How is $\partial u/\partial n$ related to the *gradient* of the temperature? In which direction (described in words) does ∇u point? Why does the left-hand side of Eq. (1) appear to make sense as a measure of the rate of flow? Now look at the right-hand side of Eq. (1): If ΔV is a small volume element in D, what does $\delta\,\Delta V$ represent? If the temperature of this element changes by an amount Δu in time Δt, what is

a) the amount, b) the average rate

of change of heat in the element? In words, what does the right-hand side of Eq. (1) represent physically? Is it reasonable to interpret Eq. (1) as saying that the rate at which heat flows out through the boundary of D is equal to the rate at which heat is being supplied from D?

22. Assuming Eq. (1), Problem 21, and assuming that there is no heat source or sink in D, derive the equation

$$\nabla \cdot (K\, \nabla u) = c\delta \frac{\partial u}{\partial t} \tag{2}$$

as the equation that must be satisfied at each point in D.

Suggestion. Apply the divergence theorem to the left-hand side of Eq. (1), and make D be a sphere of radius ε; then let $\varepsilon \to 0$.

23. Assuming the result of Problem 22, and assuming that K, c, and δ are constants, deduce that the condition for steady-state temperature in D is Laplace's equation

$$\nabla^2 u = 0, \quad \text{or} \quad \text{div (grad } u) = 0.$$

In higher mathematics, the symbol Δ is used for the *Laplace operator:*

$$\Delta u = \frac{\partial^2 u}{\partial x^2} + \frac{\partial^2 u}{\partial y^2} + \frac{\partial^2 u}{\partial z^2}.$$

Thus, in this notation,

$$\Delta u = \nabla^2 u = \nabla \cdot \nabla u = \text{div (grad } u).$$

Using the divergence theorem, and assuming that the functions u and v and their first and second order partial derivatives are continuous in the regions considered, verify the formulas in Problems 24–27. Assume that S is the boundary surface of the simply connected region D.

24. $\displaystyle\iint_S u\, \nabla v \cdot d\sigma = \iiint_D [u\, \Delta v + (\nabla u) \cdot (\nabla v)]\, dV$

25. $\displaystyle\iint_S \left(u \frac{\partial v}{\partial n} - v \frac{\partial u}{\partial n}\right) d\sigma = \iiint_D (u\, \Delta v - v\, \Delta u)\, dV$

Suggestion. Use the result of Problem 24 as given and in the form you get by interchanging u and v.

26. $\displaystyle\iint_S u \frac{\partial u}{\partial n}\, d\sigma = \iiint_D [u\, \Delta u + |\nabla u|^2]\, dV$

Suggestion. Use the result of Problem 24 with $v = u$.

27. $\displaystyle\iint_S \frac{\partial u}{\partial n}\, d\sigma = \iiint_D \Delta u\, dV$

Suggestion. Use $v = -1$ in the result of Problem 25.

28. A function u is *harmonic* in a region D if and only if it satisfies *Laplace's equation* $\Delta u = 0$ throughout D. Use the identity in Problem 26 to deduce that if u is harmonic in D and either $u = 0$ or $\partial u/\partial n = 0$ at all points on the surface S that is the boundary of D, then $\nabla u = \mathbf{0}$ throughout D, and therefore u is constant throughout D.

29. The result of Problem 28 can be used to establish the uniqueness of solutions of Laplace's equation in D, provided that either (i) the value of u is prescribed at each point on S, or (ii) the value of $\partial u/\partial n$ is prescribed at each point of S. This is done by supposing that u_1 and u_2 are harmonic in D and that both satisfy the same boundary conditions, and then letting $u = u_1 - u_2$. Complete this uniqueness proof.

30. Problems 21 through 23 deal with heat flow. Assume that K, c, and δ are constant and that the temperature $u = u(x, y, z)$ does not vary with time. Use the results of Problems 23 and 27 to conclude that the net rate of outflow of heat through the surface S is zero.

Note. This result might apply, for example, to the region D between two concentric spheres if the inner one were maintained at 100° and the outer one at 0°, so that heat would flow into D through the inner surface and out through the outer surface at the same rate.

31. Let $\rho = (x^2 + y^2 + z^2)^{1/2}$. Show that

$$u = C_1 + \frac{C_2}{\rho}$$

is harmonic where $\rho > 0$, if C_1 and C_2 are constants. Find values of C_1 and C_2 so that the following boundary conditions are satisfied:

a) $u = 100$ when $\rho = 1$, and $u = 0$ when $\rho = 2$;

b) $u = 100$ when $\rho = 1$, and $\partial u/\partial n = 0$ when $\rho = 2$.

Note. Part (a) refers to a steady-state heat flow problem that is like the one discussed at the end of Problem 30; part (b) refers to an insulated boundary on the sphere $\rho = 2$.

18 Differential Equations

18.1 Introduction

A differential equation is an equation that involves one or more derivatives, or differentials. Differential equations are classified by

a) type (namely, *ordinary* or *partial*),

b) order (that of the highest order derivative that occurs in the equation), and

c) degree (the exponent of the highest power of the highest order derivative, after the equation has been cleared of fractions and radicals in the dependent variable and its derivatives).

For example,

$$\left(\frac{d^3y}{dx^3}\right)^2 + \left(\frac{d^2y}{dx^2}\right)^5 + \frac{y}{x^2+1} = e^x \tag{1}$$

is an ordinary differential equation, of order three and degree two.

Only "ordinary" derivatives occur when the dependent variable y is a function of a single independent variable x. On the other hand, if the dependent variable y is a function of two or more independent variables, say

$$y = f(x, t),$$

where x and t are independent variables, then partial derivatives of y may occur. For example, the wave equation

$$\frac{\partial^2 y}{\partial t^2} = a^2 \frac{\partial^2 y}{\partial x^2} \tag{2}$$

(Article 15.7, Problems 32–38) is a partial differential equation, of order two and degree one. (A systematic treatment of partial differential equations lies beyond the scope of this book. For a discussion of partial differential equations, including the wave equation, and solutions of associated physical problems, see Kaplan, *Advanced Calculus,* Chapter 10.)

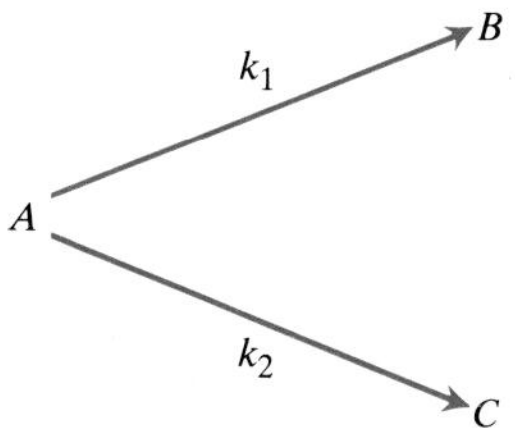

18.1 The reactant A yields products B and C at rates proportional to the amount of A present.

Many physical problems, when formulated in mathematical terms, lead to differential equations. In Article 13.2, for example, we discussed and solved the system of differential equations

$$m\frac{d^2x}{dt^2} = 0, \qquad m\frac{d^2y}{dt^2} = -mg \tag{3}$$

which described the motion of a projectile (neglecting air resistance). Indeed, one of the chief sources of differential equations in mechanics is Newton's second law:

$$\mathbf{F} = \frac{d}{dt}(m\mathbf{v}), \tag{4}$$

where $\mathbf{F}$ is the resultant of the forces acting on a body of mass m and $\mathbf{v}$ is its velocity.

An example from the field of chemical kinetics may be represented by a reactant A that undergoes parallel transformations into products B and C at rates that are proportional to the amount of A that is present at time t. (See Fig. 18.1.) If x, y, and z denote the amounts of A, B, and C present at time t, then the differential equations that describe the process are

$$\frac{dx}{dt} = -(k_1 + k_2)x, \qquad \frac{dy}{dt} = k_1x, \qquad \frac{dz}{dt} = k_2x. \tag{5}$$

If, at time $t = 0$, only A is present, the initial conditions are

$$x(0) = x_0, \qquad y(0) = 0, \qquad z(0) = 0. \tag{6}$$

Equations (5) and (6) together provide an example of an *initial value problem*. The first of Eqs. (5) can easily be solved for x to yield

$$x = x_0e^{-(k_1+k_2)t}. \tag{7a}$$

When this is substituted into the remaining Eqs. (5), they can be integrated also, to give

$$y = \frac{k_1}{k_1 + k_2}(1 - e^{-(k_1+k_2)t})\,x_0, \tag{7b}$$

$$z = \frac{k_2}{k_1 + k_2}(1 - e^{-(k_1+k_2)t})\,x_0. \tag{7c}$$

Together, Eqs. (7*a*, *b*, *c*) give a solution of the initial value problem of Eqs. (5) and (6), as can be verified by direct substitition.

Differential equations enter naturally as models for many other phenomena in physics, chemistry, biology, economics, and engineering. Many of these phenomena are complex and require more detailed knowledge of the specific subjects than we can assume here. The models that we shall study will be simple, including examples of exponential growth and decay, simple electrical circuits, vibrations, and motion.

present at that time. This leads to the equation

$$\frac{dy}{dt} = ky, \tag{1}$$

where k is a constant that is positive if the population is increasing and negative if it is decreasing. To solve Eq. (1), we separate variables and integrate, to obtain

$$\int \frac{dy}{y} = \int k\,dt,$$

or

$$\ln y = kt + C_1$$

(remember that y is positive). It follows that

$$y = e^{kt+C_1} = Ce^{kt},$$

where $C = e^{C_1}$. If y_0 denotes the population when $t = 0$, then $C = y_0$ and

$$y = y_0e^{kt}. \tag{2}$$

This equation is called the *law of exponential growth.*

PROBLEMS

Separate the variables and solve the following differential equations.

1. $x(2y - 3)\,dx + (x^2 + 1)\,dy = 0$
2. $x^2(y^2 + 1)\,dx + y\sqrt{x^3 + 1}\,dy = 0$
3. $\dfrac{dy}{dx} = e^{x-y}$
4. $\sqrt{2xy}\,\dfrac{dy}{dx} = 1$
5. $\sin x\,\dfrac{dx}{dy} + \cosh 2y = 0$
6. $\ln x\,\dfrac{dx}{dy} = \dfrac{x}{y}$
7. $xe^y\,dy + \dfrac{x^2 + 1}{y}\,dx = 0$
8. $y\sqrt{2x^2 + 3}\,dy + x\sqrt{4 - y^2}\,dx = 0$
9. $\sqrt{1 + x^2}\,dy + \sqrt{y^2 - 1}\,dx = 0$
10. $x^2y\dfrac{dy}{dx} = (1 + x)\csc y$
11. *Continuous compounding.* An investor has \$1000 with which to open an account and plans to add \$1000 per year. All funds in the account will earn 10% interest per year, continuously compounded. Assuming that the added deposits are also made continuously, show that $dx/dt = 1000 + 0.10x$, and $x(0) = 1000$, where x is the number of dollars in the account at time t. How many years would it take for the account to reach \$100,000? For more on compound interest, see Article 6.10.
12. *The snow plow problem.* One morning in January it began to snow and kept on snowing at a constant rate. A snow plow began to plow at noon. It plowed clean and at a constant rate (volume per unit time). From one o'clock until two o'clock it went only half as far as it had gone between noon and one o'clock. When did the snow begin falling?
13. *Newton's law of cooling* assumes that the temperature T of a small hot object placed in a medium of temperature T_a decreases at a rate proportional to $T - T_a$. If an object cools from 100 °C to 80 °C in twenty minutes when the surrounding temperature is 20 °C, how long does it take to cool from 100 °C to 60 °C?

(For additional problems, see Article 6.9.)

18.4
First Order: Homogeneous

Occasionally a differential equation whose variables cannot be separated can be transformed by a change of variable into an equation whose variables can be separated. This is the case with any equation that can be put into the form

$$\frac{dy}{dx} = F\left(\frac{y}{x}\right). \tag{1}$$

Such an equation is called *homogeneous*.

To transform Eq. (1) into an equation whose variables may be separated, we introduce the new variable

$$v = \frac{y}{x}. \tag{2}$$

Then

$$y = vx, \qquad \frac{dy}{dx} = v + x\frac{dv}{dx},$$

and (1) becomes

$$v + x\frac{dv}{dx} = F(v). \tag{3}$$

Equation (3) can be solved by separation of variables:

$$\frac{dx}{x} + \frac{dv}{v - F(v)} = 0. \tag{4}$$

After (4) is solved, the solution of the original equation is obtained when we replace v by y/x.

EXAMPLE Show that the equation

$$(x^2 + y^2)\,dx + 2xy\,dy = 0$$

is homogeneous, and solve it.

Solution From the given equation, we have

$$\frac{dy}{dx} = -\frac{x^2 + y^2}{2xy} = -\frac{1 + (y/x)^2}{2(y/x)}.$$

This has the form of Eq. (1), with

$$F(v) = -\frac{1 + v^2}{2v}, \qquad \text{where} \quad v = \frac{y}{x}.$$

Then Eq. (4) becomes

$$\frac{dx}{x} + \frac{dv}{v + \dfrac{1 + v^2}{2v}} = 0, \qquad \text{or} \qquad \frac{dx}{x} + \frac{2v\,dv}{1 + 3v^2} = 0.$$

The solution of this is

$$\ln|x| + \frac{1}{3}\ln(1 + 3v^2) = \frac{1}{3}\ln C,$$

so that

$$x^3(1 + 3v^2) = \pm C.$$

In terms of y and x, the solution is

$$x(x^2 + 3y^2) = C. \ \square$$

PROBLEMS

Show that the following equations are homogeneous, and solve.

1. $(x^2 + y^2)\,dx + xy\,dy = 0$
2. $x^2\,dy + (y^2 - xy)\,dx = 0$
3. $(xe^{y/x} + y)dx - x\,dy = 0$
4. $(x + y)\,dy + (x - y)\,dx = 0$
5. $y' = \frac{y}{x} + \cos\frac{y - x}{x}$
6. $\left(x \sin\frac{y}{x} - y\cos\frac{y}{x}\right) dx + x\cos\frac{y}{x}\,dy = 0$
7. Solve the equation

$$(x + y + 1)\,dx + (y - x - 3)\,dy = 0$$

by making a change of variable of the form

$$x = r + a, \qquad y = s + b,$$

and choosing the constants a and b so that the resulting equation is

$$(r + s)\,dr + (r - s)\,ds = 0.$$

Then solve this equation and express its solution in terms of x and y.

8. Use the substitution $u = x + y$ to solve the equation $y' = (x + y)^2$.

Orthogonal trajectories. If every member of a family of curves is a solution of the differential equation

$$M(x, y)\,dx + N(x, y)\,dy = 0,$$

while every member of a second family of curves is a solution of the related equation

$$N(x, y)\,dx - M(x, y)\,dy = 0,$$

then each curve of the one family is orthogonal to every curve of the other family. Each family is said to be a family of *orthogonal trajectories* of the other. In Problems 9 and 10, find the family of solutions of the given differential equation and the family of orthogonal trajectories. Sketch both families.

9. $x\,dy - 2y\,dx = 0$
10. $2xy\,dy + (x^2 - y^2)\,dx = 0$
11. Find the orthogonal trajectories of the family of curves $xy = c$.

18.5
First Order: Linear

The complexity of a differential equation depends primarily upon the way in which the *dependent* variable and its derivatives occur. Of particular importance are those equations that are linear. In a linear differential equation, each term of the equation is of degree one or zero, where, in computing the degree of a term, we add the exponents of the dependent variable and of any of its derivatives that occur. Thus, for example, (d^2y/dx^2) is of the first degree, while $y(dy/dx)$ is of the second degree because we must add 1 for the exponent of y, and 1 for the exponent of dy/dx.

A differential equation of first order, which is also linear, can always be put into the standard form

$$\frac{dy}{dx} + Py = Q, \tag{1}$$

where P and Q are functions of x.

One method for solving Eq. (1) is to find a function $\rho = \rho(x)$ such that if the equation is multiplied by ρ, the left side becomes the derivative of the product ρy. That is, we multiply (1) by ρ,

$$\rho\frac{dy}{dx} + \rho P y = \rho Q, \tag{1'}$$

and then try to impose upon ρ the condition that

$$\rho \frac{dy}{dx} + \rho P y = \frac{d}{dx}(\rho y). \tag{2}$$

When we expand the right-hand side of (2) and cancel terms, we obtain, as the condition to be satisfied by ρ,

$$\frac{d\rho}{dx} = \rho P. \tag{3}$$

In Eq. (3), $P = P(x)$ is a known function, so we can separate the variables and solve for ρ:

$$\frac{d\rho}{\rho} = P\,dx, \qquad \ln|\rho| = \int P\,dx + \ln C,$$

$$\rho = \pm C e^{\int P\,dx}. \tag{4}$$

Since we do not require the most general function ρ, we may take $\pm C = 1$ in (4) and use

$$\boxed{\rho = e^{\int P\,dx}.} \tag{5}$$

This function is called an *integrating factor* for Eq. (1). With its help, (1′) becomes

$$\frac{d}{dx}(\rho y) = \rho Q,$$

whose solution is

$$\boxed{\rho y = \int \rho Q\,dx + C.} \tag{6}$$

Since ρ is given by (5), while P and Q are known from the given differential equation (1), we have, in Eqs. (5) and (6), a summary of all that is required to solve (1).

EXAMPLE 1 Solve the equation

$$\frac{dy}{dx} + y = e^x.$$

Solution

$$P = 1, \qquad Q = e^x,$$

$$\rho = e^{\int dx} = e^x,$$

$$e^x y = \int e^{2x}\,dx + C = \frac{1}{2}e^{2x} + C,$$

$$y = \frac{1}{2}e^x + Ce^{-x}. \ \square$$

EXAMPLE 2 Solve the equation

$$x\frac{dy}{dx} - 3y = x^2.$$

Solution We put the equation in standard form,

$$\frac{dy}{dx} - \frac{3}{x}y = x,$$

and then read off

$$P = -\frac{3}{x}, \qquad Q = x.$$

Hence

$$\rho = e^{\int -(3/x)\,dx} = e^{-3\ln x} = \frac{1}{e^{3\ln x}} = \frac{1}{x^3},$$

and

$$\frac{1}{x^3}y = \int \frac{x}{x^3}\,dx + C = -\frac{1}{x} + C,$$

$$y = -x^2 + Cx^3. \ \square$$

Note that whenever $\int P\,dx$ involves logarithms, as in Example 2, it is profitable to simplify the expression for $e^{\int P\,dx}$ before substituting into Eq. (6). To simplify the expression, we use the properties of the logarithmic and exponential functions:

$$e^{\ln A} = A, \qquad e^{m\ln A} = A^m, \qquad e^{n+m\ln A} = A^m e^n.$$

A differential equation that is linear in y and dy/dx may also be separable, or homogeneous. In such cases, we have a choice of methods of solution. Observe also that an equation that is linear in x and dx/dy can be solved by the technique of this article; one need only interchange the roles of x and y in Eqs. (1), (5), and (6).

PROBLEMS

Solve Problems 1–10.

1. $\dfrac{dy}{dx} + 2y = e^{-x}$
2. $2\dfrac{dy}{dx} - y = e^{x/2}$
3. $x\dfrac{dy}{dx} + 3y = \dfrac{\sin x}{x^2}$
4. $x\,dy + y\,dx = \sin x\,dx$
5. $x\,dy + y\,dx = y\,dy$
6. $(x-1)^3\dfrac{dy}{dx} + 4(x-1)^2y = x + 1$
7. $\cosh x\,dy + (y\sinh x + e^x)\,dx = 0$
8. $e^{2y}\,dx + 2(xe^{2y} - y)\,dy = 0$
9. $(x - 2y)\,dy + y\,dx = 0$
10. $(y^2 + 1)\,dx + (2xy + 1)\,dy = 0$
11. *Blood sugar.* If glucose is fed intravenously at a constant rate, the change in the overall concentration $c(t)$ of glucose in the blood with respect to time may be described by the differential equation

$$\frac{dc}{dt} = \frac{G}{100V} - kc.$$

In this equation, G, V, and k are positive constants, G being the rate at which glucose is admitted, in milligrams per minute, and V the volume of blood in the body, in liters (around 5 liters for an adult). The concentration $c(t)$ is measured in milligrams per centiliter. The term $-kc$ is included because the glucose is assumed to be changing continually into other molecules at a rate proportional to its concentration.

a) Solve the equation for $c(t)$, using c_0 to denote $c(0)$.
b) Find the steady state concentration, $\lim_{t\to\infty} c(t)$.

12. *An investment program.* An investor has a salary of \$20,000 per year with expected increases of \$1000 per year. If an initial deposit of \$1000 is invested in a program that pays 8% per annum, compounded continuously, and additional deposits are continuously added at the rate of 5% of salary, a model for the investment x at the end of t years is of the form $dx/dt = a + bx + ct$, where a, b, and c are positive constants, and $x(0) = 1000$:

$$\frac{dx}{dt} = 0.08x + 0.05(20{,}000 + 1000t)$$

$$= 1000 + 0.08x + 50t.$$

What is x at time t? (If you have a programmable calculator you can easily compute $x(5)$, $x(10)$, $x(20)$, and $x(40)$ to see how the investment grows.)

18.6 First Order: Exact

An equation that can be written in the form

$$M(x, y)\,dx + N(x, y)\,dy = 0, \tag{1}$$

and having the property that

$$\frac{\partial M}{\partial y} = \frac{\partial N}{\partial x}, \tag{2}$$

is said to be *exact*, because its left-hand side is an exact differential. The technique of solving an exact equation consists in finding a function $f(x, y)$ such that

$$df = M\,dx + N\,dy. \tag{3}$$

Then (1) becomes

$$df = 0$$

and the solution is

$$f(x, y) = C,$$

where C is an arbitrary constant. The method of finding $f(x, y)$ to satisfy (3) is discussed and illustrated in Article 15.11.

It can be proved that every first order differential equation

$$P(x, y)\,dx + Q(x, y)\,dy = 0$$

can be made exact by multiplication by a suitable *integrating factor* $\rho(x, y)$. Such an integrating factor has the property that

$$\frac{\partial}{\partial y}[\rho(x, y)P(x, y)] = \frac{\partial}{\partial x}[\rho(x, y)Q(x, y)].$$

Unfortunately, it is not easy to determine ρ from this equation. In fact, there is no general technique by which even a single integrating factor can be produced for an arbitrary differential equation, and the search for one can be a frustrating experience. However, one can often recognize certain combinations of differentials that can be made exact by "ingenious devices."

EXAMPLE Solve the equation

$$x\,dy - y\,dx = xy^2\,dx.$$

Solution The combination $x\,dy - y\,dx$ may ring a bell in our memories and cause us to recall the formula

$$d\left(\frac{u}{v}\right) = \frac{v\,du - u\,dv}{v^2} = -\left[\frac{u\,dv - v\,du}{v^2}\right].$$

Therefore we might divide the given equation by x^2, or change signs and divide by y^2. Clearly, the latter approach will be more profitable, so we proceed as follows:

$$x\,dy - y\,dx = xy^2\,dx, \qquad \frac{y\,dx - x\,dy}{y^2} = -x\,dx,$$

$$d\left(\frac{x}{y}\right) + x\,dx = 0, \qquad \frac{x}{y} + \frac{x^2}{2} = C.$$

The same result would be obtained if we wrote our equation in the form

$$(xy^2 + y)\,dx - x\,dy = 0$$

and multiplied by the integrating factor $1/y^2$. This would give

$$\left(x + \frac{1}{y}\right)dx - \left(\frac{x}{y^2}\right)dy = 0,$$

which is exact, since

$$\frac{\partial}{\partial y}\left(x + \frac{1}{y}\right) = \frac{\partial}{\partial x}\left(-\frac{x}{y^2}\right). \ \square$$

PROBLEMS

In Problems 1–3, use the given integrating factors to make differential equations exact. Then solve the equations.

1. $(x + 2y)\,dx - x\,dy = 0$, $1/x^3$
2. $y\,dx + x\,dy = 0$,

   a) $\dfrac{1}{xy}$ b) $\dfrac{1}{(xy)^2}$
3. $y\,dx - x\,dy = 0$,

   a) $\dfrac{1}{y^2}$ b) $\dfrac{1}{x^2}$

   c) $\dfrac{1}{xy}$ d) $\dfrac{1}{x^2 + y^2}$

Solve Problems 4–13.

4. $(x + y)\,dx + (x + y^2)\,dy = 0$
5. $(2xe^y + e^x)\,dx + (x^2 + 1)e^y\,dy = 0$
6. $(2xy + y^2)\,dx + (x^2 + 2xy - y)\,dy = 0$
7. $(x + \sqrt{y^2 + 1})\,dx - \left(y - \dfrac{xy}{\sqrt{y^2 + 1}}\right)dy = 0$
8. $x\,dy - y\,dx + x^3\,dx = 0$
9. $x\,dy - y\,dx = (x^2 + y^2)\,dx$
10. $(x^2 + x - y)\,dx + x\,dy = 0$
11. $\left(e^x + \ln y + \dfrac{y}{x}\right)dx + \left(\dfrac{x}{y} + \ln x + \sin y\right)dy = 0$
12. $\left(\dfrac{y^2}{1 + x^2} - 2y\right)dx$

    $+ (2y\tan^{-1}x - 2x + \sinh y)\,dy = 0$
13. $dy + \dfrac{y - \sin x}{x}\,dx = 0$
14. If a, b, and c are constants and

    $$(ax^2 + by^2)\,dx + cxy\,dy = 0$$

    is an exact equation, what relation(s) among the constants must hold? Solve the equation subject to such conditions.
15. If a, b, and c are constants and

    $$(ax^2 + by^2)\,dx + cxy\,dy = 0$$

    is not exact, but has the integrating factor $1/x^2$, what relation(s) among the constants must hold? Solve the equation subject to such conditions.

18.7 Special Types of Second Order Equations

Certain types of second order differential equations, of which the general form is

$$F\left(x, y, \frac{dy}{dx}, \frac{d^2y}{dx^2}\right) = 0, \tag{1}$$

can be reduced to first order equations by a suitable change of variables.

Type 1. *Equations with dependent variable missing.* When Eq. (1) has the special form

$$F\left(x, \frac{dy}{dx}, \frac{d^2y}{dx^2}\right) = 0, \tag{2}$$

we can reduce it to a first order equation by substituting

$$p = \frac{dy}{dx}, \qquad \frac{d^2y}{dx^2} = \frac{dp}{dx}.$$

Then Eq. (2) takes the form

$$F\left(x, p, \frac{dp}{dx}\right) = 0,$$

which is of the first order in p. If this can be solved for p as a function of x, say

$$p = \phi(x, C_1),$$

then we can find y by one additional integration:

$$y = \int (dy/dx)\, dx = \int p\, dx = \int \phi(x, C_1)\, dx + C_2.$$

The differential equation

$$\frac{d^2y}{dx^2} = \frac{w}{H}\sqrt{1 + \left(\frac{dy}{dx}\right)^2}$$

was solved by this technique in Article 9.5.

Type 2. *Equations with independent variable missing.* When Eq. (1) does not contain x explicitly but has the form

$$F\left(y, \frac{dy}{dx}, \frac{d^2y}{dx^2}\right) = 0, \tag{3}$$

the substitutions to use are

$$p = \frac{dy}{dx} \quad \text{and} \quad \frac{d^2y}{dx^2} = p\frac{dp}{dy}.$$

Then Eq. (3) takes the form

$$F\left(y, p, p\frac{dp}{dy}\right) = 0,$$

which is of the first order in p. Its solution gives p in terms of y, and then a further integration gives the solution of Eq. (3).

EXAMPLE Solve the equation

$$\frac{d^2y}{dx^2} + y = 0.$$

Solution Let

$$\frac{dy}{dx} = p, \qquad \frac{d^2y}{dx^2} = \frac{dp}{dx} = \frac{dp}{dy}\frac{dy}{dx} = \frac{dp}{dy}p.$$

Then we proceed as follows:

$$p\frac{dp}{dy} + y = 0, \qquad p\,dp + y\,dy = 0,$$

$$\frac{p^2}{2} + \frac{y^2}{2} = \frac{C_1^2}{2}, \qquad p = \frac{dy}{dx} = \pm\sqrt{C_1^2 - y^2},$$

$$\frac{dy}{\sqrt{C_1^2 - y^2}} = \pm dx, \qquad \sin^{-1}\frac{y}{C_1} = \pm(x + C_2),$$

$$y = C_1 \sin[\pm(x + C_2)] = \pm C_1 \sin(x + C_2).$$

Since C_1 is arbitrary, there is no need for the $\pm$ sign, and we have

$$y = C_1 \sin(x + C_2)$$

as the general solution. □

PROBLEMS

Solve Problems 1–8.

1. $\dfrac{d^2y}{dx^2} + \dfrac{dy}{dx} = 0$

2. $\dfrac{d^2y}{dx^2} + y\dfrac{dy}{dx} = 0$

3. $\dfrac{d^2y}{dx^2} + x\dfrac{dy}{dx} = 0$

4. $x\dfrac{d^2y}{dx^2} + \dfrac{dy}{dx} = 0$

5. $\dfrac{d^2y}{dx^2} - y = 0$

6. $\dfrac{d^2y}{dx^2} + \omega^2 y = 0 \quad (\omega = \text{constant} \neq 0)$

7. $xy''' - 2y'' = 0$. (*Hint:* Substitute $y'' = q$.)

8. $2y'' - (y')^2 + 1 = 0$

9. *Hooke's law.* A mass m is suspended from one end of a vertical spring whose other end is attached to a rigid support. The body is allowed to come to rest and is then pulled down an additional slight amount A and released. Find its motion. (*Hint:* Assume Newton's second law of motion and Hooke's law, which says that the tension in the spring is proportional to the amount it is stretched. Let x denote the displacement of the body at time t, measured from the equilibrium position. Then $m(d^2x/dt^2) = -kx$, where k, the "spring constant," is the proportionality factor in Hooke's law.)

10. A jeep suspended from a parachute falls through space under the pull of gravity. If air resistance produces a retarding force proportional to the jeep's velocity and the jeep starts from rest at time $t = 0$, find the distance it falls in time t.

18.8 Linear Equations with Constant Coefficients

An equation of the form

$$\frac{d^n y}{dx^n} + a_1\frac{d^{n-1}y}{dx^{n-1}} + a_2\frac{d^{n-2}y}{dx^{n-2}} + \cdots + a_{n-1}\frac{dy}{dx} + a_n y = F(x), \tag{1}$$

which is linear in y and its derivatives, is called a *linear* equation of order n. If $F(x)$ is identically zero, the equation is said to be *homogeneous;* otherwise it is called *nonhomogeneous.* The equation is called linear even when the coefficients $a_1, a_2, \ldots, a_n$ are functions of x. However, we shall consider only the case where these coefficients are *constants.*

It is convenient to introduce the symbol D to represent the operation of differentiation with respect to x. That is, we write $Df(x)$ to mean $(d/dx)f(x)$. Furthermore, we define powers of D to mean taking successive derivatives:

$$D^2f(x) = D\{Df(x)\} = \frac{d^2f(x)}{dx^2},$$

$$D^3f(x) = D\{D^2f(x)\} = \frac{d^3f(x)}{dx^3},$$

and so on. A polynomial in D is to be interpreted as an operator that, when applied to $f(x)$, produces a linear combination of f and its successive derivatives. For example,

$$(D^2 + D - 2)f(x) = D^2f(x) + Df(x) - 2f(x)$$
$$= \frac{d^2f(x)}{dx^2} + \frac{df(x)}{dx} - 2f(x).$$

Such a polynomial in D is called a *linear differential operator* and may be denoted by the single letter L. If L_1 and L_2 are two such linear operators, their sum and product are defined by the equations

$$(L_1 + L_2)f(x) = L_1f(x) + L_2f(x),$$
$$L_1L_2f(x) = L_1(L_2f(x)).$$

Linear differential operators that are polynomials in D with constant coefficients satisfy basic algebraic laws that make it possible to treat them like ordinary polynomials so far as addition, multiplication, and factoring are concerned. Thus,

$$(D^2 + D - 2)f(x) = (D + 2)(D - 1)f(x)$$
$$= (D - 1)(D + 2)f(x). \tag{2}$$

Since Eq. (2) holds for any twice-differentiable function f, we also write the equality between operators:

$$D^2 + D - 2 = (D + 2)(D - 1) = (D - 1)(D + 2). \tag{3}$$

18.9 Linear Second Order Homogeneous Equations with Constant Coefficients

Suppose, now, we wish to solve a differential equation of order two, say

$$\frac{d^2y}{dx^2} + 2a\frac{dy}{dx} + by = 0, \tag{1}$$

where a and b are constants. In operator notation, this becomes

$$(D^2 + 2aD + b)y = 0. \tag{1'}$$

Associated with this differential equation is the algebraic equation

$$r^2 + 2ar + b = 0, \tag{1''}$$

which we get by replacing D by r and suppressing y. This is called the *characteristic equation* of the differential equation. Suppose the roots of (1″) are r_1 and r_2. Then

$$r^2 + 2ar + b = (r - r_1)(r - r_2)$$

and

$$D^2 + 2aD + b = (D - r_1)(D - r_2).$$

Hence Eq. (1′) is equivalent to

$$(D - r_1)(D - r_2)y = 0. \tag{2}$$

If we now let

$$(D - r_2)y = u \tag{3a}$$

and

$$(D - r_1)u = 0, \tag{3b}$$

we can solve Eq. (1′) in two steps. From Eq. (3b), which is separable, we find

$$u = C_1e^{r_1x}.$$

We substitute this into (3a), which becomes

$$(D - r_2)y = C_1e^{r_1x}$$

or

$$\frac{dy}{dx} - r_2y = C_1e^{r_1x}.$$

This equation is linear. Its integrating factor is

$$\rho = e^{-r_2x},$$

(see Article 18.5), and its solution is

$$e^{-r_2x}y = C_1 \int e^{(r_1-r_2)x}\,dx + C_2. \tag{4}$$

How we proceed at this point depends on whether r_1 and r_2 are equal.

CASE 1. If $r_1 \neq r_2$, the evaluation of the integral in Eq. (4) leads to

$$e^{-r_2x}y = \frac{C_1}{r_1 - r_2}e^{(r_1-r_2)x} + C_2$$

or

$$y = \frac{C_1}{r_1 - r_2}e^{r_1x} + C_2e^{r_2x}.$$

Since C_1 is an arbitrary constant, so is $C_1/(r_1 - r_2)$, and the solution of Eq. (2) can be written simply as

$$y = C_1e^{r_1x} + C_2e^{r_2x}, \qquad \text{if} \qquad r_1 \neq r_2. \tag{5}$$

CASE 2. If $r_1 = r_2$, then $e^{(r_1 - r_2)x} = e^0 = 1$, and Eq. (4) reduces to

$$e^{-r_2 x} y = C_1 x + C_2$$

or

$$y = (C_1 x + C_2) e^{r_2 x}, \qquad \text{if} \qquad r_1 = r_2. \tag{6}$$

EXAMPLE 1 Solve the equation

$$\frac{d^2y}{dx^2} + \frac{dy}{dx} - 2y = 0.$$

Solution $r^2 + r - 2 = 0$ has roots $r_1 = 1$, $r_2 = -2$. Hence, by Eq. (5), the solution of the differential equation is

$$y = C_1 e^x + C_2 e^{-2x}. \ \square$$

EXAMPLE 2 Solve the equation

$$\frac{d^2y}{dx^2} + 4\frac{dy}{dx} + 4y = 0.$$

Solution

$$\begin{aligned} r^2 + 4r + 4 &= (r + 2)^2, \\ r_1 = r_2 &= -2, \\ y &= (C_1 x + C_2) e^{-2x}. \ \square \end{aligned}$$

Imaginary Roots

If the coefficients a and b in Eq. (1) are real, the roots of the characteristic Eq. (1″) either will be real, or will be a pair of complex conjugate numbers:

$$r_1 = \alpha + i\beta, \qquad r_2 = \alpha - i\beta. \tag{7}$$

If $\beta \neq 0$, Eq. (5) applies, with the result

$$\begin{aligned} y &= c_1 e^{(\alpha + i\beta)x} + c_2 e^{(\alpha - i\beta)x} \\ &= e^{\alpha x}[c_1 e^{i\beta x} + c_2 e^{-i\beta x}]. \end{aligned} \tag{8}$$

By Euler's formula,

$$\begin{aligned} e^{i\beta x} &= \cos \beta x + i \sin \beta x, \\ e^{-i\beta x} &= \cos \beta x - i \sin \beta x. \end{aligned}$$

Hence, Eq. (8) may be replaced by

$$y = e^{\alpha x}[(c_1 + c_2) \cos \beta x + i(c_1 - c_2) \sin \beta x]. \tag{9}$$

Finally, if we introduce new arbitrary constants

$$\begin{aligned} C_1 &= c_1 + c_2, \\ C_2 &= i(c_1 - c_2), \end{aligned}$$

Eq. (9) takes the form

$$y = e^{\alpha x}[C_1 \cos \beta x + C_2 \sin \beta x], \quad \text{if} \quad r_1 = \alpha + i\beta, \quad r_2 = \alpha - i\beta. \tag{9'}$$

The arbitrary constants C_1 and C_2 in (9′) will be real provided the constants c_1 and c_2 in (9) are complex conjugates:

$$c_1 = \frac{1}{2}(C_1 - iC_2), \qquad c_2 = \frac{1}{2}(C_1 + iC_2).$$

To solve a problem where the roots of the characteristic equation are complex conjugates, we simply write down the appropriate version of Eq. (9′).

EXAMPLE 3 Solve the equation

$$\frac{d^2y}{dx^2} + 2\frac{dy}{dx} + 2y = 0.$$

Solution The characteristic equation $r^2 + 2r + 2 = 0$ has roots $r_1 = -1 + i$, $r_2 = -1 - i$. Hence, in Eq. (9′), we take

$$\alpha = -1, \qquad \beta = 1,$$

and obtain the solution

$$y = e^{-x}[C_1 \cos x + C_2 \sin x]. \ \square$$

EXAMPLE 4 Solve the equation

$$\frac{d^2y}{dx^2} + \omega^2 y = 0, \qquad \omega \neq 0.$$

Solution The characteristic equation $r^2 + \omega^2 = 0$ has roots $r_1 = i\omega$, $r_2 = -i\omega$. Hence we take $\alpha = 0$, $\beta = \omega$ in Eq. (9′), and obtain the solution

$$y = C_1 \cos \omega x + C_2 \sin \omega x. \ \square$$

PROBLEMS

Solve the following equations.

1. $\frac{d^2y}{dx^2} + 2\frac{dy}{dx} = 0$
2. $\frac{d^2y}{dx^2} + 5\frac{dy}{dx} + 6y = 0$
3. $\frac{d^2y}{dx^2} + 6\frac{dy}{dx} + 5y = 0$
4. $\frac{d^2y}{dx^2} - 2\frac{dy}{dx} - 3y = 0$
5. $\frac{d^2y}{dx^2} + \frac{dy}{dx} + y = 0$
6. $\frac{d^2y}{dx^2} - 4\frac{dy}{dx} + 4y = 0$
7. $\frac{d^2y}{dx^2} + 6\frac{dy}{dx} + 9y = 0$
8. $\frac{d^2y}{dx^2} - 6\frac{dy}{dx} + 10y = 0$
9. $\frac{d^2y}{dx^2} - 2\frac{dy}{dx} + 4y = 0$
10. $\frac{d^2y}{dx^2} - 10\frac{dy}{dx} + 16y = 0$

18.10

Linear Second Order Nonhomogeneous Equations with Constant Coefficients

In Article 18.9, we learned how to solve the homogeneous equation

$$\frac{d^2y}{dx^2} + 2a\frac{dy}{dx} + by = 0. \tag{1}$$

We can now describe a method for solving the nonhomogeneous equation

$$\frac{d^2y}{dx^2} + 2a\frac{dy}{dx} + by = F(x). \tag{2}$$

To solve Eq. (2), we first obtain the general solution of the related homogeneous Eq. (1) obtained by replacing $F(x)$ by zero. Denote this solution by

$$y_h = C_1u_1(x) + C_2u_2(x), \tag{3}$$

where C_1 and C_2 are arbitrary constants and $u_1(x)$, $u_2(x)$ are functions of one or more of the following forms:

$$e^{rx}, \qquad xe^{rx}, \qquad e^{\alpha x}\cos\beta x, \qquad e^{\alpha x}\sin\beta x.$$

Now we might, by inspection or by inspired guesswork, be able to discover *one* particular function $y = y_p(x)$ that satisfies Eq. (2). In this case, we would be able to solve Eq. (2) completely, as

$$\boxed{y = y_h(x) + y_p(x).}$$

REMARK. If $y_1(x)$ and $y_2(x)$ are any solutions of the same nonhomogeneous Eq. (2), then their difference $y(x) = y_1(x) - y_2(x)$ satisfies the homogeneous Eq. (1). It is for this reason that every solution of Eq. (2) is included if we add the general solution of the homogeneous equation to any particular solution of the nonhomogeneous equation. In this regard, the general homogeneous solution is analogous to the constant of integration that we add to a particular solution of $y' = f(x)$ to get the general solution.

EXAMPLE 1 Solve the equation

$$\frac{d^2y}{dx^2} + 2\frac{dy}{dx} - 3y = 6.$$

Solution y_h satisfies

$$\frac{d^2y_h}{dx^2} + 2\frac{dy_h}{dx} - 3y_h = 0.$$

The characteristic equation is

$$r^2 + 2r - 3 = 0,$$

and its roots are

$$r_1 = -3, \qquad r_2 = 1.$$

Hence

$$y_h = C_1e^{-3x} + C_2e^{x}.$$

Now, to find a particular solution of the original equation, observe that $y =$ constant would do, provided $-3y = 6$. Hence,

$$y_p = -2$$

is one particular solution. The complete solution is

$$y = y_p + y_h = -2 + C_1e^{-3x} + C_2e^{x}. \ \square$$

Variation of Parameters

Fortunately, there is a general method for finding the solution of the nonhomogeneous Eq. (2) once the general solution of the corresponding homogeneous equation is known. The method is known as the method of *variation of parameters*. It consists in replacing the constants C_1 and C_2 in Eq. (3) by functions $v_1 = v_1(x)$ and $v_2 = v_2(x)$, and requiring (in a way to be explained) that the resulting expression satisfy Eq. (2). There are two functions to be determined, and requiring that Eq. (2) be satisfied is only one condition. As a second condition, we also require that

$$v_1'u_1 + v_2'u_2 = 0. \tag{4}$$

Then we have

$$y = v_1u_1 + v_2u_2,$$

$$\frac{dy}{dx} = v_1u_1' + v_2u_2',$$

$$\frac{d^2y}{dx^2} = v_1u_1'' + v_2u_2'' + v_1'u_1' + v_2'u_2'.$$

If we substitute these expressions into the left-hand side of Eq. (2), we obtain

$$v_1\left[\frac{d^2u_1}{dx^2} + 2a\frac{du_1}{dx} + bu_1\right] + v_2\left[\frac{d^2u_2}{dx^2} + 2a\frac{du_2}{dx} + bu_2\right] + v_1'u_1' + v_2'u_2' = F(x).$$

The two bracketed terms are zero, since u_1 and u_2 are solutions of the homogeneous Eq. (1). Hence Eq. (2) is satisfied if, in addition to Eq. (4), we require that

$$v_1'u_1' + v_2'u_2' = F(x). \tag{5}$$

Equations (4) and (5) may be solved together as a pair,

$$v_1'u_1 + v_2'u_2 = 0,$$
$$v_1'u_1' + v_2'u_2' = F(x),$$

for the unknown functions v_1' and v_2'. Cramer's rule gives

$$v_1' = \frac{\begin{vmatrix} 0 & u_2 \\ F(x) & u_2' \end{vmatrix}}{\begin{vmatrix} u_1 & u_2 \\ u_1' & u_2' \end{vmatrix}} = \frac{-u_2F(x)}{D},$$

$$v_2' = \frac{\begin{vmatrix} u_1 & 0 \\ u_1' & F(x) \end{vmatrix}}{\begin{vmatrix} u_1 & u_2 \\ u_1' & u_2' \end{vmatrix}} = \frac{u_1F(x)}{D}, \tag{6}$$

where

$$D = \begin{vmatrix} u_1 & u_2 \\ u_1' & u_2' \end{vmatrix}.$$

Now v_1 and v_2 can be found by integration.

In applying the method of variation of parameters to solve the equation

$$\frac{d^2y}{dx^2} + 2a\frac{dy}{dx} + by = F(x), \tag{2}$$

we can work directly with the equations in (6). It is not necessary to rederive them. The steps are as follows:

1. Solve the associated homogeneous equation,

$$\frac{d^2y}{dx^2} + 2a\frac{dy}{dx} + by = 0,$$

   to find the functions u_1 and u_2.
2. Calculate D, v_1', and v_2' from (6).
3. Integrate v_1' and v_2' to find v_1 and v_2.
4. Write down the general solution of (2) as

$$y = v_1u_1 + v_2u_2.$$

EXAMPLE 2 Solve the equation

$$\frac{d^2y}{dx^2} + 2\frac{dy}{dx} - 3y = 6$$

of Example 1 by variation of parameters.

Solution We first solve the associated homogeneous equation

$$\frac{d^2y}{dx^2} + 2\frac{dy}{dx} - 3y = 0$$

as in Example 1 to find

$$u_1(x) = e^{-3x}, \qquad u_2(x) = e^x.$$

Then, from the equations in (6), we have

$$D = \begin{vmatrix} e^{-3x} & e^{x} \\ -3e^{-3x} & e^{x} \end{vmatrix} = e^{-2x} + 3e^{-2x} = 4e^{-2x},$$

$$v_1' = \frac{-6e^{x}}{4e^{-2x}} = -\frac{3}{2}e^{3x}, \qquad v_2' = \frac{6e^{-3x}}{4e^{-2x}} = \frac{3}{2}e^{-x}. \tag{7}$$

Hence

$$v_1 = \int -\frac{3}{2}e^{3x}\,dx = -\frac{1}{2}e^{3x} + C_1,$$

$$v_2 = \int \frac{3}{2}e^{-x}\,dx = -\frac{3}{2}e^{-x} + C_2,$$

and

$$\begin{aligned} y &= v_1u_1 + v_2u_2 \\ &= (-\tfrac{1}{2}e^{3x} + C_1)e^{-3x} + (-\tfrac{3}{2}e^{-x} + C_2)e^{x} \\ &= -2 + C_1e^{-3x} + C_2e^{x}. \ \square \end{aligned}$$

Undetermined Coefficients

The method of variation of parameters is completely general, for it will produce a particular solution of the nonhomogeneous equation (2) for any continuous function $F(x)$. But the calculations involved can be complicated, and in special cases there may be easier methods to use. For instance, we do not need to use variation of parameters to find a particular solution of

$$\frac{d^2y}{dx^2} - \frac{dy}{dx} + 5y = 3, \tag{8}$$

if we can find the particular solution $y_p = 3/5$ by inspection first. And even for an equation like

$$\frac{d^2y}{dx^2} + 3y = e^{x}, \tag{9}$$

we might guess that there is a solution of the form

$$y_p = Ae^{x}$$

and substitute $y_p = Ae^{x}$ and its second derivative into Eq. (9) to find A. If we do so, we find that the function $y_p = \frac{1}{4}e^{x}$ is a solution of Eq. (9).

Again, we might guess that the equation

$$\frac{d^2y}{dx^2} + y = 3x^2 + 4 \tag{10}$$

has a particular solution of the form

$$y_p = Cx^2 + Dx + E.$$

When we substitute this polynomial and its second derivative into Eq. (10) to see whether appropriate values for the constants C, D, and E can be found, Eq. (10) becomes

$$\begin{aligned} 2C + (Cx^2 + Dx + E) &= 3x^2 + 4, \\ Cx^2 + Dx + 2C + E &= 3x^2 + 4. \end{aligned} \tag{11}$$

This latter equation will hold for all values of x if its two sides are identical as polynomials in x; that is, if

$$C = 3, \qquad D = 0, \qquad \text{and} \qquad 2C + E = 4, \tag{12}$$

or,

$$C = 3, \qquad D = 0, \qquad E = -2.$$

We conclude that

$$y_p = 3x^2 + 0x - 2 = 3x^2 - 2 \tag{13}$$

is a solution of Eq. (10).

In each of the foregoing examples, the particular solution we found resembled the function $F(x)$ on the right side of the given differential equation. This was no accident, for we guessed the form of the particular solution by looking at $F(x)$ first. The method of first guessing the form of the solution up to certain undetermined constants, and then determining the values of these constants by using the differential equation, is known as the *method of undetermined coefficients*. It depends for its success upon our ability to recognize the form of a particular solution, and for this reason, among others, it lacks the generality of the method of variation of parameters. Nevertheless, its simplicity makes it the method of choice in a number of special cases.

We shall limit our discussion of the method of undetermined coefficients to selected equations in which the function $F(x)$ in Eq. (2) is the sum of one or more terms like

$$e^{rx}, \qquad \cos kx, \qquad \sin kx, \qquad ax^2 + bx + c.$$

Even so, we will find that the particular solutions of Eq. (2) do not always resemble $F(x)$ as closely as the ones we have seen.

EXAMPLE 3 Find a particular solution of

$$\frac{d^2y}{dx^2} - \frac{dy}{dx} = 2 \sin x. \tag{14}$$

Solution If we try to find a particular solution of the form

$$y_p = A \sin x,$$

and substitute the derivatives of y_p in the given equation, we find that A must satisfy the equation

$$-A \sin x + A \cos x = 2 \sin x \tag{15}$$

for all values of x. Since this requires A to be equal to -2 and 0 at the same time, we conclude that Eq. (14) has no solution of the form $A \sin x$.

It turns out that the required form is the sum

$$y_p = A \sin x + B \cos x. \tag{16}$$

The result of substituting the derivatives of this new candidate into Eq. (14) is

$$-A \sin x - B \cos x - (A \cos x - B \sin x) = 2 \sin x.$$
$$(B - A) \sin x - (A + B) \cos x = 2 \sin x. \qquad (17)$$

Equation (17) will be an identity if

$$B - A = 2 \quad \text{and} \quad A + B = 0,$$

or,

$$A = -1, \quad B = 1.$$

Our particular solution is

$$y_p = \cos x - \sin x. \; \square$$

EXAMPLE 4 Find a particular solution of

$$\frac{d^2y}{dx^2} - 3\frac{dy}{dx} + 2y = 5e^x. \qquad (18)$$

Solution If we substitute

$$y_p = Ae^x$$

and its derivatives in Eq. (18), we find that

$$Ae^x - 3Ae^x + 2Ae^x = 5e^x,$$
$$0 = 5e^x.$$

The trouble can be traced to the fact that $y = e^x$ is already a solution of the related homogeneous equation,

$$\frac{d^2y}{dx^2} - 3\frac{dy}{dx} + 2y = 0. \qquad (19)$$

The characteristic equation of Eq. (19) is

$$r^2 - 3r + 2 = (r - 1)(r - 2) = 0,$$

which has $r = 1$ as a simple root. We may therefore expect Ae^x to "vanish" when substituted into the left-hand side of (18).

The appropriate way to modify the trial solution in this case is to replace Ae^x by Axe^x. When we substitute

$$y_p = Axe^x$$

and its derivatives into (18), we obtain

$$(Axe^x + 2Ae^x) - 3(Axe^x + Ae^x) + 2Axe^x = 5e^x$$
$$-Ae^x = 5e^x$$
$$A = -5.$$

The function

$$y_p = -5xe^x$$

is a particular solution of Eq. (19). $\square$

EXAMPLE 5 Find a particular solution of

$$\frac{d^2y}{dx^2} - 6\frac{dy}{dx} + 9y = e^{3x}. \tag{20}$$

Solution The characteristic equation,

$$r^2 - 6r + 9 = (r - 3)^2 = 0,$$

has $r = 3$ as a *double* root. The appropriate choice for y_p in this case is neither Ae^{3x} nor Axe^{3x}, but Ax^2e^{3x}. When we substitute

$$y_p = Ax^2e^{3x}$$

and its derivatives in the given differential equation, we get

$$\begin{aligned}(9Ax^2e^{3x} + 12Axe^{3x} + 2Ae^{3x}) - 6(3Ax^2e^{3x} + 2Axe^{3x}) + 9Ax^2e^{3x} &= e^{3x}.\\ 2Ae^{3x} &= e^{3x}\\ A &= \tfrac{1}{2}.\end{aligned}$$

The function

$$y_p = \tfrac{1}{2}x^2e^{3x}$$

is a solution of Eq. (20). □

When we wish to find a particular solution of Eq. (2), and the function $F(x)$ is the sum of two or more terms, we choose a trial function for each term in $F(x)$, and add them.

EXAMPLE 6 Solve the equation

$$\frac{d^2y}{dx^2} - \frac{dy}{dx} = 5e^{x} - \sin 2x. \tag{21}$$

Solution We first check the characteristic equation, which is

$$r^2 - r = 0.$$

Its roots are

$$r_1 = 1, \qquad r_2 = 0,$$

so the general solution of the related homogeneous equation is

$$y_h = C_1e^{x} + C_2.$$

We now seek a particular solution y_p. That is, we seek a function that will produce $5e^x - \sin 2x$ when substituted into the left side of Eq. (21). One part of y_p is to produce $5e^x$, the other $-\sin 2x$.

Since any function of the form $C_1 e^x$ is a solution of the related homogeneous equation, we choose our trial y_p to be the sum

$$y_p = Axe^{x} + B\cos 2x + C\sin 2x,$$

including xe^x where we might otherwise have included e^x. When the derivatives of y_p are substituted in Eq. (21), the resulting equations are

$$\begin{aligned}&(Axe^{x} + 2Ae^{x} - 4B\cos 2x - 4C\sin 2x)\\ &\qquad - (Axe^{x} + Ae^{x} - 2B\sin 2x + 2C\cos 2x) = 5e^{x} - \sin 2x,\\ &Ae^{x} - (4B + 2C)\cos 2x + (2B - 4C)\sin 2x = 5e^{x} - \sin 2x.\end{aligned}$$

These equations will hold if

$$A = 5, \qquad (4B + 2C) = 0, \qquad (2B - 4C) = -1,$$

or,

$$A = 5, \qquad B = -\tfrac{1}{10}, \qquad C = \tfrac{1}{5}.$$

Our particular solution is

$$y_p = 5xe^x - \tfrac{1}{10}\cos 2x + \tfrac{1}{5}\sin 2x.$$

The complete solution of Eq. (21) is

$$y = y_h + y_p = C_1e^x + C_2 + 5xe^x - \tfrac{1}{10}\cos 2x + \tfrac{1}{5}\sin 2x. \quad \square$$

You may find the following table helpful in solving the problems at the end of this article.

Table 18.1 The method of undetermined coefficients for selected equations of the form

$$\frac{d^2y}{dx^2} + 2a\frac{dy}{dx} + by = F(x)$$

If $F(x)$ has a term that is a constant multiple of . . .	And if	Then include this expression in the trial function for y_p.
e^{rx}	r is not a root of the characteristic equation	Ae^{rx}
	r is a single root of the characteristic equation	Axe^{rx}
	r is a double root of the characteristic equation	Ax^2e^{rx}
$\sin kx$, $\cos kx$	ki is not a root of the characteristic equation	$B\cos kx + C\sin kx$
$ax^2 + bx + c$	0 is not a root of the characteristic equation	$Dx^2 + Ex + F$ (chosen to match the degree of $ax^2 + bx + c$)
	0 is a single root of the characteristic equation	$Dx^3 + Ex^2 + Fx$ (degree one higher than the degree of $ax^2 + bx + c$)
	0 is a double root of the characteristic equation	$Dx^4 + Ex^3 + Fx^2$ (degree two higher than the degree of $ax^2 + bx + c$)

PROBLEMS

Solve the equations in Problems 1–12 by variation of parameters.

1. $\dfrac{d^2y}{dx^2} + \dfrac{dy}{dx} = x$

2. $\dfrac{d^2y}{dx^2} + y = \tan x, \quad -\dfrac{\pi}{2} < x < \dfrac{\pi}{2}$

3. $\dfrac{d^2y}{dx^2} + y = \sin x$

4. $\dfrac{d^2y}{dx^2} + 2\dfrac{dy}{dx} + y = e^x$

5. $\dfrac{d^2y}{dx^2} + 2\dfrac{dy}{dx} + y = e^{-x}$

6. $\dfrac{d^2y}{dx^2} - y = x$

7. $\dfrac{d^2y}{dx^2} - y = e^x$

8. $\dfrac{d^2y}{dx^2} - y = \sin x$

9. $\dfrac{d^2y}{dx^2} + 4\dfrac{dy}{dx} + 5y = 10$

10. $\dfrac{d^2y}{dx^2} - \dfrac{dy}{dx} = 2^x$

11. $\dfrac{d^2y}{dx^2} + y = \sec x, \quad -\dfrac{\pi}{2} < x < \dfrac{\pi}{2}$

12. $\dfrac{d^2y}{dx^2} - \dfrac{dy}{dx} = e^x \cos x, \quad x > 0$

Solve the equations in Problems 13–28 by the method of undetermined coefficients.

13. $\dfrac{d^2y}{dx^2} - 3\dfrac{dy}{dx} - 10y = -3$

14. $\dfrac{d^2y}{dx^2} - 3\dfrac{dy}{dx} - 10y = 2x - 3$

15. $\dfrac{d^2y}{dx^2} - \dfrac{dy}{dx} = \sin x$

16. $\dfrac{d^2y}{dx^2} + 2\dfrac{dy}{dx} + y = x^2$

17. $\dfrac{d^2y}{dx^2} + y = \cos 3x$

18. $\dfrac{d^2y}{dx^2} + y = e^{2x}$

19. $\dfrac{d^2y}{dx^2} - \dfrac{dy}{dx} - 2y = 20 \cos x$

20. $\dfrac{d^2y}{dx^2} + y = 2x + 3e^x$

21. $\dfrac{d^2y}{dx^2} - y = e^x + x^2$

22. $\dfrac{d^2y}{dx^2} + 2\dfrac{dy}{dx} + y = 6 \sin 2x$

23. $\dfrac{d^2y}{dx^2} - \dfrac{dy}{dx} - 6y = e^{-x} - 7 \cos x$

24. $\dfrac{d^2y}{dx^2} + 3\dfrac{dy}{dx} + 2y = e^{-x} + e^{-2x} - x$

25. $\dfrac{d^2y}{dx^2} + 5\dfrac{dy}{dx} = 15x^2$

26. $\dfrac{d^2y}{dx^2} - \dfrac{dy}{dx} = -8x + 3$

27. $\dfrac{d^2y}{dx^2} - 3\dfrac{dy}{dx} = e^{3x} - 12x$

28. $\dfrac{d^2y}{dx^2} + 7\dfrac{dy}{dx} = 42x^2 + 5x + 1$

In each of Problems 29–31, the given differential equation has a particular solution y_p of the form given. Determine the coefficients in y_p. Then solve the differential equation.

29. $\dfrac{d^2y}{dx^2} - 5\dfrac{dy}{dx} = xe^{5x}, \quad y_p = Ax^2e^{5x} + Bxe^{5x}$

30. $\dfrac{d^2y}{dx^2} - \dfrac{dy}{dx} = \cos x + \sin x, \quad y_p = A \cos x + B \sin x$

31. $\dfrac{d^2y}{dx^2} + y = 2 \cos x + \sin x, \quad y_p = Ax \cos x + Bx \sin x$

In Problems 32–35, solve the given differential equations (a) by variation of parameters, and (b) by the method of undetermined coefficients.

32. $\dfrac{d^2y}{dx^2} - 4\dfrac{dy}{dx} + 4y = 2e^{2x}$

33. $\dfrac{d^2y}{dx^2} - \dfrac{dy}{dx} = e^x + e^{-x}$

34. $\dfrac{d^2y}{dx^2} - 9\dfrac{dy}{dx} = 9e^{9x}$

35. $\dfrac{d^2y}{dx^2} - 4\dfrac{dy}{dx} - 5y = e^x + 4$

Solve the differential equations in Problems 36–45. Some of the equations can be solved by the method of undetermined coefficients, but others cannot.

36. $\dfrac{d^2y}{dx^2} + y = \csc x, \quad 0 < x < \pi$

37. $\dfrac{d^2y}{dx^2} + y = \cot x, \quad 0 < x < \pi$

38. $\dfrac{d^2y}{dx^2} + 4y = \sin x$

39. $\dfrac{d^2y}{dx^2} - 8\dfrac{dy}{dx} = e^{8x}$

40. $\dfrac{d^2y}{dx^2} + 4\dfrac{dy}{dx} + 5y = x + 2$

41. $\dfrac{d^2y}{dx^2} - \dfrac{dy}{dx} = x^3$

42. $\dfrac{d^2y}{dx^2} + 9y = 9x - \cos x$

43. $\dfrac{d^2y}{dx^2} + 2\dfrac{dy}{dx} = x^2 - e^x$

44. $\dfrac{d^2y}{dx^2} - 3\dfrac{dy}{dx} + 2y = e^x - e^{2x}$

45. $\dfrac{d^2y}{dx^2} + y = \sec x \tan x, \quad -\dfrac{\pi}{2} < x < \dfrac{\pi}{2}$

The method of undetermined coefficients can sometimes be used to solve first order ordinary differential equations. Use the method to solve the equations in Problems 46–49.

46. $\dfrac{dy}{dx} + 4y = x$

47. $\dfrac{dy}{dx} - 3y = e^x$

48. $\dfrac{dy}{dx} + y = \sin x$

49. $\dfrac{dy}{dx} - 3y = 5e^{3x}$

Solve the differential equations in Problems 50 and 51 subject to the given initial conditions.

50. $\dfrac{d^2y}{dx^2} + y = e^{2x}; \quad y(0) = 0, \quad y'(0) = \dfrac{2}{5}$

51. $\dfrac{d^2y}{dx^2} + y = \sec^2 x, \quad -\dfrac{\pi}{2} < x < \dfrac{\pi}{2}; \quad y(0) = y'(0) = 1$

52. *Bernoulli's equation of order 2.* Solve the equation

$$\frac{dy}{dx} + y = (xy)^2$$

by carrying out the following steps: (1) divide both sides of the equation by y^2; (2) make the change of variable $u = y^{-1}$; (3) solve the resulting equation for u in terms of x; (4) let $y = u^{-1}$.

53. Solve the integral equation

$$y(x) + \int_0^x y(t)\, dt = x.$$

(*Hint:* Differentiate.)

18.11 Higher Order Linear Equations with Constant Coefficients

The methods of Articles 18.9 and 18.10 can be extended to equations of higher order. The characteristic algebraic equation associated with the differential equation

$$(D^n + a_1 D^{n-1} + \cdots + a_{n-1}D + a_n)y = F(x) \tag{1}$$

is

$$r^n + a_1 r^{n-1} + \cdots + a_{n-1}r + a_n = 0. \tag{2}$$

If its roots $r_1, r_2, \ldots, r_n$ are all distinct, the solution of the homogeneous equation obtained by replacing $F(x)$ by 0 in Eq. (1) is

$$y_h = c_1 e^{r_1 x} + c_2 e^{r_2 x} + \cdots + c_n e^{r_n x}.$$

Pairs of complex conjugate roots $\alpha \pm i\beta$ can be grouped together, and the corresponding part of y_h can be written in terms of the functions

$$e^{\alpha x} \cos \beta x \qquad \text{and} \qquad e^{\alpha x} \sin \beta x.$$

In case the roots of Eq. (2) are not all distinct, the portion of y_h which corresponds to a root r of multiplicity m is to be replaced by

$$(C_1 x^{m-1} + C_2 x^{m-2} + \cdots + C_m)e^{rx}.$$

Note that the polynomial in parentheses contains m arbitrary constants.

EXAMPLE Solve the equation

$$\frac{d^4y}{dx^4} - 3\frac{d^3y}{dx^3} + 3\frac{d^2y}{dx^2} - \frac{dy}{dx} = 0.$$

Solution $r^4 - 3r^3 + 3r^2 - r = r(r-1)^3$. The roots of the characteristic equation are

$$r_1 = 0, \qquad r_2 = r_3 = r_4 = 1.$$

The solution is

$$y = C_1 + (C_2x^2 + C_3x + C_4)e^x. \; \square$$

Variation of Parameters If the general solution of the homogeneous equation is

$$y_h = C_1u_1 + C_2u_2 + \cdots + C_nu_n,$$

then

$$y = v_1u_1 + v_2u_2 + \cdots + v_nu_n$$

will be a solution of the nonhomogeneous Eq. (1), provided

$$\begin{aligned}
v_1'u_1 &+ v_2'u_2 &+ \cdots + v_n'u_n &= 0,\\
v_1'u_1' &+ v_2'u_2' &+ \cdots + v_n'u_n' &= 0,\\
&\vdots\\
v_1'u_1^{(n-2)} &+ v_2'u_2^{(n-2)} &+ \cdots + v_n'u_n^{(n-2)} &= 0,\\
v_1'u_1^{(n-1)} &+ v_2'u_2^{(n-1)} &+ \cdots + v_n'u_n^{(n-1)} &= F(x).
\end{aligned}$$

These equations may be solved for $v_1', v_2', \ldots, v_n'$ by Cramer's rule, and the results integrated to give $v_1, v_2, \ldots, v_n$.

PROBLEMS

Solve the following equations.

1. $\dfrac{d^3y}{dx^3} - 3\dfrac{d^2y}{dx^2} + 2\dfrac{dy}{dx} = 0$
2. $\dfrac{d^3y}{dx^3} - y = 0$
3. $\dfrac{d^4y}{dx^4} - 4\dfrac{d^2y}{dx^2} + 4y = 0$
4. $\dfrac{d^4y}{dx^4} - 16y = 0$
5. $\dfrac{d^4y}{dx^4} + 16y = 0$
6. $\dfrac{d^3y}{dx^3} - 3\dfrac{dy}{dx} + 2y = e^x$
7. $\dfrac{d^4y}{dx^4} - 4\dfrac{d^3y}{dx^3} + 6\dfrac{d^2y}{dx^2} - 4\dfrac{dy}{dx} + y = 7$
8. $\dfrac{d^4y}{dx^4} + y = x + 1$

18.12

Vibrations

A spring of natural length L has its upper end fastened to a rigid support at A (Fig. 18.2). A weight W, of mass m, is suspended from the spring. The weight stretches the spring to a length $L + s$ when the system is allowed to come to rest in a new equilibrium position. By Hooke's law, the tension in the spring is ks, where k is the so-called spring constant. The force of gravity pulling down on the weight is $W = mg$. Equilibrium requires

$$ks = mg. \tag{1}$$

Suppose now that the weight is pulled down an additional amount a beyond the equilibrium position, and released. We shall discuss its motion.

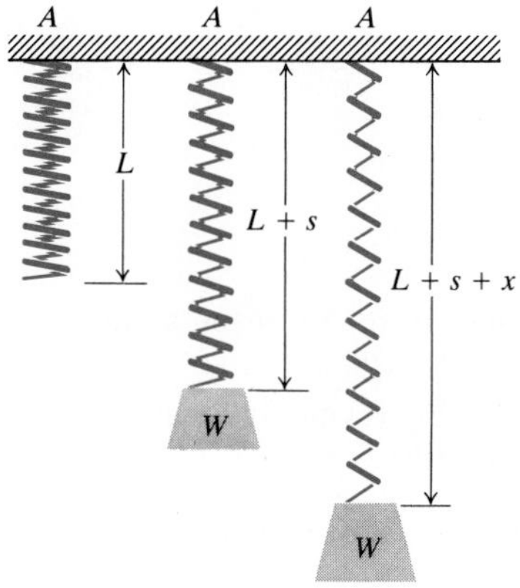

18.2 A spring stretched beyond its natural length by a weight.

Let x, positive direction downward, denote the displacement of the weight away from equilibrium at any time t after the motion has started. Then the forces acting upon the weight are

$+mg$, due to gravity,

$-k(s + x)$, due to the spring tension.

The resultant of these forces is, by Newton's second law, also equal to

$$m\frac{d^2x}{dt^2}.$$

Therefore

$$m\frac{d^2x}{dt^2} = mg - ks - kx. \tag{2}$$

By Eq. (1), $mg - ks = 0$, so that (2) becomes

$$m\frac{d^2x}{dt^2} + kx = 0. \tag{3}$$

In addition to the differential equation (3), the motion satisfies the initial conditions:

$$\text{At } t = 0: \qquad x = a \qquad \text{and} \qquad \frac{dx}{dt} = 0. \tag{4}$$

Let $\omega = \sqrt{k/m}$. Then Eq. (3) becomes

$$\frac{d^2x}{dt^2} + \omega^2 x = 0$$

or

$$(D^2 + \omega^2)x = 0,$$

where

$$D = \frac{d}{dt}.$$

The roots of the characteristic equation

$$r^2 + \omega^2 = 0$$

are complex conjugates

$$r = \pm\omega i.$$

Hence

$$x = c_1 \cos \omega t + c_2 \sin \omega t \tag{5}$$

is the general solution of the differential equation. To fit the initial conditions, we also compute

$$\frac{dx}{dt} = -c_1\omega \sin \omega t + c_2\omega \cos \omega t,$$

and then substitute from (4). This yields

$$a = c_1, \qquad 0 = c_2\omega.$$

Therefore

$$c_1 = a, \qquad c_2 = 0,$$

and

$$x = a \cos \omega t \tag{6}$$

describes the motion of the weight. Equation (6) represents simple harmonic motion of amplitude a and period $T = 2\pi/\omega$.

The two terms on the right-hand side of Eq. (5) can be combined into a single term by using the trigonometric identity

$$\sin(\omega t + \phi) = \cos \omega t \sin \phi + \sin \omega t \cos \phi.$$

To apply the identity, we take

$$c_1 = C \sin \phi, \qquad c_2 = C \cos \phi, \tag{7a}$$

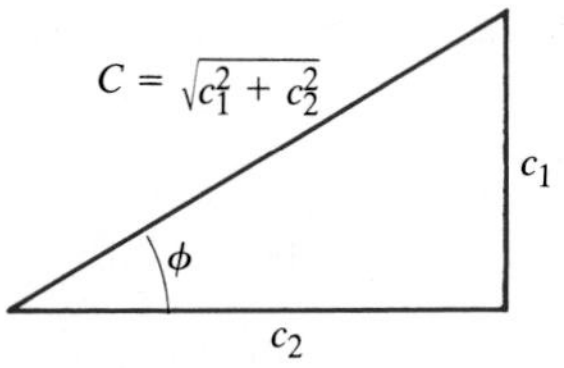

18.3 $c_1 = C \sin \phi$ and $c_2 = C \cos \phi$.

where

$$C = \sqrt{c_1^2 + c_2^2}, \qquad \phi = \tan^{-1} \frac{c_1}{c_2}, \tag{7b}$$

as in Fig. 18.3. Then Eq. (5) can be written in the alternative form

$$x = C \sin (\omega t + \phi). \tag{8}$$

Here C and ϕ may be taken as two new arbitrary constants, replacing the two constants c_1 and c_2 of Eq. (5). Equation (8) represents simple harmonic motion of amplitude C and period $T = 2\pi/\omega$. The angle $\omega t + \phi$ is called the *phase angle,* and ϕ may be interpreted as the initial value of the phase angle. A graph of Eq. (8) is given in Fig. 18.4.

Equation (3) assumes that there is no friction in the system. Next, consider the case where the motion of the weight is retarded by a friction force $c(dx/dt)$ proportional to the velocity, where c is a positive constant. Then the differential equation is

$$m \frac{d^2x}{dt^2} = -kx - c \frac{dx}{dt},$$

or

$$\frac{d^2x}{dt^2} + 2b \frac{dx}{dt} + \omega^2 x = 0, \tag{9}$$

where

$$2b = \frac{c}{m} \qquad \text{and} \qquad \omega = \sqrt{\frac{k}{m}}.$$

If we introduce the operator $D = d/dt$, Eq. (9) becomes

$$(D^2 + 2bD + \omega^2)x = 0.$$

The characteristic equation is

$$r^2 + 2br + \omega^2 = 0$$

with roots

$$r = -b \pm \sqrt{b^2 - \omega^2}. \tag{10}$$

Three cases now present themselves, depending upon the relative sizes of b and ω.

18.4 Undamped vibration.

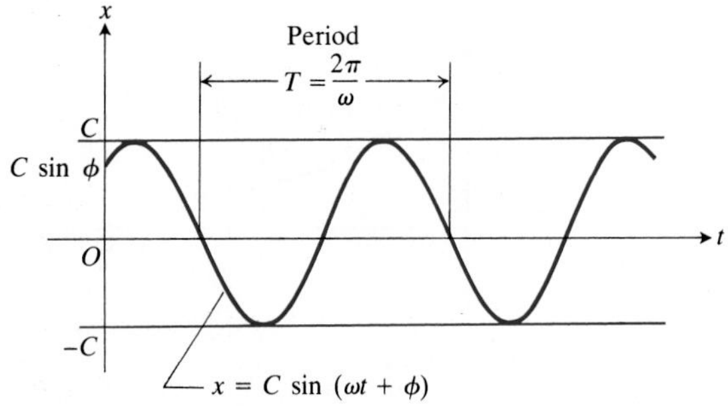

CASE 1. If $b = \omega$, the two roots in Eq. (10) are equal, and the solution of (9) is

$$x = (c_1 + c_2 t)e^{-\omega t}. \tag{11}$$

As time goes on, x approaches zero. The motion is not oscillatory.

CASE 2. If $b > \omega$, then the roots (10) are both real but unequal, and

$$x = c_1 e^{r_1 t} + c_2 e^{r_2 t}, \tag{12}$$

where

$$r_1 = -b + \sqrt{b^2 - \omega^2} \quad \text{and} \quad r_2 = -b - \sqrt{b^2 - \omega^2}.$$

Here again the motion is not oscillatory. Both r_1 and r_2 are negative, and x approaches zero as time goes on.

CASE 3. If $b < \omega$, let

$$\omega^2 - b^2 = \alpha^2.$$

Then

$$r_1 = -b + \alpha i, \qquad r_2 = -b - \alpha i$$

and

$$x = e^{-bt}[c_1 \cos \alpha t + c_2 \sin \alpha t]. \tag{13a}$$

If we introduce the substitutions (7), we may also write Eq. (13a) in the equivalent form

$$x = Ce^{-bt} \sin (\alpha t + \phi). \tag{13b}$$

This equation represents damped vibratory motion. It is analogous to simple harmonic motion, of period $T = 2\pi/\alpha$, except that the amplitude is not constant but is given by Ce^{-bt}. Since this tends to zero as t increases, the vibrations tend to die out as time goes on. Observe, however, that Eq. (13b) reduces to Eq. (8) in the absence of friction. The effect of friction is twofold:

1. $b = c/(2m)$ appears as a coefficient in the exponential *damping factor* e^{-bt}. The larger b is, the more quickly do the vibrations tend to become unnoticeable.
2. The period $T = 2\pi/\alpha = 2\pi/\sqrt{\omega^2 - b^2}$ is greater than the period $T_0 = 2\pi/\omega$ in the friction-free system.

Curves representing solutions of Eq. (9) in typical cases are shown in Figs. 18.4 and 18.5. The size of b, relative to ω, determines the kind of

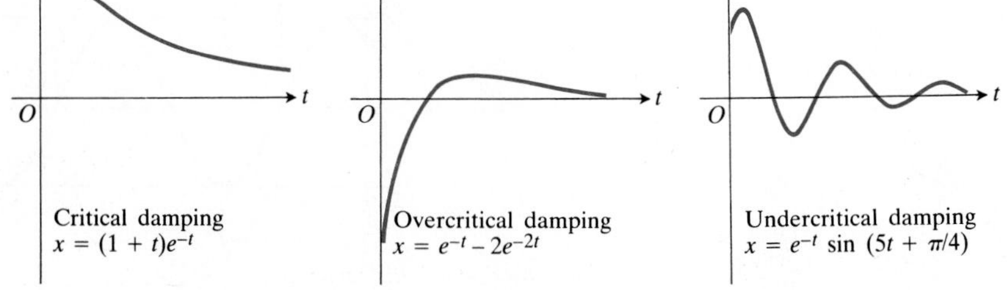

18.5 Three kinds of damping of vibration.

solution, and b also determines the rate of damping. It is therefore customary to say that there is

a) critical damping if $b = \omega$;

b) overcritical damping if $b > \omega$;

c) undercritical damping if $0 < b < \omega$;

d) no damping if $b = 0$.

PROBLEMS

1. Suppose the motion of the weight in Fig. 18.2 is described by the differential equation (3). Find the motion if $x = x_0$ and $dx/dt = v_0$ at $t = 0$. Express the answer in two equivalent forms (Eqs. 5 and 6).

2. A 5-lb weight is suspended from the lower end of a spring whose upper end is attached to a rigid support. The weight extends the spring by 2 in. If, after the weight has come to rest in its new equilibrium position, it is struck a sharp blow that starts it downward with a velocity of 4 ft/s, find its subsequent motion, assuming there is no friction.

3. A simple electrical circuit shown in Fig. 18.6 contains a capacitor of capacitance C farads, a coil of inductance L henrys, a resistance of R ohms, and a generator that produces an electromotive force E volts, in series. If the current intensity at time t at some point of the circuit is I amperes, the differential equation describing the current I is

$$L\frac{d^2I}{dt^2} + R\frac{dI}{dt} + \frac{1}{C}I = \frac{dE}{dt}.$$

Find I as a function of t if

a) $R = 0, \quad 1/(LC) = \omega^2, \quad E = \text{constant},$

b) $R = 0, \quad 1/(LC) = \omega^2, \quad E = A \sin \alpha t;$
$\alpha = \text{constant} \neq \omega,$

c) $R = 0, \quad 1/(LC) = \omega^2, \quad E = A \sin \omega t,$

d) $R = 50, \quad L = 5, \quad C = 9 \times 10^{-6}, \quad E = \text{constant}.$

4. A simple pendulum of length l makes an angle θ with the vertical. As it swings back and forth, its motion, neglecting friction, is described by the differential equation

$$\frac{d^2\theta}{dt^2} = -\frac{g}{l}\sin\theta,$$

where g (the acceleration due to gravity, $g \approx 32\ \text{ft/s}^2$) is a constant. Solve the differential equation of motion, under the assumption that θ is so small that $\sin\theta$ may be replaced by θ without appreciable error. Assume that $\theta = \theta_0$ and $d\theta/dt = 0$ when $t = 0$.

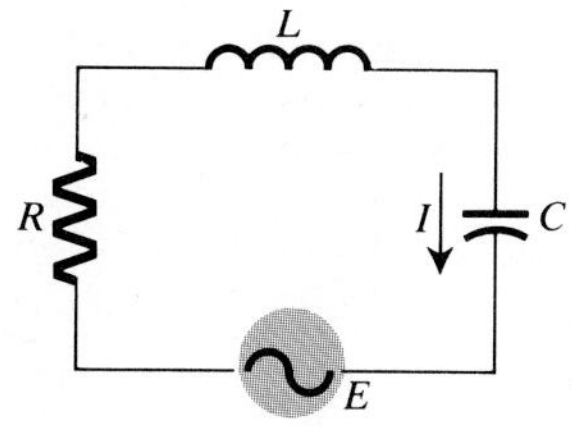

18.6 An R–L–C series circuit.

5. A circular disc of mass m and radius r is suspended by a thin wire attached to the center of one of its flat faces. If the disc is twisted through an angle θ, torsion in the wire tends to turn the disc back in the opposite direction. The differential equation for the motion is

$$\frac{1}{2}mr^2\frac{d^2\theta}{dt^2} = -k\theta,$$

where k is the coefficient of torsion of the wire. Find the motion if $\theta = \theta_0$ and $d\theta/dt = v_0$ at $t = 0$.

6. A cylindrical spar buoy, diameter 1 foot, weight 100 lb, floats partially submerged, in an upright position. When it is depressed slightly from its equilibrium position and released, it bobs up and down according to the differential equation

$$\frac{100}{g}\frac{d^2x}{dt^2} = -16\pi x - c\frac{dx}{dt}.$$

Here $c(dx/dt)$ is the frictional resistance of the water. Find c if the period of oscillation is observed to be 1.6 s. (Take $g = 32\ \text{ft/s}^2$.)

7. Suppose the upper end of the spring in Fig. 18.2 is attached, not to a rigid support at A, but to a member that itself undergoes up and down motion given by a function of the time t, say $y = f(t)$. If the positive direction of y is downward, the differential equation of motion is

$$m\frac{d^2x}{dt^2} + kx = kf(t).$$

Let $x = x_0$ and $dx/dt = 0$ when $t = 0$, and solve for x

a) if $f(t) = A \sin \alpha t$ and $\alpha \neq \sqrt{k/m}$,

b) if $f(t) = A \sin \alpha t$ and $\alpha = \sqrt{k/m}$.

***Toolkit* programs**

Damped Oscillator

Forced Damped Oscillator

18.13

Approximation Methods: Power Series

What can we do if a given differential equation fails to fit any of the categories we know how to handle? We may go to more advanced textbooks, or treatises on the subject, but even this won't always help because some simple-looking equations have no solution in terms of finite combinations of the elementary functions that we know. The next example illustrates how we might be able to solve such an equation by a power series. (3)

EXAMPLE 1 Find a power series solution for

$$u'' + x^2u = 0. \tag{1}$$

Solution Assume that there is a solution of the form

$$u = a_0 + a_1x + a_2x^2 + \cdots + a_nx^n + \cdots \tag{2}$$

where the coefficients a_k are to be determined to satisfy Eq. (1). If we differentiate the series twice, we get

$$u'' = 2a_2 + 3 \cdot 2a_3x + \cdots + n(n-1)a_nx^{n-2} + \cdots. \tag{3}$$

The series for x^2u is just x^2 times the right-hand side of Eq. (2):

$$x^2u = a_0x^2 + a_1x^3 + a_2x^4 + \cdots + a_nx^{n+2} + \cdots. \tag{4}$$

The series for $u'' + x^2u$ is the sum of the series in Eqs. (3) and (4):

$$u'' + x^2u = 2a_2 + 6a_3x + (12a_4 + a_0)x^2 + (20a_5 + a_1)x^3 + \cdots$$
$$+ [n(n-1)a_n + a_{n-4}]x^{n-2} + \cdots. \tag{5}$$

Notice that the coefficient of x^{n-2} in Eq. (4) is a_{n-4}. To satisfy Eq. (1), the coefficients of the individual powers of x in Eq. (5) must all be zero:

$$2a_2 = 0, \qquad 6a_3 = 0, \qquad 12a_4 + a_0 = 0, \qquad 20a_5 + a_1 = 0,$$

and for all $n \geq 4$,

$$n(n-1)a_n + a_{n-4} = 0. \tag{7}$$

We can see immediately from Eq. (2) that

$$a_0 = u(0), \qquad a_1 = u'(0). \tag{8}$$

In other words, the first two coefficients are the values of u and of u' at $x = 0$. The recursion formulas (6) and (7) enable us to evaluate all the other coefficients in terms of these two.

The first two of Eqs. (6) give

$$a_2 = 0, \qquad a_3 = 0.$$

Equation (7) shows that if $a_{n-4} = 0$, then $a_n = 0$; so we conclude that

$$a_6 = 0, \qquad a_7 = 0, \qquad a_{10} = 0, \qquad a_{11} = 0,$$

and whenever $n = 4k + 2$ or $4k + 3$, a_n is zero. For the other coefficients we have

$$a_n = \frac{-a_{n-4}}{n(n-1)}$$

so that

$$a_4 = \frac{-a_0}{4\cdot 3}, \qquad a_8 = \frac{-a_4}{8\cdot 7} = \frac{a_0}{3\cdot 4\cdot 7\cdot 8},$$

$$a_{12} = \frac{-a_8}{11\cdot 12} = \frac{-a_0}{3\cdot 4\cdot 7\cdot 8\cdot 11\cdot 12};$$

and

$$a_5 = \frac{-a_1}{5\cdot 4}, \qquad a_9 = \frac{-a_5}{9\cdot 8} = \frac{a_1}{4\cdot 5\cdot 8\cdot 9},$$

$$a_{13} = \frac{-a_9}{12\cdot 13} = \frac{-a_1}{4\cdot 5\cdot 8\cdot 9\cdot 12\cdot 13}.$$

The answer is best expressed as the sum of two separate series—one multiplied by a_0, the other by a_1:

$$\begin{aligned} u = a_0&\left(1 - \frac{x^4}{3\cdot 4} + \frac{x^8}{3\cdot 4\cdot 7\cdot 8} - \frac{x^{12}}{3\cdot 4\cdot 7\cdot 8\cdot 11\cdot 12} + \cdots\right) \\ + a_1&\left(x - \frac{x^5}{4\cdot 5} + \frac{x^9}{4\cdot 5\cdot 8\cdot 9} - \frac{x^{13}}{4\cdot 5\cdot 8\cdot 9\cdot 12\cdot 13} + \cdots\right). \end{aligned} \tag{9}$$

Both series converge absolutely for all values of x as is readily seen by the ratio test. □

REMARK. If the differential equation $u'' + x^2u = 0$ were of sufficient importance to justify the effort, one could tabulate values of the two series for selected values of x, or program the computations for a calculator or computer.

PROBLEMS

Find power series solutions for the following differential equations. (Some of the equations can be solved directly without series, but they are included for practice.) Find at least four nonzero terms in each series.

1. $y' = y$

2. $y' + y = 0$

3. $y' = 2y$

4. $y' + 2y = 0$

5. $y'' = y$

6. $y'' + y = 0$

7. $y'' + y = x$

8. $y'' - y = x$

9. $y'' + x = y$

10. $y'' + y = 2x$

11. $y'' + y = \sin x$

12. $y'' - x^2y = 0$

13. $y'' + x^2y = x$

14. $y'' - x^2y = e^x$

18.14 Direction Fields and Picard's Theorem

Direction Fields and Isoclines

Associated with the differential equation $y' = f(x, y)$, for a given function f, is its *direction field*. Figure 18.7 shows a part of such a field for the equation

$$y' = x + y. \tag{1}$$

This figure was done by a computer, but it can be visualized as a pattern of iron filings that have been sprinkled onto a piece of paper and have

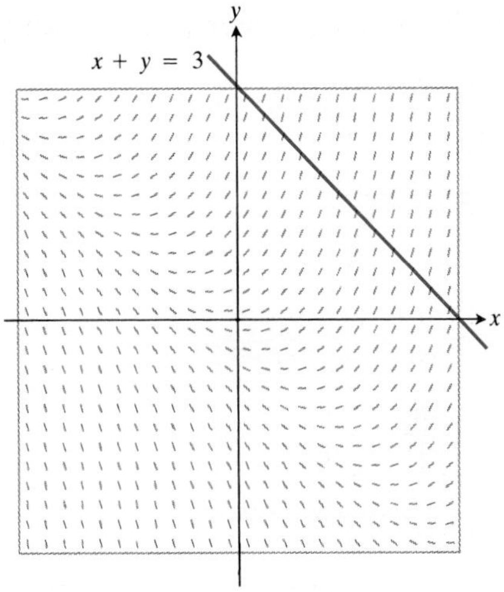

18.7 Direction field for $y' = x + y$, $-3 \leq x \leq 3$, $-3 \leq y \leq 3$.

arranged themselves so that the bit at the point $P(x, y)$ has slope $x + y$. Since iron filings won't actually do that for us, we resort to other approaches.

One simple approach is to first construct a set of *isoclines*. As the name implies, an isocline is a curve (in this case a straight line) at each point of which $f(x, y)$ is a fixed number.

For Eq. (1), each isocline has the form

$$x + y = C,$$

where C is a constant. Through some point on the line $x + y = 3$, for example, we draw a short line segment of slope 3. Having drawn one such segment, we then draw others parallel to it. Fig. 18.7 shows twelve such segments on that line. We repeat this process on each of the isoclines, starting with a short segment of slope C at a point on the line $x + y = C$. The resulting pattern is a portion of the direction field for the differential equation $y' = x + y$.

Notice one special line—the line $x + y = -1$. This line itself has slope -1 and satisfies the equation $x + y = y'$. At each point of this line, the direction field follows the line itself. Above this line, any solution curve is concave upward because $y'' = 1 + y' = 1 + x + y$ is positive if $x + y > -1$. Below the line, any solution curve must be concave downward.

Fig. 18.8 is a composite that shows two solution curves as well as a set of isoclines and part of the direction field for $y' = x + y$, in the region $-3 \leq x \leq 3$, $-3 \leq y \leq 3$. The upper solution curve is for the initial value problem

$$y' = x + y, \qquad y(0) = 1; \qquad y = -1 - x + 2e^{x}$$

and the lower curve is for

$$y' = x + y, \qquad y(-2) = 0; \qquad y = -1 - x - e^{x+2}.$$

In this example it is easy to find the general solution of

$$y' = x + y, \qquad y(x_0) = y_0.$$

Problem 15 asks you to show that the general solution is

$$y = -1 - x + (1 + x_0 + y_0)e^{x - x_0}.$$

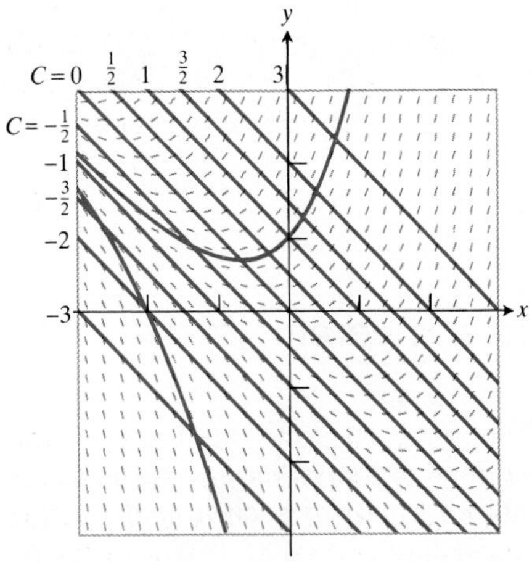

18.8 Direction field, isoclines $x + y = C$, and specific solutions for the equation $y' = x + y$, $-3 \leq x \leq 3$, $-3 \leq y \leq 3$.

Picard's Theorem

Suppose we are given an initial value problem

$$y' = f(x, y), \qquad y(x_0) = y_0 \tag{2}$$

where f is defined and continuous inside a rectangle R in the xy-plane. Does the initial value problem always have at least one solution if (x_0, y_0) is inside R? (See Fig. 18.9.) Might it have more than one solution? The next example shows that the answer to the second question is yes, unless something more is required of f.

EXAMPLE 1 The initial value problem

$$y' = y^{4/5}, \qquad y(0) = 0, \tag{3}$$

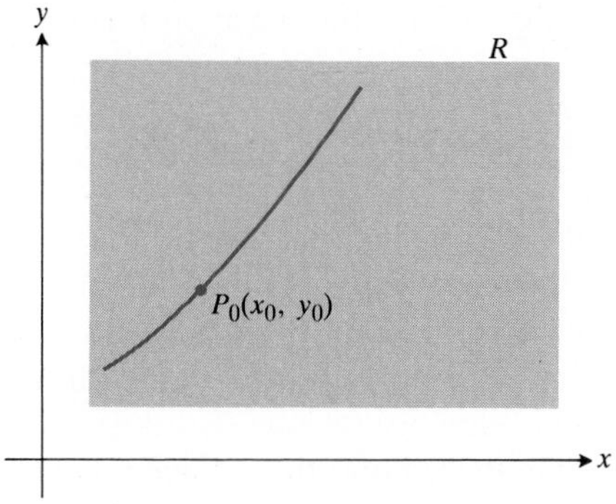

18.9 When is there exactly one solution of $y' = f(x, y)$, with $y(x_0) = y_0$? Ans. When f and f_y are continuous throughout a rectangular region R that contains $P_0(x_0, y_0)$ in its interior.

has the obvious solution $y \equiv 0$. Another solution is found by separating the variables and integrating:

$$y = \left(\frac{x}{5}\right)^5.$$

There are two more solutions (see Fig. 18.10):

$$y = \begin{cases} 0 & \text{for } x \leq 0, \\ \left(\frac{x}{5}\right)^5 & \text{for } x > 0, \end{cases}$$

and

$$y = \begin{cases} \left(\frac{x}{5}\right)^5 & \text{for } x \leq 0, \\ 0 & \text{for } x > 0. \end{cases}$$

In this example, $f(x, y) = y^{4/5}$ is continuous in the entire xy-plane. But, its partial derivative $f_y = \frac{4}{5}y^{-1/5}$ is not continuous where $y = 0$. □

There is a useful theorem, due to Picard, that gives a sufficient condition for the existence and uniqueness of a solution of the initial value problem of Eqs. (2).

THEOREM

Picard's Theorem

Let $f(x, y)$ and its first order partial derivative f_y be continuous throughout the interior and on the boundary of a rectangle R. If $P(x_0, y_0)$ is any point in the interior of R, then there exists a positive number r such that the initial value problem

$$y' = f(x, y), \quad y(x_0) = y_0 \tag{4}$$

has a unique solution, $y = y(x)$, for $x_0 - r \leq x \leq x_0 + r$.

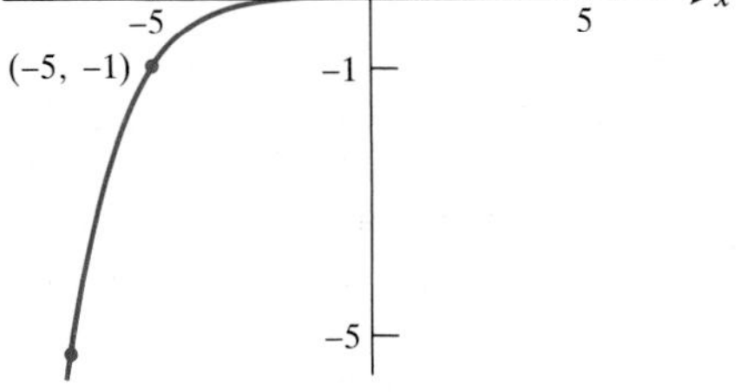

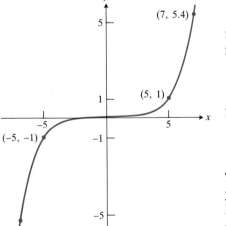

18.10 Part of graph of $y = (x/5)^5$, one of the solutions of $y' = y^{4/5}$, $y(0) = 0$. Another solution is $y = 0$.

Although we shall not prove this theorem, we relate it to Picard's iteration scheme that was mentioned in Article 2.10. In the first place, any function $y(x)$ that satisfies Eqs. (4) must also satisfy the integral equation

$$y(x) = y_0 + \int_{x_0}^{x} f(t, y(t))\, dt \tag{5}$$

because

$$\int_{x_0}^{x} \frac{dy}{dt}\, dt = y(x) - y(x_0).$$

The converse is also true: if $y(x)$ satisfies Eq. (5), then $y' = f(x, y(x))$ and $y(x_0) = y_0$. So Eqs. (4) may be replaced by Eq. (5). This sets the stage for Picard's iteration method: In the integrand in Eq. (5), replace $y(t)$ by the constant y_0, then integrate and call the resulting right-hand side of Eq. (5) $y_1(x)$:

$$y_1(x) = y_0 + \int_{x_0}^{x} f(t, y_0)\, dt. \tag{6}$$

This starts the process. To keep it going, we use the iterative formulas:

$$y_{n+1}(x) = y_0 + \int_{x_0}^{x} f(t, y_n(t))\, dt. \tag{7}$$

The proof of Picard's theorem consists in showing that this process produces a sequence of functions $\{y_n(x)\}$ that converge to a function $y(x)$ that satisfies Eqs. (4) and (5), for values of x sufficiently near x_0. (And, that the solution is unique: that is, no other method will lead to a different solution.)

The following example illustrates the Picard iteration scheme, but the computations soon become too burdensome to continue.

EXAMPLE 2 Illustrate the Picard iteration scheme for the initial value problem

$$y' = x - y, \qquad y(0) = 1. \tag{8}$$

Solution For the problem at hand, Eq. (6) becomes

$$\begin{aligned} y_1(x) &= 1 + \int_0^x (t - 1)\, dt \\ &= 1 + \frac{x^2}{2} - x. \end{aligned} \tag{9a}$$

If we now use Eq. (7) with $n = 1$, we get

$$\begin{aligned} y_2(x) &= 1 + \int_0^x \left(t - 1 - \frac{t^2}{2} + t\right) dt \\ &= 1 - x + x^2 - \frac{x^3}{6}. \end{aligned} \tag{9b}$$

The next iteration, with $n = 2$, gives

$$\begin{aligned} y_3(x) &= 1 + \int_0^x \left(t - 1 + t - t^2 + \frac{t^3}{6}\right) dt \\ &= 1 - x + x^2 - \frac{x^3}{3} + \frac{x^4}{4!}. \end{aligned} \tag{9c}$$

In this example, it is possible to find the exact solution, because

$$\frac{dy}{dx} + y = x$$

is a first order equation that is linear in y. It has an integrating factor e^x and the general solution is

$$y = x - 1 + Ce^{-x}.$$

The solution of our particular initial value problem is

$$y = x - 1 + 2e^{-x}. \tag{10}$$

If we substitute the Maclaurin series for e^{-x} in Eq. (10), we get

$$y = x - 1 + 2\left(1 - x + \frac{x^2}{2!} - \frac{x^3}{3!} + \frac{x^4}{4!} - \cdots\right)$$

$$= 1 - x + x^2 - \frac{x^3}{3} + 2\left(\frac{x^4}{4!} - \frac{x^5}{5!} + \cdots\right),$$

and we see that the Picard method has given us the first four terms of this expansion in Eq. (9c) in $y_3(x)$. □

In the next example, we cannot find a solution in terms of elementary functions. The Picard scheme is one way we could get an idea of how the solution behaves near the initial point.

EXAMPLE 3 Find $y_n(x)$ for $n = 0, 1, 2,$ and 3 for the initial value problem

$$y' = x^2 + y^2, \qquad y(0) = 0.$$

Solution By definition, $y_0(x) = y(0) = 0$. The other functions $y_n(x)$ are generated by the integral representation

$$y_{n+1}(x) = 0 + \int_0^x [t^2 + (y_n(t))^2]\,dt$$

$$= \frac{x^3}{3} + \int_0^x (y_n(t))^2\,dt.$$

We successively calculate

$$y_1(x) = \frac{x^3}{3}, \qquad y_2(x) = \frac{x^3}{3} + \frac{x^7}{63},$$

$$y_3(x) = \frac{x^3}{3} + \frac{x^7}{63} + \frac{2x^{11}}{2079} + \frac{x^{15}}{59535}. \quad \square$$

REMARK. In the next article we shall introduce numerical methods for solving initial value problems like Examples 2 and 3 above. If we program such a numerical solution for a calculator or computer, it is helpful to have an independent check on the program. For x near zero, we would expect $y_2(x)$, or $y_3(x)$, to provide such a check.

PROBLEMS

In Problems 1–6, sketch some of the isoclines and part of the direction field. Using the direction field, sketch a portion of the graph of the solution that includes the point $P(1, -1)$.

1. $y' = x$ **2.** $y' = y$ **3.** $y' = 1/x$

4. $y' = 1/y$ **5.** $y' = xy$ **6.** $y' = x^2 + y^2$

7. Sketch some of the direction field for the equation $y' = (x + y)^2$. Show isoclines where the slope is 0, $\frac{1}{4}$, 1, 4. Sketch (roughly) the solutions of the equation that include (a) $P(0, 0)$, (b) $Q(-1, 1)$, and (c) $R(1, 0)$. Now make the substitution $u = x + y$ and find the general solution of $y' = (x + y)^2$. What is the equation of the solution that passes through the origin? Sketch that solution more accurately.

8. Show that every solution of the equation $y' = (x + y)^2$ has a graph that has a point of inflection, but no maximum or minimum. (The graph also has vertical asymptotes as you may discover by letting $x + y = u$.)

9. Use isoclines to sketch part of the direction field for $y' = x - y$. Include some part of all four quadrants.
  a) Which isocline is also a solution of the differential equation?
  b) Use the differential equation to determine where the solution curves are concave upward; concave downward.

Use Picard's iteration scheme to find $y_n(x)$ for $n = 0, 1, 2, 3$ in Problems 10–15.

10. $y' = x, \quad y(1) = 2$

11. $y' = y, \quad y(0) = 1$

12. $y' = xy, \quad y(1) = 1$

13. $y' = x + y, \quad y(0) = 0$

14. $y' = x + y, \quad y(0) = 1$

15. $y' = 2x - y, \quad y(-1) = 1$

16. Show that the general solution of

$$y' = x + y, \quad y(x_0) = y_0$$

is

$$y = -1 - x + (1 + x_0 + y_0)e^{x - x_0}.$$

17. In Example 3, verify the correctness of the equation for $y_3(x)$. (Find the mistake if there is one.)

***Toolkit* programs**

Antiderivatives and Direction Fields

18.15 Numerical Methods

The Euler Method

The initial value problem

$$y' = f(x, y), \qquad y(x_0) = y_0, \tag{1}$$

provides us with a starting point, $P(x_0, y_0)$, and a slope $f(x_0, y_0)$. We know that the graph of the solution must be a curve through P with that slope. If we use the tangent through P to approximate the actual solution curve, the approximation may be fairly good from $x_0 - h$ to $x_0 + h$, for small values of h. Thus, we might choose $h = 0.1$, say, and move along the tangent line from P to $P'(x_1, y_1)$ where $x_1 = x_0 + h$ and $y_1 = y_0 + hf(x_0, y_0)$. If we think of P' as a new starting point, we can move from P' to $P''(x_2, y_2)$ where $x_2 = x_1 + h$ and $y_2 = y_1 + hf(x_1, y_1)$. (This method is due to Euler.) If we replace h by $-h$, we move to the left from P instead of to the right. The process can be continued, but the errors are likely to accumulate as we take more steps. In order to see how the process works and to gain some idea of the errors, we illustrate for the initial value problem $y' = 1 + y$, $y(0) = 1$.

EXAMPLE 1 Take $h = 0.1$ and investigate the accuracy of the Euler approximation method for the initial value problem

$$y' = 1 + y, \qquad y(0) = 1, \tag{2}$$

by letting

$$x_{n+1} = x_n + h, \quad y_{n+1} = y_n + hf(x_n, y_n). \tag{3}$$

Continue to $x = 1$.

Solution The exact solution of Eqs. (2) is

$$y = 2e^x - 1. \tag{4}$$

The following table shows the results using Eqs. (3) and the exact results rounded to four decimals for comparison.

x	y(approx)	y(exact)	Error = y(exact) − y(approx)
0	1	1	0
0.1	1.2	1.2103	0.0103
0.2	1.42	1.4428	0.0228
0.3	1.662	1.6997	0.0377
0.4	1.9282	1.9836	0.0554
0.5	2.22102	2.2974	0.0764
0.6	2.543122	2.6442	0.1011
0.7	2.8974	3.0275	0.1301
0.8	3.2872	3.4511	0.1639
0.9	3.7159	3.9192	0.2033
1.0	4.1875	4.4366	0.2491

By the time we get to $x = 1$, the error is about 5.6%. □

The Improved Euler Method

This method first gets an estimate of y_{n+1}, as in the original Euler method, but calls the result z_{n+1} and then takes the average of $f(x_n, y_n)$ and $f(x_n, z_{n+1})$ in place of $f(x_n, y_n)$. Thus

$$z_{n+1} = y_n + hf(x_n, y_n), \tag{5a}$$

$$y_{n+1} = y_n + \frac{h}{2}[f(x_n, y_n) + f(x_n, z_{n+1})]. \tag{5b}$$

If we apply this improved method to Example 1, again with $h = 0.1$, we get the following results at $x = 1$:

$$\begin{aligned} y(\text{approx}) &= 4.428161693, \\ y(\text{exact}) &= 4.436563656, \\ \text{error} = y(\text{exact}) - y(\text{approx}) &= 0.008401963, \end{aligned}$$

and the error is less than 2/10 of 1%.

The Runge-Kutta Method

The improved Euler method just described corresponds to approximating an integral by the trapezoidal formula. The method now to be described is analogous to using Simpson's approximation to an integral. It requires four intermediate calculations, as given in the following equations:

$$\begin{aligned} k_1 &= hf(x_n, y_n) \\ k_2 &= hf\left(x_n + \frac{h}{2}, y_n + \frac{k_1}{2}\right) \\ k_3 &= hf\left(x_n + \frac{h}{2}, y_n + \frac{k_2}{2}\right) \\ k_4 &= hf(x_n + h, y_n + k_3). \end{aligned} \tag{6}$$

We then calculate y_{n+1} from the formula

$$y_{n+1} = y_n + \frac{1}{6}(k_1 + 2k_2 + 2k_3 + k_4). \tag{7}$$

When we apply this method to the problem of estimating $y(1)$ for the problem $y' = 1 + y$, $y(0) = 1$, still using $h = 0.1$, we get

$$y(\text{approx}) = 4.436559490$$

with an error 0.000004166, which is less than 1/10,000 of 1%. This is clearly the most accurate of the three methods and is not difficult to program for a calculator or computer.

The next example shows that the error in the Runge-Kutta approximation did not continue to increase as the process was continued. In fact, with $h = 0.1$, the difference between the exact solutions and the approximations remained less than 10^{-6} for the two initial value problems:

a) $y' = x - y, \quad y(0) = 1,$

b) $y' = x - y, \quad y(0) = -2.$

REMARK. The fact that the differential equation is linear in y is significant. Such accuracy is not attained for the initial value problem

$$y' = x^2 + y^2, \quad y(0) = 0,$$

which becomes infinite at $x = 2^+$. (See Problem 11.) The Runge-Kutta approximation to $y(2.1)$, using $h = 0.1$, is 1.47×10^{11}.

EXAMPLE 2 The following table shows the comparison of $y(x)$ as estimated by the Runge-Kutta method with $h = 0.1$ and the true value, for solutions of $y' = x - y$ (a) with $y(0) = 1$ and (b) with $y(0) = -2$.

x	y(Runge-Kutta)	y(true value)	Difference
0	1	1	0
0.5	0.713061869	0.713061319	5.50×10^{-7}
1.0	0.735759549	0.735758882	6.67×10^{-7}
1.5	0.946260927	0.946260320	6.07×10^{-7}
2.0	1.270671057	1.270670566	4.91×10^{-7}
2.5	1.664170370	1.664169997	3.73×10^{-7}
3.0	2.099574407	2.099574137	2.70×10^{-7}
0	−2	−2	0
0.5	−1.106530935	−1.106530660	-2.75×10^{-7}
1.0	−0.367879775	−0.367879441	-3.34×10^{-7}
1.5	+0.276869537	+0.276869840	-3.03×10^{-7}
2.0	0.864664472	0.864664717	-2.46×10^{-7}
2.5	1.417914816	1.417915001	-1.85×10^{-7}
3.0	1.950212796	1.950212932	-1.36×10^{-7}

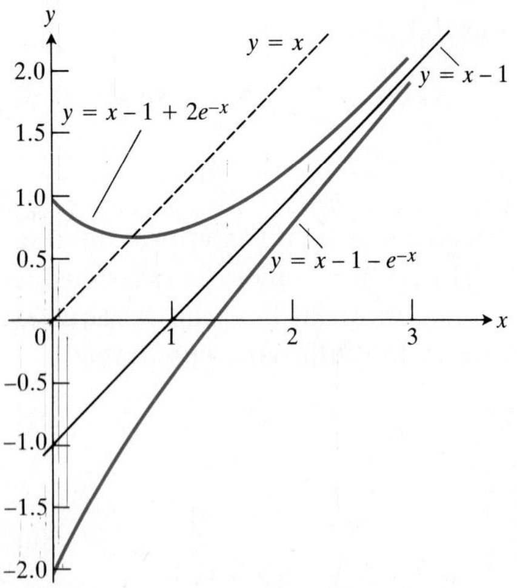

18.11 Two solutions of $y' = x - y$.
a) $y(0) = 1$, $y = x - 1 + 2e^{-x}$
b) $y(0) = -2$, $y = x - 1 - e^{-x}$

More points were actually computed and plotted to give the graphs shown in Fig. 18.11. Notice that the upper curve, $y = x - 1 + 2e^{-x}$ is concave upward and has a minimum when $x = y = \ln 2$. The lower curve is concave downward, is always rising as x increases, and crosses the x-axis

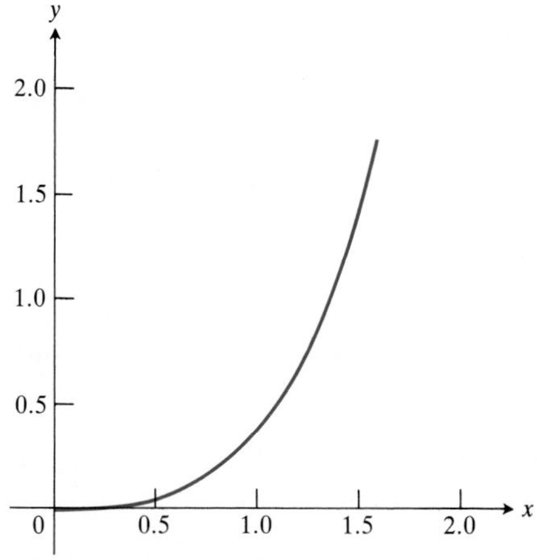

18.12 Approximate solution of $y' = x^2+y^2$, $y(0) = 0$.

at a value of x near 1.3. Both curves approach the line $y = x - 1$ as an asymptote as $x \to \infty$. □

For the initial value problem

$$y' = x^2 + y^2, \quad y(0) = 0, \tag{8}$$

some of the Runge-Kutta approximations we got with $h = 0.1$ are shown below. Part of the graph is shown in Fig. 18.12.

x	y(Runge-Kutta)	y(actual)
0	0	0
0.5	0.041791	0.041791146
1.0	0.350234	0.350231844
1.5	1.517473	1.517447544
2.0	71.578996	317.224400
2.1	1.470011×10^{11}	
2.2	9.999999×10^{99}	(meaning, "you broke the bank!")

Although there is no elementary solution of the equation $y' = x^2 + y^2$, it is possible to get an answer in terms of series. In the first place, the Maclaurin series that corresponds to the initial value $y(0) = 0$ begins this way:

$$y = \frac{x^3}{3} + \frac{x^7}{63} + \frac{2x^{11}}{6237} + \cdots. \tag{9}$$

In theory, one can get many more terms of the series, but the coefficients do not follow a simple pattern.

A more productive method involves an ingenious substitution

$$y = -\frac{u'}{u}, \tag{10a}$$

which transforms the equation $y' = x^2 + y^2$ into the second order equation

$$u'' + x^2u = 0. \tag{10b}$$

In Article 18.13, Eq. (9), we found the solution of Eq. (10b) in the form of two power series:

$$u_1(x) = 1 - \frac{x^4}{3 \cdot 4} + \frac{x^8}{3 \cdot 4 \cdot 7 \cdot 8} - \frac{x^{12}}{3 \cdot 4 \cdot 7 \cdot 8 \cdot 11 \cdot 12} + \cdots \tag{11a}$$

and

$$u_2(x) = x - \frac{x^5}{4 \cdot 5} + \frac{x^9}{4 \cdot 5 \cdot 8 \cdot 9} - \frac{x^{13}}{4 \cdot 5 \cdot 8 \cdot 9 \cdot 12 \cdot 13} + \cdots. \tag{11b}$$

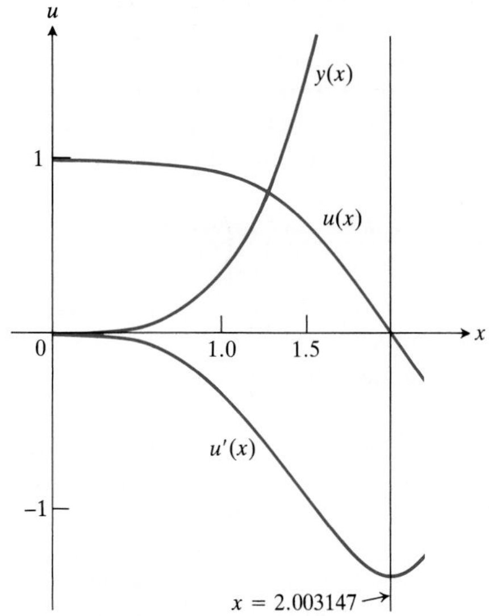

18.13 Graphs of $y(x)$, $u(x)$, and $u'(x)$ for the initial value problems

$$y' = x^2 + y^2, \; y(0) = 0$$
$$u'' + x^2u = 0, \; u(0) = 1, \; u'(0) = 0$$
$$y(x) = -u'/u.$$

The y-curve has an asymptote at $x =$ 2.003147 where $u = 0$.

The general solution of Eq. (10b) is

$$u = C_1u_1 + C_2u_2.$$

Notice that

$$u_1(0) = 1, \qquad u_2(0) = 0, \qquad u_1'(0) = 0, \qquad u_2'(0) = 1.$$

Therefore, if we wish to solve the general initial value problem

$$y' = x^2 + y^2, \qquad y(0) = y_0, \tag{12a}$$

using

$$y(x) = -\frac{u'(x)}{u(x)} = -\frac{C_1u_1'(x) + C_2u_2'(x)}{C_1u_1(x) + C_2u_2(x)}, \tag{12b}$$

we need to have

$$y(0) = -\frac{C_2}{C_1} = y_0.$$

We therefore take $C_1 = 1$, $C_2 = -y_0$. For specific values of y_0, it is relatively easy to compute values of the four series for $u_1(x)$, $u_2(x)$, $u_1'(x)$, and $u_2'(x)$ and combine them to obtain

$$y(x) = -\frac{u_1'(x) - y_0u_2'(x)}{u_1(x) - y_0u_2(x)}. \tag{12c}$$

We used this approach with $y_0 = 0$ to get the actual values of y to compare with the values obtained by using the Runge-Kutta method in the table following Eq. (8). (See Fig. 18.13.)

Problems 11 and 12 invite you to consider how you could predict that the solution of Eqs. (12a) becomes infinite near $x = 2$ if $y_0 = 0$, and before $x = 1$ if $y_0 = 1$.

PROBLEMS

1. Use the Euler method with $h = 1/5$ to estimate $y(1)$ if $y' = y$ and $y(0) = 1$. What is the exact value of $y(1)$?

2. Show that the Euler method leads to the estimate $(1 + (1/n))^n$ for $y(1)$ if $h = 1/n$, $y' = y$, and $y(0) = 1$. What is the limit as $n \to \infty$?

3. Use the improved Euler method with $h = 1/5$ to estimate $y(1)$ if $y' = y$ and $y(0) = 1$.

4. Use the Runge-Kutta method with $h = 1/5$ to estimate $y(1)$ if $y' = y$ and $y(0) = 1$.

In Problems 5–8, write an equivalent first order differential equation and initial condition for y.

5. $y = 1 + \int_0^x y(t)\,dt$

6. $y = -1 + \int_1^x t - y(t)\,dt$

7. $y = \int_1^x \frac{1}{t}\,dt$

8. $y = 2 - \int_0^x (1 + y(t))\sin t\,dt$

9. What integral equation is equivalent to the initial value problem $y' = f(x)$, $y(x_0) = y_0$?

10. If $f(x, y)$ does not depend upon y so that $f(x, y) = F(x)$, show that the Runge-Kutta method for approximating y_{n+1} is equivalent to using Simpson's approximation for

$$\int_{x_n}^{x_{n+1}} F(x)\,dx.$$

11. For the initial value problem $y' = x^2 + y^2$, $y(0) = 0$, let $y = f(a)$ be the value obtained at $x = a$, where a is positive. (For example, we could take $a = 1.5$, $f(a) = 1.517$.) As x increases beyond $x = a$, $x^2 \geq a^2$ and $y' = x^2 + y^2 \geq a^2 + y^2$. Therefore, $y(x)$ increases faster than the solution of the simpler problem $y' = a^2 + y^2$, $y(a) = f(a)$. Solve this simpler problem and thereby show that the solution of the original problem becomes infinite at a value of x no greater than

$$x^* = a + \frac{1}{a}\left(\frac{\pi}{2} - \tan^{-1}\left(\frac{f(a)}{a}\right)\right).$$

If you have a calculator, or access to a computer, show that $x^* = 2.0198$ if $a = 1.5$, $f(a) = 1.517$, and calculate x^* for $a = 2.0$, $f(a) = 317.2244$.

12. For the initial value problem $y' = x^2 + y^2$, $y(0) = y_0 = 1$, show that y increases faster than the solution of the simpler problem $y' = y^2$, $y(0) = 1$. Solve this simpler problem and thus show that the solution of the original problem becomes infinite at a value of x not greater than 1. (If you have a calculator or computer, you can use Eq. (12c) to find that y becomes infinite as x approaches 0.969810654, where $u_1 = u_2 = 0.927439876$.)

13. Find the first five nonzero terms in the Maclaurin series for the solution of the initial value problem $y' = x^2 + y^2$, $y(0) = 1$. Recall from Chapter 12 that if

$$y = a_0 + a_1x + a_2x^2 + \cdots + a_nx^n + \cdots$$

and

$$y^2 = c_0 + c_1x + c_2x^2 + \cdots + c_nx^n + \cdots,$$

then

$$c_n = \sum_{k=0}^{n} a_k a_{n-k},$$

and

$$y' = a_1 + 2a_2x + 3a_3x^2 + \cdots + na_nx^{n-1} + \cdots.$$

To satisfy the differential equation, we must have $a_1 = c_0$, $2a_2 = c_1$, $3a_3 = 1 + c_2$, and $na_n = c_{n-1}$ for $n \geq 4$. You can now determine a_0 (from the initial value), c_0, a_1, c_1, a_2, c_2, a_3, and so on.

14. Solve the problem $y' = 1 + y^2$, $y(0) = 0$, (a) by separating the variables, and (b) by using the substitution $y = -u'/u$ and solving the equivalent problem for u. (Note the similarity with the problem $y' = x^2 + y^2$, $y(0) = 0$. In the present problem we can find $u(x)$ without using series.)

***Toolkit* programs**	
First Order Initial Value Problem	Second Order Initial Value Problem

REVIEW QUESTIONS AND EXERCISES

1. List some differential equations (having physical interpretations) that you have come across in your courses in chemistry, physics, engineering, or life and social sciences; or look for some in the articles on differential equations, dynamics, electromagnetic waves, hydromechanics, quantum mechanics, or thermodynamics in the *Encyclopaedia Britannica*.

2. How are differential equations classified?

3. What is meant by a "solution" of a differential equation?

4. Review methods for solving ordinary, first order, and first degree differential equations
a) when the variables are separable,
b) when the equation is homogeneous,
c) when the equation is linear in one variable,
d) when the equation is exact.
Illustrate each with an example.

5. Review methods of solving second order equations
a) with dependent variable missing,
b) with independent variable missing.
Illustrate each with an example.

6. Review methods for solving linear differential equations with constant coefficients
a) in the homogeneous case,
b) in the nonhomogeneous case.
Illustrate each with an example.

7. If an external force F acts upon a system whose mass varies with time, Newton's law of motion is

$$\frac{d(mv)}{dt} = F + (v + u)\frac{dm}{dt}.$$

In this equation, m is the mass of the system at time t, v its velocity, and $v + u$ is the velocity of the mass that is entering (or leaving) the system at the rate dm/dt. Suppose that a rocket of initial mass m_0 starts from rest, but is driven upward by firing some of its mass directly backward at the constant rate of $dm/dt = -b$ units per second and at constant speed relative to the rocket $u = -c$. The only external force acting on the rocket is $F = -mg$ due to gravity. Under these assumptions, show that the height of the rocket above the ground at the end of t seconds (t small compared to m_0/b) is

$$y = c\left[t + \frac{m_0 - bt}{b}\ln\frac{m_0 - bt}{m_0}\right] - \frac{1}{2}gt^2.$$

8. Which of the following differential equations *cannot* be solved as a first order, exact differential equation?
a) $y^2\,dx + 2xy\,dy = 0$
b) $(2x \sin y + y^3e^x)\,dx + (x^2 \cos y + 3y^2e^x)\,dy = 0$
c) $(2x \cos y + 3x^2y)\,dx + (x^3 - x^2 \sin y - y)\,dy = 0$
d) $(y \ln y - e^{-xy})\,dx + \left(\frac{1}{y} + x \ln y\right)dy = 0$

9. The differential equation $(x^2 - y^3)y' = 2xy$ has an integrating factor of the form y^n. Find n and solve the equation.

10. If y is a solution of $(D^2 + 4)y = e^x$, show that it satisfies $(D - 1)(D^2 + 4)y = 0$, because $(D - 1)e^x = 0$, where $D = d/dx$. Find the general solution of $(D - 1)(D^2 + 4)y = 0$ and use the result to solve $(D^2 + 4)y = e^x$.

11. If you were to solve $y' = y^2$, $y(0) = 1$, by finding the Maclaurin series for $y(x)$, for what values of x would you expect the series to converge? Why?

Suggestion. Solve the initial value problem without the use of series, then expand your answer in a series.

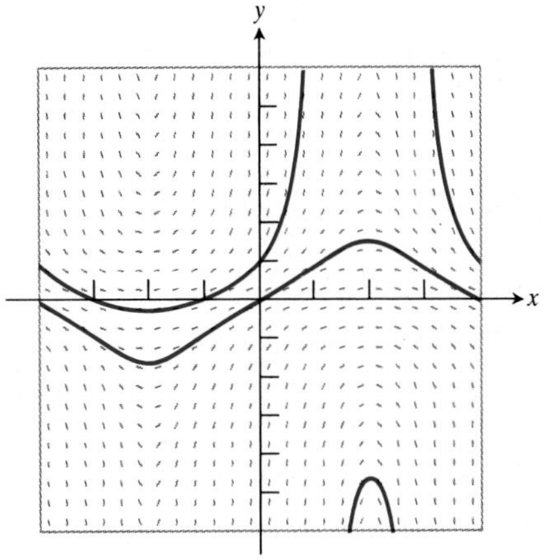

18.14 Solutions: $y = \tan(\sin x + D)$ for $D = 0, \pi/4$

12. A student of mathematics and computer science used a computer to graph solutions of

$$y' = (1 + y^2) \cos x,$$

a) for $y(0) = 0$, and
b) for $y(0) = 1$.

One curve was continuous and very well behaved. The other blew up! By solving the equation for an arbitrary initial value $y(0) = y_0$, find the set of values of y_0 for which the solution remains bounded. If a solution does not remain bounded, where are the asymptotes of its graph? See Fig. 18.14.

MISCELLANEOUS PROBLEMS

Solve the following differential equations.

1. $y \ln y \, dx + (1 + x^2)\, dy = 0$

2. $\dfrac{dy}{dx} = \dfrac{y^2 - y - 2}{x^2 + x}$

3. $e^{x+2y}\, dy - e^{y-2}\, dx = 0$

4. $\sqrt{1 + \left(\dfrac{dy}{dx}\right)^2} = ky$

5. $y\, dy = \sqrt{1 + y^4}\, dx$

6. $(2x + y)\, dx + (x - 2y)\, dy = 0$

7. $\dfrac{dy}{dx} = \dfrac{x^2 + y^2}{2xy}$

8. $x\dfrac{dy}{dx} = y + \sqrt{x^2 + y^2}$

9. $x\, dy = \left(y + x \cos^2 \dfrac{y}{x}\right) dx$

10. $x(\ln y - \ln x)\, dy = y(1 + \ln y - \ln x)\, dx$

11. $x\, dy + (2y - x^2 - 1)\, dx = 0$

12. $\cos y\, dx + (x \sin y - \cos^2 y)\, dy = 0$

13. $\cosh x\, dy - (y + \cosh x) \sinh x\, dx = 0$

14. $(x + 1)\, dy + (2y - x)\, dx = 0$

15. $(1 + y^2)\, dx + (2xy + y^2 + 1)\, dy = 0$

16. $(x^2 + y)\, dx + (e^y + x)\, dy = 0$

17. $(x^2 + y^2)\, dx + (2xy + \cosh y)\, dy = 0$

18. $(e^x + \ln y)\, dx + \dfrac{x + y}{y}\, dy = 0$

19. $x(1 + e^y)\, dx + \frac{1}{2}(x^2 + y^2)e^y\, dy = 0$

20. $\left(\sin x + \tan^{-1} \dfrac{y}{x}\right) dx - (y - \ln \sqrt{x^2 + y^2})\, dy = 0$

21. $\dfrac{d^2y}{dx^2} - 2y\dfrac{dy}{dx} = 0$

22. $\dfrac{d^2x}{dy^2} + 4x = 0$

23. $\dfrac{d^2y}{dx^2} = 1 + \left(\dfrac{dy}{dx}\right)^2$

24. $\dfrac{d^2x}{dy^2} = 1 - \left(\dfrac{dx}{dy}\right)^2$

25. $x^2\dfrac{d^2y}{dx^2} + x\dfrac{dy}{dx} = 1$

26. $\dfrac{d^2y}{dx^2} - 4\dfrac{dy}{dx} + 3y = 0$

27. $\dfrac{d^3y}{dx^3} - 2\dfrac{d^2y}{dx^2} + \dfrac{dy}{dx} = 0$

28. $\dfrac{d^2y}{dx^2} + 4y = \sec 2x$

29. $\dfrac{d^2y}{dx^2} - \dfrac{dy}{dx} - 2y = e^{2x}$

30. $\dfrac{d^2y}{dx^2} - 2\dfrac{dy}{dx} + 5y = e^{-x}$

31. Find the *general solution* of the differential equation $4x^2y'' + 4xy' - y = 0$, given that there is a particular solution of the form $y = x^c$ for some constant c.

32. Show that the only plane curves that have constant curvature are circles and straight lines. (Assume the appropriate derivatives exist and are continuous.)

33. Find the orthogonal trajectories of the family of curves $x^2 = Cy^3$. (*Caution:* The differential equation should not contain the arbitrary constant C.)

34. Find the orthogonal trajectories of the family of circles $(x - C)^2 + y^2 = C^2$.

35. Find the orthogonal trajectories of the family of parabolas $y^2 = 4C(C - x)$.

36. The equation $d^2y/dt^2 + 100y = 0$ represents a simple harmonic motion. Find the general solution of the equation and determine the constants of integration if $y = 10$, $dy/dt = 50$, when $t = 0$. Find the period and the amplitude of the motion.

Problems 37–39 refer to the differential equation $dy/dx = x + \sin y$. (See Fig. 18.15.)

37. For solutions near $y = 0$, we might use the approximation $\sin y \approx y$ and replace the original problem by $dy/dx = x + y$. Solve this equation with $y(0) = 0$. What is the Maclaurin series for your answer?

38. (*Series*) Given the initial value problem $y' = x + \sin y$, $y(0) = 0$. By implicit differentiation we can calculate successive derivatives. For example, $y'' = 1 + (\cos y)y'$, $y''' = -(\sin y)(y')^2 + (\cos y)y''$. These can, in turn, be evaluated at $x = 0$ by using the given initial value $y(0) = 0$. Using this procedure, find the terms of the Maclaurin series for $y(x)$ through x^4. (Compare the answer to Problem 37.)

39. Using the method of the preceding problem, find the terms through x^4 in the Maclaurin series for $y(x)$ if y satisfies $y' = x + \sin y$, and (a) $y(0) = \pi/2$. (b) Repeat for $y(0) = -\pi/2$.

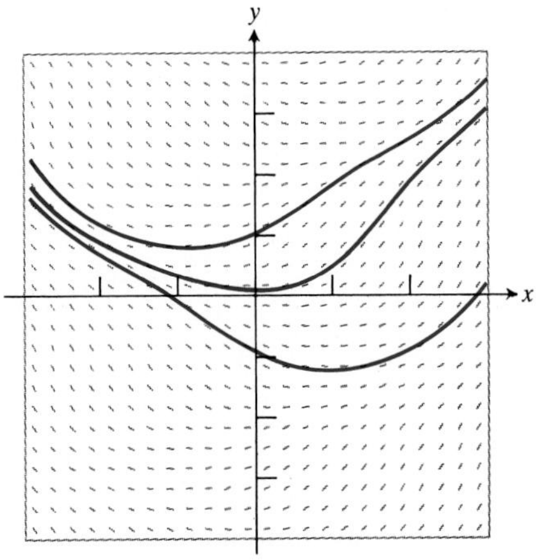

18.15 Solutions to: $dy/dx = x + \sin y$ passing through: $(0, 0)$, $(0, \pi/2)$, $(0, -\pi/2)$

APPENDIXES

A.1

Determinants and Cramer's Rule

A rectangular array of numbers like

$$A = \begin{bmatrix} 2 & 1 & 3 \\ 1 & 0 & -2 \end{bmatrix}$$

is called a *matrix*. We call A a "2 by 3" matrix because it has two rows and three columns. An "m by n" matrix has m rows and n columns, and the *entry* or *element* (number) in the ith row and jth column is often denoted by a_{ij}:

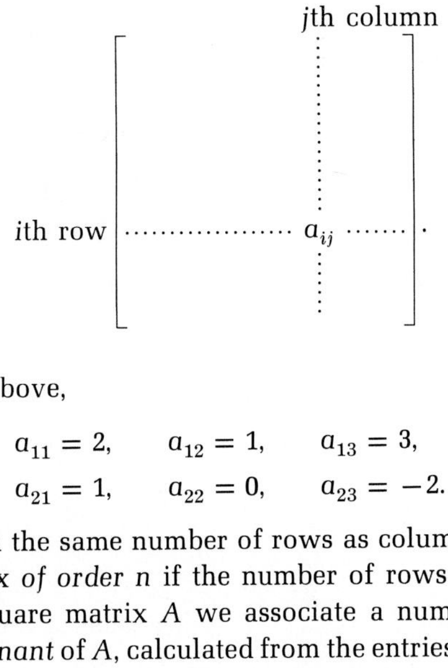

In the matrix A above,

$$a_{11} = 2, \quad a_{12} = 1, \quad a_{13} = 3,$$
$$a_{21} = 1, \quad a_{22} = 0, \quad a_{23} = -2.$$

A matrix with the same number of rows as columns is a *square* matrix. It is a *matrix of order n* if the number of rows and columns is n.

With each square matrix A we associate a number det (A) or $|a_{ij}|$ called the *determinant* of A, calculated from the entries of A in the following way. (The vertical bars in the notation $|a_{ij}|$ do not mean absolute value.) For $n = 1$ and $n = 2$ we define

$$\det\,[a] = a, \tag{1}$$

$$\det \begin{bmatrix} a_{11} & a_{12} \\ a_{21} & a_{22} \end{bmatrix} = a_{11}a_{22} - a_{21}a_{12}. \tag{2}$$

For a matrix of order 3, we define

$$\det(A) = \det\begin{bmatrix} a_{11} & a_{12} & a_{13} \\ a_{21} & a_{22} & a_{23} \\ a_{31} & a_{32} & a_{33} \end{bmatrix} = \begin{matrix}\text{Sum of all signed products} \\ \text{of the form } \pm a_{1i}a_{2j}a_{3k},\end{matrix} \tag{3}$$

where i, j, k is a permutation of 1, 2, 3 in some order. There are $3! = 6$ such permutations, so there are six terms in the sum. Half of these have plus signs and the other half have minus signs, according to the index of the permutation, where the index is a number we next define.

DEFINITION

Index of a Permutation

Given any permutation of the numbers 1, 2, 3, . . . , n, denote the permutation by $i_1, i_2, i_3, \ldots, i_n$. In this arrangement, some of the numbers following i_1 may be less than i_1, and however many of these there are is called the *number of inversions* in the arrangement pertaining to i_1. Likewise, there is a number of inversions pertaining to each of the other i's; it is the number of indices that come after that particular one in the arrangement and are less than it. The *index* of the permutation is the sum of all of the numbers of inversions pertaining to the separate indices.

EXAMPLE 1 For $n = 5$, the permutation

$$5\ 3\ 1\ 2\ 4$$

has

4 inversions pertaining to the first element, 5,

2 inversions pertaining to the second element, 3,

and no further inversions, so the index is $4 + 2 = 6$. □

The table below shows the permutations of 1, 2, 3 and the index of each permutation. The signed product in the determinant of Eq. (3) is also shown.

Permutation	*Index*	*Signed product*
1 2 3	0	$+a_{11}a_{22}a_{33}$
1 3 2	1	$-a_{11}a_{23}a_{32}$
2 1 3	1	$-a_{12}a_{21}a_{33}$
2 3 1	2	$+a_{12}a_{23}a_{31}$
3 1 2	2	$+a_{13}a_{21}a_{32}$
3 2 1	3	$-a_{13}a_{22}a_{31}$

(4)

The sum of the six signed products is

$$a_{11}(a_{22}a_{33} - a_{23}a_{32}) - a_{12}(a_{21}a_{33} - a_{23}a_{31}) + a_{13}(a_{21}a_{32} - a_{22}a_{31})$$

$$= a_{11}\begin{vmatrix} a_{22} & a_{23} \\ a_{32} & a_{33} \end{vmatrix} - a_{12}\begin{vmatrix} a_{21} & a_{23} \\ a_{31} & a_{33} \end{vmatrix} + a_{13}\begin{vmatrix} a_{21} & a_{22} \\ a_{31} & a_{32} \end{vmatrix} \tag{5}$$

$$= \begin{vmatrix} a_{11} & a_{12} & a_{13} \\ a_{21} & a_{22} & a_{23} \\ a_{31} & a_{32} & a_{33} \end{vmatrix},$$

and the formula

$$\begin{vmatrix} a_{11} & a_{12} & a_{13} \\ a_{21} & a_{22} & a_{23} \\ a_{31} & a_{32} & a_{33} \end{vmatrix} = a_{11}\begin{vmatrix} a_{22} & a_{23} \\ a_{32} & a_{33} \end{vmatrix} - a_{12}\begin{vmatrix} a_{21} & a_{23} \\ a_{31} & a_{33} \end{vmatrix} + a_{13}\begin{vmatrix} a_{21} & a_{22} \\ a_{31} & a_{32} \end{vmatrix} \tag{6}$$

is often used to calculate 3 by 3 determinants.

Equation (6) reduces the calculation of a 3 by 3 determinant to the calculation of three 2 by 2 determinants.

Many people prefer to remember the following scheme for calculating the six signed products in the determinant of a 3 by 3 matrix:

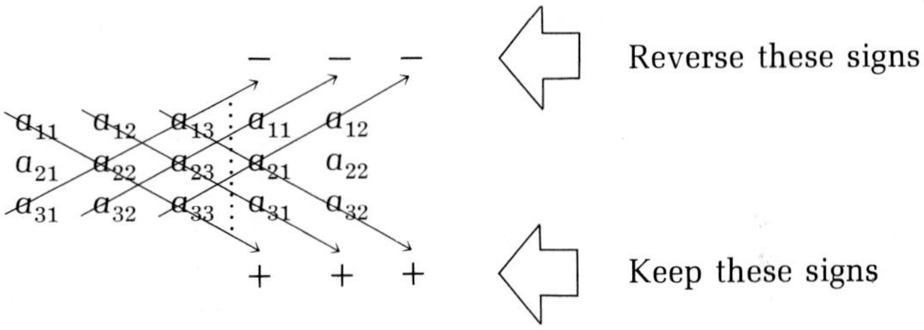

(7)

Minors and Cofactors

The second order determinants on the right-hand side of Eq. (6) are called the *minors* (short for minor determinant) of the entries they multiply. Thus,

$$\begin{vmatrix} a_{22} & a_{23} \\ a_{32} & a_{33} \end{vmatrix} \text{ is the minor of } a_{11},$$

$$\begin{vmatrix} a_{21} & a_{23} \\ a_{31} & a_{33} \end{vmatrix} \text{ is the minor of } a_{12},$$

and so on. The minor of the element a_{ij} in a matrix A is the determinant of the matrix that remains when the row and column containing a_{ij} are deleted:

$$\begin{vmatrix} a_{11} & \cancel{a_{12}} & a_{13} \\ \cancel{a_{21}} & \cancel{a_{22}} & \cancel{a_{23}} \\ a_{31} & \cancel{a_{32}} & a_{33} \end{vmatrix}. \quad \text{The minor of } a_{22} \text{ is } \begin{vmatrix} a_{11} & a_{13} \\ a_{31} & a_{33} \end{vmatrix}.$$

$$\begin{vmatrix} a_{11} & a_{12} & \cancel{a_{13}} \\ \cancel{a_{21}} & \cancel{a_{22}} & \cancel{a_{23}} \\ a_{31} & a_{32} & \cancel{a_{33}} \end{vmatrix}. \quad \text{The minor of } a_{23} \text{ is } \begin{vmatrix} a_{11} & a_{12} \\ a_{31} & a_{32} \end{vmatrix}.$$

The *cofactor* of a_{ij} is the determinant A_{ij} that is $(-1)^{i+j}$ times the minor of a_{ij}. Thus,

$$A_{22} = (-1)^{2+2}\begin{vmatrix} a_{11} & a_{13} \\ a_{31} & a_{33} \end{vmatrix} = \begin{vmatrix} a_{11} & a_{13} \\ a_{31} & a_{33} \end{vmatrix},$$

$$A_{23} = (-1)^{2+3}\begin{vmatrix} a_{11} & a_{12} \\ a_{31} & a_{32} \end{vmatrix} = -\begin{vmatrix} a_{11} & a_{12} \\ a_{31} & a_{32} \end{vmatrix}.$$

The effect of the factor $(-1)^{i+j}$ is to change the sign of the minor when the sum $i + j$ is odd. There is a checkerboard pattern for remembering these sign changes:

$$\begin{matrix} + & - & + \\ - & + & - \\ + & - & + \end{matrix}$$

In the upper left corner, $i = 1$, $j = 1$ and $(-1)^{1+1} = +1$. In going from any cell to an adjacent cell in the same row or column, we change i by 1 or j by 1, but not both, so we change the exponent from even to odd or from odd to even, which changes the sign from + to − or from − to +.

When we rewrite Eq. (6) in terms of cofactors we get

$$\det(A) = a_{11}A_{11} + a_{12}A_{12} + a_{13}A_{13}. \tag{8}$$

EXAMPLE 2 Find the determinant of the matrix

$$A = \begin{bmatrix} 2 & 1 & 3 \\ 3 & -1 & -2 \\ 2 & 3 & 1 \end{bmatrix}.$$

Solution 1 The cofactors are

$$A_{11} = (-1)^{1+1}\begin{vmatrix} -1 & -2 \\ 3 & 1 \end{vmatrix}, \qquad A_{12} = (-1)^{1+2}\begin{vmatrix} 3 & -2 \\ 2 & 1 \end{vmatrix},$$

$$A_{13} = (-1)^{1+3}\begin{vmatrix} 3 & -1 \\ 2 & 3 \end{vmatrix}.$$

To find $\det(A)$, we multiply each element of the first row of A by its cofactor and add:

$$\det(A) = 2\begin{vmatrix} -1 & -2 \\ 3 & 1 \end{vmatrix} + (-1)\begin{vmatrix} 3 & -2 \\ 2 & 1 \end{vmatrix} + 3\begin{vmatrix} 3 & -1 \\ 2 & 3 \end{vmatrix}$$

$$= 2(-1 + 6) - 1(3 + 4) + 3(9 + 2) = 10 - 7 + 33 = 36.$$

Solution 2 From Eq. (7) we find

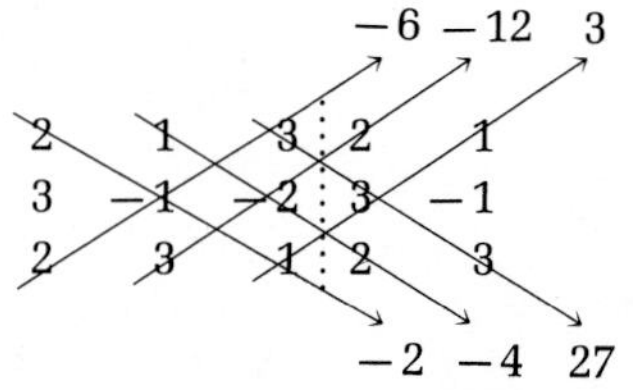

$$\det(A) = -(-6) - (-12) - 3 + (-2) + (-4) + 27 = 36. \ \square$$

Expanding by Columns or by Other Rows

The determinant of a square matrix can be calculated from the cofactors of any row or any column.

If we were to expand the determinant in Example 2 by cofactors according to elements of its third column, say, we would get

$$+3\begin{vmatrix} 3 & -1 \\ 2 & 3 \end{vmatrix} - (-2)\begin{vmatrix} 2 & 1 \\ 2 & 3 \end{vmatrix} + 1\begin{vmatrix} 2 & 1 \\ 3 & -1 \end{vmatrix}$$

$$= 3(9 + 2) + 2(6 - 2) + 1(-2 - 3) = 33 + 8 - 5 = 36.$$

Helpful Properties of Determinants

We now state some facts about determinants. You should know and be able to use these facts, but we omit the proofs.

FACT 1. If two rows of a matrix are identical, the determinant is zero.

FACT 2. If two rows of a matrix are interchanged, the determinant changes sign.

FACT 3. The determinant of a matrix is the sum of the products of the elements of the ith row (or column) by their cofactors, for any i.

FACT 4. The determinant of the transpose of a matrix is equal to the original determinant. ("Transpose" means write the rows as columns.)

FACT 5. If each element of some row (or column) of a matrix is multiplied by a constant c, the determinant is multiplied by c.

FACT 6. If all elements of a matrix above the main diagonal (or all below it) are zero, the determinant of the matrix is the product of the elements on the main diagonal.

EXAMPLE 3

$$\begin{vmatrix} 3 & 4 & 7 \\ 0 & -2 & 5 \\ 0 & 0 & 5 \end{vmatrix} = (3)(-2)(5) = -30. \quad \square$$

FACT 7. If the elements of any row of a matrix are multiplied by the cofactors of the corresponding elements of a different row and these products are summed, the sum is zero.

EXAMPLE 4 If A_{11}, A_{12}, A_{13} are the cofactors of the elements of the first row of $A = (a_{ij})$, then the sums

$$a_{21}A_{11} + a_{22}A_{12} + a_{23}A_{13}$$

(elements of second row times cofactors of elements of first row) and

$$a_{31}A_{11} + a_{32}A_{12} + a_{33}A_{13}$$

are both zero. A similar result holds for columns. □

FACT 8. If the elements of any column of a matrix are multiplied by the cofactors of the corresponding elements of a different column and these products are summed, the sum is zero.

FACT 9. If each element of a row of a matrix is multiplied by a constant c and the results added to a different row, the determinant is not changed.

EXAMPLE 5 Evaluate the fourth order determinant

$$\begin{vmatrix} 1 & -2 & 3 & 1 \\ 2 & 1 & 0 & 2 \\ -1 & 2 & 1 & -2 \\ 0 & 1 & 2 & 1 \end{vmatrix}.$$

Solution By adding appropriate multiples of row 1 to rows 2 and 3, we can get the equal determinant

$$\begin{vmatrix} 1 & -2 & 3 & 1 \\ 0 & 5 & -6 & 0 \\ 0 & 0 & 4 & -1 \\ 0 & 1 & 2 & 1 \end{vmatrix}.$$

We could further reduce this to a triangular matrix (one with zeros below the main diagonal), or expand it by cofactors of its first column. By multiplying the elements of the first column by their respective cofactors, we get the third order determinant

$$\begin{vmatrix} 5 & -6 & 0 \\ 0 & 4 & -1 \\ 1 & 2 & 1 \end{vmatrix} = 5(4+2) - (-6)(0+1) + 0 = 36. \ \square$$

Cramer's Rule

Cramer's rule is a rule for solving a system of linear equations like

$$\begin{aligned} a_{11}x + a_{12}y &= b_1, \\ a_{21}x + a_{22}y &= b_2, \end{aligned} \tag{9}$$

when the determinant of the coefficient matrix

$$D = \det(A) = \begin{vmatrix} a_{11} & a_{12} \\ a_{21} & a_{22} \end{vmatrix}$$

is different from zero. The solution of (9) is unique when $D \neq 0$ and may be calculated from the formulas

$$x = \frac{\begin{vmatrix} b_1 & a_{12} \\ b_2 & a_{22} \end{vmatrix}}{D}, \qquad y = \frac{\begin{vmatrix} a_{11} & b_1 \\ a_{21} & b_2 \end{vmatrix}}{D}. \tag{10}$$

The numerator in the formula for x comes from replacing the first column in A (the x-column) by the column of constants b_1 and b_2 (the b-column). Replacing the y-column by the b-column gives the numerator of the y-solution.

For a system of three equations in three unknowns,

$$\begin{aligned} a_{11}x + a_{12}y + a_{13}z &= b_1, \\ a_{21}x + a_{22}y + a_{23}z &= b_2, \\ a_{31}x + a_{32}y + a_{33}z &= b_3, \end{aligned}$$

Cramer's rule gives

$$\begin{aligned} x &= \frac{1}{D}\begin{vmatrix} b_1 & a_{12} & a_{13} \\ b_2 & a_{22} & a_{23} \\ b_3 & a_{32} & a_{33} \end{vmatrix}, \\ y &= \frac{1}{D}\begin{vmatrix} a_{11} & b_1 & a_{13} \\ a_{21} & b_2 & a_{23} \\ a_{31} & b_3 & a_{33} \end{vmatrix}, \\ z &= \frac{1}{D}\begin{vmatrix} a_{11} & a_{12} & b_1 \\ a_{21} & a_{22} & b_2 \\ a_{31} & a_{32} & b_3 \end{vmatrix}, \end{aligned} \tag{11}$$

where

$$D = \det(A) = \begin{vmatrix} a_{11} & a_{12} & a_{13} \\ a_{21} & a_{22} & a_{23} \\ a_{31} & a_{32} & a_{33} \end{vmatrix}.$$

The pattern continues in all higher dimensions: If $AX = B$, and $\det(A) \neq 0$, then

$$x_i = \frac{|U_i|}{|A|}, \qquad i = 1, 2, \ldots, n \tag{12}$$

where U_i is the matrix obtained from A by replacing the ith column in A by the b-column.

Other Formulas that Use Determinants

1. The equation of a line through two points (x_1, y_1) and (x_2, y_2) in the plane is

$$\begin{vmatrix} x & y & 1 \\ x_1 & y_1 & 1 \\ x_2 & y_2 & 1 \end{vmatrix} = 0.$$

2. The area of a triangle with vertices (x_1, y_1), (x_2, y_2), (x_3, y_3) is

$$\pm\frac{1}{2}\begin{vmatrix} x_1 & y_1 & 1 \\ x_2 & y_2 & 1 \\ x_3 & y_3 & 1 \end{vmatrix}.$$

3. The cross product of $\mathbf{A} = a_1\mathbf{i} + a_2\mathbf{j} + a_3\mathbf{k}$ and $\mathbf{B} = b_1\mathbf{i} + b_2\mathbf{j} + b_3\mathbf{k}$ is

$$\mathbf{A} \times \mathbf{B} = \begin{vmatrix} \mathbf{i} & \mathbf{j} & \mathbf{k} \\ a_1 & a_2 & a_3 \\ b_1 & b_2 & b_3 \end{vmatrix}.$$

4. The volume of the parallelepiped spanned by the vectors **A, B,** and **C** is

$$\pm\mathbf{A} \cdot (\mathbf{B} \times \mathbf{C}) = \pm\begin{vmatrix} a_1 & a_2 & a_3 \\ b_1 & b_2 & b_3 \\ c_1 & c_2 & c_3 \end{vmatrix}.$$

5. If $F = M(x, y, z)\mathbf{i} + N(x, y, z)\mathbf{j} + P(x, y, z)\mathbf{k}$, then

$$\text{curl}\,\mathbf{F} = \nabla \times \mathbf{F} = \begin{vmatrix} \mathbf{i} & \mathbf{j} & \mathbf{k} \\ \dfrac{\partial}{\partial x} & \dfrac{\partial}{\partial y} & \dfrac{\partial}{\partial z} \\ M & N & P \end{vmatrix}.$$

A Reduction Formula for Evaluating Determinants

The following reduction formula is derived in E. Miller's article "Evaluating an nth order determinant in n easy steps," *MATYC Journal* 12 (1978), 123–128. Evaluating determinants with this formula is relatively fast and readily programmable. For a short FORTRAN IV program, see Alban J. Rogues's article, "Determinants: A Short Program," *Two-Year College Mathematics Journal* Vol. 10, No. 5 (November 1979), pp. 340–342.

The formula for the determinant of an n by n matrix $A = (a_{ij})$ is

$$\det(A) = \left(\frac{1}{a_{11}}\right)^{n-2} \begin{vmatrix} \begin{vmatrix} a_{11} & a_{12} \\ a_{21} & a_{22} \end{vmatrix} & \begin{vmatrix} a_{11} & a_{13} \\ a_{21} & a_{23} \end{vmatrix} & \begin{vmatrix} a_{11} & a_{14} \\ a_{21} & a_{24} \end{vmatrix} & \cdots & \begin{vmatrix} a_{11} & a_{1n} \\ a_{21} & a_{2n} \end{vmatrix} \\ \begin{vmatrix} a_{11} & a_{12} \\ a_{31} & a_{32} \end{vmatrix} & \begin{vmatrix} a_{11} & a_{13} \\ a_{31} & a_{33} \end{vmatrix} & \begin{vmatrix} a_{11} & a_{14} \\ a_{31} & a_{34} \end{vmatrix} & \cdots & \begin{vmatrix} a_{11} & a_{1n} \\ a_{31} & a_{3n} \end{vmatrix} \\ \begin{vmatrix} a_{11} & a_{12} \\ a_{41} & a_{42} \end{vmatrix} & \begin{vmatrix} a_{11} & a_{13} \\ a_{41} & a_{43} \end{vmatrix} & \begin{vmatrix} a_{11} & a_{14} \\ a_{41} & a_{44} \end{vmatrix} & \cdots & \begin{vmatrix} a_{11} & a_{1n} \\ a_{41} & a_{4n} \end{vmatrix} \\ \vdots & \vdots & \vdots & & \vdots \\ \begin{vmatrix} a_{11} & a_{12} \\ a_{n1} & a_{n2} \end{vmatrix} & \begin{vmatrix} a_{11} & a_{13} \\ a_{n1} & a_{n3} \end{vmatrix} & \begin{vmatrix} a_{11} & a_{14} \\ a_{n1} & a_{n4} \end{vmatrix} & \cdots & \begin{vmatrix} a_{11} & a_{1n} \\ a_{n1} & a_{nn} \end{vmatrix} \end{vmatrix}$$

EXAMPLE 6

$$\begin{vmatrix} 1 & 0 & 2 & -1 \\ 3 & -2 & 6 & 4 \\ 5 & 4 & 3 & 0 \\ 2 & 2 & -5 & 6 \end{vmatrix} = 1^2 \begin{vmatrix} \begin{vmatrix} 1 & 0 \\ 3 & -2 \end{vmatrix} & \begin{vmatrix} 1 & 2 \\ 3 & 6 \end{vmatrix} & \begin{vmatrix} 1 & -1 \\ 3 & 4 \end{vmatrix} \\ \begin{vmatrix} 1 & 0 \\ 5 & 4 \end{vmatrix} & \begin{vmatrix} 1 & 2 \\ 5 & 3 \end{vmatrix} & \begin{vmatrix} 1 & -1 \\ 5 & 0 \end{vmatrix} \\ \begin{vmatrix} 1 & 0 \\ 2 & 2 \end{vmatrix} & \begin{vmatrix} 1 & 2 \\ 2 & -5 \end{vmatrix} & \begin{vmatrix} 1 & -1 \\ 2 & 6 \end{vmatrix} \end{vmatrix}$$

$$= \begin{vmatrix} -2 & 0 & 7 \\ 4 & -7 & 5 \\ 2 & -9 & 8 \end{vmatrix} = \left(-\frac{1}{2}\right) \begin{vmatrix} 14 & -38 \\ 18 & -30 \end{vmatrix} = -132. \ \square$$

PROBLEMS

Evaluate the following determinants.

1. $\begin{vmatrix} 2 & 3 & 1 \\ 4 & 5 & 2 \\ 1 & 2 & 3 \end{vmatrix}$

2. $\begin{vmatrix} 2 & -1 & -2 \\ -1 & 2 & 1 \\ 3 & 0 & -3 \end{vmatrix}$

3. $\begin{vmatrix} 1 & 2 & 3 & 4 \\ 0 & 1 & 2 & 3 \\ 0 & 0 & 2 & 1 \\ 0 & 0 & 3 & 2 \end{vmatrix}$

4. $\begin{vmatrix} 1 & -1 & 2 & 3 \\ 2 & 1 & 2 & 6 \\ 1 & 0 & 2 & 3 \\ -2 & 2 & 0 & -5 \end{vmatrix}$

Evaluate the following determinants by expanding according to the cofactors of (a) the third row, and (b) the the second column.

5. $\begin{vmatrix} 2 & -1 & 2 \\ 1 & 0 & 3 \\ 0 & 2 & 1 \end{vmatrix}$

6. $\begin{vmatrix} 1 & 0 & -1 \\ 0 & 2 & -2 \\ 2 & 0 & 1 \end{vmatrix}$

7. $\begin{vmatrix} 1 & 1 & 0 & 0 \\ 0 & 0 & -2 & 1 \\ 0 & -1 & 0 & 7 \\ 3 & 0 & 2 & 1 \end{vmatrix}$

8. $\begin{vmatrix} 0 & 1 & 0 & 0 \\ 0 & 1 & 1 & 0 \\ 1 & 1 & 1 & 1 \\ 1 & 1 & 0 & 0 \end{vmatrix}$

Solve the following systems of equations by Cramer's rule.

9. $\begin{aligned} x + 8y &= 4 \\ 3x - y &= -13 \end{aligned}$

10. $\begin{aligned} 2x + 3y &= 5 \\ 3x - y &= 2 \end{aligned}$

11. $\begin{aligned} 4x - 3y &= 6 \\ 3x - 2y &= 5 \end{aligned}$

12. $\begin{aligned} x + y + z &= 2 \\ 2x - y + z &= 0 \\ x + 2y - z &= 4 \end{aligned}$

13. $\begin{aligned} 2x + y - z &= 2 \\ x - y + z &= 7 \\ 2x + 2y + z &= 4 \end{aligned}$

14. $\begin{aligned} 2x - 4y &= 6 \\ x + y + z &= 1 \\ 5y + 7z &= 10 \end{aligned}$

15. $\begin{aligned} x - z &= 3 \\ 2y - 2z &= 2 \\ 2x + z &= 3 \end{aligned}$

16. $\begin{aligned} x_1 + x_2 - x_3 + x_4 &= 2 \\ x_1 - x_2 + x_3 + x_4 &= -1 \\ x_1 + x_2 + x_3 - x_4 &= 2 \\ x_1 + x_3 + x_4 &= -1 \end{aligned}$

A.2
Matrices and Linear Equations

A rectangular array of numbers like

$$A = \begin{bmatrix} 2 & 1 & 3 \\ 1 & 0 & -2 \end{bmatrix}$$

is called a *matrix*. We call A a "2 by 3" matrix because it has two rows and three columns. An "m by n" matrix has m rows and n columns, and the *entry* or *element* (number) in the ith row and jth column is often denoted by a_{ij}:

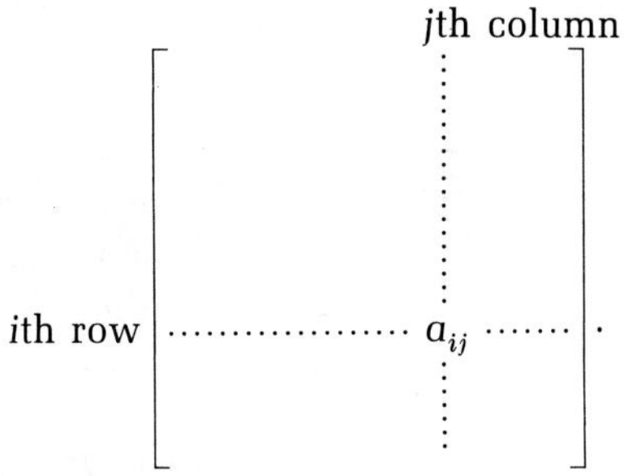

In the matrix A above,

$$a_{11} = 2, \quad a_{12} = 1, \quad a_{13} = 3,$$
$$a_{21} = 1, \quad a_{22} = 0, \quad a_{23} = -2.$$

Matrix Addition and Multiplication

Two matrices are equal if (and only if) they have the same numbers in the same positions. For example,

$$B = \begin{bmatrix} b_{11} & b_{12} & b_{13} \\ b_{21} & b_{22} & b_{23} \end{bmatrix} = \begin{bmatrix} 2 & 1 & 3 \\ 1 & 0 & -2 \end{bmatrix}$$

if and only if

$$b_{11} = 2, \quad b_{12} = 1, \quad b_{13} = 3,$$
$$b_{21} = 1, \quad b_{22} = 0, \quad b_{23} = -2.$$

For matrices A and B to be equal they must have the same *shape* (same number of rows and same number of columns) and must have

$$a_{ij} = b_{ij} \qquad \text{for all } i \text{ and } j.$$

Two matrices with the same shape can be added by adding corresponding elements. For example,

$$\begin{bmatrix} 2 & 1 & 3 \\ 1 & 0 & -2 \end{bmatrix} + \begin{bmatrix} 1 & -2 & 2 \\ 2 & 3 & -1 \end{bmatrix} = \begin{bmatrix} 3 & -1 & 5 \\ 3 & 3 & -3 \end{bmatrix}.$$

To multiply a matrix by a number c, we multiply each element by c. For example,

$$7\begin{bmatrix} 2 & 1 & 3 \\ 1 & 0 & -2 \end{bmatrix} = \begin{bmatrix} 14 & 7 & 21 \\ 7 & 0 & -14 \end{bmatrix}.$$

A system of simultaneous linear equations

$$\begin{aligned} a_{11}x + a_{12}y + a_{13}z &= b_1, \\ a_{21}x + a_{22}y + a_{23}z &= b_2. \end{aligned} \tag{1}$$

can be written in matrix form as

$$\begin{bmatrix} a_{11} & a_{12} & a_{13} \\ a_{21} & a_{22} & a_{23} \end{bmatrix} \begin{bmatrix} x \\ y \\ z \end{bmatrix} = \begin{bmatrix} b_1 \\ b_2 \end{bmatrix}, \tag{2}$$

or, more compactly, as

$$AX = B, \tag{3}$$

where

$$A = \begin{bmatrix} a_{11} & a_{12} & a_{13} \\ a_{21} & a_{22} & a_{23} \end{bmatrix}, \quad X = \begin{bmatrix} x \\ y \\ z \end{bmatrix}, \quad B = \begin{bmatrix} b_1 \\ b_2 \end{bmatrix}. \tag{4}$$

To form the product indicated by AX in Eq. (3), we take the elements of the first row of A in order from left to right and multiply by the corresponding elements of X from the top down, and add these products to get

$$a_{11}x + a_{12}y + a_{13}z,$$

which we set equal to b_1. The result is the first equation in (1). We then repeat the process with the second row in (2) to obtain the second equation in (1).

An m by n matrix A can multiply an n by p matrix B from the left to give an m by p matrix $C = AB$:

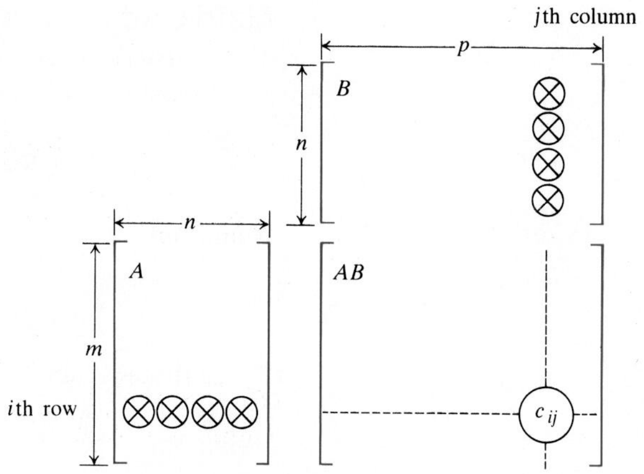

The element in the ith row and jth column of AB is the sum

$$c_{ij} = a_{i1}b_{1j} + a_{i2}b_{2j} + \cdots + a_{in}b_{nj} = \sum_{k=1}^{n} a_{ik}b_{kj}, \quad i = 1, 2, \ldots, m \quad \text{and} \quad j = 1, 2, \ldots, p. \tag{6}$$

In words, Eq. (6) says "to get the element in the ith row and jth column of AB, multiply the individual entries in the ith row of A, one after the other from left to right, by the corresponding entries in the jth column of B from top to bottom, and add these products: their sum is a single number c_{ij}."

For example,

$$\begin{array}{c} \begin{bmatrix} a & b & c \\ d & e & f \\ u & v & w \end{bmatrix} \\ \begin{bmatrix} 2 & -1 & 3 \\ 3 & 2 & 2 \end{bmatrix} \begin{bmatrix} 2a - d + 3u & 2b - e + 3v & 2c - f + 3w \\ 3a + 2d + 2u & 3b + 2e + 2v & 3c + 2f + 2w \end{bmatrix}. \end{array} \tag{7}$$

In most cases it saves space to write the matrices in a product side by side. Thus we usually write the multiplication in (7) as

$$\begin{bmatrix} 2 & -1 & 3 \\ 3 & 2 & 2 \end{bmatrix} \begin{bmatrix} a & b & c \\ d & e & f \\ u & v & w \end{bmatrix} = \begin{bmatrix} 2a - d + 3u & 2b - e + 3v & 2c - f + 3w \\ 3a + 2d + 2u & 3b + 2e + 2v & 3c + 2f + 2w \end{bmatrix}.$$

Elementary Row Operations and Row Reduction

Two systems of linear equations are called *equivalent* if they have the same set of solutions. To solve a system of linear equations it is often possible to transform it step by step into an equivalent system of equations that is so simple it can be solved by inspection. We shall illustrate such a sequence of steps by transforming the system

$$\begin{aligned} 2x + 3y - 4z &= -3, \\ x + 2y + 3z &= 3, \\ 3x - y - z &= 6, \end{aligned} \tag{8}$$

into the equivalent system

$$\begin{aligned} x \qquad\qquad &= 2, \\ y \qquad &= -1, \\ z &= 1. \end{aligned} \tag{9}$$

EXAMPLE 1 Solve the system of equations (8).

Solution The system (8) is the same as

$$AX = B, \qquad A = \begin{bmatrix} 2 & 3 & -4 \\ 1 & 2 & 3 \\ 3 & -1 & -1 \end{bmatrix}, \quad B = \begin{bmatrix} -3 \\ 3 \\ 6 \end{bmatrix}. \tag{10}$$

We start with the 3 by 4 matrix $[A \mid B]$ whose first three columns are the columns of A, and whose fourth column is B. That is,

$$[A \mid B] = \left[\begin{array}{ccc|c} 2 & 3 & -4 & -3 \\ 1 & 2 & 3 & 3 \\ 3 & -1 & -1 & 6 \end{array}\right]. \tag{11}$$

We are going to transform this augmented matrix with a sequence of so-called *elementary row operations*. These operations, which are to be performed on the rows of the matrix, are of three kinds:

1. Multiply any row by a constant different from 0.
2. Add a constant multiple of any row to another row.
3. Interchange two rows.

Our goal is to replace the matrix $[A\,|\,B]$ by the matrix $[I\,|\,S]$, where

$$I = \begin{bmatrix} 1 & 0 & 0 \\ 0 & 1 & 0 \\ 0 & 0 & 1 \end{bmatrix} \quad \text{and} \quad S = \begin{bmatrix} s_1 \\ s_2 \\ s_3 \end{bmatrix}. \tag{12}$$

If we succeed, the matrix $[I\,|\,S]$ will be the matrix of the system

$$\begin{aligned} x \qquad\qquad &= s_1, \\ y \qquad &= s_2, \\ z &= s_3. \end{aligned} \tag{13}$$

The virtue of this system is that its solution, $x = s_1$, $y = s_2$, $z = s_3$, is the same as the solution of (8).

Our systematic approach will be to get a 1 in the upper left corner and use Type 2 operations to get zeros elsewhere in the first column. That will make the first column the same as the first column of I. Then we shall use Type 1 or Type 3 operations to get a 1 in the second position in the second row, and follow that by Type 2 operations to get the second column to be what we want: namely, like the second column of I. Then we will work on the third column.

STEP 1. Interchange rows 1 and 2 and get

$$\left[\begin{array}{ccc|c} 1 & 2 & 3 & 3 \\ 2 & 3 & -4 & -3 \\ 3 & -1 & -1 & 6 \end{array}\right]. \tag{14}$$

STEP 2. Add -2 times row 1 to row 2.

STEP 3. Add -3 times row 1 to row 3.

The result of steps 2 and 3 is

$$\left[\begin{array}{ccc|c} 1 & 2 & 3 & 3 \\ 0 & -1 & -10 & -9 \\ 0 & -7 & -10 & -3 \end{array}\right]. \tag{15}$$

STEP 4. Multiply row 2 by -1; then

STEP 5. Add -2 times row 2 to row 1, and

STEP 6. Add 7 times row 2 to row 3.

The combined result of these steps is

$$\left[\begin{array}{ccc|c} 1 & 0 & -17 & -15 \\ 0 & 1 & 10 & 9 \\ 0 & 0 & 60 & 60 \end{array}\right]. \tag{16}$$

STEP 7. Multiply row 3 by $\frac{1}{60}$.

STEP 8. Add 17 times row 3 to row 1.

STEP 9. Add -10 times row 3 to row 2.

The final result is

$$[I\,|\,S] = \left[\begin{array}{ccc|c} 1 & 0 & 0 & 2 \\ 0 & 1 & 0 & -1 \\ 0 & 0 & 1 & 1 \end{array}\right]. \tag{17}$$

This represents the system (9). The solution of this system, and therefore of the system (8), is $x = 2$, $y = -1$, $x = 1$. To check the solution, we substitute these values in (8) and find that

$$\begin{aligned} 2(2) + 3(-1) - 4(1) &= -3, \\ (2) + 2(-1) + 3(1) &= 3, \\ 3(2) - (-1) - (1) &= 6. \ \square \end{aligned} \tag{18}$$

The method of using elementary row operations to reduce the augmented matrix of a system of linear equations to a simpler form is sometimes called the *method of row reduction*. It works because at each step the system of equations represented by the transformed matrix is equivalent to the original system. Thus, in Example 1, when we finally arrived at the matrix (17), which represented the system (9) whose solution could be found by inspection, we knew that this solution was also the solution of (8). Note that we checked the solution anyhow. It is always a good idea to do that.

Inverses

The matrix I in Eq. (12) is the *multiplicative identity* for all 3 by 3 matrices. That is, if M is any 3 by 3 matrix, then

$$IM = MI = M. \tag{19}$$

If P is a matrix with the property that

$$PM = I,$$

then we call P the *inverse* of M, and use the alternative notation

$$P = M^{-1},$$

pronounced "M inverse."

The sequence of row operations that we used to find the solution of the system $AX = B$ in Example 1 can be used to find the inverse of the matrix A. We start with the 3 by 6 matrix $[A \mid I]$ whose first three columns are the columns of A and whose last three columns are the columns of I, namely

$$[A \mid I] = \left[\begin{array}{rrr|rrr} 2 & 3 & -4 & 1 & 0 & 0 \\ 1 & 2 & 3 & 0 & 1 & 0 \\ 3 & -1 & -1 & 0 & 0 & 1 \end{array}\right]. \tag{20}$$

We then carry out Steps 1 through 9 of Example 1 on the augmented matrix $[A \mid I]$. The final result is

$$[I \mid A^{-1}] = \left[\begin{array}{rrr|rrr} 1 & 0 & 0 & \frac{1}{60} & \frac{7}{60} & \frac{17}{60} \\ 0 & 1 & 0 & \frac{10}{60} & \frac{10}{60} & -\frac{10}{60} \\ 0 & 0 & 1 & -\frac{7}{60} & \frac{11}{60} & \frac{1}{60} \end{array}\right]. \tag{21}$$

The 3 by 3 matrix in the last three columns is

$$A^{-1} = \frac{1}{60}\begin{bmatrix} 1 & 7 & 17 \\ 10 & 10 & -10 \\ -7 & 11 & 1 \end{bmatrix}. \tag{22}$$

By direct matrix multiplication, we verify our answer:

$$A = \frac{1}{60}\begin{bmatrix} 1 & 7 & 17 \\ 10 & 10 & -10 \\ -7 & 11 & 1 \end{bmatrix}\begin{bmatrix} 2 & 3 & -4 \\ 1 & 2 & 3 \\ 3 & -1 & -1 \end{bmatrix}$$

$$= \frac{1}{60}\begin{bmatrix} 2 + 7 + 51 & 3 + 14 - 17 & -4 + 21 - 17 \\ 20 + 10 - 30 & 30 + 20 + 10 & -40 + 30 + 10 \\ -14 + 11 + 3 & -21 + 22 - 1 & 28 + 33 - 1 \end{bmatrix} \qquad (23)$$

$$= \frac{1}{60}\begin{bmatrix} 60 & 0 & 0 \\ 0 & 60 & 0 \\ 0 & 0 & 60 \end{bmatrix} = \begin{bmatrix} 1 & 0 & 0 \\ 0 & 1 & 0 \\ 0 & 0 & 1 \end{bmatrix}.$$

Knowing A^{-1} provides a second way to solve the system of equations in (8). We write the system in the form given in (10), and then multiply B on the left by A^{-1} to find the solution matrix X. Thus,

$$X = IX = (A^{-1}A)X = A^{-1}(AX) = A^{-1}B = \frac{1}{60}\begin{bmatrix} 1 & 7 & 17 \\ 10 & 10 & -10 \\ -7 & 11 & 1 \end{bmatrix}\begin{bmatrix} -3 \\ 3 \\ 6 \end{bmatrix}$$

$$= \frac{1}{60}\begin{bmatrix} 120 \\ -60 \\ 60 \end{bmatrix} = \begin{bmatrix} 2 \\ -1 \\ 1 \end{bmatrix}.$$

If the coefficient matrix A of a system

$$AX = B$$

of n equations and n unknowns has an inverse, then the solution of the system is

$$X = A^{-1}B.$$

Only square matrices can have inverses. If an n by n matrix A has an inverse, the method shown above for $n = 3$ will give it: Put the n by n identity matrix alongside A and use the row operations to get an n by n identity matrix in place of A. The n by n matrix that is then beside that is A^{-1}.

Not every n by n matrix has an inverse. For example, the 2 by 2 matrix

$$\begin{bmatrix} 1 & 1 \\ a & a \end{bmatrix}$$

has no inverse. The method we outlined above would reduce this to

$$\begin{bmatrix} 1 & 1 \\ 0 & 0 \end{bmatrix},$$

which cannot be further changed by elementary row operations into the 2 by 2 identity matrix. We shall have more to say about inverses in a moment.

A system of m equations in n unknowns may have no solutions, only one solution, or infinitely many solutions. If there are any solutions, they can be found by the method of row reduction described above.

Determinants and the Inverse of a Matrix

Another way to find the inverse of a square matrix A (assuming A has one) depends on the fact that A has an inverse if and only if $\det A \neq 0$. We describe the method, give an example, and then indicate why the method works. We illustrate with 3 by 3 matrices, but the method works for any square matrix whose determinant is not zero.

> To find the inverse of a matrix whose determinant is not zero:
>
> 1. Construct the matrix of cofactors of A:
>
> $$\operatorname{cof} A = \begin{bmatrix} A_{11} & A_{12} & A_{13} \\ A_{21} & A_{22} & A_{23} \\ A_{31} & A_{32} & A_{33} \end{bmatrix}.$$
>
> 2. Construct the transposed matrix of cofactors (called the *adjoint* of A):
>
> $$\operatorname{adj} A = (\operatorname{cof} A)^T = \begin{bmatrix} A_{11} & A_{21} & A_{31} \\ A_{12} & A_{22} & A_{32} \\ A_{13} & A_{23} & A_{33} \end{bmatrix}.$$
>
> ("Transpose" means write the rows as columns.)
>
> 3. Then,
>
> $$A^{-1} = \frac{1}{\det A} \operatorname{adj} A.$$

EXAMPLE 2 Let us take the same matrix A that we used in illustrating the method of elementary row operations:

$$A = \begin{bmatrix} 2 & 3 & -4 \\ 1 & 2 & 3 \\ 3 & -1 & -1 \end{bmatrix}.$$

You can verify that the matrix of minors is

$$\begin{bmatrix} 1 & -10 & -7 \\ -7 & 10 & -11 \\ 17 & 10 & 1 \end{bmatrix}.$$

We next apply the sign corrections according to the checkerboard pattern $(-1)^{i+j}$ to get the matrix of cofactors

$$\operatorname{cof} A = \begin{bmatrix} 1 & 10 & -7 \\ 7 & 10 & 11 \\ 17 & -10 & 1 \end{bmatrix}.$$

The adjoint of A is the transposed cofactor matrix

$$\operatorname{adj} A = \begin{bmatrix} 1 & 7 & 17 \\ 10 & 10 & -10 \\ -7 & 11 & 1 \end{bmatrix}.$$

We get the determinant of A by multiplying the first row of A and the first column of adj A (which is the first row of the matrix of cofactors, so we are multiplying the elements of the first row of A by their own cofactors):

$$\det A = 2(1) + 3(10) + (-4)(-7) = 2 + 30 + 28 = 60.$$

Therefore, when we divide adj A by det A, we get

$$A^{-1} = \frac{1}{60}\begin{bmatrix} 1 & 7 & 17 \\ 10 & 10 & -10 \\ -7 & 11 & 1 \end{bmatrix},$$

which agrees with our previous work. □

Why does the method work? Let us take a closer look at the products A adj A and (adj A) A for $n = 3$. Because adj A is the transposed cofactor matrix, we get

$$A\,(\text{adj } A) = \begin{bmatrix} a_{11} & a_{12} & a_{13} \\ a_{21} & a_{22} & a_{23} \\ a_{31} & a_{32} & a_{33} \end{bmatrix}\begin{bmatrix} A_{11} & A_{21} & A_{31} \\ A_{12} & A_{22} & A_{32} \\ A_{13} & A_{23} & A_{33} \end{bmatrix} = \begin{bmatrix} \det A & 0 & 0 \\ 0 & \det A & 0 \\ 0 & 0 & \det A \end{bmatrix}.$$

(Here A_{ij} = cofactor of a_{ij}.)

An element on the main diagonal in the final product is the product of a row of A and the corresponding column of adj A. This is the same as the sum of the products of the elements of a row of A and the cofactors of the same row, which is just det A. For those elements not on the main diagonal in the product, we are adding products of elements of some row of A by the cofactors of the corresponding elements of a different row of A, and that sum is zero.

If we were to multiply in the other order, (adj A) A, we would again get

$$(\text{adj } A)\,A = \begin{bmatrix} \det A & 0 & 0 \\ 0 & \det A & 0 \\ 0 & 0 & \det A \end{bmatrix},$$

because we are multiplying the elements of the jth column of A, say, by the cofactors of the corresponding elements of the ith column, in order to get the entry in the ith row and jth column in the product. The result is det A when $i = j$ and is 0 when $i \neq j$.

PROBLEMS

1. a) Write the following system of linear equations in matrix form $AX = B$.

$$\begin{aligned} 2x - 3y + 4z &= -19 \\ 6x + 4y - 2z &= 8 \\ x + 5y + 4z &= 23 \end{aligned}$$

b) Show that

$$X = \begin{bmatrix} -2 \\ 5 \\ 0 \end{bmatrix}$$

is a solution of the system in part (a).

2. Let A be an arbitrary matrix with 3 rows and 3 columns and let I be the 3 by 3 matrix that has 1's on the main diagonal and zeros elsewhere:

$$I = \begin{bmatrix} 1 & 0 & 0 \\ 0 & 1 & 0 \\ 0 & 0 & 1 \end{bmatrix}.$$

Show that $IA = A$ and also that $AI = A$. This will show that I is the multiplicative identity matrix for all 3 by 3 matrices.

3. Let A be an arbitrary 3 by 3 matrix and let R_{12} be the matrix that is obtained from the 3 by 3 identity matrix by interchanging rows 1 and 2:

$$R_{12} = \begin{bmatrix} 0 & 1 & 0 \\ 1 & 0 & 0 \\ 0 & 0 & 1 \end{bmatrix}.$$

Compute $R_{12}A$ and show that you would get the same result by interchanging rows 1 and 2 of A.

4. Let A and R_{12} be as in Problem 3 above. Compute AR_{12} and show that the result is what you would get by interchanging columns 1 and 2 of A. (Note that R_{12} is also the result of interchanging columns 1 and 2 of the 3 by 3 identity matrix I.)

Solve the following systems of equations by row reduction.

5. $\begin{aligned} x + 8y &= 4 \\ 3x - y &= -13 \end{aligned}$

6. $\begin{aligned} 2x + 3y &= 5 \\ 3x - y &= 2 \end{aligned}$

7. $\begin{aligned} 4x - 3y &= 6 \\ 3x - 2y &= 5 \end{aligned}$

8. $\begin{aligned} x + y + z &= 2 \\ 2x - y + z &= 0 \\ x + 2y - z &= 4 \end{aligned}$

9. $\begin{aligned} 2x + y - z &= 2 \\ x - y + z &= 7 \\ 2x + 2y + z &= 4 \end{aligned}$

10. $\begin{aligned} 2x - 4y \quad &= 6 \\ x + y + z &= 1 \\ 5y + 7z &= 10 \end{aligned}$

11. $\begin{aligned} x \quad - z &= 3 \\ 2y - 2z &= 2 \\ 2x \quad + z &= 3 \end{aligned}$

12. $\begin{aligned} x_1 + x_2 - x_3 + x_4 &= 2 \\ x_1 - x_2 + x_3 + x_4 &= -1 \\ x_1 + x_2 + x_3 - x_4 &= 2 \\ x_1 \quad + x_3 + x_4 &= -1 \end{aligned}$

13. Verify that the inverse of

$$A = \begin{bmatrix} a & b \\ c & d \end{bmatrix}$$

is

$$A^{-1} = \frac{1}{ad - bc}\begin{bmatrix} d & -b \\ -c & a \end{bmatrix}.$$

That is, show that

$$AA^{-1} = A^{-1}A = \begin{bmatrix} 1 & 0 \\ 0 & 1 \end{bmatrix}.$$

14. a) Use the result in Problem 13 to write down the inverses of

$$A = \begin{bmatrix} 2 & -1 \\ 3 & 1 \end{bmatrix} \quad \text{and} \quad B = \begin{bmatrix} 2 & 3 \\ -1 & 1 \end{bmatrix}.$$

b) In part (a), B is the transpose of A. Is B^{-1} the transpose of A^{-1}?

15. Use the result in Problem 13 to solve the system of equations in Problem 5.

16. Given

$$A = \begin{bmatrix} 1 & 0 & -1 \\ 0 & 2 & -2 \\ 2 & 0 & 1 \end{bmatrix} \quad \text{and} \quad A^{-1} = \begin{bmatrix} \frac{1}{3} & 0 & \frac{1}{3} \\ -\frac{2}{3} & \frac{1}{2} & \frac{1}{3} \\ -\frac{2}{3} & 0 & \frac{1}{3} \end{bmatrix},$$

solve the system

$$A\begin{bmatrix} x \\ y \\ z \end{bmatrix} = \begin{bmatrix} 3 \\ 2 \\ 3 \end{bmatrix}.$$

17. a) Find the inverse of the matrix

$$\begin{bmatrix} 1 & 8 & 9 \\ 0 & 4 & 6 \\ 0 & 0 & 3 \end{bmatrix}.$$

b) Solve the following system of equations.

$$\begin{aligned} x + 8y + 9z &= 10 \\ 4y + 6z &= 10 \\ 3z &= -10 \end{aligned}$$

18. Solve the system

$$\begin{aligned} 2x - y + 2z &= 5 \\ 3x + y - 3z &= 7 \end{aligned} \quad \text{or} \quad \begin{bmatrix} 2 & -1 \\ 3 & 1 \end{bmatrix}\begin{bmatrix} x \\ y \end{bmatrix} = \begin{bmatrix} 5 - 2z \\ 7 + 3z \end{bmatrix}$$

for x and y in terms of z. Each equation on the left represents a plane. In how many points do the planes intersect?

19. Expanding the quotient

$$\frac{ax + b}{(x - r_1)(x - r_2)}$$

by partial fractions calls for finding the values of C and D that make the equation

$$\frac{ax + b}{(x - r_1)(x - r_2)} = \frac{C}{x - r_1} + \frac{D}{x - r_2}$$

hold for all x.

a) Find a system of linear equations that determines C and D.

b) Under what circumstances does the system of equations in part (a) have a unique solution? That is, when is the determinant of the coefficient matrix of the system different from zero?

A.3

Proofs of the Limit Theorems of Article 1.9

We now prove Theorems 1 and 4 of Article 1.9. Proofs of Theorems 2 and 3 of Article 1.9 can be found in Problems 1–5 at the end of this appendix.

THEOREM 1

If $\lim_{t\to c} F_1(t) = L_1$ and $\lim_{t\to c} F_2(t) = L_2$, then

i) $\lim\,[F_1(t) + F_2(t)] = L_1 + L_2$

ii) $\lim\,[F_1(t) - F_2(t)] = L_1 - L_2$

iii) $\lim F_1(t)F_2(t) = L_1L_2$

iv) $\lim kF_2(t) = kL_2$, (any number k)

v) $\lim \dfrac{F_1(t)}{F_2(t)} = \dfrac{L_1}{L_2}$, provided $L_2 \neq 0$.

The limits are all to be taken as $t \to c$.

i) $\lim\,[F_1(t) + F_2(t)] = L_1 + L_2$

To prove that the sum $F_1(t) + F_2(t)$ has the limit $L_1 + L_2$ as $t \to c$ we must show that for any $\varepsilon > 0$ there exists a $\delta > 0$ such that for all t

$$0 < |t - c| < \delta \Rightarrow |F_1(t) + F_2(t) - (L_1 + L_2)| < \varepsilon. \tag{1a}$$

If ε is positive, then so is $\varepsilon/2$, and because $\lim_{t\to c} F_1(t) = L_1$ we know that there is a $\delta_1 > 0$ such that for all t

$$0 < |t - c| < \delta_1 \Rightarrow |F_1(t) - L_1| < \varepsilon/2. \tag{1b}$$

Likewise, there is a $\delta_2 > 0$ such that for all t

$$0 < |t - c| < \delta_2 \Rightarrow |F_2(t) - L_2| < \varepsilon/2. \tag{1c}$$

Now let δ be the minimum of δ_1 and δ_2. Then δ is a positive number, and the ε inequalities in (1b) and (1c) both hold when $0 < |t - c| < \delta$. Thus, for all t, the inequality $0 < |t - c| < \delta$ implies

$$\begin{aligned}|F_1(t) + F_2(t) - (L_1 + L_2)| &= |F_1(t) - L_1 + F_2(t) - L_2| \\ &\leq |F_1(t) - L_1| + |F_2(t) - L_2| \\ &< \frac{\varepsilon}{2} + \frac{\varepsilon}{2} = \varepsilon.\end{aligned}$$

This establishes (1a) and proves part (i) of the theorem.

iii) $\lim\,[F_1(t) \cdot F_2(t)] = L_1 \cdot L_2$

Let ε be an arbitrary positive number, and write

$$F_1(t) = L_1 + (F_1(t) - L_1), \qquad F_2(t) = L_2 + (F_2(t) - L_2).$$

When we multiply these expressions and subtract L_1L_2, we get

$$\begin{aligned}F_1(t) \cdot F_2(t) - L_1L_2 = {} & L_1(F_2(t) - L_2) + L_2(F_1(t) - L_1) \\ & + (F_1(t) - L_1) \cdot (F_2(t) - L_2).\end{aligned} \tag{3a}$$

The numbers $\sqrt{\varepsilon/3}$, $\varepsilon/[3(1 + |L_1|)]$, and $\varepsilon/[3(1 + |L_2|)]$ are all positive, and, because $F_1(t)$ has limit L_1 and $F_2(t)$ has limit L_2, there are positive numbers δ_1, δ_2, δ_3, and δ_4 such that

$$\begin{aligned}
|F_1(t) - L_1| &< \sqrt{\varepsilon/3} && \text{when } 0 < |t - c| < \delta_1,\\
|F_2(t) - L_2| &< \sqrt{\varepsilon/3} && \text{when } 0 < |t - c| < \delta_2,\\
|F_1(t) - L_1| &< \varepsilon/[3(1 + |L_2|)] && \text{when } 0 < |t - c| < \delta_3,\\
|F_2(t) - L_2| &< \varepsilon/[3(1 + |L_1|)] && \text{when } 0 < |t - c| < \delta_4,
\end{aligned}$$

We now let δ be the minimum of the four positive numbers δ_1, δ_2, δ_3, δ_4. Then δ is a positive number and all four of the inequalities above are satisfied when $0 < |t - c| < \delta$. By taking absolute values in Eq. (3a) and applying the triangle inequality, we get

$$\begin{aligned}
|F_1(t)F_2(t) - L_1L_2| &\le |L_1| \cdot |F_2(t) - L_2| + |L_2| \cdot |F_1(t) - L_1|\\
&\quad + |F_1(t) - L_1| \cdot |F_2(t) - L_2|\\
&< \frac{\varepsilon}{3} + \frac{\varepsilon}{3} + \frac{\varepsilon}{3} = \varepsilon \qquad \text{when} \quad 0 < |t - c| < \delta. \qquad (3b)
\end{aligned}$$

This completes the proof of part (iii).

iv) $\lim kF_2(t) = kL_2$

This is a special case of (iii) with $F_1(t) = k$, the function whose output value is the constant k for all values of t.

ii) $\lim [F_1(t) - F_2(t)] = L_1 - L_2$

This can be deduced from (i) and (iv) in the following way:

$$\begin{aligned}
\lim [F_1(t) - F_2(t)] &= \lim [F_1(t) + (-1)F_2(t)]\\
&= \lim F_1(t) + \lim (-1)F_2(t)\\
&= \lim F_1(t) + (-1)\lim F_2(t)\\
&= \lim F_1(t) - \lim F_2(t)\\
&= L_1 - L_2.
\end{aligned}$$

v) $\lim \dfrac{F_1(t)}{F_2(t)} = \dfrac{L_1}{L_2}$ **if** $L_2 \neq 0$

Since L_2 is not zero, $|L_2|$ is a positive number, and because $F_2(t)$ has L_2 as limit when t approaches c, we know that there is a positive number δ_1 such that

$$|F_2(t) - L_2| < \frac{|L_2|}{2} \qquad \text{when} \quad 0 < |t - c| < \delta_1. \qquad (5a)$$

Now, for any numbers A and B, it can be shown that

$$|A| - |B| \le |A - B| \qquad \text{and} \qquad |B| - |A| \le |A - B|,$$

from which it follows that

$$||A| - |B|| \le |A - B|. \qquad (5b)$$

Taking $A = F_2(t)$ and $B = L_2$ in (5b), we can deduce from (5a) that

$$-\frac{1}{2}|L_2| < |F_2(t)| - |L_2| < \frac{1}{2}|L_2| \qquad \text{when} \quad 0 < |t - c| < \delta_1.$$

By adding $|L_2|$ to the three terms of the foregoing inequality we get

$$\frac{1}{2}|L_2| < |F_2(t)| < \frac{3}{2}|L_2|,$$

from which it follows that

$$\left|\frac{1}{F_2(t)} - \frac{1}{L_2}\right| = \left|\frac{L_2 - F_2(t)}{L_2F_2(t)}\right| \leq \frac{2}{|L_2|^2}|L_2 - F_2(t)| \tag{5c}$$

when $0 < |t - c| < \delta_1$.

All that we have done so far is to show that the difference between the reciprocals of $F_2(t)$ and L_2, at the left-hand side of (5c) is no greater in absolute value than a constant times $|L_2 - F_2(t)|$, when t is close enough to c. The fact that L_2 is the limit of $F_2(t)$ has not yet been used with full force.

But now let ε be an arbitrary positive number. Then $\frac{1}{2}|L_2|^2\varepsilon$ is also a positive number and there is a positive number δ_2 such that

$$|L_2 - F_2(t)| < \frac{\varepsilon}{2}|L_2|^2 \qquad \text{when} \quad 0 < |t - c| < \delta_2. \tag{5d}$$

We now let $\delta = \min\{\delta_1, \delta_2\}$ and have a positive number δ such that the inequalities in (5c) and (5d) combine to produce the result

$$\left|\frac{1}{F_2(t)} - \frac{1}{L_2}\right| < \varepsilon \qquad \text{when} \quad 0 < |t - c| < \delta.$$

What we have just shown is that

If $\lim F_2(t) = L_2$ as t approaches c, and $L_2 \neq 0$, then

$$\lim \frac{1}{F_2(t)} = \frac{1}{L_2}.$$

Having already proved the product law, we get the final quotient law by applying the product law to $F_1(t)$ and $1/F_2(t)$ as follows:

$$\lim \frac{F_1(t)}{F_2(t)} = \lim\left[F_1(t)\cdot\frac{1}{F_2(t)}\right] = [\lim F_1(t)]\cdot\left[\lim \frac{1}{F_2(t)}\right] = L_1\cdot\frac{1}{L_2}. \quad ■$$

THEOREM 4

The Sandwich Theorem

Suppose that

$$f(t) \leq g(t) \leq h(t)$$

for all values of $t \neq c$ in some interval about c, and that $f(t)$ and $h(t)$ approach the same limit L as t approaches c. Then,

$$\lim_{t\to c} g(t) = L.$$

Proof for right limits. If

$$\lim_{t\to c^+} f(t) = \lim_{t\to c^+} h(t) = L,$$

then for any $\varepsilon > 0$ there exists a $\delta > 0$ such that for all t

$$c < t < c + \delta \Rightarrow L - \varepsilon < f(t) < L + \varepsilon \qquad \text{and} \qquad L - \varepsilon < h(t) < L + \varepsilon.$$

We combine the ε-inequalities on the right with the inequality $f(t) \le g(t) \le h(t)$ to obtain

$$L - \varepsilon < f(t) \le g(t) \le h(t) < L + \varepsilon$$

and thereby

$$L - \varepsilon < g(t) < L + \varepsilon.$$

Therefore, for all t,

$$c < t < c + \delta \Rightarrow |g(t) - L| < \varepsilon.$$

This shows that

$$\lim_{t \to c^+} g(t) = L. \blacksquare$$

Proof for left limits. Given $\varepsilon > 0$, there exists a $\delta > 0$ such that for all t

$$c - \delta < t < c \Rightarrow L - \varepsilon < f(t) < L + \varepsilon \quad \text{and} \quad L - \varepsilon < h(t) < L + \varepsilon.$$

As before, we conclude that for all t

$$c - \delta < t < c \Rightarrow L - \varepsilon < g(t) < L + \varepsilon$$

and therefore that

$$\lim_{t \to c^-} g(t) = L. \blacksquare$$

Proof for two-sided limits. If $\lim_{t\to c} f(t) = L$ and $\lim_{t \to c} h(t) = L$, then f and h approach L as $t \to c^-$ and as $t \to c^+$. Therefore, as we have shown above, $g(t) \to L$ as $t \to c^-$ and $t \to c^+$. Since the right and left limits of g as t approaches c exist and are both equal to L,

$$\lim_{t \to c} g(t) = L. \blacksquare$$

PROBLEMS

1. If $F_1(t)$, $F_2(t)$, and $F_3(t)$ have limits L_1, L_2, and L_3, respectively, as t approaches c, prove that their sum has limit $L_1 + L_2 + L_3$. Generalize the result to the sum of any finite number of functions.

2. If n is any positive integer greater than 1, and $F_1(t)$, $F_2(t), \ldots, F_n(t)$ have the finite limits $L_1, L_2, \ldots, L_n$, respectively, as t approaches c, prove that the product of the n functions has limit $L_1 \cdot L_2 \cdot \cdots \cdot L_n$. (Use part (iii) of Theorem 1, and induction on n.)

3. Use the results of Article 1.9, Example 2, and Problem 2 above to deduce that $\lim_{t\to c} t^n = c^n$ for any positive integer n.

4. Use the result of Example 3 in Article 1.9 and the results of Problems 1 and 3 above to prove that $\lim_{t\to c} f(t) = f(c)$ for any polynomial function

$$f(t) = a_0 t^n + a_1 t^{n-1} + \cdots + a_n.$$

5. Use Theorem 1 and the result of Problem 4 to prove that if $f(t)$ and $g(t)$ are polynomials, and if $g(c) \neq 0$, then

$$\lim_{t \to c} \frac{f(t)}{g(t)} = \frac{f(c)}{g(c)}.$$

6. Figure A.1 (p. A-22) gives a diagram for a proof that the composite of two continuous functions is continuous. Reconstruct the proof from the diagram. The statement to be proved is this:

If $f(t)$ is continuous at $t = c$ and $g(x)$ is continuous at $x = f(c)$, then the composite $g(f(t))$ is continuous at $t = c$.

Assume also c is an interior point of the domain of f, and that $f(c)$ is an interior point of the domain of g. This will make all the limits involved two-sided. (The proofs for the cases in which one or both of c and $f(c)$ are endpoints of the domains of f and g are similar to the argument that assumes both to be interior points.)

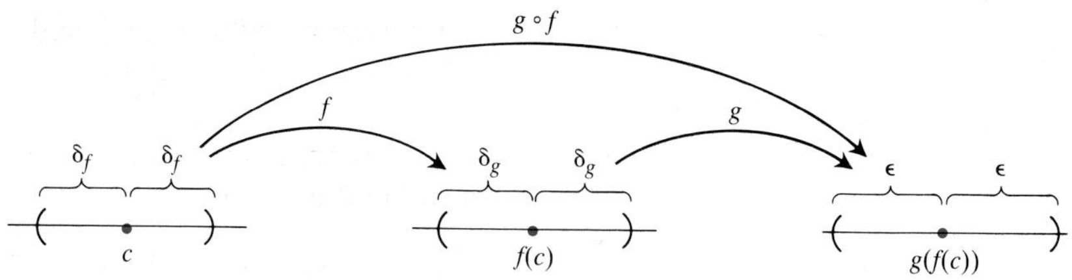

A.1 The diagram for a proof that the composite of two continuous functions is continuous. The continuity of composites holds for any finite number of functions. The only requirement is that each function be continuous where it is applied. In the diagram above, f is to be continuous at c, and g is to be continuous at $f(c)$.

A.4 The Increment and Mixed-Derivative Theorems

This appendix gives proofs for the *Increment Theorem* (Article 15.3, Theorem 2) and the *Mixed-derivative Theorem* (Article 15.7, Theorem 4) for functions of two variables.

THEOREM 2

Increment Theorem for Functions of Two Variables

Suppose that $z = f(x, y)$ is continuous and has partial derivatives throughout a region

$$R: \quad |x - x_0| < h, \qquad |y - y_0| < k$$

in the xy-plane. Suppose also that Δx *and* Δy are small enough for the point $(x_0 + \Delta x, y_0 + \Delta y)$ to lie in R. If f_x and f_y are continuous at (x_0, y_0), then the increment

$$\Delta z = f(x_0 + \Delta x, y_0 + \Delta y) - f(x_0, y_0) \tag{1}$$

can be written as

$$\Delta z = f_x(x_0, y_0)\,\Delta x + f_y(x_0, y_0)\,\Delta y + \varepsilon_1\,\Delta x + \varepsilon_2\,\Delta y, \tag{2}$$

where

$$\varepsilon_1, \varepsilon_2 \to 0 \qquad \text{as} \qquad \Delta x, \Delta y \to 0.$$

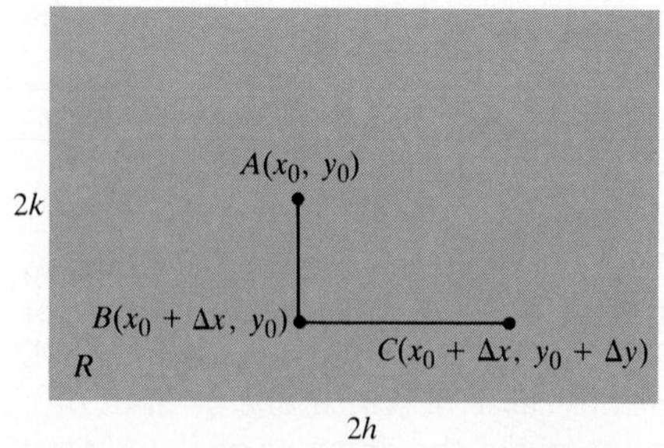

A.2 The rectangular region R in the proof of Theorem 2.

Proof. The region R (Fig. A.2) is a rectangle centered at $A(x_0, y_0)$ with dimensions $2h$ by $2k$. Since $C(x_0 + \Delta x, y_0 + \Delta y)$ lies in R, the point $B(x_0 + \Delta x, y_0)$ and the line segments AB and BC also lie in R. Thus f is continuous and has partial derivatives f_x and f_y at each point of these segments.

We may think of Δz as the sum

$$\Delta z = \Delta z_1 + \Delta z_2 \tag{3}$$

of two increments, where

$$\Delta z_1 = f(x_0 + \Delta x, y_0) - f(x_0, y_0) \tag{4}$$

is the change from A to B and

$$\Delta z_2 = f(x_0 + \Delta x, y_0 + \Delta y) - f(x_0 + \Delta x, y_0) \tag{5}$$

is the change from B to C (Fig. A.3). Note that the sum $\Delta z_1 + \Delta z_2$ equals Δz as it should:

$$\begin{aligned}\Delta z_1 + \Delta z_2 &= [f(x_0 + \Delta x, y_0) - f(x_0, y_0)] \\ &\quad + [f(x_0 + \Delta x, y_0 + \Delta y) - f(x_0 + \Delta x, y_0)] \\ &= f(x_0 + \Delta x, y_0 + \Delta y) - f(x_0, y_0) \\ &= \Delta z.\end{aligned} \tag{6}$$

The function

$$F(x) = f(x, y_0)$$

is a continuous and differentiable function of x on the closed interval joining x_0 and $x_0 + \Delta x$, with derivative

$$F'(x) = f_x(x, y_0).$$

Hence, by the Mean Value Theorem of Article 3.8, there is a point c between x and $x + \Delta x$ such that

$$F(x_0 + \Delta x) - F(x_0) = F'(c)\, \Delta x$$

A.3 Part of the surface $z = f(x, y)$ near $P_0(x_0, y_0, f(x_0, y_0))$. The points P_0, P', and P'' have the same height $z_0 = f(x_0, y_0)$ above the xy-plane. The change in z is $\Delta z = P'S$. The change

$$\Delta z_2 = f(x_0 + \Delta x, y_0) - f(x_0, y_0),$$

shown as $P''Q = P'Q'$, is caused by changing x from x_0 to $x_0 + \Delta x$ while holding y equal to y_0. Then, with x held equal to $x_0 + \Delta x$,

$$\Delta z_2 = f(x_0 + \Delta x, y_0 + \Delta y) - f(x_0 + \Delta x, y_0)$$

is the change in z caused by changing y from y_0 to $y_0 + \Delta y$. This is represented by $Q'S$. The total change in z is the sum of Δz_1 and Δz_2.

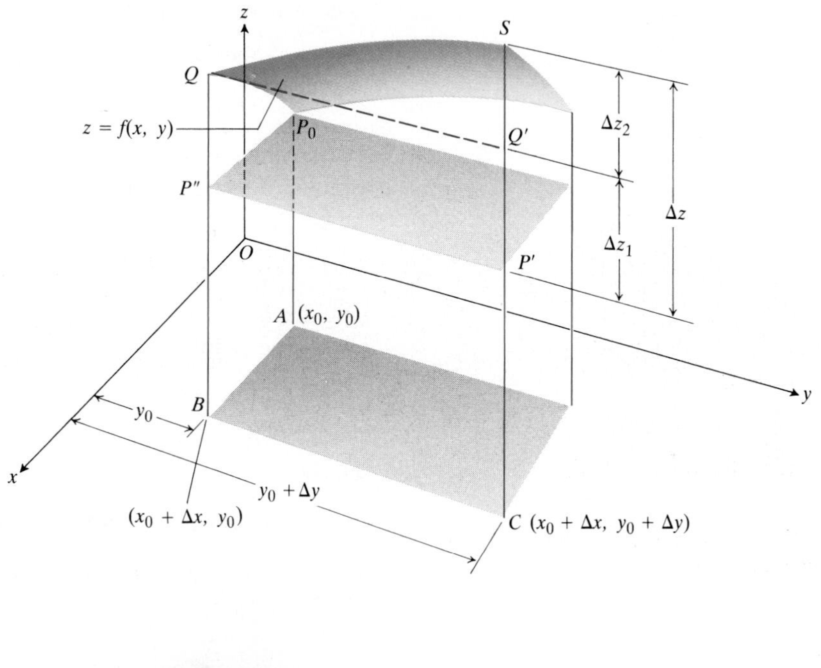

or

$$f(x_0 + \Delta x, y_0) - f(x_0, y_0) = f_x(c, y_0)\,\Delta x$$

or

$$\Delta z_1 = f_x(c, y_0)\,\Delta x. \tag{7}$$

Similarly, the function

$$G(y) = f(x_0 + \Delta x, y)$$

is a continuous and differentiable function of y on the closed interval joining y_0 and $y_0 + \Delta y$, with derivative

$$G'(y) = f_y(x_0 + \Delta x, y).$$

Hence there is a point d between y_0 and $y_0 + \Delta y$ such that

$$G(y_0 + \Delta y) - G(y_0) = G'(d)\,\Delta y$$

or

$$f(x_0 + \Delta x, y_0 + \Delta y) - f(x_0 + \Delta x, y_0) = f_y(x_0 + \Delta x, d)\,\Delta y$$

or

$$\Delta z_2 = f_y(x_0 + \Delta x, d)\,\Delta y. \tag{8}$$

Now, as Δx and $\Delta y \to 0$, we know $c \to x_0$ and $d \to y_0$. Therefore, if f_x and f_y are continuous at (x_0, y_0), the quantities

$$\begin{aligned} \varepsilon_1 &= f_x(c, y_0) - f_x(x_0, y_0), \\ \varepsilon_2 &= f_y(x_0 + \Delta x, d) - f_y(x_0, y_0) \end{aligned} \tag{9}$$

both approach zero as Δx and $\Delta y \to 0$.

Finally,

$$\begin{aligned} \Delta z &= \Delta z_1 + \Delta z_2 \\ &= f_x(c, y_0)\,\Delta x + f_y(x_0 + \Delta x, d)\,\Delta y \qquad \text{(from (7) and (8))} \\ &= [f_x(x_0, y_0) + \varepsilon_1]\,\Delta x + [f_y(x_0, y_0) + \varepsilon_2]\,\Delta y \qquad \text{(from (9))} \\ &= f_x(x_0, y_0)\,\Delta x + f_y(x_0, y_0)\,\Delta y + \varepsilon_1\,\Delta x + \varepsilon_2\,\Delta y, \end{aligned}$$

where ε_1 and $\varepsilon_2 \to 0$ as Δx and $\Delta y \to 0$. This is what we set out to prove. ■

Analogous results hold for functions of any finite number of independent variables. For a function of three variables

$$w = f(x, y, z),$$

that is continuous and has partial derivatives f_x, f_y, f_z at and in some neighborhood of the point (x_0, y_0, z_0), and whose derivatives are continuous at the point, we have

$$\begin{aligned} \Delta w &= f(x_0 + \Delta x, y_0 + \Delta y, z_0 + \Delta z) - f(x_0, y_0, z_0) \\ &= f_x\,\Delta x + f_y\,\Delta y + f_z\,\Delta z + \varepsilon_1\,\Delta x + \varepsilon_2\,\Delta y + \varepsilon_3\,\Delta z, \end{aligned} \tag{10}$$

where

$$\varepsilon_1, \varepsilon_2, \varepsilon_3 \to 0 \qquad \text{when} \quad \Delta x, \Delta y, \text{ and } \Delta z \to 0.$$

The partial derivatives f_x, f_y, f_z in this formula are to be evaluated at the point (x_0, y_0, z_0).

Note. The result (10) may be proved by treating Δw as the sum of three increments,

$$\Delta w_1 = f(x_0 + \Delta x, y_0, z_0) - f(x_0, y_0, z_0), \tag{11a}$$

$$\Delta w_2 = f(x_0 + \Delta x, y_0 + \Delta y, z_0) - f(x_0 + \Delta x, y_0, z_0), \tag{11b}$$

$$\Delta w_3 = f(x_0 + \Delta x, y_0 + \Delta y, z_0 + \Delta z) - f(x_0 + \Delta x, y_0 + \Delta y, z_0), \tag{11c}$$

and applying the Mean Value Theorem to each of these separately. Note that two coordinates remain constant and only one varies in each of these partial increments Δw_1, Δw_2, Δw_3. For example, in (11b), only y varies, since x is held equal to $x_0 + \Delta x$ and z is held equal to z_0. Since the function $f(x_0 + \Delta x, y, z_0)$ is a continuous function of y with a derivative f_y, it is subject to the Mean Value Theorem, and we have

$$\Delta w_2 = f_y(x_0 + \Delta x, y_1, z_0)\,\Delta y$$

for some y_1 between y_0 and $y_0 + \Delta y$.

THEOREM 4

The Mixed-Derivative Theorem

If $f(x, y)$ and its partial derivatives f_x, f_y, f_{xy}, and f_{yx} are defined in a region containing a point (a, b) and are all continuous at (a, b), then

$$f_{xy}(a, b) = f_{yx}(a, b). \tag{12}$$

Proof. The equality of $f_{xy}\,(a, b)$ and $f_{yx}\,(a, b)$ can be established by four applications of the Mean Value Theorem. By hypothesis, the point (a, b) lies in the interior of a rectangle R in the xy-plane on which f, f_x, f_y, f_{xy}, and f_{yx} are all continuous. We let h and k be numbers such that the point $(a + h, b + k)$ also lies in the rectangle R, and we consider the difference

$$\Delta = F(a + h) - F(a), \tag{13}$$

where we define $F(x)$ in terms of $f(x, y)$ by the equation

$$F(x) = f(x, b + k) - f(x, b). \tag{14}$$

We apply the Mean Value Theorem to the function $F(x)$ (which is continuous because it is differentiable), and Eq. (13) becomes

$$\Delta = hF'(c_1), \tag{15}$$

where c_1 lies between a and $a + h$. From Eq. (14),

$$F'(x) = f_x(x, b + k) - f_x(x, b),$$

so Eq. (15) becomes

$$\Delta = h[f_x(c_1, b + k) - f_x(c_1, b)]. \tag{16}$$

Now we apply the Mean Value Theorem to the function $g(y) = f_x(c_1, y)$ and have

$$g(b + k) - g(b) = kg'(d_1) \tag{17a}$$

or

$$f_x(c_1, b + k) - f_x(c_1, b) = kf_{xy}(c_1, d_1), \tag{17b}$$

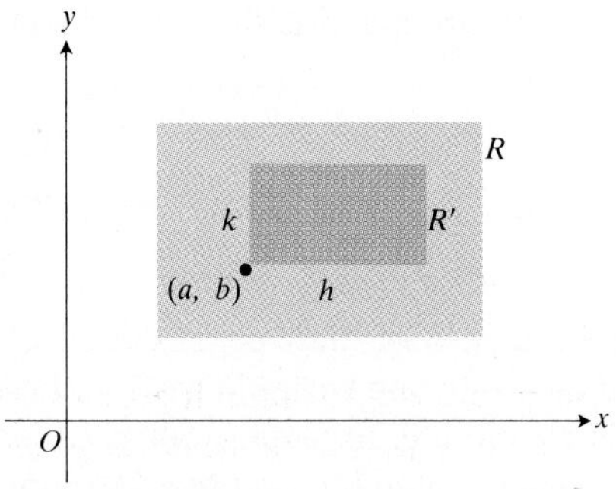

A.4 The key to proving $f_{xy}(a, b) = f_{yx}(a, b)$ is the fact that no matter how small R' is, f_{xy} and f_{yx} take on equal values somewhere inside R' (although not necessarily at the same point of R').

for some d_1 between b and $b + k$. By substituting this into Eq. (16), we get

$$\Delta = hkf_{xy}(c_1, d_1), \tag{18}$$

for some point (c_1, d_1) in the rectangle R' whose vertices are the four points (a, b), $(a + h, b)$, $(a + h, b + k)$, and $(a, b + k)$. (See Fig. A.4.)

On the other hand, by substituting from Eq. (14) into Eq. (13), we may also write

$$\begin{aligned}\Delta &= f(a + h, b + k) - f(a + h, b) - f(a, b + k) + f(a, b)\\ &= [f(a + h, b + k) - f(a, b + k)] - [f(a + h, b) - f(a, b)]\\ &= \phi(b + k) - \phi(b),\end{aligned} \tag{19}$$

where

$$\phi(y) = f(a + h, y) - f(a, y). \tag{20}$$

The Mean Value Theorem applied to Eq. (19) now gives

$$\Delta = k\phi'(d_2), \tag{21}$$

for some d_2 between b and $b + k$. By Eq. (20),

$$\phi'(y) = f_y(a + h, y) - f_y(a, y). \tag{22}$$

Substituting from Eq. (22) into Eq. (21), we have

$$\Delta = k[f_y(a + h, d_2) - f_y(a, d_2)]. \tag{23}$$

Finally, we apply the Mean Value Theorem to the expression in brackets and get

$$\Delta = khf_{yx}(c_2, d_2), \tag{24}$$

for some c_2 between a and $a + h$.

A comparison of Eqs. (18) and (24) shows that

$$f_{xy}(c_1, d_1) = f_{yx}(c_2, d_2), \tag{25}$$

where (c_1, d_1) and (c_2, d_2) both lie in the rectangle R' (Fig. A.4). Equation (25) is not quite the result we want, since it says only that the mixed derivative f_{xy} has the same value at (c_1, d_1) that the derivative f_{yx} has at (c_2, d_2). But the numbers h and k in our discussion may be made as small as we wish. The hypothesis that f_{xy} and f_{yz} are both continuous at (a, b) then means that $f_{xy}(c_1, d_1) = f_{xy}(a, b) + \varepsilon_1$ and $f_{yx}(c_2, d_2) = f_{yx}(a, b) + \varepsilon_2$, where $\varepsilon_1, \varepsilon_2 \to 0$ as $h, k \to 0$. Hence, if we let h and $k \to 0$, we have $f_{xy}(a, b) = f_{yx}(a, b)$. ■

REMARK. The equality of $f_{xy}(a, b)$ and $f_{yx}(a, b)$ can be proved with weaker hypotheses than the ones we assumed in Theorem 4. For example, it is enough for f, f_x, and f_y to exist in R and for f_{xy} to be continuous at (a, b). Then f_{yx} will exist at (a, b) and be equal to f_{xy} at that point.

A.5 Mathematical Induction

Many formulas, like

$$1 + 2 + \cdots + n = \frac{n(n+1)}{2},$$

that hold for all positive integers n can be proved by applying an axiom called *the mathematical induction principle*. A proof that uses this axiom is called a proof by mathematical induction, or a proof by induction.

The steps in proving a formula by induction are

1. check that it holds for $n = 1$, and
2. prove that if it holds for any positive integer $n = k$, then it also holds for $n = k + 1$.

Once these steps are completed (the axiom says), we know the formula holds for all positive integers n. By step 1 it holds for $n = 1$. By step 2 it holds for $n = 2$, and therefore by step 2 also for $n = 3$, and by step 2 again for $n = 4$, and so on. If the first domino falls, and the kth domino always knocks over the $(k + 1)$-st when it falls, all the dominoes fall.

From another point of view, suppose we have a sequence of statements

$$S_1, S_2, \ldots, S_n, \ldots,$$

one for each positive integer. Suppose we can show that assuming any one of the statements to be true implies that the next statement in line is true. Suppose that we can also show that S_1 is true. Then we may conclude that all the statements are true from S_1 on.

EXAMPLE 1 Show that

$$1 + 2 + \cdots + n = \frac{n(n+1)}{2}$$

for all positive integers n.

Solution We carry out the two steps of mathematical induction.

1. The formula holds for $n = 1$ because

$$1 = \frac{1(1+1)}{2}.$$

2. If

$$1 + 2 + \cdots + k = \frac{k(k+1)}{2},$$

then

$$\begin{aligned} 1 + 2 + \cdots + k + (k+1) &= \frac{k(k+1)}{2} + (k+1) \\ &= \frac{k^2 + k + 2k + 2}{2} \\ &= \frac{(k+1)(k+2)}{2} \\ &= \frac{(k+1)((k+1)+1)}{2}. \end{aligned}$$

The last expression in this string of equalities is the expression $n(n + 1)/2$ for $n = (k + 1)$.

The mathematical induction principle now guarantees the formula for all positive integers n.

Note that all we have to do here is carry out steps 1 and 2. The mathematical induction principle does the rest. □

EXAMPLE 2 Show that

$$\frac{1}{2^1} + \frac{1}{2^2} + \cdots + \frac{1}{2^n} = 1 - \frac{1}{2^n}$$

for all positive integers n.

Solution We carry out the two steps of mathematical induction.

1. The formula holds for $n = 1$ because
$$\frac{1}{2^1} = 1 - \frac{1}{2^1}.$$
2. If
$$\frac{1}{2^1} + \frac{1}{2^2} + \cdots + \frac{1}{2^k} = 1 - \frac{1}{2^k},$$
then
$$\begin{aligned}\frac{1}{2^1} + \frac{1}{2^2} + \cdots + \frac{1}{2^k} + \frac{1}{2^{k+1}} &= 1 - \frac{1}{2^k} + \frac{1}{2^{k+1}}\\ &= 1 - \frac{1 \cdot 2}{2^k \cdot 2} + \frac{1}{2^{k+1}}\\ &= 1 - \frac{2}{2^{k+1}} + \frac{1}{2^{k+1}}\\ &= 1 - \frac{1}{2^{k+1}}.\end{aligned}$$

The mathematical induction principle now guarantees the formula for all positive integers n. □

Instead of starting at $n = 1$, some induction arguments start at another integer. The steps for such an argument are

1. check that the formula holds for $n = n_1$ (whatever the appropriate first integer is), and
2. prove that if it holds for any integer $n = k \geq n_1$, then it also holds for $n = k + 1$.

Once these steps are completed, the mathematical induction principle will guarantee the formula for all $n \geq n_1$.

EXAMPLE 3 Show that $n! > 3^n$ for n sufficiently large.

Solution How large? We experiment:

n	1	2	3	4	5	6	7
$n!$	1	2	6	24	120	720	5040
3^n	3	9	27	81	243	729	2187

It looks as if $n! > 3^n$ for $n \geq 7$. To be sure, we apply mathematical induction. We take $n_1 = 7$ in step 1, and try for step 2.

Suppose $k! > 3^k$ for some $k \geq 7$. Then

$$(k+1)! = (k+1)(k!) > (k+1)3^k > 8 \cdot 3^k > 3^{k+1}.$$

Thus, for $k \geq 7$,

$$k! > 3^k \Rightarrow (k+1)! > 3^{k+1}.$$

The mathematical induction principle now guarantees $n! \geq 3^n$ for all $n \geq 7$. □

PROBLEMS

1. Assuming the triangle inequality $|a + b| \leq |a| + |b|$, show that

$$|x_1 + x_2 + \cdots + x_n| \leq |x_1| + |x_2| + \cdots + |x_n|$$

for any n numbers.

2. Show that if $r \neq 1$, then

$$1 + r + r^2 + \cdots + r^n = \frac{1 - r^{n+1}}{1 - r}$$

for all positive integers n.

3. Use the product rule

$$\frac{d}{dx}(uv) = u\frac{dv}{dx} + v\frac{du}{dx}$$

and the fact that

$$\frac{d}{dx}(x) = 1$$

to show that

$$\frac{d}{dx}(x^n) = nx^{n-1}$$

for all positive integers n.

4. Suppose that a function $f(x)$ has the property that $f(x_1x_2) = f(x_1) + f(x_2)$ for any two positive numbers x_1 and x_2. Show that

$$f(x_1x_2 \cdots x_n) = f(x_1) + f(x_2) + \cdots + f(x_n)$$

for the product of any n positive numbers x_1, x_2, $\ldots$, x_n.

5. Show that

$$\frac{2}{3^1} + \frac{2}{3^2} + \cdots + \frac{2}{3^n} = 1 - \frac{1}{3^n}$$

for all positive integers n.

6. Show that $n! > n^3$ for n sufficiently large.
7. Show that $2^n > n^2$ for n sufficiently large.
8. Show that $2^n \geq \frac{1}{8}$ for $n \geq -3$.

A.6 The Law of Cosines and the Addition Formulas from Trigonometry

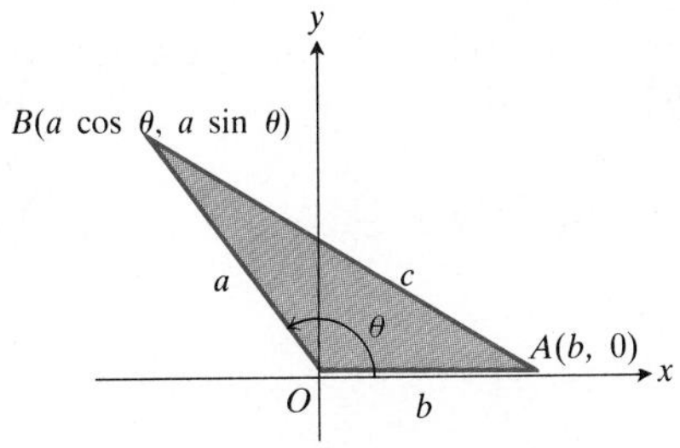

A.5 To derive the law of cosines, compute the distance between A and B, and square.

In Fig. A.5 triangle OAB has been placed with O at the origin and A on the x-axis at $A(b, 0)$. The third vertex B has coordinates

$$x = a\cos\theta, \qquad y = a\sin\theta. \tag{1}$$

The angle AOB has measure θ. By the formula for the distance between two points, the square of the distance c between A and B is

$$c^2 = (a\cos\theta - b)^2 + (a\sin\theta)^2 = a^2(\cos^2\theta + \sin^2\theta) + b^2 - 2ab\cos\theta.$$

or

$$c^2 = a^2 + b^2 - 2ab\cos\theta. \tag{2}$$

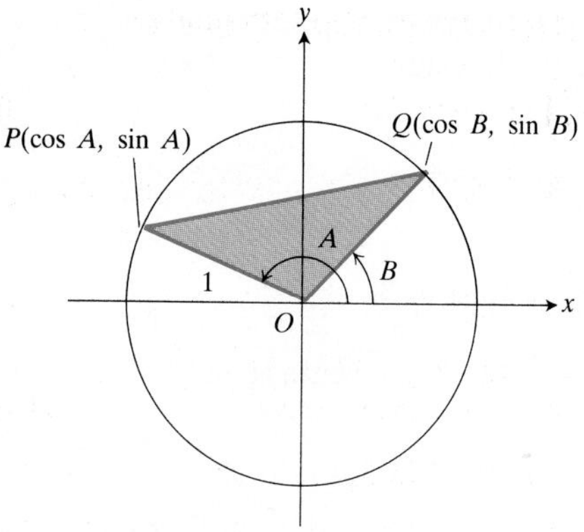

A.6 Diagram for cos $(A - B)$.

Equation (2) is called the *law of cosines*. In words, it says: "The square of any side of a triangle is equal to the sum of the squares of the other two sides minus twice the product of those two sides and the cosine of the angle between them." When the angle θ is a right angle, its cosine is zero, and Eq. (2) reduces to the theorem of Pythagoras. Equation (2) holds for a general angle θ, since it is based solely on the distance formula and on Eqs. (1). The same equation works with the exterior angle $(2\pi - \theta)$, or the opposite of $(2\pi - \theta)$, in place of θ, because

$$\cos(2\pi - \theta) = \cos(\theta - 2\pi) = \cos\theta.$$

It is still a valid formula when B is on the x-axis and $\theta = \pi$ or $\theta = 0$, as we can easily verify if we remember that $\cos 0 = 1$ and $\cos\pi = -1$. In these special cases, the right-hand side of Eq. (2) becomes $(a - b)^2$ or $(a + b)^2$.

Proofs of the Addition Formulas

Equation (11e) of Article 2.7 follows from the law of cosines applied to the triangle OPQ in Fig. A.6. We take $OP = OQ = r = 1$. Then the coordinates of P are

$$x_P = \cos A, \qquad y_P = \sin A$$

and of Q,

$$x_Q = \cos B, \qquad y_Q = \sin B.$$

Hence the square of the distance between P and Q is

$$\begin{aligned}(PQ)^2 &= (x_Q - x_P)^2 + (y_Q - y_P)^2\\ &= (x_Q^2 + y_Q^2) + (x_P^2 + y_P^2) - 2(x_Qx_P + y_Qy_P)\\ &= 2 - 2(\cos A\cos B + \sin A\sin B).\end{aligned}$$

But angle $QOP = A - B$, and the law of cosines gives

$$(PQ)^2 = (OP)^2 + (OQ)^2 - 2(OP)(OQ)\cos(A - B) = 2 - 2\cos(A - B).$$

When we equate these two expressions for $(PQ)^2$ we obtain

$$\cos(A - B) = \cos A\cos B + \sin A\sin B, \tag{11e}$$

which is Eq. (11e) of Article 2.7.

We now deduce Eqs. (a, b, d) of Article 2.7 from Eq. (11e). We shall also need the results

$$\begin{matrix}\sin 0 = 0, & \sin(\pi/2) = 1, & \sin(-\pi/2) = -1,\\ \cos 0 = 1, & \cos(\pi/2) = 0, & \cos(-\pi/2) = 0.\end{matrix} \tag{3}$$

1. In Eq. (11e), we put $A = \pi/2$ and use Eqs. (3) to get

$$\cos\left(\frac{\pi}{2} - B\right) = \sin B. \tag{4}$$

In this equation, if we replace B by $\pi/2 - B$ and $\pi/2 - B$ by $\pi/2 - (\pi/2 - B)$, we get

$$\cos B = \sin\left(\frac{\pi}{2} - B\right). \tag{5}$$

Equations (4) and (5) express the familiar results that the sine and cosine of an angle are the cosine and sine, respectively, of the complementary angle.

2. We next put $B = -\pi/2$ in Eq. (11e) and use Eqs. (3) to get

$$\cos\left(A + \frac{\pi}{2}\right) = -\sin A. \tag{6}$$

3. We can get the formula for $\cos(A + B)$ from Eq. (11e) by substituting $-B$ for B everywhere:

$$\begin{aligned}\cos(A + B) &= \cos[A - (-B)]\\ &= \cos A \cos(-B) + \sin A \sin(-B)\\ &= \cos A \cos B - \sin A \sin B.\end{aligned} \tag{11b}$$

where the final equality also uses Eqs. (11c) in Article 2.7.

4. To derive formulas for $\sin(A \pm B)$, we use Eq. (4) with B replaced by $A + B$, and then use Eq. (11e). Thus we have

$$\begin{aligned}\sin(A + B) &= \cos[\pi/2 - (A + B)]\\ &= \cos[(\pi/2 - A) - B]\\ &= \cos(\pi/2 - A)\cos B + \sin(\pi/2 - A)\sin B\\ &= \sin A \cos B + \cos A \sin B.\end{aligned} \tag{11a}$$

Equation (11d) of Article 2.7 follows from this if we replace B by $-B$:

$$\sin(A - B) = \sin A \cos B - \cos A \sin B. \tag{11d}$$

A.7

Formulas from Elementary Mathematics

ALGEBRA

1. Laws of Exponents

$$a^m a^n = a^{m+n}, \quad (ab)^m = a^m b^m, \quad (a^m)^n = a^{mn}, \quad a^{m/n} = \sqrt[n]{a^m}.$$

If $a \neq 0$,

$$\frac{a^m}{a^n} = a^{m-n}, \quad a^0 = 1, \quad a^{-m} = \frac{1}{a^m}.$$

2. Zero

$$a \cdot 0 = 0 \cdot a = 0 \text{ for any finite number } a.$$

If $a \neq 0$,

$$\frac{0}{a} = 0, \quad 0^a = 0, \quad a^0 = 1.$$

Division by zero is not defined.

3. Fractions

$$\frac{a}{b} + \frac{c}{d} = \frac{ad + bc}{bd}, \quad \frac{a}{b} \cdot \frac{c}{d} = \frac{ac}{bd}, \quad \frac{a/b}{c/d} = \frac{a}{b} \cdot \frac{d}{c}, \quad \frac{-a}{b} = -\frac{a}{b} = \frac{a}{-b}.$$

$$\frac{(a/b) + (c/d)}{(e/f) + (g/h)} = \frac{(a/b) + (c/d)}{(e/f) + (g/h)} \cdot \frac{bdfh}{bdfh} = \frac{(ad + bc)fh}{(eh + fg)bd}.$$

4. Binomial Theorem, for n = Positive Integer

$$(a+b)^n = a^n + na^{n-1}b + \frac{n(n-1)}{1\cdot 2}a^{n-2}b^2 + \frac{n(n-1)(n-2)}{1\cdot 2\cdot 3}a^{n-3}b^3 + \cdots + nab^{n-1} + b^n.$$

For instance:

$$(a+b)^1 = a+b,$$
$$(a+b)^2 = a^2 + 2ab + b^2,$$
$$(a+b)^3 = a^3 + 3a^2b + 3ab^2 + b^3,$$
$$(a+b)^4 = a^4 + 4a^3b + 6a^2b^2 + 4ab^3 + b^4.$$

5. Difference of Like Integer Powers, $n > 1$

$$a^n - b^n = (a-b)(a^{n-1} + a^{n-2}b + a^{n-3}b^2 + \cdots + ab^{n-2} + b^{n-1}),$$

For instance:

$$a^2 - b^2 = (a-b)(a+b),$$
$$a^3 - b^3 = (a-b)(a^2 + ab + b^2)$$
$$a^4 - b^4 = (a-b)(a^3 + a^2b + ab^2 + b^3).$$

6. Proportionality

The statements "y varies directly as x" and "y is directly proportional to x" mean the same thing, namely that

$$y = kx$$

for some constant k. This constant is called the *proportionality factor* of the equation.

Similarly, "y varies inversely as x" and "y is inversely proportional to x" both mean that

$$y = k\frac{1}{x}$$

for some constant k. Again, k is the porportionality factor of the equation.

7. Remainder Theorem and Factor Theorem

If the polynomial $f(x)$ is divided by $x - r$ until a remainder R independent of x is obtained, then $R = f(r)$. In particular, $x - r$ is a *factor* of $f(x)$ if and only if r is a *root* of the equation $f(x) = 0$.

8. Completing the Square

The equation

$$\begin{aligned} ax^2 + bx &= a\left(x^2 + \frac{b}{a}x\right) \\ &= a\left(x^2 + \frac{b}{a}x + \left(\frac{b}{2a}\right)^2 - \left(\frac{b}{2a}\right)^2\right) \\ &= a\left(\left(x + \frac{b}{2a}\right)^2 - \left(\frac{b}{2a}\right)^2\right) \end{aligned}$$

shows how to write $ax^2 + bx$ as a constant times the difference of two squares when $a \neq 0$.

9. Quadratic Formula

By completing the square on the first two terms of the equation

$$ax^2 + bx + c = 0$$

and solving the resulting equation for x (details omitted), one obtains the formula

$$x = \frac{-b \pm \sqrt{b^2 - 4ac}}{2a}.$$

10. Horner's Method

The fastest way (usually) to calculate the value of a polynomial

$$p(x) = a_0 + a_1x + a_2x^2 + \cdots a_nx^n$$

at $x = c$ is to calculate

$$p(c) = a_0 + c(a_1 + c(a_2 + c(a_3 + \cdots + c(a_{n-1} + ca_n))) \cdots),$$

working from the inside out. Arranged this way, the calculation does not require parenthesis keys. All we do is multiply and add n times, for a total of $2n$ operations (fewer if any of the coefficients are zero).

For example, the value of

$$p(x) = 3x^3 - 6x^2 + 4x + 5$$

at $x = 2$ is

$$p(2) = 5 + 2(4 + 2(-6 + 2 \cdot 3)) = 13.$$

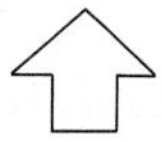

Start here,
work back

GEOMETRY (A = area, B = area of base, C = circumference, S = lateral area or surface area, V = volume.)

1. Triangle

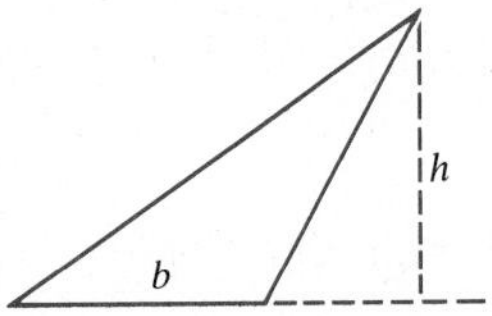

$$A = \tfrac{1}{2}bh$$

2. Similar Triangles

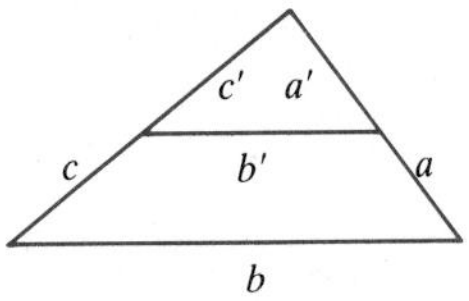

$$\frac{a'}{a} = \frac{b'}{b} = \frac{c'}{c}$$

3. Theorem of Pythagoras

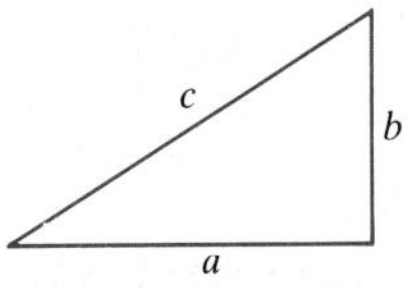

$$c^2 = a^2 + b^2$$

4. Parallelogram

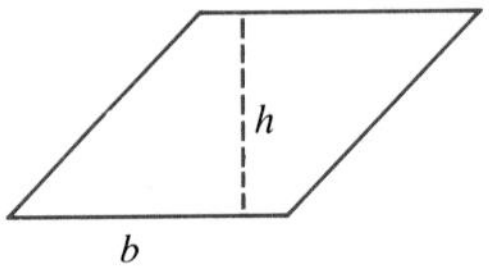

$$A = bh$$

5. Trapezoid

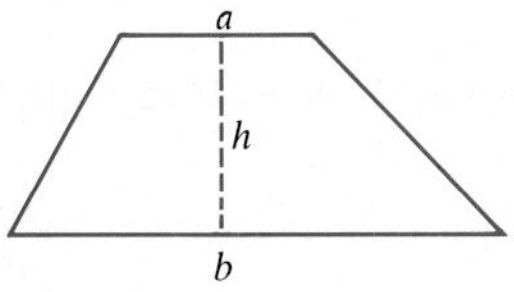

$A = \frac{1}{2}(a + b)h$

6. Circle

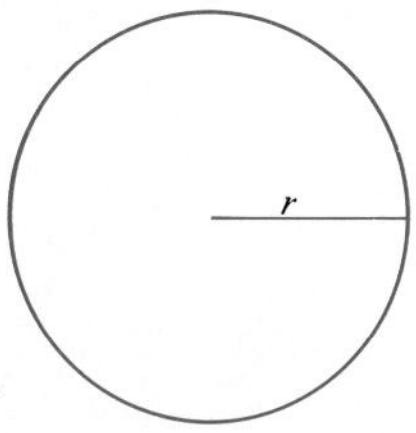

$A = \pi r^2, \qquad C = 2\pi r$

7. Any Cylinder or Prism with Parallel Bases

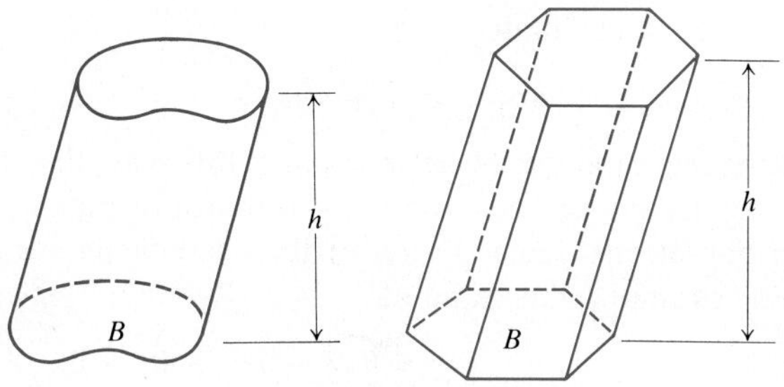

$V = Bh$

8. Right Circular Cylinder

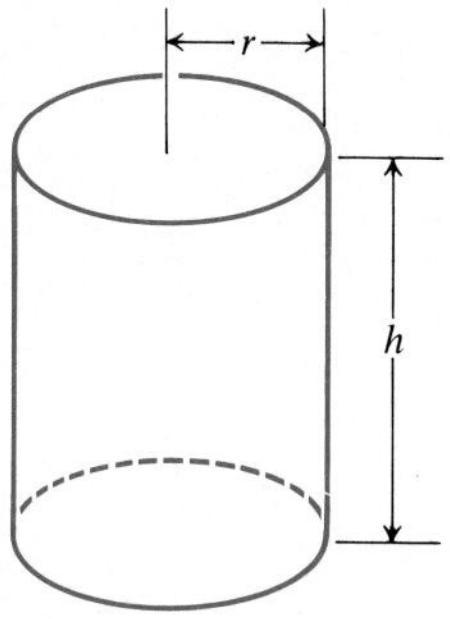

$V = \pi r^2 h, \qquad S = 2\pi rh$

9. Any Cone or Pyramid

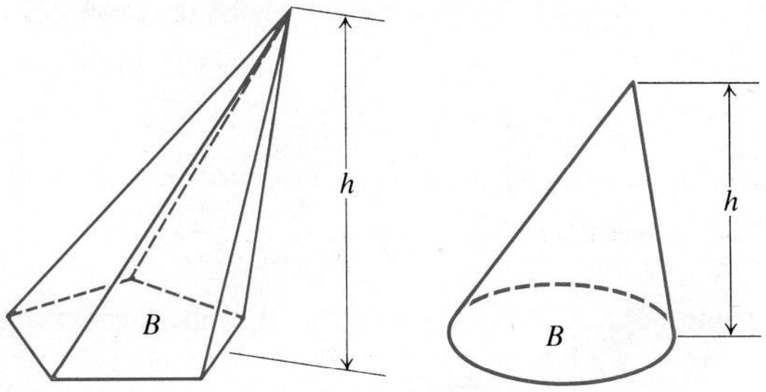

$V = \frac{1}{3}Bh$

10. Right Circular Cone

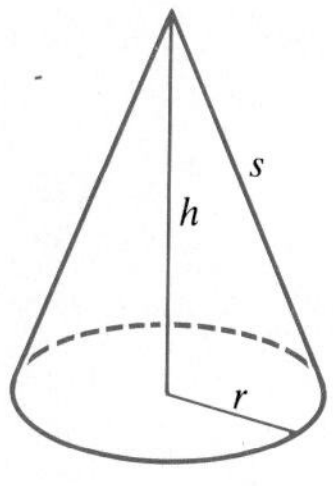

$V = \frac{1}{3}\pi r^2 h, \qquad S = \pi rs$

11. Sphere

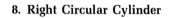

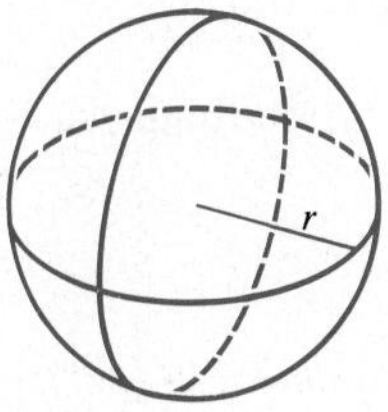

$V = \frac{4}{3}\pi r^3, \qquad S = 4\pi r^2$

TRIGONOMETRY

1. Definitions and Fundamental Identities

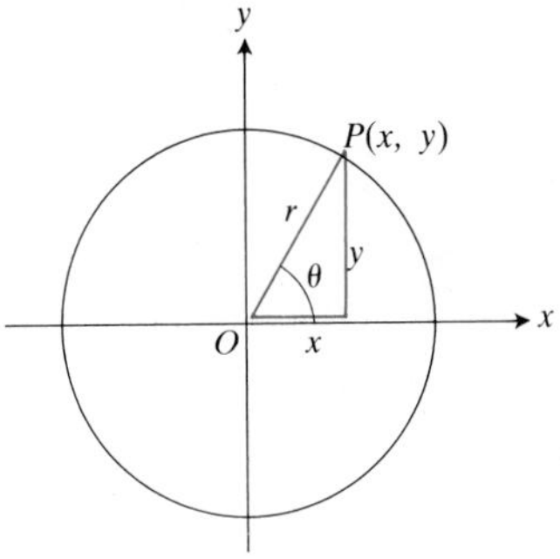

Sine: $\sin\theta = \frac{y}{r} = \frac{1}{\csc\theta}$

Cosine: $\cos\theta = \frac{x}{r} = \frac{1}{\sec\theta}$

Tangent: $\tan\theta = \frac{y}{x} = \frac{1}{\cot\theta}$

$\sin(-\theta) = -\sin\theta,$

$\cos(-\theta) = \cos\theta$

$\sin^2\theta + \cos^2\theta = 1$

$\sec^2\theta = 1 + \tan^2\theta$

$\csc^2\theta = 1 + \cot^2\theta$

$\sin 2\theta = 2\sin\theta\cos\theta,$

$\cos 2\theta = \cos^2\theta - \sin^2\theta$

$$\cos^2\theta = \frac{1+\cos 2\theta}{2},$$

$$\sin^2\theta = \frac{1-\cos 2\theta}{2}$$

$\sin(A+B) = \sin A\cos B + \cos A\sin B$

$\sin(A-B) = \sin A\cos B - \cos A\sin B$

$\cos(A+B) = \cos A\cos B - \sin A\sin B$

$\cos(A-B) = \cos A\cos B + \sin A\sin B$

$$\tan(A+B) = \frac{\tan A + \tan B}{1 - \tan A\tan B}$$

$$\tan(A-B) = \frac{\tan A - \tan B}{1 + \tan A\tan B}$$

$$\sin\left(A - \frac{\pi}{2}\right) = -\cos A, \qquad \cos\left(A - \frac{\pi}{2}\right) = \sin A$$

$$\sin\left(A + \frac{\pi}{2}\right) = \cos A, \qquad \cos\left(A + \frac{\pi}{2}\right) = -\sin A$$

$\sin A\sin B = \frac{1}{2}\cos(A-B) - \frac{1}{2}\cos(A+B)$

$\cos A\cos B = \frac{1}{2}\cos(A-B) + \frac{1}{2}\cos(A+B)$

$\sin A\cos B = \frac{1}{2}\sin(A-B) + \frac{1}{2}\sin(A+B)$

$\sin A + \sin B = 2\sin\frac{1}{2}(A+B)\cos\frac{1}{2}(A-B)$

$\sin A - \sin B = 2\cos\frac{1}{2}(A+B)\sin\frac{1}{2}(A-B)$

$\cos A + \cos B = 2\cos\frac{1}{2}(A+B)\cos\frac{1}{2}(A-B)$

$\cos A - \cos B = -2\sin\frac{1}{2}(A+B)\sin\frac{1}{2}(A-B)$

2. Common Reference Triangles

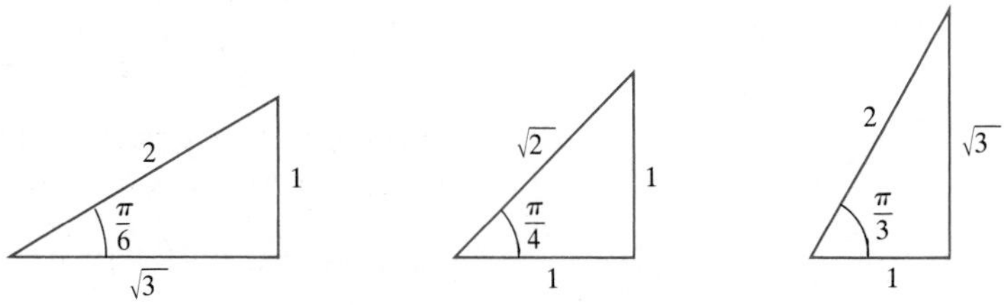

3. Angles and Sides of a Triangle

Law of cosines: $c^2 = a^2 + b^2 - 2ab \cos C$

Law of sines: $\dfrac{\sin A}{a} = \dfrac{\sin B}{b} = \dfrac{\sin C}{c}$

Area $= \frac{1}{2}bc \sin A = \frac{1}{2}ac \sin B = \frac{1}{2}ab \sin C$

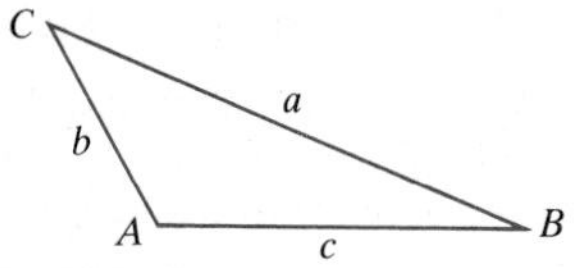

***Toolkit* programs**	
Function Evaluator	Super * Grapher

A.8

Invented Number Systems. Complex Numbers

In this appendix we shall discuss complex numbers. These are expressions of the form $a + ib$, where a and b are "real" numbers and i is a symbol for $\sqrt{-1}$. Unfortunately, the words "real" and "imaginary" have connotations that somehow place $\sqrt{-1}$ in a less favorable position than $\sqrt{2}$ in our minds. As a matter of fact, a good deal of imagination, in the sense of *inventiveness,* has been required to construct the *real* number system, which forms the basis of the calculus. In this article we shall review the various stages of this invention. The further invention of a complex number system will then not seem so strange. It is fitting for us to study such a system, since modern engineering has found therein a convenient language for describing vibratory motion, harmonic oscillation, damped vibrations, alternating currents, and other wave phenomena.

The earliest stage of number development was the recognition of the *counting numbers* 1, 2, 3, . . . , which we now call the *natural numbers,* or the *positive integers.* Certain simple arithmetical operations can be performed with these numbers without getting outside the system. That is, the system of positive integers is *closed* with respect to the operations of *addition* and *multiplication*. By this we mean that if m and n are any positive integers, then

$$m + n = p \qquad \text{and} \qquad mn = q \tag{1}$$

are also positive integers. Given the two positive integers on the *left-hand side* of either equation in (1), we can find the corresponding positive integer on the right. More than this, we may sometimes specify the positive integers m and p and find a positive integer n such that $m + n = p$. For instance, $3 + n = 7$ can be *solved* when the only numbers we know are the positive integers. But the equation $7 + n = 3$ cannot be solved unless the number system is enlarged. The number concepts that we denote by zero and the *negative* integers were invented to solve equations like that. In a civilization that recognizes all the integers

$$\ldots, -3, -2, -1, 0, 1, 2, 3, \ldots, \tag{2}$$

an educated person may always find the missing integer that solves the equation $m + n = p$ when given the other two integers in the equation.

Suppose our educated people also know how to multiply any two integers of the set in (2). If, in Eqs. (1), they are given m and q, they discover that sometimes they can find n and sometimes they can't. If their *imagination* is still in good working order, they may be inspired to invent still more numbers and introduce fractions, which are just ordered pairs m/n of integers m and n. The number zero has special properties that may bother them for a while, but they ultimately discover that it is handy to have all ratios of integers m/n, excluding only those having zero in the denominator. This system, called the set of *rational numbers,* is now rich enough for them to perform the so-called *rational operations* of arithmetic:

1. a) addition
   b) subtraction
2. a) multiplication
   b) division

on any two numbers in the system, *except that they cannot divide by zero.*

The geometry of the unit square (Fig. A.7) and the Pythagorean Theorem showed that they could construct a geometric line segment which, in terms of some basic unit of length, has length equal to $\sqrt{2}$. Thus they could solve the equation

$$x^2 = 2$$

by a geometric construction. But then they discovered that the line segment representing $\sqrt{2}$ and the line segment representing the unit of length 1 were incommensurable quantities. This means that the ratio $\sqrt{2}/1$ cannot be expressed as the ratio of two *integral* multiples of some other, presumably more fundamental, unit of length. That is, our educated people could not find a rational number solution of the equation $x^2 = 2$.

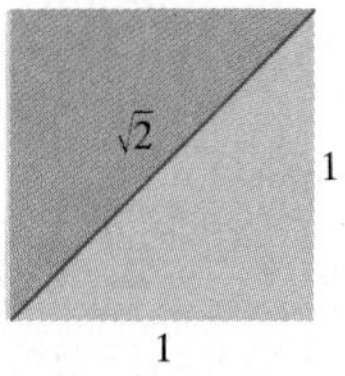

A.7 Segments of irrational length can be constructed with straightedge and compass.

There is a nice algebraic argument that there is no rational number whose square is 2. Suppose that there were such a rational number. Then we could find integers p and q with no common factor other than 1, and such that

$$p^2 = 2q^2. \tag{3}$$

Since p and q are integers, p must then be even, say $p = 2p_1$, where p_1 is an integer. This leads to $2p_1^2 = q^2$, which says that q must also be even, say $q = 2q_1$, where q_1 is also an integer. But this is contrary to our choice of p and q as integers having no common factor other than 1. Hence there is no rational number whose square is 2.

Our educated people *could*, however, get a *sequence* of rational numbers

$$\frac{1}{1}, \frac{7}{5}, \frac{41}{29}, \frac{239}{169}, \ldots, \tag{4}$$

whose squares form a sequence

$$\frac{1}{1}, \frac{49}{25}, \frac{1681}{841}, \frac{57{,}121}{28{,}561}, \ldots, \tag{5}$$

that converges to 2 as its *limit*. This time their *imagination* suggested that they needed the concept of a *limit of a sequence* of rational numbers. If we accept the fact that an increasing sequence that is bounded from above always approaches a limit, and observe that the sequence in (4) has these properties, then we want it to have a limit L. This would also mean, from (5), that $L^2 = 2$, and hence L is *not* one of our rational numbers. If to the *rational* numbers we further add the *limits* of all bounded increasing sequences of rational numbers, we arrive at the system of all "real" numbers. The word *real* is placed in quotes because there is nothing that is either "more real" or "less real" about this system than there is about any other well-defined mathematical system.

Imagination was called upon at many stages during the development of the real number system from the system of positive integers. In fact, the art of invention was needed at least three times in constructing the systems we have discussed so far:

1. The *first invented* system; the set of *all integers* as constructed from the counting numbers.
2. The *second invented* system; the set of *rational* numbers m/n as constructed from the integers.
3. The *third invented* system; the set of all "*real*" numbers x as constructed from the rational numbers.

These invented systems form a hierarchy in which each system contains the previous system. Each system is also richer than its predecessor in that it permits additional operations to be performed without going outside the system. Expressed in algebraic terms, we may say that

1. In the system of all integers, we can solve all equations of the form

$$x + a = 0, \tag{6}$$

where a may be any integer.

2. In the system of all rational numbers, we can solve all equations of the form

$$ax + b = 0 \tag{7}$$

provided a and b are rational numbers and $a \neq 0$.

3. In the system of all real numbers, we can solve all of the Eqs. (6) and (7) and, in addition, all quadratic equations

$$ax^2 + bx + c = 0 \qquad \text{having} \qquad a \neq 0 \quad \text{and} \quad b^2 - 4ac \geq 0. \tag{8}$$

Every student of algebra is familiar with the formula that gives the solutions of (8), namely,

$$x = \frac{-b \pm \sqrt{b^2 - 4ac}}{2a}, \tag{9}$$

and familiar with the further fact that when the discriminant, $d = b^2 - 4ac$, is *negative,* the solutions in (9) do *not* belong to any of the systems discussed above. In fact, the very simple quadratic equation

$$x^2 + 1 = 0 \tag{10}$$

is impossible to solve if the only number systems that can be used are the three invented systems mentioned so far.

Thus we come to the *fourth invented* system, the set of all complex numbers $a + ib$. We could, in fact, dispense entirely with the symbol i and use a notation like (a, b). We would then speak simply of a pair of real numbers a and b. Since, under algebraic operations, the numbers a and b are treated somewhat differently, it is essential to keep the *order* straight. We therefore might say that *the complex number system consists of the set of all ordered pairs of real numbers* (a, b), together with the rules by which they are to be equated, added, multiplied, and so on, listed below. We shall use both the (a, b) notation and the notation $a + ib$. We call a the "real part" and b the "imaginary part" of (a, b). We make the following definitions.

Equality

$a + ib = c + id$ if and only if $a = c$ and $b = d$.	Two complex numbers (a, b) and (c, d) are *equal* if and only if $a = c$ and $b = d$.

Addition

$(a + ib) + (c + id)$ $= (a + c) + i(b + d)$	The sum of the two complex numbers (a, b) and (c, d) is the complex number $(a + c, b + d)$.

Multiplication

$(a + ib)(c + id)$ $= (ac - bd) + i(ad + bc)$	The product of two complex numbers (a, b) and (c, d) is the complex number $(ac - bd, ad + bc)$.
$c(a + ib) = ac + i(bc)$	The product of a real number c and the complex number (a, b) is the complex number (ac, bc).

The set of all complex numbers (a, b) in which the second number is zero has all the properties of the set of ordinary "real" numbers a. For example, addition and multiplication of $(a, 0)$ and $(c, 0)$ give

$$(a, 0) + (c, 0) = (a + c, 0),$$
$$(a, 0) \cdot (c, 0) = (ac, 0),$$

which are numbers of the same type with "imaginary part" equal to zero. Also, if we multiply a "real number" $(a, 0)$ and the "complex number" (c, d), we get

$$(a, 0) \cdot (c, d) = (ac, cd) = a(c, d).$$

In particular, the complex number $(0, 0)$ plays the role of zero in the complex number system and the complex number $(1, 0)$ plays the role of unity.

The number pair $(0, 1)$, which has "real part" equal to zero and "imaginary part" equal to one has the property that its square,

$$(0, 1)(0, 1) = (-1, 0),$$

has "real part" equal to minus one and "imaginary part" equal to zero. Therefore, in the system of complex numbers (a, b), there is a number $x = (0, 1)$ whose square can be added to unity $= (1, 0)$ to produce zero $= (0, 0)$; that is,

$$(0, 1)^2 + (1, 0) = (0, 0).$$

The equation

$$x^2 + 1 = 0$$

therefore has a solution $x = (0, 1)$ in this new number system.

You are probably more familiar with the $a + ib$ notation than you are with the notation (a, b). And since the laws of algebra for the ordered pairs enable us to write

$$(a, b) = (a, 0) + (0, b) = a(1, 0) + b(0, 1),$$

while $(1, 0)$ behaves like unity and $(0, 1)$ behaves like a square root of minus one, we need not hesitate to write $a + ib$ in place of (a, b). The i associated with b is like a tracer element that tags the "imaginary part" of $a + ib$. We can pass at will from the realm of ordered pairs (a, b) to the realm of expressions $a + ib$, and conversely. But there is nothing less "real" about the symbol $(0, 1) = i$ than there is about the symbol $(1, 0) = 1$, once we have learned the laws of algebra in the complex number system (a, b).

To reduce any rational combination of complex numbers to a single complex number, we need only apply the laws of elementary algebra, replacing i^2 wherever it appears by -1. Of course, we cannot divide by the complex number $(0, 0) = 0 + i0$. But if $a + ib \neq 0$, then we may carry out a division as follows:

$$\frac{c + id}{a + ib} = \frac{(c + id)(a - ib)}{(a + ib)(a - ib)} = \frac{(ac + bd) + i(ad - bc)}{a^2 + b^2}.$$

The result is a complex number $x + iy$ with

$$x = \frac{ac + bd}{a^2 + b^2}, \qquad y = \frac{ad - bc}{a^2 + b^2},$$

and $a^2 + b^2 \neq 0$, since $a + ib = (a, b) \neq (0, 0)$.

The number $a - ib$ that is used as multiplier to clear the i out of the denominator is called the *complex conjugate* of $a + ib$. It is customary to use $\bar{z}$ (read "z bar") to denote the complex conjugate of z; thus

$$z = a + ib, \qquad \bar{z} = a - ib.$$

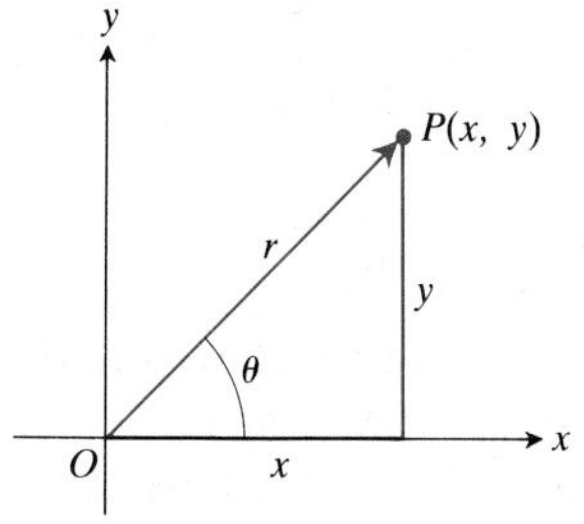

A.8 An Argand diagram representing $z = x + iy$ both as a point $P(x, y)$ and as a vector $\overrightarrow{OP}$.

Thus, we multiplied the numerator and denominator of the complex fraction $(c + id)/(a + ib)$ by the complex conjugate of the denominator. This will always replace the denominator by a real number.

Argand Diagrams There are two geometric representations of the complex number $z = x + iy$:

a) as the point $P(x, y)$ in the xy-plane, and

b) as the vector $\overrightarrow{OP}$ from the origin to P.

In each representation, the x-axis is called the "axis of reals" and the y-axis is the "imaginary axis." Both representations are called *Argand diagrams* (Fig. A.8).

In terms of the polar coordinates of x and y, we have

$$x = r\cos\theta, \qquad y = r\sin\theta,$$

and

$$z = x + iy = r(\cos\theta + i\sin\theta). \tag{11}$$

We define the *absolute value* of a complex number $x + iy$ to be the length r of a vector $\overrightarrow{OP}$ from the origin to $P(x, y)$. We denote the absolute value by vertical bars, thus:

$$|x + iy| = \sqrt{x^2 + y^2}. \tag{12a}$$

If we always choose the polar coordinates r and θ so that r is nonnegative, then we have

$$r = |x + iy|. \tag{12b}$$

The polar angle θ is called the *argument* of z and written $\theta = \arg z$. Of course, any integral multiple of 2π may be added to θ to produce another appropriate angle. The *principal value* of the argument will, in this book, be taken to be that value of θ for which $-\pi < \theta \leq +\pi$.

The following equation gives a useful formula connecting a complex number z, its conjugate $\bar{z}$, and its absolute value $|z|$, namely,

$$z \cdot \bar{z} = |z|^2. \tag{13}$$

The identity

$$e^{i\theta} = \cos\theta + i\sin\theta \tag{14}$$

introduced in Article 12.2 leads to the following rules for calculating products, quotients, powers, and roots of complex numbers. It also leads to Argand diagrams for $e^{i\theta}$. Since $\cos\theta + i\sin\theta$ is what we get from Eq. (11) by taking $r = 1$, we can say that $e^{i\theta}$ is represented by a unit vector that makes an angle θ with the positive x-axis, as shown in Fig. A.9.

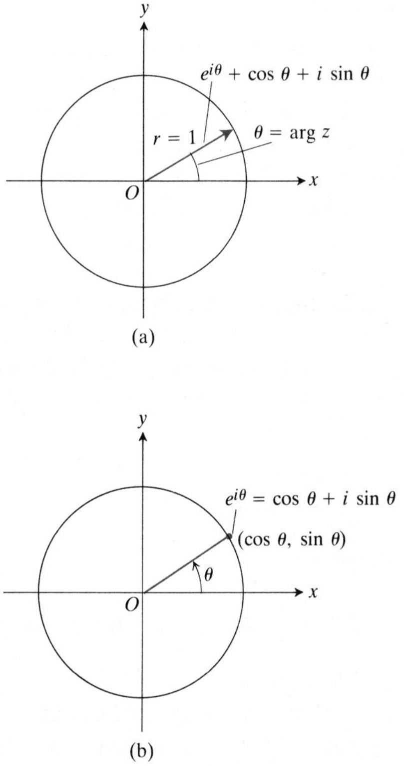

A.9 Argand diagrams for $e^{i\theta} = \cos\theta + i\sin\theta$ (a) as a vector, (b) as a point.

Products Let

$$z_1 = r_1 e^{i\theta_1} \qquad z_2 = r_2 e^{i\theta_2} \tag{15}$$

so that

$$|z_1| = r_1, \quad \arg z_1 = \theta_1; \qquad |z_2| = r_2, \quad \arg z_2 = \theta_2. \tag{16}$$

Then

$$z_1 z_2 = r_1 e^{i\theta_1} \cdot r_2 e^{i\theta_2} = r_1 r_2 e^{i(\theta_1 + \theta_2)}$$

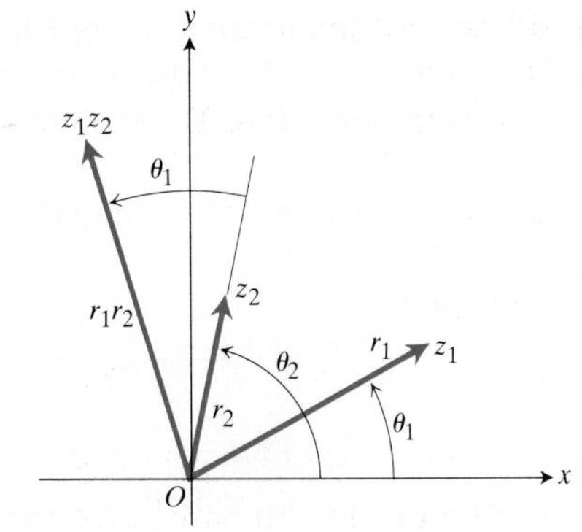

A.10 When z_1 and z_2 are multiplied, $|z_1z_2| = r_1 \cdot r_2$, and $\arg(z_1z_2) = \theta_1 + \theta_2$.

and hence

$$|z_1z_2| = r_1r_2 = |z_1| \cdot |z_2|,$$
$$\arg(z_1z_2) = \theta_1 + \theta_2 = \arg z_1 + \arg z_2. \tag{17}$$

Thus the product of two complex numbers is represented by a vector whose length is the product of the lengths of the two factors and whose argument is the sum of their arguments (Fig. A.10). In particular, a vector may be rotated in the counterclockwise direction through an angle θ by simply multiplying it by $e^{i\theta}$. Multiplication by i rotates 90°, by -1 rotates 180°, by $-i$ rotates 270°, etc.

EXAMPLE 1 Let

$$z_1 = 1 + i, \qquad z_2 = \sqrt{3} - i.$$

We plot these complex numbers in an Argand diagram (Fig. A.11) from which we read off the polar representations

$$z_1 = \sqrt{2}e^{i\pi/4}, \qquad z_2 = 2e^{-i\pi/6}.$$

Then

$$z_1z_2 = 2\sqrt{2}\exp\left(\frac{i\pi}{4} - \frac{i\pi}{6}\right) = 2\sqrt{2}\exp\left(\frac{i\pi}{12}\right)$$
$$= 2\sqrt{2}\left(\cos\frac{\pi}{12} + i\sin\frac{\pi}{12}\right) \approx 2.73 + 0.73i. \ \square$$

Quotients Suppose $r_2 \neq 0$ in Eq. (15). Then

$$\frac{z_1}{z_2} = \frac{r_1e^{i\theta_1}}{r_2e^{i\theta_2}} = \frac{r_1}{r_2}e^{i(\theta_1-\theta_2)}.$$

Hence

$$\left|\frac{z_1}{z_2}\right| = \frac{r_1}{r_2} = \frac{|z_1|}{|z_2|},$$
$$\arg\left(\frac{z_1}{z_2}\right) = \theta_1 - \theta_2 = \arg z_1 - \arg z_2.$$

That is, we divide lengths and subtract angles.

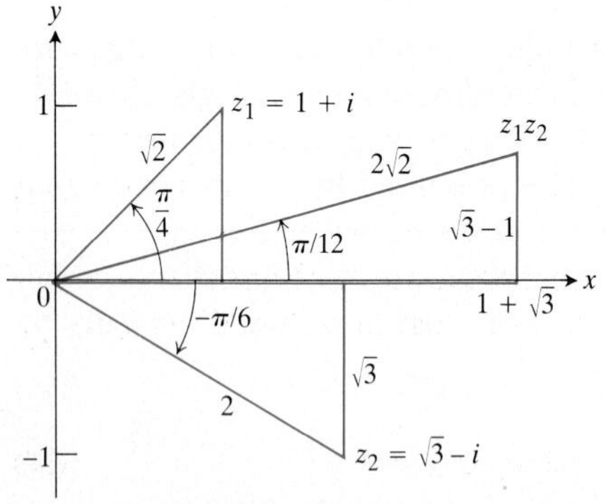

A.11 To multiply two complex numbers, one multiplies their absolute values, and adds their arguments.

EXAMPLE 2 Let $z_1 = 1 + i$ and $z_2 = \sqrt{3} - i$, as in Example 1. Then,

$$\frac{1+i}{\sqrt{3}-i} = \frac{\sqrt{2}e^{i\pi/4}}{2e^{-i\pi/6}} = \frac{\sqrt{2}}{2}e^{5\pi i/12}$$
$$\approx 0.707\left(\cos\frac{5\pi}{12} + i\sin\frac{5\pi}{12}\right)$$
$$\approx 0.183 + 0.683i. \ \square$$

Powers If n is a positive integer, we may apply the product formulas in (17) to find

$$z^n = z \cdot z \cdot \cdots \cdot z \qquad (n \text{ factors}).$$

With $z = re^{i\theta}$, we obtain

$$z^n = (re^{i\theta})^n = r^ne^{i(\theta+\theta+\cdots+\theta)} \qquad (n \text{ summands})$$
$$= r^ne^{in\theta}. \tag{18}$$

The length $r = |z|$ is raised to the nth power and the angle $\theta = \arg z$ is multiplied by n.

In particular, if we place $r = 1$ in Eq. (18), we obtain De Moivre's theorem.

THEOREM

De Moivre's Theorem

$$(\cos\theta + i\sin\theta)^n = \cos n\theta + i\sin n\theta. \tag{19}$$

If we expand the left-hand side of this equation by the binomial theorem and reduce it to the form $a + ib$, we obtain formulas for $\cos n\theta$ and $\sin n\theta$ as polynomials of degree n in $\cos\theta$ and $\sin\theta$.

EXAMPLE 4 If $n = 3$ in Eq. (19), we have

$$(\cos\theta + i\sin\theta)^3 = \cos 3\theta + i\sin 3\theta.$$

The left-hand side of this equation is

$$\cos^3\theta + 3i\cos^2\theta\sin\theta - 3\cos\theta\sin^2\theta - i\sin^3\theta.$$

The real part of this must equal $\cos 3\theta$ and the imaginary part must equal $\sin 3\theta$. Therefore,

$$\cos 3\theta = \cos^3\theta - 3\cos\theta\sin^2\theta,$$
$$\sin 3\theta = 3\cos^2\theta\sin\theta - \sin^3\theta. \ \square$$

Roots If $z = re^{i\theta}$ is a complex number different from zero and n is a positive integer, then there are precisely n different complex numbers w_0, $w_1, \ldots, w_{n-1}$, that are nth roots of z. To see why, let $w = \rho e^{i\alpha}$ be an nth root of $z = re^{i\theta}$, so that

$$w^n = z$$

or

$$\rho^n e^{in\alpha} = re^{i\theta}. \tag{20}$$

Then

$$\rho = \sqrt[n]{r} \tag{21}$$

is the real, positive, nth root of r. As regards the angle, although we cannot say that $n\alpha$ and θ must be equal, we can say that they may differ only by an integral multiple of 2π. That is,

$$n\alpha = \theta + 2k\pi, \qquad k = 0, \pm 1, \pm 2, \ldots \tag{22}$$

Therefore

$$\alpha = \frac{\theta}{n} + k\frac{2\pi}{n}.$$

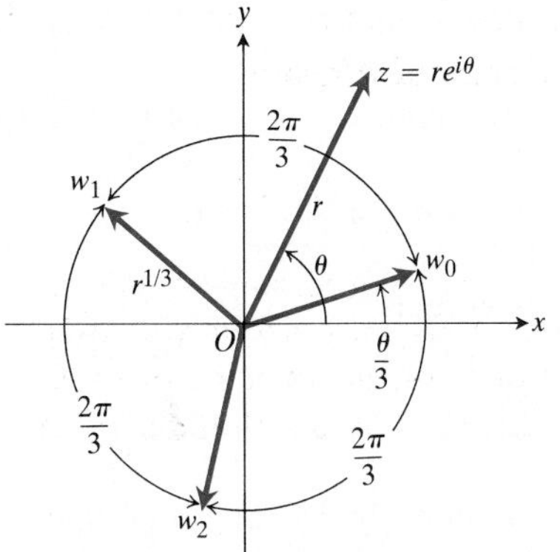

A.12 The three cube roots of $z = re^{i\theta}$.

Hence all the nth roots of $z = re^{i\theta}$ are given by

$$\sqrt[n]{re^{i\theta}} = \sqrt[n]{r}\exp\left(\frac{\theta}{n} + k\frac{2\pi}{n}\right), \qquad k = 0, \pm 1, \pm 2, \ldots . \tag{23}$$

There might appear to be infinitely many different answers corresponding to the infinitely many possible values of k. But one readily sees that $k = n + m$ gives the same answer as $k = m$ in Eq. (23). Thus we need only take n consecutive values for k to obtain all the different nth roots of z. For convenience, we may take

$$k = 0, 1, 2, \ldots, n - 1.$$

All the nth roots of $re^{i\theta}$ lie on a circle centered at the origin O and having radius equal to the real, positive nth root of r. One of them has argument $\alpha = \theta/n$. The others are uniformly spaced around the circle, each being separated from its neighbors by an angle equal to $2\pi/n$. Figure A.12 illustrates the placement of the three cube roots, w_0, w_1, w_2, of the complex number $z = re^{i\theta}$,

EXAMPLE 5 Find the four fourth roots of -16.

Solution As our first step, we plot the given number in an Argand diagram (Fig. A.13) and determine its polar representation $re^{i\theta}$. Here,

$$z = -16, \qquad r = +16, \qquad \theta = \pi.$$

One of the fourth roots of $16\,e^{i\pi}$ is $2\,e^{i\pi/4}$. We obtain others by successive additions of $2\pi/4 = \pi/2$ to the argument of this first one. Hence

$$\sqrt[4]{16\exp i\pi} = 2\exp i\left(\frac{\pi}{4}, \frac{3\pi}{4}, \frac{5\pi}{4}, \frac{7\pi}{4}\right),$$

and the four roots are

$$w_0 = 2\left[\cos\frac{\pi}{4} + i\sin\frac{\pi}{4}\right] = \sqrt{2}(1 + i),$$

$$w_1 = 2\left[\cos\frac{3\pi}{4} + i\sin\frac{3\pi}{4}\right] = \sqrt{2}(-1 + i),$$

$$w_2 = 2\left[\cos\frac{5\pi}{4} + i\sin\frac{5\pi}{4}\right] = \sqrt{2}(-1 - i),$$

$$w_3 = 2\left[\cos\frac{7\pi}{4} + i\sin\frac{7\pi}{4}\right] = \sqrt{2}(1 - i). \ \square$$

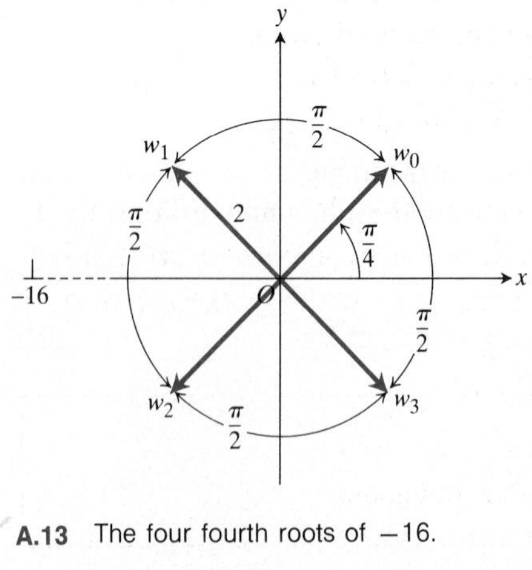

A.13 The four fourth roots of -16.

The Fundamental Theorem of Algebra One may well say that the invention of $\sqrt{-1}$ is all well and good and leads to a number system that is richer than the real number system alone; but where will this process end? Are we also going to invent still more systems so as to obtain $\sqrt[4]{-1}$, $\sqrt[6]{-1}$, and so on? By now it should be clear that this is not necessary. These numbers are already expressible in terms of the complex number system $a + ib$. In fact, the Fundamental Theorem of Algebra says that with the introduction of the complex numbers we now have enough numbers to factor every polynomial into a product of linear factors and hence enough numbers to solve every possible polynomial equation.

THEOREM

The Fundamental Theorem of Algebra

Every polynomial equation of the form

$$a_0 z^n + a_1 z^{n-1} + a_2 z^{n-2} + \cdots + a_{n-1} z + a_n = 0,$$

in which the coefficients $a_0, a_1, \ldots, a_n$ are any complex numbers, whose degree n is greater than or equal to one, and whose leading coefficient a_0 is not zero, has exactly n roots in the complex number system, provided each multiple root of multiplicity m is counted as m roots.

This theorem is rather difficult to prove, and we can only state it here.

PROBLEMS

1. Find $(a, b) \cdot (c, d) = (ac - bd, ad + bc)$.
a) $(2, 3) \cdot (4, -2)$ b) $(2, -1) \cdot (-2, 3)$
c) $(-1, -2) \cdot (2, 1)$

(*Note:* This is how complex numbers are multiplied by computers.)

2. Solve the following equations for the real numbers x and y.
a) $(3 + 4i)^2 - 2(x - iy) = x + iy$
b) $\left(\frac{1+i}{1-i}\right)^2 + \frac{1}{x + iy} = 1 + i$
c) $(3 - 2i)(x + iy) = 2(x - 2iy) + 2i - 1$

3. Show with an Argand diagram that the law for adding complex numbers is the same as the parallelogram law for adding vectors.

4. How may the following complex numbers be obtained from $z = x + iy$ geometrically? Sketch.
a) $\bar{z}$ b) $\overline{(-z)}$
c) $-z$ d) $1/z$

5. Show that the conjugate of the sum (product, or quotient) of two complex numbers z_1 and z_2 is the same as the sum (product, or quotient) of their conjugates.

6. a) Extend the results of Problem 5 to show that

$$f(\bar{z}) = \overline{f(z)} \quad \text{if}$$
$$f(z) = a_0 z^n + a_1 z^{n-1} + \cdots + a_{n-1} z + a_n$$

is a polynomial with real coefficients $a_0, \ldots, a_n$.
b) If z is a root of the equation $f(z) = 0$, where $f(z)$ is a polynomial with real coefficients as in part (a) above, show that the conjugate $\bar{z}$ is also a root of the equation. (*Hint:* Let $f(z) = u + iv = 0$; then both u and v are zero. Now use the fact that $f(\bar{z}) = \overline{f(z)} = u - iv$.)

7. Show that $|\bar{z}| = |z|$.

8. If z and $\bar{z}$ are equal, what can you say about the location of the point z in the complex plane?

9. Let $R(z)$, $I(z)$ denote respectively the real and imaginary parts of z. Show that
a) $z + \bar{z} = 2R(z)$, b) $z - \bar{z} = 2iI(z)$,
c) $|R(z)| \leq |z|$,
d) $|z_1 + z_2|^2 = |z_1|^2 + |z_2|^2 + 2R(z_1\bar{z}_2)$,
e) $|z_1 + z_2| \leq |z_1| + |z_2|$.

10. Show that the distance between the two points z_1 and z_2 in an Argand diagram is equal to $|z_1 - z_2|$.

In Problems 11–15, graph the points $z = x + iy$ that satisfy the given conditions.

11. a) $|z| = 2$ b) $|z| < 2$ c) $|z| > 2$

12. $|z - 1| = 2$ **13.** $|z + 1| = 1$

14. $|z + 1| = |z - 1|$ **15.** $|z + i| = |z - 1|$

Express the answer to each of the Problems 16–19 in the form $re^{i\theta}$, with $r \geq 0$ and $-\pi < \theta \leq \pi$. Sketch.

16. $(1 + \sqrt{-3})^2$ **17.** $\frac{1+i}{1-i}$

18. $\frac{1 + i\sqrt{3}}{1 - i\sqrt{3}}$ **19.** $(2 + 3i)(1 - 2i)$

20. Use De Moivre's theorem (Eq. 19) to express $\cos 4\theta$ and $\sin 4\theta$ as polynomials in $\cos\theta$ and $\sin\theta$.

21. Find the three cube roots of unity.

22. Find the two square roots of i.

23. Find the three cube roots of $-8i$.

24. Find the six sixth roots of 64.

25. Find the four roots of the equation $z^4 - 2z^2 + 4 = 0$.

26. Find the six roots of the equation $z^6 + 2z^3 + 2 = 0$.

27. Find all roots of the equation $x^4 + 4x^2 + 16 = 0$.

28. Solve: $x^4 + 1 = 0$.

***Toolkit* programs**

Complex Number Calculator

A.9 Tables

Table 1 Natural trigonometric functions

Angle					Angle				
Degree	Radian	Sine	Cosine	Tangent	Degree	Radian	Sine	Cosine	Tangent
0°	0.000	0.000	1.000	0.000					
1°	0.017	0.017	1.000	0.017	46°	0.803	0.719	0.695	1.036
2°	0.035	0.035	0.999	0.035	47°	0.820	0.731	0.682	1.072
3°	0.052	0.052	0.999	0.052	48°	0.838	0.743	0.669	1.111
4°	0.070	0.070	0.998	0.070	49°	0.855	0.755	0.656	1.150
5°	0.087	0.087	0.996	0.087	50°	0.873	0.766	0.643	1.192
6°	0.105	0.105	0.995	0.105	51°	0.890	0.777	0.629	1.235
7°	0.122	0.122	0.993	0.123	52°	0.908	0.788	0.616	1.280
8°	0.140	0.139	0.990	0.141	53°	0.925	0.799	0.602	1.327
9°	0.157	0.156	0.988	0.158	54°	0.942	0.809	0.588	1.376
10°	0.175	0.174	0.985	0.176	55°	0.960	0.819	0.574	1.428
11°	0.192	0.191	0.982	0.194	56°	0.977	0.829	0.559	1.483
12°	0.209	0.208	0.978	0.213	57°	0.995	0.839	0.545	1.540
13°	0.227	0.225	0.974	0.231	58°	1.012	0.848	0.530	1.600
14°	0.244	0.242	0.970	0.249	59°	1.030	0.857	0.515	1.664
15°	0.262	0.259	0.966	0.268	60°	1.047	0.866	0.500	1.732
16°	0.279	0.276	0.961	0.287	61°	1.065	0.875	0.485	1.804
17°	0.297	0.292	0.956	0.306	62°	1.082	0.883	0.469	1.881
18°	0.314	0.309	0.951	0.325	63°	1.100	0.891	0.454	1.963
19°	0.332	0.326	0.946	0.344	64°	1.117	0.899	0.438	2.050
20°	0.349	0.342	0.940	0.364	65°	1.134	0.906	0.423	2.145
21°	0.367	0.358	0.934	0.384	66°	1.152	0.914	0.407	2.246
22°	0.384	0.375	0.927	0.404	67°	1.169	0.921	0.391	2.356
23°	0.401	0.391	0.921	0.424	68°	1.187	0.927	0.375	2.475
24°	0.419	0.407	0.914	0.445	69°	1.204	0.934	0.358	2.605
25°	0.436	0.423	0.906	0.466	70°	1.222	0.940	0.342	2.748
26°	0.454	0.438	0.899	0.488	71°	1.239	0.946	0.326	2.904
27°	0.471	0.454	0.891	0.510	72°	1.257	0.951	0.309	3.078
28°	0.489	0.469	0.883	0.532	73°	1.274	0.956	0.292	3.271
29°	0.506	0.485	0.875	0.554	74°	1.292	0.961	0.276	3.487
30°	0.524	0.500	0.866	0.577	75°	1.309	0.966	0.259	3.732
31°	0.541	0.515	0.857	0.601	76°	1.326	0.970	0.242	4.011
32°	0.559	0.530	0.848	0.625	77°	1.344	0.974	0.225	4.332
33°	0.576	0.545	0.839	0.649	78°	1.361	0.978	0.208	4.705
34°	0.593	0.559	0.829	0.675	79°	1.379	0.982	0.191	5.145
35°	0.611	0.574	0.819	0.700	80°	1.396	0.985	0.174	5.671
36°	0.628	0.588	0.809	0.727	81°	1.414	0.988	0.156	6.314
37°	0.646	0.602	0.799	0.754	82°	1.431	0.990	0.139	7.115
38°	0.663	0.616	0.788	0.781	83°	1.449	0.993	0.122	8.144
39°	0.681	0.629	0.777	0.810	84°	1.466	0.995	0.105	9.514
40°	0.698	0.643	0.766	0.839	85°	1.484	0.996	0.087	11.43
41°	0.716	0.656	0.755	0.869	86°	1.501	0.998	0.070	14.30
42°	0.733	0.669	0.743	0.900	87°	1.518	0.999	0.052	19.08
43°	0.750	0.682	0.731	0.933	88°	1.536	0.999	0.035	28.64
44°	0.768	0.695	0.719	0.966	89°	1.553	1.000	0.017	57.29
45°	0.785	0.707	0.707	1.000	90°	1.571	1.000	0.000	

Table 2 Exponential functions

x	e^x	e^{-x}	x	e^x	e^{-x}
0.00	1.0000	1.0000	2.5	12.182	0.0821
0.05	1.0513	0.9512	2.6	13.464	0.0743
0.10	1.1052	0.9048	2.7	14.880	0.0672
0.15	1.1618	0.8607	2.8	16.445	0.0608
0.20	1.2214	0.8187	2.9	18.174	0.0550
0.25	1.2840	0.7788	3.0	20.086	0.0498
0.30	1.3499	0.7408	3.1	22.198	0.0450
0.35	1.4191	0.7047	3.2	24.533	0.0408
0.40	1.4918	0.6703	3.3	27.113	0.0369
0.45	1.5683	0.6376	3.4	29.964	0.0334
0.50	1.6487	0.6065	3.5	33.115	0.0302
0.55	1.7333	0.5769	3.6	36.598	0.0273
0.60	1.8221	0.5488	3.7	40.447	0.0247
0.65	1.9155	0.5220	3.8	44.701	0.0224
0.70	2.0138	0.4966	3.9	49.402	0.0202
0.75	2.1170	0.4724	4.0	54.598	0.0183
0.80	2.2255	0.4493	4.1	60.340	0.0166
0.85	2.3396	0.4274	4.2	66.686	0.0150
0.90	2.4596	0.4066	4.3	73.700	0.0136
0.95	2.5857	0.3867	4.4	81.451	0.0123
1.0	2.7183	0.3679	4.5	90.017	0.0111
1.1	3.0042	0.3329	4.6	99.484	0.0101
1.2	3.3201	0.3012	4.7	109.95	0.0091
1.3	3.6693	0.2725	4.8	121.51	0.0082
1.4	4.0552	0.2466	4.9	134.29	0.0074
1.5	4.4817	0.2231	5	148.41	0.0067
1.6	4.9530	0.2019	6	403.43	0.0025
1.7	5.4739	0.1827	7	1096.6	0.0009
1.8	6.0496	0.1653	8	2981.0	0.0003
1.9	6.6859	0.1496	9	8103.1	0.0001
2.0	7.3891	0.1353	10	22026	0.00005
2.1	8.1662	0.1225			
2.2	9.0250	0.1108			
2.3	9.9742	0.1003			
2.4	11.023	0.0907			

Table 3 Natural logarithms of numbers

n	$\log_e n$	n	$\log_e n$	n	$\log_e n$	n	$\log_e n$
0.0	*	3.5	1.2528	7.0	1.9459	15	2.7081
0.1	7.6974	3.6	1.2809	7.1	1.9601	16	2.7726
0.2	8.3906	3.7	1.3083	7.2	1.9741	17	2.8332
0.3	8.7960	3.8	1.3350	7.3	1.9879	18	2.8904
0.4	9.0837	3.9	1.3610	7.4	2.0015	19	2.9444
0.5	9.3069	4.0	1.3863	7.5	2.0149	20	2.9957
0.6	9.4892	4.1	1.4110	7.6	2.0281	25	3.2189
0.7	9.6433	4.2	1.4351	7.7	2.0412	30	3.4012
0.8	9.7769	4.3	1.4586	7.8	2.0541	35	3.5553
0.9	9.8946	4.4	1.4816	7.9	2.0669	40	3.6889
1.0	0.0000	4.5	1.5041	8.0	2.0794	45	3.8067
1.1	0.0953	4.6	1.5261	8.1	2.0919	50	3.9120
1.2	0.1823	4.7	1.5476	8.2	2.1041	55	4.0073
1.3	0.2624	4.8	1.5686	8.3	2.1163	60	4.0943
1.4	0.3365	4.9	1.5892	8.4	2.1282	65	4.1744
1.5	0.4055	5.0	1.6094	8.5	2.1401	70	4.2485
1.6	0.4700	5.1	1.6292	8.6	2.1518	75	4.3175
1.7	0.5306	5.2	1.6487	8.7	2.1633	80	4.3820
1.8	0.5878	5.3	1.6677	8.8	2.1748	85	4.4427
1.9	0.6419	5.4	1.6864	8.9	2.1861	90	4.4998
2.0	0.6931	5.5	1.7047	9.0	2.1972	95	4.5539
2.1	0.7419	5.6	1.7228	9.1	2.2083	100	4.6052
2.2	0.7885	5.7	1.7405	9.2	2.2192		
2.3	0.8329	5.8	1.7579	9.3	2.2300		
2.4	0.8755	5.9	1.7750	9.4	2.2407		
2.5	0.9163	6.0	1.7918	9.5	2.2513		
2.6	0.9555	6.1	1.8083	9.6	2.2618		
2.7	0.9933	6.2	1.8245	9.7	2.2721		
2.8	1.0296	6.3	1.8405	9.8	2.2824		
2.9	1.0647	6.4	1.8563	9.9	2.2925		
3.0	1.0986	6.5	1.8718	10	2.3026		
3.1	1.1314	6.6	1.8871	11	2.3979		
3.2	1.1632	6.7	1.9021	12	2.4849		
3.3	1.1939	6.8	1.9169	13	2.5649		
3.4	1.2238	6.9	1.9315	14	2.6391		

*Subtract 10 from $\log_e n$ entries for $n < 1.0$.

***Toolkit* programs**

Function Evaluator

A.10
The Distributive Law for Vector Cross Products

In this appendix we prove the distributive law

$$\mathbf{A} \times (\mathbf{B} + \mathbf{C}) = \mathbf{A} \times \mathbf{B} + \mathbf{A} \times \mathbf{C} \tag{1}$$

from Article 13.7.

Proof. To see that Eq. (1) is valid, we interpret the cross product $\mathbf{A} \times \mathbf{B}$ in a slightly different way. The vectors $\mathbf{A}$ and $\mathbf{B}$ are drawn from the common point O and a plane M is constructed perpendicular to $\mathbf{A}$ at O (Fig. A.14). Vector $\mathbf{B}$ is now projected orthogonally onto M, yielding a vector $\mathbf{B}'$ whose length is $|\mathbf{B}| \sin \theta$. The vector $\mathbf{B}'$ is then rotated $90°$ about $\mathbf{A}$ in the positive sense to produce a vector $\mathbf{B}''$. Finally, $\mathbf{B}''$ is multiplied by the length of $\mathbf{A}$. The resulting vector $|\mathbf{A}|\mathbf{B}''$ is equal to $\mathbf{A} \times \mathbf{B}$ since $\mathbf{B}''$ has the same direction as $\mathbf{A} \times \mathbf{B}$ by its construction (Fig. A.14) and

$$|\mathbf{A}||\mathbf{B}''| = |\mathbf{A}||\mathbf{B}'| = |\mathbf{A}||\mathbf{B}| \sin \theta = |\mathbf{A} \times \mathbf{B}|.$$

Now each of these three operations, namely,

1. projection onto M,
2. rotation about $\mathbf{A}$ through $90°$,
3. multiplication by the scalar $|\mathbf{A}|$,

when applied to a triangle whose plane is not parallel to $\mathbf{A}$, will produce another triangle. If we start with the triangle whose sides are $\mathbf{B}$, $\mathbf{C}$, and $\mathbf{B} + \mathbf{C}$ (Fig. A.15) and apply these three steps, we successively obtain:

1. a triangle whose sides are $\mathbf{B}'$, $\mathbf{C}'$, and $(\mathbf{B} + \mathbf{C})'$ satisfying the vector equation

$$\mathbf{B}' + \mathbf{C}' = (\mathbf{B} + \mathbf{C})';$$

2. a triangle whose sides are $\mathbf{B}''$, $\mathbf{C}''$, and $(\mathbf{B} + \mathbf{C})''$ satisfying the vector equation

$$\mathbf{B}'' + \mathbf{C}'' = (\mathbf{B} + \mathbf{C})'';$$

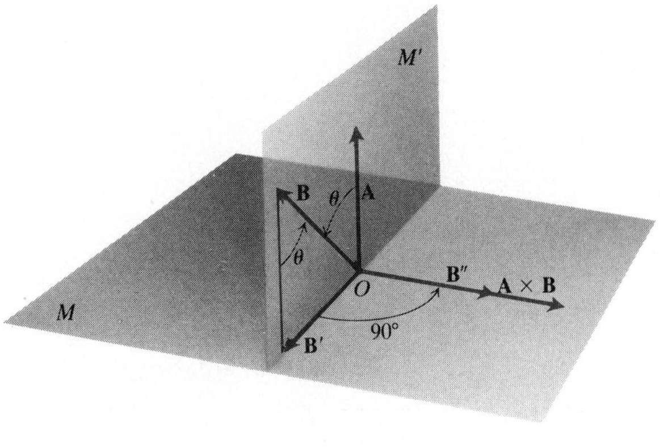

A.14 For reasons explained above. $\mathbf{A} \times \mathbf{B} = |\mathbf{A}|\mathbf{B}''$.

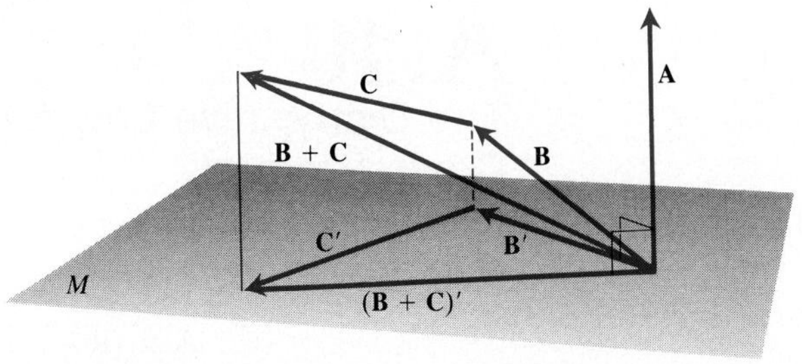

A.15 The vectors **B, C,** and **B** + **C** projected onto a plane perpendicular to **A.**

(the double-prime on each vector has the same meaning as in Fig. A.14); and finally,

3. a triangle whose sides are $|\mathbf{A}|\mathbf{B}''$, $|\mathbf{A}|\mathbf{C}''$, and $|\mathbf{A}|(\mathbf{B}+\mathbf{C})''$ satisfying the vector equation

$$|\mathbf{A}|\mathbf{B}'' + |\mathbf{A}|\mathbf{C}'' = |\mathbf{A}|(\mathbf{B}+\mathbf{C})''. \tag{2}$$

When we use the equations $|\mathbf{A}|\mathbf{B}'' = \mathbf{A}\times\mathbf{B}$, $|\mathbf{A}|\mathbf{C}'' = \mathbf{A}\times\mathbf{C}$ and $|\mathbf{A}|(\mathbf{B}+\mathbf{C})'' = \mathbf{A}\times(\mathbf{B}+\mathbf{C})$, which result from our discussion above, Eq. (2) becomes

$$\mathbf{A}\times\mathbf{B} + \mathbf{A}\times\mathbf{C} = \mathbf{A}\times(\mathbf{B}+\mathbf{C}),$$

which is the distributive law, (1), that we wanted to establish. ■

ANSWERS

CHAPTER 1

Article 1.1, pp. 3–4

	Q	R	S	T
1.	(1, 2)	(−1, −2)	(−1, 2)	(−2, 1)
3.	(−2, 2)	(2, 2)	(2, −2)	(2, −2)
5.	(0, −1)	(0, 1)	(0, −1)	(1, 0)
7.	(−2, 0)	(2, 0)	(2, 0)	(0, −2)
9.	(−1, 3)	(1, −3)	(1, 3)	(−3, −1)
11.	$(-\pi, \pi)$	$(\pi, -\pi)$	(π, π)	$(-\pi, -\pi)$

13. If P is the point (x, y), then the other points are: $Q(x, -y)$, $R(-x, y)$, $S(-x, -y)$, $T(y, x)$.
15. Missing vertices: (−1, 4), (−1, −2), (5, 2).

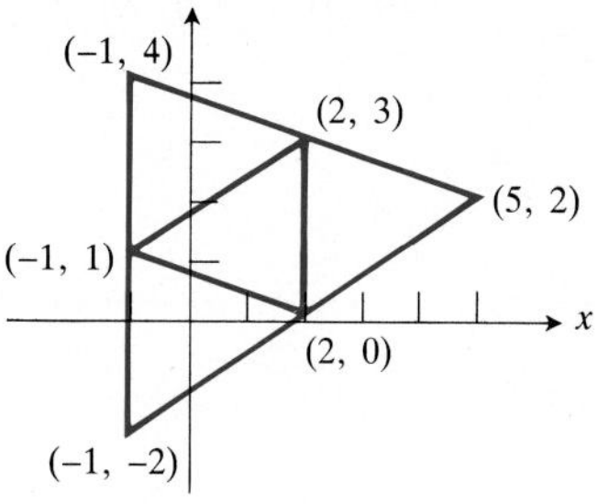

17. $A(-12, 2)$, $B(-12, -5)$, $C(9, -5)$
19. $b = 2$

Article 1.2, p. 7

1. a) $\Delta x = 57$, $\Delta y = -10$ b) $\Delta x = -2$, $\Delta y = -12$ c) $\Delta x = 1$, $\Delta y = -8$ **3.** a) $C(1, 1)$ b) $\Delta x = 2$ c) $\Delta y = 1$ d) $AB = \sqrt{5}$ **5.** a) $C(-1, 2)$ b) $\Delta x = 2$ c) $\Delta y = -4$ d) $AB = 2\sqrt{5}$ **7.** a) $C(-8, 1)$ b) $\Delta x = -5$ c) $\Delta y = 0$ d) $AB = 5$ **9.** Above and to the right of P_1. **11.** To the left of P_1. **13.** Below and to the right of P_1. **15.** Below P_1. **17.** $x^2 + y^2 = 25$ **19.** $x^2 + (y - 5)^2 = 25$, or $x^2 + y^2 - 10y = 0$ **21.** $x^2 + y^2 = 2$ **23.** $(3, -3)$ **25.** $(x, y) = (-2, -9)$ **27.** $(u - h, v - k)$ **29.** The five triangles have equal bases and equal altitudes.

Article 1.3, pp. 12–13

1. $m_{AB} = 3$, $m_\perp = -\frac{1}{3}$ **3.** $m_{AB} = -\frac{1}{3}$, $m_\perp = 3$ **5.** $m_{AB} = -1$, $m_\perp = 1$ **7.** $m_{AB} = 0$, $m_\perp$ is undefined **9.** $m_{AB} = -1$, $m_\perp = 1$ **11.** $m_{AB} = 2$, $m_\perp = -\frac{1}{2}$ **13.** $m_{AB} = y/x$, $m_\perp = -x/y$ **15.** m_{AB} is undefined, $m_\perp = 0$ **17.** $ABCD$ is a rectangle. **19.** $ABCD$ is a rectangle. **21.** $ABCD$, though not a parallelogram, is a trapezoid. **23.** 7.55 inches **25.** a) $-1/0.35 \approx -3$ deg/in. b) $(9 - 68)/(4 - 0.4) \approx -16$ deg/in. c) $(4 - 9)/(4.7 - 4) \approx -7$ deg/in. **27.** (1, 2)
29. $y/x = 1/-2 \Rightarrow y = -x/2$

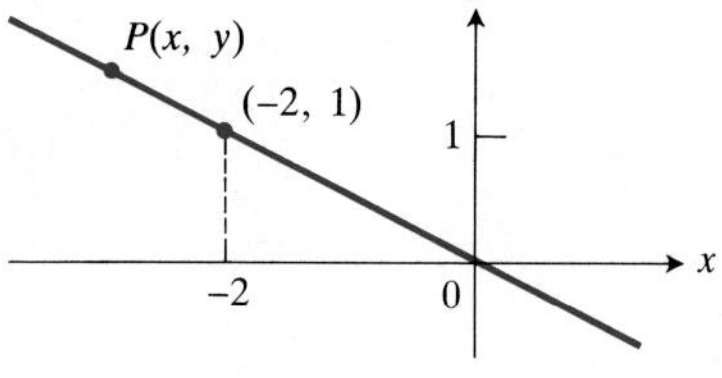

31. Not collinear **33.** Not collinear
35. Center: $(\frac{3}{2}, \frac{3}{2})$, radius is $\frac{1}{2}\sqrt{34}$ **37.** $(\frac{20}{3}, 0)$

Article 1.4, pp. 19–20

1. a) $x = 2$ b) $y = 3$ **3.** a) $x = 0$ b) $y = 0$ **5.** a) $x = -4$ b) $y = 0$ **7.** a) $x = 0$ b) $y = b$ **9.** $y = x$ **11.** $y = x + 2$ **13.** $y = 2x + b$ **15.** $y = \frac{3}{2}x$ **17.** $x = 1$ **19.** $x = -2$ **21.** $y = (-F_0/T)x + F_0$ **23.** $x = 0$ **25.** $y - 0.8 = -\frac{1}{3}(x - 1.4)$, or $y = -\frac{1}{3}x + \frac{114}{90}$ **27.** $y - y_0 = [(y_1 - y_0)/(x_1 - x_0)](x - x_0)$ or $y - y_1 = [(y_1 - y_0)/(x_1 - x_0)](x - x_1)$ **29.** $y = 3x - 2$ **31.** $y = x + \sqrt{2}$ **33.** $y = -5x + 2.5$ **35.** $m = 3$, $b = 5$ **37.** $m = -1$, $b = 2$ **39.** $m = 0.5$, $b = -2$ **41.** $m = \frac{4}{3}$, $b = -4$ **43.** $m = -\frac{4}{3}$, $b = 4$ **45.** $m = \frac{3}{2}$, $b = 3$ **47.** $m = 3$, $b = -20$ **49.** $m = -b/a$, y-intercept is $(0, b)$, x-intercept is $(a, 0)$ **51.** a) $y = 3x - 1$ b) $y = -(1/\sqrt{3})x + 1$ **53.** a) $y = x - 1$ b) $y = 3 - x$ c) $3\sqrt{2}/2$ **55.** a) $y = -x/\sqrt{3}$ b) $y = x\sqrt{3}$ c) $\frac{3}{2}$ **57.** a) $y = -2x - 2$ b) $y = \frac{1}{2}x + 3$ c) $6\sqrt{5}/5$ **59.** a) $y = 2x - 2$ b) $y = -\frac{1}{2}x + \frac{1}{2}$ c) $4\sqrt{5}/5$ **61.** a) $x = 3$ b) $y = 2$ c) 8 **63.** a) $x = a$ b) $y = b$ c) $|a + 1|$ **65.** a) $y = -\frac{4}{3}x + \frac{34}{3}$ b) $y = \frac{3}{4}x + 3$ c) 4.4 **67.** 45° **69.** 150° **71.** 116.6° **73.** 126.9° **75.** $y = \sqrt{3}x + (4 - \sqrt{3})$ **77.** $x = -2$ **79.** 5.97 atm **81.** $s = 0.0023t + 34.85$

Article 1.5, pp. 35–38

1. $x \geq -4$; $y \geq 0$ **3.** $x \neq 2$; $y \neq 0$ **5.** $x \geq 0$; $y \geq 0$ **7.** All x; $-2 \leq y \leq 2$ **9.** All x; $-3 \leq y \leq 3$ **11.** All x; all y

13. a) All x b) $y \geq 1$ c)

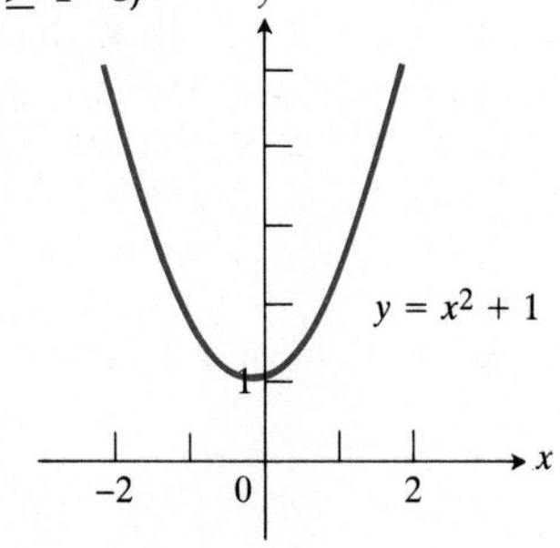

15. a) All x b) $y \leq 0$ c)

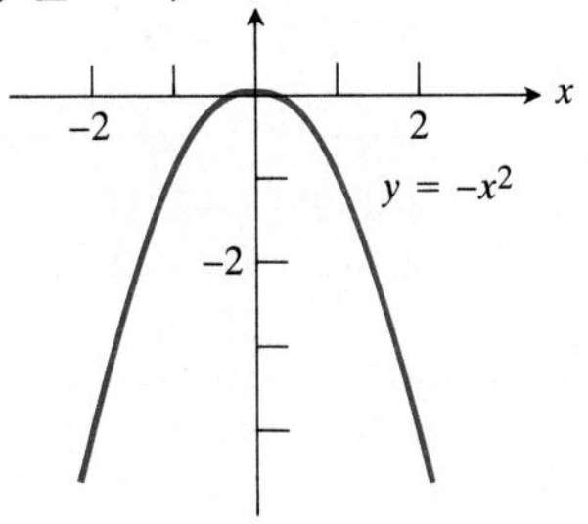

17. a) $x \geq -1$ b) $y \geq 0$ c)

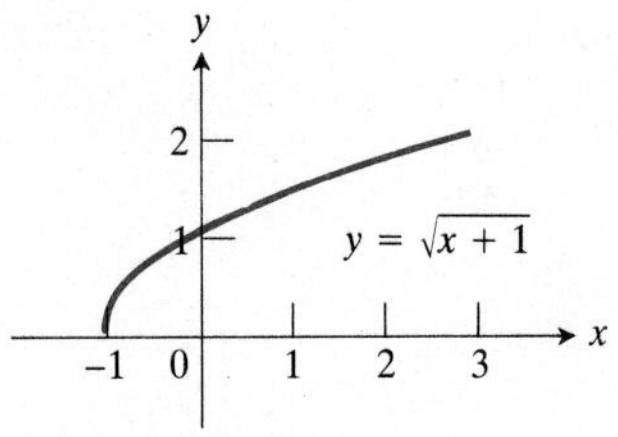

19. a) $x \geq 0$ b) $y \geq 1$ c)

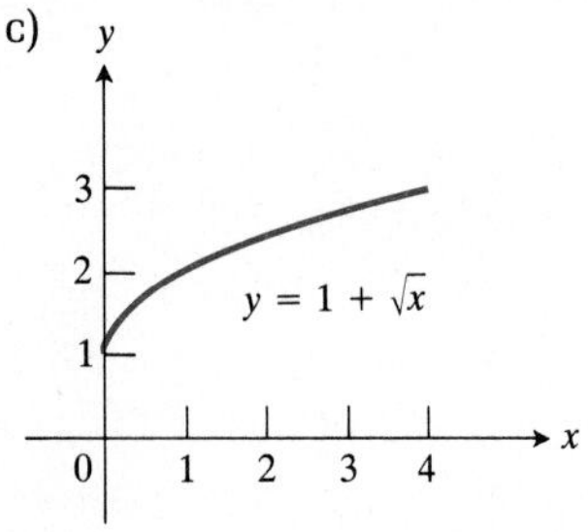

21. a) $x \geq 0$ b) $y \geq 0$ c)

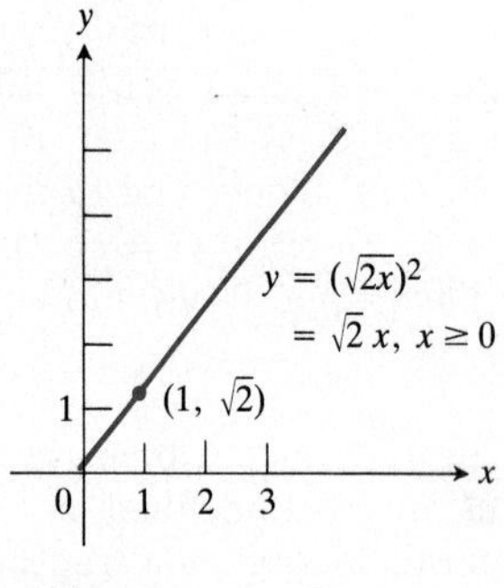

23. a) $x \neq 0$ b) $y \neq 0$ c)

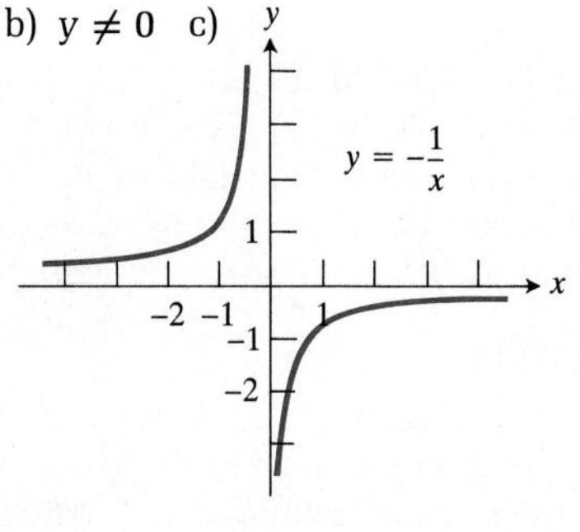

25. a) All x b) $-1 \leq y \leq 1$ c)

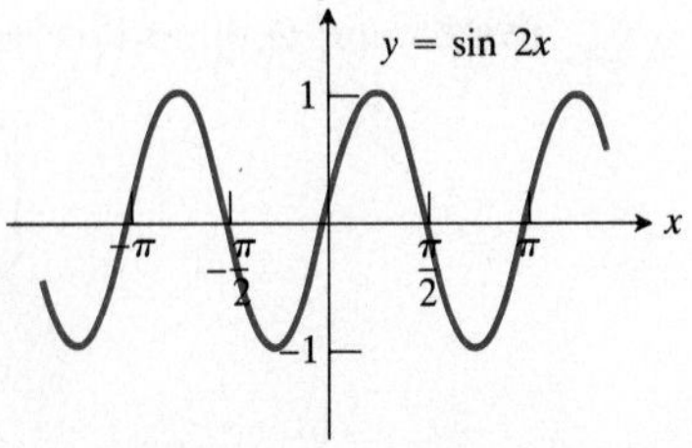

27. a) All x b) $0 \leq y \leq 1$ c)

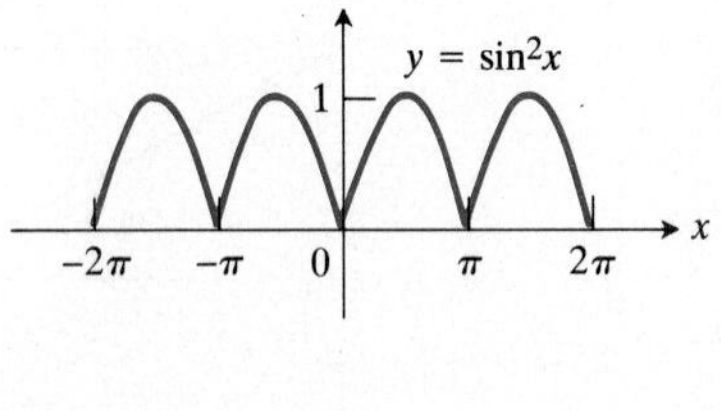

29. a) All x b) $0 \leq y \leq 2$ c)

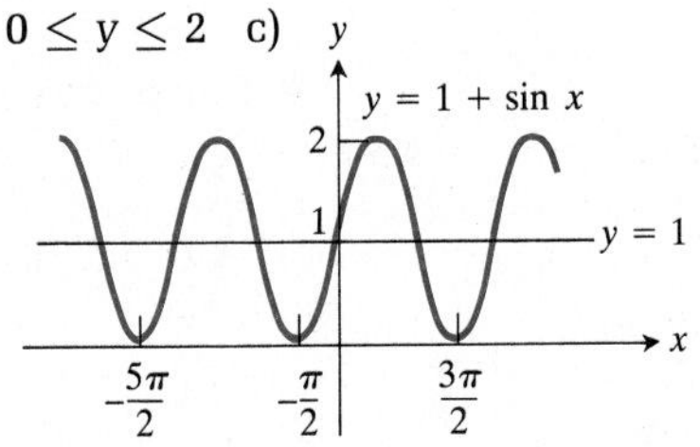

31. a) No b) No c) $x > 0$

33. a) No b) No c) No d) $0 < x \leq 1$

35. i, iii, and iv. The graph must pass through (0, 0).

37.

x	y
0	0
$\frac{1}{2}$	$\frac{1}{2}$
1	1
$\frac{3}{2}$	$\frac{1}{2}$
2	0

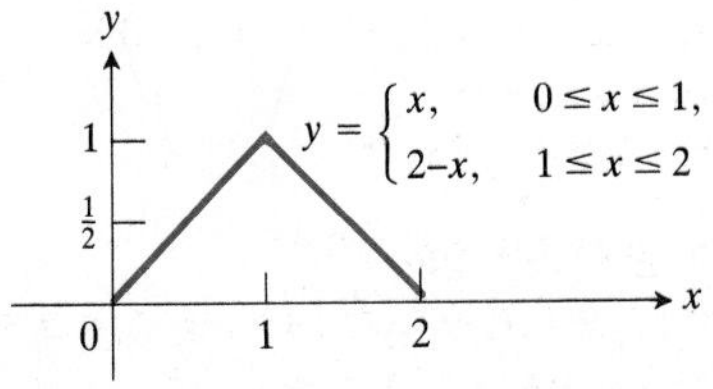

39.

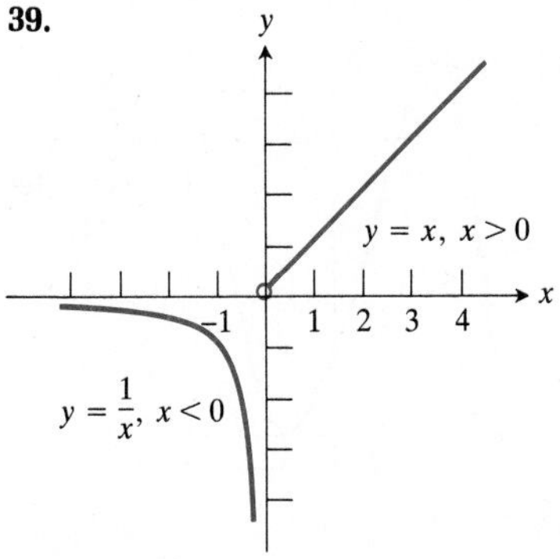

41.

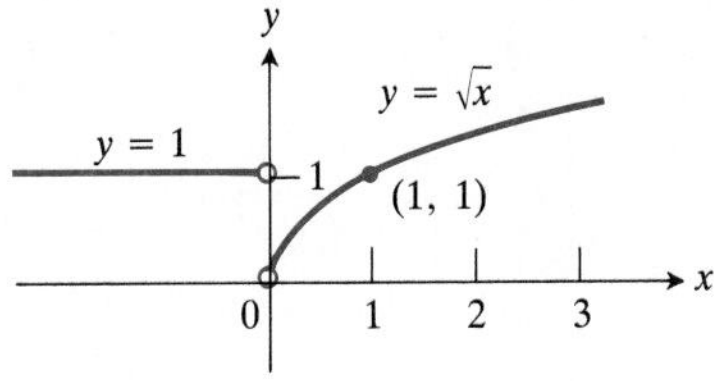

43. Domain of f is all x; domain of g is $x \geq 1$. Domains of $f + g$, $f - g$, $f \cdot g$, g/f: $x \geq 1$. Domain of f/g: $x > 1$. **45.** Domain of f is $x \geq 0$; domain of g is $x \geq -1$. Domains of $f + g$, $f - g$, $f \cdot g$, f/g: $x \geq 0$. Domain of g/f: $x > 0$. **47.** a) -4 b) 11 c) 2 d) $(x + 1)/x$ e) $(2x + 1)/2x = 1 + (1/2x)$ f) $1 + 5x$

49. $f(1 - x) = -x/(1 - x) = x/(x - 1)$ **51.** 4 **53.** $f(x) = \sin^2 x$, $g(x) = \sqrt{x}$ **55.** $x \leq -2$ or $x \geq 2$ **57.** $x < 0$ or $x > 6$ **59.** $x \leq -1$ or $x \geq 1$ **61.** $|x| < 8$ **63.** $|x + 2| < 3$ **65.** $|y| < a$ **67.** $|y - L| < \varepsilon$ **69.** $|x - x_0| < 5$ **71.** g **73.** e **75.** h **77.** b **79.** i **81.** $|1 - x| = 1 - x$ if $x \leq 1$; $|1 - x| = x - 1$ if $x \geq 1$

83.

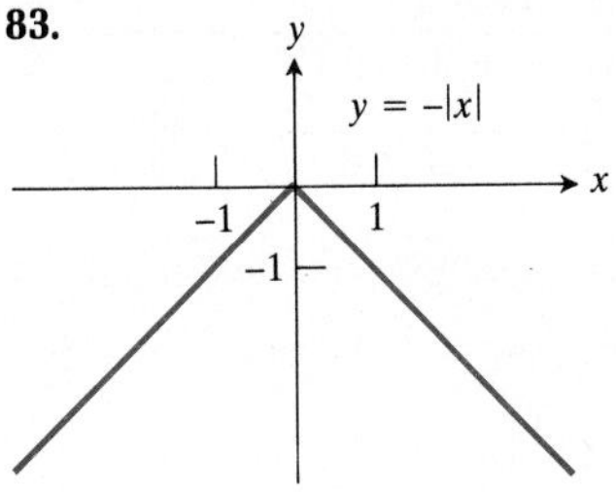

85.

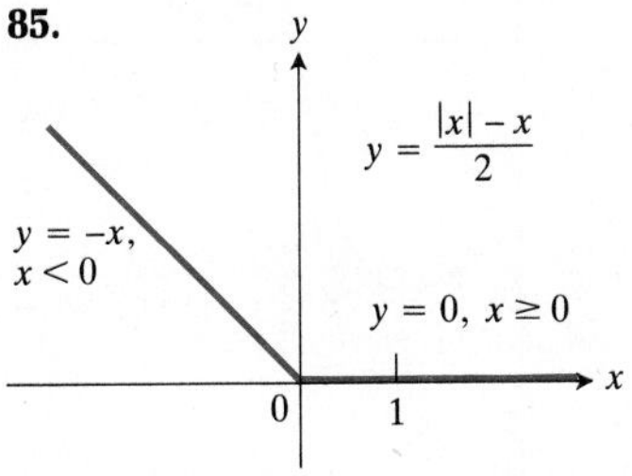

87.

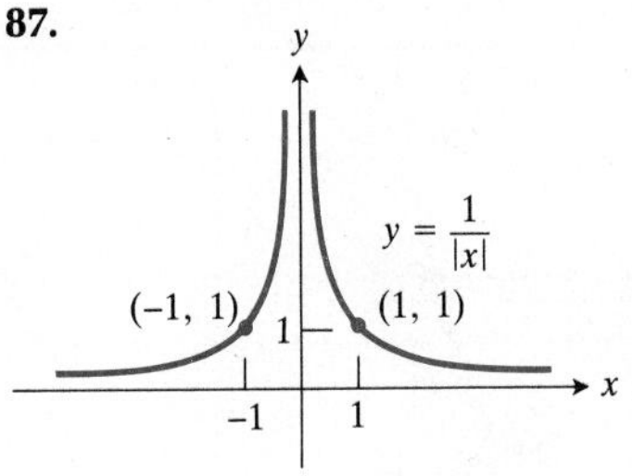

89. a) minimum $y = -2$; no maximum b) minimum $y = 1$; no maximum **91.** a) $x - y = 1$ b) $-x + y = 1$

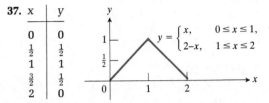

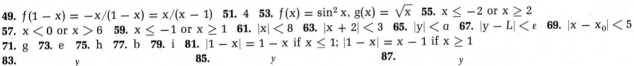

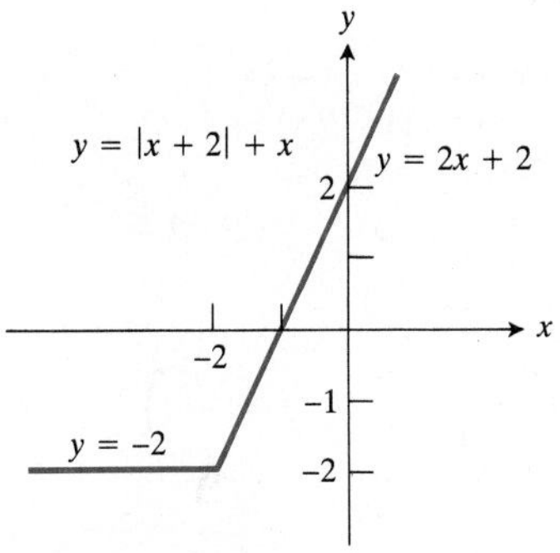

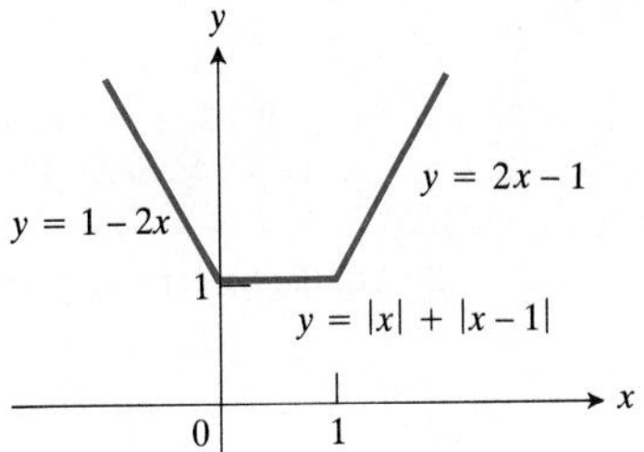

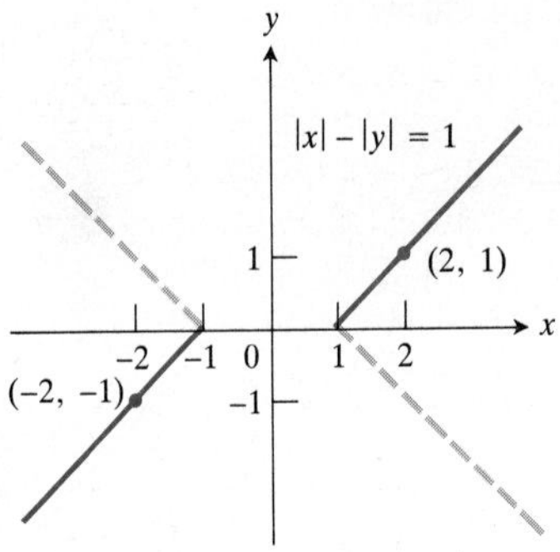

93. $[x] = 0$ for $0 \leq x < 1$ **95.** $\approx$62.4 km **97.** Odd

Article 1.6, p. 44

1. Slope is $2x - 2$

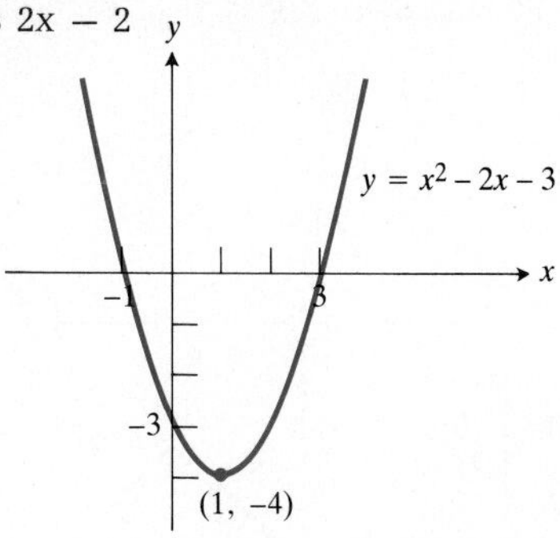

3. Slope is $-2x$

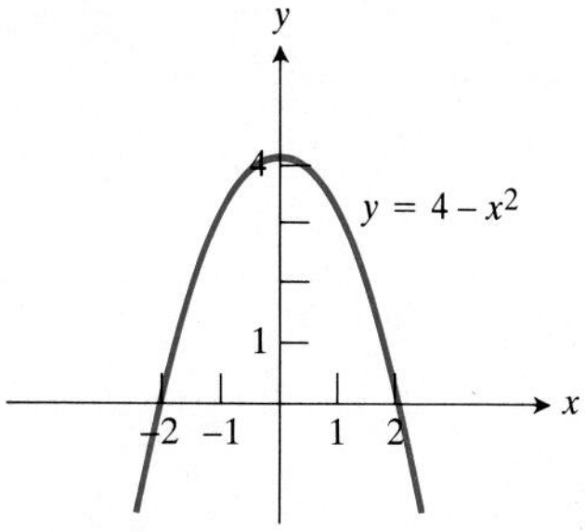

5. Slope is $2x - 4$

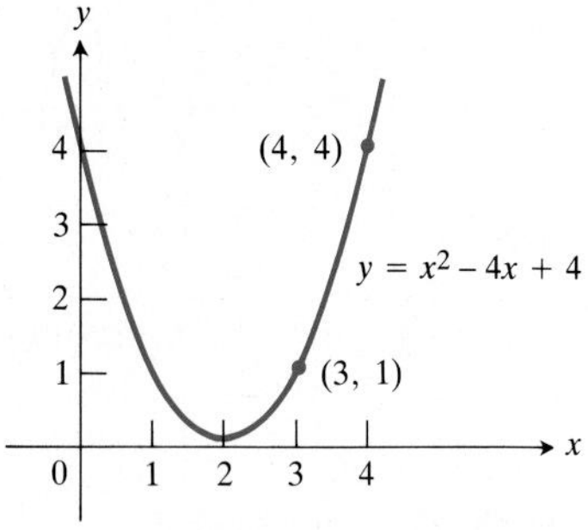

7. Slope is $1 - 2x$

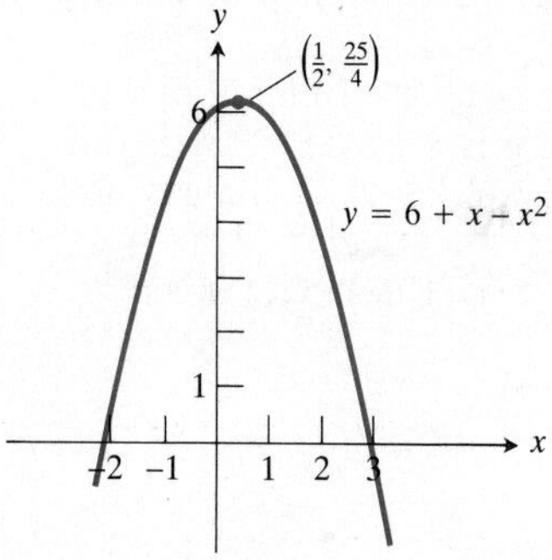

9. Slope is $2x + 3$

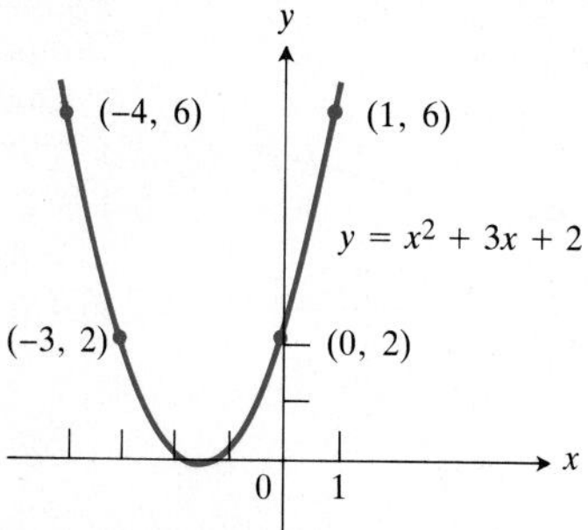

11. Slope is $6x^2 + 6x - 12$

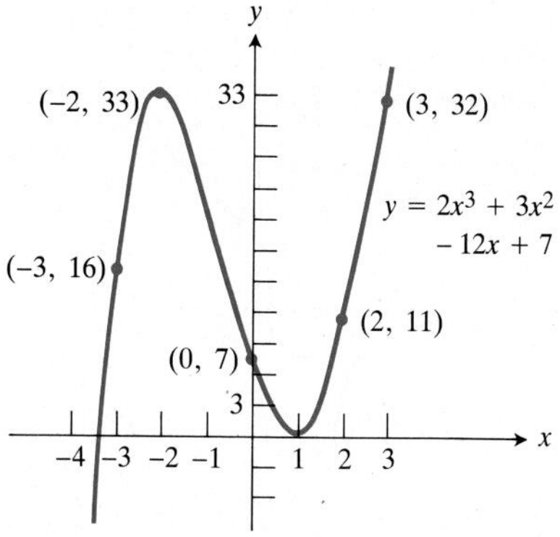

13. Slope is $3x^2 - 12$

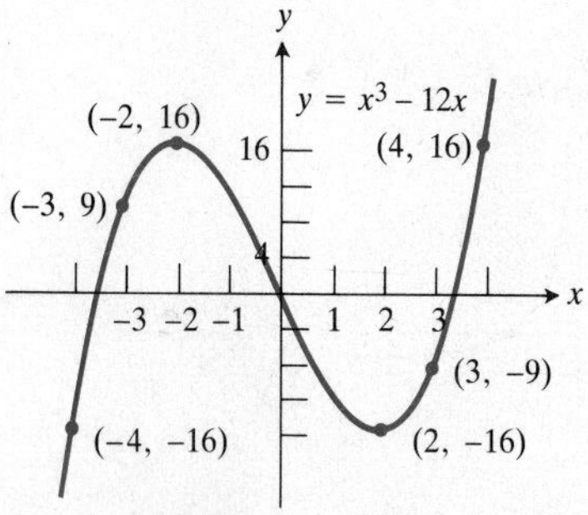

15. Slope is $3x^2 - 6x$

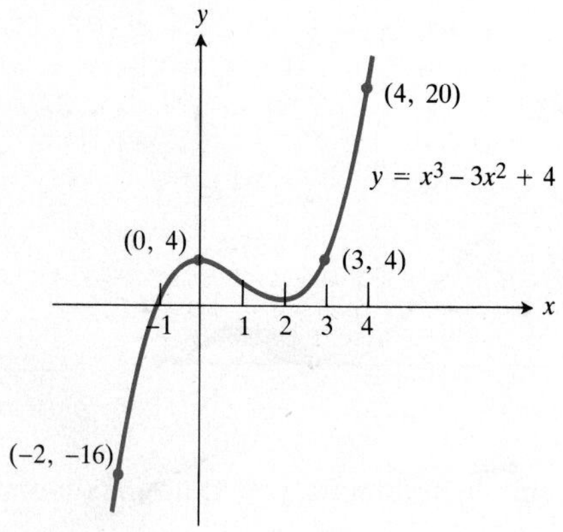

17. b) $y = 2x - 1$ c) -1; 0; 1; $\frac{3}{2}$
d) $2 - \Delta x$; 2

19. a) $y = 24x - 51$
b) $y = 24x + 57$
c) $y = 3x + 3 - 4\sqrt{2}$

Article 1.7, pp. 50–51

1. $f'(x) = 2x$; $m = 6$; $y = 6x - 9$ **3.** $f'(x) = 2$; $m = 2$; $y = 2x + 3$ **5.** $f'(x) = -1/x^2$; $m = -\frac{1}{9}$; $y = -\frac{1}{9}x + \frac{2}{3}$ **7.** $f'(x) = -2/(2x + 1)^2$; $m = -\frac{2}{49}$; $y = -\frac{2}{49}x + \frac{13}{49}$ **9.** $f'(x) = 4x - 1$; $m = 11$; $y = 11x - 13$ **11.** $f'(x) = 4x^3$; $m = 108$; $y = 108x - 243$ **13.** $f'(x) = 1 + (1/x^2)$; $m = \frac{10}{9}$; $y = \frac{10}{9}x - \frac{2}{3}$ **15.** $f'(x) = 1/\sqrt{2x}$; $m = 1/\sqrt{6}$; $\sqrt{6}y = x + 3$ **17.** $f'(x) = 1/\sqrt{2x + 3}$; $m = \frac{1}{3}$; $y = \frac{1}{3}x + 2$ **19.** $f'(x) = -(2x + 3)^{-3/2}$; $m = -\frac{1}{27}$; $y = -\frac{1}{27}x + \frac{4}{9}$

21. The two graphs are identical. The derivative of $f(x) = |x|$, $x \neq 0$ is $f'(x) = |x|/x$, $x \neq 0$.

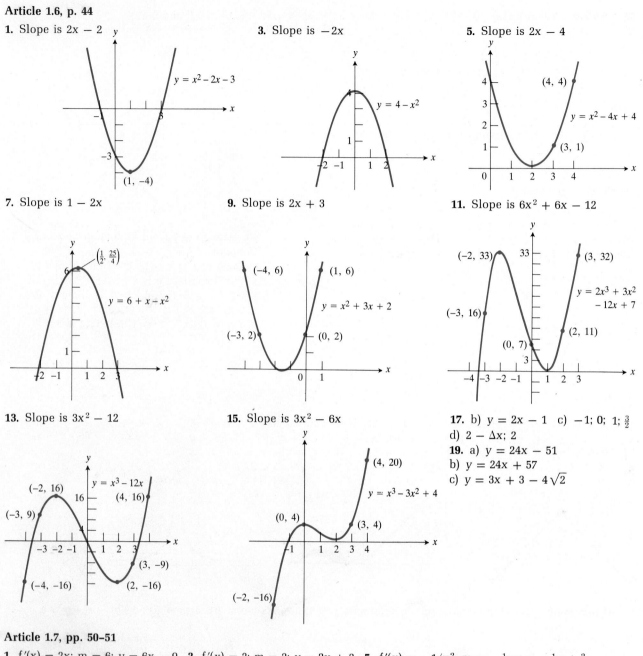

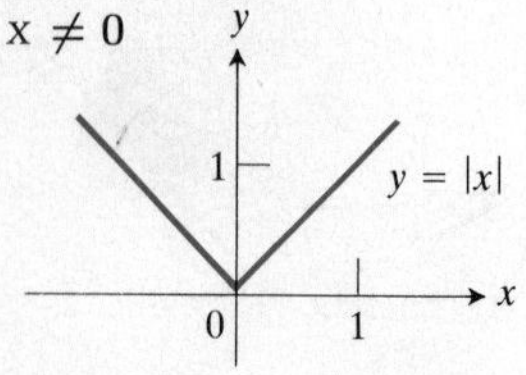

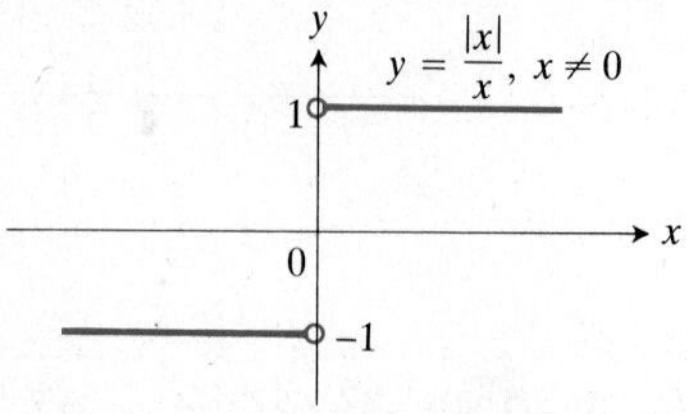

23. Greatest: $t \approx 20$, rabbit population about 1700. Least: $t \approx 65$, rabbit population about 1300.
27.

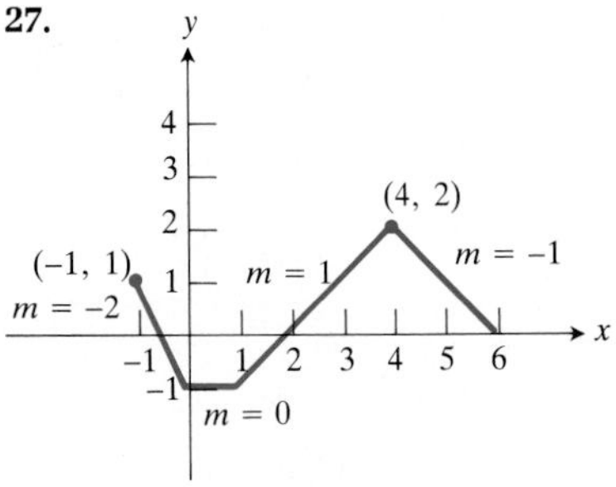

Article 1.8, pp. 56–58

3. $\Delta s = 2v_0 + 2g$; $v_{av} = v_0 + g$; $v(t) = v_0 + gt$; $v(2) = v_0 + 2g$ **5.** $\Delta s = -2$ m; $v_{av} = -1$ m/s; $v(t) = 2t - 3$; $v(2) = 1$ m/s **7.** $\Delta s = 40$ m; $v_{av} = 20$ m/s; $v(t) = 8t + 12$; $v(2) = 28$ m/s **9.** $\Delta s = -8$ m; $v_{av} = -4$ m/s; $v(t) = -4t$; $v(2) = -8$ m/s **11.** a) Eq. (3) b) $t = \frac{4}{7}$ sec; av. vel. $= 160 \div (\frac{4}{7}) = 280$ cm/sec c) The balls fall 160 cm in 17 flash intervals. There is a flash every $(\frac{4}{7}) \div 17 \approx 0.034$ sec.

13.

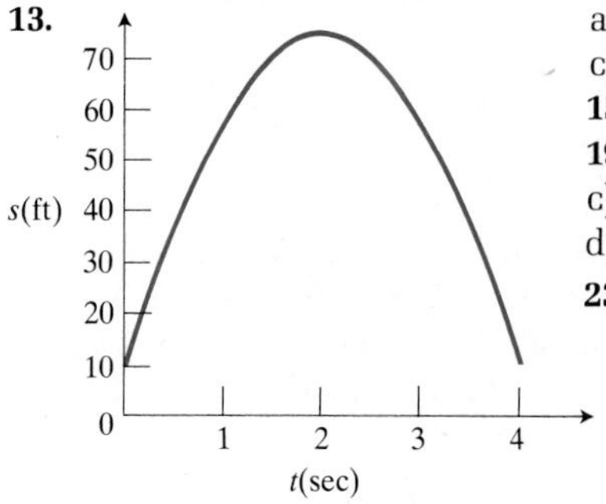

a) 32 ft/sec b) -16 ft/sec c) 8 sec; 0 ft/sec **15.** 190 ft/sec **17.** 8 seconds; 0 **19.** 2.8 seconds **21.** a) 10,000 b) 0 c) $-10{,}000$ (Units are bacteria/hour) d) ≈ 37 hr **23.** a) $2\pi r + \pi(\Delta r)$ b) $2\pi r$ **25.** πh^2

Article 1.9, pp. 71–73

1. 4 **3.** 4 **5.** 2 **7.** 25 **9.** 1 **11.** $2x$ **13.** 0 **15.** 0 **17.** 18 **19.** 36 **21.** 2 **23.** 4 **25.** 1 **27.** a) -10 b) -20 **29.** a) 4 b) -21 c) -12 d) $-\frac{7}{3}$ **31.** $\frac{5}{4}$ **33.** 0 **35.** Does not exist. **37.** $-\frac{1}{2}$ **39.** -1 **41.** $\frac{1}{3}$ **43.** $-\frac{1}{2}$ **45.** 0 **47.** $\frac{3}{4}$ **49.** -3 **51.** $3/4a$ **53.** One possibility: $f(x) = \begin{cases} 1/x & x \neq 0 \\ 0 & x = 0 \end{cases}$, $g(x) = \begin{cases} -1/x & x \neq 0 \\ 0 & x = 0 \end{cases}$ **55.** One possibility: $f(x) = x$, $g(x) = \begin{cases} 1 & x \geq 0 \\ -1 & x < 0 \end{cases}$; $f/g = |x|$ **57.** $f(x) = |-1 + x| = |x - 1|$ **59.** a) All c except 0, 1, 2. b) At $c = 2$, only the left-hand limit exists. c) At $c = 0$, only the right-hand limit exists.

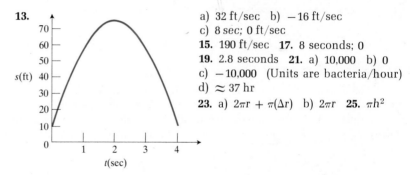

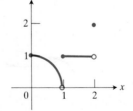

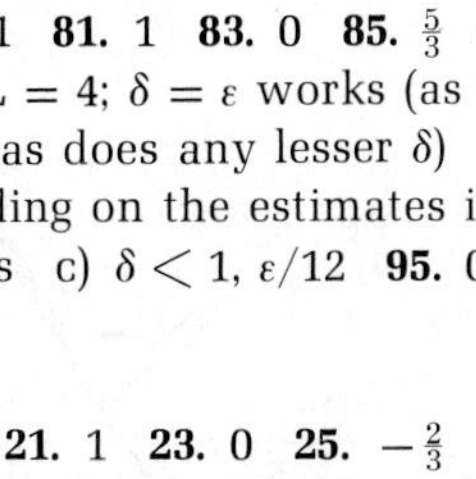

61. 0 **63.** 0

65. 1 **67.** 6 **69.** -16 **71.** All $c \neq 0$ **73.** 1 **75.** 0 **77.** 2 **79.** 1 **81.** 1 **83.** 0 **85.** $\frac{5}{3}$ **87.** 2 **91.** a) $L = -1$; $\delta = \varepsilon/3$ works (as does any lesser δ) b) $L = 7$; any $\delta > 0$ works c) $L = 4$; $\delta = \varepsilon$ works (as does any lesser δ) d) $L = 6$; $\delta = \varepsilon$ works (as does any lesser δ) e) $L = -5$; $\delta = \frac{2}{3}\varepsilon$ works (as does any lesser δ) f) $L = 2$; $\delta < 2$, $2\varepsilon/(2 + \varepsilon)$ g) $L = -\frac{1}{9}$; $\delta < 3$, $27\varepsilon/(1 + 9\varepsilon)$ **93.** Answers will vary, depending on the estimates involved. The following values work, but so do others: a) $\delta = \frac{1}{120}$ or less b) $\delta = \frac{1}{1{,}200}$ or less c) $\delta < 1$, $\varepsilon/12$ **95.** 0

Article 1.10, pp. 79–80

1. $\frac{2}{5}$ **3.** 0 **5.** 0 **7.** $\frac{1}{2}$ **9.** 1 **11.** ∞ **13.** 1 **15.** $\frac{3}{10}$ **17.** 1 **19.** 2 **21.** 1 **23.** 0 **25.** $-\frac{2}{3}$ **27.** 1 **29.** 0 **31.** 2 **33.** ∞ **35.** ∞ **37.** 4 **39.** ∞ **41.** $-\infty$ **43.** ∞ **45.** 0 **47.** $-\infty$ **49.** -7 **51.** a) $-\frac{1}{5}$ b) 0 c) $-\frac{1}{3}$

53.

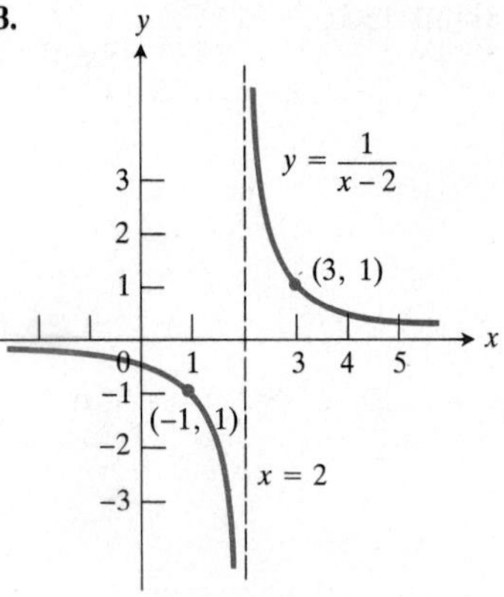

55.

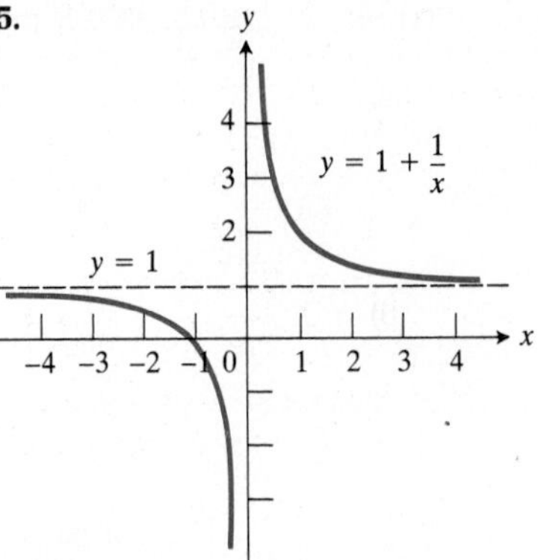

57.

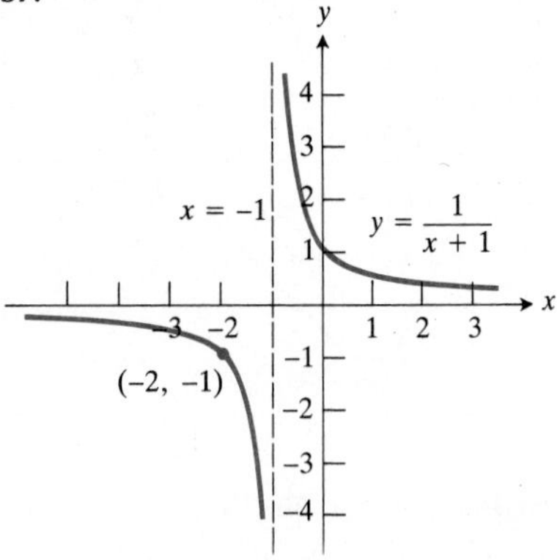

59.

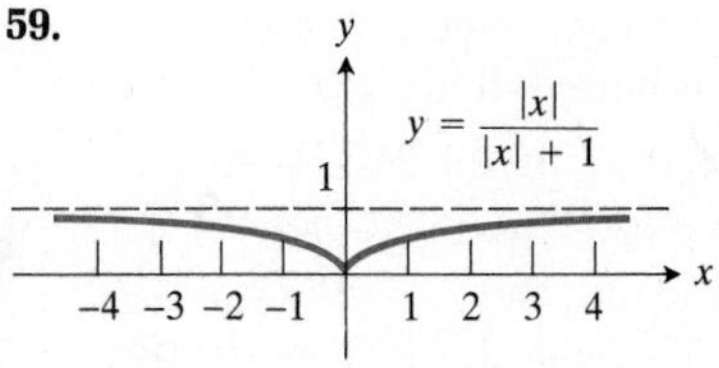

61. a) As $x \to 0^+$, $1/x^2$ increases most rapidly; $1/\sqrt{x}$ increases least rapidly. b) As $x \to +\infty$, $1/x^2$ decreases most rapidly; $1/\sqrt{x}$ decreases least rapidly.

Article 1.11, pp. 89–90

1. a) Yes, $f(-1) = 0$ b) Yes, 0 c) Yes d) Yes **3.** a) No b) No **5.** a) 0 b) Set $f(2) = 0$ **7.** All x except 0, 1 **9.** All x in the domain of f except 1, 2 **11.** All x except -1, 0, 1 **13.** Not continuous at $x = -2$ **15.** Not continuous at 1, 3 **17.** Not continuous at 5, -2 **19.** Continuous for all x **21.** Not continuous at 0 **23.** 6

25. a)

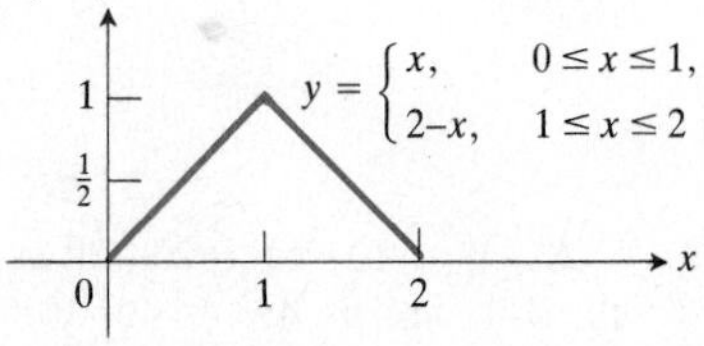

b) Yes c) No **27.** $\frac{4}{3}$ **29.** No; left- and right-hand limits disagree

31. Discontinuous for $0 \le x < 1$, and at all integers **33.** 1 **35.** Maximum is 1, minimum is 0 **37.** Minimum value of 0; no maximum **39.** Note $f(0) < 0 < f(1)$ and then use the Intermediate Value Theorem.

43.

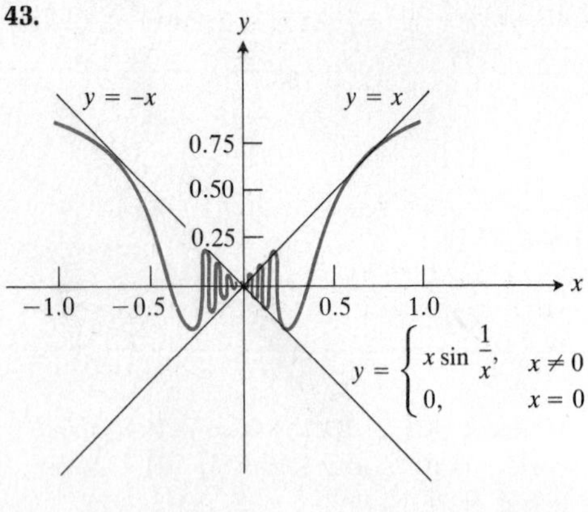

Miscellaneous Problems Chapter 1, pp. 92–94

1. b) $m_{AB} = -\frac{3}{2}$, $m_{BC} = \frac{2}{3}$, $m_{CD} = -\frac{3}{2}$, $m_{DA} = \frac{2}{3}$, $m_{CE} = 0$, m_{BD} is undefined c) *ABCD* is a parallelogram (its opposite sides are parallel). d) Yes, *AEB* e) Yes, *CD* f) *AB*: $y = -\frac{3}{2}x + 13$; *CD*: $y = -\frac{3}{2}x$; *AD*: $y = \frac{2}{3}x - \frac{13}{3}$; *CE*: $y = 6$; *BD*: $x = 2$ g) *AB*: x-intercept $\frac{26}{3}$, y-intercept 13; *CD*: x-intercept 0, y-intercept 0; *AD*: x-intercept $\frac{13}{2}$, y-intercept $-\frac{13}{3}$; *CE*: No x-intercept, y-intercept 6; *BD*: x-intercept 2, no y-intercept **3.** a) *ABC* is not a right triangle. b) *ABC* is not isosceles. c) Outside d) $y = -\frac{9}{7}$ **5.** Midpoint $(a, b) = [(x_1 + x_2)/2, (y_1 + y_2)/2]$

7. a) $-A/B$ b) $(0, C/B)$ c) $(C/A, 0)$ d) $y = (B/A)x$ **9.** Four; centers $(0, \sqrt{5} - 2)$, $(0, -\sqrt{5} - 2)$, $(-\sqrt{5} - 1, 1)$, $(\sqrt{5} - 1, 1)$. Radii (in same order): $(3\sqrt{2} - \sqrt{10})/2$, $(3\sqrt{2} + \sqrt{10})/2$, $(\sqrt{10} + \sqrt{2})/2$, $(\sqrt{10} - \sqrt{2})/2$. **11.** Straight line through point of intersection of the two given lines. **13.** $x + 5y = \pm 5\sqrt{2}$ **15.** $A = \pi r^2$, $C = 2\pi r$, $A = C^2/4\pi$ **17.** $-5 < x < 3$ **19.** $x = (y + 2)/2$, $y > 2$ **21.** a) $f(-2) = -3$, $f(-1) = -4$, $f(x_1) = x_1^2 + 2x_1 - 3$, $f(x_1 + \Delta x) = (\Delta x)^2 + (\Delta x)(2x_1 + 2) + (x_1^2 + 2x_1 - 3)$ b) $f(1/x) = -f(x) = f(-x) = (1/x) - x$
23. Range: $-2 \leq y \leq 6$

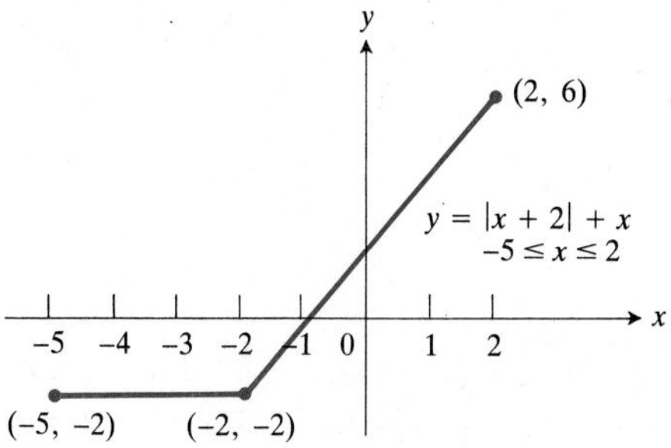

27. $ad + b = bc + d$, or $f(d) = g(b)$ **29.** a) $1/(1 - x)$ b) $x/(x + 1)$ c) $f(f(x)) = x$ d) $1 - x$ **31.** a) $2x - 3$ b) $2 - (1/3x^2)$ c) $1/2\sqrt{t - 4}$ **33.** a) $f(0) = 0$, $f(-1) = 1$, $f(1/x) = 2/(1 - x)$ b) $-2/(x - 1)(x + \Delta x - 1)$ c) $-2/(x - 1)^2$ **35.** $v(t) = 180 - 32t$; $v = 0$ at $t = \frac{45}{8}$ **37.** a) $\Delta P/\Delta V = -1/V(V + \Delta V)$ b) $P'(2) = -\frac{1}{4}$ **39.** $\frac{1}{2}$ **41.** $\frac{3}{7}$ **43.** Limit does not exist. **45.** 0 **47.** $1/2\sqrt{x}$ **49.** ∞ **51.** $-\frac{1}{10}$ **53.** $-1/2\sqrt{2}$ **55.** -1

57. 0 **59.** ∞ **61.** 4 **63.** Point of intersection is $[(c + 1)/(3c + 2), -1/5(3c + 2)]$. Limit as $c \to 1$ is $(\frac{2}{5}, -\frac{1}{25})$. **65.** $\delta \leq \min\{1, \varepsilon^2/3\}$ works. **67.** a) Roots: $-2, 2, 3$ b) $k = 5$ c) h is even **69.** Use the Sandwich Theorem.
73. I: $y = -x + 1$, $0 \leq x \leq 1$; II: $y = x + 1$, $-1 \leq x \leq 0$; III: $y = -x - 1$, $-1 \leq x \leq 0$; IV: $y = x - 1$, $0 \leq x \leq 1$.

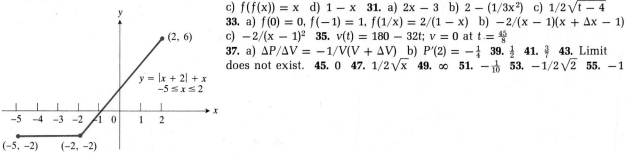

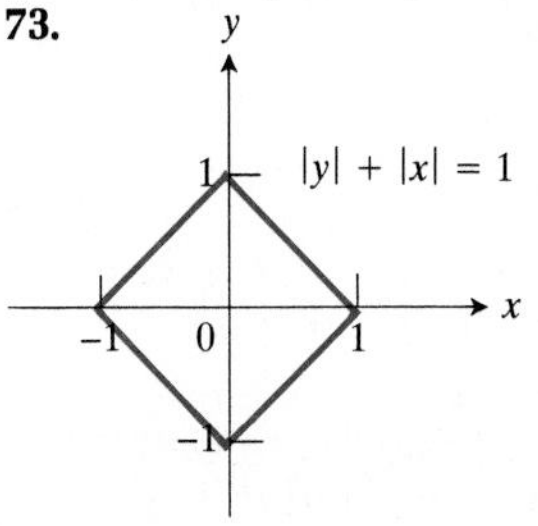

CHAPTER 2

Article 2.2, pp. 102–103

1. $v = 2t - 4$, $a = 2$ **3.** $v = gt + v_0$, $a = g$ **5.** $v = 8t + 12$, $a = 8$ **7.** $y' = 15x^2 - 15x^4$, $y'' = 30x - 60x^3$ **9.** $y' = x^3 - x^2 + x - 1$, $y'' = 3x^2 - 2x + 1$ **11.** $y' = 2x^3 - 3x - 1$, $y'' = 6x^2 - 3$ **13.** $y' = 5x^4 - 2x$, $y'' = 20x^3 - 2$ **15.** $y' = 12x + 13$, $y'' = 12$ **17.** (c) **19.** $y = -\frac{1}{9}x + \frac{29}{9}$ **21.** 29.4 m **23.** $t_1 = -\frac{1}{3}$, $t_2 = 3$, $a_1 = -10$, $a_2 = 10$ **25.** $(-1, 27)$, $(2, 0)$ **27.** $t = x_1(1 - 1/n)$; draw line through $(t, 0)$ and (x_1, x_1^n) **29.** $(2, 6)$ **31.** $a = b = 1$, $c = 0$ **33.** $\frac{1}{4}$

Article 2.3, pp. 113–114

1. $x^2 - x + 1$ **3.** $10x(x^2 + 1)^4$ **5.** $2(x + 1)(x^2 + 1)^{-3} - 6x(x + 1)^2(x^2 + 1)^{-4}$ **7.** $-19(3x - 2)^{-2}$ **9.** $(x^2 - 2x - 1)(1 + x^2)^{-1}$ **11.** $-40(2x - 3)^{-5}$ **13.** $4x - 14$ **15.** $45(2x - 1)^2(x + 7)^{-4}$ **17.** $-30(2x^3 - 3x^2 + 6x)^{-6}(x^2 - x + 1)$ **19.** $2x^{-3}(x - 1)$ **21.** $12(-(1/x^2) + (1/x^4) - (1/x^5))$ **23.** $[(x - 1)/(x^2 + x - 2)]^2 = 1/(x + 2)^2$ **25.** $(1 - t^2)/(1 + t^2)^2$ **27.** $(2 - 4t)(t^2 - t)^{-3}$ **29.** $(2 - 6t^2)/(3t^2 + 1)^2$ **31.** $3(2t + 3)(t^2 + 3t)^2$ **33.** a) 13 b) -7 c) $\frac{7}{25}$ d) 20 e) -225 f) -30 **35.** $4y - 3x = 4$ **37.** $8(3 - 2x)^{-3}$ **41.** $dy/dt = 0.09y$ **43.** a) $dP/dt = a^2k/(akt + 1)^2$ b) $t = 0$, $dP/dt|_0 = a^2k$

Article 2.4, pp. 122–124

1. $-x/y$ **3.** $-(y/x) - 2$ **5.** $(\frac{3}{2})x^{1/2}$ **7.** $-(-y/x)^{1/2}$ **9.** $y/x - (x + y)^2$ **11.** $(2x^2 + 1)(x^2 + 1)^{-1/2}$ **13.** $(1 - 2y)/(2x + 2y - 1)$ **15.** $2x/y(x^2 + 1)^2$ **17.** $5(3x + 7)^4/2y^2 = (5/\sqrt[3]{2})(3x + 7)^{2/3}$ **19.** $-(y/x)^2$ **21.** $(2x - 1)/2y$ **23.** $-\frac{1}{3}(x^2 + 9)/(x^2(x^2 + 3)^{2/3})$ **25.** $(x^2 - 6y)/(6x - y^2)$ **27.** $-\frac{2}{5}(2x + 5)^{-6/5}$ **29.** $-\frac{1}{4}(x - x^{3/2})^{-1/2}$ **31.** $\pi/\sqrt{Lg}$ **33.** (b), (c), and (d) **37.** $y' = -x^2/y^2$, $y'' = -2(xy^3 + x^4)/y^5 = -2x/y^5$ **39.** $y' = -y/(2y + x)$, $y'' = 2/(2y + x)^3$ **43.** a) $y = \frac{3}{4}x - \frac{25}{4}$ or $4y - 3x = -25$ b) $3y + 4x = 0$ **45.** a) $(x - y)/(x - 2y) = 2$, or $3y = x$ b) $y + 3x = 10$ **47.** a) $x + y = -1$ b) $x - y = -3$ **49.** $y + 2x = \pm 3$ **51.** $(-\frac{13}{4}, \frac{17}{16})$ **53.** $(\pm\sqrt{7}, 0)$; slope $= -2$ **55.** $y = [-x \pm (28 - 3x^2)^{1/2}]/2$, $y' = -\frac{4}{5}$

Article 2.5, pp. 131–133

1. a) $f(x) \approx 2x$ b) 0.20 c) 0.21 d) 0.01 e)

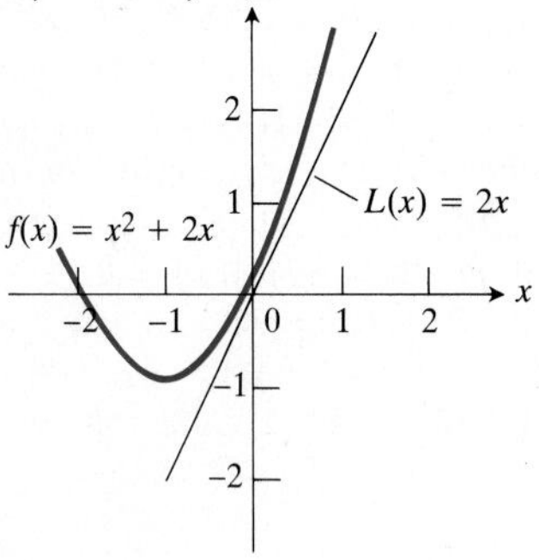

3. a) $f(x) \approx 2x - 2$ b) 0.200 c) 0.331 d) 0.131 e)

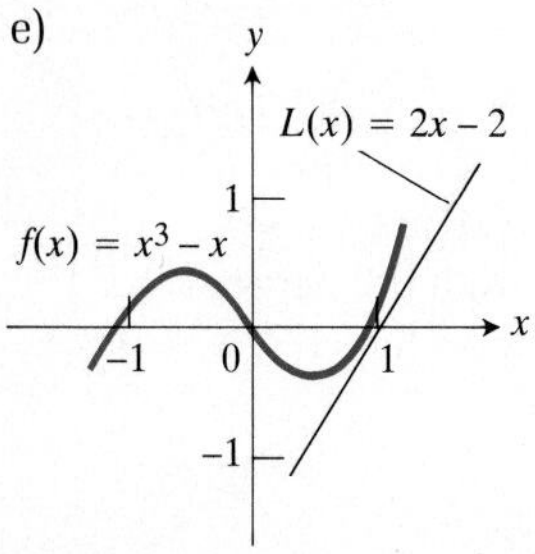

5. a) $f(x) \approx 4 - 4x$ b) -0.400 c) -0.333 d) 0.067 e)

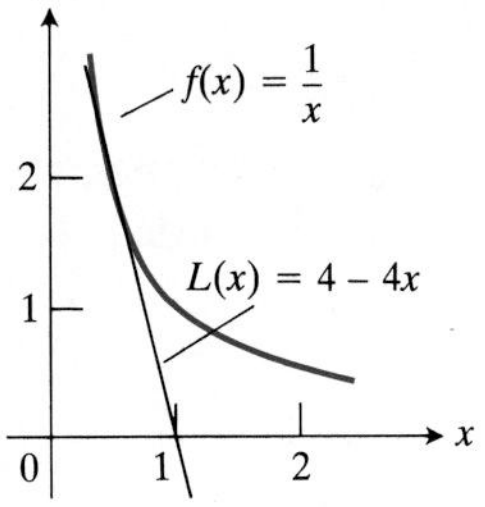

7. $\Delta V \approx 4\pi r^2\,\Delta r$ **9.** $\Delta V \approx 3x^2\,\Delta x$ **11.** $\Delta V \approx 2\pi rh\,\Delta r$ **13.** $\Delta V \approx \frac{2}{3}\pi rh\,\Delta r$ **15.** $L(x) = 3 + \frac{1}{6}(x - 8)$, $L(9.1) = 3.1833$, $f(9.1) = 3.1780$ **17.** $L(x) = 2 + \frac{1}{12}(x - 8)$, $L(8.5) = 2.0417$, $f(8.5) = 2.0408$ **19.** $L(x) = 5 - 0.8(x + 4)$, $L(-4.2) = 5.1600$, $f(-4.2) = 5.1614$ **21.** a) $\Delta A \approx 0.08\,\pi$ b) 2% **23.** 1.5% **25.** a) Within 1 percent, $|\Delta E| < (0.01)E$ b) 3 percent **27.** a) 4% b) 8% c) 12% **29.** a) Within a half percent b) Within five percent **31.** a) $L = g/(\Delta\pi^2) = 0.815$ b) $\Delta T = \pi(Lg)^{-1/2}(0.01) = 0.00613$ s c) Loses 529.6 s/day or 8 min, 50 s **33.** $1 + 2x$ **35.** $1 - 5x$ **37.** $2 + 8x$ **39.** $2 + x$ **41.** $1 - x$ **43.** a) 1.02 b) 1.003 c) 1.001 **45.** a) $g(3) = \sqrt{3} - 2 < 0$, $g(4) = \sqrt{5} - 2 > 0$ b) $x^* = x_0 - (y_0/m) = 3 - (\sqrt{3} - 2)/((1/2\sqrt{3}) + \frac{1}{4}) \approx 3.4974$ c) $g(3.4974) = -0.00914$ **49.** $v \approx 0.14c$

Article 2.6, pp. 140–141

1. $8t - 20$ **3.** $-t^2$ **5.** $-\frac{1}{9}$ **7.** $-\sqrt{t}/t^2\sqrt{2 + 2t}$ **9.** $8t + 10$ **11.** $18x + \frac{1}{3}x^{-2}$ **13.** $(16t^2 - 20t + 1)^{-1/2}(16t - 10)$ **15.** $(1 - 2v)/(v^2 - v)^2$ **17.** $y = x^{3/2}$, $dy/dx = \frac{3}{2}x^{1/2} = \frac{3}{2}t$, $dx/dt = 2t$, $dy/dt = 3t^2$ **19.** $y = x^2/(1 - x)$, $dy/dx = (2x - x^2)/(1 - x)^2 = t^2 + 2t$, $dx/dt = 1/(1 + t)^2$, $dy/dt = (t^2 + 2t)/(1 + t)^2$

21.

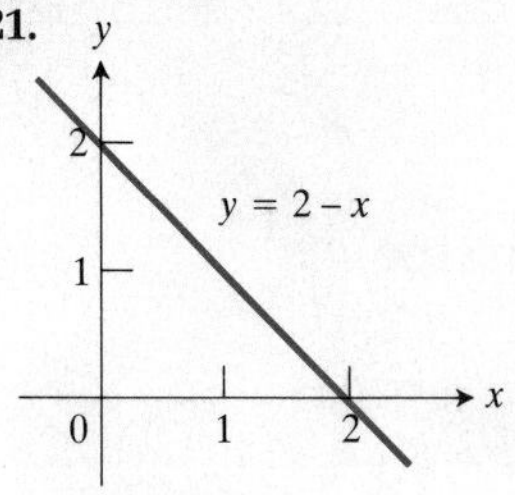

23.

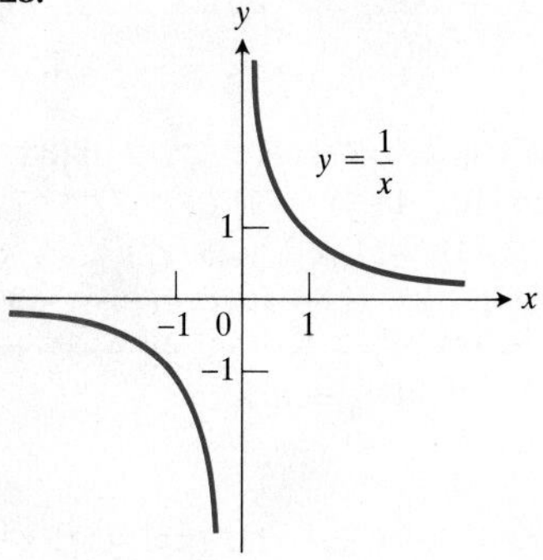

25.

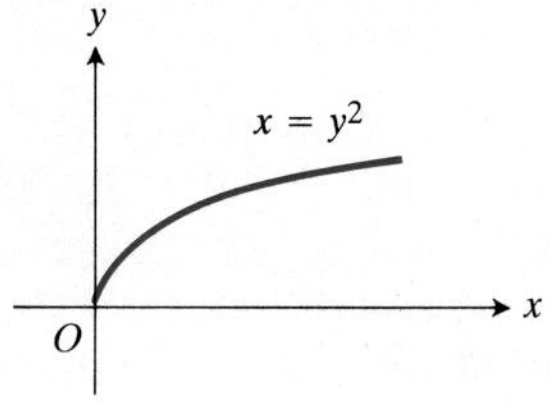

27. (c) **29.** a) $\frac{5}{3}$ b) $3y = 5x - 8$ **31.** a) $\frac{1}{11}$ b) $11y = x + 16$ **33.** a) -2 b) $2y + 4x = 9$ **35.** 2 **37.** -3 **39.** $f \circ g(x) = g \circ f(x) = x^2$, $d(x^2)/dx = 2x$, defined even when $x = 0$. Since $g'(0)$ does not exist, the assumptions of the Chain Rule Theorem are not all satisfied, but the theorem is not being contradicted. **41.** 0 **43.** $2/x^3$ **45.** a) $-4t^3/(t^2 - 1)^3$ b) $y'' = 2x^{-3}$ **47.** a) $\Delta T \approx -\frac{1}{2}(T/g)\,\Delta g$ b) T will decrease, pendulum will speed up. c) With $L = 100$ and $g = 980$, $\Delta g \approx -(2g/T)\,\Delta T \approx -0.98$ cm/s².

Article 2.7, pp. 152–154

1.

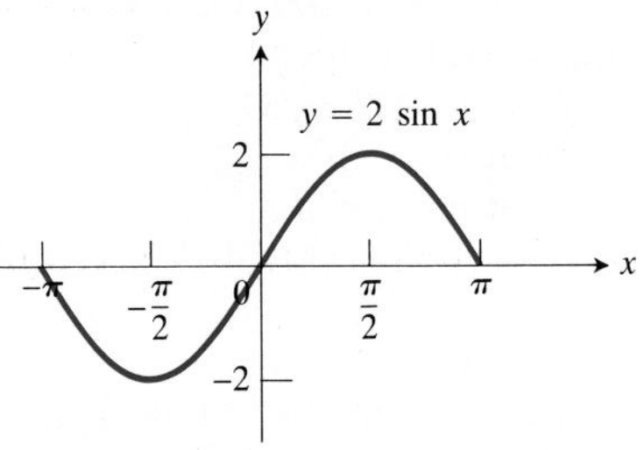

3.

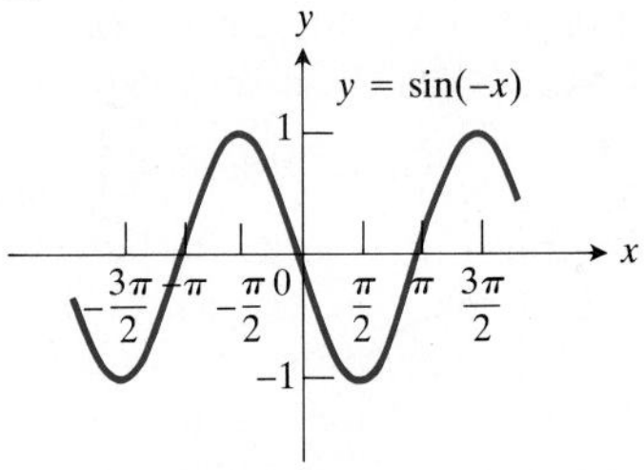

5.

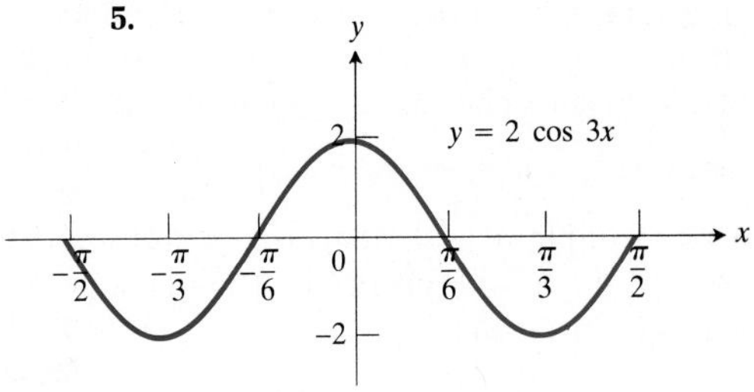

7.

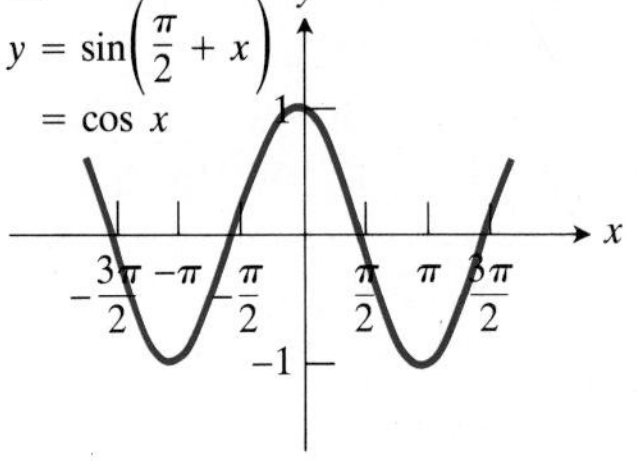

9.

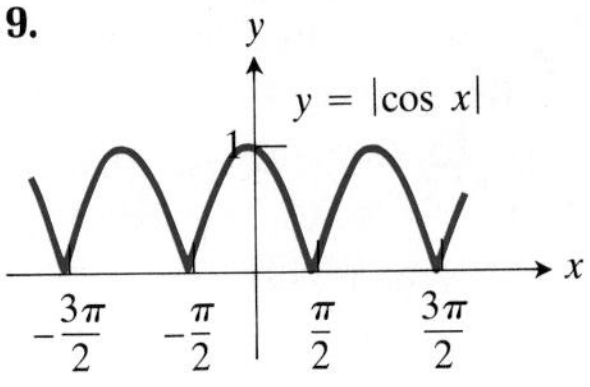

11.

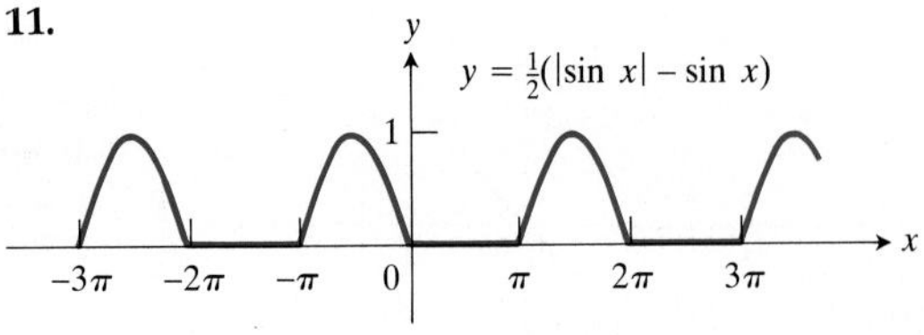

13.

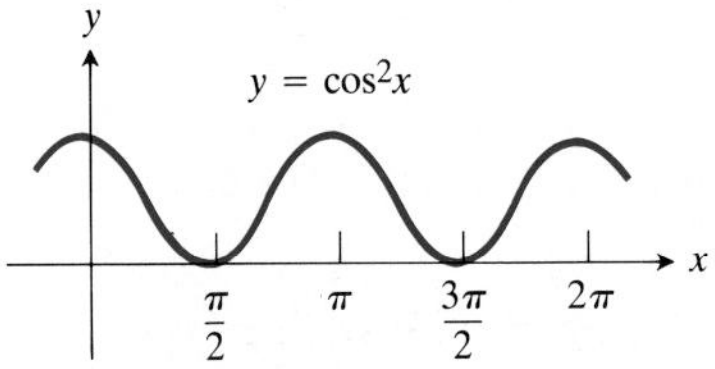

15.

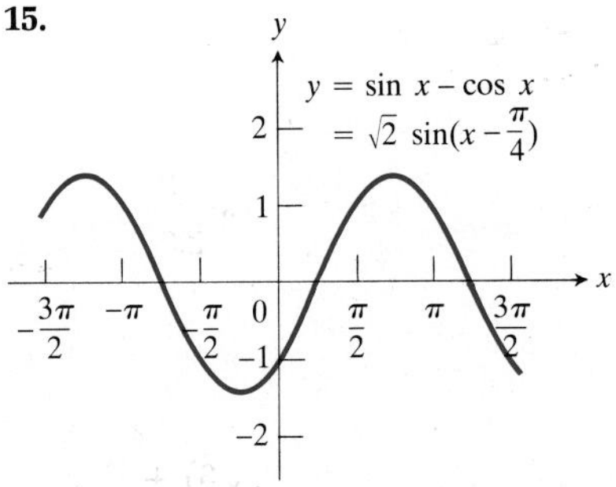

17.

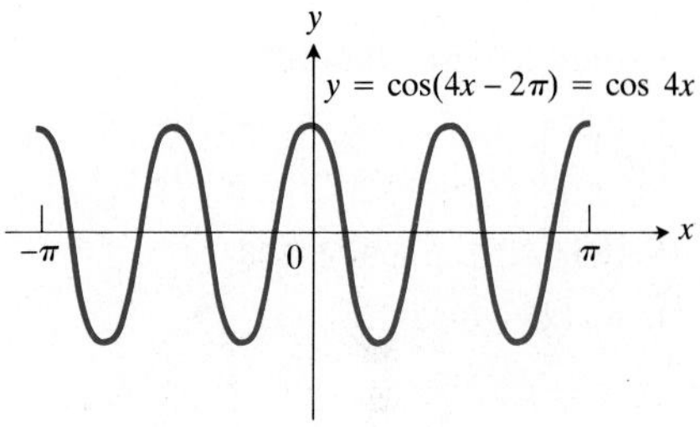

19.

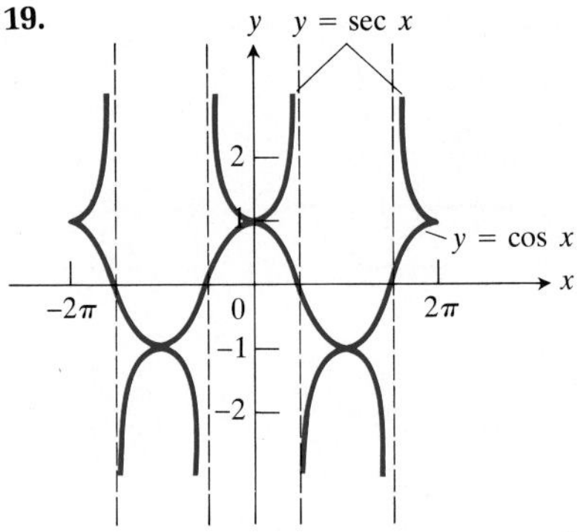

21.

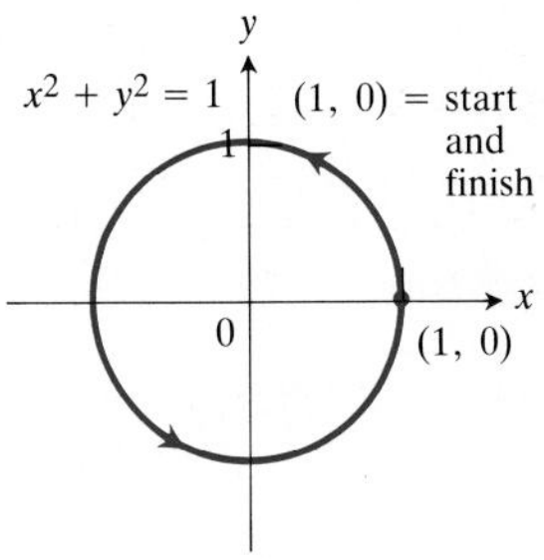

23. Same as 21. **25.**

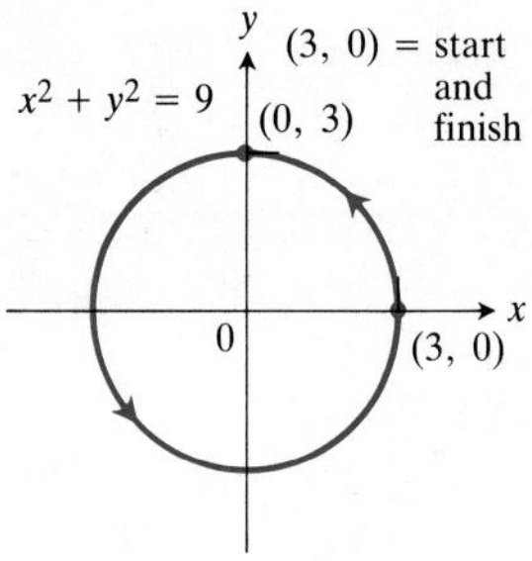

27.

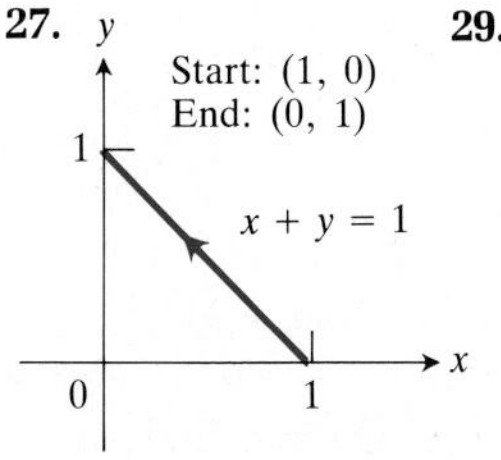

29. a) $x = 2\cos t$, $y = -2\sin t$ b) $x = 2\cos t$, $y = 2\sin t$ **31.** a) 37 b) 365 c) 101 d) 25

33. 0 **35.** 0 **37.** $\cos^2 A + \sin^2 A = \cos(0) = 1$ **39.** $\pi/4$ **41.** $(0, \sqrt{3}/2)$, $\theta = \pi/3$ **43.** $(0, 0)$, $\theta = \pi/2$; $(1, 1)$, $\theta = \pi/4$ **45.** $y = \frac{1}{2}$ **47.** $A = \cos^{-1}(3/\sqrt{34}) = \tan^{-1}(\frac{5}{3}) \approx 59°$, $B = \cos^{-1}(-1/\sqrt{26}) = \tan^{-1}(5) \approx 78.7°$, $C = \cos^{-1}(11/\sqrt{221}) = \tan^{-1}(10/11) \approx 42.3°$

Article 2.8, pp. 160–161

1. $\cos(x+1)$ **3.** $\frac{1}{2}\cos(x/2)$ **5.** $-5\sin 5x$ **7.** $-2\sin(2x)$ **9.** $3\cos(3x+4)$ **11.** $\sin x + x\cos x$ **13.** $x\cos x$ **15.** $\tan x \sec x$ **17.** $\tan x \sec x$ **19.** $-\tan(1-x)\sec(1-x)$ **21.** $2\sec^2(2x)$ **23.** $2\sin x\cos x = \sin 2x$ **25.** $-10\sin 5x\cos 5x = -5\sin 10x$ **27.** $5\sec^2(5x-1)$ **29.** $2\cos(2x)$ **31.** $(-\sin x)/y$ **33.** $-(1/(2\sqrt{x}))\sin\sqrt{x}$ **35.** $y = \sqrt{\cos^2 x} = |\cos x| = \begin{cases} \cos x, & \text{when } \cos x \geq 0 \\ -\cos x, & \text{when } \cos x \leq 0. \end{cases}$ Therefore, $y' = \begin{cases} -\sin x, & \text{when } \cos x > 0 \\ \sin x, & \text{when } \cos x < 0. \end{cases}$ When $\cos x = 0$ (for example at $x = \pm\pi/2$), y' is not defined; the graph of $y = |\cos x|$ has cusps at these points. **37.** $dy/dx = \cos^2 y$ **39.** $dy/dx = -(1/y)\sin 8x$ **41.** $-(1/x)(y + \cos^2(xy))$ **43.** $y = 2x$ **45.** 1 **47.** 1 **49.** 1 **51.** $-\sin a$ **53.** a) -1 b) -1 c) 0 **55.**

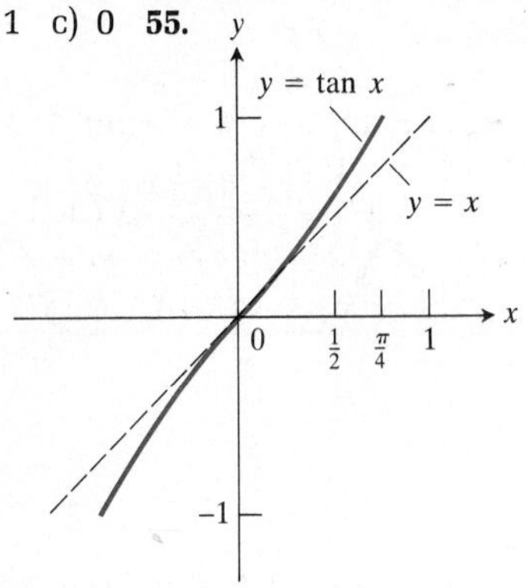

57. If $b = 1$, f is continuous, but not differentiable, at $x = 0$. **59.** a) $f(0) = -1$, $f(2) = 1 - \frac{1}{2}\sin 2 > \frac{1}{2} > 0$, $\sin x \leq 1$ always. Since f is continuous, $f(x) = 0$ for some x in $(0, 2)$. b) $f(x) \approx x - 1 - \frac{1}{2}x$, implies $x = 2$ c) $f(2) = 1 - \frac{1}{2}\sin 2 = 0.545$. (A better approximation might take x_0 near $\pi/2$.) **61.** $f(x) \approx 1 + \frac{3}{2}x$, which is the sum of the linear approximations to $\sin x$ and $[1+x]^{1/2}$. **63.** b) $\sqrt{1+x} \approx 1.3 + (x - 0.69)/2.6$, $2\cos x \approx \sqrt{2} - \sqrt{2}(x - \pi/4)$ c) $x \approx 0.82849$, at which point $\sqrt{1+x} \approx 1.3522$ and $2\cos x \approx 1.3520$.

Article 2.9, pp. 165–166

1. 0.618 **3.** 1.1640 **5.** -1.189 **7.** $x_n = x_1 = x_0 - f(x_0)/f'(x_0) = x_0$ if $f(x_0) = 0$ **9.** $f(3) = -21 < 0$ and $f(4) = 117 > 0$, so $f(c) = 0$ for some c in $(3, 4)$; $x = 3.22858$ **11.** $\pi = 3.141592654$ **13.** 0.7390851 **15.** 0.6301154, 2.57327

Article 2.10, pp. 174–175

1. a) $(x-3)/2$ b)

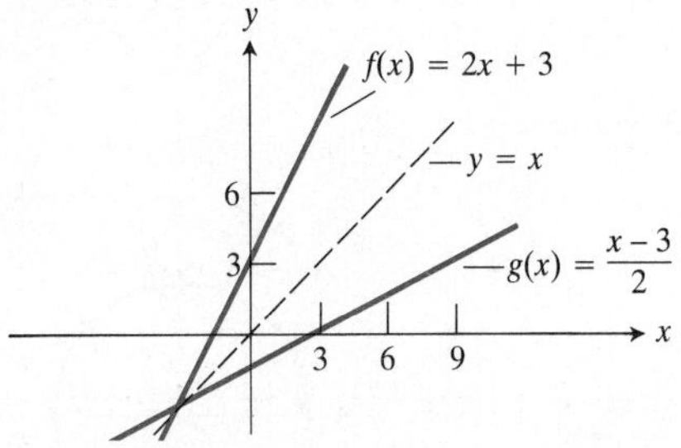

c) $f' = 2$, $g' = \frac{1}{2}$ **3.** a) $5x - 35$ b)

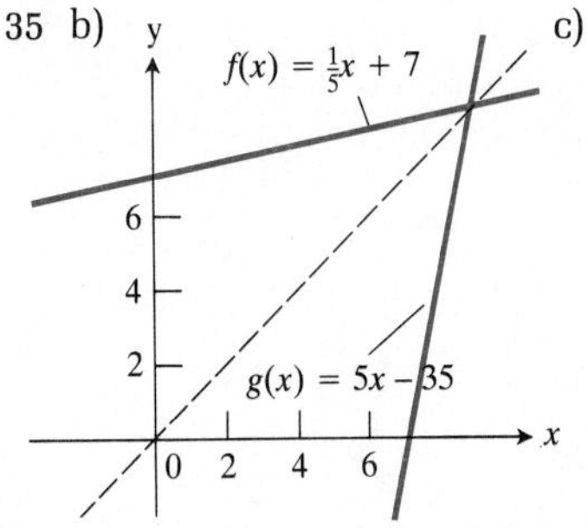

c) $f' = \frac{1}{5}$, $g' = 5$

5. a) $[x^2 - 1]^{1/2}$ b)

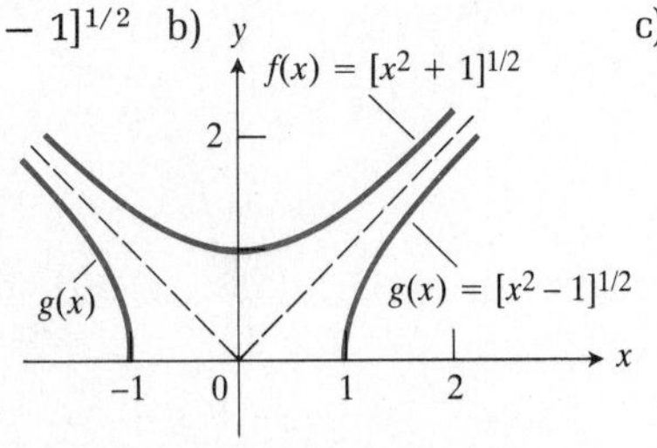

c) $f'(\sqrt{3}) = \sqrt{3}/2$, $g'(2) = 2/\sqrt{3}$

7. a) $g(x) = x - 1$ b)

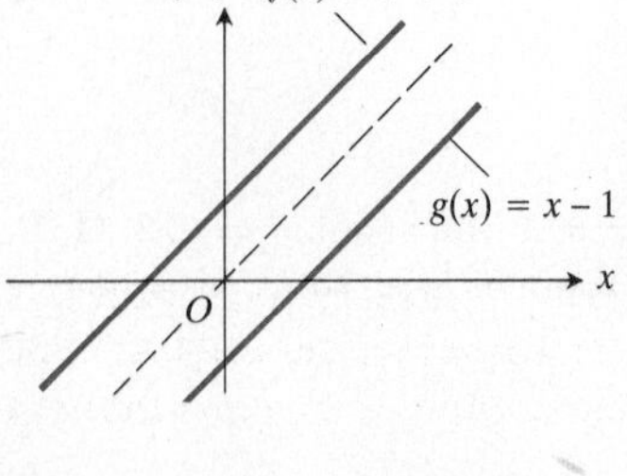

c) $f' = 1$, $g' = 1$ **9.** $x^{1/5}$ **11.** $x^{3/2}$ **13.** If $x \geq 1$, then $f^{-1} = x^{1/2} + 1$. If $x < 1$, then $f^{-1} = 1 - x^{1/2}$. **15.** $f^{-1} = x^{-1/2}$ **17.** $2y^{1/2}/(1 + 4y^{3/2})$ **19.** $-(1-y^2)^{1/2}/5y$ **21.** $6y/(3y^{1/2} - 2y^{1/3})$ **23.** b) $-1/a^2$, $-a^2$, 1 c) $1/x$ **25.** $(\frac{3}{4}, -\frac{1}{2})$ **27.** $g(x) = 1/(x-1)$, $g'(1 + 1/x) = -x^2$, $f'(x) = -1/x^2$ **29.** $x = 1$ (Picard's method directly will not yield the other root $x = 0$.) **31.** ± 1.026738 **33.** $x = 0$ **35.** Slope is greater than 1 at $x = 1$. **37.** $g(x) = 1 + \frac{1}{2}\sin x$, $x = 1.4987011$ **39.** 0.682328

Miscellaneous Problems Chapter 2, pp. 177–179

1. $-4(x^2-4)^{-3/2}$ **3.** $-y/(x+2y)$ **5.** $-y(2x+y)/(x(2y+x))$ **7.** $2\sin(1-2x)$ **9.** $(x+1)^{-2}$ **11.** $(3x^3-2a^2x)/\sqrt{x^2-a^2}$ **13.** $2x(1-x^2)^{-2}$ **15.** $10\sin(5x)/\cos^3(5x)$ **17.** $[(4x+5)/2]\sqrt{2x^2+5x}$ **19.** $-(y/x)[(2y\sqrt{xy}+1)/(4y\sqrt{xy}+1)]$ **21.** $-(y/x)^{1/3}$ **23.** $-y/x$ **25.** $-\frac{1}{2}[(x+2y+y^2)/(x+2y+xy)]$ **27.** $1/(2y(x+1)^2)$ **29.** $-(y+2)/(x+3)$ **31.** $(y-3x^2)/(3y^2-x)$ **33.** $1/(y(1-x)^2)$ **35.** $x^2/(x^3+1)^{2/3}$ **37.** $-2\csc^2 x$ **39.** $\sec x\,(2\tan^2 x+1)$ **41.** $2x\cos 8x-8x^2\sin 8x$ **43.** $(1+\cos x)^{-1}$ **45.** $-(\cot x)\csc x$ **47.** $-(\sin 2x)\sin(\sin^2 x)$ **49.** $2\tan x\sec^2 x$ **51.** $-3(\sin 6x)\sin(\sin^2 3x)$ **53.** $(4x+8)/(4x^2+16x+15)^{1/2}$ **55.** $2x[1+((1-x^2)/(1+x^2)^3)]$ **57.** $3(y+1)^{2/3}/(2(y+1)^{1/3}+1)=3t^2/(2t+1)$ **59.** $m=1,\ y=x$ **61.** (d) **63.** $8x+5y=18,\ 3x-5y=18$ **65.** 3 ft **67.** $m=2,\ y-2x=-5$ **69.** a) $10(x^2+2x)^4(x+1)$ b) $(3t-1)/\sqrt{3t^2-2t}$ c) $r[(r^2+5)^{-1/2}+(r^2-5)^{-1/2}]$ d) $4x/(x^2+1)^2$ **71.** $x+y=3$ **73.** $-6(2-3x)$, $f(x)=(2-3x)^2$ **75.** $6\pi(\Delta r)^2$; a shell around the can with thickness Δr. **77.** Revenue is $px=x(3-(x/40))^2$. $x_1=40$, $p=4(=20¢)$ **79.** $(x-1)(2x-1)$ **81.** 56 ft/s **83.** a) $h+2k=5$ b) $h=-4,\ k=\frac{9}{2},\ r=5\sqrt{5}/2$ **85.** $2/[3(2y+1)(2x+1)\sqrt{x^2+x}]$ **87.** $(3/(x+1)^2)\sin[((2x-1)/(x+1))^2]$ **91.** $-3/32$ **97.** a) Length of each side $=2r\sin(\pi/n)$ b) $2\pi r$, yes **99.** 14 ± 0.044 ft **101.** $-x$; clockwise **103.** a) all m, b such that $m=-b/\pi$ b) $m=-1,\ b=\pi$ **105.** a) $f'(0)=\lim_{x\to 0} x^2\sin(1/x)=0$ b) $2x\sin(1/x)-\cos(1/x)$ c) Yes; $f'(x)\to 0$ as $x\to 0$

CHAPTER 3

Article 3.1, p. 185

1. $-1<x<2$ **3.** $0<x<1$ **5.** $x<-2$ or $0<x<2$ **7.** $-1<x<1$ or $x>5$ **9.** $y'=2x-1$, minimum at $x=\frac{1}{2}$, increasing if $x>\frac{1}{2}$, decreasing if $x<\frac{1}{2}$. **11.** $y'=x^2-x-2$, increasing if $x<-1$ or $x>2$, decreasing if $-1<x<2$, local maximum at $x=-1$, local minimum at $x=2$. **13.** $y'=3x^2-27$, increasing for $x>3$ or $x<-3$, decreasing for $-3<x<3$, local maximum at $x=-3$, local minimum at $x=3$. **15.** $y'=6x(1-x)$, increasing for $0<x<1$, decreasing for $x<0$ or $x>1$, local maximum at $x=1$, local minimum at $x=0$. **17.** $y'=4x^3$, increasing for $x>0$, decreasing for $x<0$, local minimum at $x=0$, no local maxima. **19.** $y'=-3/x^4$ for $x>0$, decreasing over domain, no local maxima or minima. **21.** $y'=-2/(x+1)^3$, increasing for $x<-1$, decreasing for $x>-1$, no local maxima or minima. **23.** $y'=-\sin x$, increasing for $-\pi<x<0$ and $\pi<x<\frac{3}{2}\pi$, decreasing for $-\frac{3}{2}\pi<x<-\pi$ and $0<x<\pi$, local maximum at $x=0$, local minimum at $x=\pm\pi$. **25.** $y'=2x$ for $x\geq 0$, $y'=-2x$ for $x<0$, increasing everywhere, no local maxima or minima. **27.**

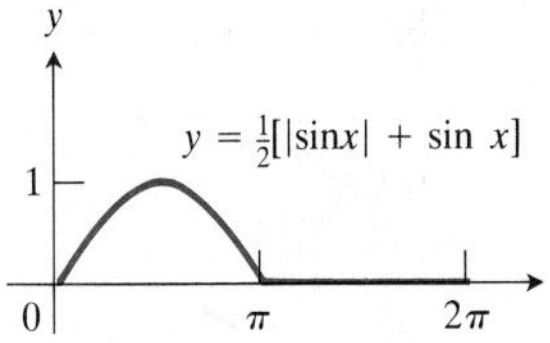

Maximum $=1$ at $x=\pi/2$, minimum $=0$ at $x=0$ and $\pi\leq x\leq 2\pi$ **29.** $y=-2x$ for $x<0$, $y=-x$ for $x\geq 0$ **31.** a) $y'<0$ when $0\leq x<\pi/3$, $y'=0$ at $x=\pi/3$, $y'>0$ for $\pi/3<x\leq\pi$ b)

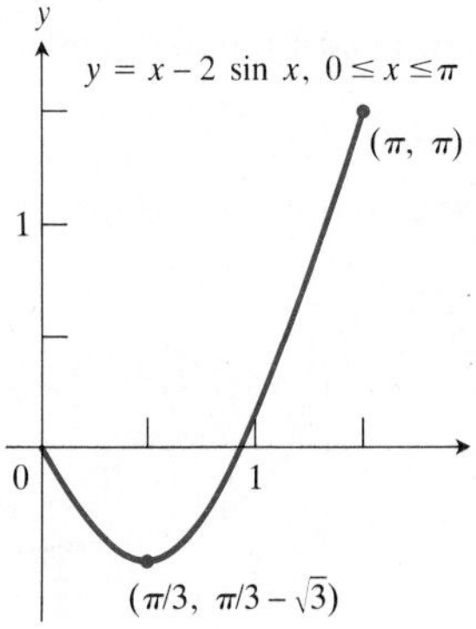

c) 1.8955

Article 3.2, pp. 189–190

1. a) $x>2$ b) $x<2$ c) Everywhere d) Nowhere. Local minimum at $(2,-1)$, no inflection points. **3.** a) $x>1$ or $x<-1$ b) $-1<x<1$ c) $x>0$ d) $x<0$. Local maximum at $(-1,5)$, local minimum at $(1,1)$, inflection at $(0,3)$. **5.** a) $x<1$ or $x>3$ b) $1<x<3$ c) $x>2$ d) $x<2$. Local maximum at $(1,5)$, local minimum at $(3,1)$, inflection at (2, 3).

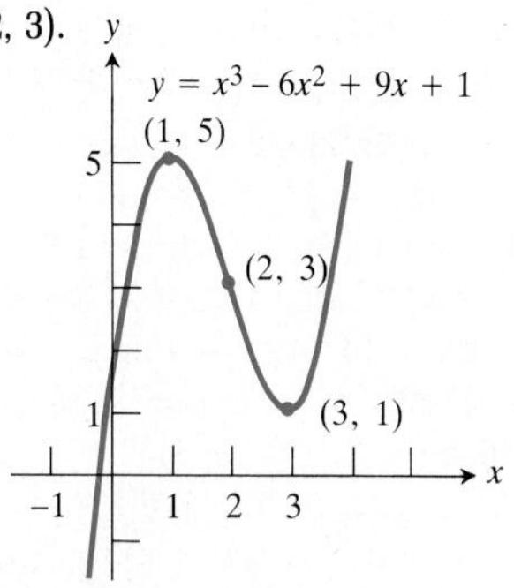

7. a) Everywhere b) Nowhere c) $x > 2$ d) $x < 2$. No maxima or minima, inflection at (2, 1). **9.** a) $-\pi/2 < x < \pi/2$ b) Nowhere c) $0 < x < \pi/2$ d) $-\pi/2 < x < 0$. No local maxima or minima, inflection at (0, 0). **11.** a) $x < 0$ b) $x > 0$ c) Nowhere d) Everywhere. Local maxima at (0, 0), no inflection points. **13.** Use $y = (x - 1)^2$, for instance. **15.** Local maximum at $(-1, 7)$, no inflection points. **17.** Minimum at (3, 0), maximum at (1, 16), inflection at (2, 8). **19.** Maximum at $(-2, 28)$, minimum at $(2, -4)$, inflection at (0, 12). **21.** Minimum at $(9, -972)$, maximum at $(-5, 400)$, inflection at $(2, -286)$. **23.** Minimum at (2, 0), no maxima or inflection points. **25.** Maxima at $(m\pi + \frac{1}{3}\pi, m\pi + \frac{1}{3}\pi + \sqrt{3}/2)$, minima at $(n\pi - \frac{1}{3}\pi, n\pi - \frac{1}{3}\pi - \sqrt{3}/2)$, inflection points at $(k\pi/2, k\pi/2)$. (m, n, k are integers.)

27. $dx/dy = 3y^2 + 6y + 3 = 3(y + 1)^2$

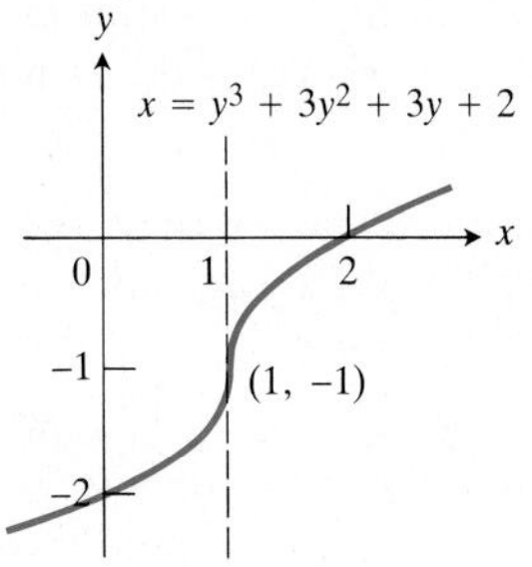

29. $dx/dy = 4y^3 - 4y = 4y(y + 1)(y - 1)$

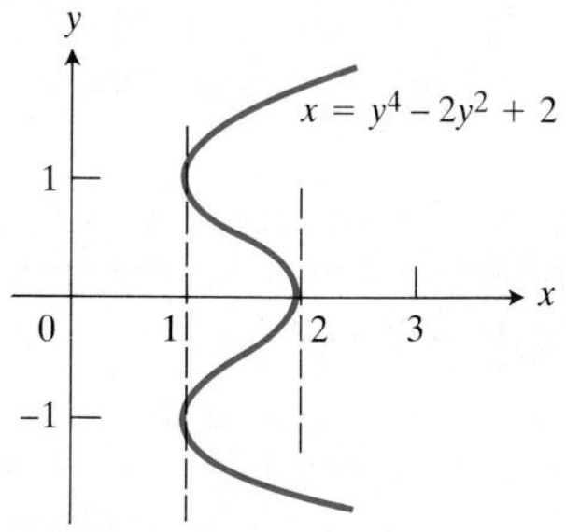

31. a) T b) P **33.**

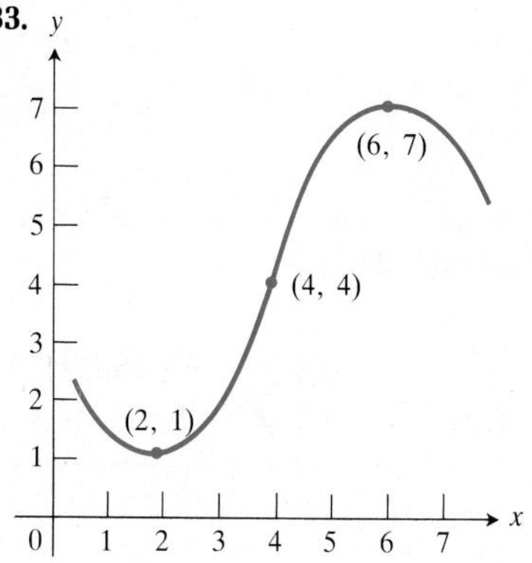

35. $f' < 0$, so graph is concave down.

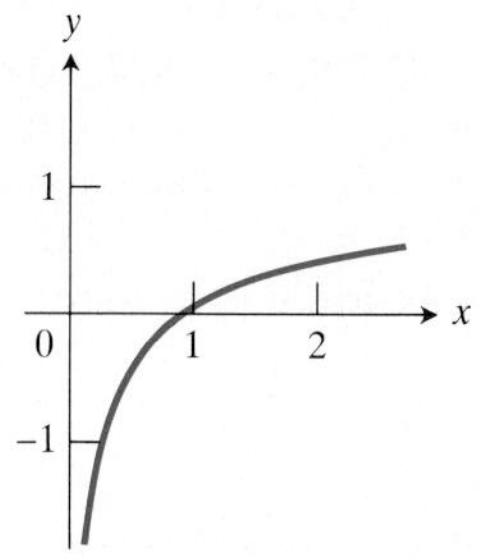

37. a) Near $x = -1.2, -0.4, +0.8$ b) Near $x = -1.8$ c) Near $x = 1.2$ **39.** Minima at $(9, -9477)$, $(-15, -37125)$, local maximum at (0, 0), inflections at $(5, -5125)$, $(-9, 21141)$. **41.** a) $x = 2$ b) $x = 4$

Article 3.3, p. 198

1. a) Symmetric about lines $y = x - \frac{3}{2}$, $y = -x + \frac{3}{2}$ and point $(\frac{3}{2}, 0)$ b) $x \neq \frac{3}{2}$, $y \neq 0$ c) $(0, -\frac{1}{3})$ d) $x = \frac{3}{2}$, $y = 0$ e) $-\frac{2}{9}$

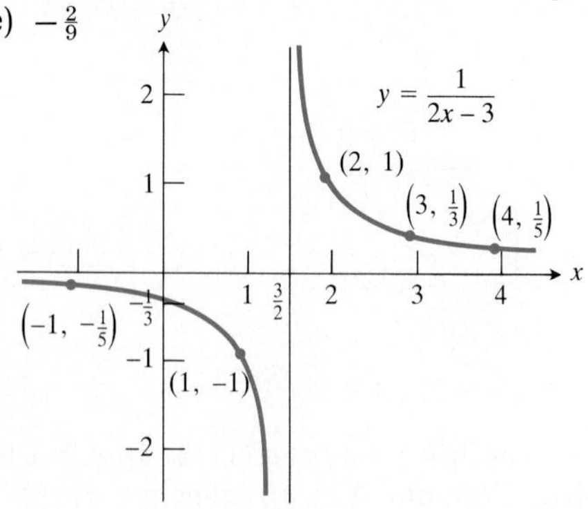

3. a) Symmetric about y-axis b) $x \neq \pm 2$, $y \geq 3$ or $y < 2$ c) (0, 3), $(\pm\sqrt{6}, 0)$ d) $x = \pm 2$, $y = 2$ e) $y' = 0$ at (0, 3), $y' = \pm 2\sqrt{6}$ at $(\pm\sqrt{6}, 0)$.

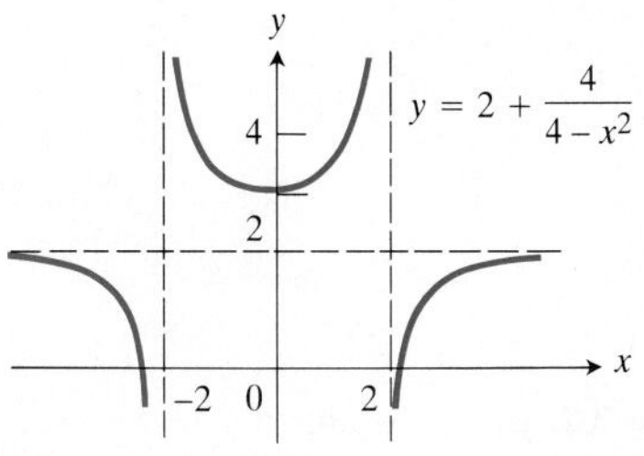

5. a) Symmetrical about origin b) $x \neq 0$ c) $(\pm 1, 0)$ d) $y = x$, $x = 0$ e) $y' = 2$ at $(\pm 1, 0)$

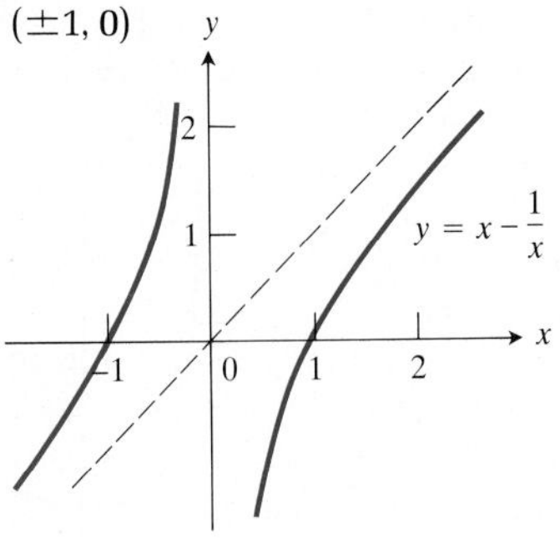

7. a) Symmetrical about $y = -x$ and $y = x + 2$ b) $x \neq -1$, $y \neq 1$ c) $(0, 0)$ d) $x = -1$, $y = 1$ e) $y' = 1$ at $(0, 0)$

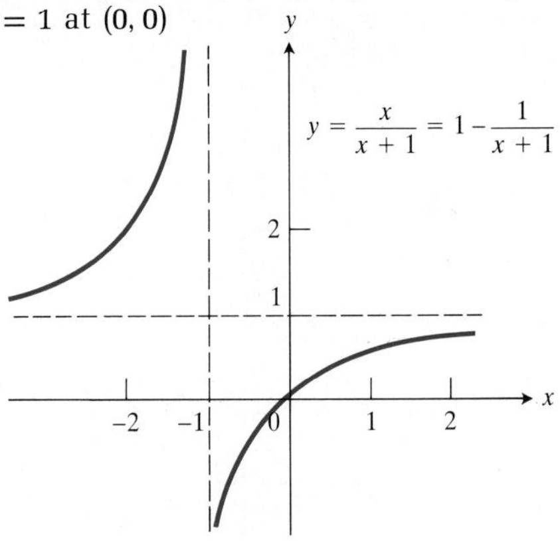

9. a) Symmetrical about y-axis b) $y \geq 1$ c) $(0, 1)$ d) No asymptotes e) $y' = 0$ at $(0, 1)$

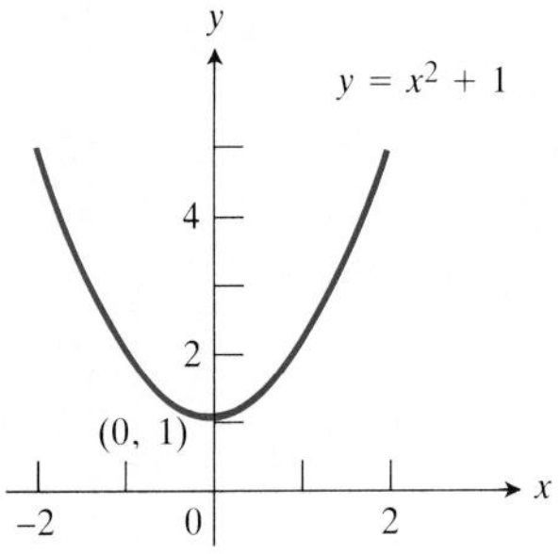

11. a) Symmetrical about y-axis b) $y \geq -1$ c) $(0, -1)$ d) No asymptotes e) $y' = 0$ at $(0, 1)$

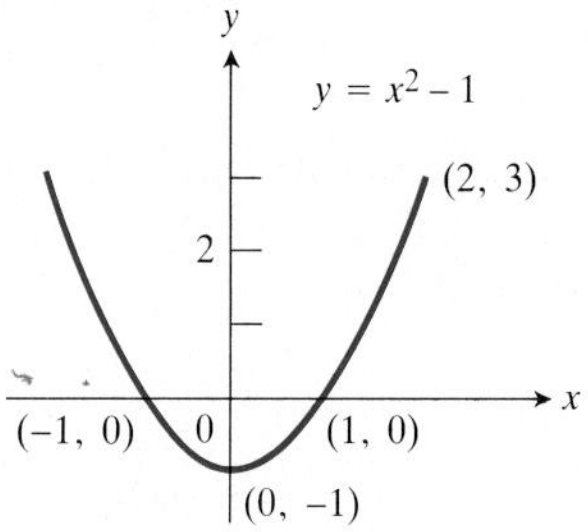

13. a) No symmetry about the origin or lines through the origin b) $x \neq 1$, $y \geq 4$ or $y \leq 0$ c) $(0, 0)$ d) $x = 1$, $y = x + 1$ e) $y' = 0$ at $(0, 0)$

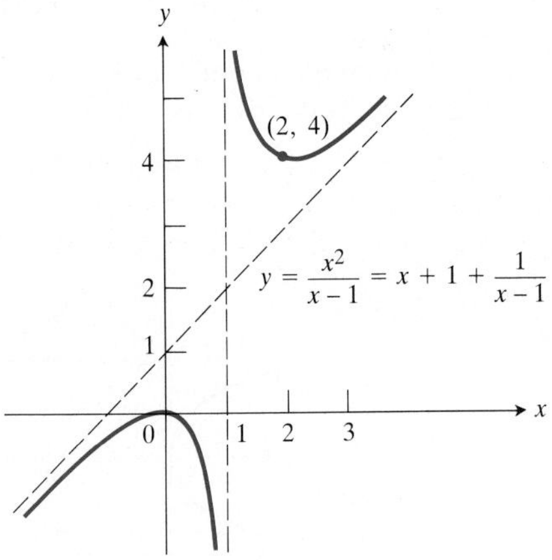

15. a) No symmetry about the origin or lines through the origin b) $x \neq 1$, $x \neq -1$, $y < 2 - 2\sqrt{2}$ or $y \geq 2 + 2\sqrt{2}$ c) $(0, -1)$, no x-intercept d) $x = 1$, $y = x + 1$ e) $y'(0) = -1$

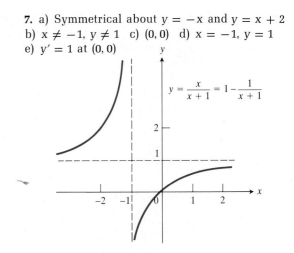

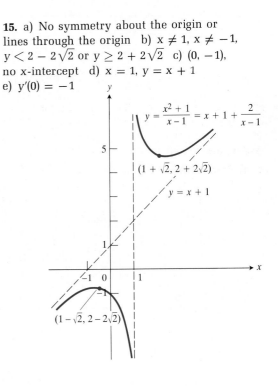

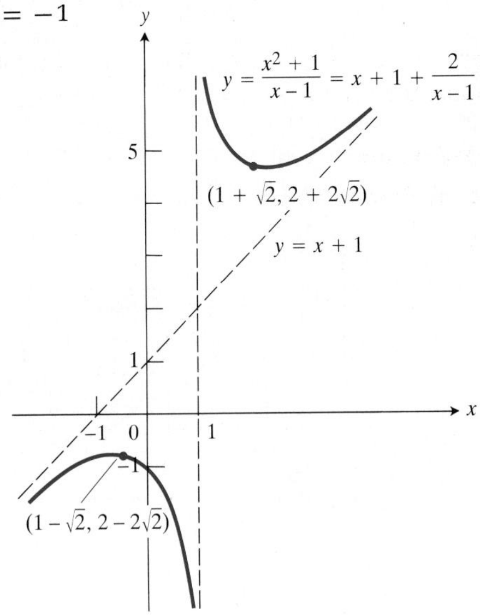

17. a) No symmetry about the origin or lines through the origin b) $x \neq 1$ c) $(0, 4)$, $(\pm 2, 0)$ d) $y = x + 1$, $x = 1$ e) $y' = 4$ at $(0, 4)$, $y' = 4$ at $(2, 0)$, $y' = \frac{4}{3}$ at $(-2, 0)$.

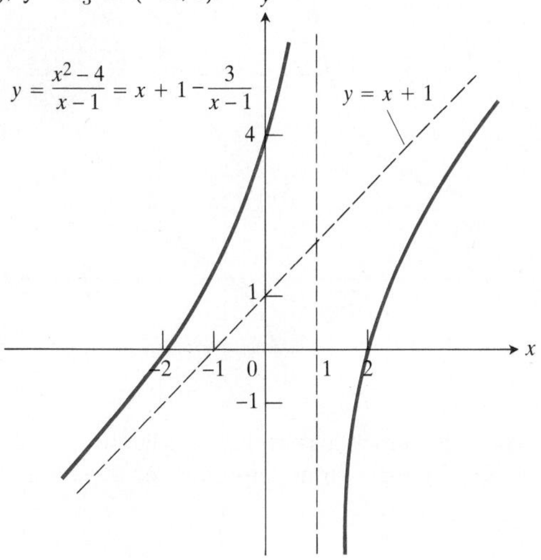

19. a) No symmetry about the origin or lines through the origin b) $x \neq -2$, $y \geq -2 + \sqrt{3}$ or $y \leq -2 - \sqrt{3}$ c) $(\pm 1, 0)$, $(0, -\frac{1}{4})$ d) $(\frac{1}{2})x - 1$, $x = -2$ e) $y' = \frac{1}{8}$ at $(0, -\frac{1}{4})$, $y' = -1$ at $(-1, 0)$, $y' = \frac{1}{3}$ at $(1, 0)$.

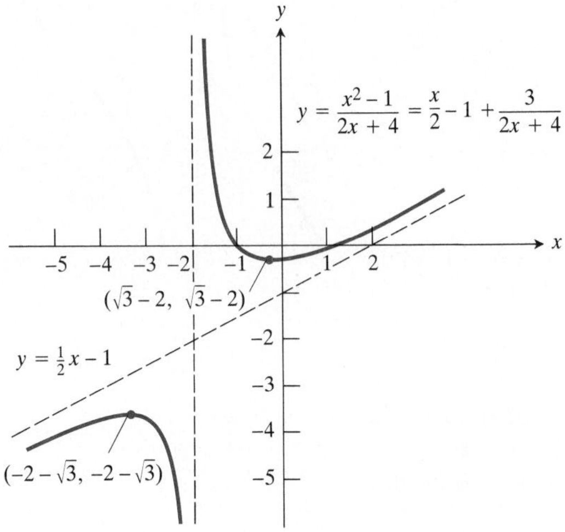

21. a) No symmetry b) $x \neq 0$ c) $(-1/\sqrt[3]{3}, 0)$ d) $x = 0$ e) $y' = -\frac{2}{3}\sqrt[3]{3} - 4\sqrt[3]{9}$

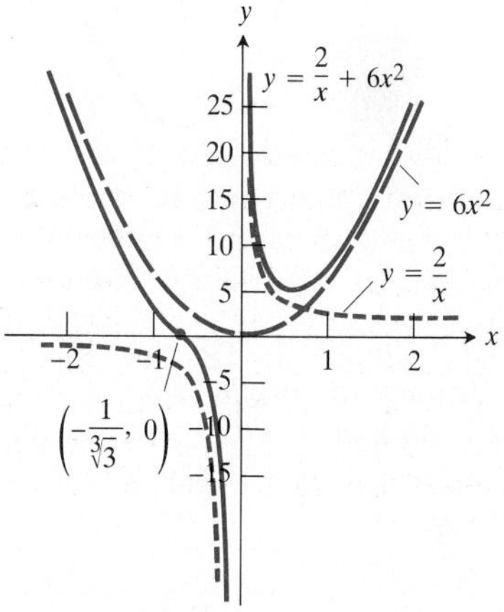

23. a) Symmetry about y-axis b) $y \geq 4$ or $y < 1$, $x \neq \pm 1$ c) $(0, 4)$, $(\pm 2, 0)$ d) $x = \pm 1$, $y = 1$ e) $y' = 0$ at $(0, 4)$, $y' = \pm \frac{4}{3}$ at$(\pm 2, 0)$.

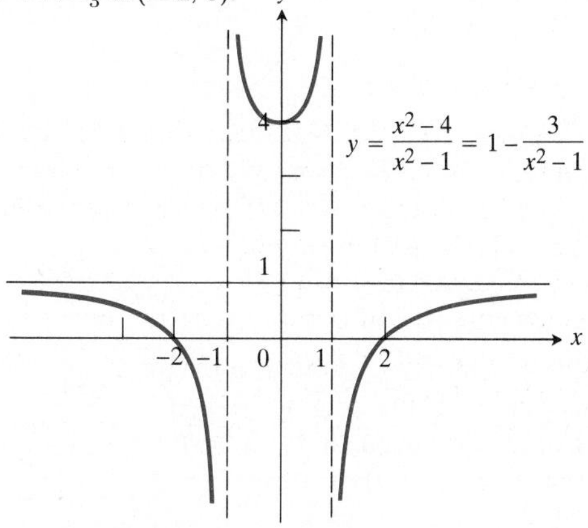

25. a) No symmetry b) $y \leq -2 - \sqrt{5}$ or $y \geq -2 + \sqrt{5}$ c) $(0, \frac{1}{3})$ d) $y = 1$, $x = 1$, $x = 3$ e) $y' = \frac{4}{9}$

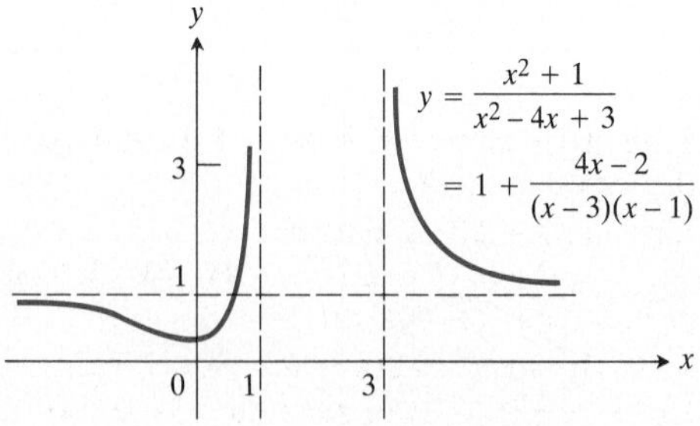

27. a) Symmetry about x-axis b) $x \geq 2$ or $x \leq 0$, $y \neq \pm 1$ c) $(0, 0)$ d) $y = \pm 1$, $x = 2$ e) $y' = \infty$ at $(0, 0)$

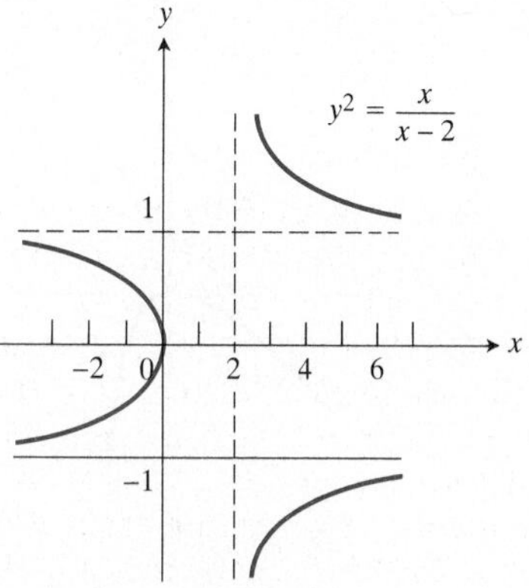

29. a) Symmetric about x-axis, y-axis, origin b) $x \geq 1$ or $x \leq -1$, $-1 < y < 1$ c) $(\pm 1, 0)$ d) $y = \pm 1$ e) $y' = \infty$ at $(\pm 1, 0)$

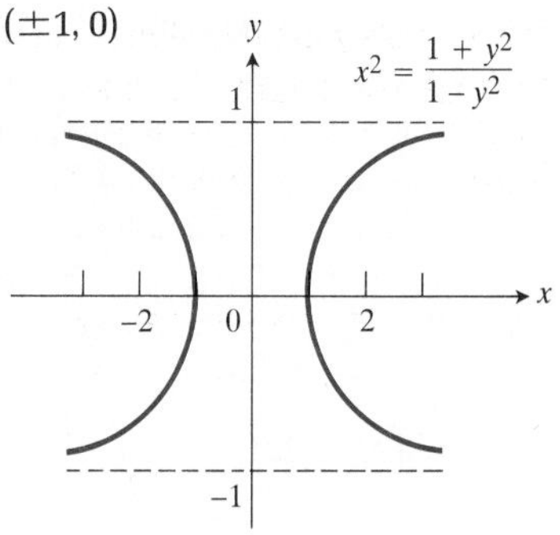

31. a) No symmetry b) $x \neq 0$, $x \neq 2$ c) $(1, 0)$ d) $y = 0$, $x = 0$, $x = 2$ e) $y' = -1$

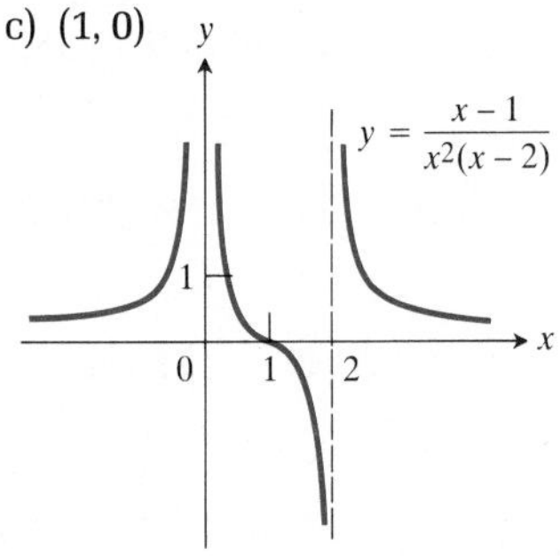

33. $y' = (x \cos x - \sin x)/x^2 \to 0$ as $x \to \infty$ **35.** a) Symmetrical about y-axis b) Symmetrical about origin c) Both symmetrical about origin

Article 3.4, pp. 204–205

1. $(\frac{1}{2}, \frac{1}{4})$ gives absolute maximum, absolute minimum = 0 **3.** $(1/\sqrt{3}, 2\sqrt{3}/9)$ gives absolute maximum, absolute minimum = 0 **5.** Local maximum at $x = -7$, local minimum at $x = 7$, no absolute maxima or minima **7.** $(\frac{1}{2}, \frac{1}{4})$ gives local maximum, absolute maximum = 6, absolute minimum = 0 **9.** No critical points or absolute maxima or minima **11.** $(1, -3)$ gives absolute minimum, absolute maximum = 8 **13.** $(0, 1)$ gives absolute minimum, no absolute maximum **15.** $(0, 0)$ gives local maximum, $(15, -37125)$ gives absolute minimum, $(-9, -9477)$ gives local minimum, no absolute maximum **17.** No critical points, no absolute maximum or minimum **19.** No critical points, no absolute maximum, absolute minimum = 0 **21.** No critical points, absolute maximum = 27, absolute minimum = 0. With domain $[-2, 3)$, there is no absolute maximum. **23.** No critical points, absolute minimum = 1, absolute maximum = $\frac{9}{2}$ **25.** Critical points are $(1, \frac{1}{2})$, $(-1, -\frac{1}{2})$, inflection points are $(0, 0)$, $(+\sqrt{3}, \sqrt{3}/4)$, $(-\sqrt{3}, -\sqrt{3}/4)$ **27.** (b) **29.** 4

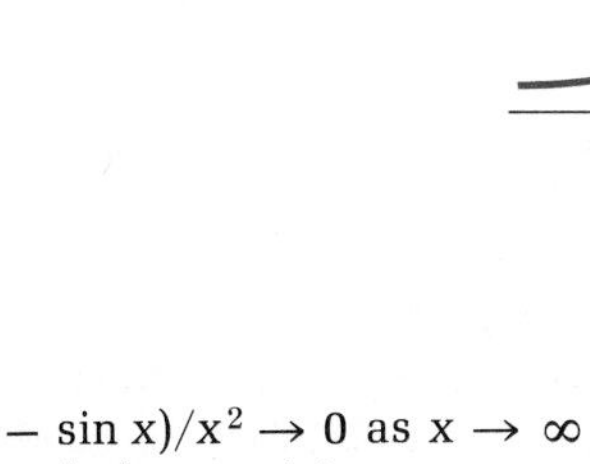

Article 3.5, pp. 212–215

3. 32 **5.** The side parallel to the river is twice as long as each of the other two sides. **7.** Base is 4 in. × 4 in., height is 2 in. **9.** Height = 18 in., width = 9 in. **11.** 12 m by 18 m, total = 3(12) + 2(18) = 72 m **13.** Radius = height = $10/\sqrt[3]{\pi}$ cm **15.** Length = height = $\sqrt{2}$ m. Least cost is $20 + 20\sqrt{2}$ dollars. **19.** $V_{max} = 2\pi h^3/(9\sqrt{3})$. ($h$ = hypotenuse.) **21.** $0.58\,L$ **23.** a) $a = -3$, $b = -9$ b) $a = -3$, $b = -24$ **25.** $(\frac{1}{2}, -\frac{1}{2})$ and $(-\frac{1}{2}, \frac{1}{2})$ **27.** $\frac{32}{81}\pi r^3$ **31.** Breadth = $\frac{1}{2}d$, depth = $(\sqrt{3}/2)d$. (d = diameter of log.) **33.** Height of rectangle = $(4 + \pi)/8$ times diameter of semicircle. **35.** $r = h = \sqrt{A/5\pi}$. (A is surface area.) **37.** No; $y = 3 + 4\cos\theta + (2\cos^2\theta - 1) = 2(1 + \cos\theta)^2 \geq 0$. **39.** Maxima = $\sqrt{2}$ at $x = \pi/4 + 2m\pi$, minima = $-\sqrt{2}$ at $x = -\frac{3}{4}\pi + 2n\pi$. ($m$, n are integers.) **41.** Diameter = altitude of cylinder = $\sqrt[3]{3V/\pi}$ ft, V = total volume. **47.** $\sqrt{2197} = 46.87$ ft (beam touches building 26 feet above ground) **49.** Maximum value is $ka^2/4$ at $x = a/2$.

Article 3.6, pp. 220–222

1. $dA/dt = 2\pi r(dr/dt)$ **3.** $dV/dt = 3s^2(ds/dt)$ **5.** a) 1 b) $-\frac{1}{3}$ c) $dV/dt = R(dI/dt) + I(dR/dt)$ d) $dR/dt = \frac{3}{2}$, increasing **7.** a) D = distance from player to third base $= 10\sqrt{117}$ ft, S = distance from player to second base = 60 ft b) $D = [90 + S^2]^{1/2}$ c) $dD/dt = S[90 + S^2]^{-1/2}(dS/dt)$ d) $-96/\sqrt{117} = -8.88$ ft/sec **9.** 119/3 ft^2/s, $t = 13\sqrt{2}/4$ s **11.** For all spheres $dV/dt = 4\pi r^2(dr/dt)$. If $dV/dt = Sk = 4\pi r^2 k$, then $dr/dt = k$. (S is surface area, k is a constant.) **13.** $(25/9\pi)$ ft/min **15.** Increasing 33.7 ft/s **17.** ± 3 m/s, $y' = \pm 2$ **19.** 8 ft/s, decreasing 3 ft/s **21.** 1500 ft/s **23.** Thickness decreasing at $5/72\pi$ in/min. Area decreasing at 10/3 in^2/min. **25.** At about 53.67 ft/hr. **27.** $260/\sqrt{37}$ mi/hr.

Article 3.7, p. 224

1. $f(-2) = 11$, $f(-1) = -1$, $f'(x) = 4x^3 + 3 < 0$ on $(-2, -1)$ **3.** $f(-1) = 1$, $f(0) = -6$, $f'(x) = 6x^2 - 6x - 12 < 0$ on $(-1, 0)$ **5.** a) i) $y = 0$ at $-2, 2$, $y' = 0$ at 0; ii) $y = 0$ at $-3, -5$, $y' = 0$ at -4; iii) $y = 0$ at $-1, 2, 2$, $y' = 0$ at 0, 2; iv) $y = 0$ at 0, 9, 24, $y' = 0$ at 4, 18. If $y = 0$ at x_1 and x_2, then $y' = 0$ somewhere between x_1 and x_2.

Article 3.8, pp. 230–231

1. $\frac{1}{2}$ **3.** $\frac{8}{27}$ **5.** $\frac{3}{2}$ **9.** Yes; all values on (0, 1) **11.** No contradiction. **17.** These are the only ones.

Article 3.9, pp. 236–237

1. $\frac{1}{4}$ **3.** $\frac{5}{7}$ **5.** 0 **7.** 5 **9.** 1 **11.** $\sqrt{2}$ **13.** +1 **15.** -1 **17.** $\frac{1}{2}$ **19.** 3 **21.** na **23.** $-\frac{1}{2}$ **25.** $\sqrt{10}$ **27.** a) 0 b) 1 c) ∞ **29.** -1 **31.** $(-1 + \sqrt{37})/3 \approx 1.694$

Article 3.10, p. 244

1. x^2 **3.** $\frac{1}{2}(1 - x)$ **5.** $\frac{1}{8}(3 + 6x - x^2)$ **7.** a) $1 + \frac{1}{2}x - \frac{1}{8}x^2$ b) $(0.9)^{-5/2}/16{,}000 = 8.13 \times 10^{-5}$ **9.** a) $1 - x^2/2$ b) $(0.001)/6 = 1.67 \times 10^{-4}$ **11.** a) $\pi - x$ b) $(0.001)/6 = 1.67 \times 10^{-4}$ **13.** a) $-1 + (x - \pi)^2/2$ b) $(0.001)/6 = 1.67 \times 10^{-4}$

Miscellaneous Problems Chapter 3, pp. 245–249

1. a) $x < \frac{9}{2}$ b) $x > \frac{9}{2}$ c) No values d) All values **3.** a) $x < 3$ b) $x > 3$ c) $0 < x < 2$ d) $x < 0$, $x > 2$ **5.** $x > \sqrt[3]{2}$ b) $x < \sqrt[3]{2}$ c) $x > 0$, $x < -\sqrt[3]{4}$ d) $-\sqrt[3]{4} < x < 0$ **7.** a) $x < 0$ b) $x > 0$ c) $x \neq 0$ d) Never **9.** a) $x \neq 0$ b) Never c) $x < 0$ d) $x > 0$ **11.** a) $x < -2b/a$, $x > 0$ b) $-2b/a < x < 0$ c) $x > -b/a$ d) $x < -b/a$ **13.** a) $x < -1$, $x > \frac{1}{3}$ b) $-1 < x < \frac{1}{3}$ c) $x > -\frac{1}{3}$ d) $x < -\frac{1}{3}$ **15.** a) $x < 1$, $x > 2$ b) $1 < x < 2$ c) $x > \frac{3}{2}$ d) $x < \frac{3}{2}$ **17.** a) $0 < x \leq 1$, $y < 0$ b) $0 < x \leq 1$, $y > 0$ c) $0 < x \leq 1$, $y > 0$ d) $0 < x \leq 1$, $y < 0$ **19.** a) A. Symmetric about x-axis; B. $0 \leq x \leq 4$, $-2 \leq y \leq 2$; C. (0, 0), (4, 0); D. None; E. ∞ at both intercepts b) A. Symmetric about x-axis; B. $x \leq 0$, $x \geq 4$, all y; C. (0, 0), (4, 0); D. $y = \pm(x - 2)$; E. ∞ at both intercepts c) A. Symmetric about x-axis; B. $0 \leq x < 4$, all y; C. (0, 0); D. $x = 4$; E. ∞ at (0, 0) **21.** a) A. None; B. All y, all x; C. (0, 0), $(-1, 0)$, (2, 0); D. None; E. -2 at (0, 0), 3 at $(-1, 0)$, 6 at (2, 0) b) A. Symmetric about x-axis; B. All y, $-1 \leq x \leq 0$, $x \geq 2$; C. (0, 0), $(-1, 0)$, (2, 0); D. None; E. ∞ at all intercepts **23.** a) A. Symmetric about the origin; B. $|y| \geq 2$, $x \neq 0$; C. None; D. $y = x$, $x = 0$; E. No intercepts b) A. About (1, 2); B. $y \leq 0$, $y \geq 4$, $x \neq 1$; C. (0, 0); D. $x = 1$, $y = x + 1$; E. 0 *Remark.* Translate axes to new origin at (1, 2); 5(b) becomes same as 5(a). **25.** A. None; B. $y \leq 4 - 2\sqrt{3}$, $y \geq 4 + 2\sqrt{3}$, $x \neq \pm 1$; C. (0, 8), (2, 0); D. $x = \pm 1$, $y = 0$; E. -4 at (0, 8), $\frac{4}{3}$ at (2, 0) **27.** A. Symmetric about the origin; B. $|x| \leq 2$, $|y| \leq 2$; C. $(0, \pm\sqrt{3})$, $(\pm\sqrt{3}, 0)$; D. None; E. $-\frac{1}{2}$ at $(0, \pm\sqrt{3})$ -2 at $(\pm\sqrt{3}, 0)$ **29.** a) $x = 1$; y' goes from + to −. b) $x = 3$; because y' goes from − to +. **31.** If $ds/dt = ks^{-1/2}$, then $d^2s/dt^2 = -\frac{1}{2}ks^{-3/2}\,ds/dt = (-k^2/2)s^{-2}$. ($k$ is a constant.) **33.** $(3/400\,\pi)$ ft/min **35.** a) Approx. 606 mi/hr b) Approx. 0.83 mi **37.** $-480\sqrt{2}/7$ **39.** $a = -gR^2/s^2$ **41.** $(\sqrt{3}/30)r$, increasing **43.** $(8/9\pi)$ ft/min **45.** 10 **47.** $a = 1$, $b = -3$, $c = -9$, $d = 5$ **49.** $r = 25$ ft **51.** $r = 4$ ft, $h = 4$ ft **53.** $4\sqrt{3}$ **55.** a) $(2, \pm\sqrt{3})$ b) (1, 0) c) (1, 0) **57.** $x = aw/\sqrt{w^2 - l^2}$, $y = b - (al/\sqrt{w^2 - l^2})$ **59.** Each $= (P - b)/2$ **71.** 6×18 ft **73.** $(h^{2/3} + w^{2/3})^{3/2}$ m **75.** $r = \sqrt{A}$, $\theta = 2$ rad **77.** a) A decreasing 0.04π cm^2/s b) $t = (5b - 3a)/(a^2 - b^2)$ **81.** $x = (1/n)\sum_{i=1}^{n} c_i$ **87.** -0.67 **89.** a) $x = (c - b)/2e$ b) $P = \frac{1}{2}(b + c)$ c) $-a + (c - b)^2/4e$ d) $P = \frac{1}{2}(b + t + c)$, it adds $\frac{1}{2}t$ to its previous price. **91.** a) 1 b) 2

CHAPTER 4

Article 4.2, p. 256

1. $y = x^2 - 7x + C$ **3.** $y = (x^3/3) + x + C$ **5.** $y = (5/x) + C$ **7.** $y = (x - 2)^5/5 + C$ **9.** $y = \frac{1}{2}x^2 - (1/x) + C$ **11.** $y^2 - x^2 = C$ **13.** $(y - 1)^2/2 = ((x + 1)^2/2) + C$, or $y = 1 + \sqrt{x^2 + 2x + D}$ **15.** $y^{1/2} = (x^{3/2}/3) + C$, or $y = ((x^{3/2}/3) + C)^2$ **17.** $y = -1/(x^2 + C)$ **19.** $y = (1/x^2) + C$ **21.** $s = t^3 + 2t^2 - 6t + C$

23. $-1/u = 2v^4 - (4/v^2) + C$ or $u = -v^2/[2(v^6 + Cv^2 - 2)]$ **25.** $y = \frac{4}{3}t^3 + 4t - (1/t) + C$ **27.** $x^2 + 3x + C$ **29.** $((x - 1)^{244}/244) + C$ **31.** $(-2(1 - x)^{3/2}/3) + C$ **33.** $-\frac{3}{5}(2 - t)^{5/3} + C$ **35.** $x + (x^4/2) + (x^7/7) + C$ **37.** $\frac{2}{3}(1 + x^3)^{3/2} + C$ **39.** $\frac{2}{15}(2 + 5y)^{3/2} + C$ **41.** $-3\sqrt{1 - r^2} + C$ **43.** $\frac{1}{2}(1 + 2t^3)^{1/3} + C$ **45.** $\frac{2}{3}x^{3/2} + 2x^{1/2} + C$ **47.** 48 m/s

Article 4.3, pp. 259–260

1. $s = t^3 + s_0$ **3.** $s = ((t + 1)^3/3) + s_0 - \frac{1}{3}$ **5.** $s = -1/(t + 1) + s_0 + 1$ **7.** $v(t) = v_0 + gt,\ s(t) = s_0 + v_0t + \frac{1}{2}gt^2$ **9.** $v(t) = \frac{3}{8}(2t + 1)^{4/3} + v_0 - \frac{3}{8},\quad s(t) = \frac{9}{112}(2t + 1)^{7/3} + v_0t - \frac{3}{8}t + s_0 - \frac{9}{112}$ **11.** $v(t) = (t^5/5) + (2t^3/3) + t + v_0$, $s(t) = (t^6/30) + (t^4/6) + (t^2/2) + v_0t + s_0$ **13.** $y = (x - 7)^4 + 9$ **15.** $y = x - (1/x) + 1$ **17.** $y = -1/(x^2 - 2)$ **19.** $y = (x^3/3) + 2x - (1/x) - \frac{1}{3}$ **21.** $y = \frac{1}{3}(1 + x^2)^{3/2} - \frac{10}{3}$ **23.** $y = x^2 - x^3 + 4x + 1$ **25.** $y = x^3/16$ **27.** $f(x) = 2x^{3/2} - 50$ **29.** 24.2 m/s

Article 4.4, pp. 265–266

1. $-\frac{1}{3}\cos 3x + C$ **3.** $\tan(x + 2) + C$ **5.** $-\csc(x + \pi/2) + C$ **7.** $-\frac{1}{4}\cos(2x^2) + C$ **9.** $-\frac{1}{2}\cos 2t + C$ **11.** $\frac{4}{3}\sin 3y + C$ **13.** $(\sin^3 x)/3 + C$ **15.** $\frac{1}{2}\tan 2\theta + C$ **17.** $2\sec(x/2) + C$ **19.** $-\cot\theta\, d\theta$ **21.** $(y/2) + (\sin 2y)/4 + C$ **23.** $(\sin 3t)/3 - (\sin^3 3t)/9 + C$ **25.** $-\csc x + C$ **27.** $\sqrt{2 - \cos 2t} + C$ **29.** $(-\cos^3(2x/3))/2 + C$ **31.** $\tan\theta + \sec\theta + C$ **33.** $\tan y - \cot y + C$ **35.** $-\frac{3}{2}\cos 2x + \frac{4}{3}\sin 3x + C$ **37.** (d) **39.** $y = \sqrt{(5x^2/2) + 3\cos x - 3}$ **41.** $y = \cos x + x^2 + 2x + 2$ **43.** 10 m

Article 4.5, pp. 275–276

1. a) $\frac{7}{4}$ b) $\frac{9}{4}$

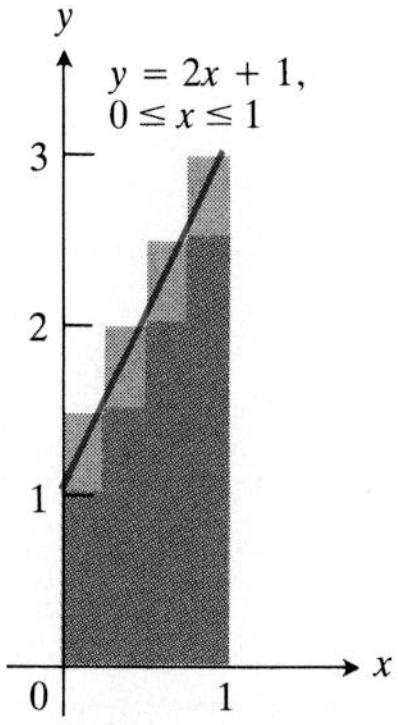

3. a) $\pi/(2\sqrt{2})$ b) $(\pi/2)(1 + (1/\sqrt{2}))$ **5.** a) $1 + \sqrt{2} + \sqrt{3}$ b) $3 + \sqrt{2} + \sqrt{3}$

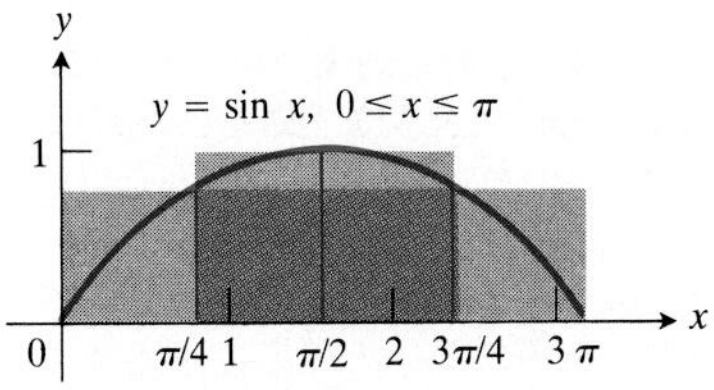

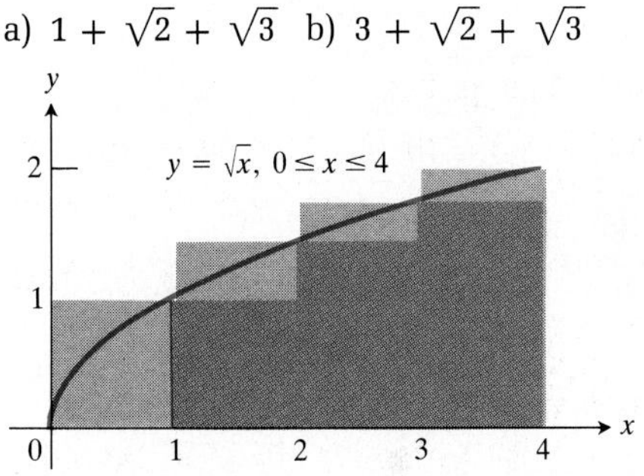

7. $\frac{1}{2} + 1 + 2 + 4 + 8$ **9.** $\frac{5}{2}$ **11.** 1 **13.** 34 **15.** All of them **17.** a) -8 b) 6 c) -5 d) 0

Article 4.6, pp. 280–281

3. $(m/2)(b^2 - a^2)$ **9.** Area under $y = 1 + x^2$ over the interval $0 \le x \le 1$, $\frac{4}{3}$ **11.** $\frac{1}{3}$ **13.** b) 1

Article 4.7, pp. 292–294

1. 12 **3.** $\frac{26}{3}$ **5.** $\frac{1}{15}$ **7.** 10 **9.** 1 **11.** $\frac{1}{2}$ **13.** 2 **15.** $\frac{4}{3}$ **17.** $\pi/6$ **19.** 8 **21.** $\frac{8}{3}$ **23.** 2 **25.** $\sqrt{2} - 1$ **27.** $\pi/2$ **29.** $\frac{2}{9}$ **31.** $\frac{38}{15}$ **33.** 0 **35.** $\frac{3}{2}$ **37.** $\sqrt{3}/8$ **39.** 16 **41.** $\frac{35}{4}$ **43.** a) π b) 3π **45.** $1/x$ **47.** $1/(1 + x^2)$ **49.** $2x/(1 + \sqrt{1 - x^2})$ **51.** $2/x$ **53.** $(\sin 4x)/\sqrt{x} - (\sin x)/2\sqrt{x}$ **55.** $\frac{1}{3}$ **59.** a) 1 b) $\frac{1}{4}$ c) $\sqrt[3]{12}$ **61.** b) 0 **63.** $\frac{1}{6}$

Article 4.8, pp. 302–303

1. $\frac{2}{3}(x^2 - 1)^{3/2} + C$ **3.** $\frac{52}{9}$ **5.** 2 **7.** $((x^2 + 1)^{11}/22) + C$ **9.** $\sqrt{1 + t} + C$ **11.** $(-3/4(1 + x^{4/3})) + C$ **13.** $\frac{1}{4}$ **15.** $2\sqrt{3}$ **17.** $(-\cos^3 2x)/6 + C$ **19.** $\sqrt{2} - 1$ **21.** $\frac{1}{3}(1 + x^4)^{3/4} + C$ **23.** 0 **25.** $-1/(x + 2) + C$ **27.** $a^3/3$ **29.** $-2/(1 + \sqrt{x}) + C$ **31.** $dy = (3x^2 - 6x + 5)\,dx$ **33.** $dy = -[(2xy + y^2)/(x^2 + 2xy)]\,dx$ **35.** $dy = [(1 - 2x^2)/(\sqrt{1 - x^2})]\,dx$ **37.** $dy = [(3(1 - x)^2(2x - 1)/(2 - 3x)^2)]\,dx$ **39.** $y = 5\ln(\sin x) + 10$ **41.** $y = \sqrt{(\sin x + 2)^2 - 1}$ **43.** Yes, $\sec^2 x = \tan^2 x + 1$. **45.** Let $x = 1 - t$. **47.** a) -2 b) -2 **51.** a) -3 b) 3

Article 4.9, pp. 310–312

1. a) 2 b) 2 c) 2 d) 1 (smallest possible for trapezoidal method) e) 2 (smallest possible for Simpson's method) **3.** a) $\frac{17}{4}$ b) 4 c) 4 d) 895 e) 2 **5.** a) ≈ 4.65509 b) ≈ 4.66622 c) $\frac{14}{3} \approx 4.66667$ d) 238 e) 20 **7.** a) $\frac{1}{4800} \le |E_T| \le \frac{1}{600}$ b) $4.16 \times 10^{-7} \le |E_S| \le 1.3 \times 10^{-5}$ **9.** a) Concave upward: $f''(x) > 0$, trapezoidal estimate too high. b) Concave downward: $f''(x) < 0$, trapezoidal estimate too low. **11.** $|E_T| \le \frac{1}{96}$, $|E_S| \le \frac{1}{1920}$ **13.** a) 40.25 lumen-milliseconds b) 55.50 lumen-milliseconds **15.** a) ≈ 0.8413 b) $|E_S| \le (\frac{2}{3}) \times 10^{-5} = 6.66667 \times 10^{-6}$

Miscellaneous Problems Chapter 4, pp. 313–314

1. $y = 1/(c - (x^2/2))$ **3.** $y^3 + 3y = x^3 - 3x + C$ **5.** $(2 + x)^{-1} = -(3 - y)^{-1} + C$ **7.** Yes; $y = x$ is the curve. **9.** $(2\sqrt{2}/3)b^{3/4}$ **11.** $v = t^2 + 3t + 4$, $\Delta s = 61\frac{1}{3}$ **13.** $h(10) = h(0) = 5$ **15.** a) $k = 38.72$ ft/s^2 b) 25 ft **17.** $h = 96t - 16t^2$, 144 ft **19.** a) ≈ 1.76 b) ≈ 1.73 **21.** $(x^2/2) - (1/x) + C$ **23.** $-\frac{1}{8}(1 + t^{4/3})^{-6} + C$ **25.** $-\frac{3}{5}(7 - 5r)^{1/3} + C$ **27.** $\frac{1}{9}\sin^3 3x + C$ **29.** $\frac{1}{2}\sin(2x - 1) + C$ **31.** $-\sqrt{2/t} + C$ **33.** $\frac{1}{3}(2 - 3x)^{-1} + C$ **35.** $r\sqrt{2}/2$ ft **37.** $|a|$ **39.** 480π mi/hr **41.** $f(x) = x/\sqrt{x^2 + 1}$ **43.** a) $\frac{1}{16}$ b) $\frac{2}{3}$ c) $2/\pi$ d) 0 e) ∞ **45.** $d^2y/dx^2 = 4y$

CHAPTER 5

Article 5.2, pp. 318–319

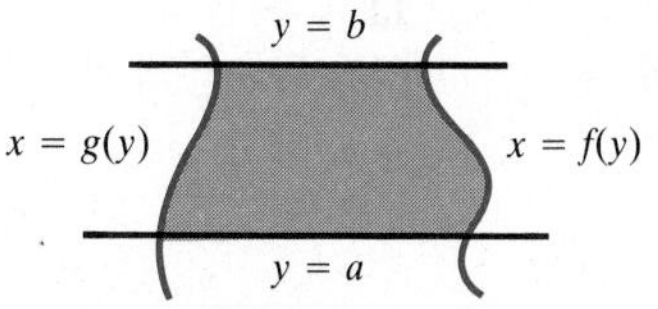

1. a) $\lim_{\Delta y \to 0} \Sigma_a^b (f(y) - g(y))\,\Delta y$ b) $\int_a^b (f(y) - g(y))\,dy$ **3.** $\frac{1}{12}$ **5.** $\frac{32}{3}$ **7.** $\frac{4}{3}$ **9.** $\frac{9}{2}$ **11.** $\frac{1}{12}$ **13.** $(4/\pi) - 1$ **15.** $\frac{9}{2}$ **17.** 4 **19.** $\frac{1}{6}$ **21.** $\sqrt{2} - 1$ **23.** $\frac{5}{6}$ **25.** $\frac{32}{3}$ **27.** $\frac{1}{2}$ **29.** $\frac{3}{4}$ **31.** $A \approx 3.141591$ (actual area $= \pi$)

Article 5.3, pp. 323–324

1. a) $0 \le t \le 2$ b) None c) 6 m d) 6 m **3.** a) $2 \le t \le 3$ b) $1 \le t \le 2$ c) $-\frac{8}{3}$ m d) $\frac{11}{3}$ m **5.** a) $0 \le t \le (\pi/3)$ b) $(\pi/3) \le t \le (\pi/2)$ c) 2 m d) 6 m **7.** a) $0 \le t \le (3\pi/4)$ b) $(3\pi/4) \le t \le \pi$ c) -2 d) $2\sqrt{2}$ **9.** $6 - \sin 2$ **11.** 2g **3.** $3\frac{1}{6}$ **15.** a) $s = \cos t$ meters b) 4 meters c) 0 meters

Article 5.4, pp. 329–330

1. $8\pi/3$ **3.** $\pi/30$ **5.** $16\pi/15$ **7.** $\pi/9$ **9.** $\pi(1 + \sqrt{3})$ **11.** $32\pi/3$ **13.** $16\pi/15$ **15.** $\pi/2$ **17.** $\pi\sqrt{3} - (\pi^2/3)$ **19.** $512\pi/15$ **21.** a) $\pi h^2[a - (h/3)]$ b) $0.1/12\pi$ ft/s $= 6/\pi$ in/min **23.** $512\pi/15$ **25.** $8a^3/3$ **27.** $\pi\sqrt{3}/16$

Article 5.5, pp. 336–337

1. $8\pi/3$ **3.** $56\pi/15$ **5.** $256\pi/5$ **7.** $117\pi/5$ **9.** $\pi/3$ **11.** a) $5\pi/3$ b) $4\pi/3$ **13.** $11\pi/15$ **15.** $512\pi/21$ **17.** $\pi/6$ **19.** $\pi/2$ **21.** $32\pi/3$ **23.** $2\pi^2$ **25.** $2\pi^2a^2b$

Article 5.6, pp. 342–343

1. 12 **3.** $\frac{14}{3}$ **5.** $\frac{123}{32}$ **7.** $\sqrt{1 + m^2}$ **9.** 4 **11.** 8 **13.** $\frac{21}{2}$ **15.** $S_{10} = 21.06792$ in.

Article 5.7, pp. 348–349

1. $(\pi/27)(10\sqrt{10} - 1)$ **3.** $1823\pi/18$ **5.** $(2\pi/3)(2\sqrt{2} - 1)$ **7.** $99\pi/2$ **9.** a) $4\pi^2$; identical with the result in Example 4 with $r = 1$. b)

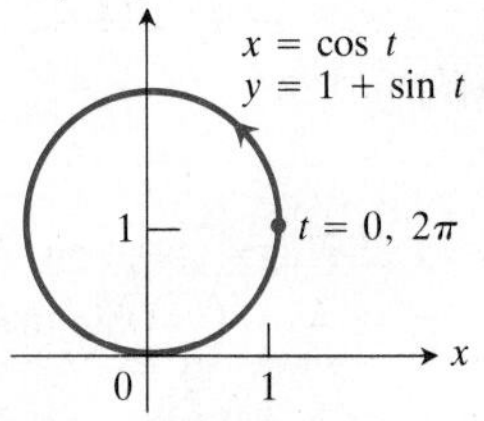

11. $12\pi a^2/5$ **13.** $56\pi\sqrt{3}/5$

Article 5.8, pp. 353–354

1. a) $2/\pi$ b) 0 **3.** $\frac{49}{12}$ **5.** $\alpha((a + b)/2) + \beta$ **9.** 600 **11.** $I_{av} = 200$; 1 dollar/day **13.** a) Average $= 67.25/60.00 \approx 1.12$ million lumens b) $t \approx 87.91$ seconds **17.** $4a/\pi$ **19.** $2a^2$

Article 5.9, p. 362

1. $(4a/3\pi, 4a/3\pi)$ **3.** $(2a/(3(4 - \pi)), 2a/(3(4 - \pi)))$ **5.** $(\frac{2}{5}, 1)$ **7.** $\frac{3}{5}h$ **9.** $5L/6$, or $L/6$ units from the heavy end. **11.** On the axis, $(3h/5)$ units from the vertex **13.** $(0, \pi r/4)$

Article 5.10, p. 365

1. $(0, 2c^2/5)$ **3.** $(0, \frac{12}{5})$ **5.** $(\frac{3}{5}, 1)$ **7.** $(0, \frac{8}{3})$ **9.** On the axis, $(2h/3)$ from the vertex **11.** $(17\sqrt{17} - 1)/12$ **13.** 47.14 in.

Article 5.11, p. 368

1. $4\pi^2$ **3.** $(0, 2r/\pi)$ **5.** $[(3\pi + 4)/3]\pi r^3$ **7.** $\pi\sqrt{2}(\pi + 2)r^2$ **9.** $(4 + 3\pi)r^3/6$

Article 5.12, p. 372

1. 375 lb **3.** $1666\frac{2}{3}$ lb **5.** $41\frac{2}{3}$ lb **7.** 8,450 ft-tons **9.** a) $93\frac{1}{3}$ lb b) To a depth of 3 ft only

Article 5.13, pp. 376–378

1. $c = 5$ lb/in; 80 inch-lb **3.** a) 1250 inch-lb b) 3750 inch-lb **5.** $18\frac{3}{4}$ inch-lb; 300 lb **7.** $\frac{1}{2}k$, where k is the proportionality constant. **9.** $\frac{1}{2}mgR$ **13.** $(200\pi/3)$ ft-tons **15.** Using 1 gal = 8 lb, we get 22,800,000 ft-lb = 11,400 ft-tons **17.** 30 hours 47 minutes

Miscellaneous Problems Chapter 5, pp. 378–380

1. $\frac{9}{2}$ **3.** 1 **5.** $\frac{9}{2}$ **7.** 18 **9.** $a^2/6$ **11.** 8 **13.** $\frac{13}{3}$ **15.** $\frac{64}{3}$ **17.** a) $\frac{3}{7}$ b) $\frac{3}{8}$ c) $y = 3/(2\sqrt{1+3x})$ d) $\frac{3}{7}$ **19.** a) $243\pi/5$ b) $189\pi/2$ c) $81\pi/2$ d) $512\pi/15$ e) $128\pi/3$ **21.** $32\pi/3$ **23.** $V = 2\pi\int_0^2 [x^2/(\sqrt{x^3+8})]\,dx = (8\pi/3)(2-\sqrt{2})$ **25.** $112\pi a^3/15$ **27.** On the axis, $3h/4$ units from the vertex **29.** $28\pi/3$ **31.** $2\sqrt{3}a^3$ **33.** $424\pi/15$ **35.** $153\pi/40$ **37.** a) $2b^2/3$ b) $2b/3$ **39.** a) $\bar{v} = 72$ b) $82\frac{2}{3}$ **41.** $(\frac{9}{10}, \frac{9}{5})$ **43.** $(1, \frac{12}{5})$ **47.** a) $\bar{x} = \bar{y} = 4(a^2+ab+b^2)/(3\pi(a+b))$ b) $(2a/\pi, 2a/\pi)$ **49.** 5 in., $7\frac{1}{9}$ in., $4\frac{8}{9}$ in. **51.** $5 \times 10^6/3$ lb **53.** $504\,d_1 + 72\,d_2$ **55.** $k(b-a)/ab$ **57.** $(320\,w/3)(15\pi + 8)$ ft-lb

CHAPTER 6

Article 6.2, pp. 386–388

1. $\sqrt{3}/2$, $\sqrt{3}/3$, $2\sqrt{3}/3$, 2 **3.** $\sqrt{2}/2$ **5.** 2 **7.** $2\sqrt{3}/3$ **9.** $\sqrt{2}/2$ **11.** 0 **13.** 0 **15.** π **17.** $-\pi/3$ **19.** 0.6 **21.** $\pi/3$ **23.** $+30°$ **25.** π **27.** $-\pi/2$ **29.** $\pi/2$ **31.** 0 **37.** $\pi/4$ **39.** b) 61°

Article 6.3, p. 393

5. $-2x/\sqrt{1-x^4}$ **7.** $1/2\sqrt{x}(1+x)$ **9.** $15/(1+9x^2)$ **11.** $1/\sqrt{4-x^2}$ **13.** $1/(|x|\sqrt{25x^2-1})$ **15.** $-2x/(x^2+1)\sqrt{x^4+2x^2}$ or $-2/[(x^2+1)\sqrt{x^2+2}]$ **17.** $-1/(2(x+1)\sqrt{x})$, $x > 0$ **19.** $2\sqrt{1-x^2}$ **21.** $4/(4+x^2)$ **23.** $1/(x^2+1)$ **25.** $(\sin^{-1}x)^2$ **27.** $\pi/6$ **29.** $\pi/12$ **31.** $\frac{1}{2}\sin^{-1}2x + C$ **33.** $\pi/8$ **35.** π **37.** $2\sin^{-1}\sqrt{x} + C$ **39.** $\sin^{-1}(x-1)^2 + C$ **41.** 2 **43.** $\frac{1}{6}$ **45.** $\pi/2$ **47.** $\sqrt{b(a+b)}$ ft **49.** a) 0 b) $\pi/2$ **51.** Yes; $\sin^{-1}x + C_1 = -\cos^{-1}x + C_2$ or $\sin^{-1}x + \cos^{-1}x = C_2 - C_1 = \pi/2$ **53.** $(\pi^2/4) - (\pi/6)$

Article 6.4, pp. 399–400

1. $1/x$ **3.** $1/x$ **5.** $-1/x$ **7.** $3(\ln x)^2/x$ **9.** $\sec x$ **11.** $3x^2\ln 2x + x^2$ **13.** $1/(x + x(\ln x)^2)$ **15.** $2x(1+\ln x^2)$ **17.** $-\tan^{-1}(x/2)$ **19.** $3(\ln x)^2 + (\ln x)^3$ **21.** $\sec^{-1}x$ **23.** $\ln|x| + C$ **25.** $\frac{1}{2}\ln|x| + C$ **27.** $\ln 2$ **29.** $\frac{1}{2}\ln|2x+3| + C$ **31.** $\frac{1}{8}\ln(4x^2+1) + C$ **33.** $\ln|\sin x| + C$ **35.** $x - \ln|x+1| + C$ **37.** $-\frac{1}{2}\ln|1-x^2| + C$ **39.** $\frac{1}{3}(\ln x)^3 + C$ **41.** $-\ln|\cos x| + C$ **43.** $\frac{1}{2}(\ln x)^2 + C$ **45.** $\frac{1}{2}[x^2\ln x - (x^2/2)] + C$ **47.** 0 **49.** 0 **51.** $\pi\ln 2$ **53.** $y + \ln y = x + \ln x$ **55.** b) $\ln 1.16 \approx 0.16$ (linear), 0.1472 (quad); $\ln 0.84 \approx -0.16$ (linear), -0.1728 (quad) c) $\ln 1.16 \approx 0.14842$, $\ln 0.84 \approx -0.17435$ d) $\ln 1.16 = 0.148420$; $\ln 0.84 = -0.174353$ **57.** f) $0.18226 < \ln 1.2 < 0.18267$

Article 6.5, pp. 405–406

1. $4a$ **3.** $\frac{3}{2}a$ **5.** $2(a-b)$ **7.** $2b - 3a$ **9.** $2b - a$ **11.** $x/(x^2+5)$ **13.** $-(3x+2)/(2x^2+2x)$ **15.** $\cot x + 2\cot 2x$ **17.** $(3x+4)/(2x^2+4x)$ **19.** $1/(x^4+x)$ **21.** $10x/(x^2+1) + 1/(2-2x)$ **23.** $(2x+1)/(2y)$ **25.** $y[1/(2(x+3)) + \cot x - \tan x]$ **27.** $(y/3)[(2x-2)/(x^2-2x) - 2x/(x^2+1)]$ **29.** $(y/3)[1/x + 1/(x+1) + 1/(x-2) - 2x/(x^2+1) - 2/(2x+3)]$ **31.** $2y[5/x + 1/(1+x^2)\tan^{-1}x + 2/(3-2x) - 1/3x]$ **33.** $\ln 2$ **35.** $\frac{1}{2}\ln 2$ **39.** $\ln 2$ **41.** a) $\ln 2$ b) $\pi(\sqrt{3} - \pi/3)$ **43.** $\bar{x} = (\ln 4 - \frac{3}{4})/(\ln 4 - 1) \simeq 1.647$, $\bar{y} = (1 + (\ln 2)^2 - \ln 4)/(\ln 4 - 1) \approx 0.2437$

45.

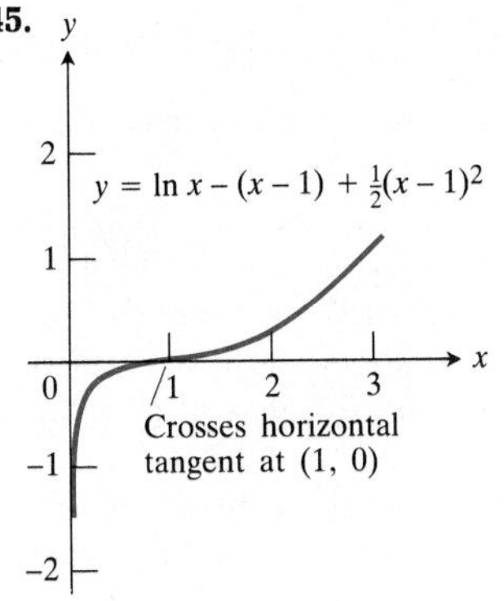

47. No inflection point; $y'' = 1/x$ is positive since x is positive.

Article 6.6, pp. 413–415

1. x **3.** $1/x^2$ **5.** $1/x$ **7.** $1/x$ **9.** $2x$ **11.** $x - x^2$ **13.** xe^x **15.** $4(\ln x)^2$ **17.** $(x+1)^2$ **19.** $2 + e^{-x}\sin x$ **21.** $3e^{3x}$ **23.** $-7e^{5-7x}$ **25.** $e^x(x^2+2x)$ **27.** $(\cos x)e^{\sin x}$ **29.** $-1 + (1/x)$ **31.** $1/(1+e^x)$ **33.** $\frac{1}{2}(e^x - e^{-x})$ **35.** $e^{\sin^{-1}x}/\sqrt{1-x^2}$ **37.** $27x^2e^{3x}$ **39.** $2e^{-x^2}(x - x^3)$ **41.** $e^x/(1+e^{2x})$ **43.** $-e^{1/x}/x^2$ **45.** $(\sin x)/x$ **47.** $((1/x) - e^{x+y})/(e^{x+y} - (1/y))$ **49.** $(e^x + (1/x))/(1 + (e^x + \ln x)^2) = (\cos^2 y)(e^x + (1/x))$ **51.** $\frac{1}{2}e^{x^2} + C$ **53.** $3e^{x/3} + C$ **55.** $2(e-1)$ **57.** $1 - (1/e)$ **59.** $\ln((1+e)/2)$ **61.** $\ln 2$ **63.** $\tan^{-1} 2 - (\pi/4)$ **65.** $\frac{1}{2}$ **67.** 1 **69.** 0 **71.** At $x = 0$. The limit is 1.

73. Max $(e, 0.36788)$; I $(4.48, 0.335)$

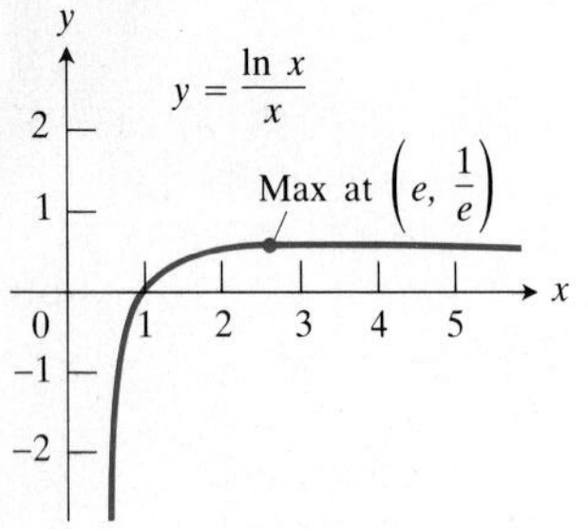

75. Abs. max at $(0, 1)$; abs. min at $(3\pi/4, -0.067)$; rel. max at $(7\pi/4, 0.003)$; rel. min at $(2\pi, 0.002)$; inflection at $(\pi, 0.043)$.

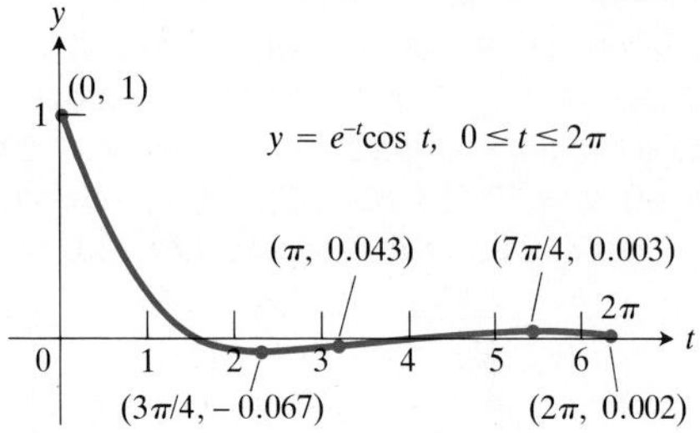

77. x_3 equals e to 9 significant digits. **79.** a) Minima at $(-\pi/2, 1/e)$, $(3\pi/2, 1/e)$, maximum at $(\pi/2, e)$. These are local and absolute maxima and minima. **81.** $1/(2e)$ **83.** $(\pi/2)(e^{-2} - 0.01) \approx 0.1969$ **85.** b) $y = 3e^{-2t}$ **87.** $y = -e^{-x} + e^{-4}$ **89.** $\ln(y+1) = \frac{1}{2}\ln x + \frac{1}{2}\ln 2$, or $y = \sqrt{2x} - 1$ **91.** a) $x = \ln(t+2)$, $y = t^2 + 1$ b) $y = (e^x - 2)^2 + 1$ c) $x = \ln(2 + \sqrt{y-1})$ d) $(y(2) - y(0))/(x(2) - x(0)) = 4/\ln 2 \approx 5.77$ e) 6 **93.** $2x/\sqrt{\pi}$

Article 6.7, pp. 422–423

1. $(\ln 2)2^x$ **3.** $(\ln 8)8^x$ **5.** $(\ln 9)9^x$ **7.** $y[(\sin x)/x + \cos x \ln x]$ **9.** $y(\sec x \tan x)\ln 2$ **11.** $y[\ln(\cos x) - x\tan x]$ **13.** $y[(1/x) + 2x\ln 2]$ **15.** $y[\ln\cos x)/2\sqrt{x} - \sqrt{x}\tan x]$ **17.** $3^{2x}/2\ln 3 + C$ **19.** $-(1/\ln 5)5^{-x} + C$ **21.** $1/\ln 4$ **23.** $\frac{3}{8}$ **25.** $(4\ln 4)^{-1} \approx 0.180$ **27.** 4 **29.** $(2\ln 4)^{-1} \approx 0.360$ **31.** a) 0 b) 2 c) $\ln 2/\ln(\frac{3}{2})$ d) $-2\ln 3/\ln 12$ **33.** a) If $f(x) = e^x$ and $g(x) = (\ln 3)x$, then $3^x = f \circ g(x)$. b) 27 **35.** 0 **37.** e **39.** 1 **43.** c) 0.485, 3.2124 **45.** a) $e^{1/e} \approx 1.445$ b) $e^{1/2e} \approx 1.202$ c) $e^{1/(ne)}$ **49.** 1, 2

Article 6.8, pp. 429–430

1. a) 2 b) $\frac{5}{3}$ c) -2 d) -2 **3.** 12 **5.** $1/(x\ln 2)$ **7.** $(2x\ln 5)^{-1}$ **9.** $-(x\ln 2)^{-1}$ **11.** $\ln 10$ **13.** $[x(\ln 2)(\ln x)]^{-1}$ **15.** $(1/x\ln 10)e^{\log_{10} x}$ **17.** a) 1.30103, 2.30103, 9.30103 – 10, 8.30103 – 10 b) 2.99574, 5.29833, 8.39056 – 10, 6.08797 – 10 **19.** $\ln 3/\ln 2$ **21.** Note: "Faster" means faster than $x^2 - 1$; likewise for "Slower." a) Same b) Faster c) Faster d) Slower e) Same f) Same g) Slower h) Slower i) Slower j) Faster **23.** The order is (b), (c), (a), (d).

Article 6.9, pp. 435–436

1. a) $10\ln(\frac{3}{14})/\ln(\frac{4}{7}) - 10 \approx 17.53$ min b) $10\ln(\frac{10}{21})/\ln(\frac{4}{7}) \approx 13.26$ min **3.** -3 °C **5.** $2^{48} \approx 281$ trillion **7.** a) $-(\ln 2)/5700$ b) $(5700\ln 10)/\ln 2 \approx 18{,}935$ years c) $-(5700\ln 0.445)/\ln 2 \approx 6{,}658$ years **9.** $-(140\ln 0.9)/\ln 2 \approx 21.2$ days **11.** $(\ln 0.1)/\ln 0.9 \approx 22$ years **13.** $(L\ln 2)/R$ **15.** a) $i = i_0e^{-(R/L)t}$, (i_0 is initial current.) b) $(L/R)\ln 2$ c) i_0/e

Article 6.10, pp. 439–440

1. $(\ln 2)/\ln 1.04 \approx 17.67$ years **5.** a) $A = A_0e^t$, where t is measured in years. The effective annual interest is 172% $(A_0e = A_0(1 + 1.72))$. b) $\ln 3 \approx 1.1$ years. c) $A_0(e - 1)$ if A_0 is the initial amount.

Miscellaneous Problems Chapter 6, pp. 440–443

1. $2\cos 2x$ **3.** $2(\cos x + \cos 2x)$ **5.** $-\csc y$ **7.** $-2\cos(x/2)$ **9.** $1 - \cos x$ **11.** $[(x\cos x - 2\sin x)/x^3]$ where $y \to \infty$ as $x \to 0$ and $y \to 0$ as $x \to \infty$. **13.** $2x\sin(1/x) - \cos(1/x)$ where $y \to 0$ as $x \to 0$ and $y \to \infty$ as $x \to \infty$. **15.** $4(e^x + e^{-x})^{-2}$ **17.** a) $1 + \ln x$ b) $\frac{1}{2}x^{-1/2}(2 + \ln x)$ **19.** $e^{-x}(2\cos 2x - \sin 2x)$ **21.** a) $1 - e^x$ b) $(1 - e^x)e^{(x - e^x)}$ **23.** $(1 - x^2)^{-1}$ **25.** $\tan^{-1}(x/2) + 2x/(x^2 + 4)$ **27.** $2\sec x$ **29.** $2x(1 - x^4)^{-1/2} - (2x^2 + 1)e^{x^2}$ **31.** $2x(2 - x)(x^2 + 2)^{1-x} - (x^2 + 2)^{2-x}\ln(x^2 + 2)$ **33.** $1/(x(x^2 + 1))$ **35.** $(6x + 4)/(3x^2 + 4x)$ **37.** $3(1 + \ln x)$ **39.** 1 **41.** e^x **43.** $2 + 1/x$ **45.** $y[3\sec^2(3x)\ln x + (\tan(3x))/x]$ **47.** $(\sec x)(\tan x)e^{\sec x}$ **49.** $[1 - (x - y)ye^{xy}]/[1 + (x - y)xe^{xy}]$

51. a) $(1-x^2)/((1+x^2)^2)$ b) $\frac{2}{3}(x^2+x-1)x^{-(2/3)}(x-2)^{-(2/3)}(x^2+1)^{-(4/3)}$ c) $y[(2/(2x-5))+(8x/(8x^2+1))-(6x^2/(x^3+2))]$ **53.** a) $a^{x^2-x}(2x-1)\ln a$ b) $(1+e^x)^{-1}$ c) $\ln x$ d) $x^{(1-2x)/x}(1-\ln x)$ **57.** $y''=4(e^x+e^{-x})^{-2}$, $y=\ln(e^x+e^{-x})+C$ **61.** $E=E_0e^{-t/40}$, $t=92.1$ sec **63.** $y=1-\frac{1}{2}(e^x+e^{-x})$ **65.** $1+4\ln\frac{3}{2}\approx 2.62$ **67.** $y=2e^x/(2-e^x)$ **71.** $\pi\sqrt{3}$ **73.** $\pi(e^4-1)/2\approx 84.2$ **75.** $y^2=4x$ **77.** $5\ln|x-3|+C$ **79.** $\frac{1}{6}$ **81.** $-\frac{1}{4}\ln|2x+1|+(x/2)+C$ **83.** $\frac{1}{2}(e^8-e^{-1})\approx 1490$ **85.** $\frac{1}{2}\ln 26\approx 1.63$ **87.** $\tan x+e^{\sin x}+C$ **89.** $-\frac{1}{4}$ **91.** e^2 **95.** 2 **97.** $2000\,\pi(\frac{9}{10})^t$ ft^3 **99.** a) No, x could be 4. b) Yes; consider the graph of $y=(\ln x)/x$ for $x>0$. **105.** $e-1$

CHAPTER 7

Article 7.1, pp. 450–451

1. $\frac{2}{3}(3x^2+5)^{3/2}+C$ **3.** $\frac{1}{2}$ **5.** $\frac{1}{4}$ **7.** $\tan(e^x)+C$ **9.** $\frac{1}{6}[3+4e^x]^{3/2}+C$ **11.** $\ln|\ln x|+C$ **13.** 2 **15.** $\frac{1}{3}(2x+3)^{3/2}+C$ **17.** $1/(14-4x)+C$ **19.** $-\ln|2+\cos x|+C$ **21.** $-\frac{1}{4}\sqrt{1-4x^2}+C$ **23.** $\pi/4$ **25.** $\frac{3}{2}\ln 2$ **27.** $(\pi/4)-\frac{1}{2}$ **29.** $\sin^{-1}(2x)+C$ **31.** $-1/(12(3x^2+4)^2)+C$ **33.** $\frac{2}{3}(x^3+5)^{1/2}+C$ **35.** $\frac{1}{2}e^{2x}+C$ **37.** $(-1/3e^{3x})+C$ **39.** $\tan^{-1}(e^x)+C$ **41.** $-\frac{1}{3}\cos^3 x+C$ **43.** $-\frac{1}{4}\cot^4 x+C$ **45.** $\frac{1}{2}\ln|e^{2x}-e^{-2x}|+C$ **47.** $-\frac{1}{4}(1+\cos\theta)^4+C$ **49.** $\sec^{-1}|2t|+C$ **51.** $\ln|\sin x|+C$ **53.** $\frac{1}{3}\sec^3 x+C$ **55.** $\frac{1}{3}e^{\tan 3x}+C$ **57.** $-\frac{1}{2}(\csc 2x+\cot 2x)+C$ **59.** $-\sqrt{1+\cot 2t}+C$ **61.** $\frac{1}{2}e^{\tan^{-1}2t}+C$ **63.** $3^x/\ln 3+C$ **65.** $\frac{1}{2}(\ln x)^2+C$ **67.** $2\ln(1+\sqrt{x})+C$ **69.** $\frac{1}{2}e^{x^2}+C$ **71.** a) $\pi/2$ b) $\pi/2$

Article 7.2, pp. 457–458

1. $-x\cos x+\sin x+C$ **3.** $-x^2\cos x+2x\sin x-2\cos x+C$ **5.** $((x^2\ln x)/2)-(x^2/4)+C$ **7.** $((x^{n+1})/(n+1))[\ln ax-1/(n+1)]+C$ **9.** $(x^3-3x^2+6x-6)e^x+C$ **11.** $(x^5-5x^4+20x^3-60x^2+120x-120)e^x+C$ **13.** $((x^2/\ln 2)-(2x/(\ln 2)^2)+(2/(\ln 2)^3))2^x+C$ **15.** $(12-3\pi^2)/16$ **17.** $(-x^5+20x^3-120x)\cos x+(5x^4-60x^2+120)\sin x+C$ **19.** $(x/2)-(\sin 2x)/4+C$ **21.** $x\sin^{-1}(ax)-(1/a)\sqrt{1-(ax)^2}+C$ **23.** $((x^2/a)-(2/a^3))\sin ax+(2x/a^2)\cos ax+C$ **25.** $2\sec^{-1}2-\sqrt{3}/2=(4\pi-3\sqrt{3})/6$ **27.** $(\ln x)^2/2+C$ **29.** $e^{2x}[\frac{2}{13}\cos 3x+\frac{3}{13}\sin 3x]+C$ **31.** $(a/(a^2+b^2))e^{ax}\sin bx-(b/(a^2+b^2))e^{ax}\cos bx+C$ **33.** $(x/2)[\cos(\ln x)+\sin(\ln x)]+C$ **35.** $\pi^2-2\pi$ **37.** $(\bar{x},\bar{y})=((e^2-3)/4(e-2),1/2(e-2))\approx(1.53,0.696)$ **39.** $\pi^2+\pi+4$ **41.** 2 **43.** $(x^4/32)[8(\ln x)^2-4\ln x+1]+C$ **45.** $(x^3-3x^2+6x-6)e^x+C$

Article 7.3, pp. 467–468

1. $\frac{1}{48}$ **3.** $-\frac{2}{3}(\cos x)^{3/2}+C$ **5.** $\ln|2-\cos\theta|+C$ **7.** $-\frac{1}{3}\ln|\cos 3x|+C$ **9.** $\sin x-(\sin^3 x)/3+C$ **11.** $\frac{1}{4}\tan 4\theta-\theta+C$ **13.** $\frac{8}{9}+\ln(\sqrt{3})$ **15.** a) $\frac{1}{5}\cos^5 x-\frac{1}{3}\cos^3 x+C$ b) $\cos x+\sec x+C$ **17.** $((\tan^{n+1}x)/(n+1))+C$ if $n\neq-1$; $\ln|\tan x|+C$ if $n=-1$ **19.** $((-\cos^{n+1}x)/(n+1))+C$ if $n\neq-1$; $-\ln|\cos x|+C$ if $n=-1$ **21.** $-\frac{1}{8}\cos^4 2x+C$ **23.** $\frac{4}{15}$ **25.** 0 **27.** $(-\cot^4 x/4)+C$ **29.** $\frac{1}{9}\tan^3 3x+\frac{1}{3}\tan 3x+C$ **31.** $\frac{1}{4}\tan^2 2x+\frac{1}{2}\ln|\cos 2x|+C$ **33.** $\frac{1}{3}\cos^3 x-\cos x+C$ **35.** $\ln|2+\tan x|+C$ **37.** $-\ln|\sin x|-\frac{1}{2}\cot^2 x+C$ **39.** $-\cot x-\frac{1}{3}\cot^3 x+C$ **41.** $\ln(1+\sqrt{2})$ **43.** $\sqrt{3}-1$ **45.** 0 **47.** π **49.** $-\frac{1}{4}\ln|\csc 4x+\cot 4x|+C$ **51.** $\ln(2+\sqrt{3})$ **53.** $(0,1/(\ln(3+2\sqrt{2}))$ **57.** a) $\int\cot^n ax\,dx=(-1/a(n-1))\cot^{n-1}ax-\int\cot^{n-2}ax\,dx$, $n\neq 1$. b) $-\frac{1}{9}\cot^3 3x+\frac{1}{3}\cot 3x-x+C$

Article 7.4, p. 472

1. $(x/8)-(\sin 4x)/32+C$ **3.** $(t/2)-(\sin 4t)/8+C$ **5.** $\frac{3}{8}x-(\sin 2ax)/4a+(\sin 4ax)/32a+C$ **7.** $\tan x+C$ **9.** $\ln|\sec z+\tan z|-\sin z+C$ **11.** $(1/\cos z)+\cos z+C$ **13.** $-\frac{1}{6}\csc^3 2t+C$ **15.** 4 **17.** $\frac{2}{5}$ **19.** $2\sqrt{2}(2\sin(\theta/2)-\theta\cos(\theta/2))$ **21.** $\ln(3+2\sqrt{2})=2\ln(\sqrt{2}+1)$ **23.** 2 **25.** $64\pi/3$ **27.** $2\sqrt{2}$

Article 7.5, pp. 479–480

1. $\pi/4$ **3.** $\frac{1}{2}\tan^{-1}2x+C$ **5.** $\sin^{-1}2x+C$ **7.** $\ln|y+\sqrt{25+y^2}|+C$ **9.** $\frac{1}{3}\ln|3y+\sqrt{25+9y^2}|+C$ **11.** $\ln|3z+\sqrt{9z^2-1}|+C$ **13.** $\frac{1}{4}\sec^{-1}|x/4|+C$ **15.** $\frac{1}{2}\sec^{-1}|2x|+C$ **17.** $2\ln 2$ **19.** $-\sqrt{1-x^2}+((1-x^2)^{3/2}/3)+C$ **21.** $\ln((1+\sqrt{5})/2)$ **23.** $(\sqrt{x^2-1}/x)+C$ **25.** $\frac{1}{2}\sin^{-1}2x+C$ **27.** $\sin^{-1}((x-1)/2)+C$ **29.** $\ln(1+\sqrt{2})$ **31.** $\sqrt{4+x^2}+C$ **33.** $(2-\sqrt{2})/3$ **35.** $-\sqrt{4-x^2}+\sin^{-1}(x/2)+C$ **37.** $\cos^{-1}((\cos\theta)/\sqrt{2})+C$ **39.** $(1/a)\ln|x/(a+\sqrt{a^2-x^2})|+C$ **41.** $(1/2a^3)(\tan^{-1}(x/a)+(ax)/(a^2+x^2))+C$ **43.** $((2x^2-1)/4)\sin^{-1}x+(x\sqrt{1-x^2}/4)+C$ **45.** $-\sqrt{1-x^2}/x+C$ **47.** $-\sqrt{16-y^2}+C$ **49.** $(-1/\sqrt{x^2-1})+C$ **51.** $3\pi/4$ **53.** $y=1/(\sqrt{2}-x+\sin^{-1}(x/2))$ **55.** a) $u/\sqrt{a^2+u^2}$ b) a/u

Article 7.6, p. 483

1. $\pi/8$ **3.** $\frac{1}{2}[\ln(x^2-2x+5)+\tan^{-1}((x-1)/2)]+C$ **5.** $\pi/6\sqrt{3}$ **7.** $\frac{1}{3}\ln|\sqrt{9x^2-6x+5}+3x-1|+C$
9. $\frac{1}{9}[\sqrt{9x^2-6x+5}+\ln|\sqrt{9x^2-6x+5}+3x-1|]+C$ **11.** $\ln|x+1+\sqrt{x^2+2x}|+C$
13. $\sqrt{x^2+4x+13}-2\ln|\sqrt{x^2+4x+13}+x+2|+C$ **15.** $\ln|x-1+\sqrt{x^2-2x-3}|+C$
17. $2\sin^{-1}(x-1)-\sqrt{2x-x^2}+C$ **19.** $2\sin^{-1}((x-2)/3)-\sqrt{5+4x-x^2}+C$ **21.** $\sqrt{8+2x-x^2}+C$
23. $\frac{1}{2}\ln(x^2+4x+5)-2\tan^{-1}(x+2)+C$ **25.** $\frac{1}{2}\sec^{-1}2=\pi/6$

Article 7.7, pp. 490–491

1. $3/(x-2)+2/(x-3)$ **3.** $1/(x+1)+(3/(x+1)^2)$ **5.** $(-2/x)-(1/x^2)+2/(x-1)$
7. $1-12/(x-2)+17/(x-3)$ **9.** $-\frac{1}{3}/(x^2+9)+(\frac{1}{3}/x^2)$ **11.** $\frac{1}{2}\ln 3$ **13.** $4-\ln 3$ **15.** $\ln 9$
17. $\ln|(x-1)^{5/7}(x+6)^{2/7}|+C$ **19.** $(\frac{9}{2})-6\ln 2$ **21.** $\ln|(x-1)^{1/6}(x+5)^{5/6}|+C$ **23.** $\ln|(x-1)^{1/3}(x+5)^{2/3}|+C$
25. $\ln 3-\ln 2-\frac{1}{4}$ **27.** $-\frac{3}{8}\ln|x|+\frac{1}{16}\ln|x+2|+\frac{5}{16}\ln|x-2|+C$ **29.** $5(\sqrt{3}-\pi/3)$
31. $\frac{1}{2}\ln(x^2/(x^2+x+1))-(1/\sqrt{3})\tan^{-1}((2x+1)/\sqrt{3})+C$ **33.** $4\ln|x|-\frac{1}{2}\ln(x^2+1)+\tan^{-1}x+C$
35. $(x^2/2)+\frac{9}{2}\ln|x+3|-\frac{3}{2}\ln|x+1|+C$ **37.** $\tan^{-1}x-(1/(x^2+1))+C$ **39.** $-\tan^{-1}x-(1/(x-1))+C$
41. $\ln 9-\ln 8$ **43.** $x-\frac{3}{2}\tan^{-1}x+(x/2(1+x^2))+C$ **45.** $-x+4\sqrt{x}-4\ln(1+\sqrt{x})+C$
47. $2\sqrt{x}-3\sqrt[3]{x}+6\sqrt[6]{x}-6\ln(\sqrt[6]{x}+1)+C$ **49.** $x\ln(x^2+1)-2x+2\tan^{-1}x+C$ **51.** a) $x=a^2kt/(1+akt)$
b) $x=b(1-e^{kt/(b-a)})/(1-(b/a)e^{kt/(b-a)})$

Article 7.8, p. 493

1. 2 **3.** $\tan x+\sec x+C$ **5.** $(4/\sqrt{3})\tan^{-1}(\sqrt{3}\tan(x/2))-x+C$
7. $(1/\sqrt{2})\ln|(\tan(x/2)+1-\sqrt{2})/(\tan(x/2)+1+\sqrt{2})|+C$ or $(1/\sqrt{2})\ln|\csc(x-\pi/4)-\cot(x-\pi/4)|+C$

Article 7.9, pp. 502–503

1. $\pi/2$ **3.** 6 **5.** 4 **7.** $\frac{1}{2}$ **9.** Diverges **11.** ln 2 **13.** Diverges **15.** Converges **17.** Converges **19.** Diverges
21. Converges **23.** Diverges **25.** Diverges **27.** Converges **29.** Converges **31.** Diverges **33.** Diverges
35. Converges **37.** Diverges **39.** Converges **41.** Converges **43.** π **45.** a) $p<1$ b) $p>1$ **47.** a) $\pi/2$ b) 2π
49. ln 2

Article 7.10, pp. 505–506

7. $8\cdot 6\cdot 4\cdot 2/9\cdot 7\cdot 5\cdot 3=\frac{128}{315}$ **9.** $(x^2/2)\cos^{-1}x+\frac{1}{4}\sin^{-1}x-\frac{1}{4}x\sqrt{1-x^2}+C$ **11.** $\frac{1}{3}\sec^2 x\tan x+\frac{2}{3}\tan x+C$
13. $(2/\sqrt{3})\tan^{-1}\sqrt{(x-3)/3}+C$ **15.** $(x/18(9-x^2))+\frac{1}{108}\ln|(x+3)/(x-3)|+C$ **17.** $-(\sqrt{7+x^2}/7x)+C$
19. $\sqrt{x^2-2}-\sqrt{2}\sec^{-1}|x/\sqrt{2}|+C$ **21.** $-\frac{1}{6}\ln|(5+4\sin 2x+3\cos 2x)/(4+5\sin 2x)|+C$
23. $\tan^2(x/2)-2\ln|\cos(x/2)|+C$ **25.** $[(ax+b)^{3/2}/a^2][(2(ax+b)/5)-(2b/3)]+C$ **27.** $3^{e^x}/e\ln 3$
29. $265e-720\approx 0.3447$

Miscellaneous Problems Chapter 7, pp. 506–509

1. $2\sqrt{1+\sin x}+C$ **3.** $(\tan^2 x)/2+C$ **5.** $\frac{2}{3}x^{3/2}+C$ **7.** $\ln|x+1+\sqrt{x^2+2x+2}|+C$ **9.** $(x^2-2x+2)e^x+C$
11. $\tan^{-1}(e^t)+C$ **13.** $2\tan^{-1}\sqrt{x}+C$ **15.** $\frac{9}{25}(t^{5/3}+1)^{5/3}+C$ **17.** $\ln[(\sqrt{1+e^t}-1)/(\sqrt{1+e^t}+1)]+C$
19. $e^{\sec x}+C$ **21.** $\sin^{-1}(x-1)+C$ **23.** $\frac{1}{2}\ln|1+\sin 2t|+C$ **25.** $(-2\cos x)/\sqrt{1+\sin x}+C$
27. $(x/a^2\sqrt{a^2-x^2})+C$ **29.** $\frac{1}{3}\ln[(1-\cos x)/(4-\cos x)]+C$ **31.** $\Sigma_{k=0}^{m}[(-1)^k/k!(m-k)!]\ln|x+k|+C$
33. $\frac{1}{3}[1/(2y^3+1)+\ln|2y^3/(2y^3+1)|]+C$ **35.** $\frac{1}{2}[\ln x^2-\ln(1+x^2)-(x^2/(1+x^2))]+C$ **37.** $\ln|e^x-1|-x+C$
39. $(1/x)+2\ln|1-(1/x)|+C$ **41.** $1/2(1-e^{2u})+C$ **43.** $\frac{1}{3}\tan^{-1}[(5x+4)/3]+C$ **45.** $e^x(\cos 2x+2\sin 2x)/5+C$
47. $\ln|x|-3\ln|1+\sqrt[3]{x}|+C$ **49.** $\frac{1}{15}(8-4z^2+3z^4)(1+z^2)^{1/2}+C$ **51.** $\frac{5}{2}(1+x^{4/5})^{1/2}+C$
53. $\frac{1}{4}[(2x^2-1)\sin^{-1}x+x\sqrt{1-x^2}]+C$ **55.** $x+\ln|(x-1)/x|+C$
57. $\frac{2}{3}(3e^{2x}-6e^x-1)^{1/2}+(1/\sqrt{3})\ln|e^x-1+\sqrt{e^{2x}-2e^x-\frac{1}{3}}|+C$ **59.** $\tan^{-1}(2\sqrt{y^2+y})+C$
61. $\frac{3}{8}\sin^{-1}x+\frac{1}{8}(5x-2x^3)\sqrt{1-x^2}+C$ **63.** $x\tan x-(x^2/2)+\ln|\cos x|+C$
65. $(x^2/4)+[(x\sin 2x)/4]+[(\cos 2x)/8]+C$ **67.** $(x^2/4)-[(x\sin 2x)/4]-[(\cos 2x)/8]+C$
69. $\frac{1}{12}[4u-3\ln(1+e^{2u})+\ln(3+e^{2u})]+C$ **71.** $(x^2+1)e^x+C$ **73.** $(-2/x^2)+(2/x)+\ln|x/(x+2)|+C$
75. $2-\sqrt{2}\ln(1+\sqrt{2})$ **77.** $-[1/2(e^{2u}+1)]+C$ **79.** $\ln|\tan t-2|-\ln|\tan t-1|+C$
81. $(1/\sqrt{2})\tan^{-1}((\tan x)/\sqrt{2})+C$ **83.** $x\ln\sqrt{x^2+1}-x+\tan^{-1}x+C$ **85.** $\frac{1}{2}(x^2-1)e^{x^2}+C$
87. $\sin^{-1}(\tan x/\sqrt{3})+C$ **89.** $\ln|2+\tan^{-1}x|+C$ **91.** $-\frac{1}{2}\cot x\csc x+\frac{1}{4}\ln[(1-\cos x)/(1+\cos x)]+C$
93. $x(\sin^{-1}x)^2-2x+2\sqrt{1-x^2}\sin^{-1}x+C$ **95.** $\frac{1}{2}\ln(x^2+1)+[1/2(x^2+1)]+C$ **97.** $\frac{1}{15}(3x-1)(2x+1)^{3/2}+C$
99. $x\ln(x-\sqrt{x^2-1})+\sqrt{x^2-1}+C$ **101.** $x-e^{-x}\tan^{-1}(e^x)-\frac{1}{2}\ln(1+e^{2x})+C$
103. $x\ln|x+\sqrt{x}|-x+\sqrt{x}-\ln(1+\sqrt{x})+C$ **105.** $x\ln(x^2+x)+\ln(x+1)-2x+C$
107. $2\sqrt{x}\sin\sqrt{x}+2\cos\sqrt{x}+C$ **109.** $(x+2)\tan^{-1}\sqrt{x+1}-\sqrt{x+1}+C$ **111.** $\frac{1}{4}x^2-\frac{1}{8}x\sin 4x-\frac{1}{32}\cos 4x+C$
113. $\ln(\sqrt{e^{2t}+1}-1)-t+C$ **115.** $(1/ac)\ln|e^{ct}/(a+be^{ct})|+K$ **117.** $-(1/x)-3\tan^{-1}(3x)+C$
119. $x\ln(2x^2+4)+2\sqrt{2}\tan^{-1}(x/\sqrt{2})-2x+C$ **121.** $\ln|2+\ln x|+C$

123. $\frac{1}{3}\ln|x+1| - \frac{1}{6}\ln(x^2 - x + 1) + (1/\sqrt{3})\tan^{-1}[(2x-1)/\sqrt{3}] + C$ **125.** $-\cos(2\sqrt{x}) + C$ **127.** $-2\sqrt{x+1}\cos\sqrt{x+1} + 2\sin\sqrt{x+1} + C$ **129.** $\ln 2$ **131.** $2/\pi$ **133.** $\pi/4$ **135.** $\pi/2$ **137.** -1 **139.** e^{-4} **141.** Area: $(\pi - 2)r^2/4$; centroid: $2r\sqrt{2}/(3\pi - 6)$ units from center **143.** $2\pi^2 kr^2$ **145.** Integrals for A, V, and S converge. $A = 1$, $V = \pi/2$, $S = \pi[\sqrt{2} + \ln(1 + \sqrt{2})]$ **147.** About 23 **149.** About 175 **151.** 1.9100989 **153.** 2π **155.** $12\pi a^2/5$ **157.** Converges; compare with $\int_1^\infty x^{-(3/2)}\,dx$ **161.** $x = a\cos kt$

CHAPTER 8

Article 8.2, p. 513

1. $x^2 = 8y + 16$ **3.** $x - y = 1$ **5.** $(x^2 + y^2)^2 + 8(y^2 - x^2) = 0$ **7.** $3x^2 + 4y^2 - 20x + 12 = 0$ **9.** $5x^2 - 4y^2 = 20$ $(x \geq 2)$ **11.** $x^2 + y^2 - 4x - 6y + 4 = 0$ **13.** $(2, 2)$, $\sqrt{5}$

Article 8.3, pp. 515–516

1. $x^2 + y^2 - 4y = 0$ **3.** $x^2 + y^2 - 6x + 8y = 0$ **5.** $x^2 + y^2 + 4x + 2y < 1$ **7.** $(0, 0)$, $r = 4$ **9.** $(0, 1)$, $r = 2$ **11.** $(-2, 0)$, $r = 4$ **13.** $(-1, -1)$, $r = 1$ **15.** $(-\frac{1}{4}, -\frac{1}{4})$, $r = \sqrt{2}/4$ **17.** $(-1, 2)$ **19.** $x^2 + y^2 - 4x - 4y = 5$ **21.** $(x - 0.7)^2 + (y - 1.3)^2 = 12.58$ **23.** $x^2 + y^2 + 2x + 2y - 23 = 0$ **25.** Outside; center is $(1, 2)$, radius is $\sqrt{2}$. Distance of $(0.1, 3.1)$ from center is $\sqrt{2.02}$

Article 8.4, pp. 521–522

1. $y = \frac{1}{8}x^2$; directrix is line $y = -2$ **3.** $x^2 + 16y = 0$; directrix is line $y = 4$ **5.** $(x + 2)^2 = 4(y - 3)$; directrix is line $y = 2$ **7.** $(y - 1)^2 = 12(x + 3)$; directrix is line $y = -6$ **9.** $y^2 = 8(x - 2)$; $F(4, 0)$ **11.** $(y - 1)^2 = -16(x + 3)$; $F(-7, 1)$ **13.** $(y - 1)^2 = 4x$; $F(1, 1)$ **15.** $V(0, 0)$; $F(2, 0)$; axis: $y = 0$; directrix: $x = -2$ **17.** $V(0, 0)$; $F(0, 25)$; axis: $x = 0$; directrix: $y = -25$ **19.** $V(1, 1)$; $x = 1$; $F(1, -1)$; $y = 3$ **21.** $V(2, 0)$; $y = 0$; $F(1, 0)$; $x = 3$ **23.** $V(-1, -1)$; $x = -1$; $F(-1, 0)$; $y = -2$ **25.** $V(2, 1)$; $y = 1$; $F(\frac{29}{16}, 1)$; $x = \frac{35}{16}$ **27.** $V(2, -2)$; $x = 2$; $F(2, -\frac{4}{3})$; $y = -\frac{8}{3}$ **29.** Points above parabola $x^2 = 8y$ **31.** $b^2 y = 4hx(b - x)$

Article 8.5, pp. 529–530

1. $(x^2/12) + (y^2/16) = 1$; $e = \frac{1}{2}$ **3.** $9x^2 + 5(y - 2)^2 = 45$; $e = \frac{2}{3}$ **5.** $(x - 2)^2 + 10(y - 2)^2 = 10$; $e = 3/\sqrt{10}$ **7.** $C = (7, 5)$; $V = (7, 5 \pm 5)$; $F = (7, 5 \pm (\sqrt{21}/5))$ **9.** $C = (-1, -4)$; $V = (-1, -4 \pm 5)$; $F = (-1, -4 \pm 4)$ **11.** $C = (3, 1)$; $V = (3, 1 \pm 5)$; $F = (3, 1 \pm \sqrt{21})$ **13.** $C = (-5, 0)$; $V = (-5 \pm 5, 0)$; $F = (-5 \pm 2\sqrt{6}, 0)$ **15.** $C = (2, -1)$; $V = (2 \pm 3, -1)$; $F = (2 \pm 2\sqrt{2}, -1)$ **17.** $C = (2, -2)$; $V = (2, -2 \pm 2)$; $F = (2, -2 \pm \sqrt{3})$ **19.** $C = (-1, 3)$; $V = (-1 \pm 4, 3)$; $F = (-1 \pm \sqrt{7}, 3)$ **23.** $e = \frac{3}{4}$; $y = \pm\frac{16}{3}$ **25.** $\frac{4}{3}\pi ab^2$ **29.** $5x^2 + 9y^2 = 180$ **33.** Graph consists of parabola $x^2 + 4y = 0$, straight line $2x - y - 3 = 0$, circle $x^2 + y^2 - 25 = 0$, ellipse $x^2 + 4y^2 - 4 = 0$

35.

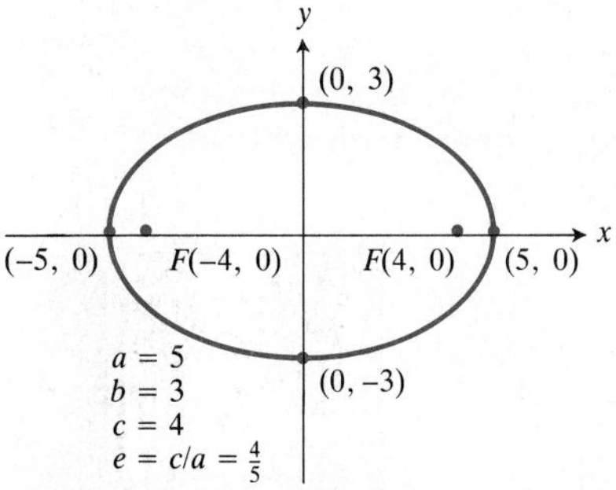

37. a) $(x - 17.5)^2/(18.09)^2 + y^2/(4.56)^2 = 1$ b) 0.584 AU, or 5.41×10^7 mi c) 35.60 AU, or 3.30×10^9 mi; January (winter) 1947–48 d) 1.67×10^9 mi or 18.0 AU

Article 8.6, pp. 537–538

1. a)

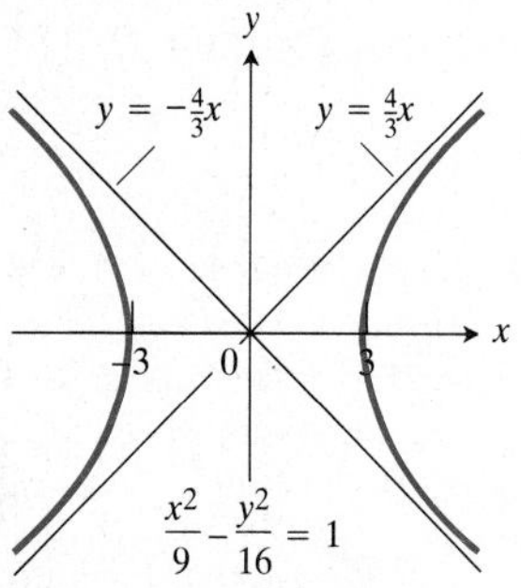

b)

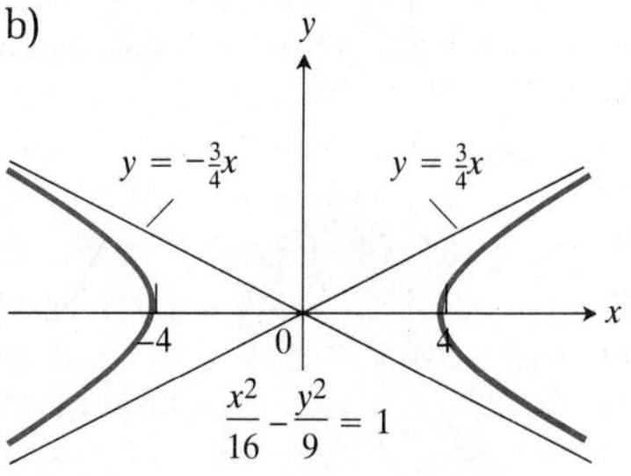

c)

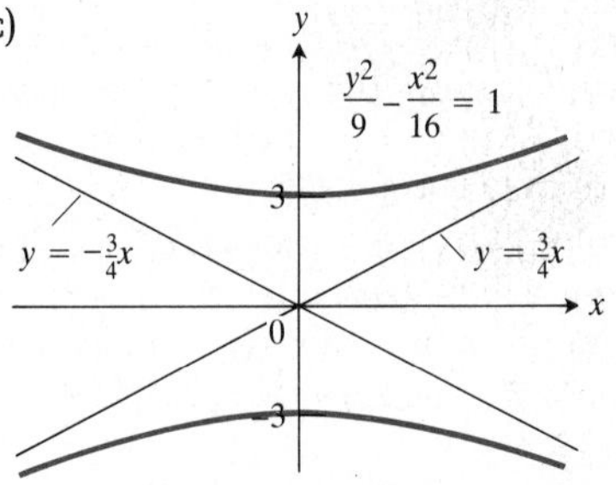

d)

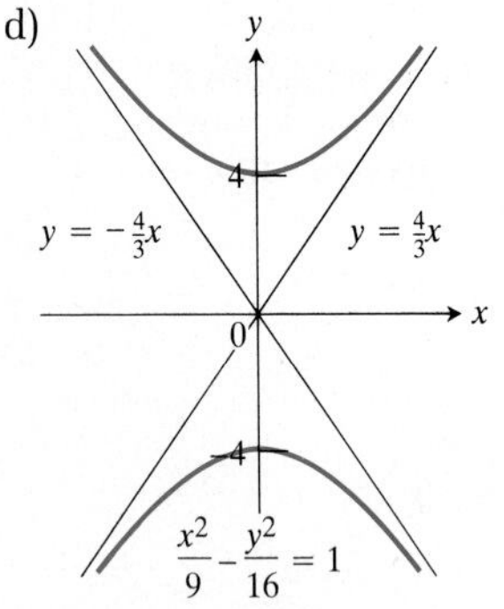

3. $C(2, -3)$; $V(2 \pm 3, -3)$; $F(2 \pm \sqrt{13}, -3)$; $e = \sqrt{13}/3$; A: $2(x - 2) = \pm 3(y + 3)$ **5.** $C(-2, 1)$; $V(-2 \pm 2, 1)$; $F(-2 \pm 3, 1)$; $e = \frac{3}{2}$; A: $\sqrt{5}(x + 2) = \pm 2(y - 1)$ **7.** $C(2, 0)$; $V(2 \pm 2, 0)$; $F(2 \pm \sqrt{5}, 0)$; $e = \sqrt{5}/2$; A: $x - 2 = \pm 2y$ **13.** $4x^2 - 5y^2 - 8x + 60y + 4 = 0$ **15.** a) It lies on the branch nearest A of the hyperbola with foci at A, B and with $a = 6.86 \times 10^5$ ft.

Article 8.7, pp. 542–543

Note: The answers to Problems 1–7 will vary, depending on the size and sense of the rotation used.

1. $x = (x' - y')/\sqrt{2}$, $y = (x' + y')/\sqrt{2}$; $(x')^2 - (y')^2 = 4$ **3.** $\alpha = 5\pi/12$; $(\frac{1}{2} - \sqrt{3})(x')^2 + (\frac{1}{2} + \sqrt{3})(y')^2 = 2$ or $-1.232(x')^2 + 2.32(y')^2 = 2$ **5.** $\alpha = \pi/4$; $5(y')^2 - (x')^2 = 10$ **7.** $\sin\theta = 1/\sqrt{5}$; $\cos\theta = 2/\sqrt{5}$ **13.** $\alpha = -\pi/4$; $2(y')^2 = 1$

Article 8.8, p. 545

1. Hyperbola **2.** Parabola **5.** Ellipse **7.** Hyperbola **9.** Circle **11.** Parabola **13.** Ellipse **15.** Hyperbola **17.** Parabola

Article 8.9, pp. 547–548

3. Line L and the point D do not exist.

Miscellaneous Problems Chapter 8, pp. 548–551

1. $x^2 + xy + y^2 = 3k^2$ **3.** a) Left of the line $x = 3$ b) Left of the line $y = x$ c) Above the parabola $y = x^2$ d) Outside the circle $x^2 + y^2 = 4$ e) Inside the ellipse $x^2 + xy + y^2 = 3$ f) Outside the ellipse $x^2 + xy + y^2 = 3$ **11.** $5x^2 + 5y^2 = 4x + 8y$ **13.** $x = 2pm$ **15.** $(\pm\frac{8}{9}\sqrt{3}, \frac{4}{3})$ **17.** $(-2, 1)$, $(2, -1)$ **21.** $(-\frac{16}{5}, \frac{53}{10})$ **23.** 10^7 mi **25.** $y^2 = 2x - 7$ **31.** No; for points inside the circle, $x^2 + (y^2 - 9)$ is negative, while $x^2 - (y^2 - 9)$ is positive, so their product cannot equal 1. **33.** $5x^2 + 9y^2 - 30x = 0$ **35.** $2ab$ **37.** $x^2 + 4xy + 5y^2 = 1$ **41.** $C = (\frac{1}{2}\sqrt{2}, 0)$; $e = \sqrt{2}$ **43.** Square, vertices at $(\pm 1, 0)$ and $(0, \pm 1)$; area = 2 **45.** a) Center $(2, -1)$, foci $((4 \pm 3\sqrt{3})/2, -1)$, major axis 6; minor axis 3; $e = \sqrt{3}/2$ b) Center $(-2, -\frac{3}{2})$, foci $(-2 \pm \sqrt{13}, -\frac{3}{2})$, major axis $= \sqrt{13}$, conjugate axis $= \sqrt{39}$, $e = 2$, asymptotes: $2y \pm 2\sqrt{3}x + (3 \pm 4\sqrt{3}) = 0$ **47.** b) Bounded; both $9x^2 + 4y^2 - 36$ and $4x^2 + 9y^2 - 36$ exceed one when $4(x^2 + y^2) > 37$. **49.** $\sqrt{2}$ **53.** $9x^2 - y^2 = 9$ or $9y^2 - x^2 = 9$ **55.** a) $C = (3, 2)$, $e = \sqrt{\frac{11}{12}}$ b) $V = (3, 0)$ **57.** With $A = (0, 0)$ and $B = (c, 0)$, curve is left branch of hyperbola $3x^2 - y^2 - 4cx + c^2 = 0$. **59.** Square and add both sides of both equations in Eq. (4), Article 8.7; the result is $x^2 + y^2 = (x')^2 + (y')^2$. **61.** a) Circle, radius a b) Points lie between the circle and the square with vertices $(\pm a, \pm a)$ c) Nearly a square with vertices at $(\pm a, \pm a)$

CHAPTER 9

Article 9.3, p. 561

3. $10\cosh 5x \sinh 5x = 5\sinh 10x$ **5.** $2\operatorname{sech}^2 2x$ **7.** $-3\operatorname{sech}^3 x \tanh x$ **9.** $\sec^2 x \operatorname{sech} y$ **11.** $-2\operatorname{csch}^2 u \coth u$ **13.** $1 + \operatorname{csch}^2 4x$ **15.** $4x^3 \sinh x + x^4 \cosh x$ **17.** $x\cosh x$ **19.** $\ln(\cosh x) + C$ **21.** $\tanh x + C$ **23.** $x - \tanh x + C$ **25.** $\frac{1}{2}x + \frac{1}{12}\sinh 6x + C$ **27.** $\frac{1}{2}x + \frac{1}{20}\sinh 10x + C$ **29.** $\sinh x + \frac{1}{3}\sinh^3 x + C$ **31.** $-\ln|\operatorname{csch} x + \coth x| + C = \ln|\tanh(x/2)| + C$ **33.** $(\sinh^3 3x \cosh 3x)/12 - (\sinh 6x)/16 + (3x/8) + C$ **35.** $(e^{5x}/10) + (e^x/2) + C$ **37.** $u/2$

Article 9.4, p. 566

7. $-\csc x$ **9.** $-\csc x$ **11.** $2x/\sqrt{x^4 - 1}$ **13.** $-|\csc x|$ **15.** $1/(2\sqrt{x^2 + x})$ **17.** $\sec x$ **19.** $x^2/(2\sqrt{x^2 - 1})$ **21.** $\frac{1}{2}\sinh^{-1} 2x + C$ **23.** 0.5493 **25.** $-\frac{1}{2}\operatorname{csch}^{-1}(|x|/2) + C = -\frac{1}{2}\sinh^{-1}(2/|x|) + C$ **27.** $x\sinh^{-1} x - \sqrt{1 + x^2} + C$ **29.** $\cosh^{-1} u + C$ **31.** $-\ln|(1 + \sqrt{u^2 + 1})/u| + C = \ln|(\sqrt{u^2 + 1} - 1)/u| + C$ **33.** $\sqrt{x^2 - 25} - 5\sec^{-1}|x/5| + C$ **35.** $\frac{25}{2}\cosh^{-1}(x/5) + (x/2)\sqrt{x^2 - 25} + C$ **37.** $(x/2)\sqrt{x^2 - 4} - 2\cosh^{-1}(x/2) + C$

Article 9.5, pp. 568–569

1. $a \sinh (x_1/a)$ **3.** $(\pi a/2)(2x_1 + a \sinh (2x_1/a))$ **5.** $(\pi a^2/4)(2x_1 + a \sinh (2x_1/a))$ **7.** c) $a = 25$ ft, sag ≈ 4.6 ft d) $H = 50$ lb

Miscellaneous Problems Chapter 9, p. 569

3. $\lim PQ = 0$
5. $\cosh x = \frac{41}{9}$, $\tanh x = -\frac{40}{41}$
9. $y = 1$
13. $\sec^2 x/(2 \tanh y \operatorname{sech}^2 y)$
15. $\tanh y \tan x$
17. $x^{-1} \operatorname{sech}^2 (\ln x) = 4x(x^2 + 1)^{-2}$
19. $3e^{3x} \cosh (\tan^{-1} e^{3x})/(1 + e^{6x})$
21. $-(\cosh y + \sinh 2x)/(2y + x \sinh y)$
23. $(2\theta - e^{-2\theta})/4 + C$
25. $(e^{3x} + 3e^{-x})/6 + C$
27. $\frac{1}{2} \tanh^{-1} (\frac{1}{2})$
29. $\sinh^{-1} (e^t) + C$
31. $\cosh^{-1} (\tan \theta) + C$

33.

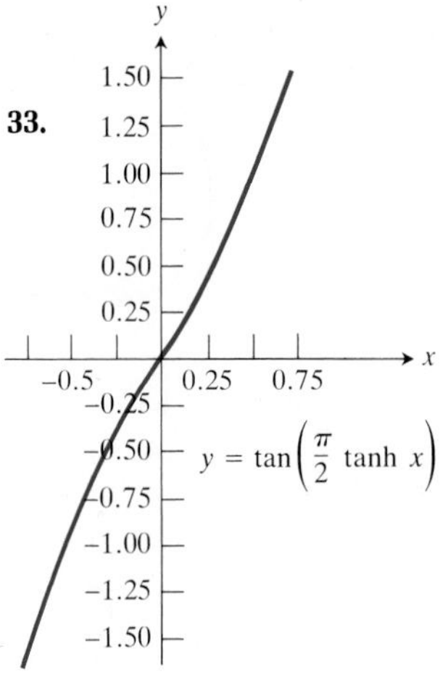

37. $\ln (2/(1 + \sqrt{2}))$

CHAPTER 10

Article 10.1, pp. 576–577

1. The cartesian coordinates of the points are a) (0, 2) b) (2, 0) c) (0, −2) d) (−2, 0). The polar coordinates are a) (2, 90° + n · 360°), (−2, −90° + n · 360°) b) (2, n · 360°), (−2, 180 + n · 360°) c) (−2, 90° + n · 360°), (2, −90° + n · 360°) d) (2, 180° + n · 360°), (−2, n · 360°) (In these formulas, n can be any integer.)

3.

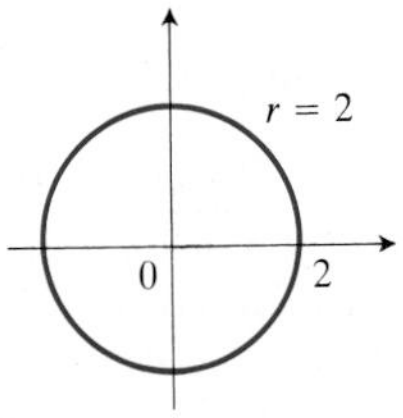

5.

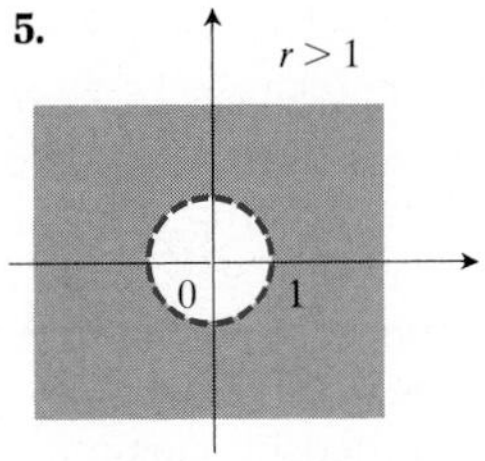

7.

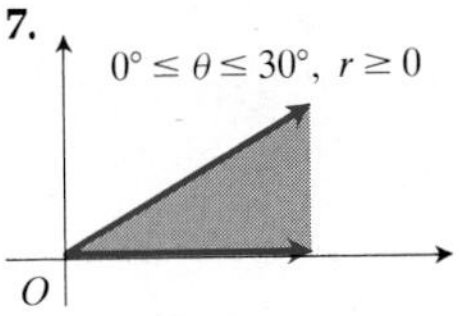

9.

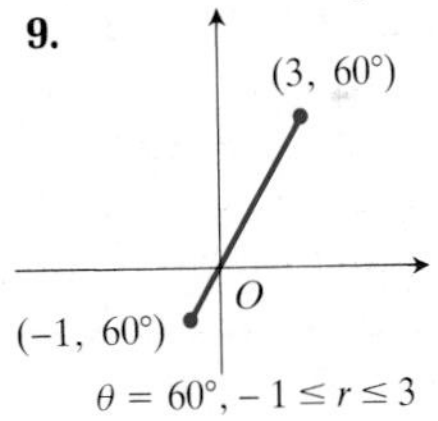

11.

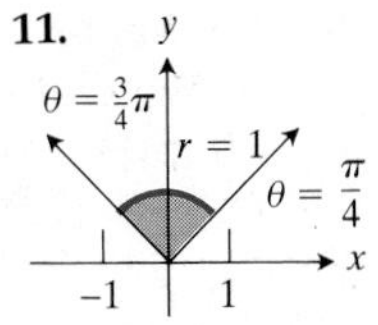

13.

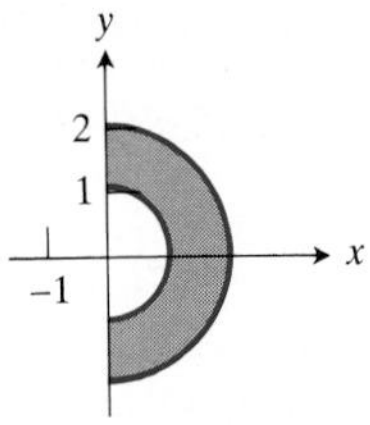

15. $x = 2$ **17.** $y = 4$ **19.** $y = 0$ **21.** $x + y = 1$ **23.** $x^2 + y^2 = 1$ **25.** $y = e^x$ **27.** $y - 2x = 5$ **29.** $x^2 = 4y$ **31.** $r \cos \theta = 7$ **33.** $\tan \theta = 1$, or $\theta = \pi/4$ **35.** $r = 2$ **37.** $(r^2 \cos^2 \theta)/9 + (r^2 \sin^2 \theta)/4 = 1$ **39.** $r \sin^2 \theta = 4 \cos \theta$, or $r = 4 \cot \theta \csc \theta$ **41.** $x + \sqrt{3}y = 6$ **43.** $x - y = 2$ **45.** $(2, 3\pi/4)$ is the same as $(-2, -\pi/4)$ and $2 \sin (-\pi/2) = -2$ **47.** If (r_1, θ_1) satisfies $r_1 = \cos \theta_1 + 1$, then the same point has coordinates $(-r_1, \theta_1 + \pi)$, which satisfy the second equation. Also, if (r_2, θ_2) satisfies the second equation, the coordinates $(-r_2, \theta_2 + \pi)$ of the same point satisfy the first equation. **49.** $(0, 0)$, $(a/\sqrt{2}, \pi/4)$ **51.** $(0, 0)$, $(\frac{8}{5}a, \sin^{-1} \frac{3}{5})$ **53.** b) $\theta = \pi/2$ in the first equation; $\theta = 0$ for the second.

Article 10.2, p. 581

1. $r[3 \cos \theta + 4 \sin \theta] = 5$ **3.** Symmetric about x-axis. **5.** Symmetric about origin, x-axis, y-axis.

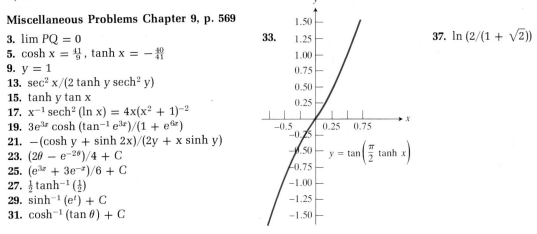

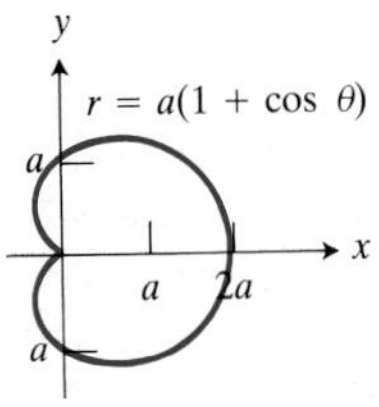

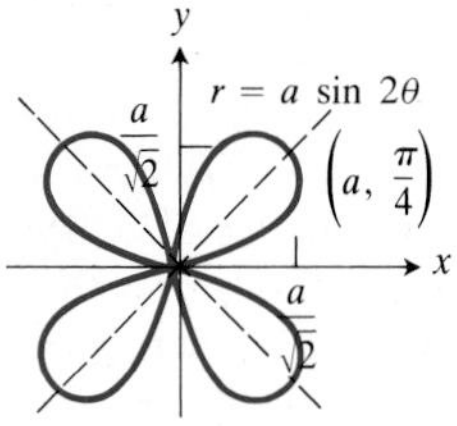

7. Symmetric about y-axis.

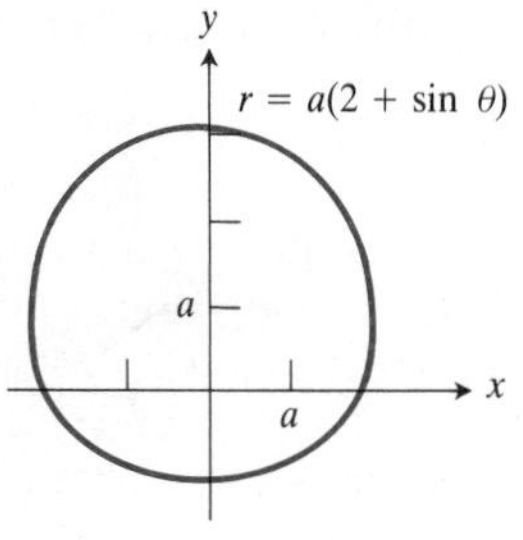

9.

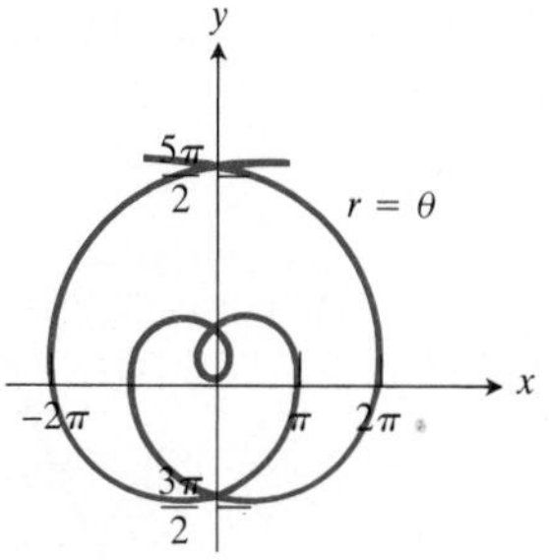

11. a) Same as answer 7, rotated $-\pi/4$ radians. b) Same as answer 3, with $a = \frac{3}{2}$. c)

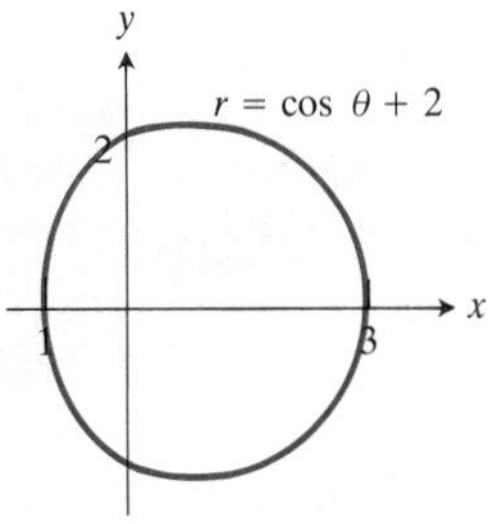

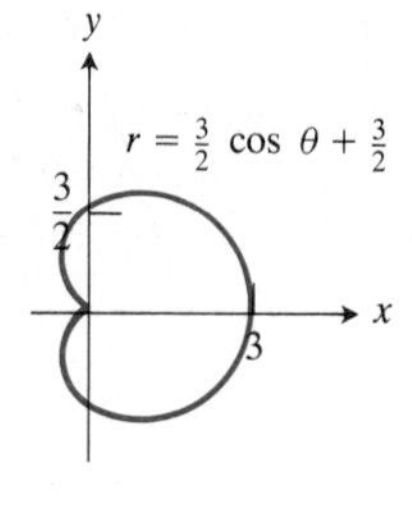

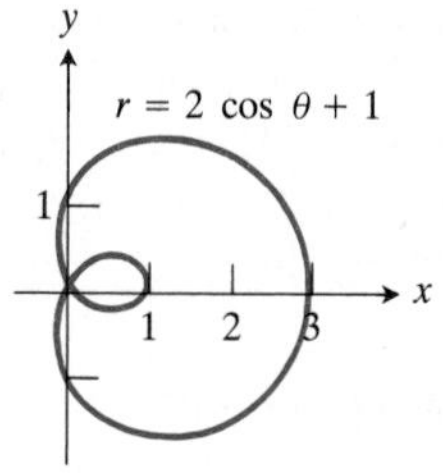

13. 8 **15.** 30°

Article 10.3, pp. 585–586

1. $(x-2)^2 + y^2 = 4$; circle with center (2, 0), radius 2 **3.** $(x+1)^2 + y^2 = 1$; circle with center $(-1, 0)$, radius 1 **5.** $[x^2 + y^2]^3 = (xy)^2$; graph is same as in Article 10.2, Problem 5, with $a = 1$ (four-leaved rose). **7.** $(x^2 + y^2)^2 = 8(x^2 - y^2)$; lemniscate with axis $y = 0$. **9.** $(x^2 + y^2 + 4x)^2 = 64(x^2 + y^2)$; graph is the graph of Article 10.2, Problem 11(c), inverted about y-axis. **11.** $(x^2 + y^2 + 4x)^2 = 16(x^2 + y^2)$; graph is the graph of Article 10.2, Problem 3, inverted about the y-axis. **13.** $y^2 = 3x^2 + 16x + 16$; graph is hyperbola, center at $(-8/\sqrt{3}, 0)$. **15.** a) Circle, center at $(1/\sqrt{2}, 1/\sqrt{2})$ b) Straight line, $\sqrt{3}y - x = 8$ c) Straight line, $x + \sqrt{3}y = 1.0$ d) Circle, center $(\frac{3}{4}, 3\sqrt{3}/4)$, radius $\frac{3}{2}$ e) Cardioid, axis at 30° f) Closed region bounded by cartioid, cusped side pointing right **17.** (e) **19.** (f) **21.** (h) **23.** (j) **25.** a) Circle $(x-1)^2 + y^2 = 1$ and line $x = 1$ b) $(1, \pm 1)$, $(1/\sqrt{2}, \pm\pi/4)$ **27.** $r = 4/(1 - \cos\theta)$ **29.** $(-16/15, 0°)$, $r = 8/(4 + \cos\theta)$ **31.** ± 1 **33.** b) $\pm\sqrt{e^2 - 1}$ **35.** $r = 2a \sin\theta$, circle with center $(0, a)$, radius $|a|$ **37.** $r \cos(\theta - \alpha) = p$ (see Fig. 10.22). **39.** $r^2 = \cos 2\theta$ (see Fig. 10.26).

Article 10.4, pp. 590–591

1. $(3\pi/2)a^2$ **3.** $2a^2$ **5.** $\frac{9}{2}\pi a^2$ **7.** πa^2 **9.** $[(2\pi/3) - (\sqrt{3}/2)]a^2$ **11.** $\pi/8$ **13.** 18π **15.** a) $2\pi + (3\sqrt{3}/2)$ b) $\pi + 3\sqrt{3}$ **17.** $(\frac{5}{6}a, 0)$ **19.** a) $2\pi a$ b) πa c) πa **21.** $2a$. The curve is the cardioid $r = (a/2)(1 - \cos\theta)$ **23.** $(a/8)(4\pi - 3\sqrt{3})$ **25.** $4\pi a^2(2 - \sqrt{2})$ **27.** $(32 - 4\sqrt{2})\pi/5$

Miscellaneous Problems Chapter 10, pp. 591–593

1. Spiral (see Article 10.3, Problem 17). **3.** a) Line: $x = a$ b) Line: $y = a$ c) Curve: $y = ax/(x - a)$; hyperbola with asymptotes $x = a$, $y = a$ **5.** Circle, radius 3, center at $(-1, -1)$ **7.** $x^2y^2 - 4a^2(x^2 + y^2) + 4a^2 = 0$; nearly hyperbolic **9.** $(x^2 + y^2)^2 = 4a^2xy$; lemniscate with axis along $\theta = \pi/4$. **11.** a) $(x^2 + y^2)^{3/2} = x^2 - y^2$; four-leaved rose with axes along coordinate axes b) $(x^2 + y^2)^2 = x^2 - y^2$; two-leaved rose with axis along $y = 0$ **13.** a) $y^2 = 4x + 4$; parabola, vertex at $(-1, 0)$ b) $x^2 + 4y = 4$; parabola, vertex at (0, 1) **15.** $(a, \pi/6)$ and $(a, 5\pi/6)$ **17.** $(a\sqrt{2}, \pi/4)$ **19.** $r = 0$; $(a/2, \theta)$ when $\theta = \pm\pi/3, \pm 2\pi/3$ **21.** $r = 2/(1 + \cos\theta)$ **23.** $r = -2a\cos\theta$ **25.** $r = 8/(3 - \cos\theta)$ **27.** Approx. $(0.75a, -3.44°)$ **29.** $p = 3\pi a/2$ **31.** $(3\pi - 8)a^2/2$ **33.** $9a^2\pi/2$ **35.** πa^2 **37.** $4a^2$ **39.** $5\pi a^2/4$ **41.** $\tan\phi = \tan(\theta/4)$ **45.** $(2, \pm 60°)$; 90° **49.** $\pi/2$ **51.** $\pi/2$ **53.** a) $(\frac{1}{4})\tan\alpha$ b) $\sec\alpha$

CHAPTER 11

Article 11.1, p. 595

1. $\frac{3}{2}, \frac{7}{4}, \frac{15}{8}, \frac{31}{16}, \frac{63}{32}, \frac{127}{64}$ **3.** $1, \frac{1}{2}, \frac{1}{4}, \frac{1}{8}, \frac{1}{16}, \frac{1}{32}$ **5.** 1, 2, 3, 5, 8, 13

Article 11.2, pp. 603–604

1. $0, -\frac{1}{4}, -\frac{2}{9}, -\frac{3}{16}$; converges to 0 **3.** $\frac{1}{3}, \frac{1}{9}, \frac{1}{27}, \frac{1}{81}$; converges to 0 **5.** $1, -\frac{1}{3}, \frac{1}{5}, -\frac{1}{7}$; converges to 0 **7.** 0, −1, 0, 1; diverges **9.** $1, -1/\sqrt{2}, 1/\sqrt{3}, -\frac{1}{2}$; converges to 0 **11.** Converges to 0 **13.** Converges to 1 **15.** Diverges **17.** Converges to $-\frac{2}{3}$ **19.** Converges to $\sqrt{2}$ **21.** Converges to 0 **23.** Diverges **25.** Converges to 1 **27.** Converges to −5 **29.** Converges to 1 **31.** Converges to 1 **33.** Converges to 5 **35.** Converges to 0 **37.** Converges to 0. **39.** Converges to $\sqrt{2}$ **41.** Diverges **43.** Converges to 0 **45.** Diverges **47.** Converges to 0 **49.** Converges to $\pi/2$ **51.** Converges to 1 **53.** Converges to $\frac{1}{2}$ **55.** Use $0 < n!/n^n \le 1/n$ **57.** Yes. No; the limit is 1 for any $x > 0$. **61.** 1

Article 11.3, pp. 607–608

1. Converges to 0 **3.** Converges to 0 **5.** Converges to 0 **7.** Converges to e^7 **9.** Converges to 0 **11.** Diverges **13.** Converges to 1 **15.** Converges to 0 **17.** Converges to $1/e$ **19.** Converges to 0 **21.** Converges to x **23.** Converges to 1 **25.** Converges to 1 **27.** Converges to $1/e$ **29.** Converges to 0 **31.** Diverges **33.** Converges to $1/(p-1)$ **35.** $N \ge 9124$ **37.** 10^{75}

Article 11.4, pp. 618–619

1. $s_n = 3(1 - (\frac{1}{3})^n)$; $\lim_{n\to\infty} s_n = 3$ **3.** $s_n = (1 - (1/e)^n)/(1 - (1/e))$; $\lim_{n\to\infty} s_n = e/(e-1)$ **5.** $s_n = \frac{1}{3}(1 - (-2)^n)$, diverges **7.** $s_n = -\ln(n+1)$, diverges **9.** a) $\Sigma_{n=-2}^{\infty} 1/[(n+4)(n+5)]$ b) $\Sigma_{n=0}^{\infty} 1/[(n+2)(n+3)]$ c) $\Sigma_{n=5}^{\infty} 1/[(n-3)(n-2)]$ **11.** $\frac{4}{3}$ **13.** $\frac{7}{3}$ **15.** 11.5 **17.** $\frac{5}{3}$ **19.** 1 **21.** $\frac{1}{9}$ **23.** a) $0.234234... = 234\,\Sigma_{n=1}^{\infty}(\frac{1}{1000})^n = \frac{26}{111}$ b) Yes. If $x = 0.a_1a_2 \ldots a_n\,\overline{a_1a_2 \ldots a_n}$, then $(10^n - 1)x = a_1a_2 \ldots a_n = p$ is an integer and $x = p/q$ with $q = 10^n - 1$. **25.** $2 + \sqrt{2}$ **27.** 1 **29.** Diverges **31.** $e^2/(e^2 - 1)$ **33.** Diverges **35.** $\frac{3}{2}$ **37.** Diverges **39.** $a = 1$, $r = -x$ **41.** 8

Article 11.5, pp. 639–641

Note. The tests mentioned in Problems 1–30 may not be the only ones that apply.

1. Converges; ratio test; geometric series, $r < 1$ **3.** Converges; compare with $\Sigma(1/2^n)$ **5.** Converges; ratio test **7.** Diverges; compare with $\Sigma(1/n)$ **9.** Converges; ratio test **11.** Diverges; compare with $\Sigma(1/n)$ **13.** Diverges; ratio test **15.** Converges; ratio test, root test **17.** Converges; compare with $\Sigma(1/n^{3/2})$ **19.** Converges; ratio test **21.** Diverges; integral test **23.** Diverges; n^{th}-term test, $a_n \to e$ **25.** Converges, ratio test **27.** Converges; ratio test or comparison with $\Sigma(1/n!)$ **29.** Converges; ratio test **31.** $\Sigma(1/(2n-1)) > \Sigma(1/2n) = \frac{1}{2}\Sigma(1/n)$; the last series diverges **33.** $x^2 \le 1$ U **35.** 1.649767731 **37.** $n = 27$, $s_n + n^{-3}/3 = 1.082324151$; $\pi^2/90 = 1.082323234$; difference $\approx 0.9 \times 10^{-6}$ **39.** $\Sigma(1/nx) = (1/x)(\Sigma(1/n))$. Since $\Sigma(1/n)$ diverges, its nonzero multiple $(1/x)(\Sigma(1/n))$ also diverges.

Article 11.6, pp. 647–648

Note. In the answers provided for this article, "Yes" means that the series converges absolutely. "No" means that the series does not converge absolutely. (The series may still converge.) The reasons given may not be the only appropriate ones.

1. Yes; p-series, $p = 2$ **3.** No; harmonic series, divergence **5.** Yes; compare with p-series, $p = 2$ **7.** Yes; compare with p-series, $p = \frac{3}{2}$ **9.** Yes; comparison test **11.** No; n^{th} term test **13.** Yes; geometric series, $r < 1$ **15.** Yes; ratio test **17.** Yes; compare with $(2+n)/n^3$

Article 11.7, pp. 654–655

1. Converges **3.** Diverges **5.** Converges **7.** Converges **9.** Converges **11.** Absolutely convergent **13.** Absolutely convergent **15.** Conditionally convergent **17.** Divergent **19.** Conditionally convergent **21.** Absolutely convergent **23.** Absolutely convergent **25.** Divergent **27.** Conditionally convergent **29.** $\frac{1}{5} = 0.2$ **31.** $(0.01)^5/5 = 2 \times 10^{-11}$ **33.** 0.54030 (actual: $\cos 1 = 0.540302306\ldots$) **35.** a) The a_n are not decreasing (condition 2) b) $-\frac{1}{2}$

Miscellaneous Problems Chapter 11, pp. 655–656

1. $s_n = \ln[(n+1)/(2n)]$; series converges to $-\ln 2$. **7.** Diverges, since n^{th} term doesn't approach 0 **9.** Diverges **11.** Converges **13.** Diverges **15.** Converges **17.** Diverges **19.** Converges

CHAPTER 12

Article 12.1, p. 662

1. $1 - x + (x^2/2!) - (x^3/3!)$; $1 - x + (x^2/2!) - (x^3/3!) + (x^4/4!)$ **3.** $1 - (x^2/2!)$; $1 - (x^2/2!) + (x^4/4!)$ **5.** $x + (x^3/3!)$; $x + (x^3/3!)$ **7.** $-2x + 1$; $x^4 - 2x + 1$ **9.** $x^2 - 2x + 1$; $x^2 - 2x + 1$ **11.** x^2 **13.** $1 + [3/2]x + [3/(4 \cdot 2!)]x^2 - [3/(8 \cdot 3!)]x^3 + [(3 \cdot 3)/(16 \cdot 4!)]x^4 - [(3 \cdot 3 \cdot 5)/(32 \cdot 5!)]x^5 + \cdots + [(\frac{3}{2})(\frac{1}{2})(-\frac{1}{2}) \cdots (\frac{3}{2} - n + 1)/n!]x^n + \cdots$ **15.** $e^{10} + e^{10}(x - 10) + (e^{10}/2!)(x - 10)^2 + (e^{10}/3!)(x - 10)^3 + \cdots = e^{10} \sum_{n=0}^{\infty} [(x - 10)^n/n!]$ **17.** $(x - 1) - [(x - 1)^2/2] + [(x - 1)^3/3] - [(x - 1)^4/4] + \cdots \sum_{n=1}^{\infty}[(-1)^{n+1}(x - 1)^n]/n$ **19.** $-1 - (x + 1) - (x + 1)^2 - (x + 1)^3 - \cdots = -\sum_{n=0}^{\infty} (x + 1)^n$ **21.** $1 + 2(x - (\pi/4)) + 2(x - (\pi/4))^2$ **23.** 1.0100 with error less than 0.0001 (use two terms)

Article 12.2, pp. 671–672

1. $1 + (x/2) + \frac{1}{4}(x^2/2!) + \frac{1}{8}(x^3/3!) + \frac{1}{16}(x^4/4!) + \cdots$ **3.** $5 - (5/\pi^2)(x^2/2!) + (5/\pi^4)(x^4/4!) - (5/\pi^6)(x^6/6!) + \cdots$ **5.** $(x^4/4!) - (x^6/6!) + (x^8/8!) - (x^{10}/10!) + \cdots$ **9.** $1 - x + x^2 + R_2(x, 0)$ **11.** $1 + (x/2) - (x^2/8) + R_2(x, 0)$ **13.** $|x| < \sqrt[5]{0.06} \approx 0.57$ **15.** $|R| < (10^{-9}/6)$; $x < \sin x$ for $x < 0$ **17.** $|R| < (e^{0.1}/6000) \approx 1.84 \times 10^{-4}$ **19.** $R_4(x, 0) < 0.0003$ **23.** a) $-1 + 0i$ b) $(\sqrt{2}/2) + (i\sqrt{2}/2)$ c) $-i = 0 - i$ d) $-1 + i$ **27.** $\int e^{ax} \cos bx\, dx = [e^{ax}/(a^2 + b^2)](a \cos bx + b \sin bx) + \text{const.}$; $\int e^{ax} \sin bx\, dx = [e^{ax}/(a^2 + b^2)](a \sin bx - b \cos bx) + \text{const.}$

Article 12.3, pp. 678–679

1. Use $\cos x \approx (\sqrt{3}/2) - \frac{1}{2}(x - (\pi/6))$; $\cos 31° \approx 0.857299$. Actual value: $\cos 31° = 0.857167\ldots$ **3.** Use $\sin x \approx x - 2\pi$; $\sin 6.3 \approx 0.0168147$. Actual value: $\sin 6.3 = 0.0168139\ldots$ **5.** Use $\ln x \approx (x - 1) - [(x - 1)^2/2] + [(x - 1)^3/3] - [(x - 1)^4/4]$; $\ln 1.25 \approx 0.222982$. Actual value: $\ln 1.25 = 0.223143\ldots$ **7.** $\ln(1 + 2x) = 2x - [(2x)^2/2] + [(2x)^3/3] - [(2x)^4/4] + \cdots = \sum_{n=1}^{\infty} [(-1)^{n+1}(2x)^n/n]$; converges for $|x| < \frac{1}{2}$ **9.** $\int_0^{0.1} [(\sin x)/x]\, dx \approx x - (x^3/18)\big|_0^{0.1} = 0.0999\overline{4}$ **13.** $\ln 1.5 \approx 0.405465$ **17.** $c_3 = 3.141592665$ **21.** c) 6 d) $1/q$

Article 12.4, p. 681

1. 1 **3.** $-\frac{1}{24}$ **5.** -2 **7.** 0 **9.** $-\frac{1}{3}$ **11.** $\frac{1}{120}$ **13.** $\frac{1}{3}$ **15.** -1 **17.** 3 **19.** 0 **21.** b) $\frac{1}{2}$
23. This approximation gives better results than the approximation $\sin x \approx x$. See table.

x	±1.0	±0.1	±0.01
$6x/(6 + x^2)$	±0.857142857	±0.099833611	±0.009999833
$\sin x$	±0.841470985	±0.099833417	±0.009999833

Article 12.5, pp. 693–694

1. $|x| < 1$. Diverges at $x = \pm 1$ **3.** $|x| < 2$. Diverges at $x = \pm 2$ **5.** $-\infty < x < \infty$ **7.** $-4 < x < 0$. Diverges at $x = -4, 0$ **9.** $|x| < 1$. Converges at $x = \pm 1$ **11.** $-\infty < x < \infty$ **13.** $|x| < (1/e)$. Converges at $x = -(1/e)$; diverges at $x = (1/e)$ **15.** $x = 0$ **17.** $\frac{1}{2} < x < \frac{3}{2}$. Diverges at $x = \frac{1}{2}, \frac{3}{2}$ **19.** $|x| < \sqrt{3}$. Diverges at $x = \pm\sqrt{3}$ **21.** e^{3x+6} **25.** $\sum_{n=1}^{\infty} 2nx^{2n-1}$ **27.** 0.002666 **29.** 0.000331 **31.** 0.363305 **33.** 0.099999 **35.** a) $\sinh^{-1} x = x - \frac{1}{2}(x^3/3!) + [(1 \cdot 3)/(2 \cdot 4)](x^5/5!) - [(1 \cdot 3 \cdot 5)/(2 \cdot 4 \cdot 6)](x^7/7!) + \cdots$ b) $\sinh^{-1} 0.25 \approx 0.247$. Actual value: $\sinh^{-1} 0.25 = 0.247466\ldots$ **37.** $\sum_{n=1}^{\infty} n^n x^n$ **39.** By Theorem 1, the series converges absolutely for $|x| < d$, for all positive $d < r$. This means that the series converges absolutely for all $|x| < r$. [Take $d = \frac{1}{2}(r + |x|)$.] **41.** $x + x^2 + (x^3/3) - (x^5/30) + \cdots$ **43.** $\ln(\sec x) = (x^2/2) + (x^4/12) + (x^6/45) + \cdots$

Miscellaneous Problems Chapter 12, pp. 694–696

1. a) $\sum_{n=2}^{\infty} (-1)^n x^n$ b) No, because it will converge in an interval symmetric about $x = 0$, and it cannot converge when $x = -1$. **3.** $e^{\sin x} = 1 + x + (x^2/2!) - (3x^4/4!)$ **5.** a) $\ln(\cos x) = -(x^2/2) - (x^4/12) - (x^6/45) - \cdots$ b) -0.00017 **7.** 0.747 **9.** $\sum_{n=0}^{\infty} (-1)^{n+1}(x - 2)^n$, $1 < x < 3$ **11.** $\cos x = \frac{1}{2}\sum_{n=0}^{\infty} (-1)^n[(1/(2n)!)(x - (\pi/3))^{2n} + (\sqrt{3}/(2n + 1)!)(x - (\pi/3))^{2n+1}]$ **13.** 1.543 **15.** $e^{(e^x)} = e(1 + x + (2x^2/2!) + (5x^3/3!) + \cdots)$ **17.** -0.0011 **19.** $e^{-1/6}$ **21.** $-5 < x < 1$; converges at $x = -5$, diverges at $x = 1$ **23.** $-\infty < x < \infty$ **25.** $1 < x < 5$; diverges at $x = 1, 5$ **27.** $0 \le x \le 2$ **29.** $-\infty < x < \infty$ **31.** $x \ge \frac{1}{2}$ **37.** $\tan^{-1} x/(1 - x) = x + x^2 + \frac{2}{3}x^3 + \frac{2}{3}x^4 + \frac{13}{15}x^5 + \cdots$

CHAPTER 13

Article 13.1, p. 703

1. a) b) c) d) e)

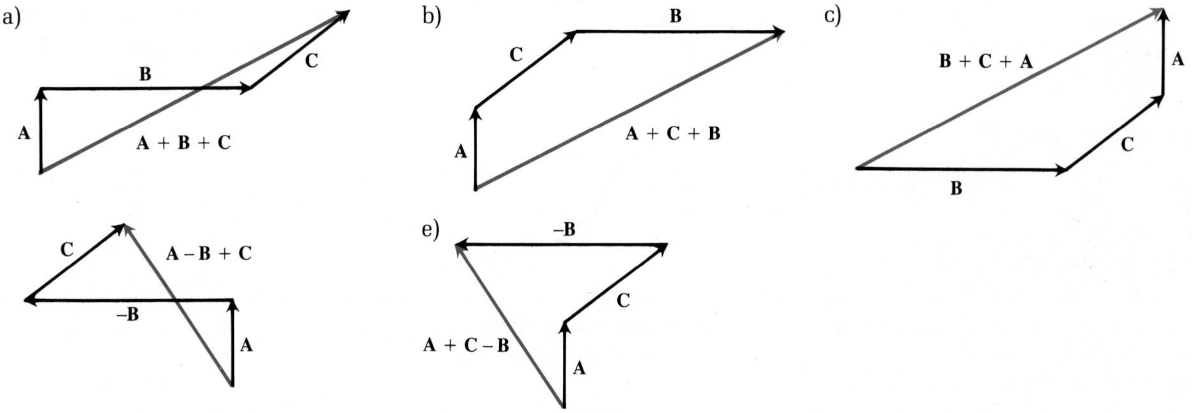

3. $-\mathbf{i}+\mathbf{j}$ **5.** 0 **7.** $\frac{1}{2}\sqrt{3}\mathbf{i}-\frac{1}{2}\mathbf{j}$ **9.** $\pm(\mathbf{i}+4\mathbf{j})/\sqrt{17}$ (either one) **11.** a) $-(\mathbf{i}+4\mathbf{j})/\sqrt{17}$ b) $(4\mathbf{i}-\mathbf{j})/\sqrt{17}$ **13.** $\sqrt{2}$, $(\mathbf{i}+\mathbf{j})/\sqrt{2}$, $\pi/4$ **15.** 2, $(\sqrt{3}\mathbf{i}+\mathbf{j})/2$, $\pi/6$ **17.** 13, $(5\mathbf{i}+12\mathbf{j})/13$, $\tan^{-1}(12/5)\approx 67.4°$ **19.** $2\sqrt{5}$

Article 13.2, p. 708

1. $\alpha_1=\frac{1}{2}\sin^{-1}0.8=26°34'$, flight time 28.0 seconds; $\alpha_2=90°-\alpha_1=63°26'$ flight time 55.9 seconds **5.** Yes; if $\alpha=\frac{1}{2}\sin^{-1}(\frac{9}{10})\approx 32°5'$, y max = 31.3 ft and range = 200 ft **7.** $x=e^t$, $y=2e^{-t}$ **9.** $x=-\ln(1-t)$, $y=1/(1-t)$ **11.** a) $x=\sin t$, $y=-\cos t$ b) $x^2+y^2=1$ c) (1, 0), (0, 1)

Article 13.3, pp. 713–715

1. $x^2+y^2=1$ **3.** $(x/4)^2+(y/2)^2=1$ **5.** $x=1-2y^2$ **7.** Right-hand branch of the hyperbola $x^2-y^2=1$ (right-hand branch because $\sec t>0$ for $-\pi/2<t<\pi/2$). **9.** One arch of the cycloid above the interval $0\le x\le 2\pi$ (see Fig. 13.16). The points (x, y) on this curve satisfy the equation $x=\cos^{-1}(1-y)-\sqrt{2y-y^2}$. **11.** $y=x^{2/3}$ **13.** $4x^2-y^2=4$ **15.** $y=x^2-2x+5$; $x\ge 1$ **17.** $[(x-3)^2/4]+[(y-4)^2/9]=1$; $x>3$ **19.** $x=a\tanh\theta$; $y=a\,\text{sech}\,\theta$ **21.** $x=a\sinh^{-1}(s/a)$; $y=\sqrt{a^2+s^2}$ **23.** $x=(a+b)\cos\theta-b\cos([(a+b)/b]\theta)$; $y=(a+b)\sin\theta-b\sin([(a+b)/b]\theta)$ **25.** $8a$ **29.** 4 **31.** $16\pi/3$ **33.** $x=2\cos t+1$; $y=\sin t$; ellipse, center at (1, 0), vertices at (−1, 0) and (3, 0), semimajor axis 2, semiminor axis 1.

Article 13.4, pp. 718–719

1. Line parallel to z-axis through point (1, 1, 0) **3.** Straight line, 5 units above and parallel to the line $y=x$ in the xy-plane **5.** Circle of radius 2 with center (0, 0, −2) and parallel to the xy-plane **7.** Circle of radius 1 in yz-plane with center at the origin **9.** Straight line in yz-plane that passes through the origin and forms an angle of $\pi/4$ radians with y and z axes

Note. In Problems 11–19, entries that are completely arbitrary are denoted by "*."

	(x, y, z)	(r, θ, z)	(ρ, ϕ, θ)
11.	(1, 0, 0)	(1, 0, 0)	$(1, \pi/2, 0)$
13.	(0, 0, 1)	(0, *, 1)	(1, 0, *)
15.	$(\sqrt{2}, 0, 1)$	$(\sqrt{2}, 0, 1)$	$(\sqrt{3}, \tan^{-1}\sqrt{2}, 0)$
17.	$(0, -\frac{3}{2}, \sqrt{3}/2)$	$(\frac{3}{2}, -\pi/2, \sqrt{3}/2)$	$(\sqrt{3}, \pi/3, -\pi/2)$
19.	$(0, 0, -\sqrt{2})$	$(0, *, -\sqrt{2})$	$(\sqrt{2}, \pi, 3\pi/2)$

21. Straight line in the plane $\theta=\pi/6$ forming an angle of 45° with the z-axis **23.** Two spirals emanating from the origin and situated on the cone formed by rotating the line $z=\frac{3}{2}x$ about the z-axis **25.** The intersection of the sphere $\rho=5$ and the cone $\phi=\pi/4$ is a circle, radius $5/\sqrt{2}$ centered at $(0, 0, 5/\sqrt{2})$ and parallel to xy-plane **27.** A semicircle in the yz-plane, center at (0, 0, 2), radius 2 **29.** $\rho=2$; $r^2+z^2=4$ **31.** $\phi=\pi/4$ or $\phi=\frac{3}{4}\pi$; $x^2+y^2=z^2$ **33.** $\theta=\pi/4$; $\theta=\pi/4$ **35.** As an equation in cylindrical coordinates, $\theta=0$ defines the cartesian plane $y=0$. To get this plane in spherical coordinates, we piece together the two half-planes $\theta = 0$ and $\theta = \pi$. As an equation in spherical coordinates, the equation $\theta = 0$ defines the cartesian half-plane $y = 0$, $x \geq 0$. To describe this half-plane in cylindrical coordinates we may write $\theta = 0$, $r \geq 0$. **37.** $x^2 + y^2 + z^2 = 1$; $r^2 + z^2 = 1$ **39.** $x = 2$, $r \cos \theta = 2$ **41.** Shell and its boundaries formed between spheres of radii 3 and 5 centered at the origin. **43.** Intersection of the solid wedge bounded by planes $y = 0$ and $y = x$, and the solid cone bounded by surface formed by rotating line $z = x$ in the xz-plane about z-axis, all in the first octant.

Article 13.5, p. 722

1. $(-2, 0, 2)$, $\sqrt{8}$ **3.** $(0, 0, a)$, $|a|$ **5.** a) $\sqrt{y^2 + z^2}$ b) $\sqrt{z^2 + x^2}$ c) $\sqrt{x^2 + y^2}$ d) $|z|$ **7.** 3 **9.** 9 **11.** $(2^2 + 3^2 + 7^2)^{1/2}/\sqrt{62} = \sqrt{62}/\sqrt{62} = 1$ **13.** $(2\mathbf{i} + 2\mathbf{j} + 4\mathbf{k})/3$

Article 13.6, pp. 728–729

1. 10, $\sqrt{13}$, $\sqrt{26}$, $(5\sqrt{2})/13$ **3.** 4, $\sqrt{14}$, 2, $\sqrt{\frac{2}{7}}$ **5.** 2, $\sqrt{34}$, $\sqrt{3}$, $\sqrt{\frac{2}{51}}$ **7.** $\sqrt{3} - \sqrt{2}$, $\sqrt{2}$, 3, $(\sqrt{3} - \sqrt{2})/(3\sqrt{2})$ **9.** -25, 5, 5, -1 **11.** $B = \cos^{-1}(\frac{15}{19}) \approx 37.9°$, $A = C = \cos^{-1}(\sqrt{\frac{2}{19}}) \approx 71.1°$ **13.** $\cos^{-1}(\sqrt{\frac{2}{3}}) \approx 35.3°$ **15.** a) $(\frac{7}{45})(10\mathbf{i} + 11\mathbf{j} - 2\mathbf{k}) = \frac{7}{45}\mathbf{A}$ b) $\frac{7}{3}$ **17.** $(3, 3, 0)$ **19.** $\mathbf{B} = [(\frac{3}{2})\mathbf{i} + (\frac{3}{2})\mathbf{j}] + [-(\frac{3}{2})\mathbf{i} + (\frac{3}{2})\mathbf{j} + 4\mathbf{k}]$ **21.** a) $2\sqrt{10}$ b) $6\sqrt{10}$ **23.** a) $\frac{1}{3}(\mathbf{i} - \mathbf{j} + \mathbf{k})$ b) $\mathbf{j} + \mathbf{k}$ c) $-\frac{1}{3}(-2\mathbf{i} - \mathbf{j} + \mathbf{k})$ **25.** $\cos \alpha = 1$, $\cos \beta = 0$, $\cos \gamma = 0$ **27.** $\cos \alpha = -2/\sqrt{53}$, $\cos \beta = 7/\sqrt{53}$, $\cos \gamma = 0$ **29.** $\cos \alpha = 1/\sqrt{2}$, $\cos \beta = 1/\sqrt{3}$, $\cos \gamma = 1/\sqrt{6}$ **31.** $\cos \alpha = 1/\sqrt{2}$, $\cos \beta = 0$, $\cos \gamma = 1/\sqrt{2}$ **33.** -5 newton-meters **35.** $2000\sqrt{3}$ newton-meters **37.** If $|\mathbf{v}_1| = |\mathbf{v}_2|$, then $(\mathbf{v}_1 + \mathbf{v}_2)$ and $(\mathbf{v}_1 - \mathbf{v}_2)$ are orthogonol. **39.** $z^2 = x^2 + y^2$

Article 13.7, pp. 731–732

1. $-\mathbf{i}$ w $3\mathbf{j} + 4\mathbf{k}$ **3.** $c(2\mathbf{i} + \mathbf{j} + \mathbf{k})$, c = scalar **5.** $\sqrt{6}/2$ **7.** 3 **9.** $2\mathbf{i} + 4\mathbf{j} + 7\mathbf{k}$; 0; 0 **11.** a) No pairs b) **A** and **C** **13.** a) $\cos^{-1}(2\sqrt{2}/\sqrt{33}) \approx 60.6°$ b) Yes; the normal vectors are not parallel, so the planes are not parallel. c) $\mathbf{A} \times \mathbf{B} = -5\mathbf{j} - 5\mathbf{k}$, or $\mathbf{j} + \mathbf{k}$ **15.** $\pm[2\sqrt{2}/3, 1/(3\sqrt{2}), 1/(3\sqrt{2})]$ **17.** $12/\sqrt{62}$

Article 13.8, pp. 738–739

1. $(9, -5, 12)$ **3.** $x = 1$, $y = 1 + t$, $z = t - 1$ **5.** $(5, 15, 10)$ **7.** $3x + y + z = 5$ **9.** $x + 3y - z = 9$ **11.** $x + 3y + 4z = 34$ **13.** $x - 2y + z = 6$ **15.** $x + 6y + z = 16$ **17.** a) $6x - 2y + 3z = 23$ b) $23/7$ c) $25/7$ d) $(x - 4)/6 = (2 - y)/2 = (z + 1)/3$ **19.** a) $z = -1$, $y = 2 - x$ b) $(x - 1)/14 = y/2 = z/15$ c) $x = 4$, $y = 2z + 1$ **21.** a) The complement of the angle between the line and a normal to the plane b) $\sin^{-1}(\frac{8}{21}) \approx 22.4°$ **23.** 3 **25.** $(\cos \alpha)(x - x_0) + (\cos \beta)(y - y_0) + (\cos \gamma)(z - z_0) = 0$ **31.** Both lines contain the point $(2, 7, 5)$, so they intersect. $5x - 7y - 4z = -59$

Article 13.9, p. 745

1. $-2\mathbf{C} = -6\mathbf{i} + 8\mathbf{j} - 24\mathbf{k}$; $-22\mathbf{A} = -88\mathbf{i} + 176\mathbf{j} - 22\mathbf{k}$ **3.** a) $\mathbf{A} \times \mathbf{B} = 15\mathbf{i} + 10\mathbf{j} + 20\mathbf{k}$; $(\mathbf{A} \times \mathbf{B}) \times \mathbf{C} = 200\mathbf{i} - 120\mathbf{j} - 90\mathbf{k}$ b) $(\mathbf{A} \cdot \mathbf{C})\mathbf{B} - (\mathbf{B} \cdot \mathbf{C})\mathbf{A} = 56\mathbf{B} + 22\mathbf{A} = 200\mathbf{i} - 120\mathbf{j} - 90\mathbf{k}$ **5.** $\frac{2}{3}$ **7.** $\frac{4}{3}$ **9.** a) 0 b) 2 c) $2\mathbf{A}$ **11.** $a = (\mathbf{A} \times \mathbf{B}) \cdot \mathbf{D}$, $b = -(\mathbf{A} \times \mathbf{B}) \cdot \mathbf{C}$ **15.** $7 - x = (3y - 19)/12 = (4 - 3z)/30$

Article 13.10, p. 748

1. Right circular cylinder of radius 1, axis along y-axis. **3.** Parabolic cylinder, containing the z-axis, and meeting the xy-plane in the parabola $x = -y^2$. **5.** Plane determined by the x-axis and the line $z = -y$ in the yz-plane. **7.** Hyperbolic cylinder, axis along the y-axis and meeting the xz-plane in the hyperbola $x^2 - z^2 = 1$. **9.** Right circular cylinder, radius 4, axis along the z-axis. **11.** Right circular cylinder, radius $\frac{1}{2}$, axis is the line $x = \frac{1}{2}$, $y = 0$. **13.** Right circular cylinder, radius a, axis along z-axis. **15.** Elliptic cylinder $x^2 + 4(z - \frac{1}{2})^2 = 1$; axis is the line $x = 0$, $z = \frac{1}{2}$ parallel to the y-axis. The cylinder lies on top of the y-axis.

Article 13.11, p. 753

1. Paraboloid of revolution, vertex $(0, 0, 1)$ opening upward. **3.** Ellipsoid, center $(0, 0, 0)$, semiaxes $a = 2$, $b = 1$, $c = 2$. **5.** Sphere, center $(0, 1, 0)$, radius 1. **7.** Two-sheeted hyperboloid of revolution, axis parallel to x-axis, center $(-2, -3, 0)$ **9.** Rotate the xy-plane through 45° and have $z^2 + 2y'^2 = 2x'^2$; elliptic cone with vertex O, axis along x′-axis. **11.** Paraboloid of revolution obtained by rotating the parabola $z = x^2$ about the z-axis. **13.** Rotate $z^2 = x$ about the z-axis. **15.** Elliptic cone, vertex O, axis is the x-axis. **17.** $(x - 1)^2 + 4(y + 1)^2 - (z - 2)^2 = 1$. One-sheeted hyperboloid, center $(1, -1, 2)$, axis $x - 1 = y + 1 = 0$. **19.** Elliptic cylinder, axis along y-axis. **21.** Plane, $z = x$. **23.** "Propeller"-shaped surface generated by rotating about the z-axis from $\theta = 0$ to $\theta = \pi/2$ a line through the z-axis and parallel to the xy-plane whose height above the xy-plane is $z = \sin\theta$. **25.** a) $A(z_1) = \pi ab(1 - (z_1^2/c^2))$ b) $V = \frac{4}{3}\pi abc$ **27.** a) $\pi abh(1 + (h^2/3c^2))$ b) $A_0 = \pi ab$, $A_h = \pi ab(1 + (h^2/c^2))$, $V = (h/3)(2A_0 + A_h)$ **29.** When $\theta = 0$, $\rho = F(\phi)$ describes a curve in the xz-plane. When this curve is rotated about the z-axis, it generates the surface $\rho = F(\phi)$. **31.** A cardioid of revolution.

Miscellaneous Problems Chapter 13, pp. 754–757

1. $x = (1-t)^{-1}$, $y = 1 - \ln(1-t)$, $t < 1$ **3.** $x = e^t$, $y = e^{(e^t)} - e$ **5.** $x = t - \sin t$, $y = 1 - \cos t$ **7.** $x = \sinh^{-1} t$, $y = t\sinh^{-1} t - \sqrt{1+t^2} + 1$ **9.** $x = 4\sin t$, $y = 4\cos t$ **13.** $x = 2a\sin^2\theta\tan\theta$, $y = 2a\sin^2\theta$ **15.** a) $3\pi a^2$ b) $8a$ c) $64\pi a^2/3$ d) $(\pi a, 5a/6)$ **17.** 0 **25.** $-\frac{1}{10}$ **27.** $7\sqrt{2}/10$ **29.** $\mathbf{j} - \mathbf{k}$ **31.** $(\mathbf{j} + \mathbf{k})/\sqrt{2}$ **33.** $(10\mathbf{i} - 2\mathbf{j} - 6\mathbf{k})/\sqrt{35}$ **35.** $\cos^{-1}(3/\sqrt{35})$ **39.** $(1, -2, -1)$; $(x-1)/-5 = (y+2)/3 = (z+1)/4$ **43.** $\frac{1}{3}\sqrt{78}$ **45.** a) $2x + 7y + 2z + 10 = 0$ b) $9/(5\sqrt{57})$ **49.** a) 1 b) $-10\mathbf{i} - 2\mathbf{j} - 12\mathbf{k}$ **57.** $(x-1)^2 + (y-2)^2 + (z-3)^2 = 9$

CHAPTER 14

Article 14.1, pp. 762–763

1. $\mathbf{i}$ **3.** $\mathbf{j} - \mathbf{k}$ **5.** $\mathbf{i} + \mathbf{k}$ **7.** All t **9.** $t \neq -1$ **11.** All t **13.** $2e^{2t}\mathbf{i} + (1-\varepsilon)e^{-t}\mathbf{j}$; domain is all t. **15.** 0; domain is all t. **17.** $[1/(|x|\sqrt{9x^2-1})]\mathbf{i} + (2\sinh 2x)\mathbf{j} + (4\,\text{sech}\,4x)\mathbf{k}$, for $|3x| > 1$. **19.** $\mathbf{F}'(s) = -\mathbf{i}/4(s-1)^{5/4} + \mathbf{j}(\cos\sqrt{s-1})/2\sqrt{s-1} + \mathbf{k}/s$ **21.** For $x + 1 > 0$, $\mathbf{F}'(x) = -\mathbf{i}\sin(x+1) + \mathbf{j}(2x+2)/[1 + (x+1)^4] + \mathbf{k}/(x+2)^2$ **23.** $\mathbf{F}'(s) = -\sqrt{2}(\mathbf{i}\cos t + \mathbf{j}\sin t)$ **25.** $\mathbf{F}'(t) = 4t(\mathbf{i}\cos 2\theta - \mathbf{j}\sin 2\theta)$

Article 14.2, pp. 766–767

1. $\mathbf{v} = (-a\omega\sin\omega t)\mathbf{i} + (a\omega\cos\omega t)\mathbf{j}$, $\mathbf{a} = -(a\omega^2\cos\omega t)\mathbf{i} + -(a\omega^2\sin\omega t)\mathbf{j} = -\omega^2\mathbf{R}$; when $\omega t = \pi/3$, $\mathbf{v} = a\omega(-\frac{1}{2}\sqrt{3}\mathbf{i} + \frac{1}{2}\mathbf{j})$, speed $= a\omega$, $\mathbf{a} = -a\omega^2(\frac{1}{2}\mathbf{i} + \frac{1}{2}\sqrt{3}\mathbf{j})$ **3.** $\mathbf{v} = \mathbf{i} + 2t\mathbf{j}$, $\mathbf{a} = 2\mathbf{j}$; at $t = 2$, $\mathbf{v} = \mathbf{i} + 4\mathbf{j}$, $\mathbf{a} = 2\mathbf{j}$, speed $= \sqrt{17}$ **5.** $\mathbf{v} = e^t\mathbf{i} - 2e^{-2t}\mathbf{j}$, $\mathbf{a} = e^t\mathbf{i} + 4e^{-2t}\mathbf{j}$; at $t = \ln 3$, $\mathbf{v} = 3\mathbf{i} - \frac{2}{9}\mathbf{j}$, $\mathbf{a} = 3\mathbf{i} + \frac{4}{9}\mathbf{j}$ speed $= \frac{1}{9}\sqrt{733}$ **7.** $\mathbf{v} = (3\sinh 3t)\mathbf{i} + (2\cosh t)\mathbf{j}$, $\mathbf{a} = (9\cosh 3t)\mathbf{i} + (2\sinh t)\mathbf{j}$; at $t = 0$, $\mathbf{v} = 2\mathbf{j}$, $\mathbf{a} = 9\mathbf{i}$, speed $= 2$ **9.** $\mathbf{v} = -e^{-t}\mathbf{i} - (6\sin 3t)\mathbf{j} + (6\cos 3t)\mathbf{k}$, $\mathbf{a} = e^{-t}\mathbf{i} - 18(\cos 3t)\mathbf{j} - 18(\sin 3t)\mathbf{k}$; at $t = 0$, $\mathbf{v} = -\mathbf{i} + 6\mathbf{k}$, $\mathbf{a} = \mathbf{i} - 18\mathbf{j}$, speed $= \sqrt{37}$ **11.** $\mathbf{v} = -(4\sin t)\mathbf{i} - (3\cos t)\mathbf{j} + 2\mathbf{k}$, $\mathbf{a} = (-4\cos t)\mathbf{i} + (3\sin t)\mathbf{j}$; at $t = \pi/3$, $\mathbf{v} = -2\sqrt{3}\mathbf{i} - \frac{3}{2}\mathbf{j} + 2\mathbf{k}$, $\mathbf{a} = -2\mathbf{i} + (3\sqrt{3}/2)\mathbf{j}$, speed $= \sqrt{73}/2$ **13.** $t = 1/\sqrt[6]{72}$ **15.** All t **17.** For $t = 0$, $\theta = \pi/2$ **19.** a) $\mathbf{v} = \mathbf{i} + 2t\mathbf{j}$, $\mathbf{a} = 2\mathbf{j}$ b) At $t = 2$, $\mathbf{v} = \mathbf{i} + 4\mathbf{j}$, $\mathbf{a} = 2\mathbf{j}$ **21.** No, it would take about 38.7 seconds to reach the hill, longer than the 31 seconds the round is airborne. **23.** $\mathbf{v} = (\frac{3}{2}t^2 + 4)\mathbf{i} + 4t\mathbf{j} + t\mathbf{k}$; $\mathbf{R} = (\frac{1}{2}t^3 + 4t)\mathbf{i} + (2t^2 + 5)\mathbf{j} + (t^2/2)\mathbf{k}$

Article 14.3, p. 771

1. $-\mathbf{i}\sin t + \mathbf{j}\cos t$ **3.** $-\mathbf{i}\cos t + \mathbf{j}\sin t$ **5.** $-(2\mathbf{i}\cos t + \mathbf{j})/\sqrt{1 + 4\cos^2 t}$ **7.** $[(\cos t - \sin t)/\sqrt{2}]\mathbf{i} + [(\cos t + \sin t)/\sqrt{2}]\mathbf{j}$ **9.** $\frac{52}{3}$ **11.** $\mathbf{j}$ **13.** $\mathbf{T} = \sqrt{\frac{1}{3}}[(\cos t - \sin t)\mathbf{i} + (\cos t + \sin t)\mathbf{j} + \mathbf{k}]$; $|\mathbf{v}| = \sqrt{3}e^t$; $\sqrt{3}(e^{\pi} - 1)$ **15.** $\mathbf{T} = (9t^2 + 25)^{(-1/2)}[3(\cos t - t\sin t)\mathbf{i} + 3(\sin t + t\cos t)\mathbf{j} + 4\mathbf{k}]$; $|\mathbf{v}| = [9t^2 + 25]^{1/2}$; $[(\pi/2)\sqrt{9\pi^2 + 25} + (\frac{25}{6})\sinh^{-1}(3\pi/5)]$ **17.** $\mathbf{v}(0) = 6\mathbf{j}$; $\mathbf{T} = \mathbf{j}$; speed $= 6$; $\mathbf{a} = 6\mathbf{i}$ **19.** $\frac{1}{4}[2\sqrt{5} + \sinh^{-1} 2]$

Article 14.4, pp. 778–779

1. $|\cos x|$ **3.** $|\cos y|$ **5.** $48y^5/[(4y^6 + 1)^2]$ **7.** $1/[2(1 + t^2)^{3/2}]$ **9.** $\mathbf{T} = (\cos t)\mathbf{i} + (\sin t)\mathbf{j}$; $\kappa = 1/t$; $\mathbf{N} = -\sin t\mathbf{i} + \cos t\mathbf{j}$ **11.** $\mathbf{T} = \sqrt{\frac{1}{2}}[(\cos t - \sin t)\mathbf{i} + (\sin t + \cos t)\mathbf{j}]$; $\kappa = \sqrt{\frac{1}{2}}e^{-t}$; $\mathbf{N} = \sqrt{\frac{1}{2}}[-(\sin t + \cos t)\mathbf{i} + (\cos t - \sin t)\mathbf{j}]$ **13.** $\mathbf{T} = \frac{1}{13}[(12\cos 2t)\mathbf{i} - (12\sin 2t)\mathbf{j} + 5\mathbf{k}]$; $\kappa = \frac{24}{169}$; $\mathbf{N} = -(\sin 2t)\mathbf{i} - (\cos 2t)\mathbf{j}$; $\mathbf{B} = \frac{1}{13}[(5\cos 2t)\mathbf{i} - (5\sin 2t)\mathbf{j} - 12\mathbf{k}]$ **15.** $\mathbf{T} = (1/\sqrt{3})[(1+t)^{1/2}\mathbf{i} - (1-t)^{1/2}\mathbf{j} + \mathbf{k}]$; $\kappa = (\sqrt{2}/3)(1 - t^2)^{-1/2}$; $\mathbf{N} = (1/\sqrt{2})[(1-t)^{1/2}\mathbf{i} + (1+t)^{1/2}\mathbf{j}]$; $\mathbf{B} = (\sqrt{2}/2)[-(1+t)^{1/2}\mathbf{i} + (1-t)^{1/2}\mathbf{j} + (2\sqrt{3}/3)\mathbf{k}]$ **17.** $\mathbf{T} = (\cos t)\mathbf{i} + (\sin t)\mathbf{j}$; $\kappa = 1/t$; $\mathbf{N} = -(\sin t)\mathbf{i} + (\cos t)\mathbf{j}$; $\mathbf{B} = \mathbf{k}$. Equations for tangent line: $x = \pi/2$, $z = 3$. **19.** $\mathbf{T} = (1/\sqrt{5})(-2(\sin t)\mathbf{i} + 2(\cos t)\mathbf{j} + \mathbf{k})$; $\kappa = \frac{2}{5}$; $\mathbf{N} = -(\mathbf{i}\cos t + \mathbf{j}\sin t)$; $\mathbf{B} = (1/\sqrt{5})[\mathbf{i}\sin t - \mathbf{j}\cos t + 2\mathbf{k}]$; $s(t) = \sqrt{5}t$ **21.** a) $\mathbf{R}(t) = (1 + (2/\sqrt{17})t)\mathbf{i} + (-2 + (2/\sqrt{17})t)\mathbf{j} + (3 - (3/\sqrt{17})t)\mathbf{k}$ b) $\mathbf{v}(t) = (1/\sqrt{17})(2\mathbf{i} + 2\mathbf{j} - 3\mathbf{k})$ c) $\mathbf{T}(t) = (1/\sqrt{17})(2\mathbf{i} + 2\mathbf{j} - 3\mathbf{k})$ d) $\mathbf{a}(t) = 0$ e) $\kappa = 0$ **23.** $[x - x_c(t)]^2 + [y - y_c(t)]^2 = \frac{1}{4}(t + t^{-1})^4$ where $x_c(t) = 2\ln t - \frac{1}{2}(t^2 - t^{-2})$, $y_c(t) = -2(t + t^{-1})$ **25.** $(x+2)^2 + (y-3)^2 = 8$ for the circle; $y' = -(x+2)/(y-3)$, $y'' = -8/(y-3)^3$. Both of these $= +1$ at $(0, 1)$. **29.** $d\mathbf{R}/dt = \mathbf{T}(ds/dt)$, $d^2\mathbf{R}/dt^2 = \mathbf{T}(d^2s/dt^2) + \mathbf{N}\kappa(ds/dt)^2$

Article 14.5, pp. 784–785

1. $\mathbf{v} = 2\mathbf{i} + 2t\mathbf{j}$; $\mathbf{a} = 2\mathbf{j}$; $ds/dt = 2(1 + t^2)^{1/2}$; $a_T = 2t/(1 + t^2)^{1/2}$; $a_N = 2[1 - t^2/(1 + t^2)]^{1/2} = 2/(1 + t^2)^{1/2}$; $\mathbf{a} = [2t/(1 + t^2)^{1/2}]\mathbf{T} + 2/(1 + t^2)^{1/2}\mathbf{N}$ **3.** $\mathbf{v} = (-2\sin t)\mathbf{i} + (2\cos t)\mathbf{j}$; $ds/dt = 2$; $a_T = 0$; $a_N = 2$;

$\mathbf{a} = -(2\cos t)\mathbf{i} - (2\sin t)\mathbf{j} = 2\mathbf{N}$ **5.** $\mathbf{v} = e^t[(\cos t - \sin t)\mathbf{i} + (\sin t + \cos t)\mathbf{j}]$; $ds/dt = \sqrt{2}e^t$; $a_T = a_N = \sqrt{2}e^t$; $\mathbf{a} = \sqrt{2}e^t(\mathbf{T} + \mathbf{N})$ **7.** a) $\mathbf{v} = -(\sin t)\mathbf{i} + (\cos t)\mathbf{j} + b\mathbf{k}$; $ds/dt = [1 + b^2]^{1/2}$; $a_T = 0$; $a_N = 1$, $\mathbf{a} = \mathbf{N}$ b) $\kappa = (b^2 + 1)^{-1}$; increasing b reduces curvature **9.** $\mathbf{v} = \mathbf{i} + 2\mathbf{j} + 2\mathbf{k}$; $\mathbf{a} = 2\mathbf{k}$; $ds/dt = 3$; $a_T = \frac{4}{3}$; $a_N = 2\sqrt{5}/3$ **11.** $\mathbf{v} = \mathbf{j} + \mathbf{k}$; $\mathbf{a} = 2\mathbf{i}$; $ds/dt = \sqrt{2}$; $a_T = 0$; $a_N = 2$ **13.** $\mathbf{v} = 3\mathbf{i} + 4\mathbf{j}$; $\mathbf{a} = 6\mathbf{i} + 8\mathbf{j} + 6\mathbf{k}$; $ds/dt = 5$; $a_T = 10$; $a_N = 6$ **15.** a) $\sqrt{3}\ 1\ k$ b) $z = 1$, $y = x - 2$ c) $x - y + 2z = 4$ **17.** If ds/dt = constant, then $d^2s/dt^2 = a_T = 0$ so there is no tangential acceleration. Thus all acceleration and hence all force is directed along the normal.

Article 14.6, pp. 789–790

1. $\mathbf{v} = (3a\sin\theta)\mathbf{u}_r + 3a(1 - \cos\theta)\mathbf{u}_\theta$, $a = 9a(2\cos\theta - 1)\mathbf{u}_r + 18a(\sin\theta)\mathbf{u}_\theta$ **3.** $\mathbf{v} = 2ae^{a\theta}\mathbf{u}_r + 2e^{a\theta}\mathbf{u}_\theta$; $\mathbf{a} = 4e^{a\theta}(a^2 - 1)\mathbf{u}_r + 8ae^{a\theta}\mathbf{u}_\theta$ **5.** $\mathbf{v} = -(8\sin 4t)\mathbf{u}_r + (4\cos 4t)\mathbf{u}_\theta$; $\mathbf{a} = -(40\cos 4t)\mathbf{u}_r - (32\sin 4t)\mathbf{u}_\theta$ **7.** a) $\mathbf{v} = (\cos t - t\sin t)\mathbf{i} + (\sin t + t\cos t)\mathbf{j} + \mathbf{k}$; $\mathbf{a} = -(2\sin t + t\cos t)\mathbf{i} + (2\cos t - t\sin t)\mathbf{j}$ b) $\mathbf{v} = \sqrt{2}\mathbf{u}_r + r\mathbf{u}_\theta = \sqrt{2}\mathbf{u}_r + \sqrt{2}\,t\,\mathbf{u}_\theta$; $\mathbf{a} = -r\mathbf{u}_r + 2\sqrt{2}\mathbf{u}_\theta = -\sqrt{2}t\mathbf{u}_r + 2\sqrt{2}\mathbf{u}_\theta$ **11.** $2a = 13.5046 \times 10^8$ cm = 8391.3 mi **17.** $v_0 = 3.01 \times 10^6$ cm/sec = 18.7 mi/sec

Miscellaneous Problems Chapter 14, pp. 790–793

1. a) $\mathbf{v} = 2^{-(3/2)}(-\mathbf{i} + \mathbf{j})$, $\mathbf{a} = 2^{-(5/2)}(\mathbf{i} - 3\mathbf{j})$ b) $t = 0$ **3.** Speed $= a\sqrt{1 + t^2}$, $a_T = at/\sqrt{1 + t^2}$, $a_N = a(t^2 + 2)/\sqrt{1 + t^2}$ **5.** b) $\pi/2$ **7.** $x = a\theta + a\sin\theta$, $y = -a(1 - \cos\theta)$ **9.** b) πab, πa^2 **11.** $y = \pm\sqrt{1 - x^2} \pm \ln[(1 - \sqrt{1 - x^2})/x] + C$. Set $C = 0$, then $x = e^{-s}[s = -\ln x$, measured from $(1, 0)]$; $y = \pm\sqrt{1 - e^{-2s}} \pm \ln(e^s - \sqrt{e^{2s} - 1})$ **13.** a) $320\sqrt{10}$ b) $16[\sqrt{2} + \ln(\sqrt{2} + 1)]$ **15.** a) $a_T = 0$, $a_N = 4$ b) 1 c) $r = 2\cos\theta$ **17.** $\kappa = |f^2 + 2f'^2 - ff''|/(f^2 + f'^2)^{3/2}$ **19.** $\sqrt{2}\mathbf{u}_r$ **21.** a) $\mathbf{v} = -\mathbf{u}_r + 3\mathbf{u}_\theta$, $\mathbf{a} = -9\mathbf{u}_r - 6\mathbf{u}_\theta$ b) $\sqrt{37} + \frac{1}{6}\ln(6 + \sqrt{37})$ **23.** a) $2\sqrt{\pi bg/(a^2 + b^2)}$ b) $\theta = [b/(a^2 + b^2)](\frac{1}{2}gt^2)$, $z = [b^2/(a^2 + b^2)]\frac{1}{2}gt^2$ c) $d\mathbf{R}/dt = gt[b/\sqrt{a^2 + b^2}]\mathbf{T}$; $d^2\mathbf{R}/dt^2 = g[b/\sqrt{a^2 + b^2}]\mathbf{T} + (gt)^2[ab^2/(a^2 + b^2)^2]\mathbf{N}$. There is never any component of acceleration in the direction of the binormal. **25.** a) $\mathbf{u}_\rho = \mathbf{i}\sin\phi\cos\theta + \mathbf{j}\sin\phi\sin\theta + \mathbf{k}\cos\phi$; $\mathbf{u}_\phi = \mathbf{i}\cos\phi\cos\theta + \mathbf{j}\cos\phi\sin\theta - \mathbf{k}\sin\phi$; $\mathbf{u}_\theta = -\mathbf{i}\sin\theta + \mathbf{j}\cos\theta$ d) Yes, they form a right-handed system of mutually orthogonal vectors because of (b) and (c). **27.** a) $ds^2 = dr^2 + r^2d\theta^2 + dz^2$ b) $ds^2 = d\rho^2 + \rho^2d\phi^2 + \rho^2\sin^2\phi\,d\theta^2$ **29.** $x = (a\cos\theta)/\sqrt{1 + \sin^2\theta}$, $y = (a\sin\theta/\sqrt{1 + \sin^2\theta}$, $z = -(a\sin\theta)/\sqrt{1 + \sin^2\theta}$; $L = 2\pi a$ **31.** $2\sqrt{3}(\mathbf{i} + \mathbf{j} - 2\mathbf{k})$ **33.** a) $\mathbf{T}(\frac{2}{3}, \frac{1}{3}, \frac{2}{3})$, $\mathbf{N}(1/\sqrt{5}, -2/\sqrt{5}, 0)$, $\mathbf{B}(4/3\sqrt{5}, 2/3\sqrt{5}, -\sqrt{5}/3)$ b) $2\sqrt{5}/9$ **35.** $\frac{1}{3}(t^2 + 1)^{-2}$ **37.** $x^2/4 + y^2/9 + z^2 = 1$, an ellipsoid

CHAPTER 15

Article 15.1, p. 799

1. Domain: $\{(x, y): xy \neq 0\}$. The xy-plane excluding the axes. Range: $(-\infty, 0) \cup (0, \infty)$. Level curves: hyperbolas $xy = k$, $k \neq 0$. **3.** Domain: $\{(x, y): y > -x\}$. Points above the line $y = -x$. Range: $(0, \infty)$. Level curves: lines $x + y = k$, $k > 0$. **5.** Domain: $\{(x, y) \neq (0, 0)\}$. xy-plane excluding the origin. Range: $(-\infty, \infty)$. Level curves: circles $x^2 + y^2 = k$. **7.** Domain: $\{(x, y): x \neq 0, y \geq 0\}$. Upper half-plane excluding y-axis. Range: $(-\infty, \infty)$. Level curves: the "punctured" parabolas $y = kx^2$, $k > 0$, $x \neq 0$, and the "punctured" x-axis, $y = 0$, $x \neq 0$. **9.** Domain: $\{(x, y): x \neq 0\}$. xy-plane excluding y-axis. Range: $(-\pi/2, \pi/2)$. Level curves: "punctured" lines through the origin $y = mx$, $x \neq 0$. **11.** Domain: xy-plane. Range: $[-1, 1]$. Level curves: hyperbolas $xy = k$. **13.** Domain: $\{(x, y): y \neq \pm x\}$. xy-plane excluding the lines $y = \pm x$. Range: $(-\infty, \infty)$. **15.** Domain: $\{(x, y, z): z \geq 0\}$. Range: $[0, \infty)$. **17.** Domain: All (x, y, z) except the origin. Range: $[0, 1]$. **19.** Spheres centered at the origin. **21.** Spheres centered at the origin. **23.** Planes x = constant, parallel to yz-plane. **25.** Planes $x + y + z$ = constant. **27.** Cylinders, with axis the x-axis. **29.** Paraboloids $z = x^2 + y^2 + k$. **31.** Parabolic cylinders $y = x^2 + k$. **33.** The surface $z = x$ is the plane containing the y-axis that makes a 45° angle with the positive x-axis. The level curves are the lines x = const. in the xy-plane. **35.** The surface $z = x^2 + y^2$ is the circular paraboloid whose axis is the z-axis and that opens upward from the origin above the xy-plane. The level curves are circles in the xy-plane centered at the origin. **37.** The surface $z = \sqrt{y - x^2}$ consists of the upper half $(z \geq 0)$ of the circular paraboloid $z^2 = y - x^2$ or $y = x^2 + z^2$ opening to the right of the origin and lying on and above the xy-plane. The level curves are the curves $\sqrt{y - x^2} = k$ or $y = x^2 + k^2$, the family of parabolas in the xy-plane, with axis the y-axis, opening upward away from the x-axis, and congruent to $y = x^2$. These level curves nest to fill the domain $y \geq x^2$ shown in Fig. 15.1.

39. Setting $f(x, y) = k$ gives $-(x-1)^2 - y^2 + 1 = k$, or $1 - k = (x-1)^2 + y^2$. From this we see that the level curves of f are the circles in the xy-plane centered at (1, 0) and with radius $\sqrt{1-k}$, $k \le 1$.

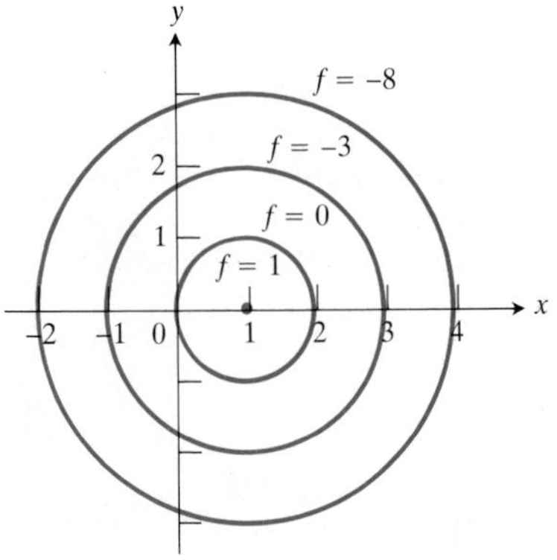

41. $x^2 + y^2 = 10$ **43.** The domain of $z = \sqrt{\sin(\pi xy)}$, $xy \ge 0$ consists of the set of points in the first and third quadrants for which $\sin(\pi xy) \ge 0$. These points make up the closed region between the coordinate axes and the hyperbola $xy = 1$, together with the alternate bands between the hyperbolas $xy = n$, n a positive integer: $2 \le xy \le 3$, $4 \le xy \le 5$, $\cdots$, $2k \le xy \le 2k+1$, $\cdots$. See the figure.

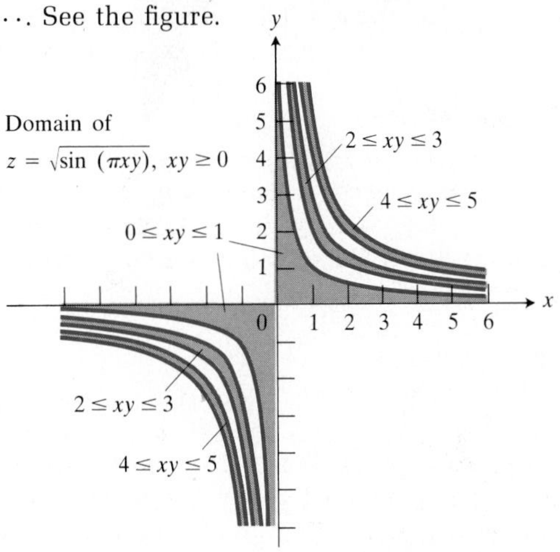

Article 15.2, p. 805

1. $\frac{5}{2}$ **3.** $\frac{1}{2}$ **5.** 5 **7.** 1 **9.** $\pi/2$ **11.** 1 **13.** No limit **15.** 0 **17.** 2 **25.** $\sqrt{x^2+y^2} < 0.1$ (inside the disc of radius 0.1 around (0, 0)) **27.** $|x| < 0.005$, $|y| < 0.005$ works **29.** $\sqrt{x^2+y^2+z^2} < 0.1$ (inside the ball of radius 0.1, centered at origin) **31.** $|x| < 0.005$, $|y| < 0.005$, $|z| < 0.005$ works **33.** Yes.

Article 15.3, pp. 810–811

1. $f_x = f_y = 1$ **3.** $f_x = ye^x$, $f_y = e^x$ **5.** $f_x = e^x \sin y$, $f_y = e^x \cos y$ **7.** $f_x = x/(x^2+y^2)$, $f_y = y/(x^2+y^2)$ **9.** $f_x = 1$, $f_y = 0$ **11.** $f_x = f_y = 0$ **13.** $f_x = 2x - y$, $f_y = 2y - x$ **15.** $f_x = f_y = -1/(x+y)^2$ **17.** $f_x = 1/(y+3)$, $f_y = -(x+2)/(y+3)^2$ **19.** $f_x = (-1-y^2)/(xy-1)^2$, $f_y = (-1-x^2)/(xy-1)^2$ **21.** $f_x = f_y = \sec(x+y)\tan(x+y)$ **23.** $f_x = ex^{e-1}y^e$, $f_y = ex^e y^{e-1}$ **25.** $f_x = 2xe^{2y+3z}\cos(4w)$, $f_y = 2x^2e^{2y+3z}\cos(4w)$, $f_z = 3x^2e^{2y+3z}\cos(4w)$, $f_w = -4x^2e^{2y+3z}\sin(4w)$ **27.** $f_u = 2u/(v^2+w^2)$, $f_v = -2v(u^2+w^2)/(v^2+w^2)^2$, $f_w = -2w(u^2-v^2)/(v^2+w^2)^2$ **29.** $f_x = 2x/(u^2+v^2)$, $f_y = 2y/(u^2+v^2)$, $f_u = -2u(x^2+y^2)/(u^2+v^2)^2$, $f_v = -2v(x^2+y^2)/(u^2+v^2)^2$ **31.** $f_x = y+z$, $f_y = z+x$, $f_z = x+y$ **33.** $f_x = 0$, $f_y = 2y$, $f_z = 4z$ **35.** $f_P = QR$, $f_Q = PR$, $f_R = PQ$ **37.** By the law of cosines, $\cos A = (b^2+c^2-a^2)/2bc$; $\partial A/\partial a = a/bc \sin A$, $\partial A/\partial b = (c^2-a^2-b^2)/2b^2c \sin A$ **39.** $x/\rho = x/\sqrt{x^2+y^2+z^2}$ **41.** $x/(x^2+y^2)$ **43.** $xz/\rho^3 \sin\phi = xz/(x^2+y^2+z^2)\sqrt{x^2+y^2}$ **45.** $\mathbf{i}\sin\phi\cos\theta + \mathbf{j}\sin\phi\sin\theta + \mathbf{k}\cos\phi$ **47.** $\rho\sin\phi(-\mathbf{i}\sin\theta + \mathbf{j}\cos\theta)$

Article 15.4, pp. 817–818

1. $4e^{2t}$ **3.** $2t(t^2+1)^2[(4t^2+1)\cos 4t - 2t(t^2+1)\sin 4t]$ **5.** a, b **11.** $2(x^2-xy-y^2)$ **13.** $e^x + y^2(1+2x)$ **15.** 2 **17.** $\partial w/\partial r = e^{2r}/\sqrt{e^{2r}+e^{2s}}$; $\partial w/\partial s = e^{2s}/\sqrt{e^{2r}+e^{2s}}$ **19.** 0

Article 15.5, p. 822

1. a) 0 b) $1+2z$ c) $1+2z$ **3.** a) $\partial U/\partial p + (v/nR)(\partial U/\partial T)$ b) $\partial U/\partial T + (nR/v)(\partial U/\partial p)$ **5.** $\partial w/\partial x = \cos\theta(\partial w/\partial r) - (\sin\theta/r)(\partial w/\partial\theta)$; $\partial w/\partial y = \sin\theta(\partial w/\partial r) + (\cos\theta/r)(\partial w/\partial\theta)$ **9.** Use $(\partial x/\partial y)_z = -f_y/f_x$, etc.

Article 15.6, pp. 834–837

1. $\mathbf{E} = -2\mathbf{i} - 2\mathbf{j} + 4\mathbf{k}$ **3.** $\mathbf{E} = 2\mathbf{i}$ **5.** $\mathbf{E} = (\mathbf{i} + 2\mathbf{j} - 2\mathbf{k})/27$ **7.** $\mathbf{E} = 5\mathbf{k}$ **9.** $-\pi/\sqrt{2}$ **11.** $\frac{2}{13}$ **13.** $(\pi-3)/2\sqrt{5} \approx 0.032$ **15.** $4\sqrt{3}$ **17.** $\frac{2}{3}$ **19.** a) $[1/\sqrt{1+(2-\pi/2)^2}][(2-(\pi/2))\mathbf{i} - \mathbf{j}]$ b) $\sqrt{1+(2-(\pi/2))^2}$ **21.** a) $(1/\sqrt{10})(-3\mathbf{i}+\mathbf{j})$ b) $8\sqrt{10}$ **23.** a) $-(10\mathbf{i}+4\mathbf{j}+10\mathbf{k})$ b) $-6\sqrt{6}$ **25.** $0.3/\sqrt{2} = 0.15\sqrt{2}$ **27.** 0.2/3 **29.** $\pm(-\frac{12}{13}\mathbf{i} + \frac{5}{13}\mathbf{j})$ **31.** $\mathbf{n} = 4\mathbf{i} + 2\mathbf{j}$ **33.** $\mathbf{n} = 2\sqrt{2}\mathbf{i} - \mathbf{j}$ **35.** $\mathbf{n} = 4\mathbf{i} - 4\mathbf{j}$ **37.** a) $6x + 10y + 8z = 36$ b) $(x-3)/6 = (y-5)/10 = (z+4)/8$ **39.** a) $z - 2x = -2$ b) $(x-2)/-2 = z-2$, $y = 0$ **41.** a) $3x + 3y + 4z = 25$

b) $(x - 1)/3 = (y - 2)/3 = (x - 4)/4$ **43.** a) $2x + 2y + z = 4$ b) $x/2 = (y - 1)/2 = z - 2$ **45.** a) $x - 2y + 2z = 9$ b) $2x = -y = z$ **47.** a) $x - y + 2z = \frac{\pi}{2}$ b) $x - 1 = 1 - y = (z - (\pi/4))/2$ **49.** a) $z = 0$ b) $x = y = 0$ **51.** a) $x + y + z = 1$ b) $x = y - 1 = z$ **53.** a) $9x - 7y - z = 21$ b) $(x - 2)/9 = (y + 3)/-7 = (z - 18)/-1$ **55.** a) $x + 2z = 2$ b) $x = (z - 1)/2,\ y = 0$ **57.** a) $x - y = 0$ b) $x = -y,\ z = 0$ **59.** $\mathbf{n} = \mathbf{i} + 2\mathbf{j} + \mathbf{k}$ **61.** $\mathbf{n} = 2\mathbf{j} + \mathbf{k}$ **63.** $\mathbf{n} = \mathbf{i} + 2\mathbf{j} + 2\mathbf{k}$ **65.** $\mathbf{n} = 2\mathbf{i} + 4\mathbf{j} - \mathbf{k}$ **67.** a) $2\mathbf{i} + 2\mathbf{j} - 2\mathbf{k}$ b) $2\sqrt{2}$ **69.** Two lines: $x = z = -y \pm 4$ **73.** a) $24\mathbf{i} + 13\mathbf{j}$ b) $\sqrt{745}$ c) $-24\mathbf{i} - 13\mathbf{j}$ d) $\pm(13\mathbf{i} - 24\mathbf{j})$ e) $24x + 13y - z = 42$ **75.** $(x - \sqrt{2})/-1 = y - \sqrt{2},\ z = 4$ **77.** $x - 1 = (z - \frac{1}{2})/-1,\ y = 1$ **79.** $y - 1 = 1 - z,\ x = 1$

Article 15.7, pp. 842–844

1. $f_{xx} = 2(y^2 - x^2)/(x^2 + y^2)^2$, $f_{yy} = 2(x^2 - y^2)/(x^2 + y^2)^2$, $f_{xy} = f_{yx} = -4xy/(x^2 + y^2)^2$ **3.** $f_{xx} = 2y - y \sin x$, $f_{yy} = -\cos y$, $f_{yx} = 2x + \cos x$ **5.** $f_{xx} = f_{yy} = 0$, $f_{xy} = z$ **11.** a) x first b) y first c) x first d) x first e) y first f) y first **13.** $w_{xy} = f_{uu} + uf_{uv} + vf_{vv}$ **15.** $2f_x + 8rsf_{xy} + 4s^2f_{xx} + 4r^2f_{yy}$ **17.** $-f_x \sin t + 2f_y + f_{xx} \cos^2 t + 4t^2f_{yy} + 4t \cos t f_{xy}$ **19.** 2 **31.** a) $f(x, y) = bxy + c^2$; b, c arbitrary b) $f(x, y) = ax^3 + bx^2y - 3axy^2 - (b/3)y^3$; a, b arbitrary **39.** $w = e^{-c^2\pi^2 t} \sin \pi x$

Article 15.8, pp. 852–854

1. a) $T = 1$ b) $T = 2(x - 1) + 2(y - 1) + 3 = 2x + 2y - 1$ **3.** a) $T = 1 + x$ b) $T = (\pi/2) - y$ **5.** a) $T = 3x - 4y + 5$ b) $T = 3x - 4y + 5$ **7.** a) $T = x$ b) $T = (x + y)/\sqrt{2}$ c) $T = (x + y + z)/\sqrt{3}$ **9.** a) $T = 1 + x$ b) $T = (\pi/2) - y - z$ **11.** $T = 7x - 3$; $|E| \le 0.02$ **13.** $T = x + y - 2$; $|E| \le \frac{1}{8}$ **15.** $T = 4x + 5y - 8$; $|E| \le \frac{1}{2}(5.2)(0.2)^2 = 0.104$ **17.** a) $T = 3x + 3y + 2z - 5$, $|E| \le 0.01$ b) $T = x + y - z - 1$, $|E| \le (3/2)(0.03)^2 = 0.00135$ c) $T = y + z + 1 - \pi/4$, $|E| \le (\sqrt{2}/2)(0.03)^2 < 0.00064$ **19.** a) $S_0[\Delta p/100 + \Delta x - 5\,\Delta w - 30\,\Delta h]$ b) A 1-cm increase in height decreases sag 6 times as much as a 1-cm increase in width. **21.** Approximately 340 ft^2 **23.** 0.14 **25.** Approximately 5% **27.** $|x - 1| \le \frac{1}{70}$, $|y - 1| \le \frac{1}{70}$. **29.** a) r: 0.014, θ: 0.003 b) r more sensitive to changes in y, θ more sensitive to changes in x **33.** a) More sensitive to changes in x b) $\Delta y/\Delta x = -2$ **35.** $h = f_0/(f_x^2 + f_y^2 + f_z^2)_0$. (Smaller values of h should probably be used in calculations.) For the three equations given, we might use $h = 0.001$. One approximate solution is $(x, y, z) = (1.059, 1.944, 3.886)$. For any solution (a, b, c), there are five other solutions: (c, b, a), $(-b, -a, c)$, $(c, -a, -b)$, $(a, -c, -b)$, and $(-b, -c, a)$.

Article 15.9, pp. 864–865

1. Minimum at $(-3, 3, -5)$ **3.** Saddle point at $(\frac{6}{5}, \frac{69}{25}, -\frac{112}{25})$ **5.** Saddle point at $(-2, 1, 3)$ **7.** Maximum at $(\frac{4}{9}, \frac{2}{9}, -3\frac{1}{9})$ **9.** Minimum at $(1, -2, 0)$ **11.** Minimum at $(2, -1, -6)$ **13.** Saddle point at $(\frac{1}{6}, \frac{4}{3}, -\frac{11}{12})$ **15.** Saddle point at $(1, 2, 9)$ **17.** Saddle point at $(0, 0, 0)$ **19.** Minimum at $(-2, 3, -2)$ **21.** Minimum at $(1, -2, -3)$ **23.** Saddle point at $(0, 0, 6)$; saddle point at $(\frac{2}{3}, \frac{2}{3}, \frac{46}{9})$ **25.** Minimum at $(0, 0, 0)$; saddle point at $(1, -1, 1)$ **27.** Saddle point at $(0, 0, 15)$; minimum at $(1, 1, 14)$ **29.** Maximum at $(1, 1, 2)$; maximum at $(-1, -1, 2)$; saddle point at $(0, 0, 0)$ **31.** Minimum of 1 at $(0, 0)$; minimum of -5 at $(1, 2)$ **33.** Minimum of 0 at $(0, 0)$; maximum of 4 at $(0, 2)$ **35.** Maximum of 11 at $(0, -3)$; minimum of $-\frac{37}{4}$ at $(5, -\frac{5}{2})$ and at $(\frac{9}{2}, -3)$ **37.** Minimum of -4 at $(2, 0)$; maximum of -3 at $(1, 0)$ and $(3, 0)$ **39.** Minimum of $-\frac{1}{4}$ at $(\frac{1}{2}, 0)$; maximum of $\frac{9}{4}$ at $(-\frac{1}{2}, \sqrt{3}/2)$ and $(-\frac{1}{2}, -\sqrt{3}/2)$ **41.** Absolute maximum of 4 at $(1, 1)$; no minima **43.** Low point: $(0, 0, 0)$; high points at the corners: $(\pm 1, \pm 1, 1)$; $\partial z/\partial x$, $\partial z/\partial y$ do not exist at $(0, 0, 0)$ but do exist at the other points and equal (in absolute value) $1/\sqrt{2}$.

Article 15.10, pp. 874–876

1. Maximum at $\pm(1/\sqrt{2}, \frac{1}{2})$; minimum at $(1/\sqrt{2}, -\frac{1}{2})$ and $(-1/\sqrt{2}, \frac{1}{2})$ **3.** Maximum value of -5.4 at $(1.2, 3.6)$ **5.** Local minimum of 0 at $(0, 3)$; local maximum of 4 at $(2, 1)$; no absolute minimum, maximum **7.** a) Minimum value -8 at $(-4, -4)$ b) Maximum value 64 at $(8, 8)$ **9.** $r = 2$ cm, $h = 4$ cm **11.** Maximum $T = 125$ at $\pm(2\sqrt{5}, -\sqrt{5})$; minimum $T = 0$ at $\pm(\sqrt{5}, 2\sqrt{5})$ **13.** Maximum of 20 at $(2, 4)$; minimum of 0 at $(0, 0)$ **15.** $(\frac{3}{2}, 2, \frac{5}{2})$ **17.** Minimum distance is 1 **19.** $(2, -2, 0)$ and $(-2, 2, 0)$ **21.** 50 **23.** $4096\sqrt{5}/125$ **25.** Minimum value $-\frac{1}{3}$ at $(0, -\frac{1}{3})$; maximum value 5 at $(0, 1)$ **27.** Hottest points are $(\frac{4}{3}, -\frac{4}{3}, -\frac{4}{3})$ and $(-\frac{4}{3}, -\frac{4}{3}, -\frac{4}{3})$ **29.** $\sqrt{a_1^2 + \cdots + a_n^2}$ **31.** $\frac{4}{3}$ **33.** Cone $|z| = r$, plane $z = 1 + r(\cos\theta + \sin\theta) = 1 + r\sqrt{2}\sin(\theta + \pi/4)$; curve of intersection lies on the hyperbolic cylinder $r = g(\theta) = 1/[\pm 1 - \sqrt{2}\sin(\theta + \pi/4)]$, $|\overrightarrow{OP}| = \sqrt{2}\, r$ with $r = g(\theta)$. The minimum value of $|g(\theta)|$ is $1/(1 + \sqrt{2}) = \sqrt{2} - 1$, and $|\overrightarrow{OB}| = \sqrt{2}(\sqrt{2} - 1) = 2 - \sqrt{2} = \sqrt{(6 - 4\sqrt{2})}$. B is the point on the cone nearest the origin; A is the point nearest the origin on the other branch of the hyperbola; there is no point farthest from the origin. **35.** a) $2xy + 2x + 2y + 1 = 0$ is the hyperbola $2(x + 1)(y + 1) = 1$ in the xy-plane. b) In space, $\{(x, y, z): 2xy + 2x + 2y + 1 = 0\}$ is a hyperbolic cylinder. Minimum for $y = x = -1 + \sqrt{\frac{1}{2}}$; no maximum.

Article 15.11, pp. 883–884

1. Yes; $f = \frac{2}{5}x^5 + x^2y^3 + \frac{3}{5}y^5 + C$ **3.** Yes; $f = x^2 + xy + y^2 + C$ **5.** No **7.** No **9.** Yes; $f = y \sin x + x \sin y + \cos y + C$ **11.** Yes; $f = 3x^2y^5 + 6xy - 5y^2 + C$ **13.** Yes; $f = 5x^2 + e^{xy} + C$ **15.** General solution: $6x^2y + 2xy^2 = C$; solution through $(1, -3)$: $6x^2y + 2xy^2 = 0$ **17.** $y \ln x + xe^y + y^2 = C$; solution through $(1, 0)$: $y \ln x + xe^y + y^2 = 1$ **19.** $x^2 - e^{-y} \cos x + e^y = 0$ **21.** a) $\alpha = 2$ b) $g = xy^5 + 2x^{3/2}y^3 + y^2 + 2$

Article 15.12, pp. 887–888

3. $y = \frac{3}{4}x + \frac{5}{3}$; $y = \frac{14}{3}$ at $x = 4$ **5.** $y = \frac{5}{14}x + \frac{15}{14}$; $y = \frac{5}{2}$ at $x = 4$ **7.** a) $y = -0.369x$ b) $I/I_0 = e^{-0.369x}$ **9.** $H \approx 17$

Miscellaneous Problems Chapter 15, pp. 889–893

1. No, $\lim_{x\to 0} f(x, 0) = 1$, $\lim_{x\to 0} f(x, x) = 0$ **5.** a) $(\partial/\partial x)(\sin xy)^2 = 2y \sin xy \cos xy = y \sin 2xy$, $(\partial/\partial y)(\sin xy)^2 = 2x \sin xy \cos xy = x \sin 2xy$ b) $(\partial/\partial x) \sin (xy)^2 = 2xy^2 \cos (xy)^2$, $(\partial/\partial y) \sin (xy)^2 = 2x^2y \cos (xy)^2$ **9.** a) Hyperboloid of one sheet b) $2\mathbf{i} + 3\mathbf{j} + 3\mathbf{k}$ c) $2(x - 2) + 3(y + 3) + 3(z - 3) = 0$, $(x - 2)/2 = (y + 3)/3 = (z - 3)/3$ **11.** $\partial f/\partial x = 1$, $\partial f/\partial y = 3$. The derivative of f at $P_0(1, 2)$ in the direction of $\mathbf{u} = \frac{3}{5}\mathbf{i} + \frac{4}{5}\mathbf{j}$ is 3. **15.** $\sqrt{3}$, maximum $= \sqrt{3}$ **17.** 7 **19.** $-\sqrt{2/3}$
21. a) $\mathbf{N}(x, y, z) = (x^2 + y^2)^{-1/2}[x(1 + 3x^2 + 3y^2)\mathbf{i} + y(1 + 3x^2 + 3y^2)\mathbf{j} - \sqrt{x^2 + y^2}\mathbf{k}]$ b) $[(1 + 3x^2 + 3y^2)^2 + 1]^{-1/2}$, $1/\sqrt{2}$ **23.** a) $4\mathbf{j} + 6\mathbf{k}$ b) $4(y + 1) + 6(z - 3) = 0$ **31.** Direction makes an angle of $\theta + \pi/2$ with the positive x-axis; magnitude is $1/r$ **35.** $h'(x) = f_x(x, y) + f_y(x, y)[-g_x(x, y)/g_y(x, y)]$
37. $dx/dz = (\sin y + \sin z - 2y^2 \cos z)(\sin y + \sin z)^{-1}(\sin x + x \cos x)^{-1}$ **39.** $\partial F/\partial x = \frac{1}{2}(\partial f/\partial u + \partial f/\partial v)$, $\partial F/\partial y = \frac{1}{2}(\partial f/\partial v - \partial f/\partial u)$ **43.** $a = b = 1$ **47.** $d^2y/dx^2 = -(1/f_y^3)[f_{xx}f_y^2 - 2f_{xy}f_xf_y + f_{yy}f_x^2]$ **49.** $x_t = x - v + ax_{vv}/x_v^2$, $0 \le v \le 1$, $t \ge 0$, $x(0, t) = 0$, $x(1, t) = 1$ **55.** $(1, 1, 1)$; $(1, -1, -1)$; $(-1, 1, -1)$; $(-1, -1, 1)$ **57.** $e^{-2}/6$ **59.** $\sqrt{3}abc/2$
61. y^2 **65.** Yes; $f = \frac{2}{5}x^5 + x^2y^3 + \frac{3}{5}y^5 + C$

CHAPTER 16

Article 16.2, pp. 903–904

1. 16

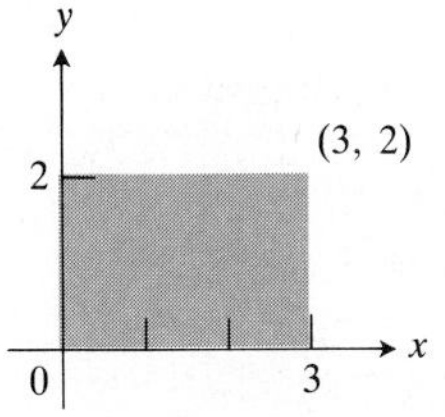

3. 0

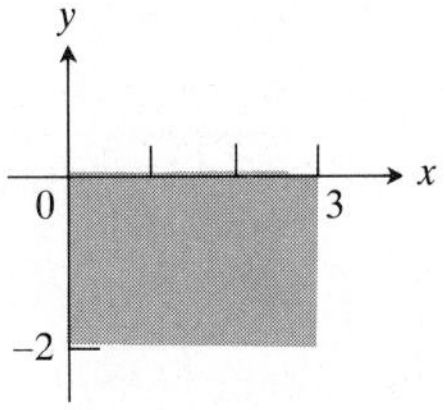

5. $(4 + \pi^2)/2$

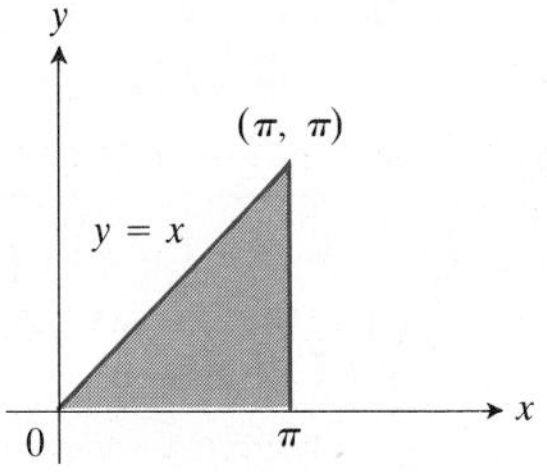

7. $\pi/4$

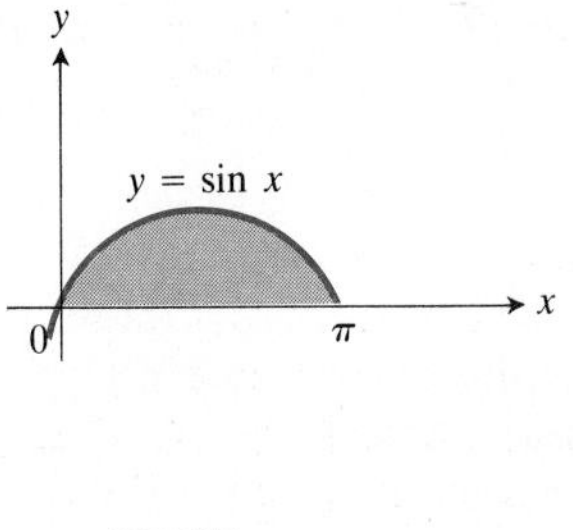

9. $9 - 9e$ **11.** $(3 \ln 2)/2$ **13.** $-\frac{1}{10}$ **15.** $(\ln 2)^2$ **17.** $\int_1^{e^2} \int_{\ln y}^{2} dx\, dy = e^2 - 3$

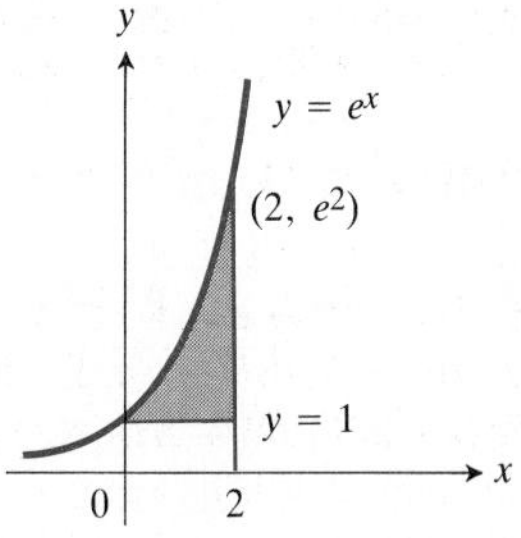

19. $\int_{-2}^{2} \int_0^{\sqrt{(4-x^2)/2}} y\, dy\, dx = \frac{8}{3}$

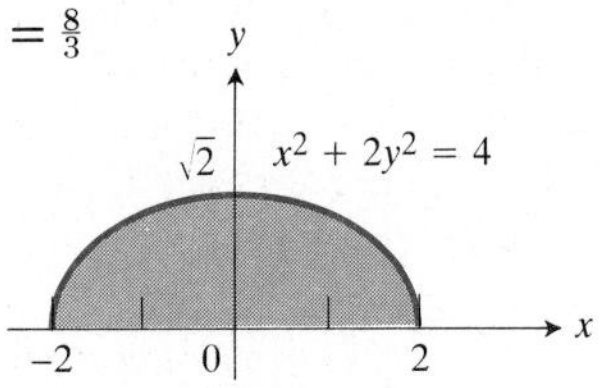

21. $\int_0^1 \int_y^{\sqrt{y}} f(x, y)\, dx\, dy$ **23.** $\int_1^e \int_{\ln y}^{1} dx\, dy$ **25.** $\int_0^8 \int_{\sqrt[3]{y}}^{2} f(x, y)\, dx\, dy$ **27.** 2 **29.** $(e - 2)/2$ **31.** $\frac{1}{4} \ln 17$ **33.** $\frac{625}{12}$ **35.** $(9\pi - 8)/3$ **37.** 32 **39.** $e - 1$ **41.** $[(\pi - 1)/2\pi] \ln 5 - (1/2\pi) \ln [(1 + 4\pi^2)/5] + 2 [\tan^{-1} 2\pi - \tan^{-1} 2]$

Article 16.3, pp. 905–906

1. $a^2/2$ **3.** 4 **5.** $\frac{1}{3}$ **7.** $\frac{4}{3}$ **9.**

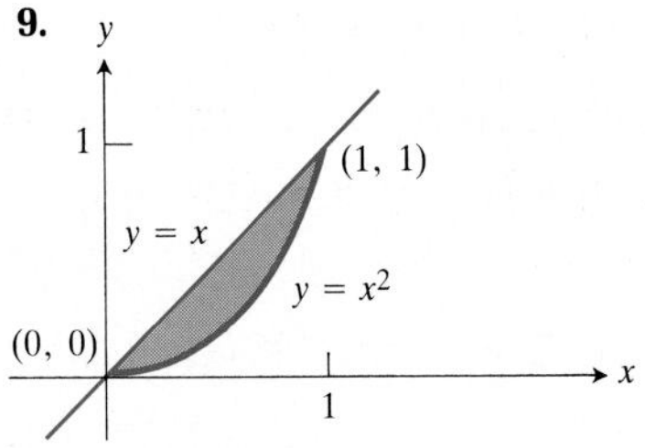

11.

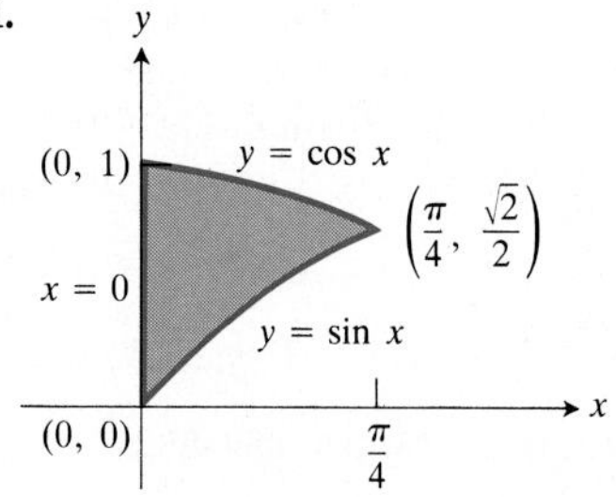

13.

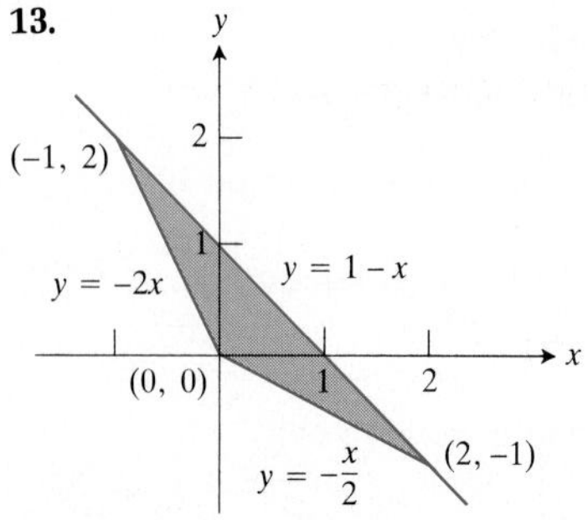

Article 16.4, pp. 909–910

1. $\frac{5}{12}$ **3.** $(a/3, a/3)$ **5.** $(0, a(8 + 3\pi)/(40 + 12\pi))$ **7.** $\bar{x} = \frac{3}{2}$ **9.** $(4a/3\pi, 4a/3\pi)$ **11.** $\frac{1}{9}[e^3 - 1]$ **13.** $\frac{64}{105}$ **15.** $\frac{104}{3}$ **17.** a) $I_x = bh^3/3$, $R_x = h/\sqrt{3}$ b) $I_x = bh^3/12$, $R_x = h/\sqrt{6}$ c) $I_x = \pi a^4/4$, $R_x = a/2$ d) $I_x = \pi a^4/16$, $R_x = a/2$ e) $I_x = \pi ab^3/4$, $R_x = b/2$ **19.** 1 **21.** $1000/3r$ lb/ft^2

Article 16.5, p. 916

1. $\int_0^{2\pi}\int_0^a r\,dr\,d\theta = \pi a^2$ **3.** $\int_0^{\pi/4}\int_0^a r^2\cos\theta\,dr\,d\theta = a^3\sqrt{2}/6$ **5.** $\int_0^{\pi/4}\int_0^{2\sec\theta} r^2\sin\theta\,dr\,d\theta = \frac{4}{3}$ **7.** $\int_0^{\pi/6}\int_0^{3\sec\theta} dr\,d\theta = \frac{3}{2}\ln 3$ **9.** $(3\pi + 8)/8$ **11.** $(\pi - 1)/2$ **13.** $\frac{4}{3}$ **15.** $((32 + 15\pi)a/(48 + 6\pi), 0)$ **17.** $(a^3/24)[32 + 15\pi]$ **19.** $(3\pi + 20 - 16\sqrt{2})2\sqrt{2}\,a^3/9$ **21.** $\int_1^3\int_1^2 (2u/v)\,dv\,du = 8\ln 2$ **23.** $\pi ab(a^2 + b^2)/4$

Article 16.6, pp. 920–921

1. $\int_0^1\int_0^2\int_0^3 dz\,dy\,dx$, $\int_0^2\int_0^1\int_0^3 dz\,dx\,dy$, $\int_0^1\int_0^3\int_0^2 dy\,dz\,dx$, $\int_0^3\int_0^2\int_0^1 dx\,dy\,dz$, $\int_0^3\int_0^1\int_0^2 dy\,dx\,dz$. The value of each integral is 6. **3.** $\int_0^3\int_0^2\int_0^{\sqrt{4-x^2}} dz\,dx\,dy$, $\int_0^2\int_0^3\int_0^{\sqrt{4-x^2}} dz\,dy\,dx$, $\int_0^2\int_0^{\sqrt{4-x^2}}\int_0^3 dy\,dz\,dx$, $\int_0^2\int_0^{\sqrt{4-z^2}}\int_0^3 dy\,dx\,dz$, $\int_0^3\int_0^2\int_0^{\sqrt{4-z^2}} dx\,dz\,dy$, $\int_0^2\int_0^3\int_0^{\sqrt{4-x^2}} dx\,dy\,dz$. The value of each integral is 12π. **5.** $4\int_0^2\int_0^{\sqrt{4-x^2}}\int_0^{x^2+y^2} dz\,dy\,dx$

7. $\frac{1}{6}|abc|$ **9.** $\frac{2}{3}$ **11.** $\frac{2}{3}$ **13.** $\frac{2}{7}$ **15.** 32π **17.** $16a^3/3$ **19.** $\frac{4}{3}\pi abc$ **21.** a) $\int_{-1}^1\int_0^{1-x^2}\int_{x^2}^{1-z} dy\,dz\,dx$ b) $\int_0^1\int_{-1}^1\int_{x^2}^{1-z} dy\,dx\,dz$ c) $\int_0^1\int_0^{1-z}\int_{-\sqrt{y}}^{\sqrt{y}} dx\,dy\,dz$ d) $\int_0^1\int_0^{1-y}\int_{-\sqrt{y}}^{\sqrt{y}} dx\,dz\,dy$ e) $\int_0^1\int_{-\sqrt{y}}^{\sqrt{y}}\int_0^{1-y} dz\,dx\,dy$

Article 16.7, pp. 923–924

1. $R_x = \sqrt{(b^2 + c^2)/12}$, $R_y = \sqrt{(a^2 + c^2)/12}$, $R_z = \sqrt{(a^2 + b^2)/12}$ **3.** $I_x = (m/3)(b^2 + c^2)$, $I_y = (m/3)(a^2 + c^2)$, $I_z = (m/3)(a^2 + b^2)$ **5.** Center of mass: $(0, 0, \frac{12}{5})$; $I_x = 3952/105$; $I_y = 2416/63$; $I_z = 128/45$ **7.** $32\,k/15$

9. $m = \int_{-2}^{2}\int_{-\sqrt{9-(9/4)x^2}}^{\sqrt{9-(9/4)x^2}}\int_{-\sqrt{1-(x^2/4)-(y^2/9)}}^{\sqrt{1-(x^2/4)-(y^2/9)}} kx\,dz\,dy\,dx$. To get M_{yz} and I_y, integrate kx^2 and $kx(x^2 + z^2)$, respectively, with the same limits and order of integration. **11.** $m(3a^2 + 4b^2)/4$

Article 16.8, pp. 928–930

1. a) $\int_0^{2\pi}\int_0^{\pi}\int_0^{2} \rho^2\sin\phi\,d\rho\,d\phi\,d\theta$ b) $\int_0^{2\pi}\int_0^{2}\int_{-\sqrt{4-r^2}}^{\sqrt{4-r^2}} r\,dz\,dr\,d\theta$ c) $\int_{-2}^{2}\int_{-\sqrt{4-x^2}}^{\sqrt{4-x^2}}\int_{-\sqrt{4-x^2-y^2}}^{\sqrt{4-x^2-y^2}} dz\,dy\,dx$

3. a) $\int_{-1}^{1}\int_{-\sqrt{1-x^2}}^{\sqrt{1-x^2}}\int_{0}^{\sqrt{4-x^2-y^2}} dz\,dy\,dx$ b) $\int_{0}^{2\pi}\int_{0}^{\pi/6}\int_{0}^{2} \rho^2 \sin\phi\, d\rho\, d\phi\, d\theta$ **5.** $\int_{-\pi/2}^{\pi/2}\int_{0}^{\cos\theta}\int_{0}^{3r^2} f(r,\theta,z)\cdot r\,dz\,dr\,d\theta$

7. a) $\int_{0}^{2\pi}\int_{0}^{1}\int_{0}^{\sqrt{1-r^2}} r^3\,dz\,dr\,d\theta$ b) $\int_0^{2\pi}\int_0^{\pi/2}\int_0^1 \rho^4 \sin^3\phi\, dp\, d\phi\, d\theta$ **9.** $\pi/2$ **11.** 8π **13.** $5\pi/2$ **15.** $(4\pi a^3/3)(8-3\sqrt{3})$
17. $(8\sqrt{2}-7)\pi a^3/6$ **19.** $100\pi/3$ **21.** $(2\pi a^5/15)(128-51\sqrt{3})$ **23.** $\bar{x}=\bar{y}=0, \bar{z}=7a/(16\sqrt{2}-14)\approx 0.8114a$
25. $\frac{2}{5}ma^2$ **27.** $64\pi a^5/35$ **29.** $\frac{7}{5}ma^2$ **31.** $\frac{15}{8}$ **33.** $\pi a^3/3$ **35.** $\pi a^3/9$ **37.** $(16-10\sqrt{2})\pi/3$ **39.** 8π **41.** $(0,0,3a/8)$
43. $a\sqrt{1270/651}\approx 1.4a$ **45.** $\pi r^4 h/10$

Article 16.9, pp. 933–934

1. $(\pi/6)[17\sqrt{17}-1]$ **3.** $(\pi/6)[17\sqrt{17}-1]$ **5.** $8\sqrt{3}/3$ **7.** $\frac{1}{3}(2\sqrt{2}-1)$ **9.** $(\pi/6)(37\sqrt{37}-5\sqrt{5})$

11. $\int_{-3}^{3}\int_{-\sqrt{(9-x^2)/4}}^{\sqrt{(9-x^2)/4}} 3\,dy\,dx/\sqrt{9-x^2-y^2}$ **15.** $\frac{1}{2}\sqrt{a^2b^2+a^2c^2+b^2c^2}$ **17.** $\pi a^2(1+c^2)$ **19.** $(\pi a^2/6)(5\sqrt{5}-1)$
21. a) $\pi a^2/2$ b) $\pi a^2/2$

Miscellaneous Problems Chapter 16, pp. 934–937

1. $\int_{-2}^{0}\int_{2x+4}^{4-x^2} dy\,dx = \frac{4}{3}$ **3.** $\int_0^1\int_{-\sqrt{y}}^{\sqrt{y}} dx\,dy$ **5.** $\int_0^1\int_x^{2-x}(x^2+y^2)\,dy\,dx=\frac{4}{3}$ **9.** $(4a/3\pi, 4a/3\pi)$

11. $(28\sqrt{2}/9\pi, 28(2-\sqrt{2})/9\pi)$ **17.** $b\sqrt{\frac{2}{5}}$ **21.** $\int_{-\pi/2}^{\pi/2}\int_0^a r^2\cos\theta\,dr\,d\theta$ **25.** $A=2a^2, I_y=(3\pi+8)a^4/12$
27. $K(\alpha)=\int_0^{\alpha\cos\beta}\int_0^{x\tan\beta}\ln(x^2+y^2)\,dy\,dx+\int_{\alpha\cos\beta}^{\alpha}\int_0^{\sqrt{\alpha^2-x^2}}\ln(x^2+y^2)\,dy\,dx$ **29.** $(0,0,\frac{1}{4})$ **31.** $(\pi/4)$ **33.** $49\pi a^5\delta/15$
35. $a^2b^2c^2/6$ **37.** $4\pi a^3/3+\frac{16}{3}[(b/2)(3a^2-b^2)\sin^{-1}(b/\sqrt{a^2-b^2})+(b^2/2)\sqrt{a^2-2b^2}-a^3\tan^{-1}(a/\sqrt{a^2-2b^2})]$

39. $\int_0^1\int_{\sqrt{1-x^2}}^{\sqrt{3-x^2}}\int_1^{\sqrt{4-x^2-y^2}} xyz^2\,dz\,dy\,dx+\int_1^{\sqrt{3}}\int_0^{\sqrt{3-x^2}}\int_1^{\sqrt{4-x^2-y^2}} xyz^2\,dz\,dy\,dx$ **41.** $2a^3(3\pi-4)/9$ **43.** $\pi^2a^3/4$

47. $2\pi(\sqrt{3}-1)$ **49.** $\pi\sqrt{2}\,a^2$ **51.** $4\int_0^a\sqrt{a^2-y^2}\,\sqrt{4y^2+1}\,dy$ **53.** $8\pi^2$ **55.** $8r^2$ **57.** $(6M/a^2)[1-h/\sqrt{a^2+h^2}]$

CHAPTER 17

Article 17.1, p. 942

1. a) At the center of the cylinder b) At the outer surface **5.** Replace x, y, z by $x-x_0$, $y-y_0$, $z-z_0$
7. $\vec{\mathbf{F}}=2(x^2+y^2+z^2)^{-1}[x\mathbf{i}+y\mathbf{j}+z\mathbf{k}]$ **9.** $\mathbf{F}=2\mathbf{i}-3\mathbf{j}+5\mathbf{k}$ **11.** $\pi a^4\delta/2$

Article 17.2, pp. 947–948

1. 9 **3.** $\frac{1}{4}abc(ab+ac+bc)$ **5.** $-\pi a^3$ **7.** $\frac{28}{3}\sqrt{14}$ **9.** $(0,0,a/2)$ **11.** $3\pi\sqrt{2}/2$ **13.** $\pi a^2/2$ **15.** $\pi a^3/6$ **17.** $\pi a^2/4$
19. $-4(e-1)$ **21.** $\frac{14}{3}$ **23.** πa^3

Article 17.3, pp. 958–960

1. $\frac{5}{6}\sqrt{2}$ **3.** π **5.** a) 0 b) $-\frac{1}{3}$. The endpoints are the same, yet the integral along different paths is different.
7. a) $\frac{9}{2}$ b) $\frac{9}{2}$ c) $\frac{9}{2}$ **9.** a) $\pi/4$ b) ln 2 c) $\frac{1}{2}$ **11.** a) 1 b) $\frac{17}{18}$ c) $\frac{1}{2}$ **13.** a) 2.143 b) 2.538 c) 2 + sin 1 **15.** a) e^3
b) e^3 c) e^3 **17.** $\frac{1}{2}$ **19.** $-\pi$ **21.** 48 **23.** $f=x^2+\frac{3}{2}y^2+2z^2$ **25.** $f=xe^{(y+2z)}$ **29.** 24 **31.** -1 **33.** 16 **35.** -2
37. 8 + ln 2 **43.** $n=-1$

Article 17.4, pp. 966–967

3. $2\pi\delta$ **5.** a) 0 b) 4π **7.** $\int_C \mathbf{F}\cdot\mathbf{n}\,ds$ represents flux, $\int_C \mathbf{G}\cdot\mathbf{T}\,ds$ represents work.

Article 17.5, pp. 974–975

1. 0 **3.** $\pi a^4/2$ **5.** $-\pi a^2$ **7.** $-\frac{1}{12}$ **9.** $\frac{3}{8}\pi a^2$ **11.** 0 **13.** -16π **15.** $\frac{2}{33}$ **21.** $x^2/4+y^2=1$

Article 17.6, pp. 982–984

1. 0 **3.** 0 **5.** 0 **7.** 48π **9.** $4\pi a^3$ **11.** 32π **13.** $128+8\pi$ **15.** 0 **17.** 5 **21.** Flux $=76\pi$ **25.** $\pi a/2$

Article 17.7, pp. 991–992

1. a) 0 b) $-\pi$ **3.** $-\frac{5}{6}$ **5.** 0 **7.** $-\pi/4$ **9.** 4π **11.** 0

Miscellaneous Problems Chapter 17, pp. 993–995

5. 0 **7.** $(\pi/60)(17^{5/2} - 41)$ **9.** $8\pi a^3$ **11.** $-\frac{25}{3}$ **13.** 0 **15.** 2π **17.** 0 **19.** 0 **31.** a) $C_1 = -100$, $C_2 = 200$ b) $C_1 = 100$, $C_2 = 0$

CHAPTER 18

Article 18.3, p. 1000

1. $(x^2 + 1)(2y - 3) = C$ **3.** $e^y = e^x + C$ **5.** $\sinh 2y - 2\cos x = C$ **7.** $(y - 1)e^y + (x^2/2) + \ln|x| = C$ **9.** $\cosh^{-1} y + \sinh^{-1} x = C$ **11.** 10 ln 10 years **13.** 48.2 minutes

Article 18.4, p. 1002

1. $x^2(x^2 + 2y^2) = C$ **3.** $\ln|x| + e^{-y/x} = C$ **5.** $\ln|x| - \ln|\sec(y/x - 1) + \tan(y/x - 1)| = C$ **7.** $\ln[(x + 2)^2 + (y - 1)^2] + 2\tan^{-1}[(x + 2)/(y - 1)] = C$, $a = -2$, $b = 1$ **9.** a) $y = kx^2$ b) $x^2 + 2y^2 = C$ **11.** $x^2 - y^2 = C$

Article 18.5, p. 1004

1. $y = e^{-x} + Ce^{-2x}$ **3.** $x^3y = C - \cos x$ **5.** $x = y/2 + C/y$ **7.** $y\cosh x = C - e^x$ **9.** $xy = y^2 + C$ **11.** a) $C = G/100\,kV + (C_0 - G/100\,kV)e^{-kt}$ b) $G/100\,kV$

Article 18.6, p. 1006

1. $y = Cx^2 - x$ **3.** $y = Cx$ **5.** $e^x + e^y(x^2 + 1) = C$ **7.** $x^2 + 2x\sqrt{y^2 + 1} - y^2 = C$ **9.** $y = x\tan(x + C)$ **11.** $e^x + x\ln y + y\ln x - \cos y = C$ **13.** $xy + \cos x = C$ **15.** $c = -2b$, a is arbitrary; $ax - b(y^2/x) = \text{const.}$

Article 18.7, p. 1008

1. $y = C_1e^{-x} + C_2$ **3.** $y = C_1\int e^{-x^2/2}\,dx + C_2$. (The integral cannot be calculated explicitly as a finite combination of elementary functions.) **5.** $y = C_1\sinh(x + C_2)$ **7.** $y = C_1x^4 + C_2x + C_3$ **9.** $x = A\sin(t\sqrt{k/m} + \pi/2) = A\cos(t\sqrt{k/m})$

Article 18.9, p. 1012

1. $y = C_1 + C_2e^{-2x}$ **3.** $y = C_1e^{-x} + C_2e^{-5x}$ **5.** $y = e^{-x/2}[C_1\cos(\sqrt{3}x/2) + C_2\sin(\sqrt{3}x/2)]$ **7.** $y = (C_1 + C_2x)e^{-3x}$ **9.** $y = e^x(C_1\cos(\sqrt{3}x) + C_2\sin(\sqrt{3}x))$

Article 18.10, pp. 1020–1021

1. $y = C_1 + C_2e^{-x} + \frac{1}{2}x^2 - x$ **3.** $y = C_1\cos x + C_2\sin x - \frac{1}{2}x\cos x$ **5.** $y = (C_1 + C_2x)e^{-x} + \frac{1}{2}x^2e^{-x}$ **7.** $y = C_1e^x + C_2e^{-x} + \frac{1}{2}xe^x$ **9.** $y = e^{-2x}(C_1\cos x + C_2\sin x) + 2$ **11.** $y = A\cos x + B\sin x + x\sin x + \cos x\ln(\cos x)$ **13.** $y = C_1e^{5x} + C_2e^{-2x} + \frac{3}{10}$ **15.** $y = C_1 + C_2e^x + \frac{1}{2}\cos x - \frac{1}{2}\sin x$ **17.** $y = C_1\cos x + C_2\sin x - \frac{1}{8}\cos 3x$ **19.** $y = C_1e^{2x} + C_2e^{-x} - 6\cos x - 2\sin x$ **21.** $y = C_1e^x + C_2e^{-x} - x^2 - 2 + \frac{1}{2}xe^x$ **23.** $y = C_1e^{3x} + C_2e^{-2x} - (e^{-x}/4) + \frac{49}{50}\cos x + \frac{7}{50}\sin x$ **25.** $y = C_1 + C_2e^{-5x} + x^3 + \frac{3}{5}x^2 - \frac{6}{25}x$ **27.** $y = C_1 + C_2e^{3x} + 2x^2 + \frac{4}{3}x + \frac{1}{3}xe^{3x}$ **29.** $y = C_1 + C_2e^{5x} + \frac{1}{10}x^2e^{5x} - \frac{1}{25}xe^{5x}$ **31.** $y = C_1\cos x + C_2\sin x - \frac{1}{2}x\cos x + x\sin x$ **33.** $y = C_1 + C_2e^x + \frac{1}{2}e^{-x} + xe^x$ **35.** $y = C_1e^{5x} + C_2e^{-x} - \frac{1}{8}e^x - \frac{4}{5}$ **37.** $y = C_1\cos x + C_2\sin x - (\sin x)[\ln(\csc x + \cot x)]$ **39.** $y = C_1 + C_2e^{8x} + \frac{1}{8}xe^{8x}$ **41.** $y = C_1 + C_2e^x - x^4/4 - x^3 - 3x^2 - 6x$ **43.** $y = C_1 + C_2e^{-2x} - \frac{1}{3}e^x + x^3/6 - x^2/4 + x/4$ **45.** $y = C_1\cos x + C_2\sin x + (x - \tan x)\cos x - \sin x\ln(\cos x) = C_1\cos x + C_2'\sin x + x\cos x - (\sin x)\ln(\cos x)$ **47.** $y = Ce^{3x} - \frac{1}{2}e^x$ **49.** $y = Ce^{3x} + 5xe^{3x}$ **51.** $y = 2\cos x + \sin x - 1 + \sin x\ln(\sec x + \tan x)$ **53.** $y = 1 - e^{-x}$

Article 18.11, p. 1023

1. $y = C_1 + C_2e^x + C_3e^{2x}$ **3.** $y = (C_1 + C_2x)e^{x\sqrt{2}} + (C_3 + C_4x)e^{-x\sqrt{2}}$ **5.** $y = e^{x\sqrt{2}}(C_1\cos\sqrt{2}x + C_2\sin\sqrt{2}x) + e^{-x\sqrt{2}}(C_3\cos\sqrt{2}x + C_4\sin\sqrt{2}x)$ **7.** $y = (C_1 + C_2x + C_3x^2 + C_4x^3)e^x + 7$

Article 18.12, p. 1027

1. $x = x_0\cos\omega t + (v_0/\omega)\sin\omega t = \sqrt{x_0^2 + v_0^2/\omega^2}\sin(\omega t + \phi)$, where $\omega = \sqrt{k/m}$, $\phi = \tan^{-1}(\omega x_0/v_0)$ **3.** a) $I = C_1\cos\omega t + C_2\sin\omega t$ b) $I = C_1\cos\omega t + C_2\sin\omega t + (A\alpha/L(\omega^2 - \alpha^2))\cos\alpha t$ c) $I = C_1\cos\omega t + C_2\sin\omega t + (A/2L)t\sin\omega t$ d) $I = e^{-5t}(C_1\cos 149t + C_2\sin 149t)$ **5.** $\theta = \theta_0\cos\omega t + (v_0/\omega)\sin\omega t$, where $\omega = \sqrt{2k/mr^2}$ **7.** a) $x = C_1\cos\omega t + C_2\sin\omega t + (\omega^2A/(\omega^2 - \alpha^2))\sin\alpha t$ where $\omega = \sqrt{k/m}$ b) $x = C_1\cos\omega t + C_2\sin\omega t - (\omega A/2)t\cos\omega t$ where $\omega = \sqrt{k/m}$

Article 18.13, p. 1029

1. $y = C\sum_{n=0}^{\infty}[x^n/n!]$ **3.** $y = C\sum_{n=0}^{\infty}[(2x)^n/n!]$ **5.** $y = C_1\sum_{n=0}^{\infty}[x^n/n!] + C_2\sum_{n=0}^{\infty}[(-1)^n x^n/n!]$ **7.** $y = x + C_1\sum_{n=0}^{\infty}[(-1)^n x^{2n}/(2n)!] + C_2\sum_{n=0}^{\infty}[(-1)^n x^{2n+1}/(2n+1)!]$ **9.** $y = x + C_1\sum_{n=0}^{\infty}[x^n/n!] + C_2\sum_{n=0}^{\infty}[(-1)^n x^n/n!]$ **11.** $y = C_1(1 - (x^2/2) + (x^4/24) - (x^6/720) + \cdots) + C_2(x - (x^5/120) - (x^9/362{,}880) + \cdots)$ **13.** $y = a_0(1 - (x^4/12) + (x^8/672) - (x^{12}/88{,}704) + \cdots) + a_1(x - (x^5/20) + (x^9/1440) - (x^{13}/224{,}640) + \cdots) + (x^3 - (x^7/42) + (x^{11}/4620) - (x^{15}/970{,}200) + \cdots)$

Article 18.14, pp. 1033–1034

7. $\tan^{-1}(x + y) = x + C$. The solution passes through (0, 0) when $C = 0$. **9.** a) $y = x - 1$ b) Concave upward for $y > x - 1$; concave downward for $y < x - 1$ **11.** $y_0(x) = 1$; $y_1(x) = 1 + x$; $y_2(x) = 1 + x + (x^2/2)$; $y_3(x) = 1 + x + (x^2/2) + (x^3/6)$ **13.** $y_0(x) = 0$; $y_1(x) = x^2/2$; $y_2(x) = (x^2/2) + (x^3/6)$; $y_3(x) = (x^2/2) + (x^3/6) + (x^4/24)$ **15.** $y_0(x) = 1$; $y_1(x) = -1 - x + x^2$; $y_2(x) = \frac{1}{6} + x + \frac{3}{2}x^2 - (x^3/3)$; $y_3(x) = -\frac{1}{4} - (x/6) + (x^2/2) - (x^3/2) + (x^4/12)$

Article 18.15, pp. 1038–1039

1. $y(1) \approx 2.488320$; exact value is $e = 2.718281$ **3.** $y(1) \approx 2.702708$ **5.** $dy/dx = y$; $y(0) = 1$ **7.** $dy/dx = 1/x$; $y(1) = 0$ **9.** $y = y_0 + \int_{x_0}^{x} f(t)\,dt$ **11.** The general solution of $y' = a^2 + y^2$ is $(1/a)\tan^{-1}(y/a) = x + C$. If $y(a) = f(a)$, then $x = a + (1/a)\tan^{-1}(y/a) - (1/a)\tan^{-1}(f(a)/a)$. **13.** $y = 1 + x + x^2 + \frac{4}{3}x^3 + \frac{5}{6}x^4$

Miscellaneous Problems Chapter 18, pp. 1040–1041

1. $\ln(C\ln y) = -\tan^{-1}x$ **3.** $y = \ln(C - \frac{1}{3}e^{-3x})$ **5.** $y^2 + \sqrt{y^4+1} = Ce^{2x}$ **7.** $y^2 = x^2 + Cx$ **9.** $y = x\tan^{-1}(\ln Cx)$ **11.** $y = (x^2+2)/4 + C/x^2$ **13.** $y = \cosh x \ln(C\cosh x)$ **15.** $y^3 + 3xy^2 + 3(x+y) = C$ **17.** $x^3 + 3xy^2 + 3\sinh y = C$ **19.** $x^2 + e^y(x^2 + y^2 - 2y + 2) = C$ **21.** $y = C$, or $y = a\tan(ax + C)$, or $y = -a\tanh(ax + C)$, or $y = -1/(x + C)$ **23.** $y = -\ln|\cos(x + C)| + D$ **25.** $y = \frac{1}{2}(\ln Cx)^2 + D$ **27.** $y = C_1 + (C_2x + C_3)e^x$ **29.** $y = C_1e^{-x} + (C_2 + x/3)e^{2x}$ **31.** $y = C_1\sqrt{x} + C_2/\sqrt{x}$ **33.** $\frac{3}{2}x^2 + y^2 = D$ **35.** $y^2 = D^2 + 2Dx$ **37.** $y = e^x - x - 1$; $y = \sum_{n=2}^{\infty}(x^n/n!)$ **39.** a) $y = (\pi/2) + x + (x^2/2!) - (x^3/3!) - (3x^4/4!)$ b) $y = -(\pi/2) - x + (x^2/2!) + (x^3/3!) - (3x^4/4!)$

APPENDIXES

Appendix A.1, p. A-8

1. -5 **3.** 1 **5.** a) -7 b) -7 **7.** a) 38 b) 38 **9.** $(-4, 1)$ **11.** $(3, 2)$ **13.** $(3, -2, 2)$ **15.** $(2, 0, -1)$

Appendix A.2, pp. A-16–A-17

1. a) $\begin{bmatrix} 2 & -3 & 4 \\ 6 & 4 & -2 \\ 1 & 5 & 4 \end{bmatrix}\begin{bmatrix} x \\ y \\ z \end{bmatrix} = \begin{bmatrix} -19 \\ 8 \\ 23 \end{bmatrix}$ **5.** $(-4, 1)$ **7.** $(3, 2)$ **9.** $(3, -2, 2)$ **11.** $(2, 0, -1)$ **15.** $(-4, 1)$

17. a) $\begin{bmatrix} 1 & -2 & 1 \\ 0 & \frac{1}{4} & -\frac{1}{2} \\ 0 & 0 & \frac{1}{3} \end{bmatrix}$ b) $(-20, \frac{15}{2}, -\frac{10}{3})$ **19.** a) $\begin{bmatrix} 1 & 1 \\ -r_2 & -r_1 \end{bmatrix}\begin{bmatrix} C \\ D \end{bmatrix} = \begin{bmatrix} a \\ b \end{bmatrix}$ b) Whenever $r_1 \neq r_2$

Appendix A.3, pp. A-21–A-22

1. Generalization: If $F_1(t) \to L_1$, $F_2(t) \to L_2$, . . . , $F_n(t) \to L_n$ as $t \to c$, then $F_1(t) + F_2(t) + \cdots + F_n(t) \to L_1 + L_2 + \cdots + L_n$ as $t \to c$, for any positive integer n.

Appendix A.8, p. A-45

1. a) (14, 8) b) $(-1, 8)$ c) $(0, -5)$ **11.** a) Circle, radius 2, center at origin b) Interior of circle in (a) c) Exterior of circle in (a) **13.** Circle, radius 1, center at $(-1, 0)$ **15.** Points on line $y = -x$ **17.** $e^{(\pi/2)i}$ **19.** $\sqrt{65}e^{i\tan^{-1}(-0.125)}$ **21.** $1, \frac{1}{2}(-1 + i\sqrt{3}), -\frac{1}{2}(1 + i\sqrt{3})$ **23.** $-(\sqrt{3} + i), (\sqrt{3} - i), 2i$ **25.** $\pm(1/\sqrt{2})(\sqrt{3} + i), \pm(1/\sqrt{2})(\sqrt{3} - i)$ **27.** $\pm(1 + i\sqrt{3}), \pm(1 - i\sqrt{3})$

INDEX

Note. Numbers in parentheses refer to problems on the pages indicated.

$d \sin^2 = 2 \sin\theta \cos\theta$

65. $\displaystyle\int \frac{\cos ax}{\sin ax}\,dx = \frac{1}{a}\ln|\sin ax| + C$

66. $\displaystyle\int \cos^n ax \sin ax\,dx = -\frac{\cos^{n+1} ax}{(n+1)a} + C, \qquad n \neq -1$

67. $\displaystyle\int \frac{\sin ax}{\cos ax}\,dx = -\frac{1}{a}\ln|\cos ax| + C$

68. $\displaystyle\int \sin^n ax \cos^m ax\,dx = -\frac{\sin^{n-1} ax \cos^{m+1} ax}{a(m+n)} + \frac{n-1}{m+n}\int \sin^{n-2} ax \cos^m ax\,dx,$

$n \neq -m \qquad$ (If $n = -m$, use No. 86.)

69. $\displaystyle\int \sin^n ax \cos^m ax\,dx = \frac{\sin^{n+1} ax \cos^{m-1} ax}{a(m+n)} + \frac{m-1}{m+n}\int \sin^n ax \cos^{m-2} ax\,dx,$

$m \neq -n \qquad$ (If $m = -n$, use No. 87.)

70. $\displaystyle\int \frac{dx}{b + c\sin ax} = \frac{-2}{a\sqrt{b^2 - c^2}}\tan^{-1}\left[\sqrt{\frac{b-c}{b+c}}\tan\left(\frac{\pi}{4} - \frac{ax}{2}\right)\right] + C, \qquad b^2 > c^2$

71. $\displaystyle\int \frac{dx}{b + c\sin ax} = \frac{-1}{a\sqrt{c^2 - b^2}}\ln\left|\frac{c + b\sin ax + \sqrt{c^2 - b^2}\cos ax}{b + c\sin ax}\right| + C, \qquad b^2 < c^2$

72. $\displaystyle\int \frac{dx}{1 + \sin ax} = -\frac{1}{a}\tan\left(\frac{\pi}{4} - \frac{ax}{2}\right) + C$

73. $\displaystyle\int \frac{dx}{1 - \sin ax} = \frac{1}{a}\tan\left(\frac{\pi}{4} + \frac{ax}{2}\right) + C$

74. $\displaystyle\int \frac{dx}{b + c\cos ax} = \frac{2}{a\sqrt{b^2 - c^2}}\tan^{-1}\left[\sqrt{\frac{b-c}{b+c}}\tan\frac{ax}{2}\right] + C, \qquad b^2 > c^2$

75. $\displaystyle\int \frac{dx}{b + c\cos ax} = \frac{1}{a\sqrt{c^2 - b^2}}\ln\left|\frac{c + b\cos ax + \sqrt{c^2 - b^2}\sin ax}{b + c\cos ax}\right| + C, \qquad b^2 < c^2$

76. $\displaystyle\int \frac{dx}{1 + \cos ax} = \frac{1}{a}\tan\frac{ax}{2} + C$

77. $\displaystyle\int \frac{dx}{1 - \cos ax} = -\frac{1}{a}\cot\frac{ax}{2} + C$

78. $\displaystyle\int x\sin ax\,dx = \frac{1}{a^2}\sin ax - \frac{x}{a}\cos ax + C$

79. $\displaystyle\int x\cos ax\,dx = \frac{1}{a^2}\cos ax + \frac{x}{a}\sin ax + C$

80. $\displaystyle\int x^n \sin ax\,dx = -\frac{x^n}{a}\cos ax + \frac{n}{a}\int x^{n-1}\cos ax\,dx$

81. $\displaystyle\int x^n \cos ax\,dx = \frac{x^n}{a}\sin ax - \frac{n}{a}\int x^{n-1}\sin ax\,dx$

82. $\displaystyle\int \tan ax\,dx = -\frac{1}{a}\ln|\cos ax| + C$

83. $\displaystyle\int \cot ax\,dx = \frac{1}{a}\ln|\sin ax| + C$

84. $\displaystyle\int \tan^2 ax\,dx = \frac{1}{a}\tan ax - x + C$

85. $\displaystyle\int \cot^2 ax\,dx = -\frac{1}{a}\cot ax - x + C$

86. $\displaystyle\int \tan^n ax\,dx = \frac{\tan^{n-1} ax}{a(n-1)} - \int \tan^{n-2} ax\,dx, \qquad n \neq 1$

87. $\displaystyle\int \cot^n ax\,dx = -\frac{\cot^{n-1} ax}{a(n-1)} - \int \cot^{n-2} ax\,dx, \qquad n \neq 1$

88. $\displaystyle\int \sec ax\,dx = \frac{1}{a}\ln|\sec ax + \tan ax| + C$

89. $\displaystyle\int \csc ax\,dx = -\frac{1}{a}\ln|\csc ax + \cot ax| + C$

Continued overleaf.

90. $\int \sec^2 ax\,dx = \frac{1}{a}\tan ax + C$

91. $\int \csc^2 ax\,dx = -\frac{1}{a}\cot ax + C$

92. $\int \sec^n ax\,dx = \frac{\sec^{n-2} ax \tan ax}{a(n-1)} + \frac{n-2}{n-1}\int \sec^{n-2} ax\,dx, \quad n \neq 1$

93. $\int \csc^n ax\,dx = -\frac{\csc^{n-2} ax \cot ax}{a(n-1)} + \frac{n-2}{n-1}\int \csc^{n-2} ax\,dx, \quad n \neq 1$

94. $\int \sec^n ax \tan ax\,dx = \frac{\sec^n ax}{na} + C, \quad n \neq 0$

95. $\int \csc^n ax \cot ax\,dx = -\frac{\csc^n ax}{na} + C, \quad n \neq 0$

96. $\int \sin^{-1} ax\,dx = x\sin^{-1} ax + \frac{1}{a}\sqrt{1-a^2x^2} + C$

97. $\int \cos^{-1} ax\,dx = x\cos^{-1} ax - \frac{1}{a}\sqrt{1-a^2x^2} + C$

98. $\int \tan^{-1} ax\,dx = x\tan^{-1} ax - \frac{1}{2a}\ln(1+a^2x^2) + C$

99. $\int x^n \sin^{-1} ax\,dx = \frac{x^{n+1}}{n+1}\sin^{-1} ax - \frac{a}{n+1}\int \frac{x^{n+1}\,dx}{\sqrt{1-a^2x^2}}, \quad n \neq -1$

100. $\int x^n \cos^{-1} ax\,dx = \frac{x^{n+1}}{n+1}\cos^{-1} ax + \frac{a}{n+1}\int \frac{x^{n+1}\,dx}{\sqrt{1-a^2x^2}}, \quad n \neq -1$

101. $\int x^n \tan^{-1} ax\,dx = \frac{x^{n+1}}{n+1}\tan^{-1} ax - \frac{a}{n+1}\int \frac{x^{n+1}\,dx}{1+a^2x^2}, \quad n \neq -1$

102. $\int e^{ax}\,dx = \frac{1}{a}e^{ax} + C$

103. $\int b^{ax}\,dx = \frac{1}{a}\frac{b^{ax}}{\ln b} + C, \quad b > 0,\ b \neq 1$

104. $\int xe^{ax}\,dx = \frac{e^{ax}}{a^2}(ax-1) + C$

105. $\int x^n e^{ax}\,dx = \frac{1}{a}x^n e^{ax} - \frac{n}{a}\int x^{n-1}e^{ax}\,dx$

106. $\int x^n b^{ax}\,dx = \frac{x^n b^{ax}}{a\ln b} - \frac{n}{a\ln b}\int x^{n-1}b^{ax}\,dx, \quad b > 0,\ b \neq 1$

107. $\int e^{ax}\sin bx\,dx = \frac{e^{ax}}{a^2+b^2}(a\sin bx - b\cos bx) + C$

108. $\int e^{ax}\cos bx\,dx = \frac{e^{ax}}{a^2+b^2}(a\cos bx + b\sin bx) + C$

109. $\int \ln ax\,dx = x\ln ax - x + C$

110. $\int x^n \ln ax\,dx = \frac{x^{n+1}}{n+1}\ln ax - \frac{x^{n+1}}{(n+1)^2} + C, \quad n \neq -1$

111. $\int x^{-1}\ln ax\,dx = \frac{1}{2}(\ln ax)^2 + C$

112. $\int \frac{dx}{x\ln ax} = \ln|\ln ax| + C$

113. $\int \sinh ax\,dx = \frac{1}{a}\cosh ax + C$

114. $\int \cosh ax\,dx = \frac{1}{a}\sinh ax + C$

115. $\int \sinh^2 ax\,dx = \frac{\sinh 2ax}{4a} - \frac{x}{2} + C$

116. $\int \cosh^2 ax\,dx = \frac{\sinh 2ax}{4a} + \frac{x}{2} + C$

117. $\int \sinh^n ax\,dx = \frac{\sinh^{n-1} ax \cosh ax}{na} - \frac{n-1}{n}\int \sinh^{n-2} ax\,dx, \quad n \neq 0$